2025 IEEE Applied Power Electronics Conference and Exposition (APEC 2025)

Atlanta, Georgia, USA
16-20 March 2025

Pages 1274-1958

IEEE Catalog Number: CFP25APE-POD
ISBN: 979-8-3315-1612-3

Copyright © 2025 by the Institute of Electrical and Electronics Engineers, Inc.
All Rights Reserved

Copyright and Reprint Permissions: Abstracting is permitted with credit to the source. Libraries are permitted to photocopy beyond the limit of U.S. copyright law for private use of patrons those articles in this volume that carry a code at the bottom of the first page, provided the per-copy fee indicated in the code is paid through Copyright Clearance Center, 222 Rosewood Drive, Danvers, MA 01923.

For other copying, reprint or republication permission, write to IEEE Copyrights Manager, IEEE Service Center, 445 Hoes Lane, Piscataway, NJ 08854. All rights reserved.

*** *This is a print representation of what appears in the IEEE Digital Library. Some format issues inherent in the e-media version may also appear in this print version.*

IEEE Catalog Number: CFP25APE-POD
ISBN (Print-On-Demand):   979-8-3315-1612-3
ISBN (Online):   979-8-3315-1611-6
ISSN: 1048-2334

Additional Copies of This Publication Are Available From:

Curran Associates, Inc
57 Morehouse Lane
Red Hook, NY 12571 USA
Phone: (845) 758-0400
Fax: (845) 758-2633
E-mail: curran@proceedings.com
Web: www.proceedings.com

TABLE OF CONTENTS

Versatile Controller Architecture for a Universal DC Fast Charging Front-End 1
Anurag Singh, Sayan Paul, Tejas Bhuse, Trent Martin, Hien Nguyen, Inder Vedula, Nikola Milivojeviœ, Dragan Maksimoviœ, Luca Corradini

A 10 kV SiC MOSFET based Three-Phase Single-Stage Isolated MVAC/LVDC Converter for Solid State Transformer Applications 9
Anup Anurag, Chi Zhang, Rudy Wang, Peter Barbosa

Direct Digital Control Applied to T-Type Vienna Rectifiers for Power Factor Correction 16
Jun-Yang Chang, Tsai-Fu Wu, Chien-Chih Hung, Jui-Yang Chiu

Active Power Decoupling Method based on Dual Active Bridge Converter without Additional Components 21
Kosuke Takeuchi, Takashi Ohno, Hiroki Watanabe, Yuki Nakata, Jun-Ichi Itoh

An ANPC-Based Building Block for Medium-Voltage Applications 27
Ahmed Rahouma, Hui Cao, David A. Porras, Zhuxuan Ma, Yue Zhao, Juan C. Balda

Analog Control of a 2.5 kW GaN based CRM PFC with Input Filter Optimization 34
Naveed Ishraq, Ayan Mallik

An iTHD and Efficiency Optimized Control Method for Triangular Conduction Mode Totem Pole Bridgeless PFC with Zero Current Detection 42
Brent McDonald, Sheng-Yang Yu

Resonance Current Suppression for AC-DC Active-Clamp Flyback Converter by Triangular Current Mode 48
Yasuo Uchida, Hiroki Watanabe, Jun-Ichi Itoh

A Universal DC Fast Charging Front-End with Optimized Film Capacitor Design 54
Sayan Paul, Anurag Singh, Tejas Bhuse, Trent Martin, Hien Nguyen, Inder Vedula, Nikola Milivojeviœ, Dragan Maksimoviœ, Luca Corradini

Power Characterization of a 1200-V/800-V 22-kVA 30-kHz Unity-Gain Dual-Active-Bridge Converter Prototype 62
Radhika Sarda, Abishek Sethupandi, Madasamy Palavesha Thevar, Howe Li Yeo, Praveenkumar Palani, Vaisambhayana B. Sriram, Anshuman Tripathi

Design of Fully Soft-Switched Semi-Dual Active DC-DC Converter for Battery Charging Application 69
Siva Prabhakar, Shiladri Chakraborty, Sandeep Anand

A ZCS-ZVS Strategy for Low Impedance Dual Active Bridges in MHz Range 77
Pushkar Saraf, Michael Solomentsev, Alex Hanson

A 6.6 kW Highly Efficient Reconfigurable Dual Active Bridge Converter Designed using Planar Transformer, SiC-Fets and Monolithic Bidirectional Devices 90
Reza Barzegarkhoo, Fabian Groon, Arkadeb Sengupta, Marco Liserre

Interleaved Switched-Inductor-Based SIPO Partial Power Converter Module for Battery Management Systems 98
Fengwang Lu, Henry Shu-Hung Chung

Single Sensor-Based Fault Localization and Detection in GaN Three-Phase Dual Active Bridge Converters 103

Satyam Sa, Yi Han, Cheng Feng Wang, Olivier Trescases

Enhanced Cocharge Operation Scheme in Bidirectional PhaseShift Full-Bridge Converters with Eliminated Voltage Overshoot and Reduced Freewheeling Current 111

Tien-Sheng Li, Minh Ngo, Rolando Burgos, Dong Dong

DC Bias Elimination in Isolated DC-DC Converters using Fundamental-Frequency Ripple 118

Arkadeb Sengupta, Thiago Antonio Pereira, Marco Liserre

Tunable Matching Network with Dual Phase-Switched Impedance Modulation Actuators 124

Alexander Jurkov, David Perreault

Soft-Switched Pulsed Bias Plasma Supply System 132

Julia Estrin, Alexander Jurkov, David J. Perreault

Analysis and Design of a Cyclo-Active-Bridge Inverter for Single-Stage Three-Phase Grid Interface 139

Mian Liao, Tanuj Sen, Yang Wu, Minjie Chen

Modular Nanosecond Pulse Generator Leveraging GaN and SiC for Versatility and Performance 147

P. Briz, H. Sarnago, O. Lucía

A Variable Frequency Technique for EMI and Efficiency Improvements in High-Level Count Flying Capacitor Multilevel Converters 151

Francesca Giardine, Sahana Krishnan, Logan Horowitz, Robert C. N. Pilawa-Podgurski

Analysis and Implementation of Minimum-Sensor Capacitor Voltage Estimators for Flying Capacitor Multilevel Converters 157

S. Tahmid Mahbub, Rahul K. Iyer, Ivan Z. Petriœ, Robert C. N. Pilawa-Podgurski

Single-Stage Bidirectional High-Frequency Link DC to Three-Phase AC (4-Wire) Grid-Tied Microinverter 164

Aniruddh Marellapudi, Satish Belkhode, Joseph Benzaquen Sune, Deepak Divan

Analysis and Design of a Constant Current LCC Class-E Inverter 171

Ju Gao, Ziheng Liu, Jiayin He, Hongjie Peng, Chengkang Ao, Jinyan Wang

Series Connected Class-E Push-Pull Converters using GaN HEMT for High-Efficiency RF Generators in Float Zone Silicon Production 175

Faheem Ahmad, Thore Stig Aunsborg, Jannick Kjær Jørgensen, Stig Munk-Nielsen

State of the Art 1.7kV Lateral GaN HEMTs, an Alternative to SiC 180

Karthick Murukesan, Robert Yang, Kamal Varadarajan, Sorin Georgescu, Doug Kang

Modeling and Characterization of Current and Future 1.2 kV Wide Bandgap Semiconductor-Based MOSFETs 185

Sushanta Gautam, Austin M. Szczublewski, Samuel K. Atwimah, Aidan P. Fox, William M. Collings, Tolen Nelson, Daniel G. Georgiev, Raghav Khanna, Andrew D. Koehler, Karl D. Hobart

2.5-kV 6.4-ns 100-kHz Repetitive GaN Marx Generator 192

Ruize Sun, Ci Pan, Wanjun Chen, Bo Zhang

Novel Dual Output LDO Architecture in 650-V GaN Technology for Power ICs 195

Plinio Bau, Thanh Hai Phung, Deniz Aygun, Bart Coomans, Mike Wens

Impact of Substrate Bias on the Stability of Bidirectional GaN HEMT in Hard- and Soft-Switching............ 202
Qihao Song, Hongchang Cui, Qiang Li, Yuhao Zhang

Characterization of LED Driven GaN-Based Photoconductive Switches 207
Samuel K. Atwimah, Tolen M. Nelson, Geoffrey M. Foster, Daniel G. Georgiev, Andrew D.
Koehler, Alan G. Jacobs, Karl D. Hobart, Micheal R. Hontz, Raghav Khanna

Development and Validation of Repetitive Transient Gate Overvoltage Rating for GaN HEMTs 214
Ricardo Garcia, Angel Espinoza, Siddhesh Gajare, Shengke Zhang

Junction Temperature Monitoring of GaN HEMT by using On-Resistance with Voltage Clamp and
Current Shunt 219
Xiao Wang, Mingrui Zou, Jiakun Gong, Yulei Wang, Zheng Zeng

False Turn-On Failure and Protection of p-Gate GaN HEMT in MHz Class-E Resonant Inverter 225
Ziheng Liu, Ju Gao, Hongjie Peng, Jiayin He, Jinyan Wang, Maojun Wang

Heat Extraction from Ferrite Cores using Metallic Laminations 231
Alyssa Brown, Duy T. Nguyen, Alex J. Hanson

Folded Flex-PCB Winding Planar Transformer for High-Frequency Isolated DC-DC Converters................. 238
Soundhariya G. Soundararajan, Hans Wouters, Wout Vanderwegen, Wilmar Martinez

Winding Strategy Analysis and Optimization for High-Current Matrix Transformer 246
Bima Nugraha Sanusi, Pinhe Wang, Michael A. E. Andersen, Ziwei Ouyang

Investigation on Impact of Transformer Parasitic Capacitance on Standby Power Consumption in
Power Converters 252
Kamran Kamran, Andrea Russo, Federica Cammarata, Claudia Malannino, S. Yuri Ciardo,
Ziwei Ouyang

PCB-Winding Integrated Transformer for 800-V Dual Active Bridge Converter using 1.2-kV GaN
Devices 258
Hans Wouters, Wei-Ren Lin, Nicolas Pirson, Thomas Jochmans, Yu Zuo, Wilmar Martinez

Comparative Assessment of Inductance Modeling for PCB-Based Circular Spiral Coils in Inductive
Power Transfer Systems 266
Gaia Petrillo, Drazen Dujic

Compact Air-Core Inductors for Variable Frequency Soft-Switching in 3 Phase Inverters.................... 272
Youssef A. Fahmy, Matthias Preindl

Simulation and Experimental Research on Cooling Performance of Fully-Immersed Evaporative
Cooling High-Frequency Transformer 278
Zhanlei Liu, Lingyu Zhu, Yuntian Gao, Yongliang Dang, Cao Zhan, Shengchang Ji

High-Efficiency PCB-Embeddable Inductor for Vertical Power IVR Applications 285
Youssef Kandeel, Liang Ye, John Flannery, Cian Ó Mathúna, Ranajit Sai, Seamus O'Driscoll,
Takayuki Tsuchida, Naoya Terauchi, Sumiaki Kishimoto, Toshio Hiraoka, Masanori Nagano

An Adaptive Zero Current Switching Control Technique for Multi-Resonant Switched-Capacitor
Converters 291
Haifah B. Sambo, Rose A. Abramson, Sahana Krishnan, Robert C. N. Pilawa-Podgurski

Small-Signal Analysis and External Ramp Design for Multiphase Current-Mode Constant On-Time
Control with Phase Overlapping... 299
Sundaramoorthy Sridhar, Qiang Li

Multiphase Constant-on-Time Minimum-Deviation Controller for Modern Processors 307
    Duo Li, Gianluca Roberts, Aleksandar Prodić, Alan Wu

Closed-Loop Control of a Dual-Side Series/Parallel Piezoelectric-Resonator-Based DC-DC
Converter 315
    Wen-Chin B. Liu, Gaël Pillonnet, Patrick P. Mercier

High-Bandwidth Embedded Rogowski Coil on Multilayer Substrate with Minimal Contribution to
Power Loop Inductance 321
    Takahiro Okamoto, Masataka Ishihara, Kazuhiro Umetani, Eiji Hiraki

Operating and Switching Frequency Circulating Current Control in Paralleled High Power
Adjustable Speed Drives with Common DC Link.. 327
    Kevin Lee, Zhihao Song, Wenxi Yao, Bo Wei

Mixed-Signal Sliding Mode Controller for Non-Inverting Buck-Boost Photovoltaic DC Optimizers............. 334
    Anurag Singh, Sayan Paul, Dragan Maksimović, Luca Corradini

A Current Sensorless Output Voltage Tracking Controller-Observer for a Boost Inverter using
Feedback Linearization 342
    *Ion Leandro Dos Santos, Tailan Orlando, Yohannes Amilcar Tekle Scherer, Telles Brunelli
Lazzarin, Hector Bessa Silveira*

Modeling and Control of a Cyclo-Active-Bridge Inverter for Single-Stage Three-Phase Grid
Interface.. 349
    Tanuj Sen, Mian Liao, Yang Wu, Minjie Chen

Turn-On Transient Modeling of 10 kV SiC MOSFET Half-Bridge Power Module in LTspice 357
    *Nianzun Qi, Jannick Kjær Jørgensen, Gao Liu, Zhixing Yan, Morten Rahr Nielsen, Asger
Bjørn Jørgensen, Hongbo Zhao, Stig Munk-Nielsen*

A Compact, Automated Sawyer-Tower System for Characterization of the High-Frequency, Soft-
Switching C_{oss} Loss of Wide Bandgap Devices 363
    Katherine Liang, Malachi Hornbuckle, Juan Rivas-Davila

Enhancing Behind-the-Meter Visibility of Grid Edge PV Systems and Electric Vehicle Charging
Loads Through Integration of Compact Low-Cost Sensors 370
    Mehrnaz Madadi, Paul Ohodnicki, Subhashish Bhattacharya

Supercapacitor based TMS Pulse Generator Design-Experimental Results Versus MATLAB
MOSFET Simulation Model 378
    Soniya Raju, Nihal Kularatna, Marcus Wilson, Alistair Steyn-Ross

Application of Artificial Intelligence for Modeling SiC Power MOSFETs.. 385
    Fredo Chavez, Danial Bavi, Sourabh Khandelwal

Multi-Objective Design Automation in Power Electronics using Bayesian Optimization Techniques 389
    Tung-Tan Nguyen, Man-Hay Pong, Huang-Jen Chiu

Reduced Order Thermal Modelling of Multi-Chip Silicon Carbide Power Modules.................................... 395
    Aamir Rafiq, Blake Nelson, Marshal Olimmah

Design and Evaluation of Dual-Resolver Emulation for Control System Verification in Aerospace
Actuation Applications 401
    *Tomas Sadilek, Julian Opificius, Jason Wright, Alec Leslie, Jeremie Tuzizila, Cesar Alzate,
Hunter Burnett, Joshua Atkinson, Justin Stricula*

Un-Terminated Blackbox Modeling for Electric Machines.. 409
   Xinliang Yang, Vladimir Mitrovic, Qing Lin, Rolando Burgos

7.2 kW GaN-Based DAB Converter with 37 kW/L Power Density and High Efficiency................................ 416
   Esmaeil Jalalabadi, Xiaoyu Wang, Jaksa Rubinic, Yang Jiao, Lucas Lu

A Novel Interleaving Method for High Power Integrated Electric Vehicle Charger with Three-Phase
Permanent Magnet Synchronous Motor 423
   Ryota Tanaka, Toshihiro Kai, Kenta Takishima, Yoshiyuki Nagai, Tetsuya Hayashi, Kantaro
   Yoshimoto

A Three-Phase CLLC Resonant Converter with Integrated Planar Magnetics for 22-kW On-Board
Chargers.. 429
   Tianlong Yuan, Zhangwei Xiang, Abdelrahman Ali, Feng Jin, Qiang Li, Wendell Da-Cunha-
   Alves, Xiaoshan Liu

Reconfigurable LLC Resonant Converter for Wide Voltage Range and Reduced Voltage Stress in
DC-Connected EV Charging Stations 436
   Yu Zuo, Xiaobing Shen, Bangli Du, Qingcheng Sui, Tim Geboers, Wilmar Martinez

Design and Control of GaN based Three-Phase / Single-Phase Combo Three-Level Flying
Capacitor PFC for OBC Applications.. 442
   Nidhi Haryani, Laszlo Huber, Anup Anurag, Juan Ruiz, Peter Barbosa

Optimization Strategy for Battery Electric Vehicle (BEV) DC Fast Charging (FC) in Cold
Environments.. 449
   Seif Sarofim, Cheng Feng Wang, Satyam Sa, Avram Kachura, Isaac Muscat, Olivier Trescases

DC-Link Voltage Reduction with Synergetic Common-Mode Voltage Control of Single-Phase Two-
Stage Non-Isolated EV Chargers.. 457
   Dongsu Lee, Juwon Lee, Jung-Ik Ha

DC-DC Converter Architecture for Fast Electric Vehicle (EV) Battery Charging Applications 464
   Shibaji Basu, Arjun Ivimey, Praveen Jain

Fast Simulator for the Estimation of Inverter DC-Link Temperature in e-Drives Subjected to Highly
Variable Working Cycles 472
   Simone Giuffrida, Fabio Mandrile, Radu Bojoi

A Monolithic Regulated 160 MHz Resonant DC-DC Converter 479
   Giacomo Ripamonti, Stefano Michelis, Georgios Bantemits, Pablo Daniel Antoszczuk, Khalil
   Khalife, Nils Hans Van Der Blij, Sokratis Koseoglou, Mattia Balutto, Francesco Driussi,
   Stefano Saggini

Reconfigurable Trans-Inductor Voltage Regulator with Improved Light Load Efficiency in Data
Center Applications 485
   Ziyao Wang, Zehui Li, Haoyu Wang

Fully Integrated Voltage Regulators (FIVRs) with Package In-Situ Coupled CoaxMIL Inductor for
High Power Density Microprocessor Applications 491
   Jaeil Baek, Beomseok Choi, Siddharth Kulasekaran, Huong Do, Brandon Marin, Jose
   Chavarria, Leigh Wojewoda, Kaladhar Radhakrishnan

Multiphase Lateral Flux Indirect Coupled Inductor for Vertical Power Delivery Voltage Regulator
Module 498
   Adhistira M. Naradhipa, Qiong Wang, Qiang Li

A High Density Three-Level Quadratic Buck Hybrid Converter for 48V-to-PoL Conversion..........505
Kejia Wang, Si Yuan Sim, Yin Quen Choong, Xin Zhang, Sriharsh Pakala, Cheng Huang

Air-LEGO: A Magnetic-Free Ultra-Thin 24V-to-1V 120A VRM with Air-Coupled Inductors..........510
Haoran Li, Wenliang Zeng, Youssef Elasser, Minjie Chen

A 15A 48V-Input Dual-Path Hybrid Dickson Converter with 6 mm³ Low Saturation Current
Inductors for Point-of-Load Conversion 518
Hua Chen, Young-Seok Noh, Minxiang Gong, Vivek De, Arijit Raychowdhury

An Ultra-Fast Control Strategy and Pre-Current-Balancing Measures Prepared for Rapid Transients
in Constant On-Time Controllers 524
Yijie Qian, Yuan Gao, Wenze Shu, Lingyun Li, Shen Xu, Weifeng Sun

Loosely Coupled Trans-Inductor Voltage Regulator (LC-TLVR) Inductor as Compensation Inductor
(Lc)..........530
Pavan Kumar, Arturo Sanchez Hernandez

Novel Complex Permeability Model of Powder Magnetic Materials..........538
Lukas Mueller, James Cox, Jun Wang, Enrique Garcia

Design Study Evaluating Impact of Gap Loss on Nanocrystalline Inductor Cores with Experimental
Validation 544
Maurice Sturdivant, Brandon Grainger, Christopher Bracken, Paul R. Ohodnicki

A Permanent Magnet Variable Inductor for DC Fault Current Limiting Applications 552
Mark Nations, Subhashish Bhattacharya

Design-Oriented Modeling and Multi-Objective Optimization of Two-Phase Coupled Inductors in
Multiphase PWM Converters 558
Yicheng Zhu, Jiarui Zou, Robert C. N. Pilawa-Podgurski

MagNetX: Extending the Magnet Database for Modeling Power Magnetics in Transient 566
*Hyukjae Kwon, Shukai Wang, Haoran Li, Youssef Elasser, Gyeong-Gu Kang, Daniel Zhou,
Davit Grigoryan, Minjie Chen*

Non-Monotonic Influence of DC Bias on Ferrite Core Loss Up to 10 MHz with Sine Wave
Excitation 573
Bohua Zhang, Martin Pfost

Comprehensive SPICE Model for Inductors Considering Magnetic Losses Under DC Bias Current 579
Yuki Sato, Hirokazu Matsumoto, Junichi Kotani, Shohei Tomioka, Kenichiro Tanaka

Indented Core to Reduce and Desensitize Inductor's Fringing Losses without Increasing Volume..........586
Rajaie Nassar, Promit Datta, Guo-Quan Lu, Christina DiMarino, Khai Ngo

Coupled Inductor Analysis and Finite Element Modeling Assisted Design for Boost Extender
Topology..........594
Vikas Kumar Rathore, Michael Evzelman, Mor Mordechai Peretz

Stability Analysis of Current-Limited Grid-Forming Inverters with Frequency Stabilization: An
Equivalent Impedance Approach..........602
Bowen Yang, Gab-Su Seo

Revisit Active Power Oscillation in Multi-Virtual Synchronous Generators Gride 609
Junjie Xiao, Pavol Bauer, Zian Qin

A Novel Current Control Technique for Off-Grid Single-Phase Inverters 616
Arpan Laha, Abirami Kalathy, Praveen Jain, Majid Pahlevani

Intelligent Low-Bandwidth Frequency Controller for VSGs at Economic Dispatch in Islanded
Microgrid.. 622
*Shraf Eldin Sati, Ahmed Al-Durra, Hatem H. Zeineldin, Tarek H.M. El-Fouly, Ehab F. El-
Saadany*

Hardware-in-the-Loop of a Grid Forming Control Strategy Applied to a DC Off-Grid Green
Hydrogen Production System... 629
*Diego Montoya-Acevedo, René Contreras-Barrios, Ángel Maureira-Riquelme, Esteban
Ibáñez-Muñoz, Catalina Gonzalez-Castaño, Carlos Restrepo*

Experimental Validation of a 40kW, 480V Point-to-Point DC Interlinks for Controller-Agnostic,
Interoperable Networked Microgrids 637
Maximiliano Ferrari, Michael Starke, John Smith, Joao Pereira, Misael Montejano

Andronov-Hopf Oscillator-Based Grid-Forming Converters with Embedded Disturbance Rejection
for Non-Ideal Loading Condition 645
Vikram Roy Chowdhury, Gab-Su Seo, Barry Mather

Estimation of Rectifier Output Current of the LLC Converter.. 651
Xin Wu, Yi Zhou, Haihong Long, Dehong Xu

A 100kHz Digitally Controlled 10kW, 2-Channel Solar MPPT Converter using 3-Level Topology
with >75W/in³ Power Density and >98.5% Peak Efficiency... 658
Ranajay Mallik, Akshat Jain

A Bootstrapless KY-S-Hybrid Buck-Boost Converter with Full Range iLs Reduction and 400%
Line Transient Response Acceleration for AI-Mobile Application... 664
Chuan-En Chang, Cheng-Ta Chuang, Hao-Ran Huang, Chieh-Ju Tsai, Ching-Jan Chen

Digital Control of a 600-V to 28-V 20-kW Two-Stage DC-DC Converter 670
*Shreyas B. Shah, Rachit Pradhan, Jiaqi Yuan, Mohamed Ibrahim, Ahmed Elezab, Samuel
Hemming, Giorgio Pietrini, Piranavan Suntharalingam, Mario F. Cruz, Ali Emadi*

Self-Calibrated Digital Current Emulation for High-Frequency Hysteretic Current-Mode Control in
GaN PFC Converters.. 676
Mohammad Shawkat Zaman, Olivier Trescases

High-Frequency Flying Capacitor Four-Level Drain Supply Modulator 682
Audrey Cheshire, Paul Flaten, Zoya Popoviœ, Dragan Maksimoviœ

Discontinuous Modulation Strategy for Voltage and Temperature Balancing of MMCs 689
Davide D'Amato, Stayner Nóbrega Barros, Jun-Hyung Jung, Marco Liserre

Damping Control and Improvement of Grid-Forming Inverter from a Wideband Stability
Perspective... 696
Rui Kong, Subham Sahoo, Yubo Song, Frede Blaabjerg

A Grid-Forming Split-Phase Three-Leg Inverter with Unbalanced Loading and Active Power
Decoupling 703
Namwon Kim, Renata Kimpara, Michael Starke

Completely Decentralized Active and Reactive Power Control of Grid-Connected Cascaded H-
Bridge Inverters with Integrated Battery Storage 711
Soham Dutta, Brian Johnson

Small-Signal Modeling and Damping Design of Unfolding-Based Single Stage AC-DC Converter using the Extra Element Theorem 719
Dakota Goodrich, Aditya Zade, Shubhangi Gurudiwan, Mahmoud Mansour, Regan Zane, Hongjie Wang

Methods to Enhance Cybersecurity of Multiple Inverters in Large Grid Connected PV / Battery Energy Storage Systems 727
Hasan Ibrahim, Jaewon Kim, Peng-Hao Huang, Vishwam Raval, Prasad Enjeti

Optimal DC-DC Converter Topology and Control Algorithm for Fuel Cell Electric Vehicle with Series-Connected Supercapacitor 733
Hyeon Soo Kim, Yun Seong Hwang, Seung Hyun Kang, Man Jae Kwon, Byoung Kuk Lee

Reliability-Constrained Design of a High-Gain Power Optimizer based on a Real Mission Profile 738
Stefano Cerutti, Francesco Iannuzzo, Ariya Sangwongwanich, Tamás Kerekes, Mario Giuseppe Pavone, Francesco Gennaro, Natale Aiello, Francesco Musolino, Paolo Stefano Crovetti

Submodule Voltage Balancing Technique of Solar MMC for Firing the Switches using Integrated PWM Modules 746
Ahmed Elsanabary, Saad Mekhilef, Mokhtar Aly, José Rodriguez

Single-Stage High-Frequency-Link Split-Phase Microinverter with High Voltage Gain based on Buck-Boost AC Chopper 751
Xuewen Li, Jia Liu, Jinjun Liu

Fault Diagnosis and Tolerant Strategy for Triple-Port Hydrogen Converter using SSA-Optimized Random Forest Algorithm 757
Shiqi Zhang, Yiyina Teng, Naizhe Diao, Xiaoqiang Guo, Vladimir Terzija, Lichong Wang

Resilient Operation for Grid-Connected Cascaded H-Bridge Multilevel Inverter with Improving PV Source Stress 761
Jinli Zhu, Yuan Li, Hector Akuta, Jeonghun Kim, Uthandi Selvarasu, Shumeng Wang, Vikram Roy Chowdhury, Brad Lehman, Fang Z. Peng

A Medium Voltage Grid-Connected PV Inverter with a New Modular High Voltage Gain Converter Featuring Internal Modified Voltage Doubling Balancers 768
Kajanan Kanathipan, Muhammad Ali Masood Cheema, John Lam

Split-Source Common-Ground Inverter for Photovoltaic Applications 775
Mahmoud A. Gaafar, Mohamed Orabi, Samir Kouro, Ahmed Ibrahim, Eltaib Abdeen D. Ibrahim

Comprehensive Investigation and Proposal of a New Wireless Charging Road Structure using Low-Environmental-Impact Magnetic Concrete 782
Shuntaro Inoue, Yuko Kano, Shin Tajima

Design of a Bidirectional High Power Inductive Power Transfer System with Auxiliary Winding for Automotive Applications 788
Luis Ruiz Chamorro, Nikola Mirkoviœ, Alberto Delgado Expósito, Pedro Alou Cervera, Miroslav Vasiœ

Mutual Inductance and Load Identification Method based on the Voltage Transients of WPT Systems 795
Xiaosheng Wang, C.Q. Jiang, Yibo Wang, Liping Mo

Digitally Controlled Misalignment-Tolerant Inductive Power Transfer System with Adaptive Hybrid Compensation for CC/CV Charging of E-Scooter 801
    Niranjan Shrestha, V.S.R. Varaprasad Oruganti, Sheldon Williamson

On/Off Control of Modular Inductive Power Transfer System 809
    Kunxiao Zhou, Guangdong Ning, Heyuan Li, Xinlin Wang, Minfan Fu

Receiver Side Regulation of LCC Wireless Power Transfer System with Variable Notch Filter 815
    Hsin-Che Hsieh, Jih-Sheng Lai

84.7 Percent Peak Efficiency Stress Tolerant DC DC Buck Converter for Li Ion Battery Driven Standby Circuits in 18nm FDSOI.................... 821
    Gautam Dey Kanungo, Pijush Kanti Panja, Vikas Bugade, Kallol Chatterjee

Leveraging Ultrasound and Neural Networks for Non-Invasive Power Converter Efficiency Estimation.................... 828
    Youssof Fassi, Vincent Heiries, Jérôme Boutet, Julien Marianne, Sébastien Martin, Mathilde Chareyron, Clément Chambon, Sébastien Boisseau

A Load-Independent Multi-Relays Wireless Power Transfer with Self-Regulation and Single Compensation Network.................... 834
    Jong-Hun Kim, Najam Ul Hassan, Seogyong Jeong, Myeong-Ho Kim, Min-Sik Kim, Jee-Hoon Jung, Byunghun Lee, Se-Un Shin

A GaN-Based Single-Stage Solid-State Transformer Replacement for 40 VA Class 2 Line-Frequency Transformers.................... 840
    Allen T. Nguyen, Charles R. Sullivan

Survey of Components and Topologies for High-Efficiency and High-Power Density 48V DC-DC Converters 848
    Joseph Winkler, Niklas Deneke, Bernhard Wicht

A Novel Solid-State Circuit Breaker using B-TRAN™ 854
    Mudit Khanna, Ruiyang Yu, Milad Tayebi, Jiankang Bu, Jeffrey Knapp

Development of a Supercritical Fluid-Insulated Fast Mechanical Switch for MVDC Hybrid Circuit Breakers.................... 860
    Zhiyang Jin, Qichen Yang, Alfonso Cruz, Lukas Graber

Dynamic Impedance Matching for a Variable Reluctance Energy Harvesting Application with Constrained Space 868
    Fernando Pérez, Alejandro Redondo, Airán Francés, Gabriel Mujica

Renewable Energy-Powered DC-Converted Refrigerator based on a Supercapacitor-Assisted Technique 874
    Nirashi Polwaththa Gallage, Nihal Kularatna, Alistair Steyn-Ross, Dulsha Kularatna-Abeywardana

Design and Evaluation of Flexible Inductors for Wearable Power Electronics.................... 880
    Sean Logi, F. Selin Bagci, Katherine A. Kim

Design of Boost Power Factor Corrector and Asymmetrical Half-Bridge Flyback Converter for USB-PD Applications.................... 887
    Yun-Keng Cheng, Tsorng-Juu Liang, Kai-Hui Chen, Ming-Chang Tsou

Computationally Efficient Current Sensorless Predictive Control for PMSM Drive Fed by a Matrix Converter with CMV-Free Operation 895
Ali Sarajian, Ibrahim Harbi, Quanxue Guan, Davood Arab Khaburi, Ralph Kennel, José Rodriguez, Patrick Wheeler, Mokhtar Aly

PMSM Motor Drive with Current Direct Digital Control and Near 1st-Order Speed Control 900
Po-Chang Lee, Tsai-Fu Wu, Han Ku, Chien-Chih Hung, Jui-Yang Chiu

Fault-Tolerant Multilevel Converter for Multiphase Switched Reluctance Motor Drives based on q+2 Converter 906
Mahmoud A. Gaafar, Mohamed Orabi, Hao Chen, Mostafa Dardeer

Uncertainty-Aware Artificial Intelligence for Gear Fault Diagnosis in Motor Drives 912
Subham Sahoo, Huai Wang, Frede Blaabjerg

Neural Network based Digital Twin Health Monitoring of BLDC Motor Drives for Robots 919
Mohamed Y. Metwly, Benjamin Luckett, Landon Clark, JiangBiao He, Biyun Xie

MTPA Control using Predictive P&O Method for Dual Parallel Surface-Mounted Permanent Magnet Synchronous Motor Drives Fed by a Single Inverter 925
Jae-Seong Kim, Kyo-Beum Lee

A Novel I-f Startup Strategy with Smooth Transition to Sensorless Control for CSI-Fed PMSM Drives used in Submersible Pumps 930
Milad Bahrami-Fard, Majid Ghasemi Korrani, Babak Fahimi

Simulation-Assisted Design and Implementation of an Electrically Excited Synchronous Motor Drive System 938
Shih-Gang Chen, Jun-Ming Hsu, Chun-Yen Chen, Ming-Shi Huang

Implementation and Analysis of Direct Torque Control on High-Speed PMSMs: A Comparative Study of Commercial and Laboratory-Developed Motors 943
Md Moniruzzaman, Kishor Joshi, Md Rashedur Rahman, Md Khurshedul Islam, Seungdeog Choi, Masoud Karimi Ghartemani

A Ferrite based Carbon Reinforced Composite Wrapped IPM Rotor Design for High-Speed Traction Applications 951
Md Rashedur Rahman, Md Khurshedul Islam, Md Moniruzzaman, Seungdeog Choi, Han-Gyu Kim, Andrew Walters

A Novel Phase-Mode Controller for Resonant Converters 958
Claudio Adragna, Daniele Cazzaniga, Stefano Manzoni

A Regulated 36V-60V-Input VIN-Insensitive Resonant Switched-Capacitor Converter with Large Voltage Conversion Ratio 966
Yichao Ji, Jingyi Yuan, Lin Cheng

A Hybrid Switched Capacitor Converter Enabling Capacitive-Based Wireless Power Transfer for Battery Charging Applications 971
Jade Sund, Samantha Coday

A 48V to 50-110V Resonant Power-Bus Charger with Reduced Conduction Loss for MHz-Frequency Long-Range LiDAR Driver 978
Hangxiao Ma, Xuchu Mu, Yang Jiang, Weihang Zhang, Jincheng Zhang, Rui P. Martins, Pui-In Mak

A Trajectory Controlled 48-to-24 V Resonant Switched Capacitor Converter with 98.7% Efficiency and Ultrafast Dynamic Response 983
Hélène T.W. Ma Yang, Liang Wang, Haoyu Wang, Wai Tung Ng

Low Power, Non-Isolated, Extremely-High Step-Up, Quasi-Resonant Hybrid DC–DC Converter 990
Kumar Joy Nag, Aleksandar Prodić

Isolated Soft-Switching Flying-Capacitor based Quasi-Resonant Step-Up Converter..................................... 997
Kumar Joy Nag, Aleksandar Prodić

Accurate Small-Signal Phasor Transformation-Based Modeling of Secondary-Side Diode-Bridge Rectifiers for Battery Charging Applications 1004
Aditya Zade, Regan Zane

High-Efficiency Isolated Piezoelectric Transformers for Magnetic-Less DC-DC Power Conversion 1012
Sourav Naval, Wentao Xu, Mustapha Touhami, Jessica D. Boles

First Characterization of GaN Power Device and IC at Deep Cryogenic Temperatures Down to 100 mK 1020
Xin Yang, Matthew Porter, Zineng Yang, Zichen Xi, Liyang Jin, Liyan Zhu, Linbo Shao, Yuhao Zhang

Dynamic Environment-Aware Lifetime Prediction of SiC MOSFET Modules Through LSTM 1026
Md Zakir Hasan, Seungdeog Choi, Youssef Aider, Prashant Singh, Chun-Hung Liu

Guarding-Based C-V Characterization of 10 kV SiC MOSFET in Half-Bridge Module Configuration.. 1034
Nianzun Qi, Gao Liu, Zhixing Yan, Shaokang Luan, Pawel Piotr Kubulus, Yuan Gao, Stefan Meyer, Hongbo Zhao, Asger Bjørn Jørgensen, Stig Munk-Nielsen

Automated Characterization Platform for Comprehensive Dynamic R_{dson} Assessment of GaN HEMTs from 50 K to 400 K.. 1040
Tian Qiu, Zheyu Zhang, Purushottam Khadka, Ahmed Siraj, Dilip Rana

A Gate Driving Scheme for GaN Git with Enhanced Short Circuit Capability for Motor Drive Application 1047
Zongjie Zhou, Yan Cheng, Kevin J. Chen

Online Detection and Reduction of the Influence of Parameter Tolerance of Paralleled SiC MOSFETs in an EV Inverter Environment.. 1051
Hadiuzzaman Syed, Jochen Streit, Robert Kragl, Muhammad Muneeb Alam, Alberto Martinez-Limia, Karl Oberdieck, Ertuðrul Sönmez

Dynamic Current Sharing Issues with Paralleling SiC Power MOSFETs 1058
Ching-Yao Liu, Chen-Chan Lee, Jih-Sheng Lai

Integrated Short-Circuit Protection Design based on Dual-Channel Gate Driver for Series Connected Medium-Voltage SiC MOSFETs 1063
Rui Wang, Drazen Dujic

Long-Term High-Temperature Dynamic Gate Stress Reliability of a Last-Generation, Automotive-Grade, Planar 1200 V SiC MOSFET.. 1070
Giuseppe Mauromicale, Alessandro Sitta, Michele Fiore, Michele Calabretta, Francesco Iannuzzo

Innovative Gate Driver Structure Achieving Low Time Skew Across Isolation Barrier for Parallel Connected SiC Modules 1076
Louison Gouy, Anne-Sophie Descamps, Nicolas Ginot, Christophe Batard

Fully Integrated Closed-Loop Active Gate Driver IC with Real-Time Control of Gate Current Change Timing by Gate Current Sensing 1084
Yaogan Liang, Katsuhiro Hata, Makoto Takamiya

Analyze and Design of Digitally Load Current Modulated Active Gate Driver for GaN HEMTs based Buck DC-DC 1090
Wentao Liu, Zhina Lian, Taotao Wu, Xiaochuan Peng, Hao Min

Impact of Real-Time Variable Gate-Drive Strength on Drive Cycle Efficiency in SiC Inverter-Fed PMSM Traction Drives... 1096
Matteo Pizzuto, Aiswarya Balamurali, Aniket Anand, Narayan C. Kar

Demonstration of Efficiency Increase of 350 V-to-13.3 V Isolated DC-DC Converters for Electric Vehicles by Active Gate Driving 1102
Yohei Sukita, Katsuhiro Hata, Hiroki Kondo, Kenichi Watanabe, Kenichi Nagayoshi, Makoto Takamiya

A Multi-Level Active Gate Driver for Achieving Thermal Balance in Parallel Connected Power MOSFETs.. 1108
Jingyuan Liang, Lingwei Sun, Wen Tao Cui, Wai Tung Ng, Motomitsu Iwamoto, Haruhiko Nishio

A Fast Short-Circuit Protection Method for Ohmic Gate P-GaN HEMT based on Gate Charge 1114
Yue Wu, Xi Jiang, Song Yuan, Xiaowu Gong, Zhaoheng Yan, Jiahong Chen, Yun Xu, Jinjie Liu

Comparison of Ultrafast-Rise-Time Gate Drivers for Wide-Bandgap Devices in Sub-Microsecond Pulsed Power Applications 1121
Soham Roy, Duy T. Nguyen, Neeraj Anantha, Alex J. Hanson

A Discrete Multilevel Active Gate Driver for GaN HEMTs to Optimize the Switching Behavior 1129
Celine Lawniczak, Martin Pfost

Attenuation of Fundamental Component of Differential Mode Noise using Active EMI Filter..................... 1135
Guru Abhilash Mulumudi, Naveed Ishraq, Ayan Mallik

Graph Neural Network based Performance Modeling for the Dual Active Bridge Converter with Operational Generalization... 1143
Weihao Lei, Fanfan Lin, Xinze Li, Xiaokun Bao, Xin Zhang

An Augmented State Space Modelling Approach for DC-DC Converter Start-Up in Closed Loop 1148
Waseah Anjum, Arkadeb Sengupta, Marco Liserre

The Utilization of a Parallel Computing Algorithm for Accelerating Switching-Level Modeling of Power Electronics Simulations in a T-Type PV Inverter 1153
Buck F. Brown III, Liwei Wang, Zheyu Zhang, Johan Enslin, Yi Li

A New Reduced Order Analytical Switching Model for eGaN HEMTs 1159
Ruqi Li, Douglas Arduini, Phen Lumod, Shobhana Punjabi, River Lin, Harold Gutierrez

Proposal of an Alternative Reverse Recovery Calculation Method... 1167
Brian Deboi, Blake Nelson, Austin Curbow

Improvement of CM EMI Attenuation Ability of Transformer with Negative Capacitor 1173
    Qinghui Huang, Yiming Li, Yirui Yang, Shuo Wang, Yanwen Lai, Zhedong Ma

Damping Factor based PCB Parasitic Inductance Value Optimization to Minimize Voltage
Overshoot and Settling Time of Semiconductors 1179
    Reza Shahbazi, Yunting Liu

Hardware Implementation of Virtual Resistance based FRT Logic in Programmable 3-Level ANPC
Inverters...1184
    Mohammad Safayet Hossain, Shuvangkar Chandra Das, Paychuda Kritprajun, Amin Banaie,
    Tapas Barik, Deepak Ramasubramanian, Aboutaleb Haddadi, Evangelos Farantatos, Ulrich
    Muenz

Rad-Hard PSFB Controller for High-Voltage Space Applications 1190
    Reynaldo S. Gonzalez, Robert E. Bolaños

Modeling, Control and Digital Implementation of a Buck Converter Operating in Triangular
Current Mode for a Wide Output Voltage Range Space Application...1197
    Regina Ramos, Sara Pérez, Guillermo Núñez, Pedro Alou, Javier Torres

Thermal Model and Optimization of a Multi-Winding Transformer for Lunar Surface Power
Transmission...1203
    Zhining Zhang, Yuzhou Yao, Junchong Fan, Juchen Yang, Robert Guenther, Pengyu Fu, Jin
    Wang

Active Gate Driver Power Supply for High-Reliability Applications 1211
    Joseph P. Kozak, Juan Ramirez, Jesse Lin, Allison Orr, Alexander Martin, Hala Tomey

A Hybrid Energy Storage System for eVTOL Unmanned Aerial Vehicles using Supercapacitors.................1217
    Ali Alenezi, PengHao Huang, Prasad Enjeti

Evaluation of Retired Lithium-Ion Batteries for Second-Life Applications Through Electrochemical
Impedance Spectroscopy 1224
    Latha Anekal, Sheldon Williamson

Uninterruptable Non-Isolated Integrated Power Electronics Converter (UNIPEC) for Commercial
Truck Auxiliary Power Unit 1230
    Pouya Zolfi, Ahmad Alzahrani, Ayman El-Refaie

Investigation of Electrical Safety for Non-Isolated Single-Phase On-Board Chargers used in
BEV/PHEV 1237
    Soya Kataoka, Shohei Funatsu, Hiroaki Matsumori, Takashi Kosaka, Keisuke Nakamura,
    Subrata Saha

An 8-Level Flying Capacitor Multilevel Converter for Electric Aircraft Pulse Deicing 1242
    Nicole Stokowski, Andrew Freeman, Aidan Rodgers, Aria Delmar, Jonathan Sengstock, Alex
    Solecki, Andrew Stillwell

Impact of Position Measurement Delay Angle on Performance of PMSM Drives for Electric Power
Steering in a Wide Speed Range...1248
    Yingzhe Wu, Hengbin Zhang, Yuxiang Xue, Lisheng Wang, Hui Li, Shan Yin

Physical Parameter Estimation for a Two-Level VSI Three-Phase PMSM Electric Drivetrain 1255
    Bernard Steyaert, Ananda Tjakra Adisurja, Matthias Preindl

A Novel Two-Dimensional Random Switching Frequency PWM Method for Variable Frequency Drives 1261

    Mostafa Abarzadeh, Kevin Lee

Optimized Maximum Torque and Minimum Loss Fault-Tolerant Control Schemes for Dual Three-Phase PMSM 1267

    Syed Mohammad Maaz, Dong-Choon Lee

Wireless Actuation of Magnetic Robots with a Modular 60 mT 3-D Helmholtz Coil System 1274

    Konstantinos Manos, Yifan Rao, Tuo Zhao, Kevin Liu, Daniel Zhou, Calvin Nguyen, Eric Chen, Glaucio H. Paulino, Minjie Chen

A Versatile PHIL based Motor Emulator Testbench using a High-Performance Power Amplifier Testbench 1279

    Seyedeh Nazanin Afrasiabi, Rajendra Thike, Mathews Boby, K. S. Amitkumar

A 450V Three Phase GaN IPM Achieving 99.1% Efficiency in Smallest 12mm x12mm Package for 250W Power Delivery without Heatsink 1286

    Maik Kaufmann, Manu Balakrishnan, Stefan Herzer, Anand Chellamuthu, Hely Zhang

FEA based High-Frequency Synthesis for the Design and Optimization of GaN-Based Dual Three Phase Motor Drive System 1294

    Syed Imam Hasan, Alper Uzum, Ashraf Siddiquee, Yilmaz Sozer, Krishna Namburi

Evaluation of Passive Common-Noise Canceller Considering Both of Thermal Equilibrium and Common-Mode Noise Cancellation 1299

    Koji Mitsui, Kenshiro Katsura, Koki Notake, Koji Yamaguchi

Performance Evaluation of Isolated DC/DC Converters in Modularized Bridge Rectifier Solid-State Transformer 1305

    Zhenchao Li, Andrea Cervone, Drazen Dujic

Active and Reactive Power Flow Control of the Dual Active Bridge Converter 1311

    Lauryn Morris, Thomas W. Francois, Jonathan Saelens, Oroghene Oboreh-Snapps, Arnold Fernandes, Praneeth Uddarraju, Sophia A. Strathman, Jonathan W. Kimball

Comparative Analysis of Carbon Footprints and Material Usage of Solid-State Transformers and Low-Frequency-Transformer-Based MVac-LVdc Interfaces for High-Power EV Charging 1318

    Luc Imperiali, Rudy Wang, Anup Anurag, Peter Barbosa, Johann W. Kolar, Jonas Huber

Trade Study of Isolation Requirements and Magnetic Core Selection for Medium Frequency-Medium Voltage Transformers 1326

    Mohendro Kumar Ghosh, Mark A. Juds, Brandon Grainger, Ahmad El Shafei, Bogdan S. Borowy, Paul Ohodnicki

Comparative Evaluation of a Multilevel LLC Resonant Converter for a Modular DC/DC Stage in a Electrolyzer Power Supply 1334

    Samuel S. Queiroz, Levy F. Costa

Cost-Effectiveness Assessment of SiC MOSFET and Si IGBT Semiconductors in a Three-Level Resonant Converter for Solid-State Transformer 1341

    Samuel S. Queiroz, Levy F. Costa

Comparative Performance Analysis of Medium Voltage 3L-ANPC and 3L-DNPC Pole Enabled by Series-Connection of 10kV SiC MOSFETs and 10kV SiC JBS Diodes for Sine Triangle PWM Operation 1347
    Sanket Parashar, Shubham Rawat, Nithin Kolli, Raj Kumar Kokkonda, Subhashish Bhattacharya

A Zero Harmonic Distortion Master Converter for Medium Voltage Microgrids 1355
    Gabriel V. Ramos, Dener A. de L. Brandão, Thiago M. Parreiras, Danilo I. Brandão, Braz de J.C. Filho

An MILP Approach for Modeling and Analyzing the BESS for Smoothing Renewable Fluctuations Considering BESS Capacity Attenuation in the Bulk Power System with High Inverter-Based Resource Penetration 1363
    Hualong Liu, Wenyuan Tang

Thermal and Efficiency Characterization of Immersion Cooled SiC Traction Inverter 1368
    Yiju Wang, Reza Ilka, JiangBiao He

FPGA-Based Hybrid Simulator for Real-Time 3-D Temperature Monitoring of Power Converters 1375
    Xianghao Mo, Daniel Ríos Linares, Regina Ramos, Miroslav Vasiœ

A New Subassembly Concept for Enhanced Heat Dissipation and Reliability of Power Module 1383
    Yosuke Nakata, Yuji Sato, Shin Uegaki, Jun Fujita, Akihiko Furukawa, Masayoshi Tarutani

Stand-Alone R_{DS-ON} Sensor for In-Situ Prognostic, Protection and Reliability Enhancement of Power Converters 1388
    Zaheen Mustakin, Qiang Mu, Lucas Pereira, Jiale Zhou, Tiefu Zhao, Babak Parkhideh

Electrical Evaluation of a Modular High Voltage 3D Power Module using Direct Dielectric Liquid Cooling 1396
    Omar Sanjakdar, Yvan Avenas, Rachelle Hanna, Guillaume Piquet Boisson, Emmanuel Marcault, Antoine Philippe

Board Level Reliability of Gull-Wing, Micro-Leaded and Lead-Less Packaged MOSFETs in Automotive Environments.. 1403
    Christopher Liu, Vijayakrishna Satyamsetti, Xuanjing Wei, Christian Radici, Peter Vines, Wayne Lawson

Cost Effective and High Noise Immunity Methodology for Aging Evaluation of DC-Link Capacitors in Traction Inverters 1408
    Seyed Hossein Aleyasin, Fausto Stella, Radu Bojoi, Enrico Vico

A 3D Structure of Single-Sided Cooling Power Module with Low Thermal Resistance and Low Inductance 1414
    Hirofumi Hisamochi, Koki Notake, Yoshiaki Takahashi, Koji Yamaguchi

Aging of Y-Capacitor in an EMI Filter and Its Impact on Common-Mode Noises 1420
    Tahmid Ibne Mannan, Seungdeog Choi, Subarto Kumar Ghosh, Md Moniruzzaman

2200A/48V-to-1V Low-Profile Direct Power Converter with Standard PCB Transformer 1427
    Alejandro Figueroa, Pablo Mazariegos, Álvaro Cobos, Javier Goicoechea, Alejandro Castro, José A. Cobos

Single-Stage 48V-to-1V Regulator with a Half-Turn Transformer and Current-Doubler Rectifier................. 1433
    Xinmiao Xu, Qiang Li

Ultra-Low-Profile Single-Stage Voltage Regulator Module (VRM) for Next-Generation AI Accelerators 1439
    Xufu Ren, Jinfeng Zhang, Zhenshuai Rong, Borong Hu, Teng Long

Novel TLVR Operation in Multi-Stage Voltage Regulator Module with Current Multipliers 1444
    Kevin Zufferli, Roberto Rizzolatti, Mario Ursino, Simone Mazzer, Gerald Deboy, Stefano Saggini

Interphase LC-Oscillation Suppression with Fast Line-Transient Response in 48-V Series-Capacitor Buck Converters for Automotive Applications 1451
    W.L. Jiang, Y. Liu, N. Khan, J. Pigott, H.J. Bergveld, V. Chaturvedi, O. Trescases

An Approach to Compensate for Low Frequency DC-Link Voltage Ripple in High Power ANPC Inverter 1459
    Shaozhe Wang, Ankit Vivek Deshpande, Rolando Sandoval, Erick Pool-Mazun, Enrique Garza-Arias, Prasad Enjeti

A Cascaded Multilevel Inverter System with Hot-Swapping and Fault Isolation Capability for Improved Resiliency 1465
    Uthandi Selvarasu, Vikram Roy Chowdhury, Shumeng Wang, Jinli Zhu, Mahshid Amirabadi, Yuan Li, Brad Lehman

Layout Optimization for Parasitic Inductance Reduction of GaN-Based NPL.X Multilevel Inverter 1473
    Ali Halawa, Jinyeong Moon, Woongkul Lee

Topology Selection and Design Methodology for SiC based Solar Photovoltaic Medium Voltage Direct Grid Connect Inverters 1481
    Jenson Joseph C. Attukadavil, Baylon G. Fernandes

EMI Modeling of PCB-Based Three-Level Active Neutral-Point-Clamped GaN Converter 1489
    Mohammad Hassan Adeli, Necmi Altin, Erkan Deniz, Adel Nasiri

A Novel Layout for Improving Current Sharing of Paralleled SiC MOSFETs with TO-247 Package 1495
    Che-Wei Chang, Matthias Spieler, Rolando Burgos, Ayman El-Refaie, Renato Amorim Torres, Dong Dong

A Sensor-Less IGBT On-State Voltage Estimation Method using Inverter Control Variables 1501
    Shuyu Ou, Subham Sahoo, Ariya Sangwongwanich, Yongjie Liu, Frede Blaabjerg

A Novel Non-Intrusive Online Monitoring Method for Diagnosing the Lift-Off of Bonding Wires in SiC MOSFETs 1507
    Keqi Song, Henry Shu-Hung Chung, Ho-Tin Tang

Optimizing MOSFET Selection for EMC-Critical Automotive Applications 1512
    Sacha J. Cazzitti, Christian Radici, Andrew J. Forsyth, Cheng Zhang, Peter Vines

Improving Dynamic Current Sharing Between Parallel MOSFETs by Optimizing Device Parameters 1519
    Kunal Jha, Kapil Kelkar, Marina Hedenik, David Penof

A 21.6 kW/L Two-Phase Immersion-Cooled Isolated DC-DC Converter 1529
    Aleksandar Ristic-Smith, Kawsar Ali, Daniel Rogers

Extraction of Common Mode Parasitic Capacitance in Balance Filter for the Prediction of EMI Noise Suppression 1537
    Qiuzhe Yang, Xingyu Chen, Zijian Wang, Qiang Li

A 660W, 96% Efficiency 3D Heterogeneously Integrated Digital DC/DC Power Module for Vertical Power Delivery 1544
Haoyu Wang, Xuliang Wang, Yan Wang, Xiaosen Liu

Planar Rogowski Coil-Based Switch Current Measurement for a 1.2 kV SiC MOSFET Embedded Die PCB 1551
Matthias Spieler, Che-Wei Chang, Ayman M. El-Refaie, Dong Dong, Rolando Burgos

Effect of Magnetic Couplings on Conducted EMI of GaN-Based PFC Converter 1557
Tyler McGrew, Qiang Li

Optically-Controlled 3.3 kV SiC MOSFET with Fast Switching Speed and Low Optical Power 1564
Xin Yang, Guannan Shi, Liyang Jin, Yuan Qin, Matthew Porter, Che-Wei Chang, Xiaoting Jia, Dong Dong, Linbo Shao, Yuhao Zhang

Optimization Techniques for Parallel-Connected Devices in IPMs for Consumer Use 1569
Keisuke Kawamoto, Haruhiko Murakami, Teruaki Nagahara, Michael Rogers, Akiko Goto, Shoji Saito, Koichiro Noguchi

Investigating the Temperature Dependency and Operating Parameters of a Self-Driving Active Gate Driver 1576
Vin Loong Choo, Martin Pfost

Use of Switched-Capacitor Circuit to Generate Negative Gate-Source Voltage Pulses 1582
Ho-Tin Tang, Henry Shu-Hung Chung

An Optically Isolated Gate Driver with Simultaneous Data and Power Transmission Through a Miniaturized, Efficient Photonic Platform 1590
Jiajun Li, Mariia Klymenko, Yanqiao Li, William Scheideler, Jason T. Stauth

Optimal Shared Energy Storage Capacity Configuration in Multi-Energy Microgrids Considering Battery Lifetime Loss based on Relaxation Techniques 1597
Hualong Liu, Wenyuan Tang

Virtual Resistance Control for an Active Battery Management System 1602
Alastair P. Thurlbeck, Ashraf Siddiquee, Mithat John Kisacikoglu, Yilmaz Sozer

Internal Voltage Source Saturation Impact on Stability Limits of Grid Forming Converter 1610
Divyanshu Bansal, Aravind G., L. Umanand

A Zero Harmonic Distortion Grid-Connected Grid-Forming Converter for Battery Energy Storage System Applications 1615
Gabriel V. Ramos, Thiago M. Parreiras, Fangzhou Zhao, Xiongfei Wang, Braz de J.C. Filho

Single Cell Energy Router Justification for Three Phase Near Zero Energy Buildings 1622
Hossein Nourollahi Hokmabad, Tala Hemmati Shahsavar, Oleksandr Matiushkin, Tanel Jalakas, Oleksandr Husev, Juri Belikov

A Multi-UAV Charging Station Enabling Free Landing by Grid Pattern Transmitter 1629
Jungho Kim, Hyunkyeong Jo, Seoktae Seo, Bonyoung Lee, Hyungki Min, Franklin Bien

Capacitor Design for Self-Resonant Coils for Long-Distance Wireless Power Transfer System 1635
Mostak Mohammad, Vandana Rallabandi, Omer C. Onar, Gui-Jia Su

A 10.4-kW High-Power-Transfer-Density Multi-MHz Capacitive Wireless Power Transfer System for EV Charging Utilizing Stacked-Inverter Stacked-Rectifier Architecture 1640

    Dheeraj Etta, Miguel Alvarez Dominguez, Sounak Maji, Syed Saeed Rashid, Khurram K. Afridi

Reduced-Fringing-Field Multi-MHz Capacitive Wireless Power Transfer System using Metasurface-Based Couplers with Active Field Cancellation 1646

    Syed Saeed Rashid, Dheeraj Etta, Matteo Ciabattoni, Francesco Monticone, Khurram K. Afridi

Living Object Detection in Wireless Power Transfer Systems using Remote Capacitive Bio-Signals Monitoring.. 1653

    Bruno M.G. Rosa, Paul D. Mitcheson

Modified N:1 Switched Capacitor Converter with Reduced Capacitor DC Bias Voltage for High Power Density 1659

    Taewoo Lee, Dam Yun, Sunghyuk Choi, Jung-Ik Ha

Wide Range Digital Control for Three-Level Buck Converters with Sensorless Flying-Cap Voltage Balancing.. 1666

    Hossein Hajisadeghian, Giovanni Bonanno

A Comparative Investigation of a New Continuous Voltage Conversion Ratio Approach in a Zero-Inductor Voltage Converter.. 1673

    Sina Salehi Dobakhshari, Aamna Nasir Hameed, Binghui He, Mojtaba Forouzesh, Yan-Fei Liu

A 96.1% Peak Efficiency, 6.8 kW/in³, 48V-to-6V On-Package Intermediate Bus Converter with LV-GaN Power Transistors.. 1681

    Mausamjeet Khatua, Nachiket Desai, Harish Krishnamurthy, Sheldon Weng, Jingshu Yu, Huong Do, Samuel Bader, Han Wui Then, Krishnan Ravichandran, James Tschanz, Kaladhar Radhakrishnan, Vivek De

A 48V to 2.4V-5V 95.8%-Peak-Efficiency 869W/in³-Power-Density Fibonacci Dual-Path Hybrid DC-DC Converter with Inductor Current Reduction and Low Output Resistance.. 1687

    Yichao Ji, Zeguo Liu, Lin Cheng

An Ultra-Fast Very Large Scale Interleaved Li-Fi Transmitter 1693

    Daniel H. Zhou, Konstantinos Manos, Minjie Chen

Isolated PWM DC-DC Converter with Single Magnetic Component, ZVS and Self-Balanced Switched-Capacitor Voltage 1701

    Pablo M. Gil, Juan Rodríguez, Diego G. Lamar

Analysis and Design of a Low-Complexity ZVS Buck-Boost Converter 1707

    Burkhard Ulrich

A High Conversion-Ratio Hybrid Series-Parallel DC-DC Converter with Pseudo-Soft-Charging and Inductor Current Frequency Multiplication.. 1715

    Avinash Maddela, Kishalay Datta, Jason T. Stauth

A Real-Time Variation Control of Deadtime in GaN-Based Bidirectional Buck-Boost Converter for Lithium-Ion Battery Formation System.. 1723

    Jong-Hun Lim, Go Woon Heo, Je-Yeong Lim, Dong Hwan Kim, Byoung Kuk Lee

A Space Vector PWM Strategy for Charging of Bootstrap Capacitor in Three-Level Neutral-Point-Clamped Inverter 1728
Anantha Hegde, Asamira Suzuki, Hirokazu Nakamura, Takamune Kabashima, Koji Higashiyama, Keiji Akamatsu

A Complementary Carrier based PWM Strategy for Average Current Sampling of Three-Phase Inverter using Single Current Sensor 1734
Byeong-Il Kim, Joon-Seok Kim, Yeongsu Bak, June-Seok Lee

Short-Circuit Ride-Through for a CRM-Based Soft-Switching Three-Phase Inverter 1741
Xingyu Chen, Gibong Son, Qiang Li

Modified Space Vector Modulation with Low Bandwidth Sensor to Reduce Losses in Soft Switching Three-Phase Inverters 1746
Md Didarul Alam, Nazmul Hassan, Iqbal Husain, Liming Liu, Hongrae Kim

A Feedforward Ripple Reduction Control Strategy based on a Hybrid GaN/Si Interleaved Inverter 1754
Mowei Lu, Jurgis Reinotas, Xiaoyang Tian, Stefan M. Goetz

IGBT Comparison for Optimized Switching Behavior in the SiC/Si-Hybrid Switch 1759
Adrian Amler, Thomas Heckel, Daniel Ruppert, Cornelius Rettner, Martin März

Forward Recovery and its Mitigation in Hybrid Si/SiC-Based DC–AC Converters 1767
Yan Zhou, Thomas Lehmeier, Adrian Amler, Martin März

Real-Time IGBT Module Ageing Characterization Through Temperature Monitoring 1774
Quirc Perez-Farre, Luis F. Gomez-Rivera, Carlos Lopez-Torres, Kai Dannehl, Antoni García-Espinosa, Alejandro Paredes-Camacho

Experimental Validation of Triangular SOA via Infrared Thermography of a MOSFET Die Operating in the Thermally Unstable Linear-Mode for Automotive Applications 1781
Yacine Ayachi Amor, Christian Radici, Kerry J. Abrams, Philip Ellis, Peter Vines, Wayne Lawson

Feasibility Study of the SuperIGBT: A Series-Connected High Voltage IGBT with a Single Gate 1786
Junhong Tong, Alex Q. Huang, Huanghaohe Zou, Zhiyuan Ma

Low Profile, Laminated Nife Transformers for Flyback Converters 1791
Xuan Wang, Reza Mounesi, Matthew Catanoso, Matthew Fox, Adel Nasiri, Mark G. Allen

Comprehensive Demonstration of New Magnetic Designs Utilizing Magnetic Anisotropy of the Cores for Integrated Magnetics 1797
Yota Takamura, Honami Nitta, Tatsuya Miyazaki, Kimito Yamanaka, Ryosuke Ishido, Akira Namba, Keisuke Fujisaki, Shigeki Nakagawa

A Two – Stage Artificial Neural Network (ANN) – based Design and Optimization of High Frequency Transformers for Dual Active Bridge Converter 1803
Lufan Zhou, Alberto Delgado Expósito, Adam Ruszczyk, Simon Round, Miroslav Vasiœ

Modeling and Optimizing Winding Arrangement for Gapped Planar Magnetics based on Artificial Neural Network 1810
Hanqing Cao, Bima Nugraha Sanusi, Ziwei Ouyang

Free-Shape Optimization of VHF Air-Core Inductors using a Constraint-Aware Genetic Algorithm 1816
Thomas Guillod, Charles R. Sullivan

Organic Direct Bonded Copper-Based Rapid Prototyping for Silicon Carbide Power Module Packaging 1824
   Shuofeng Zhao, Joshua Major, Douglas DeVoto, Sarwar Islam, Xiaoling Li, Mike Tant, Faisal Khan, Sreekant Narumanchi

Discrete Power Device Packaging with Integrated Direct Two-Phase Cooling 1832
   Jinpeng Cheng, Jinxiao Wei, Hao Feng, Li Ran

Investigation of Die Top-Side Re-Metallization for SiC-Based Double-Side Cooled Power Modules 1836
   Narayanan Rajagopal, Christina DiMarino

Design of Low Parasitic Inductance GaN HEMT Flip-Chip Power Module 1844
   Mohammad Dehan Rahman, Tanzila Akter, Abu Shahir Md Khalid Hasan, H. Alan Mantooth, Xiaoqing Song

A Scalable Dual-Orthogonal-Cooling Packaging Concept for Parallel-Series SiC Chips 1850
   Ekaterina Muravleva, Youssef Abotaleb, Blake Anderson, Zichen Zhang, Boyi Zhang, Jerry Hudgins, Jun Wang

Parasitic Impact Analysis and Design of Hybrid EMI Filter for Active Clamp Flyback SMPS 1858
   Tahmid Ibne Mannan, Seungdeog Choi, Masoud Karimi-Ghartemani

Overview of Dynamic Characterization of Switches for Three Phase Voltage Source, Current Source, and Matrix Converter Applications 1866
   Sneha Narasimhan, Sathya Rupan Thirumoorthi, Subhashish Bhattacharya

Advanced Modeling Technique of Class-E Inverter Considering Low R_{on} of eGaN FETs and Different Design Procedures 1874
   Manas Palmal, Jungwon Choi

PiezoNet and Data-Driven Models for Time-Domain Characterization of Piezoelectric Resonators 1882
   Davit Grigoryan, Mian Liao, Haoran Li, Shukai Wang, Tanuj Sen, Matthew Tan, Minjie Chen

A New Gate Charge De-Embedding Method for Accurate On-Wafer Characterization of HV MOSFET Devices 1889
   João R.R.O. Martins, Rachid Hamani, Vincent Quenette, Joerg Gessner

4 kW Auxiliary Power Module for Electric Vehicles Utilizing a Dual-Phase LLC DC-DC Converter 1892
   Mojtaba Forouzesh, Xiang Yu, Yan-Fei Liu, Paresh C. Sen

New Reverse Mode Control Method of Phase-Shift Full-Bridge Converter for Bidirectional Auxiliary Power Module 1899
   Jongyoon Chae, Dongmin Kim, Dongmin Choi, Gun-Woo Moon

In-Situ EV EIS with a High-Density Flying Capacitor Multi-Level Converter Supercapacitor System 1905
   Avram Kachura, Gaël Vergès, Samantha K. Murray, Olivier Trescases

A Novel 500-kHz LLC-T Resonant Converter with Wide Output Range 1913
   Zhengming Hou, Dong Jiao, Jih-Sheng Lai

High Efficiency Traction Drive Operation with a Partial Load Three-Phase Triangular Current Mode Modulation Concept 1919
   Bhaskar Chatterjee, Jan Allgeier, Thomas Plum, Marc Hiller

Analysis of Maximum Power Transfer Limit for Linear Operation of Dual-Active-Bridge Converters 1927

Radhika Sarda, Ezequiel Ramos Rodriguez, Gaowen Liang, Glen G. Farivar, Josep Pou, Vaisambhayana B. Sriram, Anshuman Tripathi

Enhanced Control for Integrated Active Power Decoupling in Single-Phase Three-Level Flying Capacitor PFC Converter 1935

Gleisson Balen, Cristian Blanco, Ángel Navarro-Rodríguez, Pablo García, Rafael Peña-Alzola

Improving Transient Stability of PLL-Synchronized Grid-Following Inverters 1940

Surya Prakash, Kalpana Beura, Mohamed Alkhatib, Omar Al Zaabi, Khalifa Al Hosani, Utkal Ranjan Muduli

Online Impedance-Based Analysis for Power System Stability Assessment using Transformer-Less and Filter-Less Switch-Mode Perturbation Generator 1946

Tomoya Ide, Yuko Hirase, Cheng Huang, Takanori Isobe

PIR-R Control for Three-Phase Grid-Connected Inverter with Unbalanced Grid Current Correction 1953

Haneen Ghanayem, Xingyu Yang, Mohammad Alathamneh, R.M. Nelms

Design and Placement of a Passive Clamp Snubber for Isolated SEPIC and Cuk Converters Working as Automatic Power Factor Correctors 1959

Abraham López, Juan Rodríguez, Duberney Murillo-Yarce, Javier Sebastián, Diego G. Lamar

Current Sensorless Control Strategy for Single-Phase T-Type PFC Converter 1967

Che-Yu Lu, Jia-En Zeng

Three-Phase Single-Stage Multiport AC-DC Converter with Integrated DC-DC Conversion Stages 1972

Asad Hameed, Gerry Moschopoulos

High Efficiency AC-Adapter Realized by Voltage-Clamper with Mid-Voltage AHB Converter using Synchronous Rectification 1977

Shuichiro Motoori, Akihiro Kawano, Toshiyuki Zaitsu, Riku Tatetsu, Kohei Sebata, Kazuki Miyanjou, Kimihiro Nishijima

Active Soft Switching Technique for Single Phase Series Capacitive Link Universal Rectifier 1983

Anran Wei, Brad Lehman, Mahshid Amirabadi

A Multi Mode Control Algorithm for Totem-Pole Bridgeless PFC 1990

Bosheng Sun, Sheng-Yang Yu, Amir Hussain

Protection Strategy for Flying Capacitor Totem-Pole PFC Under the AC Drop Transient 1995

Yanqing Wu, Wending Zhao, Zhenhai Zhu, Xinke Wu

Three-Phase with Three Single-Phase Single-Stage Isolated AC-DC Converters for EV Charging Station Applications 2002

Misha Kumar, Peter M. Barbosa, Juan M. Ruiz

400V SiC in Next-Generation 3-Level Flying Capacitor Bridgeless Totem-Pole PFC 2009

Rytis Beinarys, Seamus O'Driscoll

Extended Smart-Link Quasi-Single-Stage 3-Phase AC-DC Power Supply Module for AI-Driving Data Centers 2014

D. Biadene, J. Huber, J.W. Kolar, P. Mattavelli

A New Three-Phase Multi-Mode AC/DC LLC Converter with Output-Controlled Active Rectifier (with V2G and G2V Functions) for Fast DC Charging Application.. 2022
Xiaoyi Xia, John Lam

Capacitorless Notch Resonant Converters for Miniaturized LLCLC Resonant Converters in Electric Vehicle Charging Applications 2029
Haitham Kanakri, Euzeli Cipriano Dos Santos Jr., Maher Rizkalla

Multiple-Core Transformer Design based on Half-Turn Structure in Two-Stage DC-DC Converter for Battery Storage System.. 2035
Yilei Li, Bima Nugraha Sanusi, Pinhe Wang, Tianming Luo

Bidirectional DC-DC Converter Utilizing Coupled Inductors for Energy Storage System.......................... 2043
Wen-Hsuan Lee, Jiann-Fuh Chen, Hsuan Liao, Kuo Fu Liao

Comparison of 2-Level and Quasi-2-Level Topologies in a Bidirectional Isolated DC-DC Converter for MVDC Networks.. 2051
José Andrés Aguilar Croston, Jean-Yves Gauthier, Cyril Buttay, Maryam Saeedifard, Besar Asllani, Piotr Dworakowski

Sling Forward Converter for Offline Operation: Achieving High Efficiency and Wide Voltage Range Performance 2059
Nasherul Islam, Guozhu Chen, Honglei Miao, Fuxing Zhang

A Pulse Width Alternating Modulation Strategy for Three-Level Buck-Boost Converter 2066
Xinlong He, Caifeng Liu, Xudong Zou, Jiaao Zou, Tianyi Zhang, Yong Kang

ZVT Circuit Applied for Wide Input Range Isolated Converters 2070
Linguo Wang, Zhongyin Guo, Junjie Zhu, Bing Zhang, Zhiling Zuo, Xiaoguang Gao, Guangji Ma

Impact of Asymmetrical Leakage Inductance on a 380 V-12 V LLC Converter with Synchronous Rectifier for DCX Application 2075
Jinshu Lin, Shan Yin, Chen Song, Honglang Zhang, Minhai Dong, Limei Xu, Hui Li

Start-Up Techniques and Universal Closed Loop Control of Immittance Network based Resonant Converter 2082
Ripun Phukan, Misha Kumar, Randy Beckemeyer, Juan Ruiz, Peter Barbosa

Multi-Objective Efficiency-Oriented Optimization for DAB Converters Minimizing Current Stress and Backflow Power with Soft-Switching Assurance 2088
Kun Wang, Ian Laird, Jun Wang

An ISOP-PSFB PWM Converter based on Coupled Output Inductors and Phase-Shifted Modulation with Full ZVZCS Range 2096
Kang Hong, Guo Xu, Guangfu Ning, Mei Su

Design and Implementation of a GaN-Based Soft-Switched Series-Capacitor Buck Converter Operating at the CCM-DCM Boundary for High-Performance Computing Systems 2101
Ramin Rahimzadeh Khorasani, Kolman Puterman Ghitelman, Madhavan Swaminathan

Intrinsic Feedback Model for Coupled-Damped Self-Balancing of General Multiphase Hybrid Converters 2109
Haoran Xu, Weijia Hao, Desheng Zhang, Run Min, Qiaoling Tong, Xuecheng Zou

A High-Efficiency Switching Oscillation Suppression Strategy based on Damped Oscillation for Synchronous DC-DC Converter 2117
    Hao Yuan, Chuan Ni, Zhengyu Ye, Wei Lu, Hui Xue, Ting Qian

Efficient and Streamlined Demodulation Strategy for High-Frequency Talkative DC-DC Converters 2125
    Abdelmoumin Allioua, Hendrik Gockel, Gerd Griepentrog

A 90.9% Peak Efficiency KY Single-Inductor Bipolar-Output Converter with Conductance Modulation Controller for Active-Matrix Organic Light Emitting Diode Power Supply 2131
    Sheng-Han Yu, Chieh-Ju Tsai, Hao-Ran Huang, Ching-Jan Chen

Constant-on-Time Control for Zero-Bias Trans-Inductor Voltage Regulators 2138
    Hank Zeng, Justin Lee, Rixin Lai, Hang Shao

An Improved PFM Control Scheme for Three-Level Buck Converter based on Ton Extension Achieving an 810% Frequency Reduction 2143
    Yi-Chun Chang, Chieh-Ju Tsai, Ting-Lun Lee, Ching-Jan Chen

A Concept for Current Ripple or Transient Improvements in Multiphase Converters 2149
    Alexandr Ikriannikov, Alex Gao

System Solutions and Design Trade-Offs to the Input Filter Interactions with Battery Chargers 2157
    Xigen Zhou, Dan Mavencamp, Kuang-Yao Cheng

Modeling and Implementation of a Zero Bias TLVR 2162
    Lei Wang, Travis Guthrie, Peyman Asadi, Mark Alexander, Kunrong Wang, Brandon Howell

cGANET-Enhanced Voltage Gain Modeling: Elevating CLLC Converter Accuracy 2167
    Yu Zuo, Xiaobing Shen, Fanghao Tian, Jiaze Kong, Hans Wouters, Wilmar Martinez

Capacitive vs Inductive Coupling based DC-DC Converter Operating in MHz Switching Frequency Range 2173
    Saeid Pourjafar, Parham Mohseni, Oleksandr Husev, Ryszard Strzelecki, Oleksandr Matiushkin

LLC Converter Main Transformer Losses: Eliminating Air Gaps and Integrating Parallel External Inductors 2179
    Yu-Chen Liu, Shang-Syun Wu

Small-Signal Phasor Modeling of T-Type Bridge-Based Single-Sided and Double-Sided LCC Resonant Converters for WPT Applications 2194
    Aditya Zade, Shubhangi Gurudiwan, Regan Zane

A Hybrid Three-Port Topology for Urban Charging Stations 2202
    Mohammadreza Khodaparast Klidbari, Naser Souri, Zahra Sadat Habibolahi, Hamid Montazeri Hedeshi

Reconfigurable H5-Bridge based LLC-DAB Sigma Converter for EV Fast Charging Stations 2207
    Huangsheng Xu, Mingde Zhou, Qishan Pan, Haoyu Wang

A Resonant Reset Forward Converter with Ultra-High Conversion Gain using Differential Transformation Technique (DTT) 2213
    Shubham Srivastava, Mandeep S. Rana, Santanu K. Mishra

Full-Range ZVS Modulation of Switched Capacitor Converter for Sensorless Voltage Balancing 2220
    Md Tanvir Ahammed, Wensong Yu

Dimensional Parasitics Absorption in Capacitively-Isolated Æuk Converter for Medium-Voltage High Step-Down Converters 2228

Aakash Kamalapur, Jung-Soo Bae, Mark Cairnie, Rajaie Nassar, Jack Knoll, Dushan Boroyevich, Guo-Quan Lu, Christina DiMarino, Qiang Li, Khai D. T. Ngo

A 36-to-60V Input Dual-Phase 2MHz 93%-Efficiency ZVT Series-Parallel Hybrid Buck Converter using Single Auxiliary Inductor and Adaptive Time Multiplexing Control 2236

Qi Cheng, Hoi Lee

Improved Efficiency in a 10 W Class-Φ_2. Converter Utilizing a Resonant Gate Drive 2241

Malachi Hornbuckle, Katherine Liang, Juan Rivas-Davila

The Analysis and Design of a Resonant Capacitively-Isolated Cockcroft-Walton Converter 2249

Elizabeth Rabenold, Raiphy Jerez, Samantha Coday

SHSC: Non-Isolated High-Density 4:1 IBC for 48 V Applications 2254

Mario Ursino, Roberto Rizzolatti, Simone Mazzer

High-Performance Current Multiplier: A Hybrid Switched Capacitor Solution for High-Current Applications 2260

Kevin Zufferli, Roberto Rizzolatti, Mario Ursino, Simone Mazzer, Gerald Deboy, Stefano Saggini

Representation and Design Methodology for Generalized Switched-Capacitor Converter Topologies 2268

Seokwon Choi, Dam Yun, Jung-Ik Ha

A 48-V-to-1-V Gallium Nitride Switching Bus Converter for Processor Vertical Power Delivery with 2.7 mm Thickness and 3048 W/in³ Power Density 2276

Jiarui Zou, Yicheng Zhu, Nathan M. Ellis, Logan Horowitz, Robert C. N. Pilawa-Podgurski

Ripple Reduction and Efficiency Improvement of Always-Dual-Path Hybrid DC-DC Converter based on Phase Shift Operation 2284

Katsuhiro Hata, Shinsaku Tanaka, Toru Ashikaga, Yasuhiro Rikiishi

Ultralocal PQ Theory: A New Approach for Model-Free Predictive Direct Power Control of Shunt Active Power Filters 2290

Mahdi S. Mousavi, Abolfazl Nassaji, Ibrahim Harbi, Behnam Nikmaram, S. Alireza Davari, Mokhtar Aly, José Rodriguez

Symmetrical Balanced Circuit for Common-Mode Noise Mitigation in LCL-T Resonant Converter 2296

Ripun Phukan, Boyi Zhang, Juan Ruiz, Peter Barbosa

A Single-Phase Soft-Switching Buck-Boost Inverter 2303

Lukas Wipprecht, Burkhard Ulrich

Low-Complexity Model Predictive Control Method for Driving Dual Induction Motors Fed by Five-Leg Inverter 2311

Jun Young Lee, Eun Woo Lee, Dongho Choi, June-Seok Lee

Overvoltage Mitigation Filter using High-Frequency Cable Modeling in Long Transmission Lines for Silicon Carbide Inverter Systems 2317

Yun-Jin Lee, Kyo-Beum Lee

Power Delivery Network (PDN) Design and Analysis to Achieve Low Impedance in Fast Edge Rate DC-DC Converters for EMI Compliance 2322

Manraj Singh Ladhar, Sheldon Williamson

Enhancing the Performance of Dual Input Split Source Inverters using an Advanced Modulation Strategy...............2327

Mustafa Abu-Zaher, Fang Zhuo, Mokhtar Aly, Mahmoud A. Gaafar, Mohamed Orabi, José Rodriguez, Alaaeldien Hassan, Jiachen Tian, Samir Kouro

A Novel GaN-HEMT Single-Phase Single-Stage Buck-Boost Micro-Inverter Topology for PV Applications...............2332

Pengwei Li, Uiliam Kutrolli, Ali Bazzi

A Dynamic Current Sharing Method using Novel Clip Considering Mutual Inductance Coupling...............2343

Zexiang Zheng, Jianwei Lv, Yiyang Yan, Baihan Liu, Yifan Zhang, Linhao Ren, Jiaxin Liu, Cai Chen, Yong Kang, Xiong Zhang, Hao Yu, Wei Jiang

Application-Oriented Test Setup for Measuring Dynamic Output and Transfer Characteristics of GaN-HEMTs...............2348

Philipp Swoboda, Martin Fein, Simon Frank, Andreas Liske, Marc Hiller

Mitigating Gate Voltage Oscillation in Parallel SiC Power Modules for xEV...............2356

Hideo Komo, Michael Rogers, Mark Steiner, Eric Motto, Koichi Taguchi, Chihiro Kawahara, Junichi Nakashima, Yasushige Mukunoki, Seiichiro Inokuchi, Rei Yoneyama

Switching Performance Comparison of Low-Voltage GaN and Si Devices...............2361

Tianxiao Chen, Haoyang Liu, Pedro A.M. Bezerra, Eckart Hoene, Sibylle Dieckerhoff

Modeling of Switching Transients for Frequency-Domain CM EMI Analysis in Double Sided Cooling Power Modules...............2369

Sijia Liu, Liu Yang, Heng Zhang, Yifan Zhang, Zexiang Zheng, Jianwei Lv, Jiaxin Liu, Cai Chen, Yong Kang, Yuebin Zhou, Daming Wang, Shuang Zhao

Leakage Current Detection Scheme for Aging Test of 10kV SiC MOSFET Power Module...............2375

Peiyang Ding, Hong Zhang, Tianshu Yuan, Qiling Chen, Jiacheng Guo, Dingkun Ma, Peiyuan Sun, Ting Hou, Laili Wang

Physics-Informed Neural Network Approach for Early Degradation Trajectory Prediction of Power Semiconductor Modules...............2380

Jie Kong, Yi Zhang, Yichi Zhang, Lukas Wick, Frederik Lillebæk Hansen, Dao Zhou, Huai Wang

Nonlinear Output Capacitance of Bidirectional Gallium Nitride Power Switches...............2387

Michael Bosch, Jeremy Nuzzo, Dominik Koch, Mathias C.J. Weiser, Ingmar Kallfass

Novel Approach of Determining and Predicting SiC MOSFET's on Resistance from Device Case Temperature using Machine Learning...............2393

Paul Bradford, Conner Deppe, Hongjie Wang

Comparison of Static Characteristics in GaN HEMTs Across 50K to 400K Considering Diverse Techniques and Statistical Variation...............2400

Purushottam Khadka, Saumil Shivdikar, Zheyu Zhang, Tian Qiu, Ahmed Siraj

Compact Model of β-Ga$_2$O$_3$ Schottky Barrier Diode...............2407

Abu Shahir Md Khalid Hasan, Mohammad Dehan Rahman, Tanzila Akter, Md Majharul Islam, Md Maksudul Hossain, Xiaoqing Song, H. Alan Mantooth

DC-Link Capacitor Board Design for Low Parasitic Inductance...............2413

Mikayla Benson, Lifang Yi, Kangbeen Lee, Jinyeong Moon, Woongkul Lee

First Demonstration of a Gallium Oxide Power Converter 2419
    Joshua J. Piel, Elizabeth A. Sowers, Daniel M. Dryden, Thaddeus J. Asel, Adam T. Neal,
    Brenton A. Noesges, Shin Mou, Andrew J. Green

Optimized Integrated EMI Filter Design in SiC Power Modules with Terminal Inductor for Better
High-Frequency EMI Suppression 2426
    Yifan Zhang, Wenzhe Xu, Jianwei Lv, Yiyang Yan, Baihan Liu, Sijia Liu, Jiaxin Liu, Cai Chen,
    Yong Kang, Xiong Zhang, Hao Yu, Wei Jiang

Balanced Technique using Integrated Winding Coupled Inductor for High-Power Density Two-
Phase Interleaved Boost Converter 2431
    Yuta Imaeda, Jun Imaoka, Masayoshi Yamamoto, Hiroyuki Onishi

MagNetX: Foundation Neural Network Models for Simulating Power Magnetics in Transient 2438
    Shukai Wang, Hyukjae Kwon, Haoran Li, Youssef Elasser, Gyeong-Gu Kang, Daniel Zhou,
    Davit Grigoryan, Minjie Chen

Revisiting Models of Common Mode Inductors to Include the Magnetized Capacitance Effect 2446
    Rafael Bogo Portal Chagas, Marcelo Lobo Heldwein

A High Frequency Coupled Inductor with Distributed Air Gap for High Power DC-DC Converters 2453
    Muhammad Fasih Uddin, Ahmed H. Ismail, Peyman Darvish, Baher Abu Sba, Yue Zhao

High-Power Planar Transformer Design for Four-Port Converters 2461
    Arya Sadasivan, Behrooz Mirafzal

Optimal Design of Inductors with Aluminum Litz Wire for Inductive Power Transfer Systems 2468
    Jesús Acero, Claudio Carretero, Ignacio Lope, Óscar Lahuerta, José-Miguel Burdío

Analytic Design of Flat-Wire Inductors for High-Current and Compact DC-DC Converters 2474
    Sajjad Mohammadi, James L. Kirtley, Alireza Namadmalan

Insulation Dielectric Loss of High-Frequency Transformer Under Square Voltage Excitation with
Edge Oscillation 2482
    Zhanlei Liu, Lingyu Zhu, Yuntian Gao, Yongliang Dang, Cao Zhan, Shengchang Ji

Improved High-Speed Thermal Analysis based on Two-Step Simulation for High-Frequency
Transformers 2488
    Zheyuan Yi, Kai Sun, Qiang Li, Zengyang Liu

Core Material Characterization Under DC Bias Conditions 2495
    Jonas Mühlethaler, Fabrice Locher, Frédéric Mathieu, Edward Herbert

A Low-Cost Setup and Procedure for Measuring Losses in Inductors 2502
    Burkhard Ulrich

Effect of Temperature of Additively Manufactured Cores 2510
    Ken Johnson, Ali Bazzi

Extreme Temperature Permeability Engineered Soft Magnetics 2516
    Tyler W. Paplham, Alex M. Leary, Paul R. Ohodnicki Jr.

An Isolated RF Power Combining Approach with Multiple Decoupled Input Coils 2521
    Ziyang Xu, Yifan Zhao, Zhan Liu, Alex J. Hanson, Ming Liu

Simulation of a Custom Core, 15kV Isolated Gap Transformer Optimized for High Power Density 2527
    Andrew Galamb, Fei Teng, Srdjan Lukic

Low Interwinding Capacitance Design for PCB-Winding based Transformer in Self-Powered Gate Drive Power Supply for High-Voltage SiC MOSFET 2535

Yuan Zhou, Li Zhang, Yilun Chen, Tianxiang Yin, Lei Lin

Integrated 4-Level Dual-Phase Superimposed Quadratic Power Converter for High-Density Direct 48V/1V Conversion 2541

Prosenjit Ghosh, Jin Woong Kwak, Fei Zhou, D. Brian Ma

Compensation Method for Unbalance of the Multi-Channel Class E Power Amplifier using the Closed Loop Frequency Control 2547

Kyungmin Lee, Sungku Yeo

High Temperature Operation of Digital Gate Driver Integrated Into a Power Module 2551

Kazuma Saiga, Shohei Zaizen, Satoshi Nakano, Shigeru Kusunoki, Kiyoto Watabe, Katsuhiro Hata, Makoto Takamiya, Shin-Ichi Nishizawa, Wataru Saito

Evaluation Index-Based Multiphysics Coupling Model and Analysis Methodology for High-Reliable Power Supply Module 2556

Haoyu Wang, Xuliang Wang, Yan Wang, Xiaosen Liu

Electrical Characterization of Modular 3D Packaging Assembled with Compressed Metal Foams 2562

Paul Bruyere, Alexis Derbey, Betina Zynger-Capaverde, Yvan Avenas, Eric Vagnon, Jean-Luc Schanen, Jean-Michel Guichon, Omar Sanjakdar

Improvement in Short-Circuit Robustness of SiC-MOSFETs based Power Modules using Two-Level Turn-On (2LTO) 2569

Muhammad Muneeb Alam, Saad Khalid, Nisar Ahmed Khan, Ngoc Ho Tran, Sebastian Strache

GaN-Based Two Stage Point-of-Load (PoL) Converter with 2.5D Embedded Substrate Implementation 2576

Samuel Defaz, Yang Li, Fang Luo

Near-Field Coupling Mitigation of the Noise from High Voltage DC-Link Decoupling Capacitors in Voltage Source Converters 2582

Yuxuan Wu, Kushan Choksi, Samuel Defaz, Fang Luo

Advantages of Paralleling SiC MOSFETs in High-Performance Power Modules 2589

Steffen Beushausen, Dominik Alexander Ruoff, Wenqi Zhou, Karl Oberdieck

A SiC Half-Bridge Power Module based on Liquid Metal Packaging for High Performance and Low Thermal Stress 2597

Wei Mu, Ameer Janabi, Luke Shillaber, Borong Hu, Teng Long

Analysis and Modeling of Radiated EMI Considering Coupling Between Power Converter and Power Cable with LC-Type EMI Filter 2603

Qinghui Huang, Yingjie Zhang, Shuo Wang, Yirui Yang, Zhedong Ma, Yanwen Lai

Simple Prediction Method for Impacts of Switching Characteristics on EMI Noise of a Three-Phase PWM Inverter 2610

Shinobu Nagasawa, Toshiya Tadakuma, Keita Takahashi

Coaxially Nested 3.3 kV SiC MOSFET Packages with Uniform Interpackage Electric Field Distribution 2616

Jack Knoll, Mark Cairnie, Christina DiMarino

Thermal Modeling and Performance of a Bare-Die Embedded PCB for High Power Density Converters Design 2624
    Shahid Aziz Khan, Feng Zhou, Mengqi Wang, DucDung Le, Shivam Chaturvedi

Research on the Voltage Fluctuation Suppression Strategy in Weak Grid Under Pulsed Power Load Integration 2628
    Xi Chen, Jiazheng Zhang, Mingjun Bao

An Optimized Firmware-Based Cycle-by-Cycle Current Limiting Method for Power Electronic Converters in UPS 2634
    Teng Wu, Hong Liu

Frequency Stop-Band Management System for DC-DC Converters 2640
    Alessandro Bertolini, Alberto Cattani, Claudio Luise, Alessandro Gasparini

Multi-Stage Model Predictive Control with Enhanced Discrete-Time Models for Multilevel Inverters 2647
    Hoang Le, Apparao Dekka, Deepak Ronanki, Abdul R. Beig

Direct Effective Power Control (D-EPC) for LLC Resonant Converters Operating in Boost Mode using Event-Driven-Timer based Digital Controller 2654
    Yuto Yoshimura, Kenji Funatani, Kazuhiro Umetani, Toshiyuki Zaitsu, Akito Nakagaki, Masataka Ishihara, Eiji Hiraki

Mitigation Method of Resonance Between Paralleled On-Line UPS 2660
    Teng Wu, Zhenguo Huo, Shangxian Ning

An Extra-Element Small-Signal Model for a Current-Fed Resonant Dual-Active-Bridge Converter 2667
    Paolo Sbabo, Paolo Mattavelli, Giorgio Spiazzi, Andrea Petucco

Concurrent Charge Distribution and Time-Optimal Control for Unordered Single-Inductor Dual-Output Converter 2675
    Xuliang Wang, Haoyu Wang, Yang Liu, Yunxin Wang, Boran Zhang, Hongru Liu, Yan Wang, Xiaosen Liu

Circulating Current Control with Loss Reduction for Parallel Connected Inverters 2681
    Shun Endo, Takae Shimada, Masato Ando, Yuuichi Mabuchi, Masaki Miyamae, Naoki Takayama, Yohei Matsumoto, Naoto Onuma

Analysis of Power and Power Spectral Density for Quaternary Random Pulse Position Modulation 2687
    Hung-Chi Chen, Hsiang-Kai Wu, Chih-Chiang Wu

Bidirectional CLLC Converter using a Hybrid Control Method for Wide Voltage Range Applications 2692
    Jhih-Cheng Hu, Hong-Xuan Liao, Chien-Lung Liu, Wei Wang, Ming-Shi Huang

Design and Control of a High-Bandwidth Dual Active Bridge DC-DC Converter 2698
    Alper Uzum, Syed Imam Hasan, Yilmaz Sozer, Kenneth A. Loparo

Unified Model Predictive Control for DC-DC Buck Converters: From Start-Up to Steady-State Operation 2703
    Zhengchen Guo, R.M. Nelms

A Novel IPPC Method for Precise Overload Protection and Burst Mode Operation in LLC Resonant Converters 2708
    Manikanta Pallantla, Ramkumar S

An Improved Current-Sensorless Model Predictive Voltage Control for Four-Leg Voltage Source Inverters 2713
Heng Guo, Yuxin Wei, Mengmeng Jing, Wenlong Ding, Bin Duan, Chenghui Zhang

A Highly Integrable, Modular and Multi-Functional Fault Monitoring Active Gate Driver with Parallel Buffers for a Global Enhanced Reliability of Gen. 3 SiC Power MOSFETs 2718
Mathis Picot-Digoix, Léo Seugnet, Frédéric Richardeau, Jean-Marc Blaquière, Sébastien Vinnac, Thanh-Long Le, Stéphane Azzopardi

A 24 – 16 V to 0.8 – 1.2 V Merged 4-Stage Hybrid-SC-SL Converter with 96.5% Peak Efficiency and Larger Than 50% iL Reduction 2725
Chien-Hao Tseng, Cheng-Ta Chuang, Chieh-Ju Tsai, Ching-Jan Chen

Innovation Active Gate Drive Method (Named TriC3™) for MOSFET Heat Reduction and EMI 2730
Hisashi Sugie

A KY Buck-Boost Converter with Extended Ramp Control Achieving 1500% Output Variation Reduction for Smooth Mode Transition 2735
Yu-Ting Hung, Chieh-Ju Tsai, Ching-Jan Chen, Chun-Yu Hsieh

An USB Cable based Extended Conversion Range L-First Hybrid-Converter using Valley-Virtual-Inductor-Current-Mode Control with Auto-Tracking Slope Compensation Against ±50% Inductance Variation 2741
Chun-I Li, Chieh-Ju Tsai, Ching-Jan Chen

Impact of Gate Resistor Configurations on Current Balancing in Paralleled SiC MOSFETs 2746
Yifu Zhang, Shashank Karanth, Emanuel Eni

Exploring the Potential of FPGA in High-Frequency Switching DC-DC Boost Converters using Model Predictive Control 2752
Qingcheng Sui, Bangli Du, Yu Zuo, Wilmar Martinez

A 7 Bit 5A 6.7 GHz Gate-Shaping Digital Gate Driver with Burst-Sampling ADC for Iterative Switching Optimization of SiC Power MOSFETs 2757
Tobias Zekorn, Kenny Vohl, Erik Wehr, Leon Weihs, Michael Hanhart, Ralf Wunderlich, Stefan Heinen

Decentralized Interleaving of Series-Stacked DC-DC Converters via Extremum-Seeking Control 2764
Ivan Petriæ, Vignesh Iyer, Shoudong Hu, Chirayu Rajpurohit, Bailey Sauter, Milan Iliæ, Luca Corradini, Dragan Maksimoviæ

Online Dead-Time Control for Half Bridges without Preliminary Training based on Switching Transient Steepness 2772
Lukas Knappstein, Niklas Falkenberg, Martin Pfost

Impedance-Based State-of-Health Estimation for Lithium-Ion Battery Management Systems 2779
Mohammad K. Al-Smadi, Jaber A. Abu Qahouq

Stability Analysis and Resonance Damping of LC Filter-Based Voltage Source Converter with Single-Loop Voltage Control 2785
Aravind G., Divyanshu Bansal, L. Umanand

Finite Control Set Model Predictive Control Combined with Online Junction Temperature Estimation for Reliability Enhancement of Voltage Source Inverters 2790
Qiang Mu, Jiale Zhou, Zaheen Mustakin, Lucas Pereira, Babak Parkhideh, Tiefu Zhao

Framework for Dynamic Control and Operation of Power Electronics Interfaces 2797
Radha Sree Krishna Moorthy, Steven Campbell

Achieving Soft-Charging and Over 20% Input Current Ripple Reduction in a 48-to-6 V Dickson
Converter using 3-Phase Split-Phase Control 2805
*Nagesh Patle, Rose A. Abramson, Sahana Krishnan, Jiarui Zou, Robert C. N. Pilawa-
Podgurski*

Experimental Verification of Circuit-Losses Analysis-Model of DC-Output Converter Developed
using Approximated Equations from Measurement Data and Datasheet Data 2813
Ryota Kondo, Tsuyoshi Funaki

Scattering Parameter Measurement System using Probes for Surface Mount Devices Operating in
the Frequency Range from 50 kHz to 1 GHz 2821
*Ryoko Kishikawa, Masahiro Horibe, Tomokazu Shoji, Shigenori Yabuta, Toshi Ohi, Ryo
Takeda, Takamasa Arai*

Optical Transformer Design with Additional Common-Mode Noise Reduction Winding for Flyback
DC-DC Converters 2828
*Yusuke Irie, Shinichiro Eguchi, Yoichi Ishizuka, Toshiro Takeuchi, Akio Iwabuchi, Takahiro
Koga, Toshiyuki Tanaka*

Enhanced Bus Voltage Stability Through Digital Twin-Enabled Adaptive Controller Tuning 2833
Matthew Belanger, Andy Wong, Kerry Sado, Enrico Santi

Modeling and Performance Characterization of Lithium-Ion Capacitor at Different Temperature
and Voltage Values 2840
Mohammad K. Al-Smadi, Jaber A. Abu Qahouq, Sajad Saberi

Conveniently Identify Coils in Inductive Power Transfer System using Machine Learning 2846
*Yifan Zhao, Mowei Lu, Ting Chen, Heyuan Li, Xiang Gao, Zhenbin Zhang, Minfan Fu, Stefan
M. Goetz*

Accurate Modeling of LLC Resonant Converters with Enhanced Analytical Approach Considering
of Parasitic Capacitance 2851
Dong Jiao, Zhengming Hou, Jih-Sheng Lai

High-Frequency Conditioning Circuits for Power-Related Information Extraction in Non-
Sinusoidal Power Electronic Systems 2857
Haoyu Wang, Yuanxin Zhang, Di Mou, Alex Hanson, Shiqi Ji

Transconductance Model of the Dual Active Bridge Converter Under Single and Dual Phase Shift
Control 2865
Jared Cronin, Andrew Wunderlich, Enrico Santi

Lumped Parameter Modeling for Real-Time Thermal Regulation of Li-Ion Battery Packs 2871
Utkal Ranjan Muduli, Mohamed Shawky El Moursi, Khalifa Al Hosani, Ahmed Al-Durra

A Physics-Based Temperature Dependent Analytical Model for 2DEG Density in AlGaN/GaN
HEMT Devices 2877
Kashfia Tajmim Nabila, Jerry L. Hudgins

Comparative Analysis of Stator-PM Machines: Design Optimization and Electromagnetic
Performance Evaluation 2883
Maryam Salehi, Madhav Manjrekar

Elimination of Deadtime Effect on Resolver Offset Estimation using the Pulsating Current Command for Electric Vehicle Application 2889
Yingfeng Ji, Nurani Chandrasekhar

A Generic Load Emulator for Testing Motor Drives of E-Mobility 2894
Qingzheng Zhang, Kaiyuan Feng, Changsheng Hu, Dehong Xu

Design and Implementation of Power Assisted Control System for E-Bikes 2900
Che-Yu Lu, Tzu-Ping Cheng

A Hybrid PWM Strategy with Reduced Common-Mode Voltage and Extended Output Voltage Linearity for Adjustable Speed Drives 2907
Zhe Zhang, Kevin Lee

Single-Phase Open-Circuit Fault-Tolerant Control of Three-Phase PMSM Drives 2913
Yuichiro Minato, Yuki Nakata, Jun-Ichi Itoh

Multi-Vendor Encoder Position Sensing Interface using Programmable IP based Solution 2920
Rajul Bhambay, Dhaval Khandla, Pratheesh Gangadhar, Thomas Leyrer, Achala Ram, Manoj Koppolu, Archit Dev

Sensorless Control Method at Low-Speed Range using High-Frequency Voltage Injection for Synchronous Reluctance Motors Considering to Nonlinear Characteristic Due to Magnetic Saturation 2924
Sota Takizawa, Sari Maekawa

Hybrid Control Scheme for Permanent Magnet Gear Motor 2932
Bing Li, Takayoshi Matsuo, Ahmed Sayed-Ahmed, Yujia Cui, Jiangang Hu

Cost-Effective Fault Diagnosis for Motor and Inverter using Bootstrap Charging and Single DC Link Current Sensor 2937
Gyu Cheol Lim, Won Hyo Jeong, Kahyun Lee, Jung-Ik Ha

Improved PWM to Suppress Motor Overvoltage Caused by Voltage Reflection 2943
Sung-Oh Kim, Kyo-Beum Lee

Analysis of Double Pulsing Effect in Motor Drives based on Vector Diagram 2948
Byeong-Woo Kang, Kyo-Beum Lee

A Novel Speed Sensor-Less Control of a Solar-Powered PMSM Drive 2953
Abirami Kalathy, Arpan Laha, Praveen Jain, Majid Pahlevani

Design of a Compact Low-Loss MMC Double Submodule for MVDC and HVDC Applications 2960
Ali Sharaf Addin, Rainer Marquardt, Thomas Brückner

A Series-Type Dynamic Voltage Restorer Control Strategy to Cope with Voltage Swell 2968
Jiazheng Zhang, Hongyu Chen, Xi Chen, Mingjun Bao

Machine Learning Approach for Accurate Lithium-Ion Battery Temperature Prediction using Electrochemical Features Independent of Battery SOC and SOH 2973
Vincent Masabiar Tingbari, Oluwaseun Isaiah Ekuewa, Anshul Nagar, Asad Abbas, Jamil Umar, Yuxin Zhang, Woonki Na, Jonghoon Kim

A Battery Strings Circulating Current Blocking Method for Battery Energy Storage Systems 2981
Haihong Long, Ziang Sun, Yucheng Fan, Xin Wu, Dehong Xu

A Hybrid Multilevel Converter-Based High-Gain Isolated DC/DC Converter for Grid-Tied Energy Storage Applications.. 2986

    Pengyu Fu, Yizhou Cong, Jin Wang, Anant Agarwal

LCL Filter Parameter Selection using Graphical Method for a 13.8 kV ac 1.1 MVA 7-Level Flying Capacitor Grid-Connected Converter Utilizing Variable Switching Frequency.. 2992

    Arthur Mendes, David Nam, Mingze Gao, Thimothy Thacker, Dong Dong, Rolando Burgos

Online Extraction of Electrochemical Impedance Spectroscopy Pattern based on EV Load Profile and Short Time Fourier Transform for Diagnosis of Lithium-Ion Battery Safety 3000

    Miyoung Lee, Dongcheol Lee, Youngmin Bae, Jongchan An, Garam Yang, Woonki Na, Jonghoon Kim

Enhanced Incremental Capacity Analysis for Evaluating Battery Degradation Mechanisms of Optimized Fast Charging Methods... 3006

    Taehyeon Gong, Jaehyeong Lee, Sungjun Lee, Yura Kim, Bomyeong Ko, Woonki Na, Sungjin Choi, Jonghoon Kim

Co-Estimation of SOC and SOT in Lithium-Ion Batteries using an RLS-Based Heat Generation Model 3012

    Seongkyu Lee, Eunjin Kang, Minhyeok Kim, Seunghyun Lee, Minwoo Song, Jaea Lee, Woonki Na, Jonghoon Kim

Three-Stage Adaptive Control Strategy for Stability Improvement of Grid-Connected Inverter in Weak Grid... 3018

    Longxiang You, Sicong Jin, Xin Zhang, Zuoshuai Wang, Sunqing Wang

Degradation Analysis of Offshore Bifacial PV Modules Under Multiple Climatic Stressors 3024

    Aidha Muhammad Ajmal, Yongheng Yang

A Flexible Energy Management System for Solar Powered Electric-Bus Charging Stations 3030

    Supun Amarathunga, Pasan Gunawardena, Xiaoting Wang, Yunwei Li

A Vienna Rectifier based Grid-Connected Powertrain for Hydrokinetic Turbine Systems 3036

    Peidong Li, Md Tariquzzaman, Yue Cao

Condition Monitoring for DC-Link Capacitors and PV Arrays based on the Start-Up Process of the PV System 3042

    Yongjie Liu, Ariya Sangwongwanich, Chen Liu, Xing Wei, Shuyu Ou, Tamás Kerekes, Jiahong Liu, Huai Wang

Electrically and Thermally Efficient Reliable Power Converter Design for Micro–Hydrokinetic Turbine 3048

    Md Tariquzzaman, Peidong Li, Yue Cao

Comprehensive Evaluation of Cyber Attacks on Grid-Connected Smart Inverters.................................... 3054

    Rishabh Singla, Vishwam Raval, Hasan Ibrahim, Jaewon Kim, Prasad Enjeti, Narsimha Reddy

Parallel Operation of Grid-Forming Converters based on Kuramoto Oscillators with Virtual Cable Emulation for Improved Power Sharing... 3059

    Vikram Roy Chowdhury, Gab-Su Seo, Barry Mather

Enhancing Hydrogen Production in Hybrid Standalone Microgrids... 3064

    Utkal Ranjan Muduli, Mohamed Shawky El Moursi, Khalifa Al Hosani, Ahmed Al-Durra

LSTM-Based Sub-Synchronous Oscillation Detection Scheme for Type 4 Wind Farm Interfaced with Weak AC Grid 3071
Omar Abu-Rub, Muhammad F. Umar, Jana A. Sheikh Ali, Yazan Qiblawey, Abdulrahman Alassi, Maryam Saeedifard, Mohammad B. Shadmand

A Study of Module Design Method to Suppress the Oscillation Occurs Between Parallel-Connected Power Devices 3077
Shinji Yato, Hiroto Sakai, Hideo Araki, Shumei Shimosako

A High-Efficient Hybrid Traction Inverter in Electric Vehicle Applications 3083
Yousefreza Jafarian, Omid Salari, Praveen Jain, Alireza Bakhshai, Mohamed Z. Youssef

Dual-Use of Onboard Chargers to Achieve Controllable DC Bus Voltage for Electric Vehicles 3089
Anuj Maheshwari, Elie Libbos, Arijit Banerjee

Isolated Single-Phase Onboard Chargers for BEV/PHEV using Active Power Decoupling Technology 3096
Yoshiki Amano, Keigo Nishimura, Hiroaki Matsumori, Takashi Kosaka, Kenichi Nagayoshi, Kenichi Watanabe

A Practical Use of xEVCap: The Modular and Standard DC-Link Capacitor Solution for the Main EV Powertrain Inverter 3100
David Olalla, Tomas Wagner, Fernando Rodriguez, Alberto Espinar

Optimized Bidirectional On-Board Charger using a Novel Unfolder-DAB Topology 3109
Héctor Sarnago, Ignacio Álvarez, Pablo Briz, Óscar Lucía

Critical Thermal Characterization of Next-Generation Solid-State Batteries for Automotive Battery Management Systems 3114
Chandan Chetri, Sheldon Williamson

Nanocrystalline CMC Inductors for EV Charging: Trade Studies and Testing Standardization 3119
Christopher Bracken, Mark A. Juds, Paul R. Ohodnicki, Bharadwaj Reddy Andapally, Jose Gato

Predicting Efficiency of On-Board and Off-Board EV Charging Systems using Machine Learning 3124
Mohamed Yasko, Fanghao Tian, Wilmar Martinez, Johan Driesen

High-Power and High-Speed Multi-Channel VCSEL Arrays with GaN Driver for Automotive LiDAR 3129
Yifu Liu, Sichao Li, Junlei He, Changyu Hu, Bill He, Karthik Krishnamurthy, Andy Shen

Double Pulse Test Platform for Hybrid SiC-IGBT Switch Characterization and Optimal Gate Control Strategy for EV Traction Inverters 3133
Rosario Attanasio, Harsha Ademane, Ryan Satterlee, Gianni Vitale

Critical Role of Individual Cell Temperature Monitoring in Mitigating Thermal Runaway and Reducing Accelerated Degradation in Lithium-Ion Batteries 3141
Mohit Sharma, Akash Samanta, William Locke, Sheldon Williamson

Loss-Optimized Design of a Triple Active Bridge DC-DC Converter for an Electric Vehicle Application 3147
Sreejith Chakkalakkal, Kyle Kozielski, Wesam Taha, Yicheng Wang, Aniket Anand, Ali Emadi

A Magnetic-Less DC/DC Converter with Pulse Charging for 800 V Powertrains from 400 V DC Fast Chargers 3155
Duc Dung Le, Shivam Chaturvedi, Shahid Aziz Khan, Mengqi Wang, Mohamed Elshaer

Boosting Charger Efficiency: A GaN-Based Flyback Converter with Energy Recycling 3160
    Ahmad Nabizadah, Majid Ghasemi Korrani, Babak Fahimi

A Hybrid Three-Level Buck Converter with Flying Supercapacitor for High Load Current Surge
Capability using Peak Current Mode Control 3167
    *Finlay Lodge, Rafael Peña-Alzola, Martin MacFadyen, Patrick Norman, Mark Sweet,
    Graeme Burt*

Supercritical Carbon Dioxide (sCO.)-Cooled Current Source Inverter-based Integrated Motor Drive
for MW-Scale Electric Aviation Applications 3174
    *Hang Dai, John Yagielski, Thomas Jahns, Kum-Kang Huh, Vandana Rallabandi, Libing
    Wang, Tarak Saha, Wenda Feng, Bulent Sarlioglu*

The Challenge of Thermal Runaway in Soft Magnetic Materials for Inductive Power Transfer 3181
    Yibo Wang, Ben Zhang, Weisheng Guo, Tianlu Ma, Sheng Ren, C.Q. Jiang

A Capacitively Coupled Alternative Electric Field Control for Freeze-Free based High Quality
Food Preservation 3187
    Jaeyong Cho, Junhyeong Park, Sung-Bum Park, Daehyun Kim, Jinsoo Choi

The Characteristics of the Long Length Primary Loop and the Power Supply for the SCMaglev's
DWPT System 3194
    Keisuke Yamamoto, Jun Enomoto, Shunsaku Koga, Junichi Kitano

A Wireless EV Charging System with a Double-Sided LCC Network using Variable Switching
Frequency and DC-Link Voltage Control 3200
    Chae-Lyn Kim, Hyeonu Jo, Ju-A Lee, Dong Hyeon Sim, Byoung Kuk Lee

Class E/EF Inductive Power Transfer to Achieve Stable Output Under Variable Low Coupling 3206
    Yifan Zhao, Mowei Lu, Heyuan Li, Zhenbin Zhang, Minfan Fu, Stefan M. Goetz

A Motorized Air-Core Variable Inductance Winding Structure 3212
    Xindong Li, Sampath Jayalath, Cheng Zhang

Wireless Power Transfer System with Automatic Tuning Capability in Metallic Environment 3220
    *Renjie Zhang, Yue Wu, Delin Zhao, Yaohua Li, Yongbin Jiang, Yi Tang, Huan Yuan, Xiaohua
    Wang, Mingzhe Rong*

Design of Wireless Power Transmitters for Enhanced Transmission Distance and Output Power 3227
    Kaiyuan Wang, Shuang Zhao, Shuye Shang, Eric Ka-Wai Cheng, Siew-Chong Tan, Yun Yang

Optimization of Wireless Power Transfer Waveforms and In-Vivo Receivers for Implantable
Medical Devices 3232
    Hanbing Liu, Xin Zan

Comparison of Compact Power Amplifier Designs for High Frequency Resonant Wireless Power
Transfer Systems at 6.78 MHz using High-Q Resonators 3241
    Manuel Rueß, Kilian Müller, Mathias C.J. Weiser, Ingmar Kallfass

Analysis and Design of Capacitive Coupling Wireless Power Transfer System using Load-
Independent Class-EF Inverter 3248
    *Takumi Kobayashi, Yutaro Komiyama, Akihiro Konishi, Hiroaki Ota, Yuki Ito, Taichi Mishima,
    Takeshi Uematsu, Kien Nguyen, Hiroo Sekiya*

Design and Optimization of a 600 W Wireless Drone Charger for High Gravimetric Power Density 3253
    Arka Basu, Daniel Costinett

Stabilization Method for DC-Bus Oscillation in Dynamic Wireless Power Transfer Systems...... 3261
Yuki Ochiai, Keisuke Kusaka

Unveiling Aliasing Effect on Resonant Pole Locations in Wireless Battery Chargers 3267
Anwesha Mukhopadhyay, Daniel Costinett

Integrated Hybrid Inductive and Capacitive Power Transfer System with Asymmetrical PCB Self-Resonator... 3275
Yao Wang, Zhen Sun, Xiangrong Zhang, Yun Yang, Shu Yuen Ron Hui

High Frequency Noise Reduction Method of the Class E Power Amplifier.................................... 3281
Kyungmin Lee, Sungku Yeo

Single-Stage Three-Phase Buck-Matrix Rectifier with Series-Parallel Connected Transformers for High-Power 48 V Data Center Power Supplies... 3285
Yuki Ishikura, Chinmay Bhagat

Sector Transition PWM Modulation Scheme for a Three-Phase Isolated Buck-Matrix Rectifier.................. 3291
Chinmay Bhagat, Yuki Ishikura

Adaptive Capacitance Circuit for Optimal Dynamic Impedance Matching in Variable Reluctance Energy Harvesting Applications 3298
Alejandro Redondo, Fernando Pérez, Sofia García, Gabriel Mujica, Airán Francés

Gallium Nitride (GaN) based Topology Comparison for Low Power Battery Charging Applications 3304
Jai Aditya Chaudhary, Rosario Attanasio, Gianni Vitale

Server Motherboard Power Performance Study Under Immersion Cooling Environment........................... 3312
Meng Wang, Haiyan Wang, Pavan Kumar, Haijin Zhang, Xiang Li, Fengwei Bian, Jianting Deng, Jiaqi Zhu, Yiming Lei

Practical PCB Design Considerations for GaN HEMTs based Isolated DC-DC Converter.......................... 3316
Gaureej Gauttam, Harish S. Krishnamoorthy, Sai Sushma Pasupuleti

Data-Driven Characterization and Forecasting of Metal-Oxide Varistor Degradation in DC Circuit Breakers.. 3321
Zhi Jin Zhang, Yang Liu, Lukas Graber, Maryam Saeedifard

A Thyristor-Based Fault Current Bypass Solid-State Circuit Breaker for DC Microgrid Applications 3328
Jiale Zhou, Xiuhu Sun, Qiang Mu, Tiefu Zhao

Single-Stage Three-Phase AC-AC Isolated Inertialess Converter (IIC) for Industrial Drives...................... 3334
Brad Houska, Decheng Yan, Aniruddh Marellapudi, Satish Belkhode, Joseph Benzaquen Sune, Deepak Divan

Author Index

Wireless Actuation of Magnetic Robots with a Modular 60 mT 3-D Helmholtz Coil System

Konstantinos Manos, Yifan Rao, Tuo Zhao, Kevin Liu, Daniel Zhou, Calvin Nguyen, Eric Chen,
Glaucio H. Paulino, and Minjie Chen

Princeton University, Princeton, NJ, United States
Email: {km4382, minjie}@princeton.edu

Abstract—This paper presents the modeling and design of a three-dimensional (3-D) Helmholtz coil system for wireless control of magnetically actuated robots. Instead of using moving permanent magnets to generate the external magnetic field needed for magnetic actuation, driving a set of stationary Helmholtz coils with power electronics offers enhanced control flexibility and precision through active feedback. A modular drive is developed to operate the custom pulsed 15 kW 60 mT 3-D Helmholtz coil setup. A method for modeling the generated magnetic field is developed to enable the precise motion control of untethered robots. The prototype system is verified by precisely actuating a magnetized cubic robot.

Index Terms—3-D Helmholtz coil system, modular drive, wireless actuation, magnetically actuated robot

I. INTRODUCTION

THE inclusion of magnetic dipole elements in a robotic structure enables its remote manipulation under external magnetic fields. The dipoles tend to align with the applied field, generating torques, while magnetic field gradients produce forces that induce translational motion. When a dipole of magnetic moment $\mathbf{m}$ is subjected to an externally imposed flux density $\mathbf{b}$, the torque and the translational force exerted on the dipole are given [1] by

$$\boldsymbol{\tau} = \mathbf{m} \times \mathbf{b}, \tag{1}$$

$$\mathbf{f} = (\mathbf{m} \cdot \nabla)\mathbf{b}. \tag{2}$$

By leveraging the aforementioned interaction between the dipole elements embedded in the robot and the external magnetic field, sophisticated robotic motion patterns can be realized.

Owing to their simplicity – stemming from the absence of complex moving parts, external wires, and an onboard energy source – magnetically actuated robots are well-suited for critical, confined-space applications such as biomedical settings and harsh environments. These robots can take various forms, ranging from digestible capsules [2]–[4] to sophisticated robotic arms [5], [6]. While some works have relied on moving permanent magnets for generating the required external magnetic field [2], [5], this approach entails limited control speed and modulation capability and it increases the system complexity requiring an additional robot to move the permanent magnet.

In this paper we focus on using a set of stationary current-carrying coils for generating the required magnetic field. A

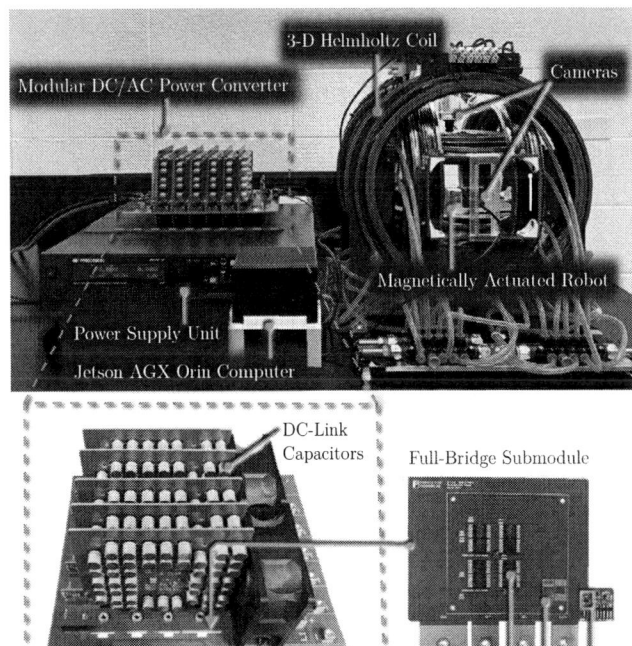

Fig. 1. Wireless magnetic robotic actuation system comprising a dc power source, a scalable power converter, and a set of copper coils. Cameras are mounted around the coil to monitor the robot motion. The video data is analyzed in real time on an NVIDIA Jetson AGX Orin computer.

modular drive system is designed for controlling the coil currents to the desired values resulting in the precise actuation of a magnetically actuated robot. A complete wireless actuation setup is presented in Fig. 1. Such a system enables fully untethered operation of the magnetic robots. Rapid and precise control of the magnetic field enables the robot to perform sophisticated tasks with high agility and a large number of control degrees of freedom.

This paper is organized as follows: Section II introduces the 3-D Helmholtz coil structure and presents an analytical model enabling the mapping between the flux density and the coil currents in 3-D space and the sizing of the coil system. It then elaborates on the design of a modular dc-ac power

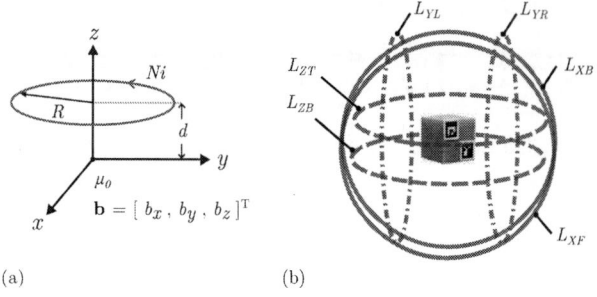

Fig. 2. (a) A filamentary circular loop of radius R carrying a current Ni placed at an offset d from the xy plane is used to model each winding of the coil system; (b) 3-D Helmholtz coil structure made of 6 individual circular windings. The first subscript indicates the axial direction, while the second denotes the relative position.: left/right (L/R), front/back (F/B), and top/bottom (T/B). Same-axis windings are identical.

converter used for the precise control of the individual winding currents. Section III presents experimental results highlighting the field generation capabilities of the setup and demonstrates the locomotion of a cubic robot as an example of magnetic robot actuation. Finally, Section IV concludes this paper.

II. SYSTEM MODELING AND DESIGN

To precisely control a magnetically actuated robot, the mapping between the coil currents and the magnetic field at the robot location needs to be determined first. The individual winding currents are then independently controlled by a drive system maintaining the desired reference values irrespective of the operating conditions.

A. 3-D Helmholtz Coil System

The fundamental building block of the coil system is a circular current-carrying winding. We model an N-turn winding by an equivalent filamentary circular loop of radius R corresponding to the mean radius of the winding carrying a current Ni, as shown in Fig. 2-(a). To maximize the robot actuation freedom, the magnetic field needs to be controlled in all 3 dimensions. A pair of coaxial loops, known as a Helmholtz coil system, can generate either an approximately uniform or a high-gradient field along the axial direction when the current directions are equal or opposite respectively. Arranging 3 such pairs perpendicular to each other, as illustrated in Fig. 2-(b), results in a coil system that allows control of the field in any direction. Due to symmetry, the windings of a coaxial pair should be identical and placed equally away from the origin. However, the windings of different axes need not have the same radii, number turns, and offset from the origin. In fact, it is convenient that their geometrical characteristics are different so that a compact 3-D coil system can be formed.

The flux density $\mathbf{b}(x, y, z)$ at any point $P(x, y, z)$ in a 3-D space produced by the circular loop of Fig. 2-(a) can be

expressed [7] as

$$b_x = \frac{C \cdot x \cdot (z - d)}{2 \cdot A^2 \cdot B \cdot \rho^2} \cdot \left[(R^2 + r^2) \cdot E(k^2) - A^2 \cdot K(k^2) \right],$$
(3)

$$b_z = \frac{C}{2 \cdot A^2 \cdot B} \cdot \left[(R^2 - r^2) \cdot E(k^2) + A^2 \cdot K(k^2) \right],$$
(4)

$$b_y = \frac{y}{x} \cdot b_x,$$
(5)

where R is the radius of the loop and

$$\rho = \sqrt{x^2 + y^2},$$
$$r^2 = x^2 + y^2 + (z - d)^2,$$
$$k^2 = 1 - \frac{A^2}{B^2},$$
$$A^2 = R^2 + r^2 - 2R\rho,$$
$$B^2 = R^2 + r^2 + 2R\rho,$$
$$C = \frac{1}{\pi} \cdot \mu_0 \cdot N \cdot i,$$

where μ_0 is the permeability of free space. The functions $K(m)$ and $E(m)$ are the complete elliptic integrals of the 1st and 2nd kinds respectively, defined [8] as

$$K(m) = \int_0^{\frac{\pi}{2}} \frac{1}{\sqrt{1 - m \sin^2(w)}} dw,$$

$$E(m) = \int_0^{\frac{\pi}{2}} \sqrt{1 - m \sin^2(w)} \, dw.$$

The magnetic field produced by the entire 3-D Helmholtz coil system can be expressed as the sum of the fields produced by the individual coils. This is because air is a linear medium and the superposition principle holds. While (3), (4), and (5) describe the flux density produced by a loop in the axial z direction, the field generated by coils positioned along the x and y axes can be easily found by swapping the spatial variables x, y, z. Noting from (3), (4), (5) that the flux density is a linear function of the loop current, the field $\mathbf{b}(x, y, z)$ due to all six windings is a linear combination of the six driving currents and it can be expressed as

$$\mathbf{b}(x, y, z) = \mathbf{A}(x, y, z) \cdot \mathbf{i},$$
(6)

where $\mathbf{i}$ is the current vector

$$\mathbf{i} = \begin{bmatrix} i_{XF} & i_{XB} & i_{YL} & i_{YR} & i_{ZT} & i_{ZB} \end{bmatrix}^T.$$

The matrix $\mathbf{A}$ is a function of the point of interest $P(x, y, z)$ and of the geometrical properties of the coils as dictated by the field equations (3), (4), and (5). Each column of $\mathbf{A}$ represents the magnetic field when the current of the respective coil is unity and all other coils are turned off. Eq. (6) allows to map the coil driving currents to the magnetic flux density at any point in free space.

When dimensioning a coil design, a simpler expression for the field along the axial direction can be useful. Setting $x = y = 0$, it is $b_x = b_y = 0$ due to symmetry and (5) becomes

$$b_z = \frac{1}{2}\mu_0 Ni \cdot \frac{R^2}{[R^2 + (z - d)^2]^{3/2}}.$$

TABLE I
COIL-SYSTEM CHARACTERISTICS

Wdg*	N	R [cm]	d [cm]	$R_{cu}^{\dagger}$ [mΩ]	L [mH]	$L_s^{\ddagger}$ [mH]
X_o	56	15.2	9.7	154.4	1.71	6.70
X_i	64	16.5	7.2	188.3	2.38	
Y_o	48	12.2	8.5	107.1	0.94	3.78
Y_i	54	13.5	6.3	133.5	1.35	
Z_o	40	10.0	6.6	74.8	0.50	1.78
Z_i	40	11.1	4.8	79.5	0.54	

*subscripts o and i for the outer and inner windings respectively
†at room temperature 25°C
‡measured series inductance reflecting magnetic coupling

The flux density due to a pair of identical coils carrying current i is thus

$$b_z = \frac{\mu_0 N i R^2}{2} \cdot \left[\frac{1}{[R^2 + (z-d)^2]^{3/2}} + \frac{1}{[R^2 + (z+d)^2]^{3/2}} \right], \quad (7)$$

and the field at the origin is

$$b_z = \frac{\mu_0 N i R^2}{(R^2 + d^2)^{3/2}}. \quad (8)$$

Each loop of the coil system of Fig. 1 is made of 10 AWG square copper magnet wire and it is composed of 2 individual windings, an inner one and an outer one that are connected in series. The characteristics of the system are summarized in Table I.

Applying the field equations (3), (4), (5) for the specific coil system yields the following matrix $\mathbf{A}$ in [mT/A] at the origin:

$$\mathbf{A}(0,0,0) = \begin{bmatrix} 0.326 & 0.326 & 0 & 0 & 0 & 0 \\ 0 & 0 & 0.322 & 0.322 & 0 & 0 \\ 0 & 0 & 0 & 0 & 0.322 & 0.322 \end{bmatrix}. \quad (9)$$

The maximum possible flux density is practically limited by the ohmic resistance of the windings. While the field scales linearly with the driving current, the heat dissipation has a quadratic dependence. Fig. 3 shows the calculated magnetic field at the origin as a function of the coil input-power requirement at 25°C when both coils of each coaxial pair are driven with the same current.

B. Drive System

The inherent current-filtering nature of the coils allows the windings to be energized by a pulsed voltage source. A modular dc-ac converter topology is implemented, as shown in Fig. 4, to precisely control the currents of the individual windings. The coils are driven independently by six full-bridge submodules operating from a common dc bus.

The implemented power converter, shown in Fig. 1, employs 100 V top-cooled Si power MOSFETs switching at 10 kHz. It can operate at a maximum input voltage of 80 V, supplying a current up to 100 A per submodule for robotic control. The main components of the drive system are presented in Table II. To reduce the conduction loss which is dominant at this switching frequency, each switch is implemented by connecting two power MOSFETs in parallel. A copper heat sink is mounted on the top side of each submodule and two

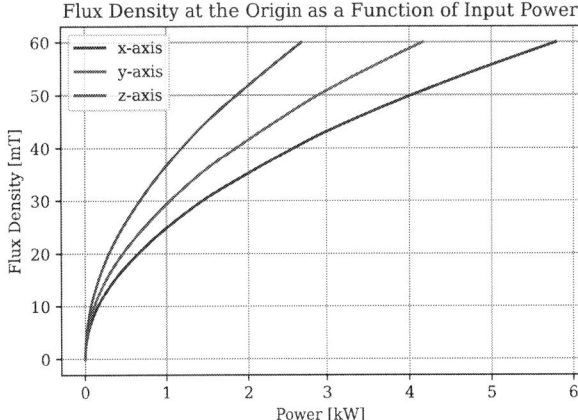

Fig. 3. Flux density at origin for the x, y, and z axis coil pairs as a function of the coil input power. Higher radius coils require more power to produce the same magnetic field.

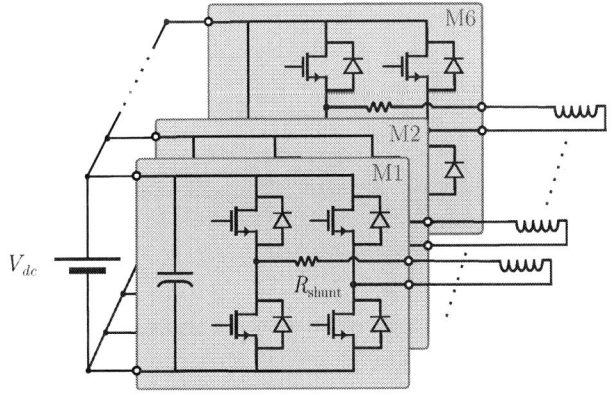

Fig. 4. Modular power-converter topology consisting of six full-bridge submodules to precisely control the current of each winding. The winding currents are sensed locally at each submodule using shunt resistors. A closed loop current-control scheme regulates the currents to the desired values.

fans provide forced convection cooling for the entire six-unit stack, as shown in Fig. 5. A single-input half-bridge gate driver with programmable dead time and a bootstrap supply circuit is used for simplicity and minimal PWM signal count requirement from the control hardware. An input dc-bus capacitance of 6.2 mF is evenly distributed among the submodules by using surface mount solid polymer aluminum capacitors. To reduce the harmonic content of the magnetic field, a unipolar PWM switching scheme [9] was implemented resulting in an effective output switching frequency of 20 kHz.

To ensure a predictable transient response and to keep the coil currents stable at the desired setpoint irrespective of the temperature conditions affecting the resistance of the copper windings, a proportional-integral current-control scheme is

TABLE II
POWER CONVERTER COMPONENTS

Component	Description
Power Switch	Infineon IPTC014N10NM5, 100 V
Gate Driver	TI LM5106 with bootstrap supply
DC-link Capacitor	Kemet A767MU226M2A, 22 μF, 100 V
Shunt Resistor	Bourns CSS2H-5930K-1L00FE, 1 mΩ
Current Sense Amplifier	TI INA241A2IDDFR
Control Hardware	TMS320F2837 control card

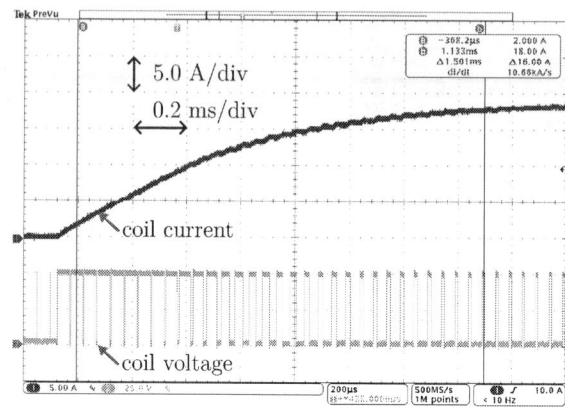

Fig. 6. Current step response from 0 to 20 A (blue) and pulse-width-modulated voltage (cyan) for $t_r = 1.5$ ms and $V_{dc} = 48$ V.

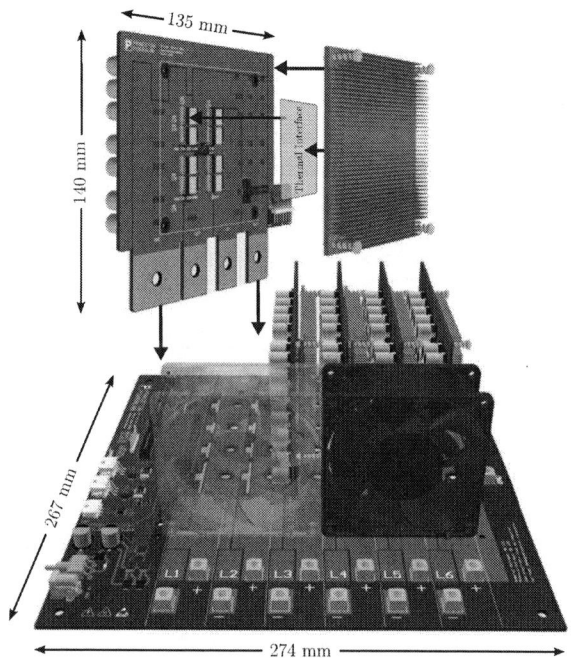

Fig. 5. The proposed modular drive is scalable allowing the use of an arbitrary number of full-bridge submodules to meet the coil-system requirements.

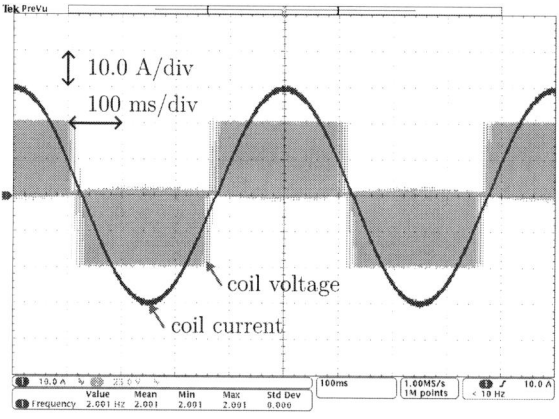

Fig. 7. Winding current (blue) and pulse-width-modulated voltage (cyan) for the Helmholtz coils. The system is capable of modulating an arbitrary signal from dc to 25 Hz with a maximum di/dt of 10 A/ms.

implemented. Selecting the controller gains as

$$k_p = \frac{L_s}{t_r} \ln 9, \qquad (10)$$

$$k_i = \frac{R_s}{t_r} \ln 9, \qquad (11)$$

results in a first order reference-to-output system with a rise time t_r. The rise time corresponds to the time it takes for the current to go from 10% to 90% of its final value. In the expressions above, L_s and R_s are the equivalent series inductance and the resistance of each inner-outer coil pair, as provided in Table I.

III. EXPERIMENTAL RESULTS & APPLICATIONS

The current tracking performance of the drive is first verified by triggering a step response, as shown in Fig. 6, where a rise time of 1.5 ms was chosen. An arbitrary current waveform can subsequently be generated by providing a sequence of reference values to the controller, as illustrated in Fig. 7.

The wireless actuation capability of the system is demonstrated by precisely controlling the motion of simple cubic robot structure. A nickel-plated N52-grade neodymium permanent magnet is embedded at the center of a 3D-printed block, as shown in Fig. 8. The robot motion is monitored in real time by motion capture cameras (e-con Systems NileCAM25CUOAGX) mounted around the Helmholtz-coil structure. Fiducial markers [10] are attached on the sides of the cubic block and a script employing the OpenCV [11] library runs in real time on a Jetson AGX Orin computer to detect the robot position and orientation. The parameters of the robotic structure are summarized in Table III.

When the same-axis coils are energized with identical current values, an approximately uniform magnetic field is generated inside the coil system. By manipulating the direction and magnitude of the flux density, the torque exerted on the magnetized robot can be precisely controlled, resulting in a well defined motion. This is illustrated in Fig. 9, where by imposing appropriate current waveforms on the coils, the robot can move by rolling in any direction as the induced electromagnetic torque, given by (1), tends to align the dipole moment **m** with the externally imposed magnetic field **b**.

979-8-3315-1612-3/25 $31.00 © 2025 IEEE

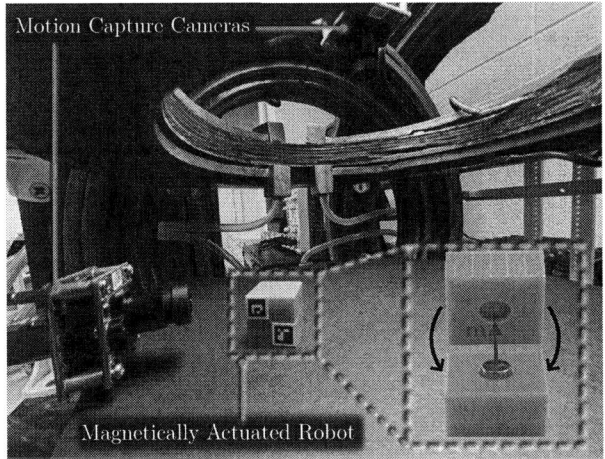

Fig. 8. Magnetically actuated robot inside the coil system. A neodymium permanent-magnet disc with an out-of-plane magnetization is placed inside a cubic-block structure.

TABLE III
ROBOTIC BLOCK PARAMETERS

Description	Value
Permanent magnet diameter	6.35 mm
Permanent magnet thickness	3.18 mm
Permanent magnet flux density	1.48 T
Cubic block material	UV-curable resins
Cubic block side length	20 mm
Assembled robot weight	10.3 g

IV. CONCLUSIONS

This paper discusses the modeling and design of a 3-D Helmholtz coil system for the wireless control of magnetically actuated robots. A modular dc-ac converter is designed to drive the 3-D Helmholtz coil resulting in a system capable of generating magnetic fields from dc to 25 Hz up to 60 mT in any direction. The developed hardware is employed to precisely control the motion of a magnetized cubic robot.

REFERENCES

[1] J. J. Abbott, E. Diller, and A. J. Petruska, "Magnetic methods in robotics," *Annual Review of Control, Robotics, and Autonomous Systems*, vol. 3, no. Volume 3, 2020, pp. 57–90, 2020.

[2] J. A. Levy, M. A. Straker, J. M. Stine, L. A. Beardslee, and R. Ghodssi, "Magnetically triggered ingestible capsule for localized microneedle drug delivery," *Device*, p. 100438, 2024.

[3] M. Sitti, *Mobile microrobotics*. Mit Press, 2017.

[4] S. Yim and M. Sitti, "Design and rolling locomotion of a magnetically actuated soft capsule endoscope," *IEEE Transactions on Robotics*, vol. 28, no. 1, pp. 183–194, 2012.

[5] Y. Kim, E. Geneviere, P. Harker, J. Choe, M. Balicki, R. W. Regenhardt, J. E. Vranic, A. A. Dmytriw, A. B. Patel, and X. Zhao, "Telerobotic neurovascular interventions with magnetic manipulation," *Science Robotics*, vol. 7, no. 65, p. eabg9907, 2022. [Online]. Available: https://www.science.org/doi/abs/10.1126/scirobotics.abg9907

[6] S. Wu, Q. Ze, J. Dai, N. Udipi, G. H. Paulino, and R. Zhao, "Stretchable origami robotic arm with omnidirectional bending and twisting," *Proceedings of the National Academy of Sciences*, vol. 118, no. 36, p. e2110023118, 2021. [Online]. Available: https://www.pnas.org/doi/abs/10.1073/pnas.2110023118

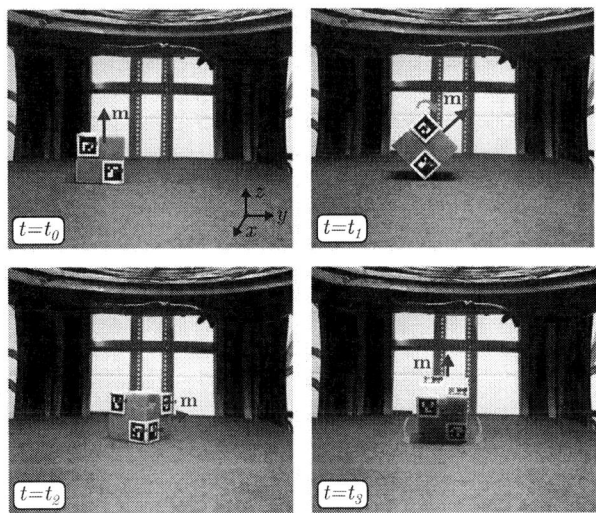

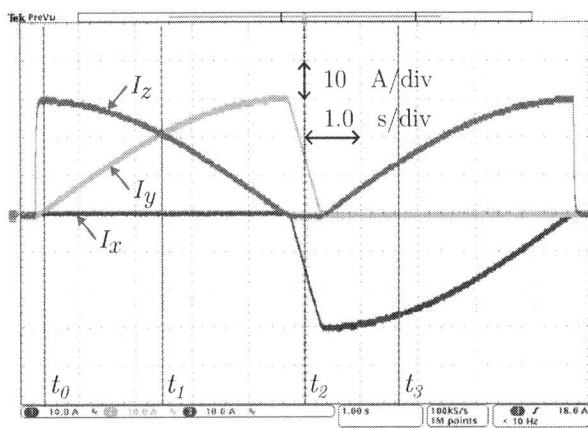

Fig. 9. The motion of the cube can be precisely controlled by generating an appropriate magnetic field. Initially, the robot rests on its face, with its magnetization **m** pointing upwards along the positive z direction. The externally imposed flux density **b** begins to rotate clockwise on the yz plane ($t = t_0$). The resulting electromagnetic torque causes the robot to move to the right by rolling clockwise around its edge ($t = t_1$). To change the direction of motion, the heading of the robot is adjusted by controlling the direction of **b** on the xy plane. As a result, the robot spins around its center of mass ($t = t_2$), allowing it to subsequently move along the x axis ($t = t_3$).

[7] J. C. Simpson, J. E. Lane, C. D. Immer, R. C. Youngquist, and T. Steinrock, "Simple analytic expressions for the magnetic field of a circular current loop," 2001.

[8] M. Abramowitz, *Handbook of Mathematical Functions, With Formulas, Graphs, and Mathematical Tables,*. USA: United States Department of Commerce, National Bureau of Standards, 1972.

[9] D. G. Holmes and T. A. Lipo, *Modulation of Single-Phase Voltage Source Inverters*. Wiley-IEEE Press, 2003, pp. 155–213. [Online]. Available: https://doi.org/10.1109/9780470546284.ch4

[10] S. Garrido-Jurado, R. Muñoz-Salinas, F. J. Madrid-Cuevas, and M. J. Marín-Jiménez, "Automatic generation and detection of highly reliable fiducial markers under occlusion," *Pattern Recognition*, vol. 47, no. 6, pp. 2280–2292, 2014. [Online]. Available: https://doi.org/10.1016/j.patcog.2014.01.005

[11] G. Bradski, "The OpenCV Library," *Dr. Dobb's Journal of Software Tools*, 2000.

A Versatile PHIL Based Motor Emulator Testbench Using a High-Performance Power Amplifier Testbench

Seyedeh Nazanin Afrasiabi
Department of Electrical and Computer Engineering
Concordia University
Montréal, QC, Canada
n_afrasi@encs.concordia.ca

Rajendra Thike
Opal-RT Technologies Inc.
Montréal, QC, Canada
rajendra.thike@opal-rt.com

Mathews Boby
Opal-RT Technologies Inc.
Montréal, QC, Canada
mathews.boby@opal-rt.com

K. S. Amitkumar
Opal-RT Technologies Inc.
Montréal, QC, Canada
amitkumar.ks@opal-rt.com

Abstract— **The advancement in Power Hardware-in-the-Loop (PHIL) based AC motor emulation plays a vital role in the development, testing, and validation of motor drive systems across various industries. This paper presents a versatile PHIL-based motor emulator test bench capable of replicating a wide range of AC motors, including permanent magnet synchronous machines (PMSMs), induction machines (IMs), and brushless DC machines (BLDCs). The proposed test bench provides flexibility in motor drive testing. Comprehensive descriptions of the system configuration and control methodologies, which allow the testing of different types of motors, are provided. Experimental implementation and results are presented to validate the operation of the motor emulator test bench under various operating conditions, demonstrating its accuracy and utility.**

Keywords— *power hardware-in-the-loop (PHIL), permanent magnet synchronous machine, induction machine, Brushless DC, real-time simulations*

I. INTRODUCTION

The continuous evolution of motor drive systems necessitates advanced tools for their development and testing. AC motors, including permanent magnet synchronous machines (PMSMs), induction machines (IMs), and brushless DC machines (BLDCs), are integral components in a wide range of industrial and commercial applications [1]-[4]. Due to the growing demand for these motors across various industrial sectors, there is a need for high-performance, reliable, and efficient motor control systems[1],[5].

Traditional methods for testing drive inverters and controls typically require physical prototypes of motors and dynamometers, which can be both costly and time-consuming to develop and operate. These methods also have limitations in testing the wide range of dynamic behaviors that occur during transient and steady-state conditions for different motor types, as there is a risk of damaging the expensive prototype motors [6]. The increasing complexity and diversity of motor applications make it challenging to address all potential operating scenarios using physical prototypes alone. Furthermore, the need to test and validate control strategies for motors across various sectors, such as transportation, aerospace, and energy, adds to the demand for more efficient and cost-effective testing solutions.

To overcome these challenges, there is a growing interest in the development of hardware-in-the-loop (HIL) and power hardware-in-the-loop (PHIL) based motor emulators. HIL simulation integrates real hardware components with virtual simulations to assess system performance under realistic operating conditions. This approach enables researchers to evaluate control algorithms in a controlled environment while maintaining the ability to interact with physical components [7],[8]. Similarly, PHIL extends this concept by incorporating power electronics into the loop, allowing for real-time testing of power devices and systems alongside the simulated motor models. These systems provide a flexible, scalable, and realistic testing environment for motor drive inverters [9]-[13]. By using real-time simulations (RTS) coupled with power amplifiers, motor emulators can replicate the electrical and mechanical behavior of various motor types without the need for physical prototypes. This allows engineers and researchers to evaluate control algorithms, optimize drive system performance, and explore different motor configurations under a wide range of operating conditions, all while minimizing cost and development time.

The accuracy of motor emulation in replicating machine transient behavior is influenced by three key factors: 1) the accuracy of the machine model used to control the emulator. Prior research [14] has addressed this by utilizing look-up table-based machine models that account for variations in parameters such as current magnitude, rotor position, and current advance angle. This approach ensures that the model reflects the dynamic changes experienced by real motors under different operating conditions. 2) The emulator's accuracy depends largely on the converter used in the emulator system. If a switching converter is employed, the emulator bandwidth is primarily determined by the converter's switching frequency. A higher switching frequency results in greater emulator bandwidth [15], [16] and using power amplifiers with high dynamic performance can significantly enhance the accuracy and utility of the emulation. 3) The control methodology implemented for the emulator, which is determined by the selected mathematical model of the machine, also plays a critical role in achieving accurate motor behavior replication. 4) PHIL test benches usually consist of two voltage source converters, each requiring a separate DC power supply. Recently, test bench costs have been reduced by utilizing a single DC power supply [17], [18].

979-8-3315-1612-3/25 $31.00 © 2025 IEEE

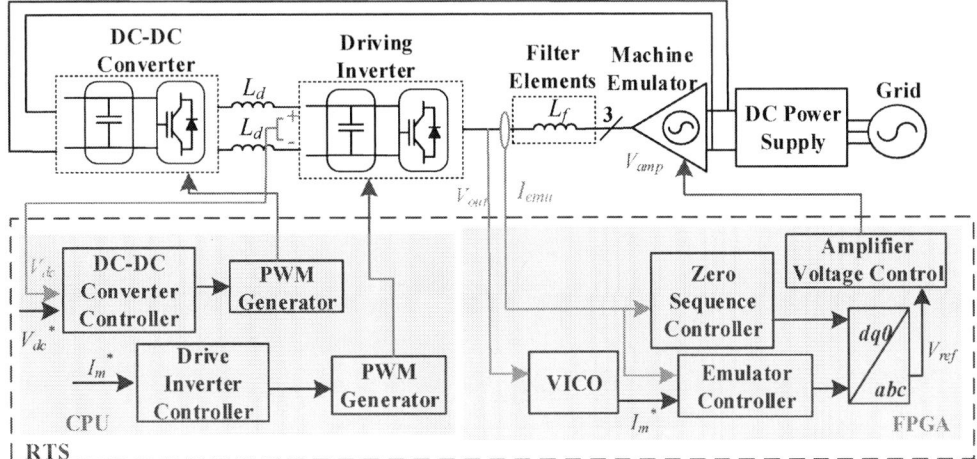

Fig. 1 Schematic of the proposed AC motor emulation

This paper presents a versatile PHIL-based motor emulator testbench capable of emulating various motor topologies, including PMSM, IM, and BLDC machines, across a wide range of operating conditions. By using a configurable real-time simulation environment, the proposed emulator can switch between different motor models, offering researchers and engineers the flexibility to study and test numerous motor types with a single platform. This versatility is critical in modern applications where multiple motor technologies are often employed, and testing needs to accommodate the unique characteristics of each motor type. Moreover, the emulator can simulate complex transient scenarios, providing a comprehensive and flexible solution for motor drive testing, development, and optimization.

The organization of this paper is as follows. Section II provides a description of the machine emulator system, including details on the control configuration of the universal setup. Section III outlines the experimental setup of the PHIL machine emulator system and presents the experimental results. These results demonstrate the emulator system's utility and accuracy in replicating different types of machine behaviors. Finally, Section IV concludes the paper.

II. PHIL BASED MACHINE EMULATOR SYSTEM

This section provides a description of the proposed motor emulation system, along with the mathematical models used for the PMSM, IM, and BLDC emulation systems.

A. System description

Fig. 1 shows a schematic of the proposed PHIL-based machine emulator. This emulator system includes the emulating power converter, drive inverter, real time simulator (RTS), and emulator control. The coupling inductor (L_f) connects the machine emulator to the drive inverter. The motor emulator is powered by a bidirectional DC power supply. The same DC power supply is used to supply the drive inverter with an intermediate DC-DC converter in between. The DC-DC converter is is controlled to generate the required DC link for the drive inverter, and its control is implemented in the RTS. A power amplifier from Opal-RT, OP8110-6 is used which provides the necessary high dynamic performance needed for motor emulation [14]. The primary function of the power amplifier is to control current flow with the drive inverter. To achieve this, the output voltages of the drive inverter (V_{out}) are measured and applied to the mathematical

model of the machine. The voltage-in, current-out mathematical model is used for IM, PMSM, and BLDC. This model generates reference currents (I_m) for the emulator current controller. These reference currents are compared with the actual measured output current (I_{emu}). The error between the machine stator currents and the measured currents are then fed to the emulator controller. Based on this error signal, the emulator controller generates a voltage reference (V_{ref}) for the power amplifier. As a result, the power amplifier ensures that the motor stator current closely follows the emulated current (I_{emu}). This process allows the power amplifier to accurately replicate the behavior of the motor.

B. PMSM mathematical model

To achieve high accuracy, the PMSM machine model is implemented on an FPGA, allowing it to run at a high sampling rate. The equation for the PMSM model can then be written as follows:

$$I_{abc} = [L_{abc}]^{-1}\left\{\int (V_{abc} - R_{abc}I_{abc})\,dt - \psi_{abc}(\theta_e)\right\} \quad (1)$$

where L_{abc} is the phase inductance matrix, I_{abc} represents the current in the stator windings, R_{abc} denotes the stator resistances, and V_{abc} is the voltage applied across the stator windings. ψ_{abc} indicates the magnetic flux linked with the stator windings. The electrical angle, denoted as θ_e. The electrical angle is expressed as follows:

$$\theta_e = pp * \theta_m \quad (2)$$

Here, pp represents the number of pole pairs and θ_m is mechanical angle. the Electromagnetic Torque can be expressed by

$$T_e = pp\left[I_{abc} \cdot \frac{\partial \psi_{abc}}{\partial \theta_e} + (L_d - L_q)i_d i_q\right] \quad (3)$$

Where is $\frac{\partial \psi_{abc}}{\partial \theta_e}$ the partial derivative of the instantaneous permanent magnet flux. If the back EMF of the machine is sinusoidal, as is the default case for PMSM machines, the torque can be further simplified into (4)

$$T_e = pp\left[\sqrt{\frac{3}{2}}\psi_M i_q + (L_d - L_q)i_d i_q\right] \quad (4)$$

979-8-3315-1612-3/25 $31.00 ©2025 IEEE

C. IM mathematical model

The output currents of the induction machine windings, I, are computed using the machine's state-space representation, where the magnetic flux linkages Ψ serve as the state variables. The forward Euler discretization of this state-space model can be expressed as follows:

$$\psi[n+1] = A_d[n]\psi(n) + B_d[n]u[n]$$
$$I[n+1] = C[n+1]\psi[n+1] \tag{5}$$

where n represents the timestep index, and the coefficient matrices A_d, B_d, and C are defined as follows:

$$A_d = T_s(-RL^{-1} - \Omega) + I_d$$
$$B_d = T_s \times I_d \tag{6}$$
$$C = L^{-1}$$

where T_s is the applied solver timestep, and I_d the identity matrix. The state variable matrices and output matrices referenced in equation (6). Their specific definitions is below:

$$u = [V_{sq} \quad V_{sd} \quad V_{s0} \quad V'_{rq} \quad V'_{rd}]^t$$
$$I = [I_{sq} \quad I_{sd} \quad I_{s0} \quad I'_{rq} \quad I'_{rd}]^t$$
$$\psi = [\psi_{sq} \quad \psi_{sd} \quad \psi_{s0} \quad \psi'_{rq} \quad \psi'_{rd}]^t$$
$$R = [R_s \quad R_s \quad R_0 \quad R'_r \quad R'_r]^t$$
$$L = \begin{bmatrix} L_{ls} & 0 & 0 & L_m & 0 \\ 0 & L_{ls} & 0 & 0 & L_m \\ 0 & 0 & L_0 & 0 & 0 \\ L_m & 0 & 0 & L'_{lr} & 0 \\ 0 & L_m & 0 & 0 & L'_{lr} \end{bmatrix} \tag{7}$$
$$\Omega = \begin{bmatrix} 0 & 0 & 0 & 0 \\ -\omega & 0 & 0 & 0 & 0 \\ 0 & 0 & 0 & 0 & 0 \\ 0 & 0 & 0 & 0 & \omega - \omega_r \\ 0 & 0 & 0 & -(\omega - \omega_r) & 0 \end{bmatrix}$$

where ω is the rotational speed of the reference frame, and ω_r is the rotational speed of the rotor's electrical frame. The Squirrel-Cage Induction Machine is modeled in the rotor reference frame, implying that $\omega = \omega_r$. Additionally, since the machine's rotor is not externally energized, it follows that $V'_{rq} = V'_{rd} = 0$ at all times.

The electromagnetic torque is defined by the following equation, where Ψ represents the flux linkage.

$$T_e = \frac{3}{2}PP * (\psi_d I_q - \psi_q I_d) \tag{8}$$

D. BLDC mathematical model

The BLDC (Brushless DC) model generates the machine currents as outputs by solving the following equation:

$$I_{abc} = [L_{abc}]^{-1}\left\{\int (V_{abc} - R_{abc}I_{abc})\,dt - \psi_{abc}(\theta_e)\right\} \tag{9}$$

where V_{abc} represents the voltage applied across the stator windings, I_{abc} denotes the phase currents, R_{abc} and L_{abc} represent the stator phase resistance and inductance, respectively, ψ_{abc} is the magnetic flux linked to the stator windings, and θ_e is the electrical angle. The generated electromagnetic torque is given by:

$$T_e = pp\left[I_{abc} \cdot \frac{\partial \psi_{abc}}{\partial \theta_e}\right] \tag{10}$$

where pp is the number of pole pairs, and $\frac{\partial \psi_{abc}}{\partial \theta_e}$ is the partial derivative of the instantaneous permanent magnet flux with respect to the electrical angle. For BLDC machines, the back EMFs have trapezoidal waveforms. This trapezoidal shape is generated using a cosine lookup table, which is created using the following equation:

$$\frac{\partial \psi_a}{\partial \theta_e} * \frac{1}{\psi_m} = \max\left(\min\left(\frac{\cos(\theta_e)}{\cos\left(\frac{H}{2}\right)}, 1\right), -1\right) \tag{11}$$

where H is the back-emf flat area in degrees and ψ_m is the PM flux. The mechanical dynamics of the machine are governed by the equation: where H represents the flat area of the back-EMF in degrees, and ψ_m is the flux of the permanent magnet. The machine mechanical dynamics are described by the following equation:

$$T_e = J\frac{d\omega_m}{dt} + B\omega_m + T_L \tag{12}$$

where T_L is the load torque, J is the moment of inertia, B is the coefficient of viscous friction, and ω_m is the rotor speed. The machine model also generates position sensor information in the form of three Hall sensor outputs (H_{abc}). These outputs are derived from the electrical angle θ_e and are used as feedback for the drive control.

III. EXPERIMENTAL SETUP AND RESULTS

This section presents the developed experimental setups and results of the PMSM, IM, and BLDC emulator systems, aimed at evaluating their functionality under various operating conditions. Detailed descriptions of the experimental setups and the corresponding results are provided in the following subsections.

A. Experimental setup

The proposed PHIL-based emulator system is illustrated in Fig. 2 (a). A four-quadrant power amplifier is utilized to replicate the machine's behavior. The OP8110 power amplifier can sink or source up to 5 kW power, with a voltage control bandwidth of 10 kHz [20]. Control of this amplifier is facilitated through a high-speed fiber optic (SFP) link, with the emulator controller implemented on the RTS (OP4610) FPGA board [19]. A bidirectional DC power supply maintains a 500V DC for this module [21]. The drive inverter connects to the power amplifier through a 2.5 mH coupling inductor for each phase [22]. A standard two-level IGBT switch-based inverter is chosen for the drive inverter, while a bidirectional DC-DC converter is controlled to maintain a 350V DC voltage across the inverter DC bus. The DC-DC converter is connected to the same 500V supply powering the OP8110 power amplifier. The drive inverter switching frequency is set to 20 kHz, and its controller is executed on the RTS CPU, with a sampling time of 20 µs. The mathematical models for the induction motor (IM), permanent magnet synchronous motor (PMSM), and brushless DC motor (BLDC) are implemented on the FPGA board of the RTS.

(1) RTS

(2) Power amplifier

(3) Drive Inverter/ DC-DC converter

(4) DC supply for the power amplifier

(a)

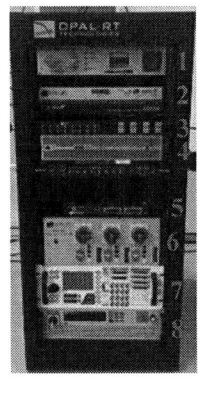

(1) Protection interface

(2) RTS

(3) Measurement panel

(4) Power amplifier

(5) OP866

(6) microgrid interface

(7) Drive Inverter/
DC-DC converter

(8) DC supply for the
power amplifier

(b)

Fig .2 (a) The experimental setup (b) The prototype motor emulation test bench

B. Testbench cpability

The proposed test bench is engineered to facilitate comprehensive testing and evaluation of drive inverters and motor controls. It is equipped with advanced features that allow for detailed analysis of motor performance. Table. I lists the key capabilities of the proposed test bench.

Table I. TEST BENCH CAPABILITY

Parameter	Value
Emulator output power	5 kW
Maximum voltage of the emulated motor	120Vrms
Maximum current of the emulated motor	14 Arms
Drive DC link	Up to 350 VDC
Encoder Type	Resolver, Quadrature
Emulator current controller bandwidth	2.5 kHz
Emulator voltage controller bandwidth	10kHz
Emulator mode of operation	Voltage and current mode
Motor type	PMSM, IM, BLDC
Motor model type	Current in voltage out/ voltage in current out
Motor Back EMF	Sinusoidal, trapezoidal
Motor number of poles	2 and higher
Motor maximum speed	120000 rpm (for 2 pole)
Link filter	2.5 mH
Test inverter switching frequency	20kHz
DC power supply maximum power	6 kW
DC power supply maximum DC voltage	500V
Motor Programmable Parameters	Direct inductance, quadrative inductance, Magnet flux linkage Winding resistance

This test bench supports a maximum emulator output power of 5 kW and operates at a maximum emulator voltage of 120 Vrms, with the capability to handle a maximum current of 14 Arms. The drive DC link is rated for up to 350 V, providing flexibility for various testing scenarios.

The encoder system supports both resolver and quadrature types, allowing for precise feedback in motor control applications. With an emulator current controller bandwidth of 2.5 kHz and a voltage controller bandwidth of 10 kHz, the test bench ensures rapid response and accurate control during motor performance evaluations.

The system can operate in both voltage and current control modes, accommodating a variety of motor types, including Permanent Magnet Synchronous Motors (PMSM), Induction Motors (IM), and Brushless DC Motors (BLDC). Additionally, the motor model type supports configurations for current-in-voltage-out and voltage-in-current-out scenarios, enhancing testing versatility.

The motor emulation capability includes various back-EMF profiles—sinusoidal, trapezoidal and custom user defined—and accommodates motors with two or more poles, achieving a maximum speed of 120,000 rpm for two-pole motors.

Additional features include a 2.5 mH link filter to reduce noise, a test inverter with a switching frequency of 20 kHz, and a DC power supply with a maximum output of 6 kW and a peak DC voltage of 500 V. The system also allows for programmable motor parameters, including direct and quadrature inductance, magnetic flux, and winding resistance, enabling thorough examination and validation of motor controls. The following sections examine the proposed motor emulator test bench in detail.

C. PMSM emulation results

Fig. 3 presents experimental results obtained from the PMSM emulator system under various operational conditions.

Fig. 3 (a) shows the speed change from 0 to 1000 rpm. The motor accelerates smoothly, with a close match between the reference current (green line) and the emulated current (magenta line), indicating accurate current tracking during startup.

Fig. 3 (b) demonstrates the torque change from 5 Nm to -5 Nm at 1000 rpm. The system successfully responds to torque variations, with the blue line showing the change in torque. The emulated current closely follows the reference current during these torque transitions, demonstrating effective control.

Fig. 3 (c) presents the speed reversal from 1000 rpm to -1000 rpm. These results confirm that the PMSM emulator system provides reliable tracking of the reference currents across different dynamic conditions, validating the accuracy and robustness of the implemented control strategy.

D. Induction motor emulation results

Fig. 4 illustrates the experimental results obtained from the induction motor (IM) emulator system under three different operational scenarios.

In Fig.4 (a), the speed is increased from 0 to 1000 rpm, where the close alignment between the reference current (green line) and the emulated current (magenta line) demonstrates precise current tracking during acceleration.

In Fig.4 (b), the torque is varied from 5 Nm to -5 Nm while the motor operates at 1000 rpm. The system responds effectively, with the torque (blue line) changing accordingly, and the emulated current maintaining close synchronization with the reference current, confirming accurate torque control.

In Fig.4 (c), the motor undergoes a speed reversal from 1000 rpm to -1000 rpm. The motor speed (red line) reverses direction, while the emulated current continues to follow the reference current, validating the system's ability to maintain precise control during dynamic transitions. These results highlight the robustness and reliability of the IM emulator

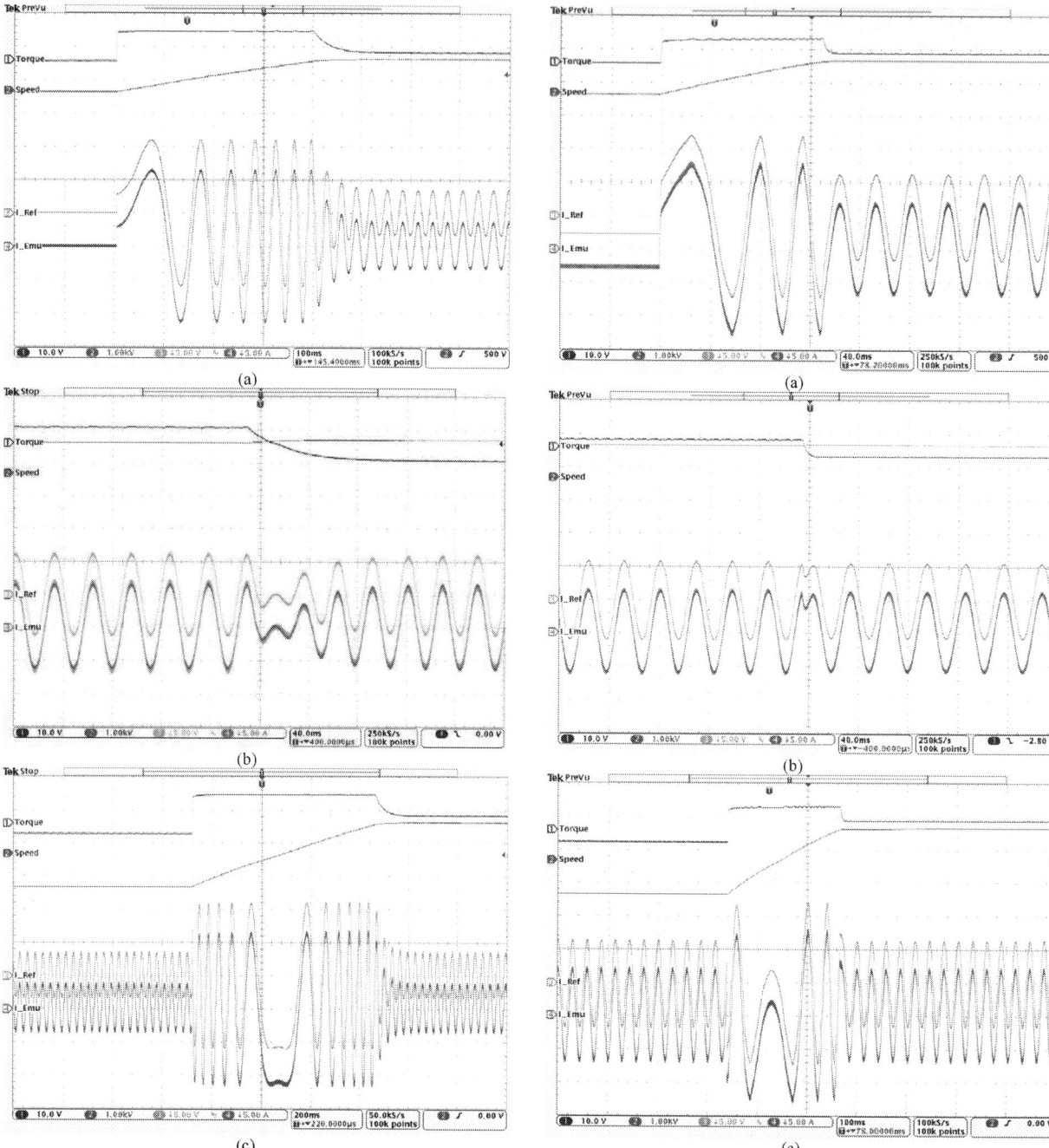

Fig. 3 Experimental results obtained from the PMSM emulator system: (a) speed change from 0 to 1000 rpm, (b) torque change from 5 Nm to -5 Nm at 1000 rpm, (c) speed reversal from 1000 rpm to -1000 rpm. The blue line represents the torque (scale: 10 Nm/div), the red line represents the motor speed (scale: 1000 rpm/div), the green line represents the reference current (scale: 5 A/div), and the magenta line represents the emulated current (scale: 5 A/div).

system in accurately replicating motor performance under varying operational conditions.

E. BLDC motor emulation results

Fig. 5 shows the experimental results obtained from the BLDC motor emulator system, demonstrating its performance under various conditions.

Fig. 4 Experimental results obtained from the IM emulator system: (a) speed change from 0 to 1000 rpm, (b) torque change from 5 Nm to -5 Nm at 1000 rpm, (c) speed reversal from 1000 rpm to -1000 rpm. The blue line represents the torque (scale: 10 Nm/div), the red line represents the motor speed (scale: 1000 rpm/div), the green line represents the reference current (scale: 5 A/div), and the magenta line represents the emulated current (scale: 5 A/div).

In Fig.5 (a), the motor undergoes a speed increase from 0 to 1000 rpm, where the reference current (green line) closely matches the emulated current (magenta line), confirming accurate current tracking during the acceleration phase.Fig.5 (b) depicts a torque variation from 5 Nm to -5 Nm while the motor operates at a constant speed of 1000 rpm. The system successfully adjusts to these torque changes, as indicated by the blue line, and maintains precise control, with the emulated current closely following the reference current.

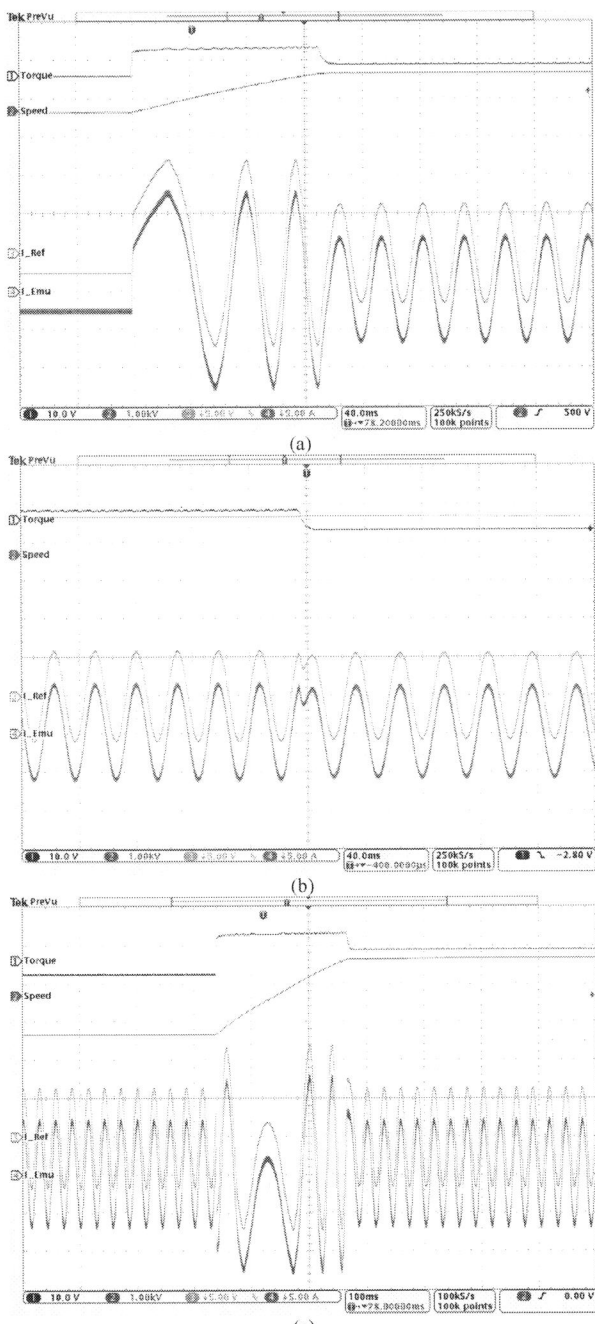

(a)

(b)

(c)

Fig. 5 Experimental results obtained from the BLDC emulator system: (a) speed change from 0 to 1000 rpm, (b) torque change from 5 Nm to -5 Nm at 1000 rpm, (c) speed reversal from 1000 rpm to -1000 rpm. The blue line represents the torque (scale: 10 Nm/div), the red line represents the motor speed (scale: 1000 rpm/div), the green line represents the reference current (scale: 5 A/div), and the magenta line represents the emulated current (scale: 5 A/div).

In Fig.5 (c), the motor performs a speed reversal from 1000 rpm to -1000 rpm. The motor speed (red line) reverses smoothly, and the emulator system continues to accurately track the reference current during the transition, reflecting its ability to manage dynamic changes effectively. Overall, the results illustrate the BLDC emulator system's capability to provide precise control and current tracking in response to speed and torque variations, demonstrating its effectiveness in replicating BLDC motor behavior.

IV. CONCLUSION

In conclusion, this paper presents a comprehensive power hardware-in-the-loop (PHIL) testbench capable of emulating three different types of AC motors: induction motors (IMs), permanent magnet synchronous motors (PMSMs), and brushless DC motors (BLDCMs). The detailed design and implementation of these motor models and their corresponding drive systems have been successfully demonstrated. Experimental results validate the emulator's ability to accurately replicate both the dynamic and steady-state responses of each motor type across various operating conditions. The flexibility of the emulator, combined with its robust control strategies, makes it a powerful tool for real-time testing, development, and optimization of motor drive systems. This versatile testbench contributes to advancing motor drive technology by providing an effective solution for system evaluation and performance enhancement under realistic conditions.

REFERENCES

[1] S. Sakunthala, R. Kiranmayi and P. N. Mandadi, "A study on industrial motor drives: Comparison and applications of PMSM and BLDC motor drives," 2017 International Conference on Energy, Communication, Data Analytics and Soft Computing (ICECDS), Chennai, India, 2017, pp. 537-540.

[2] A. M. El-Refaie, "Motors/generators for traction/propulsion applications: A review," in IEEE Vehicular Technology Magazine, vol. 8, no. 1, pp. 90-99, March 2013.

[3] A. Kermanizadeh, M. T. Mercan and P. Pillay, "Thermal Model of a Soft Magnetic Composite EV Permanent Magnet Traction Motor," 2024 IEEE Transportation Electrification Conference and Expo (ITEC), Chicago, IL, USA, 2024, pp. 1-5.

[4] S. N. Afrasiabi and C. Lai, "Investigation of LC Filter Unbalance in an Inverter-Fed Permanent Magnet Synchronous Machine Drive," 2020 23rd International Conference on Electrical Machines and Systems (ICEMS), Hamamatsu, Japan, 2020, pp. 1961-1966.

[5] K. Matsuse and D. Matsuhashi, "New technical trends on adjustable speed AC motor drives," in Chinese Journal of Electrical Engineering, vol. 3, no. 1, pp. 1-9, 2017.

[6] K. Ma and Y. Song, "Power-Electronic-Based Electric Machine Emulator Using Direct Impedance Regulation," in IEEE Transactions on Power Electronics, vol. 35, no. 10, pp. 10673-10680, Oct. 2020.

[7] J. Noon et al., "Design and Evaluation of a Power Hardware-in-the-Loop Machine Emulator," 2020 IEEE Energy Conversion Congress and Exposition (ECCE), Detroit, MI, USA, 2020.

[8] F. Alvarez-Gonzalez, A. Griffo, B. Sen and J. Wang, "Real-Time Hardware-in-the-Loop Simulation of Permanent-Magnet Synchronous Motor Drives Under Stator Faults," in IEEE Transactions on Industrial Electronics, vol. 64, no. 9, pp. 6960-6969, Sept. 2017.

[9] H. Song et al., "Development and Evaluation of a Power Hardware-in-the-Loop (PHIL) Emulator Testbench for Aerospace Applications," 2021 IEEE Energy Conversion Congress and Exposition (ECCE), Vancouver, BC, Canada, 2021.

[10] S. N. Afrasiabi, C. Lai and P. Pillay, "Dead Time Analysis of a Power-Hardware-in-the-Loop Emulator for Induction Machines," IECON 2021 – 47th Annual Conference of the IEEE Industrial Electronics Society, Toronto, ON, Canada, 2021, pp. 1-6.

[11] W. Yu and I. Husain, "FPGA-Based High-Bandwidth Motor Emulator for Interior Permanent Magnet Machine Utilizing SiC Power Converter," in IEEE Journal of Emerging and Selected Topics in Power Electronics, vol. 9, no. 4, pp. 4340-4353, Aug. 2021.

[12] C. -A. Cheng, C. -C. Chang, T. -S. Li, Z. -J. Chen and Y. -M. Chen, "Initial Rotor Position Startup Process Emulation Based on Electric Motor Emulator," 2020 IEEE 9th International Power Electronics and Motion Control Conference (IPEMC2020-ECCE Asia), Nanjing, China, 2020, pp. 605-610.

[13] D. Wang, S. Ge, Q. Li, H. Sun and P. Xu, "A Novel Control Scheme for the Electric Motor Emulator to Improve the Voltage Emulation Accuracy," in IEEE Transactions on Power Electronics, vol. 39, no. 1, pp. 374-383, Jan. 2024.

[14] M. A. Masadeh and P. Pillay, "Induction motor emulation including main and leakage flux saturation effects," 2017 IEEE International

979-8-3315-1612-3/25 $31.00 © 2025 IEEE

Electric Machines and Drives Conference (IEMDC), Miami, FL, USA, 2017, pp. 1-7.

[15] S. N. Afrasiabi, R. Thike, K. S. Amitkumar, C. Lai and P. Pillay, "Comparative Study on Machine Emulation Systems Based on Switching Converters and Linear Power Amplifier Configurations," 2023 IEEE 14th International Conference on Power Electronics and Drive Systems (PEDS), Montreal, QC, Canada, 2023, pp. 1-6.

[16] S. Xia et al., "Bandwidth Enhancement of Electric Machine Emulator With Virtual-Impedance Control," in IEEE Transactions on Power Electronics, vol. 39, no. 12, pp. 15609-15621, Dec. 2024.

[17] A. H. Kadam and S. S. Williamson, "Single Loop Control of a Common DC-Bus-Configured Traction Motor Emulator Using State Feedback Linearization Method," 2021 IEEE Southern Power Electronics Conference (SPEC), Kigali, Rwanda, 2021, pp. 1-6.

[18] G. Mademlis, N. Sharma, Y. Liu and J. Tang, "Zero-Sequence Current Reduction Technique for Electrical Machine Emulators With DC Coupling by Regulating the SVM Zero States," in IEEE Transactions on Industrial Electronics, vol. 69, no. 11, pp. 10947-10957, Nov. 2022.

[19] OPAL-RT, "Op4610 datasheet," OPAL-RT, Montreal, QC, Canada. [Online]. Available: https://blob.opal-rt.com/medias/L00161_0125.pdf

[20] OPAL-RT, "Op8110 datasheet," OPAL-RT, Montreal, QC, Canada. [Online]. Available: https://www.opal-rt.com/wpcontent/uploads/2019/08/OP8110_onepager_Mars2019_Final.pdf

[21] https://www.itechate.com/en/info_128.aspx?lcid=16&itemid=115

[22] https://www.taraztechnologies.com/products/power-electronics

A 450V Three Phase GaN IPM Achieving 99.1% Efficiency in Smallest 12mm x12mm Package for 250W Power Delivery without Heatsink

Maik Kaufmann
KilbyLabs
Texas Instruments
Freising, Germany
ORCID: 0000-0002-9954-132X

Manu Balakrishnan
Motor Drive Business Unit
Texas Instruments
Bangalore, India

Stefan Herzer
KilbyLabs
Texas Instruments
Freising, Germany

Anand Chellamuthu
Motor Drives Business Unit
Texas Instruments
Dallas, TX, USA

Hely Zhang
Motor Drives Business Unit
Texas Instruments
Shanghai, China

Abstract— This work demonstrates a three-phase motor drive Intelligent Power Module (IPM) based on Gallium-Nitride (GaN) power transistors for major increase in efficiency and power density compared to today's Industry's best in class. Thanks to the superior characteristics of GaN (lower specific resistance, best-in-class figure-of-merit for switching: Qoss*RDSon, zero reverse-recovery) and intelligent gate driver design, this work reduces switching and conduction loss and eliminates reverse-recovery achieving inverter efficiency up to 99.1% while maintaining safe operation for motors. The high efficiency enables 250W power delivery without heat sink in a 65% smaller package; a 12mm*12mm quad-flat no leads (QFN) compared to state-of-the-art silicon-based implementations in > 29mm*12mm leaded packages. High level of integration enables intelligent gate driving unleashing the full potential of GaN technology achieving high efficiency and power density while keeping the GaN transistors protected at all times. This design integrates adaptive dead-time control based on in-situ power stage sensing to minimize third-quadrant conduction loss and reduce total harmonic distortion (THD) of the motor current by 2.5x compared to silicon based implementations. GaN specific challenges such as slew-rate control despite the highly non-linear capacitances are handled by package-integrated gate drivers. The technology-inherent lower short-circuit ruggedness of GaN transistors is addressed using a fast, integrated over current protection (OCP). Measurements of this work are presented and compared to existing silicon-based IPMs to quantify the benefit in terms of efficiency, power density and THD enabled by GaN.

Keywords—GaN, Motor Drive, brushless DC motor (BLDC), intelligent power module (IPM), power density, high voltage, three phase motor, slew rate control, dead time control, over current protection (OCP), total harmonic distortion (THD)

I. INTRODUCTION

Commercially available Intelligent Power Modules (IPMs) for motor drives integrate three half bridges with power transistors, gate drivers and level shifters in one package. They are widely used for driving motors across a wide power range found in applications such as washing machines, refrigerator pumps and ceiling fans. Traditionally, silicon-based IGBT [1],[2] or MOSFET [3],[4] transistors are used and inverter efficiency in the range 97-98% at 250W output power is achieved. To dissipate the power loss of 5W-7.5W of the IPM, they are assembled in relatively large DIP packages >29mm x 12mm and require a heat sink. This work proposes the use of Gallium-Nitride (GaN) transistors to improve efficiency and power density by reducing the package size and eliminating the need for a heat sink.

II. BENEFITS FOR GaN IN MOTOR DRIVES

A. Loss Mechanisms in Motor Drive Inverters

Motor drive inverters is a hard-switching application where switching losses contribute a significant portion to the overall inverter losses. Fig. 1 depicts exemplary waveforms of a hard-switched turn-on of the low-side (LS) transistor with the load current (I_L) flowing into the switching node (V_{sw}). Until time point t_1, the high-side transistor Q_{HS} is on and conducts the current I_L. The switching event begins at t_1 with turning off Q_{HS}. The time frame $t_1 - t_2$ is the dead time where both transistors Q_{HS} and Q_{LS} are turned off to prevent shoot-through and potential destruction of the half bridge. During this dead-time phase, Q_{HS} is conducting in third quadrant leading to increased conduction loss. At t_2, the gate voltage $V_{GS,LS}$ of Q_{LS} ramps up and Q_{LS} slowly takes over the load current I_L from the high-side FET. In this dI/dt phase, maximum V_{DS} is present at Q_{LS} while the drain current $I_{D,LS}$ increases until $I_{D,LS} > I_L + I_{slew}$ where I_{slew} is the additional current in Q_{LS} discharging the parasitic capacitances at the switching node V_{sw}. Power loss in this phase can easily exceed several kilowatt ($V_{in} = 400V$, $I_D > 2.5A$).

979-8-3315-1612-3/25 $31.00 © 2025 IEEE

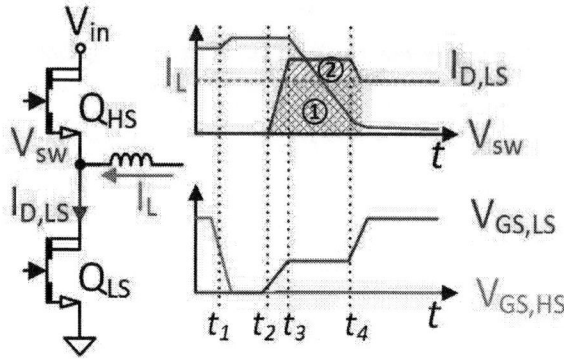

Fig. 1: Exemplary waveforms for hard-switching half bridge.

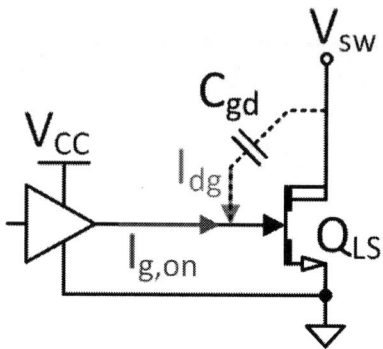

Fig. 2: Simplified schematic for slew rate control

During the dV/dt phase ($t_3 - t_4$) the switching node voltage V_{sw} reduces from maximum value to <1V ($R_{DS,on} * I_L$). The slew rate and the duration of this phase is defined by dV/dt = I_{slew} / $C_{sw}(V)$. The area ① below the IL value corresponds to I-V overlap losses and can be reduced by increasing slew rate. The area ② above IL corresponds to the charge at the switching node

$$Q_{sw} = \int C_{sw}(V_{sw})dV_{sw} \qquad (1)$$

and is independent of the slew rate.

After t_4, Q_{LS} is fully turned on, $I_{D,LS} = I_L$ and only conduction losses $I_{D,LS} * R_{DS,on}$ are present. The proposed design of this work reduces all losses of the hard-switching half bridge by exploiting the superior technology properties of GaN compared to silicon.

B. Inherent GaN-Technology Benefits

In recent years, GaN based transistors became increasingly popular for power conversion in the 50V-800V space due to their superior characteristics:

- Specific on-resistance (R_{sp} ($\Omega*mm^2$)) for 600V GaN transistors is more than one decade lower compared to 600V silicon technologies since almost 15 years [5]. This allows to integrate smaller on-resistance in a small footprint package. Thus, the power density is increased and conduction losses are reduced for high efficiency.

- GaN shows low specific capacitances/charges ($R_{on}*Q_g$, $R_{on}*Q_{oss}$). While Q_g is negligible from an efficiency point-of-view for >600V transistors, Q_{oss} is a significant part of Q_{sw} related switching losses.

- The GaN-device is a unipolar semiconductor without any junction in the power path. Thus, no reverse recovery charge exists allowing for short dead time and fast dI/dt phase reducing losses.

- Thanks to high specific current I_{sat} / Q_{oss} and low Miller capacitance C_{gd}, GaN transistors can achieve fast dV/dt slew rates in the range 150V/ns … 200V/ns leading to reduced I-V overlap losses.

III. CHALLENGES FOR GaN IN MOTOR DRIVE APPLICATIONS

Besides the significant benefits of GaN for high-voltage hard-switching applications (see Section II.B), there are several challenges considering the specifics of motor drives.

A. Challenges for Slew Rate Control with GaN FETs

While GaN allows dV/dt slew rates larger than 150V/ns, this is typically not tolerated by today's motors and can lead motor insulation degradation and then to destruction. 400V motors nowadays can typically handle 5V/ns … 15V/ns, some even support slew rates up to 50V/ns with increased winding insulation. Thus, slew rate control has to be implemented for a GaN-based motor drive in order to protect the motor.

Fig. 2 illustrates a simplified schematic of a slew-rate controlled gate driver. Traditionally, slew rate control relies on Miller feedback: the parasitic gate-drain capacitance C_{gd} of the transistor Q_{LS} causes a current flow

$$I_{dg} = C_{gd}(V_{sw}) \cdot \frac{dV_{sw}}{dt} \qquad (2)$$

I_{dg} = $C_{gd}(V_{sw})$ * dV_{sw}/dt into the gate of Q_{LS} while V_{sw} is slewing. If V_{sw} changes from a high potential to a lower, dV_{sw}/dt and subsequently I_{dg} are negative, trying to reduce the gate-source voltage $V_{GS,LS}$ and hence the slew rate dV_{sw}/dt. Thus, slew rate control can be achieved by controlling the gate current $I_{g,on}$ forming an equilibrium $I_{g,on} = I_{dg}$, leading to

$$\frac{dV_{sw}}{dt} = \frac{I_{g,on}}{C_{gd}} \qquad (3)$$

Implementing slew-rate control for GaN transistors is challenging: Due to the lateral device structure with field plates, as illustrated in Fig. 3, all capacitances have a strong voltage dependence. Especially the gate-drain capacitance (Miller-capacitance), which provides negative feedback during slewing, can change by two orders of magnitude over the operating voltage range [6]. For higher V_{DS}, the two-dimensional electron gas (2DEG) channel beneath the source-connected field plates depletes. Thereby, the distance d between the gate and the drain-connected channel increases, leading to lower capacitance $C_{dg} \sim 1/d$.

979-8-3315-1612-3/25 $31.00 © 2025 IEEE   1287

Similarly, the distance between the source connected field-plate and the drain-connected channel increase the further the channel depletes with increasing V_{DS}. Subsequently, also the drain-source capacitance C_{ds} reduces with increasing voltage. Since C_{ds} is a mature part of the GaN output capacitance C_{oss}, also C_{oss} shows a strong voltage dependence. This is depicted in the measured typical characteristic for a commercially available GaN transistor [7] drawn in Fig. 4 ("$C_{oss,LS}$").

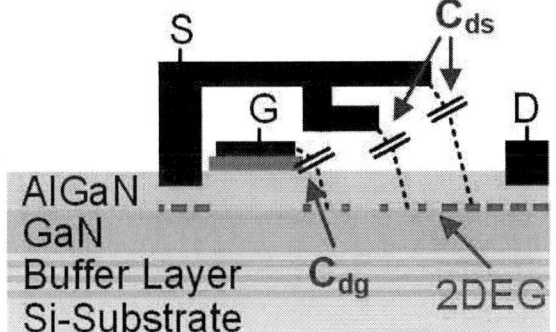

Fig. 3: Illustrated GaN device cross-section with V_{DS}-dependent capacitances

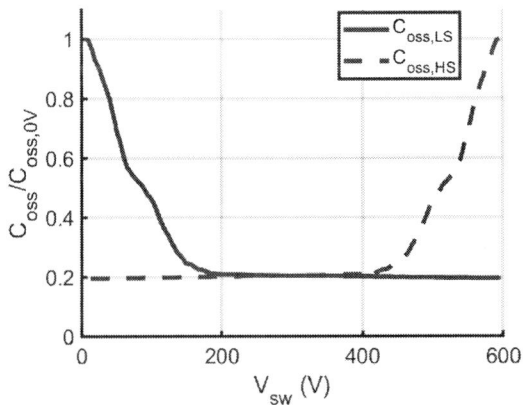

Fig. 4: Measured typical C_{oss} vs. V_{sw} for $V_{in} = 600\,V$ as published in [6]

In a half-bridge configuration, the total capacitance at the switching node is the sum of the output capacitance of both transistors, $C_{sw} = C_{oss,LS} + C_{oss,HS}$. In a general manner, the slew rate can be expressed as function of the transistor drain current I_D and the output capacitance

$$\frac{dV_{sw}}{dt} = \frac{I_D - I_L}{C_{sw}(V_{sw})} \tag{4}$$

For motor safety consideration, the maximum slew rate of V_{sw} is important. The maximum slew rate occurs during the time when $C_{sw}(V_{sw})$ is lowest, roughly between $V_{sw} = 100\,V$ and $V_{sw} = V_{in} - 100\,V$. In this voltage range, the Miller capacitance is very low and provides only negligible feedback to the gate voltage V_{GS} of the transistor, thus the relation of equation (3) is not applicable for GaN transistors.

The following section derives the relation between gate current $I_{g,on}$ and the slew rate dV_{sw}/dt without Miler feedback.

In order to determine the maximum slew rate, the drain current I_D at the time with lowest C_{sw} capacitance needs to be identified.

The discharging of C_{sw} starts once $I_D > I_L$. For the following equations, we assume the threshold voltage $V_{th} = V_{GS}$ @ $I_D = I_L$. We consider the turn-on of the Q_{LS} while similar relations are true for turn-on of Q_{HS}. Assuming a constant gate current $I_{g,on}$ for turning on the transistor, the gate-source voltage gets overcharged while C_{sw} gets discharges from V_{in} to $V_{in} - 100\,V$. The gate-source voltage as function of time and gate capacitance C_{gate} can be expressed as

$$V_{GS} - V_{th} = \frac{I_{g,on}}{C_{gate}} \cdot t \tag{5}$$

Assuming constant C_{gate}, which is a fair assumption for $V_{GS} > V_{th}$. Increasing V_{GS} leads to increasing I_D defined by fundamental transistor equations in saturation with the transconductance g_m

$$I_D(t) = (V_{GS}(t) - V_{th}) \cdot g_m \tag{6}$$

Inserting (5) in (6) yields

$$I_D(t) = \frac{I_{g,on}}{C_{gate}} \cdot t \cdot g_m \tag{7}$$

The charge Q_{100V} moved away from C_{sw} until V_{sw} drops by $100\,V$ can be expressed as integral of the drain current using equation (7)

$$Q_{100V} = \int I_D(t) dt = \frac{1}{2} \frac{I_{g,on}}{C_{gate}} \cdot g_m \cdot t^2 \tag{8}$$

Solving this equation for the time it takes to discharge C_{sw} by $100\,V$ yields

$$t_{100V} = \sqrt{\frac{2 \cdot C_{gate} \cdot Q_{100V}}{I_{g,on} \cdot g_m}} \tag{9}$$

During this time, C_{gate} gets continuously charged by $I_{g,on}$, further increasing VGS

$$(V_{GS}(t_{100V}) - V_{th}) = \frac{I_{g,on}}{C_{gate}} \cdot t_{100V} \tag{10}$$

Using expression (9) for t_{100V} yields

$$(V_{GS}(t_{100V}) - V_{th}) = \frac{I_{g,on}}{C_{gate}} \cdot \sqrt{\frac{2 \cdot C_{gate} \cdot Q_{100V}}{I_{g,on} \cdot g_m}} \tag{11}$$

which can be simplified to

$$(V_{GS}(t_{100V}) - V_{th}) = \sqrt{\frac{2 \cdot I_{g,on} \cdot Q_{100V}}{C_{gate} \cdot g_m}} \tag{12}$$

979-8-3315-1612-3/25 $31.00 © 2025 IEEE

Inserting (12) in (6) gives the drain current I_{D100V} at the time point when C_{sw} approaches its minimum value

$$I_D(t_{100V}) = \sqrt{\frac{2 \cdot I_{g,on} \cdot Q_{100V} \cdot g_m}{C_{gate}}} \qquad (13)$$

In (13) it can be seen that the drain current at lowest C_{sw} is proportional to the square root of $I_{g,on}$. Since the slew rate dV_{sw}/dt is proportional to I_D (4), it can be derived

$$\frac{dV_{sw,max}}{dt} \sim \sqrt{I_{g,on}} \qquad (14)$$

This assumes that the duration t_{100V} for initial discharge of C_{sw} is much larger than the time during the slewing from $V_{sw} = V_{in} - 100V$ to $V_{sw} = 100V$. This is fair since during the initial discharge, I_D ramps up and C_{sw} reduces by ~5x.

Comparing (3) with (14) shows that without Miller feedback, the slew rate of a transistor is no longer proportional to the gate current, but to the square root of the gate current. When trying to slow down the slew rate of the GaN transistor from its 150V/ns capability to ~10V/ns for safe operation of the motor, this would require 225x lower $I_{g,on}$. Since the turn-on delay for charging C_{gate} from 0V to V_{th} is proportional to $I_{g,on}$, this would also lead to 225x longer gate turn-on delay which is not acceptable. Thus, this work implements a dynamic gate current control with high pre-charge current and low slewing current. This keeps the turn-on delay short while controlling the slew rate to a safe value for common motors. The dynamic current adjustment with the required low delay of only few nanoseconds is enabled by very small parasitics of the gate loop, which is a benefit of co-packaging of driver and power transistor.

B. Short-Circuit Ruggedness of GaN Transistors

Another challenge for using GaN in motor drive applications is the short-circuit ruggedness. In case of a short circuit from V_{sw} to V_{in} or to ground, high instantaneous power loss upto 450V * 10A = 4.5kW can occur. GaN transistors are typically lateral power devices, thus the high short-circuit power happens in a very small volume at the surface of the die. This is illustrated in Fig. 5 (left).

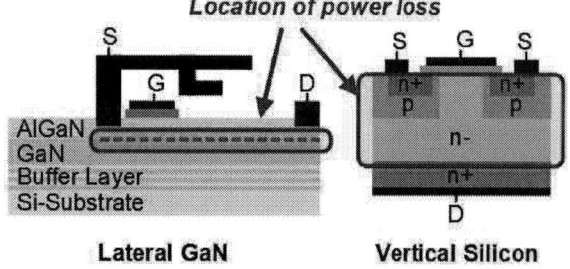

Fig. 5: Illustrated cross section of lateral GaN transistor (left) and vertical silicon transistor (right) indicating the area of high power-loss during a short-circuit event

In contrast to that, in vertical silicon transistors the short-circuit power occurs within a larger volume in the n- doped region. Hence, vertical transistors are more rugged towards short circuits. [1] specifies a short-circuit withstand time of 5µs at

V_{in} =400V for an IGBT based IPM, [4] claims upto 3µs short-circuit safe-operating-area (SOA) for a MOSFET IPM. In contrast to that, the GaN FETs of this work are turned off within less than 500ns in order to stay within SOA. Therefore, this work implements in-situ sensing of drain currents and drain voltages reliably detecting operating conditions outside SOA and shutting of the GaN transistor within typically than 200ns [8], which is confirmed by measurements as depicted in Fig. 7.

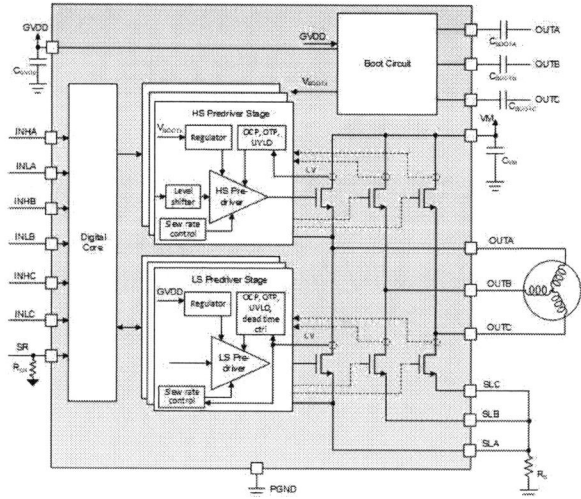

Fig. 6: Block diagram of the GaN based IPM presented in this work

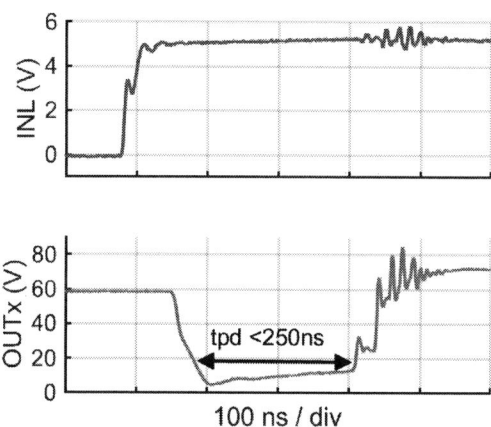

Fig. 7: Measured OUTx waveform when turning on into a short from switching node OUTx to high-voltage DC input VM showing a protection shut-off delay of less than 250ns.

C. Dead Time Losses for GaN

One more challenge in using GaN transistors in motor drive is the relatively high third quadrant conduction loss during dead time. Since there is no body diode, the channel needs to turn on during the dead time while $V_{GS} = 0V$ in order to conduct a load current forced by the inductance. This is achieved by a drain voltage lower than gate and source. However, the source-drain voltage drop is in the range of 2.5 − 3.5V for GaN at typical currents [8], which is ~3x higher than typical body-diode forward voltage of silicon FETs (0.7V − 1.5V). Thus, the dead time needs to be minimized in order to achieve high efficiency.

Thanks to the package integration of drivers and GaN transistors, all information required for adaptive dead time control is available. The drain voltage of the GaN transistor is already sensed for the short-circuit protection (Section III *B.*) and can be used to control the gate turn-on speed.

Usually if INHx goes low and INLx goes high (primary inputs in the block diagram Fig. 6) demanding the high-side to turn off and the low side to turn on, a non-overlap is inserted in the driver to prevent shoot-trough in the half bridge. If a fixed dead time is used, it needs to be designed to cover all possible variations of delays, slew time, input voltage… However, in this work, the dead time is minimized based on VDS sensing and fast VGS charging. As a result, a significant reduction of dead time is achieved and the third-quadrant conduction phase is minimized effectively to less than 200ns. This is lower than 20% of silicon-based motor drive IPMs [1]-[4] achieving lower dead time loss for this IPM despite the 3x larger voltage drop in third quadrant conduction.

IV. PREVIOUS EXAMPLES OF GaN IN MOTOR DRIVES

Despite the challenges for GaN in motor drive, there are some attempts for implementing motor drive inverters with GaN transistors. A board-level 3-phase motor drive using GaN-based half-bridges is presented in [9]. While this demonstrates the use of GaN transistors in motor drive applications, it applies only relatively fast slew rates 20V/ns ... 40V/ns which might bring some reliability concern of motor winding insulation depending on motor types [10],[11]. The board-level assembly of the three half-bridges leads to significant high-voltage routing on the PCB as well as a separate high-voltage DC BUS ceramic capacitor for each half-bridge in the distributed design. Additionally, a heat sink is required further increasing the space required for this solution.

A first concept study for a GaN-based 3-phase IPM in a 12mm x 12mm package was presented at APEC 2020 [10]. There, only few switching transitions are published but no top-level measurements such as efficiency curves or motor currents are provided. Slew rate control is achieved with Miller feedback. Therefore, the very non-linear gate-drain capacitance is linearized by adding ~1.2pF high-voltage metal-insulator-metal (MIM) capacitor, which increases the die size by ~50% of the HV power FET (estimated based on published layout picture). The provided output power for this GaN-based IPM is claimed as 226W with 80°C temperature increase, which is typically not tolerable in applications with ambient temperature above 25°C.

V. CHARACTERIZATION OF THE GaN IPM

To characterize the performance of this work and benchmark it with a silicon-based solution, two test boards are assembled, shown in Fig. 8. Both are used to drive a reference motor with tunable load to set the output power of the motor drive IPMs.

A. Measure Characterization Results

Fig. 9 shows the measured top-level waveforms of the three PWM switching nodes as well as the resulting sine-wave motor

12 * 12mm GaN IPM   33 * 19mm IGBT IPM
+ 40 * 40 * 15mm heat sink

Fig. 8 Left: Test board with GaN IPM of this work [8], right: test board with silicon IGBT IPM and heat sink [2].

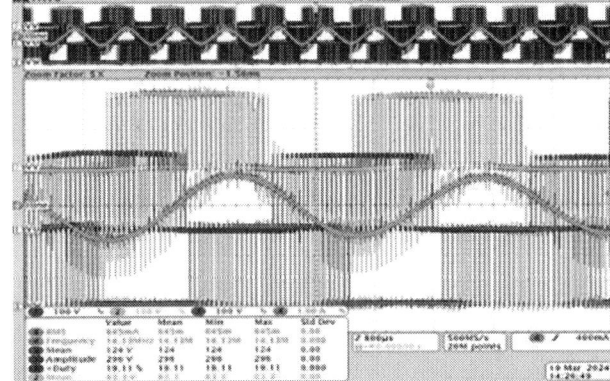

Fig. 9: Oscilloscope screenshot of top-level waveforms for the GaN IPM with all three switching nodes and the sine-wave modulated motor current.

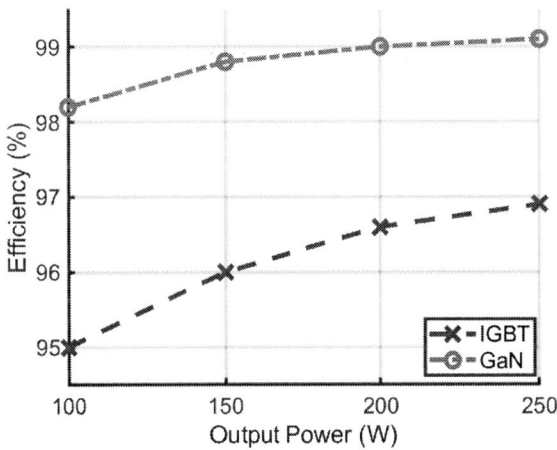

Fig. 10: Measured efficiency over output power for IGBT and GaN based IPM.

current of one phase for the GaN-based IPM of this work. The measured efficiency over output power for both, GaN and silicon-IGBT based IPMs is depicted in Fig. 10. The measurement was taken at 300V DC input and 20kHz PWM frequency. The motor current changes between 0.33A_{RMS} and 0.84A_{RMS} to get measurements at different output power levels.

979-8-3315-1612-3/25 $31.00 © 2025 IEEE

The measurements show that the GaN based implementation achieves much higher efficiency, effectively reducing the power loss by a factor of three at full output power. Despite the relatively low switching frequency for GaN, the technology benefits (lower specific on-resistance, faster slewing, zero reverse recovery) as well as the intelligent gate driver design (short propagation delays, minimized dead time) enable unprecedented efficiency in the 250W motor drive application space.

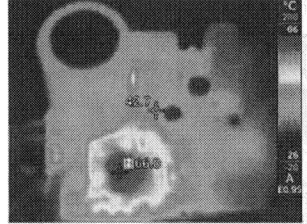 

GaN IPM without heat sink: ΔT_{case}-T_{amb} = 41°C   IGBT IPM with heat sink: ΔT_{sink}-T_{amb} = 38°C

Fig. 11: Thermal image of the two PCBs from Fig. 8 operating at 250W. Left: GaN based IPM without heat sink, right: silicon IGBT IPM with heat sink.

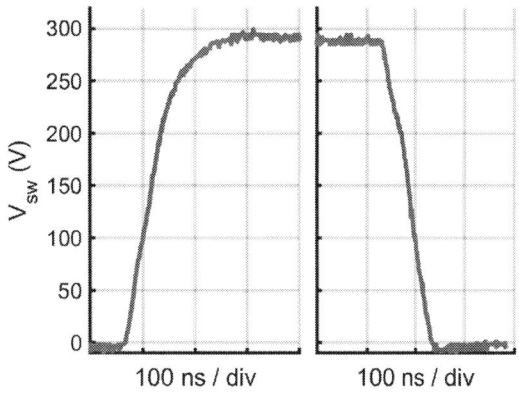

Fig 12: Measured switching transitions for the GaN IPM with 5V/ns slew rate.

The superior efficiency and lower power loss of the GaN IPM in this work is emphasized by the thermal images in Fig. 11. For these images, the two boards from Fig. 8 are operated with 250W output power at ambient temperature of 25°C. The left part shows the GaN IPM with a maximum temperature at the case top of 66°C. This aligns well with the measured power loss of ~2.2W (based of the efficiency at 250W in Fig. 10) and the thermal resistance of the package of 21.2°C/W [8]. The right part of Fig. 11 depicts the thermal image of the IGBT-based IPM with a 15mm x 40mm x 40mm aluminum heat sink, occupying a total volume of 24000mm³. The hottest point in this image is on the heat sink with comparable 63°C indicating a temperature rise of 38°C. In conclusion, at same output power the GaN IPM achieves similar temperature rise in a small 12mm x 12mm x 1mm QFN package without heat sink occupying only 144mm³. It can thus deliver the same output power in less than 1/165 of the volume, effectively increasing the inverter power density including cooling by 165x.

Fig. 12 shows measured transient V_{sw} waveforms during the switching transitions. Despite the challenge of very nonlinear capacitances of the GaN FETs (see Section III.A), the package integrated gate drivers of this work achieve a well-controlled 5V/ns slew rate (20% – 80%) as accepted by almost all motors in this application space. For further loss reduction, this work also offers slew rate settings of 10V/ns, 20V/ns and 40V/ns in case the motor accepts faster slew rates.

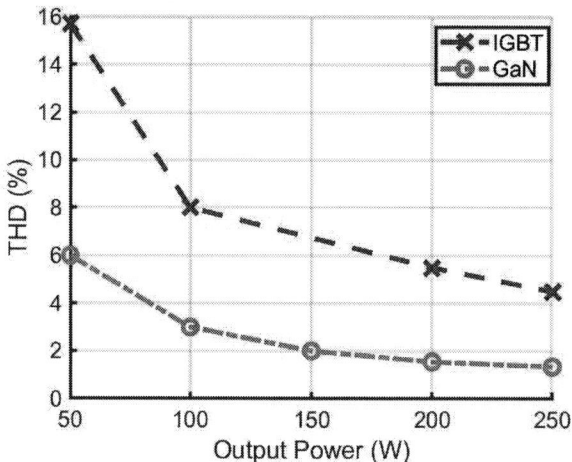

Fig 13: Measured THD of motor current of IGBT based IPM and the GaN based IPM of this work.

With low dead time, short propagation delays and wide duty cycle range, the IPM in this work achieves very good sine wave modulation of the motor phase currents. This is confirmed by the low measured total harmonic distortion (THD) of the motor phase currents (Fig. 13). In contrast, the IGBT IPM with 10x larger dead time of 2µs shows ~2.5 times larger THD. This leads to higher audible noise during operation as well as more mechanical wear of the motor bearings.

B. Comparison with State-of-the-Art

The performance and critical parameters of the GaN-based IPM of this work is compared with recently released silicon-based IPMs for benchmarking. Table I. summarizes the comparison of critical parameters. All featured devices are commercially available part released in 2023 or 2024. Since there is no other GaN IPM for this application space (offline operation supporting upto 450-500V input and ~5A phase current) available, the comparison can only be done with silicon-based IPMs using MOSFETs or IGBTs as power transistors. As extension of the comparison, a silicon MOSFET based half-bridge [13] is added, which is also suitable for motor drive applications due to a slew rate of around 3V/ns.

This work achieves lowest on-resistance in the smallest package based on the comparison overview, leading to low conduction loss of ~1.2W at 250W output power. Due to lack of reverse recovery, low dead time and faster but controlled slew rate, also switching losses are minimized to about 1W. This results in a total power loss of only 2.2W at 250W output power. Closest MOFET competition [4] shows more than double the on-resistance, indicating more than twice the conduction loss of

979-8-3315-1612-3/25 $31.00 © 2025 IEEE

TABLE I.   COMPARISON TO STATE-OF-THE-ART

Parameter \ Device, Year	GaN IPM This Work 2024	Silicon IGBT 2023 [1]	Silicon IGBT 2024 [4]	Silicon IGBT 2023 [12]	Silicon IGBT 2023 [2]	Silicon MOSFET 2023 [3]	Silicon MOSFET 2024 [4]	Silicon MOSFET Half-Bridge 2024 [13]
Technology	GaN	IGBT	IGBT	IGBT	IGBT	MOS	MOS	MOS (1/2-H)
Max op. voltage (V)	450	450	450	500	450	500	450	480
Current rating	5A @T_J=125C	4A @TC=100C	5A @TC=25C	6A @TC=25C	5A @TC=25C	5A	5A @TC=25C	5.5A
Package (Body size)	12x12 mm (QFN)	36x21mm (DIP)	36.0×14.8 mm (DIP)	32x12.5mm (DIP)	32.8x18.8mm (DIP)	32x12.5mm (DIP)	36.0×14.8 mm (DIP)	(10.8x9.4mm) x3 leaded
Dead time (us)	< 0.2	> 1	1.5	1.5	> 1	1-2	1.5	0.6
Slew rate (V/ns)	5-40	6	~ 2-4	>3	1-3	>3	~ 1-3	2.8-3.4
RDSon (Ω)	0.205	n/a	n/a	n/a	n/a	0.8	0.45	0.83
$V_{DS,1A}$ / $V_{CE,sat}$ (V)	0.205	1.6	1.75	1.6	1.85	0.8	0.45	0.83
Switching Energy 1 cycle, 6 transistors, no reverse recovery (uJ)	n/a	1170	n/a	870	n/a	582	n/a	n/a
Rth,JunctionCase (°C /W)	5.5	5.3	3.6 (IGBT) 4.2 (Diode)	10 (IGBT) 15 (Diode)	5.4	9.2	3.6	3
Rth,JunctionAmbient (°C /W)	21.2	n/a	25 (IGBT) 29 (Diode)	n/a	n/a	n/a	25	53
OC/SC protection	400ns latch 1-2us comp w/50u retry	Comparator w/ 40 us retry	Comp with 10ms retry	Integrated comparator	Integrated comparator (20us)	Integrated comparator	Comp with 10ms retry	CBC CL, DC bus UV/OV, dual OT
Dead time interlock	adaptive	Yes	No	Yes	No	Yes	No	Yes

this work. The low power loss enables the use of a small 12 x 12mm QFN package boosting the power density to unprecedented level in the motor drive IPM space. With a junction-to-ambient thermal resistance of 21.2°C/W of the package alone, this work can support 250W with only 41°C temperature increase eliminating the need for a heat sink. This also enables operation at 70°C ambient temperature while maintaining a junction temperature around 125°C. The high efficiency of above 99% allows system designs to easily meet energy efficiency standards.

The 12x12mm QFN package used for this work shows similar junction-to-case-top thermal resistance as existing IPMs in DIP packages. This indicates that if a heat sink is added, the power delivery can be enhanced much beyond 250W.

VI. CONCLUSION

This work elegantly solves all challenges of using GaN power transistors for motor drive, such as slew rate control, dead time reduction and short-circuit protection. It is achieved by high level of integration, enabling real-time measurement of all relevant signals inside the power stage. Furthermore, these measurements are used to implement intelligent gate driving with controlled slew rate and minimum dead. Thanks to the driver design fully exploiting the superior technology properties of GaN, best-in-class efficiency and concurrently highest power density are achieved. On top of that, no heat sink is required for 250W operation leading to even higher system power density, cleaner PCB layout and smaller Bill-of-Materials. Furthermore,

lowest total harmonic distortion is achieved by the wide duty cycle range and small dead time, reducing audible noise and mechanical wear of the motor. The performance and all key parameters of this work are compared to recently released silicon-based IPMs in Table I, highlighting that the design of this work is able to reduce all losses and achieve highest efficiency and smallest package footprint.

Future work can include extension of the power and voltage range to support the broad variety of motors, such as e-scooters, power tools, higher power appliances and more.

ACKNOWLEDGMENT

The authors would like to thank Cetin Kaya from Texas Instruments for valuable discussions and guidance for this work.

REFERENCES

[1] Infineon, "IGCM06F60GA Datasheet CIPOSTM MINI," *Datasheet*, URL: https://www.infineon.com/dgdl/Infineon-IGCM06F60GA-DataSheet-v03_01-EN.pdf?fileId=5546d4624fb7fef2014fcafa55727821 January 2023.

[2] Mitsubishi Electric, "SLIMDIP-S," *Datasheet*, URL: https://www.mitsubishielectric.com/semiconductors/powerdevices/datasheets/dipipm/slimdip/slimdip-s_e.pdf , June 2023.

[3] STMicroelectronics, "SLLIMM nano 2nd series IPM, 3-phase inverter, 5 A, 1.0 Ω max., 600 V, N-channel MDmesh DM2 Power MOSFET," *Datasheet*, URL: https://www.st.com/resource/en/datasheet/stipq5m60t-hz.pdf , September 2023.

979-8-3315-1612-3/25 $31.00 © 2025 IEEE

[4] Sanken Electric Co. LTD., "600 V High Voltage 3-phase Motor Drivers SIM689xM Series," *Datasheet*, URL: https://www.semicon.sanken-ele.co.jp/sk_content/sim6890m_ds_en.pdf, June 2024.

[5] M. Germain et al., "GaN-on-Si power field effect transistors," Proceedings of 2010 International Symposium on VLSI Technology, System and Application, Hsinchu, Taiwan, 2010, pp. 171-172, doi: 10.1109/VTSA.2010.5488899.

[6] X. Li et al., "GaN-on-SOI: Monolithically Integrated All-GaN ICs for Power Conversion," 2019 IEEE International Electron Devices Meeting (IEDM), San Francisco, CA, USA, 2019, pp. 4.4.1-4.4.4, doi: 10.1109/IEDM19573.2019.8993572.

[7] Texas Instruments, "LMG2610 Integrated 650-V GaN Half Bridge for Active-Clamp Flyback Converters," *Datasheet*, URL: https://www.ti.com/lit/ds/symlink/lmg2610.pdf, December 2022.

[8] Texas Instruments, "DRV7308 Three Phase 650V, 5A, GaN Intelligent Power Module," *Datasheet*, URL: https://www.ti.com/lit/ds/symlink/drv7308.pdf, May 2024

[9] Alfred Hesener, "Fully-Protected Half-Bridge Power ICs Enable Motor-Integrated Inverters," *Technical Article*, URL: https://eepower.com/technical-articles/fully-protected-half-bridge-power-ics-enable-motor-integrated-inverters/, December 21, 2022.

[10] Eric Persson, Infineon Technologies, "Gate driver solution for GaN-based low-power motor control applications," in IEEE 2020 Applied Power Electronics Conference Industry Session IS30-4, 2020.

[11] P. Wang, A. Cavallini and G. C. Montanari, "The effect of impulsive voltage rise time on insulation endurance of inverter-fed motors," 2015 IEEE 11th International Conference on the Properties and Applications of Dielectric Materials (ICPADM), Sydney, NSW, Australia, 2015, pp. 84-87, doi: 10.1109/ICPADM.2015.7295214.

[12] STMicroelectronics, "SLLIMM nano 2nd series IPM, 3-phase inverter, 6 A, 600 V, short-circuit rugged IGBT," *Datasheet*, URL: https://www.st.com/resource/en/datasheet/stgipq4c60t-hz.pdf, September 2023.

[13] Power Integrations, "BridgeSwitch-2 Family," *Datasheet*, URL: https://www.power.com/sites/default/files/documents/bridgeswitch-2_family_data_sheet.pdf, May 2024.

FEA Based High-Frequency Synthesis for The Design and Optimization of GaN-Based Dual Three Phase Motor Drive System

Syed Imam Hasan
Dept. of Electrical and Computer Engineering
The University of Akron
Akron, OH, USA
sh328@uakron.edu

Alper Uzum
Dept. of Electrical and Computer Engineering
The University of Akron
Akron, OH, USA
au25@uakron.edu

Ashraf Siddiquee
Dept. of Electrical and Computer Engineering
The University of Akron
Akron, OH, USA
as802@uakron.edu

Yilmaz Sozer
Dept. of Electrical and Computer Engineering
The University of Akron
Akron, OH, USA
ys@uakron.edu

Krishna MPK Namburi
Research and Development
Nexteer Automotive
Auburn Hills, MI, USA
krishna.namburi@nexteer.com

Abstract—**Multiphase electric machines, particularly permanent magnet synchronous motors, are increasingly used across various industries for their high torque, power density, and low torque ripple. This paper presents the modeling of a high-frequency impedance loop and the optimization of loop inductance using finite element analysis. The derived model shows the impact of parasitic loop inductances on voltage overshoot and ringing across the switch. Besides, the importance of minimizing the parasitic loop inductance for Gallium Nitride (GaN)-based dual three-phase permanent magnet synchronous motor inverter are evaluated from the switching characteristics. The conventional strategies are employed to minimize the loop inductances and estimated inductances are incorporated in the high-frequency modeling. These findings are validated through simulation and experimental results which offer valuable insights into the optimization of switching behaviors in advanced motor drive systems.**

Keywords— PMSM, GaN, Dual Three phase Inverter, FEA, Wide bandgap device, Parasitic inductance.

I. INTRODUCTION

Wide bandgap (WBG) semiconductors have emerged as a transformative technology in power electronics, enabling the development of power converters that can operate at higher voltages and frequencies, while offering superior efficiency and compactness compared to traditional silicon (Si) devices [1]. Despite these advantages, the adoption of WBG devices in motor drive applications has been limited due to constraints related to the rate of voltage change (dv/dt), as excessive dv/dt can have detrimental effects on motor winding insulation and generate harmful bearing currents [2]-[4]. This challenge has slowed the widespread use of WBG devices in motor drives. However, ongoing research is focused on overcoming these limitations, as higher switching frequencies enabled by WBG devices significantly reduce the size and cost of passive filters, presenting a compelling opportunity for their integration into motor drive systems in the future.

To effectively utilize WBG devices in power converters, the design of the Printed Circuit Board (PCB) becomes crucial, as the fast switching characteristics of WBG devices lead to significant increases in both the di/dt and dv/dt, which can greatly affect the parasitic elements of the PCB [5]. Recently, GaN FETs have gained popularity for high-frequency applications due to their fast switching capabilities, absence of reverse recovery loss, and chip-scale packaging, which minimizes stray inductance and contributes to a reduction in the overall power loop inductance. This, in turn, helps enhance system efficiency and performance by minimizing parasitic effects. Although parasitic inductance from the PCB traces can induce undesirable gate-source and drain-source voltage ringing during switching transients, which can have severe implications for the overall performance and reliability of the system [6]-[8]. These effects necessitate careful design considerations for both the gate drive and power board circuits. In some cases, to achieve higher power ratings, paralleling GaN FETs become essential, which further requires layout optimization to minimize and match the common-source loop inductance between parallel switches and mitigate overshoot across the switch [9]. Consequently, layout optimization is a critical aspect of WBG switch applications.

In this paper, the analytical models are developed that incorporates both PCB and switch parasitics to provide valuable insights into the switching behavior influenced by these parasitic elements. To reduce the overall loop inductance in a dual three-phase motor drive system, mutual flux cancellation technique is used. Besides, to match the loop inductance between two parallel GaN FETs, symmetric layout design is employed. The impact of parasitic inductance on gate-source voltage ringing is demonstrated through simulations. A Finite Element Analysis (FEA)-based high-frequency synthesis is utilized to accurately estimate the power and gate loop inductance, providing a comprehensive understanding of the system's parasitic characteristics. From the derived model, the relationship between gate loop inductance and gate resistance is established, allowing for the estimation of the required external gate resistance for an optimized layout. Finally, the proposed design is validated through detailed testing and experimentation using a 2.5 kW dual three-phase inverter, confirming the effectiveness of the layout optimization in mitigating the adverse effects of parasitics and improving system performance.

979-8-3315-1612-3/25 $31.00 © 2025 IEEE

II. ANALYTICAL MODELING OF HIGH-FREQUENCY PARASITIC IMPEDANCE LOOPS

A. Power Loop Parasitic Impedance

An analytical model has been developed to identify the parasitic elements of power loop and gate loop, responsible for the predominant adverse effects induced by high di/dt (rate of change of current). This model facilitates the determination of the critical components to avoid high-frequency oscillations that can degrade the system performance or totally damage [3]. Due to the required rated current, the power loop formed by paralleling two GaN FETs, and high frequency parasitics are represented in Fig. 1(a). Only considering high frequency power loop with lumped inductances and capacitances, model can be simplified by single-loop LC circuit in Fig. 1(b). Transfer function for the voltages across the switches with respect to high-frequency ac current in the loop is derived (3). The PCB inductances (L_{PCB2}, L_{PCB4}), switch node inductance (L_{mid}) along with the switch stray inductances (L_{GA_T}, L_{GA_B}) lumped as loop inductance (L_{loop}), and resistances are neglected.

$$V_{DS_T}(s) = V_{DS_B}(s) = I_{ac_1}(s)\left(\frac{1}{s^2 C_{oss}} + s L_{GaN}\right) \quad (1)$$

$$I_{ac_1}(s) = \frac{C_{oss} I_{ac}(s)}{s^2 C_{oss} L_{loop} C_{HF} + C_{oss} + C_{HF}} \quad (2)$$

$$\frac{V_{DS}(s)}{I_{ac}(s)} = \frac{2s^2 C_{oss} L_{GaN} + 1}{2s(s^2 C_{HF} L_{loop} C_{oss} + C_{oss} + C_{HF})} \quad (3)$$

$$\omega_n = \sqrt{\frac{C_{oss} + C_{HF}}{C_{HF} L_{loop} C_{oss}}} \cong \sqrt{\frac{1}{L_{loop} C_{oss}}} \quad (4)$$

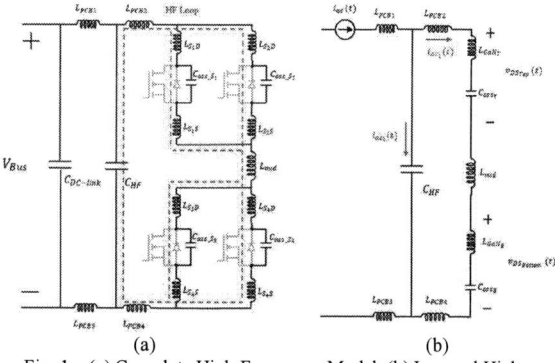

(a)   (b)

Fig. 1. (a) Complete High-Frequency Model; (b) Lumped High-Frequency Model

The natural frequency of the ringing voltage across the switches can be defined in (4), Loop inductance (L_{loop}), switch output capacitance (C_{oss}), and high-frequency capacitance (C_{HF}) selection plays a vital role in the safe operation of the device. However, the simplicity of the model and the effect of the high-frequency capacitor are not considered in (4). In generic second-order series RLC systems, the damping ratio (ζ) can be defined in (5) by using (4).

$$\zeta = \frac{R_{loop}}{2 L_{loop}} \frac{1}{w_n} = \frac{R_{loop}}{2} \sqrt{\frac{C_{oss}}{L_{loop}}} \quad (5)$$

Since output capacitances are part of the device characteristics, loop inductance minimization is essential to make system critically damped (ζ=1) without increasing the loop resistance which leads the conduction losses.

B. Gate Loop Parasitic Impedance

For the gate loop, during the turn-on process, the gate current (I_G) for the gate driver charges the gate capacitor ($C_{iss} = C_{GS} + C_{GD}$) through the gate resistor and the parasitics. The relation between gate current, input capacitance, and gate-source voltage can be represented by (6). Fig. 2 presents the parasitics at the gate loop where $L_{G1S1} + L_{S1S1} = L_{G1}$ and $R_{G1s1} + R_{s1s1} = R_{G1}$ are the total gate loop inductance and resistance, respectively. Using KVL, the gate-source voltage of the switch, S1 can be calculated by (7). Equations (6) and (7) can be used to derive the transfer function shown in (8) between gate drive voltage (V_{DR}) and gate-source voltage (V_{GS}).

$$I_G = C_{iss} \frac{dV_{GS}}{dt} \quad (6)$$

$$V_{GS} = V_{DR} - L_{G1} \frac{di_c}{dt} - R_{G1} I_c \quad (7)$$

$$\frac{V_{GS}}{V_{DR}} = \frac{\frac{1}{L_{G1} C_{iss}}}{s^2 + \frac{R_{G1}}{L_{G1}} s + \frac{1}{L_{G1} C_{iss}}} \quad (8)$$

From (8), the damping ratio (ξ) and the natural frequency (ω_n) of the at the gate loop can be calculated using the following relations.

$$\xi = \frac{R_{G1}}{2} \sqrt{\frac{C_{iss}}{L_{G1}}} \quad (9)$$

$$\omega_n = \frac{1}{\sqrt{L_{G1} C_{iss}}} \quad (10)$$

From (9), with the increase of the gate resistance and/or capacitance (paralleling GaN) the damping ratio increases whereas, with the increase of the gate loop inductance the damping decreases. However, as the gate capacitance (C_{iss}) is fixed and increase of the gate resistance (R_{G1}) will increase the switching losses, reducing gate loop inductance (L_{G1}) is an effective way to increase damping and reduce the overshoot. Besides, to meet RMS current requirement, in design, two parallel GaN FETs arrangement is used that increases C_{iss} and improves damping as presented in equation (8). However, paralleling GaN FETs can create other challenges. The formation of the common source inductance loop (Fig. 2), due to the paralleling of GaN FETs, the small mismatch between quasi-common source inductance (L_{QS}, L_{QS2}) will cause a voltage spike between the source of the two parallel GaN FETs as shown in (11).

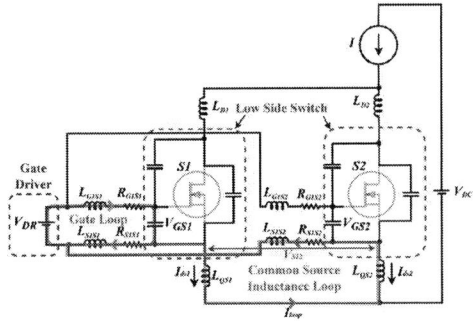

Fig. 2. Gate loop parasitics shown only at the low voltage side GaN FETs.

$$L_{QS1} \frac{dI_{ds1}}{dt} - L_{QS2} \frac{dI_{ds2}}{dt} =$$
$$L_{S1S1} \frac{dI_{loop}}{dt} + L_{S1S2} \frac{dI_{loop}}{dt} + R_{S1S1} I_{loop} \qquad (11)$$
$$+ R_{S1S2} I_{loop} = V_{S12}$$

Where, V_{S12} is the voltage between the two sources of parallel GaN FETs. These voltage drops between two sources can cause an overshoot at the gate to source voltage of the switch. To avoid this issue, the gate loop inductance between two parallel GaN FETs need to be same which validates the necessity of power and gate loop layout optimization.

III. HARDWARE DESIGN AND LAYOUT OPTIMIZATION USING FEA SIMULATIONS

The stray inductance associated with a GaN FET is provided by the manufacturer as 0.05 nH at a typical ringing frequency and is to be incorporated into FEA tools for modeling purposes. Additionally, high-frequency capacitors are represented within the model as a single copper bond wire. This is achieved by adjusting the dimensions of the bond wire to match with the equivalent series inductance (ESL) value specified in the datasheet. Furthermore, gate loop inductance is reduced by optimizing gate paths from the gate driver board to the power board. Each section undergoes individual optimization before being integrated to collectively reduce overall gate loop inductance. For the gate path, three different path structures are formed and analyzed in FEA tool. In Routing-1 (Fig. 4), the current flows through a single via, causing an increase in flux density and consequently raising the inductance along the path. Similarly, Routing-2 (Fig. 4), adopts the same strategy but incorporates a larger conductor from the bottom of the PCB. However, this leads to a longer gate path and increased inductance. In contrast, the proposed design Routing-3 (Fig. 4), aims to minimize inductance by offering multiple paths for the current, thereby reducing the skin effect.

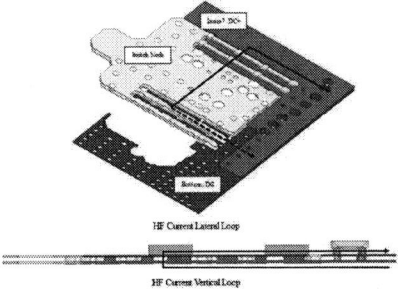

Fig. 3. High-frequency current paths of the power loop

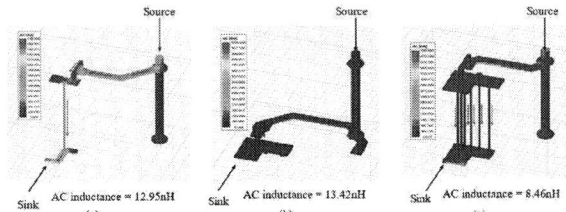

Fig. 4. Different gate path design strategies in the power board. (a) Routing-1, (b) Routing-2, (c) Routing-3

For the gate drivers, a double layer layout has been used which reduces the loop inductance more than the single-layer layout [4]. In the double-layer layout, the source, and the sink current flow through two adjacent layers in the PCB which ensures maximum mutual inductance between the trace (Fig. 5 (b)). From the relation between total loop, mutual and self-inductance as shown in (12), with the increase of the mutual inductance the overall loop inductance decreases.

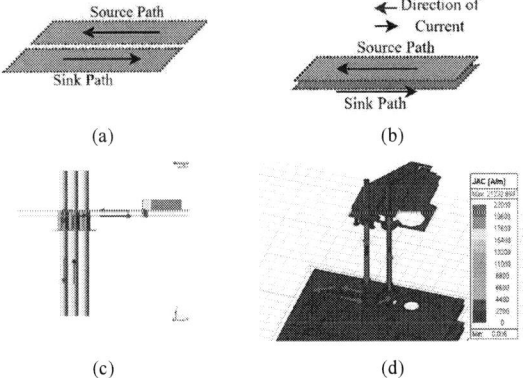

Fig. 5. Layout: (a) Single layer (b) Double layer (c) Double layer layout for Gate Driver (d) Total gate loop FEA simulation

$$L_G = L_{self_{source\_path}} + L_{self_{sink\_path}} - 2M \qquad (12)$$

Where, L_G is the total gate loop inductance, $L_{self_{source\_path}}$ and $L_{self_{sink\_path}}$ represent the self-inductance in the source and the sink path respectively and M is the mutual inductance between the source and the sink path. The optimized loop inductance of the power and the gate loop are shown in Table I.

TABLE I. DESIGNED INVERTER'S SPECIFICATIONS

Topic	Parameter	VALUE
PCB	Power loop Inductance, L_{loop}	3.4 nH
	Gate loop Inductance, L_G	11.6 nH
Inverter	DC Bus Voltage, V_{dc}	12 V
	Output Power, P_{out}	2.5 kW
	Switching frequency, f_{sw}	80 kHz
GaN FET	Part number	EPC2302
	Input Capacitance, C_{iss}	3200 pF
	Output Capacitance, C_{oss}	1000 pF
	Internal Gate Resistance, R_{gan}	0.5 Ω
Gate Driver	Part number	LMG1205
	Internal Resistance (Source Path)	2.1 Ω
	Internal Resistance (Sink Path)	0.6 Ω

By incorporating the inverter's parameters into the derived transfer function (8), the influence of gate loop inductance on the gate-to-source voltage ringing is demonstrated in Fig. 6(a) using MATLAB simulations. As shown in the figure, during the turn-on process, the gate-to-source voltage overshoot increases by approximately 16% for every 10 nH increase in the gate loop inductance, assuming no external gate resistance. Similarly, during the turn-off process, a lower internal gate resistance results in a higher voltage overshoot, as illustrated in Fig. 6(b).

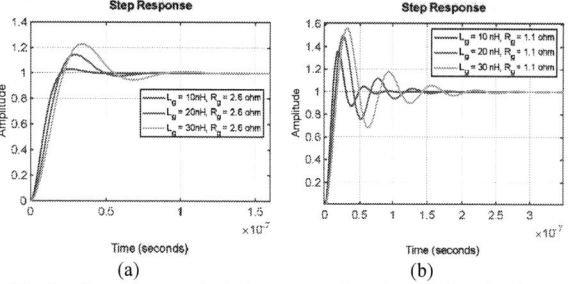

(a)   (b)

Fig. 6. Step response of gate to source voltage transfer function for different gate loop inductance (a) during switch turn on; (b) during switch turn off.

To mitigate the voltage overshoot, it is necessary to make the system approximately critically damped ($\zeta \approx 1$). This can be achieved by increasing the gate resistance. From equation (9), the required value of the external gate resistance can be calculated. Table II presents the values of the external gate resistance introduced to reduce the gate-to-source voltage ripple in the GaN FET. As shown in Fig. 7, the introduction of the external gate resistance results in a significant reduction in the voltage overshoot..

TABLE II.   SELECTED GATE RESISTANCE AND HIGH FREQUENCY CAPACITANCE

Parameter	VALUE
External Gate Resistance. R_G	2 Ω
Total Gate Resistance (during turn on)	4.6 Ω
Total Gate Resistance (during turn off)	3.1 Ω
High Frequency Capacitance, C_{HF}	400 nF

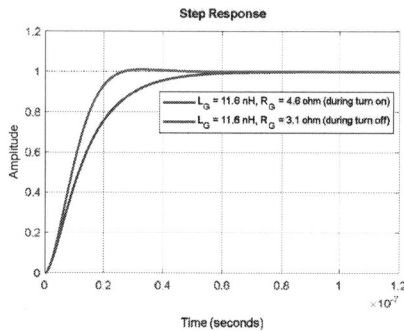

Fig. 7. Step response of gate to source voltage transfer function for the optimized gate loop inductance.

IV. EXPERIMENTAL VALIDATION

The design is validated through experiments conducted on the 2.5 kW dual three-phase inverter. Testing included a combination of double-pulse testing (DPT) for transient response analysis and RL load testing for steady-state performance. The experiments were performed under 25 A peak current condition, and the major system parameters are detailed in Tables I and II. The results are presented alongside simulation data to verify the accuracy of the proposed design

and layout optimization. The experimental setup, as shown in Fig. 8, includes a 2.5 kW dual three-phase inverter driving an RL load. The setup is equipped with measurement instruments to capture key waveforms, including line currents and line-to-line voltages. The hardware implementation integrates the optimized layout design, featuring minimized power and gate loop inductances, to mitigate parasitic effects.

The transient performance of the inverter was evaluated using a double-pulse test (DPT). The test results, presented in Fig. 9, illustrate the drain-to-source voltage waveforms during switching events. The results show excellent alignment with LTspice simulation data, confirming the effectiveness of the FEA-based modeling in predicting transient behavior. The optimized layout design achieved a power loop inductance of 3.4nH and a gate loop inductance of 11.6nH, which significantly reduced voltage overshoot and ringing. These improvements ensure stable operation under high di/dt conditions.

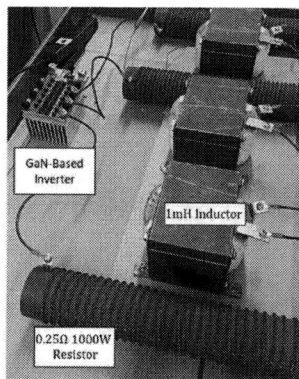

Fig. 8. GaN-based dual three-phase motor driver experimental setup

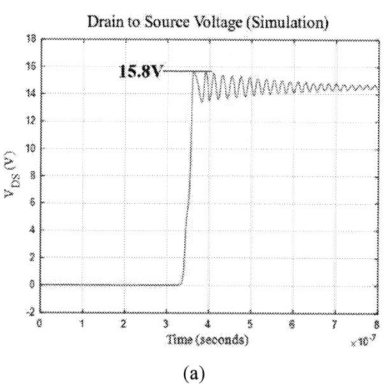

(a)

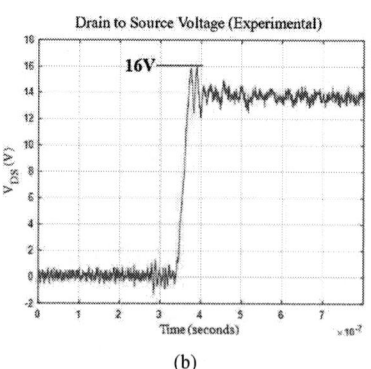

(b)

Fig. 9. Drain to Source voltage during switching transition (a) Simulation Result; (b) Experimental Result

The steady-state performance of the inverter was validated through RL load testing. The results, as shown in Fig. 10, highlight the sinusoidal three-phase line current waveforms and corresponding line-to-line voltage. The system delivered clean and symmetric line currents, demonstrating balanced operation across all phases. The inverter achieved an efficiency of 95.8%, with an input power of 216.96 W and an output power of 207.85 W. These results underscore the effectiveness of the design in maintaining high efficiency and low harmonic distortion.

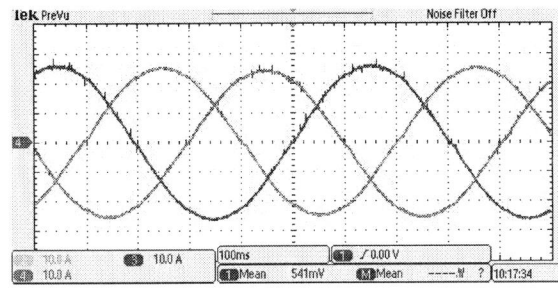

(a)

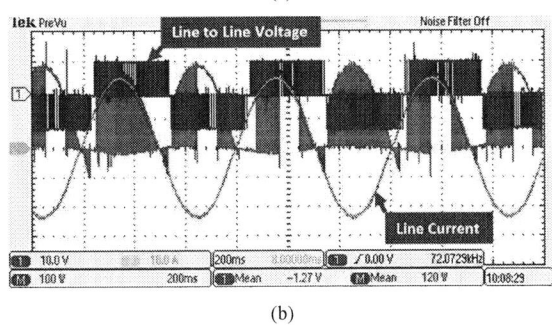

(b)

Fig. 10. Experimental results (a) three phase line currents; (b) line to line voltage and line current

The experimental validation of the proposed GaN-based dual three-phase inverter highlights its robust design and optimized layout for high-performance applications. The DPT demonstrated effective suppression of parasitic effects, including voltage overshoot and ringing, through the reduction of power and gate loop inductances. The RL load testing further validated the inverter's capability to deliver clean sinusoidal three-phase currents with high efficiency (95.8%) and balanced operation across all phases, ensuring stable and reliable performance. The close alignment between experimental results and simulation data confirms the accuracy of the FEA-based modeling and the efficacy of the proposed layout optimization, establishing the design's suitability for advanced motor drive systems.

V. CONCLUSIONS

In conclusion, this paper addresses the challenges associated with the use of GaN FETs in motor drive systems and proposes effective solutions. The effect of parallel GaN structure is addressed and symmetric layout structure is proposed to match the common source loop inductance. The non-linear models were derived to analyze the impact of parasitics on switching characteristics. To mitigate these effects, conventional layout optimization technique such as mutual flux cancellation and trace length reduction are used. FEA-based high-frequency synthesis was employed to predict accurate loop inductance for hardware design and layout optimization. Utilizing the developed model, the value of the external gate resistor for the optimized gate loop inductance was determined to achieve critical damping in the gate-to-source voltage. The results of LTspice simulations, incorporating non-linear models of the components, were validated against actual hardware test outcomes. Finally, the designed inverter was tested with an RL load, demonstrating low harmonic distortion and high efficiency, thereby confirming the efficacy of the proposed approach.

REFERENCES

[1] Y. Nakayama, Y. Kanazawa, F. Kondo, M. Inoue, K. Ohta, S. Doki, K. Shiozaki, "Efficiency Improvement of Motor Drive System by using a GaN Three Phase Inverter," *2019 IEEE International Conference on Industrial Technology (ICIT)*, Melbourne, VIC, Australia, 2019, pp. 1599-1604.

[2] Y. Xu, X. Yuan, F. Ye, Z. Wang, Y. Zhang, M. Diab, W. Zhou, "Impact of High Switching Speed and High Switching Frequency of Wide-Bandgap Motor Drives on Electric Machines," in *IEEE Access*, vol. 9, pp. 82866-82880, 2021.

[3] A. K. Morya, M. C. Gardner, B. Anvari, L. Liu, A. G. Yepes, J. Doval-Gandoy, H. A. Toliyat, "Wide Bandgap Devices in AC Electric Drives: Opportunities and Challenges," in *IEEE Transactions on Transportation Electrification*, vol. 5, no. 1, pp. 3-20, March 2019.

[4] A. Morya, M. Moosavi, M. C. Gardner and H. A. Toliyat, "Applications of Wide Bandgap (WBG) devices in AC electric drives: A technology status review," *2017 IEEE International Electric Machines and Drives Conference (IEMDC)*, Miami, FL, USA, 2017, pp. 1-8.

[5] M. E. Haque, S. Das, M. A. Rahman, M. Fesli, A. Chowdhury, A. Siddiquee, Y. Sozer, "FEA-Based Design, Optimization, and Validation of SiC Dual-Three Phase PMSM Inverter," *2023 IEEE Energy Conversion Congress and Exposition (ECCE)*, Nashville, TN, USA, 2023, pp. 1817-182.

[6] Y. Chen, W. Chen, D. Tong and X. Yang, "An Optimized PCB High Frequency Parasitic Parameter Elimination Method for GaN Based Driving Circuit," *2021 IEEE 1st International Power Electronics and Application Symposium (PEAS)*, Shanghai, China, 2021, pp. 1-6.

[7] B. Sun, K. L. Jørgensen, Z. Zhang and M. A. E. Andersen, "Research of Power Loop Layout and Parasitic Inductance in GaN Transistor Implementation," in *IEEE Transactions on Industry Applications*, vol. 57, no. 2, pp. 1677-1687, March-April 2021.

[8] K. Wang, L. Wang, X. Yang, X. Zeng, W. Chen and H. Li, "A Multiloop Method for Minimization of Parasitic Inductance in GaN-Based High-Frequency DC–DC Converter," in *IEEE Transactions on Power Electronics*, vol. 32, no. 6, pp. 4728-4740, June 2017.

[9] D. Reusch and J. Strydom, "Effectively paralleling gallium nitride transistors for high current and high frequency applications," *2015 IEEE Applied Power Electronics Conference and Exposition (APEC)*, Charlotte, NC, USA, 2015, pp. 745-751.

Evaluation of Passive Common-noise Canceller Considering Both of Thermal Equilibrium and Common-mode Noise Cancellation

Koji Mitsui
IHI Corporation
Yokohama, Japan
mitsui4110@ihi-g.com

Kenshiro Katsura
IHI Corporation
Yokohama, Japan
katsura6384@ihi-g.com

Koki Notake
IHI Corporation
Yokohama, Japan
notake3617@ihi-g.com

Koji Yamaguchi
IHI Corporation
Yokohama, Japan
yamaguchi2284@ihi-g.com

Abstract— This paper focuses on the design and evaluation of a Passive Common-noise Canceller (PCC) intended to mitigate Common Mode (CM) noise in large-capacity inverters and establish cooling structures of PCC for continuous operation. Electrification in large mobility such as aircraft for CO2 reduction requires large-capacity inverters and safety, making Electromagnetic Interference (EMI) reduction critical, and hence, this paper addresses this issue. The authors proposed cooling structures of the PCC with two different heat flow paths and applied to a motor drive system for evaluation. The effectiveness of the PCC was confirmed through significant cancellation in CM voltage and current. Moreover, the authors demonstrated that the cooling structures of the PCC are effective and operated continuously with natural air cooling, resulting in numerous advantages such as high environmental durability. This finding is critical as it proves the feasibility of the PCC, indicating its potential for practical application in large-scale inverters.

Keywords—Motor drive, Passive common-noise canceller, Pulse width modulation inverters, Noise Filters, Common-mode voltage, Electromagnetic interference

I. INTRODUCTION

Inverters, particularly those utilized in motor drive systems, inherently generate Electromagnetic Interference (EMI) that can potentially disrupt the operation of other equipment. Common Mode (CM) voltage is one of the causes of this EMI, and various reduction methods have been reported [1-4].

Contrary to the conventional methods, a Passive Common-noise Canceller (PCC) that can theoretically completely cancel the CM voltage, a CM current source, has been proposed [5]. Cancelling the CM voltage mitigates not only the noise but also electric erosion stress in motor bearings [6]. The PCC circuit configuration, characterized by its simplicity, consists of three magnetic cores and a Y capacitor. This study explores its compact and lightweight use in large-capacity inverters exceeding several hundred kVA such as aircraft applications for fuel efficiency improvement. Increasing capacity results in larger three power lines (UVW) sizes, hindering to wind the power lines multiple turns on the magnetic core. This paper, thus, focuses on a through-type PCC whose three magnetic cores are penetrated by the three power lines.

Since the maximum magnetic flux density in the through-type PCC core is much higher than in PCC with multiple winding of the power lines, it has been mentioned that the through-type PCC core suffers significant heat generation (iron loss) [5]. This hinders continuous operation with natural air

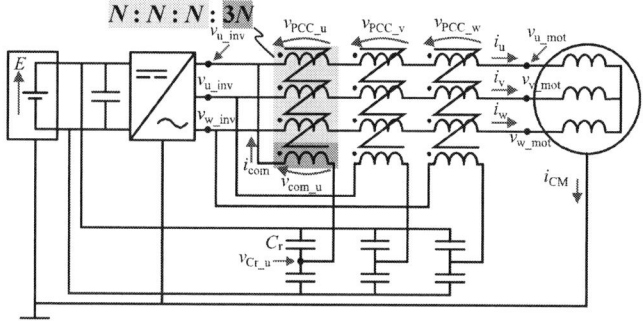

Fig. 1. Motor drive system with a through-type PCC.

cooling, which is desirable for numerous advantages such as high environmental durability required for aircraft applications, etc. To reduce the iron loss, increasing the core size and weight is necessary to reduce the magnetic flux density inside the core but increase of size and weight is not desirable. Currently, no known papers presented evaluations of cooling and continuous operation for the though-type PCC.

This paper evaluates that the performance of CM voltage cancellation of an inverter, and achievement of thermal equilibrium of the through-type PCC. The authors designed and fabricated two different shapes of magnetic core with different heat conduction paths to ambient air via aluminum cooling fins. These cores exhibit different iron losses for the same specification of an inverter and motor system as discussed later. The performance was evaluated by applying it to a 300-V/400-V 5-Arms 200-Hz motor drive system, focusing on the CM voltage removal performance and thermal equilibrium. This paper highlights major achievements in the through-type PCC design and the experimentation. First, the property of though-type PCC is introduced, followed by the design of the PCC with cooling fins for continuous operation with natural air cooling. Experimental results and conclusions, as well as future work, are then presented.

II. PROPERTY OF THE THROUGH-TYPE PCC WITH COOLING FINS

Fig. 1 shows the motor-drive system circuit diagram with a through-type PCC. There are three cores to impose an inverse of CM voltage for CM voltage cancellation. The winding turn ratio is $N : N : N : 3N$, with N is a turn number of the core. The through-type PCC has the advantage of not needing to wind on a magnetic core because the winding number $N=1$.

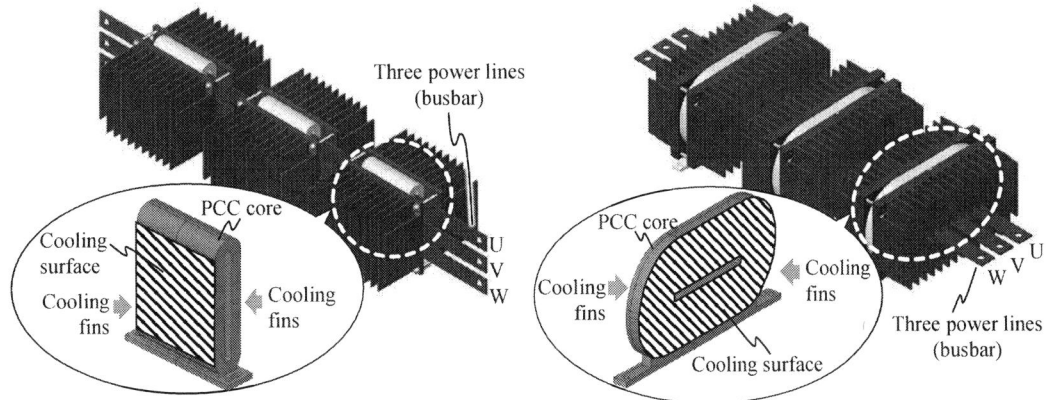

(a) Outer circumference cooling (Outer-cooling) (b) Window cross-section-cooling (Cross-section-cooling)
Fig. 2. Two cooling structures of the through-type PCC.

TABLE I. Parameters of the test condition.

	DC input voltage	Switching frequency	Inverter control method	Inverter output fundamental frequency	Y capacitor Cr
Value	400 V/ 300V	73 kHz	Open-loop	200 Hz	0.15 μF

The operating principle of the PCC is shown in this paragraph. The PCC is composed of three sets of CM transformers and a DC-side Y capacitor. This interprets the CM voltage v_{CM_inv} generated by the inverter as follows, and can cancel the CM voltage of the inverter using each of the three CM transformers [5].

$$v_{CM\_inv} = \frac{v_{u\_inv}}{3} + \frac{v_{v\_inv}}{3} + \frac{v_{w\_inv}}{3} \qquad (1)$$

Equation (1) shows that when cancelling the CM voltage, it does not cancel all at once with a   magnetic core, but cancels each phase separately. The voltage of the u phase on the inverter output side can be expressed by (2).

$$v_{u\_inv} = v_{com\_u} + v_{Cr\_u} \qquad (2)$$

where v_{com_u} is the compensation winding voltage of the u phase of the PCC, and v_{cr_u} is the voltage of the Y capacitor in the u phase. Since the impedane of the Y capacitor is to be negligible for the switching frequency, it can be assumed that v_{com_u} becomes the same voltage as v_{u_inv}. The power line voltage of the u phase of the PCC in that case is expressed by (3).

$$v_{PCC\_u} = \frac{v_{com\_u}}{3} \qquad (3)$$

The CM voltage on the motor side v_{cm_mot} is expressed by (4).

$$v_{CM\_mot} = v_{CM\_inv} - (v_{PCC\_u} + v_{PCC\_v} + v_{PCC\_w}) \qquad (4)$$

Since the v and w phases also have the power line voltage of the PCC in the same way, the CM voltage is theoretically cancelled from (1).

Because of the winding number $N=1$, the magnetic flux in the magnetic core becomes larger compared to multiple winding $N \geq 2$ resulting in a disadvantage of increased iron loss. Therefore, cooling the magnetic core in the through-type PCC is desired for its continuous operation.

Thus, the authors propose two PCC cooling structures as shown in Fig.2. The common point of these two structures is that the two cooling fins are attached both side of each magnetic

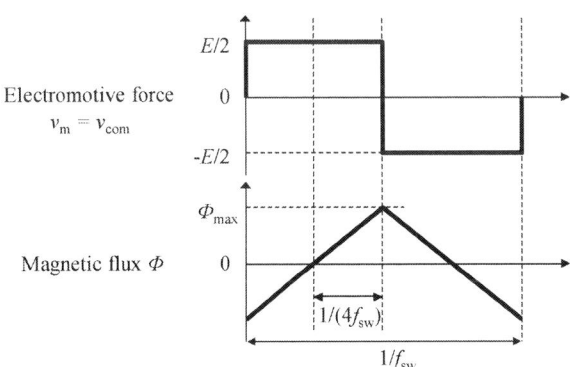

Fig. 3. Theoretical waveforms of electromotive force and flux linkage of CM transformer when flux linkage is maximum.

core, and the difference is the cooling surface of magnetic cores. Fig. 2(a) is the Outer-cooling structure (Outer-cooling), which is long in the axial direction (the direction in which the bus bar penetrates) and short in the circumferential direction, making it easy to dissipate heat from the outer surface of the magnetic core. Fig. 2(b) is the Cross-section-cooling structure (Cross-section-cooling), which is short in the axial direction and long in the circumferential direction, making it easy to dissipate heat from the cross section of the magnetic core.

III. DESIGN OF THE THROUGH-TYPE PCC FOR CONTINUOUS OPERATION

This section shows the design of the through-type PCC for continuous operation. First, the magnetic property design is shown followed by the thermal design of the PCC through the measurement of iron loss and a thermal circuit network of the PCC. Finally, the design of the cut-off frequency as an electrical filter is shown.

A. Magnetic design

Table I shows the electrical specifications of the inverter. The DC voltage is 300 V, the output voltage level is 2LV, the switching frequency is 73 kHz, the output fundamental

979-8-3315-1612-3/25 $31.00 © 2025 IEEE

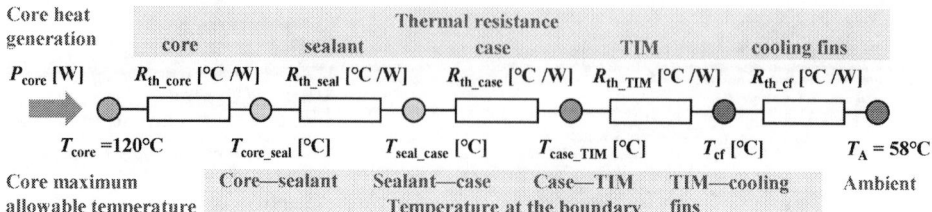

Fig. 4. Thermal circuit network of the PCC.

frequency is 200 Hz, and the output current is 5 Arms. As will be explained later, the test environment for noise evaluation and thermal evaluation is different, so the DC voltage was set to 300 V for thermal design. Fig. 3 shows the theoretical waveform of the induced electromotive force v_m and magnetic flux linkage when the magnetic flux linkage of the CM transformer is maximized. The induced electromotive force v_m is equivalent to the compensation winding voltage v_{com}. In the case of a 2-level three-phase inverter, the magnetic flux linkage is maximized when the duty ratio of the PWM signal is 50%. The maximum value of the magnetic flux linkage Φ_{max} is expressed by (5).

$$\Phi_{max} = \frac{E}{2} \cdot \frac{1}{4f_{sw}} = \frac{E}{8f_{sw}} \qquad (5)$$

where E is the input dc voltage of the inverter and f_{sw} is the switching frequency. Other expression of the maximum flux linkage shows the relation among the winding turns of the compensation winding, maximum magnetic flux density B_{max}, and the effective cross-sectional area of the core A_e in (6).

$$\Phi_{max} = 3NB_{max}A_e \qquad (6)$$

From (5) and (6), A_e is calculated as follows.

$$A_e = \frac{E}{24NB_{max}f_{sw}} \qquad (7)$$

The winding turn ratio N is set to 1 since this paper targets the through-type PCC. Under the experimental conditions $E = 300$ V, $f_{sw} = 73$ kHz, $B_{max} = 0.43$ T which is 36% of the saturation flux density of 1.2 T, the effective cross-sectional area Ae is designed to 401 mm^2 from (7). Considering the fill factor of the magnetic core material of 0.78, the actual core cross-sectional area A_{ea} becomes 514 mm^2.

In order to reduce the excitation current in the frequency range above the switching frequency of 73 kHz, the authors selected CATECH's nanocrystalline soft magnetic material 1K107B, which has a high relative permeability of 34000 at 70 kHz. The saturation magnetic flux density is 1.2 T.

B. Thermal and structure design

Fig. 4 shows the thermal circuit network of the core. The thermal circuit network is common for Outer-cooling and Cross-section-cooling structures. Under the conditions of an ambient temperature of 58 °C and an allowable maximum core temperature of 120 °C, the authors will explain the method of calculating the requirement of the heat sink thermal resistance that is thermally established in two patterns: Cross-section-cooling and outer-cooling structures. First, the iron loss P_{core} is estimated. Second, the thermal resistance of each material are calculated. Finally, the requirement of the thermal resistance for the cooling fins are achieved for the continuous operation.

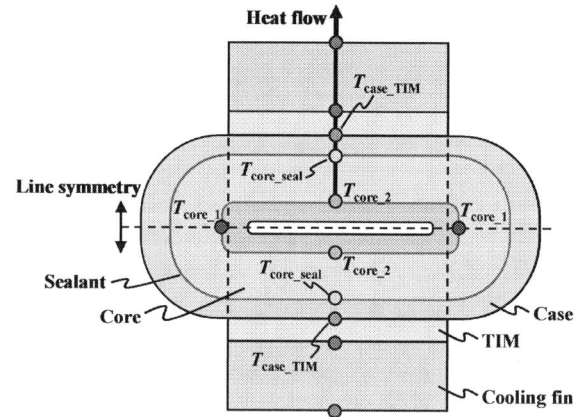

(a) Outer circumference cooling (Outer-cooling)

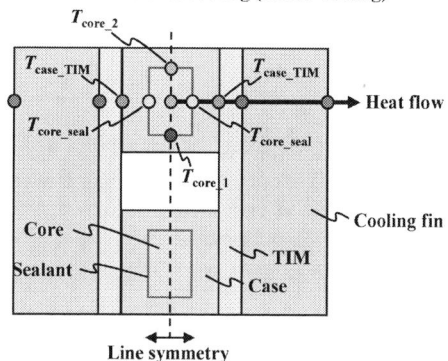

(b) Window cross-section-cooling (Cross-section-cooling)

Fig. 5. Material of two cooling structures and simplified heat flow paths.

To design the through-type PCC for continuous operation, the authors measured heat generation (iron loss) of the magnetic core using a PCC prototype made from the same magnetic material. The Steinmetz model is used to estimate the iron loss of the PCC core per weight p_{core} expressed as (8).

$$p_{core} = kf_{sw}{}^a B_{max}{}^b \qquad (8)$$

where k, a and b are constants to be fitted, a function of DC input voltage and switching frequency of the inverter [5]. The experimentation identified the constants to calculate the iron loss from the switching frequency and the DC input voltage. The switching frequency characteristic matches well with the approximate straight line, and the core iron loss is proportional to the 1.56 power of the switching frequency, as shown in (9).

$$P_{core} = k_1 f_{sw}{}^a = 0.0773 \, f_{sw}{}^{1.56} \qquad (9)$$

The maximum magnetic flux density characteristic matches well with the approximate straight line, and the core iron loss is proportional to the 2.12 power of the maximum magnetic flux density, as shown in (10).

TABLE II. Thermal resistance of each material of PCC components.

Material of the PCC components	Common values		Outer-cooling structure		Cross-section-cooling structure	
	Thermal resistance [K/W]	Thermal conductivity [W/(m·K)]	Length in the direction of heat flow [mm]	Cross-sectional area orthogonal to the heat flow [mm²]	Length in the direction of heat flow [mm]	Cross-sectional area orthogonal to the heat flow [mm²]
Magnetic core	$R_{\text{th\_core}} = \dfrac{l_{\text{core}}}{\lambda_{\text{core}} S_{\text{core-seal}}}$	$\lambda_{\text{core}} = 0.57$	$l_{\text{core}} = 5.7$	$S_{\text{core-seal}} = 191$	$l_{\text{core}} = 10/2 = 5$	$S_{\text{core-seal}} = 423$
Sealant	$R_{\text{th\_seal}} = \dfrac{l_{\text{seal}}}{\lambda_{\text{seal}} S_{\text{seal-case}}}$	$\lambda_{\text{seal}} = 0.8$	$l_{\text{seal}} = 1$	$S_{\text{seal-case}} = 195$	$l_{\text{seal}} = 1$	$S_{\text{seal-case}} = 440$
Case	$R_{\text{th\_case}} = \dfrac{l_{\text{case}}}{\lambda_{\text{case}} S_{\text{case-TIM}}}$	$\lambda_{\text{case}} = 0.16$	$l_{\text{case}} = 2$	$S_{\text{case-TIM}} = 204$	$l_{\text{case}} = 2$	$S_{\text{case-TIM}} = 473$
TIM	$R_{\text{th\_TIM}} = \dfrac{l_{\text{TIM}}}{\lambda_{\text{TIM}} S_{\text{TIM-cf}}}$	$\lambda_{\text{TIM}} = 13$	$l_{\text{TIM}} = 0.5$	$S_{\text{TIM-cf}} = 204$	$l_{\text{TIM}} = 0.5$	$S_{\text{TIM-cf}} = 473$

$$P_{\text{core}} = k_2 B_{\max}^{\,b} = 203\, B_{\max}^{2.12} \qquad (10)$$

The proportionality constant k in the (8) can be calculated by (11) from (9) and (10).

$$k = \frac{k_1}{B_{\max}^{\,b} W_{\text{core}}} = 0.316 \qquad (11)$$

where W_{core} is the weight of the core. If the material property of the core remains the same, this function can be used universally for PCC design.

Next, this paragraph shows the calculation method of the thermal circuit network of the core to ambient air to achieve the required thermal resistance $R_{\text{th\_hs}}$ of the cooling fins.

$$R_{\text{th\_cf}} = \frac{T_{\text{core}} - T_A}{P_{\text{core}}} - (R_{\text{th\_core}} + R_{\text{th\_seal}} + R_{\text{th\_case}} + R_{\text{th\_TIM}}) \qquad (12)$$

where T_{core} is the allowable maximum core temperature, T_A is the ambient air temperature, and $R_{\text{th\_x}}$ is the thermal resistance of the PCC components. Equation (12) is utilized to calculate the lower limit of the thermal resistance of the cooling fins to air which is required to keep the maximum core temperature below the maximum allowable temperature.

To achieve the thermal equilibrium, it is necessary to simultaneously consider the dimensions of the core determined by the dimension of the window (the three power lines of the inverter to be penetrated) and magnetic cross-sectional area of the core (determined by switching frequency and DC input voltage of the inverter) and the thermal resistance of the core calculated from its material property.

Fig. 5 shows the material of the PCC and simplified heat flow paths considered in the two cooling structures, it is assumed that heat is dissipated only from the surface where the core (case) and the cooling fins are in contact, and all other surfaces are insulated. Also, in the Outer-cooling structure, only the area indicated in green as the thermal path inside the core is considered. The heat flow paths are considered in only one side of the PCC because the heat flow paths are line symmetry.

Table II shows the thermal resistance of each material of the PCC components. Thermal resistance is calculated from the thermal conductivity of the material, the length of the heat flow path, and the cross-sectional area through which the heat is transmitted. Each shape parameter of the PCC components was determined considering the heat dissipation area and the feasible range.

TABLE III. Required and selected thermal resistance of the cooling fins.

	Outer-cooling structure	Cross-section-cooling structure
Required thermal resistance [W/(m·K)]	0.67	0.72
Designed thermal resistance [W/(m·K)]	0.625	0.584

TABLE IV: shape parameters of the fabricated PCC and the CM choke for experimental comparison.

	Outer-cooling	Cross-section-cooling	CM choke
Effective cross-sectional area A_e [mm²]		401	
Core weight W_{core} [kg]	0.79	1.21	0.8
Excitation inductance L_m [mH]	0.54	0.35	0.56

Finally, the required thermal resistance was calculated and designed thermal resistance was selected as shown in Table III. The thermal interface material (TIM) is the silicon-gel sheet (GR130A, Fuji Polymer Industries). Table IV shows the shape parameters of the fabricated PCC and the CM choke for experimental comparison. As will be mentioned later, unlike the PCC, the CM choke does not cancel the CM voltage. Therefore, it is described here for comparison as a filter that is generally applied.

C. Design of cut-off frequency

This demonstrates the design method for the resonance frequency of the through-type PCC. The excitation inductance L_m of the CM transformer core is calculated by (13).

$$L_m = \frac{9\mu A_e}{l_e} \qquad (13)$$

where μ is the permeability and l_e is the average magnetic path length. The following relationship holds between the maximum excitation current $I_{\text{m\_max}}$ and the maximum chain magnetic flux $\Phi_{\max}$.

$$I_{\text{m\_max}} = \frac{\Phi_{\max}}{L_{\max}} = \frac{B_{\max} l_e}{3\mu} \qquad (14)$$

The excitation current decreases as the average magnetic path length becomes shorter under the condition that the maximum magnetic flux density is constant. The cut-off frequency f_c composed of the excitation inductance of the CM transformer and the Y capacitor is set between the inverter output frequency f_{out} and the switching frequency f_{sw}. This allows only the high-frequency components above the switching frequency to be

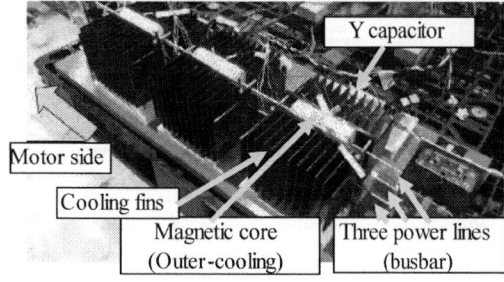

Fig. 6. Experimental environment.

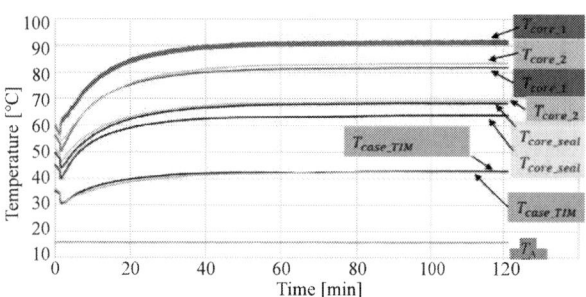

Fig. 7. Thermal equilibrium of Outer-cooling PCC.

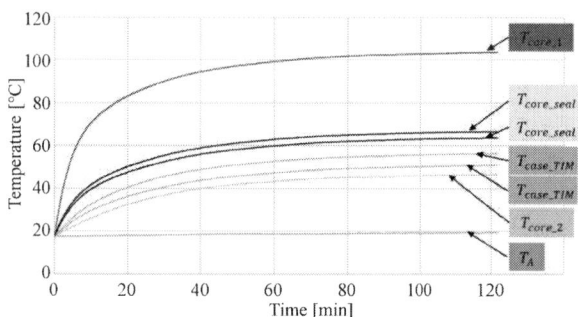

Fig. 8. Thermal equilibrium of Cross-section-cooling PCC.

removed. The capacitance C_r of the Y capacitor can be calculated by (15).

$$C_r = \frac{1}{2L_m(2\pi f_c)^2} \tag{15}$$

IV. EXPERIMENTAL EVALUATION

A. Experimental configuration

To evaluate the thermal equilibrium of the PCC, the thermal equilibrium test for continuous operation is conducted. Also, to evaluate the CM noise reduction, the CM voltage and CM current are measured by the experimentation.

The authors assembled both Outer-cooling and Cross-section-cooling test equipment composed of the PCC core and cooling fins, designed using the proposed method, as shown in Fig. 6. The Outer-cooling type is described for representative in this paper. In order to compare the noise reduction rate of the PCC, measurements were also taken using the CM choke as a conventional method.

Table I lists the test conditions used in the experiment. The DC input voltage of 400 V for the validation of CM noise cancelling effect, and 300 V for the validation of thermal

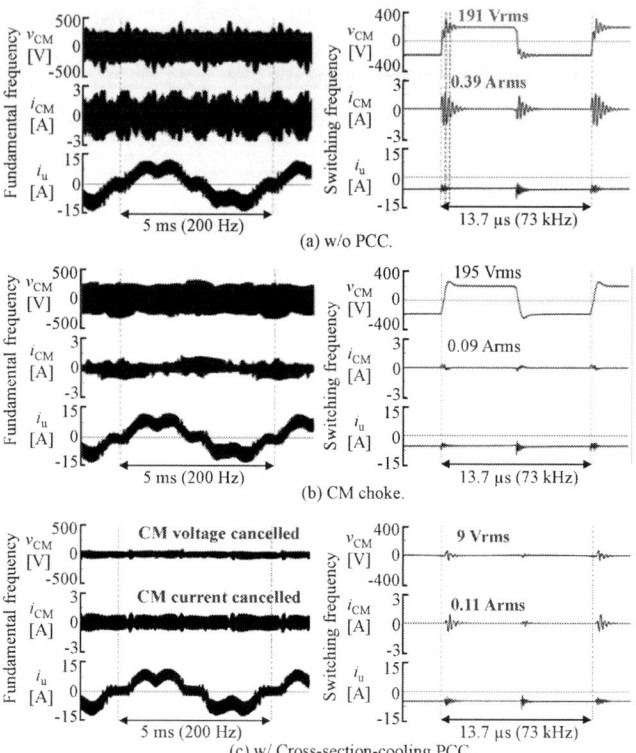

Fig. 9. CM voltage, CM current, and u-phase current comparison w/o and w/ PCC.

equilibrium due to the circumstance of the experimental environment.

A virtual common mode voltage measurement point was extracted by a star connection capacitors in to the motor terminals to measure the common mode voltage. The common mode current was measured by clamping a current probe around the 3-phase cables between the motor and PCC. The cooling fins are not grounded in this experimental setup.

B. Experimental results

Figs. 7 and 8 illustrate the thermal equilibrium temperature of the Outer-cooling and Cross-section-cooling PCCs. The surface temperature of the PCC core was observed to stabilize under natural air cooling, demonstrating the capability for continuous operation. The measured temperatures, which represent the average values in the case of line symmetry, of each part of the PCC core roughly align with the predictions of the thermal circuit network design. Nevertheless, certain temperatures measured were higher than those projected by the thermal circuit network calculations. This discrepancy can likely be attributed to the design's assumption of uniform core heat generation, while in reality, heat generation may not be uniform due to the presence of non-uniform magnetic flux inside the core [7]. The variation in dimensions of the core casing, sealant, and the other components could also contribute to the difference between the measured and calculated values. Moreover, the possibility of higher temperatures at the vertical upper part due to air convection is suggested. These findings underscore the importance of achieving a balance between thermal and electromagnetic design considerations.

979-8-3315-1612-3/25 $31.00 © 2025 IEEE

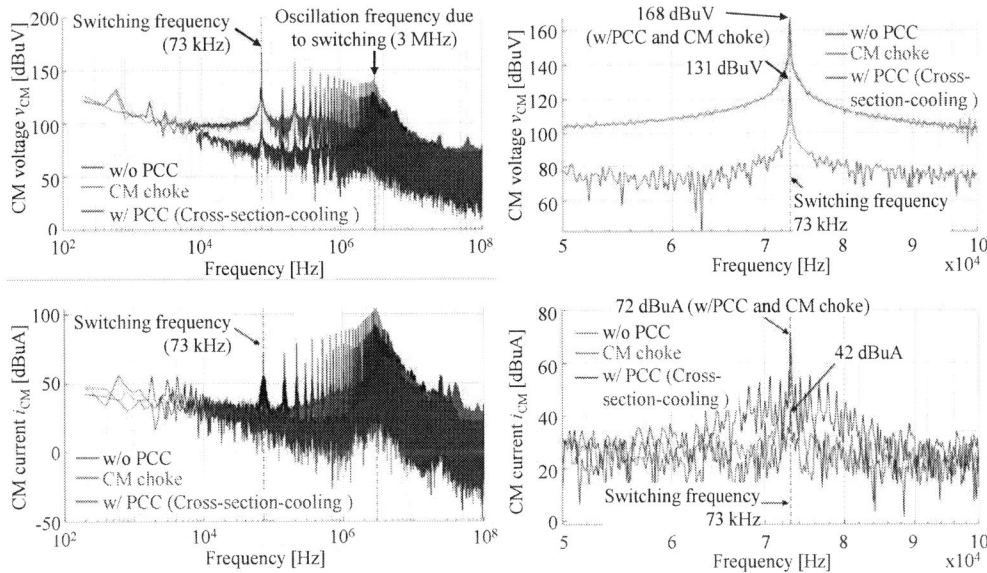

Fig. 10. CM voltage, CM current, and u-phase current comparison w/o and w/ PCC.

Fig. 9 compares the CM voltage, CM current, and u-phase current with and without the evaluated PCC. The PCCs significantly cancel both CM voltage and current, with the CM voltage decreasing by 28 dB at the switching frequency for the Cross-section-cooling PCC. Correspondingly, the CM current also experiences a decrease, with reductions of 42 dB. On the other hand, the conventional method, the CM choke, has no function to reduce the CM voltage, so it did not reduce at all.

Fig. 10 shows the frequency characteristics of the CM voltage and current when there is no PCC, when the CM choke is applied, and when the PCC (Cross-section-cooling) is applied. Using the PCC, whose thermal stability was confirmed, it was verified that the effect of significantly reducing CM noise was confirmed. The results of outer-cooling are similar in trend to those of cross-section-cooling, with the relative error in the reduction rate in dB being within 4%. Thus, the results of outer-cooling are omitted in this paper.

V. CONCLUSIONS AND FUTURE WORK

In this paper, the authors have pioneered the development of a Passive Common-noise Canceller (PCC) that has significant cancellation potential of CM noise and proposed a thermal design method of the PCC for continuous operation with natural air cooling. This achievement results in numerous advantages such as no need for power for cooling and high environmental durability.

This study is currently in the phase of establishing a design methodology. The dimensions and weight of the prototype developed for the purpose of validating the design's appropriateness have not yet been optimized for specific applications. Hence, there is considerable scope for comparative examination of various materials for the core. Similarly, various materials and cooling methods for the cooling fins, such as forced air cooling, water cooling, and oil cooling, also warrant comparative study. Further research in these areas will contribute to refining the design process.

REFERENCES

[1] E. Zhong and T. A. Lipo, "Improvements in EMC performance of inverter fed motor drives," IEEE Transactions on Industry Applications, vol. 31, no. 6, pp. 1247—1256, 1995.

[2] D. A. Rendusara and P. N. Enjeti, "An improved inverter output filter configuration reduces common and differential modes dv/dt at the motor terminals in PWM drive systems," IEEE Transactions on Power Electronics, vol. 13, no. 6, pp. 1135—1143, 1998.

[3] S. Ogasawara, H. Ayano, and H. Akagi, "An active circuit for cancellation of common-mode voltage generated by a PWM inverter,", IEEE Transactions on Power Electronics, vol. 13, No. 5, pp. 835—841, 1998.

[4] M. M. Swamy, K. Yamada, T. Kume, "Common mode current attenuation techniques for use with PWM drives," IEEE Transactions on Power Electronics, vol. 16, no. 2, pp. 248—255, 2001.

[5] S. Ohara, Y. Kawada, S. Ogasawara, and K. Orikawa, "Passive common-noise canceller for cancelling the common-mode voltage generated by inverters," IEEJ Transactions on Industry Applications, vol. 142, no. 11, pp. 825—834, 2022. (In Japanese)

[6] D. Hyypio, "Mitigation of bearing electro-erosion of inverter-fed motors through passive common-mode voltage suppression," IEEE Transactions on Industry Applications, vol. 41, no. 2, pp. 576-583, 2005.

[7] M. K. acki, M. S. Ryłko, J. G. Hayes, C. R. Sullivan, "Analysis and experimental investigation of high-frequency magnetic flux distribution in Mn-Zn ferrite cores," IEEE Transactions on Power Electronics, vol. 38, no. 1, pp. 703—716, 2023.

Performance Evaluation of Isolated DC/DC Converters in Modularized Bridge Rectifier Solid-State Transformer

Zhenchao Li, Andrea Cervone, and Drazen Dujic

Power Electronics Laboratory - PEL

École Polytechnique Fédérale de Lausanne - EPFL

Lausanne, CH-1015, Switzerland

zhenchao.li@epfl.ch, andrea.cervone@epfl.ch, drazen.dujic@epfl.ch

Abstract—**Modularized Bridge Rectifier (mBR) is a promising architecture of single-stage solid-state transformer. Without the presence of AC/DC active rectifier, the system performance solely depends on the control and optimization of isolated DC/DC converters. This paper presents a comprehensive analysis of DC/DC converters in mBR, covering topology selection, efficiency optimization, and design considerations. Additionally, under the specifications of $750\,\text{kW}$ and $10\,\text{kVac}/760\,\text{Vdc}$, two existing control methods of mBR are compared with a focus on their impact on the performance of DC/DC converters.**

Index Terms—**solid-state transformer, isolated DC/DC converter, dual active bridge converter, efficiency**

I. INTRODUCTION

Solid-State Transformer (SST) is a high-power converter with integrated high-frequency galvanic isolation. Due to its enhanced power density and functionality, SSTs are promising for applications such as data centers, EV charging, and distribution grids. To improve scalability and flexibility, many SST designs adopt the modular architecture [1], typically with input-series output-parallel (ISOP) configuration for medium-to-low voltage conversion. Among these, the cascaded H-bridge (CHB) based topology, depicted in Fig 1a, is commonly used. The CHB, composed by active switches, acts as a front-end rectifier to convert medium-voltage AC (MVAC) into separate and constant medium-voltage DC (MVDC). Under the fixed input and output voltage, the following isolated DC/DC converters can then efficiently transfer MVDC into low-voltage DC (LVDC). However, this two-stage architecture suffers from the semiconductor losses, because of the hard-switched CHB, which limit the achievable efficiency. [2].

To address this issue, a new SST family named modularized Diode Rectifier (mDR) was recently proposed by [3] and analyzed by [4]–[6]. As one type of mDR, the modularized Star Rectifier (mSR) [4], shown in Fig. 1b, uses only one diode to replace the active H-bridge in CHB-based topology, significantly reducing system complexity, losses and cost. When mSR is operating, the diodes in the branch with minimum phase-to-natural AC voltage are conducting, and the corresponding isolated DC/DC converters are deactivated. The other diodes are in blocking state and provide pulsed unipolar

voltage to the corresponding isolated DC/DC converters. However, to maintain continuous current flow on the grid side, the isolated DC/DC converters in mSR have to absorb energy from the load during some intervals. This increases the magnitude of positive power flow to ensure certain average power required by the load.

As an extended version, the modularized Bridge Rectifier (mBR) shown in Fig 1c provides additional paths for grid currents by reconfiguring half of the submodules in mSR into another cluster [3]. The control degrees of freedom increase from four to six due to more independent branches. Four are dedicated to ensuring symmetrical grid currents with the desired phase angle and magnitude, while the remaining two can be utilized for performance optimization. With different optimization targets, two different control methods were separately proposed by [5] and [6]. Method 1 proposed by [5] aims to eliminate energy backflow in the isolated DC/DC converters throughout the entire grid cycle. Method 2 introduced by [6] reduces the peak power of isolated DC/DC converters by 22.5% compared to method 1, but results in a wider operating voltage range. Fig.2 compares the operation waveforms of isolated DC/DC converters under these two control methods based on the system specifications listed in Table I.

Since mBR has only one power conversion stage, the system's performance relies entirely on the isolated DC/DC converters. Therefore, a quantitative comparison of how different control methods affect the performance of these converters is essential, and is presented in this work. Additionally, the topology selection, design, and operation of the isolated DC/DC converters must be adapted to the unique operation waveforms of mBR, which is also addressed in this paper.

II. TOPOLOGY SELECTION AND MODULATION STRATEGY OF ISOLATED DC/DC CONVERTER

A. Topology Selection

Similarly to the standard CHB-based SST, the isolated DC/DC converters in mBR can be designed to operate at frequencies significantly higher than the AC grid, thereby reducing the volume and weight of passive components. However, the biggest challenge is that these DC/DC converters

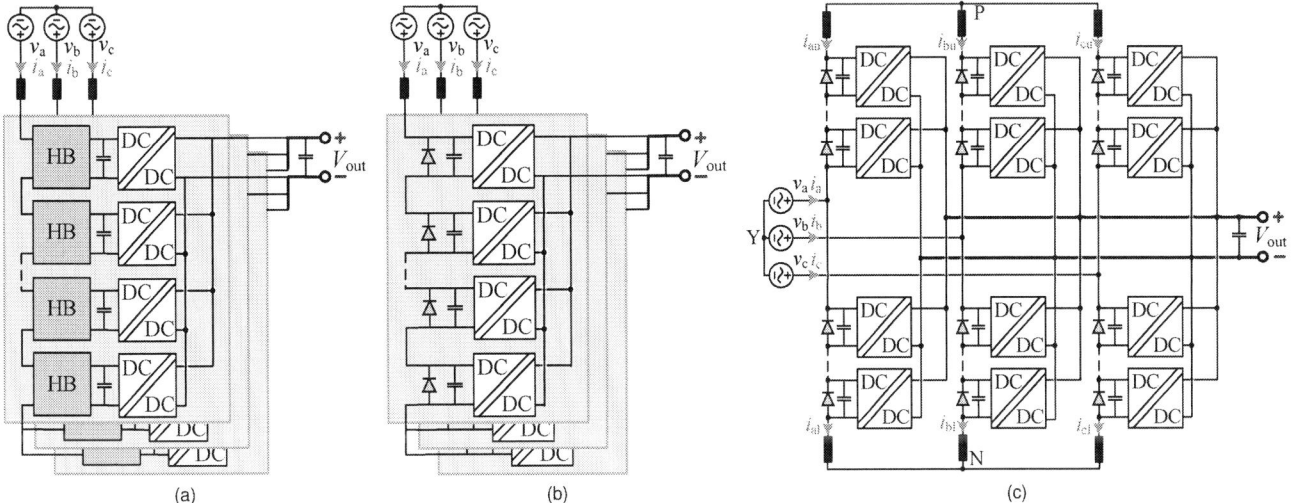

Fig. 1. Topologies of Solid-state Transformer. (a) CHB-based SST. (b) mSR SST. (c) mBR SST

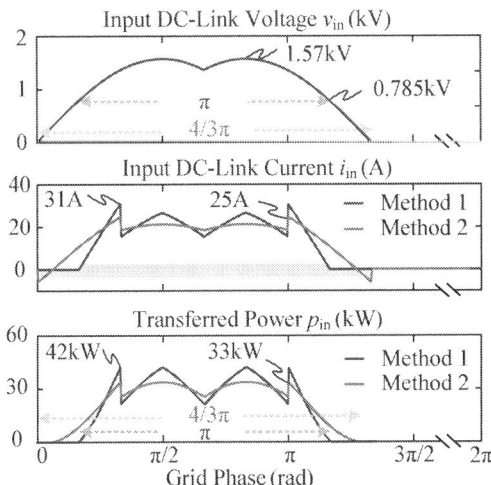

Fig. 2. Waveforms of DC/DC converter under different control methods

TABLE I
SYSTEM SPECIFICATIONS OF MBR SST

Parameter	Meaning	Value
P_{sys}	Rated Power	750kW
U_{AC}	Line-to-Line Grid Voltage	10kV (RMS)
I_{AC}	Grid Current	43.3A (RMS)
ω_{AC}	Grid Frequency	$2\pi \times 50$ rad/s
V_{out}	Output LVDC Voltage	760V
N	Submodule per Arm	9
S_{pri}	Primary Switches	G2R50MT33K
S_{ecs}	Secondary Switches	G3R20MT12K
f_{sw}	Switching Frequency	50kHz
T	MFT Core Material	Nanocrystalline

must be able to operate efficiently with significantly varying input voltage and power flow. Hence, resonant converters such as LLC and CLLC are unsuitable for mBR applications due to their limited capability to handle large input voltage variations while maintaining soft-switching operation. Besides, to achieve low-distortion grid currents, DC/DC converters must rapidly track input current commands that fluctuate at the grid frequency and may experience sudden changes. Current-source converters, such as the current-fed full-bridge converter, typically have limited bandwidth for current tracking due to the presence of input inductor. Dual-active-bridge converter (DAB) is a good candidate for that requirement since its input current can be regulated within one switching cycle. In addition, the highly efficient operation of DAB can be guaranteed by combining different phase-shift modulation schemes, such as extended-phase-shift (EPS) and dual-phase-shift (DPS). More importantly, the bidirectional power flow can be easily

achieved by applying positive or negative phase-shift between H-bridges, and this feature is necessary for mBR's four-quadrant operation [6]. Converters with secondary-side diode rectifiers, such as single-active bridge converter (SAB) cannot meet this requirement. In summary, the selection criteria and comparison of the isolated DC/DC converters in mBR are summarized in Table II, and the DAB converter is ultimately selected for further investigation.

B. Modulation Strategy

Given the unique input voltage and power profile of the DAB converter in mBR, a dedicated modulation strategy is necessary to enhance its efficiency. According to [7], twelve switching patterns can be generated by DAB, regarding the different sequence of rising and falling edge of two-side midpoint voltage v_{AB} and v_{CD}. Six of them are selected depending on the voltage conversion ratio (i.e. operation as

TABLE II
SELECTION CRITERIA AND COMPARISON OF ISOLATED DC/DC CONVERTERS IN MBR

Topology		Soft Switching	Dynamic Response	Bidirectional	Low Input Capacitance	Simplicity
Current Source		★★★	★	★★★	★★★	★
Voltage Source	Resonant	★★	★	★★	★	★
	DAB	★★★	★★★	★★★	★★	★★★
	SAB	★	★★★	No	★★	★★★

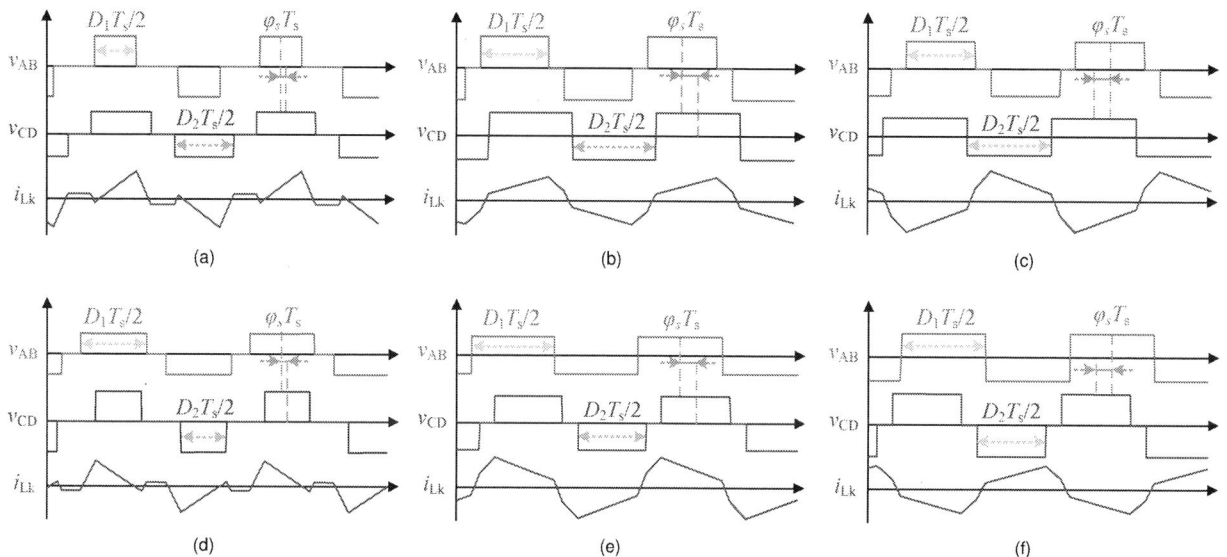

Fig. 3. Selected Switching Patterns for DAB Converter. (a) Mode **BuL**: Pos/Neg Buck Low Power. (b) Mode **BuHP**: Pos Buck High Power. (c) Mode **BuHN**: Neg Buck High Power. (d) Mode **BoL**: Pos/Neg Boost Low Power. (e) Mode **BoHP**: Pos Buck High Power. (f) Mode **BoHN**: Neg Boost High Power

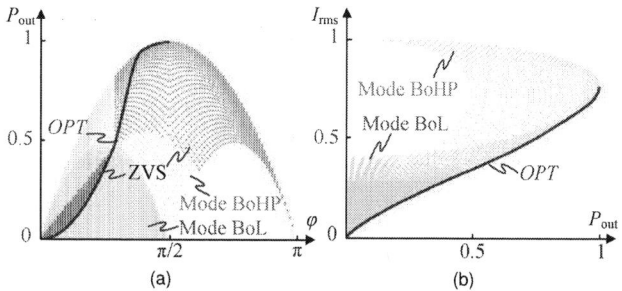

Fig. 4. Operation Trajectories of Optimal Modulation Strategy with $M = V_2/(nV_1) = 1.67$. (a) Normalized Power P_{out} and Phase-shift φ. (b) Normalized RMS Current I_{rms} and Normalized Power P_{out}

Buck or as Boost) and the transferred power (i.e. operation with positive/negative or low/high power), as shown in Fig 3. Mode **BuL** and **BoL** are particularly advantageous, achieving zero-voltage soft switching (ZVS) even with low transferred power. The remaining four modes utilize typical EPS modulation, which is suitable for heavy-load operation.

The different modulation patterns depend on three parameters: φ_s, D_1 and D_2. φ_s is the phase-shift ratio between the primary and secondary side, and is determined by the transferred power. D_1 and D_2 are the duty cycles of primary and secondary side midpoint voltage. The selection of optimal

D_1 and D_2 is aimed at minimizing the RMS current while preserving zero-voltage soft switching (ZVS) in the widest operation range, which is a compromise between conduction loss and switching loss. In the low-power mode, ZVS is prioritized due to the relatively low conduction loss. While in the high-power mode, the minimized RMS current is ensured by using the Lagrange Multiplier Method [10]. For instance, the analytical expressions for D_1 and D_2 in the positive Boost mode can be separately derived for low-power and high-power conditions, as shown in (1) and (2).

$$\begin{cases} D_2 = \frac{1}{M-1}\left(\varphi_s + \frac{4L_k I_{ZVS2}}{V_1 T_s}\right) \\ D_1 = MD_2 + \frac{4L_k I_{ZVS1}}{V_1 T_s} \end{cases} \quad (1)$$

$$\begin{cases} D_1 = 1 + \frac{\varphi_s - 1}{M} + \sqrt{\left(1 + \frac{1}{M^2}\right)(\varphi_s - 1)^2 + \frac{2(\varphi_s-1)}{M} + 1} \\ D_2 = 1 \end{cases} \quad (2)$$

n is the turns ratio (sec/pri) of high-frequency transformer. L_k is the auxiliary inductance. T_s is the switching period. $I_{ZVS1,2}$ is the minimum current amplitude for primary and secondary side ZVS transition, and can be evaluated as Q_{oss}/t_{dead} [9], where Q_{oss} is the charge stored at the output capacitance of switches and t_{dead} is the dead time between

979-8-3315-1612-3/25 $31.00 © 2025 IEEE

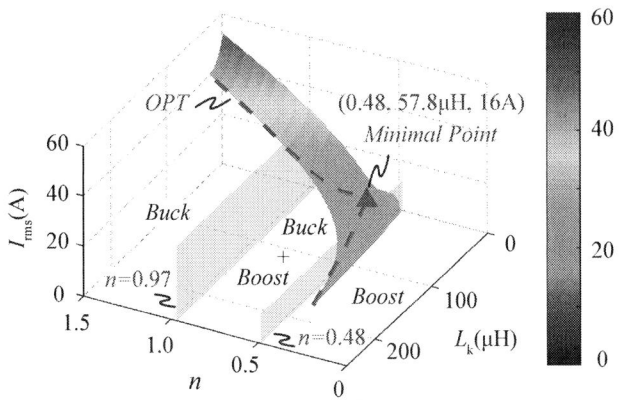

Fig. 5. Circuit parameters optimization for control method 1 with minimized RMS current of auxiliary inductor

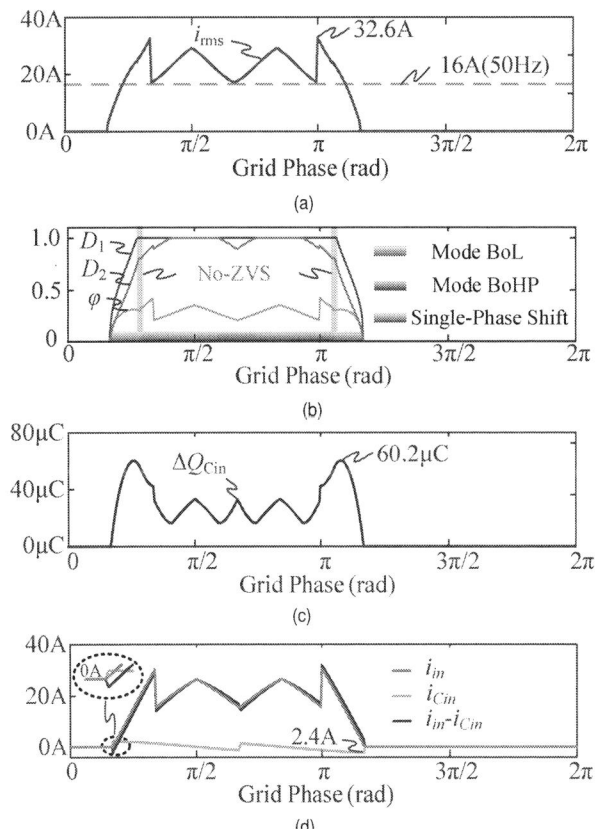

Fig. 6. Operation Waveforms of optimized DAB with control method 1. (a) RMS current of auxiliary inductor over one switching cycle. (b) Phase-shift angle φ and duty cycles D_1, D_2. (c) Amplitude of charge flowing through the input capacitor over one switching cycle. (d) Input current before capacitor i_{in}, input capacitor current i_{Cin}, input current after capacitor $i_{in} - i_{Cin}$

upper and lower switches. M is the voltage conversion ratio $V_2/(nV_1)$ of DAB converter.

To illustrate the effectiveness of the above modulation strategy, Fig 4 shows the corresponding operation trajectories at positive Boost mode. As seen, ZVS is achieved in most conditions except the boundary between Mode **BoL** and Mode **BoHP**, and the minimized RMS current is well maintained throughout the entire power range. Similarly, the optimal D_1 and D_2 can be obtained for other switching modes. The detailed implementation process of this modulation strategy can be found in [8], and is omitted here. The only modification is that φ_s should be generated by the close-loop of input current. In the following sections, this modulation strategy is implemented for further analysis.

III. DC/DC Converter Design Considering Selected Control Methods

A. Control Method 1

According to the operation waveforms shown in Fig 2, no negative power flow is required under control method 1, and the power transmission starts from half of the peak voltage. Hence, DAB converters only need to operate in positive modes (Buck/Boost High/Low). The boundary between Boost and Buck modes is determined by the selection of HFT's turns ratio n. For instance, under the system specifications listed in Table I, if n is less than or equal to 0.48, DAB will only operate in Boost mode. Similarly, the selection of auxiliary inductance L_k influences the boundary between high-power and low-power modes. As introduced by section II, these working modes have different priorities for loss minimization. Hence, it is necessary to allocate the percentage of different working modes by selecting n and L_k. Here, the optimization target for selection is to minimize the RMS current of auxiliary inductor over one grid cycle at rated power. The optimization results are shown in Fig 5. The minimal RMS current is reached with $n = 0.48$ and $L_k = 57.8\,\mu H$.

Under these optimal parameters, the time-domain waveforms of DAB converter are calculated theoretically and shown in Fig 6. Fig 6a shows the RMS current of auxiliary inductor over one switching cycle. Although the RMS current over one grid cycle is minimized as $16\,A$, its instantaneous peak value is still distinct as $32.6\,A$, which leads to the oversizing design of DC/DC converters. The reason for that is the intermittent operation of DC/DC converters in mBR SST. In other words, the operation period of DC/DC converters cannot cover the whole grid cycle. This weakness is inherent with mBR topology, and can be relieved or eliminated through the topology modification, which is out of the scope of this paper. Fig 6b shows the phase-shift angle and duty cycles of DAB. Throughout the entire operation period, DAB always works in voltage boosting. As predicted by section II, ZVS is lost only during the transition between low- and high-power modes.

In mBR SST, the input capacitors are only required to filter the high-frequency ripple generated by the switching of DC/DC converter, but do not need to buffer the low-frequency ripple as in the CHB-based SST. Hence, the minimum input capacitance can be calculated according to the desired content of high-frequency voltage ripple, as shown in (3).

$$C_{in} = \max\left[\frac{\Delta Q_{Cin}}{\Delta V_{in}}\right] = \max\left[\frac{\Delta Q_{Cin}}{1\% \times v_{in}}\right] \tag{3}$$

ΔQ_{Cin} is the amplitude of charge flowing through the input capacitor over one switching cycle. According to the ΔQ_{Cin}

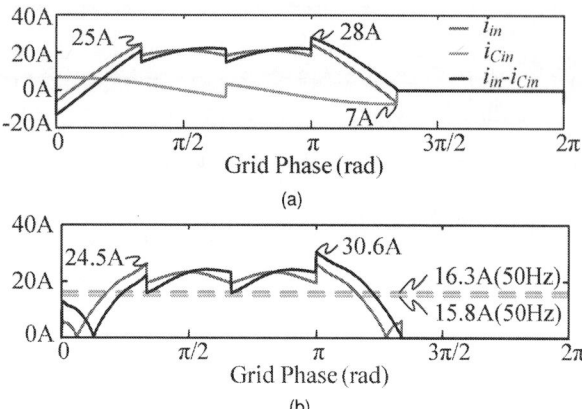

(a)

(b)

Fig. 7. Operation Waveforms of optimized DAB with control method 2. (a) Input current before capacitor i_{in}, input capacitor current i_{Cin}, input current after capacitor $i_{in} - i_{Cin}$. (b) RMS current of auxiliary inductor over one switching cycle with/without considering the influence of input capacitor

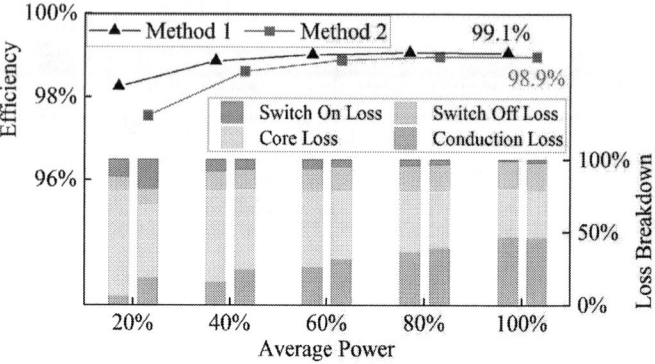

Fig. 8. Efficiency comparison and loss breakdown of DAB converter under different control methods

TABLE III
PERFORMANCE COMPARISON OF ISOLATED DC/DC CONVERTER UNDER
TWO CONTROL METHODS

Control Method	Input Capacitance	Optimal Parameters (n/L$_k$)	Efficiency	Reactive Compensation
Method 1	5.5μF	0.48 / 57.8μH	98.2%~99.1%	[-30°, +30°]
Method 2	14.3μF	0.48 / 76.8μH	97.5%~98.9%	[-180°, +180°]

waveform shown in Fig 6c, the required input capacitance under control method 1 can be obtained as 5.5 μF. The above analysis of DC/DC converters is all based on the ideal case, neglecting the influence of input capacitors. However, the voltage across diodes varying with grid frequency will induce significant low-frequency ripple current in the input capacitors. This low-frequency ripple current needs to be compensated by the input-current loop of the following DC/DC converter to meet the current target set by the AC side. Fig 6d shows the actual input current waveform of DAB after considering the input capacitor's influence. The amplitude of capacitor's current is around 10% of the total input current. And the negative power flow occurs after the starting point, which illustrates the necessity of bi-directional DC/DC converters.

B. Control Method 2

In the ideal case, control method 2 can reduce the instantaneous peak current by 22.5% [6] at the expense of wider operation range. Due to the negative power flow in some periods, DAB converters need to operate at both positive and negative modulation modes. Following the same design procedure, the optimal main circuit parameters for method 2 can be determined as $n = 0.48$ and $L_k = 76.8\,\mu\text{H}$. Since the operation range of input voltage is extended to 0V, the input capacitance requirement becomes infinitely high according to (3). Here, the required capacitance when v_{in} is equal to 100 V is chosen as the final value as 14.3 μF. Under the selected parameters and rated power level, Fig.7 compares the current waveforms with/without considering the influence of input capacitors. 25% increase in instantaneous peak RMS current is caused by the input capacitors. At lower power levels, current stress increase would become more pronounced, because the capacitor's current is only decided by applied voltage and independent of transferred power.

IV. DESIGN COMPARISON

Based on the main circuit parameters obtained from above sections, the efficiency of isolated DC/DC converters under different control methods can be estimated. In detail, the

conduction losses and turn-off losses of primary-side and secondary-side semiconductors are calculated in PLECS using the loss model provided by the manufacturers. The components' part numbers can be found in Table I. The soft-switching criterion for turn-on is based on the sign and magnitude of turn-on current, and the non-ZVS turn-on loss is evaluated by the method proposed by [9]. The core loss is obtained using the Steinmetz equation where the maximum magnetic flux density B_m is calculated based on the secondary-side duty cycle D_2 over each switching period.

Fig 8 shows the efficiency curves under different control methods. Since the circuit optimization in above sections is all based on the rated power, the maximum efficiency of DAB can achieve 99% under both control methods. However, when the power is reduced, the efficiency degradation of method 2 becomes more pronounced, mainly because of the additional current stress introduced by the input capacitors as analyzed in section III.

The comprehensive comparison of two control methods is shown in Table III. It needs to be mentioned that all the above analysis is based on the unity power factor operation of the mBR SST, which fits applications such as data centers and EV charging. If the bidirectional power flow or reactive power compensation is required, only method 2 is applicable.

V. CONCLUSION

As an emerging topology, mBR SST integrates isolated DC/DC converters into the standard three-phase diode rectifier, offering promising advantages in cost, efficiency, and power density. As the only active stage, the isolated DC/DC converters in mBR SST operate under unique waveforms. This paper first analyzed the selection criteria for suitable DC/DC topology, and identified DAB converter as the optimal

979-8-3315-1612-3/25 $31.00 © 2025 IEEE

choice. To accommodate the wide input voltage and power range, a modulation strategy with six switching modes was implemented to minimize switching losses in the low-power region and conduction losses in the high-power region. The main circuit parameters, including the turns ratio of HFT and auxiliary inductance, were also optimized to achieve minimal RMS current. Additionally, the impact of input capacitors on current stress and energy backflow was investigated for the first time. Furthermore, two mBR control methods were compared with a focus on their influence on the performance of DC/DC converters.

VI. ACKNOWLEDGMENT

The results presented in this paper are part of the project "Heating Bits", funded by the École Polytechnique Fédérale de Lausanne (EPFL) through a call Solutions4Sustainability.

REFERENCES

[1] D. Dong, M. Agamy, J. Z. Bebic, Q. Chen and G. Mandrusiak, "A Modular SiC High-Frequency Solid-State Transformer for Medium-Voltage Applications: Design, Implementation, and Testing," in IEEE Journal of Emerging and Selected Topics in Power Electronics, vol. 7, no. 2, pp. 768-778, June 2019.

[2] H. Long, J. Deng, X. Wu, Y. Zhou, Y. Wu and D. Xu, "Evaluation of the Front-End AC/DC Converter Circuits for Medium-Voltage-Connected Power Supply Systems," 2023 11th International Conference on Power Electronics and ECCE Asia (ICPE 2023 - ECCE Asia), Jeju Island, Korea.

[3] S. Götz, "Charging apparatus with a phase unit having multiple strands," U.S. Patent 10 439 407 B2, Oct., 2019.

[4] R. Raju and J. Leonard, "AC solid-state transformer using DC-DC converters and without added rectifier and inverter stages," in Proc. IEEE Appl. Power Electron. Conf. Expo. (APEC), Long Beach, CA, USA, Feb. 2024, pp. 528–532.

[5] Andrioli, G., Calligaro, S., Petrella, R., Kolar, J. W., and Huber, J, "Analysis and Comparative Evaluation of a Modularized Bridge Rectifier MVAC-LVDC Solid-State Transformer," 2024 10th International Power Electronics and Motion Control Conference (IPEMC/ECCE Asia), Chengdu, China.

[6] A. Cervone, D. Dujic, "Unlocking the potential of the Modularized Bridge Rectifier Solid-State Transformer," 2024 Energy Conversion Congress & Expo (ECCE Europe), Darmstadt, Germany.

[7] F. Krismer and J. W. Kolar, "Closed Form Solution for Minimum Conduction Loss Modulation of DAB Converters," in IEEE Transactions on Power Electronics, vol. 27, no. 1, pp. 174-188, Jan. 2012, doi: 10.1109/TPEL.2011.2157976.

[8] Z. Guo, "Modulation Scheme of Dual Active Bridge Converter for Seamless Transitions in Multiworking Modes Compromising ZVS and Conduction Loss," in IEEE Transactions on Industrial Electronics, vol. 67, no. 9, pp. 7399-7409, Sept. 2020, doi: 10.1109/TIE.2019.2945270.

[9] M. Kasper, R. M. Burkart, G. Deboy and J. W. Kolar, "ZVS of Power MOSFETs Revisited," in IEEE Transactions on Power Electronics, vol. 31, no. 12, pp. 8063-8067, Dec. 2016, doi: 10.1109/TPEL.2016.2574998.

[10] N. Hou, W. Song and M. Wu, "Minimum-Current-Stress Scheme of Dual Active Bridge DC–DC Converter With Unified Phase-Shift Control," in IEEE Transactions on Power Electronics, vol. 31, no. 12, pp. 8552-8561, Dec. 2016, doi: 10.1109/TPEL.2016.2521410.

Active and Reactive Power Flow Control of the Dual Active Bridge Converter

Lauryn Morris, Thomas W. Francois, Jonathan Saelens, Oroghene Oboreh-Snapps, Arnold Fernandes, Praneeth Uddarraju, Sophia A. Strathman, and Jonathan W. Kimball

Department of Electrical and Computer Engineering - Missouri University of Science and Technology

{lrmdhf@mst.edu, twfth9@mst.edu, jhskrw@mst.edu, oogdq@mst.edu, af2vc@mst.edu, pu5mb@mst.edu, ss6k4@mst.edu, kimballjw@mst.edu}

Abstract—The Dual Active Bridge (DAB) is a reliable and efficient converter capable of providing bi-directional power transfer and galvanic isolation. An ac-ac DAB can control both active and reactive power flow. The present work introduces a combined feedback/feed-forward current control system, utilizing the calculated and measured converter currents translated into the *dq* reference frame, to control the output power. The system was simulated in PLECS to demonstrate the control algorithm's ability to track the *dq* currents and provide the necessary output power.

Index Terms—Dual Active Bridge, AC-AC Converter, Bi-directional, Power Electronics, Reactive Power, Active Power.

I. INTRODUCTION

THE solid-state transformer (SST) is a growing topic of research as SSTs have the capability to provide better controllability. SSTs have four configurations that can provide isolation as well as DC links to tap off of in applications such as electric vehicle charging [1]–[3]. The single-stage SST solely consists of the ac-to-ac conversion stage and is shown in Fig. 1a. The two-stage SST is an ac-dc-dc-ac topology with two different variants, the low voltage DC link (LVDC) or the high voltage DC link (HVDC) with these links being accessible between the dc-dc stage. The schematic for the LVDC and HVDC SST topologies are shown in Fig. 1b and Fig. 1c. The three-stage SST is an ac-dc dc-dc dc-ac system and has both the LVDC and HVDC links accessible before and after the middle dc-dc stage. The schematic for the three-stage SST is shown in Fig. 1d [1]–[3].

The single-stage SST is simpler to construct, control and is cheaper as it does not need to access the HVDC or LVDC links. The dual active bridge (DAB) converter is a popular topology for single-stage SST configurations for its bidirectional power flow capabilities, galvanic isolation, simplicity, and single-phase and three-phase configurations [5].

The ac-ac DAB is constructed utilizing two H-bridge converters connected by a high frequency transformer, and is depicted in Fig. 2. Power transferred between the two bridges is dictated by the phase shift between the primary and secondary side. Additionally, the ac-ac DAB does support Zero Voltage Switching (ZVS) and Zero Current Switching (ZCS) [4], [6].

The design and operating principles of the DAB are well understood, but there are still many challenges associated with the control. Controlling active and reactive power is desired for a wide number of applications, and feedback

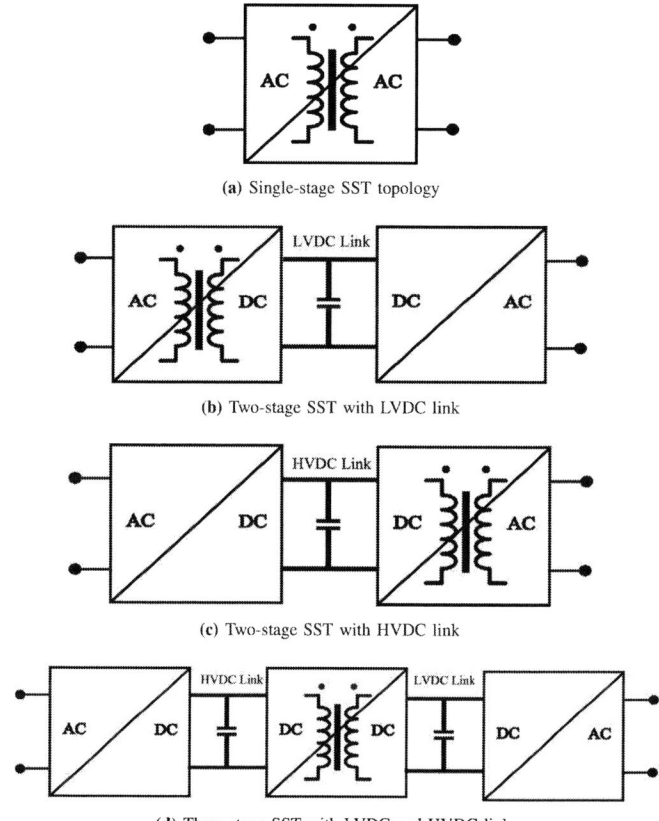

(a) Single-stage SST topology

(b) Two-stage SST with LVDC link

(c) Two-stage SST with HVDC link

(d) Three-stage SST with LVDC and HVDC links

Figure 1: SST Topologies [1]–[4]

control is desired to be able to provide stability within the system. Several methods have been proposed with one such method being described in [7] which does utilize a closed-loop system, but the controller is manipulating voltage and lacks the ability to control active and reactive power. In [8], the developed control system is utilizing a proportional-resonant (PR) controller which are more complex than a proportional-integral (PI) controller.

Another approach was developed in [6] that demonstrates the ability to control active and reactive power. In [6], the rotated reference frame generated by the Park transformation is used to calculate voltage and current, but the controller fails to accurately track the *dq* currents utilized in the system.

979-8-3315-1612-3/25 $31.00 © 2025 IEEE

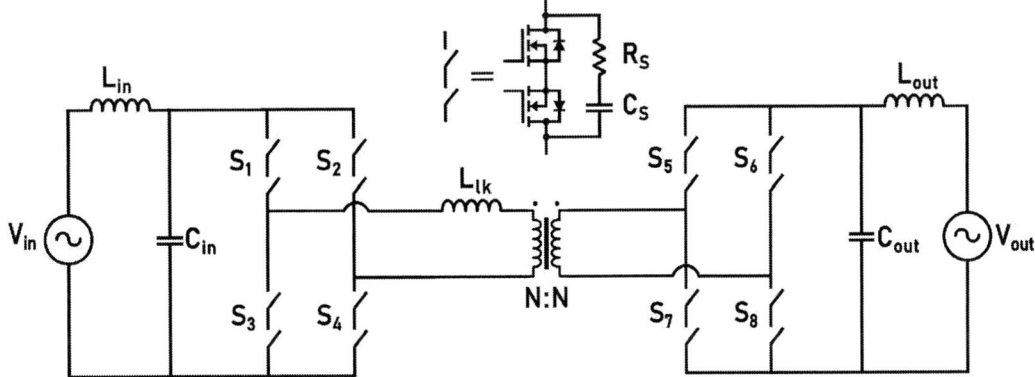

Figure 2: AC-AC dual active bridge topology schematic [4].

This work corrects the deficiencies and inaccuracies in [6] and provides accurate *dq* current tracking with a combination of feed-forward and feedback control and compares this with results from [6].

In Section II, an overview of the operating principles of the DAB is provided. Section III describes the fundamental control strategies utilized as well as the P-Q controller and its working operations. Section IV provides simulation results and compares these results to those found in [6] before Section V concludes the paper.

II. AC-AC DAB CONVERTER TOPOLOGY

The dual active bridge converter is a two-port system using two H-Bridge converters interconnected by a high-frequency transformer that provides galvanic isolation and bi-directional power flow. The ac-ac DAB system is depicted in Fig. 2, and uses anti-series switches. The anti-series switches are to manage the ac voltages and currents [4], [9]. Instantaneous power flow between the two bridges is given by

$$p(t) = \frac{v_i(t)v_o(t)}{X_L}\left(|\phi(t)| - \frac{\phi^2(t)}{\pi}\right)sgn(\phi(t)) \quad (1)$$

with $v_i(t)$ and $v_o(t)$ being the input and output voltages, and the leakage reactance, $X_L = 2\pi f_{sw}L_{lk}$, with f_{sw} being the switching frequency and L_{lk} being the leakage inductance of the transformer [4], [6], [9].

With the primary side of the DAB serving as bridge 1 and the secondary side serving as bridge 2, the phase shift, $\phi(t)$, is controlled to deliver power from bridge 1 to bridge 2 or vice versa. The desired $\phi(t)$ is found by solving (1) for $\phi(t)$ using the quadratic equation and yields [4], [6], [9]

$$\phi(t) = \frac{\pi}{2}\left(\pm 1 \pm \sqrt{1 \pm \frac{4p(t)X_L}{\pi v_i(t)v_o(t)}}\right) \quad (2)$$

where $p(t)$ is the demanded instantaneous total power. To maintain stability in the system, $\phi(t)$ must not exceed $\pm\frac{\pi}{2}$ [4]. The instantaneous active, reactive, and total power equations are are described as [4], [6], [9]

$$p_d(t) = P_d - P_d\cos(2\omega_g t) \quad (3)$$

$$q_d(t) = Q_d\sin(2\omega_g t) \quad (4)$$

$$p(t) = p_d(t) + q_d(t) \quad (5)$$

In the ac-ac DAB system, bridge 1 is the reference bridge and has a phase shift of $0°$. The calculated phase shift based on the demanded instantaneous power is then utilized in the pulse width modulation (PWM) switching scheme. The PWM uses a sawtooth based carrier and generates a switching pattern that accounts for the switching times as well as the anti-series switches [6]. A full description of the design and operating principles of the ac-ac DAB is available in [4].

III. P-Q FEEDBACK/ FEED-FORWARD CONTROLLER

A. Fundamental Control Strategies

1) Second-Order Generalized Integrator Phase-Locked Loop

A Second-Order Generalized Integrator Phase-Locked Loop (SOGI-PLL) is used in ac systems and takes either input voltage or current and generates the phase of the system as well as utilizes the Clarke and Park transformations to obtain the stationary reference frame ($\alpha\beta$) and then the rotated reference frame (dq) values for voltages and currents [10], [11]. This operation is necessary to lock onto the signal for synchronization and thus effective control.

2) Clarke and Park Transformations

In essence, the $\alpha\beta$ and dq reference frames are utilized to make an ac signal look dc. The Clarke transformation generates the $\alpha\beta$ reference frame and then the Park transformation generates the dq reference frame. More information on the operation and applications of the SOGI-PLL, $\alpha\beta$, and dq reference frames can be found in [10] and [11] respectively.

3) SOGI-PLL and Park Transformation Architecture

The ac-ac DAB described is a single-phase system with the control algorithm begining with the SOGI-PLL taking the output voltage and current, v_o and i_o, and obtaining the dq voltages and currents [6]. Since this is a single-phase system, the Clarke transformation is not needed as the system is already in the $\alpha\beta$ reference frame. The transfer function, with the gain k_s fixed at unity, for the SOGI-PLL is [6], [10]

$$G_\alpha(s) = \frac{v_\alpha(s)}{v_o(s)} = \frac{i_\alpha(s)}{i_o(s)} = \frac{k_s\omega_g s}{s^2 + k_s\omega_g s + \omega_g^2} \quad (6)$$

$$G_\beta(s) = \frac{v_\beta(s)}{v_o(s)} = \frac{i_\beta(s)}{i_o(s)} = \frac{k_s \omega_g^2}{s^2 + k_s \omega_g s + \omega_g^2} \qquad (7)$$

The v_α, v_β, i_α, and i_β from the SOGI are then passed to the Park transformation to generate v_d, v_q, i_d, and i_q [6], [10].

B. Feedback/Feed-Forward Controller Architecture

With bridge 1 of the ac-ac DAB serving as the reference bridge and bridge 2 serving as the output bridge, the control algorithm developed in [6] utilizes current control with proportional-integral (PI) controllers, as well as feed-forward and feedback components to achieve the desired output active and reactive power.

1) Feedback Architecture

The feedback portion of the controller utilizes the demanded active, P_d, and reactive, Q_d, power as well as the measured output voltage and current, v_o, and i_o, from bridge 2 [6]. The SOGI-PLL and Park transformation takes the measured voltage and current and generates the reference currents, i_d and i_q [10]. P_d, Q_d, and the peak voltage of the system are used to calculate the demanded currents, i_d^* and i_q^*, by

$$i_d^* = \frac{P_d}{V_{pk}} \qquad (8)$$

$$i_q^* = \frac{Q_d}{V_{pk}} \qquad (9)$$

The reference frame currents are subtracted from the calculated currents to produce the current error. These errors are then passed to a PI controllers to generate increments to the commanded currents [10]. The active power portion of the controller uses the i_d and i_d^* currents with $K_p = 5$ and $K_i = 10$ and generates Δi_d. The reactive power portion of the controller uses the i_q and i_q^* currents with $K_p = 10$ and $K_i = 20$ and generates Δi_q. The block diagram of this portion of the controller is shown in Fig. 3.

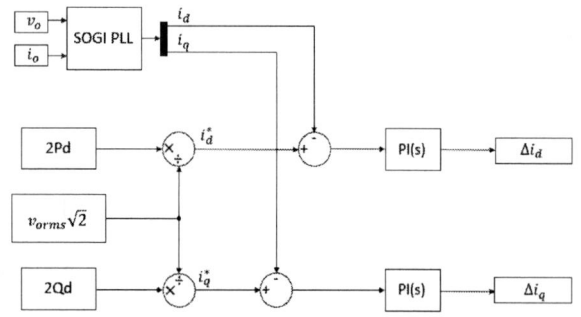

Figure 3: Feedback PQ controller block diagram [6].

2) Feed-Forward Architecture

The feedback controller *alone* is insufficient for accurate current tracking, so the feed-forward portion of the P-Q controller was developed and added to the system. This feed-forward portion takes the delta values calculated from the PI controller, Δi_d and Δi_q, and adds this to the demanded currents, i_d^* and i_q^* producing the desired current. This addition between the delta currents and the demanded currents is used

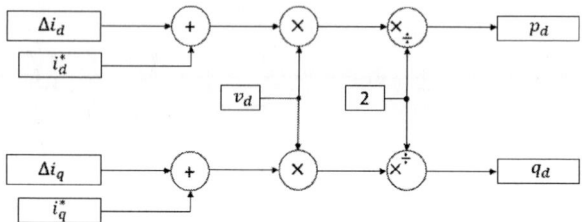

Figure 4: Feed-forward PQ controller block diagram.

to even further achieve the correct current values for the demanded active and reactive power. The desired current is then multiplied by v_d and calculates active and reactive power in the dq reference frame given by [6]

$$P_d = v_d(\Delta i_d + i_d^*) + v_q i_q \qquad (10)$$

$$Q_d = v_q i_d - v_d(\Delta i_q + i_q^*) \qquad (11)$$

As described in [10], v_q is controlled to zero leaving

$$P_d = v_d(\Delta i_d + i_d^*) \qquad (12)$$

$$Q_d = -v_d(\Delta i_q + i_q^*) \qquad (13)$$

The block diagram for this portion of the controller is shown in Fig. 4. Using (12) and (13) the instantaneous power is then calculated and utilized in (2) to calculate $\phi(t)$ and then the PWM signals to drive the switches [4], [9]. Simulation results of the developed control algorithm and a performance comparison to what was developed in [6] are provided in the next section.

IV. SIMULATION RESULTS

A. PLECS Simulation Results

To simulate the ac-ac DAB and the feedback/feed-forward control system, the system was constructed in PLECS Standalone and simulated with the parameters given below in Table I [6].

Table I: System parameters [6]

Switching Frequency	30 kHz
Duty Cycle	50%
Input Voltage ($V_{in,rms}$)	120 V
Output Voltage ($V_{out,rms}$)	120 V
Filter Inductors(L_{in}, L_{out})	$10\mu H$
Leakage Inductance(L_{lk})	$10\mu H$
Filter Capacitors(C_{in}, C_{out})	$30\mu F$
Turns Ratio (N_1:N_2)	1:1
Total Simulation Time	1 s
Step Time	0.5 s

To test the dynamic reponse of the controller, a step input for both active and reactive power was used. The total simulation time was 0.5 s and the step input was applied at 0.25 s. Below in Table II is the active and reactive power demanded for the first 0.25 s of the simulation. Table III is the active and reactive power demanded at the step input applied at 0.25 s and for the remainder of the simulation time.

To validate the controller's performance, the current tracking of the i_d and i_q currents was analyzed in PLECS as well as the output power and $\phi(t)$ values.

979-8-3315-1612-3/25 $31.00 © 2025 IEEE

Table II: Demanded active and reactive power for the first 0.25 s of simulation time

Simulation Time	0 - 0.25 s
Demanded Active Power (P_d)	1000 W
Demanded Reactive Power (Q_d)	300 VAR

Table III: Demanded active and reactive power after the demanded step input at 0.25 s

Simulation Time	0.25 - 0.5 s
Demanded Active Power (P_d)	3600 W
Demanded Reactive Power (Q_d)	-720 VAR

1) i_d and i_q Current Tracking

The simulation results for i_d, from the SOGI-PLL, and i_d^*, which is calculated from the demanded active power, are shown in Fig. 5. The full simulation of the i_d and i_d^* currents shows that over the course of the full simulation time, the i_d current is able to track the i_d^* current. In Fig. 5b, the step response that occurs at 0.25 s is shown. This shows that even when a step response to the active power is introduced, the feedback/feed-forward control system is able to respond and track the new demanded current quickly.

In Fig. 5c, the detailed results of the i_d^* and i_d current tracking is shown for the initial demanded active power of 1000 W. In Fig. 5d, the detailed results of the i_d^* and i_d current tracking is shown after the step for a demanded active power of 3600 W.

In Fig. 6 , the full simulation results for i_q, also from the SOGI PLL, and i_q^*, which is calculated from the demanded reactive power, is shown. The full simulation of the i_q and i_q^* currents also shows that over the course of the full simulation time, the i_q current is able to track the i_q^* current as seen in Fig. 6a. In Fig. 6b, the step response that occurs at 0.25 s is shown. This shows that even when a step response to the reactive power is introduced into the system, the feedback/feed-forward control system is also able to respond quickly and continue to have the i_q current track the i_q^* current.

In Fig. 6c, the detailed results of the i_q^* and i_q current tracking is also shown for the initial demanded reactive power of 300 VAR. In Fig. 6d, the detailed results of the i_q^* and i_q current tracking for -720 VAR of demanded reactive power is shown.

The results in Figs. 5 and 6 show that the feedback/feed-forward control system is able to quickly respond and track the i_d and i_q currents even when a change in the demanded i_d and i_q currents is introduced validating the designed control system.

2) Output Power and $\phi(t)$

Along with the i_d and i_q currents, the output power on bridge 2 and the $\phi(t)$ values generated are also needed to further validate the system. To measure output power on bridge 2, the measured output voltage and currents are used to find output power as shown by

$$P_o = v_o i_o \qquad (14)$$

The output power computed using (14) is then compared against the total power value found in (5) to validate that the system is able to produce the output power being demanded.

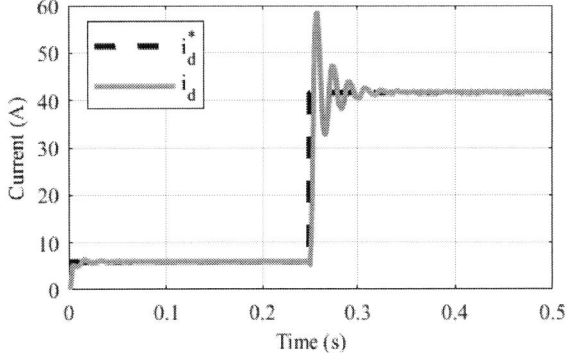

(a) Full simulation results of i_d^* and i_d current tracking

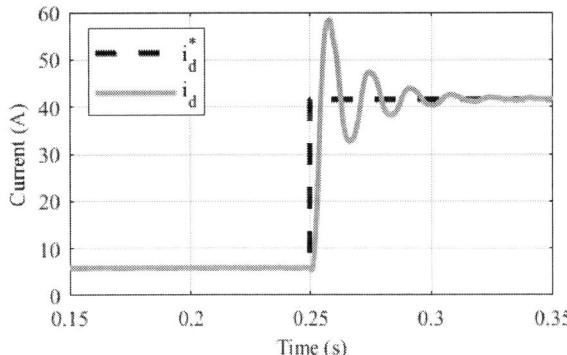

(b) Zoomed view of step response

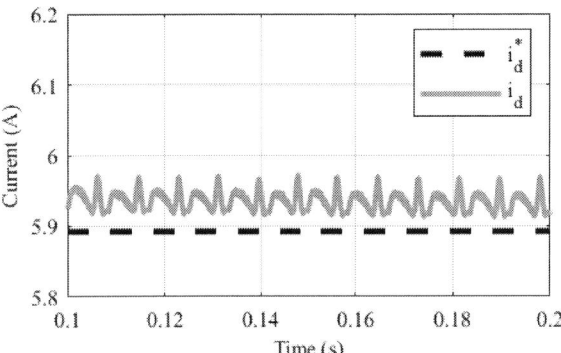

(c) Zoomed view during constant command of 1000 W

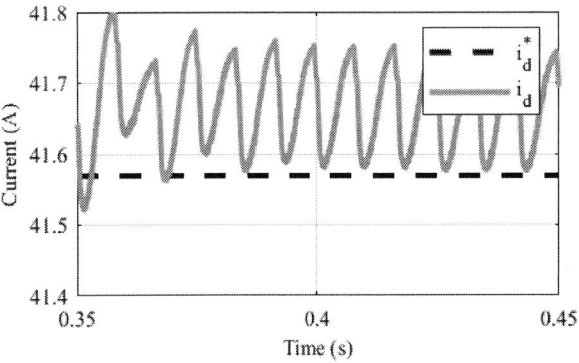

(d) Zoomed view during constant command of 3600 W

Figure 5: i_d and i_d^* current tracking

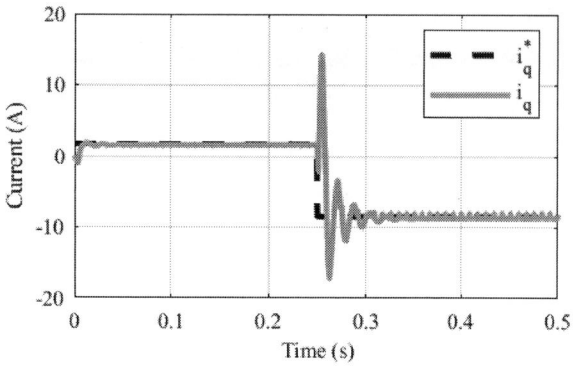

(a) Full simulation results of i_q^* and i_q current tracking

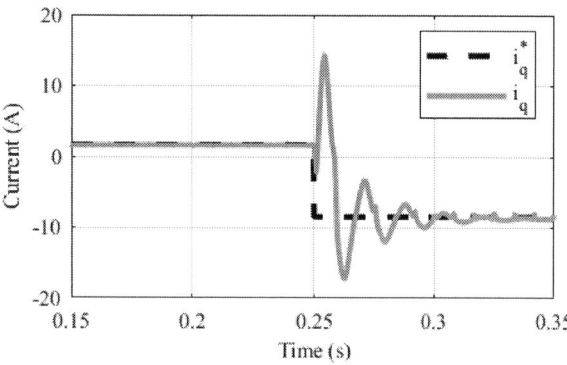

(b) Zoomed view of step response

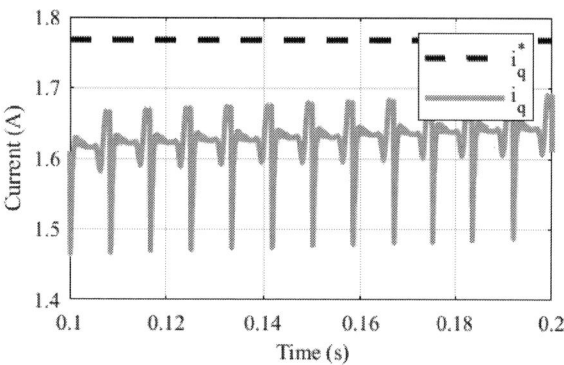

(c) Zoomed view of constant 300 VAR command

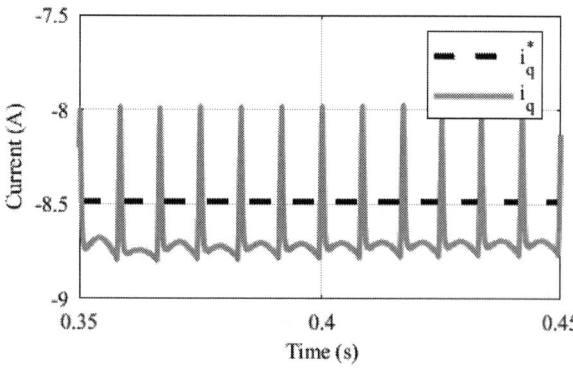

(d) Zoomed view of constant -720 VAR command

Figure 6: i_q and i_q^* current tracking

The output power of the system is shown below in Fig. 7. The output power generated by (14) is able to track the demanded output power from (5). Also in Fig. 7, the $\phi(t)$ values generated from (2) are shown and are bounded by $\pm\frac{\pi}{2}$. When the step is introduced, the $\phi(t)$ values are still able to maintain the $\pm\frac{\pi}{2}$ bound, and the measured output power is still able to track the demanded output power.

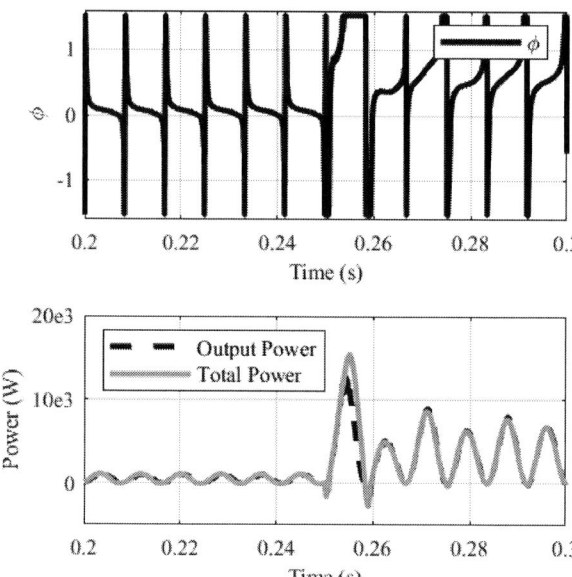

Figure 7: Output power and $\phi(t)$ for the feedback/feed-forward control system

The PLECS simulation results, including the tracking of the i_d and i_q currents, the output power, and the phase angle $\phi(t)$, collectively demonstrate the validity of the control system. The detailed analysis of these parameters confirms that the control system performs as expected even when subjected to a step change in input values.

3) Performance Comparison

The work described in [6] lacks the feed-forward architecture described in Section III as well as demonstrates poor current tracking. The control system in [6] was simulated in PLECS Standalone under the same conditions as was described in Section IV. The i_d and i_q current tracking, $\phi(t)$ values, and output power results from Section IV are compared against the results from [6].

As seen in Fig. 8a, the i_d current tracking capabilities of the results presented in Section IV is able to respond to the demanded i_d^* and accurately and quickly track the current. The i_d current tracking from [6] is in Fig. 8b. These results do not accurately track the i_d^* current and the system lacks an accurate response when the step is introduced.

The i_q current tracking results from Section IV and [6] are shown in Fig. 8c and 8d. The i_q current tracking performance from [6] is significantly better than the i_d current tracking performance from [6]. However, the step response from the system could still be improved and the current tracking could be more accurate.

979-8-3315-1612-3/25 $31.00 © 2025 IEEE

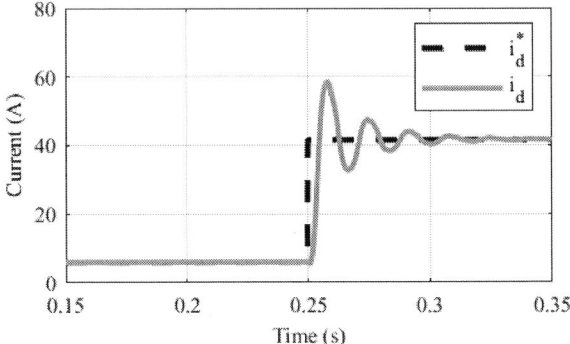

(a) Step response results of i_d^* and i_d current tracking from Section IV

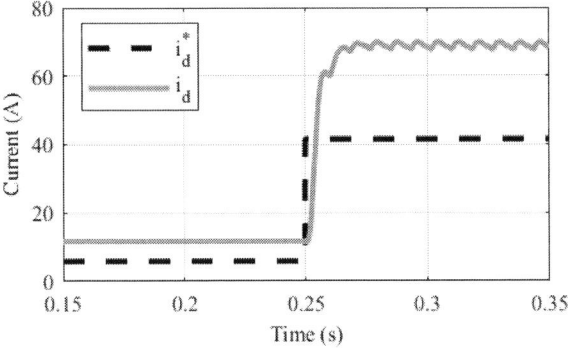

(b) Step response results of i_d^* and i_d current tracking from [6]

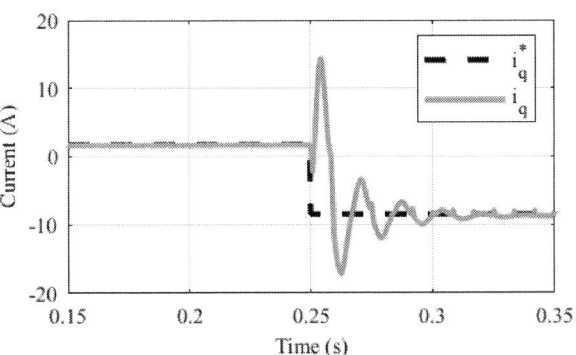

(c) Step response results of i_q^* and i_q current tracking from Section IV

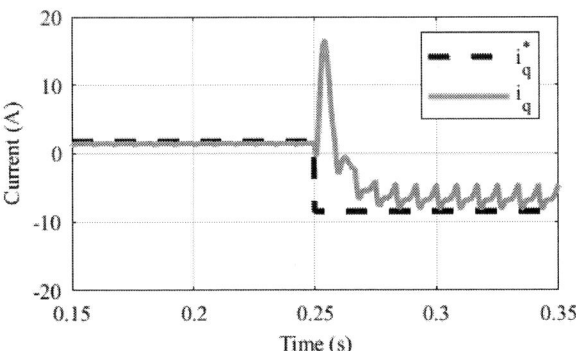

(d) Step response results of i_q^* and i_q current tracking from [6]

Figure 8: i_d and i_q current tracking comparison between results in Section IV and [6]

Lastly, the output power and $\phi(t)$ results from Section IV and [6] are shown in Fig. 9. The $\phi(t)$ values for both systems are within the $\pm\frac{\pi}{2}$ bound and also respond to the step input. The output power and demanded power results presented in Fig. 9a and Fig. 9b show that the systems are capable of producing the desired power, but the output power in Fig. 9b is slightly larger in magnitude.

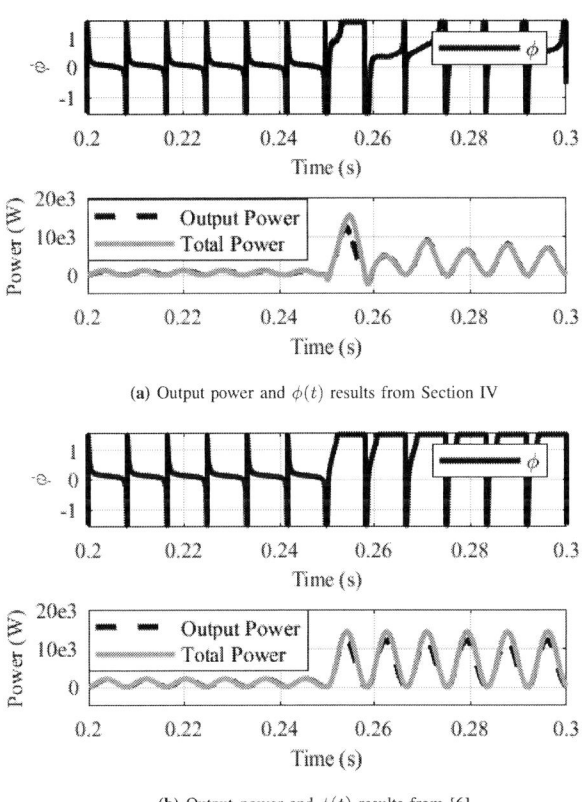

(a) Output power and $\phi(t)$ results from Section IV

(b) Output power and $\phi(t)$ results from [6]

Figure 9: Output power and $\phi(t)$ comparison between results in Section IV and [6]

Throughout the various results presented in this paper, the improvements made in the addition of the feed-forward portion of the controller and tuning the gains of the two PI controllers was able to drastically improve the current tracking performance of the controller introduce in [6].

V. CONCLUSION

This paper presents active and reactive power flow control utilizing a feedback/feed-forward method for the ac-ac dual active bridge converter. The controller utilizes the dq axis current and voltage for calculating the instantaneous active and reactive power that are then used to calculate the necessary $\phi(t)$ value for the phase shift. This was validated in PLECS using a step input from 1000 W + 300 VAR to 3600 W - 720 VAR. The results indicate that the controller has the ability to improve and track the dq currents first presented in [6] and provide the desired output power and $\phi(t)$ waveforms. Future work includes applying this control system to the dc-dc-dc and dc-ac-ac triple active bridge converters.

REFERENCES

[1] Mahammad A. Hannan, Pin Jern Ker, Molla S. Hossain Lipu, Zhen Hang Choi, M. Safwan Abd. Rahman, Kashem M. Muttaqi, and Frede Blaabjerg. State of the art of solid-state transformers: Advanced topologies, implementation issues, recent progress and improvements. *IEEE Access*, 8:19113–19132, 2020.

[2] Nasiru B. Kadandani, Mohamed Dahidah, Salaheddine Ethni, and James Yu. Solid state transformer: An overview of circuit configurations and applications. In *15th IET International Conference on AC and DC Power Transmission (ACDC 2019)*, pages 1–6, 2019.

[3] Xu She, Alex Q. Huang, and Rolando Burgos. Review of solid-state transformer technologies and their application in power distribution systems. *IEEE Journal of Emerging and Selected Topics in Power Electronics*, 1(3):186–198, 2013.

[4] Kartikeya Jp Veeramraju, Angshuman Sharma, and Jonathan W. Kimball. A comprehensive analysis on complex power flow mechanism in an ac-ac dual active bridge. In *2022 IEEE Power and Energy Conference at Illinois (PECI)*, pages 1–6, 2022.

[5] Jacob A. Mueller, Jack Flicker, Andrew Dow, Luciano Garcia Rodriguez, and Felipe Palacios. Isolated three-phase ac-ac converter with phase shift modulation. In *2024 IEEE Applied Power Electronics Conference and Exposition (APEC)*, pages 506–513, 2024.

[6] Angshuman Sharma, Kartikeya Jp Veeramraju, and Jonathan W. Kimball. Power flow control of a single-stage ac-ac solid-state transformer for ac distribution system. In *2022 IEEE Power and Energy Conference at Illinois (PECI)*, pages 1–6, 2022.

[7] Raeed Rahmoun and Michael Patt. High efficiency single-phase dual-active-bridge ac/ac converter. In *2019 21st European Conference on Power Electronics and Applications (EPE '19 ECCE Europe)*, pages P.1–P.10, 2019.

[8] Akhil Chambayil and Souvik Chattopadhyay. A single-stage single phase bidirectional ac-ac converter for solid state transformer application. In *2022 IEEE International Conference on Power Electronics, Drives and Energy Systems (PEDES)*, pages 1–6, 2022.

[9] Martin Jagau and Michael Patt. Reactive power operation of a single phase ac-ac dab converter. In *PCIM Europe 2017; International Exhibition and Conference for Power Electronics, Intelligent Motion, Renewable Energy and Energy Management*, pages 1–4, 2017.

[10] Jinming Xu, Hao Qian, Yuan Hu, Shenyiyang Bian, and Shaojun Xie. Overview of sogi-based single-phase phase-locked loops for grid synchronization under complex grid conditions. *IEEE Access*, 9:39275–39291, 2021.

[11] Ioan C. Damian, Mircea Eremia, and Lucian Toma. Advanced control of a modular multilevel high voltage direct current converter. In *2017 International Conference on ENERGY and ENVIRONMENT (CIEM)*, pages 1–5, 2017.

Comparative Analysis of Carbon Footprints and Material Usage of Solid-State Transformers and Low-Frequency-Transformer-Based MVac-LVdc Interfaces for High-Power EV Charging

Luc Imperiali*, Rudy Wang†, Anup Anurag†, Peter Barbosa†, Johann W. Kolar*, and Jonas Huber*

*Advanced Mechatronic Systems Group, ETH Zurich, Switzerland
†Delta Electronics, Raleigh, USA
imperiali@ams.ee.ethz.ch

Abstract—Medium-voltage (MV) ac to low-voltage (LV) dc conversion for, e.g., high-power EV charging can be realized either with a low-frequency transformer (LFT) and a downstream LV ac-dc converter, or, alternatively, as a modular solid-state transformer (SST) that employs high-frequency (HF) transformers to provide galvanic isolation. Both solutions achieve similar power conversion efficiencies in the order of 98% but differ significantly regarding complexity and the types and amounts of employed components and materials. Thus, this paper compares the two approaches regarding the embodied carbon footprint and the material usage, on the basis of industrial 400 kW first-generation and 1200 kW second-generation SST demonstrators, highlighting the potential of SST technology to benefit from improvements in power electronics whereas the LFT-based solutions remain constrained by high material usage for the transformer. Specifically, the 1200 kW second-generation SST demonstrator features only about 40% of the mass (2 kg/kW) and about 2/3 of the embodied carbon footprint (13.7 kg CO$_2$eq/kW) of an equally rated LFT-based solution.

Index Terms—Carbon footprint, low-frequency transformer, LFT, material usage, medium voltage, MVac-LVdc, solid-state transformer, SST, transformer.

I. INTRODUCTION

High-power low-voltage (LV) dc loads, e.g., datacenters or EV fast chargers, are supplied from the medium-voltage (MV) ac mains. Conventionally, such MVac-LVdc interfaces are conventionally realized with a low-frequency (i.e., mains frequency of 50 Hz or 60 Hz) transformer (LFT) providing galvanic separation and voltage step-down, and a downstream LV ac-dc converter, see **Fig. 1a**. Aiming at a reduced weight and size, solid-state transformers (SSTs), where high-frequency (HF) transformers (HFTs) provide the galvanic separation, e.g., as shown in **Fig. 1b**, have been considered since the 1970s, in particular for weight-/space-constrained applications like traction [1]. However, given the opposing long-term trends of increasing raw material prices (in particular for copper) and declining prices for high-volume electronics production indicated in **Fig. 2**, more recently, SSTs have been also proposed for various stationary applications like datacenters [2], electrolysis [3], PV inverters [4], and high-power EV charging [5]–[8], for which industry has demonstrated full-scale prototype systems [9]–[11].

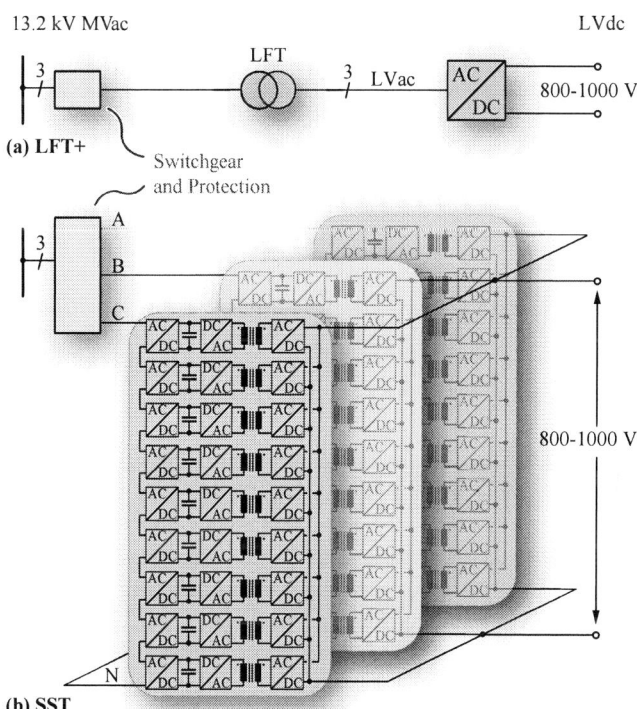

Fig. 1. MVac-LVdc interfaces realized **(a)** conventionally with a low-frequency transformer (LFT) and a LV-side ac-dc converter (denominated as "LFT+" in this paper) or **(b)** as a fully modular solid-state transformer (SST), where the galvanic isolation is provided by high-frequency transformers (HFTs) in the converter cells.

While economical considerations thus drive the interest in SST technology, the evermore visible signs of climate change and the certainty that resources are limited have lead to an increased environmental awareness in society. Thus, corresponding requirements for products are being codified in regulations like the European Union's Green Deal [15] and Circular Economy Action Plan [16] or standards like IEC 62430 (Environmentally conscious design for electric and electronic products), etc. Such considerations are relevant for power electronics, too, and there are already manufacturers

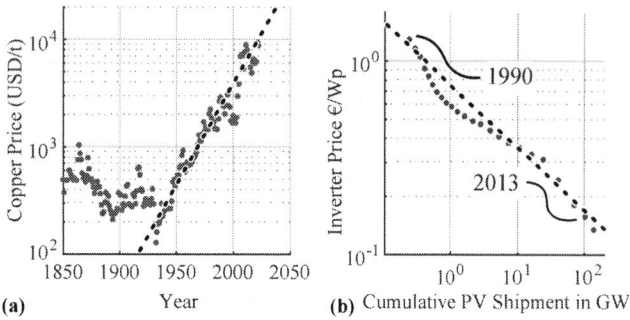

(a) Year **(b)** Cumulative PV Shipment in GW

Fig. 2. (a) Increasing copper prices over the last 175 years [12], [13] and **(b)** learning curve of LV power electronics, specifically PV inverters [14]. Thus, costs of LFT-based solutions must be expected to increase whereas costs of modular SSTs potentially decrease in the future.

publishing life-cycle assessment (LCA) studies for their power converters [17]–[20], quantifying the environmental impacts resulting over the life cycle, i.e, from raw material sourcing through production, use phase, and finally waste disposal and/or recycling. Similarly, there is a growing body of scientific literature concerning LCAs, material efficiency, circular economy compatibility, etc. of power electronic converters [17], [21]–[37]; [31], [36] give an overview.

Whereas industrial MVac-LVdc SSTs have been found to be on par with LFT-based solutions regarding efficiency and volume [2], this paper evaluates the life-cycle environmental impacts, focusing on the embodied carbon footprint and the raw material usage, of MVac-LVdc interfaces realized either conventionally with an LFT and a LV ac-dc stage (**Fig. 1a**) or as an SST (**Fig. 1b**). The evaluation of the embodied carbon footprint and material usage is based on actually available/built systems/components as far as possible; specifically, a fully rated 13.2 kV, 400 kW first-generation industrial SST prototype [10] and a 1200 kVA second-generation SST demonstrator [38] are considered. First, **Section II** details the evaluated systems and the modeling approach, before **Section III** presents the comparative evaluation results. **Section IV** closes the paper with a discussion of the results and provides an outlook on further research.

II. Modeling

In general, LFT-based MVac-LVdc conversion, i.e., an LFT plus an LV ac-dc converter and hence referred to as "LFT+" in the following, and MVac-LVdc SSTs can achieve similar conversion efficiencies [2]. In the following, we consider LFT+ solutions with two different LFTs, i.e., a dry-type LFT compatible with Tier 2 efficiency requirements defined by the EU in [39] (Legrand GREEN T.HE series [40], [41]) as shown in **Fig 3a** and a high-efficiency dry-type transformer with an amorphous core, i.e., an amorphous metal distribution transformer (AMDT) [42] (ABB EcoDry Ultra series [43], [44]). The LV ac-dc conversion is modeled by scaling a 150 kW PV inverter (SMA Sunny High Power Peak 3) for which a detailed LCA study is available [19] to the rated power, e.g., of the considered 400 kW first-generation SST prototype [10] shown in **Fig 3b**.

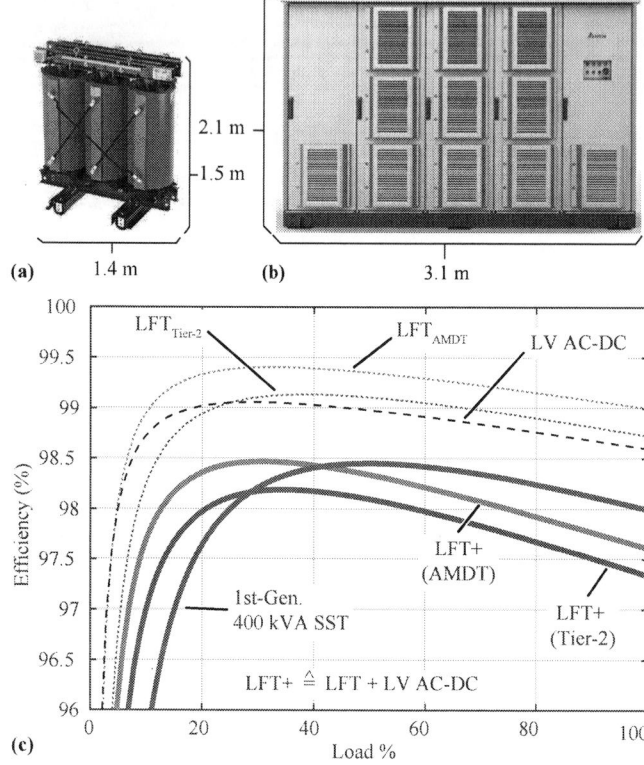

(a) 1.4 m **(b)** 3.1 m

(c) Load %

Fig. 3. Key elements of the considered 400 kW MVac-LVdc interfaces from **Fig. 1**. **(a)** 400 kVA dry-type LFT (image source: [40]) and **(b)** 400 kVA first-generation SST prototype [10] with the topology shown in **Fig. 1b**. **(c)** Efficiency curves of the considered MVac-LVdc interfaces from **Fig. 1** using directly measured values [10] of the 400 kW first-generation SST prototype shown in **(b)**. The LFT-based solutions (referred to as "LFT+" in the following) are modeled by combining an LFT and a representative LV ac-dc converter [45], whereby two different dry-type LFT options are considered (Tier 2 efficiency according to EU regulations [39] or, alternatively, an ultra-efficient amorphous metal distribution transformer (AMDT) [43], [44]).

Fig. 3c shows the corresponding efficiency curves of the considered LFTs and the resulting LFT+ solutions and of the 400 kW first-generation SST prototype. Overall, both solutions can achieve similar efficiencies, especially if the LFT+ solution is realized using high-efficiency AMDTs. Note further that the efficiencies of LFTs increase with the rated power. Therefore, the carbon emissions during the use-phase due to the conversion losses, which are covered using the grid electricity mix with a non-zero carbon footprint per kWh, are similar and depend on the use case; therefore, the use phase is not considered further here but the focus is set on the embodied carbon footprint and the material usage.

In the following, the corresponding modeling approaches and assumptions are discussed on the example of the considered the 400 kW LFT+ and SST solutions, respectively. The embodied carbon footprints are either taken from available LCA studies and/or estimated based on the generic component models introduced in [32], which in turn are based on the literature and LCA databases like ecoinvent [46]; the ecoinvent database is also directly consulted for certain parts, materials, and processing steps.

979-8-3315-1612-3/25 $31.00 © 2025 IEEE

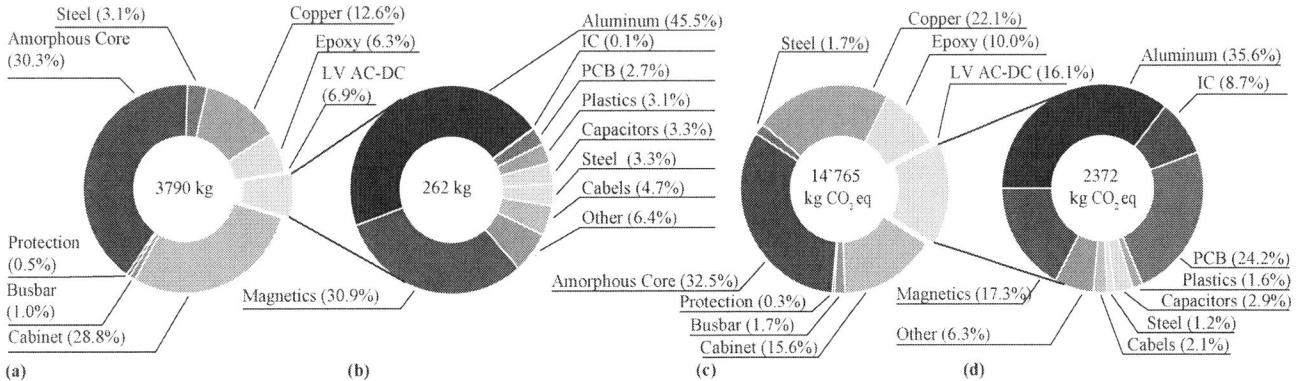

Fig. 4. (a) Weight and **(b)** embodied carbon footprint breakdown of the 400 kW LFT+ solution (see **Fig. 1a**) considering a high-efficiency AMDT LFT. The callouts give more details on the LV ac-dc converter using data from [19].

TABLE I
ESTIMATED MASS BREAKDOWNS OF DRY-TYPE TIER 2 AND AMDT LFTS
(WITHOUT CABINET); THE ABSOLUTE VALUES ARE FOR 400 kVA UNITS.

Material	LFT Tier 2		LFT AMDT	
	Share	Mass	Share	Mass
Amorphous Core			65%	1550 kg
Steel	65.4%	1150 kg	5%	120 kg
Copper	< 0.1%	< 3 kg	20%	480, kg
Aluminum	22.3%	480 kg		
Epoxy	9.8%	130 kg	10%	240 kg
PET	2.5%	40 kg		
Total	100%	1350 kg	100%	2380 kg
η_{max}	99.1%		99.4%	
$\eta_{400\,kW}$	98.7%		99.0%	

A. 400 kW LFT-based MVac-LVdc Interface (LFT+)

As mentioned, two realization options of a dry-type LFT are considered, i.e., an LFT with EU Tier 2 [39] efficiency levels and a high-efficiency AMDT with higher efficiency (see **Fig. 3**) which, however, is larger and heavier. **Tab. I** shows the corresponding material breakdowns, which are the basis for estimating the carbon footprint. For the considered Tier 2 dry-type LFT, detailed information on material composition is available from a product environmental profile (PEP) [41], whereas the mass breakdown of the AMDT is estimated based on [47], [48]. The overall weight of the AMDT is found via slight linear extrapolation of data of the ABB EcoDry Ultra series [43], [44]. As the amorphous core shows a lower saturation flux density compared to standard grain-oriented electrical steel (about 1.5 T instead of 2 T), the better efficiency of an AMDT comes at the price of higher material content and, for 400 kVA units, leads to an about 75% higher mass. The carbon footprint for the amorphous core material is assumed at 3.1 kg CO_2eq/kg [49] (i.e., about 30% higher than steel due to the required processing steps) and the ecoinvent database is consulted for all other material fractions.

The same MV-side input protection elements (except for the series inductor) used for the 400 kW first-generation SST prototype discussed below are considered. Similarly, an outdoor

cabinet housing the LFT and the protection elements is considered, and the corresponding amount of steel is obtained by scaling the data of the cabinet used for a 1200 kW second-generation SST (see **Section III**) via the required surface area (for the LFTs, the necessary cabinet size ensuring sufficient isolation distances etc. is available from the data sheets). Note that the more efficient but larger AMDT consequently requires a larger cabinet than the Tier 2 LFT.

The LV ac-dc converter is modeled based on an SMA Sunny High Power Peak 3 PV inverter [50] with a power rating of 150 kW, which includes a housing for outdoor mounting. The embodied carbon footprint is found by scaling the results of a detailed LCA study [19] to fit a 400 kW system. The LCA results were verified as far as possible using the material breakdowns provided therein and the generic component models introduced in [32], whereby an overall match within 10% has been found. Note that a PV inverter features additional power circuitry such as a maximum power point (MPP) tracker; i.e., a conservative estimate for the LV-side ac-dc converter of the LFT+ solution results.

Fig. 4a shows the resulting weight breakdown for a 400 kW LFT+ solution using a high-efficiency AMDT. The LFT and the cabinet (for the LFT and the input protection) contribute most of the total weight. For the ac-dc converter a more detailed weight breakdown is provided using the LCA data from [19]. In terms of weight, aluminum (heat sink etc.) and the magnetic components contribute around 75%. The weight of the copper bus bars in the LFT+ solution is assumed to be identical as in the SST where accurate numbers are provided.

The weight/material information is then translated into the embodied carbon footprint using the component models from [32] and available LCA data for the LV ac-dc converter. **Fig. 4cd** shows the resulting embodied carbon footprint breakdown. It is clearly visible that, e.g., the weight of printed circuit boards (PCBs) is almost negligible but contributes significantly to the carbon footprint. Similar observations can be made for integrated circuits (ICs; here including power semiconductors) due to the high energy intensity of the production. Consequently, the share of the LV ac-dc converter in the overall LFT+ carbon footprint is higher (16%) than its mass share (7%).

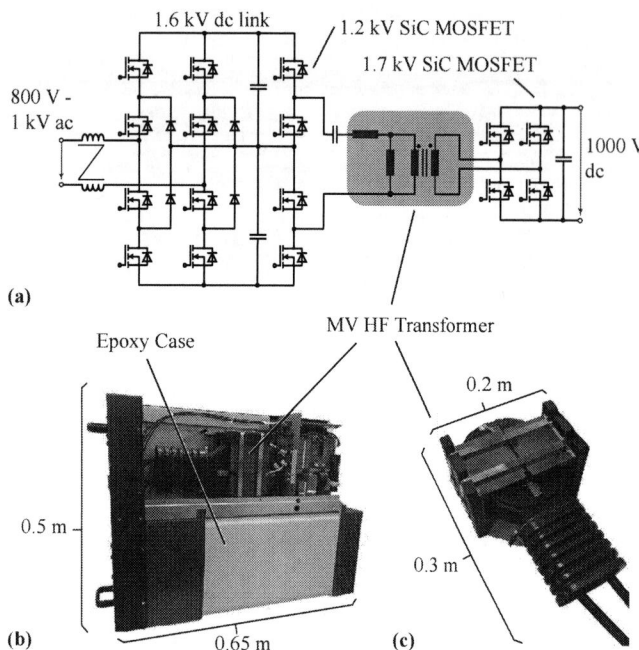

Fig. 5. (a) Power circuit of one 15 kW converter cell of the 400 kW first-generation SST demonstrator reported in [10] and shown in **Fig. 3b**, and photos of (b) the converter cell and (c) the HF transformer.

B. 400 kW First-Generation MVac-LVdc SST

The full-scale 400 kW first-generation SST prototype [10] shown in **Fig. 3b** features the power circuit topology from **Fig. 1b** and consists of 27 converter modules arranged in an input-series output-parallel (ISOP) configuration. As 9 modules share the MV grid phase voltage, each module only requires an ac input voltage of 800 V to 1 kV and a nominal power of 15 kW; the output dc voltage is 1 kV. The power circuit topology of a module is shown in **Fig. 5a** and consists of a 3-level neutral-point-clamped (NPC) single-phase ac-dc conversion stage followed by an isolated dc-dc converter. The dc-dc converter employs a 4-switch symmetric half-bridge (SHB) stage on the primary side of the HF transformer and a full-bridge on the secondary side. The module uses 1.2 kV SiC power transistors on the MV side and 1.7 kV SiC power transistors on the dc side. As the cells are connected in series on the MVac side and in parallel on the LVdc side, the HFT has to provide the galvanic isolation between MV and LV and thus must withstand the corresponding lightning impulse test voltages, hence the large bushings (see **Fig. 5c**). Similarly, the MV-side power electronics are encased in an epoxy box as shown in **Fig. 5b**. Finally, the overall SST features an MVac input section containing protection devices (fuses, surge arresters) and an input filter inductor. All 27 converter modules are interconnected by copper bus bars and, together with the the input protection, installed in a steel cabinet for outdoor use. Further implementation details are given in [10].

The material/weight breakdown of the SST prototype is found directly from the bill of materials (BOM). The carbon footprint of each converter cell is estimated using component

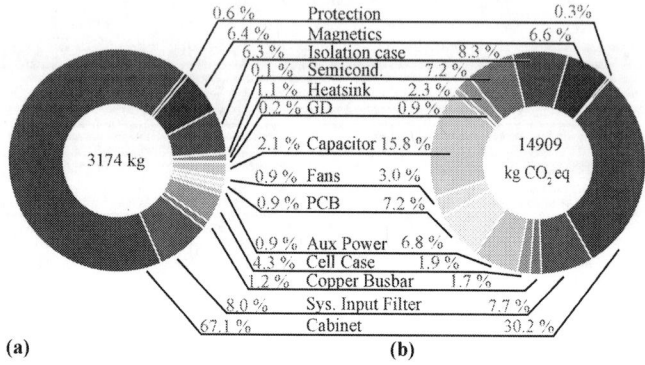

Fig. 6. (a) Mass and (b) embodied carbon footprint breakdowns of the 400 kW first-generation SST prototype reported in [10] and shown in **Fig. 3b**.

models described in [32], either via the electrical properties of the components (e.g., capacitance and voltage rating) or via measuring the mass of certain materials (e.g., copper winding of the HFT). The contributions of copper bus bars, protection devices, and the steel cabinet (for outdoor use) are considered via the respective materials and masses.

Fig. 6a presents the weight breakdown and **Fig. 6b** the breakdown of the embodied carbon footprint. It is evident that the cabinet / structural parts contribute significantly, accounting for approximately 67% of the total weight. Note that the magnetic components, i.e., the HFTs (see **Fig. 3c**) which are built using copper windings, ferrite cores, and epoxy isolation, only contribute about 6.4% of the total mass. Again, note that components such as power semiconductors have a minimal impact on the weight but substantially contribute to the overall carbon footprint, primarily due to the high energy intensity of their manufacturing process. Similar effects are observed for the PCBs, the auxiliary power units of the converter modules, and the capacitors.

III. COMPARATIVE EVALUATION

This section presents a comparative evaluation of the 400 kW LFT+ system and the first-generation SST discussed above, but also considers a 1200 kW second-generation SST demonstrator [38]. This optimized SST is shown in **Fig. 7** and consists of 30 modules in an ISOP arrangement. Each module has a nominal power rating of 40 kW and is depicted in **Fig. 7b**. Note that the overall volume is almost the same as that of the 400 kW first-generation SST prototype from **Fig. 3b**, i.e., the (volumetric) power density has been increased threefold, while similar efficiencies are achieved. Again, the mass and carbon footprint breakdowns of this SST are found based on the BOM and using the same procedure as discussed in **Section II-B**. Similarly, the results for a corresponding 1200 kW LFT+ solution are obtained as discussed above in **Section II-A**. In the following, all realization options are comparatively evaluated regarding mass, embodied carbon footprint, and raw material usage.

A. Mass

Fig. 8a presents the mass breakdowns of all considered MVac-LVdc interfaces, i.e., the LFT+ solution evaluated with

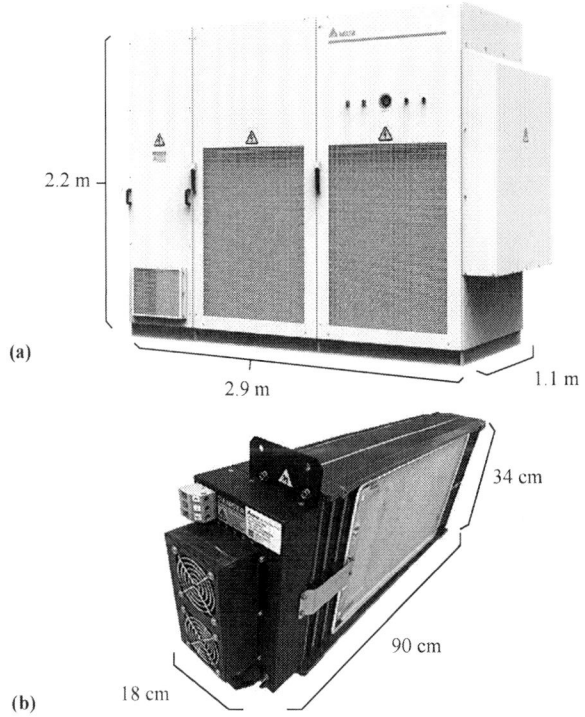

Fig. 7. (a) Photo of a 1200 kW second-generation SST demonstrator [38] and (b) detail view of one 40 kW converter cell. Note the SST's more than three times higher volumetric power density (6.9 MW/m³) compared to the 400 kW first-generation SST prototype (2 MW/m³) shown in **Fig. 3b**.

both, an ultra-efficient AMDT LFT and a standard Tier 2 LFT, for rated power levels of 400 kW and 1200 kW, as well as the 400 kW fist-generation and the 1200 kW second-generation SSTs. As previously discussed, the weight of the LFT+ systems is dominated by the LFT and the cabinet. When comparing the 400 kW and 1200 kW LFT+ realizations, note that the absolute masses of the cabinets and the LFTs do not increase in proportion to the rated power because the size of an LFT and in particular its surface area do not scale linearly with power. Interestingly, the overall weight of the 1200 kW second-generation SST is *lower* than that of the 400 kW first-generation SST despite the three times higher power rating; clearly, the 400 kW first-generation SST was a prototype designed primarily to demonstrate the viability of the concept for applications such as EV charging, but without significant optimization. For example, the cabinet contributes around two-thirds of the total weight of the 400 kW SST whereas the cabinet contribution is significantly reduced in the optimized 1200 kW second-generation SST.

Fig. 8b displays the mass breakdowns in terms of specific weight, i.e., in kilograms per kilowatt of rated power, which confirms the expectation that systems with higher nominal power feature an improved specific weight (less kg/kW). This is intuitive, as certain components such as cabinets, protective elements, and mechanical support structures do not scale with power, and in case of the LFT+ solutions also because of the favorable scaling of LFT volume/weight with power rating

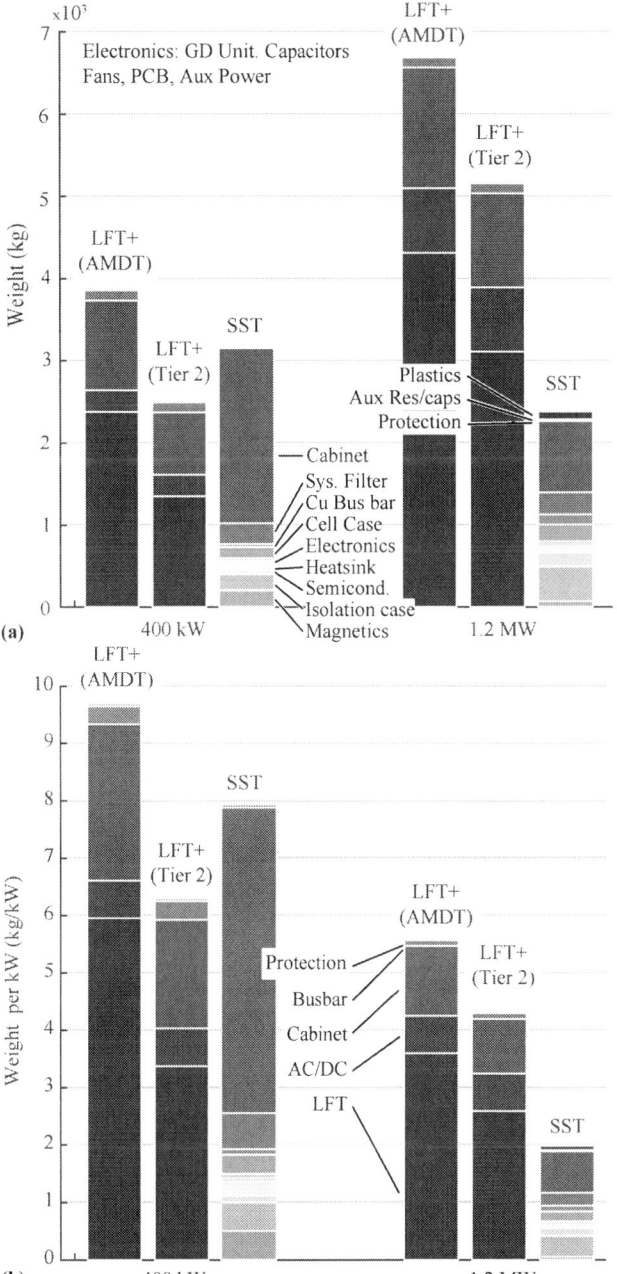

Fig. 8. Weight comparison of the considered MVac-LVdc conversion systems in (a) absolute terms and (b) normalized to the rated power (specific weight). Two different realization options are considered for the dry-type LFT, whereby the AMDT variant is more efficient but heavier compared to an LFT with Tier-2 efficiency levels according to EU regulations [39]. Further, the specific weight of the LFTs improve with the rated power. Note the significant improvement in material efficiency of the 1200 kW second-generation SST demonstrator from **Fig. 7** compared to the 400 kW first-generation SST prototype from **Fig. 3b**. Further, note that in the 1200 kVA class, the SST weight is only about 1/3 of that of the LFT+ solutions.

mentioned above. For a given power rating, the two LFT+ solutions differ in terms of specific weight depending on whether a Tier 2 LFT or a heavier (but more efficient, see **Fig. 3**) AMDT LFT is used. Finally, the 1200 kW second-generation

SST demonstrator only shows about 1/3 of the weight of the LFT+ solutions.

B. Embodied Carbon Footprint

Fig. 9a shows the embodied carbon footprints of all six MVac-LVdc interfaces. As discussed previously, the contributions of the components/subsystems to the carbon footprints is quite different than the respective weight contributions; e.g., because 1 kg of steel has a much lower carbon footprint than 1 kg of PCB. The LFTs account for about 2/3 of the LFT+ solutions' carbon footprints due to the high raw material content / large mass. In contrast, the contribution of the SSTs' HFTs to the embodied carbon footprint is comparably low but the large amount of electronic components (power semiconductors, capacitors for the buffering of the single-phase power flow in the phase-modular SST topology from **Fig. 1b**, PCBs, etc.) dominate.

Fig. 9b then shows the specific carbon footprints, i.e., the embodied carbon footprint normalized to the rated power. Similar trends as for the mass can be observed: Systems with higher rated power generally require less material (better material efficiency / lower mass), which translates into a lower carbon footprint. Still, the LFT (and its cabinet) contributes a significant part of the LFT+ solutions' carbon footprints, and there is little room for further improvement of that share; thus even if the LV ac-dc converter's specific carbon footprint could be reduced in the future, the LFT imposes a lower bound. In contrast, the clear improvement of the 1200 kW second-generation SST demonstrator compared to the 400 kW first-generation SST prototype confirms the potential of SST technology to benefit from improvements in power electronics, whereas the LFT+ solutions ultimately are remain constrained by the material usage.

C. Resource Usage

Efficient use of materials is crucial for reducing environmental impact and conserving finite resources. Prioritizing resource efficiency also lowers costs, supports sustainability, and ensures compliance with environmental regulations. **Fig. 10a** illustrates the copper usage in all three 1200 kW MVac-LVdc interfaces. The high-efficiency AMDT LFT uses copper windings; otherwise, mainly the busbars contribute to the copper usage. **Fig. 10b** shows the aluminum usage. Note that the Tier 2 LFT employs aluminum windings. Both, the LFT+ solution's LV ac-dc converter and the SST employ heat sinks. **Fig. 10c** shows that the steel usage of the LFT+ solutions is much higher than that of the SST because of the LFTs' magnetic cores.[1] Further, in all three systems, the cabinet remains a dominant contributor to the steel usage. Finally, **Fig. 10d** presents the usage of epoxy, which serves primarily as an insulation material, e.g., in the LFTs and the SST's HFTs and isolation covers of the cells. Interestingly, all three systems require similar total amounts of epoxy.

---

[1]Note that the amorphous metal core of the AMDT is considered as steel here even though further processing steps are needed compared to conventional grain-oriented electrical steel.

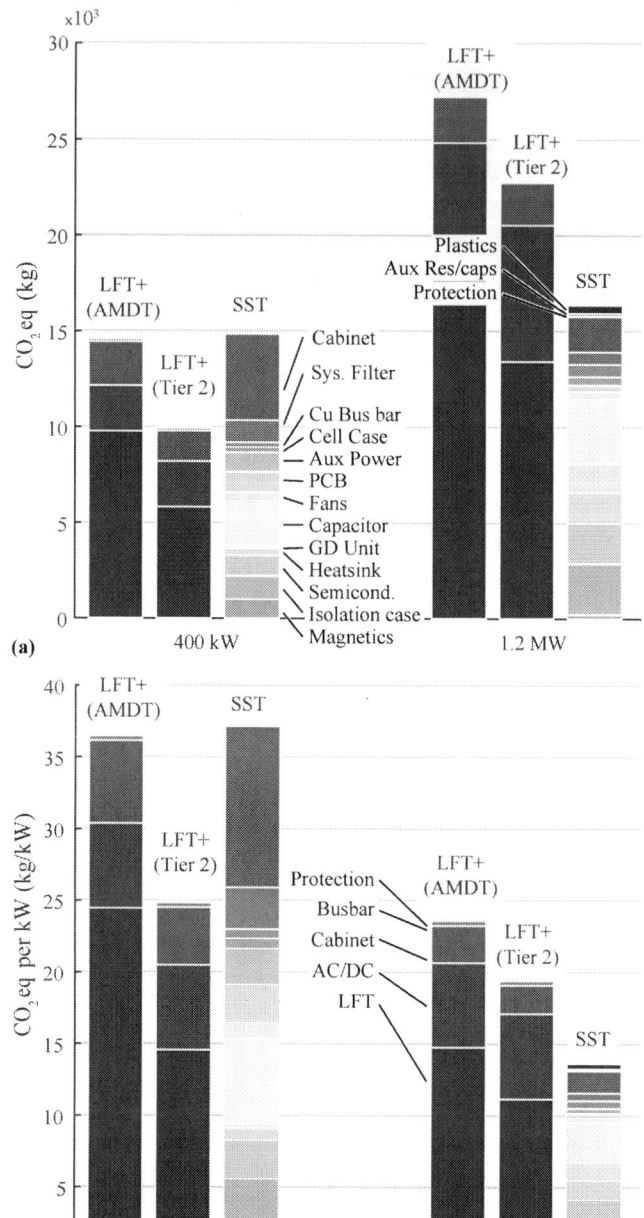

Fig. 9. Comparison of the embodied carbon footprints of the considered MVac-LVdc conversion systems in **(a)** absolute terms and **(b)** normalized to the rated power (specific carbon footprint). Two different realization options are considered for the dry-type LFT, whereby the AMDT LFT is more efficient but heavier compared to an LFT with Tier-2 efficiency levels according to EU regulations [39]; the higher weight translates to a higher embodied carbon footprint. Note the significant improvement in the specific embodied carbon footprint of the 1200 kW second-generation SST demonstrator from **Fig. 7** compared to the first-generation 400 kW first-generation SST prototype from **Fig. 3b**. Further, note that in the 1200 kVA class the embodied carbon footprint of the SST is only about half that of the LFT+ solution with the high-efficiency AMDT transformer, which is expected to show comparable efficiencies in operation (see **Fig. 3**).

IV. DISCUSSION AND CONCLUSION

MVac-LVdc SSTs, e.g., for high-power EV charging, achieve similar efficiencies and power densities as LFT-based solutions.

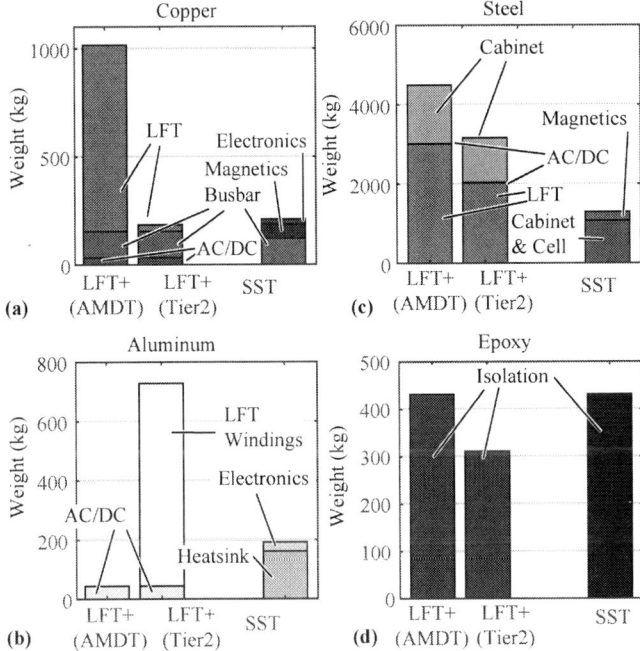

Fig. 10. Material usage of 1200 kW LFT+ MVac-LVdc interfaces and the 1200 kW second-generation SST demonstrator from **Fig. 7**. (a) Copper; (b) steel; (c) aluminum; (d) epoxy. Note that the Tier 2 LFT uses aluminum windings and the AMDT LFT employs copper windings.

higher than that of the two SSTs (27.3 kg CO_2eq/kW). Further, the modular structure of the SST simplifies maintenance and repair, i.e., only faulty modules could be replaced instead of the entire system.

On the other hand, LFTs may show a high material usage, but their relatively simply construction facilitates very high end-of-life recycling rates of 80...90% [41]. Thus, aspects such as reliability, reuse of components, recyclability and, in general, the compatibility with future circular economy concepts should be targeted by further research. Furthermore, also hybrid solutions with partial power processing (e.g., an LFT with a 12-pulse thyristor rectifier and a small LV ac-dc converter acting as an active filter [2], [51]) should be included in the comparison.

ACKNOWLEDGMENT

The Advanced Mechatronic Systems Group at ETH Zürich is generously supported by the *Else und Friedrich Hugel Fonds* via the *ETH Zurich Foundation*, for which the authors are most grateful. The authors would also like to thank the *European Center for Power Electronics e.V. (ECPE)* for the financial support. Further, the authors would like to thank Eddie Huang and Johnny Yeh from the Delta Taiwan team for providing the details of the data used in this paper.

This paper, for the first time, provides a comparative evaluation of the embodied carbon footprints and material usage, based on the BOMs of actually built industrial SST demonstrator systems.

LFTs have been optimized during decades of research and development, leaving little room for further improvements. Hence, there are lower bounds for the material usage and the embodied carbon footprint of LFT-based solutions. In contrast, SSTs are still in an evolving phase, which provides ample opportunities to enhance performance while minimizing material consumption and embodied carbon footprint, e.g, advances in semiconductor technology, high-frequency magnetics, and cooling systems offer pathways to reduce the mass and energy intensity of future SST designs. Additionally, the modular nature of SSTs allows for greater flexibility and leverages economies of scale. This is reflected by the massive reduction of the specific carbon footprint of the considered 1200 kW second-generation SST demonstrator (13.7 kg CO_2eq/kW, i.e., about 2/3 of an equivalent LFT-based solution) compared to its first-generation counterpart (37.3 kg CO_2eq/kW).

Whereas SSTs thus have the potential to outperform LFT-based solutions of similar efficiency in terms of material usage and embodied carbon footprint, LFTs typically achieve very long service lifes of 40 years and beyond, i.e., more than the power electronics (typ. 20 years). However, for the considered 1200 kW systems, even if the SST and the LV ac-dc converter of the LFT-based solution must be replaced once to reach a lifetime of 40 years, the overall embodied carbon footprint of the LFT-based solution (29.6 kg CO_2eq/kW) is still slightly

REFERENCES

[1] C. Zhao, D. Dujic, A. Mester, J. K. Steinke, M. Weiss, S. Lewdeni-Schmid, T. Chaudhuri, and P. Stefanutti, "Power electronic traction transformer—Medium voltage prototype," *IEEE Trans. Ind. Electron.*, vol. 61, no. 7, pp. 3257–3268, Jul. 2014.

[2] J. Huber, P. Wallmeier, R. Pieper, F. Schafmeister, and J. W. Kolar, "Comparative evaluation of MVAC-LVDC SST and hybrid transformer concepts for future datacenters," in *Proc. Int. Power Electron. Conf. (IPEC/ECCE Asia)*, Himeji, Japan, May 2022, pp. 2027–2034.

[3] R. Unruh, F. Schafmeister, N. Froehleke, and J. Boecker, "1-MW full-bridge MMC for high-current low-voltage (100V-400V) DC-applications," in *Proc. Power Convers. Intelligent Motion Conf. (PCIM)*, Nuremberg, Germany, Jul. 2020.

[4] "Milestones of Sungrow: 2022," 2022. [Online]. Available: https://en.sungrowpower.com/AboutSungrow/1/introduction

[5] S. Srdic and S. Lukic, "Toward extreme fast charging: Challenges and opportunities in directly connecting to medium-voltage line," *IEEE Electrific. Mag.*, vol. 7, no. 1, pp. 22–31, Mar. 2019.

[6] H. Tu, H. Feng, S. Srdic, and S. Lukic, "Extreme fast charging of electric vehicles: A technology overview," *IEEE Trans. Transp. Electrific.*, vol. 5, no. 4, pp. 861–878, Dec. 2019.

[7] A. Meintz, M. Starke, and T. Bohn, "Charging infrastructure technologies: Development of a multiport, >1 MW charging system for medium- and heavy-duty electric vehicles," presented at the DOE Vehicle Techn. Program Annu. Merit Rev. Peer Eval. Meet., Jan. 2021. [Online]. Available: https://www.nrel.gov/docs/fy22osti/79988.pdf

[8] K. Pouresmaeil, J. Duarte, K. Wijnands, M. Roes, and N. Baars, "Single-phase bidirectional ZVZCS AC-DC converter for MV-connected ultra-fast chargers," in *Proc. Power Convers. Intelligent Motion Conf. (PCIM)*, Nuremberg, Germany, May 2022.

[9] K. Nakatsu, T. Kumazaki, A. Kanouda, K. Ide, and T. Tsukishima, "Development of smart power management for achieving carbon neutrality by 2050: Energy ecosystems for widespread EV adoption," *Hitachi Rev.*, vol. 71, no. 1, pp. 51–57, 2022.

[10] C. Zhu, "High-efficiency, medium-voltage-input, solid-state-transformer-based 400-kW/1000V/400A extreme fast charger for electric vehicles," Tech. Rep. EE-0008361, Jul. 2023. [Online]. Available: https://www.osti.gov/servlets/purl/1987553/

979-8-3315-1612-3/25 $31.00 © 2025 IEEE   1324

[11] J. Yu, C. Luo, J. Duan, C. Wang, R. Lu, A. Trujillo, C. Li, and W. Li, "Design and validation of a 2MW 10kV medium-voltage solid state transformer," in *Proc. Energy Convers. Congr. Expo (ECCE USA)*, Phoenix, AZ, USA, Oct. 2024.

[12] Statista, "Copper annual market price 2022." [Online]. Available: https://www.statista.com/statistics/533292/average-price-of-copper/

[13] U.S. Geological Survey (USGS), "Metal prices in the United States through 2010," Scientific Investigations Report 2012–5188, 2013. [Online]. Available: https://pubs.usgs.gov/sir/2012/5188/

[14] Fraunhofer ISE, "Current and future cost of photovoltaics. long-term scenarios for market development, system prices and LCOE of utility-scale PV systems," Study on behalf of Agora Energiewende, Feb. 2015. [Online]. Available: https://www.agora-energiewende.org/publications/current-and-future-cost-of-photovoltaics

[15] European Commission, "The European green deal." [Online]. Available: https://commission.europa.eu/strategy-and-policy/priorities-2019-2024/european-green-deal_en

[16] ——, "Circular economy action plan." [Online]. Available: https://environment.ec.europa.eu/strategy/circular-economy-action-plan_en

[17] F. Musil, C. Harringer, A. Hiesmayr, and D. Schönmayr, "How life cycle analyses are influencing power electronics converter design," in *Proc. Power Convers. Intelligent Motion Conf. (PCIM Europe)*, Nuremberg, Germany, May 2023.

[18] Fronius International GmbH, "Fronius Tauro – Nachhaltige Technologie für eine grüne Zukunft: Lebenszyklusanalyse, (in German)," 2022. [Online]. Available: https://www.fronius.com/~/downloads/Solar%20Energy/Whitepaper/SE_WP_LCA_Tauro_DE.pdf

[19] SMA Solar Technology AG, "Sunny Highpower PEAK3 life cycle assessmenet (LCA)," Jun. 2023. [Online]. Available: https://files.sma.de/assets/280662.pdf

[20] ——, "Sunny Central 4600 UP life cycle assessmenet (LCA)," Oct. 2024. [Online]. Available: https://www.sma.de/en/newsroom/news-details/life-cycle-assessment-sunny-central-up-sustainability-performance

[21] J. Popović-Gerber, J. A. Ferreira, and J. D. van Wyk, "Quantifying the value of power electronics in sustainable electrical energy systems," *IEEE Trans. Power Electron.*, vol. 26, no. 12, pp. 3534–3544, Dec. 2011.

[22] A. Nordelöf, M. Alatalo, and M. L. Söderman, "A scalable life cycle inventory of an automotive power electronic inverter unit—Part I: Design and composition," *Int. J. Life Cycle Assess.*, vol. 24, no. 1, pp. 78–92, Jan. 2019. [Online]. Available: https://doi.org/10.1007/s11367-018-1503-3

[23] G. A. Quintana-Pedraza, S. C. Vieira-Agudelo, and N. Muñoz-Galeano, "A cradle-to-grave multi-pronged methodology to obtain the carbon footprint of electro-intensive power electronic products," *Energies*, vol. 12, no. 17, p. 3347, Jan. 2019.

[24] C. Hunziker, J. Lehmann, T. Keller, T. Heim, and N. Schulz, "Sustainability assessment of novel transformer technologies in distribution grid applications," *Sustain. Energy Grids Netw.*, vol. 21, p. 100314, Mar. 2020.

[25] M. Patra, T. Cormenier, A. Louise, and P. Mourlon, "European ecodesign material efficiency standardization supporting circular economy aspects of power drive systems for sustainability," in *Proc. 12th Int. Conf. Energy Efficiency Motor Driven Syst. (EEMODS)*, Stuttgart, Germany, May 2022.

[26] T. T. Romano, T. Alix, Y. Lembeye, N. Perry, and J.-C. Crebier, "Towards circular power electronics in the perspective of modularity," *Procedia CIRP*, vol. 116, pp. 588–593, Jan. 2023.

[27] B. Baudais, H. Ben Ahmed, G. Jodin, N. Degrenne, and S. Lefebvre, "Life cycle assessment of a 150 kW electronic power inverter," *Energies*, vol. 16, no. 5, p. 2192, Jan. 2023.

[28] S. Glaser, P. Feuchter, and A. Diaz, "Looking beyond energy efficiency - Environmental aspects and impacts of WBG devices and applications over their life cycle," in *Proc. Europ. Power Electron. Conf. (EPE)*, Aalborg, Denmark, Sep. 2023.

[29] L. B. Spejo, I. Akor, M. Rahimo, and R. A. Minamisawa, "Life-cycle energy demand comparison of medium voltage silicon IGBT and silicon carbide MOSFET power semiconductor modules in railway traction applications," *Power Electron. Dev. Comp.*, vol. 6, p. 100050, Oct. 2023.

[30] L. Vauche, G. Guillemaud, J.-C. Lopes Barbosa, and L. Di Cioccio, "Cradle-to-gate life cycle assessment (LCA) of GaN power semiconductor device," *Sustainability*, vol. 16, no. 2, p. 901, Jan. 2024.

[31] J. Huber, L. Imperiali, D. Menzi, F. Musil, and J. W. Kolar, "Energy efficiency is not enough!" *IEEE Power Electron. Mag.*, vol. 11, no. 1, pp. 18–31, Mar. 2024.

[32] L. Imperiali, D. Menzi, J. W. Kolar, and J. Huber, "Multi-objective minimization of life-cycle environmental impacts of three-phase AC-DC converter building blocks," in *Proc. IEEE Appl. Power Electron. Conf. Expo. (APEC)*, Long Beach, CA, USA, Feb. 2024.

[33] C. Minke, R. Mallwitz, P. Burfeind, and D. Hu, "Recycling potential of power electronics solutions - An exemplary study about on-board chargers," in *Proc. 13th Int. Conf. Integrated Power Systems (CIPS)*, Düsseldorf, Germany, Mar. 2024.

[34] K. Ribeiro de Faria, J.-R. Capounda, V. Rajagopal, P. Menegazzi, B. Paul, N. Kamil, S. Galeshi, and N. Messi, "Efficiency, volume and CO_2 emissions impact in a PFC converter with an active filter solution for OBC application," in *Proc. Power Convers. Intelligent Motion Conf. (PCIM Europe)*, Nuremberg, Germany, Jun. 2024.

[35] L. Fang, E. Quisbert-Trujillo, P. Lefranc, and M. Rio, "Leading LCA result interpretation towards efficient ecodesign strategies for power electronics: The case of DC-DC buck converters," *Procedia CIRP*, vol. 122, pp. 731–736, Jan. 2024.

[36] F. Salomez, H. Helbling, M. Almanza, U. Soupremanien, G. Viné, A. Voldoire, B. Allard, H. Ben-Ahmed, A. Chatroux, A. Cizeron, M. Delhommais, M. Fayolle-Lecocq, V. Grennerat, P.-O. Jeannin, L. Laudebat, B. Rahmani, P.-E. Vidal, L. Villa, L. Dupont, and J.-C. Crébier, "State of the art of research towards sustainable power electronics," *Sustainability*, vol. 16, no. 5, p. 2221, Jan. 2024.

[37] A. Campos, A. Jasi, K. Vershinin, and P. Dworakowski, "CO_2 footprint of medium voltage DC solid state transformer," in *Proc. Power Convers. Intelligent Motion Conf. (PCIM Europe)*, Nuremberg, Germany, Jun. 2024.

[38] Delta Electronics, Inc., "Solid-state transformers (SST): Enabling smarter and more efficient grid," 2024. [Online]. Available: https://www.deltaww.com/en-US/products/solid-state-transformer/ALL/

[39] The European Commission, "Commisison regulation (EU) 548/2014," May 2014. [Online]. Available: https://eur-lex.europa.eu/eli/reg/2014/548/oj

[40] Legrand SA, "Green T.HE cast resin transformers," 2021. [Online]. Available: https://datacenter.legrandwebfactory.com/sites/g/files/ocwmcr716/files/2023-06/Green_T.HE_Transformer_Brochure_16770.pdf

[41] ——, "Product environmental profile: Green transformers high efficiency TIER2," Apr. 2021. [Online]. Available: https://www.legrand.co.uk/sites/g/files/ocwmcr866/files/2023-07/bticino_green_he_transformer_tier2_lgrp-01351-v01.01-en.pdf

[42] M. Carlen, D. Xu, J. Clausen, T. Nunn, V. R. Ramanan, and D. M. Getson, "Ultra high efficiency distribution transformers," in *Proc. IEEE PES Transm. Distr. Conf. Expo. (PES T&D)*, New Orleans, LA, USA, Apr. 2010.

[43] ABB, "Technical data EcoDry dry-type transformers," 2013.

[44] ABB Ltd., "EcoDry: Ultra-efficient dry-type transformers—Lower environmental impact while cutting your costs," 2011. [Online]. Available: https://library.e.abb.com/public/e6debf4eb2025aa348257c000016fc6f/1de000076%20en%20ecodry%20dry-type%20transformers%20final.pdf

[45] SMA Solar Technology AG, "Sunny Boy US / Sunny Tripower US / Sunny Highpower US: Efficiency and derating," Oct. 2018. [Online]. Available: https://files.sma.de/downloads/WKG-Derating-US-TI-en-20.pdf

[46] ecoinvent, "Ecoinvent Database 3.9.1." [Online]. Available: https://ecoinvent.org/

[47] M. Carlen, U. Överstam, V. V. Ramanan, J. Tepper, L. Swanström, P. Klys, and E. Stryken, "Life cycle assessment of dry-type and oil-immersed distribution transformers with amorphous metal core," in *Proc. 21st Int. Conf. Electricity Distrib.*, Frankfurt, Germany, Jun. 2011.

[48] L. de Oliveira, M. Özerten, G. Kablouti, and A. Nogués, "The sustainability benefits of liquid cooled dry-type transformers in renewable energy and vent-closed applications," in *Proc. CIGRE Session*, Paris, France, Aug. 2024.

[49] Proterial, Ltd., "Issuance of a life cycle assessment (LCA) report for amorphous alloy MaDC-a™," Oct. 2024. [Online]. Available: https://www.proterial.com/e/press/2024/n1030b.html

[50] SMA Solar Technology AG, "Suny Highpower Peak3 - Customized for tomorrow today," Datasheet. [Online]. Available: https://files.sma.de/downloads/SHP-20-DS-en-22.pdf

[51] M. Schulz, D. Hoffmann, and M. Ketterer, "The resurrection of GTOs and thyristors as core components in MW-charger-application and railway/mining refurbishment," in *Proc. Power Conversion and Intelligent Motion Conf. (PCIM Europe)*, Nuremberg, Germany, May 2022.

Trade Study of Isolation Requirements and Magnetic Core Selection for Medium Frequency-Medium Voltage Transformers

Mohendro Kumar Ghosh
Department of Mechanical Engineering and Material Science
University of Pittsburgh
Pittsburgh, USA
ghosh.mk@yahoo.com

Mark A. Juds
Department of Mechanical Engineering and Material Science
University of Pittsburgh
Pittsburgh, USA
MAJ253@pitt.edu

Brandon Grainger
Department of Electrical and Computer Engineering
University of Pittsburgh
Pittsburgh, USA
BMG10@pitt.edu

Ahmad El Shafei
Eaton Research Labs
Eaton Corporation
Menomonee Falls, WI, USA
AhmadElShafei@eaton.com

Bogdan S. Borowy
Eaton Research Labs
Eaton Corporation
Menomonee Falls, WI, USA
BogdanSBorowy@eaton.com

Paul Ohodnicki
Department of Mechanical Engineering and Material Science
University of Pittsburgh
Pittsburgh, USA
PRO8@pitt.edu

Abstract— A significant emphasis exists on enabling a new generation of medium voltage and medium frequency solid state transformers using ferrite and nanocrystalline alloys for grid-tied applications including renewable integration and electric vehicle charging. Isolation requirements become a critical aspect of the transformer design in such cases, with a key design requirement being the winding-to-winding and core-to-winding spacing requirements. This paper investigates the critical impacts of isolation requirements on medium frequency transformer designs, comparing nanocrystalline and ferrite core materials in such applications. It is found that leakage flux loss becomes a critical factor as isolation distance increases. To illustrate this, a reluctance-based Magnetic Equivalent Circuit (MEC) of a single-phase shell-type transformer constructed with two EE cores is used to estimate leakage flux and associated leakage loss. Experimental results populated from a 10 kHz transformer prototype validate the model and design methodology. This paper also presents the design trade-offs between leakage flux core loss and isolation space, incorporating multi-objective optimization of 50 kW, 10 kHz transformer with ferrite and nanocrystalline cores, based on the NSGA-II algorithm under specific constrained conditions.

Keywords—ferrite core, leakage core loss, magnetic equivalent circuit analysis, medium frequency-medium voltage transformers, nanocrystalline core, NSGA-II algorithm.

I. INTRODUCTION

Transformers integrated into power electronic systems for high-power applications (e.g. 10s of kW to MW scale) operate within the medium frequency range (~1-50 kHz) [1] and are recently being extended into the high-frequency domain (>> 50 kHz), offering galvanic isolation between low and high voltage sides. The fundamental operating principle of a transformer is grounded in electromagnetic induction theory, involving magnetically coupled primary and secondary windings. In the context of medium-frequency power electronics converters such as dual active bridge topologies, medium-frequency transformers play a critical role in the overall power density and system efficiency with the operating frequency playing a significant role in the design and quantifying optimization tradeoffs [2]. In general, transformers with compact size and low loss are desired, with achievable figures of merit also being interdependent upon effective utilization of the unique performance of candidate core materials [3]. In medium-frequency applications, the core material selection is a critical design consideration in addition to the detailed optimization of topology, geometry, and windings. For line frequency transformers, electrical steels show excellent combinations of performance and cost, but these alloys exhibit unacceptable eddy current losses in medium and high-frequency transformers [4]. Instead, the soft magnetic materials that are commercially available for medium-frequency operation are comprised of nanocrystalline alloys, amorphous alloys, and ferrites. Due to unique properties of nanocrystalline and ferrite-based magnetic cores, core selection has significant impacts on the overall design strategy. As a general design rule, nanocrystalline-based magnetics offer a superior combination of efficiency and power density in the lower part of the medium frequency range while ferrite-based magnetics are superior at sufficiently high frequencies. Within this range, a cross-over is typically identified in which both ferrite and nanocrystalline-based magnetics can exhibit very similar figures of performance.

Amorphous materials are used in highly efficient line frequency transformers, and they can also exhibit a higher saturation flux density than nanocrystalline alloys. Nevertheless, amorphous components have higher losses than nanocrystalline alloys and cause audible noise in the medium frequency range due to non-zero magnetostriction. As a result,

979-8-3315-1612-3/25 $31.00 © 2025 IEEE

transformer designs based on nanocrystalline alloys are generally superior to amorphous alloys in medium frequency, therefore the design details considering amorphous alloys will not be considered here. In contrast, the metal oxides that are used in manufacturing ceramic ferrite materials have higher resistivity ($\sim10^3$-10^6 $\mu\Omega$–cm) when compared to metallic core materials with substantially reduced eddy current losses in the medium frequency range [5].

In medium-frequency transformers, a fundamental trade-off exists between core losses and flux density to maintain acceptable efficiency and limit temperature rise to allowable ranges [5]. The operating frequency dictates the specifics of the allowed tradeoff between core loss and flux density for a given core material, with eddy currents playing a dominant role in metallic core materials, thereby resulting in a stronger-frequency dependence which limits peak flux density accessible as frequency increases. As a result of this limitation, a natural transition occurs between achievable figures of merit for metallic amorphous and nanocrystalline-based transformer designs when benchmarked with ferrites. As the frequency increases in the medium frequency range, ferrite-based designs ultimately become superior despite a reduced saturation flux density. However, the specifics of this trade-off and cutoff frequency for optimal ferrite vs. nanocrystalline cores involve detailed electromagnetic and thermal considerations, driven by the unique specifications.

Winding isolation space has a significant influence on leakage flux, leakage loss, and overall size [6]. The required isolation space is defined by standards (IEC-60947-1 [7], UL-508 [8]), which vary based on the operating environment, and applied voltage. These standard requirements and industry regulations provide guidelines for isolation space in transformer design. Inadequate isolation space can lead to insulation breakdown, while excessive isolation space increases the leakage flux, leakage loss, winding loss, and size. A critical trade-off in high-frequency transformer design with ferrite and nanocrystalline core lies between leakage core loss and isolation space. A higher isolation space increases leakage inductance and leakage flux which can lead to increased core loss due to leakage flux but decreased parasitic capacitance [9]. Moreover, larger isolation space means more space is required within the transformer, potentially leading to larger and heavier designs. Conversely, smaller isolation space may improve the design compactness with a reduced leakage flux but increase parasitic capacitances at the risk of potential insulation failure. So, incorporating standard isolation space is required to ensure safety and optimal performance.

In this paper, a comprehensive study is carried out aiming to quantify the sensitivity of leakage flux core loss to variations in winding isolation space in medium frequency nanocrystalline transformers and to benchmark against ferrite transformers. Section II will explain a reluctance-based magnetic equivalent circuit model for an EE core-constructed shell-type transformer to estimate leakage flux and leakage core loss considering various leakage flux paths. In section III, MEC analysis results are validated through experiments using a 50kW-10kHz shell-type transformer prototype. In section IV, an analytical trade study between core loss due to leakage flux and variation of isolation space is discussed for nanocrystalline and ferrite transformer designs. A comprehensive trade study between leakage core loss and isolation space and the impact of leakage core loss on ferrite and nanocrystalline designs is discussed in section V. Finally, section VI concludes the article.

II. ANALYTICAL ESTIMATION OF CORE LOSSES IN LEAKAGE FLUX PATHS

In this section, a conceptual reluctance-based linear magnetic equivalent circuit model for an EE core-based shell-type transformer is discussed. The magnetic flux shown in the magnetic flux paths is assumed to be symmetrical. A shell-type transformer is chosen for leakage core loss evaluation because it provides better shielding and confinement of magnetic field leading to lower leakage core loss compared to other transformer topologies [10]. Leakage flux links only the primary winding. It exits the core which produces eddy current losses in the surface for electrically conducting core materials.

Leakage flux paths are modelled using reluctances and the resulting MMF drops across the reluctances. Generated total magnetic flux splits into two components, i) primary flux linking the secondary winding at full load operation, denoted by ϕ_{p-s} (red line) and ii) primary flux linking only the primary winding, denoted by ϕ_{p-p} (blue dashed line), shown in Fig. 1(a). When the secondary winding is shorted, it blocks all the mutual flux (red line) and only the leakage flux ϕ_{leak} (blue dashed line), exist in primary windings, shown in Fig. 1(b). Fig. 1(c) shows a reluctance-based equivalent magnetic circuit with two symmetric T-networks. Total leakage reluctance, $\mathcal{R}_T$ (1), is calculated by solving two symmetric T-networks and the leakage flux is driven by the total MMF drop $N_p I_p$. Leakage flux at the top half, $\mathcal{R}_{(top)}$ (2) and bottom half, $\mathcal{R}_{(bot)}$ (3) of the two equivalent T-networks are calculated from the MMF drop across $\mathcal{R}_{h1}$ (4), $\mathcal{R}_{h2}$ (5), $\mathcal{R}_{v(top)}$ (6), and $\mathcal{R}_{v(bot)}$ (7) and doing the series-parallel transformation. The total leakage flux and total core loss due to leakage flux can be obtained from (9) and (10) [10]. Eddy current voltage, V_E (11), is produced by the leakage flux normal to the ribbon surface around the eddy current path following Faraday's law. Eddy current path resistance, R_E (12), is defined by the size of the reluctance region on the core surface, and the depth of penetration required for eddy current leakage loss calculation is calibrated based on the prototype loss measurement. Each averaged flux density over the core surface is assumed to have an eddy current path about 1/3 of the minimum ribbon width or depth. In (12), l, s, and A are defined as the length, material conductivity, and cross section area of eddy current path, respectively. Factor 1/3 is an approximation for the width of the eddy current path which allows eddy currents to enclose the leakage flux and gives a reasonable size of the path which also correlates with core size and dimensions to capture the design specific trends. Magnetizing reluctance of the core, $\mathcal{R}_{core}$ (8) in the magnetizing flux path is quantitatively very small and neglected due to high core permeability.

979-8-3315-1612-3/25 $31.00 © 2025 IEEE

This reluctance-based analytical approximation can be validated using a single-phase shell type transformer based on the experimental prototype measurement. Fig. 2 (a) and (b) represent the 2D view of the analyzed model geometry and schematic diagram showing leakage reluctances associated with vertical and horizontal leakage fluxes as a result of a shorted secondary condition. The schematic diagram is analogous to the magnetic equivalent circuit shown in Fig. 1(c) with two symmetric T-networks. Table I shows all the prototype dimensions of this specific design. Total leakage reluctance, leakage flux and core loss due to leakage flux is calculated using (1), (9) and (10) respectively, and listed in Table II. This analytical approach computes a total leakage core loss of 60.86 W (considering 4 places) with a shorted secondary excitation using a reduced primary voltage (85.6 V_{pk}), which produces the same primary current (89.1 A_{pk}) and the same leakage flux as the full load condition. The leakage loss in $\mathcal{R}_{h1}$ is large because the winding is located near the center leg of the core.

$$\mathcal{R}_T = \mathcal{R}_{top} + \mathcal{R}_v + \mathcal{R}_{bot} + \mathcal{R}_{core} \qquad (1)$$

$$\mathcal{R}_{top} = \frac{1}{1/\mathcal{R}_{h1} + 1/\mathcal{R}_{v(top)} + 1/\mathcal{R}_{h2}} \qquad (2)$$

$$\mathcal{R}_{bot} = \frac{1}{1/\mathcal{R}_{h1} + 1/\mathcal{R}_{v(bot)} + 1/\mathcal{R}_{h2}} \qquad (3)$$

$$\mathcal{R}_{h1} = \frac{L_{\mathcal{R}h1}}{\mu_0\mu_r * D_c * W_{\mathcal{R}h1}} \qquad (4)$$

$$\mathcal{R}_{h2} = \frac{L_{\mathcal{R}h2}}{\mu_0\mu_r * D_c * W_{\mathcal{R}h2}} \qquad (5)$$

$$\mathcal{R}_{v(top)} = \frac{L_{v(top)}}{\mu_0 * D_c * S_2} \qquad (6)$$

$$\mathcal{R}_{v(bot)} = \frac{L_{v(bot)}}{\mu_0 * D_c * S_2} \qquad (7)$$

$$\mathcal{R}_{core} = \frac{L_{core}}{\mu_0\mu_r * A_c} \qquad (8)$$

$$\phi_{leak} = \frac{N_p I_p}{\mathcal{R}_T} \qquad (9)$$

$$P_E = \frac{V_E^2}{2R_E} \qquad (10)$$

$$V_E = \omega * \phi_{leak} \qquad (11)$$

$$R_E = \frac{l}{\sigma A} \qquad (12)$$

$$\mathcal{R}_v = \frac{H_{sw}}{\mu_0 * D_c * S_2} \qquad (13)$$

$$I_E = \frac{V_E}{R_E} \qquad (14)$$

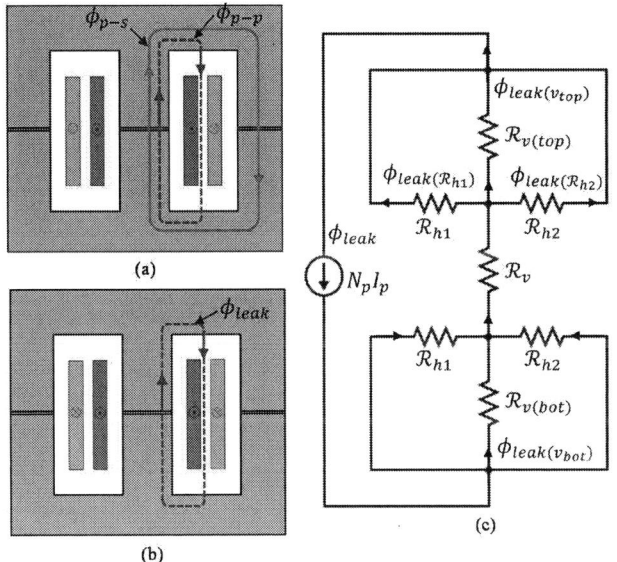

Fig. 1. Magnetic flux paths for (a) full load condition, (b) shorted secondary condition, and (c) reluctance-based equivalent magnetic circuit.

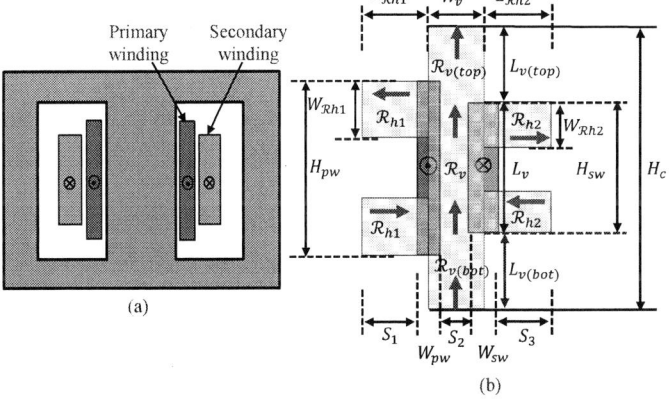

Fig. 2. Analytical validation (a) 2D view of analyzed model and (b) schematic diagram showing reluctances.

TABLE I   PROTOTYPE DESIGN DIMENSIONS

Transformer Matrix	Design Value [in]
Window height, H_c	5.51
Window width, W_c	3.15
Core depth, D_c	1.15
Eddy path depth, t	0.025
Primary winding height, H_{pw}	4.201
Primary winding width, W_{pw}	0.35
Secondary winding height, H_{sw}	2.801
Secondary winding width, W_{sw}	0.7
Isolation space, S_2	0.321

TABLE II   ANALYTICAL CALCULATIONS

Induced Reluctances in Flux Paths	Leakage Reluctances [AT/Wb]	Leakage Flux [Wb]	Leakage Core Loss [W]
Reluctance at the top or bottom, $\mathcal{R}_{v(top)}$ or $\mathcal{R}_{v(bot)}$	4.35E+07	1.12E-06	0.76
Reluctance in horizontal path (left), $\mathcal{R}_{h1}$	5.53E+06	8.83E-06	59.68
Reluctance in the horizontal path (right), $\mathcal{R}_{h2}$	5.88E+07	8.31E-07	0.44
Total (using Eq. (1))	9.51E+07	1.08E-05	60.88

III. EXPERIMENTAL VALIDATION USING DESIGN PROTOTYPE

Fig. 3 (b) shows a prototype design of 50 kW-10 kHz single phase, shell-type transformer designed with a nanocrystalline core and concentric windings with a primary RMS voltage of 800 V. Primary and secondary coils were designed with 12 turns and 15 turns, respectively. A rectangular core was chosen from the Hitachi catalog (Finemet FT-3TL) [11] and two identical cores were placed side by side. The skin effect and the proximity effect were evaluated by using 4 AWG Litz wire based on the correlation explained in [12]. Table III listed additional design specifications and parameters assumed for experimental validation.

An open secondary and shorted secondary test were performed to estimate the magnetizing core loss and leakage core loss, respectively and results are listed in Table IV. The full load test is performed at full primary voltage (1131.4 V_{pk}) with full primary and secondary current (89.1 A_{pk}, and 71.2 A_{pk}). An open secondary test was performed at full primary voltage (1131.4 V_{pk}) with zero secondary current, maximum core flux density, zero leakage flux, and a primary current equal to the magnetizing current (1.3 A_{pk}). The shorted secondary test was performed at a reduced primary voltage (85.6 V_{pk}) which produces the full load primary and secondary current (89.1 A_{pk}, and 71.2 A_{pk}) and the full load leakage flux. A relative comparison of leakage core loss obtained from a prototype shorted secondary testing and MEC analysis is given at the bottom of Table IV, which shows strong agreement.

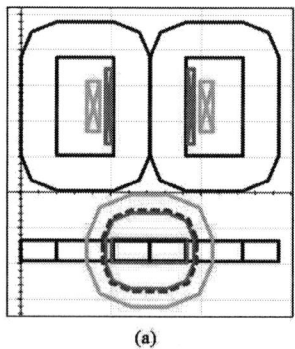 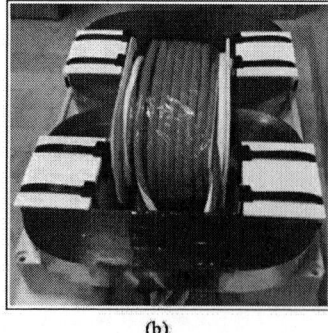

(a)   (b)

Fig. 3. (a) 2D view of design geometry (inch), and (b) 10 kHz shell-type transformer prototype.

TABLE III   DESIGN SPECIFICATION FOR EXPERIMENTAL PROTOTYPE

Parameters	Design Value
Base material	Nanocrystalline (Finemet FT-3TL)
Frequency [kHz]	10
Flux density (Peak) [T]	0.630
Core density [kg/m³] [11]	7300
Core mass [kg]	10.32
Primary voltage [V]	1131.4 (peak), 800 (RMS)
Primary current [A]	89.2 (peak), 63.1 (RMS)
Output power [kW]	50
Primary turns	12
Secondary turns	15

TABLE IV   PROTOTYPE TRANSFORMER PERFORMANCE

Parameters	Measured Results
Flux density [T]	0.630
Magnetizing core loss [W]	64.30
Leakage flux density [T]	0.047
Leakage core loss [W]	62.50
Relative Comparison in Leakage Core Loss	
Calculation Method	Leakage Core Loss [W]
Measured	62.50
MEC analytical	60.88

IV. ANALYTICAL TRADE-OFFS BETWEEN LEAKAGE FLUX CORE LOSS AND WINDING ISOLATION SPACE

In this section, an analytical trade study is carried out to explain the impact of larger winding isolation space in medium-frequency transformers. Leakage flux occupies the space between the primary and secondary windings shown in Fig. 1(b). The reluctance, $\mathcal{R}_v$ (13), of the leakage flux path between the windings can be calculated from the geometry of the flux path shown in Fig. 2(b). The largest reluctance-portion of the leakage flux path is in the center of the windings, where the space between the windings, S_2 includes the isolation space and both the winding height, H_{pw} and core depth, D_c are constant with respect to the isolation space. Leakage reluctance decreases when the isolation space increases. Reciprocally, leakage flux increases when leakage reluctance decreases. Increased leakage flux produces increased eddy current voltage, V_E, eddy current, I_E (14), and core loss due to leakage flux, P_E.

A trade study is performed for a 20 kHz-25 kW shell-type transformer at 800 V-rms with a temperature rise limit of 60°C. The potential system configurations include series connecting several transformers to achieve 13.8 kV. Isolation space is calculated from industry standards ([7] & [8]) at 800 V-rms, which gives 3 mm. Isolation space estimated with 0.10% Weibull probability (15) [13] of dielectric breakdown and using a conservative Weibull exponent of 4.5, gives an isolation space factor of 1.54 mm/kV, or 22 mm at 13.8 kV [13]. The trade study is evaluated with 4 different isolation space values (2, 9, 16, 23 mm) for both nanocrystalline and ferrite core design, and the relative performances are shown in Fig. 4.

$$P(E) = 1 - \exp\left(-\frac{E}{E_b}\right)^{\beta} \qquad (15)$$

Fig.4. (a) shows that ferrite designs have a small increase in mass with increased isolation space. This is exclusively a result of the increased isolation space between the primary and secondary windings. As shown in Fig. 4 (b), the leakage core loss is negligible (due to a very high core resistivity) and the magnetizing core loss is small, which results in a maximum core temperature rise of 8°C. Therefore, the ferrite core flux density, cross-section and magnetizing core loss remain constant with increasing isolation space. Note that the core and winding temperature rise is limited to 60°C, and it is calculated based on a cooling air temperature of 50°C at a velocity of 10 m/s. Fig. 4 (a) shows that the nanocrystalline designs have a large increase in mass with increased isolation space. As shown in Fig. 4 (c), the leakage core loss increases significantly with increased isolation space. This results in an increased core surface area to keep the core temperature-rise below the 60°C limit. The increased core surface area also results in a larger core cross-sectional area, decreased core flux density, decreased magnetizing core loss (Fig.4 (c)), and increased core mass (Fig.4 (a)). This illustrates how isolation space plays a critical role in the selection of the magnetic core material.

V. MULTI-OBJECTIVE OPTIMIZATION

A. NSGA-II Algorithm

In this paper, a multi-objective optimization is carried out based on NSGA-II algorithm [14]. The optimization framework of high-frequency transformers is an iterative process in achieving lower mass and lower loss of transformers while satisfying the specific design constraints and restrictions. The genetic algorithm, which simulates the process of biological evolution to address optimization problems, offers strong robustness and significant advantages over other methods in solving multi-objective optimization challenges. Traditionally, it simplifies multiple design goals into multi-objective functions that guide the design process. It is also possible to combine multiple objectives into a single-objective function by assigning different weighting coefficients. The expected optimal objective can be achieved if the weighting coefficients are properly selected [15]. Fig. 5 (a) shows the flow chart to create validated designs satisfying all the constraints and eligible to be used in the specified optimization algorithm. This result flows into the first modelling step of Fig. 5 (b).

Fig. 5 (b) illustrates the basic iteration steps for multi-objective optimization using NSGA-II algorithm. The algorithm starts by randomly generating an initial population of individuals (potential solutions) based on a set of decision variables. Each individual in the population is evaluated using a fitness function, which quantifies how well each solution meets the objectives or optimization criteria. Selection, crossover, and mutation operators are applied to create subpopulations. Next, the parent and offspring individuals are combined to form the next generation of the population. The combined population is reordered using fast non-dominated sorting and crowding distance calculation to create a new population. All individuals are stored in layers to maintain population diversity. These steps are repeated until the algorithm converges to the optimal solution.

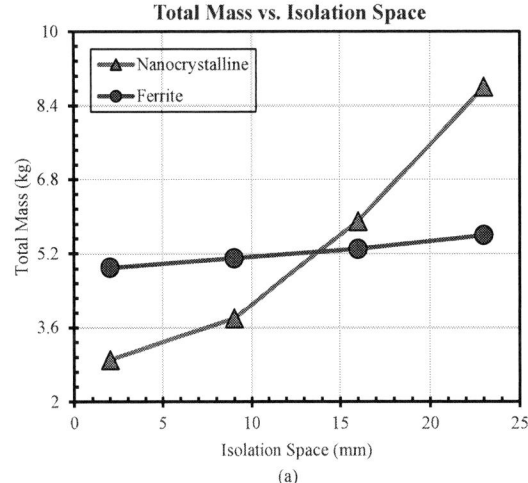

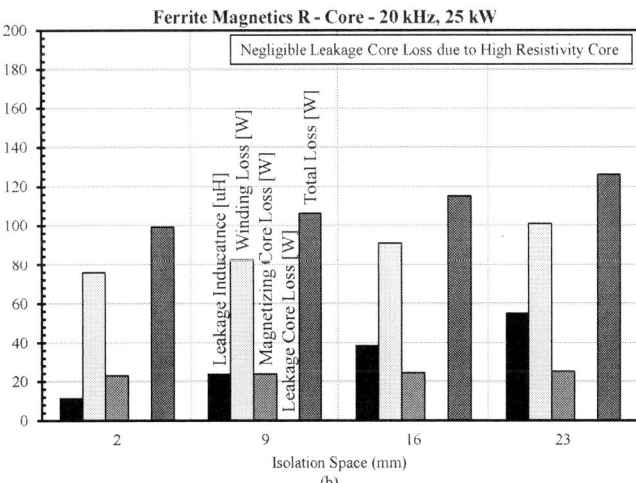

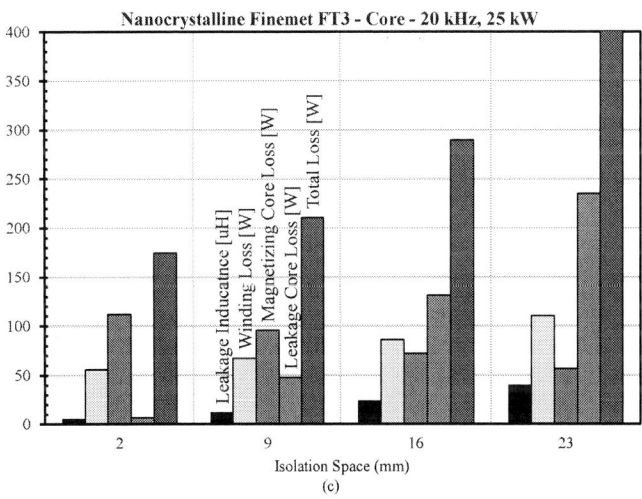

Fig. 4. Analytical trade study showing performances between (a) total mass and isolation space for both nanocrystalline and ferrite design, (b) leakage core loss and isolation space for ferrite design, and (c) leakage core loss and isolation space for nanocrystalline design.

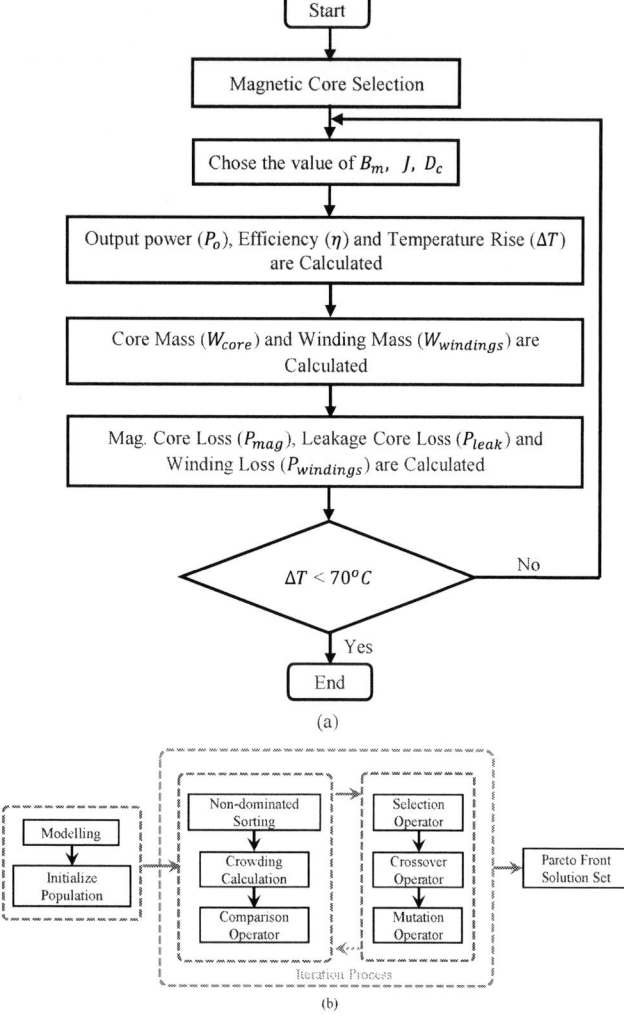

(a)

(b)

Fig. 5. (a) Steps of modeling with performance details, and (b) iteration process starting with the modeling for multi-objective optimization using NSGA-II algorithm [17].

B. Optimization Objectives and Mathematical Model

For high frequency-high power applications, the design is desired to be compact and efficient. Therefore, a reasonable selection of design parameters must be chosen to minimize the total mass and total loss. The total mass of the transformer is primarily influenced by the core size and the cross-sectional area of the window core and losses occur in the core and windings. In high-frequency transformer design, a significant amount of leakage flux is generated between the primary and secondary windings resulting in flux leaving the core normal to the surfaces. In nanocrystalline cores, this phenomenon produces a significant amount of additional core losses associated with eddy currents. In ferrite cores, this is negligible due to high resistivity. As a result, this unique loss component is also needed to be considered in the optimization.

In this section, NSGA-II algorithm is used to carry out the multi-objective optimization, taking total mass and total loss as two objectives to establish the fitness function (16). Total mass

is taken as the first objective function, $f_1(x)$, and total loss is taken as the second objective function, $f_2(x)$. In high-frequency transformer design, the ideal results can be achieved by optimizing the following parameters: magnetic flux density (B_m), winding current density (J), and core depth (D_c) [16]. The mathematical modeling to establish the fitness function for the optimization process is shown in (16), (17), and (18), respectively.

$$f(x) = \min\{ f_1(x), f_2(x)\} \tag{16}$$

$$f_1(x) = W_{core}(B_m, D_c) + W_{wind}(J) \tag{17}$$

$$f_2(x) = P_{mag}(B_m) + P_{leak}(D_c) + P_{wind}(J) \tag{18}$$

In a high-frequency transformer, the formula of the core mass, W_{core} (19), is derived from the basic relationship of voltage in a transformer, where the voltage applied in the winding is proportional to the rate of change of flux. In which, V_T, ρ_c, f, N_p and l_m are denoted as the applied primary RMS voltage of the transformer, core material density, operating frequency, number of turns at primary, and the mean magnetic path length, respectively. This relationship factors in the core material properties which influence the overall transformer mass. Similarly, (20) calculates the winding mass, W_{wind} based on the primary and secondary winding configurations, including the number of turns and the current densities. Where, ρ_{cu}, l_{wp}, l_{ws}, J_p, and J_s are denoted as, the density of copper, primary wire length, secondary wire length, primary winding current density and secondary winding current density, respectively. The winding weight derivation (20) shows the wire length, and the current densities directly impact the winding weight. Therefore, optimizing these parameters is crucial to minimizing losses and ensure transformer efficiency.

$$W_{core} = \rho_c \times \frac{\sqrt{2}V_T}{2\pi f N_p B_m} \times l_m \tag{19}$$

$$W_{wind} = \rho_{cu}\left(l_{wp} \times \frac{I_p}{J_p} + l_{ws} \times \frac{I_s}{J_s}\right) \tag{20}$$

Magnetizing core loss, P_{mag} (21) is calculated using the commonly used Steinmetz equation [10] considering a set of core loss coefficients (K, α, and β) obtained from the material database. Magnetizing core loss is influenced by both operating frequency (f) and magnetic flux density (B_m). Winding loss, P_{wind} (22) is calculated based upon the RMS current densities and the resistance (R_p, R_s) of respective windings of the transformer. Equation (22) implies, an increased current density and increased winding resistance leading to higher winding losses. Where, A_w is denoted as the cross-sectional area of the window core. Core loss due to leakage flux cannot be ignored in high-frequency transformer design and should be considered and added to the total core loss objective function, (18). This unique core loss depends on the leakage flux, the depth (D_c) and the conductivity of the core and is calculated from MEC analysis (discussed in section II).

979-8-3315-1612-3/25 $31.00 © 2025 IEEE

$$P_{mag} = K \cdot f^\alpha \cdot B_m^\beta \qquad (21)$$

$$P_{wind} = A_w^2 \left(J_{p(rms)}^2 R_p + J_{s(rms)}^2 R_s \right) \qquad (22)$$

From the above derivations, under certain frequency range and design constraints, both the transformer mass and loss is a function of magnetic flux density (B_m), winding current density (J) and core depth (D_c). Optimization results can be obtained by varying these parameters through a multi-objective algorithm.

C. Optimization Results: Impact of Isolation Space on Leakage Core Loss

In this section, the corresponding mass and loss objective functions are established for 50 kW - 10 kHz nanocrystalline and ferrite design optimization. Design parameters used in the optimization process for both nanocrystalline and ferrite design are listed in Table V. NSGA-II algorithm with a relevant mathematical model is implemented in the optimization, ensuring that the relevant parameter requirements for the high-frequency transformer design align with those of the traditional method for comparison. The parameters set for the NSGA-II algorithm are as follows: a population size of 100, a crossover rate of 0.8, a mutation rate of 0.1, 1000 iterations, and 70 offspring generated per iteration.

TABLE V   DESIGN PARAMETERS FOR 10 KHZ-50 KW TRANSFORMER

Parameters	Design Value	
Base Material	*Nanocrystalline*	*Ferrite*
Frequency [kHz]	10	
Primary voltage [V]	1131.4 (peak), 800 (RMS)	
Primary current [A]	89.2 (peak), 63.1 (RMS)	
Output power [kW]	50	
Primary turns	12	
Secondary turns	15	
Isolation space, S_2 [in]	0.321	

Fig. 6 represents the two-dimensional Pareto optimal solutions showing the trade-off between total mass and total loss for nanocrystalline and ferrite designs with an identical isolation space (0.321 in, 8.1 mm) between windings. 70 groups of optimization results with similar parameters and constraints are obtained for both simulations. Each point on the curve represents a potential design configuration that minimizes both objectives simultaneously. The Pareto analysis shows that the loss increases as the mass decreases for both designs. Example results for prototype designs have equal efficiency (99.7%) as listed in Table VI. An identical core window geometry, winding layouts, and operating at 50 kW and 10 kHz are also maintained. Under the above conditions, the nanocrystalline design has a small mass because of the high flux density, and a large leakage core loss due to low core resistivity. Whereas the ferrite design has negligible leakage core loss due to the high core resistivity. At the cross-over point, both the nanocrystalline and ferrite designs have the same weight and total loss. The nanocrystalline designs dominate to the left with higher flux density, lower mass, and higher loss. The ferrite designs dominate to the right with lower flux density, higher mass, and lower loss. At the crossover point, the nanocrystalline

designs can be reduced in mass without significantly increasing the loss, and the ferrite designs can be reduced in loss without significantly increasing the mass. The optimum design for each core material will depend upon the relative importance of mass and loss for a particular application but it is typically near the knee of each curve.

TABLE VI   PERFORMANCE RESULTS OF 10 KHZ TRANSFORMER PROTOTYPE DESIGNS

Parameters	Prototype Design Value	
Base Material	*Nanocrystalline*	*Ferrite*
Flux density [T]	0.630	0.193
Magnetizing core loss [W]	64.3	62.7
Winding loss [W]	41.6	51.08
Leakage core loss [W]	62.5	0
Total loss [W]	168.4	113.7
Total mass [kg]	10.8	23.9
Efficiency [%]	99.7	99.7

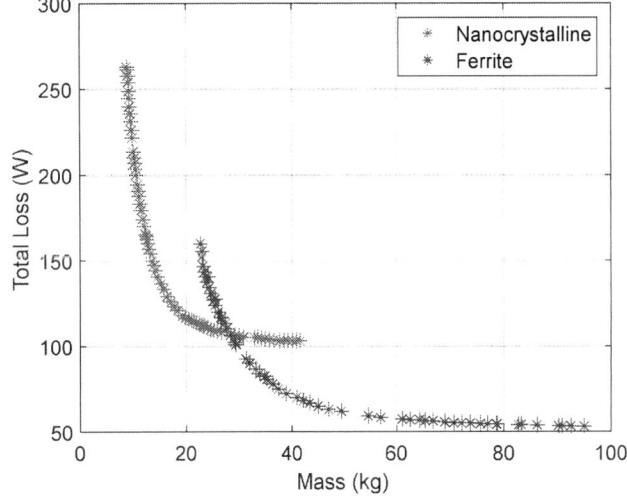

Fig. 6. Optimization results showing the trade-off between mass and loss for nanocrystalline and ferrite design at 10 kHz (with isolation space of 0.321 inch).

Fig. 7 represents the Pareto analysis showing the trade-off between the transformer mass and loss with different isolation spaces (9, 16, and 23 mm). Results are obtained from the multi-objective optimization using a 10 kHz − 50 kW transformer designed with nanocrystalline and ferrite core. This trade study indicates that increased isolation space (e.g., 23 mm), in both nanocrystalline and ferrite designs causes increase in total loss. Ferrite designs have a relatively small increment in overall total loss at increased isolation space with a negligible leakage core loss due to high core resistivity. Nanocrystalline designs have a large increment in total loss with increased isolation space due to the increased leakage flux which produces a large leakage core loss in the low resistivity core surface. With a small isolation space (e.g., 9 mm), the leakage core loss as well as the total loss is reduced, and total mass becomes relatively smaller for nanocrystalline designs in comparison to ferrite designs. Thus, isolation spacing as required by industry standards plays a critical role in the selection of magnetic material.

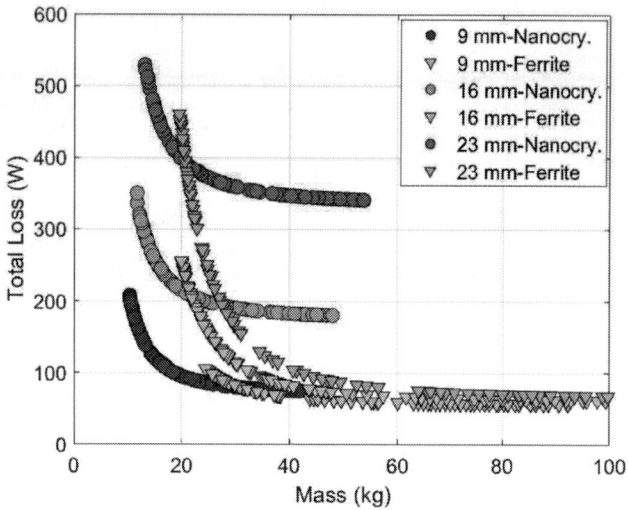

Fig. 7. Optimization results show the trade-off between mass and loss for nanocrystalline and ferrite design at 10 kHz with different isolation space.

VI. CONCLUSION

This paper investigates the impact of isolation requirements on core selection in medium frequency–medium voltage transformers. Larger isolation spaces increase leakage flux, causing significant core losses due to eddy currents in nanocrystalline cores while ferrite designs show negligible leakage core loss. For minimal isolation at medium frequencies, nanocrystalline cores achieve higher power density due to greater flux density over a significant range of operating frequencies. With increased isolation, leakage flux eddy losses of nanocrystalline cores can make ferrite cores relatively more favorable for high power density over a wider range of frequencies, balancing electrical and thermal demands as shown in our prior work [10]. Using NSGA-II, a multi-objective optimization minimizes transformer mass and loss in both core types. A trade study at 50 kW and 10 kHz reveals the crossover point (28 kg and 108 W) where nanocrystalline dominates to the left at lower mass, and ferrite dominates to the right at lower loss. Detailed trade studies at increased isolation spaces (9, 16, and 23 mm) show that nanocrystalline and ferrite designs experience increased mass and loss with larger isolation space. However, nanocrystalline designs show relatively a larger increment in total loss due to increased leakage flux in comparison to ferrite designs. It is clearly seen that (from Fig.7), the cross-over point in the pareto fronts of nanocrystalline and ferrite designs increases to larger transformer sizes as the isolation spacing requirement increases due to the increase in nanocrystalline core leakage flux losses. Future work will extend transformer modeling to explore various operating frequencies and to assess how winding isolation impacts leakage core loss and core material selection of medium frequency transformer designs. Enhanced design optimization will also explore the use of multi-objective frameworks with refined models for leakage flux-induced eddy current loss to increase design accuracy.

ACKNOWLEDGMENT

This work, supported in part by the U.S. Department of Energy's EERE under the Vehicle Technology Office (Award # DE-EE0009870), conducted in collaboration with Eaton Corporation. The work was also funded in part by the Advanced Magnetics for Power and Energy Development industry consortium.

REFERENCES

[1] A. El Shafei, S. Ozdemir, N. Altin and A. Nasiri, "Development of a high power, medium frequency transformer for medium voltage applications," in *IEEE Energy Conversion Congress and Exposition* (ECCE), pp. 891-898, Nashville, TN, USA, 2023.

[2] G. T. Nikolov, and V. C. Valchev, "Nanocrystalline magnetic materials versus ferrites in power electronics," in *Procedia Earth and Planetary Science*, pp. 1357-1361, 2009.

[3] S. D. Sudhoff, "Power Magnetic Devices-A Multi-Objective Design Approach," *IEEE Press Wiley,* 2nd Edition. 2022.

[4] R. B. Beddingfield, S. Bhattacharya and P. Ohodnicki, "Shielding of leakage flux induced losses in high power, medium frequency transformers," in *IEEE Energy Conversion Congress and Exposition* (ECCE), Baltimore, MD, USA, pp. 4154-4161, 2019.

[5] S. Balci, I. Sefa and N. Altin, "Design and analysis of a 35 kVA medium frequency power transformer with the nanocrystalline core material," in *Journal of Hydrogen Energy*, pp. 17895-17908, 2017.

[6] P. A. Dahono, Y. Sato and T. Kataoka, "Analysis and minimization of ripple components of input current and voltage of PWM inverters," in *IEEE Transactions on Industry Applications*, vol. 32, no. 4, pp. 945-950, July-Aug. 1996.

[7] Industrial Control Equipment -UL 508 [https://www.ul.com].

[8] IEC 60947-1: Low Voltage Switchgear and Control Gear – Part 1: General Rules [https://www.iecee.org].

[9] J. W. Kolar, T. Friedli, A. Looser, and J. Miniböck, "Investigation of the Effect of Coupling Capacitance on the EMI Filter Design of a 1 MHz, 10 kW Three-Phase/Level PWM Rectifier System", in *IEEE Transactions on Power Electronics, 28*(7), 3103-3115, 2013.

[10] M. K. Ghosh, C. Bracken, M. Juds, B. R. Andapally, B. Grainger and P. R. Ohodnicki, "Trade study benchmarking ferrite and nanocrystalline based medium frequency transformer technology with numerical modeling," in *IEEE Electric Ship Technologies Symposium* (ESTS), Alexandria, VA, USA, pp. 107-114, 2023.

[11] Finemet F-3TL Material Data Sheet [http://www.hitachi-metals.co.jp]

[12] M. A. Juds, *"Practical Magnetic and Electromechanical Design: If I were a flux line, where would I go?"*, Milwaukee, WI, USA, 2020.

[13] Q. Chen, R. Raju, D. Dong, and M. Agamy, "High-frequency transformer insulation in medium voltage SiC enabled air-cooled solid-state transformers," in *IEEE Energy Conversion Congress and Exposition* (ECCE), pp. 2436-2443, Portland, OR, USA, 2018.

[14] K. Deb, A. Pratap, S. Agarwal and T. Meyarivan, "A fast and elitist multiobjective genetic algorithm: NSGA-II," in *IEEE Transactions on Evolutionary Computation*, vol. 6, no. 2, pp. 182-197, April 2002.

[15] C. Yeh and J. lai, "A study on high-frequency transformer design in medium voltage solid state transformers," in *Asian Conference on Energy, Power and Transportation Electrification* (ACEPT), pp. 1-15, 2018.

[16] C. Wang, W. Han, P. Chen, J. Song and S. Yuan, "Multiobjective optimization design of high frequency transformer based on NSGA-II algorithm," in *IEEE International Conference on Mechatronics and Automation* (ICMA), Guilin, Guangxi, China, pp. 664-669, 2022.

[17] K. Zhang, W. Chen, X. Cao, Z. Song, G. Qiao and L. Sun, "Optimization Design of High-Power High-Frequency Transformer Based on Multi-Objective Genetic Algorithm," *IEEE International Power Electronics and Application Conference and Exposition* (PEAC), Shenzhen, China, pp. 1-5, 2018.

Comparative Evaluation of a Multilevel LLC Resonant Converter for a Modular DC/DC Stage in a Electrolyzer Power Supply

Samuel S. Queiroz
Faculty of Electrical Engineering
Eindhoven Universiry of Technology
Eindhoven, Netherlands
s.soares.queiroz@tue.nl

Levy F. Costa
Faculty of Electrical Engineering
Eindhoven University of Technology
Eindhoven, Netherlands
l.costa@tue.nl

Abstract—**This paper analyzes and compares 2-level and multilevel topologies for an LLC resonant converter intended for a modular DC/DC stage of a solid-state transformer (SST) in a green hydrogen system. To leverage the benefits of multilevel topologies, a design methodology is developed to evaluate the topologies concerning efficiency, power density, number of devices, and cost. A general scaling law is derived to determine the minimum number of power modules in the modular DC/DC stage. An optimization algorithm is developed to identify which number of levels achieves the best trade-off among the key performance metrics. Experimental results of an 1 kV / 5 kW three-level LLC resonant converter demonstrate an efficiency of 98.6%.**

I. INTRODUCTION

In a green hydrogen (GH$_2$) power system, the power supply interfaces the medium-voltage (MV) AC grid and the electrolyzer [1]. The most widely adopted power supply solutions for large-scale industrial water electrolyzers systems include: *i)* one-stage architecture, consisting of a three-phase low-frequency transformer (3P-LFT) with a multipulse (12 and 24-pulse) rectifier (AC/DC stage); *ii)* two-stage architecture, comprising a 3P-LFT and a multipulse rectifier connected to a non-isolated DC/DC converter [1]–[4].

These solutions offer high current capability, cost-effectiveness, and reliability [5]. However, they also introduce current harmonics into the power grid, thereby degrading power quality. As a result, additional power filters are required, which increase the overall system volume, installation area, and cost. To address these shortcomings, a solution employing a 3P-LFT combined with an IGBT-based voltage source (VSC) inverter has been investigated [3], [6]–[8]. Although this approach mitigates power quality issues, it still depends on a single power unit, i.e. the LFT, and presents disadvantages concerning volume.

To fully replace the LFT in the electrolyzer power supply, a novel architecture based on the solid-state transformer (SST) concept was introduced in [9]. The general architecture of the SST-based power supply is illustrated in Fig. 1. The SST-based architecture integrates a non-isolated AC/DC stage with a modular isolated DC/DC stage, comprising several

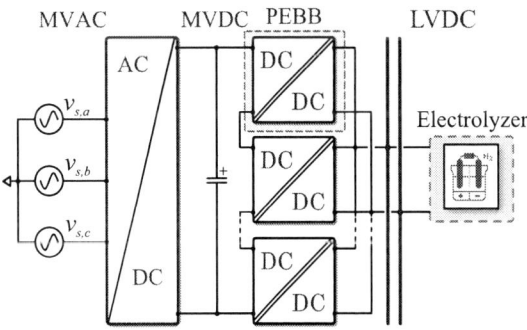

Fig. 1. General structure of the modular two-stage SST system applied the electrolyzer power-supply.

power electronic building blocks (PEBBs) in the input-series, output-parallel (ISOP) configuration. The AC/DC stage is implemented using a modular multilevel converter (MMC). The PEBBs in the modular DC/DC stage are based on LLC resonant converters, operating at the resonant frequency [10], [11]. Both 2L and multilevel topologies can be employed in the implementation of the PEBBs.

For the two-stage SST-based architecture, the benefits of multilevel topologies in the modular DC/DC stage, as well as their performance limits in terms of efficiency (η), power density (ρ), and cost, remain an open question. Despite extensive research in the SST field, no direct quantitative comparison between 2L-based and multilevel-based SSTs has not yet been presented. Furthermore, as well reported in technical literature, the DC/DC stage design represents the primary challenge in the realization of modular SST [16]. Accordingly, this paper conducts a comparative evaluation of the conventional two-level (2L) full-bridge (FB) topology [12]–[14] and the three- and five-level (3L and 5L, respectively) topologies of the series HB converter [15] (see Fig. 1(b)). The analysis is conducted using a multi-objective algorithm that incorporates key performance metrics. Additionally, semiconductors rated at 1.2 kV and 1.7 kV are considered in the design of the modular DC/DC stage to evaluate and compare their performance.

979-8-3315-1612-3/25 $31.00 © 2025 IEEE

II. DESCRIPTION OF THE SOLID-STATE TRANSFORMER SYSTEM FOR ELECTROLYZER POWER SUPPLY

The two-stage SST system is shown in Fig. 1. In the AC/DC stage, the MMC is implemented using the two-level (2L) symmetric half-bridge (HB) topology as the basic power module (see Fig. 2). Additionally, the MMC is responsible for voltage regulation at the MVDC link. The DC/DC stage consists of N_p PEBBs arranged in the ISOP configuration. The PEBB is an LLC resonant DC/DC converter, which operates as a DC transformer with constant voltage gain between V_o and V_i. Also, the PEBB provides galvanic isolation by means of the medium-frequency transformer (MFT). The primary-bridge of the PEBB can be implemented by using the 2L-FB, 3L series HB (3L-HB) and 5L series HB (5L-HB) topologies, as illustrated in Fig. 3. The secondary-bridge of the PEBB is implemented by the conventional full-wave rectifier.

By employing the SST concept, it is possible to improve the power/voltage scalability, and modularity of the system. Additionally, the SST-based architecture enables the possibility to use standard semiconductor devices with low current and low voltage ratings, along with greater flexibility in controlling voltage and power.

III. DESIGN METHODOLOGY OF THE DC/DC STAGE

A comprehensive design methodology is developed to perform the optimization of the modular DC/DC stage by taking into account key performance metrics: η, ρ, and cost per power τ. The conceptual diagram of the developed optimization tool is illustrated in Fig. 4. The system specifications for the design of the DC/DC stage are listed in Table I.

A. Modular DC/DC Stage - System Level

The most significant degrees of freedom at the system level are the number of levels m_l in the basic power module and the blocking voltage V_{ds} of the primary-bridge semiconductors. The primary-bridge of the PEBB is connected to the MVDC link. The total voltage V_{MVDC} is equally shared among the series-connected modules. For a generic topology, the minimum number of PEBBs to handle the V_{MVDC} is given by [9]

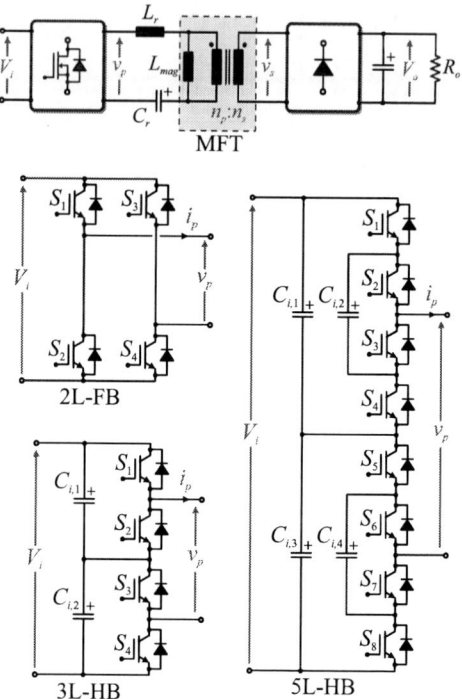

Fig. 3. (a) General structure of the modular two-stage SST system; (b) PEBB of the DC/DC stage and the main topolohies employed for the implementation of the primary-bridge - 3L-series HB and 2L-FB.

$$N_p = ceil \left[\frac{V_{MVDC}}{(k_{u,s} V_{ds})(m_l - 1)} \right]. \quad (1)$$

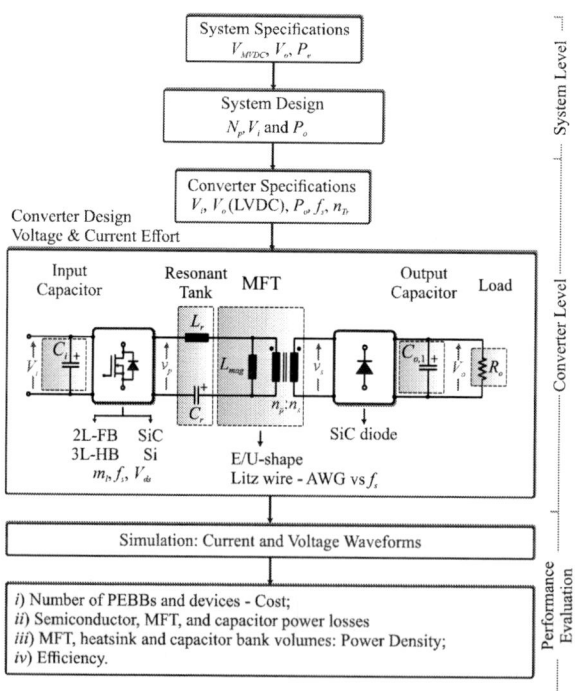

Fig. 4. Conceptual diagram of the design methodology of the modular DC/DC stage.

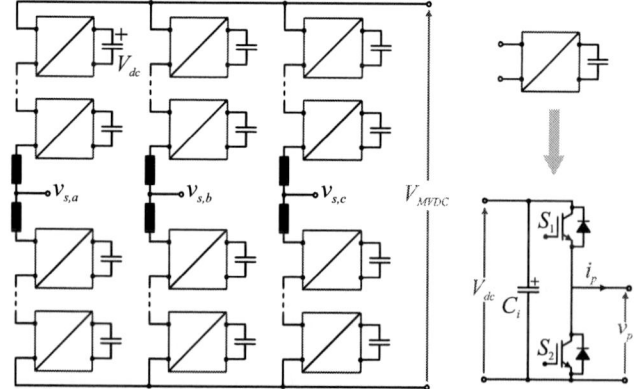

Fig. 2. General representation of the MMC in the AC/DC stage of the SST system and the basic module - 2L symmetric HB.

979-8-3315-1612-3/25 $31.00 © 2025 IEEE

TABLE I
System Specifications

Parameter	Symbol	Value
Input Voltage	V_{MVDC}	25 kV
Output Voltage - LVDC link	V_o	1 kV
Nominal Power	P_e	250 - 500 kW
Switching Frequency	f_s	20 - 50 kHz
Semiconductor Voltage Utilization	$k_{u,s}$	0.65

TABLE II
Specifications of the Selected SiC MOSFETs

SiC-MOSFET - 1.2 kV			
Part Number	I_{ds}	E_{on} / E_{off}	$R_{ds,on}$
C3M0016120K	115 A	1.1 mJ / 0.8 mJ	16 mΩ
IMZA120R020M1H	98 A	0.72 mJ / 0.47 mJ	25 mΩ
SiC-MOSFET - 1.7 kV			
C2M0045170P	67 A	0.892 mJ / 0.49 mJ	20 mΩ
G3R20MT17K	48 A	2.1 mJ / 0.86 mJ	45 mΩ
SiC Diode - 1.2 kV			
Part Number	I_D	V_F	Q_c
NDSH40120CDN	52 A	3 V	95 nC mΩ

TABLE III
Specifications of the Selected Magnetic Cores

Core	Volume	A_e	W_a
EE 100/60/28 - 3C94	202 mm^3	738 mm^2	1288 mm^2
UU 93/52/30 - 3C94	217 mm^3	840 mm^2	1737 mm^2

Therefore, the input voltage and the rated power of each PEBB are determined by $V_i = V_{MVDC}/N_p$ and $P_o = P_e/N_p$, respectively.

B. PEBB Design - Converter Level

The converter level comprises the design of the LLC resonant DCX converter. The most relevant degrees of freedom for the converter design are m_l (basic module) and the switching frequency f_s.

As shown in Fig. 4, the resonant circuit consists of the resonant inductor L_r, resonant capacitor C_r and magnetizing inductance L_{mag}. The transformer's primary-side leakage inductance is utilized as L_r. Based on transformer parameters, the inductance ratio $L_n = L_{mag}/L_r$ is calculated. The LLC resonant converter under study operates with frequency modulation (FM) [11] and f_s slightly below the resonant frequency f_o. The capacitance C_r is determined by the expression $C_r = 1/(4\pi^2 f_o^2 L_r)$.

1) PEBB - Semiconductor: For the implementation of the primary and secondary bridges of the PEBB, SiC MOSFETs and SiC diodes are employed. Table II lists the selected semiconductors and details the key electrical parameters of each device. The LLC resonant DCX converter achieves zero-voltage switching (ZVS) turn-on for the primary MOSFETs and zero-current switching (ZCS) turn-on for the secondary diodes. Therefore, the primary source of losses in the primary- and secondary-bridges are conduction losses in the MOSFETs and diodes, along with turn-off losses in the MOSFETs. The simplified conduction loss models are [17], [18]

$$P_{cond,MOSFET} = R_{ds,on}(I_{S(i),rms})^2, \qquad (2)$$

$$P_{cond,Diode} = V_F I_{D(i),avg} + R_{on,D}(I_{D(i),rms})^2, \qquad (3)$$

where $I_{S,rms}$ represents the RMS current in the MOSFET, and $I_{D,avg}$ and $I_{D,rms}$ represent the average and RMS currents in the diode, respectively. The current efforts in primary and secondary semiconductors are given by (4)-(7) [19]. For the cooling system, the heatsink is designed using the Cooling System Performance Index (CSPI) method.

$$I_{S1,rms} = \sqrt{\frac{1}{T_{sw}}\left[\frac{I_{Lr,p}^2 T_o}{4} - \left(\frac{V_o}{4L_{mag}f_o n_{Tr}}\right)^2 (T_o - t_{dt})\right]} \quad (4)$$

$$I_{S1,avg} = \frac{1}{T_o + 2t_{dt}}\left[I_{Lm,p}t_{dt} + \left(\frac{V_o n_{Tr}}{2f_{sw}R_o}\right)\right] \quad (5)$$

$$I_{D1,avg} = \frac{1}{(T_o + 2t_{dt})}\left(\frac{V_o}{2f_{sw}R_o}\right) \qquad (7)$$

The turn-off switching losses in the primary MOSFET can be modeled as [18]

$$P_{sw,MOSFET} = f_{sw}E_{off}(i_{off,(i)}), \qquad (8)$$

where i_{off} represents the turn-off current and is defined by $V_o/(4L_{mag}f_o n_{Tr})$ [19].

2) PEBB - MFT: The transformer core size A_p and the minimum number of turns of the primary side $N_{MFT,p}$ are determined as follows [20]:

$$A_p = A_e \cdot W_a = \frac{2P_o}{K_f K_u B_m J_{rms} f_s}, \qquad (9)$$

$$N_{MFT,p} = \frac{V_{p,rms}}{B_m K_f f_s A_e}. \qquad (10)$$

In (9), parameter K_f denotes the waveform coefficient, K_u is the window utilization factor, B_m is the maxium value of the flux density, and J_{rms} is the rms current density. In (10), A_e refers to the core cross-sectional area, and $V_{p,rms}$ denotes the RMS voltage applied to the primary side. For the realization of the MFT, E- and U-shape cores of N-87 and 3C94 ferrite material are taken into consideration. In Table III are listed the selected magnetic cores. To avoid skin effect, a Litz wire is adopted. Therefore, the dc losses are calculated as a function of the Lizt wire resistance and winding RMS current modeled by 11 [21]. The MFT core loss is estimated using the improved generalized Steinmetz equation (iGSE) [22]. The volume of the MFT is related to the core volume and the copper volume of the windings.

$$I_{Lr,rms} = \frac{V_o n_{Tr}}{2\sqrt{2}}\sqrt{\frac{1}{4L_{mag}^2 n_{Tr}^4 f_o^2} + \frac{\pi^2 f_o^2}{f_s^2 R_o^2}}. \qquad (11)$$

$$I_{D1,rms} = \sqrt{\frac{1}{n_{Tr}^2(T_o + 2t_{dt})} \left[\frac{I_{Lr,p}^2 T_o}{4} - \frac{T_o I_{Lm,p}}{6} + V_o T_o n_{Tr} I_{Lm,p} \left(\frac{T_o}{\pi^2 L_{mag} n_{Tr}^2} - \frac{T_s}{T_o R_o} \right) \right]} \tag{6}$$

3) PEBB - Capacitor: The capacitance of the output capacitor C_o is calculated by [23]

$$C_o = \frac{P_o \left[2f_s arcsin \left(\frac{2f_s}{\pi f_o} \right) - \pi f_s + \sqrt{(\pi f_o)^2 - (2f_s)^2} \right]}{2V_o^2 \Delta V_o \pi f_s f_o}, \tag{12}$$

where the maximum voltage ripple ΔV_o is 5%. The RMS current through C_o is given by

$$I_{Co,rms} = \frac{P_o}{V_o} \sqrt{\frac{\pi}{16 f_s \sqrt{L_r C_r}}}. \tag{13}$$

As the capacitor C_r is series connected to L_r, the rms current through is given by (11). The capacitor losses are calculated according to (14), as a function of the equivalent series resistance of the capacitor $R_{ESR(eq)}$ and the RMS current through the device. To achieve the desired capacitances, 1.25 kV film capacitors of 0.56 μF, 0.39 μF, 0.22 μF, and 0.1 μF from Kemet (R75 series) were used. The total volume and cost of the capacitor banks are estimated from the series/parallel arrangements using the selected devices.

$$P_{Cap} = R_{ESR,eq}(I_{cap,rms})^2 \tag{14}$$

IV. RESULTS OF THE DESIGN OPTIMIZATION AND PERFORMANCE EVALUATION

The developed design tool systematically explores the databases of selected components and electrical parameter vectors, evaluating all feasible combinations for designing the modular DC/DC stage. The resulting η-ρ and η-τ metrics are derived by considering the most relevant degrees of freedom of the system and the converter levels. It is worthy to mention that component costs were obtained from Mouser Electronics, based on single-unit prices.

For the system level, the results obtained from the design tool are described in Fig. 5 and Table IV. Comparing the implemented architectures, the multilevel topologies require less critical components. The 5L architecture requires the lower number of PEBBs. Advantageously, this approach achieves a reduction of 48% in the number of semiconductors and MFTs when compared to the 2L-based solution. Taking into account the number of capacitors, the 2L architecture requires 66 devices, while the 3L one requires 68. From the system perspective, the multilevel solution presents advantages in terms of number of modules and critical components, as well as implementation complexity.

The results obtained from the design tool regarding the η-ρ and η-τ limits are presented in Fig. 6. In the scatter plot, the color bar represents f_s. Overall, it is observed that all topologies can achieve efficiencies above 99%. The highest efficiencies were achieved using 1.2 kV SiC MOSFETs. Specifically, for 1.2 kV SiC MOSFETs, the study predicts a peak efficiency of 99.12% for the 2L-FB topology at $f_s = 20$

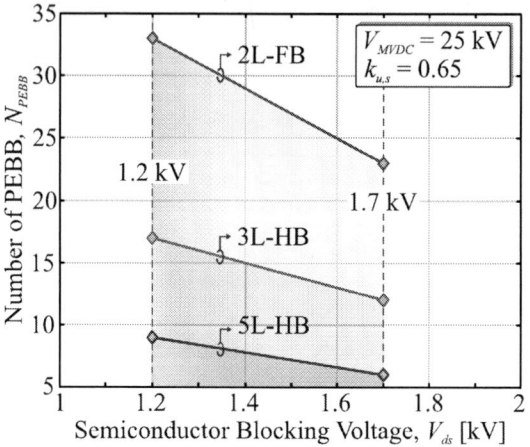

Fig. 5. Relation between the number of PEBBs and semiconductor blocking voltage.

TABLE IV
COMPARATIVE ANALYSIS OF THE 2L AND MULTILEVEL TOPOLOGIES

Semiconductor Blocking Voltage - 1.2 kV			
Parameter	2L-FB	3L-HB	5L-HB
Number of PEBBs	$N_p = 33$	$N_p = 17$	$N_p = 9$
PEEB Rated Power	$P_o = 12$ kW	$P_o = 24$ kW	$P_o = 44$ kW
Input Voltage	$V_i = 750$ V	$V_i = 1.5$ kV	$V_i = 2.7$ kV
Number of Semi.	264	136	108
Number of Cap.	66	68	54
Semiconductor Blocking Voltage - 1.7 kV			
Number of PEBBs	$N_p = 23$	$N_p = 12$	$N_p = 6$
PEEB Rated Power	$P_o = 17$ kW	$P_o = 33$ kW	$P_o = 66$ kW
Input Voltage	$V_i = 1$ kV	$V_i = 2$ kV	$V_i = 4.2$ kV
Number of Semi.	184	96	72
Number of Cap.	46	48	36

kHz. The 3L-HB and 5L-HB configurations achieved peak efficiencies of 99% and 98.88%, respectively.

In Fig. 6, the design points that achieve the best trade-off between η and τ are highlighted. For both semiconductor voltage classes, the f_s that enables the best trade-off between η-τ is 40 kHz. In the particular case of the 1.2 kV SiC MOSFET, the 5L architecture demonstrated the best trade-off between efficiency and cost. Fig. 6(c) illustrates the final η, ρ and τ for the 5L-HB solution with the 1.2 kV SiC MOSFET C3M0016120K.

With the 1.7 kV SiC MOSFETs, the 3L-HB architecture demonstrated the best trade-off between η and τ. Fig. 6(e) presents the final values of η, ρ, and τ for the 3L solution using the 1.7 kV SiC MOSFET model G3R20MT17K.

The comparison study indicates that the implementation of the traditional solutions based on 2L topologies are more complex in terms of the power circuit. The multilevel solution

979-8-3315-1612-3/25 $31.00 © 2025 IEEE

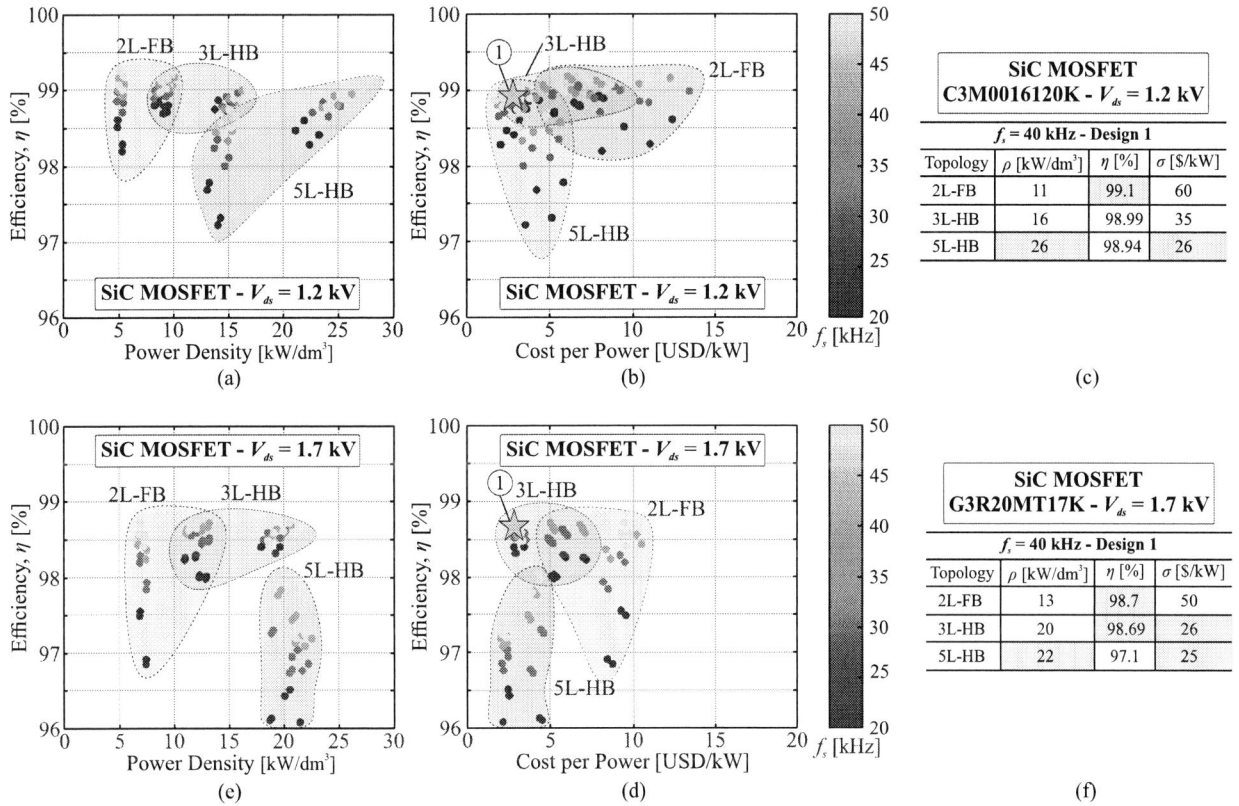

Fig. 6. Comparison study of the 2L and multilevel architectures: (a) theoretical η-ρ considering 1.2 kV SiC MOSFETs; (b) theoretical η-cost considering 1.2 kV SiC MOSFETs; (c) theoretical values of η, ρ and cost of the performed designs 1 with 1.2 kV SiC MOSFET; (d) theoretical η-ρ considering 1.7 kV SiC MOSFETs; (e) theoretical η-cost considering 1.7 kV SiC MOSFETs; (f) theoretical values of η, ρ and cost of the performed designs 1 with 1.7 kV SiC MOSFET.

is advantageous from the modularity point of view. The multilevel topologies enable improvements in power density and cost-effectiveness. The theoretical study excludes the cost of the heatsink and additional devices (such as gate drivers), leaving room for further investigations and improvements.

V. EXPERIMENTAL RESULTS

Experimental results were obtained to evaluate the performance of the 3L sereis HB LLC resonant DCX converter. The converter was operated in open-loop with FM. The modulation strategy was implemented using the TMS320F28379D microcontroller. Testing was conducted with the following specifications: V_i = 750 V and 1 kV, P_o = 1-5 kW, f_s = 48 kHz, and f_o = 50 kHz. The primary and secondary bridges were implemented using the 1.2 kV SiC MOSFET (C3M0016120K) and the 1.2 kV SiC diode (NDSH40120CDN), respectively. The specifications of the implemented MFT are presented in Table V.

The experimental results are presented in Fig. 7 and 8. The measured waveforms of the output voltage of the primary-bridge v_p, the current through the resonant circuit i_{Lr}, the voltage across the resonant capacitor v_{Cr}, and the output voltage V_o of the converter at steady-state and operating at the nominal power with V_i = 750 V are shown in Fig. 7. Fig.

8 presents the measured waveforms at the nominal power with V_i = 1 kV.

The efficiency curve in function of the output power was obtained experimentally using the power analyzer WT3000. Fig. 9 presents the efficiency curve considering two operating points: 1) V_i = 750 V and P_o = 1 - 5 kW; 2) V_i = 1 kV and P_o = 1 - 5 kW. For the operating point 1, the 3L series HB LLC resonant converter has achieved a peak efficiency of 98.56% at 4 kW. For the operating point 2, the peak efficiency was 98.6% at 5 kW. These experimental resuts demonstrate the feasibility and high performance of the 3L series HB LLC

TABLE V

TRANSFORMER SPECIFICATIONS FOR THE 3L SERIES HB LLC
RESONANT DC/DC CONVERTER

Implemented MFT		
Parameter	Leakage Inductance	Resistances
Winding - Primary	$L_{l,p} = 21.77\mu H$	$R_{w,p} = 163m\Omega$
Winding - Secondary	$L_{l,s} = 38.29\mu H$	$R_{w,s} = 164m\Omega$
Magnetizing Inductance	$L_{mag} = 5.74mH$	
Core	Cores - E 100/60/28	
N° of turns	$n_p = 34$ and $n_s = 43$	
Wires - Primary	Litz Wire - 2 x AWG41	
Wires - Secondary	Litz Wire - AWG41	

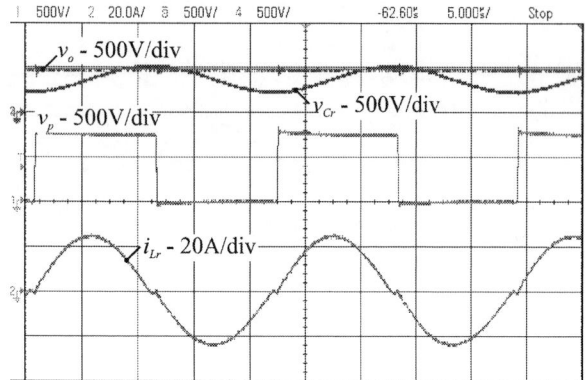

Fig. 7. Experimental results of the 3L series HB LLC resonant converter – Ch1: voltage v_p; Ch2: current i_{Lr}; Ch3: voltage v_{Cr}; Cha4: voltage V_o: (a) V_i = 750 V and P_o = 5 kW.

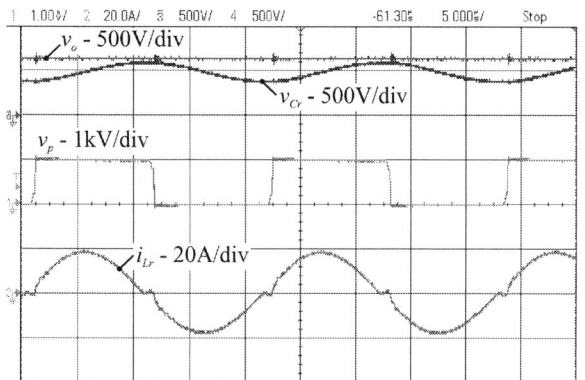

Fig. 8. Experimental results of the 3L series HB LLC resonant converter – Ch1: voltage v_p; Ch2: current i_{Lr}; Ch3: voltage v_{Cr}; Cha4: voltage V_o: (a) V_i = 1 kV and P_o = 5 kW.

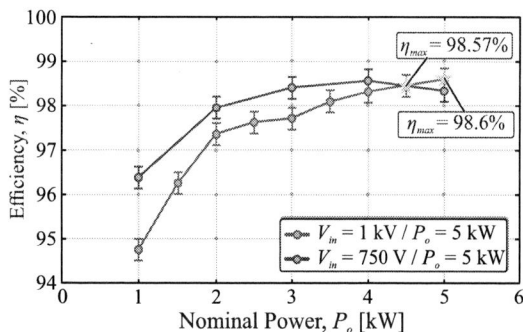

Fig. 9. Efficiency curve in function of the nominal power of the 3L series HB LLC resonant converter.

resonant converter.

VI. CONCLUSION

This paper presents the design and evaluation of an SST system's modular DC/DC stage. The DC-DC stage is an LLC resonant converter and the topologies used to implement it are the 2L-FB, 3L-HB, and 5L-HB. A comprehensive study was conducted to identify which architecture performs best in terms of efficiency, power densti and cost. The comparative study identified the 3L-HB and 5L-HB architectures as the overall best design solution since they achieved the best trade-off between efficiency and cost of critical components.

REFERENCES

[1] Chen, M.; Chou, S.-F.; Blaabjerg, F.; Davari, P. Overview of Power Electronic Converter Topologies Enabling Large-Scale Hydrogen Production via Water Electrolysis. Appl. Sci. 2022, 12, 1906. https://doi.org/10.3390/app12041906.

[2] H. Renaudineau, A. M. Llor, R. Cortés D., C. A. Rojas, C. Restrepo and S. Kouro, "Photovoltaic Green Hydrogen Challenges and Opportunities: A Power Electronics Perspective," in IEEE Industrial Electronics Magazine, vol. 16, no. 1, pp. 31-41, March 2022, doi: 10.1109/MIE.2021.3120705.

[3] Yodwong, B.; Guilbert, D.; Phattanasak, M.; Kaewmanee, W.; Hinaje, M.; Vitale, G. AC-DC Converters for Electrolyzer Applications: State of the Art and Future Challenges. Electronics 2020, 9, 912. https://doi.org/10.3390/electronics9060912.

[4] Lei, J.; Ma, H.; Qin, G.; Guo, Z.; Xia, P.; Hao, C. A Comprehensive Review on the Power Supply System of Hydrogen Production Electrolyzers for Future Integrated Energy Systems. Energies 2024, 17, 935. https://doi.org/10.3390/en17040935.

[5] J. Koponen, V. Ruuskanen, A. Kosonen, M. Niemelä and J. Ahola, "Effect of Converter Topology on the Specific Energy Consumption of Alkaline Water Electrolyzers," in IEEE Transactions on Power Electronics, vol. 34, no. 7, pp. 6171-6182, July 2019, doi: 10.1109/TPEL.2018.2876636.

[6] ABB. DC Power Supply for Electrolyzers – Safe and cost-efficient green hydrogen generation. Accessed on: August 15, 2023. Available: https://library.abb.com.

[7] Neya. Power Supplies for Large-Scale Electrolysis. Accessed on: August 15, 2023. Available: https://neya.byvonk.com/.

[8] Semikron. Power Electronics for Electrolysis. Accessed on: August 15, 2023. Available: https://www.semikron-danfoss.com/service-support/downloads/detail/semikron-danfoss-brochure-power-electronics-for-hydrogen.html.

[9] Soares Queiroz, S., and Costa, L. F. (2024). Design of a Modular Multilevel DC/DC Converter to Solid-State Transformer in a Green Hydrogen System. In 2024 IEEE Applied Power Electronics Conference and Exposition, APEC 2024 (pp. 2282-2288). Article 10509521 Institute of Electrical and Electronics Engineers. https://doi.org/10.1109/APEC48139.2024.10509521.

[10] Bo Yang, F. C. Lee, A. J. Zhang and Guisong Huang, "LLC resonant converter for front end DC/DC conversion," APEC. Seventeenth Annual IEEE Applied Power Electronics Conference and Exposition (Cat. No.02CH37335), Dallas, TX, USA, 2002, pp. 1108-1112 vol.2, doi: 10.1109/APEC.2002.989382.

[11] Y. Wei, Q. Luo and A. Mantooth, "Overview of Modulation Strategies for LLC Resonant Converter," in IEEE Transactions on Power Electronics, vol. 35, no. 10, pp. 10423-10443, Oct. 2020, doi: 10.1109/TPEL.2020.2975392.

[12] Q. Zhu, L. Wang, A. Q. Huang, K. Booth and L. Zhang, "7.2-kV Single-Stage Solid-State Transformer Based on the Current-Fed Series Resonant Converter and 15-kV SiC mosfets," in IEEE Transactions on Power Electronics, vol. 34, no. 2, pp. 1099-1112, Feb. 2019.

[13] P. Czyz, T. Guillod, F. Krismer, J. Huber and J. W. Kolar, "Design and Experimental Analysis of 166 kW Medium-Voltage Medium-Frequency Air-Core Transformer for 1:1-DCX Applications," in IEEE Journal of Emerging and Selected Topics in Power Electronics, vol. 10, no. 4, pp. 3541-3560, Aug. 2022.

[14] M. Glinka, "Prototype of multiphase modular-multilevel-converter with 2 MW power rating and 17-level-output-voltage," 2004 IEEE 35th Annual Power Electronics Specialists Conference (IEEE Cat. No.04CH37551), Aachen, Germany, 2004, pp. 2572-2576 Vol.4.

[15] I. Barbi, R. Gules, R. Redl and N. O. Sokal, "DC-DC converter: four switches V/subpk/=V/sub in//2, capacitive turn-off snubbing, ZV turn-on," in IEEE Transactions on Power Electronics, vol. 19, no. 4, pp. 918-927, July 2004, doi: 10.1109/TPEL.2004.830092.

[16] J. E. Huber and J. W. Kolar, "Solid-State Transformers: On the Origins and Evolution of Key Concepts," in IEEE Industrial Electronics Magazine, vol. 10, no. 3, pp. 19-28, Sept. 2016, doi: 10.1109/MIE.2016.2588878.

979-8-3315-1612-3/25 $31.00 © 2025 IEEE

[17] J. Biela, M. Schweizer, S. Waffler and J. W. Kolar, "SiC versus Si—Evaluation of Potentials for Performance Improvement of Inverter and DC–DC Converter Systems by SiC Power Semiconductors," in IEEE Transactions on Industrial Electronics, vol. 58, no. 7, pp. 2872-2882, July 2011, doi: 10.1109/TIE.2010.2072896.

[18] D. Dujic, G. Steinke, E. Bianda, S. Lewdeni-Schmid, C. Zhao and J. K. Steinke, "Characterization of a 6.5kV IGBT for medium-voltage high-power resonant DC-DC converter," 2013 Twenty-Eighth Annual IEEE Applied Power Electronics Conference and Exposition (APEC), Long Beach, CA, USA, 2013, pp. 1438-1444, doi: 10.1109/APEC.2013.6520487.

[19] S. S. Queiroz and L. F. Costa, "Investigation of the Influence of the Dead-Time on the Performance of an LLC Resonant Converter for High-Power Application," 2024 IEEE 15th International Symposium on Power Electronics for Distributed Generation Systems (PEDG), Luxembourg, Luxembourg, 2024, pp. 1-6, doi: 10.1109/PEDG61800.2024.10667434.

[20] M. Kazimierczuk, High-Frequency Magnetic Components. Hoboken, NJ, USA: Wiley, 2011.

[21] C. R. Sullivan and R. Y. Zhang, "Simplified design method for litz wire," 2014 IEEE Applied Power Electronics Conference and Exposition - APEC 2014, Fort Worth, TX, USA, 2014, pp. 2667-2674, doi: 10.1109/APEC.2014.6803681.

[22] K. Venkatachalam, C. R. Sullivan, T. Abdallah and H. Tacca, "Accurate prediction of ferrite core loss with nonsinusoidal waveforms using only Steinmetz parameters," 2002 IEEE Workshop on Computers in Power Electronics, 2002. Proceedings., Mayaguez, PR, USA, 2002, pp. 36-41, doi: 10.1109/CIPE.2002.1196712.

[23] D. Haake, A. Grodnichev, F. Schnabel and M. Jung, "Impact of Higher Current Harmonics on Component Current Stress and Conduction Losses of Half-Bridge-Series-Resonant-Converters in Discontinuous Conduction Mode for High-Power Applications," 2022 24th European Conference on Power Electronics and Applications (EPE`22 ECCE Europe), Hanover, Germany, 2022, pp. 01-10.

Cost-Effectiveness Assessment of SiC MOSFET and Si IGBT Semiconductors in a Three-Level Resonant Converter for Solid-State Transformer

Samuel S. Queiroz
Faculty of Electrical Engineering
Eindhoven University of Technology
Eindhoven, Netherlands
s.soares.queiroz@tue.nl

Levy F. Costa
Faculty of Electrical Engineering
Eindhoven University of Technology
Eindhoven, Netherlands
l.costa@tue.nl

Abstract—**This paper presents the analysis and experimental verification of an LLC resonant converter based on the three-level (3L) series half-bridge (HB) converter intended for a modular DC/DC stage of a two-stage solid-state transformer (SST) system. In general, efficiency and cost are key performance metrics for the SST design and implementation. Therefore, the paper focuses on the design of the modular multilevel DC/DC stage, aiming at the optimization of its efficiency and cost. An additional contribution of the paper is a comparison study considering Si IGBT and SiC MOSFET semiconductor technologies. A down-scale laboratory prototype rated at 750 V / 5 kW has shown a peak efficiency of 98.6%.**

I. INTRODUCTION

The solid-state transformer (SST) is a remarkable subject of study within the field of power electronics. Numerous studies in the technical literature have demonstrated the key features and benefits of SST technology. In the modern distribution system, the SST concept is applied as a solution in several applications, including fast-charging stations for electric vehicles and integration of renewable energy sources [1]–[3].

Typically, an SST system comprises multiple power conversion stages, as presented in [1]. Specifically, the two-stage architecture is composed by an AC/DC stage connected to an isolated DC/DC stage, which interfaces the medium-voltage (MV) AC power grid (3–30 kV) with the low-voltage (LV) DC link, as illustrated in Fig. 1(a).

In the two-stage SST architecture, the DC/DC stage connects the MVDC and LVDC and provides galvanic isolation through a medium-frequency transformer (MFT). This stage must handle high voltage and current levels in the power conversion and achieve a high efficiency. Consequently, the major challenges of the SST system emerges from the DC/DC stage [4]–[6].

The state-of-the-art two-stage SST system indicates that typically the DC/DC stage is composed by several power electronic building blocks (PEBBs) connected in a modular configuration employing the input-series, output-parallel (ISOP) configuration. A schematic of the DC/DC stage is illustrated in Fig. 1(b). The dual-active-bridge (DAB) and LLC resonant converter are the most used solution in the

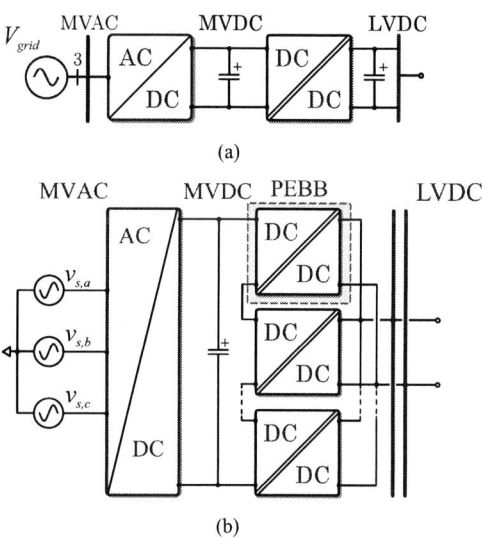

Fig. 1. SST system: (a) General representation of the two-stage architecture; (b) General representation of the system with modular DC/DC stage in the ISOP configuration.

implementation of the PEBB. Furthermore, two-level (2L) topologies, such as 2L symmetric half-bridge (HB) and 2L full-bridge (FB), are widely adopted as the basic power module (primary-bridge) for the realization of the PEBB [7]–[11]. However, the 2L-based DC/DC stage requires a large number of PEBBs to handle the voltage of the MVDC link, resulting in increased complexity [12].

Featuring higher voltage capability, multilevel topologies have been studied and evaluated for the DC/DC stage [13]–[17]. According to the literature, most LLC resonant converter-based solutions utilize the three-level (3L) neutral-point-clamped (NPC) converter and the series half-bridge (HB) converter [15]–[17].

In [17], a theoretical analysis of the series HB converter designed for the modular DC-DC stage was conducted. The study demonstrated that the 3L architecture provides the best trade-off between power losses and component count. This

979-8-3315-1612-3/25 $31.00 © 2025 IEEE   1341

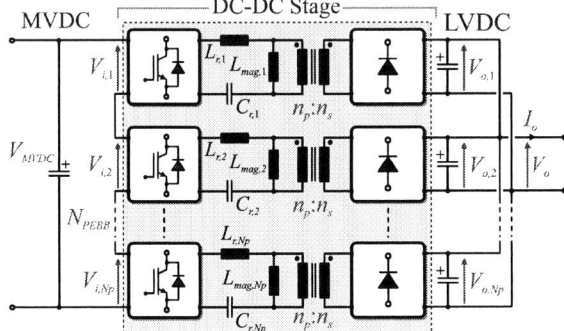

Fig. 2. Modular DC/DC stage of the SST system.

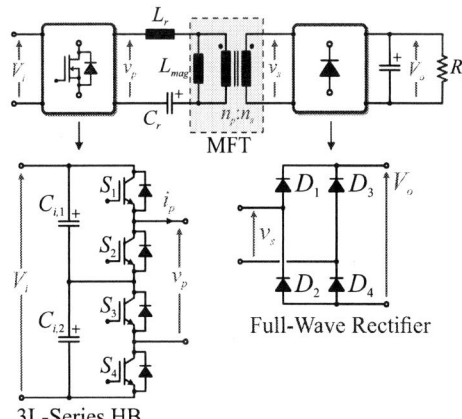

Fig. 3. PEBB: LLC resonant converter using the 3L series HB, MFT, and full-wave rectifier.

paper extends the investigation of the modular DC/DC stage, focusing on the design of the 3L series HB LLC resonant converter and optimizing the efficiency (η) of the PEBB considering the cost of critical components and power density (ρ). An in-depth comparative study of the modular DC/DC stage is also presented, evaluating the performance of Si-IGBT and SiC-MOSFET technologies. The analysis highlights the advantages of each semiconductor technology and quantifies the improvements enabled by SiC-MOSFETs in terms of power losses. Furthermore, the optimal switching frequency is identified by achieving the best trade-off between η and cost.

II. DESCRIPTION OF THE DC/DC STAGE IN THE SST SYSTEM

The generalized DC/DC stage of the SST system is illustrated in Fig. 2. The DC/DC stage consists of N_{PEBB} PEBBs arranged in the ISOP configuration. The LLC resonant converter is adopted for the PEBB due to its superior soft-switching capabilities and higher power density compared to the DAB topology [18], [19]. The 3L series HB topology is adopted as the basic submodule of the PEBB primary side [20]. For the secondary-bridge, the full-wave rectifier is used. The PEBB based on this multilevel topology is depicted in Fig. 3. The LLC resonant DC/DC converter operates using frequency modulation (FM) [19] and in discontinuous conduction mode (DCM). Its switching frequency f_s slightly below the resonant frequency f_{res} ($f_s < f_{res}$), allowing it to operate as a DC transformer [21]. As can be seen in Fig. 3, the resonant circuit is composed by the resonant inductor L_r, resonant capacitor C_r and magnetizing inductance L_{mag}. The resonant frequency is defined by

$$f_r = \frac{1}{2\pi\sqrt{L_r C_r}}. \tag{1}$$

In this operating mode, the converter achieves zero-voltage switching (ZVS) turn-on for the primary-bridge switches and zero-current switching (ZCS) for the secondary-bridge diodes. The operation of the LLC resonant converter is determined by: *i*) resonant frequency f_o; *ii*) inductor ratio L_n; *iii*) quality factor Q_e. The equation for the Q_e is given in (2). The normalized

voltage gain of the LLC resonant converter M_v is described in (3).

$$Q_e = \frac{\pi^2 n_{Tr}^2}{8R_o}\sqrt{\frac{f_r}{C_r}} \tag{2}$$

$$M_v(\omega_r) = \frac{\omega_r^2 L_n}{\sqrt{[(\omega_r^2(L_n+1))-1]^2 + [\omega_r^2 L_n Q_e(\omega_r^2-1)]^2}} \tag{3}$$

III. DESIGN METHODOLOGY

This section presents the design of the modular DC/DC stage and the 3L series HB LLC resonant DCX converter.

A. System Design

The system specifications are listed in Table I. The design of the modular DC/DC stage is realized through the methodology introduced in [3], [22]. The design process begins by determining the minimum number of PEBBs required for system operation. A semiconductor with a 1.2 kV voltage rating is initially selected. In the 3L series HB power module, the voltage on each switch is reduced by 2. The number of PEBBs is defined by

$$N_p = ceil\left[\frac{V_{MVDC}}{(k_{u,s} V_{ds}) \cdot 2}\right], \tag{4}$$

where V_{MVDC} is the voltage at the MVDC link, V_{ds} is the primary semiconductor blocking voltage and $k_{u,s}$ is the semiconductor utilization factor. Assuming the values of $k_{u,s} = 0.65$ and $V_{MVDC} = 25$ kV, it was determined that the system requires 17 PEBBs. Considering that the voltage and power are equally shared among the 17 PEBBs, the input voltage and the rated power of each PEBB are determined by $V_i = V_{MVDC}/N_p$ and $P_o = P_e/N_p$, respectively. Fig. 4 shows the shematic of the designed system, where the key parameters of the PEBB are presented.

TABLE I
SYSTEM SPECIFICATIONS

System Specifications		
Parameter	Symbol	Value
MVDC Link	V_{MVDC}	25 kV
Nominal Power	P_e	1 MW
PEBB Parameters		
Input Voltage	V_i	1.5 kV
Output Voltage - LVDC link	V_o	1 kV
Rated Power	P_o	58 kV
Switching Frequency	f_s	5 - 50 kHz
Resonant Frequency	f_{res}	$1.05 \cdot f_s$
Transf. Turns Ratio	n_{Tr}	1.33

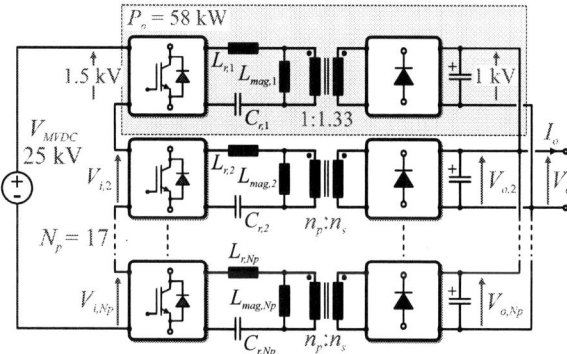

Fig. 4. Schematic of the modular DC/DC stage of the SST system.

B. Converter Design

The main parameters of the PEBB are listed in Table I. The design of the 3L series HB LLC resonant DCX converter follows the approach described in [23]. To enable its operation as a DC transformer, the converter is typically designed with L_n and Q_e set around 100 and 0.1, respectively. The capacitance C_r is determined by the expression $C_r = 1/(4\pi^2 f_o^2 L_r)$. For the topology realization, the MFT is designed using the core EE 260/91/40.

IV. RESULTS AND DISCUSSION

Firstly, the modular DC/DC stage and the 3L series HB LLC resonant DCX converter are simulated using MAT-LAB/PLECS. The simulated voltage and current waveforms of the PEBB are presented in Fig. 5. The input and output capacitances are set to 20 μF and 16 μF, respectively. Metalized film capacitors are employed for the implementation of the capacitor banks.

The power losses of the designed converter are estimated, including: *i*) primary semiconductor conduction and turn-off switching losses (P_{cond} and P_{sw}, respectively); *ii*) secondary semiconductor conduction losses ($P_{cond,diode}$); *iii*) capacitor conduction losses, and *iv*) transformer core and winding losses. Analytical models for calculating these power losses are detailed in [24]–[28]. The current efforts of each component were obtained directly from the simulated waveforms. The overall cost of the converter is evaluated by considering the primary and secondary semiconductors, capacitors, and transformer. For estimating transformer costs, the methodology presented in [29] is employed. For the other components, individual unit prices were obtained from Mouser.

The selected Si IGBT and SiC MOSFET devices are listed in Table II. It is important to mention that the 1.2 kV devices were chosen due to their wide range of current ratings. To ensure a fair comparison, Si and SiC devices with closely current ratings were selected. The full-wave rectifier on the secondary side is implemented using the 1.2 kV SiC diode CAR600M12HN6.

The results obtained from the comparison study are presented in Fig. 6. The theoretical η of the 3L series HB LLC resonant DCX converter is presented considering ρ and the overall cost of the modular DC/DC stage. In the scatter plot, the color bar represents f_s. It is observed that the Si-based and SiC-based solutions attained $\eta > 98.5\%$.

The design using the 1.2 kV Si IGBT FGY100T120SWD provides the best trade-off between η and cost. Figure 6(c).i shows the values of η, ρ, and cost for this design. The optimal f_s that achieves the best η-cost trade-off is 10 kHz. When a SiC device is used, a higher η is achieved, but at the expense of a higher overall cost. Figure 6(c).ii shows the final values

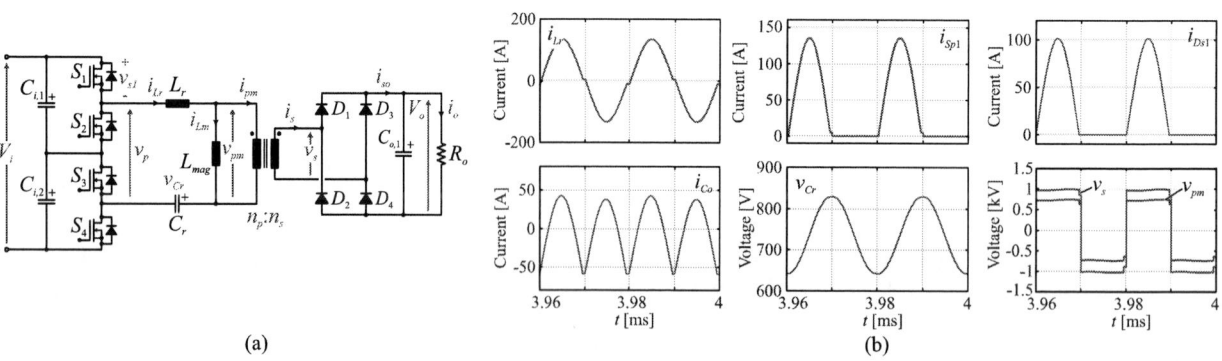

Fig. 5. (a) 3L series HB LLC resonant DC/DC converter; (b) The main waveforms of the 3L series HB LLC resonant DC/DC converter operating with FM and $f_s < f_o$.

979-8-3315-1612-3/25 $31.00 © 2025 IEEE

TABLE II
PARAMETERS OF THE SELECTED SEMICONDUCTORS FOR THE 3L SERIES HB LLC RESONANT CONVERTER

SiC MOSFET - 1.2 kV					
Part Number	Manuf.	I	$R_{ds,on}$	E_{on} / E_{off}	Cost
C3M0016120K	Cree	85 A	16 mΩ	1.1 mJ / 0.8 mJ	$ 90.13
IMZA120R020M1H	Infineon	71 A	25mΩ	0.72 mJ / 0.14 mJ	$ 26.43
STC4018KE	ROHM	57 A	18mΩ	0.52 mJ / 0.14 mJ	$ 36.77
Si IGBT - 1.2 kV					
IKY75N120CS6	Infineon	75 A	18 mΩ	3.3 mJ / 5.3 mJ	$ 8.78
IKW75N120CH7	Infineon	81 A	20 mΩ	5.8 mJ / 3.37 mJ	$ 10.0
FGY100T120SWD	Onsemi	100 A	20 mΩ	3.1 mJ / 1.6 mJ	$ 8.78

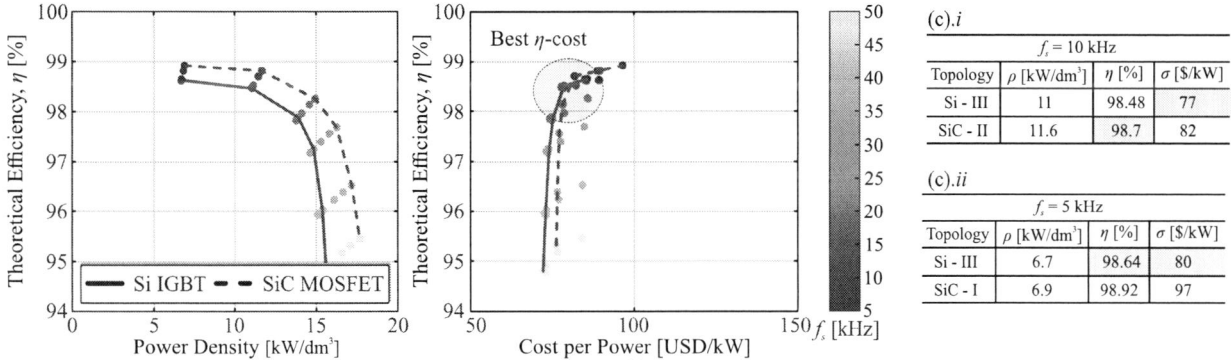

Fig. 6. Comparison study of the modular DC/DC stage based on the 3L series HB LLC resonant converter: (a) theoretical η versus cost; (b) theoretical η versus ρ; (c) theoretical results of η, ρ and cost of the performed designs.

of ρ and cost at the point the highest η. The highest efficiency was achieved using the 1.2 kV SiC MOSFET C3M0016120K, due to its lowest on-state resistance $R_{ds,on}$. However, it is the most expensive solution.

Figure 7(a) presents the breakdown of losses for the PEBB when using the Si IGBT FGY100T120SWD and the SiC MOSFET C3M0016120K, both operating at f_s = 10 kHz. In general, semiconductor power losses are the most significant contributors. As the LLC resonant converter operates under ZVS, switching losses are minimized. Consequently, Si-based solutions can achieve comparable η at f_s in range of 5 kHz - 15 kHz.

V. EXPERIMENTAL RESULTS

Experimental results were obtained from a prototype to validate the performance of the 3L series HB LLC resonant DCX converter. The main specifications of the prototype are listed in Table III. The primary side was implemented using: *i*) a 1.2 kV SiC MOSFET C3M0016120K; and *ii*) a 1.2 kV Si IGBT IKZA40N120CS7. The specifications of the implemented MFT are provided in Table IV. The capacitance C_r was calculated based on $L_r = L_{l,p}$ and $f_s = f_o = 20$ kHz, yielding $C_r = 2.4$ µF. Film capacitors of 0.56 µF were used to implement the capacitor bank.

Figure 7(b) shows the behaviour of the output voltage of the primary-bridge v_p, the current through the resonant circuit i_{Lr}

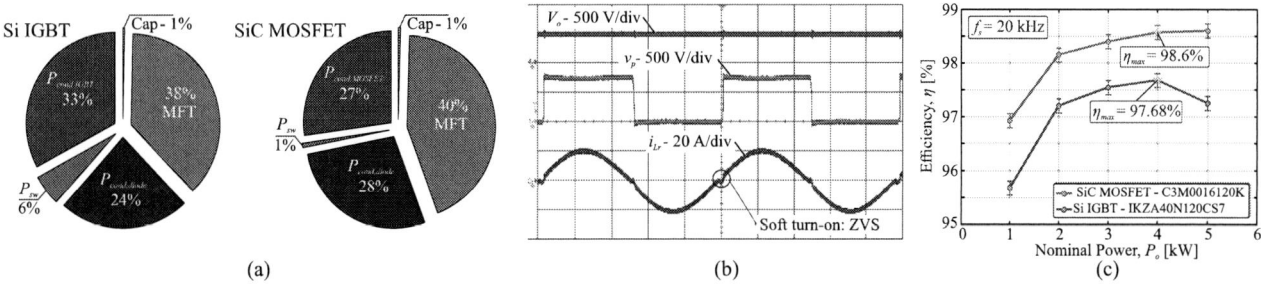

Fig. 7. Theoretical and experimental results: (a) Breakdown of losses in the 3L series HB LLC resonant DCX converter - best trade-off between η and cost with f_s = 10 kHz; (b) Experimental results of the 3L series HB LLC resonant converter – Ch1: voltage v_p; Ch3: current i_{Lr}; Ch4: voltage V_o - V_i = 750 V and P_o = 5 kW; (c) Efficiency curve in function of the nominal power of the 3L series HB LLC resonant converter.

TABLE III
SPECIFICATIONS OF THE PROTOTYPE

System Specifications		
Parameter	Symbol	Value
Input Voltage	V_i	750 V
Nominal Power	P_n	5 kW
Switching Frequency	f_s	20 kHz

TABLE IV
TRANSFORMER SPECIFICATIONS FOR THE 3L SERIES HB LLC
RESONANT DCX CONVERTER

Implemented MFT		
Parameter	Leakage Inductance	Resistances
Winding - Primary	$L_{l,p} = 21.77\mu H$	$R_{w,p} = 163m\Omega$
Winding - Secondary	$L_{l,s} = 38.29\mu H$	$R_{w,s} = 164m\Omega$
Magnetizing Inductance	$L_{mag} = 5.74mH$	
Core		Cores - E 100/60/28
N° of turns		$n_p = 34$ and $n_s = 43$

and the output voltage V_o of the 3L series HB LLC resonant DCX converter at steady-state. By the waveforms, it is demonstrated the soft switching operation (i.e., ZVS operation) at the turn-on.These experimental results were obtained from SiC-based prototype. The experimental waveforms closely match the simulated waveforms shown in Fig. 5.

The efficiency curve as a function of nominal power is shown in Fig. 7. The converter achieved a peak efficiency of 98.6% at $P_o = 5$ kW with the SiC MOSFET device. When the Si IGBT was used, the converter reached a peak efficiency of 97.68% at $P_o = 4$ kW. The SiC MOSFET semiconductor technology allows a reduction of 52% in the power losses of the converter, resulting in a η increase around 1%. These results demonstrate that SiC MOSFET semiconductor technology is crucial for minimizing power losses and enhancing overall η of the converter. Additionally, the experimental results confirm the feasibility of the 3L series HB LLC resonant DCX converter using both semiconductor technologies.

VI. CONCLUSION

This paper investigates the 3L series HB LLC resonant DCX converter as a power electronic building block (PEBB) for the modular DC/DC stage of a two-stage SST system. The design and theoretical analysis of the converter are presented, along with a comparative study of Si IGBT and SiC MOSFET technologies, using efficiency and cost as metrics of performance. Theoretical analysis indicates that the Si IGBT-based solution provides the optimal trade-off between efficiency and cost of critical components when operating at switching frequencies between 5 kHz and 10 kHz. A 750 V / 5 kW propotype of the 3L series HB LLC resonant converter was built and tested using Si IGBT and SiC MOSFET. A 750 V / 5 kW prototype of the 3L series HB LLC resonant converter was constructed and tested with both Si IGBT and SiC MOSFET technologies. Under nominal power conditions

and at a switching frequency of 20 kHz, the SiC MOSFET-based solution achieved a peak efficiency of 98.6%, while the Si IGBT-based solution reached a peak efficiency of 97.68%. The experimental results confirmed the high performance and practical feasibility of the 3L series HB LLC resonant DCX converter using both semiconductor technologies.

REFERENCES

[1] X. She, A. Q. Huang and R. Burgos, "Review of Solid-State Transformer Technologies and Their Application in Power Distribution Systems," in IEEE Journal of Emerging and Selected Topics in Power Electronics, vol. 1, no. 3, pp. 186-198, Sept. 2013, doi: 10.1109/JESTPE.2013.2277917.

[2] J. E. Huber and J. W. Kolar, "Solid-State Transformers: On the Origins and Evolution of Key Concepts," in IEEE Industrial Electronics Magazine, vol. 10, no. 3, pp. 19-28, Sept. 2016, doi: 10.1109/MIE.2016.2588878.

[3] A. Q. Huang, "Medium-Voltage Solid-State Transformer: Technology for a Smarter and Resilient Grid," in IEEE Industrial Electronics Magazine, vol. 10, no. 3, pp. 29-42, Sept. 2016, doi: 10.1109/MIE.2016.2589061.

[4] J. E. Huber and J. W. Kolar, "Solid-State Transformers: On the Origins and Evolution of Key Concepts," in IEEE Industrial Electronics Magazine, vol. 10, no. 3, pp. 19-28, Sept. 2016.

[5] X. She, A. Q. Huang and R. Burgos, "Review of Solid-State Transformer Technologies and Their Application in Power Distribution Systems," in IEEE Journal of Emerging and Selected Topics in Power Electronics, vol. 1, no. 3, pp. 186-198, Sept. 2013.

[6] M. Liserre, G. Buticchi, M. Andresen, G. De Carne, L. F. Costa and Z. -X. Zou, "The Smart Transformer: Impact on the Electric Grid and Technology Challenges," in IEEE Industrial Electronics Magazine, vol. 10, no. 2, pp. 46-58, June 2016.

[7] Q. Zhu, L. Wang, A. Q. Huang, K. Booth and L. Zhang, "7.2-kV Single-Stage Solid-State Transformer Based on the Current-Fed Series Resonant Converter and 15-kV SiC mosfets," in IEEE Transactions on Power Electronics, vol. 34, no. 2, pp. 1099-1112, Feb. 2019.

[8] P. Czyz, T. Guillod, F. Krismer, J. Huber and J. W. Kolar, "Design and Experimental Analysis of 166 kW Medium-Voltage Medium-Frequency Air-Core Transformer for 1:1-DCX Applications," in IEEE Journal of Emerging and Selected Topics in Power Electronics, vol. 10, no. 4, pp. 3541-3560, Aug. 2022.

[9] M. Steiner and H. Reinold, "Medium frequency topology in railway applications," 2007 European Conference on Power Electronics and Applications, Aalborg, Denmark, 2007, pp. 1-10.

[10] M. Glinka, "Prototype of multiphase modular-multilevel-converter with 2 MW power rating and 17-level-output-voltage," 2004 IEEE 35th Annual Power Electronics Specialists Conference (IEEE Cat. No.04CH37551), Aachen, Germany, 2004, pp. 2572-2576 Vol.4.

[11] Rothmund, Daniel, et al. "99% efficient 10 kV SiC-based 7 kV/400 V DC transformer for future data centers." IEEE Journal of Emerging and Selected Topics in Power Electronics 7.2 (2018): 753-767.

[12] C. Lu, W. Hu and F. C. Lee, "Neutral-Point Voltage Balancing Methods of Series-Half-Bridge LLC Converter for Solid State Transformer," in IEEE Transactions on Power Electronics, vol. 36, no. 6, pp. 7060-7073, June 2021, doi: 10.1109/TPEL.2020.3035150.

[13] Y. Jiao and M. M. Jovanović, "Topology Evaluation and Comparison for Isolated Multilevel DC/DC Converter for Power Cell in Solid State Transformer," 2019 IEEE Applied Power Electronics Conference and Exposition (APEC), Anaheim, CA, USA, 2019, pp. 802-809, doi: 10.1109/APEC.2019.8722253.

[14] C. Lu, W. Hu and F. C. Lee, "Neutral-Point Voltage Balancing Methods of Series-Half-Bridge LLC Converter for Solid State Transformer," in IEEE Transactions on Power Electronics, vol. 36, no. 6, pp. 7060-7073, June 2021, doi: 10.1109/TPEL.2020.3035150.

[15] C. Lu, W. Hu, W. Zhang and F. C. Lee, "Comparison of Three-Level Cell Topology and Control for Solid State Transformer," 2021 IEEE 4th International Electrical and Energy Conference (CIEEC), Wuhan, China, 2021, pp. 1-6, doi: 10.1109/CIEEC50170.2021.9510615.

[16] J. E. Huber and J. W. Kolar, "Analysis and design of fixed voltage transfer ratio DC/DC converter cells for phase-modular solid-state transformers," 2015 IEEE Energy Conversion Congress and Exposition (ECCE), Montreal, QC, Canada, 2015, pp. 5021-5029, doi: 10.1109/ECCE.2015.7310368.

979-8-3315-1612-3/25 $31.00 © 2025 IEEE

[17] S. S. Queiroz and L. F. Costa, "Evaluation and Comparison of Multilevel LLC Resonant DC-DC Converter for Solid-State Transformer," 2023 IEEE Energy Conversion Congress and Exposition (ECCE), Nashville, TN, USA, 2023, pp. 3248-3254, doi: 10.1109/ECCE53617.2023.10361987.

[18] Bo Yang, F. C. Lee, A. J. Zhang and Guisong Huang, "LLC resonant converter for front end DC/DC conversion," APEC. Seventeenth Annual IEEE Applied Power Electronics Conference and Exposition (Cat. No.02CH37335), Dallas, TX, USA, 2002, pp. 1108-1112 vol.2, doi: 10.1109/APEC.2002.989382.

[19] Y. Wei, Q. Luo and A. Mantooth, "Overview of Modulation Strategies for LLC Resonant Converter," in IEEE Transactions on Power Electronics, vol. 35, no. 10, pp. 10423-10443, Oct. 2020, doi: 10.1109/TPEL.2020.2975392.

[20] I. Barbi, R. Gules, R. Redl and N. O. Sokal, "DC-DC converter: four switches V/sub pk/=V/sub in//2, capacitive turn-off snubbing, ZV turn-on," in IEEE Transactions on Power Electronics, vol. 19, no. 4, pp. 918-927, July 2004, doi: 10.1109/TPEL.2004.830092.

[21] J. Jiao, X. Guo, C. Wang and X. You, "Time-Domain Analysis and Optimal Design of LLC-DC Transformers (LLC-DCXs) Considering Discontinuous Conduction Modes," in IEEE Transactions on Transportation Electrification, vol. 9, no. 2, pp. 2308-2323, June 2023, doi: 10.1109/TTE.2022.3205954.

[22] S. S. Queiroz and L. F. Costa, "Design of a Modular Multilevel DC/DC Converter to Solid-State Transformer in a Green Hydrogen System," 2024 IEEE Applied Power Electronics Conference and Exposition (APEC), Long Beach, CA, USA, 2024, pp. 2282-2288, doi: 10.1109/APEC48139.2024.10509521.

[23] S. S. Queiroz and L. F. Costa, "Investigation of the Influence of the Dead-Time on the Performance of an LLC Resonant Converter for High-Power Application," 2024 IEEE 15th International Symposium on Power Electronics for Distributed Generation Systems (PEDG), Luxembourg, Luxembourg, 2024, pp. 1-6, doi: 10.1109/PEDG61800.2024.10667434.

[24] J. Biela, M. Schweizer, S. Waffler and J. W. Kolar, "SiC versus Si—Evaluation of Potentials for Performance Improvement of Inverter and DC–DC Converter Systems by SiC Power Semiconductors," in IEEE Transactions on Industrial Electronics, vol. 58, no. 7, pp. 2872-2882, July 2011, doi: 10.1109/TIE.2010.2072896.

[25] D. Dujic, G. Steinke, E. Bianda, S. Lewdeni-Schmid, C. Zhao and J. K. Steinke, "Characterization of a 6.5kV IGBT for medium-voltage high-power resonant DC-DC converter," 2013 Twenty-Eighth Annual IEEE Applied Power Electronics Conference and Exposition (APEC), Long Beach, CA, USA, 2013, pp. 1438-1444, doi: 10.1109/APEC.2013.6520487.

[26] C. R. Sullivan and R. Y. Zhang, "Simplified design method for litz wire," 2014 IEEE Applied Power Electronics Conference and Exposition - APEC 2014, Fort Worth, TX, USA, 2014, pp. 2667-2674, doi: 10.1109/APEC.2014.6803681.

[27] K. Venkatachalam, C. R. Sullivan, T. Abdallah and H. Tacca, "Accurate prediction of ferrite core loss with nonsinusoidal waveforms using only Steinmetz parameters," 2002 IEEE Workshop on Computers in Power Electronics, 2002. Proceedings., Mayaguez, PR, USA, 2002, pp. 36-41, doi: 10.1109/CIPE.2002.1196712.

[28] D. Haake, A. Grodnichev, F. Schnabel and M. Jung, "Impact of Higher Current Harmonics on Component Current Stress and Conduction Losses of Half-Bridge-Series-Resonant-Converters in Discontinuous Conduction Mode for High-Power Applications," 2022 24th European Conference on Power Electronics and Applications (EPE'22 ECCE Europe), Hanover, Germany, 2022, pp. 01-10.

[29] H. Abu Bakar Siddique, A. R. Lakshminarasimhan, C. I. Odeh and R. W. De Doncker, "Comparison of modular multilevel and neutral-point-clamped converters for medium-voltage grid-connected applications," 2016 IEEE International Conference on Renewable Energy Research and Applications (ICRERA), Birmingham, UK, 2016, pp. 297-304, doi: 10.1109/ICRERA.2016.7884555.

979-8-3315-1612-3/25 $31.00 © 2025 IEEE

Comparative Performance Analysis of Medium Voltage 3L-ANPC and 3L-DNPC Pole Enabled by series-Connection of 10kV SiC MOSFETs and 10kV SiC JBS Diodes for Sine Triangle PWM Operation

Sanket Parashar,*Member,IEEE*, Shubham Rawat,Nithin Kolli. Raj Kumar Kokkonda,*Student Member,IEEE*, and Subhashish Bhattacharya,*Fellow*,IEEE
Department of ECE, North Carolina State University, Raleigh, USA
sparash1993@gmail.com

Abstract—**This paper presents a comparative performance analysis of Medium Voltage three-level Active Neutral Point Clamped (ANPC) and Diode Neutral Point Clamped (DNPC) converters, utilizing series-connected 10kV Silicon Carbide (SiC) MOSFETs and 10kV SiC Junction Barrier Schottky (JBS) diodes. Switching transitions for both topologies were mathematically modeled, accounting for critical factors such as reverse recovery current, variable device capacitance, base plate current, and parasitic bus bar inductance to accurately represent dynamic behavior. Dynamic voltage balancing and switching loss reduction were achieved using two distinct approaches: turn-off delay control strategies and R-C snubber design techniques. Validation of the proposed scheme on a three-level test bench with a 7.5kV DC bus and a 10A load current demonstrated its effectiveness in enhancing converter performance under practical conditions.**

Index Terms—**10kV SiC MOSFETs, 10kV SiC JBS Diodes, ANPC, DNPC, Dynamic Voltage Balancing**

I. INTRODUCTION

The emergence of 10 kV 4H-SiC MOSFETs has opened transformational opportunities in medium-voltage (MV) power electronics. With their fast switching capabilities, high $\frac{dv}{dt}$ tolerance, and reduced switching losses, these devices are particularly well-suited for applications such as solid-state transformers, grid connectors, and high-speed motor drives [1]-[3]. The integration of such Wide-Bandgap (WBG) devices has driven the development of high-performance converters, addressing the critical need for enhanced efficiency and reduced system complexity in modern power systems. A notable advancement in this domain is the Asynchronous Micro-grid Power Conditioning System (AsMPCS), illustrated in Fig. 1. This system integrates two Active Front End Converters (AFECs) interfacing with the 13.8 kV AC grid. By leveraging series-connected 10 kV SiC MOSFETs and SiC JBS diodes, the AsMPCS achieves remarkable power density, high efficiency, and superior reliability [4]. The series connection of SiC MOSFETs forms the core building block of the Three phase AFEC, directly influencing the overall operational

efficiency of the system. Among the viable topologies for medium-voltage (MV) applications, the Active Neutral Point Clamped (ANPC) and Diode Neutral Point Clamped (DNPC) configurations have emerged as highly promising solutions. These topologies improve power conversion efficiency, enhanced grid stability, and seamless integration with renewable energy systems. Their deployment in renewable energy integration has enabled efficient DC-to-AC power conversion for grid-connected operations and advancing sustainable energy goals [5].

To ensure efficient operation for series-connected devices , Dynamic Voltage Balancing (DVB) becomes critical, especially under high switching frequencies. A passive DVB based on R-C snubber network is commonly employed to mitigate voltage imbalances across series-connected devices. The parasitic base plate capacitance ($C_{\text{bs},Mij}$), formed between the device's drain terminal and the conductive heat sink, contributes significantly to turn off V_{ds} over shoot and Dynamic Voltage Imbalance (DVI) during switching transitions. These effects become pronounced in 3L-ANPC and 3L-DNPC converters, where asymmetry between positive and negative pole outputs results in differential turn-on and turn-off times. DVB across devices in 3L-ANPC and 3L-DNPC topologies have sharp differences due to these additional challenges [6].

This research conducts a comparative performance analysis of ANPC and DNPC topologies under hard-switching conditions, utilizing series-connected 10 kV SiC MOSFETs and JBS diodes. The primary objectives are to uncover the mechanisms driving performance differences, propose design improvements, and validate the findings through experimental testing on a three-level converter test bench. By addressing these critical challenges, this work seeks to advance the development of high-efficiency, reliable power converters for medium-voltage (MV) applications, fostering the integration of emerging wide-bandgap (WBG) technologies into practical, real-world systems [7]-[9]. Section II explores the switching

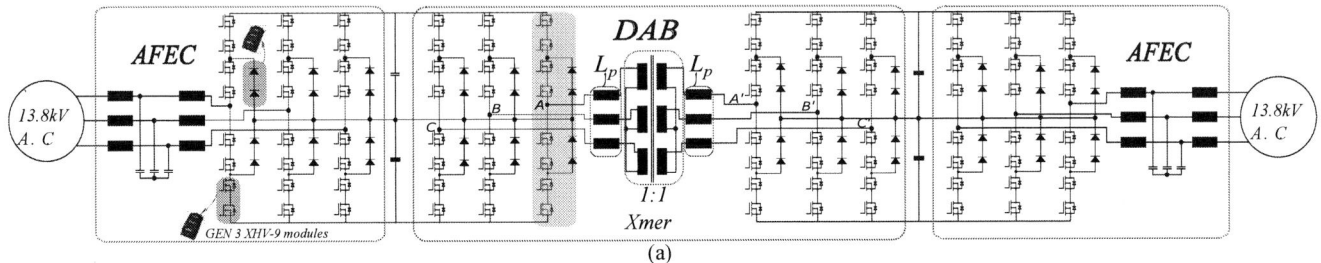

Fig. 1. Asynchronous Micro-grid Power Conditioning System used for integration of 13.8 kV AC Grid using 24 kV DC interface.

TABLE I
SWITCHING TRANSITIONS AND CONDUCTION STATE IN 3L-DNPC POLE
WITH SERIES-CONECTED MOSFETs AND SiC JBS CLAMPING DIODES

Switch → Tr. ↓	M_{11} -M_{12}	M_{21} -M_{22}	M_{31} -M_{32}	M_{41} -M_{42}	D_{11}/M_{51} -D_{12}/M_{52}	D_{21}/M_{61} -D_{22}/M_{62}
Tr.I	✓	✓	×	×	×	×
Tr.II	×	✓	×	×	✓	×
Tr.III	×	×	✓	×	×	✓
Tr.IV	×	×	✓	✓	×	×
State I	✓	✓	×	×	×	×
State II	×	✓	×	×	✓	×
State III	×	×	✓	×	×	✓
State IV	×	×	✓	✓	×	×

transitions across series-connected devices in 3L-ANPC and 3L-DNPC poles, while Section III details the algorithm developed for modeling these transitions. Section IV presents a performance analysis of both topologies, based on experimental data.

II. SWITCHING TRANSITION ANALYSIS OF 3L-DNPC AND 3L-ANPC POLE

Fig.2(a)-2(d) illustrate the detailed switching transitions occurring in a Three-Level Diode Neutral Point Clamped (3L-DNPC) pole. Similarly, Fig. 3(a)-3(d) comprehensively depict the switching dynamics in a Three-Level Active Neutral Point Clamped (3L-ANPC) pole. Additionally, Table I provides a comparative analysis of the switching characteristics observed in Active Neutral Point Clamped (ANPC) converter poles under varying operating conditions. The switching transitions in a 3L-DNPC pole are characterized by intricate interactions between complementary series-connected MOSFETs and diodes. These transitions often involve the generation of significant reverse recovery currents in the diodes, leading to considerable energy losses in the circuit.

- **Reverse Recovery Currents:** As depicted in Fig. 4, the reverse recovery currents for the 3L-DNPC pole and the 3L-ANPC converter pole are compared. The results indicate that the reverse recovery current peaks at approximately 30A during the turn-off phase of the PiN diode. This significant peak reflects the energy-intensive nature of the process, which may adversely affect system efficiency and thermal performance.
- **Diode and MOSFET Behavior:** Fig. 4 compares the reverse recovery current waveform for a 3L-DNPC pole with 3L-ANPC pole. It demonstrates that the reverse re-

covery current is negligible during the turn-off transition of the JBS (Junction Barrier Schottky) diode. The reduced reverse recovery losses in JBS diodes, compared to PiN diodes, are attributed to their intrinsic material properties.

Fig.5 illustrates the reverse recovery operation in a 3L-ANPC pole during switching transitions involving series-connected SiC MOSFETs and PiN diodes. The inclusion of an R-C snubber extends the duration of the reverse recovery current, resulting in increased switching losses. However, when a JBS diode is employed for neutral-point clamping, the reverse recovery current is significantly reduced. Fig.6 demonstrates that R-C snubber across SiC JBS diodes has no impact on losses due to reverse recovery.

III. MODELING THE SWITCHING TRANSITION IN 3L-DNPC AND ANPC POLE

The switching transition for series-connected devices in Three level pole, assumes a constant device capacitance and linear i-v_{gs} characteristics. Figures 7(a) and 7(b) illustrate a flowchart algorithm developed to evaluate the effect of variable device capacitance and the $i_{dp,Mij}$-$v_{gs,ij}$ relationship based on the switching transition model. The steps in the flowchart algorithm are described below:

- Evaluate $v_{pole}(t_p)$.
- Retrieve the $i_{dp,Mij}$-$v_{gs,ij}$ data from the Excel sheet. In the proposed scheme, the curve tracer generates 1178 data points. The time corresponding to each data point is denoted as t_p, where o ranges from 1 to 1178, and k represents the k-th iteration of the flowchart algorithm. The transconductance between two consecutive data points is assumed to be linear and is utilized to update the snubber current $i_{s,Mij}$ [10]. The transconductance at time t_p is given by $\frac{di_{dp,Mij}(t_p)}{dv_{gs,ij}(t_p)}$.
- The $i_{dp,Mij}(t_p)$ is updated in k_{th} iteration during during t_p-t_{p-1} using (1)

$$i_{dp,Mij}(t_p) = \qquad\qquad (1)$$
$$i_{dp,Mij}(t_{p-1}) - (v_{gs,ij}(t_{p-1}) - \quad v_{gs,ij}(t_{p-1})) \times \frac{di_{dp,Mij}(t_{p-1})}{dv_{gs,ij}(t_{p-1})}$$

- The $i_{s,pole}(t_p)$ is updated in k_{th} iteration during during t_p-t_{p-1} using (2) [10]:

$$i_{s,pole}(t_p) = \qquad\qquad (2)$$
$$i_{s,pole}(t_{p-1}) - v_{gs,ij}(t_{p-1}) \times \frac{di_{dp,Mij}(t_{p-1})}{dv_{gs,ij}(t_{p-1})}$$

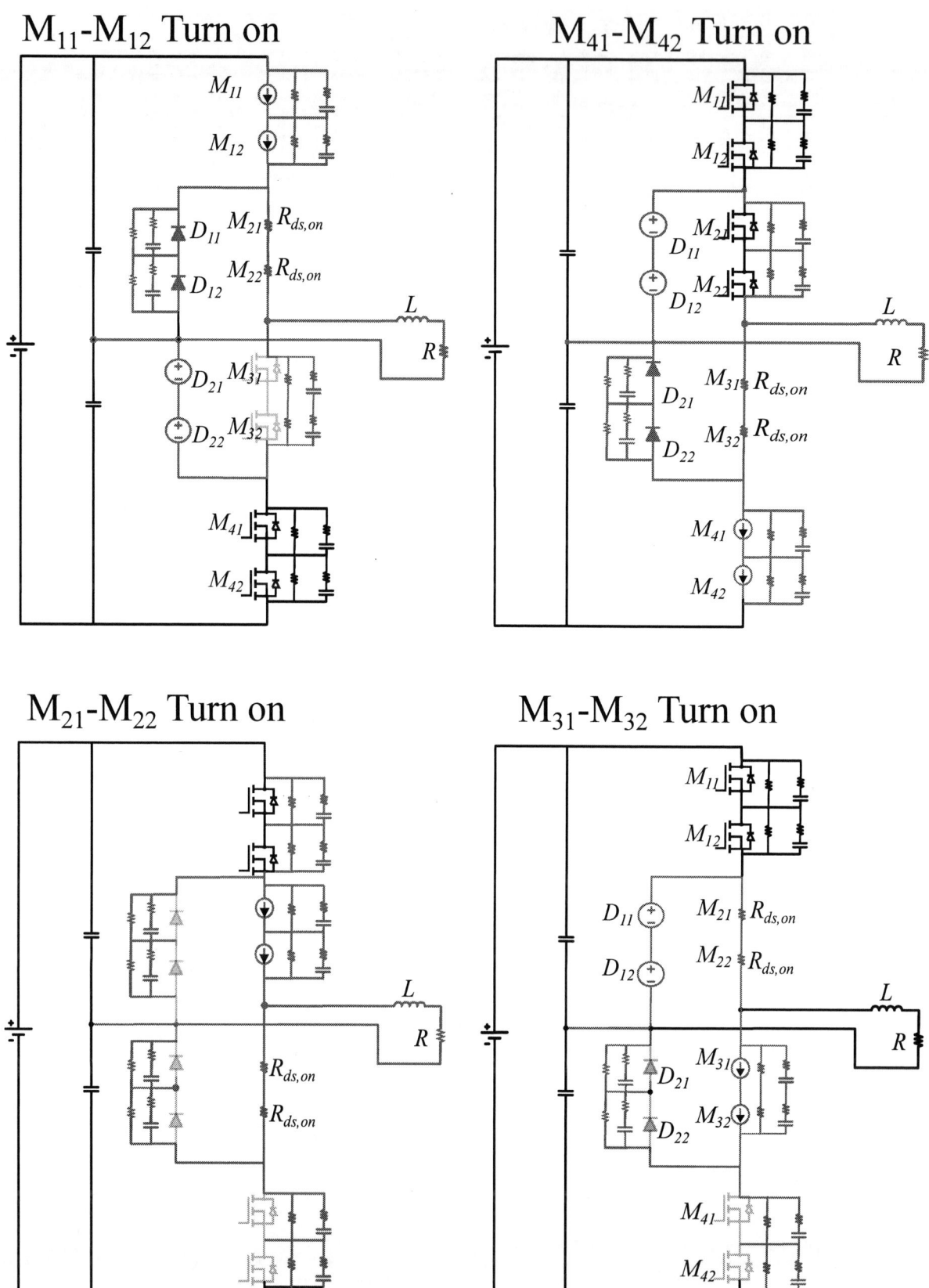

Fig. 2. a) Switching transition across M_{11}-M_{12} during device turn-on (b) Switching transition across M_{21}-M_{22} during device turn-off (c) Switching transition across M_{31}-M_{32} during device turn-off (d) Switching transition across M_{41}-M_{42} during device turn-off.

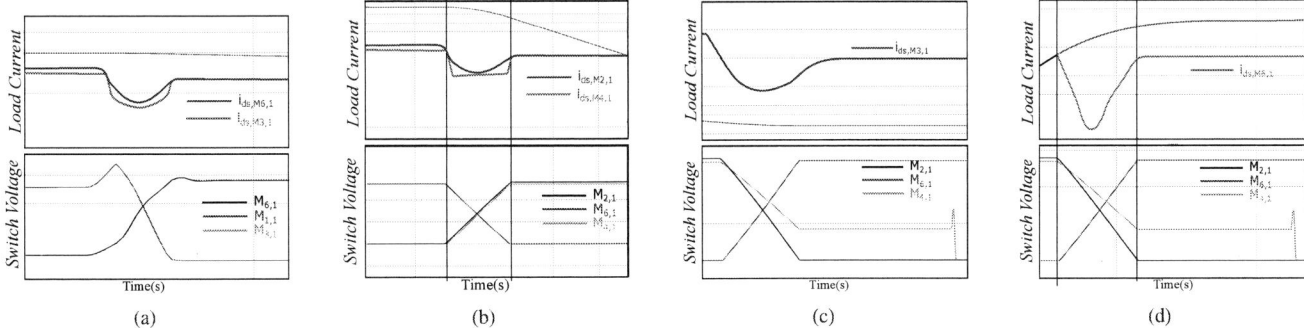

Fig. 3. a) Switching transition across M_{11}-M_{12} during device turn-on (b) Switching transition across M_{21}-M_{22} during device turn-off (c) Switching transition across M_{31}-M_{32} during device turn-off (d) Switching transition across M_{41}-M_{42} during device turn-off.

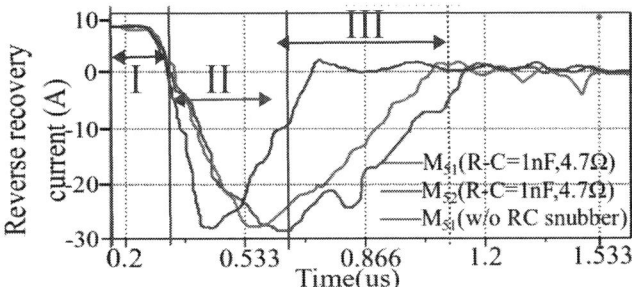

Fig. 4. Reverse recovery current flow through the MOSFET in 3L-DNPC and 3L-ANPC converter pole.

- The evaluation of Diode snubber current is followed by the comparison of the snubber current with previous value using (3) [10]:

$$\frac{i_{s,pole}(t_p) - i_{s,pole}(t_{p-1})}{i_{s,pole}(t_{p-1})} < \epsilon \quad (3)$$

A reduction in ϵ to within 5
- Eq.4 is used for correction of $v_{pole}(t_p)$ [10] :

$$v_{pole}(t_p) = v_{pole}(t_{p-1}) - \frac{\int_{t_{p-1}}^{t_p} i_{s,pole}(t_{p-1})}{C_{s,Dab}(t_{p-1})} \quad (4)$$

- After calculating $v_{pole}(t_p)$, the value of $v_{gs,ij}(t_p)$ is updated using Eq. (5) [10]:

$$v_{gs,ij}(t_p) = v_{gs,ij}(t_{p-1}) \quad (5)$$
$$- R_{g,on}(k) \times C_{gs,ij}(t_p) \times$$
$$\frac{v_{ds,Mij}(t_p) - v_{ds,Mij}(t_{o-1,m})}{t_p - t_{o-1,m}}$$

- The value of $i_{Cbs,ij}(t_{p-1})$ is computed at each iteration and compared with the previous iteration's value. This comparison ensures that the corrections to $i_{s,pole}(t_p)$ and $v_{pole}(t_p)$ converge towards the final solution. The updated value of $i_{Cbs}(t_p)$ is then calculated using Eq. (6) [10]:

$$i_{Cbs}(t_p) =$$
$$C_{bs,Mij} \frac{\sum_{o=1}^{300}(v_{ds,Mij}(t_p) - \sum V_{hs}(t_{o+1,m}))}{\Delta t} \times l_{i,j} \quad (6)$$

The convergence of $i_{Cbs}(t_p)$ at each iteration is evaluated using Eq. (7) [10]:

$$\frac{i_{Cbs}(t_p) - i_{Cbs}(t_{o-1,m})}{i_{Cbs}(t_{o-1,m})} < 0.05 \quad (7)$$

The iteration process halts once the value of ϵ meets the specified criterion, after which the next step for evaluating the impact of variable device capacitance on the switching transition is initiated. In a similar manner, the current through the base plate capacitance of D_{ab} is calculated using Eq.(8) [10]:

$$i_{Cbs,ij}(t_p) =$$
$$C_{bs} \frac{\sum_{o=1}^{300}(v_{pole}(t_p) - \sum V_{hs}(t_{o+1,m}))}{\Delta t} \times l_{i,j} \quad (8)$$

Eq.7 is employed to assess the convergence of $i_{Cbs,ij}$ at each iteration [10]:

$$\frac{i_{Cbs,ij}(t_p) - i_{Cbs,ij}(t_{o-1,m})}{i_{Cbs,ij}(t_{o-1,m})} < \epsilon \quad (9)$$

The flowchart algorithm terminates iteration for $\epsilon = 0.05$.
- Fig.7(a) illustrates the flowchart algorithm used to assess the impact of variable $C_{ds,Mij}(k)$ and $C_{jn,dn}(k)$ on the switching transition. Meanwhile, Figure 7(b) highlights the enhancement in estimating the diode junction capacitance for the theoretical evaluation of $v_{m,D1}, v_{m,D2}$, $i_{Cbs,Dij}$ and $E_{rs,Di}$. Similarly to The $v_{ds,Mij}(t_p)$ obtained from the i-$v_{gs,ij}$ calculation is updated using Eq.(10) [10]:

$$v_{ds,Mij}(t_p) = v_{ds,Mij}(t_{p-1}) -$$
$$C_{ds,Mij} \frac{v_{ds,Mij}(t_p - v_{ds,Mij}(t_{o-1,m})}{t_p - t_{o-1,m}} \quad (10)$$

- Similarly, assess the impact of the device junction capacitance on the switching transition between D_{11} and D_{22} [10].

$$v_{pole}(t_p) = v_{pole}(t_{p-1}) -$$
$$C_{jn,dn} \frac{v_{pole}(t_p) - v_{pole}(t_{o-1,m})}{t_p - t_{o-1,m}} \quad (11)$$

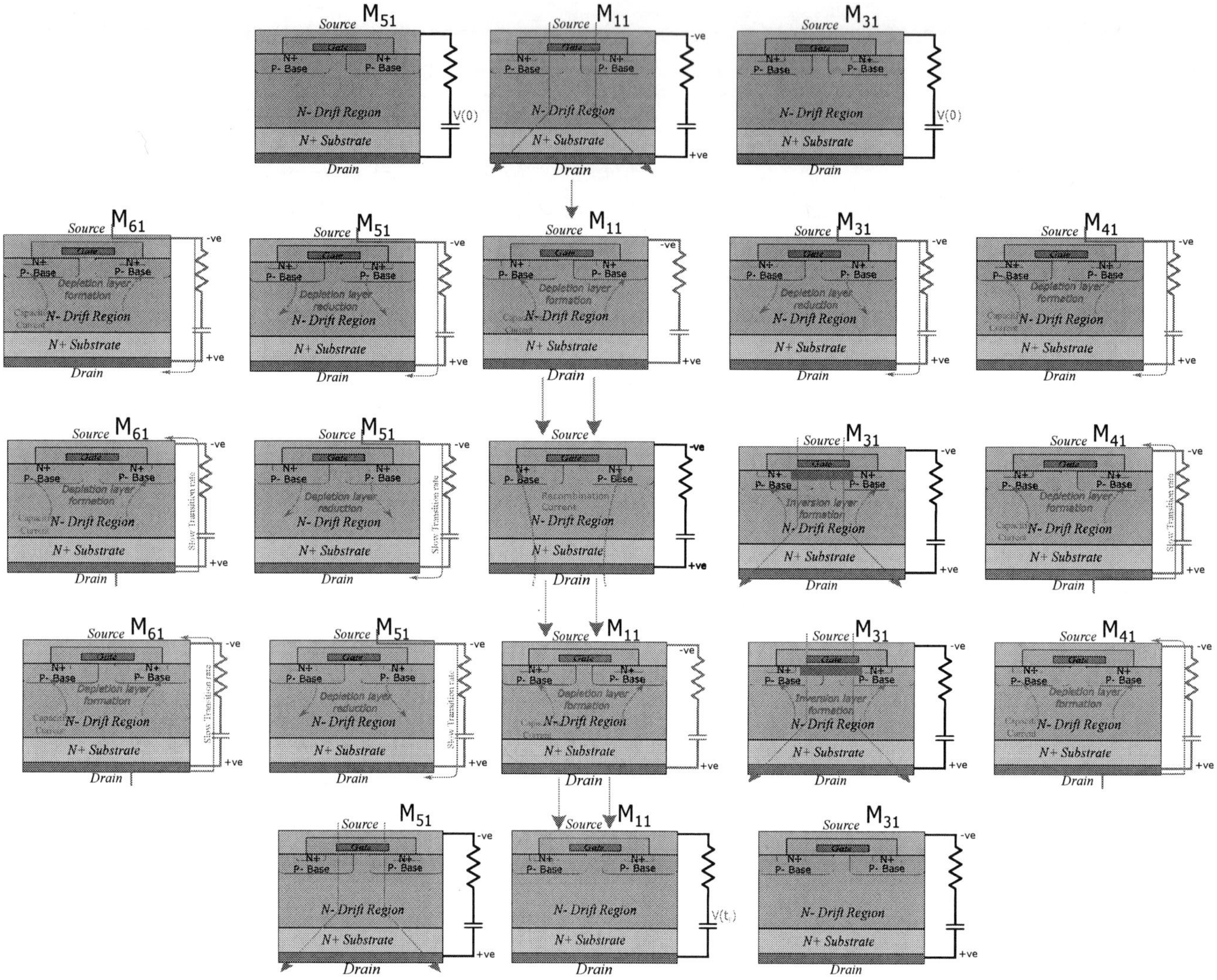

Fig. 5. Motion of holes and electrons in the 3L-ANPC poles during current commutation across the channel of the complimentary MOSFETs (M_{11}-M_{12} & M_{31}-M_{32}) and the clamping MOSFETs (M_{51}-M_{52}).

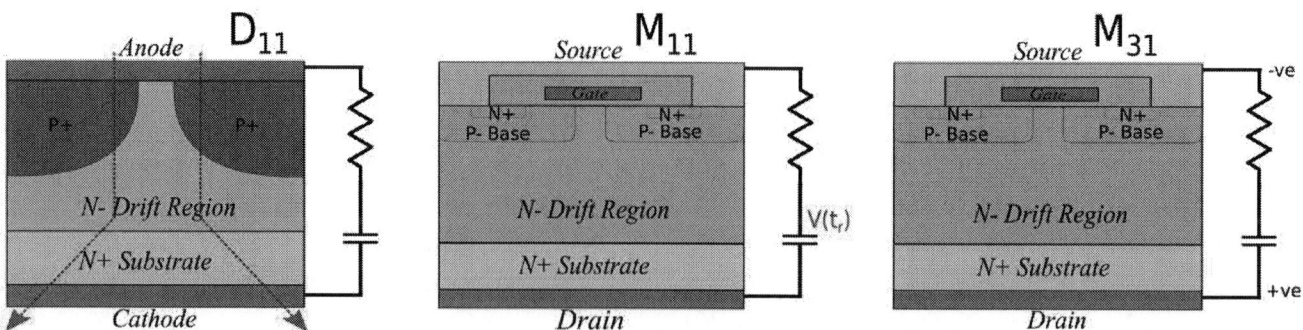

Fig. 6. Motion of holes and electrons in the 3L-DNPC poles during current commutation across the channel of the complimentary MOSFETs (M_{11}-M_{12} & M_{31}-M_{32}) and JBS diodes (D_{11}-D_{12}).

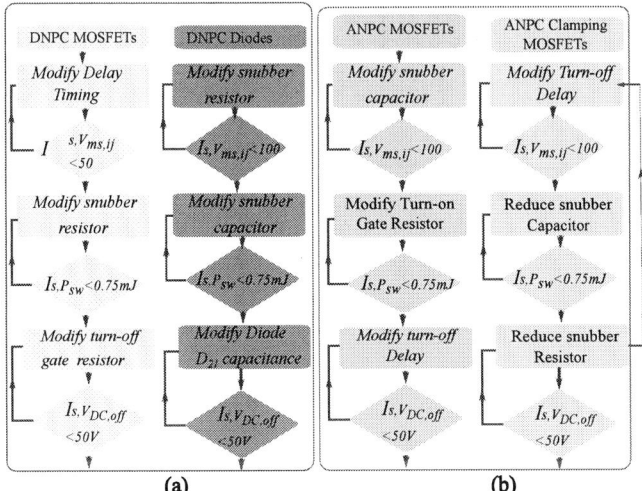

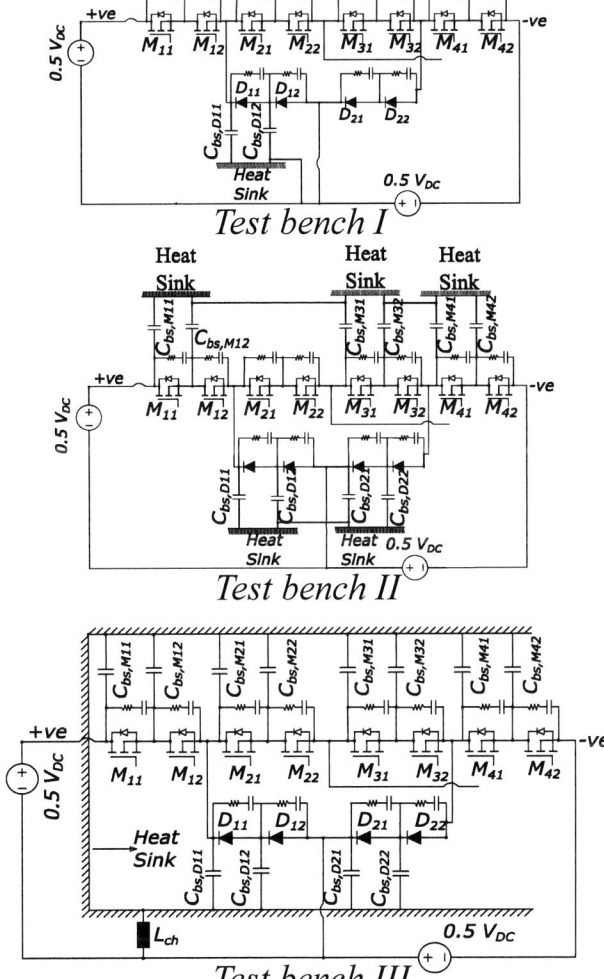

Fig. 7. (a)Flowchart algorithm to calculate R-C snubber design for DNPC pole (b) Flowchart algorithm to evaluate R-C snubber design in ANPC pole.

- The calculated v_{pole} is then compared with the value from the previous iteration using Eq.(12):

$$\frac{v_{pole}(t_p) - v_{pole}(t_{p-1})}{t_p - t_{o-1,m}} < \epsilon \qquad (12)$$

- Adjust the $C_{jn,dj}$ values using the data points obtained from the static characterization of the devices via the curve tracer. The new values of the device capacitance are updated using Eq.(15) is used to adjust new values of the device capacitance [10]:

$$C_{jn,dn}(t_{o,m+1}) = 0.5 \times (C_{jn,dn}(t_p) + C_{jn,dn}(Exp.)) \quad (13)$$

$$C_{ds,Mij}(t_{o,m+1}) = 0.5 \times (C_{ds,Mij}(t_p) + C_{ds,Mij}(Exp.)) \quad (14)$$

$$C_{gs,ij}(t_{o,m+1}) = 0.5 \times (C_{gs,ij}(t_p) + C_{gs,ij}(Exp.)) \quad (15)$$

As mentioned in the main body, $C_{rss,ij}$ is neglected due to its small size. The device capacitance approaches the experimental data from curve tracer, during each iteration. The update is repeated for all 1178 points with the corresponding value of $v_{ds,Mij}(t_p)$ during the calculation.

- Following an update of the device capacitance, the snubber current $i_{s,pole}$ is iterated using (16) [10]:

$$i_{s,pole}(t_p) =$$
$$i_{s,pole}(t_{p-1}) - \frac{\int^{t_p} t_{o+1,m} i_{jn,Dab}(t_{p-1})}{C_{s,Dab}(k)} \qquad (16)$$

- Evaluate the $L_{bs,Dab}\frac{di_{b,Dab}}{dt}$ and $L_{bs,Mij}(k)\frac{di_{b,Mij}}{dt}$ across the parasitic $L_{bs,Dab}$ and $L_{bs,Mij}$ in the power converter [10].
- Evaluate the current through $C_{bs,Mij}(k)$ and $C_{bs}(k)$ for all MOSFETs and diodes using Eq.6 and Eq.8 [10].
- If the Common-Mode Choke is connected between the heat sink and the midpoint of the DC link capacitor,

Fig. 8. Test Bench design for assessment of effect of base plate capacitance on Three level DNPC pole. Test Bench I: Performance analysis without $C_{bs,ij}$, Test Bench II: Performance analysis with $C_{bs,ij}$ of complimentary MOSFETs and Test Bench III: Performance analysis of complete system.

calculate the heat sink voltage using the expression given in Eq.17 [10]

$$V_{hs}(t_p) = L_{ch}\frac{di_{ch}(t_p)}{dt} \qquad (17)$$

Eq.18 is relates the current through CM choke to i_{Cbs} and $i_{Cbs,ij}$ [10]

$$i_{ch}(t_p) = \sum_{i=1}^{i=6} \sum_{j=1}^{j=2} i_{Cbs,ij} \qquad (18)$$

IV. EXPERIMENTAL RESULTS AND DISCUSSION

The base plate capacitance, $C_{bs,Mij}$ and $C_{bs,Dij}$, contribute to turn-off voltage mismatches and elevated switching losses in the system. Fig.8 depicts the test bench employed for analyzing the switching performance of the converter in the 3L-ANPC pole. Meanwhile, Fig.9 illustrates the test bench utilized to assess the impact of individual base plate capacitance ($C_{bs,Mij}$ and $C_{bs,Dij}$) in the 3L-DNPC pole. Fig.10

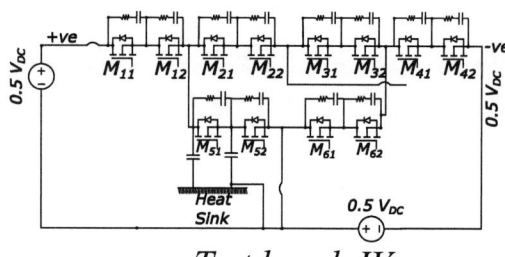

Test bench IV

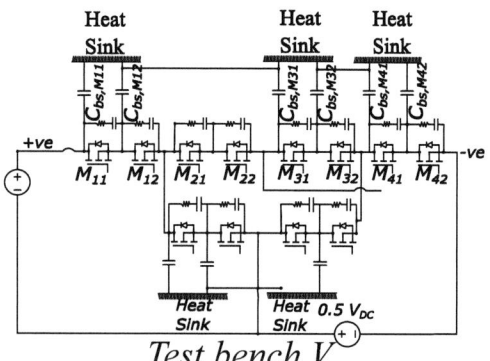

Test bench V

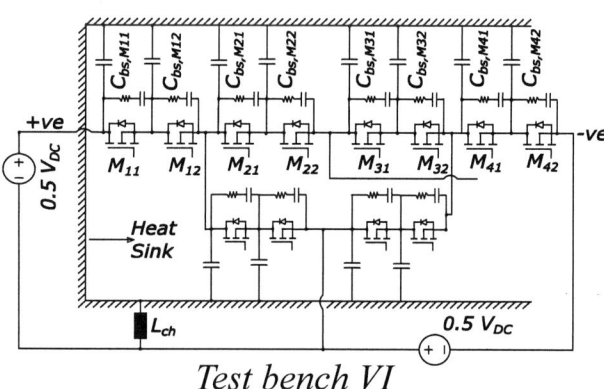

Test bench VI

Fig. 9. Test Bench design and selection for the experiment and analysis of switching loss in 3L-ANPC pole.est Bench IV: Performance analysis without $C_{bs,Mij}$, Test Bench V: Performance analysis with $C_{bs,Mij}$ of complimentary MOSFETs and Test Bench VI: Performance analysis of complete system.

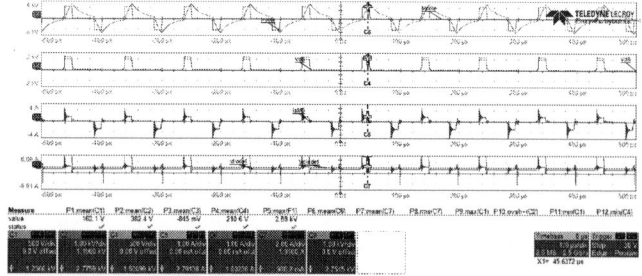

Fig. 10. Experimental result depicting operation of the Active Neutral Point Clam ped converter pole at 8kV DC bus. C1: V_{pole} (2kV/div), C2: I_{load} (10A/div), C3: v_{D22} (1.5kV/div), C4: v_{D21} (1.5kV/div), C5: Clamping MOSFET Voltage (1.5kV/div), C6: Clamping MOSFET voltage (v_{M12}).

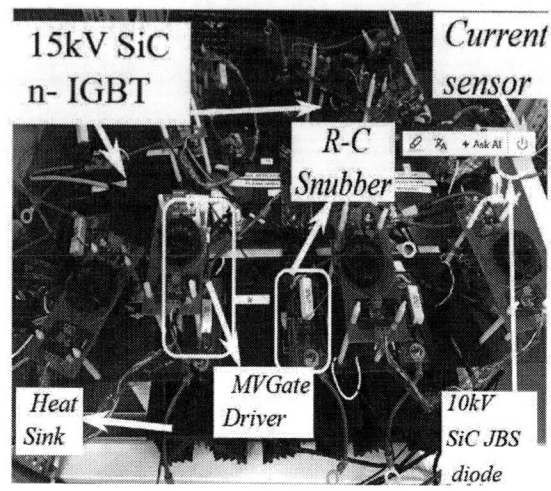

Fig. 11. Experimental results at 7kV DC bus, 10A load current for series-connected 15kV SiC IGBTs.

presents the experimental results for the 3-level Dual Active Neutral Point Clamped (DNPC) inverter, which utilizes series-connected 10 kV SiC MOSFETs and operates at a DC bus voltage of 7.5 kV. The results highlight key performance metrics, including the output voltage waveform and current waveform, offering valuable insights into the system's behavior under normal operating conditions. Fig.12 illustrates the hardware setup of the test bench for both the 3L-DNPC and 3L-Active Neutral Point Clamped (3L-ANPC) poles. The data collected from the test bench is instrumental in refining the R-C snubber network, which is subsequently utilized for design verification. This process enables the optimization of the snubber network to achieve improved performance. Fig.12(a) compares the switching losses of the 3-level DNPC and ANPC poles. The switching losses in the ANPC pole are lower compared to the DNPC pole, primarily due to the reduced R-C snubber across the neutral-point clamping MOSFETs. However, when D_{11} and D_{12} with similar snubbers have higher losses are observed in the 3L-DNPC pole. Fig.12(b) compares the snubber losses for TB3 and TB6 across different series-connected MOSFETs. The snubber losses are generally higher for the 3L-ANPC pole. A reduction in $R_{s,M51}$ and $R_{s,M61}$ significantly decreases the snubber losses in the 3L-ANPC pole. This comparison between the two configurations provides valuable insights into the overall system efficiency and reliability, particularly with regard to the impact of snubber losses and turn-off behavior during operation. Fig.13 performs a comparative analysis of turn-off voltage mismatch $V_{d,m1}$ across MOSFETs for Test Bench III (TB3) and Test Bench VI (TB6). The turn-off voltage mismatch is within specified limit, described in Table.II.

V. CONCLUSION

This paper introduces advanced design methods for R-C snubbers in series-connected 10 kV SiC MOSFETs for 3L-ANPC and 3L-DNPC poles operating under hard-switching conditions. The proposed approach focuses on three critical

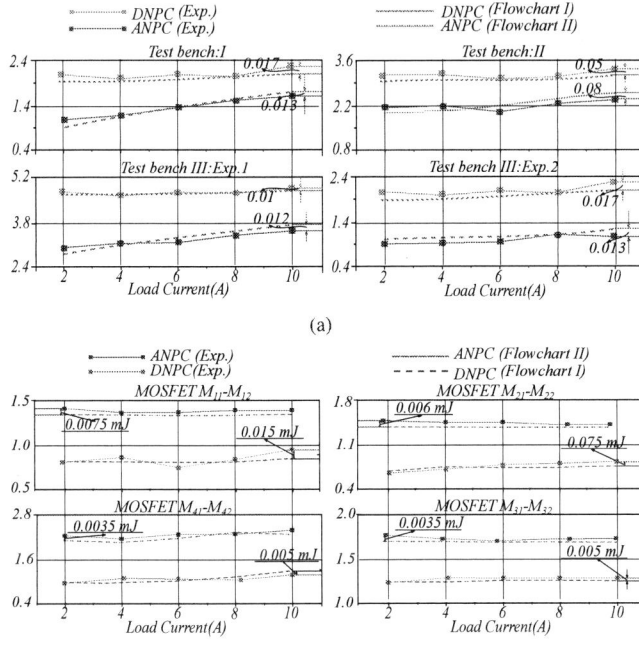

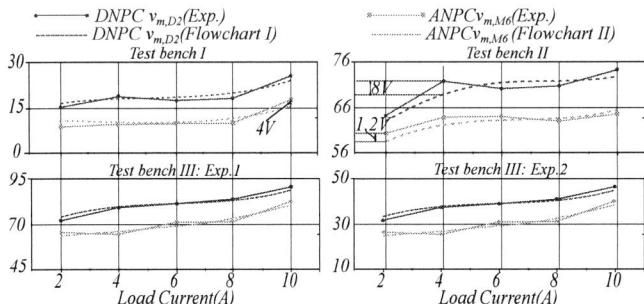

Fig. 13. Turn-off voltage mismatch in 3L-DNPC and 3L-ANPC power converter.

Fig. 12. (a) Switching loss in Three Level Active Neutral Point Clamped (ANPC) converter (b) Comparative analysis of the snubber loss in Three-level Neutral Point clamped converter.

TABLE II
PERFORMANCE METRICS FOR COMPARISON OF ANPC AND DNPC POLE

Parameter	Limit	ANPC	DNPC
Efficiency	98.5%	97.6%	98.6%
Snubber Loss	25W	18W	7W
V_{ds} Mismatch	90V	66V	74V

objectives: minimizing DC offset voltage, reducing snubber losses, and maintaining turn-off voltage mismatch within acceptable thresholds. A comparative analysis, presented in Table III, demonstrates the superior performance of the proposed solution over existing methods, particularly in mitigating voltage imbalance, DC offset voltage, and switching losses in the 3L-DNPC pole. The flowchart algorithm illustrated in Fig. 7 has been utilized to model the switching transitions in the ANPC and DNPC poles. Furthermore, the paper offers a comprehensive theoretical analysis of the reverse recovery current dynamics in the MOSFET body diode induced by the R-C snubber, along with its effect on voltage balancing

TABLE III
PERFORMANCE COMPARISON OF EXISTING LITERATURE HAVING SERIES-CONNECTION OF DEVICES

Ref.	Device	Eff.	V_{ds} Mismatch	Snubber Loss	Error
[1],[2]	Diode	93.4%	22V	100W	×
[3]	MOSFETs	95%	200V	100W	×
DNPC	MOSFETs,Diodes	97.8%	35V	7W	5%
ANPC	MOSFETs	98.8%	45V	18W	3%

across the neutral-point clamping diode. Experimental results conducted at a 7.5 kV DC bus confirm that the proposed methods effectively address turn-off voltage mismatch, improve efficiency, and maintain DC offset voltage within the specified limits outlined in Table II.

REFERENCES

[1] X. Lin, L. Ravi, Y. Zhang, D. Dong and R. Burgos, "Analysis of Parasitic Capacitors' Impact on Voltage Sharing of Series-Connected SiC MOSFETs and Body-Diodes," 2020 IEEE Applied Power Electronics Conference and Exposition (APEC), New Orleans, LA, USA, 2020, pp. 208-215, doi: 10.1109/APEC39645.2020.9124591.

[2] N. Kolli, S. Parashar, R. Kumar Kokkonda, S. Bhattacharya and V. Veliadis, "Switching Loss Analysis of Three-Phase Three- Level Neutral Point Clamped Converter Pole Enabled by Series-Connected 10 kV SiC MOSFETs," 2023 IEEE Applied Power Electronics Conference and Exposition (APEC), Orlando, FL, USA, 2023, pp. 2353-2360, doi: 10.1109/APEC43580.2023.10131392.

[3] Rudy Severns, "Design of Snubbers for Power Circuits", Cornell Dubilier, 140 Technology Place Liberty, SC. Available: https://www.cde.com/resources/technical-papers/design.pdf

[4] Y. Shi, S. Shao, X. Wang, W. Cui and J. Zhang, "Series-Connected Power Devices in a CLLC Resonant Converter for DC Transformer Applications," 2021 IEEE 1st International Power Electronics and Application Symposium (PEAS), Shanghai, China, 2021, pp. 1-6, doi: 10.1109/PEAS53589.2021.9628875.

[5] M. Hergt et al., "Modelling and Characterization of Power Semiconductors in the Frequency Domain," 2024 Energy Conversion Congress Expo Europe (ECCE Europe), Darmstadt, Germany, 2024, pp. 1-8, doi: 10.1109/ECCEEurope62508.2024.10752007.

[6] H. Riazmontazer, A. Rahnamaee, A. Mojab, S. Mehrnami, S. K. Mazumder and M. Zefran, "Closed-loop control of switching transition of SiC MOSFETs," 2015 IEEE Applied Power Electronics Conference and Exposition (APEC), Charlotte, NC, USA, 2015, pp. 782-788, doi: 10.1109/APEC.2015.7104438.

[7] V. Pala, E. V. Brunt, L. Cheng, M. O'Loughlin, J. Richmond, A. Burk, et al., "10 kV and 15 kV silicon carbide power MOSFETs for next-generation energy conversion and transmission systems", 2014 IEEE Energy Conversion Congress and Exposition (ECCE), pp. 449-454, Sep. 2014.

[8] D. Grider, M. Das, A. Agarwal and J. Palmour, "10 kV/120 A SiC DMOSFET half H-bridge power modules for 1 MVA solid state power substation", IEEE Electric Ship Technologies Symp, pp. 131-134, 2010.

[9] B.A. Hull, J. Sumakeris, M. O'Loughlin, Q. Zhang, J. Richmond, A. Powell, et al., "Performance and Stability of Large-Area 4H-SiC 10 kV Junction Barrier Schottky Rectifiers", IEEE Transactions on Electron Devices, vol. 55, pp. 1864-1870, 2008.

[10] S. Parashar, N. Kolli, R. K. Kokkonda, A. Kanale, S. Bhattacharya and B. J. Baliga, "Dynamic Voltage Balancing Across Series-Connected 10 kV SiC JBS Diodes in Medium Voltage 3L-NPC Power Converter Having Snubberless Series-Connected 10 kV SiC MOSFETs," in IEEE Open Journal of the Industrial Electronics Society, vol. 5, pp. 1058-1084, 2024, doi: 10.1109/OJIES.2024.3450509.

A Zero Harmonic Distortion Master Converter for Medium Voltage Microgrids

Gabriel V. Ramos*, Dener A. de L. Brandão*, Thiago M. Parreiras[†], Danilo I. Brandao[‡] and Braz de J. C. Filho[‡]

*Graduate Program in Electrical Engineering - Universidade Federal de Minas Gerais -
Av. Antônio Carlos 6627, 31270-901, Belo Horizonte, MG, Brazil

[†]Department of Electrical Engineering, Centro Federal de Educação Tecnológica de Minas Gerais, Belo Horizonte, Brazil

[‡]Department of Electrical Engineering, Universidade Federal de Minas Gerais, Belo Horizonte, Brazil

*gabrielvilkn@ieee.org, *dener.brandao@ieee.org, [†]thiago.parreiras@cefetmg.br,
[‡]dibrandao@ufmg.br and [‡]braz.cardoso@ieee.org

Abstract—**This paper proposes the Zero Harmonic Distortion Converter (ZHD) as a master converter (MC) for medium voltage microgrids, delivering sinusoidal waveforms in both grid-connected and islanded mode without sinusoidal capacitive filtering elements. The ZHD MC ensures seamless transitions between modes using a simple control approach that eliminates the need off a voltage closed-loop control for grid-connected, islanded, and transition modes, using only a current closed-loop control for grid-connected operation. Simulation and hardware-in-the-loop results show the performance of the ZHD MC converter.**

Index Terms—**Distributed generation, grid-connected operation, island operation, microgrid, master converter.**

I. INTRODUCTION

Electric power systems have been changed due to the rise of distributed generation (DG) driven by environmental concerns and the increasing energy demand. This transition replaces fossil fuel and nuclear power plants with renewable resources. DG provides energy closer to consumers, improving efficiency, investment, flexibility, stability, and power quality [1]–[4]. The concept of microgrids (MG) evolves with DG, integrating loads, storage, and renewable resources such as solar, wind, and biomass [5], [6]. MGs can operate in grid-connected and islanded modes: When connected to the grid, the grid dictates the voltage and frequency; while during islanded mode, at least a single one grid-forming converter with an energy storage system (ESS) is tasked with establishing the voltage and frequency of the MG [7].

The transition of an MG from grid-connected to islanded mode and back involves either a smooth transition or black-start procedures [8]. In the literature, considering conventional voltage source converters (VSCs), numerous methods can be found for seamless transition that modify inverter control loops and control the islanding switch based on droop control [9]–[15] and hybrid voltage and current mode (HVCM) control structures [16]–[19]. In droop-based control techniques the converter is always working as a voltage source showing

This study was financed in part by the Coordenação de Aperfeiçoamento de Pessoal de Nível Superior - Brasil (CAPES) - Finance Code 001, in part by Conselho Nacional de Desenvolvimento Científico e Tecnológico (CNPq), and in part by Fundação de Amparo à Pesquisa do Estado de Minas Gerais (FAPEMIG).

good performance between the transition of operation modes. However, the limitation of this approach is the low dynamic of the power loops due to bandwidth restrictions between the control loops, and since the grid current is not directly controlled, the inrush grid current during the transition modes and fault occurrence always exists [11]. In HVCM control techniques the converter operates in a current control mode in the grid-connected mode, and in voltage control mode in the islanded mode. In this way, there is a need to switch between two types of controllers depending on the operation mode, where the quality of the voltage waveforms relies on this transition [16]–[19].

This paper proposes using the Zero Harmonic Distortion (ZHD) technology as a microgrid master converter. The ZHD converter, detailed in [20] as grid-connected grid-following converter, in [21]–[24] as a islanded grid-forming converter using 2-level VSCs, and in [25], [26] using 3-level VSC converters, offers advantages such as low part count, off-the-shelf power circuitry, and no need for output capacitive filters. It generates no characteristic harmonics up to the 50th order, meeting the IEEE 519 and IEEE 1547 standards [27], [28]. This is attained with two 3-phase converters using Selective Harmonic Elimination Pulse Width Modulation (SHE PWM) connected to a three-winding transformer, creating a natural sinusoidal voltage source without sinusoidal capacitive filters that are a source of reliability problems [29], contribute negatively to possible resonance problems [30]–[34] and ferroresonance problems [35], [36] in medium voltage microgrids applications.

This paper explores the Zero Harmonic Distortion (ZHD) converter as a multifunctional master converter solution for medium voltage microgrids operating in both grid-connected and islanded modes. The ZHD converter features an inherently sinusoidal output voltage, allowing closed-loop current control in grid-connected mode and open-loop control in islanded mode and with the transitions between modes achieved by simply enabling or disabling the current regulator gains. This study evaluates the ZHD converter's performance across both operational modes, highlighting its black-start capability and seamless transitions between modes.

II. ZERO HARMONIC DISTORTION CONVERTER

A. Converter Structure

The ZHD converter is illustrated in Fig. 1, which comprises two 2-level VSCs attached to a three-winding transformer. The VSCs are connected to the wye and delta secondary windings of the transformer. The 30-degree phase shift between these secondary windings facilitates harmonic cancelation, as described in (1) [20]–[24] .

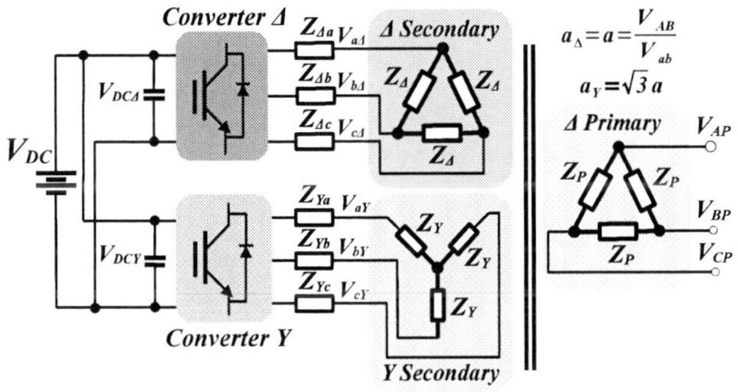

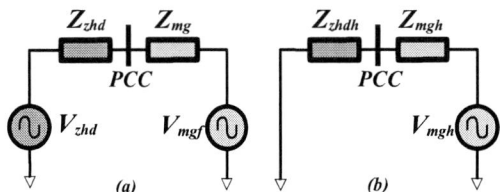

Fig. 2. Per-phase ZHD analysis: (a) Fundamental and (b) Harmonic equivalent circuits proposed in [21]–[24].

and the impedance of the delta primary winding is Z_P [21]–[24].

$$Z_{zhd} = (a^2(Z_{\Delta\phi} + {}^{Z_\Delta}/3)//a^2(Z_{Y\phi} + Z_Y)) + {}^{Z_P}/3 \quad (2)$$

B. ZHD Master Converter Control Structure

In literature, the ZHD has already been explored as a grid-connected grid-following converter in [20], later as a grid-forming converter for islanded applications in [21]–[24] considering 2-level VSCs and in [25], [26] considering 3-level VSCs. The developed current and voltage control structures for both applications can be observed in Figs. 3 (a) and (b), respectively.

1) Grid-connected operation: In grid-following grid-connected operation (Fig. 3 (a)), the ZHD converter operates following the amplitude, frequency and phase of the grid voltage of the grid voltage, as these values are necessary to control the active and reactive power delivered to the grid [20].

In this operation, Fig. 3 (a) shows a proportional-integral (PI) feedback controller where k_p and k_i are the proportional and integral gains of the controller. The voltages generated by the converter are E_d and E_q, V_d, V_q, I_d, and I_q are the output voltages and currents, respectively. V_d^* and V_q^* are the feedforward disturbance control applied to compensate the effects of external disturbances V_d and V_q. $\hat{R}_{zhd}$ and $\hat{L}_{zhd}$ are the converter estimated internal resistance and inductance also utilized as disturbance feedforward control to reject the converter resistance R_{zhd} voltage drop and axis coupling through inductances L_{zhd}. The generated current control commands signals are expressed in (3) as the sum of the feedback and feedforward control actions [20].

$$
\begin{aligned}
E_d^* &= V_d^* - \omega_0 \hat{L}_{zhd} i_q + \hat{R}_{zhd} i_d + (I_d^* - I_d)(k_p + \tfrac{k_i}{s}) \\
E_q^* &= V_q^* + \omega_0 \hat{L}_{zhd} i_d + \hat{R}_{zhd} i_q + (I_q^* - I_q)(k_p + \tfrac{k_i}{s})
\end{aligned} \quad (3)
$$

The synchronization of the ZHD converter with the grid is achieved through the double synchronous decoupled reference frame PLL (DDSRF-PLL) [39]. The angle of the grid followed by the DDSRF-PLL is used by the current control in dq synchronous reference frame allowing the decoupled control of active and reactive power as shown in (4), respectively [20].

$$
\begin{aligned}
P &= {}^3/_2 V_d I_d \\
Q &- -{}^3/_2 V_d I_q
\end{aligned} \quad (4)
$$

The gains k_p and k_i of the PI feedback controller are projected using the dynamic stiffness curve, method proposed

Fig. 1. 2-level ZHD converter.

$$h = 6k \pm 1 \ \forall \ k \ odd \ and \ integer \quad (1)$$

Therefore, it is achievable to reach a voltage waveform without all harmonics assessed according to international standards [27], [28], up to 50th order, utilizing the SHE PWM technique [37], [38] to eliminate harmonics not canceled by the transformer. Table I shows the harmonics canceled by the transformer and the ones eliminated by the SHE PWM technique. Switching angles are defined on the basis of modulation indices, calculated offline using Newton's method, and stored in Look-up Tables (LUTs) [20]–[24].

TABLE I
HARMONIC ELIMINATION AND CANCELLATION IN THE ZHD CONVERTER.

Element	Harmonic Order
Transformer:	$5^{th}, 7^{th}, 17^{th}, 19^{th}, 29^{th}, 31^{th}, 41^{th}, 43^{th}, \ldots$
SHE PWM:	$11^{th}, 13^{th}, 23^{th}, 25^{th}, 35^{th}, 37^{th}, 47^{th}, 49^{th}$

A systematic analysis of the ZHD converter in the islanded grid-forming mode is proposed in [21], allowing the equivalent fundamental and harmonic per-phase circuits shown in Fig. 2 to be obtained.

The idea behind the equivalent circuits is that the converter does not generate a harmonic until 50th is reached, providing a sinusoidal voltage waveform. In other hand, similar to the other converters, if the microgrid/grid presents harmonics, the PCC voltage waveforms deteriorate due to the voltage drop in the ZHD converter impedance (2). The VSCs impedances are given by $Z_{\Delta\phi}$ and $Z_{Y\phi}$, the delta and wye secondaries impedances by Z_Δ and Z_Y, a is the transformer winding ratio,

979-8-3315-1612-3/25 $31.00 © 2025 IEEE

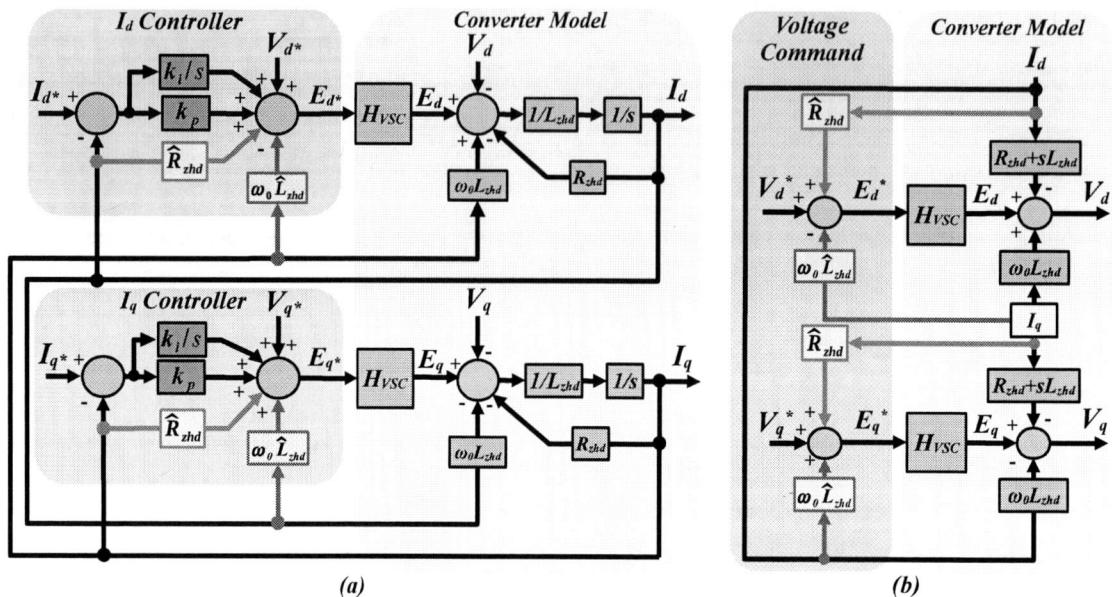

Fig. 3. (a) ZHD current control for grid-following applications proposed in [20] and (b) ZHD voltage command control for grid-forming applications proposed in [21]–[24].

in [40], of the current control shown in (5) and estimated by the segment lines shown in Fig. 4 as a function of L_{zhd}, k_p and k_i. The adjustment of the poles of the dynamic stiffness curve given in (6) is done to allow a dynamic current control without modifying the SHE PWM modulator pulse pattern with a distance of 10 times between the poles [20].

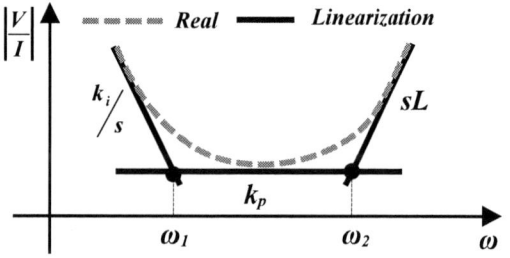

Fig. 4. Dynamic Stiffness for ZHD grid-connected current control.

$$\left| \frac{V}{I} \right| = sL_{zhd} + k_p + k_i/s \tag{5}$$

$$\begin{aligned} \omega_1 &\sim k_i/k_p \\ \omega_2 &\sim k_p/L_{zhd} \end{aligned} \tag{6}$$

2) Islanded operation: In islanded operation, the ZHD operates as a grid-forming converter that defines the voltage and frequency of the MG. The voltage command control in the dq frame shown in Fig. 3(b), proposed in [21]–[24], shows that a disturbance feedforward control is responsible for rejecting load variations, adjusting the voltage reference command, as can be shown in (7).

$$\begin{aligned} E_d^* &= V_d^* - \omega_0 \hat{L}_{zhd} I_q + \hat{R}_{zhd} I_d \\ E_q^* &= V_q^* + \omega_0 \hat{L}_{zhd} I_d + \hat{R}_{zhd} I_q \end{aligned} \tag{7}$$

3) ZHD Hybrid voltage and current mode: The present work combines the grid-connected/grid-following control structure proposed in [20] with the islanded mode/grid-forming control proposed in [21]–[24] in order to enables the ZHD converter to operate as multifunctional microgrid master converter in both grid-connected and islanded modes.

Comparing the command control signals of both current and voltage controls, (3) and (7) respectively, it is possible to observe that in the ZHD converter case, the transition between modes can be made only by enabling/disabling the current regulator gains, does not needing of a change or a additional control structures as required in conventional converters.

In this way, the overall control of the ZHD converter is shown in Fig. 5. The ZHD converter during the grid-connected operation uses the closed-loop current control in the dq rotating reference frame with the synchronizing angle generated by the DDSRF-PLL. During the islanded mode, the microgrid master controller (MGMC), that stays inside of the ZHD MC, gives a command so that the DDSRF-PLL begins to generate a reference frequency, the current regulator gains k_p and k_i are disabled through k_{pEN} and k_{iEN}, and the voltage reference is directly sent to the SHE PWM modulator by the V_d^* and V_q^* feedforward commands with the disturbance feedforward voltage drop compensation of the estimated ZHD impedance parameters $\hat{R}_{zhd}$ and $\hat{L}_{zhd}$.

The reference voltage commands generated by the control are given by E_d^* and E_q^* considering the turn ratio of the transformer a. The references are utilized to calculate the modulation indexes $m_{\Delta,Y}$ and phase compensation $\delta_{\Delta,Y}$. The

979-8-3315-1612-3/25 $31.00 © 2025 IEEE

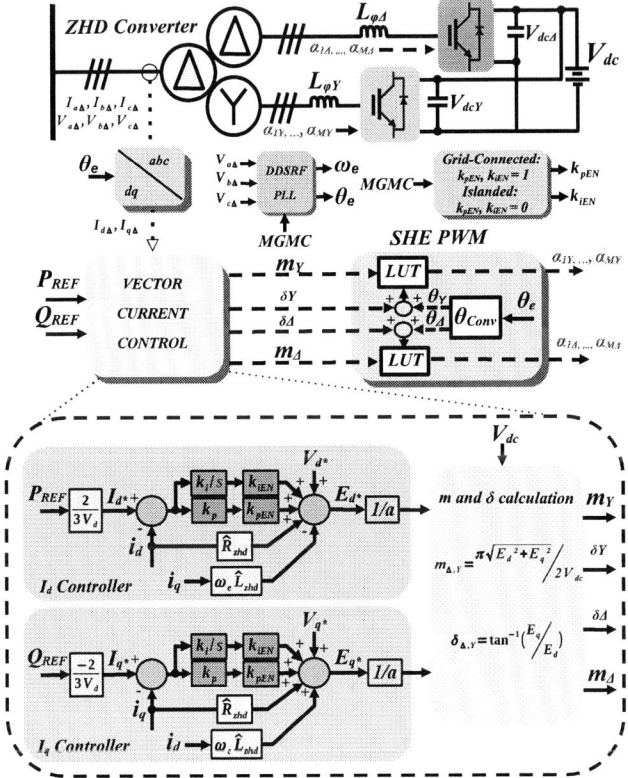

Fig. 5. Control structure of ZHD MC.

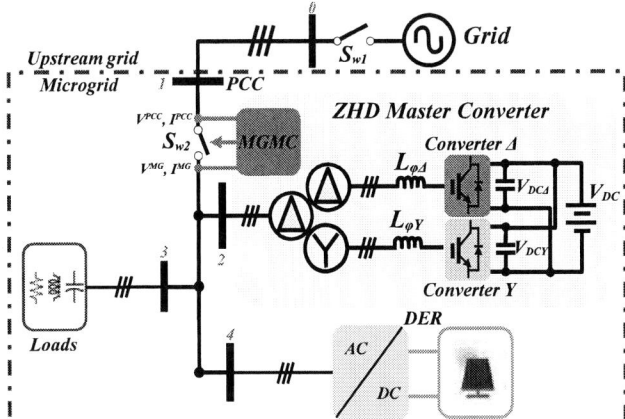

Fig. 6. Generic microgrid with the ZHD as master converter.

modulation signals and the sum of phase compensation and the angle references θ_Y and θ_Δ are compared in the LUTs generating the power switches command signals $\alpha_{1Y}, .., \alpha_{MY}$ and $\alpha_{1\Delta}, .., \alpha_{M\Delta}$ to the VSCs.

III. SIMULATION RESULTS

A. Simulation Scenarios

The ZHD MC converter based on a battery energy storage system structure (BESS) was assessed based on typical daily load variation and solar power injection scenarios. The microgrid system simulated using the MATLAB/Simulink platform is shown in Fig. 6. The simulation parameters are described in Table II.

The simulation scenarios considering, black-start, grid-connection and intentional island operations are listed:

- $t = 0$ s $\rightarrow (I) \rightarrow$ Black-start and islanded operation;
- $t = 1.55$ s $\rightarrow (II) \rightarrow$ Grid-connection and energy time shift service;
- $t = 3.1$ s $\rightarrow (III) \rightarrow$ Islanded Transition;

Fig. 7 shows the overview results of all the scenarios considering: black start, islanded operation, grid connection, and intentional islanded. Figs. 7 (a) and (b) show the active and reactive power flow, respectively, at PCC during the events. Figs. 7 (c) and (d) show the voltage and current at the ZHD MC converter and at the grid, respectively.

TABLE II
SIMULATION PARAMETERS

ZHD CONVERTER			
Parameters	**Values**	**Parameters**	**Values**
Rated Power	280 kVA	Frequency	60 Hz
Primary Voltage	13.8 kV	Δ Secondary Voltage	440 V
Winding Connections	Dd0y1	Y Secondary Voltage	440 V
$L_{\varphi\Delta}$ and $L_{\varphi Y}$ reactors	0.59/0.506 mH	DC link voltage	650 V
THREE-WINDING THREE-PHASE TRANSFORMER - Dd0y1			
Parameters	**Values**	**Parameters**	**Values**
R_m	526.7 $k\Omega$	$R_{s\Delta}$	19.4 $m\Omega$
L_m	534.5 H	$L_{s\Delta}$	3.33 μH
$R_{P\Delta}$	10.5 Ω	R_{sY}	16 $m\Omega$
$L_{P\Delta}$	108.7 mH	L_{sY}	84.26 μH
CONTROLLER PARAMETERS			
Parameters	**Values**	**Parameters**	**Values**
k_p	139.71	k_i	6320
LOAD SCENARIO 1, 2, 3 and 4			
Parameters	**Values**	**Parameters**	**Values**
Active Power (P)	100 kW	Reactive Power (Q)	0 var
Active Power (P)	148 kW	Reactive Power (Q)	36 kvar
Active Power (P)	100 kW	Reactive Power (Q)	0 var
Active Power (P)	196 kW	Reactive Power (Q)	72 kvar
DER SCENARIO 3			
Parameters	**Values**	**Parameters**	**Values**
Active Power (P)	200 kW	Reactive Power (Q)	0 kvar

B. Black-Start and Islanded operation

Considering the islanded operation time event I, Fig. 8 (a) shows the voltage and current of the ZHD MC working as a grid-forming converter. The zoomed view between 0 and 0.15 s shown in (b) shows the ZHD black-start capability, and in (c) and (d) the sinusoidal voltage waveform shape without the necessity of capacitive filters, absorbing and delivering power in Scenarios 3 and 4, respectively. Fig. 8 (e) shows the ZHD voltage magnitude and frequency regulation capability at PCC, where it is possible to verify that the voltage regulation is within an acceptable range of 5% [28]. The output voltage FFT for all phases is shown in Fig. 8 (f) with a free harmonic range until 50th order, as expected.

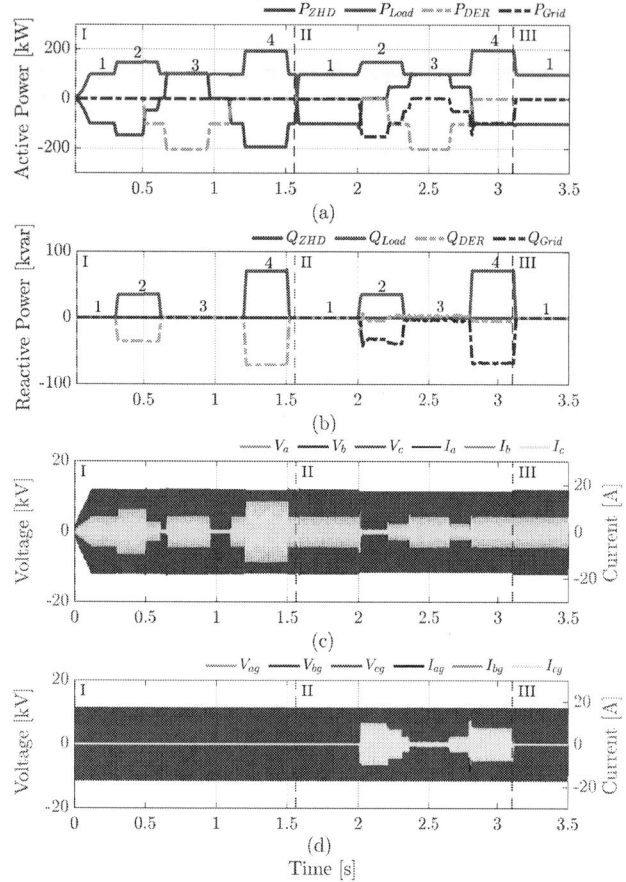

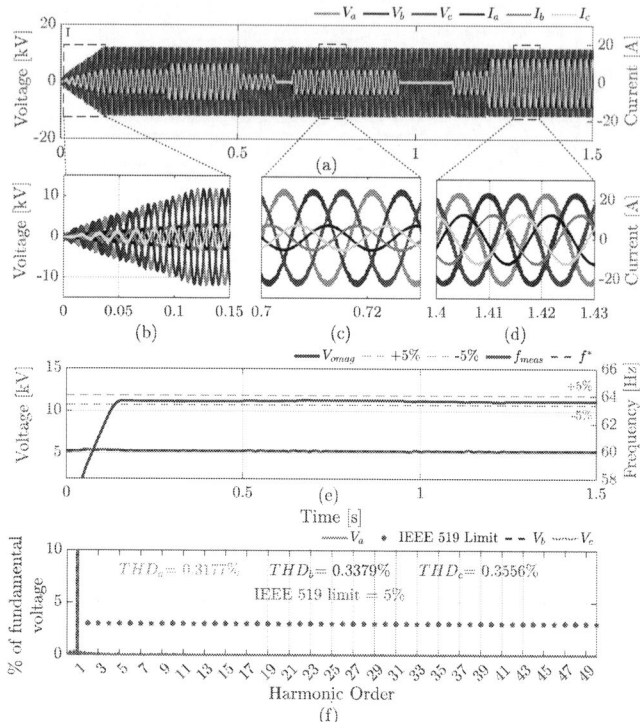

Fig. 8. Islanded operation: (a) Voltage and current at ZHD converter between: (a) 0 and 1.5 s, (b) 0 and 0.15 s, (c) 0.7 and 0.73s, (d) 1.4 and 1.43 s, (e) voltage and frequency at the PCC and (f) output voltage FFT for all the phases.

Fig. 7. (a) Active power, (b) Reactive power, (c) Voltage and current at the ZHD converter, (d) Voltage and current at the grid side.

C. Grid-Connection and Islanding Transition

The zoomed view of voltage and current at ZHD converter and grid side even as the magnitude voltage and frequency at PCC of the time events II and III are shown in Fig. 9 (a), (g) and (l), respectively. Still in the islanded mode the converter frequency is changed to 59 Hz in time event II, as can be seen in (b) and (l), in order to start the grid-connection process. The ZHD voltage magnitude and phase are compared with the grid voltage and phase in the MGMC and the connection to the grid occurs as shown in (c) and (h) the zoomed view of the voltage and current waveforms in the ZHD converter and the grid, respectively, by only enabling the current regulator gains k_{pEN} and k_{iEN}, where the ZHD MC converter changes to the grid-following mode, synchronized with the grid by DDSRF-PLL, after the signal sent by the MGMC.

The steady-state waveforms of Fig. 9 (d) and (i) show the energy time shift service realized by the ZHD MC converter absorbing power in load scenario 3 of high DER generation (Fig. 9 (d)), where the current at the grid is near zero (Fig. 9 (i)), and injecting power in scenario 4 of high load consumption (Fig. 9 (f) and (j)). These waveforms in an analogue way of the islanded case show that the ZHD converter ensures sinusoidal current injection to the MG. The output current FFT

shown in (e) shows the sinusoidal current waveform shape without harmonics until the 50th order. Finally, the same Fig. 9 shows in time event III the transition from the grid-connected to the islanded mode. Figs. 9 (f) and (k) show the zoomed views of voltage and current at the ZHD converter and in the grid, respectively, only by disabling the gains of the current regulators k_{pEN} and k_{iEN} after the signal sent by the MGMC with the DDSRF-PLL now generating the reference frequency.

IV. HARDWARE IN THE LOOP RESULTS

The ZHD MC converter with the parameters described in Table II was implemented on a real-time simulation platform to test and validate the proposed ZHD control hardware operating in both grid-connected and islanded operation modes.

Fig. 10 shows the real controller connected to the Hardware-In-the-Loop (HIL) test bench. The controller consists of a Texas Instruments TMS320F28335 Digital Signal Processor (DSP) responsible for the converter control and an Field-Programmable Gate Array (FPGA) model MAX 10 from Intel operating with sufficiently high sampling frequency (250 kHz) storing the switching angles and handling with the SHE PWM based on the modulation rates sent by the DSP. Parallel communication is performed between the DSP and the FPGA, ensuring a high data exchange rate between the devices.

Taking into account a capacitive load of 50 kVA (PF = 0.9) and a resistive load of 130 kW connected at the output of the

979-8-3315-1612-3/25 $31.00 © 2025 IEEE

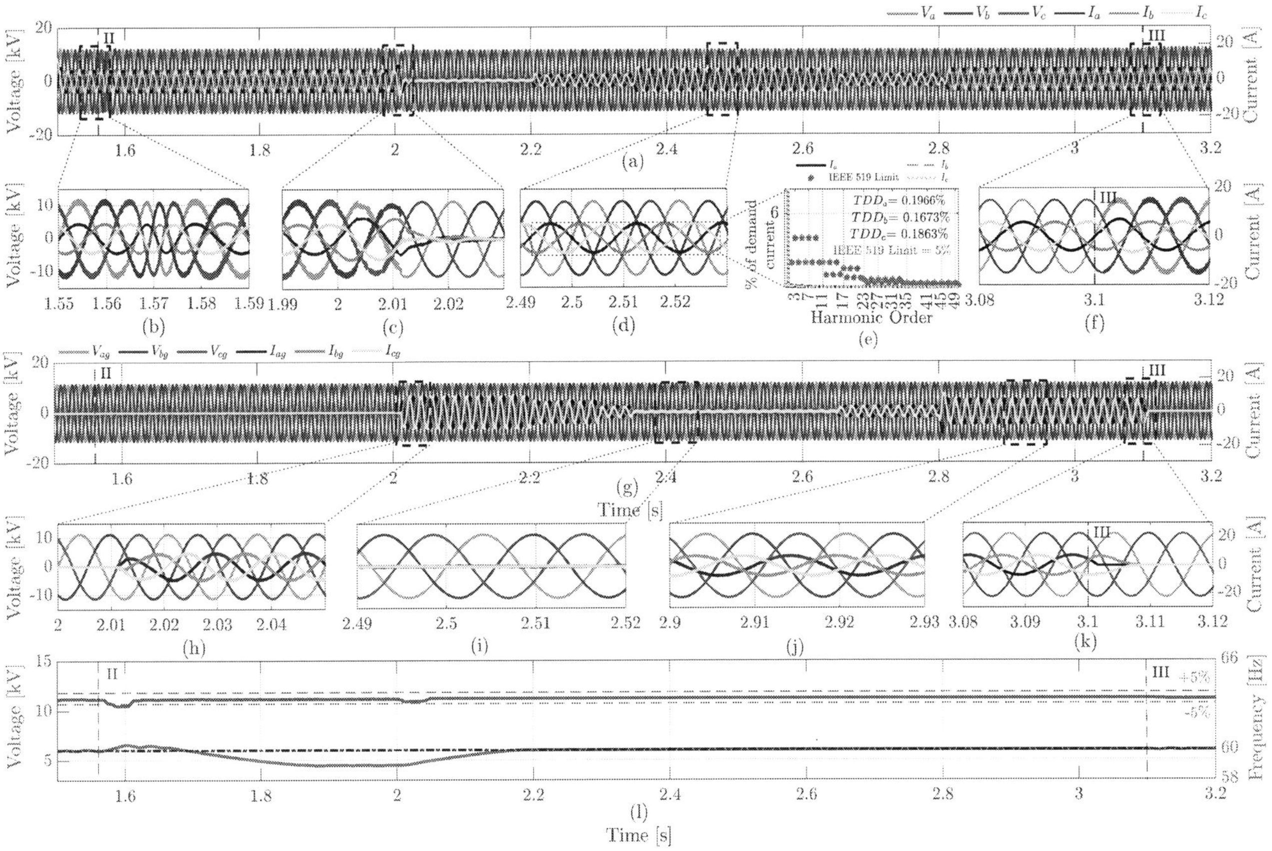

Fig. 9. Voltage and current at ZHD converter between: (a) 1.5 and 3.2 s, (b) 1.55 and 1.59 s, (c) 1.99 and 2.03 s, (d) 2.4 and 2.43 s, (e) output current FFT, (f) 3.08 and 3.12 s, voltage and current at grid side between: (h) 2 and 2.05 s, (i) 2.11 and 2.22 s, (j) 3.05 and 3.12 s, (k) voltage and frequency at PCC.

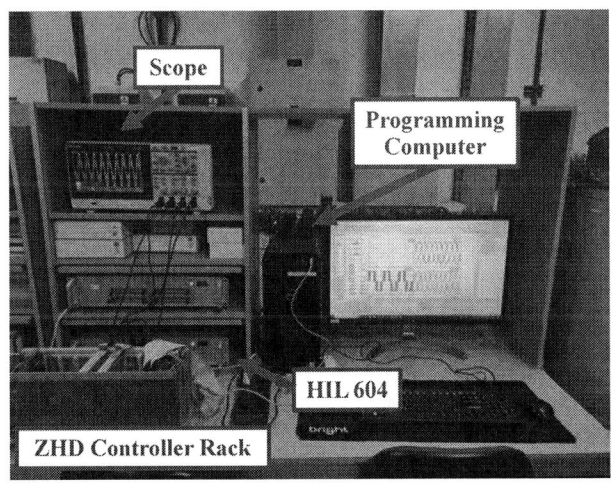

Fig. 10. Structure of tests and validation in Hardware-in-the-Loop.

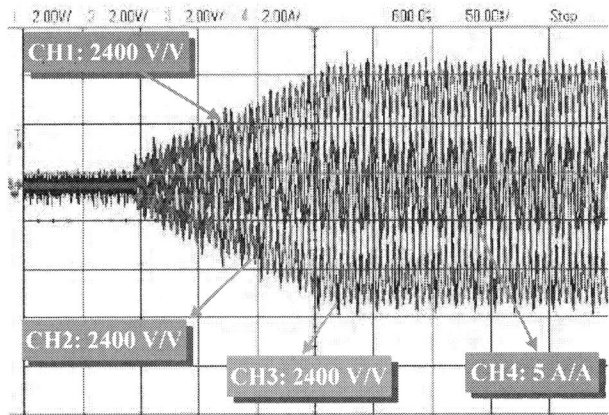

Fig. 11. ZHD converter black-start capability: phase A (CH1), B (CH2), and C (CH3) output voltages and phase A output current (CH4).

ZHD converter, Fig. 11 shows the black start capability of the converter.

Fig. 12 (a) shows the ZHD MC converter transiting from grid-connected to islanded mode by only disabling the current regulator gains, the zoomed view of the ZHD current waveform in Fig. 12 (b) shows a current harmonic FFT practically a sinusoidal current waveform.

Finally, Fig. 12 (c) shows the islanded to grid-connected mode transition, only by enabling the current regulator gains and in (d) the islanded mode zoomed view showing the converter delivering a practically sinusoidal voltage waveform.

979-8-3315-1612-3/25 $31.00 © 2025 IEEE

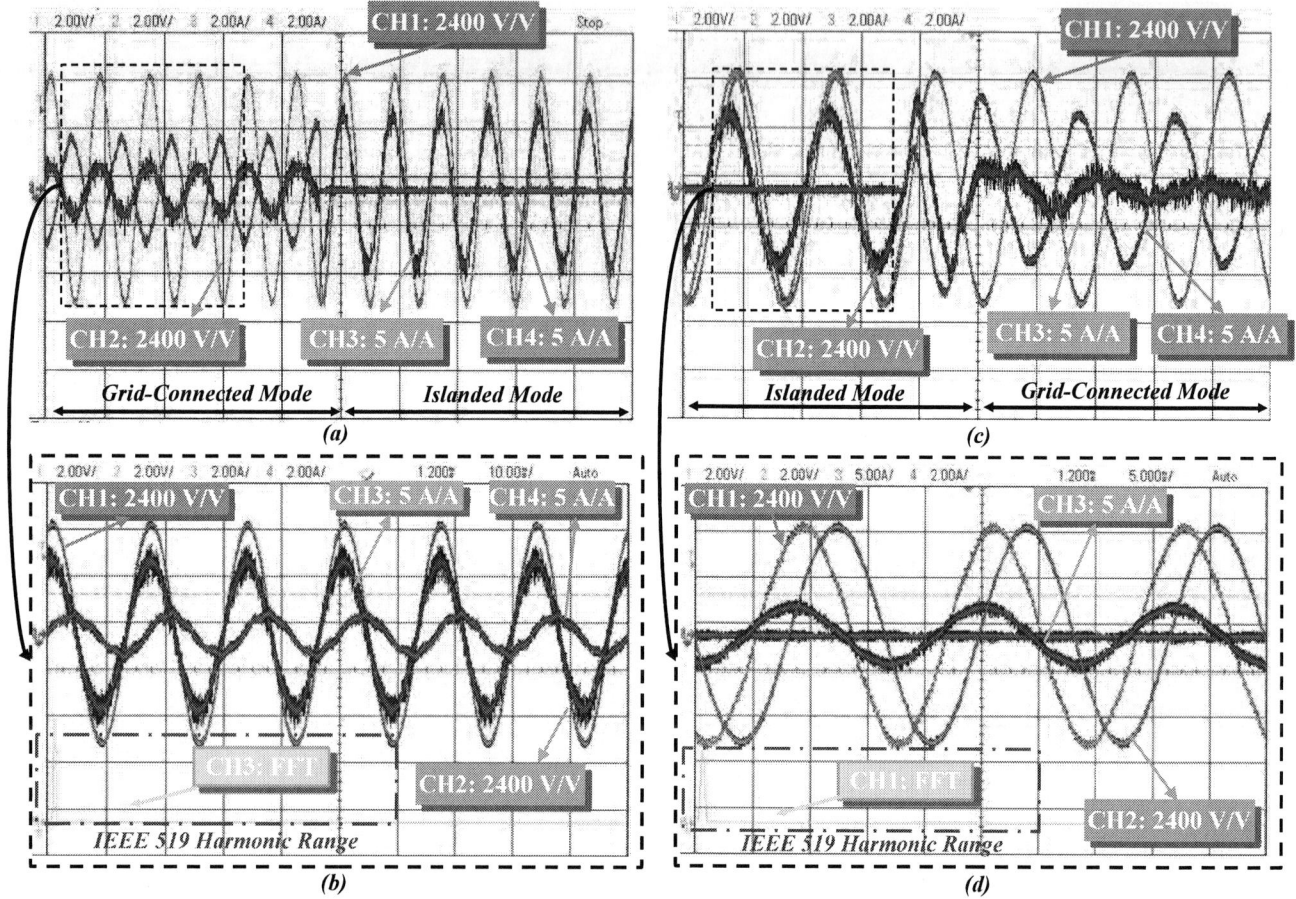

Fig. 12. (a) Intentional Islanded, (b) zoomed view of the islanded case, (c) grid-Connection, and (d) zoomed view of the grid-connection mode: ZHD converter phase A voltage (CH1), phase A grid voltage (CH2), ZHD phase A current (CH3) and phase A load current (CH4).

V. CONCLUSIONS

The Zero Harmonic Distortion (ZHD) converter presents a significant advancement in master converters for medium-voltage microgrids, offering high-quality sinusoidal waveforms and reliable performance without the need for capacitive filtering. Its simplified control approach, which eliminates the voltage closed-loop control and relies solely on a current closed-loop for grid-connected operations, enables seamless transitions between operating modes. The effectiveness of the ZHD converter is substantiated by simulation and hardware-in-the-loop results, underscoring its potential for robust integration in medium-voltage microgrid applications.

REFERENCES

[1] B. Kroposki, B. Johnson, Y. Zhang, V. Gevorgian, P. Denholm, B.-M. Hodge, and B. Hannegan, "Achieving a 100% renewable grid: Operating electric power systems with extremely high levels of variable renewable energy," *IEEE Power and Energy Magazine*, vol. 15, no. 2, pp. 61–73, 2017, doi:10.1109/MPE.2016.2637122.

[2] F. Milano, F. Dörfler, G. Hug, D. J. Hill, and G. Verbič, "Foundations and challenges of low-inertia systems (invited paper)," in *2018 Power Systems Computation Conference (PSCC)*, 2018, pp. 1–25, doi:10.23919/PSCC.2018.8450880.

[3] C. Vartanian, R. Bauer, L. Casey, C. Loutan, D. Narang, and V. Patel, "Ensuring system reliability: Distributed energy resources and bulk power system considerations," *IEEE Power and Energy Magazine*, vol. 16, no. 6, pp. 52–63, 2018, doi:10.1109/MPE.2018.2863059.

[4] M. Braun, J. Brombach, C. Hachmann, D. Lafferte, A. Klingmann, W. Heckmann, F. Welck, D. Lohmeier, and H. Becker, "The future of power system restoration: Using distributed energy resources as a force to get back online," *IEEE Power and Energy Magazine*, vol. 16, no. 6, pp. 30–41, 2018, doi:10.1109/MPE.2018.2864227.

[5] R. Lasseter and P. Paigi, "Microgrid: a conceptual solution," in *2004 IEEE 35th Annual Power Electronics Specialists Conference (IEEE Cat. No.04CH37551)*, vol. 6, 2004, pp. 4285–4290 Vol.6, doi:10.1109/PESC.2004.1354758.

[6] N. Hatziargyriou, H. Asano, R. Iravani, and C. Marnay, "Microgrids," *IEEE Power and Energy Magazine*, vol. 5, no. 4, pp. 78–94, 2007, doi:10.1109/MPAE.2007.376583.

[7] J. Rocabert, A. Luna, F. Blaabjerg, and P. Rodríguez, "Control of power converters in ac microgrids," *IEEE Transactions on Power Electronics*, vol. 27, no. 11, pp. 4734–4749, 2012, doi:10.1109/TPEL.2012.2199334.

[8] A. Vukojevic and S. Lukic, "Microgrid protection and control schemes for seamless transition to island and grid synchronization," *IEEE Transactions on Smart Grid*, vol. 11, no. 4, pp. 2845–2855, 2020, doi:10.1109/TSG.2020.2975850.

[9] Y. Li, D. Vilathgamuwa, and P. C. Loh, "Design, analysis, and real-time testing of a controller for multibus microgrid system," *IEEE Transactions on Power Electronics*, vol. 19, no. 5, pp. 1195–1204, 2004.

[10] F. Gao and M. R. Iravani, "A control strategy for a distributed generation unit in grid-connected and autonomous modes of operation," *IEEE Transactions on Power Delivery*, vol. 23, no. 2, pp. 850–859, 2008.

[11] S.-H. Hu, C.-Y. Kuo, T.-L. Lee, and J. M. Guerrero, "Droop-controlled

inverters with seamless transition between islanding and grid-connected operations," in *2011 IEEE Energy Conversion Congress and Exposition*, 2011, pp. 2196–2201, doi:10.1109/ECCE.2011.6064059.

[12] L. Arnedo, S. Dwari, V. Blasko, and S. Park, "80 kw hybrid solar inverter for standalone and grid connected applications," in *2012 Twenty-Seventh Annual IEEE Applied Power Electronics Conference and Exposition (APEC)*, 2012, pp. 270–276, doi:10.1109/APEC.2012.6165830.

[13] Y. A.-R. I. Mohamed, H. H. Zeineldin, M. M. A. Salama, and R. Seethapathy, "Seamless formation and robust control of distributed generation microgrids via direct voltage control and optimized dynamic power sharing," *IEEE Transactions on Power Electronics*, vol. 27, no. 3, pp. 1283–1294, 2012, doi:10.1109/TPEL.2011.2164939.

[14] L. S. Araujo and D. I. Brandao, "Self-adaptive control for grid-forming converter with smooth transition between microgrid operating modes," *International Journal of Electrical Power & Energy Systems*, vol. 135, p. 107479, 2022, doi:10.1016/j.ijepes.2021.107479. [Online]. Available: https://www.sciencedirect.com/science/article/pii/S0142061521007183

[15] S. Lissandron and P. Mattavelli, "A controller for the smooth transition from grid-connected to autonomous operation mode," in *2014 IEEE Energy Conversion Congress and Exposition (ECCE)*, 2014, pp. 4298–4305, doi:10.1109/ECCE.2014.6953987.

[16] Z. Liu, J. Liu, and Y. Zhao, "A unified control strategy for three-phase inverter in distributed generation," *IEEE Transactions on Power Electronics*, vol. 29, no. 3, pp. 1176–1191, 2014, doi:10.1109/TPEL.2013.2262078.

[17] O. V. Kulkarni, S. Doolla, and B. G. Fernandes, "Mode transition control strategy for multiple inverter-based distributed generators operating in grid-connected and standalone mode," *IEEE Transactions on Industry Applications*, vol. 53, no. 6, pp. 5927–5939, 2017, doi:10.1109/TIA.2017.2743682.

[18] J. Kwon, S. Yoon, and S. Choi, "Indirect current control for seamless transfer of three-phase utility interactive inverters," *IEEE Transactions on Power Electronics*, vol. 27, no. 2, pp. 773–781, 2012, doi:10.1109/TPEL.2011.2161335.

[19] K.-Y. Lo and Y.-M. Chen, "Design of a seamless grid-connected inverter for microgrid applications," *IEEE Transactions on Smart Grid*, vol. 11, no. 1, pp. 194–202, 2020, doi:10.1109/TSG.2019.2919905.

[20] T. M. Parreiras, J. C. G. Justino, and B. d. J. Cardoso Filho, "The true unity power factor converter — a practical filterless solution for sinusoidal currents," in *2015 9th International Conference on Power Electronics and ECCE Asia (ICPE-ECCE Asia)*, 2015, pp. 2557–2565, doi:10.1109/ICPE.2015.7168134.

[21] G. V. Ramos, T. M. Parreiras, D. A. d. L. Brandão, S. M. Silva, and B. d. J. C. Filho, "A zero harmonic distortion grid-forming converter for islanded microgrids," in *2023 IEEE Industry Applications Society Annual Meeting (IAS)*, 2023, pp. 1–8, doi: 10.1109/IAS54024.2023.10406502.

[22] G. V. Ramos, T. M. Parreiras, and B. de Jesus Cardoso Filho, "Control performance assessment of a zero harmonic distortion grid-forming converter in islanded microgrids," in *2023 IEEE 8th Southern Power Electronics Conference and 17th Brazilian Power Electronics Conference (SPEC/COBEP)*, 2023, pp. 1–8, doi:10.1109/SPEC56436.2023.10407128.

[23] G. V. Ramos, D. A. d. L. Brandao, T. M. Parreiras, S. M. Silva, and B. J. C. Filho, "A zero harmonic distortion grid-forming converter for medium voltage islanded microgrids," *IEEE Transactions on Industry Applications*, pp. 1–12, 2024, doi:10.1109/TIA.2024.3462913.

[24] G. V. Ramos, T. M. Parreiras, and B. J. Cardoso Filho, "Control performance assessment of a zero harmonic distortion grid-forming converter for medium voltage islanded microgrids," *Eletrônica de Potência*, vol. 29, p. e202441, Oct. 2024, doi: 10.18618/REP.e202441. [Online]. Available: https://journal.sobraep.org.br/index.php/rep/article/view/961

[25] G. V. Ramos, T. M. Parreiras, and B. d. J. C. Filho, "The three-level zero harmonic distortion grid-forming converter: A practical filterless solution for sinusoidal voltages," in *2024 IEEE 21st International Power Electronics and Motion Control Conference (PEMC)*, 2024, pp. 1–6, doi:10.1109/PEMC61721.2024.10726351.

[26] G. V. Ramos, T. M. Parreiras, and B. D. J. C. Filho, "A three-level zero harmonic distortion grid-forming converter for medium voltage islanded microgrids," in *2024 Energy Conversion Congress & Expo Europe (ECCE Europe)*, 2024, pp. 1–7, doi:10.1109/ECCEEurope62508.2024.10751867.

[27] "IEEE Standard for Harmonic Control in Electric Power Systems," *IEEE Std 519-2022 (Revision of IEEE Std 519-2014)*, pp. 1–31, 2022, doi:10.1109/IEEESTD.2022.9848440.

[28] "IEEE Standard for Interconnection and Interoperability of distributed energy Resources with Associated Electric Power Systems Interfaces," *IEEE Std 1547-2018 (Revision of IEEE Std 1547-2003)*, pp. 1–138, 2018, doi:10.1109/IEEESTD.2018.8332112.

[29] B. Yao, X. Wei, Y. Zhang, P. Correia, R. Wu, S. Song, I. Trintis, H. Wang, and H. Wang, "Accelerated degradation testing and failure mechanism analysis of metallized film capacitors for ac filtering," *IEEE Transactions on Power Electronics*, vol. 39, no. 5, pp. 6256–6270, 2024, doi:10.1109/TPEL.2024.3360373.

[30] A. Saim, A. Houari, J. M. Guerrero, A. Djerioui, M. Machmoum, and M. A. Ahmed, "Stability analysis and robust damping of multiresonances in distributed-generation-based islanded microgrids," *IEEE Transactions on Industrial Electronics*, vol. 66, no. 11, pp. 8958–8970, 2019, doi:10.1109/TIE.2019.2898611.

[31] X. Wang and F. Blaabjerg, "Harmonic stability in power electronic-based power systems: Concept, modeling, and analysis," *IEEE Transactions on Smart Grid*, vol. 10, no. 3, pp. 2858–2870, 2019, doi:10.1109/TSG.2018.2812712.

[32] R. Juntunen, J. Korhonen, T. Musikka, L. Smirnova, O. Pyrhönen, and P. Silventoinen, "Identification of resonances in parallel connected grid inverters with lc- and lcl-filters," in *2015 IEEE Applied Power Electronics Conference and Exposition (APEC)*, 2015, pp. 2122–2127.

[33] J. Enslin and P. Heskes, "Harmonic interaction between a large number of distributed power inverters and the distribution network," *IEEE Transactions on Power Electronics*, vol. 19, no. 6, pp. 1586–1593, 2004.

[34] J. He, Y. W. Li, D. Bosnjak, and B. Harris, "Investigation and active damping of multiple resonances in a parallel-inverter-based microgrid," *IEEE Transactions on Power Electronics*, vol. 28, no. 1, pp. 234–246, 2013.

[35] C. J. Mozina, "Impact of smart grid and green power generation on distribution systems," in *2012 IEEE PES Innovative Smart Grid Technologies (ISGT)*, 2012, pp. 1–13, doi:10.1109/ISGT.2012.6175625.

[36] H. Oliveira, L. H. S. Santos, J. Gomes de Matos, L. A. d. S. Ribeiro, A. Cunha Oliveira, and J. Victor Mapurunga Caracas, "Challenges in medium voltage microgrids case study: Alcântara launch center," in *2023 IEEE 8th Southern Power Electronics Conference and 17th Brazilian Power Electronics Conference (SPEC/COBEP)*, 2023, pp. 1–6.

[37] H. S. Patel and R. G. Hoft, "Generalized techniques of harmonic elimination and voltage control in thyristor inverters: Part i–harmonic elimination," *IEEE Transactions on Industry Applications*, vol. IA-9, no. 3, pp. 310–317, 1973, doi:10.1109/TIA.1973.349908.

[38] ——, "Generalized techniques of harmonic elimination and voltage control in thyristor inverters: Part ii — voltage control techniques," *IEEE Transactions on Industry Applications*, vol. IA-10, no. 5, pp. 666–673, 1974, doi:10.1109/TIA.1974.349239.

[39] P. Rodriguez, J. Pou, J. Bergas, J. I. Candela, R. P. Burgos, and D. Boroyevich, "Decoupled double synchronous reference frame pll for power converters control," *IEEE Transactions on Power Electronics*, vol. 22, no. 2, pp. 584–592, 2007, doi:10.1109/TPEL.2006.890000.

[40] R. Lorenz, T. Lipo, and D. Novotny, "Motion control with induction motors," *Proceedings of the IEEE*, vol. 82, no. 8, pp. 1215–1240, 1994, doi: 10.1109/5.301685.

An MILP Approach for Modeling and Analyzing the BESS for Smoothing Renewable Fluctuations Considering BESS Capacity Attenuation in the Bulk Power System With High Inverter-Based Resource Penetration

Hualong Liu and Wenyuan Tang
Department of Electrical and Computer Engineering
North Carolina State University
Raleigh, USA
{hliu37, wtang8}@ncsu.edu

Abstract—Analyzing the effectiveness of the battery energy storage system (BESS) in smoothing renewable fluctuations correctly is vital to the stable operation of the power system. Therefore, based on stochastic programming and attention-based time series forecasting with learnable and interpretable basis (BasisFormer), we propose a mixed integer linear programming (MILP) approach for modeling the BESS and analyzing the effectiveness of the BESS in smoothing renewable fluctuations in the bulk power system with high inverter-based resources. The approach considers the attenuation of BESS capacity and the uncertainty of renewable generation. First, we present a scenario generation and reduction method based on Latin hypercube sampling (LHS) and the Kantorovich distance. Second, we propose a state of health (SoH) prediction method of the BESS based on BasisFormer. Third, we propose an MILP approach for modeling the BESS and analyzing the effectiveness of the BESS in smoothing renewable fluctuations. Finally, we verify the proposed approach through case studies. The experimental results show that the attenuation of BESS capacity is a crucial factor in correctly analyzing the effectiveness of the BESS in smoothing renewable fluctuations.

Index Terms—Battery energy storage system, BasisFormer, mixed integer linear programming, renewable fluctuation, state of health, scenario generation and reduction.

I. INTRODUCTION

A. Research Motivation

Today's new power systems demonstrate the characteristics of "three highs", i.e., high penetration of renewables, high penetration of power electronic equipment, and high penetration of battery energy storage systems (BESSs). The "three highs" lead to a large number of inverter-based resources, which typically include wind turbines (WTs), photovoltaic (PV) panels, and others, in the power system. The high inverter-based resource penetration results in considerable differences between the dynamic behavior of the new power system and that of the traditional power system dominated by synchronous generators. These differences yield new stability problems [1], [2]. The large-scale integration of renewables

into the power system has substantial negative impacts on the stability, reliability, and resilience of the power system. These negative effects are largely due to the significant fluctuations of renewable generation. The BESS can suppress these fluctuations. Therefore, it is of great significance to exploit grid-scale BESS to smooth renewable fluctuations.

B. Literature Review

Reference [3] investigated an optimization strategy for a microgrid containing renewable energy, which takes into account the smoothing of power fluctuations. However, Reference [3] does not consider the uncertainty of renewable generation. The role of energy storage systems in suppressing the fluctuations of renewable energy was reviewed in [4]. Reference [5] reviewed the function of energy storage systems in suppressing the fluctuations of renewable energy in integrated energy systems. Reference [6] explored how to mitigate the fluctuations of renewable power in cloud data centers. Reference [7] studied how to reduce power fluctuations in integrated energy systems that include thermal energy, energy storage, wind, and PV generation. Reference [8] discussed how to conduct the optimal design of the distribution network containing wind power under the premise of considering power quality.

Although there are some studies on smoothing renewable fluctuations, there are the following problems with these state-of-the-art researches [3]–[8]. 1) Uncertainty is not considered. 2) BESS capacity attenuation is not considered. 3) These models are complex non-linear models, and focus more on microgrids.

C. Contributions

In order to fill these research gaps mentioned in Section I-B, we propose a mixed integer linear programming (MILP) approach for modeling the BESS and analyzing the effectiveness of the BESS in smoothing the fluctuations of renewable power considering the attenuation of BESS capacity and the uncertainty of renewable generation. Our key contributions are as follows.

1) We present a scenario generation and reduction method based on Latin hypercube sampling (LHS) and the Kantorovich distance.
2) We propose a state of health (SoH) prediction method of the BESS based on BasisFormer.
3) We propose an MILP approach for modeling the BESS and analyzing the effectiveness of the BESS in smoothing the fluctuations of renewable generation.
4) We verify the effectiveness and superiority of the proposed method through case studies.

D. Organization

The rest of our paper are organized as follows. Section II details a scenario generation and reduction method. A SoH prediction method of the BESS based on BasisFormer is explained minutely in Section III. An MILP approach for modeling and analyzing the BESS is derived and elaborated in Section IV. Case studies are provided in Section V. We present conclusions and future work in Section VI.

II. A SCENARIO GENERATION AND REDUCTION METHOD

Because of the uncertainties of wind and PV generation, it is unreasonable not to consider uncertainty. This paper leverages stochastic programming to handle uncertainty. In order to reduce computation burden and sampling bias, improve sampling efficiency and convergence speed, this paper selects LHS [9]–[11] instead of Monte Carlo sampling to generate scenarios. The commonly used Euclidean distance-based simultaneous backward reduction method [12] is sensitive to outliers and the scale of different dimensions of the data. However, the Kantorovich distance [13], [14] can compare two probability distributions better and is not sensitive to outliers. Thus, the Kantorovich distance is selected. The Kantorovich distance between original scenario set $\mathcal{S}$ and deleted scenario set $\mathcal{S}'$ is defined by

$$
\begin{aligned}
K\left(\mathcal{S}, \mathcal{S}'\right) = \min \Big\{ & \sum_{\mathbf{s}\in\mathcal{S}, \mathbf{s}'\in\mathcal{S}'} \|\mathbf{s}-\mathbf{s}'\|_2\, p\left(\mathbf{s},\mathbf{s}'\right) \Big| \\
& p\left(\mathbf{s},\mathbf{s}'\right) \geq 0, \forall \mathbf{s}\in\mathcal{S}, \mathbf{s}'\in\mathcal{S}'; \\
& \sum_{\mathbf{s}\in\mathcal{S}} p\left(\mathbf{s},\mathbf{s}'\right) = p_{s'}, \forall \mathbf{s}'\in\mathcal{S}'; \\
& \sum_{\mathbf{s}'\in\mathcal{S}} p\left(\mathbf{s},\mathbf{s}'\right) = p_s, \forall \mathbf{s}\in\mathcal{S} \Big\},
\end{aligned} \quad (1)
$$

where $\mathbf{s}$ and $\mathbf{s}'$ are scenarios in $\mathcal{S}$ and $\mathcal{S}'$, respectively, p_s and $p_{s'}$ are the probabilities of $\mathbf{s}$ in $\mathcal{S}$ and $\mathbf{s}'$ in $\mathcal{S}'$, respectively. $p\left(\mathbf{s},\mathbf{s}'\right)$ is the joint probability of $\mathbf{s}$ and $\mathbf{s}'$.

The proposed scenario generation and reduction method is shown in Algorithm 1.

III. A SoH PREDICTION METHOD OF THE BESS BASED ON BASISFORMER

With the use of the BESS, it deteriorates and its capacity fades, and if the environment is harsh, then the attenuation will

---

Algorithm 1 Scenario generation and reduction based on LHS and the Kantorovich distance

---

1: Generate n scenarios based on LHS.
2: **while** The number of scenarios in $\mathcal{S}$ is not equal to the preset number n_s **do**
3:     Remove scenario $\mathbf{s}'$ as per (1) from $\mathcal{S}$ and put it into $\mathcal{S}'$.
4:     Select scenario $\mathbf{s}$ closest to scenario $\mathbf{s}'$ based on (1).
5:     $p_s = p_s + p_{s'}$, $\mathcal{S} = \mathcal{S} - \{p_{s'}\}$, and $\mathcal{S}' = \mathcal{S}' + \{p_{s'}\}$.
6: **end while**

---

be more serious. Therefore, it is unreasonable not to consider the BESS attenuation [15], [16] in the design and operation of the BESS. Not taking into account BESS attenuation can lead to a significant erroneous judgement of the BESS. The SoH of the BESS is defined as the ratio of the actual capacity of the BESS at the current moment to the rated capacity. The complex physical and chemical mechanism inside the BESS is difficult to model accurately, which leads to the imprecision and poor generality of the BESS SoH prediction based on the physical model [17] and traditional machine learning [18].

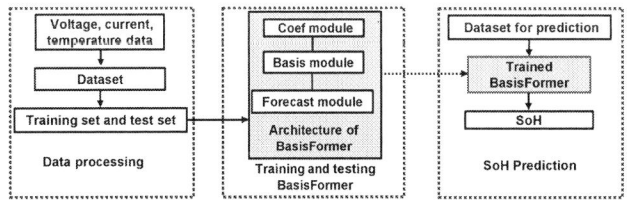

Fig. 1. Fundamental architecture of BasisFormer and framework of BasisFormer-based BESS SoH prediction.

Consequently, we propose a deep learning BESS SoH prediction method based on BasisFormer. The chief goal of BasisFormer [19] is to learn a basis $\mathbf{z}$ which can explain the behavior of all time series in the dataset. Then, we use the basis to forecast $\mathbf{y}$ given the input $\mathbf{x}$. As shown in Fig. 1, BasisFormer comprise three modules, i.e., the Coef module, the Forecast module, and the Basis module. The Coef module decides the corresponding coefficients by analyzing a time series relative to each basis vector. The Forecast module forecasts the future of the time series according to the coefficients and the future part of the basis. The Basis module learns the basis considering the view points of both history and future by adjusting the relations of the basis and the time series. Refer to [19] for details. The flowchart and framework of the BasisFormer-based BESS SoH prediction method is shown in Fig. 1.

IV. PROPOSED MILP APPROACH FOR MODELING AND ANALYZING THE BESS

A. Objective Function

The schematic diagram of the system of wind, PV generation, and the BESS is shown in Fig. 2. The objective function is to minimize the sum of the absolute value of the difference

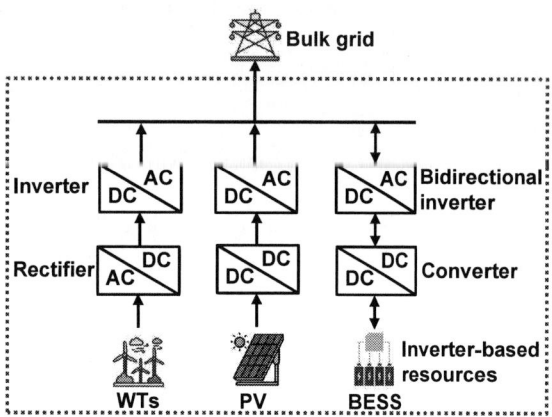

Fig. 2. System of inverter-based resources

C. An MILP Model

Function (2) contains absolute values, which are difficult and time-consuming to optimize, so we convert it to a linear function. Accordingly, an MILP model for analyzing the BESS for smoothing renewables considering BESS attenuation can be formulated as follows:

$$
\begin{aligned}
\min \quad & \sum_{s=1}^{|\mathcal{S}|} p_s \sum_{t=1}^{T} y_{s,t} \\
\text{s.t.} \quad & P_{s,t} - P_{s,t-1} \leq y_{s,t}, \\
& -(P_{s,t} - P_{s,t-1}) \leq y_{s,t}, \\
& (3)\text{–}(8).
\end{aligned}
\tag{9}
$$

where $y_{s,t}$ is the auxiliary variable.

V. CASE STUDIES

A. Description

The system shown in Fig. 2 is used to verify the validity of the proposed method. The capacities of wind and PV generation are 11 MW and 1 MW, respectively. The prediction errors of wind and PV generation follow normal distributions with the mean of 0 and the standard deviations of 9% and 7%, respectively. We first use the method in Section II to produce 1200 scenarios, and achieve 12 typical scenarios and their corresponding probabilities. Second, based on the historical and present voltage, current and temperature data of the BESS, we use the method in Section III to obtain the present BESS capacity considering BESS attenuation. Finally, we leverage the model in Section IV to analyze and evaluate the BESS. The dispatch cycle is 1 d. Each period is 15 min. The number of periods is 96. Throughout the dispatch cycle, power fluctuations are considered acceptable if they are less than or equal to 7.5% of the installed renewable capacity for any adjacent periods.

B. Results

Case 1: Based on the proposed method, we obtain the optimal BESS configuration capacity of 1 MWh. Case 2: When BESS attenuation is not considered, the system power fluctuations before and after using the BESS, the charging and discharging of the BESS, and the storied energy of the BESS are shown in Figs. 3–5. Case 3: When BESS attenuation is considered, the system power fluctuations before and after using the BESS, the charging and discharging of the BESS, and the storied energy of the BESS are shown in Fig. 6–8. As can be seen from Figs. 3–8, if BESS attenuation is not considered, the maximum power fluctuation is 0.7833 MW, which is less than 0.9 MW (12×7.5%), and thus the configured BESS capacity fully meets the requirements. However, the BESS has attenuation. After considering BESS attenuation, the maximum power fluctuation is 2.2309 MW, which is greater than 0.9 MW, and thus the effect of the BESS to smooth the power fluctuations of renewables does not meet the requirements. This is the time to consider replacing the BESS.

between the power fluctuations of all adjacent periods in the entire scheduling cycle after the BESS is used, that is,

$$
\min \sum_{s=1}^{|\mathcal{S}|} p_s \sum_{t=1}^{T} |P_{s,t} - P_{s,t-1}|,
\tag{2}
$$

$$
P_{s,t} = P_{s,t}^{\mathrm{WT}} + P_{s,t}^{\mathrm{PV}} - P_{s,t}^{\mathrm{B+}} + P_{s,t}^{\mathrm{B-}}, \quad \forall s, \forall t,
\tag{3}
$$

where $P_{s,t}$, $P_{s,t}^{\mathrm{WT}}$, $P_{s,t}^{\mathrm{PV}}$, $P_{s,t}^{\mathrm{B+}}$, and $P_{s,t}^{\mathrm{B-}}$ represent the system output power, WT, PV generation power, BESS charging power, and discharging power during period t in scenario s, respectively. p_s denotes the probability that scenario s occurs. $|\mathcal{S}|$ indicates the total number of scenarios. T signifies the number of periods in the entire scheduling cycle.

B. BESS Constraints

The BESS should satisfy the following constraints:

$$
E_{s,t}^{\mathrm{B}} = (1-\alpha)E_{s,t-1}^{\mathrm{B}} + \eta^{\mathrm{B+}} P_{s,t}^{\mathrm{B+}} - P_{s,t}^{\mathrm{B-}}/\eta^{\mathrm{B-}}, \quad \forall s, \forall t,
\tag{4}
$$

$$
0 \leq P_{s,t}^{\mathrm{B-}} \leq u_{s,t}^{\mathrm{B-}} \eta^{\mathrm{B-}} P_{\max}^{\mathrm{B}}, \quad \forall s, \forall t,
\tag{5}
$$

$$
0 \leq P_{s,t}^{\mathrm{B+}} \leq (1-u_{s,t}^{\mathrm{B-}}) P_{\max}^{\mathrm{B}}/\eta^{\mathrm{B+}}, \quad \forall s, \forall t,
\tag{6}
$$

$$
E_{\min}^{\mathrm{B}} \leq E_{s,t}^{\mathrm{B}} \leq E_{\max}^{\mathrm{B}}, \quad \forall s, \forall t,
\tag{7}
$$

$$
E_{s,1}^{\mathrm{B}} = E_{s,T}^{\mathrm{B}}, \quad \forall s, \forall t,
\tag{8}
$$

where $E_{s,t}^{\mathrm{B}}$, α, $\eta^{\mathrm{B+}}$, and $\eta^{\mathrm{B-}}$ represent the stored energy of the BESS during period t in scenario s, the self-discharging rate of the BESS, the charging efficiency of the BESS, and the discharging efficiency of the BESS, respectively. $P_{s,t}^{\mathrm{B+}}$ and $P_{s,t}^{\mathrm{B-}}$ indicate the charging power and discharging power of the BESS during period t in scenario s, respectively. $u_{s,t}^{\mathrm{B-}}$ indicates the binary decision variable of the charging and discharging of the BESS. When $u_{s,t}^{\mathrm{B-}}$ is equal to 1, it means discharging the BESS, and when it is equal to 0, it means charging the BESS. $P_{\max}^{\mathrm{B}}$, $E_{\max}^{\mathrm{B}}$, and $E_{\min}^{\mathrm{B}}$ indicate the power capacity of the BESS, the energy capacity of the BESS, and the minimum allowable stored energy of the BESS, respectively. Constraint (8) indicates that the energy stored by the BESS at the beginning of the scheduling cycle should be equal to the energy stored by the BESS at the end of the scheduling cycle in scenario s.

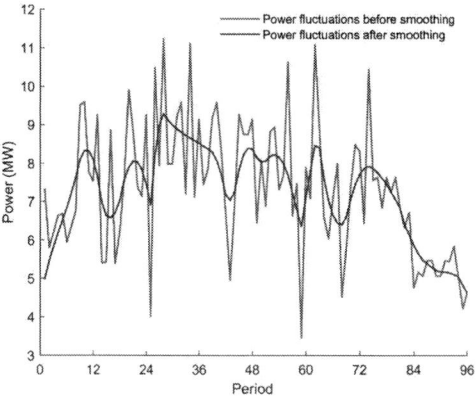

Fig. 3. Power fluctuations when BESS attenuation is not considered.

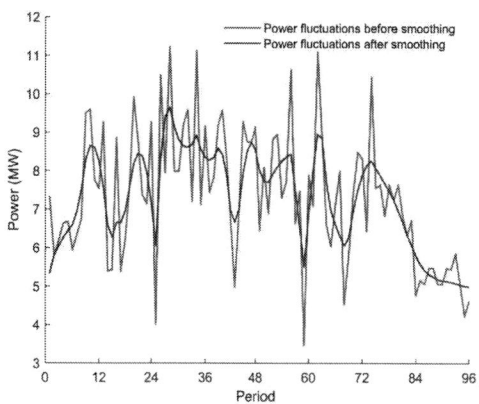

Fig. 6. Power fluctuations when BESS attenuation is considered.

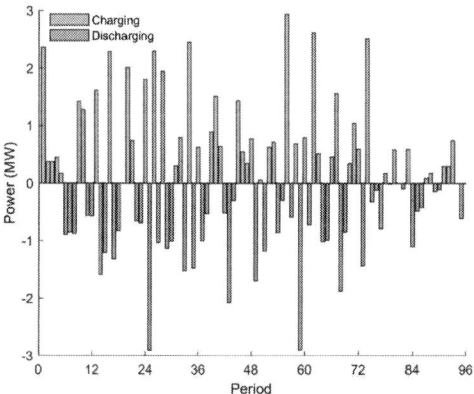

Fig. 4. BESS charging and discharging when BESS attenuation is not considered.

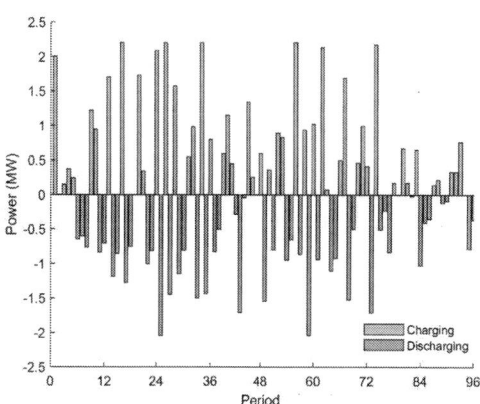

Fig. 7. BESS charging and discharging when BESS attenuation is considered.

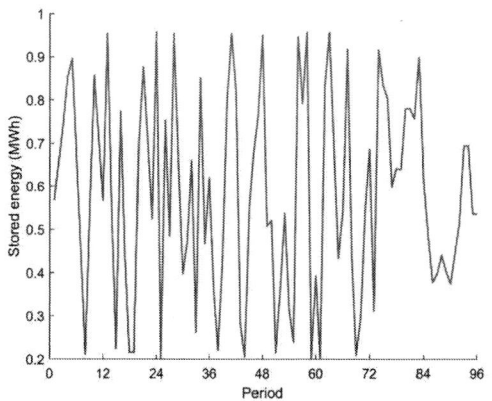

Fig. 5. Storied energy of the BESS when BESS attenuation is not considered.

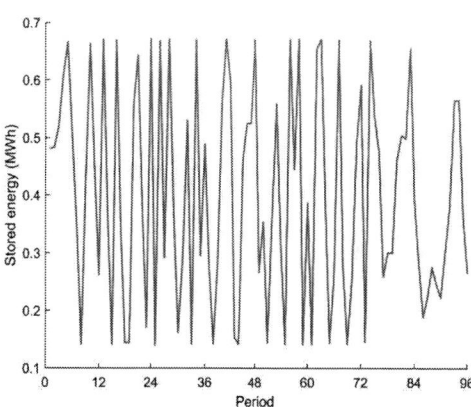

Fig. 8. Storied energy of the BESS when BESS attenuation is considered.

VI. CONCLUSIONS AND FUTURE WORK

An MILP approach for modeling and analyzing the BESS for smoothing renewable fluctuations considering BESS capacity attenuation has been proposed and validated. The proposed method can also be used to configure the optimal BESS capacity. Through conducting case studies, we can draw the following conclusions.

1) BESS attenuation is a critical factor that must be considered in BESS analysis and configuration, otherwise misjudgment will occur.
2) Under the premise of considering of the attenuation of BESS capacity, the BESS can make the fluctuations of renewable generation within reasonable ranges.
3) When the BESS capacity decays to 70%–80% of the rated capacity, the BESS should be replaced.

In the future, we will focus on the following research directions.

1) We will consider more dynamic analyses of inverters, more transient stability analyses, and the efficiencies of inverters and converters.
2) We will consider hybrid energy storage systems. We will also explore how hybrid energy storage systems can be integrated into an MILP optimization problem. This is a challenging problem, because the different forms of energy storage systems have different properties, and how to model them uniformly as an MILP problem is a difficult problem. Therefore, in order to reduce the difficulty of research, we will first study the hybrid energy storage composed of a few of these different forms of energy storage. These different forms of energy storage include battery energy storage, supercapacitor energy storage, flywheel energy storage, and compressed air energy storage.
3) We will study how to optimize power distribution in different energy storage systems.
4) In addition to electrical energy storage systems, we will also study other energy storage systems, such as hydrogen energy storage systems, thermal energy storage systems, and gas storage systems. If energy flow is considered, the integrated energy system containing these systems will be a highly complex non-convex nonlinear system, and how to deal with this system will be a problem worth studying.
5) We will also investigate the configuration of energy storage systems.

REFERENCES

[1] Y. Gu and T. C. Green, "Power system stability with a high penetration of inverter-based resources," *Proceedings of the IEEE*, vol. 111, no. 7, pp. 832–853, 2022.

[2] M. Chen, D. Zhou, and F. Blaabjerg, "High penetration of inverter-based power sources with vsg control impact on electromechanical oscillation of power system," *International Journal of Electrical Power & Energy Systems*, vol. 142, p. 108370, 2022.

[3] S. Li, J. Zhu, H. Dong, H. Zhu, and J. Fan, "A novel rolling optimization strategy considering grid-connected power fluctuations smoothing for renewable energy microgrids," *Applied Energy*, vol. 309, p. 118441, 2022.

[4] Y. Sun, Z. Zhao, M. Yang, D. Jia, W. Pei, and B. Xu, "Overview of energy storage in renewable energy power fluctuation mitigation," *CSEE Journal of Power and Energy Systems*, vol. 6, no. 1, pp. 160–173, 2019.

[5] W. Wang, B. Yuan, Q. Sun, and R. Wennersten, "Application of energy storage in integrated energy systems—a solution to fluctuation and uncertainty of renewable energy," *Journal of Energy Storage*, vol. 52, p. 104812, 2022.

[6] X. Liu, Y. Hua, X. Liu, L. Yang, and Y. Sun, "Design and implementation of smooth renewable power in cloud data centers," *IEEE Transactions on Cloud Computing*, vol. 11, no. 1, pp. 85–96, 2021.

[7] R. Hemmati, S. M. S. Ghiasi, and A. Entezariharsini, "Power fluctuation smoothing and loss reduction in grid integrated with thermal-wind-solar-storage units," *Energy*, vol. 152, pp. 759–769, 2018.

[8] A. Ghaffari, A. Askarzadeh, R. Fadaeinedjad, and G. Moschopoulos, "Optimal design of distributed network based on power quality to minimize the flicker of wind turbines," in *2024 IEEE Applied Power Electronics Conference and Exposition (APEC)*, pp. 2239–2243, IEEE, 2024.

[9] W.-L. Loh, "On latin hypercube sampling," *The annals of statistics*, vol. 24, no. 5, pp. 2058–2080, 1996.

[10] Z. Shu and P. Jirutitijaroen, "Latin hypercube sampling techniques for power systems reliability analysis with renewable energy sources," *IEEE Transactions on Power Systems*, vol. 26, no. 4, pp. 2066–2073, 2011.

[11] M. D. Shields and J. Zhang, "The generalization of latin hypercube sampling," *Reliability Engineering & System Safety*, vol. 148, pp. 96–108, 2016.

[12] H. Heitsch and W. Römisch, "Scenario reduction algorithms in stochastic programming," *Computational optimization and applications*, vol. 24, pp. 187–206, 2003.

[13] L. Kantorovitch, "On the translocation of masses," *Management science*, vol. 5, no. 1, pp. 1–4, 1958.

[14] I. Gulrajani, F. Ahmed, M. Arjovsky, V. Dumoulin, and A. C. Courville, "Improved training of wasserstein gans," *Advances in neural information processing systems*, vol. 30, 2017.

[15] Z. Wei, J. Zhao, R. Xiong, G. Dong, J. Pou, and K. J. Tseng, "Online estimation of power capacity with noise effect attenuation for lithium-ion battery," *IEEE Transactions on Industrial Electronics*, vol. 66, no. 7, pp. 5724–5735, 2018.

[16] S. K. Pradhan and B. Chakraborty, "Battery management strategies: An essential review for battery state of health monitoring techniques," *Journal of energy storage*, vol. 51, p. 104427, 2022.

[17] J.-H. Lim, G. W. Heo, J.-Y. Lim, D. H. Kim, B. Jun, and B. K. Lee, "State of charge estimation based on thermal modeling compensation considering capacity variation by internal temperature effects of lifepo 4 battery," in *2024 IEEE Applied Power Electronics Conference and Exposition (APEC)*, pp. 1800–1804, IEEE, 2024.

[18] X. Li, D. Yu, S. B. Vilsen, and D.-I. Store, "State of health estimation and remaining useful lifetime prediction of battery based on the real dynamic forklift profile," in *2024 IEEE Applied Power Electronics Conference and Exposition (APEC)*, pp. 1794–1799, IEEE, 2024.

[19] Z. Ni, H. Yu, S. Liu, J. Li, and W. Lin, "BasisFormer: Attention-based time series forecasting with learnable and interpretable basis," *Advances in Neural Information Processing Systems*, vol. 36, 2024.

Thermal and Efficiency Characterization of Immersion Cooled SiC Traction Inverter

Yiju Wang
EECS Department
University of Tennessee
Knoxville, TN, USA
yiju.wang@utk.edu

Reza Ilka
ECE Department
University of Kentucky
Lexington, KY, USA
reza.ilka@uky.edu

JiangBiao He
EECS Department
University of Tennessee
Knoxville, TN, USA
jiangbiao.he@utk.edu

Abstract—Thermal management plays a critical role in the performance of the power electronic converters. Immersion cooling, as a new promising thermal management solution for power electronics, can significantly improve the efficiency, power density, and reliability of power converters. In this paper, based on the simulation and experimental investigations, the efficiency improvement of an immersion cooled Silicon Carbide (SiC) traction inverter is characterized in comparison to the scenario based on conventional cooling solution with heatsinks and fans. It shows that the peak efficiency of the immersion cooled 5-kW 2-level SiC inverter can be improved by up to 0.16%, and the semiconductor hotspot temperature is significantly reduced by 9.6 °C, enabling much improved power cycling lifetime.

Index Terms—thermal management, SiC MOSFETs, traction inverters, immersion cooling

I. INTRODUCTION

Driven by the global trend of developing carbon-neutral economy, transportation electrification have experienced rapid progress in the recent years, especially in the area of ground vehicles and aircraft propulsion systems [1]–[3]. Pursuing high energy efficiency, high reliability, and high power density is the common goal for power converters in transportation vehicles [4]. Among the key technologies to be developed for electric vehicles, effective thermal management for power converters plays an important role in concurrently meeting these performance targets.

In most power electronic systems, heatsinks and indirect liquid cooling (e.g., with cold plate) are used to transfer heat dissipated from semiconductor devices to the surrounding air. To increase the rate of heat transfer in these systems, the heat transfer convective coefficient must be increased, which may require the installation of a fan or pump [5]–[7]. Another way is to increase the cross section surface area of heat transfer, which is achieved by attaching the heat transfer area to an external highly thermal conductive plate such as copper or aluminum heatsink or cold plate. With the increasing power rating of converters in transportation electrification applications, increasing the size of the heat sink is not a viable option as it leads to reduction in power density and rise in cost [8].

Immersion cooling is a novel thermal management method where power devices are directly submerged in a thermally-conductive dielectric-insulated liquid, as is shown in Fig. 1.

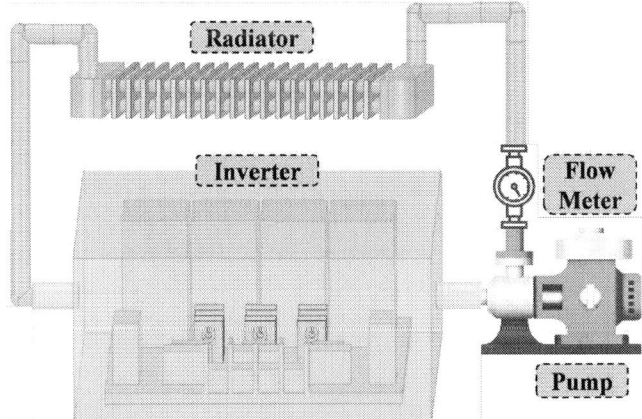

Fig. 1: Block diagram of an immersion cooled power inverter system (note: the size of the radiator, the pump, and the flow meter is magnified in this figure for illustration purpose).

This approach allows the direct transfer of heat dissipated by the devices to the cooling fluid, which circulates through the system via natural convection or pumping. Immersion cooling relies on fluids with excellent dielectric insulating properties to prevent any partial discharging risks to electronic components. The immersion cooling method presents various advantages, such as significantly reduced energy consumption, heightened power density, lower implementation costs, and uniform thermal distribution [9], [10]. Particularly in safety-critical applications such as electric transportation propulsion, immersion cooling eliminates thermal runaway risks and enhanced thermal ride-through capability. Due to its energy efficiency and space-saving features compared to conventional methods, interest in immersion cooling has rapidly increased in recent years [11].

Several studies have explored immersion cooling for power electronic devices, which demonstrate better overall performance compared to other conventional cooling methods. For example, an approach using deionized water for the immersion cooling of printed circuit boards and electronic devices is introduced in [12]. In [13], a fluid with zero global warming potential is employed, demonstrating superior cooling performance compared to forced air cooling. A dielectric

979-8-3315-1612-3/25 $31.00 © 2025 IEEE

fluid solution that increased the maximum power dissipation in high-performance electronic systems is proposed in [14]. Another study [15] presents an innovative cooling structure for a single-phase immersion cooling system using a heat sink and forced circulation. Immersion cooling for Silicon Carbide (SiC) power devices in electric vehicle applications is discussed in [16], showing lower thermal resistance than existing systems. Another study introduces a 40-kW cryogenically cooled three-level active neutral point clamped inverter with Si MOSFETs at 140 kHz switching frequency [17].

However, comprehensive research targeted on the efficiency characterization combining both simulation and experimental verification is still lacking, which is actually essential to understand and validate the thermal performance improvement of power converters. In this paper, the thermal and efficiency performance of an immersion cooled SiC power converter under various load conditions is firstly simulated, followed by experimental validations, which is achieved by monitoring the temperature variations of the SiC MOSFETs.

II. PROPOSED IMMERSION COOLED TRACTION INVERTER

A. Three-phase Two-level SiC Traction Inverter

Three-phase two-level traction inverters have been widely utilized in industrial and automotive applications due to its merits in simplicity, reliability, and cost-effectiveness. It is a dominant inverter topology used in the powertrain systems of hybrid and electric vehicles. Fig. 2 illustrates the basic circuit topology of a three-phase two-level traction inverter.

On the other hand, as is well known, SiC MOSFETs have been increasingly utilized in EVs due to its superior performance in efficiency and power density compared to traditional silicon-based devices [18]. However, due to the smaller die size of SiC MOSFETs compared to the conventional Si IGBTs, the thermal time constant with SiC MOSFETs is shorter and their junction temperature may have larger swing value during operation, exhibiting potential power cycling lifetime degradation under heavy load operating conditions. Thus, this paper will investigate the thermal and efficiency performance of a general SiC two-level traction inverter with immersion cooling.

B. Immersion Cooling System

As mentioned earlier in this paper, immersion cooling is an advanced thermal management solution designed to efficiently dissipate heat from power electronics. These systems involve submerging components, such as power inverters, into a dielectric cooling liquid, which directly absorbs the heat generated by electronics during operation. Key components of an immersion cooling system include an enclosure to house the liquid and electronics, a pump to circulate the cooling liquid, and a heat exchanger (radiator) to transfer the absorbed heat to an external cooling medium. This approach offers superior heat dissipation, improved reliability, and reduced system size and weight compared to traditional forced air cooling or indirect liquid-cooled systems, making it promising for high-power applications. Fig. 3 shows the customized lab prototype

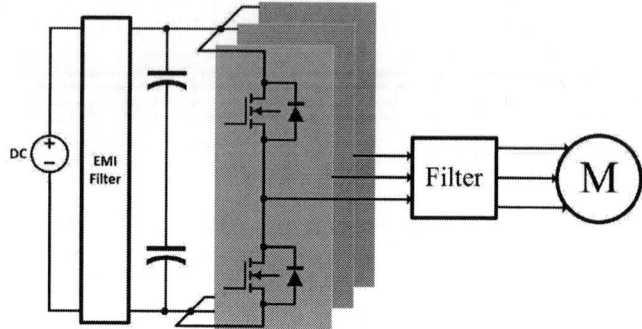

Fig. 2: Topology of a three-phase two-level traction inverter.

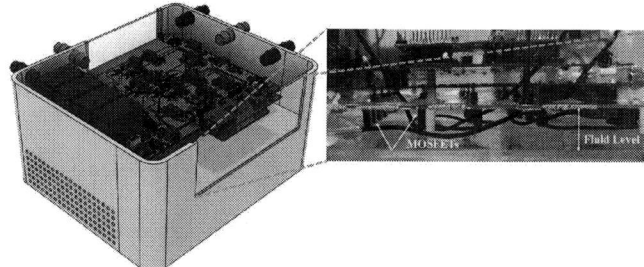

Fig. 3: The customized immersion cooled SiC inverter prototype under investigation.

of the immersion cooled SiC inverter under investigation in this paper.

C. Compare SiC Inverter with Various Cooling Solutions

To evaluate the performance of the immersion cooling against a traditional cooling design, a baseline model is established using a forced air-cooled SiC inverter with heatsinks. Before comparing the efficiency of these two cooling systems, where efficiency pertains to the total thermal losses, it is crucial to establish their respective thermal impedance network. A thermal impedance network will involve all the major components, through which the heat generated from the SiC MOSFET junction points dissipates into the ambient air. Fig. 4 illustrate the heat flow charts and the corresponding thermal models for both immersion cooling and heatsink cooling systems. The interface between the two components is represented by a thermal resistor. For all the components that are capable of both conducting and storing heat, they are treated as thermal capacitors, in which thermal capacitance is associated with the thermal capacity (specific heat) of their own materials.

III. SIMULATION RESULTS

The simulation is based on a 5-kW 2-level SiC MOSFET inverter evaluation board (part number IMZ120R045M1 by Infineon) operating under rated conditions of 480 V dc input, 50 Hz of output fundamental frequency, 25 kHz of switching frequency, a unity modulation index. The load inductance and resistance are 1.5 mH and 17 Ohm per phase, respectively. For the thermal parameters, the ambient temperature was set

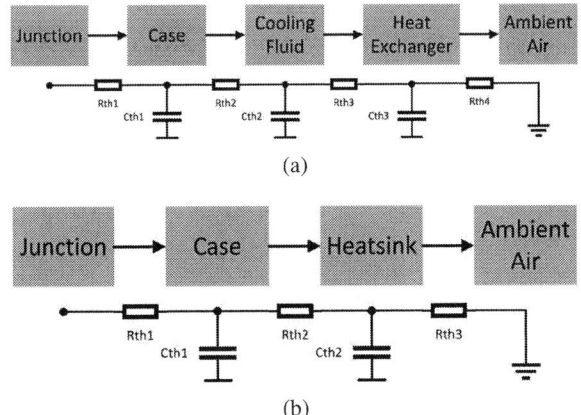

(a)

(b)

Fig. 4: Heat flow and thermal impedance network in (a) immersion cooled inverter and (b) heatsink cooled inverter.

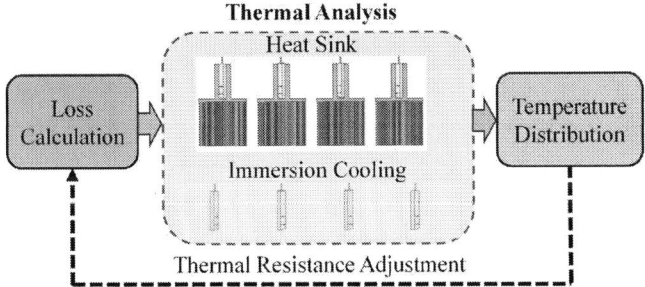

Fig. 5: Operating flowchart of the multi-physics simulation process.

at 22 °C. In [9], a multi-physics modeling approach combing both multidisciplinary modeling and electrical simulation was introduced, enabling the calculation of the values of thermal resistance and capacitance in both heatsink and immersion cooling systems. The flow chart in Fig. 5 illustrates how the multi-physics simulations converge after iterations.

A. Thermal Losses

By implementing the thermal impedance chains presented above, two groups of data of thermal losses are calculated and presented in Fig. 6. Under full load conditions (unit modulation index), the switching losses are comparable for heatsink and immersion cooling, with values of 33.8 W and 33.5 W, respectively. However, conduction losses vary more obviously between the two cooling systems, dropping from 41.8 to 37 W when immersion cooling is used.

Apart from the full load condition, thermal losses under lower output power, which is achieved by reducing the modulation index, were also examined and calculated, as presented in Table I. With increased load condition level, the thermal loss difference between these two cooling systems becomes more significant. Thus, it is necessary to focus on heavy load conditions.

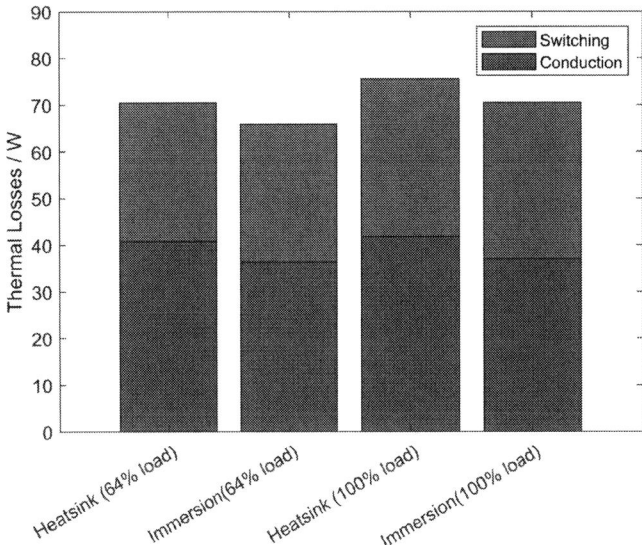

Fig. 6: SiC MOSFET thermal losses at unit modulation index with heatsink and immersion cooling.

B. Temperature Profile

Although the reduction in semiconductor thermal losses may appear modest, the corresponding decrease in semiconductor hotspot temperature is substantially more significant. The SiC MOSFET junction temperature under full load condition at steady-state is shown in Fig. 7. In the heatsink cooling scenario, the junction temperature ranges from 62.1 °C to 67.8 °C with a medium temperature of 64.3 °C. In contrast, the immersion cooling system achieves a 15.0 °C reduction in the medium temperature, bringing it down to 49.3 °C. Besides, the temperature swing in the immersion cooling system is lower, ranging from 47.2 °C to 52.7 °C. This will dramatically improve the maximum allowable output power and output frequency of the inverter, in addition to its power cycling lifetime.

In multi-physics simulations [9], the inlet and outlet velocity of the cooling fluid is also investigated to determine the optimal pump speed. Fig. 8 shows the temperature contour of the inverter board with different inlet and outlet velocity. As can be seen from the three figures of temperature distribution, with increased inlet and outlet velocity, the hotspot temperature on the whole inverter board was reduced from 72.3 °C to 59.6 °C. Thus, it is reasonable to utilize higher pump speed. However, in [9], a nonlinear relationship between the hotspot temperature and the inlet/outlet velocities was proposed, accounting for the cooling effects of both the SiC MOSFETs and the heat exchanger. This indicates that increasing the fluid velocity does not necessarily reduce the hotspot temperature. Moreover, higher fluid velocities require greater pump power, leading to increased heat dissipation from the pump itself. The optimal cooling fluid velocity for this immersion cooling system is 3 m/s, based on a trade-off study from the multi-physics analysis.

TABLE I: Thermal losses at various load conditions

Load condition (per unit)	Heatsink			Immersion		
	Conduction losses (W)	Switching losses (W)	Total losses (W)	Conduction losses (W)	Switching losses (W)	Total losses (W)
1.0	41.77	33.78	75.55	36.96	33.51	70.47
0.64	40.77	29.69	70.46	36.37	29.55	65.92
0.36	37.61	24.08	61.69	33.42	24.03	57.45
0.16	32.46	16.01	48.48	28.34	15.99	44.33

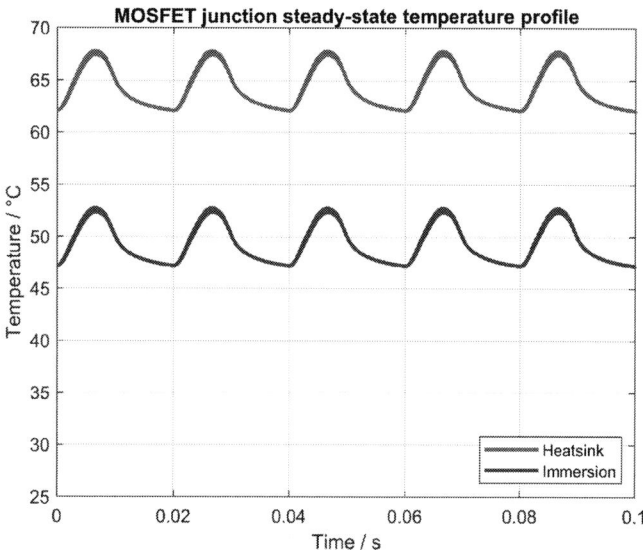

Fig. 7: Simulated SiC MOSFET junction temperature profile at steady state under full load condition.

C. Efficiency Characterization

One of the primary objectives of employing the immersion cooling is to enhance the system efficiency. The reduction in both thermal losses and operating temperatures directly contributes to this efficiency improvement. The inverter efficiency curves with both heatsink cooling and immersion cooling are presented in Fig. 9 under various load conditions. It is evident that the inverter efficiency diminishes as the loading level is decreased, whereas the efficiency deviation behaves almost inversely between the heatsink cooling and the immersion cooling system. The maximum efficiency improvement is 0.51% at the light load condition.

IV. EXPERIMENTAL VERIFICATION

A. Hardware Prototype

Experiments of the immersion-cooled inverter prototype has been carried out in lab room temperature to verify the aforementioned simulation results. Fig. 10 shows the prototyped inverter and the overall immersion cooling system. The enclosure is 3D-printed as one piece to prevent any fluid leakage during operation. The pump is driven by a stepper motor, which is controlled by high-frequency PWM pulse to achieve high-precision flow rate control. An aluminum heat

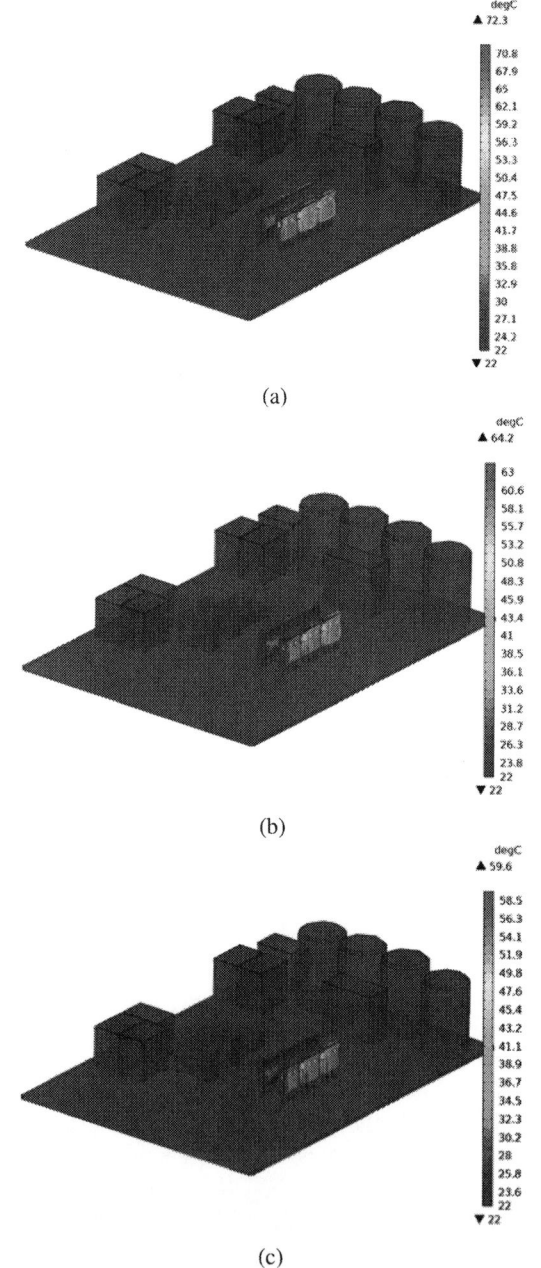

Fig. 8: Temperature contour of the immersion cooled SiC inverter with various cooling fluid velocities (a) 1 m/s, (b) 2 m/s, and (c) 3 m/s.

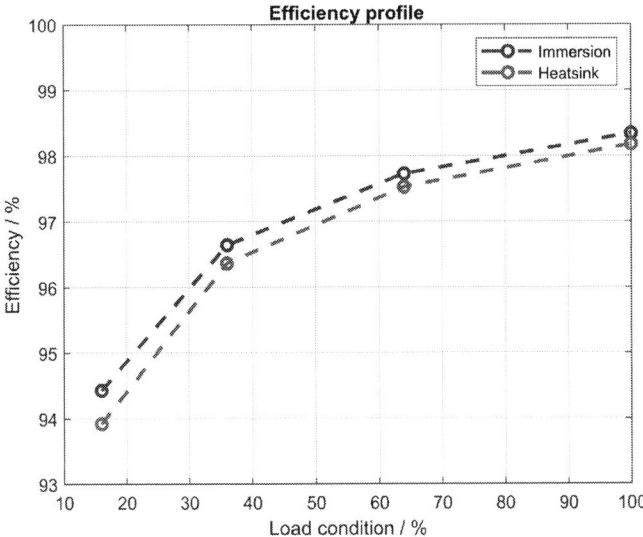

Fig. 9: Inverter efficiency under various load conditions.

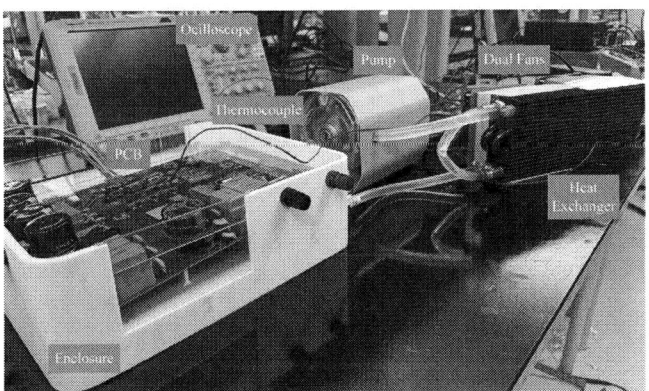

Fig. 10: Prototyped immersion cooled SiC inverter system.

exchanger is utilized to enhance the system's thermal performance and power density, leveraging aluminum's favorable properties such as low density, high ductility and malleability. Dual fans are added to further improve the performance of the heat exchanger. Thermocouples are directly attached on the thermal pad of the SiC MOSFET case. It is noted that for the heatsink testing, the utilization of the thermal interface material, which fully covers the thermal pad, leaving no space for placing a thermocouple between the heatsink and the thermal pad. Consequently, the closest measurable temperature to the semiconductor case is chosen as the hotspot case temperature of the heatsink.

B. SiC Temperature Monitoring

Two groups of experiments have been conducted in both heatsink and immersion cooling system under the load conditions of 100% and 64%. The ambient temperature was approximately maintained at 22 °C, aligning with the simulation settings. The other electrical parameters of the inverter system also remain the same as the simulation settings.

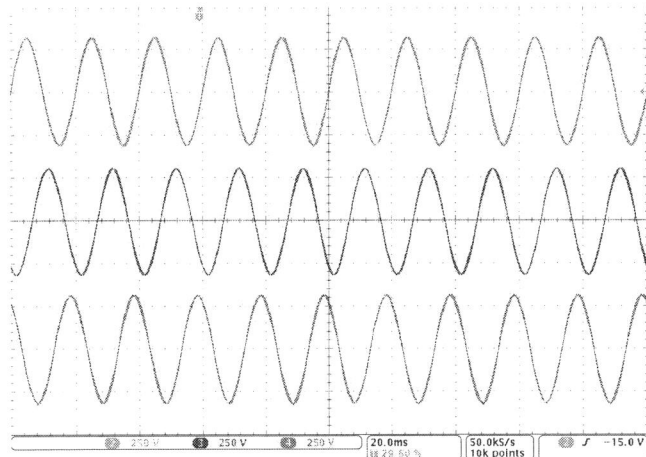

Fig. 11: Three-phase load line-to-line voltage waveform.

Before any temperature monitoring, it is essential to examine the electrical performance of the SiC inverter. The three-phase load line-to-line voltage waveform is shown in Fig. 11.

The first group of experiments was conducted in the heatsink cooling system. To better monitor the semiconductor case temperature, an infrared (IR) camera was employed. Fig. 12 demonstrate the temperature distribution of SiC MOS-FETs and heatsinks captured by the IR camera under 64% and 100% load condition after running the setup for 30 minutes to ensure a steady-state measurement. As previously mentioned, the temperatures obtained from either the IR camera or the thermocouple may not represent the actual accurate case temperature. Also, it should be noted that, the IR camera cannot capture the semiconductor temperature beneath the cooling fluid because infrared radiation emitted by an object is absorbed, reflected, or refracted by the fluid layer. Considering the direct contact between the case thermal pad and the cooling fluid without interference from any other materials, the thermocouple is able to accurately measure the case temperature. In Fig. 13, the case temperature with a time interval of 30 minutes was monitored and captured in both the heatsink and immersion cooling scenarios. At the full load condition, the case temperature has not yet reached a complete steady state, While in contrast, under 64% load condition, it is approaching.

The experimental results are quite surprising. Under 64% load condition, the heatsink cooling outperformed the immersion cooling, exhibiting roughly a 5 °C advantage, with 40 °C in heatsink cooling and 44.99 °C in immersion cooling. However, this outcome is not consistent with the simulation results. Previously, we believed that immersion cooling would outperform the heatsink cooling under all load conditions. One possible explanation is that the previous simulation did not account for the impact of the heatsink fans, which can significantly enhance the performance of the heatsink by improving heat dissipation and airflow.

Under the full load condition, although it is not the steady state yet, a significant temperature drop of 9.6 °C was ob-

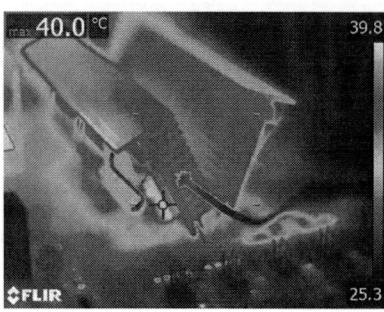

(a)

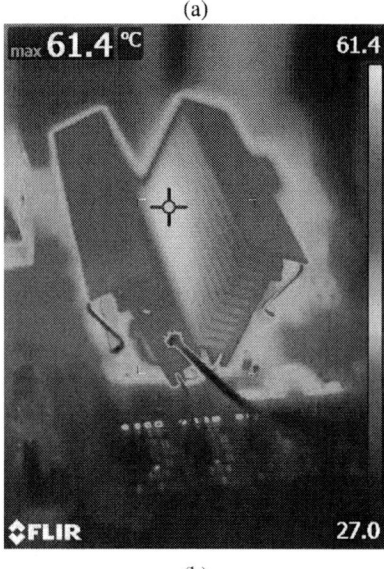

(b)

Fig. 12: Temperature distribution of the SiC inverter captured by the IR camera under (a) 64% load condition and (b) 100% load condition.

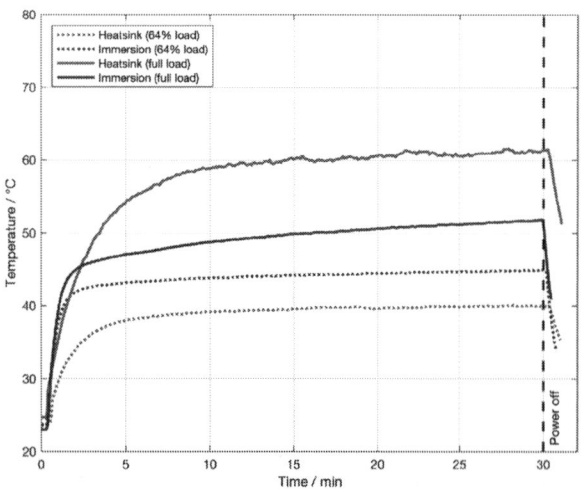

Fig. 13: Case temperature profile of the SiC MOSFETs captured by the thermocouple in testing during a time interval of 30 minutes.

served, which aligns with the simulation results. The case temperature in heatsink cooling can reach 61.7 °C, which is substantially higher than the 51.8 °C in immersion cooling.

C. Efficiency Improvement

Based on the temperature drop measured from the experiments, the value of improved efficiency can be projected. However, before calculating the efficiency, it is necessary to convert the SiC case temperature difference to the junction temperature difference.

According to the data sheet of the SiC MOSFETs used for the inverter prototype, the junction-to-case thermal resistance is 0.51 K/W, which is a shared value for both heatsink and immersion cooling scenarios. The overall equivalent thermal resistance in heatsink and immersion cooling system based on the co-simulations calculation is 4.0 K/W and 2.5 K/W, respectively. Thus, the corresponding junction temperatures of the SiC MOSFETs in each cooling scenario are calculated as follows:

Immersion cooling:

$$T_{j_1} = \frac{T_c \times 2.5}{(2.5 - 0.51)} = 65.1°C \tag{1}$$

Heatsink cooling:

$$T_{j_2} = \frac{T_c \times 4.0}{(4.0 - 0.51)} = 70.7°C \tag{2}$$

Hence, the adjusted equivalent overall junction-to-ambient thermal resistance for heatsink and immersion cooling to obtain the same temperature above in simulation is 4.7 K/W and 4.1 K/W, respectively. The corresponding inverter efficency is 98.32% and 98.16%, respectively. The eventual projected efficiency improvement is 0.16%.

V. CONCLUSION

In this paper, a novel immersion cooling system for a three-phase 2-level SiC inverter was presented to investigate the thermal and efficiency performance of the immersion cooled SiC traction inverter, with comparison to a baseline model of forced air cooled SiC inverter with heatsinks and fans. The co-simulation results show that the immersion cooling outperforms the heatsink cooling, with the highest 0.51% efficiency improvement at light load condition. The experimental results, however, demonstrate that immersion cooling Under heavy load condition exhibits a reduction of the SiC case temperature by up to 9.6 °C. The projected efficiency improvement based on the experimental results is 0.16%. To further investigate the efficiency characterization of immersion cooling, an inverter with much higher output power is under development for testing.

VI. ACKNOWLEDGMENT

This work presented in this paper is financially supported by Valvoline Global Operations.

979-8-3315-1612-3/25 $31.00 © 2025 IEEE

REFERENCES

[1] B. Sarlioglu, C. T. Morris, D. Han and S. Li, "Driving Toward Accessibility: A Review of Technological Improvements for Electric Machines, Power Electronics, and Batteries for Electric and Hybrid Vehicles," in IEEE Industry Applications Magazine, vol. 23, no. 1, pp. 14-25, Jan.-Feb. 2017.

[2] M. T. Fard, J. He, H. Huang and Y. Cao, "Aircraft Distributed Electric Propulsion Technologies—A Review," in IEEE Transactions on Transportation Electrification, vol. 8, no. 4, pp. 4067-4090, Dec. 2022.

[3] J. Benzaquen, J. He and B. Mirafzal, "Toward more electric powertrains in aircraft: Technical challenges and advancements," in CES Transactions on Electrical Machines and Systems, vol. 5, no. 3, pp. 177-193, Sept. 2021.

[4] K. P and S. Prakash, "A Review of High Efficiency Power Converters for Electric Vehicles Applications," 2024 5th International Conference on Electronics and Sustainable Communication Systems (ICESC), Coimbatore, India, 2024, pp. 26-29.

[5] S. Jones-Jackson, R. Rodriguez, Y. Yang, L. Lopera and A. Emadi, "Overview of Current Thermal Management of Automotive Power Electronics for Traction Purposes and Future Directions," in IEEE Transactions on Transportation Electrification, vol. 8, no. 2, pp. 2412-2428, June 2022.

[6] A. -C. Iradukunda, D. R. Huitink and F. Luo, "A Review of Advanced Thermal Management Solutions and the Implications for Integration in High-Voltage Packages," in IEEE Journal of Emerging and Selected Topics in Power Electronics, vol. 8, no. 1, pp. 256-271, March 2020.

[7] J. Fan, S. Shah, Z. Zhang and J. Wang, "A High Performance Liquid Metal-based Cooling System for an Ultra High Power Density Inverter," 2023 IEEE Energy Conversion Congress and Exposition (ECCE), Nashville, TN, USA, 2023, pp. 6151-6155.

[8] E. Laloya, Ó. Lucía, H. Sarnago and J. M. Burdío, "Heat Management in Power Converters: From State of the Art to Future Ultrahigh Efficiency Systems," in IEEE Transactions on Power Electronics, vol. 31, no. 11, pp. 7896-7908, Nov. 2016.

[9] R. Ilka et al., "Multi-Physics Modeling of Power Electronic Converters with Liquid Immersion Cooling," 2023 IEEE Energy Conversion Congress and Exposition (ECCE), Nashville, TN, USA, 2023, pp. 4698-4704.

[10] R. Ilka et al., "Impact of Immersion Cooling on Balancing Semiconductor Thermal Distribution in NPC Multilevel Converters for Transportation Propulsion," 2023 IEEE Energy Conversion Congress and Exposition (ECCE), Nashville, TN, USA, 2023, pp. 6547-6549.

[11] C. M. Barnes and P. E. Tuma, "Practical Considerations Relating to Immersion Cooling of Power Electronics in Traction Systems," in IEEE Trans. on Power Electronics, vol. 25, no. 9, pp. 2478-2485, Sept. 2010.

[12] P. Birbarah, T. Gebrael, T. Foulkes, A. Stillwell, A. Moore, R. Pilawa-Podgurski, and N. Miljkovic, "Water immersion cooling of high power density electronics," International Journal of Heat and Mass Transfer, vol. 147, p. 118918, 02 2020.

[13] W. Luiten, "Single Phase Passive Hydrocarbon Immersion Cooling of High-power ICs," 2021 27th International Workshop on Thermal Investigations of ICs and Systems, Berlin, Germany, 2021, pp. 1-6.

[14] J. L. Gess, S. H. Bhavnani and R. W. Johnson, "Experimental Investigation of a Direct Liquid Immersion Cooled Prototype for High Performance Electronic Systems," in IEEE Transactions on Components, Packaging and Manufacturing Technology, vol. 5, no. 10, pp. 1451-1464, Oct. 2015.

[15] C. C. Cheng, P. C. Chang, H. C. Li, and F.-I. Hsu, "Design of a single-phase immersion cooling system through experimental and numerical analysis," International Journal of Heat and Mass Transfer, vol. 160, p. 120203, 07 2020.

[16] F. Yang, C. Liu and J. Shen, "Immersion Oil Cooling Method of Discrete SiC Power Device in Electric Vehicle," 2022 IEEE Energy Conversion Congress and Exposition (ECCE), Detroit, MI, USA, 2022, pp. 1-5.

[17] H. Gui et al., "Development of High-Power High Switching Frequency Cryogenically Cooled Inverter for Aircraft Applications," in IEEE Transactions on Power Electronics, vol. 35, no. 6, pp. 5670-5682, June 2020.

[18] Y. Takatsuka, H. Hara, K. Yamada, A. Maemura and T. Kume, "A wide speed range high efficiency EV drive system using winding changeover technique and SiC devices," 2014 International Power Electronics Conference, Hiroshima, Japan, 2014, pp. 1898-1903.

FPGA-based Hybrid Simulator for Real-Time 3-D Temperature Monitoring of Power Converters

Xianghao Mo ⓘ
Centro de Electrónica Industrial
Universidad Politécnica de Madrid
Madrid, Spain
xianghao.mo@upm.es

Daniel Ríos Linares ⓘ
Centro de Electrónica Industrial
Universidad Politécnica de Madrid
Madrid, Spain
d.rios@upm.es

Regina Ramos ⓘ
Centro de Electrónica Industrial
Universidad Politécnica de Madrid
Madrid, Spain
regina.ramos@upm.es

Miroslav Vasić ⓘ
Centro de Electrónica Industrial
Universidad Politécnica de Madrid
Madrid, Spain
miroslav.vasic@upm.es

Abstract—This paper introduces an FPGA-based hybrid simulator for real-time 3-D temperature monitoring, designed as a methodology for creating digital twins of power converters. Based on the Finite-Difference Time-Domain method, derived from the physical layout, the software constructs a precise thermal digital replica with an error margin of less than 7 K. The FPGA solver accelerates the thermal simulations 50 to 200 times, providing the complete transient temperature in real-time and faster-than-real-time. A novel designed meshing technique reduces computational demands and enhances modelling flexibility. The simulator's fidelity and broad applicability are validated by monitoring the dynamic behaviour of transistors and thermal analysis of a complex three-phase LLC converter. The results illustrate the feasibility of the simulator for real time health management of power converters.

Index Terms—Finite-difference, real-time, simulations, FPGAs, temperature monitoring, power converter, thermal analysis, digital twinning

I. INTRODUCTION

Digital Twins (DTs) have recently emerged as a powerful tool for advanced monitoring in modern power electronics, showing significant potential to improve both reliability and operational stability [1]. Compact power converters, which efficiently manage power within constrained spaces, often suffer from increased component temperatures due to their densely packed design [2], [3]. Given that approximately 55% of power converter failures are directly linked to excessive thermal stress [4], the development of a power converter DT

This work was possible thanks to the ALL2GAN- SP, PCI2023-143375, funded by MICIU/AEI /10.13039/501100011033 and co-funded by the European Union. The ALL2GaN Project(Grant Agreement No 101111890) is supported by the Chips Joint Undertaking and its members including the top-up funding by Austria, Belgium, Czech Republic, Denmark, Germany, Greece, Netherlands, Norway, Slovakia, Spain, Sweden and Switzerland. View and opinions expressed are however those of author(s) only and do not necessarily reflect those of the European Union or the national granting authorities. Neither the European Union nor the national granting authorities can be held responsible for them.

seeks to facilitate coupled simulations [5], [6], enabling real-time evaluation of power converter dynamics [7], [8]. Real-time thermal analysis is crucial for enhancing power converters reliability and longevity.

To accurately replicate the thermal behaviour of a power converter and its components for Digital Twin (DT) implementation, the chosen modelling technique must be precise, efficient for its real-time implementation and flexible across varying operating conditions. Existing approaches primarily rely on the finite-element method (FEM), thermal networks, and AI-driven methods.

FEM simulations are capable of accurately capturing temperature distributions, providing a comprehensive analysis of thermal performance [9], [10]. However, due to the need for meshing, FEM simulations impose a heavy computational burden, rendering them unsuitable for real-time applications. This computational demand also hinders the ability to incorporate feedback from physical prototypes effectively.

Thermal modelling using Foster and Cauer networks has been widely used to describe power modules [11], [12], and these models can be seamlessly integrated with electrical simulations when modelling power converters [5], [6], [13]. However, the trade-off between computational burden and precision limits the scalability of these thermal models. Real-time simulations must account for prototype geometry, material properties, and dynamic boundary conditions [14], and the extraction of model parameters often requires extensive FEM simulations. These challenges constrain the broader adoption of network-based models for real-time applications.

AI-driven methods have also been explored for comprehensive transient thermal analysis [15], [16], and neural networks have been proposed for real-time 3-D thermal simulations of power converters [17]. However, these approaches are heavily dependent on extensive FEM simulations and experimental data for training purposes [18]. The demand for real-time monitoring of physical prototypes, involving detailed 3-D tem-

979-8-3315-1612-3/25 $31.00 © 2025 IEEE

perature distributions and time-dependent variations, restricts the design of these neural networks. Thus, AI-driven methods are not yet a universal solution for real-time power converter thermal simulations.

This paper presents a real-time simulation tool based on the Finite-Difference Time-Domain (FDTD) method, laying the groundwork for a power converter digital twin. Unlike existing methods, the proposed approach offers a hardware-based simulation framework with customisable conditions that can adapt to various scenarios. The thermal properties of the power converter and its components are accurately captured across both spatial and temporal domains. A key novelty of this work lies in providing a complete thermal simulation of the entire power converter, ensuring precise temperature analysis throughout all components under various operational conditions.

Compared to the FEM, the proposed method requires significantly fewer meshing points, thereby enhancing computational efficiency. In contrast to thermal network models, this approach is hardware-based, providing a flexible programming environment where simulation conditions can be tailored. Consequently, this method supports a broader range of applications, ensures computational efficiency for real-time simulations, and offers higher accuracy in predicting 3-D temperature distributions. Furthermore, compared to AI-driven methods, the proposed technique establishes the thermal model more swiftly while considering essential physical factors such as geometry, ambient conditions, and load variations. This makes it particularly advantageous for real-time temperature monitoring.

To facilitate real-time thermal simulation of power converters, this work also introduces a hybrid simulator design. Given that a real-time simulator should be embeddable within the controller or power system for Hardware-in-the-Loop (HIL) verification, or serve as a state monitor, FPGA has been identified as the ideal platform for conducting real-time simulations in power electronics due to its computational advantages and high connectivity [8], [19], [20]. Leveraging this approach, 3-D transient temperature monitoring supports real-time or faster-than-real-time thermal simulations based on the converter's physical layout, providing an advanced tool for power converter monitoring.

Section II details the thermal solver of the proposed method. Section III presents the design of the hybrid simulator. Section IV discusses the validation of the model, while Section V concludes with future work directions.

II. VOXEL-BASED THERMAL SOLVER

This section details the core of the voxel-based thermal solver, employing the FDTD method to model the 3-D temperature distribution of a power converter over time in granular cells. In this approach, the thermal model is divided into granular cells, each represented by a voxel, where each voxel functions as a distinct thermal unit within the simulation.

As illustrated in Fig. 1, the properties of each voxel are defined either at its centre or along the three directional

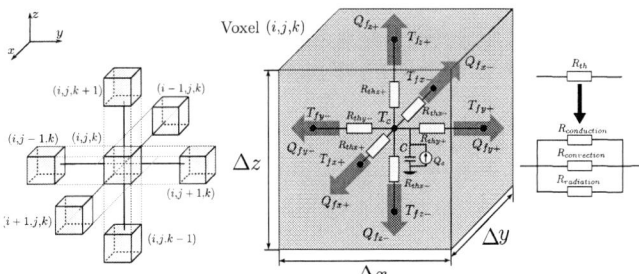

Fig. 1: Voxel thermal representation with its neighbours

axes within the dimensions Δx, Δy, and Δz. Dividing the properties in each voxel is crucial for accurately representing the thermal behaviour at a granular level. This division allows the model to capture localised variations in thermal properties such as heat capacity, conduction, and temperature gradients, which directly influence the temperature dynamics. By treating each voxel as an independent thermal unit with specific properties, the solver can more precisely track the flow of heat throughout the power converter. This level of detail is essential for real-time temperature monitoring, as it enables the simulation to quickly react to changes, predict hotspots, and accurately represent the impact of both internal heat generation and environmental conditions on each component. The centre of each voxel contains its temperature T_c, self-heat generation Q_c, and volumetric heat capacity C. Heat exchange is modelled as the interaction between a voxel and its neighbouring voxels, with heat flow across the six faces denoted as Q_f. This process involves computing the face temperatures T_{fx+}, T_{fx-}, T_{fy+}, T_{fy-}, T_{fz+}, and T_{fz-}, which are determined by the conduction resistance $R_{conduction}$, convection resistance $R_{convection}$, and radiation resistance $R_{radiation}$.

The thermal solver is based on solving the FDTD heat equation to obtain the 3-D temperature distribution over time. The heat equation for each unit is expressed in 1,

$$\rho \cdot c \cdot \Delta x \cdot \Delta y \cdot \Delta z \cdot \frac{\partial T_c}{\partial t} = \sum_i \dot{Q}_i \qquad (1)$$

where ρ is the density of the material, c is the specific heat and $\dot{Q}_i$ is the heat flow. To include the influence of external ambient conditions to volume, the heat transfer mechanism is expressed as equations (2), (3) and (4),

$$\dot{Q}_{i,\text{cond}} = \frac{k_i}{\Delta x_i/2} A_i (T_c - T_i) \qquad (2)$$

$$\dot{Q}_{i,\text{conv}} = h_i A_i (T_c - T_i) \qquad (3)$$

$$\dot{Q}_{i,\text{rad}} = \varepsilon_i \sigma A_i (T_c^4 - T_i^4) \qquad (4)$$

where the $\dot{Q}_{i,\text{cond}}$, $\dot{Q}_{i,\text{conv}}$, $\dot{Q}_{i,\text{rad}}$ model the heat transferred by conduction, convection, and radiation respectively, k_i is the thermal conductivity of the material, h_i is the convection heat transfer coefficient from Newton's linear convection law, ε_i is the emissivity of the surface, σ is the Stefan-Boltzmann constant, A_i is the surface area, and T_s is the surface temperature at the corresponding heat path direction. Finally, the forward

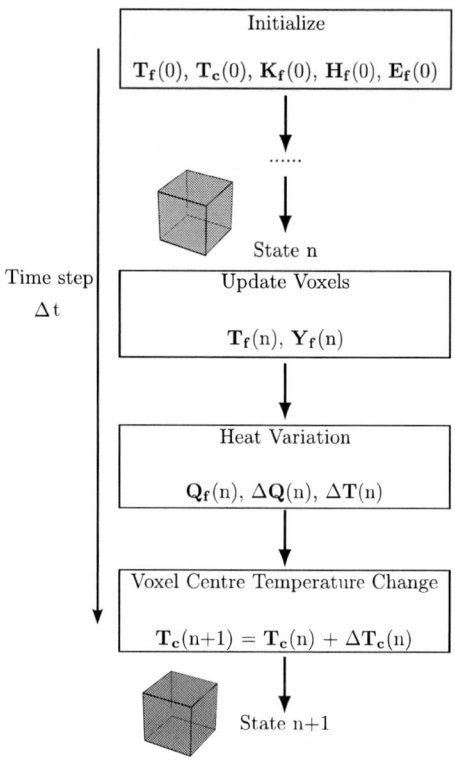

Fig. 2: Voxel thermal solver workflow

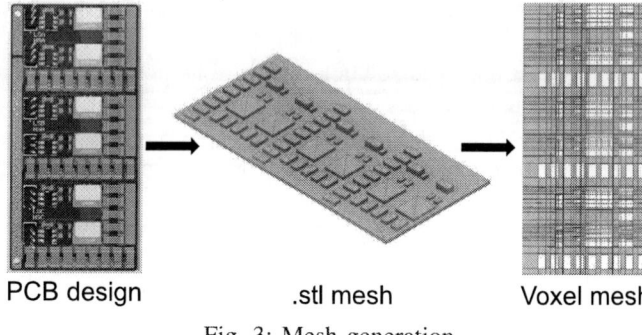

Fig. 3: Mesh generation

Euler method is applied for temporal discretisation, allowing incremental updates in voxel temperature with each time step:

$$\rho \cdot c \cdot \Delta x \cdot \Delta y \cdot \Delta z \cdot \frac{T_c^{(n+1)} - T_c^{(n)}}{\Delta t} = \sum_i \dot{Q}_i^{(n)} \quad (5)$$

$$T_c^{(n+1)} = T_c^{(n)} + \Delta T_c^{(n)} \quad (6)$$

For each instant, the total heat excess is computed as the sum of the heat flux across the six faces of the voxel along with its self-heat generation. The temperature change for the current state is determined using the time step Δt and the heat capacitance.

As shown in Fig. 2, the workflow of the thermal solver proceeds as follows:

- Face temperature calculation: at the start of each iteration, the temperature at each face of the voxel, $\mathbf{T_f}$ is calculated from the centres of two adjacent voxels, resulting in an adjustment to the voxel's thermal impedance due to temperature variations. To avoid division operations, the thermal admittance, $\mathbf{Y_f}$ is used instead.

- Heat variation: Subsequently, the heat flux, $\mathbf{Q_f}$ is determined based on the temperature difference between the voxel and its neighbouring voxels. For each voxel, the total heat excess, $\mathbf{\Delta Q}$ is computed as the sum of the heat variations across the six faces and the self-heat generation.

- Temperature update: Taking into account the time step and the heat capacitance of the voxel material, the increase in voxel temperature, $\mathbf{\Delta T_c}$, is then calculated. By

updating this change to determine the next state, a thermal iteration is completed.

III. HYBRID SIMULATOR DESIGN

In order to define the 3-D voxel array, a simulator using Python-based software was developed to handle the modelling of the power converter, including data processing and providing a graphical interface for the user. A hybrid simulator is subsequently implemented on the FPGA, seamlessly integrated for the user, enabling accelerated simulation by directly loading the compiled thermal model.

A. Software Design

This section outlines the software development process for creating a voxel-based thermal model of power converters. The software imports hardware design files, converts them into voxel meshes, and assigns physical properties, enabling accurate thermal analysis. Key aspects, including adaptive meshing, physical property assignment, and FPGA integration for real-time performance, are detailed below.

1) Voxel meshing: To ensure accurate replication of the physical prototype, the software constructs a voxel-based model of the power converter by importing the hardware design file as a polygon mesh and converting it into a voxel mesh, as illustrated in Fig. 3. A ray-intersection method is used in this process, where a given point is determined to be inside the solid by casting rays to determine the intersection with the polygon mesh. By identifying the intersections of these rays with the triangles, the point's position can be established. When the ray is cast from outside the solid, the segment starting at the first (odd) intersection point and ending at the next (even) intersection point is considered solid, while all other segments are deemed outside the bounds of the solid (i.e., air). Using this method, the voxel mesh is automatically generated by identifying the solid regions within the 3-D model.

Once the object is voxelised, each voxel serves as a discrete unit for the FDTD thermal analysis. This cellular approach allows for detailed simulation of heat transfer, accounting for variations in material properties, geometry, and boundary conditions. However, FDTD methods can result in an excessive number of voxels, increasing computational burden and thus limiting simulation speed [21]. To effectively study the

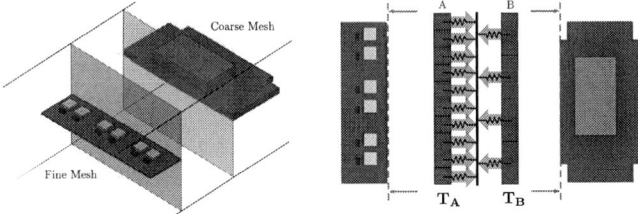

Fig. 4: Illustration of adaptive mesh

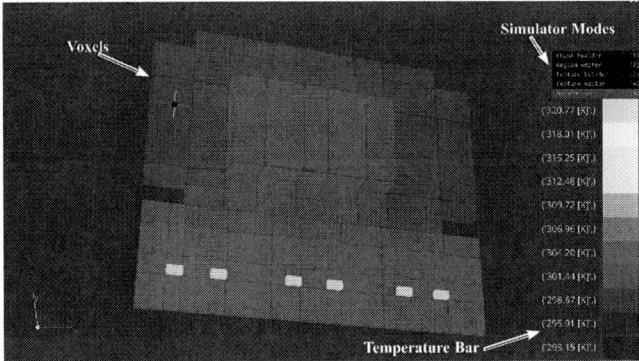

Fig. 5: Graphical user interface of the simulator

transient thermal behaviour of the power converter and its components, the simulator needs to be adaptable: thermal-critical areas of the converter, such as those where drivers and transistors are located, require fine meshing due to the small size of transistors. Conversely, less dense areas of the PCB, where components are more sparse, can be voxelised using larger elements.

We propose an adaptive mesh technique, as illustrated in Fig. 4, in which the power converter is divided into independent sections. This technique is based on dividing the power converter into regions within a rectangular grid, ensuring that the interfaces between different sections are continuous and permit heat flow. The proposed model leverages the boundary conditions of a single-region problem to enforce consistency in properties and temperature across adjacent regions. This approach has two key advantages: it facilitates flexible implementation of individual threads for the FDTD solver and allows finer discretisation of smaller components using a customised mesh.

To ensure accurate heat transfer calculations between different mesh sections, each section is simulated independently within a time step Δt. To reconcile mesh differences between sections, a set of two transformation matrices maps the temperature values, ensuring a continuous heat flow across interfaces. This matrix is calculated by the area overlap of each touching face in the interface $\mathbf{M}_i$ and $\mathbf{M}_j$ by the equation:

$$\mathbf{T}'_A = \mathbf{M}_i \cdot \mathbf{T}_B \cdot \mathbf{M}_j^T \tag{7}$$

where $\mathbf{M}_i$ and $\mathbf{M}_j^T$ are the transformation matrices which represent the projection area of the mesh to the other. As illustrated in Fig.4, heat transfer between different sections requires projecting properties from one section to another at their contact interface. For instance, in a new iteration, the left section will utilise the updated value $\mathbf{T}'_A$ calculated using equation (7).

With the proposed technique, a more flexible mesh can be generated for power converter models. The simulations can be divided into different sections, allowing the mesh for each section to be specified independently. This flexibility is particularly beneficial for modelling the complex structure of power converters, especially in the presence of heatsinks and other auxiliary components. Additionally, it offers significant computational advantages, as fine mesh grids are only allocated in thermally critical areas, reducing the overall computational burden without compromising accuracy in key regions.

2) Thermal simulation: Once the voxel mesh is constructed, initial conditions and material properties are assigned to each voxel. During this process, and to conserve computational resources, homogenising the material stack-up may be necessary to accurately represent their physical properties.

Assigning physical properties is a crucial step in the modelling process. The estimation of transient behaviours relies heavily on the heat capacitance, while the accuracy of the temperature distribution is influenced by how effectively heat conductivity is homogenised. The definition of the heat transfer coefficient is the most important aspect of assigning properties, as it determines the simulation conditions — whether the power converter operates under natural air convection, forced air convection, or without convection. This parameter can be determined empirically, as outlined in [22], either through equations or Computational Fluid Dynamics (CFD) simulations. As previously explained, the heat source is defined by self-heat generation, which is modelled as either a constant power source or a time-dependent function for dynamic scenarios.

The simulation interface, shown in Fig. 5, loads the 3-D files and generates the mesh based on the specified settings. Through this interface, users can edit the simulation region, material properties, boundary conditions, and heat sources. It also automatically calculates the time step according to the minimum voxel size and its thermal properties. The power converter is rendered with a temperature map, where each voxel is represented with its unique properties and transient thermal information.

The software is designed as a comprehensive simulation system for thermal modelling and analysis. However, the simulation speed is primarily constrained by Python's Global Interpreter Lock (GIL). To enable real-time simulation and ensure responsiveness to physical signals, an FPGA solver is proposed in this work.

B. FPGA solver design

The purpose of the FPGA solver is to accelerate simulation speed, and to be embeddable on the power converter system.

979-8-3315-1612-3/25 $31.00 © 2025 IEEE

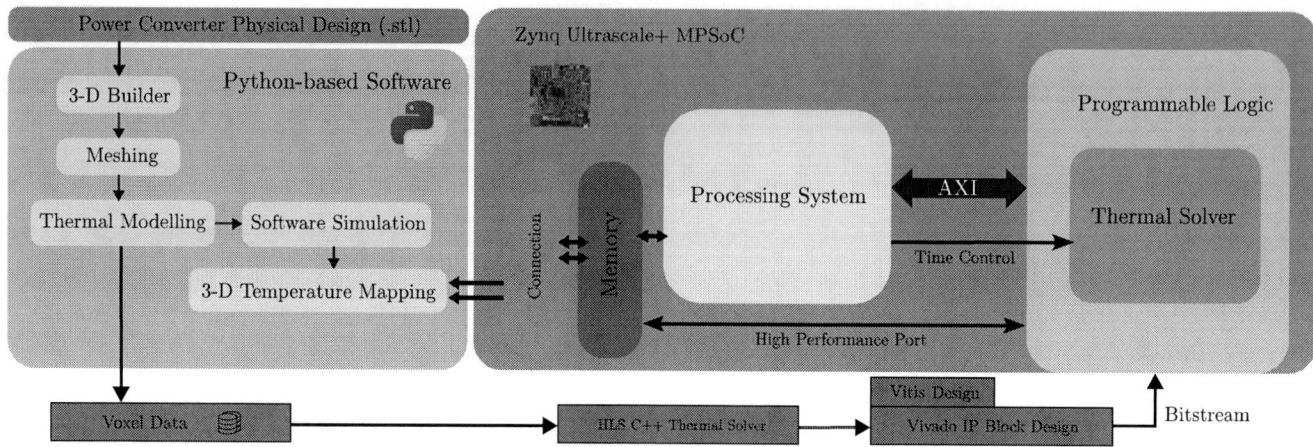

Fig. 6: Hybrid simulator design

To achieve real-time simulation, the solver must complete each iteration in less time than Δt. The thermal solver is implemented on an FPGA to achieve higher simulation speeds, and the workflow of the hybrid simulator design is shown in Fig. 6. The software, implemented in Python, provides the functionalities discussed earlier. The solver is then implemented in C++ code, which allows it to be converted to Register Transfer Level (RTL) using Vivado HLS.

The chosen platform is the Xilinx Zynq UltraScale+ MPSoC ZCU102, as it offers the ideal balance of memory capacity and computational speed for the hybrid simulator design, with the thermal solver implemented through High-Level Synthesis (HLS). This platform leverages the computational power of the Processing System (PS) alongside the massive parallelism of the Programmable Logic (PL). With the thermal solver seamlessly implemented at RTL, in the PL, the FPGA-based solver loads the mathematical arrays representing the physical properties of the power converter which is modelled by the simulation software, and performs accelerated thermal simulations.

Different from the Python solver, when the adaptive mesh is enabled, the FPGA solver can simulate various sections in parallel. The simulation speed is managed by sending time information from the PS to the PL, enabling either real-time or faster-than-real-time simulation. The management of the data and communication is designed in the PS, sending the temperature data back to the software through high-performance data communication for mapping the power converter.

C. Discussion

The simulation software proposed in this work creates a voxel-based thermal model, providing a versatile foundation for thermal analysis of power converters. The FPGA-based hybrid design offers an embedded solution for real-time monitoring. The FPGA thermal solver can also be integrated with controllers and electrical simulations, thereby constructing a powerful digital twin for the health management of power converters.

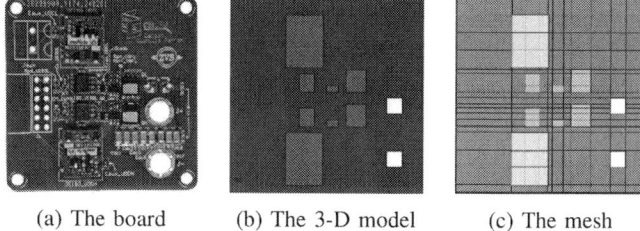

(a) The board   (b) The 3-D model   (c) The mesh

Fig. 7: The pulsating load monitoring: the board, the 3-D model and its mesh

The scalability of the voxel-based model depends on the geometrical complexity and the required level of precision. A fine mesh provides a highly accurate temperature distribution, whereas a coarser mesh enables faster thermal analysis. The limitations of real-time simulation are primarily determined by the FPGA's memory capacity and computational speed, which are influenced by the number of voxels needed. In this work, as the chosen platform is the ZCU102, the maximum of total voxels that can be processed by the PL, is 20,000.

IV. VALIDATIONS

In this work, two experimental cases are used to demonstrate the fidelity and potential of the proposed thermal simulator. The prototypes are modelled using the software to obtain 3-D transient thermal simulations, showcasing the effectiveness of the thermal modelling approach. Thermal models are also loaded onto the FPGA to perform accelerated simulations, demonstrating real-time simulation capabilities. These results are then compared with experimental data captured using the DIAS Infrared Systems PYROVIEW 640L, which provides high-resolution thermal images at a refresh rate of 1 Hz.

A. The first case: pulsating load monitoring

The first case examines the temperature tracking of transistors under dynamic load conditions (load steps), with the PCB layout illustrated in Fig.7a. The 3-D model used for simulation is depicted in Fig.7b. The model is simplified by omitting

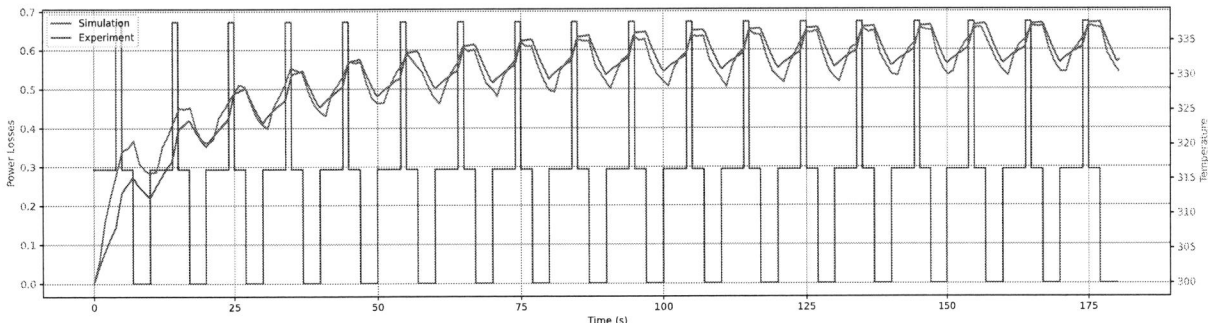

Fig. 8: The transient behaviour of the pulsating load

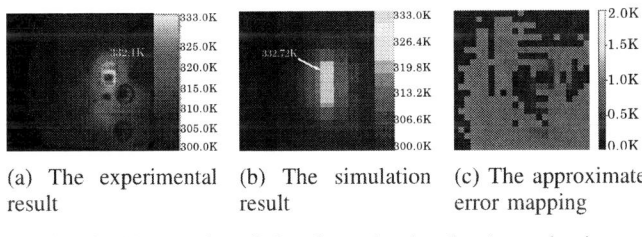

(a) The experimental result (b) The simulation result (c) The approximate error mapping

Fig. 9: The results of the the pulsating load monitoring

small components such as resistors and capacitors, as their thermal impact is negligible. Only components with significant power losses are modelled, as they serve as heat sources; these include the transistors in DSOP packages, drivers, digital isolators, and auxiliary DC/DC isolated power supplies. The final mesh, containing 1,365 voxels in total, is shown in Fig. 7c.

Power losses predominantly occur in the transistors, which are subjected to no load, nominal load, and extreme load conditions over time, as represented by the blue curve in Fig.8, repeating every 10 seconds. The thermal simulation results, experimental data, and their differences at 70 seconds are presented in Fig.9, revealing that the transistor is the hotspot with the highest temperature. In Fig.9a, the uneven temperature distribution observed in the thermal camera images is attributed to the non-uniform copper distribution, whereas in the voxel model, the board is represented with homogenised thermal conductivity. Consequently, in Fig.9b, the simulation presents a relatively symmetric distribution. Nonetheless, since the heat transfer coefficient is accurately modelled in the simulation, the maximum temperature shows only a minor error, as depicted in Fig. 9c, which approximately estimates the difference between the experimental and simulation results by mapping the average temperature of the voxels.

B. The second case: three-phase LLC converter

As shown in Fig. 10, the second case study involves temperature monitoring of a complex three-phase LLC converter. Its compact layout results in over 65% of power losses being concentrated in the transistors on both the inverter and rectifier sides, making these areas critical for monitoring due to the risk of high-temperature failures.

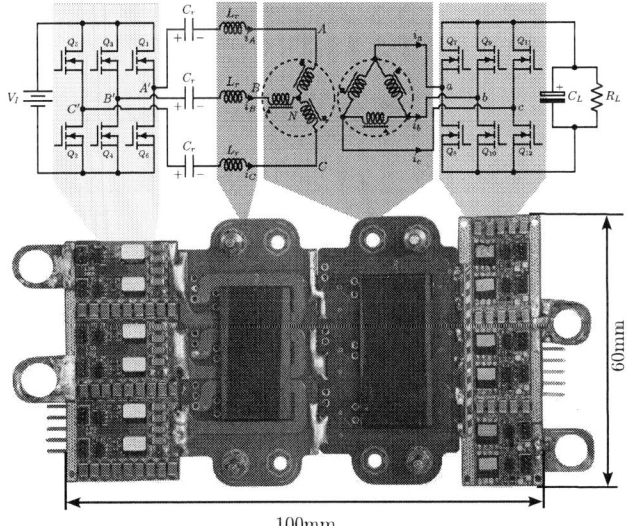

Fig. 10: The PCB board of the three phase LLC

In Fig. 11a shown the 3-D model, the small components whose thermal influence is negligible are ignored. The default mesh for the converter model consists of 14,700 voxels, owing to the complexity and misalignment of components. To reduce this, the adaptive mesh strategy proposed divides the converter into four sections: inverter, inductor, transformer, and rectifier. Fig. 11b shows the final mesh, featuring a coarser grid for magnetics and a finer mesh for transistors, resulting in four distinct regions and a total of 3,602 voxels.

Fig.12 presents the results at 720 seconds, comparing four hotspots across the four mesh sections, with an error below 2.1 K. Fig.13 illustrates the transient behaviour of these four hotspots over the 720 second period, showing a maximum error of less than 7 K. It was observed that, due to the highest power losses, the rectifier's transistors reached the highest temperatures in both simulations and experiments, identifying them as the most thermally critical components.

C. Results of hybrid simulations

The details of the hybrid simulator for the simulations of the two case studies are summarised in TABLE I. For the first

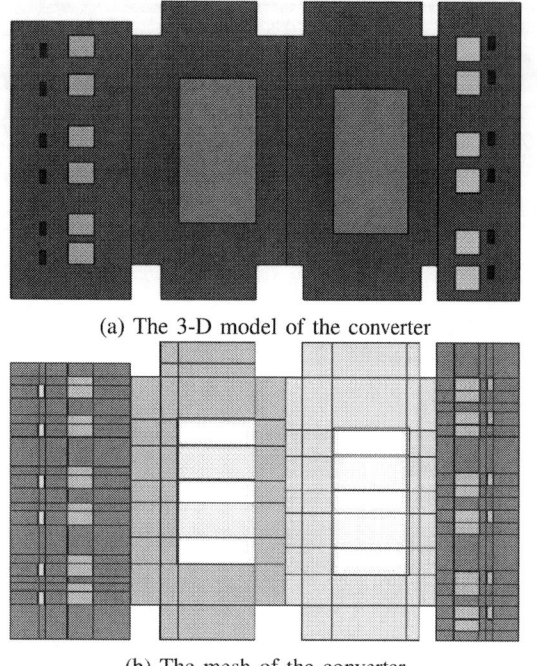

(a) The 3-D model of the converter

(b) The mesh of the converter

Fig. 11: The 3-D model and the mesh of the three-phase LLC

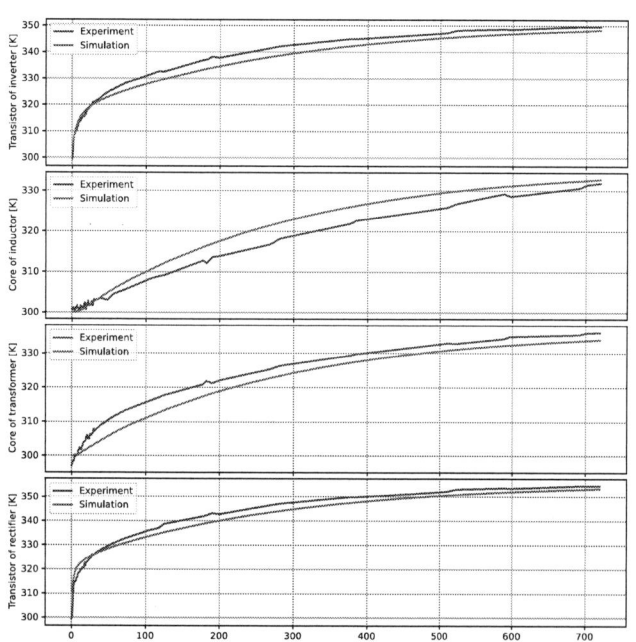

Fig. 12: The results of the three-phase LLC

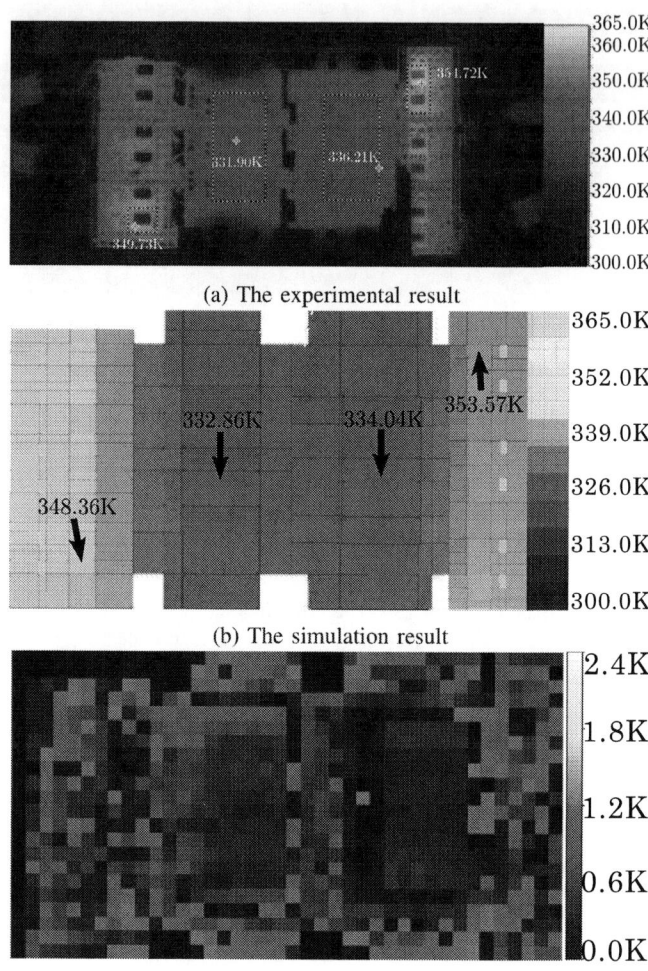

(a) The experimental result

(b) The simulation result

(c) The approximate error mapping

Fig. 13: The results of the three-phase LLC

model is divided into four chunks comprising a total of 3,602 voxels. The time step is 2.3 ms, and the software simulation speed is limited to only 7 iterations per second. The maximum FPGA simulation speed is 1,483 iterations per second, which indicates that, for this case, the thermal simulation is significantly accelerated using the FPGA, with 80% occupancy of BRAM.

Both cases demonstrate that thermal simulation can be greatly accelerated on the FPGA, enabling real-time and faster-than-real-time simulation capabilities.

V. CONCLUSION

This paper presented a hybrid simulator for thermal modelling of power converters, the FDTD method was used to predict the transient temperature distribution. A novel adaptive meshing technique ensures efficiency by applying finer meshes in critical areas, allowing for detailed thermal analysis while reducing computational costs. With the support of FPGA, simulations can be accelerated to provide real-time and faster-than-real-time results.

case, the time step is 36.75 ms, which means that 28 iterations per second are required to achieve real-time simulation. The software simulation speed is 45 iterations per second, whereas the maximum FPGA simulation speed reaches 2,374 iterations per second. This model only requires 21% of the FPGA's BRAM resources.

In the second case, involving the three-phase LLC, the

TABLE I: Summary of two models in the hybrid simulator

Study case	Pulsing power	Three-phase LLC
Mesh size (chunks of adaptive mesh)	1,365 (0 chunks)	3,602 (4 chunks)
Time step [ms]	36.75	2.3
RT requirement [iteration/s]	28	434.78
Software speed [iteration/s]	45	7
FPGA speed [iteration/s]	2,374	1,483
BRAM, DSP, FF, LUT [%]	21, 17, 3, 16	80, 85, 40, 84

The accuracy of the simulator was validated through two experimental case studies, achieving a maximum error margin of less than 7 K. Implementing these scenarios on FPGA demonstrated the feasibility of real-time simulation on embedded devices, highlighting the potential for integration into various power systems. The results further indicate the simulator's effectiveness for thermal management in modern power electronics, providing precise temperature predictions and identifying system hotspots. Its accurate, high-resolution, real-time capabilities facilitate integration with other simulators within a digital twin framework, supporting real-time monitoring and HIL testing of power converters.

Future work will focus on expanding the capabilities of the simulator, handling more specialized scenarios, such as forced cooling, hot plate conditions, and the modelling of power modules. Further improvements in resource optimisation and simulation speed will be investigated to develop digital twins of power converters that incorporate coupled simulations. The software simulator will also be extended by utilising the combined capabilities of CPUs and GPUs to accelerate simulation speed, making it capable for larger mesh counts and thus providing a more practical thermal simulation tool for real-world applications.

REFERENCES

[1] Hao Bai, Johannes Kuprat, Caio Osório, et al. "Digital Twins for Modern Power Electronics: An investigation into the enabling technologies and applications". In: *IEEE Electrification Magazine* 12.3 (2024), pp. 50–67. DOI: 10.1109/MELE.2024.3423111.

[2] Daniel Ríos Linares, Alberto Delgado Expósito, and Miroslav Vasić. "High-Gain High-Frequency Three-Phase *LLC* Resonant Converter Design Based on the Wye–Delta Transformer for Aircraft Applications". In: *IEEE Transactions on Power Electronics* 39.4 (2024), pp. 4367–4383. DOI: 10.1109/TPEL.2023.3339973.

[3] Huai Wang and Frede Blaabjerg. "Power Electronics Reliability: State of the Art and Outlook". In: *IEEE Journal of Emerging and Selected Topics in Power Electronics* 9.6 (2021), pp. 6476–6493. DOI: 10.1109/JESTPE.2020.3037161.

[4] A. Sundaram and R. Velraj. "Thermal management of electronics: A review of literature". In: *Thermal Science - THERM SCI* 12 (Jan. 2008), pp. 5–26. DOI: 10.2298/TSCI0802005A.

[5] Kerry Sado, Jarrett Peskar, Sebastian Ionita, et al. "Real-time Electro-thermal Simulations for Power Electronic Converters". In: *2024 IEEE Applied Power Electronics Conference and Exposition (APEC)*. 2024, pp. 2616–2623. DOI: 10.1109/APEC48139.2024.10509396.

[6] Yoganandam Vivekanandham Pushpalatha, Daniel Alexander Philipps, Timm Felix Baumann, et al. "Real-Time Discrete Electro-Thermal Model of Dual Active Bridge Converter for Photovoltaic Systems". In: *IEEE Journal of Emerging and Selected Topics in Power Electronics* (2024), pp. 1–1. DOI: 10.1109/JESTPE.2024.3396609.

[7] Sihui Zhang, Wensheng Song, Hu Cao, et al. "A Digital-Twin-Based Health Status Monitoring Method for Single-Phase PWM Rectifiers". In: *IEEE Transactions on Power Electronics* 38.11 (2023), pp. 14075–14087. DOI: 10.1109/TPEL.2023.3307415.

[8] Matthew Milton, Castulo De La O, Herbert L. Ginn, et al. "Controller-Embeddable Probabilistic Real-Time Digital Twins for Power Electronic Converter Diagnostics". In: *IEEE Transactions on Power Electronics* 35.9 (2020), pp. 9850–9864. DOI: 10.1109/TPEL.2020.2971775.

[9] Seyed Amir Assadi, Omri Tayyara, Joshua Palumbo, et al. "Electro-Thermal Co-design of a High-Density Power-Stage for a Reconfigurable-Battery-Assisted Electric-Vehicle Fast-Charger using Multi-Physics Co-simulation and Topology Optimization". In: *2023 IEEE Applied Power Electronics Conference and Exposition (APEC)*. 2023, pp. 1808–1815. DOI: 10.1109/APEC43580.2023.10131627.

[10] Rouhollah Shafaei, Martin Ordonez, and Mohammad Ali Saket. "Three-Dimensional Frequency-Dependent Thermal Model for Planar Transformers in LLC Resonant Converters". In: *IEEE Transactions on Power Electronics* 34.5 (2019), pp. 4641–4655. DOI: 10.1109/TPEL.2018.2859839.

[11] Mahera Musallam and C. Mark Johnson. "Real-Time Compact Thermal Models for Health Management of Power Electronics". In: *IEEE Transactions on Power Electronics* 25.6 (2010), pp. 1416–1425. DOI: 10.1109/TPEL.2010.2040634.

[12] Christoph H. van der Broeck, Robert D. Lorenz, and Rik W. De Doncker. "Monitoring 3-D Temperature Distributions and Device Losses in Power Electronic Modules". In: *IEEE Transactions on Power Electronics* 34.8 (2019), pp. 7983–7995. DOI: 10.1109/TPEL.2018.2882402.

[13] Luis Herrera, Cong Li, Xiu Yao, et al. "FPGA based real time electro-thermal modeling of power electronic converters". In: *2013 Twenty-Eighth Annual IEEE Applied Power Electronics Conference and Exposition (APEC)*. 2013, pp. 1725–1729. DOI: 10.1109/APEC.2013.6520529.

[14] Cameron Entzminger, Wei Qiao, Liyan Qu, et al. "Automated Extraction of Low-Order Thermal Model With Controllable Error Bounds for SiC MOSFET Power Modules". In: *IEEE Transactions on Power Electronics* 39.1 (2024), pp. 538–551. DOI: 10.1109/TPEL.2023.3318580.

[15] Zheng-Wei Du, Yu Zhang, Yuankui Wang, et al. "A Time Series Characterization of IGBT Junction Temperature Method Based on LSTM Network". In: *IEEE Transactions on Power Electronics* (2024), pp. 1–14. DOI: 10.1109/TPEL.2024.3459470.

[16] Md Moniruzzaman, Ahmed H. Okilly, Seungdeog Choi, et al. "A Comprehensive Study of Machine Learning Algorithms for GPU based Real-time Monitoring and Lifetime Prediction of IGBTs". In: *2024 IEEE Applied Power Electronics Conference and Exposition (APEC)*. 2024, pp. 2678–2684. DOI: 10.1109/APEC48139.2024.10509167.

[17] Hèlios Sanchis-Alepuz and Monika Stipsitz. "Towards Real Time Thermal Simulations for Design Optimization using Graph Neural Networks". In: *2022 IEEE Design Methodologies Conference (DMC)*. 2022, pp. 1–6. DOI: 10.1109/DMC55175.2022.9906469.

[18] Daniel Santamargarita, Guillermo Salinas, David Molinero, et al. "Tradeoff Between Accuracy and Computational Time for Magnetics Thermal Model Based on Artificial Neural Networks". In: *IEEE Journal of Emerging and Selected Topics in Power Electronics* 11.6 (2023), pp. 5658–5674. DOI: 10.1109/JESTPE.2022.3203934.

[19] Gard Lyng Rødal and Dimosthenis Peftitsis. "Real-Time FPGA Simulation of High-Voltage Silicon Carbide MOSFETs". In: *IEEE Transactions on Power Electronics* 38.3 (2023), pp. 3213–3234. DOI: 10.1109/TPEL.2022.3223951.

[20] Yangbin Zeng, Jialin Zheng, Zhengming Zhao, et al. "Real-Time Digital Mapped Method for Sensorless Multitimescale Operation Condition Monitoring of Power Electronics Systems". In: *IEEE Transactions on Industrial Electronics* 71.4 (2024), pp. 3628–3638. DOI: 10.1109/TIE.2023.3273259.

[21] Kang Xiong, Weixin Chen, and Jian Wang. "An Efficient FDTD Conformal Grid Mesh Method Based on Ray Tracing Algorithm". In: *2023 International Applied Computational Electromagnetics Society Symposium (ACES-China)*. 2023, pp. 1–3. DOI: 10.23919/ACES-China60289.2023.10250037.

[22] Yunus A. Cengel. *Heat Transfer: A Practical Approach*. 3rd ed. New York: McGraw-Hill, 2002.

A New Subassembly Concept for Enhanced Heat Dissipation and Reliability of Power Module

Yosuke Nakata
Power Device Works
Mitsubishi Electric Corporation
Fukuoka, Japan
Nakata.Yosuke@ds.MitsubishiEl
ectric.co.jp

Yuji Sato
Advanced Technology
R&D Center
Mitsubishi Electric Corporation
Hyogo, Japan
Sato.Yuji@dw.Mitsubishielectric
.co.jp

Shin Uegaki
Advanced Technology
R&D Center
Mitsubishi Electric Corporation
Hyogo, Japan
Uegaki.Shin@dn.Mitsubishielect
ric.co.jp

Jun Fujita
Advanced Technology
R&D Center
Mitsubishi Electric Corporation
Hyogo, Japan
Fujita.Jun@dx.Mitsubishielectric.
co.jp

Akihiko Furukawa
Power Device Works
Mitsubishi Electric Corporation
Fukuoka, Japan
Furukawa.Akihiko@df.Mitsubish
ielectric.co.jp

Masayoshi Tarutani
Power Device Works
Mitsubishi Electric Corporation
Fukuoka, Japan
Tarutani.Masayoshi@da.Mitsubis
hielectric.co.jp

Abstract— We propose a new subassembly concept that structurally enhances heat dissipation and reliability by monolithically integrating multiple silicon carbide (SiC) chips on a single plate and encapsulating them with molding resin. This subassembly is designed to be a universal solution for SiC modules that can be adapted to various module configurations. In this study, we specifically verify this concept by assuming a de facto standard module of a 6-in-1 full-bridge circuit with a pin-fin structure, which is widely used for xEVs, and partially reproducing its module shape. We validate the heat dissipation performance through detailed thermal simulations and the reliability through power cycle tests of an actual sample module. The research results demonstrate the potential of this subassembly to meet the high thermal and reliability requirements of the xEV market.

Keywords— power cycle, thermal performance, SiC module, subassembly

I. INTRODUCTION

In recent years, there has been an increased demand for power semiconductor modules, particularly for high heat dissipation and highly reliable SiC modules aimed at xEVs. Concurrently, the need for parts standardization and BCP (Business Continuity Planning), as well as the standardization of product exterior shapes, is progressing. Most SiC power modules available in the market today are packaged using standard methods used for Si-based devices, requiring innovation to fully utilize the potential advantages of SiC devices. Specifically, in large electric vehicles (EVs) equipped with high-capacity batteries, the replacement of Si with SiC is advancing. This transition enables higher efficiency and increased power output. However, merely adapting conventional packaging technologies has proven insufficient to fully exploit the performance benefits of SiC due to issues related to heat dissipation and mechanical stress. These challenges necessitate the development of innovative packaging solutions tailored to the unique properties of SiC devices and industry trends addressing these challenges are reported [1]. This paper presents an innovative subassembly concept designed to address these challenges. Through the thermal simulations and power cycle tests using actual devices, we demonstrate the effectiveness of our approach in significantly improving heat dissipation and reliability in SiC power modules.

II. SUBASSEMBLY CONCEPT

For the enhancement of heat dissipation in semiconductor modules, reducing thermal resistance is essential. To effectively reduce thermal resistance, increasing thermal diffusion is crucial. In the past, transfer molded type modules, where chips were mounted on a thick Cu heat-spreader, have been introduced to the market [2]. Especially for recent BEVs, SiC is being increasingly adopted in large-capacity modules, with multiple SiC chips being used in parallel in many cases, resulting in higher power density than ever before. Silicon nitride substrates with Cu circuit patterns have become common as high-heat dissipation insulating substrates for modules. To promote heat diffusion while avoiding thermal interference between parallel chips, it is necessary to increase the thickness of the Cu pattern. However, there are technical issues in increasing the thickness of the Cu pattern, and challenges to thickening the Cu layer have been reported [3]. There is also a growing demand for further improvements in reliability. For example, a technology has been reported that improves reliability by connecting a copper foil to the chip and then running copper wires on top of it, instead of the conventional aluminum wire bonding [4]. Further reliability improvements have been reported by molding the module to relieve thermomechanical stress on the fragile aluminum layer [5]. However, when the module is molded, mechanical issues

979-8-3315-1612-3/25 $31.00 © 2025 IEEE

often become constraints when constructing large modules. As a result, when attempting to realize high-capacity SiC modules, it often becomes necessary to use multiple modules, which may deviate from the standard form and mounting method. We propose the subassembly concept, which aims to balance improvements in reliability by molding only smaller unit subassemblies and constructing large modules of standard form using the necessary number of subassemblies. Inside the subassembly, multiple SiC chips are Ag-bonded to a Cu heat-spreader, their source electrodes are connected across each other with a Cu material, signal wirings are routed, and the entire surface is molded and sealed. As depicted in Figure 1, the development of the subassembly and its module concept achieves high heat dissipation and high reliability by mounting the subassembly on the chip mounting area of a conventional module, thereby maximizing the potential benefits of SiC devices. The following sections report on the effectiveness verification through simulation and the results of power cycle tests on a prototype.

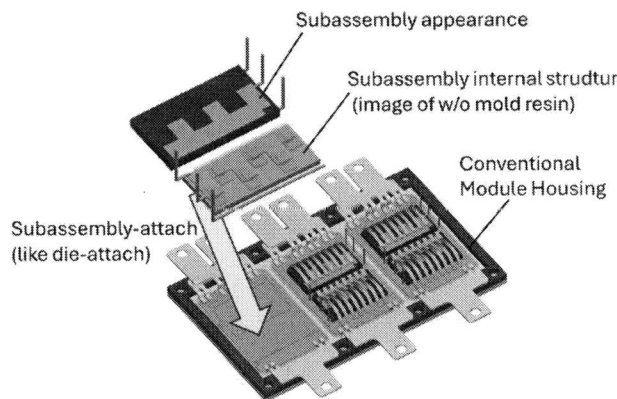

Fig. 1. Subassembly and its module concept

III. THERMAL PERFORMANCE SIMULATION

The thermal performance of a module simply equipped with a chip using methods commonly used in conventional semiconductor modules, and a module using the proposed subassemblies (hereafter referred to as the subassembly module), was verified using simulation.

A. Simulation Models

Figures 2. and 3. show the simulation models and cross-sectional diagrams for the conventional module and the subassembly module. Table 1 shows the properties of each part of the subassembly module. The subassembly module consists of a Cu base plate, lower substrate solder, active metal brazed SiN substrate, upper substrate solder, heat-spreader, lower Ag bond, SiC chip, upper Ag bond, and block from bottom to top, with the heat-spreader to block being sealed by the molding resin. The conventional module consists of a Cu base plate, lower substrate solder, active metal brazed SiN substrate, lower Ag bond, and SiC chip from bottom to top. Both modules have a case attached to the base and are sealed internally with gel. In this report, a model using multiple chips with an active area of 21.27 mm² each per arm was created, and heat flux was applied to the active area of each chip to generate heat, as indicated by the purple area in Figure 2(c). A heat transfer

coefficient of 48000 W/K (assuming water cooling by pinfins) and an ambient temperature of 65°C were applied to the back of the base plate, as indicated by the yellow area in Figure 2(d).

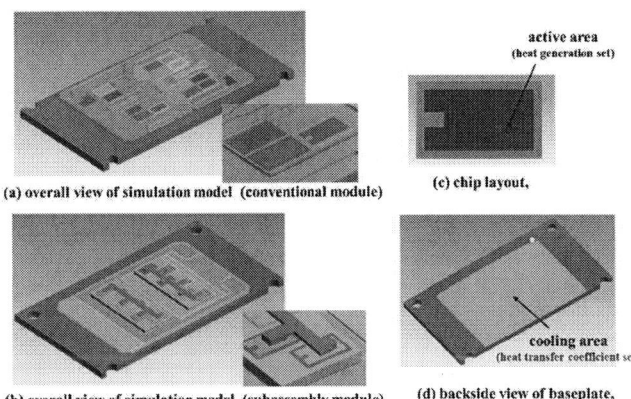

(a) overall view of simulation model (conventional module)

(b) overall view of simulation model (subassembly module)

(c) chip layout,

(d) backside view of baseplate,

Fig. 2. Simulation models

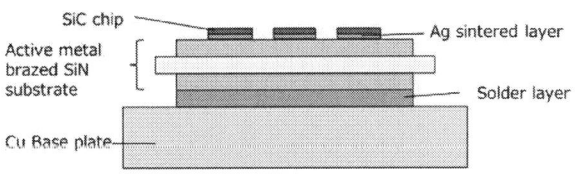

(a) cross-sectional diagram (conventional module)

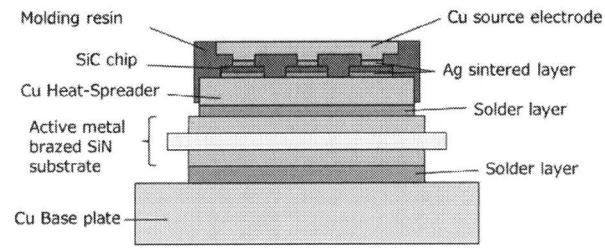

(b) cross-sectional diagram (subassembly module)

Fig. 3. Cross-sectional diagrams

TABLE I. PROPERTIES OF EACH PART

part	material	Thermal Conductivity [W/mK]
Cu source electrode	Cu	394
Ag layer (on chip)	Ag	200
SiC Chip	SiC	200
Ag layer (under chip)	Ag	200
Cu Heat-Spreader	Cu	394
Solder layer	Solder	45.3
AMB substrate	Cu/SiN/Cu	394/90/394
Solder layer	Solder	45.3
Cu base plate	Cu	394

First, the optimization of the heat-spreader thickness in the subassembly module was verified for configurations with six chips and with half the number of chips, i.e., three chips spaced apart.

B. Optimization of the Cu Heat-Spreader Thickness

The results of varying the heat-spreader thickness from 0.5 mm to 3.5 mm for each configuration are shown in Figure 4. As shown in Figure 4, it was found that the thermal resistance was minimized when the heat-spreader thickness was 1.5 mm for the configuration with six chips, and when the heat-spreader thickness was 2.5 mm for the configuration with three chips spaced apart. This suggests that as the heat-spreader thickness increases, the heat diffusion effect of the heat-spreader improves, but the thermal interference between chips also increases, resulting in an optimal thickness that varies depending on the layout. For higher output, a heat-spreader thickness of 1.5 mm, which was optimal for the configuration with six chips, was selected, and the design of the subassembly module was advanced.

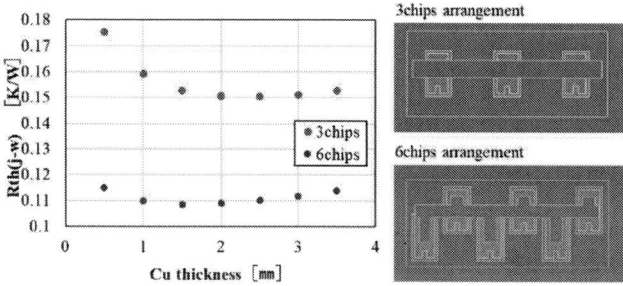

Fig. 4. Thermal resistance vs Cu thickness for different layouts

C. Comparison to the Conventional Modules

Next, the heat generation amount at which T_{jmax} becomes equal for the subassembly module and the conventional module was determined, and the superiority of output was compared when the chip active area, cooling performance, and T_{jmax} are equal. Table 2 shows the simulation result. As shown in table 2, at T_{jmax} reaches 175°C, the heat generation amount is 1011W for subassembly module, compared to 938W for the conventional module, making subassembly 7.7% larger. From this result, it was demonstrated that the subassembly module can handle a larger output compared to the conventional module when T_{jmax}, chip active area, and cooling performance are the same.

TABLE II. SIMULATION RESULT LIST

	Conventional module	Subassembly module
Themal Resstance [K/W]	0.118	0.109
Total Heat Generation [W]	938	1011 (7.7% larger)
T_{jmax} [°C]	175	175

IV. POWER CYCLE TEST ON ACTUAL SAMPLES

A. Power Cycle Test Emviroment

As shown in Figure 5, a subassembly module sample was manufactured. First, the bottom electrode of the subassembly was soldered to an insulating substrate, which was then soldered to a base plate. Next, after attaching the case, the top electrode of the subassembly was connected to the circuit pattern on the top of the insulating substrate using Cu ribbon bonds. Finally,

the interior of the case was sealed with gel to form the subassembly module. The back of the subassembly module was not made with pin fins, but with a flat base plate, and was connected to a water-cooled chiller through grease (not shown in the figure 5). Two types of power cycle tests were conducted. T_j was calculated using the value read from the temperature sense diode placed on the chip.

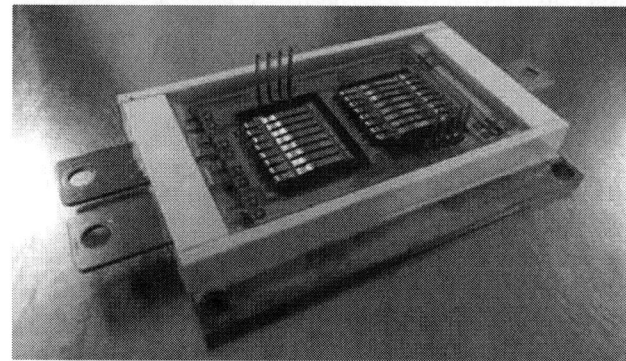

Fig. 5. Subassembly module sample

B. Power Cycle Test 1

T_j was calculated using the value read from the temperature sense diode placed on the chip. The test conditions were set at T_j=50~150°C, and the power cycle test was conducted without changing the power conditions or test environment after the start of the test (in compliance with AQG324). As shown in Figure 6, T_{jmax} increased as the test progressed, and the test was terminated when T_j reached 200°C at 30k cycles. After 30k cycles, the chip characteristics were checked, and it was confirmed that no damage or fluctuations had occurred.

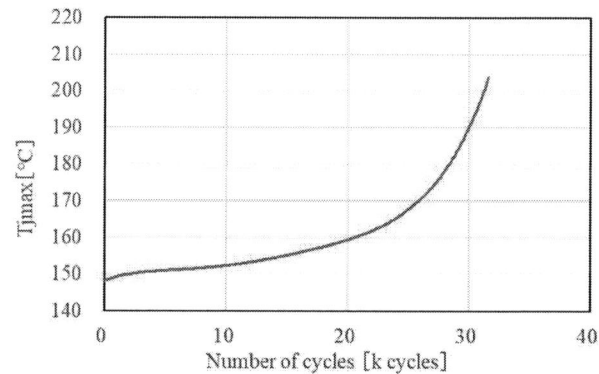

Fig. 6. Transition of T_{jmax} during power cycle test1

C. Power Cycle Test 2

This test was conducted to investigate the degradation of the subassembly itself under the assumption that no thermal resistance degradation occurs, thereby applying more long-term stress. The test conditions were adjusted to maintain a constant T_j within the range of T_j=50~150°C. Specifically, the test current was reduced each time T_{jmax} increased, in order to maintain T_{jmax}=150°C. Figure 7 shows the transition of the test current Id and T_{jmax} up to 150k cycles. After 150k cycles, the chip characteristics were checked, and it was confirmed that no damage or fluctuations had occurred.

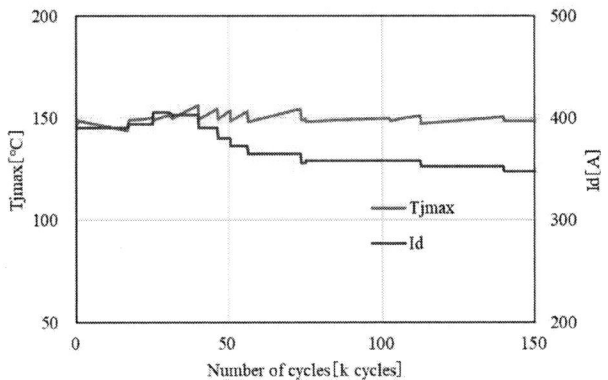

Fig. 7. Transitions of T_{jmax} and Id during power cycle test2

D. Physical Analysis of Test Samples and Discussion

In both tests, degradation was observed in the lower solder of the subassembly under Scanning Acoustic Microscope (SAM) observation as shown in Figure 8, but no degradation occurred in the lower solder of the insulating substrate.

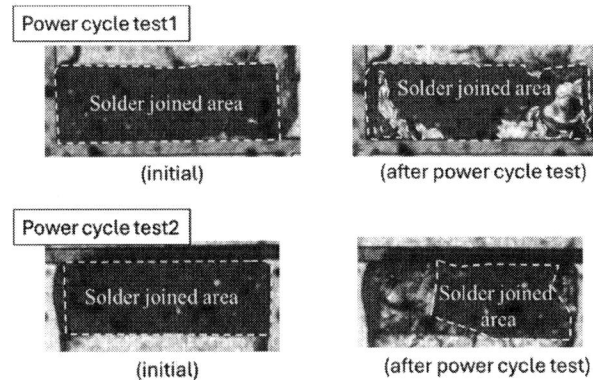

Fig. 8. Scanning Acoustic Microscope (SAM) observation images

The cross-sectional observation results, as shown in Figure 9, confirmed a lower solder crack in the subassembly, but almost no degradation was observed inside the subassembly, not even in the weak aluminum layer, nor elsewhere.

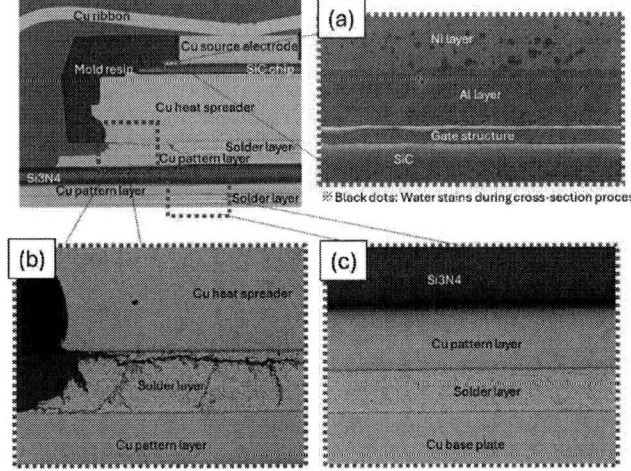

Fig. 9. The cross-sectional observation results

From these results, it is inferred that the cause of the test temperature rising in both tests was due to the degradation of the lower solder of the subassembly. It is thought that if the lower solder of the subassembly can be changed to a highly reliable bonding material to prevent degradation, it is expected that the high reliability of the subassembly module can be achieved.

V. THERMAL SIMULATION (ENHANCEMENT BY ADOPTING AG)

The power cycle tests indicated that replacing the solder on the lower connection of the subassembly with a high-reliability sintered Ag could potentially mitigate the observed degradation. To verify the thermal benefits of this replacement, we conducted an additional thermal simulation.

A. Simulation Setup

The lower connection of subassembly was modeled with Ag, and the thermal properties were adjusted accordingly. The simulation conditions were set to match those of the initial thermal simulation for consistency.

B. Results

Table 3 shows the simulation result. As shown in table 3, the thermal resistance of the subassembly with Ag was significantly reduced compared to the initial configuration. Specifically, the thermal resistance improvement was observed to increase from 7% with the conventional solder to 15% with the Ag material. This reduction in thermal resistance allows for an increase in output power while maintaining the same maximum junction temperature (T_{jmax}).

TABLE III. SIMULATION RESULT LIST

	Conventional module	*Subassembly module*
Themal Resstance [K/W]	0.118	0.103
Total Heat Generation [W]	938	1078 (15 % higher)
T_{jmax} [°C]	175	175

C. Discussion

The use of high-reliability sintered Ag as the lower connection material not only suppresses the degradation under the subassembly but also provides thermal benefits, suggesting the potential for higher output power. However, increasing the output power leads to a rise in the temperature of the solder under the insulating substrate, which could potentially limit the product's lifespan due to solder degradation. One of the failure modes during power cycle testing has been suggested to be due to degradation of the solder under the insulating substrate [6]. Table 4 shows the comparison of the heat generate and the solder temperature under the insulating board when T_{jmax} is set to 150, 175, and 200°C for the conventional module and the subassembly module. As shown in Table 4, in the conventional module, when T_{jmax} is increased to 200°C, the temperature of the solder under the insulating substrate exceeds 160°C, increasing the risk of solder deterioration. On the other hand, in the subassembly module, due to the thermal diffusion effect of the heat-spreader, even if T_{jmax} is increased to 200°C, the temperature of the solder under the insulating substrate is kept at the equivalent of $T_{jmax} = 175°C$ for the conventional module.

979-8-3315-1612-3/25 $31.00 © 2025 IEEE   1386

In addition, as shown in Figure 10, the effect of gradual temperature distribution can be obtained. This temperature reduction and uniform temperature distribution may mitigate the deterioration of the solder under the insulating substrate, making it possible to increase T_j or extend the product life.

TABLE IV.   SIMULATION RESULT LIST

	Conventional module			Subassembly module	
Total Heat Generation [W]	724	938	1151	1078	1323
T_{jmax} [°C]	150	175	200	175	200
T_{solder} [°C]	125.6	143.5	161.3	129.4	144.1

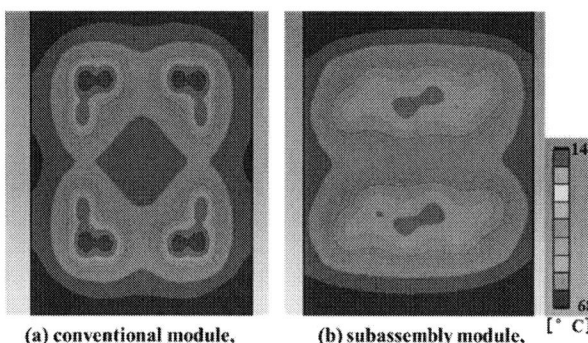

(a) conventional module,   (b) subassembly module,   [°C]

Fig. 10. Temperature of solder under insulating substrate(T_{jmax}=175°C)

VI. CONCLUSION

In this paper, the concept of subassembly and its modules was proposed, and the thermal advantages and potential for extending power cycle life were investigated.

(1) Through simulation verification, the thickness of the heat-spreader in the subassembly module was optimized, demonstrating that the thermal advantage of the subassembly module is 7.7% compared to conventional modules.

(2) A prototype of the subassembly module was manufactured, and two power cycle tests were conducted. Through the analysis of the degradation of the subassembly module, it was confirmed that the internal degradation of the subassembly was minimal, even in the weakest Al layer, although it was found that improvements were needed in the lower connection of the subassembly.

(3) Additional simulations were conducted to show that changing the lower connection of the subassembly to Ag increased the thermal advantage of the subassembly module to 15% compared to conventional modules. Furthermore, it was indicated that this subassembly module could improve the power cycle life by suppressing the temperature rise in the solder layer under the insulating substrate, thereby reducing degradation in this layer.

In conclusion, the proposed subassembly module represents a significant advancement in the packaging of SiC power modules, meeting their high thermal and reliability requirements.

REFERENCES

[1] H. Lee, V. Smet, and R. Tummala, "A Review of SiC Power Module Packaging Technologies: Challenges, Advances, and Emerging Issues," in IEEE Journal of Emerging and Selected Topics in Power Electronics, vol. 8, no. 1, pp. 239-255, Mar. 2020.

[2] T. Ueda, N. Yoshimatsu, N. Kimoto, D. Nakajima, M. Kikuchi, and T. Shinohara, "Simple, Compact, Robust and High-performance Power module T-PM (Transfer-molded Power Module)," in IEEE International Symposium on Power Semiconductor Devices & ICs, 2010, pp. 47-50.

[3] F. Ishikawa, and S. Oi, "Investigation to Improve Reliability of Substrates Having Low Thermal Resistance Using Thicker Cu Circuit Layer," in PCIM Europe, 2023, pp. 879-883.

[4] A. Syed-Khaja, "Material Solutions for High-reliability and High-temperature Power Electronics," in 2022 IEEE CPMT Symposium Japan, pp. 158-159.

[5] J. Rudzki, H. Strobel-Maier, M. Becker, P. Heimler, D. Xie, M. Alaluss, T. Basler, A. Mathew, and S. Rzepka, "Influence of Transfer Molding on the Reliability of DSM SiC Power Modules," in PCIM Europe, 2024, pp. 1937-1945.

[6] M. Junghanel, R. Schmidt, J. Strovel, and U. Scheuemann, "Investigation on Isolated Failure Mechanisms in Active Power Cycle Testing," in PCIM Europe, 2015, pp. 251-258.

Stand-alone R_{DS-ON} sensor for In-Situ Prognostic, Protection and Reliability Enhancement of Power Converters

Zaheen Mustakin
Dept. of Electrical and Computer Engineering
University of North Carolina at Charlotte
Charlotte, USA
zmustaki@charlotte.edu

Qiang Mu
Dept. of Electrical and Computer Engineering
University of North Carolina at Charlotte
Charlotte, USA
qmu1@charlotte.edu

Lucas Pereira
Dept. of Electrical and Computer Engineering
University of North Carolina at Charlotte
Charlotte, USA
lpereira@charlotte.edu

Jiale Zhou
Dept. of Electrical and Computer Engineering
University of North Carolina at Charlotte
Charlotte, USA
jzhou20@charlotte.edu

Tiefu Zhao
Dept. of Electrical and Computer Engineering
University of North Carolina at Charlotte
Charlotte, USA
Tiefu.Zhao@charlotte.edu

Babak Parkhideh
Dept. of Electrical and Computer Engineering
University of North Carolina at Charlotte
Charlotte, USA
bparkhideh@charlotte.edu

Abstract—The junction temperature is critical for evaluating the reliability of power semiconductors. However, conventional temperature-sensing methods are often inadequate for effective in situ prognostics. Research has demonstrated that the on-state resistance of a power semiconductor can be measured in situ and has a direct correlation with junction temperature. By estimating this on-state resistance, we can derive real- time junction temperature data, which can then be used to optimize converter control and enhance reliability. This study introduces a standalone on-state resistance sensor capable of measuring on-state voltage and device current. The sensor digitizes and transmits values for voltage, current, and on-state resistance. This integrated package facilitates in situ prognostics, provides robust device protection, and enables improved power converter control, ultimately leading to enhanced performance and reliability. The study focuses on evaluating the feasibility and performance of this standalone on-state resistance sensor and exploring its extended applications such as converter control and device characterization.

Index Terms—On-State Voltage V_{DS-ON}, On-State Resistance R_{DS-ON}, In-situ prognostic, Reliability, Protection

I. INTRODUCTION

The growing adoption of electric vehicles (EV) and renewable energy sources have significantly increased the reliance on power electronics in modern power systems [1], [2]. Power electronic converters play a crucial role in managing energy flow, enhancing efficiency, and integrating distributed energy sources. However, this reliance also raises the risk of

The study presented in this article builds upon research conducted by the Coastal Studies Institute (CSI) to enhance the reliability and advance the development of online health monitoring systems for a marine energy microgrid.

failures as power semiconductors are responsible for 57.1% of converter failures [3]. To address this issue, monitoring the health of power semiconductors, particularly through parameters like on-state resistance (R_{DS-ON}), has become an area of focus. R_{DS-ON} reflects the condition of a semiconductor, and its variation can indicate degradation, enabling predictive maintenance to prevent sudden failures. Studies on health and reliability monitoring of power-electronic systems have gained traction in recent years, with research emphasizing techniques to predict and mitigate failures [4], [5], [6], [7], [8]. These advancements aim to enhance the resilience of power electronics devices, ensuring reliable operation in critical applications like renewable energy integration and EV infrastructure.

Power semiconductor devices degrade over time due to electromechanical stresses [9], [10], [11], [12], leading to an increase in on-state resistance R_{DS-ON}. This degradation stems from mechanisms like gate oxide breakdown and bond wire fatigue [13], making R_{DS-ON} a key health indicator for these devices. Real-time monitoring of R_{DS-ON} enables in-situ prognostics [14], [15], [16], [17], providing valuable insights into device degradation and remaining lifespan, thereby preventing unexpected failures [18], [19], [20], [21]. Initially, health monitoring of power semiconductors focused on individual devices, namely the bottom side device, where the measurements could be referenced to system ground. However, recent research has expanded to monitor both top and bottom devices in larger systems through various indices [22], [23], such as three-phase configurations. This approach improves reliability by providing a more comprehensive view of system health. Advancements in sensor topologies and data classification techniques have further developed R_{DS-ON} monitoring. Precise sensors and advanced data analysis methods, including the application of machine learning, enable accurate and real-time health assessments. These innovations contribute to more resilient power-electronic systems, ensuring reliability in critical applications such as renewable energy, electric vehicles, and industrial automation.

No such single sensor has been presented in the studies outlined above to measure R_{DS-ON}. The potential for in-situ prognostics of R_{DS-ON} and derivation of junction temperature highlights the importance of a sensor that can perform such measurements. A standalone R_{DS-ON} sensor promises to further mature the study of power electronic reliability. This article aims to introduce a novel method for packaging the required sensors in a way that allows for the development of a standalone R_{DS-ON} sensor. This article presents a novel approach to developing a standalone R_{DS-ON} sensor by integrating the necessary measurement components into a single cohesive unit. The new sensor design aims to address key challenges in power electronics monitoring such as design motivation and challenges, evaluation methodology and results, and wider applications.

II. MOTIVATION AND BACKGROUND

On-state resistance R_{DS-ON} of a MOSFET is a combination of a couple of factors. These can be modeled into contributing equivalent resistance and the relationship can be equated as shown in Equation 1 [27],

$$R_{DS-ON} = R_{ch} + R_{JFET} + R_d + R_{package} \quad (1)$$

Among these factors, channel resistance R_{ch} is dependent on threshold voltage V_{th} as shown in Equation 2

$$R_{\mathrm{ch}} = \frac{L_{\mathrm{ch}}}{W_{\mathrm{ch}} \mu_n C_{\mathrm{ox}} (V_{GS} - V_{\mathrm{th}})}. \quad (2)$$

While, V_{th} has a relation with Oxide trap charge Q_{it} as shown in Equation 3

$$V_{\mathrm{th}} = V_{\mathrm{th0}} - \frac{Q_{\mathrm{ot}}}{C_{\mathrm{ox}}} + \frac{Q_{\mathrm{it}}}{C_{\mathrm{ox}}} \quad (3)$$

An increase in junction temperature T_j causes the Q_{it} to increase. This causes a positive shift in V_{th}. A positive shift in V_{th} will increase R_{ch} based on Equation 2. Increase in R_{ch} will increase R_{DS-ON}. All these factors combined infers that R_{DS-ON} and T_j have a predictable relationship. Here T_j is the parameter that will give us an idea of device aging;however once packaged this parameter cannot be measured. Since the relationship between T_j and R_{DS-ON} could be mapped, measuring R_{DS-ON} gives us the closest estimation of T_j.

Due to its nature, R_{DS-ON}, similar to any resistance, cannot be measured as an individual parameter as we measure voltage or current. Measuring resistance and/or impedance involves applying voltage, measuring the corresponding current through the device, and mathematically deriving it. The

study aims to understand if the associated components or measurement units can be brought together to form a stand-alone sensor.

Studies outlined in [24], [25], [26] highlight the challenges and methodologies involved in accurately measuring or deriving R_{DS-ON} for practical systems intended for grid integration. Such systems typically comprise of half-bridge sub-systems which are common building blocks of most power electronic converters. The required sensors or measurement units follow suit and are built along with the half-bridge sub-systems. Dedicated voltage sensors measure drain to source voltage (V_{DS-ON}) and current sensors can measure or estimate device current (I_{DS}). Although [25] introduces a novel method for measuring V_{DS-ON}, the basic approach remains consistent: V_{DS-ON} is measured at device terminals and I_{DS} is measured at the load. The sensors then feed data into a controller that uses the following calculation.

$$R_{DS-ON} = V_{DS-ON}/I_{DS} \quad (4)$$

If a sensor is to be designed for the sole purpose of measuring R_{DS-ON}, both the voltage sensor and the current sensor need to be in one package. Typically, to derive R_{DS-ON}, sensors are placed as shown in Fig. 1. This was the traditional way to go where measuring R_{DS-ON} was only an afterthought. Deploying the sensors in this manner does not allow us to integrate the required sensors in one package. To solve this challenge we propose that the current senor be placed directly on the output trace of the half-bridge as shown in Fig. 2. This brings the current sensor next to the voltage sensor allowing us to think about and innovate a design on a single package that houses all of the sensors and their auxiliary components. This junction point of the system is commonly called "switch-node" in the literature. The proposed configuration allows for accurate device current to be measured rather than relying on the estimation of device current. Placing the sensor in the switch node also allows us to study our current sensors in a different environment. Additionally, the current sensor in this configuration is capable of sensing shoot-through currents, which can now be built into the control system, enhancing system reliability as an added feature. We also propose that the sensor system have a dedicated controller. This allows for all the analog sensor data to be locally digitized and processed. However, there are some challenges associated with this approach. One example involves switched currents. The solution here is to have traces and provisions on the power stage to make room for the sensor placements and also to classify and shift through the data.

In contrast, the study in [27] suggests that R_{DS-ON} can be quantified without a dedicated current sensor. However, this method is not ideal for in situ prognostics. Our proposed sensor and method provide ultrafast in-situ prognostics and device protection, along with digitally quantified R_{DS-ON} data, improving reliability, performance, and the scope of further development.

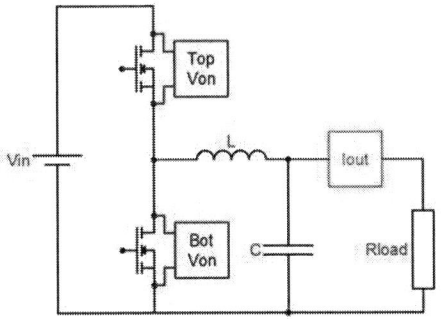

Fig. 1: Common R_{DS-ON} estimation architecture.

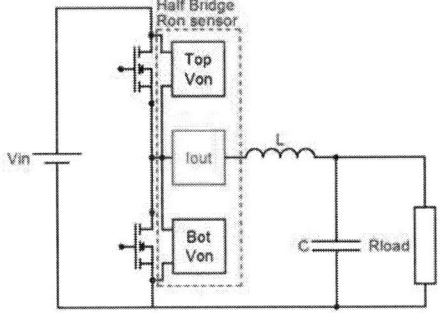

Fig. 2: Proposed R_{DS-ON} estimation architecture.

III. HARDWARE DEVELOPMENT

The sensor is designed as a plug-and-play solution for seamless integration with the converter. An evaluation platform is designed to go along with the sensor with special connectors built on it to accept the card-style sensor. The complete sensor unit architecture can be seen in Fig. 3. The real hardware can also be seen in Fig. 4.

The sensor packages include:

- 2x V_{ON} sensors.
- 2x Analog Isolation amplifiers.
- 2x Isolated power supply
- 1x On-trace contactless current sensor.
- 1x RP2040 microcontroller(connected externally)

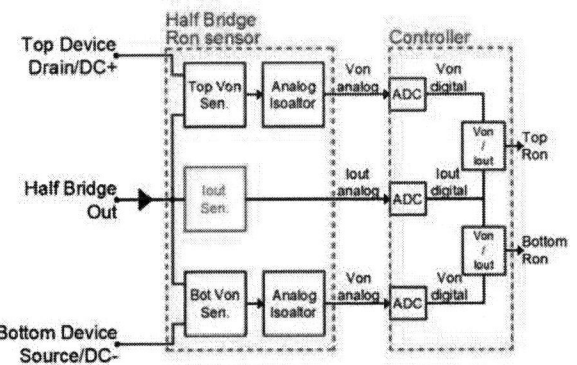

Fig. 3: Proposed R_{DS-ON} package with integrated controller.

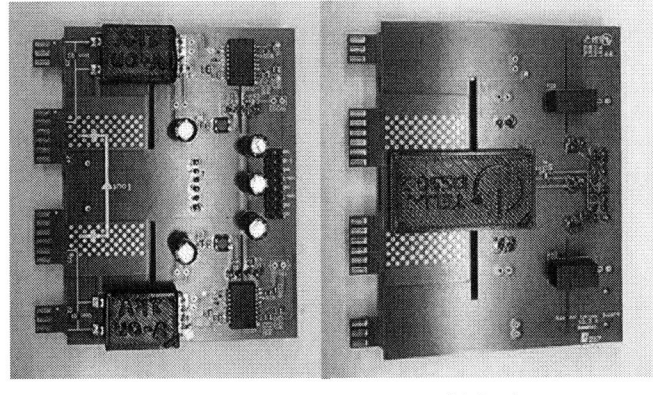

(a) Front   (b) Back

Fig. 4: Developed R_{DS-ON} sensor

- Misc. auxiliary circuitry

The sensors and crucial parts of the system are described in the following subsection.

A. V_{ON} sensor

The clamped voltage sensor, initially developed and applied in [26], has been refined and packaged into a small unit designed to be placed across the terminals of a device. The clamped voltage topology and the sensor package can be seen in Fig. 5.

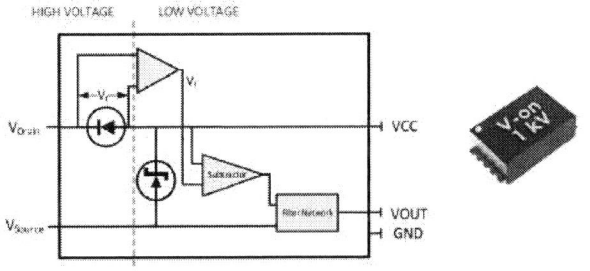

Fig. 5: V_{ON} sensor topology and package

Additional sensor specifications can be found in Table I

Attribute	Value
Blocking voltage(Max)	1000V
Settling time @1000V	50-200ns
Supply voltage	5V
MOSFET off-state voltage	2.8V
Operating temperature(Max)	80 C

TABLE I: V_{ON} sensor specifications

B. I_{DS} sensor

The current sensor is a contactless hybrid unit that combines a magneto resistor (MR) and a Rogowski coil. This combination allows for an ultra-fast, contactless response, enabling the sensor to measure both absolute current and di/dt. This capability allows for the implementation of overcurrent and di/dt protections, such as shoot-through protection.

Additional sensor specifications can be found in Table II

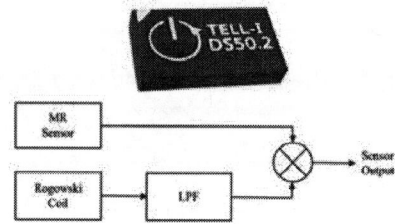

Fig. 6: I_{DS} package and architecture.

Attribute	Value
Supply voltage	5V
Bandwidth	7MHz
Mounting	On trace/ Contactless
Sensitivity	42mV/A
Operating temperature(Max)	125 C

TABLE II: I_{DS} sensor specifications

C. Analog Isolator ADUM4190

In a half-bridge system, directly connecting the controller reference to the V_{ON} sensor references (top and bottom devices) is impossible. This is due to the reference for the top side device being the signal for the bottom side device. The sensor signals need to be isolated for simultaneous measurements. To resolve this controller issue and to ensure that the high voltage of the converter does not interfere with the controller operation, an analog isolation amplifier for each sensor is used. This system requires both sensors to have their own isolated power supply so the controller can then connect to the isolated side of the isolation amplifiers. Specification of the isolator can be seen in Table III

Attribute	Value
Supply voltage	3-20V,3-20V
Bandwidth	400kHz
Gain	1.0 V/V
Isolation Voltage	5kV
Operating temperature(Max)	125 C

TABLE III: Isolator specifications

D. External Controller RP2040

Integration of a dedicated small and inexpensive yet powerful controller allows the sensor package to work as a standalone unit. The controller in question is an RP2040. Specifications of the controller can be seen in Table IV

Attribute	Value
Clock	133MHz
SRAM	246kB
ADC Channel	4
PWM Channel	16
UART	2
Operating temperature(Max)	-40 to 85 C

TABLE IV: Controller specifications

E. Half-Bridge Evaluation Platform

A half-bridge evaluation platform was designed with specific connectors for the sensor to be plugged in. The platform can be seen in Fig. 7

(a) Un-assembled　　　　　(b) Assembled

Fig. 7: Complete sensor with evaluation platform

The platform specification is tabulated in Table V.

Attribute	Value
Max DC Voltage	1200V
Max Current	50A
Max switching freq	2MHz
MOSFETs	Wolfspeed C2M0080120D
Gate Drivers	SKYWORKS SI8273GB-IS1

TABLE V: Evaluation Platform

IV. HARDWARE EVALUATION

The sensor package is comprised of multiple sensors and has multiple stages of signal translation. As a result, the package needs to be tested and evaluated by one sensor at different operating points at a time. Each sensor should be tested for,

1. Accuracy
2. Bandwidth
3. Noise immunity
4. Data loss(After Isolator)

Selected test results are presented in the sections below.

A. V_{ON} sensor test

For these test results presented below, the evaluation platform was configured as a buck converter and run at 30kHz. The sensor test was probed with a KEYSIGHT MSO-X 3034T oscilloscope. Fig. 8 and Fig. ?? show the accuracy and signal integrity of the top side and bottom side V_{ON} sensors (respectively). As it can be seen from the figures the raw V_{ON} data has a lot of dv/dt or di/dt spikes at switching states. However, the signal after the isolation amplifier is cleaner and easier to process without causing any noticeable data loss.

B. I_{DS} sensor test

For testing the current sensor the platform was again configured as a buck converter and the test was run at multiple input voltage and switching frequencies to evaluate the performance of the sensor at the switching node. Fig. 10 presents the sensor data switching at 100kHz and 22.9A ripple currents. Fig. 11 presents the sensor data switching at 500kHz and 5.2A ripple

979-8-3315-1612-3/25 $31.00 © 2025 IEEE

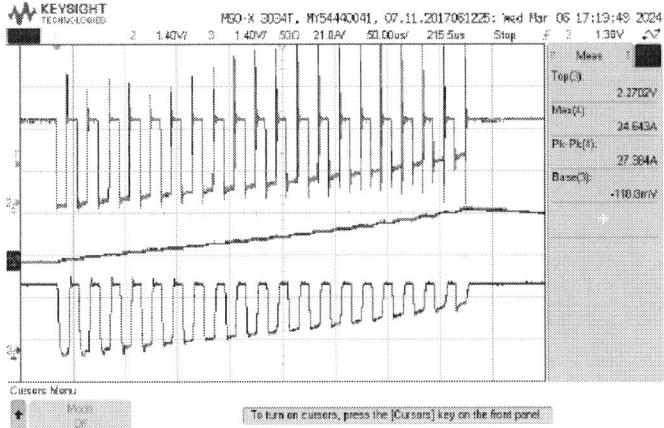

Fig. 8: Top side V_{ON} Before isolator- Green Trace, After Isolator - Blue Trace, Reference Current - Magenta Trace

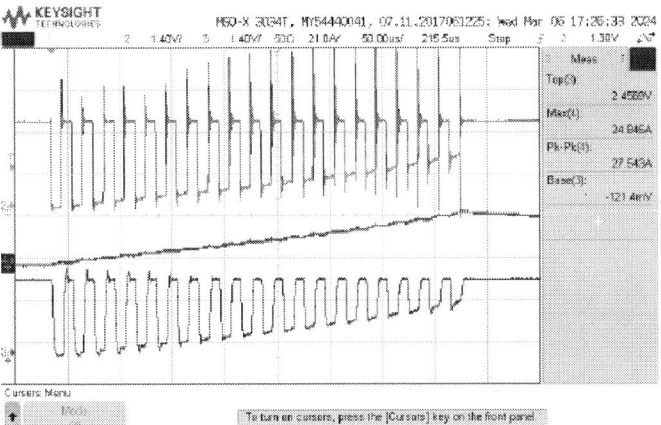

Fig. 9: Bottom side V_{ON} Before isolator - Green Trace, After Isolator - Blue Trace, Reference Current - Magenta Trace

current. Fig. 12 presents the sensor data switching at 1MhZ and 3.0A ripple current.

In all cases, there are di/dt impulses on the sensor output for switching transitions. However, it can be managed through simple analog or digital filtering. However, the output at 1Mhz seems to have lost a sufficient amount of its integrity compared to lower frequencies. This is to be expected at switched nodes. As a reference, Fig.13 presents the same 1Mhz switching condition, but this time, the sensor is now placed at the load side. This case exhibits a lot more manageable di/dt impulses. In any case, the case of noisy data can be managed numerically as long as the form of the signal is present. Meaning the sensor outputs are promising, and this plug-and-play approach can be feasible.

C. Complete R_{DS-ON} test

To evaluate the calculated value of R_{DS-ON} from the sensor, we needed to gather a reference value of the Device Under Test. In this case, we turned one of the devices on indefinitely with a stable signal while power was being supplied on the DC

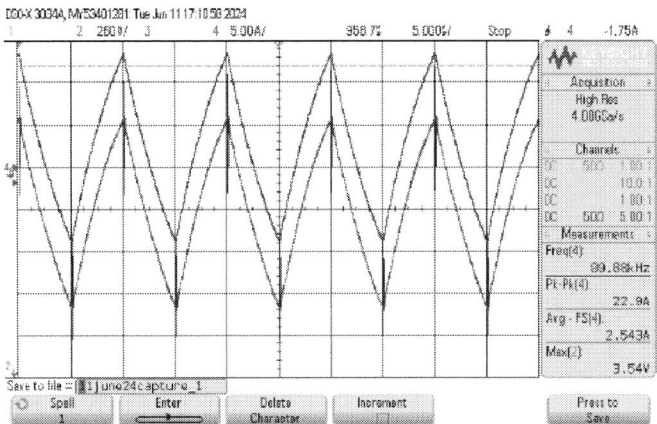

Fig. 10: 50V at 100kHz I_{DS} output at switching node. Sensor output - Green Trace, Reference Current - Magenta Trace

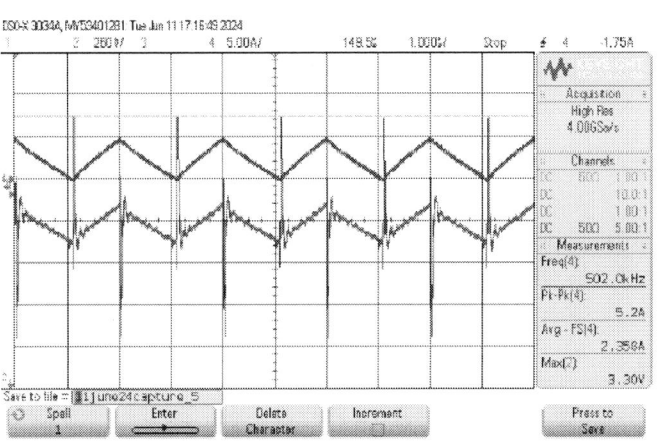

Fig. 11: 100V at 500kHz I_{DS} output at switching node. Sensor output - Green Trace, Reference Current - Magenta Trace

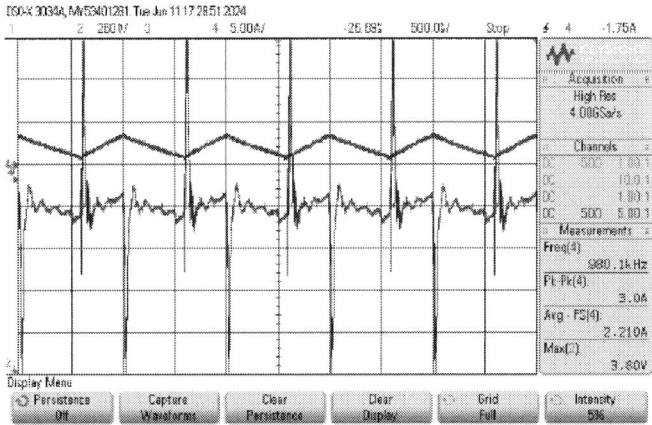

Fig. 12: 150V at 1MHz I_{DS} output at switching node. Sensor output - Green Trace, Reference Current - Magenta Trace

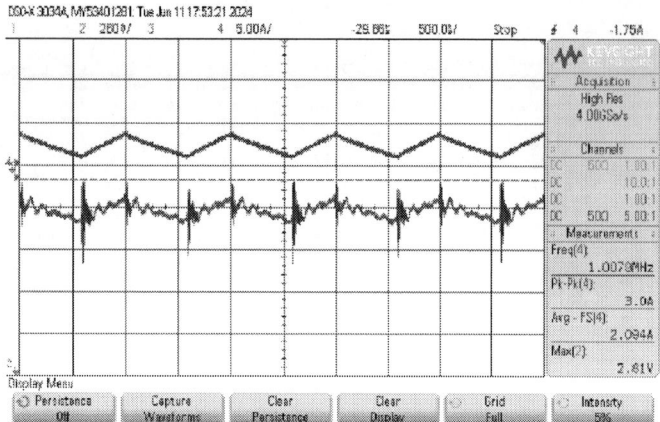

Fig. 13: 150V at 1MHz I_{DS} output at the load. Sensor output - Green Trace, Reference Current - Magenta Trace

bus and measured the R_{DS-ON} with an Agilent 34401A high-precision digital multimeter. For the sensor test, the evaluation platform was configured to a buck converter again and run at 30kHz. The calculated vs reference values can be seen in Table VI

SL#	Test	Top Dev.(mOhm)	Bottom Dev.(mOhm)
1	DC R_{DS-ON}	62.3	63.12
2	In Situ R_{DS-ON}	61.76	62.43
3	error	0.54	0.69

TABLE VI: R_{DS-ON} evaluation

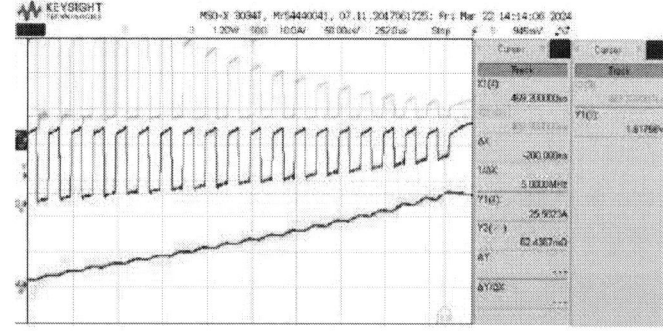

Fig. 14: Calculated R_{DS-ON} value. V_{ON} Sensor output - Blue Trace, I_{DS} sensor output - Magenta Trace, Calculated R_{DS-ON} value - pink trace

Fig. 14 presents the V_{ON}, I_{DS} and calculated R_{DS-ON} wave form. It can be seen that the scope measurements also represent the values presented in the Table. VI. The tabulated results reveal certain deviations between the DC test and the switched test. The additional impedance observed is due to the high-frequency switched operation, as elaborated in [24]. The results look promising and within the margin of error.

V. DATA SAMPLING AND R_{DS-ON} CALCULATION

Digital sampling of methods and techniques have been greatly explored and expanded on [24]. The study expands on

methods involving a central controller. The controller in this case is responsible for PWM generation and control of the converter. Information which were useful, as the data capture window and processing could be synchronized with the PWM information. This allowed for them to find a clean patch of data in a noisy environment.

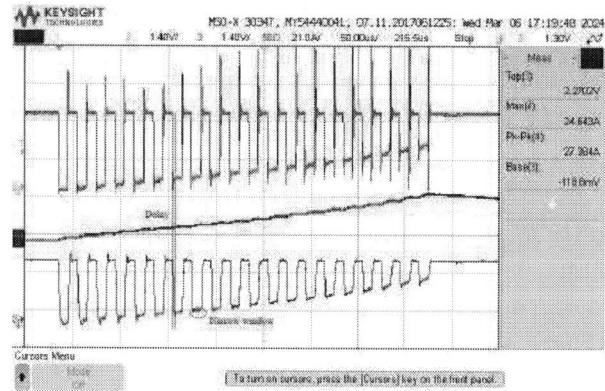

Fig. 15: Delay and narrow window issue with data capture, V_{ON} Sensor output before isolator- Green Trace, I_{DS} sensor output - Magenta Trace, V_{ON} Sensor output after isolator- Blue Trace

Fig. 15 exhibits the issues with data capture from V_{ON} and I_{DS}. Primarily the output of the isolated V_{ON} sensor has a delay induced by the isolator and the filtering network. This throws the I_{DS} data out of sync from its respective V_{ON} data. Additionally, the regions at the start and trailing end of the On-state voltage exhibits ringing from switching. Rendering this region sub-optimal for data processing barring some heavy digital filtering. Appropriate data-sampling is thus required that is suitable for our hardware approach.

The algorithm presented in Fig. 16 is currently being applied to extract and calculate the average on-cycle R_{DS-ON}. Further improvements to the process and filtering will be made as the study moves forward.

VI. APPLICATION

The sensor developed in this study is intended to be used in a marine energy microgrid. Studies presented in [28] emphasize the turbulent nature of marine energy and the impact it has on power semiconductors. It is shown that around 26% of the cost can be accounted for the operation and maintenance of the converters. Enhancing the reliability via in-situ prognostics promises to reduce the cost to a considerable degree. To conduct this study a 3-phase converter is developed, capable of housing the R_{DS-ON} sensors Fig.17

The converter specs are shown in Table VII

Experimental results are gathered and plotted in Fig.18, Fig.19, Fig.20, Fig.21. Fig. 18 shows a reduction of 20 degrees C on junction temperature while using an in-situ R_{DS-ON} based MPC. Fig.19 shows 0.005ohm reduction in dynamic R_{DS-ON} while using in-situ R_{DS-ON} based MPC. Fig.21 shows an increase in THD from 1.19% to 4.6%, however, that

979-8-3315-1612-3/25 $31.00 © 2025 IEEE

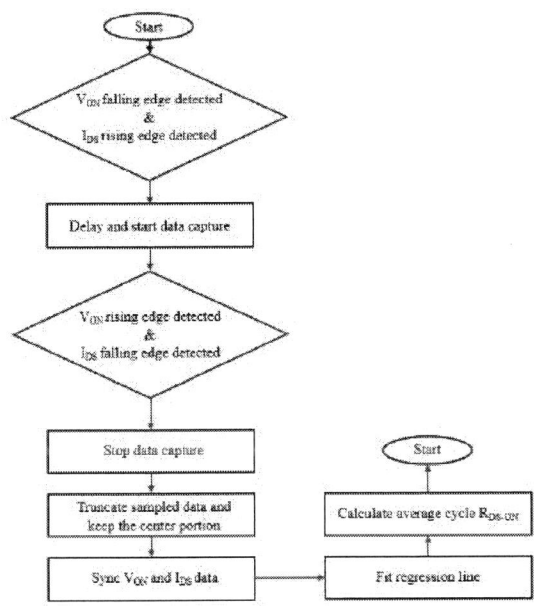

Fig. 16: Data capture and processing diagram

Fig. 17: 3 phase inverter with R_{DS-ON} sensors

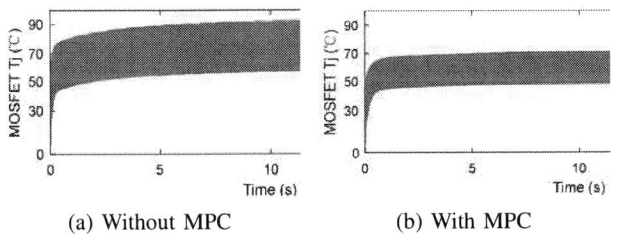

(a) Without MPC (b) With MPC

Fig. 18: Comparison of Junction temperature without and with in situ R_{DS-ON} based MPC

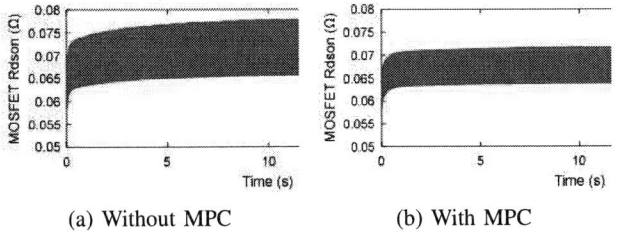

(a) Without MPC (b) With MPC

Fig. 19: Comparison of R_{DS-ON} without and with in situ R_{DS-ON} based MPC

Parameter	Value
Rated Power	20kW
DC Link Votlage	100V
AC Voltage	277/480V
AC frequency	60Hz
Current	24A
MOSFET	C2M0080120D
Ambient temperature	25 c

TABLE VII: 3-phase test parameters

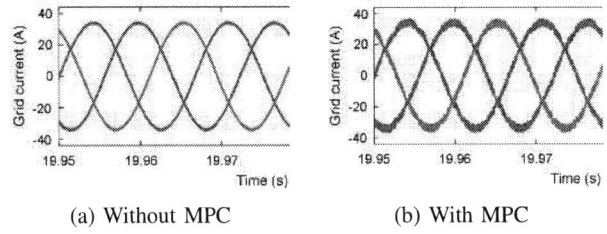

(a) Without MPC (b) With MPC

Fig. 20: Comparison of current waveforms without and with in situ R_{DS-ON} based MPC

is still within an acceptable margin and one of the compromises of Model Predictive Control. The test and simulation results are very encouraging and the team is moving forward with further refinement and field testing of the system.

VII. CONCLUSION

Presented in the earlier sections of the article we have discussed how junction temperature T_j, allows an understanding into semiconductor degradation. And how R_{DS-ON} measurements allow for T_j to be estimated. Our presented study exhibits a unique standalone method to study and extract R_{DS-ON} and apply in-situ prognostics. In addition to the applications mentioned above, in-situ R_{DS-ON} data can be utilized for several other purposes: (1) device characterization, (2) converter modeling for digital twin applications, (3) high-speed temperature sensing when mapped with junction temperature, and (4) device protection, such as di/dt or shoot-through protection. The proposed and developed sensor facilitates the study and development of R_{DS-ON} as a separate sensor. In its current form, the sensor estimates R_{DS-ON} independently of the converter controller and reports the data accordingly. The sensors long with an analysis of the sensor's dynamic performance over extended run-time. Further refinement of the sensor is already underway, with plans to remove the analog

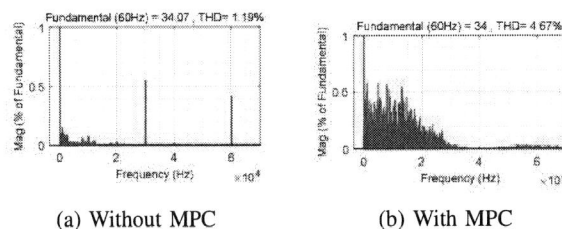

(a) Without MPC (b) With MPC

Fig. 21: Comparison of THD without and with in situ R_{DS-ON} based MPC

isolator. The sensor controller will be connected through digital isolators, thereby eliminating the analog data loss issues highlighted in [27]. Further improvements to the system is being made as we speak with the goal of developing ASIC sensor package for each discrete semiconductor devices.

REFERENCES

[1] B. K. Bose, "Global energy scenario and impact of power electronics in 21st century," IEEE Trans. Ind. Electron., vol. 60, no. 7, pp. 2638–2651, Jul. 2013.

[2] Yole Group, "Power electronics, A green revolution is giving way to new power electronics applications," Accessed: Apr. 17, 2024, [Online]. Available: https://www.yolegroup.com/thematic/power-electronics-application/

[3] L. F. Costa and M. Liserre, "Failure analysis of the dc-dc converter: A comprehensive survey of faults and solutions for improving reliability," IEEE Power Electron. Mag., vol. 5, no. 4, pp. 42– 51, Dec. 2018.

[4] J. Morroni, A. Dolgov, M. Shirazi, R. Zane, and D. Maksimovic, "Online health monitoring in digitally controlled power converters," in Proc. IEEE Power Electron. Spec. Conf., 2007, pp. 112–118.

[5] S. Pu, F. Yang, B. T. Vankayalapati, E. Ugur, C. Xu, and B. Akin, "A practical on-board SiC MOSFET condition monitoring technique for aging detection," IEEE Trans. Ind. Appl., vol. 56, no. 3, pp. 2828–2839, May/Jun. 2020.

[6] F. Gonzalez-Hernando, J. San-Sebastian, M. Arias, A. Rujas, and F. Iannuzzo, "Discontinuous PWM for online condition monitoring of SiC power modules," IEEE J. Emerg. Sel. Topics Power Electron., vol. 8, no. 1, pp. 323–330, Mar. 2020.

[7] H. You et al., "Intelligent health monitoring system hardware design for paralleled devices with fast dv/dt output," in Proc. IEEE Int. Elect. Mach. Drives Conf., 2021, pp. 1–5.

[8] E. Ugur, C. Xu, F. Yang, S. Pu, and B. Akin, "A new complete condition monitoring method for SiC power MOSFETs," IEEE Trans. Ind. Electron., vol. 68, no. 2, pp. 1654–1664, Feb. 2021.

[9] J. Millán, P. Godignon, X. Perpiñà, A. Pérez-Tomás, and J. Rebollo, "A survey of wide bandgap power semiconductor devices," IEEE Trans. Power Electron., vol. 29, no. 5, pp. 2155–2163, May 2014.

[10] A. J. Lelis, R. Green, and D. B. Habersat, "SiC MOSFET reliability and implications for qualification testing," in Proc. IEEE Int. Rel. Phys. Symp., 2017, pp. 2A–4.1–2A–4.4.

[11] J. Wei, S. Liu, J. Fang, S. Li, T. Li, and W. Sun, "Investigation on degradation mechanism and optimization for SiC power MOSFETs under long-term short-circuit stress," in Proc. IEEE 30th Int. Symp. Power Semicond. Devices ICs, 2018, pp. 399–402.

[12] X. Yang et al., "Degradation behavior and defect analysis for SiC power MOSFETs based on low-frequency noise under repetitive power-cycling stress," IEEE Trans. Electron Devices, vol. 68, no. 2, pp. 666–671, Feb. 2021.

[13] E. Ugur, F. Yang, S. Pu, S. Zhao, and B. Akin, "Degradation assessment and precursor identification for SiC MOSFETs under high temp cycling," IEEE Trans. Ind. Appl., vol. 55, no. 3, pp. 2858–2867, May/Jun. 2019.

[14] R. Li, X. Wu, S. Yang, and K. Sheng, "Dynamic on-state resistance test and evaluation of GaN power devices under hard- and soft-switching conditions by double and multiple pulses," IEEE Trans. Power Electron., vol. 34, no. 2, pp. 1044–1053, Feb. 2019.

[15] S. Dusmez, M. Bhardwaj, L. Sun, and B. Akin, "In situ condition monitoring of high-voltage discrete power MOSFET in boost converter through software frequency response analysis," IEEE Trans. Ind. Electron., vol. 63, no. 12, pp. 7693–7702, Dec. 2016.

[16] J. Chen, X. Jiang, Z. Li, H. Yu, and J. Wang, "Investigation on degradation of SiC MOSFET under accelerated stress in PFC converter," in Proc. IEEE Energy Convers. Congr. Expo., 2019, pp. 6174–6178.

[17] Y. Peng, Y. Shen, and H. Wang, "A condition monitoring method for three phase inverter based on system-level signal," in Proc. IEEE Int. Power Electron. Application Conf. Expo., 2018, pp. 1–5.

[18] M. Biglarbegian et al., "On condition monitoring of high frequency power GaN converters with adaptive prognostics," in Proc. IEEE Appl. Power Electron. Conf. Expo., 2018, pp. 1272–1279.

[19] S. Dusmez, H. Duran, and B. Akin, "Remaining useful lifetime estimation for thermally stressed power MOSFETs based on on-State resistance variation," IEEE Trans. Ind. Appl., vol. 52, no. 3, pp. 2554–2563, May/Jun. 2016.

[20] Y. Chen and D. B. Ma, "Self-Aging-prognostic GaN-based switching power converter using TJ- independent online condition monitoring and proactive temperature frequency scaling," IEEE Trans. Power Electron., vol. 36, no. 5, pp. 5022–5031, May 2021.

[21] M. Baharani, M. Biglarbegian, B. Parkhideh, and H. Tabkhi, "Real-timedeep learning at the edge for scalable reliability modeling of Si-MOSFET power electronics converters," IEEE Internet Things J., vol. 6, no. 5, pp. 7375–7385, Oct. 2019.

[22] J. Fan, D. Ma, J. Wang, M. Chinthavali, and RSK. Moorthy, "Real-time condition monitoring of power modules in grid-tied power converter," in Proc. IEEE Energy Convers. Congr. Expo., 2022, pp. 1–6.

[23] V. Mitrovic, B. Fan, Y. Cao, Y. Bai, R. Burgos, and D. Boroyevich, "Phase current reconstruction, DC link voltage and R DS-ON measurement using sensors integrated on gate drivers for SiC MOSFET," in Proc. 22nd Int. Symp. Power Electron., 2023, vol. 1, pp. 1–6.

[24] C. Roy, N. Kim, D. Evans, J. Gafford and B. Parkhideh, "Challenges and Implementation of Online In Situ RDSON Measurement in a Three-Phase Inverter," in IEEE Transactions on Power Electronics, vol. 39, no. 9, pp. 11677-11686, Sept. 2024, doi: 10.1109/TPEL.2024.3410174.

[25] C. Roy, N. Kim, D. Evans, A. P. Sirat, J. Gafford and B. Parkhideh, "A Half-Bridge On-State Voltage Sensor for In-Situ Measurements," 2022 IEEE Energy Conversion Congress and Exposition (ECCE), Detroit, MI, USA, 2022, pp. 1-7, doi: 10.1109/ECCE50734.2022.9947787.

[26] C. Roy, N. Kim, J. Gafford and B. Parkhideh, "On-State Voltage Measurement of High-Side Power Transistors in Three-Phase Four-Leg Inverter for In-Situ Prognostics," 2021 IEEE Energy Conversion Congress and Exposition (ECCE), Vancouver, BC, Canada, 2021, pp. 2770-2776, doi: 10.1109/ECCE47101.2021.9595082.

[27] F. Karakaya, A. Maheshwari, A. Banerjee and J. S. Donnal, "An Approach for Online Estimation of On-State Resistance in SiC MOSFETs Without Current Measurement," in IEEE Transactions on Power Electronics, vol. 38, no. 9, pp. 11463-11473, Sept. 2023.

[28] J. Zhou et al., "Finite Control Set Model Predictive Control Based on In-Situ Junction Temperature for Reliability Enhancement of Power Converters," 2023 IEEE 10th Workshop on Wide Bandgap Power Devices & Applications (WiPDA), Charlotte, NC, USA, 2023,

Electrical Evaluation of a Modular High Voltage 3D Power Module using Direct Dielectric Liquid Cooling

Omar Sanjakdar
Univ. Grenoble Alpes,
CEA, Liten, Campus INES
73375 Le Bourget du Lac, France
CNRS, Grenoble INP, G2Elab
38000 Grenoble, France
omar.sanjakdar@cea.fr

Yvan Avenas
Univ. Grenoble Alpes,
CNRS, Grenoble INP, G2Elab
38000 Grenoble, France
yvan.avenas@g2elab.grenoble-inp.fr

Rachelle Hanna
Univ. Grenoble Alpes,
CNRS, Grenoble INP, G2Elab
38000 Grenoble, France
rachelle.hanna@g2elab.grenoble-inp.fr

Guillaume Piquet Boisson
Univ. Grenoble Alpes,
CEA, Liten, Campus INES
73375 Le Bourget du Lac, France
guillaume.piquetboisson@cea.fr

Emmanuel Marcault
CEA, CEA Tech Occitanie
Labege, France
emmanuel.marcault@cea.fr

Antoine Philippe
Univ. Grenoble Alpes,
CEA, Liten, Campus INES
73375 Le Bourget du Lac, France
antoine.philippe@cea.fr

Abstract—In this paper, a novel 3D power module structure based on direct dielectric liquid cooling is proposed for high voltage applications. This structure allows modular assembly and puts an end to the ever-existing compromise between high electrical insulation and high thermal performance, which usually requires a trade-off in the thickness of the ceramic layer. This proposed power module assembly is illustrated and then evaluated through electromagnetic and transient electric field simulations, using Ansys Q3D Extractor and Comsol®Multiphysics respectively. Stray inductance measurements are carried out using a precision impedance analyzer. Finally, direct current (DC) breakdown voltage measurements are conducted on the dielectric liquid alone and on the structure, submerged in the liquid, in order to evaluate its voltage withstand capability.

Keywords—*3D packaging, high voltage, modular assembly, stray inductance, voltage withstand.*

I. INTRODUCTION

In multiple sectors, a wide transition to higher power and higher voltage is taking place. This is the case in applications such as photovoltaics, high voltage (HV) motor drives, or even high voltage direct current (HVDC) micro grids [1-4]. In this context, the demand for high power and HV static converters is rapidly increasing [5]. One way to develop these converters, with different topologies, is by using HV silicon carbide (SiC) MOSFETs, with voltage rating greater than 6.5 kV [6]. SiC material is required in several HV applications because it combines interesting characteristics including higher working temperature, higher electrical breakdown field, and higher switching speed when compared to silicon devices.

Although HV SiC MOSFETs receive a wide interest in research studies and industry, their packaging is also a domain of great importance for ensuring their best performances [7]. However, several technical challenges arise in the domain of packaging, especially with the existing trade-off between electrical insulation, thermal performance and parasitics (stray inductances and capacitances).

Hence, multiple researches in the literature are exploring new packaging technologies for HV power modules. In this context, several technologies are proposed in order to increase voltage withstand, such as press-pack assemblies, ceramic stacking, and sandwich power modules with metal posts [8-10]. However, it is quite clear that the achieved packages in the state-of-the-art still have room for improvement, especially when it comes to HV SiC MOSFETs applications.

In this paper, a novel packaging design, with low stray inductance values for the switching cell (SC), is proposed for these applications. The proposed design offers the possibility of increasing the voltage withstand while having a negligible impact on other physical properties. In section II, the structure of the proposed packaging is described while demonstrating its inherent modularity. The integration of the decoupling capacitor in the power module is also detailed. In section III, the packaging's performance is evaluated through stray inductance measurements; first carried out in simulation and then validated with experimental tests. Section IV describes electric transient simulations, along with breakdown voltage (BDV) experimental measurements. Finally, section V concludes with future perspectives.

II. PACKAGING CONCEPT

Multiple 3D packaging concepts exist in the literature, each having different advantages and disadvantages. For example, authors in [11] describe a stacked power module with direct die cooling where the goal is to eliminate single function components such as wirebonds and heat sinks. However, the proposed module in the present paper shows several improvements, when compared to the cited module in [11], on different levels: the implementation of heat spreaders which increases heat dissipation surface, the modularity of the structure, and the reduction of stray inductance values by integrating decoupling capacitors. The aforementioned improvements are implemented while taking into consideration voltage withstand.

979-8-3315-1612-3/25 $31.00 © 2025 IEEE

Fig. 1 shows a half-bridge circuit composed of three switching cells connected in parallel. The cross-sectional schematic of the proposed modular 3D packaging, for the electric circuit in Fig. 1, is presented in Fig. 2.

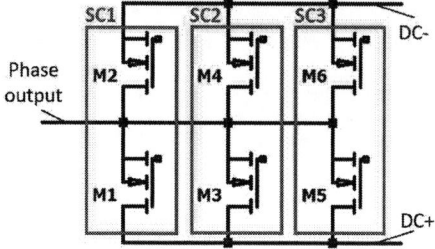

Fig. 1. Electric circuit of a half-bridge composed of three switching cells.

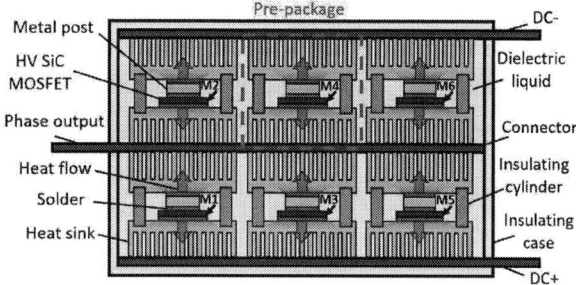

Fig. 2. Front-side view of the proposed 3D packaging.

In this structure, each SiC MOSFET die is connected, from both sides (drain and source electrodes), to copper heat sinks. The latter serve as electric connectors, in addition to their primary role of heat dissipation. Each MOSFET die is placed as in Fig. 2 with the drain (bottom side) connected directly to the heat sink; however, the source (top side) is connected through a metal post to ensure voltage withstand around voltage terminations of the chip. Furthermore, an insulating cylinder is implemented on each corner between two heat sinks; it acts as mechanical support and ensures proper spacing between these two heat sinks. Each MOSFET connected to both of its heat sinks is henceforth referred to as a "pre-package". In this example, the presented cross-sectional schematic of the packaging structure shows 6 pre-packages allowing the construction of the electric circuit of Fig. 1. All heatsinks having the same electric potential are connected to a copper plate (labelled as "DC+", "DC-", and "Phase output" in Fig. 2) in order to ensure their electrical connection with the external system. This structure is then contained in an insulating case where a dielectric liquid circulates in order to ensure electrical insulation and cooling of the semiconductor devices. The insulation with respect to the ground is hence increased when the case's thickness is increased, without having any negative impact on the thermal performance. With the proposed structure, the design of any power module in a modular way is made possible by assembling multiple pre-packages inside the module (according to the desired topology) and circulating the dielectric liquid inside the case.

In the following, the index notation 'i' will be used to refer to one of the integrated pre-packages. Regarding the control of each semi-conductor, a printed circuit board (PCB) (labelled PCB '$g_i s_i$', of light blue color in Fig. 3a) is integrated in the

associated pre-package (i). The PCB is soldered below the top heatsink of the pre-package (i) allowing the extraction of the source potential (labelled s_i, of orange color). This PCB positioning allows keeping a certain distance between the drain and the source in order to ensure voltage withstand of the structure. In addition, a contact pin is integrated in the PCB and is in contact with the gate potential (labelled g_i, of yellow color).

With each pre-package containing its own control PCB '$g_i s_i$', the control of all pre-packages connected in parallel can be achieved through only one PCB (labelled PCB 'GS' in Fig. 3b). All parallel source potentials 's_i' and gate potentials 'g_i' are hence connected to one source potential 'S' and one gate potential 'G' respectively. 'S' and 'G' are then connected, outside the insulating case, to a gate driver allowing the control of the parallel semi-conductors. Therefore, control of the whole structure (Fig. 2) is possible through only two control PCBs 'GS'.

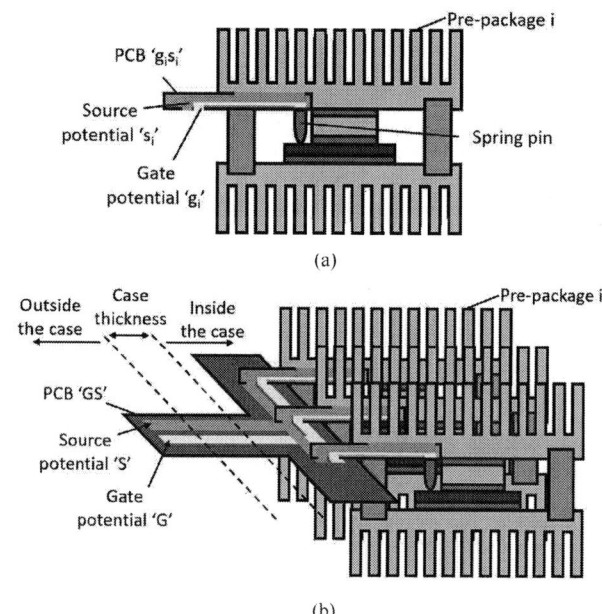

(a)

(b)

Fig. 3. Implimentation of the control PCBs for (a) a pre-package shown in right-side view (b) multiple pre-packages connected in parallel.

Furthermore, decoupling ceramic capacitors are used (represented by a purple rectangle in Fig. 4) in order to reduce the stray inductance of each switching cell.

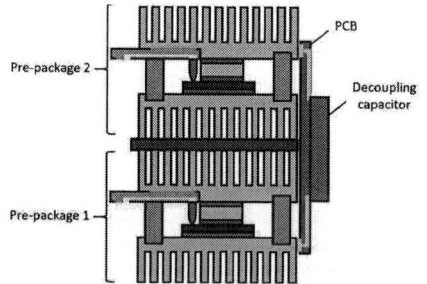

Fig. 4. Right-side view of a switching cell in the proposed 3D packaging.

Aiming to preserve the modularity of the proposed design, the decoupling ceramic capacitors are integrated on a PCB that

979-8-3315-1612-3/25 $31.00 © 2025 IEEE

is connected to both pre-packages of each switching cell as shown in Fig. 4.

In the following section, the stray inductance of this switching cell will be studied in simulation and validated through experimental tests.

III. STRAY INDUCTANCE

According to the proposed packaging concept, a switching cell composed of two pre-packages along with a PCB and its decoupling capacitor, is designed on Ansys Q3D Extractor, without the control part (see Fig. 5).

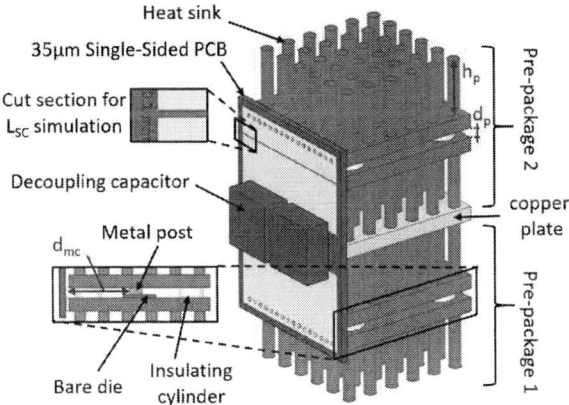

Fig. 5. 3D view of a switching cell composed of 2 pre-packages designed on Ansys Q3D Extractor.

In order to verify the proposed packaging and estimate its stray inductance, electromagnetic simulations are carried out on Ansys Q3D Extractor using finite and boundary elements methods (FEM & BEM). Table 1 presents switching cell components and their characteristics; these characteristics are implemented in Ansys Q3D Extractor for switching cell modelling.

TABLE I.   MODELLED SWITCHING CELL CHARACTERISTICS

Component	Material	Characteristics (length, width, height)
Heat sink[a]	Cu	Base plate: 21.8×21.8×2 mm Pin fins: 1.8x1.8x8.2 mm
PCB	FR4	23.8×31.2×1 mm
Insulating cylinder	RO4350B	4×4×3.2 mm
Metal post[b]	Cu	3.2×2×0.98 mm
Bare die[c]	Cu	5×5×0.42 mm
Decoupling capacitor[d]	Cu	8.4×8.4×3.8 mm
Copper plate	Cu	22×22×2 mm

[a] Dimensions correspond to the heat sink used in the experimental test.

[b] Size corresponds to the source pad of a 6.5 kV MOSFET SiC.

[c] Defined as copper to simulate the stray inductance of the SC caused by the module packaging.

[d] Defined as copper, stray inductance only relates to their geometrical shape.

Simulations are performed for stray inductance L_{SC} of the switching cell. The corresponding current path is represented in Fig. 6a; along with the inductance's terminals defined for simulations. These terminals are referred to as excitation ports, where the source and the sink have to be assigned. This exact placement of the source and sink prevents excessive deformation of the module's actual current density distribution during switching. This current deformation is also avoided by cutting a section of 0.1 mm from the PCB's copper trace (Fig. 5). This is necessary since in Q3D simulations, the excitation ports are modelled as equipotential surfaces.

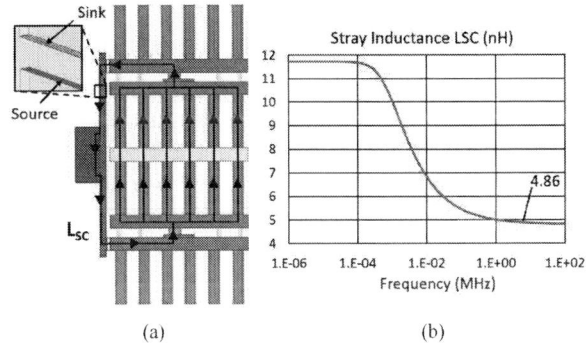

Fig. 6. (a) Right view of the switching cell designed on Ansys Q3D Extractor, and (b) its stray inductance simulation results.

For these simulations, an automatic mesh of medium accuracy is used; this guarantees good compromise between simulation time and resolution of results [13]. Simulation results indicate a stray inductance of 4.86 nH (see Fig. 6b) at 20 MHz, which is the order of magnitude of the oscillation frequency for SiC MOSFETs (associated with the device's rise and fall time). At this frequency, the skin effect will have been fully developed in the geometry [12]. Hence, all current will pass through the pin fins closest to the PCB of the decoupling capacitor. This maintains the stray inductance at a constant value for high frequencies. Results are obtained with a distance between two heatsinks d_p = 1.4 mm, a height of heat sinks' pin fins h_p = 8.2 mm and a distance between metal post and decoupling capacitors' PCB d_{mc} = 8.4 mm (in green in Fig. 5), corresponding to the fabricated structure.

It is important to mention the influence that parameters d_p, h_p, and d_{mc} have on other physical aspects in the structure under study. Increasing the parameter d_p for example, which is related to the height of the metal post, increases the voltage withstand but also causes an increase in the stray inductance and the thermal resistance.

Fig. 7 shows the impact of varying these parameters on the value of the stray inductance. For this study, parameters are varied as follows d_p: 0-10 mm, step 2 mm; h_p: 5-15 mm, step 5 mm; d_{mc}: 0-16 mm, step 4 mm. In applications where the aim is to increase the thermal performance, increasing the pin fin's height could hence be a better option (since it has a small impact on the value of Lsc) compared to increasing the heat sink's baseplate width (which is related to d_{mc} and has a higher impact on the stray inductance's value). This is of course true to a certain limit above which increasing the pin fin's height will no longer have a significant impact on the thermal performance.

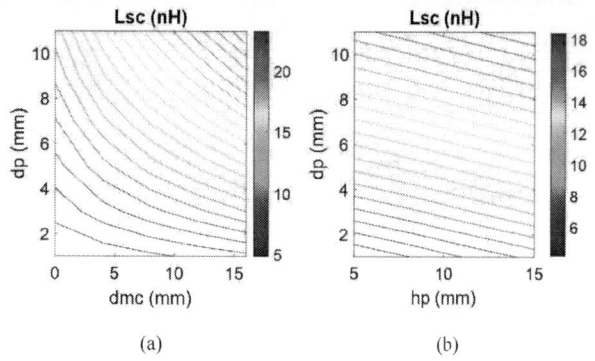

(a)   (b)

Fig. 7. Variation of the stray inductance as a function of (a) d_p and d_{mc} with $h_p = 10$ mm, and (b) d_p and h_p with $d_{mc} = 8$ mm, both for a frequency of 20 MHz.

In order to validate the stray inductance results, the switching cell modelled in Fig. 5 is fabricated and experimental measurements are carried out. Both pre-packages are fabricated, without bare dies, and assembled along a PCB (see Fig. 8) using solder paste. To compensate for the absence of bare dies, a copper post with a height of 1.4 mm is used. The rest of the components are kept as described in table 1. Since the goal is to measure only the stray inductance of the proposed packaging, the decoupling capacitor is not implemented in the fabricated SC. This is to eliminate the interference of the decoupling capacitor (ESL) in the performed measurements.

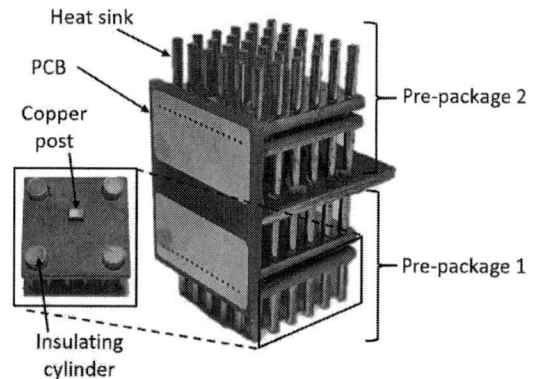

Fig. 8. Manufactured switching cell prototype composed of two pre-packages without semiconductors.

The used measurement setup is the precision impedance analyzer Keysight-E4990A, which supports analysis up to a frequency of 120 MHz. The impedance analyzer is equipped with a measurement probe Keysight-42941A. New probe placements are defined since it is impossible to keep the same placements as in the performed simulations (see Fig. 9).

Because the device's contributions in the measured impedance's value are usually not negligible, a calibration procedure is carried out to compensate for this additional impedance. Calibration is performed before any measurement is carried out and the impedance probe's jaw width is maintained after the calibration step. To limit any error that could impact the

accuracy of results, eight measurements are performed and their average is considered as the impedance value. Measurements that are not consistent with the majority are eliminated. Further details concerning this method are presented in the literature [14].

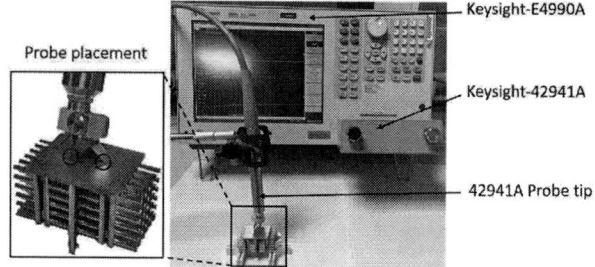

Fig. 9. Measurement configuration using an impedance analyser.

Experimental measurements indicate a stray inductance of 4.9 nH for a frequency of 20 MHz, equivalent to the value obtained in the performed simulation.

IV. VOLTAGE WITHSTAND CAPABILITY

This section aims to evaluate the maximum voltage withstand capability of the proposed structure by combining a numerical calculation of the transient electric field with DC breakdown measurements. The numerical simulation seeks to determine both the critical regions and the critical voltage frequency for the structure, assuming the model assumptions are valid. In the finite element software Comsol®Multiphysics (Fig. 10), a 2D geometry of the pre-package is implemented, regardless of the bare die, the metal post, and the control part. The geometry's dimensions are made similar to those of the fabricated structure. The upper heatsink, or electrode, (in blue in Fig. 10) is grounded, while the bottom one (in red) is connected to a high voltage source of 6.5 kV. The default boundary condition is an insulation (n.J = 0), framed in orange in Fig. 10, where the normal surface current density is set to zero. The gray part represents the liquid, while the green hatched part is the solid (insulating cylinder), their material properties are detailed in Table 2. The materials are assumed to be linear, isotropic and free of space charge, with charge injection being neglected in the calculation.

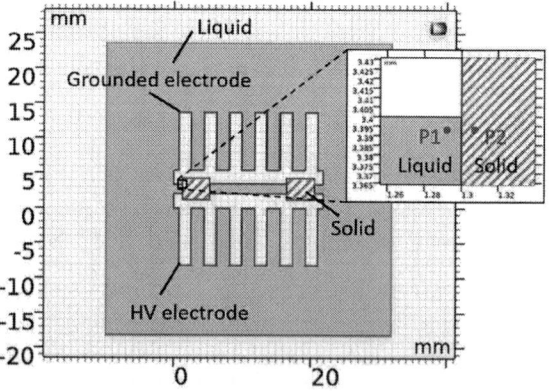

Fig. 10. Simplified 2D geometry of pre-package implemented in COMSOL®Multiphysics.

In this structure, the electric field is not fully uniformly distributed; some areas experience intensified fields, which significantly reduce the module's voltage withstand capability. This effect is especially pronounced at what is known as "triple points", where two insulating materials (solid and liquid) with different electrical properties meet alongside the metallization. In Fig. 10, two points P1 and P2 are chosen to be located 10 µm away from the triple point with a 45° angle, in both insulating materials: the liquid and the solid respectively. The location of these points is chosen after a convergence study of the electric field while changing the mesh resolution since the design is simplified; no rounding of the corners is applied. This approach helps to overcome the numerical singularities caused by the presence of the right angle.

TABLE II.   ELECTRICAL PROPERTIES OF INSULATING MATERIALS

Component		Relative permittivity at 25°C	Electric conductivity at 25°C (S/m)
Dielectric Liquid	DC Cooling Biolife 4	2.07 [e]	1E-13 [f]
Insulating Solid	Rogers 4350B	3.48	8.33E-14

[e] These values are obtained by dielectric spectroscopy measurements.

[f] Typical orders of magnitude of non-polar liquids.

Regarding the FEM simulations, triangular elements with different mesh sizes, are considered for different regions. This offers a good compromise between the accuracy of results and the required execution time and memory use—which can increase significantly if a fine mesh is applied for the whole geometry. In order to measure the quality of the applied mesh, the term skewness is used (see Fig. 11). Regular mesh elements that form an equilateral triangle have a mesh quality equal to one; degenerated elements however, have a mesh quality equal to zero [15]. In the considered simulation, skewness of the mesh is maintained between 0.51 and 1, with an average quality of 0.86.

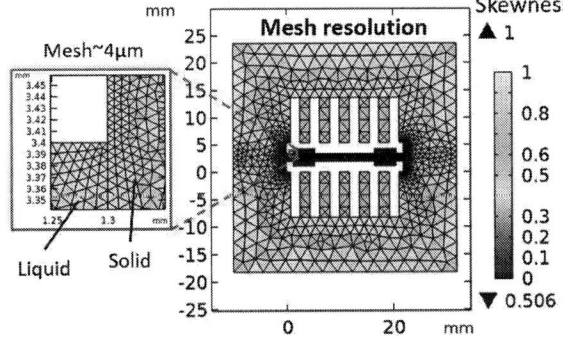

Fig. 11. Applied mesh and its resolution for the geometry implemented in COMSOL®Multiphysics.

The simulation, based on electric transient physics, shows that the electric field enhancement, whether in the dielectric liquid or within the solid cylinders, is highly time-dependent (Fig. 12). It solves the current conservation (1).

$$J = \sigma E + \frac{\partial \varepsilon E}{\partial t} \qquad (1)$$

where J is the current density (A/m^2), σ the electrical conductivity (S/m), E the electric field (V/m) and ε the absolute permittivity (F/m). For short durations of the applied voltage, a high electrostatic field appears at first. This corresponds to the early moments of impulses induced by semiconductor switching, where the electric field depends on the permittivity of the insulating materials. However, after a few seconds, the field decreases due to electrical conduction. This mode is relevant for DC applied voltage, where the field becomes dependent on the conductivity of the insulating materials.

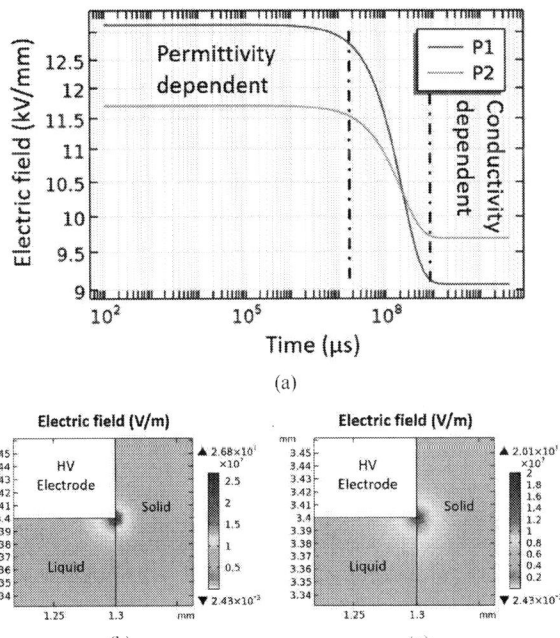

Fig. 12. (a) Transient electric field evolution at points P1 and P2 for a voltage of 6.5 kV, and its distribution during (b) field-dependent permitivity mode and (c) field-dependent conductivity mode.

Fig. 12b and Fig. 12c show the distribution of the electric field around the triple point in the modes with field-dependent permittivity and field-dependent conductivity, respectively. The electric field is higher near the triple point in the case of field-dependent permittivity mode; however, the conduction mode shows a high electric field (of $E > 7$ kV/mm) for a larger area (~60 µm) compared to the field-dependent permittivity mode.

In order to validate the results obtained in conduction mode, DC breakdown voltage measurements are conducted for both the pre-package structure, submerged in the dielectric liquid, and the dielectric liquid alone as shown respectively in Fig. 13b and Fig. 13c. The pre-package, corresponding to the geometry implemented in Comsol®Multiphysics, is fabricated without bare die, metal post, and control part.

Fig. 13a illustrates the DC breakdown voltage experimental setup, a HV DC Supply rated at 0-60 kV served as the DC generator. BDV is determined by applying DC voltage ramps at 1 kV/s until a spark occurred. Immediately after the breakdown, the high-voltage source is rapidly deactivated (within 1 ms) to minimize liquid degradation.

BDV measurements use low-energy pulses in order to minimize the impact of degradation by-products on the measurements. A minimum of 25 breakdown data points is included to ensure an accurate estimation of the breakdown voltage; a duration of 90 seconds is considered between measurements. The obtained BDV measurements provide critical insights concerning the maximum voltage the pre-package can withstand before electrical breakdown occurs.

Breakdown voltage in dielectrics typically occurs randomly and can vary even with a fixed configuration and material. Therefore, the breakdown voltage can be considered a random variable. To analyze it, the Weibull distribution is a useful statistical tool (Fig. 14).

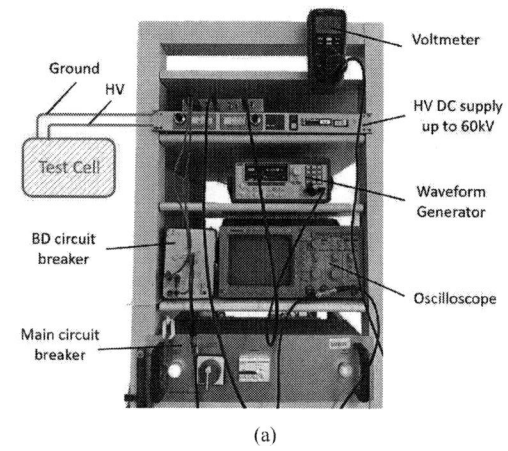

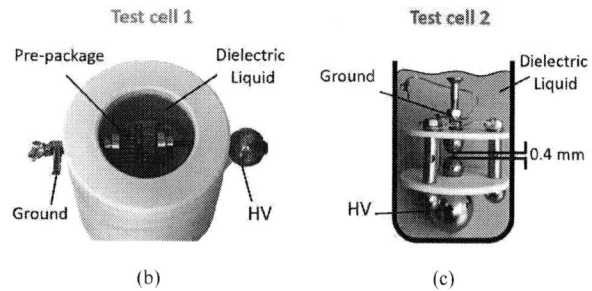

Fig. 13. (a) DC breakdown voltage experimental setup used for (b) test cell 1 and (c) test cell 2.

In this study, the breakdown data series were analyzed using the two parameters of the Weibull distribution α and β. The scale parameter α represents the breakdown voltage, where 63.2% of the samples are expected to fail. Higher values of α indicate greater durability or higher breakdown voltage. The shape parameter β, reflects the uniformity of the distribution. A larger β values correspond to a more concentrated distribution with less variability.

The probability of BDV is then linearized into the fitting of Weibull cumulative distribution function (CDF) to deduce α and β, where the CDF is defined as :

$$P(BDV) = 1 - e^{-(\frac{BDV}{\alpha})^{\beta}} \qquad (2)$$

BDV measurements are conducted for both test cells illustrated in Fig. 13b and Fig. 13c. The first test cell includes the pre-package, while the second is designed to measure the breakdown field of the liquid, placed between 2 spheres with 0.4 mm gap. This latter configuration is designed to study the BDV of the liquid alone and the influence of its impurities under uniform electric field. Comparison of results between both test cells also allows to identify the influence of the proposed structure on the voltage withstand. BDV measurement results are presented in Fig. 14.

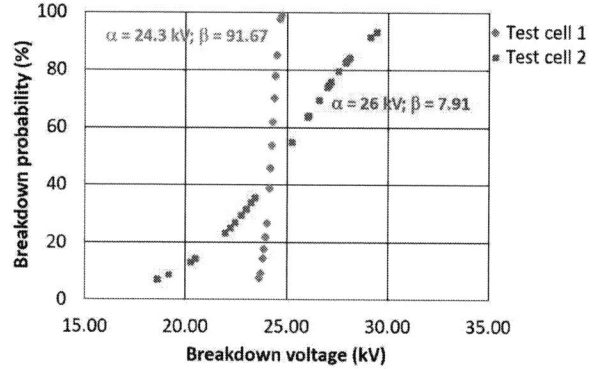

Fig. 14. Breakdown probability vs BDV of the two test cells under study at 25°C.

The values of α obtained for test cell 1 and test cell 2 are 24.3 kV and 26 kV, respectively. This indicates, while taking into consideration the difference in the distance between the electrodes (1.4 mm for test cell 1 and 0.4 mm for test cell 2), that the electric field in the proposed pre-package is not uniform; it is rather intensified in some regions of the structure as shown in the performed simulation.

The breakdown field (α = 17.3 kV/mm assuming a quasi-uniform field) in the pre-package under study is nearly twice as high as the electric field obtained in the simulation for the conduction mode case (9.1 kV/mm); hence, breakdown is unlikely to occur for a DC blocking voltage of 6.5 kV. Additional DC breakdown voltage measurements are required to validate this and confirm that breakdown begins at the interface between the solid and liquid insulators. This will enable further optimization of the structure's dielectric strength by varying the types of insulators (combinations of different solid and liquid electrical properties).

Furthermore, breakdown voltage measurements, in impulse mode, will be carried out in future work in order to validate the field-dependent permittivity part of the simulation.

V. Conclusions

In this paper the design and characterization of a novel modular 3D packaging structure targeting HV applications is presented. A prototype of the proposed packaging is evaluated on an electromagnetic and electrical level. Stray inductances related to the switching cells' layout are evaluated through electromagnetic simulations and experimental tests which show stray inductance values of 4.9 nH. These values are low compared to those of other existing high voltage packaging. Electric transient field evolution is analyzed through simulations; in addition, breakdown voltage measurements are carried out and indicate a high voltage withstand convenient for 6.5 kV SiC MOSFET.

ACKNOWLEDGMENT

The authors would like to thank Luis Gabriel Alves Rodrigues for their contributions to the development and execution of this work.

This work was supported by the French National Program "Programme d'Investissements d'Avenir – INES.2S" under Grant ANR-10-IEED-0014-01.

REFERENCES

[1] P. C. Heris, Z. Saadatizadeh, and A. Mantooth, "A Three-Port Non-Isolated High Voltage Conversion Ratio DC-DC Converter For MPPT Extraction Of Photovoltaic Systems," in 2023 IEEE Applied Power Electronics Conference and Exposition (APEC), Mar. 2023, pp. 133–136. doi: 10.1109/APEC43580.2023.10131184.

[2] X. Huang, H. Wang, Y. Zhou, X. Zhang, Y. Wang, and H. Xu, "Photovoltaic Power Plant Collection and Connection to HVDC Grid with High Voltage DC/DC Converter," Electronics, vol. 10, no. 24, Art. no. 24, Jan. 2021, doi: 10.3390/electronics10243098.

[3] X. Wang et al., "A novel Modular Multilevel Converter topology with auxiliary voltage clamping circuit for high voltage motor drive," in 2014 9th IEEE Conference on Industrial Electronics and Applications, Jun. 2014, pp. 1392–1397. doi: 10.1109/ICIEA.2014.6931386.

[4] Y. Gao et al., "Study on lightning protection scheme of multi-terminal MMC-MVDC distribution system," High Volt., vol. 5, no. 5, pp. 605–613, 2020, doi: 10.1049/hve.2019.0256.

[5] Z. Lin et al., "Research on the topology of large-scale offshore wind power based on distributed medium voltage DC collection and high voltage DC transmission," 2023 IEEE Sustainable Power and Energy Conference (iSPEC), Chongqing, China, 2023, pp. 1-6, doi: 10.1109/iSPEC58282.2023.10402895.

[6] A. V. Bilbao, J. A. Schrock, W. B. Ray, M. D. Kelley, and S. B. Bayne, "Analysis of advanced 20 KV/20 a silicon carbide power insulated gate bipolar transistor in resistive and inductive switching tests," in 2015 IEEE Pulsed Power Conference (PPC), May 2015, pp. 1–3. doi: 10.1109/PPC.2015.7296953.

[7] Z. Chen, R. Burgos, D. Boroyevich, F. Wang, and S. Leslie, "Modeling and simulation of 2 kV 50 A SiC MOSFET/JBS power modules," in 2009 IEEE Electric Ship Technologies Symposium, Apr. 2009, pp. 393–399. doi: 10.1109/ESTS.2009.4906542.

[8] N. Zhu, H. A. Mantooth, D. Xu, M. Chen, and M. D. Glover, "A Solution to Press-Pack Packaging of SiC MOSFETS," IEEE Trans. Ind. Electron., vol. 64, no. 10, pp. 8224–8234, Oct. 2017, doi: 10.1109/TIE.2017.2686365.

[9] C. DiMarino et al., "A Wire-bond-less 10 kV SiC MOSFET Power Module with Reduced Common-mode Noise and Electric Field," in PCIM Europe 2018; International Exhibition and Conference for Power Electronics, Intelligent Motion, Renewable Energy and Energy Management, Jun. 2018, pp. 1–7. Accessed: Jul. 31, 2024. [Online]. Available: https://ieeexplore.ieee.org/document/8402827

[10] Z. Zhang et al., "Packaging of an 8-kV Silicon Carbide Diode Module with Double-Side Cooling and Sintered-Silver Joints," in 2021 IEEE Electric Ship Technologies Symposium (ESTS), Aug. 2021, pp. 1–7. doi: 10.1109/ESTS49166.2021.9512339.

[11] L. M. Boteler, V. A. Niemann, D. P. Urciuoli and S. M. Miner, "Stacked power module with integrated thermal management," 2017 IEEE International Workshop On Integrated Power Packaging (IWIPP), Delft, Netherlands, 2017, pp. 1-5, doi: 10.1109/IWIPP.2017.7936764.

[12] M. S. S. Nia, S. Saadatmand, M. Altimania, P. Shamsi, and M. Ferdowsi, "Analysis of Skin Effect in High Frequency Isolation Transformers", in 2019 North American Power Symposium (NAPS), Wichita, KS, USA, Oct. 2019, pp. 1–6. doi: 10.1109/NAPS46351.2019.9000395.

[13] O. Sanjakdar, L. G. Alves Rodrigues, and Y. Avenas, "Towards a Modular Multilevel Flying Capacitor Module using SiC MOSFETs," in PCIM Europe 2023; International Exhibition and Conference for Power Electronics, Intelligent Motion, Renewable Energy and Energy Management, May 2023, pp. 1–8. doi: 10.30420/566091009.

[14] B. T. DeBoi, A. N. Lemmon, B. W. Nelson, C. D. New, and D. M. Hudson, 'Improved Methodology for Parasitic Characterization of High-Performance Power Modules', IEEE Trans. Power Electron., vol. 35, no. 12, pp. 1340013408, Dec. 2020, doi: 10.1109/TPEL.2020.2992332.

[15] C. Dorji, R. Hanna, and O. Lesaint, "Time-Resolved Non-Linear Electric Field Simulation on Liquid Embedded Substrate," in 2022 IEEE 21st International Conference on Dielectric Liquids (ICDL), May 2022, pp. 1–4. doi: 10.1109/ICDL49583.2022.9830949.

Board Level Reliability of Gull-wing, Micro-leaded and Lead-less Packaged MOSFETs in Automotive Environments

Christopher Liu
Automotive MOSFET Applications
Nexperia, Bramhall Moor Ln, Hazel Grove,
Manchester, United Kingdom
christopher.liu@nexperia.com

Vijayakrishna Satyamsetti
Automotive MOSFET Applications
Nexperia, Bramhall Moor Ln, Manchester, United Kingdom
vijayakrishna.satyamsetti@nexperia.com

Xuanjing Wei
Automotive MOSFET Applications
Nexperia, East Yunling Road, Putuo District
Shanghai, China
jae.wei@nexperia.com

Christian Radici
Automotive MOSFET Applications
Nexperia, Bramhall Moor Ln, Hazel Grove,
Manchester, United Kingdom
christian.radici@nexperia.com

Peter Vines
Automotive MOSFET Applications
Nexperia, Bramhall Moor Ln, Hazel Grove,
Manchester, United Kingdom
peter.vines@nexperia.com

Wayne Lawson
Automotive MOSFET Applications
Nexperia, Bramhall Moor Ln, Hazel Grove,
Manchester, United Kingdom
wayne.lawson@nexperia.com

Abstract—This paper explains the board level reliability of popular surface mounted power MOSFET packages which are found in automotive environments and applications. Even when AEC-Q101 standards are met, the package footprint geometry of a MOSFET inevitably has an influence on board level performance. Thermomechanical simulations were performed alongside experimental testing on micro-leaded 3x3mm² MOSFET packages, which resulted in good agreement between experimental and simulation data. A relative comparison was made through temperature cycling simulations of gull-wing, micro-leaded and lead-less packages between -55°C to 150°C as per AEC-Q101. This was done to evaluate the solder fatigue of 3x3mm² packages, which are widely used in automotive applications. The averaged accumulated creep strain per temperature cycle from simulations was then used in the double power law model derived from literature to predict MOSFET lifetime on a printed circuit board (PCB). From which, the findings of the investigation showed that gull-wing leaded MOSFETs were predicted to have higher board longevity, followed by micro-leaded and lead-less packages.

Keywords—board level reliability, solder fatigue, temperature cycling, power MOSFET.

I. INTRODUCTION

Temperature cycling is known to be a cause of failure for surface mounted MOSFETs on PCBs. In conventional automotive applications, the ambient temperature can be particularly influenced by the presence of the internal combustion engine. As the ambient temperature fluctuates, each material deforms at a different rate due to the difference in coefficient of thermal expansion (CTE), which induces deformation between various different parts of a given system. The deformation is particularly apparent in the PCB solder connecting the MOSFET to the PCB. This is because solder is vulnerable to creep strain, which is the plastic deformation from cyclic stresses over an extended time period, even if the stresses are below the material's yield stress. Creep is accelerated in materials with high homologous temperatures (above 0.4) [1]. This is the ratio of the temperature the material is operating at against its melting temperature on the Kelvin scale. As each temperature cycle is applied, the cyclic stresses from expansion and contraction result in the accumulation of creep strain in the PCB solder. This strain increases in a staircase manner with each temperature cycle, which eventually results in solder cracks that cause open circuits and thus MOSFET failure.

There are a variety of surface mounted power MOSFET packages used in automotive applications. Gull-wing packages come with pins that protrude around the mid-height of the mold compound, in which the inner pin corners are filleted. Micro-leaded packages have flat pins that protrude out low from the mold compound but not to the extent of a gull-wing package. Finally, lead-less packages have flat pins which do not protrude and sit flush with the edge of the mold compound. 3D orthographic projections of the mentioned packages are shown in Figs. 5(a), (b) and (c) respectively.

Previously, there has been a study [2] that evaluated the board level reliability between various gull-wing pin designs. In addition, another study [3] evaluated the effects of the mold compound material on board level reliability in lead-less QFN packages. This investigation aims to develop further upon the two studies by evaluating the impact of package footprint geometry on board level reliability for power MOSFETs in 3x3mm² packages.

979-8-3315-1612-3/25 $31.00 © 2025 IEEE

II. EXPERIMENTAL TESTING

Temperature cycling was performed on micro-leaded $3x3mm^2$ packages as part of gathering the experimental data. Devices were first mounted and reflowed on 1.6mm 8-layer PCB before undergoing optical and x-ray inspection. This was to ensure that each individual device was soldered correctly with no shorts or excessive PCB solder voiding (greater than 20% of pin area) prior to test. An example of a device soldered on the test PCB is given in Fig. 1.

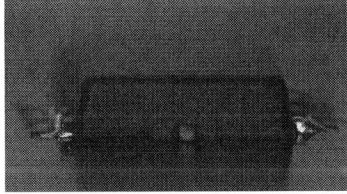

Fig. 1. Optical inspection of a $3x3mm^2$ test package from top and side angles.

To track the condition of the PCB solder, resistance monitoring was performed throughout temperature cycling as the devices were routed using a daisy chain configuration connected to an event detector. Prior to test execution, the resistance of each device in the daisy chain was measured to define the threshold resistances for failure. Temperature cycling was then carried out between -40°C to 125°C as per JESD22-A104F Test Condition G, at a rate of two cycles per hour. In addition, the duration for the test was defined to be the number of cycles which resulted in a minimum of 64% fails out of the entire test population.

However, the test was stopped after 10294 temperature cycles were performed over a total of 60 devices. The information from the failed devices was then used to generate a Weibull plot which is seen in Fig. 2.

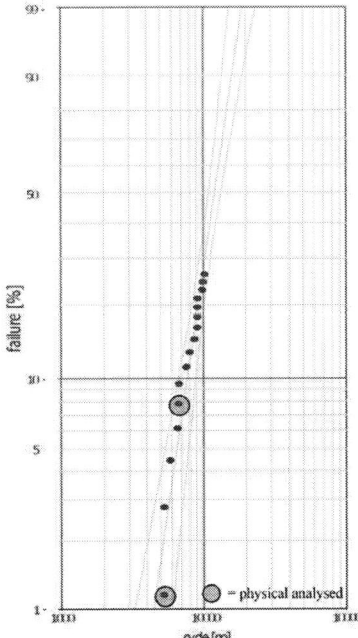

Fig. 2. Weibull plot for the micro-leaded $3x3mm^2$ derived from temperature cycling between -40°C to 125°C.

Looking at the Weibull plot produced from the temperature cycling test, the characteristic life ($N_{f,63\%}$) for the micro-leaded $3x3mm^2$ package was found to be 10373 cycles. The cause of device failures was seen to be PCB solder cracking from repeated cyclic expansion and contraction. This was identified by the event detector as significant cracks would cause the device to exceed its threshold resistance defined at the beginning of the temperature cycling test. Additionally, some devices were taken for cross section for additional confirmation of cracking as illustrated by Fig. 3.

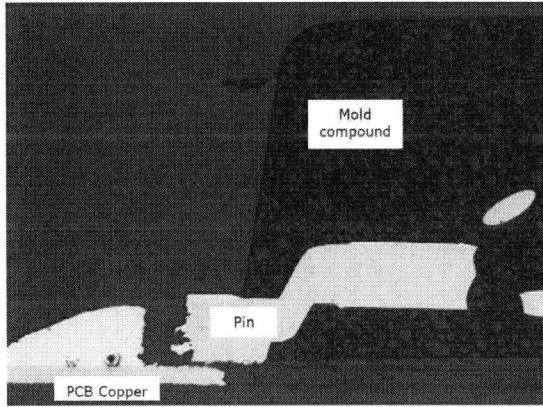

Fig. 3. Cross section of $3x3mm^2$ micro-leaded test device showing solder crack between the pin and the PCB copper pad.

III. THERMOMECHANICAL SIMULATIONS

To align with the experimental data, thermomechanical simulations were performed. The model consisted of the micro-leaded $3x3mm^2$ MOSFET on a single layer FR4 PCB measuring 40mm x 40mm x 1.6mm. No traces are present and only the manufacturer's recommended solder footprint was used for the top copper layout, with thickness of $2oz/ft^2$ as seen in Fig. 4(a). In order to focus on the effects of package footprint geometry on board level reliability, the internal MOSFET structure (wirebonds, die-attach, silicon die) was not modelled and their occupying volumes were filled and replaced with mold compound. The stress-free temperature of the model was assumed to be at 220°C before applying the temperature cycling profile used in physical testing, shown in Fig. 4(b).

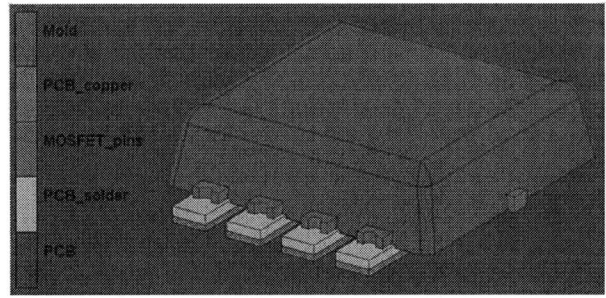

(a) Outline of micro-leaded MOSFET on PCB

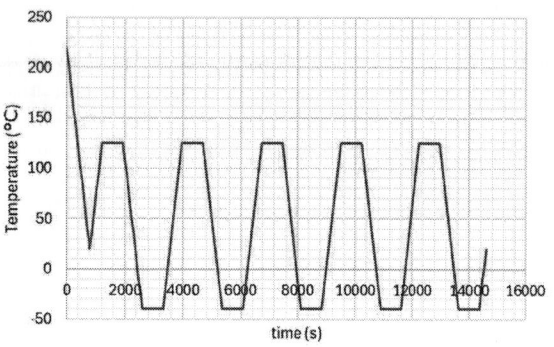

(b) Temperature cycling profile between -40°C to 125°C including an initial stress-free temperature of 220°C.

Fig. 4 Simulation setup of micro-leaded 3x3mm² MOSFET on PCB.

The Anand viscoplasticity model for SAC405 PCB solder (used in physical testing) was applied in the simulation, with parameters stated in Table I.

TABLE I. ANAND VISCOPLASTICITY PARAMETERS FOR SAC405 SOLDER [4].

Description	Symbol	unit	SAC405
Initial value of deformation resistance	S_0	MPa	1.3
Q: Activation energy. R: Universal gas constant	Q/R	K^{-1}	9000
Pre-exponential factor	A	s^{-1}	500
Stress multiplier	ξ	-	7.1
Strain rate sensitivity of stress	m	-	0.30
Hardening constant	h_0	MPa	5900
Deformation resistance saturation coefficient	$\hat{s}$	MPa	39.4
Strain rate sensitivity of saturation	N	-	0.03
Strain rate sensitivity of hardening	a	-	1.3

Five temperature cycles were performed in the simulation, in which the average creep strain per temperature cycle was calculated over the last four cycles. This was done by analyzing the nodes of the contact area between the solder joint, gate pin and the mold compound. The calculated average was then used in the double power law model [5] used to predict failure for Pb-free PCB solder alloys. This is shown in Equation (1), where $N_{f,50\%}$ is the mean number of cycles to failure and ε_{acc} is the average accumulated creep strain per temperature cycle.

$$N_{f,50\%} = (0.0468\varepsilon_{acc})^{-1} \qquad (1)$$

Table II shows the predicted mean number of cycles to failure through thermomechanical simulation using Equation (1).

TABLE II. SIMULATION RESULTS FROM TEMPERATURE CYCLING FROM -40°C TO 125°C.

Micro-leaded 3x3mm² MOSFET	
Average accumulated creep strain per temperature cycle, ε_{acc}	Mean number of cycles to failure (Double Power Law Model)
0.237%	9134

Given that the characteristic life from physical testing was shown to be 10373 cycles, this means that the mean number of cycles to failure is given to be $N_{f,50\%}$ = 8233 cycles. When

compared against the mean number of cycles from simulation, there is a 10% difference compared to that from physical testing. This shows good alignment with the experimental data, demonstrating that the simulation approach can be used to predict mean number of cycles to failure for other package footprint geometries.

IV. COMPARISON BETWEEN DIFFERENT PACKAGE GEOMETRIES USING THERMOMECHANICAL SIMULATIONS

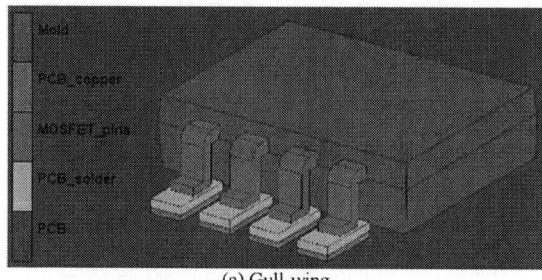

(a) Gull-wing

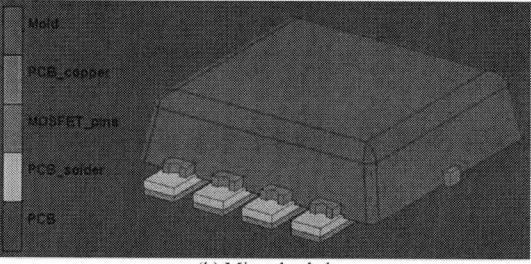

(b) Micro-leaded

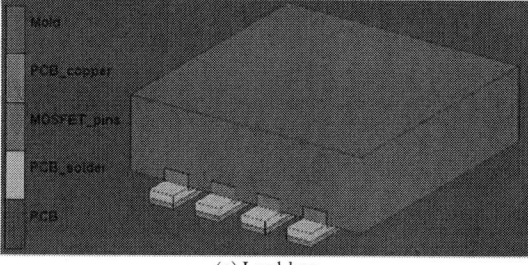

(c) Lead-less

Fig. 5. Package geometries evaluated through thermomechanical simulations using -55°C to 150°C temperature cycling profile.

Thermomechanical simulations were subsequently performed across three different 3x3mm² MOSFET package geometries, shown in Figs 5(a), (b) and (c). This time, the harsher temperature cycling profile of -55°C to 150°C was used as specified in AEC-Q101, retaining the same cycling and ramp rate per hour. The same material properties were applied for the mold compound, pins, PCB solder and PCB for all three scenarios. This was done so that a relative comparison could be made, focusing on the influence of package and pin geometry on board level reliability. As such, the contour plots showing the total accumulated creep strain after five temperature cycles is shown in Fig. 6.

The normalized contours reveal that overall, gull-wing packages accumulated less creep strain per temperature cycle, followed by micro-leaded and lead-less packages. It is known that cyclic stresses are greatest when located in areas with sharp

979-8-3315-1612-3/25 $31.00 © 2025 IEEE

edges and corners. The curved nature of the gullwing pin on the solder pad in Fig.6(a) creates a more even distribution of stress during the expansion and contraction of different parts from temperature fluctuations.

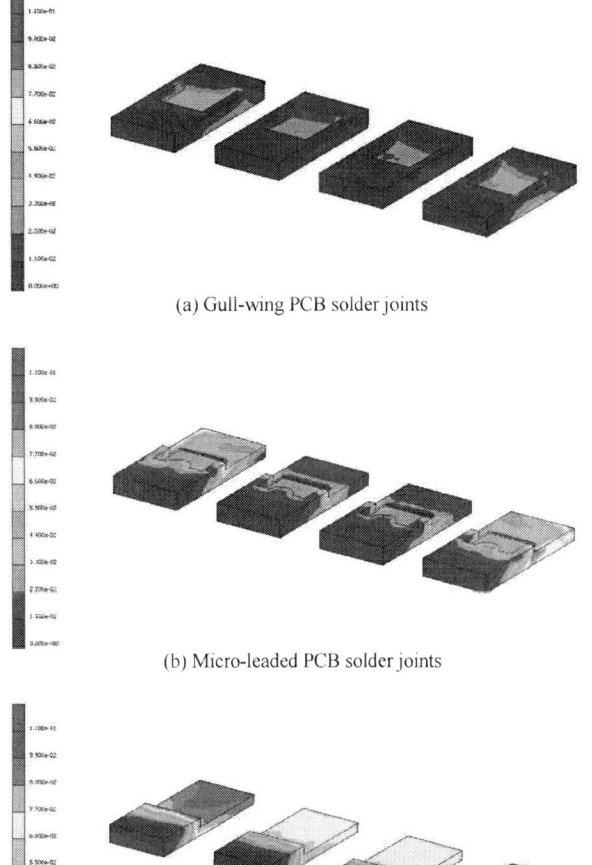

(a) Gull-wing PCB solder joints

(b) Micro-leaded PCB solder joints

(c) Lead-less PCB solder joints

Fig. 6. Normalised contour plots showing the total accumulated creep strain on PCB solder after five temperature cycles between -55°C to 150°C.

Furthermore, the mold compound is also in contact with the pins and the solder joints for the micro-leaded and lead-less packages in Fig.6(b) and 6(c) respectively.

The influence of the mold compound plays a contributing factor to the board level reliability due to the impact of its CTE on the pin and solder joint [6]. In the case of Fig.6(a), the absence of contact with the mold compound and gull-wing construction results in an even distribution of stress results in less accumulated strain per temperature cycle, ε_{acc} compared to micro-lead and lead-less over time.

As expected, the contour plots reveal creep strain is accumulated at its greatest in the corner joints across all packages [7], implying to be the first locations to encounter solder cracking. The same methodology was applied in the alignment simulation, where the ε_{acc} was calculated through averaging across all nodes in the contact area between the pin and the solder joint. The mean predicted number of cycles to failure is shown in Table III.

TABLE III. PREDICTED MEAN NUMBER OF CYCLES TO FAILURE FOR 3x3MM² PACKAGES WHEN CYCLED BETWEEN -55°C AND 150°C.

3x3mm² MOSFET lead geometry	Average accumulated creep strain per temperature cycle, ε_{acc}	Mean number of cycles to failure (Double Power Law Model)
Gull-wing lead	0.173%	12542
Micro-leaded	0.335%	6386
Lead-less	0.522%	4096

Nevertheless, all three package geometries were predicted to exceed the AEC-Q101 requirement of 1000 temperature cycles through the double power law model of Equation (1).

CONCLUSION

Despite being qualified to AEC-Q101 standards, the package footprint geometry is influential in the board level reliability of MOSFETs in automotive environments. The curved pins and the absence of contact between the mold compound and PCB solder enable gull-wing packages to have superior board longevity.

Nonetheless, the simulations do not suggest that micro-leaded and lead-less packages are unfit for purpose in automotive applications. As the automotive industry is shifting towards battery electric vehicles, there may potentially be less emphasis on the use of gull-wing packages. In the absence of an internal combustion engine, ambient temperatures can be greatly reduced. The resulting narrower temperature fluctuations would result in less expansion and contraction of electronic parts, in turn inducing less cyclic stresses and accumulating less creep strain over an automotive lifetime.

ACKNOWLEDGMENTS

The authors would like to thank Andy Lippold and Nexperia Germany for conducting the board level reliability tests on the micro-leaded 3x3mm² MOSFETs and providing the experimental data that was used in this paper.

REFERENCES

[1] Z.H. Barber, J.A. Leake and T.W. Clyne, 'The Doitpoms Project – A Web-Based Initiative For Teaching And Learning Materials Science' University of Cambridge, Introduction (doitpoms.ac.uk).

[2] R. Chao, Z. Chenhui, S. Jiang, X. Heping and X. Wenzheng, "Analysis of the influence of SOP lead shape on solder joint reliability," 2014 15th International Conference on Electronic Packaging Technology, Chengdu, China, 2014, pp. 538-541.

[3] Syed, Ahmer & Kang, Wonjoon. Board level assembly and reliability considerations for QFN type packages. SMTA International 2003.

[4] T. O. Reinikainen, P. Marjamaki and J. K. Kivilahti, "Deformation characteristics and micro- structural evolution of SnAgCu solder joints," EuroSimE 2005. Proceedings of the 6th International Conference on Thermal, Mechanial and Multi-Physics Simulation and Experiments in Micro-Electronics and Micro-Systems, 2005., Berlin, Germany, 2005, pp. 91, 98, doi: 10.1109/ESIME.2005.1502780.

[5] A. Syed, "Accumulated creep strain and energy density based thermal fatigue life prediction-Models for SnAgCu solder joints," 2004 Proceedings. 54th Electronic Components and Technology Conference

(IEEE Cat. No.04CH37546), Las Vegas, NV, USA, 2004, pp. 737-746 Vol.1, doi: 10.1109/ECTC.2004.1319419.

[6] Z. Li, H. Chen, H. Fan and J. Yang, "Optimization of Epoxy Molding Compound to Enhance the Solder Joints Robustness during Thermal Cycling for A Clip Bond Power Package," 2020 21st International Conference on Electronic Packaging Technology (ICEPT), Guangzhou, China, 2020, pp. 1-4, doi: 10.1109/ICEPT50128.2020.9201939.

[7] W. Sun et al., "Study on the Board-level SMT Assembly and Solder Joint Reliability of Different QFN Packages," 2007 International Conference on Thermal, Mechanical and Multi-Physics Simulation Experiments in Microelectronics and Micro-Systems. EuroSime 2007, London, UK, 2007, pp. 1-6, doi: 10.1109/ESIME.2007.360010.

Cost Effective and High Noise Immunity Methodology for Aging Evaluation of DC-link Capacitors in Traction Inverters

Seyed Hossein Aleyasin
Dipartimento di Energia
Politecnico di Torino
Turin, Italy
Seyed.Aleyasin@Polito.it

Fausto Stella
Dipartimento di Energia
Politecnico di Torino
Turin, Italy
Fausto.Stella@Polito.it

Radu Bojoi
Dipartimento di Energia
Politecnico di Torino
Turin, Italy
Radu.Bojoi@Polito.it

Enrico Vico
Dipartimento di Energia
Politecnico di Torino
Turin, Italy
Enrico.Vico@Polito.it

Abstract— DC-link capacitors are among the most critical components in traction inverters, with a significant potential for failure. Effective condition monitoring (CM) of these components is essential to enhance converter reliability and to prevent catastrophic failures. This paper presents a cost-effective methodology for evaluating the state-of-health of film capacitors used in traction inverters DC-link. The proposed approach is very simple and employs a controlled discharge test to measure capacitance variation over time, serving as an indirect indicator of capacitor aging. The main strengths of the proposed technique are the minimal additional component requirements, ensuring cost-efficiency, and the high noise immunity. The discharge test can be conducted during the converter turn-off and it can be fully automated, without the need to remove the capacitor or to modify the power stage of the converter. Experimental results are provided to validate the proposed methodology.

Keywords—Reliability, Condition Monitoring, DC-link Capacitors, Film Capacitors, Traction Inverters.

I. INTRODUCTION

Modern power converters used in safety-critical applications such as automotive, medical, and aerospace must guarantee high levels of reliability. In such applications, an unexpected failure may be costly or even endanger the human life. In this context, condition monitoring (CM) techniques may be implemented to evaluate the state of health of the most critical components to schedule preventive maintenance before a catastrophic failure occurs [1]. According to the literature, although there is a possibility of failure in all of the parts, semiconductors, and capacitors are among the most fragile components of power converters. Thus, designing an effective condition monitoring system to assess the health of DC-link capacitors can improve the reliability of the converter. The end of a device's useful life (EUL) can be estimated by analyzing degradation indicators collected from condition monitoring tests. This information allows for planning predictive maintenance based on condition-monitoring results. By employing this maintenance approach, downtime of the converter can be minimized, and its availability can be improved by repairing or replacing components before a failure arises [2]. According to the literature, the two main categories of lifetime indicators for capacitors are non-electrical and electrical parameters.

Non-electrical indicators, such as internal temperature, structural and weight variations, can help assess capacitor health [3]. However, evaluating these parameters often requires dedicated measurement equipment typically available only in a controlled laboratory environment. As a result, integrating such solutions into commercial applications is usually impractical, thus limiting their widespread use.

The second category of methodologies uses the variation of electrical parameters such as capacitance, ESR (equivalent series resistance), ESL (equivalent series inductance) and electrical power losses as lifetime indicators for capacitors. Different methodologies have been presented in the literature for measuring the capacitance variation, such as in [6] and [7], where high-frequency signal injection methodologies are used. However, these methods demand precise and high-bandwidth measurement sensors. In aluminum electrolytic capacitors, the ESR is predominant in the mid-frequency range, as the impedance remains capacitive at low frequencies. The research in [8] leverages this characteristic by using signal injection at low and mid frequencies to estimate the ESR value. Using converter topology to estimate capacitor current can also be applied to simpler topologies. For instance, the method from [9] aims at assessing the health state of the capacitor in a boost converter by measuring the capacitor voltage, while the current is estimated based on the current of the diode and the capacitance and the ESR are estimated using Kalman filter. A similar approach is used in [10] where the authors measured the voltage and the current of the capacitor, in order to estimate the health state. The voltage is directly measured; however, to avoid using a series current sensor, the capacitor current is calculated by subtracting the input current of the inverter (coming out of a rectifier) and the input current of the power semiconductor module. By calculating the capacitor current, the capacitance and the ESR are estimated using the Goertzel algorithm. The estimation accuracy for the capacitance in these two papers could reach to more than 10% and 3% respectively. Since the end-of-life criteria of the capacitance for film and electrolytic capacitors are respectively 95% and 80% of the initial value, these errors are considered relatively high. The average power loss can be used to estimate the capacitance, as in [11]. For this solution, samples of voltage and current are required to calculate the average power loss; therefore, a current sensor is required. The power electronic system is considered as a black-box model in [12]. Artificial neural network (ANN)-based methods are utilized to estimate capacitor parameters in this approach. The primary challenge with this method is identifying an appropriate training dataset. In [13], the capacitor discharges through motor windings and the authors use the discharging profile to assess the health state. However, this approach is prone to machine parameter variations and requires an error correction algorithm to be accurate. In [14], an external network of resistors and switches is designed to create a discharging path for the DC-link capacitor. This method however targets the ESR and capacitance in electrolytic capacitors that are not suitable for traction inverters due to low RMS current capabilities.

In addition, most of the solutions already available in the literature target electrolytic capacitors, where the ESR variation can be easily measured and used as an indirect indicator of their health state. However, electrolytic capacitors are not suited for traction converters due to their limited RMS current capability. For this reason, film capacitors are usually preferred in traction converters, due to their high RMS current capability, low ESR, good thermal stability, and their relatively high resonance frequency [4]. However, the ESR in film capacitors is too low to be measured accurately, which makes the capacitance variation the only electrical parameter that could indicate the state of health of these types of capacitors.

The end-of-life criteria for different types of capacitors is shown in TABLE I. . For film capacitors, the end-of-life condition is typically reached when capacitance decreases to 95% of its initial value C_0 [5].

TABLE I.   THE END-OF-LIFE CRITERIA OF CAPACITORS [5]

Capacitor Type	Electrolytic Capacitors	Ceramic Capacitors	Film Capacitors
End-of-life criteria	$C/C_0 < 80\%$ $R_{ESR}/R_{ESR0} > 2$	$C/C_0 < 95\%$	$C/C_0 < 95\%$

This paper introduces an innovative, cost-effective approach for assessing the health of film capacitors used in traction inverters. The proposed methodology enables to precisely evaluate the capacitance variation which can then be used as an indirect indicator of the current health state of the capacitor. The capacitance variation is evaluated by means of a simple controlled discharge test that can be performed directly on the converter with minimal additional hardware. In particular, the only additional components required by the proposed technique are a discharging resistor and a low-current power switch, while the voltage discharge profile is measured using the DC-link voltage sensor already present in standard power converters. To evaluate the effectiveness and accuracy of this technique, both simulations in PLECS and experimental results are presented in the paper. These findings show the feasibility of the proposed method, suggesting it is a viable option for real-world applications in monitoring capacitor health in traction inverters.

The structure of this paper is as follows: Section II presents the proposed methodology, Section III provides the PLECS simulation results, Section IV discusses the experimental results obtained from an H-bridge prototype using DC-link film capacitors, and Section V summarizes the results and provides the conclusions.

II. PROPOSED COMMISSIONING TEST

The proposed methodology evaluates the capacitance variation of the DC-link capacitor in a power converter to assess its health state. To achieve this, a dedicated commissioning test has been developed to measure

capacitance changes. The proposed solution estimates the capacitance by measuring the discharging voltage profile of the DC-link capacitor, thus enabling to measure the capacitance variation by repeating this test over time. The functional schematic of the proposed commissioning test is shown in Fig. 1, where the main DC-link capacitor ($C_{DC\text{-}link}$) is discharged using a dedicated resistor (R_{DSC}).

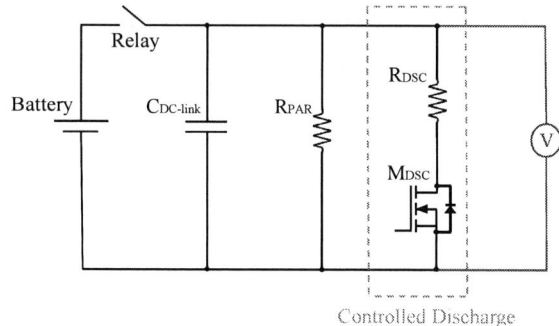

Fig. 1.   Functional electrical schematic of the proposed commissioning test circuit.

The commissioning test consists of the following steps:

1. The DC-link capacitor, initially charged at the battery voltage, is disconnected from the power source (e.g. vehicle battery) by opening the relay (usually present for safety reasons). During this phase, all the switches of the power converter (e.g. power MOSFETs of a three-phase inverter) are turned off.

2. At this point, the DC-link capacitor starts slowly discharging through R_{PAR}, which represents the equivalent parallel resistance given by the safety resistor (normally placed in parallel to the capacitor), and the parasitic resistances of converter components (e.g. power MOSFETs).

3. The MOSFET M_{DSC} is turned on and the DC-link capacitor starts discharging through R_{DSC}. The discharge voltage curve is measured using a voltage sensor (e.g. same voltage sensor already used for measuring the DC-link voltage during the normal operations of the converter).

4. The measured discharge voltage curve of the capacitor is integrated over time. **This integration starts and ends at specific voltages, respectively V_{start} and V_{end}, as shown in Fig. 2.** Here, t_{start} is the time at which the discharge voltage profile first reaches V_{start}, and t_{end} is the time when the voltage curve reaches V_{end}.

The transient discharge equation for an RC circuit is:

$$V_{new}(t) = V_0 e^{\frac{-t}{R_{DSC}C_{new}}} \; ; \; V_{aged}(t) = V_0 e^{\frac{-t}{R_{DSC}C_{aged}}} \qquad (1)$$

where C_{new} is the capacitance value of the DC-link capacitor at the beginning of the converter's life (i.e., at the first startup), C_{aged} represents the capacitance after a certain number of operational hours, R_{DSC} is the discharging resistor used for the controlled capacitor discharge (Fig. 1), while the equivalent series resistance of the MOSFET and the film capacitor can be neglected as shown in the next paragraphs.

From the integration of (1), it is possible to compute a K coefficient as shown in (2).

$$K = \int_{t_{start}}^{t_{end}} V_0 e^{\frac{-t}{R_{DSC}C}} dt \qquad (2)$$

The integration start time t_{start} and the end time t_{end} are opportunely selected so that t_{start} corresponds to the time when the capacitor voltage reaches V_{start}, and t_{end} corresponds to the time when the voltage reaches V_{end}. Assuming constant values for V_{start}, V_{end} and R_{DSC}, the integration interval t_{start} to t_{end} will vary depending on the actual capacitance value as shown in Fig. 2.

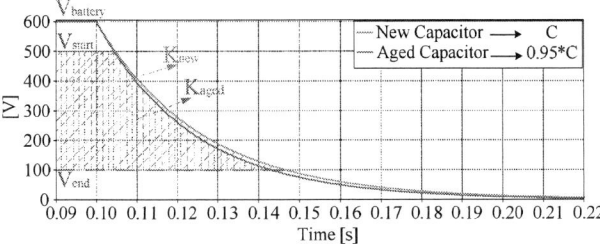

Fig. 2. Discharging voltage profiles of the new and the aged capacitor.

By solving (1) for t_{start} and t_{end}, and substituting the results in (2), the value of K coefficient is derived as follows:

$$K = V_0 R_{DSC} C \left(e^{\frac{R_{DSC}C \, ln\left(\frac{V_{start}}{V_0}\right)}{R_{DSC}C}} - e^{\frac{R_{DSC}C \, ln\left(\frac{V_{end}}{V_0}\right)}{R_{DSC}C}} \right) \qquad (3)$$

where V_0 is the initial voltage of the capacitor, V_{start} and V_{end} are the values of starting and ending point of the integral respectively.

If these two voltages V_{start} and V_{end} remain the same when the test is repeated, it is possible to obtain:

$$\frac{K_{new}}{K_{aged}} = \frac{C_{new}}{C_{aged}} \qquad (4)$$

where K_{new} is the value of K from the first test, which has been saved from the first day of the converter's operations, while K_{aged} is the value of K obtained from the latest test.

The capacitance variation over time can be easily computed by comparing the 'original' K measured at the beginning of the life of the converter with the one obtained from the latest test.

The proposed test can be conducted during the start-up or shut-down phases of the converter when it is not operating. However, during the shut-down phase, the DC-link capacitor may be at a significantly high temperature due to prior operation, potentially affecting the capacitance value. For this reason, it is preferable to perform the test during the start-up phase of the converter, when all components are at ambient temperature. The ambient temperature typically varies within a limited range, however, in cases of significant variation, this temperature can be easily measured (e.g., using the heatsink thermistor commonly present in converters). The effect of temperature on capacitance variation can then be compensated

using a model that accounts for capacitance changes with temperature.

III. SIMULATION RESULTS

Simulations were conducted in PLECS to evaluate the performance of the proposed solution in an ideal operating scenario (i.e., no measurement noise) thus computing the optimal values for V_{start}, V_{end} and R_{DSG}. In particular, selecting appropriate values for V_{start}, V_{end} and R_{DSG}, is crucial for defining the integration period. The integration period must be high enough to collect a sufficient number of samples given a fixed sampling frequency, which is typically dictated by the converter's voltage sensor used to monitor the DC-link bus voltage. The integration period can be increased either by increasing the difference between V_{start} and V_{end} or by increasing R_{DSG}.

A longer integration period would allow more precise measures, especially considering that the voltage sensor is affected by a certain percentage of noise that can be reduced by increasing the number of samples. Therefore, to increase the sensitivity of the measurement, and also to decrease the effect of noise, it is better to calculate the integral from 85-90% of the initial voltage to 10-15% of it.

In case of a traction converter, the starting voltage V_{start} must be chosen to be sufficiently below the nominal battery voltage of the vehicle, allowing the test to be conducted even when the battery is partially discharged. Additionally, V_{end} should be set high enough to avoid the non-linear measurement range of the voltage sensor, as typical power converter voltage sensors have non-linear behavior near zero voltage. Similar considerations can be made for R_{DSC} that must be sufficiently high to increase the integration time, but sufficiently low so to be able to neglect the contribution of R_{PAR}. As said, R_{PAR} represents the parallel resistor usually added in parallel to the DC-link capacitor for safety reasons, the leakage current of the power semiconductors when are commanded off, and so on. As R_{PAR} may considerably change over time (e.g., due to power semiconductor aging), it is important to select a value of R_{DSC} sufficiently low so be able to neglect R_{PAR}.

A PLECS simulation has been performed using the following parameters: $R_{DSC} = 200 \, \Omega$, $R_{PAR} = $ infinite, $C_{new} = 120 \, \mu F$, $C_{aged} = 100 \, \mu F$, and $f_{sampling} = 20 \, kHz$. The initial voltage of the DC-link capacitor is set to 600 V, while V_{start} and V_{end} are 500 V and 100 V respectively.

The discharging voltage is measured and integrated between V_{start} and V_{end} for both capacitors, thus enabling to compute K_{new} and K_{aged}. The PLECS testing circuit used during the simulation is shown in Fig. 3.

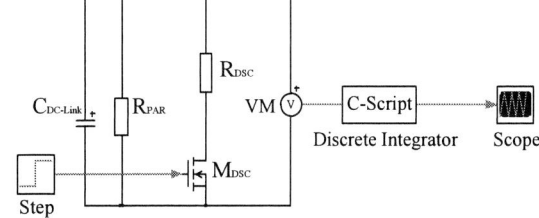

Fig. 3. PLECS implementation of the controlled discharging test.

Fig. 4 (a) shows the discharging voltage profiles in the case of new and aged capacitors, while Fig. 4(b) reports the value

979-8-3315-1612-3/25 $31.00 © 2025 IEEE

of K over time. In this case, the measured values of K_{new} and K_{aged} are respectively equal to 9.5967 and 8.0088.

According to eq. (4) the ratio between K_{new} and K_{aged} should be equal to the ratio between C_{new} and C_{aged}. In this case, the measured ratio is 1.19826, while the real one is 1.2. This corresponds to an estimation error of 0.1444%, which is due to the discretization of the integral (i.e. 20kHz). This integration error can be decreased by increasing the sampling frequency or by increasing the integration time interval (i.e., this can be achieved by increasing R_{DSC} or by increasing the difference between V_{start} and V_{end}).

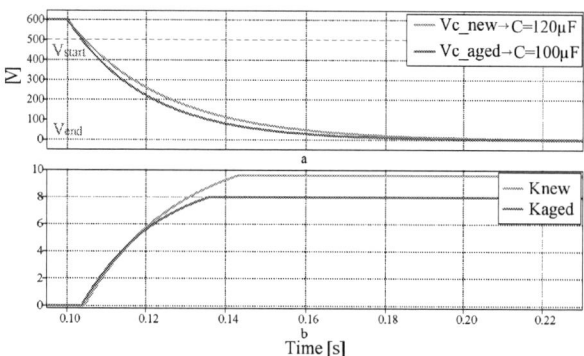

Fig. 4. Simulation results from the discharging test: (a) voltage discharging profiles over time, (b) Integrals Knew and Kaged over time.

The values of C_{new} and C_{aged} used during the simulation have been conveniently selected to clearly show the difference between the voltage discharge profiles reported in Fig. 4. However, in a real-case scenario, the end-of-life criteria for a film capacitor is met when the capacitance drops to 95% of the initial value. When the end-of-life criteria is met, necessary maintenance should be conducted to prevent possible catastrophic failures of the DC-link capacitor.

Further considerations regarding the selection of R_{DSC} can be made by analyzing the discharging current, as reported in Fig. 5. The plot shows a maximum discharging current of approximately 3 A. This relatively low current allows the use of a discharge switch with a low peak current rating, making it a cost-effective choice.

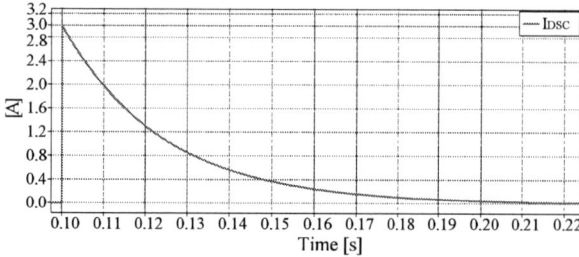

Fig. 5. Discharging current through resistor R_{DSC} in the case of new capacitor.

IV. EXPERIMENTAL RESULTS

The proposed methodology has been experimentally evaluated using an 800V, 20 kVA H-bridge SiC converter used as a rotor exciter for automotive applications that is shown in Fig. 6. For practical reasons, one of the four switches of the H-bridge converter has been used as the discharging switch for the validation of the proposed methodology as shown in Fig. 7. The discharging resistor, R_{DSC} has been attached to the output 1 to provide the discharging path through the switch M_2. The value of R_{DSC} is selected to be 200 Ω, while the DC-link capacitor is changed so that the new capacitor is 130µF, and then the capacitor is reduced to 120µF to emulate the aged capacitor.

The voltage is measured using the embedded V_{bus} measuring system of the converter, consisting of a voltage divider supplying an isolated operational amplifier whose output is sampled using the 12-bit ADC converter embedded in the microcontroller of the power converter. The ADC sampling rate is set to 20 kHz.

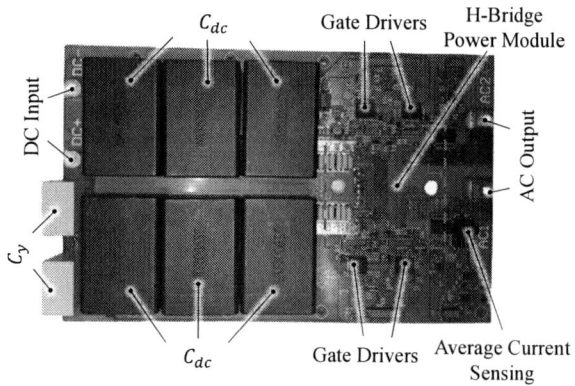

Fig. 6. H-bridge SiC power converter.

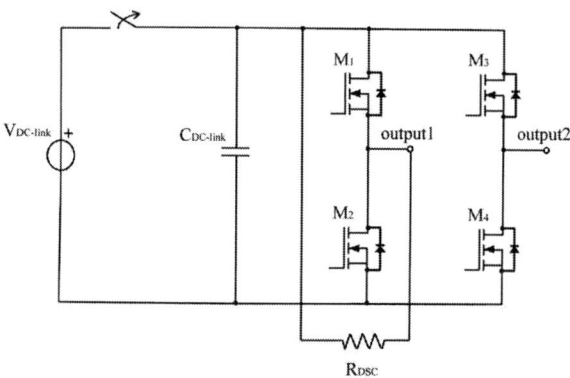

Fig. 7. Schematic of the experimental test rig, in which one of the H-bridge power MOSFETs is conveniently used as a discharging switch.

The DC-link capacitor is pre-charged up to 600 V using an external DC power supply. Once the target voltage is reached, the power supply is disconnected and the MOSFET M_2 is turn-on. At this point, the DC-link capacitor starts discharging.

The discharging voltages measured by the MCU for the 130µF and 120µF capacitors are shown in Fig. 8. A non-linearity near the zero region can be observed, which is a characteristic of the voltage sensors typically used for measuring DC-link voltage. This non-linearity explains why the capacitor voltage drops slightly below zero after complete discharge in Fig. 8. In this test, the integral was calculated between 500 V and 100 V to avoid the non-linear region. The measured, K_{new} and K_{aged} are respectively 10.39 and 9.59. The ratio of $\frac{K_{aged}}{K_{new}}$ is 0.923 the same as $\frac{C_{aged}}{C_{new}}$, which verifies the validity of (4).

979-8-3315-1612-3/25 $31.00 © 2025 IEEE   1411

Fig. 9 shows the discharging current and voltage of the 120 μF capacitor which have been monitored using an oscilloscope. Unlike measurements taken with the voltage sensor of the converter, no non-linear region is observed here due to the oscilloscope's direct measurement. The peak discharging current slightly exceeds 3A, allowing for the selection of a MOSFET with a low current rating. The value of this current is influenced by the discharging resistor, which must be designed according to the specific requirements of the application. As said, the discharging resistor should be selected such that its resistance is negligible compared to the safety parallel resistance of the capacitor and the parasitic resistances in the system (R_{PAR} in Fig. 1). Additionally, the total energy losses on the resistor (equal to the energy stored in the DC-link capacitor) must be considered, so to avoid excessive overheating of the resistor during the test.

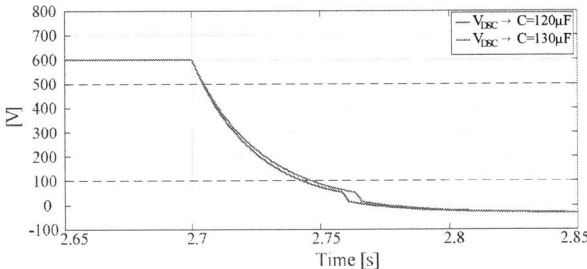

Fig. 8. Voltage discharge profiles with two different values of capacitance measured with the H-bridge embedded voltage sensor.

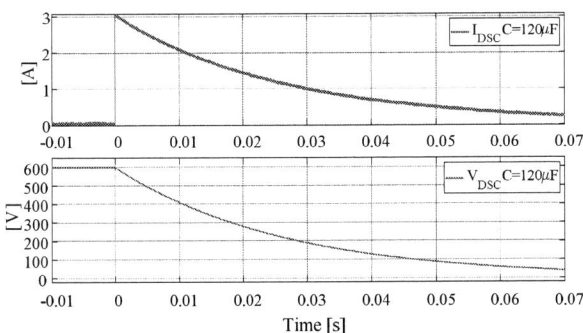

Fig. 9. Voltage and current discharge profiles of the 120μF capacitor measured with the oscilloscope.

V. CONCLUSION

In this paper, a reliable methodology for monitoring the health state of film capacitors in traction inverters has been presented and experimentally validated. The proposed approach requires only a few additional components such as a low-current switch and a resistor, making it both cost-effective and simple to implement. This technique does not require calibrated resistors and voltage sensors, as long as these components have the same characteristics over time. Even a relatively low-accuracy voltage sensor, normally used for the measurement of the DC-link voltage, is sufficient to obtain a precise estimation of the capacitance variation. By periodically repeating the commission test over time, each time at the start-up stage of the converter, it is possible to track the capacitance change.

Measuring the capacitance variation, which is the indirect indicator of the health state of the capacitor in this methodology, can be achieved with high accuracy. Once the capacitance reaches to the critical point, at 95% of the initial value, the required maintenance should be conducted to keep the converter performing in the normal way. Furthermore, this technique can be combined with other methods to assess the health of additional converter components, such as power semiconductors [15], [16], thereby enhancing the overall reliability of the converter.

Further work will consist in applying the same methodology also to power converters for traction applications having a ceramic DC-link capacitor such as [17].

REFERENCES

[1] H. Wang, M. Liserre and F. Blaabjerg, "Toward Reliable Power Electronics: Challenges, Design Tools, and Opportunities," in IEEE Industrial Electronics Magazine, vol. 7, no. 2, pp. 17-26, June 2013, doi: 10.1109/MIE.2013.2252958.

[2] M. Ghadrdan, S. Peyghami, H. Mokhtari and F. Blaabjerg, "Condition Monitoring of DC-Link Electrolytic Capacitor in Back-to-Back Converters Based on Dissipation Factor," in IEEE Transactions on Power Electronics, vol. 37, no. 8, pp. 9733-9744, Aug. 2022, doi: 10.1109/TPEL.2022.3153842.

[3] A. Shrivastava, M. H. Azarian, C. Morillo, B. Sood and M. Pecht, "Detection and Reliability Risks of Counterfeit Electrolytic Capacitors," in IEEE Transactions on Reliability, vol. 63, no. 2, pp. 468-479, June 2014, doi: 10.1109/TR.2014.2315914.

[4] H. Wang and F. Blaabjerg, "Reliability of Capacitors for DC-Link Applications in Power Electronic Converters—An Overview," in IEEE Transactions on Industry Applications, vol. 50, no. 5, pp. 3569-3578, Sept.-Oct. 2014, doi: 10.1109/TIA.2014.2308357.

[5] Z. Zhao, P. Davari, W. Lu, H. Wang and F. Blaabjerg, "An Overview of Condition Monitoring Techniques for Capacitors in DC-Link Applications," in IEEE Transactions on Power Electronics, vol. 36, no. 4, pp. 3692-3716, April 2021, doi: 10.1109/TPEL.2020.3023469.

[6] P. Sundararajan et al., "Condition Monitoring of DC-Link Capacitors Using Goertzel Algorithm for Failure Precursor Parameter and Temperature Estimation," in IEEE Transactions on Power Electronics, vol. 35, no. 6, pp. 6386-6396, June 2020, doi: 10.1109/TPEL.2019.2951859.

[7] T. H. Nguyen and D. -C. Lee, "Deterioration Monitoring of DC-Link Capacitors in AC Machine Drives by Current Injection," in IEEE Transactions on Power Electronics, vol. 30, no. 3, pp. 1126-1130, March 2015, doi: 10.1109/TPEL.2014.2339374.

[8] P. Sundararajan, M. H. M. Sathik, F. Sasongko, C. S. Tan, M. Tariq and R. Simanjorang, "Online Condition Monitoring System for DC-Link Capacitor in Industrial Power Converters," in IEEE Transactions on Industry Applications, vol. 54, no. 5, pp. 4775-4785, Sept.-Oct. 2018, doi: 10.1109/TIA.2018.2845889.

[9] K. Abdennadher, P. Venet, G. Rojat, J. -M. Rétif and C. Rosset, "A Real-Time Predictive-Maintenance System of Aluminum Electrolytic Capacitors Used in Uninterrupted Power Supplies," in IEEE Transactions on Industry Applications, vol. 46, no. 4, pp. 1644-1652, July-Aug. 2010, doi: 10.1109/TIA.2010.2049972.

[10] L. A. Rodrigues, V. Junior De Paris, A. S. Vaccari and G. Waltrich, "Real Time Measurements of Aluminum Electrolytic Capacitor Parameters in EVs Inverters," 2023 IEEE Energy Conversion Congress and Exposition (ECCE), Nashville, TN, USA, 2023, pp. 3495-3502, doi: 10.1109/ECCE53617.2023.10362689.

[11] K. Yao, W. Tang, X. Bi and J. Lyu, "An Online Monitoring Scheme of DC-Link Capacitor's ESR and C for a Boost PFC Converter," in IEEE Transactions on Power Electronics, vol. 31, no. 8, pp. 5944-5951, Aug. 2016, doi: 10.1109/TPEL.2015.2496267.

[12] H. Soliman, I. Abdelsalam, H. Wang and F. Blaabjerg, "Artificial Neural Network based DC-link capacitance estimation in a diode-bridge front-end inverter system," 2017 IEEE 3rd International Future Energy Electronics Conference and ECCE Asia (IFEEC 2017 - ECCE Asia), Kaohsiung, Taiwan, 2017, pp. 196-201, doi: 10.1109/IFEEC.2017.7992442.

[13] X. Wei, B. Yao, Y. Peng, Y. Sun, K. Wang and H. Wang, "An Improved Discharge Profile-Based DC-Link Capacitance Estimation for Traction Inverter in Electric Vehicle Applications," in IEEE Transactions on Power Electronics, vol. 39, no. 7, pp. 8696-8708, July 2024, doi: 10.1109/TPEL.2024.3383153.

979-8-3315-1612-3/25 $31.00 © 2025 IEEE

[14] Y. Wu and X. Du, "A VEN Condition Monitoring Method of DC-Link Capacitors for Power Converters," in IEEE Transactions on Industrial Electronics, vol. 66, no. 2, pp. 1296-1306, Feb. 2019, doi: 10.1109/TIE.2018.2835393.

[15] F. Stella, O. Olanrewaju, Z. Yang, A. Castellazzi, e G. Pellegrino, "Experimentally validated methodology for real-time temperature cycle tracking in SiC power modules", Microelectronics Reliability, vol. 88–90, pp. 615–619, set. 2018, doi: 10.1016/j.microrel.2018.07.072.

[16] M. Gregorio, F. Stella, R. Bojoi, e F. Pagani, «Junction Temperature Estimation via Plug-in System for the Design Validation of IGBT Industrial Power Converters», IEEE Transactions on Industry Applications, vol. 58, fasc. 5, pp. 6310–6321, set. 2022, doi: 10.1109/TIA.2022.3182320.

[17] F. Stella, E. Vico, D. Cittanti, C. Liu, J. Shen, e R. Bojoi, «Design and Testing of an Automotive Compliant 800V 550 kVA SiC Traction Inverter with Full-Ceramic DC-Link and EMI Filter», in 2022 IEEE Energy Conversion Congress and Exposition (ECCE), ott. 2022, pp. 1–8. doi: 10.1109/ECCE50734.2022.9948096.

A 3D Structure of Single-Sided Cooling Power Module with Low Thermal Resistance and Low Inductance

Hirofumi Hisamochi
Corporate research and developtment division
IHI Corporation
Yokohama, Japan
hisamochi5759@ihi-g.com

Koki Notake
Corporate research and developtment division
IHI Corporation
Yokohama, Japan
notake3617@ihi-g.com

Yoshiaki Takahashi
Corporate research and developtment division
IHI Corporation
Yokohama, Japan
takahashi1238@ihi-g.com

Koji Yamaguchi
Corporate research and developtment division
IHI Corporation
Yokohama, Japan
yamaguchi2284@ihi-g.com

Abstract— **This paper describes the three-dimensional structure of a single-sided cooling power module that uses a lead frame as a heat conduction path to cool power semiconductor dies on both sides. The proposed structure achieves low thermal resistance and low inductance, improving the efficiency of motor drive inverters. Simulation results show that the thermal resistance is reduced by 18% and the parasitic commutation loop inductance is reduced by 60% compared to the conventional module. The manufacturability of the proposed module is confirmed by fabricating prototypes, and double pulse tests (800 V_DC, turn-on current of 300 A, turn-off current of 400 A) validate its electrical performance.**

Keywords—Single-sided cooling, Three-dimensional structure, Power module packaging, Inverter, high power density, Silicon carbide

I. INTRODUCTION

As the electrification of vehicles, such as automobiles and aircraft, is progressing toward the realization of a decarbonized society, there is a need to improve the efficiency of motor drive inverters to reduce energy consumption. Recently, the replacement of silicon insulated-gate bipolar transistor (Si-IGBT) with silicon carbide metal-oxide-semiconductor field-effect transistor (SiC-MOSFET) has been advancing to achieve lower power loss through high-speed switching and low on-resistance [1]–[3]. However, high-speed switching causes high-voltage surges, which may cause dielectric breakdown and higher levels of electromagnetic interference (EMI) [4], [5]. Switching surges can be reduced by decreasing the inductance of the commutation loop in half-bridge power modules for motor drive inverters. Thus, low power loss and low surge voltage can be achieved simultaneously through high-speed switching.

Three-dimensional packaging structures have been reported to achieve the reduction of parasitic inductance in the commutation loop of half-bridge power modules [6]–[10]. As shown in Fig. 1(b), the current path between the source pad of the upper arm's MOSFET and the drain pad of the lower arm's MOSFET is much shorter than that of Fig. 1(a). As a result, the parasitic inductance between DC+ and DC- in Fig. 1(b), which is the commutation loop of the half-bridge power module, is much smaller than that in Fig. 1(a).

According to previously published studies, three-dimensional structures are commonly applied to double-sided cooling power modules. When applying three-dimensional structures to single-sided cooling half-bridge power modules, the thermal resistance between the lower arm's MOSFET and heatsink becomes higher than that of the upper arm's

MOSFET and the heatsink. This is because the size of the source pad, which is the heat conduction path, is smaller than that of the drain pad, and a conductive spacer needs to be placed between the source pad and the direct bonded copper (DBC) to ensure the electrical insulation distance.

From the perspective of the layout design of the inverter, it is desirable to place both the dc-link capacitor and the gate driver board in close proximity to the power module. However, the interface to connect the dc-link capacitor and gate driver board is very limited because the upper and lower surfaces of the double-sided cooling power module are surrounded by the heatsinks. Consequently, it is difficult to place both the dc-link capacitor and gate driver board close to the double-sided cooling power module. If the dc-link capacitor is placed far from the power module, the commutation loop inductance increases, leading to higher surge voltage. If the gate driver board is placed far from the power module, the rise and fall time of gate drive signal may increase, leading to an increase in power loss due to a decrease in switching speed.

Therefore, placing both the dc-link capacitor and the gate driver board close to the power module is more important than placing the heatsink, in order to achieve a high-efficiency motor drive inverter. This means that the area of the heat conduction path from the power module surface to the heatsink is limited because the power module for a high-efficiency motor drive inverter should be surrounded by the dc-link capacitor and the gate driver board.

Based on the above, this paper proposes a single-sided cooling half-bridge power module structure with low inductance and low thermal resistance, achieved by applying the three-dimensional structure.

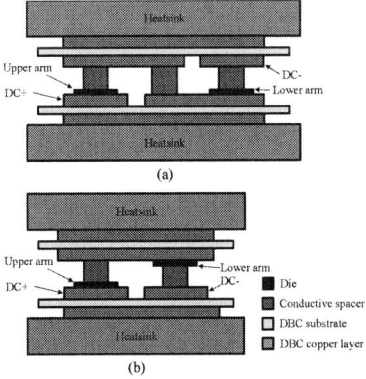

Fig. 1. Double-sided cooling power module. (a) Conventional structure. (b) Three-dimensional structure with dies in each arm attached to their respective substrates.

II. PROPOSED STRUCTURE OF SINGLE-SIDED COOLING POWER MODULE

A. Low Thermal Resistance and Low Inductance Structure

Fig. 2 shows the circuit configuration of the half-bridge power module. Fig. 3 illustrates the proposed structure of a single-sided cooling half-bridge power module. The convex part of the lead frame is directly bonded by soldering to the source pad of the upper arm's SiC-MOSFET. This convex part is provided to reduce electrical resistance and parasitic inductance between the source pad of the upper arm's SiC-MOSFET and the drain pad of the lower arm's SiC-MOSFET. This convex part also forms a heat conduction path from the source pad of the upper arm's SiC-MOSFET to the DBC substrate through the convex part in the center of the lead frame. Therefore, the heat generated by the upper arm's SiC-MOSFET is transferred from both the source and drain pads to the heatsink via the DBC substrate.

The conductive spacer is bonded by soldering between the source pad of the lower arm's SiC-MOSFET and the DBC copper to form the current path of the lower arm. Additionally, this conductive spacer forms a heat conduction path from the source pad of the lower arm's SiC-MOSFET to the DBC substrate. There is also a heat conduction path from the drain pad of the lower arm's SiC-MOSFET to the DBC substrate through the convex part in the center of the lead frame, similar to the source pad of the upper arm's SiC-MOSFET. Therefore, the heat generated by the lower arm's SiC-MOSFET is transferred from both the source and drain pads to the heatsink via the DBC substrate.

From the above features, even though it is the single-sided cooling power module with only one heatsink bonded to the bottom side of the DBC substrate, the heat generated by the upper and lower arms' SiC-MOSFETs is transferred from both the drain pads and the source pads, thereby lowering the thermal resistance.

In addition, since the lead frame forms not only the heat conduction path but also the thick and short current path, the parasitic inductance between DC+ and DC-, which forms the commutation loop of the half-bridge power module, becomes much smaller than that of the conventional structure.

B. Thermal Simulation

Thermal simulations were performed to evaluate the thermal resistance of the proposed power module using software Femtet (Murata Software). Fig. 4 shows the simulation model of the conventional single-sided cooling power module with Al wire bonds that are used for the connection of the source pads. The total Al wire cross-sectional area is 1.15 mm². Fig. 5 shows the simulation model of the proposed power module (Type A). Table I shows the material properties used in the thermal simulation of the proposed power module (Type A).

To compare the simulation results, both the conventional and proposed power module (Type A) models use the same size and shape of copper area of 1600 mm² (40 mm × 40 mm) under the DBC substrate. The number of parallel SiC-MOSFETs is four in both models. In the thermal simulation, only the heat conduction to the heatsink attached to one side (under the bottom DBC copper) is considered, and the rest of the surroundings are thermally insulated. The heat transfer coefficient of the heatsink is 5000 W/m²·K.

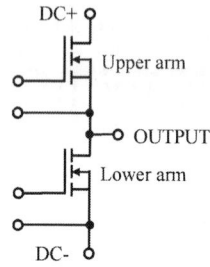

Fig. 2. Half-bridge configuration.

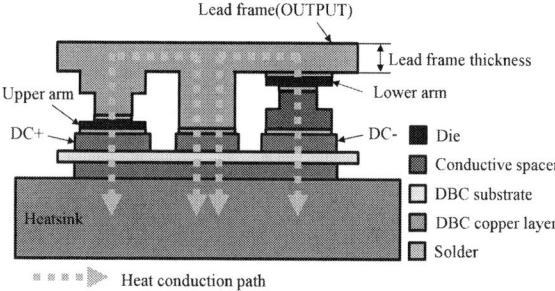

Fig. 3. Cross section of three-dimensional single-sided cooling power module.

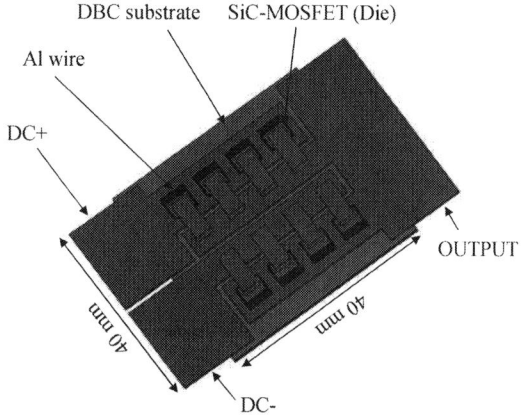

Fig. 4. Conventional single-sided cooling power module with the transparency of the lead frame changed.

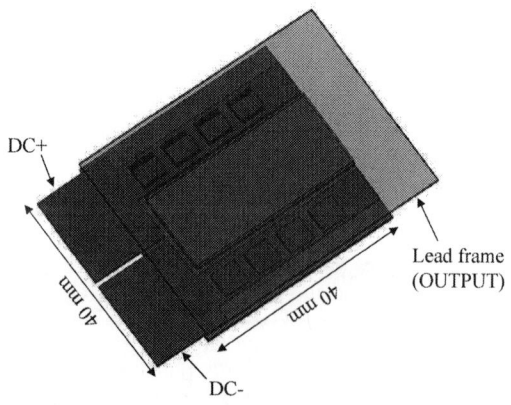

Fig. 5. Proposed power module (Type A) with the transparency of the lead frame changed.

TABLE I. MATERIAL PROPERTIES OF PROPOSED POWER MODULE
(TYPE A) IN THERMAL SIMULATION.

Items	Properties		
	Material	*Thermal conductivity [W/m/K]*	*Thickness [mm]*
MOSFET	SiC	490	–
Solder	SnAgCu	64	0.1
Substrate	Si₃N₄	90	0.32
DBC copper	Cu	402	0.5
Lead frame	Cu	402	0.5 – 7.0

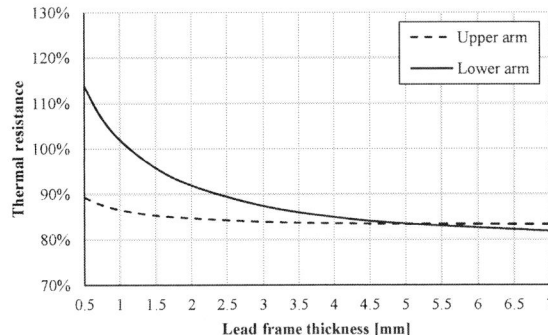

Fig. 6. Thermal simulation results.

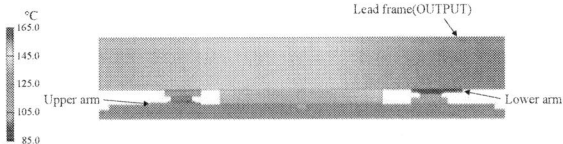

Fig. 7. Temperature distribution of the proposed power module (TypeA).

Fig. 6 shows the simulated thermal resistance Rj-c of each SiC-MOSFET in the upper and lower arms of the proposed power module (Type A), using the lead frame thickness as a parameter. For comparison with the conventional structure, the vertical axis of Fig. 6 shows the ratio of the thermal resistance of the proposed power module (Type A), based on the thermal resistance of the conventional structure. The thermal simulation results confirm an 18% reduction in thermal resistance in the proposed power module (Type A) compared to the conventional power module. Fig. 7 shows the temperature distribution of the proposed power module (Type A) when the loss per SiC-MOSFET is 50 W and the initial temperature of the power module is set to 85°C.

The thermal resistance Rj-c of the upper arm of the proposed power module (Type A) is lower than that of the conventional power module. This is due to the addition of a new heat conduction path from the source pad through the lead frame in addition to the heat conduction path similar to the conventional power module where the heat of the SiC-MOSFET is dissipated from the drain pad to the heatsink via the DBC substrate.

In addition, the thermal resistance Rj-c of the upper arm of the proposed power module (Type A) has low sensitivity to the lead frame thickness. This indicates that the heat conduction path to the heatsink via the DBC substrate is more dominant than the heat conduction path via the lead frame, as in the conventional power module.

The thermal resistance Rj-c of the lower arm of the proposed power module (Type A) is higher than that of the conventional power module at the lead frame thickness of 1.1 mm or less. This means that the thermal resistance of the lower arm is higher than 100% as shown in Fig. 6. This indicates that the heat of the lower arm SiC-MOSFET is dissipated to the DBC substrate through the conductive spacer connected to the source pad, and since the source pad has a smaller area than the drain pad, the thermal resistance of the heat conduction path to the heatsink through the DBC substrate is higher than the conventional power module.

On the other hand, the thermal resistance Rj-c of the lower arm of the proposed power module (Type A) has high sensitivity to the lead frame thickness and is lower than that of the upper arm at the lead frame thickness of 5 mm or greater. This indicates that the path of heat dissipation from the drain pad through the lead frame is more dominant than the path of heat dissipation to the heatsink through the DBC substrate. A thickness of 5 mm is quite substantial for the copper lead frame, but by applying highly anisotropic thermal conductivity materials, such as graphite, to the lead frame, it may be possible to reduce its thickness.

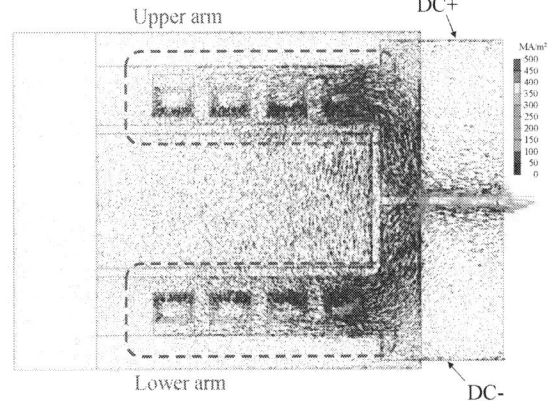

Fig. 8. Current density of proposed power module (Type A).

C. Parasitic Inductance Simulation

Parasitic inductance simulations were performed using software Femtet (Murata Software), which is also used for the thermal simulations, to evaluate the commutation loop inductance of the proposed power module (Type A). The simulation models are shown in Fig. 4 and Fig. 5.

The simulation result of the commutation loop inductance of the proposed power module (Type A) confirms a 35% reduction compared to the conventional power module, when the lead frame thickness is 2 mm. Fig. 8 shows the simulated current density of the proposed power module (Type A). The current is not evenly distributed among the four parallel SiC-MOSFETs in each of the upper and lower arms, but rather is concentrated in a single SiC-MOSFET near the DC+ terminal and another near the DC- terminal. This concentration of current is undesirable as it causes temperature imbalance among the parallel SiC-MOSFETs, especially during high-frequency switching operations.

Fig. 9 shows the proposed power module (Type B), which has the arrangement of the SiC-MOSFETs in the upper and lower arms rotated 90 degrees relative to the proposed power module (Type A). Fig. 10 illustrates the simulated current

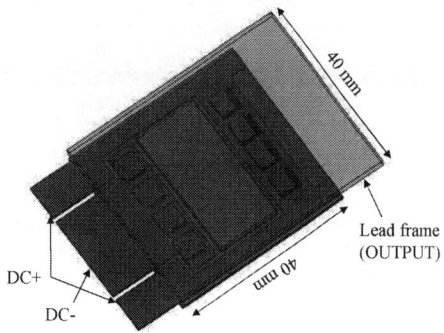

Fig. 9. Proposed power module (TypeB) with the transparency of the lead frame changed.

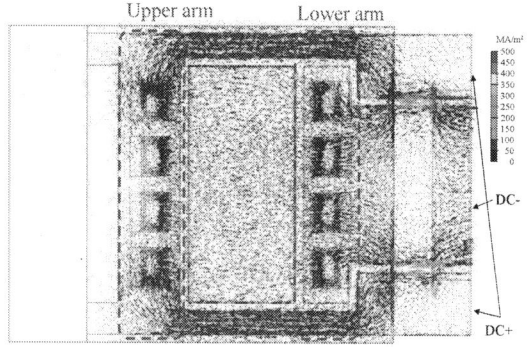

Fig. 10. Current density of proposed power module (Type B).

TABLE II. COMPARISON OF COMUTATION LOOP INDUCTANCE.

Modules	Inductance [nH]
Conventional power module	9.26
Proposed power module (Type A)	6.00
Proposed power module (Type B)	3.71

TABLE III. MAGNETIC FIELD CANCELLATION EFFECT FOR PROPOSED POWER MODULE (TYPE A).

Modules	Inductance [nH]
Proposed power module (Type A)	6.00
Proposed power module (Type A) without the bottom copper layer of the substrate	8.16

TABLE IV. MAGNETIC FIELD CANCELLATION EFFECT FOR PROPOSED POWER MODULE (TYPE B).

Modules	Inductance [nH]
Proposed power module (Type B)	3.71
Proposed power module (Type B) without the bottom copper layer of the substrate	6.18
Proposed power module (Type B) without the lead frame covering DC+ copper layer	3.96
Proposed power module (Type B) without the bottom copper layer of the substrate and the lead frame covering DC+ copper layer	7.86

density of the proposed power module (Type B). It reduces the current concentration compared to the proposed power module (Type A). The current is almost evenly distributed among the four parallel SiC-MOSFETs in the lower arm. In the upper arm, there is a slight concentration of current in the SiC-MOSFETs at both ends, but the degree of current concentration is lower than in the proposed power module (Type A).

Moreover, the magnetic field cancellation effect of the proposed power module (Type B) is larger than that of the proposed power module (Type A), resulting in a 60% reduction in commutation loop inductance compared to the conventional power module, as shown in Table II.

The evaluation results of the magnetic field cancellation effect for the proposed power module (Type A) are shown in Table III, and for the proposed power module (Type B) in Table IV. In the proposed power module (Type A), the bottom copper layer of the DBC substrate reduces the commutation loop inductance by 26.5%. Similarly, in the proposed power module (Type B), the bottom copper layer of the DBC substrate reduces the commutation loop inductance by 49.6%. The magnetic flux cancellation effect of the bottom copper layer of the DBC substrate is greater in the proposed power module (Type B) than in the proposed power module (Type A).

Additionally, in the proposed power module (Type B) as shown in Fig. 11, the magnetic field is canceled because the directions of the current flowing through the DC+ copper layer and the lead frame are opposite. Fig. 12 shows the simulation model, in which the lead frame doesn't cover the DC+ copper layer and the thickness of the lead frame is increased from 2.0 mm to 2.7 mm. This is to maintain the same

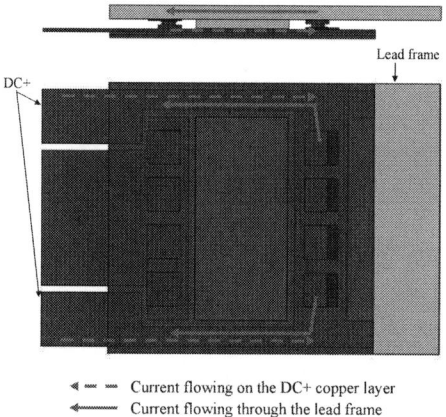

◄ ─ ── Current flowing on the DC+ copper layer
◄────── Current flowing through the lead frame

Fig. 11. Magnetic field cancellation effect by the lead frame.

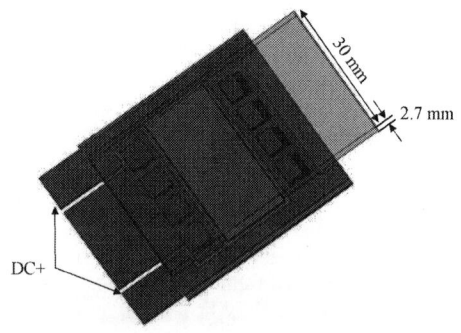

Fig. 12. Proposed power module (Type B) without the lead frame covering DC+ copper layer, with the transparency of the lead frame changed.

cross-sectional area of the lead frame. The lead frame reduces the commutation loop inductance by 21.4% in Table IV.

III. POWER MODULE FABRICATION AND TESTING

A. Fabrication of Prototypes

To validate the manufacturability of the proposed structure, prototype power modules were fabricated. Fig. 13 shows photographs of the fabricated prototype power modules. Due to the large soldering area between the convex part in the center of the lead frame and the DBC copper, solder voids are likely to occur, and voids in this area may cause a significant increase in thermal resistance. Fig. 14 shows X-ray CT images. The solder void ratio between the convex part of the lead frame and the DBC copper is 3.95% for Type A and 4.52%

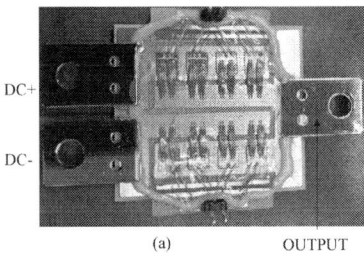

(a)

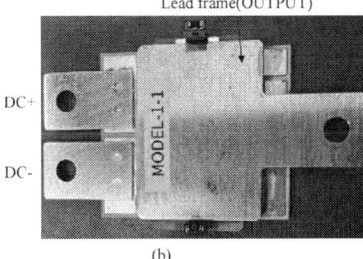

(b)

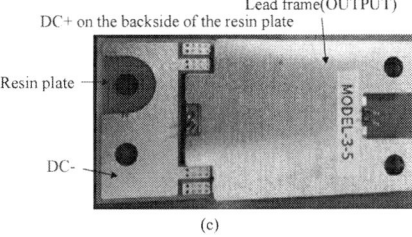

(c)

Fig. 13. Fabricated prototypes. (a) Conventional power module. (b) Proposed power module (Type A). (c) Proposed power module (TypeB).

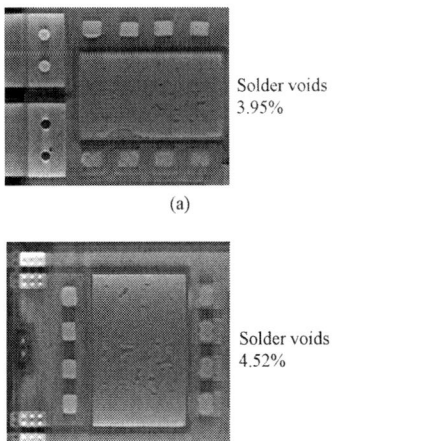

Fig. 14. Cross-sectional images taken by X-ray CT. (a) Proposed power module (Type A). (b) Proposed power module (Type B).

for Type B, confirming that the solder void does not significantly increase thermal resistance.

B. Dynamic Testing and Results

To confirm the electrical performance of the proposed power modules, double pulse tests were conducted. Fig. 15 shows the double pulse test setup. Fig. 16 shows the experimental turn-on waveforms at 800 V_{DC} and 300 A. Additionally, Fig. 17 shows the experimental turn-off waveforms at 800 V_{DC} and 400 A. Both the drain current waveforms and the drain-source voltage waveforms exhibited stable transitions during the turn-on and turn-off. The experimental results confirm that the proposed power modules, Type A and Type B, are capable of functioning with normal switching operations. Since the prototype power modules worked electrically, the proposed structure and assembly process have been confirmed.

C. Experimental Evaluation of Parasitic Inductance

According to the oscillation frequency of the experimental waveforms, the parasitic inductance can be expressed as

$$f = \frac{1}{2\pi\sqrt{LC}} \tag{1}$$

where f is the voltage oscillation frequency. C is the parasitic capacitance that includes the output capacitance of four parallel SiC-MOSFETs (209 pF for each MOSFET), the capacitance between the OUTPUT copper layer (lead frame) and the DC- copper layer, and the capacitance induced by the coupling effect between different conductor layers in the power module. The capacitance between the OUTPUT copper layer (lead frame) and the DC- copper layer is 7.6 pF for the conventional power module, 11.8 pF for the proposed power module (Type A), and 11.2 pF for the proposed power module (Type B) according to simulations using software Femtet (Murata Software). The capacitance induced by the coupling effect is 43.9 pF for the conventional power module, 51.1 pF for the proposed power module (Type A), and 44.5 pF for the proposed power module (Type B) from simulation. L is the parasitic inductance between the power module and the dc-link capacitor.

The parasitic inductance calculated in Fig. 17(b) is shown in Table V. Table V shows that the parasitic inductance is smallest in the proposed power module (Type B), followed by the proposed power module (Type A), and then the conventional power module, which is consistent with the trend of the simulation results in Table II.

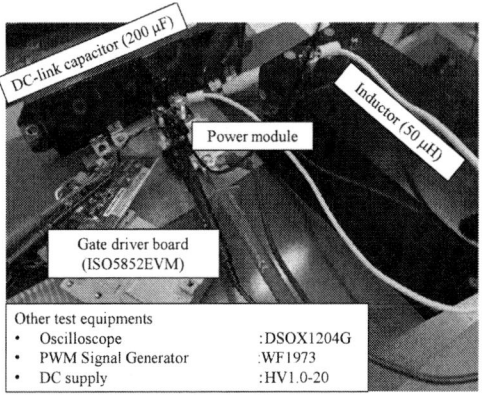

Fig. 15. Double pulse test setup.

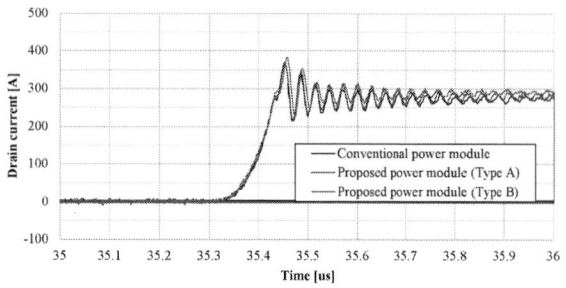

(a)

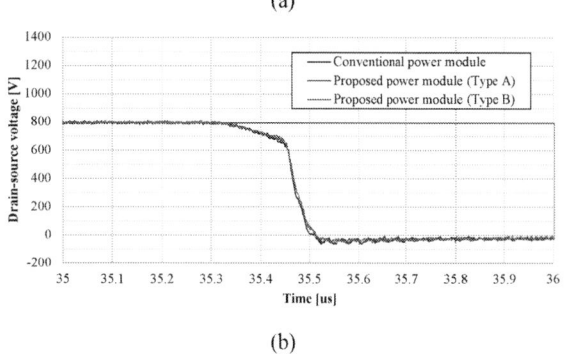

(b)

Fig. 16. Turn-on switching waveforms in double pulse test. (a) Drain current waveforms. (b) Drain-source voltage waveforms.

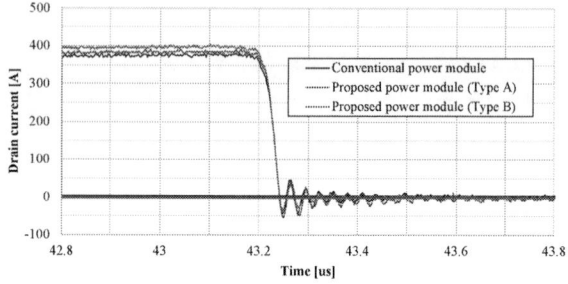

(a)

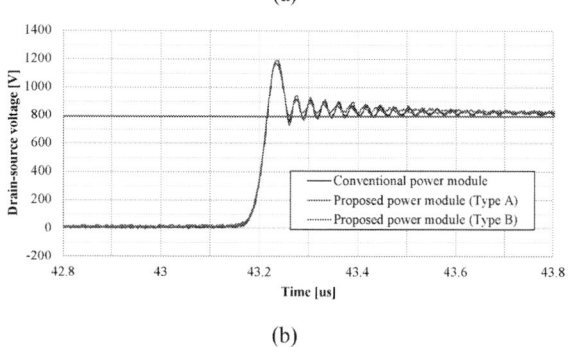

(b)

Fig. 17. Turn-off switching waveforms in double pulse test. (a) Drain current waveforms. (b) Drain-source voltage waveforms.

TABLE V.   CONPARISON OF MEASURED PARASITIC INDUCTANCE.

Modules	Parasitic capacitance [pF]	Voltage oscillation frequency [MHz]	Parasitic inductance [nH]
Conventional power module	887.5	34.1	24.6
Proposed power module (Type A)	898.9	34.9	23.1
Proposed power module (Type B)	891.7	37.4	20.3

IV. CONCLUSION AND FUTURE WORK

This paper presents two types of single-sided cooling power modules for a half-bridge circuit, featuring low thermal resistance and low inductance through a three-dimensional structure. Although the heatsink only needs to be mounted on one side, the SiC-MOSFETs in both the upper and lower arms are cooled from both the source and drain pads.

Simulation results show that the thermal resistance is reduced by 18%, and the parasitic commutation loop inductance is reduced by 35% for the proposed power module (Type A) compared to that of the conventional power module. Moreover, for the proposed power module (Type B) with improved current balance between parallel SiC-MOSFETs, the parasitic commutation loop inductance is reduced by 60% compared to that of the conventional power module. The feasibility of both proposed power modules (Type A and Type B) was confirmed by fabricating prototypes and conducting double-pulse tests (800 V_{DC}, turn-on current of 300 A, turn-off current of 400 A). The thermal resistance of the proposed power module will be validated by the experimental results in future work, and the detailed evaluation of the parasitic commutation loop inductance will also be conducted.

REFERENCES

[1] A. Elasser and T. P. "how, "Silicon carbide benefits and advantages for power electronics circuits and sy"tems," in Proceedings of the IEEE, vol. 90, no. 6, pp. 969-986, June 2002.

[2] J. W. Palmour, "Silicon carbide power device development for industrial markets," 2014 IEEE International Electron Devices Meeting, San Francisco, CA, USA, 2014, pp. 1.1.1-1.1.8

[3] L. Zhang, X. Yuan, X. Wu, C. Shi, J. Zhang and Y. Zhang, "Performance Evaluation of High-Power SiC MOSFET Modules in Comparison to Si IGBT Modules," in IEEE Transactions on Power Electronics, vol. 34, no. 2, pp. 1181-1196, Feb. 2019.

[4] J. Fabre, P. Ladoux and M. Piton, "Characterization and Implementation of Dual-SiC MOSFET Modules for Future Use in Traction Converters," in IEEE Transactions on Power Electronics, vol. 30, no. 8, pp. 4079-4090, Aug. 2015.

[5] A. K. Morya et al., "Wide Bandgap Devices in AC Electric Drives: Opportunities and Challenges," in IEEE Transactions on Transportation Electrification, vol. 5, no. 1, pp. 3-20, March 2019.

[6] Z. Liang, "Integrated double sided cooling packaging of planar SiC power modules," 2015 IEEE Energy Conversion Congress and Exposition (ECCE), pp. 4907-4912, 2015.

[7] X. Liu, Z. Wu, Y. Yan, Y. Kang and C. Chen, "A Novel Double-Sided Cooling Inverter Leg for High Power Density EV Based on Customized SiC Power Module," 2020 IEEE Energy Conversion Congress and Exposition (ECCE), pp. 3151-3154, 2020.

[8] F. Yang et al., "Interleaved Planar Packaging Method of Multichip SiC Power Module for Thermal and Electrical Performance Improvement," in IEEE Transactions on Power Electronics, vol. 37, no. 2, pp. 1615-1629, 2022.

[9] L. Ma et al., "A Double-sided Flip chip SiC Power Module with a Novel Flip Chip Method On DBC By PTFE Coating," 2023 IEEE Energy Conversion Congress and Exposition (ECCE), Nashville, TN, USA, 2023, pp. 5888-5894.

[10] F. Abed Ali, P. -O. Jeannin, Y. Avenas and P. Lefranc, "Flip-Chip Low inductive and EMC optimized PCB Power Module," 2024 IEEE Applied Power Electronics Conference and Exposition (APEC), Long Beach, CA, USA, 2024, pp. 1534-1538

Aging of Y-Capacitor in an EMI Filter and its Impact on Common-Mode Noises

Tahmid Ibne Mannan
*Department of Electrical
and Computer Engineering
Mississippi State University*
Starkville, USA
tm2445@msstate.edu

Seungdeog Choi
*Department of Electrical
and Computer Engineering
Mississippi State University*
Starkville, USA
seungdeog@ece.msstate.edu

Subarto Kumar Ghosh
*Department of Electrical
and Computer Engineering
Mississippi State University*
Starkville, USA
sg2009@msstate.edu

Md Moniruzzaman
*Department of Electrical
and Computer Engineering
Mississippi State University*
Starkville, USA
mm5111@msstate.edu

Abstract— This paper investigates the effects of Y-capacitor aging on passive electromagnetic interference (EMI) suppression filters. Due to the wide adoption of ultra-fast wide-bandgap (WBG) switches, modern power systems are subjected to severe EMI noise, risking electromagnetic compatibility (EMC). Any commercial and military power electronics system must meet EMC. Discrete passive components-based EMI filters are commonly adopted to suppress common-mode (CM) and differential-mode (DM) EMI. These filters comprise common-mode choke (L_{CM}), X-capacitors (C_X), and Y-capacitors (C_Y). Capacitors are prone to degradation due to constant voltage stress and temperature variation. Several studies have identified the changes in capacitor parameters due to aging, which can significantly impact EMI filter attenuation capabilities. This paper focuses on the aging of C_Y capacitors and its effects on CM EMI attenuation of EMI filters. Simulation modeling and experimental test results up to 500 hours are presented. The impact of C_Y degradation has been analyzed, including its effects on the CM insertion loss of the passive EMI filter and, consequently, on the CM EMI suppression capabilities of the filters.

Keywords—aging, capacitor, EMI filter, common mode, insertion loss.

I. INTRODUCTION

Modern power electronics systems have adopted fast-switching WBG switches due to their inherent high switching frequency capabilities and low-loss characteristics [1–2]. The very short switching transitions of WBG devices (in the range of nanoseconds) offer significantly reduced switching loss. However, due to the high dv/dt and di/dt characteristics, there is an increase in ground leakage current and consequently an increase in conducted EMI [3–4]. Adopting passive EMI filters with discrete passive components is a popular technique to suppress EMI along its conduction path [5–6]. Fig.1 shows a typical L-type single-stage passive EMI filter connected between a switching device and the power supply, which is the common placement of an EMI filter. The CM choke inductance (L_{CM}) works in conjunction with C_{Y1} and C_{Y2} to suppress the CM noise, whereas the leakage inductance of the choke (L_{DM}) work together with C_X to mitigate the DM EMI noise. In this paper, the effect of Y-capacitor aging on CM noise will be investigated, and therefore the effects of L_{DM} and C_X will be neglected.

Like any other system, capacitors in EMI filters can be prone to aging effects due to thermal and overvoltage stress over a long period of time. In Fig. 2, the high-frequency equivalent model of a capacitor is presented. Among the parameters, capacitance (C) and equivalent series resistance (ESR) are prone to age stress, which has been well documented in the literature [7–9]. In [7], authors report a

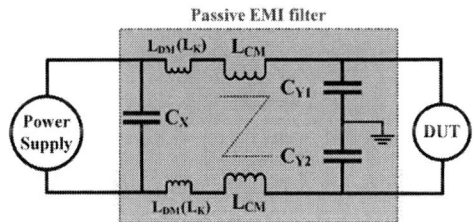

Fig. 1. Typical single-stage passive EMI filter.

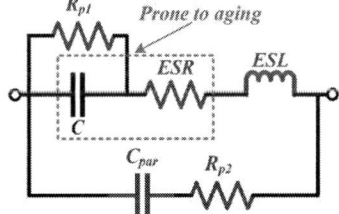

Fig. 2. High-frequency capacitor equivalent circuit model.

capacitance decrease of 20% in capacitors due to aging. [8] and [9] report that the *ESR* of the capacitor can increase up to 2 to 2.5 times with aging. Such changes in capacitors, specifically in the context of EMI filter Y-capacitors, can have crucial impacts on their CM EMI noise suppression capabilities. CM insertion loss of the EMI filter are directly related to the C_Y capacitors' impedances, a topic that has been limitedly explored in state-of-the-art until recently with the best of authors knowledge.

In this paper, the impacts of C_Y capacitor aging in EMI filters will be explored, specifically focusing on the CM EMI reduction capabilities of the EMI filter. Section II presents the design of the passive EMI filter used in this study. Section III presents the analytical and simulation analysis of Y-capacitor aging impact on the CM insertion loss of an EMI filter. Section IV describes the aging setup for the capacitors and the basic parameter measurements of the implemented filter components. Section V experimentally explores the impact of Y-capacitor aging on the CM insertion loss of the filter and the CM EMI of a SiC half-bridge inverter. Finally Section VI concludes the article.

II. EMI FILTER DESIGN

To suppress CM noise, C_{Y1} and C_{Y2} are used together with the magnetizing inductance of the CM choke, L_{CM}. The effective EMI filter equivalent circuit for CM EMI attenuation is presented in Fig. 3.

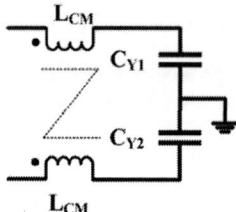

Fig. 3. EMI filter equivalent circuit for CM EMI attenuation.

For the CM attenuating circuit presented in Fig. 3, the corner frequency can be calculated as follows:

$$f_{cutoff} = \frac{1}{2\pi\sqrt{L_{CM}(C_{Y1} + C_{Y2})}} \quad (1)$$

Now, to determine the necessary values of the filter, first the required attenuation of CM EMI is quantified. Therefore, the CM EMI spectrum without the filter is measured first, and first frequency where the noise exceeds the relevant EMI standard is identified (f_{target}). The required attenuation (A) then can be calculated as follows, with a 6dB margin:

$$A\,(dB) = CM_{without\_filter}(dB) - CM_{standard}(dB) + 6dB \quad (2)$$

Once the A and f_{target} are calculated, f_{cutoff} can be calculated from the following relationship:

$$f_{cutoff} = f_{target}10^{-\frac{A}{40}} \quad (3)$$

After determining the required cut-off frequency for the EMI filter, the values for L_{CM} and C_Y capacitors can be obtained using the relationship in (1). However, it should be noted that the maximum values for the C_Y capacitors are limited by safety standards to regulate the leakage current flowing to the ground of the system and ensure safety [10]. After choosing the optimum value for C_Y, the required value of the CM choke magnetizing inductance is obtained.

To design the EMI filter for this study, a half-bridge inverter is modeled with real SiC switch parameters on PSIM. For compliance, CISPR-32, class B standard (150kHz – 30MHz) is followed. The simulation specifications are presented in Table I.

TABLE I. CM EMI SIMULATION PARAMETERS

Parameters	Specifications
Topology	Half-bridge inverter
Switching Devices	SiC MOSFETs
Switching Frequency	50kHz
DC Supply	100V
EMC Standard	CISPR 32 - Class B

From simulations, it is observed that in the concerned frequency range, the simulated CM EMI noise is around 100 dBμV, where the CISPR noise limit is around 63 dBμV. With a 6dB margin, the required f_{cutoff} is determined at around 35kHz using (2) and (3). Choosing 2.2nF capacitors for C_{Y1} and C_{Y2}, the value of L_{CM} is therefore 4.69mH. The L_{CM} value is chosen as 5mH to have some margin. Fig. 4 presents the simulated comparison between the CM EMI with and

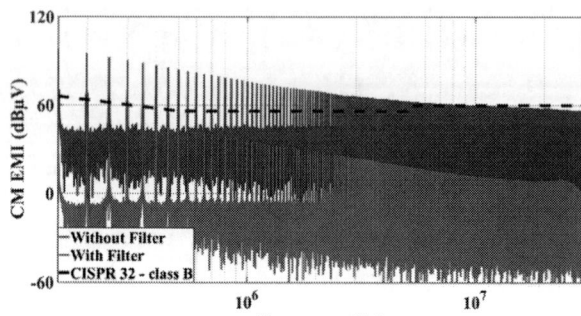

Fig. 4. Simulated CM EMI comparison for the designed filter

without the designed EMI filter. The designed EMI filter specifications are presented in Table II.

TABLE II. DESIGNED EMI FILTER SPECIFICATIONS

Parameters	Specifications
CM choke	5mH
C_{Y1}, C_{Y2}	2.2nF each
C_X	5μF

III. IMPACT OF AGING ON EMI FILTER CM INSERTION LOSS

The CM EMI attenuation capabilities of the EMI filter depend on L_{CM}, C_{Y1}, and C_{Y2}, as these components determine the cut-off frequency of the EMI filter for CM EMI attenuation. Furthermore, the parasitic parameters of these components also dictate the attenuation capabilities of the filter. The high frequency equivalent circuit of the CM choke is presented in Fig. 5 with its parasitic parameters highlighted. C_P and C_W are equivalent parallel capacitance and interwinding parasitic capacitances respectively, whereas $R_{winding}$ and R_P denote winding resistance and equivalent parallel resistance respectively. Notably, C_P can affect the high frequency suppression capabilities of the EMI filter.

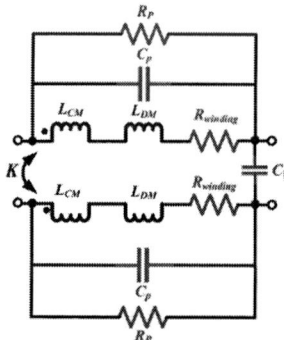

Fig. 5. High-frequency CM choke equivalent circuit model.

Insertion loss is the most important parameter to quantify and assess the noise suppression capabilities of the EMI filter. Insertion loss with respect to EMI filters is defined as the ratio between the voltage measured at the LISN from the noise source without and with the EMI filters, as illustrated in Fig. 6. Here, Z_{Lisn} and Z_S are the impedances of LISN and the switching device respectively, whereas V_{noise} is the CM noise source of the switching device. The CM filter block includes the parasitic parameters of the Y-capacitors and the CM choke that can affect the insertion loss characteristics. According to

979-8-3315-1612-3/25 $31.00 © 2025 IEEE

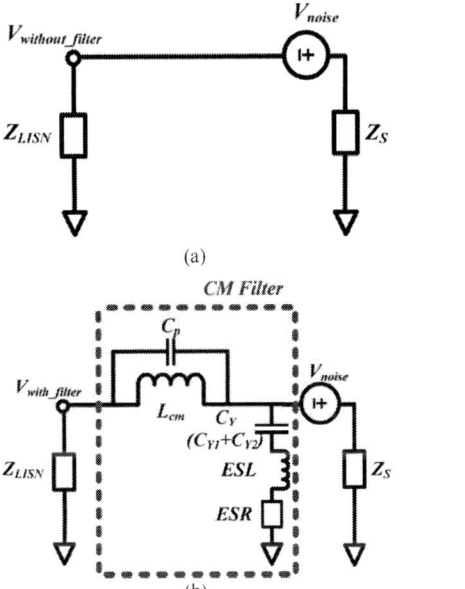

(a)

(b)

Fig. 6. Schematic diagram for CM insertion loss (a) without EMI filter (b) with EMI filter

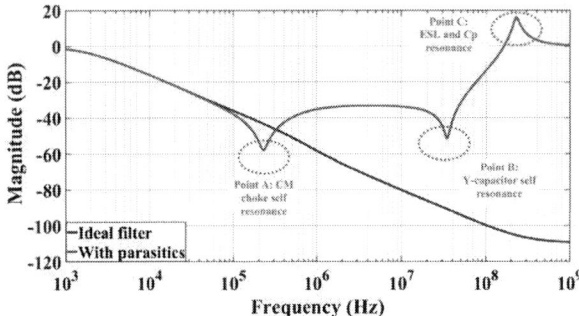

Fig. 7. IL_{CM} comparison between ideal filter components and components with parasitic parameters

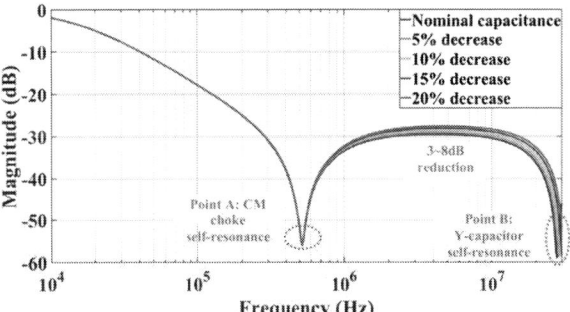

Fig. 8. Simulated changes in IL_{CM} due to changes in capacitance due to aging

CISPR-17 requirements of insertion loss characterization for EMI filters through network analyzers, Z_{Lisn} and Z_S will be assumed to be 50Ω in the following analysis for simplification [11]. From Fig. 6, V_{witho_filter} and V_{with_filter} can be expressed as follows:

$$V_{without\_filter} = V_{noise} \frac{Z_{Lisn}}{Z_{Lisn}+Z_S} \quad (4)$$

$$V_{with\_filter} =$$
$$V_{noise} \frac{Z_{Lisn}}{(Z_Y+Z_{choke}+Z_{Lisn})\left[1+\frac{Z_S(Z_Y+Z_{choke}+Z_{Lisn})}{Z_Y(Z_{choke}+Z_{Lisn})}\right]} \quad (5)$$

where Z_Y and Z_{chok} are the equivalent impedances of the Y-capacitors and the CM choke respectively. Using (4) and (5), the CM insertion loss of the EMI filter can be measured by the equation $IL_{CM} = 20\log\left|\frac{V_{withoutfilter}}{V_{withfilter}}\right| dB$. To understand the effects of Y-capacitor aging on IL_{CM} and to replicate realistic measurements, the parasitic parameters must be integrated into the equations, as they significantly affect filter performance in high frequencies [12]. Therefore, Z_Y and Z_{choke} should be replaced with $\frac{s^2 C_Y ESL+s\ _Y ESR+1}{sC_Y}$ and $\frac{sL_{CM}}{s^2 L_{CM}C_P+1}$ respectively to reflect their realistic behavior in high frequencies. A comparison of IL_{CM} magnitude is presented in Fig. 7 to demonstrate the importance of considering these parasitic elements. In these simulations, CM choke parasitic capacitance and ESL for Y-capacitors are assumed to be 40pF and 15nH, respectively, to replicate practical scenarios. As evident, with parasitic elements IL_{CM} magnitude exhibits several resonance points which are absent in the ideal scenario. The first corner frequency is the cutoff frequency of the filter. The insertion loss increases linearly after this cutoff frequency, until a resonance is observed at around 250kHz (point A), which is caused by the self-resonance of the CM choke (L_{CM} and C_P). After this resonance point, instead of increasing, the insertion loss starts to flatten with respect to frequency. After that, a second resonance point at around 25MHz (point B) is observed,

which can be attributed to the self-resonance of the Y-capacitors (C_Y and ESL). Beyond this point, the insertion loss begins to decrease and CM EMI attenuation capabilities of the filter begins diminish.

Based on these understandings, the effect of Y-capacitor aging on the CM insertion loss can be evaluated. When an EMI filter is designed, the calculations and design optimizations are done considering healthy filter components. However, as state-of-the-art suggests, the changes in capacitor parameters occur due to aging-related stress. Fig. 8 presents the changes in IL_{CM} magnitude for an EMI filter with different changes in capacitance. As evident from the figure, the changes in capacitor parameters can introduce a decrease in insertion loss after the point of CM choke's self-resonance. This is caused by the decreased capacitance of Y-capacitors, which can occur due to aging. Due to the degradation of Y-capacitors results, a 3~8dB decrease in insertion loss at all frequencies up to the self-resonance point of the Y-capacitors at point B can be expected according to the simulations. This is because the region between point A and point B is mostly dominated by the impedance of the Y-capacitors. Due to a decreased capacitance, the effective insertion loss of the filter can decrease in this region. Furthermore, at this resonance point, the filter with the degraded capacitors offers much less attenuation due to an increased ESR than the healthy capacitors. This can be detrimental to a system's CM EMI compatibility.

IV. EXPERIMENTAL SETUP

A. Experimental Setup for Capacitor Aging

For this paper, an electrical stress setup has been utilized to accelerate the aging of the capacitors under test (CUT). Fig. 9 presents the schematic and the implemented electrical stress

979-8-3315-1612-3/25 $31.00 © 2025 IEEE

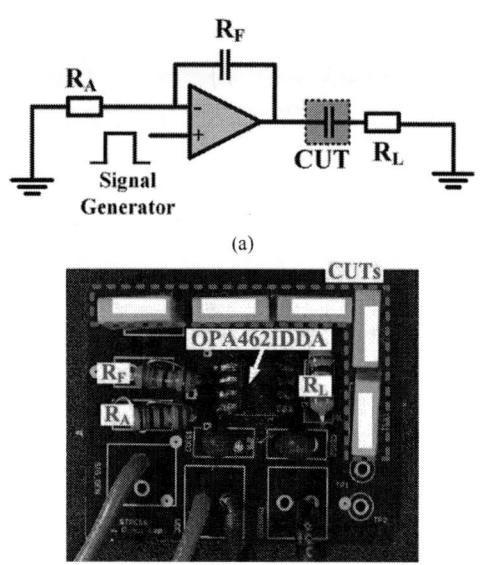

(a)

(b)

Fig. 9. Electrical stress test board for CUTs (a) schematic and (b) implemented test board

board prototype. 1Vp-p, 1Hz square wave signals from the signal generator are amplified using the op-amp, R_A and R_F. In this paper, the used op-amp is OPA462IDDA from Texas Instruments. The voltage amplification stage amplifies the signal generator input voltage to more than the rated voltage of the CUTs to repeatedly charge and discharge the CUTs. Following the CUT value, the load resistor (R_L) value is optimized to ensure complete charging and discharging during a single cycle. For this paper, 100V rated Y2 rated film capacitors are used as the CUTs. To apply electrical stress and accelarate the capacitor aging process, the applied stress voltage is 10% more than the rated voltage. Table III summarizes the electrical stress setup configurations.

TABLE III. ELECTRICAL STRESS SETUP SPECIFICATIONS

Parameters	Specifications
CUT rating	100V
Stress pulse	1Vp-p, 1Hz square wave
Amplified stress voltage	100V$_{p-p}$ + 10%
Op-amp	OPA462IDDA

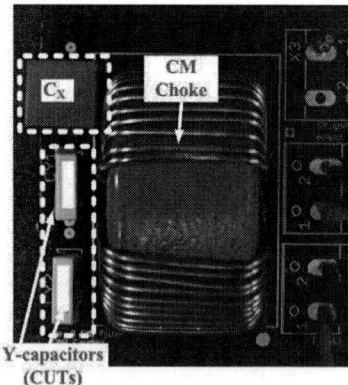

Fig. 10. Implemented passive EMI filter

B. Designed EMI Filter Prototype

A prototype passive EMI filter has been designed to test the effect of the degraded Y-capacitors on CM EMI

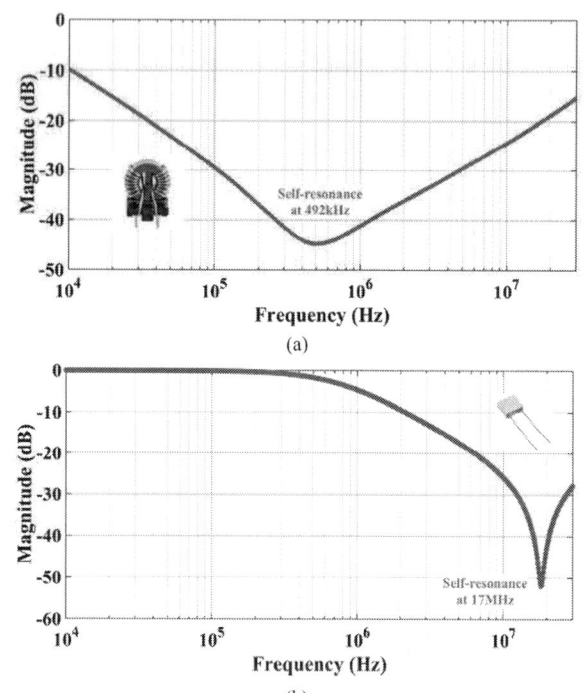

(a)

(b)

Fig. 11. Experimentally measured insertion loss magnitudes of implemented EMI filter components (a) 5mH CM choke (b) 2.2nF Y-capacitors

attenuation based on the analysis presented in Section II-A. The degraded CUTs are placed and replaced in the EMI filter after certain time intervals to extensively track the impact of capacitor aging on CM EMI attenuation. Fig. 10 presents the designed passive EMI filter prototype. Experimentally measured individual insertion loss characterisitics of the 5mH CM choke and 2.2nF Y-capacitors used in the filter are presented in Fig. 11. The choke has a self-resonance at around 492kHz, whereas for Y-capacitors, the self-resonance occurs at around 16MHz. Fig. 12 presents the experimentally measured CM insertion loss of the designed filter. As discussed in the earlier sections, the CM insertion loss magnitude plot has two notable resonances, at around 400kHz and 17MHz. The first resonance point coincides with the self-resonance of the CM choke, whereas the second resonance point is caused by the self-resonance of the Y-capacitors' shunt branch. The insertion loss magnitude in the region between these two resonance points will be affected by the aging of the Y-capacitors and their subsequent decrease in capacitances, as will be validated in the next section with experimental results.

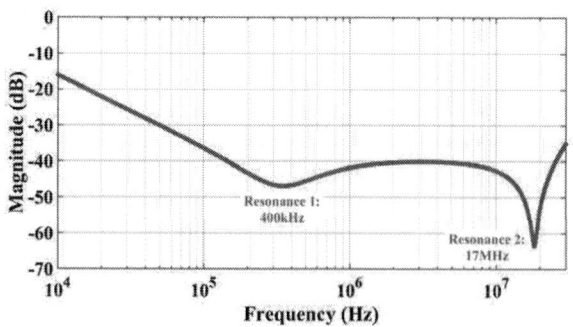

Fig. 12. Experimentally measured CM insertion loss magnitudes of implemented EMI filter with highlighted resonance points

V. Experimental Results

A. Changes in Capacitor Parameters

In this paper, 100V rated Y2 rated film capacitors are used as the CUTs. The CUTs are rated at 100V. To apply electrical stress and accelarate the capacitor aging process, the applied stress voltage is 10% more than the rated voltage. After each 100-hour interval, each CUT is removed from the electrical stress setup and characterized by an impedance analyzer. The Bode 100 network analyzer is utilized to measure and characterize crucial capacitor parameters at each stage to track the correlated changes due to degradation. This paper presents results of up to 500 hours for two CUTs. As aging indicators, several parameter changes are tracked for both CUTs. In Fig. 12, the experimentally measured changes in C, ESR, and equivalent series inductance (ESL) of both CUTs with degradation time are presented at 100kHz, 500kHz, and 25MHz, respectively. As evident, C values for both CUTs decrease almost linearly with degradation. At 500 hours, CUT-1's C decreases by 9.3%, whereas for CUT-2, the decrease in C is 9.35%. Similar linear changes are observed in ESR for both CUTs. With degradation, ESR for both CUTs

increases linearly. At 500 hours, the increase in ESR for CUT-1 and CUT-2 are 40.7% and 42.3% respectively. In the case of ESL, the changes for both CUTs with aging are not consistent, as evidenced by Fig. 12(c).

B. CM Insertion Loss

With the Bode 100 network analyzer, the CM insertion loss for the EMI filter has been experimentally measured with the CUTs at each interval. The CM insertion loss experimental measurement schematic and setup are presented in Fig. 13.

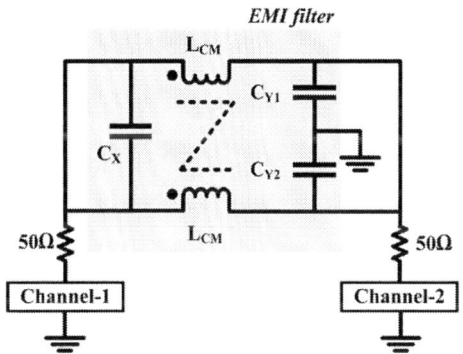

(a)

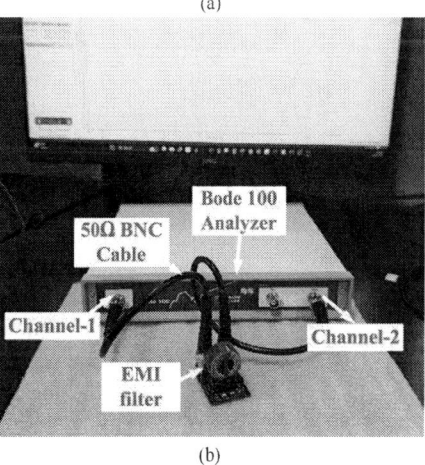

(b)

Fig. 14. Experimental CM insertion loss measurement for an EMI filter according to CISPR-17 [10] (a) schematic (b) setup

The measured CM insertion loss at different stages is presented in Fig. 14. The resonance at around 400kHz is due to the self-resonance of the CM choke, as discussed earlier. The resonance point matches with the experimentally measured resonance point of the choke presented in Fig. 11(a). The second resonance point occurs due to the equivalent capacitance of Y-capacitors and their equivalent ESL in each case. The impedance of the CM choke dominates the insertion loss magnitude before its resonance point. Therefore, despite the changes in Y-capacitor parameters, no change in insertion loss has been observed in this region. However, a decrease in insertion loss after that point can be observed for all aging time intervals up to the point of self-resonance point of the Y-

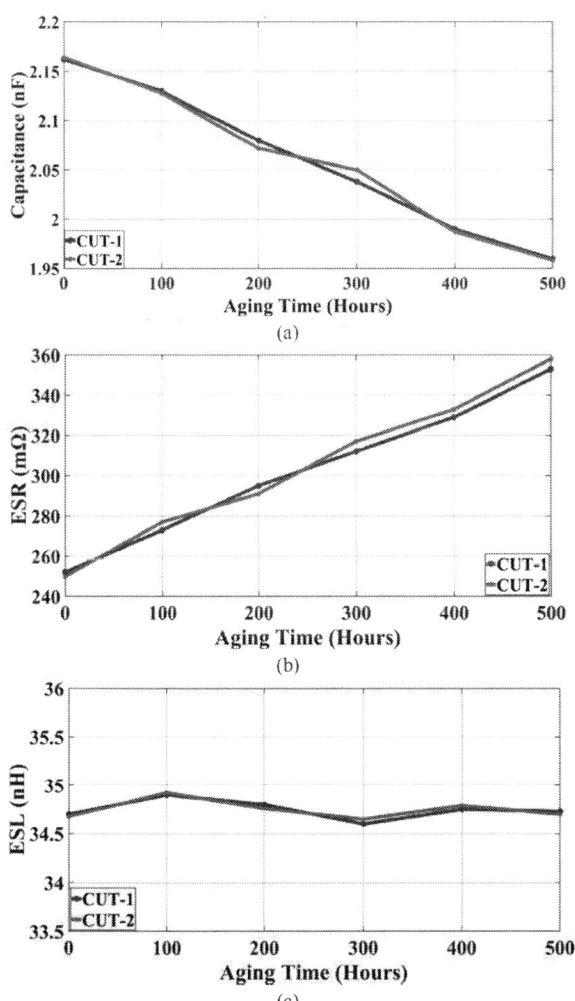

Fig. 13. Experimentally measured changes in CUT aging indicators with degradation (a) C, (b) ESR, and (c) ESL.

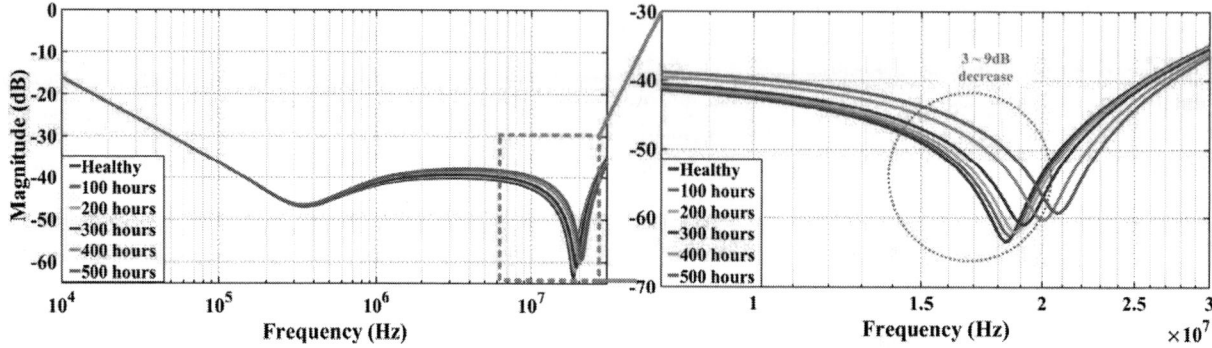

Fig. 15. (L) Experimentally measured CM insertion loss of the EMI filter with CUTs at different intervals and (R) zoomed view.

capacitors at . This is because the impedance and insertion loss in the region between these two points are dominated by the impedance of the Y-capacitors. Because of the decreased capacitances of C_{Y1} and C_{Y2}, the insertion loss incurs a decrease of 3~9dB across this region compared to the healthy state of the CUTs, as evident in the zoomed view presented in Fig. 14. Although the data is presented up to 30MHz considering CISPR standard limits, it should be noted that the attenuation observed for all cases beyond this resonance point is almost equal with increasing frequency since aging does not lead to significant changes in the *ESL* of the CUTs, as observed in Fig. 12 (c). This indicates that beyond the second resonance point, the insertion loss is dominated by the *ESL* and C_P as the insertion loss approaches 0dB.

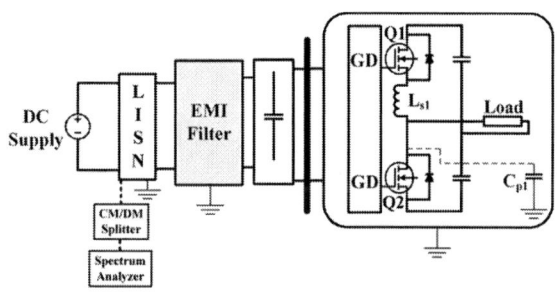

(a)

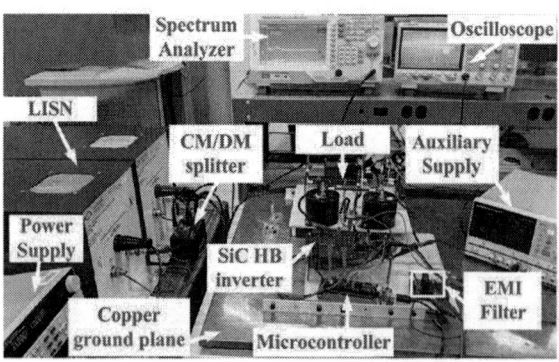

(b)

Fig. 16. CM EMI measurement testbed (a) schematic (b) experimental setup

C. CM EMI Measurements

State-of-the-art conducted EMI characterization testbed has been utilized for the experimental CM EMI measurements. The designed EMI filter with CUTs has been connected to the testbed to measure the CM EMI of a half-bridge inverter evaluation module at each interval. The testbed is presented in Fig. 15 with its schematic. The testbed specifications are detailed in Table IV.

TABLE IV.   CM EMI TESTBED SPECIFICATIONS

Parameters	Specifications
Switching frequency	50kHz
DC supply	100V
Switching device	SiC Half-bridge inverter
Spectrum Analyzer	9kHz-1.5GHz

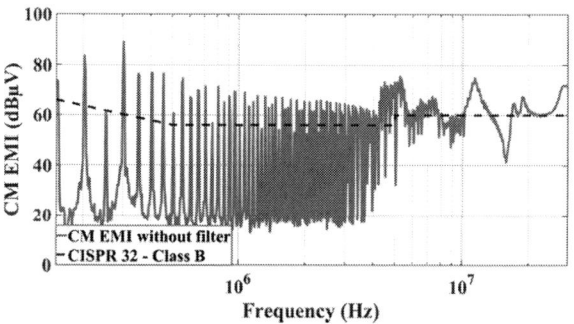

Fig. 17. Experimentally measured CM EMI spectrum of the half-bridge inverter at 100V without EMI filter

Fig. 17 presents the experimentally measured CM EMI spectrum of the half-bridge inverter without the EMI filter with CISPR-32 class B standard limit overlayed. As evident, switching frequency harmonic amplitudes at almost all frequencies exceed the specified CISPR limit.

Fig. 18 illustrates the measured CM EMI spectrum comparison across various aging stages of the Y-capacitors. The observed trends in CM EMI closely correlate with the changes in CM insertion loss as depicted in Fig. 15. This is also consistent with the theoretical and simulation analysis presented in Section III. As observed, no significant variations in CM EMI harmonic amplitudes were noted up to 400 kHz, where the CM choke resonance occurs. Beyond this frequency, however, a slight increase in CM EMI harmonics

979-8-3315-1612-3/25 $31.00 © 2025 IEEE

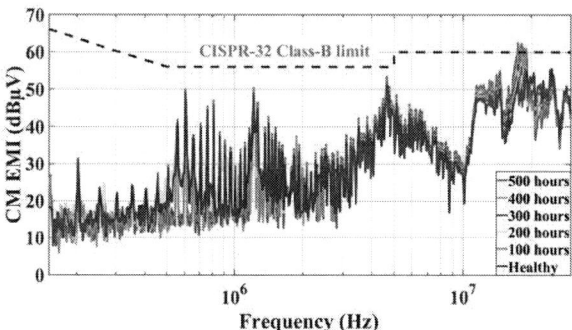

Fig. 18. Experimentally measured CM EMI with EMI filter with CUTs at different time intervals

was observed due to the gradual capacitance degradation of the Y-capacitors. This trend is in accordance with the measured CM insertion loss for the corresponding health stages. A steady increase in CM EMI between health stages was observed up to the self-resonant frequency of the Y-capacitor shunt branch, approximately 17 MHz. At this frequency, a maximum of 7-8 dB increase in CM EMI was measured for capacitors aged 500 hours compared to their healthy counterparts. Notably, at this frequency, the CM EMI levels for 500-hour degraded capacitors exceeded the CISPR 32 limit, potentially compromising electromagnetic compatibility in the system. These findings highlight the inadequacy of the conventional 6 dB margin design approach for EMI filters when considering the potential degradation of Y-capacitors over time.

VI. CONCLUSIONS

This paper presents an experimental investigation of the effects of Y-capacitor aging on a passive EMI filter's CM EMI attenuation capability. The impact of Y-capacitor aging on the CM insertion loss characteristics of an EMI filter is assessed analytically and with simulation of realistic scenarios. Experimental results are presented for up to 500 hours for two different CUTs. 9.3% and 9.35% decrease in capacitances are observed for both CUTs, whereas 40.7% and 42.3% increase in *ESR* are observed. Experimental measurement of the CM insertion loss of the EMI filter is presented for each time interval. A 3~9dB insertion loss decrease is observed with aging in the region where the Y-capacitors' impedance dominate the insertion loss of the filter. Experimental CM EMI measurement comparisons are presented for both CUTs. After 500 hours of aging, a steady increase has been observed in CM EMI after CM choke's resonant frequency point at around 400kHz. A maximum increase of 7~8dB at around 17MHz in CM EMI has been observed with degraded CUTs

after 500 hours. The experimental results match the theoretical and simulation analysis reasonably well. The experimental data demonstrates that the decrease in Y-capacitor capacitances can lead to a notable increase in CM EMI, necessitating a more robust design methodology to ensure long-term EMI compliance considering the aging effect of the capacitors.

ACKNOWLEDGMENT

This project has been supported in part by the U.S. Office of Naval Research (award #: N00014-21-1-2124). It is approved for public release (DCN# 2024-10-23-248).

REFERENCES

[1] J. Millan, P. Godignon, X. Perpina, A. Perez-Tomas, and J. Rebollo, "A survey of wide bandgap power semiconductor devices," IEEE Trans. Power Electron., vol. 29, no. 5, pp. 2155–2163, May 2014.

[2] D. Han, J. Noppakunkajorn, and B. Sarlioglu, "Comprehensive efficiency, weight, and volume comparison of SiC and Si-based bidirectional DCDC converters for hybrid electric vehicles," IEEE Trans. Veh. Technol., vol. 63, no. 7, pp. 3001–3010, Sep. 2014.

[3] D. Han, S. Li, Y. Wu, W. Choi and B. Sarlioglu, "Comparative Analysis on Conducted CM EMI Emission of Motor Drives: WBG Versus Si Devices," in IEEE Transactions on Industrial Electronics, vol. 64, no. 10, pp. 8353-8363, Oct. 2017.

[4] T. Kim, D. Feng, M. Jang, and V. G. Agelidis, "Common mode noise analysis for cascaded boost converter with silicon carbide devices," IEEE Trans. Power Electron., vol. 32, no. 3, pp. 1917–1926, Mar. 2017.

[5] H. Akagi and T. Doumoto, "A passive EMI filter for preventing high-frequency leakage current from flowing through the grounded inverter heat sink of an adjustable-speed motor drive system," IEEE Trans. Ind. Appl., vol. 41, no. 5, pp. 1215–1223, Sep. 2005.

[6] S. Ye, W. Eberle, and Y. F. Liu, "A novel EMI filter design method for switching power supplies," IEEE Trans. Power Electron., vol. 19, no. 6, pp. 1668–1678, Nov. 2004.

[7] Agarwal, N.; Ahmad, M.W.; Anand, S, "Quasi-Online Technique for Health Monitoring of Capacitor in Single-Phase Solar Inverter," in IEEE Trans. Power Electron. 2017, 33, 5283–5291.

[8] Dang, H.; Park, H.; Kwak, S.; Choi, S, "DC-Link Electrolytic Capacitors Monitoring Techniques Based on Advanced Learning Intelligence Techniques for Three-Phase Inverters," in Machines 2022, 10, 1174.

[9] Sundararajan, P.; Sathik, M.H.M.; Sasongko, F.; Tan, C.S.; Tariq, M.; Simanjorang, R, "Online Condition Monitoring System for DC-Link Capacitor in Industrial Power Converters," in IEEE Trans. Ind. Appl. 2018, 54, 4775–4785.

[10] Information Technology Equipment—Safety—Part 1: General Requirements, International Standard IEC 60950-1.

[11] CISPR, "Methods of measurement of the suppression characteristics of passive EMC filtering devices," IEC Standard CISPR-17:2011, 2011.

[12] S. Wang, R. Chen, J. D. Van Wyk, F. C. Lee and W. G. Odendaal, "Developing parasitic cancellation technologies to improve EMI filter performance for switching mode power supplies," in IEEE Transactions on Electromagnetic Compatibility, vol. 47, no. 4, pp. 921-929, Nov. 2005

2200A/48V-to-1V Low-Profile Direct Power Converter with standard PCB transformer

Alejandro Figueroa, Pablo Mazariegos, Álvaro Cobos, Javier Goicoechea, Alejandro Castro, and José A. Cobos
Differential Power S.L.
28042 Madrid, Spain
email: {alex.figueroa, pablo.mazariegos, alvaro.cobos, javier.goicoechea, alejandro.castro, jcobos}@differentialpower.com

Abstract—**This article introduces a 48V-1V high-current single-stage power converter for high performance applications. A hardware prototype is designed and built by stacking four 12:1V modules, with the primary stacks connected in series and the secondary side in parallel. This configuration provides voltage balance in the primary stacks and intrinsic current sharing between the secondary cells. The Segmented Winding Transformer is designed to achieve a 5mm height; a Planar Matrix Transformer with custom magnetic cores is implemented on a standard 3oz PCB, ensuring repeatability and simplifying manufacturing. A significant characteristic of this power converter is the high efficiency achieved, exceeding 90% for currents from 200A to 2000A, with a peak efficiency of 93.6% at 500A, including driving losses. By modifying the primary side, the power converter offers fast dynamic response and output voltage regulation, achieved through low leakage inductance and high voltage in the transformer. With a current slope exceeding 3000A/µs and interleaved phases, the power converter can handle a 440A load step with minimal output voltage deviation.**

Keywords—Low-Profile Direct Power Converter, high-current single-stage, Segmented Winding Transformer, Planar Matrix transformer, standard PCB, fast dynamic response

I. INTRODUCTION

High-performance applications are revolutionizing the power electronics market in the modern information-driven world. The need for high currents at low voltages presents challenges in achieving efficient power converters, prompting the development of innovative topologies and designs to enhance the efficiency and packaging of power converters [1], [2]. Key advancements include power distribution structures that shorten the distance between the power converter and the chip.

Future generations of artificial-intelligence chips are expected to require very high currents, likely necessitating Vertical Power Delivery architectures. As current increases, power converter losses also rise, primarily determined by the specific resistance (mΩ·mm²), which depends on several factors: the number of power switches, the footprint filling factor (percentage dedicated to power switches), RMS current, conduction losses, connections, packaging, and other losses. Addressing challenges such as efficiency, power density, and transient response is essential for managing high voltage ratios and currents. A wide range of topologies can be employed, from transformer-based designs [3], [4], [5] to switched capacitor configurations [6], [7], [8], [9]. This article presents the following strategies aimed at reducing mΩ·mm² and offering a new perspective on existing approaches in high-performance power converters:

- **Transformer-based topology**

Transformers transfer energy with minimal storage requirements, allowing for a smaller effective magnetic core area (Ae) and reduced switching frequency (fsw). A smaller Ae allows for lower profiles, and removing the energy storage gaps needed by inductors provides a substantial benefit. Alternatively, decreasing fsw improves efficiency by reducing losses such as skin, proximity, fringing, driving and switching, thereby lowering mΩ·mm² [10].

- **Split power converter with the Segmented Winding Transformer (SWT) positioned beneath the chip**

Primary side switches and controllers are positioned close to the chip, with the rectification stage located below it (Fig. 1). This configuration enhances the modularity and scalability of the rectification stage, enabling it to be adapted for larger chips. The design offers key advantages, including increased space for critical components like synchronous rectifiers, reduced losses below the chip, and enhanced thermal management.

The losses in the converter are mainly conduction losses, concentrated in the transformer and secondary side, especially in the Segmented Winding Transformer [11]. There is an N² ratio between primary and secondary losses. This led to the design concept of separating the primary and secondary sides of the converter.

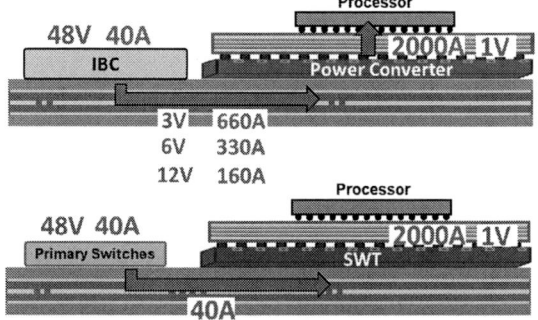

Fig. 1. Lateral-Vertical (up) and Split (down) Power Delivery.

The straightforward design of the Segmented Winding Transformer enables hundreds of balanced parallel current paths, which optimize the output current flow. The high density of synchronous rectifiers minimizes specific resistance, and the structure supports scalable vertical power delivery with intrinsic current balance, eliminating the need for additional control circuitry.

II. DIRECT POWER CONVERTER

The power converter presented in this paper is derived from the Direct Power Converter (DPx) [12] and includes additional active clamps to reduce voltage stress on the primary switches by maintaining a constant demagnetization voltage. However, neither the standard nor clamped versions support gain regulation. This design achieves a 48:1V conversion ratio by stacking four 12V to 1V primary stages and introduces a 5mm-high structure suitable for high-current applications.

A. Proposed Topology and Fundamentals

Achieving 48V-to-1V conversion in this power converter relies on a matrix transformer structure [13] with eight custom magnetic cores, arranged in four stacks with a 12:1 turns ratio per stack. The output voltage is set by the transformer turns ratio (N), calculated as the product of the primary windings per stack (N_P) and the number of primary stacks (N_{STACKS}).

The circuit in Fig. 2 is a modular, stacked power converter achieving a 48:1 step-down ratio via four identical 12:1 transformer stage, with the primary in series and the secondary in parallel. This DPx design is ideal for high-current, low-voltage applications, with each stage featuring a primary side switch network, a transformer, and a secondary side rectifier. On the secondary side, synchronous rectifiers convert AC to DC, and output capacitors smooth the signal for a stable DC output. The parallel rectified outputs share current across stages, reducing the need for extra balancing control circuitry [14], [15].

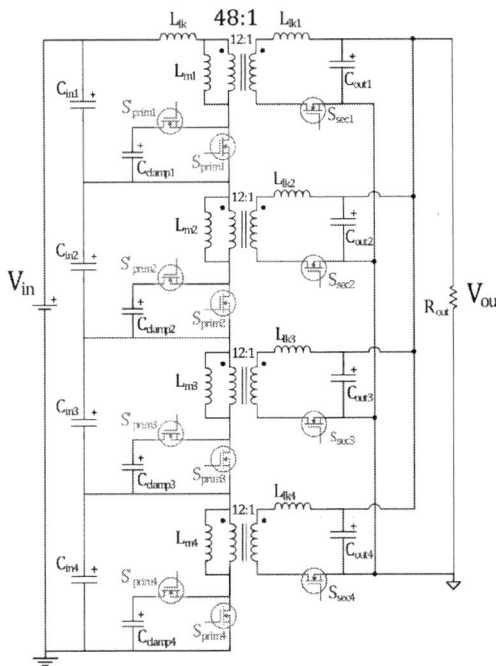

Fig. 2. Schematic of the proposed topology. Includes four stacked primaries (red GaN FETs) with their secondary stage (blue Si MOSFETs).

The power topology equivalent model includes a transformer and several switches: two primary side switches and one secondary side switch. Power transfer occurs when both main switches (Sprim and Ssec) are closed, while demagnetization takes place when they are open, and the active clamp (S'prim) is closed. During conduction, power flows from input to output, with the current waveform shaped by the leakage inductance and output capacitance. In the turn-off phase, the magnetic cores demagnetize as the active clamp switch closes. This paper advocates a 48V-to-1V single-stage Direct-to-Chip power conversion topology, highlighting the following features:

- **Transformer-Based Design:** A transformer is used for AC power processing, reducing energy storage and losses compared to inductors or capacitors.
- **Low Switching Frequency:** Operating at 200–300kHz minimizes losses in driving switches, magnetic cores, and conductors.

- **Elevated Duty Cycle:** A high duty cycle (up to 90%) reduces RMS current and conduction losses, though it increases device voltage (V_{DS}) nearing breakdown limits (V_{BV}).
- **High-Voltage Power Switches:** Discrete high-voltage switches outperform monolithic low-voltage switches in this configuration.
- **Single Rectifier Design:** A single rectifier with parallel cells minimizes footprint and losses, compared to dual rectifiers in other transformer-based converters (LLC or CDR).
- **Intrinsic Current Balance:** The topology inherently ensures current balance on the secondary side (Fig. 3), which is essential for managing numerous high-current cells injecting current into the load. This is achieved because primary current is equal across each magnetic element, providing a uniform excitation to each secondary turn.

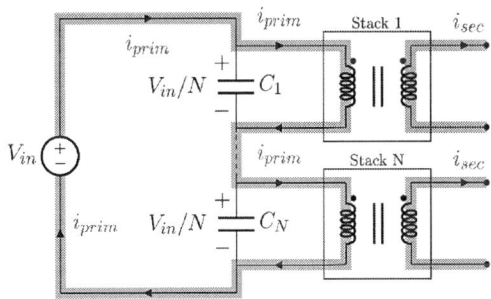

Fig. 3. Series-primary and parallel-secondary configuration for intrinsic current and voltage balance in a stacked matrix transformer.

B. Power converter structure

The DPx power converter, shown in Fig. 4, has dimensions of 90 x 80 x 5 mm and comprises four primary stacks connected in series. Each stack, controlled by independent driving circuits, includes five parallel GaN devices (EPC2215), two of which function as active clamps. The secondary side consists of Segmented Winding Transformer cells, with 192 Si devices (IQE006) connected in parallel.

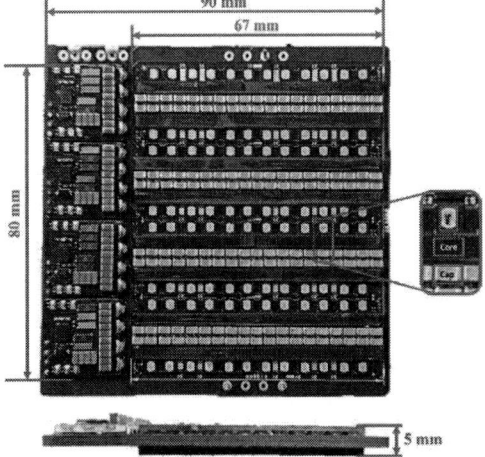

Fig. 4. DPx-Matrix-Stacked power converter with Segmented Winding implemented on a Planar Matrix Transformer.

The primary winding is connected in series, and the secondary winding is arranged in parallel to spread the current across multiple synchronous rectifiers, reducing conduction losses [16]. Since the output capacitors and secondary switches are part of the secondary turn, they should be placed close to the magnetic core to minimize termination effects. Additionally, secondary windings use the SWT concept to further lower conduction losses. The primary winding delivers equal current to each magnetic element, ensuring uniform excitation of all secondary turns and effectively balancing the parallel secondary cells.

The transformer uses a 10-layer Planar Matrix design on a 3oz. PCB, optimizing manufacturing. This design, ideal for high-current applications, reduces conduction losses by parallelizing power layers. The SWT on the secondary side is space-efficient, supports automated manufacturing processes, and enables scalability for high-power-density solutions.

III. HARDWARE DESIGN AND EXPERIMENTAL RESULTS

Hardware implementation underwent modifications throughout the testing process. During the initial tests, with output currents below 1000A, cooling the converter with fans was sufficient. As current levels increased, a cold plate was added in conjunction with fans to enhance the cooling system. Finally, a custom aluminum frame (Fig. 5) was designed to achieve optimal thermal dissipation in the converter, allowing it to reach maximum output power without thermal instability. This approach enabled an output current up to 2200A.

Fig. 5. Test set-up with custom electronic loads and cooling system.

Fig. 6 illustrates the setup connected to the power converter, which includes additional PCBs to evenly distribute thousands of amperes. The cooling system consists of a custom frame that integrates the converter with a cold plate (ATS-TCP-1001, offering 21mW/mm² cooling capacity) to efficiently dissipate heat under high load conditions.

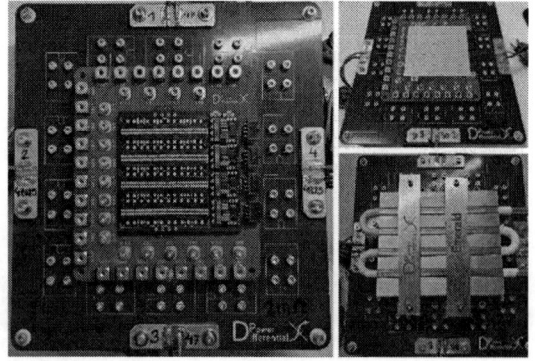

Fig. 6. Power converter assembly with aluminum frame and cold plate.

All the primary components of the converter are detailed in Table I, along with the corresponding inductance values measured from the primary side.

TABLE I. COMPONENT LIST OF THE PROPOSED POWER CONVERTER

Specification	Parameter
Primary Switch	EPC2215 (R_{ON}=8mΩ, BV=200V)
Secondary Switch	IQE006NE2LM5CGATMA1 (R_{ON}=0.65mΩ, BV=25V)
Primary Driver	LMG1210RVRR
Secondary Driver	1EDN7550UXTSA1
Input Capacitor (each stack)	C3216X5R1E476M160AC (47µF × 8 in parallel)
Output Capacitor	GRM31CR60J227ME11L (220µF × 320 in parallel)
Magnetic Material	3C95
Magnetizing Inductance (primary side)	20µH
Leakage Inductance (primary side)	30nH

The power converter waveforms for a 48V-1V and 700A output current are displayed in Fig. 7. The waveforms have soft switching and remain balanced due to the close coupling of all the secondary windings with a single primary winding. Notably, for higher output currents, the only control parameter adjusted is the duty cycle, which can reach up to 85% at full-load.

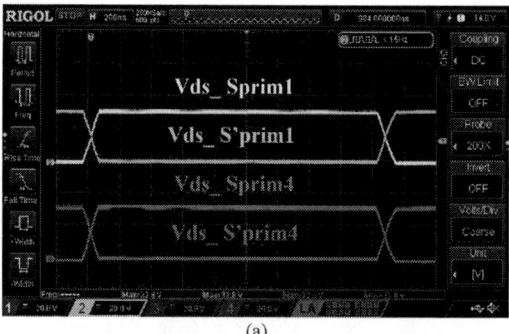

(a)

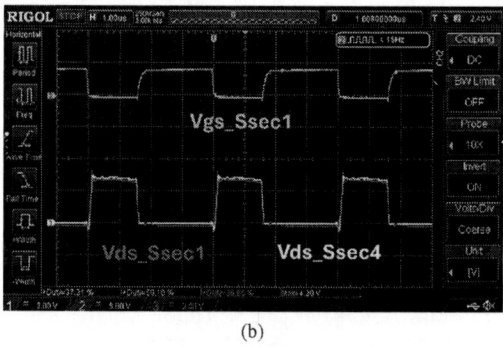

(b)

Fig. 7. Primary side (a) and secondary side (b) waveforms for a specific output current (48V-1V @ 700A).

A. Efficiency and Losses analysis

The power converter is tested for a 48V input voltage with 1V at the output for a maximum current of 2200A (Fig. 8). The switching frequency is 235kHz and the duty cycle has been taken to values above 80%. Efficiency values, including driving, align closely with the calculations from the optimization process. Peak efficiency reaches 93.6% at 500A, and efficiency values remain above 90% for currents from 200A to 2000A.

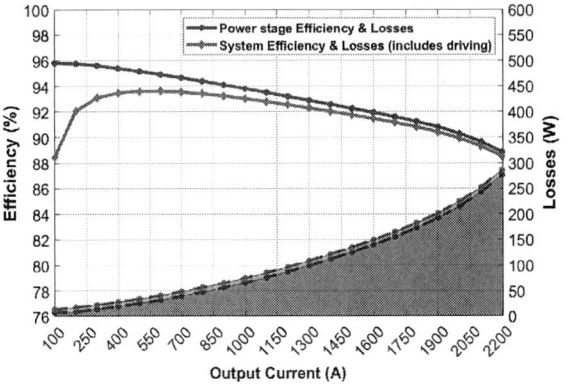

Fig. 8. Measured efficiency and losses of the power converter.

Thermal images shown in Fig. 9, depict the temperature distribution within the power converter at output currents of 500A and 700A, with cooling supplied by multiple Sanyo Denki fans (67.8CFM) under lower load conditions. The images reveal hot spots primarily around the secondary drivers, which are responsible for controlling hundreds of switches. These images demonstrate the intrinsic current balancing on the secondary side of the converter, as evidenced by the uniform temperature spread across most of the secondary components. This even temperature distribution suggests that the current is effectively balanced among the paralleled secondary elements, minimizing thermal hotspots and allowing efficient cooling under increasing loads.

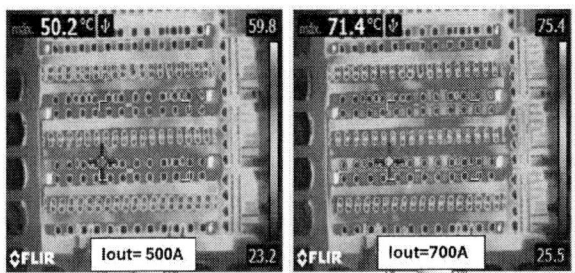

Fig. 9. Thermal images with air cooling at different low-loads.

Fig. 10 provides a detailed breakdown of power losses in the converter across different operating points and output currents, highlighting both the total losses and their distribution. Bar charts further quantify these losses by component type as output current increases. Conduction losses in both the primary and secondary windings rise considerably with increasing currents, becoming a substantial part of the total losses in power converters designed for elevated current levels. Core and driving losses are more significant at lower loads. Together, the pie and bar charts effectively illustrate how each component affects overall efficiency, with conduction and core losses emerging as primary contributors at various current levels. These insights emphasize the importance of efficient thermal management and optimized converter design for high-load conditions.

Fig. 10.b show efficiency and losses distribution at two specific operating points: at 500A, where peak efficiency is 93.35%, and at 2200A, where full-load efficiency drops to 89.1%. At lower currents, driving and core losses are the leading factors compared to other components, but as current increases up to 2200A, losses in primary and secondary conduction paths become dominant.

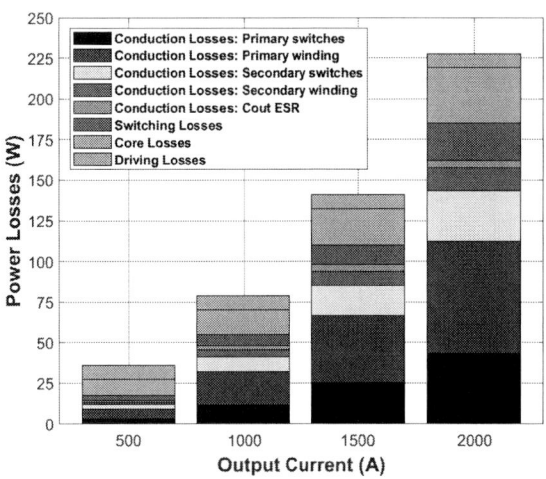

(a)

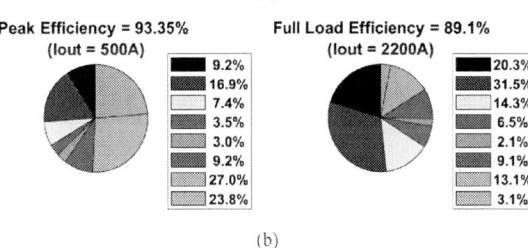

(b)

Fig. 10. Losses breakdown for different output currents (a) and distribution at peak and full-load efficiency (b).

Multiple tests were conducted at varying output voltages (0.5V, 0.8V, 1V, and 1.2V) to identify the optimal efficiency curve across different output currents. As shown in Fig. 11, efficiency improves progressively with higher output voltages; however, the current range must be limited, as the converter cannot dissipate all the heat generated at higher loads. Note that output voltage adjustments were achieved by modifying the input voltage on the power supply, as the power converter is not regulated.

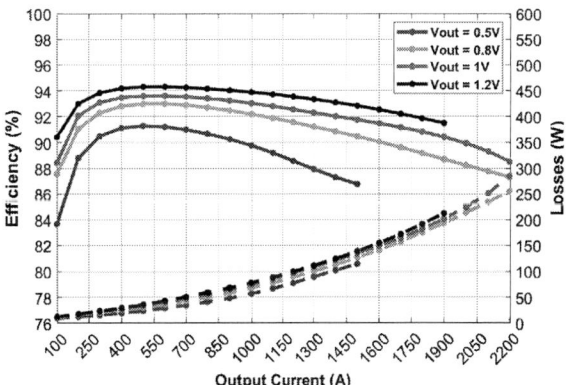

Fig. 11. Measured efficiency and losses curves including driving losses for different output voltages.

B. Load steps and Dynamic response

Achieving fast dynamics and output voltage regulation in the converter requires modifying the power topology by adding two more switches with an active clamp capacitor between the input source and the transformer. This new topology, called Direct Power Converter with Primary Controlled High Voltage (DPx-PCHV), operates similarly to the DPx for energy transfer but controls the output current using the clamped voltage through the leakage inductance. Modifying the rise time enables precise adjustment of the current injected during a single edge, facilitating the voltage regulation.

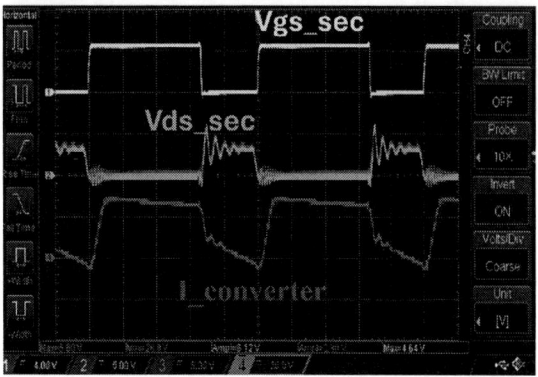

Fig. 12. Waveforms of the regulated converter (DPx-PCHV topology).

When the converter responds faster than the load, voltage drops are effectively eliminated regardless of the output capacitance. To experimentally demonstrate this, the power converter can generate an output voltage increase during a load step (Fig. 13). If the converter di/dt exceeds the load demand, the output voltage rises as the converter directly supplies the current, bypassing the need for Cout current. This results in minimal voltage deviation, showing that faster converters can step up, step down, or maintain stable output voltage even during rapid load changes. Given the structure of the power converter, it is feasible to implement interleaving of stacks to reduce ripple in the output voltage.

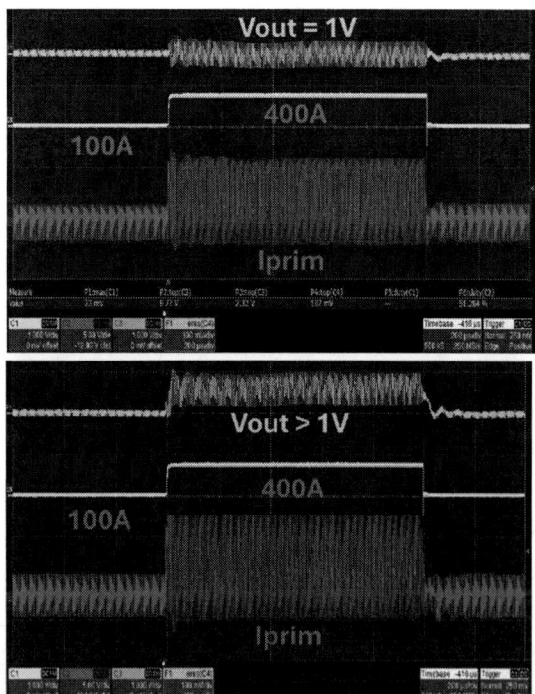

Fig. 13. Output voltage dynamics comparison under a 300A load step.

In Fig. 14, the output voltage remains at the nominal value, indicating that the power converter response is faster than the load, achieving a rate greater than 3000A/μs. The load step was conducted in an open-loop configuration, with parameters

carefully selected to ensure accurate results. Only the primary side of the prototype was modified; the secondary side remains unchanged from the previous design. Due to the fast dynamic response to load steps, the output capacitance of the power converter could be reduced and adjusted solely to meet the desired ripple value.

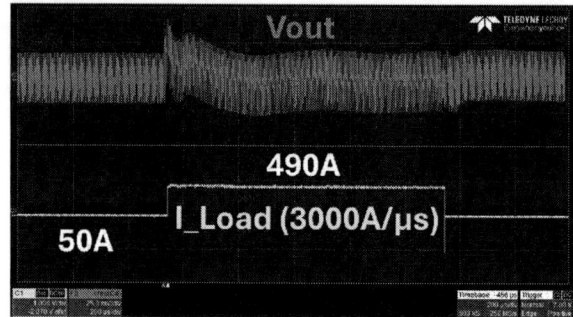

Fig. 14. Output voltage behavior to a 440A load step with a current derivative of 3000A/μs.

Transformer-based converters provide significant advantages during load steps due to their efficient energy transfer and impedance management. The transformer draws energy from the input capacitance, potentially at a less critical location, which helps optimize space and volume. The transformer conversion factor multiplies Cin by N^2, while reducing the primary side leakage inductance and resistance by N^2. This results in low impedance between the input and output, allowing high di/dt load steps.

A custom load-slammer was built and placed under the converter to test the dynamic response of the prototype (Fig. 15). This load stepper uses GaN switches (EPC2302) and includes an additional load that is briefly shorted. Each switch provides 73A at 1V with a switching speed of 500A/μs. By using six switches in parallel, the setup generates a load transient of 440A with a rate of 3000A/μs.

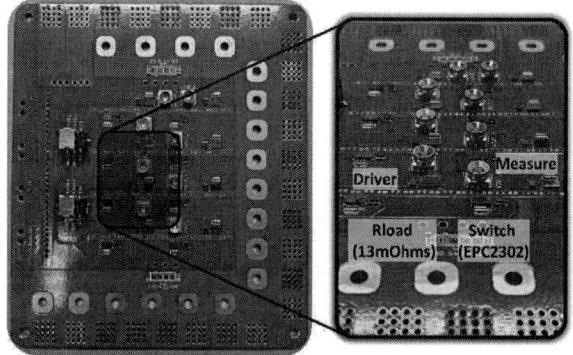

Fig. 15. Custom load slammer made up of sixteen switch-resistor modules responsible for making instantaneous load steps.

C. Performance comparison with the state of the art

For comparison with the state of the art, two widely used figures of merit (current density and power density) have been applied. While the comparison is not entirely direct, as it was made using the constant-gain prototype presented in this paper and only the SWT section (the only part of the power converter located beneath the chip), it still offers a useful reference for future designs. All the power converters shown in Fig. 16 generate 1V and high current, indicating they are all designed for the same application.

979-8-3315-1612-3/25 $31.00 © 2025 IEEE   1431

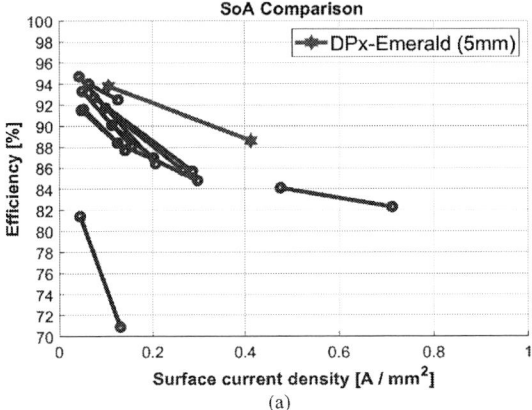

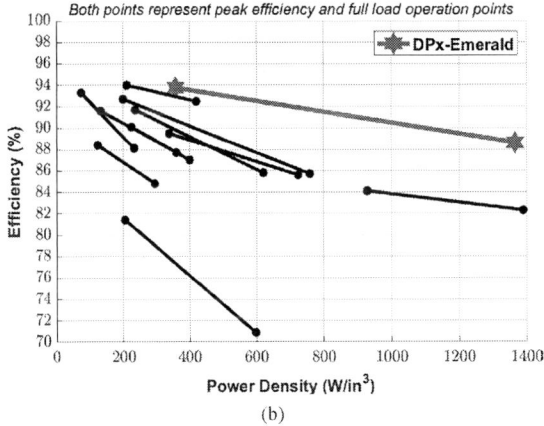

Fig. 16. Performance comparisons (Current density (a) and Power density (b)) of the DPx-Matrix-Stacked with the state-of-the-art power converters.

Full-load surface current and power density of the converter are 0.41 A/mm² and 1365 W/in³, respectively. These values demonstrate how the power converter exceeds the state-of-the-art in both graphs, marking a significant advancement in high current applications. Its 5mm height and implementation on a standard PCB make it a compact and advanced solution.

IV. CONCLUSION

This paper validates two key concepts for high-performance chips in a 48V-1V Low-Profile Direct power converter, focusing on the ability to handle high current and provide fast dynamic response. The DPx topology achieves over 90% efficiency for currents ranging from 200A to 2000A, with a full-load efficiency of 88.5% at 2200A. In terms of dynamic response, the DPx-PCHV topology demonstrates exceptional performance by maintaining no voltage deviation during a 420A load step and exceeding 3000A/μs. The main key point of the power converter lies in the integration of the updated DPx topology, which features stacking and active clamps on the primary side, combined with a Planar Matrix transformer. This combination results in a height of just 5mm, exceeding conventional solutions in power density and efficiency. These advancements surpass the current state-of-the-art (0.41A/mm² and 1365W/in³ at full-load), highlighting the designed power converter potential for Power-on-Package applications with Vertical Power

Delivery, where both space and efficiency are crucial. The scalability of the converter, featuring SWT cells on the secondary side, allows for easy integration under the chip. This design can be adapted to meet the specific requirements of high-performance applications due to its intrinsic current balance, making it an ideal solution for high-current systems that demand efficient and compact power delivery.

REFERENCES

[1] X. Li and S. Jiang, "Google 48V Rack Adaptation and Onboard Power Technology Update," in Open Compute Project (OCP) 2019 Summit, 2019.

[2] K. Radhakrishnan, M. Swaminathan, and B. K. Bhattacharyya, "Power delivery for high-performance microprocessors–challenges, solutions, and future trends," IEEE Trans. Compon. Packag. Manuf. Technol., vol. 11, no. 4, pp. 655–671, Apr. 2021.

[3] Xin Lou and Qiang Li, "Multiphase Half-Bridge Current-Doubler Rectifier: A 93.1%-Efficiency Single-Stage 48V Voltage Regulator with 1.04 kW/in3 Power Density, " 2023 IEEE Applied Power Electronics Conference and Exposition (APEC), 2023.

[4] M. H. Ahmed, C. Fei, F. C. Lee, and Q. Li, "Single-Stage High-Efficiency 48/1 V Sigma Converter With Integrated Magnetics," IEEE Transactions on Industrial Electronics, vol. 67, no. 1, pp. 192–202, Jan. 2020, doi: 10.1109/TIE.2019.2896082.

[5] B. Li, X. Huang and J. Zhang, "A High Density 400 W DC/DC Power Module with Integrated Planar Transformer and Half Bridge GaN IC," 2024 IEEE Applied Power Electronics Conference and Exposition (APEC), Long Beach, CA, USA, 2024, pp. 94-100, doi: 10.1109/APEC48139.2024.10509334.

[6] Ping Wang, David Giuliano, Stephen Allen, M. Chen. "MSC-PoL: An Ultra-Thin 220-A/48-to-1-V Hybrid GaN-Si CPU VRM with Multistack Switched Capacitor Architecture and Coupled Magnetics". 2023 IEEE Applied Power Electronics Conference and Exposition (APEC), 2023.

[7] S. Khatua, D. Kastha, and S. Kapat, "A Dual Active Bridge Derived Hybrid Switched Capacitor Converter Based Two-Stage 48 V VRM," IEEE Transactions on Power Electronics, vol. PP, Dec. 2020, doi: 10.1109/TPEL.2020.3046362.

[8] J. Baek et al., "Vertical Stacked LEGO-PoL CPU Voltage Regulator," IEEE Transactions on Power Electronics, vol. 37, no. 6, pp. 6305–6322, Jun. 2022, doi: 10.1109/TPEL.2021.3135386.

[9] Y. Zhu, J. Zou, and R. C. N. Pilawa-Podgurski, "A 1500-A/48-V-to-1-V Switching Bus Converter for Next-Generation Ultra-High-Power Processors," IEEE Transactions on Power Electronics, vol. 39, no. 9, pp. 11340–11355, Sep. 2024, doi: 10.1109/TPEL.2024.3403670.

[10] D. Feucht, "Magnetizing Current and Transformer Design Optimization," Innovatia Laboratories, published on How2Power.com, October 2023.

[11] J. A. Cobos, Á. Cobos, "Electrical power converter with segmented windings" EP4066267B1, June 28, 23. Available: https://patents.google.com/patent/EP4066267B1/en.

[12] J. A. Cobos, "Direct Electrical Power Converter," International Application PCT/EP2020/080027.

[13] E. Herbert, "Flat Matrix Transformer. US Patent 4,665,357," Patent filed in 1987.

[14] J. A. Cobos, P. Mazariegos, A. Figueroa, A. Castro and Á. Cobos, "500A Stacked Direct Power converter with standard PCB transformer," 2024 IEEE Applied Power Electronics Conference and Exposition (APEC), Long Beach, CA, USA, 2024, pp. 925-930, doi: 10.1109/APEC48139.2024.10509493.

[15] A. Figueroa, P. Mazariegos, J. Goicoechea, A. Castro and J. A. Cobos, "Low-Profile Direct Power Converter: 350A/48V-1V with Planar Matrix Transformer using standard PCB and commercial cores," 2024 IEEE Applied Power Electronics Conference and Exposition (APEC), Long Beach, CA, USA, 2024, pp. 2172-2177.

[16] Y. -C. Liu et al., "Design and Implementation of a Planar Transformer with Fractional Turns for High Power Density LLC Resonant Converters," in IEEE Transactions on Power Electronics, vol. 36, no. 5, pp. 5191-5203, May 2021.

Single-Stage 48V-to-1V Regulator with a Half-Turn Transformer and Current-Doubler Rectifier

Xinmiao Xu, Qiang Li
Center for Power Electronics Systems
Department of Electrical and Computer Engineering
Virginia Polytechnic Institute and State University
Blacksburg, VA 24061, United States
Email: {xinmiaoxu, lqvt}@vt.edu

Abstract—**Single-stage 48V-to-1V regulated solutions for high-performance processors become increasingly popular. This paper presents a full-bridge transformer-based buck converter with the current-doubler rectifier. The transformer adopts a half-turn winding structure for reduced resistance and leakage inductance. Operating principles of the converter with the half-turn transformer and the converter's optimized printed circuit board layout are illustrated. Synchronous rectifiers are integrated in high-frequency loops and are treated carefully to reduce parasitic-inductor-induced loss. With the help of recent advances in gallium nitride (GaN) and silicon transistors, a 48V-to-1.8V GaN-based prototype switching at 700 kHz is constructed, which achieves 95.34% peak efficiency, 92.79% full-load efficiency (including gating loss) and 576 W/in^3 power density with 50-A current per phase. The same hardware achieves 94.00% peak efficiency and 89.75% full-load efficiency with 55-A current per phase in 48V-to-1.0V conversion. Experimental results with 54-V input are also included in this paper, demonstrating the great potential of the single-stage solution for future data centers.**

Index Terms—**half-turn transformer, current-doubler rectifier, commutation loops, PCB layout, high switching frequency.**

I. INTRODUCTION

Data centers accounted for 4% of electric energy consumption in the United States in 2023, and this figure is projected to be more than doubled by 2030 [1]. Another noteworthy phenomenon is that the most energy-intensive hyper-scale data centers are concentrated in certain states, such as Virginia, Texas and California. According to statistics from a local utility company, data centers comprised 24% of Virginia's electric load in 2023 [2]. The significant national-level growth, coupled with uneven geographic distribution, requires more efficient and more compact energy delivery architectures.

Modern data centers have adopted 48-V buses to optimize the efficiency of power supply systems for high-performance processors. Besides two-stage solutions [3]–[9], single-stage 48V-to-1V regulators have received increased attention [10]–[15]. The vision for single-stage solutions is to have a smaller footprint without decoupling capacitors on the intermediate buses. However, for decent efficiency, most of the single-stage solutions have low switching frequencies and each output inductor conducts only 30-A current at full load. Low-frequency operation limits transient performance and current density. Also, it is not cost-effective to operate an inductor with a current that is far less than its saturation current.

This paper presents a full-bridge transformer-based buck converter that features a high switching frequency, high efficiency and a high loading capacity. Section II presents operating principles of the converter with a half-turn transformer. Section III includes a co-design of magnetic core geometry, winding arrangement and placement of rectifiers for an optimized printed circuit board (PCB) layout. In Section IV, efficiency curves and thermal images of a prototype are presented, along with a comparison to state-of-the-art solutions. Section V concludes this paper.

II. CURRENT DOUBLER RECTIFIER WITH A HALF-TURN TRANSFORMER

The multi-phase synchronous buck converter has been used extensively in the 12V-to-1V conversion for processors. However, it usually exhibits lower efficiency when it is exposed to a larger step-down ratio of input voltage to output voltage. For 48V-to-1V regulated conversion, a full-bridge transformer-based buck converter is a more appropriate topology. It uses a transformer turns ratio to assist output inductors in the voltage step-down conversion.

In point-of-load power supplies where the conduction loss dominates at heavy load, transformers with single-turn secondary windings are deemed to have a desirable trade-off between winding loss and core loss. However, a single turn might carry too much resistance and the even the single turn itself can occupy a large footprint. To further reduce the winding loss, half-turn techniques have been applied to the push-pull regulator in [16], the LLC converters with the half-bridge rectifier in [17] and the full-bridge rectifier in [18] and [19]. This paper applies the half-turn technique to the current doubler rectifier.

Compared to the half-bridge rectifier with a center-tapped secondary winding that has three wire leads to terminate, the current doubler has only two wire leads of a secondary winding for termination and doubles the number of output inductors. Also, as opposed to a center-tapped secondary winding that has only half of the winding utilized when the energy is transferred from the primary to the secondary, the current doubler uses all of the secondary winding. Therefore, it has better winding utilization and the potential to achieve a more compact design. These characteristics make the current

979-8-3315-1612-3/25 $31.00 © 2025 IEEE

doubler a proper rectifier candidate in low-voltage and high-current applications.

Fig. 1 shows a full-bridge transformer-based buck with two current-doubler rectifier units. Switch activation signals and $v_{\mathrm{pri}}(t)$, the resultant voltage applied to the transformer primary winding, are illustrated in Fig. 2. Differing from a conventional single-turn secondary winding for the current-doubler rectifier, two half-turn secondary windings are used in this paper to imitate a complete single turn. Compared to the conventional design, a half-turn transformer can achieve half of the winding resistance and a quarter of the leakage inductance with an identical turns ratio. The core idea of a half-turn transformer is that we use two individual but ideally identical currents to form a single loop with one current path being the return path of the other current. As shown in Fig. 3,

a half-turn transformer is made of a gap-less EI core. During the positive half cycle of $v_{\mathrm{pri}}(t)$ when Q_1, Q_4, SR_2 and SR_4 are turned on, both L_{out1} and L_{out3} are being charged, and their currents form a secondary loop that is electromagnetically coupled with the primary winding. Fig. 3 (b) shows the current flow of the negative half cycle when Q_2, Q_3, SR_1 and SR_3 are turned on and both L_{out2} and L_{out4} are being charged.

Switch operations of the full-bridge inverter and the current-doubler rectifier remain unchanged before and after a half-turn transformer is applied to the topology. Therefore, controllers for the multi-phase buck can be adopted for this converter except that the maximum duty ratio should be limited to 50%.

Phase-shift techniques are applicable to the full bridge for zero voltage switching if needed. This paper simply uses a symmetric (diagonal) operation for the primary switches.

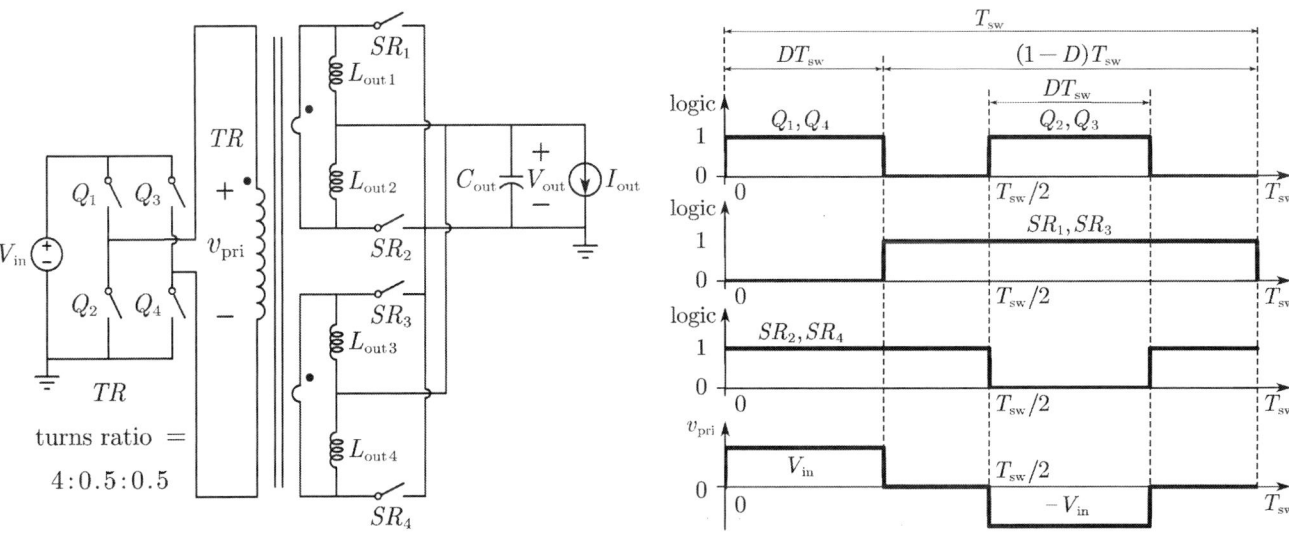

Fig. 1. Full-bridge converter with a half-turn transformer and the current doubler rectifier.

Fig. 2. Switch activation signals of the full-bridge converter and the voltage applied to the transformer primary winding.

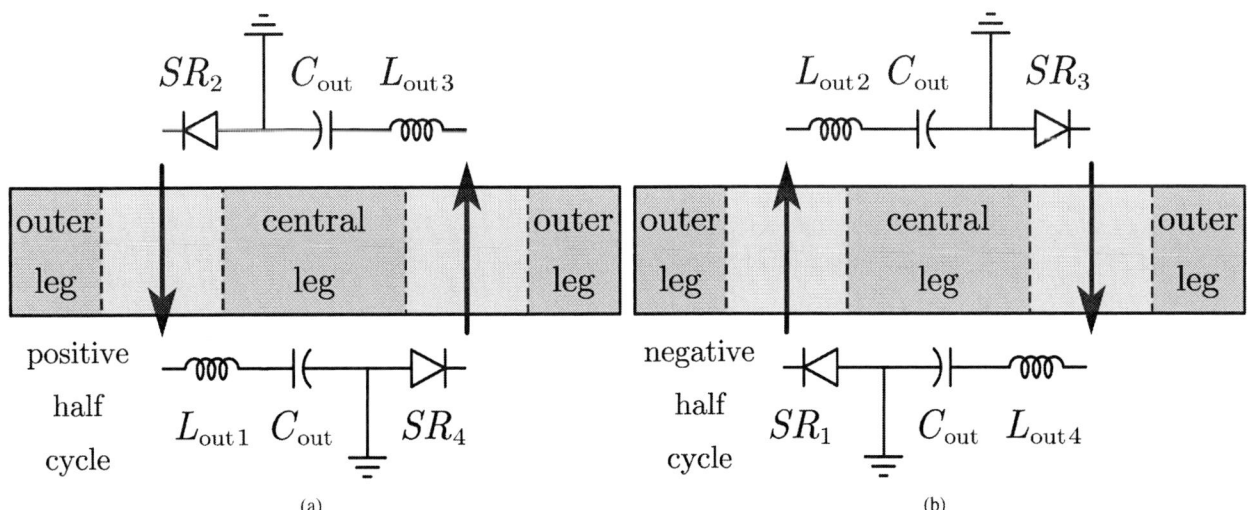

Fig. 3. Half-turn transformer's secondary currents represented by arrows during: (a) the positive half cycle of $v_{\mathrm{pri}}(t)$ when Q_1, Q_4, SR_2 and SR_4 are on, (b) the negative half cycle of $v_{\mathrm{pri}}(t)$ when Q_2, Q_3, SR_1 and SR_3 are on.

979-8-3315-1612-3/25 $31.00 © 2025 IEEE

III. Optimization of Printed Circuit Board Layout

The turn-off loss induced by commutation loop inductors theoretically proves important for step-down regulators in point-of-load applications [13]. The model of inductor-induced turn-off loss of the series-capacitor buck topology is shown in Fig. 4 (a), and this model can be extended to the current-doubler rectifier and a half-turn transformer with primary components reflected to the secondary side. In Fig. 4 (b), a dc voltage source V_{vs} shows a reflected input voltage while four dc current sources represent four inductor currents at switching instants. When we focus on $Q_1 - Q_4$ turn-off transition, switches M_1, M_2 and M_4 in Fig. 4 (b) represent primary switches (Q_1 and Q_4 together), SR_1 and SR_3 in Fig. 1, respectively. $R_3 - L_3$ and $R_5 - L_5$ symbolize parasitic resistors and inductors associated with SR_2 and SR_4. Ideally, $I_{\text{cs}1}$ and $I_{\text{cs}3}$ should be identical, and the same applies to $I_{\text{cs}2}$ and $I_{\text{cs}4}$.

Likewise, parasitic elements of secondary branches, such as $R_2 - L_2$, $R_3 - L_3$, $R_4 - L_4$ and $R_5 - L_5$, should be balanced when the converter's layout is symmetric. Therefore, according to the derivation in [13], the turn-off loss is

$$E_{\text{off}} = \frac{1}{2}\left(L_1 + L_2//L_4 + L_3//L_5\right)\left(2I_{\text{cs}1} - 2I_{\text{x}}\right)^2 \\ + \frac{1}{2}C_1\left(V_{\text{vs}} - V_{\text{x}}\right)^2 \tag{1}$$

, where I_{x} and V_{x} are the states of L_2 and C_1 when D_2 and D_4 start conducting. Evidently, L_1 and parasitic inductors of secondary branches include primary and secondary leakage inductors, respectively. According to (1) and Fig. 4 (b), we identify three critical paths shown in Fig. 5 (a) and propose an optimized layout in Fig. 5 (b). When we connect two secondary commutation paths into a loop that is tightly coupled with a primary winding, the commutation

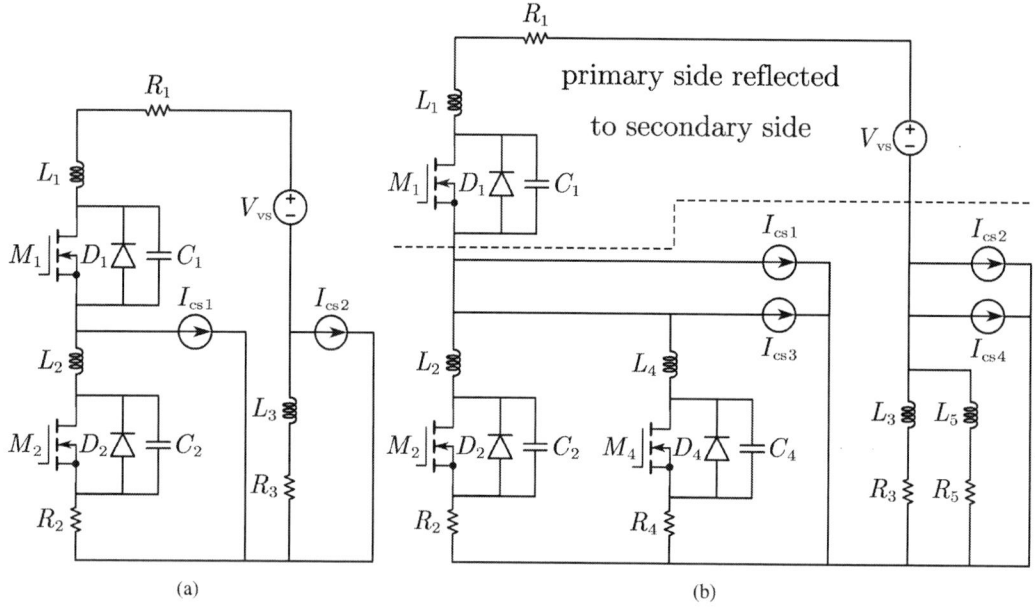

Fig. 4. Models of parasitic-element-induced loss in (a) the series-capacitor buck and (b) the current-doubler rectifier with a half-turn transformer.

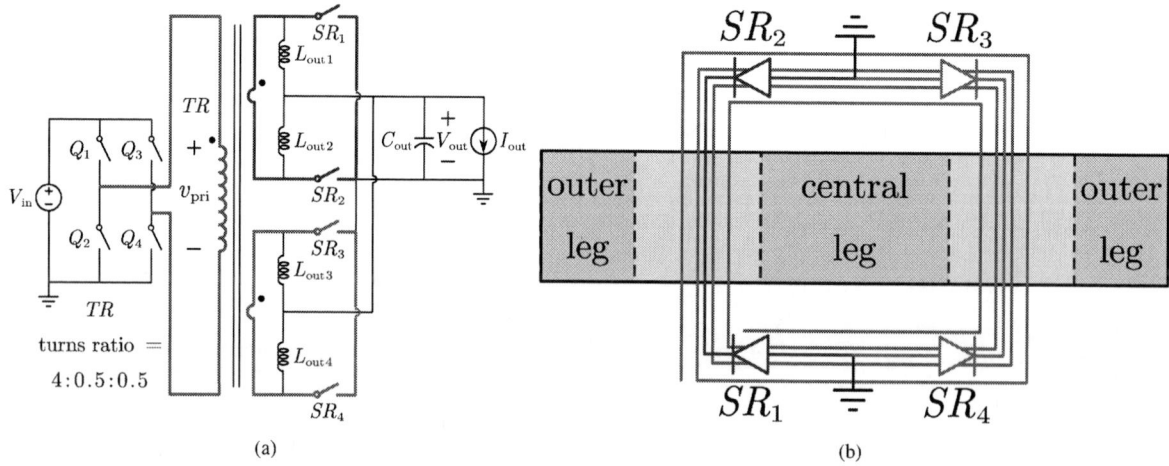

Fig. 5. Optimization of PCB layout: (a) critical paths of the commutation loop, (b) proposed optimized layout.

loop inductance, defined as $(L_1 + L_2//L_4 + L_3//L_5)$ in (1), is minimized. This is achieved by the mutual cancellation of the flux linkages on the primary and secondary sides. For example, with the specified turns ratio in Fig. 5 (a), during primary switches' turn-off transitions, when there is 1-A ac (ringing) current flowing clockwise along the red path, there will be simultaneous 4-A ac current flowing counter-clockwise in the loop formed by the blue and green paths in Fig. 5 (b). Therefore, the net of high-frequency flux linkages that correspond to the commutation loop inductors in Fig. 4 (b) will be zero ideally.

IV. HARDWARE AND EXPERIMENTAL RESULTS

Fig. 6 shows the top and bottom sides of a full-bridge converter with a half-turn transformer and the current-doubler rectifier. The upper left corner of the top side of the circuit board is reserved for sensors and feedback controllers. Main components of the prototype are listed in Table I.

This prototype is built with gallium nitride (GaN) transistors and designed for an input voltage range from 40 V to 60 V. It operates at 700 kHz and is tested at 1.8-V and 1.0-V output voltages for CPU and GPU applications, respectively. Measured efficiency curves including gating loss are shown in Fig. 7 (a) and (b). The prototype's loading capacity is more than 50 amperes per phase, which takes advantage of high saturation current of discrete inductors and makes the converter a cost-effective solution from a system perspective.

Efficiency curves of this work and state-of-the-art solutions [11], [13], [15] from academia are superimposed from Fig. 7 (c) to (f) for comparison purposes. Parameters of these 48V-to-1.8V and 48V-to-1.0V regulators are tabulated in Table II and III, respectively. For the 48V-to-1.0V regulated conversion in Table III, the prototype has simultaneously the highest switching frequency, lowest output inductance and highest efficiency. Thermal images of the prototype are shown in Fig. 8, and the thermal limit of the prototype is on rectifiers.

Despite the good performance achieved by the prototype, there are several ways for future improvement. First, instead of four non-coupled inductors, two groups of two-phase coupled inductors can be used since the currents of any two adjacent inductors are fully interleaved. The negatively coupled inductors have the potential to reduce the height of the voltage regulator module, and then the PCB side with inductors can be mounted on a sever board with copper pillars. Second, all transistors can be implemented on the same side of the PCB (e.g. bottom side in Fig. 6). This simplifies the cooling design, as transistors are the primary heat source, allowing a single cold plate or heat sink to be applied to the transistor side. Third, there are capacitor banks connected in series with the secondary windings in the prototype. These are used to help achieve current balance among the four phases but introduces unnecessary loss due to high current on the secondary side. With active feedback control implemented, these capacitor banks can be totally removed, and both efficiency and power density can be improved.

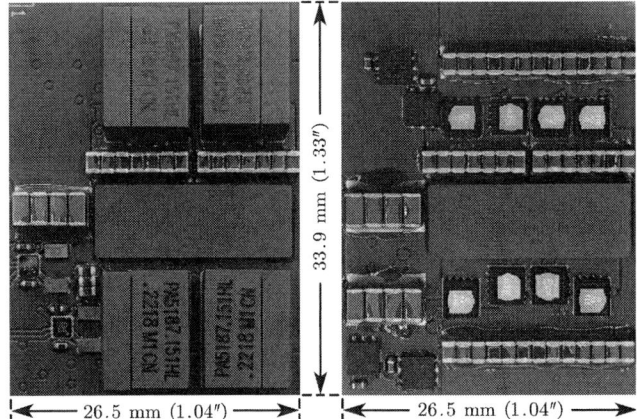

Fig. 6. Top and bottom sides of a full-bridge transformer-based buck.

TABLE I
MAIN COMPONENTS OF THE PROTOTYPE

Component	Manufacturer	Part Number
Primary switch	EPC	EPC2044
Secondary switch	Infineon	IQE004NE1LM7SC
	Infineon	IQE004NE1LM7CGSC
Primary driver	ADI	LT8418
Secondary driver	Infineon	1EDN7512G
Transformer core	DMEGC	DMR51W
Inductor	Pulse	PA5187.151HLT
Input capacitor	Murata	GRM188D72A105KE01D
	TDK	C3216X6S2A106K160AC
Output capacitor	TDK	C2012X5R0J476M125AC

V. CONCLUSION

In this paper, a half-turn winding structure is applied to a transformer in the current-doubler rectifier for the first time. Co-designed with PCB windings, layout of the current-doubler rectifier is optimized, with its theoretical connection to the series-capacitor buck pointed out. A four-phase full-bridge buck converter is experimentally tested, achieving simultaneously a higher switching frequency and higher efficiency than other single-stage solutions at both 1.8-V and 1.0-V output voltages. Also, the loading capacity is more than 50 amperes per phase, which is much larger than the prior art.

ACKNOWLEDGMENT

The authors would like to thank Murata Manufacturing for a generous sample donation of GRM188D72A105KE01D and DMEGC for manufacturing the magnetic core.

REFERENCES

[1] J. Aljbour, T. Wilson, and P. Patel, "Powering Intelligence: Analyzing Artificial Intelligence and Data Center Energy Consumption." [Online]. Available: https://www.epri.com/research/products/3002028905

[2] Dominion Energy, "Powering Your Every Day 2023 Annual Report." [Online]. Available: https://investors.dominionenergy.com/financials-and-reports/annual-materials/default.aspx

[3] S. Jiang, S. Saggini, C. Nan, X. Li, C. Chung, and M. Yazdani, "Switched Tank Converters," *IEEE Transactions on Power Electronics*, vol. 34, no. 6, pp. 5048–5062, Jun. 2019.

[4] X. Xu and Q. Li, "Symmetric Series-Capacitor Buck in 48V-to-12V Regulated Conversion for High-Performance Server Boards," in *2024 IEEE Energy Conversion Congress and Exposition (ECCE)*, Oct. 2024.

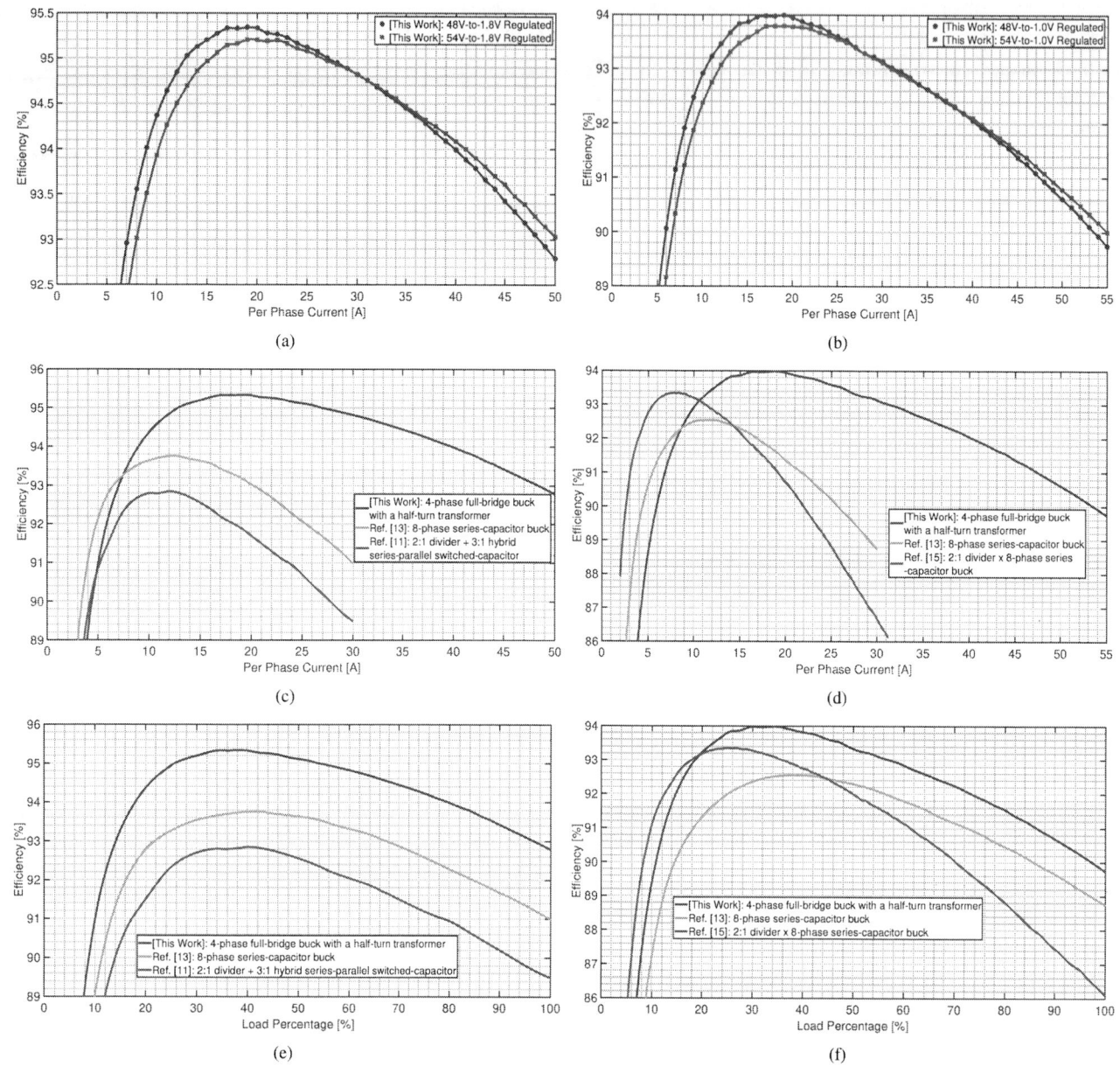

Fig. 7. Efficiency curves including gating loss: (a) this work for 1.8-V output, (b) this work for 1.0-V output. Comparison to the state-of-the-art: (c) 48V-to-1.8V [Per Phase Current], (d) 48V-to-1.0V [Per Phase Current], (e) 48V-to-1.8V [Load Percentage], (f) 48V-to-1.0V [Load Percentage].

TABLE II
PERFORMANCE COMPARISON OF 48V-TO-1.8V REGULATORS IN FIG. 7 (C)

Institute	Topology	f_{sw} [kHz]	L_{out} [nH]	η_{peak} [%]	$\eta_{full\text{-}load}$ [%]
this work	**4-phase full-bridge buck with a half-turn transformer**	**700**	**150**	**95.34% [19 A/ph]**	**92.79% [50 A/ph]**
CPES, VT [13]	8-phase series-capacitor buck	700	120	93.75% [12 A/ph]	90.99% [30 A/ph]
UC Berkeley [11]	2:1 divider + 3:1 hybrid series-parallel SC (merged)	320	360	92.8% [12 A/ph]	89.5% [30 A/ph]

TABLE III
PERFORMANCE COMPARISON OF 48V-TO-1.0V REGULATORS IN FIG. 7 (D) AND REF. [12]

Institute	Topology	f_{sw} [kHz]	L_{out} [nH]	η_{peak} [%]	$\eta_{full\text{-}load}$ [%]
this work	**4-phase full-bridge buck with a half-turn transformer**	**700**	**150**	**94.00% [19 A/ph]**	**89.75% [55 A/ph]**
CPES, VT [13]	8-phase series-capacitor buck	500	150	92.55% [12 A/ph]	88.74% [30 A/ph]
UC Berkeley [15]	2:1 divider × 8-phase series-capacitor buck (multi-stacked)	150	607†	93.4% [10 A/ph]	86.1% [31 A/ph]
Princeton [12]	2:1 divider × 4-phase series-capacitor buck (multi-stacked)	400	381†	91.65% [11 A/ph]	85.86% [28 A/ph]

† steady-state equivalent inductance [20] of coupled inductors

979-8-3315-1612-3/25 $31.00 © 2025 IEEE

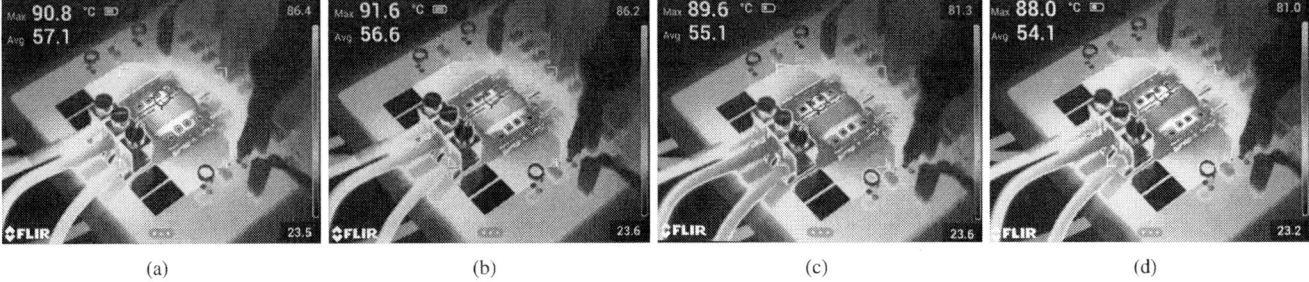

Fig. 8. Thermal images of the prototype at full load with $T_{\text{ambient}} = 25\,°\text{C}$ and 500-LFM fan cooling [air velocity measured above components]. (a) 48V-to-1.8V, $I_{\text{out}} = 200\,\text{A}$. (b) 54V-to-1.8V, $I_{\text{out}} = 200\,\text{A}$. (c) 48V-to-1.0V, $I_{\text{out}} = 220\,\text{A}$. (d) 54V-to-1.0V, $I_{\text{out}} = 220\,\text{A}$.

[5] A. M. Naradhipa, F. Zhu, and Q. Li, "Ultra-Low-Profile Twisted Core Inductor for Vertical Power Delivery Voltage Regulator," in *2024 IEEE Applied Power Electronics Conference and Exposition (APEC)*, Feb. 2024, pp. 918–924.

[6] Monolithic Power Systems, "48V Modules." [Online]. Available: https://www.monolithicpower.com/en/products/power-modules/48v-modules.html

[7] Infineon Technologies, "48 V Power Distribution." [Online]. Available: https://www.infineon.com/cms/en/applications/information-communication-technology/hyperscale-computing/48v-power-distribution/

[8] Vicor, "Power Solutions that Maximize AI, HPC and Data Center Performance." [Online]. Available: https://www.vicorpower.com/industries-and-innovations/computing

[9] Flex Power Modules, "Power Design Challenges with Unregulated IBC." [Online]. Available: https://flexpowermodules.com/power-design-challenges-with-unregulated-ibc

[10] Texas Instruments, "LMG5200POLEVM-10: LMG5200 GaN 48V to 1V Point of Load Evaluation Module." [Online]. Available: https://www.ti.com/tool/LMG5200POLEVM-10

[11] Y. Zhu, Z. Ye, T. Ge, R. Abramson, and R. C. N. Pilawa-Podgurski, "A Multi-Phase Cascaded Series-Parallel (CaSP) Hybrid Converter for Direct 48 V to Point-of-Load Applications," in *2021 IEEE Energy Conversion Congress and Exposition (ECCE)*, Oct. 2021, pp. 1973–1980.

[12] P. Wang, Y. Chen, G. Szczeszynski, S. Allen, D. M. Giuliano, and M. Chen, "MSC-PoL: Hybrid GaN–Si Multistacked Switched-Capacitor 48-V PwrSiP VRM for Chiplets," *IEEE Transactions on Power Electronics*, vol. 38, no. 10, pp. 12 815–12 833, Oct. 2023.

[13] X. Xu and Q. Li, "Analysis of Parasitic Stored Energy Loss and PCB Layout Optimization for 48V-to-1V Series-Capacitor Buck," in *2024 IEEE Applied Power Electronics Conference and Exposition (APEC)*, Feb. 2024, pp. 898–905.

[14] X. Lou and Q. Li, "Single-Stage 48 V/1.8 V Converter With a Novel Integrated Magnetics and 1000 W/in3 Power Density," *IEEE Transactions on Industrial Electronics*, vol. 71, no. 7, pp. 6601–6611, Jul. 2024.

[15] Y. Zhu, T. Ge, N. M. Ellis, L. Horowitz, and R. C. N. Pilawa-Podgurski, "The Switching Bus Converter: A High-Performance 48-V-to-1-V Architecture With Increased Switched-Capacitor Conversion Ratio," *IEEE Transactions on Power Electronics*, vol. 39, no. 7, pp. 8384–8403, Jul. 2024.

[16] D. E. Charpentier, "Low voltage high current transformer," US Patent US4 159 457A, Jun., 1979.

[17] E. Herbert, "Transformer and rectifier module with half-turn secondary windings," US Patent US5 999 078A, Dec., 1999.

[18] M. K. Ranjram, I. Moon, and D. J. Perreault, "Variable-Inverter-Rectifier-Transformer: A Hybrid Electronic and Magnetic Structure Enabling Adjustable High Step-Down Conversion Ratios," *IEEE Transactions on Power Electronics*, vol. 33, no. 8, pp. 6509–6525, Aug. 2018.

[19] D. J. Perreault, M. K. Ranjram, and I. MOON, "Variable inverter-rectifier-transformer," US Patent US11 716 030B2, Aug., 2023.

[20] P.-L. Wong, P. Xu, P. Yang, and F. Lee, "Performance improvements of interleaving VRMs with coupling inductors," *IEEE Transactions on Power Electronics*, vol. 16, no. 4, pp. 499–507, Jul. 2001.

Ultra-low-profile Single-stage Voltage Regulator Module (VRM) for Next-generation AI Accelerators

Xufu Ren, Jinfeng Zhang, Zhenshuai Rong, Borong Hu, and Teng Long
University of Cambridge, Cambridge, UK
xr222@cam.ac.uk

Abstract—**This paper presents an ultra-low-profile single-stage voltage regulator module (VRM) for next-generation AI accelerators. The proposed solution employs a cascaded transformer-coupled VRM topology, which can improve the effective duty cycle of power transistors, ensuring high-efficiency operation at large voltage transfer ratios. The proposed coupling transformer achieves a transient inductance of 13nH per phase and a steady-state inductance of 68nH per phase, which can reduce the current ripple in stable operation and provide rapid dynamic response during load transition. A prototype with 400A output current is constructed and tested. The thickness of the prototype is 3.9mm, allowing it to be mounted underneath the AI accelerator for vertical power delivery. The prototype features a peak efficiency of 89% and full-load efficiency of 81% at 54V-0.8V voltage transfer ratio. Dynamic response to fast load transition is critical for AI accelerators. In this paper, an evaluation board equipped with load steppers is designed and implemented to experimentally validate the dynamic performance of the VRM prototype.**

Keywords—*vertical power delivery, voltage regulator module (VRM), low-profile, fast transient response.*

I. INTRODUCTION

The boom in artificial intelligence (AI) applications drives a spike in electrical energy use, requiring highly efficient and compact power supply systems to energize a large number of AI accelerators such as GPUs with nearly 1kW power consumption for each core. The input voltage of the GPU motherboard, normally between 40V to 60V, is required to be transformed to low voltage (less than 1V) and high current (larger than 1000A) for powering the core by the voltage regulator modules (VRMs). The proximity to the computing cores requires the VRMs being miniaturized thus ultra-high power density and high efficiency are pivotal in such applications. In addition, the VRMs need to provide ultra-fast dynamic response to satisfy the load transition requirement (>1A/ns), which is critical for the GPU performance.

As shown in Fig. 1, the vertical power delivery (VPD) architecture is an emerging solution by mounting the VRMs underneath the microprocessor with more than 10 times reduction of power delivery network (PDN) losses. However, VPD requires ultra-low-profile VRMs with high current density to fit within the processor platform's form factor. According to state-of-the-art commercial 12V-to-1V voltage regulator for VPD, the module height is preferred to be less

than 4.0mm, which cannot be satisfied by most existing single-stage VRM solutions [1]-[7]. Additionally, the previous single-stage solutions mainly focus on the efficiency and power density improvement, while the rapid transient response is still a bottleneck for VRM.

Based on the cascaded transformer-coupled VRM topology, this paper proposes an ultra-low-profile single-stage VRM module. The 4-phase coupled transformer realizes a transient inductance of 13nH per phase and a steady-state inductance of 68nH per phase, facilitating low stable current ripple and ultra-fast dynamic response. A 400A prototype with 3.9mm thickness is constructed and tested. The dynamic performance of the VRM is experimentally evaluated using an evaluation board equipped with ultra-fast load steppers.

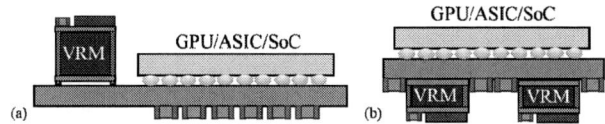

Fig. 1. Power conversion architecture. (a) Lateral power delivery. (b) Vertical power delivery.

II. TOPOLOGY

Fig. 2 shows the proposed VRM topology, comprising two half-bridge (HB) current doubler converters in interleaving operation. The control signals have 90 degree phase-shift between two HB converters. Each converter has four switching cells with an input-series-output-parallel connection. Each switching cell consists of a current doubler rectifier with two transformers with 2:1 turn-ratio, and their magnetizing inductance is integrated as the output inductance of the current doubler rectifier. The primary windings of the serially connected transformer are connected to a half-bridge circuit. Through the cascaded structure, the primary voltage applied to each switching cell can be reduced. For the topology shown in Fig. 2, the voltage conversion ratio is:

$$v_o = \frac{D}{4N} v_{in} \qquad (1)$$

where $N = 4$ is the number of cascaded switching cells, D is the duty cycle of primary transistors.

The proposed single-stage VRM topology has several advantages. Firstly, The cascaded transformer structure significantly reduces the effective duty cycle, as defined in (1).

979-8-3315-1612-3/25 $31.00 © 2025 IEEE

For 54V-0.8V voltage transfer ratio, the D is calculated as 0.237, which is much higher than the duty cycle required for a 12V-to-0.8V buck converter. Secondly, all SR power transistors have common ground, reducing the design complexity and footprint for gate drivers. The SR transistors operated in the same phase can be driven by the same gate driver. Additionally, in each switching cell, four-phase coupling is implemented, as shown in the shaded area in Fig. 2. Inversely coupling design can help to improve the stable inductance and meanwhile reduce the transient inductance, ensuring stable operation efficiency with low current ripple and minimizing transient inductance for faster dynamic performance [8]. Finally, this structure has multiple outputs in parallel, enabling high output current.

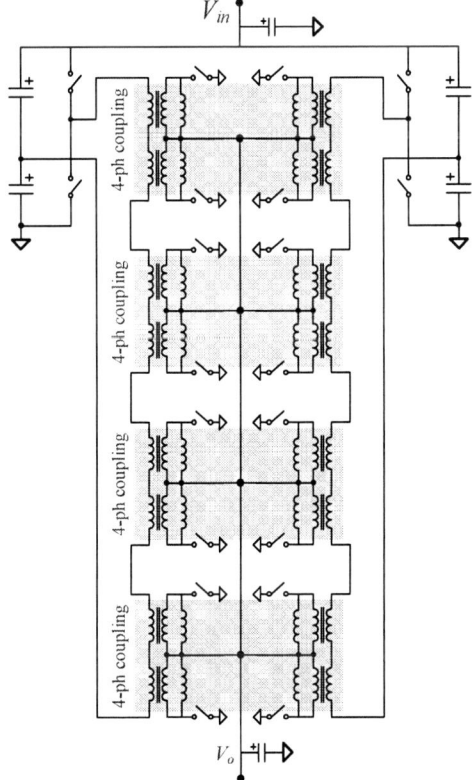

Fig. 2. The proposed single-stage VRM topology.

III. COUPLED TRANSFORMER

Inversely coupling can effectively improve the stable inductance and reduce the transient inductance [8]. As depicted in the shaded area in Fig. 2, the magnetizing inductance of adjacent cells forms 4-phase coupling. The stable inductance L_{ss} for different duty cycles is derived as:

$$\begin{cases} L_{ss} = \dfrac{(1+\alpha)(3-\alpha)}{3+(2+3\dfrac{D}{D'})\alpha}L & 0 < D < 0.25 \\[4mm] L_{ss} = \dfrac{(1+\alpha)(3-\alpha)}{3+(\dfrac{1}{2D}+\dfrac{1}{2D'}+\dfrac{D}{D'})\alpha}L & 0.25 \le D < 0.5 \end{cases} \quad (2)$$

where L is the self-inductance, $D' = 1-D$, $\alpha = 3M / L$ is the coupling coefficient, which is negative for inversely coupling, and M is the mutual inductance. The transient inductance L_{tr} is expressed as:

$$L_{tr} = (1+\alpha)L \quad (3)$$

Based on (2) and (3), the ratio between stable inductance L_{ss} and transient inductance L_{tr} can be derived as:

$$\begin{cases} \dfrac{L_{ss}}{L_{tr}} = \dfrac{3-\alpha}{3+(2+3\dfrac{D}{D'})\alpha} & 0 < D < 0.25 \\[4mm] \dfrac{L_{ss}}{L_{tr}} = \dfrac{3-\alpha}{3+(\dfrac{1}{2D}+\dfrac{1}{2D'}+\dfrac{D}{D'})\alpha} & 0.25 \le D < 0.5 \end{cases} \quad (4)$$

Fig. 3 illustrates the ratio between L_{ss} and L_{tr} as determined by (4), which should be maximized to enhance VRM performance. In a 4-phase coupling design, peak values occur when the duty cycle D is 0.25 and 0.5, providing valuable guidance for topology parameter design. In this design, with a typical voltage transfer ratio of 54V to 0.8V, the duty cycle is calculated as 0.24 based on (1), indicating that optimal performance can be achieved under this operating condition.

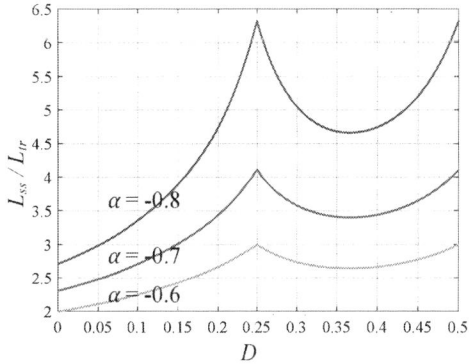

Fig. 3. The ratio between L_{ss} and L_{tr}.

The input voltage of the VRM is 40V-60V (typically 54V) and the output voltage is 0.8V-1.2V (typically 0.8V). Based on (1), the duty cycle range is calculated as: 0.21<D<0.48. In order to ensure the fast dynamic response during rapid load transition, the transient inductance is designed as 14nH. The stable inductance should be maximized to reduce the current ripple and improve efficiency. Based on Fig. 3, the coupling coefficient α is designed as -0.8 to ensure the stable inductance larger than 60nH within the duty cycle range. In this case, the self-inductance is calculated as 70nH.

A 4-phase coupled planar transformer is proposed in Fig. 4. Three key parameters including pillar radius r_l, core length l_l, and air gap length l_g should be designed. The peak flux density of the pillar cross-section is calculated as:

$$B_{l,pk} = \frac{(1+\alpha)LI}{N_s A_e} + \frac{v_o D' T_s}{2 N_s A_e} \quad (5)$$

979-8-3315-1612-3/25 $31.00 © 2025 IEEE

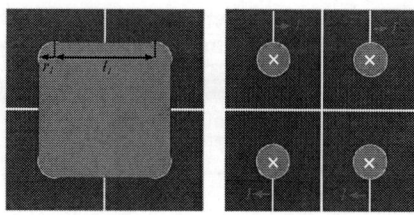

Fig. 4. The 4-phase coupled transformer structure.

where I is the DC current conducting in each winding as illustrated in Fig. 4, and $I = 25A$ for the total 400A output current; $N_s = 1$ is the turn number of the secondary winding; T_s is the switching period; $A_e = \pi r_1^2$ is the pillar cross-section area. The switching frequency is designed as 500kHz. For a given r_1, the flux density with respect to duty cycle D can be calculated. The ferrite material DMR51W is used as the transformer core material, thereby the maximum flux density is designed less than 0.2T to avoid saturation. The pillar radius r_1 is designed as 1.3mm, allowing the maximum flux density is less than 0.19T within the full duty cycle range, as depicted in Fig. 5.

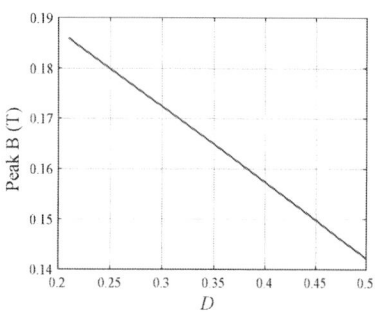

Fig. 5. The peak flux density with respect to duty cycle D.

The core length l_1 is determined by FEM simulation. According to Ansys Maxwell simulation, when $l_1 = 7.4$mm, the self-inductance is 57.1nH, transient inductance is 12.7nH, and coupling coefficient is -0.776, which are acceptable for the target design. The plate thickness of the core is designed as 1mm, ensuring an ultra-low-profile VRM.

The flux density distribution of the transformer is investigated when operating at peak current, as shown in Fig. 6. The average flux density across the pillar is about 0.18T, which accords with the theoretical analysis. Additionally, the flux density distribution is around 0.2T in the plate, which has sufficient margin to avoid saturation.

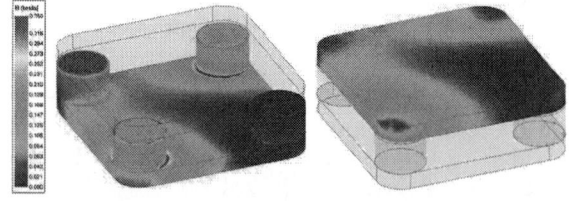

Fig. 6. Flux density distribution at peak current.

According to the simulation, the self-inductance is 57.2nH, and the transient inductance is 12.7nH. Then the stable inductance with respect to duty cycle D can be calculated, which is plotted in Fig. 7, ensuring larger than 55nH stable

inductance within the operation range. The stable inductance is 68nH for 54V-0.8V voltage transfer ratio.

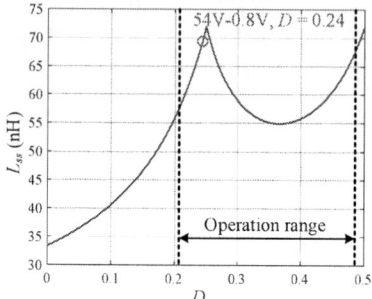

Fig. 7. The stable inductance with respect to D.

IV. PROTOTYPE AND EXPERIMENTS

A. Prototype

A VRM prototype is constructed, as shown in Fig. 8. The key parameters are listed in Table 1. For the primary switches, the GaN transistor is used. For the SR switches, the low voltage Silicon MOSFET is used. All SR switches share a common ground, hence low-side gate drivers with small footprints can be used to drive the SR switches in the same phase. In this design, 8 low-side gate drivers are utilized to drive a total of 32 SR transistors, bringing significant power density improvement. The ultra-low-profile VRM has a dimension of 67mm by 16mm by 3.9mm, achieving a power density of 1.8kW/in³.

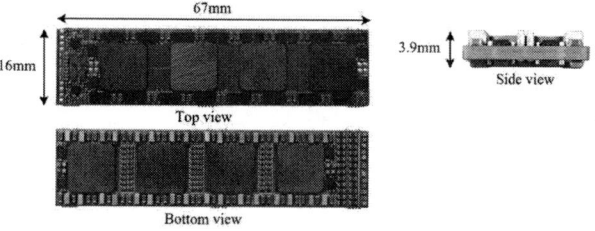

Fig. 8. The VRM topology.

Table 1. Key parameters of the prototype.

Input voltage	40V-60V (typically 54V)
Output voltage	0.8V-1.2V (typically 0.8V)
Output current	400A
Switching frequency	500kHz
Module dimension	67mm/16mm/3.9mm (L/W/H)
Stable inductance per phase	68nH
Transient inductance per phase	13nH

B. Stable Operation

Fig. 9 presents the waveform of the VRM converter under the typical operation condition (54V-0.8V). V_{sw} is the voltage between the half-bridge midpoint and GND. Following the turn-off of the primary transistors, a ringing appears on V_{sw}. This ringing occurs due to the resonance between the output capacitance of the primary transistors C_{oss} and the transformer's leakage inductance L_k. During the resonance, the energy stored in the leakage inductance is dissipated by the loop resistance, resulting energy loss.

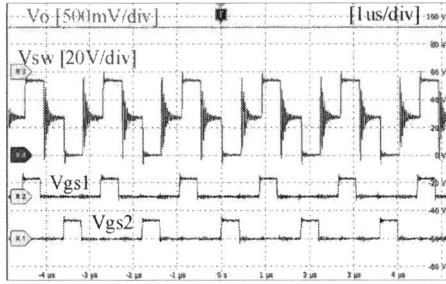

Fig. 9. The stable operation waveform. V_{sw} is the voltage between half-bridge midpoint and GND.

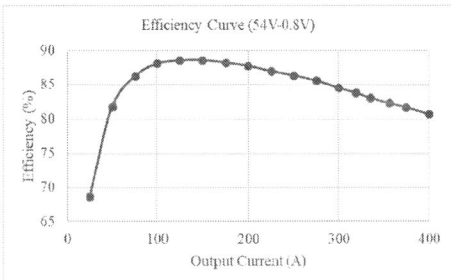

Fig. 10. Measured efficiency curve.

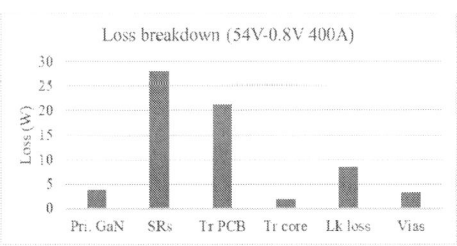

Fig. 11. Loss breakdown.

C. Efficiency and Loss Breakdown

The measured efficiency curve is illustrated in Fig. 10. At 54V-0.8V voltage transfer ratio, the prototype features a peak efficiency of 89% and full-load efficiency of 81%. The loss breakdown at 400A output is presented in Fig. 11. The transformer winding loss and SR loss dominate the loss of the overall system due to the high output current.

D. Dynamic Performance

An evaluation board is designed for dynamic transient response testing. As shown in Fig. 12, the VRM converter is positioned at the center, with load steppers arranged around it. The load stepper consists of a high-precision shunt resistor and a GaN transistor connected in series. By switching the GaN transistors on and off, current steps with an ultra-high slew rate can be achieved. As shown in Fig. 12, the circuit consists of eight resistor-switch branches. The current slew rate and step value can be controlled by adjusting the number of active parallel resistor-switch branches. The current transition is monitored by measuring the shunt voltage and output voltage. To enhance the output capacitance, numerous ceramic capacitors are mounted on the underside of the evaluation board, as depicted in Fig. 12, providing a total capacitance of approximately 3700uF.

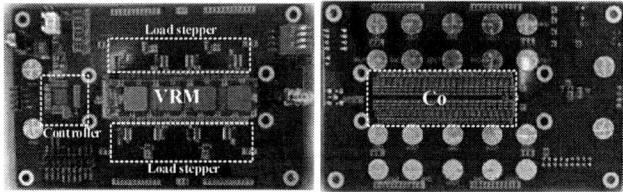

Top view   Bottom view

Fig. 12. Dynamic performance evaluation board.

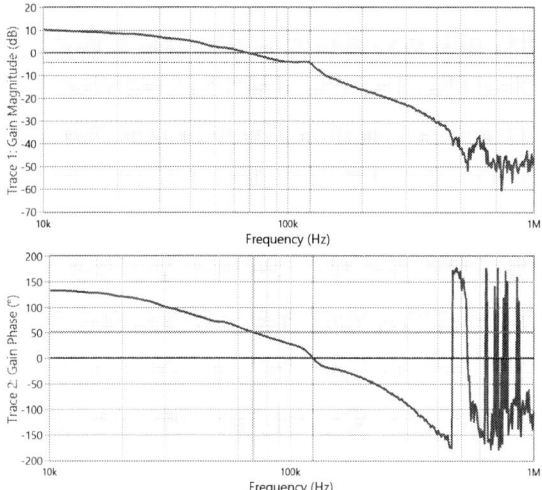

Fig. 13. The bode plot of VRM converter.

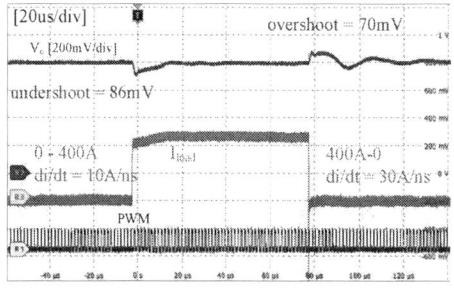

Fig. 14. Dynamic performance.

The loop analysis is performed and the bode plot is shown in Fig. 13, which reveals a crossover frequency of 70kHz, a phase margin of 51°, and a gain margin of 4.4dB. This converter supports for paralleling operation. If multiple VRM converters operate in parallel with additional interleaved phases, the equivalent switching frequency can be further increased, leading to a wider bandwidth.

In the dynamic performance test, the steady-state current is set to 400A, with a current slew rate of 10A/ns for the load step-up. Due to the faster turn-off speed of the GaN transistor compared to its turn-on speed, the current slew rate during the load step-down is significantly higher, reaching 30A/ns, which will induce oscillation during the turn-off process. Fig. 14 presents the results with fast transient response (FTR) control, which achieves an undershoot voltage of 86mV and an overshoot voltage of 70mV. FTR control is based on non-linearity algorithm and shows better performance than PID control. The rapid transient response control and dynamic performance will be further investigated in our future work.

V. CONCLUSIONS

This paper introduces the design and testing of an ultra-low-profile single-stage VRM converter, capable of providing 54V-0.8V large voltage transfer ratio and delivering 400A output current. Leveraging the coupling magnetics design, the converter features 13nH transient inductance and 68nH stable inductance, effectively minimizing current ripple during steady-state operation and enabling rapid dynamic response during load transitions. A prototype with a compact thickness of 3.9 mm is developed to support vertical power delivery for next-generation AI accelerators. According to experiments, this prototype achieves a peak efficiency of 89% and full-load efficiency of 81%. The dynamic performance of the converter is validated using a dynamic evaluation board.

REFERENCES

[1] Y. Chen et al., "Virtual Intermediate Bus CPU Voltage Regulator," in IEEE Transactions on Power Electronics, vol. 37, no. 6, pp. 6883-6898, June 2022.

[2] J. Baek et al., "Vertical Stacked LEGO-PoL CPU Voltage Regulator," in IEEE Transactions on Power Electronics, vol. 37, no. 6, pp. 6305-6322, June 2022.

[3] M. H. Ahmed, C. Fei, F. C. Lee and Q. Li, "Single-Stage High-Efficiency 48/1 V Sigma Converter With Integrated Magnetics," in IEEE Transactions on Industrial Electronics, vol. 67, no. 1, pp. 192-202, Jan. 2020.

[4] X. Lou and Q. Li, "Single-Stage 48 V/1.8 V Converter With a Novel Integrated Magnetics and 1000 W/in3 Power Density," in IEEE Transactions on Industrial Electronics, vol. 71, no. 7, pp. 6601-6611, July 2024.

[5] Y. Zhu, J. Zou and R. C. N. Pilawa-Podgurski, "A 1500-A/48-V-to-1-V Switching Bus Converter for Next-Generation Ultra-High-Power Processors," in IEEE Transactions on Power Electronics, vol. 39, no. 9, pp. 11340-11355, Sept. 2024.

[6] S. Saggini, O. Zambetti, R. Rizzolatti, M. Picca and P. Mattavelli, "An Isolated Quasi-Resonant Multiphase Single-Stage Topology for 48-V VRM Applications," in IEEE Transactions on Power Electronics, vol. 33, no. 7, pp. 6224-6237, July 2018.

[7] X. Ren, J. Zhang, Y. Jiang, X. Li and T. Long, "A 48-to-1V LLC DC Transformer," 2023 IEEE 24th Workshop on Control and Modeling for Power Electronics (COMPEL), Ann Arbor, MI, USA, 2023.

[8] Pit-Leong Wong, Peng Xu, P. Yang and F. C. Lee, "Performance improvements of interleaving VRMs with coupling inductors," in IEEE Transactions on Power Electronics, vol. 16, no. 4, pp. 499-507, July 2001.

Novel TLVR Operation In Multi-Stage Voltage Regulator Module With Current Multipliers

Kevin Zufferli, Roberto Rizzolatti, Mario Ursino,
Simone Mazzer, Gerald Deboy
Infineon Technologies, Villach, Austria
Email: kevin.zufferli, roberto.rizzolatti, mario.ursino,
simone.mazzer, gerald.deboy@infineon.com

Stefano Saggini
DPIA, University of Udine, Italy
Email: stefano.saggini@uniud.it

Abstract—**AI applications in data centers are quickly advancing, leading to a surge in the development of new hardware accelerators for Machine Learning (ML), Deep Learning (DL), and High-Performance Computing (HPC). These new devices have high power demands, thereby assigning data centers a critical role in global power consumption. Updated regulations and technological improvements, including on efficiency, will be crucial to moderate the surge in energy consumption from data centers. The power conversion chain is crucial for maximizing overall system efficiency. This paper introduces an innovative power conversion architecture that employs a multi-stage approach to ensure high efficiency and a rapid transient response, thereby enhancing Power Density and Power Delivery Network (PDN) performance.**

Index Terms—**Data Center, TLVR, Current Multiplier, AI, 48V, IBA, FPA.**

I. INTRODUCTION

The increasing demand for modern data storage has led to a significant surge in data center power consumption [1]. This has prompted a shift from traditional 12V DC server rack power distribution to 48V DC power distribution in recent years [2]. While this change has resulted in improved overall system efficiency, it also presents new challenges for voltage regulator modules (VRMs) that power multi-core processors (CPUs) and graphics processing units (GPUs) [3]. From the 48-V input one of the primary concerns for VRMs is achieving an extremely high voltage conversion ratio to a relatively low voltage range of 0.6-1.8V, while meeting the high load current requirements of CPUs and GPUs keeping high efficiency and power density. This means that VRMs must be designed to minimize energy losses while delivering high levels of current. Furthermore, they must also be capable responding quickly to high load transients, which can exceed 1A/ns. These demands pose significant technical challenges that must be addressed to ensure reliable and efficient power delivery in modern data centers. Conventional 48V architectures commonly utilize a two-stage design, comprising an unregulated stage followed by a regulated stage, connected in a cascaded configuration. One such architecture is the Intermediate Bus Architecture (IBA) [4]. It consists of a series double-stage approach, where the first stage

is unregulated, and the second stage is regulated. The first stage is named Intermediate Bus Converter (IBC), which supplies power to the intermediate bus voltage for the Voltage Regulator Module (VRM). Common IBCs used inside data centers are resonant and can be DCX converters, such as LLC [5] or Hybrid Switched Capacitor (HSC) [6] converters, or switch capacitor-based converters, such as Switched Tank Converters (STC) [7]. The VRM is typically a high-bandwidth multiphase buck converter.

The state-of-the-art of IBA is depicted in Fig. 1. The initial stage features an LLC-DCX resonant converter, which can incorporate GaN devices to elevate the switching frequency, thereby attaining high power density and efficiency. Following this, the second stage comprises a traditional multiphase buck converter, which can be realized using either silicon or GaN devices. This stage operates at high switching frequencies to ensure high power density [8].

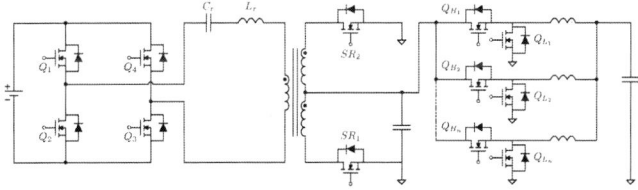

Fig. 1: State-of-the-art of Intermediate Bus Architecture.

The main advantages of IBA architecture include improved system efficiency and reduced system cost. Although PoL VRM ensures fast response to transient loads, it does not achieve high current density (A/mm^2). This limitation is increasingly becoming a bottleneck with the advent of new accelerators in data centers. As these accelerators evolve, their supply voltage continues to decrease while their power demands simultaneously rise.

Another possible way, which is also the first commercially available two-stage solution, is proposed by VICOR [2]. The Factorized Power Architecture (FPA) can be seen as IBA with the inversion stage position, connecting the regulating stages (Modular Current Driver - MCD) to the input and the fixed ratio stages (Modular Current Multipliers - MCMs) to the output. In this manner, the

---

979-8-3315-1612-3/25 $31.00 © 2025 IEEE

high current density of MCMs can be exploited, but regulating performance becomes critical.

The efficiency of the series double-stage approach, such as IBA or FPA, is calculated in a similar way. The total efficiency results from multiplying the efficiencies of the two stages as in (Eq 1), considering that the total power flows through both stages, as depicted in Fig. 2.

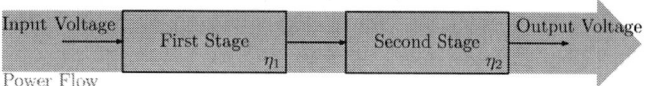

Fig. 2: Series double-stage approach.

$$\eta_{TOT} = \eta_1 \eta_2 \qquad (1)$$

It is remarkable that if the aim is to achieve the highest efficiency, the bottleneck of this solution lies in the total power processing by the regulated stage. To overcome this, Sigma architectures were proposed [9], [10] where the power is split between the two stages.

The Sigma converter is a quasi-parallel converter which performs ISOP (Input Series - Output Parallel) connection. As shown in Fig. 3, this can be useful when regulation with high efficiency is desired, especially with high step-down. In this case, the output regulation is performed maintaining high efficiency and power density, as the power is factorized between the IBC and the PoL converter.

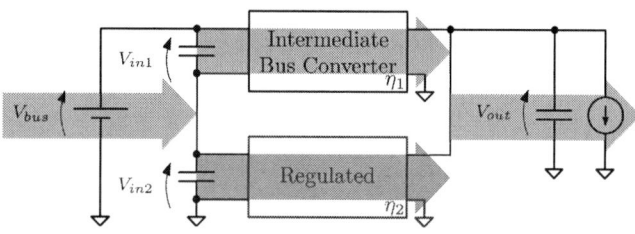

Fig. 3: Sigma (ISOP) Architecture.

The overall efficiency can be improved compared to the series double-stage approach as in (2), by using a high efficiency stage which processes most of the power, in ISOP with a low-power regulation stage.

$$\eta_{TOT} = \frac{V_{in_1}}{V_{in}}\eta_1 + \frac{V_{in_2}}{V_{in}}\eta_2 \qquad (2)$$

Sigma converters can be used for full conversion regulation from 48V to PoL. Furthermore, they can serve as a substitution for IBC or as MCD of FPA, offering semi-regulation with high power density and efficiency compared to the commonly used buck-boost topology. Consequently, MCM would be attached to the output of the Sigma converter.

In order to achieve the aforementioned critical requests in terms of power density and bandwidth, this paper presents a novel utilization of Trans-Inductor Voltage Regulator (TLVR) [11] as a hardware feed-forward between different stages during transient events and acting like a short-circuit in steady state. This achieves fast transient response, high power density and high efficiency, when applied to a Multi-Stage VRM which exploits Current Multiplier [2]. The converter performs from 48V to PoL.

II. DC-TLVR

As stated previously, multi-stage architectures are composed of two main different stages: regulated stages and unregulated stages. Unregulated stages convert the output voltage with a fixed ratio and have high efficiency and high-power density. Regulated stages can regulate the output voltage to a specific value but have low efficiency and low power density.

The performance of the entire power supply and the behavior of the unregulated stages can be improved using an electrically coupled inductor. This coupled inductor, such as a TLVR, may be placed in the circuit paths between the output of one stage and the input of another, in other words placed between *two equipotential nodes on average* (i.e. two nodes which can have a voltage ripple but do not induce an average non-zero voltage among the inductor).

Considering two series-connected stages, the output voltage of the first stage is equal to the input voltage of the second stage, so:

$$V_{out_1} = V_{in_2} \qquad (3)$$

The TLVR may be placed between these two stages, as reported in Fig. 4. In steady state, neglecting the parasitic resistance of the inductor, the DC voltage across the TLVR is zero, and the current flowing through it is determined by the total output current. In this case, the DC contribution of the TLVR is negligible, and it can be treated as a short circuit. For a better understanding, given the DC nature of TLVR nodes in steady state, it will be referred as DC-TLVR.

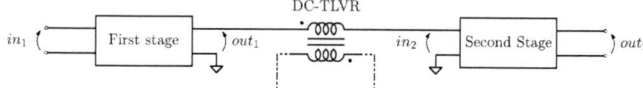

Fig. 4: DC-TLVR in the circuit path between two series stages.

The DC-TLVR is electrically coupled with the magnetic components of stages within the power supply. Assuming that during a load transient, a positive (or negative) surge of current occurs through those magnetic components, it will affect the DC-TLVR, creating a local positive or negative current boost. By managing the current slope through the DC-TLVR, it is possible to enhance the speed of the transient response of the entire converter. In steady state, the voltage across the DC-TLVR is negligible and

979-8-3315-1612-3/25 $31.00 © 2025 IEEE

that means that the ripple of the DC-TLVR is not affecting the operation of the magnetic components of stages. In a more general way, the operation of DC-TLVR involves a minimum of three different stages:

1) **DC-TLVR input stage**: regulated or unregulated stage connected to the reference input of DC-TLVR.
2) **DC-TLVR output stage**: regulated or unregulated stage connected to the reference output of DC-TLVR.
3) **DC-TLVR coupled stage**: regulated or unregulated stage electrically coupled with DC-TLVR.

Fig. 5 shows the connections for DC-TLVR operation.

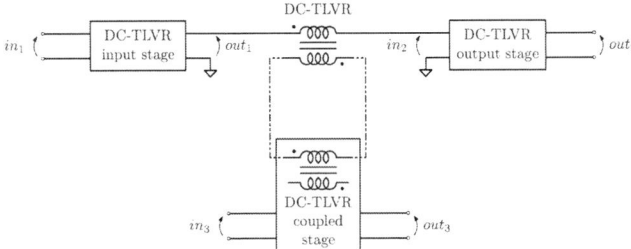

Fig. 5: General architecture with DC-TLVR implementation.

The hardware feed-forward behavior can be understood in a specific scenario where $out_2 = out_3$, and the DC-TLVR coupled stage is a TLVR multiphase buck converter with all phases coupled together and with DC-TLVR. As described in [11], when a transient event occurs, a sudden surge in current is observed within the phases and the coupling loop. This current surge is then reflected onto the DC-TLVR and directed towards the DC-TLVR output stage, which is assumed to be unregulated for simplicity. The current is subsequently amplified by the fixed ratio of the stage and supplied to the output. By carefully managing specific parameters, this configuration can enhance the transient response without requiring a high-bandwidth multiphase buck converter, unlike the approach used in IBA. As stated before, DC-TLVR operation works when its input and output voltage are of the same amount in steady state. That can be useful when considering the ISOP connection, as the output of stages are connected in parallel. Placing DC-TLVR across the output of the unregulated and the regulated stages will not affect the steady state operation of the whole Sigma converter. However, DC-TLVR can reflect a current that can be requested from the unregulated stage, making it actively responsive to the transient events.

For a better understanding, the following configuration of DC-TLVR stages will highlight the advantages of this connection:

DC-TLVR input stage: Sigma converter, composed of unregulated stages and regulated stages.

DC-TLVR output stage: Unregulated stages connected to the PoL Voltage, or Current Multipliers.

DC-TLVR coupled stage: Multiphase Buck converter.

DC-TLVR is placed between the Sigma outputs, obtaining in steady state:

$$V_{out_{11}} = V_{out_{12}} \qquad (4)$$

and the Sigma Regulated Stages are connected to Current Multipliers, as depicted in Fig. 6.

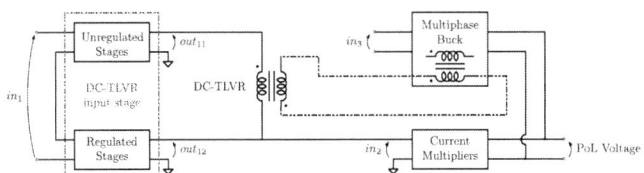

Fig. 6: DC-TLVR for ISOP converters.

The primary objective of this connection is to improve transient response affecting the operation of unregulated stages. In essence, when a transient event occurs, the boosted current flows into the unregulated stages, resembling a hardware feed-forward approach. This will enhance the transient performance of the whole converter, as the regulated and unregulated stages are responding simultaneously.

III. PROPOSED ARCHITECTURE

The novel architecture is founded on the application of DC-TLVR for ISOP connection. It comprises four distinct stages: two buck-based regulated stages and two HSC-based unregulated stages. The complete architecture is illustrated in Fig. 7.

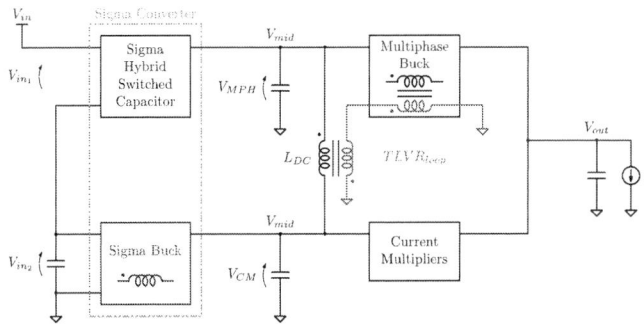

Fig. 7: MSVRM architecture.

The stages are interconnected as follows:

- The unregulated stage of the Sigma converter (S-HSC) is powered by a portion of V_{in} and is linked at the output to both the Multiphase Buck and one terminal of the DC-TLVR.
- The regulated stage of the Sigma converter (Sigma Buck - SB) is also supplied by a portion of V_{in} and is connected at the output to both the Current Multipliers and one terminal of DC-TLVR.
- The Multiphase Buck (MPH) is connected at the output to the PoL.
- The Current Multipliers (CM) are connected at the output to the PoL

979-8-3315-1612-3/25 $31.00 © 2025 IEEE   1446

At this juncture, the ISOP connection of the Sigma converter remains effective solely during steady-state conditions, when the DC-TLVR can be approximated as a short circuit, resulting in approximately equal voltages for $V_{MPH} \simeq V_{CM} \simeq V_{mid}$. However, during transient events, V_{MPH} and V_{CM} diverge. The DC-TLVR is electrically coupled to the Multiphase Buck through the $TLVR_{loop}$ (highlighted in red in Fig. 7), addressing the hardware feed forward operation.

The Sigma converter achieves a semi-regulated 10:1 down conversion, while the CMs and MPH perform a fixed 6:1 down conversion. By regulating different V_{mid}, values, various output voltages within the 1V domain can be achieved.

To gain a comprehensive understanding of the MSVRM's operation and the impact of DC-TLVR on the topology, an analysis of the control methodology is essential.

A. Control Methodology

The presence of two regulated stages enables the definition of two distinct control loops. Given the high-speed voltage regulation demands of accelerators, the high-bandwidth Multiphase Buck is specifically designed for this purpose. This is also essential for the proper operation of the DC-TLVR, as the Multiphase Buck must promptly respond to transient events. Efficiency is another critical requirement for the conversion chain in accelerator modules. If the steady-state parallel connection of the Multiphase Buck and Current Multipliers can achieve desired power sharing, most of the power will flow through the HSC-based converters, thereby enhancing overall efficiency. This objective can be realized by controlling the Sigma Buck in average current mode, comparing the current output of the Sigma Buck to a weighted value of the total output current of the Multiphase Buck converter. Considering Fig. 8, where the complete control loops are illustrated, the currents are defined as follows:

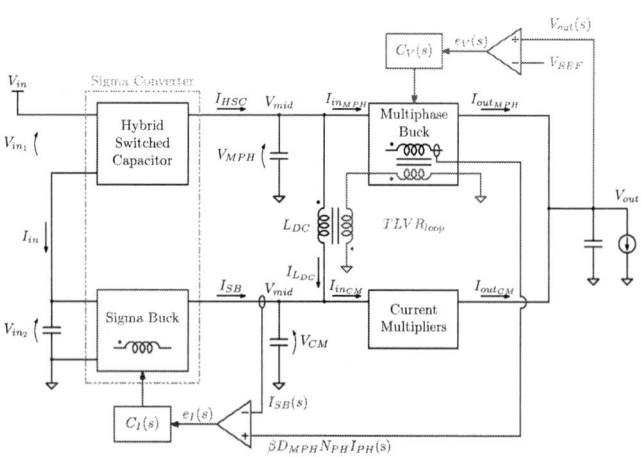

Fig. 8: MSVRM contro loops.

$$I_{out_{MPH}} = N_{PH}I_{PH} \qquad (5)$$

$$I_{in_{MPH}} = D_{MPH}I_{out_{MPH}} = D_{MPH}N_{PH}I_{PH} \qquad (6)$$

$$I_{L_{DC}} = I_{HSC} - I_{in_{MPH}} = I_{HSC} - D_{MPH}N_{PH}I_{PH} \qquad (7)$$

where N_{PH} and D_{MPH} are respectively the number of phases and Duty Cycle of the Multiphase buck.

The power splitting between Sigma Buck and Sigma Hybrid Switched Capacitor is reported in the following equations:

$$V_{in_2} = V_{in} - V_{in_1} = V_{in} - V_{mid}\left(4 + 2\frac{N_1}{N_2}\right) \qquad (8)$$

$$\frac{V_{in_2}}{V_{in}} = \frac{P_{in_2}}{P_{in}} = 1 - \frac{V_{mid}}{V_{in}}\left(4 + 2\frac{N_1}{N_2}\right) = \frac{I_{SB}}{I_{out}} \qquad (9)$$

Where V_{mid} is the mid voltage, V_{in} the total input current, N_1 and N_2 the number of primary and secondary windings of Hybrid Switched Capacitor, I_{SB} the output Sigma Buck current, I_{out} the total output current. Then, the ratio between the HSC and SB output current:

$$\frac{I_{HSC}}{I_{SB}} = \frac{\dfrac{V_{mid}}{V_{in}}\left(4 + 2\dfrac{N_1}{N_2}\right)}{1 - \dfrac{V_{mid}}{V_{in}}\left(4 + 2\dfrac{N_1}{N_2}\right)} = \alpha \qquad (10)$$

Considering that the output voltage of HSC and SB are is the same, the parameter α in (10) indirectly describes the power splitting in the Sigma converter and is imposed by its input and output voltages and the ratio of the Sigma transformer. Fig. 9 outlines the α variation based on the whole module's input and output voltages, as $V_{mid} = 6V_{out}$ imposed by the Current Multipliers.

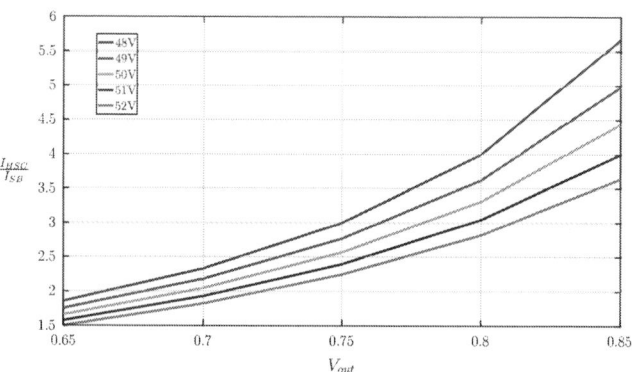

Fig. 9: α parameter in Sigma converters.

If the control imposes $I_{SB} = \beta D_{MPH}N_{PH}I_{PH}$:

$$I_{L_{DC}} = (\alpha\beta - 1)D_{MPH}N_{PH}I_{PH} \qquad (11)$$

$$I_{in_{CM}} = I_{L_{DC}} + I_{SB} = [\beta(\alpha + 1) - 1]D_{MPH}N_{PH}I_{PH} \qquad (12)$$

$$I_{out_{CM}} = K_{CM}I_{in_{CM}} = K_{CM}[\beta(\alpha + 1) - 1]D_{MPH}N_{PH}I_{PH} \qquad (13)$$

Where K_{CM} is the fixed ratio of Current Multipliers.

Finally, using (5) and (13), the ratio between output current of Multiphase Buck and total output current can be calculated as:

$$
\begin{aligned}
\frac{I_{out_{MPH}}}{I_{out}} &= \frac{I_{out_{MPH}}}{I_{out_{MPH}} + I_{out_{CM}}} = \frac{1}{1 + \dfrac{I_{out_{CM}}}{I_{out_{MPH}}}} = \\
&= \frac{1}{1 + \dfrac{K_{CM}[\beta(\alpha+1)-1)]D_{MPH}N_{PH}I_{PH}}{N_{PH}I_{PH}}} = \\
&= \frac{1}{1 + K_{CM}[\beta(\alpha+1)-1)]D_{MPH}}
\end{aligned}
\tag{14}
$$

Considering K_{CM} a fixed value, and α and D_{MPH} are fixed as depend from the steady voltages in the circuit, the power distribution between the Current Multipliers and the Multiphase Buck converter is solely influenced by the β parameter. By treating β as a customizable value that can be set through both hardware and software means, achieving the desired power distribution becomes straightforward.

B. Transient response

The control methodology described above fully leverages the DC-TLVR. Notably, the two distinct control loops operate in cascade. When a transient event occurs, the duty cycle and subsequently the output current of the Multiphase Buck converter adjust in response to load changes. This variation is then further influenced by both the DC-TLVR, coupled with the phases, and the output current of the Sigma Buck converter, which is controlled to a weighted value relative to the Multiphase Buck output current, as previously mentioned. The current behavior during a load step is illustrated in Fig. 10.

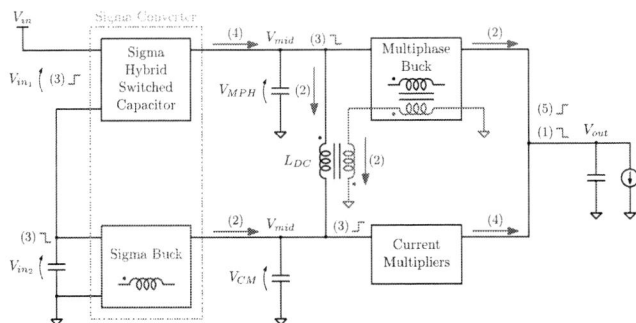

Fig. 10: Load step up transient event.

The response of the entire system can be split in five different phases:

1) Step up of load current occurs
2) Control of Multiphase Buck responds increasing the output current of the converter. The phase current is reflected in the $TLVR_{loop}$ and then reflected again in L_{DC}. Meanwhile, the control of Sigma Buck is

following the current behavior of the Multiphase Buck and then its current increases as well.
3) The current reflected in L_{DC} charges V_{CM} and discharges V_{MPH}, the increase of Sigma Buck output current charges V_{CM} and discharges V_{in_2} and, given V_{in} fixed, V_{in_1} increases.
4) Both the unregulated stages experience an increase on their input and a decrease on output voltages, then this implies that a current is driven by the voltage difference.
5) The output voltage increases and reaches the reference voltage thanks to the response of all converters inside the MSVRM.

The described operation is made possible by the DC-TLVR, which, when combined with the current boost, can perturb node voltages. Without its use, the input node of the Current Multiplier would decrease due to the Multiphase Buck converter drawing current from it. Consequently, an instantaneous current would flow in the *opposite* direction, draining from the output and resulting in a more pronounced droop in the output voltage. Fig. 11 illustrates the transient comparison between the insertion of DC-TLVR and a short circuit between V_{CM} and V_{MPH}, while maintaining the same voltage and current compensation transfer functions. Notably, during a step-up load (1kA, 2A/ns) in Fig. 11a, the DC-TLVR implementation exhibits minimal voltage drop and a faster response compared to the short circuit. Additionally, during load release in Fig. 11b, the voltage increase is similar (and even better), and once again, DC-TLVR returns to the regulation band more swiftly.

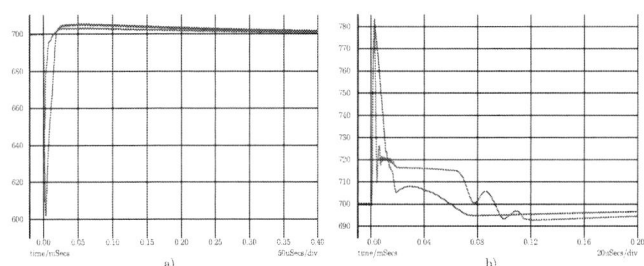

Fig. 11: Output voltage behavior during transients (in mV): a) load step-up, b) load release; in red with DC-TLVR, in green with short circuit.

IV. EXPERIMENTAL RESULTS

To validate the proposed architecture, a OAM version of MSVRM was built following OAM standards with Lateral Power Delivery (LPD) approach. The top and bottom views are depicted in Fig. 12 and Fig. 13. Converter specifications are reported in Table I.

The primary goal of this first version of MSVRM measurements is to evaluate the impact of DC-TLVR to the topology. Hence, the control of both the loops was chosen digital, linear and fixed frequency. Fig. 14

Fig. 12: MSVRM PCB top view.

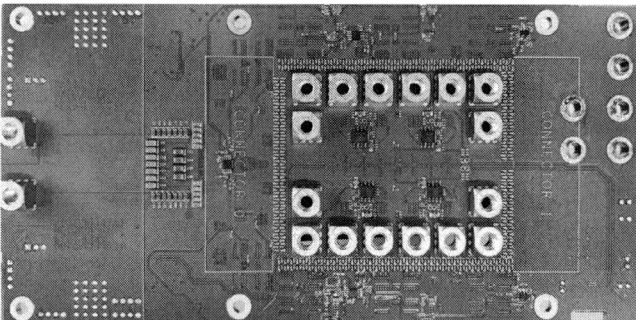

Fig. 13: MSVRM PCB bottom view.

TABLE I: OAM-based MSVRM prototype specifications

General specifications	
V_{in}	48÷52 V
V_{out}	0.65÷1 V
P_{out}	750 W
PCB technology	10 layers, 2 oz ext., 3 oz int.
Current Multipliers (8)	HSC
DC-TLVR	VLBUC12060120R07LF3, 70 nH
C_{out}	534×GRM155C80E226ME11
Down Solution RHSC	
f_{sw}	500 kHz
L_{buck} (buck)	PA5034.331HLT, 330 nH
L_{mag} (magnetizing)	1 μH
N1:N2 (turns ratio)	2:1
Q_1, Q_2, Q_4, Q_5 FETs	IQE046N08LM5CGSC, 60 V, 4.5 mΩ
Q_7, Q_8 FETs (4 per phase)	IQE004NE1LM7CGSC, 15 V, 450 $\mu\Omega$
Buck FETs	BSZ011NE2LS5I, 25 V, 1.1 mΩ
C_s (7 per phase)	GRM31CC71E226ME15
C_{res} (14 per phase)	GRM21BZ71H475KE15
$C_{out,CM}$	6×GRM21BD71A226ME44
$C_{out,MPH}$	20×GRM21BR71A106KA73
HSC drivers	4×1EDN7550B + 2×2EDN7524G
Buck drivers	2×NCP81155
PCB technology (SMD transformer)	12 layers, 2 oz ext., 3 oz int.
Core	Custom, DMR51W ferrite
Multiphase Buck	
f_{sw}	2 MHz
TLVR	VLBUC12060120R09LF3, 90 nH
Integrated power stage	TDA22590
C_{in} (6 per phase)	GRM21BR71A106KA73

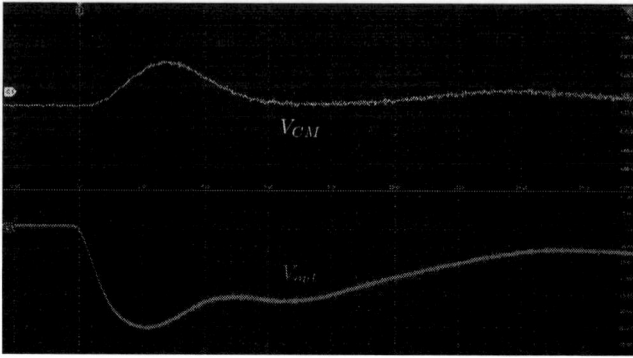

Fig. 14: Input voltage of Current Multiplier (in yellow) and output voltage of MSVRM (in red) behavior during a load step of 300 A.

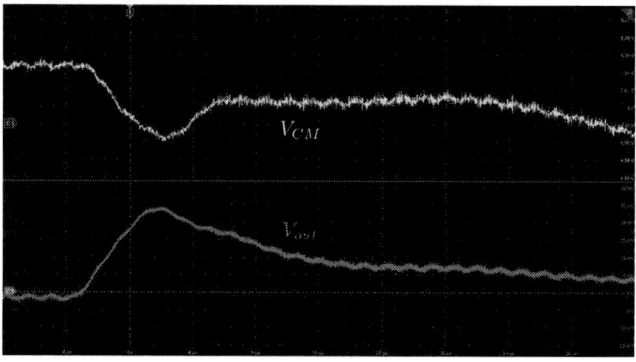

Fig. 15: Input voltage of Current Multiplier (in yellow) and output voltage of MSVRM (in red) behavior during a load release of 300 A.

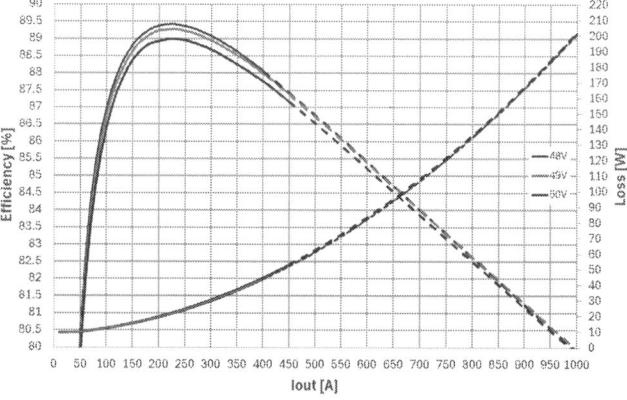

Fig. 16: MSVRM efficiency with different V_{in} and V_{out} = 0.8V.

shows the transient response to a step of 300 A load. As depicted in the image, in the first microseconds of the load step, the output voltage of the module decreases while the input voltage of Current Mutlipliers increases. Instead, Fig. 15 shows the transient response to a load release of 300 A. The behavior of waveforms is inverted, hence the output voltage of the module increases while the input voltage of Current Mutlipliers decreases. This resembles the operation described in the previous chapters, and means that the DC-TLVR is operating accordingly. Tuning the PID(s) transfer function of the two different loops the contribution of DC-TLVR can be managed, offering large possibilities of improvements with different control techniques as the Costant-On Time (COT) for the Multiphase Buck.

The overall efficiency of the converter is depicted in Fig. 16, which reveals a peak of 87% at 150A.

V. CONCLUSIONS

In this paper is proposed a novel multi-stage approach for a 48 V to PoL regulated converter for data-center applications. The architecture is composed of regulated and unregulated stages, specifically HSC and buck based. The operation of DC-TLVR ensures high slew rate for huge step loads, while the power sharing for PoL stages offers high efficiency and high current density. The board is designed in OAM standard, targeting the demands of accelerators cards for AI training in hyperscale data centers in terms of load slew rate and current capability.

REFERENCES

[1] IEA, "Electricity 2024," https://www.iea.org/reports/electricity-2024, 2024.

[2] S. Oliver, "From 48 v direct to intel vr12.0: Saving 'big data' $500,000 per data center, per year," https://www.mouser.com/pdfdocs/wp_VR12.pdf, 2012.

[3] K. Uchiyama. "Nvidia blackwell platform arrives to power a new era of computing," https://nvidianews.nvidia.com/news/nvidia-blackwell-platform-arrives-to-power-a-new-era-of-computing, 2024.

[4] R. White, "Emerging on-board power architectures," in *Eighteenth Annual IEEE Applied Power Electronics Conference and Exposition, 2003. APEC '03.*, vol. 2, 2003, pp. 799–804 vol.2.

[5] B. Yang, F. Lee, A. Zhang, and G. Huang, "Llc resonant converter for front end dc/dc conversion," in *APEC. Seventeenth Annual IEEE Applied Power Electronics Conference and Exposition (Cat. No.02CH37335)*, vol. 2, 2002, pp. 1108–1112 vol.2.

[6] R. Rizzolatti, C. Rainer, S. Saggini, and M. Ursino, "Ultra-low profile hybrid switched capacitor converter with matrix multi-tapped autostransformer," in *2021 IEEE Applied Power Electronics Conference and Exposition (APEC)*, 2021, pp. 869–874.

[7] S. Jiang, S. Saggini, C. Nan, X. Li, C. Chung, and M. Yazdani, "Switched tank converters," *IEEE Transactions on Power Electronics*, vol. 34, no. 6, pp. 5048–5062, 2019.

[8] M. H. Ahmed, C. Fei, F. C. Lee, and Q. Li, "48-v voltage regulator module with pcb winding matrix transformer for future data centers," *IEEE Transactions on Industrial Electronics*, vol. 64, no. 12, pp. 9302–9310, 2017.

[9] J. Sun, M. Xu, D. Reusch, and F. C. Lee, "High efficiency quasi-parallel voltage regulators," in *2008 Twenty-Third Annual IEEE Applied Power Electronics Conference and Exposition*, 2008, pp. 811–817.

[10] M. Ursino, R. Rizzolatti, G. Deboy, S. Saggini, and K. Zufferli, "Sigma converter family with common ground for the 48 v data center," *IEEE Transactions on Power Electronics*, vol. 38, no. 9, pp. 10 997–11 009, 2023.

[11] n/a, "Fast multi-phase trans-inductor voltage regulator," https://www.tdcommons.org/dpubs_series/2194, 2019, [Online].

Interphase *LC*-Oscillation Suppression with Fast Line-Transient Response in 48-V Series-Capacitor Buck Converters for Automotive Applications

W. L. Jiang*, Y. Liu*, N. Khan*, J. Pigott[†], H. J. Bergveld[†], V. Chaturvedi [†], and O. Trescases*
* The Edward S. Rogers Sr. Department of Electrical and Computer Engineering, University of Toronto
10 King's College Road, Toronto, ON, M5S 3G4
[†] NXP Semiconductors Inc.
E-mail: Wanlin.Jiang@mail.utoronto.ca

Abstract—This work presents a 4:1 series-capacitor buck (SCB) converter for 48 V-to-0.8 V automotive applications, designed with enhanced input transient response using current-mode control. In this work, interphase *LC* oscillations, characteristic of hybrid dc-dc converters, are damped through dynamic flying-capacitor voltage tracking. The proposed active tracking scheme is essential for current balancing in a current-mode-controlled SCB converter. Unlike existing multiphase current balancing methods, the proposed solution employs a combination of two flying-capacitor voltage regulators to adjust each phase's current command. The improved converter dynamics under a fast input voltage step are experimentally verified using a 50-W, 200-kHz prototype.

Index Terms—Hybrid converter, Series-Capacitor buck, Input transient, Transient response, Active balancing, Interphase oscillation, Current-Mode Control

I. INTRODUCTION

The adoption of 48-V distribution networks in electric vehicles (EVs) and hybrid EVs is motivated by the practical 5-kW power limit of conventional 12-V systems [1]. As power demand grows to support sub-1-V core voltages in advanced processors, state-of-the-art 48-V Point-of-Load (PoL) converters generally fall into two main categories: two-stage cascaded conversion and direct-conversion systems. The power demand of a single PoL converter in automotive power management units (PMUs) ranges from 40 W to 50 W [2]. Surveying the recent 48-V converter designs with output voltages from 0.8 V to 1.3 V, direct conversion is often used for converters rated at 100 W or less, as shown in Fig. 1 [3]–[16]. In contrast to conventional dc-dc converters, hybrid converters offer a high conversion ratio and multiphase capability for improved efficiency, making them suitable for single-stage automotive PMUs. In particular, the series-capacitor buck (SCB) converter, shown in Fig. 2, is a promising candidate with its high design modularity.

PWM-based control techniques, with equal duty cycle applied to all phases, are adopted in most SCB converters. This control approach inherently ensures steady-state current balancing. To meet the stringent output regulation requirements of automotive processors, various control techniques have been explored, including non-linear control [3], [17], [18], high-frequency auxiliary converters [2], [19], and coupled inductors [20]. Managing input transients is also critical due to load

This work was supported by NXP Semiconductors Inc. and the Natural Science and Research Council of Canada (NSERC).

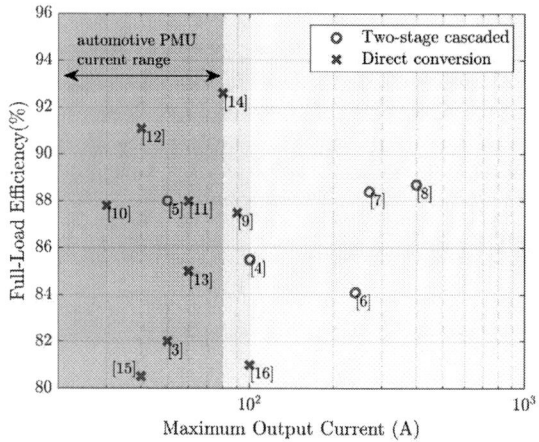

Fig. 1. The reported efficiency of 48V-to-core converters operating at full load and an output voltage nearest to 1 V [3]–[16].

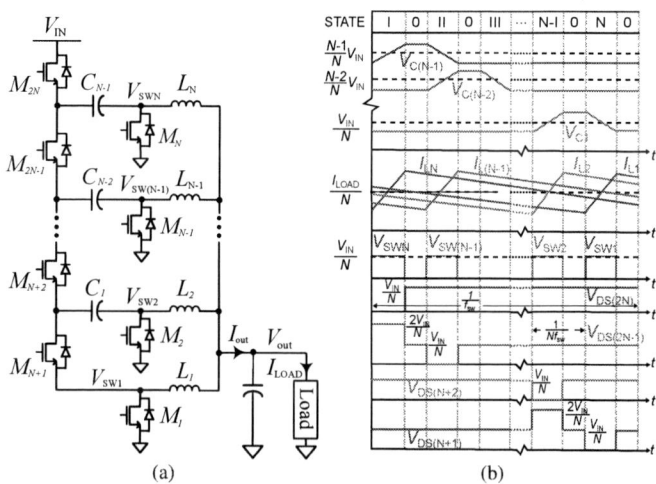

Fig. 2. (a) An *N*:1 SCB converter. (b) Steady-state waveforms.

dumps in EVs, which can cause sudden input voltage steps exceeding 200 V, unless protected by a transient voltage suppress (TVS) diode [21]. A rapid input voltage rise introduces oscillations in the *LC* network formed by the flying capacitors and power inductors in the system [22]. During such events, the inductor currents increase significantly, raising current saturation requirements and causing high voltage stress on the switches before the oscillations settle. This issue is further exacerbated in highly efficient converters with low conduction losses [23]. The impact of line transients can be mitigated by increasing the input capacitance to slow down the input voltage slew rate during transients. However, this approach comes with trade-offs, as it increases design cost and volume due to the high voltage ratings (above 60 V) and significant de-rating characteristics at high operating voltages of the capacitor.

While efforts have been made to enhance output transient response, methods for improving input step response and LC-oscillation mitigation in hybrid dc-dc converters are less frequently explored. As shown in [23], replacing individual phase inductors with coupled inductors can reduce resonance amplitudes, albeit at the cost of increased settling time and added complexity from custom magnetics. A dual-inductor-hybrid (DIH) converter design, incorporating a flying-capacitor-tapped auxiliary stage to improve output transient dynamics, is proposed in [24] and results in interphase LC oscillations during load steps. Both flying-capacitor-voltage-sensed and sensorless current balancing techniques have been proposed for the DIH converter to enhance LC dynamics. However, with only two power inductors, the DIH converter exhibits simpler LC dynamics when compared to an SCB converter with more phases.

An analogy exists between Flying Capacitor Multilevel (FCML) and SCB converters, as both use series-stacked inputs with flying capacitors. Active balancing is often applied to ensure devices and capacitors are operated within safe operating limits. However, with only one inductor connected to the output, FCML converters have fewer cross-coupled relationships between the LC elements. Additionally, a multiphase buck converter is topologically similar to an SCB converter, as both are based on the buck topology with parallel current paths to the output. The primary difference is that magnetising voltages across phase inductors in an SCB converter vary during transients due to fluctuating flying-capacitor voltages. In contrast, the power inductors in a multiphase buck converter are consistently magnetised with $V_{\text{IN}} - V_{\text{OUT}}$ and oscillations are confined to the power inductors and output capacitor.

To the best of the authors' knowledge, this work presents the first SCB converter control scheme with active interphase LC-oscillation suppression during line transients. The converter dynamics are analysed in Section II. Per-phase current commands are adjusted using outputs from two neighbouring flying-capacitor voltage-tracking regulators to achieve fast current balancing after transient events, with simulation results presented in Section III. The experimental results demonstrating the converter's improved dynamics using this control scheme are outlined in Section IV, followed by the conclusions in Section V.

II. SERIES-CAPACITOR BUCK-CONVERTER DYNAMICS

The fixed-frequency operation of an N:1 SCB converter shown in Fig. 2(b) corresponds to the switching states defined in Fig. 3. A 4:1 SCB converter, with parameters listed in

Table I, is used for simulation and experimental demonstration unless specified otherwise. The input bus line parasitic elements, L_{line} and R_{line}, were estimated based on a 1-m, 1-AWG wire.

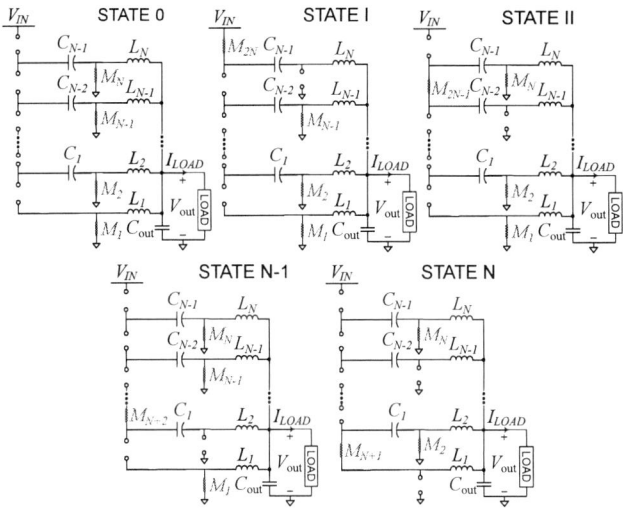

Fig. 3. Operating states of an N:1 SCB converter.

TABLE I
KEY SYSTEM PARAMETERS

Parameter	Value	Unit
Input voltage, V_{IN}	24 - 60	V
Output voltage, V_{OUT}	0.8	V
Switching frequency, f_{SW}	200	kHz
Rated load power, P_{RATED}	50	W
Line inductance, L_{line}	2.5	μH
Line resistance, R_{line}	300	mΩ
Output capacitor, C_{OUT}	800	μF
Input capacitor, C_{IN}	30	μF
High-side ON-resistance, $R_{\text{ON8,7,6,5}}$	6	mΩ
Low-side ON-resistance, $R_{\text{ON4,3,2,1}}$	0.8	mΩ
Inductors, $L_{4,3,2,1}$	1	μH
Inductor DCR	1	mΩ
Flying capacitors, $C_{3,2,1}$	30	μF

A. Interphase LC Oscillation

PWM-based control is commonly selected for an SCB converter due to its inherent steady-state current balancing characteristics [22]. In this case, underdamped interphase LC oscillations, as shown in Fig. 4, can occur when a large, rapid voltage step is applied at the input. Fig. 5 shows the propagation path of a disturbance on the input voltage, $\hat{v}_g$, across all phases. The PWM-based voltage regulation loop is decoupled from inductor currents, making it ineffective at damping the interphase LC response.

Voltage deviations in any flying capacitor, often triggered by a large load step in SCB converters using dynamically updated duty commands, can also introduce oscillations. Dynamically updated duty commands, which change after the

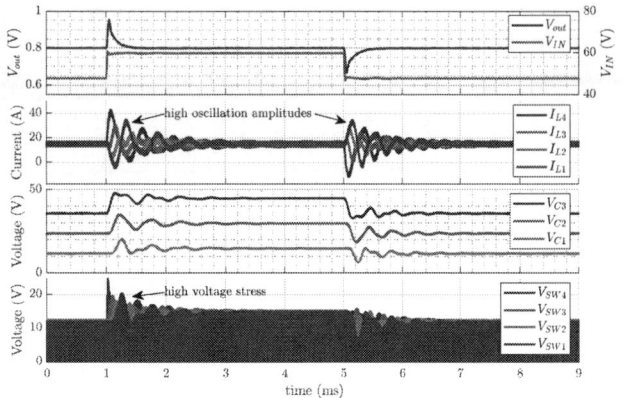

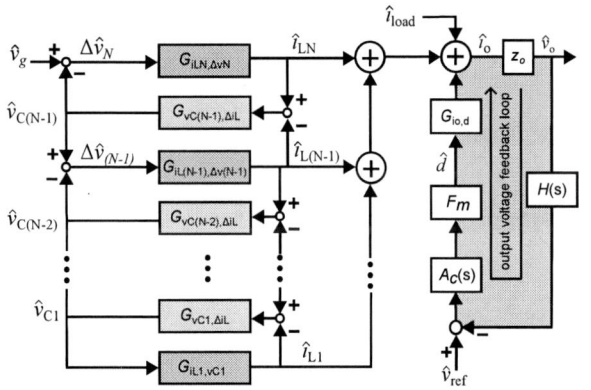

Fig. 4. Simulated response of a 4:1 SCB converter using PWM control to a line step from 48 V to 60 V with a 1-V/μs slew rate at $I_{\text{LOAD}} = 60$ A.

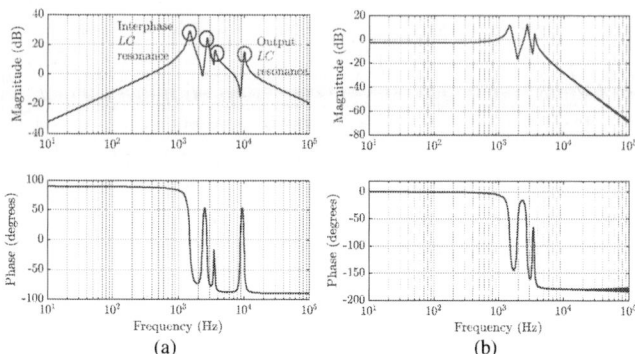

Fig. 5. Simplied small signal block diagram of an $N{:}1$ SCB converter using PWM-based control, omitting $\hat{d}$-to-$\hat{i}_{\text{L}j}$, $\hat{d}$-to-$\hat{v}_{\text{C}j}$, and $\hat{v}_o$-to-$\hat{i}_{\text{L}j}$ dynamics.

magnetization period of each phase, provide a faster closed-loop output transient response compared to using a common duty command across all phases. Nevertheless, the variation in duty can result in higher flying-capacitor voltage deviation, which has a response similar to that of an input voltage step due to the cross-coupled nature of i_L and v_C. As seen in Fig. 5, disturbances on a flying-capacitor voltage affect two inductor currents, which subsequently disturb other capacitor voltages.

For an $N{:}1$ SCB, there are N resonant frequencies in the v_g-to-$i_{\text{L}j}$ transfer functions, $G_{i\text{L}j,v_g}$, and $N-1$ resonant frequencies in the v_g-to-$v_{\text{C}j}$ transfer functions, $G_{v\text{C}j,v_g}$, as presented in Fig. 6(a) and Fig. 6(b), respectively. The damping ratio of the resonance is significantly influenced by the converter design. A more efficient and compact design reduces the damping ratio of the interphase system, as analysed in [23]. The interphase LC resonant frequencies are typically lower than the output LC pole frequency, as shown in Fig. 6(a).

B. Analysis of SCB Current-Mode Control Current Balancing

This study considers two fixed-frequency current-mode control (CMC) schemes: peak-current mode (PCM) and valley-current mode (VCM), both employing dynamically updated current commands, I_{c1} to I_{cN}. Unlike PWM-based control,

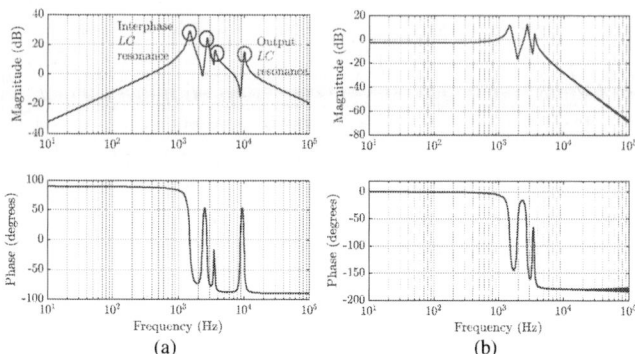

Fig. 6. The simulated open-loop transfer functions of (a) v_g-to-$i_{\text{L}4}$ and (b) the v_g-to-$v_{\text{C}3}$ using a derived small-signal model using parameters in Table I.

PCM and VCM do not provide inherent balancing [25]. A 2:1 SCB converter, which exhibits simpler dynamics than higher-level SCB converters, is analyzed using PCM control to examine the potential imbalance in response to an input voltage step-up. The timing diagram for this scenario is shown in Fig. 7, with the current command $I_c = I_{\text{L1,pk}} = I_{\text{L2,pk}}$.

A step of size $\Delta V = \alpha(V_{\text{IN}} - V_{\text{C1}})$ is applied to the input of the converter. The duration in state I, which is equivalent to the on-time of M_4 (M_N in a 2:1 SCB), t_{on2}, immediately after this input voltage step, t'_{on2}, is approximated as

$$ t'_{\text{on2}} \approx \frac{(I_c - I_{\text{val2}})L}{V_{\text{IN}} - V_{\text{C1}} + \Delta V} = \frac{t_{\text{on2}}}{\alpha + 1} , \tag{1} $$

where I_{val2} is the steady-state valley current before the transient. Given that $\Delta t_{\text{on2}} = t'_{\text{on2}} - t_{\text{on2}} < 0$, the valley current at the end of this switching cycle becomes

$$ I'_{\text{val2}} = I_c - \frac{(T_{\text{SW}} - t'_{\text{on2}})V_{\text{out}}}{L_2} , \tag{2} $$

which decreases with a shorter t_{on2}. This negative feedback to t_{on2} helps balance the currents if I'_{val2} is significantly reduced. However, due to the significant difference in positive and negative inductor current slope in the 48V-to-core application, the impact of

$$ \Delta I_{\text{val2}} = I'_{\text{val2}} - I_{\text{val2}} = \frac{\Delta t_{\text{on2}}V_{\text{out}}}{L_2} \tag{3} $$

remains insignificant compared to that of Δt_{on2}, allowing the

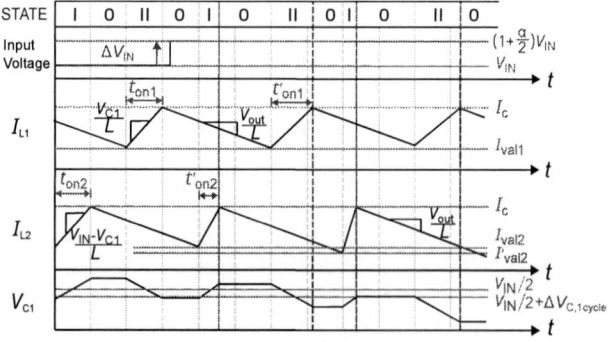

Fig. 7. Input step-up waveforms of a 2:1 SCB converter in PCM with a current command of I_c, showing a runaway capacitor voltage.

979-8-3315-1612-3/25 $31.00 © 2025 IEEE

approximation $\langle I'_{L2}\rangle_{T_{SW}} \approx \langle I_{L2}\rangle_{T_{SW}}$. Meanwhile, the change in on-time of M_4 reduces V_{C1} after a full switching period, where

$$\Delta V_{C1,cycle} = \frac{\langle I_{L2}\rangle_{T_{SW}} t'_{on2} - \langle I_{L1}\rangle_{T_{SW}} t_{on1}}{C_1}$$
$$= \frac{I_{LOAD}\Delta t_{on2}}{2C_1}, \tag{4}$$

is negative and especially noticeable at higher output currents. With the goal of increasing V_{C1} to $(V_{IN}+\Delta V)/2$, the negative value of $\Delta V_{C1,cycle}$ drives the flying capacitor away from a balanced steady-state. Furthermore, with

$$t'_{on1} \approx \frac{(I_c - I_{val1})L}{V_{C1} + \Delta V_{C1,cycle}}, \tag{5}$$

the on-time of M_3 (state II duration) extends, further reducing V_{C1} when combined with the shortened t_{on2} based on (1). The SCB converter requires non-overlapping magnetization periods between inductors from consecutive phases, adding the constraint to the maximum high-side device turn-on time with $t_{on,max} = T_{SW}/N$. Thus, t_{on1} eventually saturates at $T_{SW}/2$ as a result of the lowering V_{C1}. V_{C1} continues to drop, ultimately leading to voltage runaway.

The analysis of this imbalance condition is similar in VCM control, as the inductor current slew rate remains the same as in PCM. Here, I_c now represents I_{val1} and I_{val2}. There is a slight increase in I_{pk2} and, by association, $\langle I_{L2}\rangle_{T_{SW}}$ from ΔV due to the high inductor magnetization voltage. Although $\langle I_{L2}\rangle_{T_{SW}} > \langle I_{L1}\rangle_{T_{SW}}$ favours balancing based on (4), t_{on2} in subsequent cycles becomes

$$t'_{on2} = T_{SW} - \frac{(I'_{pk2} - I_c)L}{V_{out}}, \tag{6}$$

which reduces with higher I'_{pk2} and introduces a positive feedback that causes V_{C1} to deviate from $(V_{IN} + \Delta V)/2$. The simulated V_{C1} runaway condition of a 4:1 SCB converter under VCM control is shown in Fig. 8.

The ability to send different current commands to each phase in a CMC allows for dynamic adjustments to the commands, enabling the balancing of phase currents during

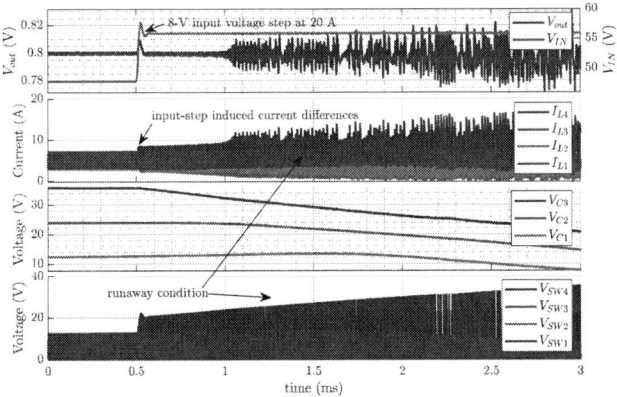

Fig. 8. Simulated 4:1 SCB using VCM control at $I_{LOAD} = 20\,\text{A}$ without active balancing.

steady-state operation and the damping of interphase LC oscillations caused by system disturbances. The right-half-plane (RHP) zero in $G_{\Delta i_L,d}$ introduces complexity in actively controlling ΔI_L resonance, particularly under light load conditions [23]. As a result, tracking V_{Cj} to their nominal steady-state values is the preferred active control technique for suppressing oscillations and balancing currents in a SCB.

III. SYSTEM ARCHITECTURE AND SIMULATION RESULTS

The proposed control architecture for an $N{:}1$ SCB converter is presented in Fig. 9(a), using adjustment current commands, $I_{Cj,adj}$, from the novel voltage-tracking regulators shown in Fig. 9(b). In fixed-frequency current-mode control, the inductor currents I_{Lj} (for $j = 1, \ldots, N$) in an SCB are equalized by maintaining uniform peak or valley currents and identical current ripple amplitudes. The current ripple of L_j on phase j, ΔI_{Lj}, can be calculated using

$$\Delta I_{Lj} = \frac{V_{Cj} - V_{C(j-1)}}{L_j t_{on}}, \tag{7}$$

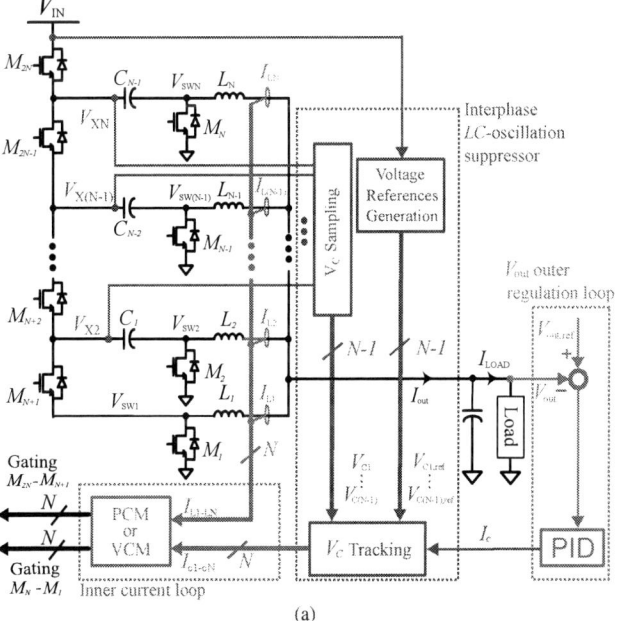

(a)

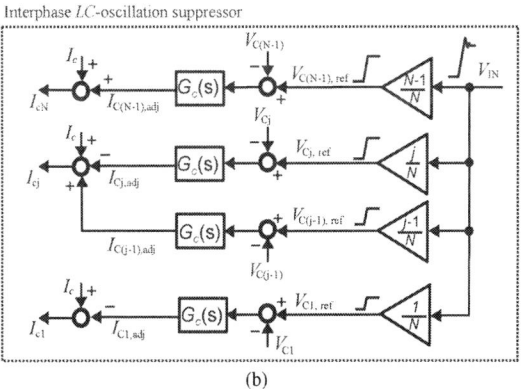

(b)

Fig. 9. (a) The proposed $N{:}1$ SCB converter control architecture using (b) the LC-oscillation suppressor with dynamic voltage-tracking.

with $V_{CN} = V_{IN}$ and $V_{C0} = 0$. For SCB with $N > 2$, I_{Lj} for $j = 2, \ldots, N-1$ are influenced by two neighbouring flying capacitors. The converter operation must satisfy $V_{Cj} - V_{C(j-1)} = V_{IN} - V_{C(N-1)} = V_{C1}$ to maintain uniform ΔV_{Cj}. Thus, the dynamic flying-capacitor voltage references, as shown in Fig. 9(b), are set based on $V_{Cj,ref} = jV_{IN}/N$ for phase-current balancing.

The control scheme employs a feedforward path from the input to modify the current command generated by the outer voltage regulation loop, thereby suppressing interphase LC oscillations when disturbances occur. Mid-capacitor voltages are computed in the V_C-sampling block by averaging the peak and valley value of $V_{C(j-1)}$, sampled using V_{Xj} on phase j, as shown in Fig. 9(a), when the low-side transistor M_j is on. With $t_{on,max} = T_{SW}/N$, slope compensation with a minimum slope, $m_{a,min}$, of

$$m_{a,min} = \frac{m_{mag}}{2} = \frac{V_{IN,max} - NV_{out}}{2NL}, \qquad (8)$$

is required to prevent sub-harmonic oscillation using VCM-based control.

Following a line transient, each V_C tracking controller immediately detects a voltage error of

$$\Delta V_{Cj,error} = \frac{j\Delta V_{IN}}{N}, \qquad (9)$$

thereby prioritising the voltage tracking of higher-level flying capacitors. This is desirable, given that I_{LN} rapidly rises and M_N experiences the largest voltage stress during a step-up transient. The adjustment commands based on the voltage of C_j, $I_{Cj,adj}$, have minimal effect on the output current. When summing the commands sent to phase j for current comparison, I_{cj}, the adjustment commands cancel. This results in

$$NI_c = \sum_{j=1}^{N} I_{cj}, \qquad (10)$$

with a reduced impact of line transients on I_{out}.

Fig. 10(a) and Fig. 10(b) show the dynamics of a 4:1 SCB converter operating in PCM and VCM (with slope compensation) at $I_{LOAD} = 60\,\text{A}$, subject to input steps between 48 V and 60 V with a slew rate of 1 V/µs. With a 32-A current reference saturation value, $I_{L,peak}$ is reduced by over 10 A during transients compared to the PWM-controlled converter shown in Fig. 4. The I_L settling time is shortened from 2 ms to 500 µs, along with a significantly lower output voltage disturbance. Except for M_4, the voltage stress on all low-side devices remains within the maximum steady-state drain-to-source voltage during line transients.

The series-input structure of an SCB converter reduces the magnetising voltage across the inductors compared to a traditional step-down converter. However, the postive inductor current slope, m_{mag}, remains significantly higher than the negative inductor current slope, espcially in SCBs with a high nominal conversion ratio. This makes achieving accurate peak sensing with minimal delay challenging. Thus, VCM

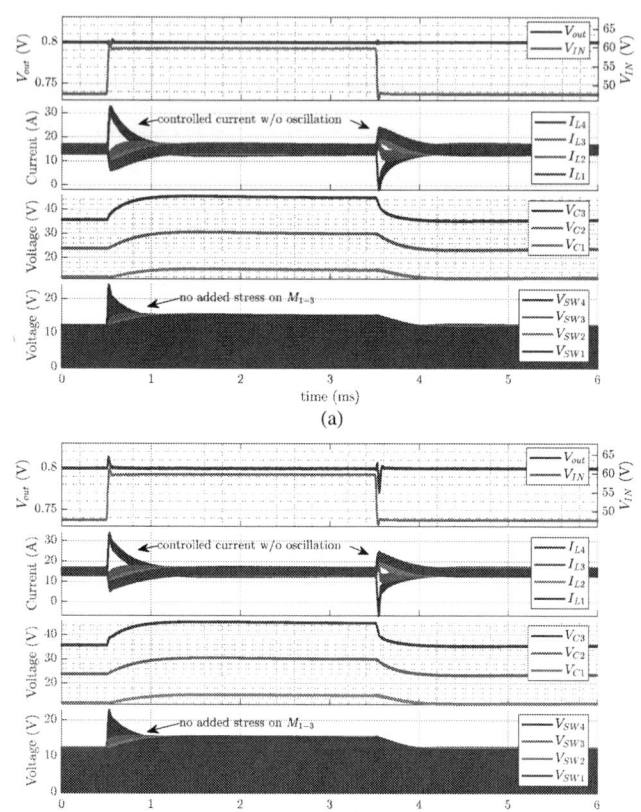

Fig. 10. The 4:1 SCB converter simulated using the proposed interphase LC-oscillation suppressor at $I_{LOAD} = 60\,\text{A}$ with (a) PCM and (b) VCM control.

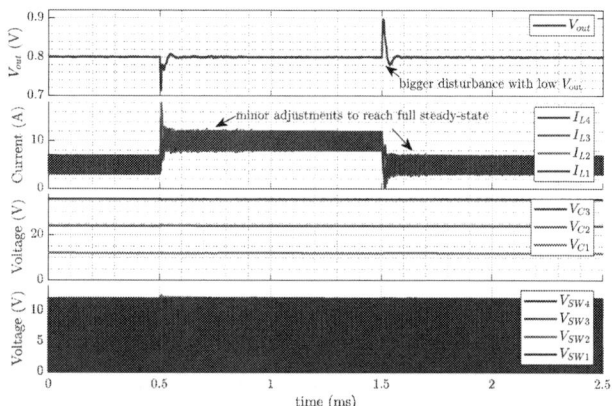

Fig. 11. Simulated response due to a load-step of I_{out} from 20 A to 40 A at 1 A/ns using VCM control with V_C-tracking, showing a regulated output voltage.

is the more practical control approach for this topology. To demonstrate the full stability of the system, a 20-A load step from $I_{out} = 20\,\text{A}$ is applied to the SCB using VCM, with the simulated results shown in Fig. 11.

IV. EXPERIMENTAL RESULTS

The 4:1 SCB converter in Fig. 12 was implemented using 60-V, 5-mΩ IAUZ40N06S5 and 40-V, 0.55-mΩ IAUC120N04S6 for switches M_{5-8} and M_{1-4}, respectively,

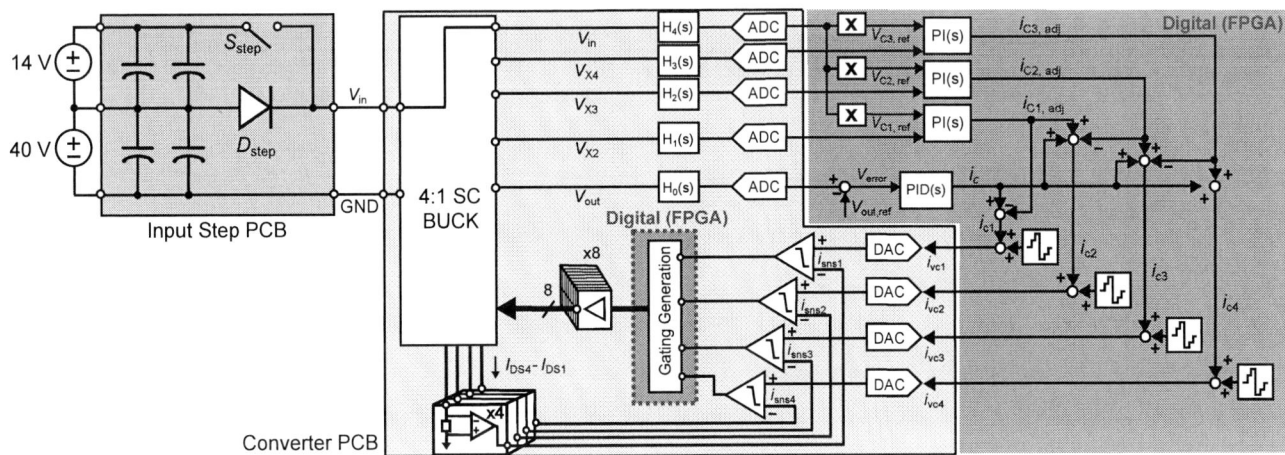

Fig. 12. System hardware implementation.

as shown in Fig. 13. An input step PCB is implemented to achieve a high input voltage step slew rate, which cannot be achieved using typical bench-top supplies. A VCM-based control with V_C-tracking is implemented using an FPGA-based digital controller. Although mid-V_{C1-3} sensing is preferable for optimal inductor current balancing, the time between magnetization periods of inductors in subsequent phases can be brief, particularly during transients. This short interval makes precise peak-voltage sensing challenging, especially when accounting for required blanking time. With a sensibly chosen flying-capacitor value based on its ripple, sensing the valley voltage is simpler to implement, with a minor current offset on the inductors. 2-mΩ current-sense resistors were connected to the sources of M_{1-4} and placed on the back side of the converter PCB. The front side contains the power stage using the same passive component values specified in Table I. During testing, the inductor current was measured only when using PWM control to avoid noise injection into the sensing circuits.

The comparison of the interphase response of the converter using a PWM-based and a VCM-based control to an input step

between 40 V and 54 V with a step-up slew rate of 1.2 V/μs at $I_{out} = 20$ A is shown in Fig. 14. The plotted switch-node voltage measurements were processed using a 2-ns moving-average filter to eliminate hard-switching switch-node ringing, making oscillations easier to observe. Comparing V_{SW} in Fig. 14(a) and Fig. 14(b), the VCM controller effectively reduces interphase LC oscillations and lowers the voltage stress on switches M_{1-3}.

The response including the output voltage deviation to the input step is shown in Fig. 15, with the valleys of V_{X2-4} reflecting the voltage behaviour of C_{1-3}. When the input voltage step is applied to the converter using PWM control, as shown in Fig. 15(a), the peak of I_{L1} quickly rises by 24 A due to the lack of direct control over I_L, resulting in a large ΔV_{out}. In comparison, the dynamics of the current-mode-controlled SCB converter with oscillation suppression are shown in Fig. 15(b). The flying-capacitor voltages are regulated to the dynamically generated references within 300 μs according to the measured V_{X1-3}. The extended oscillatory behaviour is eff-

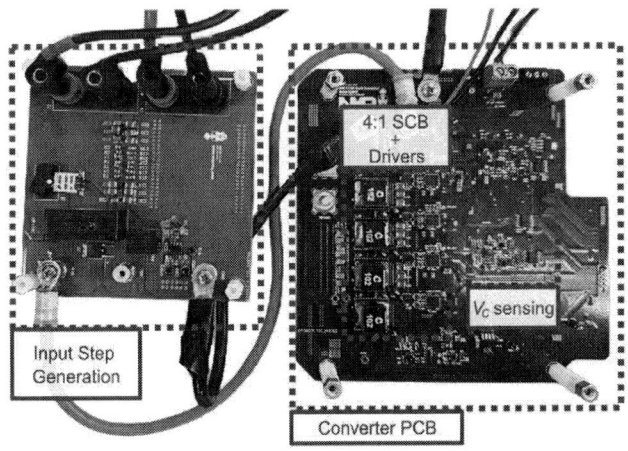

Fig. 13. Dc-dc converter experimental prototype.

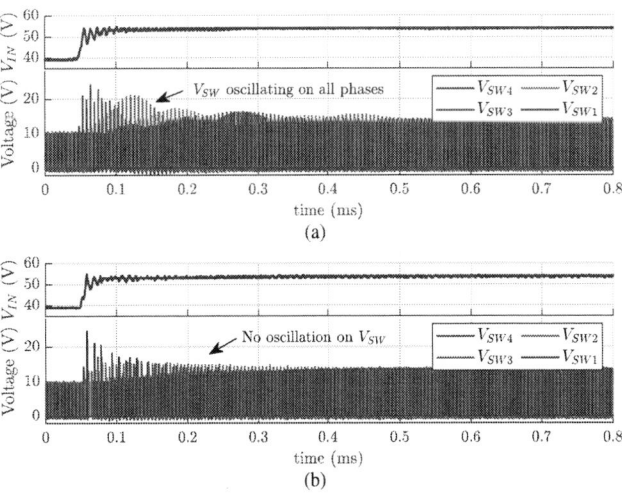

Fig. 14. Measured V_{SW} of the 4:1 SCB converter subjected to an input voltage step up from 40 V to 54 V in 10 μs at $I_{LOAD} = 20$ A using (a) PWM control and (b) VCM control with oscillation suppression.

ectively suppressed, with a $4\times$ reduction in peak ΔV_{out}. The input step-down, with a rate dictated by the converter operating current, also resulted in no major disturbance on the V_{out} when using the suppressor, as shown in Fig. 16.

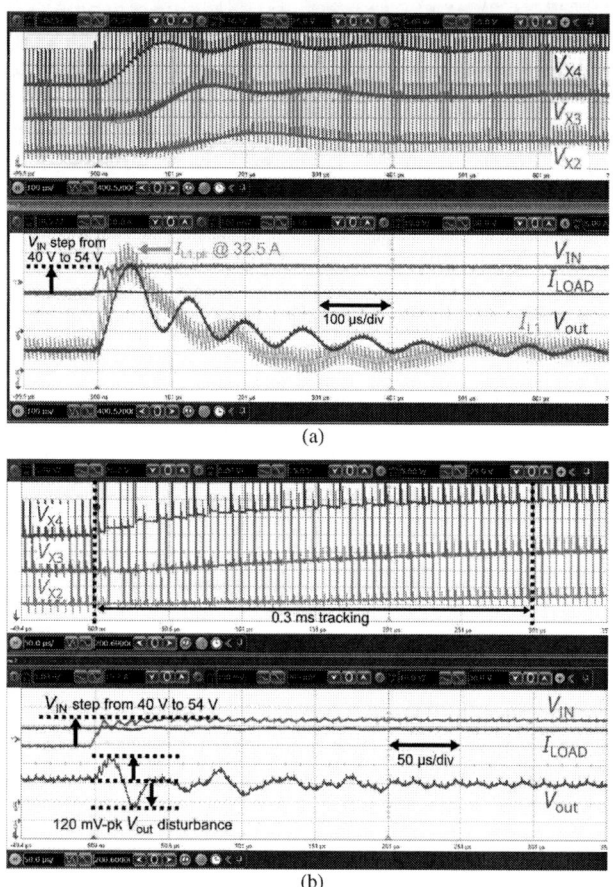

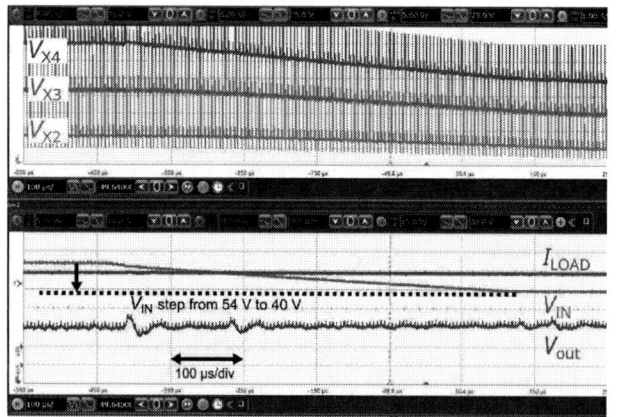

Fig. 15. Experimentally measured converter dynamics subjected to an input-voltage step up from 40 V to 54 V in 10 µs at $I_{\text{LOAD}} = 20$ A using (a) PWM control and (b) VCM control with oscillation suppression.

Fig. 16. The converter response to the input step down from 54 V to 40 V at $I_{\text{LOAD}} = 20$ A using VCM with oscillation supression. The voltage step-down slew rate is limited by I_{LOAD}.

At the nominal conversion ratio of 48 V-to-0.8 V, the peak efficiency of 88.4% was measured at 32 A, with a rated-load

efficiency of 85.6% at 62.5 A. The converter losses include the loss from the 2-mΩ sense resistors, R_{sns}, on each phase. The total sensing loss is calculated using $P_{\text{sense}} = R_{\text{sns}} I_{out}^2 / 4$, which becomes substantial at $I_{\text{out}} > 20$ A, as shown in Fig. 17. A 90.2% peak and 89.2% rated efficiency were measured when R_{sns} was removed from the power stage. This shows that by adopting near-lossless current sensing, such as using senseFETs [26], the converter can achieve efficiencies comparable to other 48-V converters [5], [10], [11].

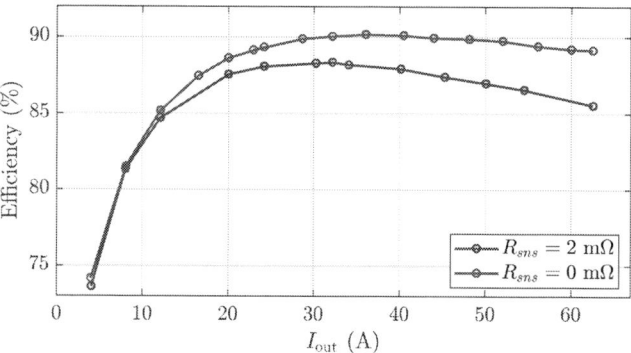

Fig. 17. Measured power stage efficiency.

V. CONCLUSIONS

The proposed control strategy for N:1 SCB converters has been experimentally demonstrated to enhance the input-transient response using a 50-W prototype. This approach satisfies the input-transient immunity requirements of emerging 48-V automotive PMUs. By incorporating dynamic flying-capacitor voltage regulation, the current-mode control scheme mitigates interphase LC oscillations, a unique condition in hybrid dc-dc converters, where the magnetising voltages of the inductors can vary significantly between phases. Leveraging this technique allows for the reduction of input capacitance and inductor saturation current requirements, thereby improving the power density of the converter.

REFERENCES

[1] M. Kane, "Tesla Confirms The Switch To 48 Volt System," 2023, accessed: 2023-08-20. [Online]. Available: https://insideevs.com/news/656775/tesla-switch-48v-voltage-system/

[2] N. Khan *et al.*, "A 40W Dual-Inductor Hybrid Converter with Flying-Capacitor-Tapped Auxiliary Stage for Fast Transient Response in 48V PoL Automotive Applications," in *2023 IEEE Applied Power Electronics Conference and Exposition (APEC)*, March 2023, pp. 40–47.

[3] S. Y. Sim, X. Zhang, J. Jiang, K. Wei, and C. Huang, "A 94.7% Efficiency Direct-Step-Down Switched-Tank-Based 48V to 1V-3.3V Hybrid Converter with Constant-Resonant-Time Closed-Loop Control," in *2024 IEEE Applied Power Electronics Conference and Exposition (APEC)*, 2024, pp. 1344–1350.

[4] H. Peng *et al.*, "Optimized Two-Stage & Two-phase 48V-1V DC-DC Converter with Large Load Current up to 100A," in *2024 IEEE 10th International Power Electronics and Motion Control Conference (IPEMC2024-ECCE Asia)*, 2024, pp. 1629–1633.

[5] N. M. Ellis, Y. Zhu, and R. C. Pilawa-Podgurski, "Gallium Nitride-based 48V-to-1V Point-of-Load (PoL) Converter for Aerospace Telecommunications and Computing Applications," in *2024 IEEE Applied Power Electronics Conference and Exposition (APEC)*, 2024, pp. 1384–1388.

[6] Y. Elasser *et al.*, "Mini-LEGO: A 1.5-MHz 240-A 48-V-to-1-V CPU VRM with 8.4-mm Height for Vertical Power Delivery," in *2023 IEEE Applied Power Electronics Conference and Exposition (APEC)*, 2023, pp. 1959–1966.

[7] Y. Zhu, T. Ge, Z. Ye, and R. C. Pilawa-Podgurski, "A Dickson-Squared Hybrid Switched-Capacitor Converter for Direct 48 V to Point-of-Load Conversion," in *2022 IEEE Applied Power Electronics Conference and Exposition (APEC)*, 2022, pp. 1272–1278.

[8] Y. Chen, H. Cheng, D. M. Giuliano, and M. Chen, "A 93.7% Efficient 400A 48V-1V Merged-Two-Stage Hybrid Switched-Capacitor Converter with 24V Virtual Intermediate Bus and Coupled Inductors," in *2021 IEEE Applied Power Electronics Conference and Exposition (APEC)*, 2021, pp. 1308–1315.

[9] Y. Zhu, Z. Ye, T. Ge, R. Abramson, and R. C. N. Pilawa-Podgurski, "A Multi-Phase Cascaded Series-Parallel (CaSP) Hybrid Converter for Direct 48 V to Point-of-Load Applications," in *2021 IEEE Energy Conversion Congress and Exposition (ECCE)*, 2021, pp. 1973–1980.

[10] C. Chen, J. Liu, and H. Lee, "A 92.7%-Efficiency 30A 48V-to-1V Dual-Path Hybrid Dickson Converter for PoL Applications," in *2021 IEEE Energy Conversion Congress and Exposition (ECCE)*, 2021, pp. 1989–1994.

[11] X. Zhang *et al.*, "A 12- or 48-V Input, 0.9-V Output Active-Clamp Forward Converter Power Block for Servers and Datacenters," *IEEE Transactions on Power Electronics*, vol. 35, no. 2, pp. 1721–1731, 2020.

[12] M. Halamicek, T. McRae, and A. Prodić, "Cross-Coupled Series-Capacitor Quadruple Step-Down Buck Converter," in *2020 IEEE Applied Power Electronics Conference and Exposition (APEC)*, 2020, pp. 1–6.

[13] M. Choi and D.-K. Jeong, "A 92.8%-Peak-Efficiency 60A 48V-to-1V 3-Level Half-Bridge DC-DC Converter with Balanced Voltage on a Flying Capacitor," in *2020 IEEE International Solid-State Circuits Conference - (ISSCC)*, 2020, pp. 296–298.

[14] M. H. Ahmed, C. Fei, F. C. Lee, and Q. Li, "Single-Stage High-Efficiency 48/1 V Sigma Converter With Integrated Magnetics," *IEEE Transactions on Industrial Electronics*, vol. 67, no. 1, pp. 192–202, 2020.

[15] R. Das, G.-S. Seo, D. Maksimovic, and H.-P. Le, "An 80-W 94.6%-Efficient Multi-Phase Multi-Inductor Hybrid Converter," in *2019 IEEE Applied Power Electronics Conference and Exposition (APEC)*, 2019, pp. 25–29.

[16] R. Das and H.-P. Le, "Regulated 48V-to-1V/100A 90.9%-Efficient Hybrid Converter for POL Applications in Data Centers and Telecommunication Systems," in *2019 IEEE Applied Power Electronics Conference and Exposition (APEC)*, 2019, pp. 1997–2001.

[17] T. Vekslender, O. Ezra, Y. Bezdenezhnykh, and M. M. Peretz, "Closed-Loop Design and Time-Optimal Control for a Series-Capacitor Buck Converter," in *2016 IEEE Applied Power Electronics Conference and Exposition (APEC)*, 2016, pp. 308–314.

[18] H. Han *et al.*, "A Monolithic 48V-to-1V 10A Quadruple Step-Down DC-DC Converter with Hysteretic Copied On-Time 4-Phase Control and 2× Slew Rate All-Hysteretic Mode," in *2022 IEEE Symposium on VLSI Technology and Circuits (VLSI Technology and Circuits)*, 2022, pp. 182–183.

[19] N. Khan *et al.*, "Ultrafast Transient Response in 48 V Automotive VRMs: An Auxiliary-Assisted Adaptive Slew-Rate Control Scheme," *IEEE Transactions on Industrial Electronics*, pp. 1–11, 2024.

[20] N. M. Ellis, R. A. Abramson, R. Mahony, and R. C. N. Pilawa-Podgurski, "The Symmetric Dual Inductor Hybrid Converter for Direct 48V-to-PoL Conversion," *IEEE Transactions on Power Electronics*, vol. 39, no. 6, pp. 7278–7289, 2024.

[21] STMicroelectronics, "Dealing with voltage surges in a 48 V automotive system and transient voltage suppressor (TVS)," Application Note, 2023, [Online]. Available: https://www.st.com/resource/en/application_note/an5958-dealing-with-voltage-surges-in-a-48-v-automotive-system-and-transient-voltage-suppressor-tvsstmicroelectronics.pdf.

[22] P. S. Shenoy *et al.*, "Automatic Current Sharing Mechanism in the Series Capacitor Buck Converter," in *2015 IEEE Energy Conversion Congress and Exposition (ECCE)*, 2015, pp. 2003–2009.

[23] P. Wang *et al.*, "Interphase *LC* Resonance and Stability Analysis of Series-Capacitor Buck Converters," *IEEE Transactions on Power Electronics*, vol. 38, no. 5, pp. 5680–5687, 2023.

[24] O. Cobani, N. Khan, J. Pigott, H. J. Bergveld, and O. Trescases, "A Sensorless Feedforward Balancing Compensator for Improved Interphase Dynamics in Auxiliary-Assisted Dual-Inductor-Hybrid Converters," in *2024 Energy Conversion Congress & Expo Europe (ECCE Europe)*, 2024, pp. 1–8.

[25] P. Majumder, S. Kapat, and D. Kastha, "Fast Transient State Feedback Digital Current Mode Control Design in Series Capacitor Buck Converters," in *2022 IEEE Applied Power Electronics Conference and Exposition (APEC)*, 2022, pp. 2080–2085.

[26] S. Zhang, M. Zhao, X. Wu, and H. Zhang, "Current-Balance Method for Multi-Phase DC–DC Buck Converters With Wide Duty Ratio Applications in Both CCM and DCM," *Electronics Letters*, vol. 53, no. 15, pp. 1062–1064, 2017.

An Approach to Compensate for Low frequency DC-Link Voltage Ripple in High Power ANPC Inverter

Shaozhe Wang
dept. Electrical & Computer Engineering
Texas A&M University
College Station, Texas
wangshaozhe1996@tamu.edu

Ankit Vivek Deshpande
dept. Electrical & Computer Engineering
Texas A&M University
College Station, Texas
ankitdesh92@tamu.edu

Rolando Sandoval
dept. Electrical & Computer Engineering
Texas A&M University
College Station, Texas
rsando18@tamu.edu

Erick Pool-Mazun
dept. Electrical & Computer Engineering
Texas A&M University
College Station, Texas
epooltamu@tamu.edu

Enrique Garza-Arias
dept. School of Engineering and Sciences
Tecnológico de Monterrey
Monterrey, Mexico
enrique.garza@tec.mx

Prasad Enjeti
dept. Electrical & Computer Engineering
Texas A&M University
College Station, Texas
enjeti@tamu.edu

Abstract— DC-link voltage ripple has an adverse impact on the performance of Electrical traction drives. Employing large DC-link capacitors may not be feasible in scenarios demanding high power density such as in Electric vehicle applications. This paper introduces a new technique to compensate for the DC-link voltage ripple in ANPC inverters employing a feedforward strategy. The proposed DC-link method is combined with hybrid space vector PWM (SVPWM) to effectively compensate for low frequency DC-link ripple as large as 20% by tracking voltage variation in real-time. The paper also includes derivation process for calculating the capacitor current in an ANPC inverter, which serves as the basis for selecting the appropriate capacitance value. Simulation and experiment results on a 14kW SiC ANPC inverter are presented to verify the effectiveness of the proposed technique, which successfully compensates for DC-link fluctuation.

Keywords — ANPC inverter, hybrid SVPWM, capacitor current, DC ripple rejection, feedforward control.

I. INTRODUCTION

Recently, electrical traction drives have gained significant attention. Electrical propulsion systems are increasingly utilized in aircraft propulsion systems. In aerospace applications, the power density requirement for inverter systems is particularly high, ranging from 30 to 40 kW/kg, with DC-link voltages around 1000 V. To address the need for lower device voltage stress and higher efficiency, a silicon carbide (SiC)-based three-level Active Neutral Point Clamped (ANPC) inverter is more suitable for these applications [1]-[2]. Compared to the PWM modulation strategy, the SVPWM method allows for better utilization of the DC bus [3]. Additionally, the SVPWM strategy can be extended to 3-level inverter, which has made it widely utilized in the control of ANPC inverter [4].

In the design of an ANPC inverter, selecting the input capacitance is crucial, as the input capacitor helps maintain a stable and constant input voltage, which is essential for achieving better output quality with reduced harmonics. The selection of capacitor value is determined by the current

variation that flows through the capacitor. [5]-[6] gives the formula to represent the DC-link capacitor current in a SVPWM controlled 2-level inverter. [7] discussed the capacitor current of a 3-level T-type inverter based on different value range of modulation index. This paper derives the capacitor current using Fourier analysis of the modulation signal and provides a detailed formula with explanations for each term.

In electrical propulsion applications, the inverter DC-link is powered by either a single-phase AC supply or a three-phase AC supply. The DC-link will include voltage ripple (typically at twice the input frequency) in case of single-phase supply or unbalanced three-phase supply conditions. Low frequency DC-link voltage ripple affects the electrical propulsion system in the following ways.

1) Generates low-frequency sidebands in the inverter output voltage, leading to torque pulsation, vibration, and overheating.

2) Contributes to higher voltage stress on the inverter switches.

3) Large size DC-link filter capacitors reduce power density.

4) Time-varying nature of the DC-link voltage ripple makes it difficult to filter and compensate.

Using larger DC-link filter components has following disadvantages, making it unsuitable for high power density applications such as aerospace systems.

1) DC-link components are heavy, bulky, and take up significant space.

2) They lead to slower response times and increased costs.

3) Low-frequency harmonics are difficult to filter using capacitors.

Consequently, it is desirable to mitigate the DC-link ripple voltage through control strategy. The elimination of DC-link ripple voltage with the programmed PWM technique and

---

This material is based upon work supported by the Department of Energy under Award Number DE-AR0001356.

associated tradeoff's is discussed in [8]. A disturbance rejection strategy based on sliding mode control for T type inverters is discussed in [9]. DC-link voltage fluctuation is predicted by a repetitive algorithm in [10].

In this paper, a real-time technique to reject DC-link voltage ripple for ANPC Inverters is discussed. The proposed technique is a feedforward approach with hybrid SVPWM method that adjusts the modulating function appropriately and is shown to adapt for time varying nature of the DC-link voltage ripple. The approach has been shown to be fast in response, easy to implement, and compatible to ANPC control algorithm. Section II of this paper describes the operation of ANPC, the DC ripple rejection, and the capacitor selection. Section III includes simulation and experiment results. Section IV is the conclusion.

II. ANALYSIS OF ANPC CONTROL AND DC RIPPLE COMPENSATION

A. Hybrid SVPWM Control for ANPC Inverter

A three phase ANPC converter is shown in Fig. 1(a). The input DC voltage is equally divided by two identical capacitors. The voltage at the middle point V_o is assumed to be zero. Each phase of the ANPC inverter consists of six switches named from Q_1 to Q_6. Point "o" is connected to the middle point of the input DC voltage link. Fig. 1 (b) gives the illustration of control signal and line-to-middle voltage of phase a. SW_i in Fig. 1(b) is the control signal for switch Q_i ($i = 1$ to 6). SW_1, SW_4, SW_5, and SW_6 are operated at fundamental frequency. SW_1 and SW_6 are the same, and complementary to SW_4 and SW_5. SW_2 and SW_3 are complementary and operated at high frequency.

The switching sequence in Fig. 1(b) is generated by the hybrid SVPWM method as shown in Fig. 2. According to Fig. 2(a), each operation period of the inverter includes six sectors, and each sector has four regions. All of the vectors can be categorized into zero vector, small vector, medium vector and large vector. Taking sector I as an example, there are six vectors in each sector, V_0 is the zero vector, V_1 and V_2 are the small

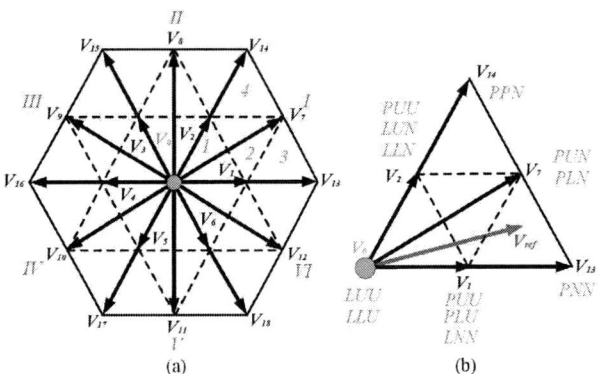

Fig. 2. (a) Hybrid SVPWM plot for a three-phase ANPC system shown in Fig. 1, (b) Vector combinations of sector I.

vector, V_7 is the medium vector, and V_{13} as well as V_{14} are the large vector. Each vector could be generated by different switching combinations of a three-phase ANPC system. In Fig. 2(b), P means output positive voltage, N means output negative voltage, L and U means output zero voltage when the output voltage is in positive or negative half cycle respectively. The order is the same as phase a, b, and c. For example, when phase a generates a positive voltage, phase b generates a zero voltage, and phase c generates a negative voltage, the indicator of the vector is "PUN". The reference vector V_{ref} consists of the three vectors that are most close to it.

B. Feedforward modulation for DC-link ripple rejection

The line-to-middle output voltage of phase a is V_{ao}, and it can be expressed by (1). V_P, V_N and V_O are the positive DC-link voltage, negative DC-link voltage and the middle point voltage of the DC-link respectively.

$$V_{ao} = \begin{cases} SW_2 \cdot (V_P - V_O) = SW_2 \cdot \frac{V_{dc}}{2}, & SW_1, SW_6 = 1 \\ SW_3 \cdot (V_N - V_O) = SW_3 \cdot (-\frac{V_{dc}}{2}), & SW_4, SW_5 = 1 \end{cases} \quad (1)$$

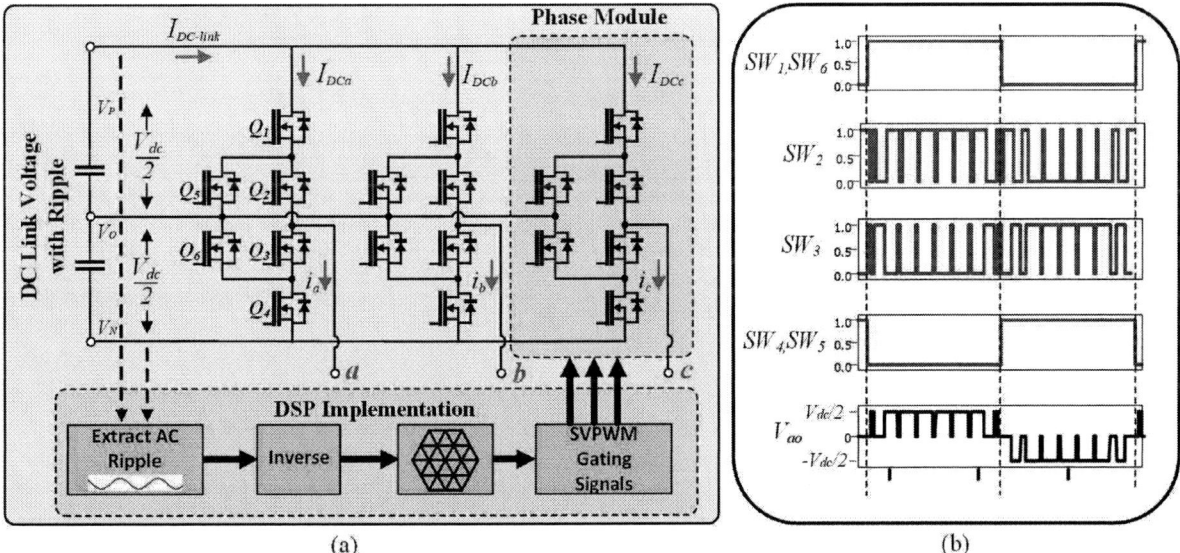

Fig. 1. (a) Three phase ANPC Inverter (b) Advanced SVPWM strategy that operates switch Q_1, Q_5, Q_6, Q_4 at line frequency and Q_2, Q_3, at high frequency.

$$\begin{cases} SW_2 = S = m_a \cdot sin(2\pi ft) + m_a \cdot \delta \cdot \sin(3 \cdot 2\pi ft) + \sum h_n \cdot \sin(2\pi f_n t), when \ Q_1, Q_6 \ ON \\ SW_3 = -S, \qquad\qquad\qquad\qquad\qquad\qquad\qquad\qquad\qquad\qquad when \ Q_4, Q_5 \ ON \end{cases} \tag{2}$$

$$V_{ao} = \frac{V_{dc}}{2} \cdot S = \frac{V_{dc}}{2} \cdot [m_a \cdot sin(2\pi ft) + m_a \cdot \delta \cdot \sin(3 \cdot 2\pi ft) + \sum h_n \cdot \sin(2\pi f_n t)] \tag{3}$$

SW_2 and SW_3 have the following Fourier expressions according to [11]. In (2), m_a is the modulation index, f is the fundamental frequency. Fourier expressions of control signals include high frequency component, the amplitude and frequency are represented by h_n and f_n respectively.

Combining Equation. 1 and 2, the line-to-middle voltage of phase a is shown in (3).The high-frequency component in (3) can be neglected, as it is easily filtered out using a small inductor. Therefore, approximately only the fundamental component and the 3rd-order harmonics remain in the equation.

Since the sum of the third-order harmonics across all three phases cancels out to zero, the third-order component does not degrade the output quality.

$$V_{ao} = \frac{V_{dc}}{2} \cdot m_a \cdot [sin(2\pi ft) + \delta \cdot \sin(3 \cdot 2\pi ft)] \tag{4}$$

When the input voltage has a low order ripple, the input voltage could be represented by the following equation.

$$V'_{dc} = V_{dc} + f(t) \times V_{dc} = V_{dc} \times (1 + f(t)) \tag{5}$$

To eliminate the DC-link ripple represented in (5), the modulation index needs to be compensated contrast to the input voltage as shown in (6).

$$m_{a\_new} = m_a \cdot \frac{1}{1+f(t)} \tag{6}$$

C. Capacitor Current and Capacitance Selection

To determine the capacitor current, the DC-link current must first be derived. As shown in Fig. 1(a), the DC-link current is represented by (7), where I_{dcx} ($x = a, b, c$) denotes the DC-link current for phase a, b, and c respectively.

$$I_{DC-link} = I_{DCa} + I_{DCb} + I_{DCc} \tag{7}$$

Taking phase a as an example, the switching signals of phase a are given in Fig. 1(b). Based on the phase current and the switching function, the DC-link current for phase a is expressed in (8). Refer to Fig. 1(a), the SW_i ($i = 1$ to 6) indicate the switching function of switches 1 to 6, and i_a is the phase current of phase a. Switch 2 is always operated at high frequency, and switch 1 is always operated at fundamental frequency.

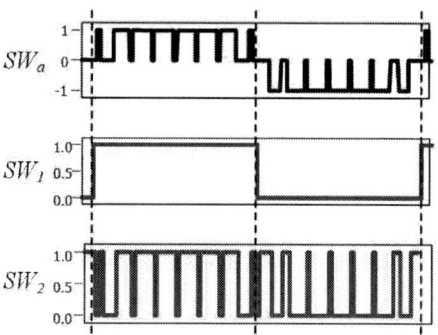

Fig. 3. Modulation signal waveforms for SW_a, SW_1, and SW_2.

$$\begin{aligned} I_{DCa} &= (SW_1 \cdot SW_2 + SW_1 \cdot SW_5 \cdot SW_6 \cdot SW_3) \cdot i_a \\ &= (SW_1 \cdot SW_2) \cdot i_a \end{aligned} \tag{8}$$

The high frequency switching signal is represented by (9), where SW_a is the modulation signal as shown in Fig. 3. With different modulation strategies, SW_a will be different. Considering the SPWM modulation is the fundamental modulation strategy, in this paper, the DC-link current for a three-phase ANPC with SPWM modulation is calculated as an example.

$$SW_2 = SW_a - SW_1 + 1 \tag{9}$$

Combine (8) and (9), I_{DCa} could be written as (10). The first six items in Fourier expansion of SW_1, and the first 5 items of Fourier expansion of SW_a are taken into consideration as shown in (11) and (12) respectively, where ω is the fundamental angular speed.

$$\begin{aligned} I_{DCa} &= [SW_1 \cdot (SW_a - SW_1 + 1)] \cdot i_a \\ &= SW_1(\omega t) \cdot SW_a(\omega t) \cdot i_a \end{aligned} \tag{10}$$

In (12), A_{h1} and A_{h2} refer to the amplitude of the high order harmonics, m is the modulation index, and m_f equals to f_{sw} / f_{ref}, which represents the frequency ratio of the inverter.

Assume that the ANPC inverter is operated on three-phase balanced conditions, then phase b and phase c have a phase shift of -120 and 120 degrees compared with phase a respectively.

$$SW_1(\omega t) = \frac{1}{2} + \frac{2}{\pi} \cdot \left[\sin(\omega t) + \frac{1}{3} \cdot \sin(3\omega t) + \frac{1}{5} \cdot \sin(5\omega t) + \frac{1}{7} \cdot \sin(7\omega t) + \frac{1}{9} \cdot \sin(9\omega t) + \frac{1}{11} \cdot \sin(11\omega t) \right] \tag{11}$$

$$SW_a(\omega t) = m_a \cdot \sin(\omega t) + A_{h1} \cdot \sin\left[(2m_f - 3) \cdot (\omega t) \right] + A_{h2} \cdot \sin\left[(2m_f - 1) \cdot (\omega t) \right] + A_{h2} \cdot \sin\left[(2m_f + 1) \cdot (\omega t) \right] +$$
$$A_{h1} \cdot \sin\left[(2m_f + 3) \cdot (\omega t) \right] \tag{12}$$

979-8-3315-1612-3/25 $31.00 © 2025 IEEE

$$I_{DC-link} = \frac{3\sqrt{2}}{4} I \cdot m \cdot \cos\varphi + \frac{2\sqrt{2}}{\pi} I \cdot m \cdot \left\{ \frac{3}{2a} \cdot \sin(a\omega t) \cdot \cos\varphi - \frac{3}{4b} \cdot \sin[(b+2)\omega t - \varphi] - \frac{3}{4c} \cdot \sin[(c-2)\omega t + \varphi] \right\} +$$

$$\begin{cases} h = 3n, & \left(\frac{3}{4b} \{\sin[(b+h-1)\omega t + \varphi] - \sin[(b-h-1)\omega t + \varphi] - \} + \frac{3}{4c} \{\sin[(c-h+1)\omega t - \varphi] - \right. \\ & \left. \sin[(c+h+1)\omega t - \varphi]\} \right) \cdot A_h \\ h = 3n+1, & \frac{3\sqrt{2}}{4} I \cdot A_h \cdot \cos[(h-1)\omega t + \varphi] + \left(\frac{3}{4a} \{\sin[(a+h-1)\omega t + \varphi] + \sin[(a-h+1)\omega t - \varphi]\} - \right. \\ & \left. \frac{3}{4b} \sin[(b+h+1)\omega t - \varphi] - \frac{3}{4c} \sin[(c-h-1)\omega t + \varphi] \right) \cdot A_h \\ h = 3n-1, & -\frac{3\sqrt{2}}{4} I \cdot A_h \cdot \cos[(h+1)\omega t - \varphi] + \left(\frac{3}{4a} \{\sin[(a+h+1)\omega t + \varphi] + \sin[(a-h-1)\omega t + \varphi]\} - \right. \\ & \left. \frac{3}{4b} \sin[(b-h+1)\omega t - \varphi] - \frac{3}{4c} \sin[(c+h-1)\omega t + \varphi] \right) \cdot A_h \end{cases} \quad (15)$$

$$a = 3, 9, 15, \ldots\ldots, b = 1, 7, 13, \ldots\ldots, c = 5, 11, 17, \ldots\ldots, h = 2m_f \pm 1, 2m_f \pm 3$$

The three phase current is represented in (13), where I is the RMS value of current, and φ represents the phase angle.

$$\begin{cases} i_a = \sqrt{2} \cdot I \cdot m \cdot \sin(\omega t - \varphi) \\ i_b = \sqrt{2} \cdot I \cdot m \cdot \sin(\omega t - 120 - \varphi) \\ i_b = \sqrt{2} \cdot I \cdot m \cdot \sin(\omega t + 120 - \varphi) \end{cases} \quad (13)$$

The equation for I_{DCb} and I_{DCc} can be expressed similarly to I_{DCa} as shown in (14).

$$\begin{cases} I_{DCb} = SW_1(\omega t + 120) \cdot SW_a(\omega t + 120) \cdot i_b \\ I_{DCc} = SW_1(\omega t - 120) \cdot SW_a(\omega t - 120) \cdot i_c \end{cases} \quad (14)$$

By integrating (10) – (14) into (7), the DC-link current can be represented by (15). The harmonics in SW_1 are all odd order harmonics, and the orders are indicated by a, b and c. The orders of the harmonics in SW_a is represented by h. When h is multiple times of three, the relative items will follow the equation after $h = 3n$; when h is an integer multiple of three plus one, the relative items will follow the equation after $h = 3n+1$; when h is an integer multiple of three minus one, the relative items will follow the equation after $h = 3n-1$. All of the harmonics in SW_a should be included, and for each harmonics of SW_a, all of the harmonics in SW_1 should be included according to (15).

The DC component in (15) is the first term $\frac{3\sqrt{2}}{4} I \cdot m \cdot \cos\varphi$. It is assumed that the DC source at the input side of the ANPC inverter supplies only DC current, while the capacitors on the DC-link absorb the current harmonics by charging and discharging. According to [11], the selection of the DC-link capacitor should satisfy the following criteria to make sure that the DC-link ripple remains within an acceptable range.

$$C \geq \frac{1}{\left(\frac{\Delta V_{max}}{2}\right)} \cdot \sum_{hc} \frac{I_{hc}}{2\pi f_{hc}} \quad (16)$$

In (16), ΔV_{max} indicates the amplitude of DC-link voltage ripple, hc represents the current harmonics as shown in (15), f_{hc} is the frequency of the corresponding current harmonic, and I_{hc} is the amplitude of it.

Another important factor in the selection of capacitor size is the equivalent series resistance (ESR) of the capacitor. A capacitor can be modeled as an ideal capacitor in series with an ESR component [12]. Consequently, the voltage ripple includes both the ripple caused by the ideal capacitor and the ripple caused by the ESR, as shown in equation (17), where ΔV, ΔV_C, and ΔV_{ESR} represent the total voltage ripple, the ripple across the ideal capacitor, and the ripple across the ESR, respectively.

$$\Delta V \leq \Delta V_C + \Delta V_{ESR} = \Delta V_C + R_{ESR} \cdot \sum_{hc} I_{hc} \quad (17)$$

In some cases, the impact of ESR on the voltage ripple can exceed that of the ideal capacitor. To reduce the voltage ripple caused by ESR, a low-ESR capacitor can be selected.

III. SIMULATION AND EXPERIMENT RESULTS

A PSIM simulation for a three-phase ANPC Inverter with the proposed DC-link ripple rejection algorithm is discussed in this section. The topology of the ANPC inverter is given in Fig. 1 (a). The basic specifications are shown in Table I. With a 600 Hz DC-link voltage ripple and a fundamental frequency of 1200 Hz in the ANPC inverter, low-frequency harmonics at 1200 ± 600, specifically, 600 Hz and 1800 Hz, are produced on the output side.

TABLE I.　　SPECIFICATIONS FOR ANPC INVERTER SHOWN IN FIG. 1.

DC-link voltage constant / V	1000V
Ripple amplitude / V	100V
DC-link capacitance / μF	96μF
Ripple frequency / Hz	600Hz
ANPC fundamental frequency / Hz	1200 Hz
High switching frequency / Hz	28800 Hz

Fig. 4 gives the simulation results of DC-link voltage and line-to-middle voltage of phase a in (a) and (b) respectively. There is a 10% ripple on the DC-link voltage.

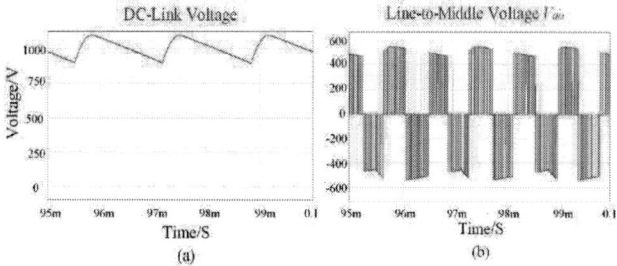

Fig. 4. When the DC input includes a 10% ripple, (a) DC-link voltage, (b) line-to-middle voltage of phase *a*.

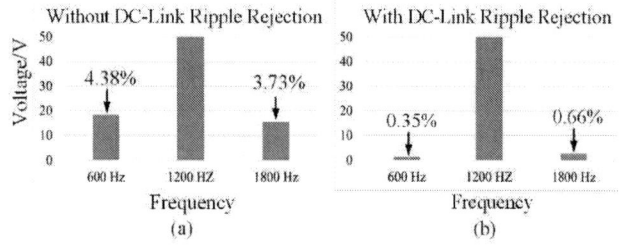

Fig. 5. FFT analysis of V_{ao} (refer to Fig. 1) based on simulation results, (a) without ripple rejection method (notice the sidebands around the fundamental because of the DC-link voltage ripple), (b) with the proposed feedforward ripple rejection algorithm.

Fig. 5 (a) and (b) shows the FFT analysis of line-to-middle voltage without and with DC ripple rejection technique respectively. The fundamental components are 420V RMS. The 600 Hz harmonic is reduced from 4.38% of the fundamental voltage to 0.35%, and the 1800 Hz harmonic is reduced from 3.73% to 0.66% with the proposed DC-link voltage rejection algorithm.

An experimental prototype employing SiC half bridge modules CAB006A12GM3 has been built and tested as shown in Fig. 6 (a). The voltage source in Fig. 6(a) generates an AC sinusoidal voltage at 300 Hz, which is firstly rectified by a full-bridge diode rectifier. This rectified voltage is then supplied to the ANPC inverter. The DC-link capacitor of the ANPC inverter also works as the filter of the diode rectifier. With two groups of 96 µF capacitors, the rectified voltage is smoothed, resulting in the DC-link voltage shown in Fig. 6(c), which

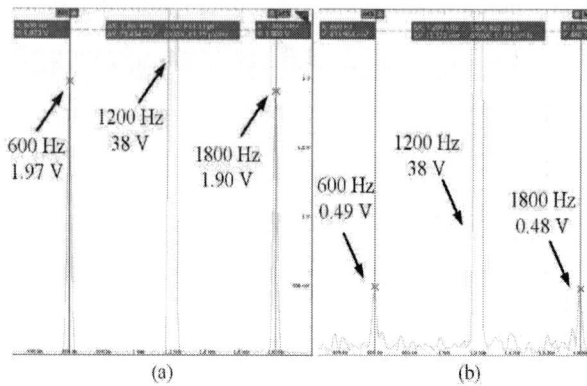

Fig. 7. FFT analysis of V_{ao} (refer to Fig. 1) based on experimental results (a) without ripple rejection, (b) with ripple rejection which shows 75% reduction of sidebands.

includes a 600 Hz component. The ANPC prototype is controlled by a DSP, and the load of the ANPC inverter is a 20 ohms resistive load.

According to Fig. 6(b), there are 3 half bridge modules in total. All of them are mounted on the cold plate, and the cold plate is placed on a heat sink. During high-power testing, the cold plate is connected to a cooling machine that circulates water in and out to maintain a low temperature. For lower-power tests, a cooling fan can be used to reduce the temperature instead.

The input voltage has a DC component of 160V with a ripple of 10.508% as shown in Fig. 5 (c), other parameters are the same as Table I. Fig. 5 (c) also shows the output voltage of the single-phase ANPC inverter.

Fig. 7 shows the FFT analysis of line-to-middle voltage without and with DC-link ripple rejection technique in (a) and (b) respectively. The fundamental components are 38V RMS. The 600 Hz harmonic is reduced from 4.38% of the fundamental to 1.29%, and the 1800 Hz harmonic is reduced from 4.2% to 1.26%. According to this result, it can be concluded that the harmonics at low frequency caused by the DC-link voltage ripple are reduced by introducing the proposed ripple rejection technique.

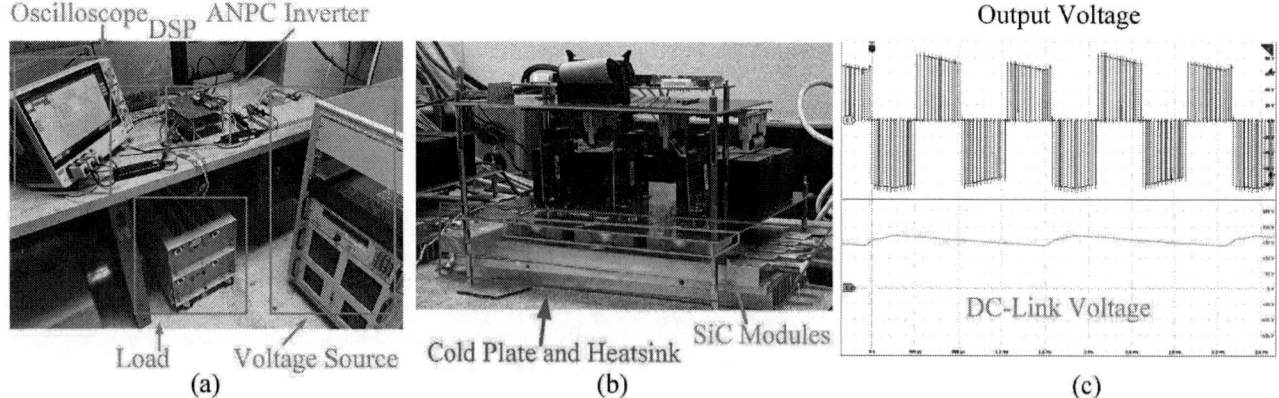

Fig. 6. (a) Experimental setup for a single-phase ANPC inverter as shown in Fig. 1, (b) Thermal management system, (c) DC-link and output voltage.

979-8-3315-1612-3/25 $31.00 © 2025 IEEE   1463

IV. CONCLUSION

In this paper a new technique to compensate for the DC-link voltage ripple in ANPC inverters employing a feedforward strategy has been implemented with a hybrid SVPWM strategy. Real-time variation of the DC-link ripple voltage is monitored via a voltage sensor, and the control logic reference is adjusted based on the DC-link voltage measurement. The proposed technique is applied to an ANPC inverter system with a 10% DC-link ripple. Simulation and experimental results demonstrate at least 75% reduction in low frequency harmonics generated as a result of DC-link voltage ripple. Employing the proposed strategy has been shown to be effective in mitigating real-time variations of the DC-link voltage ripple, which contributes to the reduction of DC-link capacitor components, and thereby increases the overall power density of the ANPC hardware. The formula for the DC-link current and guidance on selecting the appropriate capacitor size based on the capacitor current are also presented.

ACKNOWLEDGMENT

This material is based upon work supported by the Department of Energy under Award Number DE-AR0001356.

REFERENCES

[1] T. Bruckner, S. Bernet and H. Guldner, "The active NPC converter and its loss-balancing control," in IEEE Transactions on Industrial Electronics, vol. 52, no. 3, pp. 855-868, June 2005.

[2] Q. -X. Guan et al., "An Extremely High Efficient Three-Level Active Neutral-Point-Clamped Converter Comprising SiC and Si Hybrid Power Stages," in IEEE Transactions on Power Electronics, vol. 33, no. 10, pp. 8341-8352, Oct. 2018.

[3] C. Li et al., "Space Vector Modulation for SiC and Si Hybrid ANPC Converter in Medium-Voltage High-Speed Drive System," in IEEE Transactions on Power Electronics, vol. 35, no. 4, pp. 3390-3401, April 2020.

[4] A. Lewicki, Z. Krzeminski and H. Abu-Rub, "Space-Vector Pulsewidth Modulation for Three-Level NPC Converter With the Neutral Point Voltage Control," in IEEE Transactions on Industrial Electronics, vol. 58, no. 11, pp. 5076-5086, Nov. 2011.

[5] G. Narayanan, D. Zhao, H. K. Krishnamurthy, R. Ayyanar and V. T. Ranganathan, "Space Vector Based Hybrid PWM Techniques for Reduced Current Ripple," in IEEE Transactions on Industrial Electronics, vol. 55, no. 4, pp. 1614-1627, April 2008.

[6] J. Xu, Y. Wang, Y. Lu, J. Han and H. Tang, "DC-Link Capacitors RMS Current Reduction PWM Method Based High Frequency SiC Three-Phase Inverter," 2018 1st Workshop on Wide Bandgap Power Devices and Applications in Asia (WiPDA Asia), Xi'an, China, 2018, pp. 61-65.

[7] Z. Zhao, F. Diao, Y. Wu, Z. Wang and Y. Zhao, "DC-Link Capacitor Current Modeling and Analysis for Three-Level Voltage Source Inverters," 2021 IEEE Applied Power Electronics Conference and Exposition (APEC), Phoenix, AZ, USA, 2021, pp. 2434-2439.

[8] Enjeti, P.N. and Shireen, W, "A new technique to reject DC-link voltage ripple for inverters operating on programmed PWM waveforms", IEEE Transactions on Power Electronics, vol. 7, no. 1, pp. 171–180, 1992.

[9] S. A. Khan, Y. Guo, Y. P. Siwakoti, D. D.-C. Lu, and J. Zhu, "A disturbance rejection-based control strategy for five-level t-type hybrid power converters with ripple voltage estimation capability," IEEE Transactions on Industrial Electronics, vol. 67, no. 9, pp. 7364–7374, 2020.

[10] H. Ouyang, K. Zhang, P. Zhang, Y. Kang and J. Xiong, "Repetitive Compensation of Fluctuating DC Link Voltage for Railway Traction Drives," in IEEE Transactions on Power Electronics, vol. 26, no. 8, pp. 2160-2171, Aug. 2011.

[11] Suleiman M. Sharkh; Mohammad A. Abu-Sara; Georgios I. Orfanoudakis; Babar Hussain, "DC-Link Capacitor Current and Sizing in NPC and CHB Inverters," in Power Electronic Converters for Microgrids, IEEE, 2014, pp.29-50.

[12] Daniel W. Hart, Power electronics, New York : McGraw-Hill, 2011, pp. 206-207.

A Cascaded Multilevel Inverter System with Hot-Swapping and Fault Isolation Capability for Improved Resiliency

Uthandi Selvarasu
Dept. of Electrical & Computer Engineering
Northeastern University
Boston, USA
selvarasu.u@northeastern.edu

Vikram Roy Chowdhury
National Renewable Energy Laboratory
Golden, CO, USA
vikram.roychowdhury@nrel.gov

Shumeng Wang
Dept. of Electrical & Computer Engineering
Northeastern University
Boston, USA
wang.shum@northeastern.edu

Jinli zhu
Dept. of Electrical & Computer Engineering
University of Pittsburgh
Pittsburgh, USA
jinli.zhu@pitt.edu

Mahshid Amirabadi
Dept. of Electrical & Computer Engineering
Northeastern University
Boston, USA
m.amirabadi@northeastern.edu

Yuan Li
Dept. of Electrical & Computer Engineering
University of Pittsburgh
Pittsburgh, USA
yuan.li@pitt.edu

Brad Lehman
Dept. of Electrical & Computer Engineering
Northeastern University
Boston, USA
lehman@ece.neu.edu

Abstract—The cascaded multilevel inverter (CMI) system is a promising solution for reliable inverter-based resource (IBR) applications due to its inherent modular redundancy. In the event of a fault, bypassing and isolating faulty sub-modules (SMs) within the CMI mitigates total system failure and enhances fault resiliency. However, hot-swapping a SM without interrupting the CMI presents implementation challenges. In this paper, a three phase CMI system is proposed for improved resiliency by hot-swapping the faulty SM without interrupting the system operation. Controllers were developed that permit hot-swapping of both grid forming and grid following CMIs. A power electronic module called AC smart port, is proposed and can successfully isolate the faulty SM from the system. A small-scale hardware prototype of a single-phase CMI with three H-bridge sub-modules has been successfully implemented for experimental validation.

Keywords—cascaded multilevel inverter (CMI), multilevel inverters, MMC, hot-swapping, sub-module, sub-module bypass.

I. INTRODUCTION

Renewable energy resources (RES) are becoming prevalent in the modern grid system as it moves towards a carbon-free goal. Especially, the fast-responding nature of inverter-based resources (IBRs), combined with their versatility to offer multiple services to the grid through battery energy storage systems (BESS), makes them a financially viable option compared to traditional power generation methods [1]-[3]. The cascaded multilevel inverter (CMI) is a promising choice for these IBRs due to its modularity, high efficiency, and fault tolerance [4]. Its fault resilience is particularly crucial for reliable operation. To enhance the CMI's resiliency and ensure uninterrupted operation, redundant sub-modules (SMs) are often integrated into the system. In the event of a fault, the faulty SM is bypassed, allowing the rest of the system to continue operating without the defective SM. Various methods are used for this SM bypass procedure. In [5], IGBT switches in the faulty SM are

used for bypass. In [6], a dedicated bus bar mounted hot-swap switch is used to support the current at higher power level and it also utilized the faulty SM IGBT switches for smooth bypass transients. In [7], the bypass operation is solely implemented by a dedicated device. Even though the SM bypassing allows the system's uninterrupted operation, it does not eliminate the possibility of the fault disrupting the remaining system's operation which requires SM isolation from the CMI. Further, some bypass approaches have controllers designed for grid following inverters and not grid forming inverters [5] – [7]. However, the CMI controller must seamlessly switch between synchronizing with the grid in grid-following (GFL) mode and independently regulating voltage and frequency in grid-forming mode (GFM), requiring highly robust control algorithms, that regulate well during hot-swap.

In summary, this paper presents the following research contributions

- A three phase CMI system with improved reliability and fault isolation is proposed with localized SM level control for fast fault detection and isolation. The CMI fault resilience operation is simulated under grid forming as well as grid following conditions for control architecture verification.

- CMI controllers are designed and simulated for both grid-forming and grid-following mode, based on Lyapunov theory, that maintains stability during hot-swap.

- A power electronic module called AC smart port for SM bypass and isolation is prototyped. Its operation is validated with a low power single-phase CMI with three H-bridge SMs. A two-step SM bypass and isolation protocol is proposed for seamless hot-swapping of SM without interrupting the CMI operation.

- A small-scale hardware (proof-of-concept) single-phase CMI inverter is tested to validate the hot-swap bypass

---

This research was supported by Department of Energy, RACER: DE-EE0010427

procedure. The CMI utilizes a Proportional Resonant (PR) current controller.

Future research will experimentally validate the Lyapunov controllers.

II. SYSTEM DESCRIPTION

The concept of the proposed three phase CMI with multiple SMs is shown in Fig. 1. Each SM consists of DC and AC smart ports, DC-DC stage, H-bridge inverter and local microcontroller (MCU). Here DC and AC side smart ports are responsible for connecting and isolating the source on the DC side and load on the AC side respectively. The complete CMI is controlled by a master controller which is responsible for the output voltage and current control of the CMI system.

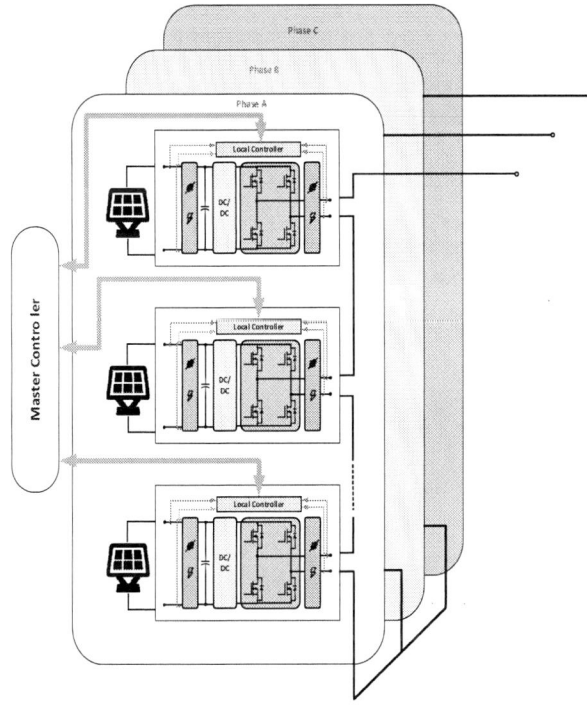

Figure 1: A three-phase cascaded multilevel inverter system for

Each SM has a local controller which is responsible for controlling the H-bridge inverter and DC-DC stage based on the commands from master controller; it is also responsible for detecting any fault in the SM based on the local parameters of the SM. In case of any faults or maintenance the local controller follows hot-swapping protocol and seamlessly bypasses the SM from the CMI and communicates it to the master controller;

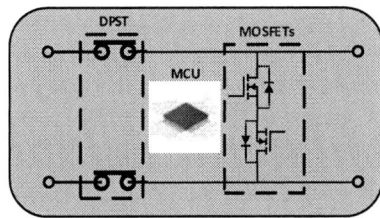

Figure 2: AC smart port

subsequently the master controller increases the output power from the other SMs to match the absence of bypassed SM.

The successful uninterrupted operation of the CMI with reduced number of SMs during and after the hot-swapping procedure is achieved by the AC/DC smart port. The DC side smart port is used to isolate the PV panel from the power conversion stage and AC side smart port is used for isolating the SM from the rest of the system and to provide a bypassed current path at the output side. The detailed circuit diagram of the AC side smart port is shown in Fig. 2. The smart port consists of a double pole single throw (DPST) relay and a pair of anti-series MOSFETs. The MOSFETs are preferred over TRIACs for better turn on/off control of the bypass switch and to avoid the false turn on due to high dv/dt [7].

This paper particularly focuses on the fault tolerant operation of the CMI system which is verified with simulation results and the small-scale hardware implementation of single-phase CMI with three SMs for experimental validation of hot-swapping ability using AC smart port. In the following section, the proposed 3-phase system's control architecture is first explained, with simulation results presented.

III. CONTROL ARCHITECTURE AND SIMULATION RESULTS

The simplified block diagram schematic of the closed loop control architecture utilized in this paper in unison with the hot-swap architecture is presented in Fig. 3 where $SM_{1,2,3}$ are SMs which only consist of a full bridge converter.

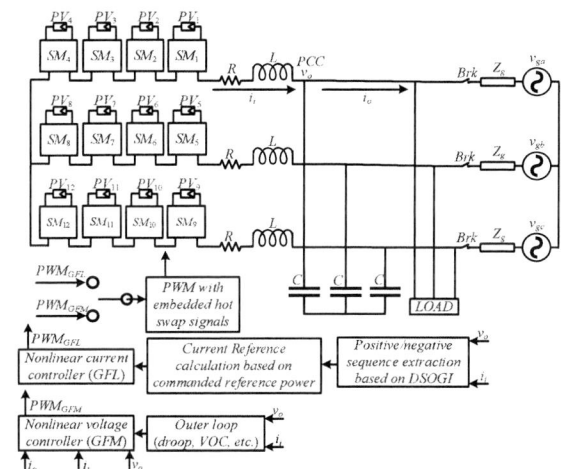

Figure 3: Simplified block diagram of the overall cascaded multilevel inverter system with control structure

In the grid-forming (GFM) mode of operation, the outer loop utilizes control strategies such as droop control or virtual oscillator control (VOC) to generate the necessary voltage references for the inner loop, based on a Lyapunov energy function architecture [8] – [9]. The dual second order generalized integrator (DSOGI) architecture [10] is utilized to extract the positive and negative sequence components of the PCC voltages. This architecture ensures system stability and effective voltage regulation. In the grid-following (GFL) mode, the current references are derived according to the methodologies presented in references [10], [11]. These references are formulated in the stationary two-phase $\alpha\beta$

domain, with the primary control objective being to ensure non-oscillatory power delivery at the point of common coupling (PCC) even under unbalanced grid voltage conditions. Both the GFM and GFL control architectures generate the modulating signals required for the system, which are then processed using pulse width modulation (*PWM*) to ensure precise signal modulation and achieve the desired control objectives. This approach facilitates seamless hot swapping and maintains the integrity of the control process during any fault condition.

A. Controller Derivation

The extraction of positive and negative sequence components of the PCC voltages are based on stationary αβ to ensure reduced number of equations as well as lesser computational burden on the microcontroller. Using [11], the power (steady state and oscillating components) is related to the generated current references as presented in (1)

$$
\begin{bmatrix} P_o \\ Q_o \\ P_{s2} \\ P_{c2} \end{bmatrix} = \frac{3}{2} \begin{bmatrix} v_{g\alpha}^p & v_{g\beta}^p & v_{g\alpha}^n & v_{g\beta}^n \\ v_{g\beta}^p & -v_{g\alpha}^p & v_{g\beta}^n & -v_{g\alpha}^n \\ v_{g\beta}^n & -v_{g\alpha}^n & -v_{g\beta}^p & v_{g\alpha}^p \\ v_{g\alpha}^n & v_{g\beta}^n & v_{g\alpha}^p & v_{g\beta}^p \end{bmatrix} \begin{bmatrix} i_{i\alpha}^{pref} \\ i_{i\beta}^{pref} \\ i_{i\alpha}^{nref} \\ i_{i\beta}^{nref} \end{bmatrix} \quad (1)
$$

where $\begin{bmatrix} P_o & Q_o & P_{s2} & P_{c2} \end{bmatrix}^T$ are the active, reactive and oscillating components of the powers during an unbalance on the PCC voltage, superscript p denotes positive sequence and n denotes negative sequence components and $D = v_{g\alpha}^{p2} + v_{g\beta}^{p2} - v_{g\alpha}^{n2} - v_{g\beta}^{n2}, D \neq 0$ as per calculation. Therefore, inverse of (1) with zero reactive power reference yields the respective current references as presented in (2).

$$
\begin{bmatrix} i_{i\alpha}^{pref} \\ i_{i\beta}^{pref} \\ i_{i\alpha}^{nref} \\ i_{i\beta}^{nref} \end{bmatrix} = \begin{bmatrix} v_{g\alpha}^p & v_{g\beta}^p & v_{g\alpha}^n & v_{g\beta}^n \\ v_{g\beta}^p & -v_{g\alpha}^p & v_{g\beta}^n & -v_{g\alpha}^n \\ v_{g\beta}^n & -v_{g\alpha}^n & -v_{g\beta}^p & v_{g\alpha}^p \\ v_{g\alpha}^n & v_{g\beta}^n & v_{g\alpha}^p & v_{g\beta}^p \end{bmatrix}^{-1} \begin{bmatrix} \frac{2}{3}P_o \\ 0 \\ 0 \\ 0 \end{bmatrix} = \frac{2P_o}{3D} \begin{bmatrix} v_{g\alpha}^p \\ v_{g\beta}^p \\ -v_{g\alpha}^n \\ -v_{g\beta}^n \end{bmatrix} \quad (2)
$$

$$
i_{i\alpha}^{tot} = i_{i\alpha}^{pref} + i_{i\alpha}^{nref}, \, i_{i\beta}^{tot} = i_{i\beta}^{pref} + i_{i\beta}^{nref}
$$

The objective of the above current reference generator is to ensure zero reactive power as well as nullifying the oscillating components of active and reactive powers denoted as P_{S2} and P_{C2}. The generated references are algebraically added instantaneously, and the current references denoted with superscript *'tot'* are generated for the inner loop controller. For (2) $P^{ref} = P_o, Q^{ref} = 0$, which is chosen depending on the converter rating and operating at minimum current (unity power factor). Next a brief discussion of implementation of the Lyapunov energy function-based controller for the inner current controller is presented. The per-phase equivalent simplified circuit diagram schematic for a converter connected to the grid (PCC/load) during either GFL or GFM mode is presented in Fig. 4.

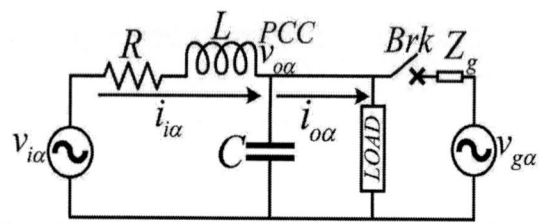

Figure 4: Per-phase equivalent circuit of a converter connected to PCC (GFL) or load (GFM)

In Fig. 4, the left side represents the per-phase equivalent for the cascaded multilevel inverter, RL are the parameters of the interfacing reactor, C is the PCC capacitor, *Brk* is the breaker to connect to the grid and Z_g is the equivalent grid impedance. Therefore, to generate the inner current control loops based on Lyapunov energy function, the error dynamics are computed with the objective of controlling the inverter side currents for GFL mode of operation. The control objective for GFM mode of operation is to ensure balanced sinusoidal voltages at the PCC despite the nature of the load balanced, unbalanced, harmonics, etc. More details of the step-by-step development of the Lyapunov energy function-based control architecture can be found in [12]-[14]. The governing dynamic equations to generate the control law(s) for the GFL inverter are (3):

$$
L\frac{d\Delta i_{i\alpha}}{dt} = -R\Delta i_{i\alpha} + \Delta m_\alpha \frac{V_{dc}}{2} \, (error \quad dynamics)
$$

$$
U = \frac{1}{2}L\Delta i_{i\alpha}^2 \, (energy \quad function)
$$

$$
\dot{U} = -R\Delta i_{i\alpha}^2 + \Delta m_\alpha \frac{V_{dc}}{2}\Delta i_{i\alpha} \, (time \quad derivative)
$$

$$
m_\alpha = m_\alpha^{ref} + \Delta m_\alpha = \begin{bmatrix} \frac{2}{V_{dc}}\left(L\frac{di_{i\alpha}^{tot}}{dt} + Ri_{i\alpha}^{tot} + v_{o\alpha} \right)(reference \quad dynamics) \\ + \frac{2}{V_{dc}}R_o\left(i_{i\alpha}^{tot} - i_{i\alpha} \right)(user \quad defined) \end{bmatrix} \quad (3)
$$

For the GFM inverter, the control laws follow (4):

$$
L\frac{d\Delta i_{i\alpha}}{dt} = -R\Delta i_{i\alpha} + \Delta m_\alpha \frac{V_{dc}}{2} - \Delta v_{o\alpha} \, (error \quad dynamics)
$$

$$
C\frac{d\Delta v_{o\alpha}}{dt} = \Delta i_{i\alpha} \, (error \quad dynamics)
$$

$$
V = \frac{1}{2}L\Delta i_{i\alpha}^2 + \frac{1}{2}C\Delta v_{o\alpha}^2 \, (energy \quad function)
$$

$$
\dot{V} = -R\Delta i_{i\alpha}^2 + \Delta m_\alpha \frac{V_{dc}}{2}\Delta i_{i\alpha} \, (time \quad derivative)
$$

$$
m_\alpha = m_\alpha^{ref} + \Delta m_\alpha = \begin{bmatrix} \frac{2}{V_{dc}}\left(L\frac{di_{i\alpha}^{ref}}{dt} + Ri_{i\alpha}^{ref} + v_{o\alpha}^{droop} \right)(reference \quad dynamics) \\ + \frac{2}{V_{dc}}K_o\left(i_{i\alpha}^{ref} - i_{i\alpha} \right)(user \quad defined) \end{bmatrix} \quad (4)
$$

where $i_{i\alpha}^{tot}$ and $i_{i\alpha}^{ref}$ are the references for the inverter side currents based on (2) or droop respectively, R_0 and K_0 are the Lyapunov energy function controller gains defined by the user, chosen to accomplish required bandwidth for the overall system. Similar control law is also implemented for the β axis and the overall control system is implemented. In Fig. 4, during GFM operation as described in (4), the dynamics of the PCC-side capacitor must be considered, as the primary control objective in this mode is to maintain a low-THD three-phase voltage waveform under unbalanced or nonlinear load conditions.

Conversely, in GFL mode, the main control objective shifts to ensuring current reference tracking based on the outer loop calculations defined in (2). As a result, the PCC capacitor dynamics are neglected in GFL mode due to the presence of a stronger grid. Thus, without loss of generality, it can be stated that (4) simplifies to (3) during the transition from GFM to GFL. Consequently, the Lyapunov control law generated for GFL mode becomes a subset of that for GFM mode. This relationship simplifies implementation and enhances computational efficiency, as only one state—the capacitor dynamics—needs to be disregarded. Using this approach, the overall system was simulated in MATLAB/Simulink. The next section presents key case study results to demonstrate the effectiveness of the proposed method.

B. Simulation Results

The Lyapunov controllers are verified with simulation, and experimental validation will occur in future research. The parameters of the overall cascaded multilevel converter are given in Table. I. Using these parameters, the results showing operation during GFL mode are first presented; the result showing the step change in current is presented in Fig. 5 under balanced *PCC* voltage condition. It is observed from this result that the proposed control architecture can achieve superior dynamic performance in terms of negligible overshoot and good reference tracking using the proposed architecture. A comparative result with traditional PI based architecture for GFL mode can be found in [14]. The result showing the step change in current with *PCC* voltage changing from balanced to unbalanced is presented in Fig.6.

TABLE I.   SIMULATION PARAMETERS OF 3 PHASE CMI

Parameter	Value
V_{DC}	75V
R_{Load}	1.0 Ω
L_{Load}	5.0 mH
C	20 μF
Z_g	$(0.05 + j0.15)$ Ω
R_0	120
K_0	70

The findings highlight the effectiveness of the proposed control architecture in proportionally commanding unbalanced grid current references, ensuring oscillation-free active power (i.e., a stable DC bus voltage). This approach reduces stress on the DC capacitors and facilitates uninterrupted MPPT.

Figure 7 illustrates the system's performance during a step change caused by an uncontrolled diode rectifier-based nonlinear load in GFM mode. It demonstrates the ability to achieve low-THD PCC voltage under heavy nonlinear loading, with a step response exhibiting negligible overshoot—underscoring the superiority of the proposed approach. Lastly, Figure 8 showcases the behavior of the three-phase converter pole voltages and load current during a submodule fault and subsequent hot-swap operation in GFM mode. The seamless bypass of the submodule and recovery of control objectives

demonstrate the robustness and efficiency of the proposed control architecture.

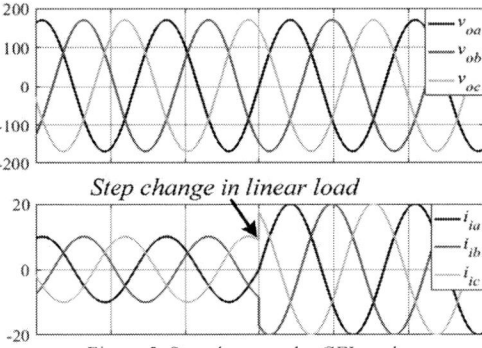

Figure 5: Step change under *GFL* mode

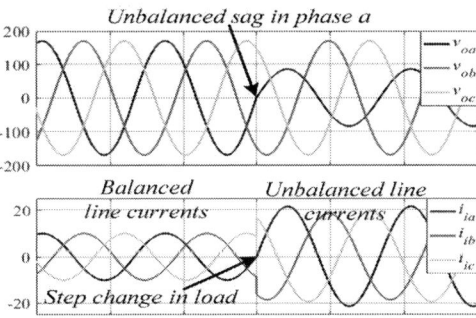

Figure 6: Line currents with asymmetrical grid voltage sag

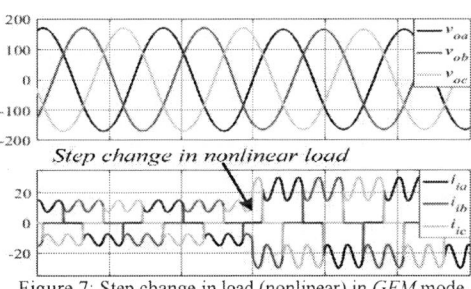

Figure 7: Step change in load (nonlinear) in *GFM* mode

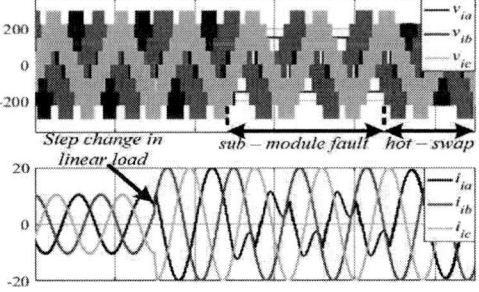

Figure 8: Hot − swap operation *GFL* mode

IV. HOT-SWAPPING PROTOCOL AND EXPERIMENTAL VALIDATION

The fault tolerant operation of a 3-phase CMI proposed in this paper primarily depends on the AC smart port. A low power (proof-of-concept) prototype of the AC smart port is developed and its operation is verified with a single phase CMI hardware setup that consists of three SMs. A single-phase CMI with 3 sets of H-bridge and AC smart port is shown in Fig. 9(a) during nominal operation. To evaluate the transient behavior of the CMI system during hot-swapping, the hot-swapping protocol is tested both under open-loop control and with closed-loop current control using a simple proportional-resonant (PR) approach. The AC smart port consists of a double pole single throw (DPST) relay and a pair of anti-series MOSFETs. This connects the H-bridge inverter to the AC side of the CMI. Each H-bridge inverter is connected to the AC side of the CMI through the AC smart port.

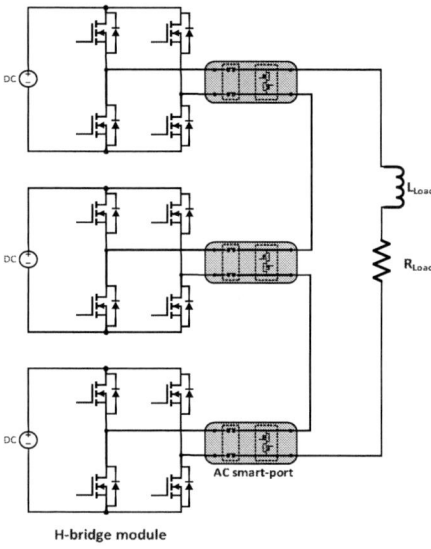

Figure 9(a): Single-phase CMI
(Nominal operation)

During the hot-swapping process when one of the SMs needs to be bypassed, the DPST relay opens and isolates the SM from the rest of the system and the anti-series MOSFETs are closed to establish a bypassed connection. However, the electromagnetic (EM) DPST relay has higher turn on/off time (10-30 ms) [15] and [16]. This causes intermittency in the CMI output voltage and current for 1-2 fundamental cycles during the hot-swapping process. To eliminate this a two-step hot-swapping protocol is followed in this paper. During the bypass process, first the higher side MOSFETs in the corresponding SM are turned off and the lower side MOSFETs are turned on. Simultaneously, the anti-series MOSFETs in the corresponding AC smart port are also turned on, and the DPST relay is turned off. Due to the slow transient of the EM DPST relay, it is not turned off for 1-2 fundamental cycles. During this time two parallel bypass paths are established for the SM which is shown in Fig. 9(b) where the bottom SM is assumed to be the faulty one. In the second step, once the DPST relay is turned off, the SM is isolated from the rest of the system, and all the MOSFETs

in the SM are deactivated. Now, the SM is ready to be removed. This is shown in Fig. 9(c).

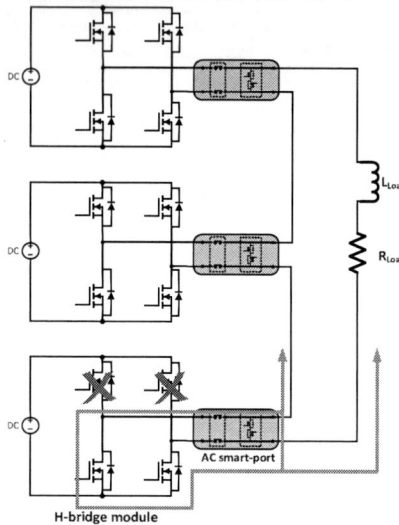

Figure 9(b): Single-phase CMI
(Hot-swap step-1)

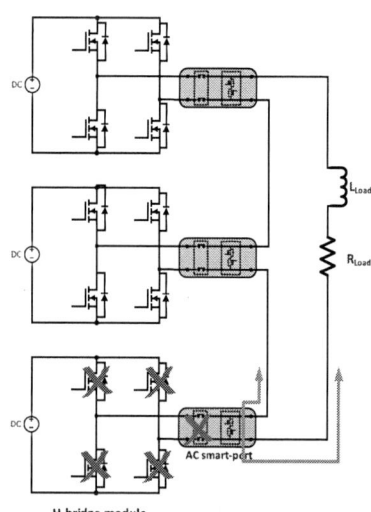

Figure 9(c): Single-phase CMI
(Hot-swap step-2)

Similarly, for reconnecting the SM into the CMI system, the 2-step hot-swapping protocol followed in the reverse order. A low power single-phase CMI hardware with 3 H-bridge SMs testbench setup is created for hot-swap experimental validation which is shown in Fig. 10. The prototype AC smart port is connected to one of the SMs in the single-phase CMI, which is assumed to be the faulty SM. The parameters of the testbench setup are given in Table. II. The single phase CMI pole voltage and current during the bypass and reconnection actions of the hot-swapping are shown in Fig. 11 and 12 respectively. The hot-swapping process first tested with open loop control to verify the change in CMI output voltage levels and current amplitude due to a SM bypass.

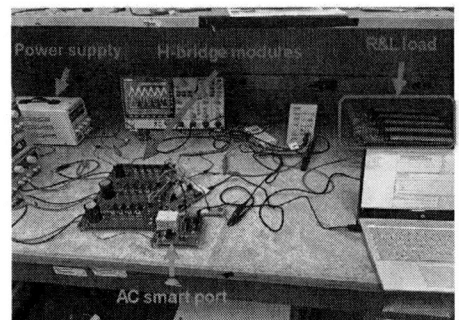

Figure 10: Low power single-phase CMI hot-swap test bench setup

TABLE II. PARAMETERS OF THE HARDWARE TESTBENCH

Parameter	Value
SM DC source voltage	10 V
Modulation index	0.8
Number of Modules	3
Switching frequency	10 kHz
Fundamental frequency	60 Hz
Modulation technique	PSPWM
Load (R&L)	25 Ohms & 2.5 mH

In Fig. 11, before bypassing the SM output voltage has 7 levels. The SM is bypassed when hot-swap step-1 is implemented, and the output voltage reduced to 5 levels with lower amplitude. Consequently, the output current amplitude also reduced after the bypass action. However, there is no intermittency in the output voltage or current. Once, the DPST relay is turned off all the PWM signals to the SM are turned off (hot-swap step-2) and the SM is safe to remove. Similarly, the reconnection of the SM is shown in Fig. 12. The same hot-swap protocol followed in reverse order.

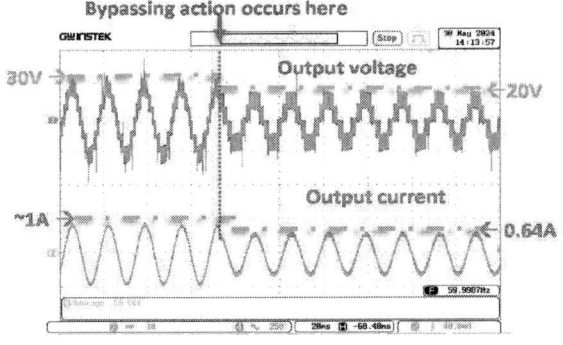

Figure 11: Sub module bypassing process

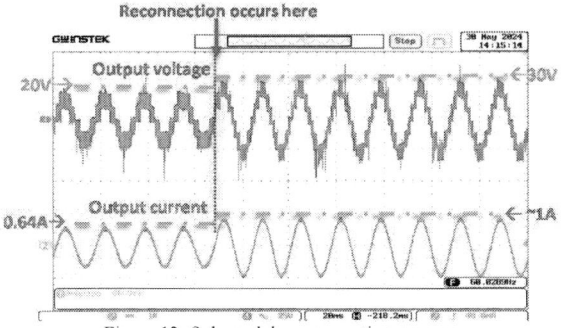

Figure 12: Sub module reconnection process

These open loop hot-swapping results shown in Figure 11 and 12 verify the uninterrupted operation of the single phase CMI. However, in a standalone mode where the CMI supplies constant power to an RL passive load, it is necessary to verify the stability of output current and voltage during hot-swap operation. In order to verify that a simple proportional resonant (PR) based current controller implemented to the single phase CMI system shown in Fig. 9(a). The block diagram of the implemented PR controller is shown in Fig. 13 where C(s) and G_p(s) are the transfer functions of the non-ideal PR controller and plant (R&L load) respectively which are given in equation (5) and (6)

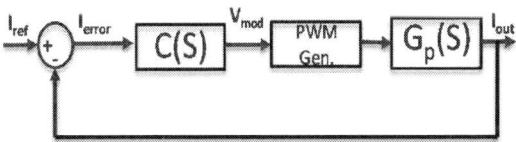

Figure 13: PR controller based closed loop current control block diagram

$$C(s) = K_P + K_R \frac{\omega_c * s}{s^2 + 2 * \omega_c * s + \omega_0^2} \qquad (5)$$

$$G_p(s) = \frac{1}{R + s * L} \qquad (6)$$

where K_P is the proportional gain, K_R is the resonant gain, ω_c is the cut off frequency and ω_0 is the natural frequency of the PR controller. The R and L in equation (6) are the load resistance and inductance values. The controller parameters are tuned for a stable and fast control based on [17]-[20]. The closed loop system parameters are given in Table. III.

TABLE III. CLOSED LOOP SYSTEM PARAMETERS

Parameter	Value
SM DC source voltage	10 V
Load (R&L)	10 Ohms & 2.5 mH
K_P	7.85
K_R	6283
W_c	0.5 radians
W_0	377 radians
I_{ref} (peak)	1.8 A

For the given parameters of Table. III, the closed loop control implemented using the single phase low power protype shown in Fig. 10. The output voltage and current of the single phase CMI protype with closed loop control is shown in Fig. 14.

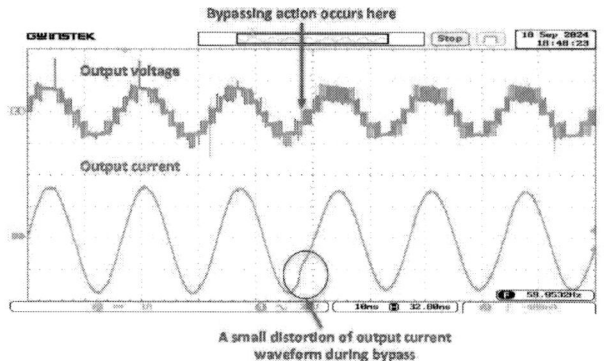

Figure 14: Sub module bypassing process with closed loop control

The output current and voltage stability during bypass action is verified by implementing the bypass action during nominal operation of the single phase CMI with 3 SMs. In Fig. 14, a small distortion can be seen in the output current during the moment of bypass action. However, the PR controller was able to withstand the disturbance during the bypass action and the current amplitude is also maintained at the desired value with one SM bypassed. Similarly, the reconnection of the SM is also implemented with the closed loop control which is shown in Fig. 15.

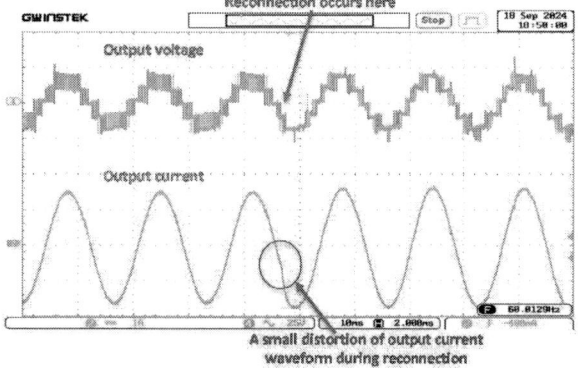

Figure 15: Sub module reconnection process with closed loop control

Except for a small distortion during the reconnection moment, the controller is able to maintain the desired output current and voltage waveforms.

V. CONCLUSION AND FUTURE WORK

The fault resilient 3-phase CMI is simulated with Lyapunov controllers, and the simulation results which show the controller effectiveness on handling non-linear disturbances and ensuring the system resiliency are presented. The key aspect of the CMI which is fault isolation and continued operation of the CMI by successful bypass of the faulty sub-module is achieved in real time by a hardware called AC smart port. The presented hot-swapping protocol and AC smart port not only bypasses the sub module but also it isolates the faulty sub module from the rest of the system without manual assistance and is valid for operation in either GFL or GFM. The presented hot-swapping protocol is also implemented with a PR based closed loop control to verify the controller ability to maintain a stable operation under the non-linear disturbances of the hot-swap transients. This differentiates the research contribution of this paper from the previous works [5] – [7]. This research is further aims to advance towards the development of fault tolerant three-phase cascaded multilevel inverter system with improved resiliency.

ACKNOWLEDGMENT

This material is based upon work supported by the U.S. Department of Energy (DOE) Solar Energy Technologies Office (SETO) Renewables Advancing Community Energy Resilience (RACER) under award number: DE-EE0010427.

ABRIDGED LEGAL DISCLAIMER

The views expressed herein do not necessarily represent the views of the U.S. Department of Energy or the United States Government.

REFERENCES

[1] S. Wang, U. Selvarasu, M. Amirabadi, Y. Li, and B. Lehman, "Enhanced Startup and Synchronization Transients for Virtual-Oscillator Grid-Tie Inverters Using Predictive Feedback-Based Method," *2024 IEEE Energy Conversion Congress and Exposition (ECCE)*, Phoenix, AZ, USA, 2024.

[2] U. Selvarasu, M. Amirabadi, Y. Li, C. Crow and B. Lehman, "DP-Based Optimization of BESS to Substitute RICE Reserves for Improved Economic Benefits," *2023 IEEE Energy Conversion Congress and Exposition (ECCE)*, Nashville, TN, USA, 2023, pp. 188-194, doi: 10.1109/ECCE53617.2023.10362257.

[3] U. Selvarasu, M. Amirabadi, Y. Li, C. Crow and B. Lehman, "BESS Sizing for PV Power Smoothing," *2024 IEEE Energy Conversion Congress and Exposition (ECCE)*, Phoenix, AZ, USA, 2024.

[4] J. Tait, S. Wang and K. H. Ahmed, "An Enhanced Series-Connected Offshore Wind Farm (SC-OWF) System Considering Fault Resiliency," in IEEE Transactions on Power Delivery, vol. 39, no. 1, pp. 352-362, Feb. 2024.

[5] Jin-Jhan Jheng, Min-Fu Shih and Kuo-Yuan Lo, "Design of a cascaded H-bridge multi-level inverter with hot swappable capability for battery energy storage systems," 2016 IEEE 8th International Power Electronics and Motion Control Conference (IPEMC-ECCE Asia), Hefei, China, 2016, pp. 2972-2977.

[6] D. Cottet et al., "Integration technologies for a medium voltage modular multi-level converter with hot swap capability," 2015 IEEE Energy Conversion Congress and Exposition (ECCE), Montreal, QC, Canada, 2015, pp. 4502-4509.

[7] S. n. vinay Mutyala, I. Cvetkovic, C. DiMarino and D. Boroyevich, "Design and Testing of a SiC-Based Solid-State Bypass Switch for 1 kV Power Electronics Building Blocks," 2022 IEEE Energy Conversion Congress and Exposition (ECCE), Detroit, MI, USA, 2022, pp. 1-8.

[8] S. Chakraborty, S. Patel, G. Saraswat, A. Maqsood and M. V. Salapaka, "Seamless Transition of Critical Infrastructures Using Droop-Controlled Grid-Forming Inverters," in IEEE Transactions on Industrial Electronics, vol. 71, no. 2, pp. 1535-1546, Feb. 2024.

[9] G. -S. Seo, M. Colombino, I. Subotic, B. Johnson, D. Groß and F. Dörfler, "Dispatchable Virtual Oscillator Control for Decentralized Inverter-dominated Power Systems: Analysis and Experiments," 2019 IEEE Applied Power Electronics Conference and Exposition (APEC), Anaheim, CA, USA, 2019, pp. 561-566.

[10] P. Rodriguez, R. Teodorescu, I. Candela, A. V. Timbus, M. Liserre and F. Blaabjerg, "New positive-sequence voltage detector for grid synchronization of power converters under faulty grid conditions," 2006

37th IEEE Power Electronics Specialists Conference, Jeju, Korea (South), 2006, pp. 1-7.

[11] Hong-Seok Song and Kwanghee Nam, "Dual current control scheme for PWM converter under unbalanced input voltage conditions," in IEEE Transactions on Industrial Electronics, vol. 46, no. 5, pp. 953-959, Oct. 1999.

[12] H. Komurcugil, N. Altin, S. Ozdemir and I. Sefa, "Lyapunov-Function and Proportional-Resonant-Based Control Strategy for Single-Phase Grid-Connected VSI With LCL Filter," in IEEE Transactions on Industrial Electronics, vol. 63, no. 5, pp. 2838-2849, May 2016.

[13] V. R. Chowdhury, "Enhanced Operation of a Virtual Synchronous Machine Under Unbalanced Grid Voltage Condition Based on Lyapunov Energy Function," in IEEE Journal of Emerging and Selected Topics in Industrial Electronics, vol. 4, no. 2, pp. 589-602, April 2023.

[14] V. R. Chowdhury and J. W. Kimball, "Robust Control Scheme for a Three Phase Grid-Tied Inverter With LCL Filter During Sensor Failures," in IEEE Transactions on Industrial Electronics, vol. 68, no. 9, pp. 8253-8264, Sept. 2021.

[15] "A High-capacity, High-dielectric-strength Relay Compatible with Momentary Voltage Drops," datasheet of G7L relays.

[16] "HE-R Relays; Power Relays (Over 2A)," datasheet of Panasonic relays .

[17] P. Hu, Z. He, S. Li and J. M. Guerrero, "Non-Ideal Proportional Resonant Control for Modular Multilevel Converters Under Sub-Module Fault Conditions," in IEEE Transactions on Energy Conversion, vol. 34, no. 4, pp. 1741-1750, Dec. 2019, doi: 10.1109/TEC.2019.2938395.

[18] A. Kuperman, "Proportional-Resonant Current Controllers Design Based on Desired Transient Performance," in IEEE Transactions on Power Electronics, vol. 30, no. 10, pp. 5341-5345, Oct. 2015, doi: 10.1109/TPEL.2015.2408053.

[19] D. G. Holmes, T. A. Lipo, B. P. McGrath and W. Y. Kong, "Optimized Design of Stationary Frame Three Phase AC Current Regulators," in IEEE Transactions on Power Electronics, vol. 24, no. 11, pp. 2417-2426, Nov. 2009, doi: 10.1109/TPEL.2009.2029548.

[20] D. N. Zmood and D. G. Holmes, "Stationary frame current regulation of PWM inverters with zero steady-state error," in IEEE Transactions on Power Electronics, vol. 18, no. 3, pp. 814-822, May 2003, doi: 10.1109/TPEL.2003.810852.

Layout Optimization for Parasitic Inductance Reduction of GaN-based NPL.X Multilevel Inverter

Ali Halawa
Elmore Family School of Electrical and Computer Engineering
Purdue University
West Lafayette, USA
ahalawa@purdue.edu

Jinyeong Moon
Department of Electrical and Computer Engineering
Florida State University
Tallahassee, FL, USA
j.moon@fsu.edu

Woongkul Lee
Elmore Family School of Electrical and '
Computer Engineering
Purdue University
West Lafayette, USA
wklee@purdue.edu

Abstract— **The increasing adoption of gallium nitride (GaN) devices has raised the bar for layout optimization techniques to achieve lower parasitic inductance design. This is driven by the high *dv/dt* characteristics of GaN devices, which make them particularly sensitive to parasitic inductances. This challenge is especially prominent in multilevel inverters (MLIs), where the neutral point connection complicates PCB layout and busbar design with low parasitic inductance. The newly proposed neutral-point-less X-type (NPL.X) MLI topology eliminates the neutral point connection, simplifying layout optimization. However, its complex structure can lead to poor switch placement, introducing undesirable parasitic elements and causing voltage overshoots. This paper proposes a novel circuit configuration for the NPL.X topology to reduce parasitic inductance from inefficient switch placement. The optimized configuration expected to significantly lowers commutation loop inductance, mitigating voltage overshoots. To validate the capability of the new configuration, a PCB layout based on the proposed configuration is analyzed, showing a 60% reduction in loop inductance. Simulation results, using extracted parasitic inductances from the PCB, show up to a 75% reduction in voltage overshoots across switches and up to 90% across diodes under varying load and *dv/dt* conditions. These results highlight the potential of the new configuration to enhance the performance and reliability of the GaN-based NPL.X.**

Keywords— *Commutation loop, gallium nitride, multilevel inverter, neutral-point-less x-type, parasitic inductance, three-level, traction inverter, voltage overshoot, wide bandgap.*

I. INTRODUCTION

Layout optimization has recently gained popularity due to the use of wide bandgap (WBG) devices, such as silicon carbide (SiC) and gallium nitride (GaN) [1]-[5]. Layout optimization is especially crucial when using GaN devices because of their high *dv/dt*. Even small parasitic inductances in the power loop can significantly impact voltage overshoot, which, in turn, affects the reliability of the switches [6]-[8] Another concern is the absence of an avalanche mode in GaN devices, which can lead to thermal runaway under high voltage overshoots [9]-[10].

In addition, GaN devices have gained attention for use in multilevel inverters (MLIs) [11]-[12]. The primary reason is that commercially available GaN devices are typically rated below 900 V, limiting the dc-link voltage in conventional two-

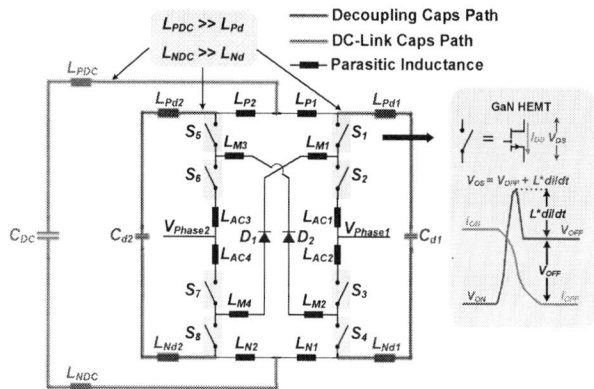

Fig. 1. Single module (two phases) of GaN-based NPL.X, including decoupling and dc-link capacitors, as well as the main parasitic inductances in the circuit that influence the voltage overshoot across the switches.

level inverters. One of the challenges associated with MLIs is the presence of a neutral point, which complicates the circuit configuration. To address this, the neutral-point-less X-type (NPL.X) three-level inverter has been recently proposed [13]. This MLI enables multilevel operation with lower voltage-rated switches without requiring a neutral point, as shown in Fig. 1. Consequently, it achieves high power density and facilitates the use of GaN devices. However, a single NPL.X module contains eight switches, making poor switch placement a common issue in conventional designs. This often leads to significant parasitic inductances that could otherwise be avoided. Hence, this article proposes a novel NPL.X configuration that optimally places switches and decoupling capacitors, minimizing parasitic inductance in the power loop.

Several approaches to reducing voltage overshoot in MLIs have been examined. One method focuses on optimizing the layout of individual switching cells [14]-[15], while another employs pulse-width modulation (PWM) techniques to minimize overshoot [16]-[17]. The voltage overshoot issue in the NPL.X inverter has been briefly explored in [18] under various operating conditions. This study indicates that the NPL.X achieves lower voltage overshoot compared to the neutral-point-clamped (NPC) inverter, primarily due to the elimination of neutral inductance. However, despite the

979-8-3315-1612-3/25 $31.00 © 2025 IEEE   1473

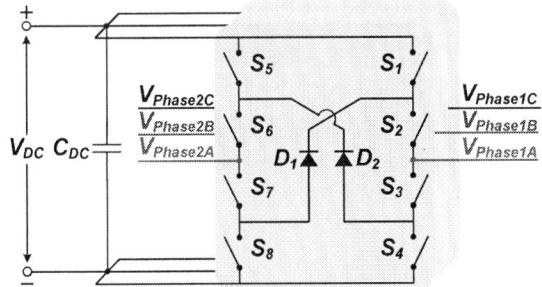

Fig. 2. Schematics of six-phase NPL.X MLI topology under investigation.

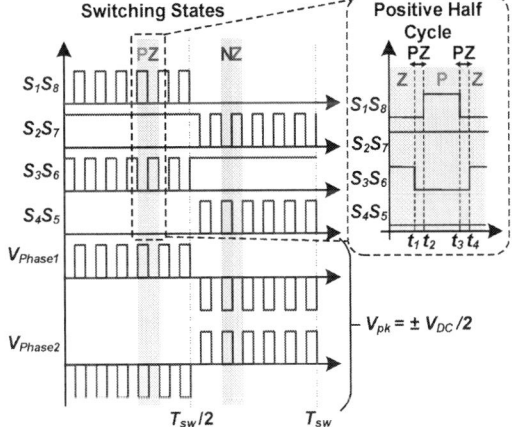

Fig. 3. Detailed switching process for a single NPL.X three-level inverter.

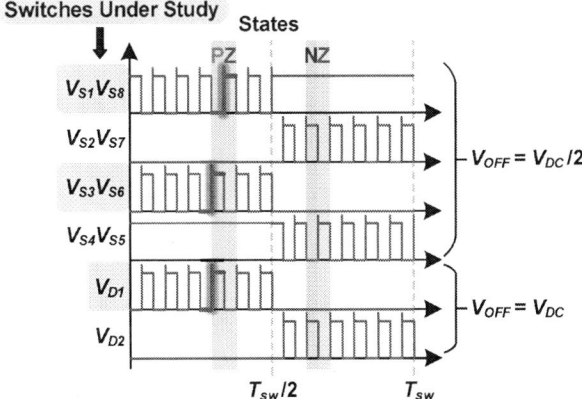

Fig. 4. Voltage waveform of each switch during the corresponding switching states.

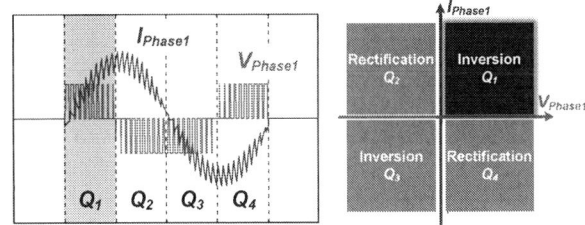

Fig. 5. Four quadrant operation of NPL.X highlighting the studied quadrant.

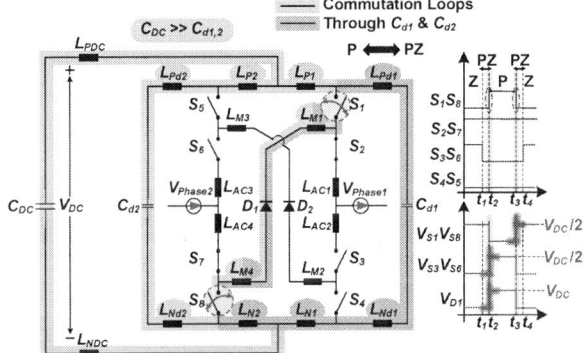

Fig. 6. Single module (two phases) of NPL.X with commutation loops formed through the decoupling capacitors in the first quadrant.

reduced overshoot in the NPL.X, it remains significant enough to impact the reliability of switching devices, largely due to the undesirable parasitic inductances that are present due to inefficient switches and decoupling capacitors placement. To address this issue, this paper proposes a novel circuit configuration for the NPL.X three-level inverter, aiming to make the NPL.X more compact by effectively placing the switching devices and decoupling capacitors on the PCB. This efficient placement of switches and decoupling capacitors will result in lower loop inductance, and therefore, lower overshoots and oscillations.

This paper begins with a brief explanation of the NPL.X MLI working principle and the commutation process in the first quadrant, highlighting the challenges associated with the conventional NPL.X configuration. Next, the proposed configuration is analyzed and compared to the conventional configuration, with particular focus on the parasitic inductances involved. A PCB layout is then presented to demonstrate how the proposed configuration achieves a reduced commutation loop inductance. Finally, simulations are performed using extracted parasitic inductance values from the layout under varying loading and *dv/dt* conditions.

II. NPL.X OPERATING PRINCIPLE AND COMMUTATION

A. Operation Principle

The NPL.X is three-level inverter that utilizes a single, compact dc-link capacitor enabling multilevel operation without requiring a neutral point connection and by using lower voltage active switches. Fig. 2 shows three NPL.X modules combined to create a balanced six-phase multilevel output. Multilevel operation is achieved by dividing the total dc-link voltage across the two outputs in a single inverter module, producing a half dc-link voltage step. To further understand the operation of the NPL.X, Fig. 3 illustrates the switching states along with the output voltage waveform. It shows that the NPL.X behaves similarly to the NPC but with an additional leg that generates an output voltage complementary to the first leg voltage, eliminating the need for a neutral point. The PZ and NZ state are transitional state are present during the deadtime which will be further analyzed in the next section. The corresponding voltages across the switches for each state is shown in Fig. 4. It should be noted that since the NPL.X have

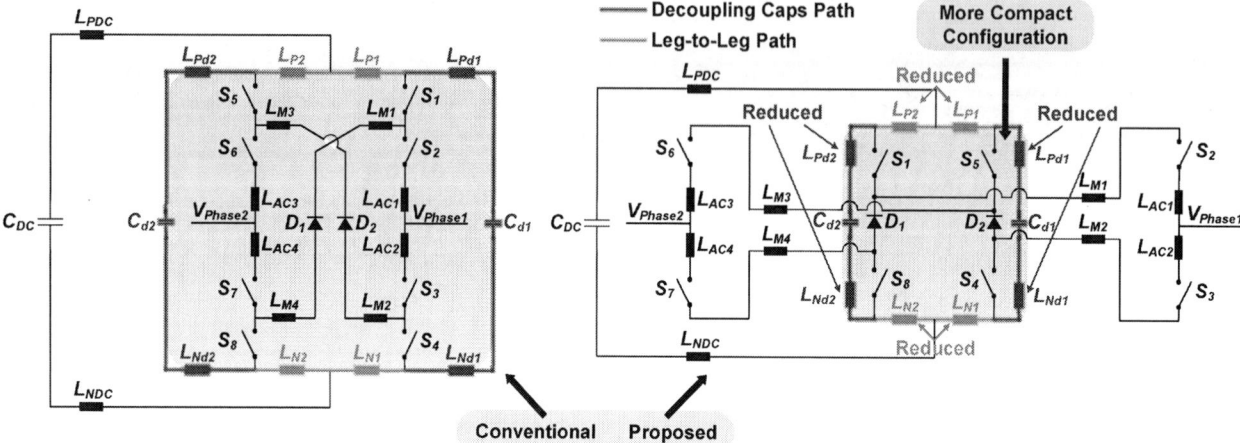

Fig. 7. Proposed NPL.X configuration B compared to the conventional configuration A highlighting the possible anticipated reduction in multiple key parasitic inductances.

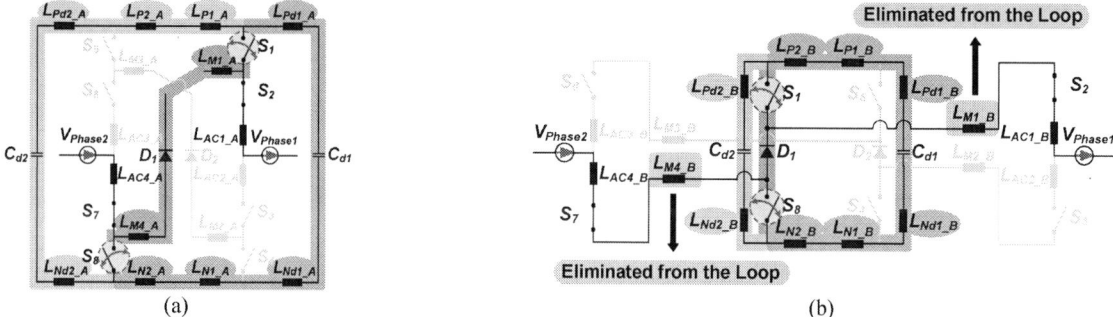

(a)   (b)

Fig. 8. Commutaion loop comparison during Q_1 between (a) conventional configuration A, and (b) proposed configuration B.

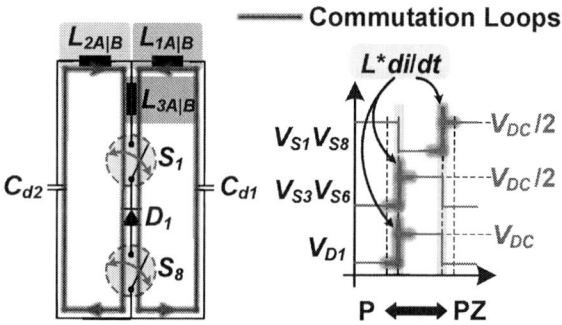

Fig. 9. Equivalent circuit illustrating the commutation loops and the switches with voltage overshoot during the commutation.

two outputs per single NPL.X module, the power rating is double of other well-known 3-level inverter topologies like the NPC or the T-Type.

B. Commutation

To address the challenges associated with the conventional NPL.X configuration, the commutation process of the NPL.X is analyzed specifically within the first quadrant (Q_1). Fig. 5 illustrates the operation of the NPL.X across all four quadrants, offering an overview of its behavior in various scenarios. In Q_1 during the zero-output voltage state (Z), switches S_2S_7 and S_3S_6 are turned on. Under this condition, the output voltages V_{phase1}

and V_{phase2} are interconnected and remain equal to zero. The current in this state flows through switches S_2 and S_7, as well as diode D_1. At t_1, when switches S_3S_6 are turned off, no commutation occurs immediately. Instead, the inverter enters an intermediate state, referred to as the PZ state, because no current flows through switch S_3 before or after this transition period. The commutation process begins when switches S_1S_8 are turned on at t_2, transitioning the inverter from the PZ state to the positive output voltage state (P). During this positive voltage state, the current path is established through switches S_1, S_8, S_2, and S_7. In this state, the output voltages V_{phase1} and V_{phase2} exhibit complementary behavior, with values of $V_{DC}/2$ and -$V_{DC}/2$, respectively. The commutation loop formed is highlighted in Fig. 6. During these transitions, the voltages of S_1S_8, S_3S_6, and D_1 will fluctuate. This loop inductance is the primary cause of the overshoot, and the conventional configuration contain excessive parasitic inductances that can be mitigated with an optimized design.

III. OPTIMIZED LAYOUT DESIGN FOR LOWER LOOP INDUCTANCE

A. Loop Inductance Analysis

The conventional NPL.X layout places decoupling capacitors around four switches and the AC link, creating a design where the path of the decoupling capacitors inherently

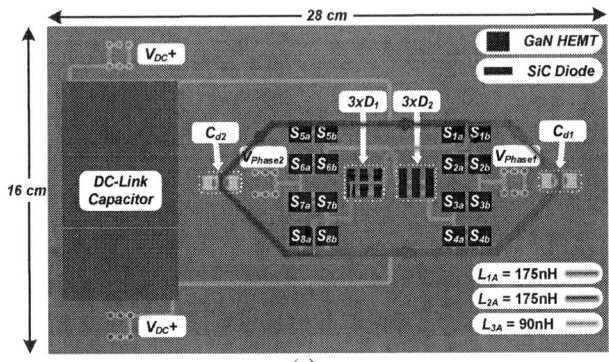 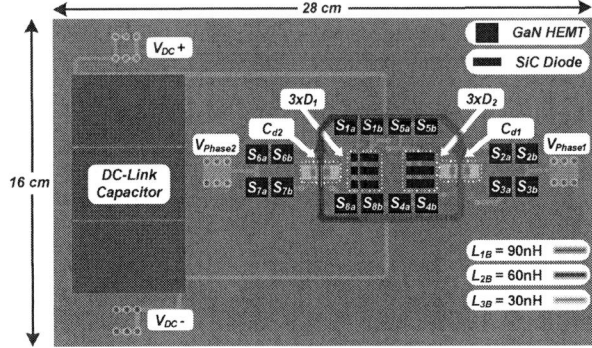

Fig. 10. Two layouts illustrating the difference in loop inductance between configurations A and B during Q_1 in (a) configuration A and (b) configuration B.

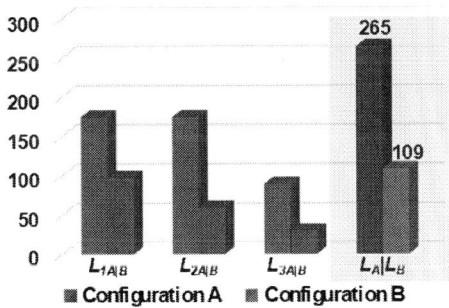

Fig. 11. Extracted loop inductances from the layout showing the reduction in configuration B loop inductance.

introduces higher-than-necessary parasitic inductance. This increased inductance can degrade the system switching performance by increasing the voltage overshoot and oscillations. This observation motivated the development of the proposed layout, illustrated in Fig. 7(b). The proposed design maintains the same steady-state functionality as the conventional configuration but optimizes the placement of the decoupling capacitors for improved dynamic performance. In this arrangement, the decoupling capacitors surround only two switches and a diode, resulting in a significantly reduced parasitic inductance along the decoupling capacitor paths. Additionally, the leg-to-leg inductance is minimized in the proposed configuration by relocating the two diodes that previously occupied the space between the legs. This adjustment reduces the distance between the two legs, thereby lowering the inductance. The proposed layout is predicted to effectively minimize the voltage overshoot across the switches.

The impact of parasitic inductances in the commutation loop shown in Fig. 8 is further analyzed using the derived equivalent circuit shown in Fig. 9. In the conventional layout, the commutation loop contains three lumped parasitic inductances: two from the paths of the decoupling capacitors (C_{d1} and C_{d2}) and one from the shared common path. These inductances contribute to higher voltage overshoots during switching transitions. In contrast, the proposed configuration addresses this issue by reconfiguring the commutation loop to minimize the contributions of these inductances. The value of each inductance can be written as follows:

$$\begin{cases} L_{1A} = L_{Pd1\_A} + L_{Nd1\_A} + L_{N1\_A} + L_{N2\_A} \\ L_{2A} = L_{Pd2\_A} + L_{Nd2\_A} + L_{P1\_A} + L_{P2\_A} \\ L_{3A} = L_{M1\_A} + L_{M4\_A} \end{cases} \quad (1)$$

$$\begin{cases} L_{1B} = L_{Pd1\_B} + L_{Nd1\_B} + L_{N1\_A} + L_{N2\_A} + L_{P1\_A} + L_{P2\_A} \\ L_{2B} = L_{Pd2\_B} + L_{Nd2\_B} \\ L_{2B} \approx 0 \end{cases} \quad (2)$$

Hence, the commutation loop inductance can be expressed as:

$$L_A = L_{Pd\_A} + L_{Nd\_A} + L_{N\_A} + L_{P\_A} + L_{M1\_A} + L_{M4\_A} \quad (3)$$

$$L_B = L_{Pd\_B} + L_{Nd\_B} + L_{N\_B} + L_{P\_B} \quad (4)$$

Such that:

$$\begin{cases} L_{Pd\_x} = L_{Pd1\_x} = L_{Pd2\_x} \\ L_{Nd\_x} = L_{Nd1\_x} = L_{Nd2\_x} \\ L_{P\_x} = L_{P1\_x} = L_{P2\_x} \\ L_{N\_x} = L_{N1\_x} = L_{N2\_x} \end{cases} \quad (5)$$

Such that x represents configurations A or B. Based on (3) and (4), it is observed that the commutation loop inductance is reduced in configuration B, as the inductances L_{M1} and L_{M2} are eliminated. Additionally, the extracted inductances L_{Pd}, L_{Nd}, L_P, and L_N in configuration B are expected to be lower due to the shorter path length of the decoupling capacitors and reduced distance between the two legs.

It worth mentioning that in the proposed layout, the common path inductance includes only the device inductance, which is negligible, often less than a nano henry for e-mode GaN devices and is therefore omitted from this analysis. Another assumption is that, in configuration A, paired switches have negligible inductance between them due to close positioning. Similarly, in configuration B, the inductances between S_1, D_1, S_8 and between S_5, D_2, S_4 are also minimal because these components can be positioned closely in the layout. These assumptions will be explained and validated during the layout analysis in the next section.

TABLE I: NPL.X PARAMETERS USED IN SIMULATION.

Parameter	Value
DC-Link Voltage, V_{DC}	800 V
Load Current, I_{Phase1}	30 A
Switching Speed, dv/dt	10 V/ns
Fundamental Frequency, f_{ref}	60 Hz
Switching Frequency, f_{sw}	5 kHz
Decoupling Capacitors, C_{d1} and C_{d2}	1 μF
Configuration A Loop Inductance, L_A	265 nH
Configuration B Loop Inductance, L_B	109 nH

B. Layout Comparison

To validate our assumptions, an example layout resembling an actual PCB design was developed. The primary objective was to create layouts that were as symmetric and similar as possible, ensuring that the same topology was represented with only variations in the placement of switches, diodes, and capacitors. This approach minimizes confounding variables, allowing for a direct comparison of parasitic inductance between configurations. The parasitic inductance for each layout was calculated using the following equation:

$$L = 5.08l \times [2.303 \log_{10}(\frac{2l}{w+t}) + \frac{1}{2} + 0.2235(\frac{w+t}{l})] \qquad (6)$$

In this context, L represents the inductance in nH, while l refers to the path length, w to the trace width, and t to the copper thickness, with each measurement expressed in inches [19].

In configuration A during quadrant Q_1, as shown in Fig. 10(a), the inductance along the blue and red current paths was found to be significantly higher compared to configuration B. Specifically, it was observed to be approximately twice the inductance of the red and blue paths in configuration B, illustrated in Fig. 10(b). For the common path highlighted in orange, the inductance in configuration B is three times lower. This reduction can be attributed to the less efficient placement of decoupling capacitors and switches in configuration A. The layout for configuration B, by contrast, effectively reduces the path lengths, leading to lower overall inductance values. These findings align well with our initial predictions based on the design principles and assumptions. The inductances present in commutation loop has been calculated using (3) and (4). A summary of these calculations is presented in Fig. 11, which provides a detailed comparison of the inductances across configurations A and B. The results highlight that the commutation loop inductance for configuration B is reduced by 60% compared to configuration A. This substantial reduction underscores the effectiveness of the proposed layout in mitigating parasitic inductances through optimized component placement.

IV. SIMULATION RESULTS

To validate our study, simulations were conducted in LTspice using the extracted parasitic inductance values. The test conditions were based on the parameters listed in Table I, unless otherwise specified. The switching device used is the Infineon GS66516B, a 650 V e-mode GaN rated for 60 A. For

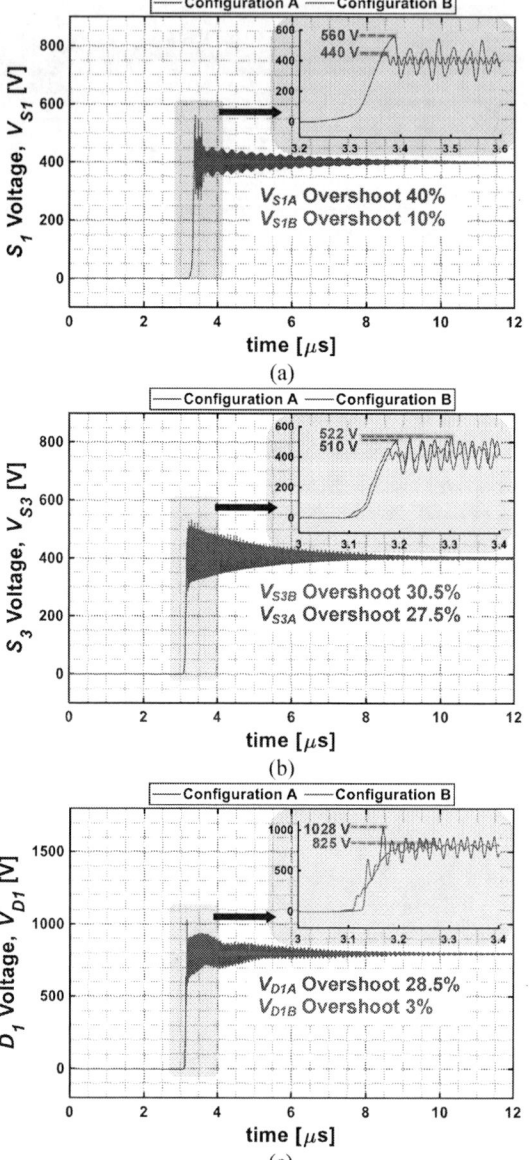

Fig. 12. Simulation results showing voltage across each switch during turn-off, highlighting voltage overshoot for the two different configurations for (a) S_1, (b) S_3, and (c) D_1 at the extracted inductance value.

the diode, the 3xC4D20120D SiC 1200 V, 20 A Schottky diode was selected. To simplify the analysis, only three switches (S_1, S_3, and D_1) were studied in detail, as the waveforms for the other switches are either similar or redundant.

When examining the turn-off waveform of S_1, as shown in Fig. 12(a), a substantial reduction in voltage overshoot is observed for Configuration B compared to Configuration A. Specifically, the overshoot across S_1 decreases by 75%, dropping from 40% in Configuration A to just 10% in Configuration B. This significant improvement highlights the effectiveness of Configuration B in mitigating voltage stress on S_1. In contrast, for switch S_3, no reduction in overshoot is

979-8-3315-1612-3/25 $31.00 © 2025 IEEE

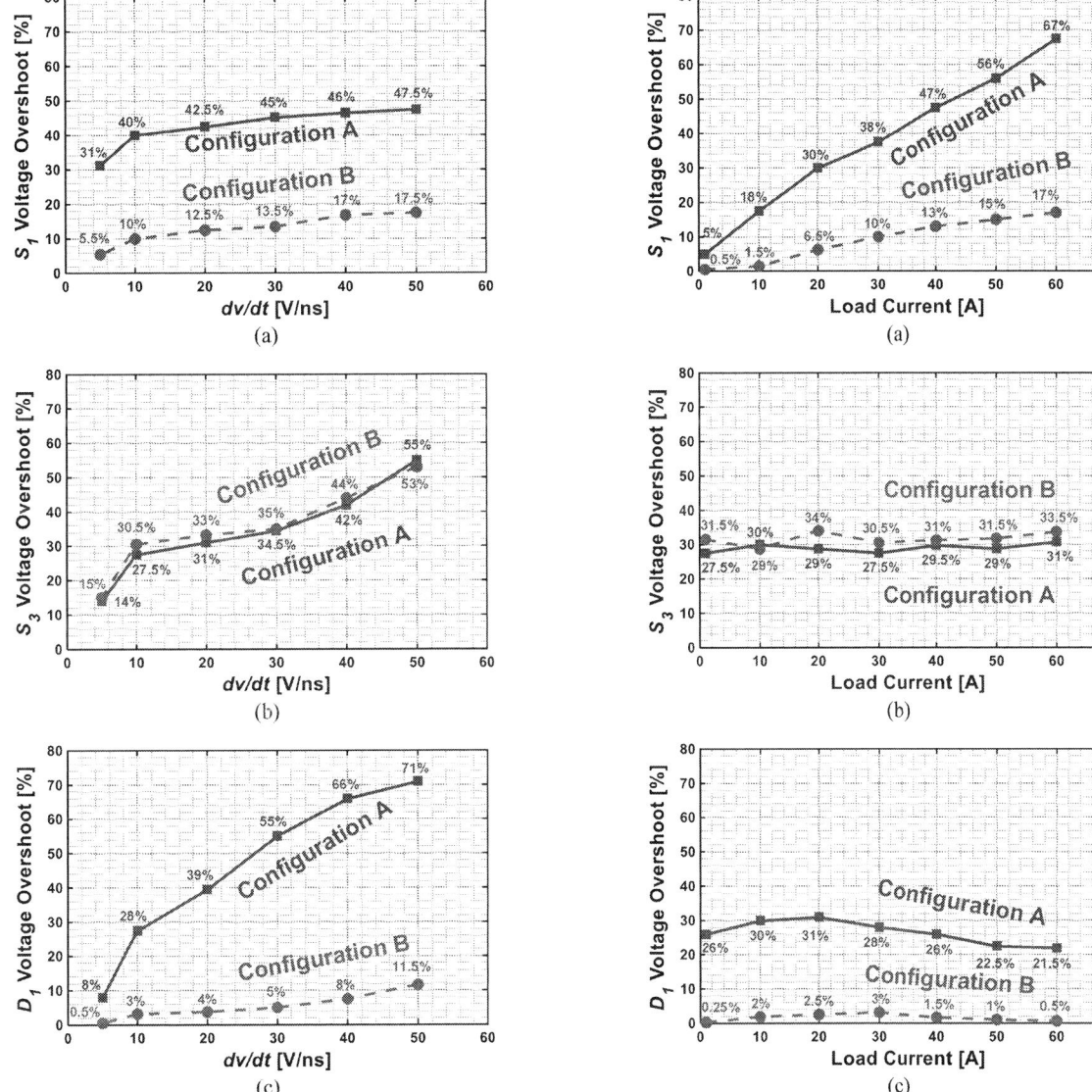

Fig. 13. Voltage overshoot percentages for Configuration A and Configuration B under varying dv/dt values for (a) S_1, (b) S_3, and (c) D_1.

Fig. 14. Voltage overshoot percentages for Configuration A and Configuration B under different loading conditions for (a) S_1, (b) S_3, and (c) D_1.

observed in Configuration B as shown in Fig. 12(b). This behavior can be attributed to the new location of the parasitic inductance, which prevents a significant improvement. However, it is important to note that Configuration B does not result in a considerable increase in overshoot for S_3, with only a modest rise of about 10% compared to Configuration A. This indicates that while Configuration B does not improve overshoot for S_3, it does not negatively impact its performance to any significant extent either. For diode D_1, Configuration B demonstrates the most significant improvement in voltage overshoot. The overshoot for D_1 decreases by more than 90%, dropping from 28.5% in Configuration A to a mere 3% in Configuration B as illustrated in Fig. 12(c). This remarkable reduction underscores the potential of Configuration B to

significantly enhance the reliability of D_1 by minimizing voltage stress.

For switch S_3, no reduction in overshoot is observed in Configuration B as shown in Fig. 12(b). This behavior can be attributed to the new location of the parasitic inductance, which prevents a significant improvement. However, it is important to note that Configuration B does not result in a considerable increase in overshoot for S_3, with only a modest rise of about 10% compared to Configuration A. This indicates that while Configuration B does not improve overshoot for S_3, it does not negatively impact its performance to any significant extent either. For diode D_1, Configuration B demonstrates the most significant improvement in voltage overshoot. The overshoot for D_1 decreases by more than 90%, dropping from 28.5% in

Configuration A to a mere 3% in Configuration B as illustrated in Fig. 12(c). This remarkable reduction underscores the potential of Configuration B to significantly enhance the reliability of D_I by minimizing voltage stress.

To explore the impact of different switching speeds, the study was expanded, as shown in Fig. 13, to analyze overshoot across S_I, S_3, and D_I for varying dv/dt. The results reveal that, across different switching speeds, Configuration B consistently maintains much lower overshoot levels for S_I and D_I compared to Configuration A. For S_3, however, the overshoot remains similar in both configurations, with only slight variations favoring one configuration over the other at specific switching speeds. This suggests that the performance of S_3 is relatively unaffected by the topology, and its reliability remains comparable in both cases.

The study was further extended to consider different loading conditions, as illustrated in Fig. 14. Under these conditions, the switches exhibited varied behavior. For S_I, increasing the load led to a corresponding increase in voltage overshoot, making Configuration B reduced overshoot even more critical for maintaining reliability. On the other hand, D_I and S_3 demonstrated minimal sensitivity to changes in loading conditions, with overshoot remaining relatively unaffected. Overall, Configuration B consistently provided lower voltage overshoot across all loading scenarios, particularly for S_I, D_I, and other similarly behaving switches. Overall, Configuration B offers substantial advantages in reducing voltage overshoot, especially for S_I, S_4, S_5, S_8, D_I, and D_2, across a wide range of loading and switching speed conditions. This reduction significantly enhances the reliability of these components by mitigating the effects of parasitic inductance. For switches such as S_2, S_3, S_6, and S_7, where overshoot remains similar in both configurations, the reliability is expected to be comparable. This makes configuration B it a more robust choice for minimizing voltage stress on the switches and diodes.

V. CONCLUSION

In conclusion, this paper presents a novel configuration for the NPL.X topology aimed at minimizing parasitic inductance within the commutation loop. The proposed configuration optimizes the placement of switches and decoupling capacitors, resulting in reduced parasitic inductance in the paths of the decoupling capacitors. Additionally, the leg-to-leg inductance is reduced by optimizing the location of the inner diodes, which helps lower the inductance between the legs. These reductions in parasitic inductance will lead to a decrease in the overall commutation loop inductance. To illustrate the effectiveness of the proposed configuration, a PCB layout representing the actual placement of switches and decoupling capacitors was analyzed. The results show that the proposed circuit configuration reduces loop inductance by 60% compared to the conventional configuration. Simulation results indicate that, with the extracted parasitic inductance values, the reduction in parasitic inductance leads to up to a 75% decrease in voltage overshoot across switches and up to a 90% reduction across

diodes under various operating conditions. These improvements demonstrate that the proposed configuration significantly enhances the reliability of the NPL.X compared to the conventional configuration, making it a more suitable choice for high dv/dt applications.

REFERENCES

[1] S. Satpathy, P. P. Das, and S. Bhattacharya, "Power Layout Design of a GaN HEMTs-Based High-Power High-Efficiency Three-Level ANPC Inverter for 800 V DC Bus System," in *IEEE Journal of Emerging and Selected Topics in Industrial Electronics*, vol. 5, no. 2, pp. 565-576, April 2024.

[2] H. Gui, R. Chen, Z. Zhang, J. Niu, L. M. Tolbert, F. Wang, D. Costinett, B. J. Blalock, and B. B. Choi, "Methodology of Low Inductance Busbar Design for Three-Level Converters," in *IEEE Journal of Emerging and Selected Topics in Power Electronics*, vol. 9, no. 3, pp. 3468-3478, June 2021.

[3] W. A. Khan, A. Ebrahimian, S. I. Hosseini S. and N. Weise, "Design of High Current, High Power Density GaN Based Motor Drive for All Electric Aircraft Application," *2022 IEEE 9th Workshop on Wide Bandgap Power Devices & Applications (WiPDA)*, Redondo Beach, CA, USA, 2022, pp. 247-253.

[4] B. Sun, Z. Zhang, and M. A. E. Andersen, "Research of Low Inductance Loop Design in GaN HEMT Application," *IECON 2018 - 44th Annual Conference of the IEEE Industrial Electronics Society*, Washington, DC, USA, 2018, pp. 1466-1470.

[5] S. -S. Yang, S. -S. Min, C. -H. Eom, R. -Y. Kim, and G. -Y. Lee, "Design Method of Vertical Lattice Loop Structure for Parasitic Inductance Reduction in a GaN HEMTs-Based Converter," in *IEEE Access*, vol. 10, pp. 117215-117224, 2022.

[6] N. Haryani, X. Zhang, R. Burgos and D. Boroyevich, "Static and dynamic characterization of GaN HEMT with low inductance vertical phase leg design for high frequency high power applications," *2016 IEEE Applied Power Electronics Conference and Exposition (APEC)*, Long Beach, CA, USA, 2016, pp. 1024-1031.

[7] N. Deneke and B. Wicht, "Overshoot Prevention in Monolithic GaN by Ultra-Low ESL Gate Loop Design Using Chip-Scale Capacitors and Gate Driver Pull-Up Path Tuning Technique," *2024 IEEE Applied Power Electronics Conference and Exposition (APEC)*, Long Beach, CA, USA, 2024, pp. 2409-2414.

[8] J. P. Kozak, R. Zhang, M. Porter, Q. Song, J. Liu, B. Wang, R. Wang, W. Saito and Y. Zhang "Stability, Reliability, and Robustness of GaN Power Devices: A Review," in *IEEE Transactions on Power Electronics*, vol. 38, no. 7, pp. 8442-8471, July 2023.

[9] R. Zhang, J. P. Kozak, M. Xiao, J. Liu and Y. Zhang, "Surge-Energy and Overvoltage Ruggedness of P-Gate GaN HEMTs," in *IEEE Transactions on Power Electronics*, vol. 35, no. 12, pp. 13409-13419, Dec. 2020.

[10] E. A. Jones, F. F. Wang, and D. Costinett, "Review of Commercial GaN Power Devices and GaN-Based Converter Design Challenges," in *IEEE Journal of Emerging and Selected Topics in Power Electronics*, vol. 4, no. 3, pp. 707-719, Sept. 2016.

[11] T. Modeer, N. Pallo, T. Foulkes, C. B. Barth, and R. C. N. Pilawa-Podgurski, "Design of a GaN-Based Interleaved Nine-Level Flying Capacitor Multilevel Inverter for Electric Aircraft Applications," in *IEEE Transactions on Power Electronics*, vol. 35, no. 11, pp. 12153-12165, Nov. 2020.

[12] G. Rohner, T. Gfrörer, P. S. Niklaus, D. Bortis, M. Schweizer, and J. W. Kolar, "Comparative Evaluation of Three-Phase Three-Level GaN and Seven-Level Si Flying Capacitor Inverters for Integrated Motor Drives Considering Overload Operation," in *IEEE Access*, vol. 12, pp. 7356-7371, 2024.

[13] M. Guven, M. Benson, X. Dong, J. Moon, and W. Lee, "Operating Principle of Neutral-Point-Less (NPL) Multilevel Inverter Topology: X-type Inverter," *2022 IEEE Transportation Electrification Conference & Expo (ITEC)*, Anaheim, CA, USA, 2022, pp. 345-350.

[14] S. Satpathy, P. P. Das, S. Bhattacharya, and V. Veliadis, "Design Considerations of a GaN-based Three-Level Traction Inverter for Electric Vehicles," *2022 IEEE 9th Workshop on Wide Bandgap Power Devices & Applications (WiPDA)*, Redondo Beach, CA, USA, 2022, pp. 192-197.

[15] E. Gurpinar, F. Iannuzzo, Y. Yang, A. Castellazzi and F. Blaabjerg, "Design of Low-Inductance Switching Power Cell for GaN HEMT Based Inverter," in *IEEE Transactions on Industry Applications*, vol. 54, no. 2, pp. 1592-1601, March-April 2018.

[16] H. Chen, S. Ai, Z. Nie, F. Cui, and P. Yuan, "Modulation Strategy for Suppressing Peak Voltage Spikes of SiC-MOSFETs During ANPC Commutation," in *IEEE Access*, vol. 11, pp. 27631-27640, 2023.

[17] J. He, D. Zhang, and D. Pan, "PWM Strategy for MW-Scale "SiC+Si" ANPC Converter in Aircraft Propulsion Applications," in *IEEE Transactions on Industry Applications*, vol. 57, no. 3, pp. 3077-3086, May-June 2021.

[18] A. Halawa, B. Schuchardt, J. Moon, and W. Lee, "Influence of High-frequency Power Loop Inductance on GaN-based Neutral-Point-Less X-type (NPL.X) Three-level Inverter," *2024 IEEE Transportation Electrification Conference and Expo (ITEC)*, Chicago, IL, USA, 2024, pp. 1-6.

[19] W. Lee and B. Sarlioglu, "Thermal Analysis of Lateral GaN HEMT Devices for High Power Density Integrated Motor Drives Considering the Effect of PCB Layout and Parasitic Parameters," *2018 IEEE Transportation Electrification Conference and Expo (ITEC)*, Long Beach, CA, USA, 2018, pp. 471-476.

Topology Selection and Design Methodology for SiC based Solar Photovoltaic Medium Voltage Direct Grid Connect Inverters

Jenson Joseph C Attukadavil, Baylon G. Fernandes

Department of Electrical Engineering, Indian Institute of Technology Bombay, Mumbai, India

jensonjoseph@iitb.ac.in, bgf@ee.iitb.ac.in

Abstract—**Solar photovoltaics (PV) is rapidly expanding as the world's leading renewable energy source by installed capacity, with utility-scale systems increasingly relying on medium voltage (MV) inverters. Recent advancements in MV silicon carbide (SiC) devices have enabled the development of direct grid-connect inverters, which offer improved efficiency over traditional central inverters. This paper provides a comprehensive design methodology for various MV inverter topologies, including three-level neutral point clamped (NPC) inverters, five-level active NPC inverters, three and five-level modular multilevel converters (MMCs), and cascaded H-bridge converter configurations. The methodology covers key design aspects such as SiC device selection, switching frequency optimization, filter design, and DC capacitor sizing. Furthermore, a refined loss model is introduced, derived from datasheet parameters and validated experimentally using 1700 V SiC MOSFETs to accurately assess converter performance. A comparative analysis of these topologies is conducted, considering component count, euro efficiency, device stress, harmonic distortion, and total volume. The results provide critical insights into the suitability of each topology for MV direct grid-connected PV applications.**

Index Terms—**Medium voltage grid connect inverter, PV central inverter, SiC MOSFET loss modelling.**

I. INTRODUCTION

Traditionally, solar PV central inverters are constructed using low-voltage IGBTs to generate low-voltage AC output, which is then stepped up to medium voltage (MV) using a low-frequency transformer. However, scaling power levels further with low-voltage IGBTs presents challenges, as their inherent limitations like high switching losses hinder efficiency at higher power ratings [1]. Additionally, low-frequency transformers due to their large size, require substantial resources for construction, transportation, and installation, making them costly and less scalable. Consequently, a direct MV grid-connected solution, utilizing isolated DC-DC converters, is an attractive alternative [2]. With recent advancements in high-voltage SiC MOSFETs, realizing MV grid-connected inverters with simpler structures is now feasible [3].

Significant research has been conducted on medium voltage inverter topologies for diverse applications [1]–[9]. The motor drive sector, in particular, holds a substantial share of the MV inverter market. Motor drive applications employ multilevel topologies such as the neutral point clamped (NPC), active neutral point clamp (ANPC), cascaded H-bridge (CHB), and modular multilevel converters (MMC). Comparative studies, such as those in [4]–[7], focus primarily on motor drives, often using low-voltage IGBTs and higher voltage levels to achieve MV output [5]. However, motor drive designs typically differ from grid-connected solar PV inverters in two ways: they rely on a transformer-rectifier front end and omit output filters, which are essential for grid interconnections.

Several studies, such as [1], [2], provide topological comparisons of MV inverters for grid-connected applications. However, these discussions are predominantly qualitative and focus on structural configurations and DC-DC conversion methods, offering limited insights into inverter design criteria. In [8], the design of NPC, flying capacitor, and MMC inverters are discussed using low-voltage Si IGBTs with a large number of voltage levels (up to 43), allowing for output line filters to be omitted. By contrast, this paper focuses on inverter topologies that leverage MV SiC MOSFETs, enabling higher operating frequencies and reducing the number of voltage levels required.

Estimating inverter losses is a critical aspect of comparing different topologies. With MV SiC MOSFETs still relatively new to the market, SPICE models for accurate loss estimation are often unavailable. Analytical loss models, such as those in [10]–[14], have been developed, yet many require data beyond what is provided in datasheets, necessitating experimental estimation. The method in [10] estimates losses based on datasheet parameters but exhibits deviations between estimated and experimental losses. This paper proposes an improved loss estimation method to address these discrepancies.

This paper provides a novel design methodology for medium voltage inverters, incorporating key elements like filter design and switching frequency selection that are underexplored in the existing literature. Unlike previous studies, which compare topologies with varying voltage levels, this paper provides a more standardized comparison by maintaining a consistent number of voltage levels. Furthermore, an improved loss model is presented based on datasheet values for SiC MOSFETs. The inverter topologies are designed and simulated in MATLAB, with modulation schemes tailored to each topology. Through simulation, losses and harmonic performance of the inverters are evaluated. Finally, a compre-

---

This work was supported by the National Centre for Photovoltaic Research and Education (NCPRE) and the Ministry of New and Renewable Energy (MNRE) of the Government of India.

979-8-3315-1612-3/25 $31.00 © 2025 IEEE

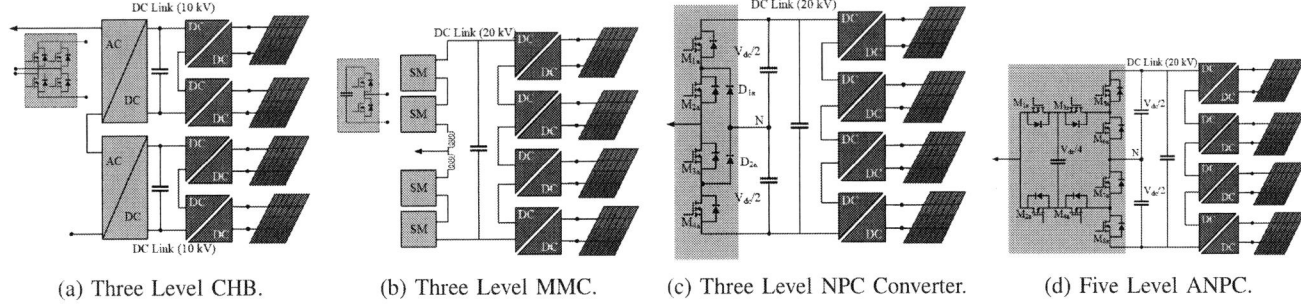

(a) Three Level CHB.	(b) Three Level MMC.	(c) Three Level NPC Converter.	(d) Five Level ANPC.

Fig. 1: Phase leg A of different medium voltage grid connect PV inverter topologies.

hensive comparison is conducted based on component count, voltage and current stress, harmonic distortion, efficiency, and total inverter volume, offering valuable insights into each topology's suitability for MV grid-connected PV applications.

II. CONVERTER DESIGN

This section discusses the design methodology of inverters for solar PV applications. The inverter ratings are 1 MVA, 11 kV output and 1500 V_{OC} (800 - 1200 V_{MPP}) as input [15]. The design of the inverter is done based on the following: topology, DC bus voltage, device ratings and number of voltage levels, sizing of filters and DC capacitors, and switching frequency. The design of inverters is discussed in detail in this section.

A. Topology selection

To choose appropriate topologies for medium voltage grid connect applications, a market survey of medium voltage drives available, with a voltage rating of 11 kV and a power rating in the range of 1-5 MVA is carried out. In low voltage levels upto 6.9 kV 3L-NPC is popular (Siemens SINAMICS GM 150 and ABBs ACS1000). At voltages in the range of 6.9 to 11 kV 5L-ANPC by ABB (ACS2000) and CHB by Siemens (Perfect Harmony GH 180), Allen Bradley (PowerFLex 6000), and Delta (MVD 1000) are popular. For further higher voltage MMC topology is primarily used and is produced by Siemens (Perfect Harmony GH 150) and Benshaw (M2L 3000).

Based on the survey the topologies chosen for realizing the MV grid connect inverter are, Neutral Point Clamped (NPC) converter, Active neutral point clamped (ANPC), Modular Multilevel Converter (MMC), and Cascaded H-Bridge (CHB) converter as shown in Fig. 1. The Solar PV plant inverters must have a long life and minimal maintenance breaks [1]; meaning simple topologies are preferred. Further using SiC MOSFETs higher frequency of operation is possible, reducing the filter requirements. Hence the topologies considered in this paper will be three-level and five-level topologies only.

B. Inverter DC Bus Design

The first step in inverter design after selecting the topology is the choice of appropriate DC bus voltage. Usually, a reserve of 10% is kept to take care of the grid transients [4]. Hence the DC bus voltage requirement is given by:

$$\frac{m V_{DC}}{2} = 1.1 V_{g_{peak}}$$
$$= \frac{1.1 \sqrt{2} V_{LL}}{\sqrt{3}} \qquad (1)$$
$$V_{DC} \approx 20.2 \; kV$$

To account for the grid transients, a 10% overrating is applied. Consequently, the DC bus voltage is set to 10 kV for the CHB topology and 20 kV for the NPC, ANPC, and MMC topologies.

C. Device Ratings and Number of Levels

The devices chosen for designing the inverters are medium voltage SiC MOSFETs with blocking voltages of 3.3 kV and 6.5 kV. The selection of switches is discussed in [3]. 3.3 kV MOSFETs (G2R120MT33J by Genesic) are chosen for designing the inverters as they provide the highest efficiency for a given operation. As the DC bus voltage rating is 20 kV, series and/or paralleling of devices is necessary as discussed in [3].

In the case of 3L-CHB, 3L-NPC, and 3L-MMC the minimum device blocking voltage requirement is 10 kV hence four 3.3 kV MOSFETs are connected in series to meet the blocking voltage requirement. In the case of 5L-CHB and 5L-MMC, the minimum device rating is only 5 kV. Hence, two 3.3 kV MOSFETs are connected in series. In case of ANPC as shown in Fig. 1d, the devices M_{1a}, M_{2a}, M_{3a} and M_{4a} are rated for 5 kV and devices M_{5a}, M_{6a}, M_{7a} and M_{8a} are rated for 10 kV.

D. Design of filters and DC capacitors

The passive components consist of the grid-side filter, DC link capacitors, floating capacitors for ANPC, and arm inductors for MMC. Selecting an appropriate filter is crucial, as its sizing depends on the filter type and switching frequency. For this application, an LCL filter is chosen due to its ability to deliver the required performance within a compact volume. The design methodology for the LCL filter is thoroughly detailed in [3]. The designed filter must comply with grid code requirements, ensuring that the output harmonics remain within the limits specified by IEEE 519 standards.

979-8-3315-1612-3/25 $31.00 © 2025 IEEE

The value of the inverter-side inductor can be determined using (2) and (3). By calculating the minimum and maximum allowable inductance, the final value can be selected based on various factors, such as the resonance frequency of the filter ($10f_{fund} \leq f_{res} \leq f_{sw}/2$), size, reactive power requirement, etc.

$$\frac{2\Delta i_L L_{in_{min}}}{M_r V_{in} T_s} = \frac{1}{6}\sin \omega t + \frac{1}{3}\sin\left(\omega t + \frac{2\pi}{3}\right) \\ - \frac{m}{2}(\sin \omega t)^2 \qquad (2)$$

where Δi_L is the maximum ripple current and is set at 30% of the rated current. V_{in} is the dc bus voltage and T_s is the switching time.

$$L_{in_{max}} = \frac{\lambda_{V_{L_{in}}} V_g}{\omega i_L} \qquad (3)$$

where, $\lambda_{V_{L_{in}}}$ is the percentage voltage drop across the inductor and is taken as 3%, V_g is the RMS value of grid voltage and i_L is the rated grid current.

Further the capacitance value of the filter can be calculated by limiting the amount of reactive power generated by the capacitor to less than 5% as given in (4)

$$C_{max} = \frac{\lambda_C P_o}{\omega_o V_g^2} \qquad (4)$$

From the above equation the value is calculated as $0.6 \,\mu F$. Therefore a capacitor of $0.5 \,\mu F$ is chosen for the filter.

Finally, the grid side inductance is calculated based on (5), limiting the switching frequency component of the grid current [16].

$$L_{grid} = \frac{1}{L_{in} C \omega_h^2 - 1}\left(L_{in} + \frac{|V_{inv}(j\omega_h)|}{\omega_h \lambda_h i_g}\right) \qquad (5)$$

where $|V_{inv}(j\omega_h)|$ is the RMS value of the dominant switching harmonic component in the inverter output voltage. λ_h is the factor for maximum switching frequency harmonics allowed, which is set to be 0.15%, and ω_h is the dominant switching frequency component.

The sizing of the DC link capacitors is crucial, particularly in MMCs, where capacitors account for approximately 70% of the converter's total size [4]. Therefore, selecting appropriately sized capacitors is essential. The DC capacitors should be designed to ensure that the ripple voltage at the DC bus does not exceed 20% of the rated voltage. Additionally, the capacitor sizing must also account for the output Total Harmonic Distortion (THD).

E. Switching frequency

Selecting an appropriate switching frequency is critical to achieving optimal system performance. This selection is based on analyzing the losses in both the converter and the filter across varying frequencies. As the switching frequency increases, the filter size and its associated losses decrease, while the inverter losses rise correspondingly. Therefore, an

optimal operational frequency exists where the total system losses are minimized. The results, as shown in Fig. 2, indicate that the minimum system losses occur in the 8–12 kHz range. Based on these findings, a switching frequency of 10 kHz is selected to ensure optimal performance.

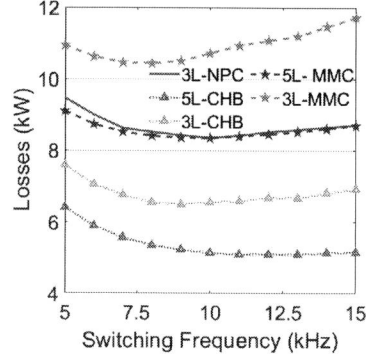

Fig. 2: Variation of losses in the inverter and filter with variation of switching frequency.

III. Modulation and Control

Control and modulation play a crucial role in comparing different inverter topologies. The chosen modulation strategy significantly affects the inverter's loss distribution and output harmonics. Additionally, the modulation scheme must prevent over-modulation, which can degrade the quality of the output waveforms. Therefore, selecting an appropriate modulation scheme is essential for optimal operation.

The control strategy, on the other hand, ensures maximum power point tracking (MPPT) and power balancing across the PV panel strings. In [1], various modulation schemes and control strategies are analyzed to achieve optimal performance in PV plants. This paper assumes uniform power generation across all PV panels (no shading) and minimal manufacturing tolerances among the panels. The front-end isolated DC-DC converter, implemented as a dual-active-bridge converter, ensures MPPT operation for the PV panels and provides a stable DC voltage at its output. These DC-DC converters are connected in series at their outputs to achieve the required DC bus voltage. Additionally, it is assumed that the DC bus voltage is sufficient to prevent the inverter from operating in over-modulation.

A. 3L-NPC

The Neutral Point Clamped (NPC) converter is typically controlled using Level Shifted Pulse Width Modulation (LS-PWM) [9]. In this paper, LS-PWM is employed to generate the gating pulses for the 3L-NPC converter. While this modulation scheme offers excellent output harmonic performance [17], its primary limitation is the unequal distribution of losses among the devices within the converter.

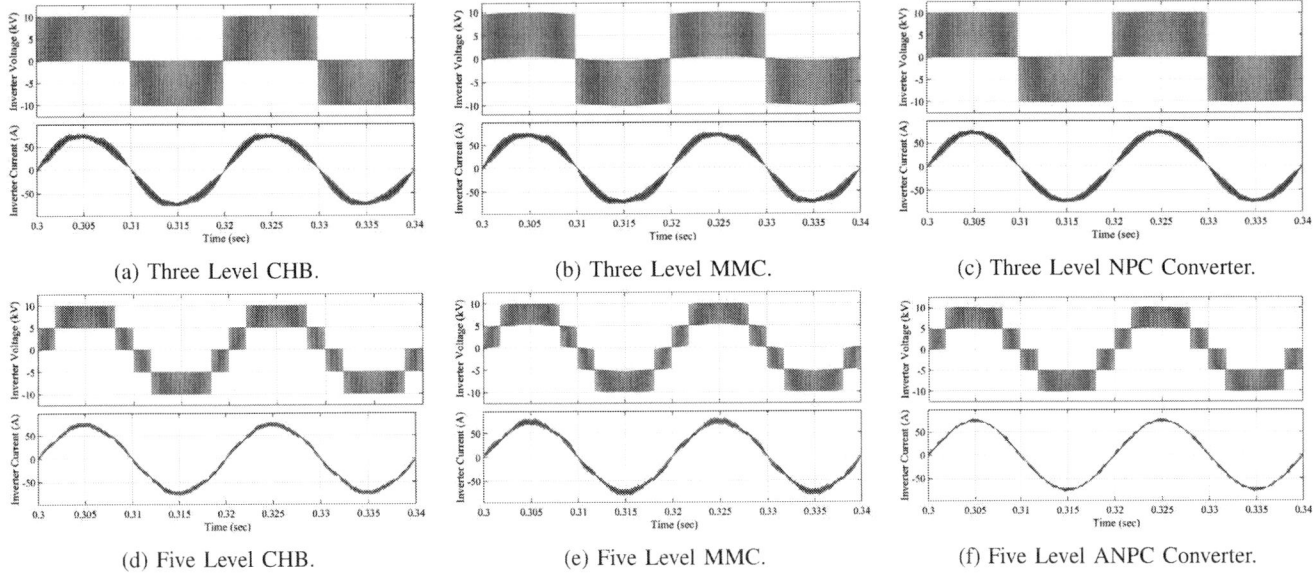

(a) Three Level CHB.   (b) Three Level MMC.   (c) Three Level NPC Converter.

(d) Five Level CHB.   (e) Five Level MMC.   (f) Five Level ANPC Converter.

Fig. 3: Simulation results of phase A inverter output voltage and current due to different topologies and modulation schemes.

B. 5L-ANPC

As noted in [18], LS PWM is a suitable modulation scheme for the 5L-ANPC topology. In this approach, devices M_{1a} to M_{4a} are modulated at the switching frequency based on the LS-PWM gating pulses, while devices M_{5a} to M_{8a} operate at the line frequency. This strategy significantly reduces the overall switching losses in the ANPC converter.

C. CHB

In this paper, Phase Shifted Carrier Pulse Width Modulation (PSC-PWM) is employed to generate the gating pulses for the CHB inverter. As highlighted in [19], PSC-PWM is among the most suitable modulation strategies for CHB inverters. This method offers several advantages, including equal power loss distribution across the inverter devices, low output voltage THD, and an effective increase in the output switching frequency by a factor of N, where N represents the number of cells in a phase leg.

D. MMC

In this paper, the Phase-Shifted Carrier Pulse Width Modulation (PSC-PWM) technique is employed to generate gating pulses for the Modular Multilevel Converter (MMC). As noted in [20], PSC-PWM is well-suited for MMCs because it inherently balances the individual submodule capacitor voltages, eliminating the need for an external cell sorting algorithm. Additionally, this method increases the effective switching frequency at the output by a factor of 2N, where N represents the number of submodules in each arm. Consequently, the requirement for output filtering is significantly reduced.

IV. LOSS MODELLING OF SiC MOSFETs

The loss modeling of SiC MOSFETs plays a critical role in designing the inverter, as the primary loss contributors in

the inverter are device losses. Traditionally, device losses in a converter are estimated using simulation tools such as PLECS or SPICE. However, medium-voltage SiC MOSFETs rated at 3.3 kV and 6.5 kV are still in the developmental stage and are not yet commercially available, which makes their SPICE models unavailable. As a result, an analytical approach for loss estimation becomes necessary.

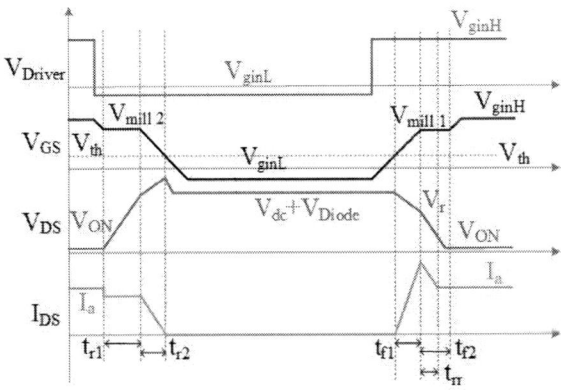

Fig. 4: Key waveforms of turn-on and turn-off transitions.

In [10], a loss model is presented based on the input and output capacitances (C_{gs}, C_{ds}, C_{gd}). The variation of these capacitances is approximated using two discrete values: one corresponding to the rated voltage (C_{ds_1} and C_{gd_1}) and another at $V_{ds} = V_{gs}$ (C_{ds_2} and C_{gd_2}). However, in practical applications, the device operates at voltages different from the rated conditions, and the capacitance values given in datasheets can vary significantly—by up to a factor of five for medium-voltage SiC MOSFETs. Thus, the capacitance values must be selected based on the actual operating voltages

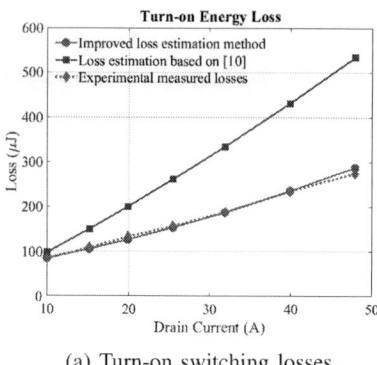

(a) Turn-on switching losses.

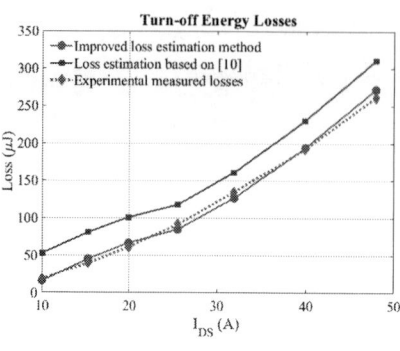

(b) Turn-off switching losses.

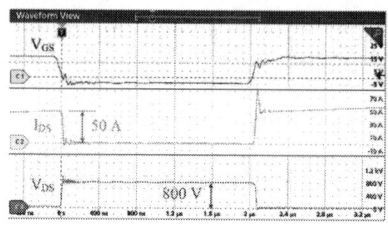

(c) DPT test results of Turn ON and OFF transition.

Fig. 5: Analytical and experimental switching energy for (a) turn-on (b) turn-off (c) DPT test results used for the loss estimation.

before incorporating them into the loss model. As illustrated in Fig. 5a and Fig. 5b, the method proposed in [10] results in inaccurate loss estimations. To address this issue, the following modifications are proposed in the loss modeling process:

1) Capacitance Estimation: The capacitance values (C_{ds} and C_{gd}) should be determined based on the applied drain-source voltage using the graphs provided in the device datasheet.

2) Reverse Recovery Loss: The reverse recovery characteristics should be modeled based on the switch current at the instant of turn-off.

Using these refinements, an enhanced loss model is developed. The switching characteristics of a SiC MOSFET, as shown in Fig. 4, reveal that turn-on and turn-off losses can be divided into specific transition segments. Each transition segment is analyzed individually, and the losses in each segment are calculated separately.

Turn-on losses consist of three distinct components (i) Current Rise: Losses during the current rise phase. (ii) Voltage Fall: Losses during the voltage fall phase. (iii) Diode Reverse Recovery: Losses associated with the reverse recovery of the anti-parallel diode. To calculate the losses for each phase, the duration of the respective transitions must be estimated. This segmented approach allows for a more accurate estimation of turn-on losses.

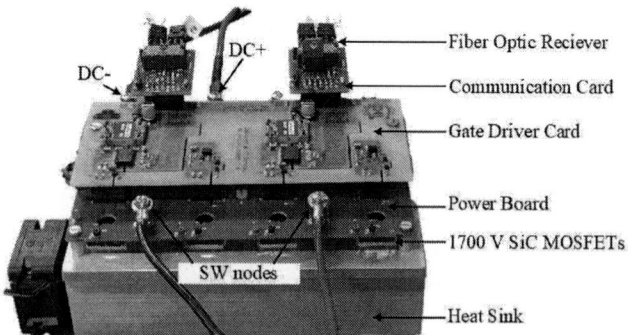

Fig. 6: Developed prototype of the 3L-CHB using 1700 V SiC MOSFETs.

The current rise time (t_{f_1}) is defined as the period for the gate-source voltage to rise from threshold voltage V_{th} to miller voltage V_{mill_1}. The threshold voltage is taken from the datasheet and the miller voltage and the duration of the current rise are calculated using the following equations.

$$V_{mill_1} = V_{th} + \frac{I_L + I_{rr}}{g_{fs}} \quad (6)$$

$$t_{f_1} = \frac{(I_L + I_{rr})C_{iss_2}R_g + (I_L + I_{rr})L_s g_{fs}}{(V_{ginH} - 0.5(V_{mill_1} + V_{th}))g_{fs}} \quad (7)$$

where L_s is the common source inductance. During this time the drain-source voltage across the device falls to recovery voltage V_r, which is the sum of operational voltage, the voltage drop across the diode of the complementary device, and the voltage drop across the source inductance due to current transition is calculated as below.

$$V_r = V_{dc} + V_{Diode} - L_s \frac{I_L + I_{rr}}{t_{f1}} \quad (8)$$

The voltage fall time (T_{f_2}) is determined as the time required for the gate-drain capacitance to discharge from the recovery voltage (V_r) to the on-state voltage (V_{on}). The value of the gate-drain capacitance used in this calculation corresponds to its value at the operational voltage (V_{dc}), as specified below:

$$t_{f_2} = \frac{(V_r - V_{on})C_{gd_1}R_g}{V_{ginH} - V_{mill_1}} \quad (9)$$

The process of reverse recovery of the diode is completed during this period, and it can be estimated based on the reverse recovery charge (Q_{rr}) and the reverse recovery current (I_{rr}). The Q_{rr} is estimated based on the load current as $Q_{rr} = Q_{rr_{datasheet}} I_L / I_{L_{datasheet}}$, where $Q_{rr_{datasheet}}$ is the reverse recovery charge at the test condition with load current $I_{L_{datasheet}}$. Hence, the time for the reverse recovery of the device (t_{rr}) is given by:

$$t_{rr} = 2Q_{rr} \frac{I_L}{I_{rr}} - t_{f_1} \frac{I_L}{I_L + I_{rr}} \quad (10)$$

The total switching energy during turn-on can be estimated by $E_{ON} = E_{Irise} + E_{Vfall} + E_{rr}$ as follows:

$$
\begin{aligned}
E_{ON} = {} & \frac{t_{f_1}}{2}(V_{dc} + V_{Diode} + V_{on})(I_L + I_{rr}) \\
& - \frac{L_s}{3}\left((I_L + I_{rr})\right)^2 + \frac{t_{f_2}}{2}(V_r + V_{on})I_L \\
& + \frac{t_{rr}}{4}(V_r + V_{on})I_{rr}
\end{aligned} \tag{11}
$$

The losses during the turn-off process consist of two components: (i) losses during the voltage rise and (ii) losses during the current fall. The first component, corresponding to the voltage rise, is calculated based on the time required for C_{gd} to charge up to the DC bus voltage plus the diode's voltage drop.

$$
t_{r_1} = \frac{(V_{dc} + V_{Diode} - V_{on})R_g C_{gd_2}}{V_{mill_2} - V_{ginL}} \tag{12}
$$

The drain current during this time consists of two parts (i) load current (I_L) and (ii) current due to the discharge of drain-source capacitor (C_{ds}). Also during this period the gate source voltage does not change and is constant at miller plateau voltage. The drain current and miller voltage is calculated as below:

$$
I_{ds} = I_L - C_{ds_1}\frac{V_{dc} + V_{Diode} + V_{on}}{t_{r_1}} \tag{13}
$$

$$
V_{mill_2} = V_{th} + \frac{I_{ds}}{g_{fs}} \tag{14}
$$

The current fall time is defined as the time taken for the gate-source voltage to fall from miller voltage to threshold voltage. This time period is calculated as below:

$$
t_{r_2} = \frac{(C_{iss_2}R_g + L_s g_{fs})I_{ds}}{(0.5(V_{mill_2} + V_{th}) - V_{ginL})g_{fs}} \tag{15}
$$

The total switching energy during turn off can be estimated by $E_{OFF} = E_{Vrise} + E_{Ifall}$ as follows:

$$
\begin{aligned}
E_{OFF} = {} & \frac{t_{r_1}}{2}I_{ds}(V_{dc} + V_{Diode} - V_{on}) \\
& + \frac{t_{r_2}}{2}(V_{dc} + V_{Diode} - V_{on})I_{ds} - \frac{L_s}{3}I_{ds}^2
\end{aligned} \tag{16}
$$

A Double Pulse Test (DPT) circuit is utilized to validate the accuracy of the developed loss model. Voltage measurements are performed using high-bandwidth, high-voltage passive probes, while device currents are captured using a Rogowski coil-based current probe. To ensure accurate loss calculations, the measured signals are de-skewed, and propagation delays across different probes are carefully matched. The results, shown in Fig. 5c, highlight the robustness of the proposed loss model. Experimental losses are determined by integrating the product of V_{ds} and I_{ds} during the turn-on and turn-off events of the MOSFETs. A comparison of the experimental and analytical losses, presented in Fig. 5, demonstrates that

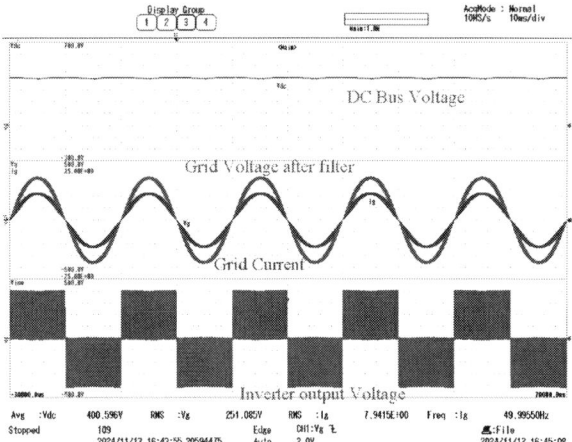

Fig. 7: Experimental waveforms of 3L-CHB inverter.

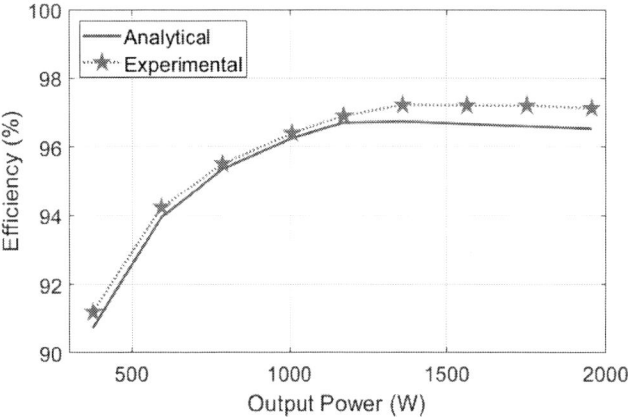

Fig. 8: Analytical and experimental efficiency of 3L-CHB with output power variation.

the analytical model aligns closely with the experimental data, affirming the model's reliability and precision.

To extend the model's validation, a three-level CHB converter prototype incorporating 1700 V SiC MOSFETs was developed, as shown in Fig. 6. Operating waveforms captured at 400 VDC and 2 kW power levels are presented in Fig. 7. The estimated efficiency derived from the loss model and the experimentally measured efficiency are compared in Fig. 8. These results demonstrate that the analytical loss model, incorporating the proposed modifications, effectively estimates converter losses, enabling its application in inverter design as detailed in subsequent sections.

V. Topological Comparison of Inverter

Based on the design methodology outlined in the preceding sections, inverters are designed for various topologies, including 3L-NPC, 5L-ANPC, 3L-MMC, 5L-MMC, and 3L-CHB and 5L-CHB configurations. Leveraging these designs and the loss model developed in Section IV, a comparative analysis of the inverter topologies is conducted. This comparison evaluates key metrics such as the number of components, voltage and current stresses, harmonic performance, efficiency, total volume, and control complexity. A detailed discussion of each metric is provided below.

A. Number of Components

Reference [3] highlights the use of 3.3 kV devices for implementing an 11 kV system, demonstrating superior performance compared to 6.5 kV SiC MOSFETs. Consequently, to realize the various inverter topologies for the 11 kV, 1 MW system, a series connection of 3.3 kV SiC MOSFETs is employed. The selected MOSFET for this implementation is the Genesic 3.3 kV, 35 A discrete device (G2R120MT33J) [21].

In three-level converters, the device blocking voltage required is a minimum of 10 kV. Hence four devices are connected in series to be able to realize the required blocking voltage. Further in 3L-NPC there is a requirement of additional clamping diodes, while the other inverters don't require any extra diodes. It is seen that the semiconductor count is the least for NPC while that of CHB and MMC is the same. However, there is an additional requirement for an arm inductor and an increased number of cell capacitors in MMC.

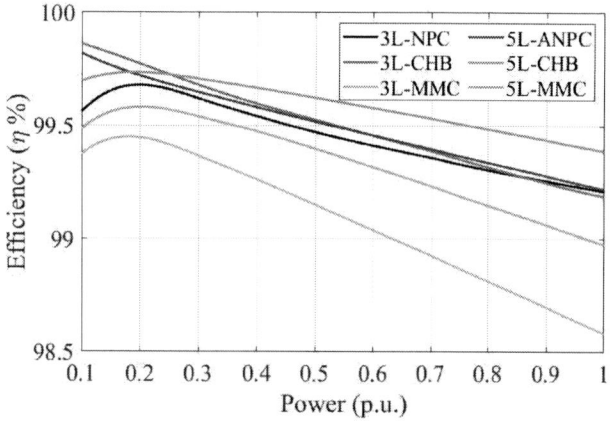

Fig. 9: Inverter efficiencies with respect to load variation.

In the five-level topology, the number of modules is increased but their voltage ratings are reduced hence the number of semiconductor devices remains the same. However,

the number of cell capacitors and auxiliary components are increased.

B. Voltage and current stress

In three-level topologies, the voltage stress on the devices is uniform at $V_{DC}/2 \approx 10$ kV. However, in the case of the 5L-ANPC topology, devices S1-S4 experience higher voltage stress compared to other five-level topologies. Specifically, S1-S4 must block a voltage of 10 kV, while S5-S8 only need to block 5kV. As a result, additional series connection of devices is required for S1-S4 in the 5L-ANPC topology.

The current stress on devices corresponds to the line current for the NPC, CHB, and ANPC topologies. In contrast, the MMC topology experiences additional circulating currents through its devices, which are necessary for balancing the cell capacitor voltages. Consequently, the devices used in the MMC require a higher current rating.

C. Harmonics

The design of the LCL filter for 3L and 5L topologies is performed using the methodology outlined in Section II-D. For the 3L-NPC and 5L-ANPC topologies, the switching frequency is equal to the output current frequency. In contrast, for CHB and MMC topologies, the output frequency increases by a factor of N and 2N, respectively. Consequently, the filters are designed based on the specific output frequency of each topology. According to the IEEE-519:2022 standards, the converter must comply with the total demand distortion (TDD) limits of less than 5%, replacing the older total harmonic distortion (THD) limits.

D. Efficiency

The efficiency of the system is determined by estimating the losses in the switches and the passive components. The variation in efficiency with load changes is illustrated in Fig. 9. For solar PV inverters, the European efficiency (Euro efficiency) is commonly used for performance comparison. The Euro efficiency is calculated as a weighted sum of the inverter efficiencies at different load levels. Based on the Euro efficiency, the 5-level Cascaded H-Bridge (5L-CHB) topology demonstrates the highest efficiency, closely followed by the 5-level Active Neutral Point Clamped (5L-ANPC) topology.

TABLE I: Medium Voltage inverter comparison for topology selection

Parameters for 1 MVA, 11 kV Inverter	Three Level			Five Level		
	NPC	*MMC*	*CHB*	*ANPC*	*MMC*	*CHB*
No. of devices	12(Q)+6(D)	24(Q)	24(Q)	36(Q)	48(Q)	48(Q)
TDD of output current (%)	1.85	1.6	1.37	1.81	3.1	3.04
Voltage stress on module (kV)	10	10	10	10,5	5	5
Current stress on module (A)	54	44 x 2	54	54	44 x 2	54
DC capacitor requirement(J/kVA)	1.25	12.5	5	2.2	8.5	3.8
Filter inductor (mH)	11	10 + 6x1	11	8	7 mH + 1 mH x 6	8
Euro efficiency (%)	99.46	99.12	99.54	99.52	99.36	99.60
Volume	Low	High	Moderate	Moderate	High	Moderate
Control complexity	Simple	Simple	Complex	Complex	Moderate	Complex

E. Total Inverter Volume

The inverter's volume is qualitatively estimated based on the sizing of passive components (DC link capacitors, filter inductors) and heat sinks. Capacitor sizing is compared using the stored energy requirement (J/kVA), as sizes vary by manufacturer for similar ratings. Heat sink sizing depends on material (e.g., aluminum, copper), technology (e.g., cold plate, phase change), and cooling method (e.g., natural/forced convection). Therefore, it is assessed based on inverter losses and their distribution. These comparisons and results are summarized in Table I.

VI. CONCLUSION

This paper presents a comprehensive design methodology for medium-voltage (MV) grid-connected solar PV inverters, leveraging advancements in silicon carbide (SiC) MOSFET technology. Key design aspects, including SiC device selection, switching frequency optimization, filter design, and DC capacitor sizing, are addressed. A refined loss estimation model, validated experimentally, provides accurate performance analysis of various inverter topologies, including 3L-NPC, 5L-ANPC, 3L- and 5L-MMC, and 3L- and 5L-CHB configurations.

The comparative analysis highlights the trade-offs among topologies. Three-level designs feature simpler control but impose higher voltage stress on devices. Five-level topologies reduce voltage stress, though they vary in current stress and capacitor requirements. Among the evaluated topologies, the 5L-CHB demonstrates the best overall performance, offering superior efficiency, manageable current stress, and optimized passive component sizing, with harmonic distortion and voltage stress comparable to the best alternatives. For lower voltage levels, the 5L-ANPC and 3L-CHB also emerge as strong candidates due to their balance of efficiency and design simplicity.

These findings establish the 5L-CHB as the most suitable topology for 11 kV MV grid-connected PV systems and provide a foundation for advancing inverter design. The proposed methodology emphasizes the importance of SiC MOSFETs and topology-specific optimization in developing efficient, scalable MV PV solutions.

REFERENCES

[1] M. Rabiul Islam, A. M. Mahfuz-Ur-Rahman, K. M. Muttaqi and D. Sutanto, "State-of-the-Art of the Medium-Voltage Power Converter Technologies for Grid Integration of Solar Photovoltaic Power Plants," in IEEE Transactions on Energy Conversion, vol. 34, no. 1, pp. 372-384, March 2019.

[2] X. Zhang, M. Wang, T. Zhao, W. Mao, Y. Hu and R. Cao, "Topological comparison and analysis of medium-voltage and high-power direct-linked PV inverter," in CES Transactions on Electrical Machines and Systems, vol. 3, no. 4, pp. 327-334, Dec. 2019.

[3] J. J. Attukadavil, S. Anand and B. G. Fernandes, "An Adaptive DC Voltage Control for SiC based Medium Voltage Photovoltaic Inverter," 2022 IEEE Energy Conversion Congress and Exposition (ECCE), Detroit, MI, USA, 2022, pp. 1-7.

[4] A. Marzoughi, R. Burgos, D. Boroyevich and Y. Xue, "Design and Comparison of Cascaded H-Bridge, Modular Multilevel Converter, and 5-L Active Neutral Point Clamped Topologies for Motor Drive Applications," in IEEE Transactions on Industry Applications, vol. 54, no. 2, pp. 1404-1413, March-April 2018.

[5] A. Marzoughi, R. Burgos and D. Boroyevich, "Investigating Impact of Emerging Medium-Voltage SiC MOSFETs on Medium-Voltage High-Power Industrial Motor Drives," in IEEE Journal of Emerging and Selected Topics in Power Electronics, vol. 7, no. 2, pp. 1371-1387, June 2019.

[6] A. Kumar, S. Bhattacharya, J. Baliga and V. Veliadis, "Performance Comparison and Demonstration of 3-L Voltage Source Inverters Using 3.3 kV SiC MOSFETs for 2.3 kV High Speed Induction Motor Drive Applications," 2021 IEEE Applied Power Electronics Conference and Exposition (APEC), Phoenix, AZ, USA, 2021, pp. 1103-1110.

[7] A. Marzoughi, R. Burgos and D. Boroyevich, "Investigating Impact of Emerging Medium-Voltage SiC MOSFETs on Medium-Voltage High-Power Industrial Motor Drives," in IEEE Journal of Emerging and Selected Topics in Power Electronics, vol. 7, no. 2, pp. 1371-1387, June 2019.

[8] Islam, M.R., Guo, Y., Zhu, J. (2014). Multilevel Converters for Step-Up-Transformer-Less Direct Integration of Renewable Generation Units with Medium Voltage Smart Microgrids. In: Hossain, J., Mahmud, A. (eds) Large Scale Renewable Power Generation. Green Energy and Technology. Springer, Singapore.

[9] S. Belkhode, P. Rao, A. Shukla and S. Doolla, "Comparative Evaluation of Silicon and Silicon-Carbide Device-Based MMC and NPC Converter for Medium-Voltage Applications," in IEEE Journal of Emerging and Selected Topics in Power Electronics, vol. 10, no. 1, pp. 856-867, Feb. 2022.

[10] S. Eskandari, K. Peng, B. Tian and E. Santi, "Accurate Analytical Switching Loss Model for High Voltage SiC MOSFETs Includes Parasitics and Body Diode Reverse Recovery Effects," 2018 IEEE Energy Conversion Congress and Exposition (ECCE), Portland, OR, USA, 2018, pp. 1867-1874.

[11] S. K. Roy and K. Basu, "Analytical Model to Study Hard Turn-off Switching Dynamics of SiC mosfet and Schottky Diode Pair," in IEEE Transactions on Power Electronics, vol. 36, no. 1, pp. 861-875, Jan. 2021.

[12] S. K. Roy and K. Basu, "Analytical Estimation of Turn on Switching Loss of SiC mosfet and Schottky Diode Pair From Datasheet Parameters," in IEEE Transactions on Power Electronics, vol. 34, no. 9, pp. 9118-9130, Sept. 2019.

[13] X. Wang, Z. Zhao, K. Li, Y. Zhu and K. Chen, "Analytical Methodology for Loss Calculation of SiC MOSFETs," in IEEE Journal of Emerging and Selected Topics in Power Electronics, vol. 7, no. 1, pp. 71-83, March 2019.

[14] M. Shen and S. Krishnamurthy, "Simplified loss analysis for high speed SiC MOSFET inverter," 2012 Twenty-Seventh Annual IEEE Applied Power Electronics Conference and Exposition (APEC), Orlando, FL, USA, 2012, pp. 1682-1687.

[15] J. He, A. Sangwongwanich, Y. Yang and F. Iannuzzo, "Performance Comparison of PV Inverter Systems Considering System Voltage Ratings and Installation Sites," 2021 IEEE Applied Power Electronics Conference and Exposition (APEC), Phoenix, AZ, USA, 2021, pp. 2620-2625.

[16] X. Ruan, X. Wang, D. Pan, D. Yang, W. Li and C. Bao, Control Techniques for LCL-Type Grid-Connected Inverters, Beijing, China:Springer Press, 2018.

[17] A. Nabae, I. Takahashi and H. Akagi, "A New Neutral-Point-Clamped PWM Inverter," in IEEE Transactions on Industry Applications, vol. IA-17, no. 5, pp. 518-523, Sept. 1981.

[18] P. Barbosa, P. Steimer, J. Steinke, M. Winkelnkemper and N. Celanovic, "Active-neutral-point-clamped (ANPC) multilevel converter technology," 2005 European Conference on Power Electronics and Applications, Dresden, Germany, 2005, pp. 10.

[19] C. D. Townsend, T. J. Summers and R. E. Betz, "Phase-Shifted Carrier Modulation Techniques for Cascaded H-Bridge Multilevel Converters," in IEEE Transactions on Industrial Electronics, vol. 62, no. 11, pp. 6684-6696, Nov. 2015

[20] S. Debnath, J. Qin, B. Bahrani, M. Saeedifard and P. Barbosa, "Operation, Control, and Applications of the Modular Multilevel Converter: A Review," in IEEE Transactions on Power Electronics, vol. 30, no. 1, pp. 37-53, Jan. 2015

[21] [Online]:https://www.genesicsemi.com/sic-mosfet/G2R120MT33J/G2R120MT33J.pdf

EMI Modeling of PCB-based Three-Level Active Neutral-Point-Clamped GaN Converter

Mohammad Hassan Adeli
Department of Electrical Engineering
Untiversity of South Carolina
Coulmbia, U.S.
madeli@email.sc.edu

Necmi Altin
Department of Electrical Engineering
Untiversity of South Carolina
Coulmbia, U.S.
altin@sc.edu

Erkan Deniz
Department of Electrical Engineering
Untiversity of South Carolina
Coulmbia, U.S.
edeniz@mailbox.sc.edu

Adel Nasiri
Department of Electrical Engineering
Untiversity of South Carolina
Coulmbia, U.S.
nasiri@sc.edu

Abstract— This paper investigates the Electromagnetic Interference (EMI) behavior of a Gallium Nitride (GaN)-based three-level Active Neutral Point Clamped (3L-ANPC) converter. GaN-based switches are utilized in the converter design to achieve high switching frequencies and to reduce overall system size and losses, but they pose challenges related to electromagnetic compatibility (EMC). Accurate modeling of the system's Common mode Equivalent circuit Model (CEM) is essential to effectively investigate EMI emissions. In this work, a PCB-integrated converter is designed to mitigate EMI emissions while enhancing system performance and reliability. The PCB structure facilitates a straightforward approach to calculating parasitic capacitances. An analytical approach is also presented for modeling system's CEM, enabling the investigation of the Common Mode (CM) current. Simulations are conducted to evaluate the performance of the CEM by comparing its emission characteristics with those obtained from a Mixed Mode (MM) model.

Keywords— Active neutral point clamped, common mode, EMI modeling, EMI current, equivalent circuit, GaN switches, inverter, parasitic capacitance.

I. INTRODUCTION

Multi-level converter topologies have been prominent in addressing the demands of high-power and high-voltage applications [1]. The three-level Neutral Point Clamped (3L-NPC) topology stands as one of the most widely favored multilevel inverters. They achieve high voltage operation and allow the use of switching devices rated for higher current and lower voltage. They improve the voltage waveform quality of a voltage source inverter's output voltage by raising waveform steps and lowering voltage stress across switches. Reduced voltage stress leads to lower dv/dt, which consequently causes fewer electromagnetic interference (EMI) problems. Moreover, The three-level Active NPC (3L-ANPC) converter is introduced to address unbalance issues and enhance performance by utilizing various PWM techniques, providing more degrees of freedom to optimize the 3L-NPC converter structure.

Wide band gap (WBG) semiconductors like Silicon Carbide (SiC) and Gallium Nitride (GaN) enable higher temperature, voltage, and switching frequencies than silicon, but this introduces EMC concerns [2]. The increased voltages, faster edge rates, and higher switching frequencies in power-dense electronic converters lead to Common Mode (CM) current

emissions, which can result in significant radiated emission problems and excessive conducted emissions [3]. This CM noise has the potential to generate radiated emissions problems in the power distribution network by finding its way through unforeseen ground routes.

For WBG devices, higher switching frequencies make busbar design crucial due to increased parasitic susceptibility. The thinner skin depth at high frequencies also impacts the busbar choice. Voltage over-shoot during turn-off depends on the commutation loop's total inductance, highlighting the need for low inductance busbar design [4]. Low inductance design aids component integration and reduces parasitic capacitances, making busbar optimization essential. Thus, the busbar design must minimize inductance and excel in magnetic flux cancellation. This also facilitates easier integration of components like sensors and resistors. Considering these aspects, opting for a PCB busbar is advantageous due to its cost-effectiveness, improved high-frequency current carrying capacity, effective minimization of loop inductance, lightweight and high-power density.

Recent EMI characterization aims to avoid over-designed filters and improve system compatibility, addressing EMC design issues effectively. Determining the precise attenuation needed to satisfy standards requires proper characterization. EMI emissions have been characterized using two fundamental methods: time-domain and frequency-domain. The modeling of EMI noise sources and noise propagation paths are components of both techniques. The modeling of the noise source is mostly where the two diverge. Simulation in time-domain may take a long time [5] for a complex circuit with several parasitic components, and convergence issues could arise. Frequency-domain analysis uses substitution theory to linearize circuits, whereby noise sources are substituted for the original switches [6]. In [7], the EMI emission of 3L-ANPC inverter for different modulation technique is investigated in frequency domain where the noise sources are modeled based on the approach introduced in [6]. The EMI modeling of an Active NPC (ANPC) is studied in [8-10], where the frequency modeling technique is deployed for modeling and the predicted model is compared with the time-domain results. The conventional open module die area is used for measurement, which can pose a risk of damage

979-8-3315-1612-3/25 $31.00 © 2025 IEEE

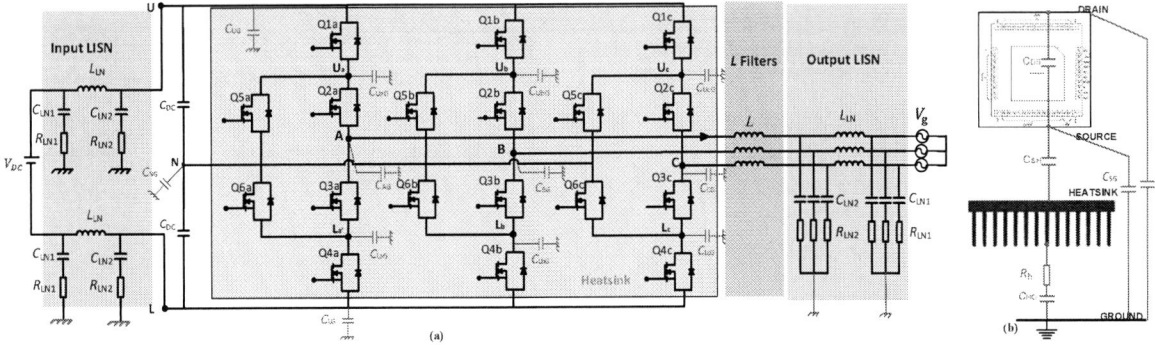

Fig. 1. a) Three-phase grid-connected 3L-ANPC converter with the filter inductor. b) each switch parasitic capacitance.

and render the module non-functional. Also, the parasitic capacitances are calculated analytically for their discrete packaging. To avoid the costly, risky, and time-consuming process of module parasitic extraction, this work employs a PCB-based EMI optimization technique to effectively calculate parasitic capacitances, ensuring both efficiency and safety while maintaining module functionality.

This paper investigates the EMI characterization of an ANPC inverter topology, where GaN switches operate at high voltage and switching frequencies, reducing system loss and size. The PCB-integrated GaN-based 3L-ANPC converter is designed to mitigate EMI emissions effectively, improving both system performance and reliability. Additionally, the PCB structure enables a straightforward method for calculating parasitic capacitances, which is a key step before proceeding with the system modeling. Using calculated capacitance, the MM model of the system is developed, offering valuable insights into its differential-mode (DM) EMI behavior. Subsequently, the Common mode Equivalent circuit Model (CEM) of the system is analytically derived from the MM model, enabling a more comprehensive evaluation of CM emissions. Moreover, the paper presents a detailed comparison between the Mixed Mode (MM) model and CEM, both in the frequency domain and time domain, using simulations.

II. COMMON MODE EQUIVALENT CIRCUIT MODELING

First, The three-phase grid-connected 25kW 3L-ANPC converter with its filter inductor is shown in Fig. 1a. To create a MM model, the dominant parasitic paths are identified and parameterized by incorporating corresponding parasitic capacitances into the Differential Mode (DM) model. Also, Line Impedance Stabilization Networks (LISNs) are incorporated into the system at the input and output, serving as terminating impedances for the equipment under test (EUT). The EUT is a converter composed of six 650V GaN switches per phase, forming an ANPC topology. Switch terminals are capacitively coupled to the inverter heatsink and ground, as illustrated in Fig. 1b. Using the obtained MM model shown in Fig. 1a, different groups of capacitors connected to the labeled super-nodes are identified. Considering an arbitrary reference point, P, distinct from the ground to define the CM voltage allows for the conversion of the MM system components into CM equivalent components. By doing so, their voltage equations are expressed relative to this arbitrary reference point as a two- or three-branch

model as CM current and voltage with respect to P for the N branches can be defined as

$$i_{\mathrm{CM}} \triangleq \sum_{k=1}^{N} i_k, \quad v_{\mathrm{CM}} \triangleq \frac{1}{N} \sum_{k=1}^{N} v_k p, \tag{1}$$

where N is a designated normalization factor. DM current and voltage are defined as

$$i_{mn} \triangleq \frac{1}{N_i}(i_m - i_n), \qquad v_{mn} \triangleq v_{mp} - v_{np}. \tag{2}$$

These definitions can then be expanded to accommodate multiple line currents and voltages as follows

$$\mathbf{i}_{\mathrm{DCM}} \triangleq \mathbf{T}_i \mathbf{i}_n, \qquad \mathbf{v}_{\mathrm{DCM}} \triangleq \mathbf{T}_v \mathbf{v}_{np}. \tag{3}$$

Here, $\mathbf{i}_n$ represents the vector of MM line currents, while $\mathbf{i}_{\mathrm{DCM}}$ denotes the vector of DM and CM currents. The same applies correspondingly to the voltage vectors of $\mathbf{v}_{np}$ and $\mathbf{v}_{\mathrm{DCM}}$. The expanded form of (3) is as follows,

$$\begin{bmatrix} i_{12} \\ i_{23} \\ \vdots \\ i_{(N-1)N} \\ I_{\mathrm{CM}} \end{bmatrix} \triangleq \begin{bmatrix} 1/N_i & -1/N_i & 0 & \cdots & 0 & 0 & 0 \\ 0 & 1/N_i & -1/N_i & & 0 & 0 & 0 \\ \vdots & & \vdots & \ddots & & \vdots & \\ 0 & 0 & 0 & \cdots & 0 & 1/N_i & -1/N_i \\ 1 & 1 & 1 & & 1 & 1 & 1 \end{bmatrix} \begin{bmatrix} i_1 \\ i_2 \\ \vdots \\ i_{(N-1)} \\ i_N \end{bmatrix} \tag{4}$$

$$\begin{bmatrix} v_{12} \\ v_{23} \\ \vdots \\ v_{(N-1)N} \\ v_{\mathrm{CM}} \end{bmatrix} \triangleq \begin{bmatrix} 1 & -1 & 0 & \cdots & 0 & 0 & 0 \\ 0 & 1 & -1 & & 0 & 0 & 0 \\ \vdots & & \vdots & \ddots & & \vdots & \\ 0 & 0 & 0 & \cdots & 0 & 1 & -1 \\ 1 & 1 & 1 & & 1 & 1 & 1 \end{bmatrix} \begin{bmatrix} v_{1P} \\ v_{2P} \\ \vdots \\ v_{(N-1)P} \\ v_{NP} \end{bmatrix}. \tag{5}$$

The matrices $\mathbf{T}_v$ and $\mathbf{T}_i$ in equations (4) and (5) define a distinct set of modes, even though these modes are not the only possible options. Additionally, there is flexibility in selecting N_i; it is not confined to a single specific choice. In this work, for a set of two super-nodes of (U, L), (U$_a$, L$_a$), (U$_b$, L$_b$), (U$_c$, L$_c$) in Fig. 1a, these transformation matrix of $\mathbf{T}_v$ and $\mathbf{T}_i$ are considered as follows

$$\mathbf{T}_v = \begin{bmatrix} 1 & -1 \\ \frac{1}{2} & \frac{1}{2} \end{bmatrix}, \qquad \mathbf{T}_i = \begin{bmatrix} \frac{1}{2} & -\frac{1}{2} \\ 1 & 1 \end{bmatrix}. \tag{6}$$

Also, for a set of three super-nodes of (U, L, N) and (A, B, C), these transformation matrix are derived as follows

$$\mathbf{T}_v = \begin{bmatrix} 1 & -1 & 0 \\ 0 & 1 & -1 \\ 1/3 & 1/3 & 1/3 \end{bmatrix}, \quad \mathbf{T}_i = \begin{bmatrix} 1/2 & -1/2 & 0 \\ 0 & 1/2 & -1/2 \\ 1 & 1 & 1 \end{bmatrix}. \tag{7}$$

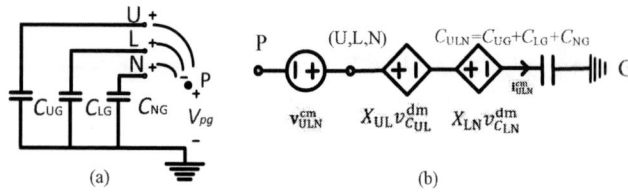

Fig. 2. a) Three-branch capacitor model, b) three-branches equivalent CM circuit.

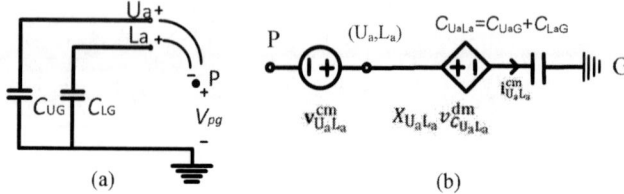

Fig. 3. a) Two-branch capacitor model, b) Two-branches equivalent CM circuit.

To model the CM, we need to analyze the voltage expression for each super-nodes of (U, L, N), (U, L), (U_a, L_a), (U_b, L_b), (U_c, L_c), and (A, B, C). This typically involves finding the relationships between the voltages at these super-nodes to an arbitrary reference point P. For the super-nodes (U, L, N) and (A, B, C), three branches are connected, including the dominant paths of parasitic capacitances. These dominant paths, characterized by high voltage slew rates (dv/dt) driven by switching operations, are the primary causes of radiated and conducted EMI. These paths are characterized by the inclusion of parasitic capacitances, which play a crucial role in defining the system's behavior under high frequency switching operations. As shown in Fig. 2a, these capacitances act as coupling elements, influencing the distribution of high-frequency currents and impacting both radiated and conducted (EMI). We can then write the voltage expression for (U, L, N) node as

$$[v_{UP} \quad v_{NP} \quad v_{LP}]^T + v_{PG}[1 \quad 1 \quad 1]^T$$

$$= [v_{C_{UG}} \quad v_{C_{NG}} \quad v_{C_{LG}}]^T = \mathbf{v}_{C_{ULN}} \quad (8)$$

where, $p = dv/dt$ is the Heaviside operator, and $\mathbf{i}_{ULN} = p\begin{bmatrix} C_{UG} & 0 & 0 \\ 0 & C_{LG} & 0 \\ 0 & 0 & C_{NG} \end{bmatrix}\mathbf{v}_{C_{ULN}}$. C_{xG} can be calculated as the equivalent capacitance between the (U, L, N) super-node and the reference point, as shown in Fig. 3a. This method can similarly be applied to other super-nodes of Fig. 3a. Left multiplying (8) by $\mathbf{T}_v$, and using (3), the common differential voltage can be expressed as

$$[v_{UL}^{dm} \quad v_{LN}^{dm} \quad v_{ULN}^{cm}]^T + v_{PG}[0 \quad 0 \quad 1]^T$$

$$= [v_{C_{UL}}^{dm} \quad v_{C_{LN}}^{dm} \quad v_{ULN}^{cm}]^T = \mathbf{v}_{C_{ULN}}^{dcm}. \quad (9)$$

Also, by left multiplying $\mathbf{i}_{ULN}$ by $\mathbf{T}_i$ and using (3), we will have

$$\mathbf{i}_{ULN}^{dcm} = [i_{UL}^{dm} \quad i_{LN}^{dm} \quad i_{ULN}^{cm}]^T$$

$$= p\mathbf{T}_i\begin{bmatrix} C_{UG} & 0 & 0 \\ 0 & C_{LG} & 0 \\ 0 & 0 & C_{NG} \end{bmatrix}(\mathbf{T}_v)^{-1}\mathbf{v}_{C_{ULN}}^{dcm}. \quad (10)$$

By substituting $\mathbf{T}_v$ and $\mathbf{T}_i$ of (7) in (10), and then inserting (10) into (9), we can simplify the common voltage expression of (9) as follows:

$$v_{ULN}^{cm} + v_{PG} = -\frac{1}{3}\left(\frac{2C_{UG} - C_{LG} - C_{NG}}{C_{UG} + C_{LG} + C_{NG}}\right)v_{C_{UL}}^{dm} \quad (11)$$

$$-\frac{1}{3}(\frac{C_{UG} + C_{LG} - 2C_{NG}}{C_{UG} + C_{LG} + C_{NG}})v_{C_{LN}}^{dm} + \frac{i_{ULN}^{cm}}{p(C_{UG} + C_{LG} + C_{NG})}.$$

A single line CM equivalent circuit of (11) can be expanded as depicted in Fig. 2b. By selecting N as the reference point for the floating point, we can calculate the dependent and independent voltage source values in Fig. 2b as

$$v_{UL}^{cm} = \frac{v_{UN} + v_{LN}}{2}, \qquad v_{ULN}^{cm} = \frac{v_{UN} + v_{LN} + v_{NN}}{3}, \quad (12)$$

$$v_{C_{UL}}^{dm} = v_{C_{UN}} - v_{C_{LN}}, \qquad v_{C_{LN}}^{dm} = v_{C_{LN}} - v_{C_{NN}}.$$

A similar analysis can be performed for the other three-path super-node (A, B, C). By conducting a linear voltage analysis for the super-nodes that include the two dominant parasitic paths in Fig. 3a, we can also model a single line CM equivalent model for these nodes. Considering (U_a, L_a) super-node, we can write the voltage expression as

$$[v_{UP} \quad v_{LP}]^T + v_{PG}[1 \quad 1]^T = [v_{C_{UG}} \quad v_{C_{LG}}]^T \quad (13)$$

$$= \mathbf{v}_{C_{UL}}$$

To calculate the common differential voltage, the $\mathbf{v}_{C_{UL}}^{dcm}$ and $\mathbf{i}_{UL}^{dcm}$ can be derived by left-multiplying equation (13) by $\mathbf{T}_v$ of (6) and left-multiplying $\mathbf{i}_{UL}$ by $\mathbf{T}_i$ of (6). After manipulating them the common differential voltage is derived as

$$v_{UL}^{cm} + v_{PG} = -\frac{1}{2}\left(\frac{C_{UG} - C_{LG}}{C_{UG} + C_{LG}}\right)v_{C_{UL}}^{dm} \quad (13)$$

$$+ \frac{i_{ULN}^{cm}}{p(C_{UG} + C_{LG} + C_{NG})}.$$

A single line CM equivalent circuit of (13) is modeled and illustrated at Fig. 3b. By selecting N as the reference point, we can calculate the voltage terms in Fig. 4b as

$$v_{U_aL_a}^{cm} = \frac{v_{U_aN} + v_{L_aN}}{2}, \quad v_{C_{U_aL}}^{dm} = v_{C_{U_aN}} - v_{C_{L_aN}} \quad (14)$$

This approach can also be applied to the other super-nodes with two parasitic paths, such as (U, L), (U_b, L_b), and (U_c, L_c), to derive the single line CM equivalent circuit for each. By connecting all of these individual single line circuits, a comprehensive equivalent CM model of the entire system can be constructed, as shown in Fig. 4b. This combined model provides a clear representation of the CM behavior across all super-nodes in the system. The capacitors in these models can be calculated using the coupling model of Fig. 1b, where just for C_{UG} as an example this capacitor can be found by the summation of four parallel paths of capacitor illustrated in Fig 4a. This can calculated as

$$C_{UG} = C_{P_1} + C_{Q_{1a}} + C_{Q_{1b}} + C_{Q_{1c}}, \quad (15)$$

979-8-3315-1612-3/25 $31.00 © 2025 IEEE

where C_{p_1} is the equivalent capacitor between two conducting layers connected to positive DC bus and the reference point, and

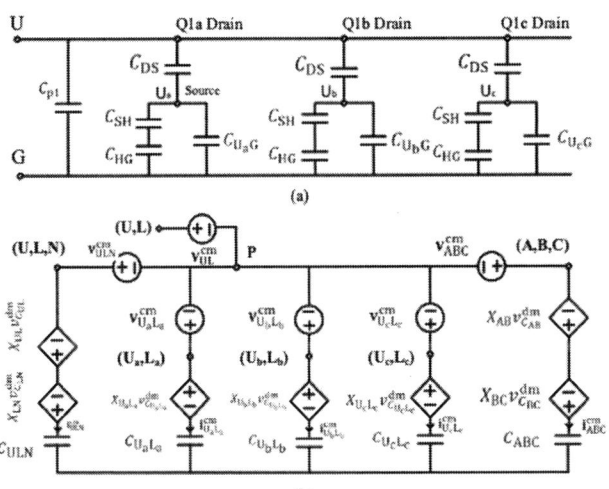

(a)

(b)

Fig. 4. a) Capacitance between the U super-node and ground, and b) Equivalent CM circuit of the ANPC converter.

$C_{Q_{1x}}$ represents the equivalent capacitance between the first switch of each phase and the reference point calculating as

$$C_{Q_{1x}} = \frac{C_{DS} \times \left(C_{U_xG} + \frac{C_{SH} \times C_{HG}}{C_{SH} + C_{HG}} \right)}{C_{DS} + C_{U_xG} + \left(\frac{C_{SH} \times C_{HG}}{C_{SH} + C_{HG}} \right)}. \tag{16}$$

With the equivalent CM circuit model, we can effectively characterize and analyze the EMI noise sources of the converter, enabling a deeper understanding of their origins and pathways. In the next section, we will discuss the PCB design approach we employed to parameterize the parasitic coupling capacitances, focusing on how layout considerations and component placement influence the coupling effects and overall EMI performance.

III. PARASITIC CAPACITANCE CALCULATION

There are several steps involved in the general process of creating an equivalent circuit from the DM model of the component. As discussed, in the first step, major CM parasitic couplings that give pathways for CM currents are extracted and added to the DM model to define the MM model of Fig. 1a. High voltage slew rates (dv/dt) caused by switching operations are the primary contributors to radiated EMI and CM noise associated with the extracted parasitics, particularly when Pulse Width Modulation (PWM) is applied. These rapid voltage changes induce strong coupling effects through parasitic capacitances, exacerbating EMI issues. Accordingly, the identified dominant parasitic paths that form the super-nodes include those associated with the drain and source terminals of the switches. These paths are critical as they establish key connections between high-frequency noise sources and the surrounding circuitry, allowing EMI and CM noise to propagate. Additionally, the DC bus has two distinct (dv/dt) versus ground at the positive and the negative sides. The packaging of

the GaN devices, along with the physical configuration and connections of the device to the heatsinks and DC bus, determines the parasitic capacitances between these nodes and

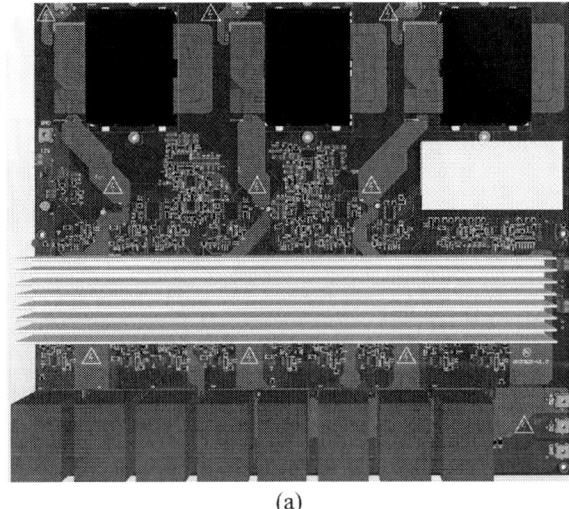

(a)

(b)

Fig. 5. a) Fabricated circuit board, and b) PCB design of the converter.

ground, Fig. 1b. In this work, a compact converter system is designed that operates at 200kHz. In this design, the DC busbar and inductors are implemented using a PCB-based approach to ensure compactness, reduced parasitic effects, and improved thermal management. The design integrates all key components, including switches, gate drivers, the control board, and sensing circuits, into a unified layout. This approach minimizes the physical footprint, enhances the system's reliability by reducing interconnect lengths and improves performance by optimizing current paths and mitigating EMI. The entire assembly is fabricated as shown in Fig. 5, highlighting the integration of these components into a cohesive and efficient design. Using the PCB design and considering the neutral point as the arbitrary reference point mentioned in Section II, we can determine the parasitic capacitances connected to each super-node. Accordingly all the intersection area of each node's conductor layer with the neutral point, as shown in Fig. 6 (for calculating C_{p1} in (14) as an example), determines the device's CM parasitic

979-8-3315-1612-3/25 $31.00 © 2025 IEEE

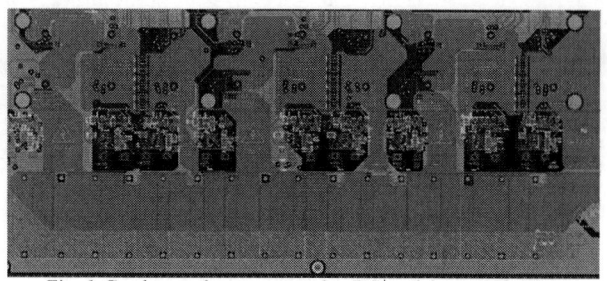

Fig. 6. Conductor plates connected to DC⁺ and the neutral point.

capacitance. The red layer in this Altium PCB design indicates the tracks connected to the positive DC bus and accordingly to the drains of the switches Q1a, Q1b, and Q1c, forming a super-node of U in Fig. 1a. The brown layer represents the neutral point of the converter, distributed across the board, and serves as the reference point for the calculation, $P = N$.

The intersection area between these two conducting layers can be determined in Altium by drawing a polygon over the affected common area. The software then calculates and provides the area of the polygon, allowing for precise measurement of the overlap. Based on the intersections between the red and brown layers, the C_{p1} can be calculated according to two parallel plates equation. This can be derived for all the added parasitic capacitances in Fig. 1a in a similar way, using $C = \epsilon_0 \epsilon_r A/d$, where $\epsilon_0 = 10^{-9}/36\pi$ is the air permittivity and $\epsilon_r = 4.3$ F/m is the permittivity of the FR4 dielectric medium between the layers of the fabricated PCB board. Also A is the area of the capacitor plates and d is the distance gap between the plates which is the thickness of the PCB board in this work. Considering the neutral point, which is different than the ground in our design, these values are calculated and deployed for CM modelling and simulating.

IV. SIMULATION RESULTS

The performance of the GaN-based 480V grid-connected 3L-ANPC converter is simulated using MATLAB/Simulink to validate its functionality and optimize key parameters. Following this, the board layout for the 25kW converter is meticulously designed in Altium. This comprehensive design integrates GaN devices, gate drivers, DC-link capacitors, measurement circuits, an output L-filter, and an FPGA controller, all consolidated onto a single PCB to ensure compactness and efficiency.

The bidirectional grid-connected converter employs Voltage Oriented Control (VOC), a control strategy that aligns the grid voltage with the direct axis of the rotating reference frame. This enables independent control of active and reactive power, facilitating precise power flow management. The converter operates seamlessly in both inverting and rectifying modes, supporting bidirectional energy transfer. Additionally, it generates three-level Space Vector PWM signals, which enhance power quality, reduce harmonic distortion, and improve overall system performance. This advanced design and control strategy make the converter highly efficient and adaptable for various grid-connected applications. From the system's steady-state DM operating point, the CM voltage inputs for CM modelling were analytically derived.

Starting from the system's steady-state DM operating point, the CM voltage inputs for CM modeling were analytically derived, ensuring accurate representation of system behavior. The MM model shown in Fig. 1a is simulated then in MATLAB/Simulink using the calculated capacitance values discussed in detail in Section III, allowing for precise incorporation of parasitic effects. Additionally, the equivalent CM model depicted in Fig. 4b was implemented in alongside the DM model to facilitate integrated analysis. In this setup, the CM circuit derives its inputs from the DM model, enabling the measurement of the corresponding CM and differential-mode voltages directly within the simulation. This dual-model approach ensures consistency between CM and MM analyses, providing deeper insights into the interaction between the modes and their impact on system performance, particularly regarding EMI and noise behavior. This comprehensive modeling strategy allows for a more robust evaluation of both DM and MM phenomena in the system.

The CM current (displacement current) through the AC-side LISN is measured and shown in Fig. 7 for both the MM and CM circuit models. To ensure a consistent comparison with the MM model results, the equivalent LISN values on the AC side are connected to the CM equivalent circuit representing the EUT. The CM current is then measured and plotted on Fig. 7b. By comparing the CM currents of Fig. 7a and Fig. 7b, we observe that the overall waveforms are similar, although the CM model exhibits larger current spikes. The larger spikes in the CM current obtained from the CM model can be attributed to several factors. The CM model accounts for CM impedance, which is generally lower than the differential-mode DM impedance considered in the MM model. This lower impedance is due to parasitic capacitances that provide low-impedance paths at high frequencies, resulting in increased current spikes, especially during switching transients.

Moreover, the simulation results for the CM current spectrum are shown in Fig. 8 for both MM model and CM model.

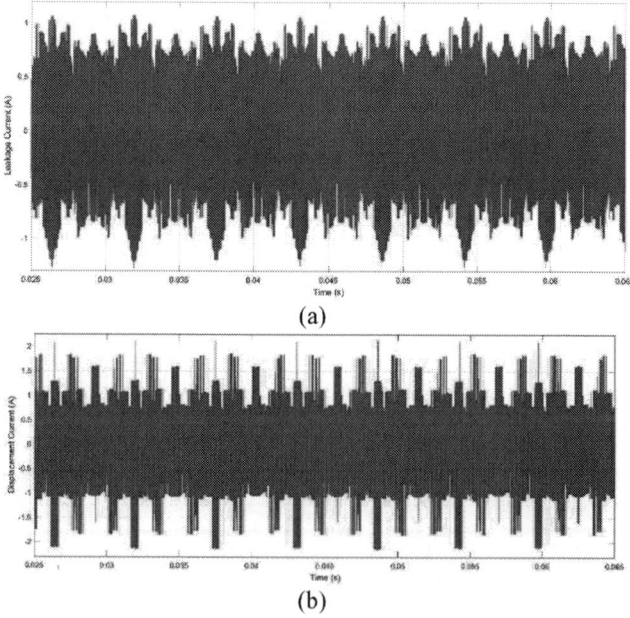

(a)

(b)

Fig. 7. The CM current in time domain a) MM model, and b) CM model.

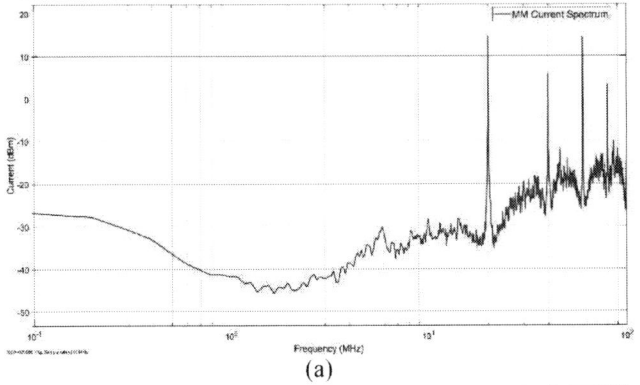

(a)

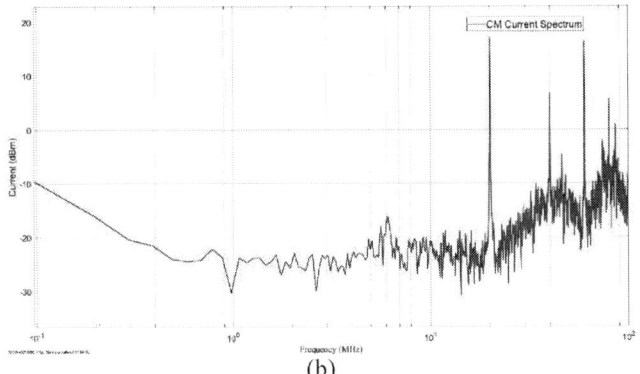

(b)

Fig. 8. The CM current in frequency domain a) MM model, and b) CM model with 200MHz sample rate.

It has been observed that they are in good agreement, with notable similarities between the two plots in terms of frequency range and the behavior of spikes at the 200kHz (switching frequency) harmonics. The CM current in the CEM of Fig. 8b is higher than in the MM model Fig. 8a, especially in the lower and mid-frequency ranges. This is consistent with the previous observations where the CM model exhibited larger spikes and overall current values.

V. CONCLUSION

In this paper, the EMI characterization of a 3L-ANPC converter topology is investigated. GaN switches are utilized to operate at high switching frequencies, aiming to reduce the overall system size and losses. The design of a PCB-integrated converter in this work mitigates EMI radiation, while also enhancing system performance and reliability. Additionally, the PCB structure enables a straightforward method for calculating parasitic capacitances, which is essential for improving the system's EMI behavior. The calculated capacitance facilitates the derivation of a MM model for the system. Besides, an analytical approach is introduced to extract the CEM from the MM model. The CEM and MM model were compared through simulations, and their results showed consistent verification, with both models exhibiting similar time and frequency spectral characteristics. The consistency between the two models affirms the validity of the approach used for EMI characterization. This approach provides a practical and accurate way to identify and analyze the CM characteristics in the design of high-frequency power converters.

ACKNOWLEDGEMENT

This material is based upon work supported by the National Science Foundation under Grant No. 1439700 and 1747757 (GRAPES I/UCRC). Any opinions, findings, and conclusions or recommendations expressed in this material are those of the author(s) and do not necessarily reflect the views of the National Science Foundation.

REFERENCES

[1] T. Brückner, S. Bernet, Loss balancing in three-level voltage source inverters applying active NPC switches", Proc. IEEE PESC, Vancouver, 2001, pp.1135-1140.

[2] J. Millán, P. Godignon, X. Perpiñà, A. Pérez-Tomás and J. Rebollo, "A Survey of Wide Bandgap Power Semiconductor Devices," in IEEE Transactions on Power Electronics, vol. 29, no. 5, pp. 2155-2163, May 2014.

[3] N. Oswald, B. H. Stark, D. Holliday, C. Hargis, and B. Drury, "Analysis of shaped pulse transitions in power electronic switching waveforms for reduced EMI generation," IEEE Transactions on Industry Applications, vol. 47, no. 5, pp. 2154–2165, 2011.

[4] Z. Wang, Y. Wu, M. H. Mahmud, Z. Yuan, Y. Zhao and H. A. Mantooth, "Busbar Design and Optimization for Voltage Overshoot Mitigation of a Silicon Carbide High-Power Three-Phase T-Type Inverter", IEEE Transactions on Power Electronics, vol. 36, no. 1, pp. 204-214, Jan. 2021.

[5] B. Revol, J. Roudet, J. L. Schanen, and P. Loizelet, "Fast EMI prediction method for three-phase inverter based on Laplace transforms," in Proc. IEEE 34th Annu. Power Electron. Spec. Conf., 2003, vol. 3, pp. 1133–1138.

[6] S. Wang and P. J. Kong, "Common mode noise reduction for boost converters using general balance technique," IEEE Trans. Power Electron., vol. 22, no. 4, pp. 1410–1416, Jul. 2007.

[7] D. Pan, M. Chen, X. Wang, H. Wang, F. Blaabjerg and W. Wang, "EMI modeling of three-level active neutral-point-clamped SiC inverter under different modulation schemes", Proc. IEEE 10th Int. Conf. Power Electron. ECCE Asia, pp. 1-6, 2019.

[8] F. A. Kharanaq, A. Emadi and B. Bilgin, "Analytical EMI Modeling of an Active Neutral Point Clamped Inverter," IECON 2021 – 47th Annual Conference of the IEEE Industrial Electronics Society, Toronto, ON, Canada, 2021, pp. 1-5.

[9] J. Wang, X. Liu, Y. Xun and S. Yu, "Common mode noise reduction of three-level active neutral point clamped inverters with uncertain parasitic capacitance of photovoltaic panels", IEEE Trans. Power Electron., vol. 35, no. 7, pp. 6974-6988, Jul. 2020.

[10] J. Wang et al., "Co-Reduction of Common Mode Noise and Loop Current of Three-Level Active Neutral Point Clamped Inverters," in IEEE Journal of Emerging and Selected Topics in Power Electronics, vol. 9, no. 1, pp. 1088-1103, Feb. 2021.

A Novel Layout for Improving Current Sharing of Paralleled SiC MOSFETs with TO-247 Package

Che-Wei Chang
Center for Power Electronics Systems (CPES)
Virginia Tech
Blacksburg, USA
cwchang@vt.edu

Matthias Spieler
Center for Power Electronics Systems (CPES)
Virginia Tech
Blacksburg, USA
mspieler@vt.edu

Rolando Burgos
Center for Power Electronics Systems (CPES)
Virginia Tech
Blacksburg, USA
rolando@vt.edu

Ayman EL-Refaie
Electrical and Computer Engineering
Marquette University
Milwaukee, USA
ayman.el-refaie@marquette.edu

Renato Amorim Torres
General Motors
Detroit, USA
renato.amorimtorres@gm.com

Dong Dong
Center for Power Electronics Systems (CPES)
Virginia Tech
Blacksburg, USA
dongd@vt.edu

Abstract — **Paralleling the silicon carbide (SiC) MOSFETs is necessary to increase the current rating in high-power applications. Moreover, the discrete device with TO-247 package features low-cost, compact, flexible to arrange, and larger cooling area per die. However, it is unclear which layout can offer the best electrical performances such as low loop inductance and parasitic capacitance. Moreover, the current sharing among paralleled devices yields non-negligible issue in device lifetime. This paper covers this gap by comparing performances of different layout options. Furthermore, to address the unbalanced current sharing issue, a novel layout concept is proposed. Both simulations and experimental double-pulse tests (DPT) verify the proposed layout concept. Compared to the conventional laminated layout technique, the proposed layout significantly improves current sharing performance while maintaining a low loop inductance.**

Keywords—Paralleled SiC MOSFETs, layout, current sharing, TO-247 package discrete devices

I. INTRODUCTION

Silicon-Carbide (SiC) devices provide a promising path for power electronics to achieve high efficiency. Combining with advanced technologies of modulation schemes, sensors, embedded die technology, gate driver, control, thermal management, and passive components, the design of power electronics can be more compact [1-15]. The SiC MOSFET with TO-247 package features low-cost, easy to arrange, and low junction to case thermal resistance. For high-power applications, it is necessary to parallel the discrete devices to increase current capability. However, the issue of unbalanced current sharing among paralleled SiC MOSFETs leads to uneven losses [16], current overshoot [17, 18], gate voltage oscillation [19], unequal junction temperatures, and thermal runaway [20, 21]. To address this issue, several techniques have been developed.

In [22], paralleling half-bridge offers a better current sharing performance than directly paralleling die. Based on this concept, cutouts have been made on the printed circuit board (PCB) to facilitate better current sharing among paralleled devices with TO-247 package [23]. With aid of distributed decoupling capacitors, a symmetric layout for paralleled TO-247 devices is proposed in [24]. Research [17] proposes a common source compensation technique for paralleled devices, requiring the overlapping of a part of gate driving loop with the power loop to achieve functionality. However, the above works result in larger commutation loop inductance L_{loop}, slower switching speed, or additional components in power loop which is undesirable for industry applications. In [25], differential mode chokes (DMC) are added in gate driver to enhance dynamic current sharing of paralleled discrete devices. Despite that, the DMC is unable to enhance static current sharing when SiC MOSFETs exhibits unequal drain to source resistance $R_{DS,on}$. Moreover, it is unclear which layout options for paralleled devices can offer best electrical performances such as low L_{loop} and low overlapping capacitance.

This paper covers this gap by comparing different layout options for paralleled devices with TO-247 package. Furthermore, a novel layout concept is proposed to improve current sharing performance.

II. COMPARISON OF DIFFERENT LAYOUT OPTIONS

A. Circuit and Layout Parasitic Components

Fig 1 shows a half-bridge circuit with 4 paralleled SiC MOSFETs in both high-side (HS) and low-side (LS) switching positions. For the layout in a half-bridge circuit, the parasitic L_{loop} and overlapping capacitances $C_{AC\,to\,DC+}$ and $C_{AC\,to\,DC-}$ should be minimized. Assuming that the inductance of DC-link capacitor C_{DC} is negligible, the L_{loop} in a half-bridge is composed of four different parasitic inductances which can be expressed as:

This work is supported by the U.S. Department of Energy's (DOE) Office of Energy Efficiency and Renewable Energy (EERE) under the Vehicle Technologies Office (VTO) Award Number DE-EE0009190.

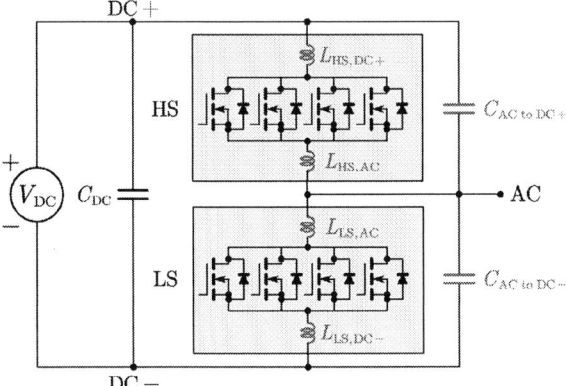

Fig. 1 The double pulse tester circuit with key parasitic components.

$$L_{\text{loop}} = L_{\text{HS,DC+}} + L_{\text{HS,AC}} + L_{\text{LS,AC}} + L_{\text{LS,DC-}} \quad (1)$$

During switching transients, due to sudden change in current direction, the induced voltage across L_{loop} results in additional voltage stress across SiC MOSFETs. The voltage stress is positively related to L_{loop} and may exceed device's rated voltage [26]. On the other hand, the layout also affects overlapping capacitances $C_{\text{AC to DC+}}$ and $C_{\text{AC to DC-}}$ according to the equation in (2) between each potential layer of the layout:

$$C = \frac{\varepsilon \cdot A}{d} \quad (2)$$

where ε is permittivity of the material, A is the overlapping area, and d is the distance. Since the distance is determined by the board thickness, the overlapping capacitances are mainly affected by overlapping area A. Both $C_{\text{AC to DC+}}$ and $C_{\text{AC to DC-}}$ are equivalent to output capacitance of switching positions which further increase the switching losses of SiC MOSFETs. Therefore, the layout parasitic L_{loop}, $C_{\text{AC to DC+}}$, and $C_{\text{AC to DC-}}$, should be minimized to reduce voltage stress and switching losses.

B. Layout Options for Paralleled Discrete Devices

Fig. 2 and Fig. 3 show two different layout options for paralleled discrete devices in a half-bridge circuit. The DC capacitors and AC output node locate at the bottom and right side of the PCB, respectively. The layout shown in Fig. 2 parallels the devices in a single row [23, 27] while those in Fig. 3 are separated into 2 rows [24, 28]. Moreover, the paralleled devices can be arranged in sequential (…, HS, HS, LS, LS, …) or interleaved (…, HS, LS, HS, LS, …).

With aid of Q3D simulation, the L_{loop} and overlapping capacitances can be extracted as shown in Fig. 4. Apparently, the interleaved arrangement always provides a lower L_{loop} than sequential arrangement because of the current direction from HS to LS between paralleled branches is opposite, generating negative mutual inductance which further decreases the total equivalent L_{loop}. Also, the 2-row layout has a shorter commutation loop than the 1-row layout.

When comparing overlapping capacitances, the 1-row layout has smaller capacitances since the layer of AC

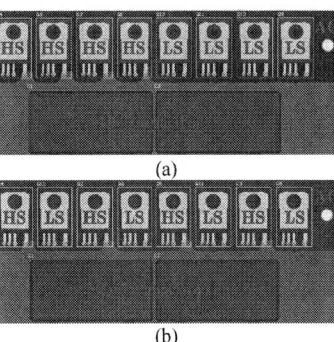

Fig. 2 The 1-row layout with devices arranged in (a) sequential arrangement (b) interleaved arrangement.

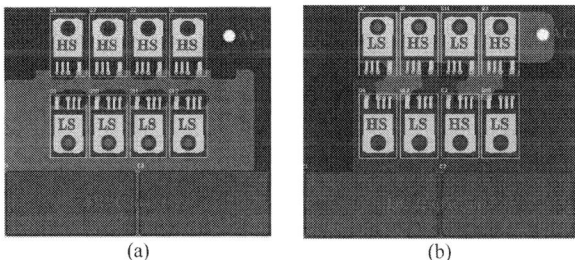

Fig. 3 The 2-row layout with devices arranged in (a) sequential arrangement (b) interleaved arrangement.

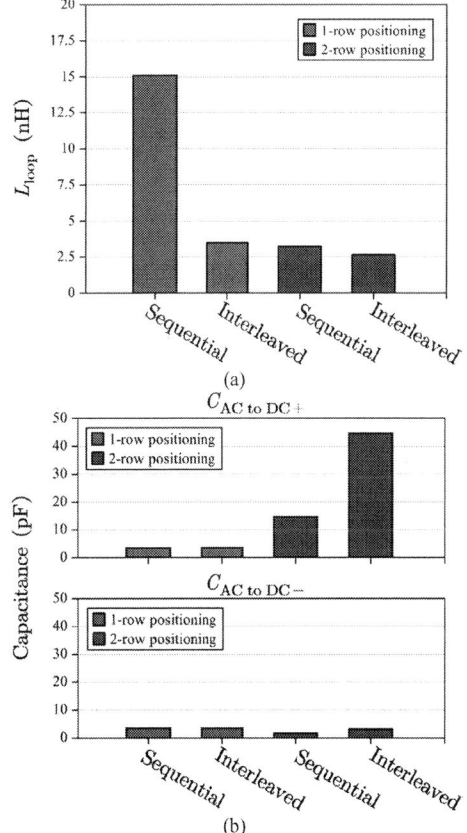

Fig. 4 Based on layout options, the comparisons of (a) L_{loop} (b) overlapping capacitances $C_{\text{AC to DC+}}$ and $C_{\text{AC to DC-}}$.

979-8-3315-1612-3/25 $31.00 © 2025 IEEE

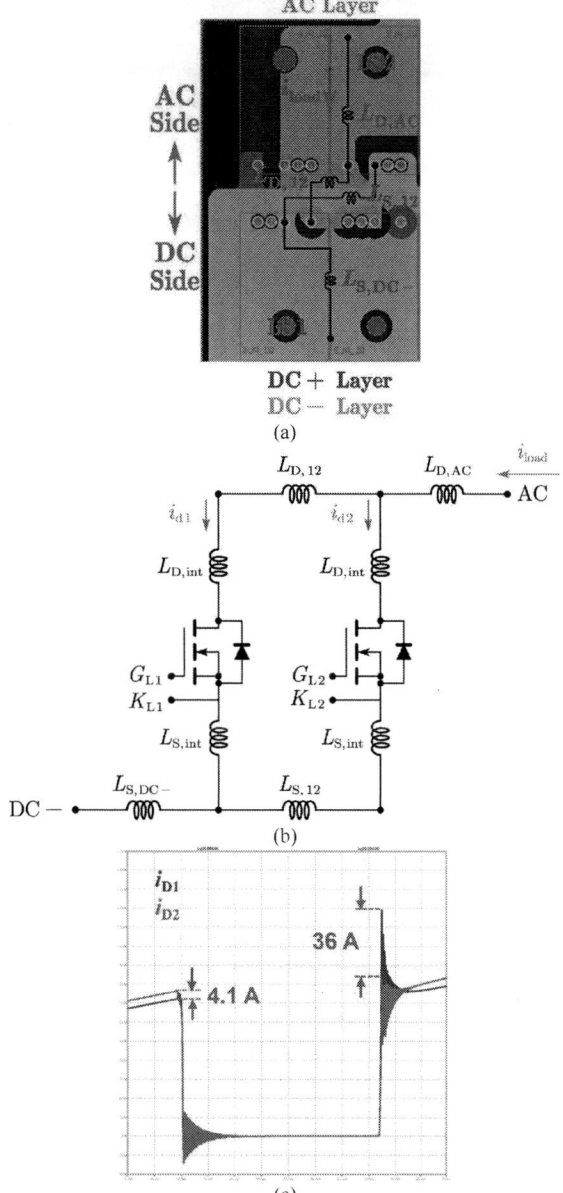

Fig. 5 The (a) conventional laminated layout technique, (b) equivalent circuit, and (c) simulated current sharing performance.

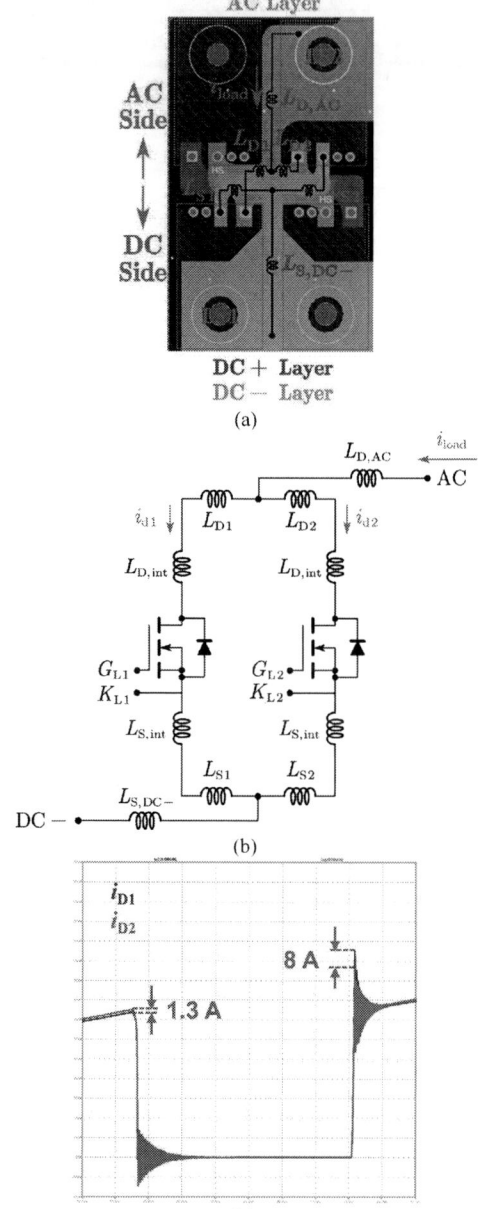

Fig. 6 The (a) proposed layout concept, (b) equivalent circuit, and (c) simulated current sharing performance.

potential barely overlaps with DC+ and DC- layers. In contrast, the 2-row layout has overlapping area in between of top and bottom row of devices, resulting in higher $C_{\text{AC to DC+}}$ and $C_{\text{AC to DC-}}$ than 1-row layout. Though 2-row layout has higher capacitances, their impact is negligible when compared to devices' C_{oss} (level of nH). Therefore, the 2-row layout with interleaved arrangement is chosen for further optimization due to low L_{loop}.

III. CURRENT SHARING ISSUE OF CONVENTIONAL LAMINATED LAYOUT TECHNIQUE

Conventionally, the laminated layout technique is adopted where the devices are directly connected by big pieces of

copper polygon as shown in Fig. 5(a). However, this layout technique inevitably results in unequal traces which further cause unbalanced current sharing. If focusing on 2 paralleled LS devices, Fig. 5(b) shows the equivalent circuit of such layout technique where L_{int} denotes internal parasitic of device's package. The drain and source inductances (L_D and L_S) between paralleled devices have mismatches of $\Delta L_D = L_{D,12}$ and $\Delta L_S = L_{S,12}$. Both ΔL_D and ΔL_S affect static current sharing, and dynamic current sharing is only sensitive to ΔL_S [25]. Fig. 5(c) shows the DPT simulation in LTspice with extracted parameters from Q3D. The total load current i_{load} is set as 150 A (75 A/each MOSFET). Even when paralleled devices exhibit equal $R_{\text{DS,on}}$ and threshold voltage V_{th}, the

979-8-3315-1612-3/25 $31.00 © 2025 IEEE

static and dynamic current differences are 4.1 A and 36 A, respectively.

IV. PROPOSED NOVEL LAYOUT CONCEPT

To enhance the current sharing performance, Fig. 6(a) shows the proposed layout concept with the same device arrangement as Fig. 5. The gaps between SiC MOSFETs are slightly enlarged to allow traces distributing to paralleled devices. Focusing on 2 paralleled LS devices, the equivalent circuit in Fig. 6(b) shows that the ΔL_D and ΔL_S are now solely depends on slight differences between L_{D1} and L_{D2}, L_{S1} and L_{S2}. The proposed layout concept significantly mitigates both ΔL_D and ΔL_S, improving both static and dynamic current sharing performances. Fig. 6(c) shows the DPT result under the same condition as that in Fig. 5(c). Obviously, the current sharing is greatly improved. While remaining the benefit of small L_{loop} due to interleaved arrangement, the static and dynamic current differences significantly decrease down to 1.3 A and 8 A, respectively. The simulation results verify the feasibility of the proposed layout concept.

V. EXPERIMENTAL VERIFICATIONS

To validate the proposed layout concept, a testing environment is built as shown in Fig. 7(a). Two half-bridge PCBs are fabricated, and both PCBs adopt 2-row layout with interleaved arrangement. The layout in Fig. 7(b) utilizes conventional laminated layout technique while that in Fig. 7(c) adopts proposed layout concept. For both PCBs, each switching position consists of 4 paralleled Infineon IMZA120R007M1H and are driven by single gate driver IC. All DPT tests are conducted under DC voltage of 800 V, i_{load} of 280 A, and gate resistance of 10 Ω (for each LS1-4).

Fig. 8 shows the comparison of static current sharing waveforms obtained from Fig. 7(b) and 7(c). The static peak current difference of conventional layout is 18 A. On the other hand, due to the minimized ΔL_D and ΔL_S, the static current difference of proposed layout concept is only 6 A.

Fig. 9 further presents the dynamic current sharing during the turn-on transient. The switching speeds of both waveforms are identical due to the same operating conditions. Apparently, the dynamic current difference in proposed layout is only 18.4 A which is much lower than that of 76 A in the conventional layout. The current difference of 18.4 A may be caused by mismatches in paralleled devices.

Similarly, for turn-off current sharing shown in Fig. 10, the peak current difference in conventional layout is around 10 A while the currents in proposed layout are almost balanced with negligible difference. Finally, obtained from Q3D simulation when 4 devices in parallel, the proposed layout has L_{loop} of 2.23 nH which is even lower than 2.53 nH from conventional layout. The experimental results verify the proposed layout in current sharing while maintaining low L_{loop} with slight penalty of 16 % larger footprint area.

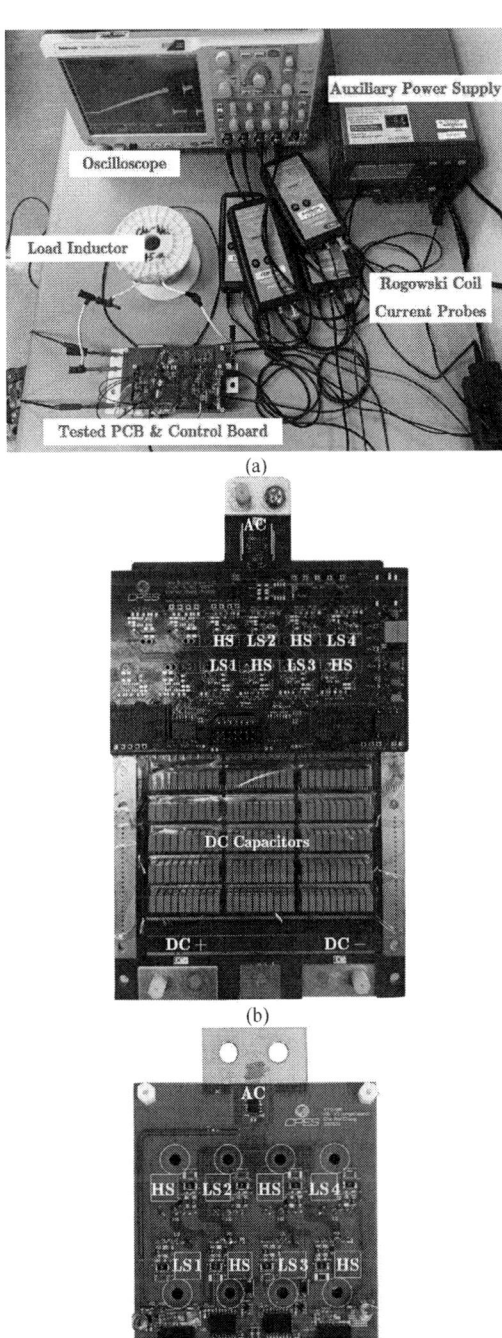

Fig. 7   The (a) testing environment, (b) PCB with conventional laminated layout, and (c) PCB with proposed layout concept.

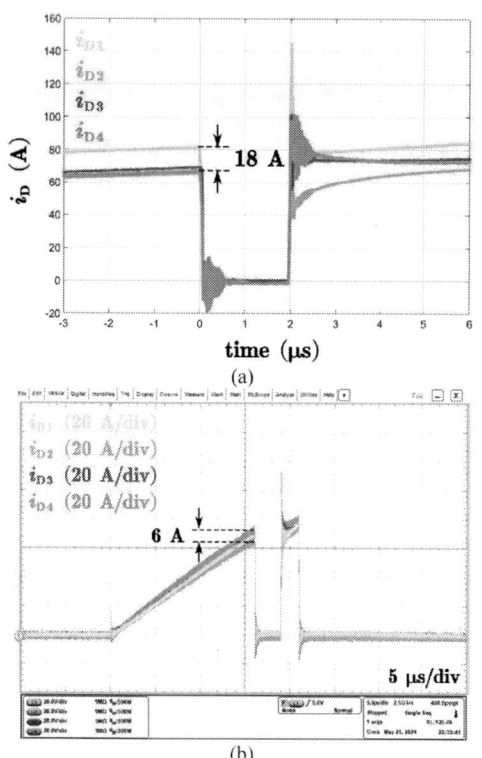

Fig. 8 Static current sharing of (a) conventional laminated layout (b) proposed layout concept.

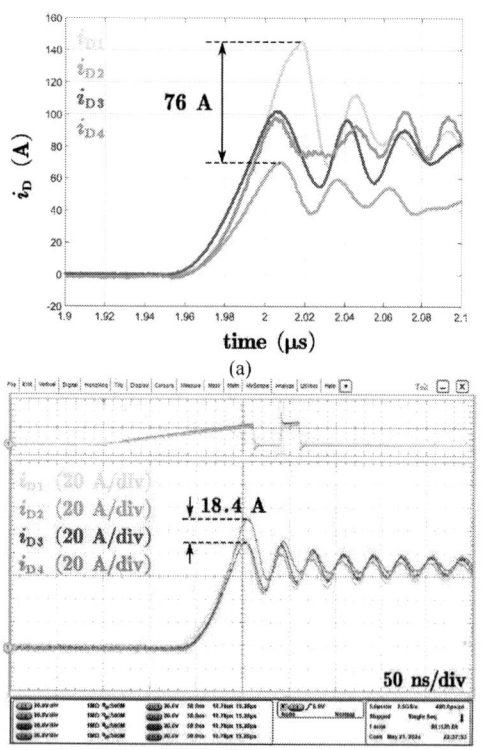

Fig. 9 Turn-on dynamic current sharing of (a) conventional laminated layout (b) proposed layout concept.

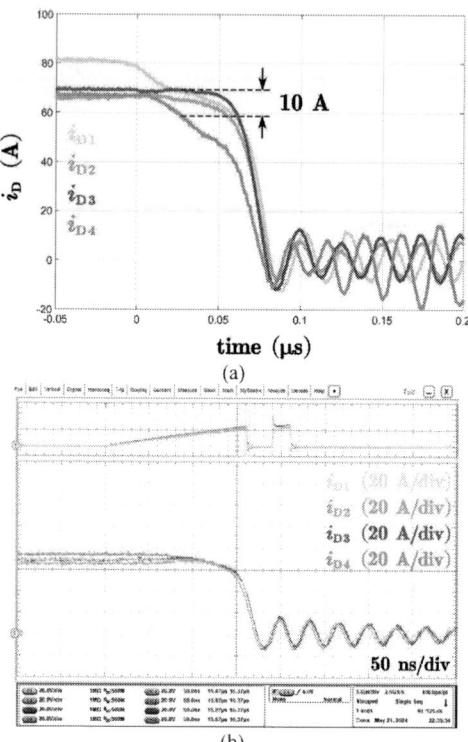

Fig. 10 Turn-off dynamic current sharing of (a) conventional laminated layout (b) proposed layout concept.

VI. CONCLUSIONS

A novel layout concept for paralleled discrete devices is proposed in this paper. This paper first explores different layout options to compare electric performances. It is discovered that the L_{loop} is smallest when the devices are interleaved arranged in 2-row layout option. Later, based on the selected layout, the current sharing performance is analyzed. To enhance the current sharing, a novel layout concept is proposed to minimize the layout mismatch. Two PCBs are fabricated to compare the proposed layout concept with conventional laminated layout technique. The experimental results well verify the proposed layout concept.

REFERENCES

[1] X. Zhao, C.-W. Chang, R. Phukan, R. Burgos, S. Uicich, P. Asfaux, and D. Dong, "An Enhanced Modulation Scheme for Multi-Level T-Type Inverter with Loss Balance and Reduction," *IEEE Transactions on Power Electronics*, 2023.

[2] X. Zhao, R. Phukan, C.-W. Chang, R. Burgos, S. Uicich, P. Asfaux, M. Debbou, A. Plat, and D. Dong, "Design and Implementation of SiC-Based 200-kW High-Density High-Speed High-Altitude Electric Propulsion AC Drive System," *IEEE Journal of Emerging and Selected Topics in Power Electronics*, 2024.

[3] M. Spieler, C.-W. Chang, A. El-Refaie, D. Dong, and R. Burgos, "Design of a High Bandwidth Compact DC-bus Embedded Planar Rogowski Coil for SiC MOSFET Current Sensing," *IEEE Transactions on Power Electronics*, 2024.

[4] M. Spieler, C.-W. Chang, A. El-Refaie, M. H. Alvi, D. Dong, and R. Burgos, "PCB Technology Comparison Enabling a 900V SiC MOSFET Half Bridge Design for Automotive Traction Inverters," in *2022 24th European Conference on Power*

979-8-3315-1612-3/25 $31.00 © 2025 IEEE

Electronics and Applications (EPE'22 ECCE Europe), 2022: IEEE, pp. 1-11.

[5] C.-W. Chang, M. Spieler, R. Burgos, A. EL-Refaie, R. A. Torres, and D. Dong, "An Improved di/dt-Based Dynamic Current RC Sensing Method for Paralleled SiC MOSFETs," in *2024 IEEE Energy Conversion Congress and Exposition (ECCE)*, 2024: IEEE.

[6] C.-W. Chang, X. Zhao, R. Phukan, R. Burgos, S. Uicich, P. Asfaux, and D. Dong, "Thermal Consideration and Design for a 200 kW SiC-Based High-Density Three-Phase Inverter in More Electric Aircraft," *IEEE Journal of Emerging and Selected Topics in Power Electronics*, 2023.

[7] T. Yuan, F. Jin, and Q. Li, "Analysis and Comparison of Integrated Planar Transformers for 22-kW On-board Chargers," *IEEE Transactions on Power Electronics*, 2024.

[8] T. Yuan, F. Jin, Z. Li, C. Zhao, and Q. Li, "Design of an integrated transformer with parallel windings for a 30-kW LLC resonant converter," *IEEE Transactions on Power Electronics*, 2023.

[9] T. Yuan, F. Jin, Z. Li, and Q. Li, "Current Sharing Analysis of a High Power Transformer with Parallel Windings," in *2023 IEEE Applied Power Electronics Conference and Exposition (APEC)*, 2023: IEEE, pp. 1551-1556.

[10] T.-S. Li, Y.-H. Yang, C.-A. Cheng, and Y.-M. Chen, "A variable DC-link voltage determination method for motor drives with SiC MOSFETs," in *2020 IEEE Workshop on Wide Bandgap Power Devices and Applications in Asia (WiPDA Asia)*, 2020: IEEE, pp. 1-6.

[11] T.-S. Li, M. Ngo, R. Burgos, and D. Dong, "Modeling and Analysis of Voltage Overshoot in Bidirectional Phase-Shift Full Bridge Converters," in *2024 IEEE Sixth International Conference on DC Microgrids (ICDCM)*, 2024: IEEE, pp. 1-7.

[12] Y.-T. Huang, C.-C. Yang, T.-S. Li, and Y.-M. Chen, "A feedforward voltage control strategy for reducing the output voltage double-line-frequency ripple in single-phase AC–DC converters," *IEEE Journal of Emerging and Selected Topics in Power Electronics*, vol. 9, no. 6, pp. 6605-6612, 2021.

[13] X. Yu, J. Feng, L. Zhu, and Q. Li, "Design and Optimization of a Planar Omnidirectional Wireless Power Transfer System for Consumer Electronics," *IEEE Open Journal of Power Electronics*, 2024.

[14] X. Yu, J. Feng, L. ZHu, and Q. Li, "Modelling of A Planar Omnidirectional Wireless Power Transfer System," in *2024 IEEE Applied Power Electronics Conference and Exposition (APEC)*, 2024: IEEE, pp. 2609-2615.

[15] X. Yu, J. Feng, and Q. Li, "A planar omnidirectional wireless power transfer platform for portable devices," in *2023 IEEE Applied Power Electronics Conference and Exposition (APEC)*, 2023: IEEE, pp. 1654-1661.

[16] C. Zhao, L. Wang, F. Zhang, and F. Yang, "A method to balance dynamic current of paralleled SiC MOSFETs with kelvin connection based on response surface model and nonlinear optimization," *IEEE Transactions on Power Electronics*, vol. 36, no. 2, pp. 2068-2079, 2020.

[17] B. Zhang, R. Wang, P. Barbosa, Q. Cheng, Y.-H. Tsai, W.-S. Wang, W.-S. Lai, and F.-Y. Shih, "Common Source Inductance Compensation Technique for Dynamic Current Balancing in SiC MOSFETs Parallel Operations," *IEEE Transactions on Power Electronics*, 2023.

[18] H. Li, S. Munk-Nielsen, X. Wang, R. Maheshwari, S. Bęczkowski, C. Uhrenfeldt, and W.-T. Franke, "Influences of device and circuit mismatches on paralleling silicon carbide MOSFETs," *IEEE Transactions on Power Electronics*, vol. 31, no. 1, pp. 621-634, 2015.

[19] Y. Zhu, H. Li, C. Luo, C. Wan, and J. Ma, "Influence of paralleled SiC MOSFET on turn-off gate voltage oscillation," in *2020 IEEE Energy Conversion Congress and Exposition (ECCE)*, 2020: IEEE, pp. 683-689.

[20] C. Zhao, L. Wang, and F. Zhang, "Effect of asymmetric layout and unequal junction temperature on current sharing of paralleled SiC MOSFETs with kelvin-source connection," *IEEE Transactions on Power Electronics*, vol. 35, no. 7, pp. 7392-7404, 2019.

[21] S. Neira, R. Mathieson, M. Parker, P. D. Judge, and S. J. Finney, "Investigation into Current Sharing of Parallel SiC MOSFET Modules using a Gate-Driver with Sub-Nanosecond Time-Skew Capability," in *2023 25th European Conference on Power Electronics and Applications (EPE'23 ECCE Europe)*, 2023: IEEE, pp. 1-8.

[22] H. Li, W. Zhou, X. Wang, S. Munk-Nielsen, D. Li, Y. Wang, and X. Dai, "Influence of paralleling dies and paralleling half-bridges on transient current distribution in multichip power modules," *IEEE Transactions on Power Electronics*, vol. 33, no. 8, pp. 6483-6487, 2018.

[23] J. Qu, Q. Zhang, X. Yuan, and S. Cui, "Design of a paralleled SiC MOSFET half-bridge unit with distributed arrangement of dc capacitors," *IEEE Transactions on Power Electronics*, vol. 35, no. 10, pp. 10879-10891, 2020.

[24] Y. He, J. Zhang, and S. Shao, "Symmetric Circuit Layout with Decoupled Modular Switching Cells for Multi-Paralleled SiC MOSFETs," *IEEE Transactions on Power Electronics*, 2023.

[25] C.-W. Chang, M. Spieler, E.-R. Ayman, R. A. Torres, R. Burgos, and D. Dong, "A Current Balancing Gate Driver for Dynamic Current Sharing of Paralleled SiC MOSFETs with Kelvin-Source Connection," *IEEE Transactions on Power Electronics*, 2024.

[26] J. Liang, B. Fan, C. Chang, R. Burgos, D. Dong, J. Tangudu, and S. Dwari, "PCB Busbar Design and Verification for a Multiphase 250 kW SiC based All-electric Aircraft Powertrain Converter," in *2023 IEEE Applied Power Electronics Conference and Exposition (APEC)*, 2023: IEEE, pp. 1031-1036.

[27] R. Bosshard and J. W. Kolar, "All-SiC 9.5 kW/dm 3 on-board power electronics for 50 kW/85 kHz automotive IPT system," *IEEE Journal of Emerging and Selected Topics in Power Electronics*, vol. 5, no. 1, pp. 419-431, 2016.

[28] D. Rothmund, T. Guillod, D. Bortis, and J. W. Kolar, "99% efficient 10 kV SiC-based 7 kV/400 V DC transformer for future data centers," *IEEE Journal of Emerging and Selected Topics in Power Electronics*, vol. 7, no. 2, pp. 753-767, 2018.

A Sensor-less IGBT On-State Voltage Estimation Method Using Inverter Control Variables

Shuyu Ou ⦿, Subham Sahoo ⦿, Ariya Sangwongwanich ⦿, Yongjie Liu ⦿, and Frede Blaabjerg ⦿

Department of Energy, Aalborg University, Aalborg, Denmark

so@energy.aau.dk, sssa@energy.aau.dk, ars@energy.aau.dk, yoli@energy.aau.dk, fbl@energy.aau.dk

Abstract—The on-state voltage is a commonly used health indicator for monitoring the health status of IGBT modules. However, deriving the on-state voltage through measurement-based methods requires additional measurement circuits, which increases both the cost and volume. Existing sensor-less methods, meanwhile, depend largely on training datasets to extract degradation features specific to particular converters, limiting their wider applications. To address these challenges, a sensor-less method uses variables collected for control purposes to estimate the on-state voltage and current is proposed. The proposed method eliminates the need for a training dataset. Additionally, the impact of noise in control variables is mitigated by averaging a series of samples collected under similar operating conditions to improve estimation accuracy. The effectiveness of the proposed method has been validated through hardware-in-the-loop experiments conducted across three case studies, each representing different degradation stages, power levels, and power factors. The estimation errors at full load are close to 1% of the initial on-state resistance.

Index Terms—IGBT, on-state voltage, sensor-less estimation, condition monitoring.

I. INTRODUCTION

The reliability of insulated-gate bipolar transistors (IGBTs) is crucial for the performance of power converters. Thermal stress can induce package-related failures, such as bond wire cracks or lift-off, which reduce the reliability of IGBTs [1]. To monitor bond wire health, the on-state voltage is commonly used as a health indicator due to its ease of online measurement, high sensitivity, and high linearity [2]–[4]. Once the on-state voltage is derived or measured, it enables end-of-life detection and remaining useful lifetime prediction [5], [6]. The derived health status also facilitates predictive maintenance, which enhances reliability while reducing maintenance costs [7].

The on-state voltage can be derived using either measurement-based or sensor-less methods. Measuring the on-state voltage of IGBT is challenging because the measurement circuit needs to meet stringent requirements: high blocking voltage, fast response, noise immunity, and high accuracy [8], [9]. These demands typically necessitate components, e.g., isolated power supplies, MOSFETs, fast recovery diodes, and high-precision resistors. To simplify the measurement circuit, recent research focuses on reducing the components in the measurement circuit. For example, reusing the isolated

This project is supported by the European Union's Horizon 2020 research and innovation program under the Marie Skłodowska-Curie grant agreement No. 955614.

power supplies with a common floating ground [10] and measuring the inverter output terminals to enable the converter level monitoring [11], [12]. Nonetheless, incorporating measurement circuits continues to increase cost, volume, and potential failure points, limiting the widespread application of measurement-based methods.

Conversely, the sensor-less methods eliminate the need for additional components by extracting degradation features from training datasets. Xiang et al. [13] established a statistical relationship between the health status of the IGBT and the magnitude of harmonics in inverter control variables, i.e., the current controller output. By incorporating the phase angle of harmonics, the method can locate the degraded component more precisely [14]. The primary drawback of these methods lies in the reliance on training datasets. Developing a training dataset that spans the entire degradation process can be costly and time-consuming. Furthermore, features extracted for a specific component and system are not easily transferable to other systems. Additionally, the impact of noises on estimation accuracy is not fully investigated, which potentially reduces the precision of these methods.

To address these challenges, a sensor-less condition monitoring method for the IGBT package is proposed. This method utilizes control variables (the current controller output v_{a}^*) and the measurement signals (including the grid voltage v_{ga}, the inverter current i_{a}, and the DC-link voltage v_{dc}) to estimate the on-state current and voltage. Then the increased on-state resistance $\Delta R_{\mathrm{CE,on}}$ is derived. As a sensor-less approach, the proposed method avoids additional components, thereby preventing increases in converter cost and volume. Compared to the existing sensor-less methods, this approach leverages Ohm's law, which eliminates the need for a training dataset and enhances its applicability across various converter systems. Additionally, noise in the control variables is mitigated by averaging estimates across multiple sampling points.

The structure of the following sections is: Section II describes the inverter system and outlines the failure mechanisms of the IGBT. Section III introduces the proposed method, while Section IV presents the hardware-in-the-loop experimental setup and results.

II. SYSTEM DESCRIPTION AND IGBT DEGRADATION

A. Inverter System

The proposed method is applied to a three-phase inverter, as illustrated in Fig. 1(a). The IGBT module consists of six

979-8-3315-1612-3/25 $31.00 © 2025 IEEE

IGBTs (S_1 - S_6) and six free-wheeling diodes (FWD) (D_1 - D_6). The DC link has a capacitor C_{bus} with a voltage level of v_{dc}. The inverter is connected to the grid v_g through a filter (with inductance L_g and parasitic resistance R_L) and the grid impedance Z_g.

The inverter is controlled using a current controller in the synchronous frame (dq-frame) [15]. Proportional-integral (PI) controller regulates the inverter output current (i_a, i_b, and i_c) by adjusting the voltage reference in the synchronous frame (v_d^*, v_q^*). Additionally, a feed-forward grid voltage (v_{gd}, v_{gq}) is added to the PI controller output v_{dqcc}^* to enable faster regulation. The voltage reference in the synchronous frame is then converted into the stationary frame as v_{abc}^*, which generates the gate signals v_{gate} through a pulse-width modulation (PWM) module. Coordinate transformation is achieved using the phase angle θ, provided by a phase-locked loop (PLL).

B. IGBT Structure and Failure Mechanism

The structure of an IGBT module is illustrated in Fig. 1(b), where the bond wire connects the chips to the direct copper bonded (DBC). Due to thermal stress and differences in coefficient of thermal expansion (CTE), the bond wire can develop cracks or lift-off [1]. This degradation in the bond wire reduces the conduction path, resulting in increased resistance and a higher on-state voltage.

To monitor the health status of bond wires, the on-state voltage is a commonly used health indicator. It comprises an initial voltage part and a resistive part [16]. The initial voltage remains stable as the IGBT degrades, while the resistive part can increase due to bond wire degradation [17].

The end-of-life criteria for on-state resistance is determined by a 5% increase in the on-state voltage [5], [18]. Once the increased on-state resistance $\Delta R_{CE,on}$ exceeds the end-of-life threshold $\Delta R_{on(eol)}$ (as illustrated in the red region of Fig. 1(c)), maintenance actions are required to prevent the functional failure of the IGBT. Failure to address this can compromise the reliability of both IGBT and the converter, potentially leading to breakdowns and higher maintenance costs.

C. Effects of On-state Voltage/ Resistance on Control Loop

In the current control loop, the on-state voltage acts as a disturbance, as illustrated in Fig. 1(d), where G_{PI}, G_{PWM}, G_{inv}, and G_L are transfer functions of the PI controller, PWM, inverter, and the filter. The disturbance $v_{CE,on}$ slightly reduces the inverter output voltage v_n and the output current i_{dq} [11]. To counteract this reduction in the output current, the current controller output v_{dq}^* and the PI controller output v_{dqcc}^* increase slightly. This compensation allows using the current controller output to indicate the increased on-state resistance $\Delta R_{CE,on}$. The transfer functions and waveforms of the disturbances are analyzed in detail in the previous work [19].

Using IGBT S_1 as an example, the on-state voltage of S_1 in Fig. 1(a) slightly reduces the inverter output voltage v_{an} and the output current i_a. The reduction of i_a is compensated by

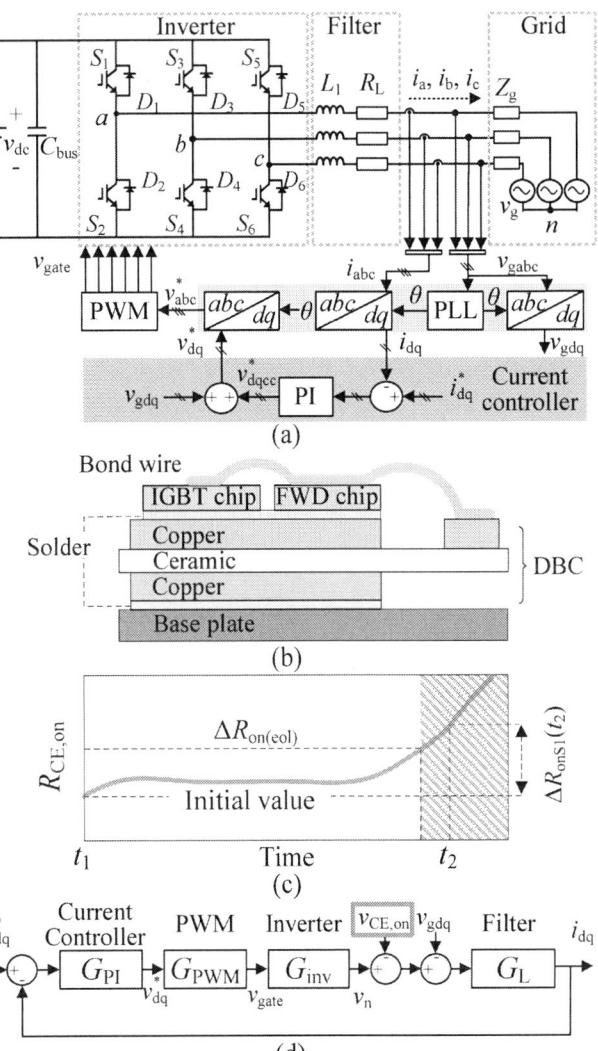

Fig. 1. System description: (a) Three-phase inverter and its controller. (b) Structure of the IGBT module. (c) Degradation leading to increment of the on-state resistance. (d) The on-state voltage adds disturbances to the current control loop.

an increase in the PI controller output in the stationary frame v_{acc}^*.

III. PROPOSED METHOD

The proposed method (see Fig. 2) monitors the on-state resistance of IGBT S_1 in Fig. 1(a), and can be applied to other IGBTs with minor modifications. The method consists of two main parts:

- Deriving the estimated on-state current and reference voltage using control variables (v_a^*) and measurement signals (v_{ga}, i_a, and v_{dc}).
- Mitigating noise effect on estimation accuracy through accumulation and averaging. The increased on-state resistance is then calculated using the average current and voltage.

The underlying hypothesis of the method is that the current controller can detect and compensate for the reduction in output current caused by an increase in on-state resistance of one of the IGBTs [19].

A. Derivation of the Equivalent Current and Voltage

To eliminate extra components and develop a sensor-less method, the low-frequency equivalent on-state current and voltage of IGBT S_1 are estimated. The estimated IGBT current and the increment in on-state voltage are derived from control variables (the current controller output $v_a^*(t)$) and the measurement signals (the grid voltage $v_{ga}(t)$, the inverter current $i_a(t)$, and the DC-link voltage $v_{dc}(t)$).

Firstly, the IGBT current $i_{S1}(t)$ is derived from the phase current $i_a(t)$, the current polarity $sign(i_a(t))$ and the modulation signal $m_{S1}(t)$ [20], [21], as illustrated in Fig. 2(b). The IGBT on-state current $i_{S1}(t)$ equals the phase current $i_a(t)$ when the phase current has a positive direction and the gate signal is high, as given in (1).

$$i_{S1}(t) = i_a(t)sign(i_a(t))m_{s1}(t) \qquad (1)$$

The current polarity $sign(i_a(t))$ is one if the current direction is positive; otherwise, it is zero, as denoted by:

$$sign(i_a(t)) = \begin{cases} 1, & i_a(t) \geq 0 \\ 0, & i_a(t) < 0 \end{cases} \qquad (2)$$

The accurate IGBT would ideally be derived using the gate signal. However, sensing the gate signal requires an additional sensing circuit with high sampling bandwidth. To simplify the proposed method, the gate signal is replaced by the reference signal $m_{S1}(t)$ as they are equivalent in the low-frequency region. The low-frequency region is defined as the frequency spectrum near the baseband and significantly lower than the switching frequency [22]. $m_{S1}(t)$ is a scalar falling within the range of [0, 1], and a 0.5 offset is added to (3).

$$m_{S1}(t) = \frac{v_a^*(t)}{v_{dc}(t)} + 0.5 \qquad (3)$$

Secondly, the increased on-state voltage is approximated by the increment in the current controller output $v_{acc}^*(t)$, since the current controller compensates for the on-state voltage of S_1 [19]. The voltage reference is multiplied by the current polarity $sign(i_a(t))$ as the IGBT S_1 conducts only when the current polarity is positive.

$$v_{accS1}^*(t) = v_{acc}^*(t)sign(i_a(t)) \qquad (4)$$

The current controller output $v_{acc}^*(t)$ is derived from the voltage reference $v_a^*(t)$ in phase A. $v_a^*(t)$ consists of two parts: the PI controller output $v_{acc}^*(t)$ which is related to the on-state voltage, and the feed-forward part which is independent of the IGBT health status:

$$v_{acc}^*(t) = v_a^*(t) - v_{ga}(t) \qquad (5)$$

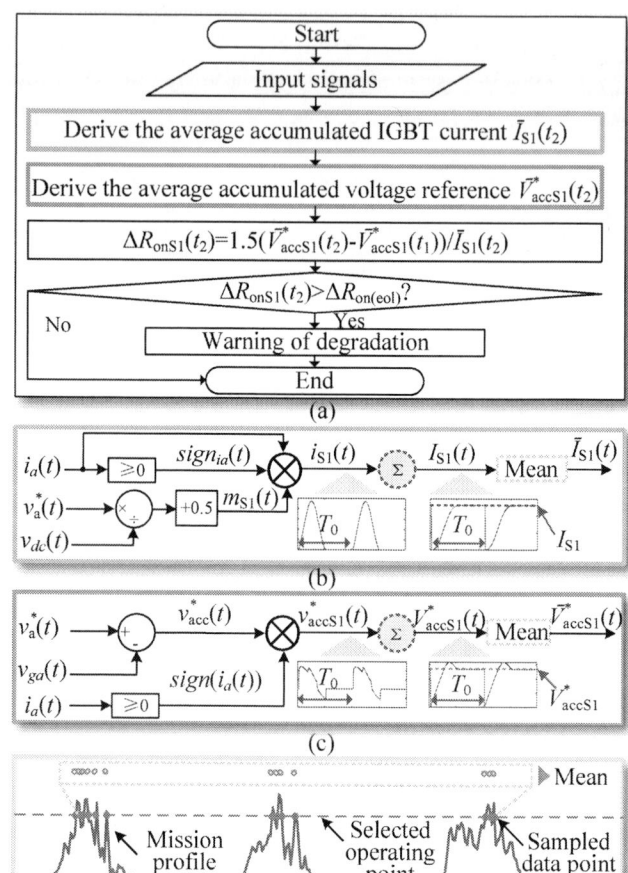

Fig. 2. Proposed method: (a) Flowchart. (b) Derive the equivalent S_1 current. (c) Derive the equivalent S_1 voltage. (d) The estimation of increased on-state resistance ΔR_{on} uses the mean value of a series of sampled points at the selected operating point.

B. Mitigate the Noise

Directly using the instantaneous current and voltage without post-processing can result in a higher estimation error due to noise and oscillations in the input signals.

To mitigate the effect of noises and obtain a stable estimation, instantaneous values are accumulated and averaged; similar methods are discussed in [23]. Specifically, the current $i_{S1}(t)$ and voltage $v_{accS1}^*(t)$, derived in (1) and (4), are accumulated over each fundamental cycle T_0 with a step size of the switching cycle T_{sw}, as given in (6) and (7), resulting in the accumulated current $I_{S1}(t)$ and voltage $V_{accS1}^*(t)$. In the equations, p is a counter and N is the number of switching cycles in T_0. The accumulation process is indicated by blue dashed circles in Figs. 2(b)-(c).

$$I_{S1}(t) = \sum_{p=1}^{N} i_{S1}(t_p) \qquad (6)$$

$$V_{accS1}^*(t) = \sum_{p=1}^{N} v_{accS1}^*(t_p) \qquad (7)$$

979-8-3315-1612-3/25 $31.00 © 2025 IEEE   1503

The accumulated current $I_{S1}(t)$ and voltage $V^*_{accS1}(t)$ are then averaged over a $200\,s$ sampling window, as derived in (8) and (9), where M represents the number of fundamental cycles within the sampling window. The averaging process is indicated with yellow dashed blocks.

$$\overline{I}_{S1}(t) = \frac{1}{M}\sum_{p=1}^{M} I_{S1}(t_p) \tag{8}$$

$$\overline{V}^*_{accS1}(t) = \frac{1}{M}\sum_{p=1}^{M} V^*_{accS1}(t_p) \tag{9}$$

Although the sampling window is relatively longer than individual switching cycles, it includes a set of shorter measurements taken at similar operating points, as illustrated in Fig. 2(d), where the mission profile encounters the selected sampling points multiple times [24]. The sampling scheme is designed to include a wide range of test waveforms, helping to mitigate noise and oscillations in the control variables. Due to the fast dynamics in the power converter, the extended sampling window is not continuous but instead consists of multiple samples collected at the same operating points.

In real applications, the sampling window duration and the number of sampling points need to be adjusted according to the mission profile, noise level, and desired accuracy. The mission profile requirement can be met in certain applications, such as a converter operating in a steady state for most of the time and the photovoltaic inverter across similar operating points as the irradiance and temperature fluctuate [25].

At last, the increased resistance is calculated following Ohm's law:

$$\Delta R_{onS1}(t_2) = 1.5\frac{\overline{V}^*_{accS1}(t_2) - \overline{V}^*_{accS1}(t_1)}{\overline{I}_{S1}(t_2)} \tag{10}$$

where a factor of 1.5 is applied as a gain for the coordinate transformation. t_1 and t_2 are illustrated in Fig. 1(c) as the initial time and the time to detect the degradation, respectively.

IV. HARDWARE-IN-THE-LOOP VALIDATION

A. Test Setup

A hardware-in-the-loop (HIL) setup is built to validate the proposed method. The test setup is illustrated in Fig. 3(a) and its parameters are listed in Table I. The inverter operates within a Typhoon HIL 404, while the controller and the proposed method are executed on a dSPACE MicroLabBox.

The IGBT module in the conventional HIL setup lacks of an adjustable on-state resistance setting; instead, the junction temperature-on-state voltage relationship is used to simulate the on-state resistance. This function was introduced in Typhoon HIL version 2020.3 [26]. The on-state resistance is implemented as a look-up table, allowing the resistance to be adjusted based on the junction temperature. The validation work only adjusts the on-state resistance of S_1, but the method for tuning on-state resistances of the IGBT is applicable to all IGBTs within the inverter.

TABLE I
PARAMETERS OF HARDWARE-IN-THE-LOOP SIMULATION.

Parameter	Symbol	Value
Input power rating	P_{in}	$20\,kW$
DC link voltage	V_{dc}	$800\,V$
Grid voltage	V_g	$325\,V$
Bus capacitance	C_{bus}	$600\,\mu F$
Bus capacitor ESR	R_C	$1\,m\Omega$
Filter inductance	L_g	$6\,mH$
Filter inductor ESR	R_L	$100\,m\Omega$
ADC voltage resolution	V_{ADC}	$50\,mV$
Initial resistance of IGBT	R_{onini}	$25\,m\Omega$
Grid frequency	f_g	$50\,Hz$
Switching frequency	f_{sw}	$20\,kHz$
Fundamental cycle	T_0	$20\,ms$
Switching cycle	T_{sw}	$50\,\mu s$
Sampling period	T_{sa}	$50\,\mu s$
Current loop P/I gain	K_p/k_i	$40/500$
Current loop bandwidth	f_{B1}	$1.05\,kHz$
ADC current resolution	I_{ADC}	$6\,mA$
End-of-life criteria	$\Delta R_{on(eol)}$	$2\,m\Omega$

The MicroLabBox records the grid voltage, v_{abc}, and the inverter output current, i_{abc}. The grid voltage is used by the PLL to determine the phase angle of the grid voltage, while the output current is used for current control. The current controller output is collected in dSPACE ControlDesk as an input signal for the proposed method. Additional signals are also gathered in ControlDesk for use in the analysis and validation process.

Typical waveforms are illustrated in Fig. 3(b), including the grid current in phase A i_a; the reference voltage v^*_{acc}; the accumulated IGBT current I_{S1}; and the accumulated reference voltage V^*_{accS1}.

v^*_{acc} is the output of the current controller converted from v^*_{dqcc} (in Fig. 1) into the stationary frame. The current controller compensates for the voltage drop across the filter and nonideal circuit behaviors, i.e., the on-state resistance on IGBTs and deadtime effects. Due to the voltage drop on the inductive filter, v^*_{acc} leads the phase current, i_a. Additionally, deadtime compensation introduces oscillations in v^*_{acc}, also as modeled in [27]. Both i_{S1} and v^*_{accS1} are accumulated over each fundamental cycle when i_a is positive (the blue period), and then held stable until the end of each fundamental cycle (the yellow period). The stable value is collected at the end of each fundamental cycle and used to derive the average values.

B. Validation Results

The proposed method is tested in three cases with different increased on-state resistances (Case I), power levels (Case II), and power factors (Case III). In Case I, different health statuses are emulated by increasing on-state resistance (Fig. 1(c)), while Cases II and III examine the effects of operating conditions on the performance of the method. The IGBT module is modeled based on Infineon FS50R12KT4 [28], which has an initial on-state resistance of $25\,m\Omega$ and the corresponding end-of-life criteria for $\Delta R_{on(eol)}$ is $2\,m\Omega$.

Case I evaluates the proposed method under full load conditions with varying resistances ΔR_{onS1} from $0\,m\Omega$ to $5\,m\Omega$

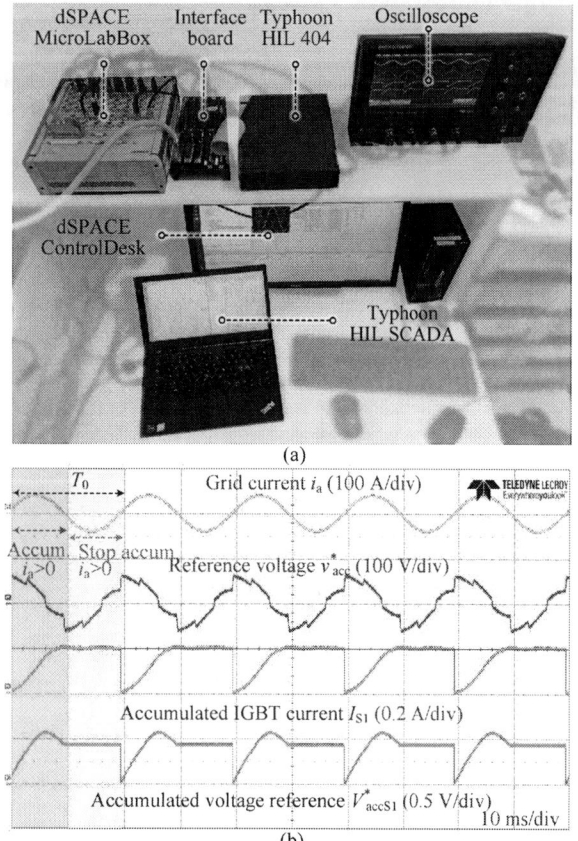

Fig. 3. Test setup: (a) Typhoon HIL and dSPACE setup. (b) Typical waveforms.

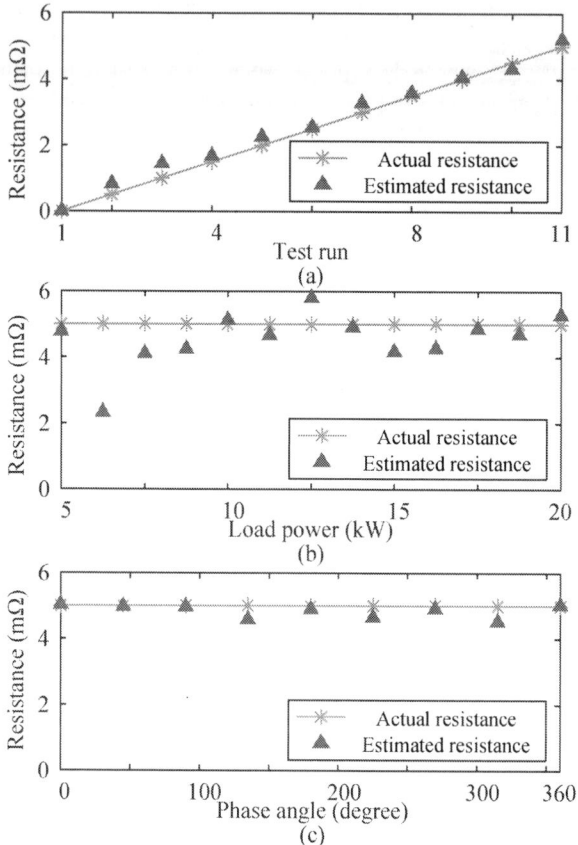

Fig. 4. HIL test results of the proposed method across (a) different on-state resistance increments. (b) different power levels. (c) different power factors.

thereby incorporating the $2\,\mathrm{m\Omega}$ end-of-life criteria within the test range. As ΔR_{onS1} increases, the mean value of v_{accS1} increases slightly as the on-state resistance increases. The estimations are illustrated in Fig. 4(a), where the estimated values (the blue triangle) closely follow the trend of the actual resistances (the green curve), with an average error of $0.29\,\mathrm{m\Omega}$ ($1.16\,\%$ of the $25\,\mathrm{m\Omega}$ initial resistance R_{onini}).

Case II examines the method across different power levels, with the output power varying from 5 to $20\,\mathrm{kW}$. The results are illustrated in Fig. 4(b). As the power increases, estimation accuracy improves, with an estimation error of $0.27\,\mathrm{m\Omega}$ ($1.08\,\%$ of R_{onini}). The improvement is attributed to the higher current flow through the IGBT at increased load levels, which corresponds to an increment in the reference voltage. Therefore, the method should be used at a higher power level to achieve an accurate and reliable estimation.

Case III examines the method across different power factors, ranging from a phase angle of $0°$ to $360°$. The results are illustrated in Fig. 4(c), demonstrating that the estimation closely tracks the actual increase in resistance with a mean error $0.17\,\mathrm{m\Omega}$ ($0.68\,\%$ of R_{onini}). Case III also indicates that the phase angle has minimum impact on resistance estimation.

V. CONCLUSION

The proposed on-state voltage estimation method is a sensorless method using only the control variable and the measurement signals. This method fully leverages the information within these variables, eliminating the need for additional sensors, so converter cost and volume remain unchanged. Compared to the existing sensor-less methods, the proposed method is generic and adaptable to different systems. Additionally, noise is addressed within the method, with a mitigation strategy integrated into the algorithm. The effectiveness of the method has been validated using HIL. The three cases demonstrate that the power level has a greater impact on the method accuracy and reliability than resistance or power factor. Thus, the method is most accurate and reliable at full load–in Case II, for example, the estimation error at full loads is $1.08\,\%$ of the initial resistance R_{onini}.

REFERENCES

[1] W. Lai, M. Chen, L. Ran, O. Alatise, S. Xu, and P. Mawby, "Low δt_j stress cycle effect in igbt power module die-attach lifetime modeling," *IEEE Trans. Power Electron.*, vol. 31, no. 9, pp. 6575–6585, 2016.

[2] F. Hosseinabadi, S. Chakraborty, S. K. Bhoi, G. Prochart, D. Hrvanovic, and O. Hegazy, "A comprehensive overview of reliability assessment strategies and testing of power electronics converters," *IEEE Open J. of Power Electron.*, vol. 5, pp. 473–512, 2024.

[3] Y. Yang and P. Zhang, "In situ junction temperature monitoring and bond wire detecting method based on igbt and fwd on-state voltage drops," *IEEE Trans. Ind. Appl.*, vol. 58, no. 1, pp. 576–587, 2022.

[4] X. Fang, S. Lin, X. Huang, F. Lin, Z. Yang, and S. Igarashi, "A review of data-driven prognostic for igbt remaining useful life," *Chin. J. Elect. Eng.*, vol. 4, no. 3, pp. 73–79, 2018.

[5] ECPE, "Aqg 324 qualification of power modules for use in power electronics converter units in motor vehicles," *ECPE Online Tutorial*, 2021.

[6] F. Qin, X. Bie, T. An, J. Dai, Y. Dai, and P. Chen, "A lifetime prediction method for igbt modules considering the self-accelerating effect of bond wire damage," *IEEE J. Emerg. Sel. Topics Power Electron.*, vol. 9, no. 2, pp. 2271–2284, 2021.

[7] C. H. Van Der Broeck, S. Kalker, and R. W. De Doncker, "Intelligent monitoring and maintenance technology for next-generation power electronic systems," *IEEE J. Emerg. Sel. Topics Power Electron.*, vol. 11, no. 4, pp. 4403–4418, 2023.

[8] M. Ghadrdan, S. Peyghami, H. Mokhtari, and F. Blaabjerg, "Floating-reference on-state voltage measurement strategy for condition monitoring application," *IEEE Trans. Power Electron.*, vol. 38, no. 2, pp. 2529–2538, 2023.

[9] M. Guacci, D. Bortis, and J. W. Kolar, "On-state voltage measurement of fast switching power semiconductors," *CPSS Trans. Power Electron. and Appl.*, vol. 3, no. 2, pp. 163–176, 2018.

[10] C. Roy, N. Kim, D. Evans, J. Gafford, and B. Parkhideh, "Challenges and implementation of online in situ rdson measurement in a three-phase inverter," *IEEE Trans. Power Electron.*, vol. 39, no. 9, pp. 11677–11686, 2024.

[11] Y. Peng, Y. Shen, and H. Wang, "A converter-level on-state voltage measurement method for power semiconductor devices," *IEEE Trans. Power Electron.*, vol. 36, no. 2, pp. 1220–1224, 2021.

[12] Y. Peng and H. Wang, "A simplified on-state voltage measurement circuit for power semiconductor devices," *IEEE Trans. Power Electron.*, vol. 36, no. 10, pp. 10993–10997, 2021.

[13] D. Xiang, L. Ran, P. Tavner, S. Yang, A. Bryant, and P. Mawby, "Condition monitoring power module solder fatigue using inverter harmonic identification," *IEEE Trans. Power Electron.*, vol. 27, no. 1, pp. 235–247, 2012.

[14] F. Yüce and M. Hiller, "Condition monitoring of power electronic systems through data analysis of measurement signals and control output variables," *IEEE J. Emerg. Sel. Topics Power Electron.*, vol. 10, no. 5, pp. 5118–5131, 2022.

[15] T.-F. Wu, A. Kumari, and C.-C. Hung, "Comparison of direct digital control and dq-based control for a three-phase three-wire inverter with lcl filter," *IEEE J. Emerg. Sel. Topics Power Electron.*, vol. 12, no. 5, pp. 5139–5151, 2024.

[16] F. Dongqiang, P. Sundararajan, M. D. Siddique, and S. K. Panda, "An improved degradation monitoring method for high power igbt modules based on on-state resistance estimation," in *in Proc. 49th Annu. Conf. IEEE Ind. Electron. Soc. (IECON)*, pp. 1–6, 2023.

[17] Y. Yang and P. Zhang, "In situ junction temperature monitoring and bond wire detecting method based on igbt and fwd on-state voltage drops," *IEEE Trans. Ind. Appl.*, vol. 58, no. 1, pp. 576–587, 2022.

[18] U.-M. Choi, F. Blaabjerg, S. Jørgensen, S. Munk-Nielsen, and B. Rannestad, "Reliability improvement of power converters by means of condition monitoring of igbt modules," *IEEE Trans. Power Electron.*, vol. 32, no. 10, pp. 7990–7997, 2017.

[19] S. Ou, A. Sangwongwanich, S. Sahoo, and F. Blaabjerg, "Semiconductor devices condition monitoring using harmonics in inverter control variables," in *2023 IEEE 25th Eur. Conf. Power Electron. Appl.*, pp. 1–8, 2023.

[20] A. Allca-Pekarovic, P. J. Kollmeyer, J. Reimers, P. Mahvelatishamsabadi, T. Mirfakhrai, P. Naghshtabrizi, and A. Emadi, "Loss modeling and testing of 800-v dc bus igbt and sic traction inverter modules," *IEEE Trans. Transp. Electrific.*, vol. 10, no. 2, pp. 2923–2935, 2024.

[21] Y. Yang, K. Zhou, H. Wang, and F. Blaabjerg, "Harmonics mitigation of dead time effects in pwm converters using a repetitive controller," in *in Proc. IEEE Appl. Power Electron. Conf. Expo. (APEC)*, pp. 1479–1486, 2015.

[22] B. P. McGrath and D. G. Holmes, "A general analytical method for calculating inverter dc-link current harmonics," *IEEE Trans. Ind. Appl.*, vol. 45, no. 5, pp. 1851–1859, 2009.

[23] C. Liu, F. Deng, Q. Yu, Y. Wang, F. Blaabjerg, and X. Cai, "Submodule capacitance monitoring strategy for phase-shifted carrier pulsewidth-modulation-based modular multilevel converters," *IEEE Trans. Ind. Electron.*, vol. 68, no. 9, pp. 8753–8767, 2021.

[24] A. Kannal, "Solar power generation data," 2020. Kaggle Dataset, [Online]. Available: https://www.kaggle.com/datasets/anikannal/solar-power-generation-data.

[25] J. Poon, P. Jain, C. Spanos, S. K. Panda, and S. R. Sanders, "Fault prognosis for power electronics systems using adaptive parameter identification," *IEEE Trans. Ind. Appl.*, vol. 53, no. 3, pp. 2862–2870, 2017.

[26] Typhoon HIL, Inc., "Igbt leg component in typhoon hil." https://www.typhoon-hil.com/documentation/typhoon-hil-software-manual/References/igbt_leg.html. Accessed: 2024-11-05.

[27] Y. Wang, W. Xie, X. Wang, and D. Gerling, "A precise voltage distortion compensation strategy for voltage source inverters," *IEEE Trans. Ind. Electron.*, vol. 65, no. 1, pp. 59–66, 2018.

[28] Infineon Technologies, *Technical information IGBT-module FS50R12KT4 B11 datasheet*, 2013. Rev. 3.0.

A Novel Non-intrusive Online Monitoring Method for Diagnosing the Lift-off of Bonding wires in SiC MOSFETs

Keqi Song
Department of Electrical Engineering
City University of Hong Kong
Hong Kong, China
kqsong2-c@my.cityu.edu.hk

Henry Shu-Hung Chung
Department of Electrical Engineering
City University of Hong Kong
Hong Kong, China
eeshc@cityu.edu.hk

Ho-Tin Tang
Department of Electrical Engineering
City University of Hong Kong
Hong Kong, China
hotintang2-c@my.cityu.edu.hk

Abstract—This article presents a real-time method for detecting bond wire lift-off in SiC MOSFETs using a transmission line impedance matching approach. By temporarily introducing a discrete transmission line (DTL) into the gate driving signal and observing energy changes on specific DTL sections, source inductance changes due to bond wire issues can be indirectly monitored. The technique was tested on a 40W, 48V/24V, 100kHz buck converter without requiring any external injection signal.

Keywords—*SiC MOSFET, bond wire lift-off, online condition monitoring, diagnosis.*

I. INTRODUCTION

High-power devices often come in wire-bonded packages that contain multiple layers of materials with different coefficients of thermal expansion, causing bond wire failure. Bond wire lift-off leads to an increase in source resistance that can result in higher conduction losses, reduced efficiency, and degraded performance [1]-[7]. Typically, 6 or 8 parallel bond wires are utilized for interconnection [8]. The failure of each bond wire leads to a significant increase in source inductance. Given that bond wire faults represent a major package-related failure, the change of source inductance serves as an indicator of the bond wire's health. The elevated source inductance leads to voltage spikes, ringing effects, and potential electromagnetic interference [8].

Ref [9] proposes a method to monitor the aging state of the bond wire using the on-state drain-source voltage separation with respect to the temperature. The package structure resistance is extracted, and a model is developed to estimate the junction temperature in a healthy state. The package structure voltage is then utilized to detect lift-off failure. However, the presence of parasitic parameters affects the accuracy of the temperature characteristics.

Ref [10] introduces an online system via acoustic signals based on morphology characteristics, which are caused by aging problems. Due to the reflected signal differences due to different propagation conditions, bond wire lift-off failure could be detected. However, these active measurement methods bring additional sensing circuits and may cause misjudgments due to other aging problems that also cause temperature rise and different radio frequency reflections.

In [11]-[12], a spread spectrum time-domain reflectometry (SSTDR) is proposed. A spread spectrum signal is introduced to analyze reflections and identify impedance mismatches at the gate-source interface. High-resolution sampling equipment and expertise in signal processing are needed.

This article introduces a real-time bond wire lift-off detection method involving a discrete DTL briefly introducing into the gate driving circuit without any extra signal injection. An analog circuit tracks high-frequency energy changes on specific DTL sections to indirectly monitor the change of the source inductance. The technique has been successfully applied to a 40W, 48V/24V, 100kHz buck converter with the DTL introduced into the gate driving circuit for 2ms.

II. PROPOSED REAL-TIME BOND WIRE LIFT-OFF DETECTION CIRCUIT

Fig. 1 shows the proposed circuit. It contains a bypass switch, S_{BP}, with a DTL connected across it via a capacitor C_C and a diode D_C. The DTL has 10 sections. Each section contains an inductor L_{DTL} and a capacitor C_{DTL}. The bypass switch is normally closed and is open upon monitoring the condition of the SiC MOSFET.

This work was supported by the Innovation and Technology Fund from the Hong Kong Special Administrative Region, China, under Project #MRP/010/21X.

979-8-3315-1612-3/25 $31.00 © 2025 IEEE

Fig. 2 shows the gate signal v_g, and the signal applied to S_{BP}, v_{bp}. The off-time duration of v_{bp} lasts 200 switching cycles to complete the entire condition monitoring.

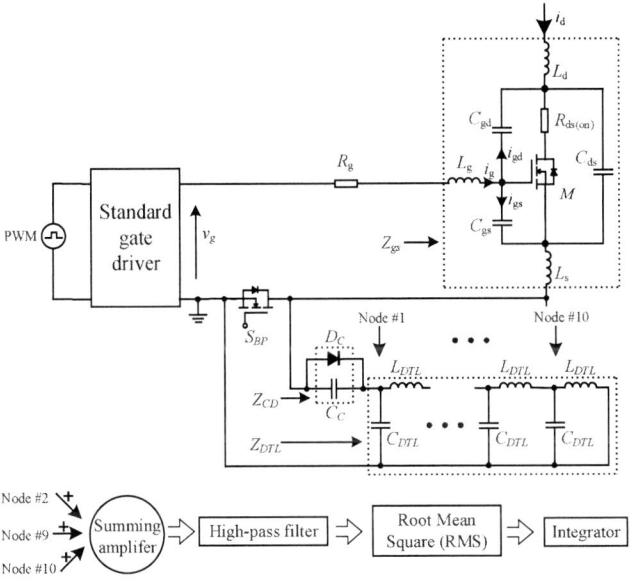

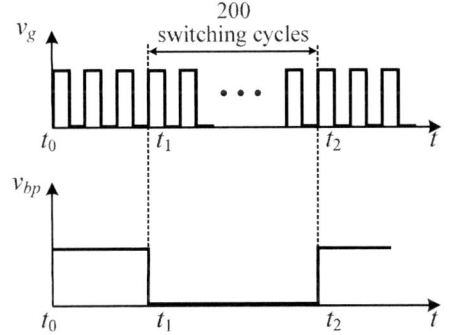

Fig. 1. Schematic of the proposed real-time bond wire lift-off detection circuit.

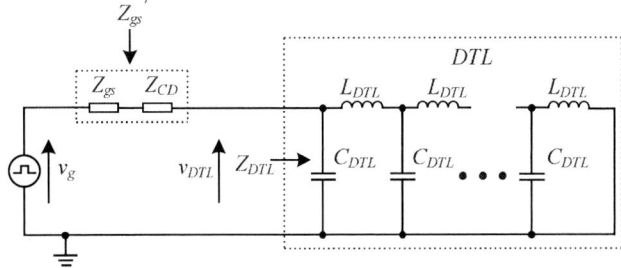

Fig. 2. Driving signals v_g and bypass switch signal v_{bp}.

By applying Thevenin's theorem, the voltages between two adjacent nodes are related by

$$V_{n-1}(s) - V_n(s) = s \cdot L_{DTL} \cdot s \cdot C_{DTL} \cdot (V_n(s) - V_{n+1}(s)) \qquad (1)$$

$$\frac{V_{n-1}(s) - V_n(s)}{s \cdot L_{DTL}} = \frac{V_n(s) - V_{n+1}(s)}{\frac{1}{s} \cdot C_{DTL}}. \qquad (2)$$

where V_n is the voltage of node n. This shows that the relationship between adjacent node voltages is determined by the parameters of L_{DTL} and C_{DTL}, and the resonant frequency variable s.

Fig. 3. Equivalent transmission line model of proposed circuits.

TABLE I. PARAMETERS OF THE EXPERIMENTAL PROTOTYPE

Parameter	Value	Parameter	Value
SiC MOSFET, M	IMW65R083 M1H	Switching frequency	100kHz
L_s	Approx. 10nH	L_{DTL}	68nH
C_{iss}	624pF	C_{DTL}	510pF
High-pass filter cutoff frequency f_c	5MHz	L_{cxt}	40nH

The equivalent transmission line model of proposed online monitoring circuit is shown in Fig. 3 with the parameters given in Table I. The impedance Z_{gs}' represents the impedance of C_C and D_C (i.e., Z_{CD} in Fig. 1), and the impedance of equivalent SiC MOSFET internal impedance (i.e., Z_{gs} in Fig. 1). After the DTL part consisting of inductances L_{DTL} and capacitors C_{DTL}, the end of the transmission line model is short-circuited. The voltages between them can be expressed by

$$V_{DTL}(s) = V_g(s) \frac{Z_{DTL}(s)}{Z_{gs}'(s) + Z_{DTL}(s)}. \qquad (3)$$

According to the transmission line model, the transmission line impedance can be represented by

$$Z_{in} = Z_0 \frac{Z_L + Z_0 \tanh(\gamma \ell)}{Z_0 + Z_L \tan(\gamma \ell)}. \qquad (4)$$

where Z_{in} is the impedance of the transmission line, Z_L is the load impedance, γ is the propagation constant, and ℓ is the length of transmission line. Since the end of DTL is short-circuited, the Z_L equals zero.

Z_0 is the characteristic impedance and can be expressed as

$$Z_0 = \sqrt{\frac{L_0}{C_0}}. \qquad (5)$$

The parameter γ can be expressed as

$$\gamma = j\omega\sqrt{L_0 C_0} \ . \tag{6}$$

Therefore, the voltage of transmission line could be expressed as

$$v_{DTL} = v_g \cdot \frac{jZ_0 \tan(\omega\sqrt{L_0 C_0}\,\ell)}{R_g' + jX_{gs}' + jZ_0 \tan(\omega\sqrt{L_0 C_0}\,\ell)} \ . \tag{7}$$

where real part R_g' and imaginary part X_{gs}' represent the impedance Z_{gs}'. The design criteria for DTL are that L_{DTL} and C_{DTL} are consistent with the characteristic inductance L_0 and characteristic capacitance C_0, and match the parameters between the gate and source of the SiC MOSFET as much as possible.

For the de V_{DTL} is maximum, when

$$X_{gs}' = -Z_0 \tan(\omega\sqrt{L_0 C_0}\,\ell) \ . \tag{8}$$

$$\frac{1}{\omega C_{gs}} + \frac{1}{\omega C_{cd}} - \omega L_s = Z_0 \tan(\omega\sqrt{L_0 C_0}\,\ell) \ . \tag{9}$$

Based on this, the resonance will occur at the frequency shown in (9).

The relationship between resonant frequency $f_{resonant}$, inductance L_s and amplitude constant α proportional to V_{DTL} is shown in Fig. 4. When L_s becomes larger, the $f_{resonant}$ will become lower and amplitude α will increase.

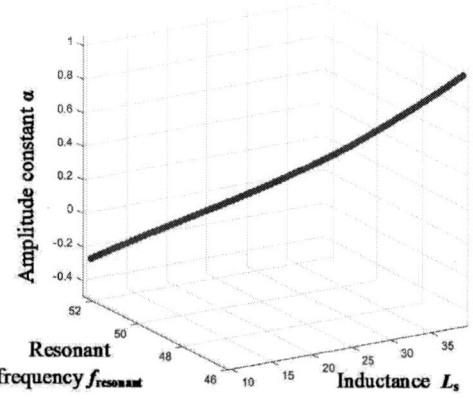

Fig. 4. Relationship between V_{DTL} Amplitude constant α and resonant frequency $f_{resonant}$, inductance L_s.

The impedance mismatching on the transmission line leading to part of the signal reflecting back and interfering with the incident signal and forming a standing wave. Due to the phase difference between the reflected wave and the incident wave, the increase in reflection and standing waves will increase the energy of high frequency components.

Three voltages of #2, #9, #10 are scaled-down, aggregated by a summer, filtered by a high-pass filter, and then rectified by a RMS module and integrated.

Based on the component values given in Table I, Fig. 5 shows the simulated voltage waveform of node #1 and gate signal applied to the SiC MOSFET. Fig. 6, Fig. 7, and Fig. 8 shows the simulated voltage waveform of node #2, node #9, and node #10.

Fig. 9 shows the frequency spectra of node #10. Fig. 10 shows output voltage of the integrator.

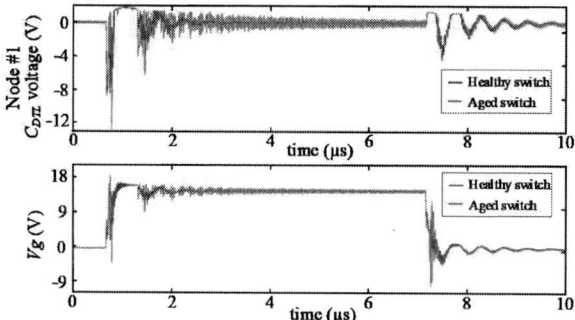

Fig. 5. Simulated voltage of node #1 and gate signal.

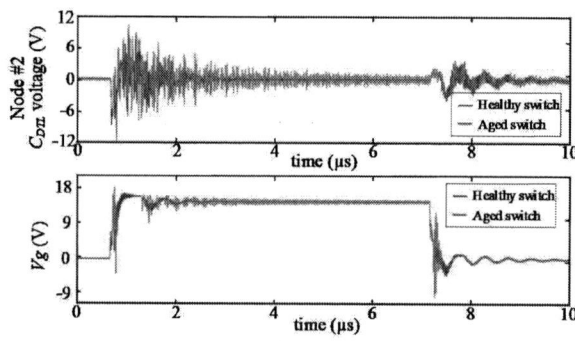

Fig. 6. Simulated voltage of node #2 and gate signal.

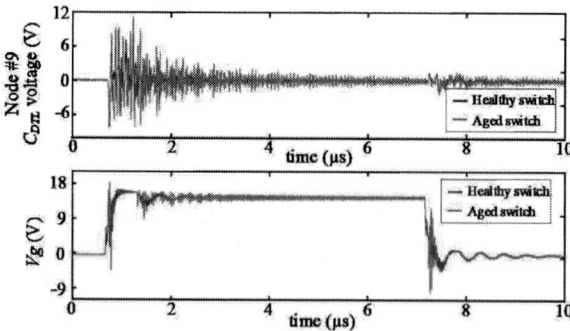

Fig. 7. Simulated voltage of node #9 and gate signal

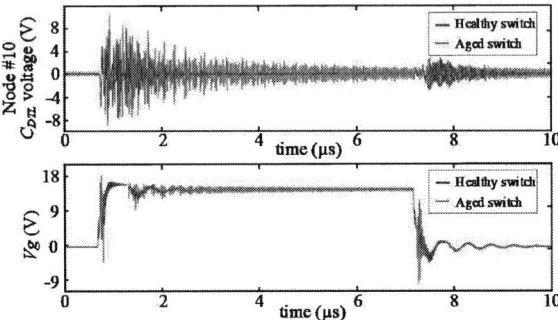

Fig. 8. Simulated voltage of node #10 and gate signal.

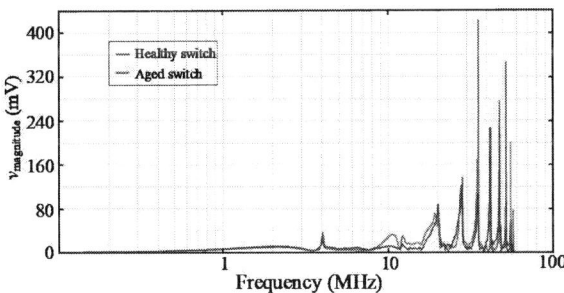

Fig. 9. Frequency spectra of node #10.

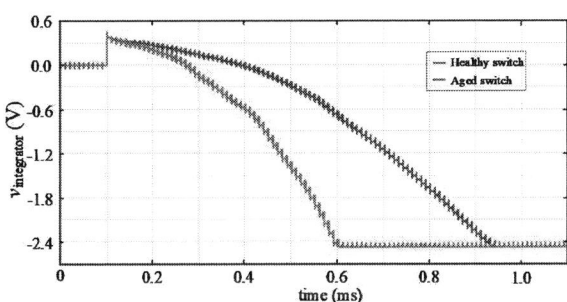

Fig. 10. Simulated output voltage of the integrator.

III. EXPERIMENTAL VERIFICATIONS

The circuit has been applied to a 40W, 48V/24V, 100kHz buck converter as shown in Fig. 11.

Fig. 11. Circuit prototype.

Fig. 12 shows the measured gate signal. The circuit does not affect the gate signal significantly and vg changes from range between -1.72V and 15.3V to range between -1.36V to 15.7V. An external inductor of 40nH is used to mimic the source inductance change.

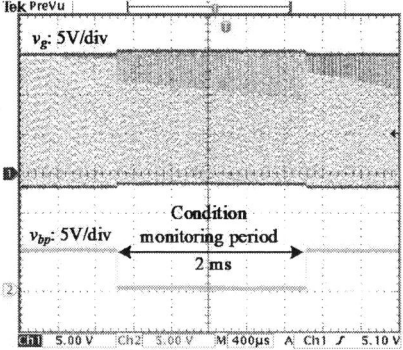

Fig. 12. Gate signal v_g and bypass switch signal v_{bp}.

Fig. 13 shows the waveform and frequency spectra of the node #10 voltage. It has more high-frequency component with the aged switch.

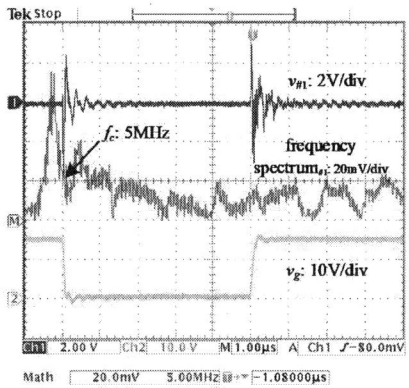

(a)

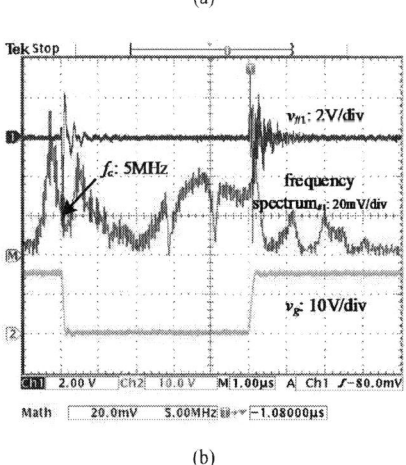

(b)

Fig. 13. Time-domain waveform and frequency spectra of the voltage of the node #1. (a) Healthy switch. (b) Aged switch.

Fig. 14 shows that during the condition monitoring period, the integrator saturation time with the healthy and aged devices are 1.30ms and 1.08ms, respectively. Such 20% difference can be used to identify the change of the source inductance.

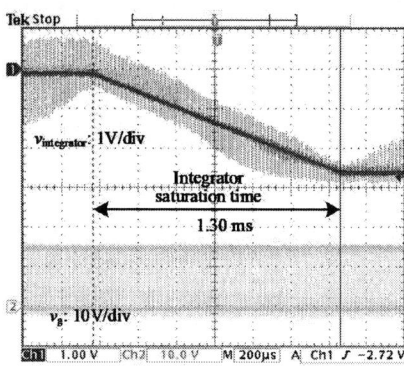

(a)

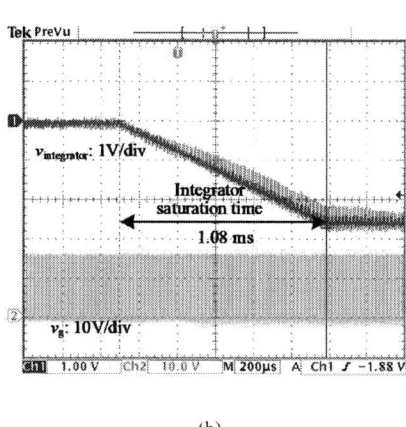

(b)

Fig. 14. Output waveforms of integrators. (a) Healthy switch. (b) Aged switch.

IV. CONCLUSION

An online SiC MOSFET monitoring circuit is proposed to detect bond wire lift-off. It does not need any other external signal injection source and does not affect the driver signal.

REFERENCES

[1] X. Deng, H. Zhu, X. Li, X. Tong, S. Gao, Y. Wen, S. Bai, W. Chen, K. Zhou, and B. Zhang, "Investigation and Failure Mode of Asymmetric and Double Trench SiC mosfets Under Avalanche Conditions," in IEEE Transactions on Power Electronics, vol. 35, no. 8, pp. 8524-8531, Aug. 2020, doi: 10.1109/TPEL.2020.2967497.

[2] H. Luo, F. Iannuzzo, N. Baker, F. Blaabjerg, W. Li and X. He, "Study of Current Density Influence on Bond Wire Degradation Rate in SiC MOSFET Modules," in IEEE Journal of Emerging and Selected Topics in Power Electronics, vol. 8, no. 2, pp. 1622-1632, June 2020, doi: 10.1109/JESTPE.2019.2920715.

[3] Y. Chen, Q. Wu, C. Li, H. Luo, Y. Xia, Q. Yin, W. Li, and X. He, "Thermal Mitigation and Optimization Via Multitier Bond Wire Layout for IGBT Modules Considering Multicellular Electro-Thermal Effect," IEEE Transactions on Power Electronics, vol. 37, no. 6, pp. 7299-7314, 2022.

[4] Y. Yang, X. Ding, and P. Zhang, "A Novel Junction Temperature Estimation Method Independent of Bond Wire Degradation for IGBT," IEEE Transactions on Power Electronics, vol. 38, no. 8, pp. 10256-10268, 2023.

[5] S. Pu, F. Yang, B. T. Vankayalapati, E. Ugur, C. Xu and B. Akin, "A Practical On-Board SiC MOSFET Condition Monitoring Technique for Aging Detection," in IEEE Transactions on Industry Applications, vol. 56, no. 3, pp. 2828-2839, May-June 2020.

[6] E. Ugur, C. Xu, F. Yang, S. Pu and B. Akin, "A New Complete Condition Monitoring Method for SiC Power MOSFETs," in IEEE Transactions on Industrial Electronics, vol. 68, no. 2, pp. 1654-1664, Feb. 2021.

[7] J. Dyer, Z. Zhang, F. Wang, D. Costinett, L. M. Tolbert and B. J. Blalock, "Online condition monitoring of SiC devices using intelligent gate drive for converter performance improvement," 2016 IEEE 4th Workshop on Wide Bandgap Power Devices and Applications (WiPDA), Fayetteville, AR, USA, 2016, pp. 182-187.

[8] Yun, M., Cai, M., Yang, D., Yang, Y., Xiao, J. and Zhang, G. "Bond wire damage detection method on discrete MOSFETs based on two-port network measurement," in Micromachines, vol. 13, no. 17, pp.1075, June 2022, doi: 10.3390/mi13071075.

[9] M. Du, J. Xin, H. Wang and Z. Ouyang, "Aging Diagnosis of Bond Wire Using On-State Drain-Source Voltage Separation for SiC MOSFET," in IEEE Transactions on Device and Materials Reliability, vol. 21, no. 1, pp. 41-47, March 2021, doi: 10.1109/TDMR.2020.3047419.

[10] Z. Zhang, C. Chen, H. Ishino, T. Endo, K. Sugiura, K. Tsuruta, and K. Suganuma, "Online Condition Monitoring of Solder Fatigue in a Clip-Bonding SiC mosfet Power Assembly via Acoustic Emission Technique," IEEE Trans. Power Electron., vol. 38, no. 2, pp. 1468-1478, 2023.

[11] A. Hanif, S. Roy and F. Khan, "Detection of gate oxide and channel degradation in SiC power MOSFETs using reflectometry," 2017 IEEE 5th Workshop on Wide Bandgap Power Devices and Applications (WiPDA), Albuquerque, NM, USA, 2017, pp. 383-387, doi: 10.1109/WiPDA.2017.8170577.

[12] A. Hanif, S. Das and F. Khan, "Active power cycling and condition monitoring of IGBT power modules using reflectometry," IEEE Applied Power Electronics Conference and Exposition (APEC), pp. 2827-2833, 2018.

Optimizing MOSFET Selection for EMC-Critical Automotive Applications

Sacha J. Cazzitti
Department of Electrical and Electronic Engineering
The University of Manchester
Manchester, United Kingdom
sacha.cazzitti@manchester.ac.uk

Christian Radici
Automotive MOSFET Applications
Nexperia
Manchester, United Kingdom
christian.radici@nexperia.com

Andrew J. Forsyth
Department of Electrical and Electronic Engineering
The University of Manchester
Manchester, United Kingdom
andrew.forsyth@manchester.ac.uk

Cheng Zhang
Department of Electrical and Electronic Engineering
The University of Manchester
Manchester, United Kingdom
cheng.zhang@manchester.ac.uk

Peter Vines
Automotive MOSFET Applications
Nexperia
Manchester, United Kingdom
peter.vines@nexperia.com

Abstract— **This paper investigates the effect of the Active Area and integrated Snubber Area of low-voltage silicon superjunction MOSFETs in a half-bridge configuration on the exhibited switching losses, ringing, and EMC performance. The damping factor is employed as a benchmark to quantify the oscillations in the drain current waveform. EMC sweeps were conducted to correlate the damping factor with electromagnetic interference levels, providing a quantitative link between oscillation behavior and EMC performance. Experimental characterization is performed on a range of devices to demonstrate the impact of the die parameters on the performance, which is then replicated through simulated testing. This research aims to identify the design choices used to select the optimal device layout for different automotive applications, including DC-DC converters, motor drives and isolation switches, based on their varying performance requirements. Alterations made to the Active Area were found to influence the parasitic capacitances of the device, resulting in a direct effect on the oscillatory behavior and associated losses. Devices featuring larger Snubber Areas demonstrated advantages in applications that necessitate strict automotive EMC standards.**

Keywords—MOSFET, EMC, Automotive Applications, Integrated Snubber, Superjunction MOSFET.

I. INTRODUCTION

Despite split-gate being the most common MOSFET structure for voltages <100 V, due to the higher cell density [1], the superjunction (SJ) structure is still used because of its higher performance in linear mode and stability under repetitive avalanche [2]. This study is driven by automotive applications requiring high-performance MOSFETs that allow for a small footprint and good efficiency. Nexperia's low voltage SJ technology addresses the minimization of the drift-resistance by implementing a form of RESURF (Reduced Surface Field) based on the introduction of p-pillars, as depicted in Fig. 1 [2][3]. This contrasts with split-gate technology which adopts another form of RESURF by splitting the trench into gate and source shield. Both technologies employ a form of integrated snubber, however, it is lumped in the SJ, while it is distributed in split-gate devices [4] - [7].

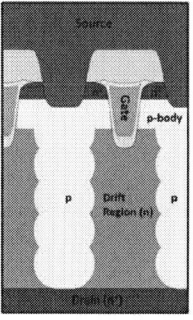

Fig. 1. Superjunction power MOSFET die structure [10].

The parasitic inductance in the switching loop in automotive application circuits has a significant influence on the device's ringing following a switching transient as it affects the frequency and damping and is a well-documented aspect of power electronics [8][9]. However, internal MOSFET parameters like the die area and layout have not been studied as extensively. The SJ integrated Snubber Area (S_A) influences the frequency and the decay of the drain current (I_D) oscillations, since the Snubber Area directly influences the snubber capacitance (C_{SNB}) which also increases the total drain to source capacitance (C_{DS}) of the MOSFET. This snubber capacitance comes from a combination of the capacitance between the gate and the p-pillar and that between the p-pillar and the drift region [10]. The snubber resistance (R_{SNB}) will impact the damping of the oscillations with a reduced effect on the switching energy losses or the frequency of the waveform as it operates in parallel with the MOSFET. The Active Area (A_A) will influence the oscillatory behavior of the drain current (I_D) as it directly impacts the device's parasitic capacitances, with the undesirable effect of slowing switching transients, indicating increased switching losses. Additional A_A will also lower the on-resistance ($R_{DS(on)}$), resulting in lower conduction losses.

This paper will focus on the performance of SJ devices in automotive applications, covering DC-DC converters, motor drives and isolation switches. Each of these applications has different requirements and will prioritize different aspects of the MOSFET performance such as conduction losses, switching losses, and parasitic ringing with its impact on EMC (Electromagnetic Compatibility). The A_A and S_A are two aspects of the device that are used to tailor the device's performance to prioritize a specific factor.

II. AUTOMOTIVE APPLICATIONS

In automotive applications, DC-DC converters are used in the management and distribution of power among diverse vehicle systems. They can be used to step-down the battery voltage to supply infotainment and navigation systems (<100 W), or from the high voltage isolated battery pack of an electric vehicle (EV) to 12 V (2.2 kW or 4.4 kW) where low voltage MOSFETs are commonly used in the secondary circuit as synchronous rectifiers or secondary regulators. Boost converters are also present, for instance, in lighting units (~50 W) and airbag systems (~300 W) to generate supply rails above 12 V.

MOSFETs employed in motor drives are used to control the speed, torque and direction of a range of motors throughout the vehicle. Brushed DC motors are used for low power applications (<200 W) such as body control (windshield wipers, windows, sunroof, seat adjustments), and various pumps across the vehicle (oil, fuel, water), the MOSFET provides on/off control without pulse width modulation. MOSFETs are also used for BLDC (Brushless DC) motor applications requiring higher power and improved reliability (>200 W) such as in electric power steering (EPS), active suspension motors, steer and brake-by-wire, and EV cooling pumps.

Isolation switches offer electrical separation between various components and systems, thereby improving safety and reliability. In automotive applications, they are generally high side switches that disconnect the load to supply, or multiplex different supplies to the same load, or multiple loads. They can be used for safety, protection of sensitive loads or to reduce energy loss during load inactivity. Isolation switches are employed to handle from 5 A to 250 A and may provide functions like reverse battery protection, fast deactivation in case of short circuit and slow activation to limit inrush current due to capacitive loads.

Table I links these applications with the design priorities of EMC, conduction losses and switching losses. Faster switching frequency applications like low power DC-DC converters and BLDC motor drives will emphasize EMC and switching losses, while applications such as brushed DC motor dives, high power DC-DC converters and isolation switches with lower switching frequencies, will have a higher priority on conduction losses.

TABLE I.   DESIGN PRIORITY IN DIFFERENT AUTOMOTIVE
APPLICATIONS

	Low power DC-DC converter	High power DC-DC converter	Brushed DC drives	BLDC motor drives	Isolation switches
EMC	✓✓	✓	✓	✓✓	✓✓
Conduction losses	✓	✓✓	✓✓✓	✓✓	✓✓✓
Switching losses	✓✓✓	✓	✓✓	✓✓	✓

To assess the EMC performance, this study employs a standardized test setup using specialized equipment on a typical automotive application, shown in Fig. 2. This involves evaluating conducted emissions using a spectrum analyzer and Elektra software, assessing conducted emissions on the input side of a DC-DC converter. The standard used is CISPR 25:2008, which regulates the testing of EMC in automotive applications. The spectrum analyzer evaluates the noise levels on the power lines of a device under test (DUT), while the Elektra software facilitates the processing and visualization of the measurement data. This method assesses the noise that is transmitted back to the power supply. The R&S ESH3-Z6 Line Impedance Stabilization Network (LISN) maintains uniform source impedance, isolates the DUT from external power line interference, and facilitates accurate measurement of conducted emissions in accordance with the CISPR 25 standards. It shunts the emissions to the spectrum analyzer, which measures the electromagnetic noise and interference.

The device under test is a buck converter, converting a 24 V input to an 11.2 V output, powering a resistive load of 2 Ω (5.6 A), and providing 62 W of power. The switching frequency was set at 300 kHz and the gate resistors of the two MOSFETs in the circuit were 0.1 Ω.

Fig. 2.   Conducted emissions testbench.

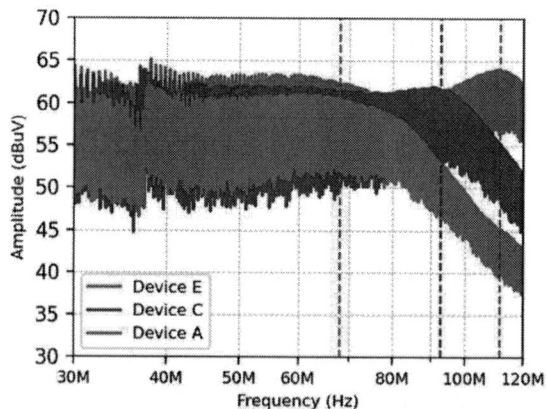

Fig. 3.   DC-DC converter input voltage conducted disturbance sweep.

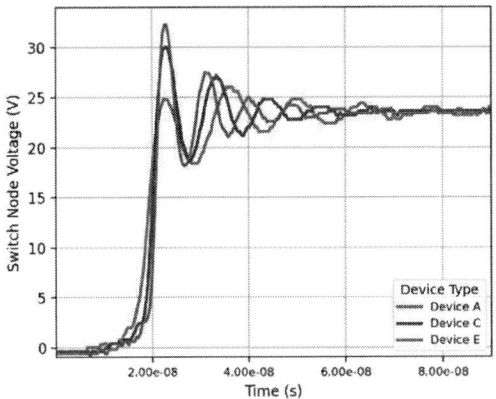

Fig. 4. MOSFET voltage waveform in buck converter operation at 300 kHz; $V_{in}= 24$ V, $R_G= 0.1$ Ω, $R_{Load}= 2$ Ω.

Measurements were made using three LFPAK56 SJ devices with different A_A and, therefore, varying C_{DS}. The impact of the capacitance on the peaks in disturbance was observed in Fig. 3 to be in the 70 to 115 MHz range. The switch node voltage rise across the low side MOSFET was also recorded and compared, as seen in Fig. 4, with notable differences in the damping and oscillation frequencies.

As detailed in Table II, the oscillation frequencies and damping factor (DF) in the converter voltage seen in Fig. 4 are correlated with the frequencies and amplitudes of the peak disturbances recorded in the EMC sweep shown in Fig. 3. Furthermore, a more heavily damped transistor voltage oscillation correlates with a reduced amplitude peak in the EMC sweep. The results are used in section IV to guide MOSFET selection according to the relative importance of EMC in the different applications, with the DF used as an indicator of likely EMC performance. The relation between MOSFET parasitic capacitances and the peaks in disturbance has also been established in previous studies [11].

TABLE II. MOSFET PARASITIC OSCILLATIONS AND EMC SWEEP CORRELATION

Device	Time Domain Oscillations		Frequency Spectrum	
	Damping factor	Frequency (MHz)	Amplitude peak (dBµV)	Frequency of peak (MHz)
A	0.123	71.8	61	66
C	0.104	94.7	62	92
E	0.097	115	64	111

$V_{in}= 24$ V, $R_G= 0.1$ Ω, $R_{Load}= 2$ Ω.

III. DEVICE CHARACTERISATION

A. Experimental Testing

A double pulse testing setup was used to test the performance of a range of SJ MOSFETs by analyzing the waveforms in the circuit. The setup employs a function generator, a DC supply with its voltage (V_{DD}) set to half of the rated voltage of the MOSFETs, an oscilloscope, a double pulse test (DPT) circuit with a 4 µH load inductor, and a personal computer (PC), as shown in Fig. 5. The printed circuit board (PCB) has a series of ceramic and electrolytic capacitors

connected in parallel with the DC supply to ensure that a low impedance source is supplied to the circuit, as illustrated in Fig. 6. The PC controls the function generator, which varies the length of the first pulse, which in turn varies the value of the drain current. The results are then recorded using the oscilloscope connected to the PC. The waveforms are subsequently plotted by the computer using a python script, as depicted in Fig. 7. The script then calculates the key parameters of the waveforms: the DF, frequency and switching energy losses. The oscilloscope records the voltage waveforms in the circuit with high impedance probes, and the drain current is measured through a 1.5mΩ co-axial shunt connected in series with the low side MOSFET. The selection of a 30 Ω gate resistance was done to achieve a balance between oscillations and switching speed, with its value within the recommended range for a half bridge circuit for automotive grade power MOSFETs [12]. To neglect its influence on the results, it remained constant throughout this study.

Fig. 5. Experimental setup for double pulse testing.

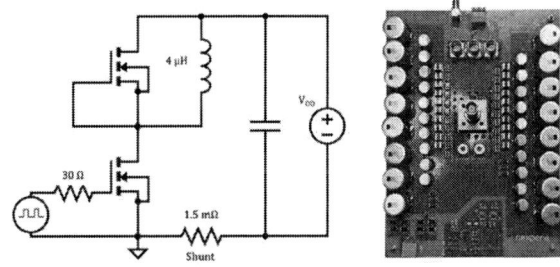

Fig. 6. Double pulse test circuit (Left) and PCB layout (Right).

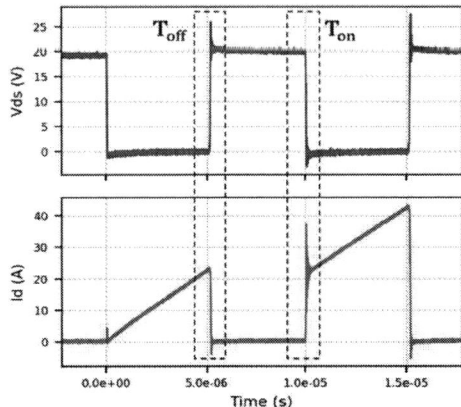

Fig. 7. Double pulse test waveforms for V_{DS} and I_D with the pulses and pause set to 5 µs; $V_{DD}= 20$V, $V_{GS}= 10$ V, $I_D= 24$ A.

979-8-3315-1612-3/25 $31.00 © 2025 IEEE

The DF is estimated using a logarithm decrement approach on the drain current rise, as described in (1) and (2), across two oscillations starting from the second peak to avoid accounting for the reverse recovery charge (Q_{rr}) present in the first peak of drain current [13]. I_D is preferred over the drain to source voltage (V_{DS}) to measure the DF as it is more sensitive to the MOSFET's parasitic capacitances and less affected by external inductances in the measurement setup, providing a more accurate representation of the device's internal switching dynamics.

$$\text{Log Decrement} = \delta = \ln\left(\frac{I_{Peak2} - I_{Trough2}}{I_{Peak3} - I_{Trough3}}\right) \quad (1)$$

$$DF = \frac{\delta}{\sqrt{4\pi^2 + \delta^2}} \quad (2)$$

The switching energy losses are calculated between when I_D or V_{DS} cross 10% of their peak values, starting with the current rise to the voltage fall for turn-on and vice-versa for turn-off. Measurements were made on five of Nexperia's LFPAK56 40 V rated MOSFETs that use the same SJ technology with variations made to their S_A and A_A, and the snubber resistance set to 1 Ω. They are named devices A to E by order of their S_A/A_A ratio, as seen in Table III.

TABLE III. 40 V SJ MOSFET DEVICES

Device:	A	B	C	D	E
A_A (mm²):	12.07	9.49	6.18	4.16	4.15
S_A (mm²):	1.24	1.23	1.00	0.79	1.00
S_A/A_A:	0.103	0.130	0.162	0.190	0.241

Fig. 8 shows a comparison of the turn-on switching waveforms for devices A and B that share the same S_A but different A_A. I_D was set to 50 A by varying the duration of the first pulse, V_{GS} was set to 10 V, and V_{DD} 20 V. As expected, the additional C_{DS} capacitance in device A decreases the frequency and improves the damping in the waveform with the consequence of a slower current rise time.

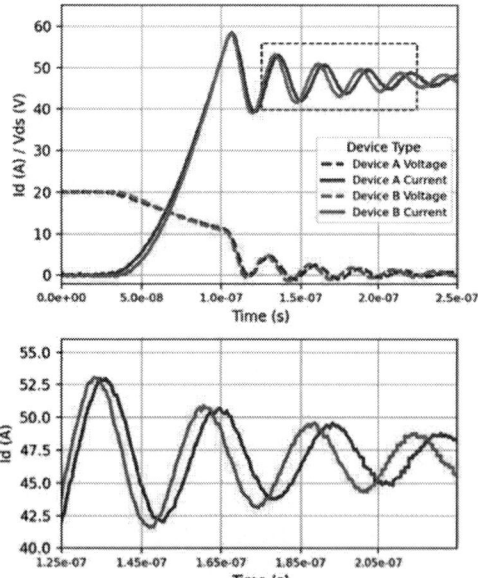

Fig. 8. Experimental DPT Waveform for LFPAK56 devices A and B. Top: Full waveform. Bottom: Magnification of oscillations. (V_{DD}= 20 V, V_{GS} = 10 V, I_D= 50 A, R_G= 30 Ω).

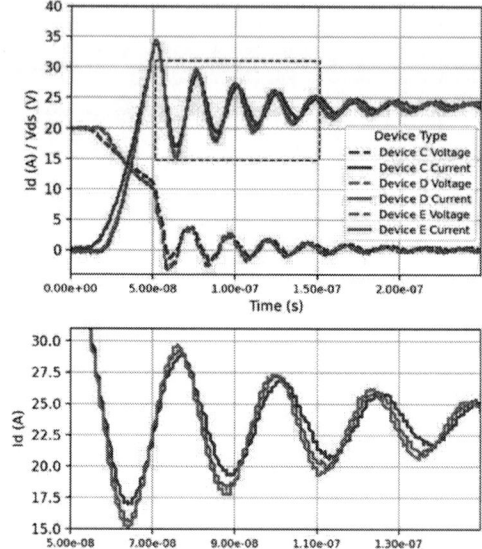

Fig. 9. Experimental DPT Waveform for LFPAK56 devices C, D and E. Top: Complete waveform. Bottom: Magnification of oscillations (V_{DD}= 20 V, V_{GS} = 10 V, I_D= 25 A, R_G= 30 Ω).

The other three devices were tested at a lower current rating of 25 A, due to their smaller die sizes. Device D and device E, share a similar A_A and snubber resistance but have a different S_A. Similarly to the higher current devices, device C and device E share the same S_A but with different A_A. The result is seen in Fig. 9. As expected, the additional snubber capacitance reduces the frequency and increases the damping in the ringing in the waveform. The device C results confirmed the effect of the A_A on the switching performance, with decreased oscillation frequency and increased DF with the additional A_A, with the downside of additional switching losses.

From these waveforms, the DF, oscillation frequency, switching losses, and the S_A/A_A ratio are compared in Figs. 10 and 11 to demonstrate the effect of A_A and S_A on the performance of the device. The DF is observed to increase and the oscillation frequency decrease when A_A or S_A is increased, while the switching losses only vary with the A_A. These additional losses are due to the increase of parasitic capacitance which slows the switching speed of the device. As the ratio of S_A/A_A is increased, the DF is noticed to decrease.

Therefore, from the outcome of this analysis, the following trends were identified:

- An increase in S_A increases the DF and has a low effect on the switching losses, which is desirable for high EMC priority applications.

- An increase in A_A increases the DF and switching losses with the desirable effect of decreasing conduction losses, which is desirable for low switching frequency and high EMC priority applications.

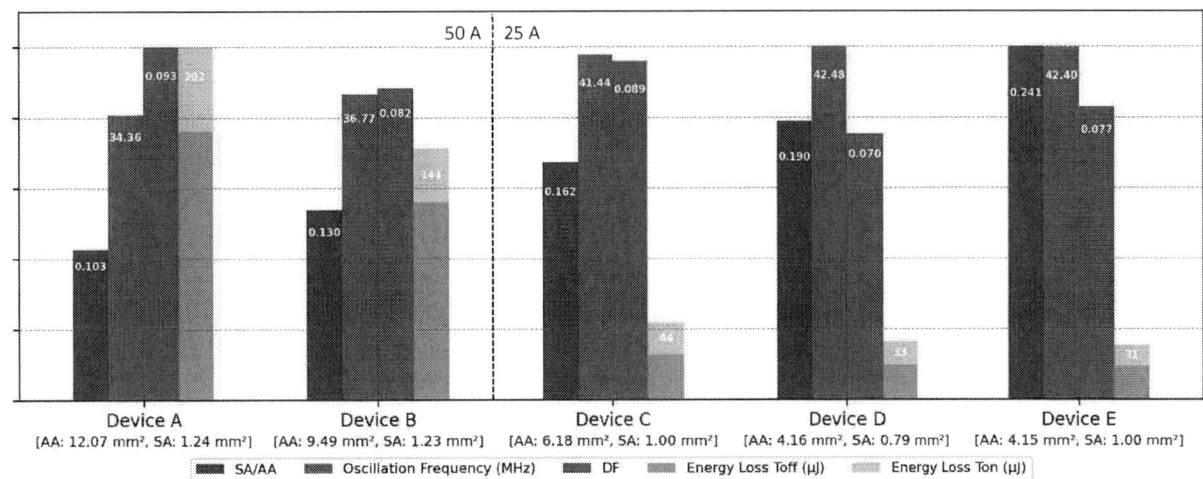

Fig. 10. Effect of Active Area and Snubber Area at high current (Device A & B) and reduced current (Device C, D & E).

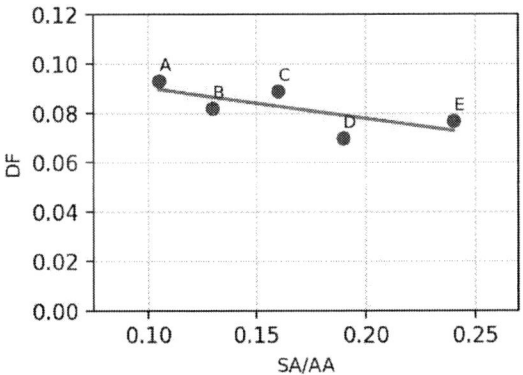

Fig. 11. Plot of damping factor against the Snubber Area / Active Area ratio.

B. SPICE Simulation

Simulations are conducted to test the effects of A_A and S_A from what has been established experimentally. A Simulation Program with Integrated Circuit Emphasis (SPICE) model is designed to replicate the physical DPT board employed for this experiment as closely as possible, with the same voltage, current and gate resistance [14]. It is achieved through the circuit analysis of the DPT board and the inclusion of the parasitic impedance of each track of the PCB using measurements made with a vector network analyzer (VNA). The DF, frequency and energy losses were also estimated using the same method as described for the experimental results.

The snubber resistance (R_{SNB}) was tested and determined that, for SJ devices, it enhances damping in the ringing without impacting the slew rate, peak and frequency. This is anticipated, as the snubber operates in parallel with the MOSFET. In this simulated study, a snubber resistance of 1 Ω was maintained for all devices, consistent with the value established in the experimental results.

For the differences in Active Area, simulations found that increasing this parameter resulted in an increase in switching losses from 72.09 µJ for device B to 86.46 µJ for device A. This was also measured at lower power ratings with the waveforms

of device C having significantly higher switching losses of 22.59 µJ than device E at 16.75 µJ. The DF was also increased with additional A_A for devices A and B, increasing from 0.088 to 0.103 with the additional 2.58 mm², as noticed from Fig. 12. The same effect was recorded for devices C and E, with the DF increasing from 0.097 to 0.106 as seen in Fig. 13.

For the variation of Snubber Area, the waveforms seen in Fig. 13 are employed to estimate the DF for devices D and E, as they share the same design, other than their Snubber Area. When the S_A was increased by 0.21 mm², the damping factor was found to also increase from 0.094 to 0.106, with a low effect on the values of switching energy loss.

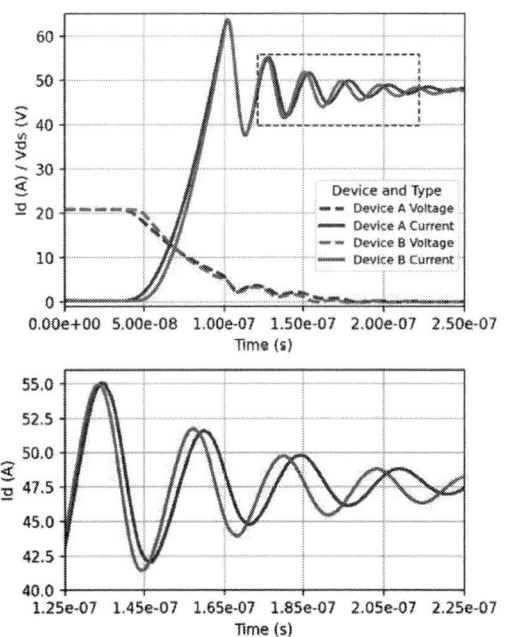

Fig. 12. Simulated DPT outcome of LFPAK56 with varying Active Area. Top: Complete waveform. Bottom: Magnification of oscillations. (V_{DD}= 20 V, V_{GS} = 10 V, I_D= 50 A, R_G= 30 Ω).

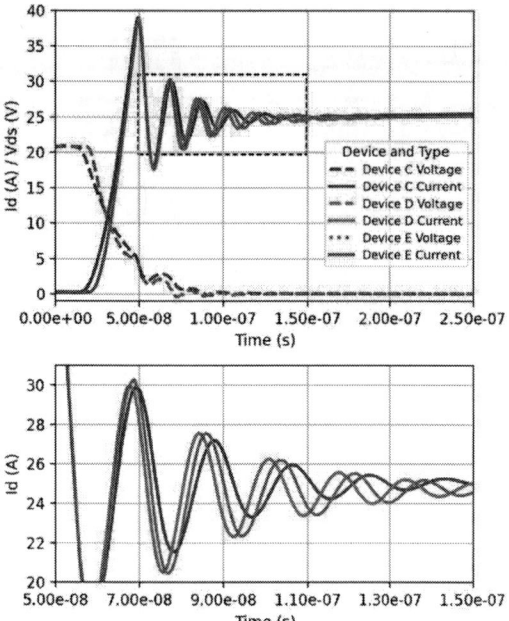

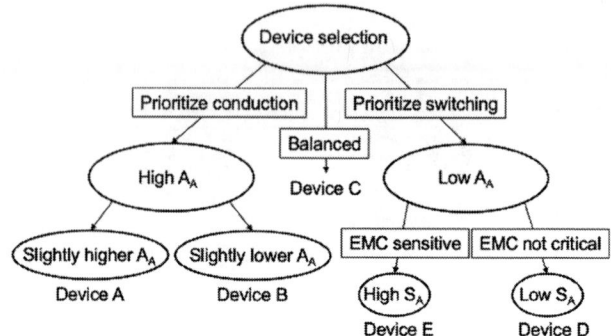

Fig. 14. MOSFET selection process: Active Area and Snubber Area considerations.

With the A_A considered, the S_A is selected based on whether the application has already mitigated EMC with filtering or not. And applications that require the lowest EMC the devices with larger snubber capacitances, as described in Fig. 14.

B. Application Selection

For low power DC-DC converters, the required switching frequency is high, sometimes surpassing 1.5 MHz. For these applications the switching speed is crucial and, therefore, conduction loss is sacrificed for a smaller Active Area. For this reason, a reduced DF might be tolerated. Moreover, at this high switching frequency a tight and compact design will reduce radiated emissions. A device type E or D would be preferable for this application.

For high power DC-DC applications, the switching frequency is lower, meaning that an increased Active Area should be more suitable. Device C would be best suited for this application.

For BLDC motor drive applications, the EMC plays a critical part in the design of the device, as the switching node is connected into the machine, allowing high frequency currents into the motor's parasitic capacitances. Larger copper areas are also required to carry higher currents, so radiated emissions have the potential to be more severe. These design considerations mean that a larger Snubber Area with a higher DF is preferable. BLDC motor drives typically operate at lower switching frequencies compared to DC-DC converters, so the Active Area of the MOSFET can be increased, particularly for larger loads. Device type C or E would be suitable for this application.

Brushed DC motor drives do not switch at high frequency and, therefore, need a large Active Area to mitigate conduction losses. The drive circuitry is simplified and is not as sensitive to EMC disturbances, as they use a constant DC supply to function with no pulse width modulation driving signal. Device type B is suitable for this application.

Isolation switches on the other hand, switch very infrequently, making their conduction performance and related thermal management a priority. In certain high power protection environments with increased sensitivity, EMC can also play an important role. Devices A or B are most likely the optimal options.

Fig. 13. Simulated DPT outcome of LFPAK56 with varying Active Area and Snubber Area. Top: Complete waveform. Bottom: Magnification of oscillations (V_{DD}= 20 V, V_{GS}= 10 V, I_D= 25 A, R_G= 30 Ω).

The pattern displayed by the A_A confirms the impact of the parasitic capacitance on the oscillations seen in the simulated output waveform. This relationship illustrates how alterations to C_{DS} directly influence the device's switching behavior, resulting in the expected changes in the frequency and amplitude of the oscillations.

The findings from the S_A analysis also correspond to the anticipated output waveforms, indicating that switching performance improves with higher integrated snubber capacitance (C_{SNB}). This demonstrates that an increase in integrated snubber capacitance enhances the damping effect on oscillations.

IV. DEVICE AND APPLICATION MATCHING

A. MOSFET Performance Optimization

The Active Area is set depending on whether switching losses or conduction losses are prioritized. This is a well-documented design process and depends on operating conditions [15]. The second step of selecting the integrated Snubber Area is primarily based on the criticality of the EMC behavior of the circuit. Larger Active Areas increase the EMC performance of the device, depending on whether this performance enhancement is sufficient, the integrated snubber may be implemented. With the range of devices put forward in this study, the MOSFET selection procedure is set out in Fig. 14.

979-8-3315-1612-3/25 $31.00 © 2025 IEEE

Different MOSFETs are used to meet different levels of EMC requirements in automotive applications. A MOSFET with a higher DF indicates it is less noisy, making it better suited for EMC sensitive applications. Additional filtering may be required depending on the application, with higher attenuation required where MOSFETs with lower DFs are used in sensitive EMC applications, resulting in more complex and expensive filtering.

CONCLUSION

This study highlights the impact of Active Area and integrated Snubber Area on the drain current oscillatory switching behavior of SJ MOSFETs by using the damping factor and energy loss as a benchmark, linking the effect these parameters have on the EMC performance of the device. It was found that increasing the die area of the device also increased the parasitic capacitances, resulting in slower switching speeds, reduced ringing, and lower frequencies in the drain current oscillations.

A device selection process was put forward to assist in making design decisions to improve particular aspects of MOSFET performance based on the specifications established by automotive application environments. Devices with reduced Active Areas were found to improve the performance for DC-DC converters through reduced switching losses, with their EMC performance further improved with additional Snubber Area. Devices with increased die areas were better suited for isolation switches and brushed DC motor drives due to their decreased conduction losses. For BLDC motor drives, increased S_A was needed to reduce ringing, and applications with a higher output power may require an adapted A_A to balance the conduction and switching losses.

ACKNOWLEDGMENT

Nexperia provided technical guidance and sponsorship for this project which significantly contributed to the research presented in this paper.

REFERENCES

[1] P. Goarin, G. Koops, R. van Dalen, C. Le Cam, and J. Saby, "Split-gate Resurf Stepped Oxide (RSO) MOSFETs for 25V applications with record low gate-to-drain charge," in *International Symposium on Power Semiconductor Devices and IC's*, May 2007, doi: https://doi.org/10.1109/ispsd.2007.4294932.

[2] P. Rutter and S. L. Peake, "Low voltage superjunction power MOSFET: An application optimized technology," in *2011 Twenty-Sixth Annual IEEE Applied Power Electronics Conference and Exposition (APEC)*, Mar. 2011, doi: https://doi.org/10.1109/apec.2011.5744642.

[3] P. Rutter and S. L. Peake, "Low voltage TrenchMOS combining low specific RDS(on) and QG FOM," in *International Symposium on Power Semiconductor Devices and IC's*, pp. 325–328, Jan. 2010.

[4] H. Yamashita et al., "Low noise superjunction MOSFET with integrated snubber structure," in *2018 IEEE 30th International Symposium on Power Semiconductor Devices and ICs (ISPSD)*, May 2018, doi: https://doi.org/10.1109/ispsd.2018.8393595.

[5] H. Afewerki, C. Lautensack, N. Bottcher, and I. Kallfass, "Design approach and analysis of a MOSFET with monolithic integrated EMI snubber for low voltage automotive applications," in *2017 International Symposium on Electromagnetic Compatibility (EMC EUROPE)*, Sep. 2017, doi: https://doi.org/10.1109/emceurope.2017.8094734.

[6] J. Roig et al., "Suitable operation conditions for different 100V trench-based power MOSFETs in 48V-input synchronous buck converters," in *European Conference on Power Electronics and Applications*, pp. 1–9, Sep. 2011.

[7] J. Chen, "Design optimal built-in snubber in trench field plate power MOSFET for superior EMI and efficiency performance," in *2015 International Conference on Simulation of Semiconductor Processes and Devices (SISPAD)*, Sep. 01, 2015, doi: https://doi.org/10.1109/SISPAD.2015.7292361.

[8] S. Zhang, "Influence of driving and parasitic parameters on the switching behaviors of the SiC MOSFET," *Frontiers in Energy Research*, vol. 10, Jan. 2023, doi: https://doi.org/10.3389/fenrg.2022.1079623.

[9] H. Qin, C. Ma, Z. Zhu, and Y. Yan, "Influence of Parasitic Parameters on Switching Characteristics and Layout Design Considerations of SiC MOSFETs," *Journal of Power Electronics*, vol. 18, no. 4, pp. 1255–1267, Jul. 2018, doi: https://doi.org/10.6113/jpe.2018.18.4.1255.

[10] P. Rutter, "Considerations in the design of a low-voltage power MOSFET technology," *IET Power Electronics*, vol. 12, no. 15, pp. 3861–3869, Dec. 2019, doi: https://doi.org/10.1049/iet-pel.2019.0284.

[11] P. Hillenbrand, M. Beltle, S. Tenbohlen, and S. Mönch, "Sensitivity analysis of behavioral MOSFET models in transient EMC simulation," in *2017 International Symposium on Electromagnetic Compatibility (EMC EUROPE)*, Sep. 2017, doi: https://doi.org/10.1109/emceurope.2017.8094762.

[12] K. Aditya, "A Comprehensive Guide for Optimizing the switching of an Automotive Grade MOSFET in a Half-Bridge Topology," in *2023 International Conference on Control, Communication and Computing (ICCC)*, May 2023, doi: https://doi.org/10.1109/iccc57789.2023.10165493.

[13] K. Prasertwong and N. Mithulananthan, "A new algorithm based on logarithm decrement to estimate the damping ratio for power system oscillation," in *2017 14th International Conference on Electrical Engineering/Electronics, Computer, Telecommunications and Information Technology (ECTI-CON)*, Jun. 2017, doi: https://doi.org/10.1109/ecticon.2017.8096288.

[14] Z. Feng, A. Berry, P. Ellis, and W. Lawson, "SPICE models for predicting EMC performance of a MOSFET based half-bridge configuration," in *2021 IEEE 1st International Power Electronics and Application Symposium (PEAS)*, Nov. 2021, doi: https://doi.org/10.1109/peas53589.2021.9628477

[15] Y. Ikegami, H. Obara, and Y. Sato, "A basic study on chip size determination of MOSFETs to minimize total power loss," in *2015 9th International Conference on Power Electronics and ECCE Asia (ICPE-ECCE)*, Jun. 2015, doi: https://doi.org/10.1109/icpe.2015.7168091.

Improving Dynamic Current Sharing between Parallel MOSFETs by Optimizing Device Parameters

Kunal Jha
Power Systems and Sensors
Infineon Technologies
Los Angeles, USA
Kunal.Jha@infineon.com

Kapil Kelkar
Power Systems and Sensors
Infineon Technologies
Los Angeles, USA
Kapil.Kelkar@infineon.com

Marina Hedenik
Power Systems and Sensors
Infineon Technologies
Villach, Austria
Marina.Hedenik@infineon.com

David Penof
Power Systems and Sensors
Infineon Technologies
Villach, Austria
David.Penof@infineon.com

Abstract—**In today's applications, there is a persistent demand for higher power capabilities, where a single MOSFET is insufficient to meet the required power needs in applications such as Light Electric Vehicles (LEVs), e-scooters, e-forklifts, etc. To address this, designers often employ multiple MOSFETs in parallel to share the currents. However, paralleling of MOSFETs can lead to dynamic and steady state current imbalance issues, primarily caused by two factors: layout asymmetries and unmatched MOSFET parameters. These imbalances will result in uneven power dissipation among the paralleled MOSFETs, which can compromise the system's overall performance and reliability. Semiconductor manufacturers are actively addressing this challenge through the development of new technologies that are designed to facilitate parallel operation and provide improved performance and reliability. This paper will demonstrate a comprehensive comparison of dynamic current sharing between parallel MOSFETs during hard switching events, and the impact of different MOSFET parameters such as gate-source threshold voltage window and transconductance on it. It will also provide designers step by step guidance on emulating different gate-source threshold voltages for evaluating dynamic current sharing, which will help them choose the best MOSFETs for their target applications.**

Index Terms—**Current Sharing, Power Electronics, MOSFETs, motor-drives, BLDC, LEVs, e-bikes, DC-DC converters, 3-phase inverters**

I. INTRODUCTION

Decarbonization initiatives are impacting every industry in various ways, and gasoline engine tools & vehicles are few of the prominent targets for electrification. These include high-power motor-drive applications such as LEVs, e-forklifts, e-scooters, e-bikes, power & gardening tools etc. which are now being designed using battery-powered Brushless DC (BLDC) motors. These applications typically use $36\,V - 84\,V$ batteries, and have $60\,V - 150\,V$ MOSFETs respectively for the motor-drive. To meet the high power requirements of up to $10\,kW$, these applications have multiple Si MOSFETs in parallel which makes the dynamic current sharing between the MOSFETs critical to ensure all the MOSFETs are experiencing similar voltage and current stresses. Today, designers de-rate the current ratings of the MOSFETs to have adequate safety margin or slow down the switching which increases

the MOSFET count and/or losses in the system. Thus, having similar current sharing during these switching events will enable better reliability, lower losses and fewer components.

This paper demonstrates the impact of MOSFET parameters on dynamic current sharing during hard-switching events and explores optimization strategies to achieve the best dynamic current sharing, thereby reducing the required safety margin and minimizing losses. The tests were conducted on four parallel MOSFETs, each with an individual co-axial current shunt to measure the currents during hard turn on and hard turn off. The test setup was based on double-pulse test methodology where the end of first pulse is a hard turn off event and start of the second pulse is a hard turn on event (when body-diode goes through reverse recovery). This paper also discusses methods to emulate extreme $V_{GS(th)}$ mismatch conditions between parallel MOSFETs, enabling designers to test worst-case scenarios for dynamic current sharing when selecting MOSFETs. Finally, it will cover the impact of improved MOSFET parameters in three-phase inverter for motor-drive to showcase improvements in switching and thermal performances.

II. DEVICE PARAMETERS THAT IMPACT DYNAMIC CURRENT SHARING

To ensure similar current sharing, various MOSFET parameters such as threshold voltage window, transconductance, and switching gate-charge need to be considered.

A. Impact of Threshold Voltage Window

When parallel MOSFETs turn on & off, the current flowing through them ramps up and down respectively. If one of the MOSFETs turns on earlier or turn off later, it will end up carrying much higher current than the other MOSFETs and have much higher losses. This uneven current sharing could lead to the MOSFET experiencing avalanche, thermal runaways, excessive voltage/current stresses, etc., potentially leading to premature failures. Therefore, it is important to have all MOSFETs turn on & off at the same time which can be achieved by having a tight threshold voltage ($V_{GS(th)}$) window.

979-8-3315-1612-3/25 $31.00 © 2025 IEEE

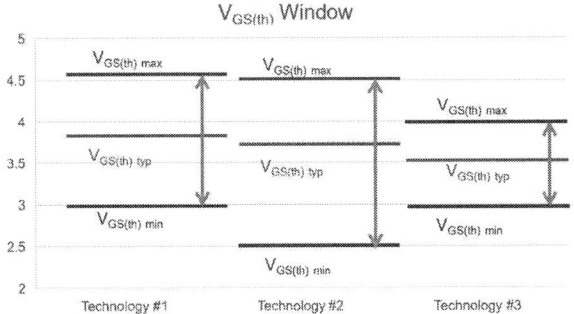

Fig. 1. Threshold voltage window comparison for different MOSFET technologies.

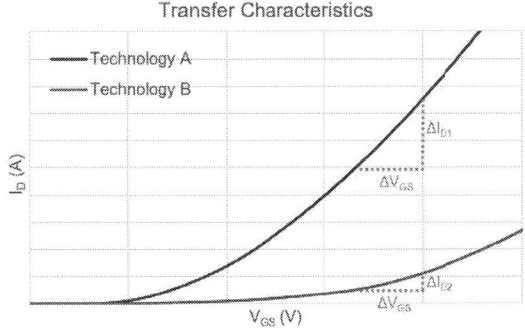

Fig. 2. Transfer characteristics comparison showing benefits of lower transconductance.

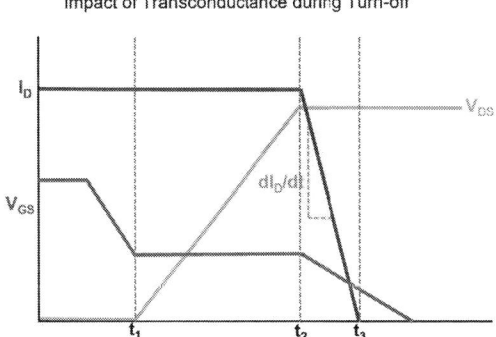

Fig. 3. turn off transition highlighting the impact of transconductance on switching performance.

$V_{GS(th)}$ window is the difference between maximum and minimum threshold voltage for any MOSFET as shown in Fig. 1. A larger window will increase the risk of having higher $V_{GS(th)}$ mismatch between parallel MOSFETs which increases the delay between the parallel switching MOSFETs. This would force the MOSFET turning on earlier or turning off later to carry a large current for longer time. In addition to tighter $V_{GS(th)}$, it is also important to have the $V_{GS(th) typ.}$ be at the center of the window to ensure the symmetry of the turn on and turn off worst-case performances, which will be discussed in Section IV.

Fig. 1 shows the $V_{GS(th)}$ window for three technologies. Technology #3 has the tightest window followed by technologies #1 and #2 respectively. While technology #1 and #3 have a symmetrical window about the typical value, technology #2 has a larger difference between typical to minimum value. This means technology #2 could have higher variation for $V_{GS(th)}$ and affect the switching performance considerably. Therefore, it is important for designers to emulate worst-case $V_{GS(th)}$ conditions to ensure the system performance is not compromised. The test methodology for emulating worst-case $V_{GS(th)}$ conditions is discussed in Section III-A.

B. Impact of Transconductance

Transconductance (g_{fs}) is the rate of change of drain current (I_D) with respect to change in gate-source voltage (V_{GS}). The g_{fs} for any device can be calculated from the transfer characteristics, where the slope of the curve is g_{fs}. Fig. 2 shows the transfer curves for two MOSFET technologies. For the same change in V_{GS}, technology B has much lower change in I_D.

When MOSFETs are paralleled, there can be V_{GS} mismatch between them due to multiple reasons such as different gate-drive loop lengths, difference in external gate resistors, etc. This V_{GS} mismatch affects the turn on & off times of the parallel MOSFETs with some switching before the others, resulting in significantly different dynamic current sharing. In such scenarios, MOSFETs with lower transconductance will ensure that the difference in I_D of the parallel MOSFETs is lower, whereas MOSFETs with higher transconductance will have higher difference in I_Ds of the parallel MOSFETs for the same difference in V_{GS}. Lower g_{fs} would also help balance the impact of difference in $V_{GS(th)}$ between parallel MOSFETs by limiting the rise in current.

Additionally, lower g_{fs} can also help reduce the turn off time. As shown in Fig. 3, the total switching time is from t_1 to t_3, where the V_{DS} of the MOSFET rises between t_1 and t_2 and the current falls between t_2 and t_3. The dI_D/dt during this fall time coupled with the parasitic inductance can cause high V_{DS} peaks and parts can avalanche.

If technologies with lower transconductance are chosen, the dI_D/dt reduces and so does the V_{DS} peak. This allows the designers to either have a lower V_{DS} peaks in the same switching time or they can reduce the external gate-resistances and reduce the turn off times to maintain the same V_{DS} peaks, while still maintaining better current sharing. This reduces the switching losses significantly and enables more efficient operation at higher frequencies, enabling designers to save costs on thermal design.

C. Impact of Switching Gate-Charge

Switching gate-charge (Q_{SW}) of a MOSFET is the total charge that the gate-source (C_{GS}) and gate-drain (C_{GD}) capacitors need during the switching transitions. Q_{SW} is the sum of Q_{GS2} and Q_{GD} [1]. The turn on transition starts once the C_{GS} capacitor charges to $V_{GS(th)}$. First, the I_D in the MOSFET

979-8-3315-1612-3/25 $31.00 © 2025 IEEE

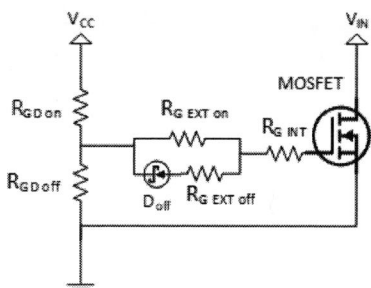

Fig. 4. Typical gate-drive circuit.

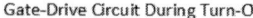

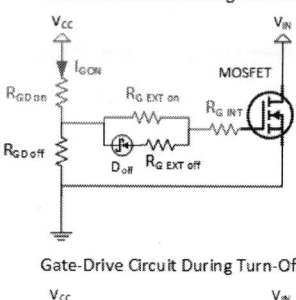

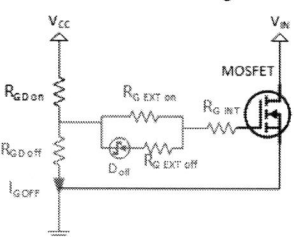

Fig. 5. Gate-drive circuit during switching transitions.

ramps up during Q_{GS2} portion (t_1 - t_2) and then the V_{DS} ramps down during Q_{GD} portion (t_2 - t_3). The total switching time therefore is from t_1 to t_3. The charging times of Q_{SW} by the gate current will dictate the total switching time and thus, if parallel MOSFETs have variations in Q_{SW}, they will turn on and off at different rates and the dynamic current sharing will be dissimilar [2]. The impact of Q_{SW} variation has been discussed in detail in [2] and not in the scope of this paper.

III. TEST METHODOLOGY AND SETUP

A. Emulating different threshold voltages

When designing new platforms, designers need to test the worst case scenarios to ensure the platform will continue to perform as per their expectations. However, the scale of the testing phase is much smaller than the scale of production where they can have MOSFETs from different wafer lots, having higher parameter deviations, that can end up on the same PCB. Therefore, it is important for designers to ensure that testing and validation can be performed for the datasheet limits without the need to purchase additional devices.

Fig. 4 shows a typical gate-drive circuit for a MOSFET, where R_{GDon} and R_{GDoff} are the internal on & off resistances of gate-driver respectively, R_{GEXTon} and $R_{GEXToff}$ are external resistors and R_{GINT} is the internal resistance of the MOSFET.

During turn on, gate-current I_{Gon} flows from V_{CC} to the gate of the MOSFET, and during turn off, gate-current I_{Goff} flows from gate of the MOSFET to ground as shown in Fig. 5. I_{Gon} and I_{Goff} can be calculated using (1) and (2).

$$I_{Gon} = \frac{V_{CC} - V_{PL}}{R_{GDon} + R_{GEXTon} + R_{GINT}} \quad (1)$$

$$I_{Goff} = \frac{V_{PL}}{R_{GDoff} + R_{GEXT} + R_{GINT}} \quad (2)$$

$$t_{on} = \frac{Q_{SW}}{I_{Gon}} \quad (3)$$

$$t_{off} = \frac{Q_{SW}}{I_{Goff}} \quad (4)$$

where V_{CC} is the gate supply voltage, V_{PL} is the plateau voltage, R_{GEXT} is the equivalent resistance of parallel combination of R_{GEXTon} and $R_{GEXToff}$, t_{on} is the turn on time and t_{off} is the turn off time.

For any MOSFET technology, the V_{PL} has variance and changes if there is change in $V_{GS(th)}$. When there is a change in $V_{GS(th)}$, the corresponding change in V_{PL} can be calculated using (5).

$$\frac{V_{GS(th)typ}}{V_{PLtyp}} = \frac{V_{GS(th)new}}{V_{PLnew}} \quad (5)$$

where $V_{GS(th)typ}$ and $V_{GS(th)new}$ are the typical and new gate-source threshold voltages respectively, and V_{PLtyp} and V_{PLnew} are the corresponding typical and new plateau voltages respectively. From (5), it is clear that the ratio of $V_{GS(th)}$ to V_{PL} is always constant, and since the MOSFET manufacturers specify the minimum and maximum values of $V_{GS(th)}$, the corresponding extreme plateau voltage values can be calculated. Using the values of $V_{PL\,min}$ and V_{PLmax} in (1) and (2) the corresponding I_{Gon} and I_{Goff} can be calculated. However, the same value of I_{Gon} and I_{Goff} can be obtained by varying the R_{GEXTon} and $R_{GEXToff}$ for a part with V_{PLtyp}, as shown in (6) to (9).

$$I_{Gon_{new}} = \frac{V_{CC} - V_{PLnew}}{R_{GDon} + R_{GEXTon} + R_{GINT}} \quad (6)$$

$$I_{Gon_{new}} = \frac{V_{CC} - V_{PLtyp}}{R_{GDon} + R_{GEXTon_{new}} + R_{GINT}} \quad (7)$$

$$I_{Goff_{new}} = \frac{V_{PLnew}}{R_{GDoff} + R_{GEXT} + R_{GINT}} \quad (8)$$

$$I_{Goff_{new}} = \frac{V_{PLtyp}}{R_{GDoff} + R_{GEXT_{new}} + R_{GINT}} \quad (9)$$

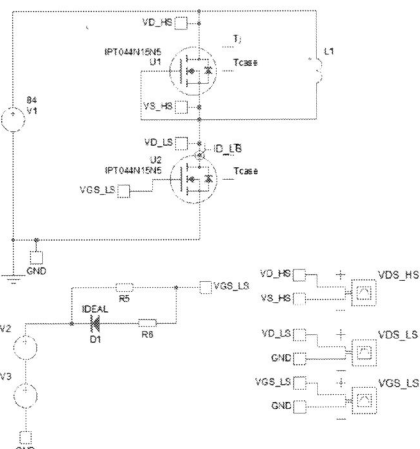

Fig. 6. SPICE simulation schematic to emulate different $V_{GS(th)}$.

TABLE I
SIMULATION RESULTS FOR $V_{GS(TH)\ MAX}$ EMULATION

Parameters	Test Condition			Units
	$V_{GS(th)\ typ}$	$V_{GS(th)\ max}$	*Emulated*	
V_{CC}	15	15	15	V
$V_{GS(th)}$	3.8	4.6	3.8	V
V_{PL}	5.4	6.54	5.4	V
R_{GDon}	0.6	0.6	0.6	Ω
R_{GDoff}	0.8	0.8	0.8	Ω
R_{GEXTon}	80	80	91	Ω
$R_{GEXToff}$	30	30	22	Ω
R_{GEXT}	21.8	21.8	17.7	Ω
R_{GINT}	0.9	0.9	0.9	Ω
I_{Gon}	118	104	104	mA
I_{Goff}	230	278	278	mA
Q_{SW}	21.3	21.3	21.3	nC
t_{on} calculated	181	205	205	ns
t_{on} simulated	183	203	205	ns
t_{off} calculated	93	77	77	ns
t_{off} simulated	95	81	77	ns

TABLE II
SIMULATION RESULTS FOR $V_{GS(TH)\ MIN}$ EMULATION

Parameters	Test Condition			Units
	$V_{GS(th)\ typ}$	$V_{GS(th)\ max}$	*Emulated*	
V_{CC}	15	15	15	V
$V_{GS(th)}$	3.8	3	3.8	V
V_{PL}	5.4	4.26	5.4	V
R_{GDon}	0.6	0.6	0.6	Ω
R_{GDoff}	0.8	0.8	0.8	Ω
R_{GEXTon}	80	80	71	Ω
$R_{GEXToff}$	30	30	46.3	Ω
R_{GEXT}	21.8	21.8	28.1	Ω
R_{GINT}	0.9	0.9	0.9	Ω
I_{Gon}	118	132	132	mA
I_{Goff}	230	181	181	mA
Q_{SW}	21.3	21.3	21.3	nC
t_{on} calculated	181	162	162	ns
t_{on} simulated	183	162	162	ns
t_{off} calculated	93	118	118	ns
t_{off} simulated	95	111	117	ns

where V_{PLnew} can be the V_{PLmin} and V_{PLmax} and $R_{GEXTon_{new}}$ and $R_{GEXT_{new}}$ are the new external resistances required to emulate the corresponding V_{PLnew} on a typical device.

Solving (6) and (7) will give the new value of R_{GEXTon} that would turn on the typical part similar to a part with minimum or maximum $V_{GS(th)}$ and solving (8) and (9) gives new value of R_{GEXT} that would turn off the typical part similar to a part with minimum or maximum $V_{GS(th)}$.

This approach was validated using a SPICE simulation for the part IPT044N15N5 using double-pulse methodology. The $V_{GS(th)\ min}$ and $V_{GS(th)\ max}$ for this part are $3\,V$ and $4.6\,V$ respectively [4]. The schematic for the simulation is shown in Fig. 6. The test was done using an input voltage of $84\,V$ and the inductor current was ramped up to $50\,A$ for the turn on and turn off. The external gate resistors R_5 and R_6 are R_{Gon} and R_{Goff} respectively. These resistors were varied to emulate the different $V_{GS(th)}$ conditions. These were compared to the waveforms that were obtained by varying the dV_{th} parameter of the model to +/-1 to represent $V_{GS(th)\ max}$ and $V_{GS(th)\ min}$ respectively. The results for each condition are compared in Tables I and II which show an error of approximately 1% for t_{on} and 5% for t_{off} values between emulation vs. varying the $V_{GS(th)}$ directly.

To use this approach for evaluating dynamic current sharing performance, designers can first tune the gate-drive circuit with the R_{GEXTon} $R_{GEXToff}$ that meets the design criteria. Once the R_{GEXTon} $R_{GEXToff}$ are finalized, then (6) to (9) can be used to find the new values of external gate resistances, $R_{GEXTon_{new}}$ and $R_{GEXT_{new}}$. These gate resistors for one or more MOSFETs in parallel can then be changed to the new value to emulate different $V_{GS(th)}$ conditions.

B. Test Setup

The tests were conducted on four parallel MOSFETs, each with an individual co-axial current shunt [5] to measure the currents during hard turn on and hard turn off. The test setup was based on double-pulse test methodology where the end

of first pulse is a hard turn off event and start of the second pulse is a hard turn on event (when body-diode goes through reverse recovery). The schematic diagram and PCB for this setup are shown in Fig. 7 and Fig. 8 respectively.

As shown in Fig. 7, $HS_1 - HS_4$ MOSFETs are the high-side MOSFETs with the gate connection shorted to the source to ensure that only the body-diode for these MOSFETs conducts during free-wheeling. The switching MOSFETs are the low-side MOSFETs, $LS_1 - LS_4$, with individual current shunts connected to the respective drains of the MOSFETs. The gate-drive circuitry for the individual low-side MOSFETs has the same scheme mentioned in Fig. 4. The inductor, L_1 is placed in parallel with the high-side MOSFETs to ramp up the current during the first pulse to the required current level.

The PCB shown in Fig. 8 was designed to ensure as symmetrical power and gate-drive loop lengths as possible between all the MOSFETs. By doing this, the impact of external factors, such as PCB parasitics, on the dynamic

979-8-3315-1612-3/25 $31.00 © 2025 IEEE

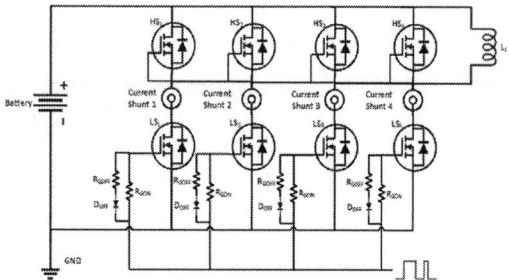

Fig. 7. Schematic diagram for measuring current sharing between four parallel MOSFETs.

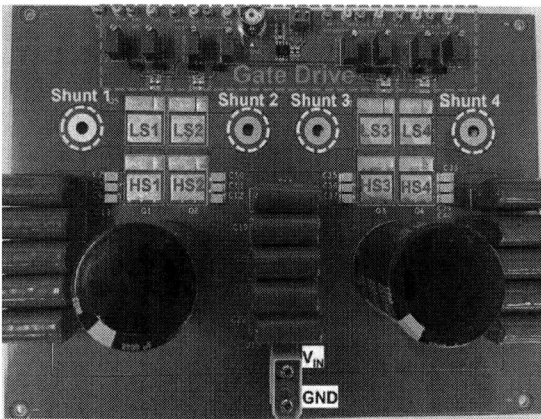

Fig. 8. PCB for measuring current sharing between four parallel MOSFETs.

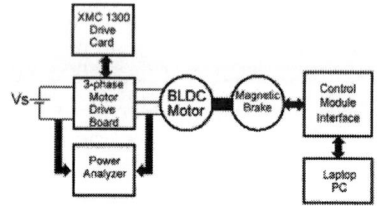

Fig. 9. Three-phase inverter with two MOSFETs in parallel for motor control.

Fig. 10. Three-phase inverter with two MOSFETs in parallel for motor control.

current sharing is minimized and the difference in current sharing will purely be due to impact of MOSFET parameters.

In addition, a three-phase inverter for motor control was also tested with two MOSFETs in parallel to compare the thermal and switching performance of different technologies at the same load conditions. The test setup block diagram is shown in Fig. 9 and the PCB is shown in Fig. 10.

C. Test Conditions

To compare the dynamic current sharing performance, two tests were performed. The first test was done using four 150 V MOSFETs in parallel and the second test was done using four 60 V MOSFETs in parallel. The three-phase inverter test for motor-drive was done using two 60 V MOSFETs in parallel to compare the impact of improved MOSFET parameters on switching and thermal performance. The test conditions for all three tests are mentioned in Table III.

The input voltage for 150 V MOSFETs is 84 V and a total current of 100 A was used for measuring the dynamic current sharing between four parallel MOSFETs. Similarly, for 60 V MOSFETs, an input voltage of 40 V was used with a total current of 100 A shared between four parallel MOSFETs. For these tests, MOSFETs from each technology were characterized and parts having the same $V_{GS(th)}$, g_{fs} and Q_{GS2} were selected to ensure that the difference in dynamic current sharing when all parallel MOSFETs have a symmetric gate drive is minimal.

For the three-phase inverter testing with two 60 V MOS-FETs in parallel, the input voltage was 40 V. A switching frequency of 20 kHz was selected since the typical switching frequency range for motor-drive applications is 10 kHz – 30 kHz. These parts were tested using block-commutation control as it is the most common control scheme used for consumer motor-drive applications. The input power level for the test was 650 W with a run time of 12 minutes to ensure the maximum case temperature for steady state operation of the MOSFETs was achieved. The two technologies were tuned using R_{GEXTon} and $R_{GEXToff}$ to ensure the V_{DS} peak is approximately 48 V (80% of breakdown voltage) as this is the standard design margin practiced by designers. For these tests, parts having a $V_{GS(th)}$ mismatch of 0.3 V and equal g_{fs} were selected to be in parallel.

D. MOSFET parameter comparison

The parameter comparisons for 150 V and 60 V MOSFETs are shown in Table IV.

Three 150 V MOSFET technologies were selected from Infineon and competitors having similar g_{fs} and Q_{GS2} but different $V_{GS(th)}$ windows so that the impact of tighter $V_{GS(th)}$ window can be understood. Competitor 1 has a $V_{GS(th)}$ window of 2 V, IPT039N15N5 has a $V_{GS(th)}$ window of 1.6 V and IPT025N15NM6, from the latest OptiMOS™ 6 family from Infineon, has a $V_{GS(th)}$ window of 1 V.

Similarly, two 60 V MOSFET technologies from Infineon were chosen, which have the same $V_{GS(th)}$ window

979-8-3315-1612-3/25 $31.00 © 2025 IEEE

TABLE III
TEST CONDITIONS

Test	Parameter	Value	Units
150 V Current Sharing	Input Voltage	84	V
	Total Current	100	A
	Parallel MOSFETs	4	
60 V Current Sharing	Input Voltage	40	V
	Total Current	100	A
	Parallel MOSFETs	4	
60 V three-phase inverter	Input Voltage	40	V
	Output Power	650	W
	Switching Frequency	20	kHz
	Parallel MOSFETs	2	
	Run Time	12	mins

TABLE IV
MOSFET PARAMETER COMPARISON

Voltage Node [V]	Part Number	$V_{GS(th)}$ Window [V]	g_{fs} [S]	Q_{GS2} [nC]
150	Competitor 1	2	175	15
	IPT039N15N5	1.6	139	9
	IPT025N15NM6	1	168	15
60	BSC014N06NS	1.2	150	10
	ISC015N06NM5LF2	1.2	56	20

of 1.2 V but different g_{fs} to understand the impact of lower g_{fs} on dynamic current sharing and overall switching performance. BSC014N06NS has a g_{fs} of 150 S whereas, ISC015N06NM5LF2 has a g_{fs} of 56 S at the same current level. These 60 V parts were also tested in three-phase inverter for motor control to show case the improvements in thermal and switching performances when a part with lower transconductance is used.

IV. EXPERIMENTAL TEST RESULTS

A. Benefits of tighter $V_{GS(th)}$ window on dynamic current sharing

During switching events, the part with the lowest $V_{GS(th)}$ will turn on first and during turn off it will turn off last. In both swiching events, this part would carry bulk of the current and increase the switching losses for it which would cause uneven heating of the parallel MOSFETs. Similarly, the part with the highest $V_{GS(th)}$ will turn on last and during turn off it will turn off first. In both swiching events, this part would carry least amount of current and the other MOSFETs in parallel will share the increased current stress.

For all three MOSFET technologies of 150 V node, the reference R_{GEXTon} and $R_{GEXToff}$ were chosen as 80 Ω and 30 Ω respectively. To test the impact of $V_{GS(th)}$ mismatch, the R_{GEXTon} and $R_{GEXToff}$ were varied for LS_1 only using (6) to (9) and the values are mentioned in Table V, while $LS_2 - LS_4$ had the same external resistance values as reference.

The dynamic current sharing comparison during turn on and turn off, when all four low-side MOSFETs have the R_{GEXTon} and $R_{GEXToff}$ as 80 Ω and 30 Ω, is shown in Fig. 13. When all MOSFETs have the same $V_{GS(th)}$, IPT025N15NM6 shows the best dynamic current sharing overall during turn-on and

TABLE V
EXTERNAL RESISTOR VALUES FOR EMULATING MINIMUM AND
MAXIMUM $V_{GS(TH)}$ FOR 150 V MOSFETs

Part	R_{GEXT} [Ω]	Reference Condition	Emulating Condition	
			$V_{GS(th) min}$	$V_{GS(th) max}$
Competitor 1	R_{GEXTon}	80	66.5	94
	$R_{GEXToff}$	30	68	21.6
IPT039N15N5	R_{GEXTon}	80	72	90.3
	$R_{GEXToff}$	30	45.7	22.3
IPT025N15NM6	R_{GEXTon}	80	74	88
	$R_{GEXToff}$	30	39.5	23.9

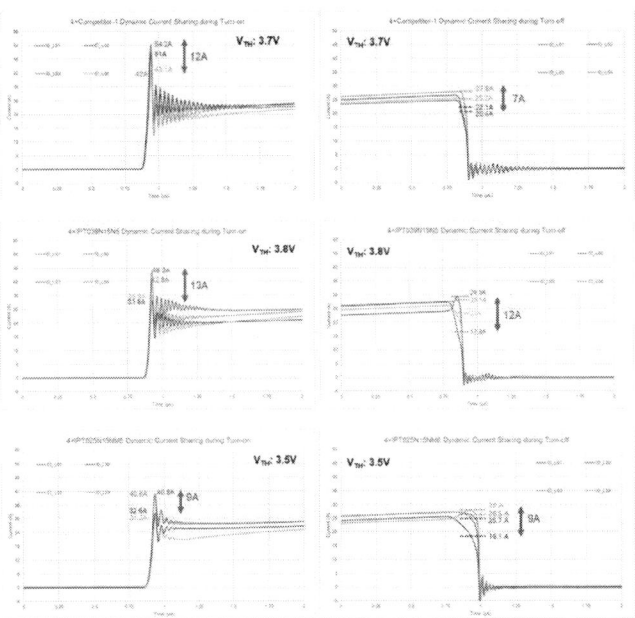

Fig. 11. Dynamic current sharing comparison during turn on and turn off for symmetric gate drive condition.

turn-off, having the maximum difference of 9 A during both transitions whereas, Competitor 1 and IPT039N15N5 show higher differences of up to 12 A in current sharing during both switching events.

To measure the dynamic current sharing when one MOSFET has lower $V_{GS(th)}$ during turn on and turn off, the R_{GEXTon} and $R_{GEXToff}$ for LS_1 are tweaked to values mentioned in Table V. The current sharing performance comparison is shown in Fig. 12. Competitor 1 part has a delta of 1.2 V between $V_{GS(th) min}$ and $V_{GS(th) typ}$, IPT039N15N5 has a delta of 0.8 V between $V_{GS(th) min}$ and $V_{GS(th) typ}$ and IPT025N15NM6 has a delta of 0.5 V between $V_{GS(th) min}$ and $V_{GS(th) typ}$.

Since the Competitor 1 part has the largest difference between $V_{GS(th) min}$ and $V_{GS(th) typ}$, LS_1 carries 95 A higher current during turn on and 144 A higher current during turn off compared to the other parallel MOSFETs. IPT025N15NM6, with lowest $V_{GS(th)}$ mismatch, has a difference in peak current of 30 A during turn on and 86 A during turn off. The smaller $V_{GS(th)}$ mismatch of IPT025N15NM6 ensures that delay between LS_1 and $LS_2 - LS_4$ during both switching transitions is smaller. As a result, the current stress on LS_1 is greatly

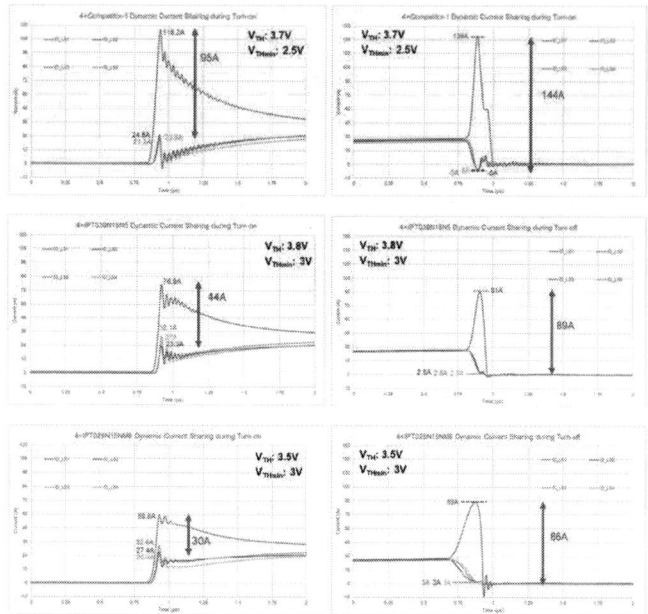

Fig. 12. Dynamic current sharing comparison during turn on and turn off when LS_1 is emulated to $V_{GS(th)\ min}$ condition.

Fig. 13. Dynamic current sharing comparison during turn on and turn off when LS_1 is emulated to $V_{GS(th)\ max}$ condition.

reduced, by up to 70%, compared to Competitor 1 parts.

To measure the dynamic current sharing when one MOSFET has higher $V_{GS(th)}$ during turn on and turn off, the R_{GEXTon} and $R_{GEXToff}$ for LS_1 are tweaked to values mentioned in Table V corresponding to $V_{GS(th)\ max}$ condition. The current sharing performance comparison is shown in Fig. 13. Competitor 1 part has a delta of $0.8\,V$ between $V_{GS(th)\ max}$ and $V_{GS(th)\ typ}$, IPT039N15N5 has a delta of $0.8\,V$ between $V_{GS(th)\ max}$ and $V_{GS(th)\ typ}$ and IPT025N15NM6 has a delta of $0.5\,V$ between $V_{GS(th)\ max}$ and $V_{GS(th)\ typ}$.

Since the Competitor 1 has the largest difference between $V_{GS(th)\ max}$ and $V_{GS(th)\ typ}$, the LS_1 carries $50\,A$ higher current during turn on and $46\,A$ higher current during turn off compared to the other parallel MOSFETs. IPT025N15NM6, with lowest $V_{GS(th)}$ mismatch, has a difference in peak current of $30\,A$ during turn on and $36\,A$ during turn off. The smaller $V_{GS(th)}$ mismatch of IPT025N15NM6 ensures that delay between LS_1 and $LS_2 - LS_4$ during both switching transitions is smaller. As a result, the current stress on LS_1 is greatly reduced, by up to 40%, compared to Competitor 1 parts.

The improvement of IPT025N15NM6 over Competitor 1 was larger when one MOSFET has lower $V_{GS(th)}$ because Competitor 1 had a higher $V_{GS(th)}$ mismatch between $V_{GS(th)\ min}$ and $V_{GS(th)\ typ}$ than it did for $V_{GS(th)\ max}$ and $V_{GS(th)\ typ}$. The asymmetry for $V_{GS(th)\ typ}$ in the $V_{GS(th)}$ window should also be avoided to ensure performance for both extreme $V_{GS(th)}$ conditions is within the design margin.

Therefore, having a tighter $V_{GS(th)}$ window improves the dynamic current sharing between parallel MOSFETs, reducing the current stress on the MOSFETs during both turn on & off conditions. Fig. 14 shows the $V_{GS(th)}$ distribution data for sixteen lots of IPT025N15NM6 showing how Infineon

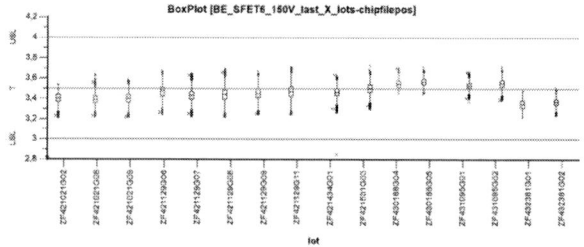

Fig. 14. $V_{GS(th)}$ distribution for IPT025N15NM6 enabling tighter $V_{GS(th)}$ window by process optimization.

was able to achieve a tighter $V_{GS(th)}$ window by process optimization.

B. Benefits of lower Transconductance on dynamic current sharing

To understand the benefits of lower g_{fs} on dynamic current sharing and switching performance, both $60\,V$ technologies were tested with symmetrical gate drive configuration as well as asymmetrical gate drive configuration where LS_1 was emulated to $V_{GS(th)\ min}$ and $V_{GS(th)\ max}$ conditions by tuning the external gate resistors. The $V_{GS(th)}$ and g_{fs} for all four parallel MOSFETs for both technologies were matched to highlight the impact of lower g_{fs} when different $V_{GS(th)}$ conditions are emulated.

The R_{GEXTon} and $R_{GEXToff}$ reference values for each technology are mentioned in Table VI. These values were selected based on typical design margin of maximum V_{DS} peak of approximately 80% - 90% of breakdown voltage. To test the impact of $V_{GS(th)}$ mismatch, the R_{GEXTon} and $R_{GEXToff}$

TABLE VI
EXTERNAL RESISTOR VALUES FOR EMULATING MINIMUM AND
MAXIMUM $V_{GS(TH)}$ FOR 60 V MOSFETs

Part	R_{GEXT} [Ω]	Reference Condition	Emulating Condition	
			$V_{GS(th)min}$	$V_{GS(th)max}$
BSC014N06NS	R_{GEXTon}	130	117.5	141
	$R_{GEXToff}$	130	380	88
ISC015N06NM5-LF2	R_{GEXTon}	62	53	69
	$R_{GEXToff}$	62	210	43

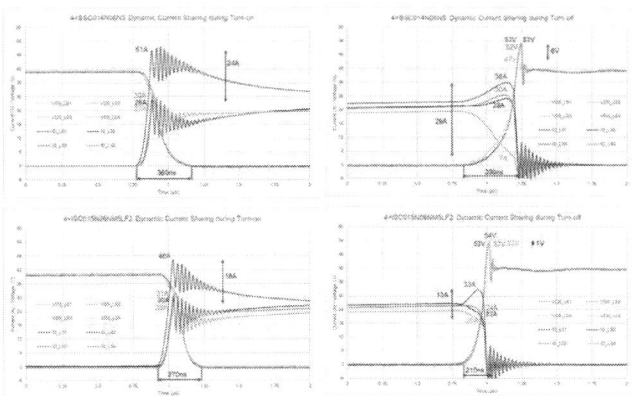

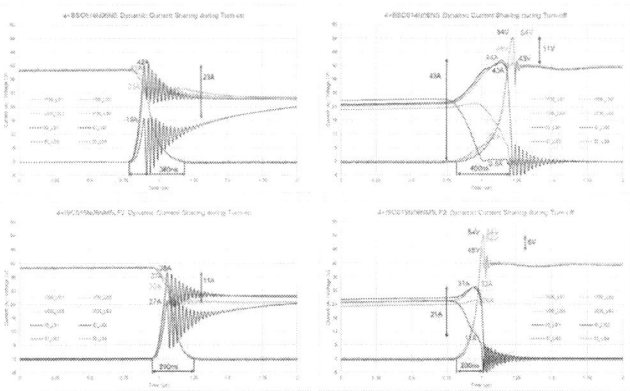

Fig. 16. Dynamic current sharing comparison during turn on and turn off when LS₁ is emulated to $V_{GS(th)\ min}$ condition showing improvements of lower g_{fs}.

Fig. 17. Dynamic current sharing comparison during turn on and turn off when LS₁ is emulated to $V_{GS(th)\ max}$ condition showing improvements of lower g_{fs}.

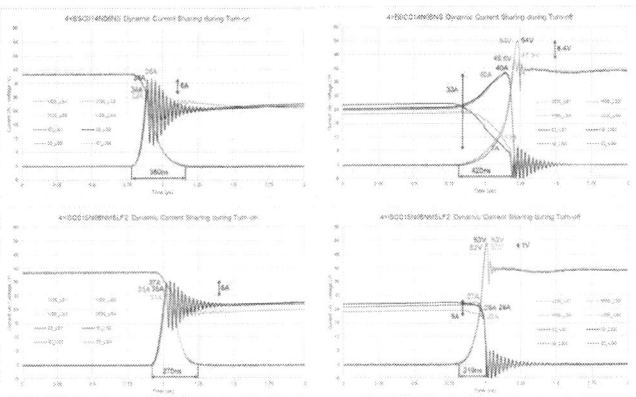

Fig. 15. Dynamic current sharing comparison during turn on and turn off for symmetric gate drive condition showing improvements of lower g_{fs}.

were varied for LS₁ only using (6) to (9) and the values are mentioned in Table VI.

The dynamic current sharing comparison during turn on and turn off, when all four low-side MOSFETs for symmetrical gate drive condition, is shown in Fig. 15. When all MOSFETs have the same $V_{GS(th)}$, ISC015N06NM5LF2 and BSC014N06NS have similar current sharing during turn on, even though ISC015N06NM5LF2 is switching 25% faster. During turn off, ISC015N06NM5LF2 switches in 210 ns which is 50% faster than BSC014N06NS and has dynamic current difference of 5 A which is 85% lower than BSC014N06NS. ISC015N06NM5LF2 also has similar V_{DS} peaks for all four parallel MOSFETs whereas, BSC014N06NS has 8 V difference due to unequal dynamic current sharing.

To measure the dynamic current sharing when one MOSFET has lower $V_{GS(th)}$ during turn on and turn off, the R_{GEXTon} and $R_{GEXToff}$ for LS₁ are tweaked to values mentioned in Table VI. The current sharing performance comparison is shown in Fig. 16. BSC014N06NS has a delta of 0.7 V between $V_{GS(th)\ min}$ and $V_{GS(th)\ typ}$ and ISC015N06NM5LF2 has a delta of 0.75 V between $V_{GS(th)\ min}$ and $V_{GS(th)\ typ}$. Both technologies have similar difference between $V_{GS(th)\ min}$ and $V_{GS(th)\ typ}$, but the the LS₁ of BSC014N06NS carries 24 A higher current during turn on and 29 A higher current during turn off compared to the other parallel MOSFETs. LS₁ of ISC015N06NM5LF2 on the other hand, has a difference in peak current of 18 A during turn on and 13 A during turn off while switching 25% and 45% faster respectively than BSC014N06NS. The lower g_{fs} helps significantly in balancing the currents even when the

$V_{GS(th)}$ mismatch is the high and the part which turns on first and turns off last doesn't have to carry the bulk of the current.

To measure the dynamic current sharing when one MOSFET has higher $V_{GS(th)}$ during turn on and turn off, the R_{GEXTon} and $R_{GEXToff}$ for LS₁ are tweaked to values mentioned in Table VI corresponding to $V_{GS(th)\ max}$ condition. The current sharing performance comparison is shown in Fig. 17. BSC014N06NS has a delta of 0.5 V and ISC015N06NM5LF2 has a delta of 0.45 V between $V_{GS(th)\ max}$ and $V_{GS(th)\ typ}$. both technologies have similar difference between $V_{GS(th)\ max}$ and $V_{GS(th)\ typ}$, but the the LS₁ of BSC014N06NS carries 23 A lower current during turn on and 43 A lower current during turn off compared to the other parallel MOSFETs. LS₁ of ISC015N06NM5LF2 on the other hand, has a difference in peak current of 11 A during turn on and 21 A during turn off while switching 25% and 50% faster respectively than BSC014N06NS. The lower g_{fs} helps significantly in balancing the currents even when there $V_{GS(th)}$ mismatch is the high. ISC015N06NM5LF2 also has similar V_{DS} peaks compared to BSC014N06NS which has 11 V difference in V_{DS} peaks due to unequal current sharing.

Therefore, having a lower g_{fs} improves the dynamic current

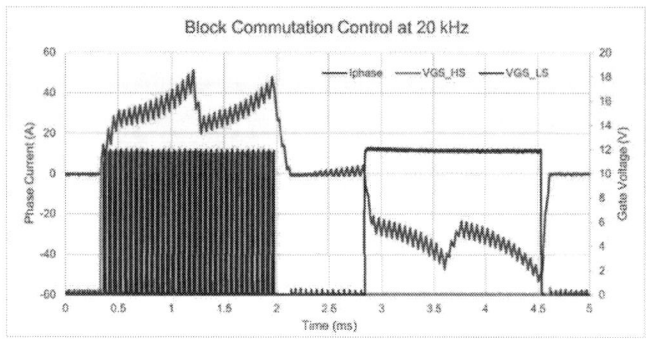

Fig. 18. Block Commutation control at 20 kHz switching frequency.

sharing between parallel MOSFETs. It balances the current and voltage stresses on the MOSFETs during both turn on & off conditions while also enabling faster switching.

C. Benefits of lower Transconductance on switching and thermal performance

To further understand the impact of lower g_{fs} on switching and thermal performance in motor-drive applications, the 60 V MOSFETs were tested in a three-phase inverter shown in Fig. 10 at test conditions mentioned in Table III. For these tests, parts having a $V_{GS(th)}$ mismatch of 0.3 V and equal g_{fs} were selected to be in parallel. To evaluate the improvement in switching performance due to lower g_{fs}, the external resistors were tuned to match V_{DS} peaks during switching transients.

The block commutation switching scheme for one period is shown in Fig. 18. The high-side MOSFETs have switching and conduction losses as the dominant loss whereas, the low-side MOSFETs have conduction and diode losses as the main contributors.Since, both 60 V MOSFETs have similar $R_{DS(on)}$ so the difference in switching performance will be the determining factor for thermal performance. In this control scheme, there are three hard switching events that will be compared which are the hard turn off of high-side MOSFETs, the hard turn off of low-side MOSFETs and hard turn on of high-side MOSFETs (when diode of low-side undergoes reverse recovery). The comparison for all three hard switching conditions is shown in Fig. 19. The target V_{DS} peak was 48 V to maintain the design margin of 80% of breakdown voltage.

From Fig. 19, during high-side hard turn off, ISC015N06NM5LF2 can switch in 135 ns whereas, BSC014N06NS has to be slowed down to 271 ns to maintain the similar V_{DS} peaks. During low-side turn off, ISC015N06NM5LF2 turns off in 127 ns while BSC014N06NS takes 272 ns to turn off while maintaining similar V_{DS} peaks. When the low-side diode undergoes reverse recovery, there is a risk of induced turn on of low-side MOSFET. To ensure, there is no induced turn-on the MOSFETs were slowed down to keep V_{GS} of low-side MOSFETs below 1.75 V. To achieve this, ISC015N06NM5LF2 high-side MOSFETs turn-on in 770 ns whereas, BSC014N06NS takes 1500 ns. In all the switching events, ISC015N06NM5LF2 can switch at least

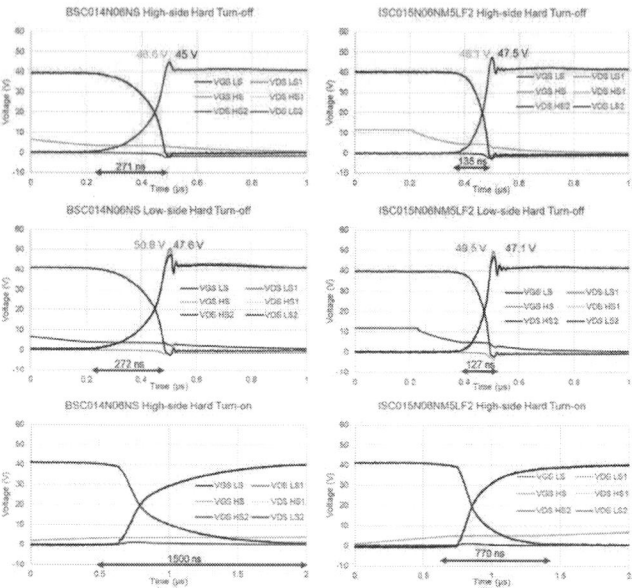

Fig. 19. Hard switching waveform comparisons for BSC014N06NS and ISC015N06NM5LF2.

50% faster than BSC014N06NS, reducing switching losses significantly.

The power loss comparison was done between BSC014N06NS and ISC015N06NM5LF2 by using the test waveforms and MATLAB [10]. Fig. 20 shows the split between total conduction losses, switching losses and diode losses for both technologies. Since both technologies have similar $R_{DS(on)}$ and were tested at the same power level, the conduction and diode losses are similar. However, since ISC015N06NM5LF2 switched at least 50% faster than BSC014N06NS in all hard switching events, there is a reduction of 50% in switching losses, going down from 16.45 W to 8.14 W. This loss reduction is primarily experienced by the high-side MOSFETs and the impact on thermal performance is shown in Fig. 21 and Fig. 22. The lower losses of ISC015N06NM5LF2 result in the high-side MOSFETs reaching temperatures 93.3 °C which is 15 °C lower than the high-side MOSFETs of BSC014N06NS.

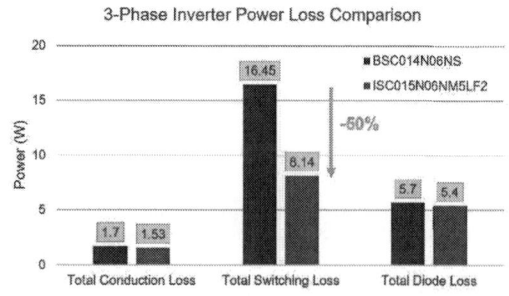

Fig. 20. Power Loss comparison between BSC014N06NS and ISC015N06NM5LF2 at 650W input power.

979-8-3315-1612-3/25 $31.00 © 2025 IEEE

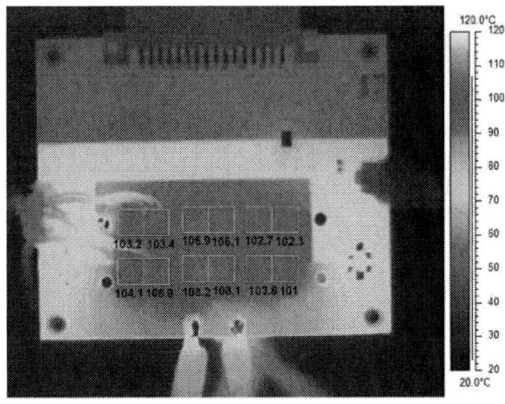

Fig. 21. BSC014N06NS thermal image at 650 W after 12 minutes.

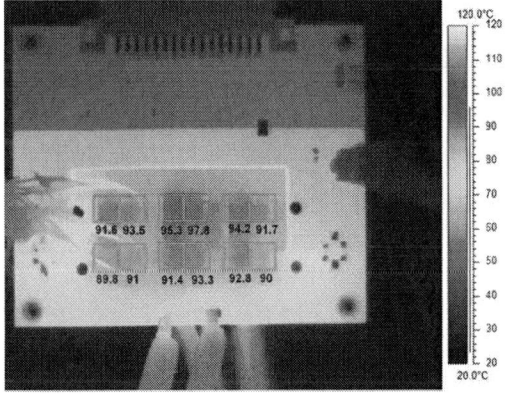

Fig. 22. ISC015N06NM5LF2 thermal image at 650 W after 12 minutes.

The higher temperature of the high-side MOSFETs of BSC014N06NS also heat up the low-side MOSFETs which result in up to $12\,^{\circ}\text{C}$ higher temperature than ISC015N06NM5LF2 low-side MOSFETs.

Therefore, having a lower g_{fs} improves not only the dynamic current sharing between parallel MOSFETs, it also improves the switching speeds greatly. This helps in improving the thermal performance of the system and reduce the MOSFET count for the same power level.

V. CONCLUSION

This paper has presented the impact of MOSFET parameters such as $V_{GS(th)}$ window and g_{fs} on dynamic current sharing between parallel MOSFETs. Tighter $V_{GS(th)}$ window reduces variation between the $V_{GS(th)}$ of parallel MOSFETs and improves the dynamic current sharing. The guidance on emulating different $V_{GS(th)}$ levels have been provided which designers can use to improve validation testing and choose the best MOSFET technology for their application. Lower g_{fs} enables better dynamic current sharing while also switching faster and maintaining similar V_{DS} peaks. This improvement in switching performance improved the thermal performance by up to $15\,^{\circ}\text{C}$ in a three-phase motor-drive at $650\,\text{W}$ input power and $20\,\text{kHz}$ switching frequency.

Future work on this will be performed using Infineon's next generation technologies for different voltage nodes, that improve MOSFET parameters further, to address the ever increasing need of MOSFET paralleling.

REFERENCES

[1] K. Jha, T. Q. Tran, K. Kelkar, S. Tegen and J. Mohammed, "OptiMOS 6 135 V for High Power Motor Drive Applications," PCIM Europe 2024; International Exhibition and Conference for Power Electronics, Intelligent Motion, Renewable Energy and Energy Management, Nürnberg, Germany, 2024, pp. 2925-2931, doi: 10.30420/566262416. keywords: Climate change;Low carbon economy;Electrification;Power electronics;Motor drives;Product development;Product design;Companies;Tools.

[2] Nexperia, "AN50005 Paralleling power mosfets in high power applications," Oct. 2024 [Online]. Available: https://assets.nexperia.com/documents/application-note/AN50005.pdf.

[3] Infineon Technologies, "AN_2309_PL51_2309_120502 MOSFET fast switching: motivation, implementation, and precautions," Oct. 2024 [Online]. Available: https://www.infineon.com/dgdl/Infineon-ApplicationNote _MOSFET_ fast _ switching_ motivation _._ implementation_ and _ precautions-ApplicationNotes-v01_00-EN.pdf?fileId=8ac78c8c8b6555fe018b66d96f1508cc.

[4] Infineon Technologies, "IPT044N15N5," Oct. 2024 [Online]. Available: https://www.infineon.com/dgdl/Infineon-IPT044N15N5-DataSheet-v02_02-EN.pdf?fileId=8ac78c8c7c9758f2017ccb22898a6c9f.

[5] T & M Research Products Inc., "A-2-01," Oct. 2024 [Online]. Available: https://www.tandmresearch.com/uploads/Product_PDFs/4-5_Watt_SBNC.PDF.

[6] Infineon Technologies, "IPT039N15N5," Oct. 2024 [Online]. Available: https://www.infineon.com/dgdl/Infineon-IPT039N15N5-DataSheet-v02_02-EN.pdf?fileId=8ac78c8c7c9758f2017ccb229dfe6ca5

[7] Infineon Technologies, "IPT025N15NM6," Oct. 2024 [Online]. Available: https://www.infineon.com/dgdl/Infineon-IPT025N15NM6-DataSheet-v02_00-EN.pdf?fileId=8ac78c8c90530b3a019077409ede701c.

[8] Infineon Technologies, "BSC014N06NS," Oct. 2024 [Online]. Available: https://www.infineon.com/dgdl/Infineon-BSC014N06NS-DataSheet-v02_05-EN.pdf?fileId=db3a3043382e837301386ab95a521dcd.

[9] Infineon Technologies, "ISC015N06NM5LF2," Oct. 2024 [Online]. Available: https://www.infineon.com/dgdl/Infineon-ISC015N06NM5LF2-DataSheet-v02_01-EN.pdf?fileId=8ac78c8c8b6555fe018be59a94313bd9.

[10] H. Amirkhanian and S. Oknaian, "Power Loss Breakdown in BLDC Drives Applications Using MATLAB," PCIM Europe 2018; International Exhibition and Conference for Power Electronics, Intelligent Motion, Renewable Energy and Energy Management, Nuremberg, Germany, 2018, pp. 1-5.

A 21.6 kW/L Two-Phase Immersion-Cooled Isolated DC-DC Converter

Aleksandar Ristic-Smith
Department of Engineering Science
University of Oxford
Oxford, UK
aleksandar.ristic-smith@eng.ox.ac.uk

Kawsar Ali
Department of Engineering Science
University of Oxford
Oxford, UK
kawsar.ali@eng.ox.ac.uk

Daniel Rogers
Department of Engineering Science
University of Oxford
Oxford, UK
daniel.rogers@eng.ox.ac.uk

Abstract—This paper introduces a sealed, passive two-phase immersion cooling concept using Novec 7000 dielectric fluid for transformer-isolated DC-DC converters with high power density. The concept relies on liquid-to-vapour phase change heat transfer from the surfaces of all converter components, supporting heat fluxes up to 40 W/cm². Experimental results are presented demonstrating the improvements in maximum winding current density and magnetic core flux density achieved by immersion in Novec 7000. This illustrates the potential for volume-minimisation of magnetic components. A novel transformer design is presented which maximises fluid contact with the core and windings while maintaining good electrical isolation between primary and secondary side. An 800 V, 18 kW CLLC resonant converter prototype with total volume of 0.81 L is designed and built. Preliminary testing demonstrates a power density of 21.6 kW/L at an input voltage of 620 V and a switching frequency of 860 kHz.

Index Terms—Isolated DC-DC converter, two-phase immersion cooling, power density, CLLC

I. INTRODUCTION

Transformer-isolated DC-DC converters are a key technology in applications like data centers, electric vehicle charging, and electrified aviation. Since available volume is typically limited, an important research challenge relates to maximisation of converter power density. Previous works in literature have focused on planar magnetics to minimise volume of isolated DC-DC converters. Air-cooled CLLC resonant prototypes with planar transformers have shown power densities of 8 kW/L [1] and 9.2 kW/L [2]. Another approach is to integrate resonant and magnetising inductances into the same magnetic component. A three phase CLLC converter with integrated magnetics for all three phases demonstrated power density of 17.6 kW/L (including the air-cooled heat sink) [3].

Comparatively few works have explored advanced cooling techniques to facilitate converter volume minimisation. In [4], single phase dielectric liquid cooling was investigated with microchannel heat sinks for the semiconductor switches. Additionally, the planar transformer featured an aluminium nitride ceramic substrate which cooled both the core and windings. The authors predicted a power density of 15 kW/L for a dual active bridge converter.

This work was supported by the U.K EPSRC and YASA Ltd as part of CASE award EP/R513295/1.

In this work a sealed passive two-phase immersion cooling concept using Novec 7000 dielectric fluid is proposed which achieves high power density due to the exceptional component-level heat transfer rates associated with liquid-to-vapour phase change. The concept is illustrated in Fig. 1. All converter components, including semiconductor switches, magnetics, capacitors and gate drivers are submerged in a pool of dielectric fluid in a sealed enclosure. The printed circuit board (PCB) extends outside the enclosure to provide signal and power connections. Heat generated by components causes the surrounding liquid to boil, vaporise, and condense in contact with a heat exchanger; liquid condensate returns to the fluid pool under gravity.

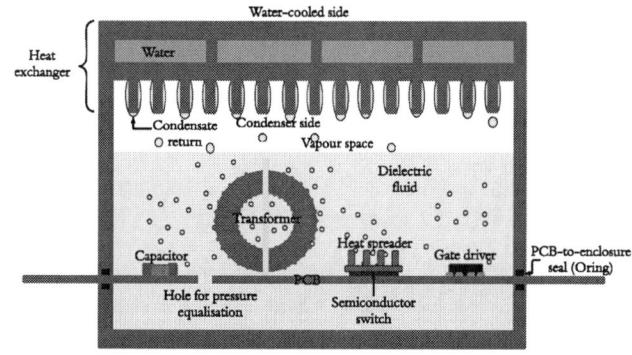

Fig. 1. Diagram of proposed two-phase immersion cooling concept for transformer isolated power converters.

There are several potential advantages of two-phase immersion cooling over alternative converter packaging techniques. Circulation of the dielectric fluid is completely passive, reducing system volume and complexity compared with pumped solutions. Dielectric fluid boiling introduces effective heat transfer from all converter components, not just the semiconductor switches. For magnetics, this allows higher winding current and core flux densities which facilitates volume minimisation. Furthermore, the dielectric fluid provides effective electrical isolation ($>10 \text{ kV/mm}$ [5]) which could enable smaller conductor spacing, dense component packing and compact circuit board layouts.

In prior literature, two-phase immersion cooled prototypes

have been demonstrated for a switched-capacitor converter circuit [6] and a transistor pulse generator [7]. To the authors' knowledge, the prototype in this work is the first example of a fully immersed transformer isolated DC-DC converter including resonant tank magnetics, capacitors, and gate drivers. Sections II and III present experiments conducted to characterise current and flux density limits for immersed copper windings and ferrite cores respectively. These results inform the design of the transformer. Section IV provides a detailed description of the bidirectional CLLC resonant converter prototype. It was succesfully tested up to 720 V and demonstrated maximum power density of 21.6 kW/L (350 W/in³).

II. TWO-PHASE IMMERSION COOLING BACKGROUND

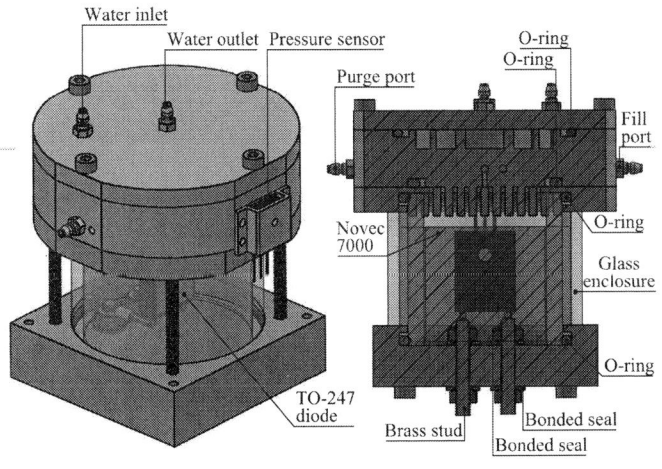

Fig. 2. Experimental apparatus for characterising critical heat flux limits of two-terminal power electronic components. See [8] for further details.

An important parameter of two-phase immersion cooled systems is the critical heat flux (CHF). This is the largest power dissipation per unit surface area which can be transferred from an electronic component via boiling. At CHF, vapour begins to blanket the component surface, accompanied by a sudden and drastic increase in temperature, leading to thermal failure. Previous experiments in the literature indicate that CHF occurs between 20 W/cm² [9] and 40 W/cm² [8] for Novec 7000.

An experimental apparatus, depicted in Fig. 2, was developed to characterise critical heat flux limits for different converter components. It consists of an open-ended cylindrical borosilicate glass enclosure sealed between a polyetheretherketone (PEEK) base and a custom aluminium water-cooled heat exchanger by an arrangement of elastomer O-rings and sealing washers. Current is passed through the component under test via a pair of brass studs inserted through the base.

To characterise the TO-247 package MOSFET used in the converter prototype, an IXYS DSEI30-06A diode was tested in the experimental apparatus under DC current. Fig. 3 shows the recorded current-voltage characteristic for the diode. A sharp decrease in diode forward voltage was observed at 45 A

due to a sudden increase in temperature following critical heat flux. This corresponds to a power dissipation of 53 W and a heat flux of 30 W/cm² when normalised to the metal tab (cathode) area. A 13.2 mm×12.1 mm×4.8 mm FIT0506 copper heat spreader from DFRobot was soldered to the metal tab, increasing surface area in contact with the fluid. A power dissipation of 166 W was sustained with no indication of critical heat flux.

The on-state resistance at 175 °C of the G3R12MT12K MOSFET in the converter prototype is 18 mΩ. If only conduction losses are considered, critical heat flux limits the RMS current through the switch to 54 A, whereas the heat spreader allows an RMS current of 96 A.

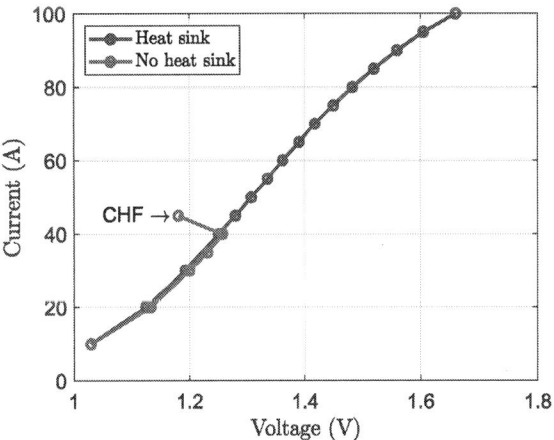

Fig. 3. Measured current-voltage characteristic for TO-247 diode with and without a heat spreader. Critical heat flux is indicated by the label CHF.

III. TWO-PHASE IMMERSION COOLED MAGNETIC COMPONENTS

A. Transformer Windings

In magnetic components, failure of the winding conductors can occur by thermal runaway arising from the increase in electrical resistance with temperature. The current just before failure is the largest value for which a single unstable thermal equilibrium exists between ohmic losses and heat transfer to the surroundings. Calculated failure currents are displayed as solid lines in Fig. 4 for bare copper wires of varying diameter under different cooling mechanisms. Interactions between the wire and surrounding medium are modelled as a constant heat transfer coefficient in all cases. The values used were 5 W/(m² K) for natural air convection [10], 50 W/(m² K) for forced air [11], 500 W/(m² K) for single-phase oil immersion [12], and 15 000 W/(m² K) for two-phase immersion in Novec 7000 [8]. In all cases, a temperature of 65 °C is assumed for the surrounding medium.

For two-phase immersion cooled components, critical heat flux introduces another winding failure mechanism. Consider a conductor with cross-section diameter d_w and electrical resistivity ρ conducting an rms current I. Equating the ohmic

TABLE I
CALCULATED AND EXPERIMENTAL CRITICAL HEAT FLUX CURRENTS FOR
DIFFERENT WIRE DIAMETERS

Wire diameter (mm)	0.315	0.4	0.5
Theoretical current (A) (Fig. 4)	35.8	49.9	68.0
Experimental current (A)	35.0	48.0	65.0

power loss to the surface heat transfer at critical heat flux q_{chf} yields the following expression for the current:

$$I = \pi \sqrt{\frac{d_w^3 q_{chf}}{4\rho}} \qquad (1)$$

The dotted line in Fig. 4 shows currents at critical heat flux, using an analytical model for horizontally-oriented cylinders to evaluate q_{chf} [13]. Critical heat flux always occurs before thermal runaway for windings immersed in Novec 7000. Despite this, the maximum current capacity is on average three times greater than single-phase oil cooling and eight times greater than forced air cooling.

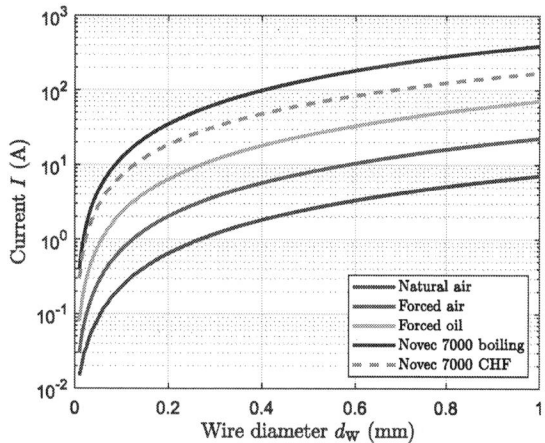

Fig. 4. Thermal runaway current limits of bare copper wires for different cooling techniques compared with critical heat flux limit for Novec 7000.

The critical heat flux currents were verified experimentally for copper wires in the experimental apparatus. Direct current was applied incrementally until critical heat flux was observed. Coil voltage was measured with separate sense leads and was used to identify critical heat flux via the sudden increase in wire resistance and voltage. Table. I compares theoretical predictions and experimental results; they agree to within 4%.

For high frequency alternating current (AC), it is typical to use Litz wire to mitigate the skin and proximity effects. Considering the skin depth of copper at 1 MHz frequency is 65 µm, Litz wires with 46 AWG strands (diameter of 40 µm) are recommended. We tested a 300x46 AWG type 2 Litz wire using our experimental apparatus. Preliminary testing with DC current of 90 A provided no indication of critical heat flux; however it was observed the insulation had melted and fused the strands together. This is to be avoided as it

could short the strands electrically, causing increased AC resistance and winding loss. The highest tested current for which strand-fusing was not observed is 62 A. Theoretical calculations predict that critical heat flux occurs at 130 A for an equivalent copper wire diameter (with higher electrical resistivity to account for the smaller fill factor of Litz wire). This indicates that when cooled in the two-phase immersion environment, the current density of Litz wire is limited by poor thermal conduction in the radial direction.

B. Ferrite Cores

The volumetric power dissipation Q''' for a magnetic core depends on the peak flux density ΔB and the excitation frequency f via the empirical Steinmetz equation below:

$$Q''' = c f^a \Delta B^b \qquad (2)$$

Due to the low thermal conductivity of ferrite materials ($\approx 0.4\,W/(m\,K)$ [14]), critical heat flux will not occur for most immersed ferrite cores. Limits on flux density are imposed either by core saturation or when portions of the core heat up beyond the Curie temperature.

An experiment was conducted to characterise the parameters a, b, and c for the Fair-rite 77 material. An inductor was constructed with 300x46 AWG type 2 Litz wire wound on a Fair-rite 5977001401 core. The inductor was immersed in Novec 7000 using the experimental apparatus and excited with square wave voltages of varying magnitude V and frequency f generated from a DC supply by a full-bridge converter. The peak flux density in response to a square wave voltage can be calculated by equation (3) for N turns wound on a core of cross-sectional area A_c.

$$\Delta B = \frac{V}{4 f N A_c} \qquad (3)$$

For each frequency and flux density, the power dissipated in the core is calculated as the product of DC supply voltage and RMS input current measured using two Keithley DMM6500 multimeters. In all tests, the input current was below 1 A, ensuring conduction losses in the converter switches and winding are negligible. The core magnetising current was monitored using a current clamp. At a certain flux density, the shape deviated from the expected triangular waveform, indicating core saturation. Once this was observed, no further voltages were tested at a given frequency. It should be noted that for 1.6 MHz, experiments were halted due to noticeable heating of the semiconductor switches, indicating converter switching losses were no longer negligible.

Fig. 5 presents volumetric power loss data for the chosen ferrite as a function of flux density at different frequencies. The Steinmetz equation in (2) was fitted to the data by multivariate linear regression and is shown as solid lines. The parameter values obtained are: $a = 2.2360$, $b = 1.8766$ and $c = 1.7962 \times 10^{-5}\,W/m^3/T^b/Hz^a$. Two-phase immersion cooling of the inductor enabled characterisation of the core material up to 0.2 T at a frequency of 1.4 MHz, a much wider range than is provided in the manufacturer data-sheet

979-8-3315-1612-3/25 $31.00 © 2025 IEEE

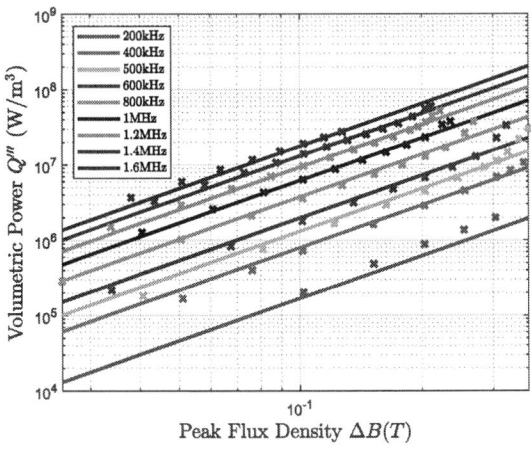

Fig. 5. Measured volumetric power dissipation as a function of peak flux density for the 5977001401 core. Steinmetz equation fits are shown as solid lines. Experimental data are shown as x.

(0.075 T at 400 kHz). It is observed that the Steinmetz model can adequately match the data up to the saturation flux density. The collected data is in close agreement with other recent works [15].

IV. CONVERTER PROTOTYPE

A. Resonant Tank Design

To demonstrate the achievable power density of two-phase immersion cooled transformer isolated DC-DC converters, a CLLC resonant converter prototype was developed. A circuit diagram of the CLLC topology is shown in Fig. 6. Full-bridge switch networks are employed for the primary and secondary sides. Depending on the power flow direction, the inverting full-bridge is switched at the switching frequency with a 50% duty cycle and a fixed dead time, while the rectifying full-bridge operates in synchronous rectification mode. The resonant tank is designed to offer symmetric operation across both power directions [16]–[18], where the resonant inductors L_{rp} and L_{rs} are integrated via the transformer leakage inductance. Converter parameters are listed in Table II. For simplicity, the voltage gain is unity and the transformer turns ratio is one. The resonant frequency of 700 kHz was selected as a compromise between power density and efficiency (efficiency decreases with frequency due to increased transformer core and MOSFET turn-off switching losses).

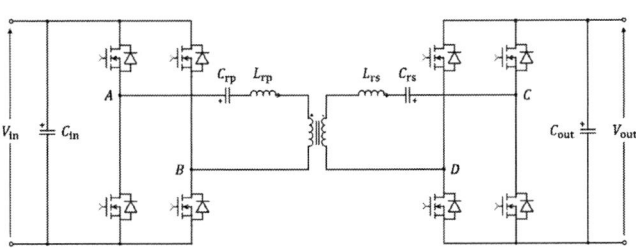

Fig. 6. Circuit diagram of the CLLC converter.

The resonant tank transformer is constructed from a Fair-rite 597700201 toroidal core with single layer, ten turn windings of 660x46 AWG type 2 Litz wire as shown in Fig. 7. This ensures the fluid can effectively contact all surfaces of the winding and core. It is assembled by cutting the core in half and adhering each piece to an FR4 interposer using epoxy adhesive. Custom 3D-printed coil formers are adhered to each core half and used to support the windings. This provides well-defined gaps between coil and core and between neighbouring turns to improve contact with and circulation of the fluid. The transformer construction introduces leakage inductance by separating the primary and secondary windings. For the chosen design, the leakage inductance measured with a N4L PSM 1735 impedance analyser was 2.8 μH. The resonant capacitance was selected as 18 nF to give the target resonant frequency.

CLLC converters are typically operated at a switching frequency slightly below the resonant frequency. Magnetising inductance should be sufficiently small such that enough magnetising current flows during the dead time to discharge the switch output capacitance, resulting in zero voltage switching (ZVS) turn-on [16]. In this work, an air gap was introduced into the magnetic path by the FR4 interposer. For the largest available interposer thickness of 1 mm, the magnetising inductance introduced proved too large to provide ZVS below the resonant frequency. Instead, the converter is operated at higher switching frequencies than resonance (840 kHz to 930 kHz) so the switch current lags the resonant tank voltage, which also guarantees ZVS [2].

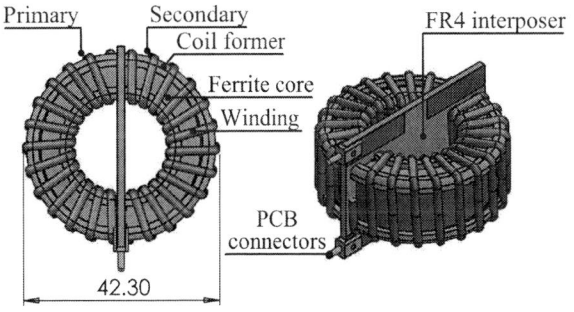

Fig. 7. Diagram of the resonant tank transformer. Dimensions are in mm.

B. Converter Design

The CLLC converter prototype is shown in Fig. 8. It is assembled on a motherboard PCB with top and bottom layer solder resist removed so the copper can be directly cooled. The components and dielectric fluid are enclosed within two aluminium 6082 chambers which seal to either side of the PCB via elastomer O-rings. Unconnected sealing traces encircle the converter active area on the top and bottom layer, providing a surface for the O-ring seals. Power and signal connections are passed into the enclosure on inner PCB layers so the sealing traces can be unbroken.

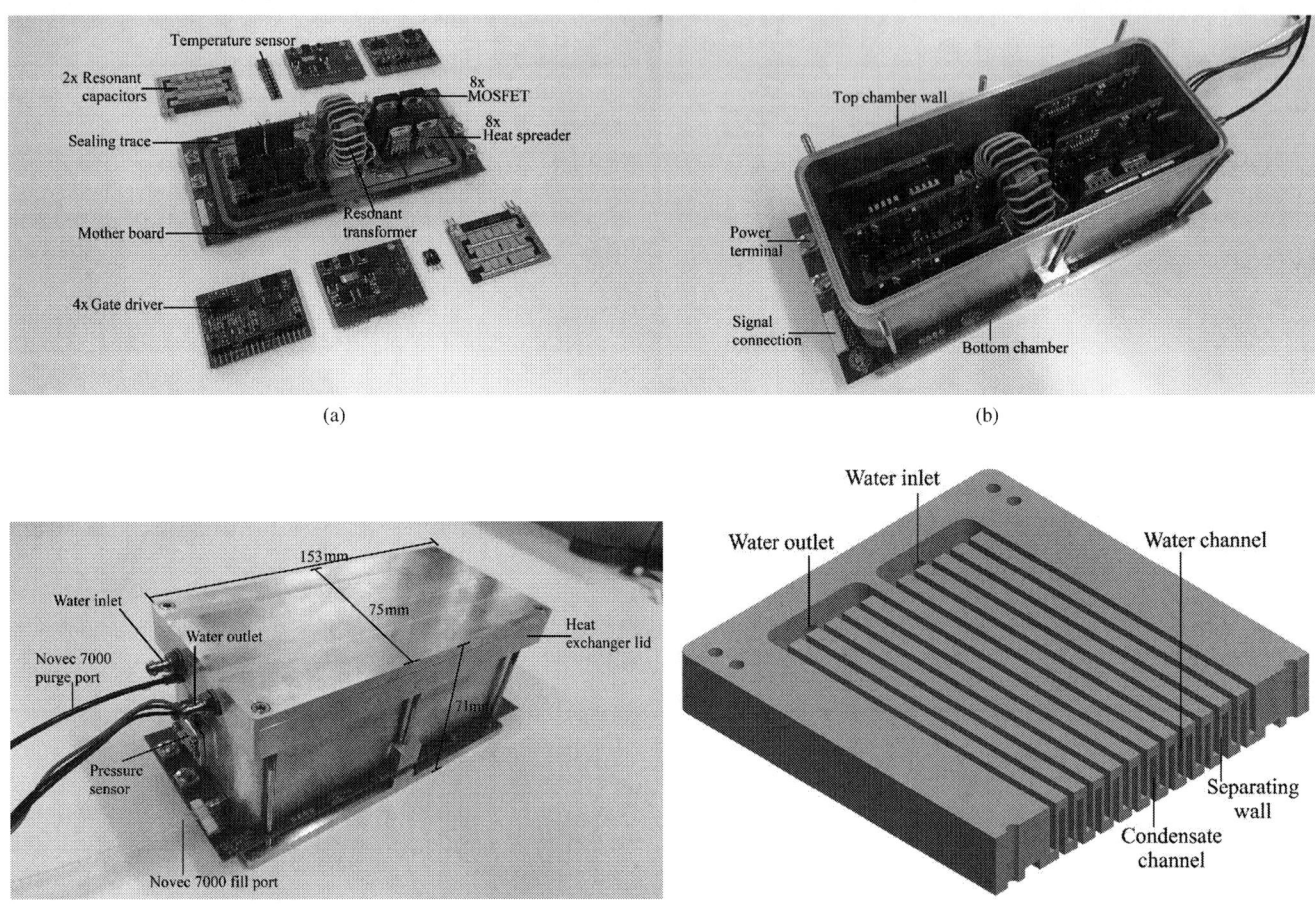

Fig. 8. Photographs of the CLLC converter prototype. (a) Motherboard and all daughter boards. (b) Partially assembled converter showing bottom chamber and top chamber wall. (c) Assembled converter with heat exchanger lid. Dimensions are in mm. (d) Cross-section view of heat exchanger active area.

TABLE II
SYMMETRICAL CLLC CONVERTER DESIGN PARAMETERS

Primary and secondary side dc voltage (V_{in}, V_{out})	800 V
Rated power (P)	20 kW
Resonant frequency (f_r)	700 kHz
Switching frequency (f_s)	840-930 kHz
Transformer turns ratio (n)	1
Resonant capacitance (C_{rp}, C_{rs})	18 nF
Resonant/Leakage inductance (L_{rp}, L_{rs})	2.8 µH
Magnetising inductance (L_m)	15.2 µH
Deadtime (t_d)	60 ns
MOSFETS	G3R12MT12K
Gate driver	NCP51561

The converter features eight G3R12MT12K SiC MOSFET switches mounted vertically on the motherboard with heat spreaders attached to increase the surface area. There are four gate driver boards, each providing control signals for two switches in a bridge leg. Each gate driver board features an Onsemi NCP51561 dual channel gate driver. It also includes custom circuitry designed to produce the 15 V and −5 V references required to gate the switches on and off. At the target switching frequency, one gate driver board dissipates 18 W. The gate driver overheats in air and its operation at such high frequency is only possible through immersion in the dielectric fluid. There are two capacitor boards, one for each side of the resonant tank. Each features eighteen 1 nF/1500 V multi-layer ceramic capacitors connected in parallel to yield the target resonant capacitance.

The fluid enclosure lid serves as a water-cooled heat exchanger to condense the fluid vapour. It features 10 mm deep and 1.4 mm wide channels on the condenser and water sides. These are interleaved to improve thermal contact between the condensing dielectric fluid and the water; they are separated by an aluminium wall which is only 0.75 mm thick. The enclosure wall features two ports used to fill and purge fluid from the enclosure. One port is positioned above the height of all components and sets the level of Novec 7000. Once the enclosure has been filled, a small quantity of fluid is boiled off to atmosphere, evacuating air present in the vapour space and dissolved in the liquid. The purging port is then shut, sealing the enclosure. A Bourns BPS130 absolute pressure sensor is mounted on the enclosure wall; the dielectric fluid temperature can then be estimated from the measured pressure using the saturation pressure-temperature relationship provided in the data-sheet [5].

V. CONVERTER TESTING

Due to limitations of available power supplies, tests were configured so the converter output power is circulated back to the input. The supply provides power equal to the total converter losses. This is illustrated in Fig. 9. Two power supplies, represented by sources V_{out} and V_{diff}, are connected in series across the CLLC input terminals with diodes preventing current flow between them. V_{out} is connected across the converter output terminals. The voltage differential between input and output is determined by V_{diff}, which was set to 20 V in all tests.

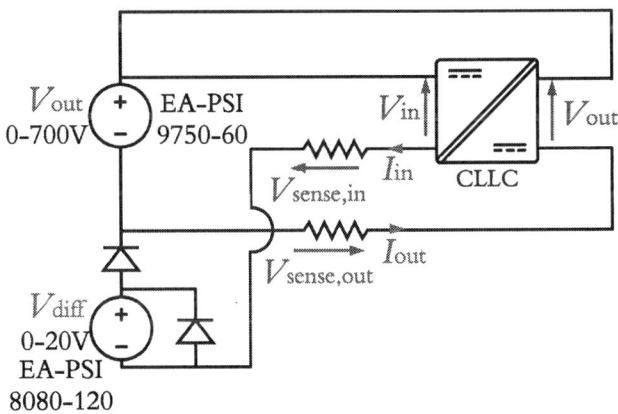

Fig. 9. Diagram of CLLC converter test set-up illustrating the power sources and voltage measurements.

For each voltage and switching frequency, the converter was allowed to reach thermal steady state observed via stabilisation of the pressure sensor voltage. The voltages V_{in} and V_{out} across the input and output were measured using two Tenma 72-774 multi-meters. The RMS input and output currents I_{in} and I_{out} were determined from the sense voltages $V_{sense,in}$ and $V_{sense,out}$ measured across $1\,m\Omega$ current shunt resistors using two Keithley DMM6500 multimeters. Due to fluctuations in the sense voltages, an RC filter (R = 2 kΩ, C = 0.33 μF) was added at the input of each multi-meter and data was recorded by averaging 100 sequential measurements.

A calorimetric approach was employed to verify the electrical measurements. In all experiments, water was circulated through the heat exchanger with a temperature set-point of 25 °C. The exposed outer surfaces of the fluid enclosure were insulated using thick polystyrene so most power loss was rejected to the water rather than to ambient air. Inlet and outlet temperatures were measured using PT1000 sensors fastened to the connecting tubing at the respective heat exchanger port. The electrically measured power losses from experiments at output voltages from 400 V to 600 V and frequencies from 840 kHz to 930 kHz are plotted against the water temperature differences from inlet to outlet in Fig. 10. The points follow a line of constant slope passing close to the origin, as expected when the water flow rate is constant, indicating the electrical

measurements are sufficiently accurate. The water flow rate predicted from the slope is 0.7 L/min.

The highest power operating point tested was for an input voltage of 620 V at a switching frequency of 860 kHz. The total power throughput was 17.6 kW, corresponding to a power density of 21.6 kW/L, with total power loss of 1127 W and efficiency of 93.6 %. The pressure inside the enclosure was 410 kPa which corresponds to a fluid temperature of 80 °C. It is estimated the surfaces of all components are maintained below 100 °C. This thermal performance was achieved with a low water flow rate, leading to a high water outlet temperature of 51 °C.

The prototype was further tested at an input voltage of 720 V with switching frequency 930 kHz, though only a transient test could be conducted as the power dissipation exceeded the heat-removal capacity of the water circulator.

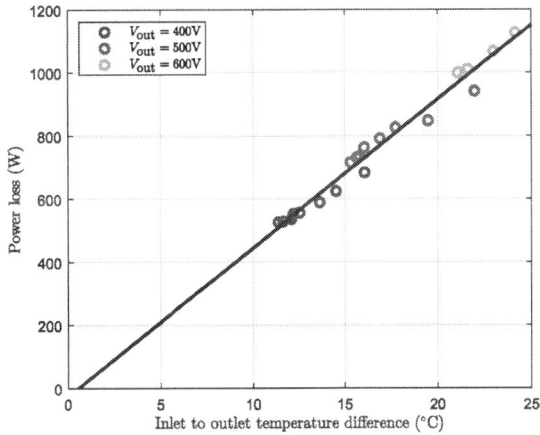

Fig. 10. Calorimetric verification of electrical power loss measurements by correlation against the water temperature difference from inlet to outlet. Data-points are shown as circles; the least-squares fit is shown as a solid line.

A. Discussion

Fig. 11 shows the measured efficiency of the converter across all voltages and frequencies tested. A trend towards decreasing efficiency with higher frequency is apparent. Fig. 12 shows oscilloscope captures of the bridge voltages V_{AB}, V_{CD} and resonant capacitor voltages $V_{Cr,p}$ and $V_{Cr,s}$ of primary and secondary side for the operating point with input voltage of 620 V and frequency of 890 kHz. The LTSPICE simulation wavefoms are overlaid and a close match is observed between experiment and simulation for the resonant capacitor voltages. We believe the mismatch in the time axis for the bridge voltages is due to not modelling the transformer core loss in the simulation and inadequate model of the body diode behaviour in the SPICE model of the MOSFET provided by the manufacturer [19].

With a total power loss of 1000 W the converter has an efficiency of 93.8 % at the 620 V 890 kHz operating point. An estimated power loss distribution is shown in Fig. 13, which accounts for 840 W. The dominant component is the turn-off

979-8-3315-1612-3/25 $31.00 © 2025 IEEE

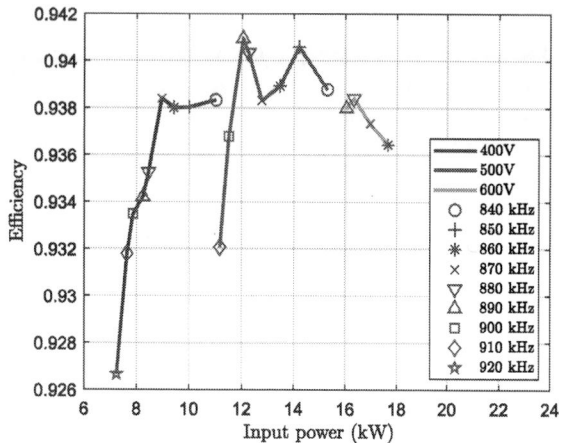

Fig. 11. Measured converter prototype efficiencies for all output voltages and frequencies tested

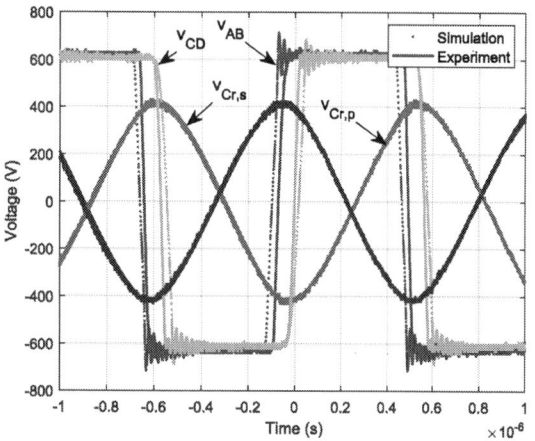

Fig. 12. Key voltage waveforms at 620 V, 890 kHz operation.

switching loss of the primary side MOSFETs. This is due to the choice of a switching frequency which is much higher than the resonant frequency, causing high turn-off current for the primary side MOSFETs. This also explains the overall decrease in converter efficiency with frequency. Considering the turn-on energy is more than 1.5 times the turn-off energy for the G3R12MT12K MOSFET, such high switching frequency was chosen to ensure zero-voltage turn-on transitions. Further research may consider aspects of the resonant tank transformer design with the aim of reducing magnetising inductance. This would allow for ZVS at switching frequencies close to the resonant frequency, improving efficiency due to a reduction in switch turn-off and core losses. The secondary side MOSFETs do not incur any switching loss because of the natural commutation of currents through the body diodes before and after the switching occurs.

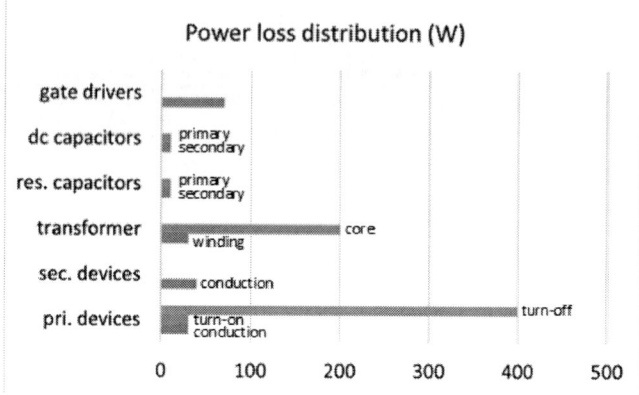

Fig. 13. Calculated power loss distribution at 620 V and 890 kHz.

VI. CONCLUSIONS

A sealed, passive two-phase immersion cooling approach for transformer-isolated DC-DC power converters was proposed with the aim of achieving high power density through volume minimisation of magnetic components. Experiments indicate that copper windings can operate at current densities up to eight times higher than in air and ferrite cores at flux densities approaching magnetic saturation, even at frequencies above 1 MHz.

A prototype two-phase immersion-cooled CLLC converter was designed, built, and tested up to a maximum power density of 21.6 kW at input voltage of 620 V and frequency of 860 kHz. This is competitive with the best power densities previously presented in the literature and demonstrates the potential for high-power density immersion cooled DC-DC converters. Further research in this area should focus on optimisation of the resonant tank transformer, incorporating design aspects which yield favourable cooling and excellent electrical isolation while ensuring ZVS circuit operation.

REFERENCES

[1] B. Li, Q. Li, and F. C. Lee, "High-frequency PCB winding transformer with integrated inductors for a bi-directional resonant converter," *IEEE Transactions on Power Electronics*, vol. 34, no. 7, pp. 6123–6135, 2019.

[2] P. He, A. Mallik, A. Sankar, and A. Khaligh, "Design of a 1-MHz high-efficiency high-power-density bidirectional GaN-based CLLC converter for electric vehicles," *IEEE Transactions on Vehicular Technology*, vol. 68, no. 1, pp. 213–223, 2019.

[3] F. Jin, A. Nabih, T. Yuan, and Q. Li, "A high-efficiency high-density three-phase CLLC resonant converter with a universally derived three-phase integrated transformer for on-board-charger application," *IEEE Transactions on Power Electronics*, vol. 39, no. 4, pp. 4350–4366, 2024.

[4] S. U. Yuruker, R. K. Mandel, P. McCluskey, M. M. Ohadi, S. Chakraborty, Y. Park, H. Yun, A. Khaligh, L. Boteler, and M. Hinojosa, "Advanced packaging and thermal management of high-power DC-DC converters," ser. International Electronic Packaging Technical Conference and Exhibition, vol. ASME 2019 International Technical Conference and Exhibition on Packaging and Integration of Electronic and Photonic Microsystems, 10 2019, p. V001T06A026. [Online]. Available: https://doi.org/10.1115/IPACK2019-6559

[5] 3M, "3M Novec7000 engineered fluid," *Data-sheet*, 9 2021.

[6] P. Birbarah, T. Gebrael, T. Foulkes, A. Stillwell, A. Moore, R. Pilawa-Podgurski, and N. Miljkovic, "Water immersion cooling of high power density electronics," *International Journal of Heat and Mass Transfer*, vol. 147, p. 118918, 2020. [Online]. Available: https://www.sciencedirect.com/science/article/pii/S0017931019336002

[7] Y. Wang, L. Ren, Z. Yang, Z. Deng, and W. Ding, "Application of two-phase immersion cooling technique for performance improvement of high power and high repetition avalanche transistorized subnanosecond pulse generators," *IEEE Transactions on Power Electronics*, vol. 37, no. 3, pp. 3024–3039, 2022.

[8] A. Ristic-Smith and D. J. Rogers, "Compact two-phase immersion cooling with dielectric fluid for PCB-based power electronics," *IEEE Open Journal of Power Electronics*, vol. 5, pp. 1107–1118, 2024.

[9] M. S. El-Genk and M. Pourghasemi, "Experimental investigation of saturation boiling of HFE-7000 dielectric liquid on rough copper surfaces," *Thermal Science and Engineering Progress*, vol. 15, p. 100428, 2020. [Online]. Available: https://www.sciencedirect.com/science/article/pii/S2451904919302343

[10] I. Villar, U. Viscarret, I. Etxeberria-Otadui, and A. Rufer, "Transient thermal model of a medium frequency power transformer," in *2008 34th Annual Conference of IEEE Industrial Electronics*, 2008, pp. 1033–1038.

[11] S. Balci, I. Sefa, and N. Altin, "Thermal behavior of a medium-frequency ferrite-core power transformer," *Journal of Electronic Materials*, vol. 45, no. 8, pp. 3978–3988, Aug 2016. [Online]. Available: https://doi.org/10.1007/s11664-016-4567-5

[12] D.-E. A. Mansour and A. M. Elsaeed, "Heat transfer properties of transformer oil-based nanofluids filled with Al2O3 nanoparticles," in *2014 IEEE International Conference on Power and Energy (PECon)*, 2014, pp. 123–127.

[13] J. H. Lienhard and V. K. Dhir, "Hydrodynamic prediction of peak pool-boiling heat fluxes from finite bodies," *Journal of Heat Transfer-transactions of The Asme*, vol. 95, pp. 152–158, 1973. [Online]. Available: https://api.semanticscholar.org/CorpusID:120647310

[14] Fair-rite. (2023) Toroids (5977001401. [Online]. Available: https://fair-rite.com/product/toroids-5977001401/

[15] D. Serrano, H. Li, S. Wang, T. Guillod, M. Luo, V. Bansal, N. K. Jha, Y. Chen, C. R. Sullivan, and M. Chen, "Why MagNet: Quantifying the complexity of modeling power magnetic material characteristics," *IEEE Transactions on Power Electronics*, vol. 38, no. 11, pp. 14 292–14 316, 2023.

[16] Y. Cao, M. Ngo, R. Burgos, A. Ismail, and D. Dong, "Switching transition analysis and optimization for bidirectional CLLC resonant DC transformer," *IEEE Transactions on Power Electronics*, vol. 37, no. 4, pp. 3786–3800, 2022.

[17] A. M. Ammar, K. Ali, and D. J. Rogers, "A bidirectional gan-based CLLC converter for plug-in electric vehicles on-board chargers," in *IECON 2020 The 46th Annual Conference of the IEEE Industrial Electronics Society*, 2020, pp. 1129–1135.

[18] B. Li, Q. Li, F. C. Lee, Z. Liu, and Y. Yang, "A high-efficiency high-density wide-bandgap device-based bidirectional on-board charger," *IEEE Journal of Emerging and Selected Topics in Power Electronics*, vol. 6, no. 3, pp. 1627–1636, 2018.

[19] "SiC MOSFET spice model usage instructions, Rev 1.0," *Application note*, 10 2020.

Extraction of Common Mode Parasitic Capacitance in Balance Filter for The Prediction of EMI Noise Suppression

Qiuzhe Yang, Xingyu Chen, Zijian Wang, Qiang Li

Center for Power Electronics Systems
Bradley Department of Electrical and Computer Engineering
Virginia Polytechnic and State University
Blacksburg, VA 24061, USA
phoenixyqz@vt.edu

Abstract—**This paper proposes a new method to extract the common mode parasitic capacitance in the balance network for the totem-pole converter. Filter concept is applied to the balance network to predict the EMI noise with the extracted parasitic capacitance. With the proposed method, the EMI noise prediction matches the experiment very well within 30MHz.**

Keywords—Common mode noise, parasitic capacitance, EMI, modeling

I. INTRODUCTION

Wide bandgap (WBG) device[1-7] brings many new opportunities[8-16] for the contemporary power electronics circuits. Their ultra-fast switching speed can greatly shrink the size of the converters[17-29] by lowering the switching loss[30-31] and pushing the operation frequency to a higher level[32-37]. Compact components like planar transformers can be deployed for a higher power density. Except for the multiple merits of WBG devices, however, they also produce many issues in the design of converter. EMI problem is one of the most severe issues[38-45] due to their fast switching speed, especially in the area of common mode (CM) noise. Many methods have been proposed to address the tricky EMI problems. Balance technique [46] is deemed as an effective method to suppress the CM noise in PFC applications like boost or totem-pole rectifier. Its simplicity and low-incurred loss make it a perfect candidate to preclude the CM noise from the source – the switching transition of the devices.

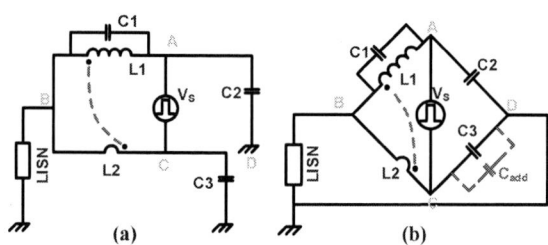

Figure 2: Simplified balance network

Fig.1 shows the 1kW single-phase totem-pole PFC converter in which balance technique is applied. There are two inductors in the converter. Except for the main inductor L1, one additional small inductor L2 is tightly coupled with L1 to form the balance bridge and serve as an adjustment factor of balance network impedance. L1 and L2 are PCB winding coupled inductors. C1 is the parasitic capacitance of L1. The structure shows on the left top of Fig.1. Only one turn is used in L2 to keep it as small as possible and no parasitic capacitance in L2. The basic principle of the balance technique is a Wheatstone bridge. Fig.2 shows the CM noise equivalent circuit. A is the switching node of the converter. B is the input AC or DC voltage source. C is the PCB ground of the converter. D is the earth ground which is usually connected to the grid earth. Except for the coupled inductor half bridge, the parasitic capacitance C2 and C3 in Fig. 2(a) form the other half bridge. C2 is the parasitic capacitance of the converter switching node referring to the earth ground. C3 is the capacitance of the converter PCB ground referring to the earth ground. To block the CM noise, the following equation should

Figure 1: Balance technique applied in Totem-Pole PFC

be satisfied throughout 150kHz to 30MHz to maintain equal voltage potential at point B and D by adding an extra small capacitor Cadd in parallel with C2 or C3, as Fig.2(b) shows.

$$\frac{V_{AB}}{V_{BC}} = \frac{V_{AD}}{V_{DC}} \tag{1}$$

The concept of the balance is simple since it's a Wheatstone bridge. The difficulty is to extract the parameters in the bridge and predict the noise attenuation with the parameters. For the coupled inductor bridge, the self-resonant frequency of L1 caused by its parasitic capacitance C1 is different from L2 due to their different number of turns and thus, different parasitic capacitances. The ratio of VAB and VBC is not a constant value throughout 30MHz. In [47], it is pointed out that a sufficiently high coupling coefficient between L1 and L2 can solve this problem. In [48], a new method is proposed to analyze and measure the inductor bridge voltage ratio. Identifying the capacitance bridge is another challenge for the design of balance network [49]. The exact value of C2 and C3 is essential to the selection of Cadd. Previously the selection of Cadd is an trail-and-error process which mainly depends on the guess. This is time-consuming and the value for the best noise attenuation remains unknown. Impedance analyzer is most commonly used for parasitic capacitance extraction. However, the self-parasitic parameters of the impedance analyzer probe introduce extra resonance and severely impact the accuracy of these PCB capacitance. Besides, the boundary of each part on PCB is vague. Except for the PCB trace, other parts like devices, PCB winding and capacitors mounted on the converter also have impact to the total capacitance. The separation of switching node or ground part on the converter cannot be identified by experience.

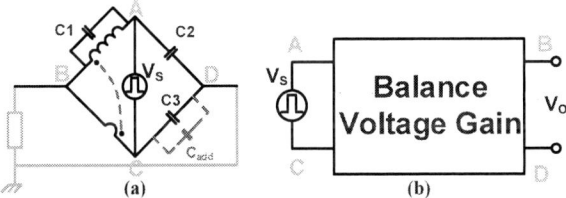

Figure 3: Balance filter model

On the other hand, the balance network is intrinsically a filter even without adding any capacitor to make the bridge balanced. Theoretically, the noise from the source could be totally attenuated if (1) were maintained. However, due to the imperfect coupling of L1 and L2 and the limited selectable capacitance value of added capacitor Cadd, the effect of balance network is compromised. Therefore, the prediction of the noise attenuation in balance network is important to meet the EMI standard and simplify the input filter design. Considering balance network in Fig.3(a) as a filter, then the noise current flowing through the LISN is the load. The insertion voltage gain of the balance filter in Fig.3(b) can be identified if all the parameters are correctly extracted. In the other word, balance filter can be used to verify the correctness of the parameters. In Fig.3, A-C is the noise input which is the switching transition of the main devices and B-D is output. Disconnecting the 50Ω LISN network, B-D will have a very high impedance and can be considered as open circuit. In

this way, the modeling of the balance filter will be greatly simplified.

In this paper, a new method to extract the common mode parasitic capacitance is proposed in the balance network. This method utilizes the spectrum of the sampled C2 and C3 voltage running in the switching mode. Filter concept is applied to verify the method and to predict the noise attenuation. Based on the analysis, the value of CM parasitic capacitance is verified in the totem-pole converter running in the DCDC mode and the balance filter model predicts the noise attenuation very well.

II. METHODOLOGY

A. Basic idea of the extraction of parasitic CM capacitance

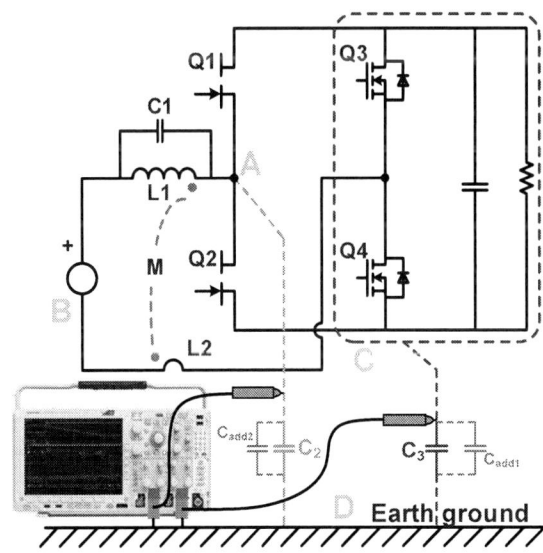

Figure 4: CM capacitance measurement method

C2 and C3 are parasitic capacitance between PCB and the earth ground. To design the balance network, the value of C2 and C3 need to be determined in order to select the added extra capacitor Cadd. C2 and C3 are essentially parallel plate capacitors and thus, are linear components. The impedance ratio of C2 and C3 is the same as the voltage ratio of C2 and C3, as (2) stated in the s domain.

$$\frac{V_{C2}(s)}{V_{C3}(s)} = \frac{VAD(s)}{VCD(s)} = \frac{sC_3}{sC_2} = m \tag{2}$$

The basic idea is utilizing the switching voltage in the running converter. To get the voltage ratio and thus, the impedance ratio of C2 and C3, the switching waveforms of AD and CD are sampled using oscilloscope probe as Fig.4 shows. Fig.5(a) is the sampled voltage. The switching waveforms include all the information. However, direct comparison of the voltage waveform to get the ratio is impossible. By doing the fast Fourier transform (FFT) in MATLAB using extracted waveform, the noise spectrum of VAD (blue) and VCD (red) is as Fig.5(b) shows.

The spectrum shows the difference between VAD and VCD is 15.65dBμV. Therefore, it can be derived that

$$20\log_{10}(VAD) - 20\log_{10}(VCD) = 15.57dB\mu V \quad (3)$$

Combining (2) and (3), the capacitance ratio is

$$\frac{C_3}{C_2} = 6 \quad (4)$$

Only one group of capacitance ratio is unable to determine the capacitance of C2 and C3. By adding different capacitors Cadd in parallel with C2 or C3, the new impedance ratio can be measured as (5) shows.

$$\frac{V_{C2+Cadd}(s)}{V_{C3}(s)} = \frac{VAD\prime(s)}{VCD\prime(s)} = \frac{sC_3}{s(C_2+C_{add})} = k \quad (5)$$

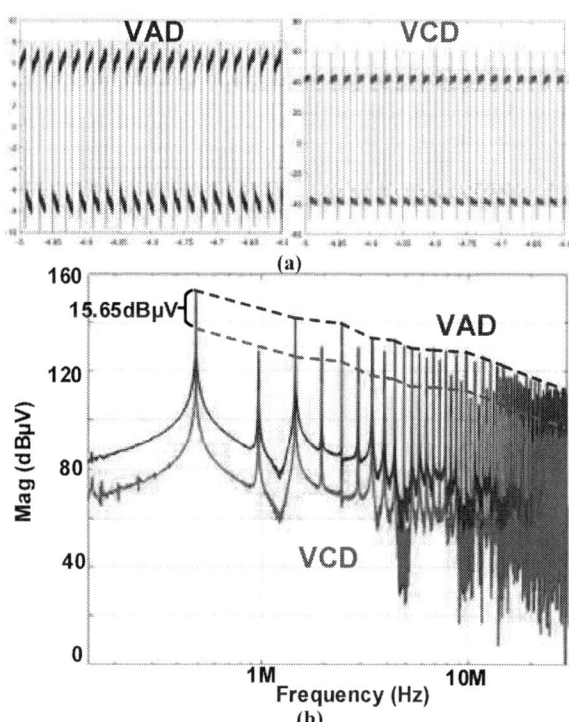

(a)

Figure 5: Waveforms and spectrums of balance capacitance

Since the value of Cadd is certain and controllable, the value of C2 can be solved with (2) and (5). And so does C3.

$$C_2 = \frac{k}{m-k}C_{add} \quad (6)$$

By using different Cadd, a relative constant C2 value can be calculated.

B. Modeling of the balance filter

Fig. 6 is the balance filter. Vs represents the noise source of the balance filter which is the switching node of the high frequency device. Vo is the output port and keeps open circuit. C1 is the intra-winding capacitance of L1. Since L2 has only one turn, no intra-winding capacitance exists in L2. M is the mutual inductance of L1 and L2. The current-voltage relationship in

balance network should maintain

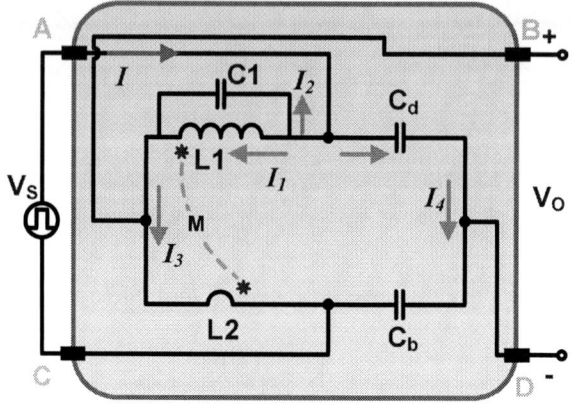

Figure 6: Balance filter structure

$$sL_1I_1 + sMI_3 = \frac{1}{sC_1}I_2 \quad (7)$$

$$V_s = \frac{1}{sC_1}I_2 + sL_2I_3 + sMI_1 \quad (8)$$

$$V_s = \left(\frac{1}{sC_d} + \frac{1}{sC_b}\right)I_4 \quad (9)$$

$$V_o = sL_2I_3 + sMI_1 - \frac{1}{sC_b}I_4 \quad (10)$$

Combining (7) – (10) together, the insertion voltage gain of balance filter is

$$\frac{V_o}{V_s} = H(s) = \frac{C_1C_b(L_1L_2-M^2)s^2+C_bL_2-C_dL_1+(C_b-C_d)M}{(C_d+C_b)(L_1L_2-M^2)s^2+(C_d+C_b)(L_1+L_2+2M)} \quad (11)$$

Substituting (2) into (11)

$$H(s) = \frac{mC_1(L_1L_2-M^2)s^2+mL_2-L_1+(m-1)M}{(1+m)(L_1L_2-M^2)s^2+(1+m)(L_1+L_2+2M)} \quad (12)$$

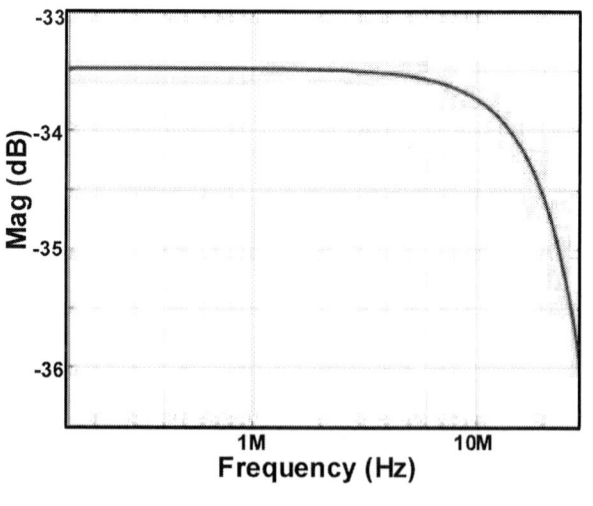

979-8-3315-1612-3/25 $31.00 © 2025 IEEE   1539

Fig.7 is the magnitude of the bode plot of (12) using the measured parameters. From the bode plot, the attenuation of the balance filter is 33.5dB at low frequency without any added capacitor to balance the bridge.

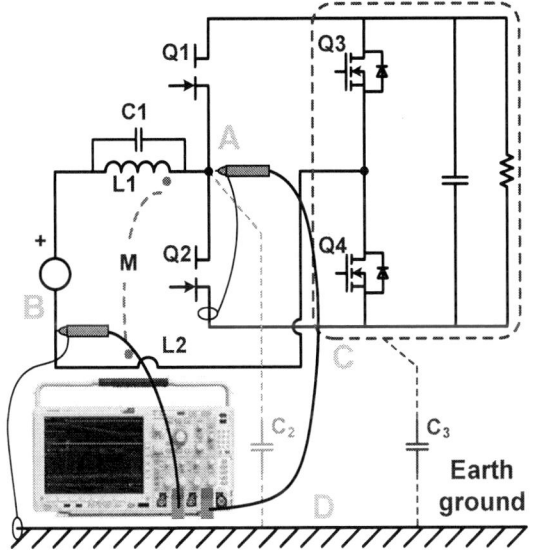

Figure 8: Input-output measurement method

III. RESULT VERIFICATION

Based on the analysis in the last section, the actual noise attenuation from the noise source to the output can be predicted using the modeled balance filter. The noise voltage VAC and VBD sampling setup is as Fig.8 shows. The converter setup is shown in Fig.9. Fig.10(a) is the sampled voltage VAC and VBD. To simplify the analysis, the converter is running in DCDC mode. After performing the FFT to the sampled voltage in MATLAB, the noise spectrum of AC (source side) and BD (output side) is shown in Fig.10(b). The blue spectrum VAC is the switching node noise spectrum, which can be considered as the 'original' noise. The red spectrum VBD is the 'filtered' noise. The noise attenuation from AC to BD is 33.5dBμV without any added capacitor at C2 or C3, which is almost the

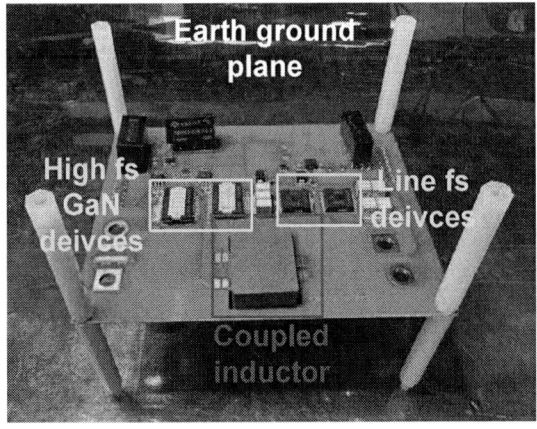

Figure 9: Converter setup

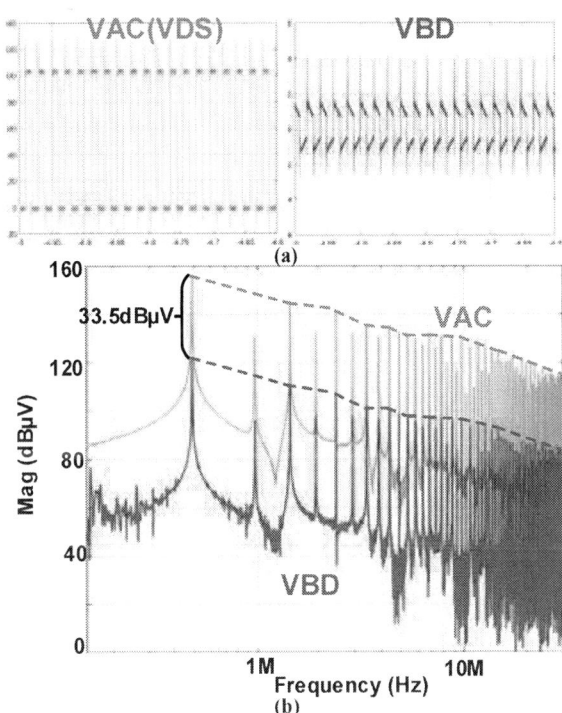

Figure 10: Input-output voltage waveforms and spectrums

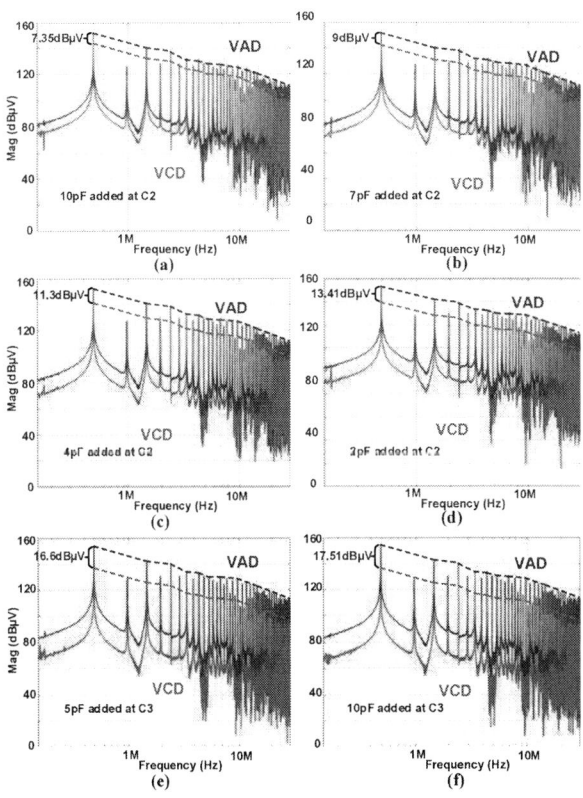

Figure 11: Balance capacitance spectrums with different Cadd

same as the insertion loss in Fig.7.

A. CM parasitic capacitance result calculation

Fig.11 is the FFT result of VAC and VCD after adding different capacitors in parallel to C2 or C3. Cadd is added to C2 in Fig.(a)(b)(c)(d) and is added to C3 in Fig.(e)(f). Based on the method proposed in last section using the exact value of added capacitor, the calculated C2 value is as Table 1 shows. It can be seen that the calculated C2 ranges from 6.3pF to 6.6pF with different Cadd, which is relatively stable within 5% error referring to 6.3pF.

TABLE I.   C2 CAPACITANCE

Fig. 11	Capacitance (pF)					
	(a)	*(b)*	*(c)*	*(d)*	*(e)*	*(f)*
Cadd	*9.987*	*7.16*	*4.015*	*1.863*	*4.911*	*9.987*
C2	6.34	6.34	6.33	6.6	6.46	6.61

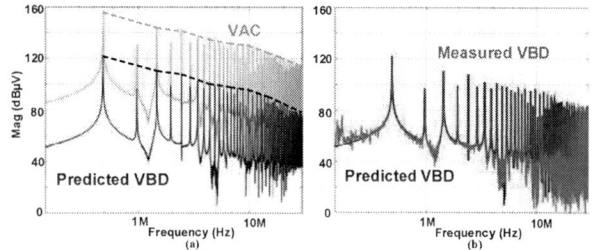

Figure 12: Prediction VS measurement

B. Balance filter attenuation prediction

With The derived insertion loss (12) and measured balance network parameters, the balance filter attenuation can be predicted. Fig.12 is the noise prediction from source (A-C) to load (B-D) without added capacitor. Fig.12(a) is the predicted noise from the noise source. The blue curve is the original noise spectrum VAC, and the black curve is the predicted noise spectrum VBD using MATLAB. Fig.12(b) is the comparison of measured VBD and predicted VBD.

Fig.13 is the comparison of spectrum measurements and predictions under different Cadds. Using the measured C2 capacitance and the inductor parameters in previous work, the noise prediction match the measurement very well.

IV. CONCLUSION

In this paper, a new method is proposed to extract the common mode parasitic PCB capacitance in balance network. By running the converter in the switching mode, the voltage spectrum of the CM parasitic capacitance is fully utilized to measure the value of CM capacitance. Filter concept is applied to both verify the correctness of the CM capacitance value and the accuracy of the predicted noise. Based on the derived balance filter model and extracted parameters, the noise is accurately predicted and the measurement fits the prediction very well throughout 30MHz.

REFERENCES

[1] X. Yang et al., "Dynamic RON free 1.2-kV vertical GaN JFET," IEEE Trans. Electron Devices, vol. 71, no. 1, pp. 720–726, Jan. 2024, doi:

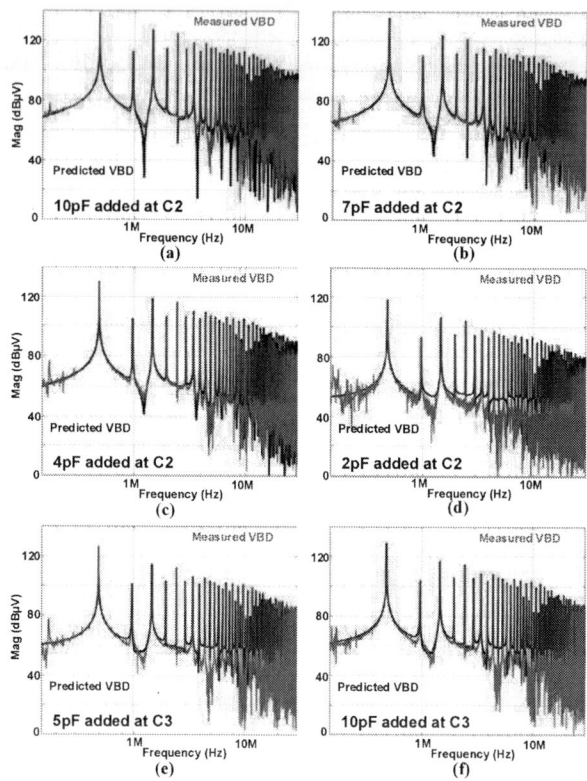

Figure 13: Prediction under different Cadd

10.1109/TED.2023.3338140.

[2] X. Yang, J. Liu, B. Wang, and G. Zhang, "Pulsed overcurrent capability of power semiconductor devices in solid-state circuit breakers: SiC MOSFET vs. Si IGBT," in Proc. IEEE Appl. Power Electron. Conf. Expo., Houston, TX, USA, 2022, pp. 966–973.

[3] Xin Yang, Guannan Shi, Liyang Jin, Yuan Qin, Matthew Porter, Xiaoting Jia, Dong Dong, Linbo Shao, Yuhao Zhang, "Ultrafast Optically Controlled Power Switch: A General Design and Demonstration With 3.3 kV SiC MOSFET," IEEE Trans. Electron Devices, Early Access, pp. 1–6, Nov. 2024, doi: 10.1109/TED.2024.3485018.

[4] X. Yang et al., " Evaluation and MHz Converter Application of 1.2-kV Vertical GaN JFET," in IEEE Transactions on Power Electronics, vol. 39, no. 12, pp. 15720-15731, Aug. 2024, doi: 10.1109/TPEL.2024.3445667.

[5] R. Zhang et al., "Switching Performance Evaluation of 650 V Vertical GaN Fin JFET," 2023 IEEE Applied Power Electronics Conference and Exposition (APEC), Orlando, FL, USA, 2023, pp. 2515-2519, doi: 10.1109/APEC43580.2023.10131473

[6] Y. Wang, M. Porter, M. Xiao, A. Lu, N. Yee, I. Kravchenko, B. Srijanto, K. Cheng, H. Y. Wong, and Y. Zhang, "Implanted guard ring edge termination with avalanche capability for vertical GaN devices," IEEE Trans. Electron Devices, vol. 71, no. 3, pp. 1481–1487, Mar. 2024, doi: 10.1109/ted.2023.3321010.

[7] Y. Wang et al, "Planar Implantation Edge Termination for Vertical GaN Power Devices," in Proc. IEEE 10th Workshop Wide Bandgap Power Devices Appl., 2023, pp. 1–5, doi: 10.1109/WiPDA58524.2023.10382233.

[8] Y. Jiang et al., "Efficiency Optimization of WPTS Based on Variable Angle Phase Shift Control for EV charging," 2020 IEEE 9th International Power Electronics and Motion Control Conference (IPEMC2020-ECCE Asia), Nanjing, China, 2020, pp. 747-752, doi: 10.1109/IPEMC-ECCEAsia48364.2020.9368148.

[9] Y. Jiang et al., "A Dynamic Efficiency Optimization Method under ZVS Conditions in the Series-Series Type Wireless Power Transfer System,"

2020 IEEE Energy Conversion Congress and Exposition (ECCE), Detroit, MI, USA, 2020, pp. 995-1001, doi: 10.1109/ECCE44975.2020.9235452.

[10] Y. Li et al., "A Universal Parameter Design Method of Resonant Coils Under Multiple Boundary Constrains for Wireless Power Transfer Systems," 2023 IEEE Energy Conversion Congress and Exposition (ECCE), Nashville, TN, USA, 2023, pp. 6489-6496, doi: 10.1109/ECCE53617.2023.10362919.

[11] L. Lin, J. Zhang, and S. Shao, "A Variable Switching Frequency Multimode Control Scheme for Single-Phase Grid-Tied Multilevel PV Microinverters," IEEE Transactions on Power Electronics, 2023.

[12] C. Li, X. Yu, Y. Li, H. Dang, J. Liu and S. Du, "A Variable Frequency Modulation Strategy for Current-Fed Dual-Active-Bridge Converter to Expand ZVS Range," 2024 IEEE Applied Power Electronics Conference and Exposition (APEC), Long Beach, CA, USA, 2024, pp. 468-473, doi: 10.1109/APEC48139.2024.10509332.

[13] M. Wu et al., "A Compact Coupler With Integrated Multiple Decoupled Coils for Wireless Power Transfer System and its Anti-Misalignment Control," in IEEE Transactions on Power Electronics, vol. 37, no. 10, pp. 12814-12827, Oct. 2022, doi: 10.1109/TPEL.2022.3166888.

[14] M. Wu et al., "Modeling of Litz-Wire DD Coil With Ferrite Core for Wireless Power Transfer System," in IEEE Transactions on Power Electronics, vol. 38, no. 5, pp. 6653-6669, May 2023, doi: 10.1109/TPEL.2022.3222228.

[15] J. Zhang, X. Huang, X. Wu, and Z. Qian, "A high efficiency flyback converter with new active clamp technique," IEEE Trans. Power Electron., vol. 25, no. 7, pp. 1775–1785, Jul. 2010.

[16] M. Wu, X. Yu, X. Yang, W. Chen and L. Wang, "An Efficiency Optimization Method for the Multiple Coils WPT System Against the Pad Misalignment," in IEEE Transactions on Transportation Electrification, doi: 10.1109/TTE.2024.3410676

[17] T. Yuan, F. Jin, Z. Li, C. Zhao and Q. Li, "Design of an Integrated Transformer With Parallel Windings for a 30-kW LLC Resonant Converter," in IEEE Transactions on Power Electronics, vol. 38, no. 11, pp. 14317-14333, Nov. 2023, doi: 10.1109/TPEL.2023.3291954.

[18] T. Yuan, F. Jin and Q. Li, "Analysis and Comparison of Integrated Planar Transformers for 22-kW On-Board Chargers," in IEEE Transactions on Power Electronics, vol. 39, no. 9, pp. 11368-11385, Sept. 2024, doi: 10.1109/TPEL.2024.3410878.

[19] L. Zhu, H. Wu, P. Xu, H. Hu, and H. Ge, "A novel high efficiency high power density three-port converter based on interleaved half-bridge converter for renewable energy applications," in Proc. IEEE Energy Convers. Congr. Expo., Pittsburgh, PA, USA, 2014, pp. 5085–5091.

[20] J. Yang, X. Wu, F. Muhammad, and Z. Deng, "External magnetic field minimization for the integrated magnetics in series resonant converter," IEEE Trans. Power Electron., vol. 37, no. 1, pp. 498–508, Jan. 2022, doi: 10.1109/TPEL.2021.3095491.

[21] T. Yuan, F. Jin, Z. Li and Q. Li, "High Frequency High Power Integrated Transformer Design for Resonant Converters with SiC Devices," 2022 IEEE 9th Workshop on Wide Bandgap Power Devices & Applications (WiPDA), Redondo Beach, CA, USA, 2022, pp. 170-175, doi: 10.1109/WiPDA56483.2022.9955265.

[22] T. Yuan, F. Jin, Z. Li and Q. Li, "Current Sharing Analysis of a High Power Transformer with Parallel Windings," 2023 IEEE Applied Power Electronics Conference and Exposition (APEC), Orlando, FL, USA, 2023, pp. 1551-1556, doi: 10.1109/APEC43580.2023.10131421.

[23] T. Yuan, F. Jin and Q. Li, "A 22-kW On-Board Charger (OBC) with an Integrated Planar Inductor and Transformer," 2024 IEEE Applied Power Electronics Conference and Exposition (APEC), Long Beach, CA, USA, 2024, pp. 1300-1304, doi: 10.1109/APEC48139.2024.10509419.

[24] P. Yao, X. Jiang, P. Xue, S. Li, S. Lu, and F. Wang, "Design optimization of medium-frequency transformer for DAB converters with DC bias capacity," IEEE J. Emerg. Sel. Topics Power Electron., vol. 9, no. 4, pp. 5043–5054, Aug. 2021.

[25] Y. Liu, H. Wu, Z. Ge, and G. Ji, "Magnetic integration for multiple resonant converters," IEEE Trans. Ind. Electron., vol. 70, no. 8, pp. 7604–7614, Aug. 2023, doi: 10.1109/TIE.2022.3229381.

[26] J. Zhang et al., "A three-port LLC resonant DC/DC converter," IEEE J. Emerg. Sel. Topics Power Electron., vol. 7, no. 4, pp. 2513–2524, Dec. 2019.

[27] Z. Zhang, J. Zhang, and S. Shao, "A novel trapezoidal wave control method for a single-phase grid-tied T-type inverter," IEEE Trans. Power Electron., vol. 36, no. 4, pp. 4711–4722, Apr. 2021.

[28] D. Gu, Z. Zhang, Y. Wu, D. Wang, H. Gui, and L. Wang, "High efficiency LLC DCX battery chargers with sinusoidal power decoupling control," in Proc. IEEE Energy Convers. Congr. Expo., Milwaukee, WI, USA, 2016, pp. 1–7.

[29] Z. Zhang, J. Zhang, S. Shao, and J. Zhang, "A high-efficiency singlephase T-type BCM microinverter," IEEE Trans. Power Electron., vol. 34, no. 1, pp. 984–995, Jan. 2019.

[30] Q. Yang, A. Nabih, R. Zhang, Q. Li and Y. Zhang, "A Converter Based Switching Loss Measurement Method for WBG Device," 2023 IEEE Applied Power Electronics Conference and Exposition (APEC), Orlando, FL, USA, 2023, pp. 8-13, doi: 10.1109/APEC43580.2023.10131509.

[31] Qiuzhe Yang, Feng Jin, Qiang Li, "An Accurate Temperature-Based Method for Fast Switching Loss Extraction of WBG Device," 2024 IEEE Applied Power Electronics Conference and Exposition (APEC), Los Angeles, CA, USA, 2024, pp. 2183-2187, doi: 10.1109/APEC48139.2024.10509222

[32] T. -S. Li, M. Ngo, R. Burgos and D. Dong, "Modeling and Analysis of Voltage Overshoot in Bidirectional Phase-Shift Full Bridge Converters," 2024 IEEE Sixth International Conference on DC Microgrids (ICDCM), Columbia, SC, USA, 2024, pp. 1-7, doi: 10.1109/ICDCM60322.2024.10665239.

[33] T. -S. Li, Y. -H. Yang, C. -A. Cheng and Y. -M. Chen, "A Variable DC-Link Voltage Determination Method for Motor Drives with SiC MOSFETs," 2020 IEEE Workshop on Wide Bandgap Power Devices and Applications in Asia (WiPDA Asia), Suita, Japan, 2020, pp. 1-6, doi: 10.1109/WiPDAAsia49671.2020.9360266.

[34] Y. -T. Huang, C. -C. Yang, T. -S. Li and Y. -M. Chen, "A Feedforward Voltage Control Strategy for Reducing the Output Voltage Double-Line-Frequency Ripple in Single-Phase AC–DC Converters," in IEEE Journal of Emerging and Selected Topics in Power Electronics, vol. 9, no. 6, pp. 6605-6612, Dec. 2021, doi: 10.1109/JESTPE.2021.3083258.

[35] X. Wu, H. Chen, and Z. Qian, "1-MHz LLC resonant dc transformer (DCX) with regulating capability," IEEE Trans. Ind. Electron., vol. 63, no. 5, pp. 2904–2912, May 2016.

[36] X. Chen, G. Son, F. Jin, and Q. Li, "A microcontroller-based high efficiency critical conduction mode control for GaN-based totem-pole PFC," in Proc. IEEE 22nd Workshop Control Model. Power Electron., 2021, pp. 1–7.

[37] G. Fan, X. Wu, T. Liu, and Y. Xu, "High-efficiency high-density MHz cellular DC/DC converter for on-board charger," IEEE Trans. Power Electron., vol. 37, no. 12, pp. 15666–15677, Dec. 2022.

[38] T. Liu, X. Wu, and S. Yang, "1 MHz 48–12 v regulated DCX with single transformer," IEEE J. Emerg. Sel. Topics Power Electron., vol. 9, no. 1, pp. 38–47, Jan. 2021.

[39] Tyler McGrew, Xingyu Chen, and Qiang Li, " PCB Inductor with Integrated Shielding to Contain Switching Electric Field and Reduce CM Noise," in IEEE Transactions on Power Electronics, Early Access, pp. 1-14, Sep. 2024, doi: 10.1109/TPEL.2024.3463486.

[40] Q. Huang, Y. Li, Z. Ma, Y. Yang, Y. Lai, and S. Wang, "RLC Balance Technique of Transformer to Reduce CM EMI for Isolated DC-DC Converters," IEEE Energy Conver. Congress and Exposition (ECCE), 2023.

[41] Q. Huang, Y. Yang, Y. Lai, Z. Ma, S. Wang, " A Survey of CM EMI Modeling and Reduction Technique of Transformer for Isolated Converters," 2024 IEEE Applied Power Electronics Conference and Exposition (APEC), Los Angeles, CA, USA, 2024, pp. 1484-1490, doi: 10.1109/APEC48139.2024.10509196

[42] Y. Lai, Y. Yang, Z. Ma, Q. Huang, S. Wang, and Z. Luo, "Development of simulation prediction techniques for low frequency emissions," in Proc. IEEE Appl. Power Electron. Conf. Expo. (APEC), Mar. 2023, pp. 2690–2696.

[43] Z. Ma, S. Wang, Q. Huang, and Y. Yang, "A review of radiated EMI research in power electronics systems," IEEE J. Emerg. Sel. Topics Power

Electron., to be published, doi: 10.1109/JESTPE.2023.3335972

[44] Q. Huang, Y. Yang, Z. Ma, Y. Lai, and S. Wang, "Transformer structure of bifilar primary winding with advanced common mode noise attenuation performance for isolated DC–DC converters," in Proc. IEEE Appl. Power Electron. Conf. Expo. (APEC), Mar. 2023, pp. 441–448.

[45] Y. Yang, Q. Huang, Y. Lai, Z. Ma, Y. Liu and S. Wang, "Analysis and Modeling of the Near Magnetic Field Distribution of Toroidal Inductors," 2023 IEEE Symposium on Electromagnetic Compatibility and Signal/Power Integrity (EMC+SIPI), Grand Rapids, MI, USA, 2023, pp. 573-578, doi: 10.1109/EMCSIPI50001.2023.10241669.

[46] S. Wang, P. Kong, and F. C. Lee, "Common mode noise reduction for boost converters using general balance technique," IEEE Trans. Power Electron., vol. 22, no. 4, pp. 1410–1416, Jul. 2007.

[47] S. Wang, F. C. Lee and Q. Li, "Improved Balance Technique for Common-Mode Noise Suppression of PCB-Based PFC," in IEEE Transactions on Power Electronics, vol. 37, no. 4, pp. 4174-4182, April 2022, doi: 10.1109/TPEL.2021.3124505.

[48] Q. Yang, S. Wang, and Q. Li, "Modeling and Analysis of The Balance Network for Common Mode EMI Noise Suppression," IEEE Energy Conver. Congress and Exposition (ECCE), 2024.

[49] T. McGrew, S. Wang, and Q. Li, "A Novel Technique to Measure Parasitic Capacitances Affecting CM Noise Emissions," in 2024 IEEE Applied Power Electronics Conference and Exposition (APEC), 2024, pp. 1498–1505.

A 660W, 96% Efficiency 3D Heterogeneously Integrated Digital DC/DC Power Module for Vertical Power Delivery

Haoyu Wang
School of Integrated Circuits
Tsinghua University
Beijing, China
hy-
wang24@mails.tsinghua.edu.cn

Xuliang Wang
School of Integrated Circuits
Tsinghua University
Beijing, China
xwangef@tsinghua.edu.cn

Yan Wang
School of Integrated Circuits
Tsinghua University
Beijing, China
wangy46@tsinghua.edu.cn

Xiaosen Liu
School of Integrated Circuits
Tsinghua University
Beijing, China
liuxiaosen@tsinghua.edu.cn

Abstract—**This work demonstrates the design, modeling and fabrication of a digital DC/DC power module featuring a record-high output power of 660 W and 96.2% efficiency for high-performance computing (HPC). To achieve vertical power delivery (VPD) in a small form factor, advanced 3DIC packaging technology is employed by heterogeneously integrating power switches, drivers and control ICs with plastic molding. To address the congested power routing in conventional lateral floorplans, modular stacking and a redistribution layer (RDL) are adopted to build a 3D high-density power delivery network (PDN) with minimal IR losses. The electro-thermal and stress properties are modeled and analyzed to ensure feasibility. Two sets of compensation algorithms are implemented to balance the tradeoff between fast response, stable output current and small voltage ripple. While dynamically adjusting the output, the power module digitally monitors the safety status and reacts with protection circuit. Experiment results from the prototype system demonstrates a power density of 1468 W/in³, output ripple less than 1%, 32.8 ms response time, and significantly improved thermal management and stress control.**

Keywords—*Vertical Power Delivery, 3DIC Packaging, Power Module, Digital Control.*

I. INTRODUCTION

Digital power supplies with high precision, stability, and modularity are widely utilized across industry, aerospace, and emerging high-performance computing (HPC) datacenters [1] [2]. However, traditional 2D board-level solutions struggle to deliver adequate power within compact form factors while maintaining the high efficiency for next-generation AI processors [3][4]. Thus, vertical power delivery (VPD) technology has been proposed to heterogeneously integrate (HI) various circuit components in three-dimensional space and package them into a monolithic 3DIC, achieving minimum IR losses, high power density and reliability simultaneously [5]. Despite significant advancements in reliable design methodologies and low-power operation, existing solutions have not yet overcome bottleneck of output power [6]-[8].

The 48 V-to-1 V power conversion paradigm has become a consensus in power supply applications for high-performance XPUs. Fig. 1 illustrates two approaches: traditional board-level converters and point-of-load (PoL) converters [9]. Board-level solutions are constrained by physical size and therefore fail to

meet the stringent requirements for conversion efficiency and response speed. Conversely, PoL converters, whether designed from on-chip or board-level, provide substantial performance improvements [10]-[16]. Furthermore, incorporating VPD technology into PoL architectures enhances the benefits of 3D integration particularly at the PoL level.

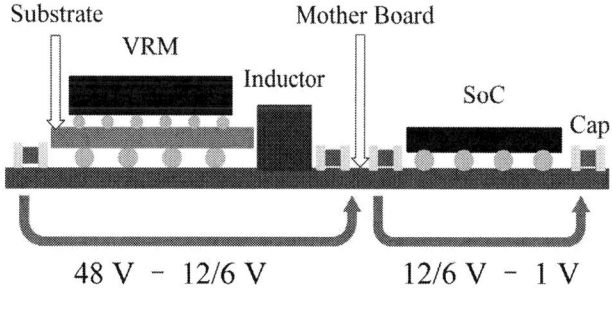

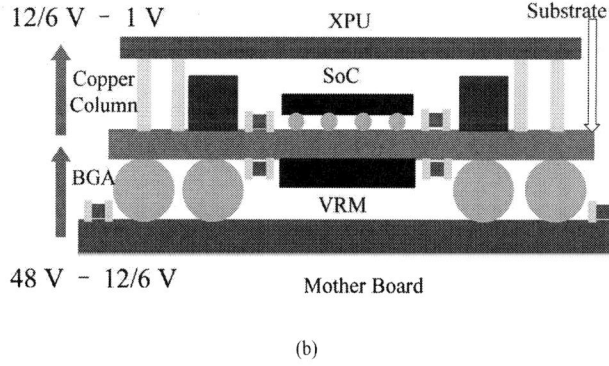

Fig. 1. Comparison of two power supply schemes: (a) traditional board-level and (b) PoL.

PoL-VPD strategies have garnered significant attention in emerging HPC datacenters applications, typically categorized into single-stage and two-stage architectures. Single-stage architectures involve onboard PoL converters that offer a highly integrated power supply solutions, rendering them promising for practical applications. However, these architectures are limited by control bandwidth. In contrast, two-stage architecture connect two DC/DC power modules, with an intermediate voltage usually being 5 V or 12 V. The first stage converts the 48 V power supply to the intermediate voltage, facilitating a seamless transition between traditional board-level solutions and single-stage architectures, provided the DC/DC module performance is sufficient.

Focusing on the first stage of the PoL architecture, this paper proposes a digital DC/DC power converter employing 3DIC technology, which heterogeneously integrates control, drive, and communication modules within a single package. This compact form factor enables the practical application of VPD technology, significantly improving the PPA metrics, particularly in terms of power density and reliability. To highlight the advantages of vertical packaging in power supply applications, this research evaluates large output voltages such as 24 V and 30 V, alongside 12 V, to determine the system's maximum power density and peak efficiency. Furthermore, high-power reliability test is conducted, including assessments of temperature, stress, and waveforms under full output conditions.

II. CIRCUIT AND PACKAGE DESIGN

This work aims to achieve extreme performance and board applicability. On the one hand, the power module has competitive performance; on the other hand, it should have a wide input voltage range and can selectively adjust the output power. The technical and physical specifications of the designed power module are shown in Table I.

TABLE I. POWER MODULE SPECIFICATIONS

Specifications	This work
Dimensions (mm³)	63.5×14.5×8
Weight (g)	<40
Power density (W/in³)	>1400
Peak efficiency (%)	>96%
Voltage ripple (%)	<1%
Input voltage range (V)	12-50
Output voltage range (V)	5-40
Output current range (A)	0-22
Max output power (W)	>650
Response time (ms)	<40
Full load temperature (℃)	<100
Working time (s)	>240

The overall architecture of the DC/DC power supply is shown in Fig. 2, which includes power stage, controller, driver, periphery bias and communication with an external digital bus. To maximize the benefits of 3D integration, most of the circuit modules are packaged together, leaving only a few larger passive components and isolated power.

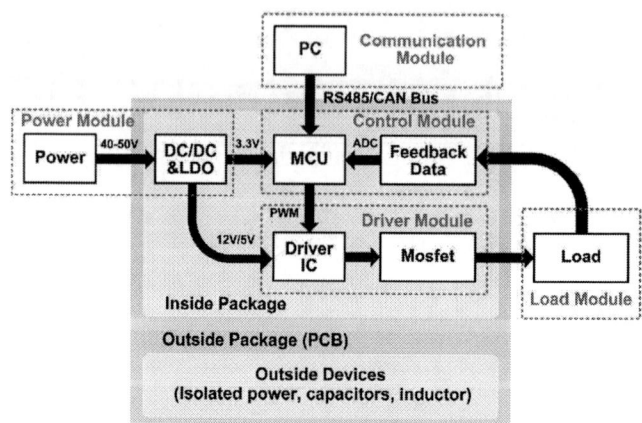

Fig. 2. Schematic of the proposed package module.

The regulation principle is illustrated in Fig. 3 and is implemented using a 12-bit ADC and MCU. The circuit is structured as a half-bridge power module with two independently operating drive signals. Closed-loop control is achieved through a double-loop feedback system for voltage and current, where the algorithms for each loop function independently. The primary and secondary relationships between the loops are determined based on command information received from the PC. The algorithm parameters strictly adhere to stability criterion conditions, ensuring a balance between dynamic response and overall system stability. To enhance the safety of the power module, a threshold calculation model for current and voltage is incorporated. This model enables the system to immediately halt output in the event of an uncontrollable increase in current or voltage, thereby protecting the module from potential damage.

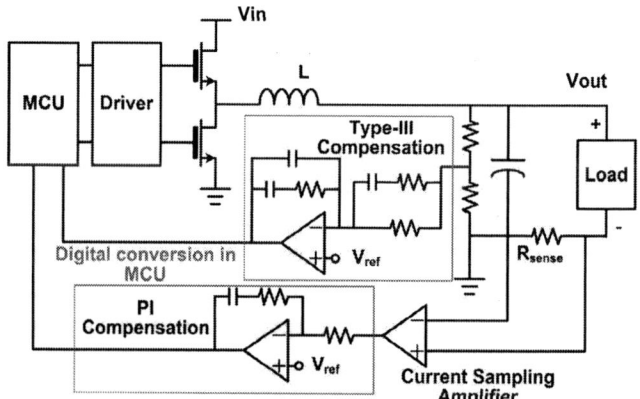

Fig. 3. Regulation principle.

The digitally calculated PWM signals are fed into the driver for controlling two MOSFETs [17] [18], and the open-loop transfer function is given by,

$$G_{vd}(s) = G_v(s)G_sG_{rc}G_{dly}(s)G_d = \frac{V_{in}}{LCs^2+\frac{L}{R}s+1} \times A_s \times \frac{1}{R_{rc}C_{rc}+1} \times e^{-s \times T_s} \times A_d \quad (1)$$

where $G_v(s)$ is the power transfer function, G_s is the sampling transfer function, A_s is related to the voltage divider or magnification, G_d is the transfer function from the sampled digital quantities to the output digital quantities, and A_d is usually obtained according to the subsequent control algorithm.

Based on the specifications in Table I and the circuit structure in Fig. 3, the chip and power components are selected as shown in Table II.

TABLE II.   MAIN COMPONENT LISTING

Component	Manufacture and Part number	Parameters
MCU	ST STM32F334C8T6	ARM, 37 I/Os
MOSFET	ONSEMI NTBLS1D5N10MC	100 V, 1.5 mΩ
Inductors	CODACA CSBX1275	6.8 µH、25.5 A
Isolated power	Taisko B1212S-1W	1000 Vdc
Driver	Silicon Labs SI8233BD-D-ISR	4 A, 8 MHz
DC/DC	XLSEMI XL7005A	400 mA, 150 kHz
LDO	ADI ADP150AUJZ-3.3	150 mA
Current-sense amplifier	TI INA180A3IDBVR	100 V/V, ±1%

The core design discussed in this paper focuses on the first stage module of the PoL power supply. This module performs the voltage conversion and supplies power through a 6/12 V SoC chip, delivering power to the HPC computer chip at 1 V or even lower. To maximize the potential of the 3DIC package technology, the entire power module is stacked within a constrained height limit. The 3D stacked structure of the package design illustrated in Fig. 4 was developed with careful consideration of wiring layout, thermal management, and the independence of power, analog, and digital domains. To optimize the fan-out capability of LDOs and amplifiers, two RDL are utilized and interconnected via wire bonding. High voltage breakdown and leakage current pose a crucial electrical overstress (EOS) risk to small-signal processing circuits. Therefore, two polymer layers are inserted to isolate different voltage domains. The connection between the substrate and the PCB is accomplished by reflow soldering, and the bump has two shapes: one is the ball grid array with a diameter of 760 µm; the other is the 3.5×3 mm² pad. Considering the warpage challenges associated with the complex vertical structure, full-chip molding is applied to encapsulate the entire module into a monolithic package as small as 63.5×14.5×8 mm³, which offers three major advantages: (1) it fully utilizes the three-dimensional space and improves the power density of the module; (2) the current sampling device can be directly connected to the MCU through lead bonding in narrowed space, avoiding interference from the power loop within the package; and (3) the plastic sealing structure not only improves the reliability against stress failure but also shields sensitive signals from influences out of package [19].

The 3D global and sectional views of the design module are shown in Fig. 5(a) and Fig. 5(b). The copper column is used as power supply inputs, and solid-state capacitors are usually added between neighboring copper column to prevent voltage overshoot from damaging the device. The connection port on the left can be used for half-duplex communication with the PC through the terminal block, and the communication method used in this paper is RS485 or CAN bus. The isolated power

can generate two drive signals in an isolated manner from a single drive signal, which is used to turn on and off the MOS. The right end of the substrate is placed with the inductor (molding choke) and output capacitor of the power loop.

Thermal management is a key focus in this paper, achieved through the following strategies: (1) Material Selection: Plastic sealing materials with high thermal conductivity are used to enhance heat dissipation. (2) Layout Optimization: The layout is designed to maintain sufficient spacing between the heat-generating components, and a large number of copper-filled vias are incorporated to create efficient thermal conductivity paths to the bottom copper layer. (3) Cooling Enhancements: A cooling fan and water-cooling system are employed to effectively mitigate temperature rise during high-power operation.

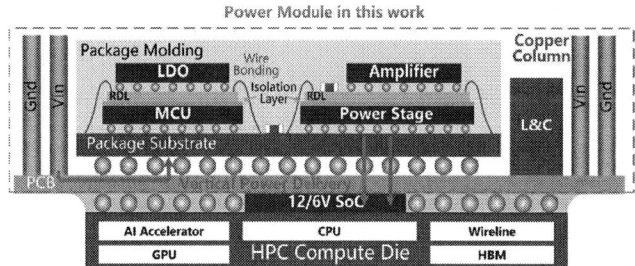

Fig. 4.   The package structure of power module.

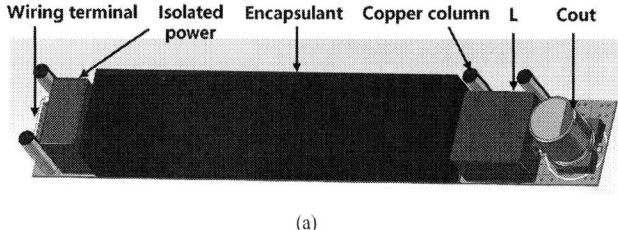

(a)

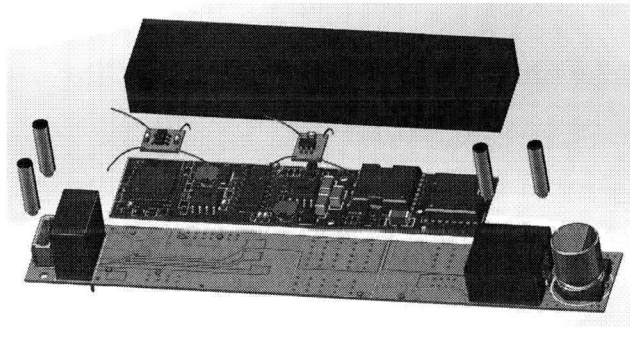

(b)

Fig. 5.   (a) The 3D overall structure and (b) the assembly diagram.

III. SIMULATION RESULTS

The simulation process is divided into two segments: loop parameters and thermal analysis, each aimed at verifying the performance and reliability of the power module.

The loop parameters are derived from the technical specifications listed in Table I, ensuring the selection of compliant power devices and determining suitable zeros and poles based on the compensation data. The parameter settings for constant voltage control and constant current control are detailed in Table III, respectively.

TABLE III.   PARAMETERS OF TRANSFER FUNCTION

Parameters	Voltage control	Current control
Input Capacitance (μF)	100 (100 V)	
Output Capacitance (μF)	220 (50 V)	
ESR (mΩ)	10	
Inductor (μF)	6.8	
DCR (mΩ)	60	
Frequency (kHz)	255	
Digital adjustment range	1-18039 (4.6 GHz/255 kHz)	
Sampling ratio	1/20-3/20	1/8-3/7
Type of compensation	Type-III	PI
Zero-polar point (Hz)	z1: 1 k; z2: 1 k; p1: 100 k; p2: 100 k	z1: 1.5 k; z2: 5 k

The zero-pole data from Table I is used to construct the compensation transfer function G_c, and the constant voltage control mode is:

$$G_c = \frac{2.243 \cdot 10^{16} \cdot s^2 + 2.817 \cdot 10^{20} \cdot s + 8.845 \cdot 10^{23}}{3.944 \cdot 10^7 s^3 + 4.956 \cdot 10^{13} \cdot s^2 + 1.557 \cdot 10^{19} \cdot s} \quad (2)$$

the constant current control mode is:

$$G_c = \frac{25000 \cdot s^2 + 1.021 \cdot 10^9 \cdot s + 7.395 \cdot 10^{12}}{2.958 \cdot 10^8} \quad (3)$$

multiplying the compensation function with Equation 1, the frequency response must have sufficient phase margins. The power module in this paper is digitally controlled, so the analog loop needs to be discretized. The S-domain can be changed to a Z-domain using the Tustin transform:

$$s = \frac{2}{T_S} \left(\frac{z-1}{z+1}\right)^m \quad (4)$$

where T_S is the sampling time.

The transfer function G_z of the discretized voltage mode is:

$$G_z = \frac{229.5 \cdot z^3 - 218.4 \cdot z^2 - 229.3 \cdot z + 218.6}{z^3 - 0.7976 \cdot z^2 - 0.1922 \cdot z - 0.01024} \quad (5)$$

the transfer function G_z of the discretized current mode is:

$$G_z = \frac{46.62 \cdot z^2 - 86.15 \cdot z + 39.72}{z^2 - 1} \quad (6)$$

The digital power supply model is built using PSIM software to verify that the compensation data meets the requirements, as shown in Fig. 6 for the voltage mode.

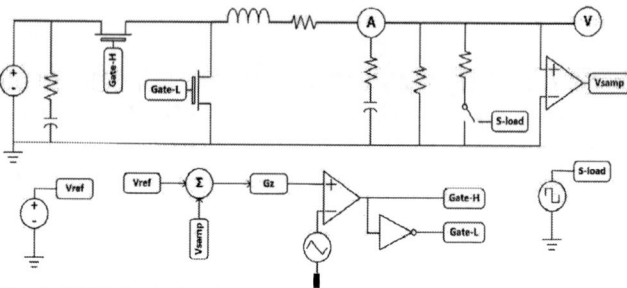

Fig. 6.   PSIM simulation circuit.

The simulation results of constant voltage control are shown in Fig. 7(a), achieving a voltage adjustment response time of 150 μs and a current adjustment accuracy of 97.8%. In the constant current mode, it is reasonable to slow down the current adjustment rate appropriately considering the hidden danger of parasitic inductance and voltage overshoot. The simulation results of constant current control are shown in Fig. 7(b), where the current can be smoothly controlled to the initial value after a sudden change with high accuracy.

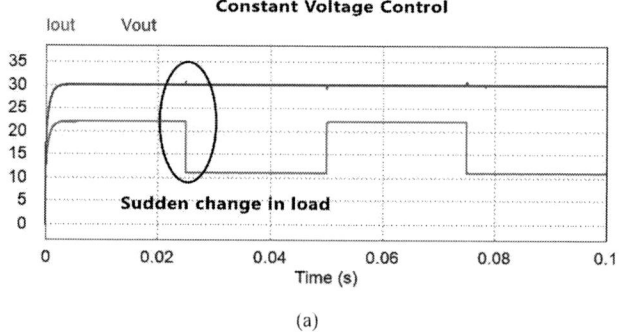

(a)

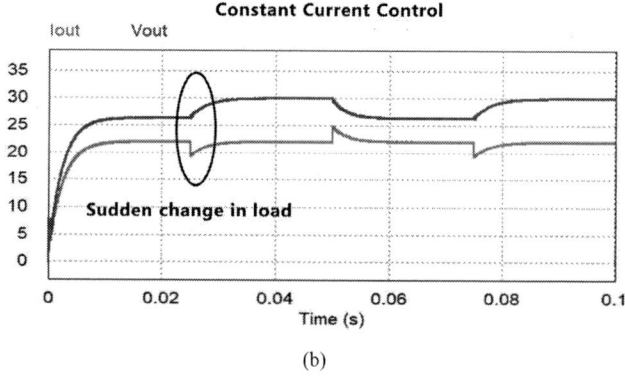

(b)

Fig. 7.   (a) Constant voltage control and (b) constant current control.

Due to the high power density, the temperature performance of the power module needs to be analyzed by constructing a hybrid electro-thermal (HET) model to address heat dissipation as follows,

$$\nabla[k(x,y,z)\nabla(x,y,z)] = -J \times E(x,y,z) \quad (7)$$

where $k(x,y,z)$ is the thermal conductivity of the solid in three directions, J is the current density and $E(x,y,z)$ is the electric field distribution.

The HET model was implemented in a numerical solution tool, yielding a theoretical maximum temperature of 85 °C. To verify the model's accuracy, a simulation was conducted using Sigrity-Power DC, which produced a highest observed temperature of 87.3 °C, as shown in Fig. 8. This corresponds to a theoretical accuracy of 97.4%. The minor discrepancy is attributed to slight inaccuracies in estimating parasitic inductance and resistance. However, the additional heat generated from these factors is negligible, confirming the reliability of the model.

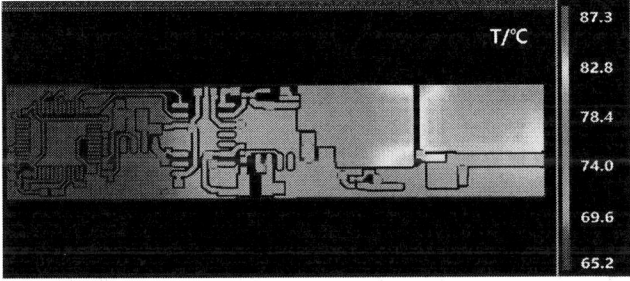

Fig. 8. Temperature simulation result.

Temperature rise and coefficient of thermal expansion (CTE) mismatch between different materials can lead to excessive deformation within the module. Precisely identifying the encapsulated stress scenario is a complex process; therefore, multilayer stacking theory is proposed to characterize the parameters of various substrates. Compared to the original stress analysis of a single geometry, the stacked model reflects the practical scenario in a 3D package.

$$\{\sigma\} = [Q]_k[\{\varepsilon\} - \{\alpha\}_k \Delta T] (k = 1,2,\cdots,n) \qquad (8)$$

where $\{\sigma\}$ is the set of stresses, $[Q]_k$ is the stiffness matrix, $\{\varepsilon\}$ is the set of strains, α is the CTE, and ΔT is the value of temperature change. The multiphysics field simulation of thermal-force follows the Eq. (8) to construct the multilayer board stacking structure as shown in Fig. 9. The overall stress value does not exceed 10 μm, and there is no significant deformation or fracture observed in the substrate of the current design size.

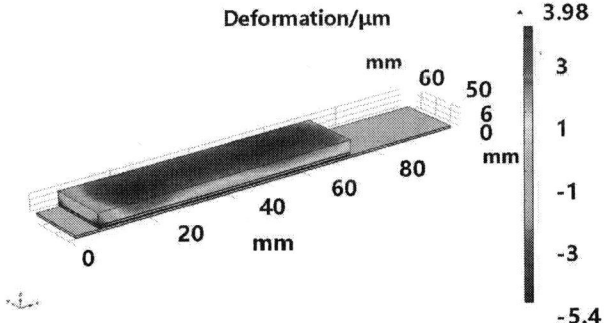

Fig. 9. Deformation simulation result.

IV. EXPERIMENTAL RESULTS

The built experimental platform is shown in Fig. 10. Commands are input from the upper computer via RS485 to select the operating mode and control the output voltage and current values.

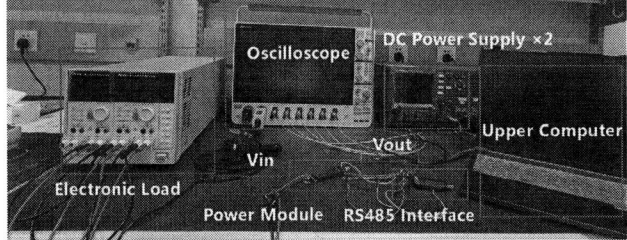

Fig. 10. Experiment platform for the high-power measurement.

First, use the oscilloscope to characterize the drive signals of the high-side and low-side MOS, as shown in Fig. 11(a). The drive signals have a certain dead time with no obvious overshoot and ringing, and the rising and falling edges are rapid, which reduces the switching losses. The response time reflects whether the power module can be adjusted quickly on command in any situation. Capturing the sending information from the PC as the starting point of calculation (usually the starting frame of valid data is low), the output voltage increases rapidly until it stabilizes as the ending point, and the measurement result is 32.8 ms as shown in Fig. 11(b). Voltage ripple not only affects the stability and efficiency of the power supply, but also affects the performance and lifetime of the entire system, and is especially important in relatively high output voltage scenarios. This paper measured the voltage ripple of the power supply module in a high power output scenario, as shown in Fig. 11(c), with a value of 270 mV, which is only 0.9% of the output value.

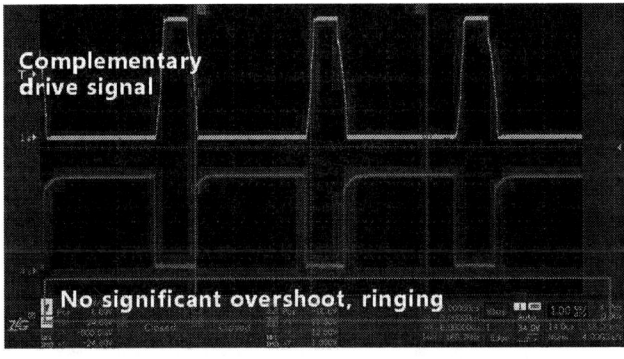

(a)

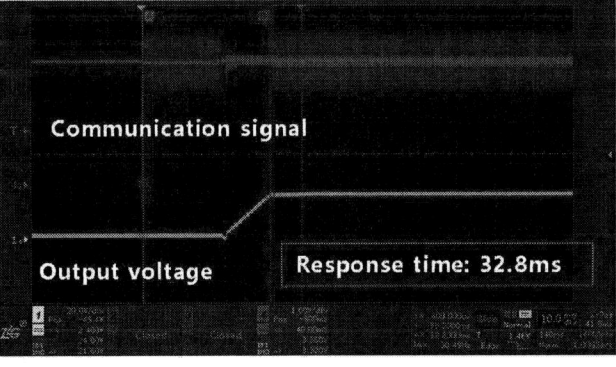

(b)

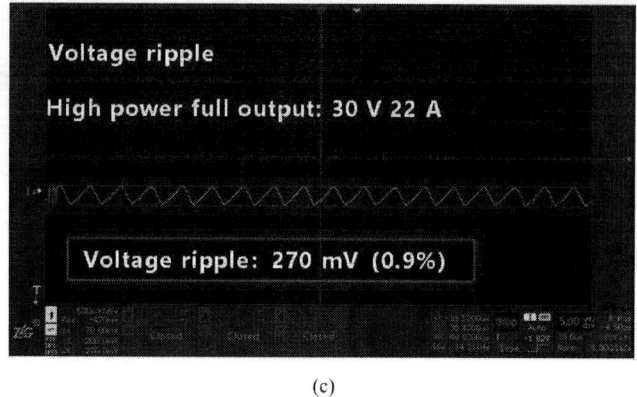

(c)

Fig. 11. Measured (a) drive signal, (b) response time, and (c) output voltage ripple.

The power module has been tested up to 22 A output current. The efficiency of output voltages of 12 V, 24 V and 30 V at 2-22 A were tested at an input voltage of 48 V with the input command, respectively. The results are shown in Table IV for the output power and efficiency results at three output voltages.

TABLE IV.　PARAMETERS OF TRANSFER FUNCTION

I_{out}	12 V Efficiency (power)	24 V Efficiency (power)	30 V Eifficiency (power)
2	74.5% (24 W)	85.6% (48 W)	88.2% (60 W)
4	77.0% (48 W)	87.1% (96 W)	89.3% (120 W)
6	83.9% (72 W)	91.3% (144 W)	92.9% (180 W)
8	87.9% (96 W)	93.4% (192 W)	94.7% (240 W)
10	90.2% (120 W)	94.7% (240 W)	95.8% (300 W)
12	91.4% (144 W)	95.5% (288 W)	96.8% (360 W)
14	92.0% (168 W)	95.8% (336 W)	96.6% (420 W)
16	92.1% (192 W)	95.8% (384 W)	96.6% (480 W)
18	91.4% (216 W)	95.2% (432 W)	96.1% (540 W)
20	90.9% (240 W)	95.2% (480 W)	96.1% (600 W)
22	90.7% (264 W)	95.0% (528 W)	96.0% (660 W)

The variation in efficiency with output current for three voltage modes is shown in Fig. 12. At the same output voltage, the efficiency increases rapidly as output current rises, peaking between 10-14 A before gradually declining. Among the three modes, the 30 V output voltage demonstrates the highest efficiency at the same output current.

The power supply operates efficiently, alleviating thermal management challenges. As shown in Fig. 13, the highest temperature at full load is 86.1 °C, primarily concentrated on the high-side MOS within the power loop. This is due to the longer on-time of the high-side MOS and the inevitable switching losses at relatively high frequencies, which contribute to localized heating. Nevertheless, the full-load temperature remains below 90 °C, meeting the requirements for continuous power output under high-performance conditions.

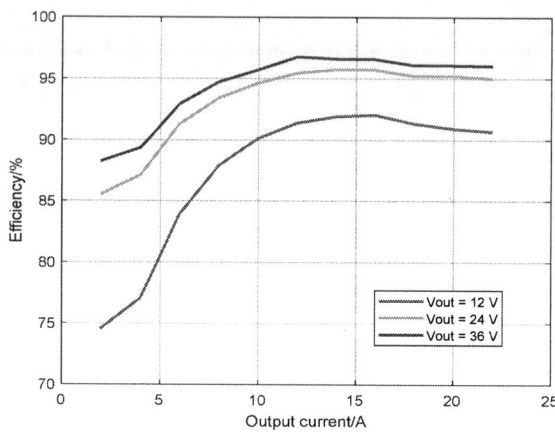

Fig. 12. The variation of efficiency with current for three output voltage modes.

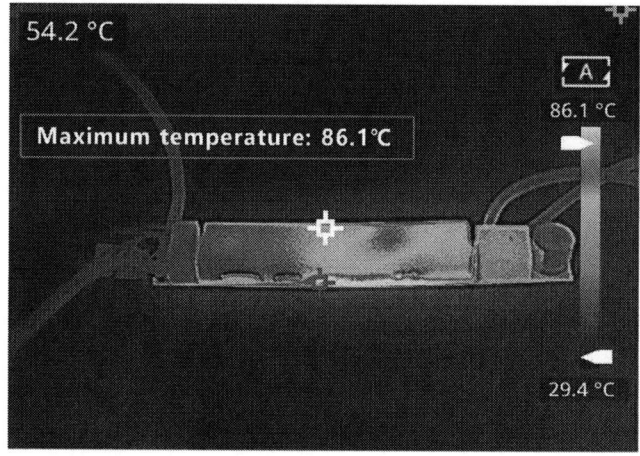

Fig. 13. Measured module temperature.

V. CONCLUSION

This work proposes a digital DC/DC power converter in the form of 3DIC with heterogeneously integrating functional modules into a monolithic package, which paves the way for VPD in HPC applications. To fully exploit the three-dimensional packaging space for maximum output power and reliability, VPD and RDL technologies are adopted to optimize the package layout. With the help of heterogenous integration and thorough physical analysis, this works achieves the highest output current up to 22 A, the highest output power up to 660 W and the best figure of merit (FoM) 2.49×10^{-5} among those references while maintaining high efficiency and minimal output ripple. The monolithic package and improved heat dissipation demonstrate the power density and reliability advantages of the proposed 3DIC approach over conventional board-level power modules.

ACKNOWLEDGMENT

This work was supported by a grant from the National Key R&D Program of China (Project No. 2022YFB4401100) and National Natural Science Foundation of China (Grant No. 62374100).

979-8-3315-1612-3/25 $31.00 © 2025 IEEE

TABLE V. COMPARISON TABLE WITH PRIOR WORKS

Parameter	Vicor PRM48B	Laser diode SF6060	APEC'24[17]	APEC'24[20]	This work
Power	600	600	400	400	660
Level	Package	Board	Package	Board	Package
Dimensions (mm³)	32.5×22.0×6.73	57.9×36.8×21	40×30×5	62.5×37.2×8	63.5×14.5×8
Density (W/in³)	2072	219.6	1091.6	352.2	1468
Efficiency (Buck)	96%	96.7%	97%	97.4%	96.2%
Ripple (%)	1.9%	0.17%	10%	>0.5%	0.9%
Max I_{out} (A)	12.5	15	16.5	16.7	22
Max T (°C)	125	80	88.3	98.5	86.1
FoM*(in³·°C/V·A²)	9.55196×10^{-5}	4.27×10^{-5}	5.05×10^{-4}	8.6×10^{-5}	2.49×10^{-5}

FoM* = (Ripple × Max T)/(Density × Efficiency × Max I_{out})

REFERENCES

[1] Yan-Fei Liu, E. Meyer, and Xiaodong Liu, "Recent Developments in Digital Control Strategies for DC/DC Switching Power Converters," *IEEE Trans. Power Electron.*, vol. 24, no. 11, pp. 2567–2577, Nov. 2009.

[2] S.-Y. Kim *et al.*, "Design of a High Efficiency DC–DC Buck Converter With Two-Step Digital PWM and Low Power Self-Tracking Zero Current Detector for IoT Applications," *IEEE Trans. Power Electron.*, vol. 33, no. 2, pp. 1428–1439, Feb. 2018.

[3] A. Smith *et al.*, "11.1 AMD InstinctTM MI300 Series Modular Chiplet Package – HPC and AI Accelerator for Exa-Class Systems," in *2024 IEEE International Solid-State Circuits Conference (ISSCC)*, San Francisco, CA, USA: IEEE, Feb. 2024, pp. 490–492.

[4] A. Song, D. Chen, and Z. Zong, "Unveiling the Truth: An Analysis of the Energy and Carbon Footprint of Training an OPT Model using DeepSpeed on the H100 GPU," in *Proceedings of the 14th International Green and Sustainable Computing Conference*, Toronto ON Canada: ACM, Oct. 2023, pp. 36–38.

[5] A. I. Emon, Mustafeez-ul-Hassan, A. B. Mirza, J. Kaplun, S. S. Vala, and F. Luo, "A Review of High-Speed GaN Power Modules: State of the Art, Challenges, and Solutions," *IEEE J. Emerg. Sel. Top. Power Electron.*, vol. 11, no. 3, pp. 2707–2729, Jun. 2023.

[6] Y. Chen *et al.*, "Thermal Mitigation and Optimization Via Multitier Bond Wire Layout for IGBT Modules Considering Multicellular Electro-Thermal Effect," *IEEE Trans. Power Electron.*, vol. 37, no. 6, pp. 7299–7314, Jun. 2022.

[7] H.-J. Lee, H. Lee, K. Lee, and J. Lee, "Small Package Size Low Power CMOS Image Sensor using Two Different Type Small Through Silicon Vias Technology for 3D Packaging," in *2022 IEEE 72nd Electronic Components and Technology Conference (ECTC)*, San Diego, CA, USA: IEEE, May 2022, pp. 967–971.

[8] W. Liu, G. Chen, B. Wang, J. X. Jiang, and E. Milligan, "Signal, Power and Thermal Co-optimization Methodology for FPGA Advanced Package," in *2024 IEEE 74th Electronic Components and Technology Conference (ECTC)*, Denver, CO, USA: IEEE, May 2024, pp. 1085–1092.

[9] P. Wang, Y. Chen, G. Szczeszynski, S. Allen, D. M. Giuliano, and M. Chen, "MSC-PoL: Hybrid GaN–Si Multistacked Switched-Capacitor 48-V PwrSiP VRM for Chiplets," *IEEE Trans. Power Electron.*, vol. 38, no. 10, pp. 12815–12833, Oct. 2023.

[10] C. Hardy *et al.*, "11.1 A Scalable Heterogeneous Integrated Two-Stage Vertical Power-Delivery Architecture for High-Performance Computing," in *2023 IEEE International Solid- State Circuits Conference (ISSCC)*, San Francisco, CA, USA: IEEE, Feb. 2023, pp. 182–184.

[11] R. Jain *et al.*, "28.6 An 87% Efficient 2V-Input, 200A Voltage Regulator Chiplet Enabling Vertical Power Delivery in Multi-kW Systems-on-Package," in *2024 IEEE International Solid-State Circuits Conference (ISSCC)*, San Francisco, CA, USA: IEEE, Feb. 2024, pp. 466–468.

[12] P. R. Prakash *et al.*, "A 2400 W/in³ 1.8 V Bus Converter Enabling Vertical Power Delivery for Next-Generation Processors," in *2024 IEEE Applied Power Electronics Conference and Exposition (APEC)*, Long Beach, CA, USA: IEEE, Feb. 2024, pp. 910–917.

[13] Y. Elasser *et al.*, "Mini-LEGO: A 1.5-MHz 240-A 48-V-to-1-V CPU VRM with 8.4-mm Height for Vertical Power Delivery," in *2023 IEEE Applied Power Electronics Conference and Exposition (APEC)*, Orlando, FL, USA: IEEE, Mar. 2023, pp. 1959–1966.

[14] A. M. Naradhipa, F. Zhu, and Q. Li, "Ultra-Low-Profile Twisted Core Inductor for Vertical Power Delivery Voltage Regulator," in *2024 IEEE Applied Power Electronics Conference and Exposition (APEC)*, Long Beach, CA, USA: IEEE, Feb. 2024, pp. 918–924.

[15] H. Gan *et al.*, "Vertical Power Delivery for 1000 Amps Machine Learning ASICs," in *2024 IEEE Applied Power Electronics Conference and Exposition (APEC)*, Long Beach, CA, USA: IEEE, Feb. 2024, pp. 906–909.

[16] Y. Elasser *et al.*, "Vertical Stacked 48V-1V LEGO-PoL CPU Voltage Regulator with 1A/mm² Current Density," in *2022 IEEE Applied Power Electronics Conference and Exposition (APEC)*, Houston, TX, USA: IEEE, Mar. 2022, pp. 1259–1266.

[17] S. S. Sinha, P. Zaghari, J. E. Ryu, and D. C. Hopkins, "A 400W, 250kHz (2kW Peak) Integrated GaN Half Bridge Power Module in a Non-Isolated Buck Converter," in *2024 IEEE Applied Power Electronics Conference and Exposition (APEC)*, Long Beach, CA, USA: IEEE, Feb. 2024, pp. 168–174.

[18] A. I. Emon *et al.*, "Design and Optimization of Gate Driver Integrated Multichip 3-D GaN Power Module," *IEEE Trans. Transp. Electrification*, vol. 8, no. 4, pp. 4391–4407, Dec. 2022.

[19] B. Bloch, V. Bar-Natan, and E. Recht, "EMI protection of composite material housings for electrooptical devices," in *2003 IEEE International Symposium on Electromagnetic Compatibility, 2003. EMC '03.*, Istanbul, Turkey: IEEE, 2003, pp. 950-953 Vol.2.

[20] B. Li, X. Huang, and J. Zhang, "A High Density 400 W DC/DC Power Module with Integrated Planar Transformer and Half Bridge GaN IC," in *2024 IEEE Applied Power Electronics Conference and Exposition (APEC)*, Long Beach, CA, USA: IEEE, Feb. 2024, pp. 94–100.

Planar Rogowski Coil-Based Switch Current Measurement for a 1.2 kV SiC MOSFET Embedded Die PCB

Matthias Spieler
Center for Power Electronics
Systems, Virginia Tech
Blacksburg, VA, USA
mspieler@vt.edu

Che-Wei Chang
Center for Power Electronics
Systems, Virginia Tech
Blacksburg, VA, USA
cwchang@vt.edu

Ayman M. EL-Refaie
Opus College of Engineering
Marquette University
Milwaukee, WI, USA
ayman.el-refaie@marquette.edu

Dong Dong
Center for Power Electronics
Systems, Virginia Tech
Blacksburg, VA, USA
dongd@vt.edu

Rolando Burgos
Center for Power Electronics
Systems, Virginia Tech
Blacksburg, VA, USA
roburgos@vt.edu

Abstract—This paper introduces the integration of a small footprint planar Rogowski coil into an 1.2 kV embedded die PCB for the switch transient current measurement thereof. Due to the high-density integration of embedded die PCBs, measuring switch transient currents has been challenging. The planar Rogowski coil enables the current measurement with a small impact on the commutation loop design and parasitic stray inductance. The geometry of the 106 mm² small coil increases the stray inductance in the commutation loop by 298 pH. The 187 MHz bandwidth current sensor allows measuring the fast switch transient currents of the 1.2 kV SiC MOSFETs. The planar Rogowski coil is first compared to alternative high-bandwidth current measurement technologies. Double pulse tests are conducted at a dc-bus voltage of 600 V and a drain current of 150 A. The overshoot voltage during the turn-off switch transient for embedded die PCBs with and without the planar Rogowski coil at a current slope of −48.2 A/ns are 646.1 V and 657.6 V, respectively. The small overshoot voltage highlights the minimal added insertion stray inductance of the current sensor into the commutation loop.

Index Terms—Current measurement, Rogowski coil, Silicon carbide, Half-bridge, Embedded die PCB design, Stray inductance

I. INTRODUCTION

Embedded die PCBs integrate one or multiple dies within their PCB layer structure. This integration approach enables a high-power dense PCB design with very small current commutation loop stray inductances. It thus allows for fast switching speeds of 70 V/ns and reduced switch transient losses. Designs with a current commutation loop inductance of below 3 nH have been achieved in literature [1]–[7].

This material is based upon work supported by the U.S. Department of Energy's Office of Energy Efficiency and Renewable Energy (EERE) under the Advanced Vehicle Technologies Office Award Number DE-EE0009190.

The current sensor can be employed for protection against overcurrent events, half-bridge output current reconstruction, and to measure the switching losses of embedded SiC MOS-FETs [8], [9]. As the current sensor must be placed within the drain or source current path of the MOSFET, its geometry will affect the design of the commutation loop and impact the parasitic stray inductance within the loop. The increase in stray inductance changes the switch transient behavior, the switch energy loss as the turn-off transient overshoot voltage increases, and a larger voltage drop over the parasitic commutation loop stray inductance during the turn-on switch is observed [8], [9]. Thus, a current sensor with little impact on the commutation loop stray inductance is beneficial.

Due to the higher-voltage application of SiC MOSFETs, an electric insulating current measurement is preferred. The sensor must have a high bandwidth to accurately measure the fast switching transients. Three potentially viable sensor options, including shunt, current transformer, and Rogowski coils, are evaluated in section two for their applicability and impact on the commutation loop design.

After comparing the three current sensor technologies for their impact on the commutation loop design, the paper focuses on the impact of a planar Rogowski current sensor on turn-off overshoot voltage. Two embedded die PCBs, one with and one without the Rogowski coil integrated. Section three outlines the design of the embedded die PCB with the integrated planar Rogowski coil. The PCB parasitics are extracted with Ansys Q3D. Section three highlights the experimental results of the double pulse tests. The double pulse tests are carried out at a dc-bus voltage of 603 V and a drain current of 150 A. The impact of the planar Rogowski coil sensor on the overshoot voltage is evaluated, and the two embedded die PCB versions are compared. The transient currents are measured,

979-8-3315-1612-3/25 $31.00 © 2025 IEEE   1551

and the commutation loop stray inductance is extracted. Lastly, conclusions are drawn about utilizing a planar Rogowski coil-based sensor to measure the transient currents of an embedded die PCB.

II. Impacts of Current Sensing Technologies on the Commutation Loop Design

The stray inductance defines, in addition to the change in current, the maximum turn-off overshoot voltage, as shown in the equation below.

$$V_{ds,os} = L_{ccl} \cdot \frac{di_d}{dt} \tag{1}$$

The stray inductance of the commutation loop (L_{ccl}) is the sum of all stray inductances within the loop, such as the power module stray inductance, dc-link/ decoupling capacitor stray inductance, PCB trace stray inductance, and the added stray inductance by the current sensor. The PCB trace stray inductance mainly depends on the trace length between the dc-link capacitors and the half-bridge, the trace width, and the mutual inductance cancelation between the current path $dc-$ and $dc+$ as highlighted in the below shown simplified loop inductance equation [10].

$$L_{ccl} = \mu_0 \cdot \frac{h}{w} l \cdot \left(\frac{1}{1 + \dfrac{h}{w}} + 0.024 \right) \tag{2}$$

Where h is the vertical height between the dc-bus traces, l is the trace length between the decoupling capacitor and the half-bridge, and w is the dc-bus trace width. The embedded die PCB design enables a commutation loop design with a very small stray inductance of 305 pH [7]. The work in this paper achieves a simulated commutation loop stray inductance of 732 pH. The commutation loop of the embedded die PCB design is depicted in Fig 1. The width of the current carrying dc-bus traces is 27 mm. A vertical commutation loop increases the mutual inductance cancellation between the dc-bus traces, and thus reduces the total commutation loop stray inductance. Fig. 2 shows a simplified cross-sectional view of the commutation loop. The decoupling capacitor is moved to the left of the half-bridge to provide enough space for a potential top-side heatsink.

Due to the small stray inductance of the commutation loop of 732 pH, the placement of the current sensor becomes crucial, as a small added stray inductance can disturb the half-bridge switch transient behavior. Furthermore, the sensor must have a high bandwidth to accurately measure the expected transient current slopes of ± 50 A/ns. The maximum expected current amplitude (I_{amp}) is 150 A. Based on the equation below, the minimum required bandwidth of the current sensor can be calculated to be 117 MHz.

$$BW_{min} = 0.35 \cdot \frac{di/dt}{I_{amp}} \tag{3}$$

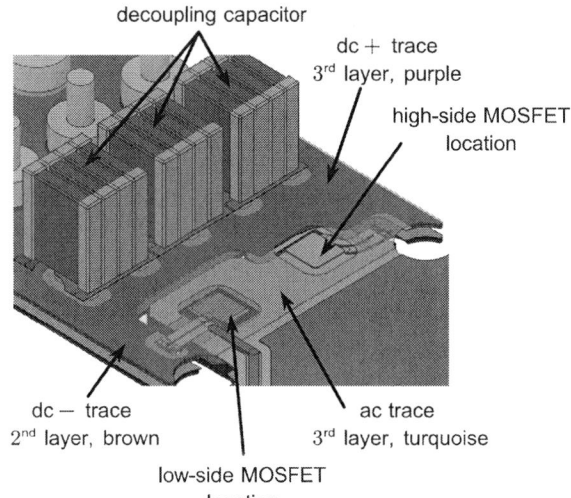

Fig. 1. Commutation loop design of the embedded die PCB without current sensor. The dc-bus traces and ac traces are highlighted. The dc- traces on the second layer are opaque to display the layer structure on layer three

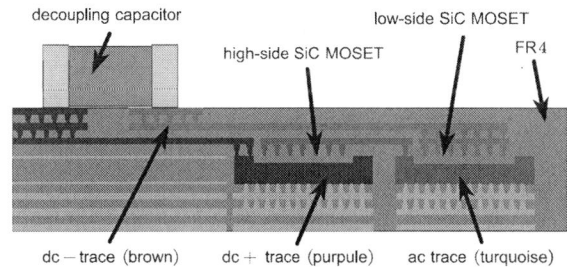

Fig. 2. Cross-sectional view of the embedded die PCB commutation loop. The loop traces are highlighted. Other traces are greyed out.

Three potentially viable sensor options: 1) shunt resistor, 2) current transformer, and 3) Rogowski coil, are compared for their applicability to measure the switch transient currents in the next subsection.

A. Current Transformer

Current transformers (CT) provide galvanic isolation between the current carrying tace and the sensor output. The to-be-measured current carrying trace is passed through the transformer core, as illustrated in Fig. 3 [11]. Commercial

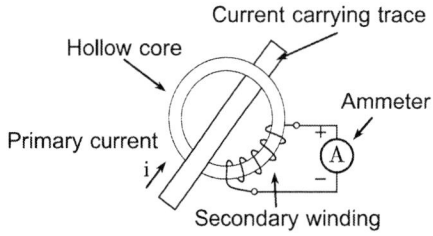

Fig. 3. Illustration of the general geometry of a current transformer.

current transformers, such as the Pearson Current Monitor 8585C, can achieve a measurement bandwidth of up to 200 MHz [12]. However, commercial current transformers are bulky, and their placement requires an extensive redesign of the commutation loop. Yin et al. addressed the bulkiness by miniaturizing the current transformer. The miniaturized sensor can be placed around a TO-247 leg [13]. Although a miniaturized sensor would not result in no distance increase between the decoupling capacitors and the SiC MOSFET half-bridge, one of the dc-bus traces would need to be narrowed to feed through the current transformer core. Following equation 2 a more narrow dc-bus trace would increase the commutation loop stray inductance.

B. Shunt Resistor

Coaxial shunt resistors have been extensively utilized to measure the switch transient currents of wide-bandgap transistors. Coaxial shunt resistors employ a high-resistivity material with minimal thermal drift for the voltage drop measurement and a highly conductive casing [14]. By minimizing the influence of parasitic inductance on the current measurement, coaxial shunt resistors can achieve very high bandwidths in the GHz range, making them desirable for measuring fast switch transients.

The sensor must be placed within the commutation loop. The physical size and the sensor's impact on current redirection result in added series stray inductance to the power loop. The total added stray inductance ranges from 2.2 nH for commercial coaxial shunt resistors to 0.52 nH for optimized coaxial shunt resistor designs [15]. The optimized low-inductance coaxial shunts are typically used for Gallium Nitride (GaN) transistors. Their small size is a disadvantage for high-current applications that require wide, heavy copper traces. According to equation 2, a small footprint shunt resistor would decrease the width of the dc-bus tracing and thus further increase the stray inductance of the commutation loop.

C. Rogowski Coil

Helical Rogowski coils encircle the current carrying trace with their windings, as displayed in Fig. 4. Helical coils achieve sensor bandwidths of up to 300 MHz by miniaturizing the sensor and thus minimizing the number of turns [16].

However, they face a similar implementation problem to that of current transformers. The helical Rogowski coil is wrapped around the current-carrying trace and thus necessitates an alteration in the commutation loop trace design. The alteration adds the sensor insertion inductance to the commutation loop's stray inductance.

In contrast to the helical Rogowski coil, the planar Rogowski coil does not circumvent the current-carrying trace but is placed in a straight line adjacent to it [17]–[21]. A planar Rogowski coil illustration is shown in Fig. 4. This implementation allows for a minimized impact on the commutation loop and can achieve bandwidths of up to 187 MHz [22]. In addition to high bandwidths, the planar Rogowski coil accommodates wide current-carrying traces, enabling current measurement in

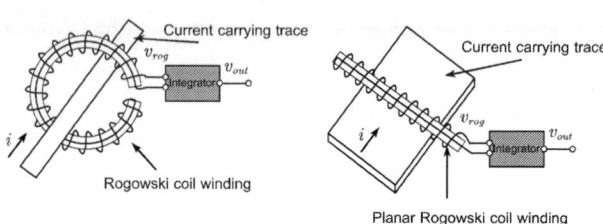

Fig. 4. General geometry comparison of a helical Rogowski coil and a planar Rogowski coil.

high-current and high-power applications without significant sensor insertion inductance and enabling the current measurement in high-current and high-power applications without a large sensor insertion inductance. The operating principle and design guidelines are extensively covered in [22], [23].

III. EMBEDDED DIE PCB DESIGN CONSIDERATIONS WITH CURRENT SENSOR

Embedded die PCBs allow for an optimized power trace design while achieving good thermal performance. Many embedded die PCB designs in the literature achieve a commutation loop stray inductance of less than 3 nH as previously highlighted in the previous section. At a constant maximum overshoot voltage, the switching speed can be increased by minimizing the stray inductance. Therefore, the switching loss of embedded die PCBs can be reduced in comparison to commercial power modules.

The low stray inductance of embedded die PCBs elevates the importance of a current sensor with minimal added parasitic stray inductance, as the additional stray inductance has an outsized impact on the converter's switch transient behavior. A planar Rogowski coil demonstrates potential for this application, as its placement is unobtrusive to the commutation loop [22].

The planar Rogowski coil is placed between the PCB layers of the dc-bus and between the dc-link capacitor and the embedded SiC MOSFET half-bridge. Fig. 5 shows the general 3D model of the embedded die PCB trace design. A simplified cross-sectional side view of the commutation loop with the integrated planar Rogowski coil is shown in Fig. 6. The $dc-$ trace is routed on the second layer. The $dc+$ trace is routed from the third layer down to the seventh layer and back to the third layer. This creates an empty area between the dc-bus traces to place the Rogowski coil windings. The Rogowski coil features 16 windings created between the third and sixth layers of the PCB, spanning the entire width of the dc-bus trace.

Two 8-layer embedded die PCB designs are created. The power loop trace design is similar, except for the planar Rogowski coil integrated into one PCB structure. Both designs utilize the same 1.2 kV SiC MOSFET die, decoupling capacitor, and current booster stage. The version with the planar Rogowski coil utilizes a smaller liquid-cooled heatsink.

979-8-3315-1612-3/25 $31.00 © 2025 IEEE

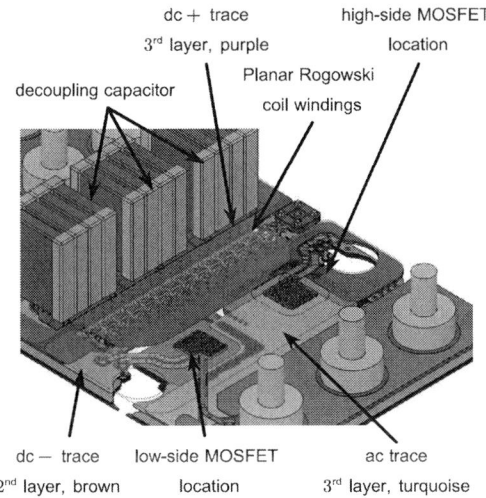

Fig. 5. The 3D model shows the inner layer structure of the embedded die PCB. The opacity of layer two is reduced to show the below layer structure.

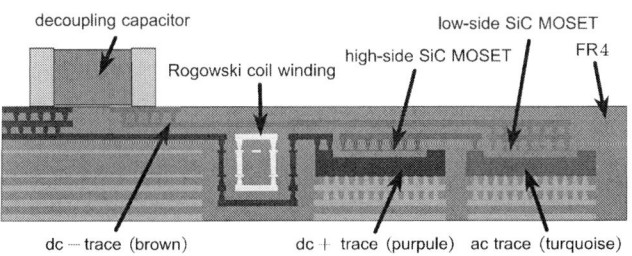

Fig. 6. Cross-sectional view of the embedded die PCB commutation loop with the planar Rogowski coil. The loop traces and Rogowski coil traces are highlighted. Other traces are greyed out.

Thus, the parasitic trace capacitances within the power loop are reduced for the embedded die PCB version with the integrated planar Rogowski coil. Ansys Q3D simulations are used to extract the stray inductance of the PCB current commutation loop and parasitic capacitances. The embedded die PCB without the planar Rogowski coil has a commutation loop stray inductance of 732 pH. The planar Rogowski coil adds a stray inductance of 298 pH to the commutation loop, resulting in a commutation loop stray inductance of 1.03 nH. The parasitic plate capacitance between the higher voltage traces causes additional switching loss of the half-bridge's SiC MOSFETs. Although the commutation loop is kept similar between the two embedded die PCB versions, slight differences in capacitance can be observed with Ansys Q3D simulations. Table I lists the results of the Ansys Q3D simulation of both embedded die PCBs.

IV. EXPERIMENTAL RESULTS

The two embedded die PCB prototypes are depicted in Fig. 7. Double pulse tests are conducted at a dc-bus voltage of 603.3 V, and a drain current of 150 A. The test setup is shown in Fig. 8. A 2.4 Ω turn-off gate resistor was used to achieve switching speeds of 67 V/ns. The same dc-link capacitor PCB

TABLE I
EMBEDDED DIE PCB PARASITIC COMPARISON

Parameter	Embedded die PCB without Rogowski coil	Embedded die PCB with Rogowski coil
Commutation loop stray inductance (L_{ccl})	732 pH	1.03 nH
Trace parasitic capacitance (dc+ to ac) ($C_{dc+,ac}$)	25.1 pF	40.94 pF
Trace parasitic capacitance (dc- to ac) ($C_{dc-,ac}$)	66.67 pF	79.64 pF
Trace parasitic capacitance (dc+ to dc-) ($C_{dc+,dc-}$)	367.84 pF	464.74 pF

and gate driver PCB were used to test both embedded die PCB versions. All test conditions were kept equal.

The drain-to-source voltage (V_{ds}) of the low-side MOSFET is measured with an 800 MHz IsoVu probe. The inductor current (I_{load}) is measured using a commercial Rogowski coil. The load current amplitude is utilized to calibrate the gain of the planar Rogowski coil. Lastly, a non-isolated passive probe directly measures the planar Rogowski coil output voltage. As the Rogowski coil's output voltage represents the current's derivative, the measured voltage must be integrated. The integration of the Rogowski coil voltage is performed in Matlab. This approach allows for the utilization of the full bandwidth of the planar Rogowski coil instead of being limited by the bandwidth of the analog integrator circuit.

The drain-to-source voltage measurements of both embedded die PCBs, the load current measurement, and the integrated planar Rogowski coil output voltage are overlaid. Fig. 9 depicts the turn-off and turn-on switch transients, respectively.

The embedded die PCB without the planar Rogowski coil has a maximum overshoot voltage of 645.1 V. In contrast, the overshoot voltage of the embedded die PCB with planar Rogowski coil increases by 12.5 V, resulting in an overshoot voltage of 657.6 V. Note that the measured low-side SiC MOSFET source current decreases during the voltage rise time. This is likely caused by the charging and discharging of the drain-source capacitance of the low-side and high-side SiC MOSFET. Thus, the current slope is measured between the time interval from $t_1 = -55.69$ ns and $t_2 = -51.94$ ns. The switching speeds are 67 V/ns and -48.2 A/ns. The 250 MHz oscillation frequency of the source current nicely matches the voltage overshoot oscillation.

Based on equation 1, the maximum overshoot voltage of 54.3 V, and a source current slope of -48.2 A/ns, the commutation loop stray inductance can be calculated to be 1.13 nH. The measured stray inductance deviates by 0.1 nH or 9.7 % from the simulated stray inductance.

979-8-3315-1612-3/25 $31.00 © 2025 IEEE   1554

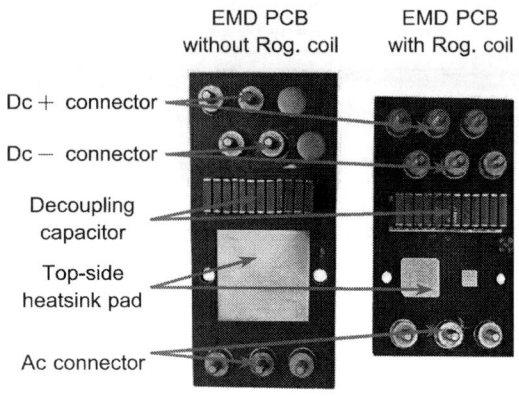

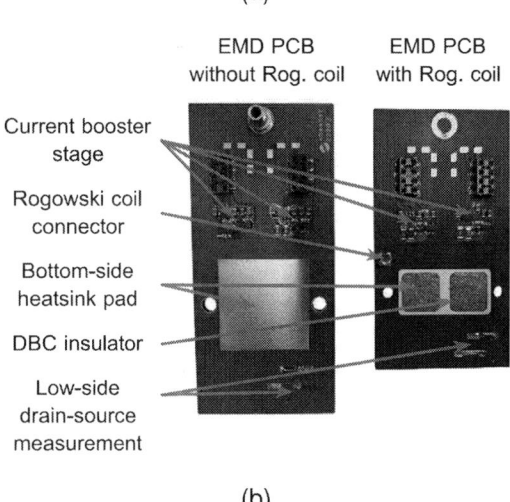

(a)

(b)

Fig. 7. (a) Top-side view of the Embedded die (EMD) PCB prototypes (b) Bottom-side view of the Embedded die (EMD) PCB prototypes

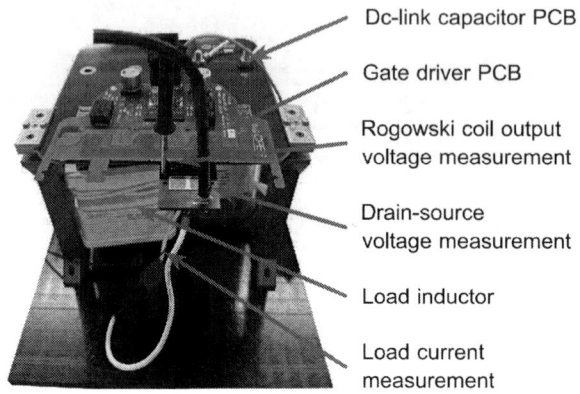

Fig. 8. Double pulse test setup to measure the drain-source voltage of the low-side embedded die PCB SiC MOSFET, load current, and Rogowski coil output voltage

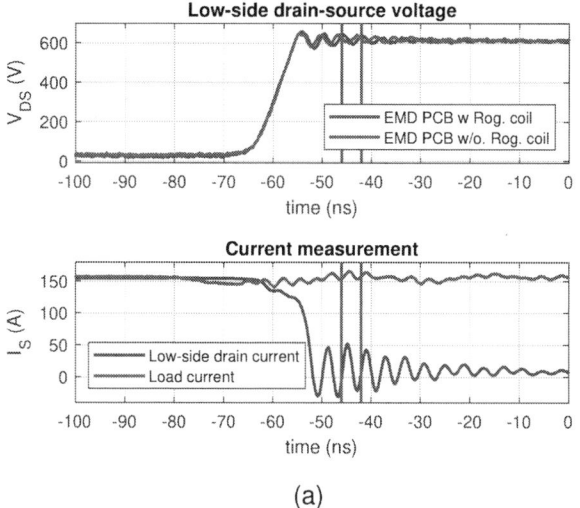

(a)

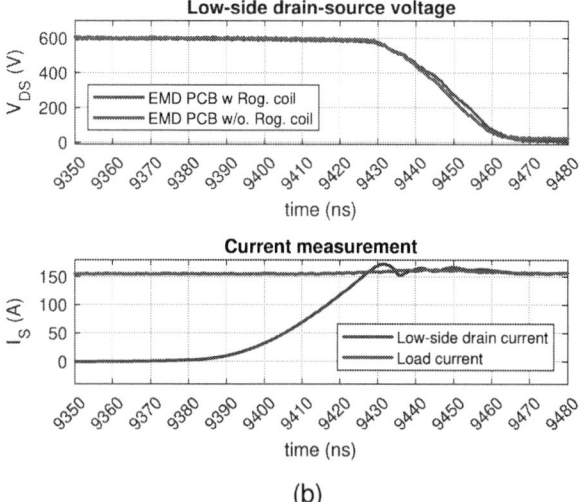

(b)

Fig. 9. (a) Turn-off switch transient waveforms (b) Turn-on switch transient waveforms

As a prototype SiC MOSFET is embedded into the PCB, no information about the die's drain-to-source capacitance is known. Thus, the commutation loop stray inductance of the embedded die PCB without the planar Rogowski cannot be derived.

The turn-on switch transient current slope is 3.98 A/ns. The current booster stage prevents the turn-on gate resistor from being reduced, as the pulsed current of the current booster cannot exceed 10 A. This hinders the use of the turn-on switch transient and the resulting voltage drop of the commutation loop inductance to evaluate the value of the stray inductance.

Overall, the planar Rogowski coil's impact and insertion inductance on the switch transient waveforms is minimal. Thus, the planar Rogowski coil is an excellent current sensor for applications with very small commutation loop stray inductances, such as embedded die PCBs.

V. CONCLUSION

The impact of a planar Rogowski coil current sensor on the switch transient behavior of a half-bridge 1.2 kV SiC

MOSFET embedded die PCB is explored. Compared to current transformers, coaxial shunt resistors, and helical Rogowski coils, the planar Rogowski coil has the lowest insertion inductance into the half-bridge's commutation loop. The parasitic insertion inductance is 298 pH. Two embedded die PCB designs, one with and one without the planar Rogowski coil, were designed and tested. Double pulse tests were carried out at a dc-bus voltage of 603 V, a drain current of 150 A, and a turn-off switching speed of 67 V/ns and −48.2 A. By adding the current sensor, the maximum drain-source overshoot voltage increased from 645.1 V to 657.6 V. The embedded die PCB's measured commutation loop stray inductance is 1.23 nH. The minimal impact of the current sensor on the switch transient behavior elevates the planar Rogowski coil as an excellent current sensor for systems with very small commutation loop stray inductances.

REFERENCES

[1] C. Neeb, J. Teichrib, R. W. D. Doncker, L. Boettcher, and A. Ostmann, "A 50 kw igbt power module for automotive applications with extremely low dc-link inductance," in *2014 16th European Conference on Power Electronics and Applications*, Conference Proceedings, pp. 1–10.

[2] E. Dechant, N. Seliger, and R. Kennel, "Performance of a gan half bridge switching cell with substrate integrated chips," in *PCIM Europe 2019; International Exhibition and Conference for Power Electronics, Intelligent Motion, Renewable Energy and Energy Management*, Conference Proceedings, pp. 1–7.

[3] K. Klein, E. Hoene, and K. Lang, "Power module design for utilizing of wbg switching performance," in *PCIM Europe 2019; International Exhibition and Conference for Power Electronics, Intelligent Motion, Renewable Energy and Energy Management*, Conference Proceedings, pp. 1–8.

[4] L. Link, A. B. Sharma, V. Polezhaev, T. Huesgen, M. Vaas, G. Stohrer, and F. Koch, "Top side isolation investigation of pcb embedded 1.2kv dies," in *2021 23rd European Microelectronics and Packaging Conference Exhibition (EMPC)*, Conference Proceedings, pp. 1–7.

[5] J. Knoll, C. DiMarino, H. Stahr, and M. Morianz, "A pcb-embedded 1.2 kv sic mosfet package with reduced manufacturing complexity," *IEEE Open Journal of Power Electronics*, vol. 4, pp. 549–560, 2023.

[6] C. Marczok, M. Martina, M. Laumen, S. Richter, A. Birkhold, B. Flieger, O. Wendt, and T. Paesler, "Sicmodul - modular high-temperature sic power electronics for fail-safe power control in electrical drive engineering," in *CIPS 2020; 11th International Conference on Integrated Power Electronics Systems*, Conference Proceedings, pp. 1–6.

[7] Z. Qi, Y. Pei, L. Wang, Q. Yang, and K. Wang, "A highly integrated pcb embedded gan full-bridge module with ultralow parasitic inductance," *IEEE Transactions on Power Electronics*, vol. 37, no. 4, pp. 4161–4173, 2022.

[8] P. Sochor, A. Huerner, and R. Elpelt, "Commutation loop design for optimized switching behavior of coolsic(exp tm) mosfets using compact models," in *PCIM Europe digital days 2020; International Exhibition and Conference for Power Electronics, Intelligent Motion, Renewable Energy and Energy Management*, Conference Proceedings, pp. 1–8.

[9] W. Teulings, J. L. Schanen, and J. Roudet, "Mosfet switching behaviour under influence of pcb stray inductance," in *IAS '96. Conference Record of the 1996 IEEE Industry Applications Conference Thirty-First IAS Annual Meeting*, vol. 3, Conference Proceedings, pp. 1449–1453 vol.3.

[10] A. Letellier, M. R. Dubois, J. P. F. Trovão, and H. Maher, "Calculation of printed circuit board power-loop stray inductance in gan or high di/dt applications," *IEEE Transactions on Power Electronics*, vol. 34, no. 1, pp. 612–623, 2019.

[11] N. Kondrath and M. K. Kazimierczuk, "Bandwidth of current transformers," *IEEE Transactions on Instrumentation and Measurement*, vol. 58, no. 6, pp. 2008–2016, 2009.

[12] I. PEARSON ELECTRONICS, "Pearson current monitor model 8585c." [Online]. Available: https://pearsonelectronics.com/pdf/8585C.pdf

[13] S. Yin, Y. Wu, M. Dong, J. Lin, and H. Li, "Design of current transformer for in situ current measurement of discrete sic power devices," *IEEE Journal of Emerging and Selected Topics in Industrial Electronics*, vol. 3, no. 4, pp. 1077–1086, 2022.

[14] W. Zhang, Z. Zhang, and F. Wang, "Review and bandwidth measurement of coaxial shunt resistors for wide-bandgap devices dynamic characterization," in *2019 IEEE Energy Conversion Congress and Exposition (ECCE)*, Conference Proceedings, pp. 3259–3264.

[15] W. Zhang, Z. Zhang, F. Wang, E. V. Brush, and N. Forcier, "High-bandwidth low-inductance current shunt for wide-bandgap devices dynamic characterization," *IEEE Transactions on Power Electronics*, vol. 36, no. 4, pp. 4522–4531, 2021.

[16] W. Zhang, S. B. Sohid, F. Wang, H. Cui, and B. Holzinger, "High-bandwidth combinational rogowski coil for sic mosfet power module," *IEEE Transactions on Power Electronics*, vol. 37, no. 4, pp. 4397–4405, 2022.

[17] P. S. Niklaus, D. Bortis, and J. W. Kolar, "Beyond 50 mhz bandwidth extension of commercial dc-current measurement sensors with ultra-compact pcb-integrated pickup coils," *IEEE Transactions on Industry Applications*, vol. 58, no. 4, pp. 5026–5041, 2022.

[18] J. Walter, J. Acuna, and I. Kallfass, "Design and implementation of an integrated current sensor for a gallium nitride half-bridge," in *PCIM Europe 2018; International Exhibition and Conference for Power Electronics, Intelligent Motion, Renewable Energy and Energy Management*, Conference Proceedings, pp. 1–8.

[19] Y. Kuwabara, K. Wada, J.-M. Guichon, J.-L. Schanen, and J. Roudet, "Design of an integrated air coil for current sensing," *IEEE Journal of Emerging and Selected Topics in Power Electronics*, vol. 8, no. 4, pp. 4122–4129, 2020. [Online]. Available: https://dx.doi.org/10.1109/jestpe.2020.2977102

[20] Y. Xue, J. Lu, Z. Wang, L. M. Tolbert, B. J. Blalock, and F. Wang, "A compact planar rogowski coil current sensor for active current balancing of parallel-connected silicon carbide mosfets," in *2014 IEEE Energy Conversion Congress and Exposition (ECCE)*, Conference Proceedings, pp. 4685–4690.

[21] A. D. Callegaro, J. Guo, M. Eull, B. Danen, J. Gibson, M. Preindl, B. Bilgin, and A. Emadi, "Bus bar design for high-power inverters," *IEEE Transactions on Power Electronics*, vol. 33, no. 3, pp. 2354–2367, 2018.

[22] M. Spieler, C. W. Chang, A. M. El-Refaie, D. Dong, and R. Burgos, "Design of a high-bandwidth compact dc-bus embedded planar rogowski coil for sic mosfet current sensing," *IEEE Transactions on Power Electronics*, vol. 39, no. 12, pp. 16 482–16 497, 2024.

[23] M. Spieler, C. W. Chang, A. El-Refaie, M. H. Alvi, D. Dong, and R. Burgos, "A dc-bus planar rogowski coil based current sensor for half-bridge applications," in *2023 IEEE 24th Workshop on Control and Modeling for Power Electronics (COMPEL)*, Conference Proceedings, pp. 1–7.

Effect of Magnetic Couplings on Conducted EMI of GaN-Based PFC Converter

Tyler McGrew and Qiang Li
Center for Power Electronic Systems
The Bradley Department of Electrical and Computer Engineering
Virginia Tech, Blacksburg, VA, USA
tymcgrew@vt.edu

Abstract—**Wide-bandgap power devices have enabled power supplies to increase in density while maintaining high efficiency. However, EMI issues remain one of the most challenging aspects when designing a power converter. These issues are often caused by parasitic near-field couplings which can degrade or bypass the attenuation of the EMI filter. This paper investigates how parasitic magnetic couplings impact the conducted noise of a high-frequency PFC converter using PCB-winding inductor. Analytical models are presented for both differential mode (DM) and common mode (CM) noise, including the effects of magnetic couplings between the PFC inductor and EMI filter. Then, three ferrite inductor core designs are evaluated to demonstrate how the core's geometry and permeability affect its leakage magnetic field. Choosing the optimal core design is shown to significantly reduce the magnetic field outside the PFC inductor, which reduces DM noise by more than 20 dB and reduces CM noise by 7 dB.**

Index Terms—**Electromagnetic interference (EMI), PCB magnetics, integrated magnetics, high-frequency, near-field, parasitics**

I. INTRODUCTION

As power conversion technology advances, electromagnetic compatability (EMC) remains one of the most challenging and least understood aspects of power supply design. After utilizing a suitable electromagnetic interference (EMI) filter, the converter's emissions can still exceed regulatory limits. In many cases, these confusing EMI behaviors are caused by parasitic near-field couplings within the power supply. Such capacitive and inductive couplings can degrade the high-frequency attenuation of the EMI filter [1]–[4] or create new noise paths which bypass the filter altogether [5]–[10].

An ideal EMI filter acts as a low-pass filter to pass line-frequency or DC power while attenuating high-frequency noise. However, all filter components are limited in bandwidth by their self-parasitics (series inductance or parallel capacitance). In addition, parasitic capacitive and magnetic couplings exist between the components within the EMI filter. These near-field couplings can further reduce the filter's bandwidth or high-frequency noise attenuation. Such effects are studied in [1]–[4], and various mitigation techniques are proposed including layout modification, cancellation, and shielding.

In addition to couplings within the EMI filter, near-field couplings also exist between the power stage and EMI filter. These parasitic couplings can create new paths for switching noise to couple to the filter or power lines and increase EMI emissions [5]–[11].

The common mode (CM) noise model of a flyback converter with filter was developed in [7]. Capacitive coupling between the switch node and filter capacitor was shown to significantly increase the converter's CM noise. Shielding and cancellation techniques were demonstrated to mitigate this coupling effect and reduce the CM noise up to 20 dB.

A differential mode (DM) noise model was developed for a boost PFC converter with EMI filter in [8]. Magnetic couplings between the boost inductor and filter were shown to increase the DM noise emissions throughout the conducted frequency range. Specifically, magnetic coupling to the CM choke increased low-frequency DM noise while coupling to the X-capacitor increased high-frequency DM noise.

The authors of [9] model how the CM choke in an EMI filter can generate both CM and DM noise when exposed to noisy magnetic fields. A stacked choke design was presented with improved immunity to external magnetic fields, and this choke was shown to significantly reduce the conducted emissions of a forward converter. With the improved choke, the converter's DM noise was reduced up to 25 dB while the converter's CM noise was reduced up to 5 dB.

This work aims to model and reduce the conducted noise of a high-frequency GaN-based bridgeless totem-pole PFC converter with PCB-winding inductor. In particular, this work focuses on design of the ferrite core for the PFC inductor, and its effect on parasitic magnetic couplings to the EMI filter.

Section II presents models for both DM and CM noise, including the effect of parasitic magnetic couplings. Section III models how the design of the PFC inductor's ferrite core affects its external/leakage magnetic field. Section IV evaluates three inductor designs to demonstrate how the air-gaps and permeability of the inductor core affect the converter's magnetic fields and conducted EMI. Section V summarizes the results and contributions of this work.

II. EMI MODEL FOR TOTEM-POLE PFC INCLUDING PARASITIC MAGNETIC COUPLINGS

A GaN-based bridgeless totem-pole PFC converter with EMI filter is shown in Fig. 1. Its components are described in Table I and its operation is detailed in Table II.

The basic structure of the PCB-winding inductor L_1 is shown in Fig. 2. This inductor integrates electrostatic shielding from [13] around the PCB windings to mitigate the effect of

---

979-8-3315-1612-3/25 $31.00 © 2025 IEEE

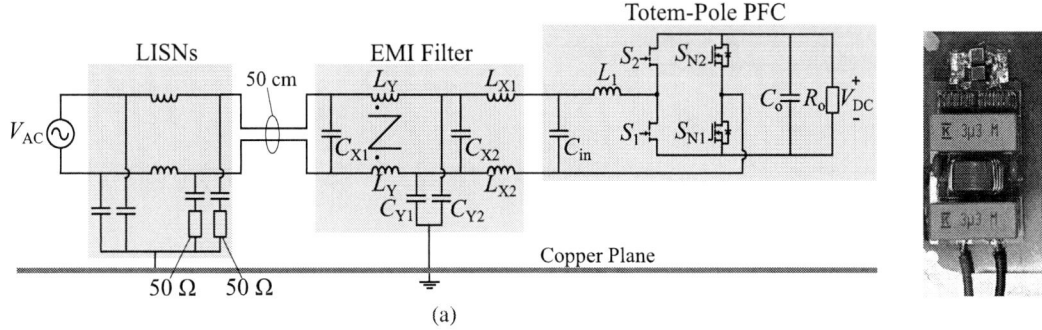
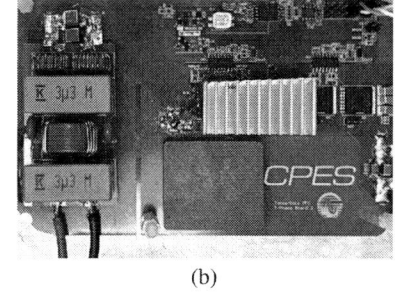

Fig. 1: GaN-based bridgeless totem-pole PFC converter with EMI filter: (a) circuit diagram, (b) hardware image.

TABLE I
DESCRIPTION OF CONVERTER COMPONENTS

Symbol	Description	Value
C_{X1}, C_{X2}	X film safety capacitors	3.3 μF
L_{X1}, L_{X2}	WE-FI 7447025 inductor	20.0 μH
C_{Y1}, C_{Y2}	Y ceramic safety capacitors	4.7 nF
L_Y	WE-CMBH 744823601 CM choke	1.0 mH
C_{in}	ceramic input capacitors	1.8 μF
L_1	PCB-winding inductor	12.6 μH
S_1, S_2	IGOT60R070D1 600 V GaN transistor	-
S_{N1}, S_{N2}	IPT60R028G7 600 V Si MOSFET	-
C_o (decoupling)	ceramic output capacitors	1.8 μF
C_o (bulk)	electrolytic output capacitors	2.0 mF
R_o	resisitive DC load	212 Ω

TABLE II
DESCRIPTION OF CONVERTER OPERATION

Symbol	Description	Value
AC line voltage	V_{AC}	230 V_{RMS}, 60 Hz
DC bus voltage	V_{DC}	400 V_{DC}
Output power	P_o	750 W
Switching frequency	f_s	320 kHz - 860 kHz
Operation mode	CRM ZVS with QSW ZVS extension [12]	

capacitive couplings on EMI. The electrostatic shield has a negligible impact on the inductor's magnetic fields or magnetic couplings, so it is not detailed here for the sake of simplicity.

The EMI filter contains one CM choke (L_Y) and symmetrical DM inductors L_{X1} and L_{X2}. This forms a single-stage CM filter and a two-stage DM filter due to the DM inductance of the CM choke ($L_{Y,DM}$). The two-stage DM filter is necessary in this design due to the large current ripple of the single-channel critical-mode PFC converter. Using a two-channel interleaved design can reduce this current ripple and allow for a single-stage DM filter [14]. However, this work uses the single-channel design as a more simple demonstration for better understanding of the EMI behaviors.

The bridgeless totem-pole PFC converter has one high-frequency switching leg with GaN devices S_1, S_2 and one line-frequency switching leg with Si MOSFETs S_{N1}, S_{N2}. The switching noise of the high frequency leg is modeled by replacing lower switch S_1 with ideal voltage source V_S and

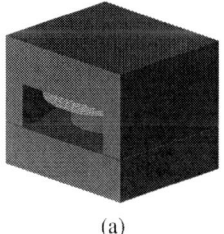
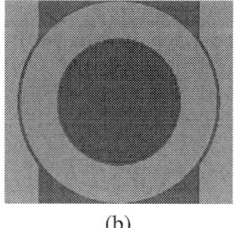

Fig. 2: Structure of PCB-winding inductor L_1 with ER/EQ ferrite core: (a) 3D view, (b) top-down view. Electrostatic shielding from [13] is integrated within the PCB windings but has negligible impact on magnetic fields.

open-circuiting the upper switch S_2 [13]. Voltage source V_S is defined as the switching voltage measured across switch S_1.

A. DM Noise Model

The PCB inductor L_1 has strong switching magnetic fields which can leak out and couple to other components in the converter [10]. There are two magnetic couplings between inductor L_1 and the EMI filter which can significantly impact the converter's DM noise emissions.

Fig. 11(a) shows the DM noise model with the effect of parasitic magnetic coupling M_{DM1} to *DM Loop 1* formed by capacitor C_{X1}, the input power cables, and the LISNs. DM currents induced in *DM Loop 1* are not attenuated by the EMI filter due to the low impedance of C_{X1}. Therefore, parasitic coupling M_{DM1} can significantly increase DM noise emissions.

Fig. 11(b) shows the DM noise model with the effect of parasitic magnetic coupling M_{DM2} to *DM Loop 2*. This DM loop is formed by capacitor C_{X2}, CM choke L_Y, the input power cables, and the LISNs. Coupling to the CM choke L_Y can be relatively strong due to the choke's magnetic core and multiple windings [9]. And, noise induced in *DM Loop 2* is only attenuated by the second stage of the DM filter. Therefore, M_{DM2} may also increase DM noise, especially at low-frequency [8].

Finally, Fig. 11(c) shows a complete, simplified DM noise model which includes the effects of both magnetic couplings M_{DM1} and M_{DM2}. In this model $L_{Y,DM}$ represents the total

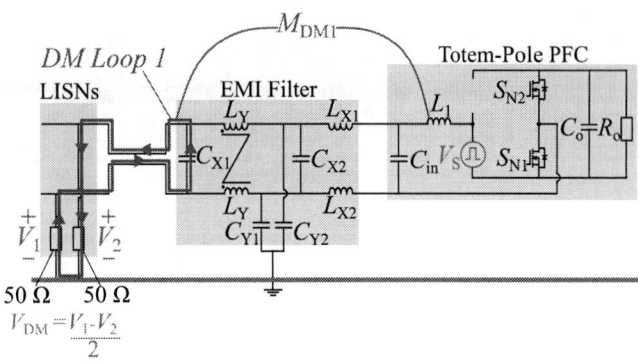

(a)

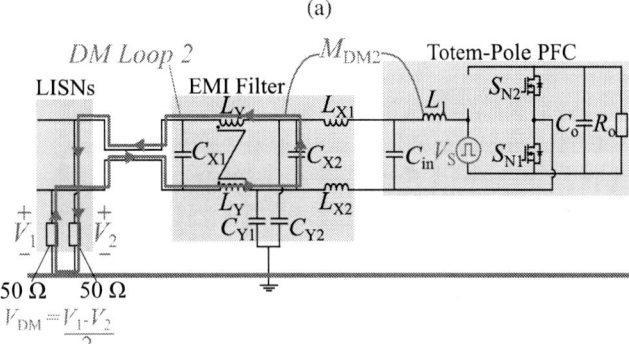

(b)

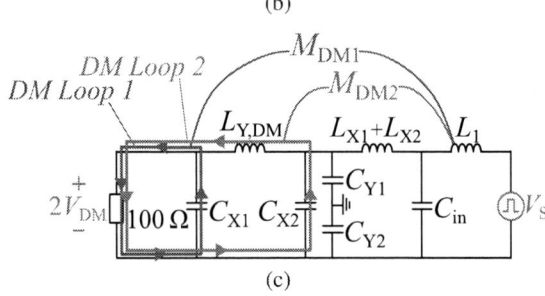

(c)

Fig. 3: DM noise model of PFC converter shown in Fig. 1 with effects of parasitic magnetic couplings: (a) with coupling M_{DM1} to *DM Loop 1*, (b) with coupling M_{DM2} to *DM Loop 2*, (c) simplified model with both couplings M_{DM1} and M_{DM2}.

DM inductive impedance of the CM choke L_{Y}, which is about 4 μH. While conducted DM noise from V_{S} must pass through a sixth-order low-pass filter, any switching noise coupling through M_{DM1} or M_{DM2} faces significantly less attenuation.

B. CM Noise Model

The CM noise of switching converters is usually associated with the effect of parasitic capacitances. The works in [6], [7], [10] study how capacitive couplings affect conducted CM noise. As shown in Fig. 4(a), there are two capacitive couplings in this converter which affect the CM noise emissions. Coupling C_{CM1} is formed between the switching voltage area and earth ground, while coupling C_{CM2} is formed between the switching voltage area and input side of the EMI filter. The CM noise through C_{CM1} is attenuated by the EMI filter. However, CM noise due to C_{CM2} bypasses the filter and can only be reduced by modifying the layout or adding shielding.

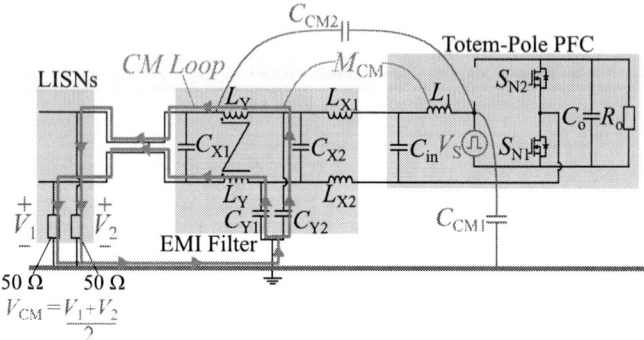

(a)

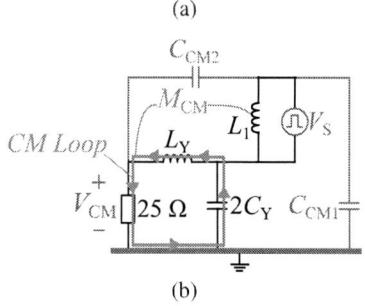

(b)

Fig. 4: CM noise model of PFC converter shown in Fig. 1 with effect of parasitic magnetic coupling M_{CM} to *CM Loop* in addition to parasitic capacitive couplings identified in [10]: (a) full circuit CM noise model, (b) simplified CM noise model.

In this work, the electrostatic shielding proposed in [13] is integrated around the windings of PCB inductor L_1 in this converter to minimize both capacitive coupling effects.

In addition to capacitive couplings, magnetic coupling can also increase the CM noise of the PFC converter. As shown in Fig. 12(a), the PFC inductor L_1 can form a magnetic coupling, M_{CM}, to the *CM Loop* comprised of the Y-capacitors, CM choke, input power cables, LISNs, and earth ground. Coupling M_{CM} to the CM choke L_{Y} can be strong due to the choke's magnetic core and multiple windings [9]. CM currents induced in the *CM Loop* are attenuated by the CM choke L_{Y}, but may still increase CM noise at low frequency where the L_{Y} impedance is smaller.

A simplified CM noise is shown in Fig. 4(b) which includes the effects of magnetic coupling M_{CM} and both capacitive couplings. While CM noise through coupling C_{CM1} faces a second-order low-pass filter, CM current induced in the *CM Loop* by M_{CM} faces only a first-order low-pass filter from L_{Y} and the 25 Ω LISN impedance. Therefore, magnetic coupling M_{CM} could increase CM noise in addition to the effect of capacitive coupling C_{CM2}.

III. Effect of Ferrite Inductor Core Design on Leakage Magnetic Field

As described in Section II, switching noise can magnetically couple from inductor L_1 to the EMI filter and increase the converter's conducted noise emissions. There are multiple approaches to mitigate the effect of these parasitic magnetic couplings, including layout modification and shielding. This

Section focuses on minimizing the stray magnetic fields generated in the air around inductor L_1 by optimizing the design of the ferrite inductor core.

The ER/EQ ferrite core for inductor L_1 is shown in Fig. 2. The PCB windings encircle the center post, and airgaps can be used on the center post and both outer core legs to achieve the required inductance and flux density within the core. The placement and size of airgaps can significantly impact the magnetic fields around the inductor, as can the permeability of the ferrite core material. These effects will be studied using an analytical magnetic reluctance model.

The magnetic reluctance model for inductor L_1 is shown in Fig. 5(a). The magnetomotive force (MMF) is generated by the inductor current according to $MMF = NI$, where $N = 6$. The reluctance of the core is divided into top plate ($\mathcal{R}_{ct}$), outer leg ($\mathcal{R}_{co}$), bottom plate ($\mathcal{R}_{cb}$), and inner post ($\mathcal{R}_{ci}$). The inner and outer airgaps have reluctance $\mathcal{R}_{gi}$ and $\mathcal{R}_{go}$, respectively. Each reluctance is given an associated reluctance $\mathcal{R}_{air,x}$ which represents the reluctance of the surrounding air. Except, it is assumed that the magnetic fields from the center post do not leak outside of the inductor, so $\mathcal{R}_{ci}$ and $\mathcal{R}_{gi}$ are not given an associated reluctance $\mathcal{R}_{air,x}$.

The reluctance model in Fig. 5(a) can then be simplified to the lumped model shown in Fig. 5(b). In this model, $\mathcal{R}_o$ is the sum of the reluctances on the outer legs and core plates

$$\mathcal{R}_o = \mathcal{R}_{ct} + \mathcal{R}_{co} + \mathcal{R}_{go} + \mathcal{R}_{cb}, \tag{1}$$

while $\mathcal{R}_i$ is the sum of the inductor's center post reluctances

$$\mathcal{R}_i = \mathcal{R}_{ci} + \mathcal{R}_{gi}. \tag{2}$$

These lumped reluctances can be further simplified when airgaps are used because the reluctance of air is so much larger than the reluctance of the ferrite core. The value of $\mathcal{R}_o$ is simplified as

$$\mathcal{R}_o \approx \begin{cases} \mathcal{R}_{go}, & \text{when } l_{g,o} > 0 \\ \mathcal{R}_{ct} + \mathcal{R}_{co} + \mathcal{R}_{cb}, & \text{when } l_{go} = 0 \end{cases} \tag{3}$$

where $l_{g,o}$ is the length of the airgap on the outer core legs. In other words, $l_{g,o} > 0$ when a non-zero airgap is used for the outer core legs. Similarly, the value of $\mathcal{R}_i$ is simplfied as

$$\mathcal{R}_i \approx \begin{cases} \mathcal{R}_{gi}, & \text{when } l_{gi} > 0 \\ \mathcal{R}_{ci}, & \text{when } l_{gi} = 0. \end{cases} \tag{4}$$

The total magnetic flux of the inductor (Φ_{total}) is determined by the L_1 inductance value and the PFC converter's operating condition (inductor current). A portion of this magnetic flux will leak into the surrounding air (Φ_{air}) and cause parasitic magnetic couplings to the EMI filter. The magnetic flux in the air on each half of the inductor is given by

$$\Phi_{air} = \frac{\mathcal{R}_o}{2\mathcal{R}_i\mathcal{R}_o + 2\mathcal{R}_i\mathcal{R}_{air} + \mathcal{R}_o\mathcal{R}_{air}} MMF \tag{5}$$

which reduces to

$$\Phi_{air} \approx \frac{\mathcal{R}_o}{\mathcal{R}_{air}(2\mathcal{R}_i + \mathcal{R}_o)} MMF, \tag{6}$$

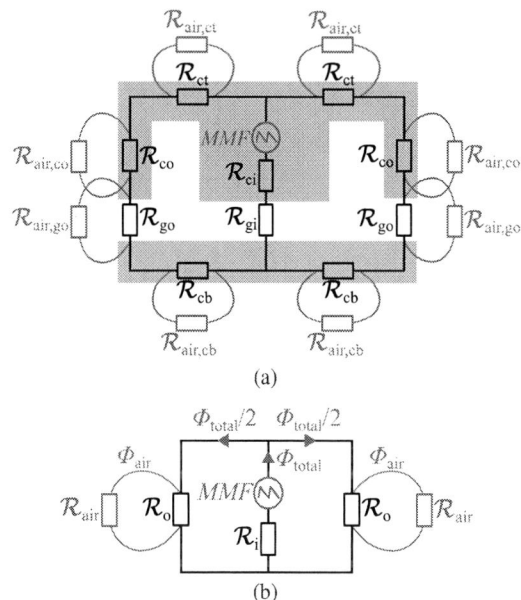

(a)

(b)

Fig. 5: Modeling stray magnetic field in air around inductor L_1: (a) magnetic reluctance model with reluctances of core, air-gaps, and surrounding air, (b) simplified model with lumped reluctances $\mathcal{R}_i$, $\mathcal{R}_o$, and $\mathcal{R}_{air}$.

by assuming that $\mathcal{R}_{air} \gg \mathcal{R}_o$.

To minimize couplings between inductor L_1 and the EMI filter, the inductor's Φ_{air} should be minimized for a given Φ_{total}. Therefore, the Φ_{air} on each half of the inductor can be normalized to the total flux on each half of the inductor ($\Phi_{total}/2$) as

$$\begin{aligned} \Phi_{air,norm} &= \frac{\Phi_{air}}{\Phi_{total}/2} \\ &= \frac{2\Phi_{air}}{MMF/\mathcal{R}_{total}} \\ &= \frac{2\Phi_{air}}{MMF/(\mathcal{R}_i + \frac{1}{2}\mathcal{R}_o)}. \end{aligned} \tag{7}$$

The expression for Φ_{air} from (6) is then substituted to derive

$$\Phi_{air,norm} \approx \frac{\mathcal{R}_o}{\mathcal{R}_{air}}. \tag{8}$$

From the derivation result in (8), it is clear that the reluctance of the outer inductor core legs $\mathcal{R}_o$ should be minimized to reduce the effect of parasitic magnetic couplings from inductor L_1. This is based on the assumption that the magnetic fields from the center post are completely contained within the inductor core.

The most effective way to reduce $\mathcal{R}_o$ is to minimize or eliminate the airgap on the outer core legs. To study the effect of the outer leg airgap, a 3D magnetostatic FEA simulation is used as shown in Fig. 6(a). The inductor model uses the same windings and core geometry as shown in Fig. 2 and the core has relative permeability $\mu_r = 2000$. For this study, a 1.0 A sinusoidal current is applied while l_{go} is varied from 0 to 1.0

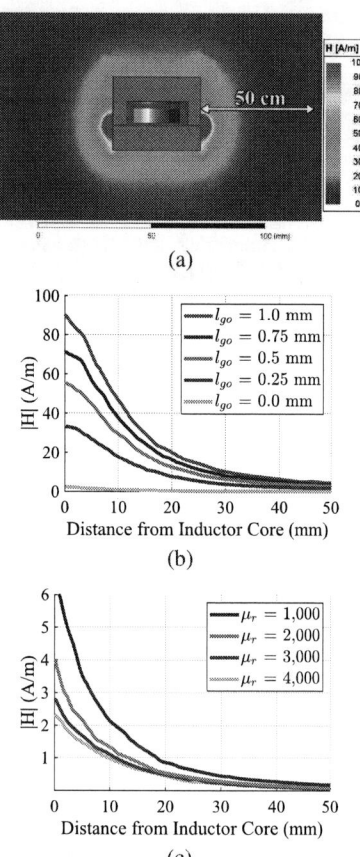

(a)

(b)

(c)

Fig. 6: Study on magnetic field outside inductor L_1 using 3D FEA simulation: (a) 3D simulation model, (b) magnetic field intensity outside core with varied outer airgap l_{go}, (c) magnetic field intensity outside core with varied core permeability μ_{r}.

mm over five steps, and the magnetic field intensity is plotted along a 50 cm line outside the core. For each value of l_{go}, the center post airgap length l_{gi} is adjusted to maintain a fixed inductance $L_1 = 12.6$ μH. As shown in Fig. 6(b), the magnetic field intensity outside the core is approximately proportional to l_{go}. This matches well to (8), where $\mathcal{R}_{\mathrm{o}} \approx \mathcal{R}_{\mathrm{go}} \propto l_{\mathrm{go}}$.

Using no airgap on the outer core legs greatly reduces the inductor's Φ_{air}, compared to the Φ_{air} with an airgap on the outer core legs. However, the remaining Φ_{air} can still couple to the EMI filter and increase noise emissions. In this case, the inductor's Φ_{air} is related to the geometry and permeability of the inductor's outer legs.

Another study is conducted with 3D FEA simulation to demonstrate the effect of the core permeability on the inductor's surrounding magnetic field. The same simulation model from Fig. 6(a) is used, except l_{go} is fixed to zero and the μ_{r} of the core material is varied from 1,000 to 4,000. The results in Fig. 6(c) show that the magnetic field intensity outside the inductor core is approximately inversely proportional to μ_{r}. Again, this matches well to (8), where $\mathcal{R}_{\mathrm{o}} = (\mathcal{R}_{\mathrm{ct}} + \mathcal{R}_{\mathrm{co}} + \mathcal{R}_{\mathrm{cb}}) \propto \frac{1}{\mu_{\mathrm{r}}}$.

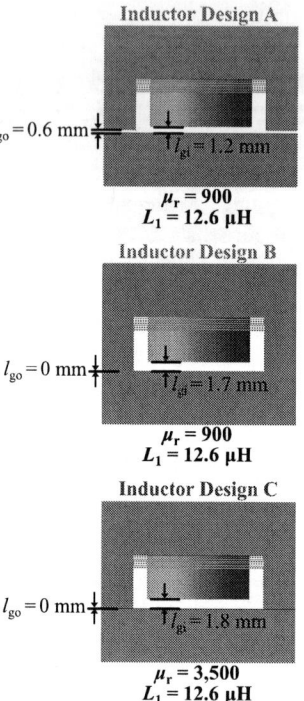

Fig. 7: Three design prototypes for inductor L_1 to evaluate the effect of air-gaps and core permeability on magnetic couplings.

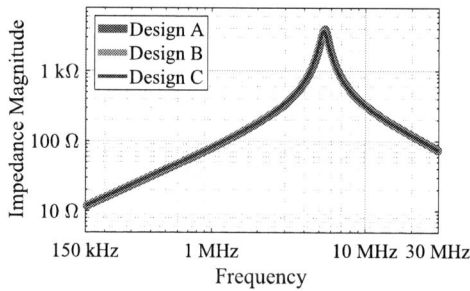

Fig. 8: Comparison of measured impedances for inductor L_1 using each of the three prototypes shown in Fig. 7.

IV. EXPERIMENTAL RESULTS

The effect of the ferrite core's outer leg air gap (l_{go}) and permeability (μ_{r}) are predicted in Section III to significantly affect the leakage magnetic field outside of the inductor. The effect of these two core design parameters are experimentally evaluated in this Section using the three inductor core prototypes shown in Fig. 7.

Inductor Design A and Design B have the same core material ($\mu_{\mathrm{r}} = 900$), but Design A has a large airgap on the outer core legs while Design B has no airgap on its outer legs. Inductor Design B and Design C both have no airgap on the outer core legs ($l_{\mathrm{go}} = 0$), but Design B uses a lower permeability ferrite material ($\mu_{\mathrm{r}} = 900$) while Design C uses a higher permeability ferrite material ($\mu_{\mathrm{r}} = 3,500$).

In each design, the inner airgap length (l_{gi}) is adjusted to

979-8-3315-1612-3/25 $31.00 © 2025 IEEE

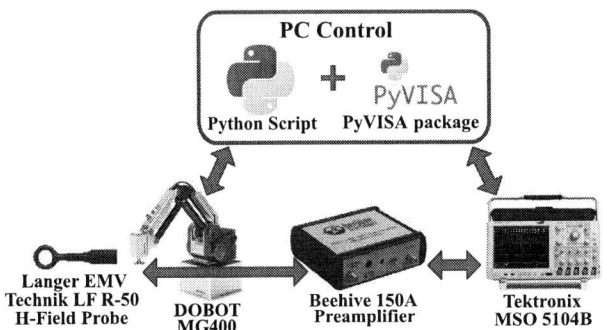

Fig. 9: Overview of automated field measurement system used to measure magnetic field.

TABLE III
AVERAGE MAGNETIC FIELD INTENSITY OVER MEASURED REGION

	Design A	Design B	Design C
Average $|H_X|$	20.6 dBA/m	1.5 dBA/m	-5.2 dBA/m
Average $|H_Y|$	9.8 dBA/m	-7.0 dBA/m	-7.9 dBA/m
Average $|H_Z|$	18.6 dBA/m	-2.3 dBA/m	-8.5 dBA/m

maintain a fixed inductance value $L_1 = 12.6$ µH. The measured impedance of inductor L_1 is compared in Fig. 8 using each of the three core prototypes shown in Fig. 7. This comparison verifies that each inductor design has the same inductance value and overall electrical characteristics.

A. Magnetic Field Measurements

An automated field measurement system, shown in Fig. 9, is used to evaluate the leakage magnetic field from each of the inductor designs. The magnetic field intensity is measured in each of the three Cartesian directions over the region where the EMI filter is located. Each converter is operated at a fixed 200 to 400 V_{dc} condition switching at 500 kHz. The measured magnetic field distributions are shown in Fig. 10, and the average magnetic field intensities for each measurement are given in Table III.

The magnetic field intensity at the location of the EMI filter is significantly reduced in all three dimensions when using Inductor Design B compared to Design A (ΔH_x = -19.1 dB,

ΔH_y = -16.8 dB, ΔH_z = -20.9 dB). This validates the prediction from Section III that having no airgap on the outer core legs significantly reduces the leakage magnetic field around the inductor.

The magnetic field intensity at the location of the EMI filter is further reduced when using Inductor Design C compared to Design B (ΔH_x = -6.7 dB, ΔH_y = -0.9 dB, ΔH_z = -6.2 dB). Again, this validates the prediction from Section III that using a core with higher permeability can further reduce the leakage magnetic field around the inductor when no airgap is used on the outer core legs.

B. Conducted EMI Measurements

The conducted DM noise of the PFC converter is shown in Fig. 11 using each of the three inductor design prototypes shown in Fig. 7. For each test, the converter is operated at the AC/DC operating condition described in Table II. These results demonstrate the strong effect of parasitic magnetic coupling on DM noise in this converter, as modeled in Section II. In addition, the changes in DM noise correspond well to the reduction in magnetic field. Changing from Design A to Design B reduces the DM noise by 20 dB at low frequency with a maximum reduction of nearly 30 dB. Changing from Design B to Design C leads to a further reduction of about 6 dB in low-frequency DM noise.

The conducted CM noise of the PFC converter is compared in Fig. 12 using each of the three inductor design prototypes

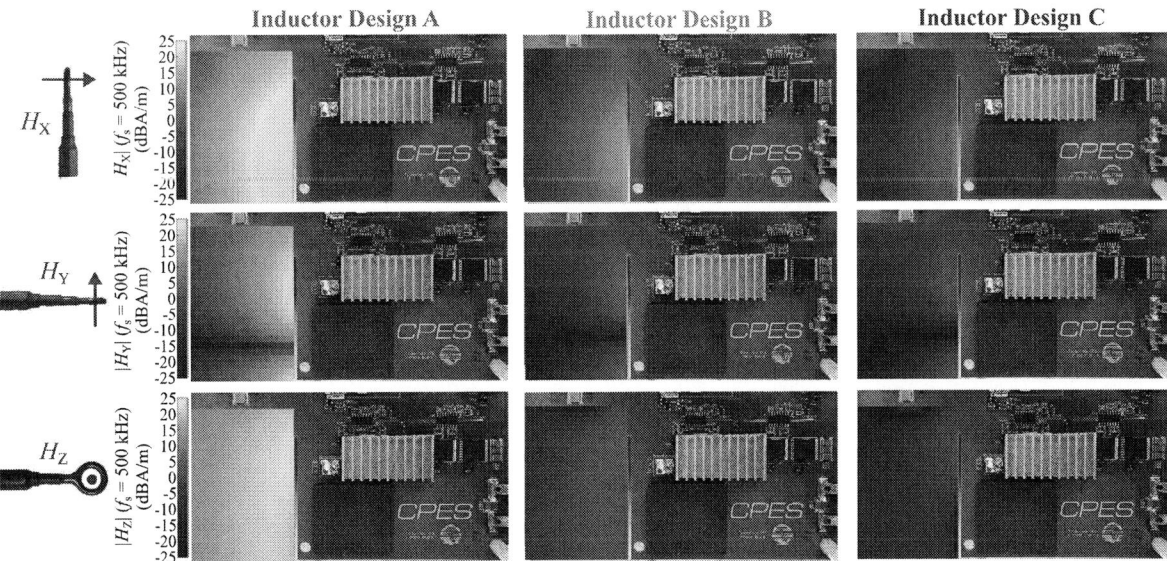

Fig. 10: Switching frequency magnetic field intensity measured at the location of the EMI filter using each of the three prototypes for inductor L_1 shown in Fig. 7.

979-8-3315-1612-3/25 $31.00 © 2025 IEEE

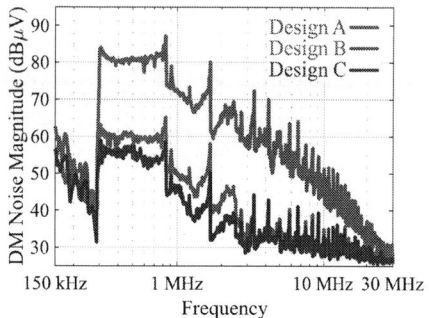

Fig. 11: Comparison of measured DM noise using each of the three prototypes for inductor L_1 shown in Fig. 7.

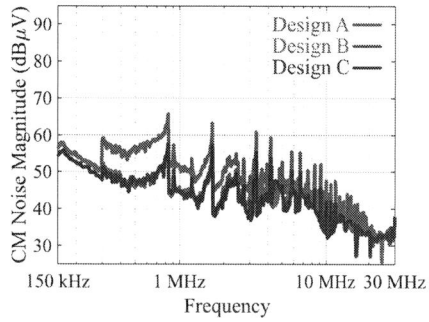

Fig. 12: Comparison of measured CM noise using each of the three prototypes for inductor L_1 shown in Fig. 7.

shown in Fig. 7. Again, the converter in each test is operated at the AC/DC operating condition described in Table II. These results demonstrate a significant but more moderate effect of parasitic magnetic couplings on the CM noise. As shown, changing from Design A to Design B reduces the low-frequency CM noise by about 7 dB but has less effect at higher frequencies. Meanwhile, changing from Design B to Design C provides no further reduction in the CM noise. This indicates that the magnetic coupling can have some effect on the CM noise in certain situations, but the CM noise is usually more related to capacitive couplings as modeled in Section II.

V. CONCLUSION

This work investigates how parasitic magnetic couplings between the PCB-winding inductor and EMI filter affect the EMI emissions of a high-frequency GaN-based PFC converter. Conducted EMI models are provided for both DM and CM noise which include these magnetic coupling effects. Then, a magnetic reluctance model is provided to relate the design of the ferrite inductor core to the leakage magnetic field outside the inductor. Three ferrite inductor core prototypes with different airgaps and permeability are then evaluated experimentally. Using an inductor with no airgaps on the outer core legs reduces the switching magnetic field intensity outside the inductor by about 20 dB, while using a high permeability ferrite core material further reduces the magnetic field by about 6 dB. This results in a 20-30 dB reduction in the converter's DM noise and about 7 dB reduction in the CM noise.

ACKNOWLEDGMENTS

This work was supported by the High Density Integration (HDI) and Power Management Consortium (PMC). The authors would also like to thank Hengdian Group DMEGC Magnetics for their in-kind donations of ferrite core samples.

REFERENCES

[1] S. Wang, F. Lee, D. Chen, and W. Odendaal, "Effects of parasitic parameters on EMI filter performance," *IEEE Transactions on Power Electronics*, vol. 19, no. 3, pp. 869–877, 2004.

[2] S. Wang, F. Lee, W. Odendaal, and J. van Wyk, "Improvement of EMI filter performance with parasitic coupling cancellation," *IEEE Transactions on Power Electronics*, vol. 20, no. 5, pp. 1221–1228, 2005.

[3] Y. Murata, K. Takahashi, T. Kanamoto, and M. Kubota, "Analysis of Parasitic Couplings in EMI Filters and Coupling Reduction Methods," *IEEE Transactions on Electromagnetic Compatibility*, vol. 59, no. 6, pp. 1880–1886, 2017.

[4] B. Liu, R. Ren, F. F. Wang, D. Costinett, and Z. Zhang, "Capacitive Coupling in EMI Filters Containing T-Shaped Joint: Mechanism, Effects, and Mitigation," *IEEE Transactions on Power Electronics*, vol. 35, no. 3, pp. 2534–2547, 2020.

[5] L. Feng, W. Chen, H. Chen, and Z. Qian, "Study on the conducted EMI due to radiated coupling in SMPS," in *Twenty-First Annual IEEE Applied Power Electronics Conference and Exposition, 2006. APEC '06.*, 2006, pp. 4 pp.–.

[6] S. Ye, X. Zheng, Z. Zheng, J. Xiao, and H. Chen, "Non-ideal electric field shielding with grounding resistor for suppressing EMI coupling in a power converter," in *2018 IEEE International Symposium on Electromagnetic Compatibility and 2018 IEEE Asia-Pacific Symposium on Electromagnetic Compatibility (EMC/APEMC)*, 2018, pp. 1089–1092.

[7] Y. Li, S. Wang, H. Sheng, and S. Lakshmikanthan, "Investigate and Reduce Capacitive Couplings in a Flyback Adapter With a DC-Bus Filter to Reduce EMI," *IEEE Transactions on Power Electronics*, vol. 35, no. 7, pp. 6963–6973, 2020.

[8] H. Chen and Z. Qian, "Modeling and Characterization of Parasitic Inductive Coupling Effects on Differential-Mode EMI Performance of a Boost Converter," *IEEE Transactions on Electromagnetic Compatibility*, vol. 53, no. 4, pp. 1072–1080, 2011.

[9] Y. Chu, S. Wang, N. Zhang, and D. Fu, "A Common Mode Inductor With External Magnetic Field Immunity, Low-Magnetic Field Emission, and High-Differential Mode Inductance," *IEEE Transactions on Power Electronics*, vol. 30, no. 12, pp. 6684–6694, 2015.

[10] T. McGrew, S. Wang, and Q. Li, "Reduction of CM Noise by Minimizing Near-Field Effects in a DC–DC Converter," *IEEE Transactions on Power Electronics*, vol. 39, no. 4, pp. 4210–4223, 2024.

[11] K. Takahashi, Y. Murata, Y. Tsubaki, T. Fujiwara, H. Maniwa, and N. Uehara, "Mechanism of Near-Field Coupling Between Noise Source and EMI Filter in Power Electronic Converter and Its Required Shielding," *IEEE Transactions on Electromagnetic Compatibility*, vol. 61, no. 5, pp. 1663–1672, 2019.

[12] Z. Liu, F. C. Lee, Q. Li, and Y. Yang, "Design of GaN-Based MHz Totem-Pole PFC Rectifier," *IEEE Journal of Emerging and Selected Topics in Power Electronics*, vol. 4, no. 3, pp. 799–807, 2016.

[13] T. McGrew, X. Chen, and Q. Li, "PCB Inductor with Integrated Shielding to Contain Switching Electric Field and Reduce CM Noise," *IEEE Transactions on Power Electronics*, pp. 1–14, 2024.

[14] Z. Liu, Z. Huang, F. C. Lee, and Q. Li, "Digital-Based Interleaving Control for GaN-Based MHz CRM Totem-Pole PFC," *IEEE Journal of Emerging and Selected Topics in Power Electronics*, vol. 4, no. 3, pp. 808–814, 2016.

Optically-Controlled 3.3 kV SiC MOSFET with Fast Switching Speed and Low Optical Power

Xin Yang
Center for Power Electronics
Systems (CPES)
Virginia Tech
Blacksburg, USA
Email: xxxyang@vt.edu

Guannan Shi
Bradley Department of Electrical
and Computer Engineering
Virginia Tech
Blacksburg, USA
Email: guannanshi@vt.edu

Liyang Jin
Bradley Department of Electrical
and Computer Engineering
Virginia Tech
Blacksburg, USA
Email: liyangjin@vt.edu

Yuan Qin
Center for Power Electronics
Systems (CPES)
Virginia Tech
Blacksburg, USA
Email: yuanqin@vt.edu

Matthew Porter
Center for Power Electronics
Systems (CPES)
Virginia Tech
Blacksburg, USA
Email: maporter@vt.edu

Che-Wei Chang
Center for Power Electronics
Systems (CPES)
Virginia Tech
Blacksburg, USA
Email: cwchang@vt.edu

Xiaoting Jia
Bradley Department of Electrical
and Computer Engineering
Virginia Tech
Blacksburg, USA
Email: xjia@vt.edu

Dong Dong
Center for Power Electronics
Systems (CPES)
Virginia Tech
Blacksburg, USA
Email: dongd@vt.edu

Linbo Shao
Bradley Department of Electrical
and Computer Engineering
Virginia Tech
Blacksburg, USA
Email: shaolb@vt.edu

Yuhao Zhang
Department of Electrical and
Electronic Engineering
The University of Hong Kong
Hong Kong, China
Email: yuhzhang@hku.hk

Abstract— Optically-controlled high-voltage power devices hold good promise for grid and renewable energy applications by providing superior electromagnetic interference (EMI) immunity and reduced switching delay. This paper proposed a novel optically-controlled gate driver architecture that applies complementary optical signals to two photodiodes (PDs) arranged in a totem-pole configuration. This configuration enables fast switching of power semiconductor devices using minimal optical power, as only low-power driver signals are optically modulated and device main current is not photogenerated. To validate this approach, we employ two InGaAs PDs to drive a 3.3 kV SiC MOSFET, the highest-voltage industrial unipolar device currently available. When each PD is illuminated by 21.7 mW optical power, the SiC MOSFET achieves hard-switching at 1500V/3A, with rise and fall times of 152 ns and 214 ns, respectively. These results set new records for switching voltage, speed, and power capacity-to-optical power ratio in optically-controlled unipolar power switches. This general optical driver design is also applicable to the future development of integrated optics for power electronics in diverse (ultra-) wide-bandgap semiconductors.[1]

Keywords— power electronics, optical driver, optical control, high voltage, switching speed, optical power, SiC MOSFET.

I. INTRODUCTION

Power devices are conventionally driven by auxiliary electrical drivers to enable easy control and fast switching [1], [2], [3], [4], [5], [6], [7], [8]. Recently, optically-driven power devices have gained increased attention due to their superior electromagnetic interference (EMI) immunity, minimized triggering delay, and ideal isolation between control and power stages [9], [10], [11], [12], [13]. These merits make them well-suited for high-voltage, high-power applications and device stacking.

Optically-driven power devices can generally be classified into two types based on their operating principles. The first type includes photoconductive semiconductor switches (PCSS) [14], [15], [16], which is primarily for pulsed power applications. The PCSS conducts the transient current at a high DC voltage, and the light illumination is usually only applied in very short pulses. The second type of optically-drive device functions as a normal switch, which include light-triggered thyristors (LTT) [17], [18]. Despite the fast progress on optically-controlled power switches (i.e., the second type), they have traditionally relied on photogenerated current, rendering a trade-off between required optical power and switching speed [19].

To mitigate this trade-off, an optically-activated gate driver has been developed, which utilize optical devices solely within the low-power driver loop. This approach enables more energy-

---

[1] This work is supported in part by the Office Naval Research monitored by Lynn Petersen (Award N000142412227) and in part by the CPES Industry Consortium.

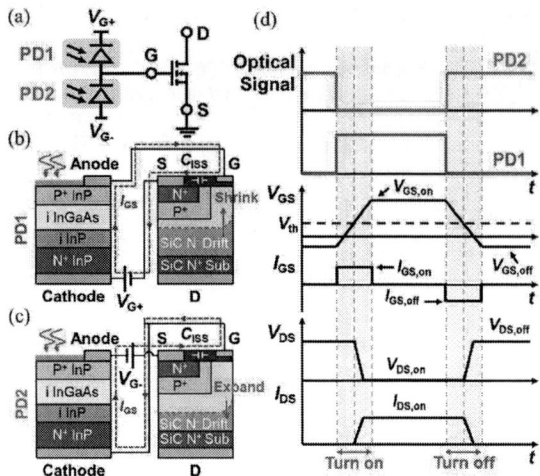

Fig. 1. (a) Circuit schematic of the proposed optically-controlled gate driver. Device dynamics at the Miller plateau in the (b) turn-on and (c) turn-off process. (d) Ideal switching waveforms of two PDs and power device.

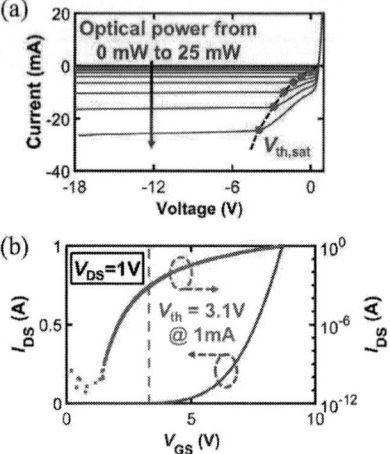

Fig. 2. (a) *I-V* characteristics of InGaAs PD at various optical powers. (b) Transfer characteristics of 3.3 kV SiC MOSFET (DUT) in the linear and semi-log scales.

efficient optical control of unipolar devices without bipolar gain. For example, in Ref. [20], the low-voltage GaAs optically triggered power transistors (OTPTs) are deployed to switch a 650 V Si Cool MOSFET or 1200 V SiC MOSFET. On the other hand, a direct optical gate control has been recently reported in unipolar devices such as vertical GaN FinFETs [21] and lateral GaN HEMTs [22], but the highest switching voltage reported in these devices is merely 30 V, which is far below the requirement in practical systems.

In this work, we propose a novel optical driving scheme using two photodiodes (PDs) triggered by complementary optical signals to drive a high-voltage unipolar power transistor. During switching, the two reverse-biased PDs are alternately illuminated to deliver current that charges and discharges the gate-drain capacitance (C_{GD}) of the power transistor during turn-on and turn-off transients. Experimental results using two InGaAs PDs to drive a 3.3 kV SiC MOSFET in a 1.5 kV hard-switching test show rise and fall times of 152 ns and 214 ns, respectively, with an optical power of only 21.7 mW on each PD. These results set several new records in the performance of optically-controlled power devices. In this conference paper, we present the main experimental results; a more detailed analysis on device physics and driver EMIs are provided in [23].

II. OPERATION PRINCIPLES AND TEST SETUP

Fig. 1(a) shows the circuit diagram of the proposed optically-controlled device. In this design, two PDs are connected in series between the positive and negative driver voltages (V_{G+} and V_{G-}), with their midpoint linked to the gate of the device under test (DUT), forming a totem-pole structure. Both PDs block the voltage bias in the absence of optical signals, which enables them to operate in the off-state with a lower junction capacitance. This lower junction capacitance can enhance the response speed under light illumination. Fig. 1(b) and (c) show the dynamic gate current flow within the PDs and the DUT at the Miller plateaus during turn-on and turn-off transitions. Fig. 1(d) shows the ideal waveforms of the complementary optical signals, along with the DUT's gate-

source voltage (V_{GS}), gate current (I_{GS}), drain-source voltage (V_{DS}), and drain current (I_{DS}) over a switching cycle.

The DUT's turn-on process initiates when PD1 is illuminated, generating photogenerated holes and electrons that migrate toward the anode and cathode of PD1, creating a gate charge current ($I_{GS,on}$). This current initially charges the DUT's input capacitance (C_{ISS}) and subsequently discharges its gate-drain capacitance (C_{GD}) at the Miller plateau. During the turn-off transition, PD2 is illuminated, producing a negative gate current ($I_{GS,off}$) that first discharges the C_{ISS} of the DUT and then charges C_{GD} at the Miller plateau.

For experimental demonstration, we adopt InGaAs PDs and trigger them with a near-infrared laser of 1550-nm wavelength [24], [25], [26]. While the ultraviolet light suffers higher propagation loss in optical fibers, the 1550 nm wavelength, which is widely used in current long-haul fiber optical communication, features a low propagation loss of <0.2 dB/km and is desired in distributed and large-scale power applications [27]. For power device, a 3.3 kV SiC MOSFET is selected for demonstration, as it has the highest voltage rating in commercially available unipolar devices.

Fig. 2(a) shows the static characteristics of the InGaAs PD (DPIN-23133) used in this study. As optical power increases from 0 mW to 25 mW, the reverse saturation threshold voltage ($V_{th,sat}$) rises to -4 V. Below $V_{th,sat}$, the saturation current (I_{sat}) increases from 7 μA to ~25 mA as optical power reaches 25 mW. For power device, a 3.3 kV/11 A rated SiC MOSFET (MSC400SMA330) is selected as the DUT [28], and its transfer and output characteristics are shown in Fig. 2(b). The DUT exhibits an on/off ratio of ~1.3×10^{10}, a threshold voltage of 3.1 V, and a ~420 mΩ on-resistance (R_{ON}) at 20 V V_{GS}.

For experimental demonstration, a test system is built to integrate a fiber-optics system and a double-pulse test (DPT) board. Fig. 3(a) shows the schematic of fiber-optics system. A tunable laser (Santec TSL) with an Erbium-doped fiber amplifier provides a 1550-nm light at desired power levels. A fiber-optic coupler splits the light equally into two paths. Polarization controllers ensure the light is at the preferred input

979-8-3315-1612-3/25 $31.00 © 2025 IEEE

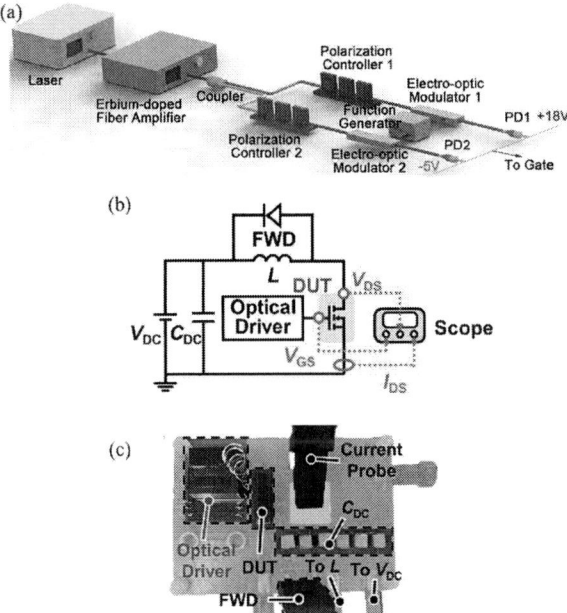

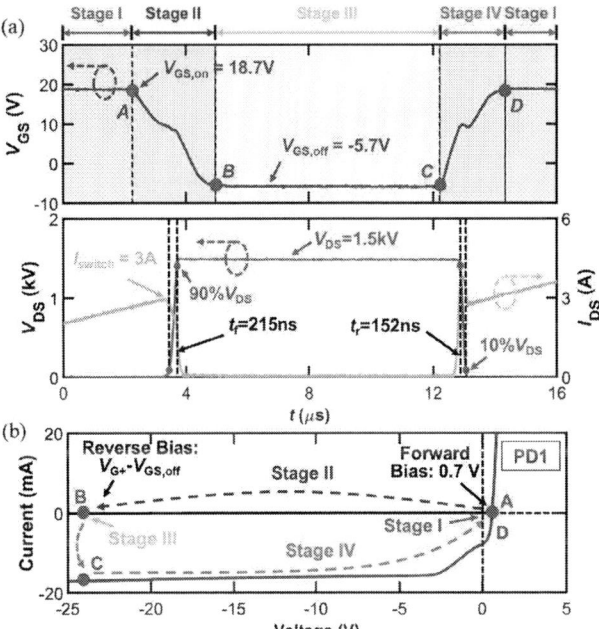

Fig. 3. Schematics of the (a) fiber-optics system and (b) DPT circuit board. (c) Photo of DPT test board.

Fig. 5. (a) 1.5kV/3A DPT waveforms under 21.7 mW optical power on each PD. (b) Switching locus of PD1. The blue curve is the PD characteristics under 21.7 mW optical power. The stages and operation points are illustrated in (a).

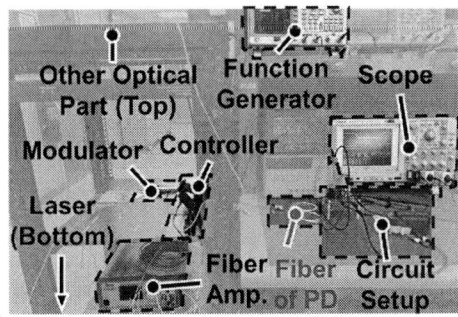

Fig. 4. Photograph of optical and electrical test system.

polarization for each electro-optic modulator [29], [30], [31]. Driven by a function generator, two electro-optic modulators provide the complementary light signals between two optical paths, which connect to two fiber-coupled InGaAs PIN PD1 and PD2. For electrical setup, as shown in Fig. 3(b), the DPT power loop includes a DC power supply (V_{DC}), DC input capacitance (C_{DC}), inductor (L), free-wheeling diode (FWD), and the DUT. The gate-to-source voltage (V_{GS}), V_{DS}, and drain current (I_{DS}) are measured using a passive probe (TPP1000), high voltage differential probe (THDP0200), and current probe (TCP0030A), respectively. Fig. 3(c) shows a photograph of test setup. The L is set as 1 mH to achieve a current rise rate of 1.5 A/μs at 1.5 kV V_{DC}. The V_{G+} and V_{G-} are set as +18 V and -5 V. And the total test setup including the optical system and electrical setup is shown in Fig. 4.

III. Switching Performance and Benchmark

Fig. 5(a) shows the switching waveforms for 1.5 kV / 3 A under 21.7 mW of optical power applied to each PD. The clean,

textbook-like hard-switching waveforms with minimal voltage overshoot confirm the effectiveness of the optical control. The fall time (t_f) and rise time (t_r), measured between 10% and 90% of V_{DS} during the transitions, are 215 ns and 152 ns, respectively. The energy loss during hard turn-on and turn-off of the DUT is extracted to be 905 μJ and 570 μJ, respectively, corresponding to a switching loss of 1.47 W at a 1 kHz frequency. The use of approximately 70 times less optical power implies that this optical system minimally affects the efficiency of practical circuits utilizing such optically-driven unipolar devices.

The waveforms also reveal a total gate-source transition time of 2.7 μs during turn-off and 2.1 μs during turn-on, which are limited by the I_{sat} of the PDs. Although this transition time is longer than t_f and t_r, it does not dramatically increase power loss, as switching loss primarily occurs at the Miller plateau. For typical applications of a 3.3 kV device operating at 10–50 kHz, this transition time is significantly shorter than the switching period, allowing ample margin for duty cycle modulation.

Interestingly, the PD operation here diverges from traditional optoelectronic applications, experiencing large-signal transitions between photoconductive (reverse-biased) and photovoltaic (forward-biased) modes. Fig. 5(b) shows the switching locus of PD1 across a full switching cycle. In Stage I (steady on-state, as labeled in Fig. 5(a)), PD1 is illuminated without current flow, resulting in a forward bias (V_F) of approximately 0.7 V, which corresponds to the observed on-state V_{GS} of 18.7 V ($V_{GS,on} = V_F + V_{G+}$). During Stage II (turn-off transient) and Stage III (steady off-state), PD1 is unilluminated. In Stage III, the off-state V_{GS} ($V_{GS,off}$) equals V_{G-} minus PD2's V_F, matching the measured -5.7 V. Under these conditions, PD1 is biased at V_{G+} - $V_{GS,off}$ = 23.7 V. In Stage IV (turn-on transient), PD1 is illuminated, initially ramping up to a higher I_{sat}, and then

979-8-3315-1612-3/25 $31.00 © 2025 IEEE

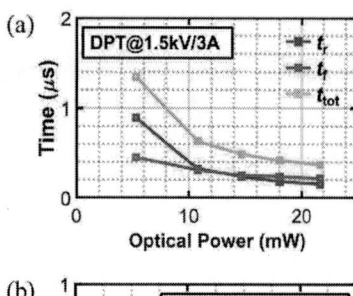

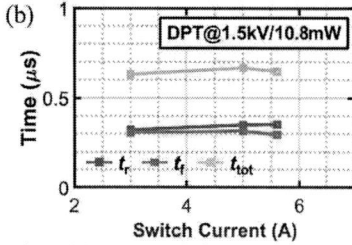

Fig. 6. Switching times (a) versus optical power at a 3 A switching current and (b) versus switching current under a 10.8 mW optical power.

transitioning to V_F along its I-V curve under 21.7 mW optical power.

Fig. 6(a) and (b) show measured t_f and t_r across varying optical powers and switching currents. Increasing optical power from 5.3 mW to 21.7 mW per PD decreases t_f by 230 ns and t_r by 740 ns due to increased charge/discharge current. At a fixed optical power of 10.8 mW, raising the switching current from 3 A to 6 A yields negligible changes in t_f and t_r, with the combined switching time ($t_r + t_f$) remaining around 640 ns. This contrasts with electrical gate drivers, where t_f and t_r are typically sensitive to switching current due to variations in gate plateau voltage at the Miller plateau [32], [33]. The detailed analysis on the device physics accounting for these phenomena is presented in [23].

The weak dependence of switching time on switching current can be advantageous for certain applications. The results suggest that both conduction and switching currents can be increased without requiring additional optical power or compromising the total switching time, highlighting the advantage of our hybrid device design over other devices that rely on photogenerated current. In these devices, optical power must increase to handle higher conduction currents.

From an application standpoint, the ideal metrics for an optically-controlled device include high power capacity (the product of conduction current and blocking voltage), low optical power consumption, low power loss, and fast switching speed. Accordingly, Table I compares these application-oriented metrics for our device with other reported optically-controlled unipolar devices [20], [21], [22], [34], despite the varied physical principles behind each device. Our hybrid device demonstrates the highest voltage, current, on/off ratio, and the best ratio of power capacity to optical power, with switching times among the shortest reported. Furthermore, our design's material-agnostic nature broadens its applicability across different materials.

TABLE I. SUMMARY AND BENCHMARK OF OPTICALLY-CONTROLLED UNIPOLAR POWER DEVICES.

Power Device	Ref	Optical Device	Optical Power (mW)	Voltage (V)	Current (A)	Total Switching Times (ns)
SiC MOS	This work	PD	44	1500	3	367
SiC MOS	[34]	OTPT	4000	500	0.77	300
Si MOS	[20]	OTPT	4000	50	1.25	410
GaN HEMT	[22]	Optical Gate	245	30	0.011	2473
GaN FinFET	[21]	Optical Gate	0.001	3	0.029	$>10^{10}$

IV. CONCLUSION

This work introduces a novel optical gate driver design to enable an optically-controlled, high-voltage unipolar power switch. The core design feature is the use of two reverse-biased PDs illuminated by complementary optical signals. Our experimental demonstration using a 3.3 kV SiC MOSFET with two InGaAs PDs establishes several performance records for optically-driven unipolar power devices. This design offers excellent scalability in terms of switching voltage, current, speed, and frequency with minimal additional optical power. It has strong potential as a foundational element for ultrafast, energy-efficient, optically-controlled power electronics across various materials and device systems.

REFERENCES

[1] Y. Zhang, F. Udrea, and H. Wang, "Multidimensional device architectures for efficient power electronics," Nat. Electron., vol. 5, no. 11, pp. 723–734, Nov. 2022, doi: 10.1038/s41928-022-00860-5.

[2] Y.-T. Huang, C.-C. Yang, T.-S. Li, and Y.-M. Chen, "A Feedforward Voltage Control Strategy for Reducing the Output Voltage Double-Line-Frequency Ripple in Single-Phase AC–DC Converters," IEEE J. Emerg. Sel. Top. Power Electron., vol. 9, no. 6, pp. 6605–6612, Dec. 2021, doi: 10.1109/JESTPE.2021.3083258.

[3] M. Wu, X. Yu, X. Yang, W. Chen, and L. Wang, "An Efficiency Optimization Method for the Multiple Coils WPT System Against the Pad Misalignment," IEEE Trans. Transp. Electrification, pp. 1–1, 2024, doi: 10.1109/TTE.2024.3410676.

[4] X. Yang et al., "Evaluation and MHz Converter Application of 1.2-kV Vertical GaN JFET," IEEE Trans. Power Electron., pp. 1–11, 2024, doi: 10.1109/TPEL.2024.3445667.

[5] Y. Jiang et al., "A Dynamic Efficiency Optimization Method under ZVS Conditions in the Series-Series Type Wireless Power Transfer System," in 2020 IEEE Energy Conversion Congress and Exposition (ECCE), Oct. 2020, pp. 995–1001. doi: 10.1109/ECCE44975.2020.9235452.

[6] M. Porter, X. Yang, H. Gong, B. Wang, Z. Yang, and Y. Zhang, "Switching figure-of-merit, optimal design, and power loss limit of (ultra-) wide bandgap power devices: A perspective," Appl. Phys. Lett., vol. 125, no. 11, p. 110501, Sep. 2024, doi: 10.1063/5.0222105.

[7] T. Yuan, F. Jin, Z. Li, C. Zhao, and Q. Li, "Design of an Integrated Transformer With Parallel Windings for a 30-kW LLC Resonant Converter," IEEE Trans. Power Electron., vol. 38, no. 11, pp. 14317–14333, Nov. 2023, doi: 10.1109/TPEL.2023.3291954.

[8] T. Yuan, F. Jin, and Q. Li, "Analysis and Comparison of Integrated Planar Transformers for 22-kW On-board Chargers," IEEE Trans. Power Electron., pp. 1–19, 2024, doi: 10.1109/TPEL.2024.3410878.

[9] S. K. Mazumder, "An Overview of Photonic Power Electronic Devices," IEEE Trans. Power Electron., vol. 31, no. 9, pp. 6562–6574, Sep. 2016, doi: 10.1109/TPEL.2015.2500903.

[10] T.-S. Li, Y.-H. Yang, C.-A. Cheng, and Y.-M. Chen, "A Variable DC-Link Voltage Determination Method for Motor Drives with SiC MOSFETs," in 2020 IEEE Workshop on Wide Bandgap Power Devices and Applications in Asia (WiPDA Asia), Sep. 2020, pp. 1–6. doi: 10.1109/WiPDAAsia49671.2020.9360266.

[11] T. Yuan, F. Jin, Z. Li, and Q. Li, "Current Sharing Analysis of a High Power Transformer with Parallel Windings," in 2023 IEEE Applied Power Electronics Conference and Exposition (APEC), Mar. 2023, pp. 1551–1556. doi: 10.1109/APEC43580.2023.10131421.

[12] Z. Xiao, W. Lei, Z. Xiang, Y. Yin, and W. Wang, "Full-state discrete-time model-based stability analysis and parameter design of dual active bridge converter in energy storage system," IET Power Electron., vol. 15, no. 13, pp. 1229–1248, 2022, doi: 10.1049/pel2.12319.

[13] C. Zhao et al., "A Load Detection Method in Multi-transmitter Dynamic Wireless Power Transfer Systems without Extra Sensors and Communication," in 2023 11th International Conference on Power Electronics and ECCE Asia (ICPE 2023 - ECCE Asia), May 2023, pp. 2573–2578. doi: 10.23919/ICPE2023-ECCEAsia54778.2023.10213968.

[14] G. M. Loubriel et al., "Photoconductive semiconductor switches," IEEE Trans. Plasma Sci., vol. 25, no. 2, pp. 124–130, Apr. 1997, doi: 10.1109/27.602482.

[15] E. Majda-Zdancewicz, M. Suproniuk, M. Pawłowski, and M. Wierzbowski, "Current state of photoconductive semiconductor switch engineering," Opto-Electron. Rev., vol. 26, no. 2, pp. 92–102, May 2018, doi: 10.1016/j.opelre.2018.02.003.

[16] J. S. Sullivan and J. R. Stanley, "Wide Bandgap Extrinsic Photoconductive Switches," IEEE Trans. Plasma Sci., vol. 36, no. 5, pp. 2528–2532, Oct. 2008, doi: 10.1109/TPS.2008.2002147.

[17] J. Vobecký et al., "Silicon Thyristors for Ultrahigh Power (GW) Applications," IEEE Trans. Electron Devices, vol. 64, no. 3, pp. 760–768, Mar. 2017, doi: 10.1109/TED.2016.2638476.

[18] X. Wang, H. Pu, Q. Liu, L. An, X. Tang, and Z. Chen, "Demonstration of 4H-SiC Thyristor Triggered by 100-mW/cm2 UV Light," IEEE Electron Device Lett., vol. 41, no. 6, pp. 824–827, Jun. 2020, doi: 10.1109/LED.2020.2988913.

[19] Y. Zhang, "Investigation and Improvement of Switching Characteristics of SiC Optically Controlled Transistor," J. Phys. Conf. Ser., vol. 2331, no. 1, p. 012006, Aug. 2022, doi: 10.1088/1742-6596/2331/1/012006.

[20] S. K. Mazumder and T. Sarkar, "Optically Activated Gate Control for Power Electronics," IEEE Trans. Power Electron., vol. 26, no. 10, pp. 2863–2886, Oct. 2011, doi: 10.1109/TPEL.2009.2034856.

[21] J.-H. Hsia, J. A. Perozek, and T. Palacios, "First Demonstration of Optically-Controlled Vertical GaN finFET for Power Applications," IEEE Electron Device Lett., vol. 45, no. 5, pp. 774–777, May 2024, doi: 10.1109/LED.2024.3375856.

[22] E. Palmese, H. Xue, D. J. Rogers, and J. J. Wierer, "Light-Triggered, Enhancement-Mode AlInN/GaN HEMTs with Sub-microsecond

Switching Times," IEEE Electron Device Lett., pp. 1–1, 2024, doi: 10.1109/LED.2024.3440177.

[23] X. Yang et al., "Ultrafast Optically Controlled Power Switch: A General Design and Demonstration With 3.3 kV SiC MOSFET," IEEE Trans. Electron Devices, pp. 1–6, 2024, doi: 10.1109/TED.2024.3485018.

[24] I. Kimukin, N. Biyikli, B. Butun, O. Aytur, S. M. Unlu, and E. Ozbay, "InGaAs-based high-performance p-i-n photodiodes," IEEE Photonics Technol. Lett., vol. 14, no. 3, pp. 366–368, Mar. 2002, doi: 10.1109/68.986815.

[25] R. H. Saul, F. S. Chen, and P. W. Shumate, "Reliability of InGaAs photodiodes for SL applications," ATT Tech. J., vol. 64, no. 3, pp. 861–882, Mar. 1985, doi: 10.1002/j.1538-7305.1985.tb00450.x.

[26] X. Wang, N. Duan, H. Chen, and J. C. Campbell, "InGaAs–InP Photodiodes With High Responsivity and High Saturation Power," IEEE Photonics Technol. Lett., vol. 19, no. 16, pp. 1272–1274, Aug. 2007, doi: 10.1109/LPT.2007.902274.

[27] Q. Lyu, H. Jiang, and K. M. Lau, "High gain and high ultraviolet/visible rejection ratio photodetectors using p-GaN/AlGaN/GaN heterostructures grown on Si," Appl. Phys. Lett., vol. 117, no. 7, p. 071101, Aug. 2020, doi: 10.1063/5.0011685.

[28] "00004433A_MSC400SMA330B4_SiC_MOSFET_Datasheet.pdf." Accessed: Dec. 16, 2024. [Online]. Available: https://ww1.microchip.com/downloads/aemDocuments/documents/PSDS/ProductDocuments/DataSheets/00004433A_MSC400SMA330B4_SiC_MOSFET_Datasheet.PDF

[29] K. Liu, C. R. Ye, S. Khan, and V. J. Sorger, "Review and perspective on ultrafast wavelength-size electro-optic modulators," Laser Photonics Rev., vol. 9, no. 2, pp. 172–194, 2015, doi: 10.1002/lpor.201400219.

[30] J. Liu, G. Xu, F. Liu, I. Kityk, X. Liu, and Z. Zhen, "Recent advances in polymer electro-optic modulators," RSC Adv., vol. 5, no. 21, pp. 15784–15794, 2015, doi: 10.1039/C4RA13250E.

[31] G. Sinatkas, T. Christopoulos, O. Tsilipakos, and E. E. Kriezis, "Electro-optic modulation in integrated photonics," J. Appl. Phys., vol. 130, no. 1, p. 010901, Jul. 2021, doi: 10.1063/5.0048712.

[32] R. Zhang et al., "Switching Performance Evaluation of 650 V Vertical GaN Fin JFET," in 2023 IEEE Applied Power Electronics Conference and Exposition (APEC), Mar. 2023, pp. 2515–2519. doi: 10.1109/APEC43580.2023.10131473.

[33] X. Yang, J. Liu, B. Wang, and G. Zhang, "Pulsed Overcurrent Capability of Power Semiconductor Devices in Solid-State Circuit Breakers: SiC MOSFET vs. Si IGBT," in 2022 IEEE Applied Power Electronics Conference and Exposition (APEC), Mar. 2022, pp. 966–973. doi: 10.1109/APEC43599.2022.9773378.

[34] A. Meyer, S. K. Mazumder, and H. Riazmontazer, "Optical control of 1200-V and 20-A SiC MOSFET," in 2012 Twenty-Seventh Annual IEEE Applied Power Electronics Conference and Exposition (APEC), Feb. 2012, pp. 2530–2533. doi: 10.1109/APEC.2012.6166179.

Optimization Techniques for Parallel-Connected Devices in IPMs for Consumer Use

Keisuke Kawamoto
Power Device Works
Mitsubishi Electric Corporation
Fukuoka, Japan
Kawamoto.Keisuke@
cw.MitsubishiElectric.co.jp

Haruhiko Murakami
Power Device Works
Mitsubishi Electric Corporation
Fukuoka, Japan
Murakami.Haruhiko@
cw.MitsubishiElectric.co.jp

Teruaki Nagahara
Power Device Works
Mitsubishi Electric Corporation
Fukuoka, Japan
Nagahara.Teruaki@
dr.MitsubishiElectric.co.jp

Michael Rogers
Semiconductor & Devices
Mitsubishi Electric US, Inc.
Export, PA USA
Michael.Rogers@meus.com

Akiko Goto
Power Device Works
Mitsubishi Electric Corporation
Fukuoka, Japan
Goto.Akiko@
bx.MitsubishiElectric.co.jp

Shoji Saito
Power Device Works
Mitsubishi Electric Corporation
Fukuoka, Japan
Saito.Shoji@
ak.MitsubishiElectric.co.jp

Koichiro Noguchi
Power Device Works
Mitsubishi Electric Corporation
Fukuoka, Japan
Noguchi.Koichiro@
cw.MitsubishiElectric.co.jp

Abstract— **Reduction of power consumption is a critical and unavoidable concern for global environmental preservation. Enhancing energy efficiency in the use of air conditioners, which consume a significant amount of energy compared to other consumer appliances, has become imperative. The consumer appliance market demands intelligent power modules (IPMs) installed on PCBs, that meet requirements such as compact size, high efficiency, and low-cost. This paper presents an optimization method for an IPM that utilizes Si devices and SiC-MOSFETs in a parallel connection for consumer applications. The active area and gate drive with delay control of the SiC-MOSFET are optimized to reduce losses in the 6in1 IPM, which is used to drive an inverter for an air conditioner compressor.**

Keywords—IGBT, SiC, Si, MOSFET, parallel, IPMs, consumer appliance, APF, SEER

I. INTRODUCTION

To enhance energy efficiency, regulations such as the Annual Performance Factor (APF) and Seasonal Energy Efficiency Ratio (SEER) have been established in various countries. These regulations are being continuously tightened over time. APF reflects total efficiency performance including steady long term operation at low current and high power rated operation at air conditioning during one year season. When losses are reduced in the low current region, higher overall efficiency are realized. In terms of inverter control, the motor leakage current can be reduced by lowering the switching frequency, so the inverter is operated at approximately 6 kHz or less. IPMs used in consumer applications require a well-balanced design to meet several requirements: reduced losses at 10% or less of rated current, appropriate control and thermal design for overloads, compact package size and reasonable cost [1]-[4]. In the past, the parallel connection of Si-IGBTs and Si-Super Junction MOSFETs (SJ-MOSFETs) with drive control by a single gate output was considered [5]. However, in this case, device miniaturization has limitations for satisfying both steady

low current and high current performance. Although there is also a possibility of parallel connection of Si-Reverse Conducting IGBTs (RC-IGBTs) and SiC-MOSFETs, meeting market demands requires using devices and operating ranges with greater advantages [6]. Furthermore, drive controls such as active gate drives cannot be integrated into consumer IPMs due to size limitations [7][8]. Additionally, the manufacturing and processing costs of SiC devices are significantly higher than Si devices.

Therefore, this paper proposes an optimization method to reduce losses in a consumer 6in1 IPM (15 A/600 V) equipped with a parallel connection of Si devices and SiC-MOSFETs for the inverter operation of an air conditioner compressor. The optimization focuses on the active area and drive control of the SiC-MOSFETs. In consideration that regulations in many countries require an efficiency improvement of about 35% per revision, the authors target 35% loss reduction compared to conventional Si IPMs [9][10].

II. INVESTIGATION OF THE ACTIVE AREA OF SiC-MOSFETs

The active area ratio in the parallel connection of Si devices and SiC-MOSFETs is limited by DC losses at 10% or less of rated current and temperature rise, so the required SiC-MOSFET active area must be clarified. First, the active area ratio of a SiC-MOSFET and DC loss reduction are explained. Fig. 1 shows the estimated current dependence of DC loss for a parallel connection of a Si-IGBT [11] and a SiC-MOSFET [12][13], normalized by the DC loss of a conventional Si-IGBT only. Each line in Fig. 1 corresponds to a different active area ratio of the SiC-MOSFET, as shown in Table I: (a) 0.7, (b) 0.54, (c) 0.4, (d) 0.27, and (e) 0.13. As shown in Fig. 1, even a SiC-MOSFET with the lowest ratio of (e) 0.13 can achieve a 35% reduction in DC loss at 10% or less of rated current. On the other hand, the smaller the active area of the SiC-MOSFET, the lower the performance of the reverse characteristic, and the temperature rise at high current operation may become an issue. Therefore, the authors next verified the current dependence of the temperature rise of SiC-MOSFET.

TABLE I. The active area ratio of Si-devices and SiC-MOSFET

	(a)	(b)	(c)	(d)	(e)
Si-IGBT	Active area (IGBT) = 1.00				
Si-RC-IGBT	Active area (IGBT) = 1.00		Active area (Diode)		
SiC-MOSFET	Active area (MOSFET) = 0.70	0.54	0.40	0.27	0.13

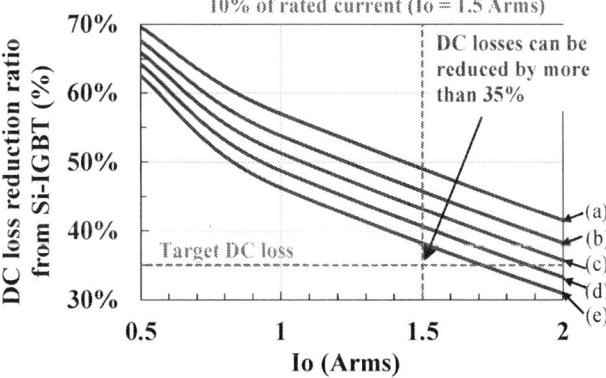

Condition: Sinusoidal, V_{cc} = 300 V, fc = 6 kHz, Power Factor = 0.8, Modulation rate = 1, Tj = 125°C
Fig. 1. Estimated: DC loss reduction ratio from Si-IGBT

Fig. 2 (a) shows estimated temperature rise Tj of several active area ratios of SiC-MOSFETs in parallel connection with a Si-IGBT. Because all recirculation current in inverter operation flows through the SiC-MOSFET in this case, only the highest ratio (a) 0.7 SiC-MOSFET can satisfy the temperature rating of the IPM at the estimated maximum rated current of air conditioner use. This situation leads to a higher cost system, so applying a Si-RC-IGBT to assist recirculation operation of the SiC-MOSFET is considered [14]. The estimated results are shown in Fig. 2 (b). When Si-RC-IGBTs are applied, almost all of the recirculation current during the dead time period at the start of recirculation flows to the Si-RC-IGBT, and during the SiC-MOSFET on period, the divided current flows to the Si-RC-IGBT and SiC-MOSFET in accordance with their DC characteristics. This reduces the recirculation current for the SiC-MOSFET, therefore a smaller area ratio SiC-MOSFET can be applied ((b) 0.54 or more). In the case of (c) 0.40, Tj of the SiC-MOSFET exceeds 125°C, making this ratio in appropraite for practical application.

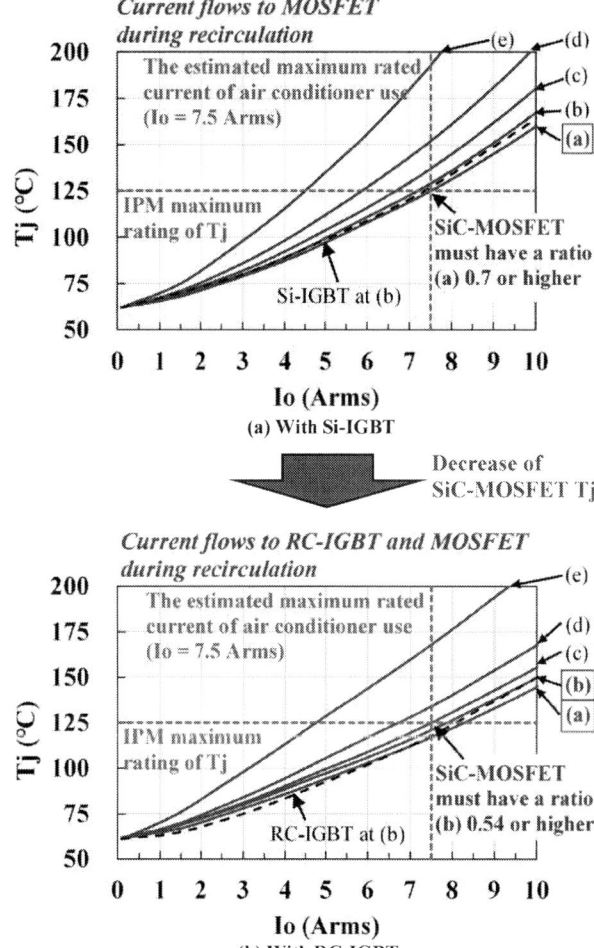

Condition: Sinusoidal, V_{cc} = 300 V, fc = 6 kHz, Power Factor = 0.8, Modulation rate = 1, Ta = 60°C
Fig. 2. Estimated: Tj of SiC-MOSFET and Si-devices

III. GATE DRIVE WITH DELAY CONTROL

As shown in the previous section, the combination of a Si-RC-IGBT and a smaller SiC-MOSFET can achieve a sufficient DC loss reduction effect. However, it may cause the thermal destruction of the SiC-MOSFET during operation, requiring optimization of the gate drive circuit. When the conventional drive circuit shown in Fig. 3 (a) is applied, the Si-RC-IGBT and SiC-MOSFET are connected by only one common gate circuit. In this case, there is a possibility that the SiC-MOSFET may turn-on faster or turn-off slower than the Si-RC-IGBT, which may cause thermal destruction of the SiC-MOSFET, especially during fault conditions at up to double rated current operation. Therefore, this combination circuit of Si-RC-IGBT and smaller SiC-MOSFET requires two gate drive circuits with proper timing delay time, as shown in Fig. 3 (b). There are multiple methods like a simple logic delay circuit or feedback by signal monitoring (e.g. gate voltage) to generate a proper delay time. In order to respond to market demand for IPMs, the authors applied the logic delay circuit shown in Fig. 3 (b), which is simple, small in circuit size, cost-effective, and operates reliably.

979-8-3315-1612-3/25 $31.00 © 2025 IEEE

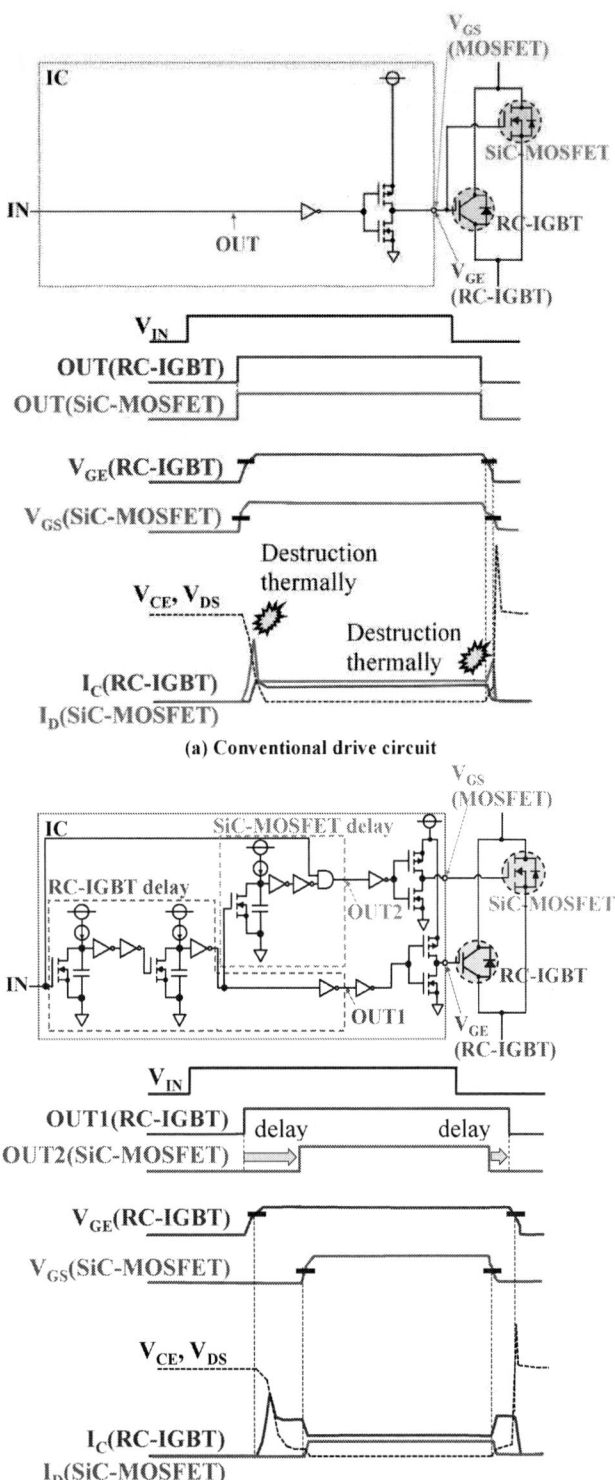

(a) Conventional drive circuit

(b) Developed drive circuit

Fig. 3. Drive circuit and timing chart

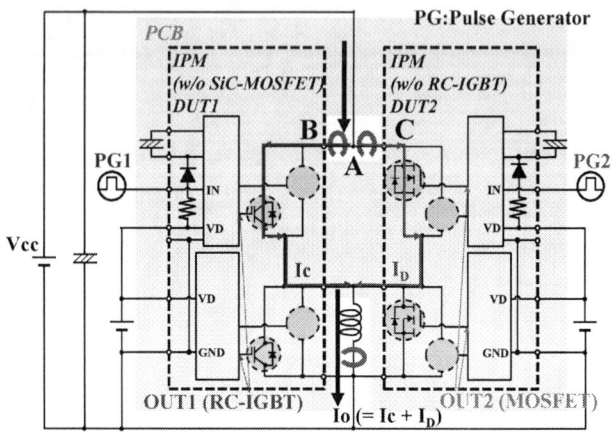

Fig. 4. Double pulse inductive test system

This circuit control allows the Si-RC-IGBTs to turn-on first followed by the SiC-MOSFETs. During turn-off, the SiC-MOSFETs turn-off first and then the Si-RC-IGBTs turn-off. When designing the delay time, it is especially important to consider the variation of the switching time of each device over all current ranges and the variation of the switching time of each device with ambient temperature. Therefore, as shown in Fig. 4, the authors evaluated the switching times of each device during parallel connection using a conventional double pulse inductive test system, where an IPM composed of Si-RC-IGBTs and an IPM composed of SiC-MOSFETs were connected parallel on the same PCB. This system is configured as follows: the three-phase inverter IPM consists of three high-side devices, three low-side devices, and ICs with built-in gate driver functions connected to each device's gate, one half-bridge circuit of each IPM is shown in Fig.4. This system can individually observe the current of each device through a parallel connection outside the IPM. Using this system, the authors measured the switching characteristics by observing the input signal (V_{IN}), voltage (V_{CE}), and each current (Si-RC-IGBT, SiC-MOSFET, and total current). Fig. 5 shows the measurement results of the switching waveforms of the Si-RC-IGBT. The turn-on switching time is defined as the time from the rise of the input signal (V_{IN}) until the collector-emitter voltage (V_{CE}) reaches 10%, and the turn-off switching time is defined as the time from the fall of the input signal (V_{IN}) until the collector current (Ic) reaches 10%. Fig. 6 shows the measurement results of the turn-on and turn-off switching times of the Si-RC-IGBT and SiC-MOSFET. The current value (Io) represents the total current, being divided between the Si-RC-IGBT and SiC-MOSFET according to their respective DC characteristics. The measurement results of the turn-on switching time indicates that a delay time of 1.3 μs is required when turning-on the Si-RC-IGBT first and then turning-on the SiC-MOSFET with delay. On the other hand, regarding turn-off, it was confirmed that the turn-off speed of the SiC-MOSFET is 0.5 μs faster than that of the Si-RC-IGBT, and no delay time setting is required. With these results and also considering the variations in drive capability and the capacitance variations of each device, a final delay time of 2 μs is determined to be necessary ensure each device operates reliably during turn-on. It should be noted that while a longer delay time prevents

979-8-3315-1612-3/25 $31.00 © 2025 IEEE   1571

thermal destruction of the SiC-MOSFET, it also shortens the on-time of the SiC-MOSFET, resulting in increased losses.

Considering the variation in switching time, a delay time of 2 µs is necessary. Fortunately, but for consumer applications with a relatively low carrier frequency of about 6 kHz, the estimated loss increases by only 0.7% as shown in Fig. 7, indicating that the proposed delay circuit is also practical.

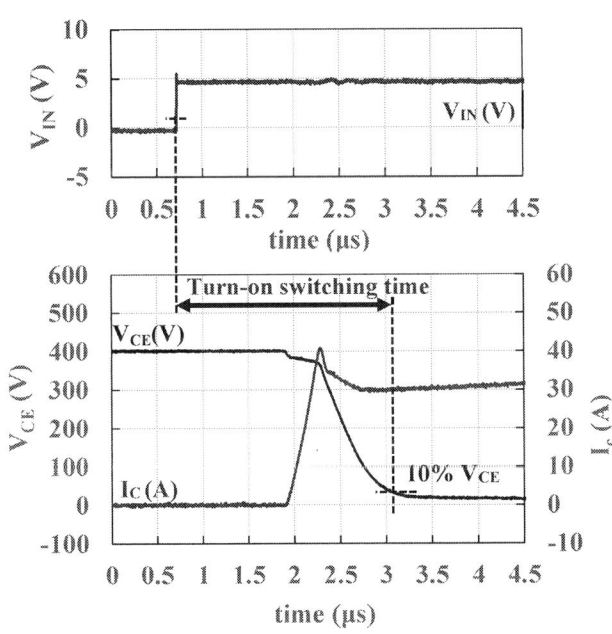

(a) turn-on

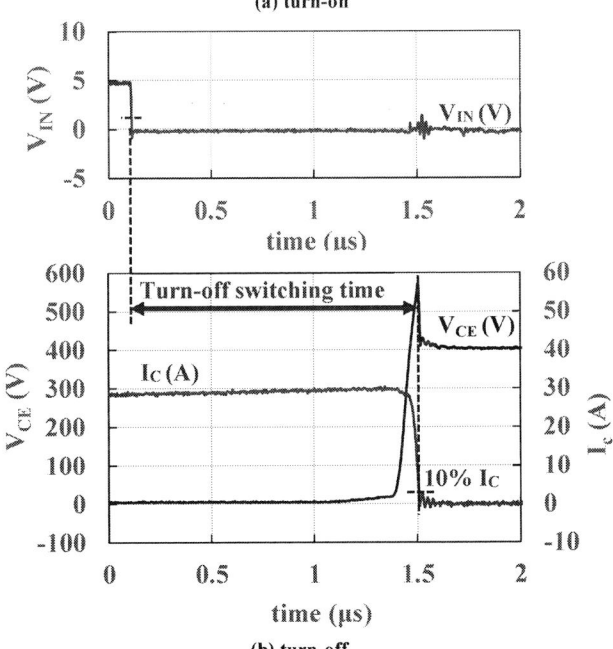

(b) turn-off

Condition: V_{cc} = 400 V, V_D = 13 V, V_{IN} = 0 ⇔ 5 V, Inductive Load, T_j = 125°C

Fig. 5. Measurement: switching waveform

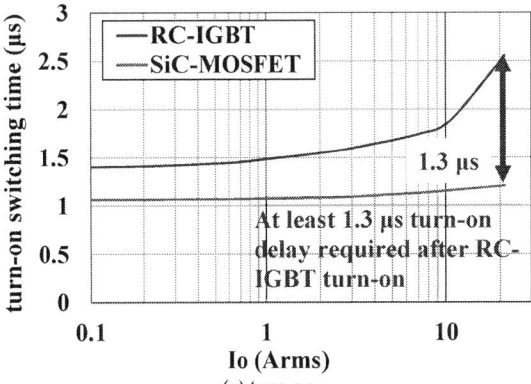

(a) turn-on

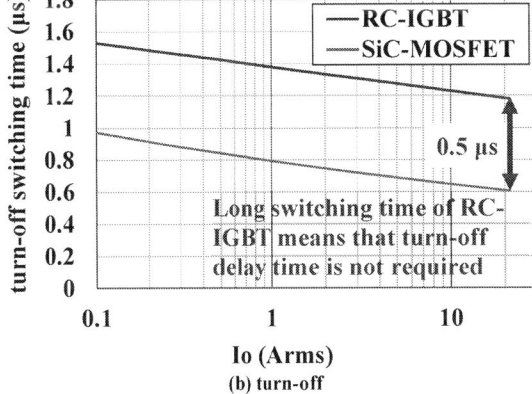

(b) turn-off

Condition: V_{cc} = 400 V, V_D = 13 V, V_{IN} = 0 ⇔ 5 V, Inductive Load, T_j = 125°C

Fig. 6. Measurement: switching time characteristics

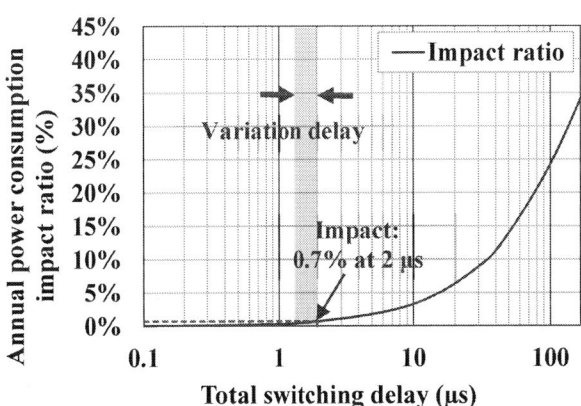

Condition: Sinusoidal, V_{cc} = 300 V, fc = 6 kHz, Power Factor = 0.8, Modulation rate = 1, fo = 60 Hz

The loss simulation results considering the inverter operating conditions of the air conditioner compressor are shown, assuming the four basic operating modes in the APF evaluation of the air conditioner (cooling rated mode, cooling intermediate mode, heating rated mode, and heating intermediate mode). The current values for each mode are as follows: cooling rated (Io = 2.5 Arms), cooling intermediate (Io = 1.5 Arms), heating rated (Io = 3 Arms), and heating intermediate (Io = 1.5 Arms).

Fig. 7. Estimated: Annual power consumption impact ratio due to total switching delay

979-8-3315-1612-3/25 $31.00 © 2025 IEEE   1572

IV. MEASUREMENT RESULTS

Fig. 8 shows the internal structure of the prototype. The active area ratio of the SiC-MOSFET is set at 0.54 relative to the Si-RC-IGBT.

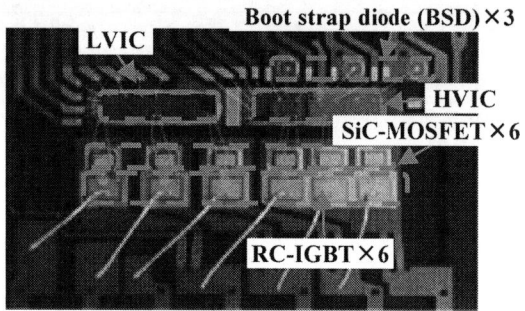

Fig. 8. Internal structure of prototype

The measurement results obtained with the prototype are shown in Figs. 9, 10, and 11. As shown in Fig. 9, the on-state voltage at Io = 1.5 Arms, which is 10% of the rated current, is dominated by the characteristics of the SiC-MOSFET and is reduced by 47% compared to the conventional Si IPM. Figs. 10 and 11 show that the turn-on switching waveform and turn-off switching waveform of Si-RC-IGBT and SiC-MOSFET are properly controlled by delay control.

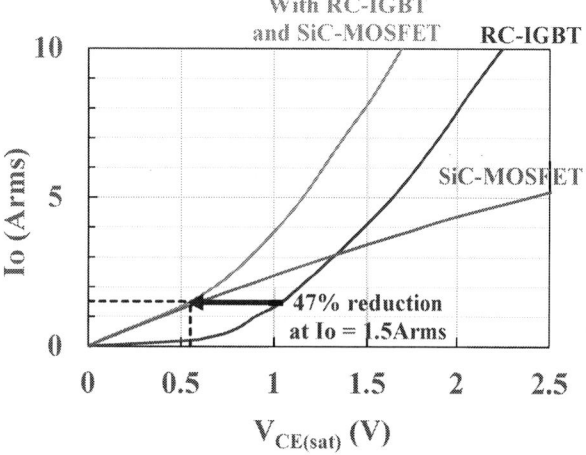

Condition: V_D = 15 V, V_{IN} = 5 V, T_j = 125℃
Fig. 9. Measurement: Output characteristic

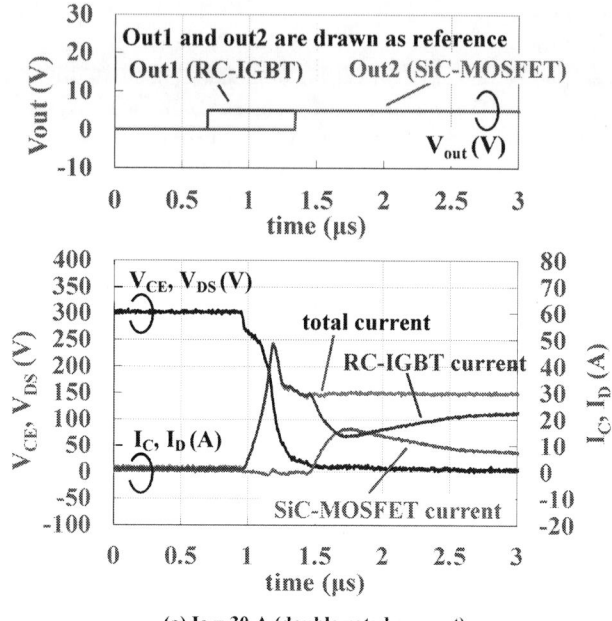

(a) Ic = 30 A (double rated current)

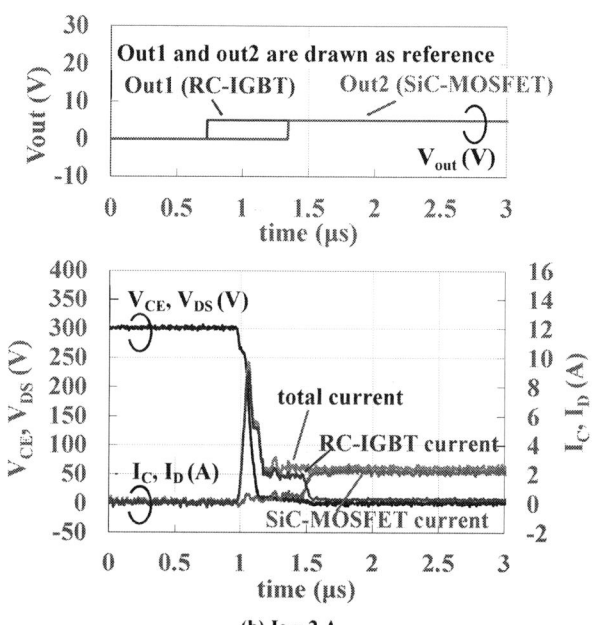

(b) Ic = 2 A

Condition: V_{cc} = 300 V, V_D = 15 V, V_{IN} = 0 $\Rightarrow$ 5 V, Inductive Load, T_j = 125℃
Fig. 10. Measurement: Turn-on switching waveform

979-8-3315-1612-3/25 $31.00 © 2025 IEEE

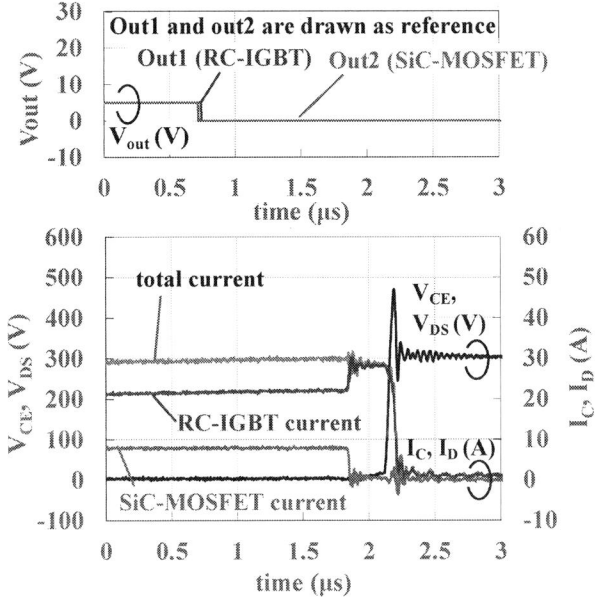

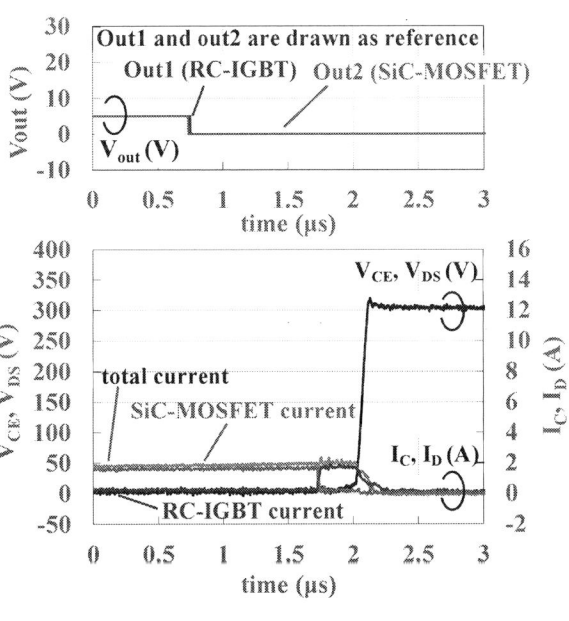

(a) Ic = 30 A (double rated current)

(b) Ic = 2 A

Condition: V_{cc} = 300 V, V_D = 15 V, V_{IN} = 5 ⇒ 0 V, Inductive Load, T_j = 125°C

Fig. 11. Measurement: Turn-off switching waveform

In addition, Fig. 12 shows the estimated total loss based on measurement results, considering the inverter operating conditions of the air conditioner compressor. Compared to the conventional IPM with only Si devices, the total loss was reduced by 40% at Io = 1.5 Arms and by 31% at Io = 3 Arms. This allows for an expected loss reduction of approximately 35% in the range of 2-3 Arms, which is a key operating point for air conditioners under 4 kW, making it possible to achieve the target loss reduction.

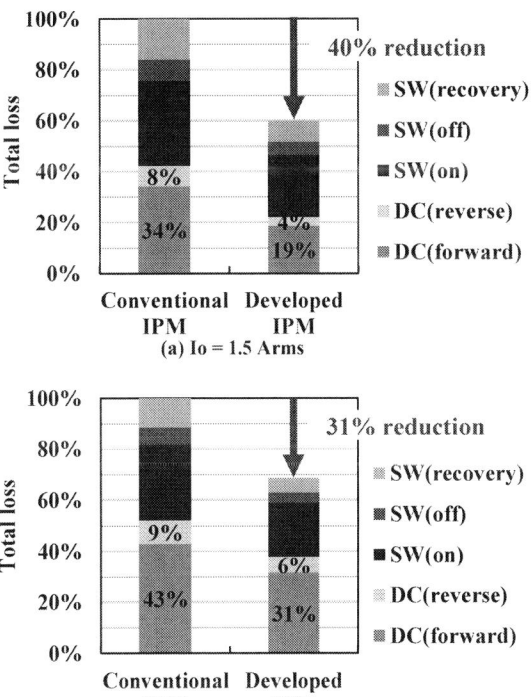

(a) Io = 1.5 Arms

(b) Io = 3.0 Arms

Condition: Sinusoidal, V_{cc} = 300 V, f_c = 6 kHz, Power Factor = 0.8, Modulation rate = 1, fo = 60 Hz, T_j = 125°C

Fig. 12. Estimated: Total loss estimation

V. CONCLUSION AND FUTURE WORK

This paper proposed three step optimization methodology for IPMs with parallel, connected Si and SiC devices as follows. First, investigation of active area of the SiC-MOSFET to optimize DC loss and temperature rise. Second, drive control for parallel connection with a suitable delay time was considered. Finally verification that the loss was reduced by 40% at 1.5 Arms by calculation at actual air conditioner compressor operating conditions. In the future, this methodology is anticipated to apply for GaN, Ga_2O_3 and the other wide band gap (WBG) devices, which make DC loss lower.

ACKNOWLEDGMENT

This paper is based on results obtained from a project, JPNP21005, subsidized by the New Energy and Industrial Technology Development Organization (NEDO).

REFERENCES

[1] N. Clark, E. Motto and S. Shibata, "New SLIM Package Intelligent Power Modules (SLIMDIP) with thin RC-IGBT for consumer goods applications," *2015 IEEE Energy Conversion Congress and Exposition (ECCE)*, Montreal, QC, Canada, 2015, pp. 4510-4512, Sept. 2015, doi: 10.1109/ECCE.2015.7310296.

[2] S. Shin et al., "Development of New 600V Smart Power Module for Home Appliances Motor Drive Application," *PCIM Europe 2018; International Exhibition and Conference for Power Electronics, Intelligent Motion, Renewable Energy and Energy Management*, Nuremberg, Germany, June. 2018, pp. 1062-1067.

[3] S. Young et al., "A new, intelligent power module with higher power density and smallest package size," *PCIM Asia 2019; International Exhibition and Conference for Power Electronics, Intelligent Motion, Renewable Energy and Energy Management*, Shanghai, China, June. 2019, pp. 227-230.

[4] T. Yamada et al., "Latest Small Intelligent Power Module For Energy-Saving," *PCIM Europe 2012; International Exhibition and Conference for Power Electronics, Intelligent Motion, Renewable Energy and Energy Management*, Nuremberg, Germany, 2012, pp. 1312-1316.

[5] M. Kato, M. Shiramizu, and T. Tanaka, "An APF oriented Transfer Mold DIPIPM utilizing MOSFET with super junction structure," *PCIM Europe 2014; International Exhibition and Conference for Power Electronics, Intelligent Motion, Renewable Energy and Energy Management*, Nuremberg, Germany, 2014, pp. 550-556.

[6] M. Rahimo et al., "The Cross Switch "XS" Silicon and Silicon Carbide Hybrid Concept," *Proceedings of PCIM Europe 2015; International Exhibition and Conference for Power Electronics, Intelligent Motion, Renewable Energy and Energy Management*, Nuremberg, Germany, 2015, pp. 382-389.

[7] Y. Wei, X. Du, D. Woldegiorgis, and A. Mantooth, "Application of An Active Gate Driver for Paralleling Operation of Si IGBT and SiC MOSFET," *2021 IEEE 12th Energy Conversion Congress & Exposition - Asia (ECCE-Asia)*, Singapore, Singapore, 2021, pp. 314-319.

[8] F. Kayser et al., "Hybrid Switch with SiC MOSFET and fast IGBT for High Power Applications," *PCIM Europe digital days 2021; International Exhibition and Conference for Power Electronics, Intelligent Motion, Renewable Energy and Energy Management*, Online, 2021, pp. 598-603.

[9] Independent Statistics and Analysis, U.S. Energy Information Administration, "Efficiency requirements for residential central AC and heat pumps to rise in 2023," [Online]. Available: https://www.eia.gov/todayinenergy/detail.php?id=40232

[10] Ministry of Economy, Trade and Industry, "New Energy Efficiency Standards Formulated for Home-use Air Conditioners," [Online]. Available: https://www.meti.go.jp/english/press/2022/0531_002.html

[11] Y. Haraguchi et al., "600V LPT-CSTBT™ on advanced thin wafer technology," *2011 IEEE 23rd International Symposium on Power Semiconductor Devices and ICs*, San Diego, CA, USA, May. 2011, pp. 68-71.

[12] M. Furuhashi et al., "Breakthrough in trade-off between threshold voltage and specific on-resistance of SiC-MOSFETs," *Proceedings of the 2013 25th International Symposium on Power Semiconductor Devices & IC's (ISPSD)*, Kanazawa, Japan, May. 2013, pp. 55-58.

[13] T. Tanioka et al., "High Performance 4H-SiC MOSFETs with Optimum Design of Active Cell and Re-Oxidation," *PCIM Europe 2018; International Exhibition and Conference for Power Electronics, Intelligent Motion, Renewable Energy and Energy Management*, Nuremberg, Germany, June. 2018, pp. 879-884.

[14] T. Yoshida, T. Takahashi, K. Suzuki, and M. Tarutani, "The second-generation 600V RC-IGBT with Optimized FWD" *Proceedings of the 2016 28th International Symposium on Power Semiconductor Devices and ICs (ISPSD)*, Prague, Czech Republic, June. 2016, pp. 159-162.

979-8-3315-1612-3/25 $31.00 © 2025 IEEE

Investigating the Temperature Dependency and Operating Parameters of a Self-Driving Active Gate Driver

1st Vin Loong Choo
Chair of Energy Conversion
TU Dortmund University
Dortmund, Germany
vinloong.choo@tu-dortmund.de

2nd Martin Pfost
Chair of Energy Conversion
TU Dortmund University
Dortmund, Germany
martin.pfost@tu-dortmund.de

Abstract—A self-driving active gate driver for silicon carbide devices is presented, which is designed to reduce current overshoot and parasitic ringing caused by high dv/dt and di/dt transients while enabling high switching speeds, with the objective of reducing switching losses. An investigation into the impact of varying operating parameters and ambient temperature measurements across a temperature range of $-40\,°C$ – $125\,°C$ was conducted to assess the influence of a V_{th} shift and the temperature dependency of the Miller plateau on the active gate driver network. The results demonstrated a reduction in turn-on switching losses compared to a conventional gate driver network across all measurements.

Index Terms—Active gate driver, switching slew rate control, multi-level gate driver, switching losses, threshold voltage, temperature characterization, Miller plateau

I. INTRODUCTION

Active gate drivers (AGDs) control the waveform of the gate-source voltage V_{GS} during turn-on/turn-off switching transients to reduce switching losses and dampen the parasitic ringing. This is achieved by either changing the gate resistance, gate-source voltage or gate current during the switching transient. Many active gate driver solutions for SiC MOSFETs use the Miller plateau of the SiC MOSFET as a reference point to adjust the aforementioned parameters that affect the gate-source voltage [1]–[4]. However, the Miller plateau is dependent on the load current, the threshold voltage V_{th} and on the ambient temperature T_{amb} [5], [6].

To ensure proper functioning of an AGD under changing operating conditions, it is important to consider the shift of the Miller plateau during operation. This requires either a feedback loop to adjust the operating parameters of the AGD or the AGD implementation must be able to operate with a shift in the Miller plateau. Systems with feedback loops require high-speed components, such as FPGAs, resulting in increased system complexity [7], [8]. In contrast, systems without a feedback loop typically operate within a limited range of conditions such as fixed delays of the active gate

driver network, and do not take the shift of the Miller plateau under consideration [9].

In this work, a self-driving 3-Level active gate driver (3L-AGD) is proposed that actively controls the gate-resistance without the use of a feedback loop or fixed delays, while improving the turn-on switching losses without increasing the overshoot. The objective of this work is to investigate the effect of the ambient temperature T_{amb} and manufacturing tolerances on the performance of the proposed 3L-AGD.

II. PROPOSED SYSTEM AND OPERATION PRINCIPLE

In Fig. 1, the double-pulse measurement with T1, a silicon carbide MOSFET, D_1 as the free-wheeling diode, L_{load} as the inductive load, C_{DC} as the DC-link capacitance and V_{DC} as the DC-link supply voltage, is shown with a simplified overview of the 3L-AGD.

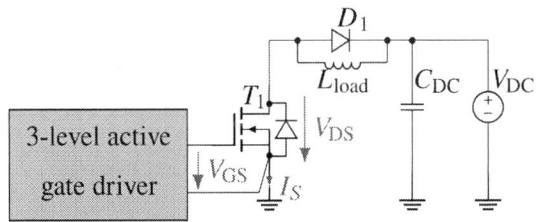

Fig. 1. Schematic of the double-pulse measurement and a simplified overview of the 3L-AGD.

The double-pulse measurement enables the system to be operated with varying load currents and can be utilized in a controlled temperature environment to examine the impact of ambient temperate T_{amb} on the Miller plateau and the threshold voltage V_{th}.

A comprehensive view of the 3L-AGD network, which is a continuation of the work shown in [2], [10], [11], is shown in Fig. 2. It comprises the gate driver IC and two auxiliary circuits in parallel to the turn-on gate resistance $R_{G,on}$, denoted as the n-AuxGD and p-AuxGD circuits. Each auxiliary circuit consists of a MOSFET connected in parallel to $R_{G,on}$. The two auxiliary circuits are not influenced by the operation of the gate driver IC and function independently of one another.

979-8-3315-1612-3/25 $31.00 © 2025 IEEE

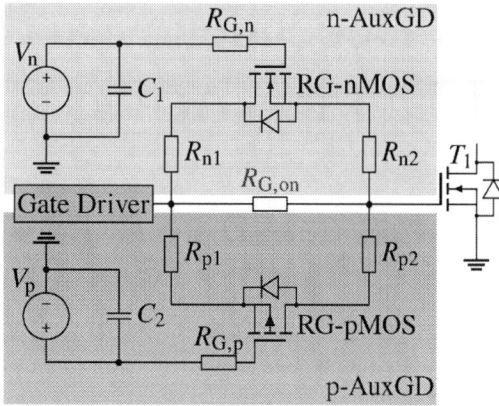

Fig. 2. A detailed view of the 3L-AGD network, consisting of the gate driver and the two circuits n-AuxGD and p-AuxGD connected in parallel to $R_{G,on}$.

The n-AuxGD contains an n-channel enhancement MOSFET, the RG-nMOS, whereas the p-AuxGD contains a p-channel enhancement MOSFET, the RG-pMOS. The resulting resistor network of the 3L-AGD can be summarized as R_{tot}, see

$$R_{tot} = R_{G,on} \parallel (R_{on,RG\text{-}nMOS} + R_{n1} + R_{n2})$$
$$\parallel (R_{on,RG\text{-}pMOS} + R_{p1} + R_{p2}) . \tag{1}$$

R_{tot} is the total turn-on gate resistance applied to T1. The total value of R_{tot} depends on the on-resistance of both silicon MOSFETs, $R_{on,RG\text{-}nMOS}$, $R_{on,RG\text{-}pMOS}$, and the gate resistor $R_{G,on}$, see (1). The resistances R_{n1}, R_{n2}, R_{p1}, and R_{p2} have a value of 1 Ω and, due to the parallel resistance network, have no significant impact on R_{tot}, as shown in (1). The buffer capacitances C_1 and C_2 have a value of 2 µF to ensure the stability of the bias voltages V_n and V_p during the switching event of the RG-nMOS and RG-pMOS. The RG-nMOS and RG-pMOS are MOSFETs with a relatively low output capacitance C_{oss} of 11 pF and 10 pF, respectively, which ensures a minimal impact on the gate driver network.

The gate terminals of the RG-nMOS and RG-pMOS are pre-biased with a voltage source V_n and V_p, see Fig. 2, and the values can be calculated with

$$V_n = V_{th,n} + V_{Mil,min} \tag{2}$$

and

$$V_p = V_{th,p} + V_{Mil,max} . \tag{3}$$

$V_{th,n}$ is the threshold voltage of the RG-nMOS, $V_{th,p}$ is the threshold voltage of the RG-pMOS, $V_{Mil,min}$ and $V_{Mil,max}$ is the range of the Miller plateau of the SiC MOSFET, all taken from their datasheets. With this gate driver circuit, the 3L-AGD dynamically changes the overall gate resistance R_{tot} during the switching transient by turning on/off the RG-nMOS and RG-pMOS. The operating principle of the 3L-AGD, the transition from low gate resistance to high gate resistance before the gate-source voltage reaches the Miller plateau and the transition back to low gate resistance after the Miller plateau, is shown for a turn-on switching event in Fig. 3(a). Fig. 3(a) shows the idealized waveform of V_{GS} of T1 for the self-driving 3L-AGD network and the changing value of R_{tot} during the turn-on switching transient, without any parasitic effects such as ringing or overshoot. The total gate resistance R_{tot}, as defined

in (1), and its dynamic resistance value are also shown in Fig. 3(a).

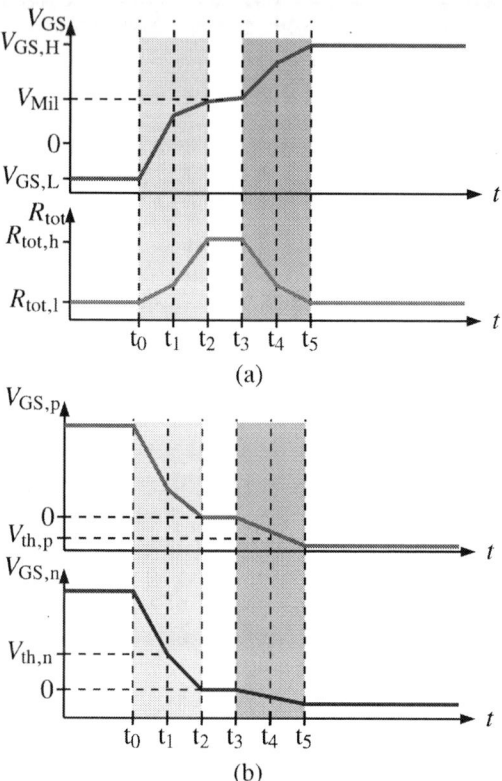

Fig. 3. Operating Principle of the 3-Level Gate Driver Network with idealized waveforms of (a) V_{GS} of T1, R_{tot} and (b) of $V_{GS,n}$ and $V_{GS,p}$ of the RG-nMOS and RG-pMOS during the turn-on switching transient.

Fig. 3(b) shows the idealized waveforms of $V_{GS,p}$ and $V_{GS,n}$, representing the gate source voltage for the RG-pMOS and RG-nMOS, respectively, within the self-driving 3L-AGD network during the turn-on switching transient. As a consequence of the parallel connection to $R_{G,on}$ and given that the voltage sources V_n and V_p are both referenced to the same ground potential as the SiC MOSFET T1, the source potentials of the RG-nMOS and RG-pMOS follow the gate source potential V_{GS} of T1, see Fig. 2. As V_n and V_p remain constant throughout the switching transient, and the source potentials of the RG-nMOS and RG-pMOS follow that of V_{GS}, the gate-source waveforms of $V_{GS,n}$ and $V_{GS,p}$ exhibit a complementary behavior.

An increase in the gate-source voltage V_{GS} of T1 results in a corresponding decrease in the gate-source voltage $V_{GS,n}$ and $V_{GS,p}$, see Fig. 3(a)-(b). This results in a change in the switching states of both the RG-nMOS and RG-pMOS without the need for a control signal. Additionally, the switching speed of both the RG-nMOS and RG-pMOS is primarily determined by the switching transient of V_{GS}. The switching states of the RG-nMOS and RG-pMOS are self-adaptive as a function of the bias voltages V_n, V_p, and V_{GS} of T1 during the turn-on switching transient.

At the start of the turn-on switching transient for $t < t_0$, the switching state of the RG-nMOS is on and off for the RG-pMOS, resulting in $R_{tot} = R_{tot,l}$ and a fast switching transient.

From $t_0 - t_2$ the total gate resistance R_{tot} transitions from $R_{\text{tot,l}}$ to the highest resistance value, reaching $R_{\text{tot}} = R_{\text{tot,h}} = R_{\text{G,on}}$ at the beginning of the Miller plateau. For $t_2 - t_3$ the switching speed is reduced during the Miller plateau as R_{tot} remains at the highest resistance value of $R_{\text{tot,h}}$, which successfully dampens the overshoot and ringing. During $t_3 - t_5$ the p-AuxGD transitions R_{tot} back to a low resistance value of $R_{\text{tot,l}}$, resulting in an increase in switching speed and a reduction in switching losses.

The switching states of RG-nMOS, RG-pMOS and the resulting total values of the gate resistance R_{tot} during the corresponding time periods can be calculated with (1) and are summarized in Tab. I.

TABLE I
OVERVIEW OF SWITCHING STATES OF RG-NMOS, RG-PMOS AND THE TOTAL GATE VALUE OF R_{tot} DURING THE TURN-ON SWITCHING TRANSIENT OF T1.

Time	RG-nMOS	RG-pMOS	R_{tot}
$t < t_0$	On	Off	$R_{\text{tot,l}}$
$t < t_2$	On→Off	Off	$R_{\text{tot,l}} \rightarrow R_{\text{tot,h}}$
$t_2 \leq t \leq t_3$	Off	Off	$R_{\text{tot,h}} = R_{\text{G,on}}$
$t \geq t_3$	Off	Off→On	$R_{\text{tot,h}} \rightarrow R_{\text{tot,l}}$

In order to achieve the desired results shown in Fig. 3 and Tab. I, the bias voltages V_n and V_p are calculated using (2) and (3) with the typical values for V_{th} and V_{Mil}, as indicated in the datasheets, resulting in $V_n = 9.6\,\text{V}$ and $V_p = 6\,\text{V}$. For the measurements with a conventional gate driver (CGD) network, the two auxiliary networks n-AuxGD and p-AuxGD were disconnected by removing the resistances R_{n1}, R_{n2}, R_{p1} and R_{p2}. Fig. 4 shows the measurement results of the gate source voltages for both the CGD and 3L-AGD network with $R_{\text{G,on}} = 10\,\Omega$ for both networks. It can be observed that before and after the Miller plateau, the transient of the gate source waveform measured with the 3L-AGD network is significantly faster in comparison to the CGD measurement, see Fig. 4.

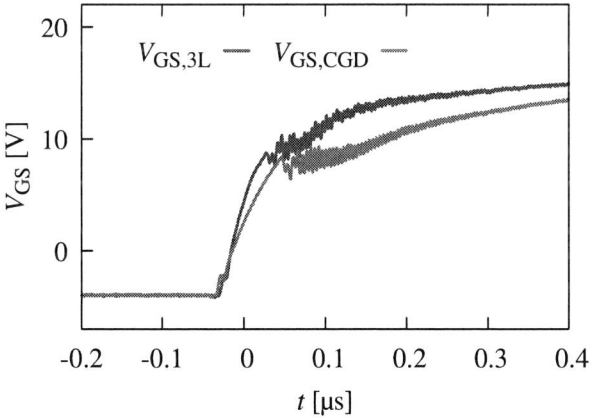

Fig. 4. Measurement of V_{GS} for CGD and 3-level AGD with $V_{\text{DC}} = 800\,\text{V}$, $I_{\text{sw}} = 20\,\text{A}$, $V_n = 9.6\,\text{V}$, $V_p = 6\,\text{V}$ and $R_{\text{G,on}} = 10\,\Omega$ in the 3L-AGD and CGD network.

This can also be observed in Fig. 5, the faster switching transients of the drain-source voltage V_{DS} and I_{S} measured with the 3L-AGD network, resulting in a reduction of switching losses. The measured turn-on switching losses can be reduced from $628.25\,\mu\text{J}$ for the CGD network to $528.39\,\mu\text{J}$ for the 3L-AGD network, representing a 15.9% reduction in turn-on switching losses. Moreover, the current overshoot for both measurements is identical despite the faster transient of the current I_{S} measured with the 3L-AGD network, as illustrated in Fig. 5. As shown in Fig. 3 and Tab. I, the highest overall gate resistance applied with the 3L-AGD network is $R_{\text{tot}} = R_{\text{G,on}}$, which in this measurement is the same gate resistance for the CGD network. Consequently, the gate resistance $R_{\text{G,on}}$ determines the maximum dampening effect on the current overshoot and is the same for the CGD and 3L-AGD network.

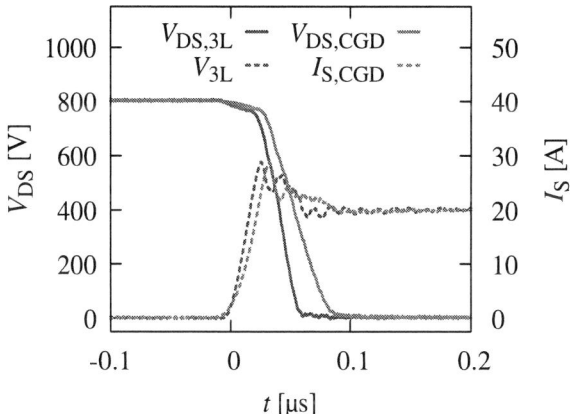

Fig. 5. Measurement of V_{DS} and I_{S} for CGD and 3-level AGD with $V_{\text{DC}} = 800\,\text{V}$, $I_{\text{sw}} = 20\,\text{A}$, $V_n = 9.6\,\text{V}$, $V_p = 6\,\text{V}$ and $R_{\text{G,on}} = 10\,\Omega$ in the 3L-AGD and CGD network.

A picture of the self-driving 3L-AGD measurement setup with the auxiliary circuits fully integrated in the PCB is shown in Fig. 6. The voltages V_n and V_p are provided by adjustable voltage regulators and are supplied from the same voltage supply as the gate driver IC [11].

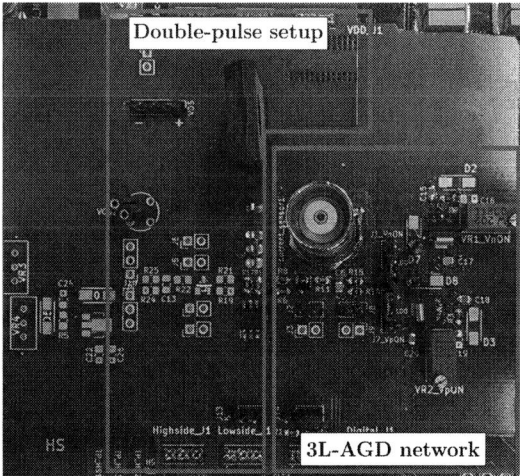

Fig. 6. An image of the double-pulse measurement setup and the 3L-AGD network, which has been subjected to investigation along with its auxiliary gate driver network n-AuxGD and p-AuxGD [11].

III. EXPERIMENTAL RESULTS

The Miller plateau of SiC MOSFETs is load and temperature dependent [5], [6]. In addition, the threshold voltage of

RG-nMOS and RG-pMOS can vary from a typical value of ±2 V. As shown in Fig. 3, (2) and (3), the operating principle of the 3L-AGD depends on the Miller plateau, bias voltages V_n, V_p, and on the threshold voltage V_{th} of the RG-nMOS and RG-pMOS. The following measurements investigate and verify the operating principle of the 3L-AGD with a deviation of V_n and V_p, calculated from (2) and (3), to simulate the voltage range of V_{th}. In addition, measurements under different ambient temperatures T_{amb} are performed to investigate the temperature dependence of the proposed 3L-AGD network.

The impact of the 3L-AGD network on the voltage overshoot and turn-off switching losses during the turn-off transient was shown to be minimal in [2]. Additionally, in [2] the influence of the load current on the operational principle on the 3L-AGD was already investigated. Therefore, the influence of a threshold voltage variation and ambient temperature is only investigated for the turn-on transient.

A. Bias Voltage Variation Measurements

The double-pulse measurements are repeated with the bias voltages V_n and V_p ranging from ±2 V from their optimal values in 0.5 V steps. This results in a voltage range of 7.6 V – 11.6 V for V_n and 4 V – 8 V for V_p. The operating paremeters V_n and V_p are varied to investigate the manufacturing tolerances of V_{th} for the RG-nMOS and RG-pMOS. As shown in equations (2) and (3), V_n and V_p depend on their respective threshold voltages $V_{th,n}$ and $V_{th,p}$, which are usually in a range of ±2 V from a typical value.

The influence of the threshold voltage V_{th} variation is investigated for each auxiliary circuit separately as well as simultaneously. The turn-on switching losses E_{on} and the current overshoot $I_{S,peak}$ with the 3L-AGD network are normalized to a CGD measurement with $R_{G,on} = 10\,\Omega$ and to the reference 3L-AGD measurement with $V_n = 9.6$ V and $V_p = 6$ V, both values calculated with (2) and (3). This approach enables an evaluation of the performance of the 3L-AGD in comparison to the CGD network, as well as an investigation of a V_{th} shift in the RG-nMOS and RG-pMOS. Fig. 7 and Fig. 8 show the measurement results for a parameter variation of $V_n = 7.6$ V – 11.6 V

It can be observed that the parameter variation of V_n results in a higher current overshoot $I_{S,peak}$ in both measurements. However, the average increase in current overshoot for the CGD normalized measurement is 7.15 %, while the average decrease in turn-on switching losses is −20.26 %, as shown in Fig. 7. Furthermore, in Fig. 8 measurements normalized to the reference 3L-AGD measurement demonstrate that the average reduction in turn-on switching losses is −2.82 %, while the average increase in current overshoot is 4.84 %.

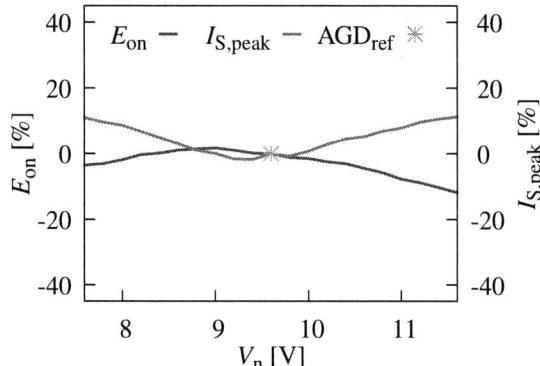

Fig. 8. Same as in Fig. 7 but now normalized to a 3L-AGD measurement with $V_n = 9.6$ V, $V_p = 6$ V and $R_{G,on} = 10\,\Omega$. The reference 3L-AGD measurement is marked as AGD$_{ref}$.

In Fig. 9 and Fig. 10 the measurement results for a parameter variation of $V_p = 4$ V – 8 V are shown. The measurements normalized to a CGD measurements show, that an increase in V_p from the calculated value of 6 V does not increase the current overshoot but the turn-on switching losses increase. On the other hand, a decrease in V_p increases the current overshoot and decreases the turn-on switching losses.

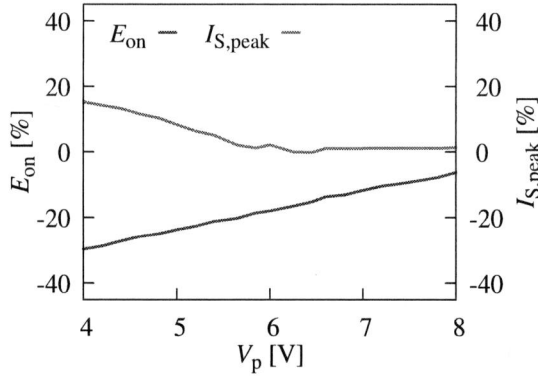

Fig. 9. Measurement of E_{on} and $I_{S,peak}$ with $V_{DC} = 800$ V, $I_{sw} = 20$ A, $V_n = 9.6$ V, $V_p = 4$ V – 8 V normalized to a CGD measurement with $R_{G,on} = 10\,\Omega$

This behavior can also be observed in the measurements normalized to a reference 3L-AGD measurement, see Fig. 10. The average increase in current overshoot for the CGD normalized measurement is 4.72 % and −17.74 % for the turn-on switching losses, while the average increase in current overshoot for the 3L-AGD normalized measurement is 2.47 % and 0.23 % for the turn-on switching losses.

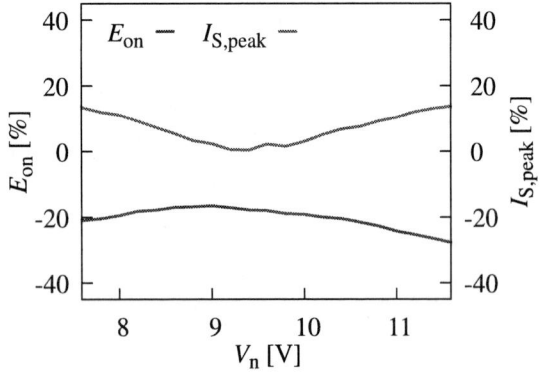

Fig. 7. Measurement of E_{on} and $I_{S,peak}$ with $V_{DC} = 800$ V, $I_{sw} = 20$ A, $V_p = 6$ V, $V_n = 7.6$ V – 11.6 V normalized to a CGD measurement with $R_{G,on} = 10\,\Omega$

979-8-3315-1612-3/25 $31.00 © 2025 IEEE   1579

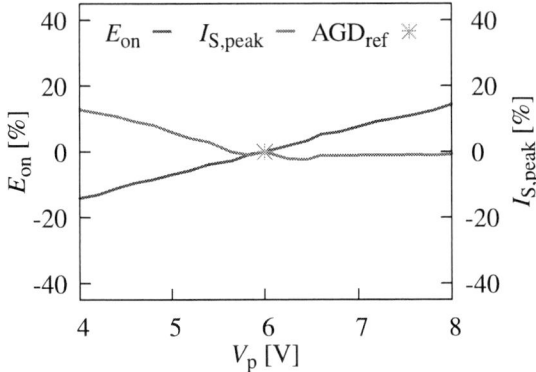

Fig. 10. Same as in Fig. 9 but now normalized to a 3L-AGD measurement with $V_n = 9.6$ V, $V_p = 6$ V and $R_{G,on} = 10\,\Omega$. The reference 3L-AGD measurement is marked as AGD_{ref}.

Fig. 11 and Fig. 12 show measurements with a simultaneous parameter variation of $V_n = 7.6$ V -11.6 V and $V_p = 4$ V -8 V. It can be observed in Fig. 11 that the peak current $I_{S,peak}$ increases by an average of 10.63 % while the turn-on switching losses decrease by -21.99 % in average. The average reduction in turn-on losses exceeds the average increase in peak current across the entire ± 2 V voltage range for both V_n and V_p.

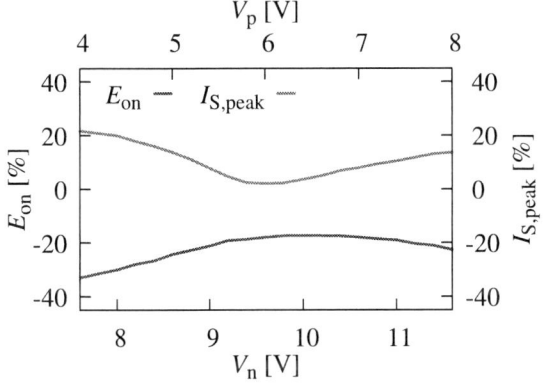

Fig. 11. Measurement of E_{on} and $I_{S,peak}$ with $V_{DC} = 800$ V, $I_{sw} = 20$ A with $V_n = 7.6$ V -11.6 V and $V_p = 4$ V -8 V normalized to a CGD measurement with $R_{G,on} = 10\,\Omega$

Fig. 12 depicts the normalized measurement results of E_{on} and $I_{S,peak}$ with the 3L-AGD network with an average increase of $I_{S,peak}$ of 8.25 % and an average decrease in E_{on} of -4.93 %.

The results of the measurements presented in Fig. 7 - Fig. 12 show that the proposed AGD network consistently results in a reduction in switching losses, regardless of the operating voltage of V_n and V_p. A wide range for V_{th}, typically of ± 2 V of the typical value of V_{th}, was simulated by varying the bias voltages V_n and V_p. This variation does not affect the operating principle of the proposed 3L-AGD network.

B. Temperature Measurements

An image of the temperature measurements is shown in Fig. 13. The ambient temperature of the measurement environment is set by a ThermoStream® thermal test system from

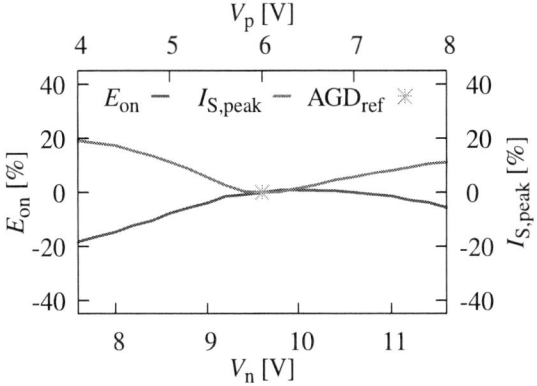

Fig. 12. Same as in Fig. 11 but now normalized to a 3L-AGD measurement with $V_n = 9.6$ V, $V_p = 6$ V and $R_{G,on} = 10\,\Omega$. The reference 3L-AGD measurement is marked as AGD_{ref}.

Temptronic. The ambient temperature T_{amb} is measured with a thermocouple and the double-pulse measurement is taken after a soak time of 2 min to ensure that the measurement setup is at the target temperature.

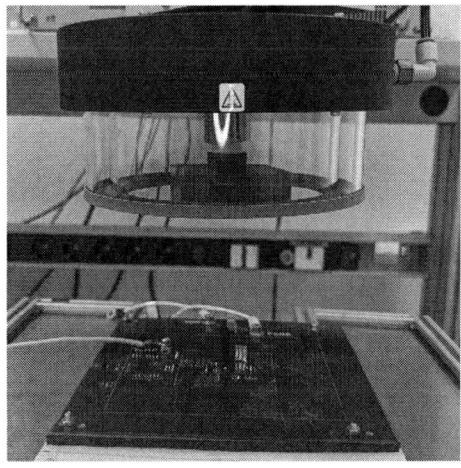

Fig. 13. Customized PCB of the double-pulse setup shown in Fig. 1 for the ambient temperature measurements.

Fig. 14 and Fig. 15 show the measurement results of E_{on} with the CGD and 3L-AGD network over an ambient temperature T_{amb} of $-40\,°C - 125\,°C$ with the specific temperatures of $-40\,°C$, $-20\,°C$, $25\,°C$, $75\,°C$, and $125\,°C$. During the measurements, the entire double-pulse setup shown in Fig. 1(a), including the 3L-AGD network shown in Fig. 1(b), is heated, see Fig. 13. This ensures that a shift of V_{th} for both the RG-nMOS and RG-pMOS as well as the SiC MOSFET T1 is accounted for. As shown in [12] and in Fig. 14, the turn-on losses with the CGD network decrease for higher T_{amb}. The turn-on losses at $125\,°C$ are 9.51% lower than the losses at $-40\,°C$, see Fig. 14.

However, the measurement results of the turn-on losses for the 3L-AGD remain unaffected by an increase in T_{amb}, see Fig. 15. The 3L-AGD network maintains a reduction of E_{on} over the entire temperature range compared to the

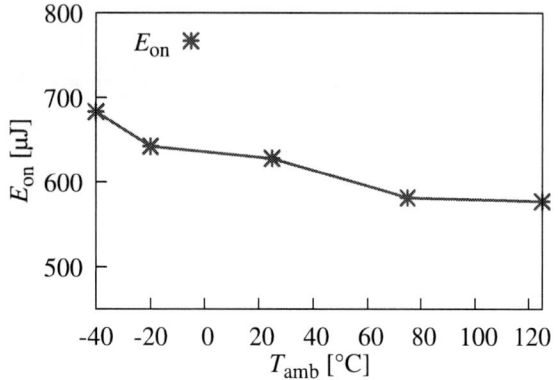

Fig. 14. Measurement of E_{on} with $V_{DC} = 800\,\text{V}$, $I_{sw} = 20\,\text{A}$ and $R_{G,on} = 10\,\Omega$ for an ambient temperature T_{amb} of $-40\,°\text{C} - 125\,°\text{C}$ with the CGD network. Lines are guide to the eye.

measurement results with the CGD network. At $T_{amb} = -40\,°\text{C}$ the reduction in E_{on} with the 3L-AGD network is 17.58% and 10% at $T_{amb} = 125\,°\text{C}$ lower compared to the measured switching losses with the CGD network, see Fig. 14.

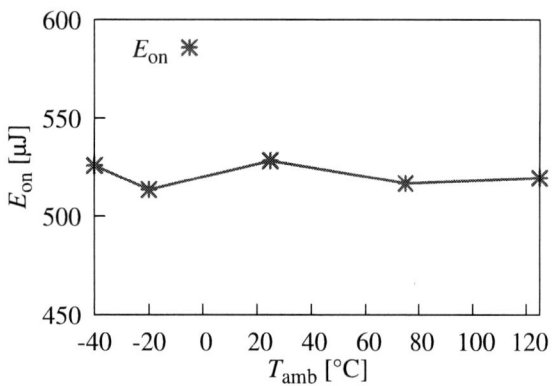

Fig. 15. Same as Fig. 14 but now with the 3L-AGD network and $V_n = 9.6\,\text{V}$ and $V_p = 6\,\text{V}$. Lines are guide to the eye.

IV. CONCLUSION

A self-driving active gate driver network for SiC MOSFETs that reduces the turn-on losses without the necessity of a complex feedback loop or specific operating parameters for certain operating conditions was presented. The impact of manufacturing tolerances on the functionality of the 3L-AGD was evaluated through a variation of the bias voltages V_n and V_p. Moreover, ambient temperature measurements were conducted on the double-pulse measurement setup, as well as the entire active gate driver circuit, from across a temperature range of $-40\,°\text{C} - 125\,°\text{C}$. This was carried out to investigate the temperature dependency of the Miller plateau and its influence on the active gate driver circuit. The results of the

measurements presented in this work demonstrate that the performance of the presented 3L-AGD network is not significantly affected by parameter variation or ambient temperature. The implemented 3L-AGD network results in a robust and efficient system that successfully increases the switching transient while reducing the switching losses without increasing the current overshoots.

ACKNOWLEDGMENT

This project has received funding from the ECSEL Joint Undertaking (JU) under grant agreement No 101007281. The JU receives support from the European Union's Horizon 2020 research and innovation programme and Austria, Germany, Slovenia, Netherlands, Belgium, Slovakia, France, Italy, Turkey.

REFERENCES

[1] S. Zhao, X. Zhao, Y. Wei, Y. Zhao, and H. A. Mantooth, "A review of switching slew rate control for silicon carbide devices using active gate drivers," vol. 9, no. 4, 2021, pp. 4096–4114.

[2] V. L. Choo and M. Pfost, "A 3-level active gate driver network for SiC MOSFETs to minimize overshoot and switching losses," in *2024 IEEE Applied Power Electronics Conference and Exposition (APEC)*, 2024, pp. 2794–2799.

[3] C. Geng, D. Zhang, X. Wu, W. Shen, and R. Dong, "A novel active gate driver with auxiliary gate current control circuit for improving switching performance of high-power SiC MOSFET modules," in *2020 IEEE 1st China International Youth Conference on Electrical Engineering (CIYCEE)*, 11 2020, pp. 1–7.

[4] Z. Gao, J. Zhang, Y. Huang, R. Guan, and Y. Zhou, "A closed-loop active gate driver of SiC MOSFET for voltage spike suppression," *IEEE Open Journal of Power Electronics*, vol. 3, pp. 723–730, 2022.

[5] J. O. Gonzalez, O. Alatise, J. Hu, L. Ran, and P. A. Mawby, "An investigation of temperature-sensitive electrical parameters for SiC power MOSFETs," *IEEE Transactions on Power Electronics*, vol. 32, no. 10, pp. 7954–7966, 2017.

[6] J. O. Gonzalez and O. Alatise, "Bias temperature instability and junction temperature measurement using electrical parameters in SiC power MOSFETs," *IEEE Transactions on Industry Applications*, vol. 57, no. 2, pp. 1664–1676, 2021.

[7] R. Li, Z. Hou, T. Liu, M. Elshazly, S. Leung, X. Peng, and W. T. Ng, "Dynamic gate drive for SiC power MOSFETs with sub-nanosecond timings," in *2023 IEEE Applied Power Electronics Conference and Exposition (APEC)*, 2023, pp. 324–330.

[8] Y. Ling, Z. Zhao, and Y. Zhu, "A self-regulating gate driver for high-power IGBTs," *IEEE Transactions on Power Electronics*, vol. 36, no. 3, pp. 3450–3461, 2021.

[9] Y. Teng, Q. Gao, Q. Zhang, J. Kou, and D. Xu, "A variable gate resistance SiC MOSFET drive circuit," in *IECON 2020 The 46th Annual Conference of the IEEE Industrial Electronics Society*, 2020, pp. 2683–2688.

[10] V. L. Choo and M. Pfost, "A variable gate resistance SiC MOSFET driver network to mitigate overshoot and parasitic ringing," in *PCIM Europe 2023; International Exhibition and Conference for Power Electronics, Intelligent Motion, Renewable Energy and Energy Management*, 2023, pp. 1–7.

[11] ——, "A self-driving 3-level active gate driver network to control the switching slew rate for SiC MOSFETs," in *PCIM Europe 2024; International Exhibition and Conference for Power Electronics, Intelligent Motion, Renewable Energy and Energy Management*, 2024.

[12] C. DiMarino, Z. Chen, M. Danilovic, D. Boroyevich, R. Burgos, and P. Mattavelli, "High-temperature characterization and comparison of 1.2 kv SiC power MOSFETs," in *2013 IEEE Energy Conversion Congress and Exposition*, 2013, pp. 3235–3242.

Use of Switched-Capacitor Circuit to Generate Negative Gate-Source Voltage Pulses

Ho-Tin Tang
Centre for Smart Energy Conversion and Utilization Research
Department of Electrical Engineering
City University of Hong Kong
Hong Kong
hotintang2-c@my.cityu.edu.hk

Henry Shu-Hung Chung
Centre for Smart Energy Conversion and Utilization Research
Department of Electrical Engineering
City University of Hong Kong
Hong Kong
eeshc@cityu.edu.hk

Abstract—A gate driver circuit that can momentarily generate negative voltage pulses to suppress positive crosstalk voltage and allow large negative crosstalk is presented. The idea is based on using a switched-capacitor circuit to generate a negative voltage from positive voltage supply when the output of the gate driver is "LOW". The proposed circuit is simple and does not require additional control signals compared with conventional gate driver. Thus, it can be implemented in conventional gate driver easily. Its performance has been evaluated on a 2600W switching circuit and 1500W synchronous buck converter. The experimental results are in close agreement with the theoretical predictions.

Keywords - Bridge-leg configuration, spurious voltage pulses, , multi-level gate driver, gate driving, switched-capacitor

I. INTRODUCTION

Bridge-leg configuration, which is formed by two series-connected switching devices, has been widely used in power converters. The two devices are alternately switched. However, due to the presence of unavoidable delay time, turn-on and turn-off times, the switching devices could be partially turned on simultaneously, leading to a shoot-through. To ensure safe operation, a practical way is to turn off both switching devices at the same time within a short time interval, known as the deadtime t_d, before turning on either one of the switches.

Despite the introduction of t_d can prevent the overlapping of the driving signal deriving from the gate drive integrated circuit (IC), switching on and off the device will result in high dv/dt on the MOSFETs and affect the gate voltage caused by the intrinsic parameters of the MOSFETs. When the control switch in the bridge-leg configuration is switched on, a positive spurious voltage occurs on the synchronous switch; if this voltage exceeds the MOSFET's threshold voltage, the MOSFET will partially turn on, leading to a large shoot-through current. This can increase the power loss of the converter or even cause device failure. Similarly, when the control switch is switched off, a negative spurious voltage occurs. If this voltage exceeds the MOSFET's maximum allowable gate-source voltage, it can also lead to device failure. Detailed analysis can

be found in [1] and [2]. This issue is even worse in wide-bandgap devices, such as silicon carbide (SiC) MOSFETs, because of their high switching speed.

A large body of literature has been devoted to address the spurious voltage issue and maximize the performance of switching devices in the bridge leg. Many solutions have been proposed and can be categorized into two main approaches. The first approach aims to reduce the magnitude of spurious voltage. Such as increasing high side gate resistance or reducing the impedance of the low side gate path, such as using a small turn-off gate resistor [3]-[4], connecting a capacitor [4]-[5] or a diode [6]-[7] across the gate resistor. Alternatively, an active approach involves inserting a transistor between the gate and source that turns on during the occurrence of spurious voltage [4],[5], [8] and [9]. However, these methods can only reduce, not eliminate, the spurious voltage due to the non-zero impedance of the low impedance path and the internal gate resistance in the MOSFET.

The second approach involves introducing a negative gate-source voltage during turn-off to ensure that the gate-source voltage is still below the threshold voltage when the spurious voltage occurs. The methods include using an additional negative voltage source for the gate driver output [6], [7], [10] and [11], passive circuit [5],[12] and [13], active circuit [6], [8], [10] and [14]-[17], or charge pump circuits [14]-[15]. As shown in Fig. 1, a simple way is to add a negative voltage source V_{GG_L} to bias the gate-source voltage to negative during the off state. However, the circuit may require an additional voltage supply to generate a negative voltage. Also, the steady negative voltage may reduce the lifetime of the MOSFET [18]-[19]. In addition, the steady negative gate-source voltage will increase the forward voltage of the body diode and reduce the allowable negative spurious voltage.

Recently, a multi-level gate driver technique has been developed [6], [8], [10]-[12] and [16]. The method is based on providing a negative gate-source voltage only within a short period before the positive spurious voltage occurs to prevent from false triggering. The gate-source voltage will be back to

---

This work was supported by the Innovation and Technology Fund from the Hong Kong Special Administrative Region, China, under Project #MRP/010/21X.

979-8-3315-1612-3/25 $31.00 © 2025 IEEE

zero so that the gate oxide's stress can be reduced. Also, the device can sustain a higher negative pulse voltage than static off-state voltage, as the off-state gate-source voltage increases to zero at the off state. This paper presents a switched-capacitor circuit that can provide a negative pulse voltage source. The proposed circuit does not require additional control signal and negative voltage source. The simple circuitry can be easily implemented by a typical gate driver IC, as shown in Fig. 2 [20]. The proposed circuit can momentarily generate a negative voltage and return to zero during off-state. It can counteract the spurious voltage and maintain the switching performance. The performance of the proposed circuit is tested on a 2600W switching circuit and 1500W synchronous buck converter.

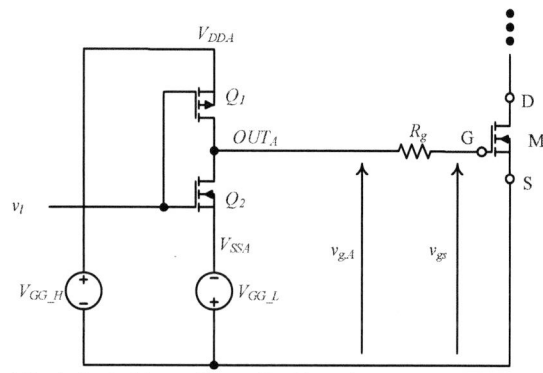

Fig.1 Typical gate driver with negative bias voltage.

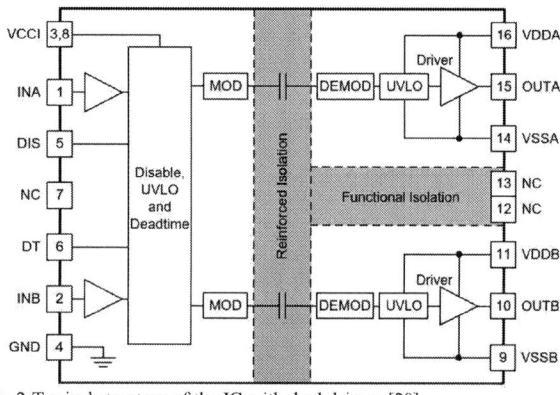

Fig.2 Typical structure of the IC with dual drivers [20].

II. PRINCIPLE OF OPERATION

The circuit schematic of the proposed circuit is shown in Fig. 3. It consists of two switches, Q_3 and Q_4, two resistors R_N and R_P, two capacitors C_N and C_P, and one diode D_P. It is powered by the same supply source V_{GG_H}. Thus, it can be easily implemented by using an IC with dual drivers as shown in Fig. 2, where the components are connected to V_{DDB}, O_{UTB}, and V_{SSB}.

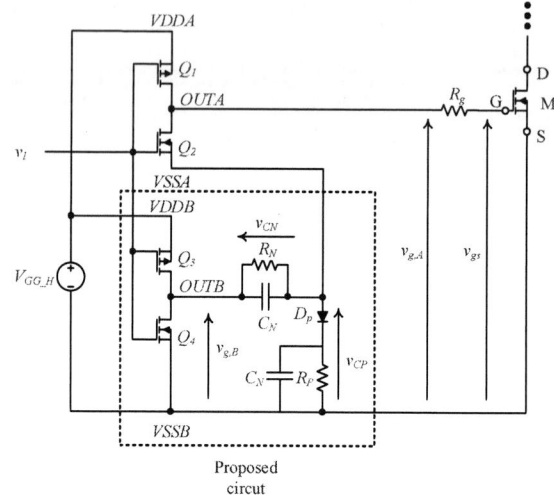

Fig. 3 Circuit schematic of the proposed circuit.

Q_1 and Q_2 form a conventional driver circuit. It is driven by the control logic v_l and its output $v_{g,A}$ is the main driver voltage, used to control the switching device M and switching between V_{DDA} and V_{SSA}. Q_3 and Q_4 form an auxiliary driver circuit, it shares the same control logic as Q_1 and Q_2. V_{DDB} is connected to the same voltage source as V_{DDA} and V_{SSB} is connected to the source of the M. Its output drives the switched capacitor circuit formed by R_N, R_P, C_N and C_P to provide a controllable negative voltage source to V_{SSA}. The ratio of R_N and R_P controlled the final state voltage of the circuit. The ratio of C_N and C_P determines the initial state voltage of the circuit. The proposed circuit can provide negative voltage when the duty cycle is small. Because the initial voltage is controlled by C_N and C_P.

When v_l is changed from logical "HIGH" to logical "LOW", Q_1 and Q_3 will turn on, and the voltage level of $v_{g,A}$ will change from V_{SSA} to V_{DDA} to turn on the MOSFET. The on-state voltage of $v_{g,A}$, $v_{g,A,ON}$, is

$$v_{g,A,ON} = V_{DDA} \qquad (1)$$

At the same time, the voltage level of $v_{g,B}$ will change from V_{SSB} to V_{DDB} which is the same as V_{DDA}, the on-state voltage of $v_{g,B}$, $v_{g,B,ON}$, is,

$$v_{g,B,ON} = V_{DDB} \qquad (2)$$

During the on state, C_N will be charged up to the voltage determined by R_N and R_P. Assuming the time constant of the switched capacitor circuit is much smaller than the period of the switching period of the switches. The voltage of v_{CN} and v_{CP}, $v_{CN,ON}$ and $v_{CP,ON}$, respectively, are expressed as

$$v_{CN,ON}(t) = \frac{(C_N + C_P)R_N + (C_P R_P - C_N R_N)e^{-\frac{t}{\tau_{ON}}}}{(C_N + C_P)(R_N + R_P)}V_{GG} \qquad (3)$$

979-8-3315-1612-3/25 $31.00 © 2025 IEEE   1583

$$v_{CP,ON}(t) = \frac{(C_N + C_P)R_P + (C_N R_N - C_P R_P)e^{-\frac{t}{\tau_{ON}}}}{(C_N + C_P)(R_N + R_P)}V_{GG} \quad (4)$$

where $t \in [0 \quad dT]$, $\tau_{on} = (C_N + C_P)\dfrac{R_N R_P}{(R_N + R_P)}$

At the end of on-state, the capacitor will be fully charged, and the final voltage of v_{CN} and v_{CP} are expressed as,

$$V_{CN,ON.f} = \frac{R_N}{R_N + R_P}V_{DDA} \quad (5)$$

$$V_{CP,ON,f} = \frac{R_P}{R_N + R_P}V_{DDA} \quad (6)$$

When v_I is changed from logical "LOW" to logical "HIGH", Q_2 and Q_4 will turn on, and the voltage level of $v_{g,A}$ will change from V_{DDA} to V_{SSA} which is connected to the midpoint of the switched capacitor network. The charge stored on C_{gs} of the MOSFET and that on C_N will be redistributed. D_P is blocking. The initial voltages of v_{CN} and v_{CP}, $V_{CN}(dT^+)$ and $V_{CP}(dT^+)$, respectively, at dT can be expressed as,

$$V_{g,A}(dT^+) = -V_{CN}(dT^+) = -\frac{C_N V_{CN,ON,f} - C_{gs}V_{DDA}}{C_n + C_{gs}} \quad (7)$$

$$V_{CP}(dT^+) = V_{CP,ON,f} \quad (8)$$

Thus, a negative voltage will be present on $v_{g,A,OFF}$ during a short period. During the off-state, C_N and C_P will be discharged and the voltages of v_{CN}, v_{CP} and $v_{g,A}$, $v_{CN,off}$, $v_{CP,off}$ and $v_{g,off}$, respectively can be expressed as,

$$v_{CN,OFF}(t) = e^{-\frac{t-dT}{\tau_{OFF}}}V_{CN}(dT^+) \quad (9)$$

$$v_{CN,OFF}(t) = -e^{-\frac{t-dT}{\tau_{CP}}}V_{CP}(dT^+) \quad (10)$$

$$v_{g,A,OFF}(t) = -v_{CN,OFF}(t) \quad (11)$$

where $t \in [dT \quad T]$, $\tau_{off} = (C_N + C_{gs})R_N$, $\tau_{CP} = C_P R_P$.

If the spurious voltage occurs, the proposed circuit can counteract the positive spurious voltage caused by turning on the control switch. The deadtime, t_d, for the control switch turns on typically within tens of nanoseconds. The maximum allowable spurious voltage, $V_{sp,max}$ can be expressed as,

$$V_{sp,max} = e^{-\frac{t_d}{\tau_{OFF}}}V_{CN}(dT^+) \quad (12)$$

Since the time constant of the switched capacitor network is much smaller than the switching period, at the end of off state, C_N will be fully discharged. The final value of $v_{g,A}$, v_{CN} and v_{CP}, at the end of off state is,

$$V_{g,A,OFF,f} = V_{CN,OFF,f} = V_{CP,OFF,f} = 0 \quad (13)$$

It allows a higher negative spurious voltage when the control switch is turned off. The timing diagram is shown in Fig. 4.

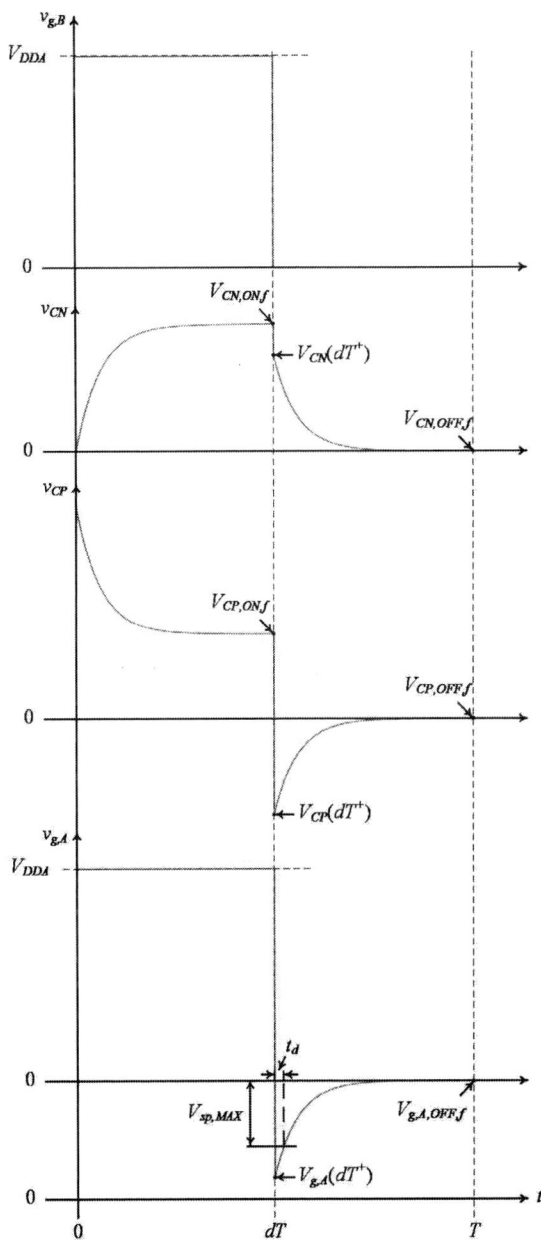

Fig. 4 Waveforms of the proposed circuit.

III. EXPERIMENTAL VERIFICATION

The proposed circuit has been tested according to the switching circuit shown in Fig. 5. The loading circuit only consists of a SiC MOSFET, M, a loading resistor, R_L and the driving circuit. $V_{CC} = 400\text{V}$ and $R_L = 60\Omega$. The proposed circuit is established by utilizing a typical gate driver IC (part no.: Si8275) with dual outputs. To evaluate the performance of the proposed circuit on negative voltage generation and the switching performance, three configurations, as shown in Fig. 6, are tested. Q_1, Q_2, Q_3 and Q_4 are implemented by using the gate driver IC's internal driving circuit. Configuration 1 is the proposed circuit. Configuration 2 uses the conventional gate driver with $V_{DDA} = 15\text{V}$ and $V_{SSA} = 0\text{V}$. Configuration 3 uses the conventional gate driver with $V_{DDA} = 15\text{V}$ and $V_{SSA} = -5\text{V}$. The part nos. and component values are given in Table I. The waveforms of primary driver voltage, $v_{g,A}$, gate-source voltage, v_{gs}, drain-source voltage, v_{ds} and drain current, i_d are studied.

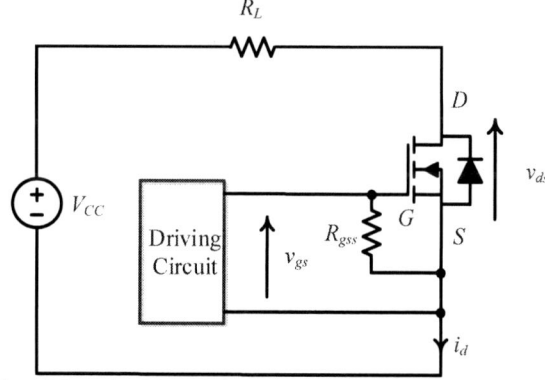

Fig. 5 Schematic of the switching circuit

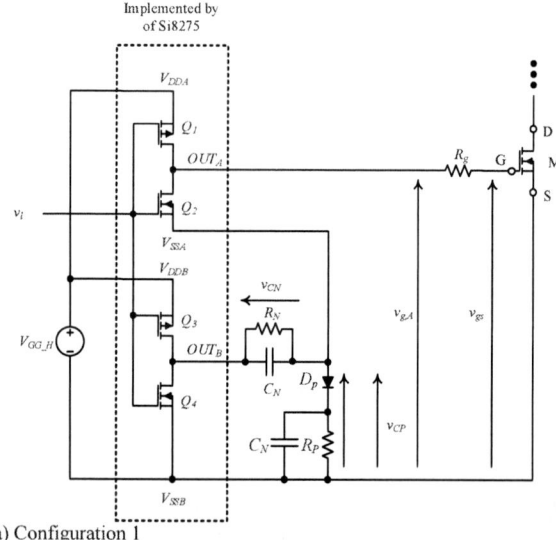

(a) Configuration 1

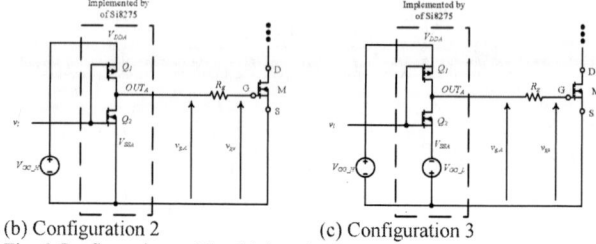

(b) Configuration 2   (c) Configuration 3
Fig. 6 Configurations of the driving circuit

TABLE I – COMPONENT PART NOS. AND VALUES

Component	Part no. / Value	Component	Part no. / Value
Configuration I			
V_{DDA}	15V	R_N	100Ω
V_{SSA}	Connected to v_{CP}	C_P	2.7nF
V_{DDB}	15V	R_P	68Ω
V_{SSB}	0V	R_g	4.7Ω
C_N	10nF	D_P	1N4148
Configuration II			
V_{DDA}	15V	R_g	4.7Ω
V_{SSA}	0V		
Configuration III			
V_{DDA}	15V	R_g	4.7Ω
V_{SSA}	-5V		

Fig. 7 shows the waveforms of $v_{g,A}$, v_{gs}, v_{ds} and i_d in a few switching cycles. Figs. 7(a), 7(b) and 7(c) show the waveforms of Configuration 1, Configuration 2 and Configuration 3, respectively. During the on state, v_{gs} follows V_{DDA} in all configurations. During the off state, v_{gs} has a momentarily short negative voltage and return to 0V in Configuration 1. In Configuration 2, v_{gs} is kept at 0V. In Configuration 3, v_{gs} is kept at -5V.

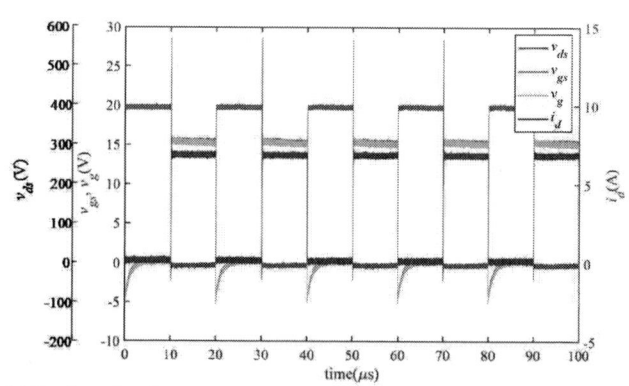

(a) Configuration 1

979-8-3315-1612-3/25 $31.00 © 2025 IEEE   1585

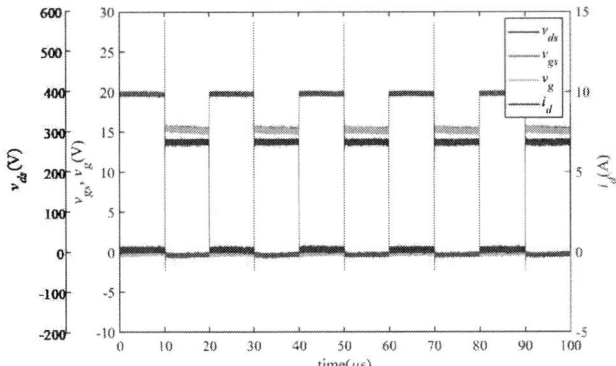

(b) Configuration 2

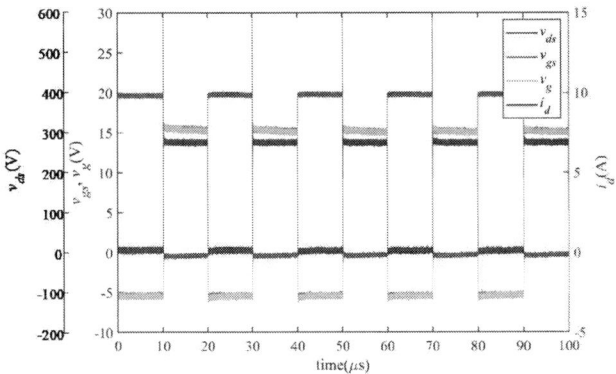

(c) Configuration 3
Fig. 7 Key waveforms of switching cycles

Fig. 8 shows the magnified waveforms upon switching off in the three configurations. The transient and steady state characteristics upon switching off are given in Table II. The nomenclature of the symbols is given as follows:

$V_{ds,OFF}$: Steady-state off-state value of v_{ds}
$t_{ds,r}$: Rise time of v_{ds} (From 10% to 90% of the steady-state value)
$V_{g,OFF}$: Steady-state off -state value of v_g
$t_{g,f}$: Fall time of v_g (From 90% to 10% of the steady-state value)
$V_{gs,OFF}$: Steady-state off -state value of v_{gs}
$t_{gs,f}$: Fall time of v_{gs} (From 90% to 10% of the steady-state value)
$I_{d,OFF}$: Steady-state off -state value of i_d
$t_{d,f}$: Fall time of i_d (From 90% to 10% of the steady-state value)

TABLE II – TRANSIENT AND STEADY-STATE CHARACTERISTICS DURING TURN OFF

Parameter	Configuration		
	I	II	III
$V_{ds,OFF}$	400V	400V	400V
$t_{ds,r}$	38.4ns	38.2ns	37.2ns
$V_{g,OFF}$	-5V	0V	-5V
$t_{g,f}$	23ns	15.6ns	17.6ns
$V_{gs,OFF}$	-5V	0V	-5V
$t_{gs,f}$	27.6ns	29.4ns	25.8ns
$I_{d,OFF}$	0A	0A	0A
$t_{d,f}$	10.2ns	8.8ns	9.8ns

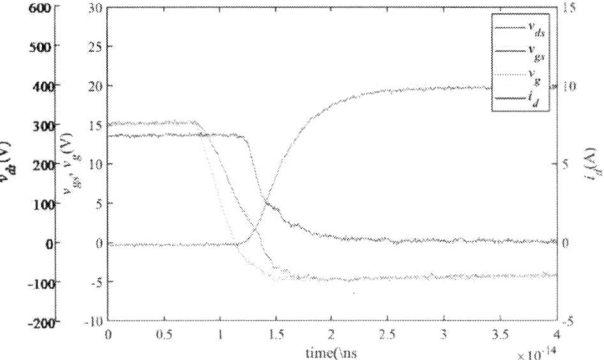

(a) Configuration 1

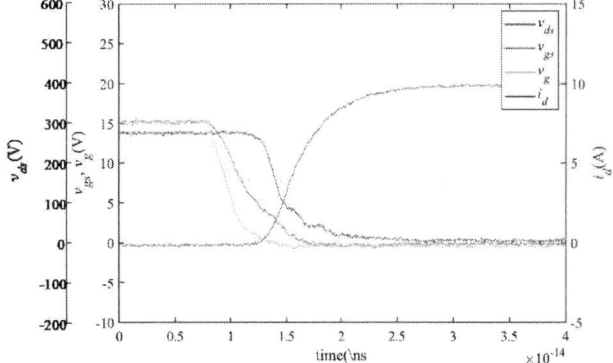

(b) Configuration 2

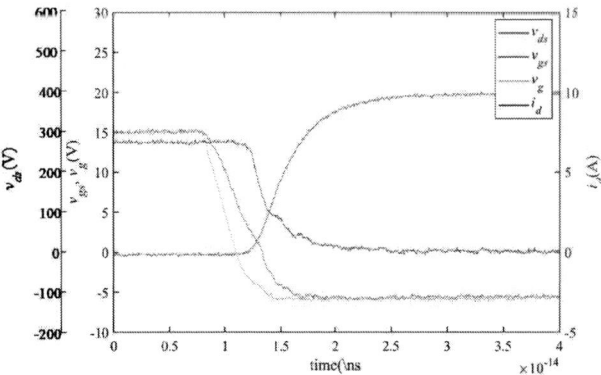

(c) Configuration 3
Fig. 8 Magnified waveforms during switching off.

With the proposed circuit, v_{CN} is temporarily negative upon switching off. V_{SSA} is also negative. Thus, when the MOSFET is switched off, v_{gs} drops from 15V to -5V. The switching speed with the proposed circuit is slightly slower than the configuration 2 and configuration 3. The reason is that the proposed circuit is implemented by a conventional gate driver IC, the resistance of the output stage of the IC is 1Ω, the total turn-off gate resistance is increased from 4.7Ω to 5.7Ω.

Fig. 9 shows the magnified waveforms upon switching on. The transient and steady state characteristics upon switching on are given in Table III. The nomenclature of the symbols is given as follows:

$V_{ds,ON}$: Steady-state on-state value of v_{ds}

$t_{ds,f}$: Fall time of v_{ds} (From 90% to 10% of the steady-state value)

$V_{g,ON}$: Steady-state on-state value of v_g

$t_{g,r}$: Rise time of v_g (From 10% to 90% of the steady-state value)

$V_{gs,ON}$: Steady-state on-state value of v_{gs}

$t_{gs,r}$: Rise time of v_{gs} (From 10% to 90% of the steady-state value)

$I_{d,ON}$: Steady-state on-state value of t_d

$t_{d,r}$: Rise time of t_d (From 10% to 90% of the steady-state value)

TABLE III – TRANSIENT AND STEADY-STATE CHARACTERISTICS DURING TURN ON

Parameter	Configuration		
	I	II	III
$V_{ds,ON}$	0.3V	0.3V	0.3V
$t_{ds,f}$	9.2ns	9ns	9ns
$V_{g,ON}$	15V	15V	15V
$t_{g,r}$	21ns	20.4ns	21.6ns
$V_{gs,ON}$	15V	15V	15V
$t_{gs,r}$	31.4ns	32.4ns	30.8ns
$I_{d,ON}$	6.7A	6.7A	6.7A
$t_{d,r}$	3.6ns	3.6ns	3.4ns

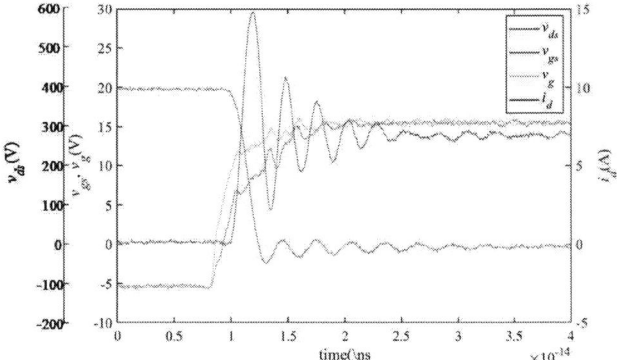

(b) Configuration 2

(c) Configuration 3

Fig. 9 Magnified waveflorms of during switching on

When the MOSFET is switched on, v_{gs} increases from 0 to 15V, the switching speed for all three configurations is close.

The proposed circuit has been further tested on a 400V$_{dc}$ / 200V$_{dc}$ synchronous buck converter shown in Fig. 10 to show a real crosstalk example for suppressing crosstalk. L_o = 0.7mH, C_o = 22μF, and R_o = 27.4Ω. M_1 is driven by a conventional gate driver IC with V_{DDA} = 15V and V_{SSA} = -5V. M_2 is driven by Configuration 1 and Configuration 3, respectively. Since the spurious voltage will occur and the positive spurious voltage which may lead to shoot through, Configuration 2 with off-state gate-source voltage = 0V is not included in this test. The part nos. and component values are same as the part nos. and component values in Table I. The waveforms of primary driver voltage, $v_{g,2}$, gate-source voltage, $v_{gs,2}$, drain-source voltage, $v_{ds,2}$ and drain current, $i_{d,2}$ are studied. Fig. 11 shows the magnified waveforms during switching off in the two configurations.

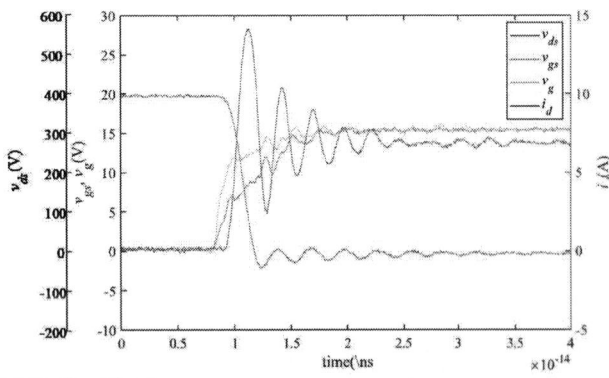

(a) Configuration 1

979-8-3315-1612-3/25 $31.00 © 2025 IEEE   1587

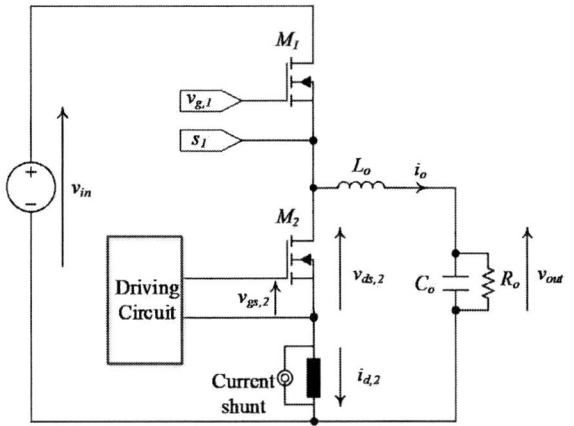

Fig. 10 Schematic of the testing setup of the synchronous buck converter

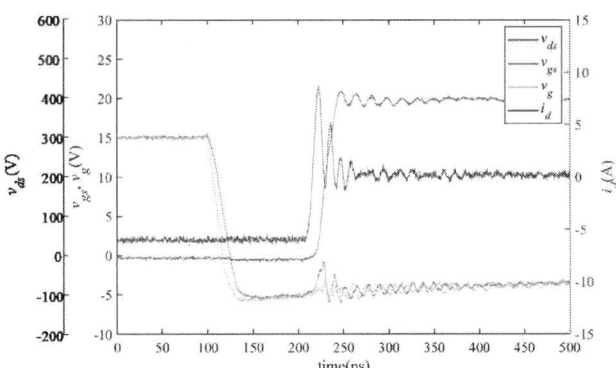

(a) Configuration 1

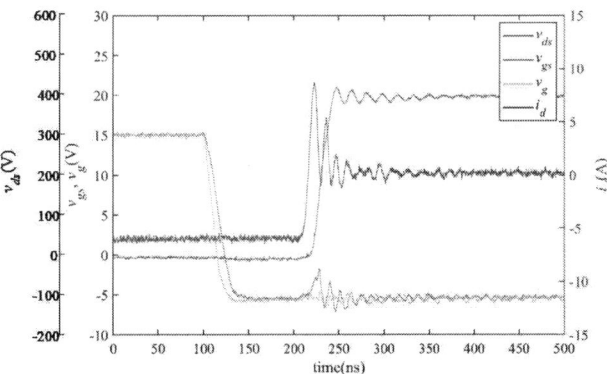

(b) Configuration 3

Fig. 11 Magnified waveforms during switching off when M_2 in synchronous mode

By observing the turn off waveforms using the proposed circuit shown in Fig. 11(a), the peak negative voltage just after the switch is turn off is -5.5V and the voltage level of $v_{gs,2}$ after 100ns of switching off, is -5.0V. The amplitude of the spurious voltage is 4.1V. By observing the turn off waveforms when configuration 3 is using in Fig. 11(b), $v_{gs,2}$ for entire off-state is -5.5V. The amplitude of the spurious voltage is 3.9V. The maximum allowable positive spurious voltage when using configuration 1 is nearly the same as that in the configuration 3.

Fig. 12 shows the magnified waveforms during switching on. As shown in Fig. 12(a), when the proposed circuit is used, at the end of the off state, C_N is fully discharged, and the voltage level of V_{SSA} is 0. When configuration 3 is using, $v_{gs,2}$ during off state is kept at -5V. Since the maximum allowable gate-source voltage of the MOSFET is -8V. Thus, the allowable negative crosstalk voltage for safe operation is 8V and 3V when configuration 1 and 3 is using, respectively.

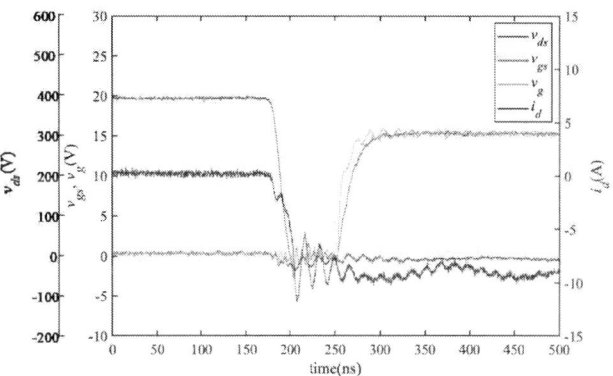

(a) Configuration 1

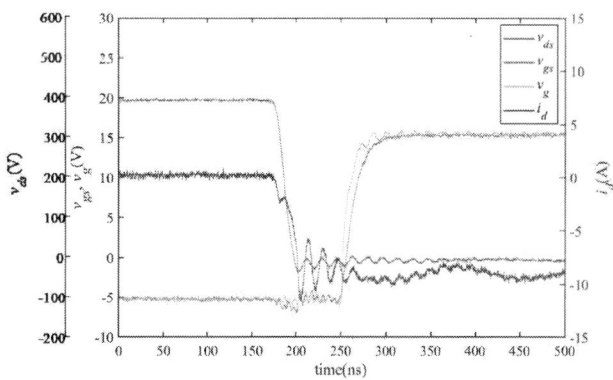

(b) Configuration 3

Fig. 12 Magnified waveforms of during switching on when M_2 in synchronous mode

To conclude, the proposed circuit have similar turn-on speed like the conventional gate driver with $V_{SSA} = 0V$ $V_{SSA} = -5V$, and switching off speed compared with conventional gate driving is slightly decreased because of the additional resistance from the IC. Also, it can provide a pulse negative voltage source to counteract the positive spurious voltage and be discharged to 0V to counteract the negative spurious voltage. Also, the circuit is easy to implement in typical gate driver IC, the control signal of the additional transistors is the same as the conventional gate driver.

IV. CONCLUSION

In this paper, a gate drive composed of a switched capacitor to provide a negative pulse voltage source has been present. The proposed circuit can generate a negative voltage by a positive voltage source to counteract the spurious voltage and improve the switching performance. Compared with the conventional gate driver, the proposed circuit only requires 2 transistors, 2 resistors, and 1 capacitor. It can easily be implemented by the typical gate driver IC with dual outputs. The proposed circuit has been tested on a 2600W switching circuit and 1500W synchronous buck converter. The experimental results show that the proposed circuit can generate the pulse negative voltage through a positive voltage source. The positive spurious voltage can then be counteracted. In addition, the switching performance is close to conventional gate drivers.

REFERENCES

[1] J. Wang, H. Chung, and T. Li, "Characterization and experimental assessment of the Effects of Parasitic Elements on the MOSFET Switching Performance," IEEE Transactions on Power Electronics, vol. 28, no. 1, pp. 573-590, Jan 2013.

[2] J. Wang and H. Chung, "Impact of Parasitic Elements on the Spurious Triggering Pulse in Synchronous Buck Converter," IEEE Transactions on Power Electronics, vol. 29, no. 12, pp. 6672-6685, Dec. 2014.

[3] Y. Li, M. Liang, J. Chen, T. Q. Zheng and H. Guo, "A Low Gate Turn-OFF Impedance Driver for Suppressing Crosstalk of SiC MOSFET Based on Different Discrete Packages," in IEEE Journal of Emerging and Selected Topics in Power Electronics, vol. 7, no. 1, pp. 353-365, March 2019

[4] K. Yamaguchi, K. Katsura, T. Yamada and Y. Sato, "Comprehensive evaluation of gate boost driver for SiC-MOSFETs," 2016 IEEE Energy Conversion Congress and Exposition (ECCE), Milwaukee, WI, 2016, pp. 1-8

[5] F. Gao, Q. Zhou, P. Wang and C. Zhang, "A Gate Driver of SiC MOSFET for Suppressing the Negative Voltage Spikes in a Bridge Circuit," in IEEE Transactions on Power Electronics, vol. 33, no. 3, pp. 2339-2353, March 2018

[6] Z. Zhang, J. Dix, F. F. Wang, B. J. Blalock, D. Costinett and L. M. Tolbert, "Intelligent Gate Drive for Fast Switching and Crosstalk Suppression of SiC Devices," in IEEE Transactions on Power Electronics, vol. 32, no. 12, pp. 9319-9332, Dec. 2017

[7] Z. Zhang, Z. Wang, F. Wang, L. M. Tolbert and B. J. Blalock, "Reliability-oriented design of gate driver for SiC devices in voltage source converter," 2015 IEEE International Workshop on Integrated Power Packaging (IWIPP), Chicago, IL, 2015, pp. 20-23

[8] Z. Zhang, F. Wang, L. M. Tolbert and B. J. Blalock, "Active Gate Driver for Crosstalk Suppression of SiC Devices in a Phase-Leg Configuration," in IEEE Transactions on Power Electronics, vol. 29, no. 4, pp. 1986-1997, April 2014

[9] S. Yin, K. J. Tseng, C. F. Tong and R. Simanjorang, "Design of high-speed gate driver to reduce switching loss and mitigate parasitic effects for SiC MOSFET," in IET Power Electronics, vol. 10, no. 10, pp. 1183-1189, 18 8 2017

[10] S. Zhao et al., "Adaptive Multi-Level Active Gate Drivers for SiC Power Devices," in IEEE Transactions on Power Electronics, vol. 35, no. 2, pp. 1882-1898, Feb. 2020

[11] Y. Yang, Y. Wen and Y. Gao, "A Novel Active Gate Driver for Improving Switching Performance of High-Power SiC MOSFET Modules," in IEEE Transactions on Power Electronics, vol. 34, no. 8, pp. 7775-7787, Aug. 2019

[12] Y. Chen, R. Wang, X. Liu and Y. Kang, "Gate-Drive Power Supply With Decayed Negative Voltage to Solve Crosstalk Problem of GaN Synchronous Buck Converter," in IEEE Transactions on Power Electronics, vol. 36, no. 1, pp. 6-11, Jan. 2021

[13] H. Tang, H. Chung, J. Fan, R. Yeung and R. Lau, "Passive Resonant Level Shifter for Suppression of Crosstalk Effect and Reduction of Body Diode Loss of SiC MOSFETs in Bridge Legs," in IEEE Transactions on Power Electronics, vol. 35, no. 7, pp. 7204-7225, July 2020

[14] F. Mo, J. Furuta and K. Kobayashi, "A low surge voltage and fast speed gate driver for SiC MOSFET with switched capacitor circuit," 2016 IEEE 4th Workshop on Wide Bandgap Power Devices and Applications (WiPDA), Fayetteville, AR, 2016, pp. 282-285

[15] H. Gui, J. Sun and L. M. Tolbert, "Charge Pump Gate Drive to Reduce Turn-ON Switching Loss of SiC MOSFETs," in IEEE Transactions on Power Electronics, vol. 35, no. 12, pp. 13136-13147, Dec. 2020

[16] Q. He, Y. Zhu, H. Zhang, A. Huang, Q. Cai and H. Kim, "A Multilevel Gate Driver of SiC mosfets for Mitigating Coupling Noise in Bridge-Leg Converter," in IEEE Transactions on Electromagnetic Compatibility, vol. 61, no. 6, pp. 1988-1996, Dec. 2019

[17] C. Li et al., "High Off-State Impedance Gate Driver of SiC MOSFETs for Crosstalk Voltage Elimination Considering Common-Source Inductance," in IEEE Transactions on Power Electronics, vol. 35, no. 3, pp. 2999-3011, March 2020

[18] A. Maerz, T. Bertelshofer, M. Bakran and M. Helsper, "A Novel Gate Drive Concept to Eliminate Parasitic Turn-on of SiC MOSFET in Low Inductance Power Modules," PCIM Europe 2017; International Exhibition and Conference for Power Electronics, Intelligent Motion, Renewable Energy and Energy Management, Nuremberg, Germany, 2017, pp. 1-7.

[19] J. Henn et al., "Intelligent Gate Drivers for Future Power Converters," in IEEE Transactions on Power Electronics, vol. 37, no. 3, pp. 3484-3503, March 2022.

[20] "Si827x Data Sheet," Datasheet of SI8275, Rev. A, SILICON LABS, Jun. 2022.

An Optically Isolated Gate Driver with Simultaneous Data and Power Transmission through a Miniaturized, Efficient Photonic Platform

Jiajun Li, Mariia Klymenko, Yanqiao Li, William Scheideler and Jason T. Stauth

Thayer School of Engineering at Dartmouth College

Hanover, NH USA

jiajun.li.th@dartmouth.edu, jason.t.stauth@dartmouth.edu

Abstract—This paper presents the design and implementation of an efficient optically-isolated gate driver that can be used for wide bandgap (WBG) devices in a wide range of high-voltage power conversion circuits where galvanically-isolated gate driving is a key challenge. The proposed isolated gate driving system achieves simultaneous signal and power transmission through a modulated LED transmitter and galvanically-isolated silicon photovoltaic receiver. Peak transmission efficiency (LED + PV combined) of the optical system was 15.1% for a single cells and 11.2% for a module comprising two series cells while achieving over 5 kV isolation and < 2.5 pF capacitance across the isolation barrier. A forward-mode receiver circuit was designed to recover a switching signal from modulated LED current. A switched-capacitor-based gate drive circuit was used to multiply the PV cell voltage in sequential steps while recovering gate charge energy to provide up to 30% total efficiency for the combined system.

I. INTRODUCTION

Recent advances in wide-bandgap (WBG) semiconductors have enabled transformational advances in power electronics for a wide range of high-voltage applications due to their ability to operate at higher voltages with better tradeoffs between conduction loss and frequency-dependent losses [1]–[3]. However, there are a number of critical yet less explored challenges related to controlling and powering WBG devices that can limit the efficiency, power-density, and reliability of modern power electronic circuits. In particular, WBG-enabled circuits increasingly require galvanic isolation for both signal and power interfaces, which motivates the need for isolated gate drivers that are both small and efficient [4], [5].

A generalized block diagram of an isolated gate driver is shown in Fig. 1. The ground-referenced control signal and power is transferred through an isolation barrier to the gate driver in the floating domain which is referenced to the source of the power field-effect transistor (FET). As the gate driver charges and discharges a predominantly capacitive load, the overall *reactive* power processed by the gate driver, P_{gg}, can be expressed as

$$P_{gg} = Q_{gg}V_{g,pp}f_{sw} = C_{gg}V_{g,pp}^2 f_{sw}, \tag{1}$$

where Q_{gg} is the total charge needed to turn on the gate, $V_{g,pp}$ is the peak-to-peak gate voltage, f_{sw} is the switching frequency of the gate driver, C_{gg} is a linear approximation of the overall gate capacitance. Therefore, here we use the metric

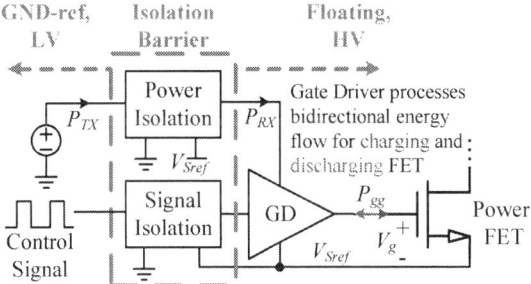

Fig. 1: A representative block diagram of isolated gate driver.

of reactive power efficiency, η_X, to quantify the performance of the gate drive circuit operating in the floating voltage domain [6]:

$$\eta_X = \frac{P_{gg}}{P_{gg} + P_{RX}}, \tag{2}$$

where P_{RX} is the total *real* power consumed by the floating gate driver (and received through the isolation barrier). We further define the efficiency of transferring P_{RX} through the isolation barrier as

$$\eta_{PT} = \frac{P_{RX}}{P_{TX}}, \tag{3}$$

where P_{TX} is the power provided from the input source or supply. These metrics combined give a total efficiency

$$\eta_{\text{total}} = \frac{P_{gg}}{P_{gg} + P_{TX}}, \tag{4}$$

which captures reactive power efficiency respective of overall transmitted input power.

Past efforts have explored isolated wireless power transfer (WPT) using magnetic transformers [7]–[9]. However, transformer-based solutions are notoriously difficult to scale down in size: magnetic components degrade both in terms of efficiency and power density as physical volume is reduced [10]. Fig. 2 provides a perspective on such approaches. For larger power modules and switches (IGBTs and large Si and SiC modules) with sizes on the order of cm^3 and power levels over 1 W, isolated DC-DC converters using cored transformers are an effective solution, *e.g.* [9]. At smaller size and lower power, high frequency air-core magnetics are needed, *e.g* [8];

979-8-3315-1612-3/25 $31.00 © 2025 IEEE

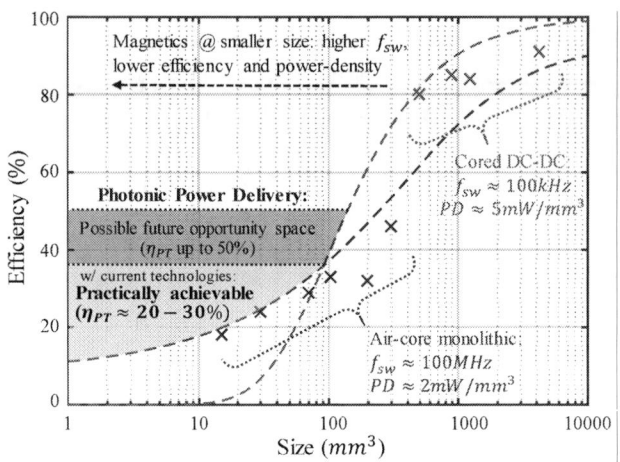

Fig. 2: A perspective on the efficiency and size of isolated power transfer via cored and coreless magnetics, highlighting a theoretical and practically achievable opportunity space for optical power transfer.

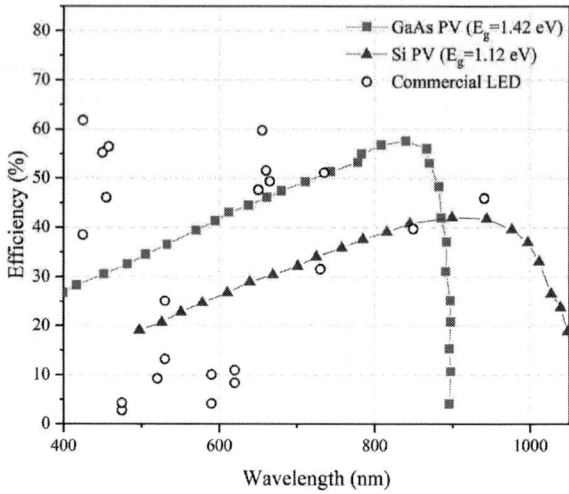

Fig. 3: Reported efficiency of the commercially available LEDs and calculated theoretical efficiency of GaAs and Si PV cells for visible and near-infrared spectrum.

however their efficiency and power density are modest and they don't scale well below $100\,mm^3$.

Other challenges with the magnetic approach that are not captured in Fig. 2 relate to galvanic isolation and parasitic capacitance. With magnetic approaches, higher isolation forces larger spacing between coils or windings; this drives lower efficiency and power density and/or larger physical size. There are similar trends with parasitic capacitance which can be a limitation in gate drive applications where high frequency transients motivate high common-mode transient immunity CMTI. However, low capacitance often correlates with larger spacing and/or weaker magnetic coupling and therefore presents a challenge in magnetic approaches [4]. Finally, electromagnetic interference and compatibility EMI/EMC can pose a constraint on magnetic or capacitive wireless power transfer which can limit frequency or power levels [4].

Another strategy is to use a photonic interface to provide optical power across a galvanic isolation barrier. Compared to conventional magnetic-based WPT, optical power transfer (OPT) has fundamentally better scaling with physical size. This makes it promising for galvanically isolated power delivery for a range of small-size and moderate to low-power applications including gate driving. Optical power delivery also has significant advantages in galvanic isolation capabilities compared to magnetic approaches. This is because optical power can be provided through insulating materials, waveguides, fiber optics, or even free space [11]–[15]. This allows small-size optical approaches achieve very high galvanic isolation suitable for high kilovolt to MV grid interface and power management electronics. Such approaches can operate with minimal parasitic capacitance, have negligible CMTI challenges, and also reduce problems with EMI/EMC [12].

Past efforts have explored a range of 'optically-activated' devices including light triggered thyristors and other junction-modulated devices used for fiber-coupled actuation [11], [16]. However, this relies on specifically-tailored devices which are difficult to scale and subject to performance tradeoffs. To date there have been examples of power-over-fiber gate drivers [12]

and optically-powered converters using fiberless assemblies [13]. However, compared to magnetic-based approaches there has been little research and development to explore the performance limits and opportunities for merged optical power delivery and communication in this area.

To outline the prospects for a future roadmap in optical power delivery, Fig. 3, shows the theoretical efficiencies of two types of PV cells, GaAs (bandgap $E_g = 1.42$ eV) and Si ($E_g = 1.12$ eV), overlaid with the reported efficiencies of commercially available LEDs [17] at drive currents of 350 mA. At lower drive currents, for example, 25–100 mA, it is possible for this LED efficiency to be higher than this nominal value [17] due to the reduced effects of series resistance and self-heating. We note that the highest efficiency commercial LEDs currently available are in the blue or red and far red parts of the visible spectrum, while most commercial near-infrared and green LEDs exhibit lower efficiency. Fig. 3 shows that due to its wider bandgap, GaAs exhibits better performance than Si for the solar cell receiver at shorter wavelengths, while Si cells are most effective at longer wavelengths in the near-infrared. Note that the monochromatic efficiency of both GaAs and Si solar cells is approximately $2\times$ higher than their often cited broadband sunlight conversion efficiency.

Fig. 4 provides a summary of the theoretical transmission efficiency $\eta_{PT} \approx \eta_{LED} \cdot \eta_{PV}$, where η_{PV} and η_{LED} are the PV and LED electrical-optical efficiencies respectively, and is thus an estimate for electrical power transmission efficiency (3). The theoretical PV cell efficiencies were estimated using

$$\eta_{PV} = EQE \cdot \frac{V_{PV}}{V_\lambda} \cdot FF, \qquad (5)$$

where EQE is the external quantum efficiency of the PV cells, V_λ photon energy at a particular wavelength expressed in Volts, and FF is the fill factor of the PV cells [18].

The theoretical optimal wavelength range for the Si cells used in this project is 900-950 nm ($\sim$1.33 eV), while for the GaAs cells, the optimal range lies between 800 and 850 nm.

979-8-3315-1612-3/25 $31.00 © 2025 IEEE

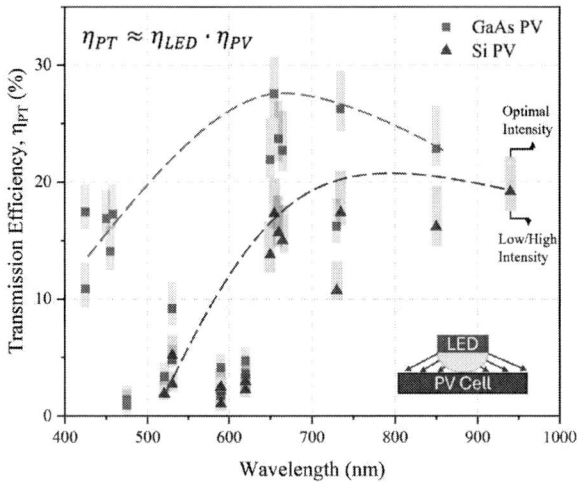

Fig. 4: Theoretical transmission efficiency (η_{PT}) is estimated using the reported efficiency of LEDs and the theoretical efficiencies of GaAs and Si PV cells across the visible and near-infrared spectrum.

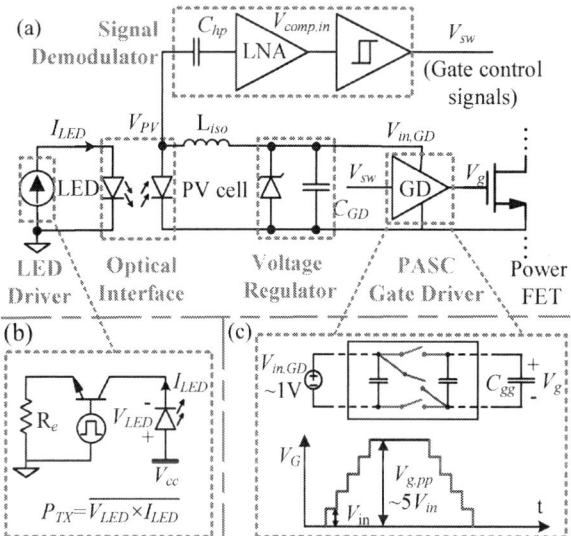

Fig. 5: (a) System diagram of proposed isolated gate driving system. (b) LED Driver Circuit (c) SC gate driver diagram and the representative gate voltage waveform.

The transmission efficiency reaches approximately 20% for Si cells coupled with commercial 940 nm LEDs, whereas the GaAs system achieves near 28% when coupled with 650 nm LEDs. The difference between the optimal optical range for the GaAs solar cell and the PV-LED module can be attributed to the fact that the transmission efficiency is determined by the product of the LED's efficiency and the PV cell's efficiency. The combined performance of the LEDs and the PV cells in Fig. 4 outlines the realistic, currently achievable performance space. Each point corresponds to a nominal reported efficiency from the manufacturer datasheet, while the shaded bands capture potential variations in efficiency related to drive power (LED current), showing the potential for exceeding 20% with Si solar cells and 30% with GaAs solar cells. This provides a basis for the *practically achievable* operating space highlighted in Fig. 2; the extended range in Fig. 2 is based on best-case reported efficiencies and advanced materials [19] that may need further validation to achieve practical viability.

To explore opportunities for wireless-optical power delivery, this work uses a custom photonic power and communication system combined with a pseudo-adiabatic switched-capacitor (PASC) gate driver [20]. The received optical power is used to operate the gate driver while the modulated optical signal is amplified and demodulated to recover the control signal in the floating domain. With the capability of recycling energy from the power device gate dielectric through PASC gate driver, the overall system efficiency is improved. The solution is able to achieve peak optical to gate-drive power efficiency over 30% with over 5kV isolation using commercially available optical components and materials, combined with the custom gate-drive integrated circuit.

II. SYSTEM ARCHITECTURE

Fig. 5 (a) shows the system diagram of the proposed optically isolated gate driving system. Both power and the gate-drive switching signal are transmitted by a LED driven by a high-speed current-mode LED driver, Fig. 5 (b). The average power is used to operate the gate driver which charges and discharges the power FET gate capacitance C_{gg}. The modulated LED current results in a small signal in the floating PV diode(s), which is detected and used to control the gate switching waveform. Note that here P_{TX} represents the average electrical power provided to the LED, the optimization of the LED driver is not within the scope of this paper.

On the receiver side, forward-biased PV cells collect both ac and dc components of the optical power and convert it into electrical voltage V_{PV}. An inductor L_{iso} is used as an ac choke to extract the ac component of V_{PV} such that the signal demodulator can recover the gate control signal V_{sw} in the floating domain. The average power from the PV cell(s) is used to power the gate drive circuit.

To improve the system efficiency, a switched-capacitor (SC) gate driver from [20] was used in this work. As shown in Fig. 5 (c), the gate driver uses multiple SC switching cells, configured in different series-parallel combinations to multiply the low-voltage ($\sim 1\,\text{V}$) PV supply input up to $\sim 5\,\text{V}$ gate swing. The gate driver switches the power FET based on the demodulated signal with all gate energy provided through the optical channel.

A. Optical Interface

The gate driver is powered by two series-connected Passivated Emitter and Rear Contact (PERC) monocrystalline Si cells. The cells were laser-diced into two sizes: 1 cm x 1 cm for initial testing and 0.5 cm x 1 cm for having 2 PV cells connected in series for the optical receiver with $V_{PV} \sim 1.2$ V.

From Fig. 3 and (5), the optimal wavelength range is predicted to be 900-950 nm. To confirm the optimal optical wavelength, initial testing was performed on PERC single 1 cm x 1 cm cells by placing the cell in direct contact with the LED lenses (Fig. 6 inset). The optimal illumination position was found by centering the LED above the cell to reach

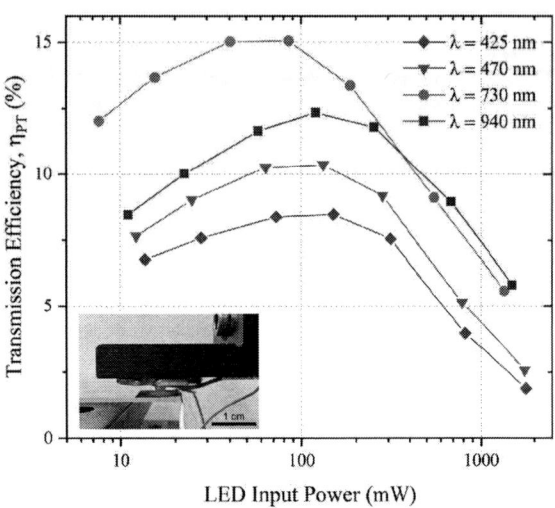

Fig. 6: Experimental transmission efficiency (η_{PT}) of a single PV cell under monochromatic illumination.

an optimum short circuit current density, J_{sc}. Fig. 6 shows the testing setup and results for monochromatic testing using LEDs from Cree (470 nm, 730 nm), Luxeon (425 nm), and New Energy (940 nm). The maximum transmission efficiency of 15.1% is observed at 50-100 mW/cm^2 for single cells under 730 nm illumination rather than the longer wavelength (940 nm). We anticipate that this is simply due to the higher performance (than the nominal datasheet value) of the 730 nm LEDs at the drive currents where the optimal system-level efficiency is achieved (around an LED input power of 100 mW).

The LED-PV module was constructed by encapsulating two series-connected PV cells with the 730 nm LED in a transparent dielectric (Crystalbond 509) (Fig. 7). The LED was driven at 5-500 mA, and the J-V measurements were taken. The optimal range of drive currents for the LEDs is determined to be 25-100 mA. At low intensities, the lower excess carrier concentrations in the solar cell result in a reduction in quasi-fermi level splitting and a lower V_{PV}, yielding lower efficiency. At higher intensity, the efficiency of the cells drops due to series resistance losses scaling quadratically with current density. Another consideration is that spacial non-uniformity of the optical power can lead to increased losses.

The maximum transmission efficiency achieved for the encapsulated PV module is $\eta_{PT} \sim 11.2\%$ for 50 mA LED drive current. The efficiency drop from 15.1% for single cells to 11.2% for the module with two series-connected cells may be explained by batch and cell-cell variability, edge effects from cell dicing, and larger overhead for interconnect for the series cells. However, as indicated in Fig. 4 higher efficiency is achievable with improved materials, processing, and assembly.

Galvanic isolation was tested using a high voltage potential tester Instek GPT-815. For the LED-PV module, the measured isolation voltage level was above the range of the instrument characterization which was 5 kV. The capacitance across the

Fig. 7: PV-LED module encapsulated in Crystalbond.

optical interface between the LED and PV cells was measured to be less than 2.5 pF using an impedance analyzer.

B. Receiver for Extracting Power and Signal

The PV module was characterized to determine its impedance and frequency response under bias to develop a small signal model used for optical signal transduction. This was done using a waveform generator to pre-bias the cells in forward-mode operation and a small-signal perturbation to determine resistance and capacitance. Fig. 8 illustrates the small signal equivalent circuit of PV cells in the receiver circuit, where I_{gd} captures the current drawn by the gate driver, R_{PV} and C_j are the equivalent resistance and capacitance of the PV cells; R_{LNA} is the input impedance of LNA stage, or the same as R_1 in Fig. 10.

The values for R_{PV} and C_j were measured to be $\sim 75\Omega$ and ~ 39nF respectively. The high relative PV capacitance illustrates a challenge in forward-mode signal transduction as this reduces the signal amplitude under optical modulation. The additional capacitance of the gate drive circuit C_{gd} also limits the ac response of the modulated signal. To reduce the impact of C_{gd} loading, an isolation inductor L_{iso} was used as a choke in series with the gate-drive circuit.

The impact of L_{iso} on the frequency response of the LED-PV module is shown in Fig. 9. The inductance results in a series and parallel resonance that can impact the circuit dynamics, but most importantly, it is seen that at high frequencies, L_{iso} shifts the capacitive signal attenuation out by removing the impact of C_{gd}. In this work we used a small 1 μH 0805 inductor with 60 mΩ dc resistance, which has negligible impact on power loss and efficiency. Nonetheless, due to the dominant effect of junction capacitance C_j, the receiver circuits for forward mode operation are different from conventional transimpedance amplifiers used in reverse-biased photodiode communication circuits in that they require significant gain (or gain×bandwidth) to amplify and detect the PV current signal.

The receiver circuits in Fig. 10 consists of a 40 dB gain stage and a comparator with adjustable hysteresis. The gain stage consists of 2 identical feedback amplifiers with 20 dB gain for each. The op-amp LTC6268 provides gain-bandwidth product of 500 MHz to minimize the latency introduced by the receiver circuitry. A 0.3 V symmetrical hysteresis window is added to eliminate chattering in the comparator which only adds 4.5 ns latency with 20 mV overdrive voltage.

Fig. 11 shows the representative waveform of the LED current (I_{LED}), PV cell voltage (V_{PV}), input voltage of

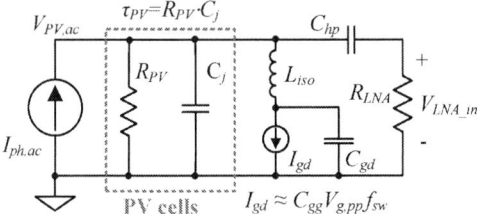

Fig. 8: Equivalent circuit of PV cells and RX circuit.

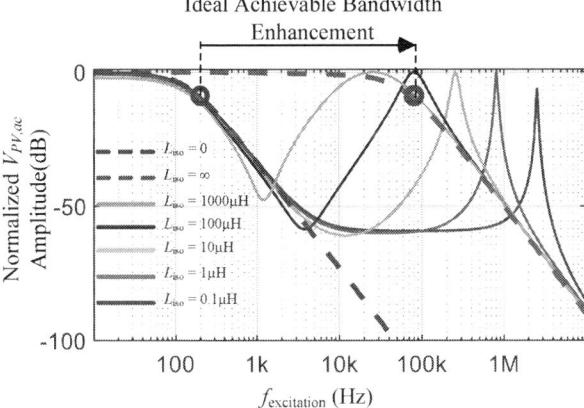

Fig. 9: Bandwidth enhancement effect by L_{iso}.

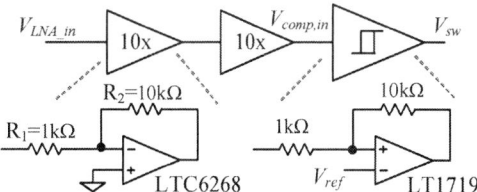

Fig. 10: Receiver detailed diagram and components value.

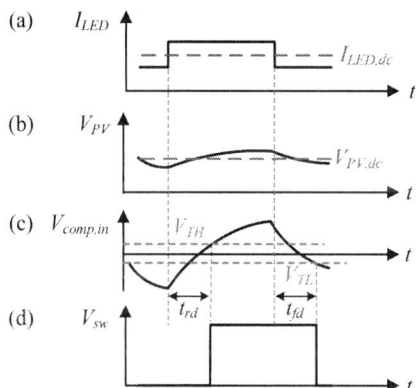

Fig. 11: Representative waveform of (a) LED Current I_{LED} (b) PV cells Voltage V_{PV} (c) Input Voltage of Comparator $V_{comp,in}$ (d) Switching Control Signal V_{sw}

comparator ($V_{comp,in}$) and switching control signal (V_{sw}). In response to square-wave ac component of LED current, PV cell voltage, V_{PV} exhibits behavior shown in Fig. 11 (b), which is exponentially rising and falling with the time constant $\tau_{RX}=R_{PV}C_j$. The signal is then coupled through the C_{hp} and amplified before being passed to the comparator.

C. Switched Capacitor Gate Driver

The top level architecture of the switched capacitor gate driver is shown in Fig. 12. The gate driver uses a 6-level series-parallel (SP) switched-capacitor (SC) circuit to sequentially multiply the input voltage V_{in}. Discussed in [20], the circuit uses 5 switching cells comprising 3 switches and 1 flying capacitor. When the parallel switches M_{pXX} are on, the flying capacitor is connected in parallel with adjacent cells; when series switch M_{sXX} is on, the capacitor is connected in series with adjacent cells.

The detailed switching sequence is shown in Fig. 13. In state S0, all capacitors are in parallel and the gate of power switch is shorted to ground. In S1, M_H turns on and V_G is connected to V_{in}. In S2, cell 1 turns to series state and then C_0 is connected in series with adjacent SC switching cell. Thus, V_g is stepped to $2V_{in}$. In the subsequent phases, the flying capacitor arrays is switched in binary segments. The gate voltage is finally stepped up to $5V_{in}$. The turn-off process is the reverse of turning-on.

By stepping up the gate voltage with multiple small voltage steps, the gate driver can reduce the $C_{gg}V_{g,pp}^2 f_{sw}$ hard-charging loss; furthermore, by using a sequential step down (gate discharge, FET turning off), the flying capacitors can

be used to recover and recycle the gate energy. In this way, the proposed isolated gate driver is able to drive a relatively large power device with a small amount optical power and/or improve the overall system efficiency. Due to five switching steps the pseudo-adiabatic gate drive process results in total gate drive loss

$$P_{GD,loss} \approx \frac{C_{gg}V_{g,pp}^2 f_{sw}}{5}, \qquad (6)$$

which is $5\times$ lower than conventional hard switching gate drivers. More details on the transistor-level circuit implementation and loss reduction mechanisms can be found in [20].

III. SYSTEM LEVEL EXPERIMENTAL RESULTS

In addition to the LED-PV optical power transfer module (Fig. 7), the optical receiver and power conditioning circuitry was implemented in a printed-circuit board PCB prototype. This was combined with the gate drive integrated circuit IC to demonstrate optically power and signal translation for the gate drive system. Fig. 14 shows annotated photo of the receiver board.

A. Receiver Functionality

Fig. 15 shows measured time domain waveforms of key receive-chain signals with the modulated LED transmitting with a 20 mA square wave at 1 MHz. With 40dB gain, the LNA can amplify the received AC signal sufficiently to be recovered through the hysteresis window of the comparator, generating a clean switching signal V_{sw}. At 1MHz the PV cell junction capacitance C_j is the dominant limitation on the receiver dynamics which has net propagation delay of 300 ns.

979-8-3315-1612-3/25 $31.00 © 2025 IEEE

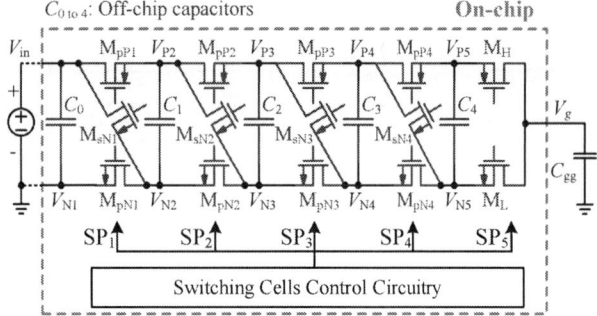

Fig. 12: Top-level architecture of SC gate driver.

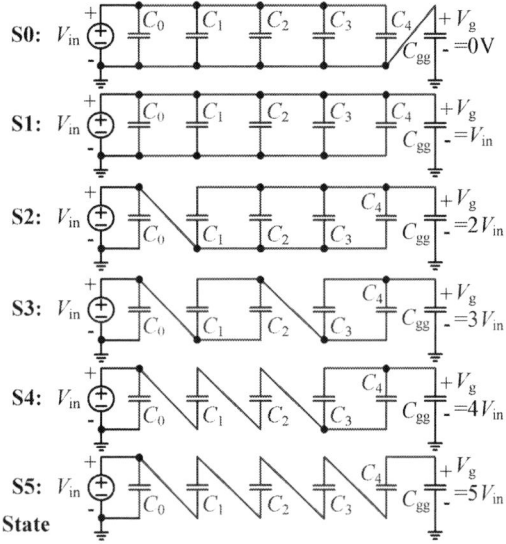

Fig. 13: 6 Circuit states when turning on the Power FET.

(a)

(b)

Fig. 14: Annotated photo of the receiver boards (a) top and (b) bottom

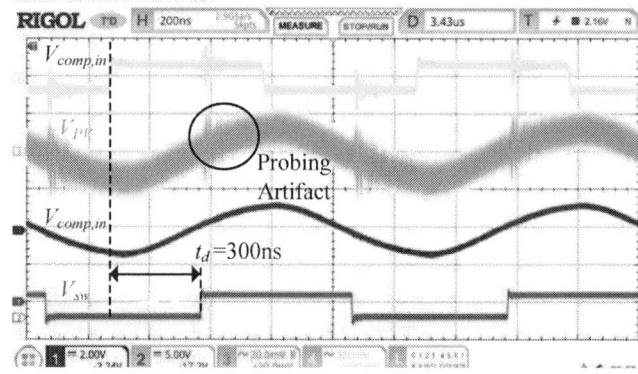

Fig. 15: Time domain waveform of LED current(I_{LED}), PV cells voltage(V_{PV}), LNA output voltage($V_{comp,in}$) and gate control signal(V_{sw}) at 1 MHz.

Fig. 16 shows the measured transient gate voltage waveform when the power FET is driven by proposed SC gate driver. The total gate voltage rise time is 15.7 ns and the $V_{g,pp} = 5$ V. The sequential charging/discharging scheme is shown in the zoom-in plot at each transition edge. The switch timing can be tuned for different scenarios to trade off rise/fall time for power savings.

B. System Efficiency

The proposed optical isolated gate driver was tested by driving a 80 V Si Power FET, ISC0602NLS, with gate charge of $\sim$10 nC. Fig. 17 (a) shows the optical power transfer efficiency, η_{PT}, and total gate-drive power efficiency, η_{total}, versus switching frequency for the peak-to-peak gate voltage($V_{g,pp} = 5V$). The gate-drive power efficiency is the ratio of the total reactive gate power ($Q_g \cdot V_{g,pp} \cdot f_{sw}$) to the total power provided to the LED. As discussed in [6], due to the efficiency advantages of the SC gate driver, the measured total gate-drive power efficiency was 27.6% at 900kHz with the gate voltage swing kept $V_{g,pp} = 5V$ by adjusting the LED drive power level. With the drive power fixed at 100 mW, $V_{g,pp}$ decreases with switching frequency as shown in Fig. 17 (b). However total gate drive efficiency is increased up to $\eta_{total} > 30\%$ as this is closer to the optimal power range for the optical interface.

IV. CONCLUSION

This paper presented a miniaturized optically isolated gate drive platform. The optical system achieved a peak efficiency (η_{PT}) of 15.1% for the LED plus single cell silicon photovoltaic cell and 11.2% for two silicon cells in series. Combined with an optical receiver circuit to recover the modulated optical transmission signal and a switched capacitor gate driver, the total efficiency (η_{total}) to optically power and drive a representative power FET was over 30%.

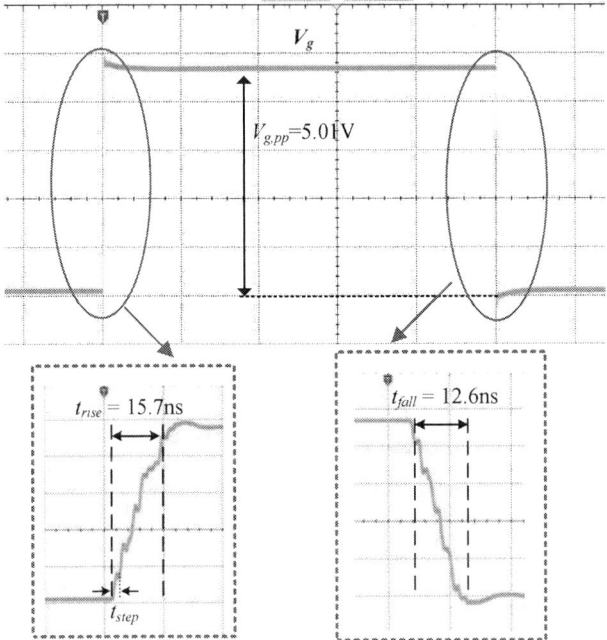

Fig. 16: Measured waveforms of Power FET gate voltage V_g

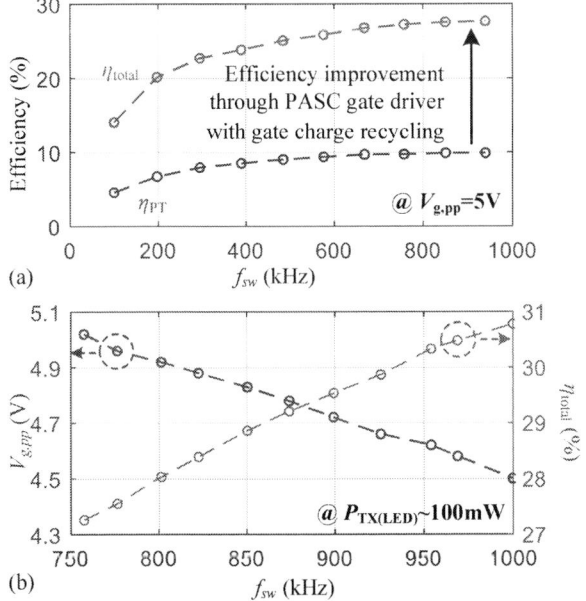

Fig. 17: Experimental results for the proposed isolated gate driver: (a) the measured power transmission efficiency, η_{PT}, and total reactive power efficiency, η_{total}, versus switching frequency, f_{sw}, for $V_{g,pp} = 5V$; (b) the $V_{g,pp}$ and η_{total} versus f_{sw} for fixed $100\,\mathrm{mW}$ TX power on LED.

V. ACKNOWLEDGEMENT

This work was funded in part by the National Science Foundation ASCENT program under award number 2328208.

REFERENCES

[1] B. Baliga, "Power semiconductor device figure of merit for high-frequency applications," *IEEE Electron Device Letters*, vol. 10, no. 10, pp. 455–457, 1989.

[2] M. Araghchini, J. Chen, V. Doan-Nguyen, D. V. Harburg, D. Jin, J. Kim, M. S. Kim, S. Lim, B. Lu, D. Piedra, J. Qiu, J. Ranson, M. Sun, X. Yu, H. Yun, M. G. Allen, J. A. Alamo, G. DesGroseilliers, F. Herrault, J. H. Lang, C. G. Levey, C. B. Murray, D. Otten, T. Palacios, D. J. Perreault, and C. R. Sullivan, "A technology overview of the powerchip development program," *IEEE Transactions on Power Electronics*, vol. 28, no. 9, pp. 4182–4201, 2013.

[3] H. A. Mantooth, M. D. Glover, and P. Shepherd, "Wide bandgap technologies and their implications on miniaturizing power electronic systems," *IEEE Journal of Emerging and Selected Topics in Power Electronics*, vol. 2, no. 3, pp. 374–385, 2014.

[4] S. Heinig, K. Jacobs, K. Ilves, S. Norrga, and H.-P. Nee, "Auxiliary power supplies for high-power converter submodules: State of the art and future prospects," *IEEE Transactions on Power Electronics*, vol. 37, no. 6, pp. 6807–6820, 2022.

[5] L. Zhang, S. Ji, S. Gu, X. Huang, J. E. Palmer, W. Giewont, F. F. Wang, and L. M. Tolbert, "Design considerations for high-voltage insulated gate drive power supply for 10-kV SiC MOSFET applied in medium-voltage converter," *IEEE Transactions on Industrial Electronics*, vol. 68, no. 7, pp. 5712–5724, 2021.

[6] Y. Li, B. Mabetha, and J. T. Stauth, "A modular switched-capacitor chip-stacking drive platform for kV-level electrostatic actuators," *IEEE Journal of Solid-State Circuits*, vol. 58, no. 12, pp. 3530–3543, 2023.

[7] B. Chen, "Air core and magnetic core transformers for isolated power conversion," in *2021 IEEE Workshop on Power Systems on Chip (PowerSOC)*, 2016.

[8] "ADUM5210 Dual-Channel Isolators with Integrated DC-to-DC Converter (2/0 Channel Directionality)," https://www.analog.com/en/products/adum5210.html, Analog Devices Inc. [Online].

[9] "Murata CRE1S0505: Isolated DC-DC converter module," https://www.murata.com/en-us/products/power/isolated-dc-dc-converter, Murata Corp. [Online].

[10] C. R. Sullivan, B. A. Reese, A. L. F. Stein, and P. A. Kyaw, "On size and magnetics: Why small efficient power inductors are rare," in *2016 International Symposium on 3D Power Electronics Integration and Manufacturing (3D-PEIM)*, 2016, pp. 1–23.

[11] S. K. Mazumder, "An overview of photonic power electronic devices," *IEEE Transactions on Power Electronics*, vol. 31, no. 9, pp. 6562–6574, 2016.

[12] X. Zhang, H. Li, J. A. Brothers, L. Fu, M. Perales, J. Wu, and J. Wang, "A gate drive with power over fiber-based isolated power supply and comprehensive protection functions for 15-kV SiC MOSFET," *IEEE Journal of Emerging and Selected Topics in Power Electronics*, vol. 4, no. 3, pp. 946–955, 2016.

[13] M. Ishigaki, S. Fafard, D. P. Masson, M. M. Wilkins, C. E. Valdivia, and K. Hinzer, "A new optically-isolated power converter for 12 V gate drive power supplies applied to high voltage and high speed switching devices," in *2017 IEEE Applied Power Electronics Conference and Exposition (APEC)*, 2017, pp. 2312–2316.

[14] H. Helmers, C. Armbruster, M. von Ravenstein, D. Derix, and C. Schöner, "6-W optical power link with integrated optical data transmission," *IEEE Transactions on Power Electronics*, vol. 35, no. 8, pp. 7904–7909, 2020.

[15] Y. Li, B. L. Dobbins, and J. T. Stauth, "An optically powered, high-voltage, switched- capacitor drive circuit for microrobotics," *IEEE Journal of Solid State Circuits*, vol. 56, no. 3, pp. 866 875, 2021.

[16] M. Ruff, H.-J. Schulze, and U. Kellner, "Progress in the development of an 8-kV light-triggered thyristor with integrated protection functions," *IEEE Transactions on Electron Devices*, vol. 46, no. 8, pp. 1768–1774, 1999.

[17] I. Cree, *Cree XLamp XP-E LED Datasheet*, Cree, Inc., Durham, NC, USA, 2015, available online: http://www.cree.com/Xlamp.

[18] P. Iles, "Nonsolar photovoltaic cells." in *IEEE Conference on Photovoltaic Specialists*, 1990, pp. 420–425 vol.1.

[19] H. Helmers, E. Lopez, O. Höhn, D. Lackner, J. Schön, M. Schauerte, M. Schachtner, F. Dimroth, and A. W. Bett, "68.9% efficient GaAs-based photonic power conversion enabled by photon recycling and optical resonance," *Phys. Status Solidi Rapid Res. Lett.*, vol. 15, no. 7, p. 2100113, Jul. 2021.

[20] Y. Li, Z. Xia, and J. T. Stauth, "A pseudo-adiabatic switched-capacitor gate driver for Si and GaN FETs achieving >5x power reduction," in *2024 IEEE Custom Integrated Circuits Conference (CICC)*, 2024, pp. 1–2.

Optimal Shared Energy Storage Capacity Configuration in Multi-energy Microgrids Considering Battery Lifetime Loss Based on Relaxation Techniques

Hualong Liu and Wenyuan Tang
Department of Electrical and Computer Engineering
North Carolina State University
Raleigh, USA
{hliu37, wtang8}@ncsu.edu

Abstract—**Installing shared battery energy storage systems (BESSs) in multi-energy microgrids (MEMGs) with the high penetration of inverter-based resources can effectively promote renewable consumption; nevertheless, how to configure the BESS capacity is an important issue. Therefore, we propose an optimal BESS capacity configuration method considering BESS lifetime loss based on stochastic programming and relaxation techniques. First, we present a basic framework of the shared BESS. Second, we consider the BESS lifetime loss based on the rain-flow counting algorithm, and the uncertainties of renewable generation and loads based on stochastic programming. Third, we propose a bi-level program for optimal shared BESS configuration in MEMGs, and then we convert the bi-level program into a single level mixed integer linear programming (MILP) model using relaxation techniques, the Karush-Kuhn-Tucker (KKT) conditions, and the Big-M method. Finally, we verify the validity of our models through conducting case studies. The results show that the shared BESS can effectively reduce the configured capacity and cost of the BESS.**

Index Terms—**Shared battery energy storage system, multi-energy microgrid, renewable, capacity configuration, uncertainty, mixed integer linear programming.**

I. INTRODUCTION

A. Research Motivation

Multi-energy microgrids (MEMGs) can realize the mutual complementarity of various energy resources, improve energy utilization efficiency, and enhance the local consumption of renewables, which is of great significance for reducing carbon emissions. Due to the strong volatility of inverter-based resources such as photovoltaic (PV) and wind generation, MEMGs are often configured with battery energy storage systems (BESSs). In MEMGs, BESS configuration plays an important role [1]–[5]. The BESS is essential to suppressing the power fluctuations of renewable generation and improving the reliability of the MEMG. However, the BESS is expensive, which limits its applications to a certain extent. The shared BESS combines the BESS technology and the sharing economy business model. Because of the disorder of the charging and discharging of each MEMG, from the overall perspective, the independent configuration of the BESS of each MEMG

will inevitably lead to the waste of BESS resources. Compared with exclusive BESS owned by each MEMG, the shared BESS by all MEMGs is a promising way to reduce the construction and operating costs of each MEMG.

Furthermore, loads and the output of renewable generation are highly uncertain. Moreover, due to the differences between the actual use environment of the BESS and the standard use environment of the BESS, as well as the frequent charging and discharging of the actual use process, the actual service lifetime of the BESS will be less than the nominal lifetime defined at the factory. Accordingly, when configuring shared energy storage capacity, we should consider uncertainty and battery lifetime loss.

In summary, it is of significant theoretical and applied value to investigate optimal shared energy storage capacity configuration in MEMGs considering uncertainty and battery lifetime loss.

B. Literature Review

At present, the research of shared BESSs is in its infancy. Reference [6] discussed the game theory control method for shared energy storage. Reference [7] studied how to distribute shared BESS within a community. How to conduct demand-side management in an intelligent grid with shared energy storage was discussed in Reference [8]. Reference [9] concluded that shared energy storage is more economical by comparing individual energy storage with shared energy storage. A distributed management method of electric vehicle charging stations with shared energy storage was studied in [10]. Reference [11] investigated how to maximize the integration rate of solar generation in microgrids containing shared energy storage. Reference [12] applied a combinatorial auction method to the distribution of shared energy storage. Reference [13] leveraged Nash bargaining theory to research peer-to-peer energy trading in conjunction with shared energy storage. Reference [14] studied peer-to-peer energy trading containing shared energy storage.

979-8-3315-1612-3/25 $31.00 © 2025 IEEE

Although there are certain research results about shared BESSs [6]–[14], the problems with the above state-of-the-art researches are summarized as follows. 1) The BESS lifetime loss is not considered. 2) The uncertainties of renewables and loads are not considered. 3) Most studies focus on microgrids containing only electricity, but seldom examine MEMGs. 4) The established models and solution methods are complicated.

C. Contributions

To address the above problems outlined in Section I-B, this paper proposes the optimal shared BESS capacity configuration in MEMGs considering the BESS lifetime loss and the uncertainties of renewables and loads. The main contributions of our work are as follows.

1) We formulate and establish a bi-level program for optimal shared BESS configuration considering BESS lifetime loss and the uncertainties of renewables and loads.
2) We propose a mixed integer linear programming (MILP) method for optimal shared BESS configuration using relaxation techniques, the Karush-Kuhn-Tucker (KKT) conditions, and the Big-M method.
3) We conduct extensive case studies, through which we validate our derivations and analyses, testify the effectiveness of our models, and provide insights for understanding and configuring shared BESSs.

D. Organization

The remainder of our paper is arranged as follows. Section II elaborates the computing method of actual BESS lifetime. A bi-level program for shared BESS capacity configuration is derived and described minutely in Section III. The proposed MILP approach for optimal shared BESS capacity and its solution process are expounded in Section IV. Case studies are illustrated in Section V. We provide the conclusions of the paper and future work in Section VI.

II. COMPUTING METHOD OF ACTUAL BESS LIFETIME

Because of the frequent charging and discharging of the BESS and the actual poor working environment compared with the standard environment, the actual BESS lifetime of the BESS is usually less than the rated lifetime set by the BESS manufacturer. Based on the rain-flow counting algorithm [15], [16], we can achieve the depth of discharge (DoD) associated with each state of charge (SoC) cycle. Then, based on the correspondence between the DoD and the number of cycles to failure of the BESS, we can obtain the number of cycles to failure of the BESS associated with each DoD. Thus, the actual lifetime of the BESS considering lifetime loss can be expressed as [17]

$$T^{\text{act}} = \left(365 \sum_{i=1}^{N} \frac{1}{N_i}\right)^{-1}, \tag{1}$$

where N_i denotes the number of cycles to failure in the i-th cycle of charging and discharging, and N indicate the total number of SoC cycles of the BESS in a day.

III. A BI-LEVEL PROGRAM FOR SHARED BESS CAPACITY CONFIGURATION

The basic framework of the shared BESS and the architecture of each MEMG are shown in Fig. 1. The MEMG encompasses the gas boiler (GB), the gas turbine (GT), the heat recovery unit (HRU), the gas boiler (GB), the electric chiller (EC), the absorption chiller (AC), PV panels, and wind turbines (WTs). The loads are composed of electrical loads (ELs), thermal loads (TLs), and cooling loads (CLs). Moreover, the MEMG is connected with the main power grid (PG) and the gas network (GN).

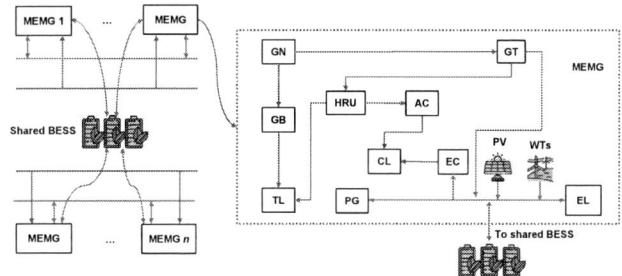

Fig. 1. Basic framework of the shared BESS and architecture of each MEMG.

A. Upper Level

The objective of the upper level is to minimize the annualized BESS investment and operating costs, which can be formulated as follows:

$$\min \sum_{s=1}^{|\Omega|} p_s \left(C_s^{\text{inv}} + 365 C_s^{\text{op}}\right), \tag{2}$$

$$C_s^{\text{inv}} = \left(a P_{s,\max}^{\text{B}} + b E_{s,\max}^{\text{B}}\right) \frac{r\left(1+r\right)^{T^{\text{act}}}}{\left(1+r\right)^{T^{\text{act}}} - 1}, \tag{3}$$

$$r = \frac{k-f}{1+f}, \tag{4}$$

$$C_s^{\text{op}} = \sum_{j=1}^{M} \sum_{t=1}^{T} \left(\pi_t^{\text{B+}} P_{j,s,t}^{\text{B+}} - \pi_t^{\text{B-}} P_{j,s,t}^{\text{B-}}\right), \tag{5}$$

where p_s indicates the probability of scenario s, and $|\Omega|$ symbolizes the total number of scenarios. C_s^{inv} and C_s^{op} signify the annualized investment cost and the daily operating cost in scenario s, respectively. a and b represent the power unit price and energy unit price of the BESS, respectively. $P_{s,\max}^{\text{B}}$ and $E_{s,\max}^{\text{B}}$ are the power capacity variable and energy capacity variable of the installed shared BESS, respectively. M, T, r, k, and f signify the total number of MEMGs, the total time steps of the scheduling cycle, the real discount rate, the nominal discount rate, and the expected inflation rate, respectively. $\pi_t^{\text{B+}}$ and $\pi_t^{\text{B-}}$ denote the unit price of charging the BESS from MEMG j during period t, and the unit price of discharging the BESS to the MEMG j during period t, respectively. $P_{j,s,t}^{\text{B+}}$ and $P_{j,s,t}^{\text{B-}}$ indicate the charging power and discharging power of MEMG j during period t in scenario s, respectively.

The shared BESS should satisfy the following constraints:

$$E_{s,t}^{B} = (1-\gamma)E_{s,t-1}^{B} + \eta^{B+}P_{s,t}^{B+} - P_{s,t}^{B-}/\eta^{B-}, \forall s, \forall t, \quad (6)$$

$$0 \leq P_{s,t}^{B-} \leq b_{s,t}^{B-}\eta^{B-}P_{s,\max}^{B}, \quad \forall s, \forall t, \quad (7)$$

$$0 \leq P_{s,t}^{B+} \leq (1-b_{s,t}^{B-})P_{s,\max}^{B}/\eta^{B+}, \quad \forall s, \forall t, \quad (8)$$

$$E_{s,\min}^{B} \leq E_{s,t}^{B} \leq E_{s,\max}^{B}, \quad \forall s, \forall t, \quad (9)$$

$$E_{s,1}^{B} = E_{s,T}^{B}, \quad \forall s, \forall t, \quad (10)$$

where $E_{s,t}^{B}$, γ, η^{B+}, and η^{B-} are the stored energy of the shared BESS during period t in scenario s, charging efficiency of the shared BESS, and discharging efficiency of the shared BESS, respectively. $P_{s,t}^{B+}$ and $P_{s,t}^{B-}$ are the charging power and discharging power of the shared BESS during period t in scenario s, respectively. $b_{s,t}^{B-}$ represents the binary charging and discharging decision variable of the shared BESS. When $b_{s,t}^{B-}$ is equal to 1, it means discharging the shared BESS; when $b_{s,t}^{B-}$ is equal to 0, it means charging the shared BESS. $E_{s,\min}^{B}$ indicates the minimum allowable stored energy of the shared BESS in scenario s. Constraint (10) ensures that the stored energy of the shared BESS at the beginning of the scheduling cycle is equal to the stored energy of the shared BESS at the end of the scheduling cycle in scenario s.

B. Lower Level

The objective of the lower level is to minimize the annualized operating cost of the system of MEMGs, which is written as

$$\min 365 \sum_{s=1}^{|\Omega|} p_s \left(C_s^{PG+} - C_s^{PG-} + C_s^{GN} - C_s^{op} \right), \quad (11)$$

where C_s^{PG+}, C_s^{PG-}, and C_s^{GN} represent the cost of buying electricity to the system in scenario s, the revenue of selling electricity from the system in scenario s, the cost of buying gas to the system in scenario s, respectively.

The constraints of the lower level are as follows:

$$P_{j,s,t}^{PV} + P_{j,s,t}^{WT} + P_{j,s,t}^{GT} + P_{j,s,t}^{PG+} - P_{j,s,t}^{PG-} + P_{j,s,t}^{B-}$$
$$- P_{j,s,t}^{B+} - P_{j,s,t}^{EC} - P_{j,s,t}^{EL} = 0, \quad \lambda_{j,s,t}^{1}; \quad (12)$$

$$P_{j,s,t}^{GB} + P_{j,s,t}^{HRU,TL} - P_{j,s,t}^{TL} = 0, \quad \lambda_{j,s,t}^{2}, \quad (13)$$

$$P_{j,s,t}^{EC} + P_{j,s,t}^{AC} - P_{j,s,t}^{CL} = 0, \quad \lambda_{j,s,t}^{3}, \quad (14)$$

$$\sum_{j=1}^{M} \left(P_{j,s,t}^{B+} - P_{j,s,t}^{B-} \right) = P_{s,t}^{B+} - P_{s,t}^{B-}, \quad \lambda_{j,s,t}^{4} \quad (15)$$

$$0 \leq P_{j,s,t}^{PG+} \leq b_{j,s,t}^{PG+}P_{j,\max}^{PG} : \lambda_{j,s,t}^{5,\min}, \lambda_{j,s,t}^{5,\max}, \quad (16)$$

$$0 \leq P_{j,s,t}^{PG-} \leq \left(1-b_{j,s,t}^{PG+}\right)P_{j,\max}^{PG} : \lambda_{j,s,t}^{6,\min}, \lambda_{j,s,t}^{6,\max}, \quad (17)$$

$$0 \leq P_{j,s,t}^{B+} \leq b_{j,s,t}^{B+}P_{j,\max}^{B} : \lambda_{j,s,t}^{7,\min}, \lambda_{j,s,t}^{7,\max}, \quad (18)$$

$$0 \leq P_{j,s,t}^{B-} \leq \left(1-b_{j,s,t}^{B+}\right)P_{j,\max}^{B} : \lambda_{j,s,t}^{8,\min}, \lambda_{j,s,t}^{8,\max}, \quad (19)$$

where $P_{j,s,t}^{PV}$, $P_{j,s,t}^{WT}$, $P_{j,s,t}^{GT}$, $P_{j,s,t}^{PG+}$, $P_{j,s,t}^{PG-}$, $P_{j,s,t}^{EC}$, and $P_{j,s,t}^{EL}$ represent the electrical power generated by the PV generation of MEMG j during period t in scenario s, the electrical power generated by the WTs of MEMG j during period t in scenario s, the electrical power generated by the GT of MEMG j

during period t in scenario s, the electrical power of MEMG j imported from the main PG during period t in scenario s, the electrical power of MEMG j exported to the main PG during period t in scenario s, the electrical power consumed by the EC during period t in scenario s, and the ELs during period t in scenario s, respectively. $P_{j,s,t}^{GB}$, $P_{j,s,t}^{HRU,TL}$, and $P_{j,s,t}^{TL}$ indicate the thermal power generated by the GB during period t in scenario s, the thermal power generated by the HRU during period t in scenario s, and the TLs during period t in scenario s, respectively. $P_{j,s,t}^{EC}$, $P_{j,s,t}^{AC}$, and $P_{j,s,t}^{CL}$ symbolize the cooling power generated by the EC during period t in scenario s, the cooling power generated by the AC during period t in scenario s, and the CLs during period t in scenario s, respectively. $P_{j,\max}^{PG}$ and $P_{j,\max}^{B}$ represent the maximum electrical power of MEMG j interacting with the main PG and the maximum electrical power of MEMG j interacting with shared BESS, respectively. $\lambda_{j,s,t}$ is the dual variable corresponding to a particular constraint. Constraint (12) imposes the electrical balance of MEMG j. Constraint (13) ensures the thermal balance of MEMG j. Constraint (14) is the cooling balance constraint of MEMG j.

IV. PROPOSED MILP APPROACH FOR OPTIMAL SHARED BESS CAPACITY AND ITS SOLUTION PROCESS

$b_{j,s,t}^{B+}$ and $b_{j,s,t}^{PG+}$ in (16)–(19) are binary variables. As a result, the bi-level model cannot be solved by strict classical mathematical methods, and can only be solved by evolutionary methods. However, the speed of the evolutionary algorithm is slow, and the results are different every time. Accordingly, we relax the binary variables in (16)–(19) into continuous variables located in $[0,1]$, and then the lower level becomes convex. This allows us to apply KKT to transforming it into a nonlinear problem, and then we convert the nonlinear problem into an MILP problem with the Big-M method. Eventually, merging the lower level with the upper level, we formulate and establish a single level MILP model for optimal shared BESS capacity configuration.

The solution process of the proposed MILP approach for optimal shared BESS capacity is shown in Algorithm 1.

Algorithm 1 Solution process of the proposed MILP approach for optimal shared BESS capacity

---

1: Relax the variables $b_{j,s,t}^{B+}$ and $b_{j,s,t}^{PG+}$ in (16)–(19).
2: Convert the lower model into a nonlinear problem using KKT.
3: Use the the Big-M method to transform the nonlinear problem into an MILP problem.
4: Merge the lower level with the upper level to obtain a single level MILP model.
5: Obtain optimal shared BESS configuration by solving this single level MILP model.

979-8-3315-1612-3/25 $31.00 © 2025 IEEE

V. CASE STUDIES

A. Description

As shown in Fig. 1, a system composed of three MEMGs is adopted in this paper to verify the effectiveness of the proposed method. The forecast errors of wind, PV generation, ELs, TLs, and CLs follow the normal distributions with the mean of 0 and the standard deviations of 9.5%, 7.5%, 2.2%, 3.2%, and 2%, respectively. The charging and discharging efficiency of the BESS is 96%. The power unit price and energy unit price of the BESS are $340/kW and $220/kWh [18], respectively. First, we generate 1000 scenarios using the Latin Hypercube method, and then reduce them to 10 scenarios using k-means. Then, we use the rain-flow counting-based method in Section II to calculate the actual BESS lifetime considering BESS lifetime loss. Finally, the MILP-based model is leveraged to calculate the optimal BESS configuration capacity.

B. Results

Case 1: The BESS is not configured for MEMGs. Case 2: Each MEMG is independently configured with BESSs. Case 3: MEMGs are configured with the shared BESS. The annualized costs, the energy and power capacities of the BESS, and the consumption rates of renewables are shown in Table I. Fig. 2 shows the operation of the shared BESS in Case 3. As can be seen from Table I and Fig. 2 , although the consumption rate of renewables in Case 2 is 27% higher than that in Case 1, the annualized cost is also 120% higher. However, the shared BESS in Case 3 can not only fully absorb renewables, but also reduce the cost by 11% compared with Case 1. Although Case 2 can also fully absorb renewables, the annualized cost, configured power capacity, and configured energy capacity of Case 2 are 2.46, 9.71, and 15 times those of Case 3, respectively.

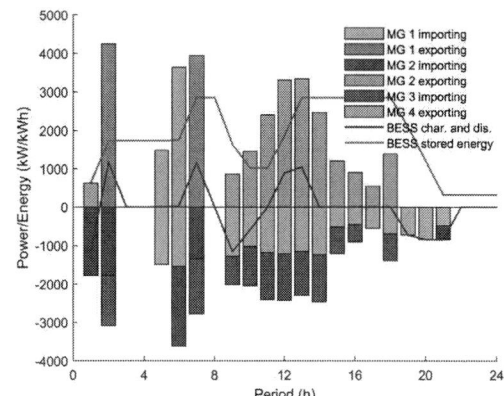

Fig. 2. Operation of the shared BESS in Case 3.

TABLE I
COSTS, BESS POWER CAPACITIES, ENERGY CAPACITIES, AND RENEWABLE CONSUMPTION RATES FOR DIFFERENT CASES

Case	Cost (k$)	Power (kW)	Energy (kWh)	Rate (%)
1	4257	/	/	73.0
2	9365	11224	47535	100
3	3807	1156	3169	100

VI. CONCLUSIONS AND FUTURE WORK

An MILP program for the optimal shared BESS capacity configuration in MEMGs considering BESS lifetime loss and uncertainties has been proposed and validated. Through the above analyses and discussions, we can obtain the following conclusions.

1) Through the shared BESS, the renewable consumption rate is significantly improved.
2) Through the shared BESS, the configuration cost and configured capacity of the BESS are greatly reduced.
3) The shared BESS can also reduce dependence of MEMGs on the main PG.

In the future, we will give attention to the following research directions.

1) We will consider electrical power flow, thermal power flow, and gas power flow. These flow constraints are non-convex nonlinear, and it is a crucial and difficult topic to study the optimization problems with these constraints.
2) We will delve into fine modeling of inverters and integrate these refined models into the optimization problems of energy systems containing inverter-based resources.
3) We will study the relationship between the BESS actual lifetime and the state of health.
4) We will study the optimal configuration and operation of shared BESS using game theory methods, and compare these methods with the method in this paper.
5) We will study the multi-time scale scheduling problem of the energy system with shared BESS.
6) We will study the dynamic performance and transient stability of energy systems with the high penetration of inverter-based resources.

REFERENCES

[1] H. Kang, S. Jung, M. Lee, and T. Hong, "How to better share energy towards a carbon-neutral city? a review on application strategies of battery energy storage system in city," *Renewable and Sustainable Energy Reviews*, vol. 157, p. 112113, 2022.

[2] Y. Zhang and Y. Li, "Energy management strategy for supercapacitor in autonomous dc microgrid using virtual impedance," in *2015 IEEE Applied Power Electronics Conference and Exposition (APEC)*, pp. 725–730, IEEE, 2015.

[3] M. Abou Houran, X. Yang, and W. Chen, "Energy management of microgrid in smart building considering air temperature impact," in *2018 IEEE Applied Power Electronics Conference and Exposition (APEC)*, pp. 2398–2404, IEEE, 2018.

[4] M. Farhadi and O. Mohammed, "Energy storage systems for high power applications," in *2015 IEEE Industry Applications Society Annual Meeting*, pp. 1–7, IEEE, 2015.

[5] A. Luna, N. Diaz, M. Savaghebi, J. C. Vasquez, J. M. Guerrero, K. Sun, G. Chen, and L. Sun, "Optimal power scheduling for a grid-connected hybrid pv-wind-battery microgrid system," in *2016 IEEE Applied Power Electronics Conference and Exposition (APEC)*, pp. 1227–1234, IEEE, 2016.

[6] X. Han, J. Li, and Z. Zhang, "Dynamic game optimization control for shared energy storage in multiple application scenarios considering energy storage economy," *Applied Energy*, vol. 350, p. 121801, 2023.

[7] H.-C. Chang, B. Ghaddar, and J. Nathwani, "Shared community energy storage allocation and optimization," *Applied Energy*, vol. 318, p. 119160, 2022.

[8] J. Jo and J. Park, "Demand-side management with shared energy storage system in smart grid," *IEEE Transactions on Smart Grid*, vol. 11, no. 5, pp. 4466–4476, 2020.

[9] A. Walker and S. Kwon, "Analysis on impact of shared energy storage in residential community: Individual versus shared energy storage," *Applied Energy*, vol. 282, p. 116172, 2021.

[10] D. Yan and Y. Chen, "Distributed coordination of charging stations with shared energy storage in a distribution network," *IEEE Transactions on Smart Grid*, vol. 14, no. 6, pp. 4666–4682, 2023.

[11] S. M. Tercan, A. Demirci, E. Gokalp, and U. Cali, "Maximizing self-consumption rates and power quality towards two-stage evaluation for solar energy and shared energy storage empowered microgrids," *Journal of Energy Storage*, vol. 51, p. 104561, 2022.

[12] W. Zhong, K. Xie, Y. Liu, C. Yang, and S. Xie, "Multi-resource allocation of shared energy storage: A distributed combinatorial auction approach," *IEEE transactions on smart grid*, vol. 11, no. 5, pp. 4105–4115, 2020.

[13] Y. Chen, W. Pei, T. Ma, and H. Xiao, "Asymmetric Nash bargaining model for peer-to-peer energy transactions combined with shared energy storage," *Energy*, vol. 278, p. 127980, 2023.

[14] B. Zheng, W. Wei, Y. Chen, Q. Wu, and S. Mei, "A peer-to-peer energy trading market embedded with residential shared energy storage units," *Applied Energy*, vol. 308, p. 118400, 2022.

[15] S. D. Downing and D. Socie, "Simple rainflow counting algorithms," *International journal of fatigue*, vol. 4, no. 1, pp. 31–40, 1982.

[16] E. Schaltz, A. Khaligh, and P. O. Rasmussen, "Influence of battery/ultra-capacitor energy-storage sizing on battery lifetime in a fuel cell hybrid electric vehicle," *IEEE Transactions on Vehicular Technology*, vol. 58, no. 8, pp. 3882–3891, 2009.

[17] H. Liu and W. Tang, "Multi-objective bi-level programs for optimal microgrid planning considering actual bess lifetime based on wgan-gp and info-gap decision theory," *Journal of Energy Storage*, vol. 89, p. 111510, 2024.

[18] X. Cao, J. Wang, and B. Zeng, "A chance constrained information-gap decision model for multi-period microgrid planning," *IEEE Transactions on Power Systems*, vol. 33, no. 3, pp. 2684–2695, 2017.

Virtual Resistance Control for an Active Battery Management System

Alastair P. Thurlbeck[*], Ashraf Siddiquee[†], Mithat John Kisacikoglu[*], and Yilmaz Sozer[†]

[*] Center for Integrated Mobility Sciences, National Renewable Energy Laboratory, Golden, CO 80401
[†] Department of Electrical and Computer Engineering, The University of Akron, Akron, Ohio 44325
E-mails: {alastair.thurlbeck, john.kisacikoglu}@nrel.gov, {as802, ys}@uakron.edu

Abstract—A virtual resistance control scheme is proposed for an active state-of-charge (SOC) balancing system within a behind-the-meter-storage (BTMS) battery pack. The proposed control enables the balancing system converters to simultaneously track a reference current while also regulating a shared 12 V bus. This enables the system to balance the cell SOCs while supporting auxiliary system loads on the 12 V bus. The advantage of the virtual resistance technique lies in its robustness to the actual resistances of the system. Without virtual resistance, the reference current tracking and 12 V bus load sharing among the converters would be dictated by the varying resistances between each converter and the 12 V bus, which are unequal and difficult to measure accurately. However, by applying the virtual resistance technique, these differences are negated such that the expected reference current tracking and equal load sharing are achieved.

I. Introduction

Renewable energy resources and their share of the electricity generation mix have rapidly increased in recent years [1]. Simultaneously, the load demand profiles are evolving to be more challenging due to electrification. In particular, DC high-power charging of electric vehicles (EVs) adds a highly variable load to grid operation. Not only is this challenging in terms of grid stability, but at a local level, requires oversizing of distribution equipment relative to the average power demand. To combat both these issues, on-site or behind-the-meter storage (BTMS) is gaining more prominence. With BTMS, the customer can control the storage to peak-shave their demand from the grid. BTMS can reduce the need for costly grid capacity upgrades by allowing for a peak charging power to be far higher than the utility connection is sized for [2]–[4].

A BTMS system, comprised of cells connected in series and parallel with an active balancing system, is shown in Fig. 1 for a DC microgrid connection. This concept, introduced in our previous study [5], allows the connection of strings of different chemistries, each comprised of series-connected modules connected to the pack bus by DC-DC converters. The voltages shown in Fig. 1 serve as an example design; and

This work was authored by the National Renewable Energy Laboratory, operated by Alliance for Sustainable Energy, LLC, for the U.S. Department of Energy (DOE) under Contract No. DE-AC36-08GO28308. Funding provided by the U.S. Department of Energy Office of Energy Efficiency and Renewable Energy Vehicles Technologies Office. The views expressed herein do not necessarily represent the views of the DOE or the U.S. Government.

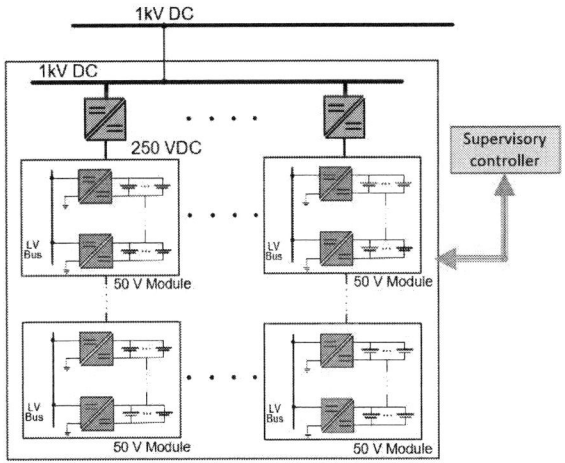

Figure 1: Overview of the BTMS system connection with DC bus.

voltage, current, and energy levels can be customized based on the application needs.

Active SOC balancing via DC-DC converters allows the healthiest cells to be more heavily utilized than the weaker cells, ensuring they charge and discharge at the same rate despite differences in capacity [6]–[11]. Dual-active-bridge (DAB) converters are connected in parallel to each series-connected cell group, allowing for energy transfer via an auxiliary 12 V low-voltage (LV) bus for balancing. This bus can also support additional system loads when needed. Measurements from the DAB converters are communicated to a supervisory controller which estimates the SOC and other critical conditions of each cell group. The supervisory controller can then command each DAB to supply current to or from the 12 V bus, such that the cells are balanced, and their SOCs converge together.

The design of this system leads to two competing control objectives. It is desired to control the current through each balancing DAB (called b-DAB in the context of this study), while it is also necessary to regulate the 12 V bus so that it can be used as a stable power supply for the auxiliary loads. In our earlier study in [5], the approach was to use paralleled PI controllers for bus voltage and b-DAB current, each attempting to vary the b-DAB phase-shift to regulate to their respective control objectives. While this was a functional solution, two independent control objectives cannot be achieved by manipu-

979-8-3315-1612-3/25 $31.00 © 2025 IEEE

lating a single control variable. Therefore, the system is prone to having a steady-state error on one or both of the control objectives.

The literature presents several control approaches for active balancing systems using DAB converters with a shared low-voltage bus [6], [7], [10], [12]. The DAB converters in [12] are regulated to maintain a fixed ratio between the cell voltages and shared LV bus voltage, resulting in autonomous balancing of the cell voltages. This limits the approach to voltage balancing and results in the shared bus voltage changing proportionally to the cell voltages. Additionally, the auxiliary load sharing is affected by differences in resistance and the cell voltages. In an earlier study, the same research group achieves active SOC balancing through distributed SOC estimation [6]. At each DAB, the local SOC estimate is mapped to a local bus voltage reference, which the controller regulates to. Droop control ensures equal load sharing of the auxiliary bus loads. However, the bus voltage still changes with the average cell SOC, since it is the means of communication between each local controller. A following study in [7] is similar to [6] but uses online estimation of cell capacity and internal resistance to modify the bus voltage reference objective mapping. Compared to these approaches, the proposed virtual resistance control regulates to a fixed 12 V bus voltage and achieves equal auxiliary load sharing among the DAB converters, while still allowing for continuous SOC balancing across the cells. However, a supervisory controller is necessary since the shared bus voltage can no longer be used to communicate the mean SOC or cell voltage.

The proposed virtual resistance control is similar to conventional droop control, in that a converter's internal voltage reference decreases with increasing output current [13]. This allows for each converter to contribute to tight regulation of the shared DC bus, and ensures a load applied to the shared DC bus splits equally among all the converters regardless of their individual output resistances. However, the proposed control scheme also shifts the internal voltage reference according to a given reference current, shifting the voltage offset of each individual droop curve. In the context of the active SOC balancing system, the proposed control allows for continuous balancing of cell SOCs, while supporting auxiliary system loads on the fixed shared 12 V bus. Simulation and extensive experimental results are presented to show the efficacy of the proposed control solution.

II. SYSTEM OVERVIEW

Fig. 2 shows the set-up of the system that will be investigated in this work. It is a scaled-down version of the concept shown in Fig. 1, with a single battery module comprised of 12 series-connected 20 Ah 3.2 V LiFeMnPO4 prismatic cells (12S1P configuration). Fig. 2 also shows a string-DAB (s-DAB) converter rated at 5 kW that converts the battery module voltage (40 V) to a higher DC bus voltage (215 V). The programmable DC supply and load on the 215 V bus operate in tandem as a bidirectional source, regulating the 215 V bus voltage as the s-DAB controls the battery pack output

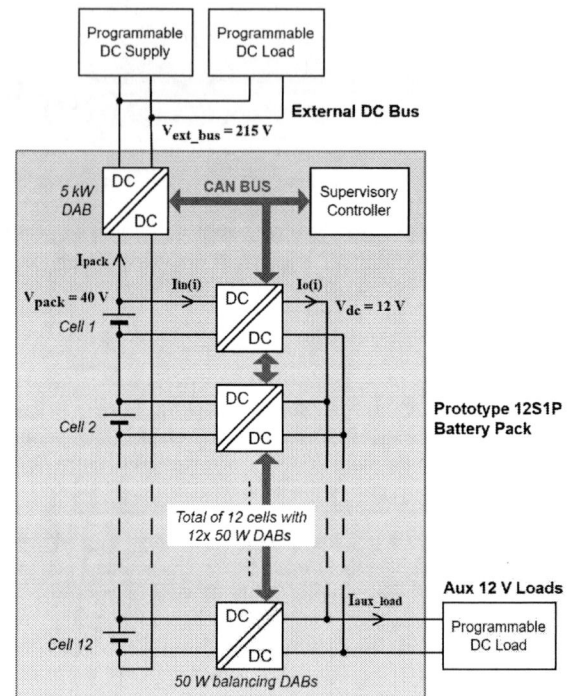

Figure 2: Overview of the 12S1P prototype battery pack with active SOC balancing system.

current I_{pack}. The programmable load connected at the LV bus emulates the behavior of auxiliary loads.

As shown in Fig. 2, each b-DAB converter uses the CAN bus to communicate cell voltage, b-DAB input current, and temperature measurements to the supervisory controller, which uses unscented Kalman filters to estimate the cell SOCs from these measurements. An unscented Kalman filter is a variation of a Kalman filter suited to the SOC estimation of Lithium-ion batteries [14]. The supervisory controller then calculates the mean SOC of the string, and generates the reference input currents for each b-DAB $I_{in}^*(i)$, proportional to each cell's SOC error from the string mean:

$$I_{in}^*(i) = K_{SOC}(SOC_{est}(i) - \overline{SOC_{est}}). \tag{1}$$

Where K_{SOC} is the SOC balancing gain, $SOC_{est}(i)$ is the estimated SOC of cell i, and $\overline{SOC_{est}}$ is the mean estimated cell SOC of the string. The supervisory controller then communicates $I_{in}^*(i)$ back to each b-DAB via the CAN bus.

III. DESIGN OF VIRTUAL RESISTANCE DROOP CONTROLLER

Both input and output current based implementations of the controller are possible. The input current is at the cell-side of the b-DAB while the output current is at the LV bus side. The output current implementation is preferred, since the voltage gain and efficiency of the b-DAB converter do not impact the controller performance. However, since the prototype b-DABs used for testing only include input current measurements, the input current version is implemented in this work. In the input

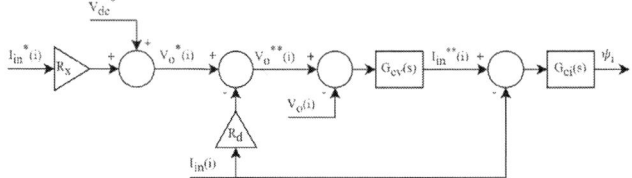

Figure 3: Proposed virtual resistance control using input current.

current implementation, there is considerable error between the reference and actual currents due to the b-DAB converter efficiency and varying voltage gain. Therefore, both input and output current based implementations are presented below for completeness.

A. Input Current based Implementation

Fig. 3 shows the proposed virtual resistance control block diagram, which is deployed as a primary controller (low-level) running on each b-DAB converter. The primary controller is a cascaded PI structure with an inner loop for input current and outer loop for output voltage. The input current reference $I_{in}^*(i)$ for each DAB is provided from the supervisory controller, calculated as shown in (1).

Fundamentally, the b-DAB converters are in voltage-control mode such that each b-DAB's measured output voltage $V_o(i)$ is tightly regulated to its reference output voltage $V_o^*(i)$. Given that each DAB converter connects to the shared LV bus with voltage V_{dc} through an output line resistance $R_o(i)$, it is possible to shift the individual references of each DAB converter $V_o^*(i)$ such that the 12 V bus regulation is maintained while each converter sources or sinks a desired current.

First, we present the case without virtual resistance. The output current of each DAB is given by:

$$I_{out}(i) = \frac{V_o^*(i) - V_{dc}}{R_o(i)}, \qquad (2)$$

where $V_{dc} \approx 12\,\text{V}$. Therefore, to achieve a desired current $I_{out}^*(i)$, the required voltage reference is:

$$V_o^*(i) - V_{dc}^* + R_{o\_nom} \cdot I_{out}^*(i), \qquad (3)$$

where $V_{dc}^* = 12\,\text{V}$, and R_{o_nom} is the nominal resistance representing all of $R_o(i)$. For the ideal case with $R_{o\_nom} = R_o(i)$, it follows that $I_{out}(i) = I_{out}^*(i)$. However, in practice, $R_o(i)$ are too small to measure accurately, temperature-dependent, and vary with the wiring length between each DAB converter and the shared 12 V bus such that $R_o(i) \neq R_{o\_nom}$. Therefore, it logically follows that $I_{out}(i) = (R_{o\_nom}/R_o(i)) \cdot I_{out}^*(i)$.

To address variations in $R_o(i)$, a virtual resistance, R_d, is employed in the controller. It is set such that $R_d \gg R_{o\_nom}$, ensuring it dominates the control behavior, negating any errors between R_{o_nom} and $R_o(i)$. The controller is developed as follows:

Firstly, implementing droop control, we define a new voltage reference for DAB output, $V_o^{**}(i)$, which is the original

voltage reference $V_o^*(i)$ minus the virtual resistance voltage drop such that:

$$V_o^{**}(i) = V_o^*(i) - I_{in}(i) \cdot R_d. \qquad (4)$$

The output current of each DAB is now different from (2) such that:

$$I_{out}(i) = \frac{V_o^{**}(i) - V_{dc}}{R_o(i)}, \qquad (5)$$

since the PI control now regulates $V_o(i)$ to $V_o^{**}(i)$ rather than $V_o^*(i)$. Substituting (4) into (5) gives:

$$I_{out}(i) = \frac{V_o^*(i) - I_{in}(i) \cdot R_d - V_{dc}}{R_o(i)}. \qquad (6)$$

Neglecting DAB losses, $I_{out}(i) = \frac{1}{G} I_{in}(i)$, where $G = V_o(i)/V_{in}(i)$. Substituting for I_{out} and rearranging gives the required voltage reference $V_o^*(i)$, which gives the desired current $I_{in}^*(i)$:

$$V_o^*(i) = V_{dc}^* + \left(\frac{1}{G}R_{o\_nom} + R_d\right) \cdot I_{in}^*(i) \qquad (7)$$

$$V_o^*(i) = V_{dc}^* + R_x \cdot I_{in}^*(i), \qquad (8)$$

where $R_x = 1/G \cdot R_{o\_nom} + R_d$. This results in the proposed controller shown in Fig. 3, which realizes (4) and (8) to obtain $V_o^{**}(i)$ and $V_o^*(i)$, respectively.

Similar to before, in the ideal case with $R_{o\_nom} = R_o(i)$, it follows that $I_{in}(i) = I_{in}^*(i)$. However, with $R_o(i) \neq R_{o\_nom}$, the actual input current is found by substituting (7) into (4), then (4) into (5), assuming that $V_{dc} = V_{dc}^*$:

$$I_{in}(i) = \frac{\frac{1}{G}R_{o\_nom} + R_d}{\frac{1}{G}R_o(i) + R_d} \cdot I_{in}^*(i). \qquad (9)$$

Provided R_d is chosen appropriately such that $R_d \gg R_{o\_nom}$, both the numerator and denominator are dominated by R_d, negating the effects of varying or mismatched $R_o(i)$.

The auxiliary load sharing can be determined by assuming the load current causes a voltage drop on the shared 12 V bus of ΔV_{dc}. First, we find the input current of a cell with $V_{dc} \neq V_{dc}^*$. Substituting (8) into (4), then (4) into (5):

$$I_{in}(i) = \frac{R_x I_{in}^*(i) + V_{dc}^* - V_{dc}}{\frac{1}{G}R_o(i) + R_d}. \qquad (10)$$

Following a load step, the output voltage drops such that $V_{dc} = V_{dc\_init} - \Delta V_{dc}$. The current before and after the load step is found by substituting $V_{dc} = V_{dc\_init}$ and $V_{dc} = V_{dc\_init} - \Delta V_{dc}$ in (10), respectively. The resulting change in input current is then:

$$\Delta I_{in}(i) = \frac{\Delta V_{dc}}{\frac{1}{G}R_o(i) + R_d}. \qquad (11)$$

Therefore each cell takes a share of the auxiliary load current $\Delta I_{in}(i)$ in proportion to the denominator of (11) (since the change in shared bus voltage ΔV_{dc} is the same for all cells). Since $R_d \gg R_o(i)$, all $\Delta I_{in}(i)$ are approximately equal, ensuring equal load sharing.

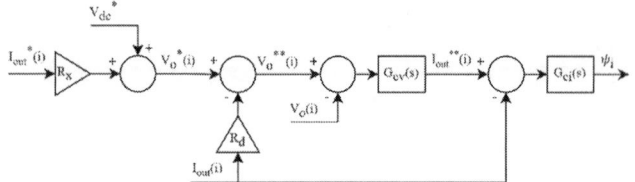

Figure 4: Proposed virtual resistance control using output current.

B. Output Current based Implementation

Fig. 4 shows the output current based implementation of the proposed controller. Using the output current measurement $I_{out}(i)$, the droop control portion becomes:

$$V_o^{**}(i) = V_o^*(i) - I_{out}(i) \cdot R_d. \qquad (12)$$

As before, the actual output current is given by:

$$I_{out}(i) = \frac{V_o^{**}(i) - V_{dc}}{R_o(i)}, \qquad (13)$$

since the PI control regulates $V_o(i)$ to $V_o^{**}(i)$ as before. Substituting (12) into (13) gives:

$$I_{out}(i) = \frac{V_o^*(i) - I_{out}(i) \cdot R_d - V_{dc}}{R_o(i)}. \qquad (14)$$

Rearranging gives the required voltage reference $V_o^*(i)$, which gives the desired current $I_{out}^*(i)$:

$$V_o^*(i) = V_{dc}^* + (R_{o\_nom} + R_d) \cdot I_{out}^*(i) \qquad (15)$$

$$V_o^*(i) = V_{dc}^* + R_x \cdot I_{out}^*(i), \qquad (16)$$

where $R_x = R_{o\_nom} + R_d$.

As before, in the ideal case with $R_{o\_nom} = R_o(i)$, it follows that $I_{out}(i) = I_{out}^*(i)$. However, with $R_o(i) \neq R_{o\_nom}$, the actual input current is found by substituting (15) into (12), then (12) into (13), assuming that $V_{dc} = V_{dc}^*$:

$$I_{out}(i) = \frac{R_{o\_nom} + R_d}{R_o(i) + R_d} \cdot I_{out}^*(i). \qquad (17)$$

As before, provided $R_d \gg R_{o\_nom}$, R_d dominates, negating the effects of the error between R_{o_nom} and $R_o(i)$.

Similarly, for the auxiliary load sharing, the output current implementation equivalents of (10) and (11) are

$$I_{out}(i) = \frac{R_x I_{out}^*(i) + V_{dc}^* - V_{dc}}{R_o(i) + R_d}, \qquad (18)$$

$$\Delta I_{out}(i) = \frac{\Delta V_{dc}}{R_o(i) + R_d}. \qquad (19)$$

Therefore, auxiliary load sharing is in proportion to the denominator of (19) and should be approximately equal if $R_d \gg R_o(i)$.

The output current implementation is preferable to the input current implementation since the output currents at the shared LV bus sum to zero (or the auxiliary load current if present). Therefore when a balanced set of reference currents is provided from the supervisory controller, these are accurately tracked (or tracked with an offset corresponding to each converter's share of the auxiliary load current). Whereas, in the input current implementation, the DAB input currents cannot sum to zero due to the converter losses, and unequal voltage gain ratios among the DAB converters:

$$I_{in}(i) = \begin{cases} \frac{1}{\mu(i)} \cdot G(i) \cdot I_{out}(i), & \text{if } I_{out} \geq 0 \\ \mu(i) \cdot G(i) \cdot I_{out}(i), & \text{otherwise} \end{cases} \qquad (20)$$

Where $\mu(i)$ and $G(i)$ are the efficiency and conversion ratio of each DAB converter, respectively. Therefore, providing a balanced set of $I_{in}^*(i)$ from the supervisory controller results in tracking errors. A possible solution is to generate a balanced set of $I_{out}^*(i)$, then convert to an unbalanced set of $I_{in}^*(i)$ accounting for all $\mu(i)$ and $G(i)$. However, $\mu(i)$ is not available without the output current measurement. If the output current measurement is available, then the output current implementation of the controller should be used for simplicity.

In practice, R_d should be selected to be several times larger than the expected R_o of the system. As R_d increases, the bus voltage drop for a given auxiliary load current increases. Therefore, R_d should be made only as large as required to dominate over $R_o(i)$ and ensure adequate load sharing performance. To eliminate the steady-state bus voltage drop and use larger R_d, each DAB's share of the auxiliary load bus current can be summed to the original reference current generated in the supervisory controller. e.g. for the output current implementation:

$$I_{out}^*(i) = K_{SOC}(SOC_{est}(i) - \overline{SOC_{est}}) + \frac{I_{aux\_load}}{n_{DABs}}. \qquad (21)$$

Where n_{DABs} is the number of DABs connected to the shared LV bus and I_{aux_load} is the auxiliary load on the LV bus.

IV. Simulation Results

The proposed virtual resistance control scheme is first simulated in MATLAB/Simulink for a four-series one-parallel (4S1P) battery pack system. Averaged DAB converter models and ideal cell models are used for simplicity. The behavior of the supervisory controller is omitted for conciseness, and instead a balanced set of reference currents are provided directly to the DAB converters. To show the advantage of the proposed virtual resistance control technique, the proposed control is compared against the control approach without virtual resistances.

To test the efficacy of the controls, each DAB converter has a different output resistance $R_o(i)$ at its connection to the shared LV bus, while a single nominal value of $R_{o\_nom}(i) = 0.01\,\Omega$ is used by all primary controllers. DAB converters 1 through 4 are given reference currents $I_{in}^*(i)$ of 7, 4, -8, and -3 A, respectively, while their respective output resistances $R_o(i)$ are 5, 8, 12, and 15 mΩ. Midway through the simulation, a load step of 4 A is applied to the 12 V bus to investigate how an auxiliary system load is shared among the four DAB converters under each control scheme.

979-8-3315-1612-3/25 $31.00 © 2025 IEEE   1605

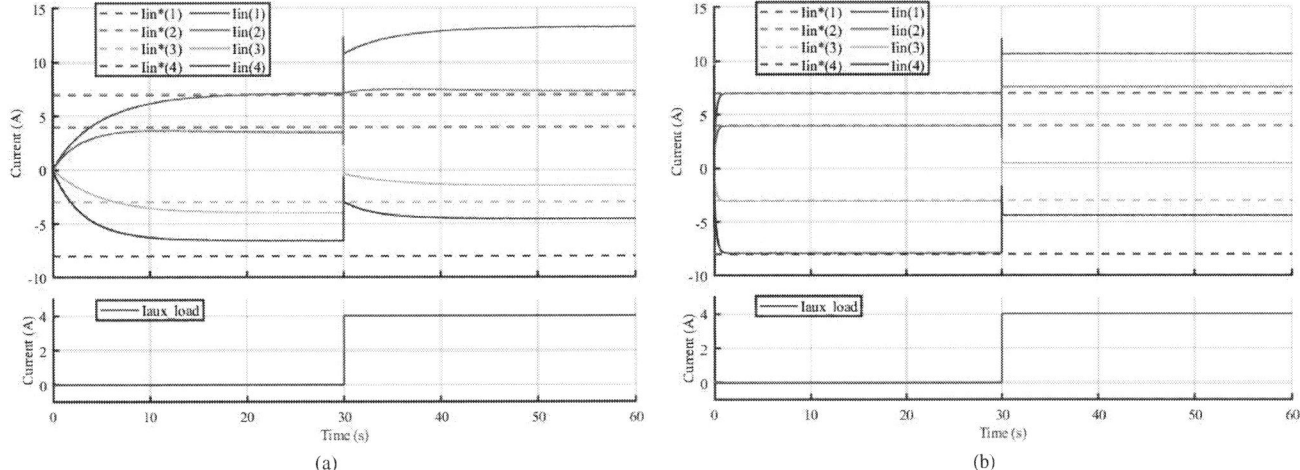

(a)   (b)

Figure 5: Simulation results for 4S1P system showing the reference and actual DAB input currents. A 5 A auxiliary load is applied to the 12 V bus at t= 30 s. a) Without virtual resistance, $R_d = 0\,\Omega$. b) With virtual resistance, $R_d = 0.05\,\Omega$.

Table I: 12V Bus Output Resistances

b-DAB i	1	2	3	4	5	6	7	8	9	10	11	12
$R_o(i)$ (mΩ)	2.98	5.69	7.76	10.7	9.40	11.5	12.0	12.0	11.2	9.30	5.66	0.03

Fig. 5a shows the simulation results without virtual resistance ($R_d = 0\,\Omega$), while Fig. 5b shows the simulation results with virtual resistance $R_d = 0.05\,\Omega$. The dashed lines show the input current references, while the solid lines show the actual input currents. Without virtual resistance, the input currents do not follow the reference currents accurately. When the auxiliary load step is applied at t = 30 s, the load does not split equally among the four DAB converters. Instead, it splits in proportion to the output resistances $R_o(i)$, with DAB 1 taking around 3x the load current share of DAB 4, since its output resistance is 3x smaller. Whereas when virtual resistance is used, the actual currents closely match the reference currents. When the auxiliary load is applied, it is observed to split equally among the four DAB currents despite their different output resistances. This comparison shows the benefit of the proposed control approach, both for accurate tracking of reference currents and equal load sharing.

V. EXPERIMENTAL TESTING

A. Experimental Setup

Fig. 6 shows the prototype 12S1P battery pack with 12 b-DAB converter active balancing system and 5 kW s-DAB connection to the 215 V high-voltage (HV) bus. Each b-DAB is situated on top of its corresponding cell, with input connected to the cell terminals, and output to the shared LV bus. A programmable DC load (Chroma 63206A-1200-240) is also connected to the shared LV bus to apply auxiliary 12 V loads during the testing. The 215 V HV bus is established by a programmable DC supply (Keysight N8930A), and DC load (second Chroma 63206A-1200-240), which together behave

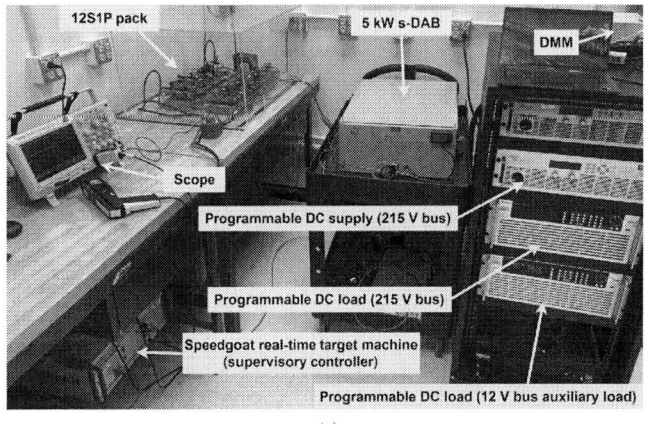

(a)

(b)

Figure 6: Experimental setup of the prototype 12S1P battery pack. (a) Full system as shown in Fig. 2 with 12S1P battery pack, s-DAB connection to HV bus, and power supplies and loads. (b) Close up view of the 12S1P battery pack with 12 b-DABs connecting each cell to the shared LV bus.

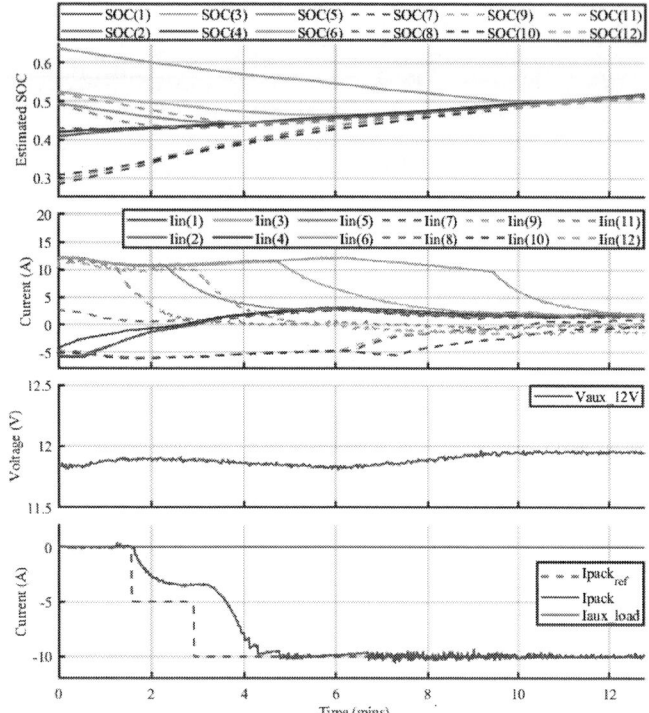

Figure 7: Initial SOC convergence and continuous balancing of the 12S1P battery pack during charging.

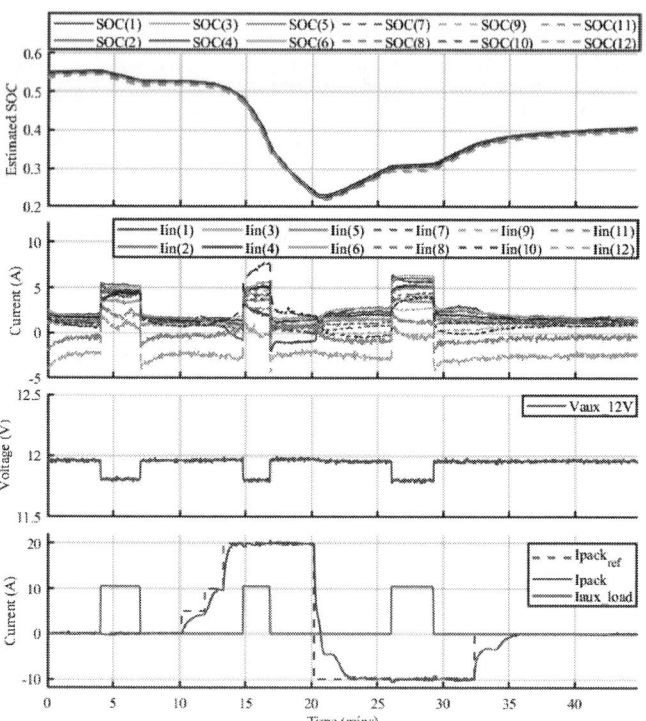

Figure 8: Continuous SOC balancing of the 12S1P battery pack with auxiliary loading on the 12V bus. The battery pack is idle from 0 to 10 minutes, discharging from 10 to 20 minutes, charging from 20 to 35 minutes, and idle from 35 minutes onwards. $1\,\Omega$ auxiliary load steps are applied to the 12V bus in each phase.

as a bidirectional source in voltage-control mode. The charging/discharging of the battery pack is controlled by current-control of the 5 kW s-DAB from the supervisory controller. A CAN bus connects the battery pack system's 12 b-DABs and the 5 kW s-DAB to a Speedgoat real-time target machine, which runs the supervisory controller. As detailed in Section II, the supervisory controller generates the reference current for each b-DAB from the estimated SOCs of each cell. The supervisory controller also sends the pack charge/discharge current commands to the s-DAB converter. The balancing gain used in the supervisory controller K_{SOC} is set to 400 and the virtual resistance R_d is set to 50 $m\Omega$ for all the b-DABs.

The measurements of the b-DAB input currents, voltages, and temperatures are logged in the Speedgoat system, along with the estimated SOCs, LV bus voltage, reference input currents, and s-DAB converter reference and measured current. A benchtop digital multimeter (DMM) measures and logs the LV bus auxiliary load current, while an oscilloscope is used for transient LV bus measurements.

Table I shows the output resistance between each b-DAB and the auxiliary load connection on the shared 12 V (LV) bus. The output resistance is measured as the sum of the resistances from the positive/negative DAB output terminals to the positive/negative auxiliary load terminals. A gaussmeter with Kelvin sensing probes is used to perform four wire resistance measurements in the milliohm range. Table I shows a wide variation in the measured output resistances. The auxiliary load is connected at the output of DAB 12, while the

other DAB outputs are connected to the DAB 12 output and auxiliary load via the shared LV bus. The LV bus connections are made via a ring bus.

B. SOC Balancing Performance

A test is performed to verify the SOC balancing system's ability to converge the cell SOCs. No auxiliary load is applied so that the balancing currents can be clearly observed, and to show the maximum speed of the balancing system. Prior to the test the 12S1P battery pack is intentionally unbalanced, resulting in an initial SOC range from 27 to 64%. The battery pack begins the test at idle (zero pack / string current) with the SOC balancing system active. The pack is commanded to charge at 5 A at t = 1.5 min, and 10 A at t = 3 min. Fig. 7 shows the experimental results with subplots for the estimated cell SOCs, b-DAB input currents $I_{in}(i)$, LV bus voltage (calculated as the mean of the b-DAB measured output voltages), and battery pack (s-DAB) and auxiliary load currents. The cell SOCs converge in around 10 minutes, remaining balanced until the end of the test. The auxiliary LV bus shows satisfactory regulation over the duration of the test, albeit with some slow-timescale fluctuations of up to 0.15 V during the initial SOC convergence period. Fig. 7 validates the performance of the virtual resistance control for balancing cell SOCs during normal operation of the battery pack.

979-8-3315-1612-3/25 $31.00 © 2025 IEEE   1607

Table II: 12V Bus Auxiliary Load Sharing

Cell / DAB i	1	2	3	4	5	6	7	8	9	10	11	12
$\Delta I_{in}(i)$ (A)	3.59	3.21	2.91	2.74	2.87	2.77	2.76	2.71	2.79	2.96	3.29	3.83
$LoadSharingError(i)$ (%)	18.2	5.56	-4.15	-9.67	-5.46	-8.73	-9.04	-10.76	-8.16	-2.51	8.45	26.3

C. LV Bus Auxiliary Loading

A second test validates the ability of the proposed control to simultaneously balance the cell SOCs while also supporting an auxiliary load of 1 Ω on the shared LV bus. This also validates the 12 V bus regulation and load sharing performance of the virtual resistance control during continuous operation of the SOC balancing system. Additionally, the battery pack is alternated through idle, discharging, and charging phases so that all functions of the battery pack and balancing system are tested simultaneously. Fig. 8 shows the experimental results with subplots for the estimated cell SOCs, b-DAB input currents $I_{in}(i)$, 12 V bus voltage (calculated as the mean of the DAB converter measured output voltages), and battery pack (s-DAB) and auxiliary load currents. This test shows the ability of the proposed control to simultaneously balance the cell SOCs while regulating the 12 V bus voltage, even under significant loading conditions. This is achieved by the virtual resistance control scheme, since it allows for each DAB converter to approximately follow an input current reference from the supervisory controller, while prioritizing the 12 V bus regulation. This control action is similar to droop control, as each DAB converter is in voltage-control mode with a droop slope R_d, while the voltage setpoint is being shifted in proportion to $I_{in}^*(i)$. Since the auxiliary load sharing is difficult to observe from Fig. 8, Table II is generated from the b-DAB input currents before and after the first auxiliary load application at t = 4 minutes.

Table II shows how the auxiliary bus load is split among the 12 cells and b-DABs. Since the SOC balancing is continually active, the auxiliary load current supplied by a single cell / b-DAB is measured as the change in a b-DAB's input current when the auxiliary load is applied. The first row of Table II shows the absolute change in input current for each b-DAB. The second row shows the load sharing percentage error for each cell / b-DAB. The load sharing percentage error is calculated as the percentage difference between each b-DAB's change in input current and the mean change in input current across the 12 b-DABs:

$$LoadSharingError(i)\,(\%) = 100 \cdot \frac{\Delta I_{in}(i) - \overline{\Delta I_{in}}}{\overline{\Delta I_{in}}}. \quad (22)$$

The maximum observed load sharing error was 26.2 % for b-DAB 12. However, the majority of b-DABs had a load sharing error of less than 10 %, demonstrating satisfactory load sharing performance. The errors appears roughly proportional to the differences in output resistances shown in Table I. This is as expected since while the virtual resistance R_d dominates over the actual output resistance $R_o(i)$ according to (10), $R_o(i)$ still introduces a small error. Increasing R_d can minimize

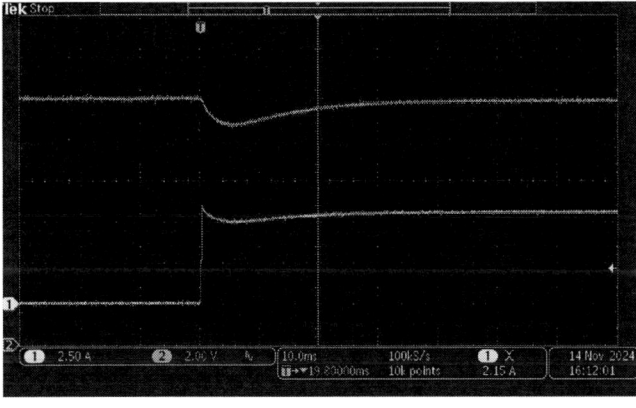

(a)

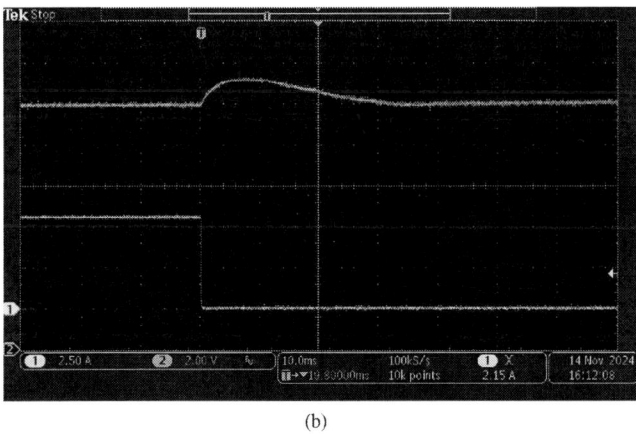

(b)

Figure 9: Transient performance of the 12V bus during auxiliary loading. Cyan = 12V bus voltage (V). Yellow = auxiliary load current (A). (a) Load step from no load to 2 Ω. (b) Load step from 2 Ω to no load.

this error further, though at the cost of increasing the steady-state LV bus voltage error when auxiliary loads are applied. However, future versions of the prototype pack may include a measurement of the auxiliary bus current communicated back to the supervisory controller. Then the supervisory controller can modify each DAB converter's reference current to include its expected share of the auxiliary load current, as shown in (21), eliminating the LV bus steady-state voltage error entirely. That said, increasing R_d further may not reduce the load sharing error significantly, since it may be predominately caused by the output voltage measurement error of each DAB converter. In this case, only improved accuracy of the voltage measurement circuitry and careful sensor calibration could further minimize the load sharing error.

D. Transient Performance

The transient performance of the virtual resistance control scheme is verified with oscilloscope measurements during positive and negative load steps applied to the 12 V bus. Similar to the test shown in Fig. 8, the auxiliary bus load is applied while the b-DABs are continuously balancing the cell SOCs. Fig. 9a shows the positive load step from no load to 2 Ω while Fig. 9b shows the negative load step from 2 Ω to no load. Both results show excellent transient responses of the control scheme, with a settling time of around 30 ms. The bus voltage experiences a brief drop or rise of less than 1.5 V due to the step change, before the voltage PI loops within each DAB's virtual resistance control implementation smoothly regulate the bus voltage back towards the 12 V setpoint.

VI. CONCLUSIONS AND FUTURE WORK

The BTMS pack uses a converter-based active SOC balancing system in which DAB converters drive a set of balancing currents between series-connected cells to converge their SOCs. To do this, a supervisory controller gives each DAB converter a reference current to follow. However, a unique challenge of this solution is that the DAB converters must also regulate a shared 12 V bus, which beyond exchanging the energy between cells, must be capable of supporting auxiliary 12 V system loads.

The control problem is solved effectively by the proposed virtual resistance control scheme, which allows each DAB converter to simultaneously track a reference current while tightly regulating the 12 V bus voltage. The experimental results show the success of the proposed control, which continually converged and balanced cell SOCs during charging and discharging operations. Further, auxiliary loads applied to the 12 V bus were split approximately equally among the 12 DAB converters, without interference to the SOC balancing operation. The virtual resistance method, which is an extension of conventional droop control, allows for these control objectives to be achieved despite differences in each DAB converter's line resistance between its output and the shared 12 V bus and auxiliary load connection point.

Future work will utilize the output current based implementation for improved reference current tracking accuracy. Additionally, to eliminate the steady-state error on the 12 V bus voltage during auxiliary loading, the supervisory controller will sum a share of the measured auxiliary load current to each DAB converters reference current. Then larger droop resistances can also be used without impacting the steady-state error on the LV bus voltage. The proposed virtual resistance control may be useful in broader applications, in which multiple converters supply a shared DC bus. Under the proposed control, all converters can participate in the DC bus voltage regulation, while still tracking a reference current from a higher-level controller.

VII. ACKNOWLEDGMENT

The authors would like to thank (in alphabetical order) Anik Chowdhury, Marco Gaxiola, Md Ehsanul Haque, Syed Imam Hasan, Jeff Holt, Mohammad Muntasir Islam, Kyle Miller, Partha Mishra, Vaibhav Pawaskar, Mohammad Arifur Rahman, and Alper Uzum for their contributions to the prototype BTMS battery pack system development.

REFERENCES

[1] "U.S. Renewable Energy Factsheet," Center for Sustainable Systems, University of Michigan, Tech. Rep., 2024, Pub. No. CSS03-12. [Online]. Available: https://css.umich.edu/sites/default/files/2024-10/Renewable%20Energy_CSS03-12.pdf

[2] A. Burrell, "Behind-the-Meter Storage," NREL, Tech. Rep., 2021, NREL/PR-5F00-79681. [Online]. Available: https://www.osti.gov/biblio/1835819

[3] M. Rezaeimozafar, R. F. Monaghan, E. Barrett, and M. Duffy, "A review of behind-the-meter energy storage systems in smart grids," Renewable and Sustainable Energy Reviews, vol. 164, p. 112573, 2022. [Online]. Available: https://www.sciencedirect.com/science/article/pii/S1364032122004695

[4] P. Carrasco Ortega, P. Durán Gómez, J. C. Mérida Sánchez, F. Echevarría Camarero, and Pardiñas, "Battery energy storage systems for the new electricity market landscape: Modeling, state diagnostics, management, and viability—a review," Energies, vol. 16, no. 17, 2023. [Online]. Available: https://www.mdpi.com/1996-1073/16/17/6334

[5] A. Chowdhury, A. Siddiquee, P. Mishra, M. E. Haque, M. J. Kisacikoglu, A. Thurlbeck, E. Watt, M. A. Rahman, Y. Sozer, and J. Holt, "Design of a multi-chemistry battery pack system for behind-the-meter storage applications," in IEEE Applied Power Electronics Conf. Expo., 2023, pp. 3044–3049.

[6] M. M. Ur Rehman, M. Evzelman, K. Hathaway, R. Zane, G. L. Plett, K. Smith, E. Wood, and D. Maksimovic, "Modular approach for continuous cell-level balancing to improve performance of large battery packs," in IEEE Energy Conversion Congr. Expo., Sep. 2014, pp. 4327–4334.

[7] M. M. U. Rehman, F. Zhang, M. Evzelman, R. Zane, K. Smith, and D. Maksimovic, "Advanced cell-level control for extending electric vehicle battery pack lifetime," in IEEE Energy Conversion Congr. Expo., Sep. 2016, pp. 1–8.

[8] S. Karmakar, T. K. Bera, and A. K. Bohre, "Review on cell balancing technologies in battery management systems in electric vehicles," in 2023 IEEE IAS Global Conference on Renewable Energy and Hydrogen Technologies (GlobConHT), 2023, pp. 1–5.

[9] M. S. Zafar, M. A. Nisar, A. Wahid, and N. A. Zaffar, "Intelligent li-ion battery management system," in 2024 IEEE Workshop on Control and Modeling for Power Electronics (COMPEL), 2024, pp. 1–7.

[10] M. S. Trimboli, A. K. de Souza, and M. A. Xavier, "Stability and control analysis for series-input/parallel-output cell balancing system for electric vehicle battery packs," IEEE Control Systems Letters, vol. 6, pp. 1388–1393, 2022.

[11] N. Ghaeminezhad, Q. Ouyang, X. Hu, G. Xu, and Z. Wang, "Active cell equalization topologies analysis for battery packs: A systematic review," IEEE Trans. Power Electron., vol. 36, no. 8, pp. 9119–9135, 2021.

[12] M. Evzelman, M. M. Ur Rehman, K. Hathaway, R. Zane, D. Costinett, and D. Maksimovic, "Active Balancing System for Electric Vehicles With Incorporated Low-Voltage Bus," IEEE Trans. Power Electron., vol. 31, no. 11, pp. 7887–7895, Nov. 2016.

[13] J. M. Guerrero, J. C. Vasquez, J. Matas, L. G. de Vicuna, and M. Castilla, "Hierarchical Control of Droop-Controlled AC and DC Microgrids—A General Approach Toward Standardization," IEEE Trans. Ind. Electron., vol. 58, no. 1, pp. 158–172, 2011.

[14] J. P. Rivera-Barrera, N. Muñoz-Galeano, and H. O. Sarmiento-Maldonado, "SoC Estimation for Lithium-ion Batteries: Review and Future Challenges," Electronics, vol. 6, no. 4, p. 102, Dec. 2017.

Internal Voltage Source Saturation Impact on Stability Limits of Grid Forming Converter

Divyanshu Bansal
Dept. of Electronic Systems Engg.
Indian Institute of Science
Karnataka, India
divyanshub@iisc.ac.in

Aravind G.
Dept. of Electronic Systems Engg.
Indian Institute of Science
Karnataka, India
aravindg@iisc.ac.in

L. Umanand
Dept. of Electronic Systems Engg.
Indian Institute of Science
Karnataka, India
lums@iisc.ac.in

Abstract—**This paper investigates the converter constraints on internal voltage source (IVS) dynamics due to weak grid disturbances. Current limit constraint affects grid synchronization and grid-forming (GFM) capability during and post fault disturbance. The effect of IVS saturation limits is shown to be significant, as it shifts the peak power capability towards increasing delta as limits increase. This provides more power throughput from the system and increased stability limits with larger critical clearing time. Along with mathematical analysis, a simulation case study on 10kW system is presented in this paper.**

Index Terms—**component, formatting, style, styling, insert.**

I. INTRODUCTION

With the focus on integrating sustainable energy source to the electrical grid and decreasing the amount of power produced by the non-renewable resources, grid connected converters are employed to tackle this challenge. Broadly, these converters are classified into two categories based on their control principle as Grid Following (GFL) and Grid Forming (GFM) converters. GFL based converters acts as current source, while being synchronized to the grid, is controlled to follow the grid conditions. As the share contributed by these GFL converter increases, the grid requirements traditionally fulfilled by the synchronous generators poses a situation where penetration limit has to be imposed to stabilize the grid, thus under-utilizing the amount of resource available.

Grid Forming (GFM) converters have emerged as a promising option in addition to Grid Following (GFL) converters in the grid. With GFM emulating synchronous machine dynamics, it provides the necessary grid support operations required to maintain the grid stability in weak grid. [1]- [3].

Grid code [4] stipulates that converter supports grid for the voltage and frequency stability under faults. But the current limit imposed on the converter, limits the capability to provide that support. There are various ways to limit the current magnitude that has been reported in the literature. One is hard-current limitation employing the circular current limit while prioritizing angle or d-q axis information. Another way is by using adaptive virtual impedance that increases its impedance as the fault increases with respect to grid voltage sag. [5]

As the converter is emulating a synchronous generator (VSG) and swing equation providing an internal voltage source (VSG), the power loops can go out of close-loop and into

saturation while current limit is active. But utilizing the angle information for the current phasor with respect to grid voltage provided still by the power loops transforms the traditional $P - \delta$ to a different curve [6]- [8].

In the reported literature on grid forming control under faults, two power loops: active power loop (APL) and reactive power loop (RPL) provides w_m and E_{int} respectively for an equivalent inductive grid. To provide the maximum power or changing controller from slow to fast dynamics during current limit condition supporting grid, APL structure is generally modified. But the effect of saturation in RPL is neglected in the literature. [8]

In this paper, we show the additional effect of RPL saturation on $P - \delta$ curve and how it enhances the stability limits provided when the E_{sat} limits are varied. The paper is organised with Section II representing the modeling of grid forming converter with its dynamic equations, Section III works on the analysis on the effect of the internal voltage source impact on stability, Section IV presents the simulation results on the analysis presented in this paper for a specific case study followed by conclusion in Section V.

II. MODELING

A three-phase grid-forming structure is represented in Fig.1. It emulates virtual synchronous machine swing equation dynamics as (1). For a weak grid with low X/R ratio, a virtual impedance loop with $X_v >> R_v$ provides the total impedance between an internal voltage source and grid source to be largely inductive in nature.

$$Jw\frac{dw}{dt} = P_o - P_{out} - D_p(w - w_o) \tag{1}$$

$$\frac{d\delta}{dt} = w - w_g = \Delta w$$
$$\frac{d\theta_m}{dt} = w \tag{2}$$

where J is system inertia, w is the virtual machine rotational speed, P_o, P_{out} is the input mechanical power and point of common coupling (PCC) power respectively. D_P is the damping constant of the system. (1)-(2) provides the power angle characteristic based active power loop (APL) dynamics as shown in Fig. 1. Since the equation represented by (1) is

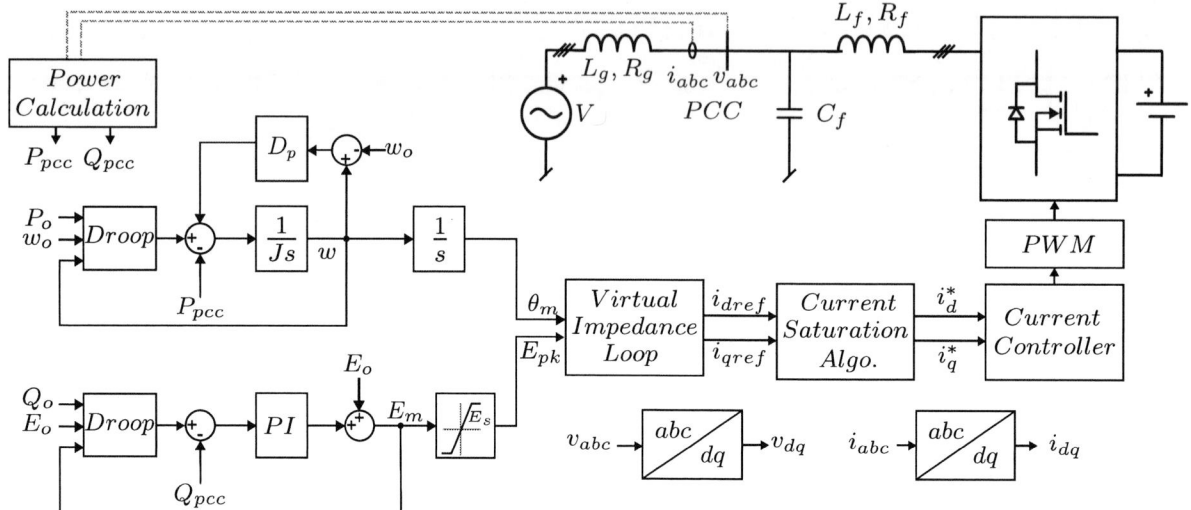

Fig. 1: System block diagram for the 3-phase GFM based Converter

non-linear in nature, an assumption is usually taken where rotational speed w is kept near to the base speed of the system, thus, $w \approx w_o$ as shown in (3)

$$Jw_o\frac{d^2\delta}{dt} = P_o - P_{out} - D_p\Delta w \qquad (3)$$

Equation (4) provides the reactive power - peak amplitude characteristic based reactive power loop dynamics where PI controller (K_{pq}, K_{iq}) and a feed-forward E_o is employed. A droop based control ($Q - E_{pk}$) could also be used that comes out to be same if $K_{iq} = 0$.

$$E_m = E_o + (K_{pq} + \int K_{iq}dt)(Q_o - Q_{out}). \qquad (4)$$

A limiter placed on the output of E_m limits the reference tracking output to $E_{pk} = E_{sat}$ when in saturated condition.

$$E_{pk} = \begin{cases} E_{sat} & \text{if } E_m > E_{sat} \\ -E_{sat} & \text{if } E_m < -E_{sat} \\ E_m & \text{otherwise} \end{cases} \qquad (5)$$

The information provided by the power loops is used to generate current references using virtual impedance loop when transformed into dq framework (6) with d-axis aligned to peak voltage of internal source (E_{pk}).

$$\begin{bmatrix} I_{refd} \\ I_{refq} \end{bmatrix} = \frac{1}{R_v^2 + X_v^2} \begin{bmatrix} R_v & X_v \\ -X_v & R_v \end{bmatrix} \begin{bmatrix} E_{pk} - V_d \\ -V_q \end{bmatrix} \qquad (6)$$

$$I_{refdq} = I_{refd} + jI_{refq} \qquad (7)$$

Current references generated are limited using angle prioritization current saturation algorithm [6] with maximum supporting current by the converter as I_{max} given by (9).

$$I = |I_{refdq}| = \sqrt{I_{refd}^2 + I_{refq}^2} \qquad (8)$$

$$I_{dq} = \begin{cases} I_{refdq} * \frac{I_{max}}{I} & \text{if } I > I_{max} \\ I_{refdq} & \text{otherwise} \end{cases} \qquad (9)$$

III. ANALYSIS

An equivalent circuit for the converter control at PCC is represented by an internal voltage source with virtual impedance as shown in Fig. 2. Thus the relationship for the current

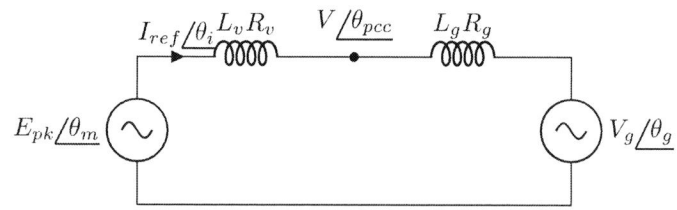

Fig. 2: Equivalent circuit diagram of GFM converter considering virtual impedance

reference information is given by,

$$I_{ref} = \frac{E_{pk}\underline{/\theta_m} - V_g\underline{/\theta_g}}{R_v + R_g + j(X_v + X_g)} \approx \frac{E_{pk}\underline{/\theta_m} - V_g\underline{/\theta_g}}{j(X_v + X_g)} \qquad (10)$$

$$V_{dq} = V_g\underline{/\theta_g - \theta_m} = V_g\underline{/-\delta} \qquad (11)$$

$$I_{dq} = I_{max}\underline{/\phi_{dq}} = I_{max}\underline{/tan^{-1}(\frac{-\frac{E_{pk}}{V_g}+cos\delta}{sin\delta})} \qquad (12)$$

$$P_{esat} = \frac{3}{2}V_gI_{max}cos(-\delta - \phi_{dq}) \qquad (13)$$

$$P_{eideal} = \frac{3}{2}\frac{V_gE_o}{(X_v + X_g)}sin(\delta) \qquad (14)$$

In (13), the maximum power during current limit depends on the grid voltage magnitude (V_g) and current limits imposed by the converter, but is independent of the IVS (E_m) magnitude. But, the angle information is dependent on IVS magnitude. Fig. 3a depicts the same when E_{sat} is varied from 1pu to 1.4pu. The observed effect shows that peak of current limit curve is constant while shifting towards increasing δ as E_{sat} is increased. It is utilized in this paper to enhance the stability

979-8-3315-1612-3/25 $31.00 © 2025 IEEE

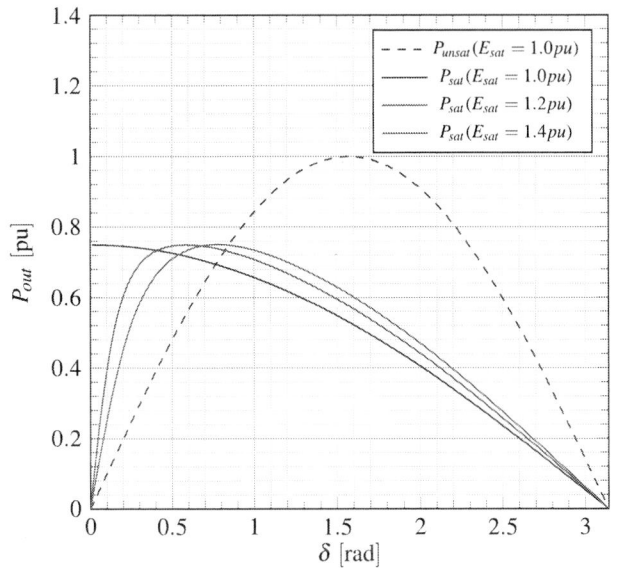

(a) P-δ curve for grid $V_{sag} = 0\%$ and $E_{sat} = \{1, 1.2, 1.4\}$pu

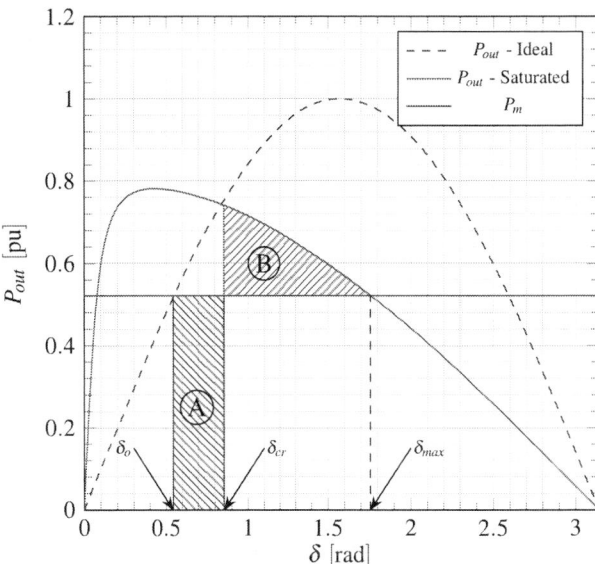

(b) EAC study for $E_{sat} = 1.1pu$ with critical clearing angle δ_{cr}

Fig. 3: P-δ curve for grid $V_{sag} = 0\%$ and $E_{sat} = \{1, 1.2, 1.4\}$pu and EAC study

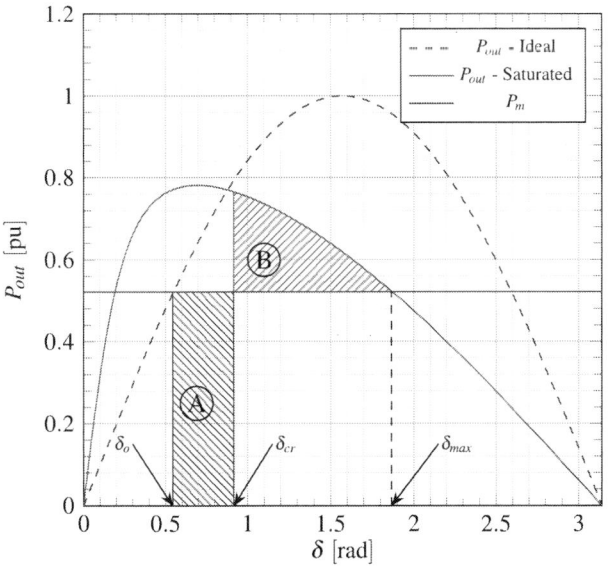

(a) EAC study for $E_{sat} = 1.3pu$ with critical clearing angle δ_{cr}

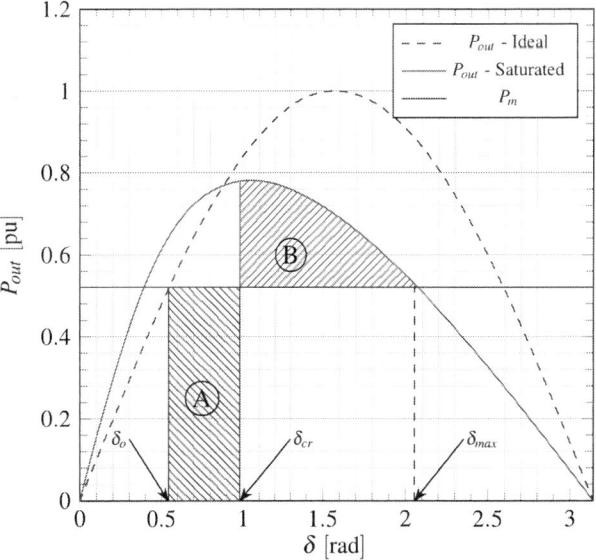

(b) EAC study for $E_{sat} = 2.0pu$ with critical clearing angle δ_{cr}

Fig. 4: Acceleration (A) and deceleration (B) area associated with symmetrical fault at $V_{gsag} = 100\%$ in EAC studies

limits in the grid where E_{sat} could be selected to provide the optimal value. In the equal area stability criterion studies, this increases the deceleration area for the converter and increases the critical clearing time for the converter.

A case study of voltage sag $V_{sag}\%$ at PCC is analyzed for the transient stability studies. Fault occurs at δ_o and cleared at δ_{cr}. For the maximum clearing point, δ_f(where P_m intersects saturated P_{out} curve in region $\delta > 90^o$) using (12) and (13).

$$P_m = P_{esat} \tag{15}$$

$$P_m = \frac{3}{2}V_g I_{max}cos(-\delta - \phi_{dq}) \tag{16}$$

On solving the above equation, we get the maximum δ_{max} supported by the system before the converter reaches unstable equilibrium point as,

$$\delta_{max} = \delta_f = x^2\frac{E_{sat}}{V_g} \pm \left(\left(x^2\frac{E_{sat}^2}{V_g^2} - 1\right)\left(x^2 - 1\right)\right)^{0.5} \tag{17}$$

$$\text{where, } x = \frac{P_m}{1.5E_{sat}I_{max}}$$

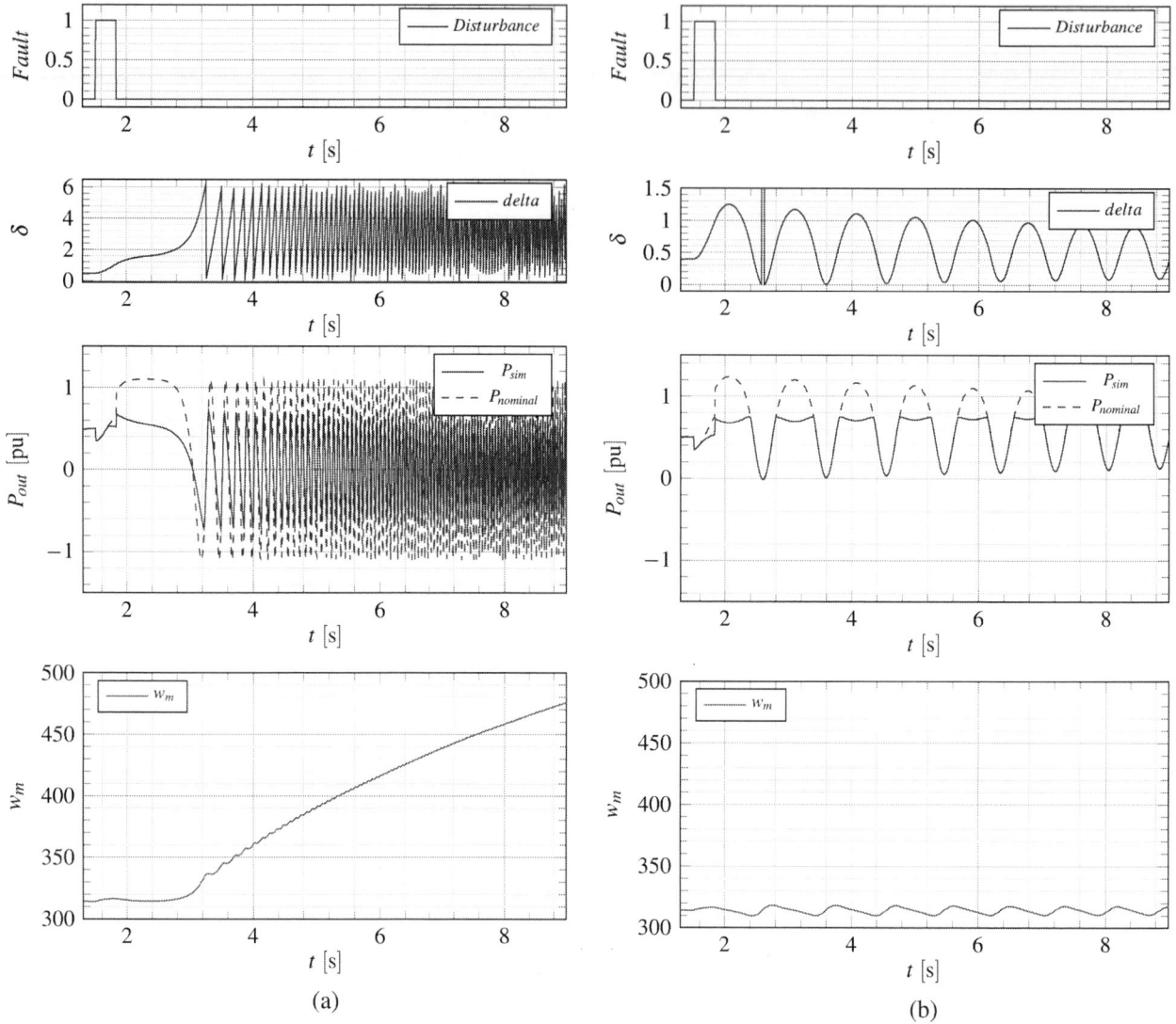

Fig. 5: Simulation case study depicting LOS for (a) when $E_{sat} = 1.1pu$ and stable state for (b) when $E_{sat} = 1.3pu$
P_{sim}: Actual power output from simulation, $P_{nominal}$: Power output if current limit is not active

Fig. 3b, 4a and 4b represents the study when E_{sat} is varied as 1.1, 1.3 and 2.0 pu respectively. In these studies, area A represents acceleration of the virtual synchronous machine's rotational speed due to symmetrical fault with 100% voltage sag and fault cleared at δ_{cr}. Also, area B represents post fault deceleration rate of the machine during recovery.

For the general case of V_{sag}, during fault, Area $A_1 = A$ and $A_2 = B$ is used as the convention in the analysis presented below. During dynamics, when it reaches the δ_{max}, (18), (19) provides the maximum fault clearing angle when $A_1 = A_2$ for equal area criterion study.

$$\text{Area } A_1: \int_{\delta_o}^{\delta_c} \left(P_m - 1.5 \frac{E_{sat} V_{gsag}}{X_v} sin\delta \right) d\delta \qquad (18)$$

$$\text{Area } A_2: \int_{\delta_c}^{\delta_f} \left(1.5 V_g I_{max} cos(-\delta - \phi_{dq}) - P_m \right) d\delta \qquad (19)$$

Solving this, we obtain when $V_{gsag} = 0$ for solid state fault, δ_{max} as (20). It can be observed that the critical clearing angle value increases with increasing E_{sat} for the same fault study. This analysis is verified with a simulation study for different values of E_{sat} in the next section.

IV. SIMULALTION RESULTS

With the following specifications for the converter, a disturbance is created at t=1.5s and cleared at t=1.84s in the system where V_{grid} is changed from 1.0pu to 0.7pu to show a voltage sag of 30% at the PCC. Table I shows the circuit parameters and control parameters used in the system.

For the same conditions in Fig. 5a and 5b, when $E_{sat} = 1.1pu$, Fig. 5a shows the unstable condition where converter has lost the synchronization with the grid when passing beyond unstable equilibrium point. But when $E_{sat} = 1.3pu$ is kept,

$$\delta_{cr} = cos^{-1}\left(\frac{1 + \left(\frac{E_{sat}}{V_g}\right)^2 - \left(1 + \frac{E_{sat}^2}{V_g^2} - 2\frac{E_{sat}}{V_g}cos\delta_f - \frac{P_m}{1.5V_g I_{lim}}(\delta_f - \delta_o)\right)}{2\frac{E_{sat}}{V_g}}\right) \quad (20)$$

Parameter	Value	Unit
P_{base}	10	kW
V_{base}	240	Vrms phase
w_n	314.16	rad/s
Z_{base}	17.225	Ω
L_g	0.0219	pu
R_g	5.80E-04	pu
L_f	0.0058	pu
V_{dc}	700	V
f_{sw}	20	kHz
J	0.8106	kgm^2
E_{sat}	1.2	pu
E_o	1	pu
K_{pq}	0.0017	$(As)^{-1}$
K_{iq}	0.04	A^{-1}
D_p	31.83	Ws/rad
K_{pc}	4.02	Ω
K_{ic}	125.66	Ω/sec
L_v	0.5	pu
R_v	0	pu

TABLE I: System specifications

we can observe that in Fig. 5b, converter post-disturbance goes to a stable state. In this study, during fault stage, to showcase the effect of swing equation as assumed in equal area criterion studies, we neglected the damping factor (D_p) for a conservative study in both conditions of E_{sat} as 1.1 and 1.3 pu.

V. CONCLUSION

This paper shows that the internal voltage source saturation limits and grid disturbance has an effect on the $P - \delta$ curve. With peak power in current limit mode shift towards increasing δ, it provides a way to transfer maximum power to the grid when internal voltage source amplitude is shifted to higher values. Using this, more deceleration area is utilized for swing in the VSG. It provides a larger critical clearing time for the disturbance to be cleared. A general analysis is presented for equal area criterion studies showing the impact on critical clearing angle. A simulation showcasing a fault disturbance of voltage sag of 30% is studied in this paper.

REFERENCES

[1] R. Rosso, X. Wang, M. Liserre, X. Lu, and S. Engelken, 'Grid-Forming Converters: Control Approaches, Grid-Synchronization, and Future Trends—A Review', IEEE Open Journal of Industry Applications, vol. 2, pp. 93–109, 2021, doi: 10.1109/OJIA.2021.3074028.

[2] Y. Gu and T. C. Green, "Power System Stability With a High Penetration of Inverter-Based Resources," Proc. IEEE, vol. 111, no. 7, pp. 832–853, Jul. 2023, doi: 10.1109/JPROC.2022.3179826.

[3] M. H. Ravanji, W. Zhou, N. Mohammed, and B. Bahrani, "Comparative Analysis of the Power Output Capabilities of Grid-Following and Grid-Forming Inverters Considering Static, Dynamic, and Thermal Limitations," IEEE Trans. Power Syst., vol. 39, no. 2, pp. 2693–2705, Mar. 2024, doi: 10.1109/TPWRS.2023.3279373.

[4] 'IEEE Standard for Interconnection and Interoperability of Inverter-Based Resources (IBRs) Interconnecting with Associated Transmission Electric Power Systems'. Apr. 2022. doi: 10.1109/IEEESTD.2022.9762253.

[5] B. Fan, T. Liu, F. Zhao, H. Wu, and X. Wang, 'A Review of Current-Limiting Control of Grid-Forming Inverters Under Symmetrical Disturbances', IEEE Open Journal of Power Electronics, vol. 3, pp. 955–969, 2022, doi: 10.1109/OJPEL.2022.3227507.

[6] K. G. Saffar, S. Driss, and F. B. Ajaei, 'Impacts of Current Limiting on the Transient Stability of the Virtual Synchronous Generator', IEEE Transactions on Power Electronics, vol. 38, no. 2, pp. 1509–1521, Feb. 2023, doi: 10.1109/TPEL.2022.3208800.

[7] J. Maeng and S. Cui, 'Enhanced Fault Ride-Through of Grid-Forming Converter with Extra Internal Voltage Source Rotation', in 2024 IEEE Applied Power Electronics Conference and Exposition (APEC), Feb. 2024, pp. 251–255. doi: 10.1109/APEC48139.2024.10509523.

[8] H. Wu and X. Wang, 'Control of Grid-Forming VSCs: A Perspective of Adaptive Fast/Slow Internal Voltage Source', IEEE Transactions on Power Electronics, vol. 38, no. 8, pp. 10151–10169, Aug. 2023, doi: 10.1109/TPEL.2023.3268374.

A Zero Harmonic Distortion Grid-Connected Grid-Forming Converter for Battery Energy Storage System Applications

Gabriel V. Ramos*, Thiago M. Parreiras[†], Fangzhou Zhao[‡], Xiongfei Wang[‡§] and Braz de J. C. Filho[¶]

*Graduate Program in Electrical Engineering - Universidade Federal de Minas Gerais -
Av. Antônio Carlos 6627, 31270-901, Belo Horizonte, MG, Brazil

[†]Department of Electrical Engineering, Centro Federal de Educação Tecnológica de Minas Gerais, Belo Horizonte, Brazil

[‡]Department of Energy, Aalborg University, Aalborg, Denmark

[§]Division of Electric Power and Energy Systems, KTH Royal Institute of Technology, Stockholm, Sweden

[¶]Department of Electrical Engineering, Universidade Federal de Minas Gerais, Belo Horizonte, Brazil

*gabrielvilkn@ieee.org, [†]thiago.parreiras@cefetmg.br, [‡]fzha@energy.aau.dk, [‡§]xiongfei@kth.se and [¶]braz.cardoso@ieee.org

Abstract—The work proposes the Zero Harmonic Distortion Converter (ZHD) as a grid-connected grid-forming (GFM) converter due to its unique sinusoidal characteristic without capacitive filters, low parts count, and off-the-shelf power circuitry with a simple open-loop control structure that does not contribute to possible control interaction and resonances in grid-connected mode. These characteristic are reached by a harmonic cancellation in a three-winding transformer along with harmonic elimination using the Selective Harmonic Elimination Pulse Width Modulation (SHE PWM). The ZHD GFM converter is a compelling alternative for GFM battery energy storage systems (BESS) in medium-voltage applications.

Index Terms—Grid-forming, voltage-source converters, SHE-PWM.

I. INTRODUCTION

The increase of renewable energy resources interfaced by voltage source converters (VSCs) has stimulated considered interest in grid-forming (GFM) control [1]–[3]. Differently from grid-following (GFL) control that handles VSCs as a current source depended of the grid voltage, the GFM-VSC is controlled as a voltage source behind an impedance, and its stable operation does not depend on the strength of the ac grid, exhibiting improved performance compared to GFL-VSC in weak grids [4]. This voltage source characteristic arises from the PCC voltage magnitude regulation, while synchronization is achieved through active power control, commonly referred to as the power synchronization loop [4].

Among the numerous advantages of GFM converters, such as black start capability, inertia, and damping, numerous studies demonstrate that they can enhance voltage stiffness, thereby improving the stability of renewable energy resources interfaced by GFL converters under weak-grid conditions, such as those found in offshore wind power plants (WPPs) and solar

This study was financed in part by the Coordenação de Aperfeiçoamento de Pessoal de Nível Superior - Brasil (CAPES) - Finance Code 001, in part by Conselho Nacional de Desenvolvimento Científico e Tecnológico (CNPq), and in part by Fundação de Amparo à Pesquisa do Estado de Minas Gerais (FAPEMIG).

farms. This improvement is particularly notable in converter structures based on battery energy storage systems (BESS) [5]–[9].

The most widely applied BESS solutions are based on 2- and 3-level (VSC) connected to line frequency transformers due to the cost-effectiveness coupled with the need to ensure the galvanic isolation required in the integration of renewable energy and in the grid-connected battery energy systems standards [10]–[12]. To achieve output sinusoidal voltage waveforms, conventional VSCs require the use of capacitive LC or LCL filters on the grid side, which contributes to resonances and instability issues [13]–[17], what is a concern in solar farms [18] and also in offshore weak grid WPPs due to long alternating current cables, transformers, shunt reactors, and converter filters [19], [20].

In addition, the GFM BESS converter control structure is generally based on closed control loops, and the necessity of a vector voltage control (VVC) may cause control interactions and oscillations in grid-connected mode, complicating the controllers parameters tuning [21]–[24].

This paper introduces the use of Zero Harmonic Distortion (ZHD) technology as a grid-connected GFM converter for GFM BESS applications. The ZHD converter, detailed in [25] as a grid-following, in [26]–[29] as an island grid-forming converter utilizing 2-level converters and in [30], [31] utilizing 3-level converters, offers advantages like low parts count, off-the-shelf power circuitry, and no need for output capacitive filters. It generates no characteristic harmonics up to the 50th order, meeting the IEEE 519 and IEEE 1547 standards [32], [33], and a desirable characteristic of recent GFM requirements [21], [34]. This is accomplished using two 3-phase converters with Selective Harmonic Elimination Pulse Width Modulation (SHE PWM), linked to a three windings transformer, creating a sinusoidal voltage source without sinusoidal filters with capacitive filter elements that are a source of reliability problems [35] and contribute negatively to possible resonance [13]–[20] and ferroresonance [36], [37] issues.

The work contributes by proposing the ZHD GFM converter for the grid-connected mode, since the converter's inherent sinusoidal voltage implies no need for output capacitive filters and consequently permits a control implementation using only a voltage open-loop structure, dismissing the commonly applied VVC control in commercial solutions, which may cause control interaction and resonances in grid-connected mode applications. This work assesses the operation of the ZHD converter highlighting its suitability for grid-connected GFM BESS in medium-voltage applications.

II. ZERO HARMONIC DISTORTION CONVERTER

A. Converter Structure

ZHD converter is depicted in Fig. 1 and is based on two 2-level VSCs integrated to a three-winding transformer. The VSCs are attached to the wye and delta secondaries of the transformer. The 30 degrees of phase shift between the secondaries implies harmonic cancelation, as described in (1) [25]–[29].

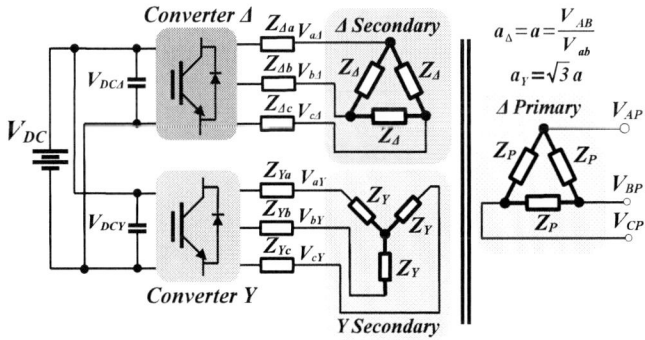

Fig. 1. 2-level ZHD technology.

$$h = 6k \pm 1 \quad \forall k \in \mathbb{Z}, \ k \text{ odd.} \tag{1}$$

Thus, it is feasible to achieve a voltage waveform that is devoid of all harmonics assessed according to international standards [32], [33], up to 50th order, using the SHE PWM technique [38], [39] to eliminate harmonics not canceled by the transformer. Table I shows the canceled and the ones eliminated in the ZHD converter structure [25]–[29].

TABLE I
HARMONIC ELIMINATION AND CANCELLATION IN THE ZHD CONVERTER.

Element	Harmonic Order
Transformer:	$5^{th}, 7^{th}, 17^{th}, 19^{th}, 29^{th}, 31^{th}, 41^{th}, 43^{th}, \ldots$
SHE PWM:	$11^{th}, 13^{th}, 23^{th}, 25^{th}, 35^{th}, 37^{th}, 47^{th}, 49^{th}$

The SHE PWM switching angles are defined based on modulation indices, calculated offline using Newton's method, and stored in Look-up Tables (LUTs). Fig. 2 presents one of the possible solutions for the commutation angles according to the modulation index [25]–[29].

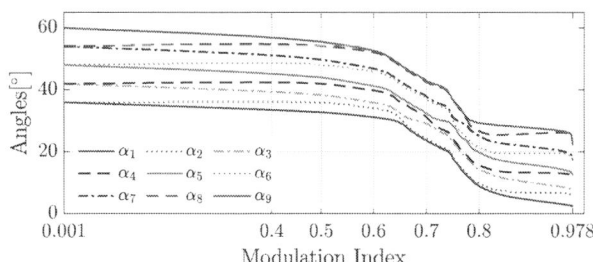

Fig. 2. 2L VSC SHE PWM switching angles stored in LUTs.

B. ZHD GFM Equivalent Circuit Model

A structured analysis of the ZHD GFM converter in islanded microgrid applications is presented in [26]–[29] and the equivalent fundamental and harmonic circuits developed are shown in Fig. 3.

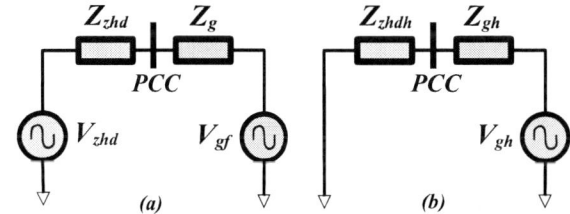

Fig. 3. (a) ZHD GFM per phase fundamental equivalent and (b) ZHD GFM per phase harmonic equivalent circuits proposed in [26]–[29].

The equivalent circuits show that the ZHD does not generate harmonics up to 50th order, providing, for practical considerations, a sinusoidal voltage waveform. Nevertheless, the presence of harmonics in the grid side, affects the PCC voltage waveforms due to the voltage drop in the series impedance of the ZHD given by (2) where $Z_{\Delta a}$, $Z_{\Delta b}$, $Z_{\Delta c}$, Z_{Ya}, Z_{Yb} and Z_{Yc} are the per-phase impedances of the converters reactors, Z_Δ and Z_Y are the impedances of the secondaries, a is the transformer winding ratio and Z_P is the transformer primary winding impedance [26]–[29].

$$Z_{zhd_a} = (a^2(Z_{\Delta a} + z_\Delta/3)//a^2(Z_{Ya} + Z_Y)) + z_P/3$$
$$Z_{zhd_b} = (a^2(Z_{\Delta b} + z_\Delta/3)//a^2(Z_{Yb} + Z_Y)) + z_P/3 \tag{2}$$
$$Z_{zhd_c} = (a^2(Z_{\Delta c} + z_\Delta/3)//a^2(Z_{Yc} + Z_Y)) + z_P/3$$

C. ZHD GFM Grid-Connected Control Structure

The intrinsic voltage source nature of the ZHD converter implies that one does not need vector voltage control (VVC), vector current control (VCC) and filter capacitors for voltage source operation, as shown [26]–[29] as GFM converter in islanded microgrid applications only a open-loop control structure. In this way, in addition to the absence of capacitors avoiding resonance and feroresonance problems, the ZHD avoids problems with respect to control resonance and control interaction problems.

The ZHD open-loop voltage control is shown in Fig. 4 in dq frame and only a disturbance feedforward control is

979-8-3315-1612-3/25 $31.00 © 2025 IEEE

responsible for rejecting load variations, thus compensating the voltage reference commands. E_d and E_q are the voltages generated by the converter and V_d, V_q, I_d, and I_q are the measured voltages and currents, respectively. R_{zhd} and L_{zhd} represent the ZHD resistance and inductance of the impedance given by (2), and ω_{ref} is the reference frequency [26]–[29].

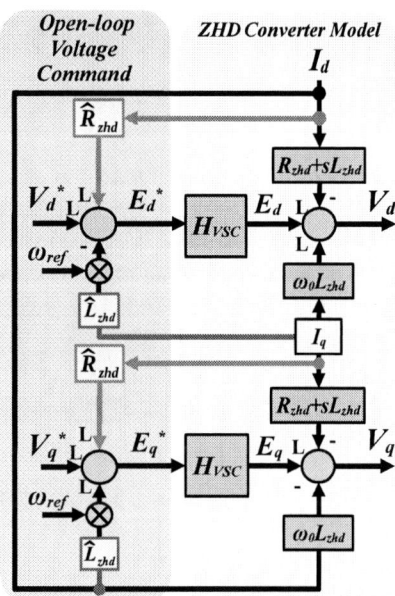

Fig. 4. Open-loop voltage control of ZHD GFM converter.

Unlike [25] that deals with the ZHD converter as a GFL converter and of [26]–[29] that deals with the ZHD as a GFM converter in islanded applications, this paper proposes the use of ZHD as a GFM grid-connected converter, where the synchronism with the grid is through the power synchronization control.

Fig. 5 shows the complete ZHD grid-connected GFM control. The measured currents in the dq reference frame, I_d and I_q, creates a corrective signal that is incorporated to the references voltages V_d^* and V_q^*. Based on a virtual impedance principle, voltage and phase compensation at the primary side is possible. $\hat{R}_{zhd}$ and $\hat{L}_{zhd}$ are the estimated resistance and inductance of ZHD converter, respectively. E_d^* and E_q^* symbolize voltage commands, and considering the transformer winding ratio a, such signals serve to determine the reference of modulation indexes $m_{\Delta,Y}$ and phase compensation $\delta_{\Delta,Y}$.

The modulation signals and the sum of phase compensation and the angle references θ_Y and θ_Δ, are compared in the LUTs generating the power switches command signals $\alpha_{1Y}, .., \alpha_{MY}$ and $\alpha_{1\Delta}, .., \alpha_{M\Delta}$ to the VSCs. The frequency ω_{ref} and angle reference θ_{ref} is generated by the active power control, responsible for by the grid synchronism, where P_{ref} is the power reference, P_o the measured output power filtered by the low-pass filter G_{LPF}, k_p is the gain of the power controller, ω_f the nominal frequency of the grid and θ_{synch} the angle of the presynchronization procedure to the grid connection.

This article focuses on the normal operation conditions of the ZHD grid-connected GFM converter. The operation under fault conditions will be our future work and the VI control will be improved in order to achieve the possibility of current limitation in the case of faults.

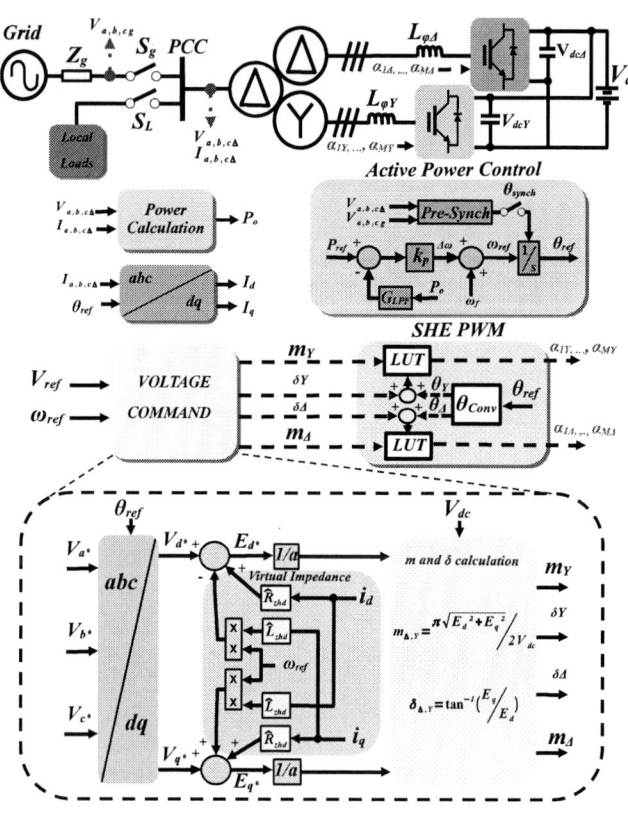

Fig. 5. Grid-connected GFM control of the ZHD converter.

III. SIMULATION RESULTS

The ZHD GFM converter in grid-connected operation was analyzed using the MATLAB/Simulink platform. The system and control parameters are described in Table II.

A. Black-Start, Synchronization and Grid-Connection

The simulation scenario considering black start, islanding operation, grid connection, and unintentional island operations are listed:

- t = 0 s → (I) → Black start of the R load;
- t = 0.5 s → (II) → Islanded Operation;
- t = 1.25 s → (III) → Grid-connection;
- t = 2 s → (IV) → Step reference of 200 kW;
- t = 2.5 s → (V) → Connection of RL and nonlinear loads;
- t = 3 s → (VI) → Unintentional island;

Figs. 6 and 7, considering weak grid scenarios (SCR = 1) and stiff grid scenarios (SCR = 5), respectively, show the results of active and reactive power in (a) and (b), voltage and current in the ZHD converter in (c) and the zoomed view of each time event in (d) - (i).

979-8-3315-1612-3/25 $31.00 © 2025 IEEE

TABLE II
SIMULATION PARAMETERS

ZHD CONVERTER PARAMETERS			
Parameters	Values	Parameters	Values
Rated Power	280 kVA	Frequency	60 Hz
Primary Voltage	13.8 kV	Δ Secondary Voltage	440 V
Winding Connections	Dd0y1	Y Secondary Voltage	440 V
$L_{\phi\Delta}$ and $L_{\phi Y}$ reactors	0.59/0.506 mH	DC link voltage	650 V
THREE-WINDING THREE-PHASE TRANSFORMER - Dd0y1			
Parameters	Values	Parameters	Values
R_m	526.7 kΩ	$R_{s\Delta}$	19.4 mΩ
L_m	534.5 H	$L_{s\Delta}$	3.33 μH
$R_{P\Delta}$	10.5 Ω	R_{sY}	16 mΩ
$L_{P\Delta}$	108.7 mH	L_{sY}	84.26 μH
LINEAR LOADS - R and RL			
Parameters	Values	Parameters	Values
Active Power	200 kW	Reactive Power	0 var
Active Power	48 kW	Reactive Power	36 kvar
NONLINEAR LOAD - THREE PHASE SIX PULSE RECTIFIER			
Parameters	Values	Parameters	Values
R_{AC}	152.35 Ω	L_{AC}	2.16 mH
C_{dc}	2.59 μH	R_L	34.56 kΩ
CONTROL PARAMETERS			
Parameters: SCR=1	Values	Parameters: SCR=5	Values
k_p	0.03 p.u.	k_p	0.01 p.u.
ω_c	10 Hz	ω_c	10 Hz

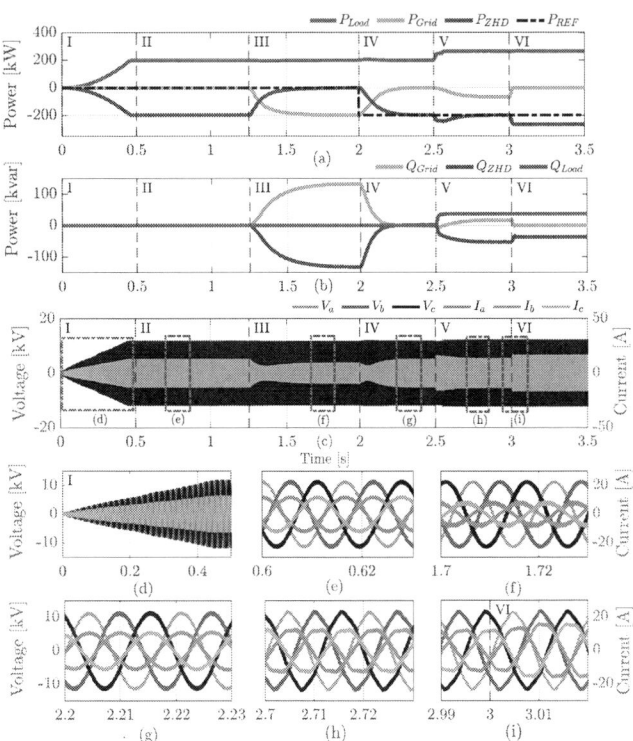

Fig. 6. SCR=1: (a) Active, and (b) Reactive power, (c) Voltage and current at the ZHD converter during different time intervals: (d) Event I, (e) Event II, (f) Event III, (g) Event IV, (h) Event V, and (i) Event VI.

In each case, the switch S_L is closed and Figs. 6 and 7 show the black-start capability of ZHD converter in (d) and operation in the island mode in (e) for time events I and II, where it is possible to perceive the converter providing a sinusoidal voltage waveform shape.

The connection to the grid occurs in time event III, where the switch to the grid S_g is closed through the presynchronization procedure that tracks the angle of the grid θ_{synch}. In both SCR conditions, after the connection to the grid, Figs. 6 and 7 (a) show the converter following the zero active power reference and in IV tracking the step power reference of 200 kW, even under load disturbance in V.

The waveforms zoomed view (f) and (g) also shows the converter capability to deliver sinusoidal waveforms in grid-connected mode, however, the delivered voltage waveforms are degraded during the operation with the nonlinear load in (h). Finally, in time event VI the grid switch S_g is open and the converter inherently assumes island operation as shown in (i).

The voltage regulation and the output phase for both grid scenarios are shown in Figs. 8 (a) and (b). The Fig. shows that in both grid strength cases the converter can regulate the voltage even in the presence of harmonics on the grid side. The voltage regulation stays well below the 5% range recommended by the standards [33], showing the converter capacity to regulate voltage in island mode, also as shown in [26]–[29], and in this work in grid-connected operation in both weak and stiff grid cases, only with an open-loop voltage command struture structure.

The voltage FFTs in Fig. 8 show the converter delivering

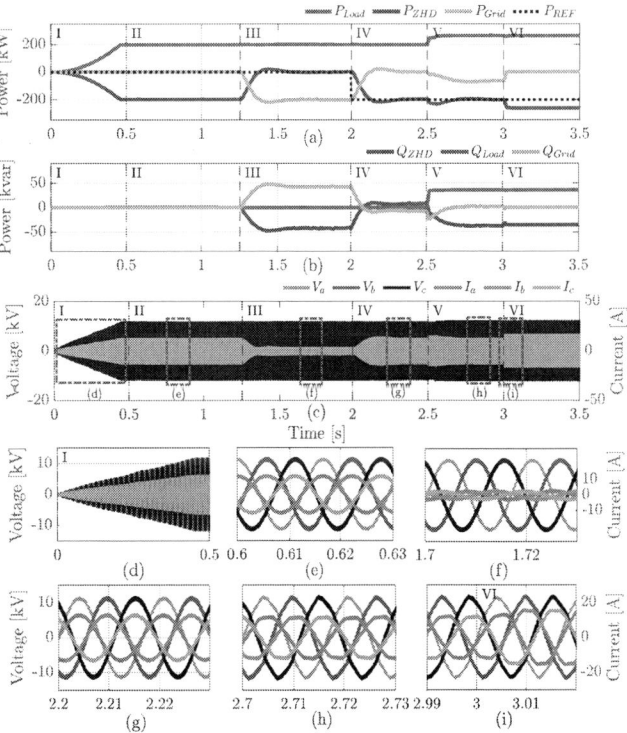

Fig. 7. SCR=5: (a) Active, and (b) Reactive power, (c) Voltage and current at the ZHD converter during distinct time events: (d) Event I, (e) Event II, (f) Event III, (g) Event IV, (h) Event V, and (i) Event VI.

979-8-3315-1612-3/25 $31.00 © 2025 IEEE

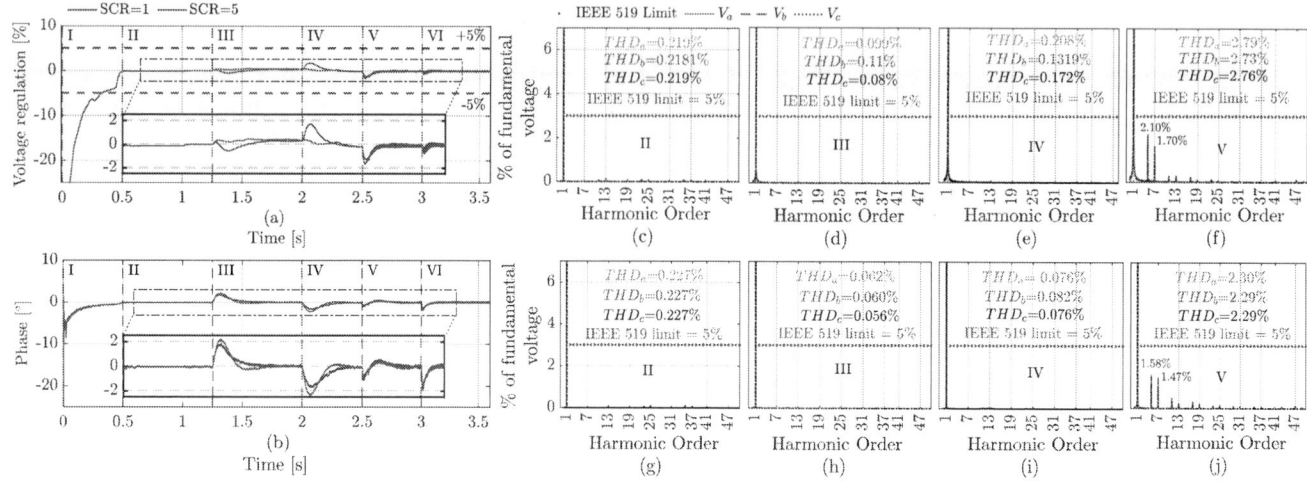

Fig. 8. (a) Voltage regulation and (b) Measured output voltage phase for SCR=1 and SCR=5 scenarios. Output voltage FFT of the ZHD converter for SCR=1 during time events: (c) II, (d) III, (e) IV, and (f) V; and for SCR=5 during time events: (g) II, (h) III, (i) IV, and (j) V.

voltage waveforms free of harmonics up to 50th order in Fig. 8 ((c)-(e)) for SCR=1, and in Fig. 8 ((g)-(i)) for SCR=5, only disturbed in the presence of harmonics at the output of the ZHD converter as shown in Fig. 8 (f) and (j).

B. Control Sensitivity Analysis

The control system sensitivity analysis was realized due to possible errors in the converter's estimated output impedance. For the same scenario considered in the last section, but considering the cases of $\pm20\%$ of estimating error in the virtual impedance parameters $\hat{R}_{zhd}$ and $\hat{L}_{zhd}$ of the open-loop voltage command, Fig. 9 shows that the converter maintains effective voltage and phase regulation, as shown in (a), within the limit $\pm5\%$ recommended by standards [33], with a maximum phase deviation of $2°$. These results show that even in the context of errors in the ZHD estimated output impedance, the effect in voltage regulation also in the grid-connected mode is acceptable.

C. Converter Secondaries Impedance Mismatches

Again, considering the same scenario simulation scenario of Section III-A, another sensitivity analysis was realized considering now possible mismatches between the secondaries reactor impedances. Taking into account the mismatch scenarios of + 20% of Z_Δ and + 20% of Z_Y, Figs. 10 (a) and (b) show the voltage regulation and the output phase voltage. It is possible to verify that even under these conditions the voltage regulation and the phase deviations remain at acceptable levels in both islanded and grid-connected modes.

Fig. 10 (c) and (d) for time event IV and Fig. 10 (e) and (f) for time event V, shows the FFT output voltage for +20% of Z_Δ and +20% of Z_Y considering an SCR = 1. Fig. 10 (g)-(j) shows the same scenarios considering an SCR=5. In these unbalance conditions the voltage FFTs (c), (d), (g), and (h) for time event IV show the presence of harmonics normally canceled in the transformer, due to the non-equal harmonic

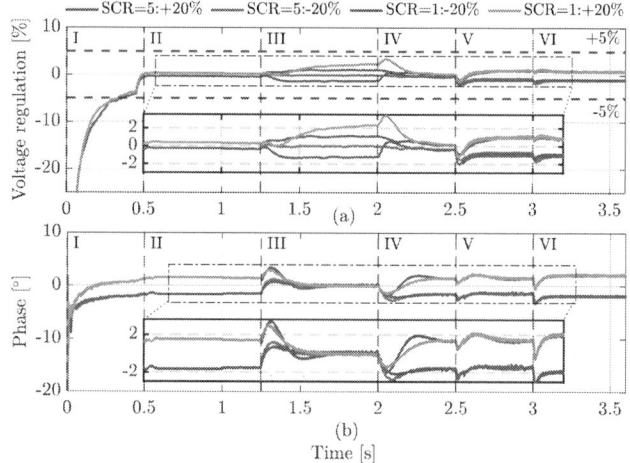

Fig. 9. (a) Voltage regulation and (b) output voltage phase under impedance estimation errors of ±20.

cancelation. However, the magnitude of these harmonics are practically negligible.

The unbalance effect was also practically negligible in time event V. Comparing Figs. 10 (e) and (g) with Figs. 8 (f) and 10 (i) and (j) with Fig. 8 (j), it is possible to verify that even with the unbalance between the reactors, the voltage harmonic deterioration is also negligible and in practice can be easily compensated with a virtual impedance.

IV. CONCLUSIONS

This paper presents the use Zero Harmonic Distortion (ZHD) converter as a promising solution for grid-forming applications in grid-connected systems. Differing from conventional converters, the ZHD dismiss capacitive sinusoidal filters, as it naturally delivers a sinusoidal voltage, resulting in lower costs, enhanced efficiency, and a more compact size. Based on a widely used industrial structure, the ZHD straightforward

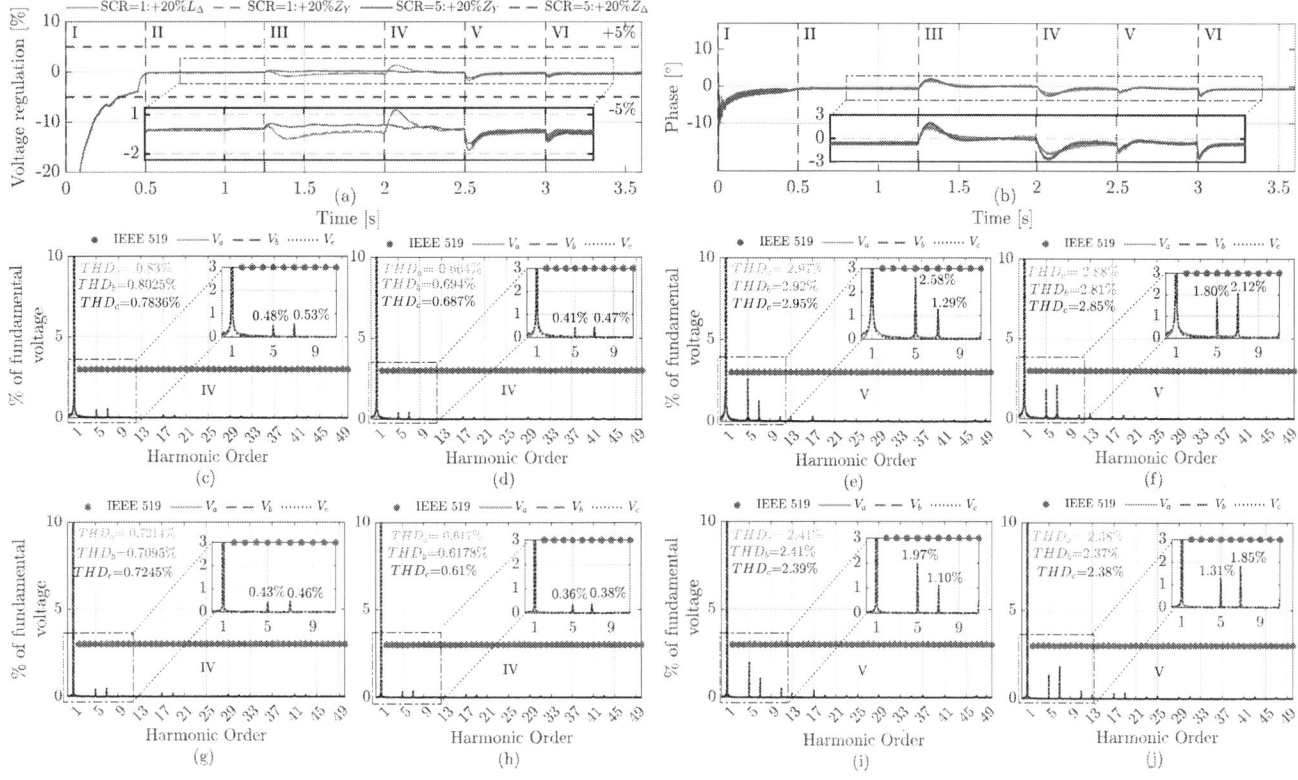

Fig. 10. (a) Voltage regulation, (b) output voltage phase, output voltage harmonic FFT considering: (c) +20% of Z_Δ, and (d) +20% of Z_Y for SCR=1 in time event IV, (e) +20% of Z_Δ, and (f) +20% of Z_Y for SCR=1 in time event V, (g) +20% of Z_Δ, and (h) +20% of Z_Y for SCR=5 in time event IV, (i) +20% of Z_Δ, and (j) +20% of Z_Y for SCR=5 in time event V.

open-loop control mitigates issues related to control interaction and resonance. Simulations results demonstrate the black-start capability, islanded operation, and the effectiveness of the converter to perform as a grid-connected grid-forming converter, in an open-loop structure even assuming, harmonics at the grid side, impedance estimating errors and unbalance between the secondaries reactors. The ZHD converter emerges as a viable and effective option for grid-forming grid-connected converter for battery energy storage systems (BESS) in medium-voltage applications.

REFERENCES

[1] J. Matevosyan, B. Badrzadeh, T. Prevost, E. Quitmann, D. Ramasubramanian, H. Urdal, S. Achilles, J. MacDowell, S. H. Huang, V. Vital, J. O'Sullivan, and R. Quint, "Grid-forming inverters: Are they the key for high renewable penetration?" *IEEE Power and Energy Magazine*, vol. 17, no. 6, pp. 89–98, 2019, doi:10.1109/MPE.2019.2933072.

[2] R. H. Lasseter, Z. Chen, and D. Pattabiraman, "Grid-forming inverters: A critical asset for the power grid," *IEEE Journal of Emerging and Selected Topics in Power Electronics*, vol. 8, no. 2, pp. 925–935, 2020, doi:10.1109/JESTPE.2019.2959271.

[3] M. Ndreko, S. Rüberg, and W. Winter, "Grid forming control scheme for power systems with up to 100% power electronic interfaced generation: a case study on great britain test system," *IET Renewable Power Generation*, vol. 14, no. 8, pp. 1268–1281, 2020, doi:10.1049/iet-rpg.2019.0700.

[4] Y. Li, Y. Gu, and T. C. Green, "Revisiting grid-forming and grid-following inverters: A duality theory," *IEEE Transactions on Power Systems*, vol. 37, no. 6, pp. 4541–4554, 2022, doi: 10.1109/TPWRS.2022.3151851.

[5] X. Quan, R. Yu, X. Zhao, Y. Lei, T. Chen, C. Li, and A. Q. Huang, "Photovoltaic synchronous generator: Architecture and control strategy for a grid-forming pv energy system," *IEEE Journal of Emerging and Selected Topics in Power Electronics*, vol. 8, no. 2, pp. 936–948, 2020, doi:10.1109/JESTPE.2019.2953178.

[6] S. Cherevatskiy, S. Zabihi, R. Korte, H. Klingenberg, S. Sproul, J. Glassmire, B. Buchholz, and H. Bitaraf, "A 30mw grid forming bess boosting reliability in south australia and providing market services on the national electricity market," in *18th Int'l Wind Integration Workshop*, Oct. 2019, pp. 1–5.

[7] G. Li, Y. Chen, A. Luo, and H. Wang, "An enhancing grid stiffness control strategy of statcom/bess for damping sub-synchronous resonance in wind farm connected to weak grid," *IEEE Transactions on Industrial Informatics*, vol. 16, no. 9, pp. 5835–5845, 2020, doi:10.1109/TII.2019.2960863.

[8] J. Fang, Y. Tang, H. Li, and X. Li, "A battery/ultracapacitor hybrid energy storage system for implementing the power management of virtual synchronous generators," *IEEE Transactions on Power Electronics*, vol. 33, no. 4, pp. 2820–2824, 2018, doi:10.1109/TPEL.2017.2759256.

[9] A. Rahmoun, A. Armstorfer, H. Biechl, and A. Rosin, "Mathematical modeling of a battery energy storage system in grid forming mode," in *2017 IEEE 58th International Scientific Conference on Power and Electrical Engineering of Riga Technical University (RTUCON)*, 2017, pp. 1–6, doi:10.1109/RTUCON.2017.8125625.

[10] F. Blaabjerg, Y. Yang, K. A. Kim, and J. Rodriguez, "Power electronics technology for large-scale renewable energy generation," *Proceedings of the IEEE*, vol. 111, no. 4, pp. 335–355, 2023, doi:10.1109/JPROC.2023.3253165.

[11] G. Wang, G. Konstantinou, C. D. Townsend, J. Pou, S. Vazquez, G. D. Demetriades, and V. G. Agelidis, "A review of power electronics for grid connection of utility-scale battery energy storage systems," *IEEE Transactions on Sustainable Energy*, vol. 7, no. 4, pp. 1778–1790, 2016, doi:10.1109/TSTE.2016.2586941.

[12] T. Engelbrecht, A. Isaacs, S. Kynev, J. Matevosyan, B. Niemann, A. J.

Owens, B. Singh, and A. Grondona, "Statcom technology evolution for tomorrow's grid: E-statcom, statcom with supercapacitor-based active power capability," *IEEE Power and Energy Magazine*, vol. 21, no. 2, pp. 30–39, 2023, doi:10.1109/MPE.2022.3230969.

[13] R. Juntunen, J. Korhonen, T. Musikka, L. Smirnova, O. Pyrhönen, and P. Silventoinen, "Identification of resonances in parallel connected grid inverters with lc- and lcl-filters," in *2015 IEEE Applied Power Electronics Conference and Exposition (APEC)*, 2015, pp. 2122–2127, doi:10.1109/APEC.2015.7104642.

[14] J. Enslin and P. Heskes, "Harmonic interaction between a large number of distributed power inverters and the distribution network," *IEEE Transactions on Power Electronics*, vol. 19, no. 6, pp. 1586–1593, 2004, doi:10.1109/TPEL.2004.836615.

[15] J. He, Y. W. Li, D. Bosnjak, and B. Harris, "Investigation and active damping of multiple resonances in a parallel-inverter-based microgrid," *IEEE Transactions on Power Electronics*, vol. 28, no. 1, pp. 234–246, 2013, doi:10.1109/TPEL.2012.2195032.

[16] P. Dang, T. Ellinger, and J. Petzoldt, "Dynamic interaction analysis of apf systems," *IEEE Transactions on Industrial Electronics*, vol. 61, no. 9, pp. 4467–4473, 2014, doi:10.1109/TIE.2013.2289896.

[17] J. L. Agorreta, M. Borrega, J. López, and L. Marroyo, "Modeling and control of n -paralleled grid-connected inverters with lcl filter coupled due to grid impedance in pv plants," *IEEE Transactions on Power Electronics*, vol. 26, no. 3, pp. 770–785, 2011, doi:10.1109/TPEL.2010.2095429.

[18] H. Hu, Q. Shi, Z. He, J. He, and S. Gao, "Potential harmonic resonance impacts of pv inverter filters on distribution systems," *IEEE Transactions on Sustainable Energy*, vol. 6, no. 1, pp. 151–161, 2015, doi:10.1109/TSTE.2014.2352931.

[19] H. Kocewiak, I. A. Aristi, B. Gustavsen, and A. Hołdyk, "Modelling of wind power plant transmission system for harmonic propagation and small-signal stability studies," *IET Renewable Power Generation*, vol. 13, no. 5, pp. 717–724, 2019, doi:10.1049/iet-rpg.2018.5077.

[20] J. Z. Zhou, H. Ding, S. Fan, Y. Zhang, and A. M. Gole, "Impact of short-circuit ratio and phase-locked-loop parameters on the small-signal behavior of a vsc-hvdc converter," *IEEE Transactions on Power Delivery*, vol. 29, no. 5, pp. 2287–2296, 2014, doi:10.1109/TPWRD.2014.2330518.

[21] F. Zhao, T. Zhu, Z. Li, and X. Wang, "Low-frequency resonances in grid-forming converters: Causes and damping control," *IEEE Transactions on Power Electronics*, vol. 39, no. 11, pp. 14 430–14 447, 2024, doi:10.1109/TPEL.2024.3424296.

[22] F. Zhao, X. Wang, Z. Zhou, L. Harnefors, J. R. Svensson, H. Kocewiak, and M. Peter Sidoroff Gryning, "Control interaction modeling and analysis of grid-forming battery energy storage system for offshore wind power plant," *IEEE Transactions on Power Systems*, vol. 37, no. 1, pp. 497–507, 2022, doi:10.1109/TPWRS.2021.3096850.

[23] T. Qoria, F. Gruson, F. Colas, X. Guillaud, M.-S. Debry, and T. Prevost, "Tuning of cascaded controllers for robust grid-forming voltage source converter," in *2018 Power Systems Computation Conference (PSCC)*, 2018, pp. 1–7, doi:10.23919/PSCC.2018.8443018.

[24] M. Dokus and A. Mertens, "On the coupling of power-related and inner inverter control loops of grid-forming converter systems," *IEEE Access*, vol. 9, pp. 16 173–16 192, 2021, doi:10.1109/ACCESS.2021.3053060.

[25] T. M. Parreiras, J. C. G. Justino, and B. d. J. Cardoso Filho, "The true unity power factor converter — a practical filterless solution for sinusoidal currents," in *2015 9th International Conference on Power Electronics and ECCE Asia (ICPE-ECCE Asia)*, 2015, pp. 2557–2565, doi:10.1109/ICPE.2015.7168134.

[26] G. V. Ramos, T. M. Parreiras, D. A. d. L. Brandão, S. M. Silva, and B. d. J. C. Filho, "A zero harmonic distortion grid-forming converter for islanded microgrids," in *2023 IEEE Industry Applications Society Annual Meeting (IAS)*, 2023, pp. 1–8, doi: 10.1109/IAS54024.2023.10406502.

[27] G. V. Ramos, T. M. Parreiras, and B. de Jesus Cardoso Filho, "Control performance assessment of a zero harmonic distortion grid-forming converter in islanded microgrids," in *2023 IEEE 8th Southern Power Electronics Conference and 17th Brazilian Power Electronics Conference (SPEC/COBEP)*, 2023, pp. 1–8, doi:10.1109/SPEC56436.2023.10407128.

[28] G. V. Ramos, D. A. d. L. Brandao, T. M. Parreiras, S. M. Silva, and B. J. C. Filho, "A zero harmonic distortion grid-forming converter for medium voltage islanded microgrids," *IEEE Transactions on Industry Applications*, pp. 1–12, 2024, doi:10.1109/TIA.2024.3462913.

[29] G. V. Ramos, T. M. Parreiras, and B. J. Cardoso Filho, "Control performance assessment of a zero harmonic distortion grid-forming converter for medium voltage islanded microgrids," *Eletrônica de Potência*, vol. 29, p. e202441, Oct. 2024, doi: 10.18618/REP.e202441. [Online]. Available: https://journal.sobraep.org.br/index.php/rep/article/view/961

[30] G. V. Ramos, T. M. Parreiras, and B. d. J. C. Filho, "The three-level zero harmonic distortion grid-forming converter: A practical filterless solution for sinusoidal voltages," in *2024 IEEE 21st International Power Electronics and Motion Control Conference (PEMC)*, 2024, pp. 1–6, doi:10.1109/PEMC61721.2024.10726351.

[31] G. V. Ramos, T. M. Parreiras, and B. D. J. C. Filho, "A three-level zero harmonic distortion grid-forming converter for medium voltage islanded microgrids," in *2024 Energy Conversion Congress & Expo Europe (ECCE Europe)*, 2024, pp. 1–7, doi:10.1109/ECCEEurope62508.2024.10751867.

[32] "IEEE Standard for Harmonic Control in Electric Power Systems," *IEEE Std 519-2022 (Revision of IEEE Std 519-2014)*, pp. 1–31, 2022, doi:10.1109/IEEESTD.2022.9848440.

[33] "IEEE Standard for Interconnection and Interoperability of distributed energy Resources with Associated Electric Power Systems Interfaces," *IEEE Std 1547-2018 (Revision of IEEE Std 1547-2003)*, pp. 1–138, 2018, doi:10.1109/IEEESTD.2018.8332112.

[34] "Unifi specifications for grid-forming inverter-based resources-version 2. unifi-2024-2-1. march 2024." *UNIFI Consortium*, 2024.

[35] B. Yao, X. Wei, Y. Zhang, P. Correia, R. Wu, S. Song, I. Trintis, H. Wang, and H. Wang, "Accelerated degradation testing and failure mechanism analysis of metallized film capacitors for ac filtering," *IEEE Transactions on Power Electronics*, vol. 39, no. 5, pp. 6256–6270, 2024, doi:10.1109/TPEL.2024.3360373.

[36] C. J. Mozina, "Impact of smart grid and green power generation on distribution systems," in *2012 IEEE PES Innovative Smart Grid Technologies (ISGT)*, 2012, pp. 1–13, doi:10.1109/ISGT.2012.6175625.

[37] M. Monadi, A. Luna, J. I. Candela, J. Rocabert, M. Fayezizadeh, and P. Rodriguez, "Analysis of ferroresonance effects in distribution networks with distributed source units," in *IECON 2013 - 39th Annual Conference of the IEEE Industrial Electronics Society*, 2013, pp. 1974–1979, doi:10.1109/IECON.2013.6699434.

[38] H. S. Patel and R. G. Hoft, "Generalized techniques of harmonic elimination and voltage control in thyristor inverters: Part i–harmonic elimination," *IEEE Transactions on Industry Applications*, vol. IA-9, no. 3, pp. 310–317, 1973, doi:10.1109/TIA.1973.349908.

[39] ——, "Generalized techniques of harmonic elimination and voltage control in thyristor inverters: Part ii — voltage control techniques," *IEEE Transactions on Industry Applications*, vol. IA-10, no. 5, pp. 666–673, 1974, doi:10.1109/TIA.1974.349239.

Single Cell Energy Router Justification for Three Phase Near Zero Energy Buildings

Hossein Nourollahi Hokmabad
Department of Electrical Power
Engineering and Mechatronics
Tallinn University of Technology
Tallinn, Estonia
hossein.nourollahi@taltech.ee

Tala Hemmati Shahsavar
Department of Electrical Power
Engineering and Mechatronics
Tallinn University of Technology
Tallinn, Estonia
talahemmati@gmail.com

Oleksandr Matiushkin
Department of Electrical Power
Engineering and Mechatronics
Tallinn University of Technology
Tallinn, Estonia
oleksandr.matiushkin @taltech.ee

Tanel Jalakas
Department of Electrical Power
Engineering and Mechatronics
Tallinn University of Technology
Tallinn, Estonia
tanel.jalakas@taltech.ee

Oleksandr Husev
Faculty of Electrical and Automation
Engineering
Gdansk University of Technology
Gdansk, Poland
oleksandr.husev@gmail.com

Juri Belikov
Department of Software Science
Tallinn University of Technology
Tallinn, Estonia
juri.belikov@taltech.ee

Abstract—**The rise of building-integrated photovoltaics and distributed electric vehicle charging has led to significant phase imbalances in utility grids, challenging service providers due to limited behind-the-meter visibility. This paper introduces a novel Single-Cell Three-Phase (SC-TP) Energy Router (ER) that accesses all phases and balances them without the complexities and costs associated with conventional three-phase systems. Our comparative analysis shows that the SC-TP ER reduces phase unbalancing by 16% and achieves cost savings, offering a viable alternative to three-phase solutions with 25% reduction in cost.**

Keywords—near zero energy building, renewable energies, grid congestion, phase imbalance, phase balancing, smart grids

I. INTRODUCTION

In recent years, the rapid integration of renewable energy sources, especially Photovoltaic (PV) systems, has significantly reshaped the topology of modern power distribution networks. This shift requires power systems to accommodate an increasing share of intermittent renewable generation while addressing new complexities in managing localized and distributed energy sources. Grid congestion and phase imbalances are intensifying due to the rise of small-scale renewable setups and Electric Vehicle (EV) charging stations, particularly in residential sectors. Additionally, phenomena like the duck curve are challenging the reliability of electricity grids [1], [2].

Phase imbalances reduce grid reliability and resiliency by decreasing power capacity and increasing losses in transformer secondaries [3], [4]. Expanding feeder capacity to address these issues would impose significant costs for Distribution System Operators (DSOs) [5], [6]. Consequently, alternative solutions which do not require large investments are preferred. One proposed solution involves clustering residential prosumers equipped with PV-Energy Storage (PV-ES) systems and high-demand buildings into evenly distributed three-phase microgrids. However, this approach faces challenges due to limited visibility of behind-the-meter installations [7], [8], [9].

Enhancing the intelligence of low-voltage equipment and integrating advanced power electronic technologies provide innovative and cost-effective solutions for phase balancing, potentially reducing or delaying the need for costly grid reinforcement [10], [11]. To this end, various solutions have been proposed. For example, some researchers have developed data-driven approaches for identifying and mitigating phase imbalances [12], [13], [14]. Others have introduced a two-stage network balancing strategy that employs phase-switch devices at terminal nodes and phase-switching soft open points across grid nodes [15]. Additionally, phase balancing challenges may also arise in highly Inverter-Based Resource (IBR) penetrated grids, where IBRs must form and stabilize the grid in the absence of a central generator. To address these scenarios, [16] proposes a solution in which inverters use droop control mechanisms to balance phases within the distribution network.

Demand-side phase balancing, proposed as a complementary solution to dispatch-side balancing in [17], involves regulating asymmetric loads. This approach integrates an optimization model that manages load imbalances from the demand side while coordinating with step voltage regulators and dispatching distributed generators to enhance overall phase stability and balance. However, with effective management and coordination, these high-demand loads could be leveraged as grid-balancing resources, offering support to DSOs [18]. Yet, there are currently very few market-ready options that autonomously address these issues without requiring end-user involvement. A promising alternative is presented in [19], which proposes an incentive-based scheme encouraging flexible consumers to assist in phase balancing. This method employs a centralized control algorithm that utilizes customers' installed converters for balancing phases at the substation. However, the paper assumes infrastructure readiness without addressing its practical implementation or scalability challenges.

While extensive research exists, practical solutions for grid phase balancing with PV, ES, and EVs often overlook the economic feasibility of the developed systems, making them less attractive to the market. This paper bridges the gap between practicality and economic efficiency by proposing a novel topology that detects unbalanced phases at terminal points and selectively directs PV or ES power to equalize

979-8-3315-1612-3/25 $31.00 © 2025 IEEE

interactions across each phase with the three-phase electricity grid. The proposed single-cell topology connects to all phases and includes an integrated non-isolated 350 V and isolated 48 - 350 V direct current (dc) bus alongside alternating current (ac) lines, enabling direct power supply to dc loads. This design reduces unnecessary energy conversion losses and minimizes harmonic injection into the ac grid.

II. PROPOSED SINGLE CELL ENERGY ROUTER CONCEPT

A. Structure of Single Cell ER:

Fig. 1 shows the proposed topology for a Single-Cell Three-Phase (SC-TP) Energy Router (ER) and its experimental realization. In this configuration, a dc bus can interact with all ac phases through an ER, but not simultaneously. Phase balance can be enhanced by detecting and reducing the power consumption of the phase with the highest demand. Since the power drawn from the three phases in a three-phase connected buildings often varies significantly, reducing the phase imbalance ratio with the proposed solution could lead to substantial economic benefits by eliminating the need for two additional converting cells. This advantage supports the rationale for adopting a single-cell approach rather than a conventional three-phase system. To achieve this goal, a Smart Energy Management Algorithm (SEMA) running beside essential low level controlling algorithm is needed to enable the ER to detect, mitigate and smooth the phase differences. Accordingly, the method proposed in [20] has been selected as the SEMA, and details regarding the energy flow optimization can be reviewed there. The technical characteristic of ER is collected in Table I. Further details related to the SC-SP ER topology is introduced in [21]. An alternative approach, which also offers dual purpose applications for both ac and dc grid is proposed in [22].

B. Test Study and Analyzed Scenarios:

To validate our hypothesis and assess the effectiveness of the proposed topology, we collected and analyzed annual load consumption data from a residential house located in Tallinn, Estonia, where the owner has an EV, and the dwelling is connected to the electricity grid via a three-phase terminal. The data were collected at a 3-second resolution; however, for simplicity and improved visualization, it was averaged and down sampled to a 1-hour resolution. Fig. 2, illustrates and compares a snapshot of initial phase imbalances and phase statuses after ideal phase balancing actions. In initial mode, it is evident that phase 3, labeled "L3", delivers less power compared to the other two phases, indicating that the EV charger should be connected to this phase. However, as observed, severe phase imbalance occurs during EV charging, regardless of which phase the EV is linked.

In this study, three ER topologies include: Single-Cell Single-Phase (SC-SP), Three-Cell Three-Phase (TC-TP), and SC-TP, are considered and their capability for phase balancing have been compared with each other. The topologies performances are first simulated and then validated with an experimental setup. The SC-TP ER is the upgraded version of the SC-SP ER. The 8 kWh Li-Ion battery pack is used as an ES. The methodology for selecting the optimal ES capacity is demonstrated in [23].

TABLE I.   ER TECHNICAL PARAMETERS

Parameters	Value
Rated power	15 kW
Grid and load side ac voltage (RMS)	230 V– 50 HZ
dc-link voltage	350 V
Nominal current of each phase	25 A
Switching frequency	65 kHz
Solar voltage input range	150 - 600 V
ES voltage input range	150 – 330 V
dc-link capacitor	3 mF

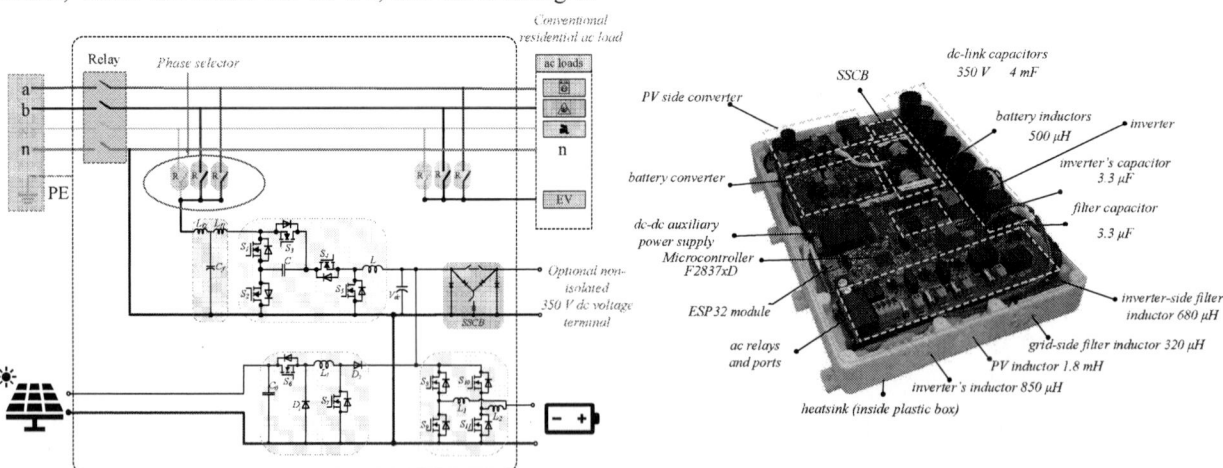

Fig. 1. Abstract view of the proposed single cell topology connected to three-phase terminal with EV charger, PV, and ES integration and its experimental realization.

979-8-3315-1612-3/25 $31.00 © 2025 IEEE   1623

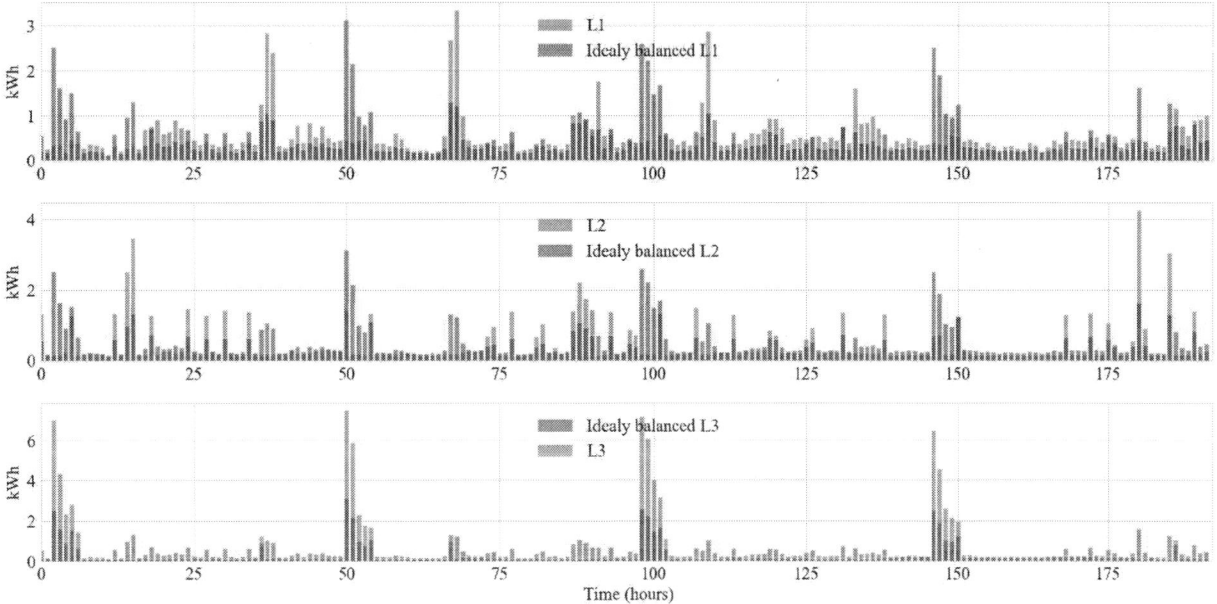

Fig. 2. Imbalanced and balanced phase power range comparison assuming ideal phase balancing possibilities.

C. Phase imbalance:

In this study, the phase imbalance ratio is defined as:

$$UB \% = \sum_{t=0}^{T} \frac{L_{unbl,t}}{L_{avg,t}} \times \frac{100}{N} \qquad (1)$$

$$L_{unbl,t} =$$

$$Max \left(\left| L_{avg,t} - L_{1,t} \right| + \left| L_{avg,t} - L_{2,t} \right| + \left| L_{avg,t} - L_{3,t} \right| \right) \qquad (2)$$

$$L_{avg,t} = (L_{1,t} + L_{2,t} + L_{3,t})/3 \qquad (3)$$

where $L_{1,t}, L_{2,t}, L_{3,t}$ is a power consumption for phase 1, 2, 3 at time t, respectively. $L_{avg,t}$ is the average power consumption from all phases in time t and $L_{unbl,t}$ is a maximum power deviation from $L_{avg,t}$ at time t. N is the total number of time steps in which the phase unbalance is calculated. Finally, $UB \%$ is the average phase imbalance. In order to minimize the UB% ratio an optimization problem is formulated for the TC-TP mode as:

$$Min \ f(x) = Min \left(\left| L''_{avg} - L''_1 \right| + \left| L''_{avg} - L''_2 \right| + \left| L''_{avg} - L''_3 \right| \right) \qquad (4)$$

subject to:

$$L''_1, L''_2, L''_3 \geq 0 \qquad (5)$$

$$PV_{l_1} + PV_{l_2} + PV_{l_3} \leq PV \qquad (6)$$

$$ES_{l_1} + ES_{l_2} + ES_{l_3} \leq ES \qquad (7)$$

$$L_i - PV_{l_i} - ES_{l_i} = L''_i, \ i = 1,2,3 \qquad (8)$$

where $L''_{avg} = (L''_1 + L''_2 + L''_3)/3$ represents the average demand and L''_y is each phase's demand from the electricity grid after allocating available renewable energy resources. Here, y denotes the phase number. PV_{l_y} and ES_{l_y} represents the allocated energy from solar energy production and energy storage, respectively. PV and ES denote the total accessible energy from renewable setups and batteries, respectively. In Eq. (4), x represents optimization factors, $x = [PV_{l_1}, PV_{l_2}, PV_{l_3}, ES_{l_1}, ES_{l_2}, ES_{l_3}]$. Finally, the ES SoC level should be updated as:

$$ES_{t+1} = \min(ES_{max}, ES - ES_{l_1} - ES_{l_2} - ES_{l_3} + PV') \qquad (9)$$

where, ES_{t+1} is the ES, State of Charge (SoC) (%) for the next time step, and ES_{max} denotes the maximum energy capacity of ES, and PV'_t denotes the remaining generated solar energy after demand responding. It should be mentioned that, in all equations, we are considering energy instead of power, since this assumption simplifies the equations.

In SC-TP mode, the optimization problem must be reformulated. Algorithm I outlines the optimization process for SC-TP mode, where the phase with the maximum load demand is identified at each time step. The ER then links this phase to the DC link and injects renewable or stored energy to meet the demand, in the selected phase. In SC-SP mode, since phase exchange is not possible and the ER remains permanently connected to phase "L1," the optimization problem simplifies into a reduced version of Algorithm I, with the index consistently set to 1. The optimization problems are solved using the Pyomo interface with the IPOPT solver, due to the nonlinear nature of the formulation. All code related to optimization process is implemented in Python 3.11.

Algorithm 1 *SC-TP Energy Optimization Algorithm*

```
1:   Input:
     Annual load profiles (⃗l₁ , ⃗l₂, ⃗l₃), Annual PV profile (PV).
2:   for i = 1 to n do:
3:       l_index, index = max (l_{1,i} , l_{2,i}, l_{3,i} )
4:       Index = [1, 2, 3]
5:       x, y = Index.drop(index)
6:       Min f(x) = Min (|l″_{avg,i} − l″_{index,i}| + |l″_{avg,i} − l″_{x,i}| +
                         |l″_{avg,i} − l″_{y,i}|)
7:       Subject to:
8:       l″_{avg,i} = avg( l″_{index,i}, l″_{x,i}, l″_{y,i})
10:      l″_{index,i} ≥ 0
11:      l″_{index,i} = l_{index,i} − PV″_i − ES″_i
12:      PV″_i ≤ PV_i
13:      ES″_i ≤ ES_i
14:      ES_{i+1} = min(ES_max , ES_i − ES″_i + PV_i − PV″_i)
22:  end for
23:  Return: optimal values PV″_i, ES″_i
```

III. CASE STUDIES EVALUATION AND DISCUSSION

In this section, the outcomes of the proposed methodology are discussed from multiple perspectives, including technical performance, operational feasibility, and economic impact. The phase balancing potential of all possible topologies is compared, followed by a discussion of the experimental realization of the SC-TP ER. Finally, a cost comparison is provided to highlight the impact of the proposed topology on cost reductions and its potential for practical adoption in building with three-phase grid-connected systems.

A. Phase Balancing Outcomes:

Fig. 3 presents a normalized radar chart comparison across all topologies, measuring variables such as UB%, PV self-consumption ratio, average SoC (%) of ES, capital costs and average load per phase during an experimental test period. Each category is normalized to its maximum observed value. For example, the average load per phase is highest when the house does not utilize any local renewable energy sources.

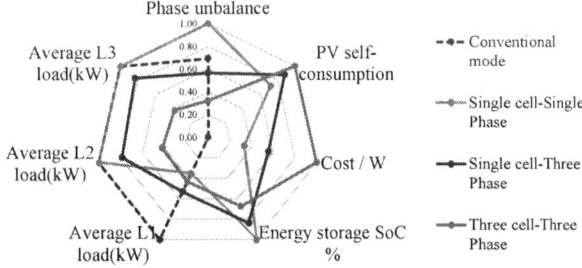

Fig. 3. Normalized radar chart comparison for all topologies.

In conventional houses, the phase balance ratio is better than in scenarios where the PV setup is connected to only one phase. The greatest phase imbalance is observed in the SC-SP topology, indicating that the presence of PV in this configuration reduces the phase balance ratio. This is because, in such operational modes, the generated renewable energy is only injected into one phase, significantly lowering the demand on that phase alone. Meanwhile, the other phases continue to draw the same demand from the grid, unable to benefit from PV production or stored energy. Considering that the SC-SP topology is a commonly used option in residential

PV setups due to its affordability [24], this underscores the necessity of proposing novel solutions for improving phase balance ratio.

The TC-TP topology outperforms other configurations in all categories, including PV self-consumption ratio, phase balance ratios, and average ES's SoC (%) levels. However, its capital cost ratio is 1.35 times higher than that of the SC-TP topology, posing a further barrier to adoption in residential buildings, where economic feasibility is a critical factor for end-users. Furthermore, the TC-TP topology can reduce grid interaction across all phases at a similar rate, whereas the SC-TP topology does not achieve the same performance, with the management algorithm primarily focused on reducing interactions on phase L1. This is because, for most of the time, the demand on L1 is higher than on the other two phases, prompting the ER to link PV and ES resources to this phase. It is important to note that phase switching frequency is constrained by numerous factors; in this study, it is set to 15-minute intervals. Improving phase-switching algorithms and increasing switching frequency will be explored in future research.

Fig. 4 compares the performance of SC-TP, SC-SP, and TC-TP topologies over a 10-day continuous operation period during spring, when PV power generation is at its moderate level. Solar power production and load demand are identical across all phases in each scenario, as shown in Fig. 4a. Additionally, Fig. 4b compares each scenario's interaction with the electricity grid. The SC-SP topology exhibits the highest energy exchange and, consequently, the lowest self-sufficiency ratio, whereas the TC-TP topology achieves the lowest energy exchange and the highest self-sufficiency ratio. Notably, phase balance is prioritized as the optimization objective across all scenarios rather than maximizing self-sufficiency, which leads to slightly lower self-sufficiency ratios than if the optimization had focused solely on maximizing self-sufficiency.

Additionally, a comparison of the ES SoC under both single-phase and multi-phase operating conditions is presented in Fig. 4c. The results indicate that when the ER distributes PV-generated power and stored energy across all phases, battery charge and discharge cycles become more frequent, which diminishes ES longevity by increasing stress on the battery cells [25]. Notably, when the ER and ES operate in single-phase mode, the SEMA has fewer opportunities to maximize the usage of locally generated energy and the self-consumption ratio. This leads to an increased need for PV curtailment or grid injection when the ES is fully charged, ultimately resulting in lower ES utilization. However, this approach helps maintain a higher battery health ratio.

A comparison of various operational modes indicates that the TC-TP topology is the least effective at protecting battery cells from rapid degradation, while the SC-SP mode achieves the highest ES health status during operation compared to other modes. Under high solar energy availability, the performance of the SC-TP mode is comparable to, though slightly better than, the TC-TP mode in terms of ES longevity. However, when solar energy is limited, the SC-TP mode outperforms TC-TP by using the ES less frequently.

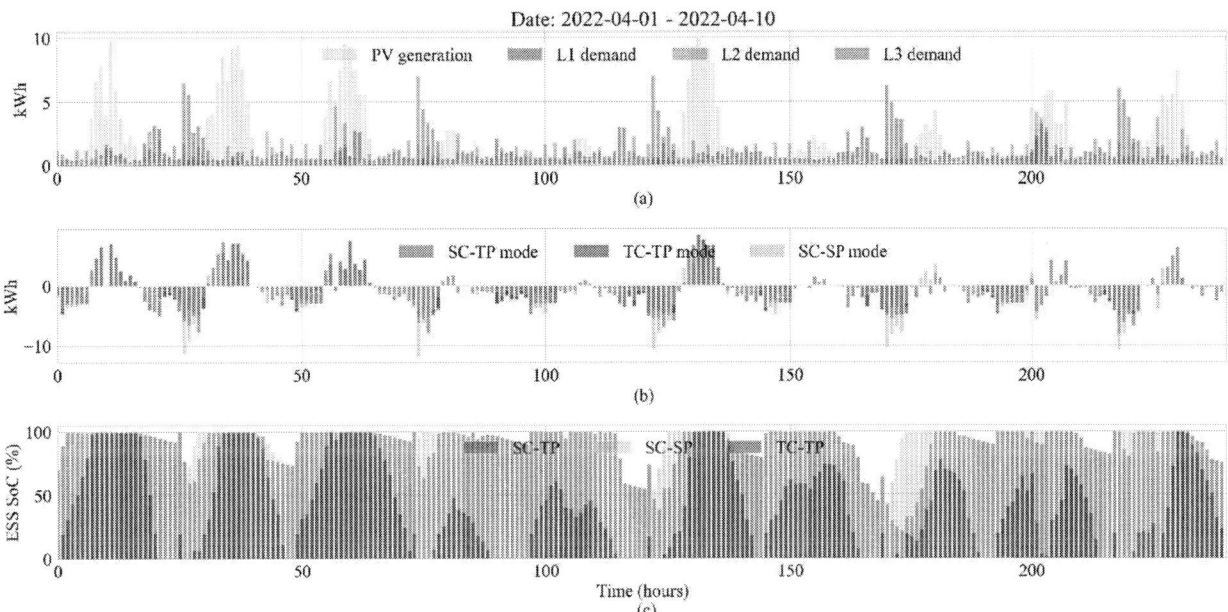

Fig. 4. Comparison of the topologies performances during first 10 days in May-2022. (a) Load demand and PV generation. (b) Amount of energy exchange with the electricity grid (negative values show imported energy and positive values show exported energy to the electricity grid). (c) ES SoC (%) level during systems' operation.

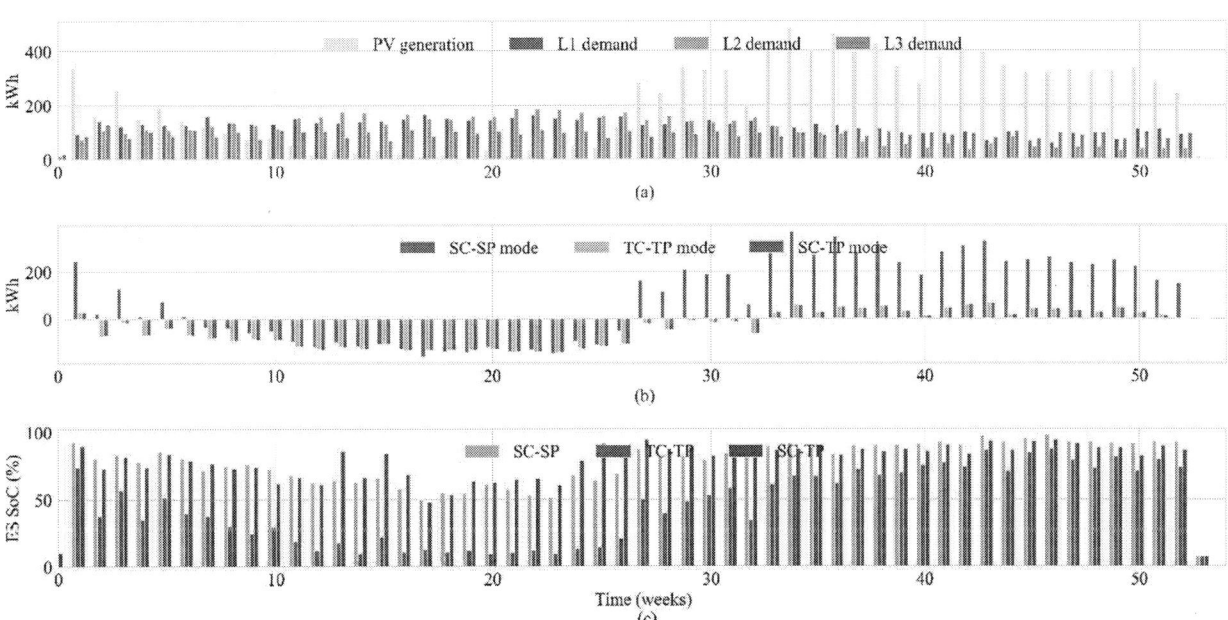

Fig. 5. Comparison of the topologies based on weekly aggregated performance over a year: (a) Total load demand and PV generation for each week; (b) Total weekly energy exchange with the electricity grid; (c) Average weekly SoC (%) ratio of the ES system.

While this improved longevity is beneficial, it may not fully meet end-user expectations, as reduced ES operation increases reliance on the electricity grid to meet demand. Therefore, it is essential for the energy management unit to account for ES degradation costs and prioritize resource allocation based on electricity tariffs.

Fig. 5 presents a comparison of the performance of the investigated operational modes over a year. From a broader perspective, the differences between topologies become clearer. For example, during winter—particularly between weeks 10 and 20—when solar generation is exceptionally low, grid interactions are quite similar across modes. However, during peak solar generation—particularly between weeks 30 and 50—the SC-SP topology shows significantly higher grid interactions, leading to increased grid power flow and potential congestion. This outcome further highlights the limitations of the SC-SP topology in maximizing PV self-consumption, as most of the on-site generated solar power is injected back into the grid. However, two other topologies have been able to utilize locally available renewable sources, to minimize their energy exchanges with the electricity network. The differences between topologies interaction with the electricity network is demonstrated in Fig. 5b.

Fig. 5c compares the average SoC (%) levels across topologies. It is evident that during periods of low solar generation, TC-TP utilizes ES more frequently than the other two topologies. However, when solar generation increases, the SoC (%) levels between topologies become more similar, though TC-TP still charges and discharges the batteries more often. Interestingly, the performance of SC-SP and SC-TP topologies is almost identical, with both benefiting from ES in an equivalent manner. Finally, Fig. 6. shows number of times the ER linked renewable resources to each phase in SC-TP topology.

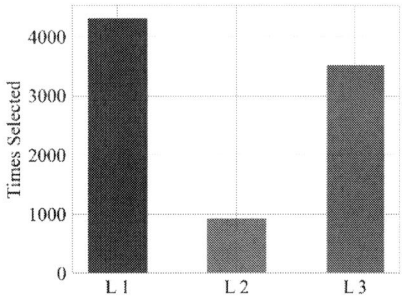

Fig. 6. A number of times, each phase is connected through ER to the local renewable energy sources.

IV. Cost Benefits Analysis

Fig. 7 presents a comparison of the cost distribution between different ER components in case of SC-TP and TC-TP.

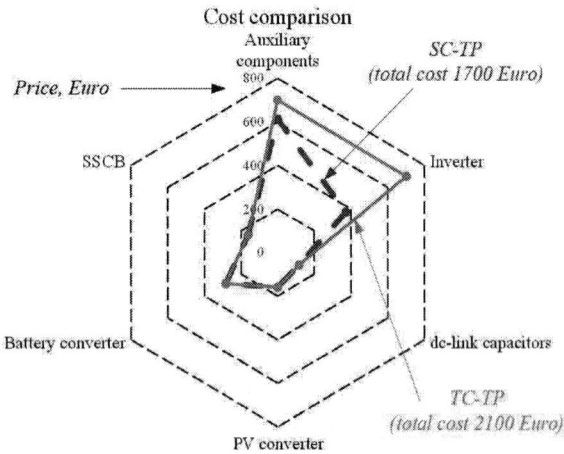

Fig. 7. Cost comparison diagram which shows the cost distribution between different ER components in case of SC-TP and TC-TP.

The cost analysis is based on the retail price of the components used for a single prototype. It includes the cost of all semiconductors and passive components, such as heatsinks and inductors. However, the cost of the printed circuit board and the enclosure used for the demonstrator are not included in the calculations, as these costs are not representative and strongly depend on scaling.

The protype of SC-TP ER is shown in Fig. 1., and the total cost of the considered components is approximately 1700 euros. The diagram shows that the most expensive part of the prototype relates to the auxiliary (common) components, such as power supply circuits, heatsinks, and relays.

The prototype of the TC-TP ER was not assembled or used in real tests. However, its cost was evaluated based on a bill of materials collected to design the prototype. The total power of the three-phase inverter, as well as the power of the single-phase inverter, was the same, following the same concept with a common ground approach. The overall cost of the components included in the analysis is around 2100 euros, which is 25% higher than that of the SC-TP ER. This cost increase is attributed to the higher number of components required for redundant circuits to connect and link PV and ESS to all three phases.

It should be noted that the absolute values presented in this work cannot be directly used for primary cost estimation, as they strongly depend on scaling, the supply chain during production, and auxiliary circuit optimization. However, the relative comparison is reliable and can be used for cost analysis.

V. Conclusions

The proposed SC-TP topology improves PV self-consumption and the phase unbalance ratio by 23% and 16% respectively, compared to the SC-SP solution. Moreover, it offers a significant cost saving of 25% compared to the TC-TP system. This achievement underscores the potential benefits of SC-TP systems as a promising solution to replace SC-SP topologies. Since the SC-TP topology offers better phase balancing, it effectively mitigates the negative effects of distributed behind-the-meter PV setups, improves grid reliability and resiliency, and better distributes the load across each phase. This highlights its value in enhancing grid stability and reducing operational challenges for utility providers.

Acknowledgments

This work was supported by the Estonian research council grants PRG675, EAG234 and PRG1463.

References

[1] I. Calero, C. A. Cañizares, K. Bhattacharya and R. Baldick, "Duck-Curve Mitigation in Power Grids With High Penetration of PV Generation," in *IEEE Transactions on Smart Grid*, vol. 13, no. 1, pp. 314-329, 2022.

[2] Hokmabad, H.N., Husev, O., Belikov, J., Vinnikov, D. and Petlenkov, E., "Energy Storage and Forecasting Error Impact Analysis In Photovoltaic Equipped Residential Nano-Grids, " in *IEEE 17th*

International Conference on Compatibility, Power Electronics and Power Engineering (CPE-POWERENG), pp. 1-6, 2023.

[3] D. Q. Hung and Y. Mishra, "Impacts of single-phase PV injection on voltage quality in 3-phase 4-wire distribution systems, " in *2018 IEEE Power & Energy Society General Meeting (PESGM)*, 2018, pp. 1–5.

[4] E. A. S. Ducoin, Y. Gu, B. Chaudhuri, and T. C. Green, "Analytical Design of Contributions of Grid-Forming and Grid-Following Inverters to Frequency Stability", in *IEEE Transactions on Power Systems*, vol. 39, no. 5, pp. 6345–6358, 2024.

[5] K. Ma, R. Li, I. Hernando-Gil and F. Li, "Quantification of Additional Reinforcement Cost From Severe Three-Phase Imbalance," in *IEEE Transactions on Power Systems*, vol. 32, no. 5, pp. 4143-4144, 2017.

[6] A. A. Mohamed, C. Sabillon, A. Golriz, M. Lavorato, M. J. Rider, and B. Venkatesh, "Capacity Market for Distribution System Operator – With Reliability Transactions – Considering Critical Loads and Microgrids, " *IEEE Transactions on Power Delivery*, vol. 38, no. 2, pp. 902–916, 2023.

[7] S. A. Raza and J. Jiang, "Mathematical Foundations for Balancing Single-Phase Residential Microgrids Connected to a Three-Phase Distribution System," in *IEEE Access*, vol. 10, pp. 5292-5303, 2022.

[8] S. Lim and J. -W. Park, "Estimation of Behind-the-Meter Power Loss of Renewables With EMS Data," in *IEEE Transactions on Power Systems*, vol. 38, no. 3, pp. 2974-2977, 2023.

[9] F. Bu, R. Cheng, and Z. Wang, "A two-layer approach for estimating behind-the-meter PV generation using smart meter data, " *IEEE Transactions on Power Systems*, vol. 38, no. 1, pp. 885–896, 2022.

[10] N. Kouvelas and R. V. Prasad, "Efficient Allocation of Harvested Energy at the Edge by Building a Tangible Micro-Grid—The Texas Case," in *IEEE Transactions on Green Communications and Networking*, vol. 5, no. 1, pp. 94-105, 2021.

[11] J. Huang, X. Wang, Y. Wang, C. Shao, G. Chen, and P. Wang, "A Game-Theoretic Approach for Electric Vehicle Aggregators Participating in Phase Balancing Considering Network Topology', "*IEEE Transactions on Smart Grid*, vol. 15, no. 1, pp. 743–756, 2024.

[12] S. Liu *et al.*, "Practical Method for Mitigating Three-Phase Unbalance Based on Data-Driven User Phase Identification," in *IEEE Transactions on Power Systems*, vol. 35, no. 2, pp. 1653-1656, 2020.

[13] P. Li, W. Wu, X. Wang, and B. Xu, "A Data-Driven Linear Optimal Power Flow Model for Distribution Networks, " *IEEE Transactions on Power Systems*, vol. 38, no. 1, pp. 956–959, 2023.

[14] D. Yousri, H. E. Z. Farag, H. Zeineldin, and E. El-Saadany, "Optimized Unsymmetrical Per-Phase Droop for Soft Line Switching of Reconfigurable Unbalanced Inverter-Based Islanded Microgrid, " *IEEE Transactions on Power Systems*, vol. 39, no. 2, pp. 3851–3866, 2024.

[15] X. Cui, G. Ruan, F. Vallée, J. -F. Toubeau and Y. Wang, "A Two-Level Coordination Strategy for Distribution Network Balancing," in *IEEE Transactions on Smart Grid*, vol. 15, no. 1, pp. 529-544, 2024.

[16] M. Lu, W. Cai, S. Dhople, and B. Johnson, "Large-signal stability of phase-balanced equilibria in single-phase grid-forming inverter systems, " *IEEE Trans. on Power Electronics*, vol. 39, no. 3, pp. 3623–3636, 2024.

[17] L. Fu, W. Wang, Z. Y. Dong and Y. Li, "Optimal Reconfiguration for Active Distribution Networks Incorporating a Phase Demand Balancing Model," *IEEE Transactions on Power Systems*, vol. 39, no. 5, pp. 6183-6195, 2024.

[18] J. Huang, X. Wang, Y. Wang, C. Shao, G. Chen and P. Wang, "A Game-Theoretic Approach for Electric Vehicle Aggregators Participating in Phase Balancing Considering Network Topology," in *IEEE Transactions on Smart Grid*, vol. 15, no. 1, pp. 743-756, 2024.

[19] L. Fang, K. Ma, F. Li and F. Xue, "Novel Incentive Scheme to Motivate Flexible Customers for Phase Balancing," in *CSEE Journal of Power and Energy Systems*, vol. 8, no. 4, pp. 1048-1059, 2022.

[20] Hokmabad HN, Husev O, Kurnitski J, Belikov J. "Optimizing size and economic feasibility assessment of photovoltaic and energy storage setup in residential applications," *Sustainable Energy. Grids and Networks*, 38:101385, 2024.

[21] S. Rahimpour, "Common-Ground Energy Router Structure with Enhanced Reliability and Protection," Ph.D. dissertation, *Tallinn University of Technology*, 2024.

[22] Shahsavar, T.H., Kurdkandi, N.V., Husev, O., Babaei, E., Sabahi, M., Khoshkbar-Sadigh, A. and Vinnikov, D., "A New Flying Capacitor-Based Buck–Boost Converter for Dual-Purpose Applications. " *IEEE Journal of Emerging and Selected Topics in Industrial Electronics*, 4(2), pp.447-459, 2023.

[23] H. N. Hokmabad, N. Shabir, V. Astapov, E. Petlenkov, O. Husev and J. Belikov, "Feasibility Study of a DC House Connected to a Conventional AC Distribution Network," 2024 IEEE 18th International Conference on Compatibility, Power Electronics and Power Engineering (CPE-POWERENG), Gdynia, Poland, pp. 1-6, 2024.

[24] Boscaino, V., Ditta, V., Marsala, G., Panzavecchia, N., Tine, G., Cosentino, V., Cataliotti, A. and Di Cara, D., "Grid-connected photovoltaic inverters: Grid codes, topologies and control techniques. " *Renewable and Sustainable Energy Reviews*, 189, p.113903, 2024.

[25] M. Amini, M. H. Nazari and S. H. Hosseinian, "Optimal Scheduling and Cost-Benefit Analysis of Lithium-Ion Batteries Based on Battery State of Health," *IEEE Access*, vol. 11, pp. 1359-1371, 2023.

A Multi-UAV Charging Station Enabling Free Landing by Grid Pattern Transmitter

Jungho Kim
School of Electrical and Computer Engineering
Ulsan National Institute of Science and Technology
Ulsan, Korea
swisshj@unist.ac.kr

Hyunkyeong Jo
School of Electrical and Computer Engineering
Ulsan National Institute of Science and Technology
Ulsan, Korea
johk@unist.ac.kr

Seoktae Seo
School of Electrical and Computer Engineering
Ulsan National Institute of Science and Technology
Ulsan, Korea
seoktae3333@gmail.com

Bonyoung Lee
School of Electrical and Computer Engineering
Ulsan National Institute of Science and Technology
Ulsan, Korea
bonyoung2@naver.com

Hyungki Min
Program in Information & Communication Technology Convergence
Ulsan National Institute of Science and Technology
Ulsan, Korea
kaimin@unist.ac.kr

Franklin Bien
School of Electrical and Computer Engineering
Ulsan National Institute of Science and Technology
Ulsan, Korea
bien@unist.ac.kr

Abstract—**This paper proposed a novel landing platform with wireless power transfer (WPT) to charge multiple unmanned aerial vehicles (UAVs) without restrictions on landing locations or alignment. A transmitter is designed using wires to charge a wide area. To overcome the disadvantages of system made by wires, a grid pattern transmitter and double power line are proposed. These structures provide multiple charging with free positioning of receivers and excite the transmitter lines in wide area properly. The optimized system is composed of 0.15m square cells, providing a 0.9 m by 0.9 m charging area. By experiment, it can efficiently charge four 0.1 m square receivers located anywhere in transmitter. The overall efficiency of four receivers is 37.5%. Also, the experimental results verify that it is unnecessary to align the receiver within the cell, as it has a 2.62% error due to rotation and a 2.94 % error due to position within the cell.**

Keywords—*wireless power transfer (WPT), UAV, grid pattern transmitter, double power line, multiple receivers*

I. INTRODUCTION

The advancement of wireless power transfer (WPT) technology has applied in various fields, such as consumer electronics, electric vehicles (EVs), and unmanned aerial vehicles (UAVs). Although, UAV are described as 'unmanned', it faces some challenges in achieving full automation. Particularly, supplying power issue is one of the challenges with automation, as manpower is needed to charge UAVs by wire

connection or replace the battery nowadays. WPT suggests a better solution for this challenge, and it can make a step forward to better UAV technology [1]. By this time, the general approaches to WPT for UAVs are charging in mid-air with tracking [2-5] or using station to charge placed along the flight path [6-9]. Due to the recent trend of operating large numbers of UAVs, station based WPT becomes important. Previous research of charging station for UAV is mainly focused on two keywords: free positioning, and wide area.

Landing perfectly on charging platform is hard, so landing that does not require accurate positioning and alignment with WPT system will help automation. In [5], helix unequal Rx coil shaped was proposed with optimized Tx coil for reducing tolerance of misalignment. With 100 W class UAV, experimental results showed 32.89% efficiency for minimum and 56.23% maximum efficiency.

UAVs are much larger than consumer electronics and usually operate in a large number. Thus, WPT for UAVs needs large area for charging many numbers instantly. Zeng et al. [6]

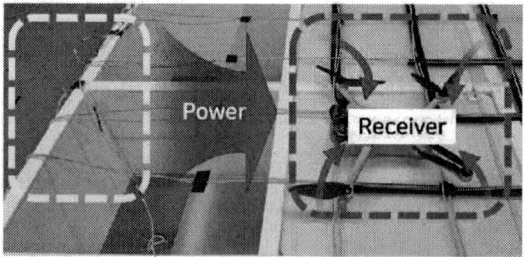

Fig 1. Principles of wireless power transfer platform for UAV. Power is delivered from double power line to grid pattern transmitter.

This work was supported by Innovative Human Resource Development for Local Intellectualization program through the Institute of Information & Communications Technology Planning & Evaluation (IITP) grant funded by the Korea government (MSIT)(IITO-2024-RS-2022-00156361, 50%)

This work was supported by Institute of Information & communications Technology Planning & Evaluation (IITP) grant funded by the Korea government (MSIT) (2022-0-00720, Development of W-band compact, high-efficiency, novel RF/power components for next-generation high-speed low-orbit satellite communications, 50%)

designed WPT system for charging multi-drones to make uniform magnetic field. Power transfer efficiency of this paper showed 85.54% and the load can receive 20 W. Main idea of paper was making uniform magnetic field for large area to charge more UAVs.

As aforementioned, misalignment tolerance and large area WPT are important characteristics for solving issues of UAVs. To satisfy these requirements, new approaches are suggested. To control wide area charging zone, traditional coil shape transmitter has limitation of covering large area. Also, it becomes hard and complex to control transmitter coils for wide area. Therefore, a new kind of system is suggested with easier to expand and simpler to control. New system replaces coils to straight wire. Additionally, this system can remove blind spots so that it can deliver power to proper position, which are covered by overlapping of coils traditionally. However, efficiency of the straight wire system is lower than coil system. Hence, making mesh shaped system with crossing two straight wires to reinforce the power delivery. Consequentially, the proposed system looks like grid pattern system when viewed on a plane. Also, the grid pattern transmitter is surrounded by two power lines so that current flows as intended by cross connecting the wires to power lines.

In this digest, novel structure of WPT system which offers advantages in misalignment tolerance and large area is proposed. To elaborate this idea, digest is organized as follows: the second section presents the theoretical analysis of the proposed system. The third section will describe simulation and experiment results. Lastly, conclusions and future work will be followed.

II. THEORETICAL ANALYSIS

A. Overall System

Fig. 2 shows the proposed system with a grid pattern transmitter and a double power line. Size of transmitter and receiver is determined by its application. Litz wire, which is composed of bundles of wire, was used for both transmitter and receiver to reduce the AC resistance caused by the skin effect. Orthogonally arranged parallel straight wires were additionally placed, so constructive interference for the single square cell is created. The receiver coil was designed to wrap around the ferrite to increase a permeability of the system.

When a UAV lands on the platform, platform automatically charges the UAV by exciting surrounded lines of UAV. The

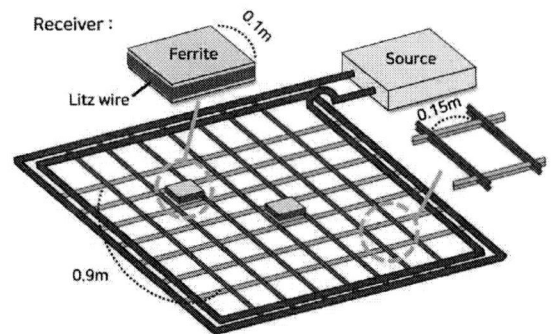

Fig 2. Proposed structure of the wireless power transfer system for UAV.

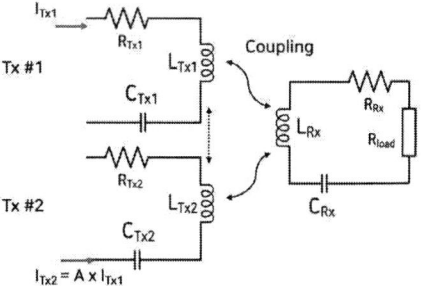

Fig 3. A general circuit model of the proposed wireless power transfer system. Parallel two wires and receiver coupled to each other.

direction of current of these lines is controlled to generate constructive magnetic flux, and that flux goes through receiver coil. Thus, the receiver can capture more magnetic flux from transmitter.

B. Efficiency Estimation

Analyzing the system's efficiency from a perspective of the circuit shows the important variables of efficiency. The receiver is affected by its surrounding four wires, but orthogonal wires are mutually decoupled. Thus, the system should be considered as two pairs of two parallel transmitter wires and single receiver as Fig. 3 described. Then, the efficiency can be written as [13]

$$\eta = \frac{|I_{Rx}|^2 R_{Rx}}{R_{Tx1}|I_{Tx}|^2 + R_{Tx2}|A \times I_{Tx}|^2 + |I_{Rx}|^2 R_{Rx}} \frac{R_{load}}{R_{Rx}} \quad (1)$$

Term A represents the current ratio between transmitter 1 (T_{x1}) and transmitter 2 (T_{x2}). This is approximated to (2)

$$\eta \approx \frac{R_{load}}{R_{Tx} \left|\frac{I_{Tx}}{I_{Rx}}\right|^2 (1 + A^2) + R_{Rx}} \quad (2)$$

(2) can be derived by using Kirchhoff's Voltage Law equation for the receiver side

$$j\omega M_1 I_{Tx} + j\omega M_2 A I_{Tx} + R_{Rx} I_{Rx} = 0 \quad (3)$$

$$\frac{I_{Tx}}{I_{Rx}} = \frac{-R_{Rx}}{j\omega(M_1 + M_2 A)} \quad (4)$$

M1 and M2 represent the mutual inductance between T_{x1}-Reciver (R_x) and T_{x2}-R_x. The unknown mutual inductance can be calculated by magnetic field density (B) by Biot-Savart Law [14] as (5). Parameters are explained at Fig. 4 and (6).

$$M \times I = B = \frac{\mu_0 I}{4\pi} \frac{1}{h}(sin\theta_1 + sin\theta_2) \quad (5)$$

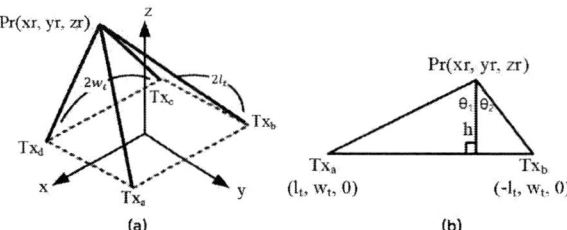

Fig 4. (a) Parameter explanation and arbitrary point Pr to calculate the mutual inductance for general case (b) separate Txa – Txb to explain parameter.

$$h = \frac{|\overrightarrow{Tx_a P_r} \times \vec{u}|}{|\vec{u}|} = \sqrt{(w_t - y_r)^2 + z_r^2}, \qquad sin\theta_1 = \frac{l_t - x_r}{\sqrt{(l_t - x_r)^2 + (w_t - y_r)^2 + z_r^2}},$$
$$sin\theta_2 = \frac{l_t + x_r}{\sqrt{(l_t + x_r)^2 + (w_t - y_r)^2 + z_r^2}} \tag{6}$$

$$B = \frac{\mu_0 I}{4\pi} \frac{1}{\sqrt{(w_t - y_r)^2 + z_r^2}} \left(\frac{l_t - x_r}{\sqrt{(l_t - x_r)^2 + (w_t - y_r)^2 + z_r^2}} + \frac{l_t + x_r}{\sqrt{(l_t + x_r)^2 + (w_t - y_r)^2 + z_r^2}} \right) \tag{7}$$

$$B_z = B \times cos\alpha = B \times \frac{w_t - y_r}{\sqrt{(w_t - y_r)^2 + z_r^2}}$$
$$= \frac{\mu_0 I}{4\pi} \frac{w_t - y_r}{(w_t - y_r)^2 + z_r^2} \left(\frac{l_t - x_r}{\sqrt{(l_t - x_r)^2 + (w_t - y_r)^2 + z_r^2}} + \frac{l_t + x_r}{\sqrt{(l_t + x_r)^2 + (w_t - y_r)^2 + z_r^2}} \right) \tag{8}$$

Rewriting B in terms of x, y, z, then (7) is derived. This system is using z component of magnetic flux, so separate z axis component of B is expressed as (8). By using this equation, the mutual inductance between the transmitter and receiver can be calculated. But it is hard to show the mutual inductance all position over the plane, so simulation result of z component of magnetic field density replaces this value. Fig. 5 shows the z component magnetic flux. The edges of the cell show higher intensity than center.

C. Double Power Line

Power line is the concentric lines that encircle the transmitter with the nodes that are connected to transmitter lines. Since delivering power to the intended place requires only a few nodes of the power line, not all lines of transmitter need to be connected. Controlling this connection helps block unwanted charging area, which can minimize energy leakage. Also, by adding this structure, each transmitter line can be excited without an individual power source. For example, a one coil needs one power source, so if ten coils are placed to charge certain area, ten power sources are needed to excite them. Plus, by using single power source, it is easy to match the phase. In this system, constructive interference is important to increase efficiency. Thus, the phase is very important to make that interference. Using the power line helps to eliminate the phase

synchronization process. Lastly, power line is easy to adapt even if the charging area become wider.

To provide power to any position for charging, the transmitter needs two paths for the current to properly excite. Thus, the transmitter needs two lines for two ways of current. There are two options that satisfying these conditions. The first option is arranging one wire in a zigzag pattern while the other wire is simply connected. The second method is connecting one end of the wire to the outside and the other end of the wire to the inside. Fig. 6 illustrates second method proved by simulation results by ANSYS Maxwell. The first method gives poor expandability because as the system becomes wider, the transmitter lines double compared to other methods. So, the system is designed with a double power line.

III. SIMULATION AND EXPERIMENTAL RESULTS

A. Simulation of Free Positioning

Delivering a power to random position of system with the proper excitation had been proved by the simulation results. Like real-world experiment condition, 0.9m by 0.9m transmitter with 36 cells was modeled using simulation tool. Each cell was surrounded by double lines to make the proper interference of magnetic flux, and a receiver was placed 0.5m above the transmitter. The receiver is made by 0.1m by 0.1m by 0.08m cuboid ferrite wound with 15 turns of litz wire. Source ports

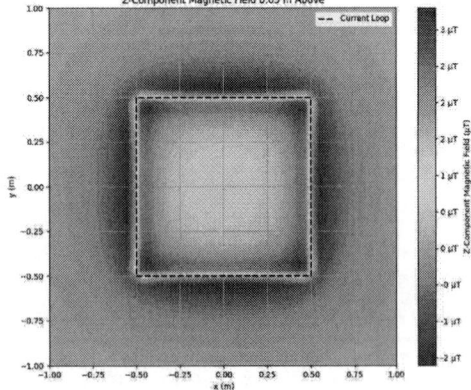

Fig 5. Calculated z component magnetic field density.

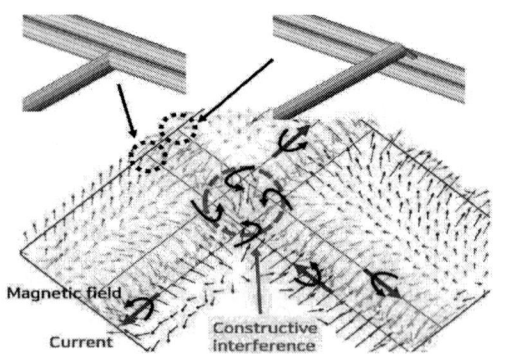

Fig 6. Description of how the power line and double line create constructive interference

979-8-3315-1612-3/25 $31.00 © 2025 IEEE

could be affect to the system so that the ports were placed far from the system.

The graph summarizing the received power of the receiver at different positions is shown in Fig. 7. The highest voltage that receiver can get was 1.62 V in 8 A condition and it was located at the cell that is farthest from the source. The lowest voltage was 1.41 V, observed at the cell closest cell to the source. The largest error within 15.5% is proved by checking all positions in the system. Also, simulation was done for the situation when the receiver was rotated. All other conditions remained the same, while the receiver was rotated by 0, 15, 30, 45 degree. The gained voltages of those receivers are 1.5683 V, 1.5687 V, 1.5622 V, 1.5585 V depending on the degree of the rotation. Only 0.6% difference happens between the maximum and the minimum values while rotating.

B. Simulation of Multiple Receivers

Adding new receivers from the previous simulation could prove that the system guarantees the ability to charge multiple receivers simultaneously. The positions of added receivers could be anywhere within the system, except in close proximity position to the original one because of hard to make constructive interference. To make constructive interference, the direction of the surrounding current around the receiver through the transmitter line is determined. It seems like small loop. If another receiver is added to a nearby cell, it is impossible to achieve the loop. However, diagonal cells can make proper loop direction of current, so there are restrictions to only up, down, left and right position.

In the results, only output voltage was provided by simulation program, but power is proportional to voltage, so comparing by adding the voltages is also valid to evaluate. Two receivers received a similar power from original one, but the power of individual become decreases when third and fourth receivers are added. Average voltages dropped 1.59 V to 0.826 V Overall voltage, however increases as presented in Fig. 8. It reaches maximum 3.30 V. When the number of receivers was

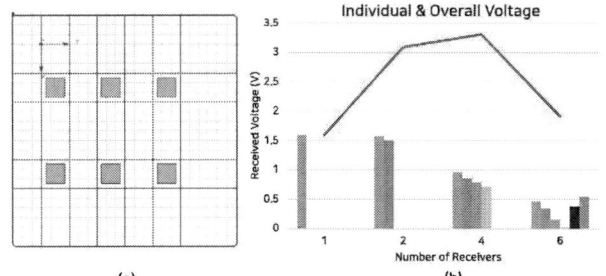

(a)   (b)

Fig 8. Simulation of multiple receivers (a) example condition of simulation and results (b) bars represent the received voltage of individual, while the red solid line indicates the total voltage of all receivers.

increased to six, system could not deliver more power than four receiver, and one of the receiver was gained nearly nothing. Average voltages of six receiver is 0.32 V.

C. Experimental Setup

Each 0.15m transmitter cell was sized based on the 0.1 m receiver which was determined by the UAV size. The 0.15 m length was selected because it is slightly longer than the diagonal length of receiver, so receiver remains in the transmitter cell even if it is rotated. And experiment situations were assuming that UAVs were landed on platform with their landing legs, so receivers were placed 0.05 m above the transmitter. Operating frequency of system was determined as 100 kHz, which is general frequency for wireless power transfer. Source was composed with signal generator and power amplifier. This could supply AC 30 W for the experiment. Since current flows more than 8 A, 18 AWG litz wire was used for transmitter. Receivers were measured in open circuit and calculate maximum efficiency. To calculate maximum efficiency, optimal load value should be calculated. From [15], optimal load to get the maximum efficiency can be determined by (9).

$$R_{L,opt} = R_{Rx} \sqrt{1 + \frac{(\omega M)^2}{R_{Tx}R_{Rx}}} \qquad (9)$$

Although the physical parameters such as receiver size and transmitter size were determined already, there were several specifications to choose. Those values are organized in the Table I. Since receivers are coils, which mean they are considered as inductor, so it is necessary to compensate. Also, transmitter is big size inductor. Simply, there are four ways to compensate the

Table I. WPT system specifications

Specification	Value
Source power	30 W
Operating frequency	100 kHz
Number of Rx turns	15 turns
Gap	0.05 m
Transmitter size	0.9 m square
Transmitter cell size	0.15 m square
	C_{rx1} = 42.4 nF
	C_{rx2} = 42.8 nF
Compensated capacitance	C_{rx3} = 43 nF
	C_{rx4} = 43.3 nF

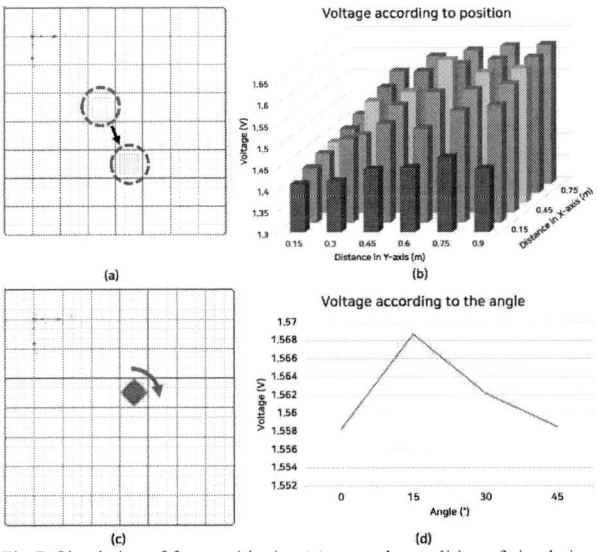

(a)   (b)

(c)   (d)

Fig 7. Simulation of free-positioning (a) example condition of simulation (b) results for all transmitter cell by position (c) example condition of simulation for rotation and (d) results for rotation of receiver

979-8-3315-1612-3/25 $31.00 © 2025 IEEE

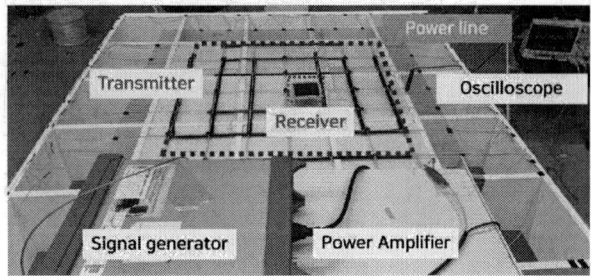

Fig 9. Experimental setup for the system.

system; series-series, series-parallel, parallel-series, parallel-parallel. In this situation, prime priority was misalignment tolerance, thus main point of choosing compensation topology was misalignment tolerance. In [16], series-series shows high mutual inductance with good misalignment tolerance, and it is independent of the coupling coefficient. Therefore, compensation of both transmitter and receiver was decided to series-series.

First experiment aimed to verify the free positioning capability of the system. Simulation already demonstrated the ability to transfer power to any desired location. In the experiment, two random positions of transmitter were selected and measured their voltage. The connections were made considering the current directions to make constructive interference. Furthermore, arrangement within the cell was proved in this experiment. Rotation and unaligned situation were tested by single receiver.

Second experiment was related to multiple receivers. The efficiency of one, two, and four receivers was measured individually, then the overall system efficiency was calculated by combining these values. Same as the previous experiment, connections were chosen to make constructive interference, maximizing power transfer efficiency.

D. Experimental Verification

The rms voltage of two random position of transmitter was measured. Measured open voltage was 1.69 V and 1.96 V. Based on this values, calculated efficiency of one receiver was 17.8%, and the other one was 20.2%. The results of experiments about rotation and placement are shown in Fig. 10 and Fig. 11. While rotation angle changed from 0° to 15°, 30°, 45°, the measured open rms voltage were 1.68 V, 1.69 V, 1.71 V, and 1.70 V. Efficiencies could be calculated, and their values were 17.7%, 17.8%, 18.0%, and 17.9%, respectively. Also, the experiment investigated whether there were a variations within the cell. The voltage of the vortex was 1.69 V with 17.8% efficiency and then edge was 1.74 V with 18.2%, and middle of the cell was 1.75 V with 18.3%. Rotation caused 1.79% error and location error in cell caused 3.55% error. First experiment verified the hypothesis of the system and simulation.

Fig. 12 presents the result of the second experiment. Measured voltage of single receiver was 1.83 V, and two receivers got 1.11 V and 1.30 V, respectively. Four receivers received 0.885 V, 1.08 V, 0.767 V, and 0.725 V. Six receivers reported 0.515 V, 0.662 V, 0.726 V, 0.837 V, 0.785 V, and 0.620 V The overall system efficiency is calculated by sum of those individual efficiency, so overall system efficiencies of four

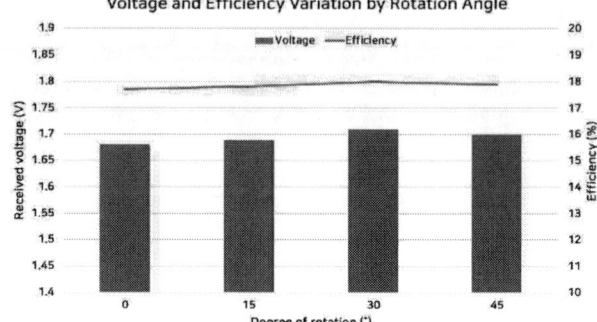

Fig 10. Result of rotation experiment.

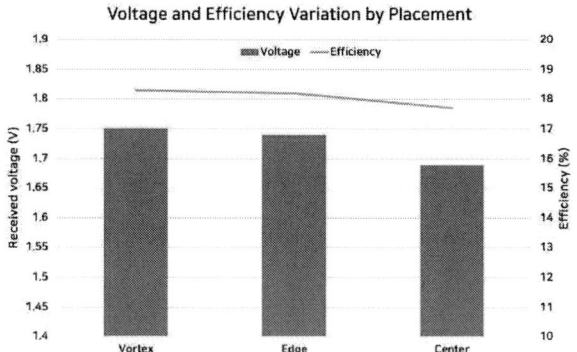

Fig 11. Result of placement experiment.

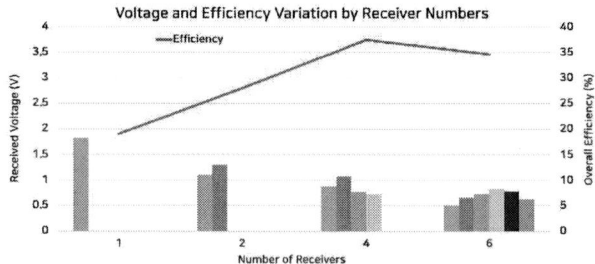

Fig 12. Result of multiple receivers experiment

condition were 19.0%, 27.9%, 37.5%, and 34.5%. The overall efficiency of each system increases by 46.9% and 34.3% compared to the previous one to the four receiver system, but decreases by 7.81% when the system expands to six receivers. Since the transmitter did not control current distribution, as the number of transmitter nodes connected to the power line increases, the current distribution became less uniform, which led to problem of system. This phenomenon makes the efficiency is not directly proportional to the number of receivers but grows at a slower rate. Unlike the simulation result, no significant efficiency dropped at six receivers, but experiment result proved that four receiver system has the highest overall efficiency.

IV. CONCLUSION

Today, the UAV operation is leading to an increase in the number of UAVs. Especially achieving true unmanned for UAVs requires WPT with free positioning and wide area. The proposed grid pattern transmitter demonstrates advantages in these characteristics. To enable the transmitter, two directions of current are required, which is achieved by two surrounding power lines. This structures also bring advantages on expandability. Simulation and experimental results confirm that system can charge anywhere over the transmitter. Even if the receiver is rotated or slightly moved within the cell, there is no problem for sending power from transmitter. Additionally, the transmitter can charge multiple receivers simultaneously. The highest efficiency of the system is achieved when charging four receivers. Even six receiver can be charged, but efficiency of system decreases and issue of dead receiver happens.

This system shows low efficiency than coil system. This is due to the limited number of turns of receiver due to the size of the transmitter cell, challenge in managing current distribution, and the poor vertical magnetic flux capture. To improve these problems, modifying the coil winding methods for the receiver and changing transmitter design to control easier will be proceeded. Furthermore, looking for solutions to block the magnetic flux leakage in areas where the receiver is not positioned, to make the system safer.

REFERENCES

[1] P. K. Chittoor, B. Chokkalingam and L. Mihet-Popa, "A Review on UAV Wireless Charging: Fundamentals, Applications, Charging Techniques and Standards," in IEEE Access, vol. 9, pp. 69235-69266, 2021, doi: 10.1109/ACCESS.2021.3077041.

[2] M. Liao et al., "Drone Charging Stations on Telecom Towers with Series-Stacked Capacitive Differential Wireless Power Transfer," 2024 IEEE Applied Power Electronics Conference and Exposition (APEC), Long Beach, CA, USA, 2024, pp. 1166-1173, doi: 10.1109/APEC48139.2024.10509159.

[3] M. A. Ali and A. Jamalipour, "Dynamic Aerial Wireless Power Transfer Optimization," in IEEE Transactions on Vehicular Technology, vol. 71, no. 4, pp. 4010-4022, April 2022, doi: 10.1109/TVT.2022.3147567.

[4] A. K. Gupta, S. Ghosh and M. R. Bhatnagar, "Pricing Based Scheme for UAV-Enabled Wireless Energy Transfer," in IEEE Transactions on Vehicular Technology, vol. 71, no. 12, pp. 13198-13209, Dec. 2022, doi: 10.1109/TVT.2022.3196551.

[5] C. Rong et al., "Optimization Design of Resonance Coils With High Misalignment Tolerance for Drone Wireless Charging Based on Genetic Algorithm," in IEEE Transactions on Industry Applications, vol. 58, no. 1, pp. 1242-1253, Jan.-Feb. 2022, doi: 10.1109/TIA.2021.3057574.

[6] Y. Zeng, C. Lu, C. Rong and M. Liu, "Optimization Design of Wireless Charging System with Uniform Magnetic Field for Multi-Drone," 2021 IEEE 2nd China International Youth Conference on Electrical Engineering (CIYCEE), Chengdu, China, 2021, pp. 1-6, doi: 10.1109/CIYCEE53554.2021.9676744.

[7] Z. Zhang and K. T. Chau, "Homogeneous Wireless Power Transfer for Move-and-Charge," in IEEE Transactions on Power Electronics, vol. 30, no. 11, pp. 6213-6220, Nov. 2015, doi: 10.1109/TPEL.2015.2414453.

[8] C. Zhang et al., "A Strong Misalignment-Tolerance Wireless Power Transfer System Based on Dynamic Diffusion Magnetic Field for Unmanned Aerial Vehicle Applications," in IEEE Transactions on Power Electronics, vol. 39, no. 11, pp. 14129-14134, Nov. 2024, doi: 10.1109/TPEL.2024.3426609.

[9] C. Cai, J. Wang, H. Nie, P. Zhang, Z. Lin and Y. -G. Zhou, "Effective-Configuration WPT Systems for Drones Charging Area Extension Featuring Quasi-Uniform Magnetic Coupling," in IEEE Transactions on Transportation Electrification, vol. 6, no. 3, pp. 920-934, Sept. 2020, doi: 10.1109/TTE.2020.2995733.

[10] J. M. Arteaga, P. D. Mitcheson and E. M. Yeatman, "Development of a Fast-Charging Platform for Buried Sensors Using High Frequency IPT for Agricultural Applications," 2022 IEEE Applied Power Electronics Conference and Exposition (APEC), Houston, TX, USA, 2022, pp. 1116-1121, doi: 10.1109/APEC43599.2022.9773730.

[11] S. A. Al Mahmud, I. Panhwar and P. Jayathurathnage, "Large-Area Free-Positioning Wireless Power Transfer to Movable Receivers," in IEEE Transactions on Industrial Electronics, vol. 69, no. 12, pp. 12807-12816, Dec. 2022, doi: 10.1109/TIE.2022.3144591.

[12] F. Iob et al., "A Novel Lightweight Wireless Charging System for UAV Applications," 2024 IEEE Applied Power Electronics Conference and Exposition (APEC), Long Beach, CA, USA, 2024, pp. 279-283, doi: 10.1109/APEC48139.2024.10509392.

[13] S. Huh and D. Ahn, "Two-Transmitter Wireless Power Transfer with Optimal Activation and Current Selection of Transmitters," in IEEE Transactions on Power Electronics, vol. 33, no. 6, pp. 4957-4967, June 2018, doi: 10.1109/TPEL.2017.2725281.

[14] D. J. Griffiths, Introduction to Electrodynamics. 4th ed. Boston, MA, USA: Pearson, 2013

[15] H. Jo, S. Seo, J. Kim and F. Bien, "A coreless track-type seamless wireless charging system using co-planar wires enabling quasi-free planar movements for mobile logistics robots," in Applied Energy, vol. 375, 2024, doi: 10.1016/j.apenergy.2024.123943.

[16] V. Shevchenko, O. Husev, R. Strzelecki, B. Pakhaliuk, N. Poliakov and N. Strzelecka, "Compensation Topologies in IPT Systems: Standards, Requirements, Classification, Analysis, Comparison and Application," in IEEE Access, vol. 7, pp. 120559-120580, 2019, doi: 10.1109/ACCESS.2019.2937891.

Capacitor Design for Self-Resonant Coils for Long-Distance Wireless Power Transfer System

Mostak Mohammad, Vandana Rallabandi, Omer C. Onar, Gui-Jia Su

Oak Ridge National Laboratory

E-mails: mohammadm@ornl.gov, rallabandivp@ornl.gov, onaroc@ornl.gov, sugj@ornl.gov

Abstract— In this paper, an integrated capacitor design is proposed for higher-order resonant tank topologies for self-resonant coils, such as series, parallel, LCC, LLC, etc. The capacitor is one of the large, lossy, and thermally vulnerable components of a high-frequency resonant tank, and designing a high-voltage, thermally stable resonant capacitor can be highly challenging. Designing the extremely high-voltage capacitor as an integral part of the coil reduces the size and complexity of the coil assembly. This paper proposes a low-loss PCB-based high-voltage capacitor design to achieve that target, which can be implemented as an integral part of the coil. The proposed capacitor designs are simulated using Multiphysics FEA and tested experimentally. A 23 kV, 133 pF capacitor prototype was built and tested as part of a 1 kW long-distance wireless charging system. The test results verify the capacitor's voltage, current, and thermal resiliency performance.

Keywords—Inductive charging, electric vehicle, leakage field, EMF, shielding effectiveness.

I. INTRODUCTION

Long-distance wireless power transfer needs a highly efficient resonant tank and coil design. An efficient resonant tank requires high precision and stable capacitor and indicator designs. The inductors are comparatively much more reliable and thermally stable than the capacitors. The challenges arise when the capacitance and the voltage of the capacitor becomes extremely high [1]. A self-resonant coil and capacitor design mitigates those challenges and provides the design flexibility, larger area for thermal dissipation, and reduces overall cost and footprint of the system.

Self-resonant coils have been proposed in other works in literature. A coil with stacked layers of conductors and dielectric materials placed inside a magnetic core with the capacitance between the layers forming a parallel LC resonance with the coil layers is proposed [2]. Two flat spiral copper layers separated by a dielectric layer forming a series capacitance resonating with the coil self-inductance is proposed in [3]. A double-sided PCB with coil layers printed on both sides uses parasitic capacitance between the turns to form a capacitor that resonates with the coil. [4]However, the common FR4 dielectric materials of the PCB board are not suitable for high-frequency capacitor design. Previously, the capacitor layers in the PCB were realized by implementing a low-loss dielectric layer instead of the common FR4 layer. While such designs are highly efficient, a large part of the dielectric may remain unused if not well designed. This work proposes an approach to utilize the entire dielectric area to achieve high voltage and low loss at target capacitance.

In contrast to the other works which focus on printed and flat coils, this work proposes a capacitance integrated with a coil formed from a copper foil. The coil is divided into two half turns, and the capacitor is realized by placing a dielectric substrate between the two half layers, as shown in Fig. 1. Different types of resonant networks can be implemented, including LC series, LC parallel, LCL, etc. This work describes the design of a self-resonant coil for high-power, long-distance wireless power transfer applications. The system's resonant frequency can be a few hundred kilohertz to a few Megahertz.

II. SELF RESONANT CAPACITOR DESIGN

The capacitors are made with either aluminum or copper conductor plates, and a dielectric separator between the plates. The properties of the dielectric material and the effective surface area between the conductors and the dielectric materials are the two most sensitive design parameters impacting the capacitance. As for the dielectric material, the relative permittivity, loss tangent, and breakdown voltages are the two key parameters. However, we can choose the best candidate from the existing materials for a specific frequency with the highest dielectric constant and lowest loss-tangent. Next, the rest of the design is focused on getting the target capacitance, breakdown voltage, and current rating within a given space.

Capacitor design needs to meet several design goals, such as capacitance, voltage rating, and current capacity. In addition, to realize a higher-order resonant tank, multiple capacitors in series or parallel within the same coil winding must be designed.

A. High-voltage capacitor design

The high frequency, high quality factor (High-Q), ceramic resonant capacitors are typically limited in voltage from 1 to 10 kV. Most of those capacitors are designed for RF and Microwave applications where the current capacity of

---

This manuscript has been authored by Oak Ridge National Laboratory, operated by UT-Battelle, LLC, under Contract No. DE-AC05-00OR22725 with the U.S. Department of Energy. The United States Government retains and the publisher, by accepting the article for publication, acknowledges that the United States Government retains a non-exclusive, paid-up, irrevocable, world-wide license to publish or reproduce the published form of this manuscript, or allow others to do so, for United States Government purposes. The Department of Energy will provide public access to these results of federally sponsored research in accordance with the DOE Public Access Plan (http://energy.gov/downloads/doe-public-access-plan).

979-8-3315-1612-3/25 $31.00 © 2025 IEEE

individual capacitors is limited. To achieve higher voltage and currents, a capacitor bank is commonly built on a PCB board using a larger number of capacitors in series and parallel. While such solutions are convenient at up to 5 to 15 kV, they can become extremely costly and thermally challenging as the required voltage goes above 15 kV and the current rating exceeds a few amps. Alternatively, thermally stable resonant capacitors, such as CELEM capacitors, are limited in their voltage rating to 1 to 1.5 kV. Making a high-voltage capacitor of more than 10 kV would make them extremely costly, heavy, and bulky. Compared to those capacitors, a self-resonant capacitor can utilize a larger area along the path of the coil, hence achieving a higher voltage rating and heat dissipation area.

The proposed design shows an extremely high-voltage capacitor using interleaving floating plates along the coil, as shown in Fig. 1, with two terminal copper plates and one overlapping floating copper plate designed and fabricated on a Roger low-loss dielectric core PCB. The Rogers PCB material has a breakdown strength exceeding 30 kV/mm [5]. With the voltage applied across 3mm, it is expected that the board is able to sustain high voltage. This type of high-voltage, high-frequency capacitor is required for long-distance, high-power wireless charging.

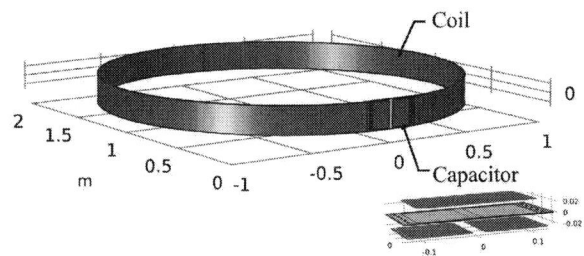

Fig. 1. A self-resonant coil made with copper trace and integrated PCB-printed high-frequency capacitor.

The voltage rating can be increased further, as shown in Fig. 2. In Fig. 2a, the full resonant capacitor voltage is applied across a dielectric material of thickness 1.54mm. The voltage rating can be increased by multiple stacking dielectric boards, which would create voids between layers, risking breakdown. Another approach is to make the coil connections such that only part of the voltage is applied across the dielectric (Fig. 2). The voltage rating of the capacitors in Fig. 2b is doubled, and that in Fig. 2c is quadrupled. This series connection method reduces the overall capacitance, which would need to be compensated by increasing the width of the overlapping area of the coil sections or by connecting multiple plates in parallel on the same layer or among different layers.

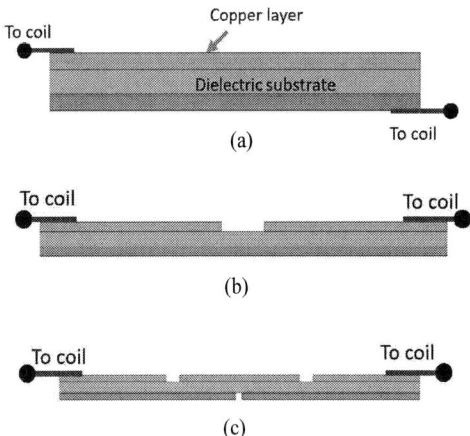

Fig. 2. Capacitor board designs with (a) the same voltage rating as the dielectric layer, (b) two times higher, and (c) four times higher voltage rating compared to the voltage rating of the dielectric layer.

B. Higher order resonant Tank design

The series and parallel capacitors are the most common topologies in resonant tank design. However, the recent development of wireless power transfer (WPT) needs higher-order resonant tanks. Especially the LCC and LLC resonant topologies are the two with the most potential, which mitigates the vulnerability of the WPT systems from wide variations of the coupling coefficient. Such higher-order resonant tanks need multiple capacitors and additional inductors in the resonant tank design. While the existing literature has presented those resonant tanks with individual component designs and partial integrations, in this paper, we present a fully integrated resonant tank design.

Series Resonant Capacitor: The series capacitor is realized by splitting the coil into two sides and connecting the dielectric board between them, as shown in Fig. 3a. The equivalent schematic diagram of the series resonant tank is shown in Fig. 3b. The capacitance can be incorporated between each turn in a multi-turn coil. If the voltage becomes significantly high, the capacitance can be distributed along the coil path to reduce localized electric field strength, following the same design method shown in Fig. 2.

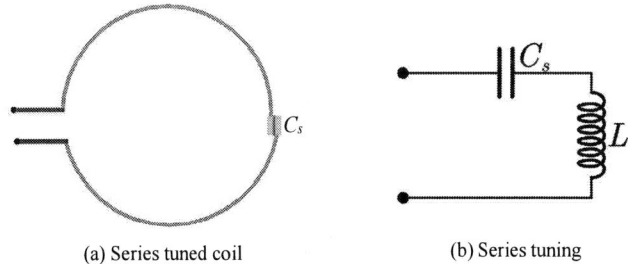

(a) Series tuned coil   (b) Series tuning

Fig. 3. Geometric configuration and equivalent schematics of the series compensated resonant coil.

Parallel Resonant Capacitor: In this case, the capacitor is integrated in parallel to the coil. The capacitor dielectric board is connected between the two terminals of the coil as seen in Fig. 4a. The equivalent schematic diagram of the parallel resonant tank is shown in Fig. 4b. While the series resonant capacitor can be anywhere along the coil path, the parallel capacitor needs to be at the terminal of the coil. The resonance frequency of the series (f_{0s}) and parallel (f_{0p}) compensation networks can be expressed as

$$f_{0s} = \frac{1}{2\pi\sqrt{(LC)}}, f_{0p} = \frac{1}{2\pi\sqrt{(LC)}} \quad (1).$$

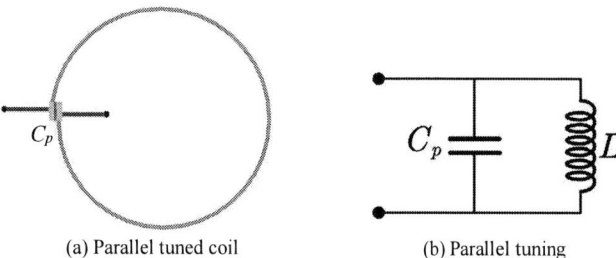

(a) Parallel tuned coil   (b) Parallel tuning

Fig. 4. Geometric configuration and equivalent schematics of the parallel compensated resonant coil.

LCC Resonant Inductor and Capacitors: The series and parallel capacitor of the LCC tuning are implemented as an integral part of the coil, as shown in Fig. 5a. The equivalent schematic diagram of the LCC resonant tank is shown in Fig. 5b. In a LCC tuning, L_f is smaller than L, which can also be implemented an integrated decoupled coil or can be made as a separate coil. The resonance frequency of the LCC compensation network can be expressed as,

$$f_{0LCC} = \frac{1}{2\pi\sqrt{((L-L_f)C_s)}} = \frac{1}{2\pi\sqrt{(L_f C_f)}} \quad (2).$$

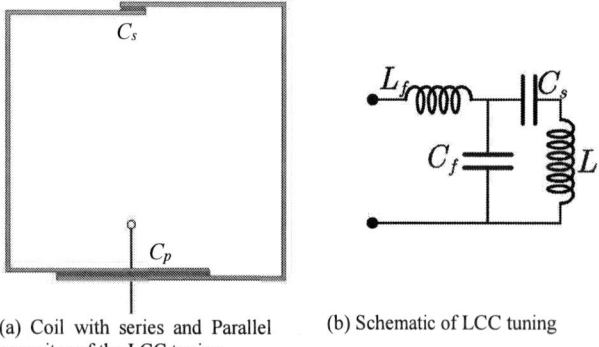

(a) Coil with series and Parallel capacitor of the LCC tuning   (b) Schematic of LCC tuning

Fig. 5. Geometric configuration and equivalent LCC compensated resonant coil schematics. The series and parallel capacitances in the case of LCC resonance are unequal, as seen in Table 1. This is achieved by incorporating different areas of dielectric material between the turns.

The method for incorporation of the series and parallel capacitances for the LCC network in the coil is shown in Fig. 5a. The series inductor L_f, as shown in Fig. 5b, is not included in the geometric diagram of Fig. 5a. Incorporation of the series inductor is a challenging task. The series inductor needs to be mutually decoupled from the wireless charging coils. One approach to implementing such a decoupled inductor can be using a DD planar inductor with a circular primary coil. Alternatively, a separate inductor can be connected in series with the self-resonant coil. The calculated values of the tuning components for different types of tuning are shown in Table 1.

Table 1: Tuning component values for different types of tuning networks.

	Series tuning	Parallel tuning	LCC tuning
Coil self-inductance [μH]	4.34	4.34	4.34
Frequency [MHz]	6.7	6.7	6.7
Series capacitance [nF]	0.13	-	0.132
Parallel capacitance [nF]	-	0.13	1.5
Series inductance [μH]	-	-	0.36

III. SIMULATION AND EXPERIMENTAL RESULTS

Three-dimensional finite element analysis (FEA) was conducted to verify the inductance and capacitance of the coils. A frequency domain eddy current model was developed to obtain the coil inductance and resistance, while an electrostatic model was developed to determine the capacitance of the coil. The designed coil has a diameter of 2m. An electrostatic 3D finite element analysis was conducted to verify the analytically calculated capacitance and the electric field distribution around the coil and the capacitor, as shown in Fig. 6.

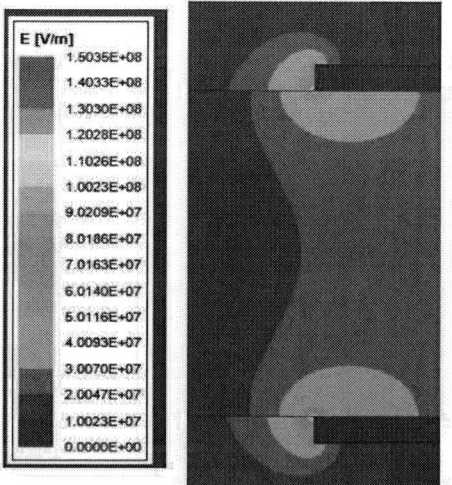

Fig. 6. FEA results of the electric field around the coil.

979-8-3315-1612-3/25 $31.00 © 2025 IEEE

Table II: Coil parameters from FEA

Self-inductance (μH)	Capacitance (pF)	Resistance (mΩ)
4.3	133	15.4

A prototype of the self-resonant coil made from 2 mils thick copper foil was built as part of a long-distance wireless power transfer system operating at a multi-MHz resonant frequency [5] and is shown in Fig. 7 and the fully integrated self-resonant coil is shown in Fig. 8. The capacitor was fabricated, and the dielectric material used was RO3003 [5]. The resonant frequency of the system was measured 6.4 MHz. The capacitance was measured using an impendence analyzer and the value 130pF was obtained, which is comparable to the value from 3D FEA (Table II). A hi-pot test was conducted to verify the voltage capability and up to 5kV DC was applied without breakdown.

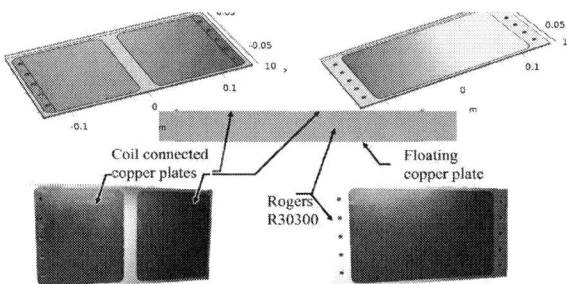

Fig. 7. High voltage series capacitor design and prototype.

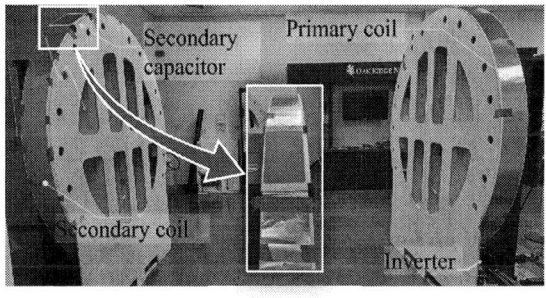

Fig. 8. Primary and secondary self-resonant coils and the secondary integrated capacitor are shown in the subfigure.

A GaN-based inverter and rectifier were built for the primary and secondary sides, as shown in Figs. 9a and 9b. With the inverter and rectifier, the self-resonant coil-based wireless power transfer system was tested up to 300W output power at a 2m air gap. The inverter output voltage and the coil current of the primary coils are shown in Fig. 10. The dc-to-dc efficiency was measured 62% at 300 W output power.

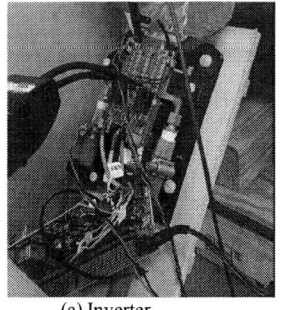 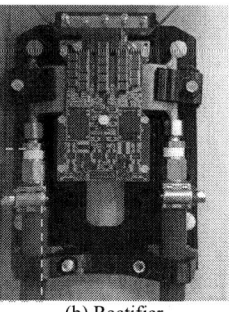

(a) Inverter   (b) Rectifier

Fig. 9. GaN-based high frequency (6.78 MHz) primary side inverter and secondary side rectifier.

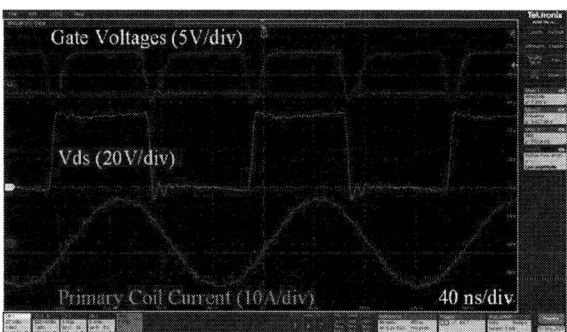

Fig. 10. Inverter side gate waveform, inverter output, and primary coil current (11ARMS) at 300 W power transfer.

IV. CONCLUSIONS

This work presents a self-resonant coil based on a copper foil winding with a dielectric substrate connected between a portion of the copper foils. Concepts for different types of resonant networks are presented - including series, parallel, and LCC. Approaches to implementing a high-voltage capacitor are also presented. Electromagnetic and electrostatic finite element analyses are conducted on the coil, and the RLC matrix for the self-resonant coil is calculated. A prototype is fabricated, and initial impedance measurements agree with the finite element analysis. This type of self-resonant coil offsets the need for discrete capacitors connected in series and parallel.

ACKNOWLEDGMENT

This research used the Power Electronics and Electric Machinery Research Center at the National Transportation Research Center, a DOE EERE User Facility operated by the Oak Ridge National Laboratory (ORNL). The authors would like to thank Dr. Burak Ozpineci (ORNL) and Lee Slezak (U.S. Department of Energy) for their support and guidance of this work.

REFERENCES

[1] M. Mohammad *et al.*, "Self-Resonant Coil Design for High-frequency High-Power Inductive Wireless Power Transfer," in *2023 IEEE Wireless Power Technology Conference and Expo (WPTCE)*, 2023.

[2] C. R. Sullivan and L. Beghou, "Design methodology for a high-Q self-resonant coil for medical and wireless-power applications," in *IEEE 14th Workshop on Control and Modeling for Power Electronics (COMPEL)*, 2013.

[3] K. Chen and Z. Zhao, "Analysis of the Double-Layer Printed Spiral Coil for Wireless Power Transfer," *IEEE Journal of Emerging and Selected Topics in Power Electronics,* vol. 1, no. 2, pp. 114-121, 2013.

[4] J. Li and D. Costinett, "Analysis and design of a series self-resonant coil for wireless power transfer," in *IEEE Applied Power Electronics Conference and Exposition (APEC)*, 2018.

[5] R. Corporation. "High-frequency circuit materials." https://www.rogerscorp.com/advanced-electronics-solutions/ro3000-series-laminates (accessed July, 2024).

A 10.4-kW High-Power-Transfer-Density Multi-MHz Capacitive Wireless Power Transfer System for EV Charging Utilizing Stacked-Inverter Stacked-Rectifier Architecture

Dheeraj Etta, Miguel Alvarez Dominguez, Sounak Maji, Syed Saeed Rashid and Khurram K. Afridi

School of Electrical and Computer Engineering
Cornell University
Ithaca, New York, USA
dke27@cornell.edu; ma2327@cornell.edu; sm2764@cornell.edu; sr944@cornell.edu; afridi@cornell.edu

Abstract—This paper introduces a high-frequency capacitive wireless power transfer (WPT) system suitable for high-power electric vehicle (EV) charging. The system uses a stacked inverter and stacked rectifier architecture, where multiple high-frequency inverters and rectifiers can be paralleled using transformers with a parallel-in series-out and series-in parallel-out configuration, respectively, thereby increasing the power transfer capability of a capacitive WPT system. Furthermore, to increase the power handling capacity of each inverter and minimize the number of parallel inverters required, a novel symmetric layout to directly parallel two gallium nitride (GaN) transistors in a full-bridge inverter at multi-MHz frequencies using six-layer printed circuit board is proposed. Experimental validation shows the proposed inverter design handles nearly twice the output current of a single GaN transistor-based inverter, supporting up to 4 kW per inverter. Finally, a 6.78-MHz 12-cm air-gap 10-kW scale capacitive WPT system prototype utilizing the proposed power scaling approaches is designed, built and tested. This system achieves record-breaking performance for a capacitive EV charging system, transferring 10.4 kW at a peak efficiency of 85.3%, corresponding to a power transfer density of 136.8 kW/m².

Keywords—wireless power transfer, capacitive WPT, high-frequency, matching network, stacked inverter, stacked rectifier, parallel-in series-out transformer, series-in parallel-out transformer, paralleling GaN transistors, current sharing.

I. INTRODUCTION

Wireless power transfer (WPT) is becoming increasingly important for EV charging and has the potential to accelerate the widespread adoption of electric vehicles [1]-[4]. EVs can be charged wirelessly using either electric fields between capacitively coupled plates [5]-[8], or magnetic fields between inductively coupled coils [9]-[12]. Capacitive WPT systems can be lighter, smaller and less expensive than inductive WPT systems due to the absence of ferrite cores [5], [6]. However, these systems typically operate at multi-MHz frequencies, presenting challenges for high-power applications. Developing reliable high-power capacitive WPT systems is difficult because paralleling semiconductor devices or inverter/rectifier modules at multi-MHz frequencies is challenging. Even slight inconsistencies in semiconductor parameters, gate signal mismatches, or design layouts can lead to imbalances in current or power sharing among the paralleled components [13]-[15].

In high-power capacitive WPT applications, the high-frequency inverters have often been the bottleneck for power scaling. This limitation arises from the undesirable dynamic on-resistance and output capacitance characteristics of commercially available gallium nitride (GaN) transistors used in them, which leads to increased losses at MHz frequencies [16]-[18]. To address inverter power scaling issues, various power-combining and frequency multiplication approaches have been proposed in [19]-[25], with transformer-based approach in [21] being particularly effective due to its scalability, low component count, and simple implementation. Though the silicon carbide (SiC) Schottky diodes used in the high frequency rectifiers have higher current carrying capabilities compared to GaN transistors, to meet the growing power demands for EV

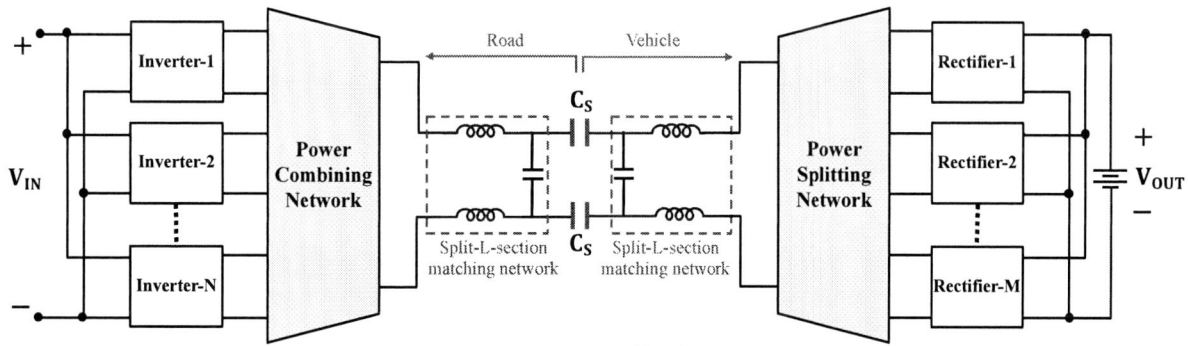

Fig. 1: Capacitive WPT system topology suitable for high power EV charging with *N* inverters and *M* rectifiers paralleled power combining and power splitting networks, respectively.

charging, there is also a need to develop power-scaling methods that enable the effective parallel operation of multiple high-frequency rectifiers.

Moreover, paralleling two or more transistors in multi-MHz inverters can enhance their power handling capability. This approach reduces the number of inverters required to achieve a specific power level and minimizes the need for additional power combining circuits. However, existing layouts in the literature and application notes lack the required symmetry for effective operation at multi-MHz frequencies [26]-[28]. These designs are limited to frequencies below a few hundreds of kHz, limiting the full advantages of device-level paralleling at multi-MHz frequencies.

To address these challenges, this paper introduces a stacked rectifier architecture, where multiple high-frequency rectifiers can be paralleled using transformers with series-in parallel-out configuration. Using this stacked rectifier approach along with stacked inverter approach from [21], a capacitive WPT system topology suitable for EV charging is proposed. A comprehensive methodology to design the proposed capacitive WPT system by leveraging magnetizing and leakage inductances of the air-core transformers used for power scaling is discussed. Furthermore, to overcome the limitations of existing layouts for direct transistor paralleling, a symmetric layout to directly parallel two GaN transistors in a full-bridge inverter at multi-MHz frequencies is also proposed and experimentally validated. Leveraging all the novel techniques proposed in this work, a 6.78-MHz 12-cm air-gap 10-kW scale capacitive WPT system prototype is designed, built and tested up to 10.4 kW at 85.3% peak dc-dc efficiency and 136.8 kW/m² power transfer density.

The remainder of this paper is organized as follows. Section II introduces the architecture of the capacitive WPT system for high-power EV charging. Section III presents the proposed high-power capacitive WPT system with stacked inverter stacked rectifier and its design. Effective layout to directly parallel two GaN transistors in a full-bridge inverter is presented in Section IV. Section V presents the

experimental results. Finally, Section VI summarizes and concludes the paper.

II. CAPACITIVE WPT SYSTEM ARCHITECTURE FOR HIGH-POWER ELECTRIC VEHICLE CHARGING

The topology of a large air-gap capacitive WPT system suitable for tens of kilowatt-scale EV charging application is illustrated in Fig. 1. In this system, power is transferred wirelessly through large air-gap by using two pairs of coupling plates (shown as C_s in Fig. 1). On the input side, each inverter converts dc input voltage to high frequency ac voltage. These inverter outputs are then fed into a power combining network, which combines the power from "N" high-frequency inverters and ensures equal power sharing among them. The output of the power combining network is then fed into a road-side L-section matching network, which steps up the voltage to enable power transfer through the air gap while maintaining relatively small displacement currents. On the vehicle side, another L-section matching network steps down the voltage and feeds it into a power splitting network, which splits the power equally among "M" high-frequency rectifiers. Finally, the rectifier outputs are paralleled and connected to the EV battery. The power combining and splitting networks enable effective parallel operation of multiple inverters and rectifiers respectively, facilitating power scaling in the WPT system. Both primary and secondary side matching networks also provide the required reactive compensation to fully compensate for the reactance of the coupling plates. To realize matching network capacitances using the parasitic capacitances present in the charging environment, the matching network inductors are divided into two equal halves - one placed in the forward path and the other in the return path, as depicted in Fig. 1.

III. HIGH-POWER CAPACITIVE WPT SYSTEM WITH STACKED INVERTER STACKED RECTIFIER

In this work, owing to its high scalability, low component count and simple implementation, parallel-in series-out transformer power combining approach is chosen for power scaling high-frequency inverters [21]. The proposed high-

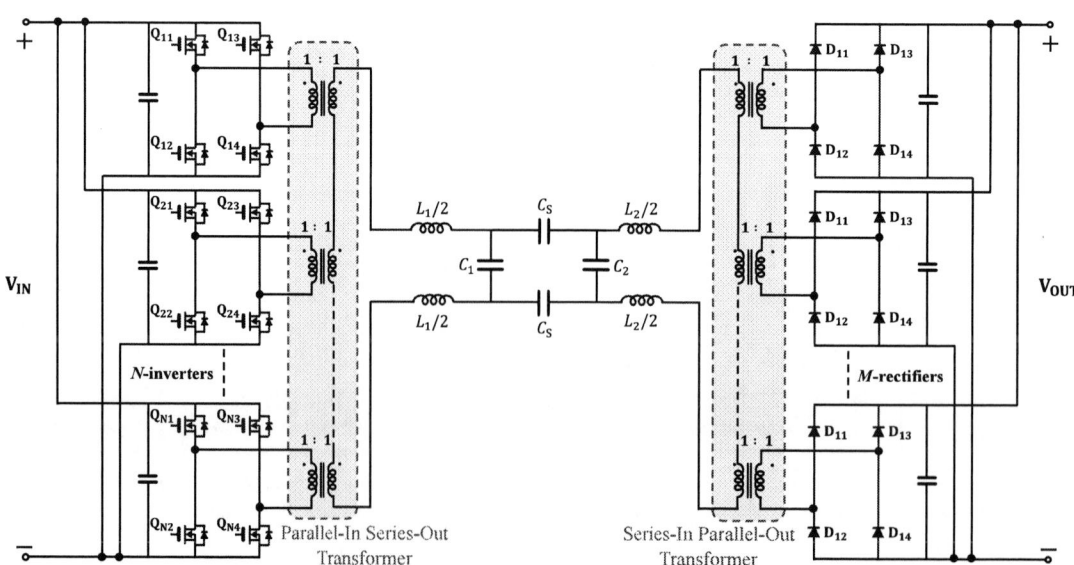

Fig. 2: High-power capacitive WPT system architecture utilizing transformer approach for power scaling.

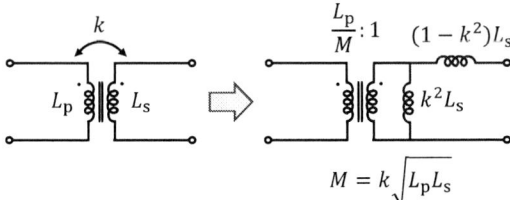

$$\frac{L_p}{M}:1 \quad (1-k^2)L_s$$

$$M = k\sqrt{L_pL_s}$$

Fig. 3: Transformer equivalent circuit model.

power capacitive WPT system architecture utilizing transformer approach for power scaling is shown in Fig. 2. In the proposed topology, to power scale high-frequency rectifiers, series-in parallel-out transformer power splitting approach is utilized. Series-in parallel-out transformer is a dual implementation of parallel-in series-out transformer, helps in splitting power equally among multiple rectifiers. The primary sides of the transformers are connected in series to distribute the voltage equally among all the stacked rectifiers, thereby enabling equal power sharing. Whereas the secondary sides of the transformers are connected across each rectifier switch nodes and the rectifier outputs are connected in parallel as shown in Fig. 2. Practically developing high quality factor high turns ratio transformer is difficult at tens of MHz frequencies, so for ease of implementation, each transformer in the proposed topology is implemented as a 1:1 transformer.

Figure 3 shows the equivalent circuit model for a general transformer for a given coupling factor (k), primary winding self-inductance (L_p) and secondary winding self-inductance (L_s). Self-inductance of each transformer winding on the inverter-side ($L_{\text{tran,inv}}$) can be chosen appropriately using the expression in [21] to achieve the zero-voltage switching (ZVS) of the inverter transistors. On the other hand, self-inductance of each transformer winding on the rectifier-side ($L_{\text{tran,rec}}$) can be designed to cancel out the effective capacitance seen across the switch nodes of each rectifier due to the junction capacitances of the rectifier diodes. Rectifier transformer winding self-inductance can be determined using

$$L_{\text{tran,rec}} = \frac{1}{\omega_s^2 C_{\text{rec}}}, \tag{1}$$

where C_{rec} is the effective capacitance seen across each rectifier switch nodes and ω_s is the angular switching frequency. Capacitance C_{rec} can be expressed in terms of

rectifier diode junction capacitance (C_j), switching frequency (f_s), output voltage (V_{OUT}) and output power (P_{OUT}) as

$$C_{\text{rec}} = \frac{2\pi(\delta - \sin(2\delta))}{((\delta - \sin(2\delta))^2 + 4\sin^4(\delta))}C_j, \tag{2}$$

where $\delta = \cos^{-1}\left(\frac{P_{\text{OUT}} - 4f_sC_jV_{\text{OUT}}^2}{P_{\text{OUT}} + 4f_sC_jV_{\text{OUT}}^2}\right)$ is the commutation angle during the diodes' transitions.

Matching networks of the WPT system can now be designed to present resistive impedance at the input of the primary side L-section matching network and at the output of the secondary-side matching network. The overall current gain (G) needed from the matching networks can be determined as

$$G = \frac{k_{\text{tran,inv}}NV_{\text{IN}}}{k_{\text{tran,rec}}MV_{\text{OUT}}}, \tag{3}$$

where $k_{\text{tran,inv}}$ and $k_{\text{tran,rec}}$ are the coupling factors of inverter-side and rectifier-side transformers, respectively. Once the optimal matching network inductor values are determined, total net inverter transformers leakage inductance, $L_{\text{leak,tran,inv}} = N(1 - k_{\text{tran,inv}}^2)L_{\text{tran,inv}}$ and total net rectifier transformers leakage inductance, $L_{\text{leak,tran,rec}} = M(1 - k_{\text{tran,rec}}^2)L_{\text{tran,rec}}$ can be absorbed into road-side and vehicle-side L-section matching network inductors L_1 and L_2, respectively.

In the proposed high-power WPT architecture, the coupling factor of the 1:1 transformer used for power scaling is critical to system performance. The system efficiency decreases significantly for lower coupling factors because the transformer network provides less voltage gain. Consequently, higher currents flow through the matching network inductors and transformer windings, leading to increased losses and reduced efficiency. Assuming equal coupling factors for the transformers on both the inverter and rectifier sides in a three-stacked inverter and three-stacked rectifier configuration, Fig. 4 shows the system efficiency relative to the transformer coupling factor for a 400 V input, 400 V output, and 10 kW output power. It can be observed that coupling factors exceeding 0.6 are needed to achieve good efficiencies.

IV. PROPOSED TWO GAN-TRANSISTOR PARALLELING LAYOUT

In this section, a layout that is effective to directly parallel two GaN systems GS66516T 650-V GaN FETs in a full-bridge inverter at multi-MHz frequencies is presented. The proposed PCB layout employs a 6-layer board, with layers 1 and 2 dedicated to laying out one full-bridge inverter composed of four GaN transistors. Concurrently, layers 5 and 6 are utilized for the layout of the other full-bridge inverter, comprising the remaining four GaN transistors. Layers 3 and 4 encompass ground planes to minimize interactions between the circuitry on the top and bottom layers. Additionally, these layers facilitate PWM signal routing from the PWM connectors to the input of the gate drive circuitry for each GaN transistor. The two paralleled transistors (Q_{xa} and Q_{xb}) are strategically positioned, with one placed on the top layer and the other positioned on the bottom layer directly beneath the top-layer transistor. The gate drive components for each

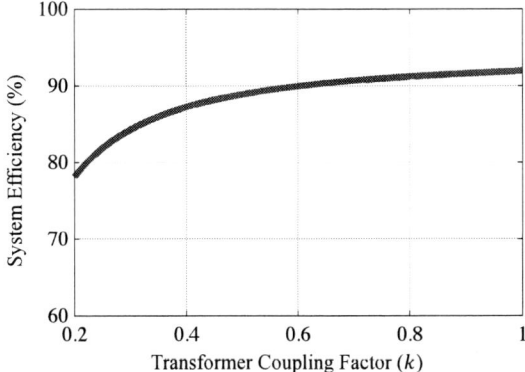

Fig. 4: Proposed system efficiency with respect to transformer coupling factor (k) for 400-V input, 400-V output and 10-kW output power.

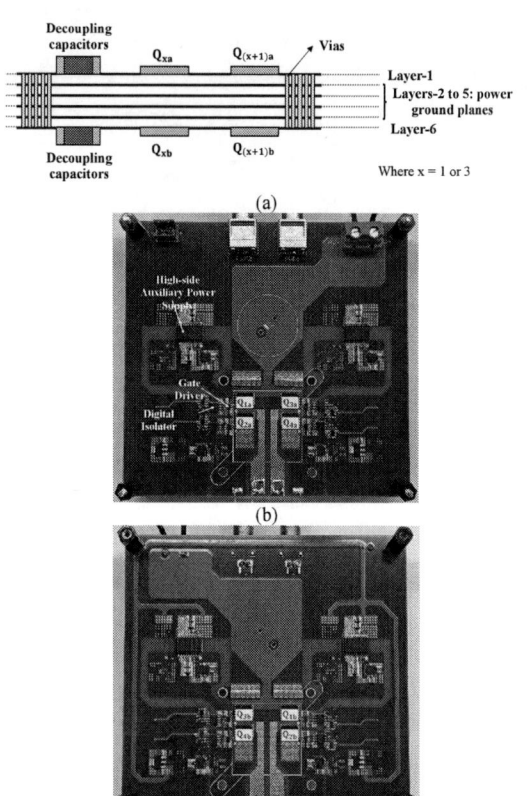

(a)

(b)

(c)

Fig. 5: Proposed two-paralleled transistor full-bridge inverter layout (a) side view of the power loop, (b) top view and (c) bottom view.

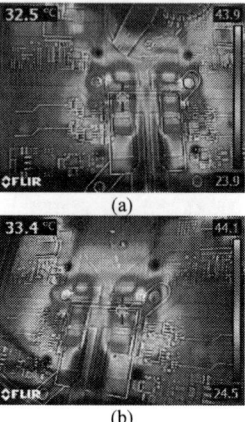

(a)

(b)

Fig. 6: Thermal image of the inverter board while delivering 400 W of output power: (a) top side, and (b) bottom side.

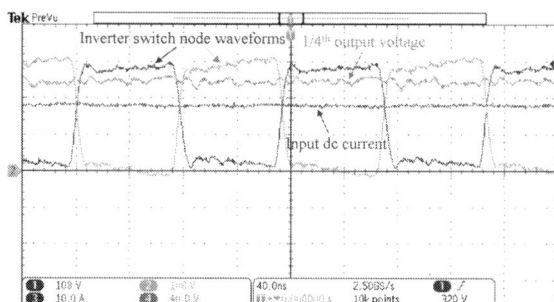

Fig. 7: Measured switch node voltage waveforms of the proposed two paralleled GaN transistor full-bridge inverter while delivering 3.95 kW output power.

transistor are laid out on the same layer as the respective transistor. To ensure symmetry in the gate drive layouts for the two parallel transistors, the digital isolators and gate drivers of the parallel transistors are also positioned underneath each other as closely as possible on both the top and bottom layers. The switch nodes of both paralleled half-bridge modules are stitched together using vias between top and bottom layers as shown in Fig. 5(b) and Fig. 5(c) from the point where it is feasible to do so after considering clearance constraints. Figure 5 shows the side view, top view and the bottom view of the proposed two-paralleled transistor inverter layout.

V. EXPERIMENTAL RESULTS

A. Experimental Validation of the Two Paralleled Transistor Layout

Two GaN systems GS66516T 650-V GaN FETs are paralleled directly in a full-bridge structure as shown in Fig. 5(b) and 5(c). This full-bridge inverter is tested in a 6.78 MHz 12-cm air-gap capacitive WPT system. Thermal images of the top and bottom sides of the PCB are presented in Fig. 6 while the inverter is processing 400 W of output power. It can be observed from Fig. 6 that the top and bottom side GaN transistors exhibit almost the same temperatures, ensuring that the current is equally shared between the paralleled transistors. Using this inverter board, the capacitive WPT system has been tested up to a maximum power of 3.95 kW, nearly ~1.9x compared to the maximum power of ~2.1 kW using a single GaN transistor full-bridge inverter, thereby proving the efficacy of proposed two transistor paralleling layout. Switch-

node voltage waveforms of the proposed two paralleled transistor full-bridge inverter obtained at 3.95 kW are shown in Fig. 7.

B. Experimental Validation of the Proposed High-Power Capacitive WPT System

A 6.78-MHz 12-cm air-gap 10-kW output power capacitive WPT prototype based on the topology shown in Fig. 2 is built and tested. Figure 8 shows a photograph of the built prototype system. In this prototype, three two-paralleled-transistor full-bridge inverters are stacked using the parallel-in series-out transformers and three full-bridge diode rectifiers are stacked using the series-in parallel-out transformers. This system is tested up to 10.4 kW at a peak dc-dc efficiency of 85.3%, achieving 136.8 kW/m^2 power transfer density. Figure

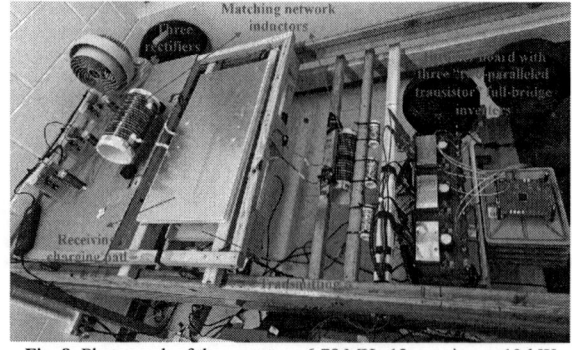

Fig. 8: Photograph of the prototype 6.78 MHz 12-cm air-gap 10-kW capacitive WPT system.

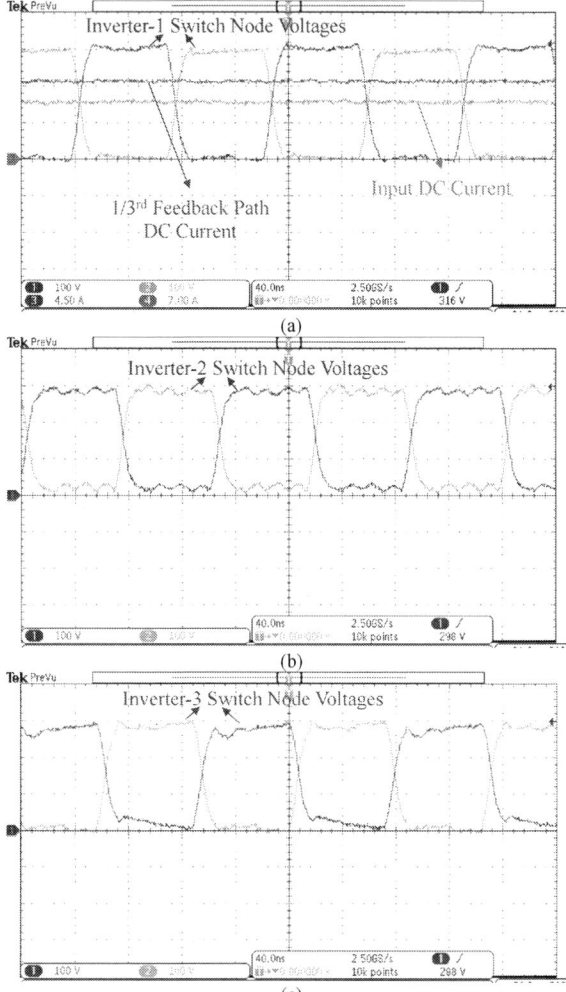

Fig. 9: Measured inverter operating waveforms of the 6.78-MHz 12-cm air-gap capacitive WPT prototype while delivering 10.4 kW output power.

9 shows the measured inverter waveforms of the prototype while delivering 10.4-kW dc output power. It can be seen from the smooth transitions of the inverters' switch-node voltages that the inverters' transistors achieve zero-voltage switching (ZVS). Also, the three inverter output currents are almost equal showing that the equal power is processed by all the inverters. Similarly equal current sharing is observed among the three stacked rectifiers. To the authors' best knowledge, this is the first demonstration of 10-kW+ capacitive WPT system and 10.4 kW at a power transfer density of 136.8 kW/m² is the highest output power and power transfer density reported for any large air-gap capacitive WPT system.

VI. SUMMARY AND CONCLUSIONS

This paper introduces a high-frequency capacitive wireless power transfer (WPT) system suitable for high-power electric vehicle (EV) charging. The system uses a stacked inverter and stacked rectifier architecture, where multiple high-frequency inverters and rectifiers can be paralleled using transformers with a parallel-in series-out and series-in parallel-out configuration, respectively, thereby increasing the power transfer capability of a capacitive WPT system. Furthermore, to increase the power handling capacity of each inverter and

minimize the number of parallel inverters required, a novel symmetric layout to directly parallel two gallium nitride (GaN) transistors in a full-bridge inverter at multi-MHz frequencies using six-layer printed circuit board is proposed. Experimental validation shows the proposed inverter design handles nearly twice the output current of a single GaN transistor-based inverter, supporting up to 4 kW per inverter. Finally, a 6.78-MHz 12-cm air-gap 10-kW scale capacitive WPT system prototype utilizing the proposed power scaling approaches is designed, built and tested. This system achieves record-breaking performance for a capacitive EV charging system, transferring 10.4 kW at a peak efficiency of 85.3%, corresponding to a power transfer density of 136.8 kW/m².

ACKNOWLEDGMENT

The authors wish to acknowledge the financial support received from the U.S. Department of Energy (DOE) through ARPA-E OPEN Program under contract no. DE-AR0001572.

REFERENCES

[1] G. A. Covic and J. T. Boys, "Inductive Power Transfer," *Proceedings of the IEEE*, vol. 101, no. 6, pp. 1276-1289, June 2013.

[2] S. Li and C. C. Mi, "Wireless Power Transfer for Electric Vehicle Applications," *IEEE Journal of Emerging and Selected Topics in Power Electronics*, vol. 3, no. 1, pp. 4-17, March 2015.

[3] G. A. Covic and J. T. Boys, "Modern Trends in Inductive Power Transfer for Transportation Applications," *IEEE Journal of Emerging and Selected Topics in Power Electronics*, vol. 1, no. 1, pp. 28-41, March 2013.

[4] S.Y.R. Hui, W. Zhong and C.K. Lee, "A Critical Review of Recent Progress in Mid-Range Wireless Power Transfer," *IEEE Transactions on Power Electronics*, vol. 29, no. 9, pp. 4500 4511, September 2014.

[5] F. Lu, H. Zhang, H. Hofmann and C. Mi, "A Double Sided LCLC-Compensated Capacitive Power Transfer System for Electric Vehicle Charging," *IEEE Transactions on Power Electronics*, vol. 30, no. 11, pp. 6011-6014, November 2015.

[6] D. Etta, S. Maji and K. K. Afridi, "High-Performance Multi-MHz Capacitive Wireless Power Transfer System with an Auxiliary ZVS Circuit," *Proceedings of the IEEE Energy Conversion Congress and Exposition (ECCE)*, Detroit, MI, October 2022.

[7] B. Luo, R. Mai, Y. Chen, Y. Zhang and Z. He, "A Voltage Stress Optimization Method of Capacitive Power Transfer Charging System," *IEEE Applied Power Electronics Conference and Exposition (APEC)*, Tampa, FL, March 2017.

[8] F. Lu, H. Zhang and C. Mi, "A Review on the Recent Development of Capacitive Wireless Power Transfer Technology," *Energies 2017 Special Issue on Wireless Power Transfer and Energy Harvesting Technologies*, art. 10, 1752, November 2017.

[9] J. M. Miller, O. C. Onar and M. Chinthavali, "Primary-Side Power Flow Control of Wireless Power Transfer for Electric Vehicle Charging," *IEEE Journal of Emerging and Selected Topics in Power Electronics*, vol. 3, no. 1, pp. 147-162, March 2015.

[10] R. Tavakoli and Z. Pantic, "Analysis, Design, and Demonstration of a 25-kW Dynamic Wireless Charging System for Roadway Electric Vehicles," *IEEE Journal of Emerging and Selected Topics in Power Electronics*, vol. 6, no. 3, pp. 1378-1393, September 2018.

[11] A. C. Bagchi, A. Kamineni, R. A. Zane and R. Carlson, "Review and Comparative Analysis of Topologies and Control Methods in Dynamic Wireless Charging of Electric Vehicles," *IEEE Journal of Emerging and Selected Topics in Power Electronics*, vol. 9, no. 4, pp. 4947-4962, August 2021.

[12] U. K. Madawala and D. J. Thrimawithana, "A Bidirectional Inductive Power Interface for Electric Vehicles in V2G Systems," *IEEE Transactions on Industrial Electronics*, vol. 58, no. 10, pp. 4789-4796, October 2011.

[13] J. B. Forsythe, "Paralleling of Power MOSFETs for Higher Power Output," *Proceedings of the Annual Meeting Industry Applications Society*, Philadelphia, PA, October 1981.

[14] Y. -F. Wu, "Paralleling high-speed GaN power HEMTs for quadrupled power output," *Proceedings of the IEEE Workshop on Applied Power Electronics Conference and Exposition (APEC)*, Long Beach, CA, March 2013.

[15] J. Lu, R. Hou and D. Chen, "Loss Distribution among Paralleled GaN HEMTs," *Proceedings of the IEEE Energy Conversion Congress and Exposition (ECCE)*, Portland, OR, September 2018.

[16] G. Zulauf, M. Guacci and J. W. Kolar, "Dynamic on-Resistance in GaN-on-Si HEMTs: Origins, Dependencies, and Future Characterization Frameworks," *IEEE Transactions on Power Electronics*, vol. 35, no. 6, pp. 5581-5588, June 2020.

[17] T. Foulkes, T. Modeer and R. C. N. Pilawa-Podgurski, "Developing a standardized method for measuring and quantifying dynamic on-state resistance via a survey of low voltage GaN HEMTs," *Proceedings of the IEEE Workshop on Applied Power Electronics Conference and Exposition (APEC)*, San Antonio, TX, March 2018.

[18] G. Zulauf, S. Park, W. Liang, K. N. Surakitbovorn and J. Rivas-Davila, "COSS Losses in 600 V GaN Power Semiconductors in Soft-Switched, High- and Very-High-Frequency Power Converters," *IEEE Transactions on Power Electronics*, vol. 33, no. 12, pp. 10748-10763, December 2018.

[19] D. Etta, S. Maji, M. Khatua and K. K. Afridi, "Impedance Control Network-Based Inverters for High-Frequency Capacitive Wireless Power Transfer Systems," *Proceedings of the IEEE 23rd Workshop on Control and Modeling for Power Electronics (COMPEL)*, Tel Aviv, Israel, June 2022.

[20] S. Maji, D. Etta and K.K. Afridi, "Design and Comparison of Power Combining Architectures for Capacitive Wireless Power Transfer Systems," *Proceedings of the IEEE Workshop on Control and Modeling for Power Electronics (COMPEL)*, Tel-Aviv, Israel, June 2022.

[21] D. Etta, S. Maji, Y. Hou and K.K. Afridi, " Stacked Inverter Architecture for High-Frequency Capacitive Wireless Power Transfer Systems," *Proceedings of the IEEE Workshop on Control and*

Modeling for Power Electronics (COMPEL), Ann Arbor, MI, June 2023.

[22] S. Maji, D. Etta and K.K. Afridi, " A High-Power Large Air-Gap Multi-MHz dc-dc Capacitive Wireless Power Transfer System for Electric Vehicle Charging," *Proceedings of the IEEE Workshop on Wireless Power Technology Conference and Expo (WPTCE)*, San Diego, CA, June 2023.

[23] K. Surakitbovorn and J. M. Rivas-Davila, "A Simple Method to Combine the Output Power from Multiple Class-E Power Amplifiers," *IEEE Journal of Emerging and Selected Topics in Power Electronics*, vol. 10, no. 2, pp. 2245-2253, April 2022.

[24] S. Maji, D. Etta and K. K. Afridi, "A Frequency Quadrupler Inverter Architecture for High-Power High-Frequency Capacitive Wireless Power Transfer Systems," *Proceedings of the IEEE Applied Power Electronics Conference and Exposition (APEC)*, Orlando, FL, March 2023.

[25] S. Maji, Y. Hou, D. Etta, K. Afridi, "A Frequency Multiplier Architecture for High-Power High-Frequency Capacitive Wireless Charging Systems," *Proceedings of the IEEE Energy Conversion Congress and Exposition (ECCE)*, Nashville, TN, October 2023.

[26] J. L. Lu and D. Chen, "Paralleling GaN E-HEMTs in 10kW–100kW systems," *Proceedings of the IEEE Workshop on Applied Power Electronics Conference and Exposition (APEC)*, Tampa, FL, March 2017.

[27] M. Wattenberg, O. Lorenz and J. Sanchez, "Efficiently Paralleling GaN-Transistors for High Current and High Frequency Applications Using a Butterfly Layout," *Proceedings of the European Conference on Power Electronics and Applications (EPE'22 ECCE Europe)*, Hanover, Germany, September 2022.

[28] H. Li *et al.*, "Paralleled Operation of High-Voltage Cascode GaN HEMTs," *IEEE Journal of Emerging and Selected Topics in Power Electronics*, vol. 4, no. 3, pp. 815-823, September 2016.

Reduced-Fringing-Field Multi-MHz Capacitive Wireless Power Transfer System Using Metasurface-based Couplers with Active Field Cancellation

Syed Saeed Rashid, Dheeraj Etta, Matteo Ciabattoni, Francesco Monticone, and Khurram K. Afridi

School of Electrical and Computer Engineering
Cornell University
Ithaca, New York, USA
sr944@cornell.edu; dke27@cornell.edu; mc2574@cornell.edu; francesco.monticone@cornell.edu; afridi@cornell.edu

Abstract— **This paper introduces a novel design for a capacitive wireless power transfer (WPT) system using metasurface-based capacitive couplers (metacouplers) with active field cancellation for electric vehicle (EV) charging application. Unlike conventional metacouplers that transfer all power through the inner plates, this approach strategically divides power transfer between the inner plates and outer rings. Moreover, active phase shift control of the inverters is done to cancel out the electric fields in the surrounding space of metacouplers. An example 6.78-MHz 12-cm air-gap 1.1-kW capacitive WPT system is proposed, designed and optimized. For experimental verification, three experimental setups (a conventional coupler system, a metacoupler system without power transfer from outer ring, and a metacoupler system transferring 10% of the output power through the outer rings) are built. The developed prototype utilizing metacouplers, with 10% of the output power transferred from outer rings, achieves an efficiency of 87.6% with a significantly improved 3.8x experimental electric field reduction compared to a conventional coupler system.**

Keywords— *wireless power transfer, capacitive WPT, electric vehicle charging, mobile robots, charging pad, reduction, high-frequency inverters, fringing electric field.*

I. INTRODUCTION

The automotive industry is undergoing a major transformation, characterized by a notable shift from traditional internal combustion engines (ICEs), which have been the standard for over a century, to electric vehicles (EVs). While this transition is accelerating, the widespread adoption of EVs still faces several challenges [1] – [5]. Chief among these are concerns about the relatively limited range of EVs compared to ICE vehicles. Furthermore, the extended charging times of EVs, in contrast to the quick refueling process of ICE vehicles, and the substantial cost associated with large-capacity batteries remain significant barriers for prospective EV adopters. By enabling the in-motion charging of vehicles from the roadway, wireless power transfer (WPT) systems offer a promising solution to these main concerns related to EV adoption [2]. Conventionally, magnetic fields are utilized to transfer power wirelessly through inductively coupled coils [6] – [19]. But such inductive WPT systems require the use of ferrites for field guidance and shielding that introduce core losses, thereby, limiting the system's operating frequency and make the system heavy and expensive [10] – [15]. Capacitive wireless power transfer (WPT) systems, unlike inductive ones, use electric fields between coupled plates to transfer power [20] – [28]. Since electric fields naturally end at these plates, there is no need for ferrites or dielectric materials to guide or shield them. This makes capacitive WPT systems thinner, lighter, cheaper, and more efficient. Additionally, the absence of fragile ferrites makes them ideal for embedding into road pavements, providing a smooth and effective charging setup for electric vehicles as they move. However, these systems are not designed to restrict the high strength high-frequency fringing electric fields in the surrounding charging premises, which is an important safety consideration [29], [30]. While designs aim to reduce these fields by limiting peak voltage across the air gap [25], this alone is often insufficient. A technique for reducing fringing electric fields in capacitive WPT systems by using metasurface-based coupling plates is introduced in [23], [26]. However, it struggles to reduce fringing fields while maintaining high efficiency in high-power applications and includes an outer ring that lowers power transfer density.

This paper addresses these challenges by utilizing metasurface-based coupling plates that not only maintain a higher efficiency than [26] but also achieves 2x more reduction

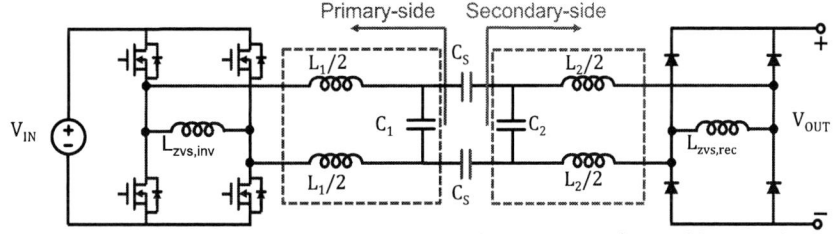

Fig. 1: Topology of a capacitive WPT system with single-stage L-section matching networks.

979-8-3315-1612-3/25 $31.00 © 2025 IEEE

in fringing electric field by active cancellation. This work also outlines the design of a capacitive WPT system based on the proposed approach, achieving an experimental 3.8x reduction in fringing electric fields when compared to a system using conventional couplers.

The remainder of this paper is organized as follows. Section II presents the architecture of the capacitive WPT system in EV charging applications. Section III details the active field cancellation using metasurface-based couplers in capacitive WPT systems. The system design of active field cancellation-enabled capacitive WPT systems is explained in Section IV. Section V presents the experimental results. Finally, Section VI summarizes and concludes the paper.

II. ARCHITECTURE OF CAPACITIVE WPT SYSTEMS FOR EV CHARGING APPLICATIONS

Figure 1 illustrates the architecture of a capacitive WPT system for EV charging application [25]. On the primary side, a full-bridge inverter transforms dc input voltage into high-frequency ac voltage, which is stepped up by an L-section matching network. This process generates a high voltage across the coupler pair, facilitating significant power transfer through a minimal displacement current over the air-gap. On the secondary side, another L-section matching network steps up the current, supplying power to the EV battery. These matching networks effectively compensate the capacitive reactance of the couplers, making the overall impedance seen by the inverter resistive. Additionally, zero voltage switching (ZVS) inductors are incorporated for the soft-switching of the inverter and rectifier transistors [31].

III. ACTIVE FIELD CANCELLATION USING METASURFACE-BASED CAPACITIVE COUPLERS

In this section, metasurface-based capacitive couplers that actively reduce the fringing electric fields in capacitive WPT systems are presented. Rectangular couplers are used for in-motion EV charging to maximize overlap time between primary and secondary pads, unlike the circular shapes in [23]. Metasurface-based couplers, as introduced in [23], use impedance sheets to reduce fringing electric fields, shown in Fig. 2(a). These sheets can be inductive or capacitive, with capacitive designs being simpler, created by adding an air gap

with a metallic ring around the coupling plate. Figures 2(b) and 2(c) show the cross-sectional and top view of one novel implementation of metasurface-based charging pads, respectively. This design is similar to the one proposed in [26] as it uses a C-shaped ring around the inner coupling plates, instead of having a full metallic ring. This approach is based on the need to restrict only the fringing electric fields outside the charging pads. By eliminating the full rings and using the C-shaped ring structure as shown in Fig. 2(b), one enhances the power transfer density.

In [26], the WPT system's configuration allows power transfer from the primary to the secondary-side exclusively through the inner plates, as illustrated in Fig. 3(a). While this setup facilitates a reduction in the electric field, its efficiency is compromised. The introduction of a gap (g) between the inner and outer plates (or rings) concentrates the electric fields within this gap, effectively minimizing the fringing electric fields beyond the metasurface-based couplers. However, this arrangement decreases the overall coupling capacitance between the primary and secondary-side by introducing a series capacitance [26].

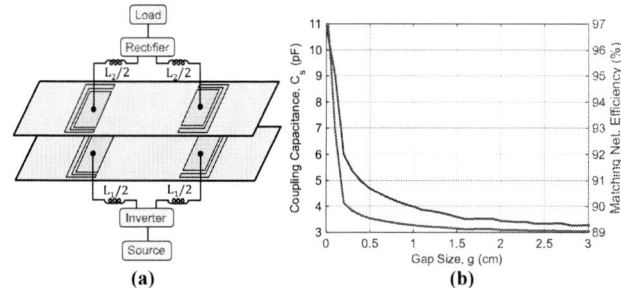

Fig. 3: (a) Conventional architecture in metasurface-based capacitive WPT systems. **(b)** Variation of coupling capacitance C_s of the metasurface-based couplers with increasing gap (g) between inner plate and outer ring. Matching network efficiency is also highlighted.

Reduction of the coupling capacitance implies that a larger inductor is needed to compensate for the coupling reactance, thus, increasing the losses. Therefore, it is natural to observe a reduction in efficiency of the WPT system with the increase in gap size (g). This exhibits a clear tradeoff between the reduction of fringing electric field and system's efficiency as depicted in Fig. 3(b).

To address this, instead of having connections just to the inner plates, one can have connections to the outer plates as well, as depicted in Fig. 4. This adaptation involves an additional inverter, matching network inductors (L'_1 and L'_2), and a rectifier. The high-frequency voltage is converted to dc output voltage by two rectifiers, each dedicated to the inner plates and outer rings. The dc output power from both is then combined to supply a common load, leveraging the ease of paralleling dc power compared to ac power.

It is intuitive to see that adding connections to the outer plates helps transfer power more efficiently by using the extra capacitance created by the outer rings. This can be seen from Fig. 5 that presents a cross-sectional view of a metacoupler

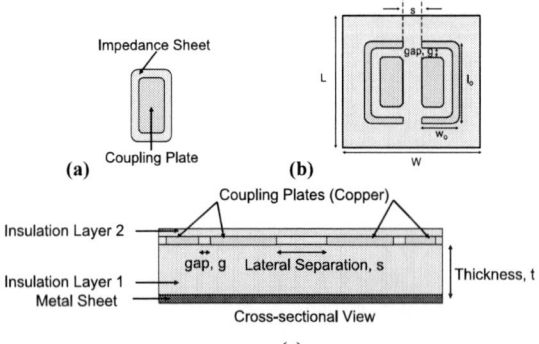

Fig. 2: (a) Conceptual metasurface-based coupling plate **(b)** A top view of the charging pads. **(c)** A cross-sectional view of the charging pads.

979-8-3315-1612-3/25 $31.00 © 2025 IEEE   1647

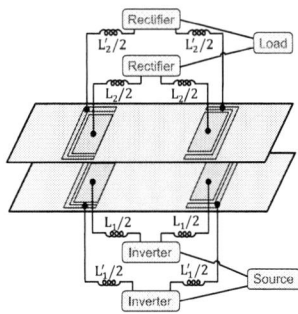

Fig. 4: Proposed architecture for metasurface-based capacitive WPT systems. (Transferring power from both outer rings and inner plates)

(inner plates and outer rings). As seen, the new arrangement of having connections with the outer plates utilizes the coupling capacitors (highlighted in green) of outer rings to transfer power as well. The diagonal capacitance C_d between an inner plate and adjacent outer C-shaped ring can absorb some power while it is being transferred. Therefore, it is imperative to maintain the gap, g between them such that the cross capacitance C_d is negligible as compared to the coupling capacitances C_s and $C_s{}'$.

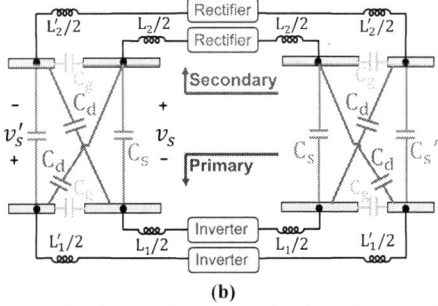

(b)

Fig. 5: Cross-sectional view of (a) conventional architecture utilizing only the inner plates for power transfer, (b) proposed architecture utilizing both inner plates and outer rings for power transfer.

Moreover, utilizing both inner plates and outer rings for power transfer could significantly increase fringing electric fields. To mitigate this, the inverters connected to the inner plates and outer rings must operate completely out-of-phase, allowing for the oppositely directed electric fields to cancel out in the surrounding free space, drawn in Fig. 6. Therefore, the proposed architecture in Fig. 4 achieves two objectives: enhancing the metacoupler-based system's overall efficiency through power transfer by outer rings and reducing fringing electric field via active phase control of the inverters.

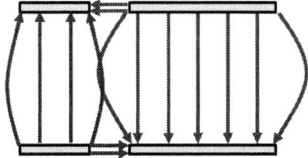

Fig. 6 A conceptual figure of a cross-sectional view of metacouplers when the inner plates and outer rings are out-of-phase. The resultant fringing electric field will reduce due to the oppositely directed electric field lines.

One might wonder if adding extra matching network inductors, an inverter, and a rectifier to use the outer ring for power transfer could lower the system's efficiency. The upcoming design section will explain how choosing the appropriate percentage of power to transfer through the outer rings can prevent any overall efficiency loss from these additional components.

IV. System Design of Active Field Cancellation-Enabled Capacitive WPT System

A methodology employed to design a capacitive WPT system for desired target specifications at chosen input (V_{IN}) and output (V_{OUT}) voltages, is described in detail in [25]. Given a target output power P_{OUT} and power transfer density constraint P_d, the outer length (l_o) and width (w_o) of the coupling plates and peak voltage across the couplers $V_{s,pk}$, are determined, using the methodology in [25]. The dimensions of the inner copper plates, i.e., scaling constants $K_l = l_i/l_o$ for length and $K_w = w_i/w_o$ for width, and the gap g are determined using the methodology presented in [26].

In the proposed architecture of Fig. 4, which utilizes outer rings for power transfer, deciding the proportion of output power transferred through these rings is critical. Let the percentage of total output power transferred through the outer rings be, $P_r = \dfrac{P_{ring}}{P_{out}}$, where P_{ring} is the output power transferred through the outer rings and P_{out} is the total output power delivered. If more power is transferred through the outer rings than the inner plates, it will mean that a higher voltage is developed across the rings, leading to higher fringing electric fields near the edges of the charging pads. As illustrated in Fig. 7, optimizing this distribution is vital for controlling fringing electric field while maintaining system efficiency.

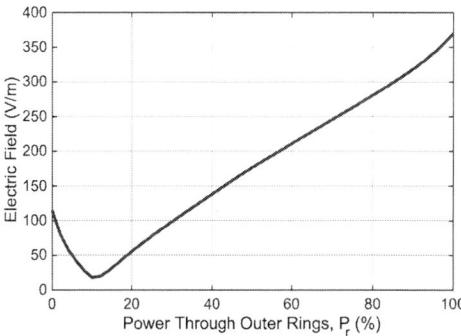

Fig. 7 Effect of increasing percentage of power transferred through outer rings on simulated electric field for an example 6.78-MHz capacitive WPT system, simulated in Ansys HFSS.

Moreover, Fig. 8 shows that system efficiency decreases as more power is allocated to the outer rings. Typically, outer rings will be smaller than inner plates to get optimum low fringing electric fields. Therefore, transferring more power through smaller coupling capacitance will lead to greater losses. Also, it is evident from Fig. 8 that having connections with the outer rings in a metacoupler is beneficial in terms of efficiency as

979-8-3315-1612-3/25 $31.00 © 2025 IEEE

compared to metacouplers with no connections to the outer ring.

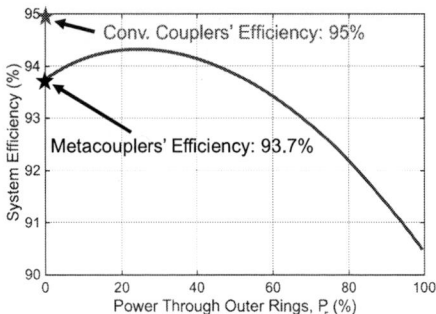

Fig. 8 Effect of increasing percentage of power transferred through outer rings on overall system efficiency for an example 6.78-MHz capacitive WPT system.

As mentioned earlier, phase difference between the two inverters is crucial for canceling fringing electric fields by ensuring out-of-phase voltages across the inner plates and outer rings. The required phase difference may not be 180°. The value of this phase difference varies with matching network inductances and capacitances. In the case where both inner plates and outer rings have identical coupling capacitances i.e., C_s, C_s' and matching network capacitances i.e., C_1, C_2, C_1', C_2'; the phase difference of the voltages developed across the inner plates (v_s) and outer rings (v_s') would be identical, therefore, inverters can be made exactly 180° out-of-phase. However, when the matching networks are not identical, the voltages developed across the plates will be phase-shifted, therefore, the inverters need to operate with a phase difference other than 180° to offset for this difference. Figure 9 represents the impact of phase difference between the actively controlled inverters. As evident, when the fringing electric fields from the inner plates and outer rings are oppositely directed, the overall resultant electric field would be the lowest. For this specific design example, when inverters are operated 220° out-of-phase, the fringing electric field is the lowest.

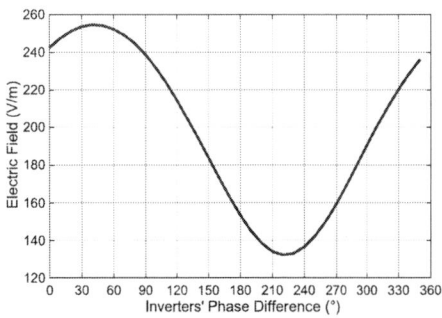

Fig. 9 Variation of simulated electric field (Ansys HFSS) with the phase-shift between two inverters, measured 79 cm away from the center of charging pads.

A 6.78-MHz, 12-cm air-gap, 1.1-kW capacitive WPT system is designed for 95% efficiency and a 4x reduction in electric fields, all within safety standards [29].

V. EXPERIMENTAL RESULTS

To verify the effectiveness of the metacoupler-based design with outer ring power transfer for reducing fringing electric fields, three 6.78-MHz, 12-cm gap, 1.1-kW capacitive WPT systems were built: a conventional coupler system, a

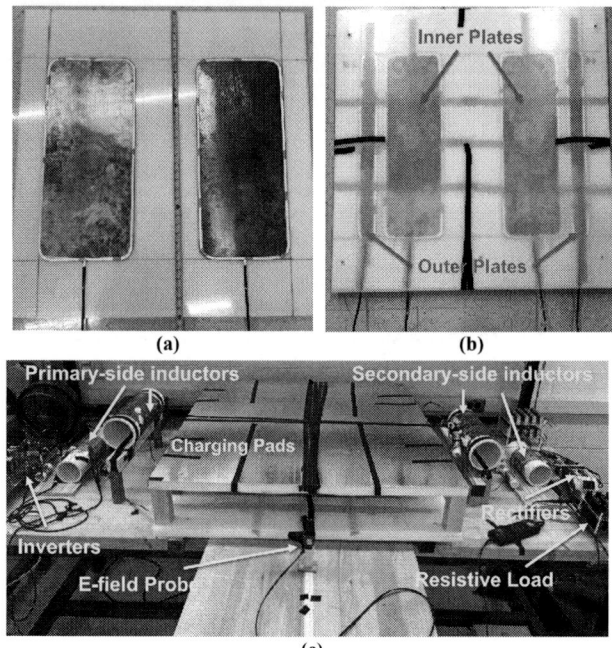

Fig. 10: Photograph of (a) conventional couplers, (b) designed metasurface-based capacitive couplers, (c) capacitive WPT experimental prototype.

metacoupler system without outer ring power transfer, and a metacoupler system transferring 10% of the output power through the outer rings. Matching networks are implemented using single-layered solenoid air-cored inductors. A 1 μH single

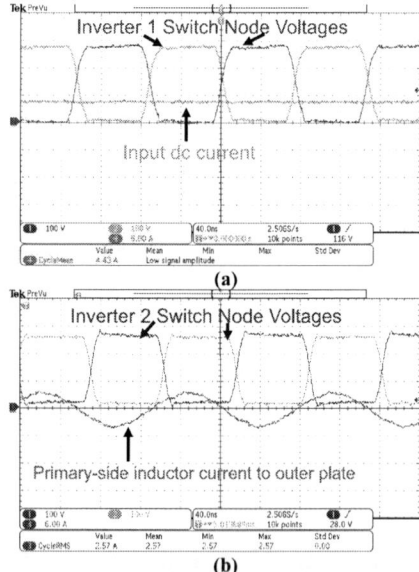

Fig. 11. Measured operating waveforms of the capacitive WPT prototype, using metacouplers, (Fig. 10(b)) while delivering 1.1-kW at 87.6% efficiency to a 40 Ω load. 10% power is processed through outer plate and the phase difference between the two inverters is 240°.

979-8-3315-1612-3/25 $31.00 © 2025 IEEE

layered solenoid air-cored inductor is used as the ZVS inductor. GS66516T 650-V GaN FETs are used as inverter transistors. On the secondary-side, 650-V STPSC12065 Schottky diodes are used in the high frequency rectifier and wideband resistors are used to emulate battery. Figure 10 illustrates the prototypes. The outer plates are not C-shaped, as electric field reduction along the car's length is unnecessary due to enough field decay over that long distance.

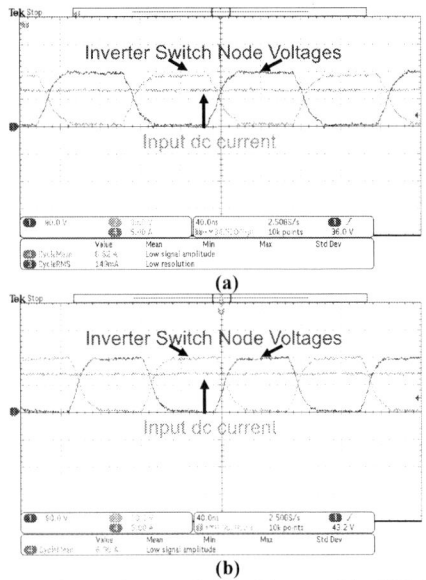

(a)

(b)

Fig. 12: Measured operating waveforms of the capacitive WPT prototype, **(a)** using conventional couplers, achieving an overall efficiency of 91%. **(b)** using metacouplers, without power transfer from outer ring, achieving an overall efficiency of 84%.

Figure 11 shows the experimental waveforms obtained from the prototype operating at 1.1 kW of output power, achieving a system efficiency of 87.6%. For comparison, capacitive WPT system using conventional couplers, as depicted in Fig. 12(a), reached an efficiency of 91%. In contrast, metacouplers that lacked connections to the outer plates delivered power at an

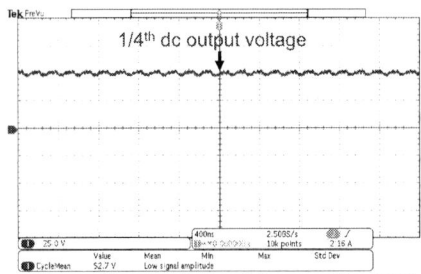

Fig. 13: Measured operating waveforms of the capacitive WPT prototype in Fig. 10(b) while delivering 1.1-kW at 87.6% efficiency to a 40 Ω load. Output voltage shown in the scope is the actual output voltage divided by 4, as it is being measured across a resistor bank out of 4 series connected resistor banks.

efficiency of 84%, as shown in Fig. 12(b). This highlights the benefits of the proposed approach, which enhances efficiency by incorporating outer plates for power transfer. Furthermore, this method achieves a significant reduction in fringing electric field —specifically, a twofold (2x) decrease compared to setups without active field cancellation and an overall 3.8x reduction when compared to conventional couplers, as evidenced from Fig. 14. Additionally, as anticipated, when the phase difference between the inverters is 0°, the electric fields from the inner and outer plates combine, resulting in higher field intensity. The resultant fields reach their minimum when the phase difference is adjusted to 240°, which aligns with the observations presented in Fig. 9. The observed system efficiencies are somewhat lower than those projected by simulations. This discrepancy can be attributed to the use of single-layered solenoid air-cored inductors within the matching networks. These inductors have a quality factor of 900, which, while high, does not match the ideal inductors assumed in simulation environments. Multi-MHz air-core inductors with quality factors reaching up to 2000 can be developed, offering the potential for enhanced overall system efficiency [32] – [34]. Additionally, for compact designs, multi-MHz core-based inductors present a viable solution, as demonstrated in [34]–[36].

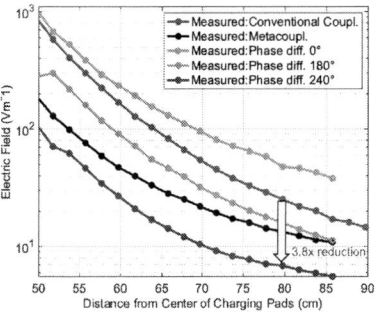

Fig. 14 Measured electric field at 130 W of output power. An experimental 3.8x reduction in electric field is observed to a conventional coupler.

Table I shows the physical dimensions of the designed 6.78-MHz 12-cm air-gap 1.1-kW capacitive WPT system.

VI. SUMMARY AND CONCLUSIONS

This paper presented a novel approach for capacitive WPT systems, utilizing metasurface-based couplers with active field cancellation. The proposed system effectively addresses challenges found in conventional capacitive WPT designs, particularly in reducing fringing electric fields while maintaining high efficiency. The design strategically distributes power transfer between inner plates and outer rings and ensures appropriate phase shift to minimize surrounding electric fields. Experimental validation on a 6.78-MHz, 12-cm air-gap WPT

TABLE I - PHYSICAL DIMENSIONS OF THE DESIGNED CHARGING PADS USING METACOUPLERS

Length, l (cm)	Width, w (cm)	Lateral Separation, s (cm)	Gap size, g (mm)	Length Scaling constant K_l	Width Scaling constant K_w	Charging Pad Total Length, L (cm)	Charging Pad Total Width, W (cm)
60	27.5	20	30	1	0.2	91	91

prototype demonstrated significant improvements. The system achieved an efficiency of 87.6%, higher than metacoupler systems without outer ring connections, which reached 84% efficiency, though it remained slightly below the 91% efficiency seen with using conventional plates. Importantly, the proposed design accomplished a 3.8x reduction in fringing electric fields compared to conventional systems, showcasing the effectiveness of the active field cancellation method. The analysis highlighted the relationship between power distribution, phase difference, and field reduction, with optimal cancellation observed at a 240° phase difference between inverters. This configuration minimized electric field intensity in surroundings.

In conclusion, the metasurface-based capacitive WPT system with active field cancellation offers a promising and efficient solution for EV charging. It enhances system efficiency and addresses safety concerns related to fringing electric fields.

ACKNOWLEDGMENT

The authors wish to acknowledge the financial support received from the U.S. Department of Energy (DOE) through ARPA-E OPEN Program under contract no. DE-AR0001572.

REFERENCES

[1] S. Y. R. Hui, W. Zhong and C. K. Lee, "A Critical Review of Recent Progress in Mid-Range Wireless Power Transfer," in *IEEE Transactions on Power Electronics*, vol. 29, no. 9, pp. 4500-4511, Sept. 2014.

[2] S. Li and C. C. Mi, "Wireless Power Transfer for Electric Vehicle Applications," *IEEE Journal of Emerging and Selected Topics in Power Electronics*, vol. 3, no. 1, pp. 4-17, March 2015.

[3] A. Mahesh, B. Chokkalingam and L. Mihet-Popa, "Inductive Wireless Power Transfer Charging for Electric Vehicles–A Review," in *IEEE Access*, vol. 9, pp. 137667-137713, 2021.

[4] Z. Zhang, H. Pang, A. Georgiadis and C. Cecati, "Wireless Power Transfer—An Overview," in *IEEE Transactions on Industrial Electronics*, vol. 66, no. 2, pp. 1044-1058, Feb. 2019.

[5] A. Ahmad, M. S. Alam and R. Chabaan, "A Comprehensive Review of Wireless Charging Technologies for Electric Vehicles," in *IEEE Transactions on Transportation Electrification*, vol. 4, no. 1, pp. 38-63, March 2018.

[6] A. Kurs, A. Karalis, R. Moffatt, J. D. Joannopoulos, P. Fisher, and M. Soljacic, "Wireless Power Transfer via Strongly Coupled Magnetic Resonances," Science, vol. 317, no. 5834, pp. 83–86, Jul. 2007

[7] M. Pinuela, D. C. Yates, S. Lucyszyn and P. D. Mitcheson, "Maximizing DC-to-Load Efficiency for Inductive Power Transfer," in *IEEE Transactions on Power Electronics*, vol. 28, no. 5, pp. 2437-2447, May 2013.

[8] M. Kesler, "Wireless Charging of Electric Vehicles," *2018 IEEE Wireless Power Transfer Conference (WPTC)*, Montreal, QC, Canada, 2018, pp. 1-4.

[9] G.A. Covic and J.T. Boys, "Modern Trends in Inductive Power Transfer for Transportation Applications," *IEEE Journal of Emerging and Selected Topics in Power Electronics*, vol. 1, no. 1, pp. 28-41, March 2013.

[10] M. Budhia, G. A. Covic and J. T. Boys, "Design and Optimization of Circular Magnetic Structures for Lumped Inductive Power Transfer Systems," *IEEE Transactions on Power Electronics*, vol. 26, no. 11, pp. 3096-3108, November 2011.

[11] B. J. Varghese, A. Kamineni, N. Roberts, M. Halling, D. J. Thrimawithana and R. A. Zane, "Design Considerations for 50 kW Dynamic Wireless Charging with Concrete-Embedded Coils," *2020 IEEE PELS Workshop on Emerging Technologies: Wireless Power Transfer (WoW)*, Seoul, Korea (South), 2020, pp. 40-44.

[12] J. B. Larsen, A. Kamineni, N. Roberts, M. Halling, P. Vaikasi and A. Barnes, "Test Platform to Evaluate Pavement Embedded Wireless Charging Pads," *2022 Wireless Power Week (WPW)*, Bordeaux, France, 2022, pp. 861-866.

[13] K. Hanawa, T. Imura, Y. Hori, H. Mashito and N. Abe, "Proposal of Coil Embedding Method in Asphalt Road Surface for Dynamic Wireless Power Transfer," *2023 IEEE Wireless Power Technology Conference and Expo (WPTCE)*, San Diego, CA, USA, 2023, pp. 1-5.

[14] J.M. Miller, O.C. Onar and M. Chinthavali, "Primary-Side Power Flow Control of Wireless Power Transfer for Electric Vehicle Charging," *IEEE Journal of Emerging and Selected Topics in Power Electronics*, vol. 3, no. 1, pp. 147-162, March 2015.

[15] V.P. Galigekere, J. Pries, O.C. Onar, G.J. Su, S. Anwar, R. Wiles, L. Seiber and J. Wilkins, "Design and Implementation of an Optimized 100 kW Stationary Wireless Charging System for EV Battery Recharging," *Proceedings of the IEEE Energy Conversion Congress and Exposition (ECCE)*, Portland, OR, September 2018.

[16] L. Gu, G. Zulauf, A. Stein, P. A. Kyaw, T. Chen and J. M. R. Davila, "6.78-MHz Wireless Power Transfer With Self-Resonant Coils at 95% DC–DC Efficiency," in *IEEE Transactions on Power Electronics*, vol. 36, no. 3, pp. 2456-2460, March 2021.

[17] A. N. Azad, A. Echols, V. A. Kulyukin, R. Zane and Z. Pantic, "Analysis, Optimization, and Demonstration of a Vehicular Detection System Intended for Dynamic Wireless Charging Applications," in *IEEE Transactions on Transportation Electrification*, vol. 5, no. 1, pp. 147-161, March 2019.

[18] C.C. Mi, G. Buja, S.Y. Choi and C.T. Rim, "Modern Advances in Wireless Power Transfer Systems for Roadway Powered Electric Vehicles," *IEEE Transactions on Industrial Electronics*, vol. 63, no. 10, pp. 6533-6545, October 2016.

[19] S.Y. Choi, B.W. Gu, S.Y. Jeon and C.T. Rim, "Advances in Wireless Power Transfer Systems for Roadway-Powered Electric Vehicles," *IEEE Journal of Emerging and Selected Topics in Power Electronics*, vol.3, no.1, pp.18-36, March 2015.

[20] C. Liu, A. P. Hu and N.K.C. Nair, "Modelling and analysis of a capacitively coupled contactless power transfer system," *IET Power Electronics*, vol. 4, no. 7, pp. 808-815, August 2011.

[21] M.P. Theodoridis, "Effective Capacitive Power Transfer," *IEEE Transactions on Power Electronics*, vol. 27, no. 12, pp. 4906-4913, December 2012.

[22] F. Lu, H. Zhang and C. Mi, "A Review on the Recent Development of Capacitive Wireless Power Transfer Technology," *Energies 2017 Special Issue on Wireless Power Transfer and Energy Harvesting Technologies*, art. 10, 1752, November 2017.

[23] S. Maji, S. Sinha, B. Regensburger, F. Monticone and K. K. Afridi, "Reduced-Fringing-Field Multi-MHz Capacitive Wireless Power Transfer System Utilizing a Metasurface-based Coupler," *2020 IEEE 21st Workshop on Control and Modeling for Power Electronics (COMPEL)*, Aalborg, Denmark, 2020, pp. 1-6

[24] S. S. Rashid, D. Etta, S. Lin and K. K. Afridi, "Pavement-Embedded Multi-MHz Capacitive Wireless Power Transfer Systems for Electric Vehicle Charging," *2024 IEEE Workshop on Control and Modeling for Power Electronics (COMPEL)*, Lahore, Pakistan, 2024, pp. 1-6.

[25] S. S. Rashid, D. Etta, S. Maji and K. K. Afridi, "Design of a High-Power Density Multi-MHz Capacitive Wireless Power Transfer System for Mobile Robots," *2023 IEEE Energy Conversion Congress and Exposition (ECCE)*, Nashville, TN, USA, 2023, pp. 1640-1645.

[26] S. S. Rashid, D. Etta, M. Ciabattoni, S. Maji, F. Monticone and K. K. Afridi, "A High-Power-Density Reduced-Fringing-Field Multi-MHz Capacitive Wireless Power Transfer System," *2024 IEEE Applied Power Electronics Conference and Exposition (APEC)*, Long Beach, CA, USA, 2024, pp. 1187-1194.

[27] D. Etta, S. S. Rashid, S. Maji and K. K. Afridi, "Design of Low-Loss Magnetic-Core Toroidal Inductor for Multi-MHz Wireless Power Transfer Systems," *2024 IEEE Workshop on Control and Modeling for Power Electronics (COMPEL)*, Lahore, Pakistan, 2024, pp. 1-6.

[28] S. S. Rashid, D. Etta, S. Maji and K. K. Afridi, "High-Performance Multi-MHz Dynamic Capacitive Wireless Power Transfer System Using Asymmetric Couplers," *2024 IEEE Wireless Power Technology Conference and Expo (WPTCE)*, Kyoto, Japan, 2024, pp. 183-188.

[29] International Commission on Non-Ionizing Radiation Protection, "ICNIRP Guidelines for limiting exposure to electromagnetic fields (100 kHz to 300 GHz)," *Health Physics*, 118(00):000–000, 2020. Pre-print. DOI: 10.1097/HP.0000000000001210.

[30] R. Ramirez-Vazquez, I. Escobar, Guy A.E. Vandenbosch, and E. Arribas, "Personal exposure to radiofrequency electromagnetic fields: A comparative analysis of international, national, and regional guidelines," *Environmental Research*, vol. 246, pp. 118124–118124, Apr. 2024.

[31] D. Etta, S. Maji and K. K. Afridi, "High-Performance Multi-MHz Capacitive Wireless Power Transfer System with an Auxiliary ZVS Circuit," *2022 IEEE Energy Conversion Congress and Exposition (ECCE)*, Detroit, MI, USA, 2022, pp. 1-6.

[32] M. Solomentsev and A. J. Hanson, "At What Frequencies Should Air-Core Magnetics Be Used?," in *IEEE Transactions on Power Electronics*, vol. 38, no. 3, pp. 3546-3558, March 2023.

[33] B. Regensburger, S. Sinha, A. Kumar, S. Maji and K. K. Afridi, "High-Performance Multi-MHz Capacitive Wireless Power Transfer System for EV Charging Utilizing Interleaved-Foil Coupled Inductors," *IEEE Journal of Emerging and Selected Topics in Power Electronics*, vol. 10, no. 1, pp. 35-51, February 2022.

[34] W. Liang, L. Raymond and J. Rivas, "3-D-Printed Air-Core Inductors for High-Frequency Power Converters," in *IEEE Transactions on Power Electronics*, vol. 31, no. 1, pp. 52-64, Jan. 2016, doi: 10.1109/TPEL.2015.2441005.

[35] D. Etta, R. Makhdoom, S. Maji, S. S. Rashid and K. K. Afridi, "High-Performance Multi-MHz Capacitive Wireless Power Transfer Systems Utilizing Magnetic-Core Coupled Inductors," *2024 IEEE Applied Power Electronics Conference and Exposition (APEC)*, Long Beach, CA, USA, 2024, pp. 2928-2932.

[36] M. V. Joisher, R. S. Bayliss, M. K. Ranjram, R. S. Yang, A. Jurkov and D. J. Perreault, "High-Performance High-Power Inductor Design for High-Frequency Applications," *2024 IEEE Applied Power Electronics Conference and Exposition (APEC)*, Long Beach, CA, USA, 2024, pp. 424-431.

[37] R. S. Yang *et al.*, "Design and Performance Comparison of Multi-Frequency Inductors for Megahertz Wireless Power Transfer," *2024 IEEE Applied Power Electronics Conference and Exposition (APEC)*, Long Beach, CA, USA, 2024, pp. 861-868.

Living object detection in wireless power transfer systems using remote capacitive bio-signals monitoring

Bruno M. G. Rosa
Department of Electrical and Electronic Engineering
Imperial College London
London, United Kingdom
b.gil-rosa@imperial.ac.uk

Paul D. Mitcheson
Department of Electrical and Electronic Engineering
Imperial College London
London, United Kingdom
paul.mitcheson@imperial.ac.uk

Abstract—Living object detection (LOD) in the context of wireless power transfer (WPT) systems can help mitigate the health risks associated with the presence of people and animals in the vicinity of high intensity magnetic fields, while assisting in the management of WPT operation in face of unforeseen link performances due to the presence of foreign objects. In this paper, we propose custom-built capacitive sensors that can detect the bio-signals for heartbeat and respiration from the body of an individual and capacitively coupled to the detection electronics during WPT operation. Performance metrics such as bio-signal amplitude and frequency content are tested for different current levels in an operating transmission coil and several distance separations from the WPT system. We hope that this study can pave the way towards the implementation of LOD systems operating in tandem with wireless power links to provide valuable feedback for safeguarding the well-being of nearby living objects, which can present high permittivity value contrasts for the typical operational frequencies in WPT.

Keywords—living object detection, capacitive sensor, wireless power transfer, bio-signals

I. INTRODUCTION

Foreign object detection (FOD) for wireless power transfer (WPT) systems remains a complex problem to solve in a robust way as systems increase their power capabilities and portfolio of applications [1], [2], ranging from small wireless chargers for mobile phones and entertainment accessories to high power units for electrical vehicles and aerial drones [3]. In any case, WPT efficiency depends on the electrical characteristics of the WPT system, including the output power of the transmitter (TX), the quality factor of the transmitting and receiving (RX) coils, the power electronics' efficiency and the coupling factor [4], [5]. While these parameters can be achieved and tested to withstand small variations in the characteristics of the TX-RX link by means of methods such as load independent tuning of the RX circuitry, frequency splitting, resonant circuit topologies

(series and parallel) and/or coil geometries adequately constrained to the dimensions of the charging area [5], factors not taken into account during the projection of the WPT system can still cause inefficiency, detuning and pose a danger to system's operation due to the presence of conductive, dielectric or ferromagnetic materials in the vicinity of the link, commonly known as foreign objects. The electrical characteristics of the WPT coils are affected by any of these objects, and their influence is stronger the larger the object and closer the proximity to the link.

In the case of high conductive objects such as metals, eddy currents can be generated from the primary magnetic field, which create their own secondary magnetic field [6], that can have a negative influence in the performance of the WPT link while causing ohmic dissipation in the surrounding medium where these currents flow, leading to a rise in temperature and possible thermal damage to the materials composing the link and foreign object itself. By its turn, ferromagnetic materials are well known concentrators of magnetic flux lines that can also negatively impact the WPT link by reducing or changing the self-inductance of the coils, thus contributing to magnetic losses in the medium [7]. Finally, human tissue, blood and other biological fluids with variable dielectric permittivity values contribute to a parasitic capacitance in the WPT system that, although negligible in terms of electrical signal at higher power levels, can still cause adverse effects to the health of people (and animals) in the vicinity of the link due to tissue heating effects, which are more pronounced for higher frequencies where WPT systems typically operate (> 100 kHz), and also to nerve stimulation effects from contact or induced currents/fields coupled to the body for frequencies in the range < 10 MHz [5]. Other physical symptoms such as dizziness, breathlessness and blood pressure changes have also been reported to occur in living objects exposed to high magnetic fields [7].

Due to the heterogeneity of materials that can pose a threat to the operation of the wireless power link, the area of foreign object detection has been traditionally divided into (foreign) metal object detection (MOD) and living object detection

This work was funded by the Engineering and Physical Sciences Research Council (EPSRC) in the United Kingdom under the grant PA1927 ("Safe Power Delivery Using a Reconfigurable Mesh of Inductive Transceivers").

979-8-3315-1612-3/25 $31.00 © 2025 IEEE

(LOD), with different solutions proposed throughout the past years to mitigate their influence in the link performance [1], [8]. While MOD solutions have been more actively researched using sensor-based methods (e.g., by deploying temperature, pressure, thermal image or radar sensors), as well as parameter detection methods (e.g., tracking self-inductance, quality factor, power transfer or power loss values in the WPT link) and sensing pattern-based methods (tracking of induced open voltages or impedance in auxiliary detection coils) [1], [4], LOD approaches have been scarcely published and applied to WPT systems, with sensor-based detection methods similar to those reported for MOD being mostly employed, but with reported low performance metrics due to impracticalities derived from the placement of either radar, optical, ultrasonic or pressure sensors inside the "field-of-view" of the WPT system without affecting the performance, or high economic costs and power requirements associated with the acquisition and deployment of such sensors, surpassing the budget allocated to build the WPT system in the first place. On the other end, other approaches for LOD using capacitive methods have come up with solutions that measure the equivalent capacitance variation of the detection system to indicate the presence of foreign objects. This is mostly achieved by placing multiple metal plates (capacitive sensors) around the transmitting coil, with different geometries and patterns [1], [5]. When a living object is close enough to be capacitively coupled to the sensor in the charging area, an extra capacitance is detected between the object and sensor, thereby causing a change in amplitude and/or phase of the original voltage signal measured between the capacitor plate and the system's ground. Nonetheless, allied to the small value of capacitance change necessary to be detected, issues associated with capacitive detection methods include the induced eddy currents in the capacitive plates of the sensor that can lead to thermal damage, the typically low signal-to-noise ratio measured and any capacitive-related perturbation not originated by living objects that degrade the detection performance.

In this paper, we propose a WPT system with living object detection capability and employing a modified version of the capacitive detection method. Unlike previous approaches that measure capacitance variations coupled within the TX circuitry, we physically separate the capacitive sensor from the transmitting circuit [5], [9], while building the former using the recent trend in distant (remote) sensing of physiological signals [10]. By monitoring bio-signals in real-time that carry important physiological information, such as the heartbeat and respiration signals wirelessly from the TX coil, we are thus more likely to correctly identify the source of capacitance disturbance in the WPT system as originating from a living object (literally by measuring the "liveliness" of the object with a periodic vital signal) than from "dead" objects. Remote sensing by capacitive sensors have been recently proposed to address the problem of direct skin contact measurements for electrocardiography (ECG) and impedance cardiography (ICG) techniques, replacing both the contact electrodes and electrolyte gel with dry (capacitive) counterparts [11] – [14]. Within this regard, AC signals generated internally in the body and carried towards the surface act like a capacitor plate connected to the electronics on the other side (or plate) of the capacitor. Electromagnetic interference (EMI) from the power grid (50/60 Hz) can also be coupled through this open connection, thus requiring different

signal amplification and conditioning circuitry relative to standard ECG measurement systems, as thoroughly described within this paper. After developing the capacitive sensor and attaching it to a magnetic coil, we deploy a class E inverter to generate a magnetic field and evaluate the effectiveness of living object detection in the presence of different field strengths, an approach not explored previously in the literature to the best knowledge of the authors of the current study. This then has the potential to solve the LOD problem in wireless power systems and, hence, associated health risks for humans using them.

II. MATERIALS AND METHODS

A. Capacitive sensor design

Capacitive sensors were designed with 3 different plates (shield, guard and active plates) on the top layer of a printed-circuit-board (PCB) to minimize electromagnetic interferences, whereas the bottom layer accommodated the electronic components, as shown in Fig. 1a. The PCB was designed in Eagle software (Autodesk, USA) and fabricated using standard PCB manufacturing technology using laminated copper foil, FR4 material and immersion silver bath for the exposed electrical pads, with a total dimension of 4 cm x 4 cm and board thickness of 1.5 mm. This electronic circuit layout followed a topology to improve performance sensitivity at larger detection distances from the signal source, as well as the reduction of induced eddy currents generated from the magnetic field, and compatibility with higher resolution acquisition systems. To that end, the active plate of the capacitive sensor was designed with different geometries (fractal and solid patterns) to test bio-signal capture in the face of strong magnetic fields and associated eddy currents induced in closed paths over the exposed metal layers. A femtoampere input-bias current amplifier (LMP7721, Texas Instruments, USA) is then connected to the active plate, to which bio-signals are capacitively coupled as shown in Fig. 1b. This signal is afterwards split into two circuit branches, one leading to signal amplification (gain of 100 V/V) while the other injects back towards the body surface a "high-pass filtered" version of the input signal through the guard plate that corresponds to the "unwanted" version of the bio-signal, whose frequency components are above 20 Hz. This signal is imposed to the guard by a buffer amplifier (OP297, Analog Devices, USA) with the common high-frequency interference (> 20 Hz) subtracted directly over the sensor plates, thus yielding a better signal quality (and increased sensitivity) at the output of a differential amplifier (AD8220, Analog Devices), followed by a 1st-order low-pass filter with cutoff frequency set around 34 Hz to attenuate the EMI signal originating from the power grid.

B. High resolution acquisition system

An 8-channel acquisition board was designed with similar PCB manufacturing parameters to acquire and digitize the analogue voltage signals originating from a maximum number

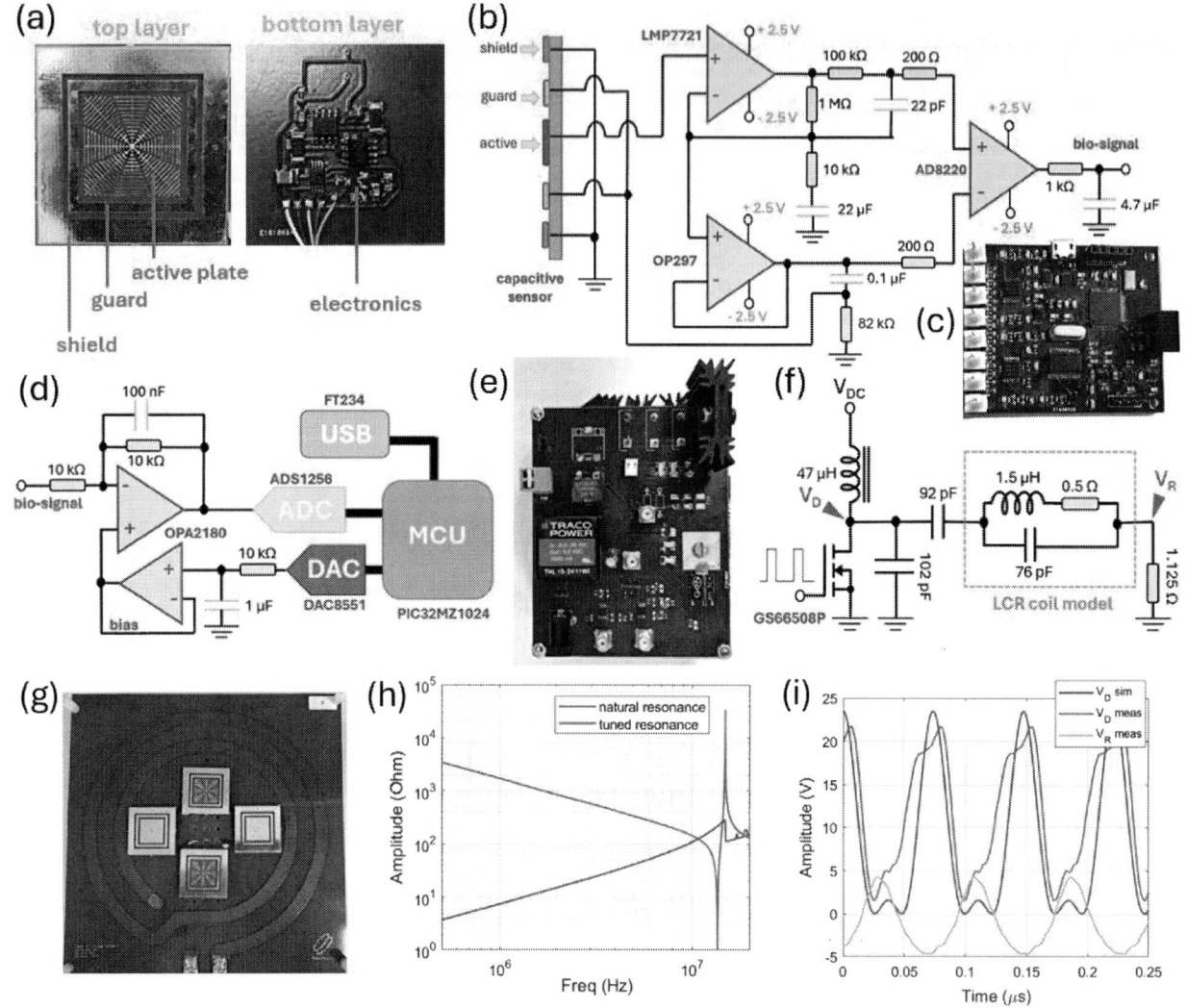

Fig. 1 – (a) Structure of the developed capacitive sensor, divided into top (signal capture) and bottom layers. (b) Electronic circuit schematic for the capacitive sensor. (c) Image of the high-resolution acquisition system designed to digitize the capacitive bio-signals. (d) Respective circuit schematic. (e) Image of the class E resonant inverter. (f) Respective circuit schematic. (g) Planar PCB containing the FR4 TX coil for inductive powering, with capacitive sensors (solid and fractal patterns) attached to the center. (h) Spectrum of the FR4 TX coil before (blue trace) and after (orange trace) tuning with a series capacitor at 13.56 MHz. (i) Voltage signals measured (V_D meas, V_R meas) and simulated (V_D sim) at different points along the circuit for the class E inverter.

of 8 capacitive sensors working simultaneously, as depicted in Figs. 1c and 1d. The high resolution (24-bit, voltage resolution ≈ 0.3 μV) and throughput analogue-to-digital converter (ADC) selected for this task (ADS1256, Analog Devices) could only accept positive inputs ($\in [0, 5]$ V, reference of + 2.5 V) and so, in order to accommodate positive and negative signal variations from the capacitive sensors (powered by rail-to-rail supplies of ± 2.5 V), a voltage level translator biased at + 2.5 V was inserted between the capacitive sensor and ADC circuitry, making use of an inverting amplifier with unitary gain and reduced feedback bandwidth (OPA2180, Texas Instruments). The biased voltage level applied to the non-inverting input of the amplifier is produced by a digital-to-analogue converter (DAC, DAC8551, Analog Devices) with digital interface to a central microcontroller (MCU, PIC32MX1024, Microchip, USA).

Digitization of the capacitive signals is performed at a rate of 100 SPS, with the samples being transferred digitally from the ADC to the MCU, before data transference to a computer using an USB transceiver (FT234, FTDIChip, UK) running at speeds of 1 Mbps. Data is finally visualized in a graphical user interface (GUI) developed in Matlab software (Mathworks Inc., USA).

C. Class E resonant inverter and FR4 TX coil

For the generation of the magnetic field, a typical class E inverter topology was adopted [15] – [19], with switching frequency set at 13.56 MHz (3 V, 50% duty cycle) for the MOSFET transistor (GS66508P, GaN Systems, Germany), as depicted in Figs. 1e and 1f. Control of the amplitude of the current signal circulating through an FR4 transmitting coil was achieved by setting different values for the DC voltage, V_{DC}, applied to input of the class E inverter. Estimation of the amplitude for the circulating current through the TX coil was performed by measuring the voltage drop produced across a resistor (1.125 Ω) connected in series with the coil and responsible for closing the current circulation path to the ground. Finally, the FR4 coil was designed as a cylindrical coil made by 2 concentric turns of copper conductor embedded on a planar PCB, with diameter of 20 cm (minor axis), conductor width of 1 cm and spacing of 1.5 cm (Fig. 1g). Measurement of the LCR model parameters for the coil was performed using an impedance analyzer (model E4990A, Keysight, USA), yielding a natural peak frequency centered at 14.930 MHz and values for L, R and C equal to 1.5 μH, 0.1 Ω and 76 pF, respectively. Tuning the coil at a frequency of 13.56 MHz was achieved experimentally by incorporating a capacitor of 92 pF in series with the coil, thus obtaining the impedance curve profile exhibited in Fig. 1h. Further theoretical validation was achieved by using the extended impedance method proposed by [20] for the simulation of the steady-state operation of a typical class E resonant inverter, with a harmonic number of K set to 8. Figure 1i shows the voltage at the transistor's drain, V_D, for the measured and simulated cases (V_{DC} = 12 V), as well as the measured voltage drop, V_R, present at the resistor (mimicking the reflected load).

III. RESULTS AND DISCUSSION

Two pairs of capacitive sensors with solid and fractal geometry patterns for the active plates were attached to the coil PCB at the center and connected to the proposed high-resolution acquisition system, both powered by USB connection to the recording computer. Power to the class E inverter was obtained from a power supply (model U8032A, Agilent Technologies, USA) with DC levels selected to generate different current amplitudes through the TX coil, with the gate signal of the transistor fed from a waveform generator (model SDG6022X, SIGLENT, China). Figures 2a and 2b then show the disturbance on the capacitive signals in real-time (computer) due to the presence of a hand and object in front of the TX coil, whereas Fig. 2c depicts thermal images (model E63900, FLIR, Sweden) captured for the distribution of the temperature profile around the capacitive sensors in the absence and presence of magnetic field, respectively, the latter produced by the class E inverter driving a TX coil current of 8 A. From the figure, a slight temperature increase ($\Delta T \approx 5.7$ °C) can be seen, specially over the solid metal area composing the active plates. The fractal pattern by contrast attenuates better the effect of circulating eddy currents in closed circuit paths and, hence, the temperature is lower.

By performing different types of body movements – walking and running – directly in front of the TX coil, exemplary capacitive signals are obtained (Figs. 2d and 2e), the former in function of the distance separation from the TX coil (without magnetic field), while the latter was recorded for a separation of 0.5 m in the presence of different magnetic field strengths according to the amplitude of the circulating current in the TX coil (from 0 A to 8 A), thus illustrating the ability of the combined WPT and capacitive systems to detect human body movements at large distances from the magnetic coil during active operation. Furthermore, the bar plot in Fig. 2f exhibits the variation of the signal amplitude detected by the tested capacitive sensors (in average) for the running exercise in function of different separation distances and current levels in the TX coil. Although not explored directly in the current study, features of the recorded capacitive signals such as amplitude, slope, frequency content and time extension and can be further explored to identify and classify different types of human activities performed around the WPT system, including movements in the upper body part (e.g., speech, respiration) using computational techniques such as machine learning and neural networks. This helps create context awareness for smart WPT systems in the presence of human individuals to adjust wireless power levels and comply with safe exposure standards.

Finally, regarding the detection of bio-signals, Figs. 2g and 2f show an extended time capture of the heartbeat and respiration signals, respectively, when the chest region of an individual is directly in front of the TX coil during active operation (coil current: 6 A). From the results, it can be seen that higher signal amplitudes are recorded for the solid plate geometry relative to the fractal pattern due to the larger capacitive-sensing area. Nonetheless, the capacitive sensors with fractal plates are still able to capture the characteristic baseline drifts and waveform patterns associated with the respiration and heartbeat, as further attested by looking at specific bands within the spectrum of these signals depicted in Fig. 2i (Fast Fourier Transform, frequency components: ≈ 0.25 and 1.25 Hz, respectively).

IV. CONCLUSION AND FUTURE WORK

A WPT system with capacitive detection capability has been presented and tested for the capture of body movements and bio-signals (heartbeat and respiration). This was achieved by building a separate detection module outside the TX circuitry, thus surpassing the drawbacks associated with the measurement of tiny equivalent capacitance variations directly over the transmitter during its operation. The proposed system helped to track capacitively-coupled body positional changes and bio-signals, which carry with them the "liveliness" mark of living objects and, thus, can help distinguish them from those originated from "dead" objects or materials. As future work, we envision to derive a mathematical model for the propagation of the electrical displacement currents from the surface of the body (cardiac current) and extended to the presence of magnetic sources in the surrounding medium. Different arrangements and combinations of capacitive sensors around the transmitting coil can also be investigated to increase the "field-of-view" for living object detection, and similar to a tomographic measurement setup. Finally, the capacitive detection capability in the presence of higher current levels circulating inside the TX coil needs to be evaluated for higher power WPT systems.

979-8-3315-1612-3/25 $31.00 © 2025 IEEE

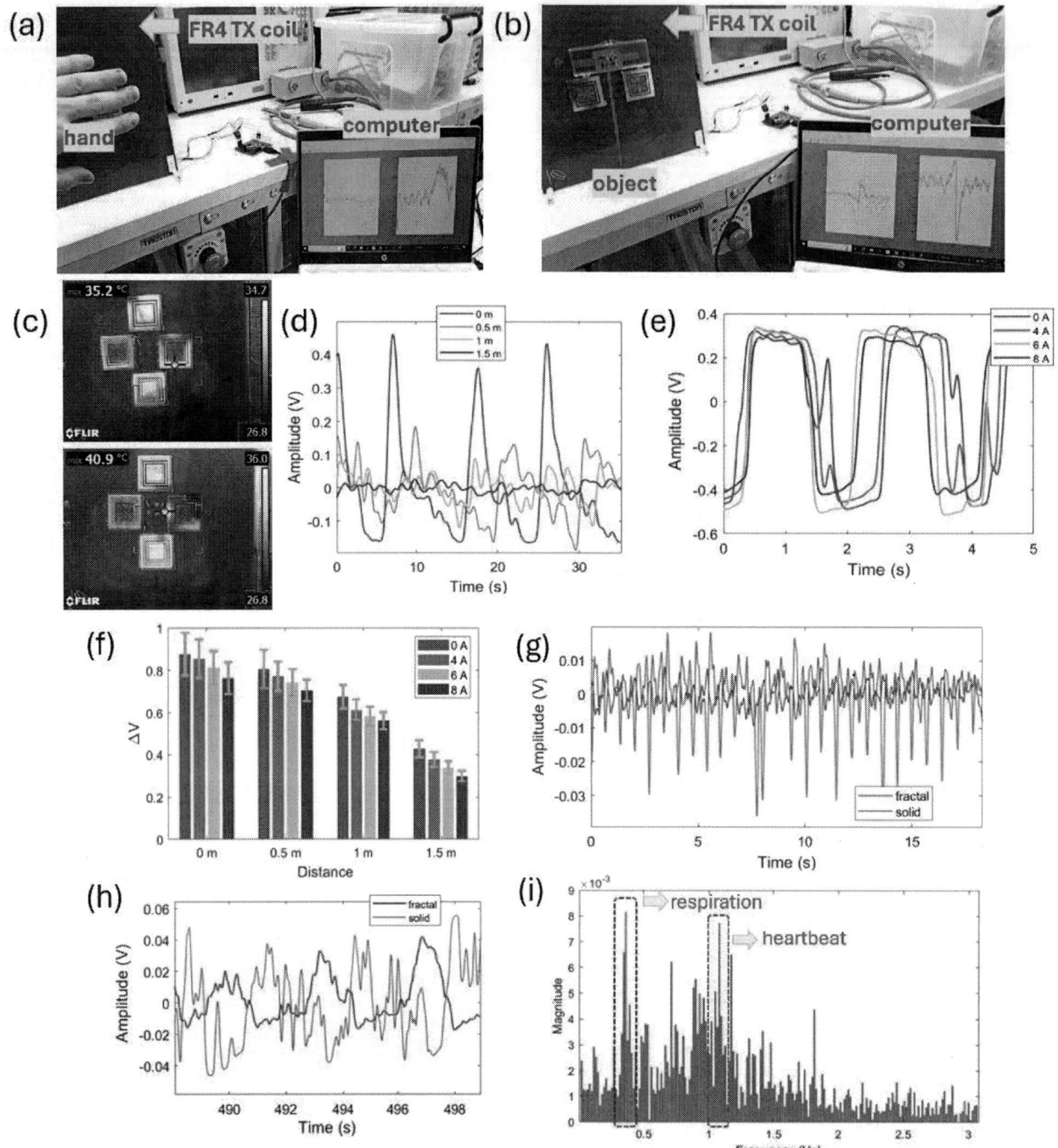

Fig. 2 – (a) – (b) Setup disturbed by a hand and electronic object placed in front of the TX coil. **(c)** Temperature profile distribution along the 4 capacitive sensors in the absence (top) and presence (bottom) of magnetic field (current level = 8 A). **(d) – (e)** Setup disturbed by performing different tasks – walking and running – by an individual in front of the WPT setup, respectively, the former for different distances from the TX coil – 0 to 1.5 m – without inductive powering (fractal geometry), while the latter is for the same distance (0.5 m) and different current levels (0 to 8 A) in the TX coil (fractal geometry). **(f)** Statistical metrics for the capacitive signal amplitude recorded amongst the tested sensors during the running task in function of the distance to the coil and current level. **(g)** Extended recording of the heartbeat signal of an individual in front of the TX coil and acquired for two types of capacitive sensors (solid and fractal geometries) at a distance of 10 cm and current level of 6 A. **(h)** Influence of breathing movements (baseline drift) in the same conditions (inhale: peaks, exhale: valleys). **(i)** Low frequency spectrum of the detected bio-signals capacitively coupled to the sensors, whose principal frequency components/bands are identified as derived from the heartbeat and respiration movements.

ACKNOWLEDGMENT

The authors would like to thank Dr. Nunzio Pucci from Imperial College London for his input in the design of the class E inverter and FR4 coil.

REFERENCES

[1] Y. Zhang, Z. Yan, J. Zhu, S. Li and C. Mi, "A review of foreign object detection (FOD) for inductive power transfer systems", *eTransportation*, vol. 1, no. 100002, 2019.

[2] L. Xiang, Z. Zhu, J. Tan and Y. Tian, "Foreign object detection in a wireless power transfer system using symmetrical coil sets", *IEEE Access*, vol. 7, pp. 44622 – 44631, 2019.

[3] J. M. Arteaga, S. Aldhaher, G. Kkelis, C. Kwan, D. C. Yates and P. D. Mitcheson, "Dynamic capabilities of multi-MHz inductive power transfer systems demosntrated with baterryless drones", *IEEE Transactions on Power Electronics*, vol. 34, no. 6, pp. 5093 – 5104, 2019.

[4] B. Zhou, Z. Z. Liu, H. X. Chen, H. Zeng and T. Hei, "A new metal detection method based on balanced coil for mobile phone wireless charging systems", *IOP Conference Series: Earth and Environmental Science*, vol. 40, no. 012029, 2016.

[5] J. Lu, G. Zhu and C. C. Mi, "Foreign object detection in wireless power transfer systems", *IEEE Transactions on Industry Applications*, vol. 58, pp. 1340 – 1354, 2022.

[6] S. Grimnes and O. G. Martinsen, *Bioimpedance & Bioelectronics Basics 2ⁿᵈ Ed.*, Elsevier Ltd., Amsterdam, 2008.

[7] K. van Schuylenbergh and R. Puers, *Inductive Powering: Basic Theory and Applications to Biomedical Systems*, Springer Nature, New York, 2009.

[8] J. Xia, X. Yuan, J. Li, S. Lu, X. Cui, S. Li and L. M. Fernandez-Ramirez, "Foreign object detection for electric vehicle wireless charging", *Electronics*, vol. 9, no. 805, 2020.

[9] V. X. Thai, J. H. Park, S. Y. Jeong and C. T. Rim, "Multiple comb pattern-based living object detection with enhanced resolution design for wireless electrical vehicle chargers", in *PCIM Europe*, 2018, pp. 1818 – 1823.

[10] F. R. Parente, M. Santonico, A. Zompanti, M. Benassai, G. Ferri, A. D'Amico and G. Pennazza, "An electronic system for the contactless reading of ECG signals", *Sensors*, vol. 17(11), no. 2474, 2017.

[11] H.-L. Peng, J.-Q. Liu, Y.-Z. Dong, B. Yang, X. Chen and C.-S. Yang, "Parylene-based flexible dry electrode for biopotential recording", *Sensors and Actuators B: Chemical*, vol. 231, pp. 1 – 11, 2016.

[12] A. A. Chlaihawi, B. B. Narakathu, S. Emamian, B. J. Bazuin and M. Z. Atashbar, "Development of printed and flexible dry ECG electrodes", *Sensing and Bio-Sensing Research*, vol. 20, pp. 9 – 15, 2018.

[13] B. Babusiak, S. Berik and L. Balogova, "Textile electrodes in capacitive signal sensing applications", *Measurement*, vol. 114, pp. 69 – 77, 2018.

[14] N. Jonassen, "Human body capacitance: static or dynamic concept?", in *IEEE Electrical Overstress/Electrostatic Discharge Symposium*, 1998, pp. 111 – 117.

[15] N. O. Sokal and A. D. Sokal, "Class E – a new class of high-efficiency tuned single-ended switching power amplifiers", *IEEE Journal of Solid-State Circuits*, vol. 10, no. 3, pp. 168 – 176, 1975.

[16] M. Acar, A. J. Annema and B. Nauta, "Analytical design equations for class-E power amplifiers", *IEEE Transactions on Circuits and Systems – I: Regular Papers*, vol. 54, no. 12, pp. 2706 – 2717, 2007.

[17] T. Suetsugu and M. K. Kazimierczuk, "Off-nominal operation of class-E amplifier at any duty cycle", *IEEE Transactions on Circuits and Systems – I: Regular Papers*, vol. 56, no. 6, pp. 1389 – 1397, 2007.

[18] F. Wen and R. Li, "Parameter analysis and optimization of class-E power amplifier used on wireless power transfer system", *Energies*, vol. 12, no. 3240, 2019.

[19] N. Pucci, J. M. Arteaga, C. H. Kwan, D. C. Yates and P. D. Mitcheson, "Induced voltage estimation from class EF switching harmonics in HF-IPT systems", *IEEE Transactions om Power Electronics*, vol. 37, no. 4, pp. 4903 – 4916, 2022.

[20] J. Liang and W.-H. Liao, "Steady-state simulation and optimization of class-E power amplifiers with extended impedance method", *IEEE Transactions on Circuits and Systems – I: Regular Papers*, vol. 58, no. 6, pp. 1433 – 1445, 2011.

Modified N:1 Switched Capacitor Converter with Reduced Capacitor DC Bias Voltage for High Power Density

Taewoo Lee
Department of Electrical and Computer Engineering
Seoul National University
Seoul, Korea
taewoo97@snu.ac.kr

Dam Yun
Department of Electrical and Computer Engineering
Seoul National University
Seoul, Korea
dam0710@snu.ac.kr

Sunghyuk Choi
Department of Electrical and Computer Engineering
Seoul National University
Seoul, Korea
sunghyukchoi@snu.ac.kr

Jung-Ik Ha
Department of Electrical and Computer Engineering
Seoul National University
Seoul, Korea
jungikha@snu.ac.kr

Abstract— **Switched capacitor converters require large capacitors due to the burden of high peak currents, which imposes a limitation on the minimum volume of passive components. This paper structurally proposes a new topology that reduces the DC voltage stress of capacitors, aiming to decrease stored energy and improve passive component volume. It is derived by modifying the connection state of the flying capacitors in the conventional Dickson converter. A consistent derivation method is applied to generalize it for a voltage conversion ratio of N:1. The improvement in passive component volume due to the reduction in DC voltage stress is confirmed through the output impedance analysis. Furthermore, the theoretical decrease of output impedance is derived through an equivalent circuit analysis based on the switch connection states. The proposed topology is experimentally verified with 48 V input to 8 V, 20 A output prototype. A full load efficiency of 96 %, which is 0.8% higher than that of the conventional Dickson converter, was achieved, with a 43.3 % reduction in passive component volume.**

Keywords—Switched capacitor converter (SCC), Dickson, DC bias voltage, Utilization, Output impedance, Fast switching limit.

I. INTRODUCTION

Switched Capacitor Converter (SCC) is composed of only capacitors and switches. Unlike PWM converters, which include bulky magnetic components, SCCs utilize high energy density capacitors to achieve high power density [1], [2]. These SCCs are applied in the design of high-speed charging systems within electronic devices, power delivery for AI semiconductors, and bus converters in data centers. SCC topologies include Doubler, Fibonacci, Series-parallel, Ladder, and Dickson [3], [4], [5]. Among these, the Dickson SCC is widely used in high-current applications due to its low switch current stress resulting from the parallel operation of capacitors and switches [6]. However, it has the disadvantage of relatively higher capacitor voltage stress compared to other topologies, leading to increased passive component volume.

Reducing the DC bias voltage leads to a decrease in capacitor volume and improves the utilization of capacitors in power transfer applications. The volume of a capacitor is proportional to its maximum energy storage capacity, which is determined by the DC bias voltage and the voltage swing. Voltage swing depends on the output current, capacitance, and switching frequency, and, assuming design specifications remain constant, the structural topology determines the voltage swing, which can be considered a fixed parameter. Thus, focusing solely on the DC bias voltage, its reduction is expected to result in reduced capacitor volume. The utilization of a capacitor refers to the ratio of the energy utilized for actual power transfer,

The utilization of a capacitor refers to the ratio of the energy utilized for actual power transfer, determined by the voltage swing, to the capacitor's maximum storable energy. Considering only the DC bias voltage, reducing the DC bias voltage enhances the utilization of the capacitor.

The output impedance of SCC is a critical indicator of the power density and efficiency of the system [7], [8], [9]. Output impedance varies with frequency and can be divided into the Slow Switching Limit (SSL) and Fast Switching Limit (FSL) regions. SSL impedances are related to capacitor losses and are determined by the capacitance and the charge flowing through the capacitors [10]. In the SSL region, the dominant factor contributing to loss is the charge-sharing loss caused by complete charging and discharging of the capacitors, resulting in output impedance inversely proportional to frequency. Conversely, in the FSL region, the dominant loss mechanism is the conduction loss of switches due to incomplete charging and discharging of the capacitors, leading to a constant minimum output impedance independent of frequency. Previous research [11] has proposed topology optimization to minimize SSL impedance under the constraint of capacitor energy storage. The SSL impedances related to capacitor charge-sharing loss are determined by the capacitance and the amount of charge flowing through the capacitor. Analytical studies exist to optimize the topology for minimizing SSL impedances under the constraint of the capacitor's energy storage capacity. Through this approach, the energy storage capacity of the capacitor can be calculated, enabling a comparison of the volume of passive components across different topologies. FSL impedances can be derived considering the charge and current flowing through the switches, as well as the voltage stress on the switches, which correspond to the conduction losses in the converter [11], [12]. Since the power flow and switch configuration vary by topology, analyzing the FSL impedances enables a comparison of converter efficiency.

In this study, a new topology is proposed for a voltage conversion ratio of N:1 by altering the positions of the capacitors in the conventional Dickson SCC. The derivation process of the proposed topology is presented, along with a comprehensive analysis of its operating principles. The SSL impedances of the proposed topology and the conventional Dickson SCC are optimized, and the differences in their capacitor energy storage are analyzed, demonstrating passive component volume reduction due to the decreased DC voltage stress on the capacitors. Additionally, the reduction in output impedance in the FSL region is theoretically derived by calculating the power loss associated with the on-resistance of the switches, considering the voltage stress applied to the switches and the equivalent series resistance (ESR) of

979-8-3315-1612-3/25 $31.00 © 2025 IEEE

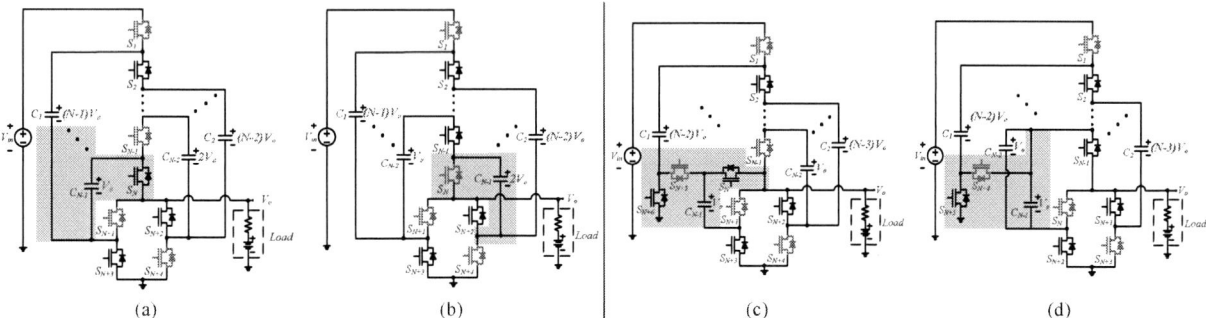

Fig. 1. Circuit diagram of N:1 Dickson SCC and proposed SCC, (a) conventional with even voltage conversion ratio, (b) conventional with odd voltage conversion ratio, (c) proposed with even voltage conversion ratio, (d) proposed with odd voltage conversion ratio.

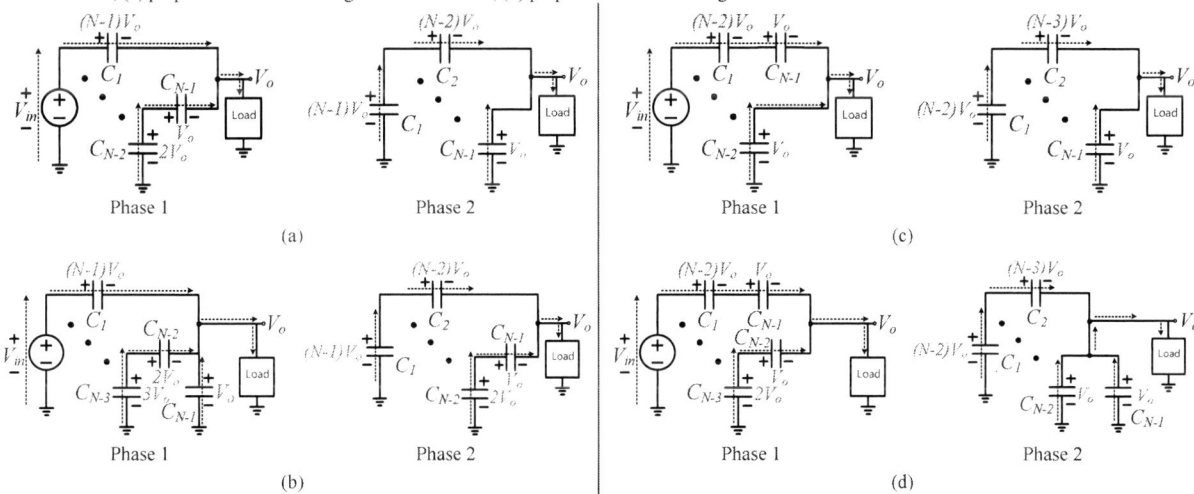

Fig. 2. Capacitor equivalent circuit, (a) conventional with even voltage conversion ratio, (b) conventional with odd voltage conversion ratio, (c) proposed with even voltage conversion ratio, (d) proposed with odd voltage conversion ratio.

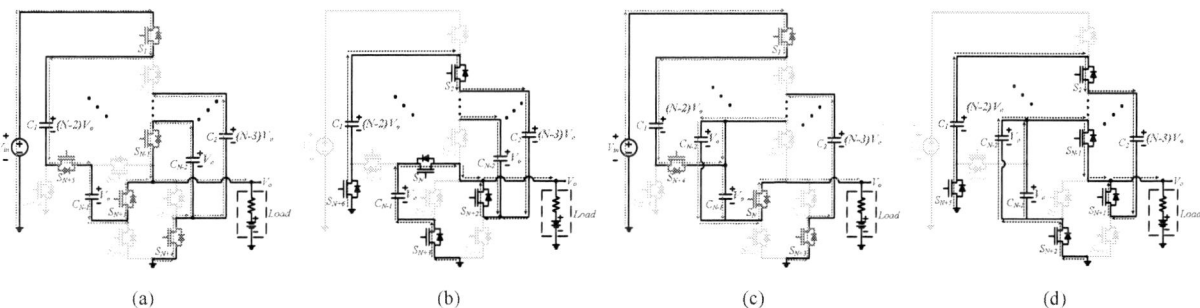

Fig. 3. Proposed circuit diagram representing the power flow in each phase, (a) phase 1 for even voltage conversion ratio, (b) phase 2 for even voltage conversion ratio, (c) phase 1 for odd voltage conversion ratio, (d) phase 2 for odd voltage conversion ratio.

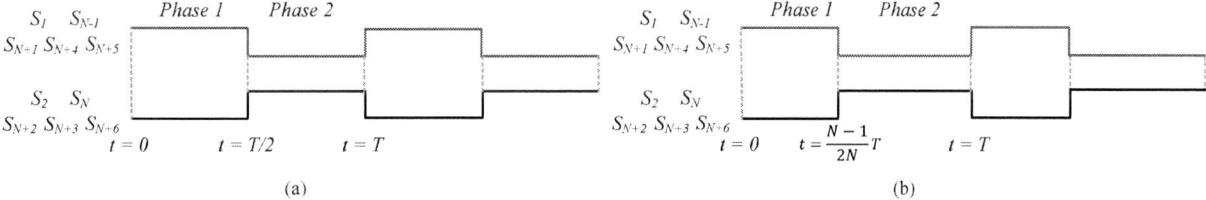

Fig. 4. Gate signals for switches in two-phase operation of proposed circuit, (a) at even voltage conversion ratio, (b) at odd voltage conversion ratio.

capacitors. Class 2 MLCCs were utilized in capacitor design to enhance power density, and a cascaded bootstrap gate driver circuit was applied to drive the switches. Through experiments on a specific voltage ratio, the validity of the proposed circuit is demonstrated by showing the reduction in passive component volume due to decreased DC bias voltage and improvements in efficiency.

II. ANALYSIS OF PROPOSED TOPOLOGY

A. Derivation of Generalized Proposed Topology

Fig. 1 shows N:1 Dickson SCC and proposed N:1 SCC for even and odd voltage conversion ratios. The switches operate in two states, distinguished by colors. Fig. 2 presents the equivalent circuits with DC voltage stress indicated for the capacitors in both the conventional and the proposed topologies for even and odd voltage ratios, respectively. The DC voltage stress of the capacitors can be expressed as an integer multiple of the output voltage V_o. In the conventional Dickson SCC, the voltage stress on the capacitor C_1 is $(N-1)V_o$. To reduce this voltage stress, the position of C_{N-1} is shifted to the upper node, placing it in series with C_1. As shown in Fig. 2(a), (b), the DC voltage stress on C_1 is distributed between C_1 and C_{N-1} in Fig. 2(c), (d), reducing the voltage stress to $(N-2)V_o$ and V_o. This adjustment lowers the DC voltage stress by V_o for all capacitors C_k ($k: 1, 2, ..., N-2$), resulting in a total reduction of $(N-2)V_o$ in capacitor DC voltage stress. The position of the capacitors is changed, but the charging state in Fig. 2(c) remains the same in the same phase. However, in Fig. 2(d), the charging and discharging states change, and the number of parallel paths for power transfer between phase 1 and phase 2 is reversed, altering the duty cycle as the charge ratio at the output changes. Therefore, when configuring the topology by connecting the switches according to the equivalent circuit, 1 or 2 additional switches are required, leading to the topologies shown in Fig. 1(c), (d).

B. Operating Principles

Fig. 3 illustrates the power flow of the generalized proposed topology under even and odd voltage conversion ratios. Fig. 4 shows the gate signals for two-phase operation, demonstrating the operation of Fig. 3 through complementary switching between the red and black switches. Fig. 3(a) and 3(b) show the current flow during Phase 1 and 2 of a circuit with an even voltage conversion ratio. During this process, capacitors C_1 and C_{N-1} are charged by the voltage source, while power transfer to the load occurs via the discharge of capacitor C_{N-2}. Then, the voltage source is disconnected, and the charging and discharging states of the capacitors are reversed to continue power transfer. Similarly, Fig. 3(c) and 3(d) show the current flow during Phase 1 and 2 of a circuit with an odd voltage conversion ratio. In this case, capacitors C_1 and C_{N-1} are charged by the voltage source, while C_{N-2} is also charged through the connected capacitors, ensuring power delivery to the load. Subsequently, the voltage source is disconnected, and the charging and discharging states of the capacitors are reversed.

C. Analysis of Capacitor Volume and Output Impedance

Capacitor volume can be evaluated by considering energy storage cost. The capacitance is optimized to minimize SSL impedances, considering the energy storage cost related to the DC voltage stress on the capacitors. As the capacitor voltage stresses differ between the conventional and proposed circuits, a difference in the optimized SSL impedances occurs. Thus, by comparing the energy storage cost under the identical SSL impedances, the reduction in capacitor volume in the proposed N:1 SCC relative to the conventional N:1 Dickson SCC is analyzed.

$$C_i = \left| \frac{a_{c,i}}{v_{c,i(rated)}} \right| \frac{2E_{tot}}{\sum_k |a_{c,k} v_{c,k(rated)}|} \quad (1)$$

Equation (1) represents the optimized capacitance that minimizes SSL impedances in two-phase operation [10]. Since the total energy constraints are satisfied and the charge vector $a_{c,i}$ is identical for both the conventional and proposed topologies, the capacitance is determined in inverse proportion to the DC voltage stress $v_{c,i(rated)}$.

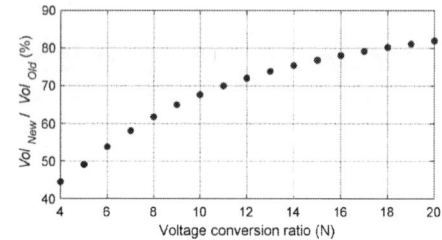

Fig. 5. Comparison of passive component volume between conventional and proposed circuits according to voltage conversion ratio.

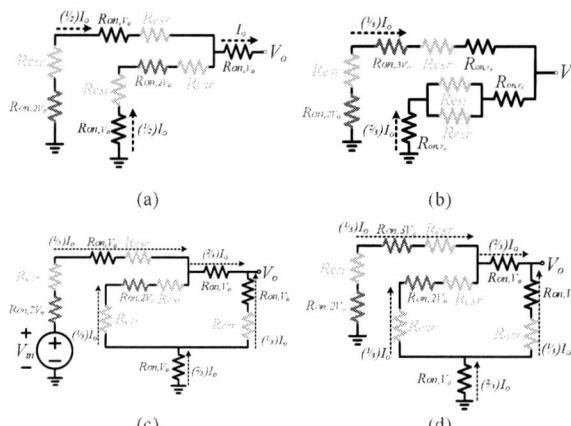

(a)

(b)

(c)

(d)

Fig. 6. Resistive component equivalent circuit of proposed SCC, (a) phase 1 (N=5), (b) phase 2 (N=5), (c) phase 1 (N=6), (d) phase 2 (N=6).

TABLE I. FSL IMPEDANCES OF N:1 SCC

N:1 SCC	FSL impedances (R_{FSL})
Dickson, even	$\frac{2N^2 - 4N + 8}{N^2} \cdot R_{on,V_o} + \frac{2N-4}{N^2} \cdot R_{on,2V_o} + \frac{4N-4}{N^2} \cdot esr$
Dickson, odd	$\frac{2N^2 - 2N + 4}{N(N+1)} \cdot R_{on,V_o} + \frac{2N-2}{N(N+1)} \cdot R_{on,2V_o} + \frac{4}{N+1} \cdot esr$
Proposed, even	$\Delta : \frac{6-4N}{N^2} \cdot R_{on,V_o} + \frac{2}{N^2} \cdot R_{on,3V_o}$
Proposed, odd	$\Delta : \left(\frac{4}{N-1} + \frac{2}{N} - \frac{10}{N+1} \right) \cdot R_{on,V_o} + \frac{2}{N(N+1)} \cdot R_{on,3V_o}$

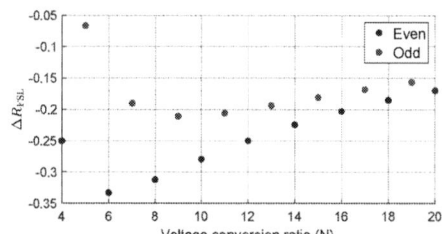

Fig. 7. Reduction of FSL impedances in the proposed topology compared to conventional Dickson SCC according to voltage conversion ratio (ΔR_{FSL}: coefficients of R_{on,V_o}, $R_{on,3V_o} = 3R_{on,V_o}$).

$$C_i = \begin{cases} \dfrac{m}{N-i} & \text{[Dickson]} \\ \dfrac{n}{N-i-1} & (i \leq N-2) \quad \text{[Proposed]} \\ n & (i = N-1) \end{cases} \quad (2)$$

TABLE II. COMPONENT LIST OF HARDWARE PROTOTYPE

Component		Part number	Parameters
Switch	Dickson	$S_1 \sim S_{12}$, Infineon, ISK024NE2LM5	25 V, 2.4 mΩ
	Proposed	$S_1, S_3 \sim S_{12}$, Infineon, ISK024NE2LM5	25 V, 2.4 mΩ
		S_2, Infineon, BSZ024N04LS6	40 V, 2.4 mΩ
Flying capacitor	Dickson	C_1, C_2 Murata, GRM31CD71H106KE11L	3216, X7T, 50 V, 10 µF* × 9 (in parallel)
			× 8 (in parallel)
		C_3, TDK, C2012X5R1V226M125AC	2012, X5R, 35 V, 22 µF* × 20 (in parallel)
		C_4, Murata, GRM188C61E226ME01W	1608, X5S, 25 V, 22 µF* × 22 (in parallel)
		C_5, Murata, GRM158R61A226ME15D	1005, X5R, 10 V, 22 µF* × 41 (in parallel)
	Proposed	C_1, Murata, GRM31CD71H106KE11L	3216, X7T, 50 V, 10 µF* × 6 (in parallel)
		C_2, TDK, C2012X5R1V226M125AC	2012, X5R, 35 V, 22 µF* × 15 (in parallel)
		C_3, Murata, GRM188C61E226ME01W	1608, X5S, 25 V, 22 µF* × 16 (in parallel)
		C_4, C_5, Murata, GRM158R61A226ME15D	1005, X5R, 10 V, 22 µF* × 30 (in parallel)
			× 30 (in parallel)
Input capacitor Output capacitor		KEMET, C1206C474KMREC7210	3216, X7R, 63 V, 0.47 µF* × 10 (in parallel)
Gate driver		Infineon, 1EDB7275FXUMA1	5 A (Peak source), 9 A (Peak sink)

* The capacitance listed in this table is the nominal value before DC derating.

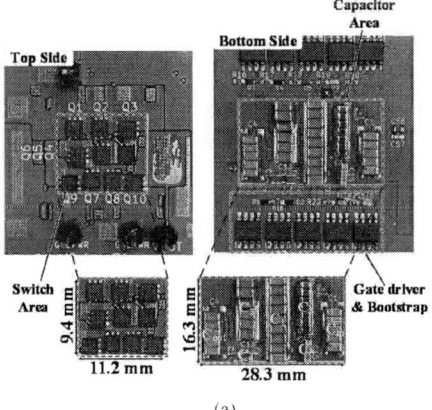
(a)

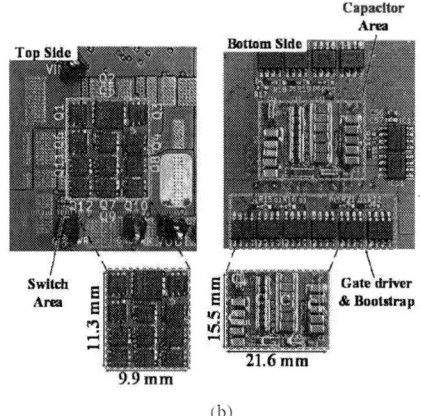
(b)

Fig. 8. Prototype of 6:1 SCC, (a) Dickson SCC, (b) proposed SCC.

Equation (2) simplifies equation (1) by expressing the inverse of the capacitor DC voltage stress $v_{c,i(rated)}$ multiplied by coefficients m and n, to generalize for a voltage conversion ratio of N:1.

$$a_{c,i} = a_{c,i}^1 = -a_{c,i}^2 = \frac{1}{N} \qquad (3)$$

$$R_{SSL} = \begin{cases} \dfrac{N-1}{2mN}/f_{sw} & \text{[Dickson]} \\ \dfrac{N^2 - 3N + 4}{2nN^2}/f_{sw} & \text{[Proposed]} \end{cases} \qquad (4)$$

$$E_{tot} = \begin{cases} \dfrac{mN(N-1)V_o^2}{4} & \text{[Dickson]} \\ \dfrac{n(N^2 - 3N + 4)V_o^2}{4} & \text{[Proposed]} \end{cases} \qquad (5)$$

$$\frac{Vol_{New}}{Vol_{Old}} = \left(\frac{N^2 - 3N + 4}{N(N-1)}\right)^2 \quad (N \geq 4) \qquad (6)$$

Here, m, n represent the charge flowing through the capacitor C_i. Since both the conventional and proposed topologies operate in two phases and the charge vector $a_{c,i}$ through the capacitors is uniformly $\frac{1}{N}$, the SSL impedances R_{SSL} are expressed as equation (4) through equations (2) and (3). Equation (5) represents the energy stored in the capacitors. Equation (6) generalizes the passive component volume of the proposed topology relative to that of the conventional Dickson

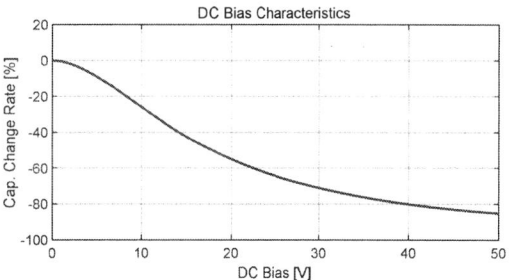

Fig. 9. Murata GRM31CD71H106KE11L capacitor dc bias characteristics.

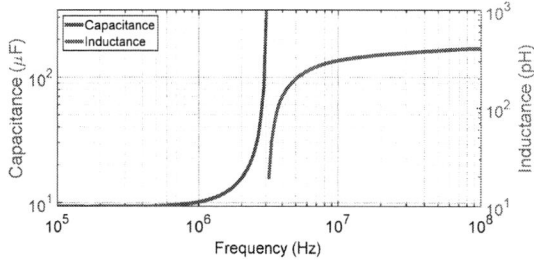

Fig. 10. Analysis of capacitance and inductance with frequency for KEMET C0805C106K3PACTU.

SCC for a voltage conversion ratio of N:1. This generalization is achieved by comparing the total energy E_{tot} from equation (5) under the assumption of equal energy density and identical SSL impedances while maintaining the optimized capacitance ratios.

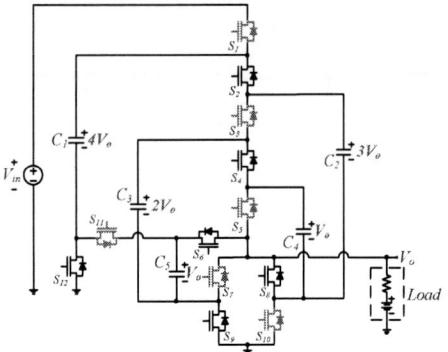

Fig. 11. Proposed 6:1 SCC.

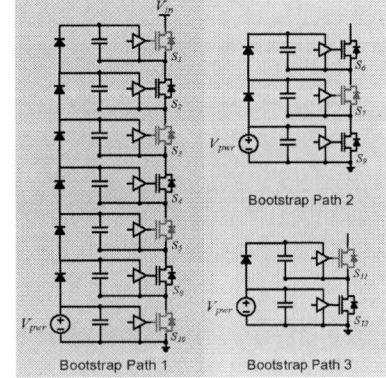

Fig. 12. Cascaded bootstrap circuits implemented in the proposed 6:1 SCC.

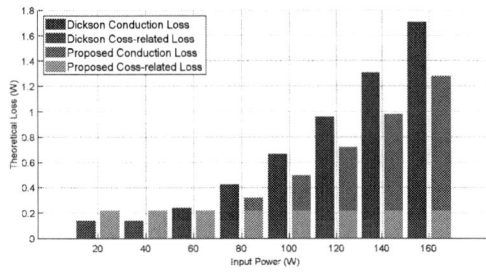

Fig. 13. Theoretical loss considering conduction and Coss elements.

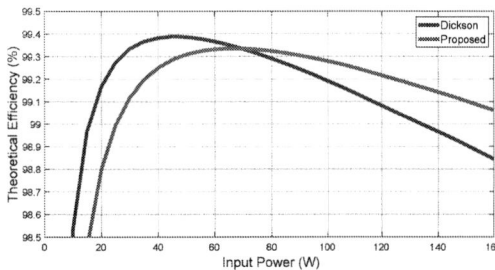

Fig. 14. Theoretical efficiency considering conduction and Coss-related loss.

Here, Vol_{Old} and Vol_{New} represent the capacitor volumes of the conventional and proposed circuits. Fig. 5 illustrates the capacitor volume ratio of the proposed N:1 SCC compared to the conventional Dickson SCC for various voltage conversion ratios N:1. This demonstrates that the proposed circuit can reduce passive component volume and indicates that the reduction rate is higher for smaller values of N.

Fig. 6 illustrates the on-resistance of the switches, considering the voltage stress in each phase within the FSL region for the proposed topology when the voltage conversion

TABLE III. System Specifications

Symbol	Parameter	Nominal Value				
P_{in}	Input Power	160 W				
V_{in}	Input Voltage	48 V				
V_{out}	Output Voltage	8 V				
I_{out}	Output Current	20 A				
$R_{on,sw}$	Switch Resistance	2.4 mΩ				
C_i	Capacitance(μF), Dickson	C_1 18	C_2 22.5	C_3 30	C_4 45	C_5 90
	Capacitance(μF), Proposed	C_1 16.5	C_2 22	C_3 33	C_4 66	C_5 66
f_{sw}	Switching Frequency	500 kHz				

ratio is 5:1 (N=5) and 6:1 (N=6), along with the equivalent series resistance (ESR) of the capacitors. The on-resistance of a switch with a voltage stress of nV_o is denoted as R_{on,nV_o}, with distinctions made using different colors. By calculating the power loss at each node and considering the duty cycle, the total power loss was determined, leading to the derivation of the FSL impedances. The FSL impedances of the Dickson SCC and the proposed circuit are generalized for a voltage conversion ratio of N:1 and presented in Table I. Δ represents the difference in FSL impedances between the proposed circuit and the Dickson SCC. Considering the voltage stress on the switches, it was confirmed that $\Delta R_{FSL} < 0$, indicating that the proposed N:1 SCC has lower FSL impedances compared to the conventional design, thereby reducing the conduction losses in the switches. Fig. 7 illustrates the reduction of FSL impedance according to voltage ratio. It is observed that the highest reduction in FSL impedance occurs at $N = 6$, and for odd voltage ratios, a significant reduction is noted at $N = 9$.

III. Hardware Implementation

The models of the selected components are summarized and presented in Table II. A hardware prototype for 48 V to 8 V conversion is designed to verify the operation and validate the theoretical analysis. Fig. 8 illustrates the 6:1 conventional and proposed prototypes.

A. Multilayer Ceramic Capacitor (MLCC) Design

Fig. 9 shows the DC derating characteristics of Class 2 MLCCs. Class 2 MLCCs are utilized to achieve high power density, making the calculation of effective capacitance essential due to the significant attenuation of capacitance under DC bias voltage. Fig. 10 illustrates the changes in capacitance and inductance with frequency. Selecting capacitors with large capacitance to account for attenuation can bring the self-resonant frequency of the capacitor closer to the operating frequency. The self-resonant frequency is determined by the resonance between the capacitance and parasitic ESL. When the self-resonant frequency is near the operating frequency, significant variations in capacitance can occur, leading to reduced reliability. Considering these constraints between frequency and capacitance and comparing the volume of passive components in the conventional and proposed circuits, capacitances are calculated with equal SSL impedances as defined in Equation (4).

979-8-3315-1612-3/25 $31.00 © 2025 IEEE

TABLE IV. VOLUME OF SWITCH, PASSIVE COMPONENTS IN POWER STAGE

6:1 Topology	Switch [Volume]	Flying capacitor (C_{fly}) [Volume]	Capacitor ($C_{fly} + C_{in} + C_{out}$) [Volume]	Capacitor + Switch [Volume]
Conventional	36 mm³	240.69 mm³	404.53 mm³	440.53 mm³
Proposed	50.49 mm³	136.41 mm³	300.25 mm³	350.74 mm³
Relative Change (%)	+ 40.25 %	− 43.33 % (Calculated*: − 46.22 %)	− 25.78 %	− 20.38 %

* : Theoretical calculation by equation (6)

TABLE V. COMPARISON OF POWER DENSITY AND EFFICIENCY

6:1 Topology	Voltage ratio	Output current	Power density (W/inch³)	Power stage efficiency
Conventional	48 V to 8 V	20 A	868.7	Full load: 95.2 % Peak: 97.7 %
Proposed			1442.5	Full load: 96.0 % Peak: 97.5 %

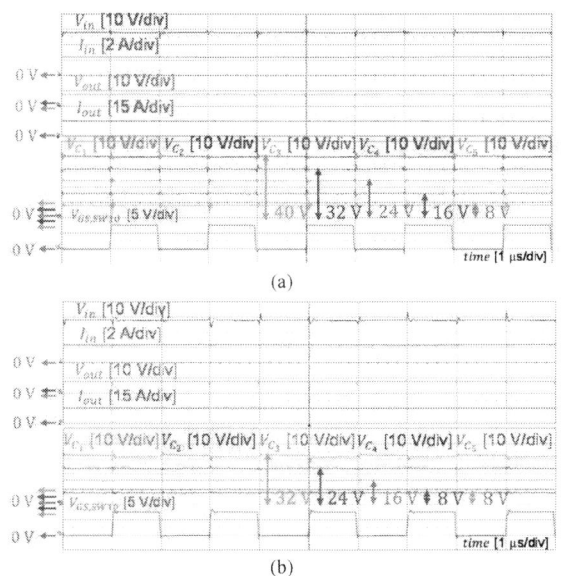

(a)

(b)

Fig. 15. Experimental results of 6:1 Dickson and proposed ($V_{in} = 48.2$ V, $V_{out} = 7.7$ V, $I_{in} = 2.5$ A, $I_{out} = 15$ A), (a) Dickson (b) proposed.

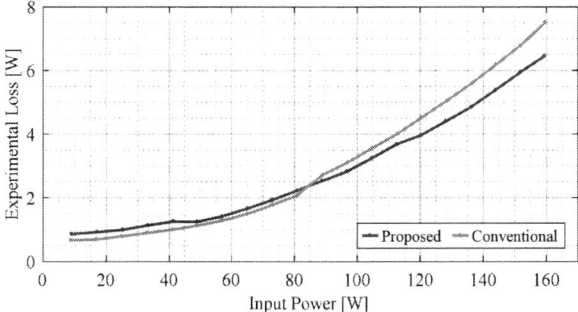

Fig. 16. Theoretical loss considering conduction and Coss elements.

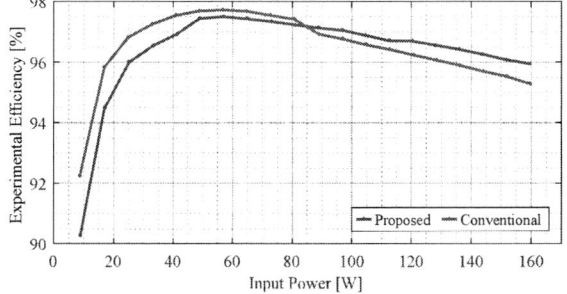

Fig. 17. Theoretical efficiency considering conduction and Coss-related loss.

B. Gate Drive Circuit

Fig. 11 illustrates the proposed 6:1 SCC. Fig. 12 shows the cascaded bootstrap circuits applied to Fig. 11. The gate driver circuit utilizes cascaded bootstrap circuits, as shown in Fig. 12, to gate all upper switches from the three lower switches connected to the ground. The circuit is configured with three paths to achieve this functionality.

C. Theoretical Loss Calculation

Fig. 13 and Fig. 14 illustrate the theoretical losses and efficiency calculated based on the conduction and Coss elements using the model data from Table II. While the light-load efficiency of the proposed circuit, which includes additional switches, was lower compared to the Dickson SCC, the proposed circuit exhibited a higher efficiency trend under rated load conditions. This was attributed to the smaller increase in conduction loss compared to the Dickson SCC as the load approached its rated value.

IV. EXPERIMENTAL RESULTS

Table III shows the prototype system specifications. Fig. 15 shows the experimental waveforms of the conventional and proposed circuits. The prototype was tested up to a maximum output of 20 A. The voltage across the capacitors confirms that the voltages of C_1, C_2, C_3, and C_4 decreases by the output voltage, validating that the proposed circuit achieves 32 V, 24 V, 16 V, and 8 V.

Fig. 16 and 17 show that at the 160 W rated load, the Dickson SCC incurs a loss of 7.52 W, while the proposed SCC incurs a loss of 6.44 W, resulting in efficiencies of 95.2 and 96 %, respectively. In the light-load region, switching losses due to the additional switches dominate, leading to lower efficiency. However, in the rated load region, the reduction in conduction losses becomes significant, resulting in higher efficiency. According to the theoretical calculations presented earlier in Table I, when N=6, a 33 % reduction in output impedance is expected, ideally leading to a proportional reduction in switch conduction losses. However, the losses measured in the experiments include not only conduction losses but also switching losses, PCB losses, and other factors combined. Given that the proposed circuit adds two additional switches, switching losses inevitably increase, partially offsetting the reduced conduction losses. The experimental results confirm that the equivalent loss, which accounts for all loss factors, is reduced in the rated load region. This validates that the reduction in switch conduction losses in the proposed circuit exceeds the rate of increase in other loss components.

Table IV summarizes the volume of the power stage, including switches, flying capacitors, and input/output capacitors. Based on Equation (6), the theoretical volume reduction rate of the flying capacitors for N=6 is calculated to be 46.22%. As shown in Table 4, the experimental hardware configuration of the flying capacitors demonstrates a volume reduction rate of 43.33%. This close agreement between the theoretically calculated and experimentally observed reduction rates validates the theoretical analysis. Furthermore, despite the volume increase caused by the two additional switches, the total power stage volume, including both switches and passive components, shows a reduction rate of 20.38%. This confirms that the reduction in flying capacitor volume outweighs the increase in switch volume. Table V compares the power density and efficiency of the conventional and proposed circuits. For a 48 V to 8 V conversion at a 20 A output, the power density is 868.7 W/inch³ for the Dickson SCC and 1442.5 W/inch³ for the proposed circuit, achieving approximately 1.7 times higher power density. However, as shown in Fig. 8, the area occupied by the switches can fit within the area occupied by the capacitors, leaving unused space. Since the calculated power density includes this remaining space, there may be room for further improvement in power density. In conclusion, the proposed circuit demonstrates a significantly higher power density compared to the conventional circuit, showcasing the potential for more compact designs when implementing this topology.

V. CONCLUSION

A new topology is proposed by reconfiguring the capacitor arrangement in the conventional Dickson SCC to reduce the DC bias voltage. This topology is extended and generalized for a voltage conversion ratio of N:1, confirming its feasibility under similar implementation methods. The reduction rates of passive component volume and output impedance in the proposed circuit compared to the conventional one were theoretically analyzed through mathematical formulations. It was observed that the reduction in passive component volume becomes more significant for smaller voltage conversion ratios, with the highest output impedance reduction occurring at N=6 for even voltage conversion ratios and N=9 for odd voltage conversion ratios.

The design of components and parameters accounted for factors essential to hardware implementation, such as Class 2 MLCC derating and self-resonant frequency. Experiments conducted for a specific voltage conversion ratio of 48 V to 8 V (N=6) at a 20 A output demonstrated the feasibility of the proposed circuit. It was verified that the capacitor DC bias voltages were reduced from 40 V, 32 V, 24 V, 16 V, and 8 V to 32 V, 24 V, 16 V, 8 V, and 8 V, respectively. A 43.33 % reduction in passive component volume was observed for the proposed circuit compared to the conventional one, closely aligning with the previously analyzed theoretical reduction rate of 46.22 %, thereby validating the theoretical analysis. Despite the volume increase due to two additional switches, the overall power stage volume showed a 20.38 % reduction, as confirmed by experiments. Furthermore, the power density of the proposed circuit was measured to be 1442.5 W/inch³, significantly higher than the 868.7 W/inch³ of the

conventional circuit, enabling more compact designs for the same power delivery.

In terms of power stage efficiency, the proposed circuit achieved a full-load efficiency of 96 % and a peak efficiency of 97.5 %, with higher efficiency in the rated load region compared to the conventional circuit. This result highlights that in the rated load region, where conduction losses dominate, the reduction in output impedance achieved by the proposed circuit outweighs the increase in switching losses caused by the additional switches. The experimentally verified output impedance reduction further validates the theoretical analysis.

REFERENCES

[1] Z. Ye, S. R. Sanders, and R. C. N. Pilawa-Podgurski, "Modeling and Comparison of Passive Component Volume of Hybrid Resonant Switched-Capacitor Converters," *IEEE Trans. Power Electron.*, vol. 37, no. 9, pp. 10903–10919, Sep. 2022, doi: 10.1109/TPEL.2022.3160675.

[2] F. Shao, K. Yao, X. Ruan, R. Feng, and L. Ren, "A 4:1 Switched Capacitor Converter With Low Flying Capacitors Voltage and High Efficiency," *IEEE Trans. Ind. Electron.*, vol. 70, no. 10, pp. 10034–10043, Oct. 2023, doi: 10.1109/TIE.2022.3224173.

[3] J.-T. Wu and K.-L. Chang, "MOS charge pumps for low-voltage operation," *IEEE J. Solid-State Circuits*, vol. 33, no. 4, pp. 592–597, Apr. 1998, doi: 10.1109/4.663564.

[4] Z. Ye, R. A. Abramson, T. Ge, and R. C. N. Pilawa-Podgurski, "Multi-Resonant Switched-Capacitor Converter: Achieving High Conversion Ratio With Reduced Component Number," *IEEE Open J. Power Electron.*, vol. 3, pp. 492–507, 2022, doi: 10.1109/OJPEL.2022.3181338.

[5] B. Veraverbeke, T. Thielemans, T. V. Daele, and F. Tavernier, "A 240V to 47.5 V Fully Integrated Switched-Capacitor Converter in GaN Achieving 62.6% Efficiency at 220 mW/mm2," in *2022 17th Conference on Ph.D Research in Microelectronics and Electronics (PRIME)*, Jun. 2022, pp. 249–252. doi: 10.1109/PRIME55000.2022.9816779.

[6] J. F. Dickson, "On-chip high-voltage generation in MNOS integrated circuits using an improved voltage multiplier technique," *IEEE J. Solid-State Circuits*, vol. 11, no. 3, pp. 374–378, Jun. 1976, doi: 10.1109/JSSC.1976.1050739.

[7] Y. Lei and R. C. N. Pilawa-Podgurski, "A General Method for Analyzing Resonant and Soft-Charging Operation of Switched-Capacitor Converters," *IEEE Trans. Power Electron.*, vol. 30, no. 10, pp. 5650–5664, Oct. 2015, doi: 10.1109/TPEL.2014.2377738.

[8] M. S. Makowski and D. Maksimovic, "Performance limits of switched-capacitor DC-DC converters," in *Proceedings of PESC '95 - Power Electronics Specialist Conference*, Atlanta, GA, USA: IEEE, 1995, pp. 1215–1221. doi: 10.1109/PESC.1995.474969.

[9] J. M. Henry and J. W. Kimball, "Practical Performance Analysis of Complex Switched-Capacitor Converters," *IEEE Trans. Power Electron.*, vol. 26, no. 1, pp. 127–136, Jan. 2011, doi: 10.1109/TPEL.2010.2052634.

[10] J. W. Kimball, P. T. Krein, and K. R. Cahill, "Modeling of capacitor impedance in switching converters," *IEEE Power Electron. Lett.*, vol. 3, no. 4, pp. 136–140, Feb. 2005, doi: 10.1109/LPEL.2005.863603.

[11] M. D. Seeman and S. R. Sanders, "Analysis and Optimization of Switched-Capacitor DC–DC Converters," *IEEE Trans. Power Electron.*, vol. 23, no. 2, pp. 841–851, Mar. 2008, doi: 10.1109/TPEL.2007.915182.

[12] I. Oota, N. Hara, and F. Ueno, "A general method for deriving output resistances of serial fixed type switched-capacitor power supplies," in *2000 IEEE International Symposium on Circuits and Systems (ISCAS)*, May 2000, pp. 503–506 vol.3. doi: 10.1109/ISCAS.2000.856107.

Wide Range Digital Control for Three-Level Buck Converters with Sensorless Flying-Cap Voltage Balancing

Hossein Hajisadeghian
Dept. of Management and Engineering (DTG)
University of Padova, Vicenza, Italy
Email: hossein.hajisadeghian@unipd.it

Giovanni Bonanno
Dept. of Information Engeneering (DEI)
University of Padova, Padova, Italy
Email:giovanni.bonanno@unipd.it

Abstract—**This paper proposes a wide-range multi-sampling peak digital predictive current mode controller (DPCMC) for a 3-level flying capacitor (3LFC) Buck DC-DC converter. The main advantage of this controller is the ability to cover the entire voltage conversion ratio range without encountering any flying capacitor (FC) voltage instability issues. The proposed control strategy maintains the FC voltage balanced without the need for measurement, estimation, and no dedicated actions of the control systems are required. The proposed modification of the multi-sampling fast-update DPCMC performs effectively also at the operating point where the output voltage is half of the input voltage. This condition is generally avoided by existing control strategies. The performance of the algorithm is demonstrated through simulations and experimental tests on a 3LFC Buck converter prototype.**

I. INTRODUCTION

In recent years hybrid topologies (i.e. switched capacitor cells + inductor) such as multi-level flying capacitor DC-DC converters has gained more attention in industry, automotive, high-power, low voltage, and point-of-load applications. These topologies offer several advantages in terms of reduced voltage stress, an equivalent frequency multiplication effect, and consequently reduced converter volume and footprint [1]–[7]. These advantages are available only when the converter works with properly balanced FC voltages. Reaching and maintaining these operating points is the most challenging issue for such topologies. Several techniques have been proposed to assess this stability and voltage balance [1], [8]–[20]. One can distinguish between active techniques, where FC voltages are maintained stable and balanced with dedicated control actions, and automatic techniques, where stability and balance are maintained with actions already embedded in the main control system. The first category uses dedicated voltage sensing circuitry to measure FC voltage or, in some cases, an analytical model to estimate FC voltage imbalances [11]. This approaches lead to a non-negligible increasing in the overall controller complexity and costs.

This paper proposes a multi-sampling digital control system that ensures *automatic* balancing of the FC voltage in 3LFC Buck converters. In a well-balanced N-level FC Buck converter, since the inductor current frequency is nominally $N-1$ times the switching rate, this opens up the possibility to *naturally* implement multi-sampling control techniques without encountering canonical issues [21], [22]. Such digital multi-sampling techniques are based on DPCMC where leading edge (LE), trailing edge (TE), and triangle trailing edge (TTE)

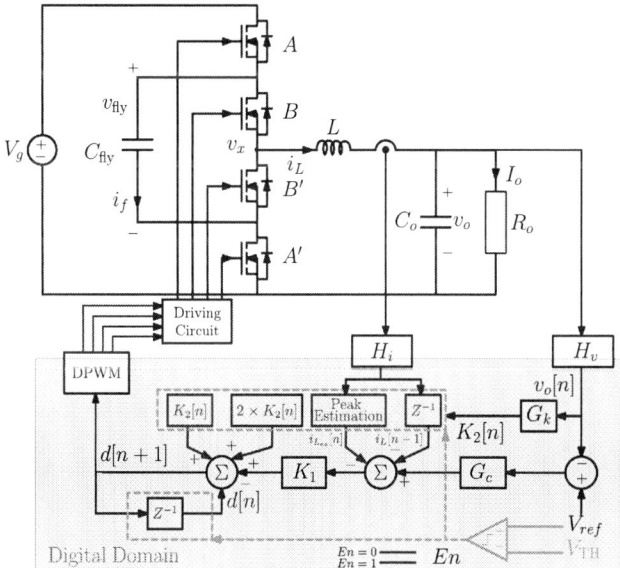

Fig. 1: Proposed Wide Range multi-sampling peak-DPCMC for 3LFC Buck converter.

carriers are used to properly assess *peak*, *valley* and *average* current-mode control (CMC) [1], [2]. The main limitations for the applications of multi-sampling DPCMC to 3LFC Buck converter is the FC voltage instability in one or both operating modes. To set the notation, one can use *mode 1* to indicate the range where the voltage conversion ratio M is less than 0.5 (i.e., $0 < M < 0.5$), conversely one can use *mode 2* when $0.5 < M < 1$.[1] Furthermore, when the 3LFC Buck converter operates in digital or analog CMC there is a stability problem when the duty cycle reaches the 50%. This corresponds to the operating point between *mode 1* and *mode 2* (i.e., $M = 0.5$). This issue prevents the use of 3LFC Buck converter in the whole operating range. A workaround is the use of a *dead-band* zone that excludes a surrounding of this operating point. This over-complicates the final design and does not solve the problem of prohibited conversion ratios and OP $M = 0.5$ is usually skipped [11], [24] in 3LFC Buck converters. A possible solution can be obtained by combining the two multi-sampling peak DPCMC and multi-sampling fast-update peak

---

[1]In general [23], for the N-level FC Buck converter one has *mode i* when $\frac{i-1}{N-1} < M < \frac{i}{N-1}$.

979-8-3315-1612-3/25 $31.00 © 2025 IEEE

DPCMC controllers analyzed in [1]. Considered separately, such control techniques provide stability of FC voltage only for $M > 0.5$ and $M < 0.5$. If it were possible to combine the two strategies and solve the problem of the operating point $M = 0.5$, it would result in a very fast digital controller that does not need FC voltage sensing to ensure stability and automatic balancing. This paper proposes a digital current-mode control algorithm that combines these multi-sampling methods in order to cover both operating modes. A second contribution is the modification of the multi-sampling fast-update DPCMC technique to ensure a proper operation also for $M = 0.5$.

The rest of the paper is organized as follows. Sec. II discusses the main limitations of multi-sampling control methods for 3LFC Buck converter and discloses the proposed algorithm. Sec. III analyzes the steady-state operation for $M = 0.5$. Sec. IV and Sec. V collect the simulation and experimental results while Sec. VI draws the conclusions.

II. WIDE RANGE MULTI SAMPLING PEAK DPCMC

Fig. 1 shows the proposed controller for 3LFC Buck converter. The inner current loop consists of the combination of the modified versions of the multi-sampling DPCMC proposed in [1]. Using the Enable (En) signal, generated by a digital comparator, the appropriate controller is activated. The outer voltage-loop regulator G_c can be used for both controllers, since the inner current-loop bandwidth of the two multi-sampling DPCMC are comparable. To sense the output voltage and the inductor current, two feedback networks with two analog-to-digital converters (ADCs) are used. The voltage and current conditioning and sensing circuitries are represented by H_v and H_i respectively. The input and output voltage are indicated as V_g and V_o, respectively, so the voltage conversion ratio can be defined as $M \triangleq \frac{V_o}{V_g}$. Note that under conditions of perfect FC voltage balancing, the conversion ratio of a 3LFC Buck converter is equal to the steady-state duty-cycle value of the controlling signals A and B. By denoting with D the steady-state duty cycle values, one has $D = M$ in all operating modes [23].

The following is the analysis of the proposed control system for the different operating modes.

A. Operating mode 1 ($M < 0.5$)

In this region, the multi-sampling fast-update DPCMC is activated. The *control equation* can be written as follows

$$d[n+1] = K_1(I_{ref} - i_L[n]) + K_2, \qquad (1)$$

where $K_1 = V_g/(2Lf_s)$ and $K_2 = M$ [1]. In the considered application the second term is not constant and depends on the output voltages V_o. Thus, K_2 is generated by sampling V_o and multiplying it by the gain $G_k \triangleq 1/V_g$. With respect to Fig. 1, the blocks enabled in this operating mode are highlighted in red color. Fig. 2(a) sketches the phase-shifted LE-DPWM signals, the controlling signals, and the inductor current waveform. Sampling, updating and controlling instances are indicated with a circle, a square, and a filled circle respectively. The time coordinate is normalized with respect the sampling rate (i.e., $t \rightarrow t/T_s$).

As stated in [1], in this controlling strategy, sampling ($i_L[n]$) and updating actions are performed in one-half of the switching period. These actions are coordinated in order to control the value of $i_L[n+1]$ to reach the current reference I_{Ref}. According to (1), this controlling strategy requires a minimum computation time Δt_{calc} to generate the new value of the duty-cycle $d[n+1]$. This time is required due to the whole delay originated in the controlling loop. Thus, one can not increase the duty-cycle above D_{max}. By indicating the time between the sampling and updating points as Δt_{calc}, this maximum can be individuated as follows

$$D_{max} = \frac{1}{2} - \frac{\Delta t_{calc}}{T_s}. \qquad (2)$$

The finite value of Δt_{calc} leads to a limitation of the application of the fast-update approach. In fact, for some operating points where $d[n]$ is close to D_{max}, the controller can face some problems. An example of this issue is highlighted in Fig. 2. In Fig. 2(a) the required action brings the reduction of the modulating signal, therefore no issues arise. The problem occurs when the error between the sampling current $i_L[n]$ and I_{ref} is negative and the current $d[n]$ is close to D_{max}. This situation is depicted in Fig. 2(b). In this condition, to reach the desired inductor current peak value, the modulating signal must be increased and the intersection with the carrier signal occurs in the red zone region caused by $\Delta t_{calc} > 0$. The update action occurs after the carrier intersection, and thus no changes in the duty cycle are obtained (i.e., $d[n+1] \leftarrow d[n]$). This issue can cause severe oscillations in the inductor current and can potentially cause damage to the converter. To address this limitation, an inductor peak current estimation process is proposed.

The issue related to $\Delta t_{calc} > 0$ can be avoided by *estimating* the peak value before reaching the real peak. Indeed, by supposing a small-ripple approximation for the FC and the output voltages, one obtains a piecewise-linear inductor current. This condition is mandatory to design the DPCMC. Under these assumptions, the estimation process can be described as follows.

The inductor current is sampled $\Delta t_{meas} > \Delta t_{calc}$ before the real peak. Fig. 2(c) shows the different situations in which Δt_{meas} is less, greater and equal to the duty-cycle. Naming $V_{(L-on)}$ and $V_{(L-off)}$ the inductor voltage during the ON and OFF phases, two states for finding the peak value can be considered.

1) Condition 1 ($d[n] \geq \Delta t_{meas}$) : In this condition, only $v_{L\text{-on}}$ is imposed on the inductor current. As a result, the peak current can be estimated with a linear equation as follows:

$$i_{L_{est}}[n] = i_L[n] + \frac{V_{L\text{-on}}}{L}\Delta t_{meas} \qquad (3)$$

The equation $v_{L\text{-on}}(t)$ of the inductor voltage during the ON phase in operating *mode 1* can be written as follows

$$v_{L\text{-on}}(t) = \begin{cases} V_g - v_{fly}(t) - v_o(t) & (A = 1, B = 0) \\ v_{fly}(t) - v_o(t) & (A = 0, B = 1) \end{cases}. \qquad (4)$$

Using the small-ripple approximation (SRA) and assuming a balanced FC voltage (i.e., $v_{fly} = \frac{V_g}{2}$) one has $V_{L\text{-on}} = \frac{V_g}{2} - V_o$

979-8-3315-1612-3/25 $31.00 © 2025 IEEE

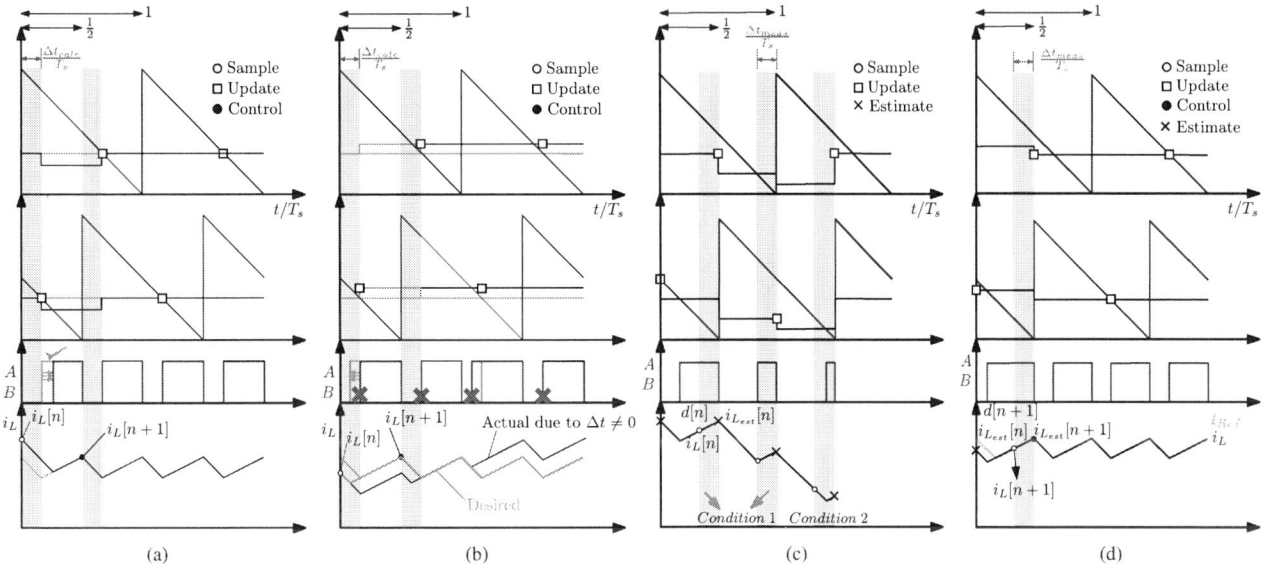

Fig. 2: multi-sampling fast-update DPCMC as in [1]: the modulation signal intersects the LE carrier (a) outside the red region (i.e., $d[n] < D_{max}$) and (b) inside the red region (i.e., $d[n] > D_{max}$). (c) Estimation of the inductor current peak value by using the sampling point $i_L[n]$ in two states and (d) by using estimation approach and removing the limitation of dead zone region.

for both $A = 1, B = 0$ and $A = 0, B = 1$. In this context, given the strong dependence on the operating point, it is necessary to evaluate the conversion ratio in the current cycle. For a 3LFC Buck converter, the voltage conversion ratio is equal to the duty-cycle of controlling signal (i.e., $M[n] = d[n]$). Thus,

$$V_{\text{L-on}} = \frac{V_g}{2} - d[n]V_g = \frac{V_g}{2}(1 - 2d[n]). \tag{5}$$

By substituting (5) in (3) the following *estimation equation* can be found

$$i_{L_{est}}[n] = i_L[n] + \frac{V_g}{2Lf_s}(1 - 2d[n])\frac{\Delta t_{\text{meas}}}{T_s} \tag{6}$$

2) Condition 2 ($d[n] < \Delta t_{meas}$) : This state involves both $v_{\text{L-on}}(t)$ and $v_{\text{L-off}}(t)$ voltages to compute the slopes of the inductor current. Based on Fig. 2(c), the estimated peak current is

$$i_{L_{est}}[n] = i_L[n] + \frac{V_{\text{L-off}}}{L}(\Delta t_{\text{meas}} - T_s d[n]) + \frac{V_{\text{L-on}}}{L}T_s d[n] \tag{7}$$

The equation $v_{\text{L-off}}(t)$ of the inductor voltage during the OFF phase in operating *mode 1* is as follows

$$v_{\text{L-off}}(t) = -v_o(t) \tag{8}$$

Using the small-ripple approximation in (8) one has $V_{\text{L-off}} = -V_o$. By substituting both ON and OFF inductor current voltages in (7) one obtains

$$i_{L_{est}}[n] = i_L[n] + \frac{V_g}{2Lf_s}\left(1 - 2\frac{\Delta t_{\text{meas}}}{T_s}\right)d[n] \tag{9}$$

The set of equations (6) and (9) for the peak current estimation together with the control equation in (1) constitute the modified version of the multi-sampling fast-update peak DPCMC proposed in this article. The modified version of the multi-sampling fast-update peak DPCMC, which hereafter is referred as *modified* fast-update DPCMC, actually eliminates the duty-cycle limitation that now, at least in theory, can reach the critical value of 50%. Sec. III details the converter operation for $M = 0.5$, which is exactly the operating point that all conventional control strategies avoid.

B. Operating mode 2 ($M > 0.5$)

In this region the multi-sample method from [1], can be directly used. As the sampling and updating actions in this control method do not occur in the same half-period, no modification in the implementation of the multi-sampling peak DPCMC is required. Thus, in operating *mode 2* the multi-sampling can cover the whole duty-cycle range. The *control equation* [1] is given by

$$d[n + 1] = K_1(I_{\text{ref}} - i_L[n - 1]) + 2K_2 - d[n] \tag{10}$$

K_1 and K_2 coefficients are the same as for the multi-sampling fast update control equation in (1). Active blocks for this operating mode are highlighted in Fig. 1 in blue color.

III. OPERATION FOR $M = 0.5$

Fig. 3 shows the DPWM signals, the inductor current and the AC component of the flying capacitor voltage for $M = 0.5$. Dead times of control signals A and B are highlighted in blue and red respectively. The AC component $\hat{v}_{\text{fly}}(t)$ of the FC voltage can be intended as the *small* residual ripple across its DC value (i.e., $\hat{v}_{\text{fly}}(t) \triangleq v_{\text{fly}}(t) - V_{\text{fly}}$). When operating at

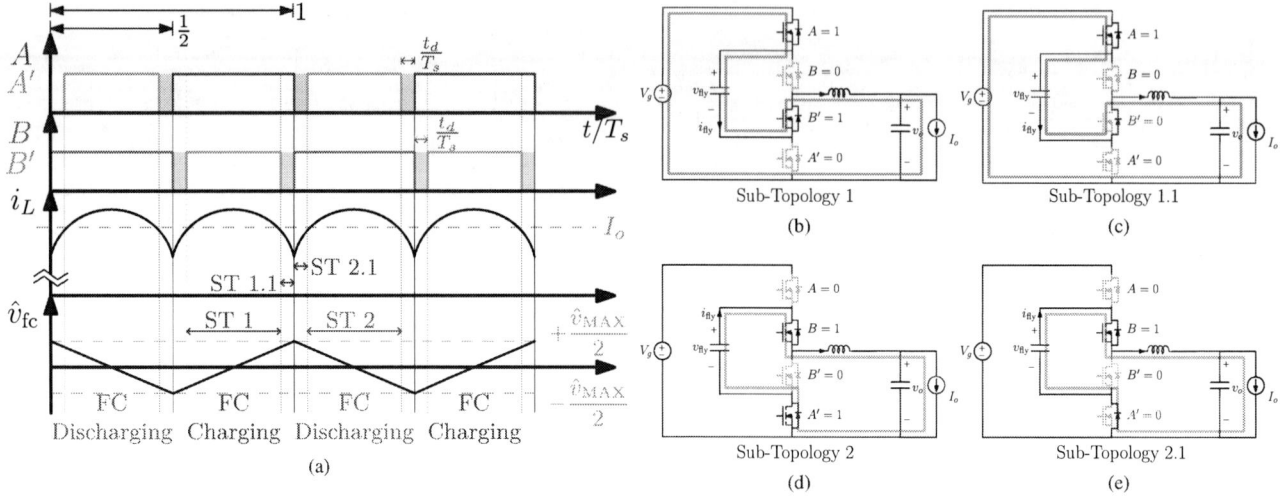

Fig. 3: (a) Operation of the 3LFC Buck converter at $M = 0.5$: (top) controlling signals A and B, (center) inductor current and (bottom) FC voltage. (b), (c), (d) and (e) report the corresponding sub-topologies (ST).

$M = 0.5$, in 3LFC Buck converters, the charge and discharge times for the FC are maximum. Therefore, the variation of its AC component is maximum. Keeping the assumption of the small ripple approximation (i.e., $\hat{v}_{\text{fly}}(t) \ll V_{\text{fly}}$), in steady-state it is possible to predict what the converter waveforms shape would be at this particular operating point and thus understand the converter operation. In the following analysis, the FC is assumed to be perfectly balanced (i.e., $V_{\text{fly}} = \frac{V_g}{2}$). Nevertheless, same qualitative results hold for the case of small residual FC voltage imbalances. Under small-ripple approximation and for $V_{\text{fly}} = \frac{V_g}{2}$ and $M = 0.5$ (i.e., $V_o = \frac{V_g}{2}$), (4) becomes

$$v_{\text{L-on}}(t) = \begin{cases} -\hat{v}_{\text{fly}}(t) & (A = 1, B = 0) \\ +\hat{v}_{\text{fly}}(t) & (A = 0, B = 1). \end{cases} \quad (11)$$

Assuming very slow changes in FC voltage (thus compatible with SRA), it can be assumed that during the charging phase the FC voltage increases and decreases linearly during the charging and discharging time, respectively. Therefore, for the i-th switching cycle, one has

$$\hat{v}_{\text{fly}}(t) = \begin{cases} -\frac{\hat{v}_{\text{MAX}}}{2} + 2\left(\frac{t}{T_s} - \left(\frac{1}{2} + i\right)\right) & (A = 1, B = 0) \\ +\frac{\hat{v}_{\text{MAX}}}{2} - 2\left(\frac{t}{T_s} - i\right) & (A = 1, B = 1). \end{cases} \quad (12)$$

As represented in Fig. 3(a) the constant $\hat{v}_{\text{MAX}}$ represents the maximum peak-to-peak FC voltage ripple. The expression (12) is compatible with a constant charge/discharge current. In fact, the inductor current ripple is very small since it is dependent only on $\hat{v}_{\text{fly}}(t)$. The residual current ripple can be formally calculated by integrating (12). The resulting waveform is reported in Fig. 3(a).

In sub-topology 1 $\hat{v}_{\text{fly}}(t)$ is increasing. Next, during all or part of the interval with sub-topology 1.1 (see Fig. 3(c)), obtained thanks to the deadtime between A and A', the body diode of the switch B' is forced in conduction, therefore

$\hat{v}_{\text{fly}}(t)$ is still increasing. The same considerations hold for sub-topology 2.1 reported in Fig. 3(d).

For $I_o = 0$ the waveform shapes are more complicated, but the qualitative results are identical. For reasons of space, this topic is not further explored and the verification of stability at zero current is devolved to simulations and experimental tests.

Now that the steady-state operation is individuated, one can verify whether this is compatible with the DPCMC (i.e., 3LFC Buck converter + DPCMC lead to a stable system). In these conditions, one has $I_{\text{ref}} - i_{\text{L}}[n] \approx 0$. Therefore, by using the modified fast-update DPCMC, detailed in Sec. II, one has $d[n+1] \approx M = 0.5$.[2] Therefore, the converter operates with constant duty-cycle with the steady-state waveforms disclosed in this section.

In more realistic cases, one can face small variation of $d[n+1]$ due to small variation of $I_{\text{ref}} - i_{\text{L}}[n] \approx 0$, mismatches in system parameters, and also due to parasitic elements. Therefore, one must verify the impact of these small fluctuations of $d[n+1]$ on the stability of the system. Verifying this with a formal analysis, as done in [23] is quite complex since one must necessarily include the deadtimes and parasitic elements. Therefore a simulation and experimental validation is used to prove the final stability of the proposed technique.

IV. SIMULATION RESULTS

The proposed wide-range control architecture is first simulated in the MATLAB/Simulink® environment using PLECS®. The converter parameters are reported in Tab. I.

First, it is necessary to verify whether the system is able to reach a steady-state for $M = V_o/V_g = 0.5$. In this operating region (nominally $0 < M \leq 0.5$), the *modified* fast-update DPCMC part is activated. The steady-state operation is analyzed in Sec. III and exemplified in Fig. 3. The simulation for this operating point, reported in Fig. 4, gives identical results. As predicted, the FC voltage is linear and the inductor

---

[2]Please note that in (1) $K_2 = M$.

current is parabolic. Since in this test I_o is high enough, during dead-times the inductor current is always positive and the sequence of the four sub-topological states is identical to the four reported in Fig. 3(b), (c), (d), and (e). For $I_o = 0$, the inductor current during the sub-topological states 1.1 and 2.1 is negative and the body diodes of A' and B' cannot be forced in conduction. Thus, the FC voltage is kept constant and the resulting inductor current is linear during dead-times. The qualitative results are identical, and the system is still able to operate stably for $M = V_o/V_g = 0.5$. For reasons of space, this simulation is not reported. In any case, the stability at zero current is verified in simulation and experimentally.

The second test, Fig. 4(b), analyzes the transition from operating mode 1 and operating mode 2. This is obtained by stepping the voltage reference V_{ref} (see Fig. 1) from 5 V to 8 V. The proposed wide-range control system correctly switches the operating mode. To avoid potential oscillations, the operating mode 1 is maintained above the operating point $M = 0.5$ ensuring $V_o = 6$ V even when parasitic elements (i.e., $R_{DS\text{-}ON}$, $R_{L\text{-}ESR}$, etc.) are introduced. In fact, under real conditions, the voltage drops in these elements lead to a slightly lower output voltage than $V_o = 6$ V when $D = M = 0.5$. This modification is also necessary to prevent continuous oscillations between the two operating modes of the control around $M = 0.5$. Thus, V_{TH} in Fig. 1 is slightly greater than $\frac{V_g}{2}$ (i.e., $V_{TH} = \frac{V_g}{2} + V_\epsilon$). Regarding the FC voltage, neither unstable nor imbalance issues are reported. The third test, Fig. 4(c), shows the full- to half-load step simulation for $M = V_o/V_g = 0.5$. Results show that the FC voltage is stable and automatically converge to $V_g/2$ after the load step. Other simulations are carried out also including randomly-generated mismatches in the controlling signal A and B. Even in these cases, the average FC voltage automatically converged to its balanced value. Therefore, the ability of the multi-sampling fast-update peak DPCMC to force the FC voltage to automatically converge to its correct value without the need to measure it and without further control action is maintained in the *modified* version proposed in this paper and also for $M = 0.5$.[3]

V. EXPERIMENTAL RESULTS

The custom prototype parameters and the experimental setup organization are shown in Tab. I. The proposed wide-range multi-sampling DPCMC is coded in VHDL using a commercial FPGA. The experimental results are shown in Fig. 5 and Fig. 6. The experimental verification is organized as follows. The first two tests investigate the stability of the 3LFC Buck converter + proposed wide-range DPCMC system around the operating point $M = 0.5$ in steady-state and in load-step changes. The other two experimental tests assess the ability of the developed control system to reach and operate properly at the critical operating point $M = 0.5$.

A. Steady-state stability and load-step responses at $M = 0.5$

Fig. 5 reports the experimental results of the proposed wide-range multi-sampling DPCMC for $V_o = 6$ V. In the first test,

---

[3]It should be noted that the original version of the fast-update control, initially proposed in [1], could not be used for $M \approx 0.5$ due to the finite computation time required before the update action. This limits the controller to operate below $M = 0.5$.

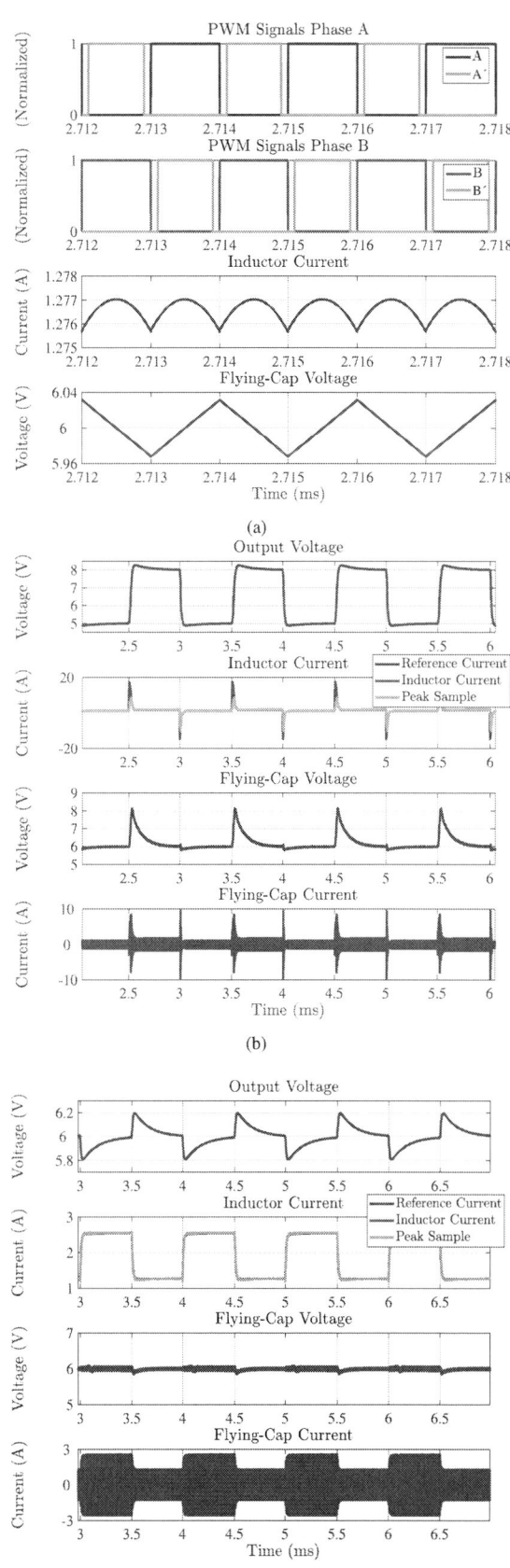

Fig. 4: Simulation results: (a) steady-state operation for $M = 0.5$, (b) step-change in V_{ref} (from 5 V to 8 V), and (c) load-step (from half to full load) at $M = \frac{6}{12} = 0.5$.

TABLE I: 3LFC Buck converter (top) main parameters and (bottom) experimental setup.

	Name and Symbol	Values
	Input Voltage (V_g)	12 V
	Output Voltage (V_o)	0.15 V to 11.5 V
Converter	Output Current (MAX) (I_o)	5 A
Parameters	Nominal Load (R_o)	4.7 Ω
	Filter Inductance (L)	5.5 μH
	Filter Capacitance (C_o)	50 μF
	Flying Capacitor (C_f)	4X5 μF
	Switching Frequency (f_s)	500 kHz
	FPGA Cyclone V	5CEBA4F23C7N
Timing	ADC Latency (t_l)	160 ns
Parameters	Calculation Time (t_c)	25 ns
	Estimation Time (Δt_{meas})	185 ns
	Dead-time (t_d)	50 ns

Custom 3LFC Buck converter prototype for the experimental validation.

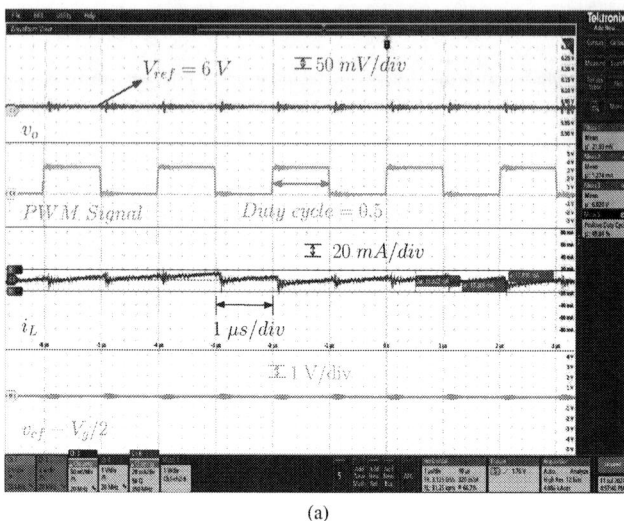

(a)

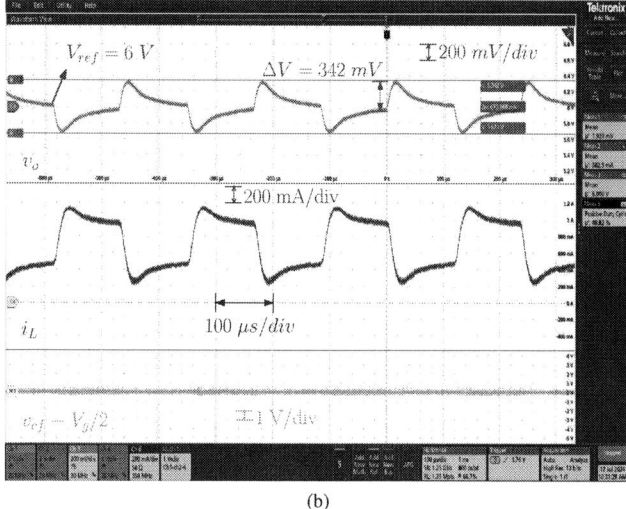

(b)

Fig. 5: Experimental results of the proposed wide-range multi-sampling DPCMC for $V_o = 6$ V: (a) no-load operation (i.e., $I_o = 0$) and (b) load-step change from half to full-load.

Fig. 5(a), the steady-state operation in no-load condition (i.e., $I_o = 0$) is verified. Because of the finite resolution of the ADC used in the sensing/conditioning circuit of the inductor current, small residual oscillations are present in the inductor current. In any case, their magnitude is very small. By defining the regulation error as $\Delta i_L = I_{ref} - i_L$ one has $\Delta i_{L\text{-MAX}} \approx 8$ mA.

The experimental tests reported in Fig. 5(b) show the ability of the proposed control method to properly react to load-step changes when operating at the critical operating point $M = 0.5$. The FC voltage remains stable and *automatically* balanced. These tests confirm the ability of the developed control system to operate without any issue at the critical operating point $M = 0.5$.

B. Ability to reach and properly operate at $M = 0.5$

These tests show how the ability of the developed control system to maintain proper operation even when the critical operating point is reached with an abrupt change in the voltage reference V_{ref}. The first test, Fig. 6(a), shows a transition between operating mode 1 and mode 2. In this case, the two multi-sampling controls from [1] and now modified and merged into the wide-range multi-sampling DPCMC, properly regulate the output voltage for both $V_{ref} = 5$ V and $V_{ref} = 7$ V. In the tests reported in Fig. 6(b) the proposed control has shown that it can operate properly under an extreme condition in which the output voltage is continuously raised to the critical value $V_o = 6$ V and then to a value far from it (i.e., $V_o = 3$ V). In all experimental tests, the FC voltage never

showed any stability or unbalance related problems, confirming the ability of the proposed wide-band control system to operate throughout the entire operating range of 3LFC Buck converter without any issue.

VI. CONCLUSION

This paper proposes a digital wide-range multi-sampling DPCMC method for 3LFC Buck converter with a full range duty cycle controllability and stable operation for $M = 0.5$. The control method guarantees the stability and proper balancing of the flying capacitor voltage in both operating modes. The FC voltage is not measured, and no extra control actions are used to regulate its value. The required controlling actions are embedded in the main CMC system. In this work, the main limitation of the multi-sampling fast-update peak DPCMC is solved by an estimation process for the peak inductor current. The ultimate performance of the proposed control method is verified in simulation and experimentally.

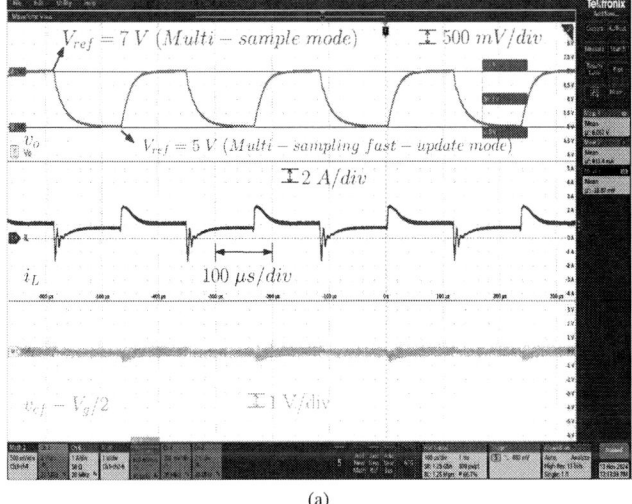

(a)

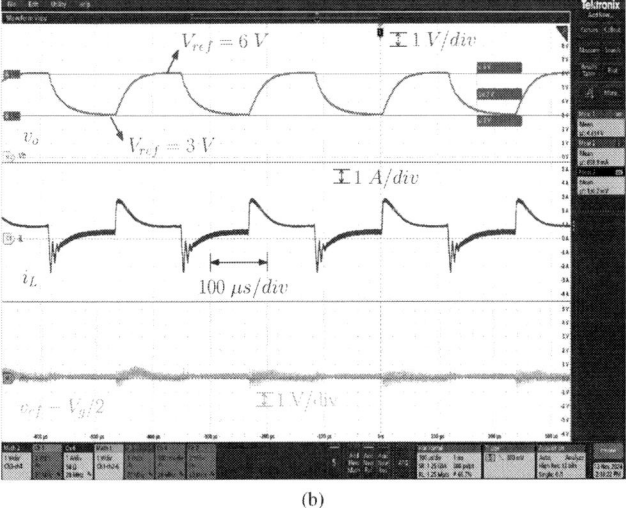

(b)

Fig. 6: Experimental results of the proposed wide-range multi-sampling DPCMC (at full-load) for step-change in the reference voltage from: (a) $V_{\text{ref}} = 5\,\text{V} \rightarrow 7\,\text{V}$ and (b) $V_{\text{ref}} = 3\,\text{V} \rightarrow 6\,\text{V}$.

REFERENCES

[1] G. Bonanno and L. Corradini, "Digital predictive current-mode control of three-level flying capacitor buck converters," *IEEE Transactions on Power Electronics*, vol. 36, no. 4, pp. 4697–4710, 2021.

[2] ——, "Stability properties of digital predictive current-mode controllers for three-level flying capacitor converters," in *2020 IEEE Applied Power Electronics Conference and Exposition (APEC)*, 2020, pp. 312–319.

[3] E. Abdelhamid, G. Bonanno, L. Corradini, P. Mattavelli, and M. Agostinelli, "Stability properties of the 3-level flying capacitor buck converter under peak or valley current programmed control," *IEEE Transactions on Power Electronics*, vol. 34, no. 8, pp. 8031–8044, 2019.

[4] T. Meynard and H. Foch, "Multi-level conversion: high voltage choppers and voltage-source inverters," in *PESC '92 Record. 23rd Annual IEEE Power Electronics Specialists Conference*, 1992, pp. 397–403 vol.1.

[5] C. F. Wang, S. Sa, R. V. Ramaraju, S. Chandra, and O. Trescases, "Seven-level flying-capacitor multi-level converter for a differential power-processing interface in multi-chemistry ev battery packs," in *2023 IEEE Applied Power Electronics Conference and Exposition (APEC)*, 2023, pp. 1–8.

[6] M. T. Elrais, M. Safayatullah, and I. Batarseh, "Generalized architecture of a gan-based modular multiport multilevel flying capacitor converter,"

IEEE Transactions on Power Electronics, vol. 38, no. 8, pp. 9818–9838, 2023.

[7] T. H. Shahsavar, N. V. Kurdkandi, O. Husev, E. Babaei, M. Sabahi, A. Khoshkbar-Sadigh, and D. Vinnikov, "A new flying capacitor-based buck–boost converter for dual-purpose applications," *IEEE Journal of Emerging and Selected Topics in Industrial Electronics*, vol. 4, no. 2, pp. 447–459, 2023.

[8] J. Celikovic, R. Das, H.-P. Le, and D. Maksimovic, "Modeling of capacitor voltage imbalance in flying capacitor multilevel dc-dc converters," in *2019 20th Workshop on Control and Modeling for Power Electronics (COMPEL)*, 2019, pp. 1–8.

[9] J. Marin, J. Gak, A. Cortes, N. Calarco, A. Oliva, E. Lindstrom, M. Miguez, A. Falcón, N. Osterman, and C. A. Rojas, "Integrated three-level flying capacitor dc-dc buck converter for cubesat applications," in *2023 Argentine Conference on Electronics (CAE)*, 2023, pp. 90–95.

[10] R. K. Iyer, I. Z. Petric, R. S. Bayliss, N. C. Brooks, and R. C. N. Pilawa-Podgurski, "A high-bandwidth parallel active balancing controller for current-controlled flying capacitor multilevel converters," *IEEE Transactions on Power Electronics*, vol. 39, no. 10, pp. 12 951–12 965, 2024.

[11] E. Abdelhamid, L. Corradini, P. Mattavelli, G. Bonanno, and M. Agostinelli, "Sensorless stabilization technique for peak current mode controlled three-level flying-capacitor converters," *IEEE Transactions on Power Electronics*, vol. 35, no. 3, pp. 3208–3220, 2020.

[12] A. El Aroudi, N. Cañas-Estrada, M. Debbat, and M. Al-Numay, "Nonlinear dynamics and stability analysis of a three-cell flying capacitor dc-dc converter," *Applied Sciences*, vol. 11, no. 4, 2021. [Online]. Available: https://www.mdpi.com/2076-3417/11/4/1395

[13] R. Garnayak, P. Majumder, S. Kapat, and C. Chakraborty, "A hybrid design framework for fast transient and voltage balancing in a three-level flying capacitor boost converter with digital current mode control," *IEEE Transactions on Power Electronics*, vol. 38, no. 11, pp. 13 674–13 685, 2023.

[14] Z. Xia, K. Datta, and J. T. Stauth, "State-space modeling and control of flying-capacitor multilevel dc–dc converters," *IEEE Transactions on Power Electronics*, vol. 38, no. 10, pp. 12 288–12 303, 2023.

[15] Y. Yang, M. Zhang, K. Shen, and D. Zhao, "Fixed frequency model predictive control of a five-level flying capacitor converters with reduced voltage ripple," in *2024 IEEE 10th International Power Electronics and Motion Control Conference (IPEMC2024-ECCE Asia)*, 2024, pp. 4354–4358.

[16] X. Zhao, Y. Zhang, and Q. Guan, "A balancing control method for flying capacitors in five-level buck/boost converter with synchronous phase shifting decoupling," in *2021 IEEE 1st International Power Electronics and Application Symposium (PEAS)*, 2021, pp. 1–6.

[17] Y. Cao, Y. Bai, V. Mitrovic, B. Fan, D. Dong, R. Burgos, D. Boroyevich, R. S. K. Moorthy, and M. Chinthavali, "A three-level buck–boost converter with planar coupled inductor and common-mode noise suppression," *IEEE Transactions on Power Electronics*, vol. 38, no. 9, pp. 10 483–10 500, 2023.

[18] C. Wang, Y. Lu, and R. P. Martins, "A highly integrated tri-path hybrid buck converter with reduced inductor current and self-balanced flying capacitor voltage," *IEEE Transactions on Circuits and Systems I: Regular Papers*, vol. 69, no. 9, pp. 3841–3850, 2022.

[19] V. Jayan and A. M. Y. M. Ghias, "A single-objective modulated model predictive control for a multilevel flying-capacitor converter in a dc microgrid," *IEEE Transactions on Power Electronics*, vol. 37, no. 2, pp. 1560–1569, 2022.

[20] S. Pan and P. K. T. Mok, "A 25 mhz fast transient adaptive-on/off-time controlled three-level buck converter," *IEEE Transactions on Circuits and Systems I: Regular Papers*, vol. 69, no. 6, pp. 2601–2613, 2022.

[21] Z. Zhou, J. Wang, Z. Liu, and J. Liu, "Accurate prediction of vertical crossings for multi-sampled digital-controlled buck converters," in *2020 IEEE Applied Power Electronics Conference and Exposition (APEC)*, 2020, pp. 292–298.

[22] G. Bonanno, A. Comacchio, and P. Mattavelli, "Multisampling digital pulse-width modulator based on asymmetric dual-edge carrier," *IEEE Transactions on Power Electronics*, vol. 39, no. 5, pp. 5121–5134, 2024.

[23] G. Bonanno, "Multi-level flying capacitor buck converters with digital-predictive current-mode control," Ph.D. dissertation, University of Padova – Department of Information Technology, 2021.

[24] N. Vukadinović, A. Prodić, B. A. Miwa, C. B. Arnold, and M. W. Baker, "Skip-duty control method for minimizing switching stress in low-power multi-level dc–dc converters," in *2015 IEEE 16th Workshop on Control and Modeling for Power Electronics (COMPEL)*, 2015, pp. 1–7.

A Comparative Investigation of a New Continuous Voltage Conversion Ratio Approach in a Zero-Inductor Voltage Converter

1st Sina Salehi Dobakhshari
Department of Electrical and Computer Engineering
Queen's University
Kingston, Canada
s.salehidobakhshari@queensu.ca

2nd Aamna Nasir Hameed
Department of Electrical and Computer Engineering
Queen's University
Kingston, Canada
21anh2@queensu.ca

3rd Binghui He
Department of Electrical and Computer Engineering
Queen's University
Kingston, Canada
binghui.he@queensu.ca

4th Mojtaba Forouzesh
Department of Electrical and Computer Engineering
Queen's University
Kingston, Canada
m.forouzesh@queensu.ca

5th Yan-Fei Liu
Department of Electrical and Computer Engineering
Queen's University
Kingston, Canada
yanfei.liu@queensu.ca

Abstract—**This paper investigates the performance of the new pulse pattern strategy for a Zero-Inductor Voltage (ZIV) converter. Unlike the other methods introduced in the literature, this method does not need additional components or sensors. Moreover, it extends the voltage conversion ratio from zero to input voltage, just like a conventional buck converter, but eliminates its drawbacks. The new method will be discussed and compared to a 3-level buck converter, and its performance as a solar battery charger will be analyzed. The results for a 250W prototype (20-60V input voltage, 12V output voltage, 21A output current) with average efficiency over the full line and load (60W to 250W) of 97.54%, are presented to justify the merit of the new method.**

Index Terms—**Full voltage conversion range, high efficiency, switched capacitor converter, zero-inductor-voltage.**

I. INTRODUCTION

Zero-Inductor Voltage (ZIV) converters are well-known for high power density and high efficiency [1]. However, as with many other capacitive-based converters (multi-level, switched capacitor converters, etc.) the inherent fixed voltage conversion ratio has limited its application. It would be desirable to have voltage conversion flexibility as a conventional buck converter without drawbacks. Some methods are proposed in the literature to extend the voltage conversion ratio in capacitive-type converters. In [2], the voltage conversion ratio of a switched-capacitor converter (SCC) has been extended, but the proposed method uses an additional switched capacitor-inductor network for the voltage regulation, and the conversion range is, however, from zero to a maximum of 0.3. In [3], a buck converter is merged with a Resonant Switched Capacitor Converter (RSCC) to regulate the output voltage. The voltage conversion range is from zero to 1/3, which is still narrow.

Another control method for an RSCC is proposed in [4]. Although it has reached very high efficiency (¿98%), the voltage conversion range is between 1/3 and 2/3. The phase shift control is introduced in [5], for a multilevel modular RSCC. Although this method does not need any extra components, the conversion range is highly limited to 0.17-0.2. One of the popular control methods in resonant switched-capacitor converters is frequency modulation.

Some RSCCs have achieved a full voltage conversion ratio through frequency control methods, as demonstrated in [6] and [7]. In [6], the RSCC attains a full voltage conversion range by using a combination of frequency and phase shift control. However, this approach requires sensing the resonant tank current, and the maximum efficiency achieved is 91%. The converter in [7] offers a wide conversion range from 0.5 to 2, allowing it to step up the input voltage. However, its maximum efficiency is 90%, which is achieved at a unity voltage ratio.

In this paper, a new control method for a ZIV converter will be discussed. The new method proposes a new gate pulse pattern strategy to fully regulate the output voltage from zero to V_{in}, without using any additional component and additional feedback signal. The converter uses seven low-voltage regular silicon switches, and the results will be compared with a Three-Level Buck Converter (TLBC) in [8] which not only uses GaN switches, but the number of switches is also less than the ZIV converter. Moreover, the converter's performance in a solar battery charger application will be investigated.

As shown in Fig. 1, the converter consists of two stages and an output filter. The first stage includes four switches (S_1, S_2, S_3, and S_4) along with a flying capacitor C_1. The

979-8-3315-1612-3/25 $31.00 © 2025 IEEE

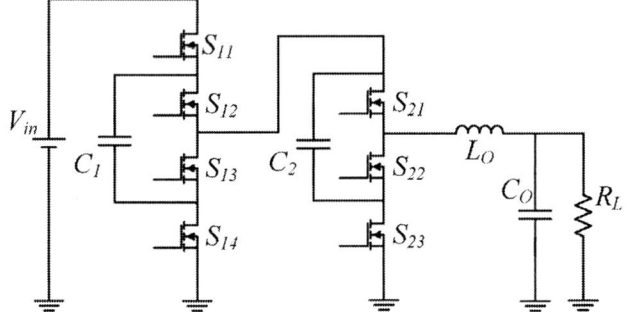

Fig. 1: The 7-Switch ZIV converter introduced in [1]

second stage comprises three switches (M_1, M_2, and M_3) and a flying capacitor C_2. Additionally, the output filter is formed by the inductor L_O and the capacitor C_O. In the diagram, the input source and output load are represented by V_{in} and R_L, respectively.

It is important to note that the original converter is referred to as a ZIV converter, as the voltage across the inductor remains at zero. However, in this control method, the voltage across the inductor is not necessarily zero in the whole voltage conversion range but is maintained at a low level, which minimizes the inductor current ripple. In this context, the term "ZIV" is used to describe the original converter topology.

II. THE PRINCIPLE OF THE NEW CONTROL METHOD

In this section, the basic principle for voltage conversion ratio extension for a seven-switch ZIV converter will be briefly explained. The converter's voltage gain inherently is ¼ as discussed in [1]. Based on the duty cycle value, the operation of the converter is divided into four modes.

A. Mode I: Duty cycle range from 0 to 1/4

The pulse pattern, inductor's current, and the equivalent circuit in a switching cycle for the $0 \leq D \leq 1/4$ are presented in Fig.2. Here the duty cycle (D) is defined as $(t_1 - t_0)/T_s$. In this mode:

1) Switches S_1 and S_3 are activated simultaneously at the start of the switching cycle and remain on for a duration of DT_S before turning off.
2) Switches S_2 and S_4 also turn on together at $t = T_S/4$ and stay on for DT_S before turning off.
3) Switch M_1 turns on at $t = T_S/2$ and stays on for $2DT_S$.
4) Switch M_2 operates in complement to M_1, so when M_1 is off, M_2 is on, and vice versa.
5) Lastly, switch M_3 turns on as S_2 turns off and stays on until the end of the switching cycle.

Fig. 2 also indicates the voltage across the inductor at each switching interval. Accordingly, the inductor current variation in each interval can be calculated as:

$$\Delta i_{Lo11} = (V_{in} - V_{C1} - V_{C2} - V_O)DT_S/L_O \qquad (1)$$

$$\Delta i_{Lo12} = (-V_O)(0.25 - DT_S)/L_O \qquad (2)$$

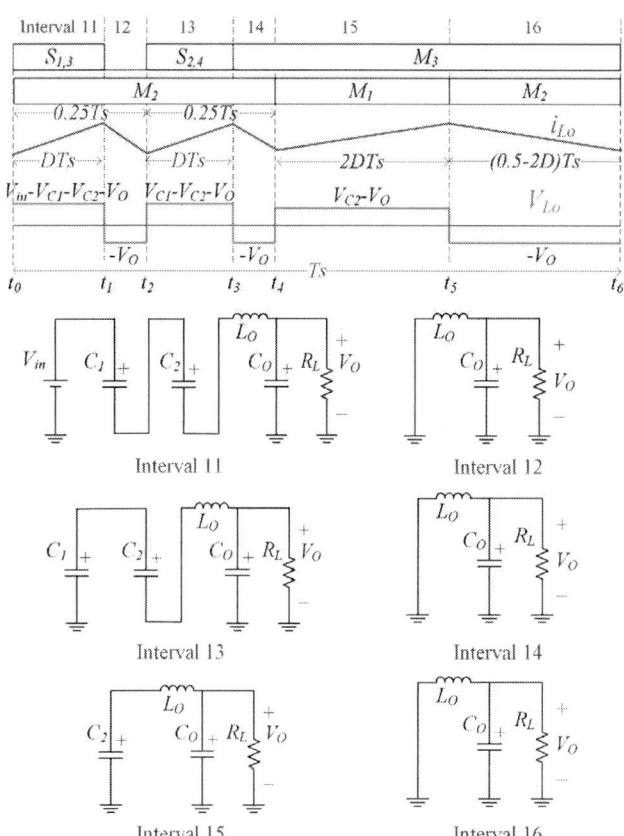

Fig. 2: The pulse pattern, inductor's current and voltage waveforms, and the equivalent circuit of the ZIV converter in each interval in Mode I.

$$\Delta i_{Lo13} = (V_{C1} - V_{C2} - V_O)DT_S/L_O \qquad (3)$$

$$\Delta i_{Lo14} = (-V_O)(0.25 - DT_S)/L_O \qquad (4)$$

$$\Delta i_{Lo15} = (V_{C2} - V_O)2DT_S/L_O \qquad (5)$$

$$\Delta i_{Lo12} = (-V_O)(0.5 - 2DT_S)/L_O \qquad (6)$$

where D in the duty cycle, and Δi_{L11}, Δi_{L12}, Δi_{L13}, Δi_{L14}, Δi_{L15}, and Δi_{L16} are inductor's current variations in interval 1 to 6 of $Mode I$, respectively. V_{C1} and V_{C2} are flying capacitors' voltage, and T_S is the switching period.

B. Mode II: Duty cycle range from 1/4 to 1/3

The pulse pattern, inductor's current, and the equivalent circuit in a switching cycle for the $1/4 \leq D \leq 1/3$ are presented in Fig.3. In this mode:

1) Switches S_1 and S_3 turn on simultaneously at the start of the switching cycle and turn off after DT_S.
2) Switches S_2 and S_4 turn on at $t = DT_S$ and turn off at $t = 2DT_S$.
3) Switch M_1 turns on at $t = 2DT_S$ and stays on for a period of $2DT_S$, extending partially into the next switching cycle.

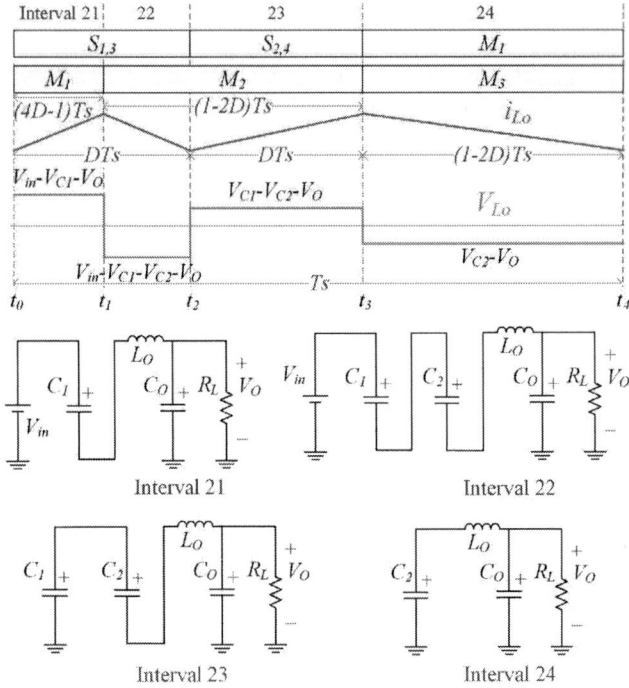

Fig. 3: The pulse pattern, inductor's current and voltage waveforms, and the equivalent circuit of the ZIV converter in each interval in Mode II.

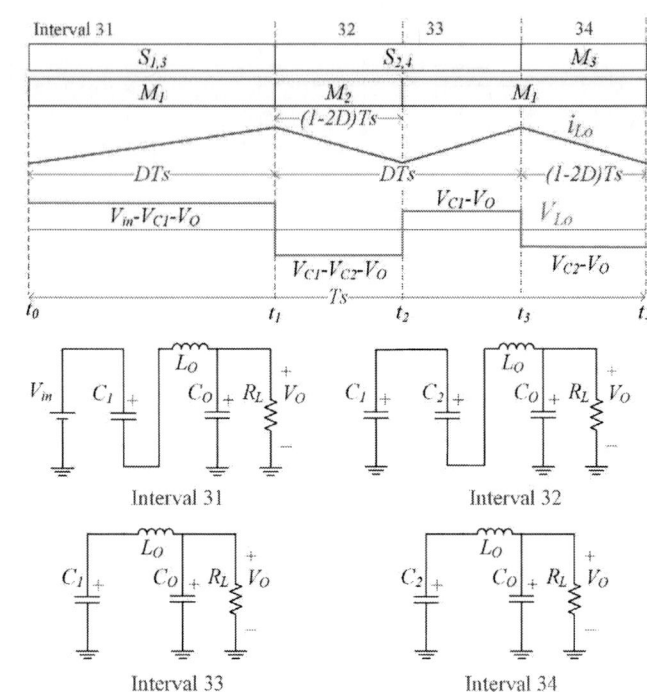

Fig. 4: The pulse pattern, inductor's current and voltage waveforms, and the equivalent circuit of the ZIV converter in each interval in Mode III.

4) Switch M_2 operates in complement to M_1, meaning it turns on whenever M_1 is off and vice versa Consequently, M_2 turns on at $t = (4D - 1)T_S$ and turns off at $t = 2DT_S$.
5) Finally, switch M_3 turns on as S_2 and S_4 turn off $t = 2DT_S$ and turns off at the end of the switching cycle.

The voltage across the inductor at each switching interval is also indicated in Fig. 3. Accordingly, the inductor current variation in each interval can be calculated as:

$$\Delta i_{Lo21} = (V_{in} - V_{C1} - V_O)(4D - 1)T_S/L_O \quad (7)$$

$$\Delta i_{Lo22} = (V_{in} - V_{C1} - V_{C2} - V_O)(1 - 3D)T_S/L_O \quad (8)$$

$$\Delta i_{Lo23} = (V_{C1} - V_{C2} - V_O)DT_S/L_O \quad (9)$$

$$\Delta i_{Lo24} = (V_{C2} - V_O)(1 - 2D)T_S/L_O \quad (10)$$

where Δi_{L21}, Δi_{L22}, Δi_{L23}, and Δi_{L24} are inductor's current variations in interval 1 to 4 of Mode II, respectively.

C. Mode III: Duty cycle range from 1/3 to 1/2

The pulse pattern, inductor's current, and the equivalent circuit in a switching cycle for the $1/3 \leq D \leq 1/2$ are presented in Fig.4. In this mode:

1) Switches S_1 and S_3 turn on together at the beginning of the switching cycle and turn off after DT_S.
2) Switches S_2 and S_4 also turn on at $t = DT_S$ and turn off at $t = 2DT_S$.

3) Switch M_1 turns on at $t = (1 - D)T_S$ and stays on for $2DT_S$, completing part of this duration in the next switching cycle. In this mode, M_1 turns off at the same time as S_1 and S_3.
4) Switch M_2 operates in a complementary manner to M_1,
5) Finally, switch M_3 turns on when S_2 and S_4 turn off ($t = DT_S$) and stays on until the end of the switching cycle.

According to the data presented in Fig. 3, the inductor current variation in each interval can be calculated as:

$$\Delta i_{Lo31} = (V_{in} - V_{C1} - V_O)DT_S/L_O \quad (11)$$

$$\Delta i_{Lo32} = (V_{C1} - V_{C2} - V_O)(1 - 2D)T_S/L_O \quad (12)$$

$$\Delta i_{Lo33} = (V_{C1} - V_O)(3D - 1)T_S/L_O \quad (13)$$

$$\Delta i_{Lo34} = (V_{C2} - V_O)(1 - 2D)T_S/L_O \quad (14)$$

where Δi_{L31}, Δi_{L32}, Δi_{L33}, and Δi_{L34} are inductor's current variations in interval 1 to 4 of Mode III, respectively.

D. Mode IV: Duty cycle range from 1/1 to 1

The pulse pattern, inductor's current, and the equivalent circuit in a switching cycle for the $1/2 \leq D \leq 1$ are presented in Fig.5. In this mode:

1) Switch S_1 turns on at the start of the switching cycle and turns off after a duration of DTS.

979-8-3315-1612-3/25 $31.00 © 2025 IEEE

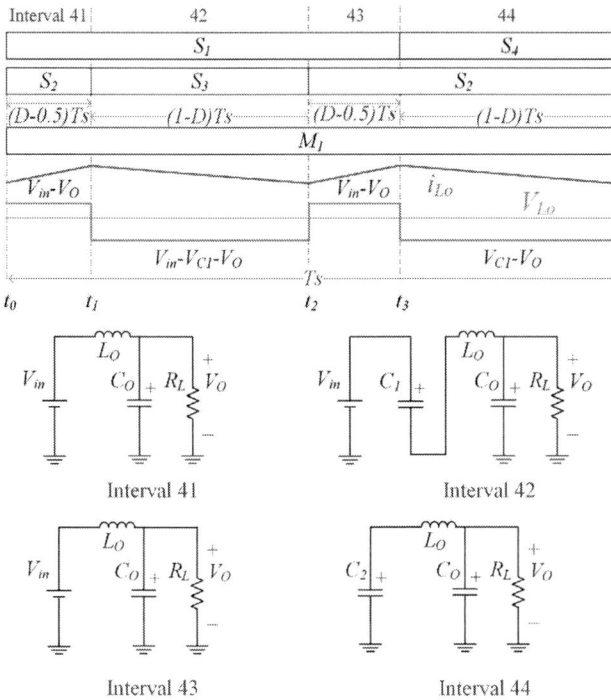

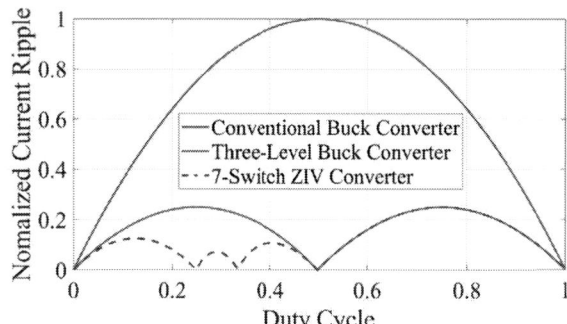

Fig. 6: Current ripple comparison among buck, TLBC, and ZIV converter.

TABLE I: Prototype specifications.

Parameters	Value
V_{in}	$20 - 60V$
V_O	$12V$
I_O	$21A$
P_O	$250W$
F_S	$100kHz$
C_1	$7 \times 10uF/50V$
C_2	$7 \times 10uF/50V$
C_O	$10 \times 10uF/50V$
L_O	$2.2uH$
$S_1 - S_4$	$BSZ025N04LS, 40V/2.5m\Omega$
$M_1 - M_3$	$BSiSA04DN, 30V/2.15m\Omega$

Fig. 5: The pulse pattern, inductor's current and voltage waveforms, and the equivalent circuit of the ZIV converter in each interval in Mode II.

2) Switch S_2 turns on at $t = DT_S/2$ and remains on for DTS, extending partially into the next switching cycle, and turns off at $t = (D - 0.5)T_S$.
3) Switch S_3 operates in complement to S_2.
4) Switch S_4 operates in complement to S_1.
5) Switch M_1 remains continuously on, bypassing the second stage.
6) Switches M_2 and M_3 are always off.

According to the data presented in Fig. 5, the inductor current variation in each interval can be calculated as:

$$\Delta i_{Lo41} = (V_{in} - V_O)(D - 0.5)T_S/L_O \qquad (15)$$

$$\Delta i_{Lo42} = (V_{in} - V_{C1} - V_O)(1 - D)T_S/L_O \qquad (16)$$

$$\Delta i_{Lo43} = (V_{C1} - V_O)(D - 0.5)T_S/L_O \qquad (17)$$

$$\Delta i_{Lo44} = (V_{C1} - V_O)(1 - D)T_S/L_O \qquad (18)$$

where Δi_{L41}, Δi_{L42}, Δi_{L43}, and Δi_{L44} are inductor's current variations in interval 1 to 4 of Mode IV, respectively.

III. PERFORMANCE ANALYSIS OF THE CONTROL METHOD

Applying volt-second balance to the inductor at each mode, the voltage gain will be achieved as:

$$V_O = DV_{in} \qquad (19)$$

Therefore, the ZIV converter can fully regulate the output voltage from zero to V_{in}, like a Buck converter. However, as mentioned earlier, it is desired to eliminate the buck converter's disadvantages, even having merits over TLBC.

Fig. 6 presents a comparison of the inductor current ripple across the conventional Buck converter, TLBC, and the ZIV converter, with each ripple normalized to the maximum ripple observed in the buck converter. It is evident that both the TLBC and ZIV converters significantly reduce the current ripple compared to the Buck. While the TLBC and ZIV converters show similar current ripple levels for $D > 0.5$, the ZIV converter demonstrates a marked improvement in ripple reduction for $D < 0.5$. This improvement is particularly notable since most operating points typically fall within this range. As a lower current ripple leads to a reduced RMS current, the ZIV converter is expected to be more efficient overall. In addition, a low current ripple enables the use of a smaller inductor, which is highly desirable in recent designs. A smaller inductor significantly increases power density, as magnetic components occupy most of the converter's volume.

IV. EXPERIMENTAL RESULTS AND COMPARISON

The experimental results based on the specifications listed in Table I, are presented in Fig. 7. The gate-pulse for S_1, S_2, and M_1 along with the inductor's current waveform are presented in this figure. The results show the transition conditions that

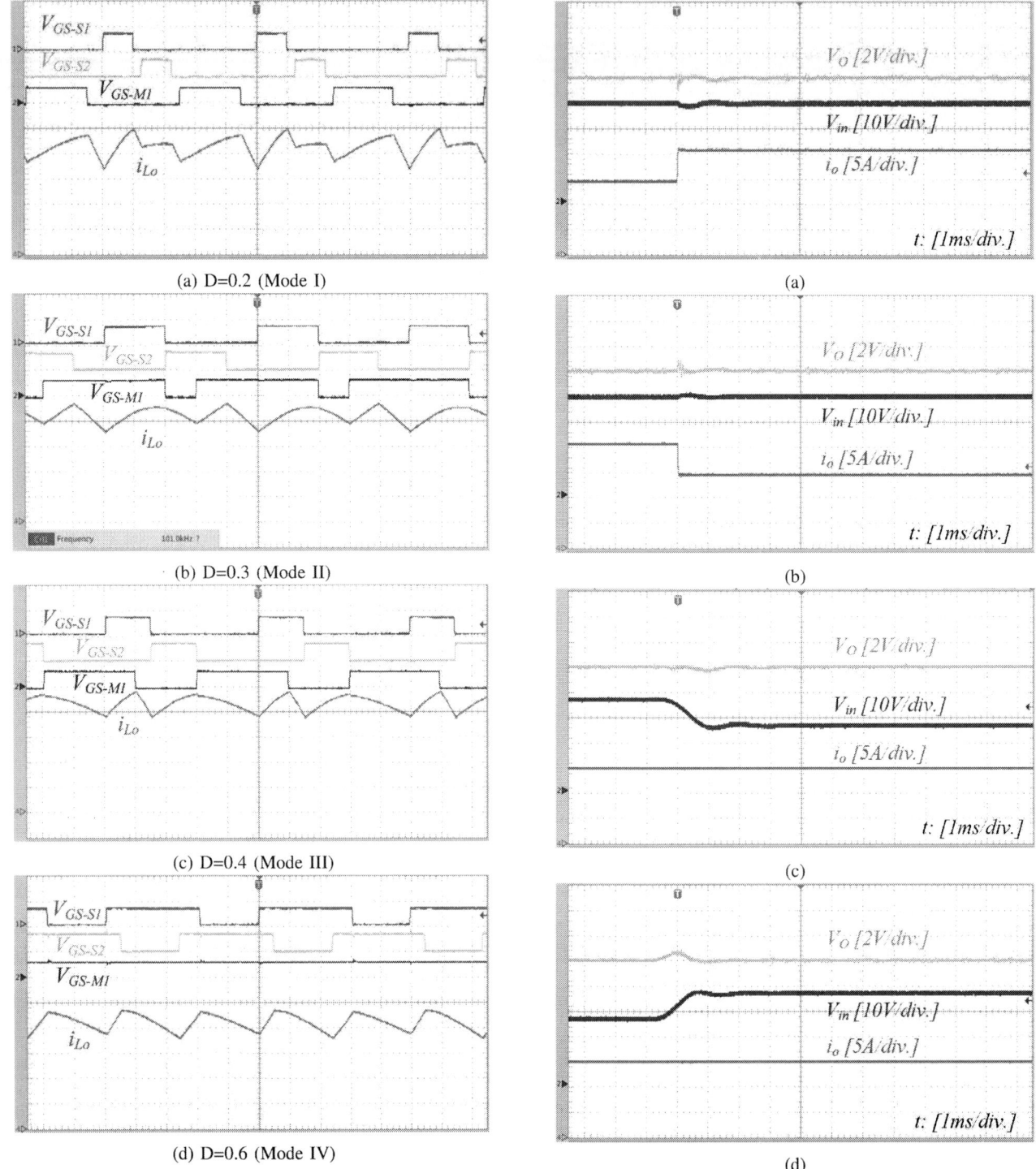

Fig. 7: Experimental waveforms for $V_{GS-S1}[5V/div.]$, $V_{GS-S2}[5V/div.]$, $V_{GS-M1}[5V/div.]$, and $i_{Lo}[5A/div.]$, ($t = 2us/div.$).

Fig. 8: Dynamic response of the ZIV converter with the new control method to (a) Load step up, (b) load step down, (c) input voltage step up, and (d) input voltage step down.

cover all four sections that were discussed before. The output current is 21A, and the converter has kept the current ripple low during the transitions.

A controller is designed for the converter to finely tune

the output voltage at line and load variations. The converter's dynamic response to these variations is shown in Fig. 8. In particular, Fig. 8(a) illustrates the response to a load increase from 15A to 21A, while Fig. 8(b) depicts the response to a load decrease from 21A to 15A, both occurring at a slew rate of 0.5A/μs. Throughout these tests, the input voltage remains constant at 40V, and the output voltage is regulated to 12V. The controller successfully stabilizes the output voltage to the new operating point within 200μs in both scenarios.

This rapid response underscores the control system's efficiency in maintaining output stability during sudden load transitions, ensuring minimal voltage deviation and fast recovery, critical for applications requiring high power accuracy and dynamic load handling.

Additionally, Fig. 8(c) and (d) illustrate the converter's dynamic response to line regulation, where the input voltage shifts from 27V to 37V and back, with a slew rate of 10V/ms and a constant load of 15A. The transient time is around 1ms. The converter's ability to quickly respond to line fluctuations ensures reliable operation in environments with unstable power supplies, making it highly adaptable for applications where both load and line stability are vital.

The efficiency of the ZIV converter with the proposed method is compared with two Bucks and a TLBC from EPC [8], [9] in Fig. 9. In Fig 9 (a), the efficiency of the Buck with GaN switches is shown. It is a product from EPC with part number EPC9205. The operation condition is $V_{in} = 48V, V_O = 12V, I_O = 2 - 15A$ (here it is named Buck I). The switching frequency is 700kHz, and the output inductor is 2.2uH. The maximum efficiency of the converter is 95.9%, and at full load, it is 95.3%.

Fig. 9 (b) shows the efficiency comparison between the TLBC and a comparative design of another Buck converter (here it is named Buck II), both from EPC. In this figure, Buck II operates at 500kHz with a 3.3uH inductor. While the TLBC operates at 320kHz, yielding an effective switching frequency of 640kHz, and uses a 1.5uH inductor with 50% less volume than the conventional two-level buck converter. At full load, TLBC achieves a 25% reduction in power loss and over 1% improvement in peak efficiency compared to Buck II, owing to reduced switching stresses and a lower effective switching frequency. The peak and full load efficiency of the TLBC are 96.8% and 96%, respectively.

The efficiency of the ZIV converter is compared with TLBC, Buck I, and Buck II in Fig. 9(c), under the same condition. The ZIV converter has significantly improved efficiency over the whole output power range from 3A to 21A. This is because the conversion ratio is ¼ which is one of the ZIV points. According to Fig. 6, at this point, the current ripple of the ZIV converter is zero, while in a TLBC and a Buck, it is 25% and 60% of the maximum current ripple in a Buck converter, respectively. In Fig. 9(b), it was mentioned that the reason for the efficiency improvement of TLBC over Buck II is the lower current ripple. Consequently, in the ZIV converter, since the current ripple is minimized, the efficiency is higher than TLBC.

The peak and full load efficiency of the four converters is listed in Table II, as well as the average efficiency over the whole output current range. The range of current over which the average efficiency is calculated is indicated in the

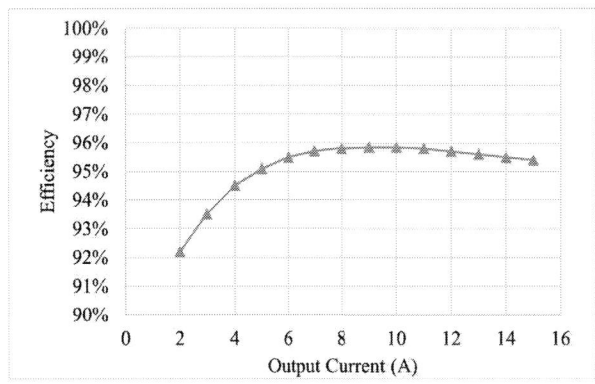

(a) Efficiency curve of Buck I [9],

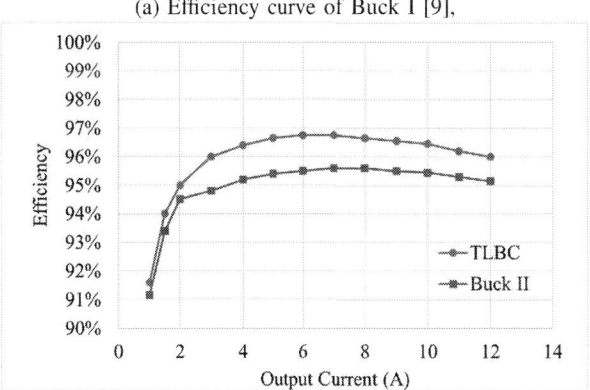

(b) Efficiency curve of TLBC [8] and Buck II [8],

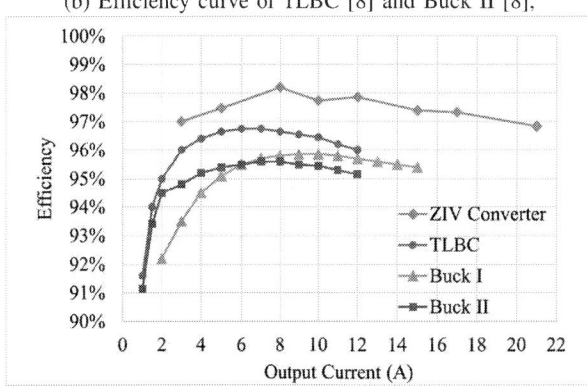

(c) Efficiency comparison of the ZIV converter, Buck I [9], Buck II [8], and TLBC [8].

Fig. 9: Efficiency curves of the converters

corresponding row. The peak and full load efficiency of the ZIV converter (even at 21A) is more than the other ones. Although the output current range of the ZIV converter is more than the others, its average efficiency is 1.9% more than Buck I, 2.2% more than Buck II, and 1.2% more than TLBC. The efficiency of the ZIV converter from light load to full load, including the control and drivers' loss is shown in Fig. 10. The average efficiency at light load is 97.7%, and at full load, it is 97.12%. The peak efficiency is 98.5% which occurs at light load. Also, the average efficiency over the whole input voltage and power range is 97.54%. Moreover, the efficiency curve is almost flat during the input voltage range and at

TABLE II: EFFICIENCY COMPARISON OF THE CONVERTERS

	Peak Efficiency	Full Load Efficiency	Average Efficiency
Buck I [9]	95.9%	95.4% @15A	95.6% [5A-15A]
Buck II [8]	95.6%	95.2% @12A	95.3% [2A-12A]
TLBC [8]	96.8%	96% @12A	96.3% [2A-12A]
This work	**98.2%**	97.4%　@15A 96.8% @21A	**97.5% [3A-21A]**

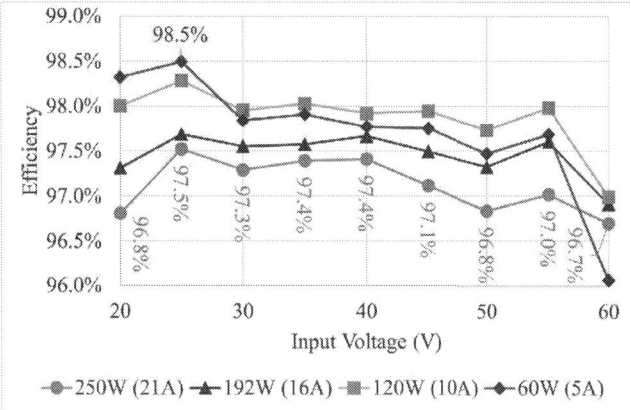

Fig. 10: The measured efficiency of the ZIV converter versus input voltage at various output power.

different output power levels. This shows that the converter preserves high and almost constant efficiency throughout the input voltage and power.

V. THE CONVERTER'S PERFORMANCE IN A SOLAR BATTERY CHARGER APPLICATION

This converter was used in a solar battery charger application, and the corresponding experimental results will be discussed in this section to validate the feasibility of the new control method further.

In this application, TerraSAS ETS 600/17 was used as a solar simulator. Also, an LC input filter ($C_f = 100uF$, $L_f = 8uH$) was used to make the input current smooth. To examine the functionality of the proposed control method in a dynamic environment, the irradiation of the solar emulator was set to change in 30-second intervals from $100W/m^2$ to $1000W/m^2$, as shown in Fig. 10. In this condition, the input voltage changes from 28.5V to 45V. The output voltage is fixed at 12V, by a constant voltage load (to mimic a battery), and the maximum power is 180W. The converter is set to do the Maximum Power Point Tracking (MPPT) with the Perturb and Observation (P&O) method. Fig. 11 shows the experimental results for the MPPT performance of the converter with the new control method. The maximum available power and the harvested power are presented in this figure. The irradiation changes every 30 seconds with a one-second transition between the levels. The irradiation changing pattern is repeated twice for the dynamic performance investigation,

and then it is kept constant to explore the static functionality. At each irradiation level, its value is indicated in this figure. The MPPT efficiency, defined as the ratio of the harvested power to the maximum available power, is illustrated in Fig. 11 for various irradiation levels. The efficiency is mostly more than 99%. This figure demonstrates that the control method effectively tracks the maximum power point under both increasing and decreasing irradiation conditions, ensuring optimal power extraction. As shown, the converter successfully adapts to changes in irradiation, maintaining high MPPT efficiency and maximizing the harvested power across a range of environmental conditions. This adaptability is essential for efficient energy capture in applications where sunlight varies frequently. Consequently, as discussed in Section IV, The converter can effectively adapt to rapid dynamic changes.

VI. CONCLUSION

This paper investigates a novel control method for the 7-switch ZIV converter that maximizes its voltage conversion range without requiring additional components or feedback signals. The converter's voltage conversion range is divided into four sections, with smooth transitions between them, ensuring that there is no sudden shift in the phase or duty cycle of any switch. The ZIV converter has the same voltage conversion ratio as a traditional buck converter but offers a significantly reduced inductor current ripple, which contributes to higher efficiency. The comparative study showed that the efficiency of the ZIV converter is higher than that of a TLBC built with GaN technology and fewer switches. The converter's dynamic response was evaluated under both line and load regulation conditions, with results indicating excellent stability and adaptability to changes. Additionally, the converter was tested in a solar battery application, where it was configured to maintain a constant load voltage while tracking maximum power at the input. Under these conditions, the MPPT efficiency proved to be outstanding, validating the controller's effectiveness in applications with variable input conditions like solar energy harvesting.

REFERENCES

[1] S. Webb, T. Liu and Y. -F. Liu, "Zero Inductor-Voltage Multilevel Bus Converter," in IEEE Transactions on Power Electronics, vol. 36, no. 10, pp. 11565-11578, Oct. 2021.

[2] M. Uno and A. Kukita, "PWM Switched Capacitor Converter With Switched-Capacitor-Inductor Cell for Adjustable High Step-Down Voltage Conversion," in IEEE Transactions on Power Electronics, vol. 34, no. 1, pp. 425-437, Jan. 2019.

[3] S. Han, Y. Wang, Y. Guan and D. Xu, "Analysis and Design of a Modular Switched Capacitor Converter with Adjustable Output Voltage in DC Microgrid," in IEEE Journal on Emerging and Selected Topics in Circuits and Systems, vol. 12, no. 1, pp. 232-241, March 2022.

[4] H. Setiadi and H. Fujita, "An Asymmetric Control Method for Switched-Capacitor-Based Resonant Converters," in IEEE Transactions on Power Electronics, vol. 36, no. 9, pp. 10729-10741, Sept. 2021.

[5] Y. Li, B. Curuvija, X. Lyu and D. Cao, "Multilevel modular switched-capacitor resonant converter with voltage regulation," 2017 IEEE Applied Power Electronics Conference and Exposition (APEC), Tampa, FL, USA, 2017, pp. 88-93.

[6] G. Ripamonti, M. Ursino, S. Saggini, S. Michelis and F. Faccio, "Regulated Resonant Switched-Capacitor Point-of-Load Converter Architecture and Modeling," in IEEE Transactions on Power Electronics, vol. 36, no. 4, pp. 4815-4827, April 2021.

979-8-3315-1612-3/25 $31.00 © 2025 IEEE

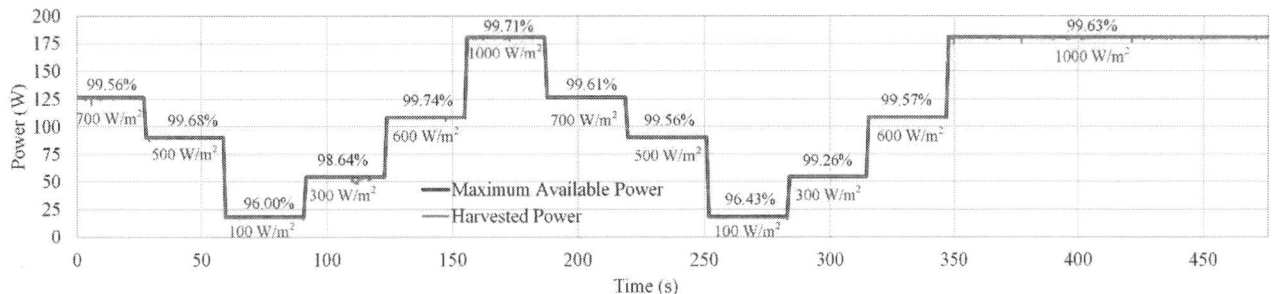

Fig. 11: The experimental results for MPPT performance of the ZIV converter with the new control method in dynamic and static conditions.

[7] A. Cervera, M. Evzelman, M. M. Peretz and S. Ben-Yaakov, "A High-Efficiency Resonant Switched Capacitor Converter with Continuous Conversion Ratio," in IEEE Transactions on Power Electronics, vol. 30, no. 3, pp. 1373-1382, March 2015.

[8] D. Reusch, S. Biswas and M. de Rooij, "GaN Based Multilevel Intermediate Bus Converter for 48 V Server Applications," PCIM Europe 2018; International Exhibition and Conference for Power Electronics, Intelligent Motion, Renewable Energy and Energy Management, Nuremberg, Germany, 2018, pp. 1-8.

[9] Efficient Power Conversion, " Development Board EPC9205 Quick Start Guide," EPC9502 Datasheet, 2018.

A 96.1% Peak Efficiency, 6.8 kW/in³, 48V-to-6V On-package Intermediate Bus Converter with LV-GaN Power Transistors

Mausamjeet Khatua, Nachiket Desai, Harish Krishnamurthy, Sheldon Weng, Jingshu Yu, Huong Do, Samuel Bader, Han Wui Then, Krishnan Ravichandran, James Tschanz, Kaladhar Radhakrishnan, and Vivek De

Intel Corporation

Abstract— **This paper presents the industry's first and only 48V, on-package, intermediate bus converter (IBC) solution with an 8:1 fixed ratio using low-voltage (LV) GaN power transistors with 10x superior figure of merit over Si laterally diffused MOS (LDMOS) for high performance computing applications. Eleven 2x2 mm GaN half-bridge dies along with the passives are co-packaged to form an 8:1 switched capacitor divider. This 1-MHz 200-W IBC prototype achieves a peak efficiency of 96.1% and a power density of 6.8 kW/in³.**

Keywords—Intermediate bus converter, 48V, GaN, switched capacitor converter, high power density

I. INTRODUCTION

With the insatiable demand for compute fueling the "AI Everywhere" future, current/power demands of future CPU/GPU platforms are trending to 1000's of Amperes putting a heavier burden on the power delivery network. 48V systems with a high efficiency, unregulated Intermediate Bus Converter (IBC) 1st stage (48V to 5-12V) and a higher frequency, high density, low z-height vertical power delivery (VPD) based 2nd stages (5-12V to sub-1V) are gaining traction to reduce I2R losses and improve E2E efficiency. Most IBCs today are typically a 4:1 divider [1], switch at 10-100's of kHz [1-3] and prioritize efficiency over density due to lack of better power devices. A 48V to 6V higher density (6.8kW/in³), higher conversion ratio (8:1), ultra-low z-height (<3mm) IBC with comparable efficiency (>95%) can facilitate better system optimization and more importantly bring the power delivery solutions closer to the CPU/GPUs thereby improving overall efficiency and is the primary focus of this paper.

Prior implementations [2-16] of the IBC rely on traditional MOSFETs with poorer FoM, forcing switching frequencies (fsw) <433kHz to maintain good efficiency. Such low frequency switching requires large capacitor footprints implemented on a PCB with larger minimum spacing rules, necessitating longer paths to and from the passives, thereby incurring larger parasitic inductances further inhibiting any chances of switching faster. This paper exploits higher frequency (up to 2MHz) to maximize power density without compromising efficiency. To deliver >96% efficiency at 1MHz (~2.5x better than SOA), the paper explores optimizations across i) power devices and their layout in conjunction with the passives to minimize resistance and inductance, ii) maximizing driver slew rates with minimal overshoot and optimal

deadtimes, and finally iii) paving the way to enable the industry's first and only 48V on-package IBC solution.

II. PROPOSED ARCHITECTURE AND CONVERTER DESIGN

The proposed architecture of the On-Package IBC is shown in Fig. 1. It uses four series-stacked 2:1 SC dividers to perform a 48V to 6V voltage conversion, with each SC divider constructed using 2mm x 2mm GaN half-bridge dice and landside MLCCs. Owing to its modular architecture, the IBC produces four intermediate bus voltages of 6V each. The follow-on multi-phase buck regulator stage can be designed to present a constant current load to the IBC to minimize the required bus capacitance thereby reducing the hard charging losses in the IBC [2-3]. The switches in the IBC block a maximum voltage of 12V, allowing the use of low-voltage (LV), enhancement-mode GaN transistors [4], which is key to maintaining high efficiency despite the higher Fsw. The 12V LV GaN transistors achieve >10× better Ron*Qg FoM compared to industry-standard Si transistors, e.g., Si extended drain MOS (EDMOS), LDMOS, and low-voltage, series-stacked CMOS transistors. Each GaN die in Fig. 1 consists of a half-bridge of switches and their clamps to prevent spurious turn-on of switches during their off state. The 1.8V-drive voltage needed to turn on the switches reduces gate drive losses.

The higher switching frequency facilitates lowering the amount of capacitance needed by almost 3x, aiding its integration directly under the die shadow of the power devices on the landside (Fig. 1) minimizing lateral routing and further reducing parasitic inductances. Lower capacitance values also enable smaller footprints (0402/0603) with ultra-low z-height profiles (<3mm) enabling tighter integration with future microprocessor systems. The flying capacitors and the bus capacitors in each module are selected using the methodology described below:

A. Flying capacitor selection methodology: The flying capacitors in each module are optimally selected to limit the voltage ripple in their corresponding bus voltage while meeting the rms current requirement. The flying capacitor in the upper modules block higher voltages and their effective capacitance is worst hit by the impact of voltage derating of the ceramic capacitors used to realize the flying capacitors. Hence, the number of discrete ceramic capacitors used to realize the flying

979-8-3315-1612-3/25 $31.00 © 2025 IEEE

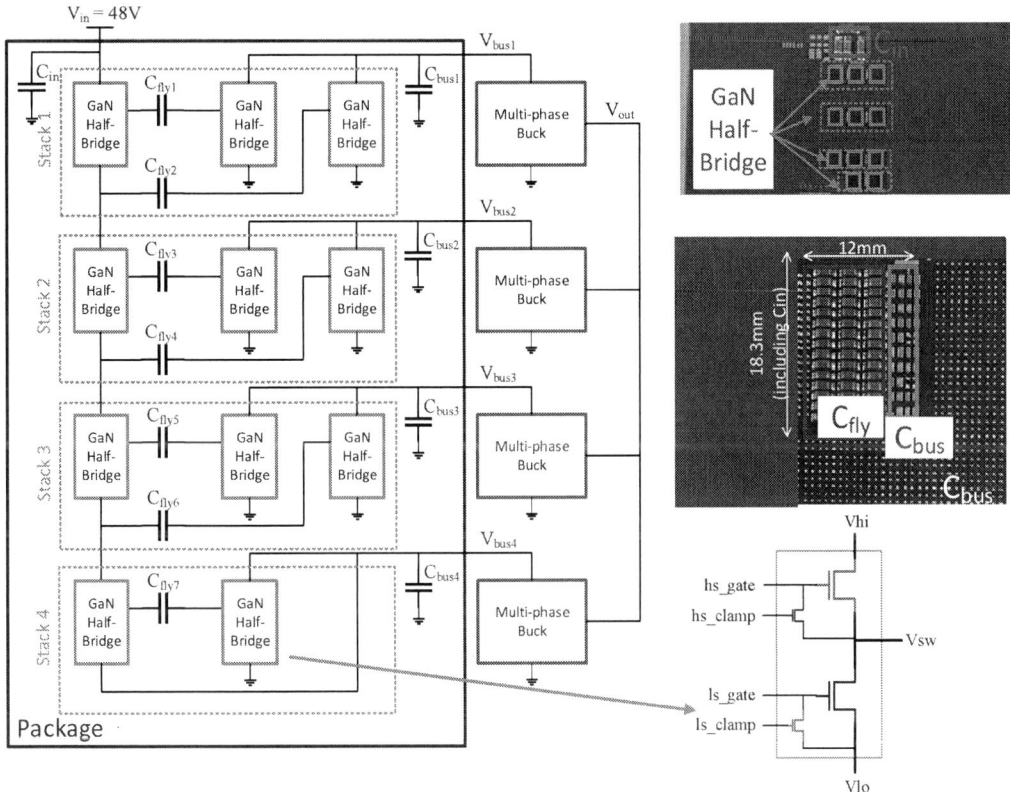

Fig. 1: Proposed architecture of the 48V-to-6V IBC with a package view of construction (top & landside).

capacitors in the upper modules is dictated by the permissible bus voltage ripple limit. On the other hand, the maximum rms current through the flying capacitor dictates the number of ceramic capacitors used to realize the flying capacitors in the lower stacks. The details of the number of capacitors used in each stack for a 200W design is shown in the table below. The top five flying capacitors (C_{fly1}-C_{fly5}) use 2.2µF, 50V capacitor (part number: GRM188R61H225ME11J) whereas the remaining two flying capacitors use 10µF, 25V capacitor (part number: GRM188R61E106KA73D). Note that these capacitors are the highest value capacitance available in 0603 footprint for the voltage rating of interest. The rms current through the flying capacitors can reach a maximum of 15A, which translates to using at least five capacitors in parallel to realize the flying capacitor as each capacitor has a rms current rating of 3.5A. The

number of capacitors in the upper stacks were chosen to limit the peak-to-peak voltage ripple at their respective bus outputs to 4V.

B. Bus capacitor selection methodology: The bus capacitors serve the funtions of being the output capacitor for the switched capacitor (SC) stage and and the input-decoupling capacitor for the follow-on buck stage. From an overall system efficiency standpoint, there is an optimal value of the bus capacitance for each stack as explained below. A larger bus capacitance would incur greater charge sharing losses in the SC stage and the effective slow switching limit output resistance of the each of the modules of the SC stage can be expressed as:

$$R_{out} = \frac{1}{4 f_s C_f \left(1 + \frac{C_f}{C_{bus}}\right)} \quad (1)$$

where, C_f is the effective flying capacitance of the stack, f_s is the switching frequency of the SC stage, and C_{bus} is the bus capacitance of the module. On the other hand, a smaller bus capacitance would result in poorer decoupling for the buck stage exacerbating its efficiency. Additionally, a smaller bus capacitance would also cause a larger drop in the bus voltage during the deadtime of the switched capacitor stage hurting its efficiency. Hence, there exists an optimal value of the bus capacitance that maximizes overall system efficiency. Figure 2 shows the result of this optimization at 50 W output power. The SC stage losses are minimized, and the efficiency is maximized when the bus capacitance is chosen to be 100 nF. The calculated hard charging loss line using (1) runs parallel to the total

TABLE. I Details of the Flying Capacitances used in the Prototype IBC

Flying capacitor	Blocking Voltage (V)	Number of Capacitors Used	Effective capacitance (µF)
C_{fly1}	42	9	1.9
C_{fly2}	36	8	1.8
C_{fly3}	30	7	2.0
C_{fly4}	24	5	1.8
C_{fly5}	18	5	2.4
C_{fly6}	12	5	7.9
C_{fly7}	6	5	17.5

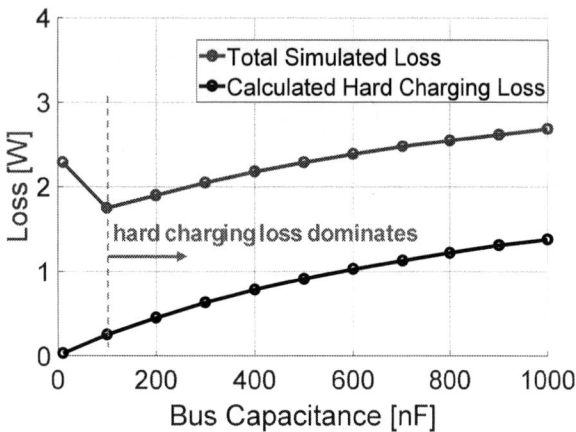

Fig. 2: Total simulated loss in the IBC and the calculated hard charging loss using (1) as a function of the bus capacitance.

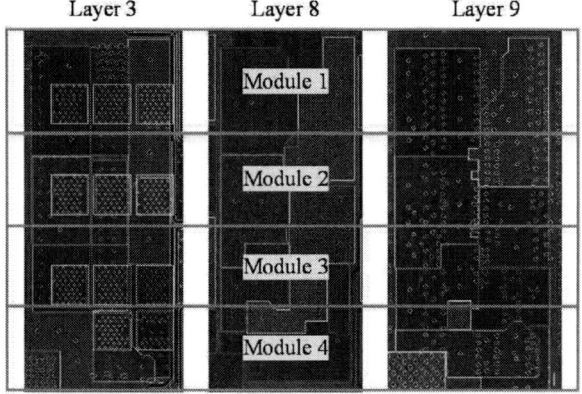

Fig. 3: Power plane layout in three different layers of the IBC package.

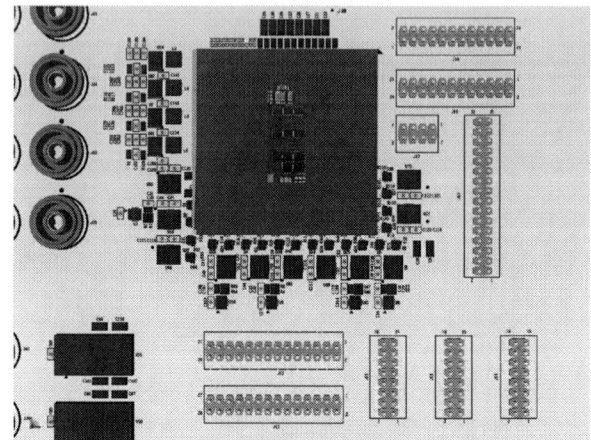

(a)

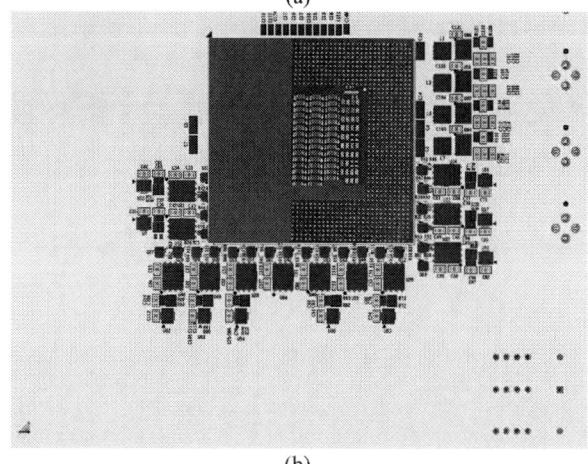

(b)

Fig. 4: Photograph of the top and bottom side of the board and the package of the prototype IBC.

simulated loss line validating the accuracy of the output resistance expression. This optimal bus capacitance of 100 nF is implemented using ten 10nF, 25V class I ceramic capacitors (part number: GRM1555C1E103JE01D). The class I capacitors have smaller effective series resistance (ESR) compared to their class II counterpart and hence a fewer of them are needed to realize the total required bus capacitance.

C. Package design: A 4-2-4 package with 200μm core thickness is used. The package layout takes advantage of the modular architecture of the IBC, ensuring that the voltage difference between neighboring traces and power planes within and across layers adhere to the technological limits of the package and prevents arcing and other undesirable effects. The power plane layout in three different layers of the package highlighting the aforementioned layout strategy is shown in Fig. 3.

III. PROTOTYPE DESIGN AND EXPERIMENTAL RESULTS

A 1MHz 200W prototype of the IBC is built and tested with a two-phase DrMOS based buck per stack loading the converter. A photograph of the prototype is shown in Fig. 4. The flying capacitors are located on the landside underneath the switches of the corresponding stack to minimize the parasitic loop inductance and resistance of the loop comprising the switches and the flying capacitors. The drivers for the power transistors are located on the top and the bottom side of the board close to the package.

Fig. 5(a) shows the four output bus voltages (Vbus1-4) of the IBC while the converter is running off an input voltage of 48V and delivering its rated 200W output power. All the bus voltages are balanced with an average value of 6V. Four of the seven flying capacitor voltages at the same operating point is shown in Fig. 5(b). In this IBC, the flying capacitors block voltages in multiples of the bus voltages (6V) and the experimental results can be seen correlating well. Flying capacitors in the upper stacks block higher voltages and are worst hit by the impact of derating, resulting in larger peak-to-peak ripple in the bus voltages of the corresponding stacks. The startup and a 0-to-200W load step transient performance of the prototype IBC is measured, and the results are shown in Fig. 6(a) and (b), respectively. It is evident that bus voltages across the four modules stay balanced both during startup and load transient events. The thermal performance of the converter

979-8-3315-1612-3/25 $31.00 © 2025 IEEE   1683

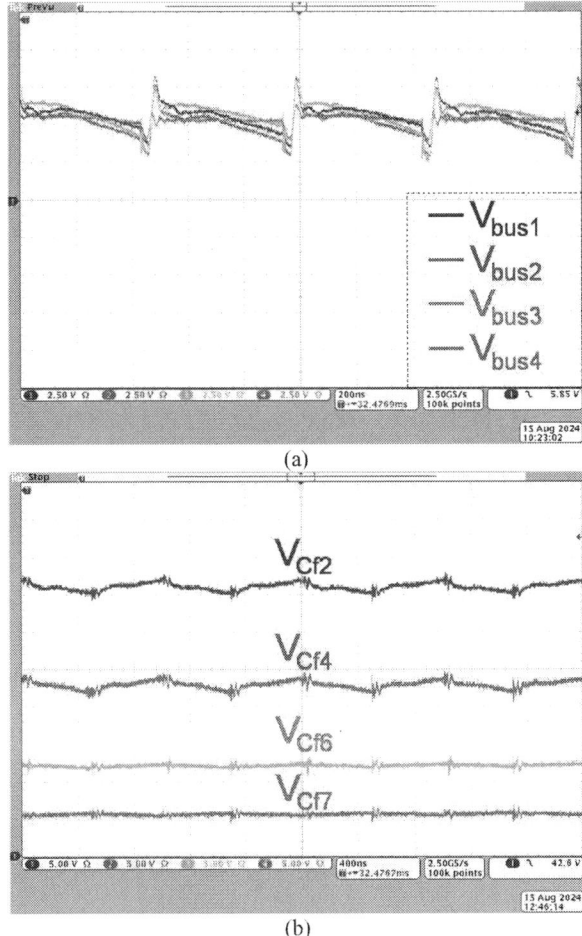

(a)

(b)

Fig. 5: Measured (a) bus voltages and (b) the four flying capacitor voltages of the prototype IBC while delivering 200W output power at 1MHz operating frequency.

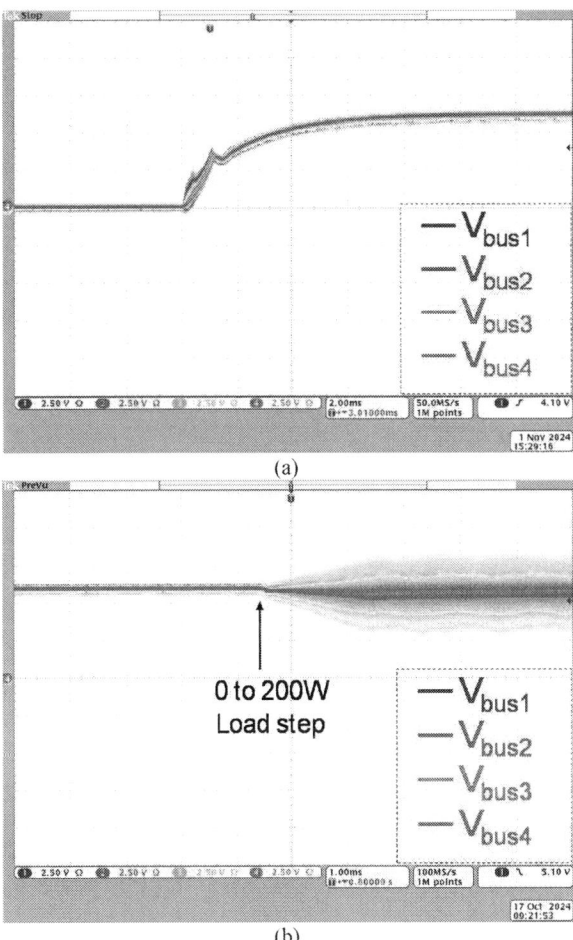

(a)

(b)

Fig. 6: Measured bus voltages of the prototype IBC at (a) startup, (b) while undergoing a 0 to 200W load step.

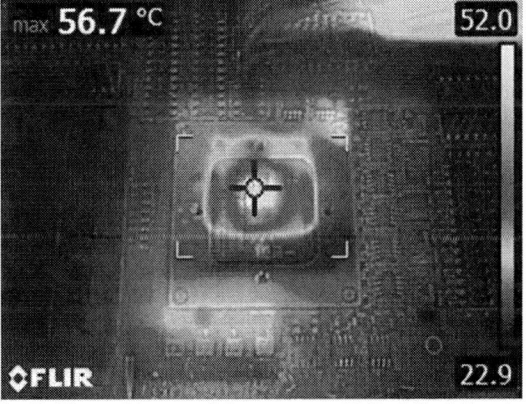

Fig. 7: Thermal image of the prototype IBC while delivering 100W output power with no active cooling.

while delivering 100W output power with no active cooling is shown in Fig. 7.

Efficiency measurement strategy: Since, the IBC is tested with a two-phase external buck converter at each of its four bus outputs, there is no direct way to measure the IBC's output current for efficiency measurement purposes. SC converters are near-perfect current converters as their output-to-input current ratio matches (after discounting the no load input current) the ideal voltage conversion ratio (8 in the presented IBC) across the load range. This key feature of the SC converters is leveraged to estimate the total bus output current and the efficiency. The expression for calculating efficiency can be expressed as:

$$\text{Efficiency} = \frac{8V_{\text{bus,avg}}I_{\text{in}} - V_{\text{in}}I_{\text{in,noload}}}{V_{\text{in}}I_{\text{in}}} \quad (2)$$

where, $V_{\text{bus,avg}}$ is the average of the four bus voltages, I_{in} is the measured input current, V_{in} is the measured input voltage. $I_{\text{in,noload}}$ is the no-load input current to the IBC measured after removing the buck stage and connecting all four buses together. The measured and estimated efficiency of the IBC is shown in Fig. 8. The peak and full load efficiencies are 96.1% and 91.8% respectively, and the estimated efficiency is correlating well

with experiment. The prototype IBC can operate at higher frequency (2 MHz: 2x faster) and the comparative measured efficiency data at the two operating frequencies (1 MHz and 2 MHz) is shown in Fig. 9. Higher frequency of operation can reduce the required flying capacitance and improve the efficiency of the follow-on buck stage, thus improving the

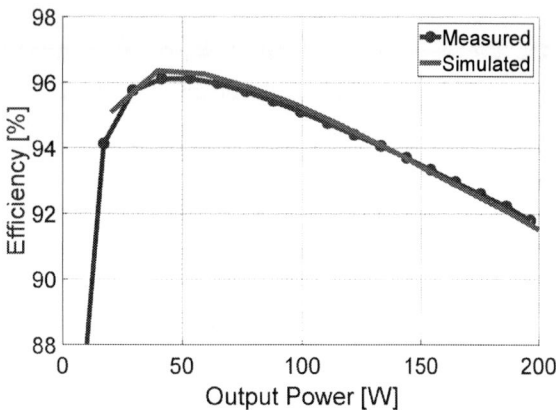

Fig. 8: Measured and Simulated efficiency of the prototype IBC across output power at 1MHz operating frequency.

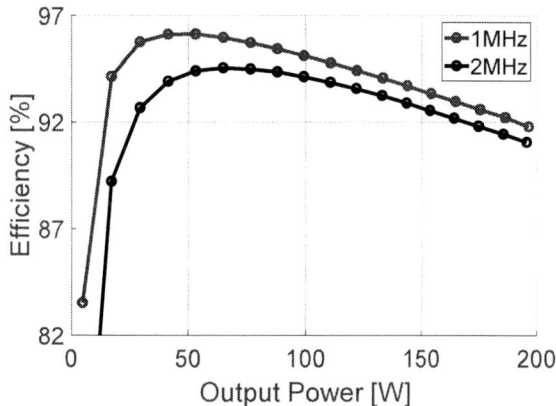

Fig. 9: Measured efficiency of the prototype IBC across output power at 1MHz and 2MHz operating frequency.

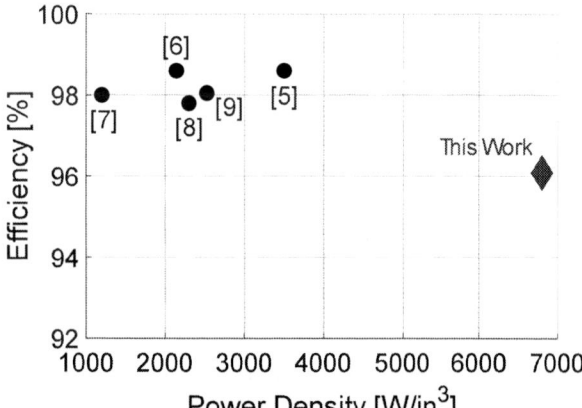

Fig. 10: Performance comparison of the presented prototype IBC with state-of-the-art 8:1 conversion ratio IBCs.

overall end-to-end efficiency and density of the power delivery solution. The efficiency and density performance of the IBC is compared with 8:1 conversion ratio state-of-the-art IBCs in Fig. 10. The proposed IBC achieves ~2x higher power density compared to its best performing counterpart while maintaining >96% peak efficiency.

IV. Conclusions and Future Work

The industry's first 48-to-6V, on-package, IBC utilizing low voltage GaN power transistors with better figure of merit is presented. This IBC uses four stacks of 2:1 SC dividers to achieve the 8:1 voltage conversion. The 1 MHz 200 W prototype IBC achieves a peak efficiency of 96.1% and a power density of 6.8 kW/in^3.

Acknowledgment

The authors would like to acknowledge Marko Radosavljevic, Pratik Koirala, Ahmad Zubair, Jason Peck, Michael Beumer, Heli Vora in Intel foundry technology research for their contributions towards developing the GaN transistors used in this work.

Appendix

This appendix provides the derivation for the slow switching limit (SSL) output resistance of a 2:1 SC converter with a finite output capacitance. Figure 11 shows the voltage across the effective flying capacitance C_f across a switching cycle under SSL operation. In phase 1, V_{in}, C_f and C_{bus} appear in series and in phase 2, C_f is in parallel with C_{bus}. In phase 1, the voltage swing and the charge received by C_f are $V_{in} - 2V_{bus,l}$ and $C_f(V_{in} - 2V_{bus,l})$, respectively, where $V_{bus,l}$ is the minimum value of the bus voltage. The same amount of charge is also delivered to the output in a half switching cycle. Equating and simplifying the charge received by C_f to the output charge delivered results in the following expression:

$$V_{bus,l} = \frac{V_{in}}{2} - \frac{I_{out}}{4f_sC_f} \tag{3}$$

where, I_{out} is the average output current. At the beginning of phase 1, the voltage across C_{bus} swings from $V_{bus,l}$ to $V_{bus,h}$, and the voltage across C_f swings from $V_{bus,l}$ to $V_{in} - V_{bus,h}$, where $V_{bus,h}$ is the maximum value of the bus voltage. The charges associated with these swings must be equal, which leads to the following charge balance expression:

$$C_{bus}(V_{bus,h} - V_{bus,l}) = C_f(V_{in} - V_{bus,h} - V_{bus,l}) \tag{4}$$

By replacing $V_{out,h}$ with $2V_{out,avg} - V_{out,l}$, (4) can be simplified to

$$V_{bus,l} = V_{bus,avg} - \frac{C_f}{C_{bus}}\left(\frac{V_{in}}{2} - V_{bus,avg}\right) \tag{5}$$

where, $V_{bus,avg}$ is the average value of the bus voltage. Replacing the expression for $V_{out,l}$ from (5) in (3) results in

$$\frac{V_{in}}{2} = V_{bus,avg} - I_{out}\left(\frac{1}{4f_sC_f\left(1 + \frac{C_f}{C_{bus}}\right)}\right) \tag{6}$$

The expression for the effective output resistance of the converter can be obtained by comparing (6) with the well-known equivalent circuit model for SC converters, and is given by:

$$R_{out} = \frac{1}{4f_sC_f\left(1 + \frac{C_f}{C_{bus}}\right)} \tag{7}$$

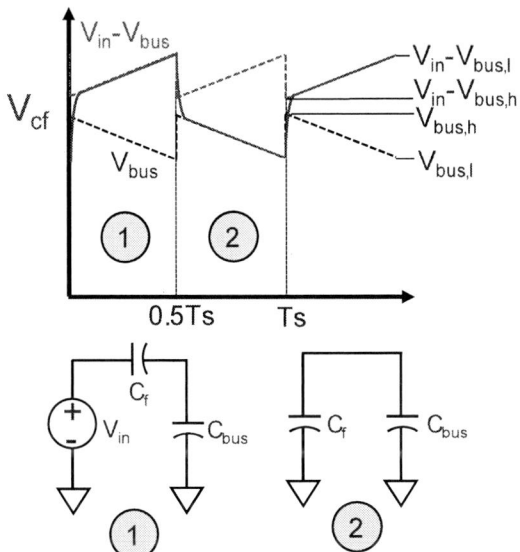

Fig. 11: Voltage waveform across the flying capacitor in a 2:1 SC converter under the assumption of finite output capacitance and SSL operation.

which is same as (1). In each stack of the presented IBC, two flying capacitors appear in series (except in the charging phase of C_{fly1} and the discharging phase of C_{fly7}) making their series combination the effective flying capacitance in (7) for these stacks. It is interesting to note that for a scenario with large output capacitance C_{bus}, (7) simplifies to the well-known expression $\frac{1}{4f_sC_f}$.

REFERENCES

[1] Z. Ye, Y. Lei and R. C. N. Pilawa-Podgurski, "A resonant switched capacitor based 4-to-1 bus converter achieving 2180 W/in3 power density and 98.9% peak efficiency," in *IEEE 2018 Applied Power Electronics Conference and Exposition*, 2018, pp. 121-126.

[2] J. Baek, Y. Elasser, K. Radhakrishnan, H. Gan, J. P. Douglas, H. K. Krishnamurthy, X. Li, S. Jiang, C. R. Sullivan, M. Chen, "Vertical Stacked LEGO-PoL CPU Voltage Regulator," in *IEEE Transactions on Power Electronics*, vol. 37, no. 6, pp. 6305-6322, Jun. 2022.

[3] Y. Elasser, J. Baek, K. Radhakrishnan, H. Gan, J. P. Douglas, H. K. Krishnamurthy, X. Li, S. Jiang, V. De, C. R. Sullivan, M. Chen, "Mini-LEGO CPU Voltage Regulator," in *IEEE Transactions on Power Electronics*, vol. 39, no. 3, pp. 3391-3410, Mar. 2024.

[4] H. W. Then et al., "Scaled Submicron Field-Plated Enhancement Mode High-K Gallium Nitride Transistors on 300mm Si(111) Wafer with Power FoM (RON xQGG) of 3.1 mohm-nC at 40V and fT/fMAX of 130/680GHz," in *IEEE 2022 International Electron Devices Meeting*, 2022, pp. 35.1.1-35.1.4.

[5] H. Wu, Y. Zhang and Z. Li, "Hybrid Resonant Converter-Based 8:1 Bus Converter With 3.5 kW/in3 and 98.6%-Efficient for 48 V Data-Center Power Systems," in *IEEE Transactions on Power Electronics*, vol. 39, no. 1, pp. 36-41, Jan. 2024, doi: 10.1109/TPEL.2023.3319312

[6] R. A. Abramson, Z. Ye, T. Ge and R. C. N. Pilawa-Podgurski, "A High Performance 48-to-6 V Multi-Resonant Cascaded Series-Parallel (CaSP) Switched-Capacitor Converter," *2021 IEEE Applied Power Electronics Conference and Exposition (APEC)*, Phoenix, AZ, USA, 2021, pp. 1328-1334, doi: 10.1109/APEC42165.2021.9487048.

[7] M. H. Ahmed, F. C. Lee and Q. Li, "Two-Stage 48-V VRM With Intermediate Bus Voltage Optimization for Data Centers," in *IEEE Journal of Emerging and Selected Topics in Power Electronics*, vol. 9, no. 1, pp. 702-715, Feb. 2021.

[8] X. Ren, J. Zhang, B. Hu, P. Xu, T. Long," Design and Testing of an 8: 1 Non-isolated Bus Converter for 48V Data Centers", *2024 IEEE Energy Conversion Conference and Expo (ECCE)*, Phoenix, AZ, USA, 2024.

[9] Flex power BMR321 750W non-isolated unregulated Intermediate Bus Converter datasheet: https://flexpowermodules.com/products/bmr321.

[10] D. Reusch, S. Biswas and Y. Zhang, "System Optimization of a High Power Density Non-Isolated Intermediate Bus Converter for 48 V Server Applications," in *IEEE Transactions on Industry Applications*, vol. 55, no. 2, pp. 1619-1627, March-April 2019, doi: 10.1109/TIA.2018.2875387.

[11] S. Webb and Y. F. Liu, "A Zero Inductor-Voltage 48V to 12V/70A Converter for Data Centers with 99.1% Peak Efficiency and 2.5kW/in3 Power Density," *2020 IEEE Applied Power Electronics Conference and Exposition (APEC)*, New Orleans, LA, USA, 2020, pp. 1858-1865, doi: 10.1109/APEC39645.2020.9124211.

[12] K. Hata, Y. Yamauchi, T. Sai, T. Sakurai and M. Takamiya, "48V-to-12V Dual-Path Hybrid DC-DC Converter," *2020 IEEE Applied Power Electronics Conference and Exposition (APEC)*, New Orleans, LA, USA, 2020, pp. 2279-2284, doi: 10.1109/APEC39645.2020.9124077.

[13] M. R. Mohammadi, M. Ebrahimi, A. Safaee and S. A. Khajehoddin, "A ZVS Bidirectional 48V/12V Converter with Magnetic Integration for 48V Mild Hybrid Vehicles," in *IEEE Transactions on Transportation Electrification*, doi: 10.1109/TTE.2024.3404864.

[14] M. Salato, "Re-architecting 48V Power Systems with a Novel Non-isolated Bus Converter," *2015 IEEE International Telecommunications Energy Conference (INTELEC)*, Osaka, Japan, 2015, pp. 1-4, doi: 10.1109/INTLEC.2015.7572441.

[15] Y. Cai, M. H. Ahmed, Q. Li and F. C. Lee, "Optimal Design of Megahertz LLC Converter for 48-V Bus Converter Application," in *IEEE Journal of Emerging and Selected Topics in Power Electronics*, vol. 8, no. 1, pp. 495-505, March 2020, doi: 10.1109/JESTPE.2019.2939469.

[16] T. Kim and S. Kwak, "A Flexible Voltage Bus Converter for the 48-/12-V Dual Supply System in Electrified Vehicles," in *IEEE Transactions on Vehicular Technology*, vol. 66, no. 3, pp. 2010-2018, March 2017, doi: 10.1109/TVT.2016.2570817.

A 48V to 2.4V-5V 95.8%-Peak-Efficiency 869W/in³-Power-Density Fibonacci Dual-Path Hybrid DC-DC Converter with Inductor Current Reduction and Low Output Resistance

Yichao Ji, Zeguo Liu, Lin Cheng

School of Microelectronics

University of Science and Technology of China, Hefei, China

Email: jyc123@mail.ustc.edu.cn, liuzeguo@mail.ustc.edu.cn, eecheng@ustc.edu.cn

Abstract—**This paper presents a 48-V-input and 2.4-V-to-5-V-output dual-path hybrid buck converter based on an interleaved Fibonacci switched-capacitor topology. This Fibonacci topology is a more effective switch strategy with well-distributed current and reduced voltage stress compared with a conventional one, and hence high efficiency and high power density can be achieved. Among classic switched-capacitor topologies, the Fibonacci configuration maximizes achievable voltage gain, allowing for a large voltage conversion ratio with fewer capacitors. Moreover, compared with conventional inductive buck converters, the dual-path structure with a capacitor always in parallel with the inductor realizes an inductor current reduction of up to 62.5%, thus significantly reducing inductor DCR loss and relieving the reliance on a bulky inductor. Topology optimization on both the capacitors and the inductor results in a lower output resistance with a minimized power loss. A hardware prototype was designed and built to validate the performance of the proposed converter. It was tested with a maximum load of 30 A, and measurement results show that the converter achieves a peak efficiency of 95.8% for 48-V-to-5-V operation and a power density (by total passive volume) of up to 869 W/in³.**

Keywords—*DC-DC converter, dual path, Fibonacci switched-capacitor topology, hybrid converter, inductor current reduction.*

I. INTRODUCTION

With the extensive use of 48V power bus in industrial and automotive applications [1], large voltage-conversion-ratio (VCR) DC-DC converters with high efficiency and high power density are in great demand to reduce power consumption and minimize system volume. Conventional inductive buck converters have to operate with a very small duty ratio, resulting in significantly increased conduction loss in the low-side switches and switching loss in the high-side switches, which are implemented by high-voltage power switches with poor figure-of-merit (FoM) [2]. Moreover, they have limited power density since the inductor with low DC resistance (DCR) and high saturation current dominates the overall size.

To accommodate a high input voltage (V_{IN}), hybrid DC-DC converters that combine inductors and capacitors have recently emerged as promising solutions. As shown in Figure 1(a), by introducing switched-capacitor (SC) networks to replace the two power switches in a conventional buck converter, voltage stress

This work was supported by the National Natural Science Foundation of China under Grant 92373203.

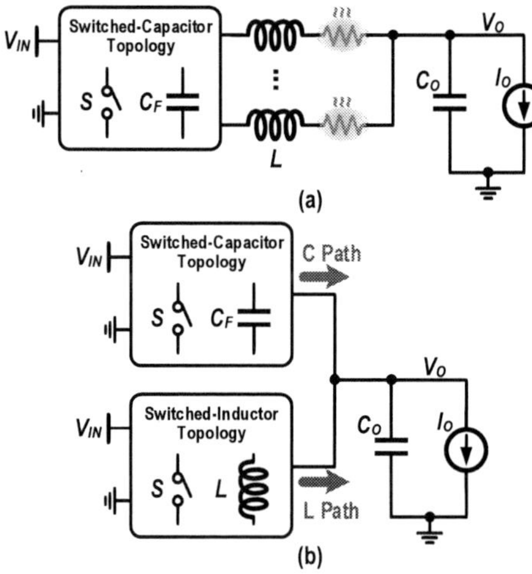

Fig. 1. Two kinds of hybrid DC-DC converters: (a) switched-capacitor hybrid converter and (b) dual-path hybrid converter.

on power switches is reduced to various levels of V_{IN}/N, thus enabling the use of low-voltage devices for higher efficiency [3]-[9]. However, bulky inductors are still required to handle the entire output current (I_O), which is similar to a conventional buck converter.

To reduce inductor DCR loss and relieve the reliance on bulky inductors, dual-path hybrid DC-DC converters have been proposed [10]-[12], featuring reduced inductor DC current (I_L) with flying capacitors (C_{FS}) providing additional current paths in parallel with the inductor, as shown in Figure 1(b). But the duty ratio is still narrow in [10] and the reduction in I_L is only 21% when converting 48 V to 5 V, while up to 6 C_{FS} are needed in [11] for a larger VCR and greater I_L reduction. Although over 50% reduction in I_L is achieved in [12] with 4 C_{FS}, it faces limitations in practical applications due to the negative switching node voltage required to energize the on-ground inductor.

To address the issues discussed above, a Fibonacci dual-path hybrid buck (FDPHB) converter is proposed in this work to realize a large VCR with fewer capacitors. The topology realizes a further reduction in I_L and has a lower output resistance. Hence,

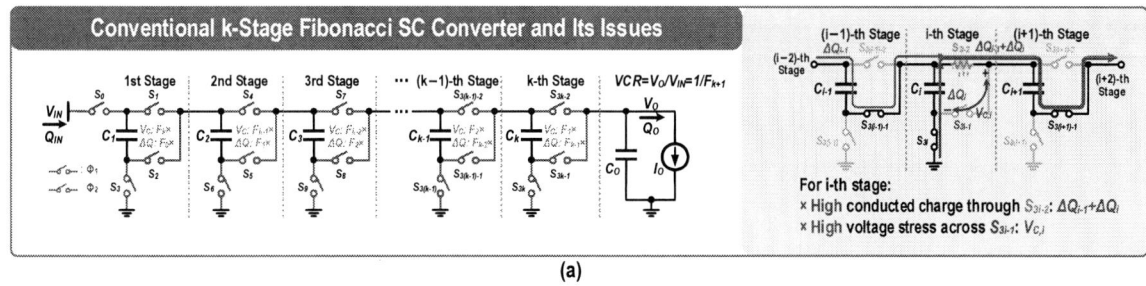

(a)

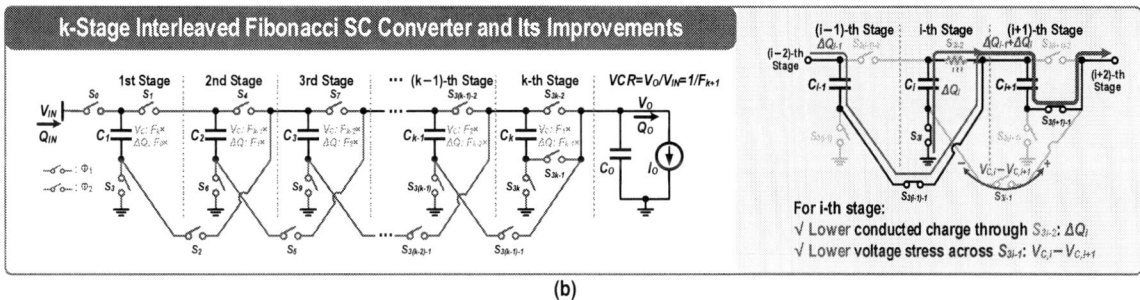

(b)

Fig. 2. (a) Conventional Fibonacci SC topology and (b) interleaved Fibonacci SC topology.

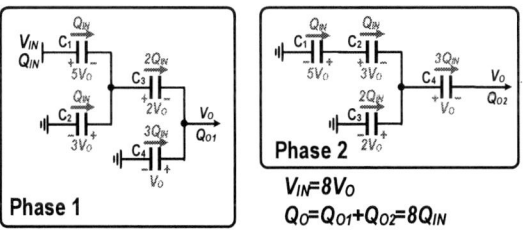

Fig. 3. Equivalent circuit model of a 4-stage step-down Fibonacci topology.

the proposed converter can simultaneously provide high efficiency and high power density.

II. PROPOSED DUAL-PATH HYBRID BUCK CONVERTER

A. Interleaved Fibonacci SC Topology

Various SC converter topologies with specific conversion ratios can be implemented by using multiple capacitors and switches. Fibonacci topology is a classic non-linear SC topology. By utilizing a given number of capacitors, it attains the highest voltage gain among all the two-phase SC topologies, as defined by the Fibonacci series F [13]. In this series, with the zeroth and the first elements defined as 1, the successive element is the sum of the preceding two elements, and hence the series can be listed as

$$F_i = \{1,1,2,3,5,8,13,...\}, i \in N . \quad (1)$$

Fig. 2(a) shows a conventional k-stage step-down Fibonacci topology achieving a voltage gain of $1/F_{k+1}$, where k is a positive integer. In this example, the even-numbered switches are turned on during Φ_1, while the odd-numbered switches are turned on during Φ_2. It could be observed that both voltages and charges of the capacitors are Fibonacci multiples of output voltage (V_O) and input charge (Q_{IN}), respectively. Fig. 3 shows the equivalent circuit model of a 4-stage step-down Fibonacci topology with 4 capacitors. During phase 1, C_3 and C_4 are connected in series to provide a boosted voltage for C_2, while V_{IN} charges C_1 and C_2

in series. The output is charged through C_3 and C_4 concurrently. During phase 2, the output and C_4 are in series to provide a boosted voltage for C_3, while C_1 and C_2 in series are discharged and charged, respectively. It realizes a voltage gain of 8, unlike a voltage gain of only 5 in linear SC topologies that also use 4 capacitors, such as Dickson and series-parallel topologies. Therefore, Fibonacci SC topology offers a better utilization of capacitors and is more suitable for achieving a large VCR with fewer capacitors.

Since the conventional Fibonacci SC topology features a string of basic SC cells, in the i-th stage, S_{3i-2} must always conduct a gathered charge from C_{i-1} and C_i, inducing large conduction loss. Besides, S_{3i-1} has a high voltage stress determined by the voltage across C_i, necessitating high-voltage power switches with large parasitic parameters. To address these issues, modified Fibonacci wiring methods were proposed in [14] and [15]. With all right nodes of S_{3i-1}s connected to the output nodes of subsequent stages, a k-stage interleaved Fibonacci SC topology is obtained, as shown in Fig. 2(b). In this configuration, S_{3i-2} only conducts a charge from C_i within its local stage, and the voltage stress on S_{3i-1} is reduced from $V_{C,i}$ to $V_{C,i}-V_{C,i+1}$, while all the characteristics of a conventional Fibonacci SC topology are still retained, and the equivalent circuit model remains the same. Therefore, this approach provides a more effective switch strategy with well-distributed current and reduced voltage stress to realize a Fibonacci SC topology with lower conduction and switching losses.

B. Proposed Topology and Operating Principle

To realize a regulated output voltage, an inductor is added to the interleaved Fibonacci SC topology, replacing one power switch. Fig. 4 shows the proposed Fibonacci dual-path hybrid buck converter, consisting of one inductor (L), four flying capacitors (C_1 to C_4), and twelve power switches (S_0 to S_{11}). It retains the benefits of the interleaved Fibonacci SC topology, achieving a large VCR with fewer capacitors and minimized switch losses. The inductor is in parallel with the final stage to

979-8-3315-1612-3/25 $31.00 © 2025 IEEE

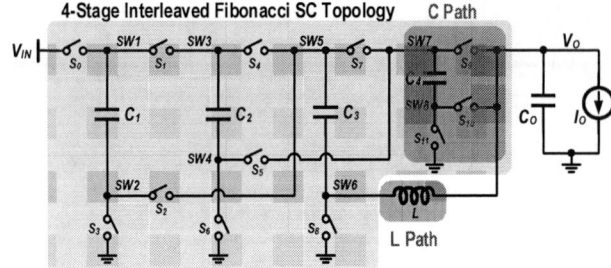

Fig. 4. The proposed Fibonacci dual-path hybrid buck converter.

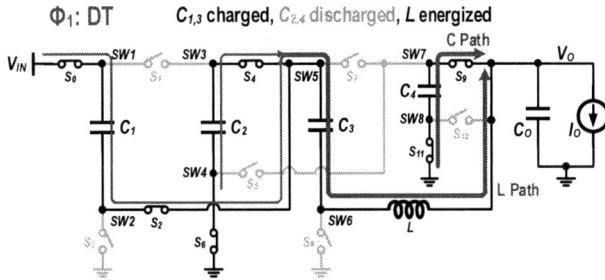

Fig. 5. Operating principle of the proposed converter.

form a dual-path structure for I_L reduction.

Fig. 5 illustrates the operating principle that contains two phases: Φ_1 and Φ_2, while Fig. 6 depicts its key operating waveforms. During Φ_1, S_0, S_2, S_4, S_6, S_9, and S_{11} are turned on, while L is energized. C_1 is directly charged by V_{IN}, and such charge, together with the discharge from C_2, constitutes the total charge transferred to the output through L. Additionally, C_3 in series with L is charged through C_1 and C_2 concurrently. C_1 to C_3 in the first three stages of the interleaved Fibonacci SC withstand a high V_{IN} and reduce the voltage swing across L, enabling the use of a smaller inductance and enlarging the duty ratio for realizing a large VCR. Moreover, the charge accumulated on C_4 in the last stage gets transferred to the output directly to share the output current. During Φ_2, S_1, S_3, S_5, S_7, S_8, and S_{10} are turned on. With SW6 switching to the ground, L is de-energized. C_1 and C_2 are in series with C_1 discharged and C_2 charged, while C_3 gets discharged. Charges from C_2 and C_3 gather in C_4 towards the output in parallel with L.

All the capacitor voltages can be self-balanced and defined by hard charging among them. Based on their connections, we can obtain $V_{C1} = (V_{IN} + 2V_O)/2$, $V_{C2} = (V_{IN} - 2V_O)/2$, $V_{C3} = 2V_O$, and $V_{C4} = V_O$. Hence, SW6 switches between $(V_{IN} - 6V_O)/2$ and 0. Performing inductor voltage-second balance principle yields the following VCR:

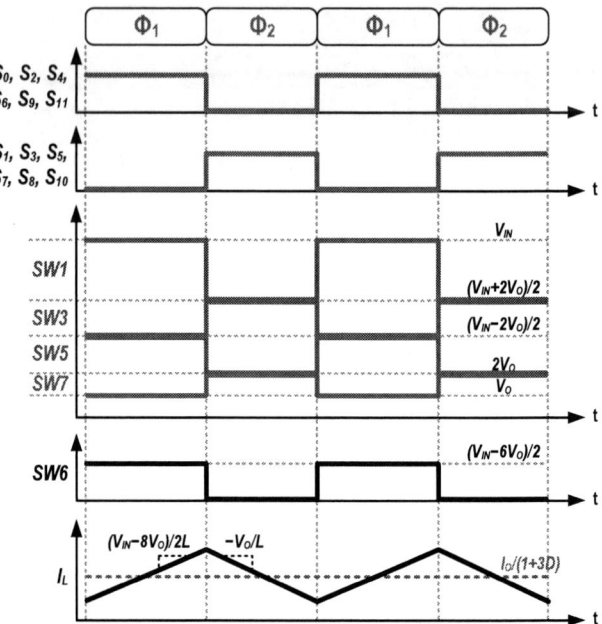

Fig. 6. Operating waveforms of the proposed converter.

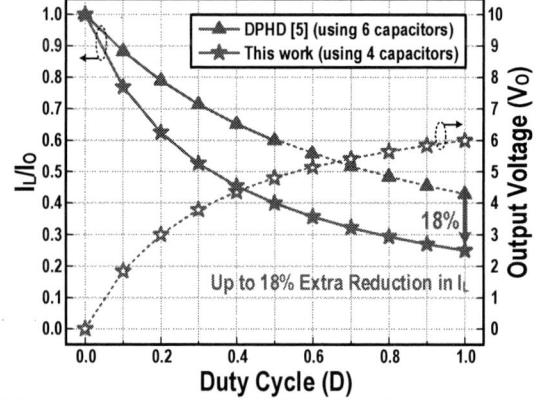

Fig. 7. Output voltage range and comparison I_L/I_O ratio.

$$\left(\frac{V_{IN} - 6V_O}{2} - V_O\right) \cdot DT = V_O \cdot (1 - D)T \tag{2}$$

$$VCR = \frac{V_O}{V_{IN}} = \frac{D}{2 + 6D} \tag{3}$$

where D represents the duty ratio of the inductor energizing phase and T is the switching period.

Notably, a capacitive path through C_4 always exists in parallel with the inductive path to share I_O, and hence the inductor DC current is further reduced. Based on the charge conservation, the total charge delivered to the output through the capacitor and the inductor over the whole T is equal to $I_O T$. Therefore, the I_L/I_O ratio can be derived as follows:

$$\frac{I_L}{I_O} = \frac{1}{1 + 3D}. \tag{4}$$

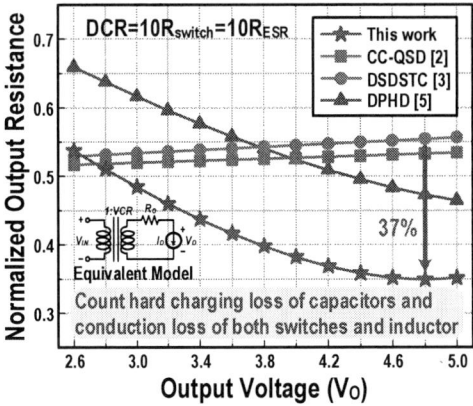

Fig. 8. Comparison of normalized output resistance.

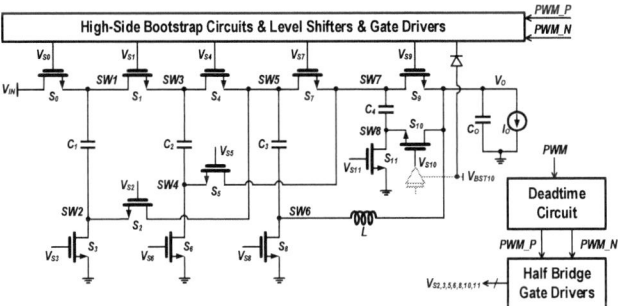

Fig. 9. System architecture of the proposed converter.

Fig. 10. Photograph of the hardware prototype.

TABLE I. CIRCUIT COMPONENTS AND SPECIFICATIONS

Components	Part Information	Parameters
S_{0-4}	Infineon BSC032N04LS	40V, 3.2mΩ
S_{5-11}	Infineon BSC009NE2LS5I	25V, 0.95mΩ
L	Bourns SRP1050WA	1μH, 2.76mΩ, 37A
C_1, C_2	Murata GRM32EC72A106KE05L	80μF (10μF×8, 1210)
C_3, C_4	Murata GRM32ER71E226KE15L	176μF (22μF×8, 1210)
C_O	Murata GRM32ER71E226KE15L Panasonic EEHZC1V271P	88μF (22μF×4, 1210) 270μF
Gate Drivers	Texas Instruments UCC27282	120V, Half Bridge

C. Topology Comparison

From (3), the proposed FDPHB converter can realize an output voltage ranging from 0 to $V_{IN}/8$, as shown in Fig. 7. The duty ratio is extended by up to 8 times that of a conventional buck converter, achieving the largest extension with four

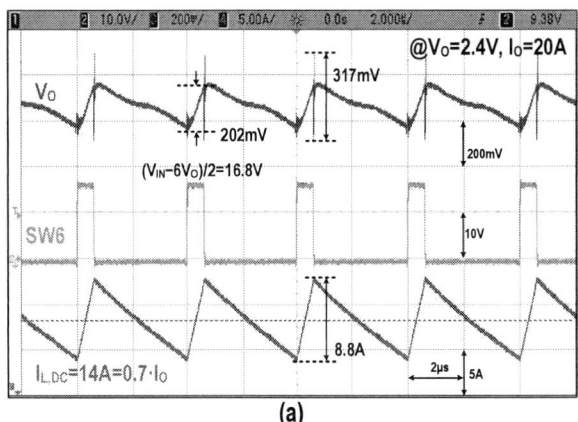

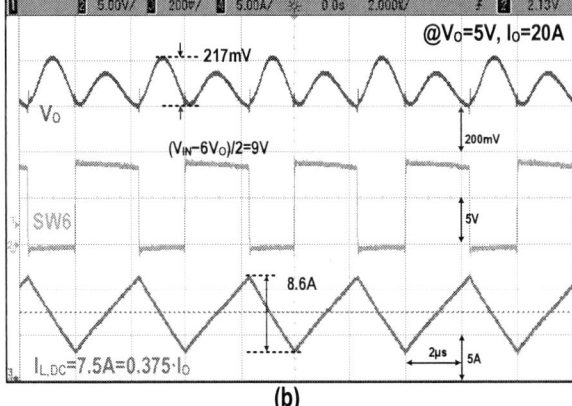

Fig. 11. Measured steady-state waveforms with $I_O = 20$ A at (a) $V_O = 3$ V and (b) $V_O = 5$ V.

capacitors thanks to the full capacitor utilization in the Fibonacci topology. With V_O between 2.4 V and 5 V, a reduction in I_L of 30% to 62.5% is achieved, and up to 18% extra reduction is obtained compared with [11].

To evaluate the performance improvement, output resistance of the converters is adopted as a metric here [16]. In the equivalent transformer model of a DC-DC converter, the output resistance (R_O) represents the power losses, which can be approximated as

$$R_O \approx \sqrt{\left(R_{SSL}\right)^2 + \left(R_{FSL}\right)^2} \qquad (5)$$

where R_{SSL} is the slow-switching limit resistance that represents hard-charging loss of capacitors, and R_{FSL} is the fast-switching limit resistance that represents the conduction loss induced by resistive elements (r_i), including switch on-resistances, parasitic equivalent series resistances (ESRs) of capacitors, and DCR of the inductor. There are

$$R_{SSL} = \sum_{C_i} \frac{a_i C_i \left(\Delta V_i\right)^2}{I_O^2 T} = \sum_i \frac{a_i Q_i^2}{I_O^2 C_i T} \qquad (6)$$

$$R_{FSL} = \sum_{r_i} \frac{I_i^2 r_i D_i}{I_O^2} \qquad (7)$$

where a_i is a coefficient of 1/2 or 1 depending on whether the

TABLE II. PERFORMANCE SUMMARY AND COMPARISON

Publication	Vicor PI3545 [17]	APEC 2019 [9]	APEC 2020 [3]	COMPEL 2020 [7]	ECCE 2021 [11]	APEC 2024 [4]	This work
Topology	Buck + ZVS	MPMIH	CC-QSD	MLB-PoL	DPHD	DSDSTC	FDPHB
V_{IN}	36 – 60V	48	48V	48V	36 – 65V	48V	48V
V_O	4 – 5.5V	1 – 2V	1 – 3.3V	1 – 2.5V	1 – 2V	1 – 3.3V	2.4 – 5V
Max. Output Power	55W	80W	132W	162.5W	60W	165W	150W
Switching Frequency	600kHz	N.A.	125kHz	250kHz	250kHz	111 – 367kHz	250kHz
Inductor (DCR)	420nH (3.15mΩ) + 65nH (0.4mΩ)	3 × 2.2μH (N.A.)	2 × 1μH (0.63mΩ)	2 × 0.6μH (0.91mΩ)	260nH (N.A.)	2μH (N.A.)	1μH (2.76mΩ)
Flying Capacitor	– –	4.4μF + 6μF + 6μF + 4μF + 2μF	60μF + 40μF + 40μF	176μF + 120μF + 400μF	6 × 6.8μF	50μF + 100μF	2 × 80μF + 2 × 176μF
Output Capacitor	282μF	N.A.	330μF	440μF	238μF	370μF	358μF
Peak Efficiency	93.5%[a]	94.6%	95.4%[a]	95.0%	92.7%	94.7%	95.8%
Power Density[b]	619W/in³	425W/in³	580W/in³	494W/in³[c]	451W/in³	595W/in³	869W/in³
I_L Reduction	No	No	No	No	Yes (7%~23% Reduction)	No	Yes (30%~62.5% Reduction)

(a) Estimated from figure. (b) Power Density = Max. output power / Total power stage component volume. (c) Power density calculated by box volume.

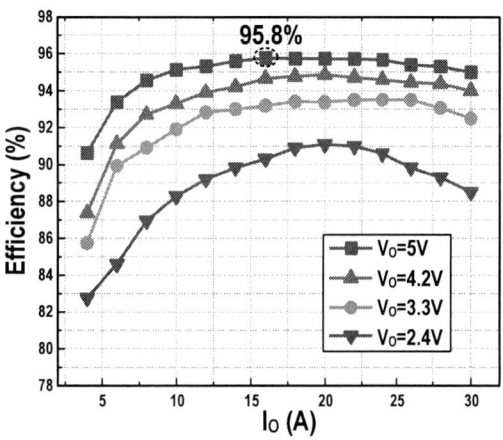

Fig. 12. Measured power efficiency.

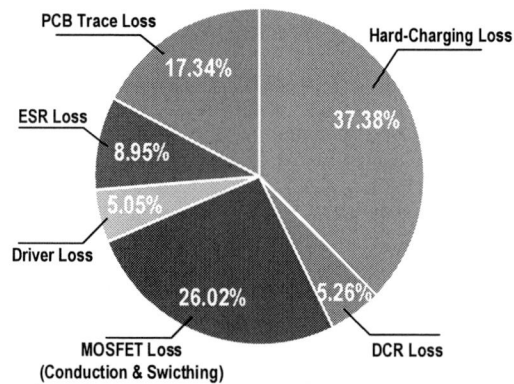

Fig. 13. Estimated power loss breakdown at V_O = 5 V and I_O = 30 A.

capacitor is soft charged by L or not, ΔV_i is the voltage ripple across C_i, Q_i is the charge transferred through it, I_i is the current through r_i, and D_i is turn-on duty ratio of r_i.

As shown in Fig. 8, when normalized by a conventional buck converter, the proposed hybrid converter achieves a reduction of up to 37% when compared with prior works. Note that not all the power switches can achieve zero-voltage switching. In the lower stages, the capacitors maintain higher DC voltages and transfer lower charges, while in the higher stages, they maintain lower DC voltages but transfer larger charges. This means that although high-voltage power switches are necessary in the lower stages, most of the output charge is conducted by low-voltage power switches in the higher stages. This arrangement capitalizes on the benefit of lower on-resistance and results in a more evenly distributed power loss.

III. IMPLEMENTATION AND EXPERIMENTAL RESULTS

Fig. 9 shows the system architecture of the proposed FDPHB converter. An external PWM signal is fed into a dead-time circuit to generate two complementary phase signals (Φ_1 and Φ_2) for each power transistor's level shifter and driver. To validate the performance, a hardware prototype with all the components implemented by discrete modules was fabricated and verified on

a 6-layer PCB with 4-oz copper, as shown in Fig. 10. All the selected key components and parameters are listed in Table I. Thanks to the reduced voltage swing across L and reduced inductor current, a compact inductor with small inductance, large DCR and low saturation current can be adopted here to realize high-density power delivery while still maintaining high efficiency. It was tested with a load capability of 30 A, operating at 250 kHz with 48-V input and 2.4-V-to-5-V output.

Fig. 11 shows the measured steady-state waveforms of V_O, SW6, and I_L at different V_O conditions (2.4 V and 5 V) when I_O is 20 A. The inductor DC current is reduced to 14 A and 7.5 A for V_O = 2.4 V and 5 V, respectively, showing a reduction in I_L from 30% to 62.5%, which aligns with our previous analysis. Notably, the hard-charging current between C_4 and C_O is the main factor influencing the V_O ripple, and hence with a narrower duty ratio, the ripple resembles a triangular waveform.

Fig. 12 shows the measured efficiency of the proposed FDPHB converter under various V_O and I_O conditions. The peak efficiency reaches 95.8% at V_O = 5 V and I_O = 16 A, benefiting from a low equivalent output resistance that minimizes power loss. Fig. 13 illustrates the estimated power loss breakdown at V_O = 5 V and I_O = 30 A, where the DCR loss is significantly reduced, making the hard-charging loss among capacitors the

979-8-3315-1612-3/25 $31.00 © 2025 IEEE

dominant factor. This indicates that using larger capacitance and a higher switching frequency could further improve efficiency.

Table II summarizes and compares the performance of the proposed converter with state-of-the-art designs. Due to the significant reduction in inductor current, a compact inductor with high DCR can be used without compromising efficiency. As a result, the power density reaches 869 W/in^3 when considering the total volume of all the power-stage components. The proposed converter achieves both the highest peak efficiency and the highest power density, with power density at least 1.4× higher than previous designs.

IV. CONCLUSIONS

This paper presents a dual-path hybrid DC-DC converter based on an interleaved Fibonacci SC topology with well-distributed current and reduced voltage stress for a large voltage conversion from 48 V to 2.4 V − 5 V without using more flying capacitors. The dual-path structure with a capacitor always in parallel with inductor realizes an inductor current reduction of up to 62.5%, thus significantly reducing inductor DCR loss and relieving the reliance on a bulky inductor. Moreover, the converter has a lower output resistance with a minimized power loss. Measurement results show that the proposed converter achieves the highest peak efficiency and the highest power density among state-of-the-art designs.

ACKNOWLEDGMENT

The authors would like to thank the Information Science Laboratory Center, University of Science and Technology of China (USTC), Hefei, China, for the hardware/software services.

Yichao Ji would like to thank Xuan Sun and Si Yuan Sim from Iowa State University for sharing submission guidelines and technical support, which were invaluable in preparing this paper as the first APEC submission from USTC.

REFERENCES

[1] X. Li and S. Jiang, "Google 48V power architecture," in *Proc. IEEE Appl. Power Electron. Conf. Expo.*, Tampa, FL, USA, 2017, Keynotes presentation.

[2] Il-Jung Kim, S. Matsumoto, T. Sakai, and T. Yachi, "New power device figure of merit for high-frequency applications," in *Proceedings of International Symposium on Power Semiconductor Devices and IC's: ISPSD '95*, Yokohama, Japan, 1995, pp. 309–314.

[3] M. Halamicek, T. McRae, and A. Prodić, "Cross-Coupled Series-Capacitor Quadruple Step-Down Buck Converter," in *2020 IEEE Applied Power Electronics Conference and Exposition (APEC)*, Mar. 2020, pp. 1–6.

[4] S. Y. Sim, X. Zhang, J. Jiang, K. Wei, and C. Huang, "A 94.7% Efficiency Direct-Step-Down Switched-Tank-Based 48V to 1V-3.3V Hybrid Converter with Constant-Resonant-Time Closed-Loop Control," in *2024 IEEE Applied Power Electronics Conference and Exposition (APEC)*, Long Beach, CA, USA, Feb. 2024, pp. 1344–1350.

[5] J. W. Kwak and D. Brian Ma, "A Conduction-Loss-Conscious 4-Level Power Converter with Tri-Path Synchronous Rectification for High Step-Down DC-DC Conversion," in *2024 IEEE Applied Power Electronics Conference and Exposition (APEC)*, Long Beach, CA, USA, Feb. 2024, pp. 2119–2123.

[6] N. M. Ellis, Y. Zhu, and R. C. N. Pilawa-Podgurski, "Gallium Nitride-based 48V-to-1V Point-of-Load (PoL) Converter for Aerospace Telecommunications and Computing Applications," in *2024 IEEE Applied Power Electronics Conference and Exposition (APEC)*, Long Beach, CA, USA, Feb. 2024, pp. 1384–1388.

[7] Z. Ye, R. A. Abramson, Y.-L. Syu, and R. C. N. Pilawa-Podgurski, "MLB-PoL: A High Performance Hybrid Converter for Direct 48 V to Point-of-Load Applications," in *2020 IEEE 21st Workshop on Control and Modeling for Power Electronics (COMPEL)*, Aalborg, Denmark, Nov. 2020, pp. 1–8.

[8] R. Das and H.-P. Le, "A Regulated 48V-to-1V/100A 90.9%-Efficient Hybrid Converter for POL Applications in Data Centers and Telecommunication Systems," in *2019 IEEE Applied Power Electronics Conference and Exposition (APEC)*, Mar. 2019, pp. 1997–2001.

[9] R. Das, G.-S. Seo, D. Maksimovic, and H.-P. Le, "An 80-W 94.6%-Efficient Multi-Phase Multi-Inductor Hybrid Converter," in *2019 IEEE Applied Power Electronics Conference and Exposition (APEC)*, Mar. 2019, pp. 25–29.

[10] K. Hata, S. Tanaka, Y. Rikiishi, and T. Matsumoto, "48 V-to-12 V Always-Dual-Path Hybrid DC-DC Converter for Inductor Current Reduction," in *2022 IEEE Energy Conversion Congress and Exposition (ECCE)*, Oct. 2022, pp. 1–6.

[11] C. Chen, J. Liu, and H. Lee, "A 92.7% Efficiency 30A 48V to 1V Dual-Path Hybrid Dickson Converter for PoL Applications," in *2021 IEEE Energy Conversion Congress and Exposition (ECCE)*, Oct. 2021, pp. 1989–1994.

[12] Y. Ji, J. Jin, and L. Cheng, "A 12V-Input 1V-1.8V-Output 94.7%-Peak-Efficiency 685A/cm^3-Current-Density Hybrid DC-DC Converter with a Charge Converging Phase," in *2024 IEEE International Solid-State Circuits Conference (ISSCC)*, San Francisco, CA, USA, Feb. 2024, pp. 458–460.

[13] M. S. Makowski and D. Maksimovic, "Performance limits of switched-capacitor DC-DC converters," in *Proceedings of PESC '95 - Power Electronics Specialist Conference*, Atlanta, GA, USA, 1995, vol. 2, pp. 1215–1221.

[14] H. Lin, W. C. Chan, W. K. Lee, Z. Chen, M. Chan, and M. Zhang, "High-Current Drivability Fibonacci Charge Pump With Connect–Point–Shift Enhancement," *IEEE Transactions on Very Large Scale Integration (VLSI) Systems*, vol. 25, no. 7, pp. 2164–2173, Jul. 2017.

[15] Z. Gu, Y. Zhang, Y. Zhao, J. Zhang, Y. Zeng, Z. Luo, and Y. Li, "Synthesizing Step-Down Switched Capacitor Power Converter Topologies," *IEEE Transactions on Circuits and Systems I: Regular Papers*, vol. 71, no. 3, pp. 1465–1479, Mar. 2024.

[16] M. D. Seeman and S. R. Sanders, "Analysis and Optimization of Switched-Capacitor DC–DC Converters," *IEEE Trans. Power Electron.*, vol. 23, no. 2, pp. 841–851, Mar. 2008.

[17] Vicor Inc., *PI354x-00 Datasheet*, "36-60V$_{IN}$ ZVS Buck Regulator & LED Driver," Oct, 2023. Available: https://www.vicorpower.com/documents/datasheets/ds_pi354x-00.pdf.

An Ultra-Fast Very Large Scale Interleaved Li-Fi Transmitter

Daniel H. Zhou, Konstantinos Manos, and Minjie Chen
Princeton University
Email: {dhzhou, minjie}@princeton.edu

Abstract—This paper combines multiphase and multilevel interleaving using distributed active switches and integrated magnetics into a unified very large scale interleaving (VLSI) technique to develop ultra-fast power electronics with outstanding large-signal tracking capability. The large-signal reference-tracking capabilities considering the fundamental sampling limit, modulator, and output filter are derived, including how reduced-amplitude, above-switching-frequency tracking is possible for highly interleaved converters. The capabilities of very large scale interleaving are demonstrated with a 64× interleaved, four-phase, 17-level FCML converter enabled by passive flying capacitor balancing provided by a four-phase tightly coupled inductor. The applicability and efficacy of the theory are verified by using the converter to directly power a 400 W Li-Fi transmitter communicating with OOK/16-QAM at 2.4× the switching frequency and 95.5% efficiency.

Index Terms—interleaving, flying capacitor multilevel converters, Li-Fi, transmitter, multiphase, multilevel, coupled inductors

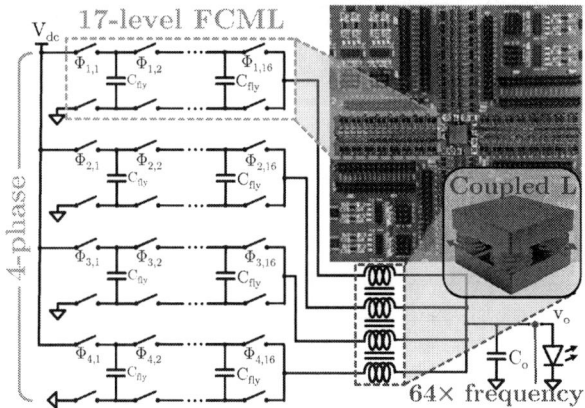

Fig. 1. Four-phase, 17-level coupled inductor FCML converter schematic.

I. INTRODUCTION

Multiphase interleaving, multilevel interleaving, and coupled magnetics [1]–[5] are important techniques that extend the capabilities of PWM (pulse-width-modulated) converters in high-speed applications such as envelope tracking [6], [7] and communication-over-power [8], [9]. In particular, interleaved power converters can be of benefit for Li-Fi (Light Fidelity) [10], [11], as they can provide high efficiency power delivery and fast modulation for LED illumination. Multiphase and multilevel interleaving multiply the effective switching frequency of the current and voltage ripples in the converter, reducing loss and the required passive component sizes [12]. Multiphase converters can also take advantage of coupling the magnetics [5], while multilevel topologies such as the FCML (flying capacitor multilevel) converter [1] can yield major efficiency and density benefits by replacing inductor volume with energy-dense capacitors and interleaved switches [12]–[14]. This has motivated FCML converters with many levels [15]–[17] and variations which also leverage multiple phases and coupled inductors [18], [19].

However, major obstacles remain with interleaved FCML converters: first, the flying capacitors must be balanced to maintain an undistorted output and appropriate switch voltage stress [20], [21], which is challenging with many levels and higher switching frequencies. Second, the output voltage tracking capabilities of interleaved converters are not fully understood beyond half the switching frequency [22]–[24], at beat-frequency harmonics [25]–[27], and with nontraditional PWM carriers [28], [29]. In particular, the relation between the switching frequency, effective switching frequency, and maximum output tracking frequency is not clear at present.

This paper presents several contributions to the theory and application of interleaved power electronics: (i) a unification of multilevel and multiphase interleaving together with coupled magnetics as a very large scale interleaving (VLSI) technique in power electronics, (ii) a complete theory on the large-signal tracking capabilities of open-loop interleaved converters, (iii) a 64× interleaved, four-phase, 17-level coupled inductor FCML converter (Fig. 1) pushing the experimental limits of interleaved switching, (iv) an application of large-scale coupled inductor FCML balancing, and (v) a demonstration of advanced communication-over-power at 2.4× the switching frequency on a directly-powered 400 W Li-Fi transmitter.

II. VERY LARGE SCALE INTERLEAVING (VLSI)

Interleaving fundamentally involves splitting one switch into more than one and driving those switches with phase-shifted gate signals. This can be done by putting multiple switches in parallel, such as with the multiphase buck converter, or in series, such as with FCML converters. In this section, we unify these two interleaving concepts under the assumption that the total switch area is fixed and all switches have the same size. Our base case is a buck converter with two equally sized switches with width W and length L. When one of the switches is on, it carries the full inductor current, which has an average of I_o. When off, it blocks the full input voltage, V_{dc}. The total switch area is $2WL$.

The voltage-blocking and current-carrying capability of a switch are proportional to its length and width respectively. Ideally, we can divide the total switch area $2WL$ into multiple narrower devices in parallel or shorter devices in series and handle the same V_{dc} and I_o so long as each carries a current and blocks a voltage proportional to its width and length. This

979-8-3315-1612-3/25 $31.00 © 2025 IEEE

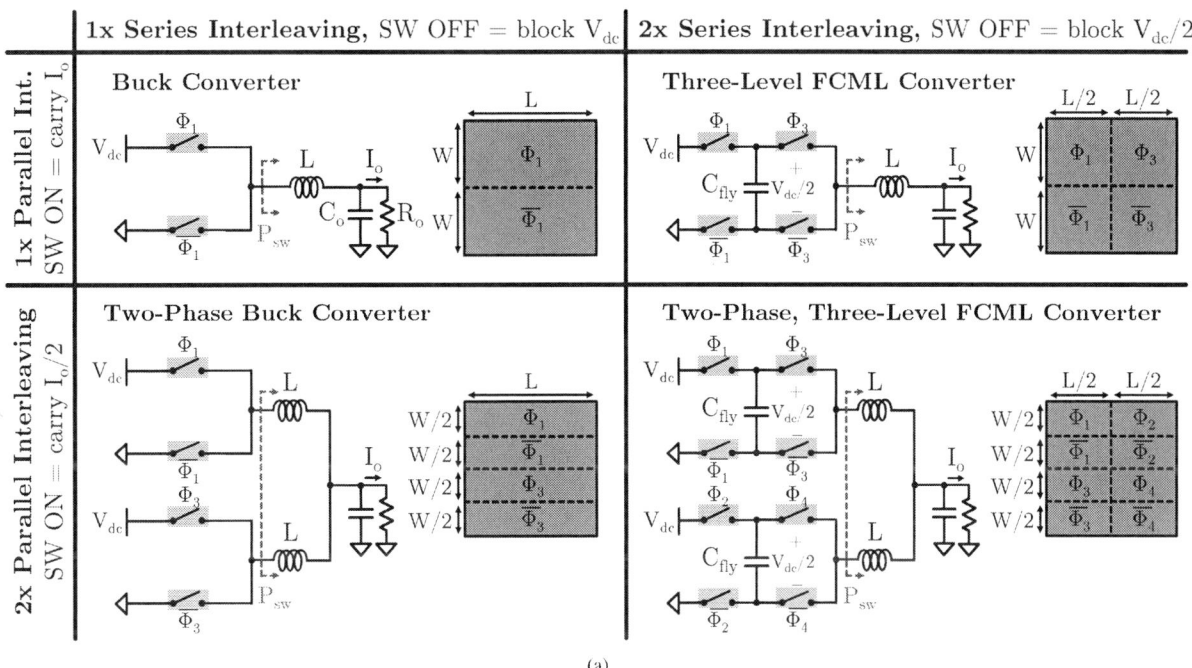

1x Series Interleaving, SW OFF = block V_{dc}	2x Series Interleaving, SW OFF = block $V_{dc}/2$
1x Parallel Int. SW ON = carry I_o — Buck Converter	**Three-Level FCML Converter**
2x Parallel Interleaving SW ON = carry $I_o/2$ — Two-Phase Buck Converter	**Two-Phase, Three-Level FCML Converter**

(a)

Fig. 2. Chart of switch areas (Φ denoting gate signals) and example schematics of converters with parallel interleaving, series interleaving, and a combination.

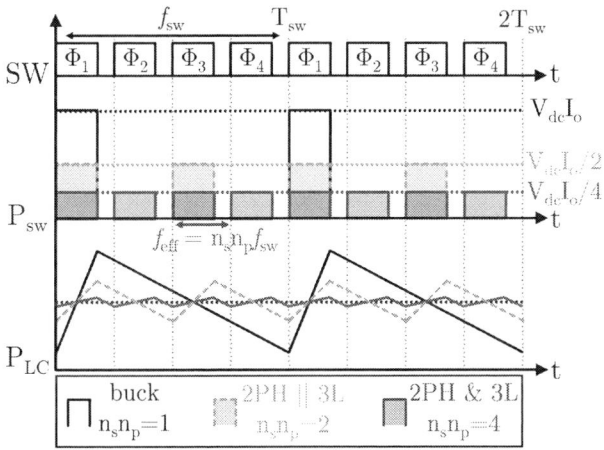

Fig. 3. Power flow waveforms of the four converters in Fig. 2. Interleaving increases the frequency of pulses and reduces their amplitude, reducing the level of energy that must be stored in the inductor and capacitor.

is shown in Fig. 2 with the two-phase buck converter and three-level FCML converter, or a combination of both multiphase and multilevel switching at once. If we then phase shift the gates, we multiply the effective switching frequency of the converter without changing the switch area. We define the series and parallel interleaving factors n_s and n_p as the number of top-side switches in series and parallel respectively. We then index the $n_s n_p$ top-side switches by their phase $p = 1, \ldots, n_p$ and their position in series $s = 1, \ldots, n_s$, where the switch closest to the input is $s = 1$. The switches are driven with uniformly phase-shifted PWM signals with phase shift

$$\theta_{sp} = \theta_k = k \frac{2\pi}{n_s n_p}, \tag{1}$$

where $k = (p - 1) + (s - 1)n_p$ indexes the switches from $k = 0$ to $k = n_s n_p - 1$ by the order of their phase shift. The effective switching frequency of an interleaved converter with uniform phase shifting is

$$f_{\text{eff}} = n_s n_p f_{\text{sw}} = N f_{\text{sw}}, \tag{2}$$

where $N = n_s n_p$ is the total interleaving factor of the converter. Since the two interleaving techniques are dividing different things (voltage or current), we now examine how interleaving divides power flow. Assuming the currents and blocking voltages are balanced, the inductor current carried by each phase is I_o/n_p and the switch node voltage of each phase is equal to the number of switches turned on in that phase multiplied by the blocking voltage. The power transferred to the output network as labeled in Fig. 2 is

$$\begin{aligned}
P_{\text{sw}} &= \sum_{p=1}^{n_p} v_{\text{sw},p} \times i_{L,p} = \sum_{p=1}^{n_p} \left(\frac{I_o}{n_p} \times \sum_{s=1}^{n_s} \Phi_{sp} \frac{V_{dc}}{n_s} \right) \\
&= \frac{I_o V_{dc}}{n_p n_s} \sum_{p=1}^{n_p} \sum_{s=1}^{n_s} \Phi_{sp} \\
&= \frac{I_o V_{dc}}{N} \sum_{k=0}^{N-1} \Phi_k = \frac{I_o V_{dc}}{N} \times n_{\Phi=1},
\end{aligned} \tag{3}$$

where $n_{\Phi=1}$ is the number of top switches that are turned on. Eq. (3) shows that interleaving divides the maximum input power, $I_o V_{dc}$, into N equal divisions controlled by the N top switches. Eq. (2) and (3) quantify the key benefit of interleaving: ideally, **we can split the same switch area into smaller switches that divide the power flow into steps controlled with greater granularity and at a higher frequency.** This is illustrated in Fig. 3. For the one-phase buck converter, $N = 1$ and there is only one switch which delivers the maximum input power $V_{dc}I_o$ when it is on, and none otherwise, with energy storage components handling the balance. An interleaved converter can switch between power levels much closer to the load at a greater frequency, reducing the required energy storage. Note that an interleaved converter only divides power flow control, but does not change the maximum or minimum power flow. Even if all the switches of an interleaved converter are turned on or off, the minimum and maximum power transferred to the load is still the same as a buck converter. Since we often wish to track a particular output voltage, we define the effective switch node voltage as

$$v_{sw,eff} = \frac{V_{dc}}{N} \times n_{\Phi=1}, \tag{4}$$

taken from eq. (3). We use $v_{sw,eff}$ to track a reference signal and filter it with the L-C output filter. The effective switch node voltage reformulates the power divisions in (3) as voltage divisions, even if there may physically be multiple switch nodes in the multiphase converter.

III. THE LARGE-SIGNAL REFERENCE-TRACKING LIMITS OF OPEN-LOOP VLSI CONVERTERS

The Nyquist sampling theorem, in its most common form, states that if a signal is sampled *"at a rate slightly higher than twice the highest significant signal frequency, then the samples contain all of the information of the original signal"* [30], making signal reconstruction possible with an ideal low-pass filter (LPF). Interleaved power converters bear many similarities to the communication systems for which this principle was originally written; we modulate a reference signal to drive PWM switches and recover the desired signal and attenuate the switching harmonics with an L-C filter. In this section, we adapt the Nyquist sampling principle and other elements of modulation theory to answer the following question: how do the switching frequency and effective switching frequency determine the signal-tracking capability of an interleaved converter? We find that interleaving **splits one large control action into smaller ones at a higher frequency, improving large-signal reference tracking resolution and range**, albeit at a reduced amplitude. We assume that the converter is balanced and has ideal phase shifts.

A. The Sampling Principle Adapted to Interleaved Converters

First, we adapt the sampling principle above while assuming an ideal modulator and LPF to derive the fundamental tracking limit. We assume that the reference signal is

$$v_{ref}(t) = \frac{A_{ref}}{2} + \frac{A_{ref}}{2}\cos(2\pi f_{ref}t), \tag{5}$$

with peak-to-peak amplitude A_{ref}, frequency f_{ref}, and period $T_{ref} = \frac{1}{f_{ref}}$, which we seek to track as closely as possible with $v_{sw,eff}$. With a fixed switching frequency f_{sw}, the N switches are turned on and off once per $T_{sw} = \frac{1}{f_{sw}}$, with each switching event increasing or decreasing the effective switch node voltage by $\frac{V_{dc}}{N}$. The two frequency ranges of interest are:

(i) $f_{ref} \leq f_{sw}$: a large-N interleaved converter can track a sub-f_{sw} reference signal with any amplitude $A_{ref} \leq V_{dc}$, as

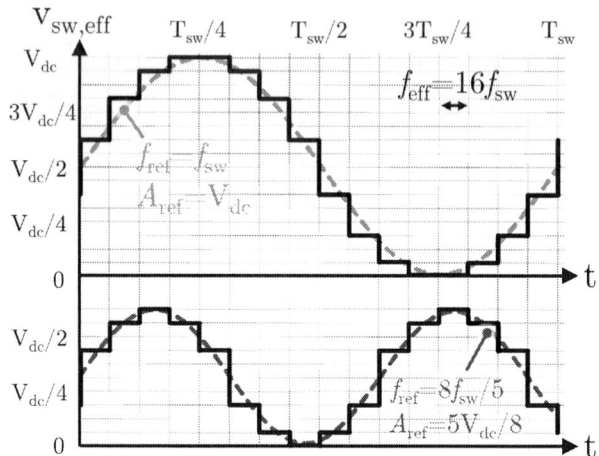

Fig. 4. A $N = 16\times$ interleaved converter with ideal PAM-approximating level-selection tracking reference signals at the output-slope limit.

illustrated in Fig. 4 assuming the switching events happen at uniform times separated by T_{eff}. This is because in time T_{sw}, the reference signal will traverse, at most, from zero to V_{dc} and back once. In the same time, we are able to turn on and off every switch once and traverse the same amplitude with $v_{sw,eff}$. If N is high, the interleaved converter can output levels very close to v_{ref} for each time segment T_{eff}, as shown in Fig. 4, allowing signal recovery from $v_{sw,eff}$ through the LPF.

(ii) $f_{sw} \leq f_{ref} \ll f_{eff}$. If we let $f_{ref} = \frac{f_{eff}}{\rho}$, where $\{\rho \in \mathbb{R} \mid 1 \ll \rho \leq N\}$, we can turn on and off at least $\lceil \text{floor}(\rho) \rceil$ switches in time T_{ref}, so the output of the converter can track a signal with peak-to-peak amplitude up to $A_{ref} \leq \frac{\text{floor}(\rho)}{N}V_{dc}$. This is illustrated in Fig. 4 for $\rho = 10$: the $N = 16$ converter can track a signal with frequency $f_{ref} = \frac{f_{eff}}{10} = \frac{8f_{sw}}{5}$ up to amplitude $A_{ref} = \frac{5V_{dc}}{8}$. If we substitute the reference frequency into the maximum amplitude expression, we derive the inequality

$$A_{ref}f_{ref} \leq \frac{\text{floor}(\rho)}{\rho}V_{dc}f_{sw}, \tag{6}$$

the fundamental output tracking limit of an interleaved converter for $f_{sw} \leq f_{ref} \ll f_{eff}$. Both sides of the inequality (6) are in units [volts/time], making it interpretable as a restriction on the maximum large-signal slope of the reference signal. Essentially, the voltage traversal or average slope of the reference signal, $A_{ref}f_{ref}$, must not be greater than what the converter can fundamentally provide, $V_{dc}f_{sw}$, with a possible reduction if $\rho \notin \mathbb{Z}$ because of the finite number of levels. This relates to section II; interleaving multiplies the frequency of control actions, but also proportionally divides their amplitude.

If the reference signal frequency and amplitude requirements are at the limits above (equality case in (6)), the maximum frequency may be limited by the quantization error introduced by the finite number of levels. An interleaved converter in this regime acts similarly to a pulse amplitude modulation (PAM) system, which samples a waveform and outputs pulses of modulated amplitude instead of width like in PWM. As shown in [30], a PAM system can be used to exactly represent a signal by taking regular samples and extending

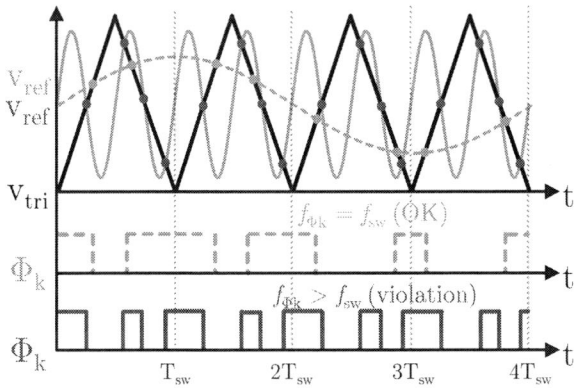

Fig. 5. PWM signal generation with reference signal below and above the slope limit, producing PWM signals with correct and excessive frequency.

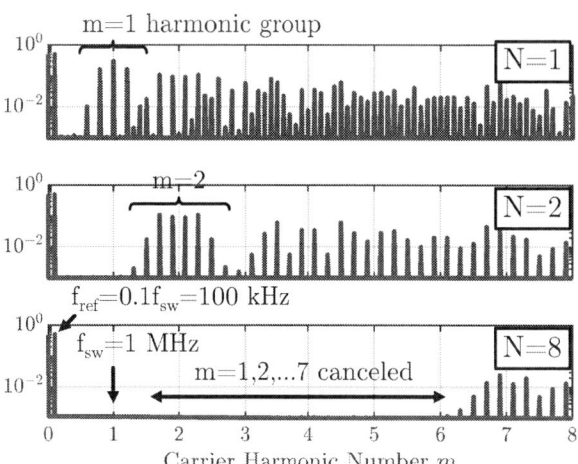

Fig. 6. Spectrum of $v_{\text{sw,eff}}$ for interleaved converters with $f_{\text{sw}} = 1$ MHz, $f_{\text{ref}} = 100$ kHz, and $V_{\text{dc}} = 1$ V. Harmonics up to $m = N - 1$ are canceled.

them into an array of flat-top pulses of duration T_{eff}, then passing the result through an ideal LPF to result in

$$v_{\text{o,LPF}} = \text{sinc}\left(\frac{\pi f_{\text{ref}}}{f_{\text{eff}}}\right) \frac{A_{\text{ref}}}{2}\left(1 + \cos\left[2\pi f_{\text{ref}}\left(t - \frac{T_{\text{eff}}}{2}\right)\right]\right),$$
(7)

adapted from eq. (4-1) of [30], meaning the original signal is recovered perfectly, albeit with a phase shift and attenuation factor. Although we do not conduct a detailed investigation of the effect of quantization on reference tracking here, we note that as long as the effective switching frequency of the converter is much higher than the reference frequency, the converter can represent a reference signal well, allowing it to be reconstructed like in a PAM system.

B. Large-Signal Slope and Harmonic Limit of the Modulator

The reference signal is typically modulated with a triangle or sawtooth carrier wave to generate the PWM waveforms, and the slope of the carrier limits the maximum slope of the reference signal. For non-interleaved converters, the reference signal frequency and slope tend to be much lower than the carrier waveform. Interleaved converters may track signals near or above the switching frequency, so we must consider the carrier slope restriction. The PWM waveform is typically generated by comparing the reference $v_{\text{ref}}(t)$ to the carrier:

$$\Phi_k(t) = \begin{cases} 1 & v_{\text{ref}}(t) \geq v_{\text{carrier},k}(t) \\ 0 & v_{\text{ref}}(t) < v_{\text{carrier},k}(t) \end{cases},$$
(8)

where $v_{\text{carrier},k}(t)$ is the carrier for the kth switch with phase shift defined in (1). The carrier wave is periodic with the switching frequency f_{sw}, ramping up and down with slope $\frac{2V_{\text{dc}}}{T_{\text{sw}}}$ with a triangle carrier or slope $\frac{V_{\text{dc}}}{T_{\text{sw}}}$ up (trailing-edge) or down (leading-edge) with a sawtooth carrier. The reference signal should intersect with the carrier twice per period, as illustrated with the dotted v_{ref} in Fig. 5, such that v_{PWM} turns on once and off once per period. This is guaranteed if the maximum slope of v_{ref} is less than the slope of the carrier. From eq. (5), the maximum slope of the signal is is $\pi A_{\text{ref}} f_{\text{ref}}$, so the modulator restricts the large-signal reference to

$$A_{\text{ref}} f_{\text{ref}} \leq \begin{cases} \frac{2}{\pi} V_{\text{dc}} f_{\text{sw}}, & \text{triangle carrier} \\ \frac{1}{\pi} V_{\text{dc}} f_{\text{sw}}, & \text{sawtooth carrier} \end{cases}$$
(9)

The modulator restricts the slope of the reference signal below the theoretical maximum (6). If the slope of the reference signal exceeds the carrier, such as with the solid reference signal in Fig. 5, the PWM signal frequency will exceed the desired switching frequency or be distorted if latched. Thus far, we have only derived the large-signal slope restriction for valid PWM, but we have not yet shown that the output will track the reference signal correctly. To do so, we study the spectrum of the effective switch node voltage from eq. (4)

$$v_{\text{sw,eff}}(t) = \frac{V_{\text{dc}}}{N}\sum_{k=0}^{N-1}\Phi_k(t) = \underbrace{\frac{A_{\text{ref}}}{2}}_{\text{dc component}} + \underbrace{\frac{A_{\text{ref}}}{2}\cos(\omega_{\text{ref}}t)}_{\text{desired reference signal}}$$

$$+ \underbrace{\sum_{m=1}^{\infty}\frac{4V_{\text{dc}}J_0\left(mM\frac{\pi}{2}\right)\sin\left(m\frac{\pi}{2}\right)}{N\pi m}\times\underbrace{\sum_{k=0}^{N-1}\cos\left(m\omega_{\text{sw}}t + m\theta_k\right)}_{\alpha}}_{\text{carrier harmonics}}$$

$$+ \underbrace{\sum_{m=1}^{\infty}\sum_{n=\pm1}^{\pm\infty}\frac{4V_{\text{dc}}\sin\left((m+n)\frac{\pi}{2}\right)}{N\pi}\times\underbrace{\sum_{k=0}^{N-1}\cos\left((m\omega_{\text{sw}}+n\omega_{\text{ref}})t + m\theta_k\right)}_{\beta}}_{\text{sideband harmonics}}$$
(10)

with a triangle carrier as adapted from (3.39) of [31], where ω_{sw} and ω_{ref} are the angular switching and reference frequencies and $M = \frac{A_{\text{ref}}}{V_{\text{dc}}}$ is the modulation ratio. The Fourier series consists of four parts: a desired dc component, a desired modulated component at the reference frequency, and the undesired carrier and sideband harmonics. The carrier harmonics occur at multiples m of the switching frequency f_{sw} and sideband harmonics occur around the carrier harmonics at multiples n of the reference frequency f_{ref}. The harmonic magnitudes decrease as m and n increase. Since there are an infinite number of sideband harmonics of a PWM signal, there is no strict boundary between acceptable and unacceptable distortion for signal reconstruction. Instead, the signal is considered well-represented by a PWM signal if the dominant sideband and carrier harmonics are far enough from f_{ref} to be filtered by the LPF to a negligible level and the smaller magnitude harmonics

979-8-3315-1612-3/25 $31.00 © 2025 IEEE

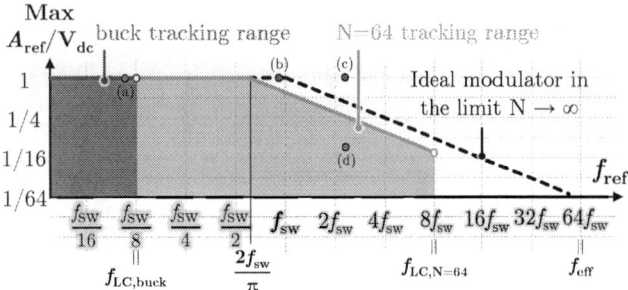

Fig. 7. Plot of maximum reference amplitude vs. reference frequency of a $N = 64\times$ interleaved converter compared to a buck converter. The points (a) through (d) denote reference signals tracked in the experimental section.

are ignored. Since the first group of harmonics occurs at and about the switching frequency, we must prove that interleaved converters cancel carrier and sideband harmonics if we wish to track reference signals around or above f_{sw}.

The expressions α and β in eq. (10) are sums of N cosines for the carrier and sideband harmonics produced by each PWM, except with a different phase shift θ_k for each term. If $\alpha = 0$ or $\beta = 0$ for a given carrier or sideband harmonic, then that harmonic will be canceled. We consider that

$$\sum_{k=0}^{N-1} \cos(x + m\theta_k) = \sum_{k=0}^{N-1} \frac{\left(e^{j(x+m\theta_k)} + e^{-j(x+m\theta_k)}\right)}{2} \quad (11)$$

$$= \frac{e^{jx}}{2} \sum_{k=0}^{N-1} \left(e^{jm\frac{2\pi}{N}}\right)^k + \frac{e^{-jx}}{2} \sum_{k=0}^{N-1} \left(e^{-jm\frac{2\pi}{N}}\right)^k \quad (12)$$

$$= \frac{e^{jx}}{2} \frac{1 - e^{jm2\pi}}{1 - e^{jm\frac{2\pi}{N}}} + \frac{e^{-jx}}{2} \frac{1 - e^{-jm2\pi}}{1 - e^{-jm\frac{2\pi}{N}}} = 0, m \notin N\mathbb{Z}, \quad (13)$$

where we expand the cosine terms with Euler's identity in (11), apply the geometric series in (12) for any non-integer ratio $\frac{m}{N}$, and note that the numerators cancel to zero for any valid m in (13). Therefore, for any m from $m = 1$ to $m = N - 1$, the sum of cosines in (11) sums to zero. Since expressions α and β of eq. (10) are in the form of eq. (11), we conclude that the carrier and sideband harmonics of an $N\times$ interleaved converter at and about the first $N - 1$ carrier harmonics all cancel out. The lowest frequency non-canceled harmonics occur at the effective switching frequency, as verified in Fig. 6.

The preceding result also shows how interleaving allows the use of an smaller L-C filter with a higher cutoff frequency. The L-C cutoff frequency, which needs to be set far below the switching frequency in a buck converter, may now be set relative to the effective switching frequency. The analysis is limited since practical phase shifting is affected by factors like propagation delay, which will lead to imperfect cancellation of undesired harmonics. Additionally, the fact that $f_{\text{eff}} = Nf_{\text{sw}}$ does not mean we can track a reference signal $N\times$ faster than a buck converter, because the sideband harmonics are located about the carrier harmonics at integer multiples of the reference frequency. Therefore, as the reference frequency increases, the sidebands (especially those with a higher order n) will approach the in-band frequencies passed by the LPF. Finally, we do not address beat-frequency harmonics or intrin-

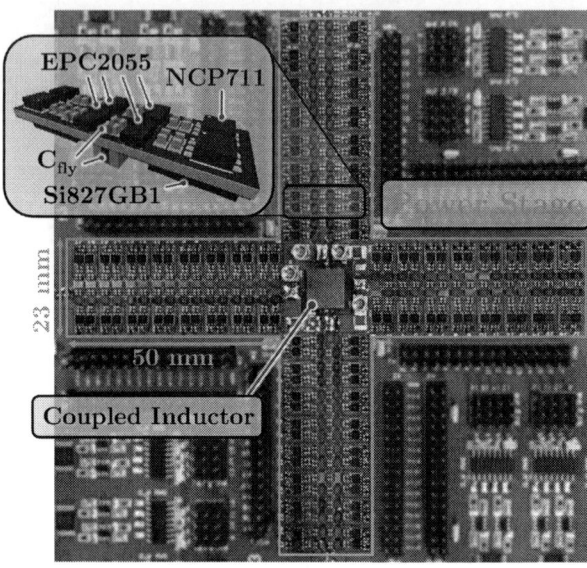

Fig. 8. Four-phase, 17-level FCML converter with coupled inductors.

TABLE I
CIRCUIT PARAMETERS OF THE FCML PROTOTYPE

f_{sw}	V_{dc}	C_{fly}	L_l	L_μ	C_o
500 kHz	48 V	10 μF	20.4 nH	230 nH	0.1 μF / 0.7 μF

sic unbalancing of interleaved converters when the reference frequency is an exact multiple of the switching frequency.

C. Summary of Large-Signal Tracking Capabilities

The large-signal tracking capabilities of interleaved converters, as derived in the preceding sections, are summarized in Fig. 7. For this plot, we assume $N = 64$ as an example, and assume that the LPF cutoff frequency is set to one-eighth of the effective switching frequency. A buck converter has a tracking range limited below the cutoff frequency. On the other hand, a highly interleaved triangle-modulated converter can track-full amplitude signals up to $f_{\text{ref}} = \frac{2f_{\text{sw}}}{\pi}$, after which the slope of the reference must be kept below the carrier wave, leading to a -6 dB/dec maximum gain roll-off with frequency. The triangle modulator restricts the reference amplitude slightly under the ideal limit (6). In a practical design, the cutoff frequency of the interleaved converter may need to be set lower to filter sidebands depending on their frequency and magnitude.

The signal tracking range of the interleaved converter is larger than a buck converter and the harmonic performance is better in their overlapping range. Fig. 7 may be likened to the Bode plot of an operational amplifier. By limiting the gain of the interleaved converter, we can dramatically increase the frequency range. For example, if we restrict $\frac{A_{\text{ref}}}{V_{\text{dc}}} \leq \frac{1}{8}$, the system will ideally have flat gain up to $8f_{\text{sw}}$, $N = 64\times$ higher than a buck converter. This is useful in many high speed applications needing fast tracking but not over a large amplitude. For example, communication-over-power technology like Li-Fi and visible light communication require a large dc signal to power a load, plus a small high frequency

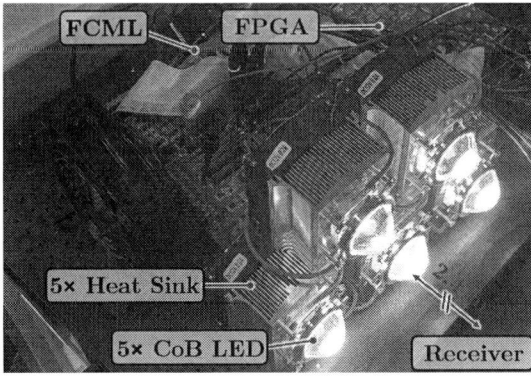

Fig. 9. 400 W Li-Fi CoB LED transmitter array.

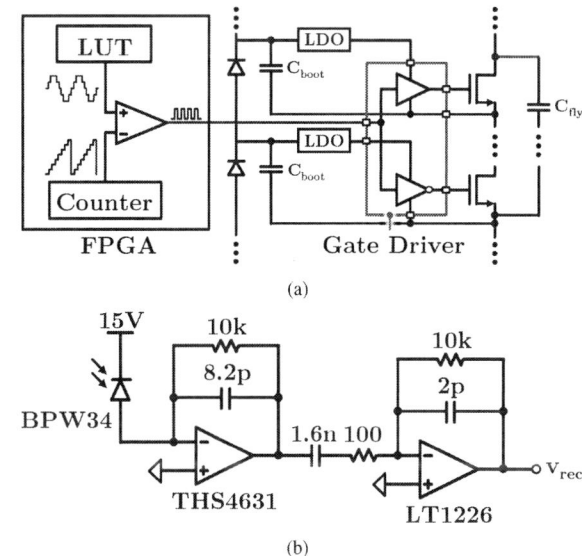

(a)

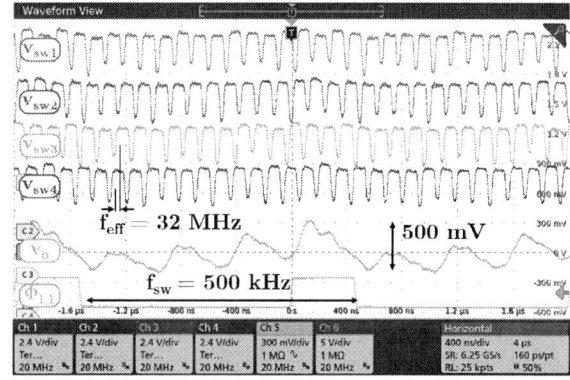

(b)

Fig. 10. Schematic of the (a) FPGA signal generation and bootstrapping and (b) light receiver filtering and amplification circuit.

component for communication. Interleaved converters are able to efficiently deliver the high power dc component, while also achieving high frequency reference-tracking performance.

IV. EXPERIMENTAL RESULTS

To verify the large-signal tracking properties of interleaved converters, we design a four-phase, 17-level FCML converter with $64\times$ interleaving and coupled inductors for ripple reduction and flying capacitor balancing. The power stage schematic is highlighted in Fig. 1, along with the physical design in Fig. 8 with key component values listed in Table I.

A. FCML Converter Design and Operation

Each FCML converter has 16 pairs of complimentary EPC2055 switches split into repeated switching cells with four switches and two half bridge drivers (Si827GB1-IS1) each. The switching cell, highlighted in Fig. 8, is designed with the low-inductance principles in [32] for maximum density and speed. There are 15 flying capacitors C_{fly} per phase with ideal dc voltages at evenly spaced fractions of the input voltage V_{dc}, starting at $\frac{15V_{\text{dc}}}{16}$ closest to the input and decreasing to $\frac{V_{\text{dc}}}{16}$ closest to the output. The flying capacitors are balanced by the coupled inductors at the output [19], which also serve to reduce the ripple and size of the magnetics. Each of the 64 gate signals are generated with phase shifted carriers according to eq. (1) and Fig. 10(a), where each complementary pair of switches is driven with an isolated half-bridge gate driver powered by a bootstrapping and regulation circuit [33]. The 64 open-loop gate signals are provided by an EP4CE15F23C8 FPGA operating with a 224 MHz internal clock to compare a digital counter and LUT to follow arbitrary reference signals. The steady-state operation at $d = 0.23$ is shown in Fig. 11 with good balancing provided by the coupled inductors. This plot does reveal one limit of large-scale interleaving: due to the differences in propagation delay (including from the PCB traces themselves), there is a phase shift variation of a few nanoseconds per phase, which means that the harmonic cancellation is not perfect; thus, the ripple is dominated by components below the effective switching frequency. Finally, Fig. 12 shows the converter tracking reference signals at points (a), (b), and (c) on Fig. 7. At low frequency, the converter

tracks the signal with very high resolution due to the $64\times$ interleaving. Fig. 12(b) and (c) show the converter tracking with distortion outside of the allowable large-signal range with trailing- and double-edge modulation.

B. FCML Powered Li-Fi Transmitter Experimental Setup

The FCML converter is used to directly power an array of five Chanzon 100 W high-brightness CoB (chip-on-board) LEDs on heat sinks, as shown in Fig. 9. The LEDs have a forward voltage of around 31 V, so an average duty cycle of $d = \frac{31}{48} = 0.65$ is used, plus a data signal on a 1.2 MHz sinusoidal carrier, $2.4\times$ higher than f_{sw}. In accordance with the theory of section III, the amplitude is limited to $A_{\text{ref}} < \frac{V_{\text{dc}}}{1.2\pi}$. Because of the high carrier frequency, the signal is imperceptible to the human eye. The LEDs are pointed at a light receiver circuit 2.4 meters away which amplifies and band-pass filters the communication signal (Fig. 10(b)).

C. Li-Fi Communication Performance

To minimize the impact of the timing mismatch ripple (section IV-A), the output capacitance is increased to

979-8-3315-1612-3/25 $31.00 © 2025 IEEE

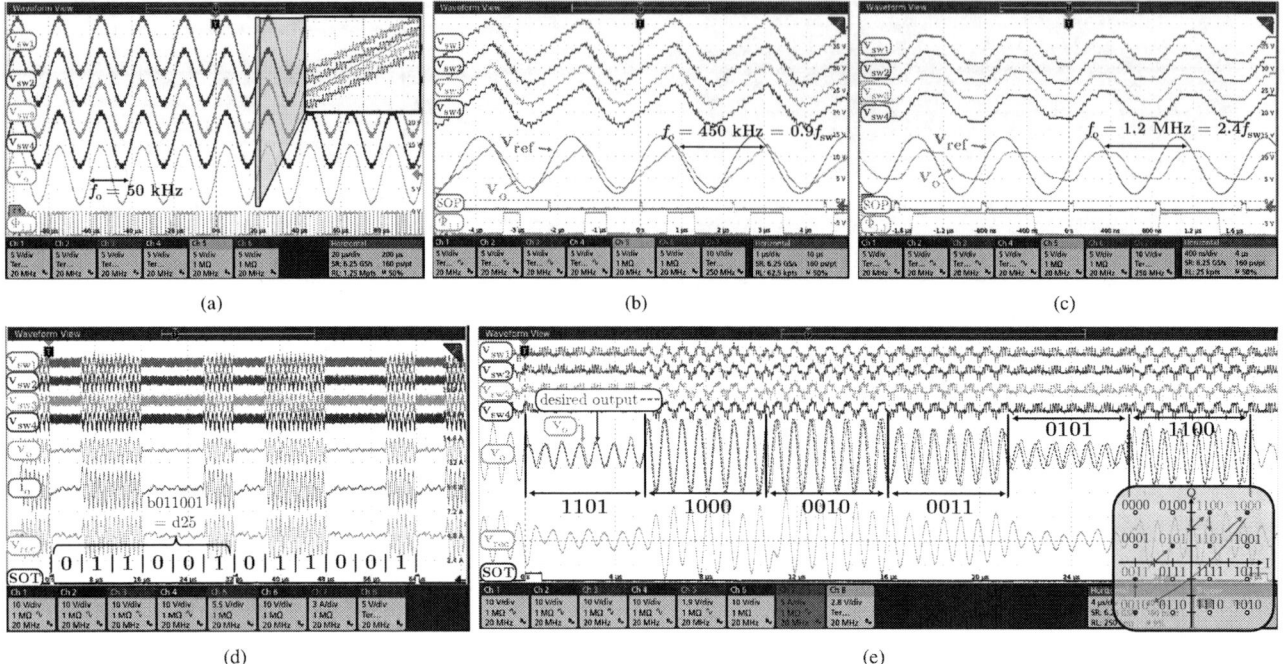

Fig. 12. Converter operation (a) well below the switching frequency, beyond large-signal limits with (b) trailing- and (c) double-edge modulation, and Li-Fi LED transmission and reception modulated at 1.2 MHz with (d) OOK, sending "25" (APEC 2025), and (e) 6 symbols of 16-QAM.

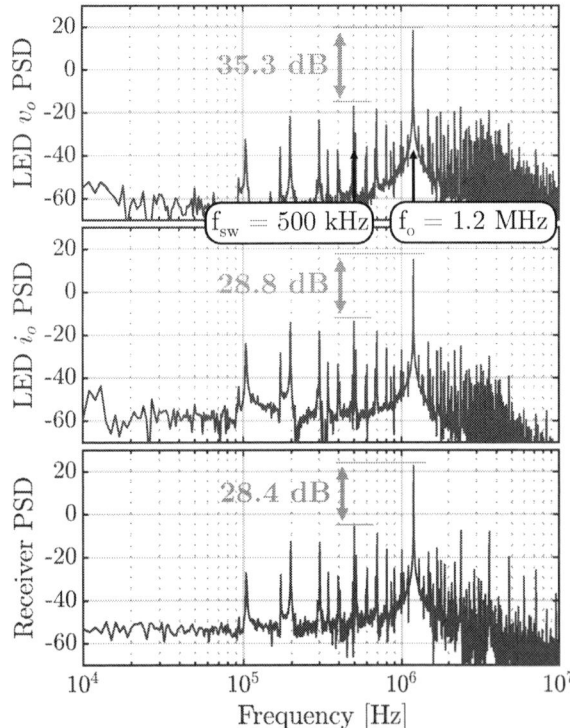

Fig. 13. PSD of single-frequency sinusoid tracking at the LED voltage, current, and amplified photodiode output.

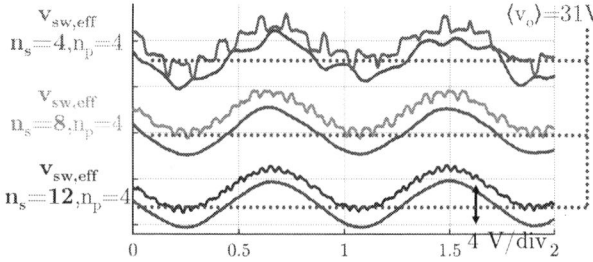

Fig. 14. LED driving performance with 5, 9, and 13 levels, showing increasing resolution with increasing levels.

density) of the LED voltage, current, and received voltage are shown in Fig. 13. The signal-to-noise ratio to the largest noise component at f_{sw} is high for all three, with some degradation due to the nonlinear LED I-V relationship which could be compensated with closed-loop control/pre-distortion. The output power is **383.4 W** and the conversion efficiency is **95.5%**, reduced negligibly from **96.0%** when no signal is transmitted. The gate drive loss and signal path loss are not included in the efficiency calculation. Fig. 12(d) and (e) shows the transmission of signals encoded with OOK (on-off keying) and 16-QAM. The converter output voltage faithfully reproduces the desired signals with a carrier frequency much higher than the switching frequency.

Practical problems remain with the optimization of interleaved converter design and with the experimental setup used here as an example. The problem of imperfect phase shifting sets a limit on the achievable harmonic cancellation. This, along with other non-idealities such as device packaging, layout space, and parasitics introduced by interleaving, may

$C_o = 0.7$ μF, making $f_{LC} = 1.3$ MHz given the leakage inductance $L_l = 20.4$ nH of the inductor. With the output carrier frequency $f_{ref} = 2.4 f_{sw} = 1.2$ MHz, amplitude $A = \frac{V_{dc}}{9.2}$ (point (d) in Fig. 7), the PSD (power spectral

979-8-3315-1612-3/25 $31.00 © 2025 IEEE

decide the optimal level of interleaving for a given application. Fig. 14 shows the LED reference signal tracking experiments repeated for the converter operated with 5-, 9-, and 13- levels, where increasing levels does improve resolution, but perhaps with an upper limit determined by practical factors. We have studied open-loop converters assuming good balancing and avoiding beat-frequency harmonics; the closed-loop behavior and stability around problematic frequency ratios, and the influence of coupled inductors are not studied here.

V. CONCLUSION

Multiphase interleaving, multilevel interleaving, and magnetic coupling multiply the effective switching frequency, reducing the required passives size and extending the signal tracking capabilities to and above the switching frequency of power electronics as long as the large-signal slope limits are not exceeded. This property is leveraged in an ultra-fast Li-Fi transmitter design using a $64\times$ interleaved four-phase, 17-level FCML converter with coupled inductors for passive balancing. This enables above-switching-frequency small-signal Li-Fi communication on top of high efficiency power for a 400 W transmitter.

ACKNOWLEDGEMENTS

This work was supported by the National Science Foundation (Award #1847365) and the Natural Sciences and Engineering Research Council of Canada (Award #557270−2021).

REFERENCES

[1] T. Meynard and H. Foch, "Multi-level conversion: high voltage choppers and voltage-source inverters," in *PESC '92 Record. 23rd Annual IEEE Power Electronics Specialists Conference*, 1992, pp. 397–403 vol.1.

[2] T. Meynard, M. Fadel, and N. Aouda, "Modeling of multilevel converters," *IEEE Transactions on Industrial Electronics*, vol. 44, no. 3, pp. 356–364, 1997.

[3] J. Rodriguez, L. G. Franquelo, S. Kouro, J. I. Leon, R. C. Portillo, M. M. Prats, and M. A. Perez, "Multilevel converters: An enabling technology for high-power applications," *Proceedings of the IEEE*, vol. 97, no. 11, pp. 1786–1817, 2009.

[4] J. Azurza Anderson, G. Zulauf, P. Papamanolis, S. Hobi, S. Mirić, and J. W. Kolar, "Three levels are not enough: Scaling laws for multilevel converters in ac/dc applications," *IEEE Transactions on Power Electronics*, vol. 36, no. 4, pp. 3967–3986, 2021.

[5] M. Chen and C. R. Sullivan, "Unified models for coupled inductors applied to multiphase pwm converters," *IEEE Transactions on Power Electronics*, vol. 36, no. 12, pp. 14 155–14 174, 2021.

[6] Y. Zhang, M. Rodríguez, and D. Maksimović, "Output filter design in high-efficiency wide-bandwidth multi-phase buck envelope amplifiers," in *2015 IEEE Applied Power Electronics Conference and Exposition (APEC)*, 2015, pp. 2026–2032.

[7] S. Yerra and H. Krishnamoorthy, "Multi-phase three-level buck converter with current self-balancing for high bandwidth envelope tracking power supply," in *2020 IEEE Applied Power Electronics Conference and Exposition (APEC)*, 2020, pp. 1872–1877.

[8] J. Chen, K. Liu, J. Wu, R. Wang, W. Weng, and X. He, "Simultaneous power and data transmission using combined three degrees of freedom modulation strategy in dc–dc converters," *IEEE Transactions on Power Electronics*, vol. 38, no. 3, pp. 3191–3200, 2023.

[9] X. He, R. Wang, J. Wu, and W. Li, "Nature of power electronics and integration of power conversion with communication for talkative power," *Nature communications*, vol. 11, no. 1, p. 2479, 2020.

[10] H. Haas, L. Yin, Y. Wang, and C. Chen, "What is lifi?" *Journal of Lightwave Technology*, vol. 34, no. 6, pp. 1533–1544, 2016.

[11] L. Teixeira, F. Loose, J. M. Alonso, C. H. Barriquello, V. Alfonso Reguera, and M. A. Dalla Costa, "A review of visible light communication led drivers," *IEEE Journal of Emerging and Selected Topics in Power Electronics*, vol. 10, no. 1, pp. 919–933, 2022.

[12] N. C. Brooks, J. Zou, S. Coday, T. Ge, N. M. Ellis, and R. C. N. Pilawa-Podgurski, "On the size and weight of passive components: Scaling trends for high-density power converter designs," *IEEE Transactions on Power Electronics*, vol. 39, no. 7, pp. 8459–8477, 2024.

[13] J. Azurza Anderson, G. Zulauf, J. W. Kolar, and G. Deboy, "New figure-of-merit combining semiconductor and multi-level converter properties," *IEEE Open Journal of Power Electronics*, vol. 1, pp. 322–338, 2020.

[14] Z. Ye, S. R. Sanders, and R. C. N. Pilawa-Podgurski, "Modeling and comparison of passive component volume of hybrid resonant switched-capacitor converters," *IEEE Transactions on Power Electronics*, vol. 37, no. 9, pp. 10 903–10 919, 2022.

[15] T. Modeer, C. B. Barth, N. Pallo, W. H. Chung, T. Foulkes, and R. C. N. Pilawa-Podgurski, "Design of a gan-based, 9-level flying capacitor multilevel inverter with low inductance layout," in *2017 IEEE Applied Power Electronics Conference and Exposition (APEC)*, 2017.

[16] S. Coday, A. Barchowsky, and R. C. Pilawa-Podgurski, "A 10-level gan-based flying capacitor multilevel boost converter for radiation-hardened operation in space applications," in *2021 IEEE Applied Power Electronics Conference and Exposition (APEC)*, 2021, pp. 2798–2803.

[17] C. B. Barth, P. Assem, T. Foulkes, W. H. Chung, T. Modeer, Y. Lei, and R. C. N. Pilawa-Podgurski, "Design and control of a gan-based, 13-level, flying capacitor multilevel inverter," *IEEE Journal of Emerging and Selected Topics in Power Electronics*, vol. 8, no. 3, 2020.

[18] D. H. Zhou and M. Chen, "Switching frequency is not the limit: Multiphase coupled inductor fcml converter tracking signals above the switching frequency," in *2023 IEEE 24th Workshop on Control and Modeling for Power Electronics (COMPEL)*, 2023, pp. 1–7.

[19] D. H. Zhou, J. Čeliković, D. Maksimović, and M. Chen, "Balancing multiphase fcml converters with coupled inductors: Modeling, analysis, limitations," *IEEE Transactions on Power Electronics*, pp. 1–24, 2024.

[20] R. Wilkinson, H. du Mouton, and T. Meynard, "Natural balance of multicell converters," in *IEEE 34th Annual Conference on Power Electronics Specialist, 2003. PESC '03.*, vol. 3, 2003, pp. 1307–1312 vol.3.

[21] B. P. McGrath and D. G. Holmes, "Analytical modelling of voltage balance dynamics for a flying capacitor multilevel converter," in *2007 IEEE Power Electronics Specialists Conference*, 2007, pp. 1810–1816.

[22] S.-F. Hsiao, C.-F. Nien, D. Chen, and C.-J. Chen, "Four-frequency small-signal model for high-bandwidth voltage regulator with current-mode control," *IEEE Access*, vol. 10, pp. 25 633–25 644, 2022.

[23] Y. Jiang, Y. Sun, J. Lin, S. Xie, M. Su, and Y. Liu, "Unified extended-frequency model of buck converters under different carriers," *IEEE Transactions on Industrial Electronics*, vol. 70, no. 4, 2022.

[24] S.-F. Hsiao, D. Chen, C.-J. Chen, and H.-S. Nien, "A new multiple-frequency small-signal model for high-bandwidth computer v-core regulator applications," *IEEE Transactions on Power Electronics*, vol. 31, no. 1, pp. 733–742, 2015.

[25] Y. Qiu, M. Xu, K. Yao, J. Sun, and F. C. Lee, "Multifrequency small-signal model for buck and multiphase buck converters," *IEEE transactions on power electronics*, vol. 21, no. 5, pp. 1185–1192, 2006.

[26] X. Yue, F. Zhuo, S. Yang, Y. Pei, and H. Yi, "A matrix-based multifrequency output impedance model for beat frequency oscillation analysis in distributed power systems," *IEEE Journal of Emerging and Selected Topics in Power Electronics*, vol. 4, no. 1, pp. 80–92, 2015.

[27] J. Sun, F. C. Lee, M. Xu, and Y. Qiu, "Modeling and analysis for beat-frequency current sharing issue in multiphase voltage regulators," in *2007 IEEE Power Electronics Specialists Conference*. IEEE, 2007.

[28] L. Mathe, F. Lungeanu, D. Sera, P. O. Rasmussen, and J. K. Pedersen, "Spread spectrum modulation by using asymmetric-carrier random pwm," *IEEE Transactions on Industrial Electronics*, vol. 59, no. 10, pp. 3710–3718, 2011.

[29] R. Wang, Z. Lin, J. Du, J. Wu, and X. He, "Direct sequence spread spectrum-based pwm strategy for harmonic reduction and communication," *IEEE Transactions on Power Electronics*, vol. 32, no. 6, pp. 4455–4465, 2016.

[30] H. S. Black, *Modulation theory*. van Nostrand, 1953.

[31] D. G. Holmes and T. A. Lipo, *Pulse width modulation for power converters: principles and practice*. John Wiley & Sons, 2003, vol. 18.

[32] N. C. Brooks, L. Horowitz, R. Abramson, and R. C. N. Pilawa-Podgurski, "Low-inductance asymmetrical hybrid gan hemt switching cell design for the fcml converter in high step-down applications," in *2021 IEEE Applied Power Electronics Conference and Exposition (APEC)*, 2021, pp. 9–15.

[33] Z. Ye, Y. Lei, W.-C. Liu, P. S. Shenoy, and R. C. Pilawa-Podgurski, "Improved bootstrap methods for powering floating gate drivers of flying capacitor multilevel converters and hybrid switched-capacitor converters," *IEEE Transactions on Power Electronics*, vol. 35, no. 6, pp. 5965–5977, 2020.

Isolated PWM DC-DC Converter with Single Magnetic Component, ZVS and Self-Balanced Switched-Capacitor Voltage

Pablo M. Gil, Juan Rodríguez and Diego G. Lamar
Electronic Power Supply Systems Group (SEA)
University of Oviedo
Gijon 33204, Spain
Email: {mateospablo, rodriguezmjuan, gonzalezdiego}@uniovi.es

Abstract— **The Single-Active Bridge (SAB) converter is a suitable candidate for dc-dc conversions that require galvanic isolation, unidirectional power flow, high efficiency and high power density. The topology can operate at constant frequency with regulation capability, thus avoiding the penalization in terms of electromagnetic compatibility of resonant converters. Moreover, the SAB converter achieves Zero Voltage Switching (ZVS) for all MOSFET and it can be implemented with a single magnetic component by properly designing the transformer leakage inductance. Unfortunately, the transformer highly penalizes the efficiency and the power density when high step-down conversions are addressed. In this paper, a novel version of the SAB converter is proposed in order to alleviate the aforementioned drawback. The novel topology incorporates a switched-capacitor structure that reduces the voltage across the transformer primary winding. Consequently, higher step-down conversions can be addressed and the integration of the series inductance into the transformer is easier. Moreover, the novel converter exhibits additional benefits, such as soft charge/discharge of the switched capacitor with self-balance mechanism of the voltage across that capacitor, reduced voltage stress across three MOSFETs and no transformer saturation problem. Experimental results of a 400V-to-48V converter prototype with a switching frequency of 100kHz and a peak output power of 500W that achieves a peak efficiency of 94.1% and a full load efficiency of 93% are provided.**

Keywords—DC-DC Power Converter, High step-down conversion, Zero Voltage Switching (ZVS), Switched Capacitor.

I. Introduction

The Single-Active-Bridge (SAB) converter, which is also referred as full-bridge dc-dc converter with capacitive output filter, is a suitable candidate for dc-dc conversions that require galvanic isolation, unidirectional power flow, high efficiency, high power density, constant switching frequency, high reliability and a relatively low cost [1]-[4]. The topology has high output impedance not only in Discontinuous Conduction Mode (DCM), but also in Continuous Conduction Mode (CCM). In other words, the behavior is similar to a current source, an unusual characteristic that makes it very attractive for some applications, such as battery chargers or LED drivers. Nowadays, these applications tend to require higher step-down conversions and, consequently, a high step-down transformer must be used in the SAB converter. Unfortunately, that transformer highly penalizes the converter in terms of efficiency and power density.

During the last years, novel isolated dc-dc converters that include switched-capacitor structures have been proposed to enable higher step-down conversions. The flying-capacitor structure is combined with a half-bridge dc-dc converter with current-doubler rectifier in [5]. Another example can be found in [6], where a stacked half-bridge dc-dc converter with current-doubler rectifier is proposed. The series-capacitor structure has been explored for the full-bridge dc-dc converter with full-bridge rectifier [7] and current-doubler rectifier [8]. However, in all previous combinations the derived topologies require two or three magnetic elements: one transformer and one or two inductors.

In this paper, a novel isolated dc-dc converter is derived (see Fig. 1) by incorporating into the SAB converter a switched-capacitor structure and by modifying the control strategy in order to address higher step-down conversions. In contrast to previous combinations of isolated dc-dc converters with switched-capacitor structures, the proposed converter could be integrated with a single magnetic element (the inductance connected in series with the transformer models its leakage inductance). It is important to note that integrating that inductance into the transformer is easier than in a conventional SAB converter. The reason is that the switched-capacitor structure reduces the magnitude of the voltage applied across the transformer primary winding to half the input voltage, which also facilitates the higher step-down conversion. Additional benefits of the proposed converter are: the leakage inductance of the transformer soft charges/discharges the switched capacitor, all MOSFETs achieve Zero Voltage Switching (ZVS), the voltage stress of three MOSFETs is half the input voltage, the switched capacitor ensures no

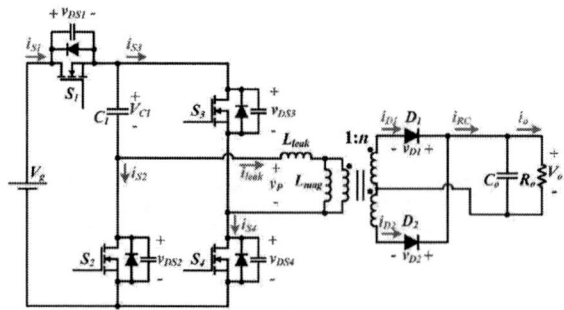

Fig. 1. Schematic of the proposed converter including the notation of the voltage and current waveforms.

transformer saturation problem, the converter operates at constant switching frequency and the voltage across the switched capacitor is naturally balanced.

In Section II, the principle of operation of the proposed converter is given, including the description of the topology, the various stages of operation within a switching period and a design guideline of the converter to operate in CCM. In Section III, the explanation of the self-balancing mechanism of the switched capacitor voltage is provided. In Section IV, experimental results verify the operation of the proposed converter in CCM. In Section V, main conclusions gathered from this work are summarized.

II. PRINCIPLE OF OPERATION IN CCM

A. Topology description

The primary side of the proposed converter (see Fig. 1) is made up of four MOSFETs (S_1, S_2, S_3 and S_4), a switched capacitor (C_1) and the transformer primary winding. As Fig. 2 shows, the control signals of S_1 and S_3 have the same duty cycle (d) value and a phase shift of 180°. Moreover, the control signals of S_2 and S_4 are complementary to those of S_1 and S_3, respectively. The switched capacitor reduces the voltage applied to the transformer's primary winding to $V_g/2$ (where V_g is the input voltage), as well as ensures that the voltage stresses in three of the switches (S_1, S_2 and S_4) are reduced to $V_g/2$. L_{leak} and L_{mag} model the leakage inductance and the magnetizing inductance of the transformer, respectively. L_{leak} fulfills three tasks: it transfers the stored energy to the load, enables ZVS for all MOSFETs and soft charges/discharges the switched capacitor. The secondary side comprises the two windings of the center-tapped transformer, two diodes (D_1 and D_2), the output capacitor (C_o) and the resistive load (R_o). Note that in Fig. 1, the output voltage is referred to as V_o and the turns ratio between the secondary (n_2) and a primary (n_1) windings is referred to as n. Some assumptions have been considered for the operation analysis in order to simplify the explanations. First, all components are ideal except for the MOSFETs, which are represented along with their parasitic output capacitors and antiparallel diodes. Second, the L_{mag} value is very large and, consequently, the magnetizing current is neglected. Third, the capacitances of C_1 and C_o are large values and, as a result, voltage ripple across them can be neglected. Fourth, the paper only considers the operation in CCM (ZVS in the turn-on of the four switches of the primary side is only possible in this conduction mode). Furthermore, it is assumed that the voltage across C_1 (V_{C1}) is naturally balanced to $V_g/2$, which will be further discussed in the following section.

B. Stages of operation

The converter operation during a switching period (T_s) can be divided into six different stages (see Fig. 3). For the sake of simplicity, dead-time intervals are not included in this description, despite they are essential for achieving ZVS in the four switches of the primary side. A brief explanation of the aforementioned stages is given below:

- At the beginning of Stage I ($t=0$), S_2 is turned-off and since the current through L_{leak} (i_{leak}) is negative (i.e., equal to $-I_{p2}$) at that moment, S_1 is turned-on with ZVS. In this way, during Stage I [$0<t<d'\cdot T_s$], both S_1 and S_4 are ON, while D_2 is conducting. C_1 is

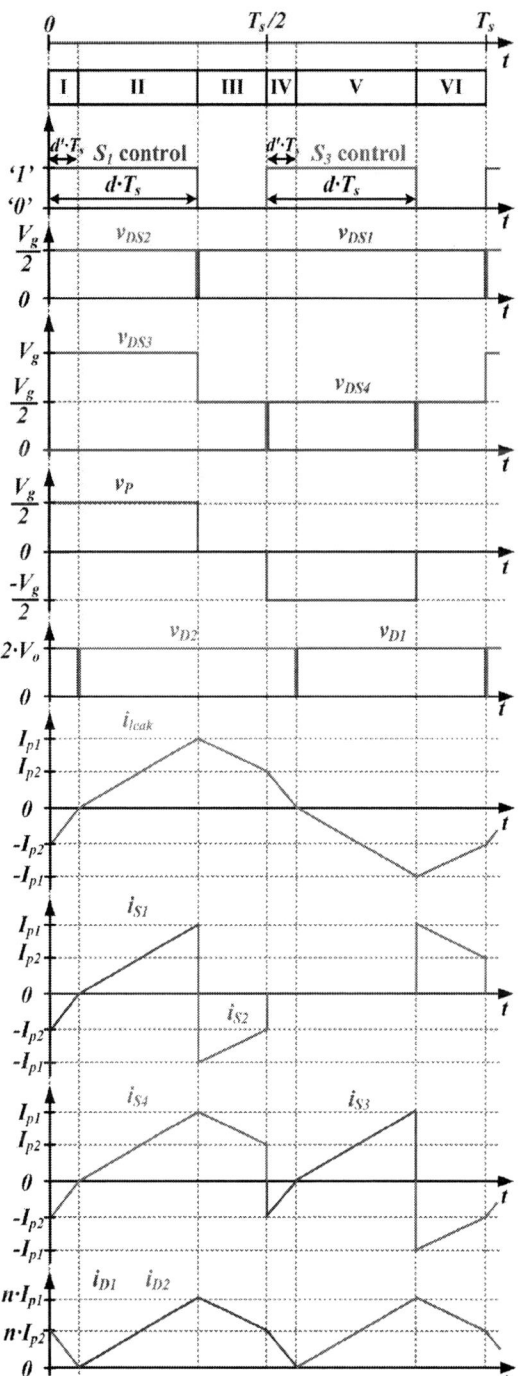

Fig. 2. Main Voltage and current waveforms of the proposed converter.

discharged and a voltage of [$V_g/2 + V_o/n$] is applied across L_{leak} (note that V_{C1} is subtracted to V_g). As a result, i_{leak} linearly rises and Stage I ends when i_{leak} reaches 0. The length of this interval is $d'\cdot T_s$, where d' is defined as:

$$d' = \frac{d}{2} - \frac{1}{2}\cdot\frac{V_o}{n\cdot V_g}, \quad (1)$$

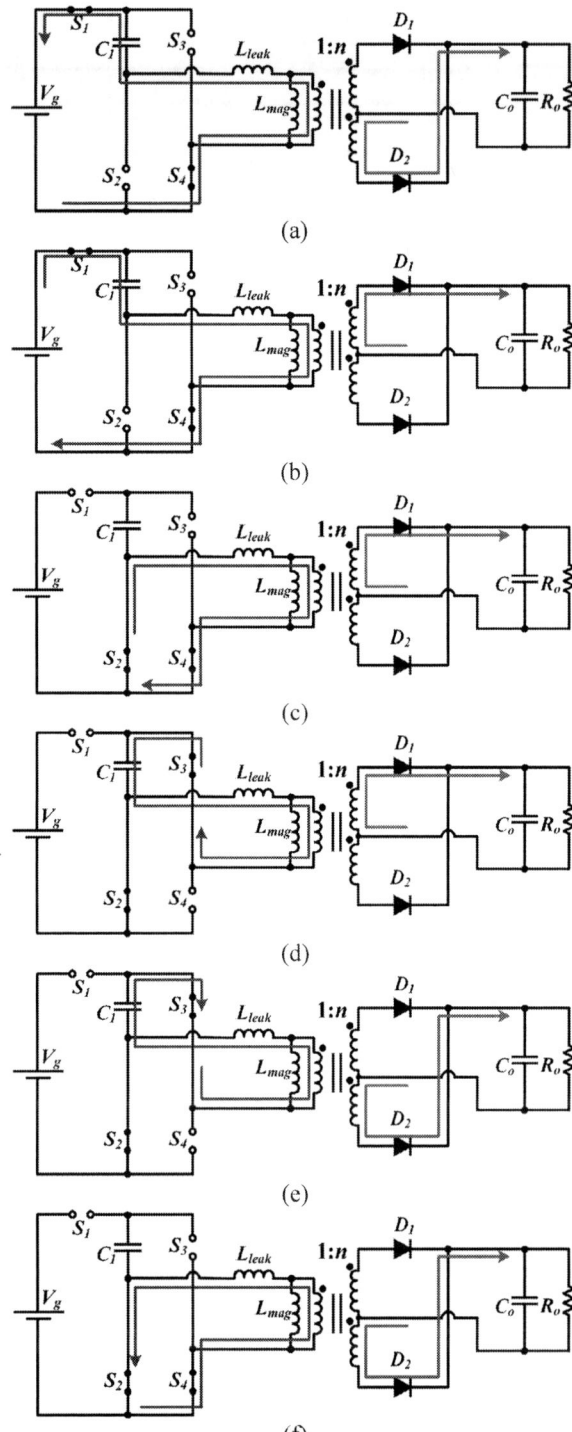

Fig. 3. Current paths of the proposed converter: a) State I. b) State II. c) State III. d) State IV. e) State V. f) State VI.

while the current value I_{P2} is given by the expression:

$$I_{p2} = \frac{T_s}{2 \cdot L_{leak}} \cdot \left(V_g + \frac{V_o}{n} \right) \cdot \left(d - \frac{1}{2} \cdot \frac{V_o}{n \cdot V_g} \right). \quad (2)$$

- During Stage II [$d' \cdot T_s < t < d \cdot T_s$], S_1 and S_4 remain ON, but in this case D_1 conducts because i_{leak} is positive. C_1 is charged and a voltage of [$V_g/2 - V_o/n$]

is applied across L_{leak} (i.e., i_{leak} rises linearly, but with a different slope). At the end of this stage, i_{leak} reaches I_{P1}, which is given by the expression:

$$I_{p1} = \frac{T_s}{2 \cdot L_{leak}} \cdot \left(V_g - \frac{V_o}{n} \right) \cdot \left(d + \frac{1}{2} \cdot \frac{V_o}{n \cdot V_g} \right). \quad (3)$$

- At the beginning of Stage III ($t=d \cdot T_s$), S_1 is turned-off and since i_{leak} is positive at that moment, S_2 is turned-on with ZVS. In this way, during Stage III [$d \cdot T_s < t < T_s/2$], S_2 and S_4 are ON. The voltage across L_{leak} is $-V_o/n$ and, consequently, i_{leak} linearly falls until it reaches I_{P2}. D_1 still conducts because i_{leak} is still positive.

- At the beginning of Stage IV ($t=T_s/2$), S_4 is turned-off and, since i_{leak} is positive at that moment, S_3 is turned-on with ZVS. Therefore, during Stage IV [$T_s/2 < t < (1/2+d') \cdot T_s$], S_2 and S_3 are ON, while diode D_1 is conducting. C_1 is charged and a voltage equal to [$-V_g/2 - V_o/n$] is applied across L_{leak}. As a result, i_{leak} linearly falls and Stage IV ends when i_{leak} reaches 0.

- During Stage V [$(1/2+d') \cdot T_s < t < (1/2+d) \cdot T_s$], S_2 and S_3 are still ON, but now D_2 conducts because i_{leak} is negative. C_1 is discharged and a voltage of [$-V_g/2 + V_o/n$] is applied across L_{leak}. Therefore, i_{leak} continues decreasing with a different slope. Stage V ends when it reaches $-I_{P1}$.

- At the beginning of Stage VI ($t=1/2+d \cdot T_s$), S_3 is turned-off and, since i_{leak} is negative at that moment, S_4 is turned-on with ZVS. Hence, during Stage VI [$(1/2+d) \cdot T_s < t < T_s$], S_2 and S_4 are ON and the voltage across L_{leak} is V_o/n. At the end of the stage, S_2 is turned-off and a new switching period begins.

The output current (I_o) can be obtained by averaging the current injected through the output RC network (i_{RC}):

$$I_o = \frac{T_s}{4 \cdot L_{leak} \cdot n} \cdot \left[V_g \cdot d \cdot (1-d) - \frac{\left(\frac{V_o}{n} \right)^2}{V_g} \right], \quad (4)$$

where, V_o can be determined by multiplying I_o and R_o, which, after some manipulations, leads to the following voltage gain:

$$M = \frac{V_o}{V_g} = \frac{2 \cdot (1-d) \cdot d}{k + \sqrt{k^2 + 4 \cdot (1-d) \cdot d}}, \quad (5)$$

where, k is an unidimensional parameter defined as:

$$k = \frac{4 \cdot L_{leak} \cdot n^2}{R_o \cdot T_s}. \quad (6)$$

As can be seen, V_o in CCM depends on the load, thus showing an unusual current source behavior. As previously explained, this characteristic makes it highly attractive for some applications, such as battery chargers or LED drivers. The provided analysis is valid if R_o is lower than the load resistance critical value, which follows this expression:

979-8-3315-1612-3/25 $31.00 © 2025 IEEE

$$R_{o-crit} = \frac{2 \cdot L_{leak} \cdot n^2}{T_s \cdot \left(\frac{1}{2} - d_{crit}\right)}, \tag{7}$$

where, d_{crit} is the value of duty cycle at which the converter operates in Boundary Conduction Mode (BCM) and can be defined as:

$$d_{crit} = \frac{V_o}{n \cdot V_g}. \tag{8}$$

While $d > d_{crit}$, the converter will operate in CCM. Otherwise, the converter would operate in DCM.

C. Design guidelines

The design goal is to ensure that the converter operates in CCM over a specified output power range while keeping a fixed output voltage at a constant frequency, f_s. Based on this requirement, the desired critical duty cycle, d_{crit}, is set. Then, by reorganizing equation (8), the turns ratio of the transformer, n, can be determined. Once n is defined, the value of L_{leak}, needed to meet the desired output power, P_o, at a specific duty cycle, d, (which must be lower than 0.5) can be calculated from the output current expression (4).

$$L_{leak} = \frac{V_g \cdot d \cdot (1-d) - \frac{V_o^2}{V_g \cdot n^2}}{4 \cdot n \cdot f_s \cdot I_o}. \tag{9}$$

In Fig. 4, a generic curve illustrating the evolution of P_o in relation to a range of d values is depicted (after having previously selected the values of f_s, n, V_o and L_{leak}). Although an analysis of the converter in DCM is beyond the scope of this paper, the portion of the d-P_o curve corresponding to this conduction mode is also included.

III. Self-Balance Mechanism of the Voltage Across the Switched Capacitor

In contrast to other hybrid switched-capacitors converters, the proposed topology does not require additional circuitry focused on balancing the voltage across the switched capacitor. This benefit can be demonstrated by applying the volt-second balance to L_{leak}. The voltage across L_{leak} in each of the intervals is shown in Table I. Since V_{C1} is unknown for the demonstration, we cannot consider the lengths of the stages indicated in the previous section (i.e., Stage I length equal to Stage IV length, Stage II length equal to Stage V length, etc.). After a large mathematical analysis,

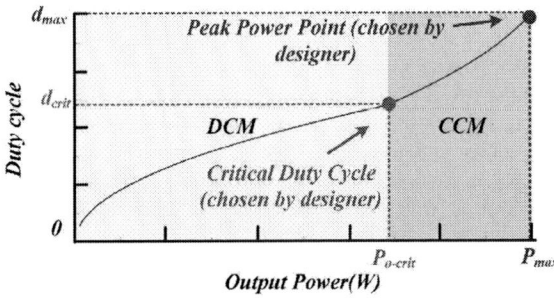

Fig. 4. Relationship between P_o and d for fixed values of f_s, n, V_o and L_{leak}.

TABLE I. Leakage Inductance Voltages

Stage	Value
I	$V_g - V_{C_1} + \dfrac{V_o}{n}$
II	$V_g - V_{C_1} - \dfrac{V_o}{n}$
III	$-\dfrac{V_o}{n}$
IV	$-V_{C_1} - \dfrac{V_o}{n}$
V	$-V_{C_1} + \dfrac{V_o}{n}$
VI	$\dfrac{V_o}{n}$

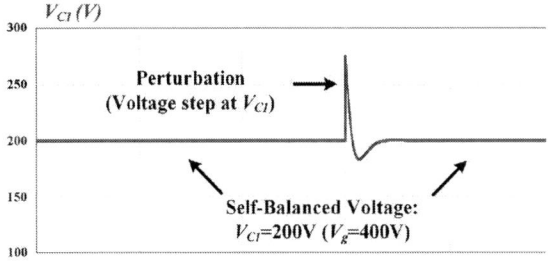

Fig. 5. Validation through simulation (PSIM) of the self-balance mechanism of V_{C1}

those stages lengths can be determined. Then, the volt-second balance can be applied, leading to:

$$V_{C1} = \frac{V_g}{2}. \tag{10}$$

As Fig. 5 shows, C_1 recovers the desired voltage (i.e., $V_g/2$) after any perturbation.

IV. Experimental Results

A. Main voltage and current waveforms

A prototype of the proposed converter was built in order to experimentally validate the described operation and benefits (Table II shows the main components). The converter is designed to handle a peak output power of 500W with a voltage step-down conversion from 400V to 48V at a switching frequency of 100kHz. Fig. 6 shows the main experimental waveforms that validate the described operation. Fig. 6(a) includes the voltage across S_3 (v_{DS3}) and S_4 (v_{DS4}), the voltage across the switched capacitor (i.e., V_{C1}) and the current through the leakage inductance (i.e., i_{leak}) at P_o=500W, whereas Fig 6(b) depicts the same waveforms for P_o=375W, situation in which the converter still operates in CCM, but very close to DCM. This figure allows us to verify the self-balance mechanism of V_{C1} and the reduction of voltage stresses on the main switches compared with the SAB converter. Figure 7(a) shows the turn-off and the turn-on of S_4, including v_{DS4} source and its gate-to-source voltage (v_{GS4}), whilst Figure 7(b) shows the same case for S_3. As can be noticed, ZVS is achieved at the turn-on of both switches.

B. Efficiency tests

The efficiency of the proposed converter was measured by setting a fixed value of the output voltage V_o of 48V for different processed power levels (different output currents) as shown in Fig.4. To meet this goal, both the load value and the converter's duty cycle were adjusted for each case. Efficiency measurements were only taken at points where the converter operates in CCM. As shown in Fig. 8, the highest efficiency was obtained at an output power level of 400W (I_o=8.34A), although the efficiency curve remains similar in every measured point. The maximum recorded efficiency is 94.1%, while at full load, the efficiency achieved is 93%.

V. CONCLUSIONS AND FUTURE WORK

In this paper, a novel switched-capacitor topology derived from the SAB converter has been proposed to address high step-down dc-dc conversions with galvanic isolation, a single magnetic component and constant frequency operation. The converter achieves ZVS in the four MOSFETs, along with reduced voltage stress in three

TABLE II. PROTOTYPE COMPONENTS.

Component	Value
S_1, S_2, S_3, S_4	IPP65R125C7 (650V, $R_{DS-ON,max}$=125mΩ)
D_1, D_2	MBRB40250G (250V)
C_1	C4AQOBU5100M12J (10uF)
C_o	B32562H1106 (10uF)
Driver	ADUM3123CRZ
Transformer	n =1/2 (n_1=16, n_2=8), L_{leak} = 35μF, L_{mag}=760uF

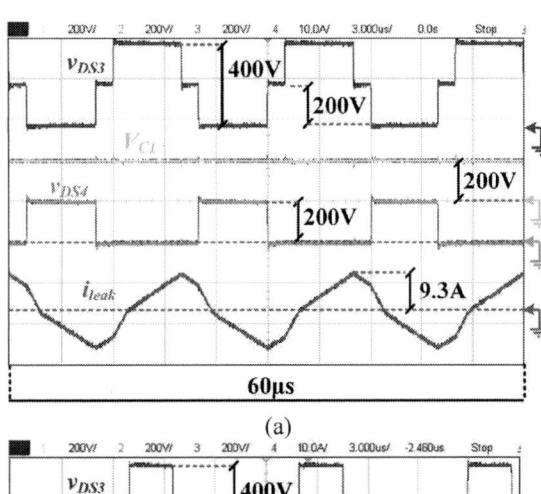

(a)

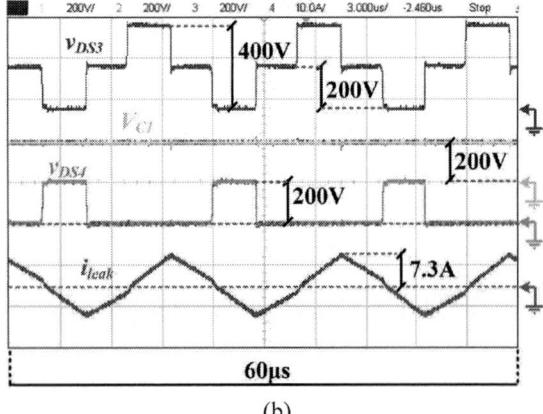

(b)

Fig. 6. Main experimental waveforms: v_{DS3}, v_{DS4}, V_{C1} and i_{leak}: a) P_o=500W. b) P_o=375W.

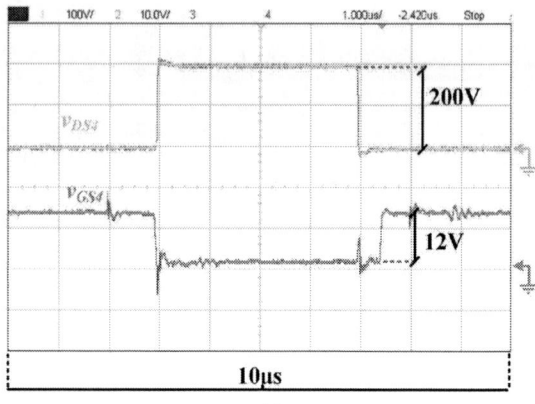

(a)

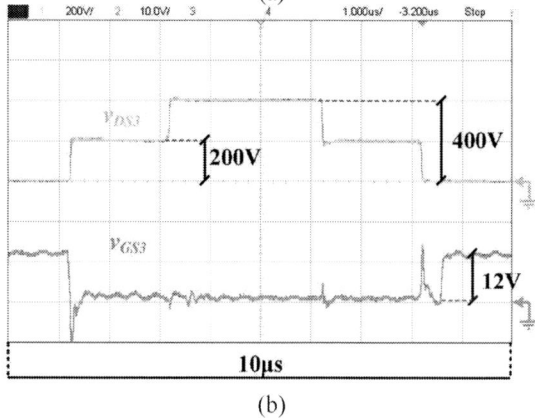

(b)

Fig. 7. ZVS experimental waveforms (P_o=500W): a) Demonstration of ZVS on S_4.(V_{DS4} and V_{GS4}) b) Demonstration of ZVS on S_3 (V_{DS3} and V_{GS3}).

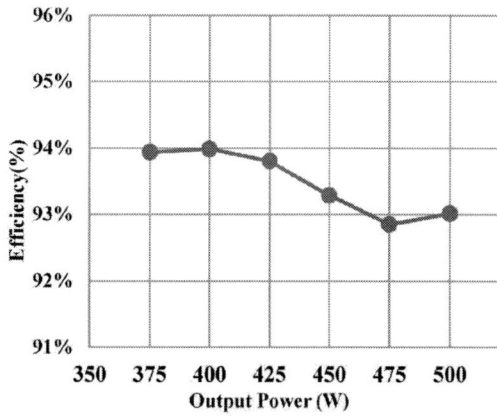

Fig. 8: Measured efficiency of the proposed converter for different output power levels (V_g=400V).

of them. A design guide of the converter to operate in CCM for a certain output power range is provided and the self-balance mechanism of the switched-capacitor voltage, which simplifies the design by not requiring any control circuitry, is demonstrated. The converter has been evaluated using a peak output power of 500W and 400V-to-48V experimental prototype that achieves a peak efficiency of 94,1% and a full load efficiency of 93% .

ACKNOWLEDGMENT

This work was supported in part by the Spanish Government under Project PID2022-136969OB-I00 and "Formación de Personal Investigador" (FPI) Program Grant no. PRE2022-000348.

REFERENCES

[1] I. D. Jitaru, "A 3 kW soft switching DC-DC converter," *APEC 2000. Fifteenth Annual IEEE Applied Power Electronics Conference and Exposition (Cat. No.00CH37058)*, New Orleans, LA, USA, 2000, pp. 86-92 vol.1.

[2] D. S. Gautam, F. Musavi, W. Eberle and W. G. Dunford, "A Zero-Voltage Switching Full-Bridge DC--DC Converter With Capacitive Output Filter for Plug-In Hybrid Electric Vehicle Battery Charging," in *IEEE Transactions on Power Electronics*, vol. 28, no. 12, pp. 5728-5735, Dec. 2013.

[3] K. Domoto, Y. Ishizuka, S. Abe and T. Ninomiya, "Output-inductor-less full-bridge converter with SiC-MOSFETs for low noise and ZVS operation," *2016 IEEE Applied Power Electronics Conference and Exposition (APEC)*, Long Beach, CA, USA, 2016, pp. 2422-2429.

[4] A. Rodríguez, A. A. Gomez, M. M. Hernando, D. G. Lamar and J. Sebastian, "A Dynamic study of the Single Active Bridge Converter," in IEEE Transactions on Power Electronics, doi: 10.1109/TPEL.2024.3426590.

[5] S. Khatua, D. Kastha and S. Kapat, "A new single-stage 48-V-input VRM topology using an isolated stacked half-bridge converter", *IEEE Trans. Power Electron.*, vol. 35, no. 11, pp. 11976-11987, Nov. 2020.

[6] S. -H. Kim, H. Cha, H. F. Ahmed, B. Choi and H. -G. Kim, "Isolated Double Step-Down DC–DC Converter With Improved ZVS Range and No Transformer Saturation Problem," in *IEEE Transactions on Power Electronics*, vol. 32, no. 3, pp. 1792-1804, March 2017.

[7] S. -H. Kim, H. Cha, H. F. Ahmed, B. Choi and H. -G. Kim, "Isolated Double Step-Down DC–DC Converter With Improved ZVS Range and No Transformer Saturation Problem," in *IEEE Transactions on Power Electronics*, vol. 32, no. 3, pp. 1792-1804, March 2017.

[8] X. Li *et al.*, "A Novel Series Capacitor Isolated DC–DC Converter With Reduced Voltage Stress of Primary Switches, Full-Range ZVS Operation, and Improved Light-Load Efficiency," in *IEEE Transactions on Power Electronics*, vol. 39, no. 1, pp. 1046-1059, Jan. 2024.

Analysis and Design of a Low-Complexity ZVS Buck-Boost Converter

Burkhard Ulrich
Electronics & Drives
Reutlingen University
Reutlingen, Germany
burkhard.ulrich@reutlingen-university.de

Abstract— A circuit structure and operating method for achieving a soft-switching operation of a non-inverting buck-boost converter is proposed. The converter is derived from the two-switch buck-boost converter. It achieves ZVS by replacing a diode with an active switch, driven with a gate signal complementary to the main transistors' control signals. In this way, the inductor current can be clamped and used to achieve soft-switching transitions of all transistors in the converter. The circuit achieves ZVS by design, and the ZVS range can be adjusted by adding a capacitor parallel to a diode or switch. The converter does not need a complex control or sensor circuit for proper operation. The operating principle and converter design are described in detail. The proposed converter operation is verified using simulations and experimental results.

Keywords—ZVS, dc-dc converter, resonant, soft-switching

I. INTRODUCTION

Applications powered by solar cells, batteries, or fuel cells require dc-dc converters with a wide input range, providing a constant output voltage where input and output voltage ranges overlap. The non-inverting buck-boost converter [1],[2] is a suitable solution for these applications, and as with other power converters, a high efficiency and a high power density are desired here. Therefore, soft-switching converters, especially zero-voltage switching (ZVS), seem favorable because they reduce switching losses and mitigate electromagnetic emissions (EMI). However, achieving ZVS can be challenging over a wide input and load range and usually requires complex control approaches or additional components. In the literature, several approaches to achieve soft-switching for the non-inverting buck-boost converter have been proposed [3]-[8], which all rely on using a four-switch buck-boost converter. In [3], ZVS is achieved by an additional coupled-inductor snubber network, which is added to both half-bridges to achieve ZVS. In [4], a four-switch converter using an additional parallel capacitance across the inductor and a dedicated control approach is used. The converters described in [4]-[8] use complex calculations, incorporating several measured electrical quantities, such as input and output voltages and currents, to derive the switch's drive signals to achieve ZVS.

In contrast, this article aims to describe a possible solution to achieve a ZVS operation by employing a modified non-inverting buck-boost converter using three active switches and a diode, achieving ZVS over a wide operating range (i.e., input voltage and load variations) while reducing the circuit and control complexity compared to existing solutions [3]-[9].

II. PROPOSED CONVERTER CIRCUIT AND OPERATION

A. Converter Circuit Structure

Fig. 1 shows the converter considered in this article. It is derived from a non-inverting buck-boost converter. In contrast to the conventional two-switch buck-boost converter, the circuit considered here uses three active controllable switches: T_1, T_2, T_{CL}, and a diode D. Therefore, one of the diodes of the two-switch converter is replaced by an active-controlled switch. The additional switch T_{CL} forms, together with the inverse diode of transistor T_2, a clamp-switch network similar to the converters discussed [9]-[11]. A capacitor C_{ZVS1} or C_{ZVS2} is also placed either across T_2 or D. This capacitor allows the ZVS range of the converter to be adjusted, as will be discussed in this article; the idea behind this is similar to the addition of such a capacitance in a boost converter introduced in [11].

B. Basic Converter Operation

This article considers a simple PWM control of the converter. The corresponding idealized control signals are depicted in **Fig. 2**. The transistors T_1 and T_2 are driven with the same PWM control signal at a constant switching frequency $f_S = 1/T_S$ and with a duty cycle D. The additional switch T_{CL} is driven by a complementary PWM signal, with inserted dead times (cf. to intervals $\Delta t_{d,1}$ and $\Delta t_{d,2}$ in **Fig. 2**). Therefore, the signal for T_{CL} is directly derived from the main control signal and together with the addition of the dead times, a ZVS operation of the converter can be achieved, without the need for additional sensors or computations. The converter is designed

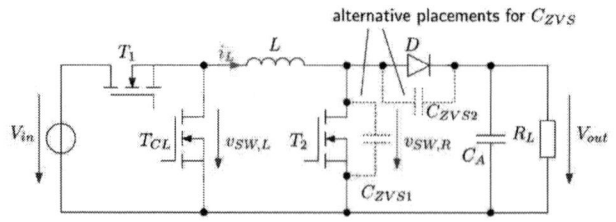

Fig. 1. Proposed three-switch non-inverting buck-boost converter

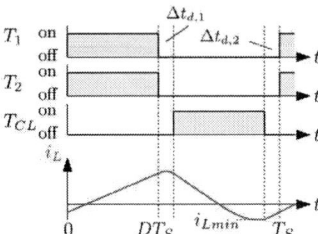

Fig. 2. Idealized control signals and inductor current

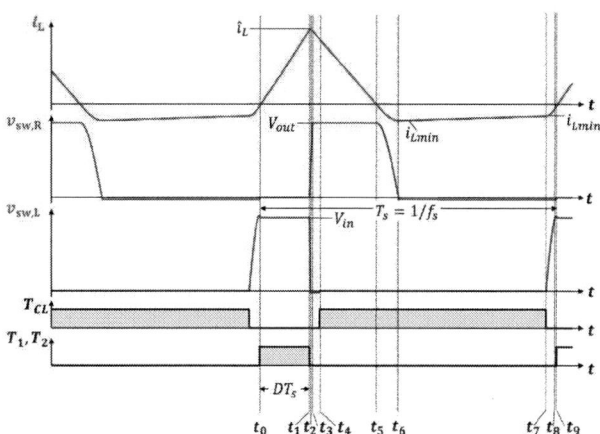

Fig. 3. Major simulated converter waveforms

with such a low inductance value that the inductor current reaches zero during the turn-on phase of T_{CL}. I.e., if a diode replaces T_{CL}, the converter will operate in discontinous conduction mode (DCM). The addition of T_{CL} allows a ZVS operation similar to converters using a clamp-switch, as discussed in [9]-[11]. In these types of converters, the negative peak inductor current (i_{Lmin}) is freewheeling through a series connection of a diode and an active switch (placed across the converter's inductor), and by turning off this clamp-switch, the stored energy is used to achieve a soft-switching ZVS transition. In the buck-boost converter considered here, the clamping action can be achieved using the power switches already present in the topology without additional components. The resulting idealized inductor current i_L is shown in the lower trace of **Fig. 2** and will be discussed in more detail in the next section.

III. CONVERTER OPERATION STATES

In this section, the converter operation in a steady state is described using **Fig. 3** and **Fig. 4**. **Fig. 3** shows selected waveforms from a simulation using ideal devices, where a parallel connection of an ideal switch, a diode, and a capacitor models all transistors. **Fig. 4** depicts the corresponding nine converter states during a switching period in a steady state.

State 1: $t_0 < t < t_1$

At t_0, the transistors T_1 and T_2 are turned on simultaneously. Prior to this, the inverse diodes of both transistors conducted the

current, leading to a ZVS turn-on of T_1 and T_2. The input voltage V_{in} is applied across L, leading to an increase in the inductor current i_L. Therefore, in state 1, energy is stored in the inductor.

State 2: $t_1 < t < t_2$

After the turn-off of T_1 and T_2, at t_1, the inductor current discharges and charges both the capacitors at the left and right switch node voltage in a resonant manner. Note that these transitions will not end simultaneously, as a different amount of charge $Q_{sw,L} = V_{in} \cdot C_{sw,L}$ and $Q_{sw,R} = V_{out} \cdot C_{sw,R}$, respectively, are needed to change the voltages on both switch nodes entirely. Here, $C_{sw,L}$ and $C_{sw,R}$ are the sum of the charge equivalent capacitance values of the semiconductor device and parasitic components acting at the left and right switch nodes, respectively.

State 3: $t_2 < t < t_3$

Assuming that a larger charge $Q_{sw,R} > Q_{sw,L}$ is required to charge the right switch node, then at first, the inverse diode of T_{CL} is

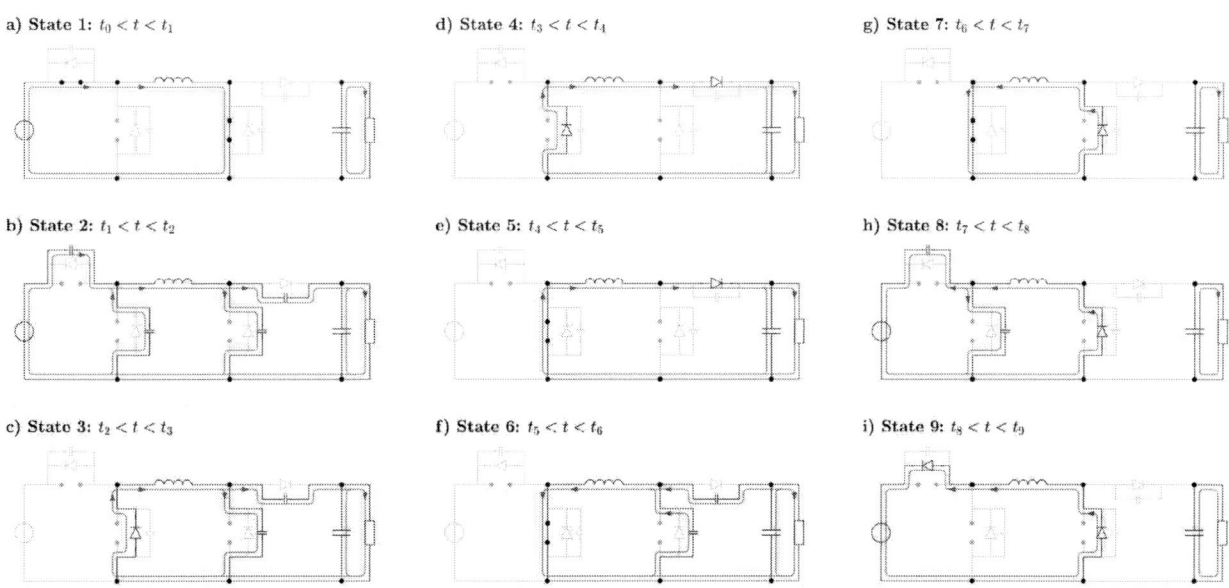

Fig. 4. Converter operating states

turned on at t_3. This turn-on action clamps the voltage $v_{sw,L}$ to a low negative value of $V_{FW,DTCL}$ (forward voltage of T_{CL}'s inverse diode). The resonant transition on the right switch node will continue until D gets forward-biased at t_3. I.e., $v_{sw,R}$ reaches a value $V_{out} + V_{FW,D}$ (forward voltage of D) at time instant t_3.

State 4: $t_3 < t < t_4$

This state is similar to the turn-off interval of a two-switch buck-boost converter. Both the output diode D and the inverse diode D_{TCL} of T_{CL} are conducting during this interval. Therefore, the negative output voltage -V_{out} is applied across the inductor, decreasing the inductor current i_L.

State 5: $t_4 < t < t_5$

At the time instant t_4, the clamp-transistor T_{CL} is turned on. Before t_4, the inverse diode of T_{CL} has already conducted the current; therefore, this turn-on occurs with ZVS. The basic operation is unchanged compared to state 4. However, this turn-on must take place before the zero crossing of the inductor current. This timing is necessary to clamp the inductor current and achieve a ZVS operation of T_1 later.

State 6: $t_5 < t < t_6$

At t_5, the inductor current i_L reaches zero. Therefore, the output diode D turns off, initiating a resonant discharge of the capacitors connected to the right switch node. The voltage on the left switch node remains unchanged, as the T_{CL} clamps it to zero volts. The resonant transition lasts until the right switch node voltages reaches zero and the inductor current reaches it's negative peak value i_{Lmin} at t_6.

State 7: $t_6 < t < t_7$

At t_6, the inverse diode of T_2 gets forward-biased. Therefore, the inductor current freewheels through the series connection of T_{CL} and the inverse diode of T_2. The voltage across the inductor is now clamped to a small value – the combined voltage drop of T_{CL} and the forward voltage of the inverse diode T_2. As a consequence, the inductor current is now freewheeling at a negative value of approximately i_{Lmin}. Actually, the current slowly increases due to the small positive voltage applied across L as the inductor demagnetizes.

State 8: $t_7 < t < t_8$

Assuming that the current i_L is, at t_7, still at a negative value i'_{Lmin}, a resonant transition can occur as T_{CL} turns off at t_7. Afterward, the capacitances connected at the left switch node are charged in a resonant manner, leading to an increase in the voltage $v_{sw,L}$. If the stored energy in L is larger than the energy needed to discharge these capacitors by V_{in}, then the inductor current i_L will still be negative when the body diode of T_1 turns on at t_8.

State 9: $t_8 < t < t_9$

Both body diodes of T_1 and T_2 will conduct simultaneously after t_8. Therefore, at t_9, transistors T_1 and T_2 can be turned on simultaneously, but only with the forward diode voltage drops across them; The turn-on process will consequently occur with ZVS. Therefore, with the proposed operating scheme, a ZVS of all transistors is possible by design and without additional sensors.

IV. Converter Design For ZVS

The discussion in the preceding section assumes that the circuit is designed correctly to achieve ZVS. This aspect will be addressed in this section. Achieving ZVS of all transistors relies basically on the following:

1) Operating the converter with such low inductance value L that the diode D turns off prior to the turn on of T_1 and T_2. I.e., designing the converter such that it would operate in discontinuous conduction mode if a diode replaced T_{CL}.
2) Having a sufficient negative current i'_{Lmin} and, therefore, enough energy stored in L at the turn-off of the clamp transistor T_{CL} to fully discharge the capacitances at the left-switch node. This current value can be adjusted by adding a proper capacitance C_{ZVS} at the right switch node.
3) Using proper timing regarding the dead times.

The following discussion assumes that the capacitances and inductances can be (simplified) modeled by ideal linear lumped elements. E.g., at the left switch node, the resultant capacitance $C_{sw,L}$ is

$$C_{sw,L} = C_{TCL} + C_{T1} + C_{par,L}, \tag{1}$$

where C_{TCL} and C_{T1} represent the capacitances of T_{CL} and T_1, respectively, and $C_{par,L}$ represents all other parasitic influences acting at the left switch node, such as layout capacitance and winding capacitance of L. Similarly, the linear equivalent capacitance $C_{sw,R}$ at the right switch node will be

$$C_{sw,R} = C_{T2} + C_D + C_{par,R} + C_{ZVS}, \tag{2}$$

where C_{T2} and C_D are the capacitances of T_2 and output diode D. $C_{par,R}$ represents the other parasitic influences and C_{ZVS} is an additional added capacitor either in position C_{ZVS1} or C_{ZVS2} as shown in **Fig. 1**.

A. Inductance Value Selection

The inductance value depends on the maximum output power P_{out}, the switching frequency f_S, and the input and output voltages V_{in} and V_{out}. Using the simplified assumption that the converter will operate at the maximum output power at the boundary between DCM and continuous conduction mode (CCM), the following equation can be derived:

$$L < \frac{1}{2 \cdot \frac{P_{out}}{\eta} \cdot f_S} \left(\frac{V_{in} V_{out}}{V_{in} + V_{out}} \right)^2. \tag{3}$$

This equation can be used to determine the maximum allowed inductance value. Assuming a constant V_{out} and varying V_{in}, the worst-case value will occur at minimum V_{in} and maximum P_{out}. In **Fig. 5**, the evaluation of (3) is depicted for parameters similar to the prototype presented in section V. As can be seen, a maximum inductance value of approximately 2.9 µH is allowed to achieve a ZVS operation across the here considered input voltage range when assuming an efficiency of $\eta = 0.9$.

979-8-3315-1612-3/25 $31.00 © 2025 IEEE   1709

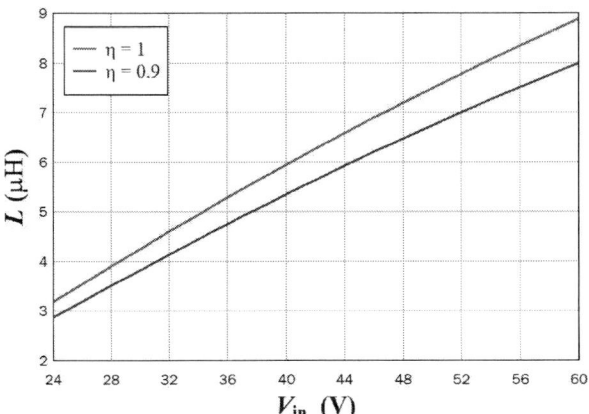

Fig. 5. Evaluation of (3) for two different efficiencies a) $\eta = 1$ (red) and b) $\eta = 0.9$ (blue) for $V_{in} = 24 \ldots 60$ V, $V_{out} = 48$ V, $P_{out} = 100$ W and $f_s = 400$ kHz.

B. ZVS Capacitor Selection

The selection of the inductance value according to (3) is a necessary but not solely sufficient condition to achieve a ZVS operation. In addition, to achieve a ZVS of T_1, the energy stored in L must be sufficient to fully discharge the capacitance $C_{sw,R}$ during state 8 by the input voltage V_{in}. Therefore, the following condition for a ZVS of T_1 can be derived

$$L \cdot i'^2_{Lmin} > C_{sw,L} \cdot V^2_{in}. \tag{4}$$

The current $i'_{Lmin} < 0$, is the current at the turn off of T_{CL} at t_7, which can be expressed as

$$i'_{Lmin} = i_{Lmin} + \frac{V_{FW,DT2} + V_{TCL}}{L} \cdot (t_7 - t_6). \tag{5}$$

Here, the second term $V_{FW,DT2} + V_{TCL}$ represents the combined voltage drop of T_{CL} and the inverse diode of T_2 during the clamping interval (cf. state 7 in **Fig. 4**), which leads to demagnetization of L. Neglecting the latter effect, the following simplified relationship can be derived

$$i'_{Lmin} \approx i_{Lmin} = -V_{out}\sqrt{\frac{C_{sw,R}}{L}} = -V_{out}\sqrt{\frac{C_{T2}+C_D+C_{par,R}+C_{ZVS}}{L}}. \tag{6}$$

From (6), it can be seen that the minimum current i_{Lmin} and, therefore, i'_{Lmin} can be adjusted (i.e., increased) by adding a capacitor C_{ZVS} to the right switch node. If the right-hand side (6) is substituted into (4) and also the definition (1) for $C_{sw,R}$ is considered, then the following design equation can be derived for selecting C_{ZVS}:

$$C_{ZVS} > \left(\frac{V_{in}}{V_{out}}\right)^2 \cdot (C_{T1} + C_{TCL} + C_{par,L}) - C_D - C_{T2} - C_{par,R}. \tag{7}$$

The design equation (7) should be evaluated for maximum V_{in} as the required energy will increase with a larger V_{in}.

Additionally, it has to be considered that the semiconductor device capacitors are strongly voltage-dependent and also different for the devices, as the maximum voltage for T_1 and T_{CL} is V_{in}, but for T_2 and D, it is V_{out}. Using the components of the prototype converter in the following section, the values of the capacitances are as follows: $C_{T1} = C_{TCL} \approx 830$ pF (evaluated at $V_{DS} = 60$ V), $C_{T2} \approx 940$ pF (at $V_{DS} = 48$ V) and $C_D \approx 81$ pF (at $V_R = 48$ V). The given values have been extracted using the curves given in the datasheets [12], [13] and calculating the charge equivalent capacitances as described by

$$C_{T,Q,eq} = \frac{\int_0^{V_{DS}} C_{oss}(v_{DS})dv_{DS}}{V_{DS}} \quad \text{and} \tag{8a}$$

$$C_{D,Q,eq} = \frac{\int_0^{V_R} C_j(v_R)dv_R}{V_R}. \tag{8b}$$

With these values, and neglecting $C_{par,L}$ and $C_{par,R}$, a value of $C_{ZVS} > 1.53$ nF has been calculated, which is required to achieve a ZVS operation for a V_{in} ranging from 24 V to 60 V at $V_{out} = 48$ V. It should be noted that, in reality, a larger value for C_{ZVS} will be required, as the demagnetizing during state 7 is neglected.

C. Setting proper dead times

Selecting the proper timing is an additional requirement to achieve a ZVS operation. In the converter here, the different transistors have different strict timing requirements. The most severe requirement regards transistor T_1. Its turn-on should be delayed by an amount $\Delta t_{d2} = t_9 - t_7$, which is, in the optimal case, equivalent to one-quarter of the resultant resonance period determined by $C_{sw,L}$, and L. Therefore, the following optimal value is obtained

$$\Delta t_{D2,opt} = \frac{\pi}{4}\sqrt{L \cdot C_{sw,L}}. \tag{9}$$

T_1 will turn on with the minimal drain-source voltage if this value is used. The timing requirements for transistors T_2 and T_{CL} are less strict than for T_1. Switch T_2 can be turned on at any time after the start of the freewheeling phase at t_6, as its inverse diode will conduct the inductor current during this phase. Therefore, T_2 achieves ZVS if condition (4) is met regardless of the timing if the inductor current is still negative at T_2's turn-on. For transistor T_{CL}, the timing requirements are also less strict as T_{CL} replaces a diode and acts as a synchronous rectifier in the original non-inverting buck-boost converter circuit. The required dead time $\Delta t_{d1} = t_4 - t_1$ will always be lower than the value given by (9) and will decrease with increasing output power and, therefore, with increasing peak inductor current. Therefore, a practical approach is to choose equal dead times:

$$\Delta t_{D1} = \Delta t_{D2} = \Delta t_{D2,opt}. \tag{10}$$

Note that this will increase conduction losses in the inverse diode of T_{CL} before the switch turns on but allows a simple implementation.

V. FURTHER DESIGN EQUATIONS

In this section some design equations are presented to allow a proper selection of the components. These equations are derived from the simplified inductor current waveform i_L presented in **Fig. 6**. The waveform is divided into four distinct phases: two linear segments ($D_1 T_S$ and $D_2 T_S$), one (quarter wave) sinusoidal segment ($D_3 T_S$), and one segment, where the current is constant at i_{Lmin} ($D_4 T_S$). The key parameters for this waveform are summarized in **Table I**.

TABLE I.   PARAMETERS INDUCTOR CURRENT

parameter	value
$\hat{i}_L$	$\sqrt{2 \cdot P_{out}/(L \cdot f_S)}$
i_{Lmin}	$-V_{out}\sqrt{C_{sw,R}/L}$
D_1	$\dfrac{V_{out}}{V_{in}}\left(\dfrac{\sqrt{2 \cdot P_{out} \cdot L \cdot f_S}}{V_{out}} + f_s\sqrt{L \cdot C_{sw,R}}\right)$
D_2	$\dfrac{\sqrt{2 \cdot P_{out} \cdot L \cdot f_S}}{V_{out}}$
D_3	$f_S \cdot \dfrac{\pi}{2}\sqrt{C_{sw,R}/L}$
D_4	$1 - D_1 - D_2 - D_3$

The current waveforms of the four semiconductor devices can be derived from the inductor current i_L, as depicted in **Fig. 7**, and their respective average and RMS values can be expressed using the analytical expressions provided in **Table I**. These calculated current values are, together with the maximum device voltage stress (V_{max}), summarized in **Table II**. Note that in the second row, the values for the inverse diode of transistor T_2 are shown separately, as this diode will conduct the inductor current when T_2 is off during the interval $D_4 T_S$ (cf. to **Fig. 6**).

TABLE II.   COMPONENT STRESSES

device	V_{max}	RMS current	AVG current
T_1, T_2	V_{in}, V_{out}	$\sqrt{\dfrac{D_1}{3}} \cdot (\hat{i}_L - i_{Lmin})$	
T_2 (diode)	V_{out}	$\sqrt{D_4} \cdot i_{Lmin}$	$D_4 \cdot i_{Lmin}$
T_{CL}	V_{in}	$\sqrt{\dfrac{D_2}{3}\hat{i}_L^2 + \left(\dfrac{D_3}{2} + D_4\right) i_{Lmin}^2}$	
D	V_{out}	$\sqrt{\dfrac{D_2}{3}} \cdot \hat{i}_L$	$\dfrac{1}{2}D_2 \cdot \hat{i}_L$

VI. PROTOTYPE AND EXPERIMENTAL RESULTS

A. Prototype Converter

A low-voltage and low-power prototype converter has been built to validate the proposed converter and its operation. The converter specifications are summarized in **Table III**, and the major components used are listed in **Table IV**. In **Fig. 8** a schematic representation of the prototype is shown, and the power stage printed circuit board (PCB) is depicted in **Fig. 9**. As capacitance C_{ZVS} a 4.7 nF ceramic capacitor with C0G dielectric was chosen to ensure ZVS up to the maximum input

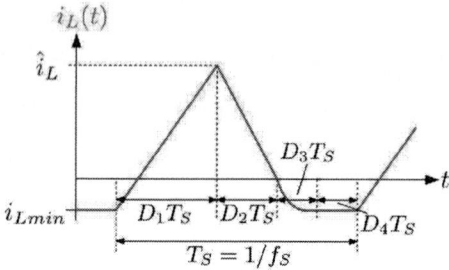

Fig. 6.   Simplified inductor current waveform, neglecting the dead time intervals and the demagnetization during the freewheeling phase.

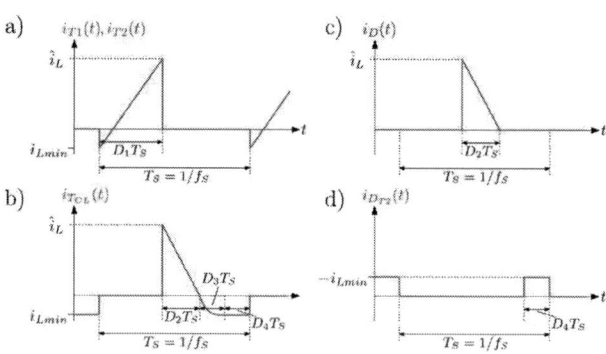

Fig. 7.   Idealized current waveforms for the semiconductor devices a) transistors T_1 and T_2, b) transistor T_{CL}, c) output diode D and d) inverse diode of T_2.

voltage of 60 V. The converter uses a voltage mode control scheme, as shown in the lower part of **Fig. 8**, to generate the complementary PWM drive signals for T_1, T_2, and T_{CL} at a constant switching frequency $f_S = 400$ kHz. The control concept is implemented using a PSoC5LP microcontroller eval board [14] but could also be implemented using analog circuitry.

TABLE III.   CONVERTER SPECIFICATIONS

parameter	value
switching frequency f_S	400 kHz
input voltage range V_{in}	24 V … 60 V
output voltage V_{out}	48 V
maximum output power P_{out}	100 W

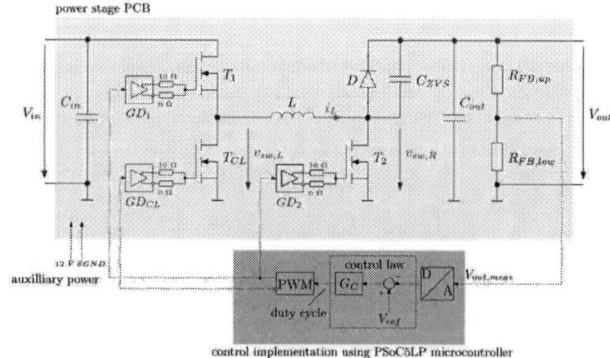

Fig. 8.   Prototype schematic

979-8-3315-1612-3/25 $31.00 © 2025 IEEE

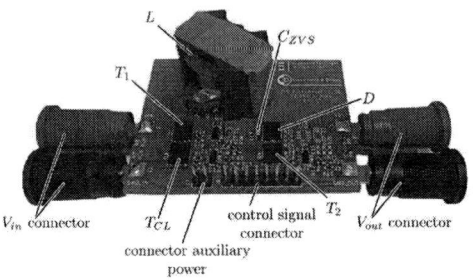

Fig. 9. Prototype converter power stage PCB (size 60 mm 45.5 mm). Not shown is the external controller board.

TABLE IV. PROTOTYPE CONVERTER COMPONENTS

component	value
transistors T_1, T_2 and T_{CL}	ISC060N10NM6 (Infineon) [12]
output diode D	MBR8H100MFS (onsemi) [13]
inductor L	2.8 uH / 2x4 windings parallel / 90x0.1 mm litz wire / core RM10 / material TDK N49
gate driver ICs: GD_1, GD_2, GD_{CL}	1EDN7550 (Infineon)
controller	PSoC5LP evaluation board [14] CY8CKIT-059
additional capacitor C_{ZVS}	4.7 nF / 100 V / ceramic C0G / size 0805
input capacitor C_{in}	4x10 uF in parallel / 100 V/ ceramic X7R / size 1206
output capacitor C_{out}	4x10 uF in parallel / 100 V/ ceramic X7R / size 1206

The microcontroller is only used for convenience to easily adapt the controller settings such as dead times, switching frequency and reference voltage, without needing to exchange components. For implementation of the voltage mode control approach the output voltage V_{out} is measured using a voltage divider and digitized using an analog to digital converter. This measured value is compared against an internal reference value V_{ref} to generate the control error signal, which is used to calculate the duty cycle value using a digital implemented control law. The duty cycle value is then used as input to a digital PWM module, which generates the complementary drive signals for all transistors and also inserts the required (symmetrical) dead times.

B. Experimental Results

Fig. 10a and **10b** present experimental waveforms that resemble the simulation results in **Fig. 3** of the converter. Here,

for the input voltage values of $V_{in} = 24$ V and $V_{in} = 60$ V, the voltages at both sides of the inductor, $v_{sw,L}$ and $v_{sw,R}$, are depicted, together with the inductor current i_L. In **Fig. 11a-c**, the switching process of the three active switches (T_1, T_2, and T_{CL}) is shown in more detail. In all cases, the gate-source voltage of the MOSFETs is depicted together with either the voltage $v_{sw,L}$ (for T_1 and T_{CL}) or $v_{sw,R}$ (for T_2). These waveforms demonstrate a successful ZVS operation for each of the transistors. As can be seen, the respective gate-source voltage is applied after the drain-source voltage has decreased to zero, and no Miller plateau is present. **Fig. 11d** shows the measured inductor current i_L and the voltage v_L across the inductor, indicating the clamping action. Note that although the voltage waveform is similar to a converter operating in DCM, the inductor current stays constant as the voltage is at zero, and the typical (low frequency) ringing in the inductor voltage and current waveforms is not present.

The measured waveforms reveal that the right switch node voltage $v_{sw,R}$ exhibits a high-frequency ringing. This effect is due to the parasitic components of the output power loop and would also occur in a hard-switched converter operating in DCM. A possible reduction for this ringing, which will also be present in the output voltage V_{out}, as the decoupling capacitor of the power loop, is the capacitor C_{out}, could be the addition of an LC filter (using a small inductor or ferrite bead) connected at the output. However, it is possible to influence the ringing by choosing the right placement of C_{ZVS}, as is briefly described in the Appendix.

Fig. 12 depicts the converter power stage's measured efficiency and power loss for different V_{in} values and varying output loads. From the curves, it can be deduced that the converter reaches a constant efficiency of over 95 %, and therefore linear increasing losses, for all V_{in} values, regardless of the input voltage. Also, it can be seen that the efficiency drops at high output power for the measurement at $V_{in} = 24$ V; this is due to the loss of ZVS at large duty cycles. In this case, the negative inductor current can no longer fully discharge the capacitors, and switching losses will increase. This effect is increased due to added C_{ZVS} capacitance, which will eventually discharge through T_2, strongly increasing its switching losses. While the inductor has been designed according to (3), this relationship is simplified as it assumes operation at the boundary between DCM and CCM. Consequently, the

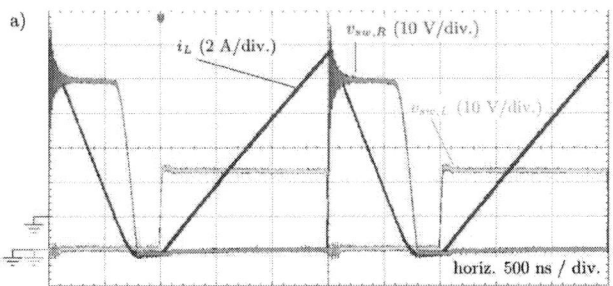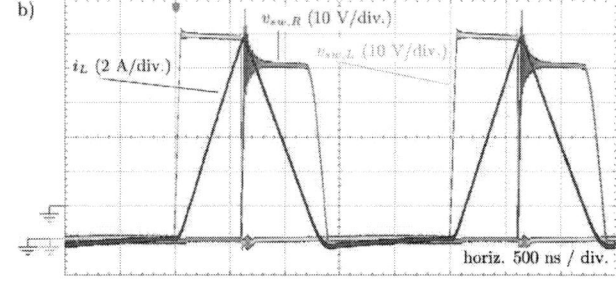

Fig. 10. Measured protototype waveforms for $v_{sw,L}$, $v_{sw,R}$ and i_L for an output current of $I_{out} = 1$ A and $V_{out} = 48$ V a) for $V_{in} = 24$ V and b) for $V_{in} = 60$ V.

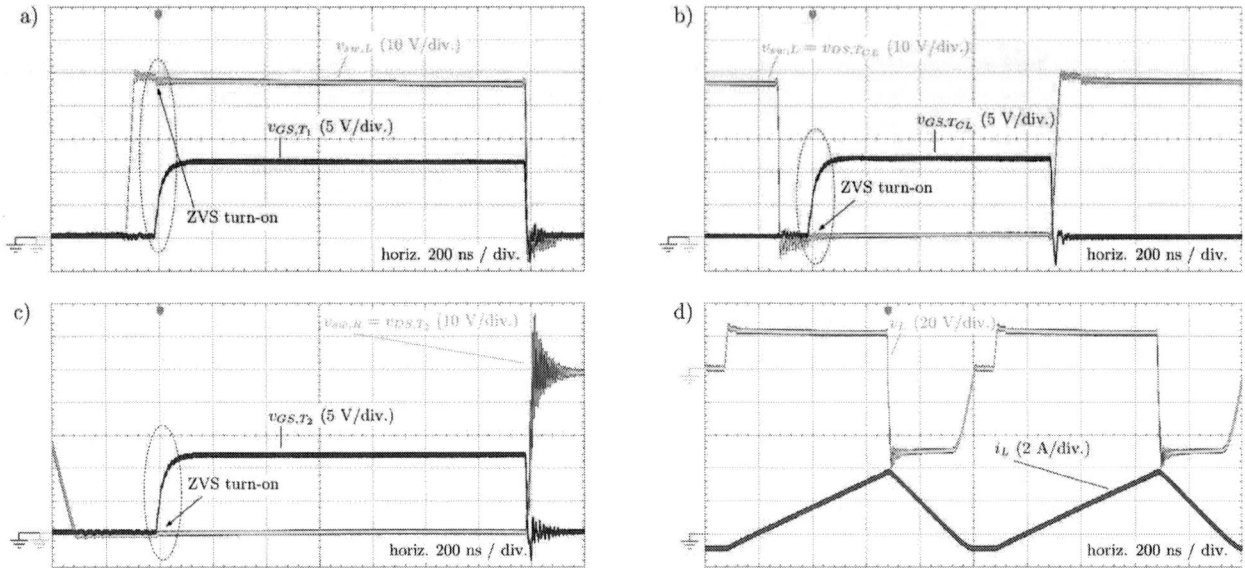

Fig. 11. Detail views of a) switching process of T_1, b) switching process of T_{CL}, c) switching process of T_2 and d) measured inductor current i_L and voltage v_L. All measurements for an output current of $I_{out} = 1$ A, $V_{out} = 48$ V and $V_{in} = 24$ V.

relationship (3) and its evaluation in **Fig. 3** neglect the required times to build up a negative inductor current sufficient for a ZVS operation. To prevent this, the inductance value should be chosen smaller than the value predicted by (3).

Another effect shown in **Fig. 12** is that the loss curve indicates that there will be a constant loss even at zero output power. This constant loss can be attributed to the conduction losses caused by the negative inductor current flowing through T_{CL} and the inverse diode of T_2 during the freewheeling phase.

Compared to commercial products (e.g.: [15]), the efficiency of the converter presented here is with a value of 94 - 95% similar at power levels of 40 W – 72 W (cf. to Fig. 8 in [15]), but lower at than the maximum efficiency of 96.7 % of the product in [15]. Also, the measurements here do not include the auxiliary power. Nevertheless, the proposed converter is scalable and offers a lower complexity. Also, the efficiency could be increased, as described in the following section.

VII. DISCUSSION

A. Comparison to other approaches

The proposed converter and operation principle are similar to the use of clamp switches across the inductor, as reported in [8]-[10]. In contrast to these references, the non-inverting buck-boost converter considered here does not need additional switches, as these are inherent in the topology. Although, for the buck-boost converter, different ZVS approaches have been proposed in the literature [3]-[8], these approaches require more controlled switches, additional circuit components, or a more complicated control. In contrast, the circuit used here only needs the generation of a simple complementary (constant frequency) PWM drive signal with inserted dead times. The addition of a capacitor to enhance the ZVS range follows a similar idea as discussed in [11], but here, a completely

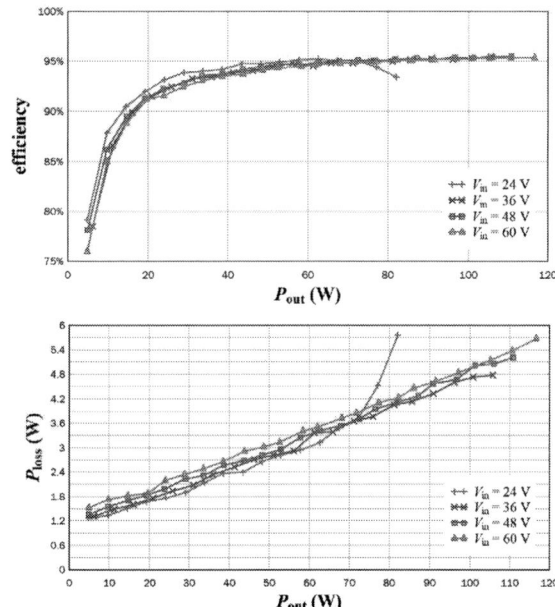

Fig. 12. Measured efficiency a) and losses b) of prototype. Values are only for the power stage, and do not include the auxiliary power for the gate drivers (≈ 0.46 W) and the controller (≈ 0.21 W).

different topology is used. In summary, the structure here allows for a reduced control system and circuit complexity compared to existing approaches.

B. Possible modifications to increase efficiency

The simple operational principle of the converter, although efficient, leaves room for improvement. The major contributors to the losses are the inductor L, output diode D, and transistor T_2. The optimization of the inductor and discussion of its losses are beyond the scope of this article. However, for diode D and switch T_2, a simple approach to reduce losses is the use of synchronous rectification. This method can significantly enhance efficiency, as the diode conduction losses dominate the losses in T_2 and D. E.g., assuming a constant forward voltage 0.8 V for D and the inverse diode of T_2, then for $V_{in} = 60$ V, $I_{out} = 2$ A and $P_{out} = 96$ W, the diode losses will be $P_{L,D} \approx 1.6$ W in D and $P_{L,DT2} \approx 0.547$ W in T_2. The calculated combined loss of 2.15 W represents about 44.3% of the total measured losses (cf. to **Fig. 12**). If both devices would use an active rectification using the same on-resistance value as the devices already in the design (≈ 6 mΩ), then the losses could be reduced to 0.105 W ($P_{L,D}$) and 0.014 W ($P_{L,DT2}$). Therefore, an overall reduction of the losses by 2.03 W at this operating point could be achieved, increasing efficiency by about 2% at the cost of a more complex control.

VIII. CONCLUSION

A novel non-inverting buck-boost converter is proposed, utilizing three active switches that achieve ZVS at a constant switching frequency. This ZVS operation is realized by design with the addition of a single capacitor, eliminating the need for complex control systems or sensor circuitry. A detailed analysis of the converter's operating mode and design criteria for ZVS is presented, with experimental results on a 100 W prototype validating its effectiveness and feasibility. The prototype achieves a peak efficiency of over 95%, underscoring its suitability for high-efficiency power applications.

APPENDIX – C_{ZVS} INFLUENCE ON PARASITIC RINGING

The additional capacitor C_{ZVS} can be placed either across output diode D or in parallel to T_2. Although the placement does not change the basic behavior, it affects the parasitic ringing, which occurs at the turn-on of diode D. The cause of this ringing is the presence of parasitic components, in particular the loop inductance of the right switching loop, which comprises of T_2, D and the output capacitor C_{out}. Depending on the position of C_{ZVS}, there will be a difference in the ringing that occurs. **Fig. 13** illustrates this effect. If C_{ZVS} is in parallel to D, the resultant oscillation starts earlier, has a higher overvoltage value, and the frequency of the oscillation will be higher (blue traces in **Fig. 13**). In contrast, placing C_{ZVS} in parallel to T_2 will

reduce the overvoltage and the frequency of the parasitic ringing. The cause is that the current charging C_{ZVS} will either flow through the output loop (in case $C_{ZVS}\|D$) or not (in case $C_{ZVS}\|T_2$).

REFERENCES

[1] M. Orellana, S. Petibon, B. Estibals and C. Alonso, "Four Switch Buck-Boost Converter for Photovoltaic DC-DC power applications," IECON 2010 - 36th Annual Conference on IEEE Industrial Electronics Society, Glendale, AZ, USA, 2010, pp. 469-474 doi: 10.1109/IECON.2010.5674983.

[2] E. Schaltz, P. O. Rasmussen and A. Khaligh, "Non-inverting buck-boost converter for fuel cell applications," 2008 34th Annual Conference of IEEE Industrial Electronics, Orlando, FL, USA, 2008, pp. 855-860, doi: 10.1109/IECON.2008.4758065.

[3] Y. Zhang, X. -F. Cheng and C. Yin, "A Soft-Switching Non-Inverting Buck–Boost Converter With Efficiency and Performance Improvement," in IEEE Tr.ansactions on Power Electronics, vol. 34, no. 12, pp. 11526-11530, Dec. 2019, doi: 10.1109/TPEL.2019.2920310.

[4] A. Wei, B. Lehman, W. Bowhers and M. Amirabadi, "A soft-switching non-inverting buck-boost converter," 2021 IEEE Applied Power Electronics Conference and Exposition (APEC), Phoenix, AZ, USA, 2021, pp. 1920-1926, doi: 10.1109/APEC42165.2021.9487051.

[5] Z. Zhou, H. Li and X. Wu, "A Constant Frequency ZVS Control System for the Four-Switch Buck–Boost DC-DC Converter With Reduced pInductor Current," in IEEE Transactions on Power Electronics, vol. 34, no. 7, pp. 5996-6003, July 2019, doi: 10.1109/TPEL.2018.2884950.

[6] F. Liu, J. Xu, Z. Chen, R. Huang and X. Chen, "A Constant Frequency ZVS Modulation Scheme for Four-Switch Buck–Boost Converter With Wide Input and Output Voltage Ranges and Reduced Inductor Current," in IEEE Transactions on Industrial Electronics, vol. 70, no. 5, pp. 4931-4941, May 2023, doi: 10.1109/TIE.2022.3187591.

[7] F. Liu, J. Xu, Z. Chen, P. Yang, K. Deng and X. Chen, "A Multi-Frequency PCCM ZVS Modulation Scheme for Optimizing Overall Efficiency of Four-Switch Buck–Boost Converter With Wide Input and Output Voltage Ranges," in IEEE Transactions on Industrial Electronics, vol. 70, no. 12, pp. 12431-12441, Dec. 2023, doi: 10.1109/TIE.2022.3232660.

[8] P. Vinciarelli, "Buck-boost dc-dc switching power conversion", US Patent US 6,788,033, 08-Aug-2002.

[9] O. Knecht, D. Bortis and J. W. Kolar, "ZVS Modulation Scheme for Reduced Complexity Clamp-Switch TCM DC–DC Boost Converter," in IEEE Transactions on Power Electronics, vol. 33, no. 5, pp. 4204-4214, May 2018, doi: 10.1109/TPEL.2017.2720729.

[10] J. Prager and P. Vinciarelli, "Loss and noise reduction in power converters", US Patent US 6,522,108, 18-Feb-2003.

[11] B. Ulrich, "Improved Clamp-Switch Boost Converter with Extended ZVS range," 2021 IEEE Applied Power Electronics Conference and Exposition (APEC), Phoenix, AZ, USA, 2021, pp. 1747-1754, doi: 10.1109/APEC42165.2021.9487229.

[12] OptiMOS 6 Power-Transistor 100V Datasheet, Rev. 2.2, Feb. 07, 2023 [online] Available: https://www.infineon.com/dgdl/Infineon-ISC060N10NM6-DataSheet-v02_02EN.pdf?fileId=8ac78c8c7bb971ed017bb9b35e3600c3. (Accessed: Oct. 29, 2024)

[13] MBR8H100MFS – Switch Mode Power Rectifiers Datasheet, Rev. 5, June 2024, [online]. Available: https://www.onsemi.com/pdf/datasheet/mbr8h100mfs-d.pdf, (Accessed: Oct. 29, 2024)

[14] CY8CKIT-059 PSoC5LP Prototyping Kit Guide, Rev. G, [online]. Available: https://www.infineon.com/cms/de/product/evaluation-boards/cy8ckit-059/, (Accessed: Oct. 29, 2024)

[15] PRMTM Regulator - PRM48AH480x200A00, Rev. 1.6, 09/2020, [online] Available: https://www.vicorpower.com/documents/datasheets/PRM48AH480T200 A00_ds.pdf (accessed: Oct. 29, 2024)

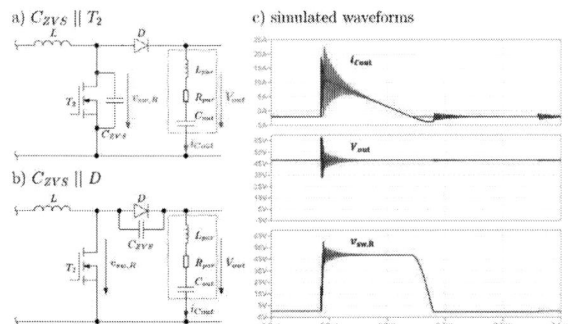

Fig. 13. Influence of C_{ZVS} placement on parasitic ringing. Left equivalent circuits considered. Right simulation results. Red traces correspond to circuit a) ($C_{ZVS}\|T_2$), and blue traces to circuit b) ($C_{ZVS}\|D$). Simulation parameters: $C_{ZVS} = 4.7$ nF, $V_{in} = 60$ V, $L_{par} = 2$ nH, $C_{out} = 40$ µF and $R_{par} = 3$ mΩ.

A High Conversion-Ratio Hybrid Series-Parallel DC-DC Converter with Pseudo-Soft-Charging and Inductor Current Frequency Multiplication

Avinash Maddela, Kishalay Datta and Jason T. Stauth

Thayer School of Engineering at Dartmouth College

Hanover, NH USA

Avinash.Maddela.TH@dartmouth.edu, Kishalay.Datta.TH@dartmouth.edu, Jason.T.Stauth@dartmouth.edu

Abstract—This work explores hybrid Series-Parallel (SP) switched capacitor (SC) DC-DC converter architectures designed for high-voltage, high-conversion-ratio applications that use inductor current ripple frequency multiplication, a property previously reported for Flying Capacitor Multilevel (FCML) converters. The design is realized using a single high-voltage GaN switch and low-voltage Si switches for the SP and PWM stages. Due to the better passive utilization of SP over other architectures, a lower switching frequency is achieved while meeting the same ripple requirements with superior flying capacitor voltage balancing. The SC stage is used to reduce C_{OSS} loss at the GaN switching node via a *pseudo-soft charging* process. A simple voltage-mode active balancing scheme provides single-cycle flying capacitor voltage regulation. The design achieves 48V:3V and 60V:3.75V step down with over 91% efficiency and peak input voltage of 170V with 92% efficiency at 64W output power.

I. INTRODUCTION

High voltage-conversion-ratio (VCR) DC-DC converters are essential for performance computing, renewable energy, and grid interface applications where significant voltage step-downs are needed to power low-voltage components efficiently [1]–[8]. Traditional buck converters, though widely used, face significant challenges in high step-down applications due to high device stress, large inductor size, short duty cycles, and high rms/dc current ratios [8]–[15].

These problems have been addressed by hybrid switched-capacitor (SC) converter topologies which use high-density SC stages to relax inductor energy storage/volume and PWM timing constraints [3]–[14]. Most past examples are based on Flying Capacitor Multilevel (FCML) [9]–[12] and Dickson-based topologies [13]–[16]. The extensive family of Dickson topologies benefits from best-achievable active device utilization, but has poor utilization of passive components and limited multi-level/mode flexibility [13]. FCML or multilevel buck converters are among the most widespread hybrid architectures due to seamless multilevel operation and frequency multiplication of inductor current ripple [9]. However, being intermediate in both active and passive component utilization, FCML converters are theoretically one of the lower performing hybrid topologies in terms of efficiency and power density [8]. Both FCML and Dickson-based architectures have challenges with flying-capacitor voltage balance, which is needed to prevent high voltage stress on active devices and provide safe, reliable operation [9], [12], [17]–[23].

This work explores the series-parallel (SP) hybrid topology [1]–[3], which is known to have the best passive component utilization, but poor active device utilization [8], [24]. The unique position of the SP converter has led to little exploration in the literature compared to other topology classes. However, as shown in this work, SP has operational and implementation advantages similar to FCML converters that may make it attractive. The limitations of active utilization may be modest for discrete sub-200V silicon and GaN devices [15]. Furthermore, as will be demonstrated here, pseudo soft-charging can be used to approximate zero-voltage switching (ZVS), which improves the active utilization.

In this paper, we will explore multi-mode/level capabilities of the SP topology that allow for frequency multiplication of inductor current ripple. Such operation has been used extensively for the FCML topology and has been reported for the exponential topology [25] in prior literature. As compared to the popular FCML, the SP topology has an improved frequency multiplication effect, which can be a major performance and operational advantage for high-voltage and high-conversion ratio applications. Such multi-mode operation is relatively easier to implement in SP which also benefits from simpler voltage balancing dynamics, a critical challenge in hybrid SC converters [17]. Consequently, the SP architecture allows a simple non-linear active balance scheme, related to *constant-switch-stress* CSS-control [19]. Finally, in this work, we will utilize a modified SP architecture [3] which requires only a single high-voltage switch, the rest being low-voltage rated. Therefore, this work seeks to position the SP topology as a competitive alternative to existing FCML and/or Dickson topologies for high efficiency, high power density, and large VCR applications. To validate this we built a prototype targeting a high step-down conversion ratios with a range of input voltages spanning 48 to 170V.

II. MODIFIED SERIES-PARALLEL HYBRID SC ARCHITECTURE

The architecture of the modified SP converter for $N = 4$ is shown in Fig. 1 (a), where N is the nominal (or resonant) conversion ratio. In normal (balanced) operation, all flying capacitors hold a voltage of V_{in}/N, where V_{in} is the input voltage. Compared to the conventional SP converter [24], the

979-8-3315-1612-3/25 $31.00 © 2025 IEEE

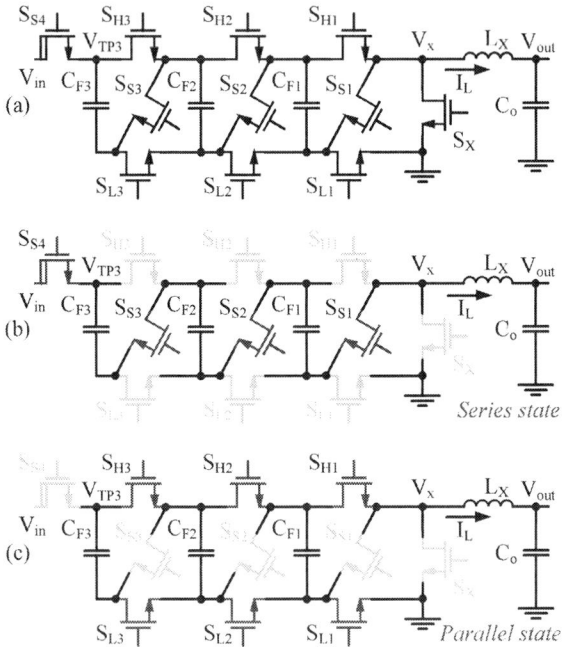

Fig. 1. (a) Schematic of $N = 4$ modified SP topology, (b) Series state and (c) Parallel state.

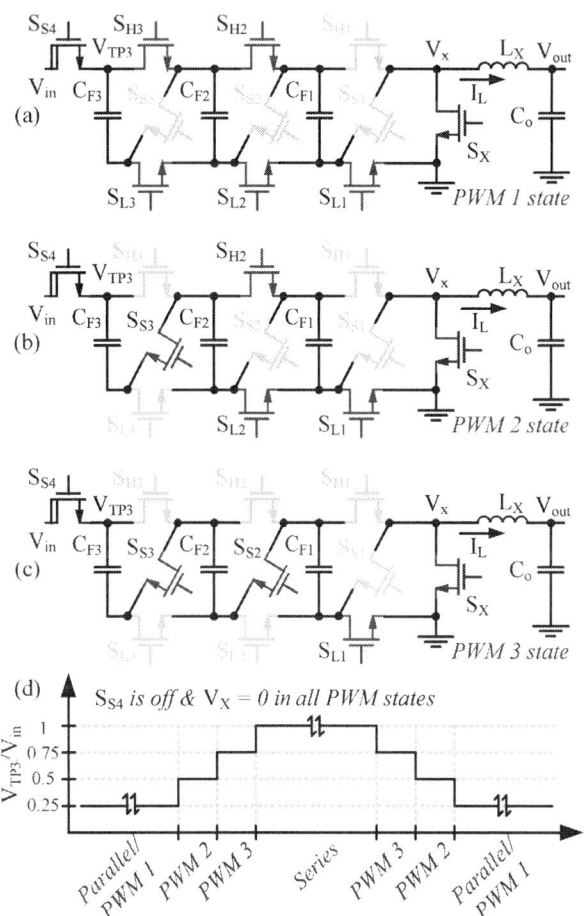

Fig. 2. PWM states of converter when (a) $V_{TP3} = V_{in}/4$ (b) $V_{TP3} = V_{in}/2$, (c) $V_{TP3} = 3V_{in}/4$. (d) Pseudo soft-charging waveform to achieve approximate ZVS for S_{S4}.

modified architecture allows all switches except S_{S4} in Fig. 1, to be realized using low-voltage devices, rated to V_{in}/N [3]. The series state as shown in Fig. 1 (b), remains the same as the conventional architecture. However, the parallel state, illustrated in Fig. 1 (c), has low-voltage switches (S_{H1}-S_{H3} and S_{L1}-S_{L3}) connected in series. This comes with the trade-off of higher current rating for some of the switches, *i.e.* S_{H1} and S_{L1} handle higher current than S_{H3} and S_{L3}. However, the overall switch $\sum VA$ of the architecture remains the same as the conventional SP, with the lower voltage rating of the parallel switches being traded off with higher current handling requirements [3].

The additional low-voltage switch S_x is analogous to the low-side switch of a buck converter, which facilitates PWM operation; while it is redundant with the switching capabilities of S_{S1} and S_{L1}, its use does not degrade the overall $\sum VA$ of the converter and it simplifies the design, layout, and optimization of the design. Besides the two SC (Series & Parallel) phases, the third category of PWM phases are illustrated in Fig. 2. In any of the possible PWM states, $V_x = 0$, with switches S_{S1}, S_{S4} and S_{H1} being always off. The decoupled SC stage can be reconfigured into any one of the three states as shown in Figs. 2 (a)-(c). Each of the states can set the voltage at the source of the switch S_{S4} (V_{TP3}) to integer multiples of V_{in}/N as depicted in Fig. 2 (d).

A. Choosing Switches and Approximating ZVS

The low voltage rating of all switches except for S_{S4}, makes them compatible with Si power switches for high step-down DC-DC conversion. For example, with $N = 4$ and $V_{in} = 170V$, switches must block nominally 42.5V; in this design 60V rated Si switches were used to provide some margin of

safety. The high-side switch S_{S4} on the other hand, is required to block a worst case voltage of $(N - 1)V_{in}/N = 3V_{in}/4$. For $V_{in} = 170V$, the nominal blocking voltage is 127.5V, but potentially up to the full 170V in a startup or fault condition. In this work, we used a 200V-rated GaN switch to provide a margin of safety and benefit from the benefits of GaN in this voltage range [26]–[28].

Due to the high voltage swing at the source of S_{S4}, there is the potential for high output capacitance C_{OSS} switching losses due to ($V_{DS,SS4}$) swing. This is a key contributor towards poor active component utilization of the SP topology. Particularly while driving lighter loads, this may limit efficiency. To circumvent such high switching losses of the high-side switch (S_{S4} in Fig. 1), the SP topology presents a mechanism to approximate zero-voltage switching ZVS. We can exploit the PWM states effectively to accomplish this as depicted in Fig. 2 (d).

Suppose we wish to turn-on S_{S4} for transitioning from parallel to series state. Instead of turning all the series switches on (and all parallel switches off) simultaneously, we can transition in a sequence of steps. The goal is to make the source of the switch S_{S4} (V_{TP3}) step up gradually from $V_{in}/4$

to $3V_{in}/4$ in two steps (*i.e.*, PWM2 and PWM3). Then S_{S4} and S_{S1} can be turned on while turning off S_{L1} and S_X to transition from PWM3 to series state. By doing so, S_{S4} is turned on, when $V_{DS,SS4}$ is a relatively small voltage of $V_{in}/4$, following the stair-case $V_{DS,SS4}$ waveform in Fig. 2 (d).

By reducing the voltage step across the equivalent capacitance C_{OSS} on the high voltage switching node V_{TP3}, the overall switching losses can be reduced by a factor of $3\times$. The same process can be followed in reverse to turn off the switch and approximate ZVS for V_{TP3}. Interestingly, unlike in conventional buck converters, this mechanism does not require negative inductor current to achieve (approximate) ZVS for high-side switches. In fact, the inductor L_X in Fig. 1 is not involved to accomplish this pseudo-ZVS mechanism. Therefore, the apparent disadvantage of poor active utilization in the SP topology can be improved by this approximate ZVS switching scheme. The practical efficiency benefits of this scheme at light loads will be demonstrated in Section V.

III. INDUCTOR CURRENT FREQUENCY MULTIPLICATION

A periodic steady-state (PSS) timing waveform for the modified SP converter is shown in Fig. 3. The waveforms are depicted assuming all flying capacitor voltages are balanced at V_{in}/N. As discussed in [3], to achieve charge balance of the flying capacitors in a hybrid SP converter, the total time duration for the parallel state must equal to $(N-1)$ times the time duration of the series state. This is because in the series state each capacitor is charged by a current which is $(N-1)$ times higher than the discharging current in the parallel state. However, instead of having a single parallel state with a longer time duration as in [3], here we implement $(N-1)$ parallel phases each with the same duration as the series state duration. To provide regulated pulse-width modulation PWM control, each of these states operates with a per-phase duty cycle D, such that the duration of the SC states is DT_{SP}/N, where $T_{SP} = 1/f_{SP}$ is the overall switching period of the series-parallel converter. A consequence of such a switching scheme is that low-voltage switches S_x and S_{H1} operate at a higher frequency than the rest of the switches.

The DC output voltage of the converter can be determined by taking the average of the switching node voltage V_x due to inductor volt-second balance. From Fig. 3, we get

$$V_{out} = \left(\frac{D}{N}\right) V_{in}. \tag{1}$$

Next, we assume the small voltage ripple approximation on V_x to determine the characteristics of the inductor current ripple. When the converter is in any of the SC phases (*i.e.*, series or parallel), the inductor current, I_L, ramps up with a positive slope of $(1-D)V_{in}/NL_X$. On the other hand, during the PWM states, I_L ramps down with a negative slope of DV_{in}/NL_X. Since the time duration of all SC phases is identical to each other, the period of I_L ripple as shown in Fig. 3, becomes T_{SP}/N. Therefore, the fundamental component (f_L) of the inductor current ripple becomes

$$f_L = N \cdot f_{SP}. \tag{2}$$

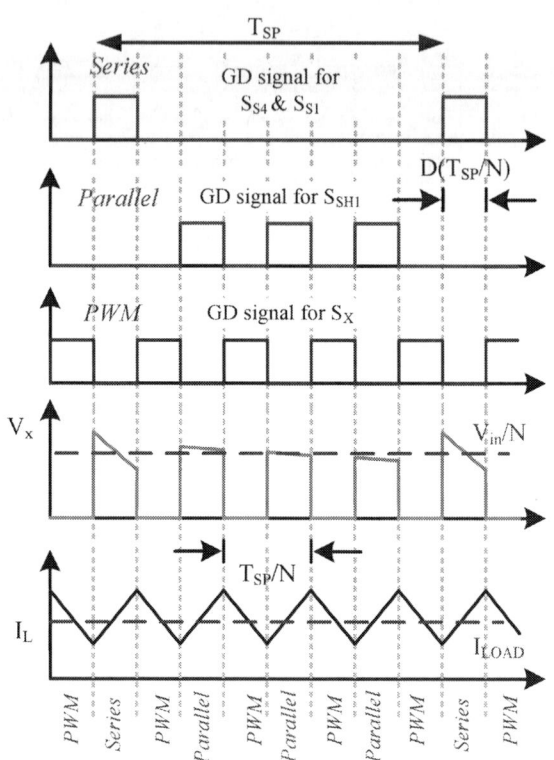

Fig. 3. Steady-state timing waveforms for a $N = 4$ modified SP hybrid SC converter. Gate-Drive (GD) signal for the relevant switches enabling the series, parallel and PWM states are shown. The inductor current ripple frequency is $N = 4$ times the converter frequency.

The current ripple frequency multiplication in (2) follows the similar concept used in FCML converters. The major benefit of the inductor current frequency multiplication is an N times reduction of the inductance required to achieve a similar current ripple in conventional buck converters. This allows for smaller or more efficient inductors and hence high converter power density or efficiency. Therefore, the SP topology can provide advantages similar to FCML topologies which have led to their widespread use.

A. Improved Frequency Multiplication Over FCML

The SP topology has the best passive utilization among all hybrid SC topologies [8], [24] and has a uniform low voltage rating across all flying capacitors at V_{in}/N. The FCML converter with a nominal VCR of N, has flying capacitors rated from V_{in}/N to $(N-1)V_{in}/N$. The higher voltage-rated capacitors lead to lower capacitance density, especially when factoring in voltage derating if class-II ceramic capacitors are used [15]. This advantage further enhances the extent of inductor current frequency multiplication in the SP over the FCML topology.

To understand this, consider a design where we need to design an SP and FCML converter to achieve a similar current waveform using the same inductor and flying capacitor volume. We also assume that the *flying capacitor voltage ripple is the same* for both topologies to maintain the same worst-case switch voltage stress. While the detailed derivation

979-8-3315-1612-3/25 $31.00 © 2025 IEEE

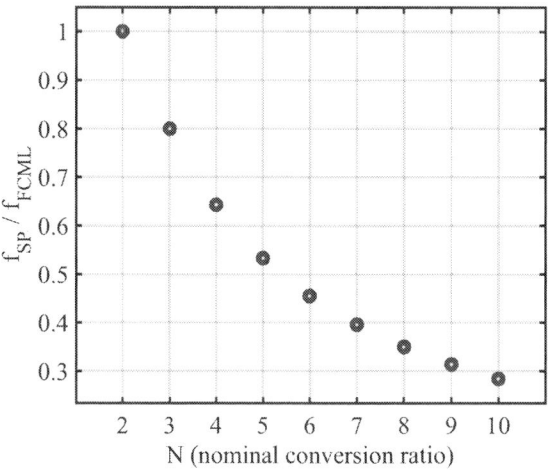

Fig. 4. Plot showing switching frequency ratio between FCML and SP with respect to nominal conversion ratio N.

is provided in Appendix A, the ratio of the converter switching frequency for the two cases is

$$\frac{f_{\mathrm{SP}}}{f_{\mathrm{FCML}}} = \frac{6(N-1)}{N(2N-1)}, \qquad (3)$$

where f_{FCML} is the switching frequency for the FCML converter.

As seen in Fig. 4, the SP converter can be operated at a lower switching frequency than the FCML converter for the same passive volume and ripple constraints. With $N = 4$ the SP can operate at $\sim 2/3$ the frequency of the FCML for the same ripple and voltage stress. With increasing nominal VCR, N, the relative benefit of the SP topology increases almost linearly. Intuitively, this is because the SP converter can provide more charge per switching phase for the same flying capacitor voltage ripple or device voltage stress, a manifestation of the better passive component utilization.

IV. FLYING CAPACITOR VOLTAGE BALANCING USING PARALLEL-STATE COUNT MODULATION

A major challenge associated with any hybrid SC architecture is the flying capacitor voltage balancing [9], [17]–[20], [29], [30]. The extremely slow dynamics of natural balancing motivate the use of active voltage balancing to accelerate the process. Detailed in [17], the balancing dynamics of the SP turns out to be 1st order, independent of N, unlike FCML or Dickson which have higher order dynamics and in some cases, uncontrollable switching states [18]. Thus the SP is simpler to control and balance, which can improve transient response and overall converter reliability.

In this work we develop a control algorithm based on the *constant-switch-stress* (CSS) control [19]. The CSS algorithm is a non-linear frequency modulation based voltage balancing scheme employed for the FCML converters. Depending on the whether flying capacitors are over- or under-charged, the pulse widths of the SC and following PWM states are decreased or increased, so as to restore balance.

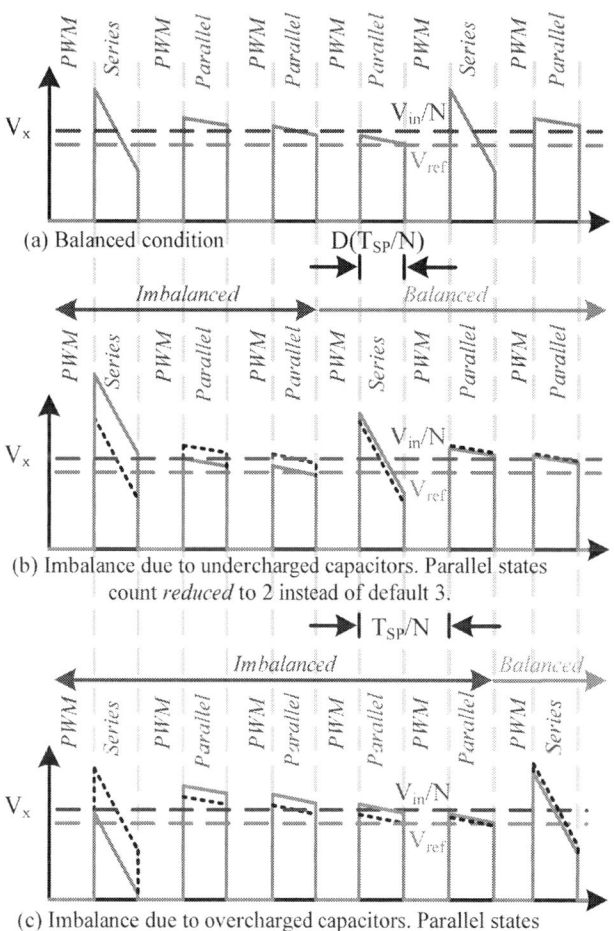

(a) Balanced condition

(b) Imbalance due to undercharged capacitors. Parallel states count *reduced* to 2 instead of default 3.

(c) Imbalance due to overcharged capacitors. Parallel states count *increased* to 4 instead of default 3.

Fig. 5. V_x node waveform when the $N = 4$ hybrid SP converter is (a) perfectly balanced and the control algorithm restores balance for imbalance caused by (b) undercharged and (c) overcharged flying capacitors.

The action of the algorithm is illustrated in Fig. 5. Just like CSS, this is another frequency modulation scheme, where instead of adjusting the pulse width of individual states, the *count of the parallel states* of a SP converter is modulated to achieve voltage balance. The series state and the PWM states that follows it, is kept the same as it would have been without any control. We want to remind that for a $N = 4$ hybrid SP converter,

$$V_x = \begin{cases} V_{in} - 3V_c & \text{in the series state,} \\ V_c & \text{in the parallel state,} \\ 0 & \text{in the PWM state,} \end{cases} \qquad (4)$$

where V_c refers to the flying capacitor voltage. Hence, in the parallel state, the $V_x \,(= V_c)$ node voltage is compared against a reference voltage V_{ref}. When V_x becomes lower than V_{ref}, the flying capacitors are considered undercharged. However, unlike CSS, the parallel state is allowed to continue for its set duration of DT_{SP}/N and the following PWM state duration is also unchanged. Following this, the SP converter is transitioned to the series state, where the flying capacitors are charged.

979-8-3315-1612-3/25 $31.00 © 2025 IEEE

Next we shall describe the mechanism of restoring balance from imbalanced scenarios by this algorithm. Suppose the imbalance occurs due to the flying capacitors being under-charged as illustrated by the example in Fig. 5 (b). Owing to the imbalance, during the second parallel state, V_x goes below V_{ref}. Since going to the third parallel state like in the balanced condition (Fig. 5 (a)) would unnecessarily discharge the capacitors further, the algorithm forces the converter to go to the series state earlier to recharge them and restore voltage balance. As a result, the number of parallel states reduce to two from the default three. Therefore, the instantaneous converter switching frequency during the imbalance event increases temporarily. Once balance is restored, the converter resumes its default behavior of one series and three parallel states.

The opposite scenario of imbalance due to overcharged flying capacitors is shown in Fig. 5 (c). Due to the overcharged capacitors, V_x remains higher than V_{ref} in the third parallel state. Consequently, the converter does not transition to the default series state and continues to a fourth parallel state to help the overcharged capacitors further discharge. In the particular case illustrated in Fig. 5 (c), at the fourth parallel state, V_x goes below V_{ref}. Hence, after the PWM state, the converter transitions to the series state with nearly balanced flying capacitors to resume the default steady-state operation. Since, the count of parallel states increases due to overcharging based imbalance, the instantaneous converter switching frequency decreases temporarily to handle the imbalance event.

This variation of a frequency modulated voltage balancing algorithm for the hybrid SP converter is hereby referred to as *parallel-state count modulation*. This is because the instantaneous converter switching frequency is adjusted by increasing or decreasing the number of parallel states for balanced operation.

We should also note that, in this particular scheme since the duration of the SC and PWM states are not changed, the fundamental component of the inductor current ripple frequency would still be $\approx N \cdot f_{SP}$. Finally, because the balanced voltage for the flying capacitors in the SP converter is V_{in}/N, the magnitude of the reference voltage, V_{ref}, is kept below V_{in}/N as shown in Fig. 5. The exact value of V_{ref} is decided based on the worst case load current, flying capacitance values and other practical timing considerations based on the implementation.

V. EXPERIMENTAL SETUP AND RESULTS

The hardware prototype of the proposed modified SP converter was designed and implemented for $N = 4$, achieving a conversion ratio of 16:1. The annotated PCB photograph of the converter is shown in Fig. 6, with the key components listed in Table I.

The top-level block diagram of the converter with voltage balancing control implementation is presented in Fig. 7. The converter was operated at a switching frequency of $f_{SP} = 250\,\text{kHz}$, resulting in an effective inductor ripple current frequency of $f_L = 1\,\text{MHz}$. The control logic monitors the switching node voltage V_x, which reflects the voltage of the flying capacitors. This voltage, scaled using a resistive divider,

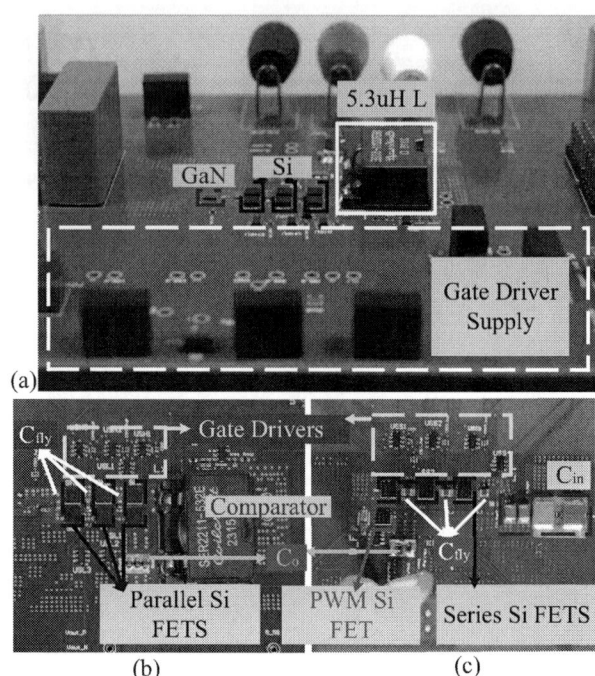

Fig. 6. Annotated PCB photograph: (a) full board, (b) top (c) bottom.

TABLE I
COMPONENT LIST OF HARDWARE PROTOTYPE

Component	Part Number	Specification
GaN FET	EPC2207	$200\,\text{V}, 22\,\text{m}\Omega$
Si FET	BSZ099N06LS5	$60\,\text{V}, 9.9\,\text{m}\Omega$
Flying capacitors (top)	GRM188R61H225	$50\,\text{V}, 2.2\,\mu\text{F}$
Flying capacitors (bottom)	GRT21BR61H475	$50\,\text{V}, 4.7\,\mu\text{F}$
Output capacitor	GRM188R61C106	$16\,\text{V}, 3 \times 10\,\mu\text{F}$
Input capacitor	C5750X7R2E105	$250\,\text{V}, 3 \times 1\,\mu\text{F}$
Inductor	SER2211-532E	$5.3\,\mu\text{H}$
Gate driver	LTC4440	-
Gate driver supply	CRE1S0505S3C	-
Comparator	LTC6752	-
FPGA module	Zybo Z7	-

was compared against an external reference voltage V_{ref} using a comparator. The comparator output served as a digital input to the FPGA, which adjusted the gate-drive signals for the converter switches to ensure flying capacitor voltage balance through a finite state machine (FSM).

The pseudo-soft charging mechanism was implemented to improve the converter's efficiency under light load conditions. This implementation was verified using oscilloscope measurements, as shown in Fig. 8. The gradual voltage step-up observed in the waveform of the source voltage of S_{S4} (V_{TP3}) matches the expected pseudo-soft charging behavior depicted in Fig. 2 (d), supporting the mechanism to approximate ZVS.

The benefits of pseudo-soft charging were validated by comparing the converter efficiency with and without the technique. As shown in Fig. 9, the converter showed a peak

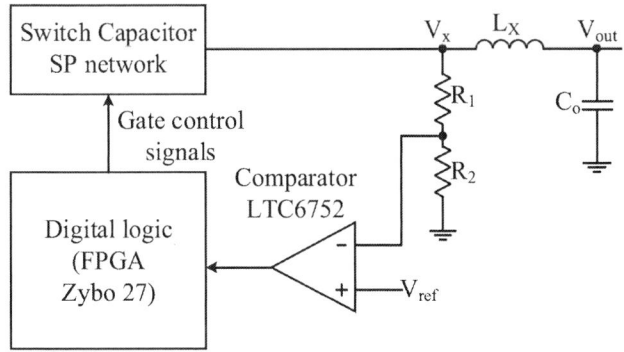

Fig. 7. The top-level block diagram of the converter with voltage balancing control.

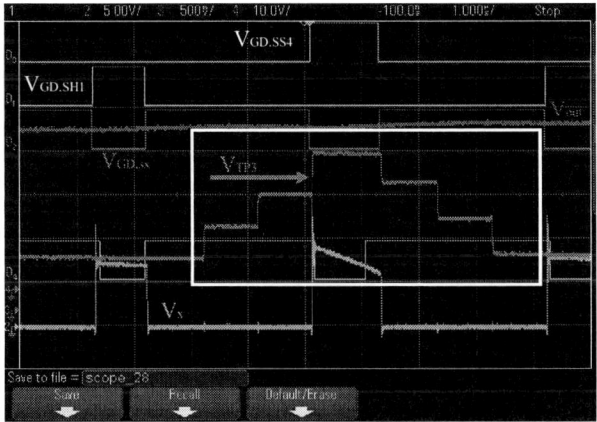

Fig. 8. Scope plot showing pseudo-soft charging of the source of S_{S4}.

efficiency improvement of 3.5% at an operating point of 60 V input, 3.75 V output, and a load current of 500 mA. This demonstrates the effectiveness of the implemented design in minimizing switching losses during low-load operation.

The efficiency of the converter was measured across a range

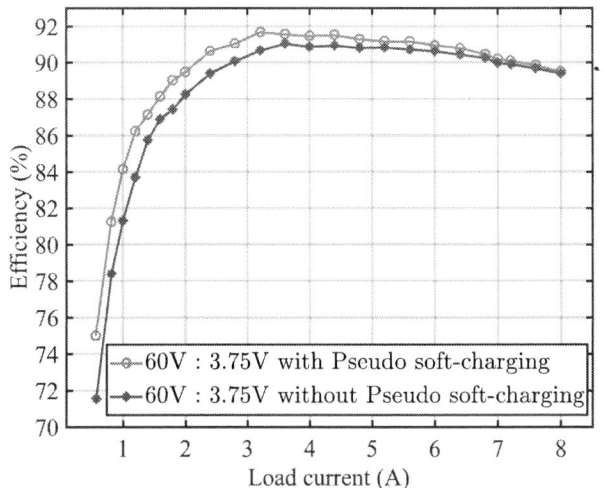

Fig. 9. Efficiency comparison with and without pseudo-soft charging.

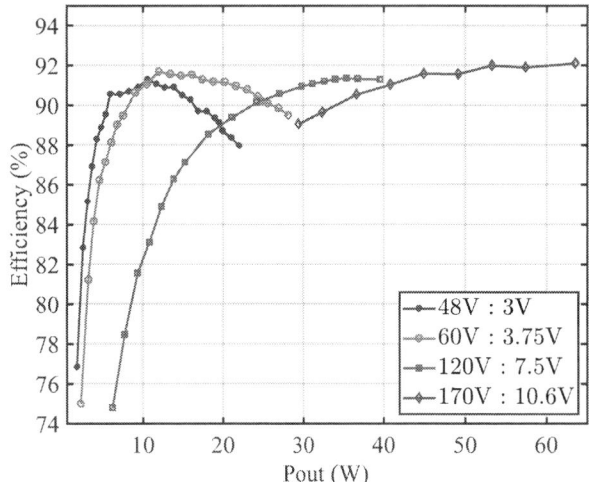

Fig. 10. Efficiency measurements across input voltages for different output power levels.

of output power levels for four different input voltages, while maintaining a fixed 16:1 conversion ratio. As shown in Fig. 10, the converter achieved a peak efficiency of 92.05% at an input voltage of 170 V, output voltage of 10.6 V, and a load current of 5 A, delivering 64 W of output power.

The converter demonstrated consistent performance across the tested input voltage range, maintaining high efficiency and stable operation. These results validate the design's suitability for applications requiring a wide input voltage range while achieving efficient power conversion.

VI. CONCLUSION

This paper presented a hybrid Series-Parallel (SP) switched capacitor DC-DC converter designed for high voltage conversion ratios. The converter is tested with 48 V to 170 V inputs with 16:1 step down ratios using a single high-voltage GaN switch and multiple Si-based switches. Owing to better passive-component utilization, the SP topology can achieve higher inductor-current frequency multiplication compared to FCML converters. An attractive feature of pseudo-ZVS for the high-side reduces switching loss and improves efficiency. A simple flying capacitor balancing scheme is introduced to ensure fast and reliable balanced operation. A prototype achieving a peak efficiency of 92.05% at a 170V input demonstrates the SP converter's potential as an efficient and compact solution for high-voltage applications.

VII. ACKNOWLEDGEMENT

This work was supported in part by the National Science Foundation (NSF) Power Management Integration Center (PMIC) under award number 2333329.

APPENDIX A
DERIVATION OF ENHANCED INDUCTOR CURRENT
FREQUENCY MULTIPLICATION OF SP OVER FCML

This section derives the relationship between flying capacitor voltage ripple and the utilization of passive components in series-parallel (SP) and flying capacitor multilevel (FCML) converters, along with the relationship between their switching frequencies under equal ripple conditions.

Referring to Fig. 3, the charge transfer in each high parallel state, Q_j, can be expressed as

$$Q_j = \frac{I_{\text{load}} \cdot D}{f_L}, \tag{5}$$

where I_{load} is the load current, D is the duty ratio, and f_L is the inductor ripple current frequency. The corresponding voltage ripple on the flying capacitor is then

$$v_{\text{fly}} = \frac{Q_j}{C_{\text{fly}}} = \frac{I_{\text{load}} \cdot D}{f_L \cdot C_{\text{fly}}}. \tag{6}$$

As established in [8], capacitance C_i of a flying capacitor is inversely proportional to the squared voltage stress across the capacitor :

$$C_i \propto \frac{1}{\sum_{i \in \text{Caps}} V_{c,i}^2}. \tag{7}$$

For SP converters, all flying capacitors are uniformly rated at $V_{c,i,\text{SP}} = \frac{V_{\text{in}}}{N}$, leading to

$$\sum_{i \in \text{Caps}} V_{c,i,\text{SP}}^2 = \frac{V_{\text{in}}^2}{N^2} \cdot (N-1). \tag{8}$$

In FCML converters, the voltage stress is non-uniform, with $V_{c,i,\text{FCML}} = \frac{(N-i) \cdot V_{\text{in}}}{N}$, resulting in

$$\sum_{i \in \text{Caps}} V_{c,i,\text{FCML}}^2 = \frac{V_{\text{in}}^2}{N^2} \cdot \frac{(N-1)N(2N-1)}{6}. \tag{9}$$

Substituting these into the capacitance ratio yields

$$\frac{C_{i,\text{FCML}}}{C_{i,\text{SP}}} = \frac{6}{N(2N-1)}. \tag{10}$$

The worst-case flying capacitor voltage ripple for FCML is

$$v_{\text{cfly,FCML}} = \frac{I_{\text{load}} \cdot D}{f_L \cdot C_{\text{fly,FCML}}}. \tag{11}$$

For SP converters, the ripple accumulates over $(N-1)$ parallel states, giving

$$v_{\text{cfly,SP}} = v_{\text{cfly,FCML}} \frac{6(N-1)}{N(2N-1)}, \tag{12}$$

using the capacitance ratio from (6).

To maintain equal ripple in both converters, the relationship between their switching frequencies becomes (3),

$$f_{\text{SP}} = \frac{6(N-1)}{N(2N-1)} f_{\text{FCML}}.$$

This final expression shows that for equal capacitor ripple, the FCML converter requires a higher switching frequency than the SP converter. The scaling factor depends on N, reflecting the differences in voltage stress distribution and passive component utilization between the two topologies.

REFERENCES

[1] R. Middlebrook, "Transformerless dc-to-dc converters with large conversion ratios," *IEEE Transactions on Power Electronics*, vol. 3, no. 4, pp. 484–488, 1988.

[2] R. C. N. Pilawa-Podgurski and D. J. Perreault, "Merged two-stage power converter with soft charging switched-capacitor stage in 180 nm cmos," *IEEE Journal of Solid-State Circuits*, vol. 47, no. 7, pp. 1557–1567, 2012.

[3] C. Schaef and J. T. Stauth, "A highly integrated series–parallel switched-capacitor converter with 12 v input and quasi-resonant voltage-mode regulation," *IEEE Journal of Emerging and Selected Topics in Power Electronics*, vol. 6, no. 2, pp. 456–464, 2018.

[4] K. Nishijima, K. Harada, T. Nakano, T. Nabeshima, and T. Sato, "Analysis of double step-down two-phase buck converter for vrm," in *INTELEC 05 - Twenty-Seventh International Telecommunications Conference*, 2005, pp. 497–502.

[5] P. S. Shenoy, M. Amaro, J. Morroni, and D. Freeman, "Comparison of a buck converter and a series capacitor buck converter for high-frequency, high-conversion-ratio voltage regulators," *IEEE Transactions on Power Electronics*, vol. 31, no. 10, pp. 7006–7015, 2016.

[6] J. Baek, Y. Elasser, K. Radhakrishnan, H. Gan, J. P. Douglas, H. K. Krishnamurthy, X. Li, S. Jiang, C. R. Sullivan, and M. Chen, "Vertical stacked lego-pol cpu voltage regulator," *IEEE Transactions on Power Electronics*, vol. 37, no. 6, pp. 6305–6322, 2022.

[7] N. Khan, O. Cobani, G. V. Piqué, J. Pigott, H. J. Bergveld, and O. Trescases, "A 48v-1v auxiliary-assisted hybrid dc–dc converter with flying-capacitor-based virtual bus for fast transient response," *IEEE Transactions on Power Electronics*, vol. 39, no. 5, pp. 5848–5861, 2024.

[8] P. H. McLaughlin, J. S. Rentmeister, M. H. Kiani, and J. T. Stauth, "Analysis and comparison of hybrid-resonant switched-capacitor dc–dc converters with passive component size constraints," *IEEE Transactions on Power Electronics*, vol. 36, no. 3, pp. 3111–3125, 2021.

[9] R. H. Wilkinson, T. A. Meynard, and H. du Toit Mouton, "Natural balance of multicell converters: The general case," *IEEE Transactions on Power Electronics*, vol. 21, no. 6, pp. 1658–1666, 2006.

[10] A. Stupar, T. McRae, N. Vukadinović, A. Prodić, and J. A. Taylor, "Multi-objective optimization of multi-level dc–dc converters using geometric programming," *IEEE Transactions on Power Electronics*, vol. 34, no. 12, pp. 11 912–11 939, 2019.

[11] M. T. Elrais, M. Safayatullah, and I. Batarseh, "Generalized architecture of a gan-based modular multiport multilevel flying capacitor converter," *IEEE Transactions on Power Electronics*, vol. 38, no. 8, pp. 9818–9838, 2023.

[12] D. H. Zhou, J. Čeliković, D. Maksimović, and M. Chen, "Balancing multiphase fcml converters with coupled inductors: Modeling, analysis, limitations," *IEEE Transactions on Power Electronics*, vol. 39, no. 8, pp. 9268–9291, 2024.

[13] N. M. Ellis and R. Amirtharajah, "Large signal analysis on variations of the hybridized dickson switched-capacitor converter," *IEEE Transactions on Power Electronics*, vol. 37, no. 12, pp. 15 005–15 019, 2022.

[14] R. Das, G.-S. Seo, and H.-P. Le, "A 120v-to-1.8v 91.5 efficient 36-w dual-inductor hybrid converter with natural soft-charging operations for direct extreme conversion ratios," in *2018 IEEE Energy Conversion Congress and Exposition (ECCE)*, 2018, pp. 1266–1271.

[15] N. C. Brooks, J. Zou, S. Coday, T. Ge, N. M. Ellis, and R. C. N. Pilawa-Podgurski, "On the size and weight of passive components: Scaling trends for high-density power converter designs," *IEEE Transactions on Power Electronics*, vol. 39, no. 7, pp. 8459–8477, 2024.

[16] S. K. Murray, A. Kachura, and O. Trescases, "A 400 v dual-phase series-capacitor buck converter gan ic with integrated closed-loop control," *IEEE Transactions on Power Electronics*, vol. 39, no. 8, pp. 9579–9590, 2024.

[17] K. Datta and J. T. Stauth, "Comparison of voltage balance and state estimation dynamics for hybrid switched-capacitor converter topologies," in *2023 IEEE 24th Workshop on Control and Modeling for Power Electronics (COMPEL)*, 2023, pp. 1–8.

[18] Z. Xia, K. Datta, and J. T. Stauth, "State-space modeling and control of flying-capacitor multilevel dc–dc converters," *IEEE Transactions on Power Electronics*, vol. 38, no. 10, pp. 12 288–12 303, 2023.

[19] Z. Xia and J. T. Stauth, "Constant switch stress control of hybrid switched capacitor dc-dc converters," in *2022 IEEE Applied Power Electronics Conference and Exposition (APEC)*, 2022, pp. 1214–1221.

[20] J. Celikovic, R. Das, H.-P. Le, and D. Maksimovic, "Modeling of capacitor voltage imbalance in flying capacitor multilevel dc-dc converters," in *2019 20th Workshop on Control and Modeling for Power Electronics (COMPEL)*, 2019, pp. 1–8.

979-8-3315-1612-3/25 $31.00 © 2025 IEEE

[21] S. Thielemans, A. Ruderman, B. Reznikov, and J. Melkebeek, "Improved natural balancing with modified phase-shifted pwm for single-leg five-level flying-capacitor converters," *IEEE Transactions on Power Electronics*, vol. 27, no. 4, pp. 1658–1667, 2012.

[22] S. d. Silva Carvalho, N. Vukadinović, and A. Prodić, "Phase-shift control of flying capacitor voltages in multilevel converters," in *2020 IEEE Applied Power Electronics Conference and Exposition (APEC)*, 2020, pp. 299–304.

[23] L. Corradini and D. Maksimović, "Steady-state indeterminacy in lossless switched-mode power converters," *IEEE Transactions on Power Electronics*, vol. 38, no. 3, pp. 3001–3013, 2023.

[24] M. D. Seeman and S. R. Sanders, "Analysis and optimization of switched-capacitor dc–dc converters," *IEEE Transactions on Power Electronics*, vol. 23, no. 2, pp. 841–851, 2008.

[25] Z. Xia and J. T. Stauth, "A cascaded hybrid switched-capacitor dc–dc converter capable of fast self startup for usb power delivery," *IEEE Journal of Solid-State Circuits*, vol. 57, no. 6, pp. 1854–1864, 2022.

[26] D. Reusch and J. Strydom, "Evaluation of gallium nitride transistors in high frequency resonant and soft-switching dc–dc converters," *IEEE Transactions on Power Electronics*, vol. 30, no. 9, pp. 5151–5158, 2015.

[27] E. A. Jones, F. F. Wang, and D. Costinett, "Review of commercial gan power devices and gan-based converter design challenges," *IEEE Journal of Emerging and Selected Topics in Power Electronics*, vol. 4, no. 3, pp. 707–719, 2016.

[28] M. Kaufmann and B. Wicht, "A monolithic gan-ic with integrated control loop for 400-v offline buck operation achieving 95.6% peak efficiency," *IEEE Journal of Solid-State Circuits*, vol. 55, no. 12, pp. 3169–3178, 2020.

[29] R. K. Iyer, I. Z. Petric, R. S. Bayliss, N. C. Brooks, and R. C. N. Pilawa-Podgurski, "A high-bandwidth parallel active balancing controller for current-controlled flying capacitor multilevel converters," in *2023 IEEE Applied Power Electronics Conference and Exposition (APEC)*, 2023, pp. 775–781.

[30] L. Lu, D. Li, and A. Prodić, "Absolute minimum deviation controller for multi-level flying capacitor direct energy transfer converters," in *2020 IEEE Applied Power Electronics Conference and Exposition (APEC)*, 2020, pp. 305–311.

979-8-3315-1612-3/25 $31.00 © 2025 IEEE

A Real-time Variation Control of Deadtime in GaN-based Bidirectional Buck-Boost Converter for Lithium-ion Battery Formation System

Jong-hun Lim
Department of Electrical and Computer Engineering
Sungkyunkwan University
Suwon, Korea
jhlim9614@g.skku.edu

Go woon Heo
Department of Electrical and Computer Engineering
Sungkyunkwan University
Suwon, Korea
heogowoon@g.skku.edu

Je-yeong Lim
Department of Electrical and Computer Engineering
Sungkyunkwan University
Suwon, Korea
wpdud9704@g.skku.edu

Dong Hwan Kim
Department of Electrical and Computer Engineering
Sungkyunkwan University
Suwon, Korea
tesla@g.skku.edu

Byoung Kuk Lee [†]
Department of Electrical and Computer Engineering
Sungkyunkwan University
Suwon, Korea
bkleeskku@skku.edu

Abstract— **In this paper, a real-time deadtime control method is proposed based on optimized switching time delay calculations, considering the dynamic characteristics of the battery formation system. The proposed technique reduces reverse conduction loss by controlling the deadtime during dynamic transitions between charging and discharging operations in both directions. Experimental verification demonstrated an average efficiency improvement of 0.87% in a 5V/100A battery formation device and a significant temperature reduction of up to 11.5°C in the synchronous switch, thereby enhancing overall system performance in charging mode operation.**

Keywords— battery formation system, Gallium Nitride high electron mobility transistor, dead time control, buck-boost converter

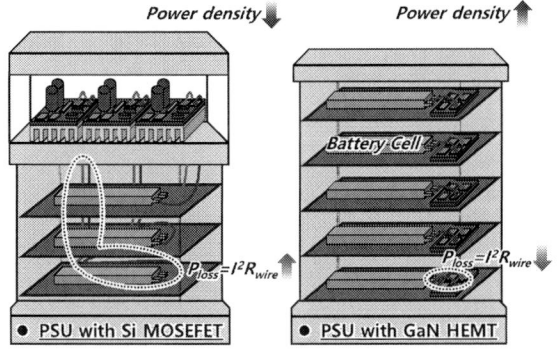

Fig. 1. Process advantages of applying GaN HEMT.

I. INTRODUCTION

The battery formation process plays a pivotal role in defining the final quality of a battery cell, as it imparts essential electrical characteristics by utilizing complex sequences of charge and discharge currents [1]. This phase ensures that each cell meets the necessary performance specifications. However, a significant challenge arises from conduction losses in the wiring between the power supply unit (PSU) and the battery, which severely impact system efficiency [2]-[3]. These losses are primarily due to the extended distance between the PSU and the battery, which increases resistance and results in greater energy dissipation. Addressing these efficiency losses requires positioning the battery and converter as close as possible, as shown in Fig. 1, to minimize conductive path lengths and thereby reduce resistive losses. Without optimizing these configurations, battery formation processes risk increased costs and energy waste, ultimately impacting large-scale battery manufacturing and quality.

To further improve this setup, considerable efforts have been made to miniaturize converters by leveraging advancements in Wide Bandgap (WBG) semiconductor devices, enhancing switching frequencies, and reducing the size of passive components. Specifically, there has been a shift toward replacing silicon-based MOSFETs with gallium nitride (GaN) high electron mobility transistors (HEMTs) for low-voltage, high-current applications [4]-[5]. GaN HEMTs offer superior electrical performance characteristics, such as faster switching speeds and lower conduction losses, which make them highly suitable for compact, efficient converter designs. However, one of the trade-offs with GaN devices is their higher losses during deadtime compared to silicon MOSFETs, which poses challenges for thermal management and further impacts system efficiency. Deadtime losses, caused by the overlap of conduction states between switches, require careful management to prevent excess heat generation and maintain high efficiency levels.

Previous research has proposed calculating optimal fixed deadtime based on the parasitic properties of GaN devices, yet this approach has proven insufficient due to the dynamic nature of switching times, which vary with load current. Fixed deadtime settings fail to adapt to changing load conditions, leading to suboptimal performance. More recent studies have suggested dynamically adjusting deadtime according to channel current, thereby reducing deadtime-associated losses more effectively [6]-[15]. This method is especially critical for high-current applications where the load varies significantly. Accurately calculating the on/off delay points for the converter switches is essential for this approach, as is implementing a real-time control mechanism that adapts to these calculations during operation.

979-8-3315-1612-3/25 $31.00 © 2025 IEEE

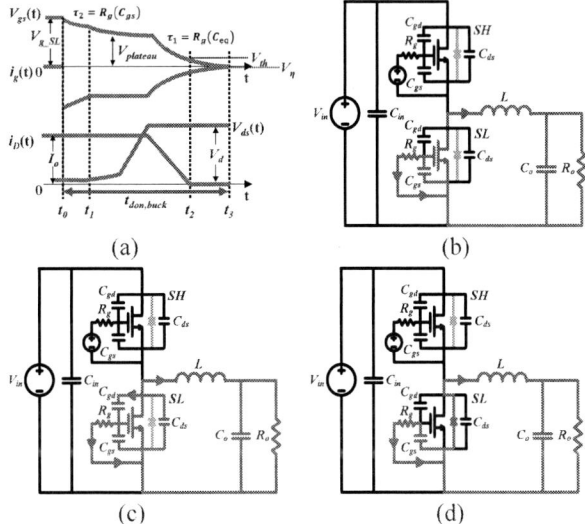

Fig. 2. Buck mode turn-on deadtime analysis (a) waveforms of SL turn-off, (b) t0-t1 operation, (c) t1-t2 operation, (d) t2-t3 operation.

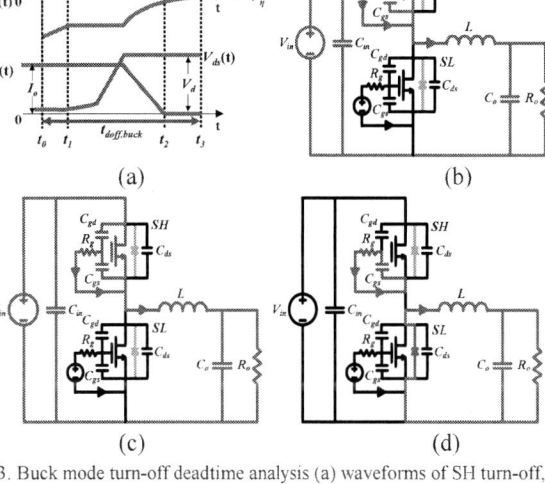

Fig. 3. Buck mode turn-off deadtime analysis (a) waveforms of SH turn-off, (b) t0-t1 operation, (c) t1-t2 operation, (d) t2-t3 operation.

TABLE I. BATTERY FORMATION SYSTEM'S SPECIFICATION

Parameters	Value	Unit
Input voltage, V_{in}	14	[V]
Switching frequency, f_{sw}	500	[kHz]
Output voltage, V_{out}	0.5-5	[V]
Output current, I_{out}	10-100	[A]

TABLE II. GAN HEMT ELECTRIC CHARACTERISTICS

Parameters		Value	Unit
Switch on resistance, R_{ds_on}		1.2	[mΩ]
Threshold voltage, V_{th}		2.2	[V]
Internal gate resistance, I_{out}		8.7	[Ω]
Parasitic capacitance	C_{gs}	2.5	[nF]
	C_{gd}	1.0	
	C_{ds}	0.6	

Experimental validation is also necessary to confirm the effectiveness of such real-time adjustments in practical applications. Real-time control in deadtime optimization not only enhances efficiency but also holds promise for broader applications in battery-intensive industries like electric vehicles and energy storage systems. Experimental validation is also necessary to confirm the effectiveness of such real-time adjustments in practical applications.

This paper proposes a method to calculate optimal deadtime based on the output current of the battery formation system and introduces a real-time deadtime control mechanism for improved system efficiency and thermal management. The proposed method calculates turn-on and turn-off deadtime control points for output currents up to 100A and applies the optimal deadtime values in real-time to a bidirectional buck-boost converter. To validate the effectiveness of this approach, a 5V/100A battery formation device is used to compare the switch temperature and

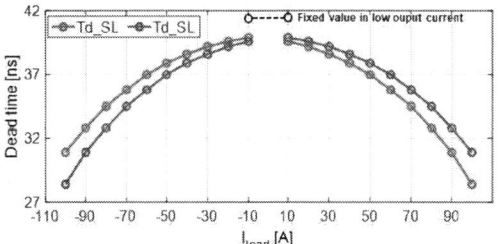

Fig. 4. SL/SH deadtime look-up table for *Ifig. 4,*.

system efficiency of the proposed method against a fixed deadtime setting, providing critical insights into the thermal and efficiency improvements achieved by this dynamic deadtime control strategy. These results are expected to offer a robust framework for enhanced control and performance in high-current battery formation processes.

II. DEADTIME DERIVATION BASED ON PARASITIC CAPACITANCE OF GAN HEMTS

The optimal deadtime for a GaN HEMT-based bidirectional buck-boost converter can be mathematically calculated using the parasitic capacitance of the switches. Therefore, turn-on/off deadtimes are derived by analyzing the dynamic characteristics of the upper switch (SH) and the lower switch (SL), the turn-on deadtime (t_{don}) can be derived from when the gate-source voltage (V_{gs}) of the lower switch decreases to zero during SL turn-off, as shown in Fig. 2 (a) When a turn-off signal is applied to the gate of the SL, the V_{gs} changes as shown in Fig. 2(a). This dynamic characteristic can be divided into three parts as shown in Fig. 2 (b)-(d) and can be derived as (1):

$$t_{don} = t_{gL} - \left(R_g C_{gs} \ln \frac{V_{Miller}}{V_{gs(t=t_0)}} + R_g C_{eq} \ln \frac{V_{th}}{V_{Miller}} + R_g C_{gs} \ln \frac{V_{g.off}}{V_{th}} \right), \quad (1)$$

where t_{gL} is the fall time of the lower gate driver, C_{gs} is the capacitance between the gate and source, V_{miller} is the Miller voltage of the switch which can be derived from channel current, C_{eq} is the equivalent capacitance during V_{gs} at V_{miller}, V_{th} is the threshold voltage, and $V_{g.off}$ is the gate-off

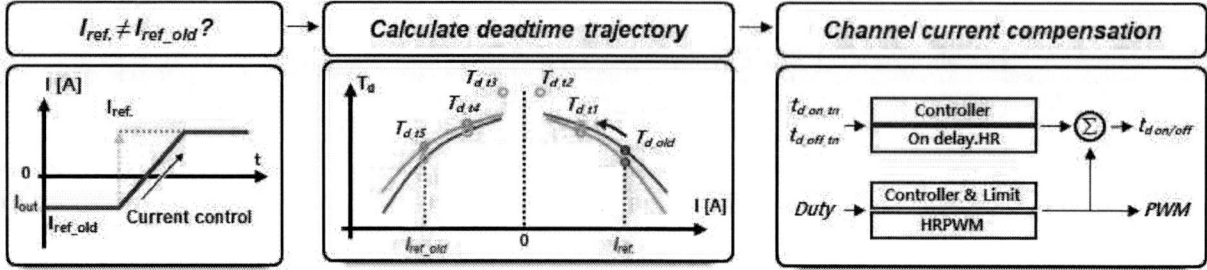

Fig. 5. Flowchart of proposed deadtime variation trajectory control method.

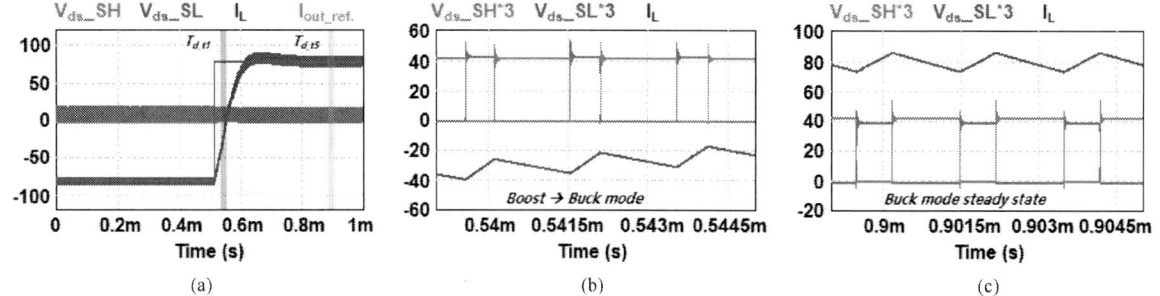

Fig. 6. Simulation result of dead time variable according to I_{ref} : (a) -80 to 80A full scale waveform (b) zoomed in waveform for T_{d_t1}, (c) zoomed in waveform for T_{d_t5}.

voltage from the experimental value. Similarly, the turn-off deadtime can be derived from the operational waveform analysis shown in Fig. 3(a)-(d), and the turn-off delay can be derived as (2), and the detailed mathematical analysis for optimizing the deadtime is described in [14],[16].

$$t_{doff} = t_{gH} - \left(R_g C_{gs} \ln \frac{V_{Miller}}{V_{gs(t=t_0)}} + R_g C_{eq} \ln \frac{V_{th}}{V_{Miller}} + R_g C_{gs} \ln \frac{V_{g\_off}}{V_{th}} \right), \quad (2)$$

The system parameters used to derive the dead time for application in the battery formation PSU, based on (1) and (2), are presented in Table I. For real-time control, the channel current can be converted into the load current (I_{load}) to derive the corresponding dead time, as shown in Fig. 4. In this study, the deadtime values were meticulously derived for varying load currents by taking into account the parasitic capacitance and switching parameters specific to the Innoscience INN040FQ015A low-voltage GaN device, as outlined in its datasheet as shown in Table II. This process involved analyzing key device characteristics, including gate resistance, parasitic capacitance, and the switching transition behavior of the GaN device, to ensure optimal deadtime intervals for different current levels. By carefully integrating these switch parameters and parasitic effects, the derived deadtime settings aim to minimize deadtime losses and enhance efficiency under dynamic load conditions.

III. REAL-TIME DEADTIME CONTROL METHOD CONSIDERING BATTERY FORMATION PROCESS

The real-time deadtime control technique designed for the battery formation process is demonstrated in Fig. 5. This approach leverages the feedback received from the battery-side current, which enables the dynamic adjustment of the deadtime variable in accordance with fluctuations in the output current. By doing so, it becomes possible to adapt deadtime settings in real-time, providing enhanced flexibility and control during the formation process. To achieve this, it

is crucial to select an optimal target reference value for deadtime based on detected changes in the reference current of the battery (I_{ref}). This selection process begins by assessing whether the polarity of I_{ref} has reversed, as a change in polarity imposes distinct requirements on deadtime handling. Specifically, when a polarity reversal occurs, the deadtime cannot simply be adjusted linearly. Instead, a deadtime that accounts for the safe operating area (SOA) must be chosen to ensure safe and efficient operation under these dynamic conditions.

The proposed technique addresses this requirement by dividing the deadtime control values based on the polarity of I_{ref}, thereby providing sufficient deadtime in the zero-crossing section. This approach helps maintain stability during the critical moment when the current changes direction. Additionally, this study develops an exponential function variable trajectory by constructing a function for each segment of the time series. This function regulates both PWM control and deadtime adjustments within a 30ms time frame, which is equivalent to the activation current response time. The relationship for this exponential trajectory can be represented as follows:

$$T_d(t + T_c) = T_d(t) + f(T_d(t)), \quad (3)$$

$$f(T_d(t)) = -T_d e^{-\frac{t}{\tau}} \begin{cases} if) I_{ref}. \ polarity \ reverse, \tau = 0.06 \\ else) \ \tau = 0.15 \end{cases}, \quad (4)$$

where $T_d(t)$ represents the current deadtime value and $T_d(t+T_c)$ signifies the next deadtime control point. Despite the adaptability of this function, challenges may arise in the transient state, particularly for current-controlled conditions. During these transient periods, there is a possibility that the variable trajectory may deviate from the SOA due to fluctuations in the duty ratio. To prevent this, an additional compensation loop is incorporated, ensuring that the transient state remains stable by continually adjusting the deadtime. This loop actively monitors and adjusts the deadtime trajectory, providing real-time stabilization.

979-8-3315-1612-3/25 $31.00 © 2025 IEEE   1725

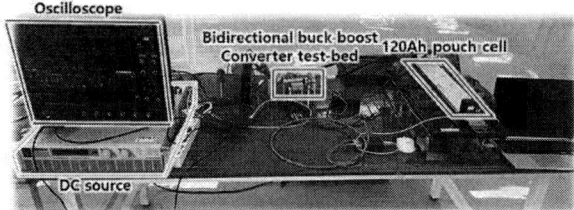

Fig. 7. Experimental set-up

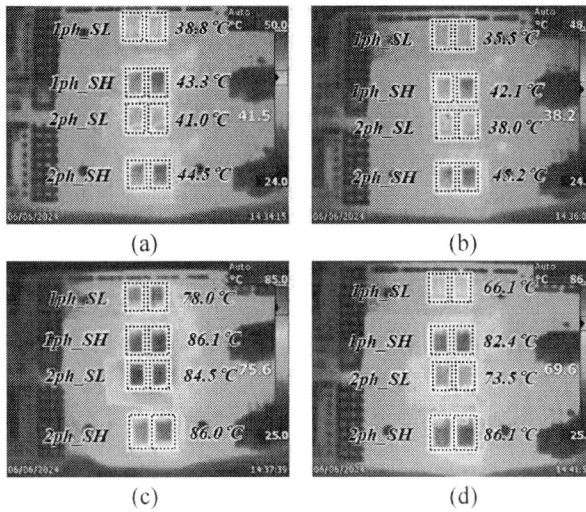

(a) (b)

(c) (d)

Fig. 8. Thermal variation of GaN HEMT under load condition: (a) 20A I_{out} with fixed deadtime, (b) 20A I_{out} with proposed method, (c) 80A th fixed deadtime, (d) 80A I_{out} with proposed method.

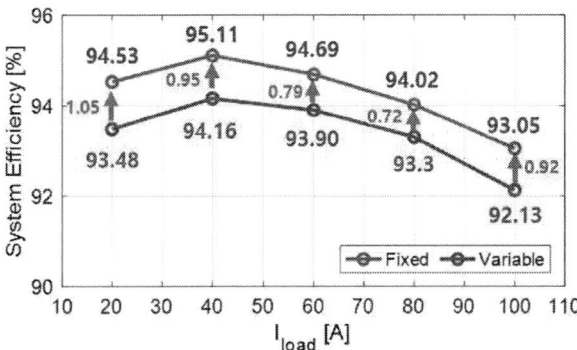

Fig. 9. Comparison of system efficiency in different control method. Red line is the system efficiency with fixed deadtime, and blue line is the system efficiency with proposed method.

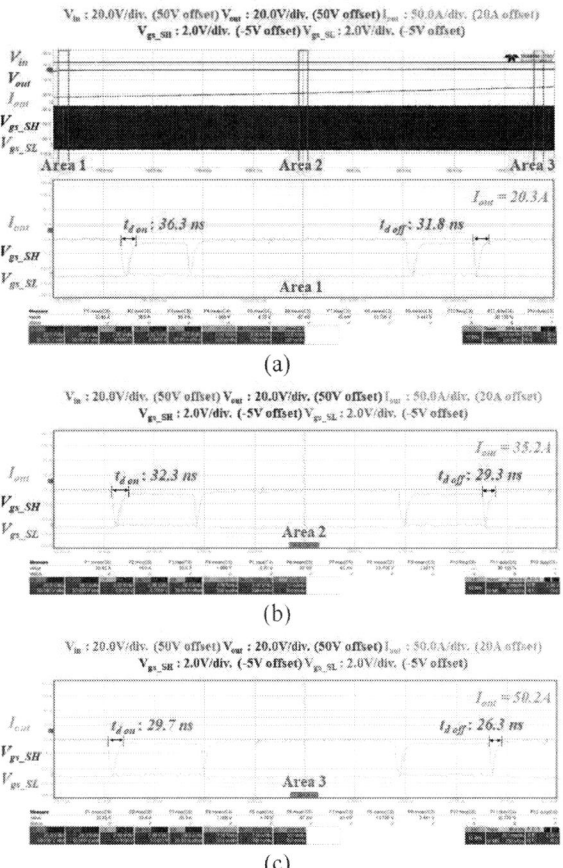

(a)

(b)

(c)

Fig. 10. Zoomed in waveform for real-time deadtime control in 20 to 50A load variation: (a) Full scale and area 1, (b) Area 2, (c) Area 3.

occur due to V_{DS} changes. This limitation affects the accuracy of the simulated deadtime trajectory under actual operating conditions. To address these limitations, the subsequent chapter will present experimental results using real battery load conditions, providing a more accurate assessment of the proposed deadtime control technique's performance in a practical setup. These experiments aim to validate the effectiveness of the technique in real-world applications and to explore potential improvements in system efficiency and thermal management achieved through precise deadtime control.

IV. VERIFICATION OF PROPOSED METHDOD

To validate the effectiveness of the proposed optimal deadtime and the real-time variable deadtime control, a series of experiments were conducted using a 120Ah capacity Lithium-ion battery, as shown in Fig. 7. These experiments aimed to confirm the normal operation of the proposed deadtime control strategy specifically during the charging mode, where efficient switching control is critical for system performance. Through this setup, the system's response was tested under varying load conditions to ensure the deadtime adjustments could adapt effectively in real-time, providing continuous optimization of switching intervals.

One primary objective of the experiment was to verify the reduction in reverse current voltage drop losses, a significant source of inefficiency in switching operations. To assess this, measurements of system efficiency and switch temperature were taken under both the proposed variable deadtime control and a traditional fixed deadtime setting. By

To verify the effectiveness of the proposed real-time deadtime control method, simulations were conducted using PSIM software, as shown in Fig. 6(a). A LEVEL2 model was implemented to examine the impact of the variable deadtime, with measurements taken of both drain-source voltages (V_{DS}) to confirm that deadtime changes dynamically in response to varying conditions. The simulation results demonstrate how the proposed method adjusts the deadtime when the battery output current transitions from a discharging mode at 80A to a charging mode at 80A. The deadtime adjustments at critical points, including sections T_{d_tl} and the steady-state deadtime, are shown in Fig. 6(b)-(c), illustrating how the real-time control adapts to significant changes in current direction. It should be noted that the LEVEL2 model utilized in these simulations does not fully capture the complex variations in parasitic capacitance that

979-8-3315-1612-3/25 $31.00 ©2025 IEEE   1726

dynamically adjusting the deadtime, the method aimed to reduce the gate turn-off reverse conduction period with certain output current (I_{out}), which, as shown in Fig. 8 (a)-(d), effectively reduce heat generation in the switch components. This reduction in the reverse conduction period directly led to a lower switch temperature, validating that real-time deadtime adjustments can positively impact the thermal profile of the system.

Further analysis of system efficiency with the optimal deadtime is presented in Fig. 9. Here, the efficiency improved by a maximum of 1.05%, a minimum of 0.72%, and an average of 0.87% over the fixed deadtime control approach. In terms of thermal performance, the variable deadtime control yielded notable temperature reductions. For instance, at load conditions of 20A and 80A, switch temperatures were reduced by 3.2°C and 11.5°C, respectively. These temperature reductions are especially valuable in battery formation systems, where circuit boards may operate under varying ambient temperatures, often subject to high thermal stresses. Lower temperatures enhance the durability and reliability of the printed circuit boards and can also reduce the demands on cooling system design, potentially leading to cost savings and simpler cooling requirements.

Lastly, Fig. 10 shows the operating waveforms of I_{out} and V_{gs} the system dynamically adjusts to reflect the optimal deadtime in real-time. The experiment tracked these waveforms as the output current varied from 20A to 50A under a variable load. These results demonstrate that the proposed deadtime control effectively adapts to fluctuating load conditions, maintaining optimal switching intervals and improving overall system stability. Such adaptability is crucial for battery formation applications, where load variations are common, and the ability to maintain efficiency across these conditions can contribute significantly to system performance and longevity.

V. CONLUSIONS

In this study, a real-time variable deadtime control technique was developed and validated to improve the efficiency and thermal management of high-current battery formation systems. By dynamically adjusting deadtime based on real-time output current feedback, the proposed method optimizes switching intervals, minimizing reverse current losses and reducing switch temperatures. Experimental results with a 120Ah Lithium-ion battery showed that the optimal deadtime control improved system efficiency by up to 1.05% compared to traditional fixed deadtime methods, with switch temperature reductions of up to 11.5°C under high-load conditions. These improvements are critical in battery formation systems, as they help mitigate the impact of varying ambient temperatures and reduce cooling system demands, enhancing the durability and reliability of circuit components. Overall, the proposed deadtime control technique demonstrates promising potential for energy savings and thermal management in industrial battery formation processes, particularly for applications in large-scale energy storage and electric vehicle manufacturing.

ACKNOWLEDGEMENT

This work was supported by the Announcement of Materials/Parts Technology Development Program (20024898, Development of 600kW Battery Emulator for dynamometer system) funded By the Ministry of Trade, Industry & Energy(MOTIE, Korea)

REFERENCES

[1] Zhao, Haichuan, et al. "A novel high-efficient lithium-ion battery serial formation system scheme based on partial power conversion." Journal of Energy Storage vol. 97, 2024.

[2] Mao, Chengyu, et al. "Balancing formation time and electrochemical performance of high energy lithium-ion batteries." Journal of Power Sources vol. 402, pp 107-115, 2018.

[3] Tat-Thang, L. E., et al. "Modular Bidirectional Differential Converter with Series Parallel Connected Output for Ultra-Wide-Voltage Applications: Control, Module Shedding, and Fail-Safe Operation." IEEE Transactions on Power Electronics vol. 37.1 pp 617-628, 2021.

[4] Tournier, Dominique, et al. "Wide band gap semiconductors benefits for high power, high voltage and high temperature applications." Advanced Materials Research vol. 324, 2011.

[5] Sørensen, Charlie, et al. "Conduction, reverse conduction and switching characteristics of GaN E-HEMT." 2015 IEEE 6th International Symposium on Power Electronics for Distributed Generation Systems (PEDG). IEEE, 2015.pp. 113-123, Feb. 2006.

[6] Reusch, David, Johan Strydom, and Alex Lidow. "Monolithic integration of GaN transistors for higher efficiency and power density in DC-DC converters." Proceedings of PCIM Europe 2015; International Exhibition and Conference for Power Electronics, Intelligent Motion, Renewable Energy and Energy Management. VDE, 2015.

[7] Khan, M. A., et al. "New developments in gallium nitride and the impact on power electronics." 2005 IEEE 36th Power Electronics Specialists Conference. IEEE, 2005.

[8] Conversion, Efficient Power. "Gallium nitride (gan) technology overview." Power 3.2, 2012.

[9] Han, Di, and Bulent Sarlioglu. "Understanding the influence of dead-time on GaN based synchronous boost converter." 2014 IEEE Workshop on Wide Bandgap Power Devices and Applications. IEEE, 2014.

[10] Conversion, EPC-Efficient Power. "Dead-time optimization for maximum efficiency.", 2014.

[11] Zhang, Haiyu, and Robert S. Balog. "Loss analysis during dead time and thermal study of gallium nitride devices." 2015 IEEE Applied Power Electronics Conference and Exposition (APEC). IEEE, 2015.

[12] Ke, Xugang, et al. "A 3-to-40-V automotive-use GaN driver with active bootstrap balancing and V SW dual-edge dead-time modulation techniques." IEEE Journal of Solid-State Circuits 56.2, 2020.

[13] Han, Di, and Bulent Sarlioglu. "Deadtime effect on GaN-based synchronous boost converter and analytical model for optimal deadtime selection." IEEE Transactions on Power Electronics vol. 31, 2015.

[14] M. Asad, A. K. Singha and R. M. S. Rao, "Dead Time Optimization in a GaN-Based Buck Converter," in IEEE Transactions on Power Electronics, vol. 37, no. 3, pp. 2830-2844, March 2022.

[15] T. LaBella, B. York, C. Hutchens, and J.-S. Lai, "Dead time optimization through loss analysis of an active-clamp flyback converter utilizing GaN devices," in Proc. IEEE Energy Convers. Congr. Expo., 2012, pp. 3882–3889.

[16] E. A. Jones, Z. Zhang, and F. Wang, "Analysis of the dv/dt transient of enhancement-mode GaN FETs," in Proc. IEEE Appl. Power Electron. Conf. Expo., 2017, pp. 2692–2699.

A Space Vector PWM Strategy for Charging of Bootstrap Capacitor in Three-Level Neutral-Point-Clamped Inverter

Anantha Hegde, Asamira Suzuki, Hirokazu Nakamura, Takamune Kabashima, Koji Higashiyama and Keiji Akamatsu

Engineering Division, Energy Solution Development Center
Panasonic Industry Co., Ltd.
Osaka 570-8501, Japan
Email: { hegde.anantha, suzuki.asamira, nakamura.hirokazu, kabashima.takamune, higashiyama.koji, akamatsu.keiji}@jp.panasonic.com

Abstract— **This paper presents a space vector PWM method to achieve a high power density three-level inverter (3LI) with a bootstrapped gate drive circuit. The objective is to shift from a two-level inverter (2LI) to a 3LI but without a significant increase in the size and cost of the inverter board. The proposed method solves an inherent issue of the charging of the bootstrap capacitors during the inverter operation. The effectiveness of the method is demonstrated through experiments carried out on a 5 kW motor drive with an input voltage of 350 V and a switching frequency of 12 kHz. Stable and efficient performance is maintained across a wide modulation index range, with a bootstrapped 3LI board approximately half the size of a conventional 3LI. Furthermore, loss comparisons between the bootstrapped 3LI board and a 2LI board with reactive load show that the power output can be increased by 1.5 times due to reduced switching losses.**

Keywords— *inverters, space vector PWM, three level inverter, bootstrap capacitor, motor drive, pulse width modulation*

I. INTRODUCTION

Recently there has been a demand for servo motors with high precision and power density to improve the productivity of equipment in the factory automation (FA) industry. To address this demand, introduction of three-level inverters (3LIs) [1] is an effective solution. Compared to conventional 2LIs, 3LIs offer advantages like improved waveform quality and lower switching losses providing a leeway to increase the total power before thermal issues become a challenge. On the other hand, 3LIs lead to an increase in the cost and size of the board. While 2LIs have six switching devices which require four isolated supplies, 3LIs have twelve switching devices which require at least ten isolated supplies for the gates. The introduction of bootstrap circuits is viable to solve these problems.

So far, there have been few reports of 3LIs using bootstrap circuits due to the difficulty of charging the bootstrap capacitors while maintaining inverter operation. Earlier approaches to operating a bootstrapped 3LI have included techniques such as using conventional PWM but sizing the capacitors for operation above a minimum speed [2] and employing carrier wave methods like dipolar modulation [3].

In this paper, we investigate the bootstrap configuration in a 3LI shown in Fig. 1 which reduces the required number of isolated gate supplies to just one. The proposed novel space vector approach allows stable operation of the bootstrapped 3LI even at very low speeds and offers more freedom when it comes to choosing switching losses and total harmonic distortion (THD) compared to carrier wave methods [4]. Initial sizing of the bootstrap capacitors for the proposed sequence is carried out using SPICE simulations. Loss comparisons between a 2LI, 3LI

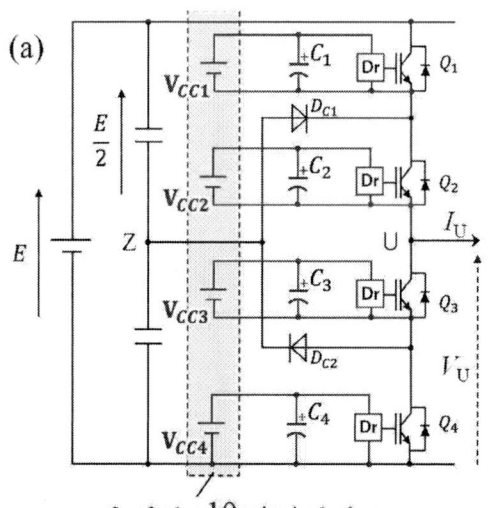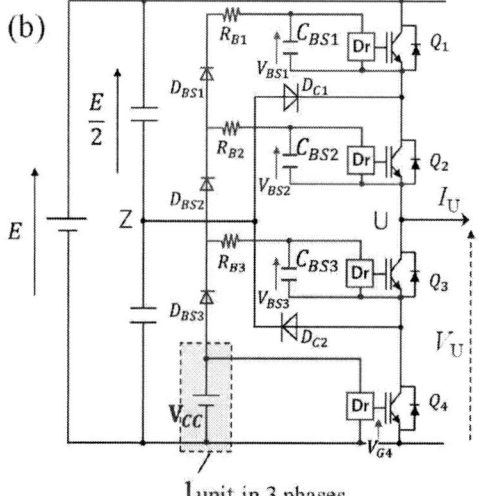

Fig. 1. One leg of a three-phase three-level Neutral Point Clamped Inverter. (a) Conventional 3LI. (b) Bootstrapped 3LI

and bootstrapped 3LI are then carried out on an inductive load with open loop control which demonstrate the improved efficiency and waveform quality of the method. Finally, the method proposed in this paper is used to demonstrate experiments on a 5 kW motor with closed loop control across a wide modulation index range.

II. OPERATION AND CONTROL METHOD

A. Inverter States and Bootstrap Operation

During the operation of a 3LI, in each leg there are three states available to us. Table I shows these states, the respective pole voltage and how the state affects the charging of the bootstrap capacitors. The charging paths during the three states are shown in Fig. 2.

In the P switching state, the two switches in the upper half, Q_1 and Q_2 are on. The pole voltage with respect to the neutral point, V_{UZ} is E/2. In this state, the bootstrap capacitor C_{BS2} can charge the bootstrap capacitor C_{BS1}.

In the O switching state, the two switches in the middle, Q_2 and Q_3 are on. The pole voltage with respect to the neutral point, V_{UZ} is 0. In this state, the bootstrap capacitor C_{BS3} can charge the bootstrap capacitors C_{BS2} and C_{BS1}. There is also a path for C_{BS1} to be charged by C_{BS2}.

TABLE I.   SWITCHING STATES AND THE CORRESPONDING BOOTSTRAP OPERATION

Switching State	Device Switching Status Q_1 Q_2 Q_3 Q_4	Pole Voltage V_{UZ}	Bootstrap Operation	
			Target	Source
P	1 1 0 0	+E/2	C_{BS1}	C_{BS2}
O	0 1 1 0	0	C_{BS1} C_{BS2}	C_{BS2} & C_{BS3} C_{BS3}
N	0 0 1 1	-E/2	C_{BS2} C_{BS3}	C_{BS3} & V_{CC} V_{CC}

In the N switching state, the two switches in the bottom half, Q_3 and Q_4 are on. The pole voltage with respect to the neutral point, V_{UZ} is -E/2. In this state, the supply V_{CC} can charge the bootstrap capacitors C_{BS3} and C_{BS2}. There is also a path for C_{BS2} to be charged by C_{BS3}.

As shown in the table, charging from the supply V_{CC} happens only in the N state. Hence, the N state in each leg is crucial for maintaining the bootstrap capacitor voltages V_{BS1}, V_{BS2}, V_{BS3} above a certain value required for efficient operation of the IGBTs. Even with the N switching state, the P and O switching states are essential to charge the bootstrap capacitor C_{BS1} and this is used in the startup sequence which is addressed later.

A more detailed explanation of the bootstrap operation can be found in [2], [3].

B. Challenges with Conventional Space Vector PWM

The space vector diagram for a 3LI is shown in Fig. 3 [5]. The diagram is divided into six sectors denoted by the dashed lines. The small vector in each sector is denoted as the pivot vector for that sector [6]. The pivot vector and the large vector for each sector are shown in Table II.

In conventional space vector PWM (CSVPWM), the three nearest vectors to a reference vector V_{ref} are considered. One of the three nearest vectors will always be the pivot vector of that sector and the time interval for each vector in a period T_P can be calculated considering the pivot vector to be the origin for that sector [6]. The three nearest vectors V_x, V_y, V_z and their time intervals T_x, T_y, T_z satisfy (1) and (2).

$$V_{ref}\, T_P = V_x\, T_x + V_y\, T_y + V_z\, T_z \qquad (1)$$

$$T_P = T_x + T_y + T_z \qquad (2)$$

For example, for the reference vector $\mathbf{V}_a$ in Fig. 3, the conventional space vector PWM sequence over one switching period $2T_P$ and the corresponding time interval for each vector is shown in Table III. The switching waveforms are shown in Fig. 3(b). (The switching period is referred to as $2T_P$ as that is the switching period of the carrier used to obtain a similar sequence using carrier wave PWM. For a carrier wave switching frequency of 10 kHz, T_P would be 50 μs.)

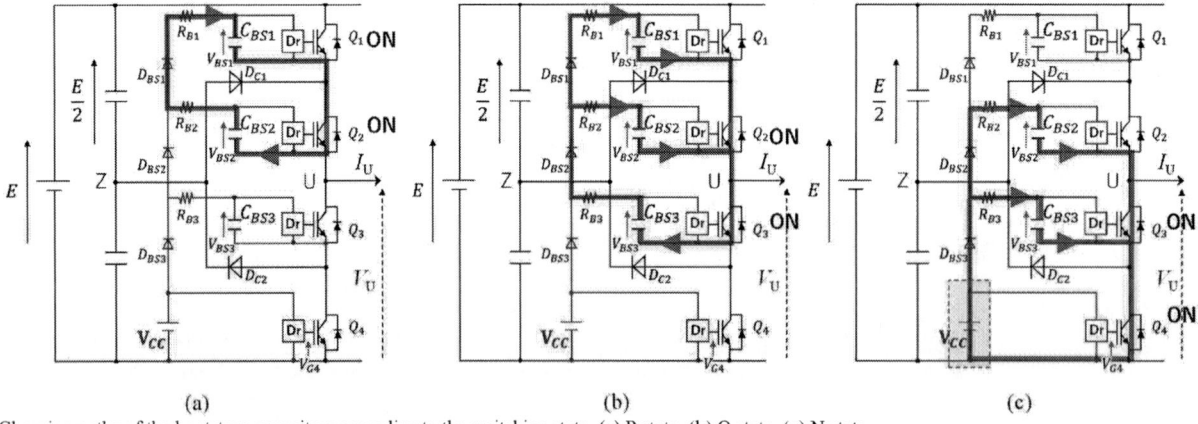

(a)   (b)   (c)

Fig. 2. Charging paths of the bootstrap capacitors according to the switching state. (a) P state. (b) O state. (c) N state

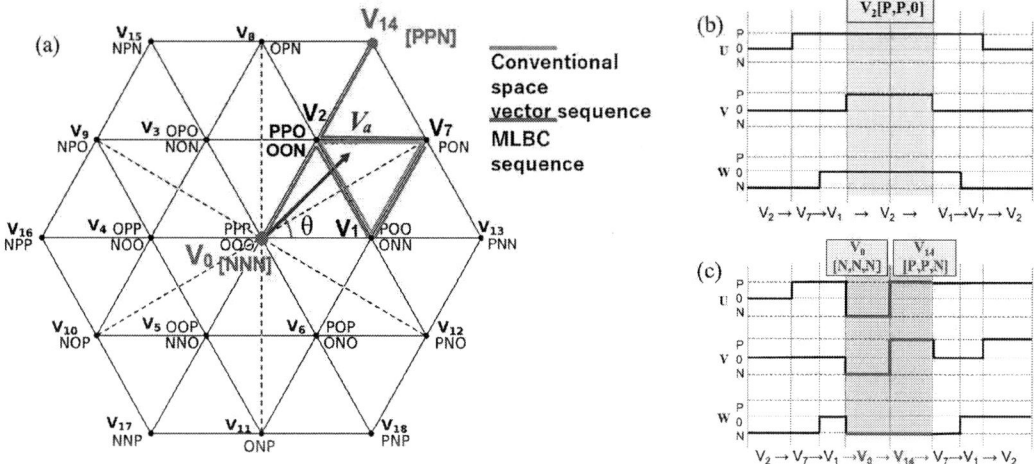

Fig. 3. (a) Space vector diagram for a 3LI. Switching states for the three legs over one switching period for (b) conventional space vector sequence and (c) MLBC sequence

TABLE II.   PIVOT VECTOR AND LARGE VECTOR FOR EACH SECTOR

Sector S	Pivot Vector V_S	Large Vector V_{LS}
1 (-30°<θ<30°)	V_1 (POO & ONN)	V_{13} (PNN)
2 (30°<θ<90°)	V_2 (PPO & OON)	V_{14} (PPN)
3 (90°<θ<150°)	V_3 (OPO & NON)	V_{15} (NPN)
4 (150°<θ<210°)	V_4 (OPP & NOO)	V_{16} (NPP)
5 (-150°<θ<-90°)	V_5 (OOP & NNO)	V_{17} (NNP)
6 (-90°<θ<-30°)	V_6 (POP & ONO)	V_{18} (PNP)

TABLE III.   CONVENTIONAL SPACE VECTOR SEQUENCE FOR V_a IN FIG. 3.

vector	V_2	V_7	V_1	V_2	V_2	V_1	V_7	V_2
state	OON	PON	POO	PPO	PPO	POO	PON	OON
time	$0.5T_2$	T_7	T_1	$0.5T_2$	$0.5T_2$	T_1	T_7	$0.5T_2$

TABLE IV.   PROPOSED SPACE VECTOR SEQUENCE (MLBC) FOR V_a IN FIG. 3.

vector	V_2	V_7	V_1	V_0	V_{14}	V_7	V_1	V_2
state	OON	PON	POO	NNN	PPN	PON	POO	PPO
time	$0.5T_2$	T_7	T_1	$0.5T_2$	$0.5T_2$	T_7	T_1	$0.5T_2$

In Table III the time interval T_2 for the pivot vector V_2 is equally divided between its two possible states. For a reference vector like V_a which lies outside the inner hexagon, in CSVPWM as in Table III, for -90°<θ<90° in Fig. 3, the N switching state for the U leg never occurs in any of the three nearest vectors. Hence, for one half of the fundamental period, the bootstrap capacitor voltages V_{BS1}, V_{BS2}, V_{BS3} continue to drop as shown in Fig. 4(a). Although the bootstrap capacitors charge in the next half of the fundamental period, this drop in one half of the fundamental period becomes more pronounced as the fundamental frequency decreases. For very low motor speeds, the bootstrap voltage could even fall below the gate threshold voltage of the IGBTs.

C. Proposed Space Vector PWM Sequence

In order to address the issue mentioned before, there is a need to have a vector with the N state for leg U in the sequence. For -90°<θ<90° the closest vector which contains the N state is the V_0 (NNN) vector. Not just the U leg but the V leg and W leg of the inverter are also in the N state for this vector. This paper proposes the Multilevel Bootstrap Charge (MLBC) sequence which makes use of this vector in the sequence. In each sector S, the pivot vector V_S can also be expressed as the vector sum of V_0 and the large vector V_{LS} as in (3).

$$\mathbf{V}_S = \mathbf{V}_0 + \mathbf{V}_{LS} \qquad (3)$$

By replacing the pivot vector with these two vectors for equal time intervals the voltage balance equation in (1) still remains satisfied. For the previous example of reference vector V_a in Fig. 3, the MLBC sequence over one switching period $2T_P$ and the corresponding time interval for each vector is shown in Table IV. The switching waveforms for the MLBC sequence are shown in Fig. 3(c).

Note that the MLBC sequence is also equivalent to the conventional sequence with respect to the current drawn from the neutral point [7], [8].

An advantage of this sequence is that it need not be used in every switching period. As a design example, a MLBC 20% sequence is one in which the MLBC sequence from Table IV is used once in every five switching periods while the conventional sequence from Table III is used for the other four switching periods.

III. SIMULATION RESULTS AND EXPERIMENTAL WORK

The MLBC sequence has more switching transitions in each switching period compared to the conventional sequence. Increasing the ratio of the MLBC sequence could charge the bootstrap capacitors more often but negatively impact the switching losses and THD. These considerations are explored in this section.

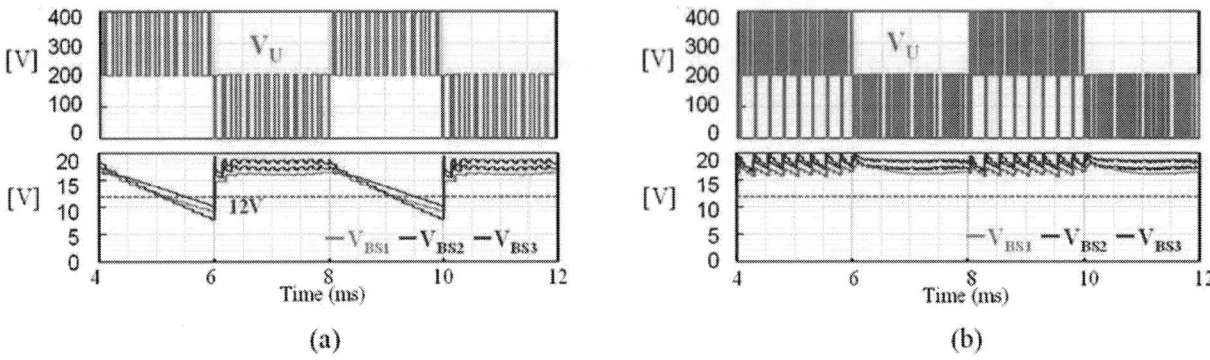

Fig. 4. Simulated waveforms for the pole voltage V_U and the bootstrap capacitor voltages for (a) conventional space vector sequence (b) MLBC 33% sequence

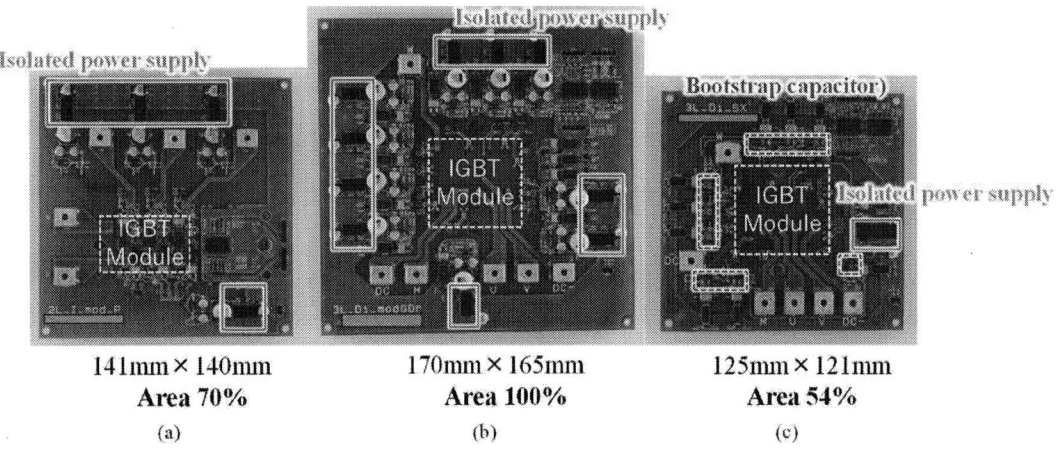

Fig. 5. Dimensions of a (a) 2LI board, (b) 3LI board and (c) bootstrapped 3LI board.

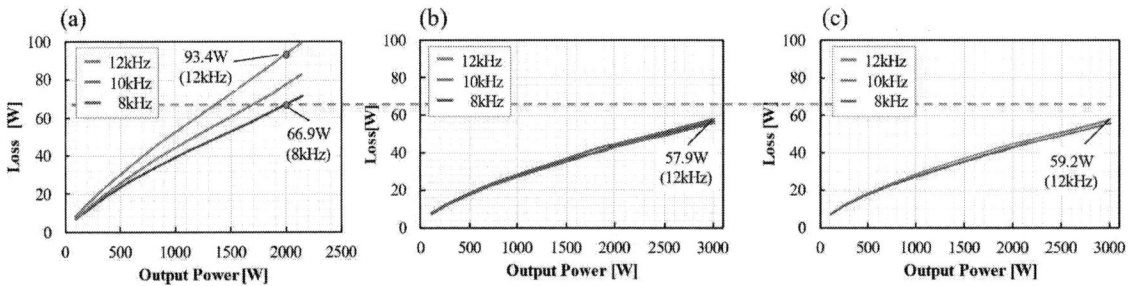

Fig. 6. Total loss against output power for various switching frequencies for (a) 2LI, (b) 3LI with conventional space vector PWM and (c) Bootstrapped 3LI with MLBC 33%

A. Sizing the Bootstrap Capacitors and Comparison of the Inverter Boards

Depending on the capacitance of the bootstrap capacitor, the sequence can be used to only charge the bootstrap capacitor when needed. SPICE simulations were carried out to estimate the size of the bootstrap capacitors. The feasibility of the operation was investigated for various capacitor sizes and MLBC ratios. Fig. 4(b) shows the simulated waveforms for the pole voltage V_U and the voltages V_{BS1}, V_{BS2}, V_{BS3} when the MLBC sequence is applied once every three switching periods (MLBC 33%). The voltages remain above a baseline of 12V.

In our bootstrapped 3LI board, we used bootstrap capacitors of 2.2 µF as confirmed by the SPICE simulations. A comparison of board sizes for the 2LI, 3LI and bootstrapped 3LI are shown in Fig. 5. The bootstrapped 3LI board leads to a board having approximately half the size of the conventional 3LI board and comparable to the 2LI board.

B. Loss and THD Comparisions with Inductive Load

In order to evaluate the bootstrapped 3LI board, we conducted comparisons between the 2LI board, conventional 3LI board and bootstrapped 3LI board.

Before performing experiments with a motor, we performed experiments with a star connected inductive load having inductance 1.2 mH and resistance 16 Ω. A fundamental frequency of 250 Hz and various switching frequencies (f_{sw}) were used. The results are shown in Fig. 6.

As seen in Fig. 6, going from a 2LI to 3LI there is a significant decrease in the switching loss as compared to the conduction loss. Therefore, increasing f_{sw} in the case of a 3LI does not lead to the total loss increasing as much. Hence, compared to conventional 3LI, although the MLBC sequence has more switch transitions per period and a slightly higher switching loss, the performance is still improved compared to the 2LI. Thermal issues due to the higher loss deter increasing the power beyond 2 kW for the 2LI board. On the other hand, even on increasing the output power by 1.5 times to 3 kW, the total losses are still less than the 2LI version.

Next, we performed simulations to see the impact of the MLBC sequence on the THD of the load current. Fig. 7(a) shows the THD for different modulation indices at a switching frequency of 10 kHz. The modulation index over here is given by the ratio of the fundamental frequency component of the pole voltage V_{UZ} to E/2. Although the MLBC 33% sequence performs slightly worse than the conventional space vector sequence, it is still better than the 2LI version.

Increasing the ratio of MLBC makes it possible to decrease the size of the bootstrap capacitor further but at the risk of increasing the switching losses and the THD. For our application, using the MLBC 33% sequence makes the performance comparable to the conventional 3LI board while still being able to charge the bootstrap capacitors.

C. Startup Sequence

There remains an issue of charging the bootstrap capacitors for the very first time when starting the motor operation. The bootstrap capacitors have no charge, so trying to operate by applying the MLBC sequence could result in an unintended voltage being applied on the motor. To address this issue, each leg was operated in phase with 50% P state and 50% N state. Although there is a phase difference of 120° in the control signals of each leg during normal operation, operating each leg with the same PWM during startup ensures that there is no voltage applied on the motor. Avoiding the O state helps to avoid drawing current from the from the neutral point and using the P and N states alternately ensures that the bootstrap capacitors C_{BS2} and C_{BS3} get charged during the N state and the bootstrap capacitor C_{BS1} gets charged during the P state. It takes a few switching periods for the capacitors to get charged and hence the entire startup sequence takes a few milli seconds with minimal impact on the motor operation.

D. Motor Experiments

Finally, the bootstrapped 2LI with the MLBC control is evaluated using a motor drive. The MLBC sequence was implemented on the TMS320F28003x microcontroller and used to control a 5 kW motor in a closed loop. The DC bus voltage was fixed at 350 V and acceleration/deceleration tests were carried out. Fig. 8(a) shows the motor current as the motor speed is increased from 20 rpm to 2000 rpm. The power output at 2000

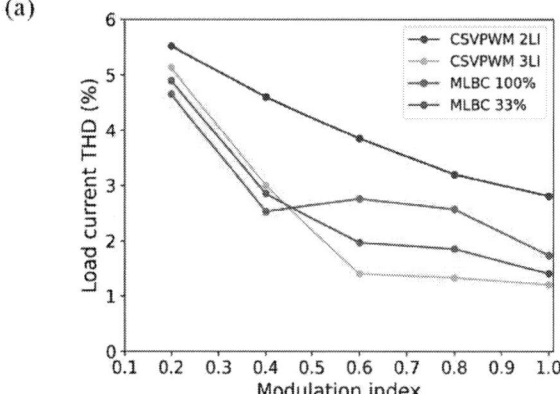

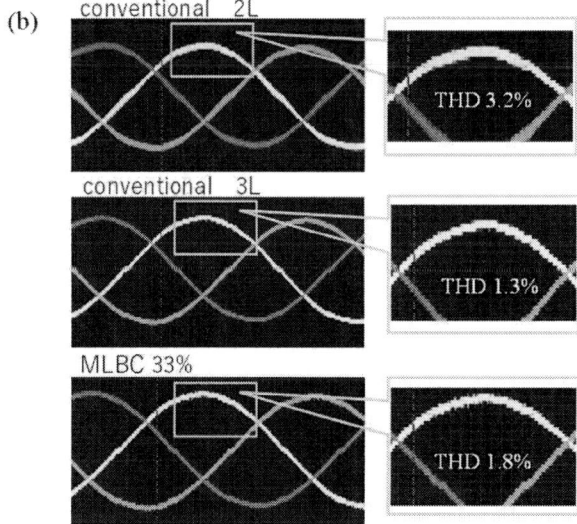

Fig. 7. (a) THD of the load current vs the modulation index for different PWM sequences. (b) Snapshot of the current waveforms for modulation index 0.8

rpm is 5 kW. Stable operation of the bootstrapped board is confirmed across the various speeds.

Fig. 8(b) shows the associated waveforms at a motor speed of 2000 rpm. The bootstrap capacitor voltages V_{BS1}, V_{BS2}, V_{BS3} and the gate signal V_{GV4} for switch Q4 of leg V are shown. It can be seen that the bootstrap voltages remain above a voltage of 14 V. Their charging coincides with the NNN pulse which appears in V_{GV4} and can be seen in Fig. 8(c). The neutral point voltage V_Z is also held stable during the entire operation.

As shown in this section, a 5 kW motor drive is successfully demonstrated using the boot strapped 3LI board with the MLBC control.

IV. CONCLUSIONS

The MLBC sequence to operate a bootstrapped 3LI is introduced and experimental verification on a 5 kW motor drive was shown. The proposed sequence is based on the averaging of the $\mathbf{V_0}$ vector and the corresponding large vector in each sector. Based on the size of the bootstrap capacitors, this sequence need

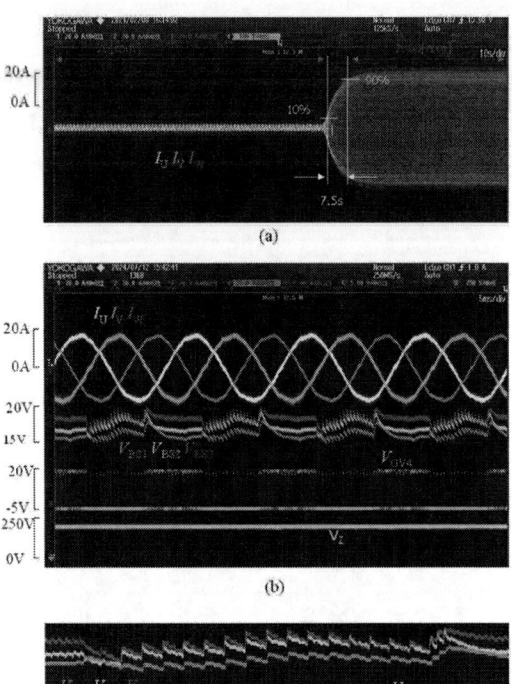

Fig. 8. (a) Motor current for the three legs as the motor speed increases from 20 rpm to 2000 rpm. (b) Motor current, bootstrap capacitor voltages, gate voltage at a motor speed of 2000 rpm. (c) A snapshot with the bootstrap capacitor voltages and gate voltage V_{GV4} for switch Q_4 of leg V.

not be used every switching cycle offering more flexibility. Furthermore, experimental analysis with an LR load showed that a bootstrapped 3LI with MLBC allows to increase the power by 1.5 times while keeping the size of the board comparable to that of a 2LI. Compared to previous approaches, with the proposed method stable motor operation and charging of the bootstrap capacitors was achieved, even at very low speeds.

Although the switching loss and THD worsen slightly compared to the conventional space vector sequence, there is still a significant improvement over a conventional 2LI. Adjusting the ratio of the MLBC sequence to the conventional sequence allows the tradeoffs to be tailored according to the application. Consequently, the proposed MLBC sequence is a

promising solution for applications looking to switch from a 2LI to a 3LI without an increase in the board size.

Another parameter that can be varied is the time interval for the vector V_0 in each switching period. This aspect has not been addressed in this paper but is planned for consideration in future work.

ACKNOWLEDGMENT

We would like to thank Mr. Yasuhiro Arai for his support with this project and Dr. Satoshi Nakazawa and Dr. Toru Yamada for their invaluable feedback and insights on this paper through various stages.

REFERENCES

[1] A. Nabea, I. Takahashi, and H. Akagi, "A new neutral-point clamped PWM inverter," IEEE Trans. Ind. Appl., vol. IA-17, no. 5, pp. 518–523, Sep./Oct. 1981.

[2] Nguyen Qui Tu Vo and Dong-Choon Lee, "Bootstrap power supply for three-level neutral-point-clamped voltage source inverters," Proceedings of the 7th International Power Electronics and Motion Control Conference, Harbin, China, 2012, pp. 1038-1043.

[3] J. -H. Jung, H. -K. Ku, W. -S. Im and J. -M. Kim, "A Carrier-Based PWM Control Strategy for Three-Level NPC Inverter Based on Bootstrap Gate Drive Circuit," in IEEE Transactions on Power Electronics, vol. 35, no. 3, pp. 2843-2860, March 2020.

[4] W. Yao, H. Hu and Z. Lu, "Comparisons of Space-Vector Modulation and Carrier-Based Modulation of Multilevel Inverter," in IEEE Transactions on Power Electronics, vol. 23, no. 1, pp. 45-51, Jan. 2008.

[5] P. F. Seixas, M. A. Severo Mendes, P. Donoso-Garcia and A. M. N. Lima, "A space vector PWM method for three-level voltage source inverters" APEC 2000. Fifteenth Annual IEEE Applied Power Electronics Conference and Exposition (Cat. No.00CH37058), New Orleans, LA, USA, 2000, pp. 549-555 vol.1

[6] Beig, A.R. & Narayanan, G. & Ranganathan, V.T. (2002). Space vector based synchronized PWM algorithm for three level voltage source inverters: principles and application to V/f drives. 1249 - 1254 vol.2. 10.1109/IECON.2002.1185453.

[7] S. Busquets-Monge, J. Bordonau, D. Boroyevich and S. Somavilla, "The nearest three virtual space vector PWM - a modulation for the comprehensive neutral-point balancing in the three-level NPC inverter," in IEEE Power Electronics Letters, vol. 2, no. 1, pp. 11-15, March 2004.

[8] C. -Q. Xiang, C. Shu, D. Han, B. -K. Mao, X. Wu and T. -J. Yu, "Improved Virtual Space Vector Modulation for Three-Level Neutral-Point-Clamped Converter With Feedback of Neutral-Point Voltage," in IEEE Transactions on Power Electronics, vol. 33, no. 6, pp. 5452-5464, June 2018.

A Complementary Carrier based PWM Strategy for Average Current Sampling of Three-Phase Inverter Using Single Current Sensor

Byeong-Il Kim
School of Electronics and Electrical Engineering
Dankook University
South Korea
kbi@dankook.ac.kr

Joon-Seok Kim
School of Electronics and Electrical Engineering
Dankook University
South Korea
KJS1702@dankook.ac.kr

Yeongsu Bak
Department of Electrical Energy Engineering
Keimyung University
South Korea
ysbak@kmu.ac.kr

June-Seok Lee
School of Electronics and Electrical Engineering
Dankook University
South Korea
ljs@dankook.ac.kr

Abstract— This paper proposes a complementary carrier-based pulse width modulation (PWM) strategy for the average current sampling of three-phase inverter using single current sensor (SCS) in dc-link. Conventional three-phase current reconstruction method for a three-phase inverter using SCS reconstructs the three-phase currents by measuring the two-phase currents corresponding to the two active voltage vectors. However, the current measured by SCS is not the average current. In addition, there is a current reconstruction dead zone (CRDZ) near the boundary of each sector. The proposed PWM strategy outputs the voltage of the inverter using two carriers that are inverted in phase with each other. This strategy compares the inverted carrier with the voltage references of one or two phases and compares the remaining phase voltage references with the noninverted carrier. This allows us to measure the average current and significantly reduces the CRDZ to the corners of the space vector hexagon. The validity of the proposed strategy is verified by simulation results.

Keywords—Single Current Sensor (SCS), DC-Link Sensor, One-Shunt Sensor Operation, Current Sensing, Inverter

I. INTRODUCTION

Vector control with symmetric space vector pulse width modulation (SVPWM) is widely used for stable, high-performance motor control. However, it requires at least two current sensors which increase the volume of the system and cost. To address these issues, recent studies have focused on a motor drive strategy using a single current sensor (SCS) that reconstructs the three-phase currents with only one current sensor in the dc-link, as shown in Fig. 1.

The conventional method for reconstructing three-phase currents uses two active voltage vectors adjacent to the reference voltage vector to obtain two phase currents over one cycle. However, the reconstructed current does not represent the average current, and a current reconstruction dead zone (CRDZ) arises where current reconstruction becomes impossible.

In [2]-[4], although a method has been proposed to enable current reconstruction even when the voltage reference is in the CRDZ, the average current still cannot be determined, and the reconstruction error may increase further in the CRDZ. In [5]-[6], methods for predicting current have been proposed, but they require additional calculation time and are sensitive to motor parameter errors.

In [7]-[10], methods to minimize or move the CRDZ were proposed. In [7], [8], and [9], the area of the CRDZ was modified by moving the location of the current sensor.

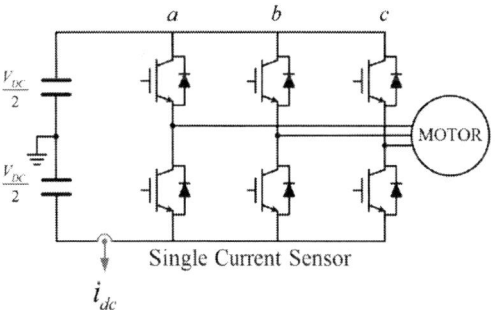

Fig. 1. Three-phase inverter system using single current sensor

However, these methods have the disadvantage of requiring additional circuit configurations that increase the volume and cost of the system or have relatively large CRDZ. In [10], the method facilitates average current measurement and reduces the CRDZ without changing the position of the current sensor. However, it is characterized by a switching imbalance between the upper and lower switches.

In this paper, we propose a novel PWM strategy that enables average current measurement even under large phase current ripples and has a narrow CRDZ. In particular, the proposed strategy has the advantage of being implemented in a very intuitive and simple way without the contribution of additional circuits and of having a very short computation time for implementation with a digital signal processor.

II. CONVENTIONAL SCS CONTROL STRATEGY

A. Basic Principle of Conventional SCS Control

In the SVPWM method, which outputs voltage using the two active voltage vectors adjacent to the reference voltage vector, these two active voltage vectors enable the SCS to measure the currents of two phases. The currents of SCS according to the active voltage vector can be seen in Table I. To accurately measure phase-current using SCS, there is a minimum time (T_{min}) during which the active voltage vector must be maintained. T_{min} can be expressed as the sum of dead time (T_{dead}), settling time of the dc-link current (T_{set}), and the analog to digital conversion time ($T_{A/D}$). This is determined as follows:

$$T_{min} = T_{stable} + T_{A/D} = T_{dead} + T_{set} + T_{A/D}, \quad (1)$$

979-8-3315-1612-3/25 $31.00 © 2025 IEEE

TABLE I.	RELATIONSHIP BETWEEN SWITCHING STATE AND PHASE CURRENT	
Voltage vector	Switching state	DC-link current i_{dc}
$\vec{V}_1$	100	$+i_a$
$\vec{V}_2$	011	$-i_c$
$\vec{V}_3$	010	$+i_b$
$\vec{V}_4$	011	$-i_a$
$\vec{V}_5$	001	$+i_c$
$\vec{V}_6$	101	$-i_b$
$\vec{V}_0$, $\vec{V}_7$	000, 111	0

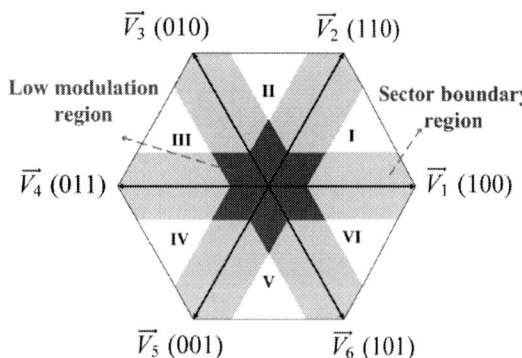

Fig. 2. Current reconstruction dead zone in conventional SCS control strategy.

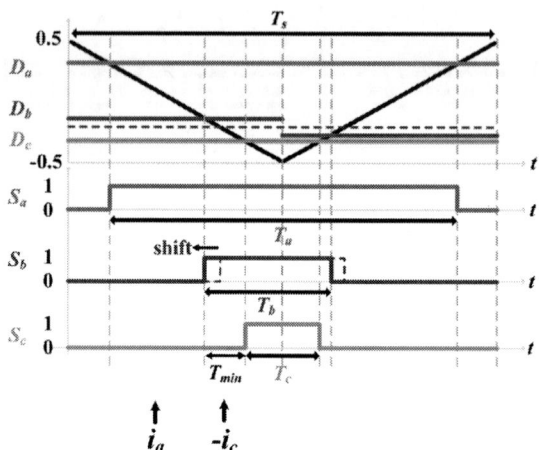

Fig. 3. Voltage injectioin in current reconstrtion dead zone

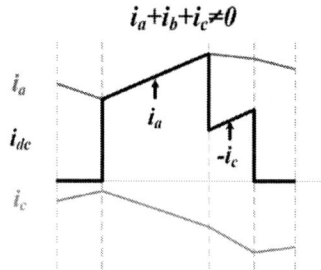

Fig. 4. Causes of current reconstruction error in conventional strategy

where, T_{stable} represents the combined time of the T_{dead} and T_{set}. The SCS must initiate current sampling only after the active voltage vector has been maintained for at least the duration of T_{stable}. If the time of active voltage vector required to output the reference voltage vector is shorter than T_{min}, the reliability of the reconstructed three-phase currents is reduced. As shown in Fig. 2, this CRDZ exists near the boundaries of the sectors. During motor operation it is inevitable for the reference voltage vector to pass through the CRDZ. Therefore, it is necessary to enable three-phase currents reconstruction even when the reference voltage vector is located within the CRDZ.

B. A Strategy For Reconstruction Three-phase Currents Within The CRDZ

If the time of active voltage vector is shorter than T_{min}, voltage is injected to secure T_{min} as shown in Fig. 3. In Fig. 3, D_a, D_b, and D_c are the values obtained by dividing the phase voltage references of phases a, b, and c by half of dc-link voltage, respectively. In addition, S_a, S_b, and S_c are switching functions that indicate the activation of the upper switches of phases a, b, and c, respectively. The average output voltage vector after voltage injection remains unchanged and is as follows:

$$\vec{V}_{ref} = (T_a\vec{V}_1 + T_b\vec{V}_3 + T_c\vec{V}_5)/T_s, \qquad (2)$$

where, T_a, T_b, and T_c represent the times at which S_a, S_b, and S_c become 1, respectively, while T_s denotes the switching period.

Although the issue of the CRDZ can be addressed through voltage injection, several problems remain. Additional calculation steps are required to determine the appropriate voltage to inject. Furthermore, as shown as Fig. 4, the currents of two phases cannot be sampled at the center of the carrier simultaneously, resulting in the inability to determine the average current. In addition, due to the differences in sampling times, the current reconstruction error i_{err}, which represents the difference between the reconstructed phase current and the actual average phase current, significantly increases. i_{err} becomes particularly severe when current ripple is significant, leading to a reduction in system stability.

III. NOVEL STRATEGY FOR AVERAGE CURRENT MEASUREMENT IN SCS SYSTEM

A. Basic Principles of Complementary Carrier Based PWM

Proposed strategy compares one or two voltage references with the inverted carrier. When a phase voltage reference is compared to an inverted carrier, the section in which the upper switch of the corresponding phase is activated moves to both ends of the PWM cycle. According to (2), the average output voltage vector does not change, regardless of which phase is compared with the inverted carrier.

The phase current is measured once per T_s, at the exact center of the carrier. By measuring different currents over $2T_s$, the average three-phase current can be reconstructed. This strategy can eliminate the presence of the CRDZ in the sector boundary region. To narrow the CRDZ as much as

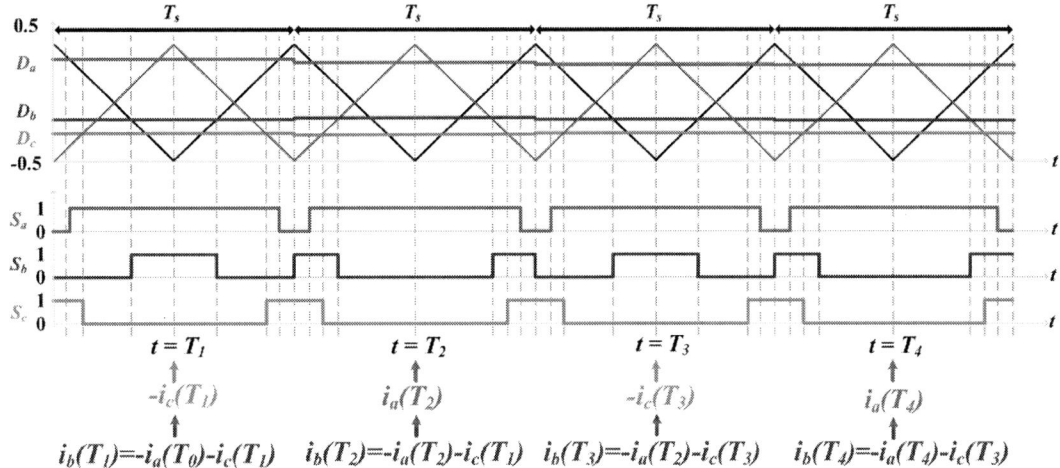

$$i_b(T_1)=-i_a(T_0)-i_c(T_1) \quad i_b(T_2)=-i_a(T_2)-i_c(T_1) \quad i_b(T_3)=-i_a(T_2)-i_c(T_3) \quad i_b(T_4)=-i_a(T_4)-i_c(T_3)$$

Fig. 5. Proposed current reconstruction strategy

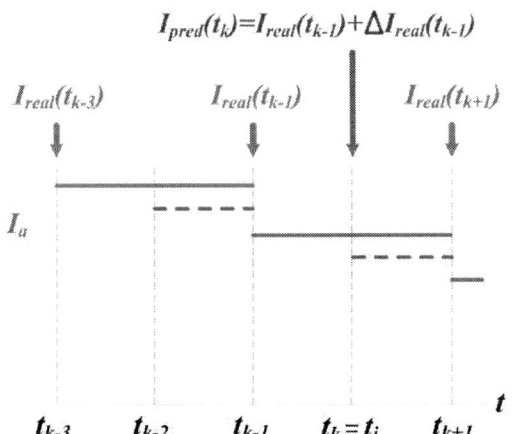

$$I_{pred}(t_k)=I_{real}(t_{k-1})+\Delta I_{real}(t_{k-1})$$

Fig. 6. Proposed Current Prediction Method

possible, the phases to be compared with the inverted carrier in each sector are predetermined.

As shown in Fig. 5, during the first PWM cycle, the minimum voltage reference D_c is compared with the inverted carrier. In the subsequent cycle, both the minimum and middle voltage references, D_c and D_b, are compared with the inverted carrier. In the following cycle, the minimum voltage reference D_c is again compared with the inverted carrier. This sequence is repeated to ensure that the SCS measures the current of a different phase in each cycle. This approach offers a significant advantage in systems with large current ripple, as it consistently allows for the accurate measurement of the average phase current.

B. Error Compensation Method for Phase Current Time Delay

As shown in Fig. 5, the phase currents directly measured for phases a and c are updated only once every two T_s. Therefore, in any cycle, i_{err} occurs in either phase a or phase c. As a result, i_{err} also consistently occur in the remaining phase current, calculated based on the assumption that the sum of the three-phase currents is zero. The method to reduce the i_{err} caused by this issue is shown in Fig. 6. If the

current of phase a is not measured at t_i, the current at that time can be predicted based on the previously measured variations and is given as follows:

$$
\begin{aligned}
I_{a\_pred}(t_i) &= I_{a\_real}(t_{i-1}) + \Delta I_{a\_real}(t_{i-1}) \\
&= I_{a\_real}(t_{i-1}) + \frac{I_{a\_real}(t_{i-1}) - I_{a\_real}(t_{i-3})}{2},
\end{aligned} \quad (3)
$$

where, t_i refers to the current measurement point in the i-th sampling period within the phase current measurement interval (PCMI), which refers to the intervals where the currents of each phase are either directly measured by SCS or predicted within one AC cycle. $I_{a\_pred}(t_i)$ and $I_{a\_real}(t_{i-1})$ denote the predicted current and the actual current at that time, respectively.

To verify the accuracy of the prediction, it is assumed that the motor operates in a steady state with a constant current frequency f and peak current I_{peak}. Under these conditions, $I_{a\_pred}(t_i)$ can be reobtained as follows:

$$
\begin{aligned}
V_a(t) &= V_m \cos(\omega t) \quad (\omega = 2\pi f) \\
I_{a\_real}(t) &= I_{peak} \cos(\omega t - \theta) \\
I_{a\_pred}(t_i) &= I_{a\_real}(t_{i-1}) + \frac{I_{a\_real}(t_{i-1}) - I_{a\_real}(t_{i-3})}{2} \\
&= I_{peak} \cos(\omega t_{i-1} - \theta) \\
&\quad + 0.5 I_{peak} \left(\cos(\omega t_{i-1} - \theta) - \cos(\omega t_{i-3} - \theta) \right)
\end{aligned} \quad (4)
$$

The error rate $R_{err}(t_i)$ between the predicted current and the actual current at t_i, is calculated as follows:

$$
\begin{aligned}
R_{err}(t_i) &= \frac{\left| I_{a\_real}(t_i) - I_{a\_pred}(t_i) \right|}{I_{peak}} \times 100\% \\
&= \left| \cos(\omega t_{t_i} - \theta) - 1.5\cos(\omega t_{i-1} - \theta) + 0.5\cos(\omega t_{i-3} - \theta) \right| \times 100\%
\end{aligned} \quad (5)
$$

If the current is measured at the t_i, the $R_{err}(t_i)$ is considered to be 0 instead of the value given in (5). As shown in Fig. 7, in the proposed PWM strategy, the PCMI corresponds to the intervals where each phase voltage is either the maximum or minimum among the three-phase voltages. The average error rate, denoted as $\overline{R_{err}}$, in the PCMI is approximated as follows:

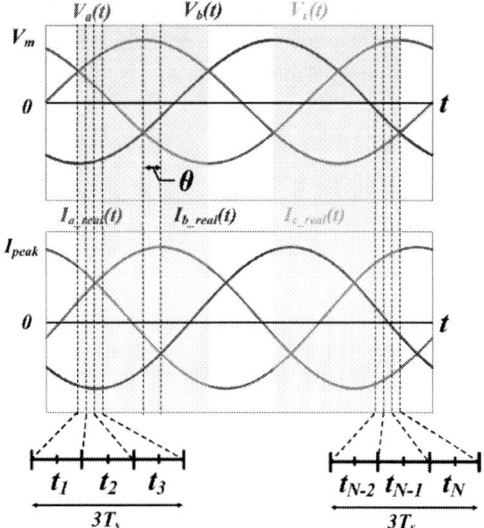

Phase Current Measurement Interval (PCMI)

Fig. 7. Interval for Calculating Error Rate

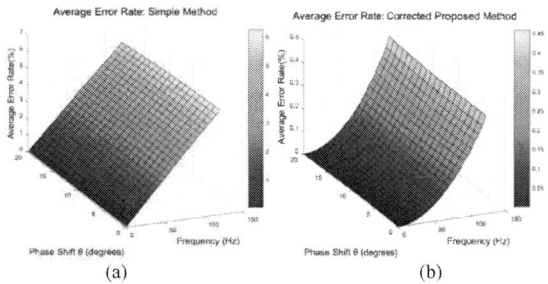

(a)   (b)

Fig. 8. Average Error Rate (a) Without Proposed Prediction Method (b) With Proposed Prediction Method

$$\overline{R_{err}} = \frac{1}{N}\sum_{i=1}^{N} R_{err}(t_i)$$
$$= \frac{1}{N}\left\{R_{err}(t_i) + R_{err}(t_{i+1}) + R_{err}(t_{i+2}) + ... + R_{err}(t_N)\right\} \quad ,(6)$$

where, N represents the total number of samples in PCMI. Since PCMI corresponds to 2/3 of one current cycle, N is calculated as follows based on the total number of samples N_{total} in one current cycle, f, and the T_s:

$$N = \left\lfloor \frac{2}{3} N_{total} \right\rfloor = \left\lfloor \frac{2}{3} \frac{1}{f \cdot T_s} \right\rfloor , \quad (7)$$

where, if N has a decimal point, it is rounded down.

To compare the results before and after applying the proposed current prediction method, a 3D graph of $\overline{R_{err}}$ was plotted, considering variations in f and θ while keeping $T_s = 100\mu s$ constant. As shown in Fig. 8(a), and Fig. 8(b), when the proposed current prediction method is used, $\overline{R_{err}}$ is significantly reduced.

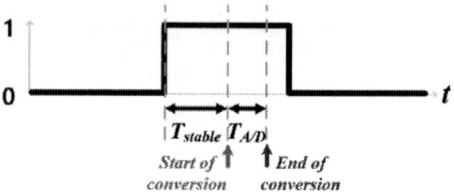

Fig. 9. Three-phase inverter system using single current sensor

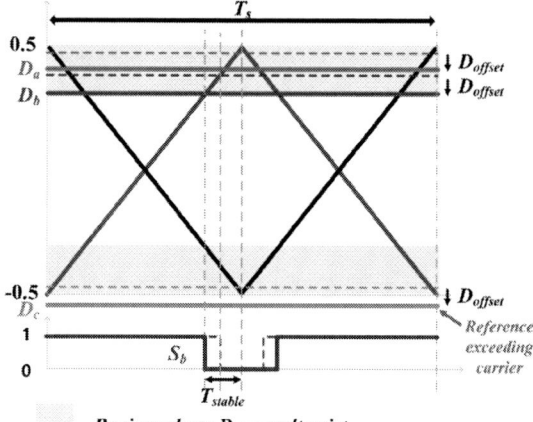

Fig. 10. Case where D_{offset} cannot be applied

C. Complementary Carrier Based PWM Strategy Considering Constraints in SCS System

To measure the current at the exact center of the carrier, the active voltage vector must be applied at least T_{stable} before the center. As shown as Fig. 9, since the A/D conversion time, $T_{A/D}$ is typically shorter than T_{stable}, it is assumed that the A/D conversion is completed before the active voltage vector ends. To satisfy constraint, the following conditions must be met.

$$-0.5 + \frac{2T_{stable}}{T_s} \le D_{mid} \le 0.5 - \frac{2T_{stable}}{T_s} , \quad (8)$$

where, D_{mid} is middle voltage reference. If D_{mid} fails to satisfy the constraint, an offset voltage that satisfies the (3) is applied equally to the three-phase voltage references. In this case, the offset voltage, D_{offset} is as follows:

$$D_{offset} = \begin{cases} 0.5 - \dfrac{2T_{stable}}{T_s} - D_{mid} & \left(when\ D_{mid} > 1 - \dfrac{2T_{stable}}{T_s}\right) \\[2mm] \dfrac{2T_{stable}}{T_s} - D_{mid} & \left(when\ D_{mid} < \dfrac{2T_{stable}}{T_s}\right) \end{cases} . \quad (9)$$

As shown in Fig. 10, since the voltage reference cannot exceed the carrier range, there are instances where the D_{offset} cannot be applied. These cases occur when the reference voltage vector is within the CRDZ, which, as shown in Fig. 12, is located at the corners of the space vector hexagon. Due to these CRDZ, the upper limit of the inverter's linear modulation range, MI_{linear_max}, is defined as follows:

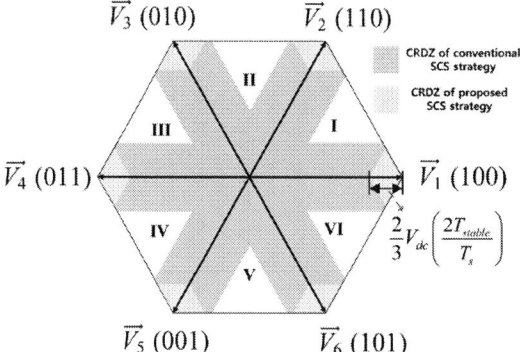

Fig. 11. Case where D_{offset} cannot be applied

TABLE II.　SIMULATION PARAMETERS

Parameters	Mark	Value	Unit
Stator resistance	R_s	0.295	Ω
Rotor resistance	R_r	0.379	Ω
Total stator inductance	L_s	60.79	mH
Total rotor inductance	L_r	60.79	mH
Magnetizing inductance	L_m	59	mH
Pole	P	4	-
Current stabilization time	T_{stable}	8	μs
Analog to digital conversion time	$T_{A/D}$	2	μs
Switching period	T_s	100	μs
DC-Link voltage	V_{dc}	200	V

$$\left|\vec{V}_{ref}\right| = \frac{2}{3}V_{dc} \cdot MI \quad (0 \le MI \le 1)$$

$$MI_{linear\_max} = \begin{cases} 1 - \dfrac{2T_{stable}}{T_s} & \left(when \ \dfrac{2T_{stable}}{T_s} > \dfrac{2-\sqrt{3}}{2}\right) \\ \dfrac{\sqrt{3}}{2} & \left(when \ \dfrac{2T_{stable}}{T_s} <= \dfrac{2-\sqrt{3}}{2}\right) \end{cases}, \quad (10)$$

where, $\left|\vec{V}_{ref}\right|$ represents the magnitude of the reference voltage vector and can be expressed in terms of the modulation index (MI) and the input dc-link voltage (V_{dc}). If the inverter operates at an MI lower than MI_{linear_max}, CRDZ can be consistently avoided.

IV. SIMULATION RESULTS

To verify whether the proposed complementary carrier-based PWM strategy can obtain the average values of three-phase currents over a wide operating range, an induction motor drive was simulated under various operating conditions using PSIM. The parameters used in the simulation are listed in Table II, and all simulations were conducted with current prediction applied in the PCMI.

In Fig. 12, a comparison between the conventional and proposed strategies is shown at MI = 0.3. As illustrated in Fig. 12(a), the conventional strategy exhibits irregular i_{err} consistently, and the reconstructed current shows significant fluctuations, especially when the sector of the voltage reference vector changes. In contrast, as shown in Fig. 12(b),

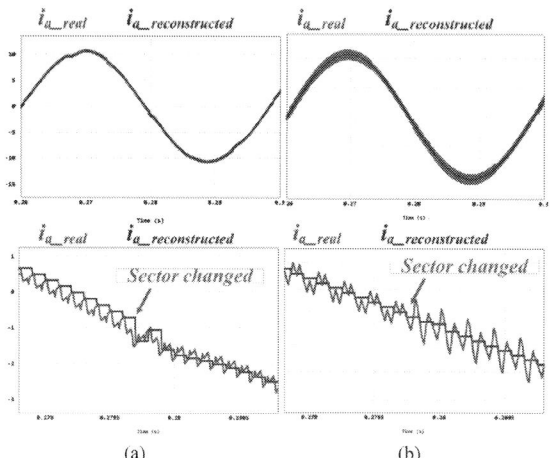

Fig. 12. Comparison of actual and reconstructed phase currnt (MI = 0.3, Operating at 700rpm) (a) Conventional current reconstruction strategy (b) Proposed current reconstruction strategy

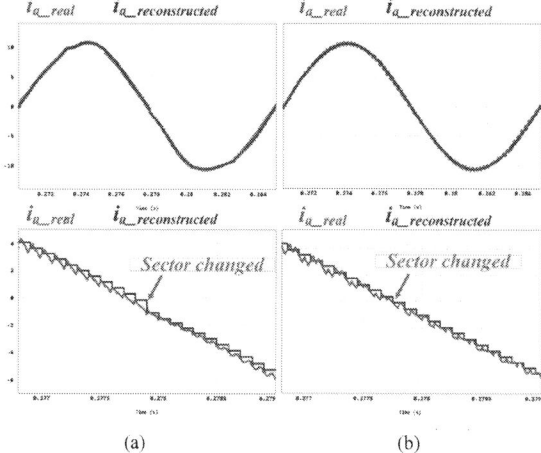

Fig. 13. Comparison of actual and reconstructed phase currnt (MI = 0.75, Operating at 2000rpm) (a) Conventional current reconstruction strategy (b) Proposed current reconstruction strategy

the proposed strategy exhibits a significantly smaller i_{err}, and the reconstructed current remains stable even when the sector changes.

In Fig. 13, a comparison between the conventional and proposed strategies is shown at MI = 0.75. As shown in Fig. 13(a) and Fig. 13(b), while the conventional strategy still has a large i_{err}, the proposed strategy maintains small i_{err} even at high MI.

The reconstructed three-phase currents of the conventional and proposed strategies are shown in Fig. 13, and the d-q axis currents in the synchronous reference frame, calculated from these, are compared with the actual values over a wide operating range. In Fig. 14(a) and Fig. 14(b), the differences between the conventional and proposed strategies at a low MI can be observed, while Fig. 14(c) and Fig. 14(d) show the differences between the two strategies at a high MI.

Fig. 15 illustrates the differences before and after applying the proposed current prediction method. Phase a and phase c currents are measured once every two sampling periods, and it can be observed that i_{err} exists when the

979-8-3315-1612-3/25 $31.00 © 2025 IEEE　　　1738

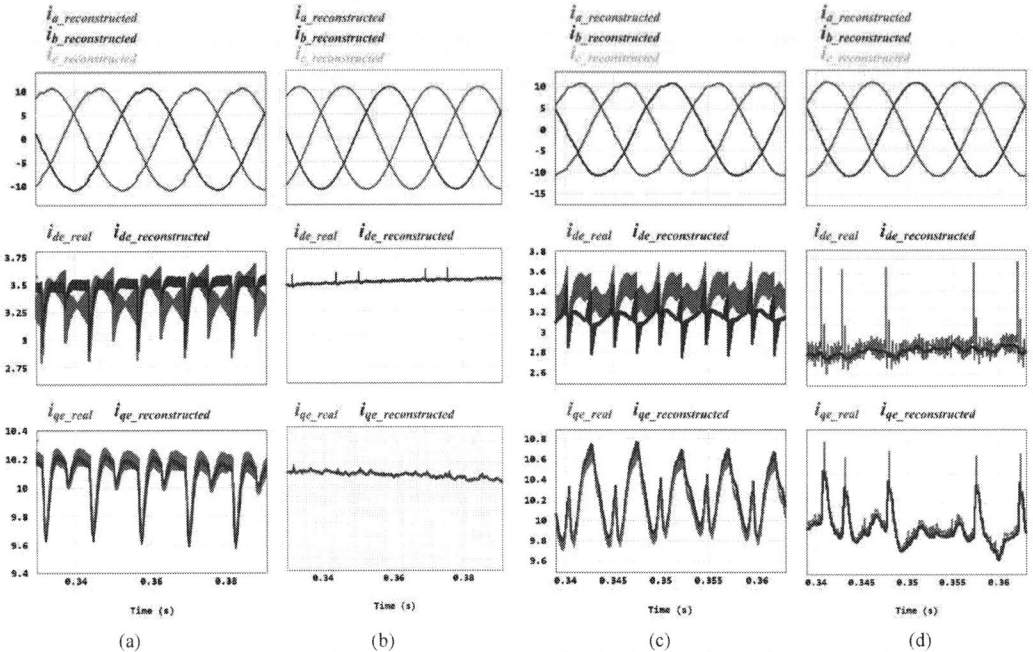

Fig. 14. Comparison of d-q axis current tracking performance between the conventional and proposed strategies (a) Conventional strategy at MI = 0.3, 700rpm (b) Proposed strategy at MI = 0.3, 700rpm (c) Conventional strategy at MI = 0.75, 2000rpm (d) Proposed strategy at MI = 0.75, 2000rpm

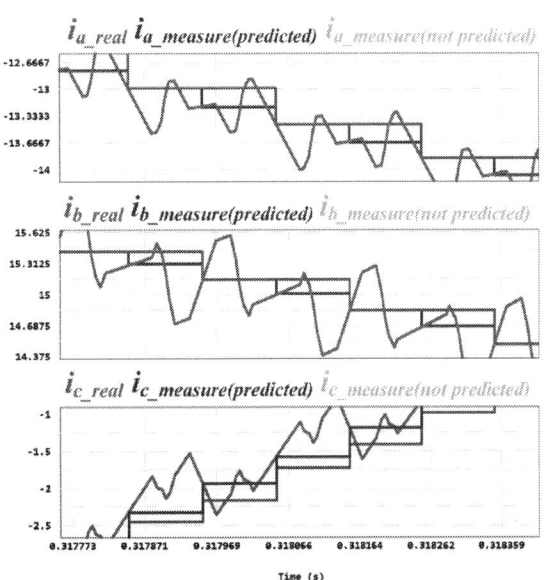

Fig. 15. Comparison of Current Prediction Method Applied and Not Applied

currents are not directly measured. When the proposed prediction method is applied, the predicted current closely matches the actual current, resulting in a significant reduction in i_{err}, which had been consistently observed in phase b.

In the proposed strategy, the reconstructed three-phase currents are closer to a pure sine wave across a wide range of MI compared to the conventional strategy. Additionally, the error between the reconstructed d-q axis currents and the actual d-q axis currents is reduced in the proposed strategy. Specifically, in the conventional strategy, the d-q axis currents fluctuate significantly when the sector changes, which may reduce stability during induction motor control. In contrast, the proposed strategy exhibits reduced fluctuationsin the d-q axis currents, resulting in improved stability.

V. CONCLUSIONS

This paper proposes a novel current reconstruction strategy for a three-phase inverter using single current sensor in dc-link. In the conventional strategy, the average current cannot be measured, and irregular current reconstruction errors consistently occur. In contrast, the proposed strategy allows for accurate measurement of the average current, with significantly smaller current reconstruction errors. In both the conventional and proposed strategies, the difference between the d-q axis currents calculated from the reconstructed three-phase currents and the actual d-q axis currents was analyzed through simulation. The results show that the proposed strategy achieves less fluctuation in the d-q axis, thereby improving motor control stability compared to the conventional strategy.

Additionally, implementing the conventional strategy requires extra computation time, potentially preventing the addition of other control algorithms. However, the proposed strategy can be implemented simply across all regions except the CRDZ, resulting in shorter computation times and making it easier to incorporate additional algorithms, such as overmodulation.

ACKNOWLEDGMENT

This work was supported by the Startup Growth Technology Development Project (RS-2024-00468283) funded by the Ministry of SMEs and Startups(MSS, Korea).

REFERENCES

[1] F. Blaabjerg, J. K. Pedersen, U. Jaeger and P. Thoegersen, "Single current sensor technique in the DC link of three-phase PWM-VS inverters: a review and a novel solution," in IEEE Transactions on Industry Applications, vol. 33, no. 5, pp. 1241-1253, Sept.-Oct. 1997.

[2] Y. Gu, F. Ni, D. Yang and H. Liu, "Switching-State Phase Shift Method for Three-Phase-Current Reconstruction With a Single DC-Link Current Sensor," in IEEE Transactions on Industrial Electronics, vol. 58, no. 11, pp. 5186-5194, Nov. 2011.

[3] J. -I. Ha, "Voltage Injection Method for Three-Phase Current Reconstruction in PWM Inverters Using a Single Sensor," in IEEE Transactions on Power Electronics, vol. 24, no. 3, pp. 767-775, March 2009.

[4] H. Kim and T. M. Jahns, "Current Control for AC Motor Drives Using a Single DC-Link Current Sensor and Measurement Voltage Vectors," in IEEE Transactions on Industry Applications, vol. 42, no. 6, pp. 1539-1547, Nov.-dec. 2006.

[5] J. -I. Ha, "Current Prediction in Vector-Controlled PWM Inverters Using Single DC-Link Current Sensor," in IEEE Transactions on Industrial Electronics, vol. 57, no. 2, pp. 716-726, Feb. 2010.

[6] B. Saritha and P. A. Janakiraman, "Sinusoidal Three-Phase Current Reconstruction and Control Using a DC-Link Current Sensor and a Curve-Fitting Observer," in IEEE Transactions on Industrial Electronics, vol. 54, no. 5, pp. 2657-2664, Oct. 2007.

[7] Y. Xu, H. Yan, J. Zou, B. Wang and Y. Li, "Zero Voltage Vector Sampling Method for PMSM Three-Phase Current Reconstruction Using Single Current Sensor," in IEEE Transactions on Power Electronics, vol. 32, no. 5, pp. 3797-3807, May 2017.

[8] Y. -S. Lai, Y. -K. Lin and C. -W. Chen, "New Hybrid Pulsewidth Modulation Technique to Reduce Current Distortion and Extend Current Reconstruction Range for a Three-Phase Inverter Using Only DC-link Sensor," in IEEE Transactions on Power Electronics, vol. 28, no. 3, pp. 1331-1337, March 2013.

[9] Y. Cho, T. LaBella and J. -S. Lai, "A Three-Phase Current Reconstruction Strategy With Online Current Offset Compensation Using a Single Current Sensor," in IEEE Transactions on Industrial Electronics, vol. 59, no. 7, pp. 2924-2933, July 2012.

[10] H. Lu, X. Cheng, W. Qu, S. Sheng, Y. Li and Z. Wang, "A Three-Phase Current Reconstruction Technique Using Single DC Current Sensor Based on TSPWM," in IEEE Transactions on Power Electronics, vol. 29, no. 3, pp. 1542-1550, March 2014.

Short-Circuit Ride-Through for a CRM-Based Soft-Switching Three-Phase Inverter

Xingyu Chen[1], Gibong Son[2] and Qiang Li[1]
[1]Center for Power Electronics Systems, Virginia Tech, Blacksburg VA USA
[2]Tesla, Palo Alto CA USA
chenxy@vt.edu

Abstract— This paper introduces the short-circuit ride-through operation for a critical conduction mode (CRM) based soft-switching three-phase three-wire inverter. A CRM - discontinuous conduction mode – discontinuous pulse width modulation hybrid modulation is implemented to a three-phase inverter to realize high efficiency and high power-density with hundreds of kilohertz operation. A fixed switching-frequency modulation is applied during short-circuit condition instead of the original hybrid modulation to enable the ride-through function. The necessity of the separate modulation is analyzed, and the parameters of the separate modulation are optimized. Finally, experiments are performed to verify the proposed short-circuit ride-through operation.

Keywords—three-phase inverter, CRM, soft switching, DPWM, short-circuit ride-through

I. INTRODUCTION

High efficiency is always desired for power converters in modern applications. As for three-phase inverters, commercial products can achieve a good efficiency around 98-99%, but the power density is always lower than $15W/in^3$ [1]. This is because the switching frequency is limited to 20-30kHz by the high switching losses of Silicon (Si)-based semiconductor devices, and the low switching frequency leads to a large volume of the passive components including the inductors and capacitors.

Wide bandgap (WBG) semiconductor devices, including Silicon-Carbide (SiC) devices and Gallium-Nitride (GaN) devices, have better performances compared to Si-based counterparts [2], [3], [4], [5], [6]. However, WBG-based three-phase inverters do not show much system-level benefit when they operate under the conventional continuous conduction mode (CCM) modulation because the switching frequency is still limited to a similar level, which leads to very little improvement in power density [7]. Soft-switching is an effective practice to push the switching frequency to a level of 10-20

times higher than the conventional converters to increase the power density without sacrifice of the efficiency [7], [8], [9], [10], [11], [12], [13]. One approach for three-phase inverter to realize soft-switching is proposed by Huang *et al.* [7]. This novel modulation combines critical conduction mode (CRM), discontinuous conduction mode (DCM) and discontinuous pulse width modulation (DPWM) to increase switching frequency, realize soft switching and limit the maximum switching frequency at the same time.

Three-phase inverters play a critical role in many applications including photovoltaic (PV) facilities, uninterruptible power supplies (UPS), inverter-based distributed generation systems, etc. [14], [15], [16], [17], [18]. In practical operations, inverters may encounter various abnormal situations such as unbalanced grid, voltage swell and short-circuit conditions. Son *et al.* studied the operation of this CRM-based soft switching inverter under unbalanced grid to realize a constant-power delivery [19]. On the other hand, the operation of such inverter under short-circuit condition has not been studied yet. During a short-circuit scenario, one common option is that the inverter may shut down itself by disabling all gate driver signals to realize self-protection, which is generally acceptable when powering a single load. However, when the inverter supplies a large network including some critical loads, shutting down is not an option. Instead, the inverter must be capable of riding through the fault condition without shutting down [20], [21], [22].

This article presents the short-circuit operation of the CRM-based three-phase inverter to realize the ride-through function. Section II reviews the CRM-based soft-switching modulation under normal operation and points out the issue under short-circuit condition. Section III discusses the modulation for short-circuit ride-through and provides the parameter optimization analysis. Section IV exhibits the experimental results. Section V concludes the article.

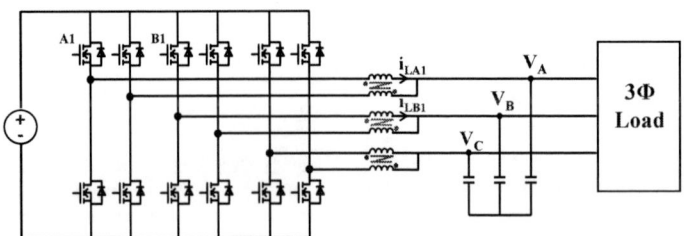

Fig. 1   Proposed 3-phase inverter stage of the auxiliary power supply for railway applications

979-8-3315-1612-3/25 $31.00 © 2025 IEEE

II. CRM-Based Inverter Modulation and Issue under Short-circuit Conditions

A. Review of CRM-Based Three-Phase Inverter Modulation[7]

Fig. 1 shows the circuit of a two-channel interleaved three-phase inverter with negative coupled inductors. A novel CRM-DCM-DPWM hybrid modulation is proposed for three-phase bidirectional rectifier/inverter to realize soft switching and increase switching frequency. Fig. 2 shows the detailed modulation principle. As shown in Fig. 2(a), the phase with the highest absolute alternate current (AC) voltage does not have high frequency PWM signal but is clamped to the direct current (DC) bus (when phase voltage is positive) or ground (when phase voltage is negative). CRM modulation is applied to the phase with the second highest voltage to realize soft switching. The last phase which has the lowest phase voltage absolute value works in DCM mode accordingly, and the active switching turn-on instant is synchronized to the CRM phase. The switching cycle gate signals V_{gs} and inductor currents i_L when phase A is at 135° are shown in Fig. 2(b).

The topology showed in Fig. 1 together with the hybrid modulation has unique advantages under high switching frequency operation. The DPWM avoids the switching loss in one AC phase. The CRM helps realize soft switching in another AC phase. If the last phase also works in CRM mode, extremely high switching frequency may happen. Instead, DCM with frequency synchronization limits the switching frequency range and simplifies the control complexity. The two-channel

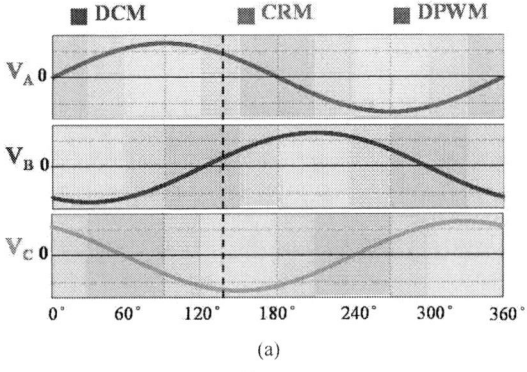

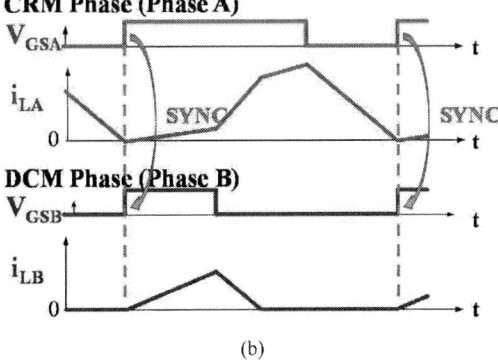

Fig. 2   CRM + DCM synchronization + DPWM modulation for 3-phase inverter (a) Line cycle operation mode distribution (b) Switching cycle waveforms when phase A is at 135°

interleaved topology reduces the current ripple and reduce the filter component sizes.

B. Issues with Proposed Hybrid Modulation under Short-circuit Condition

For stand-alone inverters and inverter-based micro-grids, the inverter should have short-circuit fault ride-through function to achieve the continuity of the power delivery[20], [21]. However, the proposed CRM-based hybrid modulation has some intrinsic issues under short-circuit situation.

Take the case of phase A voltage v_{ac} at angle $120 - 150°$ as an example. Based on the operating mode selection principle mentioned in Subsection II.A, phase A works in CRM, phase B is in DCM and phase C clamps to ground. Switches in phase A and B have different on-off modes which makes variable combinations of switching voltages. To simplify the case, non-interleaved circuit is considered in the discussion. As shown in Fig. 3, the circuit has a three-phase short-circuit condition, and phase C bottom switch S_{C2} keeps on to clamp it to ground. All possible switching modes and switching voltages, normalized by V_{dc}, are listed in Table I.

TABLE I.   Switching Status and Voltage Values (Normalized by V_{DC}) in Short-circuit Condition

Mode	Phase A(CRM)			Phase B(DCM)			Phase C(DPWM)			V_{sN}	V_{LA}	V_{LB}	V_{LC}
	S_{A1}	S_{A2}	V_{AN}	S_{B1}	S_{B2}	V_{BN}	S_{C1}	S_{C2}	V_{CN}				
I	1	0	1	1	1	0	1				2/3	1/3	1/3	-2/3
II	1	0	1	0	1	0		0	1	0	1/3	2/3	-1/3	-1/3
III	1	0	1	0	0	1/2					1/2	1/2	0	-1/2
IV	0	1	0	0	0	1/2					1/6	-1/6	1/3	-1/6

In Table I, S_{x1} and S_{x2} are top switch and bottom switch gate signals in phase X. V_{XN} is the switching node voltage of phase X normalized by V_{dc}. The resonance between the inductor and semiconductor device output capacitor is ignored. V_{sN} is the zero-sequence voltage. It is determined by switching node voltages as expressed in (1).

$$V_{sN} = \frac{1}{3}(V_{AN} + V_{BN} + V_{CN}) \quad (1)$$

And V_{LX} is the inductor voltage in each state given by (2).

$$V_{LX} = V_{XN} - V_{sN} \quad (2)$$

The inverter DC bus voltage, output AC voltage and the zero-sequence voltage are used to magnetize and demagnetize

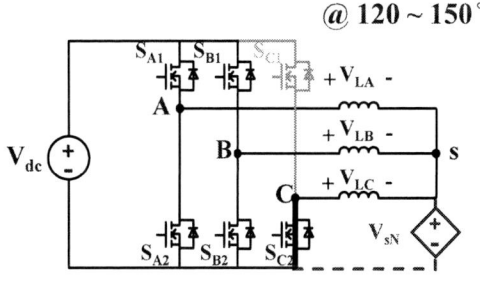

Fig.3   Three-phase inverter under three-phase short-circuit condition

979-8-3315-1612-3/25 $31.00 © 2025 IEEE

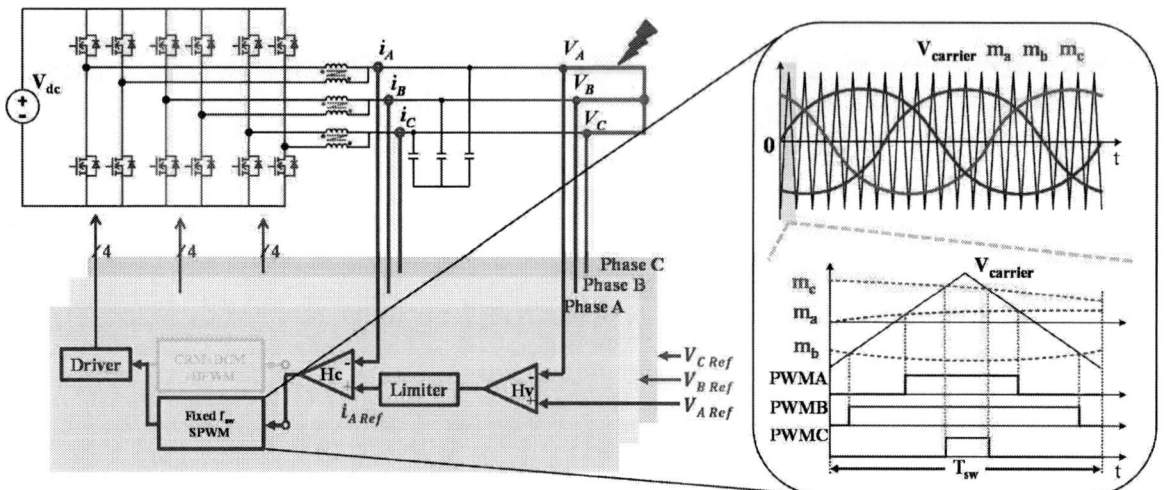

Fig. 4 Control diagram for short-circuit ride-through

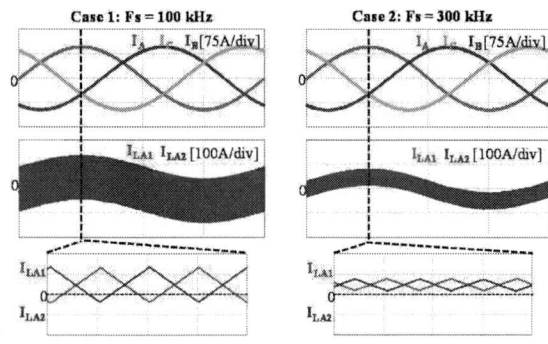

Fig. 5 Short-circuit operation simulation waveforms

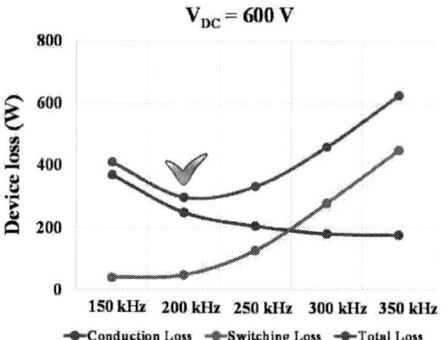

Fig. 6 Device loss breakdown under short circuit

the power inductors. During the short-circuit condition, the AC voltages become zero, so the zero-sequence voltage plays a critical role because it becomes the only voltage source which can demagnetize the inductors in each phase. By considering all possible switching statuses as listed in Table I, it reveals that phase C, the clamping mode phase, always has a negative inductor voltage. This means that in the CRM-based soft-switching modulation, there is no chance for the clamping mode phase to achieve the inductor voltage-second balance. The same principle is applicable to the two-channel interleaved three-phase inverter. In conclusion, with the CRM-based soft-switching modulation, there is no way to handle the output short-circuit. This problem does not only happens in the soft-switching modulation, but also exists in all inverter modulation strategies with DPWM. DPWM cannot be used during the output short-circuit.

III. SHORT-CIRCUIT OPERATION

A. CCM Operation for Short-circuit Ride-Through

To enable the short-circuit ride-through function, the original modulation itself must be replaced. In Fig. 4, a fixed frequency operation is depicted for short-circuit condition. Once the output short-circuit is detected, the modulation method is switched to the fixed switching frequency sinusoidal PWM (SPWM) as shown in the enlarged function block in Fig. 4. When the voltage controller is saturated, but the AC voltages

retain as 0, the inverter modulation is switched to fixed frequency SPWM. The modulation signals (m_a, m_b, m_c) are compared with a fixed frequency carrier signal $V_{carrier}$. In normal operation, the voltage controller provides the current reference values. Under the short-circuit condition, the current reference is clamped to 1.5 times of the full power rated current and the inverter works as a current source. Average current mode control is applied to generate the modulation signals.

B. Switching Frequency Optimization under Short-Circuit Ride-Through

Typically, the inverter should handle the short-circuit fault condition for no longer than 10s. If the fault condition is not cleared after 10s, the inverter will shut down to protect itself. So, the short-circuit modulation should guarantee that the inverter can handle the overcurrent and the hardware operates safely. Thus, the CCM switching frequency is optimized based on the thermal stress on the circuit components. Since the semiconductor devices have much smaller heat capability than the power inductors, the temperature rise on the semiconductor devices is much quicker. Therefore, the semiconductor device loss is selected as the optimization objective.

Fig. 5 depicts the simulation waveforms of two different switching frequencies, with case 1 100kHz and case 2 300kHz. I_A, I_B and I_C are filtered phase currents, and i_{LA1}, i_{LA2} are two inductor currents in phase A. The zoomed-in waveforms show that case 1 inductor current can cross 0, which is called quasi-square wave (QSW) mode. The benefit of QSW is that the devices can achieve zero-voltage switching (ZVS) and get rid of turn-on loss. But the drawback is that the high current ripple leads to high turn-off energy and high conduction loss. On the other hand, Case 2 has smaller current ripple, while QSW is not guaranteed with smaller current ripple, leading to high turn-on loss with partial hard switching. Basically, the choice of the switching frequency is a tradeoff between conduction loss and switching loss. Fig. 6 represents the device loss comparison under the different switching frequencies. The total device loss is the sum of conduction loss and switching loss. The final selection is 200 kHz. At this switching frequency, in one line cycle, the circuit works in CCM during part of the time and TCM during the other part of the time. Switching frequency of 200kHz gives the lowest total device loss and can minimize device thermal stress during short circuit ride-through.

IV. EXPERIMENTAL VERIFICATION

A 30kW SiC-based three-phase inverter with two-channel interleaved topology and CRM-based hybrid modulation is proposed in [23]. The hardware prototype is shown in Fig. 7. And this inverter is used to verify the proposed short-circuit ride-through strategy in this paper. The detailed circuit parameter is listed in Table II.

TABLE II. INVERTER PARAMETERS

Device	C3M0016120K
Switching frequency	≥200kHz
Inductor self inductance	8.4μH
Coupling coefficient	-0.5
Input DC voltage	600-800V
Output AC line voltage	380V
Power	30kVA
Short-circuit switching frequency	200kHz
Short circuit ride-through time	≤10s

The short-circuit ride-through test under three-phase short-circuit situation is performed on the prototype. Fig. 8 shows the test waveforms. Fig. 8(a) includes the waveforms for phase A. V_{GS_A1} (dark blue) is the gate driving signal of device A1. i_{LA1} (green) and i_{LA2} (purple) are the inductor currents. The asymmetric current sharing between two inductors is caused by

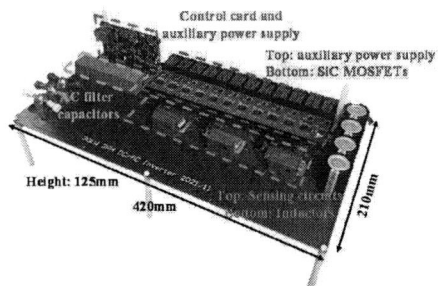

Fig. 7 Inverter prototype

the current sensing shunt resistor in channel 1. This unbalanced current sharing does not affect the controllability because the total phase current is sensed as the feedback and treated as the control target. The open-loop interleaving of the gate signals is not affected by the inductor current [24]. The light blue waveform I_A is the filtered phase A current. The waveforms prove that the inverter works under 200kHz CCM+TCM operation, and the phase current is well regulated. Fig. 8(b) shows all three phase currents I_A, I_B and I_C in green, purple and light blue. It is shown that the output AC currents are controlled as sinusoidal, and each phase has 120° shifted from each other. The red line is phase A voltage V_A, which keeps as zero due to the short circuit situation. Fig. 8(c) shows the transient response from normal operation to three-phase short-circuit. The inverter can realize a smooth transition from normal operation to short circuit operation.

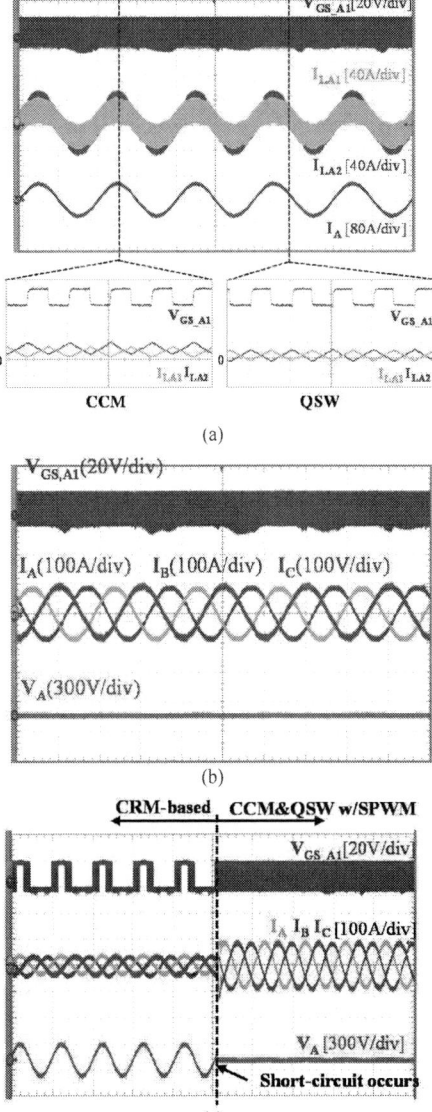

Fig. 8 Short circuit test waveform (a) Phase A in steady state (b) All phases in steady state (c) Transient response

V. SUMMARY

This paper introduces the short circuit operation of a CRM-based soft switching three-phase three-wire inverter. The issue with the original hybrid modulation is analyzed, and a fixed frequency modulation is proposed to solve the issue. The frequency value is optimized based on semiconductor device loss. Finally, experiments are performed to evaluate the effectiveness of the proposed modulation.

REFERENCES

[1] D. Aggeler, F. Canales, H. Zelaya-De La Parra, A. Coccia, N. Butcher, and O. Apeldoorn, "Ultra-fast DC-charge infrastructures for EV-mobility and future smart grids," in *2010 IEEE PES Innovative Smart Grid Technologies Conference Europe (ISGT Europe)*, Oct. 2010, pp. 1–8. doi: 10.1109/ISGTEUROPE.2010.5638899.

[2] J. P. Kozak *et al.*, "Stability, Reliability, and Robustness of GaN Power Devices: A Review," *IEEE Trans. Power Electron.*, vol. 38, no. 7, pp. 8442–8471, Jul. 2023, doi: 10.1109/TPEL.2023.3266365.

[3] Q. Huang and A. Q. Huang, "Review of GaN totem-pole bridgeless PFC," *CPSS Trans. Power Electron. Appl.*, vol. 2, no. 3, pp. 187–196, Sep. 2017, doi: 10.24295/CPSSTPEA.2017.00018.

[4] A. Marzoughi, R. Burgos, and D. Boroyevich, "Characterization and comparison of latest generation 900-V and 1.2-kV SiC MOSFETs," in *2016 IEEE Enssergy Conversion Congress and Exposition (ECCE)*, Sep. 2016, pp. 1–8. doi: 10.1109/ECCE.2016.7854911.

[5] T. Liu, C. Chen, K. Xu, Y. Zhang, and Y. Kang, "GaN-Based Megahertz Single-Phase Inverter With a Hybrid TCM Control Method for High Efficiency and High-Power Density," *IEEE Trans. Power Electron.*, vol. 36, no. 6, pp. 6797–6813, Jun. 2021, doi: 10.1109/TPEL.2020.3039386.

[6] Y. Cao, K. Ngo, and D. Dong, "A Scalable Electronic-Embedded Transformer, a New Concept Toward Ultra-High-Frequency High-Power Transformer in DC–DC Converters," *IEEE Trans. Power Electron.*, vol. 38, no. 8, pp. 9278–9293, Aug. 2023, doi: 10.1109/TPEL.2023.3279259.

[7] Z. Huang, Z. Liu, F. C. Lee, and Q. Li, "Critical-Mode-Based Soft-Switching Modulation for High-Frequency Three-Phase Bidirectional AC–DC Converters," *IEEE Trans. Power Electron.*, vol. 34, no. 4, pp. 3888–3898, Apr. 2019, doi: 10.1109/TPEL.2018.2854302.

[8] B. Fan, Q. Wang, R. Burgos, A. Ismail, and D. Boroyevich, "Adaptive Hysteresis Current Based ZVS Modulation and Voltage Gain Compensation for High-Frequency Three-Phase Converters," *IEEE Trans. Power Electron.*, vol. 36, no. 1, pp. 1143–1156, Jan. 2021, doi: 10.1109/TPEL.2020.3002894.

[9] X. Chen, G. Son, F. Jin, and Q. Li, "A Microcontroller-Based High Efficiency Critical Conduction Mode Control for GaN-Based Totem-Pole PFC," in *2021 IEEE 22nd Workshop on Control and Modelling of Power Electronics (COMPEL)*, Cartagena, Colombia: IEEE, Nov. 2021, pp. 1–7. doi: 10.1109/COMPEL52922.2021.9646009.

[10] Y. Li *et al.*, "Optimal Synergetic Operation and Experimental Evaluation of an Ultra-Compact GaN-Based Three-Phase 10 kW EV Charger," *IEEE Trans. Transp. Electrification*, pp. 1–1, 2023, doi: 10.1109/TTE.2023.3297502.

[11] J. Sun, L. Zhu, R. Qin, D. J. Costinett, and L. M. Tolbert, "Single-Phase GaN-Based T-Type Totem-Pole Rectifier With Full-Range ZVS Control and Reactive Power Regulation," *IEEE Trans. Power Electron.*, vol. 38, no. 2, pp. 2191–2201, Feb. 2023, doi: 10.1109/TPEL.2022.3215969.

[12] H. Xi, L. Li, G. Xu, and M. Su, "SiC-Based High-Frequency Soft-Switching Interleaved Totem-Pole Bridgeless PFC Converter without ZCD Circuits," in *2021 IEEE 1st International Power Electronics and Application Symposium (PEAS)*, Nov. 2021, pp. 1–5. doi: 10.1109/PEAS53589.2021.9628803.

[13] C. Marxgut, F. Krismer, D. Bortis, and J. W. Kolar, "Ultraflat Interleaved Triangular Current Mode (TCM) Single-Phase PFC Rectifier," *IEEE Trans. Power Electron.*, vol. 29, no. 2, pp. 873–882, Feb. 2014, doi: 10.1109/TPEL.2013.2258941.

[14] R. Aboelsaud, A. Ibrahim, and A. G. Garganeev, "Review of three-phase inverters control for unbalanced load compensation," *Int. J. Power Electron. Drive Syst. IJPEDS*, vol. 10, no. 1, p. 242, Mar. 2019, doi: 10.11591/ijpeds.v10.i1.pp242-255.

[15] R. Mechouma, B. Azoui, and M. Chaabane, "Three-phase grid connected inverter for photovoltaic systems, a review," in *2012 First International Conference on Renewable Energies and Vehicular Technology*, Mar. 2012, pp. 37–42. doi: 10.1109/REVET.2012.6195245.

[16] F. Rojas *et al.*, "An Overview of Four-Leg Converters: Topologies, Modulations, Control and Applications," *IEEE Access*, vol. 10, pp. 61277–61325, 2022, doi: 10.1109/ACCESS.2022.3180746.

[17] D.-E. Kim and D.-C. Lee, "Feedback Linearization Control of Three-Phase UPS Inverter Systems," *IEEE Trans. Ind. Electron.*, vol. 57, no. 3, pp. 963–968, Mar. 2010, doi: 10.1109/TIE.2009.2038404.

[18] Z. Liu, J. Liu, and Y. Zhao, "A Unified Control Strategy for Three-Phase Inverter in Distributed Generation," *IEEE Trans. Power Electron.*, vol. 29, no. 3, pp. 1176–1191, Mar. 2014, doi: 10.1109/TPEL.2013.2262078.

[19] G. Son and Q. Li, "Control Techniques for CRM-Based High-Frequency Soft-Switching Three-Phase Inverter Under Unbalanced Grid Conditions," *IEEE Trans. Power Electron.*, vol. 37, no. 6, pp. 6613–6624, Jun. 2022, doi: 10.1109/TPEL.2022.3141978.

[20] J. He, P. Liu, B. Liu, and S. Duan, "An Asymmetric Short-Circuit Fault Ride-Through Strategy Providing Current Limiting and Continuous Voltage Supply for Three-Phase Three-Wire Stand-Alone Inverters," *IEEE Access*, vol. 8, pp. 211063–211073, 2020, doi: 10.1109/ACCESS.2020.3038220.

[21] Z. Liang, X. Lin, Y. Kang, B. Gao, and H. Lei, "Short Circuit Current Characteristics Analysis and Improved Current Limiting Strategy for Three-phase Three-leg Inverter under Asymmetric Short Circuit Fault," *IEEE Trans. Power Electron.*, vol. 33, no. 8, pp. 7214–7228, Aug. 2018, doi: 10.1109/TPEL.2017.2759161.

[22] W. Qian, N. Zhou, J. Wu, Y. Li, Q. Wang, and P. Guo, "Probabilistic Short-Circuit Current in Active Distribution Networks Considering Low Voltage Ride-Through of Photovoltaic Generation," *IEEE Access*, vol. 7, pp. 140071–140083, 2019, doi: 10.1109/ACCESS.2019.2944195.

[23] X. Chen, G. Son, Z. Huang, F. Jin, and Q. Li, "High Frequency Three-Phase CRM Inverter with Integrated Magnetics for Auxiliary Power Supply in Railway Applications," in *2024 IEEE Energy Conversion Congress and Exposition (ECCE)*, Phoenix, AZ, USA: IEEE, Oct. 2024, pp. 01–06.

[24] Z. Liu, Z. Huang, F. C. Lee, and Q. Li, "Digital-Based Interleaving Control for GaN-Based MHz CRM Totem-Pole PFC," *IEEE J. Emerg. Sel. Top. Power Electron.*, vol. 4, no. 3, pp. 808–814, Sep. 2016, doi: 10.1109/JESTPE.2016.2571302.

Modified Space Vector Modulation with Low Bandwidth Sensor to Reduce Losses in Soft Switching Three-phase Inverters

Md Didarul Alam
FREEDM Systems Center
North Carolina State University
Raleigh, NC, United States
malam5@ncsu.edu

Nazmul Hassan
FREEDM Systems Center
North Carolina State University
Raleigh, NC, United States
nhassan@ncsu.edu

Iqbal Husain
FREEDM Systems Center
North Carolina State University
Raleigh, NC, United States
ihusain2@ncsu.edu

Liming Liu
Power Electronics Center of Excellence
Eaton Corporation
Raleigh, NC, United States
limingLiu@eaton.com

Hongrae Kim
Power Electronics Center of Excellence
Eaton Corporation
Raleigh, NC, United States
HongraeKim@eaton.com

Abstract—Space Vector Modulation Zero Voltage Switching Topology (SVM-ZVS) offers simple control for DC-link resonant three-phase converters but suffers from lower efficiency due to hard switching when one of the phase currents becomes zero or close to zero. Moreover, the SVM-ZVS topology does not correctly address the delay issues that come from the controller implementation and sensors, which are critical for maintaining soft-switching. This paper proposes a Modified Space Vector Modulation Zero Voltage Switching (MSVM-ZVS) to reduce the losses which is particularly beneficial for soft-switching at low levels of phase currents. Moreover, the proposed MSVM-ZVS also properly addresses the delay issues and can maintain soft-switching even if the zero crossing current is not properly detected; this relaxation of sensing requirement enables the use of low bandwidth, low cost current sensors. The MSVM-ZVS has been verified in simulation, and it is observed that the proposed topology works better than SVM-ZVS in terms of efficiency and the current stress of the switches. Simulation results shows the efficacy of the proposed modulation scheme.

Index Terms—Space Vector Modulation, Zero Voltage Switching (ZVS), Conduction loss, Switching loss, Zero Crossing Current Detection, Low bandwidth sensor

I. INTRODUCTION

Three-phase inverters are extensively utilized in industrial drives, uninterruptible power supply (UPS) systems, and traction converters [1], [2]. With growing concerns about climate change, there is an increasing focus on integrating renewable energy systems with high-performance characteristics, such as improved efficiency and higher power density. Achieving higher power density requires devices to operate at elevated switching frequencies which helps minimize the size of passive components. However, higher switching frequencies lead to increased switching losses and more complex thermal management, ultimately limiting the achievable power density [3], [4]. Soft switching offers a promising solution for achieving higher switching frequencies without adding to energy losses,

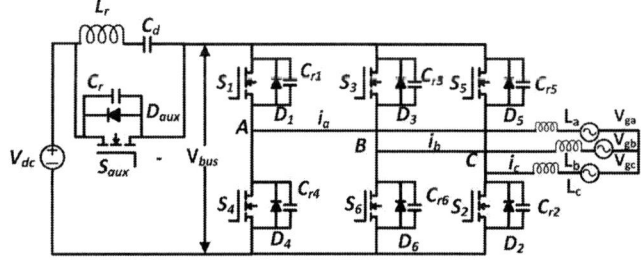

Fig. 1. Three-phase ZVS-SVM grid connected inverter topology.

which helps improve the power density compared to traditional methods. Researchers have explored several approaches to soft switching in three-phase inverters, including resonant circuits [5]–[9] and the Triangular Current Mode (TCM) technique [4], [10], [11]. TCM uses the energy stored in the output AC filter inductor to achieve zero-voltage switching (ZVS) during turn-on, giving the output current a triangular shape. The key advantage of TCM is that it doesn't require extra resonant circuits, but it comes with challenges like higher peak currents and variable switching frequencies, which can lead to more electromagnetic interference (EMI) noise. Resonant circuit-based soft switching can be categorized into AC-side and DC-side topologies. A well-known DC-side example is the resonant DC link (RDCL) inverter [12], where an auxiliary switch enables the DC bus voltage to resonate to zero. However, it comes with the drawback of high voltage stress on the switches. To address this, the active clamped resonant DC link (ACRL) inverter [13] was introduced to reduce voltage stress. Despite this improvement, both RDCL and ACRL inverters rely on DPM methods, where the resonance frequency is much higher than the PWM control frequency, resulting in subharmonic noise. To overcome these limitations, the quasi-resonant

979-8-3315-1612-3/25 $31.00 © 2025 IEEE

DC link PWM inverter (QRDCL) was proposed in [7], [14]–[16], which reduces voltage stress and supports PWM control by limiting the auxiliary circuit's operation to resonating the DC bus voltage to zero during state transitions. However, the complexity of its circuitry and the frequent operation of the auxiliary circuit make QRDCL less practical for many applications. Considering voltage stress, PWM control, and easy circuitry all together, a ZVS space vector modulation [16] scheme is proposed by Xu which is known as ZVS-SVM inverter. In this design, the main and auxiliary switches achieve ZVS turn-on in a SiC-based converter, with the auxiliary switch operating only once per switching cycle. The circuit of ZVS-SVM is shown in Fig. 1, where auxiliary circuit is placed between the DC source and the three-phase bridge legs. The ZVS-SVM inverter operates with a carefully designed vector sequence. Each switching cycle uses two adjacent active vectors and one zero vector to synthesize the reference vector [8]. The auxiliary switch turns OFF once per cycle when the current transitions from the diode to its complementary power semiconductor. This causes the bus voltage to resonate and momentarily clamp to zero, enabling ZVS turn-on for the main switches. The inverter operates in PWM mode at a fixed switching frequency, achieving ZVS for both the main and auxiliary switches. In [17]–[19] an SVM-ZVS method was proposed for achieving unity [17] and limited non-unity [18], [19] power factors with minimal auxiliary switch operations. However, the current SVM-ZVS method has some drawbacks: it experiences hard switching when one of the phase currents approaches or reaches zero. This happens because the snubber capacitor across the MOSFET cannot discharge within the dead time, requiring additional capacitors to maintain ZVS for both main and auxiliary switches. In this paper, we propose a Modified Space Vector Modulation Zero Voltage Switching (MSVM-ZVS) method that resolves this issue by maintaining soft-switching even when phase currents are very low or zero. The method also eliminates the need for precise zero-crossing detection and works effectively with low-bandwidth, cost-effective sensors.

The remainder of this paper is structured as follows: Section II provides an overview of ZVS fundamentals and evaluates the existing modulation strategy. The challenges inherent in the existing SVM-ZVS topology are critically analyzed in Section III. Section IV introduces the proposed Modified Space Vector Modulation (MSVM) technique for achieving ZVS, while Section V delves into the control structure of the proposed MSVM-ZVS topology. Simulation results are presented in Section VI, followed by a summary of key findings and conclusions in Section VII.

II. EXISTING MODULATION STRATEGY

Fig. 1 illustrates the circuit configuration of the proposed three-phase zero-voltage switching (ZVS) space vector modulation (SVM) inverter. The system operates with a constant DC voltage source (V_{dc}) and is interfaced with the grid through a three-phase LLL filter consisting of inductances L_a, L_b, and L_c. This filter minimizes harmonic distortion in

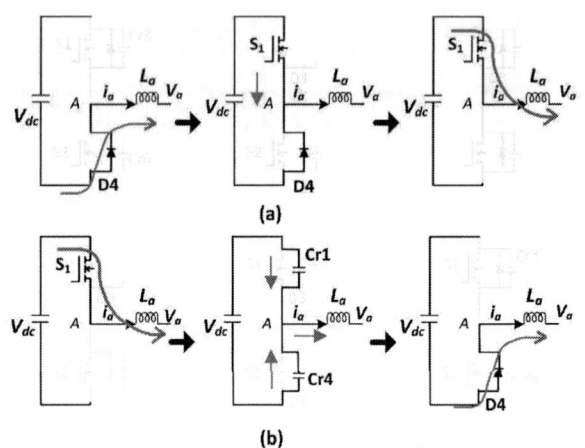

Fig. 2. Two forms of current commutation based on $phase - a$ positive current . (a) First from of commutation. (b) Second from of commutation.

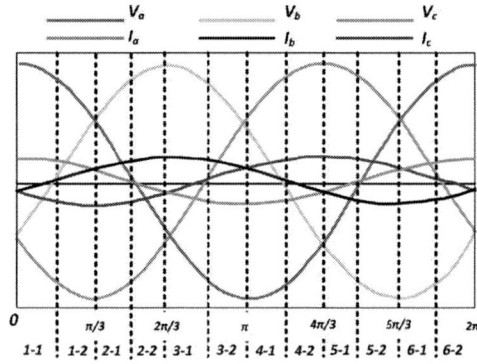

Fig. 3. Three Phase voltage and current waveforms with sector determination.

the output currents (i_a, i_b, i_c) and ensures compliance with grid standards. The grid voltages are denoted by V_a, V_b, and V_c with f_g representing the fundamental grid frequency.

To enable ZVS operation, an auxiliary resonant circuit is integrated between the DC source and the three-phase bridge inverter. This auxiliary circuit comprises an auxiliary switch (S_{aux}), a resonant capacitor (C_r), a resonant inductor (L_r), a clamping capacitor (C_d), and a diode (D_{aux}). The auxiliary circuit dynamically toggles the inverter output voltage (v_{inv}) between V_{dc} and 0 V, facilitating soft-switching transitions for the main inverter switches. The primary switching network consists of six switches (S_1 to S_6) configured in a three-phase bridge topology. Each switch is paired with an anti-parallel diode (D_1 to D_6) and a snubber capacitor (C_{r1} to C_{r6}). These components ensure efficient switching operation and limit voltage stress across the switches during transient conditions.

A. Basic understanding of ZVS Inverter during current commutation

In a zero-voltage switching (ZVS) inverter leg, current commutation can occur in two distinct forms, depending on the transitions between the switches and their antiparallel

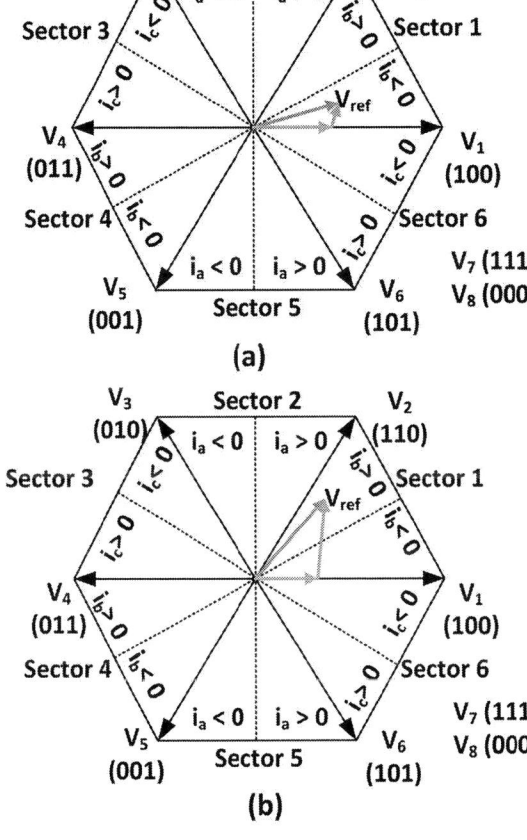

(a)

(b)

Fig. 4. Sector determination based on (a) $i_b < 0$ and (b) $i_b > 0$

diodes. The first form of current commutation occurs when the current transitions from the antiparallel diode of one switch to its complementary switch. For example, as shown in Fig. 2(a), the current commutates from the lower diode D_4 to the upper switch S_1 for a positive current in $phase - a$. During this transition, S_1 undergoes hard switching at turn-on, which results in increased turn-on losses due to the reverse recovery effect of D_4. This reverse recovery imposes additional stress on the switch, leading to significant power losses and reduced converter efficiency. The second form of commutation, illustrated in Fig. 2(b), occurs when the current transitions from the upper switch S_1 to the lower diode D_4. Under this condition, S_1 achieves ZVS during turn-off due to the snubber effect provided by the parallel capacitors across the switches. These capacitors charge and discharge during transitions, enabling a soft-switching mechanism, provided the phase current exceeds a certain threshold.

Among these two forms, the first form—current commutation from the antiparallel diode to its complementary switch—inevitably results in hard switching, contributing to higher switching losses and reduced overall efficiency. In space vector modulation (SVM), the switching state of a leg is defined as 1 when the upper switch is turned on

and 0 when it is turned off. State transitions occur in two directions: $0 \rightarrow 1$ and $1 \rightarrow 0$. The timing of the first form of commutation is closely linked to the direction of the output current. Specifically:

- For a positive output current in a given phase leg, the $0 \rightarrow 1$ transition corresponds to the first form of commutation, where current transfers from the lower diode to the upper switch.
- For a negative output current, the first form of commutation occurs during the $1 \rightarrow 0$ transition, where current transfers from the upper diode to the lower switch.

Achieving ZVS during the first form of commutation in ZVS-SVM inverters requires ensuring that the DC bus voltage resonates to zero before the commutation event. This is accomplished by controlling an auxiliary switch, S_{aux}. By turning off S_{aux}, the required resonance condition is established, enabling ZVS during the commutation process. In a three-phase inverter, each leg undergoes first and second form of commutation commutation within a single switching period. Synchronizing the second form of commutations across all three phase legs allows S_{aux} to be activated only once per pulse-width modulation (PWM) cycle. This synchronization aligns the switching frequency of S_{aux} with that of the main switches, reducing overall switching losses and simplifying the control strategy.

B. Vector Sequence Selection Based on Voltage and Current Polarities

To ensure soft-switching for both forms of commutations, the polarities of the currents are essential for selecting the appropriate vector sequences. In this approach, the sectors are determined by the voltage information, while the sub-sectors are determined by the current information. The grid voltages and currents as shown in Fig. 3 are described by the (1) and (2), respectively-

$$
\begin{aligned}
v_a(t) &= V_m \cos(2\pi f_g t), \\
v_b(t) &= V_m \cos\left(2\pi f_g t - \frac{2\pi}{3}\right), \\
v_c(t) &= V_m \cos\left(2\pi f_g t + \frac{2\pi}{3}\right),
\end{aligned} \tag{1}
$$

$$
\begin{aligned}
i_a(t) &= I_m \cos(2\pi f_g t - \theta), \\
i_b(t) &= I_m \cos\left(2\pi f_g t - \frac{2\pi}{3} - \theta\right), \\
i_c(t) &= I_m \cos\left(2\pi f_g t + \frac{2\pi}{3} - \theta\right).
\end{aligned} \tag{2}
$$

Here, V_m represents the magnitude of the grid voltage, I_m is the magnitude of the grid current, and θ denotes the power factor angle. By integrating the voltage and current information, as illustrated in Fig. 4 , the vector sequence can be dynamically adjusted to ensure effective commutation. This approach guarantees soft-switching conditions are maintained across all operating states, as summarized in Table I.

979-8-3315-1612-3/25 $31.00 © 2025 IEEE 1748

TABLE I
SWITCHING SEQUENCES OF MAIN SWITCHES

Sector	1st Vector	2nd Vector	3rd Vector	1st Vector	Zero Crossing Current
I ($i_b < 0$)	111	100	110	111	i_b
I ($i_b > 0$)	000	110	100	000	i_b
II ($i_a < 0$)	111	010	110	111	i_a
II ($i_a > 0$)	000	110	010	000	i_a
III ($i_c < 0$)	111	010	011	111	i_c
III ($i_c > 0$)	000	011	010	000	i_c
IV ($i_b < 0$)	111	011	001	111	i_b
IV ($i_b > 0$)	000	001	011	000	i_b
V ($i_a < 0$)	111	001	101	111	i_a
V ($i_a > 0$)	000	101	001	000	i_a
VI ($i_c < 0$)	111	100	101	111	i_c
VI ($i_c > 0$)	000	101	100	000	i_c

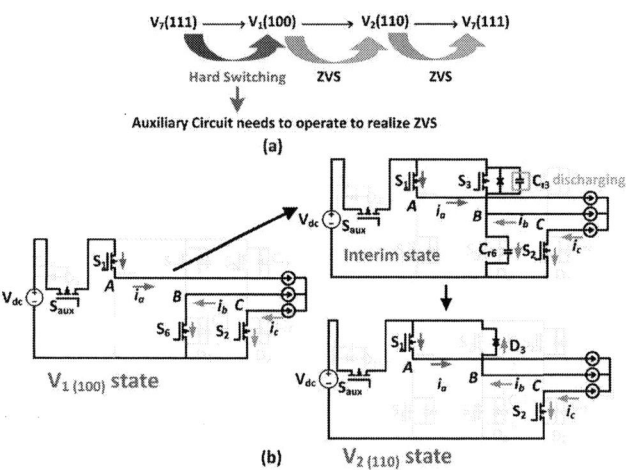

Fig. 5. SVM-ZVS for sector 1-1: (a) Switching state transition (b) Circuit transition from $V_1(100)$ to $V_2(110)$

III. ISSUES IN EXISTING SPACE VECTOR MODULATION FOR ZVS

The generalized space vector modulation (SVM) method with soft-switching for three-phase inverters, as discussed in [19], effectively maintains zero-voltage switching (ZVS) over a broad range of power factors. The detailed circuit operation of Fig. 1 is explained in [19]. However, it still encounters hard switching losses due to low current levels, system delays, and inaccurate zero-crossing current detection, which will be discussed in this section.

A. Low Current Issue in Existing SVM-ZVS

To simplify the analysis, only sector 1-1 is examined to address the issue of low or near-zero phase current for power factor angle $0 \leq \theta \leq \frac{\pi}{6}$, though it is applicable for other sector as well. In Fig. 3, with i_a positive, i_b negative and i_c negative, vector sequences are chosen to minimize auxiliary

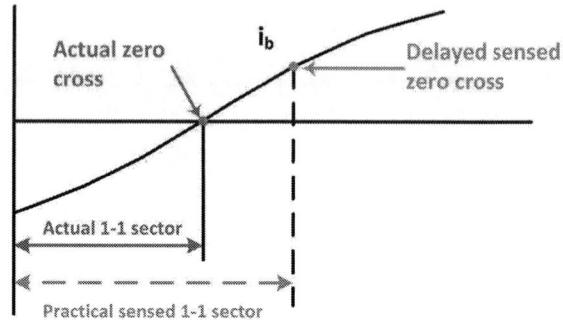

Fig. 6. Error in sector determination due to sensor delay

switch operations. Fig. 5(a) shows the existing SVM-ZVS switching transition for sector 1-1, where the auxiliary circuit ensures soft switching from $V_7(111)$ to $V_1(100)$. However, as shown in Fig. 5(b), the transitions from V_1 (100) to V_2 (110) and V_2 (110) to V_7 (111) naturally facilitate soft switching, eliminating the need for auxiliary switch operations. This is achieved when the phase i_b and i_c currents are sufficient to fully discharge the additional snubber capacitor, which is larger than the inherent C_{oss}. In sector 1- 1, $phase - b$ current (i_b) is critical because it crosses zero at the end of sector 1-1, making it the lowest current. Fig. 5(b) illustrates the state transition from V_1 (100) to V_2 (110) in Sector 1-1, highlighting the dependency of soft switching on the $phase - b$ current during low-current conditions. If i_b is insufficient to fully discharge the resonant capacitor C_{r3} during the dead time (t_{dead}), the voltage across C_{r3} ($V_{C_{r3}}$) will remain non-zero. This residual voltage results in a significant current spike when MOSFET S_3 turns on, leading to hard switching. Consequently, this increases switching losses, induces additional thermal stress on the MOSFET, and degrades system efficiency.

B. Issue of Sensor Delay and Imperfect Zero Crossing Current Detection in the Existing SVM-ZVS

Precise zero-crossing current detection poses a significant challenge in power electronic systems due to inherent delays in controllers and the limited bandwidth of sensors. These delays can lead to inaccuracies in identifying the exact moment the current crosses zero, a critical parameter in maintaining optimal switching conditions. The accuracy of zero-crossing detection heavily depends on the bandwidth of the current sensors, with higher-bandwidth sensors offering better precision but at higher cost and complexity. One of the key limitations of existing SVM-ZVS topologies is their reliance on zero-crossing current detection to determine the subsector. The identified subsector directly influences the choice of the preferred vector switching sequence, which is critical for achieving soft-switching during various state transitions. Fig. 6 illustrates a scenario where the actual i_b current crosses zero but remains below the threshold value i_{bt}. Due to delays in detection, the transition from sector 1-1 to sector 1-2 is not identified accurately. At this point, the phase currents i_a, i_b, and i_c are positive, positive, and negative, respectively.

979-8-3315-1612-3/25 $31.00 © 2025 IEEE   1749

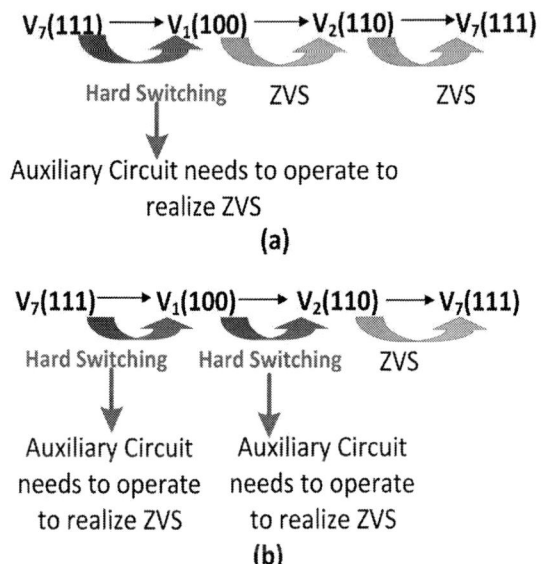

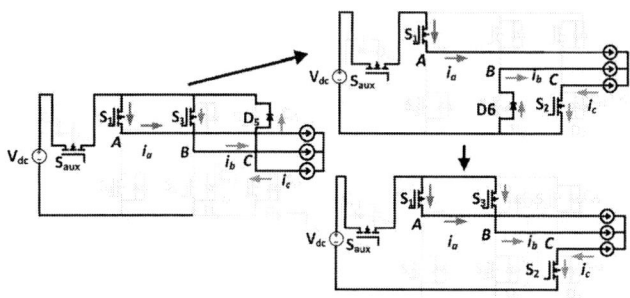

Fig. 8. Proposed MSVM-ZVS strategy for sector 1-1 while current polarities of phase current are changed due to delays: Circuit switching transition

Fig. 7. Proposed MSVM-ZVS switching transition for sector 1-1 based on i_{bt} based (a) $|i_b| > |i_{bt}|$,(b) $|i_b| < |i_{bt}|$

However, the switching sequences corresponding to sector 1-1 continue to be applied, leading to a mismatch in the intended switching vector sequence, which causes the converter to operate in hard-switching mode.

IV. PROPOSED MODIFIED SPACE VECTOR MODULATION FOR ZVS

To address the issues discussed in Section III, a Modified Space Vector Modulation with Zero-Voltage Switching (MSVM-ZVS) strategy is proposed based on the critical threshold current. This current represents the minimum value required to completely discharge the snubber capacitors within the dead time (t_{dead}), ensuring that the voltages across the capacitors reaches zero before switches turn on. It is mentionable that three-phase balanced load conditions are assumed, where the critical threshold currents for all phases are equal. This assumption simplifies implementation and ensures consistent operation across different phases

Using the method outlined in [20], the critical threshold current (i_{bt}) for Sector 1-1 can be determined using the following (3):

$$|i_{bt}| = \frac{2C_{r3}V_{dc}}{t_{\text{dead}}} \tag{3}$$

The proposed MSVM-ZVS method, illustrated in Fig. 7, is shown for a power factor angle within the range $0 \leq \theta \leq \frac{\pi}{6}$ for Sector 1-1. However, the same approach can be applied to other sectors as well, based on the preferred vector sequences. Fig. 7 depicts the sequence of inverter switching states ($V_7(111) \rightarrow V_1(100) \rightarrow V_2(110) \rightarrow V_7(111)$) and the role of the auxiliary circuit in achieving Zero Voltage Switching (ZVS). Two cases are depicted based on the relationship between the current magnitude ($|i_b|$) and a threshold current

($|i_{bt}|$). In the first case ($|i_b| > |i_{bt}|$) as shown in Fig. 7(a), ZVS is achieved during the transitions $V_1(100) \rightarrow V_2(110)$ and $V_2(110) \rightarrow V_7(111)$, while hard switching occurs during $V_7(111) \rightarrow V_1(100)$, requiring the auxiliary circuit to operate for ZVS realization. In contrast, the second case ($|i_b| < |i_{bt}|$) depicted in Fig. 7 (b) experiences hard switching in both $V_7(111) \rightarrow V_1(100)$ and $V_1(100) \rightarrow V_2(110)$, with ZVS achieved only during $V_2(110) \rightarrow V_7(111)$. The auxiliary circuit intervenes during all hard-switching transitions to facilitate ZVS, thereby reducing power losses and improving overall efficiency.

The turn-on duty ratio of the auxiliary switch, d_{aux}, plays a crucial role in maintaining Zero Voltage Switching (ZVS) for both the auxiliary and main switches.Based on the average bus voltage [19], v_{bus} in a switching cycle, this duty ratio is mathematically expressed as (4):

$$d_{\text{aux}} = 1 - \frac{\left\{\max(|i_a, i_b, i_c|) + \frac{V_{DC}}{Z_{rs}}\right\} + 2L_r}{\frac{V_{DC}}{f_s} + 2L_r\frac{V_{DC}}{Z_{rs}}} \tag{4}$$

Where, Z_{rs}, the resonant impedance, is defined as $Z_{\text{rs}} = \sqrt{\frac{L_r}{3C_r+C_{\text{aux}}}}$. Unlike existing SVM-ZVS topologies that rely on precise zero-crossing detection to avoid hard switching, the proposed MSVM-ZVS ensures soft-switching even in the presence of imperfect detection. Fig. 8 illustrates the solution to the delay issue highlighted in Fig. 6. With MSVM-ZVS, the inverter maintains ZVS up to the threshold current i_{bt}, even in the absence of precise zero-crossing detection, as demonstrated in Fig. 8. This approach eliminates the need for highly accurate zero-crossing detection, allowing the use of low-cost, low-bandwidth sensors without compromising soft-switching performance.

V. CONTROL TECHNIQUE FOR THE PROPOSED MSVM-ZVS

The control block diagram of the three-phase Minimum Voltage Active Clamping (MVAC) inverter, based on the rotating coordinate system, is shown in Fig. 9. The control system utilizes active and reactive current references, denoted as i_d^{ref} and i_q^{ref}, respectively. A Phase-Locked Loop (PLL) is used to generate the transformed current references i_α^{ref} and

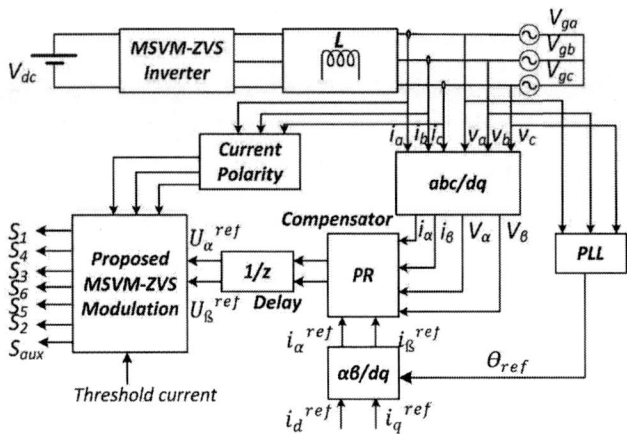

Fig. 9. Block diagram showing the control architecture of the proposed MSVM-ZVS inverter

TABLE II
PROTOTYPE PARAMETERS USED IN SIMULATION

Symbol	Parameter	Value
V_{dc}	Rated DC Voltage	750 V
S	Rated Power	5 kVA
S_i $(i = 1..6)$	Main Switches	SiC C2M0080120D
S_{aux}	Auxiliary Switch	SiC C2M0025120D
f_s	Switching frequency	100 kHz
$v_a = v_b = v_c$	Grid Phase Voltage	220 Vrms
$L_a = L_b = L_c$	Filter inductor	0.7 mH
C_{ri} $(i = 1..6)$	External cap. (Main Sw.)	0.12 nF
C_{aux}	External cap. (Aux. Sw.)	0 uF
C_{cl}	Clamping Capacitor	0 66uF
L_r	Resonant Inductor	8.6 μH
K_{pl}	Proportional Gain	8.4
T_r	Resonant time constant	0.0067
f_b	Bandwidth of Res. Com.	1 Hz

i_β^{ref}, which align with the grid voltage phase. The grid voltage and current signals are sensed in the stationary abc-frame and transformed into the rotating $\alpha\beta$-frame for easier control and modulation. The control system uses a proportional regulator, which is described by the transfer function (5):

$$ G_{\text{pr}}(s) = K_p \left[1 + \frac{1}{T_r} \frac{s}{s^2 + \omega_b s + \omega_r^2} \right]. \tag{5} $$

where K_p is the proportional gain, T_r is the resonant time constant, ω_b is the bandwidth, and ω_r is the resonance frequency. These parameters are provided in Table II and are critical for achieving effective current regulation. For digital implementation, the sampling period $T_s = 1/f_s$ is used, where f_s is the sampling frequency, which is equal to the switching frequency. The PR transfer function is discretized using the Tustin prewarping method, ensuring accurate representation in the digital domain. Additionally, to reflect real-world behavior, the simulation incorporates a controller delay of one sampling period, which accounts for the latency introduced by digital processing. Once the control loop generates the reference voltage vectors U_α^{ref} and U_β^{ref}, the control system determines the location and magnitude of the corresponding sector in the modulation sequence. Using the modulation sequences of the proposed MSVM-ZVS strategy, the switching vectors and their time durations— T_{v1} (1st Vector), T_{v2} (2nd Vector), T_{v3} (3rd Vector) —are calculated as follows (6):

$$ T_{v1} = m_a T_s \sin\left(k\frac{\pi}{3} - \theta_{ref} \right), $$
$$ T_{v2} = m_a T_s \sin\left(\theta_{ref} - (k-1)\frac{\pi}{3} \right), \tag{6} $$
$$ T_{v3} = d_{\text{aux}} T_s - T_{v1} - T_{v2}. $$

Here, the modulation index is defined as:

$$ m_a = \left[\sqrt{\left(U_\alpha^{\text{ref}}\right)^2 + \left(U_\beta^{\text{ref}}\right)^2} \right] \sqrt{\frac{3}{V_{\text{dc}}}}, \ \theta_{ref} = \arctan\left(\frac{U_\beta^{\text{ref}}}{U_\alpha^{\text{ref}}} \right), $$

and $k = 1, \ldots, 6$ denote Sectors I, ..., VI, respectively. These durations dictate the time each vector is applied, ensuring

optimal switching transitions. Furthermore, the duty ratio of the auxiliary switch, d_{aux}, is calculated using (4). This auxiliary duty cycle plays a key role in maintaining soft-switching, reducing switching losses, and enhancing overall efficiency.

VI. SIMULATION RESULTS

To validate the proposed control method, simulations were carried out using PLECS and LTspice on a three-phase MSVM-ZVS grid connected system. To compare the proposed MSVM-ZVS with the conventional SVM-ZVS at unity power factor, the same inverter parameters from [19] were adopted, as detailed in Table II. The performance improvements of MSVM-ZVS are evident in Fig. 10. Fig. 10 (a) illustrates a significant current spike (~ 145 A) in MOSFET S_3 with SVM-ZVS, which leads to increased power losses. In contrast, Fig. 10 (b) demonstrates that MSVM-ZVS drastically reduces the current spike to approximately 5 A, effectively lowering switching losses. This improvement highlights the superior current control and switching behavior of the proposed MSVM-ZVS.

The realization of ZVS for MSVM-ZVS is further validated in Fig. 11(a) and Fig. 11(b), which show the critical $phase - b$ current corresponding to switches S_6 and S_3, respectively. The MSVM-ZVS ensures near-zero voltage switching, significantly reducing energy losses associated with the switching process. Table III compares the performance of SVM-ZVS and MSVM-ZVS, demonstrating the overall higher efficiency of MSVM-ZVS at peak power. A key advantage of the proposed topology is the reduced current stress on both main and auxiliary switches. With SVM-ZVS, the current stresses on these switches are approximately nine times their rated currents, leading to greater thermal and operational challenges. Conversely, MSVM-ZVS mitigates these stresses, ensuring enhanced reliability and performance. It can be mentioned that

979-8-3315-1612-3/25 $31.00 © 2025 IEEE

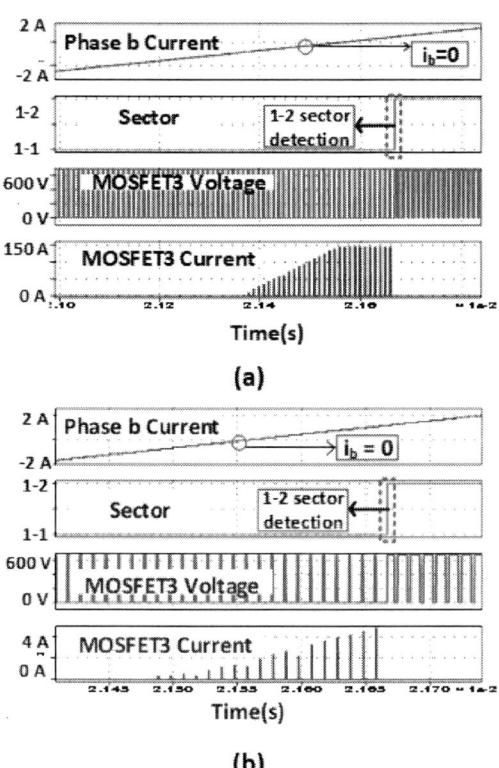

Fig. 10. MOSFET-3 (S_3) Current in sector 1-1 with delays for critical $phase - b$ current: (a) High current in SVM-ZVS;(b) Reduced current in MSVM-ZVS

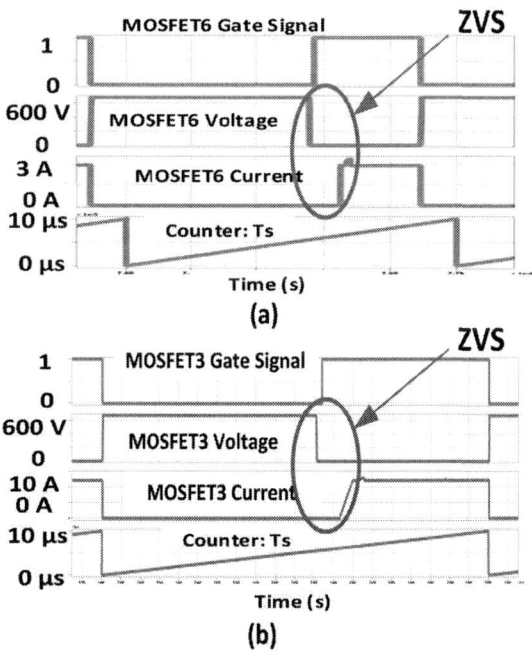

Fig. 11. ZVS turn-on realization of: (a) S_6, (b) S_3

TABLE III
COMPARISON BETWEEN SVM-ZVS AND MSVM-ZVS AT RATED CONDITION

Parameters	SVM-ZVS	MSVM-ZVS
Main Swi. Current Stress	9 times I_{rated}	Equal to I_{rated}
Aux. Swi. Current Stress	9 times I_{rated}	Equal to I_{rated}
Switching Losses	56 W	24 W
Conduction Losses	70 W	68 W
Efficiency (%) Peak	97.48	98.16

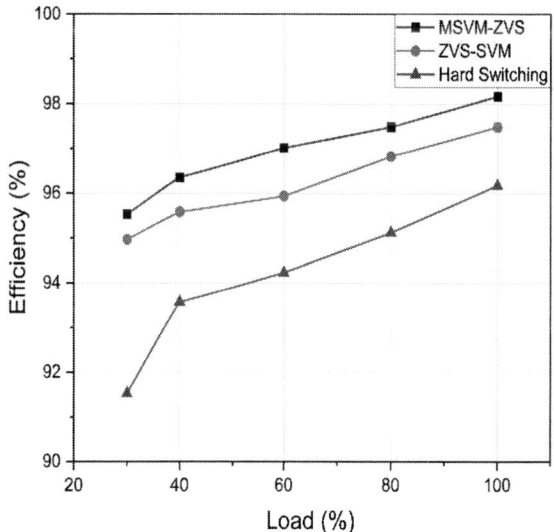

Fig. 12. Efficiency comparison among different modulation strategy

the phase RMS load current of the inverter is 7.57 A, and the current ratings of the auxiliary and main switches are different. Despite the absence of a perfect zero-crossing current due to delays, as shown in Fig. 10 (b), MSVM-ZVS achieves almost ZVS. This is enabled by the threshold current i_{bt}, which compensates for system delays, thus minimizing switching losses. In comparison, Fig. 10 (a) highlights the hard-switching behavior of SVM-ZVS due to the imprecise detection of zero-crossing currents. This results in exacerbated switching losses in the MOSFET S_3, as sensors and controller delays are not effectively addressed in the existing method. However, MSVM-ZVS maintains negligible switching losses even when system delays are considered, showcasing its effectiveness.

Fig. 12 compares the efficiency of three switching techniques: MSVM-ZVS, ZVS-SVM, and hard switching in varying load percentages (30%, 40%, 60%, 80%, and 100%). The data reveal that MSVM-ZVS consistently achieves the highest efficiency at all load levels, reaching a maximum of 98.16% at full load, followed by ZVS-SVM, which peaks at 97.48%. Hard-switching, although it improves in efficiency with increasing load, remains the least efficient, with a maximum of 96.17% at full load. The trend across all techniques indicates that efficiency improves as load increases, reflecting reduced losses or better performance under higher power levels. Notably, the results emphasize the advantages of soft-switching methods, such as MSVM-ZVS and ZVS-SVM, over traditional hard-switching, particularly in minimizing switching and conduction losses, making them more suitable for energy-efficient applications across varying operational conditions.

VII. CONCLUSIONS

The proposed MSVM-ZVS method significantly enhances inverter efficiency without requiring high-bandwidth current sensors, addressing a key limitation of conventional approaches. By optimizing the switching strategy, MSVM-ZVS reduces inverter switching losses by 57% and conduction losses by 1%, contributing to a 0.7% overall efficiency improvement, validated through simulations. A notable advancement over SVM-ZVS is its ability to achieve ZVS across the entire power factor range (-1 to +1), while SVM-ZVS is limited to a narrower range (-0.866 to +1). This comprehensive ZVS capability minimizes switching losses and current stresses, ensuring superior performance and reliability. Furthermore, MSVM-ZVS maintains its high efficiency even in the presence of system delays, providing a robust solution for a three-phase inverter.

REFERENCES

[1] J. Carrasco, L. Franquelo, J. Bialasiewicz, E. Galvan, R. PortilloGuisado, M. Prats, J. Leon, and N. Moreno-Alfonso, "Power-electronic systems for the grid integration of renewable energy sources: A survey," *IEEE Transactions on Industrial Electronics*, vol. 53, no. 4, pp. 1002–1016, 2006.

[2] S. Kim, M. Kwon, and S. Choi, "Operation and control strategy of a new hybrid ess-ups system," *IEEE Transactions on Power Electronics*, vol. 33, no. 6, pp. 4746–4755, 2018.

[3] F. Xu, B. Guo, Z. Xu, L. M. Tolbert, F. Wang, and B. J. Blalock, "Paralleled three-phase current-source rectifiers for high-efficiency power supply applications," *IEEE Transactions on Industry Applications*, vol. 51, no. 3, pp. 2388–2397, 2015.

[4] S. Safari, A. Castellazzi, and P. Wheeler, "Experimental and analytical performance evaluation of sic power devices in the matrix converter," *IEEE Transactions on Power Electronics*, vol. 29, no. 5, pp. 2584–2596, 2014.

[5] R. De Doncker and J. Lyons, "The auxiliary resonant commutated pole converter," in *Conference Record of the 1990 IEEE Industry Applications Society Annual Meeting*, 1990, pp. 1228–1235 vol.2.

[6] G. Venkataramanan, D. Divan, and T. Jahns, "Discrete pulse modulation strategies for high-frequency inverter systems," *IEEE Transactions on Power Electronics*, vol. 8, no. 3, pp. 279–287, 1993.

[7] M. R. Amini and H. Farzanehfard, "Three-phase soft-switching inverter with minimum components," *IEEE Transactions on Industrial Electronics*, vol. 58, no. 6, pp. 2258–2264, 2011.

[8] R. Li and D. Xu, "A zero-voltage switching three-phase inverter," *IEEE Transactions on Power Electronics*, vol. 29, no. 3, pp. 1200–1210, 2014.

[9] C. Du, X. Zhang, and D. Xu, "Study on an active clamping soft switching grid-connected inverter," in *2012 Twenty-Seventh Annual IEEE Applied Power Electronics Conference and Exposition (APEC)*, 2012, pp. 232–239.

[10] Z. Liu, Z. Huang, F. C. Lee, and Q. Li, "Digital-based interleaving control for gan-based mhz crm totem-pole pfc," *IEEE Journal of Emerging and Selected Topics in Power Electronics*, vol. 4, no. 3, pp. 808–814, 2016.

[11] J. Chen, D. Sha, J. Zhang, and X. Liao, "An sic mosfet based three-phase zvs inverter employing variable switching frequency space vector pwm control," *IEEE Transactions on Power Electronics*, vol. 34, no. 7, pp. 6320–6331, 2019.

[12] D. Divan, "The resonant dc link converter-a new concept in static power conversion," *IEEE Transactions on Industry Applications*, vol. 25, no. 2, pp. 317–325, 1989.

[13] D. Divan and G. Skibinski, "Zero-switching-loss inverters for high-power applications," *IEEE Transactions on Industry Applications*, vol. 25, no. 4, pp. 634–643, 1989.

[14] E. Chu, H. Xie, Z. Chen, J. Bao, Y. Zhou, and H. Zhang, "Parallel resonant dc link inverter topology and analysis of its operation principle," *IEEE Journal of Emerging and Selected Topics in Power Electronics*, vol. 8, no. 3, pp. 3124–3138, 2020.

[15] Q. Wang, G. Guo, Y. Wang, and J. Chen, "An efficient three-phase resonant dc-link inverter with low energy consumption," *IEEE Transactions on Power Electronics*, vol. 36, no. 1, pp. 702–715, 2021.

[16] D. Xu and B. Feng, "Novel zvs three-phase pfc converters and zero-voltage-switching space vector modulation (zvs-svm) control," in *Proceedings. 2004 First International Conference on Power Electronics Systems and Applications, 2004.*, 2004, pp. 30–37.

[17] N. He, M. Chen, J. Wu, N. Zhu, and D. Xu, "20-kw zero-voltage-switching sic-mosfet grid inverter with 300 khz switching frequency," *IEEE Transactions on Power Electronics*, vol. 34, no. 6, pp. 5175–5190, 2019.

[18] E. Chu, Y. Kang, P. Zhang, Z. Wang, and T. Zhang, "An svpwm method for parallel resonant dc-link inverter with the smallest loss in the auxiliary commutation circuit," *IEEE Transactions on Power Electronics*, vol. 37, no. 2, pp. 1772–1787, 2022.

[19] Y. Wu, N. He, M. Chen, and D. Xu, "Generalized space-vector-modulation method for soft-switching three-phase inverters," *IEEE Transactions on Power Electronics*, vol. 36, no. 5, pp. 6030–6045, 2021.

[20] Y. Yan, H. Gui, and H. Bai, "Complete zvs analysis in dual active bridge," *IEEE Transactions on Power Electronics*, vol. 36, no. 2, pp. 1247–1252, 2021.

979-8-3315-1612-3/25 $31.00 © 2025 IEEE

A Feedforward Ripple Reduction Control Strategy based on a Hybrid GaN/Si Interleaved Inverter

Mowei Lu, *Student Member, IEEE,* Jurgis Reinotas, *Student Member, IEEE,* Xiaoyang Tian, *Member, IEEE,*
Stefan M. Goetz, *Member, IEEE*

Abstract—While wide-bandgap (WBG) devices such as silicon carbide (SiC) and particularly gallium nitride (GaN) offer advantages such as low switching losses and high-frequency switching, they also come with higher cost per current as well as substantially lower overload and fault tolerance. Recent research suggested interleaved operation that the silicon part could take most of the load and WBG contributes only some, while it further compensates distortion. However, such a mode required fast closed-loop control with extreme demands on sensing, filtering, and control bandwidth. Additionally, previous use was limited to dc conversion, where the quasi-stationary output voltage simplifies operation. This paper proposes a novel strategy to reduce the inductor in a hybrid GaN/Si interleaved inverter and use low-bandwidth feed-back control for fundamental current distribution in combination with a feed-forward term based on ripple estimation. This approach decouples high-frequency ripple from low-frequency fundamental currents, achieving fast response and efficient ripple cancellation without the excessive demands of full bandwidth feedback control.

Index Terms—Hybrid switch, wide-bandgap transistor, interleaved switching, pulse-width modulation, ripple compensation, low-bandwidth control.

I. INTRODUCTION

POWER inverters form the backbone of modern electronics-dominated power grids, and play a crucial role in both major energy sources like wind and solar power and major loads such as electric vehicle chargers. Whereas previous research primarily aimed at power density and cost [1]–[3], the large presence of power electronics in the grid shifts the focus to stabilisation, grid control, and ancillary services [4]–[8]. For instance, Solar inverters, battery grid interfaces, and vehicle chargers are expected to provide reactive power management and also support the grid with other functions in the future [9]–[11]. Additional dedicated electronic units, such as voltage conditioners, in-line voltage regulators, soft open points (SOP), and unified power quality conditioners (UPQC) mitigate power quality problems, unwanted power flows, and voltage issues, such as voltage sag/swell or flicker [12]–[16].

In these grid-connected converters, achieving high power density and fast response is crucial. This is typically accomplished by increasing the switching frequency, which helps reduce the size of passive components, particularly the output filters. However, in conventional silicon devices such as IGBTs, which are widely used in inverters, higher switching frequencies lead to increased losses. Wide-bandgap devices such as GaN and SiC transistors exhibit low switching losses;

GaN transistors stand out for their ability to switch even at megahertz frequencies, providing superior efficiency compared to SiC and Si devices, even under hard-switching conditions [17], [18]. Additionally, with a maximum voltage rating of up to 650 V, commercial GaN HEMTs are well-suited for applications in single-phase low-voltage grids (110 V, 230 V) [1]. However, their high costs, limited current ratings, and low over-load capacity restrict widespread use. These barriers make it essential to explore new designs that capitalize on wide-bandgap devices' benefits but also compensate their limitations in high-power high-current application [19].

Hybrid wide-bandgap/silicon interleaved inverters can both achieve high-frequency switching and manage large grid currents, even fault current [20]. The topology can reduce output current ripple by high-frequency switching wide-bandgap bridges to compensate for the significant ripple produced by low-frequency switching silicon bridges. However, conventional methods rely on closed-loop control for the high-speed wide-bandgap bridge's ripple tracking, which imposes extreme demands on response time, delay management, signal processing speeds, and high-bandwidth sensors [21]. Therefore, a more efficient solution is necessary.

This paper presents a hybrid GaN/Si interleaved single-phase inverter topology discussed in our previous research [20] and introduces a novel direct feed-forward control method for ripple cancellation. The proposed method can exclusively use low-bandwidth control loops for the fundamental current share and a ripple estimator that supplies a feed-forward injection dedicated to ripple tracking and absorption. Unlike conventional full-bandwidth controllers, this strategy effectively decouples high-frequency ripple from low-frequency fundamental currents and enables fast response and efficient ripple reduction without excessive computational demands. The low bandwidth alleviates the need for extremely fast microcontrollers and eliminates the requirement for high sampling rate. Moreover, an experimental test validates both the control method and the topology, and extends the interleaved operation beyond the DC/DC converter domain.

II. OPERATION PRINCIPLE

Figure 1 presents the circuit diagram of the hybrid GaN/Si interleaved inverter. Two GaN bridges operate at a high switching frequency with low inductance, while two Si bridges switch at a lower frequency with slightly higher inductance,

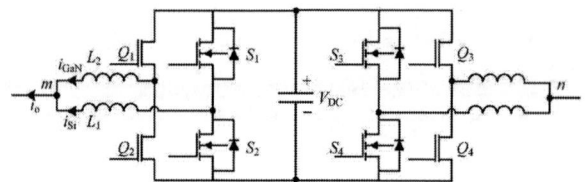

Fig. 1. Circuit diagram of the hybrid GaN/Si interleaved inverter.

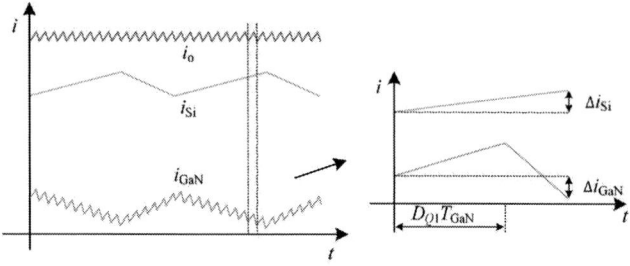

Fig. 2. Working waveform of the hybrid WBG/Si interleaved inverter in two cycles.

though still significantly lower than that of purely Si-based bridges [20]. Each bridge has two complementary switches, and can be modulated in either unipolar or bipolar modulation as common inverters. The innovative aspect of this topology is its ability to cancel out the ripple generated by the Si bridges through the fast-switching GaN bridges (Fig. 2). This design ensures that the final ripple only contains that from the GaN devices. Additionally, the total output current can be distributed between the Si and GaN bridges in a controlled ratio, which increases the current rating compared to inverters based solely on either Si or GaN devices.

To derive the relationship between the duty ratios of the two bridges, we can analyse one cycle of GaN switching. The current variation of the Si bridge is

$$\Delta i_{\mathrm{Si}} = \frac{(D_{S_1} V_{\mathrm{DC}} - V_{\mathrm{mn}})T_{\mathrm{GaN}}}{2L_1}, \tag{1}$$

where D_{S_1} is the switching state of S_1: when $S_1 = 1$, the Si device is closed and the current increases, and when $S_1 = 0$, the Si device is open and the current decreases; V_{DC} is the DC link voltage; V_{mn} is the output voltage; T_{GaN} is the period of the GaN bridge; L_1 is the filtering inductor of the Si bridge.

The difference between the increase and decrease of the GaN bridge's current ripple during one period can be obtained as

$$\Delta i_{\mathrm{GaN}} = \frac{(V_{\mathrm{DC}} - V_{\mathrm{mn}})D_{Q_1}T_{\mathrm{GaN}} - V_{\mathrm{mn}}(1 - D_{Q_1})T_{\mathrm{GaN}}}{2L_2}, \tag{2}$$

where D_{Q_1} is the swiching state of Q_1 and L_2 is the filter inductor of the GaN bridge.

To fully cancel out the large ripple produced by the Si bridge, the two variations should follow

$$\Delta i_{\mathrm{Si}} + \Delta i_{\mathrm{GaN}} = 0. \tag{3}$$

Therefore, the duty ratio of GaN device Q_1 is derive as

$$D_{Q_1} = \left(\frac{L_2}{L_1} + 1\right)\frac{V_{\mathrm{mn}}}{V_{\mathrm{DC}}} - \frac{L_2}{L_1}D_{S_1}. \tag{4}$$

Similarly, other switches of Si and GaN bridges also follow the relationship shown in equation (4).

III. PROPOSED CONTROL STRATEGY

A. Limitation of full bandwidth controller

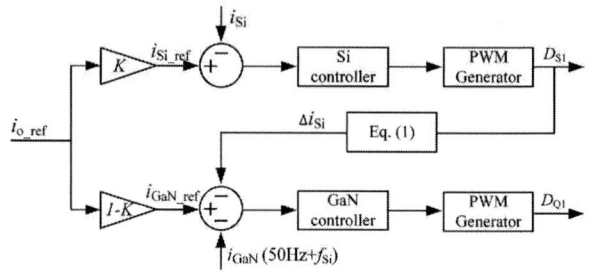

Fig. 3. The block diagram of conventional control strategy with full bandwidth.

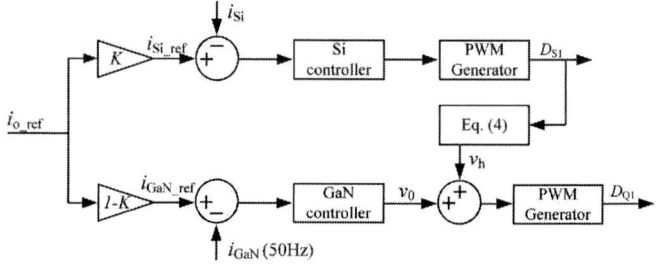

Fig. 4. The block diagram of the proposed feed-forward control.

A possible control method for reducing ripple in the Si bridge would add the variation to the GaN bridge's current reference (Fig. 3) [21]. However, this approach requires high-frequency sampling of the GaN current. The sampled data should include both the fundamental signal and the ripple reduction component, with the latter equal to the swiching rate of Si devices (e.g. 10 kHz). Therefore, the ADC sampling rate should be at least 500 kHz $\sim$ 1 MHz for accurate tracking. Moreover, the bandwidth of the controller must be designed to be some ten times that of the reference signal, i.e., on the order of 100 kHz. The large bandwidth places significant demands on the digital signal processor. Typical controllers for inverters, such as C2000, Tricore-, or ARM-M-based controllers struggle to meet such levels requirements.

B. The proposed Feed-Forward Control

The proposed feed-forward control (Fig. 4) eliminates the need for high-frequency components in the closed-loop controller and instead estimates the duty ratio required to compensate for the Si bridge's ripple with Equation 4. This approach allows the GaN controller to focus solely on the fundamental frequency, while the feed-forward term handles the modulation signal for the high-frequency components. The signal processor has direct access to the switching states of the Si devices, as it is setting them in after all, and avoids any current measurement with wide-bandwidth low-latency

979-8-3315-1612-3/25 $31.00 © 2025 IEEE

$$v_0 = \left(k_{\mathrm{p}} + \frac{2k_{\mathrm{r}}\omega_c s}{s^2 + 2\omega_c s + \omega_0{}^2} \right) \cdot \left(i_{\mathrm{GaN\_ref}} + \frac{V_{\mathrm{mn}}}{\left(k_{\mathrm{p}} + \frac{2k_{\mathrm{r}}\omega_c s}{s^2 + 2\omega_c s + \omega_0{}^2} \right) \cdot V_{\mathrm{DC}}} - i_{\mathrm{GaN}} \right) \tag{5}$$

$$v_{\mathrm{ref}} = \left(k_{\mathrm{p}} + \frac{2k_{\mathrm{r}}\omega_c s}{s^2 + 2\omega_c s + \omega_0{}^2} \right) \cdot \left(i_{\mathrm{GaN\_ref}} + \frac{V_{\mathrm{mn}}}{\left(k_{\mathrm{p}} + \frac{2k_{\mathrm{r}}\omega_c s}{s^2 + 2\omega_c s + \omega_0{}^2} \right) \cdot V_{\mathrm{DC}}} - i_{\mathrm{GaN}} \right) + \left(\frac{L_2}{L_1} + 1 \right) \cdot \frac{V_{\mathrm{mn}}}{V_{\mathrm{DC}}} - \frac{L_2}{L_1} \cdot D_{\mathrm{S}_1} \tag{6}$$

sensors. As a result, the controller only requires bandwidth that covers the fundamental frequency and significantly reduces the demands on the controller, sampling, and processing.

Figure 5 presents the detailed current control loop. We use a PR controller per $H_1(s) = k_{\mathrm{p}} + \frac{2k_{\mathrm{r}}\omega_c s}{s^2 + 2\omega_c s + \omega_0{}^2}$ to control the fundamental current, and a compensator $H_2(s) = \frac{1}{H_1(s) \cdot V_{\mathrm{DC}}}$ to compensate the disturbance of the output voltage V_{mn}. The modulation signal for the fundamental part is derived as Equation (5). Thereafter, the reference gate signal is the summation of the fundamental current term and the feed-forward ripple cancellation term as given in Equation (6).

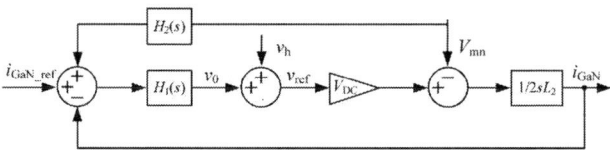

Fig. 5. The schematic of the current control loop for the GaN bridge.

The transfer function of $i_{\mathrm{GaN\_ref}}$ to i_{GaN} follows

$$\frac{i_{\mathrm{GaN}}}{i_{\mathrm{GaN\_ref}}} = \frac{\left(k_{\mathrm{p}} + \frac{2k_{\mathrm{r}}\omega_c s}{s^2 + 2\omega_c s + \omega_0{}^2} \right) \cdot \frac{V_{\mathrm{DC}}}{2sL_2}}{1 + \left(k_{\mathrm{p}} + \frac{2k_{\mathrm{r}}\omega_c s}{s^2 + 2\omega_c s + \omega_0{}^2} \right) \cdot \frac{V_{\mathrm{DC}}}{2sL_2}}. \tag{7}$$

From the above, we observe that the controller design is independent of the feed-forward ripple cancellation term. Therefore, the controller can be designed similarly to a standard pure closed-loop current control approach, without including the high-frequency ripple tracking.

With the system parameters $k_{\mathrm{p}} = 0.0045$; $k_{\mathrm{r}} = 20$; $\omega_c = 0.5\,\pi$; $\omega_0 = 100\pi$; $V_{\mathrm{DC}} = 150$ V; $L_2 = 16$ µH, we tune the controller with a sufficient phase margin of 60.7° (Fig. 6) for stable operation.

IV. EXPERIMENTAL RESULTS

TABLE I
PARAMETERS OF THE EXPERIMENTAL SETUP

Description	Value
DC link voltage	150 V
Load current (pk–pk)	32 A
Filtering inductance	L_1: 120 µH; L_2: 16 µH
Switching rate	Si bridge: 10 kHz; GaN bridge: 500 kHz
ADC sampling rate	50 kHz

We implemented an experimental system to validate the proposed method with two test scenarios: normal operation with a current ratio of Si bridge to GaN bridge of 3:1, and load variation. Here two GaN half-bridges (IGOT60R070D1) were connected in parallel with two IGBT half-bridges (1ED020I12) to

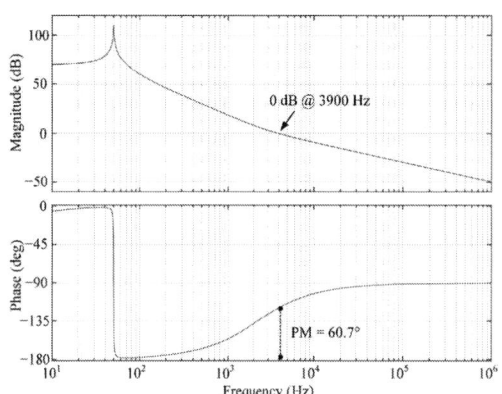

Fig. 6. Bode diagram of the current control loop gain.

form the interleaved design. A digital signal processor (C2000 F28379D) controls the setup, while two 50 kHz bandwidth LES 50-NP sensors measure the current. Further parameters are listed in TABLE I.

Figure 7 illustrates the results when the current ratio between the Si and GaN bridges is set to 3:1. The red curve represents the load, the blue curve the Si bridge, and the green curve the GaN bridge current. In Fig. 7, the GaN bridge effectively cancels out the ripple generated by the Si bridge and thus in the ripple in the load. Additionally, the fundamental current distribution is accurately controlled to 3:1. The ripple cancellation effect is further clarified in the 200- and 500-times zoomed-in graphs of Figure 8 and Figure 9. The proposed method achieves a total harmonic distortion (THD) of 2.3% and an efficiency of 90.2%, while a purely IGBT-based inverter with the same inductance yields 11.1% THD and 90.8% efficiency. Therefore, the proposed method maintains a comparable efficiency but significantly improves the output quality and enables a more compact overall design [20].

Figure 10 illustrates the control response during a step load change from 16 A to 32 A pk–pk. The current remains well-controlled throughout the transition. The ripple-tracking method follows rapidly (< 1.5 ms) so that the THD changes moderately from 2.6% to 2.3%. The measurement demonstrates the proposed method's ability to quickly follow load variations.

V. CONCLUSIONS

The proposed control strategy for the hybrid GaN/Si interleaved inverter substantially reduces current ripple and minimizes the THD (about five-fold). The GaN part contributes

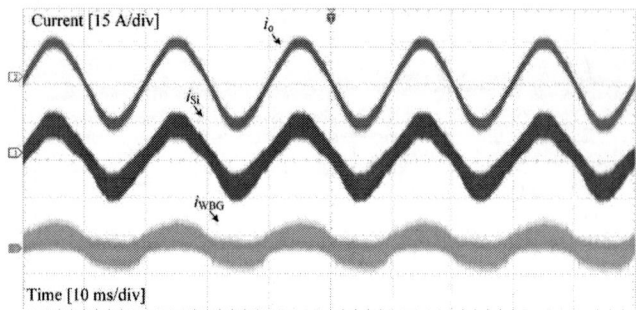

Fig. 7. Experimental results for current ratio 3:1.

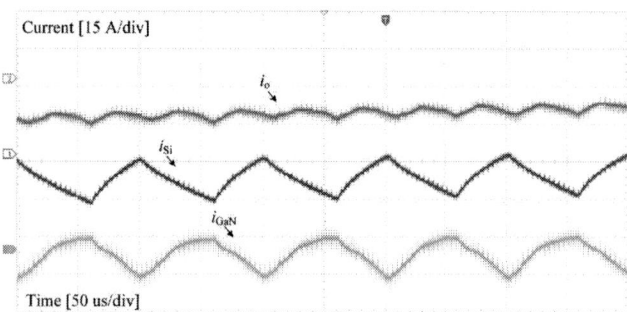

Fig. 8. Experimental results for current ratio 3:1 (200-times scaled view).

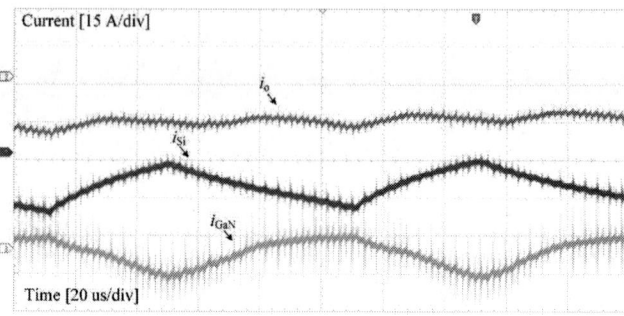

Fig. 9. Experimental results for current ratio 3:1 (500-times scaled view).

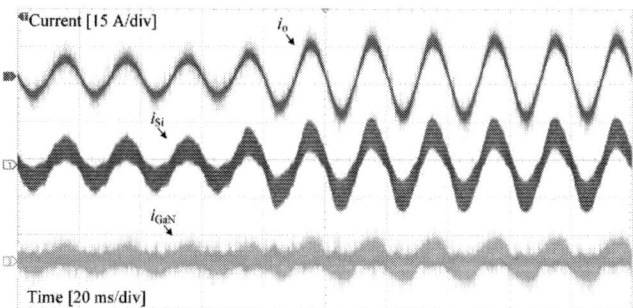

Fig. 10. Experimental results for load variation.

to the load current according to its capacity and additionally cleans up most of the ripple of the Si branch. The solution uses a ripple estimator in combination with a feed-forward loop, and thus decouples high-frequency ripple from low-frequency fundamental currents. It achieves fast response and efficient ripple reduction without imposing excessive demands on signal processing as well as control speed or requiring high-frequency sampling. Morever, the method can equally serve for SiC/Si hybrids so that it can be expanded to higher voltage ranges.

REFERENCES

[1] T. Liu, C. Chen, K. Xu, Y. Zhang, and Y. Kang, "Gan-based megahertz single-phase inverter with a hybrid tcm control method for high efficiency and high-power density," *IEEE Transactions on Power Electronics*, vol. 36, no. 6, pp. 6797–6813, 2021.

[2] M. Lu, M. Qin, J. Kacetl, E. Suresh, T. Long, and S. M. Goetz, "A novel direct-injection universal power flow and quality control circuit," *IEEE Journal of Emerging and Selected Topics in Power Electronics*, vol. 11, no. 6, pp. 6028–6041, 2023.

[3] X. Ren, Y. Jiang, H. Weng, T. Long, and D. Xu, "Design and testing of a soft-switching solid-state transformer module," *IEEE Journal of Emerging and Selected Topics in Power Electronics*, pp. 1–1, 2024.

[4] M. Lu, W. Mu, M. Qin, A. Koehler, J. Fang, and S. M. Goetz, "Differential detection of feeder and mesh impedances through a series–parallel direct-injection soft open point," *IEEE Transactions on Power Electronics*, pp. 1–10, 2024.

[5] X. Pan, L. Zhang, Y. Li, K. Li, and H. Huang, "Modulated model predictive control with branch and band scheme for unbalanced load compensation by mmcc-statcom," *IEEE Transactions on Power Electronics*, vol. 37, no. 8, pp. 8948–8962, 2022.

[6] J. Zhu, Y. Li, F. Z. Peng, B. Lehman, and H. Huang, "From active resistor to lossless and virtual resistors: A review, insights, and broader applications to energy grids," *IEEE Journal of Emerging and Selected Topics in Power Electronics*, pp. 1–1, 2024.

[7] X. Li, P. Feng, Z. Tian, M. Huang, S. Shu, X. Zha, and P. Hu, "Transient stability analysis for grid-tied VSG considering high-order nonlinear interactions between active and reactive power control loops," *IEEE Transactions on Power Electronics*, vol. 39, no. 6, pp. 6974–6988, 2024.

[8] X. Li, Z. Tian, X. Zha, P. Sun, Y. Hu, and M. Huang, "An iterative equal area criterion for transient stability analysis of grid-tied converter systems with varying damping," *IEEE Transactions on Power Systems*, vol. 39, no. 1, pp. 1771–1784, 2024.

[9] Y. Yang, F. Blaabjerg, and H. Wang, "Low-voltage ride-through of single-phase transformerless photovoltaic inverters," *IEEE Transactions on Industry Applications*, vol. 50, no. 3, pp. 1942–1952, 2014.

[10] S. Liu, C. Lu, and G. He, "Distributed electric bicycle batteries for subway station energy management as a virtual power plant," *Applied Energy*, vol. 370, p. 123094, 2024.

[11] J. Zhu, C. Mao, B. Liu, Z. Wang, D. Wang, and H. Wang, "Design and field application of dc and harmonic suppression system for neutral current in 220 kv substation," *IEEE Transactions on Industrial Electronics*, vol. 69, no. 8, pp. 7560–7570, 2022.

[12] R. M. Abdalaal and C. N. M. Ho, "System modeling and stability analysis of single-phase transformerless upqc integrated input grid voltage regulation," *IEEE Journal of Emerging and Selected Topics in Industrial Electronics*, vol. 3, no. 3, pp. 670–682, 2022.

[13] S. Rivera, S. M. Goetz, S. Kouro, P. W. Lehn, M. Pathmanathan, P. Bauer, and R. A. Mastromauro, "Charging infrastructure and grid integration for electromobility," *Proceedings of the IEEE*, vol. 111, no. 4, pp. 371–396, 2023.

[14] Y. Gu and T. C. Green, "Power system stability with a high penetration of inverter-based resources," *Proceedings of the IEEE*, vol. 111, no. 7, pp. 832–853, 2023.

[15] Z. Lai, H. Yi, Z. Wang, F. Zhuo, and H. Zhuang, "Transient analysis and overcurrent limited strategy for supply restoration-oriented hybrid soft open point," *IEEE Transactions on Power Electronics*, vol. 39, no. 2, pp. 2660–2676, 2024.

[16] H. Zhao, W. Chen, G. He, and J. Wang, "A new shared module soft open point for power distribution network," *IEEE Transactions on Power Electronics*, vol. 38, no. 3, pp. 3363–3374, 2023.

[17] W. Qian, J. Lu, H. Bai, and S. Averitt, "Hard-switching 650-v gan hemts in an 800-v dc-grid system with no-diode-clamping active-balancing three-level topology," *IEEE Journal of Emerging and Selected Topics in Power Electronics*, vol. 7, no. 2, pp. 1060–1070, 2019.

[18] S. Ni, C. Li, and Z. Zheng, "Control strategy of a hybrid sic-si traction

979-8-3315-1612-3/25 $31.00 © 2025 IEEE

inverter for direct-drive multiphase pmsms in marine propulsion," *IEEE Transactions on Power Electronics*, vol. 39, no. 12, pp. 16400–16414, 2024.

[19] C. Zhang, X. Yuan, J. Wang, W. Chen, B. Hu, and Z. J. Shen, "Adaptive power sharing and switching frequency control for power loss optimization in wbg/si hybrid half-bridge converters," *IEEE Transactions on Power Electronics*, vol. 38, no. 4, pp. 4440–4450, 2023.

[20] M. Lu, M. Qin, W. Mu, J. Fang, and S. M. Goetz, "A hybrid gallium-nitride–silicon direct-injection universal power flow and quality control circuit with reduced magnetics," *IEEE Transactions on Industrial Electronics*, pp. 1–14, 2024.

[21] C. Zhang, J. Wang, K. Qu, B. Hu, Z. Li, X. Yin, and Z. J. Shen, "Wbg and si hybrid half-bridge power processing toward optimal efficiency, power quality, and cost tradeoff," *IEEE Transactions on Power Electronics*, vol. 37, no. 6, pp. 6844–6856, 2022.

IGBT Comparison for Optimized Switching Behavior in the SiC/Si-Hybrid Switch

Adrian Amler
Institute of Power Electronics
Friedrich-Alexander-Universität
Erlangen-Nürnberg
Nürnberg, Germany
adrian.amler@fau.de

Thomas Heckel
Medium Voltage Electronics
Fraunhofer Institute for Integrated Systems
and Device Technology
Erlangen, Germany
thomas.heckel@iisb.fraunhofer.de

Daniel Ruppert
R&D / Design Electric Drives
Audi AG
Ingolstadt, Germany
daniel.ruppert@audi.de

Cornelius Rettner
R&D Power Electronics, Volkswagen Group Components
Volkswagen AG
Ingolstadt, Germany
cornelius.rettner@volkswagen.de

Martin März
Institute of Power Electronics
Friedrich-Alexander-Universität Erlangen-Nürnberg
Erlangen, Germany
martin.maerz@fau.de

Abstract—**The parallel connection of Si-IGBTs and SiC-MOSFETs in a so-called hybrid switch offers new degrees of freedom to optimize the efficiency of bridge-type inverters and converters. By separately controlling the parallel devices with a gate delay, the switching performance can be improved to resemble that of the superior MOSFET. It is known that the optimal switching strategy depends on the operating point (temperature, current) and the hybrid switch design (chip area ratio, gate resistance). Additionally, the switching and conduction behavior changes drastically when using different IGBT technologies and devices. The comprehensive experimental study performed for this paper demonstrates that the IGBT technology used has a strong influence on the system efficiency as a result of its forward characteristics and its turn-OFF performance and gives an overview of the relevant dependencies. Different established and state-of-the-art IGBTs are tested in a double pulse setup and compared regarding their switching performance in the hybrid switch. It is shown that the hybrid switch variants can achieve from up to 54 % to 71 % system loss reduction compared to a standalone IGBT system depending on the device selection. It thus performs on-par with SiC MOSFETs under comparable operating conditions.**

Keywords—SiC-MOSFET, Si-IGBT, NPT, PT, Trench, hybrid switch, delay, switching performance, switching losses, efficiency, double-pulse test, experimental study

I. Introduction

A hybrid switch (HyS) or cross switch (XS) hybrid, i.e. the parallel connection of IGBTs and MOSFETs to form a single topological switch, can be used for optimizing the conduction behavior and the switching behavior as well as the system cost of inverters and other bridge-type converters. While conduction losses can be reduced by driving both devices from the same gate circuit [1, 2], the switching can be optimized by separately controlling the IGBT and the MOSFET [3], thus, lowering the switching losses compared to IGBT-dominated switching. In literature, the focus has been on optimized control (switching

This work was supported in part by the Fraunhofer Internal Programs under Grant No. PREPARE 40-08262

patterns [4, 5], driver circuits [6–8], modelling [9]), sizing and diode selection [10–13], and applicational optimization [1, 14]. A highly efficient, MOSFET-dominated switching can be achieved by turning the IGBT OFF earlier and ON later. The optimal delay is highly dependent on the individual design (IGBT/MOSFET chip area ratio, parasitic inductance, temperature, gate control). An important aspect is the selection of the IGBT, meaning its generation, technology, optimization [15]. In conventional hard-switching conditions, IGBTs exhibit a tail current caused by bipolar device effects. Similarly, a current spike appears in the hybrid switch during the voltage transition even after an earlier turn-OFF of the IGBT. This unusual behavior warrants a detailed experimental study which was performed for this paper. It covers most importantly the IGBT device selection as well as influence factors such as current, delay, temperature, gate resistance, and clamping. Different IGBT generations along Infineon's and IXYS/Littlefuse's lineup of discrete 1200 V devices are compared by comprehensive double pulse measurements for a 750 V dc link.

This paper is organized as follows: The double pulse test setup, its properties, and the methods for evaluation are explained in Section II. The influence factors on the IGBT behavior including the different devices are explained using exemplary measurements in Section III. Section IV evaluates the usefulness of the different devices from an applicational perspective before the concluding Section V. It is shown that the IGBT selection heavily affects the turn-OFF losses while the diode selection matters for the turn-ON and the combined forward characteristics define the conduction losses.

II. Test Setup and Evaluation Methods

To precisely characterize the different hybrid switch configurations, double pulse tests are performed. Fig. 1 explains the schematic of the setup. The circuit board employs components in a TO-247 package (4-pin Kelvin variant for switches if available, 2-pin variant for diodes) to allow both an easy handling of different configurations and a moderately low parasitic inductance in the commutation loop as well as between

979-8-3315-1612-3/25 $31.00 © 2025 IEEE

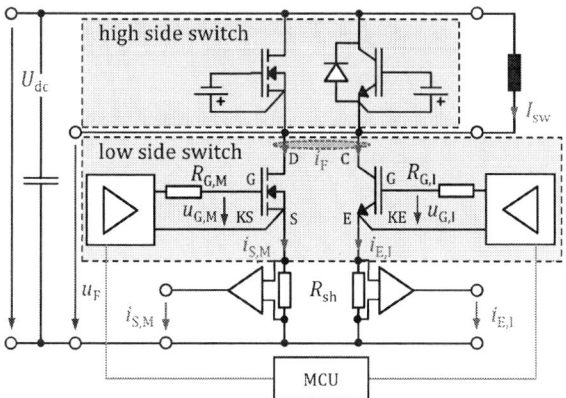

Fig. 1. Schematic for double pulse tests.

TABLE I.   IGBTs Under Research

Device	I_C	U_{CE}	Technology	Year
IKY40N120CS6 (IFX)	40 A	1.85 V	Trenchstop, Standard 6	2018
IKZA40N120CS7 (IFX)	40 A	1.65 V	Trenchstop, Standard 7	2022
IKZA40N120CH7 (IFX)	40 A	1.7 V	Trenchstop, Highspeed 7	2022
SGW25N120 (IFX)	25 A	3.1 V	NPT	2006
IXYH40N120B4H1 (LF)	40 A	1.85 V	XPT/Trench, med. freq.	2023
IXYH40N120C4H1 (LF)	40 A	2.2 V	XPT/Trench, high freq.	2023
IXGH30N120B3D1 (LF)	30 A	2.95 V	PT, med. freq.	2008

values for junc. temp. $T_j = 25\,°C$, on-state gate voltage $U_{G,I} = 15\,V$

the components of a single topological switch. The high side and low side hybrid switches can differ. IGBTs are usually paired with antiparallel (PN) diodes. To avoid their influence on the IGBT behavior, a SiC Schottky Diode (SBD) for the high side switch is used in most measurements to compare only the IGBT and eliminate the influence of its antiparallel diode. The SiC devices used are Infineon's IMZ120R060M1H SiC-MOSFET ($1200\,V$, $60\,m\Omega$) and Onsemi's FFSH40120A SiC-Schottky diode ($1200\,V$, $40\,A$). The IGBTs under consideration are listed in Table I. The orange markings are used in the following as abbreviations to identify the respective devices. The components have been chosen from Infineon's (IFX) and IXYS/Littlefuse's (LF) lineup to cover state-of-the-art Fieldstop-Trench IGBTs and older non-punch-through (NPT) and punch-through (PT) IGBTs. Of the Trench IGBTs, different devices with a balanced design between switching and conduction performance ("Standard", "medium frequency") and designs focused on switching performance ("Highspeed", "high frequency") are tested. All investigated combined switches have a nominal dc current of 60 ... 75 A according to their datasheets.

The separate drivers for the IGBT and the MOSFET are controlled by a microcontroller, which enables setting the delay with a precision of up to 7 ns. Henceforth, a positive delay $t_{d,I}$ means that the IGBT is turned ON or OFF after the MOSFET and refers to the shift in logic signals. Using an MPI ThermalAir temperature conditioning system, the full automotive temperature range from $-40\,°C$ to $175\,°C$ can be covered. If not mentioned otherwise, the turn-ON and turn-OFF is performed with an external gate resistance of $R_{G,M} = R_{G,I} = 15\,\Omega$. The internal gate resistance of the used MOSFET and the S6 IGBT have been measured at $R_{G,int} = 2.4\,\Omega$ and $0.6\,\Omega$, respectively.

The gate voltages in use are $U_{G+/-} = +15/-5\,V$. Waveforms are obtained by a Keysight MXR058A 500 MHz oscilloscope with 400 MHz differential probes for u_F and separate 100 MHz shunt measurements for the MOSFET source and IGBT emitter currents $i_{S,M}$ and $i_{E,I}$. The shunt resistors are selected to $R_{sh} = 6.8\,m\Omega$ so that $R_{sh} \ll R_{ds(ON)}$ to avoid a significant influence of the current measurement on the current distribution between the paralleled devices and in turn a change in switching behavior. Due to the separate Kelvin Source/Emitter, gate currents are excluded from the current measurements. As a compromise between noise and interference, preamplification using Analog Devices AD8009 amplifiers and passive probes are employed. The results are computationally evaluated for switching energies, voltage gradients, etc. Switching energies E_{sw} are calculated to reflect not only the voltage transition but also the additional losses caused during the delay while one transistor is OFF and by an unbalanced current sharing between the parallel devices due to the slow redistribution of current after the switching. $E_{sw,\Sigma}$ means the total turn-ON or -OFF energy, while $E_{sw,I}$ and $E_{sw,M}$ imply the IGBT or MOSFET share. Gradients (du_F/dt, di_F/dt) are defined from 20 % to 80 % of U_{dc} and I_{sw}, respectively.

III. Switching Behavior of the Hybrid Switch

The switching behavior of the hybrid switch is mainly defined by the control strategy and the properties of the involved devices. While the unipolar channel of a MOSFET acts very predictably and its behavior is well-understood, the IGBT, being a bipolar device, is dominated by the injection and sweep out of the charge carriers in the drift region of the device. This leads to forward and reverse recovery effects which manifest e.g. in the well-known tail current. They are also relevant for the hybrid switch. In the following, characteristic waveforms are shown and the IGBT influence on the switching properties is described. First, the turn-ON is explained but only briefly as it is dominated by effects covered elsewhere before. Then, the turn-OFF is investigated and its influences are analyzed in detail.

A. Turn-ON

Turn-ON waveforms are dominated by the switch which turns ON earlier. The turn-ON order is either dependent on the device properties or can be forced by introducing a delay. For the measurements seen in Fig. 2, the hybrid switch comprises an S6 IGBT and its co-packaged PN diode has been used as commutation partner.

If the IGBT turns ON simultaneously or later than the MOSFET, a MOSFET-like turn-ON can be observed, as seen in Fig. 2 for $t_{d,I} = 0\,ns$ and $50\,ns$. The voltage and current waveforms differ from when the MOSFET would be operated standalone: The voltage gradient du_F/dt is smaller due to the higher output capacitance of the combined switch. A current peak occurs due to reverse recovery of the PN diode used as commutation partner. If a SiC Schottky diode had been used instead, the MOSFET-dominated turn-ON would improve drastically. The effects of the anti-parallel diode have been investigated previously in [11] and its effects can also be seen later in Subsection G. When the IGBT performs the turn-ON (at least in part) with $t_{d,I} < 0\,ns$, the IGBT current rises initially but drops again and is conducted by the MOSFET due to a slower

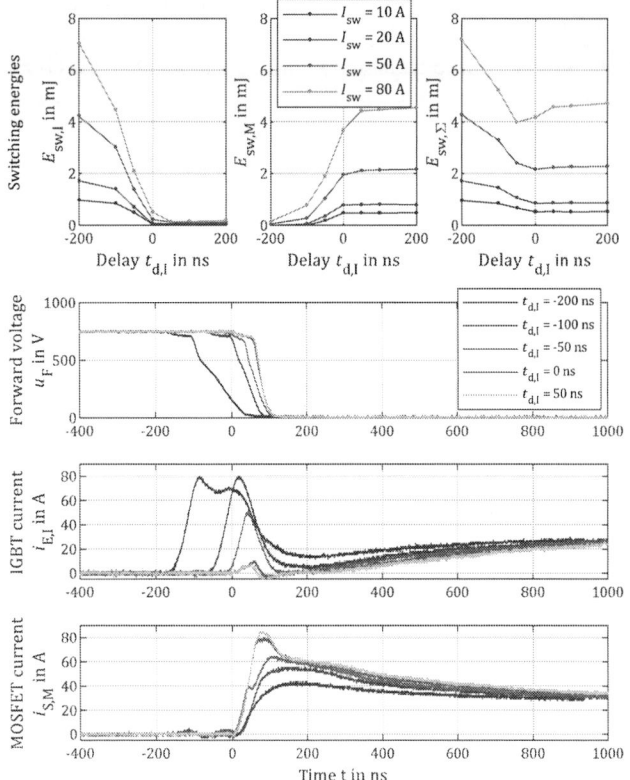

Fig. 2. Turn-ON waveforms at $I_{sw} = 50$ A and energies for different delays.

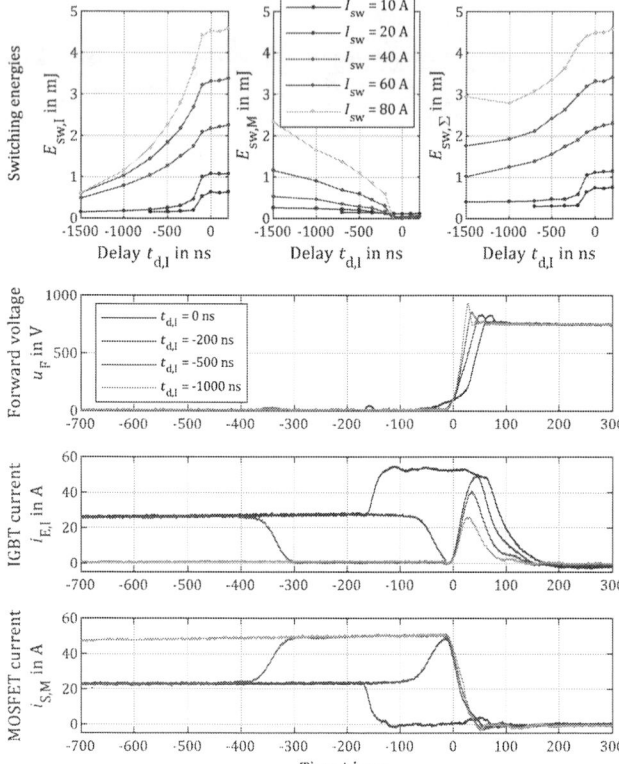

Fig. 3. Turn-OFF waveforms at $I_{sw} = 50$ A and energies for different delays.

turn-ON of the IGBT, which is caused by a higher input capacitance C_{ies} as well as the IGBT's forward recovery behavior. The change from IGBT- to MOSFET-dominated behavior by means of altering $t_{d,I}$ occurs within a narrow range of 150 ns and these few ns are sufficient for a drastic change in behavior, as can be seen for the switching energies in Fig. 2. If both devices turn ON approximately simultaneously, a minimum in the switching energies can be observed, at least for higher load currents, which can be attributed to a shorter current rise time (higher di_F/dt), and thus, a faster overall turn-ON. As a simultaneous turn-ON is difficult to achieve (e.g. using the method of [16]), after turn-ON the current distribution is out of balance compared to the steady-state according to the forward characteristics of the devices. Said steady-state is reached slowly following the principle of an L-R-low-pass, as described in [17].

B. Turn-OFF and Turn-OFF Delay

As mentioned, the turn-OFF is dominated by the IGBT's bipolar effects which in turn are affected by delay, gate resistance, clamping, current and current share, the temperature, and the IGBT selection. While the gate drive is usually determined during the design process and the other factors are beyond the system's control, the delay can be dynamically optimized, which requires a detailed knowledge of its dependencies.

In general, if the IGBT is turned-OFF later (with $t_{d,I} \gtrsim -100$ ns considering the difference in turn-OFF delay), an IGBT-like turn-OFF with a low du_F/dt and a tail current can be observed. However, contrary to the turn-ON, by turning OFF the IGBT earlier a MOSFET-like turn-OFF cannot be obtained in many operating points. For $t_{d,I}$ even beyond -1 μs, the turn-

OFF is dominated by the removal of carriers remaining in the IGBT's drift region after the turn-OFF. When the IGBT is turned-OFF at zero voltage, the missing blocking voltage prevents an extended depletion region from forming and the carriers are removed only by the recombination mechanism with a time constant corresponding to the effective carrier lifetime of the stored charge. Then, when the MOSFET turns OFF, the remaining carriers, and thus, conductivity of the IGBT allow for a current to form during the voltage rise to remove the remaining charge carriers. Thus, the characteristic current pulse occurs. It looks like a conventional parasitic turn-ON due to the Miller capacitance, but is caused by the described bipolar effect.

The IGBT behavior improves with a higher $|t_{d,I}|$ because less carriers remain, but only gradually compared to the turn-ON. The du_F/dt increases and the current pulse becomes smaller, as seen in Fig. 3. A practical measure for the current pulse is the pulse charge $Q_{pk,I}$, i.e. the integrated pulse current defined by

$$Q_{pk,I} = \int_{t_{r,u}}^{t_{tail}} i_{E,I}(t) \, dt \qquad (1)$$

with $t_{r,u}$ denoting the beginning of the voltage rise and t_{tail} the end of the IGBT current tail.

However, while the IGBT behavior improves with an earlier turn-OFF, the MOSFET has to conduct the full load current in the meantime. This results in additional conduction losses. In case of higher currents the MOSFET losses increase disproportionally as the MOSFET has to take over the full current and thus operates in a high current condition with an increased $R_{\text{ds(ON)}}$. As the MOSFET losses increase with a higher $|t_{\text{d,I}}|$, an optimum emerges for the minimum overall switching energy $E_{\text{sw,}\Sigma}$. In Fig. 3, this is visible at $t_{\text{d,I}} = -1000$ ns for $I_{\text{sw}} = 80$ A. The optimum is current and temperature dependent.

C. Gate Resistance Influence on Turn-OFF

The gate resistances need to be set during the design of the hybrid switch. When altering the MOSFET gate resistance $R_{\text{G,M}}$, no significant change in the turn-OFF waveforms of the hybrid switch can be observed unless the MOSFET is switched extremely slow, as seen in [16]. This is due to the already turned-OFF IGBT taking over the load current with its remaining conductivity, which allows the MOSFET to perform an almost zero voltage turn-OFF. This is even the case at low currents when the IGBT has no current share and therefore no recovery effect. However, it is then caused by the current commutating to the relatively large parasitic output capacitance. In summary, for the turn-OFF $R_{\text{G,M}}$ is only relevant at very large delays.

The influence of the IGBT gate resistance $R_{\text{G,I}}$ has multiple consequences. If the IGBT performs the turn-OFF alone (i.e. at $t_{\text{d,I}} \gtrsim -200$ ns), the behavior is altered toward lower du_{F}/dt and higher $E_{\text{sw,I}}$ as it would in case of a non-hybrid switch. In the exemplary operating points shown in Fig. 4, at $t_{\text{d,I}} = 0$ ns $E_{\text{sw,I}}$ increases by about 25 %, when changing $R_{\text{G,I}}$ from 15 Ω to 44 Ω, which roughly corresponds to the expected increase in E_{off} derived from the IGBT's datasheet. However, at $t_{\text{d,I}} < -200$ ns a higher $R_{\text{G,I}}$ also has the effect of prolonging the IGBT turn-OFF process which reduces the effective delay. Therefore, the reduction in switching energies starts at higher delay values. When compensating for the effective delay, as seen in Fig. 4 for the curves for $R_{\text{G,I}} = 44$ Ω shifted by $\Delta t_{\text{d,I}} = 400$ ns, a similar behavior with an identical trend for the reduction of $E_{\text{sw,}\Sigma}$ with higher $|t_{\text{d,I}}|$ can be seen.

While the effects of the gate resistances during turn-OFF influence the performance in some operating points, a better tool is the use of $t_{\text{d,I}}$ to optimize switching. The turn-OFF can then be precisely controlled by means of $t_{\text{d,I}}$ and frees $R_{\text{G,I}}$ and $R_{\text{G,M}}$ to control the turn-ON, which might eliminate the need for separate turn-ON and turn-OFF gate resistors. The principle for current balancing during turn-ON to avoid over-currents presented in [16] can then be applied easily, as well.

D. Clamping

Clamping with a lower gate resistance for the not currently turned-ON or -OFF device can slightly improve the performance in some operating points by means of gate control. For the measurements performed for Fig. 5 and Fig. 6, a clamping circuit is introduced. It operates similar to the principle of an active miller clamp by turning OFF the IGBT normally and activating a lower gate resistance ($R_{\text{cl}} = 5$ Ω) during the voltage transition. The clamping circuit is active only if $|t_{\text{d,I}}| > 200$ ns.

It can be observed in Fig. 5 that for a delay of $t_{\text{d,I}} = -350$ ns a slightly smaller current pulse $Q_{\text{pk,I}}$ and a slightly higher voltage gradient du_{F}/dt is achieved due to the clamping circuit. However, for $t_{\text{d,I}} = -1000$ ns, the waveforms are completely identical. When looking at the IGBT gate voltage

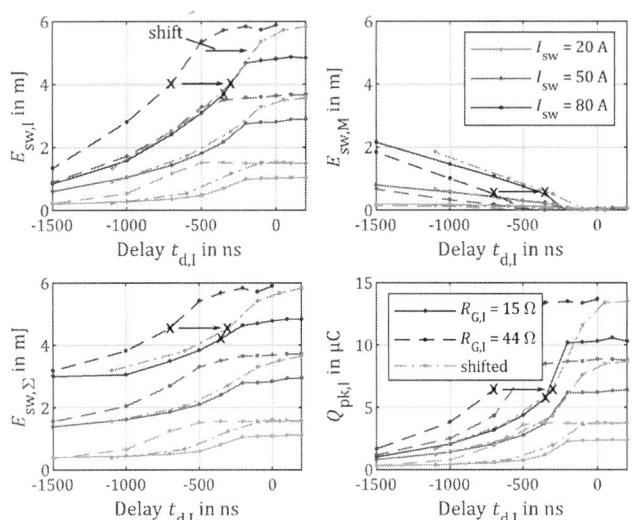

Fig. 4. Effect of gate resistance on the turn-OFF swiching energies and current pulse.

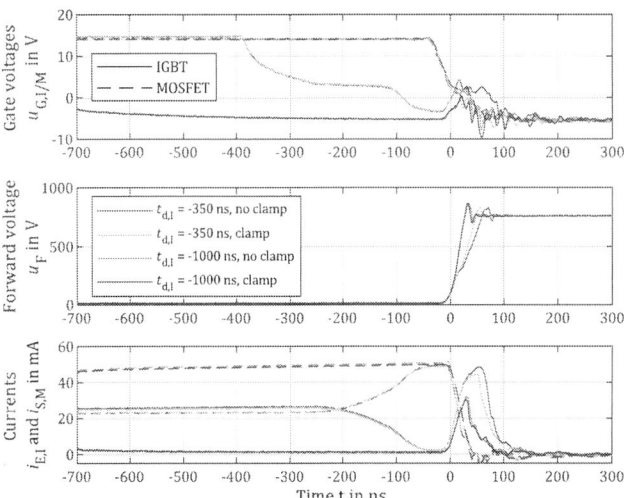

Fig. 5. Effect of clamping on the turn-OFF swiching waveforms. ($I_{\text{sw}} = 50$ A)

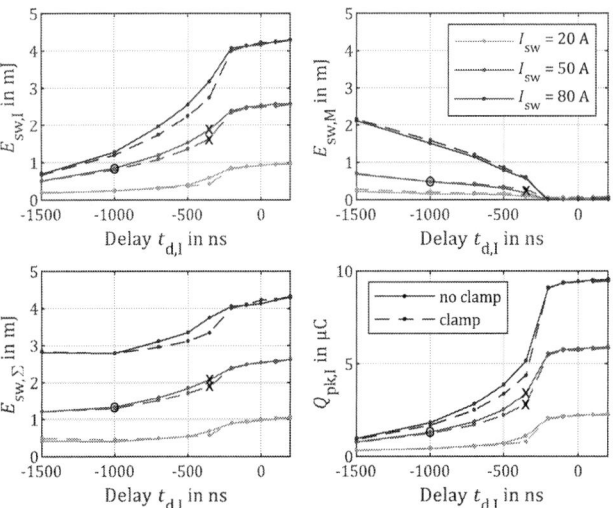

Fig. 6. Effect of clamping on the turn-OFF swiching energies and current pulse.

979-8-3315-1612-3/25 $31.00 © 2025 IEEE

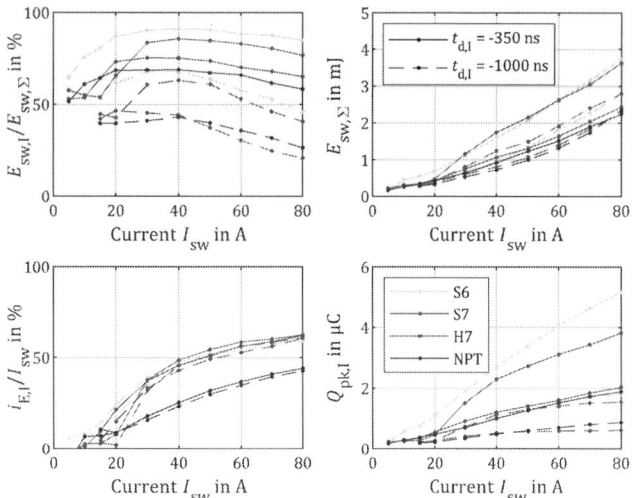

Fig. 7. Current and energy share of different IGBTs for various currents.

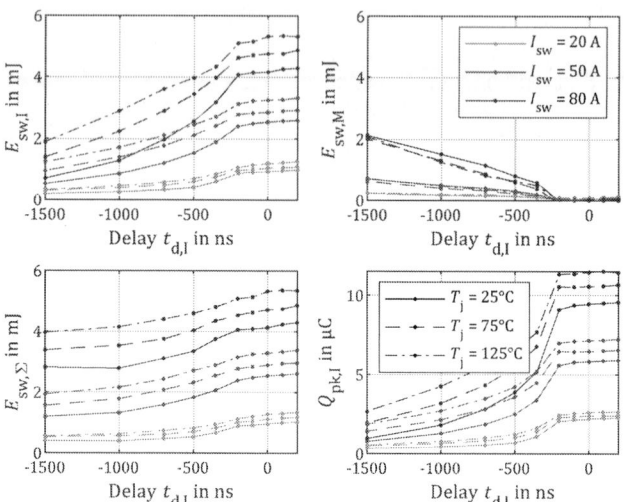

Fig. 8. Temperature dependence of switching energies and current peak.

$u_{G,I}$ during the voltage transition, a pulse corresponding to a charge of the gate is visible. Although it looks like a conventional parasitic turn-ON through charging C_{gc} during the voltage transition, it improves with higher $|t_{d,I}|$. This can be attributed to charge carriers remaining in the drift region. At $t_{d,I} = -350$ ns, the charging of the gate is sufficient to reach the threshold voltage U_{th}, and thus, more carriers are injected into the drift region. In that case, a faster discharge of the gate by means of clamping improves the switching performance. In case of $t_{d,I} = -1000$ ns, where $u_{G,I}$ never reaches U_{th} the clamping circuit has no effect. The corresponding switching energies are shown in Fig. 6 and the previously mentioned operating points marked by x where clamping is advantageous and o where not. An improvement of up to 13 % in $E_{sw,I}$ and up to 8 % in $E_{sw,\Sigma}$ as well as in $Q_{pk,I}$ is obtained within -200 ns $> t_{d,I} > -1000$ ns but it has no effect for $t_{d,I} \leq -1000$ ns. While all of the investigated Infineon IGBTs exhibit only minor charging of the gate in few operation points, the effect is more pronounced in Littlefuse's B4 and C4 IGBTs. Thus, a greater benefit of adding a clamping circuit can be expected there.

E. Current Dependence of Turn-OFF

Regarding current dependence, the hybrid switch differs from the expected behavior in non-hybrid standalone devices. Both in IGBTs and SiC-MOSFETs, a linear trend of the switching energies with increasing current is common. In the hybrid switch, however, a bend and a faster increase of the switching energies at higher currents can be observed, as visible in Fig. 7. In the diagram, switching energies $E_{sw,\Sigma}$, the IGBT share of the switching energies $E_{sw,I}/E_{sw,\Sigma}$, the current distribution before turn-OFF $i_{E,I}/I_{sw}$ and the cumulated current pulse $Q_{pk,I}$ are shown for two delay values and different IGBTs. The bend in the trend of the switching energies occurs when the forward voltage of the completely turned-ON switch is high enough for the IGBT to take over a significant part of the current. As previously mentioned, the turn-OFF waveforms and energies are mainly dependent on the IGBT carrier removal process due to the recovery current peak. For each of the devices, $Q_{pk,I}$ is not proportional to the total current I_{sw} but roughly to the current conducted by the IGBT $i_{E,I}$ at the beginning of the turn-OFF

process. This is plausible, because the carriers present in the drift region are current dependent. Of the shown variants, particularly the S7 IGBT exhibits a fast rise in the current share and a corresponding increase in E_{sw} for $I > 20$ A, especially at $t_{d,I} = -350$ ns. Both the H7 and NPT IGBTs have an overall more efficient turn-OFF, which leads to smaller E_{sw} for both delays. As seen later in Fig. 10, they also show a significantly faster decline in switching losses at increasing delays with the energy minimum at $t_{d,I,opt} \approx -700$ ns at $I_{sw} = 80$ A compared to $t_{d,I,opt} \approx -1500$ ns for the S6. For small currents ($I < 20$ A), an almost MOSFET-like turn-OFF can be expected. However, at high delays and small currents, the IGBT energy share is not zero. This is at least in part due to the capacitive charge of the parasitic setup-related and output capacitances which cannot be separated from the switching losses in the measurements. Using a coarse estimate of 250 pF for the combined switch, the stored energy amounts to approximately 70 µJ.

F. Temperature Dependence on Turn-OFF

The temperature also has an influence on the switching. Three mechanisms contribute: As with standalone IGBTs and MOSFETs, switching slows down and switching energies increase with a higher temperature in the hybrid switch. Also, when the IGBT is turned OFF earlier, the additional conduction losses in the MOSFET rise due to a higher $R_{ds(ON)}$. Simultaneously, in the IGBT the carrier mobility is reduced, a larger current peak appears and the turn-OFF slows down. From the switching energies shown in Fig. 8, a disproportional increase in switching energies with temperature is visible at higher $|t_{t,I}|$ for a given I_{sw}.

G. IGBT Selection

Of the major influences described above, most depend at least in part on the IGBT device, technology and generation selected for the hybrid switch. Various devices along Infineon's and IXYS' lineup of discrete 1200 V IGBTs have been tested. Among them are up-to-date Infineon 6th and 7th generation of standard and high speed (S6, S7, H7) "Trenchstop" IGBTs and IXYS' 4th generation medium frequency (B4) and high frequency (C4) "XPT" Trench devices. For the IGBT design, optimized conduction behavior can commonly be traded for a

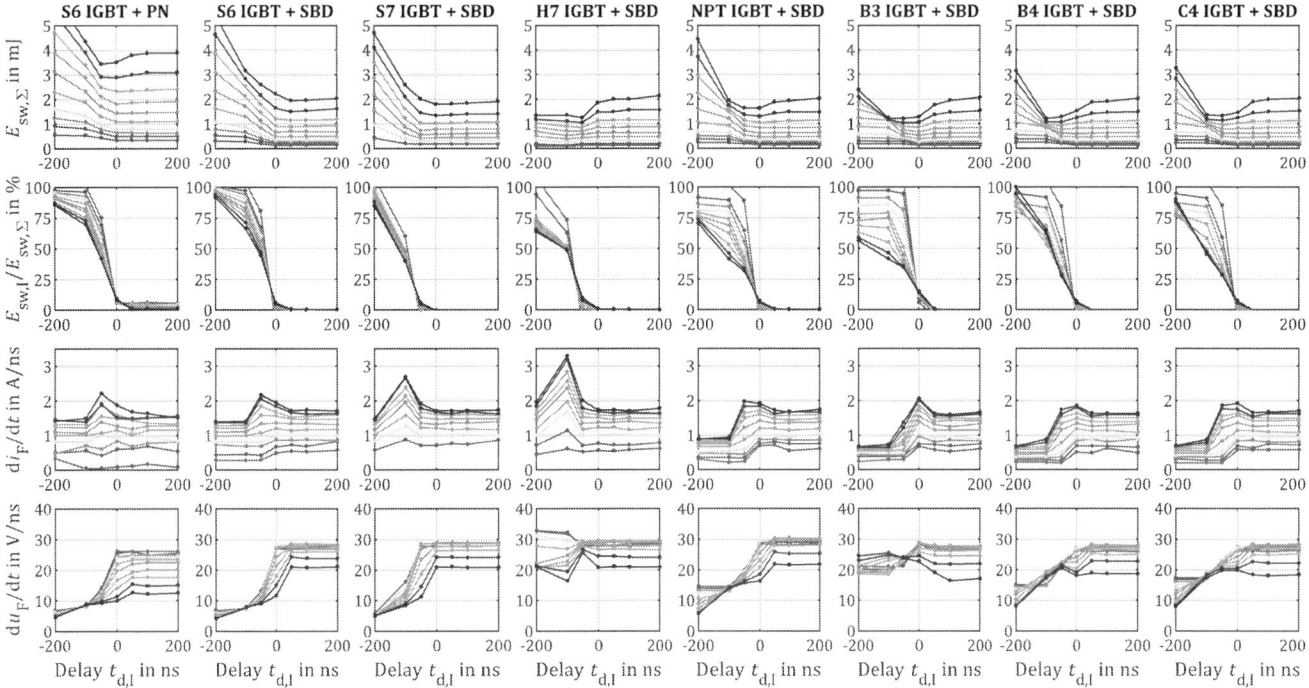

Fig. 9. Measured switching energies for a hybrid switch with different IGBTs for various operating points during turn-ON.

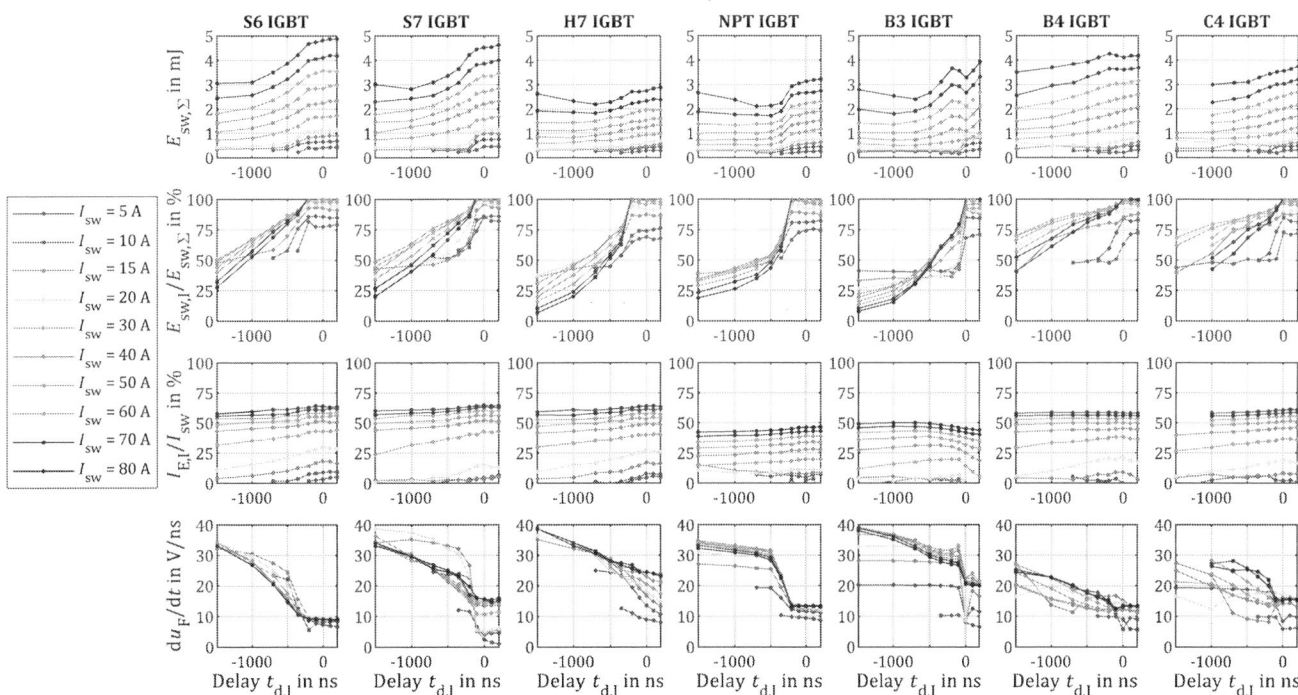

Fig. 10. Measured switching energies for a hybrid switch with different IGBTs for various operating points during turn-OFF.

better switching performance [18]. Compared to NPT IGBTs, those with a buffer or field stop layer (i.e., all mentioned modern IGBTs) improve switching performance for hard-switching at nominal voltage and reduce the tail current without the need to reduce the carrier lifetime. By not (heavily) relying on recombination for the hard-switching turn-OFF, the carrier lifetime can be increased for a lower forward voltage [19, 20].

However, in the hybrid switch, carrier removal due to expanding the depletion region during the voltage transition does not occur at the zero-voltage turn-OFF and all but the recombined carriers remain until the voltage transition during the MOSFET turn-OFF. Therefore, because a lot of modern IGBTs are optimized for their hard-switching and conduction performance, they might exhibit sub-optimal behavior under

TABLE II. COMPARISON OF SWITCHING ENERGIES AND LOSSES PER HALF BRIDGE FOR DIFFERENT DESIGNS/OPERATING POINTS

Hybrid Switch variant	Switching energies @ 20 A						Switching energies @ 50 A						System behavior @ 20 A							System behavior @ 50 A						
	$E_{on,\Sigma}$ [mJ]	$E_{off,\Sigma}$ [mJ]				E_{tot} [mJ]	$E_{on,\Sigma}$ [mJ]	$E_{off,\Sigma}$ [mJ]				E_{tot} [mJ]	$f_{sw}=15$ kHz				$f_{sw}=45$ kHz			$f_{sw}=15$ kHz				$f_{sw}=45$ kHz		
	$t_{d,I}=$ opt.	$t_{d,I}=0$ ns	$t_{d,I}=-500$ ns	$t_{d,I}=-1$ µs	$t_{d,I}=$ opt.	$t_{d,I}=$ opt.	$t_{d,I}=$ opt.	$t_{d,I}=0$ ns	$t_{d,I}=-500$ ns	$t_{d,I}=-1$ µs	$t_{d,I}=$ opt.	@$t_{d,I}=$ opt.	P_{cond} [W]	P_{sw} [W]	P_{total} [W]	P_{sw}/P_{total} [%]	P_{sw} [W]	P_{total} [W]	P_{sw}/P_{total} [%]	P_{cond} [W]	P_{sw} [W]	P_{total} [W]	P_{sw}/P_{total} [%]	P_{sw} [W]	P_{total} [W]	P_{sw}/P_{total} [%]
S6 + PN	0.77	1.10	0.61	0.46	0.42	1.19	1.81	2.85	2.01	1.63	1.41	3.23	22	18	40	45	53	75	71	83	48	132	37	145	228	64
S6 + SBD	0.35	1.10	0.61	0.46	0.42	0.76	0.91	2.85	2.01	1.63	1.41	2.33	22	11	33	34	34	56	61	83	35	118	30	105	188	56
S7 + PN	0.95	1.12	0.47	0.42	0.41	1.36	2.32	2.71	1.98	1.51	1.45	3.76	21	20	41	50	61	82	75	75	56	132	43	169	244	69
S7 + SBD	0.31	1.12	0.47	0.42	0.41	0.72	0.78	2.71	1.98	1.51	1.45	2.22	21	11	31	34	32	53	61	75	33	108	31	100	175	57
H7 + PN	0.50	0.61	0.44	0.38	0.36	0.86	1.11	1.53	1.29	1.09	1.09	2.20	21	13	34	38	39	59	65	77	33	110	30	99	176	56
H7 + SBD	0.25	0.61	0.44	0.38	0.36	0.61	0.70	1.53	1.29	1.09	1.09	1.79	21	9	30	31	27	48	57	77	27	104	26	81	157	51
NPT + SBD	0.33	0.72	0.32	0.34	0.32	0.65	0.84	1.85	1.05	1.01	0.97	1.81	28	10	37	26	29	57	51	146	27	173	16	82	227	36
B3 + SBD	0.28	0.89	0.32	0.34	0.32	0.60	0.68	1.56	1.20	1.00	1.00	1.67	30	9	39	23	27	57	47	135	25	160	16	75	211	36
B4 + SBD	0.32	0.71	0.50	0.49	0.49	0.81	0.82	2.48	2.06	1.66	1.50	2.32	23	12	35	35	36	59	61	87	35	122	29	105	192	55
C4 + SBD	0.31	0.65	0.54	0.62	0.62	0.93	0.81	2.03	1.66	1.43	1.43	2.25	26	14	40	35	42	67	62	98	34	132	26	101	199	51
SiC (fast) *	0.40				0.32	0.71	0.97				0.67	1.64	14	11	25	43	32	46	69	116	25	141	17	74	190	39
SiC (slow) *	1.05				0.44	1.50	3.00				1.11	4.11	14	22	37	61	67	82	83	116	62	178	35	185	301	61
IGBT (S6)	1.88				1.23	3.11	4.16				2.90	7.06	25	47	71	65	140	165	85	79	106	185	57	318	397	80

*: $R_{G,slow} = 3 \cdot R_{G,fast}$, $du_F/dt|_{fast} \leq 45$ V/ns   general assumptions: $f_{sw} = \{15,45\}$ kHz, $I_{load} = \{20,50\}$ A, $T_j = 25$ °C, $D = 100$ %

soft-switching/zero-voltage conditions, because carrier recombination helps to perform a loss-reduced voltage transition. Therefore, older, less optimized IGBTs are tested, as well, namely an NPT and a PT (B3) IGBT.

Fig. 9 (turn-ON) and Fig. 10 (turn-OFF) show measured total switching energies $E_{sw,\Sigma}$, the IGBT share $E_{sw,I}/E_{sw,\Sigma}$, the voltage gradient du_F/dt, and the current gradient di_F/dt or the current share $I_{E,I}/I_{sw}$ before turn-OFF.

Regarding turn-ON, the previously mentioned optima of the switching energies at the highest di_F/dt can be observed. For example, when looking at the S6 IGBT at $I_{sw} = 50$ A, MOSFET-dominated switching reduces the switching energy by roughly 50 % compared to IGBT dominated switching for a Si-PN diode as a commutation partner. When using a SiC Schottky diode, a further improvement toward 75 % can be achieved. When comparing MOSFET-dominated switching between the different tested variants using the Schottky diode, it stands out that the switching energies as well as all displayed parameters are almost identical, thus proving that the turn-ON is mainly influenced by the MOSFET and the antiparallel diode and that the IGBT selection has no significant effect. When comparing an IGBT-dominated turn-ON, it appears that the tested IXYS/Littlefuse's IGBTs as well as Infineon's H7 IGBT can achieve lower turn-ON energies than the other Infineon devices. When looking only at the variants with a SiC Schottky diodes for a fair comparison, the S6, the S7, and the NPT IGBTs behave relatively similar, although the S6 and S7 have a better conduction behavior, showing the improved properties of modern IGBTs. The H7 device forms an exception in the field of tested IGBTs. It displays higher switching speeds and lower switching losses for an IGBT-dominated turn-ON than for the MOSFET-dominated turn-ON with $E_{sw,\Sigma}$ being up to 45 % lower at the optimal delay.

Looking into the turn-OFF, the switching energies decline with a higher $|t_{d,I}|$ when the IGBT turns OFF earlier. The decline occurs fastest for the older NPT and PT (B3) IGBTs and slowest for IXYS/Littlefuse's modern IGBTs. The H7 IGBT is an exception here as well, performing significantly better than any other tested modern IGBT in the hybrid switch. For low currents and high $|t_{d,I}|$, some IGBTs are perfectly turned-OFF and the hybrid switch performs a true MOSFET-like turn-OFF. The hybrid switch can then even achieve a loss-reduced ZVS-like turn-OFF if the MOSFET is turned-OFF fast enough and the voltage rises slow enough due to the higher output capacitance. This is substantiated by the switching energies increasing only minimally with I_{sw} and being close to the energy stored in the parasitic capacitances. Also, the voltage gradient du_F/dt is proportional to I_{sw}. This effect is notable for all investigated devices and mostly when the IGBTs have conducted (almost) no current before turn-OFF. For the S7 IGBT, the effect is visible best. Therefore, a good low-load switching performance of the hybrid switch can be expected.

IV. APPLICATIONAL EVALUATION

To judge on the system level performance, switching and conduction losses are calculated on a half bridge level for part-load and near-full-load operating points and for switching frequencies of $f_{sw} = \{15, 45\}$ kHz. The results are shown in Table II and the comparative performance is highlighted by color in the respective category. Regarding the switching loss share, variants/operating points where the switching losses are dominant are colored magenta and blue color marks where conduction losses dominate. Table II also includes standalone SiC-MOSFET or Si-IGBT half bridges for comparison. In that case, the respective chip area is doubled compared to the same devices in the hybrid switch. Regarding the switching performance, operation at the optimal delay is assumed. However, for comparative purposes turn-OFF energies at various delays are shown.

Of the compared IGBTs, S7 und H7 offer the best conduction behavior with losses in the combined switch being roughly 10 % lower than the S6, B4, and C4 devices and roughly 45% lower than the older NPT and B3 devices at $I_{sw} = 50$ A. Performance is similar to the standalone IGBT but 50 % worse under part-load and 35 % better under full-load than the standalone MOSFET.

The best switching performance of the tested hybrid switches can be found in the H7 IGBT and the older NPT and PT (C3) IGBTs with performance on par with standalone SiC MOSFETs at medium switching speeds (20 ... 45 V/ns). Going from a standalone IGBT to the best of the compared SiC/Si hybrid switches (H7+SBD) saves up to 80 % of the switching losses and even the worst combination (S7+PN) saves about 50 % on average for the shown operating points. The antiparallel diode for the IGBT makes a significant difference. A Schottky diode yields on average roughly 50 % loss reduction during turn-ON, which translates to 35 % reduction in switching losses and 19 % reduction in overall losses.

Regarding the total losses, at a switching frequency of $f_{sw} = 15$ kHz, conduction losses are dominant, especially at high load for the standalone SiC MOSFET. Compared to the IGBT, the standalone SiC MOSFET improves the overall losses by 65/24 % for part-load/full-load, whereas the best hybrid switch achieves 58/44 %, thus being even more efficient than the SiC MOSFET at high load. Switching losses are dominant at 45 kHz, especially for the standalone IGBT. Starting there, the improvements are 72/52 % for the SiC MOSFET and 71/60 % for the best hybrid switch under part-load/full-load. From the worst (NPT) to the best (H7) hybrid switch variant the improvement is 36 %. It should be noted that at $f_{sw} = 45$ kHz, switching losses for the IGBT-only variant are prohibitively excessive discouraging its use at said operating frequency. Also, conduction losses in the SiC-MOSFET-only variant at $I_{sw} = 50$ A are disproportionately high, suggesting the need for more chip area for a proper design. However, this furthers the already high saving of 50 % SiC MOSFET chip area in the hybrid switch compared to a standalone SiC MOSFET.

V. Conclusion

This paper shows the importance of understanding the bipolar device effects of the IGBT in the SiC/Si hybrid switch especially during turn-OFF and that a proper design should consider the IGBT technology used. The hybrid switch can achieve from up to 54 % to 71 % system loss reduction compared to a standalone IGBT system depending on the device selection. It thus performs on-par with SiC MOSFETs under comparable operating conditions when selecting the best device combination

References

[1] M. Ippisch, T. Reiter, M. Niendorf, W. Jakobi M. Muenzer, "Maximizing Cost-Efficiency in Electric Drivetrains: A SiC/Si Fusion Switch Approach," in *PCIM Europe 2024*, Nürnberg, Germany, Jun. 2024, pp. 2633–2640.

[2] M. Rahimo *et al.*, "Characterization of a Silicon IGBT and Silicon Carbide MOSFET Cross-Switch Hybrid," *IEEE Trans. Power Electron.*, vol. 30, no. 9, pp. 4638–4642, 2015.

[3] Y. Jiang, G. C. Hua, E. Yang, and F. C. Lee, "Soft-Switching of IGBTs with the Help of MOSFETs in Bridge-Type Converters," in *Proceedings*

of IEEE Power Electronics Specialist Conference - PESC '93, Seattle, WA, USA, 1993, pp. 151–157.

[4] A. Deshpande and F. Luo, "Comprehensive Evaluation of a Silicon-WBG Hybrid Switch," in *2016 IEEE Energy Conversion Congress and Exposition (ECCE)*, Milwaukee, WI, USA, 2016, pp. 1–8.

[5] Y. Wang, M. Chen, C. Yan, and D. Xu, "Efficiency Improvement of Grid Inverters With Hybrid Devices," *IEEE Trans. Power Electron.*, vol. 34, no. 8, pp. 7558–7572, 2019.

[6] Z. Li *et al.*, "A Novel Gate Driver for Si/SiC Hybrid Switch for Multi-Objective Optimization," *IET Power Electronics*, vol. 14, no. 2, pp. 422–431, 2021.

[7] Y. Wei, D. Woldegiorgis, R. Sweeting, and A. Mantooth, "Four Control Freedoms AGD for Hybrid SiC MOSFET and Si IGBT Application," in *2021 IEEE Applied Power Electronics Conference and Exposition (APEC)*, Phoenix, AZ, USA, Jun. 2021, pp. 2211–2216.

[8] F. Kayser and H.-G. Eckel, "Event-Triggered Gate Drive for a 1.7 kV Si-SiC Hybrid Switch with IGBT-like Short-Circuit Robustness," in *2023 25th European Conference on Power Electronics and Applications (EPE'23 ECCE Europe)*, Aalborg, Denmark, Sep. 2023, pp. 1–9.

[9] A. Deshpande and F. Luo, "Design of a Silicon-WBG Hybrid Switch," in *2015 IEEE 3rd Workshop on Wide Bandgap Power Devices and Applications (WiPDA)*, Blacksburg, VA, USA, 2015, pp. 296–299.

[10] J. Yu, Z. Li, Z. He, X. Jiang, C. Zhang, and J. Wang, "Performance Comparison of Traditional and JBS Integrated SiC MOSFETs in Si/SiC Hybrid Switch," in *2019 IEEE Energy Conversion Congress and Exposition (ECCE)*, Baltimore, MD, USA, Sep. 2019, pp. 1918–1921.

[11] A. Amler, T. Heckel, D. Ruppert, C. Rettner, and M. März, "Evaluation of Current, Delay, and Temperature Influence and Diode Selection on the Switching Behavior of a SiC/Si Hybrid Switch," in *2024 IEEE Applied Power Electronics Conference and Exposition (APEC)*, Long Beach, CA, USA, Feb. 2024, pp. 1775–1782.

[12] F. Kayser, F. Pfirsch, F.-J. Niedernostheide, R. Baburske, and H.-G. Eckel, "Novel Si-SiC Hybrid Switch and its Design Optimization Path," in *2022 IEEE 34th International Symposium on Power Semiconductor Devices and ICs (ISPSD)*, Vancouver, BC, Canada, May. 2022, pp. 225–228.

[13] Z. Li, J. Wang, B. Ji, and Z. J. Shen, "Power Loss Model and Device Sizing Optimization of Si/SiC Hybrid Switches," *IEEE Trans. Power Electron.*, vol. 35, no. 8, pp. 8512–8523, 2020.

[14] C. Tan, M. Stecca, T. B. Soeiro, J. Dong, and P. Bauer, "Performance Evaluation of an Electric Vehicle Traction Drive using Si/SiC Hybrid Switches," in *2021 IEEE 19th International Power Electronics and Motion Control Conference (PEMC)*, Gliwice, Poland, Apr. 2021, pp. 278–283.

[15] S. Ueno, N. Kimura, T. Morizane, and H. Omori, "Study on Characteristics of Hybrid Switch using Si IGBT and SiC MOSFET Depending on External Parameters," in *2017 19th European Conference on Power Electronics and Applications (EPE'17 ECCE Europe)*, Warsaw, Sep. 2017, P.1-P.10.

[16] X. Jiang *et al.*, "Impact of Gate Resistance on Improving the Dynamic Overcurrent Stress of the Si/SiC Hybrid Switch," *IEEE Trans. Power Electron.*, vol. 37, no. 11, pp. 13319–13331, 2022.

[17] M. Andrade, B. Cougo, and L. M. F. Morais, "Current Sharing Dynamics During IGBT ZVS Turn-On in a Hybrid Si IGBT/SiC MOSFET Switch," in *European Conference on Power Electronics and Applications EPE'23 ECCE Europe*, Aalborg, Denmark, 2023, pp. 1–9.

[18] A. Ramamurthy, S. Sawant, and B. J. Baliga, "Modeling the [dV/dt] of the IGBT during Inductive Turn Off," *IEEE Trans. Power Electron.*, vol. 14, no. 4, pp. 601–606, 1999.

[19] T. Laska, M. Munzer, F. Pfirsch, C. Schaeffer, and T. Schmidt, "The Field Stop IGBT (FS IGBT). A new Power Device Concept with a Great Improvement Potential," in *12th International Symposium on Power Semiconductor Devices & ICs. Proceedings*, Toulouse, France, 2000, pp. 355–358.

[20] A. R. Hefner and D. L. Blackburn, "Performance Trade-off for the Insulated Gate Bipolar Transistor: Buffer Layer versus Base Lifetime Reduction," in *1986 17th Annual IEEE Power Electronics Specialists Conference*, Vancouver, Canada, 1986, pp. 27–38.

Forward Recovery and its Mitigation in Hybrid Si/SiC-based DC–AC Converters

Yan Zhou
Institute of Power Electronics
Friedrich-Alexander-Universität Erlangen-Nürnberg
Nürnberg, Germany
yan.yan.zhou@fau.de

Thomas Lehmeier
Institute of Power Electronics
Friedrich-Alexander-Universität Erlangen-Nürnberg
Nürnberg, Germany
thomas.lehmeier@fau.de

Adrian Amler
Institute of Power Electronics
Friedrich-Alexander-Universität Erlangen-Nürnberg
Nürnberg, Germany
adrian.amler@fau.de

Martin März
Institute of Power Electronics
Friedrich-Alexander-Universität Erlangen-Nürnberg
Nürnberg, Germany
martin.maerz@fau.de

Abstract—**The ongoing progress in the field of wide bandgap (WBG) devices enables new topologies and unmatched power density in high-power dc–ac converters. However, the still high cost of WBG devices, such as silicon carbide (SiC) MOSFETs, continues to hinder their widespread adoption in commercial converters. To address this challenge, emerging hybrid topologies combine fast-switching SiC MOSFETs with Si IGBTs operating at the fundamental frequency, achieving an optimal balance between cost and performance. In such topologies, the rapid switching of SiC devices induces high current slew-rates, which trigger forward recovery effects in the IGBTs. This phenomenon leads to additional losses in these bipolar devices and generates transient overvoltages, compromising the converter's reliability. This article investigates the forward recovery behavior of different IGBT technologies and explores the mitigation by introducing a snubber capacitor to decouple the different stages of the topology. The results demonstrate the effectiveness of this approach in mitigating the detrimental impacts of forward recovery in hybrid dc–ac converters.**

Index Terms—**Active neutral point clamped (ANPC), dc–ac converter, comparison, forward recovery, hybrid configuration, multilevel, silicon (Si) IGBT, silicon carbide (SiC) MOSFET, snubber capacitor.**

I. INTRODUCTION

The increasing demand for highly efficient and compact power electronic systems has led to a trend of increased switching frequencies using wide bandgap devices such as SiC MOSFETs [1]. This enables the downsizing of passive components, resulting in cost savings while maintaining compliance with electromagnetic interference standards. Furthermore, multilevel dc–ac converter topologies have become the industry standard due to their ability to handle higher voltages with standard commercial power devices and deliver higher-quality waveforms in high-power and medium-voltage applications [2], [3]. Specifically, the three-level (3L) *Active Neutral Point Clamped* (ANPC) and *T-Type Neutral Point Clamped* (T^2-NPC) topologies are widely adopted for their superior efficiency and performance [3]. However, these topologies require a substantial number of active switching devices. The still high market price of SiC MOSFETs compared to traditional Si IGBTs presents a significant barrier, limiting the widespread adoption of pure SiC-based multilevel converters [4]. To address this challenge, recent studies have explored the use of hybrid assemblies within ANPC [4]–[7] and T^2-NPC [8] converters, combining SiC and Si devices in one power converter topology. An exemplary hybrid topology consisting of a Si and SiC stage is shown in Fig. 1.

This hybrid configuration combines cost-reduction benefits with high output voltage quality by utilizing Si IGBTs operating at the fundamental frequency and SiC devices that switch at a high frequency. The studies in [4], [5], [7] demonstrated that Si/SiC assemblies can achieve efficiencies comparable to those of full SiC-based variants while offering significant cost-saving potential. Despite their benefits, a major challenge hybrid topologies face is the forward recovery (FR) effect in bipolar devices triggered by the rapid switching of SiC devices. The charge carrier density of the IGBT's drift region is insufficient to sustain the current with a high di/dt caused by the rapid switching of the SiC devices, resulting

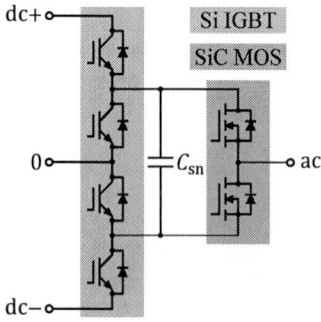

Fig. 1. Circuit diagram of an exemplary hybrid Si/SiC-based 3L-ANPC topology in single-phase configuration. Si IGBTs switching at fundamental frequency, SiC MOSFETs switching at high frequency.

979-8-3315-1612-3/25 $31.00 © 2025 IEEE

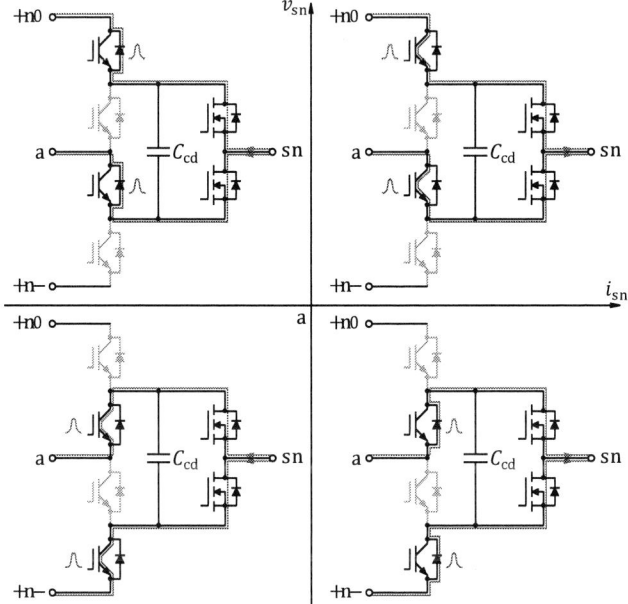

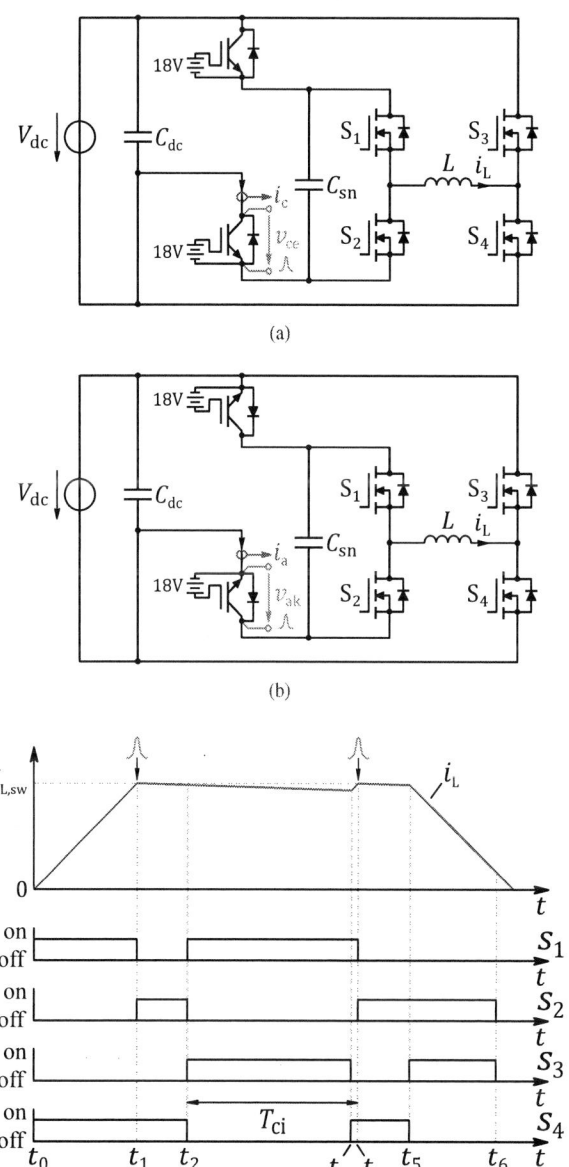

Fig. 2. Current paths and the respective semiconductor devices subjected to forward recovery in the four quadrants of a hybrid 3L-ANPC topology. The modulation is chosen so that the IGBTs remain turned-ON during the positive and negative voltage half-waves, respectively.

in a temporary lower effective conductivity of the already turned-ON Si IGBT [9], [10]. The elevated resistance leads to a transient overvoltage (superimposed voltage spike) across the IGBTs, causing significant induced losses and simultaneously increasing the voltage stress on SiC MOSFETs. The same principle applies to the i-region of a PiN-diode. The magnitude and speed of this phenomenon are significantly influenced by the technology of the bipolar components and their properties, as well as by the operating conditions, as investigated in [9], [11], [12]). Affecting key factors include:

- Switched current level of the SiC MOSFET
- Current slew-rate within the bipolar device
- Charge carrier lifetime of the bipolar device
- Junction temperature of the bipolar device

As shown in Fig. 2, the devices affected by forward recovery depend on the current paths of the respective operating quadrants. Although previous research has explored this effect in 3L-ANPC inverters, practical guidance on mitigating it remains limited. To fill these gaps, this research suggests the use of a snubber capacitor C_{sn} to reduce the FR effect. Previous studies [5], [6], [13], [14] on hybrid assemblies have recommended the use of C_{sn} only with the aim of shortening the length of long critical commutation loops by decoupling the two stages.

This research indicates that C_{sn} with a small capacity is also helpful in mitigating the forward recovery effect, which has been previously ignored. For this reason, this paper aims to thoroughly investigate the FR effect in bipolar switching devices within a hybrid topology to gain insight into the underlying mechanism and how to mitigate it using a small capacitor between the high-frequency and low-frequency stages.

Fig. 3. Modified test circuit for investigating the forward recovery effect in the (a) IGBT and (b) antiparallel diode. (c) Applied switching test sequence using first quadrant operation.

II. TEST METHODOLOGY

The test circuit depicted in Fig. 3(a) and (b) is utilized to investigate the FR effect at a dc-link voltage of 800 V. The IGBTs are permanently turned-ON using nine-volt batteries. 1 μF multilayer ceramic capacitors (MLCCs) are soldered between the gate and the kelvin-emitter to enhance gate voltage stability. The orientation of the device under test (DUT) is reversed to investigate its antiparallel diode. To eliminate the influence of the intrinsic parasitic inductance of the device during measurement, the voltage across the DUT is acquired using the four-wire measurement technique. A differential probe *Keysight DP0001A* acquires the voltage between the

TABLE I
EXPERIMENTAL PROTOTYPE AND TEST PARAMETERS.

Parameter / Component	Value
SiC MOSFETS $S_1 \sim S_4$	Sanan AMS1200016B
dc-link voltage V_{dc}	800 V
dc-link capacitance C_{dc}	120 μF
Inductance L	97.4 μH
Snubber capacitances C_{sn}	0 ∼ 70 nF [step size: 14 nF]
Switched current levels $I_{L,sw}$	10 ∼ 50 A [step size: 10 A]
Current interruption durations T_{ci}	2 ∼ 20 μs [step size: 2 μs]
Gate driver resistance (turn-OFF) $R_{g,OFF}$	0 Ω
Dead-time T_{dead} of SiC MOSFETS	120 ns
Junction temperature T_j	≈ 20°C

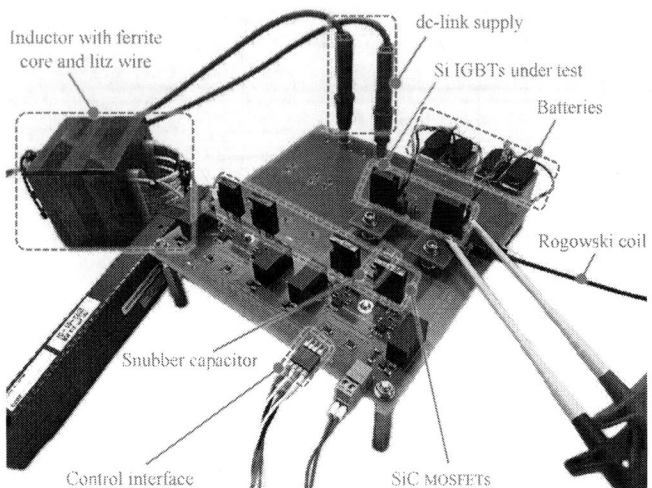

Fig. 4. Photograph of the experimental setup. The dc-link capacitors are soldered on the lower side of the PCB.

TABLE II
SPECIFICATIONS OF THE Si IGBTs UNDER EVALUATION.

Identifier	Technology / Generation	Part number
S6	IGBT 6	IKY40N120CS6
S7	IGBT 7	IKZA40N120CS7
H3	High speed 3	IKY50N120CH3
H7	High speed 7	IKZA50N120CH7

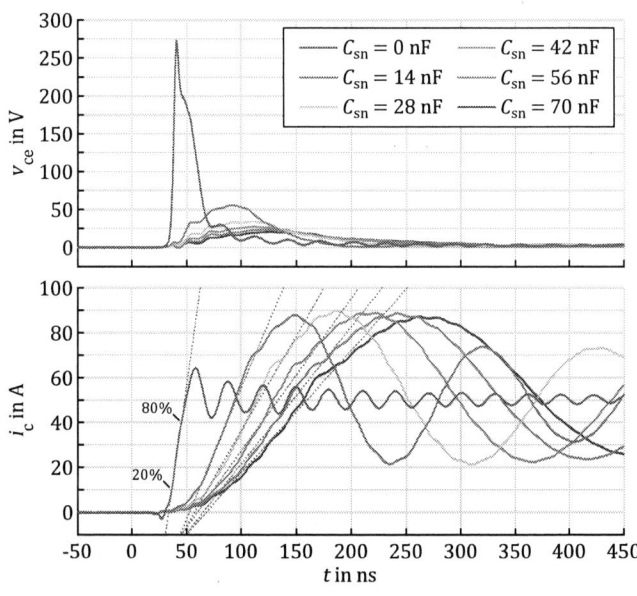

Fig. 5. Voltage and current waveforms across IGBT H7 after turning OFF the SiC MOSFET S_1 at t_1. Test parameters: $I_{L,sw} = 50$ A, single-shot measurement with $T_{ci} \to \infty$. Note that t_1 corresponds to $t = 0$ in this timescale. The current slew-rate $di_c/dt|_{20}^{80}$ in Fig. 6 and 7 is defined by the gradient between 20% and 80% of the current transition.

cooling surface (collector) and the kelvin-emitter of the IGBT while the load current flows through the collector and the main emitter pin. The DUT's collector current is measured using a Rogowski coil *Iwatsu SS-683*. The switching sequence applied in the test is illustrated in Fig. 3(c). Between interval t_0 and t_1, the inductor current ramps up to the desired switch current $I_{L,sw}$. At t_1 and t_4, the SiC MOSFET S_1 is turned-OFF, resulting in a high di/dt in the DUT during the current commutation process, which induces the forward recovery effect, manifesting itself as a transient overvoltage across the DUT. The influence of carrier recombination in IGBTs and diodes between two switching instants is investigated using different current interruption durations T_{ci} for the DUT. Minor deviations in the switched current caused by a long T_{ci} can be compensated within the interval $[t_3; t_4]$ to ensure consistent test conditions.

Table I summarizes the components and test parameters. *Sanan AMS1200016B* SiC MOSFETs were used, which were turned-OFF using an external gate resistance of 0 Ω as is usual in soft-switched applications. It should be noted that the PCB design inherently possesses a structure-related capacitance, which was optimized to 23 pF, and is therefore negligible in the analysis. Type 1 ceramic and film capacitors are preferred as snubber capacitors due to their low equivalent series resistance (ESR), which results in minimal heat generation and high stability. In addition, they can handle high current gradients without a significant piezoelectric effect. Here, C0G MLCCs are chosen.

A picture of the setup is shown in Fig. 4. Both IGBTs are soldered onto a separate small PCB, which simplifies the test process by allowing rapid exchange and reorientation.

III. EXPERIMENTAL RESULTS AND DISCUSSION

Experimental tests were conducted using various latest generation 1200 V Infineon IGBT Trenchstop™ technologies {S6, S7, H3, and H7} with comparable current ratings. An overview of these devices under test is provided in Table II. A *Keysight MXR058A* oscilloscope was used to acquire all relevant time signals simultaneously. Fig. 5 shows the voltage and current waveforms of IGBT H7 and its voltage overshoots caused by the FR effect after turning OFF the SiC MOSFET S_1 at t_1 for different snubber capacitance values.

Similar transient voltage overshoots can be observed across the upper IGBT and its antiparallel diode during the turn-OFF of the transistor S_2. According to Kirchhoff's voltage

979-8-3315-1612-3/25 $31.00 © 2025 IEEE

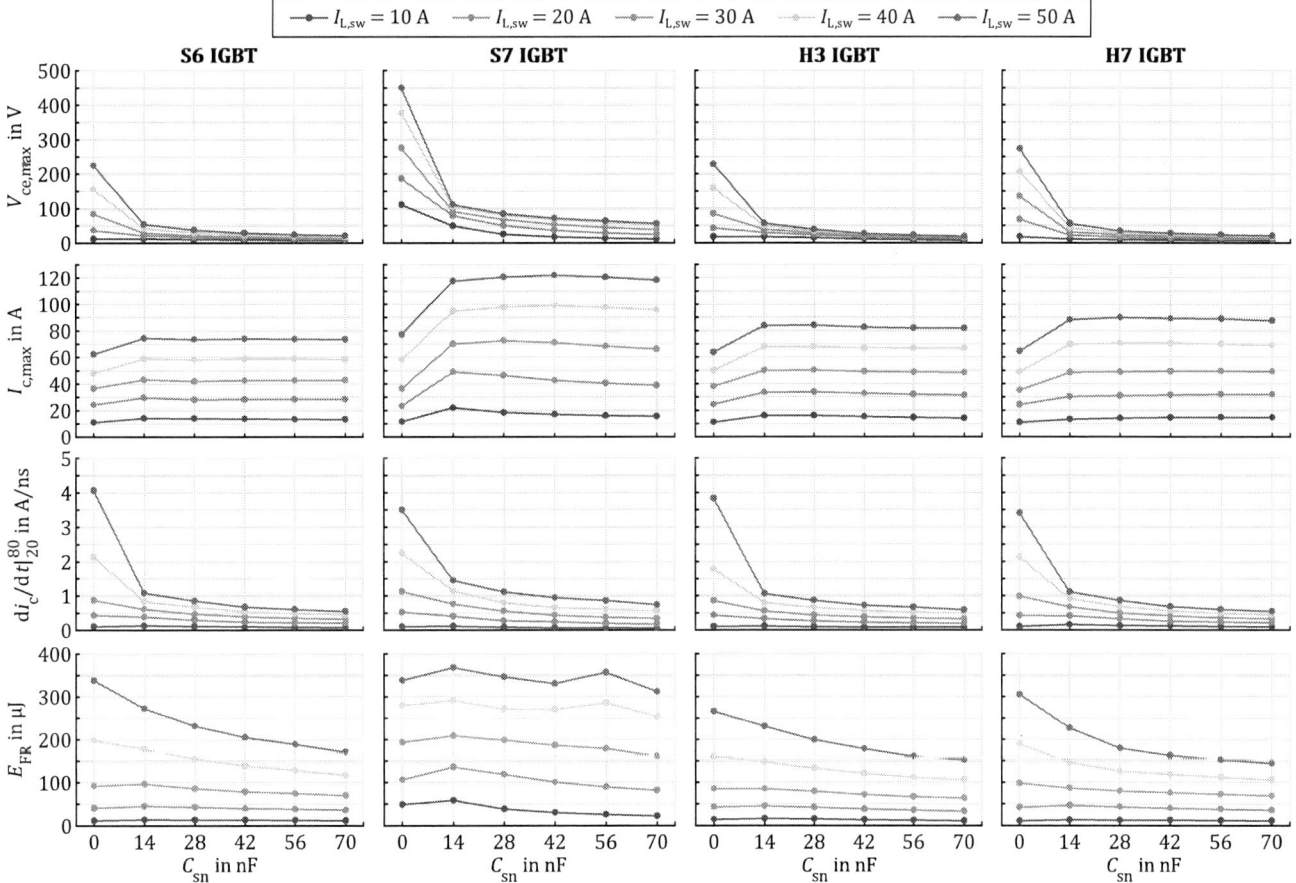

Fig. 6. Maximum transient collector-emitter overvoltage $V_{\text{ce,max}}$, maximum collector overcurrent $I_{\text{c,max}}$, current slew-rate $\mathrm{d}i_{\text{c}}/\mathrm{d}t|_{20}^{80}$ [for definition see Fig. 5], and forward recovery energy E_{FR} for the different IGBT devices under test, shown as functions of switched currents and snubber capacitance values. The single-shot results of the SiC MOSFET turn-OFF at t_1 are displayed, $T_{\text{ci}} \to \infty$.

law, the overshoots also appear across the SiC MOSFETs [9], [11], compromising the converter's reliability if no mitigation measures are taken.

A. Analysis of Single Pulse FR in the IGBT

In the following analysis, the transient overvoltage induced by the forward recovery effect and the resulting additional energy losses in the semiconductor devices using single-shot measurements ($T_{\text{ci}} \to \infty$) are examined for the different IGBT technologies. To assess the influence of the proposed measure of a snubber capacitor C_{sn}, the capacitance was varied in increments of 14 nF. The peak values of the voltage and current waveforms, as observed in Fig. 5, were plotted against the different values of C_{sn} and switched currents $I_{\text{L,sw}}$. The results are shown in Fig. 6.

Notably, the IGBT S7 exhibits a forward recovery overvoltage $V_{\text{ce,max}}$ of 450 V at a switched current of 50 A, which is twice that of the other tested devices. This elevated overvoltage could potentially lead to device failure when operating with a dc-link voltage of 800 V. Since previous tests and the findings in [11] have shown that the dc-link voltage has a negligible impact on the forward recovery effect, V_{dc} was reduced to 600 V for this particular measurement as a precaution. However, a

snubber capacitor can significantly reduce the peak voltage, even with low capacitance values. It is also observed that increasing the value of C_{sn} delays the current commutation [see also Fig. 5] and reduces the resonant frequency of the series resonant circuit composed of C_{sn} and the inductance of the commutation loop.

The transient overvoltage caused by the forward recovery effect results in additional power losses, significantly reducing system efficiency at high switching frequencies even in soft-switched applications, and therefore must be considered. The additional energy E_{FR} dissipated in the device at each switching instant is characterized by the voltage-current (V–I)-overlap during the FR event and can be calculated as follows

$$E_{\text{FR}} = \int_0^{t_\alpha} i_{\text{c}}(t) \left[u_{\text{ce}}(t) - u_{\text{ce}}(I_{\text{L,sw}}) \right] \mathrm{d}t, \qquad (1)$$

where $t = 0$ corresponds to the turn-OFF time of S_1 and t_α is defined by $\mathrm{d}u_{\text{ce}}/\mathrm{d}t \approx 0$ at the maximum of the first voltage peak. The graphs in Fig. 6 show a consistent trend: the forward recovery energy gradually decreases with increasing C_{sn}-values, especially at high switched currents. Once again, IGBT S7 shows the worst performance among the devices

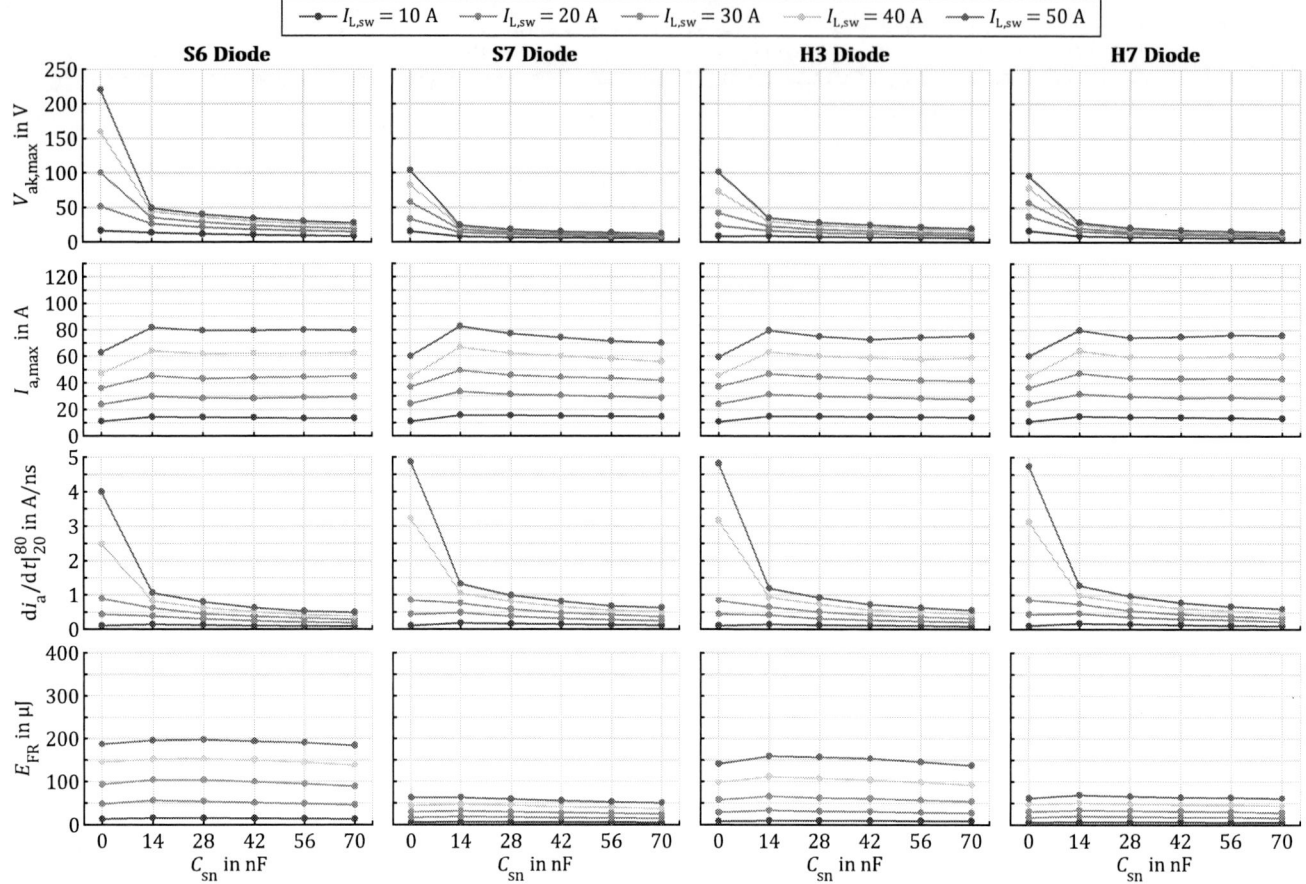

Fig. 7. Maximum transient anode-cathode overvoltage $V_{\text{ak,max}}$, maximum anode overcurrent $I_{\text{a,max}}$, current slew-rate $di_{\text{a}}/dt|_{20}^{80}$ [for definition see Fig. 5], and forward recovery energy E_{FR} for the diodes of various IGBTs under test, shown as functions of switched currents and snubber capacitance values. The single-shot results of the SiC MOSFET turn-OFF at t_1 are displayed, $T_{\text{ci}} \to \infty$.

tested, while H6, H7, and S6 behave very similar in terms of FR.

B. Analysis of Single Pulse FR in the Diode

Since the forward recovery effect also occurs in diodes, its influence on the antiparallel diodes of the tested IGBTs is investigated in the following analysis. Fig. 7 again shows the results as functions of the snubber capacitance values and switched current levels.

When examining the transient overvoltages across the diodes, it is evident that the forward recovery effect is most distinct in the diode of IGBT S6, peaking at 220 V. However, the transient overvoltage values of the diodes are less than half of those observed in the IGBTs. Both data sets, IGBT and diode, indicate that even a small capacitance of approximately 14 nF is sufficient to effectively dampen the overvoltages to less than 50 V. Further analysis of the forward recovery energies E_{FR} reveals that the diodes of IGBT S7 and H7 perform best, indicating the lowest losses. Despite decreasing overvoltages with increasing C_{sn} values, the E_{FR} reduction becomes marginal. This is because the (V–I)-overlap duration increases, resulting in similar energy dissipation even as the peak overvoltages decrease. Interestingly, the intensity of the

forward recovery effects in IGBT S7 and its diode reveals completely opposite behaviors. While IGBT S7 shows higher overvoltages and energy losses, its antiparallel diode provides the best performance among the devices tested. This inverse relationship can be attributed to different optimizations in the device design and physics implemented by the semiconductor manufacturer.

C. Memory Effect of FR during Repeating Pulses

In theory, the single-shot measurement represents the worst-case scenario because the charge carriers in the drift region have fully recombined. However, this scenario is no longer valid during the operation of a power converter with switching frequencies that exceed ≈ 50 kHz. In such cases, the carrier lifetime can exceed the current interruption duration between two switching events [9]. Therefore, to better assess the effect of forward recovery under these operating conditions, tests were carried out that involved a second consecutive SiC MOSFET turn-OFF event. Fig. 8 shows the acquired current and voltage waveforms for two consecutive turn-OFFs of IGBT H7.

The magnitude of the transient overvoltage caused by the first FR event corresponds to the results presented in Fig. 6. At

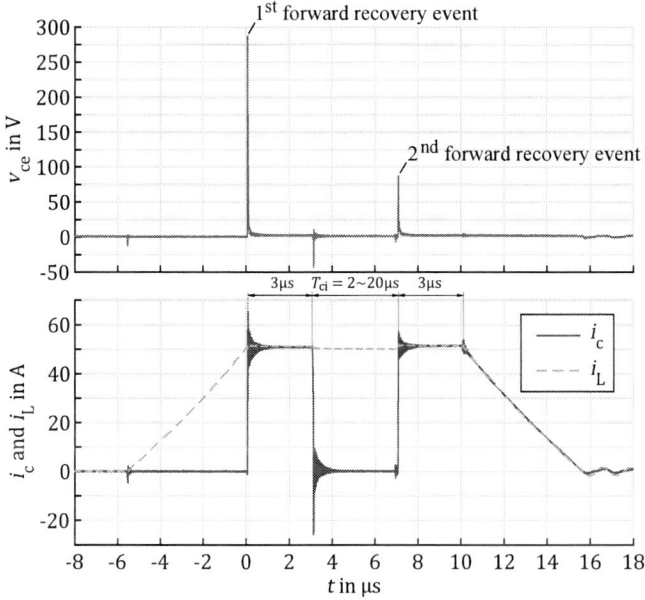

Fig. 8. Voltage and current waveforms across IGBT H7 of two consecutive SiC MOSFET S_1 turn-OFF events at $t_1 = 0$ and $t_4 = 7\,\mu s$. Test parameters: $C_{sn} = 0$, $I_{L,sw} = 50\,A$, $T_{ci} = 4\,\mu s$.

$t_2 = 3\,\mu s$, the SiC MOSFET S_1 is turned-ON. During this interval, no current flows through the IGBT under evaluation, and the charge carriers in the n-drift region gradually recombine. At $t_4 = 7\,\mu s$, the SiC MOSFET S_1 is turned-OFF again, causing the second consecutive FR event. The transient overvoltage observed during the second event is only one-third of the first FR event because many free charge carriers from the previous conduction state are still present in the n-drift region.

In the following experiments, the MOSFET turn-OFF interval was varied in 2 μs increments within the range between 2 μs to 20 μs. Fig. 9 contrasts the single-shot results with the acquired multi-shot results.

The measurement series reveals that the magnitude of the second FR events for IGBT S6 and H7 reaches the values observed during the single-shot tests after a current interruption duration of only 10 μs. This indicates that S6 and H7 have the shortest charge carrier lifetimes among the devices tested, with almost all charge carriers recombined in the n-drift region within this duration. Consequently, the IGBTs exhibit increased conductivity due to the presence of residual free charge carriers with shorter current interruption durations, allowing the collector current to rise more rapidly. This increased conductivity is demonstrated by the higher collector current slew-rate $di_c/dt|_{20}^{80}$ observed with $C_{sn} = 0$.

The experiment was repeated for the antiparallel diodes of the IGBTs [data not shown here], which revealed that different current interruption durations had minimal influence on the forward recovery effect. This indicates that the tested diodes demonstrate excellent dynamic properties and explains why the diode's FR overvoltage is approximately half that of the IGBTs.

D. Discussion

In the practical operation of hybrid converters, the FR effect occurs during the dead-time of each switching event of the fast-switching SiC MOSFETs. Since it depends on the current interruption of the IGBT—defined by the switching frequency and duty cycle—and the switched current, the energy losses and the magnitude of the transient overvoltages vary throughout the fundamental cycle.

The findings indicate that even a small-capacity snubber capacitor can effectively reduce the transient overvoltages and energy losses caused by FR. Furthermore, by shortening the commutation loops, the additional voltage across the SiC MOSFET during its turn-OFF—caused by the loop's inductance—can be reduced. C_{sn} acts as a high-frequency decoupler between the two stages and serves as energy storage during current commutations. By incorporating C_{sn}, the IGBT current slew-rate is reduced, thereby decreasing the maximum overvoltage. However, as the results also demonstrate, the maximum overcurrents through the IGBT and diode increase with the use of C_{sn}. Introducing a small resistor connected in series with C_{sn} could address this issue, although it requires a trade-off between overcurrent and current slew-rate.

Considering all results, IGBTs H3 and H7 show the best forward recovery behavior in a hybrid Si/SiC-based converter. Depending on the operating quadrants, the FR effect occurs in either the IGBT or its antiparallel diode; thus, selecting H3 or H7 may be preferable based on the specific operating conditions. Apart from that, if demand increases, the semiconductor manufacturer could enhance the FR behavior by combining the diode from IGBT S7 with the IGBT H3 or H7. However, no direct correlation could be established between the experimental results and the datasheet parameters provided by the semiconductor manufacturer. Consequently, predicting the FR behavior of a bipolar device *a priori* is not straightforward, making the selection of the optimal transistor in hybrid topologies challenging and time-consuming.

It should also be noted that the measurements were conducted at room temperature. As demonstrated in [11], the magnitude of the overvoltage increases by $\approx 50\%$ at a junction temperature of 100°C compared to room temperature. Therefore, the FR effect may impose even more significant limitations during practical operation if not addressed adequately.

IV. CONCLUSION

This article investigates and compares the forward recovery effect of different IGBTs and their antiparallel diodes employed in hybrid Si/SiC-based converters. The study reveals that the switched current, the current slew-rate, and the current interruption duration significantly influence the magnitude and duration of this effect in the various bipolar devices tested. The energy losses caused by forward recovery substantially impact the converter's overall efficiency, while the transient overvoltages induce additional voltage stress across the SiC MOSFETs. This added stress affects the reliability and lifespan of the devices employed in such hybrid topologies, necessitating careful consideration during the design phase.

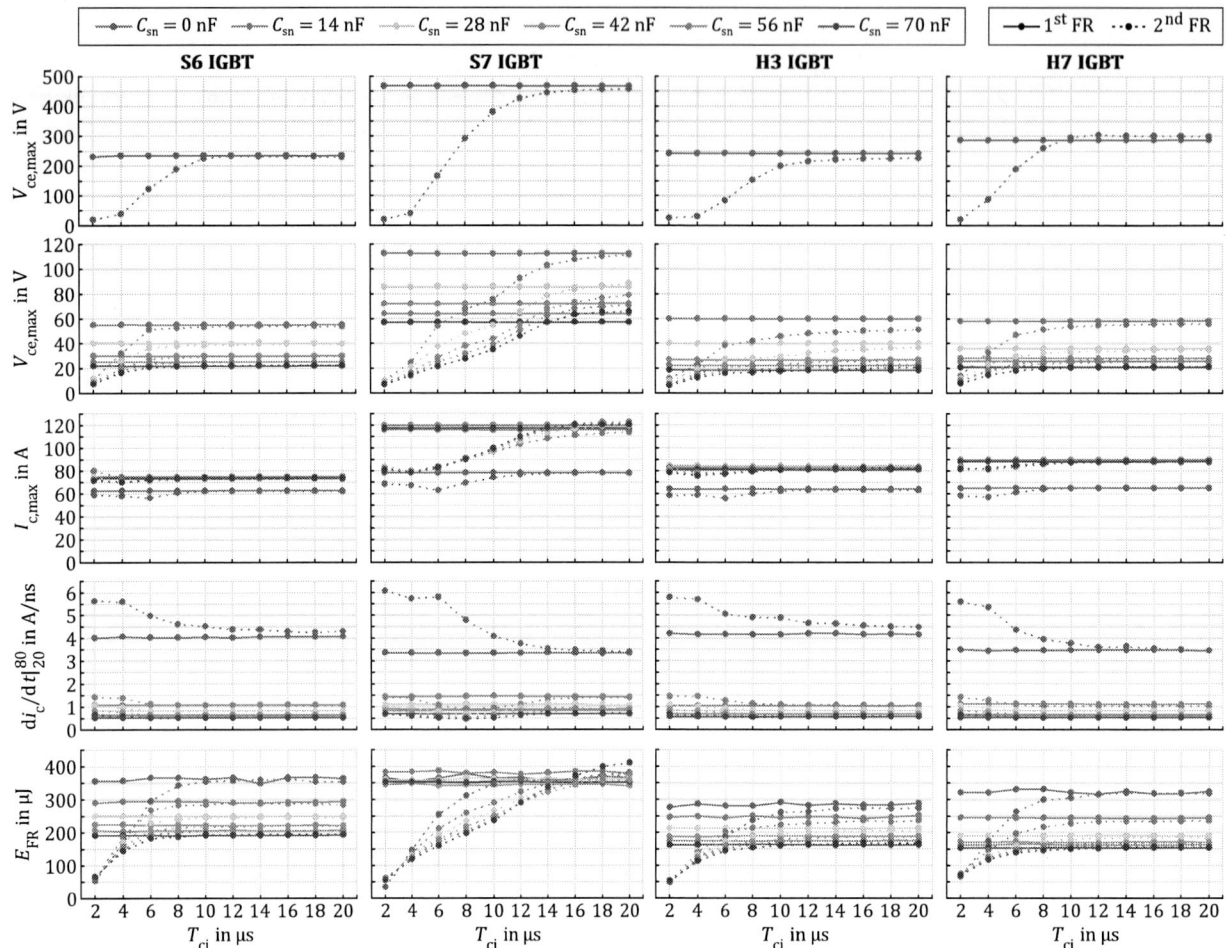

Fig. 9. Maximum transient collector-emitter overvoltage $V_{ce,max}$, maximum collector overcurrent $I_{c,max}$, current slew-rate $di_c/dt|_{20}^{80}$ [for definition see Fig. 5], and forward recovery energy E_{FR} for the different IGBT devices under test, shown as functions of current interruption durations and snubber capacitance values. The multi-shot results of two consecutive SiC MOSFET turn-OFFs are contrasted to the single-shot results.

To combat this, a snubber capacitor is employed to decouple the low-frequency from the high-frequency stage. This simple yet effective measure reduces the loop inductance of the commutation cells and effectively mitigates the detrimental effects of high di/dt on IGBTs.

REFERENCES

[1] Y. Wang, O. Lucia, Z. Zhang, *et al.*, "A review of high frequency power converters and related technologies," *IEEE Open J. Ind. Electron. Soc.*, vol. 1, pp. 247–260, Sep. 2020.

[2] L. Franquelo, J. Rodriguez, J. Leon, *et al.*, "The age of multilevel converters arrives," *EEE Ind. Electron. Mag.*, vol. 2, no. 2, pp. 28–39, Jun. 2008.

[3] I. Harbi, J. Rodriguez, A. Poorfakhraei, *et al.*, "Common DC-Link multilevel converters: Topologies, control and industrial applications," *IEEE Open J. Power Electron.*, vol. 4, pp. 512–538, Jul. 2023.

[4] Q.-X. Guan, C. Li, Y. Zhang, *et al.*, "An extremely high efficient three-level active neutral-point-clamped converter comprising SiC and Si hybrid power stages," *IEEE Trans. Power Electron.*, vol. 33, no. 10, pp. 8341–8352, Oct. 2018.

[5] Y. Zhang, C. Li, C. Li, *et al.*, "A SiC and Si hybrid five-level unidirectional rectifier for medium voltage applications," *IEEE Trans. on Ind. Applicat.*, vol. 69, no. 8, pp. 7537–7548, Aug. 2022.

[6] Di Zhang, J. He, and S. Madhusoodhanan, "Three-level two-stage decoupled active NPC converter with Si IGBT and SiC MOSFET," *IEEE Trans. Ind. Applicat.*, vol. 54, no. 6, pp. 6169–6178, Nov. 2018.

[7] L. Zhang, X. Lou, C. Li, *et al.*, "Evaluation of different Si/SiC hybrid three-level active NPC inverters for high power density," *IEEE Trans. Power Electron.*, vol. 35, no. 8, pp. 8224–8236, Aug. 2020.

[8] A. I. Emon, Z. Yuan, A. B. Mirza, *et al.*, "1200 V/650 V/160 A SiC+Si IGBT 3L hybrid T-Type NPC power module with enhanced EMI shielding," *IEEE Trans. Power Electron.*, vol. 36, no. 12, pp. 13 660–13 673, Dec. 2021.

[9] A. C. Schittler, D. Heer, and C. R. Müller, "Interaction of power-module design and modulation scheme for active neutral-point-clamped inverters," in *Proc. PCIM Europe 2019*, Nuremberg, Germany, 2019, pp. 809–815.

[10] C. L. Ma and P. O. Lauritzen, "A simple power diode model with forward and reverse recovery," *IEEE Trans. Power Electron.*, vol. 8, no. 4, pp. 342–346, Oct. 1993.

[11] S. Lakshmeesha, C. L. Kahraman, S. Rosado, and T. Wijekoon, "Switching behavior of a hybrid Si-IGBT and SiC MOSFET based ANPC topology," in *Proc. IEEE 8th Workshop Wide Bandgap Power Devices Appl.*, Redondo Beach, CA, USA, 2021, pp. 293–298.

[12] C. L. Kahraman, S. Lakshmeesha, S. Rosado, and T. Wijekoon, "Impact of forward recovery effects in different Si-IGBT technologies used in hybrid Si-IGBT, SiC-MOSFET based ANPC topology," in *Proc. IEEE Appl. Power Electron. Conf. Expo.*, Houston, TX, USA, 2022, pp. 1364–1370.

[13] C. Li, Q.-X. Guan, J. Lei, *et al.*, "An SiC MOSFET and Si diode hybrid three-phase high-power three-level rectifier," *IEEE Trans. Power Electron.*, vol. 34, no. 7, pp. 6076–6087, Jul. 2019.

[14] Pham, Ha Trieu To, Eckel, Hans-Günter, "Oscillation damping in a 500kW hybrid Si/SiC three-level ANPC inverter with decoupling capacitor," in *Proc. 24th Eur. Conf. Power Electron. Appl.*, Hanover, 2022, P1–P.10.

979-8-3315-1612-3/25 $31.00 © 2025 IEEE

Real-Time IGBT Module Ageing Characterization Through Temperature Monitoring

Quirc Perez-Farre
MCIA Research Center
Electrical Engineering Department
Universitat Politecnica de Catalunya
Terrassa, Spain
quirc.perez@upc.edu

Luis F. Gomez-Rivera
MCIA Research Center
Electronic Engineering Department
Universitat Politecnica de Catalunya
Terrassa, Spain
luis.felipe.gomez@upc.edu

Carlos Lopez-Torres
eDrive engineering & validation
SEAT, SA
Martorell, Spain
carlos.lopez3@seat.es

Kai Dannehl
E-drive, Charging Energy Systems
SEAT, SA
Martorell, Spain
kai.dannehl@seat.es

Antoni García-Espinosa
MCIA Research Center
Electrical Engineering Department
Universitat Politecnica de Catalunya
Terrassa, Spain
antoni.garcia@upc.edu

Alejandro Paredes-Camacho
MCIA Research Center
Electronic Engineering Department
Universitat Politecnica de Catalunya
Terrassa, Spain
alejandro.paredes@upc.edu

Abstract—This paper introduces a novel method for monitoring the ageing of Si insulated-gate bipolar transistors (IGBT) in real-time based on component temperature measurements. The method consists of simulating power losses and the thermal behaviour models of an IGBT module in different predefined ageing states. These simulations determine the values of expected thermal indicators for contemplated ageing conditions and establish a correlation with the IGBT module ageing state expressed on conventional indicators, such as collector-emitter on-state voltage (V_{CE-ON}) and junction-to-case thermal resistance (R_{TH-JC}). This innovative method allows the simultaneous monitoring of several ageing mechanisms and opens the door for studying their dynamics during the process. To assess the potential of this method, this manuscript presents a validation of the methodology in a virtual environment, accompanied by experimental results, making a significant advancement in the field of IGBT ageing monitoring.

Index Terms—Accelerated lifetime, IGBT ageing estimation, real-time monitoring, thermal indicators, traction inverter.

I. INTRODUCTION

Automotive inverters are subjected to rigorous operating situations due to significant and rapid load variations and changing environmental conditions. Power inverters are increasingly subject to greater energy density due to the new power requirements and existing space limitations on electric vehicle (EV) structures. Accordingly, monitoring the ageing of traction inverters is gaining increasing interest among EV manufacturers.

Power semiconductor modules, such as IGBT modules, are critical components inside the inverter. As is shown in Fig. 1, power inverters cause 31% of the failures [1]. Many environmental factors, such as ambient humidity or dust and mechanical vibrations, can affect the ageing of IGBT power

modules [1], [2]. However, these ageing factors can be considerably minimized by ensuring adequate component sealing and fastening to prevent the transmission of vibrations from the power train to the inverter.

Besides that, it has been shown that thermal stress, a factor of significant gravity, is the most determining ageing factor in the traction inverter [2], [3]. In contrast to the previously discussed factors, although thermal stress can be minimised, this is intrinsic to the operation of the component. Thermal stress translates into mechanical fatigue due to the difference in coefficients of thermal expansion (CTE) of the materials, which results in cracks and voids [4], [5]. The operating conditions and ageing factors by thermal stress also define the dominant ageing mechanisms that will affect the IGBT components, with bond-wire degradation and substrate degradation being the main ones [4], [6].

Monitoring the ageing of IGBT power modules is a complex task, as it involves measuring challenging electrical parameters

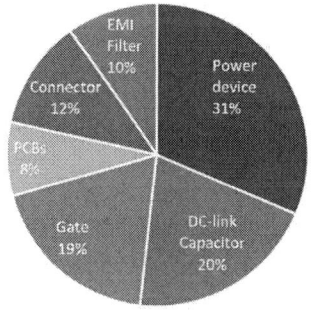

Fig. 1. Fragile components in a traction inverter for EVs.

such as collector-emitter on-state voltage (V_{CE_ON}) [7]. These variables are commonly used indicators for ageing monitoring, but their measurement is hindered by high switching voltage oscillations and very small voltage amplitudes. In practice, obtaining these indicators is complex and requires costly conditioning circuits that increase the complexity of the set. Alternatively, some authors have presented methods to determine ageing indicators indirectly by correlating electrical parameters, e.g., V_{CE_ON}, and factors such as harmonics [8], transistor switching state and gate voltages [9], [10], and thermal parameters [13]. As a consequence, these approaches eliminate the need for external auxiliary circuits.

The indicators based on thermal parameters present considerably slower transients, significantly facilitating noise filtering. Furthermore, the relative variation of these variables is more significant than that of the electrical parameters, so they are easier to detect, although still challenging, and new methods should be explored [11]. However, these proposals do not enable the distinction or characterization of the primary ageing mechanisms, such as bond-wire and substrate degradation [12].

This paper proposes a novel method for monitoring the ageing of IGBT power modules based on component temperature measurements to obtain ageing indicators indirectly. The method is based on simulating the losses and the thermal behavior of the component in predefined ageing states. Further, the main ageing mechanisms are stipulated based on the operating conditions of the component, bond-wire and substrate degradation, and they are emulated in the thermal network of the component. This allows obtaining a temperature map of different points of the component in terms of its ageing state. By contrasting these temperature maps with the real measured values of the element, a main ageing indicator forecasting of the component, i.e., an increase in V_{CE_ON} and R_{TH_JC}, will be obtained. This method also allows monitoring component ageing in real time under predefined operating conditions, permitting several ageing mechanisms to be monitored simultaneously. Therefore, the dynamics of said mechanisms can be determined by studying the different measurements. Using thermal indicators is less intrusive and does not require conditioning or acquisition circuits. Furthermore, it is not necessary to interrupt the regular operation of the component for testing or ageing monitoring.

The paper is organized as follows: Section II presents the proposed method, its description, and the ageing mechanism definitions. Section III describes the simulation power losses, thermal behaviour, and ageing analysis. Section IV includes experimental data and ageing evaluation. Finally, Section V concludes the paper.

II. PROPOSED METHOD

Fig. 2 shows the proposed method. This is based on the observation that, under identical operating and ambient conditions, component temperatures increase as the component ages due to increased heat generation and worsening heat

dissipation capacity. These changes can be directly related to the ageing mechanisms mentioned above:

- Bond wire degradation refers to damage resulting from the formation and propagation of cracks within the bond wire and its hill, which may lead to bond wire lift-off. These cracks decrease the cross-sectional area of the bond wires, resulting in increased electrical resistance and, consequently, higher losses within the component.
- Substrate degradation encompasses damage due to the initiation and propagation of cracks within the die-attached solder layer and the solder layer itself, potentially resulting in the formation of voids. This degradation reduces the effective thermal conduction area, diminish-

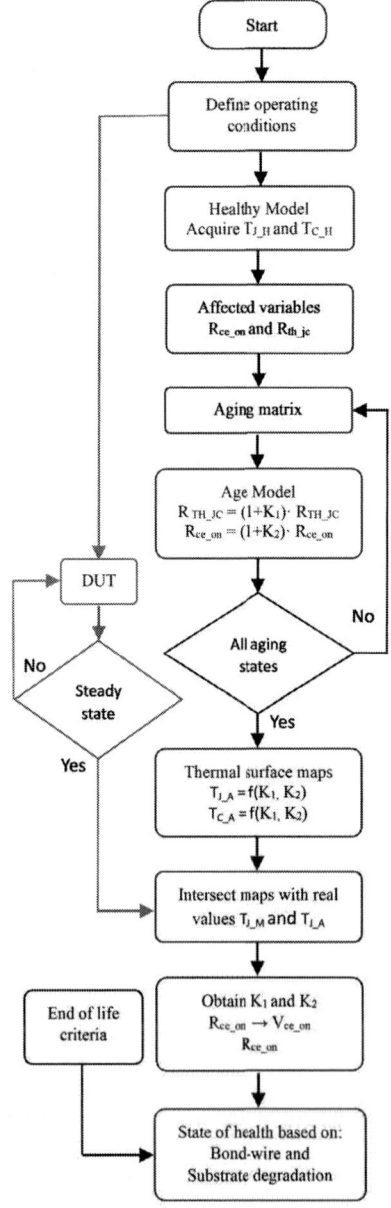

Fig. 2. Flowchart of the proposed method.

979-8-3315-1612-3/25 $31.00 © 2025 IEEE

ing the component's heat dissipation capacity, as air acts as a near-insulator for thermal conduction.

The effect of these ageing mechanisms from a thermal point of view, which means temperature increase, is not homogenous throughout the component. Depending on the point selected, one ageing mechanism will more affect the temperature increase than another.

To apply this method, it is necessary to select at least two temperature points, junction temperature (T_J) and Casing temperature (T_C), that present a differentiated behaviour in front of the selected ageing mechanisms.

Bond-wire degradation leads to a similar percentage increase in temperature at both measurement points. In contrast, substrate degradation results in a more pronounced percentage increase in junction temperature (T_J) compared to case temperature (T_C).

The thermal circuit and its effects by the mechanism can be effectively understood through comparison with an equivalent electrical circuit. The loss source is represented as an ideal current source, and the thermal resistance corresponds to electrical resistance, while temperature corresponds to voltage. In this framework, the equivalent circuit can be conceptualized as a voltage divider, as illustrated in Fig. 3.

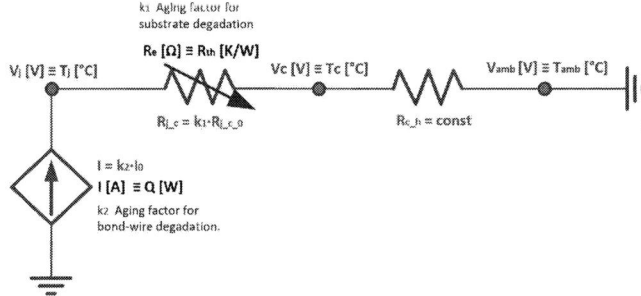

Fig. 3. Equivalent electric representation of the thermal network.

Considering the circuit in Fig. 3, if only the effect of bond-wire degradation is considered, $k_1 = 1$ (no effect) and $k_2 > 1$, then V_J and V_C can be represented as:

$$V_J = (k_2 \cdot I_0) \cdot (k_1 \cdot R_{J\_C\_0} + R_{C\_H\_0}) = k_2 \cdot V_{J\_0} \quad (1)$$

$$V_C = (k_2 \cdot I_0) \cdot (R_{C\_H\_0}) = k_2 \cdot V_{c\_0} \quad (2)$$

On the other side, if only the effect of substrate degradation is considered, $k_1 > 1$ and $k_2 = 1$ (no effect), in consequence, V_J and V_C can be represented as:

$$V_J = I_0 \cdot (k_1 \cdot R_{J\_C\_0} + R_{C\_H\_0}) \quad (3)$$

$$V_C = V_{c\_0} \quad (4)$$

The proposed equivalent model is simplified and does not account for thermal coupling effects or the temperature-dependency of thermal parameters. Nonetheless, it serves as a

helpful analogy that accurately represents the behaviour and influence of these mechanisms.

This approach enables differentiation between the two modelled ageing mechanisms, which represents the primary challenge in utilizing thermal indicators.

The methodology represented in Fig. 2 consists of the following steps:

- Define any mission profile with the sole requirement that the thermal equilibrium of the component is achieved at some point during the profile.
- Build a digital twin of the component in its healthy state and model the impact of ageing mechanisms on the system, incorporating the variables influenced by these mechanisms, in this, case thermal and electrical resistances:

$$R_{\text{TH}\_J\_C} = k_1 \cdot R_{J\_C\_0} = (1 + K_1) \cdot R_{J\_C\_0} \quad (5)$$

$$R_{\text{CE\_ON}} = k_2 \cdot R_{\text{CE\_ON}\_T} = (1 + K_2) \cdot R_{\text{CE\_ON}\_T} \quad (6)$$

- A physically realistic maximum value should be defined for the two ageing mechanisms [12], [13]. The end-of-life criterion will be determined experimentally. However, these limits are initially selected as the end-of-life criterion for the preliminary calculations as follows:

$$K_1 = [0, 0.1, 0.2, 0.3, 0.4] \quad (7)$$

$$K_2 = [0, 0.05, 0.1, 0.15, 0.2] \quad (8)$$

The growth rate of these vectors is constrained by the maximum number of simulations established, which in this case is 25. A lower growth rate of these vectors results in a higher resolution; however, this comes at the cost of an exponentially increasing simulation time.

- Simulate the different ageing states, as defined in equation (9), to obtain the junction and casing temperatures and generate the temperature maps.

$$AGEING\ MATRIX = \quad (9)$$

K_1/K_2	0	0.05	0.1	0.15	0.2
0	0/0	0/0.05	0/0.1	0/0.15	0/0.2
0.1	0.1/0	0.1/0.05	0.1/0.1	0.1/0.15	0.1/0.2
0.2	0.2/0	0.2/0.05	0.2/0.1	0.2/0.15	0.2/0.2
0.3	0.3/0	0.3/0.05	0.3/0.1	0.3/0.15	0.3/0.2
0.4	0.4/0	0.4/0.05	0.4/0.1	0.4/0.15	0.4/0.2

- Acquire the actual temperature measurements of the component and compare them with the generated temperature maps. Subsequently, the respective coefficients K_1 and K_2, along with the values for R_{TH_JC} and R_{CE_ON}, are obtained, which allows the two mechanisms to be characterized independently.
- The state of health (SOH) is a percentage derived from two ageing coefficients, K_1 and K_2. This SOH measurement cannot be directly extrapolated to time estimation

or cycle count, as the ageing of these components is non-linear and accelerates due to positive feedback between ageing mechanisms and increasing temperature. To do this, it is necessary to carry out a series of measurements throughout the life of the component and perform a trend analysis of the values obtained.

III. EVALUATED METHOD BY SIMULATIONS

The digital twin is a thermoelectric model in which the loss model is derived from empirical equations, and the thermal network employs a lumped parameter approach grounded in energy balance equations. The T_J of the transistors and diodes couples both models. Based on temperature, the R_{CE_ON} value can be recalculated as:

$$R_{CE\_ON\_T} = R_{CE\_ON\_20} + \alpha \cdot (T_J - 20) \qquad (10)$$

On the other hand, if ageing is added, R_{CE_ON} can be obtained as:

$$R_{CE\_ON} = (1 + K_2) \cdot R_{CE\_ON\_T} = \qquad (11)$$

$$(1 + K_2) \cdot (R_{CE\_ON\_20} + \alpha \cdot (T_J - 20))$$

Then, the conduction power losses (P_{COND}), and the switching power losses (P_{SWI}) can be calculated as follows:

$$P_{COND} = I_C \cdot (V_{CE\_ON} + I_C \cdot R_{CE\_ON}) \qquad (12)$$

$$P_{SWI} = P_{SWI\_ON} + P_{SWI\_OFF} = \qquad (13)$$

$$f_{sw} \cdot (E_{ON} + E_{OFF}) \cdot \frac{U_{PROFILE}}{U_{DATASHEET}} \cdot \frac{I_{PROFILE}}{I_{DATASHEET}} \cdot 10^{-3}$$

Both E_{ON} and E_{OFF} are functions of the operating conditions of the component, with a strong dependence on the collector current and the transistor's junction temperature (T_J). To account for these dependencies, look-up tables (LUTs) derived from the manufacturer's datasheets are utilised.

The thermal model is mainly defined by three nodes, representing the T_J, T_C and the heat sink surface temperature (T_H). Foster and Cauer circuit models are used in combination. With the electrothermal model of the healthy component constructed, equation (5) and (6) are applied.

Once all the ageing states predefined by the ageing matrix are simulated, the temperature maps of Fig. 4 and Fig. 5 are obtained for T_J and T_C, respectively.

The intersection of these temperature maps with the measurements obtained from the IGBT module, T_J and T_C in an undetermined ageing state produces the lines presented in Fig. 6. To validate the methodology, two values within the physically realistic range were selected, e.g., T_J = 155 ° C and T_C = 142 ° C.

The following linear equations represent all potential ageing states that independently result in the observed temperatures measured in the component.

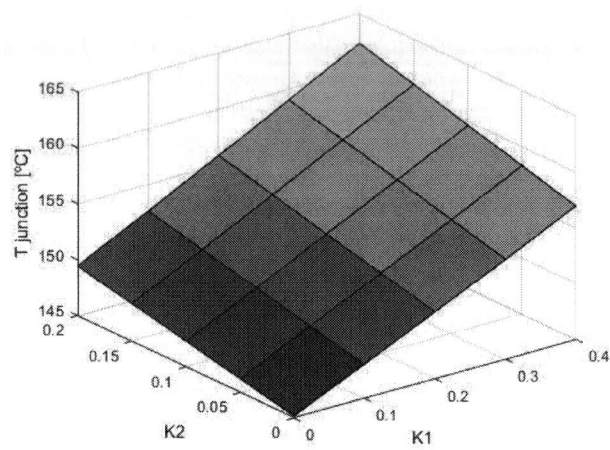

Fig. 4. Junction temperature map with ageing indicators coefficients.

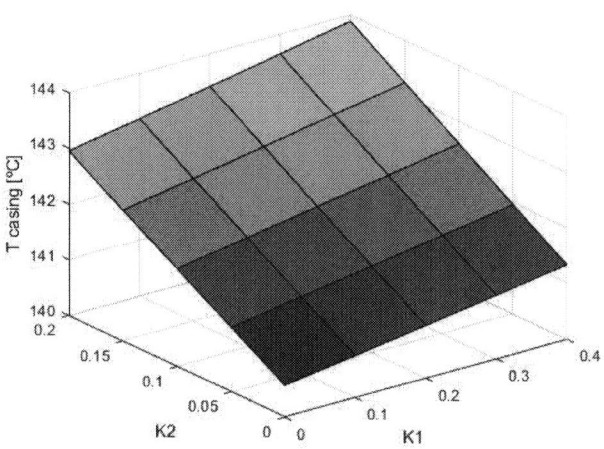

Fig. 5. Casing temperature map with ageing indicator coefficients.

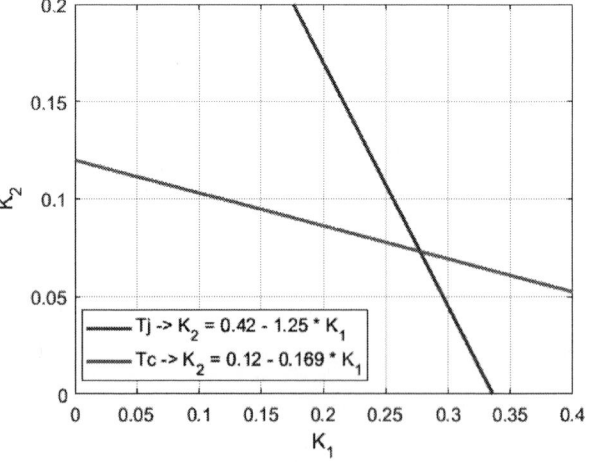

Fig. 6. Possible ageing States.

979-8-3315-1612-3/25 $31.00 © 2025 IEEE   1777

$$\text{For } T_J: \ K_2 = 0.42 - 1.25 \cdot K_1 \qquad (14)$$

$$\text{For } T_C: \ K_2 = 0.12 - 0.169 \cdot K_1 \qquad (15)$$

Since the increase in temperature and the effect of the mechanism are not homogeneous at different points of the transistor, the two straight lines intersect at one point. This point estimates the SOH model of the component based on the mechanisms studied. For this case, the percentages for substrate and bond wire degradation can be obtained as follows:

$$K_1 = 0.2759 \rightarrow \text{Substrate degradation at } \frac{K_1}{0.4} \cdot 100 = 69\%$$
$$(16)$$

$$K_2 = 0.0739 \rightarrow \text{Bondwire degradation at } \frac{K_2}{0.2} \cdot 100 = 37\%$$
$$(17)$$

IV. Experimental Evaluation Results

An experimental setup was designed to achieve two primary functions: data collection of electrical variables (V_{CE_ON} and I_C) and thermal variables (T_J, T_C and T_{AMB}) and the accelerated ageing of a commercial IGBT module using a chopper circuit. The experimental setup is shown in Fig. 7. In this case study, IGBT module SKM50GB12T4 by Semikron and an Elektro-Automatik power supply EA-PS 9750-60 were used. An ESP32 microprocessor performs pulse generation and thermal data capture. Meanwhile, the electrical data collection of V_{CE_ON} and I_{CE} is performed using the oscilloscope Tektronix MDO3024 and Tektronix probes THDP0200 and

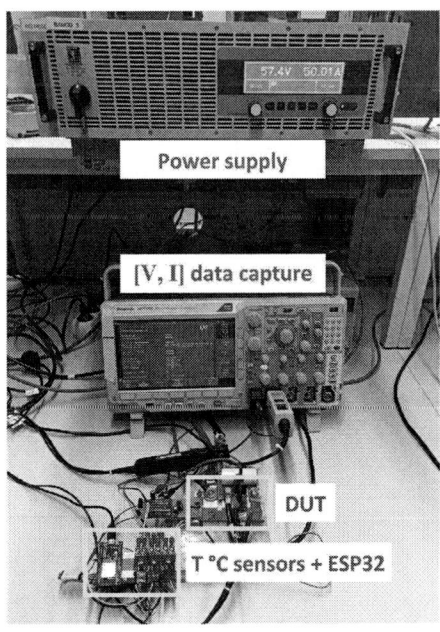

Fig. 7. Experimental setup for IGBT module ageing evaluation.

TCP0030A voltage and current, respectively. The data from the oscilloscope is sent to MATLAB via an ethernet connection.

The load and the power supply were defined constant throughout the test to the following values: 100% of I_{C_Nom} (50 A) and 4% V_{CE_Nom} (1200 V) semi-continuous operation with a frequency of 50 mHz and duty of 50%. The test was performed for 10.1 hours, and the data was processed and analysed using scripts in MATLAB. Fig. 8 shows the evolution of the V_{CE_ON}, where three regions can be distinguished.

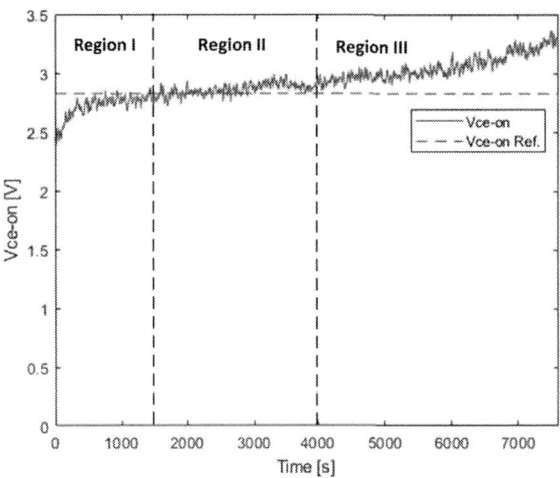

Fig. 8. Evolution of V_{CE_ON}

The regions are also reflected in Fig. 9 and Fig. 10, which depict the evolution of the R_{TH_JC} calculated by equation (18), and show the main temperatures of the IGBT module, T_J, T_C and T_{AMB}, respectively.

$$R_{\text{TH\_JC}} = \frac{T_J - T_C}{P_{\text{LOSSES}}} \qquad (18)$$

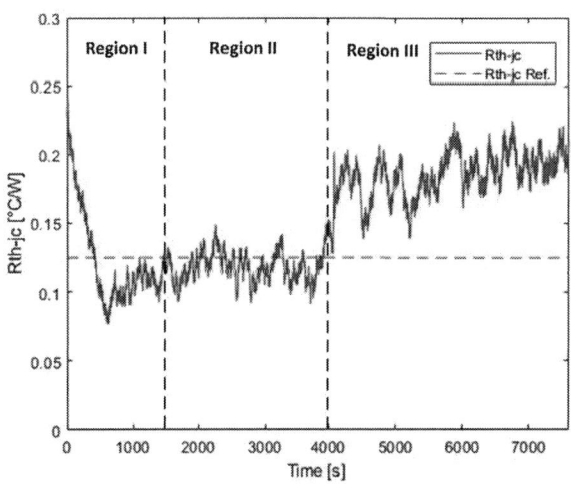

Fig. 9. Evolution of R_{TH_JC}

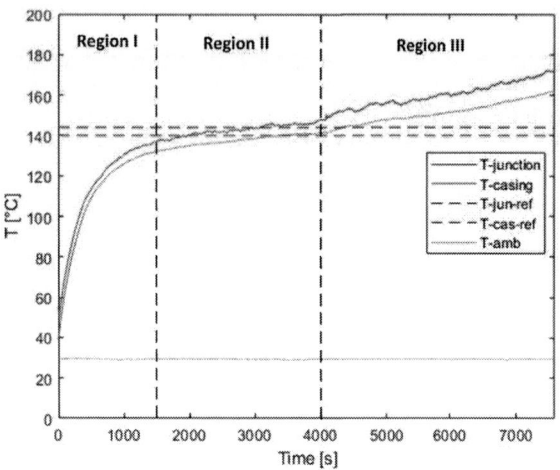

Fig. 10. Average transistor temperatures

In this case, the value of P_{LOSSES} is determined through the direct multiplication of V_{CE} and I_C, with both variables recorded using an oscilloscope. Due to the semi-continuous nature of the test (fsw = 50 mHz), the switching losses are negligible.

- **Region I: from 0 to 1500 s**

In the first region, V_{CE_ON} increases rapidly due to thermal transient. R_{TH_JC} starts from a high value and decreases during this period due to the difference in thermal inertia of the junction vs casing, $I_{TH\_J} << I_{TH\_C}$. This transient is also clearly reflected in the temperature measurements.

- **Region II: from 1500 to 4000 s**

The second region has reached thermal equilibrium. In this region, the measured value closely aligns with the reference value, which represents the average signal value obtained from prior tests.

- **Region III from 4000 to 7600 s**

Accelerated increase of all indicators, both direct (V_{CE_ON} and R_{TH_JC}) and indirect (T_J and T_C). Thus, the device under test (DUT) could be considered in the thermal runaway phase.

The test was automatically terminated when the T_J exceeded 175 ° C. Under these final conditions, the ageing coefficients K_1 and K_2 were determined using equations (19), (20) and (21).

$$K_1 = \frac{R_{\text{TH\_JC}\_T=7600s} - R_{\text{TH\_JC\_REF}}}{R_{\text{TH\_JC\_REF}}} = \quad (19)$$

$$= 0.746 \rightarrow \frac{K_1}{0.4} \cdot 100 = 7.5\%$$

$$K'_2 = \frac{V_{\text{CE\_ON}\_T=7600s} - V_{\text{CE\_ON\_REF}\_145° \text{ C}}}{V_{\text{CE\_ON\_REF}\_145° \text{ C}}} = 0.168 \quad (20)$$

The value obtained from equation (20) is inaccurate, as it does not account for the temperature increase. Therefore, the reference value of V_{CE_ON} at 175°C is estimated through simulation and applied as follows:

$$K_2 = \frac{V_{\text{CE\_ON}\_T=7600s} - V_{\text{CE\_ON\_REF}\_175° \text{ C}}}{V_{\text{CE\_ON\_REF}\_175° \text{ C}}} = \quad (21)$$

$$= 0.015 \rightarrow \frac{K_2}{0.2} \cdot 100 = 7.5\%$$

Regarding the ageing coefficient K_1 in equation (19), the impact of the temperature rise has already been incorporated into the component losses in equation (18).

For the thermal indicators, an increase in $\Delta T_J = 28°$ C of 20% and an increase in $\Delta T_C = 20°$ C of 14% are observed. The initial experimental test demonstrates a clear correlation between the thermal and electrical indicators. Regarding the results, the accelerated tendency of the indicators can be observed, which is crucial for the subsequent trend analysis.

Additionally, the results validate the hypothesis that the temperature increase of the transistor is not uniform but dependent on the dominant ageing mechanism. In this case study, the substrate degradation of the ΔT_C for Region II is 5.3 ° C, and the end of Region III is 10 ° C.

V. CONCLUSIONS

The improvements in computational resources and access to more advanced simulation tools allow this type of method to simulate many ageing states to estimate thermal behaviour. This method enables the characterisation of component ageing by considering the combined effects of multiple mechanisms. However, they must be those that have a more significant impact on the component's temperature increase. On the other hand, the evaluation of ageing must be carried out in a thermal equilibrium of the element. The first experimental results showed the hypotheses validation of accelerated ageing, the non-uniformity of ageing effects, and the correlation between electrical and thermal indicators.

Future work on this method involves defining new profiles that present a lower thermal stress, which is more similar to the operating conditions of the component and allows the unravelling of similar ageing dynamics. The test duration will be prolonged to hundreds of hours, rather than the current duration of tens of hours.

ACKNOWLEDGMENT

The authors would like to thank the support of the Generalitat de Catalunya government under the Industrial Doctorate 2023-0008.

REFERENCES

[1] S. Yang, A. Bryant, P. Mawby, D. Xiang, L. Ran and P. Tavner, "An Industry-Based Survey of Reliability in Power Electronic Converters," in IEEE Transactions on Industry Applications, vol. 47, no. 3, pp. 1441-1451, May-June 2011, doi: 10.1109/TIA.2011.2124436.

[2] S. Rahimpour, H. Tarzamni, N. V. Kurdkandi, O. Husev, D. Vinnikov and F. Tahami, "An Overview of Lifetime Management of Power Electronic Converters," in IEEE Access, vol. 10, pp. 109688-109711, 2022, doi: 10.1109/ACCESS.2022.3214320.

[3] S. M. I. Rahman et al., "Emerging Trends and Challenges in Thermal Management of Power Electronic Converters: A State of the Art Review," in IEEE Access, vol. 12, pp. 50633-50672, 2024, doi: 10.1109/ACCESS.2024.3385429.

[4] Luhong Xie, Erping Deng, Shaohua Yang, Ying Zhang, Yan Zhong, Yanhao Wang, Yongzhang Huang, State-of-the-art of the bond wire failure mechanism and power cycling lifetime in power electronics, Microelectronics Reliability, Volume 147, 2023, 115060, ISSN 0026-2714, https://doi.org/10.1016/j.microrel.2023.115060.

[5] Q. Li et al., "Review of the Failure Mechanism and Methodologies of IGBT Bonding Wire," in IEEE Transactions on Components, Packaging and Manufacturing Technology, vol. 13, no. 7, pp. 1045-1057, July 2023, doi: 10.1109/TCPMT.2023.3297224.

[6] F. Hosseinabadi, S. Chakraborty, S. K. Bhoi, G. Prochart, D. Hrvanovic and O. Hegazy, "A Comprehensive Overview of Reliability Assessment Strategies and Testing of Power Electronics Converters," in IEEE Open Journal of Power Electronics, vol. 5, pp. 473-512, 2024, doi: 10.1109/OJPEL.2024.3379294.

[7] A. Laspeyres, A. -S. Descamps, C. Batard, N. Ginot, T. -L. Le and S. Azzopardi, "Active Clamp Circuit for Online ON-State Voltage Measurement of High Voltage SiC MOSFETs Power Module," 2023 IEEE Applied Power Electronics Conference and Exposition (APEC), Orlando, FL, USA, 2023, pp. 276-282, doi: 10.1109/APEC43580.2023.10131211.

[8] S. Ou, A. Sangwongwanich, S. Sahoo and F. Blaabjerg, "Semiconductor Devices Condition Monitoring Using Harmonics in Inverter Control Variables," 2023 25th European Conference on Power Electronics and Applications (EPE'23 ECCE Europe), Aalborg, Denmark, 2023, pp. 1-8, doi: 10.23919/EPE23ECCEEurope58414.2023.10264538.

[9] S. Dusmez and B. Akin, "Comprehensive parametric analyses of thermally aged power MOSFETs for failure precursor identification and lifetime estimation based on gate threshold voltage," 2016 IEEE Applied Power Electronics Conference and Exposition (APEC), Long Beach, CA, USA, 2016, pp. 2108-2113, doi: 10.1109/APEC.2016.7468158.

[10] W. Song, P. Xu, J. Chen, H. Tan, K. Yang and T. Tang, "A Chip Open-Circuit Failure Monitoring Method in Multichip IGBT Modules Based on the Fall Time of Gate-Voltage," in IEEE Transactions on Transportation Electrification, doi: 10.1109/TTE.2024.3351697

[11] S. Kalker et al., "Reviewing Thermal-Monitoring Techniques for Smart Power Modules," in IEEE Journal of Emerging and Selected Topics in Power Electronics, vol. 10, no. 2, pp. 1326-1341, April 2022, doi: 10.1109/JESTPE.2021.3063305.

[12] Z. Wang, B. Tian, W. Qiao and L. Qu, "Real-Time Aging Monitoring for IGBT Modules Using Case Temperature," in IEEE Transactions on Industrial Electronics, vol. 63, no. 2, pp. 1168-1178, Feb. 2016, doi: 10.1109/TIE.2015.2497665.

[13] H. Huang and P. A. Mawby, "A Lifetime Estimation Technique for Voltage Source Inverters," in IEEE Transactions on Power Electronics, vol. 28, no. 8, pp. 4113-4119, Aug. 2013, doi: 10.1109/TPEL.2012.2229472.

Experimental Validation of Triangular SOA via Infrared Thermography of a MOSFET Die Operating in the Thermally Unstable Linear-mode for Automotive Applications

Yacine Ayachi Amor
Automotive MOSFET Applications
Nexperia
Manchester, United Kingdom
yacine.ayachi.amor@nexperia.com

Christian Radici
Automotive MOSFET Applications
Nexperia
Manchester, United Kingdom
christian.radici@nexperia.com

Kerry J. Abrams
Failure Analysis, Quality Labs
Nexperia
Manchester, United Kingdom
kerry.abrams@nexperia.com

Philip Ellis
Automotive MOSFET Applications
Nexperia
Manchester, United Kingdom
philip.ellis@nexperia.com

Peter Vines
Automotive MOSFET Applications
Nexperia
Manchester, United Kingdom
peter.vines@nexperia.com

Wayne Lawson
Automotive MOSFET Applications
Nexperia
Manchester, United Kingdom
wayne.lawson@nexperia.com

Abstract— This paper explores for the first time the robustness of triangular and rectangular pulses of equal energy through thermal analysis, looking at hotspot formation across the die of a power MOSFET during operation in its electro-thermally unstable linear-mode region. This research demonstrates the validity of a technique widely employed for thermally stable linear-mode operation, considering three types of power pulses corresponding to different automotive applications: i) rectangular ii) triangular - linearly decreasing current with constant voltage and iii) triangular - linearly decreasing voltage and constant current. In automotive applications, these types of triangular pulses are characteristics of active clamp and capacitive pre-charge, respectively. In this work, we use an exposed low voltage power MOSFET and present experimental data to prove the 2x SOA capability for triangular pulses uniquely from a thermal point of view. In contrast to our previous work, the same device under test (DUT) will be subjected to three different operating conditions and imaged using infrared thermography, thus avoiding the destruction of the sample for the validation of the technique.

Keywords—Automotive MOSFETs, linear-mode operation, SOA, IR thermal camera

I. INTRODUCTION

The push for increasing power density and cost reduction in modern automotive systems has led to the targeting of very low on-state resistance R_{DSon} in progressively smaller dies [1]. In general, this trend has resulted in a worsening of the linear-mode performance [2], which is directly linked to the die shrinking. MOSFETs operating in linear-mode are subjected to high thermal and electrical stresses due to their high output impedance, giving rise to high drain-to-source voltage (V_{DS}) even at high drain currents (I_D). Furthermore, for high enough

V_{DS}, a MOSFET in linear-mode might operate in a thermally unstable regime which greatly reduces its current capability [3].

The datasheet Safe Operating Area (SOA) can be used during the design phase to verify that the specific linear-mode working point lies below the corresponding SOA curve, however, it is only specified for rectangular pulses at a mounting base temperature (Tmb) of 25°C.

In automotive applications, linear-mode is employed mainly in the case of overcurrent protection and voltage or current regulation. Also, rectangular power pulses can be found in applications where MOSFETs are used as voltage-controlled current sources, such as in low-voltage dropout regulators (LDOs) and airbag "safing". Linearly decreasing current pulses are characteristic of active clamp applications, such as in fuel injection systems and solenoid drivers; while linearly decreasing voltage pulses can be found in capacitive pre-charge (in-rush current limiting or soft start), such as in isolation switches for system protection.

In case of operation within the thermally unstable linear-mode and for a non-rectangular power pulse, actual testing can be avoided by applying a technique normally used in the case of operation during thermally stable conditions, involving a power shape conversion [4]. We improve the estimation technique by way of a non-destructive technique using infrared thermography which, in contrast to previous studies, allows testing of the same device under three different operating conditions.

II. CONCEPT

During a thermally stable linear-mode and in case of non-rectangular power pulses, the operative limit of a MOSFET cannot be verified by direct comparison with the datasheet SOA graph. The actual limit can be found at a junction temperature of 175°C (for automotive MOSFETs) using a thermal model (Cauer or Foster) or precision electrothermal model; alternatively, by applying a power shape conversion at constant energy. If the MOSFET is operating in a thermally unstable linear-mode a similar approach has proven to be valid, considering that the device has a higher thermal impedance than the $Z_{th(j-mb)}$ stated in the datasheet and a different Cauer model should be used [4].

In case of triangular pulses, the actual SOA limit becomes twice that of a rectangular pulse having the same time duration [4].

$$I_{D,triangular} \approx 2 * I_{D,SOA} \qquad (1)$$

As shown in Fig. 1, the stable portion of linear-mode operation, which is inversely proportional to the thermal impedance junction-to-mounting base ($Z_{th(j-mb)}$) of the device, is indicated on the SOA by line A separated by an inflexion point (Z) from the thermally unstable one indicated by the line B. Thermal instability is caused by the uneven distribution of current across the die [5] which can lead to thermal runaway and the destruction of the device, with burn marks appearing close to the die bonding structure [1], [2], [6].

Active clamp (inductive loads) and soft start (capacitive loads) both lead to a triangular power pulse. However, as shown in Fig. 1, the main difference between the two is that in case of a soft start the operating point moves away from the thermally unstable linear-mode region, which leads to a lower risk of thermal runaway.

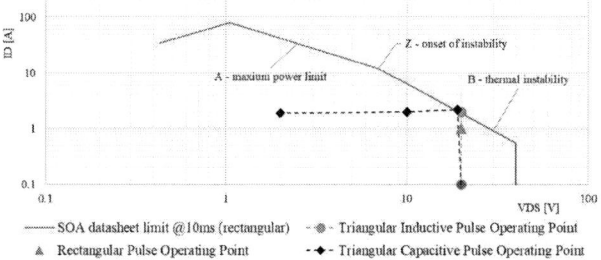

Fig. 1. SOA graph at 10ms pulse and three test operating points (OFF-ON) trajectories.

III. TESTING CAMPAIGN

In this study, three different circuits were implemented using the test setup shown in Fig. 2 with different loads to apply various operating conditions to the same MOSFET: i) a rectangular pulse (SOA pulse), ii) a linearly decreasing voltage pulse (capacitive load), and iii) a linearly decreasing current pulse (inductive load).

A 3x3mm 40V voltage automotive MOSFET with bond wires (MLPAK33) was selected for this study. The device was decapsulated by immersing into a mixture of hot fuming nitric acid and sulfuric acid (HNO3:H2SO4 = 1:2) to expose the die surface for thermal assessment using a high-resolution Infra-Red thermal camera.

The camera features a high-resolution detector with 640 x 512 pixels (each pixel is around 15um in size) and a cooled InSb (Indium Antimonide) sensor that operates within a spectral range of 7.5 to 11 μm with a thermal sensitivity of less than 20 mK, making it suitable for detailed thermal analysis of a silicon MOSFET die in dynamic conditions.

For accurate thermography, the importance of emissivity coefficient calibration is well-known [7]. The die surface is typically made from various materials such as Silicon oxide, Polysilicon, Passivation layers, etc. each with distinct emissivity coefficients. Therefore, the die surface was painted with a very thin high-temperature black matt spray with a known emissivity coefficient ε = 0.95.

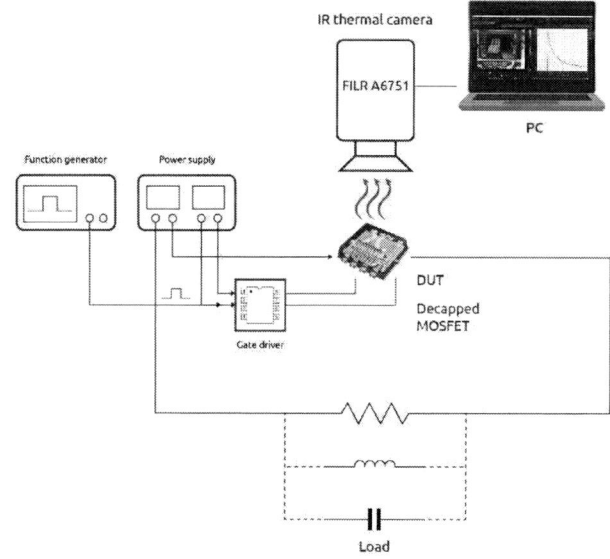

Fig. 2. Test setup diagram.

IV. EXPERIMENTAL RESULTS

In the following, experimental results are presented to prove equation (1).

Fig. 3, 4, and 5 show respectively the three 10ms pulses applied to the DUT driving the MOSFET during thermally unstable linear-mode, where the first operating condition is V_{DS} = 20V, I_D = 1A (20W peak power and 200mJ of energy), for a 10ms square pulse, the second operating condition is V_{DS} = 20V, I_D = 2.3A (46W peak), for a 10ms triangular pulse (linearly decreasing current, constant voltage), and the third pulse operating condition is V_{DS} = 20V, I_D = 2.3A (46W peak), for the second type of 10ms triangular pulse (linearly decreasing voltage, constant current). The current for both triangular pulses is set to 2.3 times that of the rectangular one to guarantee a dissipation of exactly 200mJ of energy.

Fig. 6 showcases the thermal images of the same device; the frames depict the highest peak temperature under the three different load conditions.

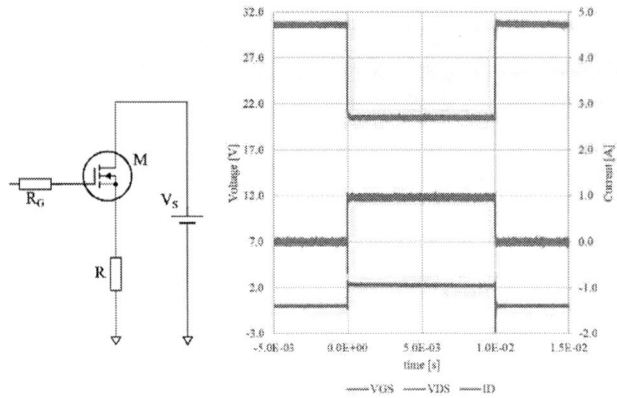

Fig. 3. Schematic for square pulse and waveforms (20V, 1A, 10ms).

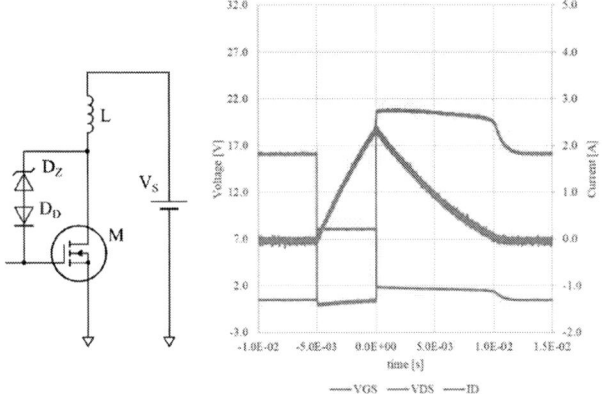

Fig. 4. Schematic for triangular pulse and waveforms - Active clamp (20V, 2.3A, 10ms).

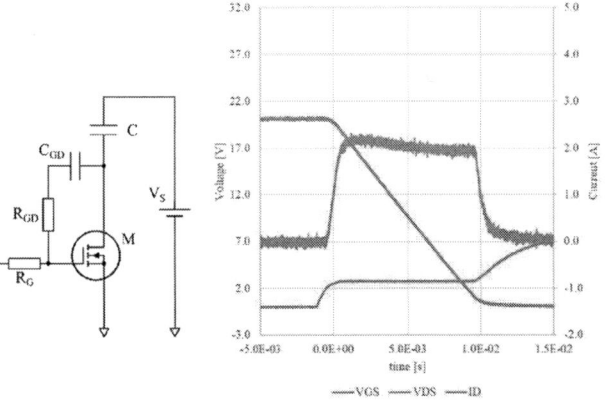

Fig. 5. Schematic for triangular pulse and waveforms – Inrush (20V, 2.3A, 10ms).

The first thing to observe is that the temperature distribution is not uniform across the die area, in fact, a hotspot in red triangle is visible in the bottom left corner of the die in all three frames. This is in line with the observations made by P. Spirito at al. [3] which attributed this phenomenon to the uneven distribution of current across the die causing hotspots. As observed, hotspot formation will be influenced by the leads-to-die bonding

technique with hotspots forming close to the source attachment points.

Comparing the three thermal images in Fig. 6 (a), (b) and (c), the hot spot location for the three different pulsations is almost in the same position and gives rise to a similar peak temperature.

The rectangular pulse with bias condition $V_{DS} = 20V$, $I_D = 1A$, tp = 10ms leads to a peak junction temperature of 80.79°C.

The triangular pulse (linearly decreasing current with constant voltage) with bias condition $V_{DS} = 20V$, $I_D = 2.3A$, tp = 10ms leads to a peak junction temperature of 78.56°C.

The next triangular pulse (linearly decreasing voltage with constant current) with bias condition $V_{DS} = 20V$, $I_D = 2.3A$, tp = 10ms leads to a peak junction temperature of 76.12°C.

The temperature distribution shown in Fig. 6 (a), (b), and (c) is further represented in Fig. 7 (a), (b), and (c). The $640_X \times 512_Y$ array data from the thermal camera are mapped in a 3-D plot, pinpointing the exact location of the hotspot and peak temperature.

Three more samples are tested in the same conditions and the result data are summarised in Table I.

This result provides empirical support for the hypothesis that the SOA current limit for a triangular pulse is around twice that of a rectangular pulse.

TABLE I.   SUMMARY OF RESULTS FOR ADDITIONAL SAMPLES

	Pulse (20V, 10ms)	Max current	Peak Temp	Hotspot location
DUT 2	Sq-pulse	1.02A	83.0 C°	●
	Tri-pulse (Active clamp)	2.27A	77.7 C°	●
	Tri-pulse (Inrush)	2.34A	81.1 C°	●
DUT 3	Sq-pulse	1.03A	74.8 C°	●
	Tri-pulse (Active clamp)	2.32A	74.5 C°	●
	Tri-pulse (Inrush)	2.40A	75.9 C°	●
DUT 4	Sq-pulse	1.05A	77.9 C°	●
	Tri-pulse (Active clamp)	2.23A	73.4 C°	●
	Tri-pulse (Inrush)	2.33A	74.5 C°	●

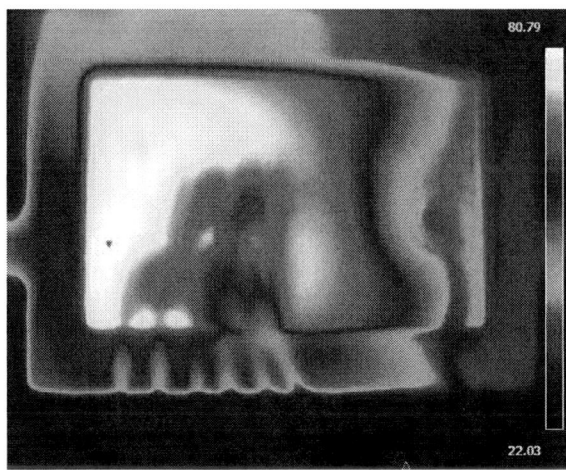

(a) Square pulse (20V, 1A, 10ms)

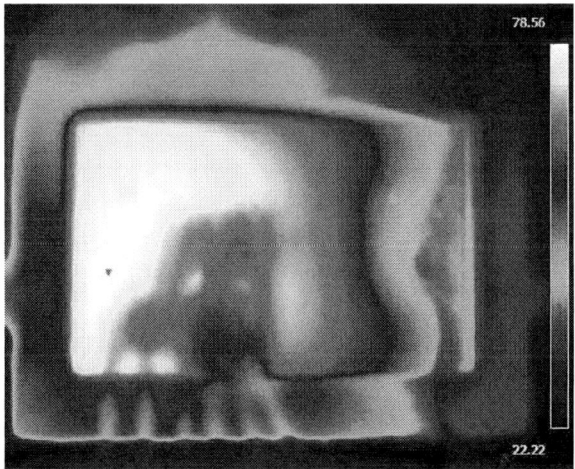

(b) Triangular pulse – Active clamp (20V, 2.3A, 10ms)

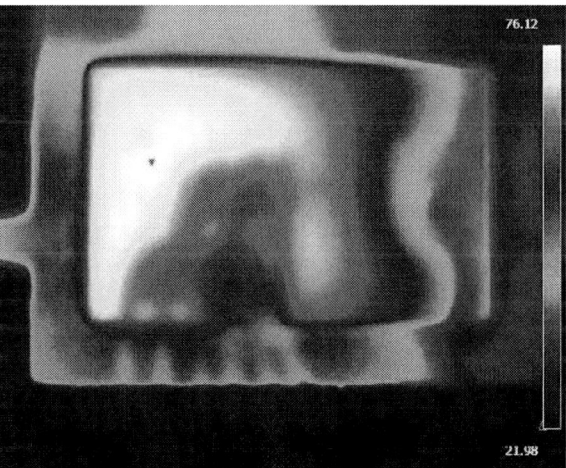

(c) Triangular pulse – Inrush (20V, 2.3A, 10ms)

Fig. 6. DUT thermal mapping using IR camera when the MOSFETs reach the peak temperature.

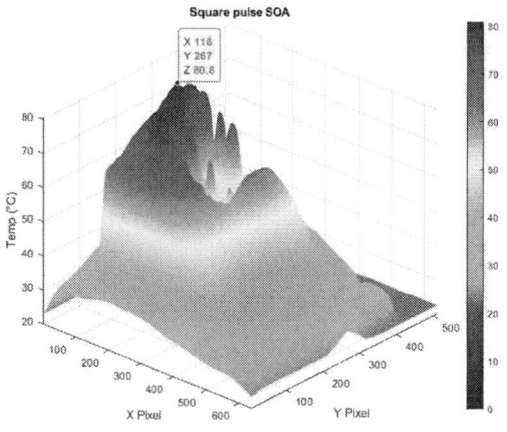

(a) Square pulse (20V, 1A, 10ms)

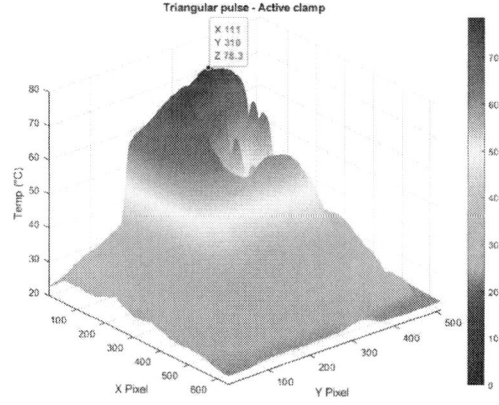

(b) Triangular pulse - Active clamp (20V, 2.3A, 10ms)

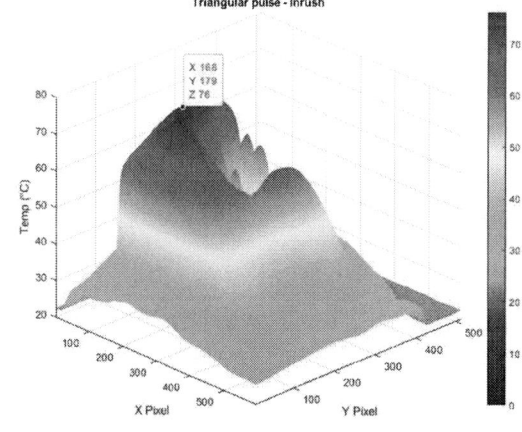

(C) Triangular pulse - Inrush (20V, 2.3A, 10ms)

Fig. 7. 3D thermal map of the die surface at peak temperature.

V. Conclusions and Future Work

This paper explores the thermal analysis of a MOSFET's die operating in a thermally unstable linear-mode. The testing has focused on comparing three different operating conditions (square pulse, triangular pulse - active clamp, and triangular pulse - inrush) proving the 2x SOA capability for the triangular pulse. The use of a high-resolution IR camera provides dynamic information on the thermal stress during operation and as the technique is non-destructive, allows the same device to be tested under different conditions, leading to a more accurate estimation of an SOA for triangular pulses.

Knowing the dynamic thermal behavior of the silicon inside the MOSFET helps to understand the physics of degradation mechanisms if reliability issues arise, especially in repetitive linear-mode operation. A possible future work could be investigating the MOSFET reliability under repetitive linear-mode operation.

References

[1] O. M. Alatise, I. Kennedy, G. Petkos, K. Khan, A. Koh and P. Rutter, "Understanding Linear-Mode Robustness in Low-Voltage Trench Power MOSFETs," in IEEE Transactions on Device and Materials Reliability, vol. 10, no. 1, pp. 123-129, March 2010, doi: 10.1109/TDMR.2009.2036001.

[2] Phil Rutter, "Considerations in the design of a low-voltage power MOSFET technology," 2018 IET Power Electron., 2019, Vol. 12 Iss. 15, pp. 3861-3869

[3] P. Spirito, G. Breglio, V. d'Alessandro and N. Rinaldi, "Thermal instabilities in high current power MOS devices: experimental evidence, electro-thermal simulations and analytical modeling," 2002 23rd International Conference on Microelectronics. Proceedings (Cat. No.02TH8595), Nis, Yugoslavia, 2002, pp. 23-30 vol.1, doi: 10.1109/MIEL.2002.1003144.

[4] C. Radici, V. Satyamsetti, P. Ellis, P. Vines and W. Lawson, "Triangular and Rectangular Power Pulses in Automotive MOSFETs Applications for Thermally Unstable Linear-mode," 2024 IEEE Applied Power Electronics Conference and Exposition (APEC), Long Beach, CA, USA, 2024, pp. 2415-2418, doi: 10.1109/APEC48139.2024.10509288.

[5] G. Breglio, F. Frisina, A. Magri and P. Spirito, "Electro-thermal instability in low voltage power MOS: Experimental characterization," 11th International Symposium on Power Semiconductor Devices and ICs. ISPSD'99 Proceedings (Cat. No.99CH36312), Toron, ON, Canada, 1999, pp. 233-236, doi: 10.1109/ISPSD.1999.764106.

[6] Failure signature of electrical overstress on power MOSFETs AN11243, Rev. 02 — 19 December 2017, https://assets.nexperia.com/documents/application-note/AN11243.pdf.

[7] Frank Liebmann "Emissivity - The Crux of Accurate Radiometric Measurement" , 2009/01/01

Feasibility Study of the SuperIGBT: A Series-Connected High Voltage IGBT With A Single Gate

Junhong Tong
Semiconductor Power Electronics Center (SPEC)
The University of Texas at Austin
Austin, Texas, USA
jredtong@utexas.edu

Alex Q. Huang
Semiconductor Power Electronics Center (SPEC)
The University of Texas at Austin
Austin, Texas, USA
aqhuang@utexas.edu

Huanghaohe Zou
Semiconductor Power Electronics Center (SPEC)
The University of Texas at Austin
Austin, Texas, USA
hzou@utexas.edu

Zhiyuan Ma
Semiconductor Power Electronics Center (SPEC)
The University of Texas at Austin
Austin, Texas, USA
zhiyuanma@utexas.edu

Abstract— **This paper introduces the design and characterization of the SuperIGBT - a series connected high voltage IGBT switch with only one gate. The SuperIGBT uses the same circuit configuration as the previously published Austin SuperMOS. A half-bridge 2.4kV SuperIGBT, which utilizes two 1.2kV Si IGBTs in each switch position, is developed to evaluate the performance. The static and dynamic performance of the 2.4 kV SuperIGBT is analyzed experimentally. Since the IGBT is a bipolar device, achieving good voltage balancing is substantially more challenging than the unipolar MOSFET. A new voltage balancing theory based on the IGBT device physics is developed.**

Keywords—IGBT, medium voltage, series connection, cascode, SuperCascode

I. INTRODUCTION

Power semiconductor devices are recognized as the fundamental elements for all power electronics systems. Si high-voltage semiconductor power devices such as the Insulated Gate Bipolar Transistors (IGBTs) are widely used in many high-power applications such as wind and solar inverters, industrial motor drives, high-voltage DC (HVDC) transmission converters, flexible AC transmission system (FACTS) devices. High voltage IGBTs from 3.3 to 6.5kV have also been developed to support medium voltage applications. However, due to the fabrication limitation and increased Terrestrial Cosmic Radiation failures, medium voltage IGBTs are expensive, and device voltage utilization is lower than for higher voltage devices. Series connection of lower voltage devices provides an alternative method to achieving high blocking voltages or providing better device utilization. The overall switch cost could be reduced. Multiple studies have already achieved series IGBT connections based on conventional individual gate control configurations [1-3]. Although this approach achieves good voltage balancing performance and is a viable option for obtaining a high-voltage switch, it needs an additional driver circuit and power supply for each additional

series connected IGBT. So the auxiliary system complexity increases substantially.

This paper introduced a novel three-terminal IGBT, the SuperIGBT, for the first time. The SuperIGBT is based on the cascode connection of several IGBTs. Only one gate driver is needed for each medium-voltage switch. 2.4kV and 3.6kV prototypes be built by series connected 1.2kV IGBTs. The test results are presented to verify the feasibility. The voltage balancing issue is identified and discussed.

II. SUPERCASCODE BASED ON JFET AND SUPERMOS BASED ON MOSFET

The cascode connection of semiconductor devices originates as a method to improve a linear amplifier. Friedrichs et al [4] extended the cascode concept into a SuperCascode by using the series connection of a number of normally on JFETs switch with a normally-off MOSFET in 2003. Biela and Li et al. [5, 6] studied the SuperCascode dynamic balance issue. 6kV and 15 kV SiC SuperCascode devices are developed and demonstrated by Ni and Song in [7-8]. Zhang et al. extended the concept to the so-called Austin SuperMOS [9,10], in which all devices used are normally off MOSFET while Sen et al.[11] Different voltages for the SiC MOSFET and JFET were used for the SuperMOS design. In addition, medium voltage SuperCascode and SuperMOS modules based on the bare die packaging [10,12] reduce the overall power switch or module footprint. In short, the SuperCascode or SuperMOS all result in a three-terminal high-voltage switch by using either SiC JFET or MOSFET. SiC JFET and MOSFET are unipolar power switches; therefore, their dynamics, especially the dynamic voltage balancing, are determined by the device's parasitic capacitance, which is well characterized and understood[13,14]. Excellent voltage balance can be achieved in these devices with careful design due to the simpler device physics during the turn-off[15,16,17].

979-8-3315-1612-3/25 $31.00 © 2025 IEEE

Table I: Specifications of the 2.4kV/75A SuperIGBT

Parameters	Value
Configuration	SuperIGBT
Number of 1200V IGBTs	2
Rated Peak Voltage	2.4kV
Recommended DC Link Voltage	≤1.8kV
Collector-Emitter on Voltage @75A	3.5V
Balancing cap C1	6.1nF

III. SUPERIGBT CONFIGURATION

Si IGBT is a widely used bipolar power switch with a MOS gate terminal. Therefore, a natural question is whether we can realize three-terminal high-voltage IGBTs by using lower-voltage IGBTs similar to a SuperMOS. The resulting device, called SuperIGBT, is shown in Fig. 1 for a 2.4 kV SuperIGBT and in Fig.2 for a 3.6 kV SuperIGBT, both based on 1200V IGBT. Two SuperIGBTs are further connected as a half-bridge for experimental evaluation. Table I summarizes the key specifications of the developed 2.4kV SuperIGBT.

A. Configuration of the SuperIGBT

As the schematic in Fig.1 shows, two Infineon 1200V TO247 Si IGBT discrete devices, IKQ75N120CS7, are configured as a SuperIGBT to achieve the 2.4kV blocking voltage. R1~R2 are

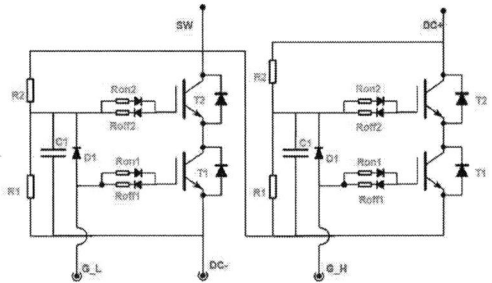

Figure 1: 2.4kV IGBT Cascode half-bridge schematic.

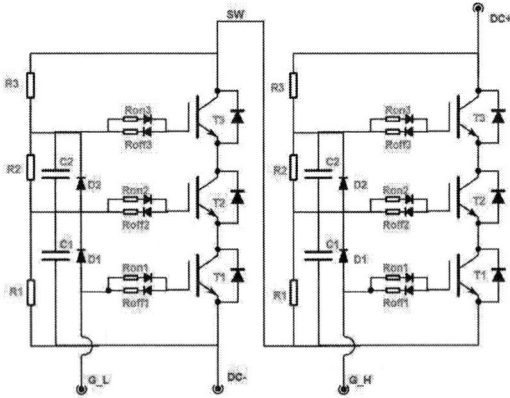

Figure 2: 3.6kV SuperIGBT half-bridge schematic

Figure 3. Half-bridge SuperIGBT test board

used for static voltage balancing, and C1 is used for dynamic voltage balancing. Rgon1, Rgon2, Rgoff1, and Rgoff2 are damping resistors that limit the gate loop charging and discharging currents in the C1 capacitor branch during the turn-on and turn-off transients. D1 is an avalanche diode for voltage clamping and enhancing the turn-on performance. More IGBTs can be used in a similar fashion to form a higher voltage SuperIGBT, as shown in Fig.2 for a 3.6kV SuperIGBT.

B. Operation principle of the SuperIGBT

As a three-terminal switch, the SuperIGBT only needs one gate to control the series of connected devices. In the on-state, a positive gate voltage is applied to all IGBTs through the diode D1, D2 etc. that are connected between the gates of the devices. To turn off the SuperIGBT, a negative gate driver signal (−5 V) is applied to the bottom IGBT (T1) gate, and T1 is turned off. Then, T1's Vce increases, boosting the potential of T2's emitter, resulting in a negative voltage across T2's Gate to Emitter. Once T2's Vge voltage reaches its threshold voltage, T2 is turned off. Similarly, this principle can be used for higher voltage SuperIGBT.

A critical issue for the proposed SuperIGBT is the voltage sharing during device turn-off. R1~R2 achieves static voltage sharing, and the design of this part is straightforward and is determined by the typical leakage current of the IGBT. The C1 is used to achieve good dynamic voltage balancing. When the T1 turns off, the inductive load current shifts to two paths: one path charges the T1 collector-emitter, which can be represented by a dynamic capacitance Cies, and the other discharges Cge of T2 in series with the C1 and Rgoff2. Similarly, for multiple IGBT series conditions, during the other IGBTs turn-off, other balance capacitors will be charged.

IV. PERFORMANCE EVALUATION AND DISCUSSION

A special half-bridge tester, shown in Fig. 3, was developed for the static and dynamic evaluation of the SuperIGBT. In this paper, we will focus on the test results for the 2.4 kV SuperIGBT.

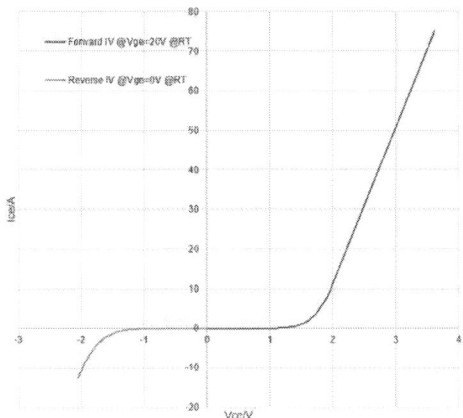

Figure 4: I-V curves of 2.4kV SuperIGBT in the first and third quadrants at room temperature

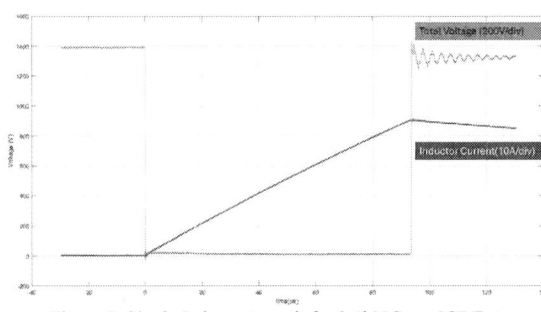

Figure 5: Single Pulse test result for 2.4kV SuperIGBT at 1.4kV/45A

A. Static performance

As a series switch, the SuperIGBT's static performance is directly related to the low-voltage discrete devices used. The effective I−V curves, including the forward and reverse conductions of the 2.4kV SuperIGBT, are characterized at room temperature, as shown in Fig.4. The forward voltage drop is around 3.5V when Ic=75A and Vge=20V. And the transconductance is around 40S. For the reverse conduction with a T1 gate bias Vgs=0V, the knee voltage is also around 1.5V, determined by the two IGBT's antiparallel diodes.

B. Dynamic performance

Fig.5 and Fig.6 show the switching performance of the 2.4kV SuperIGBT at 1.4kV/45A. The dV/dt for turning on and off from 10% to 90% voltage is 12.99V/ns and 8V/ns separately. Fig.6 shows the dynamic voltage sharing performance during the 1.4kV/45A switching. The voltages of each IGBT's collector and emitter are measured and plotted. The dynamic voltage balancing is well achieved at turn-on and turn-off transients at this current level. However, unlike the traditional JFET-based SuperCascode or MOSFET-based SuperMOS, there are a number of challenges for the SuperIGBT. One of the most critical ones is the current-dependent voltage balancing performance.

C. Current Dependent Voltage Balance

The voltage balance performance of the 2.4kV SuperIGBT is shown in Fig.7 when the load current varies. The voltage

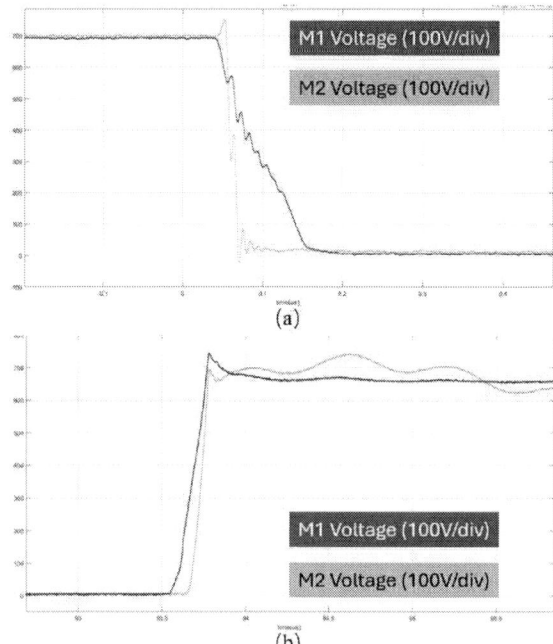

Figure 6: Dynamic voltage sharing at 1.4kV/45A (a)turn-on (b) turn-off

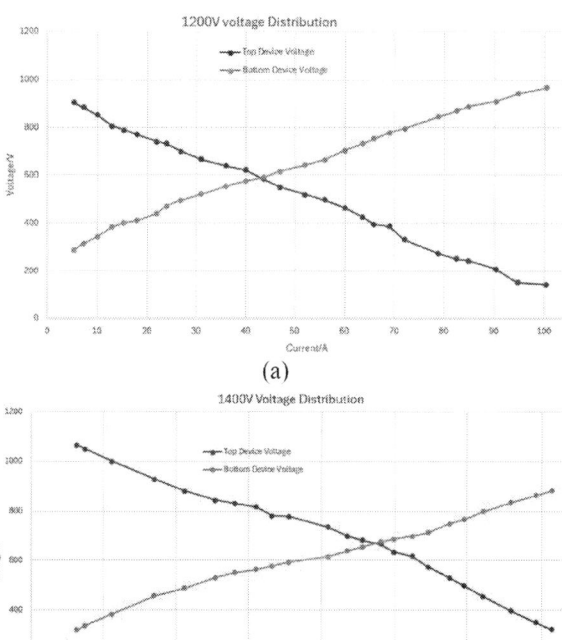

Figure 7: 2.4kV SuperIGBT dynamic voltage balance as a function of the currents at (a)1200V and (b)1400V

sharing between the two devices changes substantially when the load current increases. The voltage of the bottom device becomes higher and higher as the load current increases. For

Vdc=1200V, a perfect balance condition is reached at a load current of 45A. For Vdc=1400V, this crossover point is at 48A.

Traditional SuperCascode or SuperMOS use SiC JFET and MOSFET; when the bottom device turns off, the inductive load current shifts to two paths: one path charges the bottom device Coss, and the other one discharges Cgs of the upper device and charges C1. Therefore, the balance capacitances C1 are usually designed according to Eq. (1):

$$C1=Qg/(\text{desired balanced voltage}) \qquad (1)$$

where Qg is the gate charge of the JFET or MOSFET, which does not change with the load current. The design based on Eq. (1) works very well for voltage balance performance across a wide range of load currents. This equation's basis is that most of the load current is flowing through the C1 path instead of the T1 path.

For the IGBT, due to the bipolar current conduction mechanism during the ON state and during the turn-off, the device cannot be modeled by a simple capacitor or a traditional load-current independent Qg. To gain a better understanding, Fig.8 shows the measured Qg and Qcg of the Infineon 1200V Si IGBT IKQ75N120CS7 as a function of the load current obtained during the device inductive turn-off by measuring the differential of gate current integral. For a load current 5A, Qg is equal to 0.68uC, and Qcg is equal to 0.125uC. The IGBT's Qg decreases with the load current increase, while the IGBT's Qcg shows almost no change. Therefore, at least a load-dependent Qg will have to be used in (1) to obtain the correct C1 value. So, only one load current condition will result in a good voltage balance condition. For a given C1 value, a higher load current should result in a lower C1 voltage, which is the bottom IGBT voltage. This is clearly opposite from the result shown in Fig.7. So (1) can't be used for the SuperIGBT and a new explanation must be developed.

To understand this, one must review the fundamental difference between the IGBT and MOSFET turn-off. MOSFET is characterized as a small capacitance device determined by its junction capacitance. On the other hand, IGBT is a bipolar device with a large amount of stored charge, hence a large load current dependent capacitance. So, during the on state, a large amount of charge is stored in the device Q_{IGBT}. This charge is a function of device design, carrier lifetime, and the load current. During the turn-off, the inductive load current, or more precisely, a portion of the load current, is responsible for removing this large charge during the IGBT's voltage rise phase and determining the IGBT's dV/dt. This percentage of the current that is responsible for the removal of Q_{IGBT} is related to $(1-\alpha_{pnp})*Ic$. Here α_{pnp} is the common base current gain of the internal IGBT parasitic PNP transistor. α_{pnp} is typically governed by the collector injection efficiency and device structure (i.e. trench or planar). The current gain generally increases as the voltage rises. So, there is a certain level of complexity in modeling the current that is responsible for the IGBT charge removal during the first phase of the IGBT inductive turn-off. In modern IGBts, the second phase (current tail phase) is responsible for the removal of only a small amount of charges left in the buffer region of the field stop IGBT.

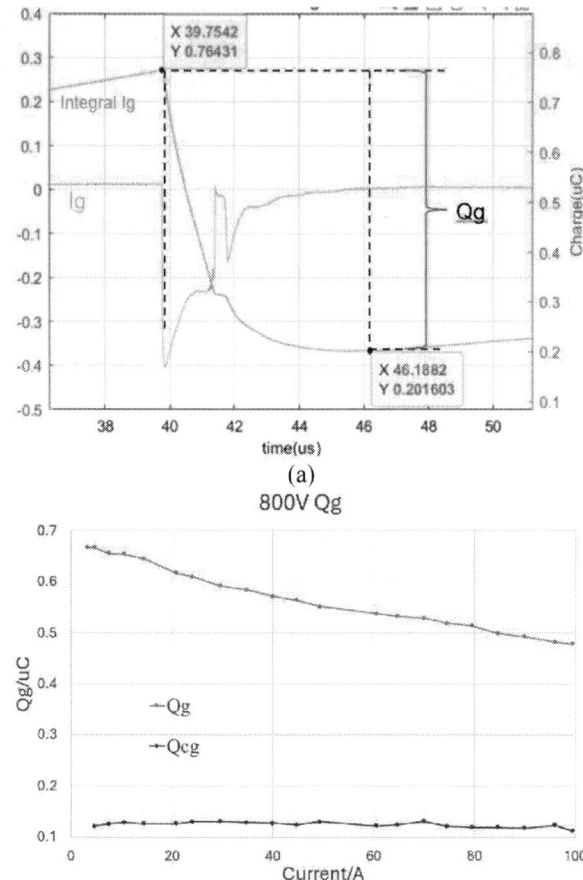

(a)
800V Qg

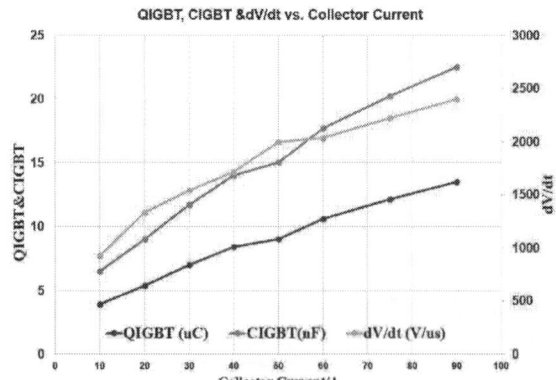

(b)

Figure 8: (a)Measured gate current during IGBT turn-off and current integral. (b) Measured Qg & Qcg of the IGBT IKQ75N120CS7 at 800V turn-off

In general, the Q_{IGBT} increases as the load current increases, but the increase is sublinear, as shown in Fig.9 for the studied Infineon IGBT. The Q_{IGBT} is extracted using the datasheet value of the voltage rise time tr, and an assumed $\alpha_{pnp}=0.4$. The slope of Q_{IGBT} vs Ic is about 0.125uC/A. On the other hand, the device turn-off dV/dt is proportional to the current $(1-\alpha_{pnp})*Ic$. So dV/dt $\propto (1-\alpha_{pnp})*Ic/Q_{IGBT}$ is actually getting higher at a higher current,

Figure 9: Q_{IGBT} and C_{IGBT} vs. collector current extracted from the datasheet.

as shown in Fig.9. An equivalent IGBT capacitance C_{IGBT} can also be extracted and is also shown in Fig.9. As can be seen, the IGBT's turn-off dynamics can be represented as a current-dependent large diffusion capacitance C_{IGBT}.

Once the capacitor C1 is chosen in the SuperIGBT, its final voltage is the same as the bottom device voltage. It is determined by the dV/dt of the bottom IGBT with C1 acting as a dV/dt snubber with this relationship $dV/dt \propto (1-\alpha_{pnp})*Ic /[C_{IGBT}+ C1]$. This is why the voltage for the bottom device in Fig.7 is higher as the load current increases, even if the C_{IGBT} also increases. This phenomenon means good balancing can only be achieved for a narrow range of SuperIGBT turn-off currents. An example of such a condition can be found in the series resonant converter, in which the turn-off current is typically equal and independent of the converter load power.

CONCLUSIONS

A novel SuperIGBT switch using series connections of 1.2kV IGBTs to achieve higher voltage has been designed and tested for the first time. Load current-dependent dynamic voltage balance is observed for the first time, and a qualitative explanation is provided. The root course is the charge storage effect of IGBT and carrier removal process, which are both strongly dependent on current.

REFERENCES

[1] P. Briff, J. Chivite-Zabalza, J. Nicholls and K. Vershinin, "Turn-Off Delay Compensation of Series-Connected IGBTs for HVDC Applications," in IEEE Transactions on Power Electronics, vol. 35, no. 11, pp. 11294-11298, Nov. 2020

[2] S. Ji, T. Lu, Z. Zhao, H. Yu and L. Yuan, "Series-Connected HV-IGBTs Using Active Voltage Balancing Control With Status Feedback Circuit," in IEEE Transactions on Power Electronics, vol. 30, no. 8, pp. 4165-4174, Aug. 2015

[3] T. C. Lim, B. W. Williams, S. J. Finney and P. R. Palmer, "Series-Connected IGBTs Using Active Voltage Control Technique," in IEEE Transactions on Power Electronics, vol. 28, no. 8, pp. 4083-4103, Aug. 2013

[4] P. Friedrichs, H. Mitlehner, R. Schorner, K. . -O. Dohnke, R. Elpelt and D. Stephani, "Stacked high voltage switch based on SiC VJFETs," ISPSD '03. 2003 IEEE 15th International Symposium on Power Semiconductor Devices and ICs, 2003. Proceedings., Cambridge, UK, 2003, pp. 139-142

[5] J. Biela, D. Aggeler, D. Bortis and J. W. Kolar, "Balancing circuit for a 5kV/50ns pulsed power switch based on SiC-JFET Super Cascode," 2009 IEEE Pulsed Power Conference, Washington, DC, USA, 2009, pp. 635-640

[6] X. Li, H. Zhang, P. Alexandrov and A. Bhalla, "Medium voltage power switch based on SiC JFETs," 2016 IEEE Applied Power Electronics Conference and Exposition (APEC), Long Beach, CA, USA, 2016, pp. 2973-2980

[7] X. Ni, R. Gao, X. Song, A. Q. Huang and W. Yu, "Development of 6kV SiC hybrid power switch based on 1200V SiC JFET and MOSFET," 2015 IEEE Energy Conversion Congress and Exposition (ECCE), Montreal, QC, Canada, 2015, pp. 4113-4118

[8] X. Song, A. Q. Huang, L. Zhang, P. Liu and X. Ni, "15kV/40A FREEDM super-cascode: A cost effective SiC high voltage and high frequency power switch," 2016 IEEE Energy Conversion Congress and Exposition (ECCE), Milwaukee, WI, USA, 2016, pp. 1-8

[9] L. Zhang, S. Sen and A. Q. Huang, "7.2-kV/60-A Austin SuperMOS: An Intelligent Medium-Voltage SiC Power Switch," in IEEE Journal of Emerging and Selected Topics in Power Electronics, vol. 8, no. 1, pp. 6-15, March 20207G.

[10] J. Tong, R. Yu, S. Sen and A. Q. Huang, "Packaging and Characterization of a Novel 7.2kV/85A SiC Austin SuperMOS Half-Bridge Intelligent Power Module (IPM)," 2022 IEEE Applied Power Electronics Conference and Exposition (APEC), Houston, TX, USA, 2022, pp. 1720-1724, doi: 10.1109/APEC43599.2022.9773740.

[11] S. Sen, J. Tong, Z. Guo and A. Q. Huang, "Design and Characterization of 4.5kV/ $15\mathrm{m}\Omega$ SiC SuperMOS Half-bridge Module," 2022 IEEE Applied Power Electronics Conference and Exposition (APEC), Houston, TX, USA, 2022, pp. 957-961, doi: 10.1109/APEC43599.2022.9773712.

[12] J. Tong, R. Yu and A. Q. Huang, "Design, Packaging, and Characterization of a 6.8kV/160A SiC SuperCascode Half-Bridge Module for Medium Voltage Applications," 2023 IEEE Energy Conversion Congress and Exposition (ECCE), Nashville, TN, USA, 2023, pp. 5538-5542, doi: 10.1109/ECCE53617.2023.10362553

[13] B. Gao, A. J. Morgan, Y. Xu, X. Zhao and D. C. Hopkins, "6.0kV, 100A, 175kHz super cascode power module for medium voltage, high power applications," 2018 IEEE Applied Power Electronics Conference and Exposition (APEC), San Antonio, TX, USA, 2018, pp. 1288-1293, doi: 10.1109/APEC.2018.8341182

[14] B. Gao, A. Morgan, Y. Xu, X. Zhao, B. Ballard and D. C. Hopkins, "6.5kV SiC JFET-based Super Cascode Power Module with High Avalanche Energy Handling Capability," 2018 IEEE 6th Workshop on Wide Bandgap Power Devices and Applications (WiPDA), Atlanta, GA, USA, 2018, pp. 319-322, doi: 10.1109/WiPDA.2018.8569146

[15] L. Pang, T. Long, K. He, Y. Huang and Q. Zhang, "A Compact Series-Connected SiC MOSFETs Module and Its Application in High Voltage Nanosecond Pulse Generator," in IEEE Transactions on Industrial Electronics, vol. 66, no. 12, pp. 9238-9247, Dec. 2019, doi: 10.1109/TIE.2019.2891441

[16] L. Gill, L. A. G. Rodriguez, J. Mueller and J. Neely, "A Comparative Study of SiC JFET Super-Cascode Topologies," 2021 IEEE Energy Conversion Congress and Exposition (ECCE), Vancouver, BC, Canada, 2021, pp. 1741-1748, doi: 10.1109/ECCE47101.2021.9595065

[17] B. Hu et al., "Characterization and evaluation of 4.5 kV 40 A SiC super-cascode device," 2017 IEEE 5th Workshop on Wide Bandgap Power Devices and Applications (WiPDA), Albuquerque, NM, USA, 2017, pp. 321-326, doi: 10.1109/WiPDA.2017.8170567.

LOW PROFILE, LAMINATED NIFE TRANSFORMERS FOR FLYBACK CONVERTERS

Xuan Wang
*Department of Electrical and
System Engineering
University of Pennsylvania*
Philadelphia, PA, USA
wxuan@seas.upenn.edu

Reza Mounesi
*Department of Electrical
Engineering
University of South Carolina*
Columbia, SC, USA
amounesi@email.sc.edu

Matthew Catanoso
*Department of Electrical and
System Engineering
University of Pennsylvania*
Philadelphia, PA, USA
mattcat@seas.upenn.edu

Matthew Fox
*Department of Electrical and
System Engineering
University of Pennsylvania*
Philadelphia, PA, USA
foxmatt@seas.upenn.edu

Adel Nasiri
*Department of Electrical
Engineering
University of South Carolina*
Columbia, SC, USA
nasiri@mailbox.sc.edu

Mark G. Allen
*Department of Electrical and
System Engineering
University of Pennsylvania*
Philadelphia, PA, USA
mallen@seas.upenn.edu

Abstract—We introduce a flyback transformer with a laminated NiFe magnetic core, fabricated using a CMOS-compatible microfabrication technique involving sequential multilayer electrodeposition, and demonstrate its application in a flyback AC/DC converter with constant output voltage control operating at 500 kHz. The core adopts a 'UT' shape to facilitate winding, as opposed to the traditional EI shape. The transformer includes a 45-turn primary winding and a 3-turn secondary winding, featuring a magnetizing inductance of 317 µH and a coupling coefficient of 0.98. The total transformer core thickness was 4.2mm. Electrical performance was evaluated using a customized flyback circuit on a printed circuit board (PCB), which provided a stable isolated 5 V, 1 A DC output from a 60 Hz sinusoidal AC input ranging from 80 V_{rms} to 220 V_{rms}. The system achieved an end-to-end efficiency of 80% with a 120 V_{rms} input and maintained over 68% efficiency across the entire input range.

Keywords—laminated magnetic core, sequential multilayer electrodeposition, flyback transformer, AC/DC converter

I. INTRODUCTION

With the rapid development of modern electronic devices, the demand for efficient universal input AC to DC adaptors and chargers is steadily growing, particularly for lightweight converters with higher power density. Flyback converters are widely favored for low-power applications due to their simple peripheral circuitry, minimal component count, control simplicity, and ability to provide electrical isolation [1]-[4]. In a flyback converter, the transformer serves as a critical component. Unlike conventional transformers, which directly transfer the energy in a forward fashion, the flyback transformer initially stores energy in the primary side's magnetic field (mainly within the airgap) during the on-cycle and subsequently releases it to the secondary side during the off-cycle [5]-[6]. This

operating mechanism is fundamental to the flyback topology, enabling effective energy transfer while maintaining isolation; however, it also imposes specific requirements for the transformer regarding energy storage.

As the main energy storage element, the transformer in a flyback converter often occupies significant space, posing a challenge to achieving a compact design. To minimize size, one effective strategy is to increase the switching frequency, which decreases the necessary inductance and enables the use of smaller transformer cores [7]-[8]. Recent advancements in wide-bandgap semiconductors, such as gallium-nitride (GaN), have enabled operation at frequencies within the MHz range, contributing to a reduction in the size of passive components [9]-[11]. Another approach focuses on the core material selection. At present, ferrite cores are commonly utilized in flyback transformers due to their low loss tangent, which effectively reduces core losses at high frequencies [12]. However, ferrite cores often face a trade-off between operating frequency and saturation flux density, with most high-frequency ferrites exhibit saturation flux densities below 500 mT. Well-known products include PC200 from TDK[13], 79 from Fair-Rite [14], and 4F1 form Ferroxcube [15]. This limitation may constrain the operating flux and power density, thereby hindering further miniaturization. In contrast, metallic alloys, such as permalloy ($Ni_{80}Fe_{20}$), offer higher achievable power density due to their higher saturation flux density. However, their considerably lower electrical resistivity can lead to significant eddy current loss at high frequencies (0.1-10 MHz), especially when the core thickness exceeds the skin depth of the material [16]-[17].

To address this issue, an alternating electroplating technique for metallic layers (NiFe) and insulation layers (polypyrrole, abbreviated as PPy) to suppress eddy current loss has been proposed [18]-[19]. The thickness of individual metallic layers can then be small compared to the skin depth, while the total thickness of the multilayer stack can be larger for higher power handling. While millimeter-scale plating molds for such stacks can be employed if desired [20], such molds can be challenging to fabricate. To streamline this process, surface tension-driven self-assembly can be used to efficiently stack individual multilayer subcomponents, facilitating the construction of thicker transformer cores that meet both the inductance and power-handling requirements while reducing overall fabrication time [21]. Moreover, by tuning electroplating parameters such as current density and plating duration, the thickness of individual NiFe layer can be customized from micrometer to nanometer scale, ensuring compatibility across a range of switching frequencies, especially as designs move toward high-frequency applications.

In this paper, we focus on the design, fabrication and performance validation of a laminated NiFe transformer, measuring only 4.2 mm in thickness and fabricated through a combined approach of alternating electroplating and self-assembly. This transformer was integrated into a universal input flyback converter operating at 500 kHz, delivering a 5 W (5V, 1A DC) output with an end-to-end efficiency of 80% at 120 V_{rms} input voltage and maintaining over 68% efficiency across the input voltage range from 80 V_{rms} to 220 V_{rms}. This technique offers a promising solution to the challenges of high-frequency operation and compact converter designs, providing insights into the applicability of laminated NiFe transformers for high power density converter applications.

II. TRANSFORMER DESIGN AND FABRICATION

Various studies [1], [5], [8], [22]-[27] provide design guidelines for flyback transformers. The transformers must achieve targeted values of magnetizing inductance L_m and a certain turn ratio. In addition, achieving a well-coupled transformer is essential, as high leakage inductance not only prevents efficient energy transfer between windings, reducing overall efficiency, but also generates voltage spikes, causing high voltage stress on the switch. As discussed above, laminated NiFe possesses high saturation flux density and the ability to be customizable to desired thicknesses for various operation frequencies. In addition, its high permeability [28] enables more effective concentration of magnetic flux, thereby enhancing coupling between the primary and secondary windings. With these design considerations in mind, the following sections describe the fabrication of the laminated NiFe transformer, covering both core construction and winding strategies to simultaneously achieve both compact dimensions and targeted electrical specifications.

A. Laminated NiFe core design and fabrication

The three-dimensional and planar schematics of the NiFe core are depicted in Fig. 1(a) and Fig. 1(b), respectively. The core was designed in a 'UT' shape instead of the common 'EI' shape to simplify winding on the central 'T' part for ease of testing. An air gap was introduced to prevent core saturation and to store the required energy for flyback operation. To mitigate additional eddy current losses resulting from electrical currents crossing the layers due to the non-zero conductivity of the

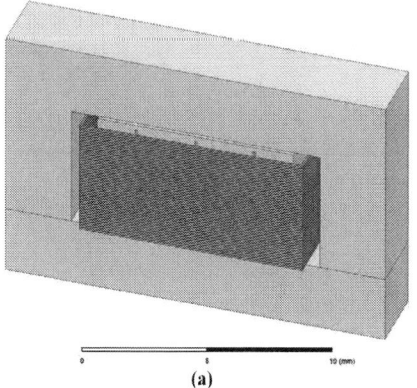

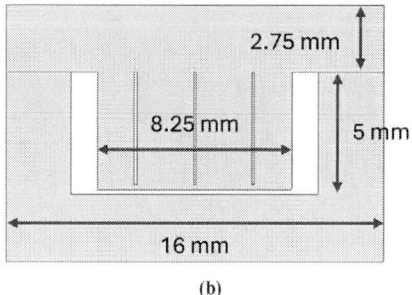

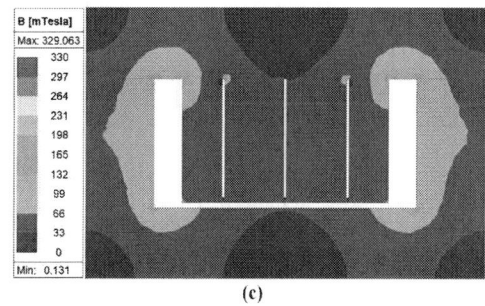

Fig. 1. Finite element simulation model of the transformer. (a) 3D representation of the transformer with windings; (b) Planar schematic of 'UT' cores with dimensions; (c) Magnetic field distribution across the cross-section with an excitation current of 150 mA applied to a 45-turn winding.

interlamination insulation, slits were incorporated in the central 'T' region. These slits segmented the region into several subregions, each with a width smaller than the critical threshold [29], thereby minimizing additional eddy current effects.

A three-dimensional finite element model, developed with Ansys Maxwell3D (Fig. 1), was used to determine the dimensions and winding turns necessary for achieving a magnetizing inductance L_m (300 μH) and a 15:1 turn ratio, tailored for the flyback circuit. This model also helped optimize the core geometry to prevent core saturation with a 150 mA input current, another desired parameter for the flyback circuit.

With the design parameters determined, the fabrication process for the laminated NiFe core was carried out as shown in Fig. 2. First, a seed layer of 200 nm Ti and 500 nm Cu was sputtered onto a 4-inch fused silica wafer, followed by the

979-8-3315-1612-3/25 $31.00 © 2025 IEEE

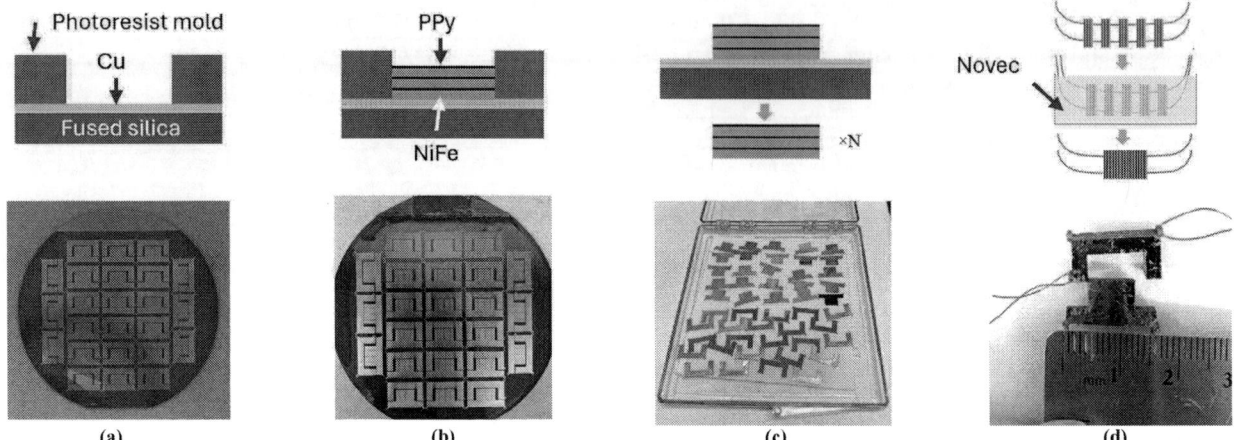

Fig. 2. Fabrication process of the laminated NiFe core. (a) Mold formation; (b) Sequential electrodeposition; (c) Mold removal and multilayer constructs release; (d) Self-assembly.

lithographic patterning of an electrically insulating photoresist mold (AZ4620, EMD Performance Materials) to define the lateral extents of the multilayer cores [see Fig. 2(a)]. Subsequently, alternating sequential electrodeposition of NiFe and interlayer polypyrrole (PPy) was performed within the mold using an automated plating robot [see Fig. 2(b)]. The electrodeposition process utilized a three-bath system, starting with the deposition of the NiFe layer, followed by the polypyrrole layer, and concluding with a thin Ni strike layer to activate the polymer surface for subsequent NiFe electrodeposition. An automated robot facilitated the transfer of the wafer between these three baths, repeating the electrodeposition cycle until the desired numbers and thicknesses of layers are achieved. Detailed electrodeposition procedures, including bath compositions and plating current densities, are provided in [18]. Afterwards, the photoresist mold was dissolved using acetone, and the NiFe/PPy multilayer constructs were detached from the wafer substrate by etching the

Cu seed layer using a commercial copper etchant (Copper Etch BTP, Transene) [see Fig. 2(c)]. Finally, a self-assembly approach was exploited to transform the multilayer constructs into a thick core [see Fig. 2(d)]: the commercial magnet wires secured the multilayer constructs in place through guiding holes as they were immersed in Novec 2704 solution; as the solvent dried, the constructs were coated with fluoropolymer insulation and self-aligned, resulting in a polymer-insulated assembly. Additional details regarding the surface tension-driven assembly can be found in [21].

The fabricated NiFe core is shown in Fig. 3, featuring a footprint of 16 mm by 10.5 mm and a thickness of 4.2 mm. The individual thickness of each NiFe layer was targeted at 5 μm, with the goal of achieving a thickness smaller than the skin depth at 500 kHz. To show the cross-sectional detail of the multilayer, a 5-layer NiFe/PPy stack was milled using a focused ion beam (FIB) and characterized through scanning electron microscopy (SEM). The FIB-SEM image indicates that actual measurements

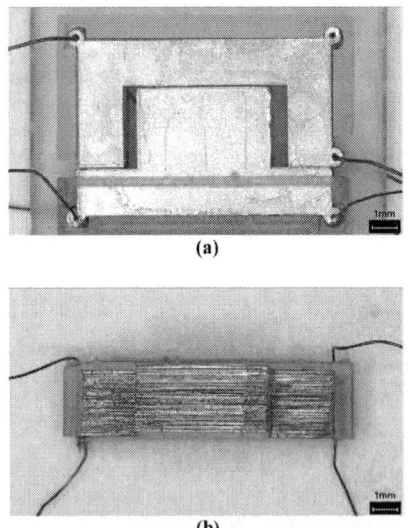

Fig. 3. Optical images of the laminated NiFe core. (a) Top-down view; (b) Side view.

Fig . 4. FIB-SEM image of a 5-layer NiFe/PPy laminated stack.

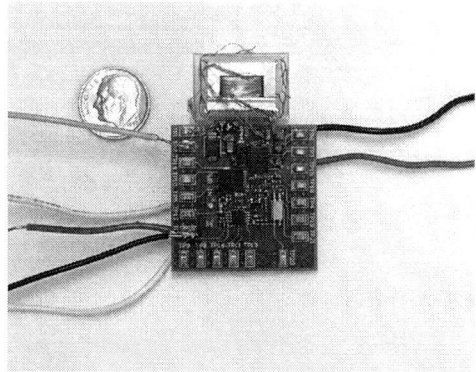

Fig. 5. The fabricated transformer soldered onto the customized PCB.

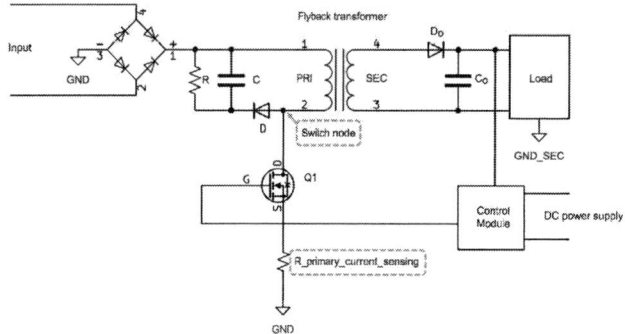

Fig. 7. Simplified circuit with switch node and primary current sensing resistor highlighted.

Fig. 6. Transformer and PCB under test with an AC power input and osilliscope traces showing various voltages and current in the circuit.

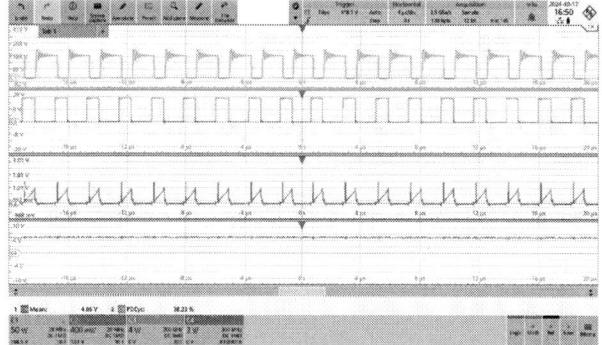

Fig. 8. Oscilloscope waveforms captured during transformer operation with a 60 Hz, 120 V_{rms} input and 5 V, 1 A output. Signals from top to bottom: voltage at the switch node (50 V/div), PWM signal from the controller (4 V/div), voltage across the 3.3 Ω sensing resistor (400 mV/div), and output voltage (2 V/div). Horizontal axis: 4 μs/div.

of the NiFe layer thickness typically ranged around 4.6 μm, as shown in Fig. 4.

B. Winding

Based on the simulation results and required turns ratio, the transformer design incorporates a primary winding of 45 turns and a secondary winding of 3 turns. Various winding and electrical isolation strategies were explored, leading to a final design where a single-layer secondary winding encircled a single-layer primary winding. The windings were wound onto a 3D-printed bobbin successively, with the primary winding placed first, followed by a thin layer of Kapton tape for isolation, and then the secondary winding wrapped closely around the primary. The assembled bobbin was slotted onto the central leg of the 'T' core, and the 'U' core was then positioned to complete the magnetic path, thus forming the transformer. The bobbin design not only protected the laminated core during the winding but also created a margin between the conductors and the air gap, reducing the effect of fringing flux on the conductors current distribution [5].

The primary winding was wound using a single strand of 39-gauge enameled copper wire, with each turn arranged in an orderly fashion to fill the winding window. Given the large 15:1 turns ratio required, achieving optimal coupling with a single wire for the secondary winding would be challenging due to the disparity in turns. To address this, the secondary winding was constructed with nine parallel strands of 38-gauge enameled copper wire, effectively increasing coverage over the primary

layer and enhancing coupling. The use of parallel strands for the secondary also lowered the DC resistance, thereby reducing conduction losses. Although a sandwiched interleaving approach (positioning the secondary layer between two primary layers) could further reduce leakage inductance, it would also significantly increase the parasitic capacitance between the windings [22]. The fabricated laminated NiFe transformer demonstrated a magnetizing inductance of 317 μH at 500 kHz and a coupling coefficient of 0.98, measured using an LCR meter (IM3536, Hioki).

III. EXPERIMENTAL RESULTS

The performance of the laminated NiFe transformer was demonstrated using a customized AC to DC flyback converter (Fig. 5). This converter was based on a commercial controller chip (UCC28C42DGK, Texas Instruments) to provide a stable 5 V, 1 A output. The experimental setup, depicted in Fig. 6, included an AC power source (IT7321, iTech) connected to the converter input, a DC power source (2231A-30-3, Keithley) connected to the controller, and a 5 Ω resistor connected to the output.

The transformer and converter provided a stable 5 W, 5 V DC output with a 60 Hz sinusoidal input ranging from 80 V_{rms} to 220 V_{rms}. The system achieved an end-to-end efficiency of 80% at 120 V_{rms} input and maintained above 68% efficiency across the entire input range. The output voltage measured at the load resistor was 4.86 V, with the slight deviation from the 5 V target attributed to lead wire resistance.

Fig.7 presents the simplified circuit schematic, and Fig. 8 shows the waveforms captured using an oscilloscope during operation with a 120 V_{rms} input. From top to bottom, the waveforms display the switch node voltage, the PWM signal from the controller, the voltage across the current-sensing resistor, and the output voltage.

In ideal flyback operation, the current in the primary winding ramps up while the switch is on, with the slope dependent on the input voltage and primary inductance; during the switch-off period in discontinuous conduction mode, this current should ramp down to zero. However, the ramp-down phase was not visible in our case because the voltage across the current sensing resistor dropped to zero when the switch turned off. In addition, the flyback converter exhibited a turn-on spike during switching-on transitions, caused by the reverse recovery current of the output diode and the parasitic capacitance of the transformer [4]. Feedback control also caused noticeable changes in the duty cycle in response to input voltage change, as expected. Overall, the measured waveform was consistent with our expectations, confirming the proper functionality of the transformer within the circuit.

IV. CONCLUSION

A CMOS-compatible microfabrication method was employed to fabricate laminated NiFe cores for a flyback transformer using sequential multilayer electrodeposition. The transformer was evaluated in a universal input AC to DC converter with constant 5 V DC output voltage control, successfully delivering 5 W output with an end-to-end efficiency of 80% at a 120 V_{rms} input voltage and maintaining over 68% efficiency across the input voltage range of 80 V_{rms} to 220 V_{rms}. Such laminated NiFe cores, developed through CMOS-compatible microfabrication, emerge as a promising alternative to traditional ferrite cores for miniaturized systems, providing significant benefits in terms of size reduction and integration capability.

To enhance the performance of laminated NiFe cores, future work can focus on reducing core losses. This can be achieved by using thinner laminations to further minimize eddy current losses and applying a magnetic field during electroplating to align magnetic domains in order to reduce hysteresis losses [30]-[31]. Additionally, integrating laminated transformer core fabrication with PCB processes can streamline the winding process and improve the arrangement of windings, potentially reducing parasitic effects [32].

ACKNOWLEDGMENT AND DISCLAIMER

This effort was sponsored in whole or in part by the Central Intelligence Agency (CIA), through CIA Federal Labs. The U.S. Government is authorized to reproduce and distribute reprints for Governmental purposes notwithstanding any copyright notation thereon. The views and conclusions contained herein are those of the authors and should not be interpreted as necessarily representing the official policies or endorsements, either expressed or implied, of the Central Intelligence Agency.

REFERENCES

[1] S. Zhao, J. Zhang, and Y. Shi, "A low cost low power flyback converter with a simple transformer," Proceedings of The 7th International Power Electronics and Motion Control Conference, Harbin, 2012, pp. 1336–1342.

[2] U. S. Padiyar and V. Kamath, "Design and implementation of a universal input flyback converter," 2016 International Conference on Electrical, Electronics, and Optimization Techniques (ICEEOT), Chennai, India, 2016, pp. 3428–3433.

[3] I. Ahmad and B. G. Fernandes, "Concept of Universal USB Charger," 2020 IEEE Industry Applications Society Annual Meeting, Detroit, MI, USA, 2020, pp. 1–5.

[4] T.-J. Liang, K.-H. Chen, and J.-F. Chen, "Primary side control for flyback converter operating in DCM and CCM," IEEE Transactions on Power Electronics, vol. 33, no. 4, pp. 3604–3612, Apr. 2018.

[5] R. Prieto, J. A. Cobos, O. Garcia, R. Asensi, and J. Uceda, "Optimizing the winding strategy of the transformer in a flyback converter," 1996 IEEE 27th Annual Power Electronics Specialists Conference (PESC Record), Baveno, Italy, 1996, vol. 2, pp. 1456–1462.

[6] J. G. Kassakian, D. J. Perreault, G. C. Verghese, and M. F. Schlecht, Principles of Power Electronics, 2nd ed., Cambridge University Press, 2023, pp. 153–154.

[7] N. A. Rahim and A. M. Omar, "Ferrite core analysis for DC-DC flyback converter," 2000 TENCON Proceedings. Intelligent Systems and Technologies for the New Millennium, Kuala Lumpur, Malaysia, 2000, vol. 3, pp. 290–294.

[8] X. Guo, S. Wu, Y. Zhang, C. Dou, and Y. Chi, "Optimal design of high frequency transformer for high power density flyback converter," IEEE Transactions on Circuits and Systems II: Express Briefs, vol. 69, no. 11, pp. 4399–4403, Nov. 2022.

[9] X. Huang, W. Du, F. C. Lee, and Q. Li, "A novel driving scheme for synchronous rectifier in MHz CRM flyback converter with GaN devices," 2015 IEEE Energy Conversion Congress and Exposition (ECCE), Montreal, QC, Canada, 2015, pp. 5089–5095.

[10] Z. Zhang, K. D. T. Ngo, and J. L. Nilles, "A 30-W flyback converter operating at 5 MHz," 2014 IEEE Applied Power Electronics Conference and Exposition (APEC), Fort Worth, TX, USA, 2014, pp. 1415–1421.

[11] X. Huang, J. Feng, W. Du, F. C. Lee, and Q. Li, "Design consideration of MHz active clamp flyback converter with GaN devices for low power adapter application," 2016 IEEE Applied Power Electronics Conference and Exposition (APEC), Long Beach, CA, USA, 2016, pp. 2334–2341.

[12] P. Kumari and S. K. Gawre, "Design of isolated flyback transformer for 50-200 kHz switching frequency range," IEEE International Students' Conference on Electrical, Electronics and Computer Science, Bhopal, India, 2024, pp. 1–6.

[13] TDK Corporation, " High-Frequency, Low-Loss Ferrite Material PC200," Available online: https://product.tdk.com/en/techlibrary/productovervie w/ferrite_pc200.html (Accessed: Oct. 30, 2024).

[14] Fair-Rite Products Corp., "Ferrite Material data sheets," Available online: https://fair-rite.com/materials/ (Accessed: Oct. 30, 2024).

[15] Ferroxcube, "Ferrite Materials for Power Conversion," Available online: https://www.ferroxcube.com/en- global/ak_material/index/power_conve rsion#6 (Accessed: Oct. 30, 2024).

[16] D. Flynn, A. Toon, L. Allen, R. Dhariwal, and M. P. Y. Desmulliez, "Characterization of core materials for microscale magnetic components operating in the megahertz frequency range," IEEE Transactions on Magnetics, vol. 43, no. 7, pp. 3171-3180, July 2007.

[17] J. Kim, M. Kim, P. Galle, F. Herrault, R. Shafer, J. Y. Park, and M. G. Allen, "Nanolaminated permalloy core for high-flux, high-frequency ultracompact power conversion," IEEE Transactions on Power Electronics, vol. 28, no. 9, pp. 4376–4383, Sept. 2013.

[18] M. Synodis, J. B. Pyo, M. Kim, H. Oh, X. Wang, and M. G. Allen, "Fully additive fabrication of electrically anisotropic multilayer materials based on sequential electrodeposition," Journal of Microelectromechanical Systems, vol. 29, no. 6, pp. 1510-1517, Dec. 2020.

[19] J. B. Pyo, X. Wang, M. Kim, H. Oh, R. Kauffman, and M. G. Allen, "Suppression of eddy current loss in multilayer NiFe-Polypyrrole magnetic cores fabricated using a continuous electrodeposition process,"

IEEE Journal of Emerging and Selected Topics in Power Electronics, vol. 10, no. 6, pp. 7433-7440, Dec. 2022.

[20] J. Kim, "UV-LED lithography for millimeter-tall high-aspect ratio 3D structures," 20th International Conference on Solid-State Sensors, Actuators and Microsystems & Eurosensors XXXIII (TRANSDUCERS & EUROSENSORS XXXIII), Berlin, Germany, 2019, pp. 100-103.

[21] J. Kim, M. Kim, and M. G. Allen, "Surface tension-driven assembly of metallic nanosheets at the liquid-air interface: application to highly laminated magnetic cores," 18th International Conference on Solid-State Sensors, Actuators and Microsystems (TRANSDUCERS), 2015, pp. 2224-2227.

[22] T. E. Salem, W. Tipton, and D. Porschet, "Fabrication and practical considerations of a flyback transformer for use in high pulsed-power applications," 2006 Proceedings of the Thirty-Eighth Southeastern Symposium on System Theory, Cookeville, TN, USA, 2006, pp. 406–409.

[23] C. R. Sullivan, T. Abdallah, and T. Fujiwara, "Optimization of a flyback transformer winding considering two-dimensional field effects, cost and loss," 2001 IEEE Applied Power Electronics Conference and Exposition (APEC), Anaheim, CA, USA, 2001, vol. 1, pp. 116–122.

[24] X. Zhang, H. Liu, and D. Xu, "Analysis and design of the flyback transformer," 2003 IEEE 29th Annual Conference of the IEEE Industrial Electronics Society (IECON'03), Roanoke, VA, USA, 2003, pp. 715–719.

[25] A. Stadler and M. Albach, "The influence of the winding layout on the core losses and the leakage inductance in high frequency transformers," IEEE Transactions on Magnetics, vol. 42, no. 4, pp. 735-738, Apr. 2006.

[26] D. Leuenberger and J. Biela, "Accurate and computationally efficient modeling of flyback transformer parasitics and their influence on converter losses," 2015 17th European Conference on Power Electronics and Applications (EPE'15 ECCE-Europe), Geneva, Switzerland, 2015, pp. 1–10.

[27] A. W. Roesler, J. M. Schare, S. J. Glass, K. G. Ewsuk, G. Slama, D. Abel, and D. Schofield, "Planar LTCC transformers for high-voltage flyback converters," IEEE Transactions on Components and Packaging Technologies, vol. 33, no. 2, pp. 359–372, June 2010.

[28] H. D. Arnold and G. W. Elmen, "Permalloy, a new magnetic material of very high permeability," The Bell System Technical Journal, vol. 2, no. 3, pp. 101–111, July 1923.

[29] M. Kim and M. G. Allen, "Interlamination insulation design considerations for laminated magnetics operating at high frequencies," IEEE Transactions on Magnetics, vol. 55, no. 8, pp. 1–11, Aug. 2019.

[30] L. Yang, L. Gao, C. Chen, and Z. Liu, "Electrodeposition of anisotropic NiFe thin films for integrated high-frequency micro-inductor," 19th International Conference on Electronic Packaging Technology (ICEPT), 2018, pp. 1241–1245.

[31] J. Kim, M. Kim, J. -K. Kim, F. Herrault, and M. G. Allen, "Anisotropic nanolaminated CoNiFe cores integrated into microinductors for high-frequency DC–DC power conversion," Journal of Physics D: Applied Physics, vol. 48, no. 46, pp. 462001, Oct. 2015.

[32] Y. Ding, X. Wang, and M. G. Allen, "A PCB-integrated inductor with an additively electrodeposited laminated NiFe core for MHz DC–DC power conversion," IEEE Transactions on Power Electronics, vol. 38, no. 12, pp. 15157–15161, Dec. 2023.

Comprehensive Demonstration Of New Magnetic Designs Utilizing Magnetic Anisotropy Of The Cores For Integrated Magnetics

Yota Takamura
Dept. of Electrical and Electronic Eng.
Institute of Science Tokyo
Tokyo, Japan
takamura.y.0404@m.isct.ac.jp

Honami Nitta
Dept. of Electrical and Electronic Eng.
Institute of Science Tokyo
Tokyo, Japan
nitta.h.ab@m.titech.ac.jp

Tatsuya Miyazaki
ROHM Research & Development Center
ROHM Co., Ltd.
Kyoto, Japan
Tatsuya.Miyazaki@dsn.rohm.co.jp

Kimito Yamanaka
Dept. of Electrical and Electronic Eng.
Institute of Science Tokyo
Tokyo, Japan
yamanaka.k.84c9@m.isct.ac.jp

Ryosuke Ishido
System Solutions Engineering Headquaters
ROHM Co., Ltd.
Kyoto, Japan
Ryosuke.Ishido@dsn.rohm.co.jp

Akira Namba
ROHM Research & Development Center
ROHM Co., Ltd.
Kyoto, Japan
Akira.Namba@dsn.rohm.co.jp

Keisuke Fujisaki
Dept. of Advanced Science and Technology
Toyota Technological Institute
Aichi, Japan
fujisaki@toyota-ti.ac.jp

Shigeki Nakagawa
Dept. of Electrical and Electronic Eng.
Institute of Science Tokyo
Tokyo, Japan
nakagawa.s.fd4e@m.isct.ac.jp

Abstract—This study presents novel magnetic core designs, magnetically anisotropic plain or tiled types, for planar spiral inductors that can effectively use magnetic anisotropy in the place of nano-granular [CoFeB-SiO₂/SiO₂] laminated magnetic cores. Inductance and resonant frequency for the proposed inductors showed significant enhancement through both finite element method calculation and experimentation. Additionally, the inductor was implemented onto a DC-DC converter successfully to demonstrate the step-down performance and showing corresponding waveforms. The efficiency of DC-DC converter with the inductor exhibited higher values under light load conditions. This comprehensive demonstration highlights the potential of our innovative design for advancing the field of integrated magnetics and developing more efficient power electronic devices.

Keywords—Magnetic inductors, integrated inductors, high frequency magnetics, anisotropic cores

I. INTRODUCTION

Magnetic inductors fabricated using semiconductor manufacturing processes [1-4] have gained significant attention due to their potential for miniaturization and high-frequency operation in power electronics circuits [5-6]. Planar-type inductors, in particular, reduce the overall height of the inductor, minimizing the volume of the entire power electronics package. Additionally, integrated inductors are highly compatible with current trends in point-of-load power delivery and vertical power delivery (VPD) for semiconductor chips.

Nano-granular magnetic films [7-10] are promising magnetic core materials for high-frequency inductors because they offer higher electrical resistivity compared to soft amorphous magnetic materials [2] and larger saturation magnetization than soft ferrite [11]. Additionally, they pose no risk of thermal runway. These nano-granular films can be formed using sputtering techniques, which are widely employed in semiconductor manufacturing processes.

Introducing uniaxial magnetic anisotropy [8,9] is a key solution for developing high frequency magnetic cores. By applying uniaxial magnetic anisotropy to the magnetic nano-granular films, hysteresis loss along the hard axis can be significantly reduced. In this configuration, the magnetic flux density (B) changes linearly with an applied magnetic field H, minimizing hysteresis loss compared to isotropic soft magnetic materials, which exhibit finite coercivity, therefore retain hysteresis loss. Additionally, the relative permeability μ_r of the magnetic material can be tuned by adjusting the magnetic anisotropic field (H_k) while maintaining the saturation magnetic flux density (B_S). This characters to fine-tune μ_r without sacrificing B_S is particularly beneficial in high-frequency inductor applications.

The effective use of anisotropic magnetic core play a crucial role for designing the performance of high frequency integrated inductors for DC-DC converters. For this use of inductor, the inductor current and the induced magnetic field are unidirectional most of cases. By such magnetic field, B is barely change along the easy axis, i.e. $\mu_r = 1$, while μ_r of the hard axis

979-8-3315-1612-3/25 $31.00 © 2025 IEEE

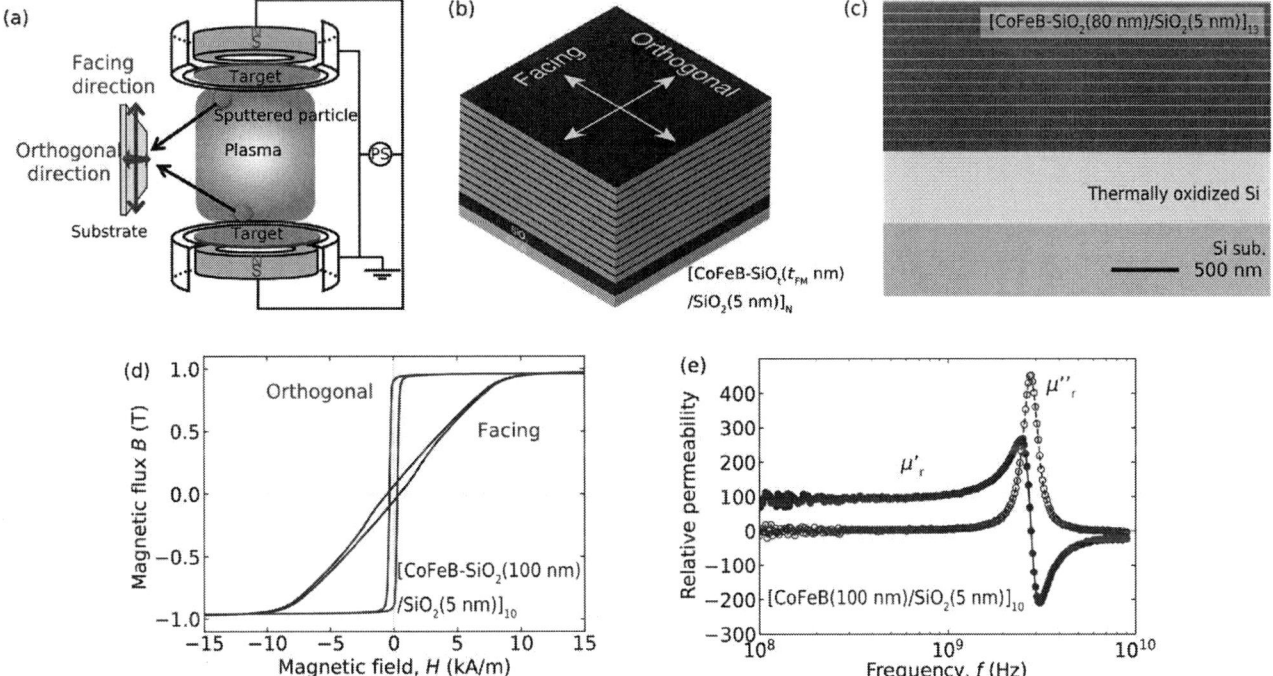

Fig. 1. (a) Facing target sputtering system (b) Stack structure of magnetic cores [CoFeB-SiO$_2$(t_{FM})/SiO$_2$(5 nm)]$_N$ (c) Corsssectional STEM image of a stack of [CoFeB-SiO$_2$(80 nm)/SiO$_2$(5 nm)]$_N$ (d) Magnetization versus magnetic field curves for [CoFeB-SiO$_2$(100 nm)/SiO$_2$(5 nm)]$_{10}$ for the facing and orthogonal directions in the plane. (e) Copmlex relative permeability for the facing direction as a function of frequency.

is high. Therefore, conventional planar-type inductors typically have a rectangular shape and only part of the coil are covered by the anisotropic magnetic core where the magnetic field aligns in the hard axis direction [12]. However, part of the coil may not fully benefit from the core's anisotropic properties.

In this study, we propose the effective integration of anisotropic magnetic film cores onto a planar spiral inductor, aligning its hard axis to interlink with all parts of the spiral coil wires, thereby enhancing the magnetic flux generated by the coil. Furthermore, we propose an improved magnetic core with low residual capacitance. The proposed inductors with magnetically anisotropic nano-granular [CoFeB-SiO$_2$/SiO$_2$]$_N$ laminated cores are successfully demonstrated by both the finite element method (FEM) and experimentation, showing enhanced inductance and a higher resonant frequency above 100 MHz. Finally, we demonstrate the successful operation of a DC-DC converter incorporating the proposed inductor, capable of MHz operation.

II. CHARACTERIZATION OF ANISOTROPIC MAGNETIC CORES

First, we present an anisotropic CoFeB-SiO$_2$ laminated core formed using the facing target sputtering (FTS) technique as illustrated in Fig. 1(a). This method induces significant in-plane magnetic anisotropy due to the oblique incident of sputtered particles [2]. Fig. 1(b) shows the stack structure of the anisotropic magnetic core formed by the FTS system. A nano-granular CoFeB-SiO$_2$ layer and electrically insulating SiO$_2$ layer were alternatively laminated until the total thickness of the CoFeB-SiO$_2$ layers reached 1 µm. The multilayered structure helps reduce eddy current losses in the magnetic layer and enhance soft-magnetic properties without forming the stripe-

shaped magnetic domains [8,13]. It is worth noting that the deposition time for this stack is approximately 15 hours, making further increases in thickness impractical. Fig. 1(c) shows a scanning transmission electron microscope (STEM) image of the [CoFeB-SiO$_2$ (70 nm)/SiO$_2$ (5 nm)]$_{13}$ sample, where a well-defined layered structure with sharp interface is observed across a wide range. The CoFeB-SiO$_2$ layers were completely isolated by the 5 nm thick SiO$_2$ layers, effectively suppressing the columnar growth of CoFeB-SiO$_2$ throughout the entire stack, with no evidence of layer mixing.

The B-H curves for the [CoFeB-SiO$_2$ (100 nm)/SiO$_2$ (5 nm)]$_{10}$ sample are shown in Fig. 1(d). The magnetic core exhibits anisotropic magnetic behavior in the plane with a hard (easy) axis formed in the facing (orthogonal) direction. B_s was close to 1 T. The anisotropic magnetic field, H_k, which is the magnetic field required to saturate magnetization along the hard axis direction, was approximately 8.2 kA/m. The relative permeability for the facing direction was determined to be 96.

Fig. 1(e) shows complex relative permeability $\mu_r (= \mu'_r - j\mu''_r)$ as a function of frequency f. In this measurement, a very small

TABLE I. PROPERTIES OF MAGNETIC CORES

Name	Value
Saturation magnetization, B_s	0.98 T
Anisotropic field, H_k	8.2 kA/m
Relative permeability, μ_r	96 (Facing, hard) 1 (Orthogonal, easy)
Ferromagentic resonance frequency, f_r	2.4 GHz
Electrical resistivity, ρ	590 µΩ-cm

AC magnetic field was applied in the facing direction, *i.e.* hard axis direction. The absolute values of the signals were calibrated using the *B-H* curves. μ'_r remained constant up to 1 GHz, and no significant increase in μ''_r were observed, showing the potential of this magnetic cores for high frequency magnetic inductors. The ferromagnetic resonance was observed at 2.4 GHz which is consistent with estimation using the Kittel's equation based on the *B-H* curves. For the orthogonal direction, no response to the the AC field was observed because the applied filed was much smaller than the coercive field.

The key material parameters of the magnetic cores are summarized in Table I. Due to the in-plane magnetic anisotropy, μ_r varies with direction: it is 96 for the hard axis and 1 for the easy axis. The electrical resistivity (ρ) was 590 $\mu\Omega$-cm and remained constant regardless of the measurement direction.

III. PROPOSAL OF SPIRAL INDCUTORS AND FEM CALCULATION

Here, we propose two new designs of magnetic spiral inductors to efficiently utilize anisotropic magnetic cores, positioned at the top and/or bottom of the spiral coils, are shown in Figs. 2(a) and 2(b). The magnetic cores are divided into two regions, shaded light and dark blue, based on the direction of the hard axis, as indicated by the arrows. In both designs, the hard axes are oriented orthogonally to the coil wires to ensure effective magnetic flux linkage generated by the coils. This configuration allows for optimal utilization of the hard axes' properties, resulting in lower hysteresis loss and higher permeability across a wide flux range. Additionally, the anisotropic magnetic core shown in Fig. 2(b) features a tiled design with slits introduced to reduce residual capacitance, thereby achieving a higher resonant frequency. Hereafter, the cores shown in Figs. 2(a) and 2(b) will be referred as the anisotropic plain type and anisotropic tiled type, respectively.

The impedance of spiral inductors with the proposed

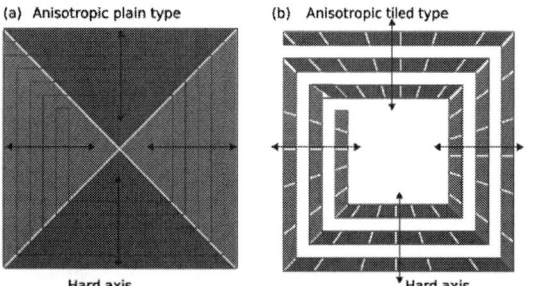

Fig. 2. Schematics of spiral inductors with the proposed magnetic core. (a) anisotropic plain type and (b) anisotropic tiled type. The arrows mean hard axes of the magnetic core layers.

TABLE II. DIMENSION OF THE MAGNETIC SPIRAL INDUCTROR

Name	Value
Outer size	9 mm x 9mm
The width of wires	250 μm
The thickness of the wires	50 μm
The gap between the wires	100 μm
The thickness of a SiO2 layer between core and coil	1.1 μm
The number of turns	6
The width of slits (The anisotropic tiled type only)	Approximately 20 μm

magnetic cores were simulated with a finite element method (FEM) using commercial software FEMTET [14] developed by Murata Software. In this simulation, magnetization flux is linear to the magnetic field, and thus no saturation behavior is included. The relative permeability for the hard and easy axes are set to 96 and 1, respectively. Note that the effect of the displacement current is not considered in this simulation and thus no residual capacitance is calculated. For a simplicity, the magnetic cores were treated as a single layer of 1 μm thick homogeneous magnetic layer. The cores are only placed one side of the coil, consistent with fabricated inductors in this paper. The cross section of the coil is 250 μm wide and 50 μm thick to reduce DC resistance. The insulating gap between the coil and cores is 1.1 μm and filled with SiO2. Other parameters for the FEM calculation are shown in Table II. Both air-core and plain magnetic core (with a single anisotropic core covering the entire coil area) were also simulated for comparison.

Figs. 3(a) and 3(b) show the simulated inductance (L_s) and AC resistance (R_s) as a function of frequency (f), respectively. The magnetic cores contribute to an increase in inductance. The contribution of the magnetic cores in L_s for the anisotropic plain type (Fig. 2(a)) is doubled compared to the plain magnetic core. The anisotropic tiled type (Fig. 2 (b)) shows a slight decrease in L_s compared to the anisotropic plain type because the volume of the magnetic cores reduces. However, it still maintains a larger L_s than the plain core over a wide frequency range. R_s also depends on the magnetic core, reflecting the reduction of eddy current loss in the cores.

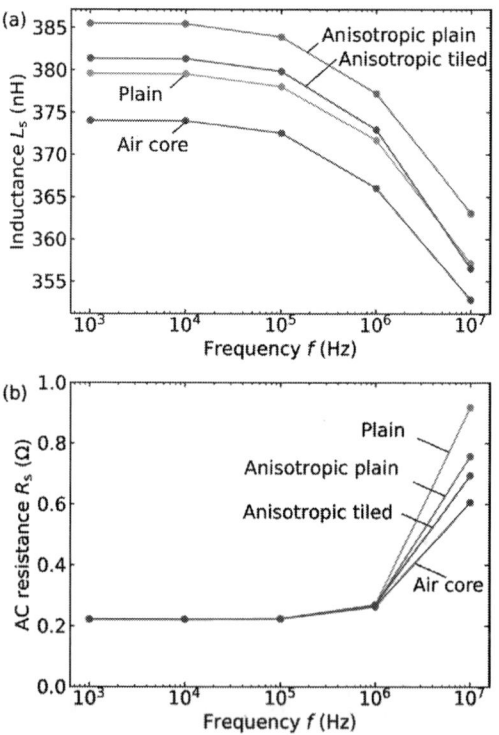

Fig. 3. FEM calculation results. (a) L_s and (b) R_s as a function of frequnecy f for various core.

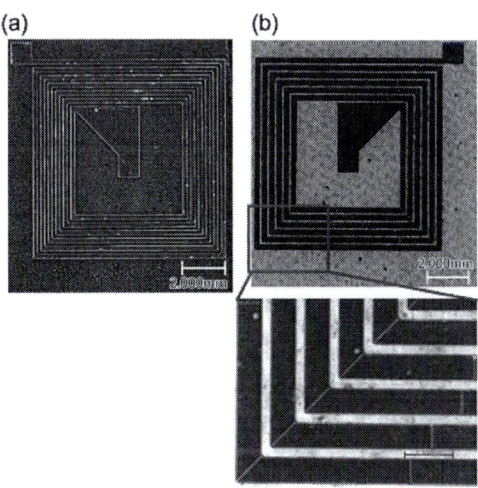

Fig. 4. Microsopic images of the fabricated anisotropic tiled type inductor taken from (a) the top and (b) bottom.

IV. FABRICATION AND CHARACTERIZATION OF PROPOSED INDUCTORS

We fabricated spiral inductors with the proposed magnetic cores. The fabrication process for anisotropic tiled type inductor is the following steps: First, a tiled patterns of the half of the magnetic cores region is formed with photoresist (S1818, micro resist technology). Then a stack of [CoFeB-SiO$_2$(100 nm)/SiO$_2$(5 nm)]$_{10}$ was deposited by FTS followed by a lift off process. After patterning the other half of the magnetic core, the samples was placed into the FTS chamber, rotated 90 degrees, and the same stack was deposited. Following the lift-off process, a 1.1 μm thick SiO$_2$ layer was sputtered to insulate the magnetic cores. Next, a Cr(30 nm)/Cr(100 nm) bilayer was deposited with

electron beam evaporation method as a seed layer for electroplating. The samples were then spin-coated with a thick photoresist (P-BK5000, Tokyo Ohka Kogyo), which was patterned into a coil. A 50 μm thick Cu layers was electroplated and subsequently dry-etched to form a spiral coil [15].

Photomicrographs of the fabricated anisotropic type spiral inductor are shown in Fig. 4. In the top view (Fig. 4(a)), continuous Cu wires patterned in to spiral coils are visible while the tiled magnetic cores can be seen in the bottom views (Fig. 4(b)).

Fig. 5(a) shows a photograph of our impedance measurement setup where a prober is used with an impedance analyzer (4294A, Keysight). The measurement system was calibrated to minimize residual impedance. Figs. 5(b) and 5(c) show L_s and R_s for the fabricated inductors up to 10 MHz, respectively. All spiral inductor with magnetic cores exhibited larger L_s than that of the air-core inductor. The increases in L_s by the proposed inductors were almost the same as the FEM results. Additionally, no sign of resonance was found for the anisotropic tiled type up to 10 MHz, while the anisotropic plain type and plain core showed the onset of resonance above 1 MHz. It is worth noting that the gradual decrease in L_s with increasing frequency, observed in all the inductors, is most likely due to the residual impedance which was not fully excluded from the measurement probes.

Further impedance measurement up to 100 MHz was conducted for the anisotropic tiled type inductor. In this measurement, the sample was mounted on a printed circuit board (PCB) as shown in Fig. 5(d) to reduce further residual impedance at higher frequency. The impedance measurement results are shown in Fig. 5(e). Note that the vertical axis ranges are identical to Figs. 5(b) and 5(c). L_s at 10 kHz is close to the previous measurement, with a 7 % margin of error, while the

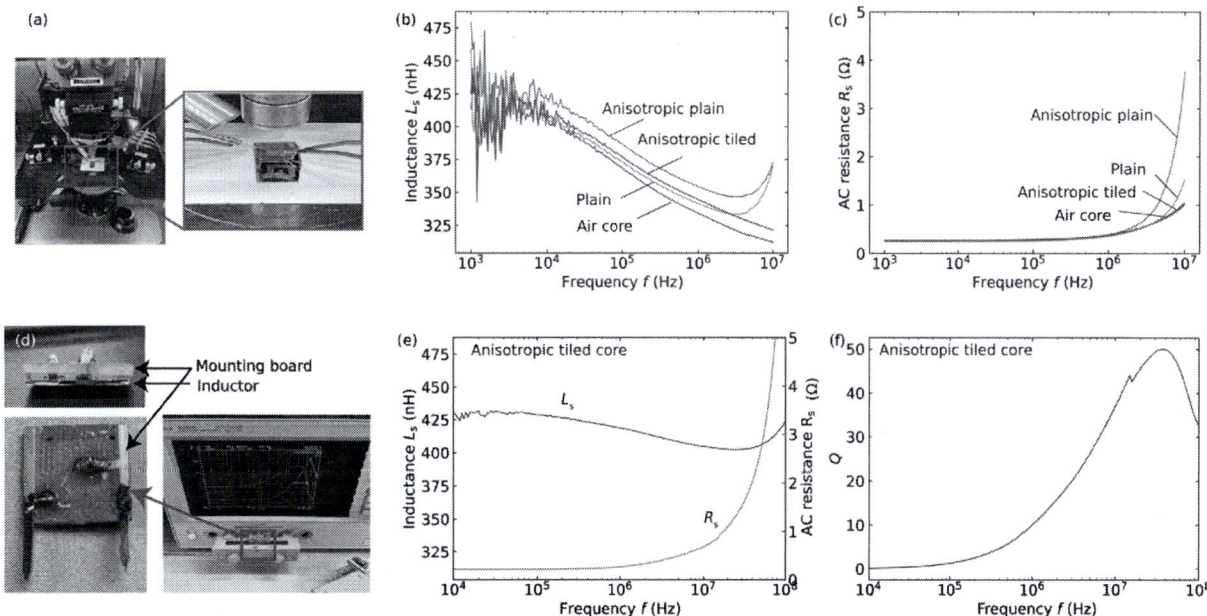

Fig. 5. (a) Impedance measurement using probers. (b) Inductance and (c) AC resistance for fabricated inductors up to 10 MHz. (d) Impedance measurmenet with a mounting board (e) Inductance and AC resistance for anisotropic tiled type inductor up to 100 MHz. (f) Q value.

significant decrease in L_s with increasing f was suppressed in this measurement. L_s was between 400 nH and 425 nH. The slight drop from 100 kHz to 5 MHz could be due to the residual impedance from the measurement system. Even up to 100 MHz, no resonance occurred thanks to the tile patterning. The change in $f > 50$ MHz may be the onset of the resonance. R_s at 100 kHz was measured as 0.2 Ω. Since this value includes residual resistance, the actual R_s is expected to be smaller.

Fig. 5(f) shows the Q factor for the anisotropic tiled core as a function of f. The Q factor reached the maximum value of nearly 50 around 30 MHz. This high Q at such high frequency indicates the effective utilization of the anisotropic core and a reduction of residual capacitance due to the slits in the magnetic core.

V. DEMONSTRATION IN A DC-DC CONVERTER

The operation of the spiral inductor with the anisotropic tiled core was tested on a DC-DC converter board. A commercially available DC-DC converter board (BU90007GWZ-E2EVK-101, ROHM Co., Ltd.)[16 with a carrier frequency of approximately 4 MHz was used in a pulse width modulation (PWM) control mode. The original inductor (MLP2520K1R0ST0S1, 1 μH, TDK)[17] was removed and then the fabricated inductor was mounted on it using connection wires to measure the inductor current as shown in Fig. 6(a). The length of the connection wires were minimized to reduce additional impedance. The current (I_L) and voltage (V_L) waveforms for the mounted inductor were measured using a multichannel oscilloscope (Tektronix MSO58) with probes cramping the wires. The output voltage (V_{out}) of the DC-DC converter was also monitored. The input voltage V_{in} was 3.4 V. The output current (I_{out}) was varied by using an electronic load. No significant side effects on output or the efficiency were observed due to the connection wires. In this paper, only the anisotropic tiled core spiral inductor from the fabricated inductors was measured.

Fig. 6(b) shows waveforms of I_L, V_L and V_{out} for the anisotropic tiled tyle spiral inductor measured under various output current I_{out} ranging from 0 mA to 500 mA. The waveforms for the original inductor mounted with the same method are shown in Fig. 6(c) as a reference. The waveforms for the fabricated inductor showed clear triangle and rectangular shape without no significant distortion and no saturation behavior. The ringing phenomena in V_L were observed for both inductors, coming from the wires not the fabricated inductor itself. V_{out} for the fabricated inductor exhibited larger ringing signals, possibly due to the mismatch in inductance compared to the original inductor. The inductance for the fabricated inductor evaluated from the waveforms were almost 440 nH and constant

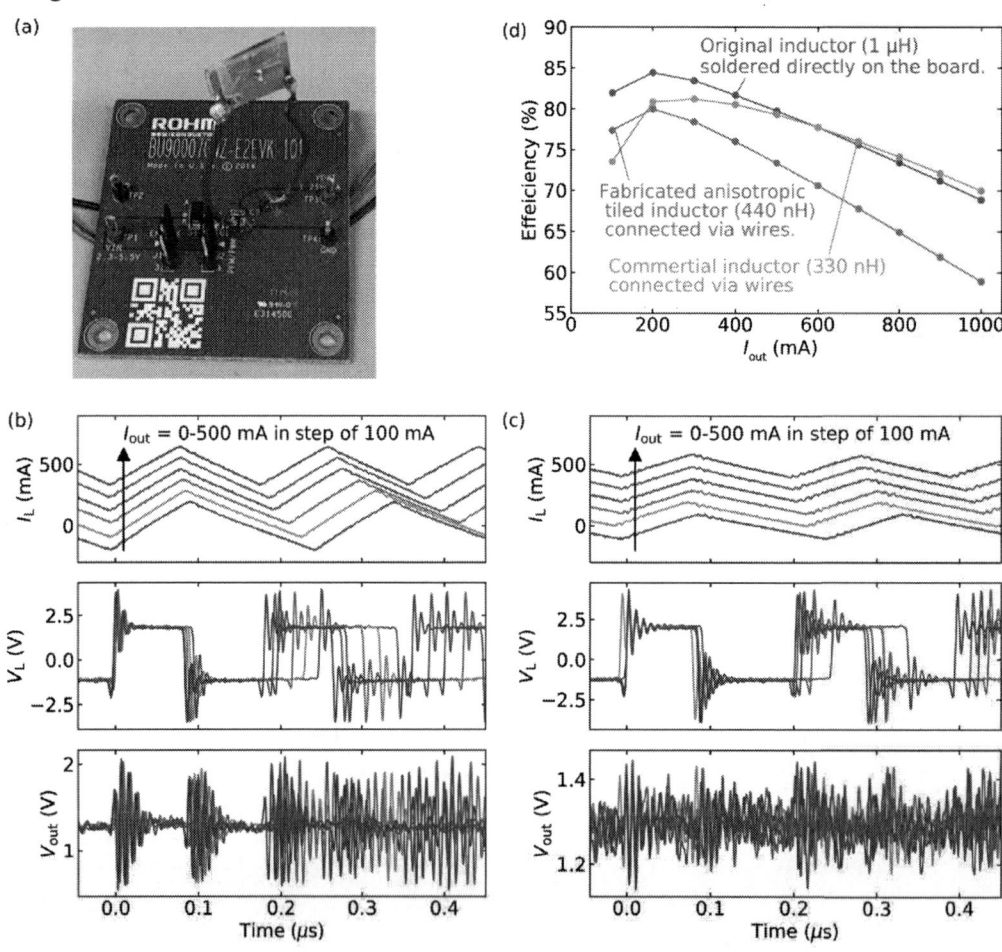

Fig. 6. (a) DCDC converter with the anisotropic tiled spiral inductor. I_L, V_L, V_{out} waveforrms with various I_{out} for (b) the fabricated anisotropic tile spiral indcutr and (c) origianl inductor, 1 μH. (d) Efficiency as a function of I_{out} for various inductors.

over all I_{out} condition. This value is consistent with the impedance measurement shown in Fig. 5(e). Considering yielded magnetic field is 0.5 kA/m for 0.5 A based on FEM calculation, this also consistent with the B-H curves in Fig. 1(d).

Fig. 6(d) presents the efficiency of the DC-DC converter for various inductors. In this measurement, both the anisotropic tiled core spiral inductor and a commercially available inductor (DFE252012F-R33M, Murata Manufacturing)[18] with a inductance of 330 nH, similar to the fabricated inductor, were mounted via the wires as shown in Fig. 6(a). For comparison, the original inductor with 1 μH inductance was also tested on the board. The efficiency for the anisotropic tiled core spiral inductor ranged from 74% to 80%, depending on I_{out}. At I_{out} = 100 mA, the anisotropic tiled core spiral inductor showed higher efficiency than the commercial 330 nH inductor. However, for $I_{out} \geq 200$ mA, efficiency decreased, especially for the fabricated inductor. The exact cause of this behavior is not addressed in this paper. R_s at 4 MHz of the fabricated inductor was 0.4 Ω, which is comparable to the other inductors; R_{ac} is 0.4 Ω and 1 Ω for the 330 nH and 1 μH inductors, respectively, according to their datasheets. This difference does not appear to be due to R_s. One possible explanation is Joule heating. As I_{out} increased further, the DC-DC converter operation became unstable. Since no thermal design is applied, it may be necessary for further improving thermal management to enhance efficiency.

VI. CONCLUSIONS

We have successfully demonstrated the potential of our anisotropic tiled magnetic granular [CoFeB-SiO$_2$/SiO$_2$] layered cores within a planar spiral inductor through both calculation and experimentation. Our results show enhanced inductance and significantly higher operational frequency up to 100 MHz, representing a dramatic improvement over previous or conventional types. Additionally, we achieved successful DC-DC converter operation implemented with our proposed spiral inductor, demonstrating higher efficiency under light load conditions such as output current of 100 mA. This paper showcases the innovative design of our spiral inductor, which is poised to advance high-frequency power electronics circuits for integrated magnetics.

ACKNOWLEDGMENT

Part of this work was supported by Japan Power Academy. The authors would like to thank Prof. Kagohashi from Institute of Science Tokyo for his guidance and advice in managing this joint project. We also thank Prof. Yamaguchi and Mr. Y. Miyazawa, from Tohoku University for his assistance with the permeability measurements. The device fabrication for this work was carried out at Nanofab, Institute of Science Tokyo, and Nanotechnology Research Center (NTRC), Waseda University, supported from "Advanced Research Infrastructure for Materials and Nanotechnology in Japan (ARIM)" of the Ministry of Education, Culture, Sports, Science and Technology (MEXT) under Grant Numbers JPMXP1224IT0004, JPMXP1223IT0030.

REFERENCES

[1] M. Yamaguchi, K. Yamada, and K. H. Kim, "Slit Design Consideration on the Ferromagnetic RF Integrated Inductor," *IEEE Trans. Magn.*, vol. 42, no. 10, pp. 3341–3343, Oct. 2006, doi: 10.1109/TMAG.2006.879636.

[2] D. V. Harburg *et al.*, "Microfabricated Racetrack Inductors With Thin-Film Magnetic Cores for On-Chip Power Conversion," *IEEE J. Emerg. Sel. Topics Power Electron.*, vol. 6, no. 3, pp. 1280–1294, Sep. 2018, doi: 10.1109/JESTPE.2018.2808375.

[3] M. Sonehara, K. Ikeda, and T. Sato, "Control of Magnetic Moment in Uniaxial Anisotropy Magnetic Thin Film Taking Leakage Flux in RF Planar Spiral Inductors with Closed-Magnetic Core into Account," *Electr Eng Jpn*, vol. 191, no. 2, pp. 1–6, Apr. 2015, doi: 10.1002/eej.22693.

[4] P. Zou, S. Wei, X. Qiang, C. Liang, C. Yue, H. Chen, W. Jiake, W. Xinyu, L. Hongwei, C. Xiaojuan, "A 100MHz IVR PMIC with On-silicon Magnetic Thin Film Inductors," *Int. Workshop on Power-Supply-on-Chip (PwrSoC), 0.3.* 2018.

[5] K. Fujisaki, "Magnetic Material Excited by Power Electronics in Electrical Engineering," presented at the TMS2017 46th Annual Meeting & Exhibition, Sandiego, California, USA, Mar. 01, 2017.

[6] D. S. Gardner *et al.*, "Integrated On-Chip Inductors with Magnetic Films," in *2006 International Electron Devices Meeting*, San Francisco, CA, USA: IEEE, 2006, pp. 1–4. doi: 10.1109/IEDM.2006.347002.

[7] M. Naoe, N. Kobayashi, S. Ohnuma, T. Iwasa, K. Arai, H. Masumoto, "Ultra-high resistive and anisotropic CoPd-CaF2 nanogranular soft magnetic films prepared by tandem-sputtering deposition," J. Magn. Magn. Mater., vol. 391, pp. 213-222, Oct. 2015.

[8] Y. Takamura *et al.*, "Fabrication of CoFeB-SiO$_2$ Films with Large Uniaxial Anisotropic by Facing Target Sputtering and its Application to High Frequency Planar Type Spiral Inductors," *IEEE Trans. Magn.*, vol. 59, no. 11, pp. 2801204/1–5, 2023, doi: 10.1109/TMAG.2023.3291879.

[9] H. Nitta, Y. Takamura, T. Kaneko, and S. Nakagawa, "Fabrication and Characterization CoZrO Films Deposited by Facing Targets Reactive Sputtering for Micromagnetic Inductors," *IEEE Magn. Lett.*, vol. 14, pp. 1–5, 2023, doi: 10.1109/LMAG.2023.3320495.

[10] M. Munakata *et al.*, "Drastically Increased Electrical Resistivity of a (CoFeB)-SiO$_2$ Magnetic Thin-Film Core in a Small GHz Range.," *J. Magn. Soc. Jpn.*, vol. 26, no. 4, pp. 509–512, 2002, doi: 10.3379/jmsjmag.26.509.

[11] W. Zhang, et al., , " Characterization of Low Temperature Sintered Ferrite Laminates for High Frequency Point-of-Load (POL) Converters," *IEEE Trans. Magn.*, vol. 49, no. 11, pp. 5454-5463, July. 2013.

[12] D. V. Harburg *et al.*, "Microfabricated Racetrack Inductors With Thin-Film Magnetic Cores for On-Chip Power Conversion," *IEEE J. Emerging and Selected Topics in Power Electronics*, vol. 6, no. 3, pp. 1280-1294, Sept. 2018.

[13] Y. Takamura, "Simulation and fabrication of planar type spiral inductors with facing target sputtered CoFeB-SiO$_2$ magnetic layers," presented at the The First International Symposium on Integrated Magnetics (iSIM) 2023, Sendai, Japan, May 14, 2023.13

[14] *FEMTET*. (2023). Murata Software Co., Ltd. [Online]. Available: https://www.muratasoftware.com/en/.

[15] C. P. Yue and S. S. Wong, "Physical modeling of spiral inductors on silicon," *IEEE Trans. Electron Devices*, vol. 47, no. 3, pp. 560–568, Mar. 2000.

[16] Step-down Switching regulators with Built-in Power MOSFET, ROHM Co., Ltd., 2015. [Online]. Avalable: https://fscdn.rohm.com/en/products/databook/datasheet/ic/power/switching_regulator/bu9000xgwz-e.pdf.

[17] SMD / SMT Inductors (Coils), TDK, [Online]. Avalable: https://product.tdk.com/en/search/inductor/inductor/smd/info?part_no=MLP2520K1R0ST0S1.

[18] DFE252012F-R33M, Murata, [Online]. Available https://www.murata.com/en-us/products/productdetail?partno=DFE252012F-R33M%23

A Two - Stage Artificial Neural Network (ANN) - Based Design and Optimization of High Frequency Transformers for Dual Active Bridge Converter

Lufan Zhou
Centro de Electrónica Industrial
Universidad Politécnica de Madrid
Madrid, Spain
lufan.zhou@upm.es

Alberto Delgado Expósito
Centro de Electrónica Industrial
Universidad Politécnica de Madrid
Madrid, Spain
a.delgado@upm.es

Adam Ruszczyk
Hitachi Energy
Kraków, Poland
adam.ruszczyk1@hitachienergy.com

Simon Round
Hitachi Energy
Zurich, Switzerland
simon.round@hitachienergy.com

Miroslav Vasić
Centro de Electrónica Industrial
Universidad Politécnica de Madrid
Madrid, Spain
miroslav.vasic@upm.es

Abstract—With the increase in high-frequency applications such as electric vehicles, finding a transformer that can handle high frequency, high power density, and high efficiency while ensuring effective thermal management is crucial. This paper presents a novel two-stage model-based design optimization methodology for high-frequency transformers based on Artificial Neural Networks (ANN), focusing on electromagnetic (EM) and thermal modelling. This methodology accelerates the design process using FEM simulations, enabling the selection of the most suitable transformer design from 70,784 design points in 5 minutes instead of 246 days. The relative errors between FEM simulations and ANN models for 99.7% of the dataset are less than 2.5% for EM modelling and 7% for thermal modelling. First, the methodology based on FEM simulations was verified experimentally with a 100 kHz Litz wire E core transformer and a 75 kHz flat bar E core transformer. Then, a 95 kHz U core-based transformer for a dual active bridge converter, selected from the Pareto front using the ANN optimization method, was built, with results matching both the electromagnetic and thermal comparisons.

Index Terms—Artificial Neural Network Optimization, Finite Element Method (FEM) Simulations, Transformer, Electromagnetic Modelling, Thermal Modelling.

I. INTRODUCTION

High-frequency transformers play a crucial role in modern power electronics, enabling efficient power conversion and isolation across various applications, such as electric vehicles and data centers. Traditional transformer designs often face challenges in optimizing multiple objectives such as efficiency, power density, and thermal performance [1]–[4]. The latter is critical for high-frequency and high-power density transformers; the high-power losses generated by the windings lead to high temperatures. The integration of advanced materials and innovative cooling methods has addressed some of these issues [5]–[7], but the design process remains complex and resource intensive. Moreover, the accurate estimation of transformer

performance in the designing part under different operating conditions is crucial but challenging due to the intricate interactions between various factors such as core and winding losses, leakage inductance, and thermal effects.

Recent advancements in artificial neural networks (ANN) offer promising solutions to these challenges by providing a robust framework for optimizing transformer designs [8]–[12]. While FEM can provide highly accurate modelling, its computational cost is often high, especially for complex geometry models that require a large number of mesh elements and may not be feasible for extensive design space explorations. Conversely, analytical models offer good computational speed, but their deviation is often too high compared to FEM. To overcome the tradeoff between computational cost and accuracy, an optimization using artificial neural network (ANN) method is proposed.

In recent years, the dual active bridge (DAB) converter has become an attractive choice due to its bidirectional power flow, galvanic isolation, simple control, wide voltage regulation range, and zero-voltage switching (ZVS) of semiconductor devices. These advantages make it an good choice for the transformer tests conducted in this study.

In this paper, we propose a two-stage ANN-based optimization framework for general high-frequency transformer design for DAB converters to provide accurate estimations of hot-spot temperatures, core and winding losses, magnetizing inductance, and leakage inductance. This framework enables rapid evaluation and optimization of design parameters. The ANN model is trained using a comprehensive dataset generated from FEM simulations, facilitating the design of transformers with higher efficiency and performance. The design specifications are summarized in Table I.

TABLE I: Summary of specifications for the Transformer for DAB

Parameter	Value	Description
V_{in}	750 V	Input voltage
V_{out}	750 V	Output voltage
P_{out}	[25, 100] kW	Output power range
f_{sw}	[50, 125] kHz	Switching frequency range
N_{Turns}	[3, 10] turns	Number of turns range
B_{max}	[20, 280] mT	Maximum magnetic flux density range
J_{max}	[2, 9] A/mm^2	Maximum current density range

II. METHODOLOGY OF TRANSFORMER DESIGN

Before addressing the optimization using the ANN method, several factors in transformer structure design are considered. When designing high-frequency transformers, it is essential to analyze geometric parameters such as window area, wire specifications, and core shape. The design process begins with selecting the core shape; in this paper, a U core is used due to its superior heat dissipation compared to an E core. A parallel winding structure is implemented to enhance transformer performance by increasing the current-carrying capacity, which reduces resistance and heat generation. This structure, shown in fig. 1, also optimizes the utilization of the core window, leading to a more compact and efficient design, where the green represents the primary winding and the orange represents the secondary winding.

As ANN optimization requires generating a sufficient dataset for input and output parameters, the framework presented in fig. 2 visualizes the complete modelling process for dataset generation in transformer design. The first step is to calculate inductances and power losses (winding losses and core losses) based on initial parameters, such as core material properties, maximum magnetic flux density, geometry parameters (width, height, and length of the transformer, window size), input and output voltages, and output power. From the calculated power losses, the temperatures of the core and winding will be obtained through thermal FEM simulations.

For the calculation of winding losses ($P_{\text{LossWinding}}$), magnetizing inductance, and leakage inductance, the process is based on the extraction of impedance matrices from FEM simulations using Ansys Maxwell, with each simulation taking 1 minute. During the construction of the transformer model, a homogenized litz wire [13] is employed to accelerate computational speed, utilizing modified conductivity and imaginary permeability to account for skin losses and proximity losses. A FEM model is built and simulated, and after extracting the frequency-dependent impedance matrices, winding losses are calculated as in (2) [14]. Magnetizing inductance, L_{mag} and leakage inductance, L_{leak} referred to primary can be calculated also based on the inductance matrix using equations (5), which k is coupling factor of inductance in (4).

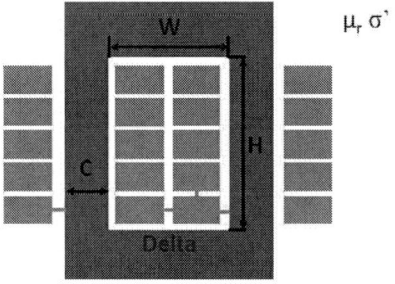

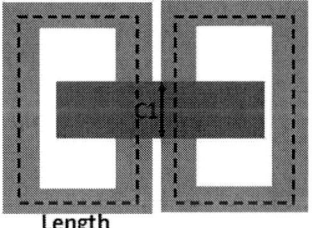

(a) Front and top view of the transformer

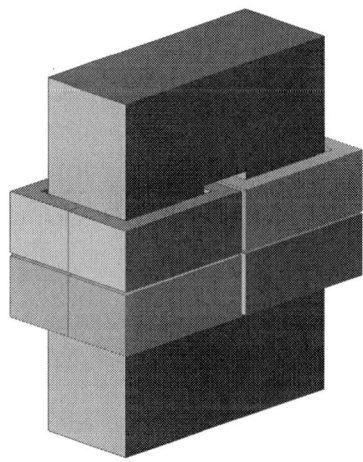

(b) U core 3D representation

Fig. 1: U core Transformer with its views and geometry parameters

$$\begin{pmatrix} V_1 \\ V_2 \end{pmatrix} = \begin{pmatrix} R_{11}(w) + j\omega L_{11} & R_{12}(w) + j\omega L_{12} \\ R_{21}(w) + j\omega L_{21} & R_{22}(w) + j\omega L_{22} \end{pmatrix} \quad (1)$$

$$P_{\text{lossWinding}} = \begin{pmatrix} Z_{11} & Z_{12} \\ Z_{21} & Z_{22} \end{pmatrix} \begin{pmatrix} I_1 \\ I_2 \end{pmatrix} \begin{pmatrix} I_1 \\ I_2 \end{pmatrix}^* \quad (2)$$

$$R_{12} = \frac{P_{\text{LossWinding}} - R_{11}I_1^2 - R_{22}I_2^2}{2I_1I_2} \quad (3)$$

$$k^2 = \frac{L_{12}}{\sqrt{L_{11}L_{22}}} \quad (4)$$

$$L_{\text{leak}} = (1 - k^2)L_{11}; \quad L_{\text{mag}} = k^2 L_{11} \quad (5)$$

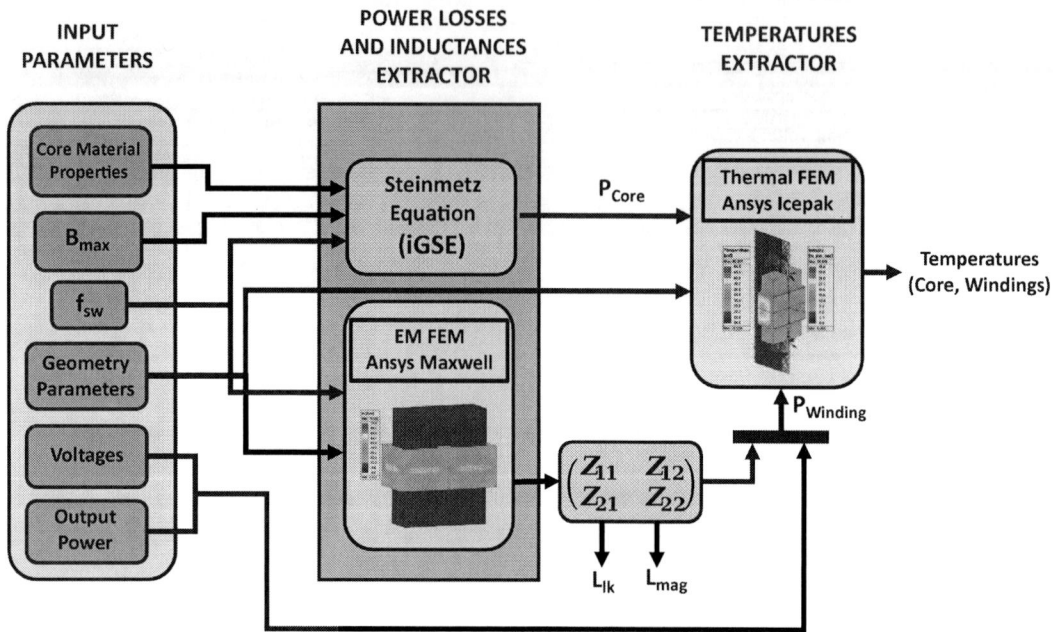

Fig. 2: Framework of methodology of transformer design

On the other hand, there are several approaches to estimating core losses in high-frequency transformers. The Improved Generalized Steinmetz Equation (IGSE) is selected due to its optimal and comprehensive approach to estimating core losses [15] with equations (6). This method stands out for its high accuracy across a wide operating range, simplicity, and use of standard datasheet. It considers the peak-to-peak magnetic induction and its instantaneous rate of change, enhancing the transformer design process by providing accurate loss predictions while reducing computational cost.

$$P_v = \frac{1}{T} \int_0^T k_i \left| \frac{dB(t)}{dt} \right|^\alpha (\Delta B)^{\beta-\alpha} \, dt; \qquad (6)$$

$$k_i = \frac{K}{(2\pi)^{\alpha-1} \int_0^{2\pi} |\cos(\theta)|^\alpha 2^{\beta-\alpha} \, d\theta} \qquad (7)$$

Where K, β, α are the coefficients of Steinmetz equation extracted from the manufacturer material datasheet, and ΔB is the peak-to-peak flux density.

The thermal modelling of the transformer is conducted using FEM simulations in Ansys Icepak, considering various conditions such as natural and forced convection, with each simulation taking 5 minutes. In these simulations, a homogenized litz wire is applied to the model using modified thermal conductivity [16], to reduce the computational cost. A thermal block is assigned to the core and windings based on the losses obtained from the electromagnetic modelling. It is crucial to consider the refinement of the geometry in FEM simulations to ensure accuracy.

III. ARTIFICIAL NEURAL NETWORK (ANN) OPTIMIZATION

After establishing the electromagnetic and thermal modelling for generation of datasets, the next step is to train the estimation of parameters for transformer design. The ANN training is divided into two stages: electromagnetic (EM) and thermal. The electromagnetic ANN model is trained to predict the inductance and resistance matrix extracted from FEM simulations using the geometry parameters shown in fig. 1 and the frequency as input parameters. The thermal ANN model is trained to predict different significant points of temperatures of the core and windings using the geometry parameters and the power losses extracted from the EM ANN model and calculated by analytical equations.

However, there is a bottleneck in the EM ANN model: predicting mutual resistance poses a challenge due to its dependency on the signal convention of the voltage variable [17], which generates proximity effect losses that can be either positive or negative. This leads to large deviations during ANN training and significantly influences the winding losses. To address this issue, instead of training on mutual resistance, winding losses ($P_{\text{LossWinding}}$) from FEM are used in (3).

The ANN method (fig. 3) is implemented in MATLAB using the Deep Learning Toolbox. The "Levenberg-Marquardt backpropagation" training function, combined with the Sigmoid activation function, is employed to handle the non-linear equations from the training data. The dataset size is 2000 for EM and 1870 for thermal modelling, which is split into 80% for training and 20% for testing. The network consists of two hidden layers, each with 20 neurons, ensuring high training performance. The cost function used to measure error

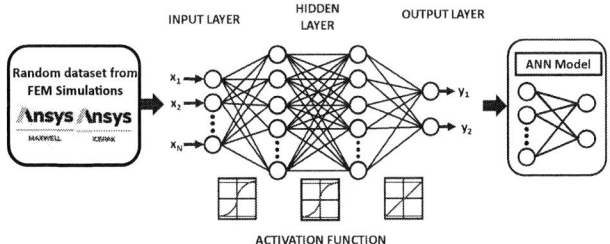

Fig. 3: ANN training framework

is the mean squared error (MSE). The output parameters in the EM ANN training, the inductance and resistance matrices, typically exhibit logarithmic behaviors, and logarithmic transformation is implemented in the output features to linearize them and improve fitting accuracy. The time required to train the ANN model depends on the previous factors, with this case taking 3 minutes for training EM ANN model and 2 minutes for thermal ANN model.

According to the results obtained from the ANN training compared with the FEM simulations shown in fig. 4, the

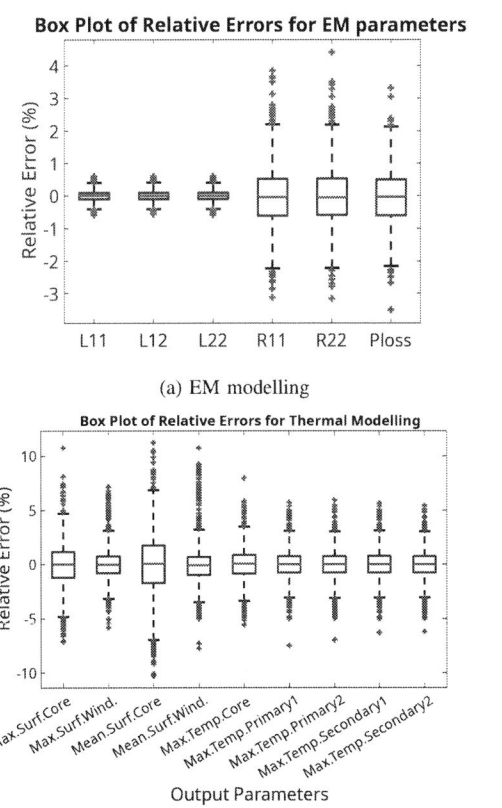

(a) EM modelling

(b) Thermal modelling

Fig. 4: Boxplot of Relative Errors for different modellings

relative error in electromagnetic (EM) modelling is less than 5% at maximum, with 99.7% of the dataset showing an error difference of within 2.5% for all output parameters. For thermal modelling, the maximum error is 10.5%, and 99.7% of the dataset has a relative error within 7% for all output parameters. The inductance predictions perform better than the resistance predictions in these ANN models. This discrepancy arises because, as the input parameters for geometry and frequency increase, the inductance changes slowly with geometry, and the changes are not very significant in this transformer structure. The thermal ANN models are trained to predict points of temperature such as maximum and mean surface of core and windings temperatures, and maximum hot spot core and winding temperatures. The surface temperature predictions have higher errors than the maximum temperature predictions because turbulence generated by air cooling affects surface temperatures more in FEM simulations, depending on the air velocity.

Once the ANN models for electromagnetic (EM) and thermal modelling are obtained, a brute force sweep algorithm is used to explore the electromagnetic and thermal parameters. This process involves using ANN models and analytical equations to generate a plot of all possible transformer designs with a Pareto front, based on the specified range of requirements. The execution time required to generate Pareto Front is 5 min for 70,784 design points. These plots are mathematically feasible designs, depicted in fig. 5, are then filtered based on key factors such as efficiency, power density, and surface temperatures of the core and windings. These designs are filtered in order to have the designs with realistic temperatures, and they are reoptimized using commercially available cores and wire values to select the most suitable components that are readily available.

TABLE II: Specifications for the Selected Transformer

Parameter	Value	Description
P_{out}	100 kW	Output Power
f_{sw}	95 kHz	Switching Frequency
N_1/N_2	7 / 7	Turn ratio
C	28 mm	Width of transformer leg
H	126 mm	Height of transformer window
W	37 mm	Width of transformer window
C_1	100 mm	Length of transformer
$\varnothing_{strand}$	0.071 mm	Litz wire diameter
$N_{strands}$	3072	Number of strands

This method offers a straightforward and intuitive approach for designers to select the optimal transformer with acceptable tradeoffs. It provides the flexibility to choose between multiple designs with similar performance and specifications. The final transformer design is chosen based on specification preferences such as available components and construction simplicity, and it is highlighted with a red star in the Pareto front, shown in fig. 5. The specifications for the selected

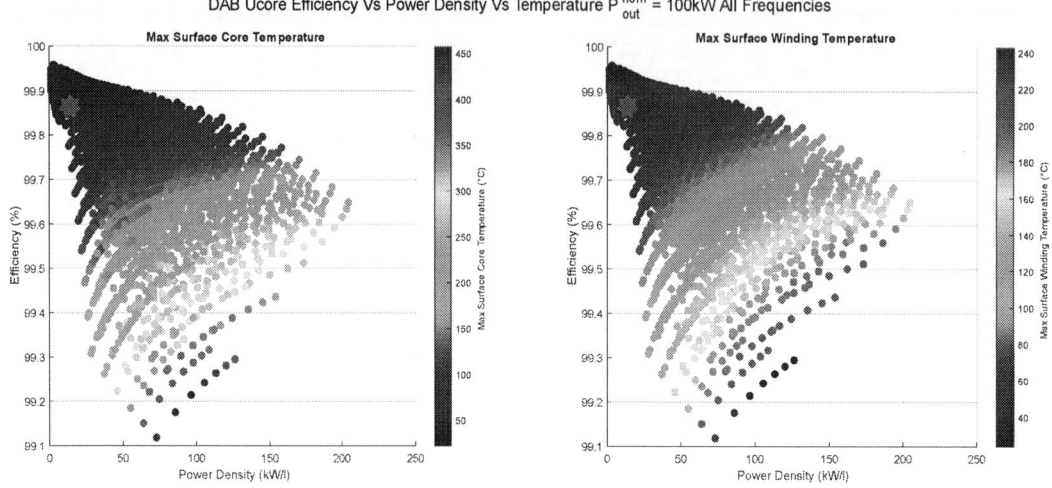

Fig. 5: Pareto Front generated by ANN models

transformer are presented in Table II.

IV. EXPERIMENTAL VERIFICATION

An E core transformer using Litz wire and flat bar, shown in figs. 6a and 7a, was first tested using electromagnetic and thermal modelling. The specifications for the Litz wire are a strand diameter of 0.2 mm, 300 strands, square shape, and a ferrite core material of 3C94 with an E71/33/32 core size. The turn ratio is 9:8. The flat bar specifications are a copper profile of 14 x 2.1 mm, a ferrite core of E80/38/20 with material 3C95, and a primary and secondary turn count of 5 each.

To verify electromagnetic performance, small-signal measurements were conducted using an impedance analyzer, "Keysight E4990." As shown in figs. 6b and 7b, the measured results for inductance and resistance in open circuit conditions accurately matched the FEM simulations. The estimation error was less than 1.5% for inductance and 8% for resistance across the entire frequency range.

To verify thermal characteristics and achieve a steady state, the Litz wire E core transformer was tested in natural convection, and the flat bar E core transformer in forced convection using the OA180AP fan. Measurements were based on resistance and frequency, reaching 100 kHz on the Litz wire and 75 kHz on the flat bar. The transformer was then set in a short-circuit configuration, and the results were compared with FEM thermal simulations, as shown in figs. 6c-d and 7c-d.

The results showed a good correlation between the experimental data and the FEM simulations, validating the proposed methodology. It can be concluded that the results obtained from measurements and simulations match in terms of resistance, inductance, and temperature. Additionally, the accuracy of the ANN depends on the accuracy of FEM simulations, so

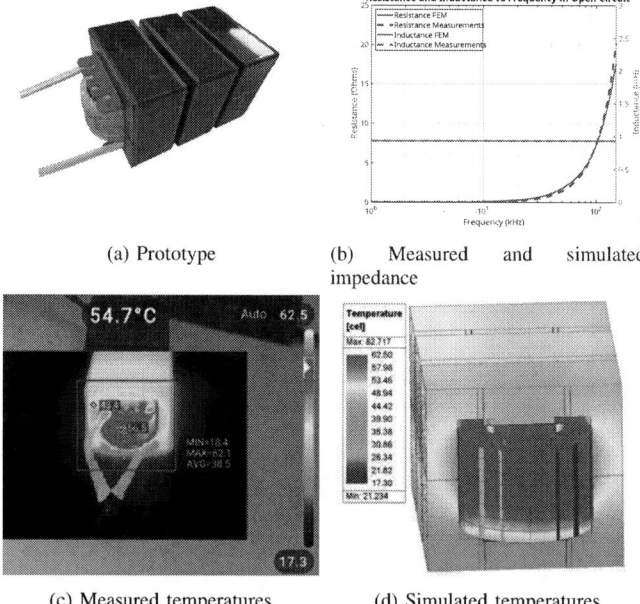

(a) Prototype

(b) Measured and simulated impedance

(c) Measured temperatures

(d) Simulated temperatures

Fig. 6: E core litz wire transformer for validations

these experiments were used to verify the transformer models, using Litz wires and flat bars. Since the simulations match the real prototypes, the ANN results will implicitly match as well, as verified and demonstrated in the previous section.

Once verified, a selected prototype from the Pareto front with specifications shown in Table II was built, as shown in fig. 8a. Since the we could not find the same dimensions of

979-8-3315-1612-3/25 $31.00 © 2025 IEEE

(a) Prototype

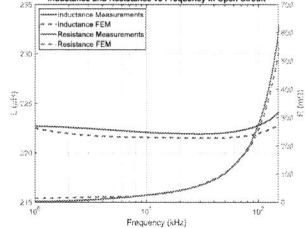

(b) Measured and simulated impedance

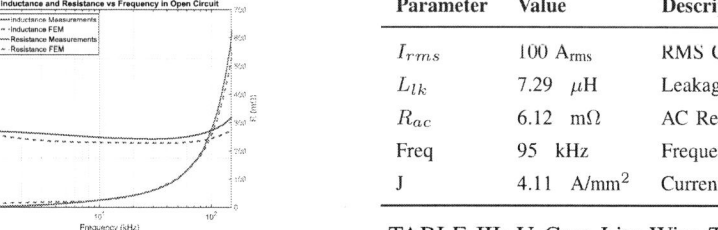

Parameter	Value	Description
I_{rms}	100 A_{rms}	RMS Current
L_{lk}	7.29 μH	Leakage Inductance
R_{ac}	6.12 mΩ	AC Resistance
Freq	95 kHz	Frequency
J	4.11 A/mm^2	Current Density

TABLE III: U Core Litz Wire Transformer

(c) Measured temperatures

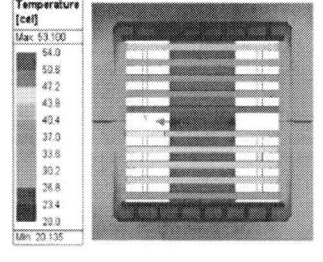

(d) Simulated temperatures

Fig. 7: E core flat bar transformer for validations.

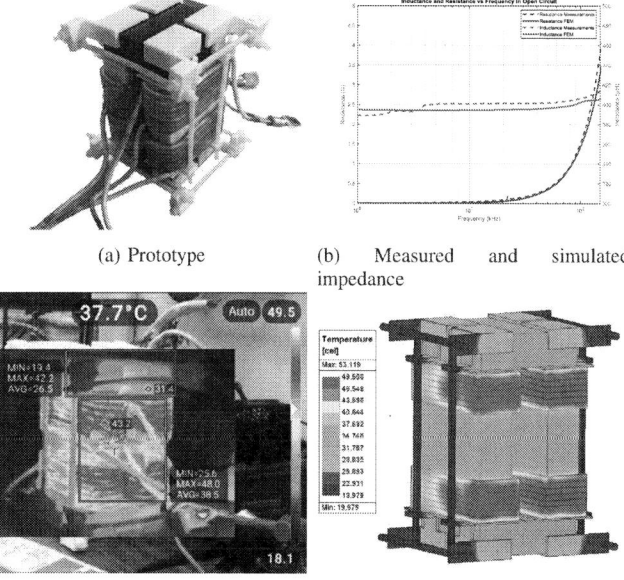

(a) Prototype

(b) Measured and simulated impedance

(c) Measured temperatures

(d) Simulated temperatures

Fig. 8: U core Litz Wire transformer for validations.

commercial U core, the transformer prototype was constructed using ferrite blocks. The material of ferrite blocks is N87. The winding type is litz wire of 3072 strands and 0.071 mm of diameter of strand. The bobbins were made with 3D printing using ASA filament material, which has good thermal resistance. The threaded rod and hex nuts were made of nylon to avoid influencing the electromagnetic and thermal characteristics, as they are non-ferromagnetic materials. The transformer is shown in fig. 8a.

The measured inductance and resistance of the U core transformer are compared to the simulated results in open circuit conditions. The estimation error was less than 2% for inductance and 8% for resistance across the entire frequency range, as shown in fig. 8b. For thermal verification, 95 kHz and 100 A were used in short circuit configurations, with 2 m/s of air velocity in average, the parameters used in the experiment are shown in table III, and the results were compared with FEM thermal simulations, as shown in figs. 8c–d. The estimation error for the maximum surface temperature of the windings was less than 3.5%, with measurements at 48°C and FEM simulations at 49.71°C. Both measurements (electromagnetic and thermal) are matched with the ANN results and the FEM simulations.

V. CONCLUSIONS

In this paper, a two-stage ANN-based design and optimization of a high-frequency transformer are presented, with the first stage focusing on electromagnetic modelling and the second stage on thermal modelling. It is demonstrated that the ANN and FEM simulations have errors of 2.5% and 7%, respectively, for 99.7% of test data. The time required to find the optimal transformer is reduced to 5 minutes instead of

246 days. To verify this methodology, two real transformer prototypes were compared with FEM simulations, and the results matched, thereby implicitly verifying the ANN. Finally, a high-frequency transformer selected by ANN optimization for a DAB converter is built, verified by electromagnetic and thermal experiments, which have less than 2% for inductance, less than 8% for resistance and 3.5% for maximum surface winding temperature.

VI. ACKNOWLEDGMENT

The author would like to acknowledge Hitachi Energy, for the support and collaboration during the design and selection of the transformer.

REFERENCES

[1] M. Mogorovic and D. Dujic. "100 kW, 10 kHz Medium-Frequency Transformer Design Optimization and Experimental Verification". In: *IEEE Transactions on Power Electronics* 34.2 (Feb. 2019), pp. 1696–1708. DOI: 10.1109/TPEL.2018.2835564.

[2] Z. Guo, R. Yu, W. Xu, et al. "Design and Optimization of a 200-kW Medium-Frequency Transformer for Medium-Voltage SiC PV Inverters". In: *IEEE Transactions on Power Electronics* 36.9 (Sept. 2021), pp. 10548–10560. DOI: 10.1109/TPEL.2021.3059879.

[3]   D. Rothmund, T. Guillod, D. Bortis, et al. "99% Efficient 10 kV SiC-Based 7 kV-400 V DC Transformer for Future Data Centers". In: *IEEE Journal of Emerging and Selected Topics in Power Electronics* 7.2 (June 2019), pp. 753–767. DOI: 10.1109/JESTPE.2018.2886139.

[4]   M. Mogorovic and D. Dujic. "FEM-Based Statistical Data-Driven Approach for MFT Design Optimization". In: *IEEE Transactions on Power Electronics* 35.10 (Nov. 2020), pp. 10863–10872. DOI: 10.1109/TPEL.2020.2982130.

[5]   M. Ngo, Y. Cao, D. Dong, et al. "Forced Air-Cooling Thermal Design Methodology for High-Density, High-Frequency, and High-Power Planar Transformers in 1U Applications". In: *IEEE Journal of Emerging and Selected Topics in Power Electronics* 11.2 (Apr. 2023), pp. 2015–2028. DOI: 10.1109/JESTPE.2022.3223849.

[6]   M. Leibl, G. Ortiz, and J. W. Kolar. "Design and Experimental Analysis of a Medium-Frequency Transformer for Solid-State Transformer Applications". In: *IEEE Journal of Emerging and Selected Topics in Power Electronics* 5.1 (Mar. 2017), pp. 110–123. DOI: 10.1109/JESTPE.2016.2623679.

[7]   Z. Guo et al. "A Novel High Insulation 100 kW Medium Frequency Transformer". In: *IEEE Transactions on Power Electronics* 38.1 (Jan. 2023), pp. 112–117. DOI: 10.1109/TPEL.2022.3205646.

[8]   T. Guillod, P. Papamanolis, and J. W. Kolar. "Artificial Neural Network (ANN) Based Fast and Accurate Inductor Modelling and Design". In: *IEEE Open Journal of Power Electronics* 1 (2020), pp. 284–299. DOI: 10.1109/OJPEL.2020.3012777.

[9]   D. Santamargarita, D. Molinero, E. Bueno, et al. "On-Line Monitoring of Maximum Temperature and Loss Distribution of a Medium Frequency Transformer Using Artificial Neural Networks". In: *IEEE Transactions on Power Electronics* 38.12 (Dec. 2023), pp. 15818–15828. DOI: 10.1109/TPEL.2023.3308613.

[10] D. Santamargarita, G. Salinas, D. Molinero, et al. "Tradeoff Between Accuracy and Computational Time for Magnetics Thermal Model Based on Artificial Neural Networks". In: *IEEE Journal of Emerging and Selected Topics in Power Electronics* 11.6 (Dec. 2023), pp. 5658–5674. DOI: 10.1109/JESTPE.2022.3203934.

[11] X. Liu et al. "Convolutional Neural Network (CNN) based Planar Inductor Evaluation and Optimization". In: *2022 IEEE Applied Power Electronics Conference and Exposition (APEC)*. Mar. 2022. DOI: 10.1109/APEC43599.2022.9773675.

[12] S. Inoue et al. "Fast Design Optimization Method Utilizing a Combination of Artificial Neural Networks and Genetic Algorithms for Dynamic Inductive Power Transfer Systems". In: *IEEE Open Journal of Power Electronics* 3 (2022), pp. 915–929. DOI: 10.1109/OJPEL.2022.3224422.

[13] A. Delgado, G. Salinas, J. A. Oliver, et al. "Equivalent Conductor Layer for Fast 3-D Finite Element Simulations of Inductive Power Transfer Coils". In: *IEEE Transactions on Power Electronics* 35.6 (June 2020), pp. 6221–6230. DOI: 10.1109/TPEL.2019.2949438.

[14] J. H. Spreen. "Electrical terminal representation of conductor loss in transformers". In: *IEEE Transactions on Power Electronics* 5.4 (Oct. 1990), pp. 424–429. DOI: 10.1109/63.60685.

[15] K. Venkatachalam, C. R. Sullivan, T. Abdallah, et al. "Accurate prediction of ferrite core loss with nonsinusoidal waveforms using only Steinmetz parameters". In: *2002 IEEE Workshop on Computers in Power Electronics, 2002. Proceedings.* Mayaguez, PR, USA, 2002, pp. 36–41. DOI: 10.1109/CIPE.2002.1196712.

[16] G. Salinas López, A. D. Expósito, J. Muñoz-Antón, et al. "Fast and Accurate Thermal Modelling of Magnetic Components by FEA-Based Homogenization". In: *IEEE Transactions on Power Electronics* 35.2 (Feb. 2020), pp. 1830–1844. DOI: 10.1109/TPEL.2019.2921160.

[17] K. Niyomsatian, J. J. C. Gyselinck, and R. V. Sabariego. "Experimental Extraction of Winding Resistance in Litz-Wire Transformers—Influence of Winding Mutual Resistance". In: *IEEE Transactions on Power Electronics* 34.7 (July 2019), pp. 6736–6746. DOI: 10.1109/TPEL.2018.2876310.

Modeling and Optimizing Winding Arrangement for Gapped Planar Magnetics based on Artificial Neural Network

Hanqing Cao, Bima Nugraha Sanusi, Ziwei Ouyang

Department of Electrical and Photonics Engineering
Technical University of Denmark
Kgs. Lyngby, 2800, Denmark
s230037@student.dtu.dk, bnusa@dtu.dk, ziou@dtu.dk

Abstract—Fringing loss dominated nonlinear winding loss in gapped inductor poses a great challenge for magnetic loss characterization due to the lack of full physical models. This paper investigates the impact of winding arrangements on the AC resistance in gapped inductors, using an EI core as a case study. It specifically examines how the distance of the conductor from the gap affects AC resistance. The study reveals that asymmetric winding arrangements exhibit lower AC resistance compared to symmetric ones. To facilitate rapid computation of eddy current losses, the generation of frequency-based AC resistance maps, and the analysis of AC resistance under various winding configurations, a software tool based on an artificial neural network (ANN) model is developed. The proposed ANN-based model uses $5.7\,\mu s$ with a deviation less than 4.29% with respect to FEA simulations.

Index Terms—Winding loss, fringing effect, data-driven method, artificial neural network, finite element analysis.

I. INTRODUCTION

High frequency (HF) magnetic components, such as HF transformers and HF inductors, are crucial for modern electronics [1]. Winding losses are significantly increased at high frequency as a result of eddy current effects. Eddy current effects, including skin effects, proximity effects, and fringing effects, significantly degrade the performance of magnetic components in high frequency power conversion [2], [3]. These effects lead to nonuniform current density across the conductor's cross section, resulting in higher winding resistance at high frequencies.

The fringing field from the air gap in magnetic components can lead to significant eddy current losses in the winding. These fringing effects create non-homogeneous magnetic field distributions, which are a primary cause of eddy current losses [4], [5]. In the design stage of magnetic components, it is essential to predict these losses to prevent overheating and the need for redesigns. So, a guide on the winding arrangement is needed to mitigate fringing effect. This paper identifies several design optimization factors that help minimize fringing losses.

This work was supported by the European Research Council (ERC) Consolidator Grant under project H3PMAG.

Fringing losses cannot be calculated using the 1-D Dowell equation [6]. Finite element analysis (FEA) is the most widely applied approach to calculate the fringing field, but it can take several hours or even days for simulation. To accurately model fringing losses compared to skin and proximity losses, it is essential to include input variables that describe the distance between the winding and the air gap. This is because the strength of the fringing field varies between different locations within the winding window [5], [7]. A layer close to the air gap will typically experience higher losses than layers farther away [8]. In addition, fringing loss models must take into account the distortion of the fringing field of the conductor due to the feedback of the conductor current on the fringing field [9]. In recent years, analytical models based on solving diffusion and Laplace equations have been developed [10]–[12]. However, these models involve approximations during the derivation process and contain numerous complex exponential terms in their solutions. This complexity makes these models challenging to apply. All of these models have known accuracy limitations for specific winding arrangements. So, a method that can quickly and accurately calculate gapped inductor eddy current losses needs to be built.

Recently machine learning has proven highly effective for solving nonlinear multivariable regression problems. Many efforts for solving nonlinear problems apply machine learning to power electronics; the most popular implementation of such methods is based on Artificial Neural Networks (ANNs) [13]–[18], which can also be applied in predicting fringing loss. In the calculation of winding loss for gapped inductors, compared to the empirical formulation, the ANNs model can integrate more correlated parameters in a unified model and establish the connection between the parameters through the neurons in hidden layers. This paper proposes a hybrid method for predicting winding losses that combines the accuracy of FEA with the flexibility of ANNs. An example based on planar inductors is selected because it is representative. The methods used for planar inductors are versatile and applicable to other power electronic components, such as transformers, and can be widely used in both industry and academia.

979-8-3315-1612-3/25 $31.00 © 2025 IEEE

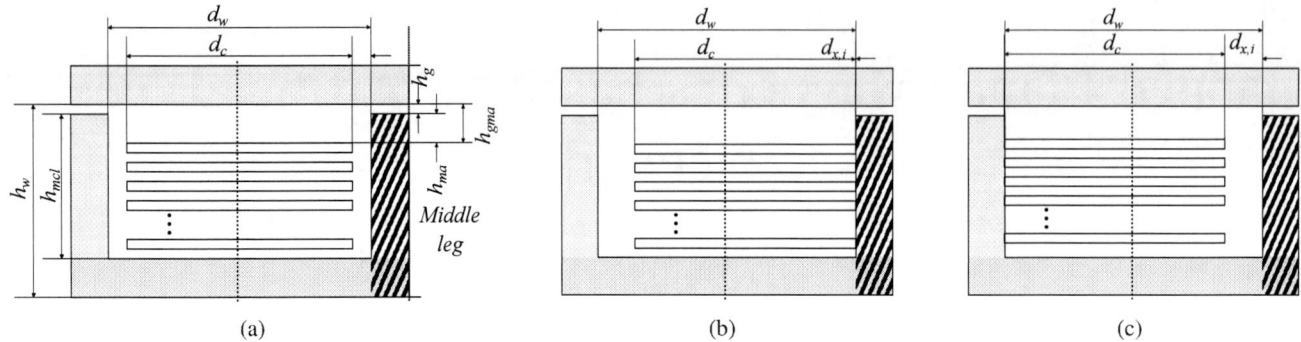

(a)   (b)   (c)

Fig. 1: Simplification of the geometrical model for planar inductor with EI core in the 2-D Cartesian coordinate system. (a) Symmetry model. (b) Asymmetric model ($k_{dx} = 0$). (b) Asymmetric model ($k_{dx} = 1$).

II. KEY PARAMETER IDENTIFICATION FOR WINDING ARRANGEMENT

In planar inductor with EI cores in Fig. 1a, the cross section demonstrates the inner winding arrangement. The model has seven variables: window width d_w, copper width d_c, total number of layers n, window height h_{mcl} and distance from top layer to gap h_{ma}, total height h_w and gap length h_g. To analyze its relationship with AC resistance R_{ac} (which can be decoupled into R_{fring}, R_{skin}, and R_{pro} in Eq. (1)), the normalized variable that defines the distance from the conductor to the gap h_{ma}/h_{mcl} is introduced.

$$R_{ac} = R_{skin} + R_{pro} + R_{fring} \qquad (1)$$

After fixing h_{ma}/h_{mcl}, k_{dx} is defined as a parameter in Eq. (2) describing the horizontal position, where the distance between the conductor and middle leg denoted as $d_{x,i}$. Besides, eddy current ratio R_{ac}/R_{dc} instead of absolute R_{ac} is chosen because it can be further exclude copper cross-sectional area effect in more general cases.

$$k_{dx} = \frac{d_{x,i}}{d_w - d_c} \qquad (2)$$

To study the effect of h_{ma}/h_{mcl} and k_{dx}, Ansys Maxwell is used to calculate the total AC loss. Fig. 2 shows a comparison of R_{ac}/R_{dc} varying with h_{ma}/h_{mcl}. The previous literature in [4] has mentioned that air gaps have a negligible influence if the spacing between the gap and the conductor is at least one-fourth larger than the window width, however, that conclusion is not sufficiently precise. By comparing different winding layers of the same ELP core, it can be seen that the trend of R_{ac}/R_{dc} does not consistently follow same h_{ma}/h_{mcl} between 8-layer and 6-layer winding results. Different h_{ma}/h_{mcl} curves can be obtained from different numbers of layers of winding. This result implies that the one-fourth is only a rough approximation rather than an optimal solution for a specific number of winding layers. A large number of simulation results in 3D and 2D can be summarized as follows:

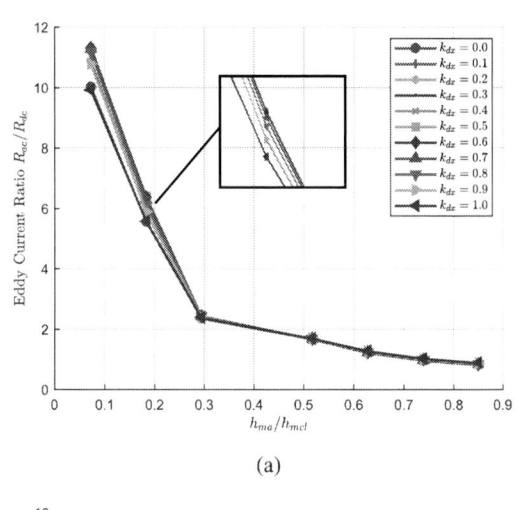

(a)

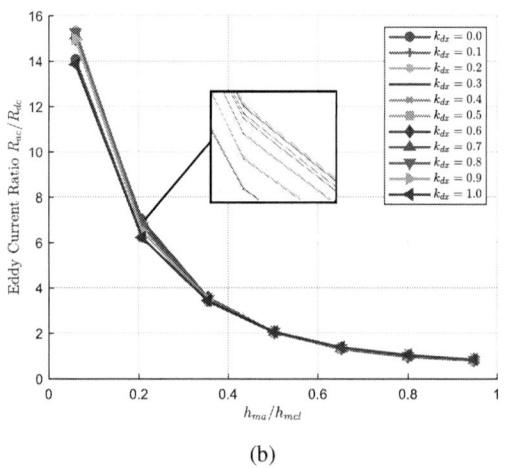

(b)

Fig. 2: Eddy current ratio R_{ac}/R_{dc} varying with h_{ma}/h_{mcl}. The frequency is $1\,\mathrm{MHz}$; the air gap is $0.1\,\mathrm{mm}$; copper thickness of each copper layers is $35\,\mu\mathrm{m}$. (a) n-layers = 6. (b) n-layers = 8.

1) As shown in Fig. 2, the ratio of the winding distance from the gap to the window height is the most critical factor that affects R_{ac}. When h_{ma}/h_{mcl} is less than 0.2,

the slope of this curve changes abruptly, which means that R_{ac} changes exponentially. Until $h_{ma}/h_{mcl} < 0.5$, the R_{ac}/R_{dc} still reduces rapidly. This fact shows that the fringing loss dominates the eddy current loss and R_{fring} is larger than R_{skin} and R_{pro}. The slope of this curve can be used as a tradeoff between core height and winding loss. i.e., larger h_{ma}/h_{mcl} leads to larger core size but less fringing loss.

2) R_{ac} is larger in a symmetric winding arrangement with respect to the axis in Fig. 1a than in an asymmetric arrangement. Counter-intuitively, when the width of the copper does not fill the d_w, R_{ac} is minimized. This means that the copper is tightly centered or sideways (asymmetric winding arrangement) instead of center the conductor in the window (symmetric winding arrangement). In other words, it is better to choose an asymmetric winding arrangement.

To minimize R_{ac} and address the fringing effect, the optimal values for k_{dx} and h_{ma}/h_{mcl} can be crucial. These parameters can significantly influence fringing losses, which vary depending on the window arrangement within the winding window. One potential solution to effectively manage this complexity is to create a comprehensive loss map via extensive simulations, although this approach demands substantial time and resources. An alternative and more efficient approach is to model winding losses in gapped inductors by using neural networks as dynamic datasheets, as described in [13], [14]. This method leverages machine learning to predict loss characteristics, offering a practical tool for design optimization without the extensive computational overhead typically associated with traditional simulations.

III. ANN WINDING LOSS MODEL

To study the effect of the winding arrangement in section II, fast access of R_{ac}/R_{dc} at different h_{ma}/h_{mcl} and k_{dx} is achieved by implementing ANN in R_{ac}/R_{dc} maps.

A. Data Acquisition

This study uses an ANN, to develop a model of AC resistance. Geometry of core and winding as input to simulation. Data concerning eddy current losses are extracted, recorded, and converted to R_{ac}. The eddy current ratio R_{ac}/R_{dc} is the output for the training of the ANN model.

Fig. 3 includes the detail of this work, where the dataset created through simulation is used as input to the ANN. The results of the simulation consist mainly of 2D results, and 3D results are used to calibrate the 2D simulation results to confirm the accuracy of the 2D simulation prior to data collection. Then the datasets are saved in CSV format.

Generally, the number of meshes significantly affects the accuracy of simulation results; using a larger number of meshes typically yields more precise outcomes. However, generating data sets with the optimal mesh configuration is time-consuming. Consequently, a balance must be struck between accuracy and simulation speed. After extensive testing, the mesh refinement study for this article has determined that

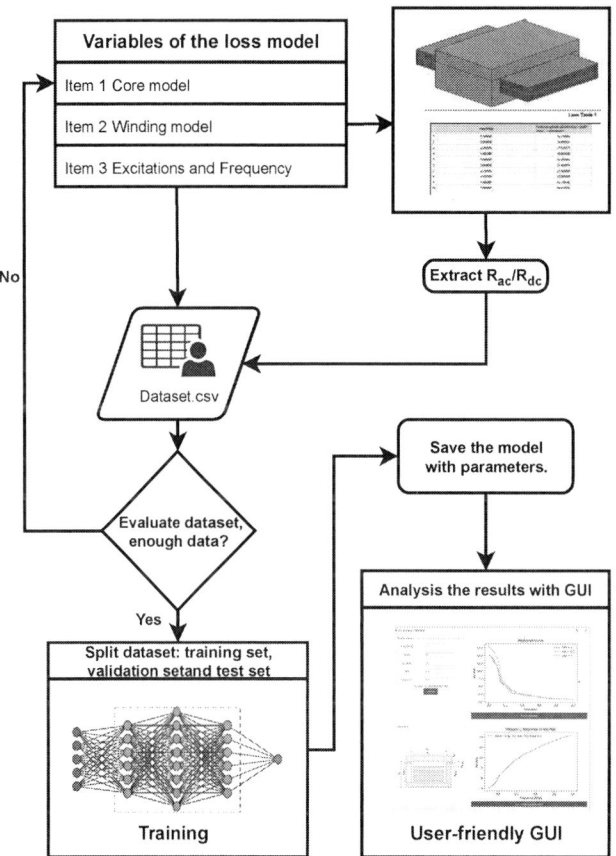

Fig. 3: Workflow for generating the datasets and training the ANN model for winding loss.

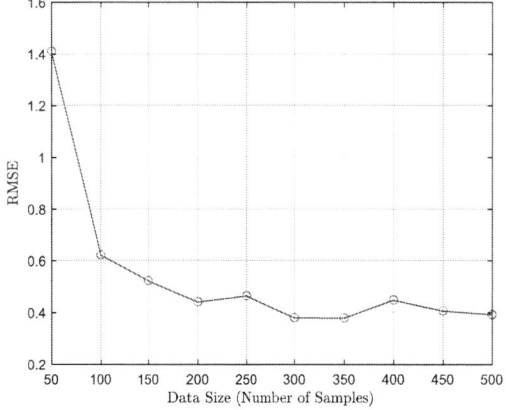

Fig. 4: ANN Performance for Varying Dataset Sizes (Neurons: 16, 32, 16; Epochs: 100).

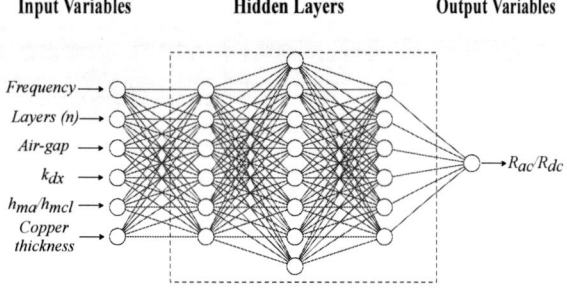

Input Variables   **Hidden Layers**   **Output Variables**

Fig. 5: Structure of an ANN with 6 inputs variables, 1 output variable, three hidden layers.

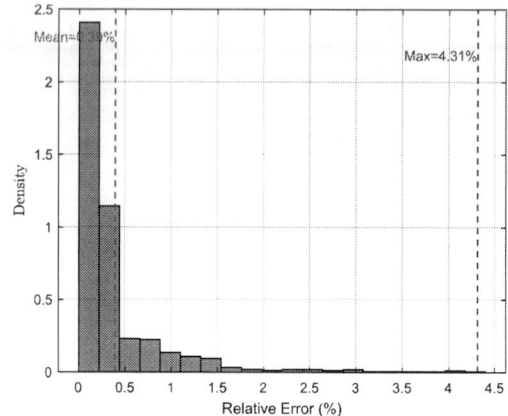

Fig. 6: RMSE error Distribution for comparison of prediction results.

the optimal number of elements is 20,000. This configuration satisfies the accuracy requirements for calculating eddy current losses, even with a copper thickness of $35\,\mu m$.

Fig. 4 presents a study aimed at finding the optimal data size to balance the accuracy of the ANN with the time required to generate the dataset. Each training session randomly splits the dataset into training, validation, and testing phases, with respective proportions of 70%, 15%, and 15%, and involves training each dataset for 200 epochs. The focus is on training specific cases with a fixed number of layers determined by the core, such as the case for core ELP 43/10/28 with 8 layers. This approach is adopted to maximize the efficiency of simulation resources, allowing for the collection of as much data as possible for different cases within limited resources. The Root Mean Square Error (RMSE) in Eq. (3) is utilized to quantify the errors. RMSE is a standard way to measure the accuracy of an ANN model's predictions. It is calculated by taking the square root of the average squared differences between the predicted values ($\hat{y}_i$) and the actual values (y_i). In the formula, n represents the number of observations. The RMSE provides a measure of the magnitude of the prediction error, with lower values indicating better model performance. It has been observed that when the data size exceeds 100, the RMSE falls below 0.6, indicating that the prediction error is within an acceptable range.

$$\text{RMSE} = \sqrt{\frac{1}{n}\sum_{i=1}^{n}(y_i - \hat{y}_i)^2} \tag{3}$$

After obtaining a sufficiently sized dataset, the trained ANN model can support fast computation of R_{ac}.

B. ANN Structure

The model implemented in PyTorch [19], an example of a 6-layer ANN model is illustrated in Fig. 5. This network comprises one input layer, one output layer, and three hidden layers. For ease of understanding and computation, the input layer has six variables: the number of winding layers n, the length of the air gap h_g, the normalized parameter for the horizontal position k_{dx}, the normalized parameter describing the distance from the top layer to the air gap h_{ma}/h_{mcl},

the copper thickness t_c and the excitation frequency f. The output layer includes one parameter: the eddy current ratio R_{ac}/R_{dc}. The total eddy current loss can be determined using R_{ac}/R_{dc}. This ANN model, by providing a measure of total resistance, offers a quick method for power electronics designers to estimate the overall winding loss, which varies with the relative position to the air gap and the intensity of the fringing field.

TABLE I: Comparsion of Different Neurons in Hidden Layers

Cases	Neurons in hidden layers	Params	Max error	RMSE
1	(8,16,8)	337	7.84%	1.03%
2	(16,32,16)	1185	6.01%	0.81%
3	(32,64,32)	4417	5.85%	0.77%
4	(64,128,64)	17025	5.30%	0.74%

For an ANN model, the number of neurons in the hidden layers directly influences the model's size and the total number of parameters. Therefore, it is essential to find an optimal balance between model size and computational accuracy. To validate the appropriate number of neurons, four different cases are compared in Tab. I. It can be seen that the number of neurons, when raised to a certain number, does not result in a significant increase in the accuracy of the prediction results. On the contrary, it leads to a larger model (more parameters). Once the size of each design case dataset and the optimal number of neurons are determined, the training of the ANN can begin. Given that the current dataset includes various cases with different winding arrangements, a total of 64 neurons was chosen for the model. Specifically, in case 2, the neuron configuration was set to (16, 32, 16). This structure allows for a balanced distribution of computational resources across layers, potentially improving the model's ability to capture complex patterns within the dataset.

Considering that core size directly influences the length and number of winding layers, and given the challenges in normalizing core windows with varying geometries and sizes, this study treats the core as a whole as a variable. Although the core type is incorporated as an input variable in the ANN model, this approach does not align well with practical physical considerations. Furthermore, adding the core type as an additional dimension to the ANN is problematic due to fitting limitations. Therefore, using a separate ANN as a datasheet for each core type presents a more viable solution.

Unlike the commonly used Sigmoid, the Rectified Linear Unit (ReLU) activation function is selected to mitigate the vanishing gradient issue, yet the Exponential Linear Unit (ELU) proves more effective, particularly because it aligns with the exponential nature of the partial differential equations used for calculating fringing loss.

Fig. 6 shows the results of the ANN model training, represented by RMSE, from the comparison of the simulated data and the predicted results, where the simulated data were randomly divided into the testset. This comparison shows that the neural network can achieve results comparable to the simulation. The maximum error is about 4.31%, the average error is 0.35%, and the largest interval of the error distribution is between 0 and 0.5%. It can be considered that the results predicted by the ANN are sufficiently accurate.

Calculations of the eddy current ratio R_{ac}/R_{dc} are swiftly performed in just 5.7 μs per point using a Geforce RTX 3050ti, enabling efficient AC resistance evaluations across various winding configurations and frequencies.

IV. PROTOTYPING AND SOFTWARE IMPLEMENTATION

With the presented method and the trained model, planar magnetics with EI core can be quickly optimized. Tab. II shows a 6 layers planar inductor with ELP 43/10/28 core. The R_{ac} measured with the Agilent 4294A precision impedance analyzer [20] is shown in Fig. 7.

TABLE II: Prototype Parameters

Parameter	Value
Copper thickness (t_c)	0.035 mm
Copper width (d_c)	12.65 mm
Layers (n)	6
Air gap length (h_g)	0.5 mm
Insulation thickness (d)	0.8 mm
Core	ELP 43/10/28
Prototype	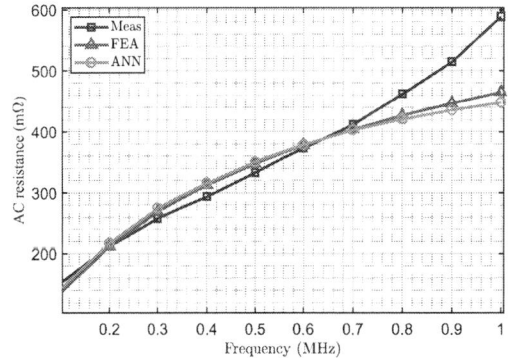

Fig. 8 compares the measured values with the simulation and ANNs-based model. The results show that the maximum deviations from 100 kHz to 1 MHz between 3D simulation and the ANN is 4.29% while the average error is 1.93%. R_{ac}/R_{dc} is 15.3 at 1 MHz where the total resistance between R_{prox} and R_{skin} is only 1.63 by Dowell equation, which means that about 89.3% of eddy current loss comes from the nonlinear fringing loss. Compared to measured results, the ANN predictions have an average error of 8.15% and a maximum error of 23.94%. FEA prediction data show an average error of 7.16% and a maximum error of 21.05%. This ANN model reduces the number of simulation hours

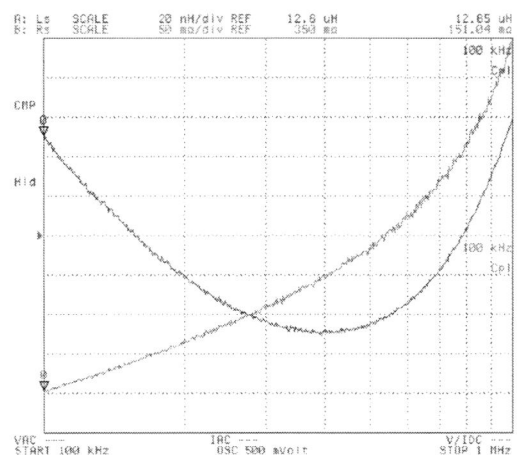

Fig. 7: Measurement results from impedance analyzer.

to less than 6 μs. The results show that the ANN results are extremely close to the FEA results and within reasonable error to the measured results. Measurement deviations are due to inconsistencies in the distance between the soft copper foil windings, as well as terminal losses.

A user-friendly graphical user interface (GUI) is developed as shown in Fig. 9. It significantly simplifies the design process. This GUI allows inductor designers to conveniently input the geometry and design constraints of the inductor into text boxes. By clicking the Run button, the software generates the optimized shape and power loss predictions. This software is now available for download on GitHub [21] and is still being updated. More than 5,000 data samples from both 3D and 2D simulations were collected.

Fig. 8: Experimental losses comparison results.

V. CONCLUSION

This article applies machine learning to model power magnetics. We first present the normalized key parameters k_{dx} and h_{ma}/h_{mcl}, which accurately describe the winding losses in gapped inductors, for data-driven winding loss modeling. Subsequently, an initial dataset was established using FEA and trained using ANN, validating the feasibility of ANN predictions for nonlinear winding losses. Experimental results

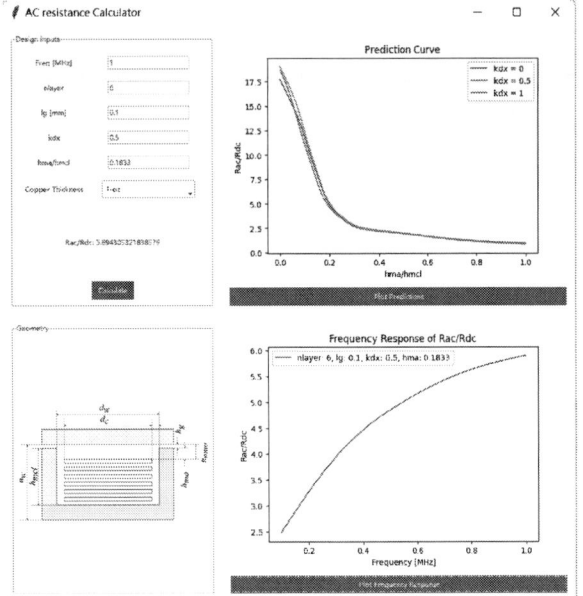

Fig. 9: Graphical interface facilitating interactive of AC resistance estimation with a specific design example

demonstrate that trained ANNs offer low computational costs and can provide real-time estimations of winding losses driven by the fringing effect, which are otherwise challenging to accurately calculate with analytical models. These findings confirm that ANNs, trained from FEA results, can serve as a reliable and practical tool for estimating winding losses in high-frequency magnetic components.

REFERENCES

[1] W. G. Hurley and W. H. Wölfle, *Transformers and Inductors for Power Electronics: Theory, Design and Applications*. Wiley, 1 ed.

[2] A. Van den Bossche and V. Valchev, *Inductors and Transformers for Power Electronics*. 01 2005.

[3] Z. Ouyang and M. A. E. Andersen, "Overview of planar magnetic technology—fundamental properties," *IEEE Transactions on Power Electronics*, vol. 29, no. 9, pp. 4888–4900, 2014.

[4] J. Hu and C. Sullivan, "Ac resistance of planar power inductors and the quasidistributed gap technique," *IEEE Transactions on Power Electronics*, vol. 16, no. 4, pp. 558–567, 2001.

[5] W. A. Roshen, "Fringing field formulas and winding loss due to an air gap," *IEEE Transactions on Magnetics*, vol. 43, no. 8, pp. 3387–3394, 2007.

[6] P. L. Dowell, "Effects of eddy currents in transformer windings," 1966.

[7] W. A. Roshen, "High-frequency fringing fields loss in thick rectangular and round wire windings," *IEEE Transactions on Magnetics*, vol. 44, no. 10, pp. 2396–2401, 2008.

[8] J. Sun and V. Mehrotra, "Orthogonal winding structures and design for planar integrated magnetics," *IEEE Transactions on Industrial Electronics*, vol. 55, no. 3, pp. 1463–1469, 2008.

[9] P. Wallmeier, "Improved analytical modeling of conductive losses in gapped high-frequency inductors," *IEEE Transactions on Industry Applications*, vol. 37, no. 4, pp. 1045–1054, 2001.

[10] T. Ewald and J. Biela, "Analytical eddy current loss model for foil conductors in gapped cores," in *2021 23rd European Conference on Power Electronics and Applications (EPE'21 ECCE Europe)*, pp. 1–10, IEEE.

[11] T. Ewald and J. Biela, "Frequency-dependent inductance and winding loss model for gapped foil inductors," *IEEE Transactions on Power Electronics*, vol. 37, no. 10, pp. 12370–12379.

[12] Z. Yu, X. Yang, Y. Xu, Q. Gao, Y. Zhou, J. Wu, K. Wang, F. Zhang, L. Wang, and W. Chen, "2-d analytical copper loss model for PCB and copper foil magnetics with arbitrary air gaps," *IEEE Transactions on Power Electronics*, vol. 38, no. 11, pp. 14274–14291.

[13] H. Li, D. Serrano, T. Guillod, E. Dogariu, A. Nadler, S. Wang, M. Luo, V. Bansal, Y. Chen, C. R. Sullivan, and M. Chen, "MagNet: An open-source database for data-driven magnetic core loss modeling," in *2022 IEEE Applied Power Electronics Conference and Exposition (APEC)*, pp. 588–595, IEEE.

[14] H. Li, D. Serrano, S. Wang, and M. Chen, "MagNet-AI: Neural network as datasheet for magnetics modeling and material recommendation," *IEEE Transactions on Power Electronics*, vol. 38, no. 12, pp. 15854–15869.

[15] B. N. Sanusi, M. Zambach, C. Frandsen, M. Beleggia, A. Michael Jørgensen, and Z. Ouyang, "Investigation and modeling of dc bias impact on core losses at high frequency," *IEEE Transactions on Power Electronics*, vol. 38, no. 6, pp. 7444–7458, 2023.

[16] J. Deng, W. Wang, Z. Ning, P. Venugopal, J. Popovic, and G. Rietveld, "High-frequency core loss modeling based on knowledge-aware artificial neural network," *IEEE Transactions on Power Electronics*, vol. 39, no. 2, pp. 1968–1973.

[17] D. Santamargarita, D. Molinero, E. Bueno, M. Marrón, and M. Vasić, "On-line monitoring of maximum temperature and loss distribution of a medium frequency transformer using artificial neural networks," *IEEE Transactions on Power Electronics*, vol. 38, no. 12, pp. 15818–15828.

[18] N. Rasekh, J. Wang, and X. Yuan, "Artificial neural network aided loss maps for inductors and transformers," *IEEE Open Journal of Power Electronics*, vol. 3, pp. 886–898.

[19] Paszke, Adam and Gross, Sam and Massa, Francisco and Lerer, Adam and Bradbury, James and Chanan, Gregory and Killeen, Trevor and Lin, Zeming and Gimelshein, Natalia and Antiga, Luca and Desmaison, Alban and Kopf, Andreas and Yang, Edward and DeVito, Zachary and Raison, Martin and Tejani, Alykhan and Chilamkurthy, Sasank and Steiner, Benoit and Fang, Lu and Bai, Junjie and Chintala, Soumith, "PyTorch: An imperative style, high-performance deep learning library."

[20] "Agilent 4294A Precision Impedance Analyzer Service Manual."

[21] "https://github.com/HanqingCao/PE-fringing-loss."

Free-Shape Optimization of VHF Air-Core Inductors using a Constraint-Aware Genetic Algorithm

Thomas Guillod and Charles R. Sullivan

Dartmouth College, Hanover NH, United States

Abstract—**This paper focuses on the optimization of air-core inductors which are widely used in Very High Frequency (VHF) integrated converters. Instead of considering classical geometries (e.g., spiral and solenoid), a free-shape optimization algorithm is implemented, i.e. any geometry respecting the design rules can be considered. The optimization is performed with a fast Partial Element Equivalent Circuit (PEEC) solver and a custom genetic algorithm that enforces the non-linear design constraints. Finally, the inductor of a 1.6 W Integrated Voltage Regulator (IVR) operated at 40.68 MHz is optimized under various conditions (minimization of the losses, placement of the terminals, footprint constraint, and magnetic near-field reduction). It is found that the shape optimizer is particularly useful for problems with complex and/or unusual constraints.**

Index Terms—**Air-core inductor, electromagnetics, integrated voltage regulator, shape optimization, topological optimization, genetic algorithms, PEEC, open-source software.**

I. INTRODUCTION

Air-core inductors feature several advantages for integrated converters: linearity, advantageous scaling at high frequency, and compatibility with various microfabrication processes [1]–[4]. Several standard coil geometries can be found in the literature (e.g., spiral, staple, solenoid, and toroid) [1], [5] and the corresponding parameters can be determined with multi-objective optimization [6], [7]. However, these standard coil geometries only cover a subset of the complete design space offered by the fabrication processes. Therefore, it is unclear if such geometries are optimal, especially with complex constraints and objectives (e.g., multi-layer process, non-standard footprint, and near-field limit).

Free-shape optimization (also called topological optimization) techniques can be used to explore the full design space. Such algorithms have been successfully applied to various electromagnetic problems such as electrical machine geometry, magnetic core shape, or magnetic field shaping [8]–[13]. More specifically, free-shape coil optimization has been demonstrated for Magnetic Resonance Imaging (MRI) and high-energy physics applications [14]–[18]. However, these methods are not fully compatible with the design constraints (design rules and objective function) of air-core inductors used in IVRs.

A fundamental question for free-shape optimization is the representation of the geometry. General descriptions, such as pixel matrices or meshes, can be used to represent the component's geometry [8], [19], [20]. However, imposing complex design rules on such arbitrary geometries is not always possible and invalid designs can be generated [8], [11]. This is particularly critical for magnetic components as the optimal designs are often located at the boundary of the design rules. For this reason, a specialized geometry description using variable-width traces and vias (similar to the GERBER format) is selected for the parametrization of air-core inductors.

A second important decision is the choice of the optimization algorithm. Free-shape optimization methods can be divided into two main categories: gradient-based and gradient-free methods. For air-core inductors, discrete variables (e.g. number of turns and layers), local minima, non-linear/non-differentiable/discontinuous objectives, and complex constraints are typical and limit the applicability of gradient-based methods [21]. In this paper, a genetic algorithm, that is able to enforce complex constraints, is selected [13], [19], [20]. A typical bottleneck for topological optimization of magnetics, is the computational cost of the magnetic field simulation, especially for 3D geometries affected by high-frequency Eddy currents. In this work, a custom FFT-accelerated PEEC solver is used and allows for the computation of thousands of 3D geometries per hour [22], [23].

This paper is organized as follows. Section II presents the design rules, the frequency-domain magnetic field solver, and the optimization algorithm. Section III applies the developed workflow to the inductor of a 1.6 W integrated Buck converter operated at 40.68 MHz. Section IV compares the optimization results with different constraints and objectives. Finally, the Python implementation of the proposed free-shape optimization workflow (including the 3D PEEC solver) is available under an open-source license [22], [24], [25].

II. OPTIMIZATION METHOD

A. Optimization Workflow

Fig. 1(a) depicts the optimization workflow. The shape optimization process is divided into two main steps: the generation of an initial pool of valid designs and the global optimization algorithm (using a constraint-aware genetic algorithm). The constraint and objective functions (see Fig. 1(b)) convert an abstract description of the component's geometry into a set of constraint and objective values.

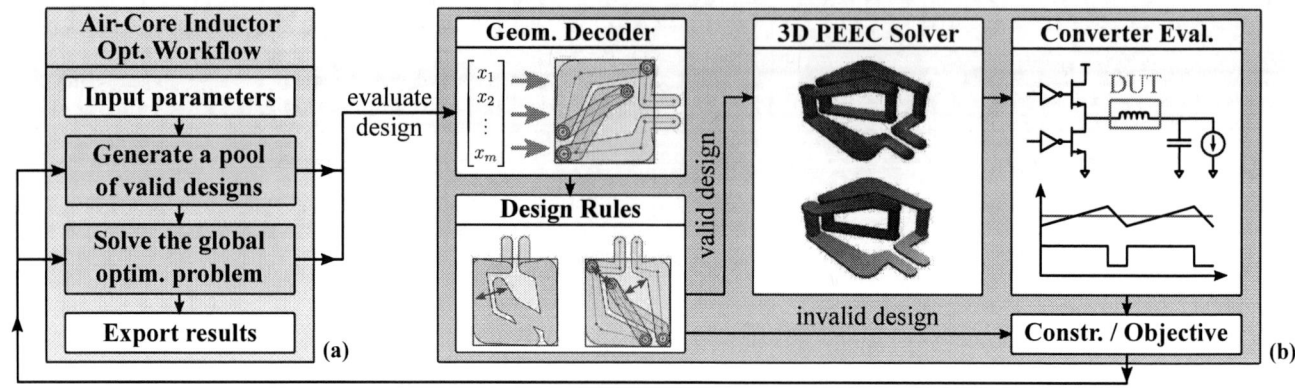

Fig. 1. (a) The shape optimization workflow is divided into two sequential steps: the generation of an initial pool of valid designs and the global optimization algorithm (using a constraint-aware genetic algorithm). (b) The constraint function evaluates the compatibility of the geometry with the design rules. The objective function computes the inductor performance within the power converter.

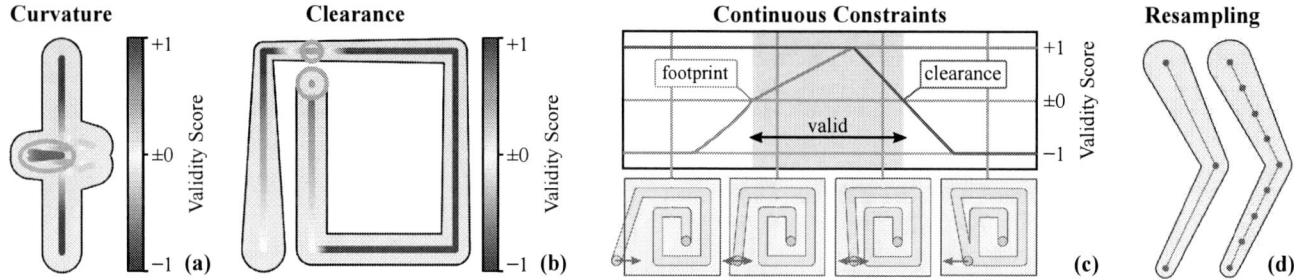

Fig. 2. (a) Design rule eliminating noisy shapes. (b) Design rule computing the clearance distances. (c) Illustration of the non-boolean constraints (footprint and clearance constraints). The validity score is normalized into $[-1, +1]$ where negative values label a constraint violation. (d) Resampling highlights that the description of a given geometry is not unique.

The geometry description consists of a vector describing the trace coordinates, the trace width, and the position of the traces in the layer stack. This geometry description is first converted into a set of polygons and the design rules are checked. With free-shape optimization techniques, it is critical that the design constraints are complete and robust as the optimizer is likely to find and exploit any loophole. For the considered multi-layer designs, the following design rules are implemented: footprint constraint, clearance distance, trace length, trace width, trace width variation, curvature radius of the traces, and angle between traces. The curvature radius constraint (see Fig. 2(a)) is computed with a convolution filter and is used to penalize noisy geometries that free-shape optimizers tend to generate [8], [19]. The clearance between distinct traces and vias can be easily computed. However, computing the clearances within the same trace (see Fig. 2(b)) is more challenging and is done by comparing the Euclidean distance and the shortest distance along the shape for the different points composing the trace. As shown in Fig. 2(c), the constraints are not implemented as boolean variables but as continuous variables determining how close the design is to the validity threshold. Additionally, it should be noted that the description of a given geometry is not unique (see Fig. 2(d)) and the design rules have to

be consistent for such scenarios. Leveraging an optimized computational geometry framework ("Shapely"), it is possible to decode and check 1620 geometries per second (with an AMD EPYC 9354 CPU) [26].

For invalid designs, the corresponding design rule violations are returned by the constraint function and the objective function value is penalized. For valid designs, a 3D frequency-domain solver is used for the extraction of the magnetic parameters (e.g., DC/AC inductance, DC/AC resistance, and near-field pattern). The PEEC method is particularly well-suited for simulating air-core inductors [27]. In this paper, a modern variant of the PEEC method, that represents the geometry with a voxel structure, is selected [23]. The selection of a voxel structure features two key advantages: most of the coefficients of the inductance matrix are repeated (reduction of the computational cost and memory footprint from $O(n^2)$ to $O(n)$) and the dense matrix multiplication can be accelerated with a FFT algorithm (reduction of the computational cost from $O(n^2)$ to $O(n \log(n))$). The open-source implementation developed by the authors ("PyPEEC") is able to solve 6.5 geometries per second (with an AMD EPYC 9354 CPU, more details in Appendix A) [22].

979-8-3315-1612-3/25 $31.00 © 2025 IEEE

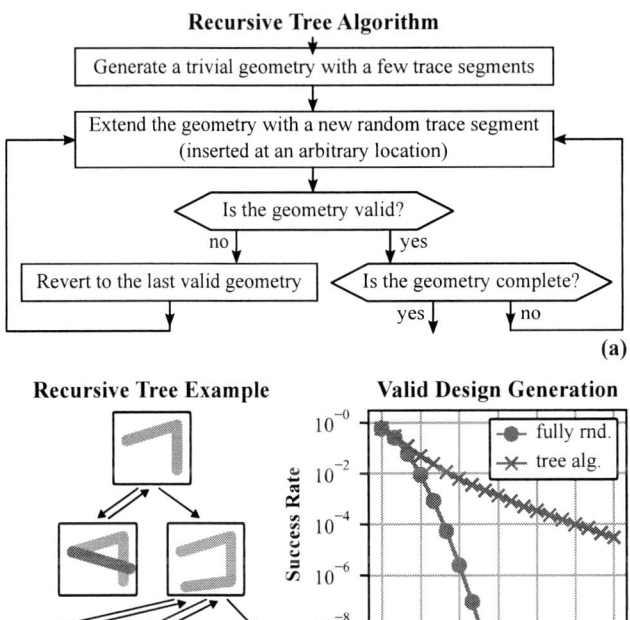

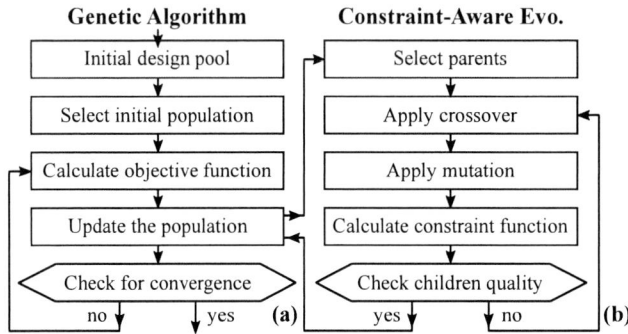

Fig. 4. (a) Optimization workflow using a genetic algorithm. (b) Constraint-aware update (through selection, crossover, and mutation) of the algorithm population between the generations.

Instead of generating the full geometry directly, the algorithm progressively constructs the coil by adding trace segments and backtracks when a design rule violation is encountered.

Fig. 3(c) shows the generation success rate for the proposed recursive tree algorithm and the fully random generation method. The success rate is computed with the average number of design rule checks required to find a valid design. It can be seen that the recursive tree algorithm allows for the generation of valid coil geometries with a few thousand trials, which is fully compatible with the computational cost of the design rule checks.

C. Genetic Algorithm

The optimization problem features a large design space, local minima, continuous variables (trace coordinates and widths), discrete variables (position of the traces in the layer stack), non-linear/non-differentiable constraints, and a non-linear/discontinuous objective function. Different global optimization methods (e.g., genetic algorithm, differential evolution, particle swarm, tree-structured Parzen estimator, CMA-ES, and NgIohTuned) have been compared and it has been found that the main difficulties are linked to the complex and restrictive design rules [28]–[31]. A custom genetic algorithm that integrates the design rules (implemented with the "PyGAD" framework) is selected as the most robust optimization method [29].

The complete optimization workflow is depicted in Fig. 4(a). After the generation of a pool of valid geometries, the best designs are selected as the initial population of the genetic algorithm. Afterward, the genetic algorithm, iteratively, computes the objective function and updates the population until the convergence criterion is reached.

Two mechanisms are used to generate new designs (or children) from existing designs (or parents): crossover and mutation. The single-point crossover process is illustrated in Fig. 5(a) where the two parent geometries are cut at a random location and recombined. The random mutation process is depicted in Fig. 5(b) where part of the parent geometry (trace coordinates, trace widths, and/or position of the traces in the layer stack) is randomly modified. These examples show how

Fig. 3. (a) Recursive tree algorithm used to generate random designs that respect the design rules. (b) Illustration of the recursive random design generation. (c) Success rate of the random design generation process. The recursive tree algorithm is compared with the fully random generation method (for different numbers of trace segments).

It should be noted that the inductor properties and the semiconductor losses are interdependent as the inductance value directly impacts the ripple current and, therefore, the current flowing through the semiconductors. This implies that the inductor with the lowest losses is not necessarily optimal at the system level. Hence, the performance of the inductor is evaluated together with the converter and the total losses (DC/AC inductor losses and semiconductor conduction/switching losses) are computed, ensuring that the inductor design is optimal at the system level (co-optimization) [6].

B. Initial Design Pool

An initial pool of inductor designs is required for selecting the initial population of the genetic algorithm. A possibility is to feed the optimizer with classical geometries (e.g., spiral and solenoid). However, the optimizer is then biased towards these geometries and the full design space will not be explored. Therefore, a random generation of the initial population is preferable. Yet, the probability that a fully random two-layer coil (composed of 10 trace segments) respects the design rules is lower than 10^{-9}. This implies that the computational cost to find a valid initial population with fully random coil geometries is unreasonable. This also highlights that a tiny fraction of the design space is valid, which represents a challenge for the optimization algorithm. This issue is mitigated by the recursive tree algorithm illustrated in Figs. 3(a)-(b).

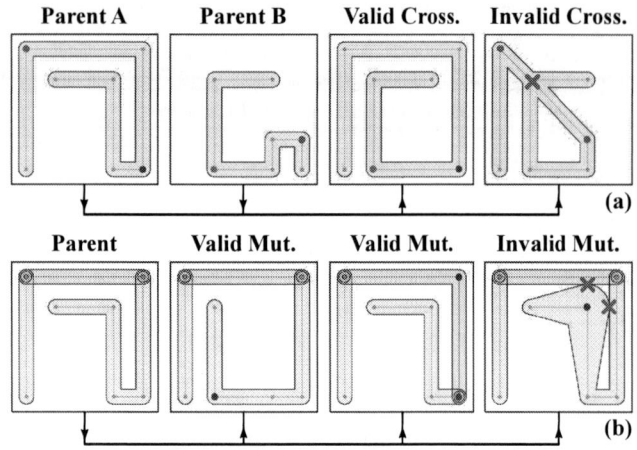

Fig. 5. (a) Single-point crossover. (b) Random mutation. The location of the crossovers and mutations are highlighted with colored dots. The design rule violations are indicated with a red cross.

crossover and mutation operations are able to generate better children geometries from the parent geometries. However, the crossover and mutation processes can also generate invalid children from valid parents. This implies that the population of the genetic algorithm is quickly dominated by invalid designs. This problem is solved by iteratively applying the genetic operation (crossover, mutation, or both) and the design rule checks until the quality of the generated children reaches a predefined threshold (see Fig. 4(b)).

Fig. 6 illustrates the constraint-aware update of the population. A parent population composed of exclusively valid random geometries (with 10 trace segments) is considered and the fraction of valid children is monitored during the population update process. Without the iterative constraint enforcement, the crossover and mutation operators produce less than 10 % and 5 % of valid children, respectively. After each iteration, the number of valid children increases and the process is repeated until the required quality threshold is attained (or a maximum number of iterations is reached). It should be noted that this iterative process is used between every generation, implying that the constraint function is called more often than the objective function (see Fig. 4). However, as the computational cost of the constraint function is much lower than the objective function, this does not represent a problem.

D. Monitoring and Post-Processing

Through the complete optimization process (initial design pool generation and genetic algorithm), all the generated designs (coil geometry, design rule checks, magnetic parameters, field patterns, converter waveforms, inductor losses, semiconductor losses, etc.) are stored in a PostgreSQL database [32]. This allows for a detailed tracking of the design space exploration and the convergence processes. The complete optimization workflow also takes advantage of parallel computing.

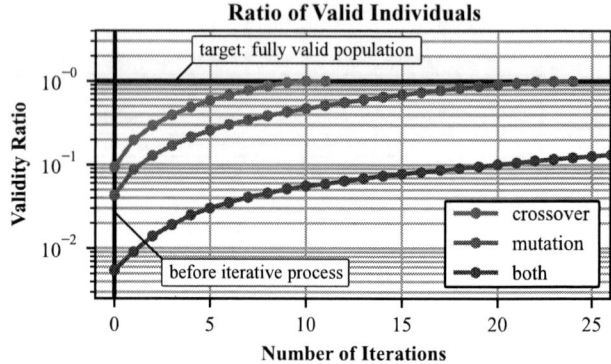

Fig. 6. Constraint-aware update of the population (crossover, mutation, or both). The parent population is composed of exclusively valid random geometries (with 10 trace segments). Single-point crossover (crossover rate of 100 %) and random mutation (mutation rate of 10 %) are used.

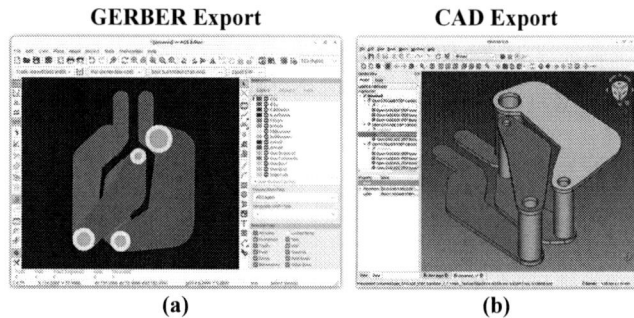

Fig. 7. (a) GERBER export. (b) CAD export.

At the end of the optimization process, the optimal designs are retrieved from the database and a comprehensive report is generated (characteristics, figures of merit, geometry plots, current density plots, and magnetic near-field plots). The coil geometry can also be exported in various formats: 2D vector shapes, 3D STL files, 3D STEP files, and GERBER files (see Fig. 7). More details on the open-source implementation of the shape optimization workflow can be found in Appendix B.

III. OPTIMIZATION RESULTS

A. Optimization Problem

The inductor uses the following air-core technology: two-layer process, 24 μm copper layer thickness, 312 μm substrate thickness, 30 μm minimum clearance distance, 80 μm minimum trace width, and 380 μm maximum trace width. Additionally, the inductor has to fit into the following footprint constraint: 1 mm × 1 mm. The number of trace segments forming the coil can vary between 6 and 12.

The air-core inductor performance metrics are evaluated within a VHF IVR [2], [4]. A two-level Buck converter operating between 3.3 V and 0.8 V with a 2.0 A output current in the 40.68 MHz ISM band is selected. For each inductor design, the optimal modulation scheme (CCM or DCM) is selected and the waveforms are computed.

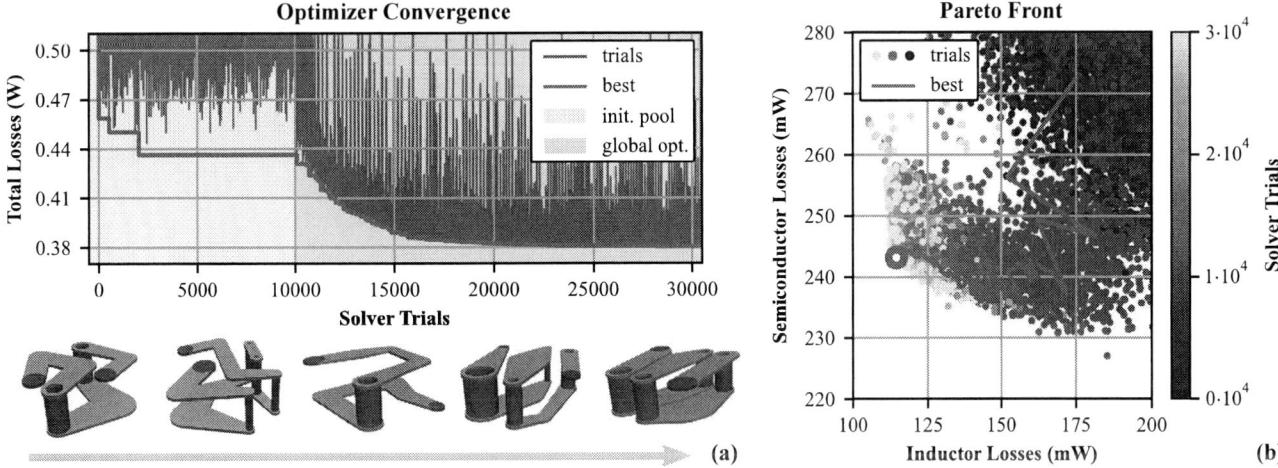

Fig. 8. (a) Convergence of the optimization workflow (initial design pool generation and genetic algorithm). (b) Pareto front representing the evolution of the trade-off between the inductor and semiconductor losses. The best candidates are highlighted in red.

For the switches, a simplified model consisting of a specific (per chip area) conduction resistance ($R_{\mathrm{on,sp}}$) and switching energy ($E_{\mathrm{sw,sp}}$) is considered. Then, the optimal switch size and losses can be computed as [33], [34]

$$A_{\mathrm{opt}} = \sqrt{\frac{I_{\mathrm{RMS}}^2 R_{\mathrm{on,sp}}}{f_{\mathrm{sw}} E_{\mathrm{sw,sp}}}},$$
$$P_{\min} = 2\sqrt{I_{\mathrm{RMS}}^2 f_{\mathrm{sw}} R_{\mathrm{on,sp}} E_{\mathrm{sw,sp}}},$$

where f_{sw} is the switching frequency and I_{RMS} the RMS current through the switch. Typical parameters for a SOI process (180 nm and 1.8 V) are considered: $R_{\mathrm{on,sp}} = 0.45\,\mathrm{mOhm} \cdot \mathrm{mm}^2$ and $E_{\mathrm{sw,sp}} = 14.22\,\mathrm{nJ/mm}^2$. The transistor voltage is scaled using a cascode arrangement with two transistors [34].

For each coil geometry, the following steps are computed: design rule checks, 3D field simulation, extraction of the DC/AC magnetic parameters, computation of the converter RMS and peak currents, DC/AC inductor losses, DC/AC inductor current density, DC/AC magnetic near-field, optimal chip area, and semiconductor losses. Finally, the total losses of the converter system are extracted and used as the objective for the optimization process.

B. Optimization Results

Fig. 8(a) shows the optimization results. During the initialization phase, the objective value quickly stops to improve, indicating that the problem is too large for a brute-force approach. Once 10000 valid designs are obtained, the initialization is stopped and 400 designs are selected as the initial population of the genetic algorithm. The following selection method is used: 300 designs are extracted for having the best performance and 100 designs are randomly chosen in order to ensure the diversity of the initial population. The genetic algorithm quickly

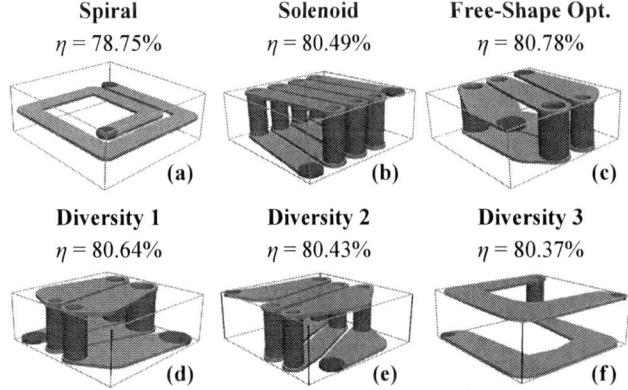

Fig. 9. (a) Optimal spiral geometry (classical optimization). (b) Optimal solenoid geometry (classical optimization). (c) Optimal free-shape geometry. (d)-(f) Free-shape designs generated during the optimization process (illustrating the design space diversity).

improves the objective value and convergence is achieved after 20000 evaluations. The optimal geometry consists of a two-layer spiral design. Interestingly, both spiral and solenoid shapes are generated during the optimization process, indicating that different concepts are competing with each other. Fig. 8(b) illustrates the trade-off between the inductor and semiconductor losses. It can be seen that the algorithm is, progressively, constructing and extending the performance space until the optimal design ($\eta = 80.67\,\%$), located on the Pareto front, is found.

It is found that 10000 initial designs and a population of 400 individuals is a good compromise between the computational speed and the risk of getting trapped in a local minimum. It should be noted that, due to the stochastic nature of the optimizer, there is no guarantee that two optimization runs will generate the same design. However, the achieved efficiency can be reproduced within a $\pm 0.2\,\%$ margin.

	Floating Terminals	Fixed Terminals	Footprint Constraint	Three-Layer Geometry	Half-Load Optimization	Three-Level Topology	Near-Field Optimization
	(a)	(b)	(c)	(d)	(e)	(f)	(g)
$\eta_{\text{Converter}}$ [%]	80.78	78.98	77.79	79.96	78.79	83.80	78.12
η_{Inductor} [%]	93.31	90.80	89.74	90.55	90.84	93.46	90.22
R_{DC} / R_{AC} [mΩ]	11.9 / 19.1	16.9 / 32.4	17.8 / 28.0	21.8 / 43.4	32.4 / 52.8	11.1 / 18.5	16.5 / 26.9
L_{DC} / L_{AC} [nH]	1.99 / 1.88	2.17 / 1.99	1.74 / 1.66	3.42 / 3.12	3.81 / 3.61	1.31 / 1.23	1.76 / 1.66
H_{DC} / H_{AC} [A/m]	733.1 / 618.9	771.5 / 616.7	677.4 / 667.0	1413.3 / 895.7	395.2 / 346.1	712.6 / 650.1	288.8 / 249.8

Fig. 10. Free-shape optimized inductors under various conditions: (a) floating terminals, (b) fixed terminals, (c) complex footprint constraint (non-square outline and keepout area), (d) three-layer geometry, (e) half-load optimization, (f) three-level converter, and (g) magnetic near-field reduction. The RMS magnetic near-field is evaluated at a distance of 0.3 mm from the coil and the maximum value of considered.

Figs. 9(a)-(b) depict the optimal spiral and solenoid inductors obtained with classical optimization (and not with free-shape optimization). It can be observed that the solenoid is more efficient than the single-layer spiral. Fig. 9(c) shows the free-shape geometry obtained with a large optimization run (50000 initial designs, 2000 individuals, and 60000 evaluations for the genetic algorithm). It can be seen that the free-shape optimized design is slightly more efficient than the solenoid (80.78 % vs. 80.49 %). However, it appears that, in this case, the advantage of free-shape optimization is marginal.

Figs. 9(d)-(f) highlight several designs generated during the optimization process. It can be seen that widely different geometries features similar performance, indicating that optimization problem features several local minima [21]. This confirms that a local optimization algorithm cannot solve this problem. However, this design space diversity offers an opportunity to include additional constraints with a limited impact on the achieved efficiency.

IV. DESIGN SPACE EXPLORATION

A. Terminal Constraint

Fig. 9 indicates that the position of the inductor terminals is variable and inconsistent between the designs. The first problem is that the interconnects between the coil and the converter are potentially difficult or even impossible to construct. Furthermore, the resistance and inductance of the interconnects are not always negligible. Finally, the inductance of an open loop is ill-defined and only the partial inductance is properly defined [35].

Fig. 10(a) depicts the optimal design without any terminal constraint (see Fig. 9(c)). In contrast, Fig. 10(b) shows the optimal design with a coplanar terminal constraint. It can be observed that the impact of the terminals on the coil geometry and the achieved efficiency is not negligible (1.8 % reduction). The coplanar terminal constraint will be used for the remainder of the paper and the design shown in Fig. 10(b) will be used as a benchmark.

B. Footprint Constraint

Fig. 10(c) shows the design obtained with a complex footprint constraint (non-square outline and keepout area). The design is only slightly less efficient than the reference design (1.2 % reduction), which indicates that the free-shape optimization is able to find good geometries with non-standard constraints.

C. Three-Layer Inductor

Fig. 10(d) depicts the obtained design with a third metal layer. The new layer is inserted in the middle of the layer stack and the total thickness of the component remains unchanged. With this additional degree of freedom, the efficiency is slightly higher (1.0 % improvement), which is explained by the higher inductance value and lower ripple current. More precisely, the additional layer allows for better trade-offs between the inductance and resistance values.

D. Half-Load Optimization

Fig. 10(e) shows the optimal coil geometry when the output current of the converter is reduced from 2.0 A to 1.0 A. In order to limit the ripple ratio, a higher inductance value is selected. The increase of the resistance value is mitigated by the reduction of the output current and the efficiency of the converter remains almost unchanged (0.2 % reduction).

E. Three-Level Converter

Fig. 10(f) shows the optimal design with a three-level flying capacitor Buck converter [36], [37]. The frequency applied to the inductor remains unchanged (40.68 MHz), implying the switching frequency of the semiconductors is reduced by a factor of two (20.36 MHz). The reduced voltage applied to the inductor shifts the optimum towards a design with low inductance and resistance values. The three-level converter is significantly more efficient (4.8 % improvement) than the two-level variant.

979-8-3315-1612-3/25 $31.00 © 2025 IEEE

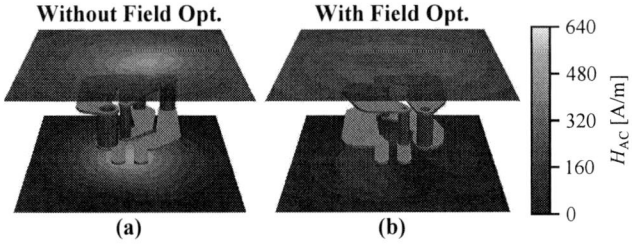

Fig. 11. (a) Magnetic near-field pattern for the reference design (see Fig. 10(b)). (a) Magnetic near-field pattern for the field-optimized design (see Fig. 10(g)). The RMS value of the AC magnetic field is considered.

F. Near-Field Constraint

Air-core inductors typically generate a large magnetic near-field as the field pattern is not confined by a high-permeability core. The magnetic near-field can produce additional losses in nearby metallic planes and/or EMI issues. For this reason, the RMS magnetic near-field is evaluated at a distance of $0.3\,\mathrm{mm}$ from the coil and the maximum value is considered. The optimizer tries to minimize the converter losses while keeping the maximum DC and AC fields below $500\,\mathrm{A/m}$ and $250\,\mathrm{A/m}$, respectively. This requires a trade-off between a priori incompatible targets: obtaining a high inductance value and reducing the generated magnetic field.

Fig. 10(g) depicts the obtained design. With a small reduction of the efficiency (0.9 % reduction), a massive reduction of the magnetic field is achieved ($2.5\times$ reduction). Fig. 11 shows the magnetic near-field pattern achieved with and without the additional field constraint. The near-field reduction is obtained with a partial compensation of the field between the layers and the turns. This indicates that the shape optimizer is able to find highly optimized geometries that are custom-tailored to the converter specifications and constraints.

V. CONCLUSION

This paper presents a free-shape optimization method for multi-layer air-core inductors. A complete design rule checker is implemented and the magnetic field problem is solved with a FFT-accelerated 3D PEEC solver. The optimization problem features continuous variables, discrete variables, and a non-linear/non-differentiable/discontinuous objective function. Moreover, due to the complex design rules, only a tiny fraction of the design space maps to valid geometries. A two-step optimization approach has been adopted. First, a recursive tree algorithm generates a pool of random designs that respect the design rules. Afterward, a constraint-aware genetic algorithm finds the optimal geometry.

This optimization workflow is successfully applied to VHF inductors used in IVRs. Without particular constraints, it is found that classical solenoid geometries lead to optimal losses. However, with additional restrictions (terminal placement, complex footprint, or magnetic near-field reduction), the optimizer finds innovative and unexpected geometries that

integrate the new constraints with a minimal impact on the performance. For example, the shape optimizer is able to reduce the generated magnetic near-field by a factor $2.5\times$ with a quasi-negligible impact on the system efficiency (0.9 % reduction).

ACKNOWLEDGEMENTS

This work was supported by the Power Management Integration Center (NSF IUCRC) at Dartmouth College under Grant No. PMIC-062. The authors would like to thank Kishalay Datta and Yue (Will) Wu for their insightful comments on the converter and semiconductor modeling. Additionally, the authors would like to show their gratitude to the Research Computing and Data Services at Dartmouth College for providing the required computational resources and support.

APPENDIX A
PEEC SOLVER VALIDATION

The 3D PEEC solver uses a regular voxel geometry which allows for FFT-accelerated matrix multiplications [22], [23]. The FFT-acceleration is particularly useful for PEEC problems as the equation systems feature dense (and not sparse) matrices. There is naturally a trade-off between the voxel size, the computational speed, and the achieved accuracy. A free-shape inductor with a $1\,\mathrm{mm}\times1\,\mathrm{mm}$ footprint is considered for the benchmark. Figs. 12(a)-(b) depict the meshed geometry for different voxel sizes. The magnetic field problem is solved at $0\,\mathrm{Hz}$ and $40.68\,\mathrm{MHz}$ (skin depth of $10.4\,\mu\mathrm{m}$).

As a comparison, the problem is also solved with a computationally intensive finite element analysis (using ANSYS Maxwell 2023R2, a 0.05 % convergence tolerance, and a boundary layer mesh for the skin depth). Fig. 12(c) shows the obtained DC and AC magnetic parameters. The deviation between the PEEC solver and FEA is below 0.5 %. A voxel size of $12\,\mu\mathrm{m}$, which represents a good trade-off between the computational speed and the achieved accuracy, is selected for the shape optimization workflow.

APPENDIX B
SOURCE CODE

The complete shape optimization workflow is implemented in Python and is available under an open-source license [24], [25]. Besides the genetic algorithm described in this paper, several other local and global optimization methods are implemented using the SciPy, PyGAD, Optuna, and Nevergrad libraries. It should be noted that the ability of the workflow to scale to more complex geometries (i.e. inductors with over 25 trace segments) remains an open-question.

PyPEEC, the custom quasi-magnetostatic 3D PEEC solver (FFT and GPU accelerated), is also distributed under an open-source license and packages are available on PyPI and conda-forge [22]. PyPEEC is a general-purpose solver that can handle arbitrary geometries (with and without magnetic cores). Therefore, PyPEEC can be used to model a large variety of components (inductors, transformers, chokes, IPT coils, busbars, interconnects, etc.).

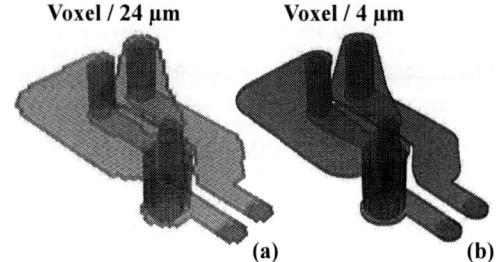

Voxel / 24 μm **Voxel / 4 μm**

Solver Name	DoFs [#]	Speed [#/s]	R_{DC} / R_{AC} [mΩ]	L_{DC} / L_{AC} [nH]
PyPEEC / 24 μm	7670	37.9	17.62 / 27.59	2.154 / 1.999
PyPEEC / 12 μm	73165	6.55	16.89 / 32.36	2.167 / 1.986
PyPEEC / 8 μm	260194	2.03	16.97 / 33.99	2.172 / 1.987
PyPEEC / 6 μm	631648	0.97	16.90 / 34.50	2.175 / 1.987
PyPEEC / 4 μm	2189956	0.21	16.90 / 34.91	2.181 / 1.990
ANSYS Maxwell	n/a	n/a	16.98 / 34.77	2.173 / 1.986

(a) (b) (c)

Fig. 12. (a) PEEC voxel mesh with a 24 μm voxel size, (b) PEEC voxel mesh with a 4 μm voxel size, (c) Convergence of the PEEC solver (at 0 Hz and 40.68 MHz) with respect to the voxel size. A finite element solution of the problem is also shown (using ANSYS Maxwell 2023R2). For the PEEC solver, the number of degrees of freedom (DoFS) and the computational cost (number of computed designs per second using an AMD EPYC 9354 CPU) is reported.

References

[1] Y. Wu and C. R. Sullivan, "Optimizations and Comparisons of Air-Core Inductors Based on a Semi-Analytical Calculation Toolkit," in *Proc. of the IEEE Workshop on Control and Modelling of Power Electronics (COMPEL)*, 2021.

[2] C. Schaef *et al.*, "A Light-Load Efficient Fully Integrated Voltage Regulator in 14-nm CMOS With 2.5-nH Package-Embedded Air-Core Inductors," *IEEE Journal of Solid-State Circuits*, vol. 54, no. 12, 2019.

[3] C. Shetty, "A Detailed Study of Qdc of 3D Micro Air-Core Inductors for Integrated Power Supplies: Power Supply in Package (PSiP) and Power Supply on Chip (PSoC)," *Power Electronic Devices and Components*, vol. 2, 2022.

[4] H. Lin, G. Van der Plas, X. Sun, D. Velenis, E. Beyne, and R. Lauwereins, "System Optimization: High-Frequency Buck Converter With 3-D In-Package Air-Core Inductor," *IEEE Transactions on Components, Packaging and Manufacturing Technology*, vol. 12, no. 3, 2022.

[5] Z. Tong, W. D. Braun, and J. M. Rivas-Davila, "Design and Fabrication of Three-Dimensional Printed Air-Core Transformers for High-Frequency Power Applications," *IEEE Transactions on Power Electronics*, vol. 35, no. 8, pp. 8472–8489, 2020.

[6] R. M. Burkart and J. W. Kolar, "Comparative Life Cycle Cost Analysis of Si and SiC PV Converter Systems Based on Advanced η- ρ-σ Multiobjective Optimization Techniques," *IEEE Transactions on Power Electronics*, vol. 32, no. 6, 2017.

[7] S. Koziel, P. Kurgan, and J. W. Bandler, "Multi-objective mixed-integer design optimization of planar inductors using surrogate modeling techniques," in *Proc. of the IEEE MTT-S Int. Microwave Symposium (IMS)*, 2017.

[8] F. Lucchini, R. Torchio, V. Cirimele, P. Alotto, and P. Bettini, "Topology Optimization for Electromagnetics: A Survey," *IEEE Access*, vol. 10, 2022.

[9] T. M. Evans *et al.*, "PowerSynth: A Power Module Layout Generation Tool," *IEEE Transactions on Power Electronics*, vol. 34, no. 6, 2019.

[10] T. Pham, P. Kwon, and S. Foster, "Additive Manufacturing and Topology Optimization of Magnetic Materials for Electrical Machines—A Review," *Energies*, vol. 14, no. 2, 2021.

[11] A. J. Mäkinen, R. Zetter, J. Iivanainen, K. C. J. Zevenhoven, L. Parkkonen, and R. J. Ilmoniemi, "Magnetic-Field Modeling with Surface Currents. Part I. Physical and Computational Principles of bfieldtools," *Journal of Applied Physics*, vol. 128, no. 6, 2020.

[12] J. Hu and C. R. Sullivan, "Optimization of Shapes for Round-Wire High-Frequency Gapped-Inductor Windings," in *Proc. of the IEEE Industry Applications Conf. (IAS)*, vol. 2, 1998.

[13] H. Sato and H. Igarashi, "Fast Multi-Objective Optimization of Electromagnetic Devices Using Adaptive Neural Network Surrogate Model," *IEEE Transactions on Magnetics*, vol. 58, no. 5, 2022.

[14] S. Pissanetzky, "Minimum Energy MRI Gradient Coils of General Geometry," *Measurement Science and Technology*, vol. 3, no. 7, jul 1992.

[15] T. Takahashi, "Shape Optimization Method for Coils Consisting of Free Curves," *IEEE Transactions on Magnetics*, vol. 29, no. 2, 1993.

[16] B. Auchmann and S. Russenschuck, "Coil end Design for Superconducting Magnets Applying Differential Geometry Methods," *IEEE Transactions on Magnetics*, vol. 40, no. 2, 2004.

[17] S. Izquierdo Bermudez *et al.*, "Coil End Optimization of the Nb3Sn Quadrupole for the High Luminosity LIIC," *IEEE Transactions on Applied Superconductivity*, vol. 25, no. 3, 2015.

[18] M. Yu *et al.*, "Coil End Parts Development Using BEND and Design for MQXF by LARP," *IEEE Transactions on Applied Superconductivity*, vol. 27, no. 4, 2017.

[19] J. Johnson and Y. Rahmat-Samii, "Genetic Algorithms and Method of Moments (GA/MOM) for the Design of Integrated Antennas," *IEEE Transactions on Antennas and Propagation*, vol. 47, no. 10, 1999.

[20] K. Watanabe *et al.*, "Optimization of Inductors Using Evolutionary Algorithms and Its Experimental Validation," *IEEE Transactions on Magnetics*, vol. 46, no. 8, 2010.

[21] T. Guillod and J. W. Kolar, "Medium-frequency transformer scaling laws: Derivation, verification, and critical analysis," *CPSS Transactions on Power Electronics and Applications*, vol. 5, no. 1, 2020.

[22] T. Guillod, "PyPEEC - 3D Quasi-Magnetostatic Solver," 2024. [Online]. Available: https://pypeec.otvam.ch

[23] R. Torchio, F. Lucchini, J. L. Schanen, O. Chadebec, and G. Meunier, "FFT-PEEC: A Fast Tool From CAD to Power Electronics Simulations," *IEEE Transactions on Power Electronics*, vol. 37, no. 1, 2022.

[24] T. Guillod. (2024) PyFreeCoil - Free-Shape Optimization of Air-Core Inductors. GitHub. [Online]. Available: https://github.com/otvam/pyfreecoil

[25] T. Guillod. (2024) PyFreeCoil - Free-Shape Optimization of Air-Core Inductors. Zenodo. [Online]. Available: https://doi.org/10.5281/zenodo.14247697

[26] S. Gillies and Shapely contributors, "The Shapely User Manual," 2024. [Online]. Available: https://shapely.readthedocs.io

[27] A. E. Ruehli, "Equivalent Circuit Models for Three-Dimensional Multiconductor Systems," *IEEE Transactions on Microwave Theory and Techniques*, vol. 22, no. 3, 1974.

[28] P. Virtanen *et al.*, "SciPy 1.0: Fundamental Algorithms for Scientific Computing in Python," *Nature Methods*, vol. 17, 2020.

[29] A. F. Gad, "Pygad: An Intuitive Genetic Algorithm Python Library," *Multimedia Tools and Applications*, vol. 83, no. 20, 2023.

[30] T. Akiba, S. Sano, T. Yanase, T. Ohta, and M. Koyama, "Optuna: A Next-generation Hyperparameter Optimization Framework," in *Proc. of the International Conference on Knowledge Discovery and Data Mining*, 2019.

[31] P. Bennet, C. Doerr, A. Moreau, J. Rapin, F. Teytaud, and O. Teytaud, "Nevergrad: Black-Box Optimization Platform," *SIGEVOlution*, vol. 14, no. 1, 2021.

[32] The PostgreSQL Global Development Group, "PostgreSQL 14.15 Documentation," 2024. [Online]. Available: https://www.postgresql.org/docs

[33] A. Endruschat, T. Heckel, H. Gerstner, C. Joffe, B. Eckardt, and M. März, "Application-related characterization and theoretical potential of wide-bandgap devices," in *Proc. of the IEEE Workshop on Wide Bandgap Power Devices and Applications (WiPDA)*, 2017.

[34] J. T. Stauth, "Pathways to mm-scale DC-DC Converters: Trends, Opportunities, and Limitations," in *Proc. of the IEEE Custom Integrated Circuits Conf. (CICC)*, 2018.

[35] A. Ruehli, C. Paul, and J. Garrett, "Inductance calculations using partial inductances and macromodels," in *Proc. of the IEEE International Symposium on Electromagnetic Compatibility*, 1995.

[36] X. Liu, X. Huang, and P. K. T. Mok, "A High-Frequency Three-Level Buck Converter With Real-Time Calibration and Wide Output Range for Fast-DVS," *IEEE Journal of Solid-State Circuits*, vol. 53, no. 2, 2018.

[37] J. Falin and A. Aguilar, "Maximize Power Density with Three-Level Buckswitching Chargers," *Analog Design Journal*, vol. 1, 2021.

Organic Direct Bonded Copper-Based Rapid Prototyping for Silicon Carbide Power Module Packaging

Shuofeng Zhao[1]
shuofeng.zhao@nrel.gov

Joshua Major[1]
joshua.major@nrel.gov

Douglas DeVoto[1]
douglas.devoto@nrel.gov

Sarwar Islam[1]
sarwar.islam@nrel.gov

Xiaoling Li[1]
xiaoling.li@nrel.gov

Mike Tant[1]
mike.tant@nrel.gov

Faisal Khan[2]
faisal.khan@nrel.gov

Sreekant Narumanchi[1]
sreekant.narumanchi@nrel.gov

[1]*Advanced Power Electronics and Electric Machines Group*
National Renewable Energy Laboratory, Golden, USA

[2]*Energy Conversion and Storage System*
National Renewable Energy Laboratory, Golden, USA

Abstract—Silicon Carbide (SiC) power devices are playing ever-growing roles in high power density power electronic (PE) converters by offering benefits such as high voltage rating, fast transient, and high thermal performance. Organic direct bonded copper (ODBC)-based packaging, due to its ductility and ease to handle, allows the possibility of more flexible layout design which may better tap the potential of SiC benefits. In this work, an ODBC-based prototyping routine is developed which accelerates the iterations of packaging layout design with low cost. The property of ODBC and its handling is briefly introduced, and tools and fabrication steps explained. Following this routine, a 1.2 kV SiC half-bridge (HB) power module is designed and fabricated with the focus on sub-nanohenry ultra-low loop inductance. Simulation and experimental validation are conducted, respectively.

Index Terms—Organic Direct Bonded Copper, Silicon Carbide MOSFETs, Packaging, Parasitics, Rapid Prototyping

I. INTRODUCTION

SiC devices, with relative maturity in high voltage, high power wide bandgap in power devices, are seeing significant growth in power conversion applications [1]. SiC metal oxide field effect transistors (MOSFETs) are being rapidly adopted in drivetrain electrification for its potential to achieve very high-power density. In high rating SiC MOSFETs, the vertical design is prevailing. Electrical contacts on both sides of the devices have implications on their packaging layout. Advanced packaging design is necessary to fully tap these potentials by providing low electromagnetic parasitics, high thermal dissipation, and structural strength. These aspects have been studied in [2]- [4], [5], and [6], respectively. In [3], a quantitative analysis and comparative experimental results are presented on SiC MOSFETs to emphasize the importance of reducing power loop parasitic inductance in mitigating voltage ring and overshoot during turn-on and turn-off transients in high di/dt scenarios. In [4], a 1.2kV, 400A SiC MOSFETs HB module is designed and validated to have low power loop inductance of no more than 1.4nH, though the design adopted

conventional ceramic-based DBC structure and showed high level of fabrication sophistication.

DBC is a very prominent structural feature in packaging designs. Conventional DBC uses various ceramics as dielectric materials for their high dielectric strength and high thermal conductivity. However, the stress coming from the mismatch of coefficient of thermal expansion (CTE) put heavy constrains on ceramic DBC such as vertical symmetry and maximum copper thickness. ODBC, on the other hand, uses ductile films as dielectric layers, lifting many of the constrains and allowing more flexibility in layout design. A polyimide material was studied in [7], whose results indicate very high dielectric strength and low effective thermal resistance when applied as a thin film. A 1.7 kV SiC MOSFETs PE converter element, comprised of multiple stages of HBs, was designed, fabricated, and tested in [8] as a validation for ODBC 1kV level packaging design.

In this work, a routine of rapid prototyping for ODBC-based packaging layout designs is presented. The key properties of the polyimide dielectric material (PDM) of concern and the experiences in handling are briefly introduced. The design of a 1.2 kV SiC MOSFETs HB module, which leverages the properties of the PDM, is then presented. The rapid prototyping routine is described next. It uses the HB module design as an example to demonstrate the tools and steps involved in the routine. This routine features fast processing, requires modest and common tool set, and has low cost and no hazard. It may serve as a technique or inspire similar techniques for fast iteration of ODBC-based packaging designs. Experimental validation of the HB module prototype is subsequently given as evidence for both the benefit of the module design and the effectiveness of the rapid prototyping routine.

II. POLYIMIDE DIELECTRIC MATERIAL PROPERTIES

Figure 1 summarizes some key properties of the PDM in this work. In general, apart from the ductility, the PDM has

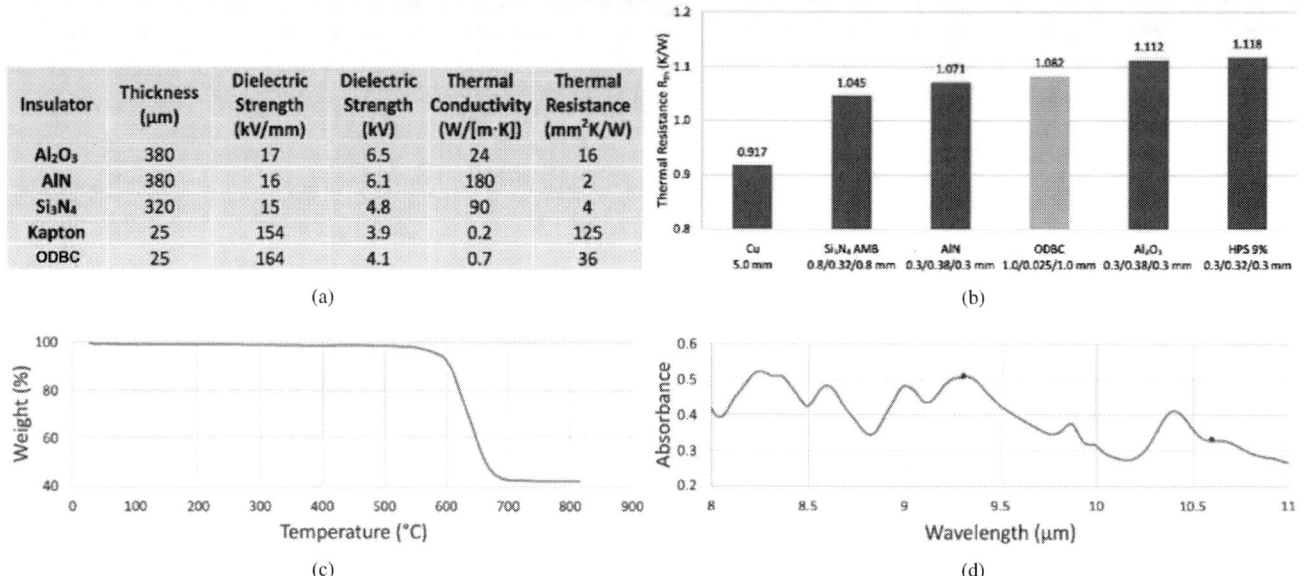

Insulator	Thickness (μm)	Dielectric Strength (kV/mm)	Dielectric Strength (kV)	Thermal Conductivity (W/[m·K])	Thermal Resistance (mm²K/W)
Al₂O₃	380	17	6.5	24	16
AlN	380	16	6.1	180	2
Si₃N₄	320	15	4.8	90	4
Kapton	25	154	3.9	0.2	125
ODBC	25	164	4.1	0.7	36

(a)

(b)

(c)

(d)

Fig. 1: Key properties of the polyimide dielectric material (a) summery of comparison (b) thermal resistance comparison (c) thermal stability (d) Fourier-transform infrared spectroscopy.

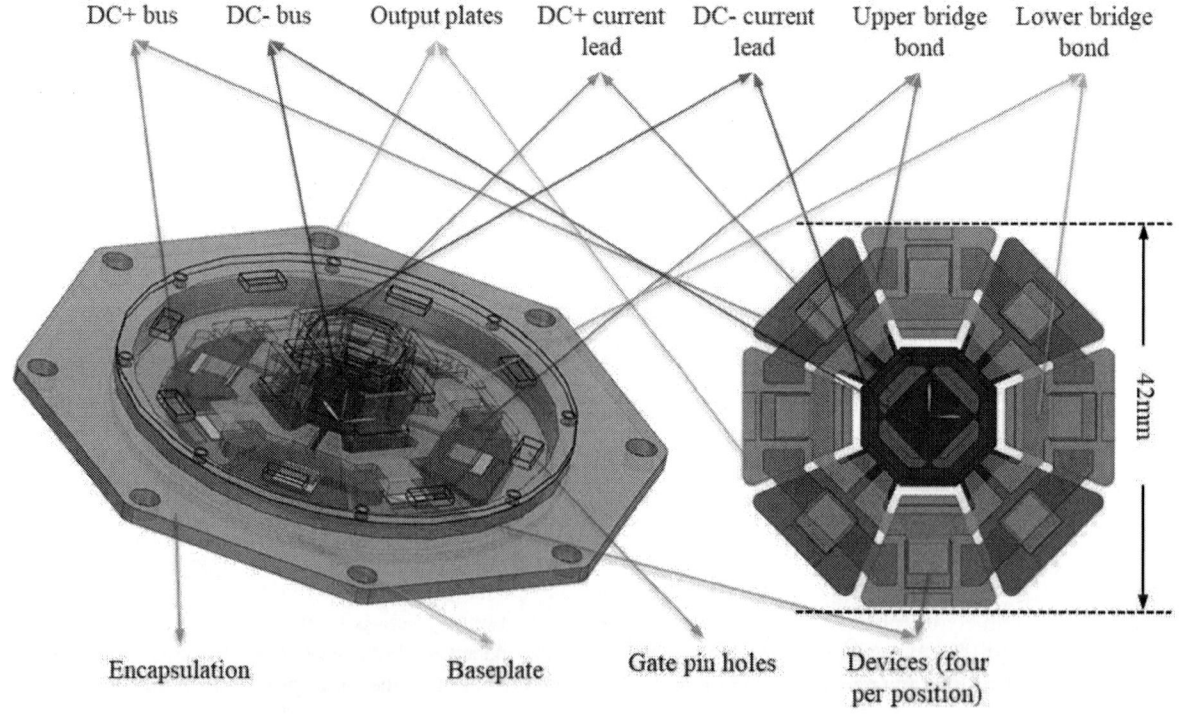

Fig. 2: Example SiC MOSFETs HB module design CAD.

extremely high specific dielectric strength and reasonable thermal resistance compared to the most popular ceramic materials in their typical thicknesses. Its thermal stability is very high below 500 °C, providing considerable fabrication processing headroom for the rapid prototyping routine in the later section. The spectroscopy chart shows two energy absorption peaks at $9.3\mu m$ and $10.6\mu m$, two wavelengths of common CO_2 laser cutters, suggesting it can be easily customized into features using common laser in large batch.

Various experiments were performed on the PDM to study its intrinsic properties and its interactions with different materials whose details are not given here for conciseness. In summary, some experiences obtained during handling the PDM are given as follows:

- The PDM bonds to metal surfaces by being subjected to high temperature and pressure at the same time. Temperature and pressure form a quasi-linear safety operation area, with 300psi and 300°C to 400 °C a comfort zoom for reliable bonding.
- The PDM does not bond to most non-metal materials including but not limited to various plastics, rubber, resin, PTFE, silicon dioxide, etc.
- The surface roughness of the object being bonded to is important for a reliable bonding. An Ra value of $10\mu m$ to $20\mu m$ is desirable. Too smooth a surface may result in weak bonding, whilst too rough a surface may puncture the PDM. Oxidation layer normally results in bonding failure. Dirt particles on the PDM surface may compromise the isolation strength and surface smoothness.
- As long as the desirable conditions above are met, the bonding process can be repeated for indefinite times without compromising the PDM.

These properties of the PDM offer a high degree of flexibility in rapid prototyping mainly by allowing deforming of already-bonded components into desirable geometries and allowing multi-step masking and bonding to form desirable dielectric patterns.

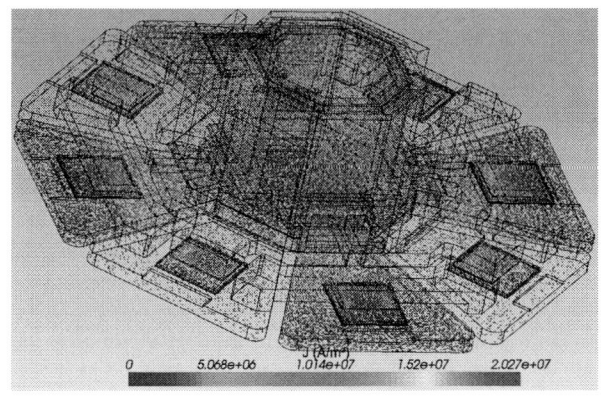

(a)

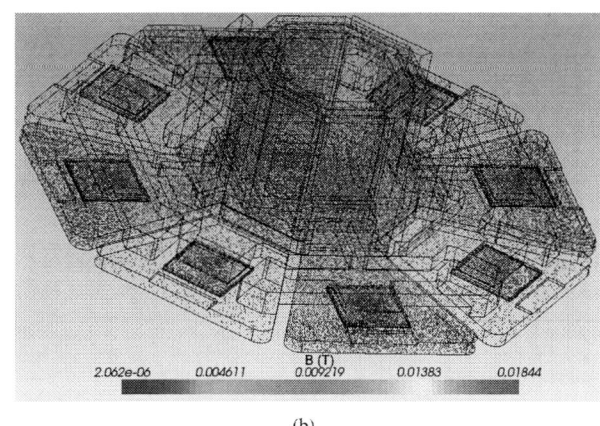

(b)

Inductance matrix (mH)	
	Coil(1)
Coil(1)	7.26091E-07

(c)

Fig. 3: Example SiC MOSFETs HB module design simulation results (a) current density field, (b) magnetic flux density field, and (c) inductance matrix value.

III. SILICON CARBIDE HALF BRIDGE MODULE DESIGN

Figure 2 shows the example design inspired by the PDM properties. There are four devices per position in parallel. It is essentially a symmetrical layout around the z-axis. Symmetrical designs have been conceived in existing research, such as in [9], due to the fact that they intuitively facilitate the inductance cancellation. Nevertheless, the potential of symmetry in current paths has rarely been fully tapped because doing so will require more sophisticated 3D layouts which may be too challenging for conventional ceramic-based DBC. However, with the properties of the PDM, a highly symmetrical design with minimized current loops such as in this example can be realized.

In this example design, the DC+ and Mid point bus bars are used as primary thermal conduction path. The loss heat goes through these bus bars and the main PDM layer to reach the copper baseplate. The DC- bus bar has a compact and symmetrical footprint and is bonded to the top surface of the DC+ bus bar with a secondary layer of PDM to allow reduced current loop area and symmetrical current paths and as the same time provide moderate DC capacitance. Four pairs of bond bridges, connecting the high-side source to the mid point and the low-side source to DC- from each die to its adjacent die position respectively, complete the compact and symmetrical current paths.

Figure 3 shows the FEA simulation results in Siemens Simcenter Magnet 3D. The simulated loop inductance with 400A current excitation is 726.1pH. From the field distributions, high degree of cancellation can be observed.

It should be noted that, increases common mode current (CMC) due to the increased coupling capacitance from the

Fig. 4: Key steps in the fabrication process of the example HB module following the proposed rapid prototyping routine.

output node to the baseplate as a result of much smaller dielectric thickness can be an issue for ODBC designs. In this example case, experiences from [10] are referred to in the design for CMC cancellation. The output node and the DC bus have 287pF and 532pF coupling capacitance to the baseplate, respectively.

IV. RAPID PROTOTYPING ROUTINE

The rapid prototyping steps are shown in Figure 4. In summary, there are four key pieces of machinery required.

- A consumer grade CNC milling machine which can work on copper at 0.1mm precision, for copper features.
- A hot-press which can be assembled from consumer grade linear stages, stepper motors, heating plate which can go above 350°C, thermal couples and temperature control, pressure sensors, Python programming for controlling temperature and pressure, for bonding PDM and copper. These items provide an example of high low-cost, fast, and flexible approach to realize the hot pressing functionality. Alternative approaches with the same functionality can also be adopted.
- A 3D printer which can print materials with temperature not less than the hot press temperature and sufficient rigidity to withstand at least 300psi pressure without major uneven deformation.
- A consumer grade laser cutter whic can cut through 50μm polyimide material sheet.

With the example HB module design as a paradigm, keys steps of the rapid prototyping routine are summarized as follows.

- Machine all copper features.
- Cut the PDM per design specs using laser cutter.
- 3D-print bonding molds per the footprints of the bonding interface and print soldering mold per the footprints of 3D overlay components. In the case of the example

HB module, SLA printing using a highly rigid resin is employed to produce molds with high infill density and homogeneity.
- Pre-bond PDM and copper using the bonding molds for reliable alignment purposes. Due to the fact that the PDM material tends to have very low friction on the metal surface-to-be-bonded, this step is often important to fix PDM and copper components in the correct positions.
- Bond PDM and copper using pressing tool and high-temperature rubber for the stable bonding and check electrical integrity using high potential tester. The high potential testing should be conducted after each bonding to guarantee consistent dielectric strength of the packaging.
- align power devices and 3D overlay features in the designed order using the soldering molds. After all alignments, solder components in a single shot to allow the highest reliability.
- Bond (or solder if pads have nickel or gold plating) source, gate, and Kelvin pads using wire bonding machine.
- Assemble gate leads and finish encapsulation.

In the example case, two types of solder, IND5.7LT and NC-SMQ80 are used. the HB module is finish per the routine and passed 1.5kV offline high potential tests. A double pulse testing (DPT)/buck converter test bench is designed and assembled accommodate the footprint of the example module as shown in Figure 5.

V. EXPERIMENTAL RESULTS

The DPT/buck converter test setup is designed and fabricated to handle the target ratings of the HB module prototype of 1.2kV and 400A, and the assembled and encapsulated prototype has passed 1.2kV high potential tests on all nodes. However, due to the limit capacity of serviceable power

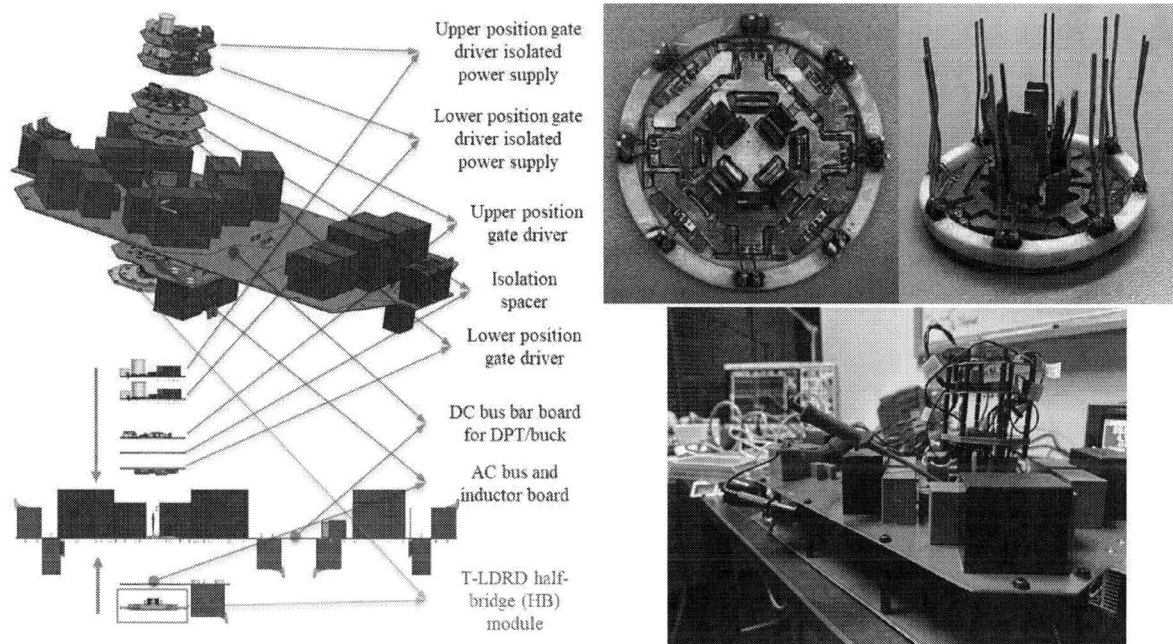

Fig. 5: Finished module and test bench setup. Left - explosion view of the mounted test setup. Upper-right - HB module prototype right before encapsulation. Lower-right - mounted setup under test.

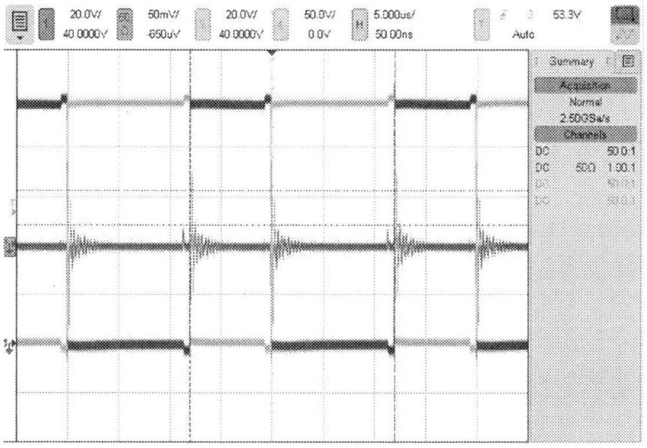

(a)

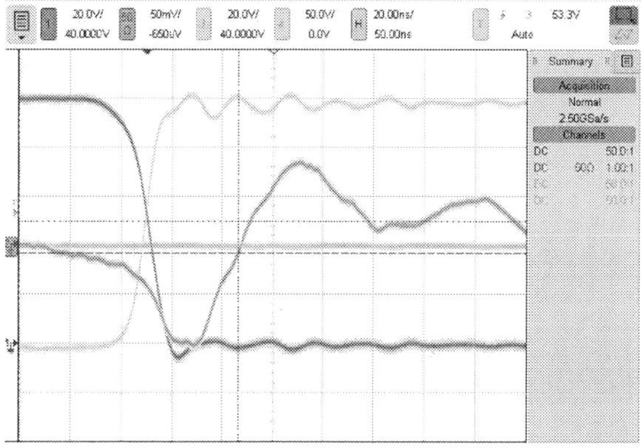

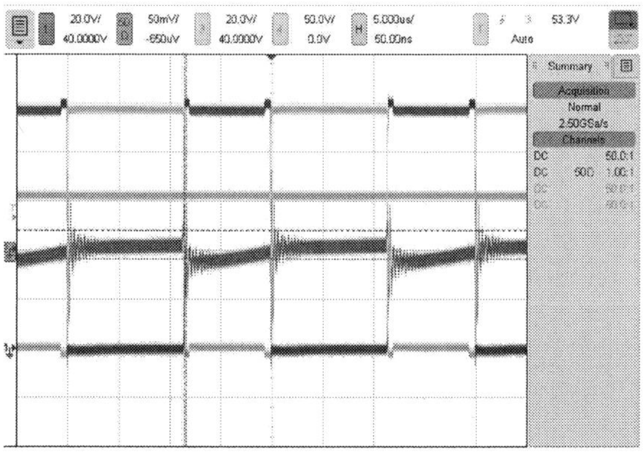

(b)

(c)

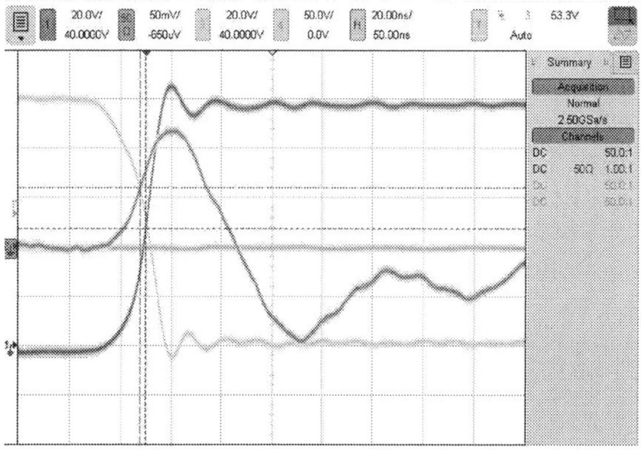

(d)

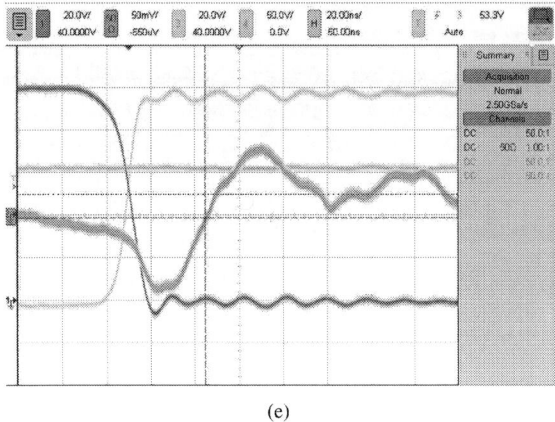

(e)

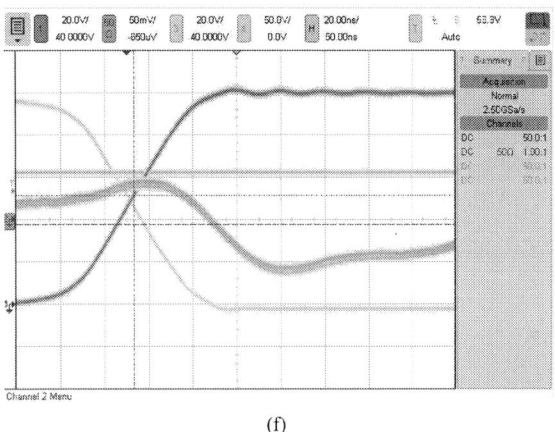

(f)

Fig. 6: Experimental evaluation of the switching transient performance of the HB module prototype. Legend: blue – high-side switch voltage, yellow – low-side switch voltage, green – load voltage, magenta – 1/4 of HB module DC-current. (a) 50kHz switching waveform overview, zero load with no output inductor connected, (b) 50kHz switching waveform overview, 10A load current, (c) high-side switch turn-on transient under zero load with no output inductor connected, (d) high-side switch turn-off transient under zero load with no output inductor connected, (e) high-side switch turn-on transient under 10A load current, and (f) high-side switch turn-off transient under 10A load current.

sources at the moment of the testing, a downgraded 100V, 10A test is conducted for a proof-of-concept.

The test setup is run in buck converter mode during the testing. On the primary DC side, the test setup has an effective $97\mu F$ DC link capacitance and 84.4nF decoupling capacitance across the mounting contacts for the DC terminal leads of the module. Both the DC link capacitors and the decoupling capacitors are arranged symmetrically corresponding to the four pairs of terminal leads of the module to achieve current symmetry. The effectively total inductance between the decoupling capacitors and the contacts of the DC board and the module terminal leads is approximately 0.5nH according to

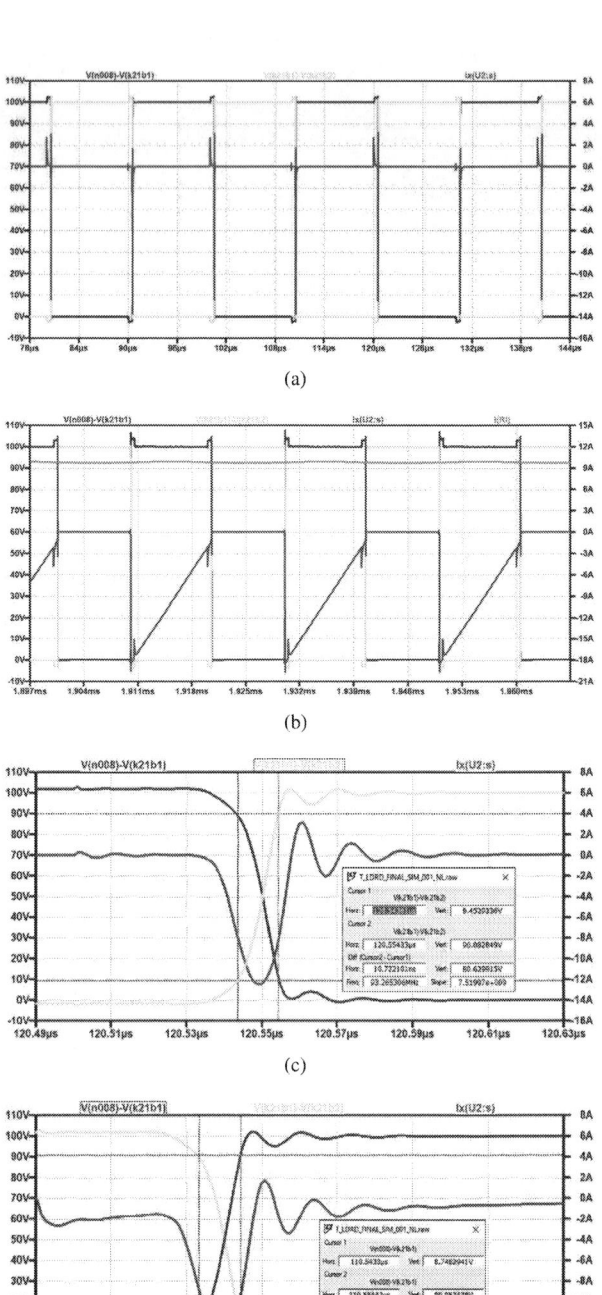

(a)

(b)

(c)

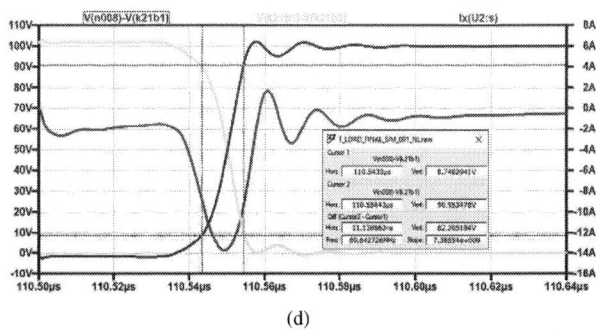

(d)

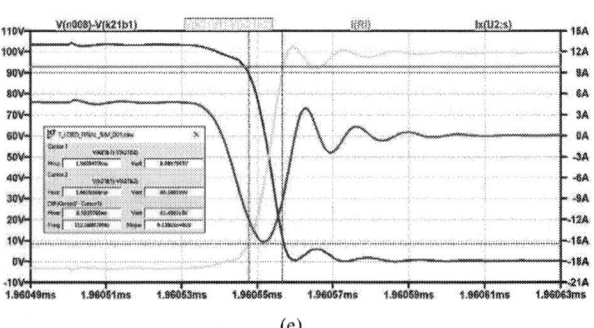

(e)

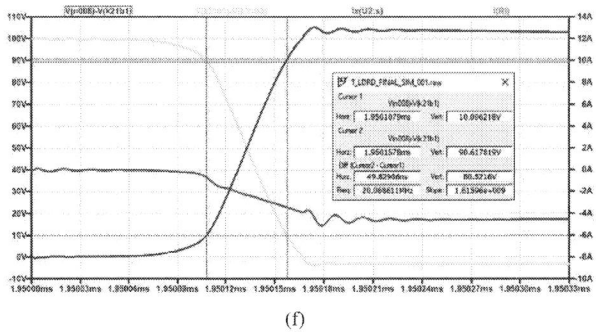

(f)

Fig. 7: Corresponding simulation results of the switching transients of the HB module prototype. Legend: blue – high-side switch voltage, yellow – low-side switch voltage, green – load current, magenta – 1/4 of HB module DC- current. (a) 50kHz switching waveform overview, zero load with no output inductor connected, (b) 50kHz switching waveform overview, 10A load current, (c) high-side switch turn-on transient under zero load with no output inductor connected, (d) high-side switch turn-off transient under zero load with no output inductor connected, (e) high-side switch turn-on transient under 10A load current, and (f) high-side switch turn-off transient under 10A load current.

FEA simulation result whose details are not shown here for conciseness. The buck converter has $45\mu F$ output capacitance. A 5Ω resistive load is used to induced as high as 10A load current condition. A Rogowsk coil current probe is attached to one of the the DC- terminal leads of the module to monitor transient current. The gate drivers for both positions have 1.0Ω output resistance.

Figure 6 gives the experimental results under both zero load condition with no output inductor and 10A load condition. The gate drivers provide maximum voltage slew rate of approximately 10V/ns. Under this voltage slew rate, the turn-on and turn-off ringing overshoot is no more than 5% of the total voltage, and the peak-to-peak voltage ringing is no more than 8% of the total voltage. These values show no noticeable changes between zero load and 10A load conditions. It is evident that low loop inductance is achieve with the test setup.

In order to quantitatively verify the sub-nH loop inductance achieved with the HB module prototype, LTspice circuit simulation models with manufacturer SPICE model of the SiC bare die MOSFETs and test bench capacitance and inductance values from the real-world measurements or FEA results. Stray inductance between the decoupling capacitors and the SiC devices terminals are adjusted such that the simulated switching transient waveforms match the experimental measurements. Figure 7 gives the simulation results under the exact settings of the experimental testing. As shown in all subfigures, the transient waveforms have no noticeable difference from the experimental measurements regarding voltage slew rates, overshoot percentage, and peak-to-peak ringing percentage. The Stray inductance between the decoupling capacitors and the

SiC devices terminals for these simulations are set to be 1.2nH with even distribution for high-side and low-side. As as result, the estimated HB module loop inductance is approximately 0.7nH which sufficiently matches the FEA results in Figure 3.

VI. CONCLUSIONS AND FUTURE WORK

This work presents a low cost, rapid prototyping routine for ODBC-based power module packaging layout design. Key properties of a polyimide dielectric material and the experiences in working with the material have been summarized. Based on these properties and experiences, details and steps in the rapid prototyping routine have been developed. An example SiC HB module with sub-nanohenry ultra-low loop inductance has been designed to leverage the ODBC properties and has been fabricated following the rapid prototyping routine, and the assembled and encapsulated module has pass high potential tests. Downgraded Experimental validation has been conducted on a correspondingly designed buck converter test setup, and sub-nH loop inductance has been verified with experimental measurements and comparing to device-level accurate circuit simulation results. Further future work will involve comprehensive electrothermal co-simulation and evaluation of the design, conducting full-scale experimental testing on the prototype HB module, and expanding the capability of the routine for more flexibility.

ACKNOWLEDGMENT

This work was authored by the National Renewable Energy Laboratory, operated by Alliance for Sustainable Energy, LLC, for the U.S. Department of Energy (DOE) under Contract No. DE-AC36-08GO28308. Funding provided by the U.S. Department of Energy Vehicle Technologies Office. The views expressed in the article do not necessarily represent the views of the DOE or the U.S. Government. The U.S. Government retains and the publisher, by accepting the article for publication, acknowledges that the U.S. Government retains a nonexclusive, paid-up, irrevocable, worldwide license to publish or reproduce the published form of this work, or allow others to do so, for U.S. Government purposes.

REFERENCES

[1] X. She, A. Q. Huang, Ó. Lucía and B. Ozpineci, "Review of Silicon Carbide Power Devices and Their Applications," in IEEE Transactions on Industrial Electronics, vol. 64, no. 10, pp. 8193-8205, Oct. 2017, doi: 10.1109/TIE.2017.2652401.

[2] R. Alizadeh and H. Alan Mantooth, "A Review of Architectural Design and System Compatibility of Power Modules and Their Impacts on Power Electronics Systems," in IEEE Transactions on Power Electronics, vol. 36, no. 10, pp. 11631-11646, Oct. 2021, doi: 10.1109/TPEL.2021.3068760.

[3] B. Zhang and S. Wang, "Parasitic Inductance Modeling and Reduction for Wire-Bonded Half-Bridge SiC Multichip Power Modules," in IEEE Transactions on Power Electronics, vol. 36, no. 5, pp. 5892-5903, May 2021, doi: 10.1109/TPEL.2020.3032521.

[4] Beckedahl, Peter, Sven Buetow, Andreas Maul, Martin Roeblitz, and Matthias Spang. "400 A, 1200 V SiC power module with 1nH commutation inductance." In CIPS 2016; 9th International Conference on Integrated Power Electronics Systems, pp. 1-6. VDE, 2016.

[5] Broughton, J., Smet, V., Tummala, R. R., and Joshi, Y. K. (August 20, 2018). "Review of Thermal Packaging Technologies for Automotive Power Electronics for Traction Purposes." ASME. J. Electron. Packag. December 2018; 140(4): 040801. https://doi.org/10.1115/1.4040828

[6] C. Chen, F. Luo and Y. Kang, "A review of SiC power module packaging: Layout, material system and integration," in CPSS Transactions on Power Electronics and Applications, vol. 2, no. 3, pp. 170-186, Sept. 2017, doi: 10.24295/CPSSTPEA.2017.00017.

[7] Rajesh Tripathi, Sejin Im, Douglas Devoto, Joshua Major, Sreekant Narumanchi, Paul Paret, and Xuhui Feng (2019) Power electronics thermal solutions using thermally conductive polyimide films. Additional Conferences (Device Packaging, HiTEC, HiTEN, and CICMT): January 2019, Vol. 2019, No. DPC, pp. 000616-000646.

[8] N. Rajagopal, C. DiMarino, R. Burgos, I. Cvetkovic and M. Shawky, "Design of a High-Density Integrated Power Electronics Building Block (iPEBB) Based on 1.7 kV SiC MOSFETs on a Common Substrate," 2021 IEEE Applied Power Electronics Conference and Exposition (APEC), Phoenix, AZ, USA, 2021, pp. 1-8, doi: 10.1109/APEC42165.2021.9487167.

[9] L. Wang, W. Wang, G. Rietveld and R. J. E. Hueting, "Development of Pressure Contact Technology for Multi-chip SiC Modules with Low Parasitics," 2024 IEEE Applied Power Electronics Conference and Exposition (APEC), Long Beach, CA, USA, 2024, pp. 202-209, doi: 10.1109/APEC48139.2024.10509116.

[10] N. Rajagopal, C. DiMarino, B. DeBoi, A. Lemmon and A. Brovont, "EMI Evaluation of a SiC MOSFET Module with Organic DBC Substrate," 2021 IEEE Applied Power Electronics Conference and Exposition (APEC), Phoenix, AZ, USA, 2021, pp. 2338-2344, doi: 10.1109/APEC42165.2021.9487439.

Discrete Power Device Packaging with Integrated Direct Two-Phase Cooling

Jinpeng Cheng
Chongqing-Warwick Joint
Laboratory in Silicon Carbide
Power Electronics
Chongqing University
Chongqing, China
Jinpeng.C@cqu.edu.cn

Jinxiao Wei
Chongqing-Warwick Joint
Laboratory in Silicon Carbide
Power Electronics
Chongqing University
Chongqing, China
jxwei@cqu.edu.cn

Hao Feng
Chongqing-Warwick Joint
Laboratory in Silicon Carbide
Power Electronics
Chongqing University
Chongqing, China
hfeng6@cqu.edu.cn

Li Ran
Chongqing-Warwick Joint
Laboratory in Silicon Carbide
Power Electronics
University of Warwick
Coventry, the UK
l.ran@warwick.ac.uk

Abstract—This paper presents a discrete device packaging with integrated direct cooling. Firstly, the fabricated three-section heat pipe meets the common 650 V, 1200 V and 1700 V devices in terms of electrical insulation, while exhibiting thermal performance comparable to commercial heat pipes. This validates the feasibility of the proposed discrete device packaging. Secondly, the proposed discrete device was fabricated, and experimental results show that it has excellent thermal performance, with a temperature difference of less than 1.5 °C between the evaporator and the condenser. Additionally, simulations reveal that, under identical conditions, the proposed design achieves a 34.5°C reduction in junction temperature compared to conventional discrete device cooling solutions. The design is expected to be widely used in high power density converters.

Keywords—*power device, packaging, two-phase cooling, heat pipe*

I. INTRODUCTION

Discrete devices have been favored by engineers as one of the most common components in today's power electronic converter systems, such as the single-pipe parallel electric drive technology applied by Tesla. The discrete devices have low packaging costs due to their mature and automated production lines. Also, the standard design allows them to be plug-and-play. Where in practice, discrete devices such as TO247 packages are bolted or clamped to a metal heat sink coated with thermal interface material (TIM), usually with a ceramic insulator between the device and the heat sink [1]. On the one hand, discrete devices and ceramic insulator have a high thermal resistance, leading to a poor thermal performance, e.g., the junction-case thermal resistance of discrete devices in TO247 packages is > 0.3 °C/W [2]. This will trouble the thermal management of the whole system. On the other hand, the heat from the chip can only be transferred to the heat sink at a certain diffusion angle and each layer of media in the heat path has poor lateral heat spreading [3]. This can lead to inefficient heat dissipation and incomplete heat sink utilization, resulting in a redundant heat sink design.

In addition, reliability issues such as deterioration of the TIMs can arise. These issues are bottlenecks to the continued

widespread use of discrete devices in higher power density converters.

Extensive work has been done to ensure safe and reliable operation of discrete devices. A repackaging method using discrete devices is proposed in [4]. By using layout optimization and integrating some peripheral circuits, achieving excellent loop inductance and thermal resistance. The researchers investigated various TIMs, including screen printed phase change material, to improve the thermal contact resistance between the ceramic and the discrete device substrate [1]. Also, some advanced thermal management solutions have been proposed to mitigate thermal concerns [5]. However, the above studies are improvements on inherent package applications, and the enhancement of the thermal performance of discrete devices packaging is not studied thoroughly.

To this end, a novel discrete device packaging design based on a three-section heat pipe is proposed in this paper. Benefiting from the three-section structure [6], the electrical isolation between the bare die and the outside is guaranteed. At the same time, the designed device has excellent thermal performance due to the superior thermal conductivity of the three-section heat pipe. The flat and flexible thermal design makes it a promising choice for the next generation of high-power electronic converters.

II. STRUCTURE DESIGN OF THREE-SECTION HEAT PIPE

A. Operating Principle of Heat Pipe

The heat pipe is a highly efficient passive device utilizing the latent heat of the vaporized working fluid rather than the sensible heat, and it can transfer heat at high rates over considerable distances with minimal temperature drops, extreme flexibility, simple structure, easy control with no external pumping power. Its equivalent thermal conductivity ranges from 10,000 to 100,000 W/m·K, much superior to thermal anisotropic graphite (~1500 W/m·K) [7].

The components of a heat pipe include a pipe wall, a wick structure and a small amount of working fluid in equilibrium with its own vapor [8]. As shown in Fig. 1, when heat is supplied to the evaporator, the working fluid begins to

---

This work was supported by the National Natural Science Foundation of China under Grant No.52107179.

979-8-3315-1612-3/25 $31.00 © 2025 IEEE

evaporate and its internal vapor-liquid balance is disrupted. The resulting vapor, which is at a higher pressure than the liquid and it passes through the vapor space into the condenser. Vapor condenses giving away its latent heat of vaporization to the heat sink. Capillary pressure in the wick, sometimes assisted by gravity, pumps the condensed liquid back to the evaporator, and so the cycle repeats itself.

The thermal resistance network of a typical heat pipe is shown in Fig. 2. R_{Oe}, R_{Oc}, R_{Be}, R_{Bc}, R_{We}, R_{Wc}, R_{ev}, R_{cd}, R_v represent the thermal resistance of the outer surface of evaporator, the outer surface of condenser, the evaporator wall, the condenser wall, the evaporator wick, the condenser wick, the evaporation process, the condensation process, and the vapor flow, respectively.

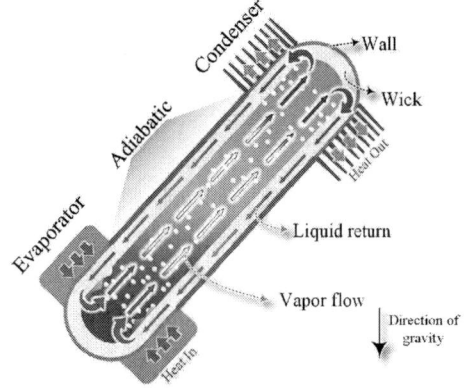

Fig. 1. Operating principle of heat pipe.

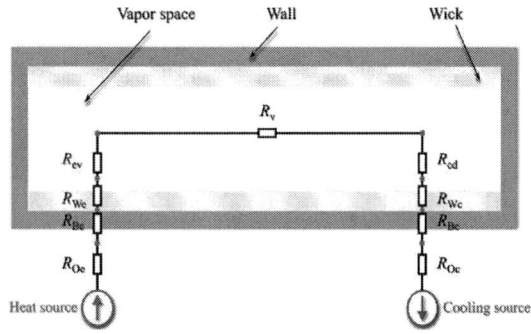

Fig. 2. The thermal resistance network of a typical heat pipe.

B. Structure Design of Three-section Heat Pipe

As previously mentioned, the latent heat of the phase change is the primary heat transfer mechanism, rather than heat conduction through the cavity. This creates the condition for replacing the middle section of the heat pipe with an insulated material. After the replacement, the R_v in the heat pipe keep constant, resulting in a three-section heat pipe that combines high heat transfer performance with electrical insulation capability.

For the insulated section, this study used Polyethylene (PE) material for 3D printing to form a cylindrical casing with a grooved structure that acts to create capillary driving force. Cylindrical copper tubes with sintered wick were used on both sides. The PE casing and copper tubing are bonded with thermosetting epoxy resin to ensure high temperature resistance and sealing. Additionally, an insulating working fluid, isopropyl alcohol (IPA) is chosen. Following the above, a three-section heat pipe was fabricated as shown in Fig. 3, where some details are presented.

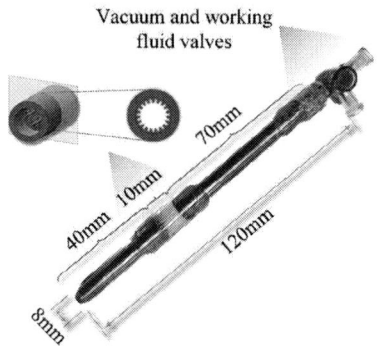

Fig. 3. The fabricated three-section heat pipe.

C. Thermal Performance and Electrical Insulation Verification Three-section Heat Pipe

The partial discharge (PD) test rig is shown in Fig. 4. During the test, a pre-heated copper block maintained the three-section heat pipe in working condition. The platform used the pulse current method to collect the local discharge signal of the DUT and adjusted the input voltage to achieve high-voltage outputs at different levels. The voltage was gradually increased in 0.5 kV steps, with each level maintained for 1 minute, until a significant PD pulse signal (>100 pC) was detected.

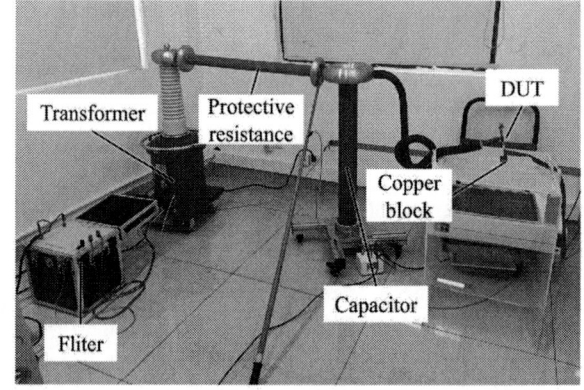

Fig. 4. Test rig of partial discharge.

As shown in Fig. 5, no PD phenomenon was observed when an AC voltage of 4 kV (50 Hz, RMS) was applied for 1 minute. This shows that the discrete device packaging based on the concept of a three-section heat pipe with an insulation length of 10mm meets the requirements of 650 V, 1200 V and even 1700 V bare dies. Fig. 6 shows the temperature response of the three-section heat pipe during operation. Initially, the

979-8-3315-1612-3/25 $31.00 © 2025 IEEE   1833

condenser temperature lags slightly behind the evaporator temperature but closely tracks it, with a steady-state difference of only 1 °C. Without natural air convection, this difference would be even smaller.

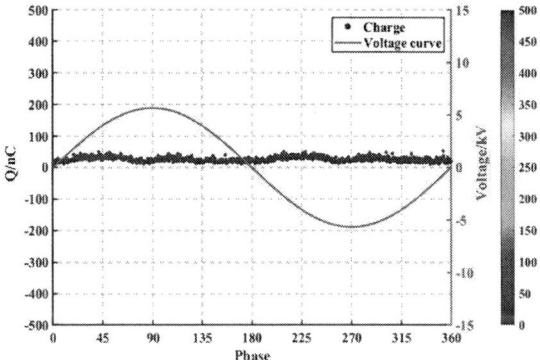

Fig. 5. Partial discharge result.

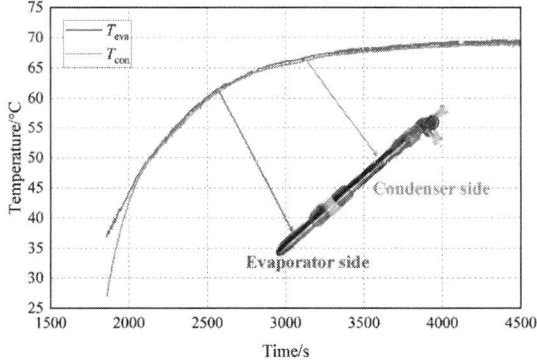

Fig. 6. Temperature response of the three-section heat pipe.

III. NOVEL DISCRETE DEVICE PACKAGING BASED ON THREE-SECTION HEAT PIPE

The proposed discrete device based on the three-section heat pipe packaging concept is illustrated in Fig. 7, and a sample of the fabricated is shown in Fig. 8. It contains a three-section heat pipe, a printed circuit board (PCB) layer and a SiC MOSFET bare die. Based on the improvement of the round three-section heat pipe in Section II, specially, a flat three-section heat pipe is utilized here for facilitating die soldering. It is worth mentioning that the operating principle and the fabricated method of the flat three-section heat pipe are unchanged as described above.

The evaporator of the flat three-section heat pipe acts as a substrate for the bare die, with a small area left for the "Drain" terminal. The bottom-layer of the PCB corresponds to the bare die is soldered to the gate and source pads of the bare die using flip-chip soldering. The rest of the PCB bottom-layer can be soldered to the substrate to ensure a large enough soldering area for better reliability. The top-layer of the PCB is divided into three regions, "Source", "Gate" and "Kelvin Source", which are connected to the corresponding pads on the bottom-layer of the PCB through via-holes.

Finally, the evaporator and insulated section are encapsulated as a whole by molding compound and thus electrically isolated from the outside environment.

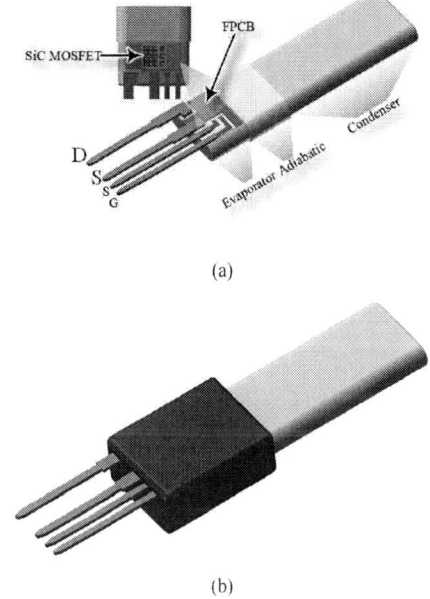

(a)

(b)

Fig. 7. (a) The proposed discrete device design; (b) The Discrete device encapsulated by molding compound.

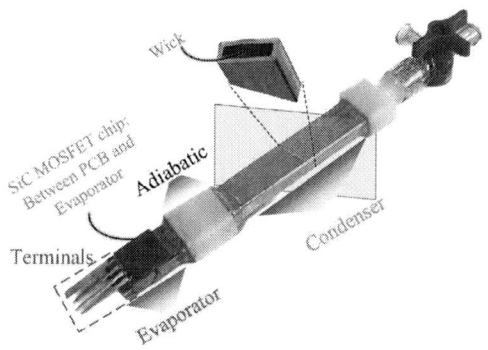

Fig. 8. The fabricated discrete device.

IV. EXPERIMENT TEST AND SIMULATION

Thermal testing was performed on the fabricated discrete device, and the temperature distribution is shown in Fig. 9. The actual temperature can be accurately measured at the black spray paint coverage, revealing that the temperature difference between the evaporator and the condenser is less than 1.5 °C. Fig. 10 shows the corresponding temperature response curve. Throughout the entire power cycle, the temperature of the condenser initially lags slightly but soon closely follows the evaporator, maintaining only a minimal temperature difference. This indicates that the designed discrete device packaging based on a three-section heat pipe has excellent heat dissipation capability.

Fig. 11 shows the discrete device in TO247 packaging is attached to a copper heat sink by a Al2O3 ceramic sheet and

TIM, alongside the proposed discrete device with integrated copper fins. Both schemes utilize the same heat sink size.

Fig. 9. Temperature distribution of the proposed device.

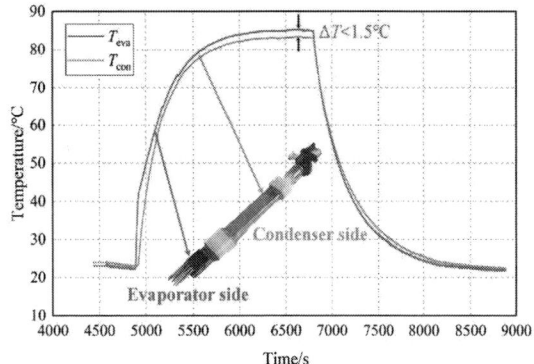

Fig. 10. Temperature response of the proposed device.

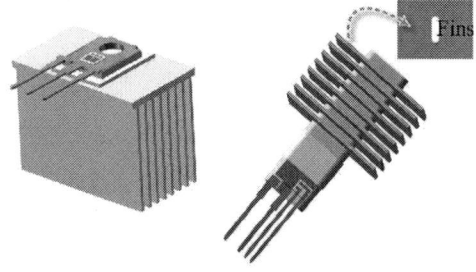

Fig. 11. The conventional and proposed discrete devices.

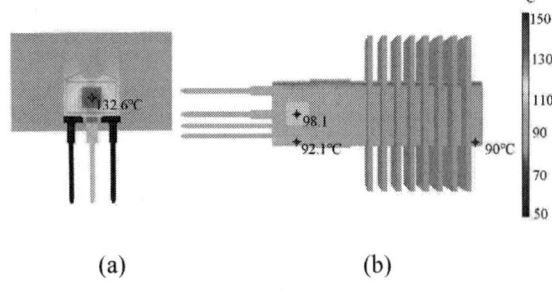

Fig. 12. The temperature results. (a) The conventional discrete device; (b) The proposed discrete device.

Both schemes use the same single SiC bare die and have the same heat sink size. Comparing the thermal performance of two schemes using FEA in COMSOL. The equivalent thermal conductivity of the three-section heat pipe is set to 16,000 W/m·K, and the temperature difference between its two sides is 2.1 °C, which is similar to the result in Fig. 9. Under the same conditions, the temperature simulation results of the two schemes are shown in Fig. 12. Due to the superior thermal conductivity of the three-section heat pipe, the junction temperature of the proposed discrete device is reduced by 34.5 °C compared to conventional discrete device.

V. CONCLUSIONS

This paper proposes a novel discrete device packaging with integrated three-section heat pipe. Firstly, a typical round three-section heat pipe was fabricated. Experimental tests show that the three-section heat pipe with 10mm insulation length fully meets the insulation requirements of 650 V, 1200 V and 1700 V devices, and has a strong heat spreading capability due to the liquid-vapor phase change process. The temperature difference between the two ends of the three-section heat pipe is only 1 °C. Based on this, a new discrete device packaging with an integrated cooling was fabricated. Experimental results showed that it has excellent thermal performance. Furthermore, compared to the cooling method of conventional TO247-packaged discrete devices, simulation results show that the proposed design reduces the junction temperature by 34.5 °C under identical conditions.

REFERENCES

[1] M. Garcia-Poulin, M. Ahmadi, M. Bahrami, E. Lau and C. Botting, "Thermal resistance of electrical insulation for bolted and clamped discrete power devices," 2018 34th Thermal Measurement, Modeling & Management Symposium (SEMI-THERM), San Jose, CA, USA, 2018, pp. 176-180.

[2] Wolfspeed of Cree Inc., "Silicon Carbide Power MOSFET C3MTM MOSFET Technology," C3M0021120K datasheet, Dec. 2023.

[3] D. Schweitzer and L. Chen, "Heat spreading revisited – effective heat spreading angle," 2015 31st Thermal Measurement, Modeling & Management Symposium (SEMI-THERM), San Jose, CA, USA, 2015, pp. 88-94.

[4] Z. Chen and A. Q. Huang, "High Performance SiC Power Module Based on Repackaging of Discrete SiC Devices," in IEEE Transactions on Power Electronics, vol. 38, no. 8, pp. 9306-9310, Aug. 2023.

[5] A.-C. Iradukunda, D. R. Huitink and F. Luo, "A Review of Advanced Thermal Management Solutions and the Implications for Integration in High-Voltage Packages," in IEEE Journal of Emerging and Selected Topics in Power Electronics, vol. 8, no. 1, pp. 256-271, March 2020.K. Elissa, "Title of paper if known," unpublished.

[6] Egli A, Feldman K T. Feasibility of an Insulting Heat Pipe for High Voltage Applications[J]. IEEE Transactions on Power Apparatus and Systems, 1982 (9): 3001-3008.

[7] J. Cheng, X. Zhang, J. Wei, H. Feng and L. Ran, "Integrated Design for Enhanced Power Module Thermal Tolerance Utilizing the Phase Change Material and Thermal Anisotropic Graphite," 2023 IEEE Energy Conversion Congress and Exposition (ECCE), Nashville, TN, USA, 2023, pp. 5491-5496.

[8] Faghri A. Heat pipes: review, opportunities and challenges[J]. Frontiers in Heat Pipes (FHP), 2014, 5(1).

Investigation of Die Top-Side Re-Metallization for SiC-Based Double-Side Cooled Power Modules

Narayanan Rajagopal
Center for Power Electronics Systems (CPES)
Virginia Tech
Arlington, Virginia, USA
nrajagopal@vt.edu

Christina DiMarino
Center for Power Electronics Systems (CPES)
Virginia Tech
Arlington, Virginia, USA
dimarino@vt.edu

Abstract—The advancement of wide-bandgap (WBG) technology has enabled a new generation of high-density inverters for automotive applications. To fully benefit from WBG devices, like silicon carbide (SiC), there is an increased focus on double-side cooled (DSC) modules with wirebond-less interconnects. To enable wirebond-less interconnects, the die top-side must be solderable or sinterable. Additionally, the quality of die top-side metallization is critical to ensure that interconnects can conduct high currents and dissipate heat. This paper investigates the thin-film deposition process to provide key insights into the challenges of re-metallizing SiC MOSFET for sintered interconnect applications. Key challenges with the re-metallization, including iridescent discoloration, low bond strength, and high contact resistance, are identified and validated. X-ray photoelectron spectroscopy analysis of the dies identifies the presence of organic residue and oxides on the aluminum source pad, which can lead to adhesion challenges. XPS scans of the shadow mask stencils identify electrolytes (sodium and calcium) as a possible source of contamination in the re-metallization fixtures.

Index Terms—Double-side Cooled, SiC, Module, Packaging, Wirebond-less, Interconnect, Metallization

I. INTRODUCTION

The advancement of wideband-gap (WBG) technology like silicon carbide (SiC) has been revolutionary in the automotive industry, where there is a growing demand for high-density inverters [1]. To fully benefit from these devices, the package becomes a critical factor in enabling good thermal, electrical, and mechanical performance. Hence, double-side cooled (DSC) packages have become popular for automotive applications due to better thermal performance and lower inductances, enabling higher power density and efficiency. In this work, a DSC module for a new high-power dense (1 kV, 100 kW/l) traction inverter is being explored for commercial vehicles to meet the Department of Energy's 2025 targets [2].

A critical subset of DSC module packaging research is on wirebond-less interconnects to enable a 3D design of power modules [3] [4]. Wirebond-less interconnects have been shown to help with improved current carrying abilities, source-side cooling of devices, and reduction in power-loop inductance [3]. Various wirebond-less interconnects have been identified in the literature, including copper posts, molybdenum posts, copper clips, lead frames, silver-porous material, and fuzz buttons [5]. While some have explored press-pack dry-contact

connections for wirebond-less interconnects, most work has focused on soldering and sintering the interconnects for low resistance, reliable connection [5] [6].

To enable these solder or sinter connections, the metallization of the device must be suitable [7]. Traditionally, the bottom side (drain) of the die is plated in a combination of nickel (Ni), palladium (Pd), and silver (Ag) or gold (Au) to solder or sinter a device onto a substrate [8]. The top side of the die is metallized with aluminum (Al), which is compatible with Al wire bonds. Hence, many researchers have focused on re-metallizing the top-side Al pads with metals compatible with soldering and sintering [8].

A. Re-Metallization Methods

A common re-metallization process and surface treatment of dies involve electroless nickel-gold immersion (ENIG) process and wet chemical etching, respectively [9] [10]. An example was demonstrated successfully in [11] where the Al pads were re-metallized in the bare die form using this process. The process involves stripping the native oxide layer of the Al pad and depositing a zinc adhesion layer and a nickel layer. Additionally, [12] reported the process for wet chemical etching and electroless plating for metallization of aluminum bond pad arrays on a CMOS chip. Chemicals like phosphoric acid, ammonium bifluoride, and butyl cellosolve are used to penetrate through the oxide barrier at room temperature [12]. The complete electroless metallization process is highlighted in [12]. Additional information on ENIG for solder bumps is also discussed in [10].

In contrast, some researchers have explored plasma cleaning and thin film deposition as an alternative to wet chemical cleaning and ENIG. In contrast to wet chemical cleaning, plasma cleaning is inexpensive and has no undesirable health and safety hazards. As SiC devices continue to shrink, plasma cleaning has become a popular way to clean device surfaces. [7] highlights some advantages of plasma cleaning over wet chemical etching, including a reduction in the residue left on the surface, higher effectiveness in removing organic contaminants, and the ability of plasma gas to penetrate small areas and surfaces. After cleaning the die surface, some researchers have used a sputtering system to deposit different metals on

979-8-3315-1612-3/25 $31.00 © 2025 IEEE

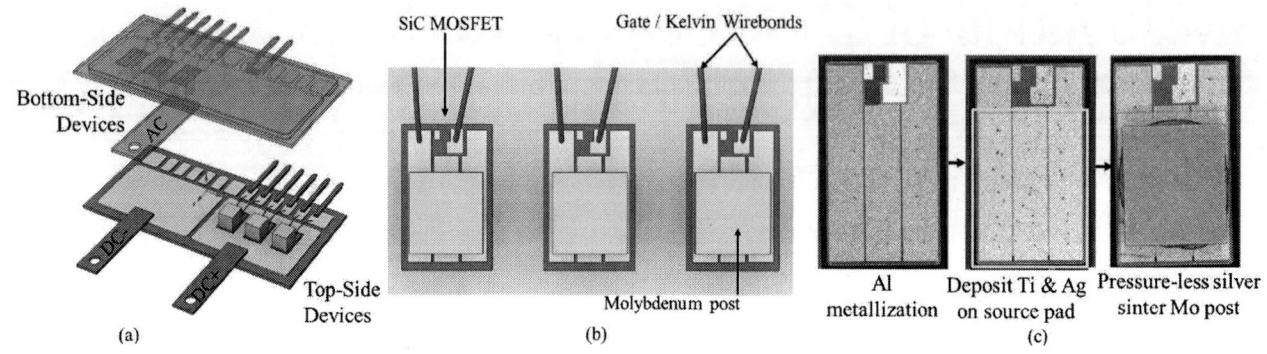

Fig. 1. (a) exploded view of the DSC module showing the layout of the module, (b) close-up of the dies and molybdenum (Mo) source interconnects, (c) the metallization and sintering process.

the Al pad surface, including Ag, Au, Ni, titanium (Ti), copper (Cu), and chromium (Cr). In [13], two types of re-metallization stackup were studied, including Ti / Ni / Cu and Cr / Cu. The thickness of the sputtered layers varies from 0.5 μm to 2 μm to enable solderable surfaces. In [8], the die surface is plasma cleaned, and then an electron beam evaporation system is used to deposit Ti / Ni / Ag at various thicknesses for solder and sinter bonding. Additionally, 10 kV SiC MOSFET devices were re-metallized with Ti and Ag deposition to enable sintered molybdenum (Mo) post interconnects for the source and gate pads [14].

The literature demonstrates successful re-metallization processes, but little detail is provided about the challenges and possible pitfalls associated with these processes. In particular, there is limited information published on the approach, the repeatability, and the yield. Hence, this paper explores the challenges of re-metallizing dies using a thin-film deposition process and provides insight into possible symptoms and diagnosing methods. The work presented in this paper will enable others to diagnose their re-metallization challenges, improve their processes, and identify probable root causes.

II. DIE RE-METALLIZATION FOR DSC POWER MODULE

The thin film re-metallization process was selected for the DSC power module in this work. Fig. 1 (a) shows an exploded view of the DSC module design where three SiC MOSFET dies are in parallel per switch position. Fig. 1 (b) shows a close-up view, where the wirebond-less Mo post interconnects are used for the source connection, and wire bonds are used for the gate and Kelvin connections. Fig. 1 (c) shows the thin-film deposition of Ti / Ag on the source pad followed by silver sintering of the Mo post. The following subsections will cover the re-metallization process.

A. Process

In this study, a re-metallization process for the 1.7 kV, 20 mΩ SiC MOSFETs was developed. Fig. 2 shows the stack up of the die being explored. The drain side of the die is unmodified and has the manufacturer's original 0.8 μm of Ni, 0.2 μm of palladium (Pd), and 0.1 μm of Au metallization.

Fig. 2. Diagram showing the die's original metallization stack up and the re-metallization Ti / Ag layers being deposited on the source pad.

In contrast, the source side has a 4 μm of Al to enable wirebonding. On top of the Al, the thin film deposition process will place 0.15 μm of Ti and 0.2 μm of Ag. The Ag layer serves as the sintering surface for the Mo interconnect, and the Ti layer is an adhesion layer between the Al and Ag. Fig. 1 (c) shows that the re-metallization area is limited to the yellow box (5 mm x 4 mm) and does not include the gate pad and some parts of the source pad.

For this study, 16 dies were studied, divided into three batches. Four dies are in Batch 1, eight are in Batch 2, and four are in Batch 3. Batch 1 and 2 underwent the re-metallization process in different runs. Batch 3 had no metallized dies and was treated as a control for surface analysis. Firstly, dies from Batch 1 and 2 are washed in acetone, isopropyl alcohol, and de-ionized water using an ultrasonic bath to clean their surface. Then, they are placed in stainless-steel fixtures and covered in a shadow mask stencil to expose only the source pad. The fixture with the dies is then inserted into an AJA ATC 1800 Sputtering unit with a vacuum of 4 x 10-06 Torr. A 25 W argon plasma cleaning for 1 minute is completed on an exposed source pad and the fixture. This is followed by the deposition of 0.15 μm of Ti and 0.2 μm of Ag on the source pad. The deposition rate for Ti was 73 Å/min and Ag was 350 Å/min at 200 W DC power, respectively.

After removing the dies from the chamber and fixture, they are washed again in acetone, isopropyl alcohol, and de-ionized water using an ultrasonic bath. The pressure-less sintering process involves using Indium QuickSinter paste to attach

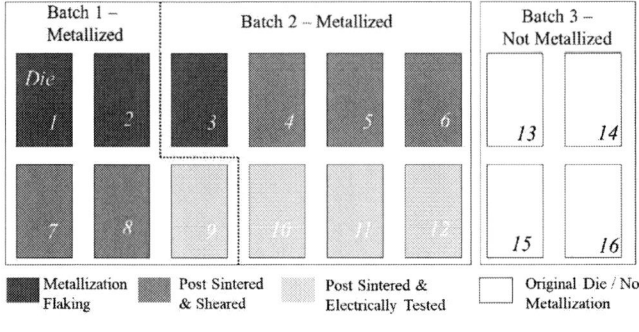

Fig. 3. A high-level diagram showing the dies used for the re-metallization study and the tests they underwent.

silver-plated Mo posts to the re-metallized surface with a 3 °C / min profile and 250 °C for 30 minutes. A similar profile is used for silver sintering the drain of the die to a substrate. The quality of the re-metallization is evaluated through visual observation, shear tests, and electrical measurements.

III. CHALLENGES AND SYMPTOMS

After re-metallizing a batch of dies, a few key challenges were observed. This included poor adhesion of the re-metallized layers, low shear strength, and high contact resistance at the re-metallized surface. Fig. 3 summarizes the dies being tested in this section.

A. Adhesion

During the ultrasonic wash process, adhesion challenges were identified when the Ag and Ti layers started flaking off two dies (die 1 and 2). Although the flaking was small, as shown in Fig. 4 (a), it indicates poor adhesion quality between the re-metallized layer and the Al pad. The dies that survived the wash process then underwent the sintering process to attach the Mo post to the re-metallized source pad. Afterward, the drain side was sintered to a substrate using the same sintering process. After the second sintering process, the post and metallization cracked off die 3 (Fig. 4 (b)). This failure, brought about by two sintering processes, indicates poor metallization and low reliability. The dies that did not experience re-metallization failure were then put through shear tests.

B. Discoloration and Bond Strength

Due to the concern of the re-metallization quality, five re-metallized dies (dies 4, 5, 6, 7, 8) that survived the sintering process underwent shear tests. Shear tests were completed on the Mo post (source-side bond), and the die's backside (drain-side bond) as a comparison. The source-side bond is between two Ag-metallized surfaces, while the drain-side bond is between an Au-metallized die and the Ag-metallized substrate.

The results are shown in Fig. 5 and highlight that the average shear strength of the source-side bond is approximately 5

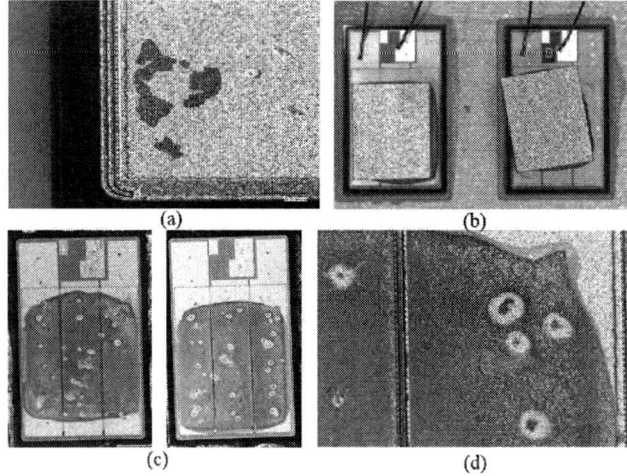

Fig. 4. The re-metallized dies experiencing (a) flaking off of the Ti and Ag layers, (b) the metallization cracking and failing after two two sintering cycles, (c) the purple surface of the die after shear tests, and (d) a close up of iridescent purple surface on the die.

MPa. As a check, the dies were also sheared to test the drain-side sinter bond to the substrate. The shear strength average was approximately 45 MPa, confirming the sintering profile and process. Greater than 40 MPa shear strengths on all bonds are desired in automotive applications to ensure good quality bonds.

Additionally, the failure point on the source-side bonds was at or near the die re-metallization layer, with most of the sinter bond and some of the metallization coming off with the Mo post. Additionally, it is noticed that the sheared region has an iridescent, purple surface (Fig. 4 (c)). The iridescent color becomes prominent towards the edge of the surface. Still, it is absent in the locations where the dies were probed during the manufacturer/vendor qualification process (Fig. 4 (d)). It is difficult to access visually and with standard measurement equipment which layer of the die the purple discoloration originates from. Still, from the shear tests, it is clear that the low shear strengths and discoloration are symptoms of a larger challenge with the re-metallization quality and process.

C. Contact Resistance

After collecting the shear test results, the remaining re-metallized dies with sintered Mo posts (Dies 9, 10, 11, 12) were electrically tested on a curve tracer to see if the re-metallization quality can be further assessed. Resistance measurements were taken at the exposed Al source pad and the re-metallized surface. Ideally, both measurements should show the same resistance, but the re-metallized surface had 3 - 4 times higher total resistance than the Al pad. The total measured resistance ranged between approximately 90 mΩ - 100 mΩ. Fig. 6 shows the resistance measurements, where $R_{DS,ON}$ is the die's internal on-resistance, $R_{DS,CONT}$ is the contact resistance across the re-metallized surface, and

979-8-3315-1612-3/25 $31.00 © 2025 IEEE

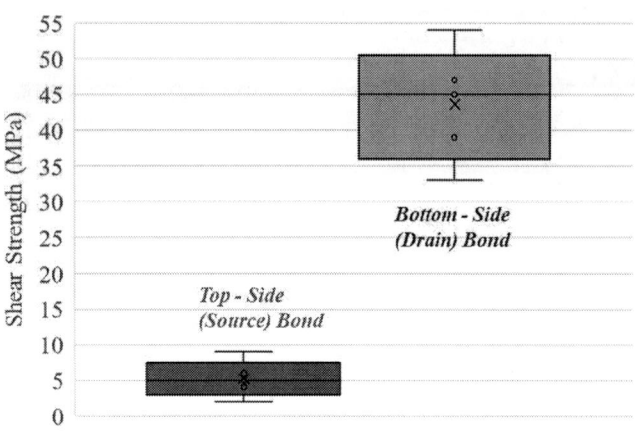

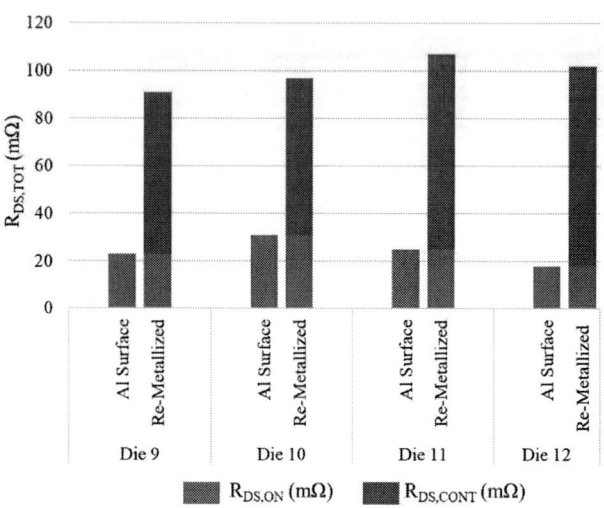

Fig. 5. Shear strengths of four re-metallized dies. The blue graph shows the shear strength of the Mo post sintered on the source-side metallization. The grey graph shows the shear strength of the drain-side die attach.

Fig. 6. Resistance measurements were taken at the aluminum source pad and at the re-metallized surface four dies. The grey bar highlights the on-resistance of the dies, and the red bar indicates contact resistance introduced by the re-metallized layers.

$R_{DS,TOT}$ is the total resistance. It can also be described as:

$$R_{DS,TOT} = R_{DS,ON} + R_{DS,CONT} \qquad (1)$$

The $R_{DS,ON}$ was measured for each die since there will be some manufacturing variation causing the value to range between 18 mΩ and 25 mΩ. Ideally, the $R_{DS,CONT}$ should negligible, such that and $R_{DS,TOT} \sim R_{DS,ON}$ if the re-metallization quality is good. The high contact resistance value shows that the adhesion is poor or that there are high-resistance layers on the die surface, causing issues. This high contact resistance is a key roadblock in re-metallization and degrades the benefits of low $R_{DS,ON}$, high-current SiC devices. The high resistance is also concentrated in a very thin contact region, so it can become a critical failure point during device operation.

IV. THEORIES

Low shear strength, discoloration, and high contact resistance make developing a high-power DSC module with good reliability difficult. Hence, based on initial findings, several theories were identified to understand the cause of these challenges. Some theories included 1) possible challenges with the sinter paste, 2) contamination of a specific batch of dies, and 3) the development of a gold-aluminum intermetallic layer in the re-metallization stack up [15]. Quick tests were completed to narrow down and eliminate these theories. For example, the dies shown in Fig. 3, were an amalgamation of different purchase orders spanning nearly a year. The results showed the same re-metallization challenges regardless of the origin of the die. The theory about forming a gold-aluminum intermetallic is unlikely due to the absence of gold in the sputtering chamber, the sintering paste, or on the top surface of the dies. Additionally, the impact of the paste was decoupled by using different types of silver sinter pastes, which yielded similar discoloration and high contact resistance.

The theory then shifted towards the possibility of leftover organic residue on the die's Al pad, stemming from the

manufacturing process, which could cause re-metallization challenges [7]. Insufficient or improper die surface preparation and cleaning could manifest as adhesion issues, weakening the bond between the Al pad and the deposited metals [16]. Factors beyond the die surface, such as contamination of the stencils and fixtures, could also contribute to the re-metallization challenges. Hence, an additional literature review on the challenges stemming from surface preparation of the dies and stencils is completed.

A. Theory: Die Surface and Contamination Challenges

A critical aspect of the top-side re-metallization process is the surface preparation of the die [16]. The surface quality will determine the adhesion and bond quality between the Al pad, the sputtered metallization, and the wirebond-less interconnect. Although the thin film deposition method has many advantages, it is highly sensitive to the surface cleanliness of the Al pad where the metallization is being deposited. Many factors could affect the Al pad's cleanliness, including handling, cleaning, and process contamination [16]. Additionally, the plasma cleaning process, including factors like the vacuum pressure, the flow rate, and the RF power, can determine the surface quality of the Al pad [7]. Other factors include possible leftover organic residue from polyimide passivation of the devices during manufacturing, which can highly impact the quality of thin film deposition.

In [16], silicon nitride and polyimide passivated insulated-gate bipolar transistor (IGBT) devices are compared using an X-ray photoelectron spectroscopy (XPS), and the findings show that the polyimide passivated devices often have higher concentrations of organic residue. It was also noticed that the cleaning processes, including plasma cleaning, have minimal

979-8-3315-1612-3/25 $31.00 © 2025 IEEE

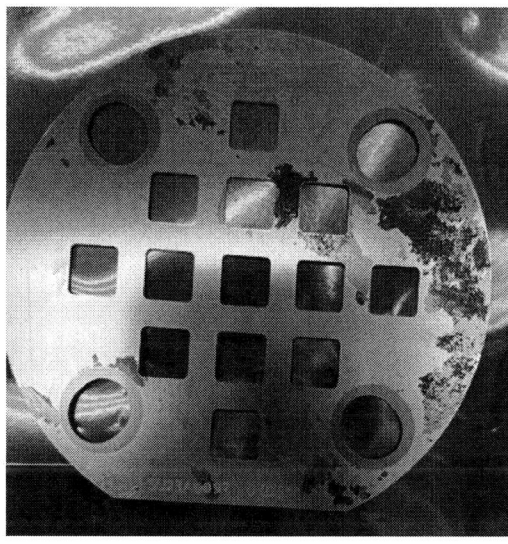

Fig. 7. Stencil used in the re-metallization process displaying purple and iridescent colors.

impact on removing hydrocarbon contamination and oxides. Similar concerns with organic contamination on the surface of other IGBT devices were discussed in [17]. In [18], the process of electroless Ni(P) under-bump metallization (UBM) was hindered by the organic surface residue from commercial devices. Fourier Transform Infrared Spectroscopy (FTIR) analysis showed that the organic contamination from polyimide passivation hindered the zinc seed deposition and growth, which reduced nickel nucleation and deposition rates [18].

Overall, it is clear that organic residue, possible oxide layers, and improper cleaning can impact the quality of the thin film deposition and manifest as poor metal adhesion to the Al pads, high surface resistance, and low shear strength. Any possible contamination from the fixtures and shadow mask stencils could add unwanted factors to the re-metallization process and exacerbate the challenges. This is an important factor since it was observed that the shadow masks started displaying similar purple and iridescent colors (Fig. 7).

This raises the possibility that multiple factors might be at play, from the die surface preparation to contamination on shadow mask stencils, which can cause the re-metallization to be poor. Thus far, the symptoms of the re-metallization challenges have been tested and documented, but to truly understand the root cause, surface chemistry analysis needs to be completed on the dies and shadow mask stencils to validate the theories.

V. SURFACE CHEMISTRY ANALYSIS

A. Surface Analysis Process

To narrow down the theories and better understand the root cause, surface analysis using a PHI Quantera Hybrid XPS was completed on dies 13-16 and on the shadow mask stencils. By using the XPS, a better understanding of the

die surface composition can be ascertained and help inform the cleaning and re-metallization process. The PHI Quantera Hybrid XPS has a scanning monochromatic Al Ka (1486.7 eV) x-ray source with a highly focused beam in selectable sizes from 9 to 200 μm. The XPS is useful in identifying surface elements by identifying electrons with respect to electron binding energies. The binding energies can be viewed as the ionization energies of atoms from a particular shell. The kinetic energy of electrons emitted from the surface helps identify the elements on the surface. By understanding the exact binding energy, the elemental and chemical composition of the surface can be identified. For all the measurements, carbon (C1s), aluminum (Al2s), and oxygen (O1s) peaks were measured, and the spectral binding energies were calibrated with respect to C1s hydrocarbons. The base pressure of the PHI Quantera was kept below 7 x 10-7 Torr.

B. Die Surfaces Scans

Prior to the XPS scanning, the dies were taken from the waffle pack and washed in acetone, isopropyl alcohol, and de-ionized water using an ultrasonic bath. Sample, wide, and multiplex scans were completed on bare dies with a depth analysis of 0.1 mm. Sample scans take a survey of different parts of the device to see how uniform the surface is in chemical composition. For example, in Fig. 8, the devices are loaded into the XPS, and three points of interest are located on the source pad for sample scans. After the sample scans, wide scans of one location are measured multiple times to achieve a high-resolution, low-noise plot. Afterward, elements of interest are selected for multiplex scans, which narrow the scan window to receive a more intense peak of elements and compounds.

Firstly, dies had their Al pad scanned, and the results show a large concentration of C1s, O1s, fluorine (F1s), and silicon (Si2p) (Fig. 9). Most samples have C1s at concentrations above 50% of the surface. The high concentration of C1s is a known issue for polyimide passivated devices due to the residue from the manufacturing process. It noted that polyimide passivated devices can have 5-6 times higher C1s on their surfaces than silicon nitride passivated devices [7]. Die 15 had the lowest C1s concentration, highlighting the surface variation between dies from the same manufacturer or vendor (Fig. 9). If not properly removed, the large concentration of C1s could cause adhesion issues between the Al surface and the deposited Ti and Ag layers [16].

Additionally, the scans reveal the presence of F1s on the surface (Fig. 9). The F1s contamination could have originated from the fabrication process during the polyimide etching process, which can use oxygen mixed with a small amount of fluorine gasses like CF_4 to increase the etch rate via enhanced radical generation [16]. It is noted that fluorine can migrate from the plasma to the surface of the polymer, where it can extract hydrogen, leaving behind a radical [16].

Al2p was also present on the surface but at lower quantities (less than 15%), indicating that XPS cannot penetrate deep

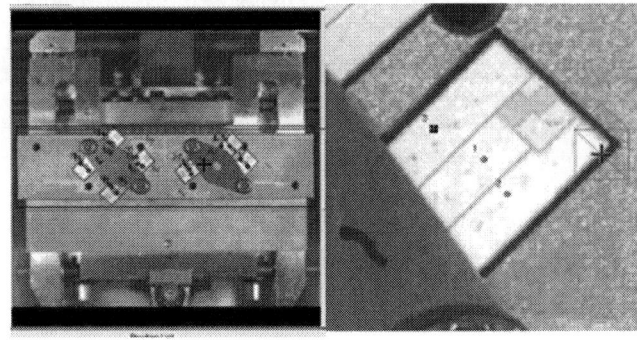

Fig. 8. PHI Quantera Hybrid XPS with devices inside the holder. The three dots show where measurements were taken on the surface.

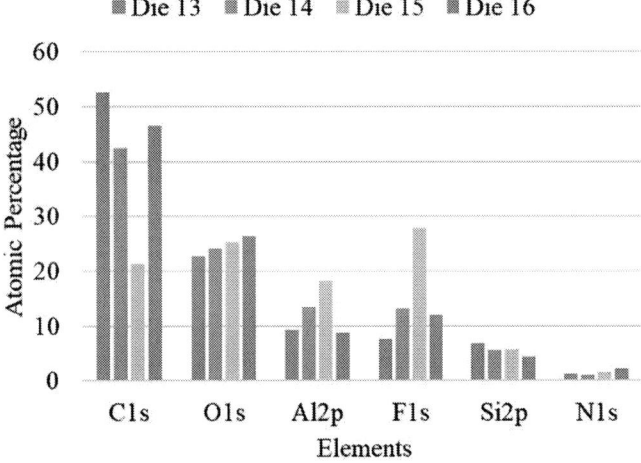

Fig. 9. XPS scanning results showing the atomic percentage of each element on the surface of four dies without any re-metallization.

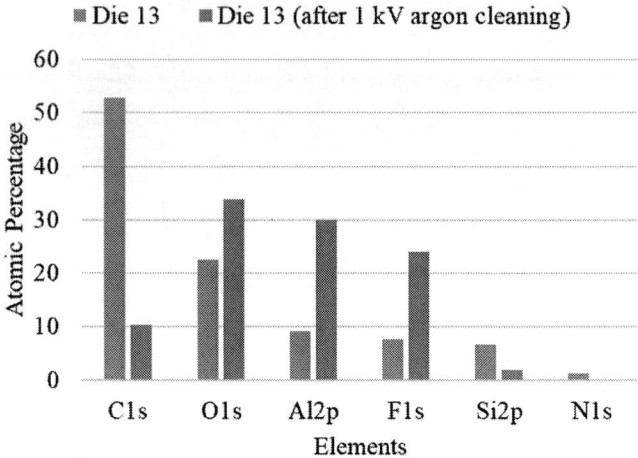

Fig. 10. XPS scanning results showing the atomic percentage of each element on the surface of die 13 before and after argon cleaning.

TABLE I
SUMMARY OF COMPOUNDS DETECTED ON THE DIES FROM XPS MULTI-PLEX SCANS

Compounds	Name	Die 13	Die 14	Die 15	Die 16
ALF_3	Aluminum Fluoride	x		x	x
Al_2O_3	Aluminum Oxide	x	x		x
$SiO_2(Al_2O_3)_2$	Alumino-silicate		x	x	

enough through the surface residue to detect clear signals from the Al pad (Fig. 9). The high concentration of C1s can mask the Al pad, contributing to high contact resistance.

Additionally, die 13 was cleaned with 1 kV argon to assess the cleaning process's impact on the surface. Fig. 10 shows the resulting atomic percentage of each element. There is a significant reduction in C1s concentration, although the argon etch has not removed all of them. Further analysis of the remaining C1s must be completed to understand whether they are single or double-bonded carbon compounds. Double-bonded carbon compounds are highly hydrophobic and can resist the adhesion of sputtered layers. Fig. 10 shows larger concentrations of F1s and Al2p are now being detected due to the removal of C1s. Interestingly, the argon cleaning has minimal impact on removing F1s from the surface. It is noted in the literature that the etching process could have exposed more of the F1s trapped in the surface due to the etching [16]. The increased percentage of Al2p detected indicates that the XPS can better detect the die's Al surface. Al2p was studied further with additional narrowed scans to understand the Al compounds on the die surface.

Multi-plex scans were completed on Al2p to identify the different binding energies on the surface. Two main binding energy peaks were 76.8 eV and 75 eV, corresponding to aluminum fluoride, AlF_3, and aluminum oxide, Al_2O_3 (Fig. 11). The blue peak is the fitted curve for AlF_3, and the green curve is the fitted curve for Al_2O_3. The black curve represents the experiment curve. The red curve is the theoretical curve added by 3 fitting curves. The main takeaway is that the Al pad has native aluminum oxide and aluminum fluoride, while no pure Al metal is seen (which should have a binding energy of 72.6 eV). From these results, it cannot be taken for granted that the source pad is just aluminum, but oxide and fluoride compounds are also on the surface, even after an argon cleaning. This reality must be considered when developing a thin-metallization process for SiC MOSFETs. Additional multi-plex scans of Al2p were completed on die 13 - 16, and the compounds detected are shown in Table I. The four scans show high concentrations of aluminum fluoride, aluminum oxide, and aluminosilicate compounds, which can lead to adhesion issues and cause poor re-metallization.

C. Shadow Mask Stencil

Additional XPS scans on the shadow masks were completed to understand better the possible cause of the stencils' discoloration (Fig. 7). The stencils were washed in acetone,

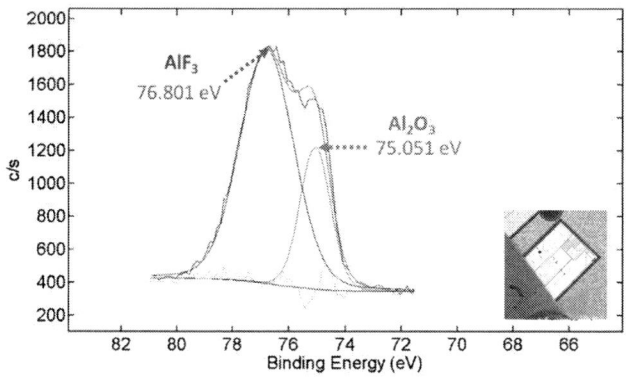

Fig. 11. Die 13 - multi-plex scan of Al2p showing the different compounds on the surface. The blue is the fitted curve for AlF_3, the green is the fitted curve for Al_2O_3, the black curve is the experimental curve, and the red is the theoretical curve of the three fitted curves.

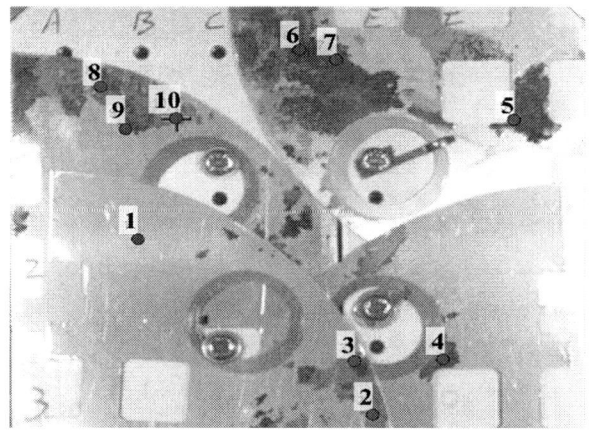

Fig. 12. Shadow mask stencil diced up and inside of the XPS. Different points that were studied on the stencil are highlighted.

isopropyl alcohol, and de-ionized water using an ultrasonic bath to clean their surface. Then, they were diced up and placed inside the XPS, and several points were measured as shown in Fig. 12. Point 1 (P1) was taken on the surface of the stencil where there was no discoloration as a control point. The other measurements were at points of discoloration and flaking on the stencil (P2 - P10). Wide scans were completed, and the results are shown in Table II. The table excludes Ti / Ag and Si since they are expected to be on the stencil due to the

TABLE II
SUMMARY OF ELEMENTS DETECTED ON THE SHADOW MASK STENCIL
FROM XPS WIDE SCANS

Elements	P1	P2	P3	P4	P5	P6	P7	P8	P9	P10
F1s	x	x	x	x	x		x	x	x	x
Ca2s		x	x	x	x		x		x	x
Na2s		x	x	x	x	x	x	x	x	x
K2s		x	x	x			x			

re-metallization and the stainless steel stencil manufacturing process, respectively.

Table II focuses on the unexpected electrolyte elements: calcium (Ca2s), sodium (Na2s), and potassium (K2s). P1, as a control, does not detect any of the electrolyte elements on the surface. On the other hand, Na, K, and Ca are detected when measurements are taken at the discolored locations like P2 - P10. Certain discoloration locations like P6 show no traces of Ca, Na, or K, but this is likely due to a high C1s value that can mask some possible electrolytes on the surface. These results indicate contamination on the stencil surface at locations of discoloration and flaking, which could point to handling, cleaning, and processing issues.

VI. DISCUSSION

Overall, the investigation indicates the possible organic residue, oxides, and fluorides on the surface of the dies, along with contamination from the stencils, might be sources of re-metallization challenges. This aligns well with the theory presented earlier in the paper. Additionally, quality of the re-metallization adhesion is highly dependent on the type and concentration of residues, oxides, and contaminants. The practical takeaway from this study is that dies must be carefully pre-screened to see the amount of unwanted residue on the surface, and a fine-tuned argon cleaning process must be implemented before re-metallizing. The XPS scans reveal that the shadow mask stencils have surface contamination, including Na, Ca, and K. The practical takeaway is that researchers need to stay vigilant against possible contamination sources during handling, and stencils should be replaced periodically to minimize the chances of contamination spreading.

Although the die and stencil surface were identified as possible causes, there is no clear-cut explanation about how the two factors are connected and the main source of the re-metallization challenges. The XPS is useful in explaining surface composition and validating the theories about insufficient cleaning, organic residue, and possible contamination. Still, it cannot describe how the dies, stencils, and sputtering chamber interact during re-metallization. Since these factors are highly interdependent, additional work is needed to develop a methodology to decouple these interdependencies through testing and verification. It is also important to acknowledge that though the XPS is very useful, it also has limitations regarding the depth and footprint it can scan, limiting our ability to understand the surfaces fully.

From an application standpoint, re-metallization remains crucial for enabling wirebond-less interconnects for DSC modules. The symptoms mentioned earlier, like the high contact resistance and poor adhesion, are key roadblocks in reliable, high-yield DSC power modules. Additionally, the sensitivity of the thin-film deposition process might lead others to consider the wet etching process (discussed earlier in the paper), which comes with its own challenges.

VII. CONCLUSION

This paper explores the quality of top-side re-metallization of SiC MOSFET dies using a thin-film deposition process. Top-side metallization is crucial to conduct high currents and dissipate heat as the need for wire bond-less interconnects grows for double-side cooled (DSC) modules. The deposition process for Ti / Ag is introduced, and the dies are tested to see the re-metallization adhesion, bond strength, and electrical performance. Testing revealed challenges, including a low average shear strength of 5 MPa, purple irradiant discoloration on the sheared surface, and a high contact resistance at the re-metallized layers (80 mΩ). A comprehensive literature review indicates that the thin-deposition process is highly sensitive to the die surface quality, the cleaning process, and unwanted contamination. XPS results highlight that the die's Al source pad has organic residue, oxides, and fluorides that could inhibit good adhesion. XPS scans on the stencils reveal possible electrolyte contamination that could impact re-metallization.

Although the thin film re-metallization process is highly attractive in its simplicity and implementation, many unexpected challenges may arise. This paper highlights the potential challenges associated with this process, hoping it informs other researchers. This paper narrows probable causes and is crucial in understanding and diagnosing the die re-metallization process for DSC power modules.

ACKNOWLEDGMENTS

This material is based upon work supported by the U.S. Department of Energy's Office of Energy Efficiency and Renewable Energy (EERE) under the Vehicle Technologies Office Award Number DE-EE0009652. This report was prepared as an account of work sponsored by an agency of the United States Government. Neither the United States Government nor any agency thereof, nor any of their employees, makes any warranty, express or implied, or assumes any legal liability or responsibility for the accuracy, completeness, or usefulness of any information, apparatus, product, or process disclosed, or represents that its use would not infringe privately owned rights. Reference herein to any specific commercial product, process, or service by trade name, trademark, manufacturer, or otherwise does not necessarily constitute or imply its endorsement, recommendation, or favoring by the United States Government or any agency thereof. The views and opinions of authors expressed herein do not necessarily state or reflect those of the United States Government or any agency thereof.

The authors acknowledge Jack Knoll and Mark Cairnie for their feedback and discussion. The authors would also acknowledge the University of Maryland NanoCenter staff for their help with the sputtering chamber. The authors would also acknowledge Virginia Tech Nanoscale Characterization and Fabrication Lab staff for their help with the XPS scans.

REFERENCES

[1] X. She, A. Q. Huang, Lucía, and B. Ozpineci, "Review of silicon carbide power devices and their applications," *IEEE Transactions on Industrial Electronics*, vol. 64, no. 10, pp. 8193–8205, 2017.

[2] A. Roy, C. DiMarino, and J. Miranda-Santos, "Design and fabrication of an inverter module co-designed with the busbar and gate driver," vol. ASME 2023 InterPACK, p. V001T02A008, 10 2023. [Online]. Available: https://doi.org/10.1115/IPACK2023-111921

[3] K. Wang, Z. Qi, F. Li, L. Wang, and X. Yang, "Review of state-of-the-art integration technologies in power electronic systems," vol. 2, no. 4, 2017, pp. 292–305.

[4] R. Selim, R. Hoofman, R. Labie, V. Sandeep, T. Drischel, and K. Torki, "Scale up of advanced packaging and system integration for hybrid technologies," in *2021 Smart Systems Integration (SSI)*, 2021, pp. 1–4.

[5] F. Hou, W. Wang, L. Cao, J. Li, M. Su, T. Lin, G. Zhang, and B. Ferreira, "Review of packaging schemes for power module," *IEEE Journal of Emerging and Selected Topics in Power Electronics*, vol. 8, no. 1, pp. 223–238, 2020.

[6] S. Seal and H. Mantooth, "High performance silicon carbide power packaging—past trends, present practices, and future directions," *Energies*, vol. 10, p. 341, 03 2017.

[7] S. Haque and G.-Q. Lu, "Effects of device passivation materials on solderable metallization of igbts," vol. 41, no. 5, 2001, pp. 639–647. [Online]. Available: https://www.sciencedirect.com/science/article/pii/S0026271401000087

[8] Y. Chen, A. Iradukunda, H. A. Mantooth, Z. Chen, and D. Huitink, "A tutorial on high-density power module packaging," vol. 11, no. 3, 2023, pp. 2469–2486.

[9] Seal, "The development of novel interconnection technologies for 3d packaging of wire bondless silicon carbide power modules." *Energies*, vol. Theses and Dissertations Retrieved from https://scholarworks.uark.edu/etd/2884, p. 12, 12 2017.

[10] A. J. Strandjord, S. Popelar, and C. Jauernig, "Interconnecting to aluminum- and copper-based semiconductors (electroless-nickel/gold for solder bumping and wire bonding)," *Microelectronics Reliability*, vol. 42, no. 2, 2002, pp. 265–283, 2002. [Online]. Available: https://www.sciencedirect.com/science/article/pii/S0026271401002360

[11] S. Seal, M. Glover, and H. Mantooth, "3-d wire bondless switching cell using flip-chip-bonded silicon carbide power devices," vol. PP, 12 2017, pp. 1–1.

[12] M. Datta, S. Merritt, and M. Dagenais, "Electroless remetallization of aluminum bond pads on cmos driver chip for flip-chip attachment to vertical cavity surface emitting lasers (vcsel's)," *IEEE Transactions on Components and Packaging Technologies*, vol. 22, no. 2, pp. 299–306, 1999.

[13] S. Haque, K. Xing, R.-L. Lin, C. Suchicital, G.-Q. Lu, D. Nelson, D. Borojevic, and F. Lee, "An innovative technique for packaging power electronic building blocks using metal posts interconnected parallel plate structures," vol. 22, no. 2, 1999, pp. 136–144.

[14] C. M. DiMarino, B. Mouawad, C. M. Johnson, D. Boroyevich, and R. Burgos, "10-kv sic mosfet power module with reduced common-mode noise and electric field," vol. 35, no. 6, 2020, pp. 6050–6060.

[15] E. Galli, G. Majni, C. Nobili, and O. Giampiero, "Gold-aluminium intermetallic compound formation," *ElectroComponent Science and Technology*, vol. 6, 01 1980.

[16] A. S. Haque, "Processing and characterization of device solder interconnection and module attachment for power electronics modules," 1999. [Online]. Available: https://api.semanticscholar.org/CorpusID:27995955

[17] D. Barry, "Design, analysis and experimental verification of a mechanically compliant interface for fabricating reliable, double-side cooled, high temperature, sintered silver interconnected power modules," 2014.

[18] W. Chen, P. McCloskey, J. F. Rohan, P. Byrne, and P. J. McNally, "Preparation and temperature cycling reliability of electroless ni(p) under bump metallization," vol. 30, no. 1, 2007, pp. 144–151.

979-8-3315-1612-3/25 $31.00 © 2025 IEEE

Design of Low Parasitic Inductance GaN HEMT Flip-Chip Power Module

Mohammad Dehan Rahman, Tanzila Akter, Abu Shahir Md Khalid Hasan, H. Alan Mantooth, Xiaoqing Song

Department of Electrical Engineering and Computer Science,
University of Arkansas,
Fayetteville, AR, United States
mr117@uark.edu, tanzilaa@uark.edu, ah162@uark.edu, mantooth@uark.edu, songx@uark.edu

Abstract— Gallium Nitride power devices, with an energy bandgap roughly three times larger than that of silicon, provide lower specific conduction resistance and faster switching speeds. These characteristics enable the design of more efficient and compact power converters. However, challenges such as high voltage overshoot during rapid switching transients continue to limit the full potential of GaN technology. This study presents an innovative alternative to the conventional wire-bonded GaN power module through the development of a flip-chip design employing solder ball bonding. A detailed finite element analysis (FEA) model is constructed to evaluate the junction temperature and parasitic inductance of power modules. Furthermore, the bonding strength is examined to ensure the mechanical integrity of the design. The proposed flip-chip design is experimentally validated by static characterization and double pulse test. The design reduces parasitics and enables compact, efficient GaN power electronics for high-frequency use.

Keywords— Flip-chip, GaN HEMT, FEA, parasitic inductance, static characterization

I. INTRODUCTION

GaN devices have demonstrated significant potential for high switching frequencies and high power density integration in electric vehicle applications. Their capability for fast switching reduces switching losses and enables high-frequency operation in power converters, thus improving overall efficiency. However, the high switching speeds of GaN devices also induce voltage overshoot during switching transients, which can lead to increased losses, concentrated hot spots, and current derating. Minimizing parasitic inductances to control switching overshoot voltage has become a critical challenge in GaN module packaging [1-4]. For power electronics in electric vehicles, achieving lightweight, high power density, and long-term reliability remains essential for meeting performance and durability requirements [5].

Wire bonding has long been one of the most widely used interconnect methods for establishing connections to and from high-power devices in bare die form. This technique offers a simple and cost-effective solution, making it suitable for a wide range of applications, from low-power MOSFETs in integrated circuits rated at just a few watts to high-power IGBT modules operating at hundreds of kilowatts. However, as system performance requirements grow, the limitations of wire bonding become increasingly evident, highlighting the need for alternative interconnect solutions in high-performance applications.

To effectively leverage the newly developed wide bandgap semiconductor devices now entering the market, it is imperative to adopt innovative packaging techniques and power module designs. Traditional packaging solutions are typically bulky, hindering the proper integration of essential gate driver and measurement circuits for power levels of several kilowatts [6-7]. The requirements for power module integration differ significantly for gallium nitride high electron mobility transistors (GaN HEMTs) compared to silicon (Si) and silicon carbide (SiC) devices due to their unique electrical and structural characteristics.

Unlike SiC chips, which have gate and source terminals on the top side, GaN chips feature drain and source terminals on the top side and the gate terminal on the bottom. This configuration often necessitates increased wire bonding for GaN HEMT power module packages. However, wire bonded architectures generally exhibit lower current carrying capabilities than other design architectures such as solder balls or flip-chip designs. Consequently, more wire bonds are required to enhance current carrying capacity [8].

Wire bonding significantly contributes to larger power loop within the power module package, which increases the parasitic inductance of the module [9-10]. The use of additional wire bonds exacerbates EMI, particularly in high-frequency applications where EMI poses a considerable concern [11]. Furthermore, reliability issues arise with wire bonded structures. Under conditions of high temperature or thermal cycling, wire bonds may fracture or lift off due to mechanical and thermal stress [12-14].

The aforementioned challenges can be alleviated through the use of flip-chip power module designs. In this approach, terminals are directly bonded to the direct bonded copper (DBC) substrate using solder balls, eliminating the need for wire bonds. The gate terminal, located on the back of the chip, is connected to another DBC or printed circuit board (PCB) to establish the gate connection. However, incorporating an additional layer of DBC or PCB increases the overall size and thermal resistance of the GaN power module design.

In this paper, both wire bond and flip-chip GaN HEMT power modules are discussed. Finite element analysis (FEA) models were developed for both architectures to evaluate their thermal performance, focusing on the junction temperatures. Additionally, the parasitic inductance of the models was examined, demonstrating the advantages of using solder balls over wire bonds. A shear test was conducted to assess the bonding strength of the die when bonded with solder balls. The static characterization of the flip-chip die using solder is also presented.

II. FLIP-CHIP PACKAGING OF GAN DIE

Flip-chip packaging, in principle, seeks to minimize the use of packaging materials. The electrical path to/from the power device is made as short as possible by using this minimalist approach. With lesser interfaces and lesser materials, the possibilities of failure are expected to reduce.

This work was supported by U.S. National Science Foundation Award 2327474, and the U.S. National Science Foundation (NSF) Center on Grid Connected Advanced Power Electronic Systems (GRAPES) under Grant 1939144.

979-8-3315-1612-3/25 $31.00 © 2025 IEEE

Fig.1: Solder ball bonded GaN die

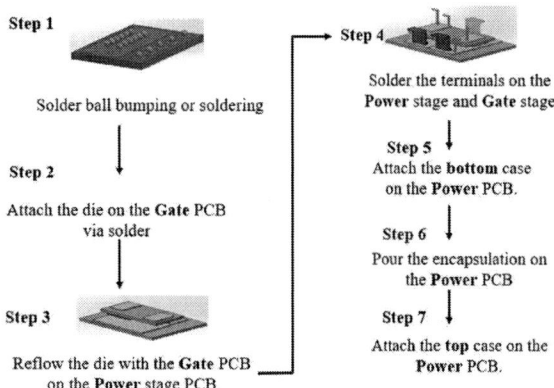

Fig.2: Reflow process of flip-chip power module

Figure 1 illustrates the proposed solder ball-bonded flip-chip configuration for a GaN die power device. In this design, the source and drain pads on the die are bonded directly with solder balls. The size of the solder balls has been optimized to maximize coverage of the drain pad, thereby minimizing current crowding and reducing parasitic inductance. The direct bonding of the GaN die to the DBC substrate eliminates the need for wire bonds, further enhancing the electrical and thermal performance of the module.

Figure 2 illustrates the fabrication process for the flip-chip module. In the initial step, solder balls are bonded to the GaN die. Following this, the die is attached to the gate PCB to provide additional support and stability for bonding with the power stage PCB. In the third step, the GaN die with the gate PCB is bonded to the power stage PCB. Next, the terminals are soldered, and the bottom casing is attached to the module. Encapsulation is then applied for additional protection, and finally, the top casing is secured to complete the module assembly.

III. THERMAL RESISTANCE ANALYSIS VIA FEA MODEL

A finite element analysis (FEA) model was developed to assess the impact of the solder ball-based flip-chip power module design. Figure 3 provides a comparison of the structures of both wire-bonded and flip-chip power modules. Wire bonded architecture is shown in Fig. 3(a) and the solder ball-based flip-chip architecture in Fig. 3(b). Both designs utilize a similar module layout to enable a direct evaluation of the electrical improvements provided by the flip-chip design. The semiconductor device has dimensions of 4.67 mm × 7.1 mm × 0.32 mm.

The finite element analysis (FEA) model was developed using the material properties listed in Table I. A heat transfer coefficient of 1000 W/m²·K was applied to simulate the junction temperature of the samples, indicating forced liquid

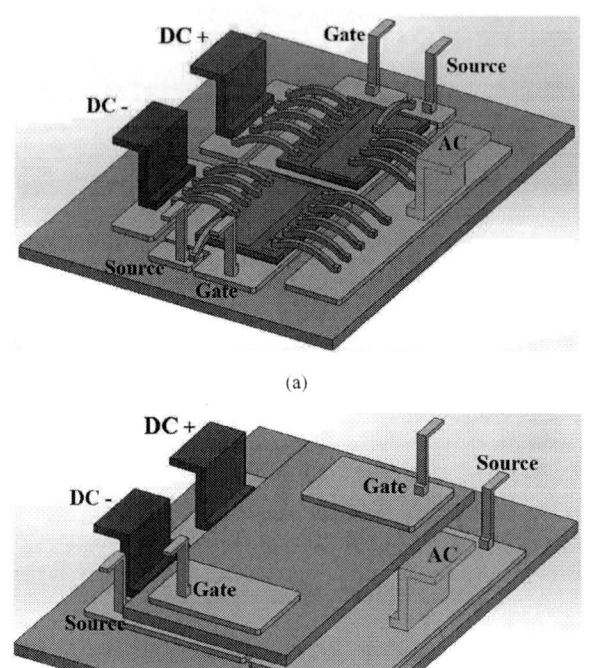

(a)

(b)

Fig.3: (a) Structure of the wire bonded GaN power module. (b) Structure of the flip-chip GaN power module

Table 1: Material parameters used in the FEA simulation.

Material	Thermal Conductivity W/(m·K)	Density (kg/m³)	Heat Capacity (J/(kg·K))
GaN	230	6095	431
Copper	398	385	8600
AlN	310	3320	780
Silver	420	8600	235
Solder balls	350	385	8600

cooling. For the simulation, the power loss of the chip was set at 20 W.

Thermal analysis of both the wire bonded and flip chip designs is shown in Fig. 4. From the figure it is observed that the flip-chip design has a significantly higher junction temperature compared to the wire-bonded design. This increase is attributed to the additional DBC layer required for establishing gate connections and the lower thermal conductivity of solder balls compared to the silver paste used in silver sintering. Moreover, the contact area of the solder balls with the GaN chip, which acts as the heat source, is considerably smaller than that of the silver bonding layer used in the wire-bonded design. Consequently, the junction temperature is higher in the flip-chip GaN half-bridge power module.

Thermal resistance of both power modules is shown in Fig. 5. Due to higher maximum temperature of the flip-chip module, the thermal resistance of the module is also higher. From the figure, it is observed that the thermal resistance of the designed module is almost 2.2 times more than wire bonded design. The relation between thermal resistance and maximum temperature is shown in the following equation. As maximum temperature of solder ball based flip-chip module is higher due to small contact area of the solder, thermal resistance is also higher. Enhanced cooling is required to reduce the thermal resistance of the flip-chip module.

$$R_{Th} = \frac{T_j - T_C}{P}$$

To improve the temperature profile of the flip-chip power module, double side cooling approach is integrated. Simulation results of double side cooling are shown in Fig.6. This cooling implementation improves the temperature profile of the flip-chip design. From the figure it is observed that double side cooling reduces the maximum temperature by almost 50% when compared with conventional bottom side cooling. Implementation of double side cooling is easier in the proposed design, due to the hybrid packaging of the flip-chip module.

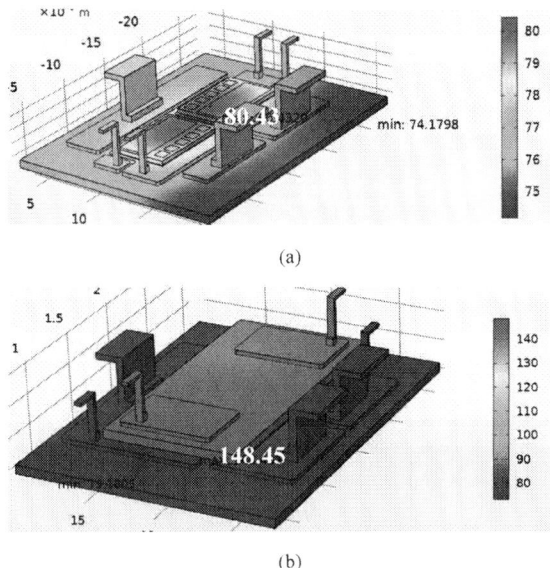

(a)

(b)

Fig.4: Junction temperature comparison between the wire bonded and flip-chip GaN half bridge power module (chip size: 7.1mm × 4.67mm, power losses: 20 W). (a) Wire bonded GaN power module maximum junction temperature: 80.4320 °C. (b) flip-chipped GaN power module maximum junction temperature: 148.457 °C.

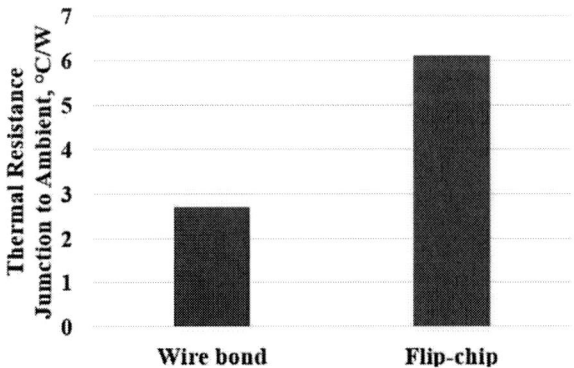

Fig. 5: Thermal resistance comparison between wire bonded and flip-chip design

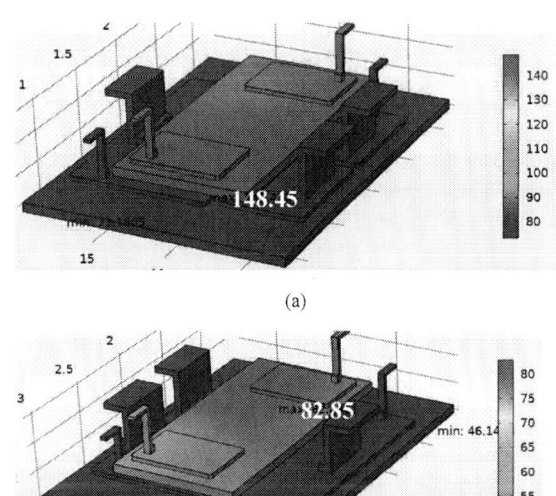

(a)

(b)

Fig.6: Junction temperature comparison: (a) Bottom side cooling of flip-chip design, (b) Double side cooling of flip-chip

IV. PARASITIC INDUCTANCE ANALYSIS

A Q3D simulation is done to investigate the parasitic inductance of both the architecture. In conventional design, wire bonds are a major source of inductance in any power module. Wire bonds increase the power loop of the module, thus increasing the loop inductance. This increase in parasitic inductance causes switching loss, voltage overshoot, reduced switching speed and high electromagnetic interference. Solder ball based flip-chip design aims at reducing the inductance in the module, by decreasing the current loop.

Parasitic inductance at various switching frequencies for both wire-bonded and flip-chip power modules was examined. As shown in Fig. 7, parasitic inductance follows an exponential decay relationship with increasing switching frequency. The flip-chip design exhibits approximately 43.38% lower parasitic inductance than the wire-bonded module. This reduction is due to the GaN die being directly bonded to the substrate, thereby minimizing the conductive path and reducing the loop area between the power semiconductor die and the substrate.

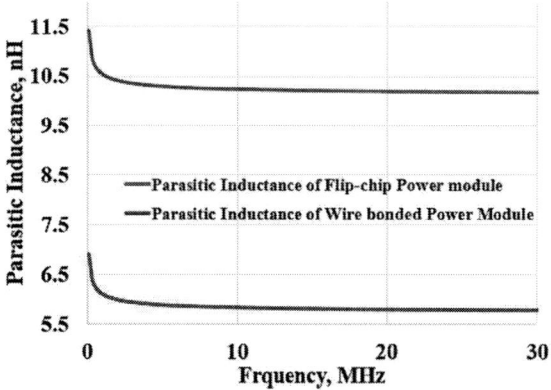

Fig. 7: Parasitic Inductance comparison between wire bonded and flip-chip design

V. EXPERIMANTAL ANALYSIS OF THE SOLDER BALL BONDED GAN HEMT DIE

An experimental setup is made to analyze the bonding strength of the die, bonded with solder balls. Conventional solder paste is used to bond the die with PCB board as shown in Fig. 8. The spacing between the solder ball pads needed to be sufficient to prevent adjacent solder balls from merging when they became molten during the reflow process. The following step in the process was to create a dependable bond.

The shear test is conducted on the samples by using the Dage DS 100 machine. The shear test results are shown in Fig. 9. The bonding strength of the sample was found to be 120 N. Even though the sample underwent bonding failure at 120 N, the result is quite encouraging. The bonding strength of the solder ball architecture can be further strengthened by using under-fill materials.

The static characterization of the solder ball connected to the GaN die was conducted using a curve tracer, as illustrated in Fig. 10. This setup allowed for a detailed assessment of several critical parameters of the GaN die, including output characteristics, transfer characteristics, and on-resistance (Rdson). These parameters validated the electrical viability of the proposed approach. Due to limitations in the testing probes, the maximum drain current for this characterization was capped at 4A, providing insight primarily into the device's behavior under low-current conditions.

Figure 11 presents the output characteristics of the GaN die, demonstrating the device's capability to block current up to V_{gs} = -16V. This observation is corroborated by the transfer characteristics curve shown in Fig. 12, where the drain

current begins to rise at V_{gs} = -16V. The R_{dson} value is relatively high (45 m Ω) due to the low current and voltage settings used during the test. This elevated R_{dson} value suggests potential areas for optimization in the die's design to enhance its low-voltage performance. Under typical operating conditions—outside the constraints of the test's limited current range—the R_{dson} is anticipated to decrease, resulting in more efficient conduction with reduced losses. This R_{dson} measurement serves as a useful baseline for future high-current testing, where improved probing capabilities can facilitate a comprehensive evaluation of the die's low-resistance characteristics.

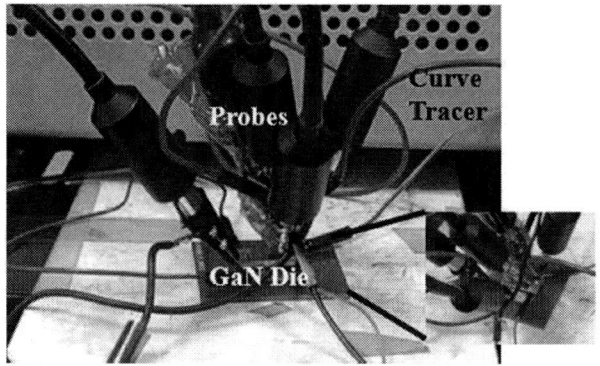

Fig. 10: Experimental setup of GaN Die characterization

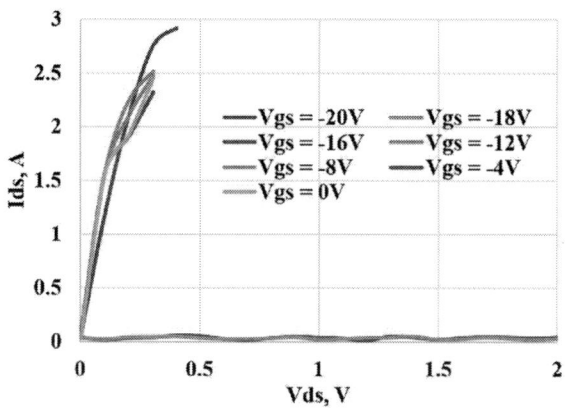

Fig. 11: Output characteristics of flip-chip GaN Die

Fig. 8: GaN die flip-chip bonded by solder balls

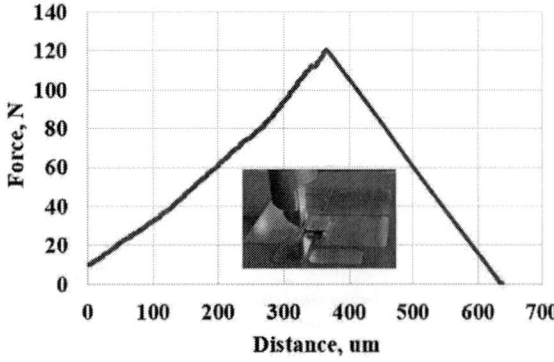

Fig. 9: Force-Displacement curve for the die shear test conducted on a flip-chip bonded sample

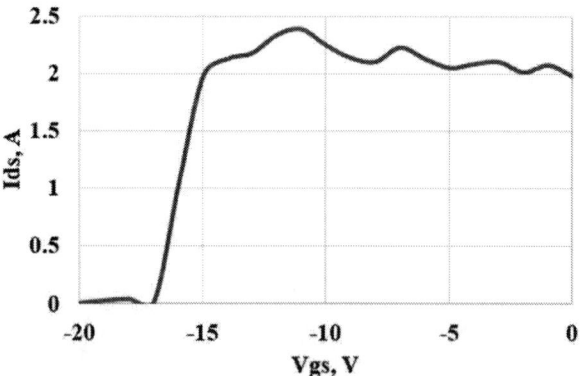

Fig. 12: Transfer Characteristics of flip-chip GaN Die

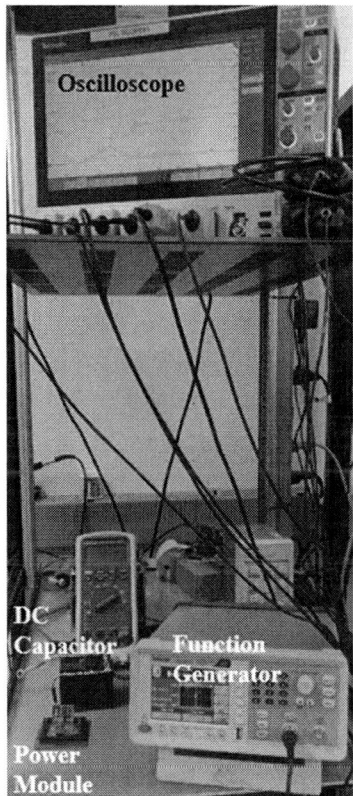

Fig. 13: Double pulse setup for the flip-chip module

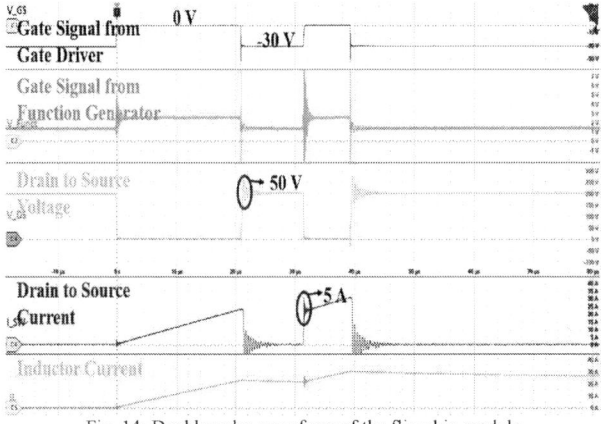

Fig. 14: Double pulse waveform of the flip-chip module

A double pulse test platform was used to evaluate the switching performance of the proposed flip-chip architecture as shown in Fig. 13. The test setup was used to characterize the switching performance of the module.

The test waveforms are shown in Fig. 14. From the figure it is observed that for 200 V DC voltage, the maximum voltage overshoot is 50 V. Gate signals from function generator is susceptible to noise as seen from Fig. 14. Thus, a gate driver was used to send clearer gate signals to GaN die, as observed from the figure. Some current ringing is observed at turn off condition due to parasitic inductance. But it is only observed for around 2us.

VI. CONCLUSION

A wire bondless, flip-chip GaN power module package utilizing solder ball bonding is proposed. Parasitic inductance of the flip-chip module is compared with conventional wire bonded power module. In the proposed architecture, parasitic inductance is reduced by 43.38%. The thermal performance of this design is analyzed and compared with that of a conventional wire-bonded design. Due to the smaller bonded area of the solder balls, the flip-chip solder ball module exhibits a higher junction temperature, which can be mitigated by implementing double-sided cooling. Double side cooling helps reduce maximum temperature of the solder ball based flip-chip module by almost 50%. The solder ball bonded sample exhibited a die bonding strength of 120 N without the use of underfill material. The electrical viability of the process is confirmed through static characterization and double pulse test of the GaN die bonded with solder balls.

REFERENCE

[1] C. Chen, Z. Huang, L. Chen, Y. Tan, Y. Kang and F. Luo, "Flexible PCB-Based 3-D Integrated SiC Half-Bridge Power Module With ThreeSided Cooling Using Ultralow Inductive Hybrid Packaging Structure," in IEEE Transactions on Power Electronics, vol. 34, no. 6, pp. 55795593, June 2019

[2] M. Wang et al., "Advanced Packaging Technology of GaN HEMT Module for High-Power and High-Frequency Applications: A Review," in IEEE Transactions on Components, Packaging and Manufacturing Technology, vol. 14, no. 9, pp. 1537-1550, Sept. 2024, doi: 10.1109/TCPMT.2024.3447079.

[3] A. I. Emon, Mustafeez-ul-Hassan, A. B. Mirza, J. Kaplun, S. S. Vala and F. Luo, "A Review of High-Speed GaN Power Modules: State of the Art, Challenges, and Solutions," in IEEE Journal of Emerging and Selected Topics in Power Electronics, vol. 11, no. 3, pp. 2707-2729, June 2023, doi: 10.1109/JESTPE.2022.3232265.

[4] K. Wang, L. Wang, X. Yang, X. Zeng, W. Chen and H. Li, "A Multiloop Method for Minimization of Parasitic Inductance in GaN-Based High-Frequency DC–DC Converter," in IEEE Transactions on Power Electronics, vol. 32, no. 6, pp. 4728-4740, June 2017, doi: 10.1109/TPEL.2016.2597183.

[5] F. Guo, T. Yang, C. Li, S. Bozhko and P. Wheeler, "Active Modulation Strategy for Capacitor Voltage Balancing of Three-Level Neutral-PointClamped Converters in High-Speed Drives," in IEEE Transactions on Industrial Electronics, vol. 69, no. 3, pp. 2276-2287, March 2022.

[6] M. D. Rahman and X. Song, "Investigation of the Impact of Low Thermal Conductivity on Gallium Oxide Power Module Packaging," 2023 IEEE 10th Workshop on Wide Bandgap Power Devices & Applications (WiPDA), Charlotte, NC, USA, 2023, pp. 1-5, doi: 10.1109/WiPDA58524.2023.10382165.

[7] M. D. Rahman and X. Song, "A Phase Change Material Based Silicon Carbide Power Module Packaging," 2024 IEEE 10th International Power Electronics and Motion Control Conference (IPEMC2024-ECCE Asia), Chengdu, China, 2024, pp. 4916-4920, doi: 10.1109/IPEMC-ECCEAsia60879.2024.10567611.

[8] Gomez, Frederick Ray I., Anthony R. Moreno, and Jonathan C. Pulido. "Wirebond solution of semiconductor IC package through modeling and simulation." Journal of Engineering Research and Reports 7.3 (2019): 1-10.

[9] H. Chen et al., "Demonstration of Wire Bondless Silicon Carbide Power Module with Integrated LTCC Jet Impingement Cooler," 2022 IEEE Energy Conversion Congress and Exposition (ECCE), Detroit, MI, USA, 2022, pp. 1-6, doi: 10.1109/ECCE50734.2022.9947694.

[10] S. Seal, M. D. Glover, A. K. Wallace and H. A. Mantooth, "Flip-chip bonded silicon carbide MOSFETs as a low parasitic alternative to wire-bonding," 2016 IEEE 4th Workshop on Wide Bandgap Power Devices and Applications (WiPDA), Fayetteville, AR, USA, 2016, pp. 194-199, doi: 10.1109/WiPDA.2016.7799936.

[11] D. Reusch and J. Strydom, "Understanding the Effect of PCB Layout on Circuit Performance in a High-Frequency Gallium-Nitride-Based

Point of Load Converter," IEEE Trans. Power Electron., vol. 29, no. 4, pp. 2008–2015, Apr. 2014.

[12] W. Liu, D. Zhou, F. Iannuzzo, M. Hartmann and F. Blaabjerg, "Separation and Validation of Bond-Wire and Solder Layer Failure Modes in IGBT Modules," in *IEEE Transactions on Industry Applications*, vol. 58, no. 2, pp. 2324-2331, March-April 2022, doi: 10.1109/TIA.2022.3141034.

[13] P. Rajaguru, T. Tilford, C. Bailey and S. Stoyanov, "Damage Mechanics-Based Failure Prediction of Wirebond in Power Electronic Module," in *IEEE Access*, vol. 12, pp. 25215-25227, 2024, doi: 10.1109/ACCESS.2023.3342689.

[14] Y. Liu, S. Irving, and T. Luk, "Thermosonic Wire Bonding Process Simulation and Bond Pad Over Active Stress Analysis," IEEE Trans. Electron.Packag. Manuf., vol. 31, no. 1, pp. 61–71, Jan. 2008.

A Scalable Dual-Orthogonal-Cooling Packaging Concept for Parallel-Series SiC Chips

1st Ekaterina Muravleva
Dept. of Elect. and Comput. Eng.
University of Nebraska-Lincoln
Lincoln, NE 68588-0511, USA
emuravleva2@huskers.unl.edu

2nd Youssef Abotaleb
Dept. of Elect. and Comput. Eng.
University of Nebraska-Lincoln
Lincoln, NE 68588-0511, USA
yabotaleb2@huskers.unl.edu

3rd Blake Anderson
Dept. of Elect. and Comput. Eng.
University of Nebraska-Lincoln
Lincoln, NE 68588-0511, USA
banderson76@unl.edu

4th Zichen Zhang
Dept. of Elect. and Comput. Eng.
Virginia Tech
Blacksburg, VA 24061-0002, USA
zichen2013@vt.edu

5th Boyi Zhang
Milan Power Electronics Lab
Delta Electronics (Americas) Ltd.
RTP, NC 27519, USA
boyi.zhang@deltaww.com

6th Jerry Hudgins
Dept. of Elect. and Comput. Eng.
University of Nebraska-Lincoln
Lincoln, NE 68588-0511, USA
jhudgins2@unl.edu

7th Jun Wang
Dept. of Elect. and Comput. Eng.
University of Nebraska-Lincoln
Lincoln, NE 68588-0511, USA
junwang@unl.edu

Abstract—Conventional high-frequency SiC power modules with isolated substrates typically bond the SiC dies directly onto the substrates. Although effective for parallel die configurations and simple package topologies, this design becomes limiting for high-voltage, high-current modules that require parallel-series die connections in the package and complex topologies. This paper introduces a novel dual-orthogonal-cooling (DOC) packaging concept that positions the dies perpendicularly to the substrate using a proposed switch bar structure. The switch bars are designed to conduct high currents horizontally; confine power loops within the ceramic layer of the top substrate and external decoupling capacitors; segment voltage stresses to limit maximum E-field and dv/dt; and spread heat to enable DOC through dual-sided heat dissipation. The new design reduces the package footprint and features high scalability to accommodate various power ratings and topologies. The proposed DOC packaging concept has been validated through FEA simulations, demonstrating reduced power-loop inductance, mitigated E-field near triple points, and enhanced thermal performance. Finally, a mechanical prototype with 2-parallel-4-series dies was fabricated to assess the manufacturability and identify manufacturing challenges.

Index Terms—SiC MOSFET package, in-package parallel-series die connection, isolated substrates, dual orthogonal cooling.

I. INTRODUCTION

The rapid growth in electricity demand driven by artificial intelligence (AI) data centers and electric vehicle (EV) fast chargers is straining the existing grid distribution infrastructure. To address this, medium-voltage ac (MVAC) to low-voltage dc (LVDC) power factor correction (PFC) rectifiers with higher power, efficiency, and density, as well as bidirectional power flow capabilities, are urgently needed. However, a critical challenge lies in the lack of higher-voltage ($> 10\,\text{kV}$), higher-current ($> 1\,\text{kA}$) SiC power semiconductor modules, which are essential for enabling simpler and more efficient grid converter topologies. SiC MOSFET or JFET dies, being small with limited voltage and current ratings but very fast switching speed, require module designs that effectively address parallel-series die connections, parasitics, partial discharge (PD) concerns, and thermal management for practical medium-voltage (MV) grid applications.

Existing SiC modules rated above 3.3 kV mainly adopt half-bridge configurations with multiple parallel dies per switch position, where the power-loop inductance (L_s) is a critical performance metric [1]. These 3.3 kV SiC modules typically offer current ratings ranging from $200\,\text{A}$–$1000\,\text{A}$, primarily targeting traction inverter applications. They exhibit $10\,\text{nH}$–$35\,\text{nH}$ L_s when using external decoupling capacitors [2]–[6], while incorporating MV MLCC decoupling capacitors directly within the module packaging can reduce L_s to as low as $6.9\,\text{nH}$ [7]. SiC modules rated higher than 3.3 kV, 1 kA are presently unavailable. Existing $6\,\text{kV}$–$7.2\,\text{kV}$ SiC modules are rated below $400\,\text{A}$ with L_s exceeding $23\,\text{nH}$ when using external decoupling capacitors [8]–[10]. By confining the power loop between two substrates, L_s can be reduced to 2.6 nH [11]; this technique is also seen in 1.2 kV SiC module designs [12]. Current 10 kV SiC modules are rated below 240 A with a 16.0 nH L_s when using external decoupling capacitors [13]. Designs with internal MLCC decoupling capacitors have achieved inductance values of 4.4 nH [14] and 5.6 nH [15], respectively. On the other hand, SiC modules featuring in-package series die configurations remain scarce. Ref. [16]

979-8-3315-1612-3/25 $31.00 © 2025 IEEE

packaged six 1.2 kV series-connected dies with respective RC snubbers to build a 6.5 kV single-switch-position module (not half bridge). The study compared metal spacer versus bond wire connections and evaluates different layouts, attaining a minimum L_s of 24.6 nH. Ref. [17] used two series-connected 1.2 kV dies per switch position (a total of four dies) to make a 2.4 kV half-bridge module, obtaining 5.8 nH with an internal MLCC capacitor. The aforementioned research efforts highlight two critical limitations in existing designs: 1) Achieving $L_s < 10$ nH without integrating internal decoupling capacitors seems impractical. However, embedding MLCC capacitors within the high-temperature module environment introduces reliability concerns, especially the short-circuit ruggedness concern due to their susceptibility to mechanical stress. 2) Configuring SiC dies in both parallel and series connections requires an extensive substrate area, making it very challenging to accommodate both configurations simultaneously for high-power MV grid applications.

Power modules with isolated substrates have junctions at the interface among the ceramic layer, copper plate, and silicone gel or epoxy, commonly referred to as triple points. The high voltage across copper plates creates a strong, distorted E-field at the triple points, increasing the occurrence of PD events. These events accelerate electrical treeing, especially under square-wave excitation, leading to premature power module breakdown [18]. PD mitigation strategies are categorized into two paths: optimizing substrate geometry and employing advanced dielectric/semiconductor materials [19]. The former include stacked substrates, ceramic geometry modifications, field plates, and electret structures. Among these, stacked substrates with dc midpoint (dc-mid) referencing, which create bipolar, symmetrical voltage stresses, offer the simplest solution without significantly complicating the manufacturing process [20]. However, this design necessitates embedding decoupling capacitors within the package, raising the same short-circuit ruggedness concern of the MLCC capacitors as previously noted. In general, maintaining the E-field below 30 kV/mm at a distance of 15 μm from the worst-case triple point is preferred [21]–[24]. For modules with in-package series-connected dies, only Ref. [16] optimized the dimensions of the substrate copper plate and copper spacer, specifically for a single-sided cooling structure.

Dual-sided cooling designs for EV power modules rated up to 1.2 kV with in-package parallel dies have been comprehensively reviewed in [25]. These designs sandwich all dies between two substrates, employing two main approaches to enable heat transfer through the die's front side: direct bonding to a top substrate or through metal spacers. In either approach, the dies for both half-bridge switch positions can be entirely bonded to the bottom substrate. Alternatively, as mentioned in [12], the dies of one switch position can be mounted on the bottom substrate, while those of the other switch position are placed on the top substrate to minimize L_s. The variations in thermal resistance with dual-sided cooling primarily depend on the die area (very different between Si IGBTs and SiC MOSFETs) and metallization, as well as the

materials and geometries of the DBC/AMB/ODBC substrates, metal spacers, Cu/AlSiC baseplates (with or without cooling channels), and all bonding mediums. The junction-to-case thermal resistance per die $R_{\text{th,j-c,die}}$ for low-voltage dies is ranged in 0.018 K/W–0.4 K/W, with the lowest result achieved by CRRC/Dynex [26]. For modules utilizing metal spacers, a method for further reducing thermal resistance involves filling the non-conductive gaps with low-temperature co-fired ceramic (LTCC) interposers (some loaded with graphene or pyrolytic graphite sheets), thereby enhancing thermal routing and mechanical support between the two substrates [12]. From the thermal perspective, these designs also apply to > 3.3 kV parallel-die modules if PD is well addressed [24]. Nevertheless, for high-power grid-oriented modules with in-package parallel-series dies, the limited substrate area remains a challenge.

In summary, scalable packaging solutions for accommodating substantial in-package parallel-series SiC dies, targeting MV grid distribution, remain unexplored. Existing package designs for L_s minimization, PD mitigation, and dual-sided cooling render valuable design guidance but may fall short of achieving optimal overall performance. As such, this work introduces a novel dual-orthogonal-cooling (DOC) packaging concept that positions the dies perpendicularly to the substrate using a proposed switch bar structure. The switch bars are designed to 1) conduct high currents horizontally; 2) confine power loops within the ceramic layer of the top substrate and external decoupling capacitors; 3) segment voltage stresses to limit maximum E-field and dv/dt; and 4) spread heat to enable DOC through dual-sided heat dissipation. The new design reduces the package footprint and features high scalability to accommodate various power ratings and topologies.

The paper is organized as follows: Section II introduces the proposed DOC packaging concept; Section III verifies power-loop inductance, E-field near triple points, and DOC through FEA simulations; Section IV demonstrates a mechanical prototype for fabrication process validation; and Section V concludes the study and discusses manufacturing challenges and future work.

II. PROPOSED DOC PACKAGING CONCEPT

The proposed DOC concept is illustrated with a half-bridge module configuration, where each switch position incorporates two series-connected dies (**Fig. 1(a)**). Building upon the existing design in [16] (a single-switch-position module rather than a half-bridge), an intuitive conventional approach would use copper spacers and a top substrate to interconnect the series dies (**Fig. 1(b)**). This design yields a lower stray inductance than using bond wires and is compatible with dual-sided cooling (not implemented, though) [16]. However, scaling this design to accommodate many parallel-series dies introduces significant challenges: 1) Placing all dies on the substrate requires a substantial substrate area and module footprint. 2) The zig-zag power loop results in high L_s; for a loop with four dies, the optimized L_s already reached 24.6 nH [16]. 3) The locations of dc+ and dc− are asymmetrical and may reside on

979-8-3315-1612-3/25 $31.00 © 2025 IEEE

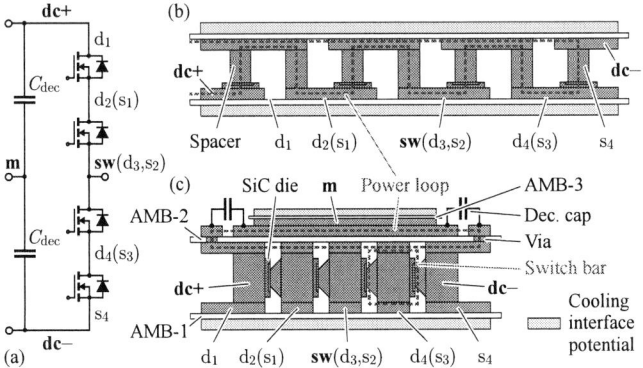

Fig. 1: (a) Half-bridge module with two series-connected dies at each switch position. (b) Existing package design using copper spacers [16]. (c) Proposed DOC package design using novel switch bars.

different substrates, leading to uneven E-field distribution and complex module termination. 4) The use of numerous metal spacers complicates the module fabrication.

To achieve a balanced performance across scalability, footprint, parasitics, E-field, and thermal management, a scalable DOC packaging concept is proposed (**Fig. 1(c)**). The key element of the DOC package is the switch bar structure. A switch bar consists of a slim copper bar that carries multiple parallel dies arranged in a line. A gate-driver distribution PCB can be mounted on the same surface as the die, connecting to the gate and source pads via bond wires. The opposite side of each die includes trapezoidal pyramid extrusions, enabling the horizontal stacking of switch bars to create series die connections. The two remaining slim sides act as cooling interfaces, bonded to the AMB-1 and AMB-2 substrates at the bottom and top, respectively. In this distinctive configuration, all dies are oriented perpendicularly to the substrates. AMB-2 is designed with connectivity between the bottom and top copper plates at the left and right edges, e.g., by means of solder-filled vias. The top middle copper plate on AMB-2 is designed to connect to the dc-mid (m) potential, facilitating the external placement of MLCC decoupling capacitors. For galvanic isolation, a third substrate, AMB-3, is bonded onto AMB-2. Since AMB-3 does not carry high current, it can be designed with minimal copper and ceramic thicknesses, enhancing top-side heat dissipation. Notably, the bottom copper plate of AMB-1 and the top copper plate of AMB-3, both at the cooling interface potential, should be connected to the dc-mid potential through high impedance for E-field management and minimal leakage current.

The DOC package provides several advantages. 1) High scalability: Extending the length of the switch bar to accommodate more dies increases the module's current rating, while laterally stacking more bars enhances the voltage rating. 2) Compact footprint: The switch bars are tightly integrated, minimizing the overall footprint regardless of die dimensions and spacing. 3) Reduced parasitics: The power loop is confined within the AMB-2 ceramic layer, with minimal gaps between the switch bars, achieving power-loop inductance below 10 nH even if external decoupling capacitors are used. 4) External

decoupling capacitors: The design lowers the ambient temperature around the MLCC capacitors, mitigates short-circuit ruggedness concerns, and allows for easy replacement of damaged capacitors without replacing the module. 5) Reduced triple-point E-field: The internal E-field is effectively graded by segmenting the switching potentials, preventing the close proximity of copper layers with high voltage differentials. Additionally, the use of the dc-mid as a reference potential halves the maximum voltage difference by creating bipolar, symmetrical voltage stresses. 6) Dual orthogonal cooling: Heat flux is generated horizontally on both sides of the dies and dissipates vertically through the two substrates, forming a DOC mechanism. While achieving the lowest per-die junction-to-case thermal resistance may be challenging compared to existing dual-sided cooling modules, a decent thermal resistance is attainable. This trade-off is acceptable given the compact and highly integrated design.

III. PACKAGE MODELING AND FEA VERIFICATION

A. Package Modeling

A conceptual 5.1 kV, 500 A DOC half-bridge module is modeled. **Fig. 2(a)** illustrates the proposed switch bar design, comprising a slim copper bar, 10 parallel SiC MOSFET dies,

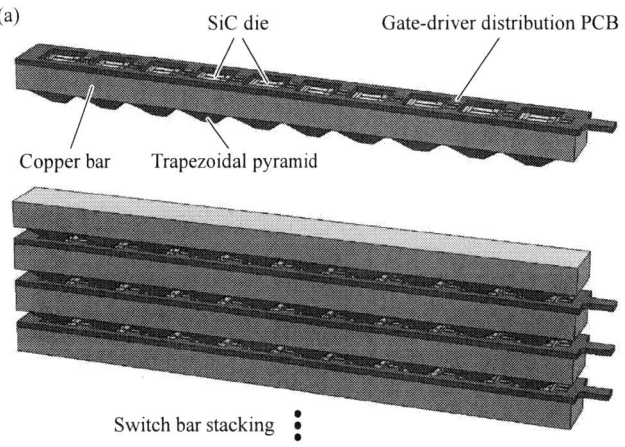

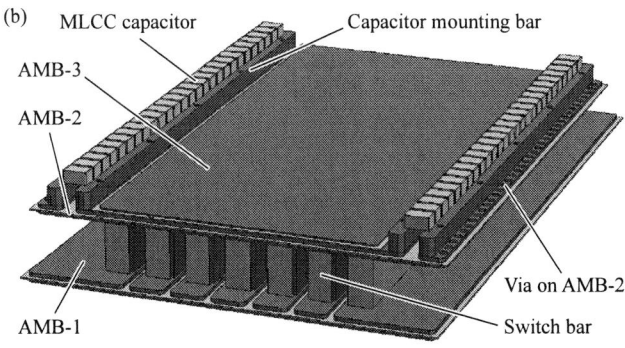

Fig. 2: (a) Conceptual switch bar with 10 parallel dies and the stacking of switch bars. (b) Conceptual 5.1 kV, 500 A DOC half-bridge module with a total of 60 dies. Each switch position consists of 10-parallel-3-series 1.7 kV, 50 A SiC dies. AMB-1 area: 85 mm × 160 mm.

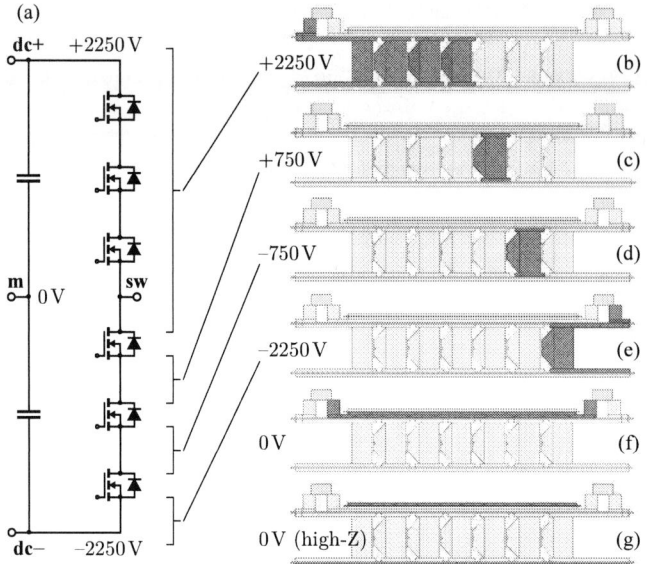

Fig. 3: Voltage potentials of all conductive parts when the top switch position is turned on at a 4.5 kV dc bus, assuming the dies at the bottom switch position share balanced blocking voltages. (a) Module circuit topology. Conductive parts are at the potentials of (b) +2250 V, (c) +750 V, (d) −750 V, (e) −2250 V, (f) dc-mid 0 V, and (g) cooling interfaces 0 V (connected to dc-mid through high impedance).

and a gate-driver distribution PCB. The back side of each copper bar features 10 trapezoidal pyramid extrusions, enabling vertical sintering of the switch bars in a stacked configuration. **Fig. 2(b)** shows the conceptual half-bridge module designed for FEA verification. Notably, four capacitor mounting bars are bonded onto AMB-2, facilitating the assembly of MLCC decoupling capacitors and separating them from the module package. The MLCC capacitor assemblies are intended to be potted for enhanced stability and insulation. With a 4.5 kV dc bus, when the top switch position is turned on, the voltage potentials of all conductive parts are denoted in **Fig. 3**. A mirrored voltage potential distribution will occur when the bottom switch position is turned on.

B. Power-loop Inductance

The commutation power loop is composed of the switch bars, AMB-2 copper plates, and external MLCC capacitors and their mounting bars. This loop tightly encloses the narrow space occupied by the AMB-2 ceramic layer (0.32 mm thick Si₃N₄) and the minimal gaps between switch bars and between capacitor mounting bars (**Fig. 4(a)**). As a result, the power-loop inductance remains exceptionally low, even with the decoupling capacitors mounted externally to the module. In the ANSYS Q3D simulation, a pair of source and sink are assigned to the top surfaces of the right-side capacitor mounting bars (**Fig. 4(b)**). The right-side MLCC capacitors are assigned as non-model elements, while the left-side capacitors are assigned as copper conductors. Under these settings, the simulated $L_s = 3.7$ nH. Furthermore, this module can be integrated as a submodule within a larger package, potentially reducing L_s even further by parallel.

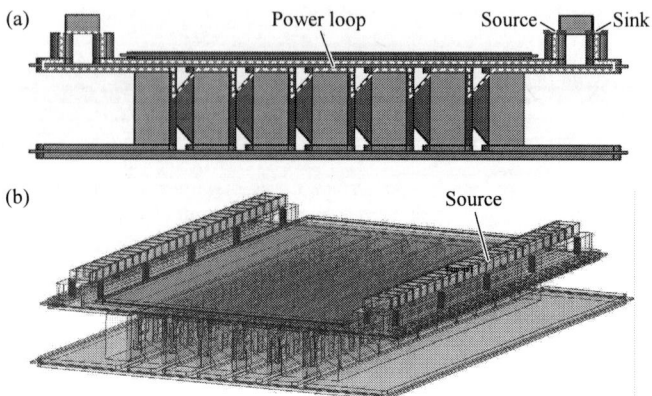

Fig. 4: (a) Power loop from the cross-section view. AMB-2: copper 0.8 mm, Si₃N₄ 0.32 mm. (b) ANSYS Q3D result: $L_s = 3.7$ nH.

C. E-field Near Triple Points

Using the voltage potential settings shown in **Fig. 3**, the E-field strength within the package was simulated using the COMSOL AC/DC Module. The E-field magnitude was found to be sensitive to the mesh size, increasing as the mesh size decreased. Simulations with upper mesh limits of 5 μm, 10 μm, and 25 μm revealed significant variation in the E-field near the critical triple point, with differences ranging from 30% to

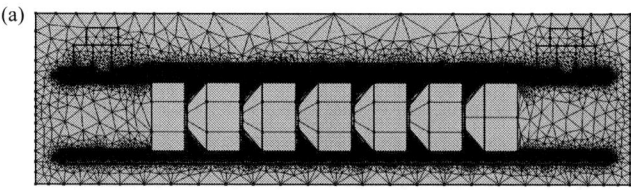

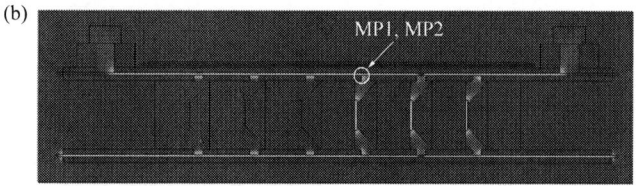

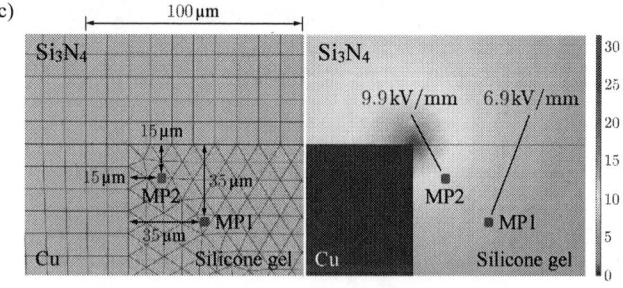

Fig. 5: (a) COMSOL mesh setting with an upper mesh size limit of 10 μm applied to critical components. (b) E-field distribution of the module. (c) Triple point mesh and E-field. AMB-1 and AMB-2: copper 0.8 mm, Si₃N₄ 0.32 mm. AMB-3: copper 0.3 mm, Si₃N₄ 0.25 mm. Material parameters: electrical conductivity: Si₃N₄ 0 S/m, copper 5.96×10^7 S/m, silicone gel 1.0×10^{-13} S/m; relative permittivity: Si₃N₄ 9.7, copper 1, and silicone gel 2.7. E-field at triple point: 36.5 kV/mm; MP1: 6.9 kV/mm; MP2: 9.9 kV/mm.

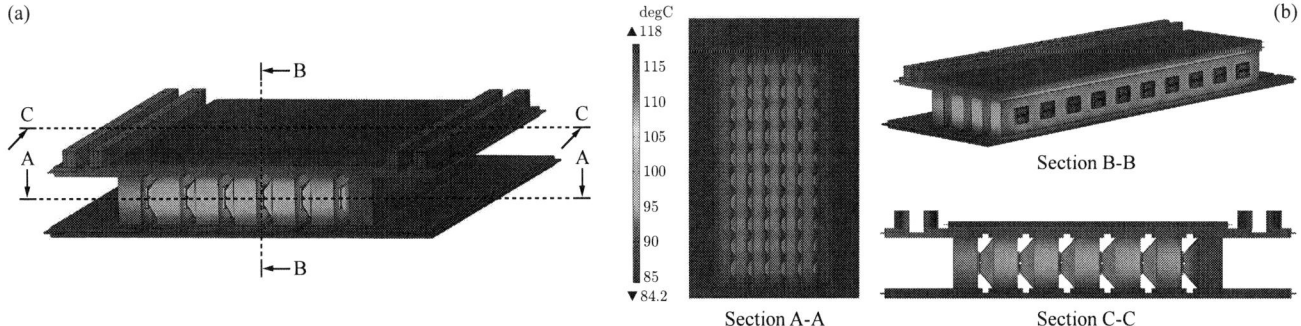

Fig. 6: (a) DOC thermal simulation with fixed substrate interface temperatures at $T_{c,fixed} = 85°C$, with a device power loss of 125 W per die, amounting to a total of 7.5 kW for the 60 dies in the half-bridge module. The simulated $T_{j,max,fixed} = 118°C$. (b) Three cross-section views.

70%. However, beyond a distance of 20 μm from the edge, the E-field variation reduced to less than 1% [21]. To mitigate the influence of mesh dependence, the E-field measurement point (MP1) was taken at a distance of 50 μm from the critical edge, as recommended in prior studies [21], [27]. In this simulation, the 50 μm diagonal distance was made by setting the horizontal and vertical distances to 35 μm. E-field at horizontal and vertical distances of 15 μm (MP2) is also measured for comparison with [22].

As seen in **Fig. 5(a)** and **Fig. 5(c)**, manual meshing was implemented on the module, with an upper mesh size limit of 10 μm applied to critical components, including the triple-point ceramic, silicone gel, and copper substrate. Following the mesh size definitions, a free triangular mesh was applied to the encapsulation material, while a mapped mesh was used for the copper and ceramic components. For the remaining components, a combination of mapped and triangular meshing was employed, with normal and extra coarse mesh sizes.

The simulation results in **Fig. 5(b)** revealed that the E-field strength peaks at the triple point, reaching 36.5 kV/mm, while the value was 6.9 kV/mm at MP1 and 9.9 kV/mm at MP2 (**Fig. 5(c)**). This should guarantee PD-free performance according to [22]. However, further encapsulation and PD testing need to be conducted for validation.

D. DOC Thermal Assessment

The DOC design has two main goals: 1) to ensure the SiC die junction temperature $T_j < 150°C$ (with a 25°C margin to 175°C) and 2) to keep the substrate cooling interface temperature $T_c < 85°C$, specifically, the bottom copper plate of AMB-1 and the top copper plate of AMB-3. Pin-fin baseplate water cooling is presumed, operating with a typical water volume flow rate of 3 LPM–24 LPM (or 0.05 L/s–0.4 L/s).

Thermal simulations in COMSOL were conducted in two steps to verify the design goals. In the first step, the cooling interface temperatures were fixed at $T_{c,fixed} = 85°C$, with a device power loss of 125 W per die, amounting to a total of 7.5 kW for the 60 dies in the half-bridge module (**Fig. 6**). The result indicates that $T_{j,max,fixed} = 118°C$, meeting the target requirements. The calculated junction-to-case thermal resistance is $R_{th,j-c} = 4.4$ K/kW, with a per-die thermal resistance

of $R_{th,j-c,die} = 0.264$ K/W. This represents a 40% reduction compared to typical discrete devices (0.45 K/W) and falls within the range of dual-sided cooling performance reported in the literature. Although the cooling interface temperature is not uniform in practice, this verification provides a reasonable estimate of the DOC's thermal performance.

In the second step, a thermal simulation coupled with CFD analysis is performed to verify that the pin-fin baseplate water

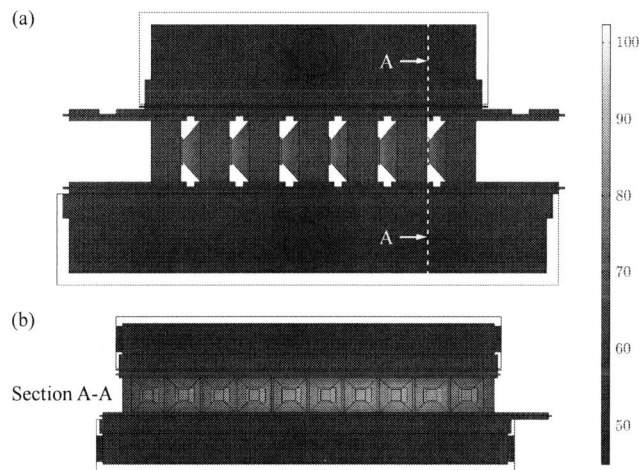

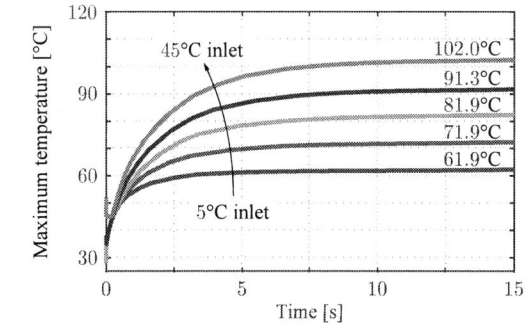

Fig. 7: (a) Simulated module temperature distribution with water cooling at a 45°C inlet temperature and a flow rate of 6 LPM, measured at $t = 15$ s. (b) Section A-A view to show the die temperatures. (c) Transient maximum temperature curves when various inlet water temperatures are applied: 5°C, 15°C, 25°C, 35°C, and 45°C.

979-8-3315-1612-3/25 $31.00 © 2025 IEEE

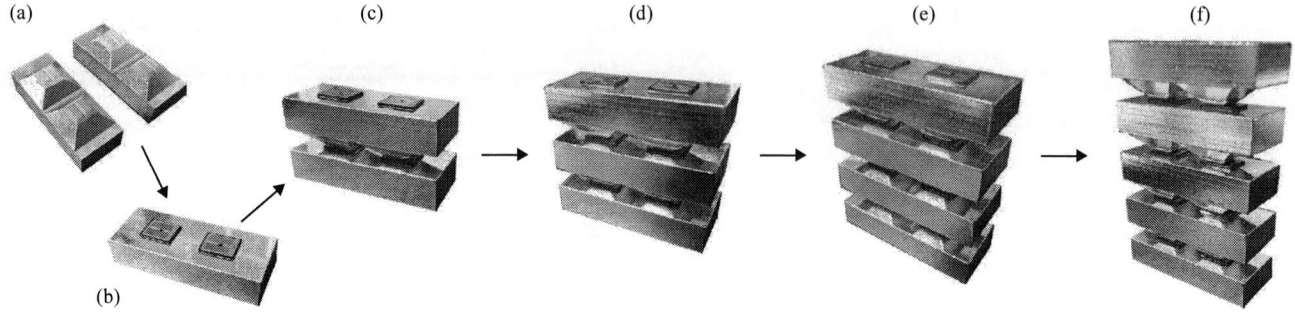

Fig. 8: Fabrication process of a mechanical prototype with 2-parallel-4-series dies. Die data: Microchip 1.7 kV, 35 mΩ MSC035SMA170D/S SiC MOSFET mechanical dies with ENEPIG top-side metallization, 5.67 mm × 4.42 mm × 0.37 mm. Copper bar machining tolerance: ±20 μm. Sintering machine: AMX X-Sinter P52. Sintering material: MacDermidAlpha Argomax 8020 film. (a) Sinter film transfer. (b) Sinter two SiC dies onto each copper bar to make a switch bar, repeated for four switch bars. (c)–(f) Layer-by-layer switch bar sintering using custom-designed sintering fixtures with precise side alignment.

cooling design maintains $T_c < 85°C$. The pin-fin baseplates used in this study are based on the Wieland MDT, a non-subtractive method for fabricating pin-fin structures. The top and bottom pin-fin arrays have effective areas of $A_{\text{eff,top}} = 0.0174\,\text{m}^2$ and $A_{\text{eff,bot}} = 0.0264\,\text{m}^2$, respectively. Using the equation $R_{\text{th}} = 1/(A_{\text{eff}} \cdot h)$ and an estimated convective heat transfer coefficient $h = 8.6\,\text{kW/m}^2\cdot\text{K}$ at a 6 LPM flow rate, the cooler thermal resistances are calculated as $R_{\text{th,c,top}} = 6.7\,\text{K/kW}$ and $R_{\text{th,c,bot}} = 4.4\,\text{K/kW}$, respectively. **Fig. 7(a)** and **Fig. 7(b)** illustrate the steady-state ($t = 15\,\text{s}$) temperature distribution of the module with an inlet water temperature of 45°C, simulated using the COMSOL CFD Module and Heat Transfer Module. The average substrate cooling interface temperature is $T_{\text{c,cfd}} = 66.9°C$, and the maximum junction temperature is $T_{\text{j,max,cfd}} = 102.0°C$. **Fig. 7(c)** shows the time-domain transient maximum temperature curves when 5°C–45°C inlet water temperatures are applied, where the steady-state temperatures are linear but the time constants increase slightly. Note that all performed DOC thermal simulations do not include bonding material layers, so the actual junction and substrate temperatures are expected moderately higher than the simulation results. Also, no thermal interface material is used.

IV. MECHANICAL PROTOTYPE AND FABRICATION PROCESS VALIDATION

A mechanical prototype was fabricated to validate the manufacturability of the proposed DOC packaging concept. The prototype uses eight Microchip 1.7 kV, 35 mΩ MSC035SMA170D/S SiC MOSFET mechanical dies with ENEPIG top-side metallization. Five copper bars with Ni/Ag plating were designed and CNC machined with a ±20 μm tolerance, each capable of accommodating two dies on the top surface. The bottom copper bar is the only one without trapezoidal pyramid extrusions.

The fabrication process involved four steps, all performed on an AMX X-Sinter P52 using MacDermid Alpha Argomax 8020 sintering film:

- Film transfer: The sintering film was transferred to the back side of the dies and the surfaces of the trapezoidal pyramids under 5 MPa, 150°C, and 30 s (**Fig. 8(a)**).

- Die sintering: Two dies were sintered onto designated locations on each copper bar under 15 MPa, 250°C, and 180 s, repeated for four switch bars (**Fig. 8(b)**).

- Layer-by-layer sintering: The switch bars were stacked and sintered layer by layer using custom-designed sintering fixtures to ensure precise side alignment (**Fig. 9(a)**–**Fig. 9(b)**), and the process was conducted under 15 MPa, 250°C, and 240 s (**Fig. 8(c)**–**Fig. 8(f)**). The extended time duration was for a better heat transfer through layers.

- Substrate soldering: The completed switch bar stack was soldered to two substrates. Due to the use of a hot plate, the soldering was performed in two separate steps.

Two tests were conducted during the fabrication process to validate the packaging quality. Before substrate soldering, the completed switch bar stack was placed horizontally on a Fujifilm Prescale film (low pressure) and subjected to a controlled

Fig. 9: (a) Custom-designed sintering fixture with precise side alignment. (b) Sintering fixtures applied to stacked switch bars. (c) Switch bar side alignment validation using Fujifilm Prescale film (LW, 350 PSI–1400 PSI). (d) Shear strength test using Nordson 4000PLUS: 1981 N (79 MPa).

(a)

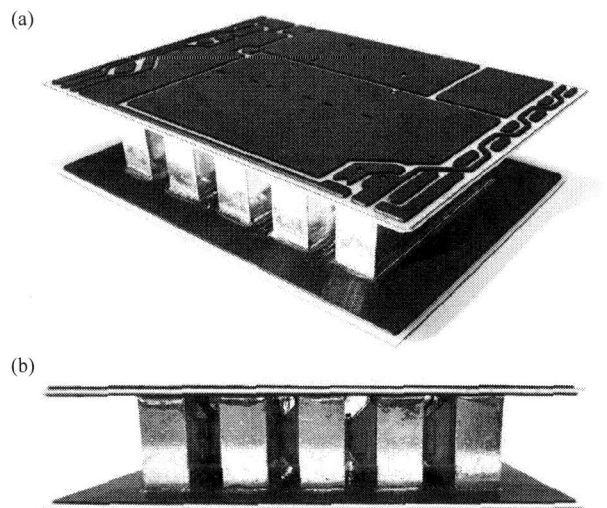

(b)

Fig. 10: (a) Completed mechanical prototype using Rogers free substrate samples. (b) Side view of the prototype.

pressure within 2.4 MPa–9.6 MPa (350 PSI–1400 PSI) using the P52 sintering machine. This test was performed to evaluate the side alignment accuracy. The result in **Fig. 9(c)** indicates satisfactory side alignment with a slight mismatch, which can be effectively compensated for with an appropriate solder bond line thickness. Additionally, shear strength test was conducted on several sintered dies using the Nordson 4000PLUS bond tester (**Fig. 9(d)**). The tested minimum strength was 1981 N (79 MPa), confirming the robustness of the sintered bonds. **Fig. 10** presents the fully assembled mechanical prototype.

V. CONCLUSION

This work presents a novel scalable DOC packaging concept for SiC modules with numerous parallel-series dies, targeting high-power MV grid applications. The proposed design incorporates a switch bar structure that positions dies perpendicularly to the substrates, enabling horizontal heat flux on both sides of the dies and vertical dissipation through the top and bottom substrates. FEA simulations demonstrated the concept's effectiveness, achieving: 1) a power-loop inductance of 3.7 nH with external MLCC decoupling capacitors, 2) an maximum E-field strength of 6.9 kV/mm at a 50 µm distance from the worst-case triple point, 3) a per-die junction-to-case thermal resistance of 0.264 K/W, and 4) a maximum junction temperature of 102°C under 7.5 kW power loss, 45°C inlet water temperature, and a 6 LPM flow rate. The manufacturability of the proposed design was validated through the fabrication of a mechanical prototype, which confirmed satisfactory switch bar alignment and robust bonding strength.

However, two manufacturing challenges were identified: 1) The reliability of the die's front-side bonding to trapezoidal pyramids requires improvement. This will be addressed by optimizing the sintering process and exploring alternative sinter materials, such as Ag or Cu sinter paste. 2) The side alignment process using the current fixture design is complex

and demands further refinement to enhance precision and ease of use. Future work will focus on advancing the concept through thermal cycling tests on mechanical prototypes, encapsulation and PD testing, as well as the packaging and electrical testing of active dies to validate the performance under operational conditions.

VI. ACKNOWLEDGMENT

We sincerely thank Microchip Technologies, Delta Electronics (Americas), AMX Automatrix, Rogers Corporation, MacDermid Alpha Electronics Solutions, and Heraeus Electronics for their invaluable support of this work.

REFERENCES

[1] Y. Li, Mustafeez-ul-Hassan, A. Mirza, Y. Xie, S. Deng, S. S. Vala, F. Luo, X. Feng, S. Narumanchi, and J. Flicker, "State-of-the-art medium- and high-voltage silicon carbide power modules, challenges and mitigation techniques: a review," *IEEE Trans. Compon. Packag. Manuf. Technol.*, Early Access, doi: 10.1109/TCPMT.2024.3391653.

[2] B. Mouawad, A. Hussein, and A. Castellazzi, "A 3.3 kV SiC MOSFET half-bridge power module," in *Proc. IEEE Int. Conf. Integr. Power Electron. Systems*, 2018, pp. 1–6.

[3] K. Yasui, S. Hayakawa, T. Ishigaki, T. Morita, T. Tabata, Y. Takayanagi, Y. Inoue, T. Murata, A. Tadano, K. Kinoshita, M. Hoshi, K. Koseki, K. Shono, K. Hamada, T. Imaizumi, H. Matsushima, H. Miki, T. Masuda, T. Kobayashi, T. Ando, A. Konno, and K. Saito, "A 3.3 kV 1000 A high power density SiC power module with sintered copper die attach technology," in *Proc. IEEE Int. Exhib. Conf. Power Electron. Intell. Motion Renewable Energy Energy Manage.*, 2019, pp. 1–6.

[4] S. Kicin, R. Burkart, J.-Y. Loisy, F. Canales, M. Nawaz, G. Stampf, P. Morin, and T. Keller, "Ultra-fast switching 3.3 kV SiC high-power module," in *Proc. IEEE Int. Exhib. Conf. Power Electron. Intell. Motion Renewable Energy Energy Manage.*, 2020, pp. 1–8.

[5] Z. Guo, L. Zhang, S. Sen, and A. Q. Huang, "A novel 3.6 kV/400A SiC intelligent power module (IPM)," in *Proc. IEEE Appl. Power Electron. Conf. Expo.*, 2021, pp. 39-43.

[6] Y. Sekino, S. Ewald, S. Yamamoto, S. Iwamoto, T. Uchida, K. Okumura, Y. Kusunoki, Y. Onozawa, H. Kimura, Y. Kobayashi, and T. Shiigi, "3.3 kV All SiC Module with 2nd Generation Trench gate SiC MOSFETs for traction," in *Proc. IEEE Int. Exhib. Conf. Power Electron. Intell. Motion Renewable Energy Energy Manage.*, 2022, pp. 1–7.

[7] Y. Chen, X. Du, L. Du, X. Du, A. S. Md Khalid Hasan, X. Li, H. Chen, R. Paul, S. Chinnaiyan, Y. Zhao, and H. A. Mantooth, "3.3 kV low-inductance full SiC power module," in *Proc. IEEE Appl. Power Electron. Conf. Expo.*, pp. 2634–2640.

[8] B. Gao, A. J. Morgan, Y. Xu, X. Zhao, and D. C. Hopkins, "6.0kV, 100A, 175kHz super cascode power module for medium voltage, high power applications," in *Proc. IEEE Appl. Power Electron. Conf. Expo.*, 2018, pp. 1288-1293.

[9] B. DeBoi, A. Lemmon, B. Nelson, C. New, and D. Hudson, "Modeling and validation of medium voltage SiC power modules," in *Proc. IEEE Appl. Power Electron. Conf. Expo.*, 2020, pp. 1964–1971.

[10] J. Nakashima, A. Fukumoto, Y. Obiraki, T. Oi, Y. Mitsui, H. Nakatake, Y. Toyoda, A. Nishizawa, K. Kawahara, S. Hino, H. Watanabe, T. Negishi, and S. -i. Iura, "6.5-kV full-SiC power module (HV100) with SBD-embedded SiC-MOSFETs," in *Proc. IEEE Int. Exhib. Conf. Power Electron. Intell. Motion Renewable Energy Energy Manage.*, 2018, pp. 1-7.

[11] L. Ma, H. Zhang, T. Yuan, D. Ma, Y. Nie, L. Li, Y. Yao, and L. Wang, "A double-sided cooling 6.5kV SiC MOSFET power module with insulation enhancement design," in *Proc. IEEE Appl. Power Electron. Conf. Expo.*, 2023, pp. 2550-2555.

[12] R. Paul, R. Alizadeh, X. Li, H. Chen, Y. Wang, and H. A. Mantooth, "A double-sided cooled SiC MOSFET power module for EV inverters," *IEEE Trans. Power Electron.*, vol. 39, no. 9, pp. 11047-11059, Sept. 2024.

[13] B. Passmore, Z. Cole, B. McGee, M. Wells, J. Stabach, J. Bradshaw, R. Shaw, D. Martin, T. McNutt, E. VanBrunt, B. Hull, and D. Grider, "The next generation of high voltage (10 kV) silicon carbide power modules," in *Proc. IEEE Workshop Wide Bandgap Power Devices Appl.*, 2016, pp. 1–4.

[14] M. Johnson, C. DiMarino, B. Mouawad, J. Li, R. Skuriat, M. Wang, Y. Tan, G. Lu, and R. Burgos, "10 kV SiC power module packaging," in *Proc. IEEE Int. Conf. Integr. Power Electron. Systems*, 2018, pp. 1–8.

[15] X. Li, Y. Chen, H. Chen, R. Paul, X. Song, and H. A. Mantooth, "A 10 kV SiC MOSFET power module with optimized system interface and electric field distribution," *IEEE Trans. Power Electron.*, vol. 39, no. 8, pp. 9540-9553, Aug. 2024.

[16] H. Shang, L. Liang, and Y. Wang, "Design and performance of high voltage chip-level series-connected SiC MOSFET module," *IEEE Trans. Power Electron.*, vol. 38, no. 2, pp. 1757–1767, Feb. 2023.

[17] T. N. Ubostad and D. Peftitsis, "Power module design with chip-level series-connected SiC MOSFETs," in *Proc. IEEE Appl. Power Electron. Conf. Expo.*, 2024, pp. 181–187.

[18] T. Do, O. Lesaint, and J.-L. Auge, "Streamers and partial discharge mechanisms in silicone gel under impulse and AC voltages," *IEEE Trans. Dielectr. Electr. Insul.*, vol. 15, no. 6, pp. 1526–1534, Dec. 2008.

[19] L. Wang, J. Gong, T. Long, Y. Wang, H. Zheng, B. Hu, W. Mu, J. Li, and Z. Zeng, "A review of partial discharge in medium voltage SiC power modules under square wave excitation: Characterization, mitigation, and detection," *IEEE J. Emerg. Sel. Topics Power Electron.*, vol. 12, no. 4, pp. 3588-3606, Aug. 2024.

[20] C. M. DiMarino, B. Mouawad, C. M. Johnson, D. Boroyevich, and R. Burgos, "10-kV SiC MOSFET power module with reduced common-mode noise and electric field," *IEEE Trans. Power Electron.*, vol. 35, no. 6, pp. 6050–6060, Jun. 2020.

[21] C. F. Bayer, E. Baer, U. Waltrich, D. Malipaard, and A. Schletz, "Simulation of the electric field strength in the vicinity of metallization edges on dielectric substrates," *IEEE Trans. Dielec. Electr. Ins.*, vol. 22, no. 1, pp. 257-265, Feb. 2015.

[22] Z. Zhang, P. Fu, J. Lynch, S. Lu, C. Nicholas, A. Morgan, W. Sung, K. D. T. Ngo, and G. -Q. Lu, "Insulation capability at 10 kV, >300 V/ns of a nonlinear resistive polymer nanocomposite field-grading coating in a 15-kV silicon carbide module," *IEEE Trans. Power Electron.*, vol. 39, no. 12, pp. 15748-15756, Dec. 2024.

[23] Z. Zhang, S. Lu, B. Wang, Y. Zhang, N. Yun, W. Sung, K. D. T. Ngo, and G. -Q. Lu, "Packaging of a 10-kV double-side cooled silicon carbide diode module with thin substrates coated by a nonlinear resistive polymer-nanoparticle composite," *IEEE Trans. Power Electron.*, vol. 37, no. 12, pp. 14462-14470, Dec. 2022.

[24] Z. Zhang, E. Arriola, C. Nicholas, J. Lynch, N. Yun, A. Morgan, W. Sung, K. D. T. Ngo, and G. -Q. Lu, "Package design and analysis of a 20-kV double-sided silicon carbide diode module with polymer nanocomposite field-grading coating," *IEEE Trans. Comp. Packaging Manu. Tech.*, vol. 14, no. 5, pp. 776-783, May 2024.

[25] M. Liu, A. Coppola, M. Alvi, and M. Anwar, "Comprehensive review and state of development of double-sided cooled package technology for automotive power modules," *IEEE Open J. Power Electron.*, vol. 3, pp. 271-289, 2022.

[26] Y. Wang, Y. Li, Y. Wu, X. Dai, Y. Ma, P. Mumby-Croft, J. Booth, M. Packwood, S. Jones, and G. Liu, "Mitigation of challenges in automotive power module packaging by dual sided cooling," in *Proc. IEEE Eur. Conf. Power Electron. Appl.*, 2016, pp. 1-8.

[27] F. Yan, L. Wang, B. Zhang, L. Yu, K. Wang, and T. Yang, "Geometrical design of ceramic substrate for high voltage power modules," in *Proc. IEEE Int. Conf. High Voltage Engineering Appl.*, 2020, pp. 1-4.

979-8-3315-1612-3/25 $31.00 © 2025 IEEE

Parasitic Impact Analysis and Design of Hybrid EMI Filter for Active Clamp Flyback SMPS

Tahmid Ibne Mannan
Department of Electrical and
Computer Engineering
Mississippi State University
Starkville, USA
tm2445@msstate.edu

Seungdeog Choi
Department of Electrical and
Computer Engineering
Mississippi State University
Starkville, USA
seungdeog@ece.msstate.edu

Masoud Karimi-Ghartemani
Department of Electrical and
Computer Engineering
Mississippi State University
Starkville, USA
karimi@ece.msstate.edu

Abstract— This paper presents the impact of parasitic elements on the design and performance of an active EMI filter (AEF) for an active clamp GaN MOSFET flyback switching mode power supply (SMPS). SMPS with wide-bandgap (WBG) power switches are increasingly used due to their high switching speed and frequency capabilities. However, extremely high-frequency operations of these switches lead to considerable common-mode (CM) electromagnetic interference (EMI) generation. Passive EMI filters (PEF) are commonly employed, comprising bulky CM chokes to mitigate CM EMI. To increase power density, Hybrid EMI Filters (HEF) have been introduced, where AEFs work in conjunction with PEFs. However, current transformerless AEFs use multiple passive components to sense and cancel noise signals and their attenuation capabilities are severely limited by the parasitic parameters of these passive elements. This is a crucial aspect in AEF and HEF design which has been limitedly explored in the existing literature. To address this research gap, , which is one of the major bottlenecks in AEF bandwidth limitation, the effect of injection and sensing capacitor parasitic parameters is mathematically presented and experimentally validated in this paper. The HEF design considerations are discussed with simulation and mathematical analysis. Experimental studies are conducted to validate the proposed filter design and show the effects of parasitic elements on the bandwidth of AEF and HEF, and their CM EMI attenuation capabilities.

Keywords—EMI filter, active EMI filter, parasitic parameters, insertion loss, common mode EMI.

I. INTRODUCTION

Gallium Nitride (GaN) devices are increasingly used in switching-mode power supplies (SMPS) applications due to their high switching frequency capabilities [1–2]. However, GaN SMPS are prone to high levels of conducted EMI due to their larger dv/dt. As a result, bulky and costly PEFs are employed in commercial flyback GaN SMPSs to meet CISPR-32 standards specified for such devices [3] – [4], as depicted in Fig. 1. Several studies have presented a hybrid EMI filter (HEF) approach for SMPS applications that integrates an AEF to decrease the PEF components' size [5] – [7], as illustrated in Fig. 2. However, these studies mostly focus on low-switching frequency systems. To further reduce the size of the AEF, voltage sensing and current compensation (VSCC) topology is becoming popular [8] – [9], as it eliminates the necessity of bulky current transformers and the filter can be realized through passive components only. However, the bandwidth of AEF is limited by the bandwidth of the op-amp and the

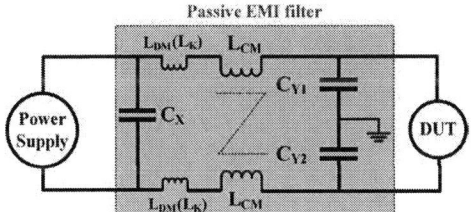

Fig. 1. Schematic of a typical single-stage PEF

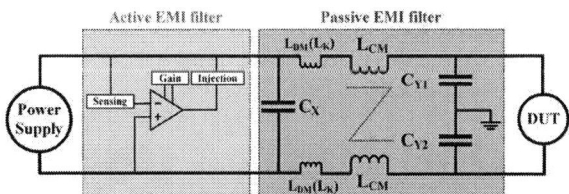

Fig. 2. Schematic of a HEF comprising FB-VSCC AEF

parasitic effects of the passive components in the sensing and injection part of the AEF [10] – [11]. Although the op-amp's issue can be fixed by using off-the-shelf high bandwidth options, the effects of passive component parasitic elements on the AEF bandwidth have not been adequately explored in the literature. Due to their inherent parasitic effects, the AEF's sensing and injection passive components can significantly limit its effective attenuation and operating bandwidth which will be explored in this paper.

In this paper, first the design of a hybrid EMI filter is presented for CM EMI mitigation of a GaN-based active-clamp flyback SMPS targeted at high switching frequency applications. An FB-VSCC AEF is presented, which reduces the required CM choke inductance as demonstrated and validated through simulation and experimental CM EMI studies. Then, detailed mathematical modeling of the factors contributing to AEF bandwidth and attenuation is presented with and without the parasitic parameters. The effects of passive component parasitic on AEF performance are analytically presented and validated through simulation and experimental results.

In this paper, a hybrid EMI filter will be designed for a flyback SMPS, then the effects of the sensing and injection capacitor parasitic parameters on the AEF attenuation characteristics will be explored. Section II presents a brief discussion on the CM EMI propagation path of an active clamp flyback SMPS. Section III presents the detailed design of the HEF and PEF for this study. Section IV explores the impact of parasitic parameters on AEF performance with

979-8-3315-1612-3/25 $31.00 © 2025 IEEE

analytical and simulation studies. In Section V, the experimental validations for the designed filters are presented, along with experimental assessment of the parasitic parameters' impact on the designed AEF's CM EMI attenuation capabilities. Finally Section VI concludes the article.

II. CM EMI PROPAGATION PATH IN ACTIVE CLAMP FLYBACK SMPS

In this section, the CM EMI generation and propagation in an active clamp flyback SMPS will be briefly discussed. Fig. 3 presents the CM EMI propagation path for an active clamp flyback SMPS, with the major sources and coupling paths highlighted. The primary switch induces high dv/dt and functions as the major source of CM EMI noise (presented as V_{CM}). The three parasitic capacitances contributing to the flow of CM EMI noise are : C_{P1} (primary switch drain to earth), $C_{P\text{-}tr}$ (transformer interwinding parasitic capacitance) and C_{P2} (secondary to earth). These parasitic capacitances function as the major coupling path for the CM noise to flow to the system ground. The total CM noise current (I_{CM_total}) flows through the LISN where CM EMI is measured. The noise then flows back to the noise source through the diode bridge at the input of the converter. The noise generated by the primary switch is majorly coupled to the ground through C_{P1}, which is formed between the pulsating node at its drain and the system ground. The noise generated on the primary side of the SMPS flows to the secondary side through the interwinding capacitances of the transformer ($C_{P\text{-}tr}$). This noise flows back to the LISN through the ground connection of C_{P2}.

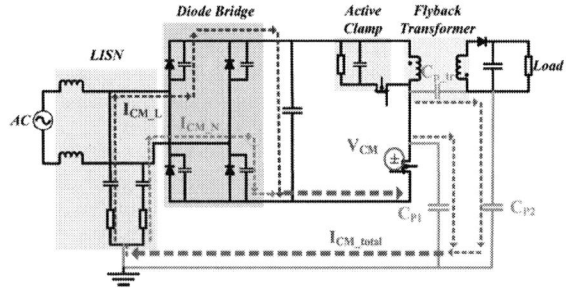

Fig. 3. CM EMI noise propagation path of the active clamp flyback converter.

For the CM EMI evaluations in this paper, an active-clamp flyback evaluation module from Texas Instruments has been utilized, rated at 45W, 20V outputs. The major specifications for the module are summarized in Table I.

TABLE I: ACTIVE CLAMP FLYBACK SMPS SPECIFICATIONS

Parameter	Value
Input voltage	90-V_{RMS} ~ 264-V_{RMS}
Input line frequency	47 ~ 63Hz
Output	20V_{DC}, 2.25A, 45W
Switching device	GaN MOSFETs
Switching frequency	325kHz

III. ANALYSIS AND SIMULATION FOR HEF DESIGN

In this section, analysis and simulation results are presented for the EMI filter design of the active clamp flyback. First, a standalone PEF is designed to meet CISPR-32, class-

B standards for comparison through simulation studies. Then, an FB-VSCC AEF is proposed that can work in conjunction with a PEF with a smaller CM choke to meet the CISPR-32 standards. Then, a finalized HEF design is presented. The CM EMI simulation parameters on PSIM for the simulations presented in this section are summarized in Table II.

TABLE II: CM EMI SIMULATION PARAMETERS

Parameter	Value
Input voltage	264-V_{RMS}
Input line frequency	60Hz
Output	20V_{DC}, 45W
Switching device	GaN MOSFETs
Switching frequency	325kHz

A. Passive EMI Filter (PEF) Design

To suppress CM noise with a PEF, C_{Y1} and C_{Y2} are used together with the magnetizing inductance of the CM choke, L_{CM}. The CM EMI attenuation corner frequency for a PEF can be presented as (1). To determine f_{cutoff}, the CM EMI spectrum without the filter is measured first. The first frequency where the noise exceeds the relevant EMI standard is identified (f_{target}). The required attenuation (A) can then be calculated using (2), with a 6dB margin.

$$f_{cutoff} = \frac{1}{2\pi\sqrt{L_{CM}(C_{Y1}+C_{Y2})}} \quad (1)$$

$$A\,(dB) = CM_{measured}(dB) - CM_{standard}(dB) + 6dB \quad (2)$$

Once the A and f_{target} are calculated, f_{cutoff} can be calculated from the following relationship:

$$f_{cutoff} = f_{target}10^{-\frac{A}{40}}. \quad (3)$$

Then, using (1), the value of L_{CM} can be calculated based on the C_{Y1} and C_{Y2} values, which are limited due to safety limits. From simulations, it is observed that in the concerned frequency range, the simulated CM EMI noise harmonic at 325kHz is around 112 dBμV, where the CISPR noise limit is around 60 dBμV. With a 6dB margin, the required f_{cutoff} is determined to be around 16kHz using (2) and (3). Choosing 2.2nF capacitors for C_{Y1} and C_{Y2}, the value of calculated L_{CM} is therefore 18mH from (1). Fig. 4 presents the simulated comparison between the CM EMI with and without the PEF. With $L_{CM} = 18$mH, $C_{Y1} = C_{Y2} = 2.2$nF and $C_X = 0.1$uF, CM EMI of the flyback converter meets the CISPR-32 limits.

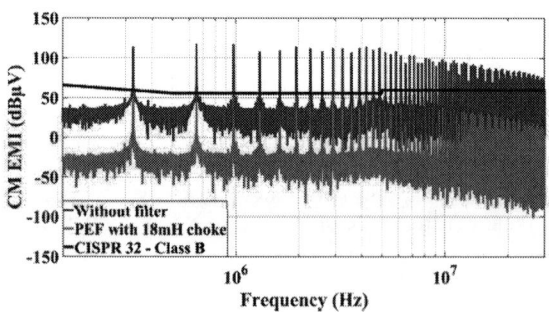

Fig. 4. CM EMI simulation with 18mH choke standalone PEF

979-8-3315-1612-3/25 $31.00 © 2025 IEEE

B. HEF Design with FB-VSCC AEF

The required CM choke size in a PEF can be reduced if an AEF is added. Fig. 5(a) illustrates the FB-VSCC topology-based AEF presented in this paper. The AEF operation is symmetric at both power lines, and Fig. 5(b) presents the AEF circuit as equivalent to the system's ground.

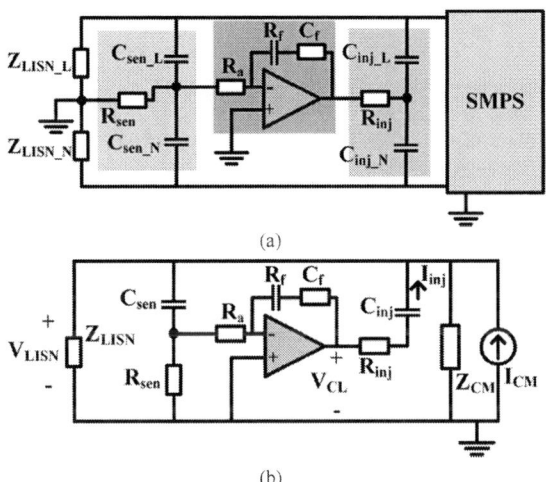

(a)

(b)

Fig. 5. FB-VSCC AEF (a) schematic and (b) equivalent circuit with reference to system ground.

C_{sen_L}, C_{sen_N}, and R_{sen} comprise the high pass sensing network and G_{sen} is the gain for this part. R_a, R_f, and C_f set the closed-loop transfer function gain (G_{CL}) of the AEF. C_{inj_L}, C_{inj_N} and R_{inj} comprise the noise-cancelling injection network of the AEF. Z_{LISN_L} and Z_{LISN_N} represent the LISN impedances, whereas Z_{CM} and I_{CM} represent the CM noise source impedance and current. The op-amp in the presented AEF is used as an inverting configuration. G_{sen}, G_{inj} and G_{CL} of the AEF can be expressed as follows:

$$G_{sen} = \frac{sC_{sen}R_{sen}}{sC_{sen}R_{sen} + 1} \tag{4}$$

$$G_{inj} = \frac{sC_{inj}R_{inj} + 1}{sC_{inj}} \tag{5}$$

$$G_{CL} = \frac{Z_f}{Z_a} \tag{6}$$

The noise cancellation current, I_{inj} and the measured noise voltage at the LISN, V_{LISN} from Fig. 5(b) can be expressed as follows:

$$I_{inj} = \frac{V_{cl} - V_{lisn}}{Z_{inj}} = \frac{G_{cl}(V_{ref} - G_{sen}V_{lisn}) - V_{lisn}}{G_{inj}} \tag{7}$$

$$V_{lisn} = \frac{Z_{lisn}}{Z_{lisn} + Z_{cm}}(V_{cm} + Z_{cm}I_{inj}) \tag{8}$$

where $V_{ref} = 0$ is the system ground voltage. Using (7) and (8), the closed-loop transfer function for the system with and without AEF can be presented as:

$$TF_{with\ AEF} = \frac{(Z_{lisn}||G_{inj})||Z_{cm}}{1 + \frac{Z_{lisn}||Z_{cm}}{Z_{lisn}||Z_{cm}+Z_{inj}}G_{CL}G_{sen}} \tag{9}$$

$$TF_{without\ AEF} = Z_{lisn}||Z_{cm} \tag{10}$$

From (9), the loop gain (G_{OP}) of the AEF can be defined as follows,

$$G_{OP} = \frac{Z_{lisn}||Z_{cm}}{Z_{lisn}||Z_{cm} + G_{inj}}G_{CL}G_{sen} \tag{11}$$

Insertion loss is the most important parameter to quantify and assess the noise suppression capabilities of the EMI filter. Insertion loss with respect to EMI filters is defined as the ratio between the voltage measured at the LISN without and with the EMI filters, as illustrated in Fig. 6.

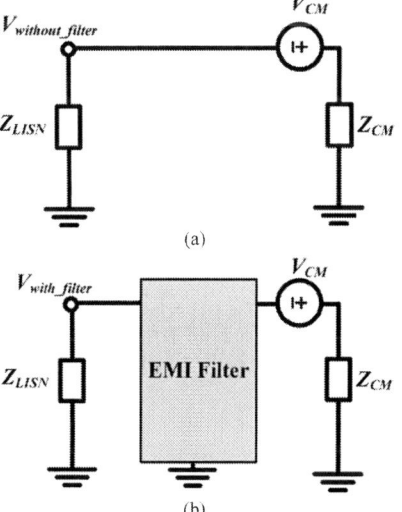

(a)

(b)

Fig. 6. Schematic diagram for filter insertion loss (a) without EMI filter (b) with EMI filter

Assuming, $Z_{CM} \gg Z_{lisn}$ in the concerned frequency range, the insertion loss provided by the AEF (IL_{AEF}) can be derived from (9) and (10) as follows:

$$IL_{AEF} = \frac{TF_{without\ AEF}}{TF_{with\ AEF}} = (1 + G_{OP})\left(1 + \frac{Z_{lisn}}{G_{inj}}\right)$$

$$= 1 + \frac{Z_{lisn}}{G_{inj}}(1 + G_{sen}G_{CL}) \tag{12}$$

If R_f and C_f is chosen such that $R_{sen}C_{sen} = R_fC_f$ to compensate the sensing circuit pole in G_{OP}, then IL_{AEF} can be expressed as (13). It is clear from (13) that the insertion loss of the AEF will depend on the feedback network (R_f and R_a) and the gain of the injection network, G_{inj}, as Z_{lisn} will be constant at (50Ω||50Ω) during EMI measurements.

$$IL_{AEF} = 1 + \frac{Z_{lisn}}{G_{inj}}\left(1 + \frac{R_f}{R_a}\right) \tag{13}$$

The sensing part used in the presented AEF topology is a high-pass RC circuit, which should have a cut-off frequency adequately lower than the switching frequency, to ensure that sufficient noise detection from the 1ˢᵗ switching harmonic can be achieved. Choosing a cut-off frequency at 150kHz and $R_{sen}= 50\Omega$, C_{sen} then can be calculated from (14).

$$C_{sen} = \frac{1}{2\pi f_{sen\_cutoff} R_{sen}} \quad (14)$$

The minimum attenuation frequency of the AEF will depend on the cut-off frequency of the injection circuit. However, a higher value of R_{inj} to gain a lower minimum cutoff frequency will lead to a decrease in the maximum attenuation of the AEF, as the cancellation current will be decreased. Therefore, as a tradeoff $R_{inj}= 10\Omega$ is chosen with a cut-off frequency at around 16kHz. C_{inj} then can be calculated from (15).

$$C_{inj} = \frac{1}{2\pi f_{inj\_cutoff} R_{inj}} \quad (15)$$

The circuit component values for the designed AEF are presented in Table I.

TABLE II: AEF COMPONENT VALUES

Parameter	Value
C_{sen}	10nF
R_{sen}	50Ω
C_{inj}	1μF
R_{inj}	10Ω
C_f	18pF
R_f	30kΩ
R_a	1kΩ

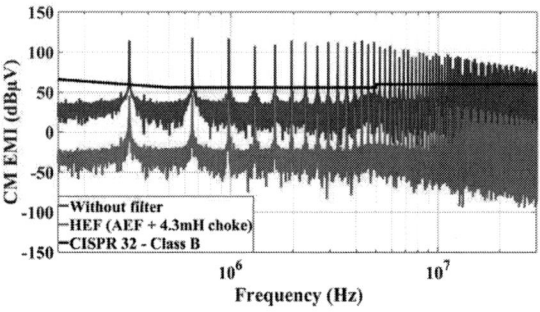

Fig. 7. Simulated CM EMI comparison with hybrid EMI filter (AEF + 4.3mH choke)

In Fig. 7, the CM EMI simulation result for the hybrid EMI filter is presented, which includes the designed AEF. In these simulations, the op-amp in AEF is modeled after AD829, which has a gain bandwidth product of 120MHz. The simulations show that the AEF provides a good reduction in CM EMI up to 6.5MHz, ranging from 5 ~ 27dB at the switching harmonics. Furthermore, it is observed that when the AEF is used, the simulated CM EMI of the DUT meets the CISPR 32 standard with a 4.3mH choke PEF, which is around 4 times smaller than the case of a standalone PEF. This will allow for a significantly smaller CM choke to be used. In Table III, the designed filter specifications are summarized.

TABLE III: DESIGNED FILTER SPECIFICATIONS BY SIMULATIONS

Component	Only PEF (Without AEF)	HEF (With AEF)
CM choke	18mH	4.3mH
C_{Y1}, C_{Y2}	2.2nF	1nF
C_X	0.1μF	0.1μF

IV. EFFECT OF CAPACITOR PARASITICS ON AEF PERFORMANCE

In the previous section, in the presented analysis the sensing and injection capacitors are assumed to be ideal. However, according to the high frequency equivalent circuit of a capacitor, a capacitor's impedance in high frequency is greatly affected by its parasitic series inductance (*ESL*), as presented in Fig. 8. Therefore, G_{sen} and G_{inj} presented in (4) – (5) must be replaced with the gain functions that integrate *ESL* for both sensing and injection capacitors, as presented in (16) – (17). Here, L_{sen} and L_{inj} denote the *ESL* of sensing and injection capacitors respectively.

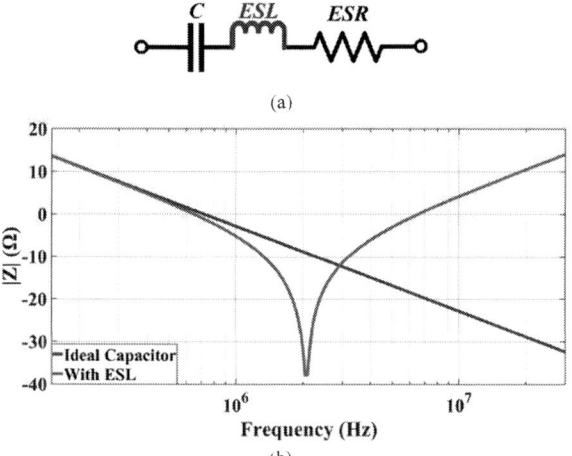

(a)

(b)

Fig. 8. Effect of *ESL* on capacitor (a) High frequency equivalent circuit (b) comparison of impedance

$$G_{sen\_parasitic} = \frac{sC_{sen}R_{sen}}{s^2 C_{sen}L_{sen} + sC_{sen}R_{sen} + 1} \quad (16)$$

$$G_{inj\_parasitic} = \frac{s^2 C_{inj}L_{inj} + sC_{inj}R_{inj} + 1}{sC_{inj}} \quad (17)$$

A. Effects on AEF Insertion Loss

As seen from (16) and (17), the integration of *ESL* into both gain functions introduce unwanted resonances that must be evaluated while designing an AEF. From the analysis in the previous section, $G_{sen\_parasitic}$ will affect the loop gain of the AEF, whereas $G_{inj\_parasitic}$ will have impacts on both the loop gain and insertion loss of the AEF. Fig. 9 presents the simulated comparison of IL_{AEF} magnitude for different values of L_{inj} and L_{sen}. As predicted, integrating L_{inj} into the injection branch has considerable impact on the magnitude of the AEF insertion loss. As much as 7 ~ 9dB decrease in noise attenuation can be expected due to the *ESL* of the injection capacitor in high frequencies, thus significantly limiting the effective bandwidth of the AEF. Also, as expected, L_{sen} has no impact on the insertion loss of the AEF.

979-8-3315-1612-3/25 $31.00 © 2025 IEEE

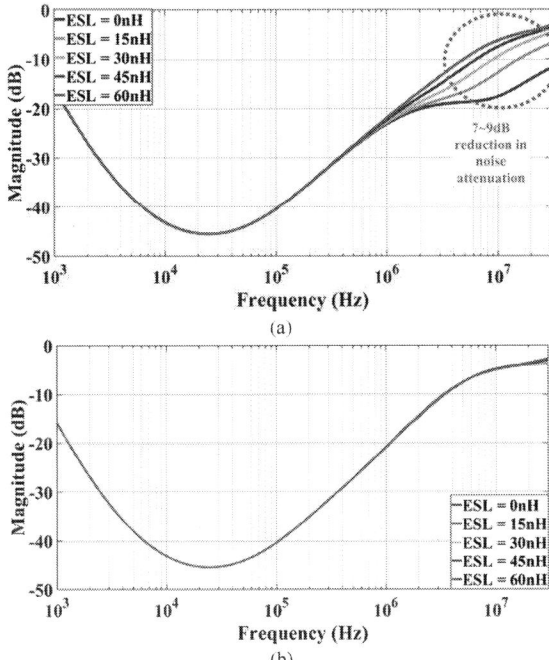

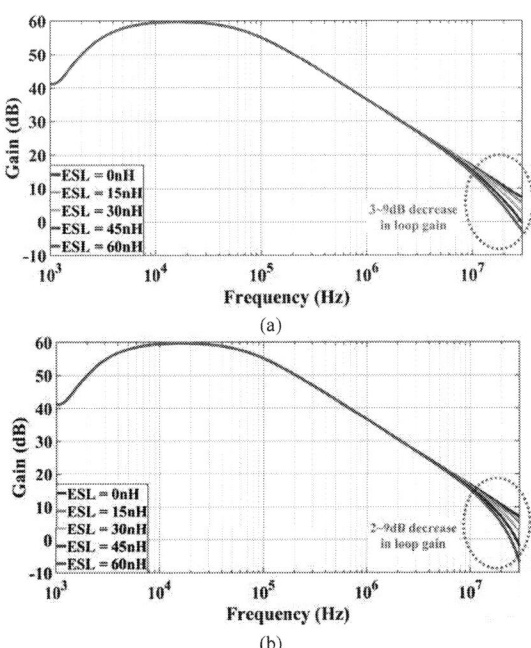

Fig. 9. IL_{AEF} comparison with (a) L_{inj} variation (b) L_{sen} variation

B. Effects on AEF Loop Gain

Fig. 10 presents the G_{OP} magnitude comparison with different values of L_{inj} and L_{sen}. It can be observed that in both cases, there is notable decrease in loop gain beyond 10MHz. In case of L_{inj} variation, around $3 \sim 9$dB decrease in the loop gain magnitude can be observed, whereas with L_{sen} variation, the observed decrease in the loop gain is around $2 \sim 9$dB in the same frequency region. These results indicate that due to the ESL of the capacitors, there can be significant deviations in AEF performance, in terms of noise attenuation and loop gain.

Fig. 10. G_{OP} comparison with (a) L_{inj} variation (b) L_{sen} variation

V. EXPERIMENTAL VALIDATIONS

State-of-the-art CISPR-32 compliant conducted EMI testbed (see Fig. 11). has been utilized to validate the designed filters and assess the impact of capacitor parasitic parameters on the AEF performance. The specifications of the testbed are presented in Table IV. A standalone PEF with a 20mH choke has been implemented, whereas the hybrid EMI filter with AEF has a 5mH choke. The designed filters are presented in Fig. 12. Table V summarizes the filter passive components for both filters.

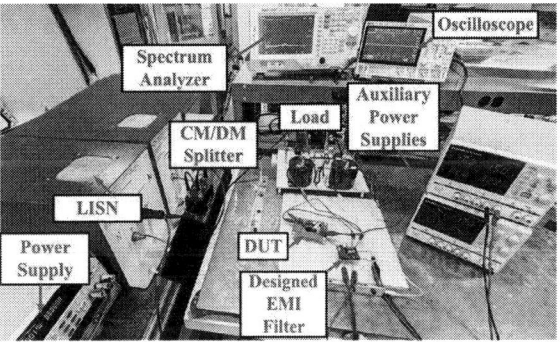

Fig. 11. CM EMI measurement testbed.

TABLE IV: CM EMI TESTBED SPECIFICATIONS

Parameter	Specifications
Supply Voltage	265Vac
DUT	TI active-clamp flyback EVM
Switching Frequency	325kHz
Spectrum Analyzer	9kHz – 1.5GHz

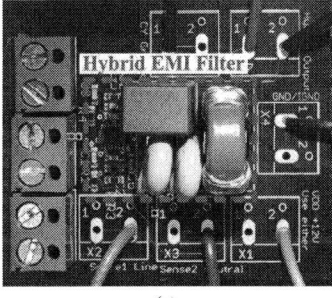
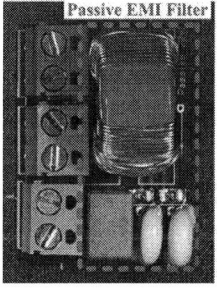

Fig. 12. Implemented EMI filters (a) HEF and (b) PEF.

TABLE V: IMPLEMENTED FILTER SPECIFICATIONS

Component	Only PEF (Without AEF)	Hybrid Filter (With AEF)
CM choke	20mH	5mH
C_{Y1}, C_{Y2}	2.2nF	1nF
C_X	0.1μF	0.1μF

In the following subsections, first, the designed filters are validated with CM EMI measurements. Then, the effect of sensing and injection capacitor *ESL* on the AEF performance is evaluated experimentally.

A. Validation of the Designed EMI Filters

Fig. 13(a) presents experimentally measured CM EMI without filter and with only AEF. The designed AEF reduces CM EMI by 5~26dB for the first ten switching harmonics, with a reasonable reduction up to 5.5MHz. The reductions for the first ten harmonics are presented in Table VI. Fig. 13(b)

979-8-3315-1612-3/25 $31.00 © 2025 IEEE

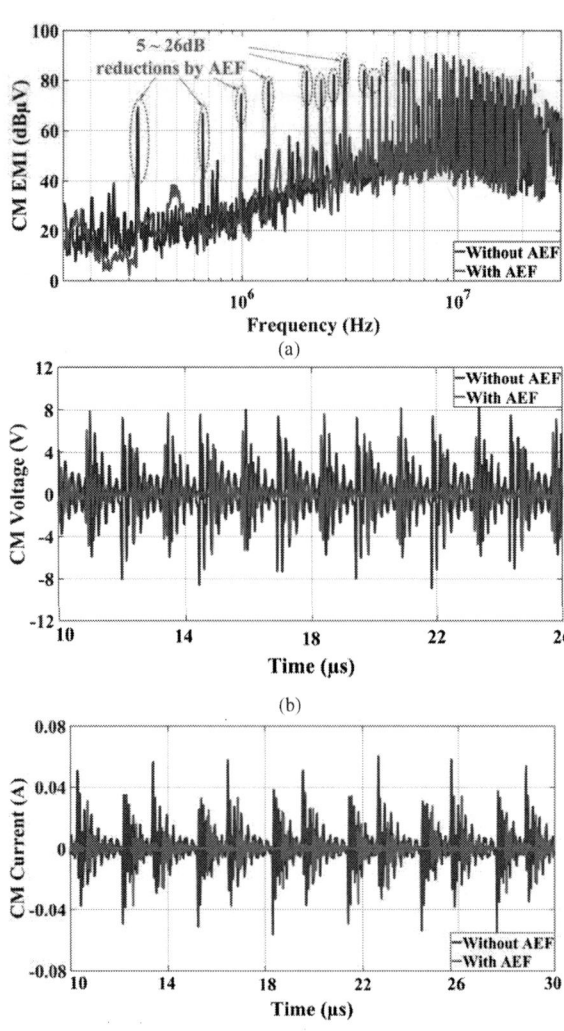

(a)

(b)

(c)

Fig. 13. CM EMI comparison with AEF (a) frequency domain (b) CM voltage and (c) CM current

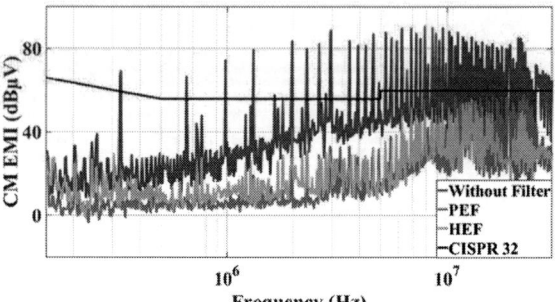

Fig. 14. CM EMI comparison between hybrid EMI filter and PEF.

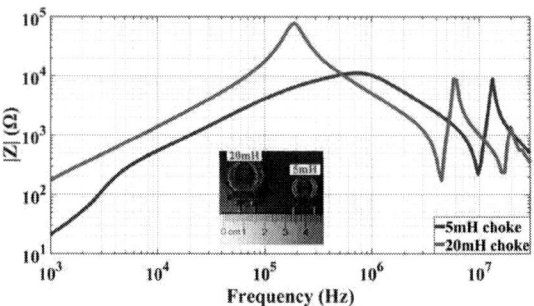

Fig. 15. 20mH and 5mH choke experimentally measured impedance magnitude comparison.

experimentally measured impedance characteristics for both CM chokes. The experimentally measured specifications for both chokes are summarized in Table VII.

TABLE VII: CM CHOKE COMPARISON

Parameter	5mH choke	20mH choke
Current rating	1A	1A
SRF	734kHz	190kHz
Parasitic Capacitance	9.42pF	35.08pF
PCB footprint	17.5 x 7.5mm²	23 x 13.2mm²

B. Effects of Parasitic Parameters on AEF Performance

To comprehensively evaluate the impact of injection and sensing capacitor *ESL* on AEF performance, three distinct capacitors are incorporated into both the sensing and injection circuits of the AEF. While all three injection capacitors were 1μF, they exhibited varying *ESL* values. Similarly, the three sensing capacitors were 10nF each, but with differing *ESL* values.

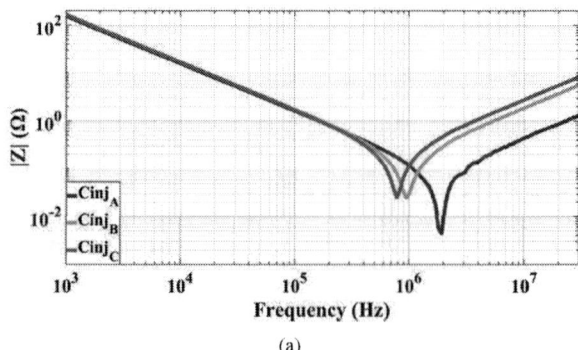

(a)

presents the time domain CM voltage comparison. A 2.86V reduction is observed in the CM voltage transient amplitude with the AEF. In case of time domain CM current presented in Fig. 13(c), it can be observed that with AEF, the noise current amplitude reduction is around 20mAp-p.

TABLE VI: CM EMI REDUCTION BY AEF

Harmonic	Reduction	Harmonic	Reduction
1st	26dB	6th	12dB
2nd	25dB	7th	11dB
3rd	14dB	8th	10dB
4th	15dB	9th	9dB
5th	15dB	10th	7dB

Fig. 14 compares the performance of the proposed hybrid EMI filter with the PEF. When only PEF is used, a 20mH CM choke is required to meet the CISPR-32 standard of CM EMI. However, with AEF, the standard can be met with a 5mH choke. Furthermore, due to the lower parasitic parameters in the smaller CM choke, the hybrid EMI filter performs better in high-frequency regions. Fig. 15 compares the size and

979-8-3315-1612-3/25 $31.00 © 2025 IEEE

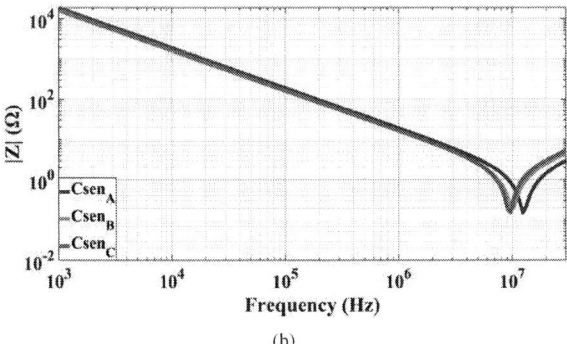

(b)

Fig. 16. Experimentally measured impedance magnitude comparison for capacitors with varying *ESL* (a) injection capacitors and (b) sensing capacitors

Fig. 16 presents the experimentally measured impedance characteristics for the considered capacitors. In Tables VIII – IX, the experimentally measured self-resonant frequency (*SRF*) and *ESL* of the capacitors are presented. Having different *ESL* values, but same capacitances will allow us to examine the effects of the *ESL* of both injection and sensing capacitors experimentally.

TABLE VIII: DIFFERENT INJECTION CAPACITORS

Capacitor	SRF	ESL
C_{inj_A}	1.93MHz	6.8nH
C_{inj_B}	949kHz	28.1nH
C_{inj_C}	782kHz	41.4nH

TABLE IX: DIFFERENT SENSING CAPACITORS

Capacitor	SRF	ESL
C_{sen_A}	12.52MHz	16.1nH
C_{sen_B}	10.32MHz	23.7nH
C_{sen_C}	9.67MHz	27.1nH

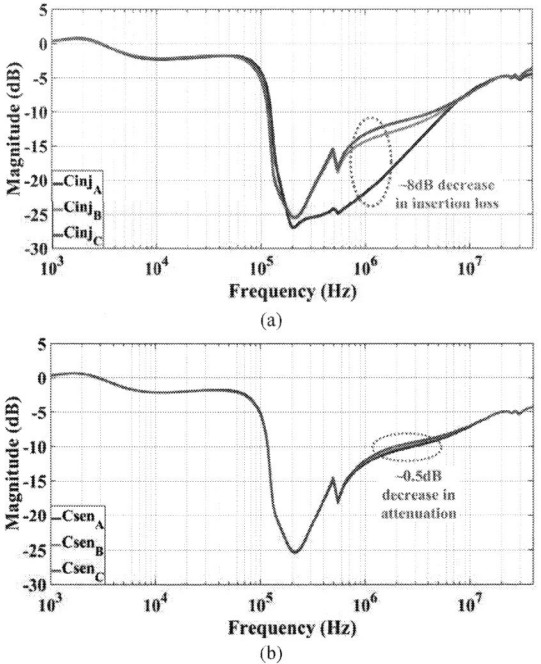

Fig. 17. Experimentally measured AEF insertion loss comparison for capacitors with varying *ESL* (a) injection capacitors and (b) sensing capacitors

Fig. 17 presents a comparative analysis of experimentally measured insertion loss for the AEF, considering different injection and sensing capacitors. As illustrated in Fig. 18(a), C_{inj_C}, which exhibits the highest *ESL* among the evaluated 1 µF injection circuit capacitors, provides the smallest insertion loss within the highlighted frequency range around 1 MHz, approximately 8 dB lower than C_{inj_A}. This region corresponds to the *SRF* of the capacitor as shown in Fig. 16(a). Within this frequency band, C_{inj_A} offers the greatest noise attenuation due to its lower *ESL* value compared to the other capacitors. These observations align with the analytical and simulation analyses presented earlier in the paper. Beyond 7 MHz, the difference in insertion loss becomes less pronounced due to the op-amp's diminishing bandwidth. However, it is anticipated that with an op-amp of higher bandwidth, the impact of the injection capacitor's *ESL* on AEF insertion loss would be more evident beyond 7 MHz. Fig. 17(b) further validates the previous analysis, demonstrating that variations in the sensing capacitor's *ESL* do not significantly affect the AEF insertion loss.

Fig. 18 presents a comparative analysis of experimentally measured loop gain for the AEF, considering different injection and sensing capacitors. As depicted in Fig. 18(a), employing C_{inj_C} results in approximately a 7 dB reduction in loop gain at frequencies exceeding 10 MHz. Conversely, the highest loop gain within this frequency range is observed with C_{inj_A}, which exhibits the lowest *ESL*. A similar trend is evident in Fig. 18(b) for sensing capacitors, where the AEF loop gain with C_{sen_A} achieves the maximum value. In contrast, utilizing C_{sen_C} leads to a 3 dB decrease in loop gain at frequencies beyond 10 MHz compared to C_{sen_A}.

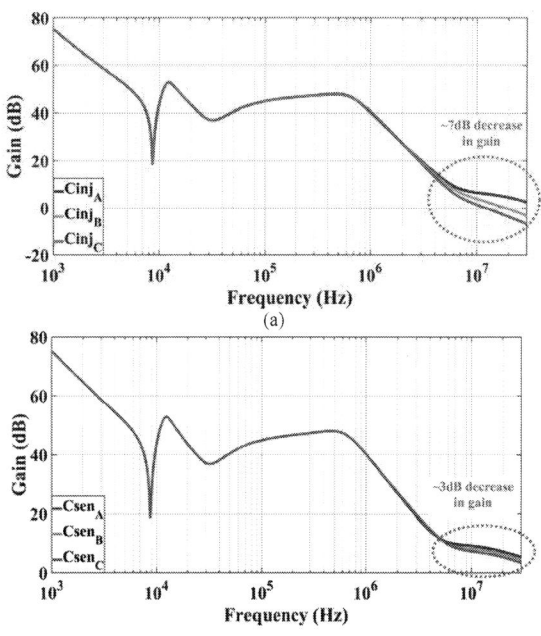

Fig. 18. Experimentally measured AEF loop gain comparison for capacitors with varying *ESL* (a) injection capacitors and (b) sensing capacitors

Fig. 19 illustrates a comparative analysis of measured CM EMI with the AEF alone to evaluate the influence of varying *ESL* in the injection capacitors C_{inj_A}, C_{inj_B} and C_{inj_C}. The

979-8-3315-1612-3/25 $31.00 © 2025 IEEE   1864

results clearly demonstrate that higher *ESL* values associated with C_{inj_B} and C_{inj_C} led to increased CM EMI levels compared to C_{inj_A}. This observation aligns with the previously presented insertion loss findings. Within the frequency range where the capacitors exhibit varying insertion losses due to differing *ESL* values, the CM EMI spectrum correspondingly demonstrates a similar increase, reflecting the impact of higher *ESL* in the injection capacitors. Table X summarizes the variations in measured CM EMI switching harmonic values for the three distinct injection capacitors. These findings underscore how the parasitic elements of passive components within the AEF can significantly constrain its effective noise attenuation bandwidth.

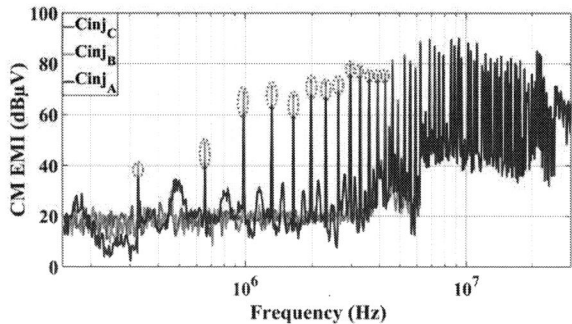

Fig. 19. Experimentally measured CM EMI comparison with different injection capacitors of varying *ESL*

TABLE X: MEASURED CM EMI WITH DIFFERENT INJECTION CAPACITORS

Harmonics	C_{inj_A}	C_{inj_B}	C_{inj_C}
1st (325kHz)	35.8 dBµV	40.1 dBµV	40.3 dBµV
2nd (650kHz)	41.1 dBµV	48.5 dBµV	49.1 dBµV
3rd (975kHz)	60.5 dBµV	68.1 dBµV	69.1 dBµV
4th (1.3MHz)	63.6 dBµV	70.2 dBµV	71.4 dBµV
5th (1.625MHz)	59.3 dBµV	66.5 dBµV	67.6 dBµV
6th (1.95MHz)	67.6 dBµV	72.6 dBµV	73.7 dBµV
7th (2.275MHz)	66.8 dBµV	71.1 dBµV	72.8 dBµV

VI. CONCLUSIONS

This paper presents the design of a hybrid EMI filter, incorporating a FB-VSCC topology based AEF, specifically tailored for a high-frequency active clamp GaN flyback SMPS. A rigorous analysis of the parasitic elements within the AEF was conducted and their consequential impact on its overall performance was evaluated. A comprehensive theoretical analysis, coupled with extensive experimental validation, is presented to substantiate the superior performance of the proposed AEF design in comparison to traditional PEFs. The investigation meticulously examines the influence of *ESL* associated with both the injection and sensing capacitors on the AEF's critical parameters, including insertion loss and loop gain. Through simulations and empirical observations, it has been observed that the *ESL* of the injection capacitor exerts a significant impact on the insertion loss of the AEF, as around 8 dB decrease in the AEF insertion loss was observed. Conversely, the *ESL* of both the injection and sensing capacitors is found to significantly affect the loop gain. Specifically, high-*ESL* injection capacitors can lead to a substantial reduction in insertion loss, resulting in a

corresponding increase in the amplitude of the initial CM EMI switching harmonics by around 6 to 8dB. Furthermore, the loop gain is observed to decrease at higher frequencies due to the combined effects of *ESL* in both the injection and sensing capacitors, having decreased by 7dB and 3dB respectively.

These findings underscore the critical role of parasitic elements in limiting the effective noise attenuation bandwidth of the AEF, even when employing operational amplifiers with higher bandwidth capabilities.

ACKNOWLEDGMENT

This effort was sponsored in whole or in part by the Central Intelligence Agency (CIA), through CIA Federal Labs. The U.S. Government is authorized to reproduce and distribute reprints for Governmental purposes notwithstanding any copyright notation thereon. The views and conclusions contained herein are those of the authors and should not be interpreted as necessarily representing the official policies or endorsements, either expressed or implied, of the Central Intelligence Agency.

REFERENCES

[1] Z. Zhang, K. D. T. Ngo, and J. L. Nilles, "A 30-W flyback converter operating at 5 MHz," in *proc. IEEE APEC*, 2014, pp 1415-142.

[2] T. Labella, B. York, C. Hutchens, and J. S. Lai, "Dead time optimization through loss analysis of an active-clamp flyback converter utilizing GaN devices," in *proc. IEEE ECCE*, 2012, pp 3882-3889.

[3] CISPR 32 Electromagnetic compatibility of multimedia equipment – Emission requirements, March 2015.

[4] I. Cadirci, B. Saka, and Y. Eristiren, "Practical EMI-filter-design procedure for high-power high-frequency SMPS according to MIL-STD 461," in *Proc. IEE Electr. Power Appl.*, 2005, vol. 152, pp. 775–782.

[5] Y. Han, Z. Wu and D. Wu, "Hybrid Common-mode EMI Filter Design for Electric Vehicle Traction Inverters," in *Chinese Journal of Electrical Engineering*, vol. 8, no. 4, pp. 52-60, December 2022.

[6] B. Narayanasamy and F. Luo, "A Survey of Active EMI Filters for Conducted EMI Noise Reduction in Power Electronic Converters," in *IEEE Transactions on Electromagnetic Compatibility*, vol. 61, no. 6, pp. 2040-2049, Dec. 2019.

[7] W. Chen, X. Yang, and Z. Wang, "An active EMI filtering technique for improving passive filter low-frequency performance," *IEEE Trans. Electromagn. Compat.*, vol. 48, no. 1, pp. 172–177, Feb. 2006.

[8] Q. Chen, R. Zhang, Z. Niu and C. Gong, "Frequency Characteristics of Insertion Loss and Loop Gain of VSCC Feedback Active EMI Filters," in IEEE Transactions on Electromagnetic Compatibility, doi: 10.1109/TEMC.2024.3454127.

[9] Y. Zhang, Q. Li and D. Jiang, "A Motor CM Impedance Based Transformerless Active EMI Filter for DC-Side Common-Mode EMI Suppression in Motor Drive System," in *IEEE Transactions on Power Electronics*, vol. 35, no. 10, pp. 10238-10248, Oct. 2020.

[10] Z. Zhang and A. M. Bazzi, "A Virtual Impedance Enhancement Based Transformer-Less Active EMI Filter for Conducted EMI Suppression in Power Converters," in IEEE Transactions on Power Electronics, vol. 37, no. 10, pp. 11962-11973, Oct. 2022.

[11] W. Chen, X. Yang and Z. Wang, "A Novel Hybrid Common-Mode EMI Filter With Active Impedance Multiplication," in IEEE Transactions on Industrial Electronics, vol. 58, no. 5, pp. 1826-1834, May 2011.

[12] CISPR, "Methods of measurement of the suppression characteristics of passive EMC filtering devices," IEC Standard CISPR-17:2011, 2011.

Overview of Dynamic Characterization of Switches for Three Phase Voltage Source, Current Source, and Matrix Converter Applications

Sneha Narasimhan
Electrical & Computer Engineering
North Carolina State University
Raleigh, USA
snarasi7@ncsu.edu

Sathya Rupan Thirumoorthi
Electrical & Computer Engineering
North Carolina State University
Raleigh, USA
sthirum5@ncsu.edu

Subhashish Bhattacharya
Electrical & Computer Engineering
North Carolina State University
Raleigh, USA
sbhatta4@ncsu.edu

Abstract—The double pulse test (DPT) is a widely accepted method to evaluate the dynamic performance of power devices. However, the results are accurate only if the DPT setup emulates the commutation loop. This paper presents an overview of DPT methods for three-phase two-level voltage source, two-level current source, and direct matrix converter applications. The design criteria for the three DPT circuits are presented. Probe selection and the impact of switching voltage-current (V-I) timing misalignment on the testing results are discussed, and a V-I alignment approach is introduced for all the DPT systems. A 1.2 kV/15 A SiC-based DPT presents the results for the three configurations. The impact of parasitic capacitances on the switching losses and the methods to minimize the effect on the DPT results are finally presented.

Index Terms—Current source, double pulse test, low voltage, matrix, medium voltage, V-I alignment, parasitic inductance, probes, voltage source, CS, CSI, DPT, LV, MV, VS, VSI

I. INTRODUCTION

A widely accepted method to obtain the power devices' turn-on, turn-off, and reverse recovery losses in a power conversion system (PCS) is the double pulse test (DPT) method. DPT uses two pulses to obtain the turn-off and turn-on losses of the power devices. Based on the first pulse, the voltage or current can be controlled to help capture the device under test (DUT) switching transients. DPT results help quantify the switching performance of power devices, offering essential insights for converter design considerations. These include the selection of optimal switching frequency and dead-time or overlap time selection, as well as strategies for effective thermal management and accurate efficiency estimation [1]–[3].

Consideration for DPT for wide bandgap (WBG) devices used for voltage source converters (VSC) is discussed in [4], [5]. However, these considerations are for low voltage (LV) systems. [6] discusses the improved DPT considerations for medium voltage (MV) systems. The increased losses and electromagnetic interference (EMI) concerns due to capacitive coupling are also presented.

Current source converters (CSC) have gained traction due to the re-emergence of WBG devices [7]–[9]. The DPT of reverse voltage blocking (RVB) switches comprising MOSFET or IGBT with a series diode is discussed in [1]. However, the DPT circuit uses the modified commutation circuit to obtain the losses. [10] discusses the DPT circuit based on the commutation of the CSI. However, concerns about the effect of parasitic capacitance, passive component sizing, and V-I misalignment are not discussed.

With the emergence of bidirectional switches, AC-AC power conversion systems are gaining traction [11], [12]. Thus, developing a commutation circuit is also important to obtain accurate losses. A 4-quadrant switch switching investigation is presented in [13]. However, the details about parasitic capacitance, device switching loss, and V-I misalignment are not presented.

The above literature review emphasizes the importance of accurate dynamic testing methodologies to ensure efficient and reliable operation of the converters in various applications. This paper provides a comprehensive overview of the DPT circuit and design for the two-level voltage source, two-level current source, and direct matrix converter applications. The paper is organized as follows: Section II introduces the DPT circuit for the three PCS. Section III provides the details of the design criteria of the DPT circuit. The probe selection and effect of probe delays are discussed in Section IV. Section V presents the experimental hardware results of the DPT for three PCS. Finally, conclusions are presented in Section VI.

II. DOUBLE PULSE TEST CONFIGURATIONS FOR THREE DIFFERENT CONVERTER APPLICATIONS

A. Voltage Source Converter

Fig. 1(a) is a three-phase two-level VSC system that shows the commutation loop of the switches. In this configuration, three switches, one per phase, are simultaneously active at any given moment. This ensures the dc-bus capacitor is not short at any instant. The commutation process occurs between the top and bottom switches within each leg of the converter. Fig. 1(a) shows the commutation between the a-phase top and bottom switches.

979-8-3315-1612-3/25 $31.00 © 2025 IEEE

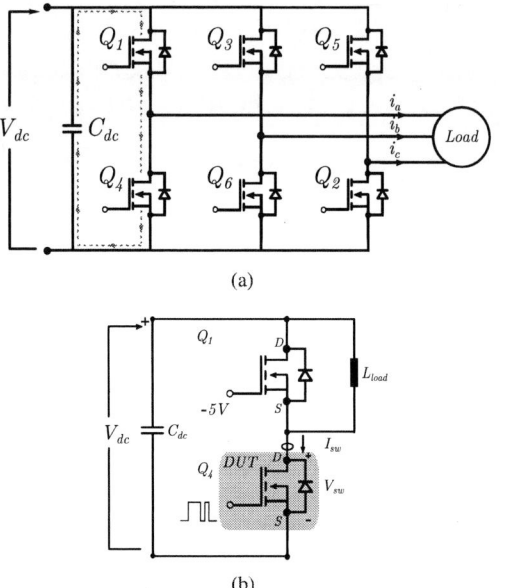

Fig. 1: a) Schematic of the VSC with the commutation paths, and b) DPT circuit for VSC.

Fig. 1(b) depicts the DPT setup for the VSC, derived from the commutation circuit. This setup is critical for evaluating the switching characteristics and performance of the VSC switches. In a VSC, a constant DC bus voltage is maintained. The modulation index of the inverter is adjusted to control the amplitude of the output current, thereby achieving the desired AC output. Hence, the input voltage is kept constant for the DPT setup, and the current is regulated by changing the first pulse width. The turn-off and turn-on losses are measured at the end of the first pulse and the beginning of the second pulse, respectively.

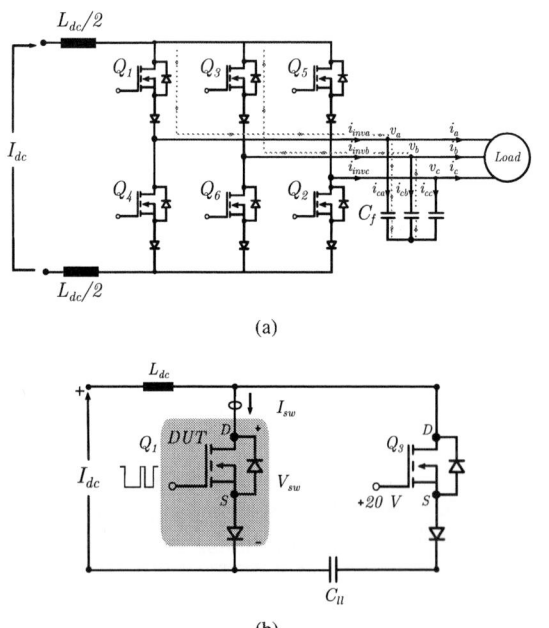

Fig. 2: a) Schematic of the CSC with the commutation paths, and b) DPT circuit for CSC.

B. Current Source Converter

The three-phase two-level CSC, depicted in Fig. 2(a), includes the commutation path for the switches. In a CSC, commutation occurs either among the three upper switches or the three lower switches. Unlike a VSC, where three switches are on at any given time, a CSC operates with only two switches on at any instance. This ensures there is a path for the dc-link current at any instant. Fig. 2(a) shows the commutation between the a-phase and b-phase top switches.

The modulation index is kept constant for the current source inverter (CSI). The output voltage is regulated by varying the DC-link current on the inverter side. On the converter side, since the input grid voltage remains constant, the modulation index is adjusted to accommodate changes in the DC-link current. This adjustment enables control over the output voltage, ensuring the system can meet varying load demands while maintaining stable operation. For the DPT setup, the input current is kept constant. The pulse pattern for the CSC DPT setup is the inverse of that used for the VSC. The required voltage is achieved by turning off the DUT during the first pulse width. The turn-on losses are measured at the end of the first pulse, while the turn-off losses are measured at the beginning of the second pulse.

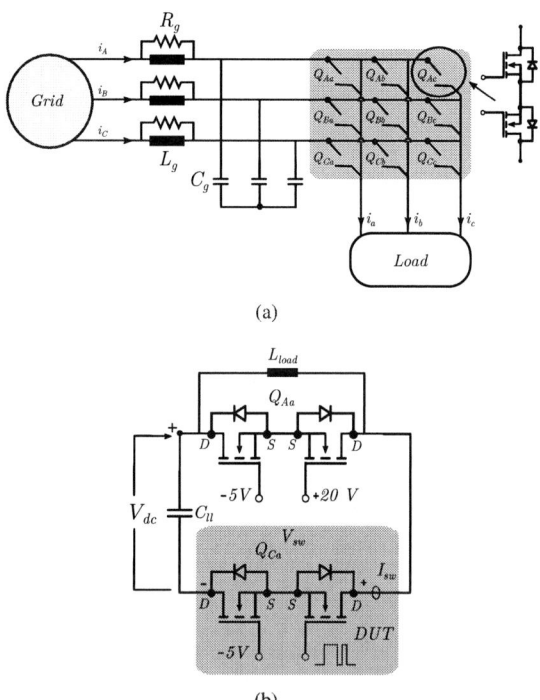

Fig. 3: a) Schematic of the DMC with the commutation paths, and b) DPT circuit for DMC.

C. Matrix Converter

A matrix converter (MC) is used for the AC-AC power conversion system. A direct matrix converter (DMC) is shown in Fig. 3(a). The input capacitor should not be shorted at any instant, and the inductive load current path should not

979-8-3315-1612-3/25 $31.00 © 2025 IEEE

be interrupted. To achieve these conditions, only one switch per output phase (a, b, and c) should be turned on.

The commutation for the DMC occurs between the switches of each phase (a, b, and c). Four-step commutation is followed for the converter [14], and switching losses were calculated using a half-bridge with an R-L load in [13]. The DPT configuration for a DMC is shown in Fig 3(b), which mirrors the VSC setup, with the switches arranged in a common-source configuration. This arrangement captures the maximum switching loss across various voltage and current operating points. A similar result will be obtained for common-drain configuration as well [1], [15]. The turn-off losses are measured at the end of the first pulse, and turn-on losses are estimated at the start of the second pulse, similar to that of a VSC.

III. Design Criteria of the DPT Circuit

A. Inductor Design

The DPT circuit for the VSC and MC consists of a load inductor, DC-link capacitor, and gate driver circuits, and the DPT circuit for the CSC consists of a DC inductor, line-line capacitor, and gate driver circuits. The design of these components, placement, and connections significantly affect the DUT's switching behavior.

The load inductor helps establish the current, and the selected inductance needs to minimize the current variation when the current flows through the non-DUT.

$$L_{load} \geq \frac{V_{dc}}{\Delta I_l} * t_{sw} \qquad (1)$$

where t_{sw} is the time to complete a switching event, i.e., the minimum time required to turn on and turn off the device.

The DC link inductor for the CSC helps maintain the DC current when the commutation occurs.

$$L_{dc} \geq \frac{C_{ll} * V_{ll}^2}{\Delta I_{dc}^2} \qquad (2)$$

The variation of the load current is set to 1-5%. The inductor is also a current source for the DUT. The maximum voltage and minimum load current should be considered to obtain the minimum load inductor. The series connection of load inductors or the DC link inductor is recommended to minimize the equivalent parallel capacitance.

B. Capacitor Design

The capacitor helps establish the voltage for a CSC, and the selected capacitance needs to minimize the voltage variation when the current flows through the DUT.

$$C_{ll} \geq \frac{I_{dc}}{\Delta V_{dc}} * t_{sw} \qquad (3)$$

The DC bus capacitor for the VSC and MC applications is designed to support a stable DC bus voltage. This capacitor provides energy to establish the inductor current during the first pulse. The capacitance can be calculated as

$$C_{dc} \geq \frac{L_{load} * I_{dc}^2}{\Delta V_{dc}^2} \qquad (4)$$

The maximum current and minimum voltage should be considered to obtain the minimum capacitance. The variation of the voltage is set to 1-5%.

Decoupling capacitors are placed across the DC bus and closer to the switches for the VSC and MC applications. These capacities have minimal equivalent series inductances and supply the transient currents during switch commutation. Additionally, they reduce the voltage overshoot across the switches by providing a low-loop inductance path. For CSC applications, the decoupling capacitors (C_{dec}) are placed across line-line as the commutation happens between the top and bottom switches. According to [16], C_{dec} should be equal to 100 times the output capacitance of the device (C_{oss}). Increasing the value beyond that does not bring a substantial benefit. Film and ceramic capacitors are preferred for C_{dec}. The decoupling capacitors should also introduce minimum parasitic inductance to the system.

C. Power and Gate Driver Boards

The power boards must mimic the converter layout to obtain accurate switching losses. Key aspects of a well-designed DPT layout include minimizing parasitic components like gate loop inductance, power loop inductance, and mutual parasitic capacitance (C_{gd}) between the gate and power loops, which can significantly impact switching performance [4]. To achieve this, a modular design is recommended where the power stage is separated from the gate driver, with each power device having its dedicated gate driver board. This approach allows for standardized gate driver layouts across different converter topologies while minimizing parasitic coupling between the power and gate loops by using separate paths and Kelvin source connections, leading to improved switching characteristics and reduced switching losses [17]. Reducing the parasitic capacitance introduced by the gate-driver board is also important. The typical capacitance of the gate-driver isolated transformer design is 1 pF to 8 pF [18].

The PCB stack up, and the layout of traces or planes are crucial in determining the loop inductance and parasitic capacitance, significantly influencing the switching transients. While optimizing the design to minimize parasitic elements is possible, they cannot be entirely eliminated. Furthermore, the layout will differ depending on the converter topology, as the di/dt loops and dv/dt nodes vary for each configuration. Experimental results on the effects of PCB parasitic elements in VSC and CSC are shown in Section V.

Furthermore, the device baseplate and the hotplate can introduce additional parasitic capacitance. The parasitic capacitance depends on the area, the distance between the device and the hotplate, and the insulation material between the device and the hotplate [6]. For this study, the hotplate is eliminated; however, using a hotplate with a smaller area and thicker insulation between the device and the hotplate will significantly reduce the parasitic capacitance.

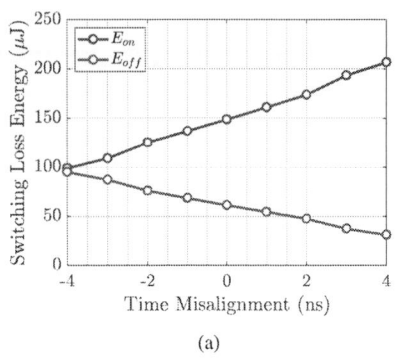 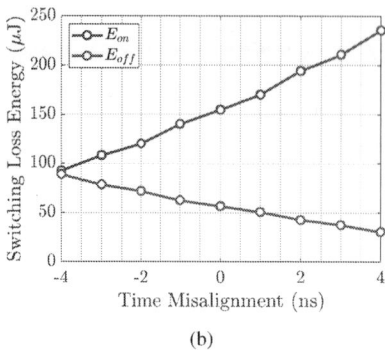 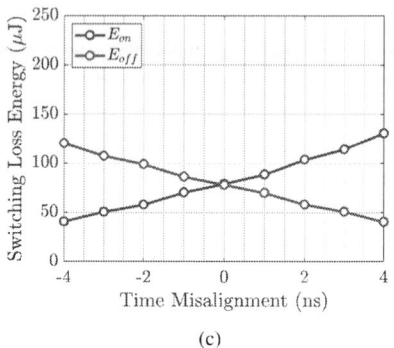

(a)   (b)   (c)

Fig. 4: a) VSC, b) CSC, and c) MC V-I misalignment concerns at 800 V, 15 A at $R_{g,on}$ = 19 Ω and $R_{g,on}$ = 14 Ω .

IV. MEASUREMENT TECHNIQUES

A. Probes

To measure the fast switching transients of the DUT, the minimum bandwidth (BW) required is:

$$BW = \frac{0.35}{min(t_r, t_f)} \quad (5)$$

where t_r and t_f are the rise and fall time of the DUT, respectively [4]. For a 1200V-rated SiC device, the typical rise and fall time is around 8-40 ns. The minimum bandwidth requirements based on Eqn. 5 is 43.75 MHz. A differential or passive probe can measure the V_{ds} of the lower switch DUT. Passive probes have higher bandwidth; however, they have no galvanic isolation.

On the other hand, differential probes have galvanic isolation. However, they have limited bandwidth and longer probe leads. Since the rise and fall time of the SiC MOSFET used for this study is around 10 ns. The differential probe HVD3206A, which has a bandwidth of 120 MHz, can be used.

The rise time (t_{rg}) of a gate-source voltage (V_{gs}) based on the gate resistance can be calculated as

$$t_{rg} = \frac{C_{gs} * R_g}{V_{step}} \quad (6)$$

where V_{step} is the voltage step applied to the gate and R_g is the gate resistance of the DUT. For this study, the total R_g is 14-19 Ω. The operating V_{gs} is +15/-5V. The gate-source capacitance C_{gs} is 6.91 nF. The minimum bandwidth required is 50 MHz. Therefore, for V_{gs} measurement, a high voltage optical isolated probe is preferred as it helps significantly reduce the common-mode noise due to high common mode rejection ratio (CMRR), minimal loading on the DUT, high safety in high voltage environments, and the ability to accurately measure floating voltages due to the optical transmission of the signal. However, they are expensive. A 1 GHz passive probe can also measure the gate voltage, which is cheaper, or a low voltage differential probe can be used.

Depending on the device package and the system layout, several current measurement techniques exist for I_{ds} measurement, including coaxial shunt, current transformer, split-core current probe, and Rogowski coil. The coaxial shunt provides the highest bandwidth, typically in the range of 1200-2000 MHz. The current transformer has a high bandwidth, such as Pearson 8590C, which has 150 MHz. These are clamped on current monitors. A split-core current probe, such as the Tektronix TCP0030A with 120 MHz bandwidth, can be considered. However, measuring at the non-switching end is recommended to avoid large dv/dt or di/dt effects, and modifications need to be made to use a split-core current probe or the current transformer. Rogowski coils have lower bandwidth, such as the Ultra MiniCWTUM/06/B, which has 30 MHz and can easily be coupled with noise.

The details of the experimental DPT instrument details are presented in Table. I. The instruments have a BW ≥ 100 MHz. The gate probe measures V_{gs}. The switch voltage V_{sw} is measured using an HVD3206A passive probe, and CP031A is used to measure the switch current I_{sw} as the gate resistance of the DUT is set to a minimum resistance of 14 Ω.

TABLE I: Details of DPT experimental platform

Instrument	Part Number	Bandwidth
Oscilloscope	MDA803A	350 MHz
Voltage Probe	HVD3206A	120 MHz
Current Probe	CP031A	100 MHz
Gate Probe	HVFO103	150 MHz

B. Probe Delays

Each voltage and current probe introduces its specific propagation delay. The mismatch between these delays, or skew, can result in inaccurate switching loss measurements, especially in fast-switching WBG devices. Fig. 4(a), Fig. 4(b) and Fig. 4(c) shows the effect of voltage-current (V-I) misalignment in the turn-on and turn-off losses for the DUT used in VSC, CSC, and MC applications, respectively. These results are for 800 V, 15 A with $R_{g,on}$ = 19 Ω a $R_{g,off}$ = 14 Ω. The sensitivity curves will vary at different operating conditions and gate resistance values. Since the magnitude of the turn-off losses is lower than the turn-on losses for VSC and CSC, the overall losses increase with a positive misalignment and decrease with a negative misalignment. For MC, the turn-on and turn-off losses are almost equal. However, the V-I misalignment is still a concern. Hence, it is important to fix

979-8-3315-1612-3/25 $31.00 © 2025 IEEE   1869

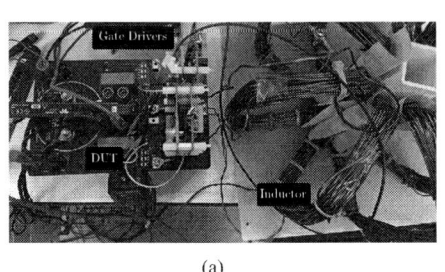 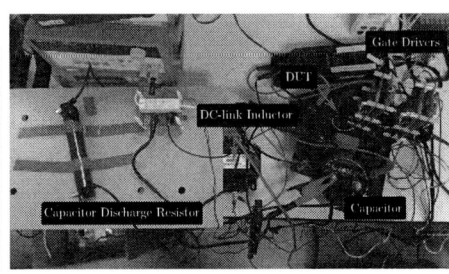 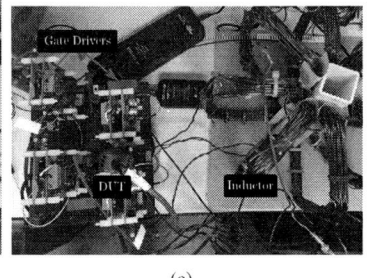

(a)	(b)	(c)

Fig. 5: a) VSC, b) CSC, and c) Matrix DPT Setup.

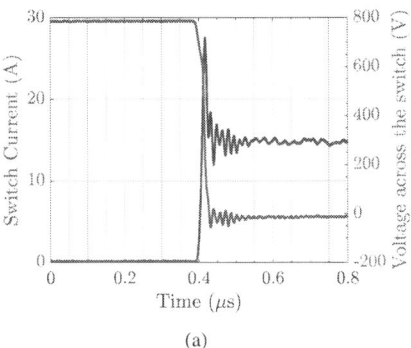 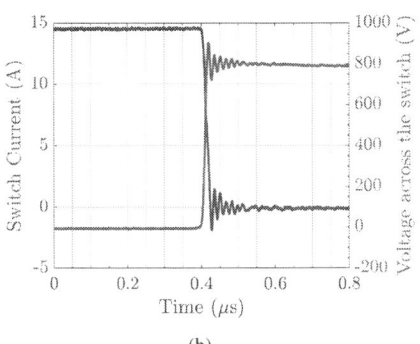 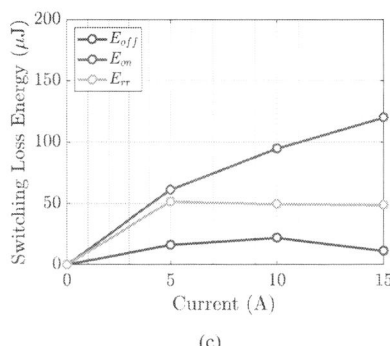

(a)	(b)	(c)

Fig. 6: Voltage Source Converter (VSC) - a) turn-on and b) turn-off losses of the DUT operating at 800 V and 15 A at room temperature. c) Total losses of the DUT operating at 800 V and various currents.

the V-I misalignment, which is essential to deciding the PCS's safe operating area (SOA). The following method can be used to fix V-I misalignment while measuring the switching losses of the DUT.

- Remove the inductor from the DPT circuit and replace it with a low parasitic inductance and capacitance 100 Ω resistor. In this resistive load DPT, the voltage across the resistor is 100 times the current, and the equivalent series inductance of the resistor contributes negligible phase shift. The resistor's power rating determines the maximum input voltage, typically reaching several hundred volts.
- Provide a single pulse to the DUT large enough to switch the device, then measure the voltage and current across it to verify they are in phase with zero phase shift.
- If a non-zero phase shift is observed, account for this value during post-processing to ensure accurate results.

For a CSC, it is important to use a voltage source at the input and turn off the non-DUT.

V. EXPERIMENTAL RESULTS

A single SiC MOSFET from CREE (C3M0075120K) and SiC Schottky diode from GeneSiC (GD30MPS12H) are used to obtain the losses. For the VSI, a single SiC MOSFET is used as the switch. The SiC MOSFET in series with the SiC Schottky diode is used as the reverse voltage blocking switch for the CSI. The SiC MOSFETs in common-source configuration act as the switch for MC. The turn-on gate resistance is 19 Ω, and the turn-off gate resistance is 13.7 Ω.

An air-core inductor is used as the load inductor L_{load} for VSC and MC DPT, as it helps to avoid the saturation of the core. The parasitic capacitance is measured to be 8 pF at 100 MHz, which is negligible compared to the junction capacitance of the DUT. A DC inductor is used for the DC-link of the CSC DPT and to help maintain a constant current. Series connecting the inductors help reduce the parasitic capacitance contributed by the inductor. Three 5 mH inductors are used for the test setup.

Decoupling capacitors are used in the experimental setup to help provide a low-impedance path for high-frequency currents. A combination of film and decoupling capacitors is used to reduce the impact of equivalent series inductances (ESL).

A. Voltage Source Converter

To obtain losses of the DUT for the VSC, the voltage is kept constant at 800 V, and the current varies. The results are obtained at 5 A, 10 A, and 15 A. When the current is negative, i.e., flows from the source of the MOSFET to the drain through the body diode, the reverse recovery losses of the switches are observed.

Fig. 5(a) shows the hardware board to obtain the losses for a VSC configuration. The turn-on and turn-off losses at 800 V, 15 A, and room temperature are presented in Fig. 6(a) and Fig. 6(b), respectively. The turn-on di/dt is 0.85 $kA/\mu S$, and the turn-off dv/dt is 18.3 $kV/\mu S$. The VSC's turn-on, turn-off, and reverse recovery losses are shown in Fig. 6(c). At a given voltage, the turn-on and turn-off losses increase as the current increases. The reverse recovery losses, however, are similar

979-8-3315-1612-3/25 $31.00 © 2025 IEEE

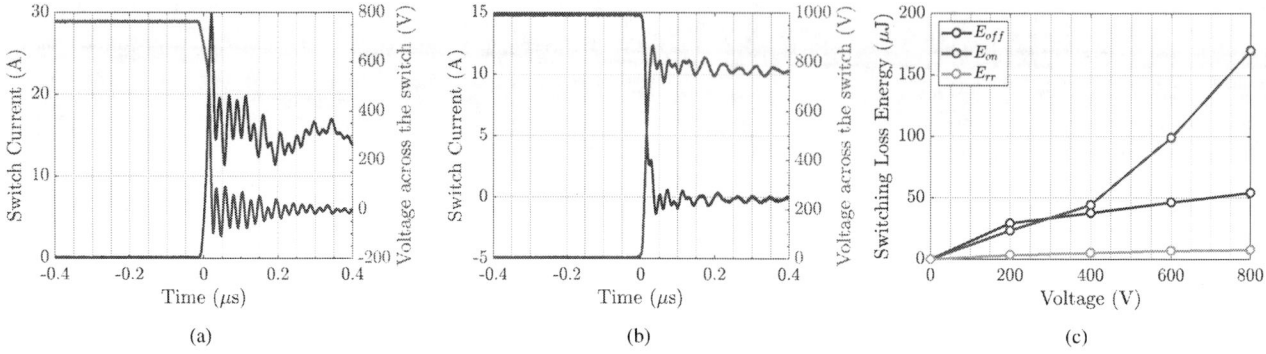

Fig. 7: Current Source Converter (CSC) - a) turn-on and b) turn-off losses of the DUT operating at 800V and 15 A at room temperature. c) Total losses of the DUT operating at 15 A and various voltages.

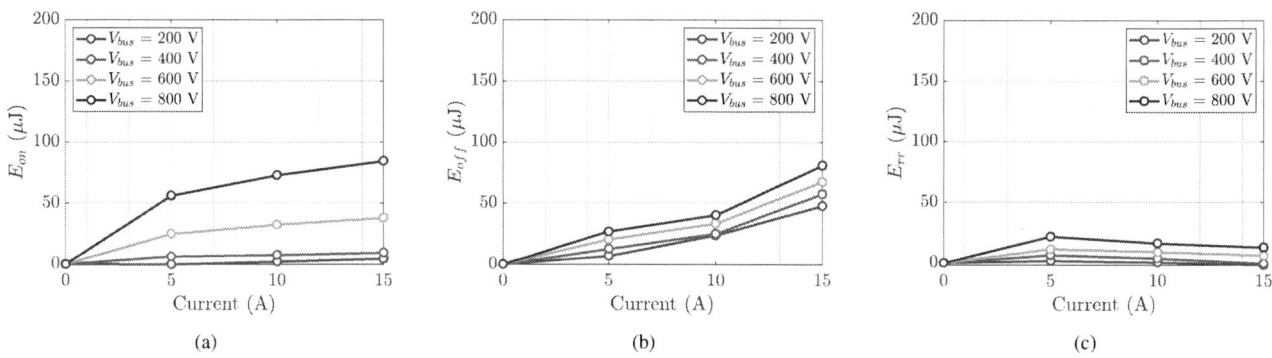

Fig. 8: Matrix Converter (MC) - a) turn-on, b) turn-off, and c) reverse recovery losses of the DUT operating at various voltages varying from 200 V to 800 V and currents 5 A to 15 A.

at various current levels. The reverse recovery losses are the losses of the DUT when the inductor is corrected across it and is attributed to the body diode. Thus, the reverse recovery losses are higher than the turn-off losses as the body diode introduces higher losses [19].

B. Current Source Converter

To obtain losses of the DUT for the CSC, the current is kept constant at 15 A, and the voltage varies from 0 to 800 V. Fig. 5(b) shows the hardware board to obtain the losses for a CSC configuration. The capacitor is precharged to the required voltage to obtain the losses of the DUT when the input acts as a voltage source with Q_1 turned off and Q_3 turned on. Then, the source is changed to a current source, and both switches Q_1 and Q_3 are turned on. Then, the source is changed to a current source, and the DUT is provided a pulse to obtain the turn-off and turn-on losses, respectively. The turn-on and turn-off losses at 800 V, 15 A, and room temperature are presented in Fig. 7(a) and Fig. 7(b), respectively. The turn-on di/dt is 0.53 $kA/\mu S$, and the turn-off dv/dt is 12.6 $kV/\mu S$. The CSC's turn-on and turn-off losses are shown in Fig. 7(c). At a given voltage, the turn-on and turn-off losses increase as the current increases. It is also important to note that the turn-on and turn-off losses are similar at lower voltages. As observed in the datasheet, the device capacitances are larger at lower voltages, leading to larger turn-off times and higher turn-off

losses. The reverse recovery losses of the CSC are the least, as they are contributed by the series SiC diode.

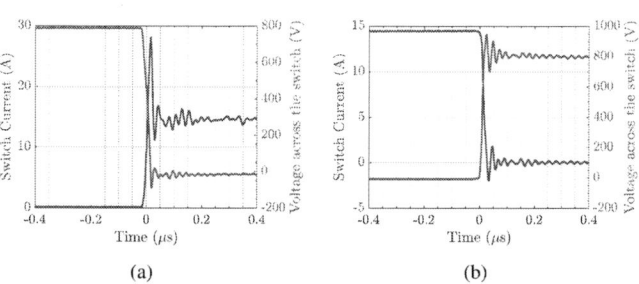

Fig. 9: Matrix Converter (MC) - a) turn-on and b) turn-off losses of the DUT operating at 800V and 15 A at room temperature.

C. Matrix Converter

To obtain losses of the DUT for the MC, the current varies from 0 to 15 A, and the voltage ranges from 0 to 800 V. The varying operating point is adapted to realize the losses during the commutation of the matrix converter at a steady state. The turn-on and turn-off losses at 800 V, 15 A, and room temperature are presented in Fig. 9(a) and Fig. 9(b), respectively. The turn-on di/dt is 0.62 $kA/\mu S$, and the turn-off dv/dt is 18.1 $kV/\mu S$. The MC DUT turn-on and turn-off losses are shown in Fig. 8(a) and Fig. 8(b), respectively. At a given voltage, the turn-on and turn-off losses increase with the increased current. As observed in the datasheet, the

979-8-3315-1612-3/25 $31.00 © 2025 IEEE

device capacitances are larger at lower voltages, leading to larger turn-off times and higher turn-off losses. The reverse recovery losses are presented in Fig. 8(c) and are high due to the losses of the body diode.

D. Effect of PCB Parasitics in the Device Switching Loss

Two scenarios were analyzed to examine the effect of PCB-induced parasitics on switching loss in a VSC. In Case 1, the DC+, DC-, and switching node layers were fully overlapped, resulting in high capacitance between the switching node and the DC+/DC-. In Case 2, the overlap with the switching node was removed, but the overlap between the DC+ and DC- was maintained, leading to lower capacitance compared to Case 1. The parameters for both cases are listed for VSC in Table II at 800 V, 10 A. $C_{Q4-DtoVdc}$ and $C_{Q4-DtoGND}$ are the parasitic capacitances from the drain of the DUT to V_{dc} and ground, respectively. During the turn-on transient, the parasitic capacitance causes significant charging and discharging currents in the DUT due to the high dv/dt at the switching node, resulting in higher losses in Case 1 compared to Case 2 [20]. Since the MC has the same derived DPT as the VSC, similar considerations should be made in the layout.

TABLE II: PCB Parasitic in VSC and Experimental results at 800 V, 10 A

Parameter	Case 1	Case 2
$C_{Q4-DtoVdc}$ (pF)	85.9	1.7
$C_{Q4-DtoGND}$ (pF)	261.1	1.7
Turn-on Loss (μJ)	305	94.5
Turn-off Loss (μJ)	69.7	21.8
Turn-on Peak Current (A)	34	21
Turn-off Peak Voltage (V)	925	846

For CSC, in Case 1, all the planes were fully overlapped, resulting in high capacitance between I_{dc} and ground, whereas in Case 2, there was no overlap between any of the planes. Table III shows the parameters for both cases for the CSC at 800 V, 10 A. $C_{IdctoGND}$ is the parasitic capacitances from the DC current input to the ground.

TABLE III: PCB Parasitic in CSC and Experimental results at 800 V, 10 A

Parameter	Case 1	Case 2
$C_{IdctoGND}$ (pF)	468.2	0.9
$V_{CaptoGND}$ (V)	763	740
Turn-on Loss (μJ)	424.2	119.3
Turn-off Loss (μJ)	31.9	34.9
Turn-on Peak Current (A)	78.6	23.4
Turn-off Peak Voltage (V)	807	787

Since the commutation of the CSC happens between the top and bottom switches, the DC link and the ground overlap can be limited. This results in lower turn-on peak currents.

The parasitic capacitance $C_{IdctoGND}$ is in parallel with the DUT switch (Q_1). During the turn-on transition of the DUT, the voltage across the parasitic capacitance decreases from $V_{CaptoGND}$ to zero. The discharge current flows into DUT, reducing loss to higher peak currents and switching loss. Thus, having a minimal capacitance between the power planes helps reduce loss.

VI. CONCLUSIONS

This paper provides an overview of the DPT circuit generated from the commutation circuits of the two-level voltage source, two-level current source, and direct matrix converters. The details of designing the DPT circuits' passive components, power, and gate boards are discussed. The effect of V-I misalignment is presented for three circuits, and the probe selection process is explained. Experimental results in the three DPT circuits to obtain the losses of the DUT are presented and analyzed. Additionally, the effects of parasitic capacitance on the switching losses of the DUT for the converters and the ways to minimize the parasitic inductances are discussed.

ACKNOWLEDGEMENT

The authors would like to acknowledge the help of Harshal Talur Lokesha for his help with the experiments.

REFERENCES

[1] S. Narasimhan, A. Kanale, S. Bhattacharya, and J. B. Baliga, "Performance evaluation of 3.3 kv sic mosfet and schottky diode based reverse voltage blocking switch for medium voltage current source inverter application," *IEEE Access*, vol. 11, pp. 89 277–89 289, 2023.

[2] H. Muhsen, S. Hiller, and J. Lutz, "Three-phase voltage source inverter using sic mosfets design and optimization," in *2015 17th European Conference on Power Electronics and Applications (EPE'15 ECCE-Europe)*, 2015, pp. 1–9.

[3] S. Madhusoodhanan, K. Mainali, A. K. Tripathi, A. Kadavelugu, D. Patel et al., "Power loss analysis of medium-voltage three-phase converters using 15-kv/40-a sic n-igbt," *IEEE Journal of Emerging and Selected Topics in Power Electronics*, vol. 4, no. 3, pp. 902–917, 2016.

[4] Z. Zhang, B. Guo, F. F. Wang, E. A. Jones, L. M. Tolbert et al., "Methodology for wide band-gap device dynamic characterization," *IEEE Transactions on Power Electronics*, vol. 32, no. 12, pp. 9307–9318, 2017.

[5] N. Haryani, X. Zhang, R. Burgos, and D. Boroyevich, "Static and dynamic characterization of gan hemt with low inductance vertical phase leg design for high frequency high power applications," in *2016 IEEE Applied Power Electronics Conference and Exposition (APEC)*, 2016, pp. 1024–1031.

[6] H. Li, Z. Gao, R. Chen, and F. Wang, "Improved double pulse test for accurate dynamic characterization of medium voltage sic devices," *IEEE Transactions on Power Electronics*, vol. 38, no. 2, pp. 1779–1790, 2023.

[7] V. Madonna, G. Migliazza, P. Giangrande, E. Lorenzani, G. Buticchi et al., "The rebirth of the current source inverter: Advantages for aerospace motor design," *IEEE Industrial Electronics Magazine*, vol. 13, no. 4, pp. 65–76, 2019.

[8] H. Dai, R. A. Torres, F. Chen, T. M. Jahns, and B. Sarlioglu, "An h8 current-source inverter using single-gate wbg bidirectional switches," *IEEE Transactions on Transportation Electrification*, vol. 9, no. 1, pp. 1311–1329, 2023.

[9] D. Zhang, M. Guacci, M. Haider, D. Bortis, J. W. Kolar et al., "Three-phase bidirectional buck-boost current dc-link ev battery charger featuring a wide output voltage range of 200 to 1000v," in *2020 IEEE Energy Conversion Congress and Exposition (ECCE)*, 2020, pp. 4555–4562.

[10] F. Chen, S. Lee, R. A. Torres, T. M. Jahns, and B. Sarlioglu, "Performance evaluation and loss modeling of wbg devices based on a novel double-pulse test method for current source inverter," in *2021 IEEE Transportation Electrification Conference & Expo (ITEC)*, 2021, pp. 219–224.

[11] A. Kanale, T.-H. Cheng, S. S. Shah, K. Han, A. Agarwal et al., "Switching characteristics of a 1.2 kv, 50 mΩ sic monolithic bidirectional field effect transistor (bidfet) with integrated jbs diodes," in *2021 IEEE Applied Power Electronics Conference and Exposition (APEC)*, 2021, pp. 1267–1274.

[12] J. Huber and J. W. Kolar, "Monolithic bidirectional power transistors," *IEEE Power Electronics Magazine*, vol. 10, no. 1, pp. 28–38, 2023.

[13] N. Anurag and S. Nath, "Switching investigation of sic mosfet based 4-quadrant switch," *IEEE Access*, vol. 11, pp. 1094–1103, 2023.

[14] S. Solemanifard, Y.-X. Chen, M. Lak, and T.-L. Lee, "A novel three-step commutation method for direct matrix converter free from inrush current and voltage error," *IEEE Access*, vol. 11, pp. 25 020–25 034, 2023.

[15] A. Kanale, S. Narasimhan, T.-H. Cheng, A. Agarwal, S. S. Shah *et al.*, "Comparison of the capacitances and switching losses of 1.2 kv common-source and common- drain bidirectional switch topologies," in *2021 IEEE 8th Workshop on Wide Bandgap Power Devices and Applications (WiPDA)*, 2021, pp. 112–117.

[16] R. Paul, A. Hassan, and H. A. Mantooth, "A double-sided cooled power module with embedded decoupling capacitors," *IEEE Journal of Emerging and Selected Topics in Power Electronics*, vol. 12, no. 2, pp. 1813–1821, 2024.

[17] H. Gui, Z. Zhang, R. Chen, J. Niu, L. M. Tolbert *et al.*, "Gate drive technology evaluation and development to maximize switching speed of

[18] sic discrete devices and power modules in hard switching applications," *IEEE Journal of Emerging and Selected Topics in Power Electronics*, vol. 8, no. 4, pp. 4160–4172, 2020.

[18] A. Anurag, S. Acharya, N. Kolli, and S. Bhattacharya, "Gate drivers for medium-voltage sic devices," *IEEE Journal of Emerging and Selected Topics in Industrial Electronics*, vol. 2, no. 1, pp. 1–12, 2021.

[19] M. R. Ahmed, R. Todd, and A. J. Forsyth, "Switching performance of a sic mosfet body diode and sic schottky diodes at different temperatures," in *2017 IEEE Energy Conversion Congress and Exposition (ECCE)*, 2017, pp. 5487–5494.

[20] R. S. Krishna Moorthy, B. Aberg, M. Olimmah, L. Yang, D. Rahman *et al.*, "Estimation, minimization, and validation of commutation loop inductance for a 135-kw sic ev traction inverter," *IEEE Journal of Emerging and Selected Topics in Power Electronics*, vol. 8, no. 1, pp. 286–297, 2020.

Advanced Modeling Technique of Class-E Inverter Considering Low R_{on} of eGaN FETs and Different Design Procedures

Manas Palmal
Dept. of Electrical and Computer Engineering
University of Washington
Seattle, USA
manas613@uw.edu

Jungwon Choi
Dept. of Electrical and Computer Engineering
University of Washington
Seattle, USA
jungchoi@uw.edu

Abstract—**Class-E inverters have been used extensively in high-frequency applications with eGaN (enhancement-mode Gallium Nitride) devices due to their high switching speed and low on-state resistance (in the order of a few $m\Omega$). However, the developed generalized averaging model to control the class-E inverter in prior literature works only for a class-E inverter with a high-Q resonant output network and high on-state resistance ($R_{on} \geq 0.5\Omega$) switches due to the presence of R_{on} in the denominator in their state equations. Therefore, this paper first eliminates R_{on} from the denominator of the state equations by incorporating $u(t)$ appropriately in the state equations, where $u(t)$ is the complementary function of the gate drive signal. Subsequently, leveraging the developed state equations, we propose a model suitable for a class-E inverter with a high-Q resonant output network. Lastly, due to the high distortion of the output current and the voltage across the eGaN FET, we develop another model for the class-E inverter with a DC-blocking output capacitor by adding more harmonic components in state variables without increasing complexity. Our simulation results from PLECS and experimental results with the 5MHz eGaN FET-based class-E inverter are provided to validate the efficacy of the proposed models.**

Index Terms—**Class-E inverter, quality factor, soft-switching, generalized modeling.**

I. INTRODUCTION

Class-E inverters have been widely used in high-frequency (HF: 3-30 MHz) applications such as RF power amplifiers, wireless power transfer, plasma generation, and induction heating due to their topological simplicity and high efficiency [1], [2]. Although a class-E inverter is a single switch converter topology and operates under zero voltage switching (ZVS) and zero dv/dt switching conditions, it is a high-order nonlinear dynamic system with ubiquitous cross-coupling dependencies among the circuit elements.

Moreover, the state variables in a class-E inverter are either DC or AC or a combination of both DC and AC components. Consequently, the extended describing function approach [3], [4] or the discrete-time state-space modeling [5] fails to provide design insight, as both methods strongly rely on numerical solutions using a computer-aided program. In addition, discrete-time state-space modeling suffers from

convergence issues. Also, the state-space averaging technique requires the small-ripple approximation to be satisfied, which is not suitable for class-E inverter modeling. Therefore, the generalized averaging method (GAM) [6], also known as multifrequency averaging (MFA) [7], must be adopted to model a class-E inverter [3].

In [8], the authors developed a model for a class-E inverter with a high-Q resonant output network using the MFA approach. However, the model's accuracy is limited to cases where R_{on} of the device is greater than or equal to 0.5Ω. Therefore, the model cannot be used for the Class-E inverter with eGaN FETs as most of the eGaN FETs have much lower R_{on} (for example, an eGaN FET, GS66508T, from GaN systems, R_{on} is 50mΩ). Moreover, a class-E converter can be designed with either a resonant output network [9] or a DC-blocking output capacitor [10]. The infinite input inductance of a class-E converter with a resonant output network can be replaced by a finite one to extend the ZVS range and improve transient performance with load modulations [11]–[15]. However, previous studies did not address the necessity of employing different models to accurately predict the dynamic behavior and steady-state performance of class-E inverters, considering different design procedures.

Therefore, this paper methodically investigates the state equations to eliminate the R_{on} from the denominator of the state equations of [8] by incorporating the control variable $u(t)$ appropriately. Subsequently, this paper proposes a generalized averaging model suitable for a class-E inverter with a high-Q (≥ 5) resonant output network by selecting suitable harmonic components in the state variables. Afterward, another model suitable for a class-E inverter with a DC-blocking output capacitor is proposed to capture the high distortion in the output current and the voltage across the eGaN FET. Finally, extensive simulation and experimental results are provided to validate the efficacy of the proposed model.

II. SWITCHING MODEL OF A CLASS-E INVERTER

A class-E inverter with four state variables corresponding to four energy storage elements is depicted in Fig. 1. In the

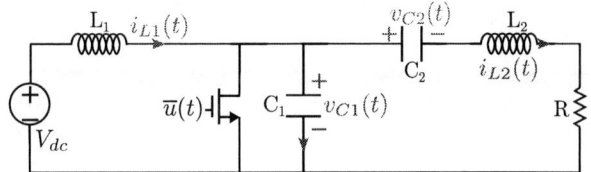

Fig. 1. Class-E inverter with four independent state variables.

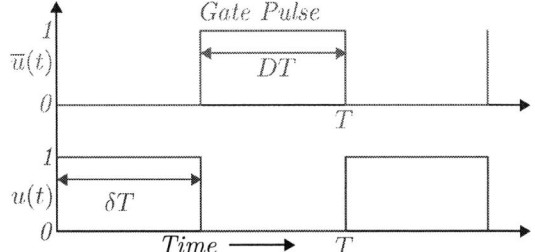

Fig. 2. Gate pulse and corresponding control variable.

previous study [8], the state equation for v_{C1} was written as given in (1) for the circuit as presented in Fig. 1, where $\overline{u}(t)$ is the gate signal, as shown in Fig. 2.

$$\frac{dv_{C1}(t)}{dt} = \frac{1}{C_1} \left\{ i_{L1}(t) - i_{L2}(t) - \frac{\overline{u}(t) \cdot v_{C1}(t)}{R_{on}} \right\}. \quad (1)$$

To eliminate R_{on} from the denominator for using eGaN FETs, we use an ideal switching device with $R_{on} = 0$. Thus, when the eGaN FET is on, the current difference, $(i_{L1} - i_{L2})$, flows through the eGaN FET. When the eGaN FET is off, the current difference, $(i_{L1} - i_{L2})$, flows through the capacitor, C_1, connected across the device. Therefore, (1) is rewritten as presented in (2) for an ideal switching device. Similarly, to describe the exact time-domain behavior of a class-E inverter when the converter is operating under ZVS switching conditions, the complete set of the state equations are updated as given in (2)-(5).

$$C_1 \frac{dv_{C1}(t)}{dt} = u(t) \cdot \{i_{L1}(t) - i_{L2}(t)\}, \quad (2)$$

$$L_1 \frac{di_{L1}(t)}{dt} = V_{dc} - u(t) \cdot v_{C1}(t), \quad (3)$$

$$L_2 \frac{di_{L2}(t)}{dt} = u(t) \cdot v_{C1}(t) - v_{C2}(t) - i_{L2}(t) \cdot R, \quad (4)$$

$$C_2 \frac{dv_{C2}(t)}{dt} = i_{L2}(t), \quad (5)$$

where $u(t) = 0$ when the eGaN FET is on, and $u(t) = 1$ when the eGaN FET is off. The actual gate signal is complementary of $u(t)$, as indicated in Fig. 2. The control variable, $u(t)$, must be used in (3) and (4) to avoid convergence issues that have not been reported in [4] despite using ideal devices for multi-resonant converters. Since the state variables of a class-E inverter have multiple harmonics, i.e. the state variables in a class-E inverter are either DC or AC or the combination of both DC and AC components, we leverage the generalized averaging approach, to model the class-E inverters with either

a resonant output network or a DC-blocking output capacitor. Therefore, a brief review of the generalized averaging approach is discussed in the next section.

III. REVIEW OF GENERALIZED AVERAGING APPROACH

Though the details of the generalized averaging approach are found in [6]–[8], [16], for the completeness of the paper, we summarize the generalized averaging approach. In the generalized averaging approach, any state variable $x(\tau)$ is represented by the complex Fourier series over the switching period interval ($\tau \in [t - T, t]$), as given below:

$$x(\tau) = \sum_{k=-\infty}^{\infty} \langle x \rangle_k e^{jk\omega\tau}, \quad (6)$$

$$\langle x \rangle_k = \frac{1}{T} \int_{t-T}^{t} x(\tau) e^{-jk\omega\tau} d\tau, \quad (7)$$

where $\langle x \rangle_k$ are the complex Fourier coefficients and are calculated using (7). The kth coefficient of the complex Fourier series is also known as *index-k* average. In other words, *index-0, 1,* and *2* refer to the average value, fundamental component, and second harmonic component, respectively.

According to [6], [7], the time derivative of the kth coefficients of complex Fourier series for a state variable x is computed using the Leibniz integral rule:

$$\frac{d\langle x \rangle_k(t)}{dt} = \left\langle \frac{dx}{dt} \right\rangle_k (t) - jk\omega\langle x \rangle_k(t). \quad (8)$$

For a circuit consisting of linear elements such as resistors, inductors, and capacitors, the time derivative of the kth coefficients can be written as follows:

$$\langle v \rangle_k = R\langle i \rangle_k, \quad (9)$$

$$\langle v \rangle_k = L \frac{d\langle i \rangle_k}{dt} + jk\omega L\langle i \rangle_k, \quad (10)$$

$$\langle i \rangle_k = C \frac{d\langle v \rangle_k}{dt} + jk\omega C\langle v \rangle_k. \quad (11)$$

The product of two variables x and u is computed by discrete convolution, as given in (12).

$$\langle ux \rangle_k = \sum_{i=-\infty}^{\infty} \langle u \rangle_{k-i}\langle x \rangle_i. \quad (12)$$

As an example, assume that the control variable, u, and the state variable, x, are approximated by *index-0, 1, 2,* and *3,* and *index-0, 1,* and *2* respectively. Therefore, *index-1* of the product of two variables, $\langle ux \rangle_1$, produces the following results:

$$\langle ux \rangle_1 = \langle u \rangle_3 \langle x \rangle_{-2} + \langle u \rangle_2 \langle x \rangle_{-1} + \langle u \rangle_1 \langle x \rangle_0 \\ + \langle u \rangle_0 \langle x \rangle_1 + \langle u \rangle_{-1} \langle x \rangle_2. \quad (13)$$

It must be noted that in (13) all components, except the average values, are the complex quantities. Moreover, positive and negative indexes are complex conjugates in nature. Thus, mathematically, we can write them down as follows:

$$\langle x \rangle_k = \langle x \rangle_k^R + j\langle x \rangle_k^I = \langle x \rangle_{-k}^* = (\langle x \rangle_{-k}^R + j\langle x \rangle_{-k}^I)^*, \quad (14)$$

$$\langle u \rangle_k = \langle u \rangle_k^R + j\langle u \rangle_k^I = \langle u \rangle_{-k}^* = (\langle u \rangle_{-k}^R + j\langle u \rangle_{-k}^I)^*, \quad (15)$$

where the superscripts R, I, and '∗' imply the real part, imaginary part, and complex conjugate of a complex quantity respectively. Substituting (14) & (15) into (13), the final relationships are obtained as presented in (16) & (17).

$$\langle ux \rangle_1^R = \langle u \rangle_1^R \langle x \rangle_0 + (\langle u \rangle_0 + \langle u \rangle_2^R) \langle x \rangle_1^R + \langle u \rangle_2^I \langle x \rangle_1^I \\ + (\langle u \rangle_1^R + \langle u \rangle_3^R) \langle x \rangle_2^R + (\langle u \rangle_1^I + \langle u \rangle_3^I) \langle x \rangle_2^I, \quad (16)$$

$$\langle ux \rangle_1^I = \langle u \rangle_1^I \langle x \rangle_0 + \langle u \rangle_2^I \langle x \rangle_1^R + (\langle u \rangle_0 - \langle u \rangle_2^R) \langle x \rangle_1^I \\ + (\langle u \rangle_3^I - \langle u \rangle_1^I) \langle x \rangle_2^R + (\langle u \rangle_1^R - \langle u \rangle_3^R) \langle x \rangle_2^I. \quad (17)$$

The key properties described in this section are utilized in the following sections to model the class-E inverter with different design conditions.

IV. MODELING OF CLASS-E INVERTER

As mentioned in the introduction, the circuit parameters of a class-E inverter, as depicted in Fig. 1, can be designed by two different procedures. The first approach [9] assumes that the output capacitor (C_2) and output inductor (L_2) are in resonance, whereas in the second approach [10], output capacitor (C_2) is a DC-blocking capacitor. Therefore, in the following subsections, we have modeled each of them.

A. Class-E Inverter with a High-Q Resonant Output Network

In a high-Q (≥ 5) design, the output current, i_{L2}, and the output capacitor voltage, v_{C2}, are presumably sinusoidal. Note that a class E inverter with a resonant output network with $Q < 5$ is not recommended due to distorted outputs. Therefore, we focus on a high-Q design in this subsection.

In a high-Q design, index-0, and 1 suffice to account for their sub-switching cycle variations for $i_{L2}(t)$, and $v_{C2}(t)$. In general, though a class-E inverter can be designed with an infinite input inductance, the infinite input inductance of a class-E converter with a resonant output network can be replaced by a finite one to extend the ZVS range and improve transient performance with load modulations. Therefore, index-0, and 1 must be employed for the input current, i_{L1}, to capture the ripple in i_{L1}. And, for the state-variable, $v_{C1}(t)$, as it is also seen from Fig. 4(c), index-0, 1, and 2 have to take into account for substantial accuracy [8]. Control variable $u(t)$ is assumed to be a pulse waveform (0 and 1), as shown in Fig. 2, and it must be approximated by the sum of index-0, 1, 2, and 3 to ensure adequate accuracy for any duty ratio. These results are summarized in (18).

$$v_{C2}(t) \approx \langle v_{C2} \rangle_0 + \langle v_{C2} \rangle_1 e^{j\omega t} + \langle v_{C2} \rangle_{-1} e^{-j\omega t}, \quad (18a)$$

$$i_{L2}(t) \approx \langle i_{L2} \rangle_0 + \langle i_{L2} \rangle_1 e^{j\omega t} + \langle i_{L2} \rangle_{-1} e^{-j\omega t}, \quad (18b)$$

$$i_{L1}(t) \approx \langle i_{L1} \rangle_0 + \langle i_{L1} \rangle_1 e^{j\omega t} + \langle i_{L1} \rangle_{-1} e^{-j\omega t}, \quad (18c)$$

$$v_{C1}(t) \approx \langle v_{C1} \rangle_0 + \langle v_{C1} \rangle_1 e^{j\omega t} + \langle v_{C1} \rangle_{-1} e^{-j\omega t} \\ + \langle v_{C1} \rangle_2 e^{j2\omega t} + \langle v_{C1} \rangle_{-2} e^{-j2\omega t}, \quad (18d)$$

$$u(t) \approx \langle u \rangle_0 + \langle u \rangle_1 e^{j\omega t} + \langle u \rangle_{-1} e^{-j\omega t} + \langle u \rangle_2 e^{j2\omega t} \\ + \langle u \rangle_{-2} e^{-j2\omega t} + \langle u \rangle_3 e^{j3\omega t} + \langle u \rangle_{-3} e^{-j3\omega t}. \quad (18e)$$

Substituting the relationships (18) into (2)-(5) and following the steps, as described in section III, we obtained fourteen

nonlinear large-signal equations. The complete set of equations is included in the appendix. But, some of the key equations have been listed below.

$$L_1 \frac{d\langle i_{L1} \rangle_0}{dt} = V_{dc} - \langle u \rangle_0 \langle v_{C1} \rangle_0 - 2 \left[\langle u \rangle_2^R \langle v_{C1} \rangle_2^R \\ + \langle u \rangle_2^I \langle v_{C1} \rangle_2^I + \langle u \rangle_1^R \langle v_{C1} \rangle_1^R + \langle u \rangle_1^I \langle v_{C1} \rangle_1^I \right], \quad (19)$$

$$L_2 \frac{d\langle i_{L2} \rangle_1^R}{dt} = \omega L_2 \langle i_{L2} \rangle_1^I - R \cdot \langle i_{L2} \rangle_1^R + \langle u \rangle_1^R \langle v_{C1} \rangle_0 \\ + (\langle u \rangle_0 + \langle u \rangle_2^R) \langle v_{C1} \rangle_1^R + \langle u \rangle_2^I \langle v_{C1} \rangle_1^I - \langle v_{C2} \rangle_1^R \\ + (\langle u \rangle_3^R + \langle u \rangle_1^R) \langle v_{C1} \rangle_2^R + (\langle u \rangle_3^I + \langle u \rangle_1^I) \langle v_{C1} \rangle_2^I, \quad (20)$$

$$L_2 \frac{d\langle i_{L2} \rangle_1^I}{dt} = -\omega L_2 \langle i_{L2} \rangle_1^R - R \cdot \langle i_{L2} \rangle_1^I + \langle u \rangle_1^I \langle v_{C1} \rangle_0 \\ + \langle u \rangle_2^I \langle v_{C1} \rangle_1^R + (\langle u \rangle_0 - \langle u \rangle_2^R) \langle v_{C1} \rangle_1^I - \langle v_{C2} \rangle_1^I \\ + (\langle u \rangle_3^I - \langle u \rangle_1^I) \langle v_{C1} \rangle_2^R + (\langle u \rangle_1^R - \langle u \rangle_3^R) \langle v_{C1} \rangle_2^I, \quad (21)$$

$$C_2 \frac{d\langle v_{C2} \rangle_0}{dt} = \langle i_{L2} \rangle_0. \quad (22)$$

It must be pointed out at this stage that though the output current, i_{L2}, does not have an average value in steady-state, index-0 i.e. $\langle i_{L2} \rangle_0$ must be considered to be non-zero in the dynamic state, as (22) indicates, to model the average voltage dynamics of the output capacitor.

B. Class-E Inverter with a DC-Blocking Output Capacitor

The output current is highly distorted in a class-E inverter with a DC-blocking output capacitor. Therefore, index-0, 1 & 2 are considered for the output current, whereas index-0, 1, 2, & 3 are taken into consideration for the voltage across the switching device and the control variable, $u(t)$. Only the average component, i.e., index-0, is sufficient for input current and output capacitor voltage because L_1 is an infinite RF choke, and C_2 is used for blocking purposes. Following the procedure as described in the section III, fourteen nonlinear large-signal equations are obtained. Some of the key equations have been listed in (23)-(26), and the remaining nonlinear large-signal equations are included in the appendix.

$$L_2 \frac{d\langle i_{L2} \rangle_1^R}{dt} = \omega L_2 \langle i_{L2} \rangle_1^I - R \cdot \langle i_{L2} \rangle_1^R + \langle u \rangle_1^R \langle v_{C1} \rangle_0 + \\ (\langle u \rangle_0 + \langle u \rangle_2^R) \langle v_{C1} \rangle_1^R + \langle u \rangle_2^I \langle v_{C1} \rangle_1^I + (\langle u \rangle_1^R + \langle u \rangle_3^R) \\ \langle v_{C1} \rangle_2^R + (\langle u \rangle_1^I + \langle u \rangle_3^I) \langle v_{C1} \rangle_2^I + \langle u \rangle_2^R \langle v_{C1} \rangle_3^R + \langle u \rangle_2^I \langle v_{C1} \rangle_3^I, \quad (23)$$

$$L_2 \frac{d\langle i_{L2} \rangle_1^I}{dt} = -\omega L_2 \langle i_{L2} \rangle_1^R - R \cdot \langle i_{L2} \rangle_1^I + \langle u \rangle_1^I \langle v_{C1} \rangle_0 + \\ \langle u \rangle_2^I \langle v_{C1} \rangle_1^R + (\langle u \rangle_0 - \langle u \rangle_2^R) \langle v_{C1} \rangle_1^I + (\langle u \rangle_3^I - \langle u \rangle_1^I) \langle v_{C1} \rangle_2^R \\ + (\langle u \rangle_1^R - \langle u \rangle_3^R) \langle v_{C1} \rangle_2^I - \langle u \rangle_2^I \langle v_{C1} \rangle_3^R + \langle u \rangle_2^R \langle v_{C1} \rangle_3^I, \quad (24)$$

$$L_2 \frac{d\langle i_{L2} \rangle_2^R}{dt} = 2\omega L_2 \langle i_{L2} \rangle_2^I - R \cdot \langle i_{L2} \rangle_2^R + \langle u \rangle_2^R \langle v_{C1} \rangle_0 \\ + (\langle u \rangle_3^R + \langle u \rangle_1^R) \langle v_{C1} \rangle_1^R + (\langle u \rangle_3^I - \langle u \rangle_1^I) \langle v_{C1} \rangle_1^I \\ + \langle u \rangle_0 \langle v_{C1} \rangle_2^R + \langle u \rangle_1^R \langle v_{C1} \rangle_3^R + \langle u \rangle_1^I \langle v_{C1} \rangle_3^I, \quad (25)$$

979-8-3315-1612-3/25 $31.00 © 2025 IEEE

$$L_2 \frac{d\langle i_{L2}\rangle_2^I}{dt} = -2\omega L_2 \langle i_{L2}\rangle_2^R - R \cdot \langle i_{L2}\rangle_2^I + \langle u\rangle_2^I \langle v_{C1}\rangle_0 +$$
$$\left(\langle u\rangle_1^I + \langle u\rangle_3^I\right)\langle v_{C1}\rangle_1^R + \left(\langle u\rangle_1^R - \langle u\rangle_3^R\right)\langle v_{C1}\rangle_1^I + \langle u\rangle_0 \langle v_{C1}\rangle_2^I$$
$$- \langle u\rangle_1^I \langle v_{C1}\rangle_3^R + \langle u\rangle_1^R \langle v_{C1}\rangle_3^I. \tag{26}$$

The complex Fourier coefficients for the control variable, $u(t)$, for the given Fig. 2, are calculated using (7) and they are summarized in (27).

$$\langle u\rangle_0 = \delta, \tag{27a}$$

$$\langle u\rangle_k^R = \frac{1}{2k\pi}\sin\left(2\pi k\delta\right), \tag{27b}$$

$$\langle u\rangle_k^I = \frac{1}{2k\pi}\left[\cos\left(2\pi k\delta\right) - 1\right]. \tag{27c}$$

V. SIMULATION AND EXPERIMENTAL RESULTS

To validate the proposed generalized averaging models, we followed the design procedures described in [9], [10] to design two infinite RF choke class-E inverters, one with a high-Q resonant output network and another one with a DC-blocking output capacitor. For the second design, the selection of the blocking capacitor needs some optimization because it has to block the DC voltage while offering minimal impedance compared to the output inductance. However, selecting a large capacitor can degrade the transient performance [17]. Subsequently, a circuit model as depicted in Fig. 1, a switch model as described by (2)-(5), and the proposed generalized averaging model, as discussed in section IV, have been implemented in PLECS using the designed parameter presented in Table. I. State variables are reconstructed from the complex Fourier coefficients ($\langle x\rangle_k^R$ & $\langle x\rangle_k^I$) using (28).

$$x(t) = \langle x\rangle_0 + 2\sum_{k=1}^{\infty}\langle x\rangle_k^R \cos k\omega t - \langle x\rangle_k^I \sin k\omega t. \tag{28}$$

After simulations, hardware prototypes for each design have been developed in the laboratory. An eGaN FET, GS66508T,

TABLE I
COMPONENT AND PARAMETER VALUES FOR DESIGNED CLASS-E INVERTER

Parameter	Values	Values
Input Voltage, V_{dc}	100Volts	
Operating Frequency	5MHz	
Quality factor Q	2.8658	5
Load Resistance R	25Ω	25Ω
Output Inductance L_2	1.4228μH	3.9789μH
Output Capacitance C_2	0.2μF	378.48pF
Ext Device Capacitance C_1	277[†] pF	233.75[‡] pF
Input RF choke L_1	30μH	

[†] 147 pF external capacitance is connected in hardware.

[‡] 100 pF external capacitance is connected in hardware.

from GaN systems, with $R_{on} = 50m\Omega$ is used as a switching device, and a gate driver IC, LM5114BMF/NOPB, from Texas Instruments is used to drive the eGaN FET at 5MHz switching frequency. An external gate resistance of 2.15Ω is also added to the hardware circuitry. It must be noted that the drain-to-source capacitance of an eGaN FET is voltage-dependent and nonlinear. However, for simplicity, we assume that the drain-to-source capacitance is linear and the value is approximated to $C_{oss} \approx 130pF$. Therefore, an external capacitance (C0G) is also soldered across the eGaN FET to match the values as presented in the Table. I.

The simulation and experimental results for the dynamic response of the voltage across the eGaN FET, v_{C1}, and input current, i_{L1}, for a class-E inverter with a high-Q (≥ 5) resonant output network are shown in Fig. 3(a) and Fig. 3(b) respectively. The generalized averaging model of a class-E inverter with a high-Q resonant output network predicts that the average input current is 1.41 Amp, which is very close to the actual current of 1.57 Amp. Since output capacitance, C_2, and output inductance, L_2, are in resonance, the voltage across the output capacitance, v_{C2}, is sinusoid. Consequently, the generalized model accurately predicts the voltage across the output capacitance, as shown in Fig. 3(c).

The simulation and experimental results for the steady-state voltage across the eGaN FET, v_{C1}, and output current, i_{L2}, are shown in Fig. 4(a) and Fig. 4(b) respectively. The peak value of the voltage across the eGaN FET from the circuit model, switching model, and experimental setup is 380 Volts, whereas the generalized averaging model predicts 340 Volts. Fourier spectrum for v_{C1} and i_{L2} is given in Fig. 4(c). Upon examination of the Fourier spectrum, it is clear that though the proposed GAM cannot precisely predict the high-frequency voltage waveform across the switching device with limited harmonic considerations of the control variable and state variables, the input current, as shown in Fig. 3(a), does not deviate from its actual value unlike the GAM previously developed in [8]. Including additional harmonics can provide a more accurate representation of the high-frequency voltage waveform across the switching device, but at the cost of increasing the complexity of the model.

Similarly, simulation and experiment have been conducted for a class-E converter with a DC-blocking output capacitor. Dynamic response of the voltage across the eGaN FET and input current from the PLECS simulation and experiment are shown in Fig. 5(a) and Fig. 5(b). The generalized averaging model provides a good approximation for the average input current with 1.3 Amp for the experimental value with 1.52 Amp. As expected, the output capacitance, C_2, blocks 100 Volts, also shown in Fig. 5(c). As shown in Fig. 6(a), and Fig. 6(b), the steady-state waveform of the voltage across the eGaN FET, v_{C1}, and output current, i_{L2}, from the generalized averaging model match with the results from the circuit model, switching model, and experimental setup. The peak values of v_{C1} and i_{L2} are 390 Volts and 4 Amp respectively. Based on the Fourier spectrum as depicted in Fig. 6(c), it can be observed that the generalized averaging model provided a

979-8-3315-1612-3/25 $31.00 © 2025 IEEE

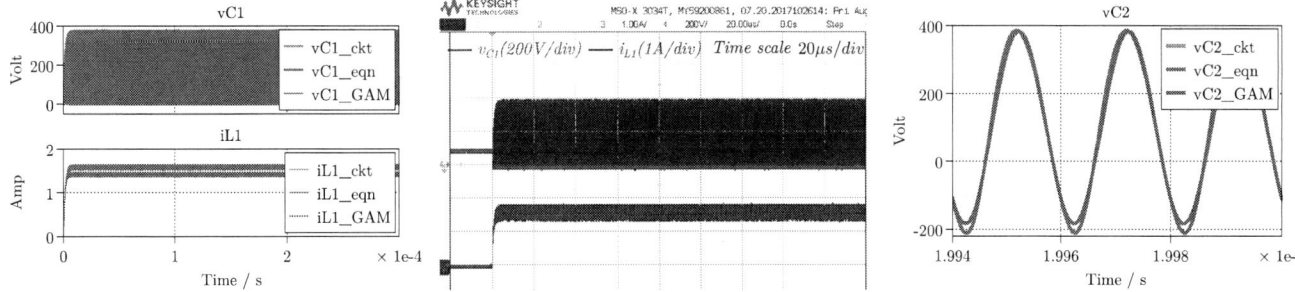

(a) Simulation results for the dynamic response of voltage across the eGaN FET and input current.

(b) Experimental results for the dynamic response of voltage across the eGaN FET and input current.

(c) Simulation results for the steady-state response of voltage across the output capacitor, v_{C2}.

Fig. 3. Simulation and experimental results of v_{C1}, i_{L1}, and v_{C2} in a class-E inverter with a high-Q (≥ 5) resonant output network. Subscripts _ckt, _eqn, and _GAM indicate that the corresponding waveform is obtained from the circuit model, as depicted in Fig. 1, switching model, as described by (2)-(5), and proposed generalized averaging model, as discussed in section IV, respectively.

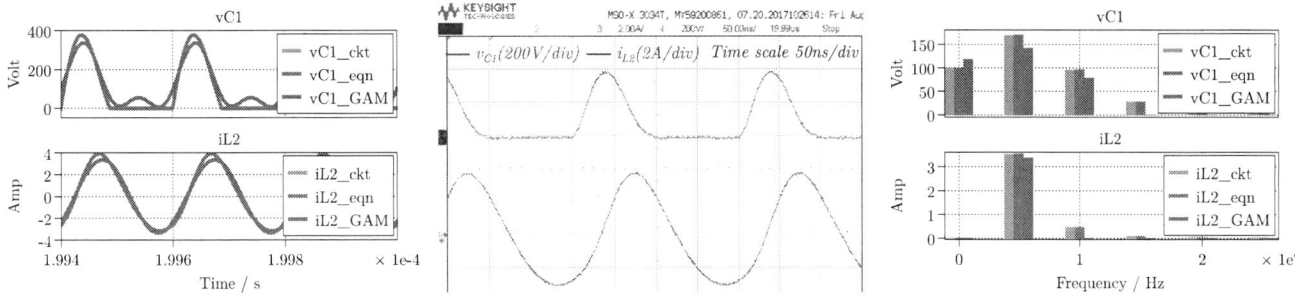

(a) Simulation results for the steady-state voltage across the eGaN FET and output current.

(b) Experimental waveform for the steady-state voltage across the eGaN FET and output current.

(c) Fourier spectrum of the steady-state voltage across the eGaN FET and output current.

Fig. 4. Simulation and experimental results of voltage across the eGaN FET and output current in a class-E inverter with a high-Q resonant output network. Subscripts _ckt, _eqn, and _GAM indicate that the corresponding waveform is obtained from the circuit model, as depicted in Fig. 1, switching model, as described by (2)-(5), and proposed generalized averaging model, as discussed in section IV, respectively.

substantial accuracy for v_{C1}. Despite a close match of the experimental and simulation results, a minor discrepancy is observed in the experimental results owing to the non-linearity in the device output capacitance and ideal device consideration in PLECS simulation. In addition to that probe capacitance and parasitics of the PCB are also ignored in the simulation.

VI. CONCLUSION

This paper presents the state equations of a class-E inverter to discard R_{on} from the state equations of [8]. Afterward, leveraging the developed state equations, two generalized averaging models are proposed for the class-E inverter, considering different design procedures. The first approach yields a fourteen-order model that can readily predict the transient and steady-state performance for a class-E inverter with a high-Q resonant output network. Another model is developed for a class-E inverter with a DC-blocking output capacitor. This model also yields a fourteen-order model that can predict the output current and voltage across the eGaN FET more accurately by considering more harmonic components for the output current and voltage across the eGaN FET. The developed models in this paper are backed up with extensive simulation and experimental results.

APPENDIX

In Section IV, we develop the generalized averaging model for the class-E inverter with a high-Q resonant output network and with a DC blocking output capacitor. Though some key equations are already presented in Section IV, the remaining nonlinear large-signal equations are presented here.

A. Nonlinear Large-signal Equations of Class E Inverter with a High-Q Resonant Output Network

As mentioned, a class-E inverter with a high-Q resonant output network can be designed either with infinite or finite input inductance. In particular, for a finite input inductance design, the input current contains the DC component and the switching frequency ripple component. The nonlinear large-signal equations of the real and imaginary parts of the input current are given in (29) and (30).

$$
\begin{aligned}
L_1 \frac{d\langle i_{L1}\rangle_1^R}{dt} &= \omega L_1 \langle i_{L1}\rangle_1^I - \langle u\rangle_1^R \langle v_{C1}\rangle_0 - \left(\langle u\rangle_0 + \langle u\rangle_2^R\right) \\
&\quad \langle v_{C1}\rangle_1^R - \langle u\rangle_2^I \langle v_{C1}\rangle_1^I - \left(\langle u\rangle_3^R + \langle u\rangle_1^R\right) \langle v_{C1}\rangle_2^R \\
&\quad - \left(\langle u\rangle_3^I + \langle u\rangle_1^I\right) \langle v_{C1}\rangle_2^I,
\end{aligned}
$$
(29)

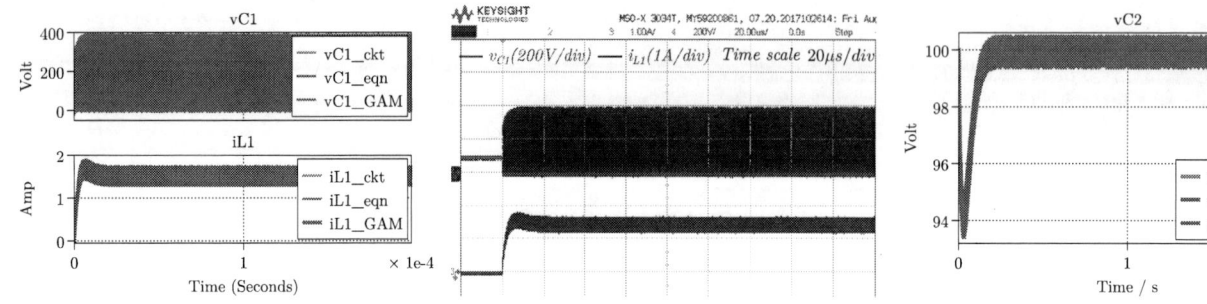

(a) Simulation results for the dynamic response of voltage across the eGaN FET and input current.
(b) Experimental results for the dynamic response of voltage across the eGaN FET and input current.
(c) Simulation results of voltage across the output capacitor, v_{C2}.

Fig. 5. Simulation and experimental results of v_{C1}, i_{L1}, and v_{C2} in a class-E inverter with a DC-blocking output capacitor. Subscripts _ckt, _eqn, and _GAM indicate that the corresponding waveform is obtained from the circuit model, as depicted in Fig. 1, switching model, as described by (2)-(5), and proposed generalized averaging model, as discussed in section IV, respectively.

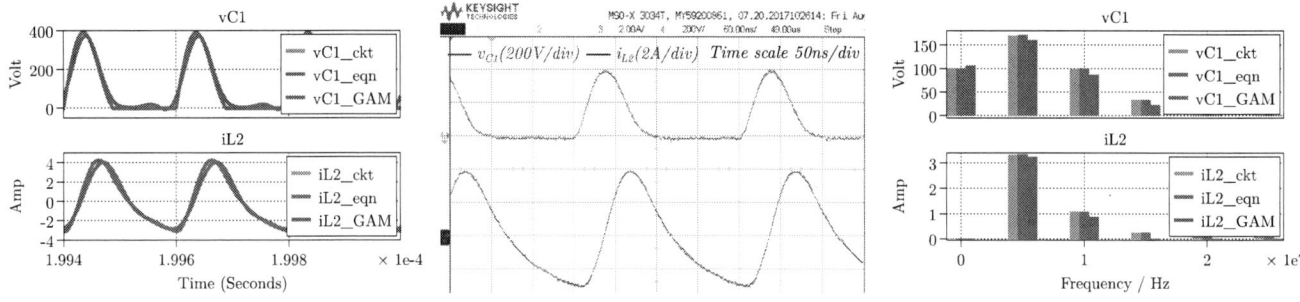

(a) Simulation results for the steady-state voltage across the eGaN FET and output current.
(b) Experimental waveform for the steady-state voltage across the eGaN FET and output current.
(c) Fourier spectrum of the steady-state voltage across the eGaN FET and output current.

Fig. 6. Simulation and experimental waveform of v_{C1}, i_{L2} and corresponding Fourier spectrum in a class-E inverter with a DC-blocking output capacitor. Subscripts _ckt, _eqn, and _GAM indicate that the corresponding waveform is obtained from the circuit model, as depicted in Fig. 1, switching model, as described by (2)-(5), and proposed generalized averaging model, as discussed in section IV, respectively.

$$
\begin{aligned}
L_1 \frac{d\langle i_{L1}\rangle_1^I}{dt} &= -\omega L_1 \langle i_{L1}\rangle_1^R - \langle u\rangle_1^I \langle v_{C1}\rangle_0 - \langle u\rangle_2^I \langle v_{C1}\rangle_1^R \\
&+ \left(\langle u\rangle_2^R - \langle u\rangle_0\right)\langle v_{C1}\rangle_1^I + \left(\langle u\rangle_1^I - \langle u\rangle_3^I\right)\langle v_{C1}\rangle_2^R \\
&+ \left(\langle u\rangle_3^R - \langle u\rangle_1^R\right)\langle v_{C1}\rangle_2^I.
\end{aligned} \tag{30}
$$

To model the average voltage dynamics of the output capacitor, the *index*-0 of the output current, $\langle i_{L2}\rangle_0$ is taken into account, and the corresponding nonlinear large-signal equation is presented in (31).

$$
\begin{aligned}
L_2 \frac{d\langle i_{L2}\rangle_0}{dt} &= \langle u\rangle_0 \langle v_{C1}\rangle_0 - R\cdot\langle i_{L2}\rangle_0 + 2\left[\langle u\rangle_2^R \langle v_{C1}\rangle_2^R + \right. \\
&\left. \langle u\rangle_2^I \langle v_{C1}\rangle_2^I + \langle u\rangle_1^R \langle v_{C1}\rangle_1^R + \langle u\rangle_1^I \langle v_{C1}\rangle_1^I\right] - \langle v_{C2}\rangle_0.
\end{aligned} \tag{31}
$$

The voltage across the eGaN FET, v_{C1}, contains higher harmonics in addition to the average component. In a class-E inverter with a high-Q resonant output network, *index*-0, 1, and 2 is considered for v_{C1}, and corresponding nonlinear large-signal relations are given in (32)-(36).

$$
\begin{aligned}
C_1 \frac{d\langle v_{C1}\rangle_0}{dt} &= \langle u\rangle_0 \langle i_{L1}\rangle_0 + 2\left[\langle u\rangle_1^R \langle i_{L1}\rangle_1^R + \langle u\rangle_1^I \langle i_{L1}\rangle_1^I\right] \\
&- \langle u\rangle_0 \langle i_{L2}\rangle_0 - 2\left[\langle u\rangle_1^R \langle i_{L2}\rangle_1^R + \langle u\rangle_1^I \langle i_{L2}\rangle_1^I\right],
\end{aligned} \tag{32}
$$

$$
\begin{aligned}
C_1 \frac{d\langle v_{C1}\rangle_1^R}{dt} &= \omega C_1 \langle v_{C1}\rangle_1^I + \langle u\rangle_0(\langle i_{L1}\rangle_1^R - \langle i_{L2}\rangle_1^R) \\
&+ \langle u\rangle_1^R(\langle i_{L1}\rangle_0 - \langle i_{L2}\rangle_0) + \langle u\rangle_2^R(\langle i_{L1}\rangle_1^R - \langle i_{L2}\rangle_1^R) \\
&+ \langle u\rangle_2^I(\langle i_{L1}\rangle_1^I - \langle i_{L2}\rangle_1^I),
\end{aligned} \tag{33}
$$

$$
\begin{aligned}
C_1 \frac{d\langle v_{C1}\rangle_1^I}{dt} &= -\omega C_1 \langle v_{C1}\rangle_1^R + \langle u\rangle_0(\langle i_{L1}\rangle_1^I - \langle i_{L2}\rangle_1^I) \\
&+ \langle u\rangle_1^I(\langle i_{L1}\rangle_0 - \langle i_{L2}\rangle_0) + \langle u\rangle_2^I(\langle i_{L1}\rangle_1^R - \langle i_{L2}\rangle_1^R) \\
&+ \langle u\rangle_2^R(\langle i_{L1}\rangle_1^I - \langle i_{L2}\rangle_1^I),
\end{aligned} \tag{34}
$$

$$
\begin{aligned}
C_1 \frac{d\langle v_{C1}\rangle_2^R}{dt} &= 2\omega C_1 \langle v_{C1}\rangle_2^I + \langle u\rangle_1^R(\langle i_{L1}\rangle_1^R - \langle i_{L2}\rangle_1^R) \\
&- \langle u\rangle_1^I(\langle i_{L1}\rangle_1^I - \langle i_{L2}\rangle_1^I) + \langle u\rangle_2^R(\langle i_{L1}\rangle_0 - \langle i_{L2}\rangle_0) \\
&+ \langle u\rangle_3^R(\langle i_{L1}\rangle_1^R - \langle i_{L2}\rangle_1^R) + \langle u\rangle_3^I(\langle i_{L1}\rangle_1^I - \langle i_{L2}\rangle_1^I),
\end{aligned} \tag{35}
$$

$$
\begin{aligned}
C_1 \frac{d\langle v_{C1}\rangle_2^I}{dt} &= -2\omega C_1 \langle v_{C1}\rangle_2^R + \langle u\rangle_1^I(\langle i_{L1}\rangle_1^R - \langle i_{L2}\rangle_1^R) \\
&+ \langle u\rangle_1^R(\langle i_{L1}\rangle_1^I - \langle i_{L2}\rangle_1^I) + \langle u\rangle_2^I(\langle i_{L1}\rangle_0 - \langle i_{L2}\rangle_0) \\
&+ \langle u\rangle_3^I(\langle i_{L1}\rangle_1^R - \langle i_{L2}\rangle_1^R) - \langle u\rangle_3^R(\langle i_{L1}\rangle_1^I - \langle i_{L2}\rangle_1^I).
\end{aligned} \tag{36}
$$

979-8-3315-1612-3/25 $31.00 © 2025 IEEE

The nonlinear large-signal equations of the real and imaginary parts of the output capacitor voltage are given in (37) and (38).

$$C_2 \frac{d\langle v_{C2}\rangle_1^R}{dt} = \langle i_{L2}\rangle_1^R + \omega_s C_2 \cdot \langle v_{C2}\rangle_1^I, \tag{37}$$

$$C_2 \frac{d\langle v_{C2}\rangle_1^I}{dt} = \langle i_{L2}\rangle_1^I - \omega_s C_2 \cdot \langle v_{C2}\rangle_1^R. \tag{38}$$

B. Nonlinear Large-signal Equations of Class-E Inverter with a DC-Blocking Output Capacitor

A class-E inverter with a DC-blocking capacitor is mostly designed with an input infinite RF choke. Consequently, the input inductor current is a constant DC current with negligible ripple component. The corresponding nonlinear large-signal equation of the input current is presented in (39).

$$L_1 \frac{d\langle i_{L1}\rangle_0}{dt} = V_{dc} - \langle u\rangle_0 \langle v_{C1}\rangle_0 - 2\left[\langle u\rangle_3^R \langle v_{C1}\rangle_3^R \right.$$
$$+ \langle u\rangle_3^I \langle v_{C1}\rangle_3^I + \langle u\rangle_2^R \langle v_{C1}\rangle_2^R + \langle u\rangle_2^I \langle v_{C1}\rangle_2^I \tag{39}$$
$$\left. + \langle u\rangle_1^R \langle v_{C1}\rangle_1^R + \langle u\rangle_1^I \langle v_{C1}\rangle_1^I\right].$$

As the output capacitor is used for DC blocking purposes only, it does not contain any switching components. The average voltage dynamic of the output capacitor is directly related to the non-zero average value of the output current as indicated in (40).

$$C_2 \frac{d\langle v_{C2}\rangle_0}{dt} = \langle i_{L2}\rangle_0. \tag{40}$$

The average output current in the steady state is zero. However, in the dynamic state, the average value of the output current is non-zero as presented in (41).

$$L_2 \frac{d\langle i_{L2}\rangle_0}{dt} = \langle u\rangle_0 \langle v_{C1}\rangle_0 - \langle v_{C2}\rangle_0 - R \cdot \langle i_{L2}\rangle_0$$
$$+ 2\left[\langle u\rangle_3^R \langle v_{C1}\rangle_3^R + \langle u\rangle_3^I \langle v_{C1}\rangle_3^I + \langle u\rangle_2^R \langle v_{C1}\rangle_2^R \right. \tag{41}$$
$$\left. + \langle u\rangle_2^I \langle v_{C1}\rangle_2^I + \langle u\rangle_1^R \langle v_{C1}\rangle_1^R + \langle u\rangle_1^I \langle v_{C1}\rangle_1^I\right].$$

The voltage across the switching device is highly distorted in a class-E inverter with a DC-blocking output capacitor, as shown in Fig. 6(c). The nonlinear large-signal equations for the voltage across the switching device are presented in (42)-(48).

$$C_1 \frac{d\langle v_{C1}\rangle_0}{dt} = \langle u\rangle_0 \left(\langle i_{L1}\rangle_0 - \langle i_{L2}\rangle_0\right) - 2\left[\langle u\rangle_1^R \langle i_{L2}\rangle_1^R \right.$$
$$\left. + \langle u\rangle_1^I \langle i_{L2}\rangle_1^I + \langle u\rangle_2^R \langle i_{L2}\rangle_2^R + \langle u\rangle_2^I \langle i_{L2}\rangle_2^I\right], \tag{42}$$

$$C_1 \frac{d\langle v_{C1}\rangle_1^R}{dt} = \omega C_1 \langle v_{C1}\rangle_1^I + \langle u\rangle_1^R \left(\langle i_{L1}\rangle_0 - \langle i_{L2}\rangle_0\right)$$
$$- \left(\langle u\rangle_0 + \langle u\rangle_2^R\right) \langle i_{L2}\rangle_1^R - \langle u\rangle_2^I \langle i_{L2}\rangle_1^I \tag{43}$$
$$- \left(\langle u\rangle_3^R + \langle u\rangle_1^R\right) \langle i_{L2}\rangle_2^R - \left(\langle u\rangle_3^I + \langle u\rangle_1^I\right) \langle i_{L2}\rangle_2^I,$$

$$C_1 \frac{d\langle v_{C1}\rangle_1^I}{dt} = -\omega C_1 \langle v_{C1}\rangle_1^R + \left(\langle u\rangle_1^I - \langle u\rangle_1^I\right) \langle i_{L2}\rangle_0 -$$
$$\langle u\rangle_2^I \langle i_{L2}\rangle_1^R + \left(\langle u\rangle_2^R - \langle u\rangle_0\right) \langle i_{L2}\rangle_1^I + \left(\langle u\rangle_1^I - \langle u\rangle_3^I\right)$$
$$\langle i_{L2}\rangle_2^R + \left(\langle u\rangle_3^R - \langle u\rangle_1^R\right) \langle i_{L2}\rangle_2^I, \tag{44}$$

$$C_1 \frac{d\langle v_{C1}\rangle_2^R}{dt} = 2\omega C_1 \langle v_{C1}\rangle_2^I + \left(\langle u\rangle_2^R - \langle u\rangle_2^R\right) \langle i_{L2}\rangle_0$$
$$- \left(\langle u\rangle_1^R + \langle u\rangle_3^R\right) \langle i_{L2}\rangle_1^R + \left(\langle u\rangle_1^I - \langle u\rangle_3^I\right) \langle i_{L2}\rangle_1^I \tag{45}$$
$$- \langle u\rangle_0 \langle i_{L2}\rangle_2^R,$$

$$C_1 \frac{d\langle v_{C1}\rangle_2^I}{dt} = -2\omega C_1 \langle v_{C1}\rangle_2^R + \left(\langle u\rangle_2^I - \langle u\rangle_2^I\right) \langle i_{L2}\rangle_0$$
$$- \left(\langle u\rangle_1^I + \langle u\rangle_3^I\right) \langle i_{L2}\rangle_1^R + \left(\langle u\rangle_3^R - \langle u\rangle_1^R\right) \langle i_{L2}\rangle_1^I \tag{46}$$
$$- \langle u\rangle_0 \langle i_{L2}\rangle_2^I,$$

$$C_1 \frac{d\langle v_{C1}\rangle_3^R}{dt} = 3\omega C_1 \langle v_{C1}\rangle_3^I + \langle u\rangle_3^R \left(\langle i_{L1}\rangle_0 - \langle i_{L2}\rangle_0\right) -$$
$$\langle u\rangle_2^R \langle i_{L2}\rangle_1^R + \langle u\rangle_2^I \langle i_{L2}\rangle_1^I - \langle u\rangle_1^R \langle i_{L2}\rangle_2^R + \langle u\rangle_1^I \langle i_{L2}\rangle_2^I, \tag{47}$$

$$C_1 \frac{d\langle v_{C1}\rangle_3^I}{dt} = -3\omega C_1 \langle v_{C1}\rangle_3^R + \langle u\rangle_3^I \left(\langle i_{L1}\rangle_0 - \langle i_{L2}\rangle_0\right) -$$
$$\langle u\rangle_2^I \langle i_{L2}\rangle_1^R - \langle u\rangle_2^R \langle i_{L2}\rangle_1^I - \langle u\rangle_1^I \langle i_{L2}\rangle_2^R - \langle u\rangle_1^R \langle i_{L2}\rangle_2^I. \tag{48}$$

References

[1] M. Madsen, A. Knott, and M. A. E. Andersen, "Low power very high frequency switch-mode power supply with 50 v input and 5 v output," *IEEE Transactions on Power Electronics*, vol. 29, no. 12, pp. 6569–6580, 2014.

[2] K. Surakitbovorn and J. M. Rivas-Davila, "A simple method to combine the output power from multiple class-e power amplifiers," *IEEE Journal of Emerging and Selected Topics in Power Electronics*, vol. 10, no. 2, pp. 2245–2253, 2022.

[3] X. Yue, X. Wang, and F. Blaabjerg, "Review of small-signal modeling methods including frequency-coupling dynamics of power converters," *IEEE Transactions on Power Electronics*, vol. 34, no. 4, pp. 3313–3328, 2019.

[4] E. X.-Q. Yang, "Extended describing function method for small-signal modeling of resonant and multi-resonant converters," Ph.D. dissertation, Virginia Tech, 1994.

[5] J. A. Baxter and D. J. Costinett, "Converter analysis using discrete time state-space modeling," in *2019 20th Workshop on Control and Modeling for Power Electronics (COMPEL)*, 2019, pp. 1–8.

[6] S. Sanders, J. Noworolski, X. Liu, and G. Verghese, "Generalized averaging method for power conversion circuits," *IEEE Transactions on Power Electronics*, vol. 6, no. 2, pp. 251–259, 1991.

[7] V. Caliskan, O. Verghese, and A. Stankovic, "Multifrequency averaging of dc/dc converters," *IEEE Transactions on Power Electronics*, vol. 14, no. 1, pp. 124–133, 1999.

[8] C. Bernal, E. Oyarbide, P. Molina, and A. Mediano, "Multi-frequency model of a single switch zvs class E inverter," in *2010 IEEE International Symposium on Industrial Electronics*, 2010, pp. 939–944.

[9] N. Sokal and A. Sokal, "Class E-a new class of high-efficiency tuned single-ended switching power amplifiers," *IEEE Journal of Solid-State Circuits*, vol. 10, no. 3, pp. 168–176, 1975.

[10] M. Kazimierczuk, "Class E tuned power amplifier with nonsinusoidal output voltage," *IEEE Journal of Solid-State Circuits*, vol. 21, no. 4, pp. 575–581, 1986.

[11] L. Roslaniec, A. S. Jurkov, A. A. Bastami, and D. J. Perreault, "Design of single-switch inverters for variable resistance/load modulation operation," *IEEE Transactions on Power Electronics*, vol. 30, no. 6, pp. 3200–3214, 2015.

[12] M. Acar, A. J. Annema, and B. Nauta, "Generalized design equations for class-e power amplifiers with finite dc feed inductance," in *2006 European Microwave Conference*, 2006, pp. 1308–1311.

[13] L. Gu, G. Zulauf, Z. Zhang, S. Chakraborty, and J. Rivas-Davila, "Push–pull class ϕ_2 rf power amplifier," *IEEE Transactions on Power Electronics*, vol. 35, no. 10, pp. 10 515–10 531, 2020.

[14] R. Zulinski and J. Steadman, "Class e power amplifiers and frequency multipliers with finite dc-feed inductance," *IEEE Transactions on Circuits and Systems*, vol. 34, no. 9, pp. 1074–1087, 1987.

[15] S. Aldhaher, D. C. Yates, and P. D. Mitcheson, "Load-Independent Class E/EF Inverters and Rectifiers for MHz-Switching Applications," *IEEE Transactions on Power Electronics*, vol. 33, no. 10, pp. 8270–8287, 2018.

[16] H. Qin and J. W. Kimball, "Generalized average modeling of dual active bridge dc–dc converter," *IEEE Transactions on Power Electronics*, vol. 27, no. 4, pp. 2078–2084, 2012.

[17] J. M. Rivas, Y. Han, O. Leitermann, A. D. Sagneri, and D. J. Perreault, "A high-frequency resonant inverter topology with low-voltage stress," *IEEE Transactions on Power Electronics*, vol. 23, no. 4, pp. 1759–1771, 2008.

PiezoNet and Data-Driven Models for Time-Domain Characterization of Piezoelectric Resonators

Davit Grigoryan, Mian Liao, Haoran Li, Shukai Wang, Tanuj Sen, Matthew Tan and Minjie Chen
Princeton University, Princeton, NJ, United States
Email: {dg1210, minjie}@princeton.edu

Abstract—This paper presents a fully automated data acquisition platform and the resulting database – PiezoNet[1] – for data-driven time-domain characterization of piezoelectric resonators used in power electronics. The platform measures the voltage and the current over piezoelectric resonators across a wide range of excitation waveforms and ambient temperatures. The power stage dynamically adjusts to efficient operating points for best operation under different load conditions. This system is mechanically versatile, accommodating crystals of diverse materials and dimensions. The platform enables comprehensive and precise characterization by automating the data collection process, thereby providing extensive datasets essential for training data-driven models (e.g., neural networks) to predict operating points and nonlinear behaviors, and to quantify the sample-to-sample variation of piezoelectric resonators. A family of sequence-to-sequence neural network models were trained and tested to validate the feasibility of time-domain data-driven models for piezoelectric resonators.

Index Terms—piezoelectric resonator, hysteresis loop, machine learning, data acquisition, data-driven methods

I. INTRODUCTION

COMMON dc-dc power converters predominantly rely on magnetics for energy storage, but this approach limits efficiency and power density at smaller scales. Despite significant advancements in semiconductors that enhance power density, the overall system's performance is still constrained by passive components, particularly magnetics [1], [2]. Piezoelectric resonators (PRs) offer an attractive alternative by storing energy through mechanical compliance and inertia, which enhances efficiency and power density in miniaturized designs [3]. PR-based converters also enable planar form factors and integration potential, achieving high efficiency through soft charging, and zero-voltage switching (ZVS) [4], [5].

However, the electromechanical nature of PRs introduces nonlinearities that are influenced by geometry, temperature, switching frequency, packaging method, and power level. Databases for frequency-domain characteristics exist [6]. Characterizing PRs in time-domain and modeling PRs with data-driven models are the main focuses of this paper. The widely used Butterworth-Van Dyke model [7] inadequately captures the nonlinear characteristics of PRs, limiting its effectiveness in high-performance designs [8]–[10]. Neural networks offer a promising alternative to equation-based models, as they can more accurately model the complex nonlinear behaviors of hysteresis energy materials if a large-enough high-quality

[1]GitHub Repository: https://github.com/Davit-Grigoryan-dev/PiezoNet.git

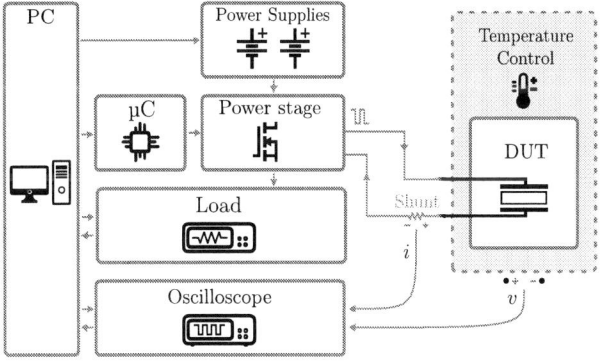

Fig. 1: The fully automated data acquisition system for PiezoNet, including a 3D-printed mechanical fixture, a power stage for actuation, and a floating oscilloscope with probes.

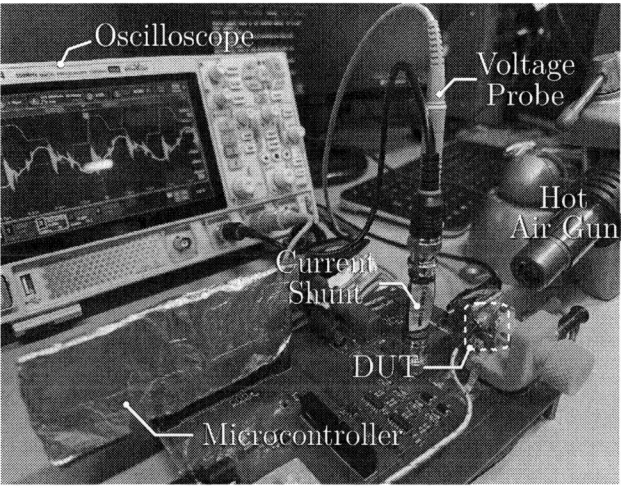

Fig. 2: The PiezoNet data acquisition system comprises driving and sensing circuits, temperature control, and a 3-D printed mechanical fixture (PETG material which is capable of handling 200 °C) which may influence the mechanical property but ensures uniform pressure across the PRs under test.

database is available [11], [12]. The Butterworth-Van Dyke model employs only 4 parameters to model a PR, whereas a neural network usually utilizes over 100 parameters. This significantly greater capacity enables neural networks to capture more intricate details and be well-suited for modeling of PRs.

Following the concept of developing a large-scale high-quality database for the study of complex power electronics materials, such as the MagNet database and the MagNet community [13], we envision a large-enough high-quality database for PRs can greatly enhance the understanding and development of PR-based power electronics, leveraging the recent advances in artificial intelligence and machine learning. The data acquisition system needs to be highly automated to enable rapid data collection and data quality control.

II. PiezoNet Data Acquisition

Fig. 1 and Fig. 2 show the fully automated V-I data acquisition system, which is highly effective in capturing nonlinearities and effectively unifying influencing factors such as temperature and sample-to-sample variation for complex power materials. Fig. 3 illustrates the principles of the PiezoNet data acquisition system. The PR is modeled using the Butterworth-Van Dyke model and represents the electromechanical behavior as a L-C-R network with the plate capacitance C_p [7]. A 6-stage switching sequence was used to ensure efficient energy transfer and soft charging [4], [8], [14]. A complete switching cycle is displayed in Fig. 4. The switching sequence determines how the PR's terminals are connected and disconnected from the source/load system throughout each cycle. These sequences must achieve ideal resonant charging and discharging ensuring efficient energy transfer. The stages are categorized into connected stages (where the PR interacts directly with the source/load system), zero stages (where the PR terminals are short-circuited), and open stages (where the PR terminals are open-circuited, allowing for resonance).

Each switching stage aims to balance energy and charge within the PR over a complete cycle determined by ensuring the switching frequency falls in the inductive region of the PR for high-performance operation. The switches in the secondary half-bridge are implemented as diodes. The control loop adjusts the timings of S_1 and S_2 to maintain the optimal operating point for the PRs and for the ZVS of the switches.

The driver features fully isolated voltage and signal channels. The signal isolators and the gate drivers are implemented as ISO7420 and STDRIVE600 respectively. For the dual half-bridges, the MOSFETs are implemented as GaN GS66504B, diodes are implemented as MURS260T3G, enabling both 4-switch and diode-based operations. The isolated channels facilitate floating measurements, allowing a 1-Ω current shunt resistor (Vishay Dale Bulk Metal) to be placed on one side for current measurement. This setup enables synchronous measurement of the current and voltage of the PR. To allow for rapid characterization of many different samples of varying sizes the measurement setup uses a 3D-printed "C" clamp (Fig. 2) to secure the samples during characterization, the samples are held in place with only the fiction of the clamp walls to minimize the mechanical damping effects.

The PRs need to operate at the optimal operating condition for high efficiency. To enable automatic tuning of the power stage, two resistive voltage dividers are used in conjugation with high-voltage buffers and a low-pass filter to measure the voltage at the switch nodes. The measured voltages in combination with the sent target voltages from the host computer are

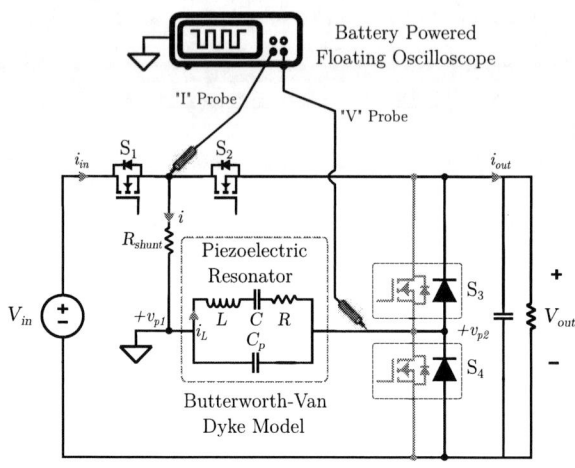

Fig. 3: An example topology for PR-based power conversion adopted as the platform for this paper. The battery-powered oscilloscope (Rigol DHO814) is floating. The Butterworth-Van Dyke model for the PR can be replaced by a neural network.

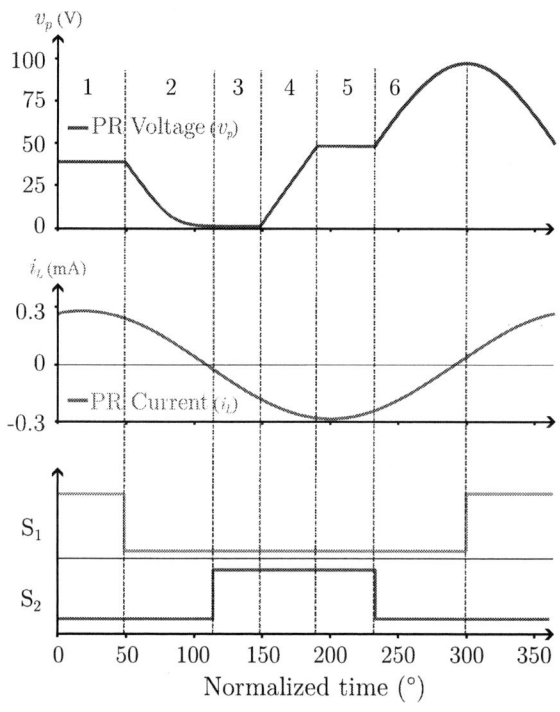

Fig. 4: Example operational waveforms for the PR-based power converter shown in Fig. 3. The converter has 6 operation stages in each switching cycle. The PR voltage v_p is measured as the voltage between v_{p1} and v_{p2} in Fig. 3.

used to ensure zero-voltage-switching (ZVS) of the switches. S_1 and S_2 are triggered by sensing voltages v_{p2} and v_{p1}, respectively (as shown in Fig. 5). S_1's on-time regulates V_{out}, while S_2's ZVS maximizes efficiency. S_1 is turned on during stage 6 when i_L crosses zero, detected by v_{p2} rising above zero. The error between the measured V_{out} and the desired

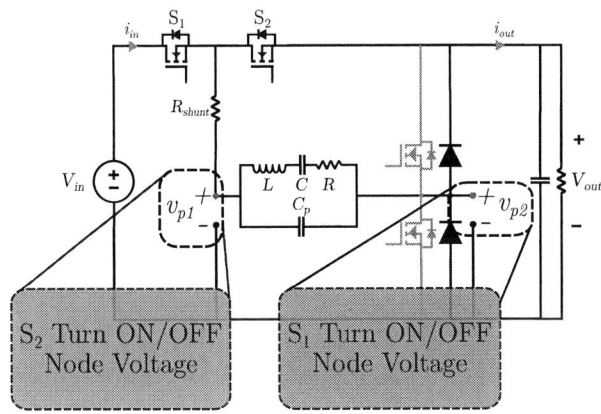

Fig. 5: Voltage Nodes v_{p1} and v_{p2} used for the control of S_1 and S_2 following the methods described in [14].

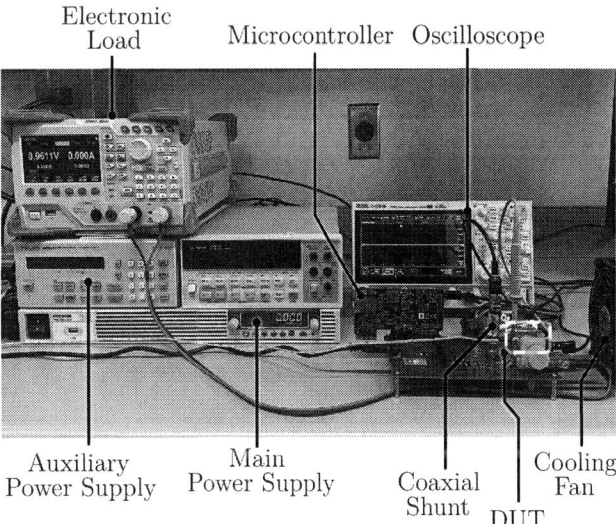

Fig. 6: Experimental setup of the data acquisition system of PiezoNet. The full setup is controlled by a personal computer and is fully automated. It takes about 2 minutes to collect the V-I loop data for the PR in one operating condition.

output (V_{set}) drives S_1's feedback loop. S_2 is turned on at the beginning of stage 3 when v_{p1} falls to V_{out} [14].

Figure 6 shows a picture of the experimental setup of the data acquisition system. While the proposed method is functional, it presents several limitations, primarily due to the restricted dynamic range of the ADC and control system (a simple PID control), which reduces adaptability across different operating conditions. This limitation can hinder precise control, especially in the application where stability is critical. To address these challenges, a refined approach has been implemented, as shown in Fig. 9. This setup leverages a PLECS server to manage control functions directly, interfacing seamlessly with the microcontroller and thus offloading complex control tasks from the embedded system. The control loop for S_2 is calculated on the PLECS server, adjusting the duty cycle at a 20 Hz refresh rate to maintain zero-voltage-switching (ZVS) at S_2.

Running in parallel with this control setup, a Python script automates equipment handling and data collection. Before acquisition begins, the Python script performs a complete sweep of the phase difference between S_1 and S_2, transmitting each phase value to the PLECS server via XML-RPC. At every phase step, the script records the efficiency of the driver, identifying the phase with the highest efficiency as the final phase setting for S_1 and S_2. The system efficiency is around 94% at its peak and drops to 75-80% during the sweep. This integrated system not only allows for precise efficiency tuning but also offers the flexibility to adapt to dynamic load or operating regimes without compromising stability. This dual-system approach brings distinct advantages. By eliminating the need for rapid response to output changes, it reduces the risk of overshoot or entrapment in local minima or maxima. Furthermore, reliance on the ADC and tight control loop precision is minimized, extending the dynamic range for high-voltage operation and enhancing system stability.

The voltage across the PR is not referenced to the ground of the excitation circuit, making voltage and current measurements challenging. A battery-powered galvanically isolated oscilloscope (Rigol DHO814) was used to ensure minimal

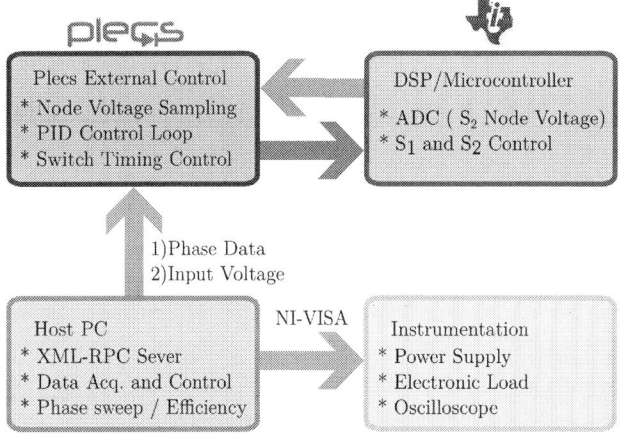

Fig. 7: Data flow of the PiezoNet data acquisition and control system, showing the communication implemented between the host computer, PLECS, and the Microcontroller.

impact on the operation of the PR (Fig. 8). LXI LAN was used to communicate with the oscilloscope, this approach eliminates the need to use a cable which would compromise the isolation of the oscilloscope.

The current version of the PiezoNet database comprises four sections: three dedicated to PR sample ITEMS 2323 (19.8 × 2.18 mm), 2324 (19.8 × 0.8 mm), and 1561 (12.9 × 2.5 mm) from APC Int., and a fourth section documenting a temperature sweep performed at 55°C, 70°C, 80°C, 90°C, and 100°C for a single sample of ITEM 1561. For each PR item, ten samples ordered from the same batch are characterized within an input voltage range of 65 V to 150 V, with the input voltage incremented in 5 V steps. Simultaneously, the output voltage is swept in 4 V increments starting at

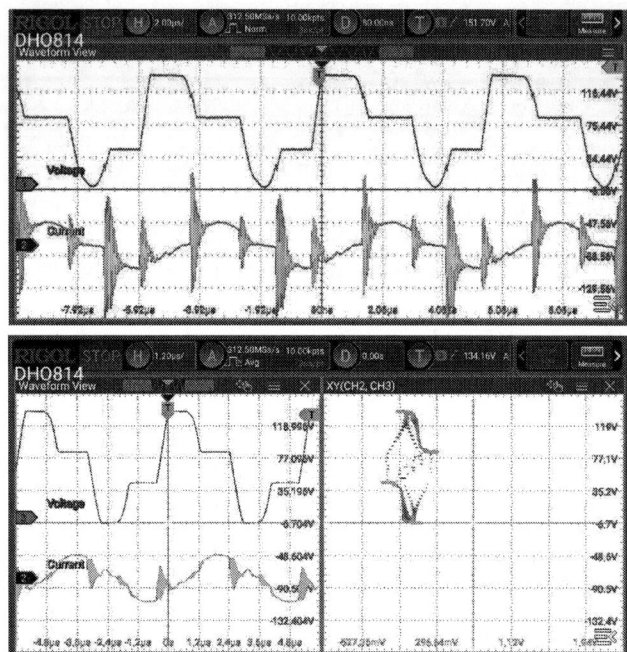

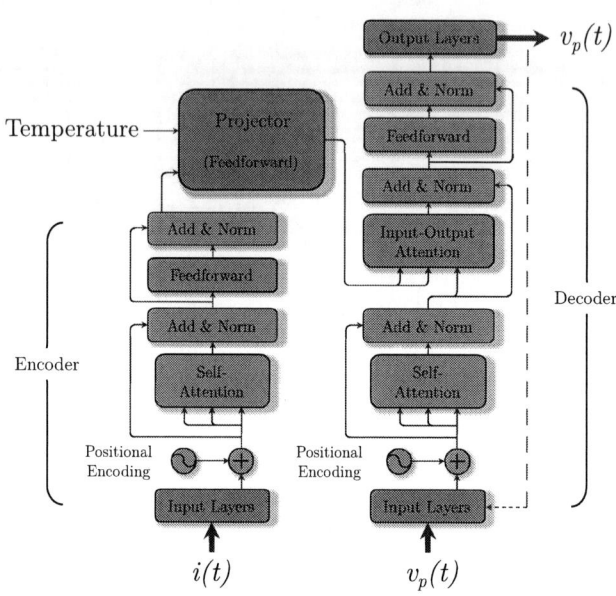

Fig. 8: Steady-State waveforms of $v_p(t)$ and $i(t)$ during data acquisition (120V Input, 40V Output, 92% Efficiency). In addition, an XY plot showing the hysteresis behavior during operation (130V Input, 45V Output, 93% Efficiency).

Fig. 9: Neural network structure of the transformer-based encoder–projector– decoder architecture. $i(t)$ waveform is the sequence input of the encoder. T is the scalar input to the projector. During the model training the target $v_p(t)$ is fed directly to the decoder as a reference.

20 V and increasing to $\frac{V_{in}}{2}$. Once this limit is reached the input voltage is incremented and the process repeats. The characterization procedure requires approximately 2.5 hours per sample for collecting about 120 V-I loops. During normal sample characterization, a large ducted fan is used to maintain the samples at room temperature. For the temperature sweep the fan is replaced with a hot air gun to reach and regulate the required temperature.

III. PIEZONET DATA-DRIVEN MODELS

Similar to other hysteresis materials in power electronics (e.g., capacitors and magnetics), the characteristics of a PR can be described by the voltage time sequence $v_p(t)$ and the current time sequence $i(t)$. The injection of $i(t)$ leads to a change of $v_p(t)$ in the PR. The voltage-current relationship in PRs is influenced by various operating conditions, including resonator type, size, geometry, and environmental factors such as temperature and mechanical pressure.

Different PR types exhibit unique impedance characteristics, and temperature variations significantly alter their behavior. To develop a neural network model capable of predicting the voltage waveform under varying operating conditions, we propose to use a transformer-based encoder–projector–decoder architecture, similar to the ones used in [15]. The general concept of this architecture is to map a time series input $i(t)$ into another time series output $v_p(t)$, while incorporating scalar inputs such as temperature T. This framework is designed to predict the voltage waveform across three distinct piezoelectric resonator types. The encoder processes the input sequence $i(t)$ and transforms it into a fixed-dimensional latent representation

by capturing sequential information and temporal correlations of the current waveform. A feedforward neural network (FNN) first maps the sequence to a d-dimensional representation, which is then combined with a positional encoding vector to retain temporal dependencies. The hidden vectors produced by the encoder are modified in the projector to incorporate scalar inputs such as temperature T. The projector employs a shallow FNN to adjust the latent representation, accounting for the influence of external operating conditions [15].

The decoder reconstructs the voltage waveform $v_p(t)$ autoregressively. It uses the modified hidden vectors from the projector and a reference sequence as inputs. During training, the reference sequence corresponds to the target $v_p(t)$, whereas during inference, it is initialized with zeros and iteratively replaced by the model's own predictions. Similar to the encoder, the reference sequence is processed through a self-attention module, where it is combined with positional encoding to generate decoder-specific hidden vectors. These are further refined using an input-output attention mechanism to incorporate the modified latent representation from the projector, followed by a final FNN that produces the predicted voltage waveform [15].

The proposed architecture effectively handles variability across different PR types and incorporates temperature as a scalar input. The attention mechanism in the transformer ensures accurate modeling of temporal dependencies in the voltage-current relationship, surpassing traditional recurrent neural network (RNN). During model inference, the autoregressive generation of the output sequence ensures the temporal causality of the prediction is preserved. As mentioned

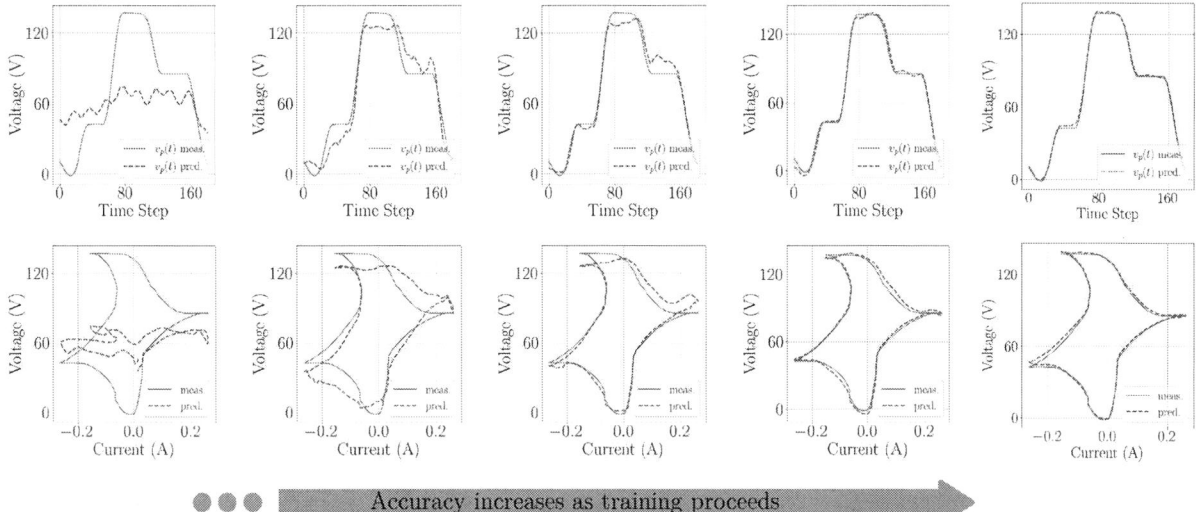

Fig. 10: Prediction results of the $v_p(t)$ waveform and the V-I loop of an example testing point (125 V Input, 44 V Output, 177 kHz, 25 °C) at different stages of the training (4 epoch, 20 epoch, 50 epoch, 100 epoch, 500 epoch).

the database consists of 4 sections and a separate model is provided for each, the 3 sections collected at room temperature omit the projector (which injects the temperature information) and provide a direct current to voltage mapping. Further details regarding the operation of the transformer model are described by the pseudo-code in Algorithm 1.

IV. TRAINING AND TESTING RESULTS

The transformer-based model is implemented using the PyTorch framework. The hyperparameters of the network are optimized based on experimental training results. In this model, the dimension of the value vector was set to 32, the number of attention heads was set to 4, and both the encoder and decoder consists of 1 layer. These configurations result in a total of 46,081 learnable parameters in the network. The model was trained for 1,000 epochs on an RTX 3060 GPU, with each training session taking approximately one hour. The training dataset comprises voltage and current waveform pairs, split into 90%, 10% for training and validation respectively. The mean-squared error (MSE) between the predicted voltage waveform $v_{\text{pred}}(t)$ and the actual waveform $v_{\text{meas}}(t)$ is used as the loss function for backpropagation. The Adam optimizer is employed for training, with an initial learning rate of 0.0001, which decays exponentially by 10% every 150 epochs to improve model convergence. Data augmentation techniques are applied such as down-sampling and averaging to make the model more robust and convergence easier. The transformer's attention mechanism is able to capture long-range dependencies in time-series data. The model's performance on the validation set indicates strong generalization capabilities, effectively handling various waveform patterns and operating conditions.

A. V-I Loop Prediction

The performance of the trained transformer models was evaluated on the test set to validate their ability to predict

---

Algorithm 1 Transformer-Based Model

Input: Current $i(t)$, Temperature T;
Output: Voltage $v_p(t)$;

1) $X \leftarrow \text{FNN}_1(i(t))$ [Mapping]
2) $X \leftarrow X + \text{Positional Encoding}$
3) $X \leftarrow \text{Norm}(X + \text{Self-Attention}_1(X))$
4) $X \leftarrow \text{Norm}(X + \text{FNN}_2(X))$ [Encoder]
5) $X' \leftarrow \text{FNN}_3(X, T)$ [Projector]
6) **if training then**
   a) $Y \leftarrow \text{FNN}_4(v_p(t))$ [Mapping]
   b) $Y \leftarrow Y + \text{Positional Encoding}$
   c) $Y \leftarrow \text{Norm}(Y + \text{Self-Attention}_2(Y))$
   d) $Y' \leftarrow \text{Norm}(Y + \text{Input-Output-Attention}(X', Y))$
   e) $Y' \leftarrow \text{Norm}(Y' + \text{FNN}_5(Y'))$ [Decoder]
   f) $v_p(t) \leftarrow \text{FNN}_6(Y')$ [Mapping]
7) **else if testing then**
   a) Initialize $v_p^0(t) \leftarrow 0$
   b) **for** $i = 1$ **to** L **do**
     i) $Y \leftarrow \text{FNN}_4(v_p^{i-1}(t))$ [Mapping]
     ii) $Y \leftarrow Y + \text{Positional Encoding}$
     iii) $Y \leftarrow \text{Norm}(Y + \text{Self-Attention}_2(Y))$
     iv) $Y' \leftarrow \text{Norm}(Y + \text{Input-Output-Attention}(X', Y))$
     v) $Y' \leftarrow \text{Norm}(Y' + \text{FNN}_5(Y'))$ [Decoder]
     vi) $v_p^i(t) \leftarrow \text{FNN}_6(Y')$ [Mapping]
   c) $v_p(t) \leftarrow v_p^L(t)$
8) **return** $v_p(t)$

---

the voltage waveform $v_{\text{pred}}(t)$ from the current waveform $i(t)$. Fig. 10 illustrates a series of example predictions generated by the transformer-based model for a specific test case (125 V input, 50 V output, 177 kHz, 25 °C) at different stages of training. As the training progresses, the model gradually converges, reducing the discrepancy between the predicted and

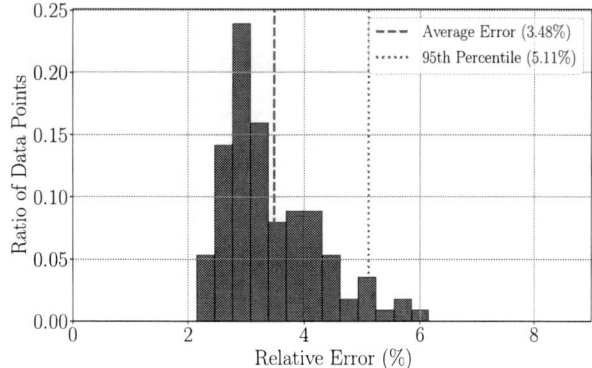

Fig. 11: Relative error distributions of the predicted $v_p(t)$ sequence generated by the transformer model.

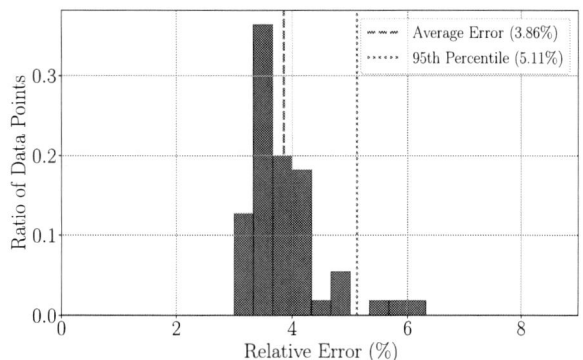

Fig. 12: Relative error distributions of the predicted $v_p(t)$ sequence at different temperatures generated by the transformer.

measured voltage waveforms, achieving a close match.

To quantitatively evaluate the prediction accuracy of the models, the relative error between the predicted sequence $v_{\mathrm{pred}}(t)$ and the measured sequence $v_{\mathrm{meas}}(t)$, as defined below, was used as the metric to assess the results:

$$\text{Relative Error of Sequence} = \frac{\text{rms}(v_{\mathrm{pred}} - v_{\mathrm{meas}})}{\text{rms}(v_{\mathrm{meas}})} = \frac{\sqrt{\frac{1}{n}\sum_{t=t_1}^{t_n}(v_{\mathrm{pred}}(t) - v_{\mathrm{meas}}(t))^2}}{\sqrt{\frac{1}{n}\sum_{t=t_1}^{t_n}(v_{\mathrm{meas}}(t))^2}}. \quad (1)$$

Fig. 11 shows the distribution of relative errors in the $v_p(t)$ predictions generated by the transformer-based model. The results demonstrate that the model can accurately predict the voltage waveforms across a wide range of operating conditions. The average relative error for the transformer-based model is 3.48%, while the median relative error is 3.17%. The 95th percentile relative error, which provides insight into the worst-case prediction accuracy, is 5.11%. Considering the sample-to-sample variations of PR.

The inference set includes data points representing various waveform shapes and spans the same ranges of frequency, temperature, and current magnitudes as the training set. These statistics confirm that the proposed model is capable of making accurate voltage waveform predictions across the defined operating range based on the current excitations to the PRs.

B. Temperature Effect Modeling

As mentioned the final section of the database is constructed by conducting the full V-I characterization at temperatures 55°C, 70°C, 80°C, 90°C, and 100°C for a single sample of ITEM 1561. As the characteristics of the PR are dependent on temperature a separate model is trained on this data to analyze and predict the degradation. Fig. 13 presents the V-I hysteresis loops while running the controller in an open-loop configuration for temperatures 25°C to 105°C showing the deviation. The plot shows that the temperature affects the PR behavior as the stages that interact with the source-load system show a small error due to the strength of the source and the load, while the stages where the PR is let to resonate show a small jump in the relative error. The deviation for the training data can be quantified by integrating the loop area, this results in a mean percentage error of 8.04% and a standard deviation of 4.94%. The transformer model trained on this data helps to generalize and predict this deviation from 25°C operation. Fig. 12 shows the distribution of relative errors in the predictions generated by this model. This demonstrates that the model generalizes well to unseen data, effectively capturing the temperature-dependent behavior of the PR.

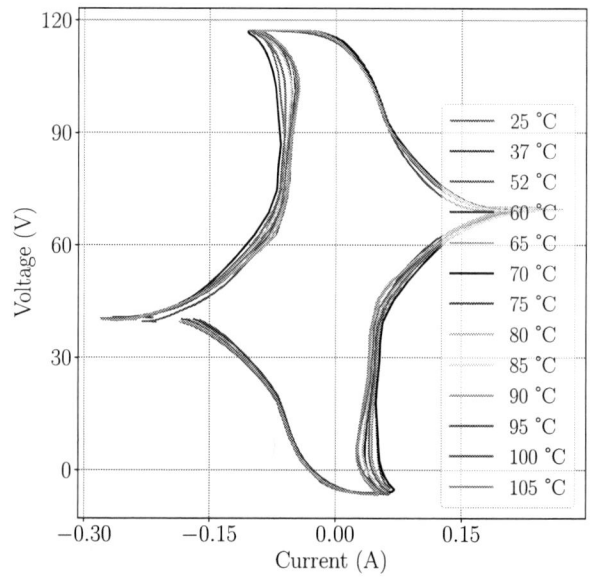

Fig. 13: V-I hysteresis measurements across a temperature range, for a single operating point (110 V input, 40 V output). For this sample (ITEM 1561) at this operating point, the V-I graphs do not show significant temperature dependence.

C. Sample Variation Analysis

The relative error analysis was conducted to study the sample-to-sample variation across samples. Voltage sequences under the same operating points where compared to a reference

979-8-3315-1612-3/25 $31.00 © 2025 IEEE

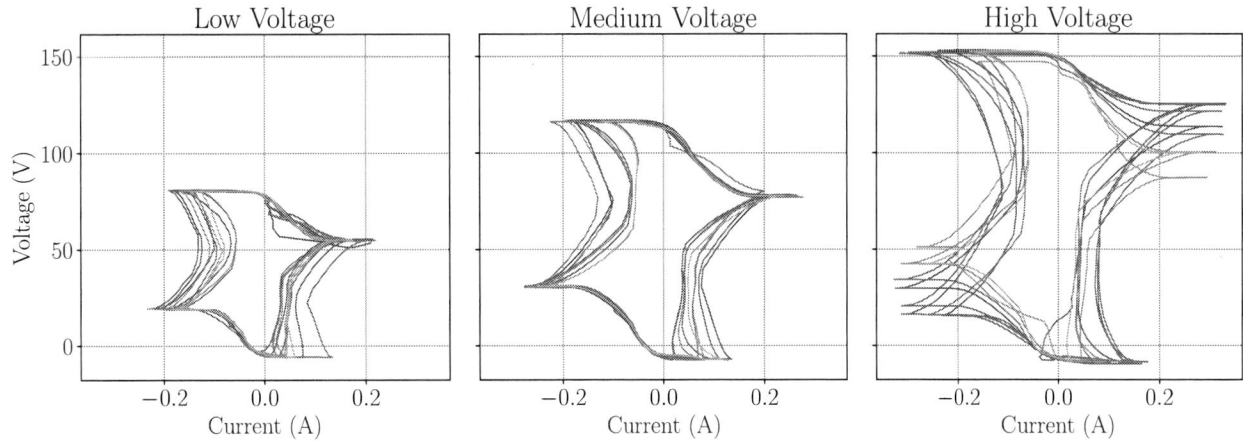

Fig. 14: Sample to sample variation between PRs with measurement data at low (80 V), medium (120 V) and high (150 V) voltage for 10 samples excited at the same operating point. A significant sample-to-sample difference was observed.

sample (Sample 1). The average relative error ranges from approximately 4% for closely matching samples (e.g., Sample 3) to over 9% for samples exhibiting higher deviations (e.g., Sample 7). This variability is also reflected in the hysteresis loops plotted for low, medium, and high voltages (Fig. 14), where deviations in loop shapes and areas suggest differences in energy dissipation and nonlinear characteristics across samples. These discrepancies have been observed as differences in the physical size and resonant frequency of the PR samples, leading to inconsistent performance under identical operating points. The observed sample-to-sample variations have implications for the accuracy of the transformer model – the accuracy of the model is bounded by the accuracy of the data. Hence the model is optimized to balance the trade-off between under-fitting and over-fitting, ensuring it captures the underlying patterns in the data, this approach maximizes predictive accuracy while enhancing the model's ability to generalize effectively across the full range.

V. CONCLUSION

This paper introduced PiezoNet, an automated platform, and the resulted database for time-domain characterization of piezoelectric resonators in power converters by employing data-driven methods. This approach enhances the predictability of PR-based power converters, enabling optimized designs and more precise simulations. The database and resulting data-driven models can also lead to deeper understanding of piezoelectric resonators in power electronics.

ACKNOWLEDGEMENTS

This work was jointly supported by the National Science Foundation under Award #2344664 and Princeton Andlinger Center for Energy and the Environment.

REFERENCES

[1] C. R. Sullivan, B. A. Reese, A. L. F. Stein, and P. A. Kyaw, "On size and magnetics: Why small efficient power inductors are rare," *2016 International Symposium on 3D Power Electronics Integration and Manufacturing (3D-PEIM)*, pp. 1–23, 2016.

[2] D. J. Perreault, J. Hu, J. M. Rivas, Y. Han, O. Leitermann, R. C. Pilawa-Podgurski, A. Sagneri, and C. R. Sullivan, "Opportunities and challenges in very high frequency power conversion," *IEEE Applied Power Electronics Conference and Exposition*, pp. 1–14, 2009.

[3] P. A. Kyaw, A. L. F. Stein, and C. R. Sullivan, "Fundamental examination of multiple potential passive component technologies for future power electronics," *IEEE Transactions on Power Electronics*, vol. 33, no. 12, pp. 10 708–10 722, 2018.

[4] J. D. Boles, J. J. Piel, and D. J. Perreault, "Enumeration and analysis of dc–dc converter implementations based on piezoelectric resonators," *IEEE Trans. on Power Electronics*, vol. 36, no. 1, pp. 129–145, 2021.

[5] J. D. Boles, J. J. Piel, E. Ng, J. E. Bonavia, B. M. Wanyeki, J. H. Lang, and D. J. Perreault, "Opportunities, progress, and challenges in piezoelectric-based power electronics," *IEEJ Journal of Industry Applications*, vol. 12, no. 3, pp. 254–263, 2023.

[6] A. K. Jackson, J. W. Perreault, J. H. Lang, and D. J. Perreault, "Large-signal characterization of piezoelectric resonators for power conversion," in *2024 IEEE Applied Power Electronics Conference and Exposition (APEC)*, 2024, pp. 2637–2643.

[7] K. Van Dyke, "The piezo-electric resonator and its equivalent network," *Proc. of the Inst. of Radio Engineers*, vol. 16, no. 6, pp. 742–764, 1928.

[8] J. D. Boles, J. E. Bonavia, J. H. Lang, and D. J. Perreault, "A piezoelectric-resonator-based dc-dc converter demonstrating 1 kw/cm resonator power density," *IEEE Transactions on Power Electronics*, vol. 38, no. 3, pp. 2811–2815, 2023.

[9] M. Hagiwara, S. Takahashi, T. Hoshina, H. Takeda, and T. Tsurumi, "Analysis of nonlinear transient responses of piezoelectric resonators," *IEEE Transactions on Ultrasonics, Ferroelectrics, and Frequency Control*, vol. 58, no. 9, pp. 1721–1729, 2011.

[10] C. Daniel, E. Stolt, W. Braun, R. Lu, and J. Rivas-Davila, "Nonlinear losses and material limits of piezoelectric resonators for dc-dc converters," in *2024 IEEE Applied Power Electronics Conference and Exposition (APEC)*, 2024, pp. 1560–1565.

[11] J. Forrester, J. N. Davidson, M. P. Foster, and D. A. Stone, "Influence of spurious modes on the efficiency of piezoelectric transformers: A sensitivity analysis," *IEEE Transactions on Power Electronics*, vol. 36, no. 1, pp. 617–629, 2021.

[12] D. Serrano, H. Li, S. Wang, T. Guillod, M. Luo, V. Bansal, N. K. Jha, Y. Chen, C. R. Sullivan, and M. Chen, "Why magnet: Quantifying the complexity of modeling power magnetic material characteristics," *IEEE Trans. on Power Electronics*, vol. 38, no. 11, pp. 14 292–14 316, 2023.

[13] M. Chen *et al.*, "Magnet challenge for data-driven power magnetics modeling," *IEEE Open Journal of Power Electronics*, pp. 1–16, 2024.

[14] J. J. Piel, J. D. Boles, J. H. Lang, and D. J. Perreault, "Feedback control for a piezoelectric-resonator-based dc-dc power converter," in *2021 IEEE 22nd Workshop on Control and Modelling of Power Electronics (COMPEL)*, 2021, pp. 1–8.

[15] H. Li, D. Serrano, T. Guillod, S. Wang, E. Dogariu, A. Nadler, M. Luo, V. Bansal, N. K. Jha, Y. Chen, C. R. Sullivan, and M. Chen, "How magnet: Machine learning framework for modeling power magnetic material characteristics," *IEEE Transactions on Power Electronics*, vol. 38, no. 12, pp. 15 829–15 853, 2023.

A New Gate Charge De-Embedding Method for Accurate On-Wafer Characterization of HV MOSFET Devices

Joao R. R. O. Martins, Member, IEEE*, Rachid Hamani*, Vincent Quenette*, and Joerg Gessner[†]
*X-FAB Semiconductor Foundries France, 224 Bd John Kennedy, 91100 Corbeil-Essonnes, France
[†]X-FAB Semiconductor Foundries Erfurt, 2 Haarbergstraße 67, 99097 Erfurt, Germany

Abstract—A new de-embedding technique for on-wafer gate charge measurements is presented. Since in the existing techniques, the interconnect capacitances between gate and source and gate and drain are not considered, our proposed approach includes all relevant parasitics and allows more accurate results of the intrinsic gate charges figures-of-merit (FOMs). To demonstrate this, experimental results obtained from 180 nm node 40 V HVMOS device are compared to three different state-of-the-art techniques for gate charge measurements. This developed method aims to accurately characterize the transient behavior at high voltages, additional to the typical CV measurements capability, and thus helps accurate modeling of high-speed switching and high-power devices.

Index Terms—Gate Charge, De-embedding, on-waffer, measurements, parasitics, switching

I. INTRODUCTION

Gate charge parameters (Q_g, Q_{gs}, Q_{gd}) are key FOMs of power MOSFET devices [1]. Additional to typical CV modeling, these quantities are essential for many power electronic applications to assess the switching performance of different devices (e.g., HVMOS, IGBT, SiC, GaN). In modern integrated power electronics, parasitic elements introduced by test equipment and measurements techniques need to be handled carefully due to their increase impact, compared to discrete devices/older technological nodes [2]. Therefore, accurate on-wafer gate charge measurements are essential to build and verify power and high voltage device models suitable for electrification and energy efficiency designs. In literature a few methods have been offered for measuring those parameters, most of them applying or measuring the current at the gate as given in [3]. However, in [4] a procedure for estimating only the total gate charge (Qg) is developed by using just capacitance (CV) measurements. In [5], a standard method is developed using a forced gate current with a current limited voltage source between drain and source without any calibration performed. In [6], Tektronix proposes an automated solution from [5] for on-wafer gate charge measurements, where a calibration is done for excluding cable charging and inherit capacitances of the measurement equipment.

This work proposes a new de-embedding technique that considers not only the parasitic capacitances of the measurement system, but also the gate to source and

gate to drain capacitances of the interconnects. This aims to accurately characterize the transient behavior at high voltages and thus helps accurate modeling of high-speed switching and high-power devices.

II. THE NEW GATE CHARGE DE-EMBEDDING METHOD

For accurate on-wafer gate charge measurements, parasitics capacitances coming from the measurement system cables, contact pads and interconnects of the test-structure must be properly handled to ensure accurate measurement of the Device Under Test (DUT) (see Fig. 1). The capacitances between gate and source and between gate and drain which come from either the system capacitance ($C_{gs_{sys}}$) or from the interconnects and connect pads (C_{gs_p}, C_{gd_p}) should be also considered [3]. Their presence will add parasitic charges to the device charge ($Q(t)$) as follows:

$$\frac{dQ(t)}{dt} = I_G + \frac{dQ_{gs_{tot}}(t)}{dt} + \frac{dQ_{gd_{tot}}(t)}{dt} \quad (1)$$

$$Q_{gd_p} = C_{gd_p} \cdot V_{gd_p}(t) = C_{gd_p} \cdot (V_D(t) - V_G(t)) \quad (2)$$

$$Q_{gs_{tot}} = C_{gs_{tot}} \cdot V_G(t) \quad (3)$$

where I_G is the input current on the gate charge, V_G and V_D are the gate to ground voltage and drain to ground voltage, respectively. To quantify the interconnect parasitics, an open structure is made by removing the contact from the DUT. By following the technique in [5], a current is forced in the High pin and the voltage is measured over time on the same pin while the Low pin is connected to the ground. The capacitance between both pins is simply calculated by the slope of the voltage waveform ($C = (dV(t)/dt)^{-1})$). The characterization of the parasitic network is performed as follows: first, the capacitance of the open test structure between gate and source $C_{gs_{tot}} = C_{gs_{sys}} + C_{gs_p}$ is measured, and then second, the capacitance of the needles on the air between gate and drain is measured, and finally the open test structure is measured between gate and drain to obtain the C_{gd_p}

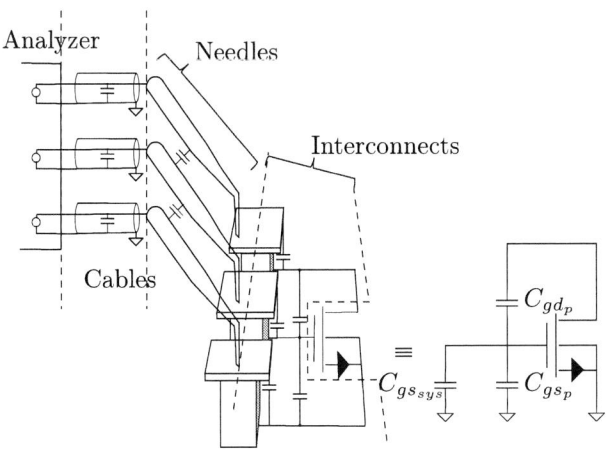

Fig. 1. Parasitic capacitances and their circuit representation on a 3 terminal 40 V high voltage MOSFET device

capacitance by excluding the capacitance of needles on-air. Once the different capacitances are measured, the de-embedded charge can be calculated using (1) and the parasitics charges from (2) and (3).

The proposed technique has been validated on a 40 V N-type HVMOS (W=100 μm, L=0.4 μm, NF=10) from X-FAB 180 nm BCD on SOI process (1.8 V for standard devices), using first the method described in [4]. On a second step, the JEDEC time-based method described in [5] is applied using the X-FAB's PDK model (HiSIM_HV [7] [8]). And finally, the old de-embedding method presented in [6] is used for final and complete comparison.

III. NEW METHOD MEASUREMENTS AND VALIDATION

The gate charge of a 40 V N-type DMOS, was measured using a Keithley 4200 SCS analyzer by applying a 5 pA gate current and a drain voltage of 3 V with current compliance of 100 μA. The gate charge results were de-embedded with the new technique and the different FOMs are calculated using the method explained [5] (see Fig.2 and Tab.1).

To compare the total gate charge, one can use the technique proposed in [4] that uses the measured total gate capacitance (C_{gdsb}) and the gate to drain capacitance (C_{gd}) to estimate the Q_g for a given transistor. The capacitance integrals were performed using trapezoidal rule and total gate charge at Vg=Vd=3 V of 8.06 pC is calculated (see Table 1). The total gate charge using our method is giving 8.33 pC and this is in accordance with the CV based method with a relative error of about 3%. As it can be seen from Table I, our new method is also providing gate to source charge (Q_{gs}) and gate to drain charge (Q_{gd}) which are also a relevant FOMs for switching performances of power devices. Both results were compared to our PDK model, where a schematic level test bench replicating the technique proposed in [5]

was simulated using Spectre [9]. The model used for this validation is extracted from a full DC, CV (de-embedded), Pulsed IV, and Reverse Recovery. As it can be noticed, not only the simulated total gate charge (Q_g= 8.22 pC) but also the Q_{gd} and Q_{gs} are comparable to the measured results using the new technique (see Tab. I).

For completeness, the technique from [6] is also included in our comparison. This technique considers only the system capacitance between gate and source where $C_{gs_{sys}}$ is first measured by doing a measurement with the chuck on separation between gate and source.

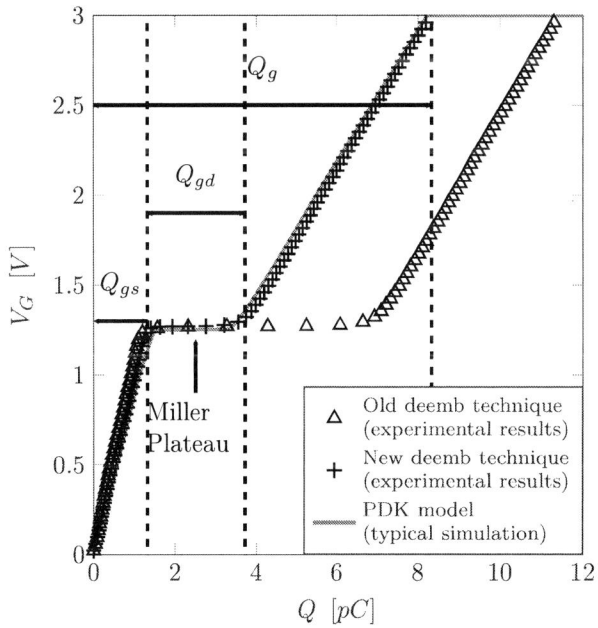

Fig. 2. Measurements and simulation of gate voltage vs. gate charge 40 V HVMOS.

The comparison of the four different techniques is presented in Tab. I, and the different time-based measurements and simulations can be seen in Fig. 2. As it can be seen, the proposed de-embedding technique agrees with [4] and simulation results within 3.5% maximum relative error.

As shown in Fig.2, in the first part of the measurements, before the Miller Plateau, both measurement techniques present similar results, due to a low metallization capacitance compared to the system capacitance ($C_{gs_p}/C_{gs_{sys}} = 5\%$). The same observation for the slope after the miller plateau is noted. The Miller Plateau itself is longer for the old technique, indicating a bigger measured C_{gd} capacitance. These observations confirm that old de-embedding technique is not sufficient to accurately exclude the parasitics on both gate do source and gate to drain capacitances. The measured Q_{gd} using the new method is higher than the simulated model. The difference is attributed to the fact that the measurement is performed

979-8-3315-1612-3/25 $31.00 © 2025 IEEE   1890

TABLE I
Comparison between different gate charge FOMs obtained from different method

	Calculated using [4]	Simulated using PDK model (Typical ± 3sigma)	Measured using new de-embedding	Measured using old de-embedding [6]
Q_g [pC]	8.06	8.22 ±1.52	8.33	12.99
Q_{gd} [pC]	N/A	2.03 ± 1.07	2.4	6.37
Q_{gs} [pC]	N/A	1.41 ± 0.53	1.32	0.71

on a single die and the measured silicon is around 1 sigma of typical value. Due to the complex parasitic network, the measurements procedure and the de-embedding algorithm requires several steps as detailed in this section. The technique has been also reproduced in a 60 V HVMOS device as it can be seen on Fig. 3

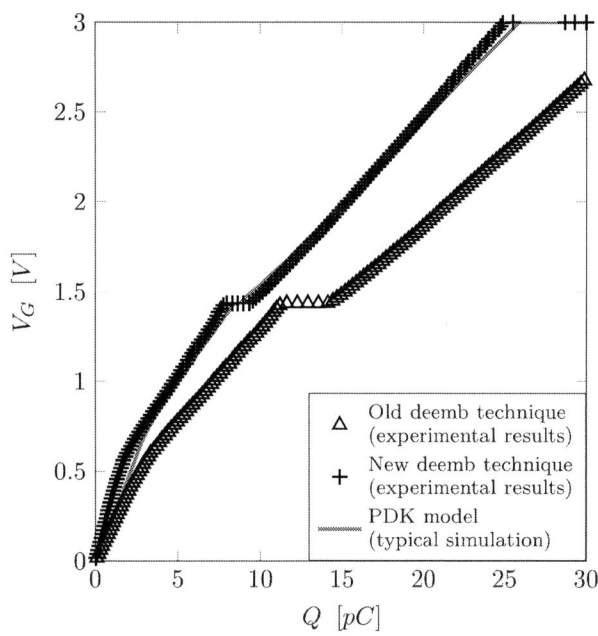

Fig. 3. Measurements and simulation of gate voltage vs. gate charge of a 60 V HVMOS.

IV. CONCLUSION

In this study, a new de-embedding technique for accurate on-wafer gate charge measurements is presented. This new technique [10] allows for correct de-embedding of all relevant parasitic capacitances on a gate charge measurement by considering the capacitances between gate and source and between gate and drain. The proposed technique has been validated on 180nm node 40 V HVMOS device and compared to three state-of-the-art methods. The technique has been also reproduced in a 60 V HVMOS device. This developed method aims to accurately characterize the transient behavior at high voltages, additional to the typical CV measurements capability, and thus helps accurate modeling for high-speed switching and high-power devices.

References

[1] B. J. Baliga, Advanced Power MOSFET Concepts. Boston, MA: Springer US, 2010.

[2] A. Kundu, "Limitation of CMOS Scaling and Effects of Parasitic Elements on the RF Performance," in Handbook of Emerging Materials for Semiconductor Industry, Y. S. Song, L. R. Thoutam, S. Tayal, S. B. Rahi, and T. S. A. Samuel, Eds. Singapore: Springer Nature Singapore, 2024, pp. 353–378.

[3] B. J. Baliga, Fundamentals of power semiconductor devices. New York, NY: Springer, 2008.

[4] V. Krishnamurthy, A. Gyure, and P. Francis, "Simple gate charge (Qg) measurement technique for on-wafer statistical monitoring and modeling of power semiconductor devices," in 2012 IEEE International Conference on Microelectronic Test Structures. San Diego, CA, USA: IEEE, Mar. 2012, pp. 98–100.

[5] J. JESD24-2, "Jesd24-2 gate charge test method," 2002.

[6] Tektronix, "Measuring gate charge of a device with acs software," 2022.

[7] H. University, HiSIM HV 2.5.1User's Manual, Hiroshima, 2023.

[8] H. J. Mattausch, N. Sadachika, M. Yokomichi, M. Miyake, T. Kajiwara, Y. Oritsuki, T. Sakuda, H. Kikuchihara, U. Feldmann, and M. Miura-Mattausch, "HiSIM-HV: A Scalable, Surface-Potential-Based Compact Model for High-Voltage MOSFETs," in POWER/HVMOS Devices Compact Modeling, W. Grabinski and T. Gneiting, Eds. Dordrecht: Springer Netherlands, 2010, pp. 33–64.

[9] Cadence, "Spectre® Circuit Simulator Reference," 2020.

[10] "Patent pending."

4 kW Auxiliary Power Module for Electric Vehicles Utilizing a Dual-Phase LLC DC-DC Converter

Mojtaba Forouzesh
Department of Electrical and Computer Engineerig
Queen's University
Kingston, ON, Canada
m.forouzesh@queensu.ca

Xiang Yu
Department of Electrical and Computer Engineerig
Queen's University
Kingston, ON, Canada
yuxiang_hfut@163.com

Yan-Fei Liu
Department of Electrical and Computer Engineerig
Queen's University
Kingston, ON, Canada
yanfei.liu@queensu.ca

Paresh C. Sen[†]
Department of Electrical and Computer Engineerig
Queen's University
Kingston, ON, Canada
paresh.sen@queensu.ca

Abstract— This paper introduces a novel resonant tank design approach for dual-phase LLC DC-DC resonant converters in Auxiliary Power Module (APM) applications. The proposed design ensures that the impedance of one phase is consistently either larger or smaller than that of the other phase across a wide range of input and output voltages. As a result, only a single Switch-Controlled Capacitor (SCC) circuit is needed to achieve current sharing across the load range. This reduces system complexity and implementation costs without compromising efficiency or current-sharing performance. The design procedure is detailed for both phases, and experimental results from a full-scale, GaN-based 4 kW APM validate the effectiveness of the proposed approach in maintaining current sharing across varying input/output voltages and load conditions. Furthermore, flat efficiency curves with calculated average of more than 95.5% were achieved for different battery voltages, with a peak efficiency surpassing 96.3%.

Keywords—Dual-phase LLC, DC-DC Resonant Converter, SCC-LLC, Auxiliary Power Module (APM), Electric Vehicle (EV).

I. INTRODUCTION

The High-Voltage (HV) to Low-Voltage (LV) DC-DC converter in Electric Vehicles (EVs) supports the 12 V battery from the HV battery to supply on-board auxiliary loads. One of the key requirements for such a system is to accommodate the larger demands of future EVs, which may require a maximum load of approximately 4 kW to 5 kW. The most significant challenge in achieving improved performance for Auxiliary Power Modules (APMs) lies in the wide input and output voltage range, due to the varying operating voltage ranges of HV and LV batteries based on their State-of-Charge (SoC) [1].

The main topologies that can meet the requirements for an APM are typically based on full-bridge configurations. The Phase-Shifted Full-Bridge (PSFB) is a topology that employs a fixed switching frequency and variable phase shift control, enabling the achievement of soft switching [2]. Proper implementation of PSFB can realize ZVS at the turn-on instant by utilizing leakage/stray inductances and the parasitic capacitances of the switching devices [3]. However, some drawbacks of the PSFB topology include the loss of ZVS under light load conditions, the load dependency of the ZVS delay time, and higher circulating and RMS currents.

When full-bridge active switch rectification is used in a FSFB converter, the topology is referred to as a Dual Active Bridge (DAB) converter [4]. The DAB is very popular in APM applications due to its ability to achieve ZVS, flexibility of control, and high efficiency [5] and [6]. However, a drawback of the DAB is the loss of ZVS across wide operating voltage and load ranges, which can deteriorate the converter's efficiency and power density. The three-phase DAB converter is another topology suitable for high-power applications with high power density requirements [7], This configuration inherently reduces current ripple, requiring smaller filtering on both sides, and offers fault tolerance capabilities, which are advantageous for automotive applications. Despite the advantages of DAB converters, the switch turn-off current in both single-phase and three-phase DAB topologies undergoes hard switching, leading to significant losses, particularly in the LV side switches, and poses severe EMI challenges.

In an approach to reduce turn-off current of LV side switches, a two-stage topology is proposed in [8], utilizing an interleaved buck in the front followed by a DC transformer (DCX) converter, where wide voltage adjustments occur in the first stage, and DCX operates as an LLC converter at the resonant frequency. This approach eliminates turn-on/-off losses of the LV side switches via Zero Current Switching (ZCS). However, while this two-stage approach reduces switching losses in the DCX stage, it introduces additional conduction losses, which can negatively impact overall efficiency and power density. To mitigate conduction losses and improve the efficiency of the LLC converter across a wide range of operating voltages, various hybrid control modes, including Discontinuous Conduction Mode (DCM), Continuous Conduction Mode (CCM) and Boundary Conduction Mode (BCM), are explored in [9].

Another aspect of achieving high-efficiency performance is the minimization of conduction losses. A common approach is to distribute conduction losses across multiple components by implementing a parallel converter design. However, resonant converters can face challenges with unbalanced loading and current sharing due to resonant tank component tolerances, even with slight impedance mismatches. In [10], a double-phase LLC converter is proposed for APM applications with dual control

---

[†] Deceased on November 22, 2023.

loops to balance the current. However, this approach results in two different switching frequencies, creating a variable beat frequency with fluctuations that can deteriorate the converter's performance. In [11], dual duty cycle control is used in a two-phase LLC converter for DCX applications to balance the load between phases while maintaining the same switching frequency for both phases. Additionally, passive current-sharing methods have been implemented in multi-phase LLC converters [12]. Despite these efforts, none of these methods can achieve accurate current sharing over a wide voltage and load range with large component tolerances. Moreover, interleaving to reduce current ripple and filtering requirements is not possible in most of these solutions.

In [13], a Switch-Controlled Capacitor (SCC) was incorporated in series with the resonant capacitors of both phases in an interleaved LCLC converter. This configuration matched the impedance of each phase at every switching frequency, ensuring effective load sharing. The method is both accurate and reliable across the entire load range, as it leverages active balancing based on sensed currents. In [14], a multi-phase LLC converter is proposed for APM applications, employing SCC circuits on all phases. The SCC circuits actively tune the impedance of the LLC tanks to ensure that the resonant tank impedances match across all phases. This approach mitigates many of the challenges associated with parallel LLC converters while effectively balancing load sharing between phases under all operating conditions. However, a notable drawback of the three-phase APM topology described in [14] is the increased cost associated with adding SCC components to each phase, which represents a significant barrier in the automotive industry.

In this paper, a novel resonant tank design is proposed for dual-phase LLC DC-DC converters. The design ensures that the voltage gains of the parallel phases are separate without any intersection over the frequency range, meaning that the voltage gain of one phase is always either larger or smaller than that of the other phase. This approach enables current sharing over a wide operating voltage range using only one SCC circuit, reducing cost and complexity while enhancing reliability. However, special care should be taken into the design of the resonant tanks to be able to achieve the latter, which will be described in the following sections.

II. THE PROPOSED DUAL-PHASE LLC DC-DC CONVERTER

As mentioned earlier, the impedances of the resonant tanks in multi-phase resonant converters are sensitive to component tolerances, which directly affect the current distribution between the phases. A phase with a resonant tank of higher impedance will carry a smaller share of the output current. On the other hand, an SCC circuit can only reduce the equivalent resonant capacitance compared to its original value. Therefore, the SCC circuit is limited to increasing the load current share in a given phase. Fig. 1 illustrates the proposed two-phase LLC converter with one SCC circuit on phase 2. It is crucial in the design procedure of the proposed converter to make sure that when the SCC circuit is removed, phase 2 always carries less output current, and when the SCC circuit is operating, the output current of phase 2 can be increased to achieve a balanced current sharing in every operating point. For the sake of manufacturing convenience, the magnetics are built with the same values (i.e.,

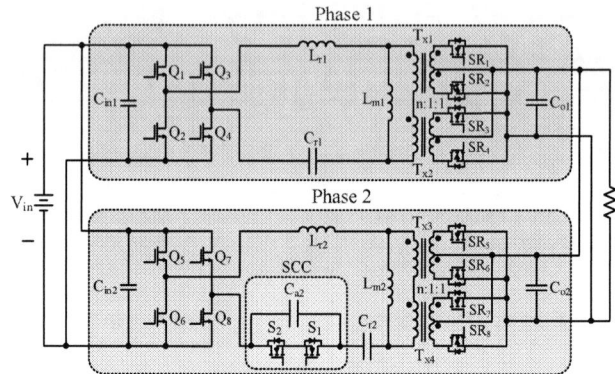

Fig. 1. The proposed dual-phase LLC DC-DC converter.

$L_{r1} \approx L_{r2}$ and $L_{m1} \approx L_{m2}$) and only the resonant capacitance of phase 2 is chosen to be larger than the resonant capacitance of phase 1 (i.e., $C_{r2} > C_{r1}$).

The first step in the design is the selection of the transformer turn ratio and then the resonant tank parameters of phase 1 for the half-load operation. As mentioned before, the resonant tank of phase 2 can be designed such that the impedance of phase 2 with SCC is always smaller than phase 1 over the whole operation conditions. Then, the SCC capacitor on phase 2 (i.e., C_{a2}) should be designed such that the impedance matching can be achieved based on the predefined component tolerances. The detailed requirement and design procedure of the proposed converter will be provided in the following of this section.

The design requirements of the DC-DC converter for EV APM are outlined in the following: the desired input voltage range is 250 V to 475 V, the output voltage range is 9 V to 16 V. The rated output continuous current is available through 320V to 450V input voltage range for 9 V to 14 V output voltage range, and from 14 V to 16 V output voltage range the output current derates to keep the output power constant at 4 kW. It should be mentioned that for the input voltage range below 320 V and above 450 V a linear derating can be implemented from 4 kW to around zero watts. The derating starts from 285 A for the output voltage of 9 V to 14 V, and it starts from 245 A for the output voltage of 16 V. It should be mentioned that when the output voltage varies between 14 V to 16 V, the output current is restricted by the rated output power. The nominal operation condition is with 320 V input voltage and 14 V output voltage at 285 A load current.

In this design, an operating frequency range between 200 kHz to 500 kHz is considered for the resonant converter to take advantage of passive component miniaturization. Moreover, 650 V GaN switches have been considered for the HV side switching bridge of each phase for maximum performance improvement over the operating range.

The turn ratio of the transformer is determined the same as that of the conventional design approaches. As it is desired to keep most of the operation range below the series resonant frequency (i.e., f_r) to achieve ZCS, the resonant frequency where the voltage gain of the LLC tank is unity should be set for the maximum input voltage and minimum output voltage. Hence, the transformer turn ratio can be found from

$$n = N_p/N_s = V_{in\_max}/V_{o\_min} \qquad (1)$$

where N_p and N_s are the transformer's primary and secondary number of turns.

As the input and output voltage range of the EV APM are too wide and we want to minimize the circulating current at the primary side of the transformers, we design the transformer turn ratio for the gain of 0.9 instead of 1 meaning that the switching frequency goes above resonant to meet highest input and lowest output voltage condition. With 450 V maximum input and 9 V minimum output, the equivalent transformer turn ratio should be $0.9 \times (450/9) = 45$. It should be mentioned that in high current applications it is usually desired to use two or more center tap transformers with series connected primary and parallel connected secondary so the large output current will be distributed as well as the conduction losses. As we want to use two center-tapped transformers in each phase, we select the final turn ratio to be 44 and each transformer has a turn ratio of $n{:}1{:}1 = 22{:}1{:}1$. In the next section, the resonant tank design of the phase without SCC is provided first (i.e., phase 1), and then resonant tank design of the phase with an SCC circuit is provided (i.e., phase 2).

III. DESIGN CONSIDERATION FOR THE PROPOSED DUAL-PHASE LLC DC-DC CONVERTER

A. Design of the Resonant Components in the Phase Without an SCC Circuit

In phase 1, the designed parameters include L_{r1}, C_{r1}, and L_{m1}. The parameters are designed the same as conventional LLC converters. The design objective is to meet the desired frequency range and all the input/output voltage and load regulations and at the same time reduce the circuiting current at the primary side of the transformers.

The worst-case scenario for the voltage gain requirement of the LLC tank happens with the highest load condition at the highest gain. The highest gain within the normal operating condition is for 320 V input to 16 V output and the hardest voltage gain to achieve is for the nominal operating condition with the highest load. The minimum voltage gain condition is for 450 V input to 9 V output and 140 A load. Hence, the following corner conditions are listed below for half of the rated power of the EV APM:

(1) The maximum voltage Gain = $44 \times 16 / 320 = 2.2$ for 320 V input, 16 V output, and 122.5 A load.

(2) The worst-case scenario with Gain = $44 \times 14 / 320 = 1.93$ for 320V input, 14 V output, and 140 A load.

(3) The minimum voltage Gain = $44 \times 9 / 450 = 0.88$ for 450 V input, 9 V output, and 140 A load.

The design is based on the well-known First Harmonic Approximation (FHA) method, and then the parameters need to be fine-tuned based on computer simulation results to consider the effect of higher-order harmonics. It should be mentioned that the voltage gain found from FHA is lower than the actual gain for the switching frequencies far away from the resonant frequency. In this design, the desired resonant frequency is between 450 kHz to 500 kHz. A relatively small quality factor

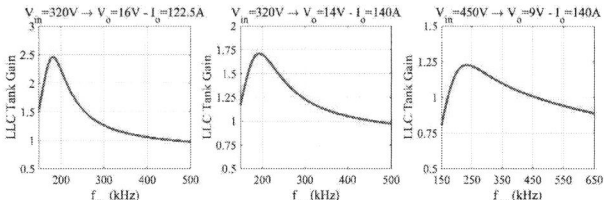

Fig. 2. The plot of voltage gain versus switching frequency at different operating conditions.

(i.e., $Q = 0.3$) is considered for the worst-case scenario, as a wide input/output voltage range and load range need to be accommodated. Moreover, the circulating currents can be reduced by using a large inductance ratio. Hence, a relatively large inductance ratio is considered in the design (i.e., $L_m/L_r = 6$). The final resonant tank parameters of phase 1 are listed in TABLE I. Fig. 2 illustrates the voltage again curves for different operating conditions. As can be observed the voltage gain of 2.2 is achievable for the first case (left curve) and the voltage gain of 0.88 is also achievable for the third case (right curve). As mentioned before, the hardest gain is for the nominal condition at rated power that is shown in the middle curve of Fig. 2. Although the required 1.93 voltage gain is not achieved with this design using FHA, the simulation results show that the achievable voltage gain using the designed resonant parameters is more than 1.93.

TABLE I.   PARAMETERS OF PHASE 1 OF THE PROPOSED CONVERTER

Parameters	Values
L_r inductance	15 µH
L_m inductance	90 µH
C_r capacitance	8 nF
Transformer turn ratio ($n{:}1{:}1$)	22:1:1

B. Design of the Resonant Components in the Phase With an SCC Circuit

In phase 2, the magnetics are identical to phase 1 (i.e., $L_{r2} = L_{r1}$ and $L_{m2} = L_{m1}$), and hence only the resonant capacitance C_{r2} and the SCC capacitance C_{a2} need to be designed. The design criteria are such that the voltage gain of phase 2 is kept below the voltage gain of phase 1 over the switching frequency range in all operating conditions. Both capacitances are designed based on the operation principles of the SCC circuit.

By modulating the SCC switches, the equivalent resonant capacitance in phase 2 is modified, which alters the impedance of phase 2. When the SCC modulation angle (i.e., α) is a certain value between the minimum and the maximum, the impedance of phase 1 matches that of phase 2, and current sharing is achieved. If α increases, the impedances of phase 2 increase, and phase 2 carry smaller currents. On the contrary, if α decreases, the impedances of phase 2 decrease, and phase 2 carry larger currents. Thus, the design criteria can be summarized as follows:

(1) When α is maximum, in any components' tolerances and operation points, phase 2 carries less current than phase 1.

(2) When α is minimum, in any components' tolerances and operation points, phase 2 carries more current than phase 1.

The components' tolerances are the main reason causing impedances unmatched and current unbalancing, which is compensated by the SCC circuit. In this design, +/-5% tolerances for all the resonant elements and the SCC capacitor are considered. The two design criteria mentioned above can be expressed by

$$I_{o2}\big(M, f_{sw}, L_{m2}, L_{r2}, C_{r-eq(\alpha=max)}\big) \quad (2)$$
$$< I_{o1}\big(M, f_{sw}, L_{m1}, L_{r1}, C_{r1}\big)$$

$$I_{o2}\big(M, f_{sw}, L_{m2}, L_{r2}, C_{r-eq(\alpha=min)}\big) \quad (3)$$
$$> I_{o1}\big(M, f_{sw}, L_{m1}, L_{r1}, C_{r1}\big)$$

where I_{o1} and I_{o2} are the output currents of phase 1 and phase 2, respectively. M is the LLC tank voltage gain, f_{sw} is the switching frequency, $C_{r-eq(\alpha=max)}$ and $C_{r-eq(\alpha=min)}$ are the equivalent capacitances when the SCC modulation angles are maximum and minimum, respectively. Note that M and f_{sw} indicate different operating conditions. The equivalent resonant capacitance C_{r-eq} satisfies,

$$\frac{C_{a2}C_{r2}}{C_{a2}+C_{r2}} < C_{r-eq(\alpha=min)} < C_{r-eq} < C_{r-eq(\alpha=max)} < C_{r2} \quad (4)$$

When components' tolerances are considered, the following inequations can be found

$$I_{o1}(M, f_{sw}, L_{m1}, L_{r1}, C_{r1}) \quad (5)$$
$$\leq I_{o1}\big(M, f_{sw}, L_{m1(min)}, L_{r1(min)}, C_{r1(min)}\big)$$

$$I_{o1}(M, f_{sw}, L_{m1}, L_{r1}, C_{r1}) \quad (6)$$
$$\geq I_{o1}\big(M, f_{sw}, L_{m1(max)}, L_{r1(max)}, C_{r1(max)}\big)$$

$$I_{o2}\big(M, f_{sw}, L_{m2}, L_{r2}, C_{r-eq(\alpha=max)}\big) \quad (7)$$
$$\leq I_{o2}\big(M, f_{sw}, L_{m2(min)}, L_{r2(min)}, C_{r-eq(\alpha=max,C_{a2}=min,C_{r2}=min)}\big)$$

$$I_{o2}\big(M, f_{sw}, L_{m2}, L_{r2}, C_{r-eq(\alpha=min)}\big) \quad (8)$$
$$\geq I_{o2}\big(M, f_{sw}, L_{m2(max)}, L_{r2(max)}, C_{r-eq(\alpha=min,C_{a2}=max,C_{r2}=max)}\big)$$

where the subscripts _min_ and _max_ indicate the minimum and maximum values of the associated variables due to components' tolerances. $C_{r-eq(\alpha=max,C_{a2}=min,C_{r2}=min)}$ is the equivalent resonant capacitance when α is maximum, C_{a2} is minimum and C_{r2} is minimum, and $C_{r-eq(\alpha=min,C_{a2}=max,C_{r2}=max)}$ is the equivalent resonant capacitance when α is minimum, C_{a2} is maximum and C_{r2} is maximum.

Reminding that increasing any component's value in a passive impedance network will increase the total impedance and a larger impedance will contribute to a smaller output current. Hence, the two criteria in equations (2) and (3) can be expressed as follows:

$$I_{o2}\big(M, f_{sw}, L_{m2(min)}, L_{r2(min)}, C_{r-eq(\alpha=max,C_{a2}=min,C_{r2}=min)}\big) \quad (9)$$
$$< I_{o1}\big(M, f_{sw}, L_{m1(max)}, L_{r1(max)}, C_{r1(max)}\big)$$

$$I_{o2}\big(M, f_{sw}, L_{m2(max)}, L_{r2(max)}, C_{r-eq(\alpha=min,C_{a2}=max,C_{r2}=max)}\big) \quad (10)$$
$$> I_{o1}\big(M, f_{sw}, L_{m1(min)}, L_{r1(min)}, C_{r1(min)}\big)$$

where $X_{max} = 1.05 \times X$ and $X_{min} = 0.95 \times X$ considering +/-5% tolerances, X represents L_{m1}, L_{m2}, L_{r1}, L_{r2}, C_{r1}, C_{r2}, and C_{a2}. The two equivalent capacitances $C_{r-eq(\alpha=max,C_{a2}=min,C_{r2}=min)}$ and $C_{r-eq(\alpha=min,C_{a2}=max,C_{r2}=max)}$ which satisfy the design criteria are determined first.

The current sharing performance becomes worse at heavy loads, and it is less influenced by the voltage gain. The two equivalent capacitances are determined first in the heaviest load operation at the nominal condition voltage gain, then are verified and adjusted in other operation points. After some iterations, the two equivalent capacitances are determined as

$$C_{r-eq(\alpha=max,C_{a2}=min,C_{r2}=min)} = 10 \ nF \quad (11)$$

$$C_{r-eq(\alpha=min,C_{a2}=max,C_{r2}=max)} = 6 \ nF \quad (12)$$

With the help of the expression of the equivalent resonant capacitance and computer simulations, and considering the minimum and maximum α angles, C_{r2} and C_{a2} are designed as $C_{r2} = 11$ nF, $C_{a2} = 9.5$ nF. Hence, the final LLC tank parameters of the proposed converter are listed in TABLE II.

TABLE II.   DESIGNED PARAMETERS OF THE PROPOSED DUAL-PHASE LLC DC-DC CONVERTER

Parameters	Phase 1	Phase 2
L_r inductance	15 µH	15 µH
L_m inductance	90 µH	90 µH
C_r capacitance	8 nF	11 nF
C_a capacitance	-	9.5 nF
Transformer turn ratio (n:1:1)	22:1:1	22:1:1

IV. SIMULATION AND EXPERIMENTAL VERIFICATIONS

A. Simulation Results

In order to verify the theoretical design of the two-phase SCC-LLC converter some computer simulations have been done in the PSIM environment. The parameters used in the simulation are the same as the resonant tank parameters designed in the previous section and a maximum of 5% tolerance is considered for the resonant tank components of both phases. The test conditions are (1) V_{in}=320 V to V_o=16 V at rated output power that is equal to I_o=245 A, (2) V_{in}=320 V to V_o=14 V at rated output power with I_o=280 A, and (3) V_{in}=450 V to V_o=9 V at the rated output current of I_o=280 A. Fig. 3 illustrates the simulation results with +5% on the resonant components of phase 1 and with -5 % on the resonant components of phase 2. Fig. 4 illustrates the simulation results with -5% on the resonant components of phase 1 and with +5 % on the resonant components of phase 2. It can be observed that in both extreme cases the SCC circuit was able to achieve current balancing for all corner cases.

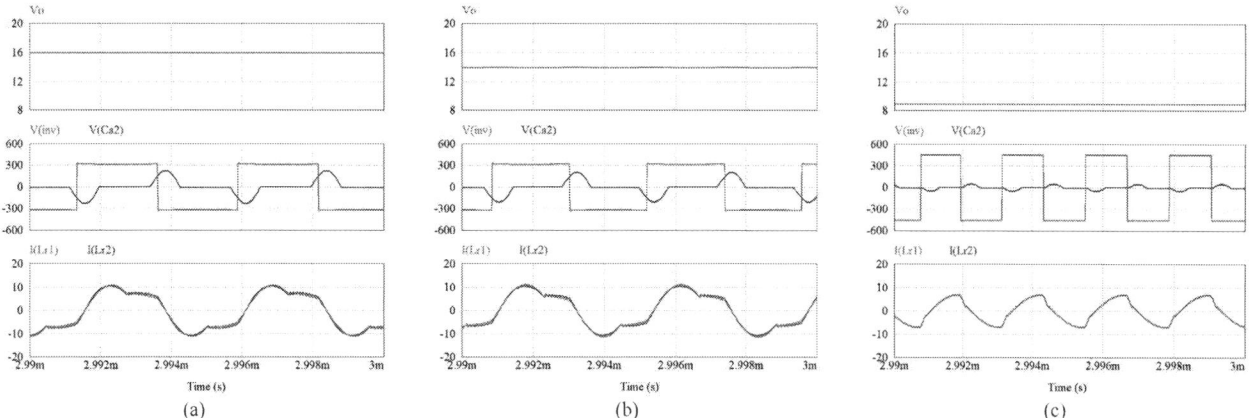

Fig. 3. Simulation results of the proposed dual-phase design for different conditions considering extreme component tolerances as L_{r1}+5%, L_{m1}+5%, C_{r1}+5%, L_{r2}-5%, L_{m2}-5%, C_{r2}-5%, C_{a2}-5%, (a) V_{in}=320V, V_o=16V, I_o=250A, f_{sw}=218.8k, α=145°, (b) V_{in}=320V, V_o=14V, I_o=280A, f_{sw}=229.2k α=144°, and (c) V_{in}=450V, V_o=9V, I_o=280A, f_{sw}=421.8k α=141°.

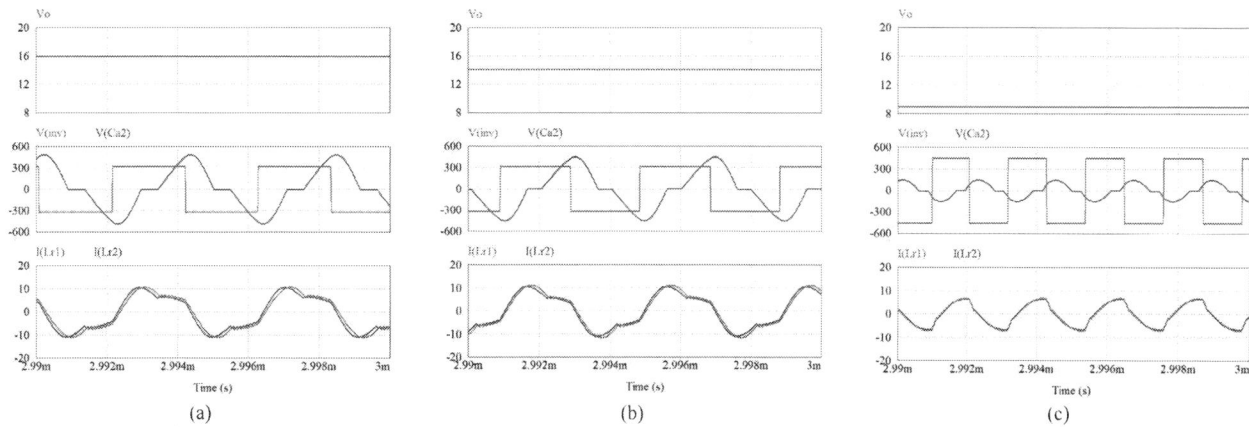

Fig. 4. Simulation results of the proposed dual-phase design for different conditions considering extreme component tolerances as L_{r1}-5%, L_{m1}-5%, C_{r1}-5%, L_{r2}+5%, L_{m2}+5%, C_{r2}+5%, C_{a2}+5%, (a) V_{in}=320V, V_o=16V, I_o=225A, f_{sw}=241.8k α=99°, (b) V_{in}=320V, V_o=14V, I_o=280A, f_{sw}=252.6k α=95°, and (c) V_{in}=450V, V_o=9V, I_o=280A, f_{sw}=454.2k α=116°

B. Experimental results

A 4 kW laboratory prototype is built to test the performance of the proposed two-phase SCC-LLC converter. The general system specifications used for the design of the prototype and the components used in the prototype are listed in TABLE III. Fig. 5 shows the 4 kW EV APM prototype that is mounted on a cold plate for proper thermal dissipation. The dimensions of the

TABLE III. SYSTEM SPECIFICATIONS AND PARAMETERS OF EACH PHASE IMPLEMENTED IN THE EXPERIMENTAL SETUP

Parameters/Descriptions	Values/Part Number
Input voltage range	250 V – 450 V
Output voltage range	10 V – 16 V
Rated output voltage	14 V
Maximum output current	285A
Maximum output power	4 kW (2 kW × 2)
Transformer	n = 22:1:1, (× 2)
Parallel inductor	L_{m1} = 85µH, L_{m2} = 85µH, (PQ32/20 - 3C97)
Series inductor	L_{r1} = 15µH, L_{r2} = 15µH, (PQ32/20 - 3C97)
Series capacitor of phase 1	C_{r1} = 8.1nF (4.7nF × 4 + 6.8nF × 2), (CGA6M1C0G3A472J200AE × 4 + CGA6M1C0G3A682J200AE × 2)
Series capacitor of phase 2	C_{r2} = 10.9nF (6.8nF × 2 + 15nF × 2), (CGA6M1C0G3A682J200AE × 2 + C3225C0G3A153J250AC × 2)
SCC capacitor of phase 2	C_{a2} = 9.4nF (4.7nF + 4.7nF), (CGA6M1C0G3A472J200AE × 2)
Primary side switches	650V, 30A, (GS66508B × 4)
Secondary side switches	40V, 250A, (IAUA250N04S6N007AUMA1 × 8)
SCC switches	650V, 69A, (IPT65R033 × 2)
Output capacitor	620µF (10µF × 31 × 2) + 180uF (10uF × 18, (12105C106K4Z2A × 80)
Micro-controller	TC375TP96F300WAAKXUMA1

979-8-3315-1612-3/25 $31.00 © 2025 IEEE

prototype are 22cm × 17.2cm × 2.6cm resulting in a 4 kW/L power density. It should be noted that phase shedding is employed to enhance efficiency across the load range. For load currents up to approximately 120 A, only phase 1 operates, while for currents between 120 A and 280 A, both phases are active to share the load.

Fig. 6 presents the experimental waveforms for two-phase operation under different input voltages and a nominal output voltage at the rated current. It can be observed that, under all operating conditions, the resonant currents in both phases are highly similar, indicating a well-balanced load distribution between the two phases. It should be mentioned that the output current of each phase is measured and compared within a 5 A hysteresis band to modulate the SCC circuit for effective current sharing. Fig. 7 illustrates the current difference between the phases across the tested load range, from 100 A to 280 A, under

different input voltage conditions. It can be observed that current sharing is successfully maintained within the hysteresis band.

For further testing with thermal performance consideration, the prototype was tested with the liquid cooling temperature set to 40°C and 65°C. The thermal images operations with V_{in}=380 V and a fixed V_o=14 V at rated current with 40°C coolant temperature are illustrated in Fig. 8. It should be mentioned that in all cases the circuit was operating for more than 10 minutes so the component temperatures were stabilized. It can be observed that the SR of phase 2 operates at a slightly higher temperature that could be due to the SCC operation. The maximum operating junction temperature of the MOSFETs used for the SR is 175°C. Therefore, the implemented EV APM prototype can reliably operate under rated power conditions with SR temperatures of up to 100°C. Additionally, the temperatures of the parallel inductors remain stable between the two phases,

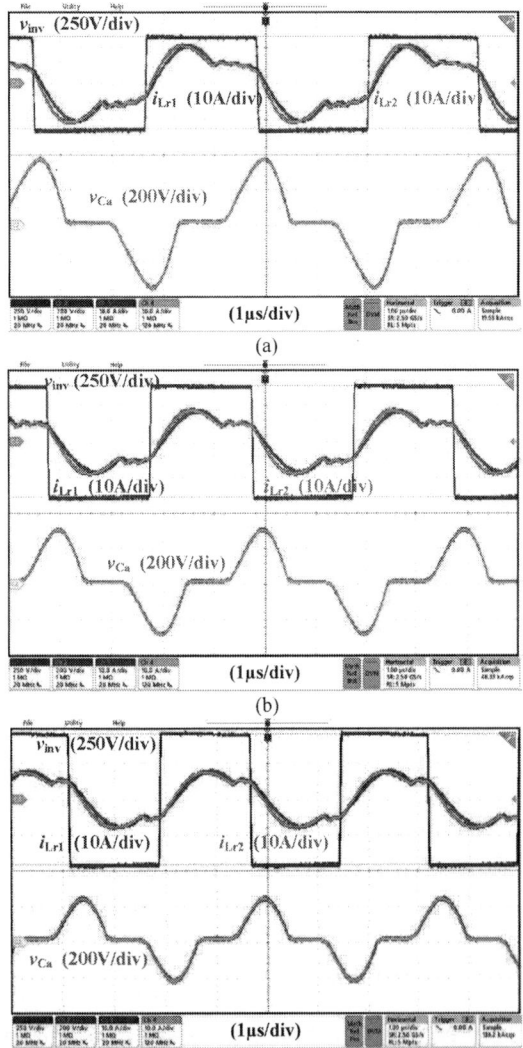

Fig. 6. Experimental results of the two-phase operation with SCC operation, (a) V_{in}=320 V to V_o=14 V and I_o=285 A, (b) V_{in}=380 V to V_o=14 V and I_o=280 A, and (c) V_{in}=450 V to V_o=14 V and I_o=285 A.

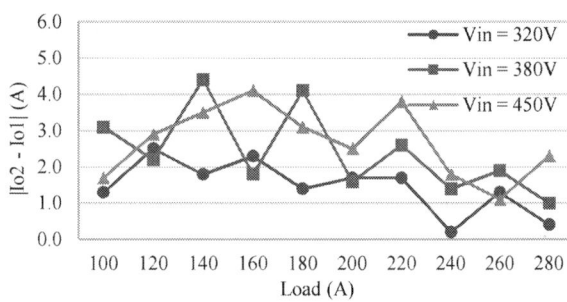

Fig. 7. Current sharing performance of the proposed dual-phase LLC converter from 100 A to 280 A load current.

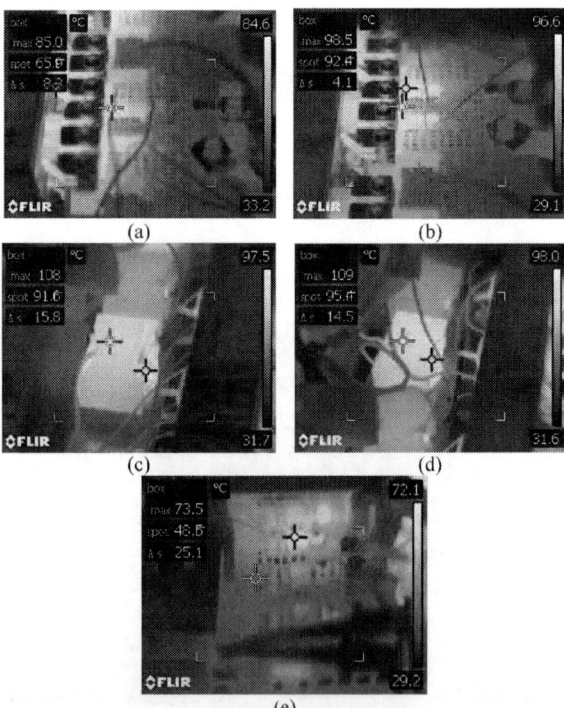

Fig. 8. Thermal images captured for dual-phase operation for V_{in}=380 V, V_o=14 V, and I_o=280 A with 40°C coolant temperature, (a) SR of phase 1, (b) SR of phase 2, (c) L_{m1}, (d) L_{m2}, and (e) GaN and SCC switches.

979-8-3315-1612-3/25 $31.00 © 2025 IEEE

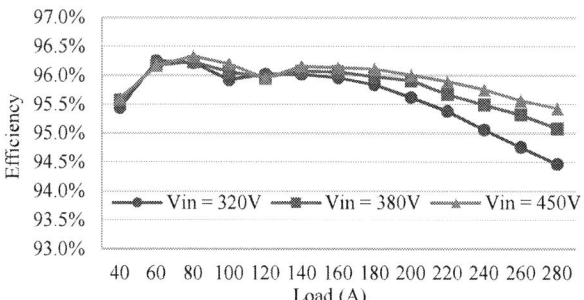

Fig. 9. Efficiency measurement results of the proposed two-phase SCC-LLC converter over the load range with 40℃ coolant temperature.

and the maximum temperature of the GaN devices and SCC MOSFETs are well below 100℃.

Fig. 9 illustrates the efficiency curves for two-phase operation at different input voltage levels and V_o=14 V over the load range, with a coolant temperature of 40°C. A flat efficiency curve exceeding 95% is observed across the load range. Up to a load current of 120 A, only phase 1 operates, while beyond 120 A, both phases operate to share the load. The average efficiencies for V_{in}=320 V, V_{in}=380 V, and V_{in}=450 V are calculated as 95.6%, 95.8%, and 95.9%, respectively.

V. Conclusion

This paper introduces a novel resonant tank design for a dual-phase LLC converter in APM applications, reducing the cost and complexity of conventional multi-phase designs. Unlike traditional approaches requiring SCC circuits on each phase, the proposed method utilized a single SCC circuit on one phase while achieving current sharing across a wide voltage range. In the design procedure the impedance of the SCC-equipped phase is deliberately kept lower to ensure effective current balancing, even under ±5% component tolerances. Computer simulations and experimental results from a 4 kW EV APM prototype validated the proposed design method in achieving accurate current sharing. Experimental measurements demonstrated successful current balancing within a 5 A limit, achieving a power density of 4 kW/L and a peak efficiency of 96.3%.

References

[1] C. Wang, P. Zheng and J. Bauman, "A Review of Electric Vehicle Auxiliary Power Modules: Challenges, Topologies, and Future Trends," *IEEE Trans. Power Electron.*, vol. 38, no. 9, pp. 11233-11244, Sept. 2023, doi: 10.1109/TPEL.2023.3288393

[2] R. L. Steigerwald and K. D. Ngo, "Full-bridge lossless switching converter," US Patent US4864479A, 1989.

[3] L. Maheshwari, A. A. Keraj, S. J. Shah, K. A. Khan, I. Makda, H. Qamar and A. Usman., "Efficient High-Frequency GaN-Based Phase Shifted Full Bridge Zero-Voltage Switching Converter for Electric Vehicle Auxiliary Power Modules," In *Proc. IEEE Workshop on Control and Modeling Power Electron.*, Lahore, Pakistan, 2024, pp. 1-6, doi: 10.1109/COMPEL57542.2024.10614037.

[4] K. Vangen, T. Melaa, A. K. Adnanes and P. E. Kristiansen, "Dual active bridge converter with large soft-switching range," in *Proc. Fifth Euro. Conf. Power Electron. Appl.*, Brighton, UK, 1993, pp. 328-333 vol.3.

[5] S. Chaurasiya and B. Singh, "A Bidirectional Fast EV Charger for Wide Voltage Range Using Three-Level DAB Based on Current and Voltage Stress Optimization," *IEEE Trans. Transp. Electrific.*, vol. 9, no. 1, pp. 1330-1340, March 2023, doi: 10.1109/TTE.2022.3201979.

[6] K. Siebke, M. Giacomazzo and R. Mallwitz, "Design of a Dual Active Bridge Converter for On-Board Vehicle Chargers using GaN and into Transformer Integrated Series Inductance," in *Proc. IEEE 22nd Euro. Conf. Power Electron. Appl. (EPE'20 ECCE Europe)*, Lyon, France, 2020, pp. 1-8, doi: 10.23919/EPE20ECCEEurope43536.2020.9215962.

[7] R. W. A. A. D. Doncker, D. M. Divan, and M. H. Kheraluwala, "A three-phase soft-switched high-power-density DC/DC converter for high-power applications," *IEEE Trans. Ind. Appl.*, vol. 27, no. 1, pp. 63-73, Jan.-Feb. 1991, doi: 10.1109/28.67533.

[8] L. Zhu, H. Bai, A. Brown, and M. McAmmond, "Design a 400 V–12 V 6 kW Bidirectional Auxiliary Power Module for Electric or Autonomous Vehicles With Fast Precharge Dynamics and Zero DC-Bias Current," *IEEE Trans. Power Electron.*, vol. 36, no. 5, pp. 5323-5335, Oct. 2021, doi: 10.1109/TPEL.2020.3028361.

[9] G. C. Knabben, J. Scähfer, J. W. Kolar, G. Zulauf, M. J. Kasper, and G. Deboy, "Wide-Input-Voltage-Range 3 kW DC-DC Converter with Hybrid LLC & Boundary / Discontinuous Mode Control," in *Proc. IEEE Appl. Power Electron. Conf. Expo.*, New Orleans, LA, USA, 15-19 March 2020 2020, pp. 1359-1366, doi: 10.1109/APEC39645.2020.9124410.

[10] G. Yang, P. Dubus, and D. Sadarnac, "Double-Phase High-Efficiency, Wide Load Range High- Voltage/Low-Voltage LLC DC/DC Converter for Electric/Hybrid Vehicles," *IEEE Trans. Power Electron.*, vol. 30, no. 4, pp. 1876-1886, Jun. 2015, doi: 10.1109/TPEL.2014.2328554.

[11] "Two-Phase interleaved LLC resonant converter design with C2000™ microcontrollers," Texas Instrument, Dallas, TX, USA, Technical Documents TIDUCT9, Dallas, TX, USA, Jan. 2017. Accessed: May 2024. [Online]. Available: https://www.ti.com/lit/pdf/tiduct9

[12] H. Wang, Y. Chen, Y. -F. Liu, J. Afsharian, and Z. Yang, "A Passive Current Sharing Method With Common Inductor Multiphase LLC Resonant Converter," *IEEE Trans. Power Electron.*, vol. 32, no. 9, pp. 6994-7010, Sep. 2017, doi: 10.1109/TPEL.2016.2626312.

[13] M. Forouzesh, B. Sheng and Y. -F. Liu, "Interleaved SCC-LCLC Converter with TO-220 GaN HEMTs and Accurate Current Sharing for Wide Operating Range in Data Center Application," in *Proc. IEEE Appl. Power Electron. Conf. Expo.*, New Orleans, LA, USA, 2020, pp. 482-489, doi: 10.1109/APEC39645.2020.9124489.

[14] X. Zhou, B. Sheng, W. Liu, Y. Chen, L. Wang, Y. -F. Liu and P. C. Sen, "A High-Efficiency High-Power-Density On-Board Low-Voltage DC-DC Converter for Electric Vehicles Application," *IEEE Trans. Power Electron.*, vol. 36, no. 11, pp. 12781-12794, Apr. 2021, doi: 10.1109/TPEL.2021.3076773.

New Reverse Mode Control Method of Phase-Shift Full-Bridge Converter for Bidirectional Auxiliary Power Module

Jongyoon Chae

Division of Future Vehicle
KAIST
Daejeon, South Korea
artalex@kaist.ac.kr

Dongmin Kim
Electronics and
Telecommunications Research
Institute (ETRI)
Daejeon, South Korea
dmkim@etri.re.kr

Dongmin Choi
Department of Electrical
Engineering
KAIST
Daejeon, South Korea
dmdm0402@kaist.ac.kr

Gun-Woo Moon
Department of Electrical
Engineering
KAIST
Daejeon, South Korea
gwmoon@kaist.ac.kr

Abstract— **This paper introduces a novel reverse mode control strategy of a phase-shift full-bridge (PSFB) converter for bidirectional auxiliary power modules (APM) in electric vehicles. The proposed method utilizes a new leading leg phase delay (LLPD) modulation to reduce excessive voltage overshoots on the current-fed side switches. Small-signal modeling and relative gain array (RGA) analysis confirm the applicability of the decentralized control structure for the proposed method. Moreover, the control method to extend the zero-voltage switching (ZVS) range for lagging leg switches is presented. The effectiveness of the proposed control method is experimentally verified through a 500-W PSFB converter prototype.**

Keywords—Electric Vehicle (EV), Auxiliary Power Module (APM), Bidirectional DC-DC Converter, Phase-Shift Full Bridge Converter, Decentralized Control, Switch Modulation Method

I. INTRODUCTION

As greenhouse gas emissions and environmental pollution have become a significant concern, various research had focused on the electrification of vehicles, since the conventional internal combustion engine vehicles take up a big portion of total CO_2 emissions. As a result, various electrified vehicles, so-called xEVs, such as fuel cell electric vehicles (FCEVs), plug-in hybrid electric vehicles (PHEVs), and battery electric vehicles (BEVs) have been commercialized and have become mainstream nowadays.

The auxiliary power module (APM) plays a key role in the abovementioned xEVs. APM converts power from a high voltage (HV) battery to charge low voltage (LV) battery, and to supply power to electronic devices inside the vehicle. At the early stage of xEVs, required power capacity of APM was only about hundreds of watts. However, nowadays, due to the advent of advanced features including autonomous driving, the required power capacity of APM has increased up to the kilo-watt range.

APM usually requires galvanic isolation, step-down voltage gain, high power capacity and high power efficiency. Among the various topologies, the phase-shift full-bridge (PSFB) converter, shown in Fig.1, is one of the most promising topologies for the APM [1], [2]. The PSFB converter is a buck-derived topology that offers galvanic isolation and a step-down voltage ratio. Output *LC* filter on the LV side guarantees low output current ripple which is advantageous in terms of efficiency and battery safety. Moreover, during forward

Jongyoon Chae and Dongmin Kim are co-first authors.

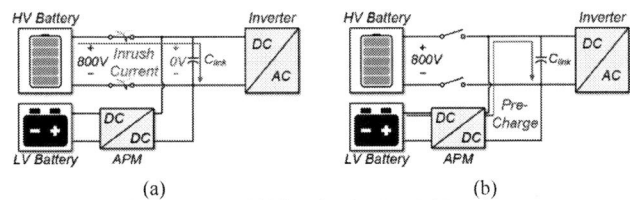

Fig.1. Circuit diagram of the phase-shift full-bridge (PSFB) converter.

Fig.2. Diagram for necessity of bidirectional APM. (a) Inrush current occurs to voltage potential difference. (b) Pre-charging of C_{link} through bidirectional operation of APM.

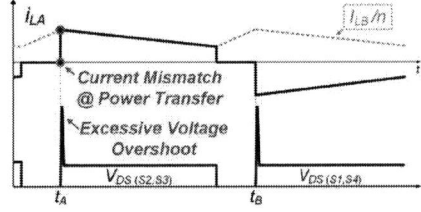

Fig.3. Voltage overshoot in conventional reverse mode PSFB converter

operation, zero-voltage switching (ZVS) of entire switches can be realized. Since more than a hundred Amps of current flows on the LV side, utilizing low on-resistance ($R_{ds(on)}$) switches rather than diodes for rectification is preferred.

Recently, the demand for the bidirectional operation of auxiliary power modules (APMs) in xEVs has increased [3]. For example, as depicted in Fig.2 (a), when the xEV turns on the relays to drive the traction motor through an inverter, an excessive inrush current flows into the inverter's link capacitor (C_{link}). This inrush current due to the severe voltage potential difference between C_{link} and HV battery may pose a significant damage to the system. A pre-charge relay and current limiting resistor [4], [5] can be a solution but this increases the overall volume, cost, and weight. If the APM is capable of pre-charging C_{link} in prior through the reverse operation, as in Fig.2 (b), it would be an effective solution for reducing the inrush current without any expenses.

979-8-3315-1612-3/25 $31.00 © 2025 IEEE

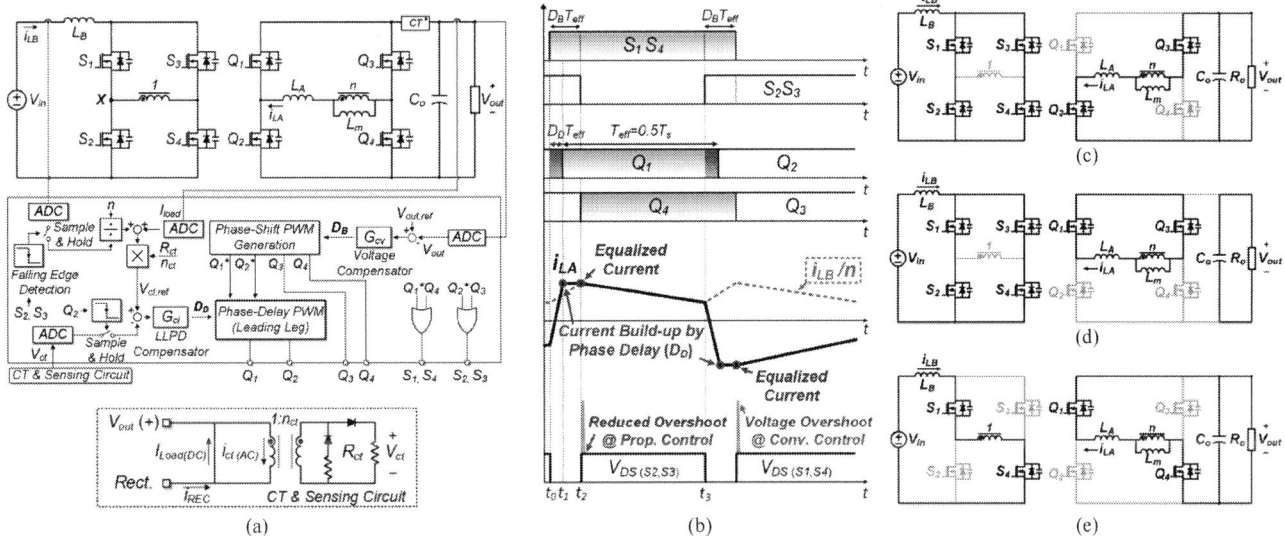

Fig. 4. Proposed reverse mode control of PSFB converter. (a) Control block diagram. (b) Key waveforms. (c) Circuit operation at Mode 1 (t_0-t_1). (d) Circuit operation at Mode 2 (t_1-t_2). (e) Circuit operation at Mode 3 (t_2-t_3).

As the PSFB converter utilizes switches (S_1~S_4) for synchronous rectification during the forward operation, reverse operation is also possible. However, if the switches are modulated the same as the forward operation, excessive voltage overshoot is induced on the current-fed side switches (S_1~S_4). As in Fig. 3, this is due to the current mismatch between the inductors at the transition (t_A, t_B) to power transfer mode [6]. RCD snubbers and clamp circuits [7], [8] can reduce voltage overshoot, but these degrade the efficiency or power density. [6] and [9] suggested a switch modulation that can reduce the voltage overshoot by equalizing the current of the voltage-fed side reflected boost inductor and other inductances prior to a power transfer. However, open loop control of [6] and [9] leads to discrepancy between currents due to the parasitic parameters, which results in circulating current or voltage overshoot.

In this work, a new reverse mode control method of the PSFB converter is proposed. Leading leg phase delay (LLPD) control through the feedback loop equalizes currents before power delivery and removes the voltage overshoot. Moreover, based on the small signal model and relative gain array (RGA) analysis, output voltage and current build-up are controlled in a decentralized structure.

II. ANALYSIS OF THE PROPOSED CONTROL METHOD

A. Operation of the Proposed Reverse Mode Control Method

Fig. 4 (a) and (b) show the control block diagram and key waveforms of the proposed control method, respectively. As the PSFB converter operates symmetrically under this control method, only the three modes in the half-cycle are analyzed. Each operating modes are depicted in Fig. 4(c), (d) and (e). The following assumptions are made to simplify the analysis:
1) Leakage inductance and series inductor are lumped together and considered a single auxiliary inductor (L_A).
2) Magnetizing inductance (L_m) is much larger than auxiliary inductor (L_A).
3) The transient time of the drain-source voltage (V_{DS}) and the dead time is short enough to be disregarded.

Mode 1 [$t_0 - t_1$]: Mode 1 starts when the S_1 and S_4 switches turn on. On the LV side, the energy is stored on the boost inductor (L_B), and no power is transferred to the HV side. Auxiliary inductor current (i_{LA}) gradually increases due to the phase delay introduced on the HV-side switch (Q_2). Q_2 turns off when i_{LA} is built up to $I_{LA,ref}$ of (1), and this mode ends.

$$I_{LA,ref} = \frac{i_{LB}(t_2)}{n} \qquad (1)$$

Mode 2 [$t_1 - t_2$]: Mode 2 begins when Q_1 turns on. The LV side operates the same as mode 1, where the energy is stored on the L_B. On the HV side, the i_{LA} circulates through Q_1 and Q_3.

Mode 3 [$t_2 - t_3$]: Mode 3 starts when the S_2 and S_3 switches turn off, transferring the energy stored in the L_B to the HV side. As i_{LA} equals i_{LB}/n at the beginning of the power transfer (t_2), voltage overshoot on S_2 and S_3 is reduced. Power transfer continues until the S_2 and S_3 switches turn back on.

In the conventional control method, i_{LA} and i_{LB}/n mismatches at the transition point (t_2) to power delivery operation. Thus, an excessive voltage overshoot is induced on the LV side switches to build up the i_{LA} and match it with i_{LB}/n in an instant. The proposed method gradually increases i_{LA} through leading leg phase delay (LLPD) control during mode 1 to match the i_{LA} and i_{LB}/n prior to power delivery. Therefore, the proposed control can successfully reduce the voltage overshoots on LV switches.

As shown in Fig. 4(a), the proposed method employs a decentralized control structure with two independent feedback loops: *output voltage control* and *LLPD control*. The output voltage is regulated by D_B, which determines the energy storage duration on the LV side. The LLPD control loop regulates i_{LA} to the reference value $i_{LA,ref}$ by controlling D_D, which determines the phase delay applied to the leading leg switches. Unlike open-loop control methods [6], [9], the proposed method can remove the voltage overshoot regardless of parasitic parameters since it is based on feedback control. The applicability of the decentralized control structure is confirmed through small-signal modeling and relative gain array (RGA) analysis throughout the rest of this section.

979-8-3315-1612-3/25 $31.00 © 2025 IEEE   1900

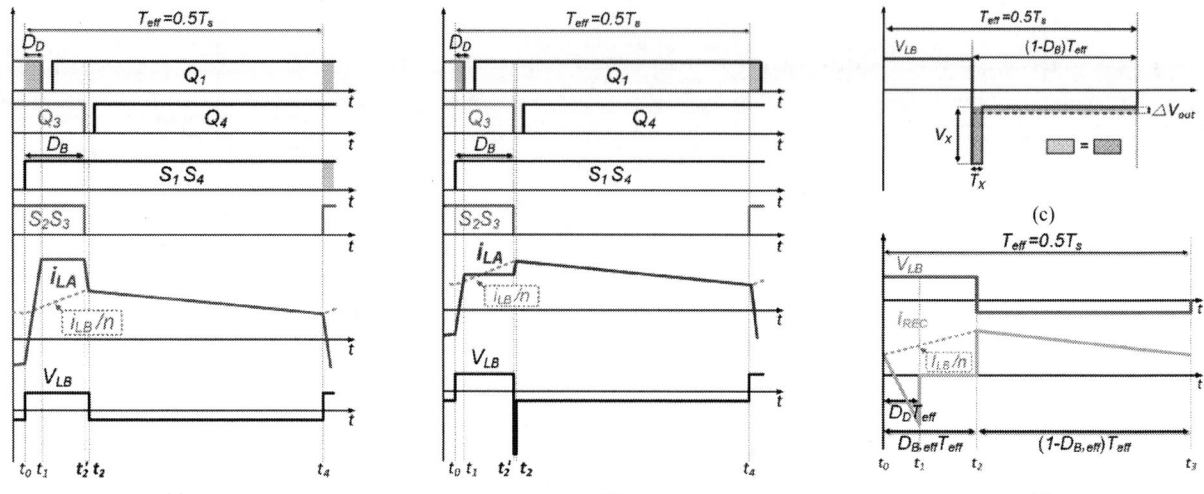

Fig. 5. Key waveforms for the small signal analysis of the proposed control method. (a) Case 1: $D_D > D_{D,opt}$ (b) Case 2: $D_D < D_{D,opt}$ (c) Inductor voltage waveform (V_{LB}) in Case 2. (d) Key waveforms based on effective boost duty ratio ($D_{B,eff}$).

B. Small Signal Modeling of the Proposed Control Method

When the phase delay control variable (D_D) is an optimal value ($D_{D,opt}$) ($i_{LA}(t_2)$ equals $i_{LB}(t_2)/n$ in steady state), the voltage gain is only determined by the boost duty ratio (D_B). However, in the case where D_D is larger or smaller than $D_{D,opt}$ ($i_{LA}(t_2)$ larger or smaller than $i_{LB}(t_2)/n$), an additional time duration to match two currents is created before the power delivery, resulting in an increase or decrease in the voltage gain. This phenomenon should be considered for a small signal model. The effective boost duty ratio ($D_{B,eff}$) reflecting the decrease and increase of voltage gain in two cases is analyzed as follows.

Fig. 5(a) shows the key waveforms of Case 1 where i_{LA} is built up larger than $i_{LB}(t_2)/n$. Since the inductor current is continuous during the steady state, an additional time duration ($t_2' \sim t_2$) is created before the power delivery to match the two currents. During this period, V_{out} is applied on L_A, resulting in zero voltage on the transformer. From the perspective of the boost inductor (L_B), the extended energy storage period by (2) increases the voltage gain, which can be expressed through $D_{B,eff}$ as (3).

Fig. 5(b) shows the key waveforms of Case 2 where i_{LA} is built up smaller than $i_{LB}(t_2)/n$. Similar to Case 1, an additional time duration (T_X) to equalize two currents is created. However, in this case, an excessive surge voltage is induced and applied to L_A. Therefore, a large negative voltage applied to L_B decreases the steady-state output voltage.

The voltage on node X (V_X) in Fig. 4(a) during T_X is as (4) [6], which is assumed to be a square wave for simplicity. T_X is expressed as (5), where C_{oss} is the output capacitance of the switches. Steady-state output voltage deviation (ΔV_{out}) is analyzed based on Fig. 5(c), where the gray dash indicates V_{LB} in the ideal case ($D_D=D_{opt}$), and the red line denotes V_{LB} in this case. As V_{in} and D_B stays the same in both cases, gray and red shaded areas are the same due to the voltage-second balance on the inductor, which is as (6). Therefore, a decrease in voltage gain can be also expressed through $D_{B,eff}$ as (3).

The effect of D_D on voltage gain is expressed through $D_{B,eff}$

$$t_2 - t_2' = \frac{L_A \cdot \left(i_{LA}(t_1) - \frac{i_{LB}(t_2)}{n}\right)}{V_{out}} \tag{2}$$

$$D_{B,eff} = D_B + \frac{L_A \cdot \left(i_{LA}(t_1) - \frac{i_{LB}(t_2)}{n}\right)}{V_{out} \cdot T_{eff}} \tag{3}$$

$$V_X = [i_{LB}(t_2) - n \cdot i_{LA}(t_1)] \cdot \sqrt{\frac{L_A}{2 \cdot n^2 \cdot C_{oss}}} \tag{4}$$

$$T_X = t_2 - t_2' = \sqrt{\frac{2 \cdot L_A \cdot C_{oss}}{n^2}} \tag{5}$$

$$T_X \cdot V_X = T_{eff} \cdot (1 - D_B) \cdot \frac{\Delta V_{out}}{n} \tag{6}$$

$$\frac{di_{LB}}{dt} = \begin{cases} \dfrac{V_{in}}{L_B} \cdots (t_0 \le t \le D_{B,eff}T_{eff}) \\ \dfrac{V_{in}}{L_B} - \dfrac{v_{out}}{nL_B} \cdots (D_{B,eff}T_{eff} \le t \le T_{eff}) \end{cases} \tag{7}$$

$$\frac{dv_{out}}{dt} = \begin{cases} \dfrac{i_{LB}(t_0)}{nC_o} - \dfrac{v_{out}}{L_A C_o}t - \dfrac{v_{out}}{R_o C_o} \cdots (t_0 \le t \le D_D T_{eff}) \\ -\dfrac{v_{out}}{R_o C_o} \cdots (D_D T_{eff} \le t \le D_{B,eff}T_{eff}) \\ \dfrac{i_{LB}}{nC_o} - \dfrac{v_{out}}{R_o C_o} \cdots (D_{B,eff}T_{eff} \le t \le T_{eff}) \end{cases} \tag{8}$$

as (3) for both cases. Therefore, the operating mode of the proposed control method can be divided into three modes based on $D_{B,eff}$, considering the effect of D_D. The key waveforms are shown in Fig. 5 (d). Differential equations of state variables during half of the switching period (T_{eff}) are as (7) and (8). Through state space averaging and perturbation, the control-to-output transfer function is derived as (9). As shown in Fig. 6, the derived small signal model matches the MATLAB simulation results well. i_{peak} equals to the $i_{LA}(t_2)$.

979-8-3315-1612-3/25 $31.00 © 2025 IEEE   1901

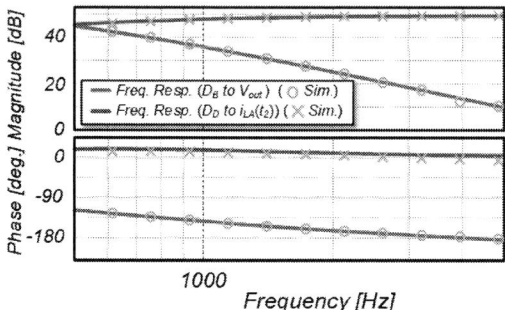

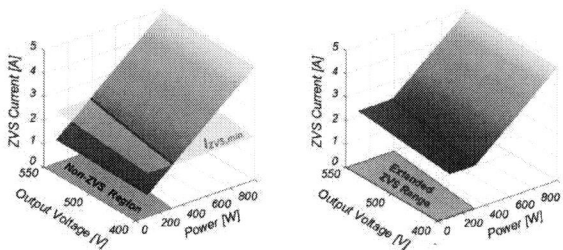

Fig.8. ZVS analysis of the lagging leg switches. (a) Without ZVS range extension method. (b) With ZVS range extension method.

Fig.6. Bode plot of the derived transfer functions (dV_{out}/d_B, dI_{peak}/d_D).

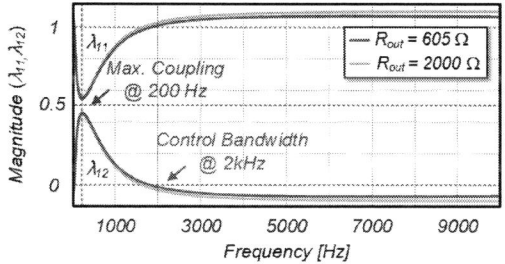

Fig.7. RGA plot showing the cross-coupling under different conditions.

C. RGA Analysis of the Proposed Control Method

Cross-coupling of two control loops in the proposed method is analyzed through relative gain array (RGA) analysis [10], which is defined as (10), where **G** is the transfer function matrix. Elements of the RGA denote the degree of cross-coupling. For example, if. one of the elements in the same row is close to 1 and the other is close to 0 at the bandwidth frequency, it indicates low cross-coupling between two control loops [11]. Oppositely, if two elements in the same row have similar values, it means strong cross-coupling. Fig. 7 shows the magnitude of λ_{11} and λ_{12} according to the frequency and the load (R_o). A strong cross-coupling exists around 200 Hz, and a low cross-coupling exists at a frequency above 1.5 kHz. Therefore, with a control bandwidth higher than 1.5 kHz, the decentralized structure can be applied in the proposed reverse mode control.

D. Zero-Voltage-Switching Range Extension Method

Zero voltage switching (ZVS) of lagging leg switches of PSFB converter is limited under light load conditions, which is due to insufficient energy stored in L_A. The proposed control method can extend the ZVS range of lagging leg switches by

$$R = \begin{bmatrix} \lambda_{11} & \lambda_{12} \\ \lambda_{21} & \lambda_{22} \end{bmatrix} \triangleq G \circ (G^{-1})^{\mathrm{T}} \qquad (10)$$

$$G = \begin{bmatrix} \left.\dfrac{\widehat{dv_{out}}}{\widehat{d_B}}\right|_{\widehat{d_D}=0} & \left.\dfrac{\widehat{dv_{out}}}{\widehat{d_D}}\right|_{\widehat{d_B}=0} \\ \left.\dfrac{\widehat{di_{peak}}}{\widehat{d_B}}\right|_{\widehat{d_D}=0} & \left.\dfrac{\widehat{di_{peak}}}{\widehat{d_D}}\right|_{\widehat{d_B}=0} \end{bmatrix}$$

$$\frac{1}{2} \cdot 2 \cdot C_{oss} V_{out}^2 \leq \frac{1}{2} \cdot 2 \cdot L_A \cdot i_{LA}(t_1)^2 \qquad (11)$$

changing the LLPD control reference. To achieve ZVS of Q_3 and Q_4, condition (11) should be fulfilled, where I_{ZVS} is the ZVS current of Q_3 and Q_4. Fig. 8 (a) shows the minimum required ZVS current ($I_{ZVS,min}$) and I_{ZVS} when LLPD control reference is (1). As I_{ZVS} is smaller than $I_{ZVS,min}$ under 25% load condition or below, Q_3 and Q_4 fail to achieve ZVS. If the LLPD control reference is set to the $I_{ZVS,min}$ at light load conditions, the ZVS boundary of lagging leg switches could be extended, as shown in Fig.8 (b). This ZVS range extension method can be utilized without affecting the voltage loop if two controllers are decoupled by utilizing the bandwidth of 1.5 kHz or above.

III. CONTROLLER DESIGN AND IMPLEMENTATION

A. Output Voltage Controller Design

The output voltage controller is designed based on the derived transfer function. The output voltage loop response suffers from large phase delays due to inherent RHP zero and double-pole effect similar to boost converters. Therefore, a type 3 compensator with transfer function ($G_{cv}(s)$) of (12) is utilized to secure a sufficient phase margin. The bode plot of

$$\begin{bmatrix} \left.\dfrac{\widehat{di_{LB}}}{\widehat{d_B}}\right|_{\widehat{d_D}=0} & \left.\dfrac{\widehat{di_{LB}}}{\widehat{d_D}}\right|_{\widehat{d_B}=0} \\ \left.\dfrac{\widehat{dv_{out}}}{\widehat{d_B}}\right|_{\widehat{d_D}=0} & \left.\dfrac{\widehat{dv_{out}}}{\widehat{d_D}}\right|_{\widehat{d_B}=0} \end{bmatrix} = = (s\mathrm{I} - \mathrm{A})^{-1} \cdot \mathrm{B} \quad \begin{bmatrix} \left.\dfrac{\widehat{di_{peak}}}{\widehat{d_B}}\right|_{\widehat{d_D}=0} \\ \left.\dfrac{\widehat{di_{peak}}}{\widehat{d_B}}\right|_{\widehat{d_B}=0} \end{bmatrix} = \begin{bmatrix} \dfrac{V_{in}T_{eff}}{2nL_B} - \dfrac{1}{n}\cdot\dfrac{\widehat{di_{LB}}}{\widehat{d_B}} + \dfrac{D_D T_{eff}}{L_A}\cdot\dfrac{\widehat{dv_{out}}}{\widehat{d_B}} \\ \dfrac{V_{out}T_{eff}}{L_A} - \dfrac{1}{n}\cdot\dfrac{\widehat{di_{LB}}}{\widehat{d_D}} + \dfrac{D_D T_{eff}}{L_A}\cdot\dfrac{\widehat{dv_{out}}}{\widehat{d_D}} \end{bmatrix}$$

$$(9)$$

$$A = \begin{bmatrix} -\dfrac{\alpha}{nL_B} & -\dfrac{2V_{out}\cdot(D_B+D_D-1)+nV_{in}-\alpha I_{LB}}{nL_B V_{out}} \\ \dfrac{4\alpha L_B I_{LB} + 2L_B V_{out}\cdot(1-D_B) + \alpha D_D V_{in}T_{eff}}{2nL_B C_o V_{out}} & \dfrac{-2L_A - R_o T_{eff} D_D^2}{R_o L_A C_o} + \dfrac{2L_B\cdot(1-D_B) - V_{in}T_{eff}D_D\cdot(D_B+D_D)}{2nL_B C_o V_{out}} \end{bmatrix}, \quad B = \begin{bmatrix} \dfrac{V_{out}}{nL_B} & \dfrac{V_{out}}{nL_B} \\ -\dfrac{I_{LB}}{nC_o} - \dfrac{V_{in}T_{eff}D_D}{2nL_B C_o} & \dfrac{-D_D T_{eff}V_{out}}{L_A C_o} + \dfrac{V_{in}T_{eff}\cdot(\alpha I_{LB} - 2D_D V_{out} - D_B V_{out})}{2nL_B C_o V_{out}} \end{bmatrix}$$

979-8-3315-1612-3/25 $31.00 © 2025 IEEE   1902

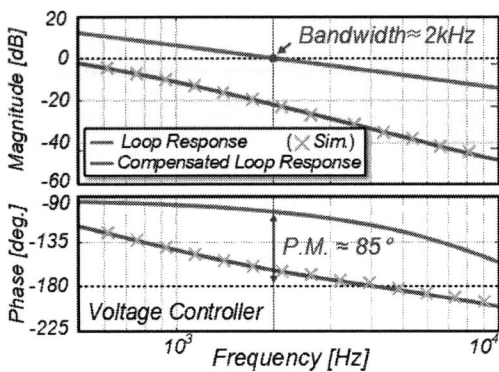

Fig.9. Bode plot of uncompensated and compensated voltage loop response.

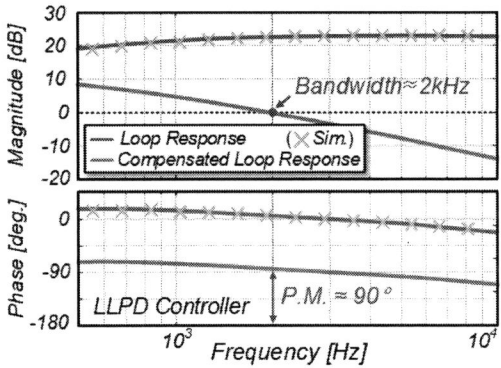

Fig.10. Bode plot of uncompensated and compensated LLPD loop response.

TABLE I
PROTOTYPE DESIGN PARAMETERS

HV side Switch (Q_1-Q_4)	Infineon AIMCQ120R040M1T (1200 V, 40 mΩ)
LV side Switch (S_1-S_4)	Onsemi NVMTSC4D3N15MC (150 V, 4.5 mΩ)
Boost Inductor (L_B)	CH330060 (1.9 µH)
Transformer	EE5555A Ferrite Core (L_m = 500 µH, n=1:20)
Auxiliary Inductor (L_A)	PQ2620 (12 µH)
Output Capacitor (C_o)	100 µF

uncompensated and compensated voltage loop response is shown in Fig. 9. The voltage control loop is designed with a 2 kHz bandwidth for minimum cross-coupling, and the phase margin is 85 degrees.

B. LLPD Control Implementation and Controller Design

The objective of the LLPD control is to control the i_{LA} as (1). Instead of an additional current sensor for i_{LA}, the voltage on the current transformer (CT) existing on the HV side is utilized for the control. Fig. 4(a) shows the sensing circuit based on CT, which is located between the positive terminal (+) of output voltage (V_{out}) and the rectifier. Since only AC components transfer through the CT, the absolute value of i_{LA} at t_1 can be expressed as (13). Therefore, the LLPD control can be

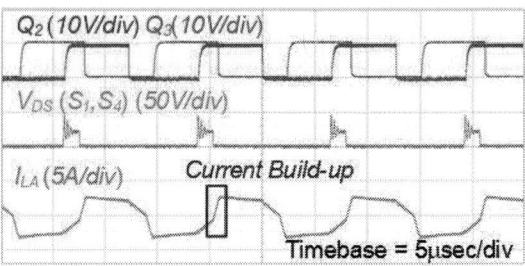

Fig.11. Experiment waveforms of the proposed control method (V_{in} = 11.5 V, V_{out} = 450 V, P_{out} = 500 W).

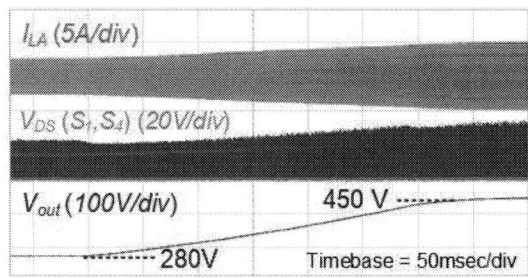

Fig.12. Key waveforms of the capacitor charging experiment with the proposed control method. (V_{out}: 280 V to 450 V)

$$G_{cv}(s) = \frac{3.4056 \cdot 10^7 \cdot (s+446.6) \cdot (s+4480),}{s \cdot (s+1.331 \cdot 10^5) \cdot (s+2.416 \cdot 10^5)} \quad (12)$$

$$|i_{LA}(t_1)| = i_{ct}(t_1) - I_{Load} \quad (13)$$

$$V_{ct,ref} = \frac{R_{ct}}{n_{ct}} \cdot \left(\frac{i_{LB}(t_2)}{n} + I_{Load}\right) \quad (14)$$

$$G_{ci}(s) = \frac{920}{s} \quad (15)$$

implemented by controlling the CT voltage to control reference ($V_{ct,ref}$) of (14).

The LLPD controller is designed based on the derived transfer function (9). An integral controller with a transfer function ($G_{ci}(s)$) of (15) is utilized for compensation. The bode plot of uncompensated and compensated LLPD control loop response is shown in Fig. 10. The bandwidth of the LLPD control loop is also designed to be 2 kHz to minimize cross-coupling, and the phase margin is 90 degrees.

IV. EXPERIMENTAL VERIFICATION

To verify the proposed reverse mode control method, a 500 W PSFB converter prototype was tested on 11.5-15.5 V input, 280-450 V output, and 78 kHz switching frequency conditions. The PSFB converter prototype is designed based on Table I. Fig. 11 shows the key waveforms of the proposed method under 11.5 V input, 450 V output, and maximum power condition. LLPD control ramps up the i_{LA} current to match the voltage-fed side reflected boost inductor current and reduces the voltage overshoot. Remained voltage oscillation is due to the resonance between the junction capacitor of the switches and L_A, which also exists in forward operation. The experiment result of capacitor charging is shown in Fig. 12. During the capacitor charging from 280 V to 450 V, the proposed method operates stably without any drain-source voltage overshoot. It is verified that the voltage overshoot is removed by the proposed method both in a steady-state operation and dynamic situation.

V. CONCLUSION

In this paper, a new reverse mode control strategy for the PSFB converter is proposed. The proposed leading leg phase delay (LLPD) control ramps up the inductor current (i_{LA}) to match the voltage-fed side reflected boost inductor current, consequently mitigating voltage overshoot across the current-fed side switches. Furthermore, by adopting a decentralized control structure, the two control loops can operate independently. This decentralized control enables the extension of the zero-voltage switching (ZVS) range for lagging leg switches by adjusting the LLPD control reference, potentially leading to improved efficiency under light load conditions. As a result, the proposed control method can eliminate voltage overshoots and improve efficiency without any additional components, making it a promising solution for bidirectional auxiliary power modules (APMs) in electric vehicles.

ACKNOWLEDGEMENT

This work was supported by the National Research Foundation of Korea(NRF) grant funded by the Korea government(MSIT) (No. 2022M3I8A1077243)

REFERENCES

[1] I. Aghabali, J. Bauman, P. J. Kollmeyer, Y. Wang, B. Bilgin and A. Emadi, "800-V Electric Vehicle Powertrains: Review and Analysis of Benefits, Challenges, and Future Trends," in IEEE Transactions on Transportation Electrification, vol. 7, no. 3, pp. 927-948, Sept. 2021.

[2] K. -H. Park, M. Lee and G. -W. Moon, "A new Phase Shift Full Bridge DC/DC Converter with Integrated Inter-module Battery Equalization Circuit (IBEC)," in IEEE Transactions on Transportation Electrification, Early Access, doi: 10.1109/TTE.2023.3332859.

[3] C. Wang, P. Zheng and J. Bauman, "A Review of Electric Vehicle Auxiliary Power Modules: Challenges, Topologies, and Future Trends," in IEEE Transactions on Power Electronics, vol. 38, no. 9, pp. 11233 11244, Sept. 2023.

[4] L. Zhu, H. Bai, A. Brown and M. McAmmond, "Design a 400 V–12 V 6 kW Bidirectional Auxiliary Power Module for Electric or Autonomous Vehicles With Fast Precharge Dynamics and Zero DC-Bias Current," in IEEE Transactions on Power Electronics, vol. 36, no. 5, pp. 5323-5335, May 2021.

[5] D. Y. Jung, K. S. Park, S. In Kim, H. G. Jang, J. Won, Y. H. Lee, and J.-W. Lim, Precharge switch based on metal–oxide–semiconductor-controlled thyristor for power relay assembly of battery electric vehicles, ETRI Journal (2024), 1–12,

[6] M. Escudero, D. Meneses, N. Rodriguez and D. P. Morales, "Modulation Scheme for the Bidirectional Operation of the Phase-Shift Full-Bridge Power Converter," in IEEE Transactions on Power Electronics, vol. 35, no. 2, pp. 1377-1391, Feb. 2020.

[7] Lizhi Zhu, Kunrong Wang, F. C. Lee and Jih-Sheng Lai, "New start-up schemes for isolated full-bridge boost converters," in IEEE Transactions on Power Electronics, vol. 18, no. 4, pp. 946-951, July 2003.

[8] M. Lee, D. Choi, J. Bae, J. Chae and G. -W. Moon, "A New Secondary Clamp Diode for Phase-Shift Full Bridge Converter," 2022 International Power Electronics Conference (IPEC-Himeji 2022- ECCE Asia), Himeji, Japan, 2022, pp. 1596-1600.

[9] Y. Yang et al., "An Improved Control Scheme for Reducing Circulating Current and Reverse Power of Bidirectional Phase-Shifted Full-Bridge Converter," in IEEE Transactions on Power Electronics, vol. 37, no. 10, pp. 11620-11635, Oct. 2022.

[10] Skogestad, Sigurd & Postlethwaite, I. (2005). Multivariable Feedback Control: Analysis and Design.

[11] A. Sarkar, N. Deshmukh and S. Anand, "Relative Gain Array Based Decoupled Controller design for GaN Based Multiple Output Flyback Converter," 2023 IEEE Applied Power Electronics Conference and Exposition (APEC), Orlando, FL, USA, 2023, pp. 721-728.

In-Situ EV EIS with a High-Density Flying Capacitor Multi-Level Converter Supercapacitor System

Avram Kachura, Gaël Vergès, Samantha K. Murray, and Olivier Trescases

The Edward S. Rogers Sr. Department of Electrical & Computer Engineering, University of Toronto, Canada

E-mail: avram.kachura@mail.utoronto.ca

Abstract—**This paper presents the implementation of a novel Electrochemical Impedance Spectroscopy (EIS) system for battery pack-level perturbation, which leverages an optimized bidirectional hybrid switched-capacitor dc-dc converter. The system consists of a 400 V, 2 kW, 6-Level Flying Capacitor Multi-Level (FCML) dc-dc converter and a 209 V Supercapacitor (SC) module, all integrated within a total volume of 2.1 L. Simulations of the EIS measurement model are used to quantify the trade-offs between the number of perturbation cycles and the accuracy of the measured impedance. The air-cooled FCML prototype utilizes closed-loop control to apply sinusoidal current perturbations for a 400 V battery without a dc offset. Experimental results validate the operation of the SC module and the bidirectional FCML converter together for a battery voltage of 400 V with 2 kW peak perturbations across frequencies ranging from 40 mHz to 2 kHz. EIS measurements are performed on 44 Ah lithium Nickel-Manganese-Cobalt-oxide pouch cells at varying state-of-charge.**

I. INTRODUCTION

In 2023, the number of Direct-Current Fast-Charging (DCFC) stations reached 1.4 million globally, marking an increase of 0.5 million from the previous year [1]. DCFC stations have ratings of 50 kW or higher and can fast-charge Electric Vehicles (EVs) in 20 minutes to an hour [2]. Accurate predictions of the EV battery voltage response during fast charging are crucial for optimizing charging performance and ensuring battery longevity. The prediction accuracy of battery voltage response to high current transients improves with the use of impedance-informed Equivalent Circuit Models (ECMs) [3], [4]. Aged EV batteries exhibit higher impedance compared to unaged cells, which limits both their maximum charging rate and the power available for vehicle propulsion [5]. Battery impedance can be accurately measured with Electrochemical Impedance Spectroscopy (EIS), which is performed by applying a sinusoidal current to the battery and measuring the voltage response over a wide frequency range [6].

A battery ECM typically consists of a state-of-charge (SOC) dependent voltage source in series with various impedance elements, as shown in Fig. 1. A representative impedance Nyquist plot for a lithium-ion battery is shown in Fig. 1, highlighting several key battery impedance elements [7]. The R_1 parameter represents the Solid-Electrolyte-Interface (SEI) resistance, which manifests at middle frequencies above 10 Hz and increases significantly with battery aging [8]. The R_2 parameter represents the charge-transfer resistance related to ion intercalation, affecting the impedance at frequencies between 10 Hz and 0.1 Hz. Both R_1 and R_2 increase significantly at low temperatures ($< 10\,°C$) with R_2 increasing exponentially, causing the capacitive semi-circles associated with the

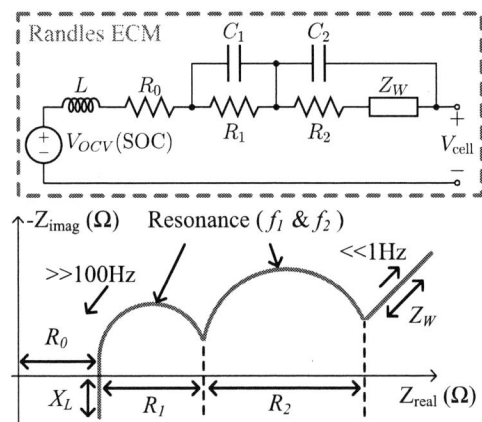

Fig. 1. The Randles ECM and the lithium-ion impedance representative Nyquist plot.

SEI and charge-transfer to enlarge by more than two times [8], [9]. Additionally, R_2 increases linearly with decreasing state-of-health (SOH) [10] and shrinks significantly as the dc current increases [9]. The ohmic resistance, R_0, includes the resistances of the battery terminals, electrode connections, and bulk electrolyte, which can be directly extracted at the imaginary impedance zero-crossing. The R_0 resistance slightly increases with decreasing temperatures. Both R_0 and R_2 values increase at low ($< 20\%$) and high ($> 80\%$) SOCs [11]. The non-linear Warburg impedance $\left(Z_W = \sigma(1-j)/\sqrt{\omega}\right)$ accounts for the slow transport processes of the lithium ions, predominantly impacting the impedance at frequencies below 1 Hz [7]. The reactance, X_L, models the battery's inductance at frequencies above 200 Hz. Given that battery impedance varies significantly under different conditions of SOC, SOH, temperature, and current, these variations emphasize the necessity for an in-situ EIS device to provide accurate and dynamic impedance measurements across operating conditions.

To perform accurate EIS measurements in noisy environments, perturbation currents exceeding 1 A_{pk} are needed to obtain measurable 1 mV voltage responses, due to the sub-milli-ohm impedance of EV batteries [12]. Capturing the full battery impedance spectrum requires frequencies lower than 1 mHz [13]. Measuring multiple cycles at this frequency can extend the measurement time beyond 30 minutes. Limiting the EIS frequency range decreases the measurement time at the expense of omitting battery features in Fig. 1. At low frequencies (< 1 mHz), the measurement time can approach

979-8-3315-1612-3/25 $31.00 © 2025 IEEE

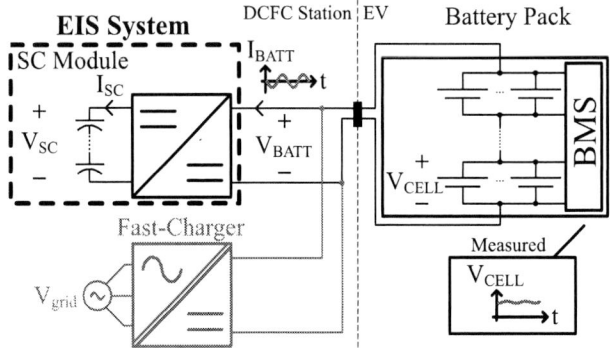

Fig. 2. EV DCFC architecture with proposed EIS add-on system.

the EV charge time, significantly affecting the charging time. Broadband frequency methods, such as multi-sine EIS [14], can reduce the EIS measurement time by over 30% by perturbing the battery with multiple sinusoids simultaneously [15]. The multi-sine method reduces measurement accuracy due to a lower signal-to-noise ratio (SNR) from overlapping signal responses. Broadband techniques, as with single-frequency methods, are constrained by the time required to complete the cycles at the lowest perturbation frequency. Minimizing the number of perturbation cycles reduces the total measurement time, while existing findings indicate that multiple cycles are needed to capture accurate data [14], [16]. One of the objectives of this work is to analyze the number of single-frequency EIS perturbation cycles required to obtain precise steady-state battery impedance data, thereby reducing the energy storage requirements of the EIS perturbation system.

EV integrated EIS perturbation hardware typically relies on an energy storage supply, such as the vehicle's battery or an additional energy storage module, along with a method to process the perturbation power. The hardware design requirements are primarily determined by the battery voltage, minimum perturbation frequency, and current needed to achieve a sufficient SNR for accurately measuring the cell voltage response in EVs. Several EIS perturbation architectures for EVs have been presented in the literature [17]–[22]. In [17], [18], cell-level EIS perturbation hardware is integrated within the EV's on-board BMS, utilizing dc-dc converters at each cell to generate currents below 200 mA_{pk} for active balancing and EIS. These converters require over-design of the balancing current to achieve the necessary SNR for accurate EIS measurements. In [19], an on-board 400 V:12 V bidirectional triple-active-bridge dc-dc converter performed EV half-pack-level perturbations. This system requires a connection to the 200 V midpoint of the series-connected EV battery modules, using each half of the battery for energy storage and bidirectional power flow. The work in [20] utilized a 250 V:325 V dual-active-bridge dc-dc converter within the on-board charger to superimpose a 1 A_{pk} perturbation current on the dc charging current. In [21], the on-board inverter was utilized to create a variable ac voltage source for cell-level perturbations. In [22], an off-board pack-level perturbation system at the DCFC station utilized a MOSFET current regulator in the charging path.

This architecture introduced a dc offset in the charging path by combining the fast-charger output with the perturbation current. The perturbation architectures demonstrated in the literature are challenging to deploy in existing EV and DCFC systems, as they require the replacement or redesign of existing hardware, highlighting the need for a compact, standalone EIS perturbation system.

The primary objective of this work is to demonstrate a modular system for pack-level EIS that integrates seamlessly with existing DCFC hardware as an optional add-on, as shown in Fig. 2. Alternatively, the system can function as a standalone battery state-of-health diagnostic device. A bidirectional EIS system eliminates the dc offset that introduces errors in the extracted ECM parameters [9] and reduces energy storage requirements within the EIS architecture. The proposed system includes its own converter and energy storage module, making it compatible with DCFC stations that have limited control of power flow or only support unidirectional power flow. Minimizing system volume is essential for optimizing the use of limited space at DCFC stations and the portability of the device. The system architecture consists of a bidirectional Flying Capacitor Multi-Level (FCML) dc-dc converter and a Supercapacitor (SC) module, as illustrated in Fig. 3.

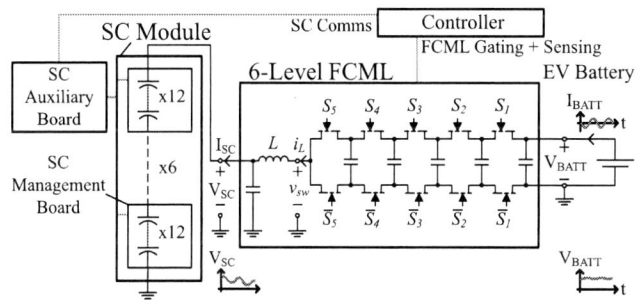

Fig. 3. EIS system architecture.

II. PROPOSED SYSTEM DESIGN

The targeted frequency range of this work is 0.1-2000 Hz, with two EIS cycles per frequency, using a 5 A_{pk} perturbation current required to measure the battery cell voltage response from a 400 V battery system. The dc-dc converter perturbs the battery with 2 kW_{pk} power with a maximum energy transfer of 1.8 Wh. An SC energy storage system is suited for the targeted high power and energy EIS requirements because of its high power density [23]. Hybrid dc-dc converters, such as the FCML converter, combine flying capacitors with inductors to achieve high-density designs, leveraging the 1000× higher energy density of capacitors compared to inductors [24]. The designed EIS system, as shown in Fig. 3, was selected from the volume-optimized design space shown in Fig. 4, generated using the algorithm proposed in [25]. The selected design, as shown in Fig. 4, though not the smallest, offers higher safety margins than the volume-optimum design for SC energy storage capacity and power device drain voltages. The algorithm generates optimized SC module and FCML dc-dc converter designs by accounting for trade-offs in energy

979-8-3315-1612-3/25 $31.00 © 2025 IEEE

storage and converter volumes. The algorithm sweeps through different SC cells and initial voltages to create modules with the minimum number of cells (n). The energy rating of the SC module scales linearly with system volume, as shown in Fig. 5, due to the proportional relationship between the SC module volume and the number of cells ($\text{Vol}_{SC} \propto n$). The relationship between the SC module energy and the number of cells is given by

$$E_{SC,max} = 0.5 C_{cell} V_{cell,0}^2 n \tag{1}$$

where C_{cell} and $V_{cell,0}$ are the SC cell capacitance and initial voltage, respectively. The algorithm sweeps SC module designs and converter levels to determine the optimal switching frequency, parallel phases, and components, to minimize overall converter volume. The energy storage requirement is set by the lowest frequency perturbation at 0.1 Hz, while the FCML converter design is dictated by the maximum inductor current, simulated with the algorithm considering bidirectional power losses. Only even-level converters are considered for their stronger immunity to flying capacitor imbalance [26]. The 6-Level FCML converter topology facilitates high power density dc-dc converter designs by requiring lower blocking voltages and achieving an effective switching frequency that is five times higher at the inductor compared to the power devices. An example 340 V, 650 W, 6-Level FCML achieved a high power density within a compact 45 mL volume [27]. In this work, an air-cooled single-phase 400 V, 2 kW, 6-Level FCML was designed and fabricated with a volume of 135 mL.

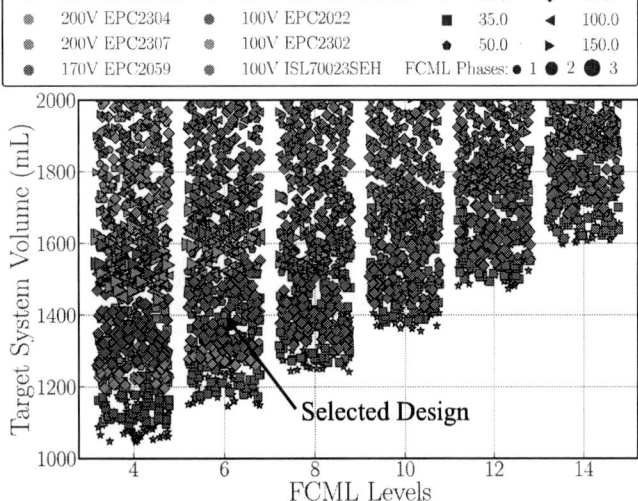

Fig. 4. Generated design space of optimized 400 V EIS systems with selection, showing system volumes versus FCML levels.

III. SIMULATED RESULTS

A. EIS Measurement Model

A battery ECM based on Tesla Model 3 sub-module parameters was simulated to evaluate the minimum number of EIS cycles needed to achieve measurement accuracy within 2% of the ideal case under modeled noise conditions. Minimizing the

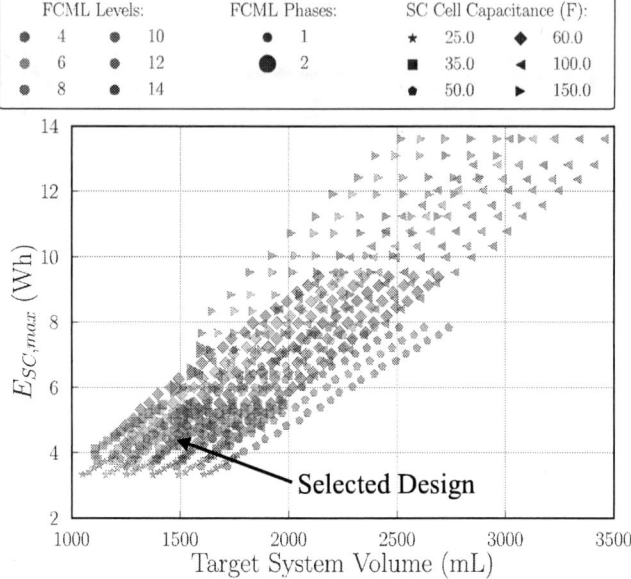

Fig. 5. Optimized design space for 400 V EIS systems, highlighting maximum SC module energy versus system volume. For clarity, only $V_{SC}(0)$ values are shown in 10 V increments.

number of EIS cycles per perturbation frequency reduces both measurement time and system volume. The system volume is directly correlated with the energy storage rating for the lowest frequency EIS measurement, as shown in Fig. 5. The EIS signal processing model shown in Fig. 6 was used to determine that more than one cycle is required to achieve an impedance magnitude error of less than 2% as shown in Fig. 7. The baseline battery model is implemented as a linear 2nd-order resistor-capacitor ECM with resonances at 18.96 Hz and 0.14 Hz according to (2).

$$f_i = \frac{1}{2\pi R_i C_i}, \; i = 1, 2 \tag{2}$$

The battery ECM uses parameters $R_0 = 0.406\,\text{m}\Omega$, $R_1 = 0.317\,\text{m}\Omega$, $C_1 = 265\,\text{F}$, $R_2 = 0.159\,\text{m}\Omega$, $C_2 = 6.96\,\text{kF}$ from [28], and is used to calculate the error for the measured impedance. The EIS measurement model also includes signal processing blocks that capture a single cycle of the battery current, I_{BATT}, and voltage, V_{CELL}, as well as blocks to compute the voltage and current phasors. The battery voltage response during the first perturbation cycle contains additional frequency components beyond the EIS frequency due to the initial voltage transients. The initial transients lead to impedance inaccuracies near the ECM resonances, as highlighted in Fig. 7. Applying a Hamming Window (HW) function to the sampled data improves the accuracy of impedance measurements by reducing spectral leakage, as shown in Fig. 7. In the model, the perturbation current has an amplitude of 5 A, with constant injected noise amplitudes of 0.1 A and 0.5 mV, consistent with harsh EV environments observed in [29]. The noise is modeled as additive white Gaussian noise. Increasing the number of cycles beyond five results in diminishing returns for error reduction due to the injected noise in the model. At EIS frequencies above 600 Hz, the impedance magnitude is low, and the phase of the impedance, ϕ_Z, approaches $0°$. In this frequency range, the injected noise significantly impacts the measured current and voltage phases relative to ϕ_Z, resulting in large errors in the measured impedance phase, as shown in Fig. 7.

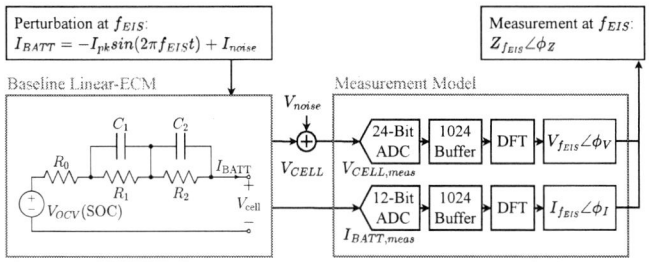

Fig. 6. Time-domain EIS signal processing model, implemented in Simulink and simulated for each perturbation frequency.

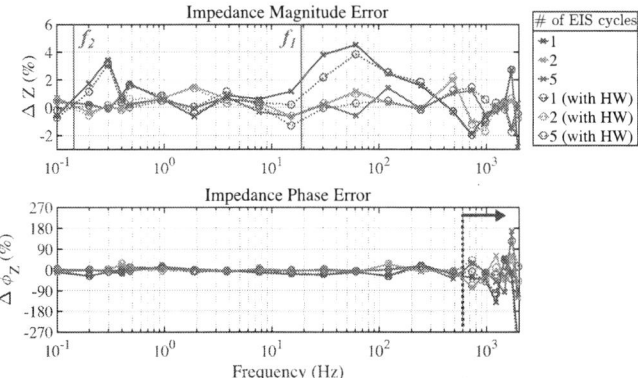

Fig. 7. Magnitude and phase error in impedance using different numbers of EIS cycles captured from 0.1 Hz to 2 kHz.

B. Design Verification

To verify the energy storage and current regulation of the selected design from the optimization in Section II, the system was simulated at the lowest frequency of 0.1 Hz over two cycles as shown in Fig. 8. To maintain a sinusoidal battery current, I_{BATT}, throughout the low-frequency cycles,

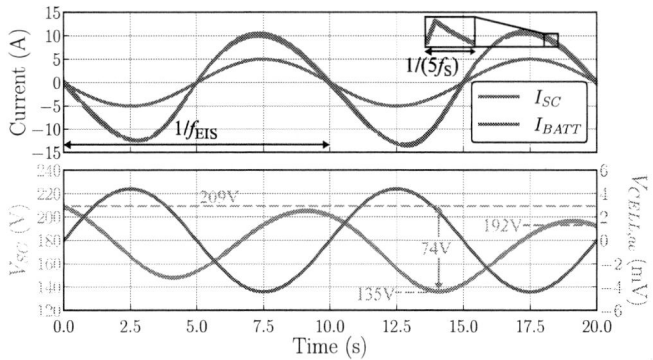

Fig. 8. Verification of the 6-Level FCML converter simulated in PLECS using 72 50 F SC series-connected cells showing current and voltage waveforms at the minimum EIS frequency of 0.1 Hz.

the SC module current, I_{SC}, becomes non-sinusoidal due to the decreasing voltage of the SC modules and power losses in the system.

IV. HARDWARE PROTOTYPE

A. SC Module

The SC module consists of 72 50 F cells connected in series, arranged in a compact 6×12 array to minimize the footprint as shown in Fig. 9. The SC module is configured with 6 SC management boards, each handling 12 cells with passive balancing and overvoltage detection. The SC auxiliary board interfaces with the management boards to monitor the SC module voltage and process overvoltage signals to the controller. A total SC module capacitance of 0.77 F was measured with an equivalent-series-resistance of $0.90\,\Omega$. The specifications for the SC module are listed in Table I.

TABLE I
SC MODULE SPECIFICATIONS

Specification	Value
SC cell	BCAP0050
Number of cells	72
Maximum voltage	209 V
Total capacitance	0.77 F
Total equivalent-series-resistance	$0.90\,\Omega$
Energy rating	4.7 Wh
Volume (without auxiliary board)	2 L

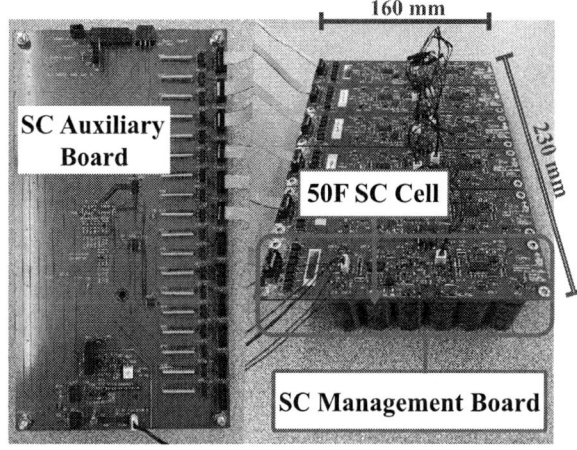

Fig. 9. The 4.7 Wh 209 V SC module with management and auxiliary boards processing the SC cell over-voltage logic for safe system shutdown.

B. FCML DC-DC Converter

To implement the proposed EIS architecture, a 400 V bidirectional 6-Level FCML dc-dc converter was designed and fabricated, as shown in Fig. 12. The converter operates in buck and boost modes to sink and source 2 kW from the battery, and was designed specifically for the worst-case operating point at the perturbation frequency of 0.1 Hz. In buck mode, the battery discharges ($I_{BATT} > 0$) and the SC module charges, while in boost mode, the SC module discharges and the battery charges ($I_{BATT} < 0$). The specifications for the prototype are listed in Table II.

TABLE II
DC-DC CONVERTER SPECIFICATIONS

Specification	Value
Operating voltage at battery side, V_{BATT}	400 V
Operating voltage at SC side, V_{SC}	209 V–123 V
Switching frequency, f_s	89.6 kHz

The 6-Level FCML topology reduces the inductor volt-second product by a factor of $1/25$ and minimizes switching losses by lowering the blocking voltage by a factor of $1/5$ compared to a traditional buck converter. The effective inductor frequency is 448 kHz, five times the switching frequency of the power devices. The reduced power device voltage stress enables the use of low-voltage GaN devices with an excellent switching figure-of-merit. The converter uses a phase-shifted pulse-width modulation (PSPWM) scheme, creating 5 fixed phase shifts for the 5 complementary switch pairs. Each complementary switch pair gating signal, $(S_i, \bar{S}_i)$, is phase-shifted by $72°$ with respect to the adjacent switch pairs, $(S_{i-1}, \bar{S}_{i-1})$. The FCML topology supports a wide conversion ratio given by $V_{SC}/V_{BATT} = D$, where D is the duty ratio. The FCML operation was demonstrated in both buck and boost modes over a wide range of conversion ratios at 400 V, displaying self-balancing of the flying capacitors. The converter steady-state waveforms at 1 kW with conversion ratios of 210 V:402 V and 406 V:130 V are shown in Figs. 10 and 11, respectively.

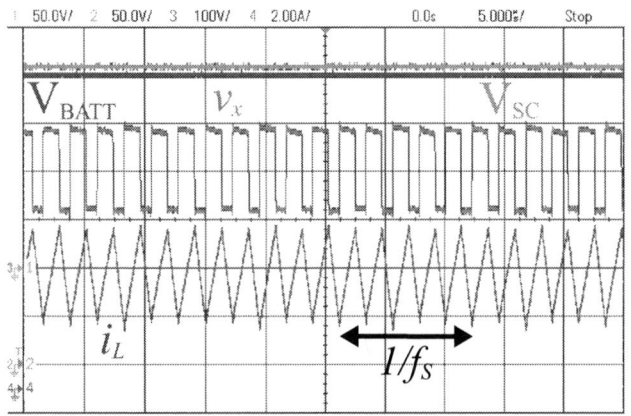

Fig. 10. Measured 6-Level FCML inductor current and switch node operating in buck mode at 210 V:402 V, 1 kW.

The converter was designed for the worst-case operating point in boost mode, with a minimum SC module current of -15 A at an SC module voltage of 143 V. In steady-state operation, each flying capacitor is biased to a different nominal voltage and designed with derating to maintain a maximum ripple of 5%. Local decoupling capacitors are positioned close to the power devices to reduce commutation loop inductance, following the optimized power-loop PCB layout presented in [30]. Blind vias between the top and bottom layers enable the placement of the gate drivers directly under the power devices on the bottom side of the PCB to minimize the gate loop inductance. The gate driver provides a 7.4 V gate-source swing, with the EZdrive^{SM} [31] circuit enabling negative

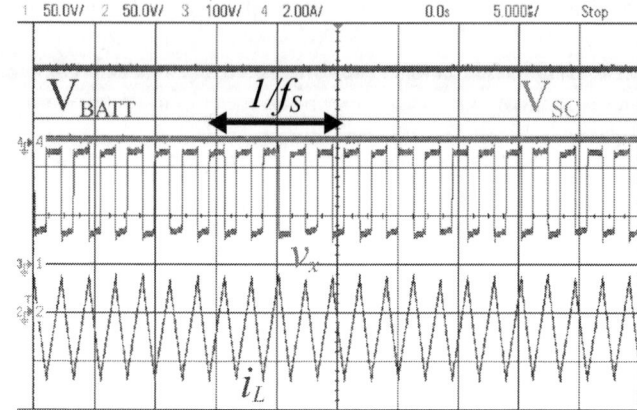

Fig. 11. Measured 6-Level FCML inductor current and switch node operating in boost mode at 406 V:130 V, 1 kW.

drive to reduce the risk of self-turn-on. The 6-Level FCML prototype components are detailed in Table III.

TABLE III
6-LEVEL FCML COMPONENT LIST

Component	Part Number	Value
Power devices	EPC2304	200 V, 5 mΩ
Gate drivers	AHV85110KNHTR	
Inductors	IHLP6767GZER5R6M11	5.6 µH
Flying capacitors	C5750X6S2W225KA250KA	450 V, 2.2 µF
Decoupling capacitors	C1210C154KCRACTU	500 V, 0.15 µF
FPGA controller	10M16SCE144I7G	

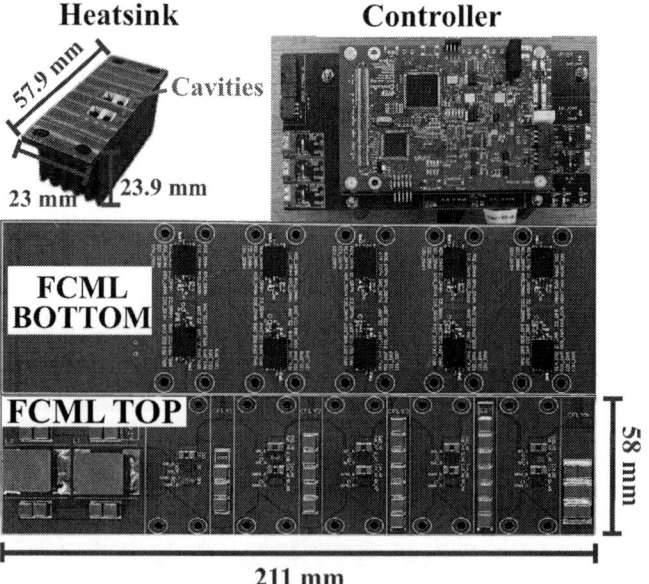

Fig. 12. The 2 kW 6-Level Flying Capacitor Multi-Level dc-dc converter with controller board and custom machined heatsink.

Five heatsinks, each with a volume of 31.8 mL and thermal resistance of 2.2 °C/W when used with forced air, are mounted on the complementary switch pairs between the bulk flying capacitors. The heatsinks are custom machined with cavities, as shown in Fig. 12, to provide space for the decoupling capacitors, ensuring proper contact with the thermal interface material on the power devices. The design allows precise adjustment for even compression, minimizing the thermal resistance of the thermal interface material. The FCML prototype volume with the heatsinks mounted is 345 mL.

C. FCML Control System

The FCML uses current mode control to achieve sinusoidal perturbations across frequencies ranging from 0.1 Hz to 2034 Hz. The closed-loop control scheme, shown in Fig. 13, measures the filtered battery current using a shunt sensor with an effective sampling frequency of 1.67 MHz. The battery current measurement, used as feedback in the control loop, ensures a symmetric sinusoidal shape by actively compensating for non-sinusoidal inductor currents caused by the varying voltage of the SC module. The proportional integral (PI) controller that tracks the sinusoidal current uses different P and I gains depending on the perturbation frequency. The controller uses a constant feedforward duty ratio, D_{FF}, where $V_{out,0}$ is the FCML output voltage measured at V_{out} before enabling the closed-loop control and closing the contactor, S_{SC}, to limit the inrush current.

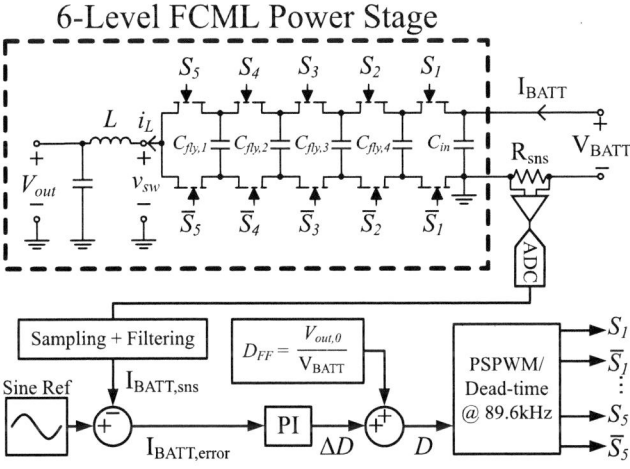

Fig. 13. Closed-loop control architecture of the 6-Level FCML for EIS.

V. Experimental Results

The EIS test setup circuit is shown in Fig. 14, with the battery cell under test connected below the 400 V battery emulator. The perturbation frequencies and PI parameters are controlled via a PC, which also collects data from the oscilloscope. The battery ac cell voltage response, $v_{CELL,ac}$, is measured using a differential amplifier to offset the dc cell voltage with a reference set to $V_{CELL,DC}$. Measurements of $v_{CELL,ac}$ and I_{BATT} are triggered by the control signal, V_{trig}, to ensure an integer number of cycles are captured.

The EIS measurement accuracy was verified to be within 10 μΩ using a known resistance of 0.5 mΩ for frequencies below 100 Hz. The system measured the impedance across 24 perturbation frequencies on two 44 Ah lithium Nickel-Manganese-Cobalt-oxide (NMC) pouch cells. 2 kW$_{pk}$

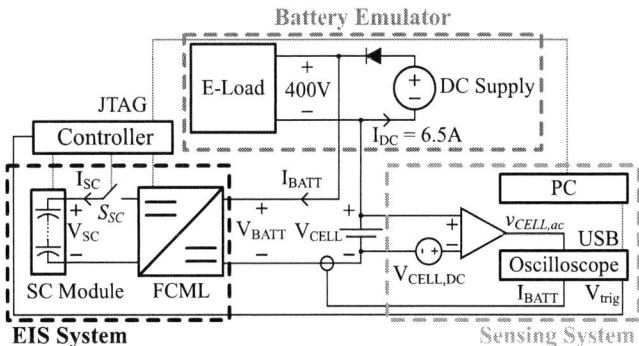

Fig. 14. Battery cell EIS measurement test setup with the sensing system and battery emulator stacked on top of a real lithium NMC cell.

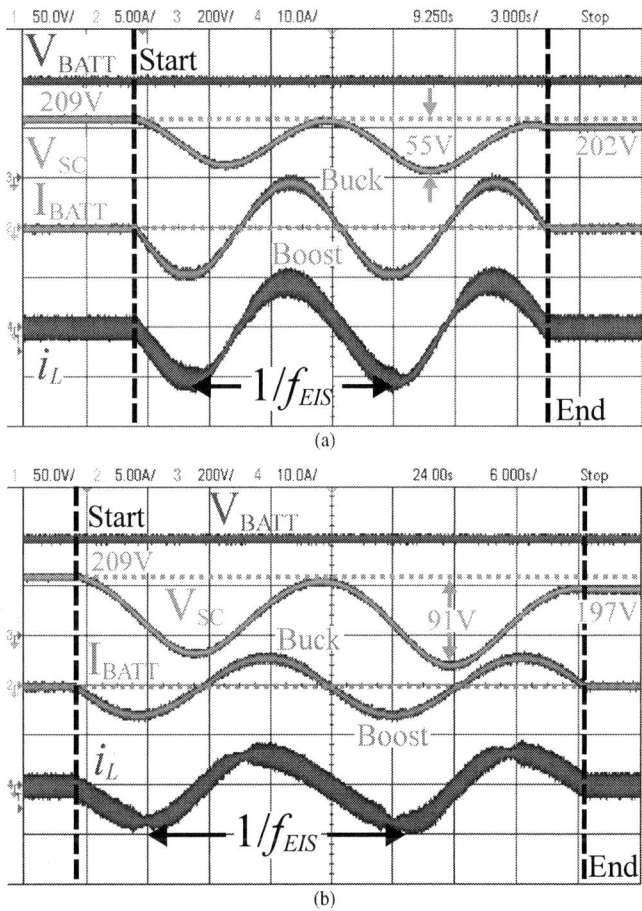

Fig. 15. 6-Level FCML current waveforms in EIS mode with V_{BATT} = 400 V showing (a) 0.1 Hz and (b) 40 mHz perturbations with the varying SC voltage.

perturbations were applied at the design's targeted frequencies ranging from 0.1 Hz to 2034 Hz. Additional perturbations were performed at 80 mHz, 60 mHz, and 40 mHz with a lower power to meet energy storage constraints by limiting the energy transfer from the SC module. These lower frequencies capture the Warburg feature of the lithium-ion battery, with the lower current providing sufficient SNR as battery impedance magnitudes increase at lower frequencies. When performing EIS perturbations at frequencies below 0.3 Hz, the SC module was fully charged to 209 V to meet the energy requirements. A 0.1 Hz, 5 A_{pk} perturbation with two cycles, based on the analysis in Section III, results in a maximum SC voltage swing of 55 V, as shown in Fig. 15(a). A 40 mHz, 3 A_{pk} perturbation corresponds to a maximum energy transfer of 2.7 Wh and results in a SC module voltage swing of 91 V, as shown in Fig. 15(b). Despite the higher energy transfer at 40 mHz, the lower perturbation power reduces the peak conduction losses compared to the 0.1 Hz case, preventing SC voltage collapse. The perturbations at 0.1 Hz and 40 mHz show non-sinusoidal inductor currents and ripple variations caused by the changing FCML duty ratios.

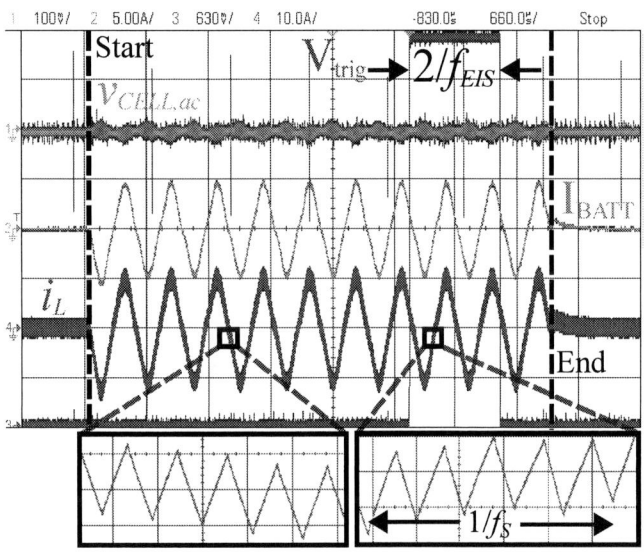

Fig. 16. 6-Level FCML current waveforms in EIS mode with $V_{BATT} = 400$ V and an initial SC module voltage of $V_{SC}(0) = 180$ V showing the 2034 Hz perturbation with the battery cell ac response and measurement trigger.

For frequencies above 100 Hz, ten perturbation cycles were applied to allow the current controller to settle, with two cycles captured for measurement using the V_{trig} signal. The measured perturbation and cell voltage response at the highest EIS frequency of 2034 Hz is shown in Fig. 16. For perturbation frequencies greater than 0.5 Hz, the SC module voltage is precharged to 180 V to prevent overvoltage resulting from high perturbation power under low energy loss conditions.

TABLE IV
EXTRACTED ECM PARAMETERS FOR 44 AH LITHIUM NMC CELL

Parameter	Cell C	Cell C	Cell B	Cell B
V_{CELL} (V)	3.05	3.51	3.07	3.51
R_0 (mΩ)	0.685	0.637	0.688	0.660
R_1 (mΩ)	0.213	0.137	0.192	0.130
C_1 (F)	10.458	7.324	12.568	9.149
R_2 (mΩ)	1.131	0.446	1.080	0.411
C_2 (F)	58.42	47.98	60.12	46.55
σ	0.000382	0.000107	0.000361	0.000099
L (nH)	38.8	37.2	37.0	42.0

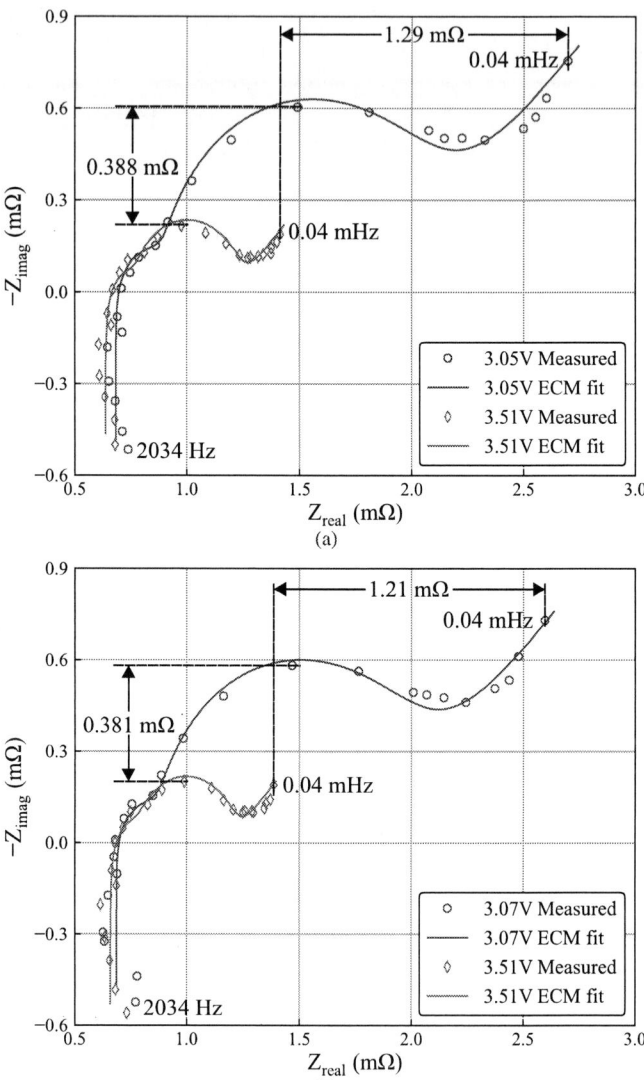

Fig. 17. Measured EIS spectra from the 44 Ah lithium NMC (a) Cell C and (b) Cell B at 23 °C and two different cell voltages.

The EIS spectra for the two different cells at two SOCs are shown in Figs. 17(a) and 17(b) and were measured using the system shown in Fig. 14. The spectra show higher impedance at 3.1 V compared to 3.5 V, consistent with the literature [11]. The Randles ECM, shown in Fig. 1, was fitted with the measured EIS data using the non-linear least squares algorithm, achieving a good fit with the parameters listed in Table IV. The R_2 parameter more than doubled between Cell C and B, while R_0 and R_1 also increased. Variations in ECM parameters between cells highlight the need for cell-level modeling, as the battery pack is constrained by the worst-performing cell.

VI. CONCLUSIONS

This work demonstrates the first experimental demonstration of a hybrid switched-capacitor bidirectional dc-dc converter for a modular EIS system designed for EV pack-level perturbation. The EIS measurement analysis reveals that using two or more perturbation cycles reduces the impedance measurement error by over 50%. The custom FCML and SC module system achieved a broad operational frequency range from 40 mHz to 2 kHz. The GaN-based 6-Level FCML prototype operated reliably at 400 V, delivering up to 2 kW$_{pk}$ power under closed-loop control. The compact 2.3 L EIS system, which includes heatsinks, effectively measured the impedance of a 44 Ah lithium NMC cell across various SOCs. Results highlighted the sensitivity of impedance and fitted circuit parameters to SOC variations, underscoring the system's utility for advanced battery diagnostics. The system can be used as a standalone diagnostic device or integrated into existing DCFC stations, where the extracted battery circuit parameters can inform and optimize fast charging strategies, as suggested in [32].

ACKNOWLEDGMENT

This research was supported by the Natural Sciences and Engineering Research Council (NSERC) of Canada.

REFERENCES

[1] Eaton, "Global EV Data Explorer." [Online]. Available: https://www.iea.org/data-and-statistics/data-tools/global-ev-data-explorer

[2] U.S. Department of Transportation, "Charger Types and Speeds." [Online]. Available: https://www.transportation.gov/rural/ev/toolkit/ev-basics/charging-speeds

[3] M. Dubarry and B. Y. Liaw, "Development of a universal modeling tool for rechargeable lithium batteries," *Journal of Power Sources*, vol. 174, no. 2, pp. 856–860, 2007.

[4] J. Zhang, P. Wang, Y. Liu, and Z. Cheng, "Variable-Order Equivalent Circuit Modeling and State of Charge Estimation of Lithium-Ion Battery Based on Electrochemical Impedance Spectroscopy," *Energies*, vol. 14, no. 3, 2021.

[5] S. F. Schuster et al., "Nonlinear aging characteristics of lithium-ion cells under different operational conditions," *Journal of Energy Storage*, vol. 1, pp. 44–53, 2015.

[6] M. Koseoglou, E. Tsioumas, D. Papagiannis, N. Jabbour, and C. Mademlis, "A Novel On-Board Electrochemical Impedance Spectroscopy System for Real-Time Battery Impedance Estimation," *IEEE Transactions on Power Electronics*, vol. 36, no. 9, pp. 10 776–10 787, 2021.

[7] P. Iurilli, C. Brivio, and V. Wood, "On the use of electrochemical impedance spectroscopy to characterize and model the aging phenomena of lithium-ion batteries: a critical review," *Journal of Power Sources*, vol. 505, p. 229860, 2021.

[8] M. Steinhauer, S. Risse, N. Wagner, and K. A. Friedrich, "Investigation of the solid electrolyte interphase formation at graphite anodes in lithium-ion batteries with electrochemical impedance spectroscopy," *Electrochimica Acta*, vol. 228, pp. 652–658, 2017.

[9] L. W. Juang, P. J. Kollmeyer, R. Zhao, T. M. Jahns, and R. D. Lorenz, "The impact of DC bias current on the modeling of lithium iron phosphate and lead-acid batteries observed using electrochemical impedance spectroscopy," in *Energy Conversion Congress and Exposition*, 2014.

[10] Q. Zhang, C.-G. Huang, H. Li, G. Feng, and W. Peng, "Electrochemical Impedance Spectroscopy Based State-of-Health Estimation for Lithium-Ion Battery Considering Temperature and State-of-Charge Effect," *IEEE Transactions on Transportation Electrification*, vol. 8, no. 4, pp. 4633–4645, 2022.

[11] H. Xiao et al., "State of Charge Effects on the Parameters of Electrochemical Impedance Spectroscopy Equivalent Circuit Model for Lithium Ion Batteries," *IOP Conference Series: Earth and Environmental Science*, vol. 474, no. 5, p. 052038, 2020.

[12] L. Lu, X. Han, J. Li, J. Hua, and M. Ouyang, "A review on the key issues for lithium-ion battery management in electric vehicles," *Journal of Power Sources*, vol. 226, pp. 272–288, 2013.

[13] C. Dunn and J. Scott, "Achieving Reliable and Repeatable Electrochemical Impedance Spectroscopy of Rechargeable Batteries at Extra-Low Frequencies," *IEEE Transactions on Instrumentation and Measurement*, vol. 71, pp. 1–8, 2022.

[14] B. Ulgut, "Methods-Employing Multisine Electrochemical Impedance Spectroscopy for Batteries In Galvanostatic Mode," *Journal of The Electrochemical Society*, vol. 169, no. 11, p. 110510, 2022.

[15] Ivium Technologies, "MultiSine EIS Time Reduction Trade-off Against Accuracy Loss," 2024, Electrochemical Notes, IEN7.1. [Online]. Available: https://www.ivium.com/wp-content/uploads/2023/01/IEN7-1-MultiSine-EIS-time-reduction-trade-off-against-accuracy-loss.pdf

[16] G. S. Popkirov and R. N. Schindler, "Optimization of the perturbation signal for electrochemical impedance spectroscopy in the time domain," *Review of Scientific Instruments*, vol. 64, no. 11, pp. 3111–3115, 11 1993.

[17] E. Din, C. Schaef, K. Moffat, and J. T. Stauth, "A Scalable Active Battery Management System With Embedded Real-Time Electrochemical Impedance Spectroscopy," *IEEE Transactions on Power Electronics*, vol. 32, no. 7, pp. 5688–5698, 2017.

[18] Y. Elasser, Y. Chen, M. Liu, and M. Chen, "A Multiway Bidirectional Multiport-Ac-Coupled (MAC) Battery Balancer with Online Electrochemical Impedance Spectroscopy," in *2020 IEEE Applied Power Electronics Conference and Exposition (APEC)*, 2020, pp. 1475–1482.

[19] S. A. Assadi et al., "In-Situ EV Battery Electrochemical Impedance Spectroscopy with Pack-Level Current Perturbation from a 400V-to-12V Triple-Active-Bridge," in *2022 IEEE Applied Power Electronics Conference and Exposition (APEC)*, 2022, pp. 1056–1063.

[20] Y.-D. Lee, S.-Y. Park, and S.-B. Han, "Online Embedded Impedance Measurement Using High-Power Battery Charger," *IEEE Transactions on Industry Applications*, vol. 51, no. 1, pp. 498–508, 2015.

[21] A. Kersten et al., "Online and On-Board Battery Impedance Estimation of Battery Cells, Modules or Packs in a Reconfigurable Battery System or Multilevel Inverter," in *IECON 2020 The 46th Annual Conference of the IEEE Industrial Electronics Society*, 2020, pp. 1884–1891.

[22] Z. Gong et al., "Pack-Level Electrochemical Impedance Spectroscopy in EV Batteries Enabled by a DC Fast Charger," in *Annual Conference of the IEEE Industrial Electronics Society*, 2021.

[23] IEC Market Strategy Board, "Electrical energy storage," International Electrotechnical Commission, White Paper, 2011.

[24] N. C. Brooks et al., "On the Size and Weight of Passive Components: Scaling Trends for High-Density Power Converter Designs," *IEEE Transactions on Power Electronics*, vol. 39, no. 7, pp. 8459–8477, 2024.

[25] A. Kachura, M. S. Zaman, and O. Trescases, "Design Optimization of a Multi-Level Converter Supercapacitor System for Electrochemical Impedance Spectroscopy in EV Fast-Charging Stations," in *2024 IEEE Transportation Electrification Conference and Expo (ITEC)*, 2024, pp. 1–6.

[26] Z. Ye, Y. Lei, Z. Liao, and R. C. N. Pilawa-Podgurski, "Investigation of capacitor voltage balancing in practical implementations of flying capacitor multilevel converters," in *2017 IEEE 18th Workshop on Control and Modeling for Power Electronics (COMPEL)*, 2017, pp. 1–7.

[27] N. C. Brooks, L. Horowitz, R. Abramson, and R. C. N. Pilawa-Podgurski, "Low-Inductance Asymmetrical Hybrid GaN HEMT Switching Cell Design for the FCML Converter in High Step-Down Applications," in *2021 IEEE Applied Power Electronics Conference and Exposition (APEC)*, 2021, pp. 9–15.

[28] Z. Gong et al., "An EV-Scale Demonstration of In-Situ Battery Electrochemical Impedance Spectroscopy and BMS-Limited Pack Performance Analysis," *IEEE Transactions on Industrial Electronics*, vol. 70, no. 9, pp. 9112–9122, Oct. 2023.

[29] S. A. Assadi et al., "In-Situ EV Battery Electrochemical Impedance Spectroscopy with Pack-Level Current Perturbation from a 400V-to-12V Triple-Active-Bridge," in *2022 IEEE Applied Power Electronics Conference and Exposition (APEC)*, 2022, pp. 1056–1063.

[30] David Reusch, "Optimizing PCB Layout," 2019, White Paper. [Online]. Available: https://epc-co.com/epc/Portals/0/epc/documents/papers/Optimizing%20PCB%20Layout%20with%20eGaN%20FETs.pdf

[31] GaN Systems, "EZDrive^{SM} Solution for GaN Systems' E-HEMT," 2018, GN010 Application Note. [Online]. Available: https://gansystems.com/wp-content/uploads/2018/12/GN010-EZDrive-Solution-for-GaN-Systems-E-HEMTs-_20181221.pdf

[32] S. Sarofim et al., "Optimization Strategy for Battery Electric Vehicle (BEV) DC Fast Charging (FC) in Cold Environments," in *2025 IEEE Applied Power Electronics Conference and Exposition (APEC)*, 2025.

A Novel 500-kHz LLC-T Resonant Converter with Wide Output Range

Zhengming Hou
Future Energy Electronics Center
Virginia Tech
Blacksburg, Virginia, USA
hzhengm@vt.edu

Dong Jiao
Future Energy Electronics Center
Virginia Tech
Blacksburg, Virginia, USA
jdong@vt.edu

Jih-Sheng Lai
Future Energy Electronics Center
Virginia Tech
Blacksburg, Virginia, USA
ORCID: 0000-0003-2315-8460

Abstract—The LLC resonant converter is attractive for the electric vehicle charging application due to its high efficiency and high power-density. Nevertheless, the efficiency of the conventional LLC converter suffers from the wide switching frequency for the wide output range applications. In this paper, a novel LLC-T resonant converter, an extra auxiliary transformer in parallel with the L-C resonant tank, is proposed to narrow the switching frequency range for the EV charging application. A 500-kHz, 1.35-kW, 400-V/250-450-V prototype is built to verify the merits of the proposed converter and demonstrates 97.82% peak efficiency.

Index Terms—Isolated dc-dc converter, LLC resonant converter, wide output range, high efficiency, narrow switching frequency range, EV charging, high frequency.

I. INTRODUCTION

As the development in the electric vehicle (EV) industry, it has gained more interests in how to charge EV battery more efficiently. The EV batteries charging typically involves with the constant current (CC) charging and constant voltage (CV) charging as depicted in Fig.1. A wide battery voltage range exhibits in the CC region. The conventional PWM converter can be implemented for the battery charging in [1] [2] [3], but the leakage current raises the safety concerns. Hence, a galvanic isolated topology is preferred in the EV charging application. The LLC resonant converter is a popular topology widely implemented for the charging applications in [4] [5] [6] [7] [8] due to its soft switching capability. Nevertheless, the wide output range leads to a wide switching frequency range.

The two-stage structure was proposed in [9] to meet the voltage regulation requirement via the PWM converter and the galvanic isolation via the LLC-DCX. Although, this structure resolves the issues, it greatly raises the cost. A variable dc-link voltage structure was proposed in [10] to regulate the voltage through the power factor correction (PFC) stage to eliminate the extra stage, whereas, the voltage stress is doubled on the dc-link rail. Consequently, it is critical to realize the wide output range through the LLC stage while minimizing the frequency range. Several techniques have been proposed in the past to reduce the frequency range of the LLC resonant converter. It is possible to divide the entire output range into different operational modes to reduce the frequency range, named as multi-mode technique. This can be realized by 1)

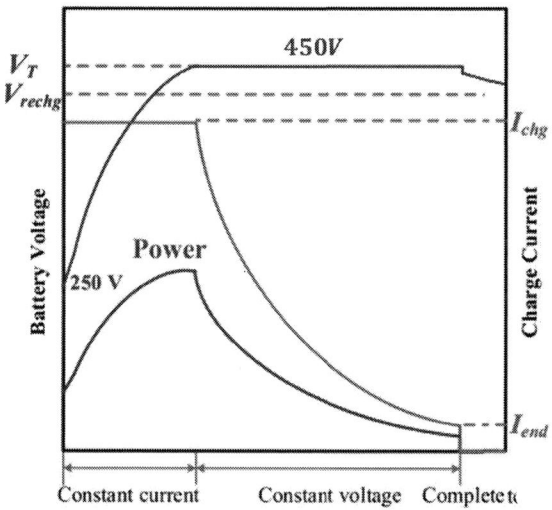

Fig. 1: Typical EV battery charging profile.

adjusting the resonant tank parameters as proposed in [11] [12] [13] [14]; 2) reconfiguring the inverter and rectifier structures in [15] [16] [17] [18] [19]; 3) changing the output connections in [20] [21] [22]. Nevertheless, the multimode operation increases the control complexity and the number of the semiconductors. The partial power technique was implemented to reduce the control complexity while operating under fixed frequency. The major power is proposed via the LLC-DCX and the partial power is utilized for the voltage regulation. The partial power resonant converter proposed in [23] [24] [25] demonstrated a good efficiency and high power density. They are considered as a single stage topologies, whereas a large number of semiconductors is still inevitable. The PWM controlled LLC converters with wide output voltage range were also proposed in [26] [27] [28] [29] [30]. However, they suffer from the high turn-off current which leads to the high switching losses. Furthermore, the asymmetrical current, generally found in the PWM controlled LLC converter, causes the difficulty of the magnetic component design. The multi-resonant converter were also investigated in [31] [32] [33] [34] to achieve a wide range within a narrow switching frequency range, but the injected higher order harmonics also lead to the

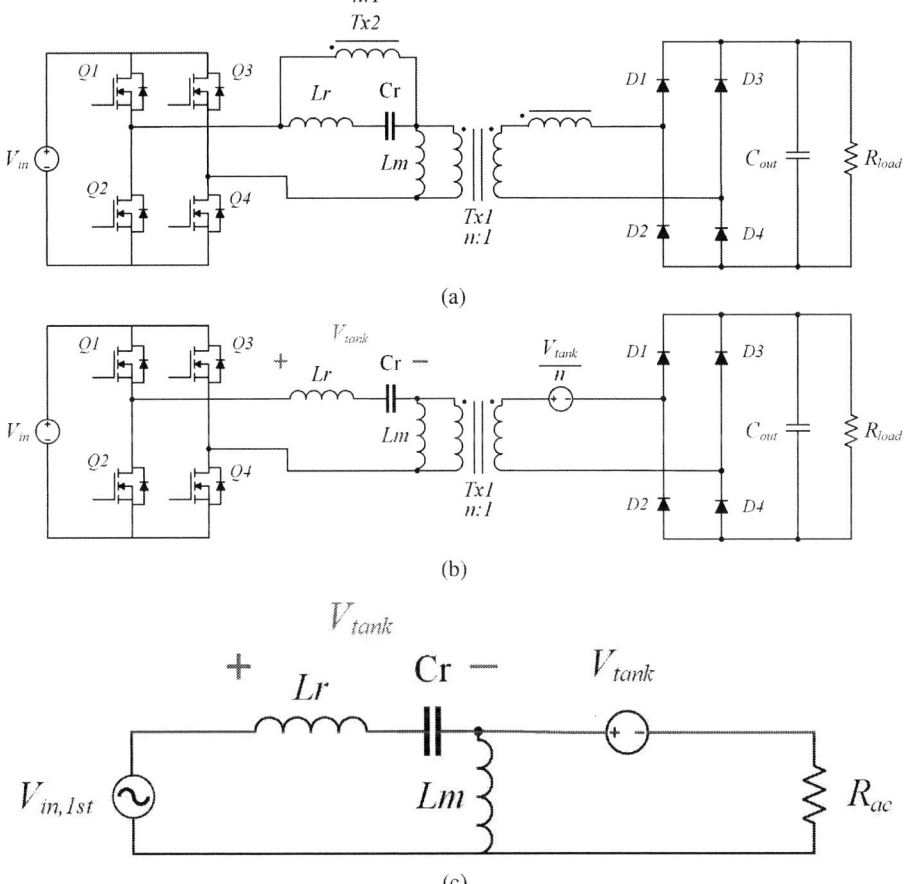

Fig. 2: (a) LLC-T resonant converter topology. (b) Equivalent circuit of the LLC-T resonant converter. (c) FHA equivalent model of the LLC-T resonant converter.

difficulty of the magnetic design. In this paper, a novel LLC resonant converter with an auxiliary transformer in parallel with the L-C resonant tank, named LLC-T, is proposed to resolve the drawbacks of the conventional LLC resonant converter. Compared to the conventional LLC resonant converter, the proposed LLC-T resonant converter can achieve the same gain range within a narrow frequency range. Furthermore, a FHA model is derived to model the voltage gain characteristic of the proposed topology.

II. OPERATION PRINCIPLE OF LLC-T RESONANT CONVERTER

The LLC-T resonant converter is presented in Fig.2 (a). The full-bridges are implemented on primary and secondary. Akin to the LLC resonant converter, the zero-voltage switching (ZVS) is realized by the magnetizing current. An unlike-wound auxiliary transformer, Tx2, is placed in parallel with the L-C resonant tank and its secondary winding is in series with the main transformer, Tx1. The auxiliary transformer reflects the voltage across the resonant tank, V_{tank}, to the secondary, while the main transformer transfers voltage, V_{LLC}, to the secondary side. The auxiliary transformer is treated as an

ideal transformer so that the secondary can be modeled as an dependent voltage source. The equivalent circuit is derived in Fig.2 (b). Hence, the LLC-T topology can achieve a wide voltage range within a narrow frequency range compared to the conventional LLC resonant converter, because the voltage across the resonant tank is counted twice. The relationships are expressed in (1) and (2).

$$V_{\text{out,LLC}} = \frac{V_{\text{in}} - V_{\text{tank}}}{n} \qquad (1)$$

$$V_{\text{out,LLC-T}} = \frac{V_{\text{in}} - 2V_{\text{tank}}}{n} \qquad (2)$$

A. LLC-T resonant converter gain curve

As a PFM resonant converter, the LLC-T topology contains three operation regions: 1) At resonant frequency; 2) Above resonant frequency; 3) Below resonant frequency. The proposed converter achieves a higher boost gain in the below resonant frequency region, a lower buck gain in the above resonant frequency region, and achieves a unity gain at the resonant frequency according to (2). The first harmonic approximation (FHA) analysis can be utilized to derive the gain curve from the equivalent circuit shown in Fig.2 (c). First of all, the voltage across the resonant tank shall be derived from the conventional LLC resonant converter as expressed from (3)-(8). Then substitute the derived resonant tank voltage back into (2) to resolve the voltage gain expressed in (9).

$$R_{\mathrm{ac}} = \frac{8n^2}{\pi^2} R_{\mathrm{load}} \qquad (3)$$

$$Z_{\mathrm{Lm}} = j\omega L_m \qquad (4)$$

$$Z_{\mathrm{Lr}} = j\omega L_r \qquad (5)$$

$$Z_{\mathrm{Cr}} = \frac{1}{j\omega C_r} \qquad (6)$$

$$V_{\mathrm{out,LLC}} = \left| \frac{Z_{\mathrm{Lm}} + R_{\mathrm{ac}}}{R_{\mathrm{ac}} + Z_{\mathrm{Lm}} + Z_{\mathrm{Lm}} R_{\mathrm{ac}}(Z_{\mathrm{Lr}} + Z_{\mathrm{Cr}})} \right| V_{\mathrm{in}} \qquad (7)$$

$$V_{\mathrm{tank}} = V_{\mathrm{in}} - nV_{\mathrm{out,LLC}} \qquad (8)$$

$$G_{\mathrm{LLC\text{-}T}} = 2 \left| \frac{Z_{\mathrm{Lm}} + R_{\mathrm{ac}}}{R_{\mathrm{ac}} + Z_{\mathrm{Lm}} + Z_{\mathrm{Lm}} R_{\mathrm{ac}}(Z_{\mathrm{Lr}} + Z_{\mathrm{Cr}})} \right| - 1 \qquad (9)$$

The final voltage gain curve is plotted in the Fig.3 for a 400V input/250V-450V output with 3A CC charging rate design. It can be seen that with the same resonant tank design, the LLC resonant converter is not able to step down the voltage till 250V. Compared with the conventional LLC resonant converter, the proposed topology shows a steeper

voltage gain with the same resonant tank. Hence, it LLC-T resonant converter was demonstrated to achieve a narrow switching frequency range for a wide output voltage range.

B. Design considerations

One advantage of the proposed LLC-T resonant converter is that it can be designed in the same procedure as for an LLC resonant converter. First of all, the transformer turns ratio is determined based on ratio between the input voltage and the nominal battery pack voltage.

$$n = \frac{V_{\mathrm{in}}}{V_{\mathrm{bat,nomial}}} \qquad (10)$$

As LLC-T resonant converter can operate at a higher frequency due to 1) the small value of the resonant inductance; 2) a narrow frequency range. A 500-kHz switching frequency is determined to minimize the volume of the magnetic components. The resonant frequency can be calculated in (11).

$$f_r = \frac{1}{2\pi\sqrt{L_r C_r}} \qquad (11)$$

The magnetizing inductance is of the main transformer is designed based on the lowest output voltage condition due to its minimum voltage across the transformer and the highest switching frequency.

$$L_m \le \frac{nV_{\mathrm{bat,min}}}{8Q_{\mathrm{oss}}f_{\mathrm{s,max}}} t_d \qquad (12)$$

The inductance ratio, L_n, is defined as the ratio between the magnetizing inductance and the resonant inductance. Different L_n values are swept for the shape of the gain curves. In general, a small L_n requires a narrow frequency range to step down the voltage, but it limits the maximum boost gain. After determining L_n, the resonant inductance, L_r, can be derived based on the determined magnetizing inductance, L_m.

TABLE I: LLC-T Converter Prototype

Parameters	Values
Input voltage V_{in}	400 V
Output voltage V_{out}	250V-450 V
Constant current I_{cc}	3 A
Constant voltage, V_{cv}	450 V
Peak power, P_{pk}	1.35 kW
Magnetizing inductance L_{m1}	35 μH
Magnetizing inductance L_{m2}	500 μH
Resonant inductance L_r	7 μH
Resonant capacitor C_r	14.1 nF
Main transformer turns	16:14
Auxiliary transformer turns	8:7
Dead time	50 ns
Frequency range	380 kHz - 630 kHz

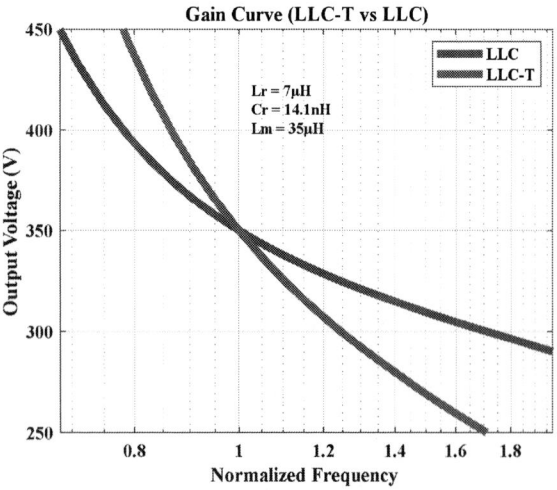

Fig. 3: Voltage gains comparison between LLC-T and LLC.

III. EXPERIMENTAL RESULTS

A 400V input, 250V-450V output prototype is built to verify the merits of the proposed LLC-T resonant converter. The prototype follows the battery charging profile shown in Fig.1 with a maximum 3A charging current. The design parameters are concluded in Table I.

The waveforms are captured in CC region first. The converter is operating above the switching frequency, which is 626.5-kHz at the lowest output voltage, 250V. Channel 1 is the gate-source waveform and channel 2 is the drain-source waveform as shown in Fig.4. The ZVS is achieved for the primary side which can be observed from the waveforms. Fig.5 are the waveforms captured when it is operating at the resonant frequency. The resonant current is in a pure sinusoidal waveform and the ZVS is also achieved. The waveforms operating at its maximum voltage, 450V, are captured in Fig.6. The switching frequency is 382.3-kHz, which is below the resonant frequency. The resonant current clearly shows that it is clamped by the magnetizing current at the end of each cycles. The waveforms also indicates the ZVS is achieves.

The measured voltage gain is compared with the FHA derived voltage gain as shown in Fig.7. The experimental results reveal a steeper gain curve of the LLC-T resonant converter. It can be seen that the voltage gain matches well when it is operating around the resonant frequency. Some mismatch is observed when the operation point deviates from the resonant point. It is a common limitation in the FHA model because the current waveform is assumed in sinusoidal

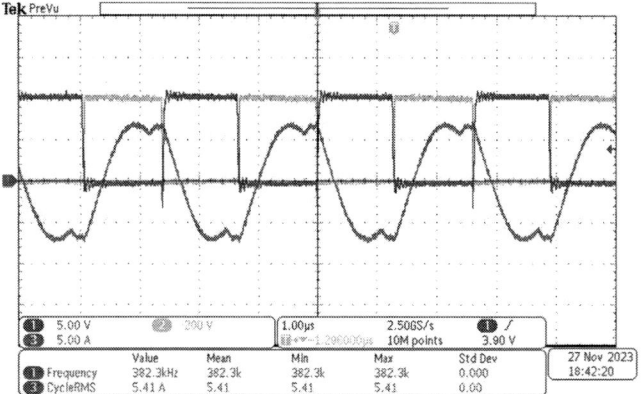

Fig. 6: Waveforms captured at 450V/3A conditions.

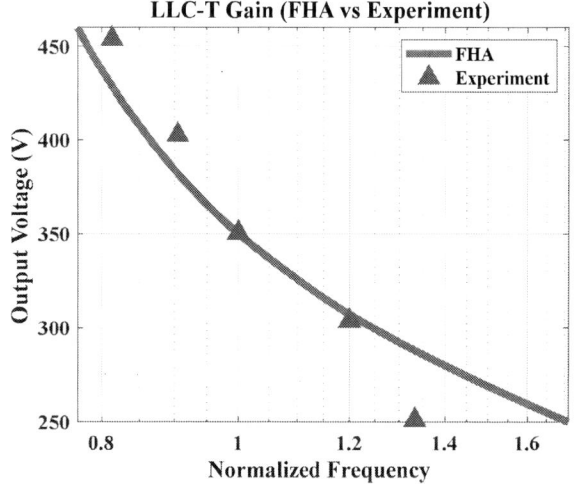

Fig. 7: LLC-T voltage gain comparison between FHA and experimental results.

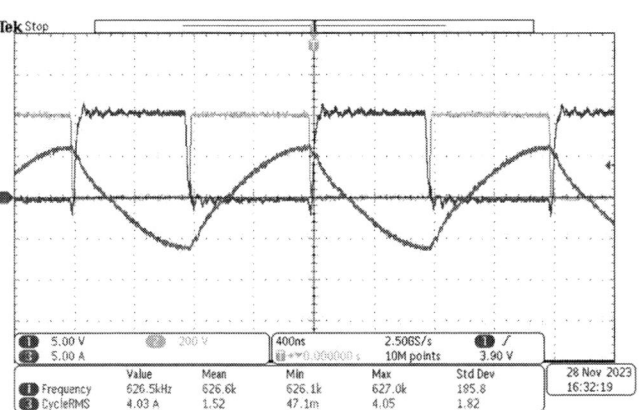

Fig. 4: Waveforms captured at 250V/3A conditions.

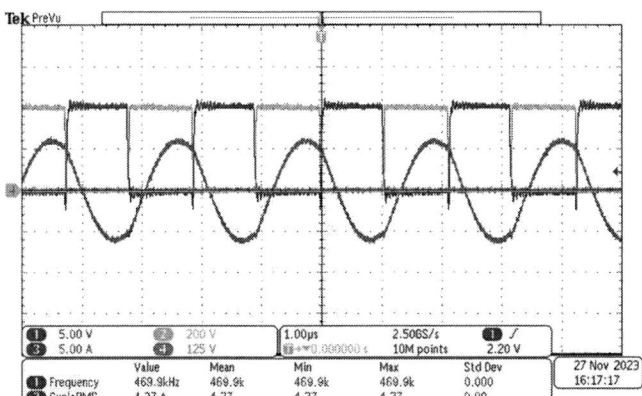

Fig. 5: Waveforms captured at 350V/3A conditions.

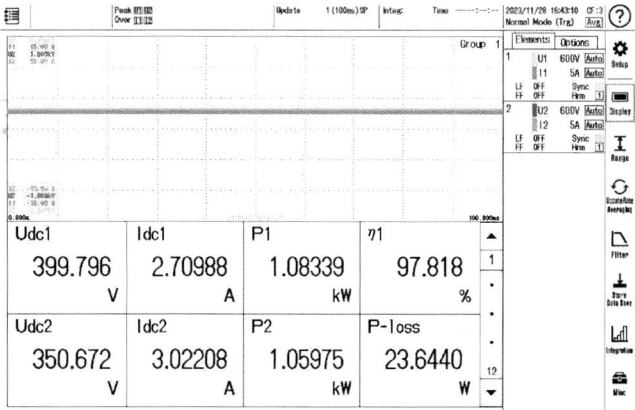

Fig. 8: Measured 97.82% peak efficiency at resonant frequency.

shapes. However, the waveforms clearly show that the resonant currents are no longer in the sinusoidal waveforms when the frequency is not at the its resonant frequency. Furthermore, the parasitic parameters also have impacts on the voltage gain which is not considered in the FHA model. According to the experimental results, the waveforms of the LLC-T resonant converter are similar to the conventional LLC resonant converter. The peak efficiency is 97.82% measured under 350V/3A output condition as shown in Fig.8.

IV. CONCLUSION

A novel LLC resonant converter with wide output range is proposed in this paper. An auxiliary transformer is placed in parallel with the L-C resonant tank to narrow the frequency range requirement. The operation principles and the FHA analysis were explored. The design procedure was presented which is similar to the conventional LLC resonant converter. A 1.35kW, 500-kHz prototype verified the proposed LLC-T resonant converter can achieve 250V-450V within a narrow frequency range. The experiment results demonstrated 97.82% peak efficiency at resonant frequency.

REFERENCES

[1] Z. Hou, D. Jiao, B. C. Gutierrez, J. -S. Lai and P. -L. Chen, "Design of a 15kW High-Efficiency and High Power Density Bidirectional TCM Buck/Boost Converter," 2024 IEEE Applied Power Electronics Conference and Exposition (APEC), Long Beach, CA, USA, 2024, pp. 341-347, doi: 10.1109/APEC48139.2024.10509033.

[2] M. Jahnes, L. Zhou, M. Eull, W. Wang and M. Preindl, "Design of a 22-kW Transformerless EV Charger With V2G Capabilities and Peak 99.5% Efficiency," in IEEE Transactions on Industrial Electronics, vol. 70, no. 6, pp. 5862-5871, June 2023, doi: 10.1109/TIE.2022.3192697.

[3] Y. Cao et al., "A Three-Level Buck–Boost Converter With Planar Coupled Inductor and Common-Mode Noise Suppression," in IEEE Transactions on Power Electronics, vol. 38, no. 9, pp. 10483-10500, Sept. 2023, doi: 10.1109/TPEL.2023.3279987.

[4] Z. Hou, S. C. Kao, J. -S. Lai, Z. Xu, C. Chen and C. -L. Wang, "A Cost-Effective Winding Structure On Modular Matrix Transformer LLC Application," 2022 IEEE Energy Conversion Congress and Exposition (ECCE), Detroit, MI, USA, 2022, pp. 1-7, doi: 10.1109/ECCE50734.2022.9948049.

[5] J. Li, W. -H. Lee, J. -F. Chen and T. -J. Liang, "Design and Implementation of CLLLC Bidirectional Converter with Wide Voltage Range on Low Side," 2023 IEEE Energy Conversion Congress and Exposition (ECCE), Nashville, TN, USA, 2023, pp. 2610-2617, doi: 10.1109/ECCE53617.2023.10362785.

[6] J. Rocha, S. Amin, G. Rego and V. Monteiro, "Step-by-Step Design of a LLC Resonant Converter for EV Fast Charging Applications," 2024 8th International Young Engineers Forum on Electrical and Computer Engineering (YEF-ECE), Caparica / Lisbon, Portugal, 2024, pp. 45-50, doi: 10.1109/YEF-ECE62614.2024.10625506.

[7] T. Yuan, F. Jin and Q. Li, "Analysis and Comparison of Integrated Planar Transformers for 22-kW On-Board Chargers," in IEEE Transactions on Power Electronics, vol. 39, no. 9, pp. 11368-11385, Sept. 2024, doi: 10.1109/TPEL.2024.3410878.

[8] S. -Q. Chen et al., "Design and Implementation of 3.5 kW Digital Controlled Battery Charge System," 2023 IEEE International Future Energy Electronics Conference (IFEEC), Sydney, Australia, 2023, pp. 62-66, doi: 10.1109/IFEEC58486.2023.10458467.

[9] Z. Hou, S. C. Kao, D. Jiao and J. -S. Lai, "Variable Turns-Ratio Matrix Transformer based LLC Converter for Two-Stage Electric Vehicle Auxiliary Power Module Applications," 2023 IEEE Energy Conversion Congress and Exposition (ECCE), Nashville, TN, USA, 2023, pp. 5859-5865, doi: 10.1109/ECCE53617.2023.10362741.

[10] S. Zhao, A. Kempitiya, W. T. Chou, V. Palija and C. Bonfiglio, "Variable DC-Link Voltage LLC Resonant DC/DC Converter With Wide Bandgap Power Devices," in IEEE Transactions on Industry Applications, vol. 58, no. 3, pp. 2965-2977, May-June 2022, doi: 10.1109/TIA.2022.3151867.

[11] J. -H. Teng, S. -S. Chen, Z. -X. Chou and B. -H. Liu, "Novel Half-Bridge LLC Resonant Converter With Variable Resonant Inductor," in IEEE Transactions on Industry Applications, vol. 59, no. 6, pp. 6952-6962, Nov.-Dec. 2023, doi: 10.1109/TIA.2023.3299912.

[12] H. -W. Jo, D. H. Sim, J. -A. Lee, W. -J. Son and B. K. Lee, "Design and Control of the Adjustable Turn-ratio LLC Converter for High-Efficiency Operation of Wired/Wireless Integrated EV Charging System," 2022 International Power Electronics Conference (IPEC-Himeji 2022- ECCE Asia), Himeji, Japan, 2022, pp. 2104-2110, doi: 10.23919/IPEC-Himeji2022-ECCE53331.2022.9806931.

[13] N. G. Feliciani dos Santos and M. L. da Silva Martins, "A Reconfigurable Partial Power Converter with Adjustable Transformer Turns Ratio for a 6.6-kW Integrated On-Board Charger," 2023 IEEE 8th Southern Power Electronics Conference and 17th Brazilian Power Electronics Conference (SPEC/COBEP), Florianopolis, Brazil, 2023, pp. 1-6, doi: 10.1109/SPEC56436.2023.10408639.

[14] Y. Wei, D. Woldegiorgis and A. Mantooth, "Variable Resonant and Magnetizing Inductor Control for LLC Resonant Converter," 2020 IEEE 11th International Symposium on Power Electronics for Distributed Generation Systems (PEDG), Dubrovnik, Croatia, 2020, pp. 149-153, doi: 10.1109/PEDG48541.2020.9244466.

[15] H. Zhang, J. Nie, X. Sun, Y. Deng and Z. Shu, "A Dual-Transformer-Based LLC Resonant Converter With Ultrawide Output Voltage Range," in IEEE Transactions on Power Electronics, doi: 10.1109/TPEL.2024.3468288.

[16] Y. Zuo, X. Pan and C. Wang, "A Reconfigurable Bidirectional Isolated LLC Resonant Converter For Ultra-Wide Voltage-Gain Range Applications," in IEEE Transactions on Industrial Electronics, vol. 69, no. 6, pp. 5713-5723, June 2022, doi: 10.1109/TIE.2021.3088355.

[17] C. -H. Chou, C. -Y. Hsiao and Y. -H. Liu, "Half-Bridge LLC Series-Resonant Converter With Hybrid Rectifier for LED Signage Backlighting Systems," in IEEE Transactions on Circuits and Systems II: Express Briefs, vol. 70, no. 2, pp. 566-570, Feb. 2023, doi: 10.1109/TCSII.2022.3172482.

[18] R. Gu, D. Zhang, J. Duan, W. Zhang and A. Li, "A Reconfigurable Current-Fed LLC Resonant Converter With Circulant PWM Control and Wide Gain Range," in IEEE Transactions on Industrial Electronics, doi: 10.1109/TIE.2024.3453849.

[19] J. -W. Kim, B. Kim, D. Lee, Y. Cho, S. K. Ji and D. Ryu, "LLC Resonant Converter for Fast Electric Vehicle Charging Module with a Reconfigurable Bi-Directional Switch," in IEEE Transactions on Transportation Electrification, doi: 10.1109/TTE.2024.3372993.

[20] A. Elezab, O. Zayed, A. Abuelnaga and M. Narimani, "High Efficiency LLC Resonant Converter With Wide Output Range of 200–1000 V for DC-Connected EVs Ultra-Fast Charging Stations," in IEEE Access, vol. 11, pp. 33037-33048, 2023, doi: 10.1109/ACCESS.2023.3263486.

[21] B. O. Aarninkhof, D. Lyu, T. B. Soeiro and P. Bauer, "A Reconfigurable Two-Stage 11 kW DC-DC Resonant Converter for EV Charging With a 150–1000 V Output Voltage Range," in IEEE Transactions on Transportation Electrification, vol. 10, no. 1, pp. 509-522, March 2024, doi: 10.1109/TTE.2023.3279211.

[22] S. Qazi, P. Venugopal, A. J. Watson, P. Wheeler and T. B. Soeiro, "Design and Analysis of Reconfigurable Resonant Converter With Ultrawide Output Voltage Range," in IEEE Transactions on Power Electronics, vol. 39, no. 5, pp. 5750-5763, May 2024, doi: 10.1109/TPEL.2024.3365391.

[23] Y. Cao, M. Ngo, N. Yan, D. Dong, R. Burgos and A. Ismail, "Design and Implementation of an 18-kW 500-kHz 98.8% Efficiency High-Density Battery Charger With Partial Power Processing," in IEEE Journal of Emerging and Selected Topics in Power Electronics, vol. 10, no. 6, pp. 7963-7975, Dec. 2022, doi: 10.1109/JESTPE.2021.3108717.

[24] W. Xiong, M. Wang, G. Ning, Y. Sun and M. Su, "A ZVS Branch-Sharing Partial Power Converter With Bipolar Voltage Regulation Capability," in IEEE Transactions on Industrial Electronics, vol. 71, no. 2, pp. 1572-1582, Feb. 2024, doi: 10.1109/TIE.2023.3262890.

[25] B. Hu et al., "Hybrid LLC Resonant Converter With Partial-Power Auxiliary Unit for Improved Performance," in IEEE Transactions on Industry Applications, vol. 60, no. 1, pp. 1255-1267, Jan.-Feb. 2024, doi: 10.1109/TIA.2023.3281299.

[26] Y. Zuo, X. Shen, B. Du, D. B. Cobaleda, H. Wouters and W. Martinez, "Fixed-frequency Dual PWM Interleaved Boost LLC Resonant Con-

verter for A Wide Input Voltage for Photovoltaic Applications," 2024 IEEE 10th International Power Electronics and Motion Control Conference (IPEMC2024-ECCE Asia), Chengdu, China, 2024, pp. 4172-4177, doi: 10.1109/IPEMC-ECCEAsia60879.2024.10567987.

[27] Y. Zuo, D. B. Cobaleda, X. Shen and W. Martinez, "High Step-up Ratio Interleaved Boost L-LLC Resonant Converter with PWM and PFM Control for Wide Input and Output Voltage Range," 2024 IEEE Applied Power Electronics Conference and Exposition (APEC), Long Beach, CA, USA, 2024, pp. 1396-1402, doi: 10.1109/APEC48139.2024.10509079.

[28] M. Abbasi, R. Emamalipour, M. A. Masood Cheema and J. Lam, "A Constant Frequency Step-Up Resonant Converter With A Re-Structural Feature And A PWM-Controlled Voltage Multiplier," 2021 IEEE Applied Power Electronics Conference and Exposition (APEC), Phoenix, AZ, USA, 2021, pp. 343-348, doi: 10.1109/APEC42165.2021.9487394.

[29] Y. Zuo, X. Pan, H. Pervaiz, F. Tian and W. Martinez, "Fixed-frequency PWM LLC resonant converter with 8-type rectifier for wide input voltage applications," 2023 IEEE Energy Conversion Congress and Exposition (ECCE), Nashville, TN, USA, 2023, pp. 59-64, doi: 10.1109/ECCE53617.2023.10362590.

[30] S. Li, C. Chen, Z. Liu, Y. Liu, Z. Luo and F. Liu, "Wide Output Range PWM Controlled Dual Resonant Capacitor LLC Resonant Converter," 2024 IEEE 10th International Power Electronics and Motion Control Conference (IPEMC2024-ECCE Asia), Chengdu, China, 2024, pp. 99-104, doi: 10.1109/IPEMC-ECCEAsia60879.2024.10567855.

[31] H. Wen, D. Jiao, J. -S. Lai, J. Strydom and B. Lu, "A MHz LCLCL Resonant Converter Based Single-Stage Soft-Switching Isolated Inverter with Variable Frequency Modulation," 2022 IEEE Applied Power Electronics Conference and Exposition (APEC), Houston, TX, USA, 2022, pp. 848-854, doi: 10.1109/APEC43599.2022.9773541.

[32] D. Jiao, H. Wen and J. -S. Lai, "LLC Resonant Converter Based Single-stage Inverter with Multi-resonant Branches using Variable Frequency Modulation," 2023 IEEE Applied Power Electronics Conference and Exposition (APEC), Orlando, FL, USA, 2023, pp. 263-270, doi: 10.1109/APEC43580.2023.10131127.

[33] D. Jiao, Z. Hou and J. -S. Lai, "LLC Type Resonant Converter Adopting Peak Current Shaving with Third Harmonics Injection for Wide Output Voltage Range Application," 2024 IEEE Applied Power Electronics Conference and Exposition (APEC), Long Beach, CA, USA, 2024, pp. 2232-2238, doi: 10.1109/APEC48139.2024.10509270.

[34] R. M. Reddy, A. K. Jana and M. Das, "Novel Wide Voltage Range Multi-Resonant Bidirectional DC-DC Converter," 2020 IEEE International Conference on Power Electronics, Drives and Energy Systems (PEDES), Jaipur, India, 2020, pp. 1-6, doi: 10.1109/PEDES49360.2020.9379888.

979-8-3315-1612-3/25 $31.00 © 2025 IEEE

High Efficiency Traction Drive Operation with a Partial Load Three-Phase Triangular Current Mode Modulation Concept

Bhaskar Chatterjee[1,2], Jan Allgeier[1], Thomas Plum[1] and Marc Hiller[2]

[1]Corporate Research, Robert Bosch GmbH, Renningen, Germany

[2]Institute of Electrical Engineering, Karlsruhe Institute of Technology, Karlsruhe, Germany

Email: bhaskar.chatterjee@de.bosch.com

Abstract—**Improving drive cycle efficiency in electrical vehicles is one of the key design criteria in traction drives, pushed by a need for lower total energy consumption and reduction of battery capacity costs. Additionally, in passenger cars the peak performance requirements also need to be respected for an adequate driving experience. This paper proposes an operating-point dependent inverter voltage modulation concept termed as Partial-Load Triangular Current Mode modulation, which operates either under Triangular Current Mode modulation or Space Vector modulation for low or high loads respectively, targeting higher cycle-efficiency without over-rating inverter components at full load operation. This work elaborates on the implementation of such a modulation scheme for a 300 kW, 800 V synchronous machine drive with Silicon Carbide switches. It then proposes subsequent modifications to the scheme with an optimal common mode voltage injection to achieve further efficiency gains. Finally, the loss validation is conducted in two steps. First the power module, filter inductor and capacitor are individually characterized and then the inverter and filter system losses are measured with triangular currents. This concept aims to save up to 23% drive losses (4.5% gain in driving range) in the WLTP drive cycle when compared with space vector modulation at a constant 10 kHz switching frequency, a saving comparable to gains made during the transition from Silicon to Silicon Carbide devices for traction inverters.**

Index Terms—**partial load triangular current mode (PL-TCM), high efficiency traction drive, inductor saturation**

I. INTRODUCTION

Energy efficiency is one of the more important benchmarks for electric vehicles mainly due to battery capacity costs which can be the most expensive single component in an electric power train [1]. The situation is further exacerbated in passenger cars with the ongoing trend of ever larger battery capacities to fulfill the demand for higher driving range, leading to added environmental stresses due to mining. To increase the energy efficiency in passenger cars, this work targets drive cycle consumption by reducing the power losses in the electrical traction drive. First, two salient observations are highlighted that are characterized by the vehicle usage trends in partial load operation and by machine PWM harmonic losses due to the current voltage modulation schemes (section I), which are then exploited to develop the Partial Load Triangular Current Mode (PL-TCM) modulation concept. Next, a general PL-TCM design guideline is provided with methods to further improve drive efficiency (section II) and lastly, a description of

component & system loss validation measurements is provided (section III).

A. Partial Load Operation

In passenger cars it is observed from real driving data [2] that the relative duration of low and partial loads is exponentially higher than high and peak load operation, as can be seen in Fig. 1 which shows a histogram of cumulative time duration spend at a speed & torque operating point. Here a power-law like relative distribution can be observed where more that 90% of time duration is spent under 30% load demand and the relative duration of peak load is less than 1%. It can be argued that for passenger cars the power required to accelerate the mass of the vehicle is much higher than required to maintain speed working against drag and frictional forces. Lastly, also noteworthy is that this trend is observed across all vehicle classes and types, as seen in Fig. 1 for (a) a small electric vehicle, (b) a mid size combustion engine sedan and (c) a large combustion engine SUV, with the vehicle consumption test cycle WLTP [3] roughly aligning over the partial load region.

B. PWM Harmonic Losses in Machine

PWM excitation of the machine adds to significant eddy current and hysteresis losses in the machine iron [5] and magnets [6]. Moreover, as these losses occur due to the shape of the applied voltage waveform they persist over the entire operating region, contributing significantly to the energy consumption in partial loads. The Fig. 2 below outlines the various energy loss contributions for a simulated 300 kW, 800 V, 600 A_{rms} drive [7] with 1200 V SiC devices operated with a constant 10 kHz switching frequency and 20 V/ns switching in the space vector based modulation scheme (SVPWM) for a large SUV class vehicle under WLTP driving. It can be seen in Fig. 2 that approximately a third of total drive losses occur in the machine due to PWM voltage harmonics of the applied voltage. These harmonic losses also lead to a higher rotor temperature affecting the continuous performance of the machine [8]. Additionally the pulsed voltage waveform might also increasingly stress the machine insulation [9] and cause high transient bearing currents damaging the machine shaft and race [10].

979-8-3315-1612-3/25 $31.00 © 2025 IEEE

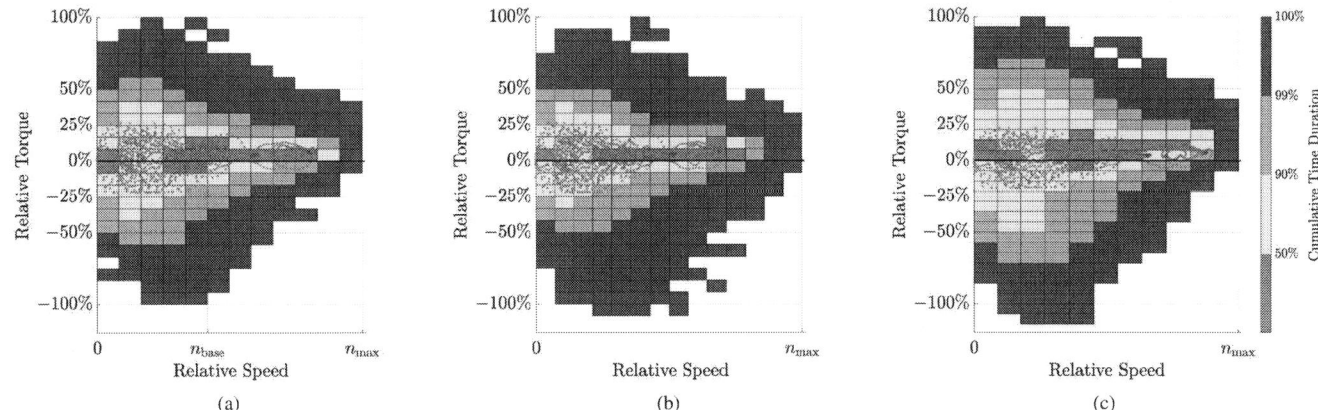

Fig. 1. Histogram of cumulative time duration spent over speed & torque operating point for various vehicle classes from the *2012 California Household Travel Survey* [2]. Trend shows significant relative duration of partial loads for real driving speed data with an assumed vehicle weight for (a) *NISSAN* Electric Vehicle with 300 hours of driving (1500 kg, 80 kW) (b) *HONDA* 4-cylinder Sedan with 1000 hours driving (1500 kg, 85 kW) and (c) *FORD* 8-cylinder SUV with 80 hours driving (2500 kg, 180 kW). Superimposed on top (*blue* scattered) is the WLTP [3] cycle sampled at 1 s intervals, simulated for the same vehicle.

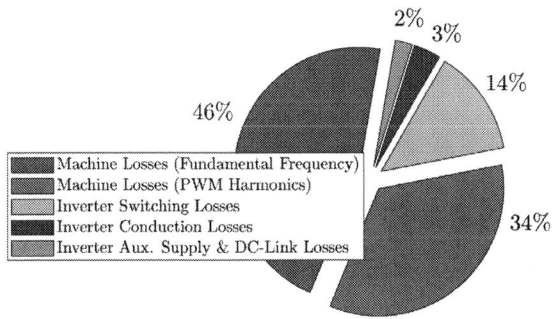

Fig. 2. Relative distribution of energy losses as shown in [7] for a 800 V, 600 A_{rms} 300 kW electric drive during WLTP at constant switching frequency of 10 kHz for 2500 kg vehicle powered by a 4 pole-pair Permanent Magnet Synchronous Machine & 1200 V SiC switches

To mitigate the voltage PWM harmonic losses in the machine, a partial load triangular current mode modulation (PL-TCM) concept is proposed in [7] which utilizes an additional low pass filter (as seen in Fig. 3) to operate under the triangular current mode (TCM) modulation [11] [12] in partial loads. The filter inductor facilitates firstly, zero voltage switching (ZVS) in the inverter allowing higher switching frequencies (up to 140 kHz) and also acts as part of a low pass filter for attenuating PWM harmonics. Any over design of the filter and inverter switches is avoided by saturating the filter inductor (as seen in Fig. 4) and disconnecting the filter capacitors at high loads, then operating under standard SVPWM. However at very low speeds where the machine fundamental voltage amplitude is near zero, TCM operation leads to maximal switching frequency and high envelop current (Fig. 5) leading to higher losses in both the filter and semiconductor (Fig. 6a). This work proposes further methods to improve PL-TCM efficiency, and allow PL-TCM to be operated for the entire speed region at partial loads.

II. PARTIAL LOAD TRIANGULAR CURRENT MODE MODULATION

The PL-TCM operation [7] can be shortly described from Fig 3. The schematic consists of an additional low pass filter L_{123}, C_{123} connect via auxiliary switches A_{123}. The filter inductor is designed to saturate at higher load operation (Fig. 4). For low and partial loads the system is operated under TCM modulation. At higher load demand, the auxiliary switches are opened at the nearest zero crossing voltage, and the drive is operated under standard SVPWM. The designated partial load boundary current i_B is less than half of the saturation limit of the inductor for a feasible TCM operation. The peak load rating is remains unchanged for the inverter.

From [7] it can be shown that the TCM switching frequency f_{Swt} can be defined as

$$f_{Swt}(t) = \frac{U_{DC}}{4L} \cdot \frac{1 - m(t)^2}{\Delta i_{Node}(t)} \tag{1}$$

where the normalized voltage m can be defined as

$$v_{Pha,T-}(t) = \frac{U_{DC}}{2}\big(1 + m(t)\big). \tag{2}$$

Here U_{DC}, L, Δi_{Node} and $v_{Pha,T-}$ are respectively the DC-link voltage, the filter inductance, the absolute inverter node current ripple and the machine terminal to negative DC-link rail voltage respectively, as seen in Fig. 3a. Here m is the general normalized voltage containing common mode components and can be written as,

$$m(t) = m_0 + m_1 \sin(\omega t + \phi_1) \\ + m_3 \sin(3\omega t + \phi_3) + m_n(t) \tag{3}$$

with ω, m_1 and ϕ_1 are the fundamental frequency, machine operating voltage amplitude and power factor angle defined by the desired operating point. The quantities m_0 and m_3 are the dc and 3rd order common mode voltage amplitudes

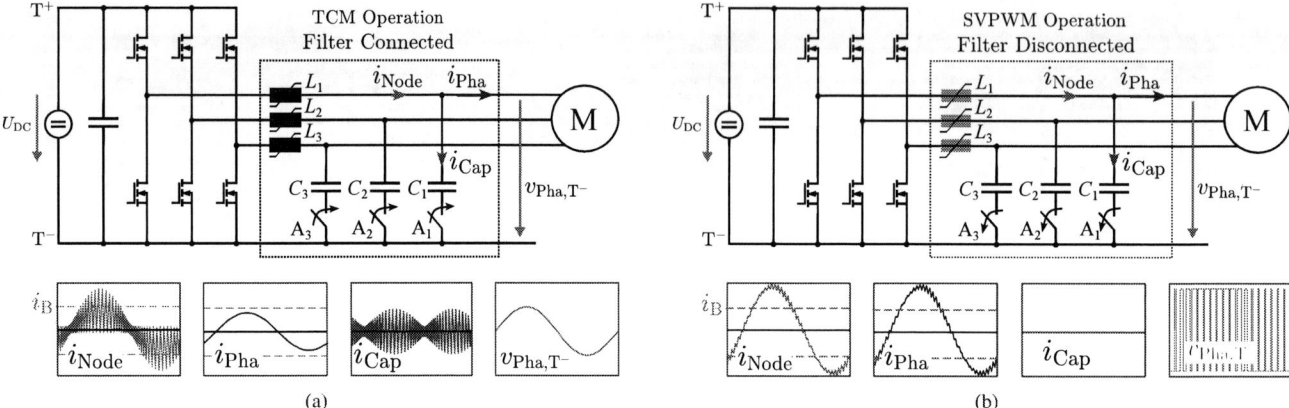

Fig. 3. PL-TCM [7] circuit schematic for (a) low & partial loads under TCM operation with a connected filter (inductors L_{123} unsaturated & auxiliary switches A_{123} closed), (b) high loads under SVPWM operation with the filter effectively 'disconnected' (inductors L_{123} saturated & auxiliary switches A_{123} open) and (c) the characteristic behavior of the filter inductor (L_{123}) over current.

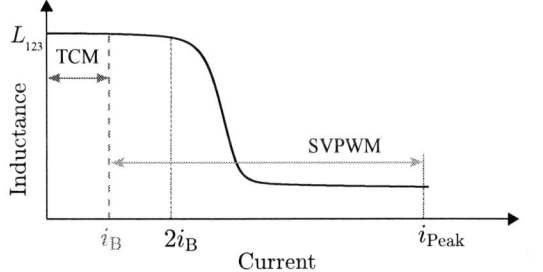

Fig. 4. Characteristic behavior diagram of the saturable filter inductor (L_{123} in Fig. 3) over current, with TCM modulation operated till the boundary current i_{B} (with an inductor saturation limit at $2i_{\mathrm{B}}$) and SVPWM for higher currents.

respectively for a three phase system with ϕ_3 being the injection angle of the 3rd harmonic and $m_n(t)$ is a general higher order common mode voltage.

Equation (1) can be rearranged to form a TCM relevant loss figure of merit (FOM),

$$f_{\mathrm{Swt}}(t) \cdot \Delta i_{\mathrm{Node}}(t) = \frac{U_{\mathrm{DC}}}{4L} \cdot (1 - m(t)^2) \qquad (4)$$

with both f_{Swt} and Δi_{Node} being constraint variables leading to high switching and conduction losses. Equation (4) shows that the absolute value of instantaneous voltage $m(t)$ is inversely proportional to TCM loss figure of merit. In TCM, lower instantaneous voltage $m(t)$ would first lead to an increasing switching frequency, which upon reaching its limit $f_{\mathrm{Swt}}^{\max}$ will require the increase of inverter node current Δi_{Node}. In TCM mode it is required that maximum current never reaches the designated saturation limit of the inductor (Fig. 4), with the filter inductance value chosen to safeguard this criteria even during voltage zero crossings ($m(t) = 0$)

with the highest node currents,

$$\Delta i_{\mathrm{Node}}(t) \leq 2i_{\mathrm{B}} - |i_{\mathrm{ZVS}}^{\max}| \qquad (5)$$

$$L \geq \frac{U_{\mathrm{DC}}}{8 f_{\mathrm{Swt}}^{\max} i_{\mathrm{B}}} \qquad (6)$$

Here $i_{\mathrm{ZVS}}^{\max}$ is the maximum reverse current required for the full zero voltage switching operation. The boundary current i_{B} is determined for the particular reference drive by accounting for lifetime load profiles, overall inductor loss characteristics and the saturation limit of core material and can be found over an multi-objective optimization as in [7].

Now, for a high value of $m(t)$ the voltage across the inductor reduces to a point where the switching frequency hits a lower limit $f_{\mathrm{Swt}}^{\min}$ as Δi_{Node} cannot be reduced indefinitely without violating valid TCM ZVS operation. The minimum switching frequency $f_{\mathrm{Swt}}^{\min}$ is defined based on TCM filter cut-off frequency. This leads to the requirement of a minimum inductor voltage drop $v_L^{\min}$ lowering the DC-link voltage utilization. The instantaneous voltage limit can hence be defined as

$$m^{\mathrm{lim}}(t) = \pm \sqrt{1 - \frac{4 L f_{\mathrm{Swt}}^{\min}}{U_{\mathrm{DC}}} \Delta i_{\mathrm{Node}}(t)} \qquad (7)$$

with

$$v_L^{\min}(t) = \frac{U_{\mathrm{DC}}}{2}\left(1 - m^{\mathrm{lim}}(t)\right). \qquad (8)$$

This reduction in voltage utilization leads to field weakening at a lower speeds increasing fundamental frequency conduction losses in the machine and a lowest possible inductance value L is chosen in [7] to lessen the effect.

Equations (4) and (7) can now be transformed to form a set of developmental guidelines to further reduce the PL-TCM drive losses,

1) Maximizing the absolute value of instantaneous voltage $|m(t)|$ when it is low.

2) Maximizing DC-link voltage utilization when $|m(t)|$ is high.

Both points above can be implemented by injecting various common mode voltage waveforms.

A. Common Mode Voltage Injection at Low Speeds

At very low speeds when $m(t)$ hovers close to zero the entire voltage signal can be raised (or lowered) with the addition of a DC common mode voltage. This is a feasible solution for traction drives as the machine is at a floating potential as seen in Fig. 3 and $v_{\mathrm{Pha,T-}}(t)$ can be raised (or lowered) simultaneously for all three phases. The following analysis is made for a positive common mode DC voltage but a similar process can be followed for negative DC voltage injection. The optimal common mode DC voltage can be calculated from the limit when $v_{\mathrm{Pha,T-}}$ contacts the maximal voltage utilization waveform at contact time t_c,

$$v_{\mathrm{Pha,T-}}(t_c) = U_{\mathrm{DC}} - v_L^{\min}(t_c). \tag{9}$$

Additionally the smooth analytical voltage waveforms have equal slope at the point of contact as seen in Fig. 5 giving,

$$\frac{\mathrm{d}v_{\mathrm{Pha,T-}}}{\mathrm{d}t} = -\frac{\mathrm{d}v_L^{\min}}{\mathrm{d}t} \tag{10}$$

At the contact point t_c the system has maximum voltage utilization and therefore has the minimum switching frequency and minimum required inverter node current. Using (2), (3), (7) and (9) and rearranging terms,

$$m_0 + m_1 \sin(\omega t_c + \phi_1) = \sqrt{1 - \frac{8Lf_{\mathrm{Swt}}^{\min}}{U_{\mathrm{DC}}} \hat{i}_{\mathrm{Pha}} |\sin(\omega t_c)|}. \tag{11}$$

The absolute value of $\sin(\omega t_c)$ in (11) expresses the fact that $v_L^{\min}$ always lowers voltage utilization. In the PL-TCM concept for traction drives the inductance L is chosen to be small for a minimal inductive voltage drop and therefore the above expression can be linearized and rearranged as,

$$m_0 + m_1 \sin(\omega t_c + \phi_1) = 1 - \frac{4Lf_{\mathrm{Swt}}^{\min}}{U_{\mathrm{DC}}} \hat{i}_{\mathrm{Pha}} |\sin(\omega t_c)|. \tag{12}$$

Two cases can now be formed based on the direction of current during contact time t_c. For the case when $\omega t_c \in [0, \pi]$, (12) can be rewritten and a similar treatment can be made for (10) to form,

$$m_1 \sin(\omega t_c + \phi_1) + i_{\mathrm{Norm}} \sin(\omega t_c) = 1 - m_0. \tag{13}$$
$$m_1 \cos(\omega t_c + \phi_1) + i_{\mathrm{Norm}} \cos(\omega t_c) = 0. \tag{14}$$

with the term i_{Norm} defined as a normalized current,

$$i_{\mathrm{Norm}} = \frac{4Lf_{\mathrm{Swt}}^{\min}}{U_{\mathrm{DC}}} \hat{i}_{\mathrm{Pha}}. \tag{15}$$

By exploiting the symmetry in (13) and (14) a closed form for the contact time t_c and DC injected voltage m_0 can be obtained as

$$m_0 = 1 - \sqrt{m_1^2 + i_{\mathrm{Norm}}^2 + 2m_1 i_{\mathrm{Norm}} \cos(\phi_1)} \tag{16}$$
$$t_c = \frac{1}{\omega} \tan^{-1} \left(\frac{m_1 \sin(\phi_1)}{m_1 \cos(\phi_1) + i_{\mathrm{Norm}}} \right). \tag{17}$$

A similar treatment can be made for the case when $\omega t_c \in [\pi, 2\pi]$ to obtain,

$$m_0 = 1 - \sqrt{m_1^2 + i_{\mathrm{Norm}}^2 - 2m_1 i_{\mathrm{Norm}} \cos(\phi_1)} \tag{18}$$
$$t_c = \frac{\pi}{\omega} + \frac{1}{\omega} \tan^{-1} \left(\frac{m_1 \sin(\phi_1)}{m_1 \cos(\phi_1) - i_{\mathrm{Norm}}} \right). \tag{19}$$

The added DC voltage can be raised further by increasing the voltage utilization factor with the injection of a 3rd order harmonic. The constraints of contact remain identical with (13) and (14) to have

$$m_1 \sin(\omega t_c + \phi_1) + m_3 \sin(3\omega t_c + \phi_3) + i_{\mathrm{Norm}} \sin(\omega t_c)$$
$$= 1 - m_0. \tag{20}$$
$$m_1 \cos(\omega t_c + \phi_1) + 3m_3 \cos(3\omega t_c + \phi_3) + i_{\mathrm{Norm}} \cos(\omega t_c)$$
$$= 0. \tag{21}$$

It can be seen that (20) and (21) are not enough not uniquely determine m_3, ϕ_3, m_0 and t_c. An approximate analytical solution could be formed by focusing only on the voltage utilization using (16) and (17) and adding the 3rd order harmonic similar to [13] [14] to obtain,

$$m_3 = \frac{1}{6} \sqrt{m_1^2 + i_{\mathrm{Norm}}^2 + 2m_1 i_{\mathrm{Norm}} \cos(\phi_1)} \tag{22}$$
$$\phi_3 = \tan^{-1} \left(\frac{m_1 \sin(\phi_1)}{m_1 \cos(\phi_1) + i_{\mathrm{Norm}}} \right) \tag{23}$$
$$m_0 = 1 - |m_{\mathrm{AC}}| \tag{24}$$

with m_{AC} defined as

$$m_{\mathrm{AC}}^2 = m_1^2 + m_3^2 + i_{\mathrm{Norm}}^2 + 2m_1 i_{\mathrm{Norm}} \cos(\phi_1) +$$
$$+ 2m_3 i_{\mathrm{Norm}} \cos(\phi_3) + 2m_1 m_3 \cos(\phi_1) \cos(\phi_3) \tag{25}$$

The exact solution can be found by adding the constraint that, the lowest point of the voltage signal should be maximally away from $\frac{uDc}{2}$ (seen in Fig. 5). This constraint directly targets the figure of merit (4) and can be expressed by first defining a term A for a given drive operating point using the general normalized voltage $m(t)$ (3) as,

$$A = \int_0^{\frac{2\pi}{\omega}} m(t)^2 \cdot dt \tag{26}$$

and then finding the optimum by,

$$\frac{\partial A}{\partial m_0} + \frac{\partial A}{\partial m_3} + \frac{\partial A}{\partial \phi_3} = 0 \tag{27}$$

The above expression can be solved numerically and pre-computed values for each operating point can be used from a look-up table for controlling the drive online.

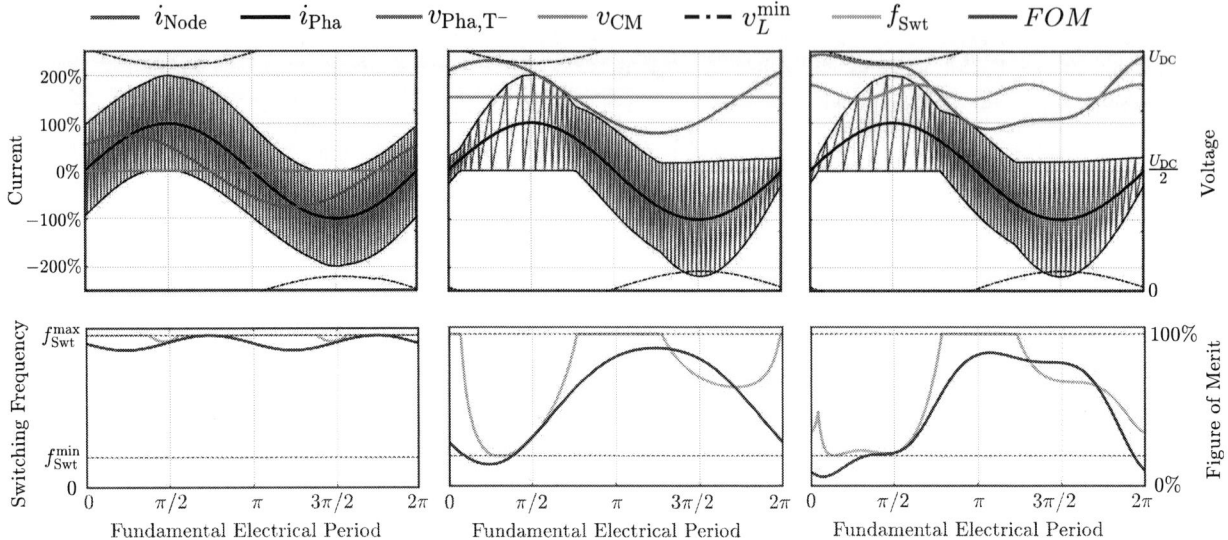

Fig. 5. Current, voltage, switching frequency and figure of merit (4) waveforms for a low speed operating point with successive addition of DC & 3rd common mode voltages.

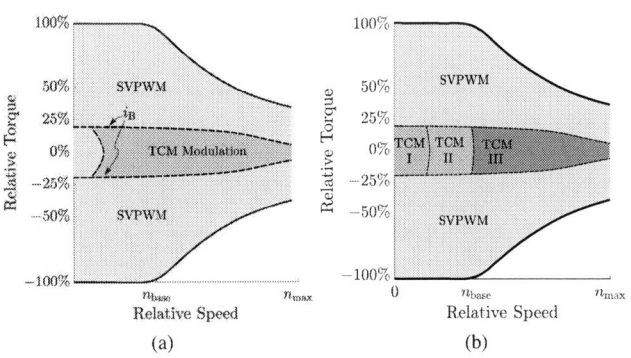

Fig. 6. Traction drive speed-torque operating maps showing (a) PL-TCM [7] modulation strategy with a gap in TCM operating at low speeds in partial load region and (b) Improved PL-TCM modulation strategy with common mode voltage injection with three sub-operating region with TCM-I at very low speeds ($n < n_{\mathrm{base}}/2$) with added DC bias and 3rd harmonic, TCM-II at low speeds till field-weakening region ($n_{\mathrm{base}}/2 \leq n < n_{\mathrm{base}}$) with no additional common mode voltage and TCM-III in field-weakening region ($n \geq n_{\mathrm{base}}$) with 3rd harmonic injection.

TABLE I
PL-TCM REFERENCE DRIVE PARAMETERS

Parameter	Name*	Value
Max. Rated Power		300 kW
DC-Link Voltage	U_{DC}	800 V
Max. Phase/Node Current	$i_{\mathrm{Pha}}^{\max}$, $i_{\mathrm{Node}}^{\max}$	850 A (600 A_{rms})
Max. TCM Phase Current	i_{B}	141 A (100 A_{rms})
Max. TCM Node Current	i_{Node}	330 A (130 A_{rms})
Max. TCM Capacitor Current	i_{Cap}	160 A (80 A_{rms})
Max. ZVS Reverse Current	i_{ZVS}	29 A (at 800 V)
Max. Fundamental Frequency		1200 Hz
TCM Filter Cut-Off Frequency		13 kHz
Min. TCM Switching Frequency	$f_{\mathrm{Swt}}^{\min}$	40 kHz
Max. TCM Switching Frequency	$f_{\mathrm{Swt}}^{\max}$	140 kHz

*in reference to Fig. 3.

TABLE II
INVERTER COMPONENTS

Components	Name*	Description
DC-Link Capacitor		450 µF, 1200 V, foil wound
Inverter Main Switches		3*240 mm², 1200 V SiC
Filter Aux. Switches	A_{123}	3*16 mm², 1200 V SiC
Filter Inductor	L_{123}	5.3 µH, ferrite, solid copper
Filter Capacitor	C_{123}	27 µF, 1200 V, foil wound

*in reference to Fig. 3.

B. Common Mode Voltage Injection at High Speeds

At high speed when the AC voltage is near its maximum limit, only a 3rd order harmonic needs to be added with the amplitude m_3 and injection angle ϕ_3 evaluated similar to process shown above and with implementations as shown in [13] [14].

III. OPERATIONAL PARAMETERS AND LOSS VALIDATION

Tab. I describes the system level design and operational parameters of the PL-TCM concept for the considered electrical drive with Tab. II defining the inverter component values. Some key highlights to be noted here are that firstly, the partial load TCM operating boundary (i_{B}) is defined at one sixth

of the peak load current at 100 A_{rms} with the inductor core saturating at roughly 350 A at 140 °C, moreover the inductor design has to be made such that peak current of 600 A_{rms} still flows through the winding (as shown in Fig. 3). Also the filter inductance value of 5.3 µH is approximately two orders of magnitude smaller than the machine inductance, designed to

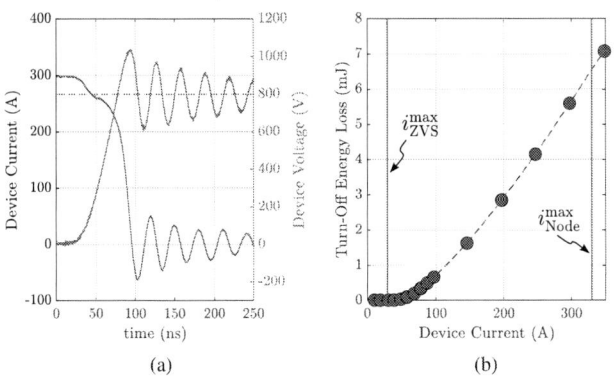

Fig. 7. Switching losses at 800 V for 8*30 mm^2 1200 V SiC devices with an R_g^{off} of 2.5 Ω. (a) Switching waveform for turn-off event at 300 A with a voltage slope of 20 V/ns (b) Turn-off energy losses during zero voltage switching operation.

avoid excessive voltage drop both during TCM operation but also during SVPWM operating region when it is not saturated.

The loss validation measurements for the inverter side components are conducted in parts with the results collected to compute the total inverter losses. The machine losses are simulated as in [7] and together with inverter losses a prognostic calculation is made for the energy consumption during TCM operation.

A. Switching Losses

Active turn-off losses dominate zero voltage switching (ZVS) events, especially for long switching periods either due to slower switching speeds or large load currents [15]. A double pulse test setup is made for the designated power module to firstly ascertain the active turn-off losses but also to find a minimum possible external gate turn-off resistance R_g^{off} which allows for a small enough switching duration in TCM mode to achieve near lossless switching [16] without exceeding the over voltage limit at peak current loads.

Fig. 7a shows the turn-off event switching waveform for a chosen R_g^{off} of 2.5 Ω for a power module consisting of 8 parallel 30 mm^2 SiC chips at 800 V which allows near lossless switching upto 48 A which is higher than the ZVS reverse current i_{ZVS} of 29 A (Tab. I) as seen in Fig. 7b. It is to be noted here that the energy stored in the MOSFET output capacitor E_{oss} which is recovered during the resonance period of ZVS cannot be directly determined in the double pulse test and therefore is calculated from the data sheet and removed.

B. Filter Inductor & Capacitor Loss Characterization

The PL-TCM inductor is required to be operational at moderately high currents upto 129 A$_{\text{rms}}$ at a high frequency of 140 kHz and upon saturation beyond the designated operational current of 330 A conduct up to 600 A$_{\text{rms}}$ at lower fundamental frequency of up to 1.2 kHz. As it is imperative to have a low loss design and high performance requirements at high temperatures, a *Ferroxcube-3C97* ferrite pot core is designed. Two air-gaps tare chosen n either side of the winding

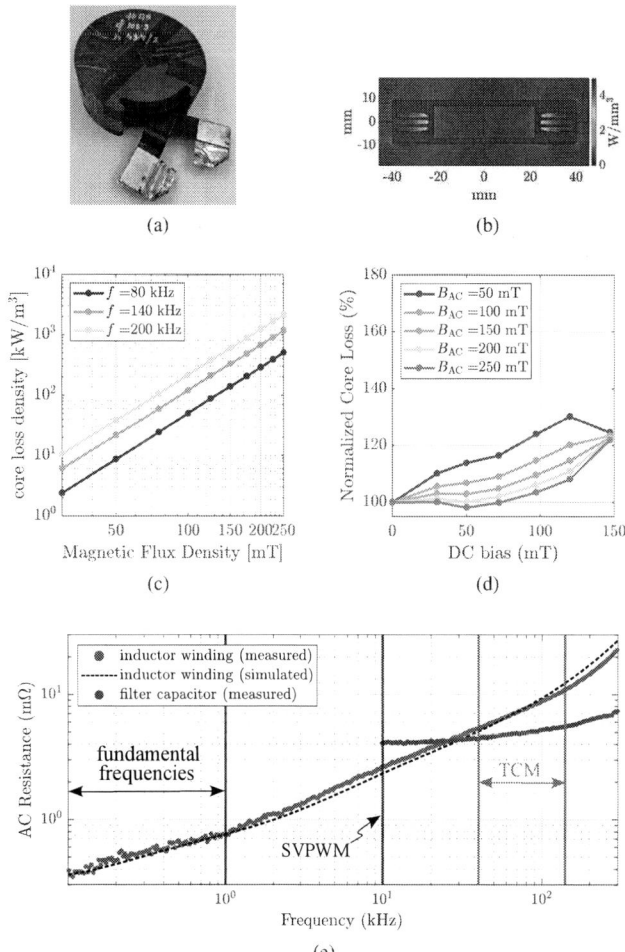

Fig. 8. Filter loss characterization with (a) the constructed saturable inductor, (b) loss density simulation of inductor in FEMM [17] for a conservative operating point (141 A at 1.2 kHz and 141 A at 140 kHz with *Ferroxcube-3C97* core characteristics at 140 °C), (c) core material characterization measurements with sinusoidal excitation (25 °C with 0 mT DC-bias), (d) normalized increase in core losses for added DC-bias (25 °C and 140 kHz) and (e) small signal impedance measurements for the inductor winding and filter capacitor (measured with the *Keysight-E4990A*).

(as seen in Fig. 8) to avoid as far as possible any intersection of the winding and the air-gap fringing flux. From [7] it can also seen that a solid copper winding for such a core provides a similar loss profile compared to a litz design with the same current carrying capacity but with a higher power density.

The characterization of the ferrite core material for sinusoidal excitation with DC-bias can be made similar to [7] by measuring active power in a resonant circuit with an additional DC excitation. The extracted Steinmetz model parameters [18] and a loss scaling factor for DC-bias can be modeled like [19] as seen in Fig. 8c and 8d. For triangular current waveforms, the model can be extended to the 'improved generalized Steinmetz equation' [20] and in the following section III-C measurements are conducted for its validation.

The inductor winding losses can be directly measured over a small signal impedance analyzer (as seen in Fig. 8e,

979-8-3315-1612-3/25 $31.00 © 2025 IEEE

measured on a *Keysight-E4990A*), the losses in the air-gapped core for small voltages can be neglected. Additionally the filter capacitor can be similarly characterized, also shown in Fig. 8e, noteworthy is that the filter capacitor losses only conducts during TCM operation with approximately 80 % of the TCM node rms current at the worse operating point. The filter capacitor is identical to the DC-link capacitors with the latter being a set of 16 of these foil wounds, the DC-link capacitor losses therefore can be characterized with the above measurement, with roughly 8 % of the losses in the filter capacitor for a given operating point.

For the semiconductor conduction losses, no explicit measurements are made but datasheet values are relied upon assuming that any deviation in the device drain source resistance R_{DS}^{on} is negligible. For the auxiliary switches (A_{123} from Fig. 3) with a custom chip area a scaled R_{DS}^{on} is used for loss calculation.

C. Inverter & Filter Losses

To realistically stress the PL-TCM filter inductor, measurements with triangular currents are made in a buck converter supplying reactive power as seen in Fig. 9a and 9b. This is done to avoid the requirement of a high active power source and sink in the test laboratory. Apart from measuring the effect of DC-bias which is already tested earlier, this setup follows the real operation closely. The half-bridge switches are in ZVS mode and the power losses of the inverter and filter are measured together as part of the DC supply power. Measurements are made over varying switching frequency, DC-link voltage and duty cycle with Fig. 9c showing the current waveforms at 800 V and 80 kHz. The different inductor current slopes and range forms the basis of validating the 'improved generalized Steinmetz equation' (iGSE) over different slope and range of magnetic flux density. After removing the various losses discussed above, the remaining core losses are compared with the model as seen in Fig. 9e showing agreement.

Using the characterized losses and the common voltage technique discussed in the previous proceeding sections, a prognosis can be made for the loss energy consumption for the WLTP cycle for the 300 kW, 600 A_{rms}, 800 V traction drive similar to [7] as shown in Fig. 10. Due to the inclusion of the low speed region in TCM for the WLTP cycle (Fig. 6), PWM harmonic losses in the machine are removed. Further as both the node current and switching frequency are reduced the switching losses & filter losses are also reduced considerable in the lower speed region (Fig. 7b & 8). The resulting WLTP loss consumption (scaled for 100 km) is subsequently reduced by approximately 0.5 kWh per 100 km from a value of 0.3 kWh without any common mode voltage injection [7]. This corresponds to roughly a 4.5 % gain in range for a large SUV class vehicle when compared to only SVPWM operation.

IV. CONCLUSION

This work, firstly highlights that a very large portion of passenger car driving occurs in partial & low loads and then argues for the possibility of substantial savings by mitigating

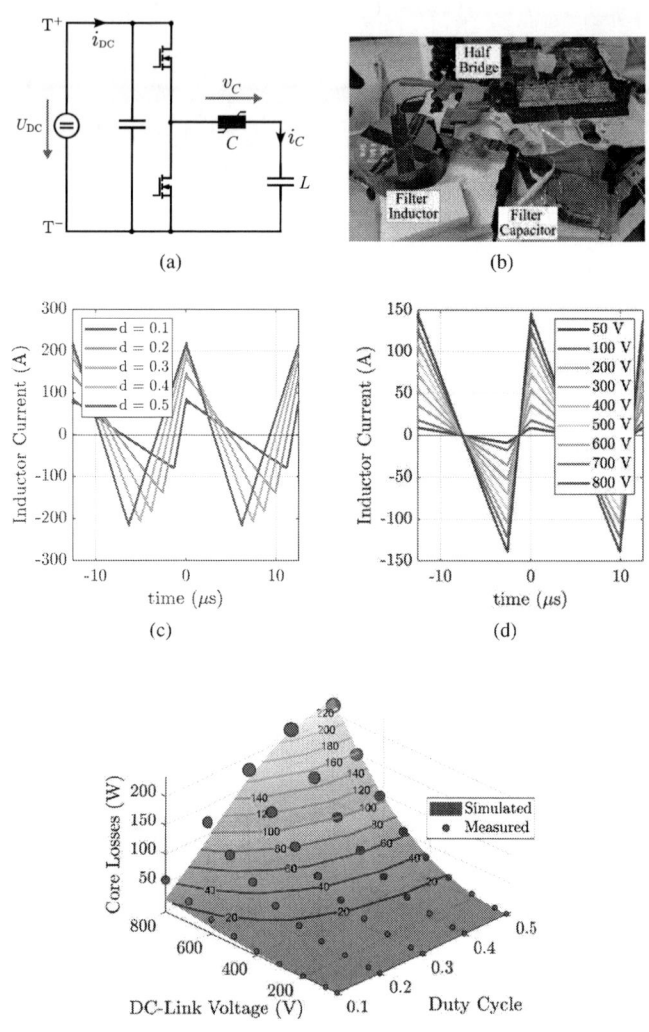

Fig. 9. Core characterization test over triangular current with (a) Circuit schematic with the measured vairables i_L, v_L, i_{DC} and U_{DC}, (b) the hardware setup, (c) measured i_L waveforms at 80 kHz and 800 V over different duty cycles, (d) measured i_L waveforms at 80 kHz and 0.2 duty cycle for different DC-link voltages and (e) core loss comparison between measurements and simulation at 80 kHz.

the PWM harmonic losses in the machine. It then further develops the PL-TCM concept from [7] by first defining a TCM loss figure of merit (FOM), consisting of node current and switching frequency, and then substantially improving it in TCM operation at very low speeds by the injection of DC & AC common mode voltages. Next, the PL-TCM inverter is characterized by measurements of the semiconductor switches, filter inductor and capacitor and the collective loss models are validated over an inverter-filter system, tested over triangular currents emulating TCM operation.

Validation tests with an actual drive system at rated power and real driving loads is considered for future work along with control aspects. Response of the system during fault events is also an important aspect to consider. Additionally, further work on the development of gate drivers capable of load dependent

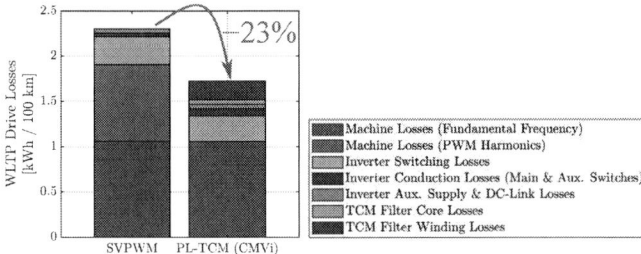

Fig. 10. Comparison of energy loss consumption in a WLTP cycle (scaled to 100 km) for a 300 kW, 600 A_{rms}, 800 V traction drive for between 10 kHz SVPWM and PL-TCM (CMVi) having selective DC & 3rd harmonic common mode voltage injection.

switching speeds and low parasitic inductance power module could further push the gains made by the PL-TCM concept..

ACKNOWLEDGMENT

The authors would like to thank Mr. Jonathan Winkler, Mr. Nisar Ahmed Khan & Mr. Eric Jacob from Robert Bosch GmbH for their support in arranging the power module infrastructure, Mr. Christoph Bantel from Robert Bosch GmbH for his support in making copper laser welding and Mr. Dennis Bura & Mr. Johannes Rickert from Robert Bosch GmbH for their kind support during laboratory measurements.

REFERENCES

[1] "Lithium-ion battery price worldwide from 2013 to 2023 (in 2023 u.s. dollars per kilowatt-hour) [graph]," BloombergNEF. [Online]. Available: https://www.statista.com/statistics/883118/global-lithium-ion-battery-pack-costs/

[2] "2012 California Household Travel Survey, Transportation Secure Data Center," National Renewable Energy Laboratory, accessed Jan. 15, 2017. [Online]. Available: www.nrel.gov/tsdc

[3] "Proposal for a new global technical regulation on the Worldwide harmonized Light vehicles Test Procedure (WLTP)," World Forum for Harmonization of Vehicle Regulations, Inland Transport Committee, United Nations Economic Commission for Europe. [Online]. Available: http://www.unece.org/DAM/trans/doc/2014/wp29/ECE-TRANS-WP29-2014-027e.pdf

[4] I. Husain, B. Ozpineci, M. S. Islam, E. Gurpinar, G.-J. Su, W. Yu, S. Chowdhury, L. Xue, D. Rahman, and R. Sahu, "Electric drive technology trends, challenges, and opportunities for future electric vehicles," *Proceedings of the IEEE*, vol. 109, no. 6, pp. 1039–1059, 2021.

[5] K. Yamazaki and Y. Seto, "Iron loss analysis of interior permanent-magnet synchronous motors-variation of main loss factors due to driving condition," *IEEE Transactions on Industry Applications*, vol. 42, no. 4, pp. 1045–1052, Jul. 2006. [Online]. Available: http://ieeexplore.ieee.org/document/1658335/

[6] Y. Miyama, M. Hazeyama, S. Hanioka, N. Watanabe, A. Daikoku, and M. Inoue, "PWM Carrier Harmonic Iron Loss Reduction Technique of Permanent-Magnet Motors for Electric Vehicles," *IEEE Transactions on Industry Applications*, vol. 52, no. 4, pp. 2865–2871, Jul. 2016. [Online]. Available: http://ieeexplore.ieee.org/document/7416178/

[7] B. Chatterjee, J. Allgeier, T. Plum, and M. Hiller, "A partial load three-phase triangular current mode modulation concept with an optimized filter inductor for high efficiency traction drives," in *PCIM Europe 2024; International Exhibition and Conference for Power Electronics, Intelligent Motion, Renewable Energy and Energy Management*, 2024, pp. 774–783.

[8] N. Zhao, Z. Q. Zhu, and W. Liu, "Rotor eddy current loss calculation and thermal analysis of permanent magnet motor and generator," *IEEE Transactions on Magnetics*, vol. 47, no. 10, pp. 4199–4202, 2011.

[9] T. Petri, M. Keller, and N. Parspour, "The influence of voltage form on the insulation resilience of inverter-fed low voltage traction machines with hairpin windings," in *2023 IEEE International Electric Machines & Drives Conference (IEMDC)*, 2023, pp. 1–6.

[10] M. Asefi and J. Nazarzadeh, "Survey on high frequency models of PWM electric drives for shaft voltage and bearing current analysis," *IET Electrical Systems in Transportation*, vol. 7, no. 3, pp. 179–189, Sep. 2017. [Online]. Available: https://onlinelibrary.wiley.com/doi/10.1049/iet-est.2016.0051

[11] J. Cho, D. Hu, and G. Cho, "Three phase sine wave voltage source inverter using the soft switched resonant poles," in *15th Annual Conference of IEEE Industrial Electronics Society*, 1989, pp. 48–53 vol.1.

[12] C. Marxgut, J. Biela, and J. W. Kolar, "Interleaved triangular current mode (tcm) resonant transition, single phase pfc rectifier with high efficiency and high power density," in *The 2010 International Power Electronics Conference - ECCE ASIA -*, 2010, pp. 1725–1732.

[13] M. Haider, J. A. Anderson, S. Miric, N. Nain, G. Zulauf, J. W. Kolar, D. Xu, and G. Deboy, "Novel ZVS s-TCM modulation of three-phase AC/DC converters," vol. 1, pp. 529–543. [Online]. Available: https://ieeexplore.ieee.org/document/9268159/

[14] Q. Wang and R. Burgos, "A method for increasing modulation index of three phase triangular conduction mode converter," in *2018 IEEE 19th Workshop on Control and Modeling for Power Electronics (COMPEL)*, 2018, pp. 1–5.

[15] M. Kasper, R. Burkat, F. Deboy, and J. Kolar, "ZVS of Power MOSFETs Revisited," *IEEE Transactions on Power Electronics*, pp. 1–1, 2016. [Online]. Available: http://ieeexplore.ieee.org/document/7482851/

[16] X. Li, X. Li, P. Liu, S. Guo, L. Zhang, A. Q. Huang, X. Deng, and B. Zhang, "Achieving zero switching loss in silicon carbide mosfet," *IEEE Transactions on Power Electronics*, vol. 34, no. 12, pp. 12 193–12 199, 2019.

[17] D. Meeker, "Finite Element Method Magnetics," version 4.2. [Online]. Available: https://www.femm.info/Archives/doc/manual42.pdf

[18] C. Steinmetz, "On the law of hysteresis," *Proceedings of the IEEE*, vol. 72, no. 2, pp. 197–221, 1984.

[19] J. Muhlethaler, J. Biela, J. W. Kolar, and A. Ecklebe, "Core losses under the dc bias condition based on steinmetz parameters," *IEEE Transactions on Power Electronics*, vol. 27, no. 2, pp. 953–963, 2012.

[20] J. Reinert, A. Brockmeyer, and R. De Doncker, "Calculation of losses in ferro- and ferrimagnetic materials based on the modified Steinmetz equation," *IEEE Transactions on Industry Applications*, vol. 37, no. 4, pp. 1055–1061, aug 2001. [Online]. Available: http://ieeexplore.ieee.org/document/936396/

Analysis of Maximum Power Transfer Limit for Linear Operation of Dual-Active-Bridge Converters

Radhika Sarda[*], Ezequiel Ramos Rodriguez[†], Gaowen Liang[†], Glen G. Farivar[‡], Josep Pou[§],
Vaisambhayana B. Sriram[†], and Anshuman Tripathi[†]

[*]Energy Research Institute at NTU, Interdisciplinary Graduate Programme, Nanyang Technological University, Singapore
Email: radhika010@e.ntu.edu.sg

[†]Energy Research Institute at NTU, Nanyang Technological University, Singapore
Email: ezequiel.rr@ntu.edu.sg, gaowen.liang@ntu.edu.sg, vsriram@ntu.edu.sg, antri@ntu.edu.sg

[‡]Department of Electrical and Electronic Engineering, University of Melbourne, Melbourne, Australia
Email: gfarivar@unimelb.edu.au

[§]Department of Electrical Engineering, City University of Hong Kong, Hong Kong
Email: josep.pou@ieee.org

Abstract—**In power electronics applications where the dual-active-bridge (DAB) converter topology serves as an isolation stage between high-impedance input and output ports, significant voltage fluctuations can arise during transient conditions. These variations restrict the DAB converter's maximum power transfer capacity and narrow its linear operating region, a limitation further exacerbated when low-capacitance capacitors are used at the DAB converter terminals. This paper introduces a comprehensive control design framework tailored to maximize the DAB converter's power transfer capacity, accommodating a range of input and output capacitor sizes. Additionally, we examine multiple control strategies that directly influence voltage variations at the DAB converter terminals, demonstrating their critical role when trying to maximize the DAB converter's power capacity.**

Index Terms—**Energy distribution, dual-active bridges, solid-state transformers, stability, maximum power.**

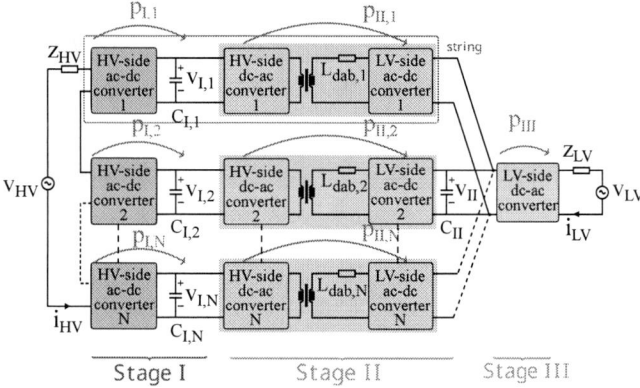

Fig. 1. Single-phase three-stage architecture of a solid-state transformer.

I. INTRODUCTION

With the rapid growth in the use of renewable energy sources in the power distribution network, the need for smarter technologies to regulate the power flow at critical nodes is increasing. Two prominent examples of such technologies that provide flexible energy flow control are the solid state transformer (SST) and DC microgrids [1]–[4]. A key requirement of flexible energy routers is to provide isolation among various sources and loads. High-frequency dual-active-bridge (DAB) converters are a widely studied and a suited topology as an interface providing isolation [4]. Notable research has been done on DAB converters with controlled voltage sources and energy storages on either input, output, or both-sides of the converter [4]–[7]. However, DAB converters, particularly those within SSTs that are connected to capacitors at both input and output sides, face the challenge of dealing with varying capacitor voltages during transient events. Varying input and output capacitor voltages affect the power flow capacity of DAB converters.

As an illustration, Fig. 1 shows a commonly used three-stage SST with a modular architecture with N strings [8], [9].

It consists of a cascaded H-bridge (CHB) ac-dc converter in Stage I. Each floating high-voltage (HV) dc-link capacitor of Stage I is connected to a DAB converter in Stage II. The outputs of all DAB converters are connected in parallel to form a low-voltage (LV) dc-link capacitor. Finally, a two-level inverter converts LVdc to LVac in Stage III. Stages I and III are tightly regulated to ensure precise ac grid current tracking on both the HV and LV sides. Consequently, the input and output ports of the DAB converters in Fig. 1 can be considered as power-controlled sources and sinks. Note that, during transients, the input and output voltages at the DAB converter terminals are impacted and thus, the power transfer capacity. When load variations exceed this capacity, the DAB converters will saturate, resulting in overshoots in high-frequency currents flowing through the transformer.

Different SST control strategies directly influence voltage variations at the DAB converter terminals. For example, the outer control loop of Stage I can be designed to regulate the total energy (voltage) of the HV-side capacitors [5], the energy (voltage) of the LV-side capacitors [10], or the combined energy (voltages) of both HV and LV capacitors [11]. Likewise, Stage II can adopt various control targets beyond

979-8-3315-1612-3/25 $31.00 © 2025 IEEE

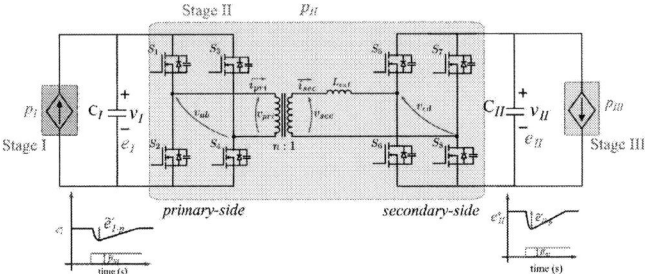

Fig. 2. Single-string equivalent model of three-stage cascaded SST.

the conventional energy (voltage) balance of the floating HV-side capacitors [5], [12]. A summary of the most common SST control strategies is presented in Table I. The voltage variations at the DAB converter terminals, and consequently its power transfer capacity, are affected by the control strategy, bandwidth of the control loops, and the size of the capacitors. This paper offers a comprehensive framework for assessing the transient power capacity of the DAB converters through input-output energy distribution charts, providing a valuable foundation for the selection of control strategies, bandwidths, and capacitor sizes in SST applications.

This paper is structured as follows. Section II provides the background of the power flow in a DAB converter. Section III analyses the energy fluctuations with different control strategies for the SSTs. Section IV analyses the effect of energy fluctuations on maximum power transfer capacity using various control strategies and with different dc-link capacitors. Section V provides the time-domain simulations of the analysis presented in Section IV. Section VI concludes the paper.

II. DAB CONVERTER POWER FLOW

Fig. 2 shows a reduced-order model of the N-string SST architecture, depicted in Fig. 1. As it can be observed, the input and output terminals of the DAB converter have been modeled as power sources and sinks, p_I and p_{III}, to facilitate the subsequent analysis. Besides, the floating HV-side capacitors are modeled using equivalent capacitor C_I holding equivalent voltage v_I (energy e_I). Similarly, the LV dc-link capacitor C_{II} holds voltage v_{II} (energy e_{II}). Variables p_I, p_{II} and p_{III} represent the averaged power (over a grid frequency period) flowing through Stages I, II and III, respectively. The parameter L_{ext} represents the external DAB converter inductance of the $n : 1$ transformer operating at f Hz. The average power flow through the DAB converter using single-phase shift is given by [4],

$$p_{II} = \frac{nv_I v_{II} d(1-d)}{2fL_{ext}}, \tag{1}$$

where d is the phase-shift between the primary and secondary converter voltage waveforms ($-0.5 \leq d \leq 0.5$). Equation (1) reveals the dependence of the power flow on both v_I and v_{II} (or equivalently, e_I and e_{II}). The maximum positive power flow from Stage I to Stage III, occurs with $d = 0.5$, given by,

$$p_{II-max} = \frac{nv_I v_{II}}{8fL_{ext}}. \tag{2}$$

TABLE I
CONTROL LAWS FOR DIFFERENT SST CONTROL STRATEGIES.

Control Strategy	Stage I	Stage II
Conventional [5]	$\tilde{p}_I = -\alpha\tilde{e}_I$	$\tilde{p}_{II} = -\beta\tilde{e}_{II}$
Decoupled [11]	$\tilde{p}_I = -\alpha\tilde{e}_\Sigma$	$\tilde{p}_{II} = -\beta\tilde{e}_{II}$
Balanced [12]	$\tilde{p}_I = -\alpha\tilde{e}_\Sigma$	$\tilde{p}_{II} = -\beta\tilde{e}_\Delta$

TABLE II
RELATIONSHIP BETWEEN PEAK TRANSIENT ENERGIES USING DIFFERENT CONTROL STRATEGIES.

Control Strategy	Stage I	Stage II
Conventional [5]	$\tilde{e}_{I-p} = \frac{\tilde{p}_{III}}{\alpha}$	$\tilde{e}_{II-p} = \frac{\tilde{p}_{III}}{\beta}$
Decoupled [11]	$\tilde{e}_{I-p} = \tilde{p}_{III}(\frac{1}{\alpha} - \frac{1}{\beta})$	$\tilde{e}_{II-p} = \frac{\tilde{p}_{III}}{\beta}$
Balanced [12]	$\tilde{e}_{I-p} = \frac{\tilde{p}_{III}}{2\alpha}$	$\tilde{e}_{II-p} = \frac{\tilde{p}_{III}}{2\alpha}$

During load transients, the DAB converter terminal voltages deviate from the references (v_I^* and v_{II}^*), thus impacting p_{II-max}. This analysis is particularly important for step-up load changes (p_{III}), as a sudden increase in load demand will cause the DAB converter terminal voltages to drop, reducing the power transfer capacity. If the required load surpasses this reduced DAB converter power transfer capacity during the transient, saturation will occur. For subsequent consistency with energy variables, (2) is rewritten considering $e_I = C_I v_I^2/2$, and similarly for e_{II}, resulting in,

$$p_{II-max} = \frac{n}{4fL_{ext}}\sqrt{\frac{e_I e_{II}}{C_I C_{II}}}. \tag{3}$$

Note that, the energy variables (e_I and e_{II}) are used instead of voltage variables (v_I and v_{II}) as the system's dynamics (relationship between capacitor energy and power) becomes a simple integrator independent of the operating point (v_I^* and v_{II}^*) [13]. Besides the analysis can be easily extended to systems with different stored energies.

III. ENERGY DISTRIBUTION DURING LOAD TRANSIENTS

The dynamic model of the system in Fig. 2 corresponds to,

$$\dot{e}_I = p_I - p_{II} \tag{4}$$
$$\dot{e}_{II} = p_{II} - p_{III} \tag{5}$$

or, equivalently, in terms of incremental variables (where the equilibrium is shifted from (e_I^*, e_{II}^*) to $(0, 0)$ via $\tilde{x} = x - x^*$),

$$\dot{\tilde{e}}_I = \tilde{p}_I - \tilde{p}_{II} \tag{6}$$
$$\dot{\tilde{e}}_{II} = \tilde{p}_{II} - \tilde{p}_{III}. \tag{7}$$

The existing control strategies for SSTs can be categorized into conventional, decoupled, and balanced control, as outlined by the control laws in Table I [5], [10]–[12]. In the conventional strategy, Stage I manages $\tilde{e}_I$ while Stage II manages $\tilde{e}_{II}$ [14]. In the decoupled strategy, Stage I handles the total sum of energies, $\tilde{e}_\Sigma = \tilde{e}_I + \tilde{e}_{II}$, and Stage II controls $\tilde{e}_{II}$ [11]. Lastly, in the balanced strategy, Stage I is responsible for $\tilde{e}_\Sigma$ and Stage II regulates the energy difference, $\tilde{e}_\Delta = \tilde{e}_{II} - \tilde{e}_I$ [12]. Positive constants α and β in Table I denote the control gains for Stages I and II, respectively. Note that integral terms have been omitted in Table I's control laws to facilitate the following analysis. Substituting the control laws from Table I into (6) and (7), the following closed-loop state-space representations are obtained for the different SST control strategies considered,

$$\left[\begin{array}{c} \dot{\tilde{e}}_I \\ \dot{\tilde{e}}_{II} \end{array} \right]_{CC} = \left[\begin{array}{cc} -\alpha & \beta \\ 0 & -\beta \end{array} \right] \left[\begin{array}{c} \tilde{e}_I \\ \tilde{e}_{II} \end{array} \right] + \left[\begin{array}{c} 0 \\ -1 \end{array} \right] \tilde{p}_{III}, \quad (8)$$

$$\left[\begin{array}{c} \dot{\tilde{e}}_I \\ \dot{\tilde{e}}_{II} \end{array} \right]_{DC} = \left[\begin{array}{cc} -\alpha & \beta - \alpha \\ 0 & -\beta \end{array} \right] \left[\begin{array}{c} \tilde{e}_I \\ \tilde{e}_{II} \end{array} \right] + \left[\begin{array}{c} 0 \\ -1 \end{array} \right] \tilde{p}_{III}, \quad (9)$$

$$\left[\begin{array}{c} \dot{\tilde{e}}_I \\ \dot{\tilde{e}}_{II} \end{array} \right]_{BC} = \left[\begin{array}{cc} -\beta & \beta \\ -\alpha + \beta & -\alpha - \beta \end{array} \right] \left[\begin{array}{c} \tilde{e}_I \\ \tilde{e}_{II} \end{array} \right] + \left[\begin{array}{c} 0 \\ -1 \end{array} \right] \tilde{p}_{III}, \quad (10)$$

where subscripts CC, DC, and BC refer to the conventional, decoupled, and balanced controls, respectively. Solving for the time-domain solutions of the state-space representations in (8)-(10), for zero initial condition and for a constant load disturbance $\tilde{p}_{III}$, the peak value of transient energies in the capacitors C_I and C_{II}, referred to as $\tilde{e}_{I-p}$ and $\tilde{e}_{II-p}$, respectively, can be readily derived. Specifically, Table II shows the peak value of the solutions with different control strategies. The peak value of the solutions are selected for the proposed analysis since they represent the worst-case scenario, i.e. when the maximum power transfer capacity of the DAB converter reaches its minimum, which can be obtained by evaluating (3) with the values in Table II. Fig. 3 illustrates how these peak values vary with different $k = \frac{\beta}{\alpha}$ ratios in the range [1,100], for a given $\tilde{p}_{III} = 1$ and $\alpha = 50$, for the different control strategies. Specifically, Fig. 3 compares the theoretical results from Table II with simulation results. As it can be observed, theoretical predictions and simulation results are in agreement, with slight discrepancies due to the integral variables considered in the simulation to enforce the steady-state error to zero.

In Fig. 3(a), where the feasible peak energy distributions with conventional control are shown, it can be inferred that with different k values, $\tilde{e}_{I-p}$ remains constant while $\tilde{e}_{\Sigma-p}$ changes, where $\tilde{e}_{\Sigma-p} = \tilde{e}_{I-p} + \tilde{e}_{II-p}$. Using the decoupled control strategy, $\tilde{e}_{\Sigma-p}$ remains constant and distributes between $\tilde{e}_{I-p}$ and $\tilde{e}_{II-p}$ with different k values, as seen in Fig. 3(b). Finally, using the balanced control, feasible peak energy distributions remain approximately equal, i.e., $\tilde{e}_{I-p} \approx \tilde{e}_{II-p}$ irrespective of k, as seen in Fig. 3(c).

IV. EFFECT OF ENERGY DISTRIBUTION ON DAB CONVERTER POWER TRANSFER CAPACITY

Note that the peak transient energy distribution charts in Fig. 3 can be readily mapped to peak transient voltage (v_{I-p} and v_{II-p}) distribution charts using the following relationships:

TABLE III
SIMULATION PARAMETERS.

Parameters	Values
Input voltage, v_I	250 V
Output voltage, v_{II}	350 V
External inductance, L_{ext}	100 μH
Switching frequency, f	50 kHz
Turns ratio, $n : 1$	1:1
HV-side dc-link capacitor, C_I	800 μF, 250 μF
LV-side dc-link capacitor, C_{II}	800 μF, 250 μF
Control gain, α	50
Relative control gain, k	[1, 100]

$$\frac{1}{2} C_I v_{I-p}^2 = e_I^* - \tilde{e}_{I-p}$$
$$\frac{1}{2} C_{II} v_{II-p}^2 = e_{II}^* - \tilde{e}_{II-p}, \quad (11)$$

where the effect of different capacitor sizes C_I and C_{II} can be analysed.

To illustrate this, the DAB converter shown in Fig. 2 with parameters listed in Table III is considered. Two case studies are presented, featuring high-capacitance (HC) and low-capacitance (LC) configurations for the dc-link capacitors C_I and C_{II}. The HC configuration considers $C_I = C_{II} = 800\mu F$, while the LC configuration considers $C_I = C_{II} = 250\mu F$. As in Fig. 3, the value of α is kept constant to 50, whereas gain β is varied with the ratio $k = [1, 100]$. Figs. 4(a)-(c) present the peak transient energy distribution charts, analogous to those in Fig. 3, but with the load power scaled from a unit-step change to 750 W. These energy distribution charts are transformed into voltage distribution charts using (11), resulting in Figs. 4(d)–(f) for the HC scenario and Figs. 4(g)–(i) for the LC scenario. In these voltage distribution charts, the maximum power transfer capacity of the DAB converter, p_{II-max}, derived from (2), is depicted with a solid pink line, while the load transient requirement of $p_{III} = 750$ W is represented by a dashed pink line. Under any control strategy employed, should $p_{III} \geq p_{II-max}$, the DAB converter's operation enters saturation, potentially undermining stability.

1) Conventional Control: Note that, $\tilde{e}_{I-p}$ remains fairly constant, thus agreeing with Table II, and hence, v_{I-p} remains constant, as seen in Figs. 4(d) and (g), depending on the value of capacitance. Specifically, for the HC case, v_{I-p} corresponds to approximately 150 V, whereas it becomes zero for the LC case. Contrarily, v_{II-p} changes with k, approaching a straight line for the HC case and a parabola for the LC case. Two different operating points, X_1 and X_2, are considered, corresponding to $k = 10$ and $k = 1.33$, respectively. As it can be observed in Fig. 4(d), X_1 and X_2 satisfy $p_{III} < p_{II-max}$ for the HC case, thus having a favorable operation. However, using LC system in Fig. 4(g), the conventional control fails to provide a favorable distribution of energies among C_I and

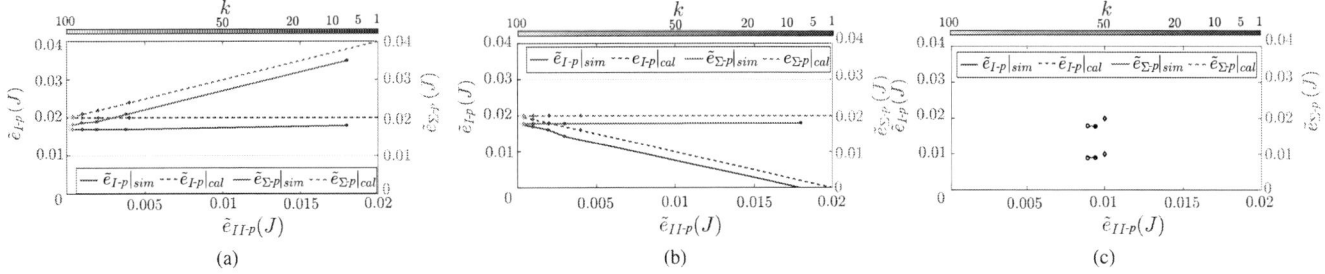

Fig. 3. Peak energy distribution among input and output DAB converter terminals for relative bandwidth ratios k between the energy controls of Stages I and II in the range [1, 100], and a unit load step, using (a) conventional control, (b) decoupled control, and (c) balanced control.

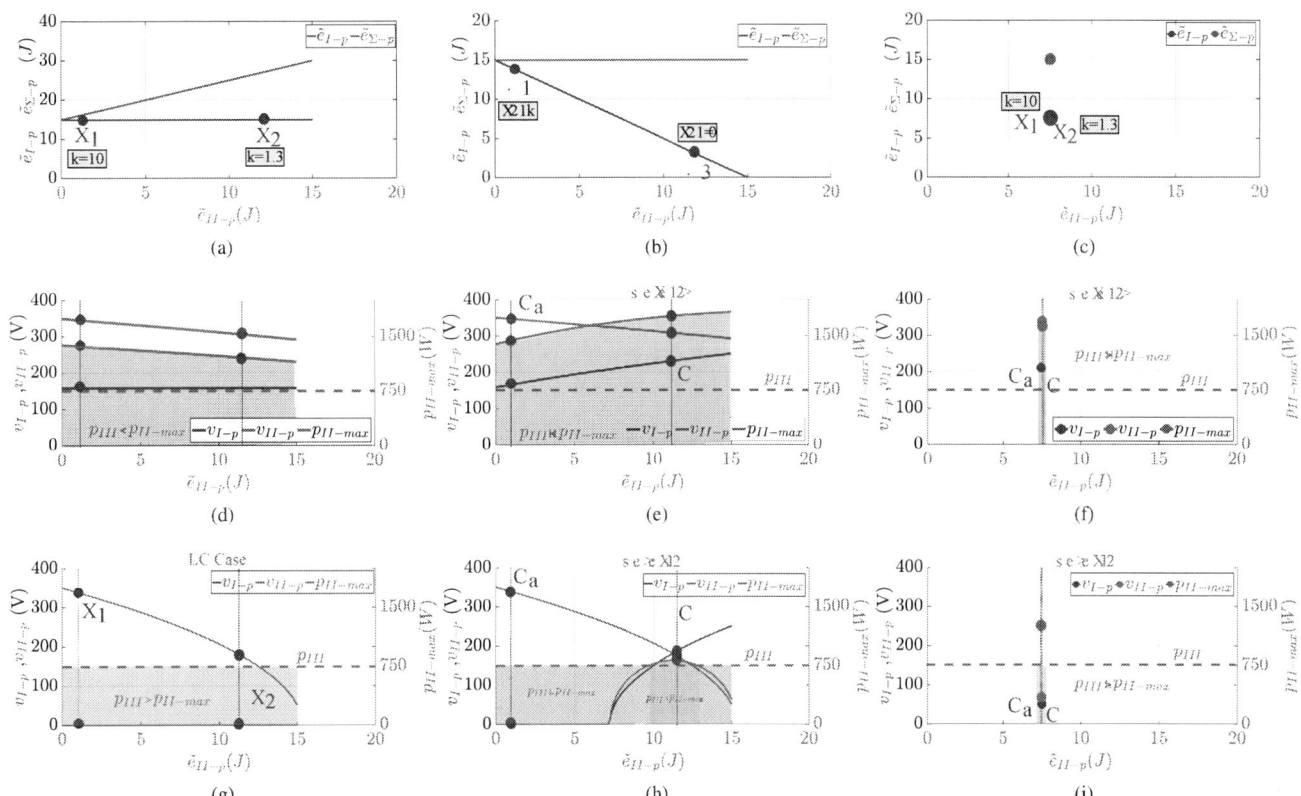

Fig. 4. Peak transient energy and voltage distribution charts among the HV and LV dc-link capacitors during a 750-W load transient. From left to right, the columns correspond to conventional control, decoupled control, and balanced control. From top to bottom, the rows display: the energy distribution charts, the voltage distribution charts for the HC scenario, and the voltage distribution charts for the LC scenario.

C_{II}, thus the saturated operation of the DAB converter cannot be avoided by manipulating k.

2) Decoupled Control: Unlike the conventional control, $\tilde{e}_{\Sigma-p}$ remains constant in case of decoupled control. Thus, as $\tilde{e}_{II-p}$ increases by reducing k (similar to the conventional case), $\tilde{e}_{I-p}$ decreases, preserving $\tilde{e}_{I-p} + \tilde{e}_{II-p} = \tilde{e}_{\Sigma-p}$. The corresponding energy changes are reflected in the respective voltage profiles.

For the HC case in Fig. 4(e), v_{II-p} reduces, while v_{I-p} increases in an approximately straight line fashion. In Fig. 4(e), operating points X_1 and X_2 satisfy $p_{III} < p_{II-max}$, thus having favorable operations. Compared to the conventional

case for the HC case in Fig. 4(d), the range $p_{II-max} - p_{III}$ in Fig. 4(e) increases by 50 W and 500 W for the X_1 and X_2 operating points, respectively.

For the LC case in Fig. 4(h), v_{II-p} reduces in an approximately parabolic fashion similar to the conventional control case. However, v_{I-p} does not remain zero for the entire range of $\tilde{e}_{II-p}$ unlike the conventional control. Importantly, v_{I-p} is nonzero for some regions of the LC case, thus enabling a nonzero DAB converter maximum power transfer capacity. In Fig. 4(h), operating point X_1 results in $p_{III} > p_{II-max}$, pushing the DAB converter operation towards saturation. However, operating point X_2 provides a favorable distribution of tran-

TABLE IV
SUMMARY OF RESULTS FROM FIG. 4: IDENTIFYING SST CONTROL STRATEGIES THAT ENSURE LINEAR DAB CONVERTER OPERATION FOR TWO DIFFERENT OPERATING POINTS (CONTROL BANDWIDTH RATIOS) WITH A GIVEN LOAD TRANSIENT.

Control Strategy	HC X_1	HC X_2	LC X_1	LC X_2
Conventional (Fig. 4(d) and 4(g))	✓	✓	x	x
Decoupled (Fig. 4(e) and 4(h))	✓	✓	x	✓
Balanced (Fig. 4(f) and 4(i))	✓	✓	x	x

sient energies ($\tilde{e}_{I-p}$ and $\tilde{e}_{II-p}$) retaining the linear operation of DAB converter (i.e., $p_{III} < p_{II-max}$). The operating point X_2 increases the range by 775 W with decoupled control, while X_1 being unfeasible for both conventional and decoupled controls. Thus, by appropriately distributing the transient energies among C_I and C_{II}, via the k relative bandwidth (or β gain equivalently), the power transfer capacity can be maximized avoiding saturated DAB converter operation.

3) Balanced Control: Unlike the conventional and decoupled controls, where the energy distribution chart depends on the bandwidth ratio k, the balanced control yields an invariant energy distribution chart. Hence, Figs. 4(c), (f) and (i) represent single points. For this reason, only the operating point X_1 is considered. For the case of HC, the results are presented in Fig. 4(f) and as $p_{III} < p_{II-max}$, the linear region of operation is retained. However, for the case of LC in Fig. 4(i), the balanced distribution of energy presents saturated condition as $p_{II-max} = 350$ W which is lower than given $p_{III} = 750$ W. In this case, unlike the conventional and decoupled controls, since the distribution is independent of k, the control design lacks flexibility to distribute the transient energy appropriately to maximize the power delivery range of DAB converter.

The results are summarized in Table IV, where it can be concluded that the three considered SST controls preserve DAB converter linear operation for the considered load transient under the HC scenario, whereas only the decoupled control can achieve linear DAB converter operation for the LC case. The distribution charts vary based on the load transient, the $C_I - C_{II}$ pair, and the α gain, but despite these variations, the proposed theoretical framework remains general to account for these changes and offer a solid foundation for SST design. The results from Table IV are validated in the following section, where time-domain simulation results support the aforementioned analysis.

V. SIMULATIONS

The switched model DAB converter shown in Fig. 2, with parameters listed in Table III, is considered in this section for simulations using PLECS. The simulations consider the same conditions as Section IV analysis, namely, a 750-W load change, HC and LC scenarios, three different SST control strategies, and two operating points for each. Specifically, operating points X_1 and X_2 correspond to $k = 10$ and $k = 1.3$, respectively. Note that for the balanced control case, only $k = 10$ is considered due to the overlapping between X_1 and X_2.

1) Conventional Control: According to Table IV, for conventional control, both operating points X_1 and X_2 yield favorable results for the HC system; however, they lead to saturated operation for the LC system. The time-domain simulations are presented in Fig. 5 for the HC case. As inferred from Fig. 4(a), in Figs. 5(a) and (e), $\tilde{e}_{I-p}$ remains constant at operating points X_1 and X_2, to approximately 14.5 J. In contrast, $\tilde{e}_{II-p}$ for X_1 is lower in Fig. 5(a) and higher for X_2 in Fig. 5(e). Consequently, the total energy disturbance ($\tilde{e}_{\Sigma-p}$) increases from 15 J to 25 J as the operating point shifts from X_1 to X_2. Thus, the voltage dip at C_I remains unchanged, irrespective of k, as inferred in Fig. 4(d) and corroborated in Figs. 5(b) and (f), with a constant voltage dip of 36% for v_I irrespective of the operating point X_1, X_2. However, the energy transient on the LV dc-link capacitor (C_{II}) increases with decrease in k, i.e., 1.5% voltage drop with X_1 and 11% voltage drop with X_2. Nonetheless, since $p_{III} = 750$ W is lower than $p_{II-max} = \{1425$ W,1250 W$\}$, for X_1 and X_2 respectively, DAB converter saturated operation is avoided, as shown in Figs. 5(c) and (g). The impact of distribution of energies among C_I and C_{II} is also visible in the primary-side high-frequency transformer current, which results in 60% and 40% overshoots with X_1 and X_2, respectively.

For the LC case in Fig. 6, operating points X_1 and X_2 produce unfavorable results as $p_{III} > p_{II-max}$. The peak voltage dip $v_{I-p} = 0$ inferred from Fig. 4(g), does not reach zero in Fig. 6(b) and (f) due to saturated DAB converter operation. Thus, the linear analysis from Figs. 4(a) and (g) does not align with the time-domain simulations, except in predicting the saturation. The phase-shift in saturated operation ($d = 0.5$), is shown in Figs. 6(c) and (g), with obvious negative effects on the high-frequency current overshoots in Figs. 6(d) and (h).

2) Decoupled Control: From Table IV, for the decoupled control, both operating points X_1 and X_2 yield favorable results for the HC system, however only X_2 yields favorable operation for the LC system. The time-domain simulations are presented in Fig. 7 for the HC case. As inferred from Fig. 4(b), $\tilde{e}_{\Sigma-p}$ remains constant at operating points X_1 and X_2, to approximately 14.5 J in Figs. 7(a) and (e). However, $\tilde{e}_{I-p}$ decreases as $\tilde{e}_{II-p}$ increases with the movement of the operating point from X_1 to X_2. Thus, $\{\tilde{e}_{I-p}, \tilde{e}_{II-p}\}$ corresponds to $\{12.5$ J, 2 J$\}$ for X_1 in Fig. 7(a) and $\{4$ J, 10.5 J$\}$ for X_2 in Fig. 7(e). The voltage dips at C_I and C_{II} follows the similar analysis inferred from Fig. 4(e). The voltage profile at C_{II}, similar to the conventional control results, observes a 1.5% and 32% drop, seen in Figs. 7(b) and (f) for X_1 and X_2, respectively. However, unlike the conventional control, where v_{I-p} is constant, a 32% and 8% voltage dip is observed at C_I for X_1 and X_2, respectively, presented in Figs. 7(b) and (f). Though, DAB converter saturated operation is avoided for both

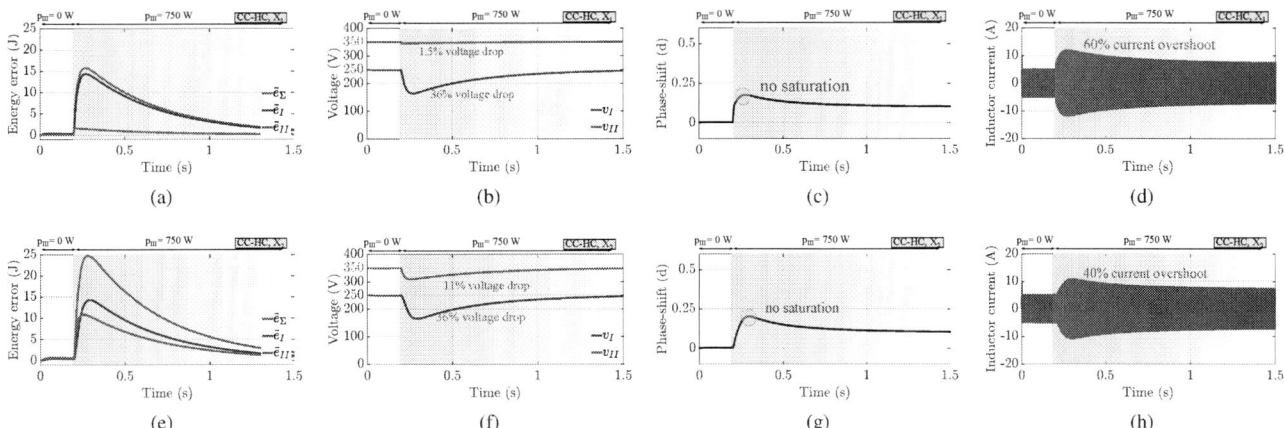

Fig. 5. Simulation results for conventional control (CC) under a load step-up change from 0 W to 750 W in the HC-DAB converter scenario. The left column shows the incremental energy variables ($\tilde{e}_I$, $\tilde{e}_{II}$ and $\tilde{e}_\Sigma$), the middle-left column presents the dc-link capacitor voltages (v_I and v_{II}), the middle-right column displays the DAB converter normalised phase-shift (d), and the right column shows the primary-side high-frequency transformer current (i_{pri}). The top row corresponds to operating point X_1, and the bottom row corresponds to operating point X_2.

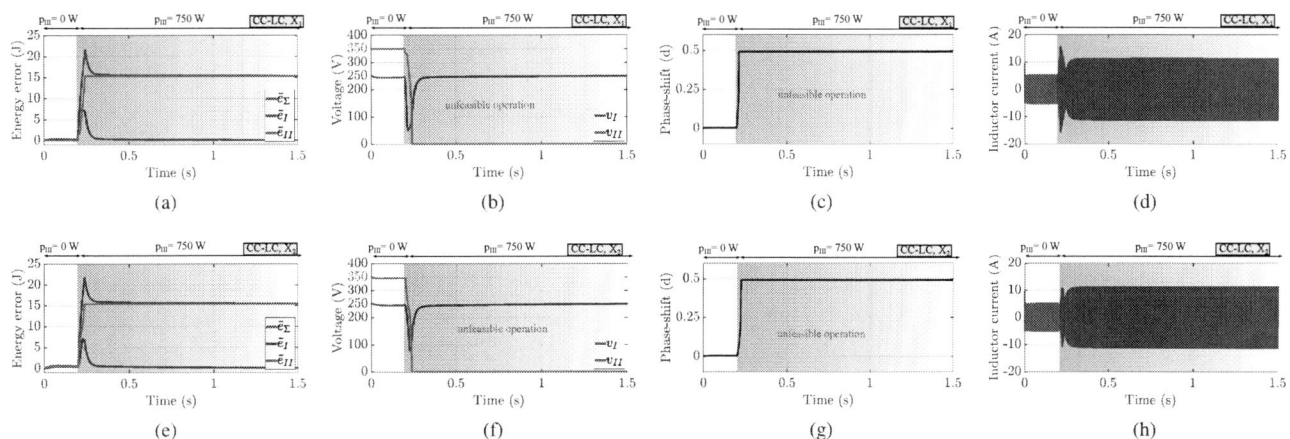

Fig. 6. Simulation results for conventional control (CC) under a load step-up change from 0 W to 750 W in the LC-DAB converter scenario. The left column shows the incremental energy variables ($\tilde{e}_I$, $\tilde{e}_{II}$ and $\tilde{e}_\Sigma$), the middle-left column presents the dc-link capacitor voltages (v_I and v_{II}), the middle-right column displays the DAB converter normalised phase-shift (d), and the right column shows the primary-side high-frequency transformer current (i_{pri}). The top row corresponds to operating point X_1, and the bottom row corresponds to operating point X_2.

X_1 and X_2, as shown in Figs. 7(c) and (g), a 33% overshoot is observed in the phase-shift for X_1 in Fig 7(c), whereas no overshoot for X_2 in Fig. 7(h). This is because the operating point X_2 provides a higher range of p_{II-max} in comparison to X_1, corroborating the results discussed for Fig. 4(e). The impact of distribution of energies among C_I and C_{II} is also visible in the primary-side high-frequency transformer current in Figs. 7(d) and (h), resulting in 47% and no overshoots with X_1 and X_2, respectively.

Extending a similar analysis for Fig. 8 with the LC case, the operating point X_1 yield unfavorable results as $p_{III} > p_{II-max}$, whereas X_2 yields a favorable result. For results with the LC case and X_1, in Figs. 8(a)-(d), the phase-shift results in saturated operation as shown in Fig. 8(c). The negative effects of saturated operation is visible in the high-frequency current overshoots presented in Fig. 8(d). Similar to Fig. 6, as the system operates in saturated region with X_1, the energy and voltage curves in Figs. 8(a) and (b) cannot be

directly verified by comparing it to Fig. 4(b). However, with X_2, $p_{II-max}(X_2)$ value from Fig. 4(h) is 775 W, which is 25 W higher than the given load change. Thus, preserving the linear operating range of the DAB converter as seen in Fig. 8(e)-(h). The peak total energy variation ($\tilde{e}_{\Sigma-p}$) in Fig. 8(e) is 14.5 J and $\{\tilde{e}_{I-p}, \tilde{e}_{II-p}\}$ corresponds to $\{3.5$ J, 11 J$\}$, very close to the HC case. Though, due to reduced capacitance, the peak energy changes correspond to 25% and 43% voltage dips at C_I and C_{II} respectively, the phase-shift is not pushed to saturation, as shown in Fig. 8(g). The high-frequency transformer current in Fig. 8(h) also does not experience overshoot as 80% of the transient energy demand is compensated by C_{II}. However, the range of operation where $p_{II-max} > p_{III}$ is narrower for the LC case, when compared with the HC case.

3) Balanced Control: From Table IV, using the balanced control, X_1 is a favourable energy distribution for the HC case, however with the LC case, X_1 pushes the system to saturated

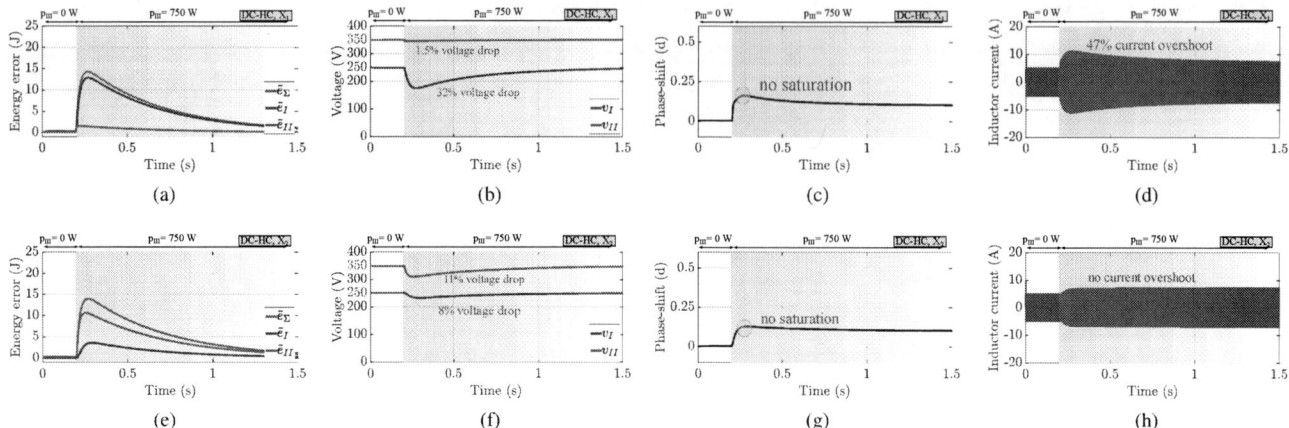

Fig. 7. Simulation results for decoupled control (DC) under a load step-up change from 0 W to 750 W in the HC-DAB converter scenario. The left column shows the incremental energy variables ($\tilde{e}_I$, $\tilde{e}_{II}$ and $\tilde{e}_\Sigma$), the middle-left column presents the dc-link capacitor voltages (v_I and v_{II}), the middle-right column displays the DAB converter normalised phase-shift (d), and the right column shows the primary-side high-frequency transformer current (i_{pri}). The top row corresponds to operating point X_1, and the bottom row corresponds to operating point X_2.

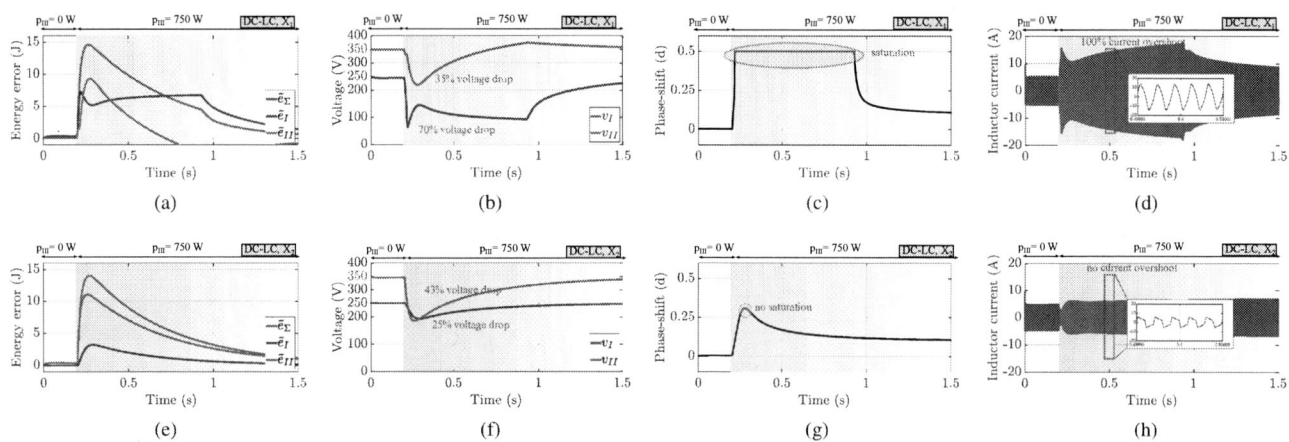

Fig. 8. Simulation results for decoupled control (DC) under a load step-up change from 0 W to 750 W in the LC-DAB converter scenario. The left column shows the incremental energy variables ($\tilde{e}_I$, $\tilde{e}_{II}$ and $\tilde{e}_\Sigma$), the middle-left column presents the dc-link capacitor voltages (v_I and v_{II}), the middle-right column displays the DAB converter normalised phase-shift (d), and the right column shows the primary-side high-frequency transformer current (i_{pri}). The top row corresponds to operating point X_1, and the bottom row corresponds to operating point X_2.

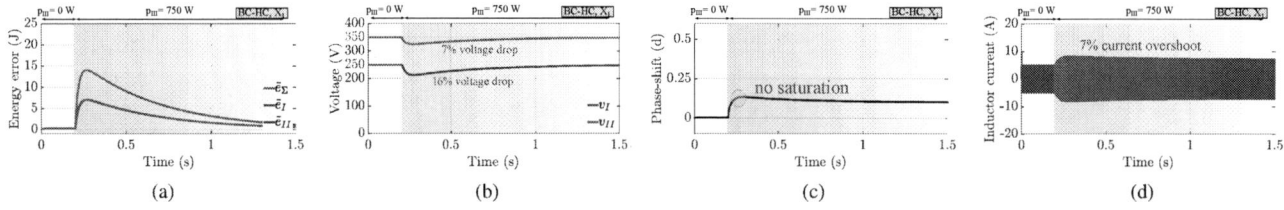

Fig. 9. Simulation results for balanced control (BC) under a load step-up change from 0 W to 750 W in the HC-DAB converter scenario. The left column shows the incremental energy variables ($\tilde{e}_I$, $\tilde{e}_{II}$ and $\tilde{e}_\Sigma$), the middle-left column presents the dc-link capacitor voltages (v_I and v_{II}), the middle-right column displays the DAB converter normalised phase-shift (d), and the right column shows the primary-side high-frequency transformer current (i_{pri}).

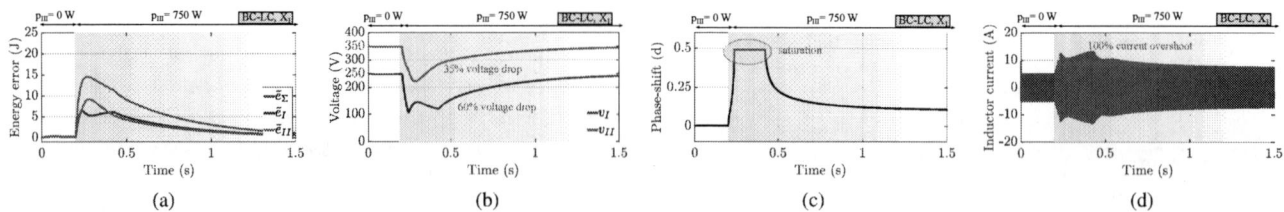

Fig. 10. Simulation results for balanced control (BC) under a load step-up change from 0 W to 750 W in the LC-DAB converter scenario. The left column shows the incremental energy variables ($\tilde{e}_I$, $\tilde{e}_{II}$ and $\tilde{e}_\Sigma$), the middle-left column presents the dc-link capacitor voltages (v_I and v_{II}), the middle-right column displays the DAB converter normalised phase-shift (d), and the right column shows the primary-side high-frequency transformer current (i_{pri}).

operation. The time-domain simulation results for the HC case is presented in Fig. 9. In Fig. 9(a), $\tilde{e}_{I-p}$ is approximately equal to $\tilde{e}_{II-p}$ at 7.25 J using the balanced control, corroborating the analysis in Fig. 4(c). As the voltage reference values are different for C_I and C_{II}, the energy variation of 7.5 J results in 16% and 7% voltage drop, respectively. The DAB converter operation is not saturated as seen in Fig. 9(c), with a 7% current overshoot experienced by the high-frequency transformer, as seen in Fig. 9(d).

The time-domain simulation results for the LC case is presented in Fig. 10. Here, p_{II-max} value is 350 W during the transient, thus lower than 750 W. The DAB converter operation enters saturation as observed with the phase-shift in Fig. 10(c) and 100% current overshoot in the transformer as shown in Fig. 10(d). Unlike the decoupled control, the control bandwidth (k) cannot be manipulated using this control strategy.

VI. CONCLUSION

This paper has analyzed the energy distribution among the input and output capacitors of a DAB converter during load transient conditions with various control strategies applied to SSTs. The paper highlights the impact of control strategies on the power transfer capacity of the DAB converters and discusses the control parameter tuning to extend the operational range. Among the conventional, decoupled and balanced control strategies studied for SSTs, the decoupled control provides the highest flexibility in distribution of transient energy among the input and output capacitors, thus an opportunity to maximize the power transfer capacity during transients without incurring DAB converter saturation. Simulation results validate the developed analysis and the proposed method to tune the control parameters.

ACKNOWLEDGMENT

This work was supported in part by the Energy Research Institute, Nanyang Technological University, Singapore. The authors extend their gratitude to the institute for its financial support.

REFERENCES

[1] B. Liu, W. Song, Y. Li, and B. Zhan, "Performance improvement of dc capacitor voltage balancing control for cascaded H-bridge multilevel converters," *IEEE Trans. Power Electron.*, vol. 36, no. 3, pp. 3354–3366, Mar. 2021.

[2] X. She, A. Q. Huang, and R. Burgos, "Review of solid-state transformer technologies and their application in power distribution systems," *IEEE J. Emerg. Sel. Topics Power Electron.*, vol. 1, no. 3, pp. 186–198, Sep. 2013.

[3] J. E. Huber and J. W. Kolar, "Solid-state transformers: On the origins and evolution of key concepts," *IEEE Ind. Electron. Mag.*, vol. 10, no. 3, pp. 19–28, Sep. 2016.

[4] T. Zhao, G. Wang, S. Bhattacharya, and A. Q. Huang, "Voltage and power balance control for a cascaded H-bridge converter-based solid-state transformer," *IEEE Trans. Power Electron.*, vol. 28, no. 4, pp. 1523–1532, Apr. 2013.

[5] S. Pugliese, M. Andresen, R. A. Mastromauro, G. Buticchi, S. Stasi, and M. Liserre, "A new voltage balancing technique for a three-stage modular smart transformer interfacing a dc multibus," *IEEE Trans. Power Electron.*, vol. 34, no. 3, pp. 2829–2840, Mar. 2019.

[6] N. Zhao, J. Liu, Y. Ai, J. Yang, J. Zhang, and X. You, "Power-linked predictive control strategy for power electronic traction transformer," *IEEE Trans. Power Electron.*, vol. 35, no. 6, pp. 6559–6571, Jun. 2020.

[7] S. Shao, L. Chen, Z. Shan, F. Gao, H. Chen, D. Sha, and T. Dragicevic, "Modeling and advanced control of dual-active-bridge dc-dc converters: A review," *IEEE Trans. Power Electron.*, vol. 37, no. 2, pp. 1524–1547, Feb. 2022.

[8] S. Falcones, X. Mao, and R. Ayyanar, "Topology comparison for solid state transformer implementation," in *Proc. IEEE PES General Meeting*, Jul. 2010, pp. 1–8.

[9] R. Pena-Alzola, G. Gohil, L. Mathe, M. Liserre, and F. Blaabjerg, "Review of modular power converters solutions for smart transformer in distribution system," in *Proc. IEEE Energy Conversion Congress and Exposition*, pp. 380–387, Sep. 2013.

[10] R. Sarda, E. Rodriguez, G. G. Farivar, J. Pou, V. B. Sriram, and A. Tripathi, "Disturbance rejection ability comparison for different solid-state transformer control strategies," in *Proc. 2023 IEEE Energy Conversion Congress and Exposition (ECCE)*, Dec. 2023, pp. 869–876.

[11] X. Mao, S. Falcones, and R. Ayyanar, "Energy-based control design for a solid state transformer," in *Proc. IEEE PES General Meeting*, Sep. 2010, pp. 1–7.

[12] R. Sarda, E. R. Ramos, N. B. Y. Gorla, G. G. Farivar, J. Pou, Y. H. Li, S. B. Vaisambhayana, and A. Tripathi, "A control strategy for solid-state transformers with coupled load disturbance attenuation ability," *IEEE Trans. Power Electron.*, vol. 39, no. 4, pp. 4029–4041, Apr. 2024.

[13] G. Farivar, B. Hredzak, and V. G. Agelidis, "Decoupled control cystem for cascaded H-bridge multilevel converter based STATCOM," *IEEE Trans. Ind. Electron.*, vol. 63, no. 1, pp. 322–331, Jan. 2016.

[14] S. Pugliese, M. Andresen, R. A. Mastromauro, G. Buticchi, S. Stasi, and M. Liserre, "A new voltage balancing technique for a three-stage modular smart transformer interfacing a dc multibus," *IEEE Trans. Power Electron.*, vol. 34, no. 3, pp. 2829–2840, Mar. 2019.

Enhanced Control for Integrated Active Power Decoupling in Single-Phase Three-Level Flying Capacitor PFC Converter

Gleisson Balen ⊚, Cristian Blanco⊚,
Ángel Navarro-Rodríguez ⊚, Pablo García ⊚
LEMUR Research Group
University of Oviedo
Gijon, Spain
{balengleisson, blancocristian,
navarroangel, garciafpablo}@uniovi.es

Rafael Peña-Alzola ⊚
Department of Electronic and Electrical Engineering
University of Strathclyde
Glasgow, United Kingdom
rafael.pena-alzola@strath.ac.uk

Abstract—This paper proposes an improved control for an integrated active power decoupling in a single-phase power factor correction flying capacitor topology. In a bridge rectifier, the distorted current is a major concern. Besides, the double-line frequency DC voltage ripple may require a bulky capacitor, resulting in extra volume. This paper presents a control strategy to use the intrinsic film flying capacitors available at the three-level power factor correction (PFC) converter to reduce the DC side voltage ripple without adding additional components, while the input current is controlled to achieve a unit PF. To mitigate the output voltage oscillations, a flying capacitor (FC) voltage control takes advantage of the FC tolerance for a higher voltage oscillation near the operational voltage. The proposed control is straightforward and has been verified through simulations. The proposed method achieved 30% DC bus ripple reduction, reducing the DC output filter capacitor and integrating the FC PFC converter with an internal active power decoupling.

Index Terms—Power Factor Correction (PFC), Active Power Decoupling (APD), Flying Capacitor (FC), Integrated

I. INTRODUCTION

The growing demand for battery chargers, such as in electric vehicles, is driving the need for more efficient converters with higher power density. Also, higher PF and lower EMI are desired for ac-dc converters [1], [2]. In a full-bridge rectifier, the input current is highly distorted, requiring some circuitry and control to improve it. To achieve a unit PF and sinusoidal current, several topologies and control schemes have been proposed [2], [3]. The boost PFC rectifier is an uncomplicated and inexpensive implementation, in which the input current is controlled, to achieve a high PF, and the output DC voltage is controlled [4], [5].

Between the topologies, high efficiencies are achieved with bridgeless interleaved PFC boost converter and totem pole designs [2]–[5]. Besides, regarding the mentioned converters, the FC topology has advantages such as the need for low on-resistance switches and, consequently, a low-voltage rating device, leading to higher efficiency and power density [6].

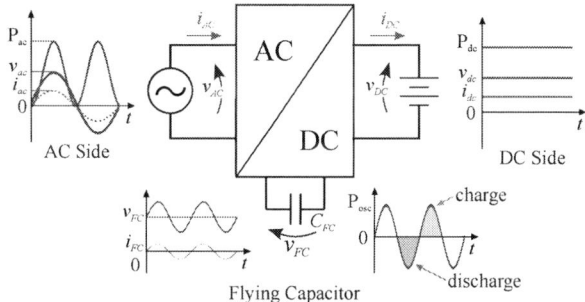

Fig. 1. Active power decoupling principle.

The FC converter, with PWM phase displacement, achieves a higher harmonic cancellation, compared to a boost PFC [7].

However, the DC output voltage will face oscillations at the double line frequency, resulting in the need for an output filter [8]. The simplest approach is to add a large capacitor in parallel with the DC bus [9]–[11]. The passive filters using capacitors and inductors add volume to the converter. Also, it is known that the electrolytic capacitors can limit the converter's lifetime [8], [12]. A possible approach to this problem is to add an active power decoupling (APD) circuit in series or parallel [9]–[11]. The FC is highlighted as a highly efficient power decoupling topology. In [6], [13], the FC is applied as a boost dc-dc and APD converter. The APD circuit uses the FC to absorb the oscillating power, taking advantage of the FC tolerance for a larger voltage ripple. In [8], the FC APD is improved, using a virtual impedance control. The PFC correction with an integrated FC APD is presented in [14]. However, the adopted control strategy relies on a non-linear bang-bang controller with variable frequency, which is difficult to analyse.

Considering the reduction of power ripple oscillation in the DC output and the PF correction, this paper proposes a new control strategy, using a proportional resonant (PR) with anti-windup to mitigate the pulsating power on the DC side using

979-8-3315-1612-3/25 $31.00 © 2025 IEEE

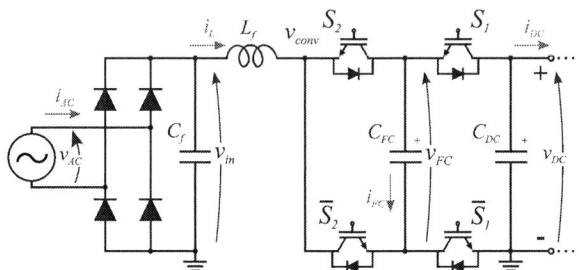

Fig. 2. Single-phase 3L flying capacitor PFC converter with integrated active power decoupling.

TABLE I
SWITCHING MODES

Mode	S_1	S_2	v_{conv}	i_{FC}	FC
00	0	0	0	0	-
01	0	1	v_{FC}	i_L	charge
10	1	0	$V_{dc} - v_{FC}$	$-i_L$	discharge
11	1	1	V_{dc}	0	-

the integrated film capacitor presented at the FC converter. The proposed control regulates the FC voltage according to the processed power. A slightly larger film FC replaces the bulky electrolytic capacitor used to reduce the DC bus ripple.

II. ACTIVE POWER DECOUPLING PRINCIPLE

The single-phase three-level (3L) PFC FC converter with integrated APD is presented in Fig. 2. The converter has in the input a full-bridge rectifier with a C_f and L_f filter. Acting as an APD, the converter C_{FC} operates under a voltage oscillation near to the rated value. In doing that, the DC bus capacitor C_{DC}, normally bulky, is reduced, leading to the adoption of C_{FC} and C_{DC} film capacitors. Fig. 1 highlights the power decoupling principle. The AC side voltage and current are:

$$v_{AC}(t) = V_{AC_p} \cos(\omega t + \theta_v) \quad (1)$$

$$i_{AC}(t) = I_{AC_p} \cos(\omega t + \theta_i) \quad (2)$$

in which, V_{AC_p} and I_{AC_p} are peak values and θ_v and θ_i phase angles of load voltage and current, respectively; ω is angular grid AC frequency. From (1) and (2) multiplication, and for unit PF, the instantaneous power, $p_{AC}(t) = v_{AC}(t)i_{AC}(t)$, is obtained:

$$p_{AC}(t) = \underbrace{\frac{1}{2}V_{AC_p}I_{AC_p}}_{P_{DC} \ (DC \ power)} + \underbrace{\frac{1}{2}V_{AC_p}I_{AC_p}\cos(2\omega t)}_{P_{osc} \ (Double \ Frequency)} \quad (3)$$

The P_{DC} is the DC component of eq. (3), while the double-frequency oscillation in eq. (3) is the part needed to be drained in the FC, to reduce C_{DC} capacitor. The positive semi-cycle of the oscillating part from P_{AC} is forced to charge the FC, while the negative part discharges the FC. The FC voltage oscillates at 2ω around the rated DC voltage. The operation

modes, corresponding voltages, and currents are presented in Fig. 3 and Tab. I, respectively.

III. FC PFC AND INTEGRATED ACTIVE POWER DECOUPLING CONTROL

The FC PFC operates normally with the same duty-cycle reference for both half-bridges $S_1/\bar{S}_1$ and $S_2/\bar{S}_2$ legs. Each half-bridge leg PWM carrier is shifted phased in $180°$. The APD control is possible by adding a slight duty-cycle variation to control the C_{FC}.

The control scheme is presented in Fig. 4. In Fig. 4 (a), the PFC control is divided into an inner faster i_L current controller and an outer slower V_{DC} voltage controller. The gains are obtained by an Internal Model Control design [15] for the current loop, and Symmetrical Optimum Criteria for the voltage loop. The corresponding gains and bandwidth are presented in Table II. The SOGI PLL tracks the grid voltage to synchronize the θ, which will be used to generate a current reference [16].

The proposed FC APD control is presented in Fig. 4 (b). The measured P_{AC} power is filtered by a high-pass filter, selecting only the oscillating AC power, which is subtracted from 0 (the desired reference). A PR control with anti-windup is designed at the oscillating $2\omega_o$ frequency, according to [17], [18]. The output is an P_{FC} action, which needs to be converted to the desired adjusted duty-cycle (d_{adj}) responsible for slightly linearly altering the duty cycle of the FC legs. In case of no adjustment, $d_{adj} = 0$, the duty cycles will be the same, $d = d_1 = d_2$, as for the normal FC operation. Thus, $S_1/\bar{S}_1$ and $S_2/\bar{S}_2$ will face $d1 = d - d_{adj}$ and $d2 = d - d_{adj}$. Moreover, according to the operation modes, the FC current is $i_{FC} = i_L(d_1 - d_2)$. Substituting d_1 and d_2, i_{FC} becomes $i_{FC} = 2d_{adj}$, leading to $P_{FC} = 2d_{adj}V_{FC}$. With the P_{FC} equation, the d_{adj} is obtained. In addition, a PI control is added to keep the V_{FC} around the rated voltage ($0.5V_{DC}$). The low-pass filter, at V_{FC}, selects only the constant part needed for a proper FC operation at the rated voltage. The duty cycle is limited due to the divisions by values equal to zero, from the sinusoidal inputs [14]. In Fig. 4 (c), a control decoupler is added to avoid interferences between the converter's PFC and APD controls [6], [8]. The converter parameters are initially obtained for the normal operation of the PFC FC converter. The inductor L_f is determined in (4) for an allowed ripple of 20% [19].

$$L_f = \frac{1}{\%Ripple} \frac{V_{ac}^2}{P_{conv}} \left(1 - \frac{\sqrt{2}V_{ac}}{V_o}\right) T_{sw} = 617\,\mu H \quad (4)$$

Also on the AC side, the filter capacitor C_f is designed to form a pole with the inductor at a frequency approximately one decade below the switching frequency. The required capacitance of the filter capacitor is determined using the equation (5).

$$C_f = \left(\frac{10}{f_{sw}}\right)^2 \frac{1}{4\pi^2 L_f} = 2.56\mu F \quad (5)$$

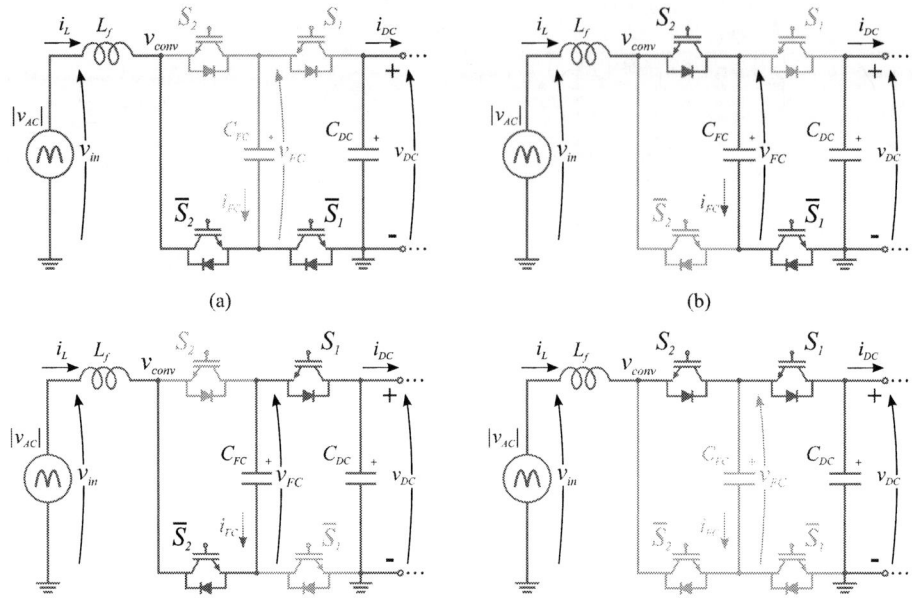

(a)   (b)

(c)   (d)

Fig. 3. FC converter with integrated APD operation modes. The input rectifier diode bridge is represented by a full-wave semi-cycle voltage source. In (a), both S_1 and S_2 are off, or 00 mode, the V_{conv} is 0. The FC is charged in (b), at mode 01, when $V_{conv} = v_{FC}$. In (c), mode 10, the V_{conv} is $V_{dc} - v_{FC}$, the FC is discharged. In the last stage (d), both S_1 and S_2 are at on, or mode 11, the V_{dc} is connected to the inductor, and nothing happens with the FC.

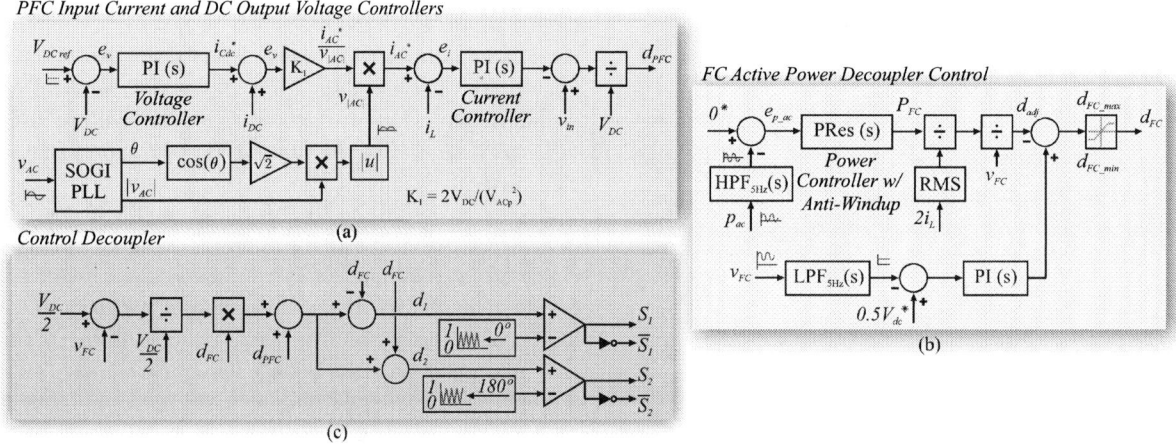

Fig. 4. FC PFC and integrated APD control scheme. In (a), the PFC input current and DC output voltage controls are detailed. The proposed FC with integrated APD control blocks is shown in (b). In (c), the duty cycle decoupling scheme is presented. Controllers' gains are presented in Tab. II.

Moreover, the DC side capacitor C_{DC} for a ripple $\Delta V_{DC} = 20$ V needs to be calculated. In equation (6), the DC side capacitor is found for a nominal operation of the PFC FC converter without the integrated APD. The ΔV_{DC} ripple is set to 20V.

$$C_{DC_0} = \frac{P_{conv}}{2\pi f_{ac} \Delta V_{DC} V_o} = 800\mu F \quad (6)$$

The flying capacitor for normal converter operation can be determined by substituting equation (7) into (8)

$$I_{peak} = \frac{\sqrt{2}P_{conv}}{V_{ac}}\left(1 + \frac{\%\text{Ripple}}{2}\right) \approx 13.53A \quad (7)$$

$$C_{FC_0} = \frac{I_{\text{peak}}}{2f_{\text{sw}}\Delta V_{FC}} \approx 30\mu F \quad (8)$$

in which $\Delta V_{FC} = 5V$ is the allowed voltage ripple in the flying capacitor, under no APD operation mode.

Adding the ADP function, C_{FC} is recalculated to absorb part of the output voltage ripple, reducing the output capacitor C_{DC}. The C_{FC} is calculated by the equation (9), to achieve a voltage fluctuation of 100 V.

$$C_{FC} = \frac{P_{conv}}{2\omega_o V_{FC}\Delta_{FC}} = \frac{2000}{2\omega_o 200 \cdot 100} \approx 150\mu F \quad (9)$$

in which Δ_{FC} is the allowed oscillation, and $V_{FC} = 0.5V_{DC}$.

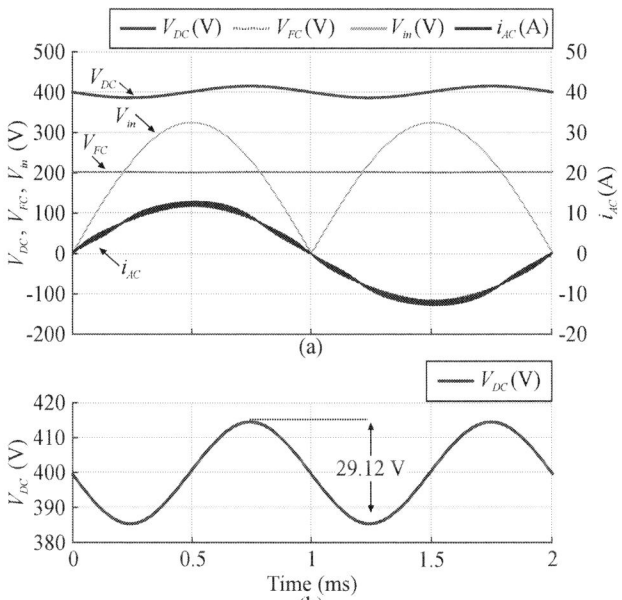

Fig. 5. Simulation results for the FC PFC without APD: (a) DC-side voltage, input voltage, flying capacitor voltage, and input current; (b) zoomed view of V_{DC}.

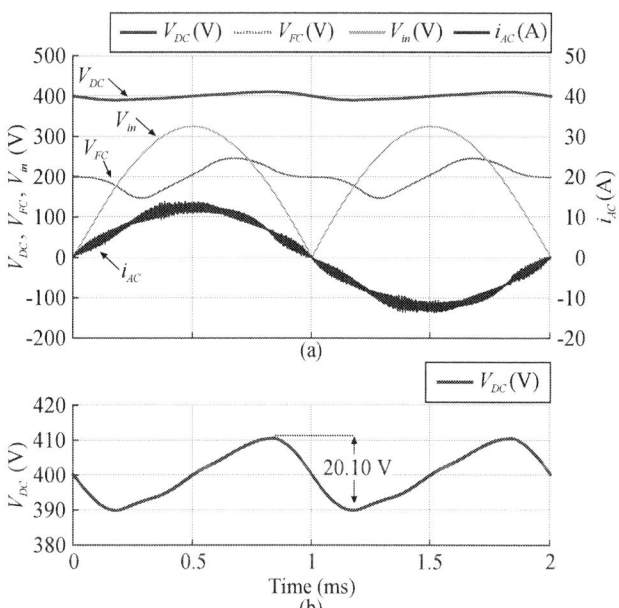

Fig. 6. Simulation results for the FC PFC with APD: (a) DC-side voltage, input voltage, flying capacitor voltage, and input current; (b) zoomed view of V_{DC}, highlighting the reduced V_{DC} ripple.

TABLE II
CONVERTER AND CONTROL PARAMETERS

Parameter	Var.	Value
Converter Power	P_{conv}	$2\ kW$
Input AC Freq.	f_o	$50\ Hz$
DC Bus Output	V_{DC}	$400\ V$
Grid RMS Voltage	V_{ac}	$230\ V$
Filter Inductor	L_f	$617\ \mu H$
Filter Capacitor	C_f	$2.56\ \mu F$
Flying Capacitor	C_{FC_0}	$30\ \mu F$
DC Bus Capacitor	C_{DC_0}	$800\ \mu F$
Flying Capacitor with APD	C_{FC}	$150\ \mu F$
DC Bus Capacitor with APD	C_{DC}	$550\ \mu F$
Switching Frequency	f_{sw}	$40\ kHz$
Current Loop BW	BW_i	10 kHz
PFC Current P Gain	K_p	7.48
PFC Current I Gain	K_i	164.43
PFC Voltage BW	BW_v	100 Hz
PFC Voltage P Gain	K_p	$15e-3$
PFC Voltage I Gain	K_i	$375e-3$
FC P Gain	K_p	0.0001
FC I Gain	K_i	0.02
PR K_p and K_r	$K_p = K_r$	1
PR Resonant Freq	f_r	100 Hz

IV. SIMULATION VALIDATION

Simulations are performed to validate the proposed control strategy. Table II presents the converter and the proposed control parameters.

Two scenarios are considered. In the first, the flying capacitor operates without the APD control, utilizing the standard calculated values for the FC PFC. In the second scenario,

the flying capacitor is configured to absorb part of the output ripple, working as APD, resulting in a corresponding reduction in output capacitance. Figures 5 and 6 represents the second scenario.

In the first scenario, without APD and with the calculated capacitances, $C_{FC_0} = 30\mu F$ and $C_{DC_0} = 800\mu F$, the desired voltage ripple $\Delta V_{DC} = 20.55$ V in the output, is obtained. Moreover, the input current distortion is $THD_i = 4.47\%$. In this case, the total circuit capacitance is $C_{FC} + C_{DC} = 830\mu F$.

In the second scenario, the capacitor C_{FC} is modified for the calculated $C_{FC} = 150\mu F$, to absorb part of the output voltage ripple with the proposed APD operation. With the PFC and the integrated APD proposed control working together, the C_{DC} is reduced to $550\mu F$, keeping the output ripple at 20.10 V. Fig. 6 illustrates the operation of the FC with both PFC and APD. The oscillating power is partially absorbed by the FC due to the V_{FC} oscillating around $0.5V_{DC}$. The voltage oscillates between 246.31 V and 146.37 V. The input current exhibits a drawback of slightly higher distortion, with $THD_i = 7.50\%$. However, the total circuit capacitance is reduced to $C_{FC} + C_{DC} = 700\mu F$.

In contrast, Fig. 5 presents the operation of the single-phase FC PFC, without APD. The capacitors remain the same as in the APD scenario, $C_{FC} = 150\mu F$ and $C_{DC} = 550\mu F$, but the resulting ripple increases to 29.12 V, as highlighted in Fig.5 (b). The input current i_{AC} is sinusoidal, demonstrating an effective controller performance from the PFC regulation. The input current distortion is $THD_i = 4.57\%$, similar to the distortion observed with the calculated values for the no APD condition. Additionally, the V_{FC} stays constant at $0.5V_{DC}$.

In summary, the proposed FC PFC with integrated APD control working results in a 30% reduction in V_{DC} ripple, decreasing it from 29.12 V to 20.10 V. This improvement is attributed to the oscillating power being partially absorbed by the FC, facilitated by the V_{FC} oscillation around $0.5V_{DC}$.

In addition, it is worth noting that no direct equation exists to determine the proportional reduction in DC-side capacitance, due to the FC APD control. Therefore, iterative simulations were conducted to determine the minimum C_{DC} value that maintains the same voltage ripple as calculated for standard operation ($\Delta V_{DC} = 20$ V). For the scenario without the APD and aiming the same output voltage ripple, the total capacitance of $C_{FC} + C_{DC} = 830\mu F$ is required. However, with the oscillating power being partially absorbed by the FC, the total capacitance is reduced to $C_{FC} + C_{DC} = 700\mu F$. As a result, the total capacitance is reduced by 15.7%. A further reduction in the flying capacitor size is possible at the trade-off of an increase in THD_i.

V. CONCLUSIONS

A single-phase 3L FC converter for a PFC with an active power decoupling is an attractive solution for AC-DC converter, increasing efficiency and reducing converter volume. The proposed APD control enables a 30% reduction in output voltage ripple, leading to a 15% decrease in the overall capacitor size and making it possible to use film capacitors for a longer converter lifetime.

ACKNOWLEDGMENT

The present work has been partially supported by the Spanish Ministry of Economy and Competitiveness funded by MCIN/AEI/10.13039/501100011033 under Grants MCINN-23-PID2022-139479OB-C22, MCINN-22-TED2021-129796B-C21, MCINN-24-PLEC2023-010252, and in part by the European Union NextGenerationEU/PRTR.

REFERENCES

[1] H. Karneddi and D. Ronanki, "Universal bridgeless nonisolated battery charger with wide-output voltage range," *IEEE Transactions on Power Electronics*, vol. 38, no. 3, pp. 2816–2820, 2023.

[2] S. S. Sayed and A. M. Massoud, "Review on state-of-the-art unidirectional non-isolated power factor correction converters for short-/long-distance electric vehicles," *IEEE Access*, vol. 10, pp. 11 308–11 340, 2022.

[3] A. D. E. Dutra, M. A. Vitorino, and M. B. D. R. Correa, "A survey on multilevel rectifiers with reduced switch count," *IEEE Access*, vol. 11, pp. 56 098–56 141, 2023.

[4] F. Musavi, M. Edington, W. Eberle, and W. G. Dunford, "Evaluation and efficiency comparison of front end ac-dc plug-in hybrid charger topologies," *IEEE Transactions on Smart Grid*, vol. 3, no. 1, pp. 413–421, 2012.

[5] L. Huber, Y. Jang, and M. M. Jovanovic, "Performance evaluation of bridgeless pfc boost rectifiers," *IEEE Transactions on Power Electronics*, vol. 23, no. 3, pp. 1381–1390, 2008.

[6] H. Watanabe, T. Sakuraba, K. Furukawa, K. Kusaka, and J.-i. Itoh, "Development of dc to single-phase ac voltage source inverter with active power decoupling based on flying capacitor dc/dc converter," *IEEE Transactions on Power Electronics*, vol. 33, no. 6, pp. 4992–5004, 2018.

[7] C. A. Teixeira, D. G. Holmes, and B. P. McGrath, "Single-phase semi-bridge five-level flying-capacitor rectifier," *IEEE Transactions on Industry Applications*, vol. 49, no. 5, pp. 2158–2166, 2013.

[8] S. Kan, X. Ruan, X. Huang, and H. Dang, "Second harmonic current reduction for flying capacitor clamped boost three-level converter in photovoltaic grid-connected inverter," *IEEE Transactions on Power Electronics*, vol. 36, no. 2, pp. 1669–1679, 2021.

[9] J. Jiang, H. Sun, and H. Wang, "A comprehensive review of single-phase converter topologies with 2w-ripple suppress," *Electrical Engineering*, vol. 106, pp. 225–262, 2 2024.

[10] H. Zhang, X. Li, B. Ge, and R. S. Balog, "Capacitance, dc voltage utilizaton, and current stress: Comparison of double-line frequency ripple power decoupling for single-phase systems," *IEEE Industrial Electronics Magazine*, vol. 11, pp. 37–49, 9 2017.

[11] Y. Liu, W. Zhang, Y. Sun, M. Su, G. Xu, and H. Dan, "Review and comparison of control strategies in active power decoupling," *IEEE Transactions on Power Electronics*, vol. 36, pp. 14 436–14 455, 12 2021.

[12] H. Wang, H. Wang, G. Zhu, and F. Blaabjerg, "An overview of capacitive dc-links-topology derivation and scalability analysis," *IEEE Transactions on Power Electronics*, vol. 35, no. 2, 2020.

[13] X. Luo, F. Yu, C. Ma, L. Ding, and Y. Peng, "A novel compensation power-decoupling strategy for single-phase three-level flying capacitor pv micro-inverters," *International Journal of Circuit Theory and Applications*, vol. 50, no. 5, pp. 1667–1685, 2022. [Online]. Available: https://onlinelibrary.wiley.com/doi/abs/10.1002/cta.3233

[14] D. Menzi, S. Weihe, J. A. Anderson, J. Everts, and J. W. Kolar, "Single-phase pfc rectifier with integrated flying capacitor power pulsation buffer," *IEEE Open Journal of Power Electronics*, vol. 3, pp. 866–875, 2022.

[15] R. Ottersten, *On control of back-to-back converters and sensorless induction machine drives*. Department of Electric Power Engineering, Chalmers University of Technology and Goteborg University, 2003.

[16] R. Teodorescu, M. Liserre, and P. Rodriiguez, *Grid converters for photovoltaic and wind power systems*. IEEE ; Wiley, 2011, book SOGI PLL completo com isso.

[17] S. Buso and P. Mattavelli, *Digital control in power electronics*, 1st ed., J. Hudgins, Ed. Italy: Morgan and Claypool Publishers, 2006, vol. 1.

[18] S. A. Richter and R. W. De Doncker, "Digital proportional-resonant (pr) control with anti-windup applied to a voltage-source inverter," in *Proceedings of the 2011 14th European Conference on Power Electronics and Applications*, 2011, pp. 1–10.

[19] I. Technologies, "Pfc boost converter design guide," 2016, application Note AN-v02_00-EN.

Improving Transient Stability of PLL-Synchronized Grid-Following Inverters

Surya Prakash
Dept. of EE
NIT Jamshedpur
Jharkhand, India
suryaprakash.ee@nitjsr.ac.in

Kalpana Beura
Dept. of ECE, COE
UAE University
Al Ain, Abu Dhabi, UAE
kalpanabehura1995@gmail.com

Mohamed Alkhatib
Dept. of ECE, COE
UAE University
Al Ain, United Arab Emirates
mohdalkhatib@uaeu.ac.ae

Omar Al Zaabi
APEC, EE Department
Khalifa University
Abu Dhabi, UAE
omar.alzaabi@ku.ac.ae

Khalifa Al Hosani
APEC, EE Department
Khalifa University
Abu Dhabi, UAE
khalifa.halhosani@ku.ac.ae

Utkal Ranjan Muduli
APEC, EE Department
Khalifa University
Abu Dhabi, UAE
utkal.muduli@ku.ac.ae

Abstract—This paper examines the issues related to synchronization of utility network through applications of three-phase grid-following inverters (GFIs). There are problems when trying to capture the positive sequence (PS) components of distorted and unbalanced grid voltage. From the grid signal, the PS components are extracted with the help of a dual second-order generalized integrator phase-locked loop (DSOGI PLL). When the grid voltage has lower order harmonics it fluctuates in both frequency and phase. To overcome these limitations, as suggested in the paper, notch filtering technique available in SOGI architecture, low-pass filter and modification of symmetrical components in type-3 PLL are recommended. This method preserves the performance of the DSOGI-GFI system. The paper also uses the GFI system with the Lyapunov-function-based current controller described above. Current controllers use PLLs for phase and frequency estimation of three-phase systems in a dq-transformed frame. The proposed PLL is compared with other existing PLLs based on stability and transient characteristics. It also includes an experimental result that demonstrates the effectiveness of PLL.

Index Terms—Lyapunov stability, phase-locked loop, grid-following inverters.

I. INTRODUCTION

Carbon reduction and decentralized power generation catalyze renewable energy generation (REG) [1], [2]. REG systems are standalone or grid-connected depending on the application. REG power is intermittent and requires batteries to maintain power independently. The high cost of batteries makes standalone applications uneconomical. According to [3], grid-following inverters (GFI) are a more economical option. GFIs connect the REG to the power grid. Synchronization unit and controllers for inner loop are required for GFI system control. dq To control the GFI system in the synchronous domain, the inner-loop controllers use grid power

"This work was supported in part by AUA-UAEU Joint Research Grant *12N146*, UAE University, Abudhabi, United Arab Emirates; in part by UAEU Startup Grant *12N158*, UAE University, Abudhabi, United Arab Emirates; and in part by the Advanced Technology Research Council ASPIRE Virtual Research Institute (VRI) Program, Abu Dhabi, United Arab Emirates, under Award *VRI20-07*."

from the synchronization unit [4]. Synchronization of phase-locked loops (PLLs) using SRF is prevalent. The performance of the PLL, an essential part of GFI systems, determines the stability, reliability and power quality of power conversion systems [5], [6].

The grid voltage in real circumstances may be distorted and unbalanced due to fluctuating values of the voltage sensors, non-linearity electronic switching devices, voltage and load unbalance, faults, voltage dips, and other disturbances [6]–[8]. Therefore, the distorted form of voltage signals makes its measurement and control by the SRF-PLL difficult, and hence there are errors. To address this challenge, PS and NS components extraction from the grid unbalance voltage and uses the Instantaneous symmetrical component (ISC) technique [9]. Based on the OSG-based PLLs, particularly the second-order generalized integrator (SOGI), which has a comparatively simpler structure than the DDSRF filtering-based PLL, a significant amount of work has been carried out because of their excellent performance. Based on the type of signals and secondary path selection as well as response and computation time, the most suitable solution is again the type-2 PLL that relies on the dual SOGI (DSOGI) [10], [11]. Nevertheless, the accuracy of grid parameter estimation may be adversely affected by the presence of a DC-offset component within the grid voltage signal.

Several state-of-the-art type-2 FAPSs have been proposed in recent years to address the presence of harmonics, and DC offset in input signals [9], [12]–[14]. Third-order generalized integrator-based type-2 PLL reduces the low-order harmonics as it estimates the PS voltage components [15], [16]. However, type-2 PLLs have steady-state phase angle errors when asked to track inputs that have ramp frequency [17]. These PLLs utilize higher-order filters, which limit their bandwidth. However, if grid disturbances are to be handled, then pre-filters such as SOGI can be utilized.

To address this issue for obtain higher immunity to the

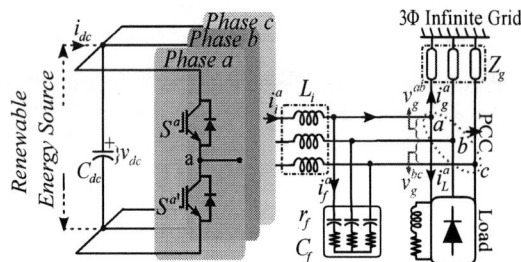

Fig. 1. GFI system configuration for REG application

grid disturbances and much faster dynamic response, type-3 PLL algorithms are proposed [18]–[23]; however, these PLLs cannot work effectively for voltage sag disturbances and are very sensitive to unexpected changes in DC offset. Techniques such as phase compensation and gain compensation have been created for single-phase applications to overcome this problem, but using these techniques is challenging [21]–[23]. Nevertheless, the extraction of PS components and distorted unbalanced grid conditions is not discussed in type-3 PLL.

This paper introduces modified ISC techniques in type-3 PLL to handle grid interruptions. It also extracts the PS component from the unbalanced distorted grid voltage. The proposed method can attenuate the effect of the low-order harmonics in the estimated frequency and phase angle. The performance of the proposed PLL is compared with its counterpart in terms of settling time, overshoot, steady-state frequency and phase error oscillation. Furthermore, the proposed PLL incorporates a GFI mechanism to increase the PCC grid's reliability and power quality. It also ensures that the total harmonic distortion of the grid current is less than 5% as per IEEE-519 standard.

II. SYSTEM CONFIGURATION

To connect the GFI system to the electric grid, three-wire connections are used at the point of common coupling (PCC), as seen in Fig. 1. In the GFI system, an inductor (L_i) is employed to reduce the grid currents ripple. In addition, ripple filter of R_f and C_f in a shunt decrease switching noises during operation. GFI regulates DC bus for active and reactive power interchange and improves the PCC power quality from load current harmonics and unbalanced grid voltage distortion. For this purpose, the inner-loop controller utilizes the LFBC technique to produce the gate pulses for the GFI system where the information about load currents, inverter currents, and grid voltages is essential. Measured signal is used for control signal establishment (s_i^a, s_i^b, s_i^c). Park's transformation creates instantaneously rotating d-q variables of the load currents, currents through the inverter, and voltages of the grid. This transformation requires a positive sequence phase angle, which can be achieved by using the proposed PLL explained in the next Section. The proposed PLL enables the estimation of the positive sequence phase angle ($\hat{\theta}_{gp}$), frequency ($\hat{\omega}_p$), and amplitude (v_g^{dp}) of the PCC voltage (v_g^{abc}). It has to be noted that superscripts $\{a, b, c\}$ and $\{d, q\}$ refer to three phase and dq-plane, respectively [24], [25].

In the synchronous rotating d-q frames, the filter capacitor voltage v_c^{dq} and inverter currents i_i^{dq} can be represented as follows:

$$\dot{v}_c^{dq} = -\left(\tau_g^{-1} - j\hat{\omega}_p\right) v_c^{dq} + \tau_g^{-1} v_g^{dq} \tag{1}$$

$$\dot{i}_i^{dq} = -\left(\tau_i^{-1} - j\hat{\omega}_p\right) i_i^{dq} + L_i^{-1}\left(v_g^{dq} - 0.5 v_{dc} s_i^{dq}\right) \tag{2}$$

The time constants for the ripple filter and interfacing components are denoted by τ_g ($=1/R_f C_f$) and τ_i ($=L_i/R_i$), respectively. The superscript $[\,]^{dq}$ denotes the complex variable $[\,]^d + j[\,]^q$, where j is considered as a complex operator. Let $x \in \left\{i_i^d, i_i^q, v_c^d, v_c^q, s_i^d, s_i^q\right\}$ denote the state variable of the PV-GFI system, where $x = \hat{x} + \tilde{x}$ with $\hat{x}$ and $\tilde{x}$ representing its steady-state and perturbed value. At the equilibrium point, the steady-state value of control signal in dq-frame ($\hat{s}_i^d$, $\hat{s}_i^q$) can be derived as

$$\hat{s}_i^d = \frac{2}{\hat{v}_{dc}}\left(\hat{v}_g^d - L_i \tau_i^{-1}\hat{i}_i^d - \hat{\omega}_p\hat{i}_i^q\right) \tag{3}$$

$$\hat{s}_i^q = \frac{2}{\hat{v}_{dc}}\left(\hat{v}_g^q - L_i \tau_i^{-1}\hat{i}_i^q + \hat{\omega}_p\hat{i}_i^d\right). \tag{4}$$

The linearized averaged state-space model of the GFI system can be derived by substituting the state variables in (2) as

$$\frac{d}{dt}\underbrace{\begin{bmatrix} \tilde{i}_i^d \\ \tilde{i}_i^q \\ \tilde{v}_c^d \\ \tilde{v}_c^q \end{bmatrix}}_{X} = \underbrace{\begin{bmatrix} -\tau_i^{-1} & \omega_p & 0 & 0 \\ \omega_p & -\tau_i^{-1} & 0 & 0 \\ 0 & 0 & -\tau_c^{-1} & \omega_p \\ 0 & 0 & \omega_p & -\tau_c^{-1} \end{bmatrix}}_{A} \underbrace{\begin{bmatrix} \tilde{i}_i^d \\ \tilde{i}_i^q \\ \tilde{v}_c^d \\ \tilde{v}_c^q \end{bmatrix}}_{X}$$

$$+ \underbrace{\begin{bmatrix} 0.5\hat{v}_{dc}L_i^{-1} & 0 \\ 0 & 0.5\hat{v}_{dc}L_i^{-1} \\ 0 & 0 \\ 0 & 0 \end{bmatrix}}_{B} \underbrace{\begin{bmatrix} \tilde{s}_i^d \\ \tilde{s}_i^q \end{bmatrix}}_{U}. \tag{5}$$

III. CONTROL STRATEGY

The proposed PLL and the inner loop of the GFI system are considered as components that form the control strategy of the GFI system. To improve the control performance of the GFI system, it is essential to have an accurate estimate of the frequency, phase angle and amplitude of the utility grid. This is especially useful in cases where grid conditions are difficult. Regardless of any grid interruptions, the proposed PLL generates the reference current for the GFI system based on the estimated grid parameters. This reference current is transmitted to the GFI's inner-loop controller, which allows efficient management of power flow and maintenance of ideal performance levels, thereby ensuring reliable performance.

A. Proposed PLL

For the detection of the positive sequence components of the grid voltage, the PLL control technique is employed, which provides a means of synchronized vital even when the faults are unbalanced. To separate the impact of Negative Sequence components in v_g^α and v_g^β, Dual Second-Order Generalized Integrator (DSOGI) blocks are used. Further, the modified DSOGIs is use the notch filtering property of $v_n(s)$ to enhance

979-8-3315-1612-3/25 $31.00 © 2025 IEEE

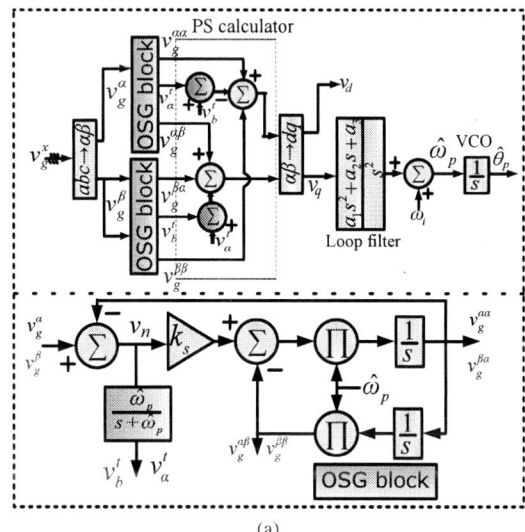

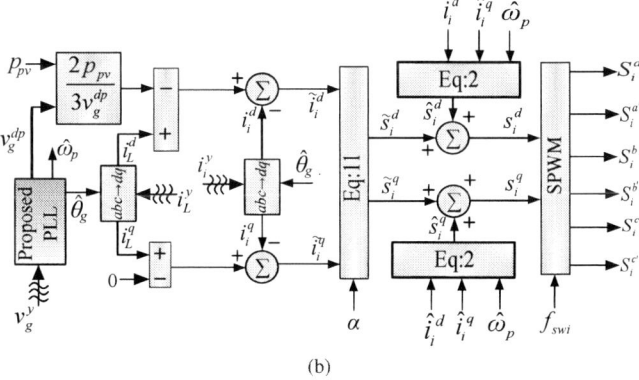

(a)

(b)

Fig. 2. control circuit(b) Proposed PLL (c) Inner loop controller

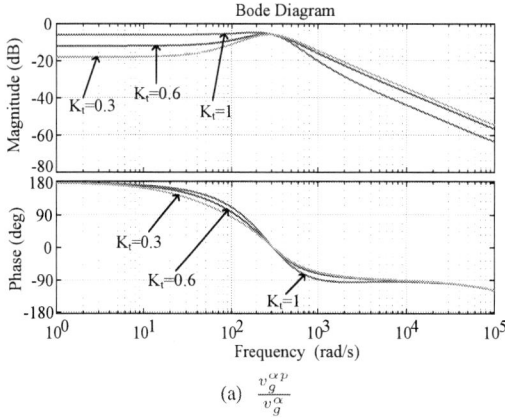

(a) $\dfrac{v_g^{\alpha p}}{v_g^{\alpha}}$

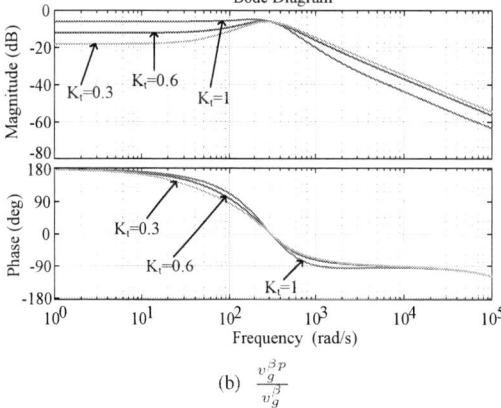

(b) $\dfrac{v_g^{\beta p}}{v_g^{\beta}}$

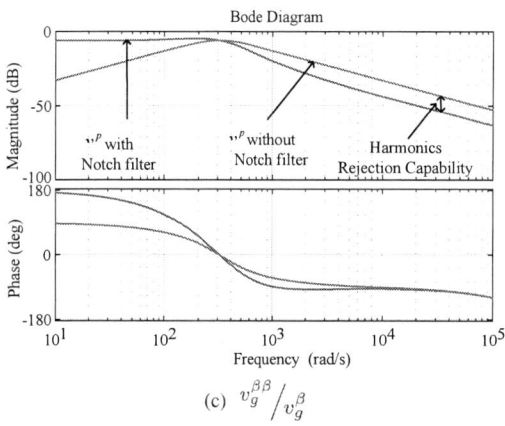

(c) $v_g^{\beta\beta}\Big/ v_g^{\beta}$

Fig. 3. Frequency response of modified symmetrical components techniques with different values of k_t.

the lower-order harmonics components of the input signals. To mitigate undesired oscillations resulting from low-order harmonics, it is possible to send $T_n(s)$ via a low-pass filter, as seen in Fig. 1. The transfer function of the modified DSOGI block is expressed as:

$$T_\alpha(s) = \frac{v_g^{\alpha\alpha}(s)}{v_g^{\alpha}(s)} = \frac{v_g^{\beta\alpha}(s)}{v_g^{\beta}(s)} = \frac{k\hat\omega_p s}{D}; \qquad (6)$$

$$T_\beta(s) = \frac{v_g^{\beta\alpha}(s)}{v_g^{\beta}(s)} = \frac{v_g^{\beta\beta}(s)}{v_g^{\beta}(s)} = \frac{k\hat\omega_p^2}{D}; \qquad (7)$$

$$T_{nl}(s) = \frac{v_\alpha^t}{v_n} = \frac{v_\beta^t}{v_n} = \underbrace{\frac{s^2 + \hat\omega_p^2}{D}}_{} \underbrace{\frac{\hat\omega_p}{s + \hat\omega_p}}_{}; \qquad (8)$$

where $D = s^2 + k\hat\omega_p s + \hat\omega_p^2$. The PS components of orthogonal signals ($v_g^{\alpha p}$, $v_g^{\beta p}$) can be obtained by the proposed symmetrical components method as shown in Fig. 1(b) in the s-domain as

$$v_g^{\alpha p}(s) = \frac{1}{2}\left(v_g^{\alpha\alpha}(s) - v_g^{\beta\beta}(s) - k_t\left[v_\alpha^t(s) + v_\alpha^t(s)\right]\right) \qquad (9)$$

$$v_g^{\alpha p}(s) = \frac{T_\alpha(s)}{2}\left(\begin{array}{l}\left[1 - k_t\frac{s^2+\hat\omega_p^2}{s(s+\hat\omega_p)}\right]v_\alpha(s) \\ + \left[\frac{s-\hat\omega_p}{s+\hat\omega_p} - k_t\frac{s^2+\hat\omega_p^2}{s(s+\hat\omega_p)}\right]v_g^{\alpha}(s)\end{array}\right) \qquad (10)$$

$$v_g^{\beta p}(s) = \frac{1}{2}\left(v_g^{\alpha\beta}(s) - v_g^{\alpha\beta}(s) - k_t\left[v_\alpha^t(s) + v_\alpha^t(s)\right]\right) \qquad (11)$$

$$v_g^{\beta p}(s) = \frac{T_\alpha(s)}{2}\left(\begin{array}{l}\left[1 - k_t\frac{s^2+\hat\omega_p^2}{s(s+\hat\omega_p)}\right]v_\beta(s) \\ + \left[\frac{s-\hat\omega_p}{s+\hat\omega_p} - k_t\frac{s^2+\hat\omega_p^2}{s(s+\hat\omega_p)}\right]v_g^{\beta}(s)\end{array}\right) \qquad (12)$$

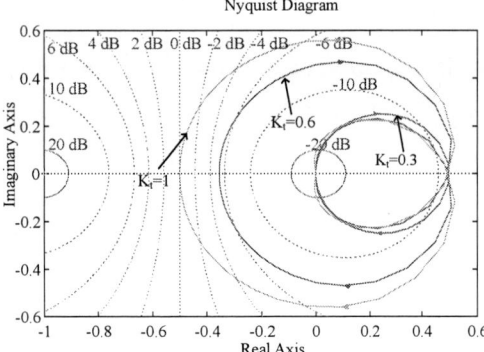

Fig. 4. Stability analysis of modified ISC techniques with different values of k_t.

where k_t is the gain factor. Frequency domain analysis is performed for $v_g^{\alpha p}/v_g^{\alpha}$, $v_g^{\beta p}/v_g^{\beta}$ with k_t varying from 0.3 to 1 as shown in Fig. 3 (a) and Fig. 3 (b). It is observed that as k_t increases, harmonics attenuation capability increases. however, it degrades the DC rejection capability. Moreover, k_t affects the bandwidth, harmonics attenuation and the magnitude gain of the filter. Therefore, k_t is chosen as a trade-off between these factors.

The effectiveness of the OSG block with and without a notch filter is observed in Fig 3 (c). It is evident from the plots that the OSG block with a notch filter exhibits more attenuated harmonics. Furthermore, the proposed ISC techniques in OSG filter in type-3 PLL demonstrate superior performance in attenuating low-frequency components below the fundamental frequency while also providing enhanced suppression of harmonics compared to the conventional ISC method. Nyquist stability analysis for modified ISC techniques with different values of k_t as shown in Fig 3. It can be observed that $\frac{v_g^{\alpha p}}{v_g^{\alpha}}$ is far from $(-1 + j0)$. It indicates that no poles of $\frac{v_g^{\alpha p}}{v_g^{\alpha}}$ in the right half of the s-plane. therefore, the system is stable.

B. inner-loop controller

The GFI inner-loop controller ensures global stability through the use of LFBC. Lyapunov stability theory decreases the energy of the GFI and achieves global stabilization. The energy of the interface inductor and ripple capacitor may be mathematically represented using the dq form [25].

$$E = \frac{3}{2}L_i\left((\tilde{i}_i^d)^2 + (\tilde{i}_i^q)^2\right) + \frac{3}{2}C_f\left((\tilde{v}_c^d)^2 + (\tilde{v}_c^q)^2\right) + \frac{3}{2}C_{dc}(\tilde{v}_{dc})^2 \tag{13}$$

To achieve global system stability, the derivative of equation (13) must be negative. By substituting (5) in derivative of E, which may be derived as

$$\dot{E} = \underbrace{-3\tau_i^{-1}\left((\tilde{i}_i^d)^2 + (\tilde{i}_i^q)^2\right) - 3\tau_g^{-1}\left((\tilde{v}_c^d)^2 + (\tilde{v}_c^q)^2\right)}_{term1}$$
$$\underbrace{-\frac{3}{2}\tilde{s}_i^d\hat{v}_{dc}\tilde{i}_i^d}_{term2}\underbrace{-\frac{3}{2}\tilde{s}_i^q\hat{v}_{dc}\tilde{i}_i^q}_{term3} \tag{14}$$

TABLE I
PROTOTYPE SYSTEM PARAMETERS.

Parameter	Value
Inductor L_g	5mH
Shunt parameter r_f, C_f,	5 Ω, 10 μF,
Inverter Switching frequency f_{inv}	10 kHz
Grid impedance Z_g	0.01 Ω, 0.15 mH
Grid voltage V_g^{ab}	100 V
frequency f_g	50 Hz

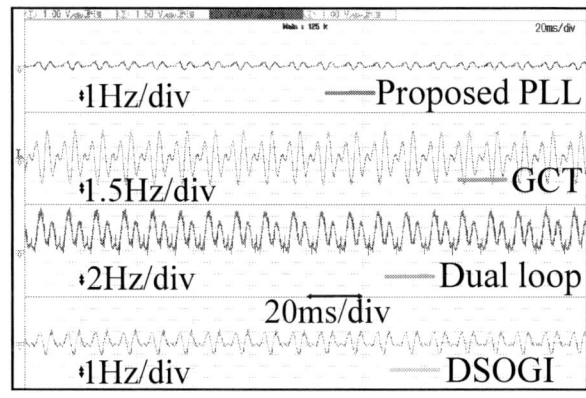

(a) frequency

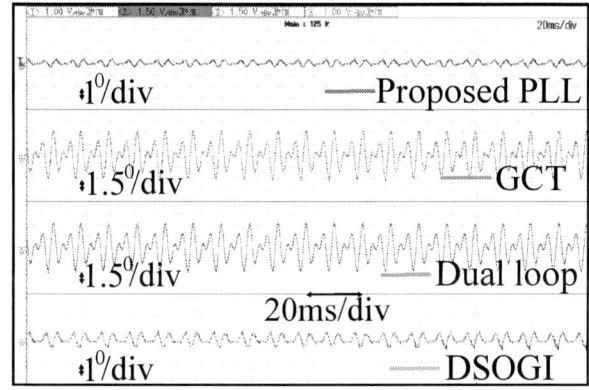

(b) phase error

Fig. 5. Comparison of type-2 and type-3 PLL performance with respect to harmonics

Based on equation (14), the first term is consistently negative. In order to maintain global stability, the perturbations of the switching function ($\tilde{s}_i^d$ and $\tilde{s}_i^q$) might be chosen as follows:

$$\tilde{s}_i^d = \alpha\left(\hat{v}_{dc}\tilde{i}_i^d\right); \tilde{s}_i^q = \alpha\left(\hat{v}_{dc}\tilde{i}_i^q\right) \tag{15}$$

where α denotes the gain of the controller, which should be greater than zero. The overall switching function outputs (s_i^d and s_i^q) are converted to the abc-frame. To generate the gate signals for GFI switches, the sinusoidal pulse width modulation (SPWM) is applied.

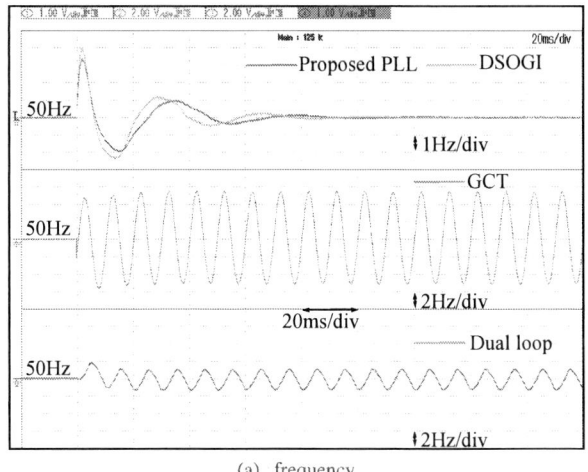

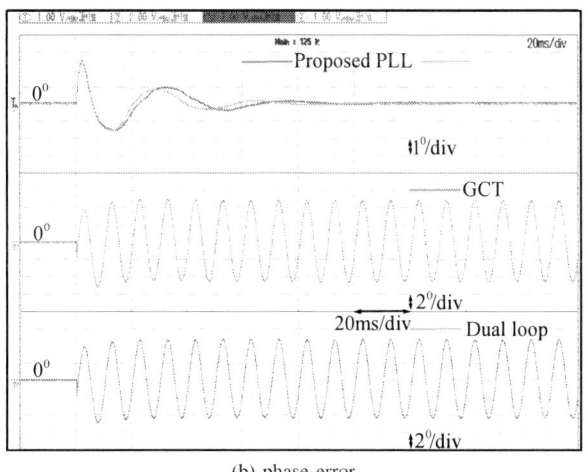

(a) frequency

(b) phase error

Fig. 6. Comparison of type-2 and type-3 PLL performance with respect to unbalanced grid voltage

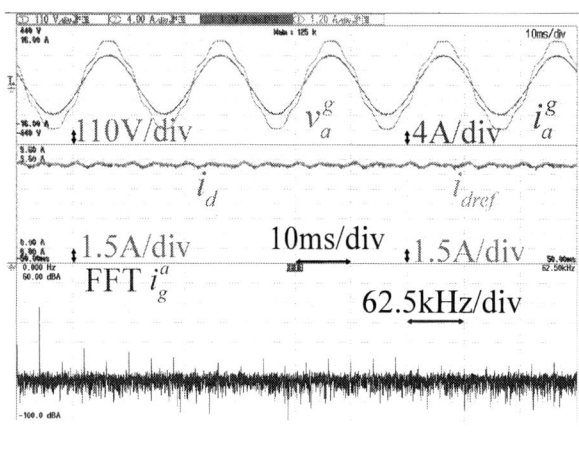

(a) frequency

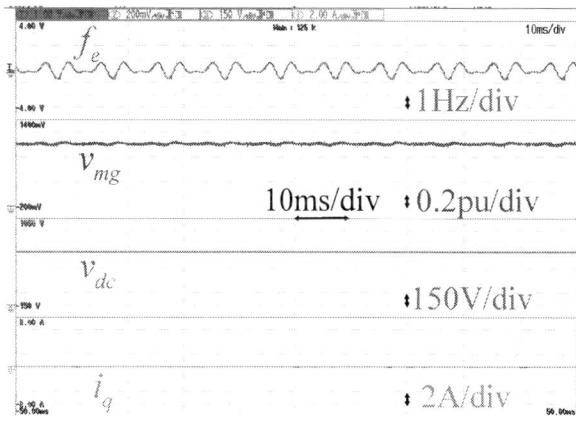

(b) phase error

Fig. 7. Performances of GFI system with a proposed PLL under distorted grid unbalance scenario

IV. RESULTS AND DISCUSSION

The performance of the PLL was assessed experimentally during three-phase grid tests and through current and voltage signals' MATLAB/Simulink simulation. The proposed controller is evaluated with two platforms: one is the eFPGAsim hardware-synchronized platform, and the other is the OPAL-RT 5700 real-time simulator platform. As for the experimental device, its frequency is 50 Hz; at the same time, the frequency with which the circuit samples its inputs is 10 microseconds. When using the proposed PLL the GFI system performance depends on grid disturbances. Furthermore, the efficacy of the proposed PLL with the GFI system has utilized the parameters provided in Table I and prototype testing during grid disturbances.

Steady-state performances of estimated frequency and phase error for various PLLs are depicted in Fig. 5 with the presence of harmonics in the grid voltage. The grid voltage is distorted by 5th and 7th-order harmonics with a magnitude of 0.05 p.u. The dual-loop type -3 PLL and type 3-GCT -PLL show more oscillations in the estimated frequency and and phase error

due to the lack of filter. By comparison, the oscillations of type-2 DSOGI-PLL, dual-loop type -3 PLL and type 3-GCT -PLL are much larger, with a frequency error of about 1.5 Hz. However, there is harmonic distortion though a recommended PLL has low oscillation around the fundamental frequency in estimated frequency and phase error.

Transient performance of phase error and frequency responses in type-2, and type-3 PLLs under the unbalanced grid voltage are observed in the Fig. 6. In this scenario, unbalanced grid voltage occur as $1p.u$, $0.5p.u$, and $0.7p.u$. It can be observed that proposed PLL has less overshoot and minimum settling time in estimated frequency compared type-2 DSOGI-PLL. Even though the grid voltages are unequal, the steady-sate phase error is zero in type-3 proposed PLL and type-2 DSOGI-PLL. The dual-loop type-3 PLL and the type-3 GCT-PLL have phase inaccuracy in steady-state and frequency oscillations produced by the unbalanced grid voltages.

As illustrated in Fig. 7, the proposed PLL is employed to assess the performance of a three-phase PV-GFI system based on the disturbance of unbalanced grid voltage distortion.

979-8-3315-1612-3/25 $31.00 © 2025 IEEE   1944

It can be observed from Fig. 7 (a), active components of inject grid i_d track the active components of reference grid current i_{dref}. Moreover, estimated frequency, magnitude and phase angle for grid voltage are obtained by the proposed PLL as shown in Fig. 7 (b). It can be observed that the estimated frequency has less oscillation since the proposed PLL extract the PS components from unbalanced grid voltage distortion with the help of modified ISC techniques. Thus, it can be recognized that the grid currents (i_g^a) show positive and sinusoidal behavior and more importantly satisfy the limits set by the IEEE-519 in terms of total harmonic distortion. In addition, grid current is within phase with grid voltage.

V. CONCLUSION

A modified ISC control algorithm in type-3 PLL has been proposed to control the GFI system. The required grid voltage characteristics are obtained from the proposed PLL with high accuracy, even in the conditions of grid disturbances. Moreover, it extracts the positive sequence component from the unbalanced distortion grid voltage and, at the same time, minimises the effect of the harmonic in estimated frequency and phase. Another advantage is that the proposed PLL does not affect the dynamics performance of the DSOGI-GFI system. The mathematical analysis for extracting the PS component from unbalanced distorted grid voltage are also discussed. Furthermore, studies have been conducted to assess and compare the proposed PLL to existing type-2 and type-3 PLLs in terms of steady-state and dynamic responses. The proposed approach has been proven to be superior in terms of precision and speed of phase locking. Furthermore, the suggested PLL is simple and easy to implement, making it a suitable PLL technique contender in reality. In addition, the effectiveness and superiority of the proposed PLL are examined in the GFI system under challenges posed by unbalanced distorted grid voltage. It effectively injects the grid current within THD limits.

REFERENCES

[1] M. K. Senapati, O. Al Zaabi, K. Al Hosani, K. Al Jaafari, C. Pradhan, and U. Ranjan Muduli, "Advancing electric vehicle charging ecosystems with intelligent control of dc microgrid stability," *IEEE Transactions on Industry Applications*, vol. 60, no. 5, pp. 7264–7278, 2024.

[2] A. Sahoo, J. Ravishankar, and C. Jones, "Phase-locked loop independent second-order generalized integrator for single-phase grid synchronization," *IEEE Transactions on Instrumentation and Measurement*, vol. 70, pp. 1–9, 2021.

[3] P. Montero-Robina, K. Rouzbehi, F. Gordillo, and J. Pou, "Grid-following voltage source converters: Basic schemes and current control techniques to operate with unbalanced voltage conditions," *IEEE Open Journal of the Industrial Electronics Society*, vol. 2, pp. 528–544, 2021.

[4] IEEE, "Standard for interconnecting distributed resources with electric power systems," *IEEE Std 1547-2003*, pp. 1–28, 2003.

[5] F. Blaabjerg, R. Teodorescu, M. Liserre, and A. Timbus, "Overview of control and grid synchronization for distributed power generation systems," *IEEE Transactions on Industrial Electronics*, vol. 53, no. 5, pp. 1398–1409, 2006.

[6] Y. Wu, H. Wu, F. Zhao, Z. Li, and X. Wang, "Influence of pll on stability of interconnected grid-forming and grid-following converters," *IEEE Transactions on Power Electronics*, vol. 39, no. 10, pp. 11 980–11 985, 2024.

[7] P. K. Chamarthi, U. R. Muduli, M. S. E. Moursi, A. Al-Durra, V. Khadkikar, K. A. Hosani, and T. H. M. EL-Fouly, "Novel 1-φ high-

voltage boosting transformerless inverter topology with optimal power components and negligible leakage currents," *IEEE Transactions on Industry Applications*, vol. 59, no. 5, pp. 6273–6287, 2023.

[8] P. K. Chamarthi, U. R. Muduli, M. S. E. Moursi, A. Al-Durra, A. S. Al-Sumaiti, and K. A. Hosani, "Improved pwm approach for cascaded five-level npc h-bridge configurations in multilevel inverter," *IEEE Transactions on Industry Applications*, vol. 60, no. 5, pp. 7048–7060, 2024.

[9] S. Golestan, J. M. Guerrero, and J. C. Vasquez, "Three-phase plls: A review of recent advances," *IEEE Transactions on Power Electronics*, vol. 32, no. 3, pp. 1894–1907, 2017.

[10] S. Golestan, M. Monfared, and F. D. Freijedo, "Design-oriented study of advanced synchronous reference frame phase-locked loops," *IEEE Transactions on Power Electronics*, vol. 28, no. 2, pp. 765–778, 2013.

[11] J. Lei, X. Quan, S. Feng, J. Zhao, and W. Chen, "Accurate modeling of pll with frequency-adaptive prefilter: On the positive feedback effect," *IEEE Transactions on Power Electronics*, vol. 37, no. 4, pp. 3747–3752, 2022.

[12] Y. Han, M. Luo, X. Zhao, J. M. Guerrero, and L. Xu, "Comparative performance evaluation of orthogonal-signal-generators-based single-phase pll algorithms—a survey," *IEEE Transactions on Power Electronics*, vol. 31, no. 5, pp. 3932–3944, 2016.

[13] K.-H. Nguyen, A. A. Nazeri, X. Yu, and P. Zacharias, "A novel modified-togi based pll for the three-phase unbalanced and distorted grid conditions," in *2022 24th European Conference on Power Electronics and Applications (EPE'22 ECCE Europe)*, 2022, pp. 1–10.

[14] S. Golestan, M. Monfared, F. D. Freijedo, and J. M. Guerrero, "Dynamics assessment of advanced single-phase PLL structures," *IEEE Trans. Ind. Electron.*, vol. 60, no. 6, pp. 2167–2177, 2013.

[15] V. N. Giotopoulos and G. N. Korres, "A laboratory pmu based on third-order generalized integrator phase-locked loop," *IEEE Transactions on Instrumentation and Measurement*, vol. 73, pp. 1–11, 2024.

[16] C. Zhang, X. Zhao, X. Wang, X. Chai, Z. Zhang, and X. Guo, "A grid synchronization PLL method based on mixed second- and third-order generalized integrator for DC offset elimination and frequency adaptability," *IEEE Journal of Emerging and Selected Topics in Power Electronics*, vol. 6, no. 3, pp. 1517–1526, 2018.

[17] J. Xu, Z. Ling, Y. Luo, J. Kan, H. Diao, and S. Xie, "Synchronization stability analysis and parameter design of grid-following inverters considering the interactions of current control and phase-locked loop," *IEEE Journal of Emerging and Selected Topics in Power Electronics*, vol. 12, no. 5, pp. 5013–5027, 2024.

[18] A. Bamigbade, V. Khadkikar, M. Al Hosani, H. H. Zeineldin, and M. S. El Moursi, "Gain compensation approach for low-voltage ride-through and dynamic performance improvement of three-phase type-3 PLL," *IET Power Electronics*, vol. 13, no. 8, pp. 1613–1621, 2020.

[19] S. Golestan, M. Ramezani, and J. M. Guerrero, "An analysis of the plls with secondary control path," *IEEE Transactions on Industrial Electronics*, vol. 61, no. 9, pp. 4824–4828, 2014.

[20] S. Golestan, M. Monfared, F. D. Freijedo, and J. M. Guerrero, "Advantages and challenges of a type-3 PLL," *IEEE Trans. Power Electron.*, vol. 28, no. 11, pp. 4985–4997, 2013.

[21] A. Bamigbade, V. Khadkikar, and M. A. Hosani, "A type-3 PLL for single-phase applications," *IEEE Transactions on Industry Applications*, vol. 56, no. 5, pp. 5533–5542, 2020.

[22] S. Prakash, J. K. Singh, R. K. Behera, and A. Mondal, "A type-3 modified SOGI-PLL with grid disturbance rejection capability for single-phase grid-tied converters," *IEEE Transactions on Industry Applications*, vol. 57, no. 4, pp. 4242–4252, 2021.

[23] S. Prakash, M. Alkhatib, O. A. Zaabi, K. A. Hosani, R. K. Behera, and U. R. Muduli, "Adaptive phase synthesizer with in-loop variable gain control for reliable grid integration," *IEEE Journal of Emerging and Selected Topics in Power Electronics*, pp. 1–1, 2024.

[24] J. K. Singh, S. Prakash, K. Al Jaafari, O. Al Zaabi, K. Al Hosani, R. K. Behera, and U. R. Muduli, "Active disturbance rejection control of photovoltaic three-phase grid following inverters under uncertainty and grid voltage variations," *IEEE Transactions on Power Delivery*, vol. 38, no. 5, pp. 3155–3168, 2023.

[25] M. Rezkallah, S. K. Sharma, A. Chandra, B. Singh, and D. R. Rousse, "Lyapunov function and sliding mode control approach for the solar-PV grid interface system," *IEEE Transactions on Industrial Electronics*, vol. 64, no. 1, pp. 785–795, Jan. 2017.

Online Impedance-based Analysis for Power System Stability Assessment Using Transformer-less and Filter-less Switch-Mode Perturbation Generator

Tomoya Ide
Dept. of Electrical, Electronic and Communications Engineering
Toyo University
Saitama, Japan
s36c02400015@toyo.jp

Yuko Hirase
Dept. of Electrical, Electronic and Communications Engineering
Toyo University
Saitama, Japan
ORCID:0000-0002-6217-6437

Cheng Huang
Institute of Pure and Applied Sciences
University of Tsukuba
Ibaraki, Japan

ORCID: 0000-0001-9951-8560

Takanori Isobe
Institute of Pure and Applied Sciences
University of Tsukuba
Ibaraki, Japan

ORCID: 0000-0003-4253-5567

Abstract— This paper proposes an online impedance-based analysis method for power systems, utilizing a switch-mode perturbation generator. The proposed approach employs frequency scanning with rectangular perturbations, which is achieved through single pulse switching of the dedicated inverter for perturbation generating. This method eliminates the need for filter inductors, which often limit the applicability of conventional techniques. Additionally, the inverter is designed to enable transformer-less operation. This paper presents comparative results from simulations and experiments using both sinusoidal and rectangular perturbations. The results show good agreement, demonstrating the basic feasibility of the proposed method.

Keywords—distributed energy resource, impedance-base analysis, perturbation, power converter, stability analysis

I. INTRODUCTION

The integration of numerous inverter-based resources (IBRs) into modern power systems has introduced challenges, including resonance phenomena caused by the high penetration of distributed energy resources (DERs) [1,2]. References [3,4] reported various cases of sub-synchronous oscillations happen in practical power systems, which are caused by PLL and others related to inverter control. To ensure power system stability, it is essential to analyze the system and eliminate instability factors before connecting DERs.

An impedance-based analysis can evaluate system stability by only using measured voltage or current data [5–8], and does not require the information about control algorithms and configurations of the power converters and the system. Therefore, it is a promising method for stability analysis and a practical tool for tuning power converters in DERs within actual power systems. For examples, references [9, 10] evaluated the stability of high-voltage direct current (HVDC) transmission systems by this method. Reference [11] proposed an impedance reshaping mitigation, which reshapes the impedance characteristics of a power system in frequency domain, and demonstrated its effectiveness in suppressing oscillations through a test. Reference [12, 13] illustrates how each element

of the system affects impedance characteristics through sensitivity analysis and participation factors.

For the dynamic nature of today's power systems, where IBRs and loads (LDs) are frequently connected and disconnected, continuous stability assessments over time and at multiple locations in the system can provide valuable insights and opportunities to enhance system stability further. Achieving such capabilities would require the installation of numerous sets of measurement equipment. However, the impedance-based analysis relies on high-performance perturbation sources, and linear amplifiers are typically used in laboratory environments. When this method is applied across multiple points in the power system, linear amplifiers become impractical due to their size and cost [14]. Reference [15] proposed using the inverter itself to generate perturbations for grid-side impedance measurements; however, this approach cannot measure the inverter's own impedance, necessitating additional simulations. On the other hand, using a separate setup enables direct measurement of both grid-side and inverter-side impedances, and/or retrofit, snap-in applications.

This study proposes the use of a dedicated switch-mode inverter as the perturbation generator. Output filters are usually required to generate sinusoidal perturbation voltage using pulse width modulation (PWM), and they limit the output bandwidth and increases the size of the equipment. Perturbations are typically required across a wide frequency range, extending up to several tens of kilohertz. To meet this requirement, single pulse switching, where the semiconductor switches are turned on and off at the same frequency as the perturbation, is preferred for the dedicated inverter. This approach eliminates the need for filter inductors, reduces switching losses, and enables the development of compact measurement equipment. However, the perturbation waveforms cannot be sinusoidal.

To realize this concept, we propose extending impedance-based analysis to utilize rectangular perturbation voltages for accurate impedance measurement. Using rectangular perturbation waveforms itself has been reported in [16] for the different purpose, which is an optimization of perturbation spectrum; however, it is not used for injecting perturbation on dq-synchronous reference frames. This paper extends an existing method of synthesizing sinusoidal perturbation for the synchronous reference frames to synthesizing the rectangular perturbation, which is achieved by the single pulse switching perturbation generator. This paper also presents some experimental verifications, including comparative discussions of the proposed method and the conventional sinusoidal perturbation technique.

II. THEORETICAL FOUNDATIONS OF IMPEDANCE-BASED ANALSIS

A. Stability Assessment Method

In impedance-based stability analysis for power systems, the entire system is divided into source and load subsystems, and small perturbation power is injected at the split points. Fig. 1 shows the equivalent block diagrams of the source and load subsystems, with impedances $Z_S(s)$ and $Z_L(s)$ represented by the Thévenin and Norton equivalent circuits, respectively, where $V_S(s)$ is the voltage and $I_L(s)$ is the current at the split point, and s is Laplace's differential operator. These voltages, currents, and impedances are scalar quantities for DC systems; however, they are expressed as 2×2 matrices for three-phase AC system with dq-synchronous reference frames [17–19].

The injected small perturbation power is shared between the source and load subsystems. For example, if the perturbation power is injected as voltage, the perturbation currents flowing through both systems are common but with opposite directions. Similarly, if it is injected as current, the perturbation voltage is common. $Z_S(s)$ and $Z_L(s)$ at the frequency of the injected perturbation can be extracted from the voltage and current waveforms measured on each side of the split point. In a DC system, $V_S(s)$ and $I(s)$ are easily expressed as

$$V(s) = \left(V_S(s) + I_L(s)Z_S(s)\right)\{Z_L(s)/(Z_S(s) + Z_L(s))\},$$
$$\cdots\cdots\cdots\cdots\cdots\cdots\cdots\cdots\cdots\cdots\cdots\cdots\cdots\cdots\cdots\cdots\cdots\cdots\cdots\cdots\cdots\cdots\cdots\cdots (1)$$

$$I(s) = \left(-I_L(s) + V_S(s)/Z_L(s)\right)\{Z_L(s)/(Z_S(s) + Z_L(s))\},$$
$$\cdots\cdots\cdots\cdots\cdots\cdots\cdots\cdots\cdots\cdots\cdots\cdots\cdots\cdots\cdots\cdots\cdots\cdots\cdots\cdots\cdots\cdots\cdots (2)$$

by using the parameters of the Thévenin and Norton equivalent circuits [20].

Given that $V_S(s)$ and $I_L(s)$ are stable, it follows from Eqs. (1) and (2) that the system's stability is determined by the term $Z_L(s)/(Z_S(s) + Z_L(s))$. When expressed in terms of an open transfer function, the loop gain is given by $L(s) = Z_S(s)/Z_L(s)$. For a three-phase AC system, the loop gain becomes a 2×2 matrix, $L(s) = Z_S(s)Z_L^{-1}$. According to the Nyquist criterion,

a system is considered stable if and only if all eigenvalue loci of $L(s)$ encircle the point $(-1, j0)$ in the complex plane counterclockwise a number of times equal to the number of right half-plane poles in the system.

However, determining the number of right half-plane poles in a complex power system with many inverter-based DERs is challenging. Moreover, it requires to obtain control algorithms and parameters of inverter-based DERs, and this requirement drastically diminishes the advantages of a data-driven, impedance-based analysis. On the other hand, if the power system is represented by a Thévenin-Norton equivalent circuit, as described above, it is evident that the voltage or current becomes unstable at frequencies where the eigenvalues of $L(s)$ approach $(-1, j0)$, as shown in Eqs. (1) and (2). Therefore, in impedance-based analysis of power systems, it is standard practice to infer stability from the load-side and source-side impedances at frequencies approaching $(-1, j0)$. Therefore, the stability can be discussed from the impedances, which can be measured in practice. If resonance is anticipated, corrective measures are taken to increase both the phase margin and gain margin [21].

B. Synthesizing Sinusoidal Perturbation

The control system of a three-phase AC system is typically constructed using dq-synchronous reference frames, therefore, the stability should be discussed based on their respective dq impedance matrices. To calculate the impedance matrices, $Z_S(s)$ and $Z_L(s)$, two independent perturbation powers are injected at the system's split points. In this case, it is considered that the perturbations are injected along the d-axis and q-axis in the synchronous reference frame. Applying the Clarke-Park (dq) transformation and its inverse introduces interference between the d-axis and q-axis control systems, and the off-diagonal elements of the 2×2 impedance matrix represent the interference. It is well known that in many power systems, resonance is observed in these off-diagonal elements [18].

To inject a perturbation with the angular frequency of ω_p in the synchronous reference frame along the d- or q-axis, a positive sequence component with its frequency of $\omega_p + \omega_n$ and a negative sequence component with its frequency of $\omega_p - \omega_n$ are injected in the stationary reference frame, where ω_n is the system frequency. For example, a positive sequence component with 300 Hz and a negative sequence component with 200 Hz are injected for the perturbation of 250 Hz and the system frequency of 50 Hz. When those are added, the perturbation voltages in the stationary reference frame are expressed as

$$\begin{bmatrix} \tilde{v}_a \\ \tilde{v}_b \\ \tilde{v}_c \end{bmatrix} = \begin{bmatrix} \sin\left(\theta_p + (\theta_n + \alpha)\right) + \sin\left(\theta_p - (\theta_n + \alpha)\right) \\ \sin\left(\theta_p + (\theta_n + \alpha) - \frac{2\pi}{3}\right) + \sin\left(\theta_p - (\theta_n + \alpha) + \frac{2\pi}{3}\right) \\ \sin\left(\theta_p + (\theta_n + \alpha) + \frac{2\pi}{3}\right) + \sin\left(\theta_p - (\theta_n - \alpha) - \frac{2\pi}{3}\right) \end{bmatrix}, \cdots (3)$$

where θ_p is the phase of the synthetic perturbation voltage, θ_n is the phase of the system voltage, and α rad is the phase difference between these voltages. On the other hand, when their difference is taken, the perturbation voltages are expressed as

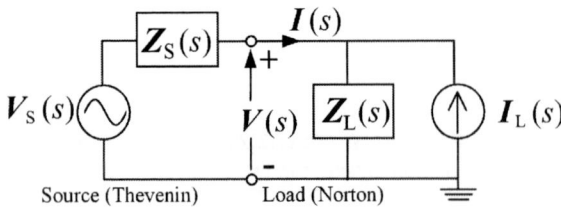

Fig. 1. System diagram split by Thevenin and Norton equivalent circuits.

$$
\begin{bmatrix} \tilde{v}_a \\ \tilde{v}_b \\ \tilde{v}_c \end{bmatrix} = \begin{bmatrix} \sin\left(\theta_p + (\theta_n + \alpha)\right) - \sin\left(\theta_p - (\theta_n + \alpha)\right) \\ \sin\left(\theta_p + (\theta_n + \alpha) - \frac{2\pi}{3}\right) - \sin\left(\theta_p - (\theta_n + \alpha) + \frac{2\pi}{3}\right) \\ \sin\left(\theta_p + (\theta_n + \alpha) + \frac{2\pi}{3}\right) - \sin\left(\theta_p - (\theta_n - \alpha) - \frac{2\pi}{3}\right) \end{bmatrix}, \cdots (4)
$$

By applying dq-transformation to (3), the perturbation voltages in the synchronous reference frame for the sum of the positive and negative sequence components are obtained as

$$
\begin{bmatrix} \tilde{v}_d \\ \tilde{v}_q \end{bmatrix} = \frac{1}{3} \begin{bmatrix} 3\sin(\theta_p)\cos(\alpha) \\ +\sin(\theta_p + 2\theta_n + \alpha) + \sin\left(\theta_p + 2\theta_n + \alpha - \frac{4\pi}{3}\right) \\ +\sin\left(\theta_p + 2\theta_n + \alpha + \frac{4\pi}{3}\right) + \sin\left(\theta_p - (2\theta_n + \alpha)\right) \\ +\sin\left(\theta_p - (2\theta_n + \alpha) + \frac{4\pi}{3}\right) + \sin\left(\theta_p - (2\theta_n + \alpha) - \frac{4\pi}{3}\right) \\ 3\sin(\theta_p)\sin(\alpha) \\ +\cos(\theta_p + 2\theta_n + \alpha) + \cos\left(\theta_p + 2\theta_n + \alpha - \frac{4\pi}{3}\right) \\ +\cos\left(\theta_p + 2\theta_n + \alpha + \frac{4\pi}{3}\right) - \cos\left(\theta_p - (2\theta_n + \alpha)\right) \\ -\cos\left(\theta_p - (2\theta_n + \alpha) + \frac{4\pi}{3}\right) - \cos\left(\theta_p - (2\theta_n + \alpha) - \frac{4\pi}{3}\right) \end{bmatrix}. \quad (5)
$$

By assuming the system voltage is three-phase balanced, the second and subsequent terms of each element in (5) become zero; therefore it can be simplified as

$$
\begin{bmatrix} \tilde{v}_d \\ \tilde{v}_q \end{bmatrix} = \begin{bmatrix} \sin(\theta_p)\cos(\alpha) \\ \sin(\theta_p)\sin(\alpha) \end{bmatrix}. \quad\quad\quad\quad\quad\quad (5)
$$

It indicates that the perturbation voltage with the angular frequency of ω_p is superimposed in the synchronus reference frame and be along the d-axis when $\alpha = 0$. In the same way, the perturbation voltages in the synchronous reference frame for the difference of the positive and negative sequence components is expressed as

$$
\begin{bmatrix} \tilde{v}_d \\ \tilde{v}_q \end{bmatrix} = \begin{bmatrix} \cos(\theta_p)\sin(\alpha) \\ \cos(\theta_p)\cos(\alpha) \end{bmatrix}. \quad\quad\quad\quad\quad\quad (6)
$$

With $\alpha = 0$, this case injects the perturbation along the q-axis.

Fig. 2 illustrates a schematic view of the perturbation in the stationary and synchronous reference frames in a case that the sinusoidal perturbation of 250 Hz is injected along the d-axis of the synchronous reference frame for a 50 Hz system voltage. Figs. 2(a)–2(c) show the perturbations in the stationary reference frame. As can be seen from Figs. 2(a) and 2(b), the positive sequence component of 300 Hz in the stationary reference frame, and the negative sequence component of 200 Hz are injected. Fig. 2(c) shows the sum of the two components. Fig. 2(d) illustrates the perturbation voltages superimposed on the system voltages in the stationary reference frame. Fig. 2(e) shows the transformed values of them in the synchronous reference frame by the dq-transformation. It can be confirmed that the perturbation is only injected into the d-axis component.

C. Synthesizing Rectangular Perturbation

Useful rectangular-based perturbation voltages can be derived based on the discussion for the sinusoidal perturbation in the previous subsection. As same as for the sinusoidal perturbation synthesis, a positive sequence and negative sequence components, but with rectangular waveforms, are superimposed on the system voltage.

The positive sequence component is expressed as

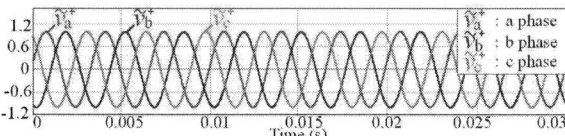

(a) Positive sequence voltage in the stationary frame.

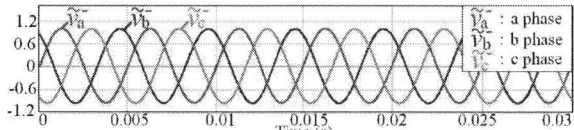

(b) Negative sequence voltage in the stationary frame.

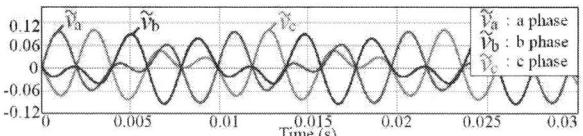

(c) Synthetic sinusoidal perturbation voltage in the stationary frame.

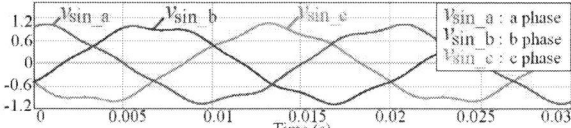

(d) System and perturbation voltages in the stationary frame.

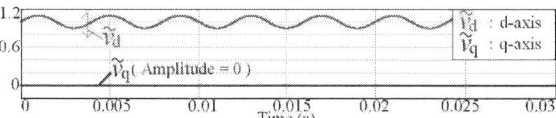

(e) System and perturbation voltages in the synchronous frame.

Fig. 2. Example of synthetic sinusoidal perturbation waveforms in stationary and synchronous reference frames.

$$
\begin{cases} \bar{v}_a^+ = \begin{cases} +1, & 0 \le t \le T^+ \\ -1, & T^+ t \le 2T^+ \end{cases} \\ \bar{v}_b^+ = \begin{cases} +1, & -2T^+/3 \le t \le T^+/3 \\ -1, & T^+/3 \le t \le 4T^+/3 \end{cases} \\ \bar{v}_c^+ = \begin{cases} +1, & -4T^+/3 \le t \le -T^+/3 \\ -1, & -T^+/3 \le t \le 2T^+/3 \end{cases} \end{cases}, \quad\cdots\cdots\cdots (7)
$$

where T^+ represents the half cycle of the positive sequence component in the stationary reference frames as $T^+ = 1/2(\omega_p + \omega_n)$. The negative sequence component is expressed as

$$
\begin{cases} \bar{v}_a^- = \begin{cases} +1, & 0 \le t \le T^- \\ -1, & T^- t \le 2T^- \end{cases} \\ \bar{v}_b^- = \begin{cases} +1, & -4T^-/3 \le t \le -T^-/3 \\ -1, & -T^-/3 \le t \le 2T^-/3 \end{cases} \\ \bar{v}_c^- = \begin{cases} +1, & -2T^-/3 \le t \le T^-/3 \\ -1, & T^-/3 \le t \le 4T^-/3 \end{cases} \end{cases} \quad\cdots\cdots\cdots (8)
$$

where T^- represents the half cycle of the negative sequence component in the stationary reference frames as $T^- = 1/2(\omega_p - \omega_n)$. Those are then added or subtracted, and superimposed on the system voltage as similar way with the sinusoidal perturbation case.

Fig. 3 illustrates the schematic view of the perturbation in the stationary and synchronous reference frames in the same case with that shown in Fig. 2. Figs. 3(a)–3(c) show the

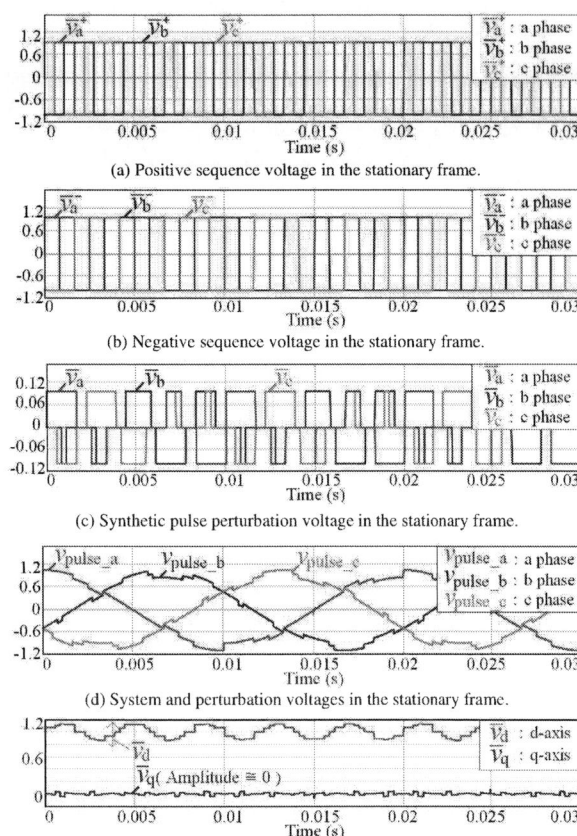

(a) Positive sequence voltage in the stationary frame.

(b) Negative sequence voltage in the stationary frame.

(c) Synthetic pulse perturbation voltage in the stationary frame.

(d) System and perturbation voltages in the stationary frame.

(e) System and perturbation voltages in the synchronous frame.

Fig. 3.   Example of synthetic pulse perturbation waveforms in stationary and synchronous reference frames.

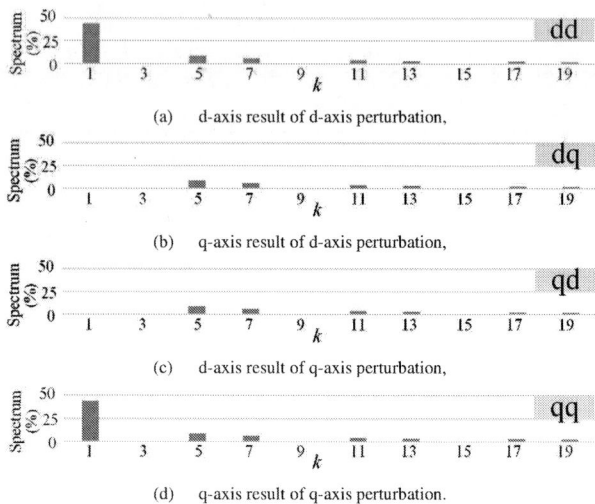

(a)   d-axis result of d-axis perturbation,

(b)   q-axis result of d-axis perturbation,

(c)   d-axis result of q-axis perturbation,

(d)   q-axis result of q-axis perturbation.

Fig. 4.   Spectrum of perturbation in a synchronous reference frame when 250 Hz is injected.

perturbation voltages in the stationary reference frame. Fig. 3(c) shows the sum of the positive and negative phase components, which are referred to as rectangular perturbation voltages in this study. Fig. 3(e) shows the system voltage with the perturbation in the synchronous reference frame. As can be seen from the figure, the perturbation is only injected into the d-axis component; however, harmonic components remain both in the d- and q- axis components.

The dq impedance is calculated by extracting components of the perturbation frequency by applying the discrete Fourier transformation (DFT) from the measured waveforms of voltage and current in synchronous reference frame. Although the d- and q-axis components contain harmonic components coming from the rectangular waveforms, those do not affect the resulting dq impedance. By applying dq-transformation and Fourier series expansion to the rectangular perturbation voltages, $\bar{v}_a$, $\bar{v}_b$, and $\bar{v}_c$, expressed as (7) and (8), the perturbation voltages in the synchronous reference frames for the sum of the positive and negative sequence components, $\bar{v}_d$ and $\bar{v}_q$, can be obtained as

$$\begin{bmatrix} \bar{v}_d \\ \bar{v}_q \end{bmatrix} = \frac{2\sqrt{6}}{\pi} \begin{bmatrix} \sin(\omega_p t) + \sum_{k=1}^{\infty} \frac{1}{6k \mp 1} \sin\left((6k \mp 1)\omega_p t\right) \cos(6k\omega_n t) \\ \pm \sum_{k=1}^{\infty} \frac{1}{6k \mp 1} \sin\left((6k \mp 1)\omega_p t\right) \sin(6k\omega_n t) \end{bmatrix},$$

.... (9)

and those for the difference of them can be expressed as

$$\begin{bmatrix} \bar{v}_d \\ \bar{v}_q \end{bmatrix} =$$
$$\frac{2\sqrt{6}}{\pi} \begin{bmatrix} \sum_{k=1}^{\infty} \frac{1}{6k \mp 1} \cos\left((6k \mp 1)\omega_p t\right) \sin(6k\omega_n t) \\ -\cos(\omega_p t) \pm \sum_{k=1}^{\infty} \frac{1}{6k \mp 1} \cos\left((6k \mp 1)\omega_p t\right) \cos(6k\omega_n t) \end{bmatrix}.$$

.... (10)

It can be derived that (9) corresponds to the case where the d-axis perturbation voltage is injected, while (10) corresponds to the case where the q-axis perturbation voltage is injected. Note that k is a natural number.

Fig. 4 shows the spectra of $\bar{v}_d$ and $\bar{v}_q$ of the case shown in Fig. 3, which is calculated from Eqs. (9) and (10), where the horizontal axis represents k in Eqs. (9) and (10). $k = 1$ represents the perturbation frequency to be injected (250 Hz in this case), and $k > 1$ represents the harmonics of the injected perturbation. Figs. 4(a) and 4(b) show the d- and q-axis voltage spectra for the injected d-axis perturbation voltage, respectively, while Figs. 4(c) and 4(d) show the d- and q-axis voltage spectra for the injected q-axis perturbation voltage. As can be seen from the figures, the perturbations at the target frequency ($k = 1$) are injected along the d- or q-axis as expected even though harmonic components can be found. It can be calculated that the amplitude of the fundamental component ($k = 1$) is 45% of the amplitude of the total components. This implies that when the amplitude of the perturbation voltage is set to 5% of the system voltage amplitude, the component of the target frequency becomes approximately 2.25% of the system voltage.

III. VERIFICATION OF IMPEDANCE ANALYSIS WITH SWITCHIMODE RECTANGURAR PERTURBATION GENERATOR

A. Using Commercially Availabe Inverter with Transformer

Using the proposed synthetic rectangular perturbation scheme, the stability of the system shown in Fig. 5 was analyzed based on the measured impedance. As a preliminary

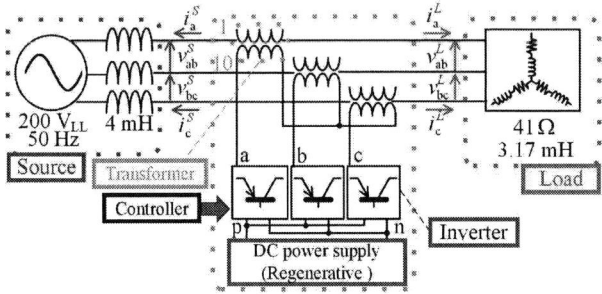

Fig. 5. System configuration.

Table 1. Magnetizing inductance of transformer for perturbation injection.

Frequency (Hz)	Primary (mH)	Secondary (mH)
5	1091.74	11.12
50	919.02	10.98
500	582.54	9.98
1,000	498.46	8.61
5,000	227.90	3.50
10,000	123.18	1.87
20,000	40.35	0.54

Table 2. Leakage inductance of transformer for perturbation injection.

Frequency (Hz)	Primary (mH)	Secondary (mH)
5	4.15004	0.0454085
50	2.39169	0.0365949
500	2.36245	0.0363825
1,000	2.34673	0.0361335
5,000	2.12743	0.0327232

demonstration, a commercially availabe inverter, which has several handreds of the voltage rating, with three single-phase line-frequency transformers was used. The winding ratio of the transformers is 10:1. The inductance values of the transformers with various frequencies are shown in Tables 1 and 2.

The line voltages and phase currents were measured both on the source and load sides. Measurements were performed with varying perturbation frequencies from 0.1 Hz to 5 kHz. The amplitude of the perturbation voltage was set to 5% in the low-frequency band ($\omega_p < \omega_n$), where the source-side impedance is low and measurement becomes difficult. On the other hand, it is set to 4% in the high-frequency band ($\omega_p > \omega_n$), where the perturbation current tends to be increased. Furthermore, a special arrangement was applied for the frequency around 100 Hz in the synchronous reference frame, which corresponds to 0 Hz in the negative sequence in the stationary reference frame. Since it can lead to magnetic saturation in the transformer, a ramp rate of −0.16 %/Hz on the low-frequency side of 50 Hz and +0.13 %/Hz on the high frequency side of 50 Hz was applied.

To avoid fluctuations in system frequency and voltage during measurements at a specific perturbation frequency, three switching amplifiers (DP015RS) were used as a source subsystem instead of the commercial grid. The output voltage was set to 200 V in line-to-line, and three 4 mH inductors were connected in series to emurate source side impedance. The load subsystem consisted of a simple system with a 41 Ω resistive load and a 3.17 mH inductive load connected in series in each phase.

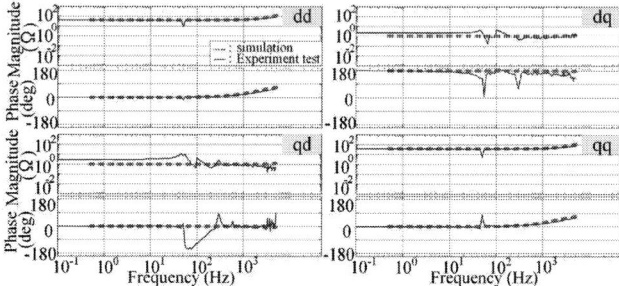

Fig. 6. Load-side impedance by simulation and experimental tests.

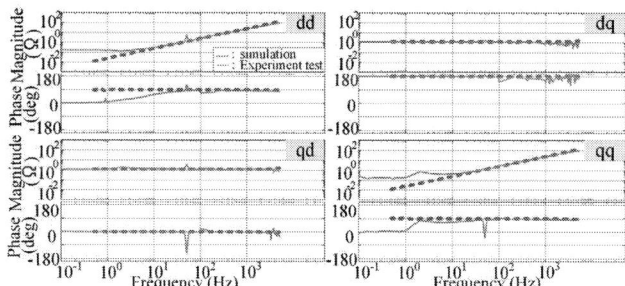

Fig. 7. Source-side impedance by simulation and experimental tests.

Figs. 6 and 7 show the Bode diagrams of the load-side and source-side impedances, respectively. In these figures, the upper graphs display the results when the perturbation voltages were injected on the d-axis, and the lower graphs show the results when they were injected on the q-axis. The left column presents the impedances on the d-axis, while the right column presents the impedances on the q-axis. The solid lines represent the results from the experimental tests using the synthetic rectangular perturbation method, and the dashed lines represent the results from MATLAB/Simulink simulations using the synthetic sinusoidal perturbation. The measured impedance values around 50 Hz are considered to have significant error due to the extremely small amplitude of the perturbation voltage with the applied ramp rate. The discrepancies between the values of simulations and experimental tests in other frequency bands suggest that the characteristics of the load-side and the switching amplifier for the source-side may not have been modeled accurately in the simulations. Despite these differences, the impedance characteristics obtained by both methods were in close agreement over the entire perturbation frequency range.

B. Using Dedicated Switch-Mode Perturbation Generator without Transformer

In conventional systems employing transformers, challenges such as increased device size and magnetic saturation of the transformer degrade the usefulness and availability of the on-line impedance measurement. To overcome these limitations, this study introduces an approach that eliminates the need for a transformer by directly connecting the switch mode perturbation generator to the system in series. Fig. 8 illustrates the circuit configuration of the experimental setup to test this approach. The source-side and load-side configurations are the same with those depicted in Fig. 5.

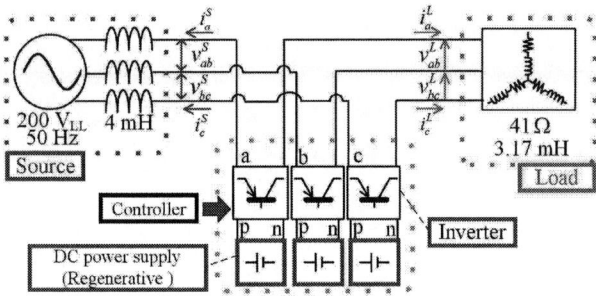

Fig. 8. System configuration without transformer.

Fig. 9. Experimental setup of perturbation generator.

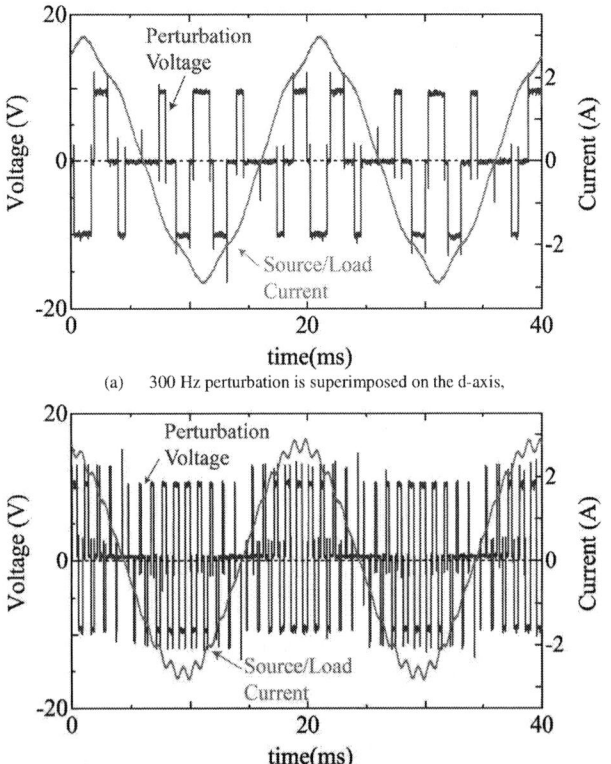

(a) 300 Hz perturbation is superimposed on the d-axis,

(b) 1k Hz perturbation is superimposed on the d-axis,

Fig. 10. Perturbation voltages and source/load currents.

Fig. 9 shows an overview of the fabricated perturbation generator in the experimental setup. It consists of six half-bridge boards to configure three full-bridge converters connected to the

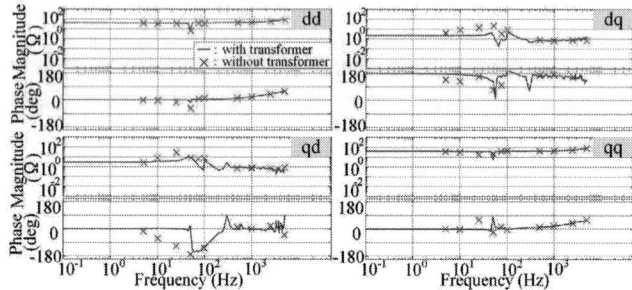

Fig. 11. Load-side impedance with and without transformer.

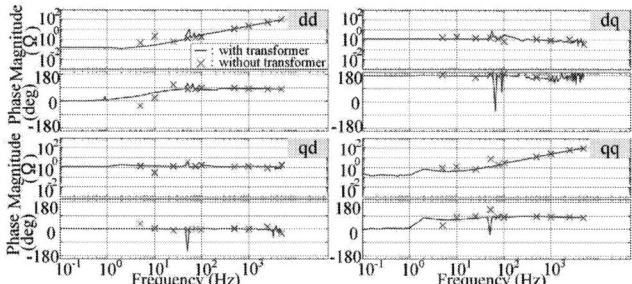

Fig. 12. Source-side impedance with and without transformer.

lines in series. Separated three DC power sources were connected. SiC-MOSFETs with 750 V of voltage rating are used. By using four devices connected in parallel, the equivalent resistance of an arm in the bridge was 15 mΩ, which is sufficiently low not to affect the perturbation voltage with the relatively low amplitude. The fabricated boards are actually capable to inject several hundreds of voltage and several tens ampere, which can be used for high power applications; however, they were used for the laboratory scale experiments this time.

Figs. 10(a) and 10(b) show the perturbation voltage and load current waveforms for perturbation frequencies of 300 Hz and 1 kHz, respectively, applied in the d-axis direction using the fabricated perturbation generator. To make a clear view of the concept, relatively high perturbation voltage, which was 10 V in peak, was injected. No ramp rate was applied for this transformer-less configuration.

Figs. 11 and 12 show the impedances calculated from the measured voltage and current waveforms on the load and source sides, respectively. The solid blue line represents the results obtained using transformers, as previously shown in Figs. 6 and 7, while the red crosses indicate the measurements performed with the developed compact, filter-less perturbation generator directly connected to the system. The close agreement between the two methods validates the feasibility of the proposed approach.

IV. CONCUSION

In this study, the feasibility of the impedance-based analysis using rectangular perturbations using single pulse switching was successfully verified. It is motivated by the requirement for achieving a compact, filter-less, transformer-less switch-mode perturbation generator to be installed at numerous sites. Although the proposed synthetic rectangular perturbation generates harmonic components, the impedances calculated by the proposed methods were found to agree with those by simulation; therefore, it could be confirmed that the algorithm of the proposed method is valid. The compact perturbation generator, when installed in DERs in the future, will enable impedance-based stability analysis of the power system at any arbitrary point and time, significantly contributing to the improvement and maintenance of system-wide stability.

This study focused only on verifying the concept using the rectangular perturbation; therefore, amplitude of the perturbation to measure the impedance effectively was not discussed. As a grid-connecting converter, high perturbation amplitude cannot be allowed for the grid connecting regulation including harmonics. On the other hand, too small perturbation voltage makes it difficult to measure the perturbation current because it is superimposed on the line currents, which have much larger amplitude. Therefore, an optimum perturbation amplitude in accordance with the capability of the measurement equipment should be discussed. In addition, its amplitude may need to be changed according to the perturbation frequency or grid situation. Control procedure for frequency scanning is also a remaining task to be discussed.

ACKNOWLEDGEMENTS

Funding: This study was supported by the Japan Society for the Promotion of Science (JSPS) KAKENHI Grants JP22K14246.

REFERENCES

[1] S. Shah, P. Koralewicz, V. Gevorgian, H. Liu, and J. Fu, "Impedance Methods for Analyzing Stability Impacts of Inverter-Based Resources: Stability Analysis Tools for Modern Power Systems," *IEEE Electrification Magazine*, Vol.9, No.1, pp.53–65 (2021).

[2] C. Buchhagen, C. Rauscher, A. Menze, and J. Jung, "BorWin1 - First Experiences with harmonic interactions in converter dominated grids," International ETG Congress 2015; Die Energiewende - Blueprints for the new energy age, pp.1–7 (2015).

[3] Y. Cheng, L. Fan, J. Rose, S. -H. Huang, J. Schmall, X. Wang, et al. "Real-World Subsynchronous Oscillation Events in Power Grids with High Penetrations of Inverter-Based Resources," *IEEE Transactions on Power Electronics*, Vol.38, No.1, pp.316–330 (2023).

[4] X. Lin, J. Yu, R. Yu, J. Zhang, Z. Yan, and H. Wen, "Improving Small-Signal Stability of Grid-Connected Inverter Under Weak Grid by Decoupling Phase-Lock Loop and Grid Impedance," *IEEE Transactions on Industrial Electronics*, Vol.69, No.7, pp.7040–7053 (2022).

[5] B. Wen, D. Boroyevich, R. Burgos, P. Mattavelli, and Z. Shen, "Analysis of DQ small-signal impedance of grid-tied inverters," *IEEE Transactions on Power Electronics*, Vol.31, No.1, pp.675–687 (2015).

[6] A. Rygg, M. Molinas, C. Zhang, and X. Cai, "A modified sequence-domain impedance definition and its equivalence to the dq-domain impedance definition for the stability analysis of AC power electronic systems," *IEEE Journal of Emerging and Selected Topics in Power Electronics*, Vol.4, No.4, pp. 1383–1396 (2016).

[7] S. Gao, H. Zhao, Y. Gui, J. Luo, and F. Blaabjerg, "Impedance Analysis of Voltage Source Converter Using Direct Power Control," *IEEE Transactions on Energy Conversion*, Vol.36, No.2, pp.831–840 (2021).

[8] C. Zhang, M. Molinas, S. Føyen, J. A. Suul, and T. Isobe, "Harmonic-domain SISO equivalent impedance modeling and stability analysis of a single-phase grid-connected VSC," *IEEE Transactions on Power Electronics*, Vol.35, No.9, pp.9770–9783 (2020).

[9] B. Liu, Z. Li, X. Zhang, X. Dong, and X. Liu, "Impedance-Based Analysis of Control Interactions in Weak-Grid-Tied PMSG Wind Turbines," *IEEE*

Journal on Emerging and Selected Topics in Circuits and Systems, Vol.11, No.1, pp.90–98 (2021).

[10] K. Ji, S. Liu, H. Pang, J. Yang, Z. Xu, Z. He, et al. "Generalized Impedance Analysis and New Sight at Damping Controls for Wind Farm Connected MMC–HVdc," IEEE Journal of Emerging and Selected Topics in Power Electronics, Vol.9, No.6, pp.7278–7295 (2021).

[11] X. Xie, W. Liu, J. Shair, C. Dai, and X. Liu, "Subsynchronous Control Interaction: Real-World Events and Practical Impedance Reshaping Controls," *The 10th Renewable Power Generation Conference (RPG 2021)*, pp.282–288, 14-15 October (2021).

[12] Z. Yang, C. Shah, T. Chen, L. Yu, P. Joebges, and R. W. Doncker, "Stability Investigation of Three-Phase Grid-Tied PV Inverter Systems Using Impedance Models," *IEEE Journal of Emerging and Selected Topics in Power Electronics*, Vol.10, No.3, pp.2672–2684 (2022).

[13] N. Cifuentes, M. Sun, R. Gupta, and B. Pal, "Black-Box Impedance-Based Stability Assessment of Dynamic Interactions Between Converters and Grid," *IEEE Transactions on Power Electronics*, Vol.37, No.4, pp.2976–2987 (2022).

[14] Y. Hirase, K. Ohuchi, T. Ide, K. Ljøkelsøy, and S. D'Arco, "Impedance-Based Stability Analysis of Power Systems Incorporating Inverter-Based Resources: Techniques, Methods, and Applications," *2023 10th International Conference on Power and Energy Systems Engineering (CPESE)*, Nagoya, Japan, pp.107–114 (2023)

[15] Q. Lin, B. Wen, and R. Burgos, "RHP Poles Trajectory Study for D–Q Impedance-Based Stability Monitoring Using a Power-Hardware-in-the-Loop Testbed," *IEEE Journal of Emerging and Selected Topics in Power Electronics*, vol. 12, no. 2, pp. 1560–1572, Apil. 2024.

[16] T. Roimila, M. Vilkko, and J. Sun, "Online Grid Impedance Measurement Using Discrete-Interval Binary Sequence Injection," *IEEE Journal of Emerging and Selected Topics in Power Electronics*, Vol.2, No.4, pp.985– 993 (2014)

[17] M. Amin, C. Zhang, A. Rygg, M. Molinas, E. Unamuno, and M. Belkhayat, "Nyquist stability criterion and its application to power electronics systems," *Wiley Encyclopedia of Electrical and Electronics Engineering.*, pp. 1–22, May. 2019 .

[18] K. Ohuchi, Y. Hirase, and M. Molinas, "MIMO and SISO Impedance-based Stability Assessment of a Distributed Power Source Dominated System," *IEEJ Journal of Industry Applications*, Vol.12, No.4, pp.653–653 (2023)

[19] S. Zhu, K. Liu, X. Liao, L. Qin, Q. Huai, and Y. Xu, "D-Q Frame Impedance Modeling of Modular Multilevel Converter and Its Application in High-Frequency Resonance Analysis," *IEEE Transactions on Power Delivery*, Vol.36, No.3, pp.1517– 1530 (2021)

[20] J. Sun, "Impedance-Based Stability Criterion for Grid-Connected Inverters," *IEEE Transactions on Power Electronics*, Vol.26, No.11, pp.3075– 3078 (2011)

[21] M. Amin, C. Zhang, A. Rygg, M. Molinas, E. Unamuno, and M. Belkhayat, "Nyquist stability criterion and its application to power electronics systems," *Wiley Encyclopedia of Electrical and Electronics Engineering*, pp.1-22 (2019)

PIR-R Control for Three-phase Grid-connected Inverter with Unbalanced Grid Current Correction

Haneen Ghanayem
Dept. of Electrical Engineering
Al-Huson University College
Al-Balqa Applied University
Irbid, Jordan
highanayem@bau.edu.jo

Xingyu Yang
Dept. of Electrical and
Computer Engineering
Auburn University
Auburn, Alabama
xzy0020@auburn.edu

Mohammad Alathamneh
Dept. of Electrical Engineering
Al-Huson University College
Al-Balqa Applied University
Irbid, Jordan
mqalathamneh@bau.edu.jo

R. M. Nelms
Dept. of Electrical and
Computer Engineering
Auburn University
Auburn, Alabama
nelmsrm@auburn.edu

Abstract—**Proposed in this paper a proportional-integral-resonant-resonant (PIR-R) control method for a three-phase grid-connected inverter. Unbalanced loads at the point of common coupling (PCC) can disrupt grid current balance and adversely impact other PCC connections, impairing the inverter's ability to accurately track active and reactive power commands. The PIR-R control integrates PI and PR components for fundamental reference current tracking and second-order harmonic rejection. The first resonant controller component targets the grid frequency, while the second addresses the double-frequency oscillations caused by unbalanced grid currents. Without the need for additional correction devices such as shunt active power filters at the PCC, the PIR-R method ensures effective reference command tracking and corrects unbalanced grid currents. Simulation and experimental results confirm that the PIR-R control provides improved transient response and reduced steady-state error compared to conventional methods.**

Index Terms—**Grid-connected three-phase inverter, proportional-integral-resonant-resonant controller, PIR-R controller, unbalanced three-phase.**

I. INTRODUCTION

The integration of renewable energy sources (RES) with the electrical grid is central to the future of sustainable energy systems. When combined with battery storage, RES can connect to the grid via inverters to operate as distributed generators (DGs). This configuration not only supports environmental sustainability by reducing reliance on fossil fuels and decreasing global carbon emissions, but it also enhances grid performance. In addition to generating cleaner energy, DGs can support grid stability by addressing power quality issues, including reactive power compensation and active filtering [1].

In the past decade, the overall capacity of wind and photovoltaic energy has grown substantially. Moreover, interest in transitioning to carbon-free, zero-emission energy systems has surged [2]. However, integrating distributed generation (DG) into the grid presents challenges due to non-ideal conditions, such as asymmetrical grid states [3]. In an unbalanced three-phase grid, the waveforms of both real and reactive power become sinusoidal with a double-frequency component [4]. This instability negatively affects the performance of three-phase loads connected to the grid, degrading the efficiency of three-phase induction motors and leading to unwanted ripples and vibrations, which can disrupt speed control [5].

Various control algorithms have been applied to regulate power and current in grid-connected three-phase inverters, typically relying on conventional controllers [6]. These methods, however, generally assume balanced grid conditions. When the grid is unbalanced, a PI controller induces double-frequency oscillations in the current, leading to fluctuations in the real and reactive power delivered to the grid.

In reference [7], a PR-controller was implemented to stabilize three-phase output voltages, reducing system complexity by avoiding the need for a phase-locked loop (PLL).

The proposed method leverages a PIR-R controller to eliminate double-frequency ripples in real and reactive power, achieving a smoother power injection. Experimental validation demonstrates the method's performance compared to PI and PIR controllers. The grid-following inverter injects adjusted currents to counteract grid imbalances while simultaneously delivering power, achieving both grid current balancing and power delivery.

The organization of this paper is as follows: Section II introduces and discusses the proposed PIR-R controller. Section III presents a comparison of simulation and experimental results among the three controllers. Finally, Section IV provides a summary and concludes the paper.

II. THREE TYPES OF CONTROLLERS

Fig.1 shows a three-phase grid-connected inverter system, which will inject current i_{inv} into the grid following active and reactive power commands. However, an unbalanced load connected at the PCC, will generate the unbalanced load current i_{ubl} and eventually imbalance the grid current i_g [8]. A conventional PI controller in the dq reference frame often fails to maintain accurate reference power tracking under these conditions, resulting in double-frequency (120 Hz) oscillations in the instantaneous power waveform. Applying a shunt active power filter at the PCC will correct the imbalance but requires extra equipment [9]. The proposed PIR-R controller will minimize the effect caused by the unbalanced load and protect the other balanced load at the PCC. The proposed PIR-R is compared with PI and PIR controllers in the dq reference frame.

979-8-3315-1612-3/25 $31.00 © 2025 IEEE

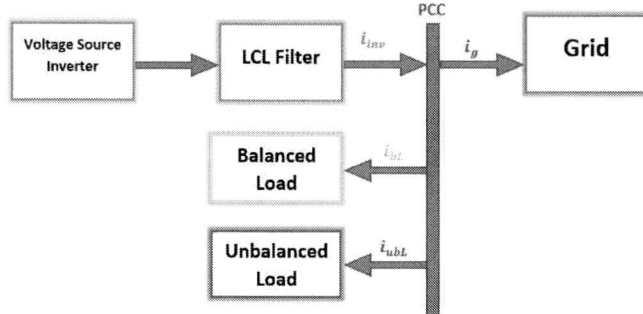

Fig. 1. Grid-connected System with Unbalanced Load.

The current control of the grid-connected inverter follows Eq. (1) to calculate the reference current and then achieve active power and reactive power tracking [10].

$$\begin{bmatrix} i_{ref_d} \\ i_{ref_q} \end{bmatrix} = \frac{1}{V_d^2 + V_q^2} \begin{bmatrix} V_d & V_q \\ V_q & -V_d \end{bmatrix} \begin{bmatrix} P_{ref} \\ Q_{ref} \end{bmatrix} \quad (1)$$

A. PI Controller

The PI controller is a widely used control strategy in power systems, particularly for maintaining stable and accurate control of grid-connected inverters. The PI controller combines proportional and integral actions to minimize the error between the desired reference value and actual value, improving steady-state performance and reducing offset. The proportional component modifies the control signal according to the current error, while the integral component considers accumulated past errors to ensure elimination of any steady-state error. The PI controller's transfer function is expressed as shown in (2).

$$G_{\text{PI}}(s) = K_p + \frac{K_i}{s} \quad (2)$$

where K_p is the proportional gain, K_i is the integral gain, and s is the complex frequency variable. This controller is effective in systems with well-balanced conditions, but may struggle with unbalanced grid scenarios.

B. PIR Controller

The PIR controller is commonly used in three-phase grid-connected inverter applications to regulate current and the injected power. By combining proportional (P), integral (I), and resonant (R) components, the PIR controller ensures robust tracking of the reference current while effectively rejecting specific harmonic components, particularly at the grid's fundamental frequency. This enhances steady-state performance and minimizes tracking error. The PIR controller's transfer function is expressed as shown in (3).

$$G_{\text{PIR}}(s) = K_p + \frac{K_i}{s} + \frac{K_r \cdot s}{s^2 + \omega_0^2} \quad (3)$$

where K_p and K_i are the proportional and integral gains, respectively, K_r is the resonant gain, and ω_0 represents the fundamental angular frequency. The resonant term effectively amplifies the response at the fundamental frequency, improving the controller's ability to maintain current tracking despite grid disturbances.

C. PIR-R Controller

Many control methods have been applied, such as PI or PIR. However, these controllers only work well under balanced conditions. To handle an unbalanced load connected at the PCC, a PIR-R is utilized with the transfer function in Eq. (4), where ω_0 is the fundamental angular frequency and ω_2 is the double fundamental angular frequency.

The block diagram of the PIR-R controller is illustrated in Fig. 2. The current control is in the dq reference frame. The PI controller will provide good steady-state error performance, while the two resonant controllers will decrease the oscillation in the power waveform.

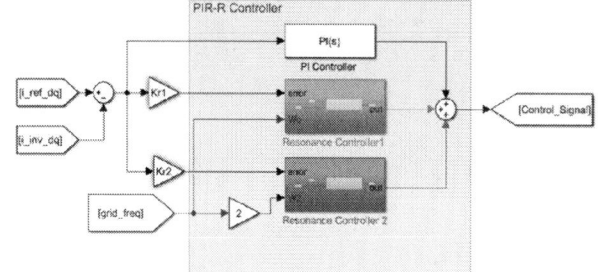

Fig. 2. Diagram of the PIR-R Control System.

$$G_{PIR-R}(s) = K_p + \frac{K_i}{s} + \frac{K_{r1}s}{s^2 + \omega_0^2} + \frac{K_{r2}s}{s^2 + \omega_2^2} \quad (4)$$

III. CASE STUDIES

In this section, the proposed PIR-R approach is validated using a prototype system of a grid-connected three-phase inverter. The performance of the proposed PIR-R controller method in the dq reference frame was tested and compared with the PI controller and PIR controller in the dq reference frame. The system parameters are shown in Table I. An unbalanced load is connected to the PCC, which precipitates the imbalance in the grid current. The performance of the three methods (PI, PIR, and PIR-R) was assessed through both simulation and hardware experimentation.

TABLE I
TEST SYSTEM PARAMETERS.

f	Grid frequency	60 Hz
V_g	Grid phase voltage	120 V
P_B	Rated Power	5 kW
V_{dc}	DC source voltage	400 V
f_{sw}	Switching frequency	5000 Hz
L_1	Inverter side inductance	2.3 mH
L_2	Grid side inductance	0.58 mH
C	Filter capacitance	15 μF
R_f	Damping resistor	1.5 Ω

979-8-3315-1612-3/25 $31.00 © 2025 IEEE   1954

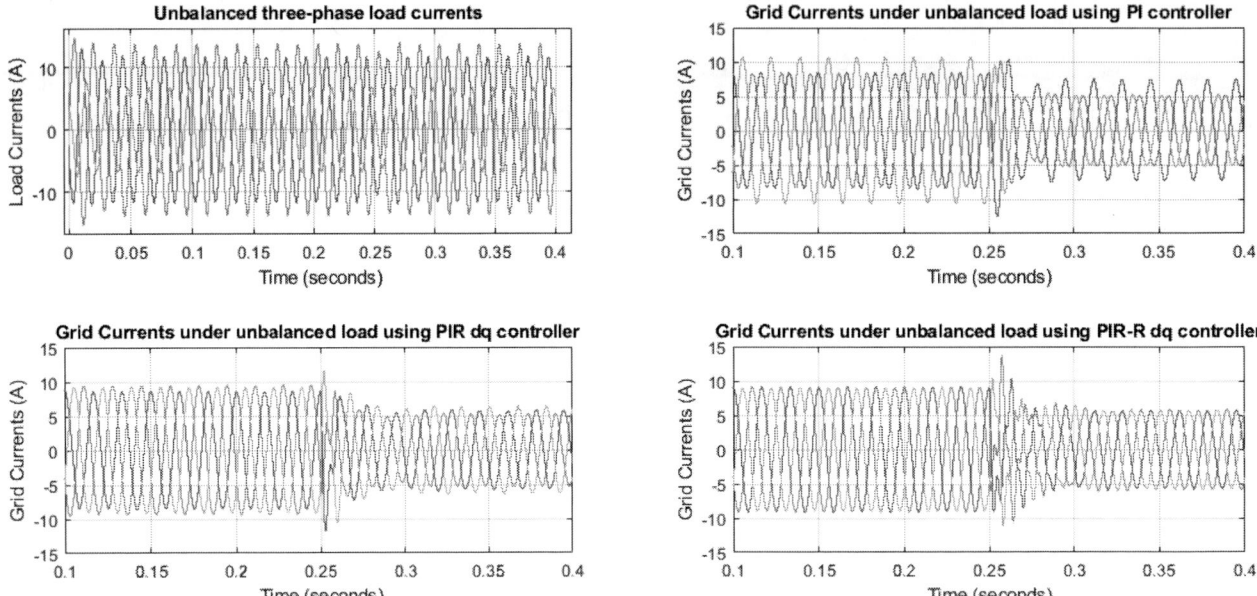

Fig. 3. Grid Current with Unbalanced Load at PCC.

A. Simulation Results

The simulation comparison between PI, PIR, and PIR-R was conducted using SIMULINK. The three-phase inverter system is connected at the PCC and will react to active and reactive power commands based on the control algorithm in the dq reference frame. An unbalanced three-phase load is also connected at the PCC, which will generate unbalanced load currents and compromise the power control of the three-phase inverter.

As shown in Fig. 3, the unbalanced load current will also make the grid current unbalanced. This scenario will downgrade the inverter's power tracking ability because its controllers are not optimized for it. In the typical case, an extra shunt active power filter (APF) applied at the PCC will balance the grid current back and guarantee the inverter's power tracking performance. However, this method requires extra equipment at the PCC, which might not be accessible for all scenarios and definitely increases the system's complexity and cost.

The proposed PIR-R controller does not require an APF at the PCC. The grid current has been consistently balanced thus the unbalanced load will not affect the inverter.

Fig. 4 confirms that the proposed PIR-R controller has the best reference power tracking ability, the least steady-state error among the three controllers, and minimizes the double-frequency oscillation caused by the unbalanced load.

B. Hardware Results

The experimental setup includes an NHR 9410 grid simulator, which simulates a 120V AC system, and an AgileSwitch 100 kW DC-AC inverter operating with a switching frequency of 5 kHz. The control algorithm is implemented and run on a dSPACE DS1202 real-time interface (RTI) platform. A

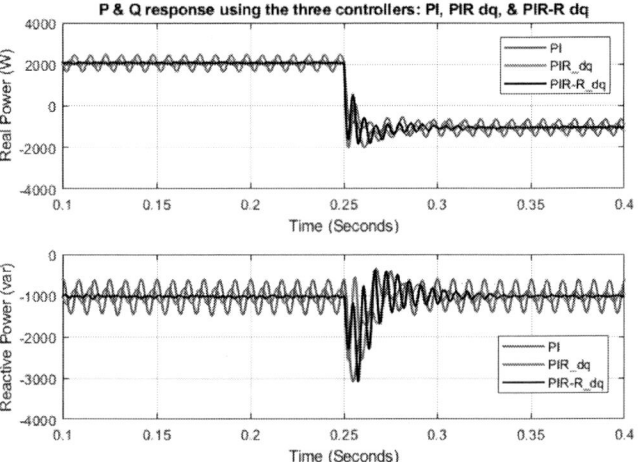

Fig. 4. Reference Active and Reactive Power Tracking.

NHR 9210 battery test system provides the DC power source, with the DC link voltage maintained at 400 V. The hardware experiments were conducted using the same settings as the previous simulation section. The test system setup is shown in Fig. 5.

The main focus of the hardware experiments is active and reactive power command tracking. The grid current will also be investigated to check if the proposed PIR-R controller can guarantee a balanced grid current. In the end, the THD of the grid current will be compared to verify that the PIR-R will not generate extra harmonics. The experimental results of three types of controllers are shown below.

Fig. 6 to Fig. 8 show the instantaneous active and reactive power waveform by the dSPACE control desk when the

979-8-3315-1612-3/25 $31.00 © 2025 IEEE

Fig. 5. Test system setup.

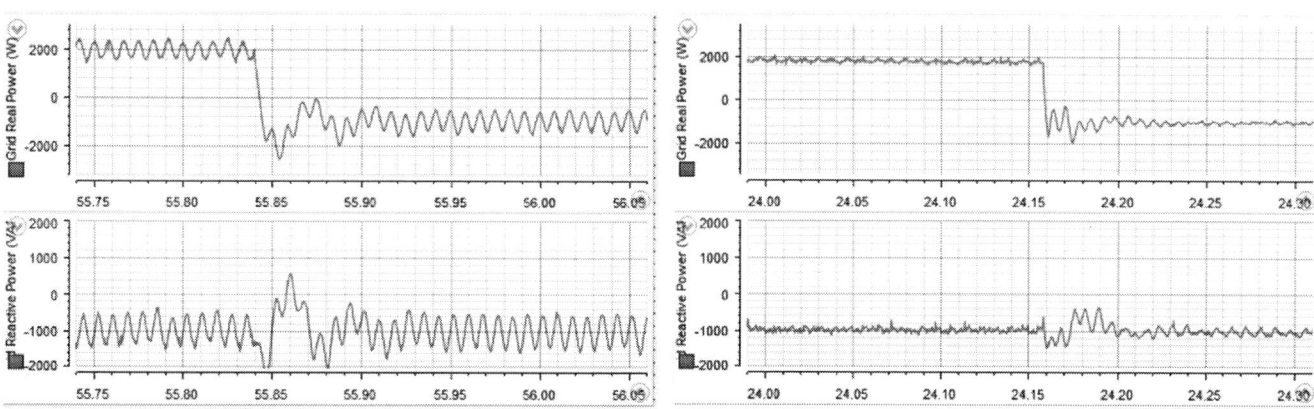

Fig. 6. Reference Power Tracking with PI.

Fig. 8. Reference Power Tracking with PIR-R.

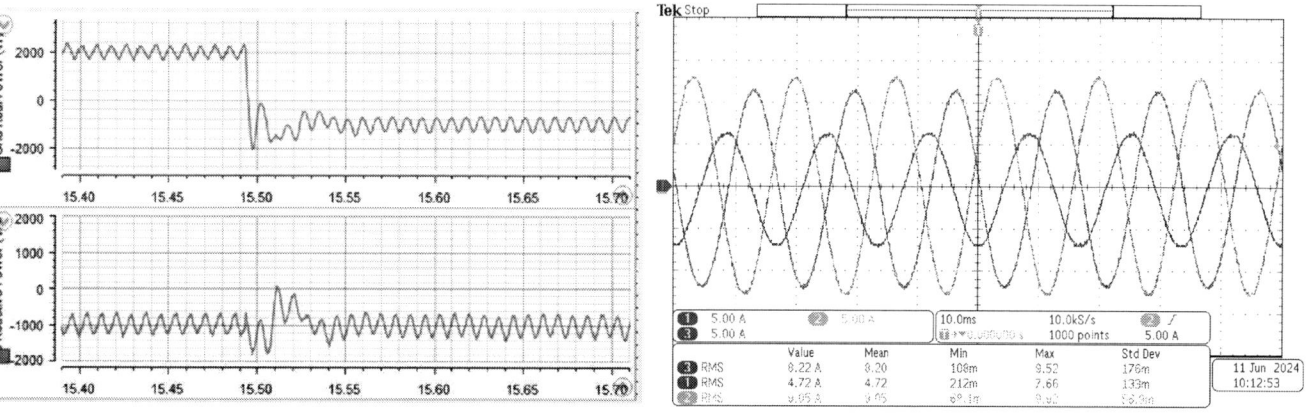

Fig. 7. Reference Power Tracking with PIR.

Fig. 9. Unbalanced Load Current (5A/div).

active power reference changes from 2000 W to 1000 W and the reactive power reference remains at 1000 var. The PIR-R controller achieves the best performance with minimum steady-state error.

Fig. 9 shows the current of a three-phase unbalanced load at the PCC. Fig. 10 to Fig. 12 show the gird currents with different controllers, only the proposed PIR-R controller can

achieve good grid current correction.

Fig. 13 to Fig. 14 show that the THD of the gird-current has a very minimum increment with the proposed PIR-R controller.

The experimental results verify and indicate that the pro-

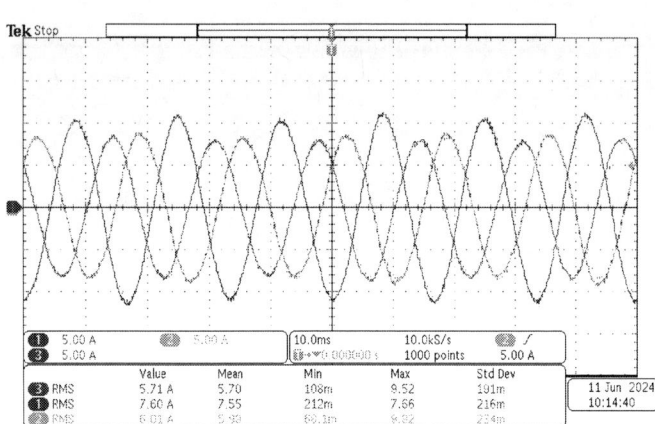

Fig. 10. Grid Current with PI Controller (5A/div).

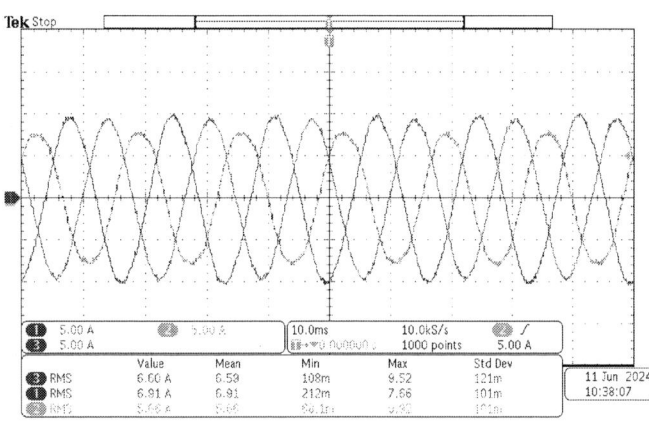

Fig. 11. Grid Current with PIR Controller (5A/div).

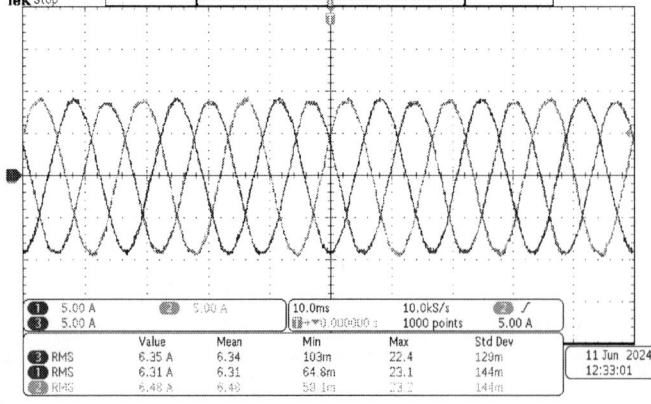

Fig. 12. Grid Current with PIR-R Controller (5A/div).

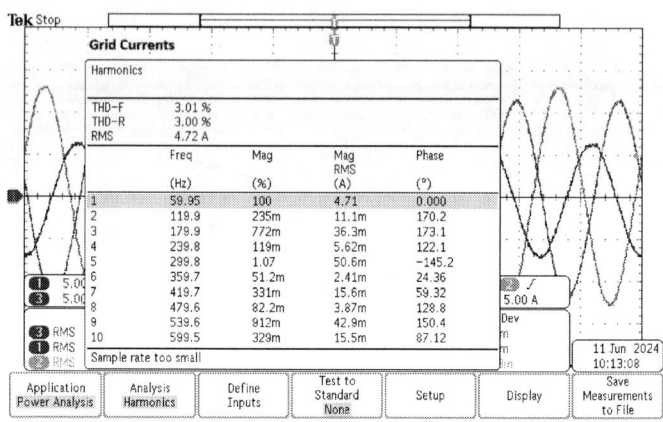

Fig. 13. THD of Unbalanced Load Current (5A/div).

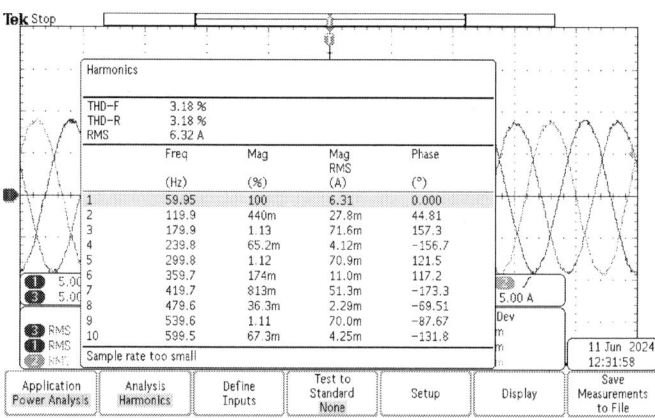

Fig. 14. Grid Current with PIR-R Controller (5A/div).

posed PIR-R controller is a practical and cost-effective solution that can correct the grid current and maintain good power tracking with the unbalanced load connected at the PCC.

IV. CONCLUSIONS

This paper presents an unbalanced grid correction method based on a PIR-R control for a three-phase grid-connected inverter. Compared to previous works, the proposed control method, the inverter can achieve better active and reactive power delivery and simultaneously correct the unbalanced grid current caused by unbalanced loads connected at the PCC. Both simulation and experimental findings validate the effectiveness of the proposed method. The future full paper submission will demonstrate that the proposed method will also apply to non-linear loads and extend with the adaptive frequency control.

979-8-3315-1612-3/25 $31.00 © 2025 IEEE

REFERENCES

[1] X. Guo, Y. Yang, and X. Zhang, "Advanced control of grid-connected current source converter under unbalanced grid voltage conditions," IEEE Trans. Ind. Electron., vol. 65, no. 12, pp. 9225-9233, 2018.

[2] "Renewables 2017 global status report," Global Wind Energy Council (GWEC), Paris: REN21, ISBN 978-3-9818107-6-9, 2018.

[3] Y. Hu, Z. Q. Zhu, and M. Odavic, "An improved method of DC bus voltage pulsation suppression for asymmetric wind power PMSG systems with a compensation unit in parallel," IEEE Trans. Energy Convers., vol. 32, no. 3, pp. 1231-1239, 2017.

[4] D. Zhou, P. Tu, and Y. Tang, "Multivector model predictive power control of three-phase rectifiers with reduced power ripples under nonideal grid conditions," IEEE Trans. Ind. Electron., vol. 65, no. 9, pp. 6850-6859, 2018.

[5] A. Yazdani and R. Iravani, Voltage-sourced converters in power systems: modeling, control, and applications: John Wiley & Sons, 2010.

[6] Wooyoung Choi, C. Morris and B. Sarlioglu, "Modeling three-phase grid-connected inverter system using complex vector in synchronous dq reference frame and analysis on the influence of tuning parameters of synchronous frame PI controller," 2016 IEEE Power and Energy Conference at Illinois (PECI), Urbana, IL, 2016, pp. 1-8.

[7] H. Cai, P. Zhang, H. Zhao, J. Shi, W. Yao and X. He, "Controller design for three-phase inverter with power unbalanced loads applied in microgrids," 2015 IEEE Energy Conversion Congress and Exposition (ECCE), 2015, pp. 4588-4593.

[8] F. Göthner, E. Tedeschi and D. I. Brandao, "Unbalanced Load Compensation by Power-Based Control in the Synchronous Reference Frame," 2019 10th International Conference on Power Electronics and ECCE Asia (ICPE 2019 - ECCE Asia), Busan, Korea (South), 2019, pp. 1383-1388, doi: 10.23919/ICPE2019-ECCEAsia42246.2019.8797154.

[9] M. Alathamneh, H. Ghanayem and R. M. Nelms, "A Robust Three-Phase Shunt Active Power Filter with Frequency Adaptive PR Controller and Sensorless Voltage Control," 2023 IEEE Industry Applications Society Annual Meeting (IAS), Nashville, TN, USA, 2023, pp. 1-8, doi: 10.1109/IAS54024.2023.10406769.

[10] H. Akagi, Y. Kanazawa and A. Nabae, "Instantaneous Reactive Power Compensators Comprising Switching Devices without Energy Storage Components," in IEEE Transactions on Industry Applications, vol. IA-20, no. 3, pp. 625-630, May 1984, doi: 10.1109/TIA.1984.4504460

[11] H. Ghanayem, M. Alathamneh and R. M. Nelms, "Adaptive PIR Current Controllers for Torque Ripple Reduction of PMSM using Decoupled Speed and Flux Control," 2023 IEEE 14th Annual Ubiquitous Computing, Electronics & Mobile Communication Conference (UEMCON), New York, NY, USA, 2023, pp. 0440-0444, doi: 10.1109/UEMCON59035.2023.10316059.

AUTHOR INDEX

Abarzadeh, Mostafa 1261
Abbas, Asad 2973
Abotaleb, Youssef 1850
Abrams, Kerry J. 1781
Abramson, Rose A. 291, 2805
Abu-Rub, Omar 3071
Abu-Zaher, Mustafa 2327
Acero, Jesús 2468
Addin, Ali Sharaf 2960
Adeli, Mohammad Hassan 1489
Ademane, Harsha 3133
Adisurja, Ananda Tjakra 1255
Adragna, Claudio 958
Afrasiabi, Seyedeh Nazanin 1279
Afridi, Khurram K. 1640, 1646
Agarwal, Anant 2986
Ahammed, Md Tanvir 2220
Ahmad, Faheem 175
Aider, Youssef 1026
Aiello, Natale 738
Ajmal, Aidha Muhammad 3024
Akamatsu, Keiji 1728
Akter, Tanzila 1844, 2407
Akuta, Hector 761
Alam, Md Didarul 1746
Alam, Muhammad Muneeb 1051, 2569
Alassi, Abdulrahman 3071
Alathamneh, Mohammad 1953
Al-Durra, Ahmed 622, 2871, 3064
Alenezi, Ali 1217
Alexander, Mark 2162
Aleyasin, Seyed Hossein 1408
Ali, Abdelrahman 429
Ali, Jana A. Sheikh 3071
Ali, Kawsar 1529
Alkhatib, Mohamed 1940
Allen, Mark G. 1791
Allgeier, Jan 1919
Allioua, Abdelmoumin 2125
Alou, Pedro 1197
Al-Smadi, Mohammad K. 2779, 2840
Altin, Necmi 1489
Álvarez, Ignacio 3109
Aly, Mokhtar 746, 895, 2290, 2327
Alzahrani, Ahmad 1230
Alzate, Cesar 401
Amano, Yoshiki 3096
Amarathunga, Supun 3030

Amirabadi, Mahshid 1465, 1983
Amitkumar, K. S. 1279
Amler, Adrian 1759, 1767
Amor, Yacine Ayachi 1781
An, Jongchan 3000
Anand, Aniket 1096, 3147
Anand, Sandeep 69
Anantha, Neeraj 1121
Andapally, Bharadwaj Reddy 3119
Andersen, Michael A. E. 246
Anderson, Blake 1850
Ando, Masato 2681
Anekal, Latha 1224
Anjum, Waseah 1148
Antoszczuk, Pablo Daniel 479
Anurag, Anup 9, 442, 1318
Ao, Chengkang 171
Arai, Takamasa 2821
Araki, Hideo 3077
Aravind, G. 1610, 2785
Arduini, Douglas 1159
Asadi, Peyman 2162
Asel, Thaddeus J. 2419
Ashikaga, Toru 2284
Asllani, Besar 2051
Atkinson, Joshua 401
Attanasio, Rosario 3133, 3304
Attukadavil, Jenson Joseph C. 1481
Atwimah, Samuel K. 185, 207
Aunsborg, Thore Stig 175
Avenas, Yvan 1396, 2562
Aygun, Deniz 195
Azzopardi, Stéphane 2718
Bader, Samuel 1681
Bae, Jung-Soo 2228
Bae, Youngmin 3000
Baek, Jaeil 491
Bagci, F. Selin 880
Bahrami-Fard, Milad 930
Bak, Yeongsu 1734
Bakhshai, Alireza 3083
Balakrishnan, Manu 1286
Balamurali, Aiswarya 1096
Balda, Juan C. 27
Balen, Gleisson 1935
Balutto, Mattia 479
Banaie, Amin 1184
Banerjee, Arijit 3089

Bansal, Divyanshu	1610, 2785
Bantemits, Georgios	479
Bao, Mingjun	2628, 2968
Bao, Xiaokun	1143
Barbosa, Peter M.	2002
Barbosa, Peter	9, 442, 1318, 2082, 2296
Barik, Tapas	1184
Barros, Stayner Nóbrega	689
Barzegarkhoo, Reza	90
Basu, Arka	3253
Basu, Shibaji	464
Batard, Christophe	1076
Bau, Plinio	195
Bauer, Pavol	609
Bavi, Danial	385
Bazzi, Ali	2332, 2510
Beckemeyer, Randy	2082
Beig, Abdul R.	2647
Beinarys, Rytis	2009
Belanger, Matthew	2833
Belikov, Juri	1622
Belkhode, Satish	164, 3334
Benson, Mikayla	2413
Bergveld, H.J.	1451
Bertolini, Alessandro	2640
Beura, Kalpana	1940
Beushausen, Steffen	2589
Bezerra, Pedro A.M.	2361
Bhagat, Chinmay	3285, 3291
Bhambay, Rajul	2920
Bhattacharya, Subhashish	370, 552, 1347, 1866
Bhuse, Tejas	1, 54
Biadene, D.	2014
Bian, Fengwei	3312
Bien, Franklin	1629
Blaabjerg, Frede	696, 912, 1501
Blanco, Cristian	1935
Blaquière, Jean-Marc	2718
Blij, Nils Hans Van Der	479
Boby, Mathews	1279
Boisseau, Sébastien	828
Boisson, Guillaume Piquet	1396
Bojoi, Radu	472, 1408
Bolaños, Robert E.	1190
Boles, Jessica D.	1012
Bonanno, Giovanni	1666
Borowy, Bogdan S.	1326
Boroyevich, Dushan	2228
Bosch, Michael	2387
Boutet, Jérôme	828
Bracken, Christopher	544, 3119
Bradford, Paul	2393
Brandão, Danilo I.	1355
Brandão, Dener A. de L.	1355
Briz, P.	147
Briz, Pablo	3109
Brown, Alyssa	231
Brown III, Buck F.	1153
Brückner, Thomas	2960
Bruyere, Paul	2562
Bu, Jiankang	854
Bugade, Vikas	821
Burdío, José-Miguel	2468
Burgos, Rolando	111, 409, 1495, 1551, 2992
Burnett, Hunter	401
Burt, Graeme	3167
Buttay, Cyril	2051
Cairnie, Mark	2228, 2616
Calabretta, Michele	1070
Cammarata, Federica	252
Campbell, Steven	2797
Cao, Hanqing	1810
Cao, Hui	27
Cao, Yue	3036, 3048
Carretero, Claudio	2468
Castro, Alejandro	1427
Catanoso, Matthew	1791
Cattani, Alberto	2640
Cazzaniga, Daniele	958
Cazzitti, Sacha J.	1512
Cerutti, Stefano	738
Cervera, Pedro Alou	788
Cervone, Andrea	1305
Chae, Jongyoon	1899
Chagas, Rafael Bogo Portal	2446
Chakkalakkal, Sreejith	3147
Chakraborty, Shiladri	69
Chambon, Clément	828
Chamorro, Luis Ruiz	788
Chandrasekhar, Nurani	2889
Chang, Che-Wei	1495, 1551, 1564
Chang, Chuan-En	664
Chang, Jun-Yang	16
Chang, Yi-Chun	2143
Chareyron, Mathilde	828
Chatterjee, Bhaskar	1919
Chatterjee, Kallol	821
Chaturvedi, Shivam	2624, 3155
Chaturvedi, V.	1451
Chaudhary, Jai Aditya	3304
Chavarria, Jose	491
Chavez, Fredo	385
Cheema, Muhammad Ali Masood	768
Chellamuthu, Anand	1286

Chen, Cai 2343, 2369, 2426
Chen, Ching-Jan 664, 2131, 2143, 2725, 2735, 2741
Chen, Chun-Yen...938
Chen, Eric 1274
Chen, Guozhu 2059
Chen, Hao 906
Chen, Hongyu... 2968
Chen, Hua 518
Chen, Hung-Chi.. 2687
Chen, Jiahong 1114
Chen, Jiann-Fuh... 2043
Chen, Kai-Hui..887
Chen, Kevin J. 1047
Chen, Minjie 139, 349, 510, 566, 1274, 1693, 1882, 2438
Chen, Qiling.. 2375
Chen, Shih-Gang...938
Chen, Tianxiao.. 2361
Chen, Ting 2846
Chen, Wanjun... 192
Chen, Xi... 2628, 2968
Chen, Xingyu 1537, 1741
Chen, Yilun... 2535
Cheng, Eric Ka-Wai.. 3227
Cheng, Jinpeng.. 1832
Cheng, Kuang-Yao... 2157
Cheng, Lin 966, 1687
Cheng, Qi.. 2236
Cheng, Tzu-Ping.. 2900
Cheng, Yan.. 1047
Cheng, Yun-Keng..887
Cheshire, Audrey 682
Chetri, Chandan.. 3114
Chiu, Huang-Jen..389
Chiu, Jui-Yang 16, 900
Cho, Jaeyong 3187
Choi, Beomseok..491
Choi, Dongho.. 2311
Choi, Dongmin 1899
Choi, Jinsoo 3187
Choi, Jungwon 1874
Choi, Seokwon.. 2268
Choi, Seungdeog........................... 943, 951, 1026, 1420, 1858
Choi, Sunghyuk 1659
Choi, Sungjin.. 3006
Choksi, Kushan.. 2582
Choo, Vin Loong.. 1576
Choong, Yin Quen 505
Chowdhury, Vikram Roy 645, 761, 1465, 3059
Chuang, Cheng-Ta 664, 2725
Chung, Henry Shu-Hung 98, 1507, 1582
Ciabattoni, Matteo 1646
Ciardo, S. Yuri 252

Clark, Landon... 919
Cobos, Álvaro 1427
Cobos, José A. 1427
Coday, Samantha...971, 2249
Collings, William M.. 185
Cong, Yizhou...2986
Contreras-Barrios, René.. 629
Coomans, Bart.. 195
Corradini, Luca 1, 54, 334, 2764
Costa, Levy F. 1334, 1341
Costinett, Daniel...3253, 3267
Cox, James 538
Cronin, Jared 2865
Croston, José Andrés Aguilar 2051
Crovetti, Paolo Stefano 738
Cruz, Alfonso 860
Cruz, Mario F. 670
Cui, Hongchang 202
Cui, Wen Tao..1108
Cui, Yujia..2932
Curbow, Austin..1167
D'Amato, Davide 689
Da-Cunha-Alves, Wendell 429
Dai, Hang 3174
Dang, Yongliang..278, 2482
Dannehl, Kai..1774
Dardeer, Mostafa 906
Darvish, Peyman..2453
Das, Shuvangkar Chandra 1184
Datta, Kishalay 1715
Datta, Promit 586
Davari, S. Alireza..2290
De, Vivek 518, 1681
Deboi, Brian..1167
Deboy, Gerald 1444, 2260
Defaz, Samuel 2576, 2582
Dekka, Apparao...2647
Delmar, Aria...1242
Deneke, Niklas 848
Deng, Jianting...3312
Deniz, Erkan..1489
Deppe, Conner 2393
Derbey, Alexis..2562
Desai, Nachiket...1681
Descamps, Anne-Sophie..1076
Deshpande, Ankit Vivek...1459
Dev, Archit...2920
DeVoto, Douglas 1824
Diao, Naizhe 757
Dieckerhoff, Sibylle..2361
DiMarino, Christina 586, 1836, 2228, 2616
Ding, Peiyang...2375

Ding, Wenlong 2713
Divan, Deepak 164, 3334
Do, Huong 491, 1681
Dobakhshari, Sina Salehi 1673
Dominguez, Miguel Alvarez 1640
Dong, Minhai 2075
Dong, ...111, 1495, 1551, 1564, 2992
Driesen, Johan 3124
Driussi, Francesco..479
Dryden, Daniel M. 2419
Du, Bangli 436, 2752
Duan, Bin..2713
Dujic, Drazen 266, 1063, 1305
Dutta, Soham 711
Dworakowski, Piotr 2051
Eguchi, Shinichiro 2828
Ekuewa, Oluwaseun Isaiah 2973
Elasser, Youssef 510, 566, 2438
Elezab, Ahmed 670
El-Fouly, Tarek H.M. 622
Ellis, Nathan M. 2276
Ellis, Philip 1781
El-Refaie, Ayman M. 1551
El-Refaie, Ayman 1230, 1495
El-Saadany, Ehab F. 622
Elsanabary, Ahmed 746
Elshaer, Mohamed 3155
Emadi, Ali... 670, 3147
Endo, Shun..2681
Eni, Emanuel 2746
Enjeti, Prasad 727, 1217, 1459, 3054
Enomoto, Jun 3194
Enslin, Johan...1153
Espinar, Alberto 3100
Espinoza, Angel 214
Estrin, Julia 132
Etta, Dheeraj 1640, 1646
Evzelman, Michael 594
Expósito, Alberto Delgado...................... 788, 1803
Fahimi, Babak.. 930, 3160
Fahmy, Youssef A. 272
Falkenberg, Niklas 2772
Fan, Junchong.. 1203
Fan, Yucheng..2981
Farantatos, Evangelos 1184
Farivar, Glen G. 1927
Fassi, Youssof 828
Fein, Martin 2348
Feng, Hao 1832
Feng, Kaiyuan..2894
Feng, Wenda 3174
Fernandes, Arnold... 1311

Fernandes, Baylon G. 1481
Ferrari, Maximiliano 637
Figueroa, Alejandro...1427
Filho, Braz de J.C..................................... 1355, 1615
Fiore, Michele..1070
Flannery, John 285
Flaten, Paul 682
Forouzesh, Mojtaba.................................. 1673, 1892
Forsyth, Andrew J. 1512
Foster, Geoffrey M. 207
Fox, Aidan P. 185
Fox, Matthew 1791
Francés, Airán 868, 3298
Francois, Thomas W..1311
Frank, Simon 2348
Freeman, Andrew 1242
Fu, Minfan 809, 2846, 3206
Fu, Pengyu 1203, 2986
Fujisaki, Keisuke...1797
Fujita, Jun 1383
Funaki, Tsuyoshi 2813
Funatani, Kenji 2654
Funatsu, Shohei 1237
Furukawa, Akihiko 1383
Gaafar, Mahmoud A. 775, 906, 2327
Gajare, Siddhesh 214
Galamb, Andrew 2527
Gallage, Nirashi Polwaththa 874
Gangadhar, Pratheesh 2920
Gao, Alex 2149
Gao, Ju 171, 225
Gao, Mingze 2992
Gao, Xiang 2846
Gao, Xiaoguang 2070
Gao, Yuan 524, 1034
Gao, Yuntian..278, 2482
Garcia, Enrique 538
García, Pablo..1935
Garcia, Ricardo 214
García, Sofía 3298
García-Espinosa, Antoni....................................1774
Garza-Arias, Enrique..1459
Gasparini, Alessandro 2640
Gato, Jose..3119
Gautam, Sushanta.. 185
Gauthier, Jean-Yves..2051
Gauttam, Gaureej 3316
Geboers, Tim 436
Gennaro, Francesco 738
Georgescu, Sorin 180
Georgiev, Daniel G..................................185, 207
Gessner, Joerg 1889

Ghanayem, Haneen 1953
Ghartemani, Masoud Karimi 943
Ghitelman, Kolman Puterman 2101
Ghosh, Mohendro Kumar 1326
Ghosh, Prosenjit 2541
Ghosh, Subarto Kumar 1420
Giardine, Francesca 151
Gil, Pablo M. 1701
Ginot, Nicolas 1076
Giuffrida, Simone 472
Gockel, Hendrik 2125
Goetz, Stefan M. 1754, 2846, 3206
Goicoechea, Javier 1427
Gomez-Rivera, Luis F. 1774
Gong, Jiakun 219
Gong, Minxiang 518
Gong, Taehyeon 3006
Gong, Xiaowu 1114
Gonzalez, Reynaldo S. 1190
Gonzalez-Castaño, Catalina 629
Goodrich, Dakota 719
Goto, Akiko 1569
Gouy, Louison 1076
Graber, Lukas 860, 3321
Grainger, Brandon 544, 1326
Green, Andrew J. 2419
Griepentrog, Gerd 2125
Grigoryan, Davit 566, 1882, 2438
Groon, Fabian 90
Guan, Quanxue 895
Guenther, Robert 1203
Guichon, Jean-Michel 2562
Guillod, Thomas 1816
Gunawardena, Pasan 3030
Guo, Heng 2713
Guo, Jiacheng 2375
Guo, Weisheng 3181
Guo, Xiaoqiang 757
Guo, Zhengchen 2703
Guo, Zhongyin 2070
Gurudiwan, Shubhangi 719, 2194
Guthrie, Travis 2162
Gutierrez, Harold 1159
Ha, Jung-Ik 457, 1659, 2268, 2937
Habibolahi, Zahra Sadat 2202
Haddadi, Aboutaleb 1184
Hajisadeghian, Hossein 1666
Halawa, Ali 1473
Hamani, Rachid 1889
Hameed, Aamna Nasir 1673
Hameed, Asad 1972
Han, Yi 103

Hanhart, Michael 2757
Hanna, Rachelle 1396
Hansen, Frederik Lillebæk 2380
Hanson, Alex J. 231, 1121, 2521
Hanson, Alex 77, 2857
Hao, Weijia 2109
Harbi, Ibrahim 895, 2290
Haryani, Nidhi 442
Hasan, Abu Shahir Md Khalid 1844, 2407
Hasan, Md Zakir 1026
Hasan, Syed Imam 1294, 2698
Hassan, Alaaeldien 2327
Hassan, Najam Ul 834
Hassan, Nazmul 1746
Hata, Katsuhiro 1084, 1102, 2284, 2551
Hayashi, Tetsuya 423
He, Bill 3129
He, Binghui 1673
He, JiangBiao 919, 1368
He, Jiayin 171, 225
He, Junlei 3129
He, Xinlong 2066
Heckel, Thomas 1759
Hedenik, Marina 1519
Hedeshi, Hamid Montazeri 2202
Hegde, Anantha 1728
Heinen, Stefan 2757
Heiries, Vincent 828
Heldwein, Marcelo Lobo 2446
Hemming, Samuel 670
Heo, Go Woon 1723
Herbert, Edward 2495
Hernandez, Arturo Sanchez 530
Herzer, Stefan 1286
Higashiyama, Koji 1728
Hiller, Marc 1919, 2348
Hiraki, Eiji 321, 2654
Hiraoka, Toshio 285
Hirase, Yuko 1946
Hisamochi, Hirofumi 1414
Hobart, Karl D. 185, 207
Hoene, Eckart 2361
Hokmabad, Hossein Nourollahi 1622
Hong, Kang 2096
Hontz, Micheal R. 207
Horibe, Masahiro 2821
Hornbuckle, Malachi 363, 2241
Horowitz, Logan 151, 2276
Hosani, Khalifa Al 1940, 2871, 3064
Hossain, Md Maksudul 2407
Hossain, Mohammad Safayet 1184
Hou, Ting 2375

Hou, Zhengming	1913, 2851
Houska, Brad	3334
Howell, Brandon	2162
Hsieh, Chun-Yu	2735
Hsieh, Hsin-Che	815
Hsu, Jun-Ming	938
Hu, Borong	1439, 2597
Hu, Changsheng	2894
Hu, Changyu	3129
Hu, Jhih-Cheng	2692
Hu, Jiangang	2932
Hu, Shoudong	2764
Huang, Alex Q.	1786
Huang, Cheng	505, 1946
Huang, Hao-Ran	664, 2131
Huang, Ming-Shi	938, 2692
Huang, PengHao	1217
Huang, Peng-Hao	727
Huang, Qinghui	1173, 2603
Huber, J.	2014
Huber, Jonas	1318
Huber, Laszlo	442
Hudgins, Jerry L.	2877
Hudgins, Jerry	1850
Huh, Kum-Kang	3174
Hui, Shu Yuen Ron	3275
Hung, Chien-Chih	16, 900
Hung, Yu-Ting	2735
Huo, Zhenguo	2660
Husain, Iqbal	1746
Husev, Oleksandr	1622, 2173
Hussain, Amir	1990
Hwang, Yun Seong	733
Iannuzzo, Francesco	738, 1070
Ibáñez-Muñoz, Esteban	629
Ibrahim, Ahmed	775
Ibrahim, Eltaib Abdeen D.	775
Ibrahim, Hasan	727, 3054
Ibrahim, Mohamed	670
Ide, Tomoya	1946
Ikriannikov, Alexandr	2149
Iliæ, Milan	2764
Ilka, Reza	1368
Imaeda, Yuta	2431
Imaoka, Jun	2431
Imperiali, Luc	1318
Inokuchi, Seiichiro	2356
Inoue, Shuntaro	782
Irie, Yusuke	2828
Ishido, Ryosuke	1797
Ishihara, Masataka	321, 2654
Ishikura, Yuki	3285, 3291

Ishizuka, Yoichi	2828
Ishraq, Naveed	34, 1135
Islam, Md Khurshedul	943, 951
Islam, Md Majharul	2407
Islam, Nasherul	2059
Islam, Sarwar	1824
Ismail, Ahmed H.	2453
Isobe, Takanori	1946
Ito, Yuki	3248
Itoh, Jun-Ichi	21, 48, 2913
Ivimey, Arjun	464
Iwabuchi, Akio	2828
Iwamoto, Motomitsu	1108
Iyer, Rahul K.	157
Iyer, Vignesh	2764
Jacobs, Alan G.	207
Jafarian, Yousefreza	3083
Jahns, Thomas	3174
Jain, Akshat	658
Jain, Praveen	464, 616, 2953, 3083
Jalakas, Tanel	1622
Jalalabadi, Esmaeil	416
Janabi, Ameer	2597
Jayalath, Sampath	3212
Jeong, Seogyong	834
Jeong, Won Hyo	2937
Jerez, Raiphy	2249
Jha, Kunal	1519
Ji, Shengchang	278, 2482
Ji, Shiqi	2857
Ji, Yichao	966, 1687
Ji, Yingfeng	2889
Jia, Xiaoting	1564
Jiang, C.Q.	795, 3181
Jiang, W.L.	1451
Jiang, Wei	2343, 2426
Jiang, Xi	1114
Jiang, Yang	978
Jiang, Yongbin	3220
Jiao, Dong	1913, 2851
Jiao, Yang	416
Jin, Feng	429
Jin, Liyang	1020, 1564
Jin, Sicong	3018
Jin, Zhiyang	860
Jing, Mengmeng	2713
Jo, Hyeonu	3200
Jo, Hyunkyeong	1629
Jochmans, Thomas	258
Johnson, Brian	711
Johnson, Ken	2510
Jørgensen, Asger Bjørn	357, 1034

Jørgensen, Jannick Kjær 175, 357
Joshi, Kishor 943
Juds, Mark A. 1326, 3119
Jung, Jee-Hoon 834
Jung, Jun-Hyung 689
Jurkov, Alexander 124, 132
Kabashima, Takamune 1728
Kachura, Avram 449, 1905
Kai, Toshihiro 423
Kalathy, Abirami 616, 2953
Kallfass, Ingmar 2387, 3241
Kamalapur, Aakash 2228
Kamran, ... 252
Kanakri, Haitham 2029
Kanathipan, Kajanan 768
Kandeel, Youssef 285
Kang, Byeong-Woo 2948
Kang, Doug. 180
Kang, Eunjin 3012
Kang, Gyeong-Gu 566, 2438
Kang, Seung Hyun 733
Kang, Yong 2066, 2313, 2369, 2426
Kano, Yuko 782
Kanungo, Gautam Dey 821
Kar, Narayan C. 1096
Karanth, Shashank 2746
Karimi-Ghartemani, Masoud 1858
Kataoka, Soya 1237
Katsura, Kenshiro 1299
Kaufmann, Maik 1286
Kawahara, Chihiro 2356
Kawamoto, Keisuke 1569
Kawano, Akihiro 1977
Kelkar, Kapil. 1519
Kennel, Ralph 895
Kerekes, Tamás 738, 3042
Khaburi, Davood Arab 895
Khadka, Purushottam 1040, 2400
Khalid, Saad. 2569
Khalife, Khalil 479
Khan, Faisal 1824
Khan, N. 1451
Khan, Nisar Ahmed 2569
Khan, Shahid Aziz 2624, 3155
Khandelwal, Sourabh 385
Khandla, Dhaval 2920
Khanna, Mudit 854
Khanna, Raghav 185, 207
Khatua, Mausamjeet 1681
Khorasani, Ramin Rahimzadeh 2101
Kim, Byeong-Il 1734
Kim, Chae-Lyn 3200

Kim, Daehyun 3187
Kim, Dong Hwan 1723
Kim, Dongmin 1899
Kim, Han-Gyu 951
Kim, Hongrae 1746
Kim, Hyeon Soo 733
Kim, Jae-Seong 925
Kim, Jaewon 727, 3054
Kim, Jeonghun 761
Kim, Jonghoon 2973, 3000, 3006, 3012
Kim, Jong-Hun 834
Kim, Joon-Seok 1734
Kim, Jungho 1629
Kim, Katherine A. 880
Kim, Minhyeok 3012
Kim, Min-Sik 834
Kim, Myeong-Ho 834
Kim, Namwon 703
Kim, Sung-Oh 2943
Kim, Yura 3006
Kimball, Jonathan W. 1311
Kimpara, Renata 703
Kirtley, James L. 2474
Kisacikoglu, Mithat John 1602
Kishikawa, Ryoko 2821
Kishimoto, Sumiaki 285
Kitano, Junichi 3194
Klidbari, Mohammadreza Khodaparast 2202
Klymenko, Mariia 1590
Knapp, Jeffrey 854
Knappstein, Lukas 2772
Knoll, Jack 2228, 2616
Ko, Bomyeong 3006
Kobayashi, Takumi 3248
Koch, Dominik 2387
Koehler, Andrew D. 185, 207
Koga, Shunsaku 3194
Koga, Takahiro 2828
Kokkonda, Raj Kumar 1347
Kolar, J.W. 2014
Kolar, Johann W. 1318
Kolli, Nithin 1347
Komiyama, Yutaro 3248
Komo, Hideo 2356
Kondo, Hiroki 1102
Kondo, Ryota 2813
Kong, Jiaze 2167
Kong, Jie 2380
Kong, Rui 696
Konishi, Akihiro 3248
Koppolu, Manoj 2920
Korrani, Majid Ghasemi 930, 3160

Kosaka, Takashi	1237, 3096
Koseoglou, Sokratis	479
Kotani, Junichi	579
Kouro, Samir	775, 2327
Kozak, Joseph P.	1211
Kozielski, Kyle	3147
Kragl, Robert	1051
Krishnamoorthy, Harish S.	3316
Krishnamurthy, Harish	1681
Krishnamurthy, Karthik	3129
Krishnan, Sahana	151, 291, 2805
Kritprajun, Paychuda	1184
Ku, Han	900
Kubulus, Pawel Piotr	1034
Kularatna, Nihal	378, 874
Kularatna-Abeywardana, Dulsha	874
Kulasekaran, Siddharth	491
Kumar, Misha	2002, 2082
Kumar, Pavan	530, 3312
Kusaka, Keisuke	3261
Kusunoki, Shigeru	2551
Kutrolli, Uiliam	2332
Kwak, Jin Woong	2541
Kwon, Hyukjae	566, 2438
Kwon, Man Jae	733
Ladhar, Manraj Singh	2322
Laha, Arpan	616, 2953
Lahuerta, Óscar	2468
Lai, Jih-Sheng	815, 1058, 1913, 2851
Lai, Rixin	2138
Lai, Yanwen	1173, 2603
Laird, Ian	2088
Lam, John	768, 2022
Lamar, Diego G.	1701, 1959
Lawniczak, Celine	1129
Lawson, Wayne	1403, 1781
Lazzarin, Telles Brunelli	342
Le, Duc Dung	3155
Le, DucDung	2624
Le, Hoang	2647
Le, Thanh-Long	2718
Leary, Alex M.	2516
Lee, Bonyoung	1629
Lee, Byoung Kuk	733, 1723, 3200
Lee, Byunghun	834
Lee, Chen-Chan	1058
Lee, Dongcheol	3000
Lee, Dong-Choon	1267
Lee, Dongsu	457
Lee, Eun Woo	2311
Lee, Hoi	2236
Lee, Jaea	3012

Lee, Jaehyeong	3006
Lee, Ju-A	3200
Lee, Jun Young	2311
Lee, June-Seok	1734, 2311
Lee, Justin	2138
Lee, Juwon	457
Lee, Kahyun	2937
Lee, Kangbeen	2413
Lee, Kevin	327, 1261, 2907
Lee, Kyo-Beum	925, 2317, 2943, 2948
Lee, Kyungmin	2547, 3281
Lee, Miyoung	3000
Lee, Po-Chang	900
Lee, Seongkyu	3012
Lee, Seunghyun	3012
Lee, Sungjun	3006
Lee, Taewoo	1659
Lee, Ting-Lun	2143
Lee, Wen-Hsuan	2043
Lee, Woongkul	1473, 2413
Lee, Yun-Jin	2317
Lehman, Brad	761, 1465, 1983
Lehmeier, Thomas	1767
Lei, Weihao	1143
Lei, Yiming	3312
Leslie, Alec	401
Leyrer, Thomas	2920
Li, Bing	2932
Li, Chun-I	2741
Li, Duo	307
Li, Haoran	510, 566, 1882, 2438
Li, Heyuan	809, 2846, 3206
Li, Hui	1248, 2075
Li, Jiajun	1590
Li, Lingyun	524
Li, Peidong	3036, 3048
Li, Pengwei	2332
Li, Qiang	202, 299, 429, 498, 1433, 1537, 1557, 1741, 2228, 2488
Li, Ruqi	1159
Li, Sichao	3129
Li, Tien-Sheng	111
Li, Xiang	3312
Li, Xiaoling	1824
Li, Xindong	3212
Li, Xinze	1143
Li, Xuewen	751
Li, Yang	2576
Li, Yanqiao	1590
Li, Yaohua	3220
Li, Yi	1153
Li, Yilei	2035

Li, Yiming 1173
Li, Yuan 761, 1465
Li, Yunwei 3030
Li, Zehui 485
Li, Zhenchao 1305
Lian, Zhina 1090
Liang, Gaowen 1927
Liang, Jingyuan 1108
Liang, Katherine 363, 2241
Liang, Tsorng-Juu 887
Liang, Yaogan 1084
Liao, Hong-Xuan 2692
Liao, Hsuan 2043
Liao, Kuo Fu 2043
Liao, Mian 139, 349, 1882
Libbos, Elie 3089
Lim, Gyu Cheol 2937
Lim, Je-Yeong 1723
Lim, Jong-Hun 1723
Lin, Fanfan 1143
Lin, Jesse 1211
Lin, Jinshu 2075
Lin, Lei 2535
Lin, Qing 409
Lin, River 1159
Lin, Wei-Ren 258
Linares, Daniel Ríos 1375
Liserre, Marco 90, 118, 689, 1148
Liske, Andreas 2348
Liu, Baihan 2343, 2426
Liu, Caifeng 2066
Liu, Chen 3042
Liu, Chien-Lung 2692
Liu, Ching-Yao 1058
Liu, Christopher 1403
Liu, Chun-Hung 1026
Liu, Gao 357, 1034
Liu, Hanbing 3232
Liu, Haoyang 2361
Liu, Hong 2634
Liu, Hongru 2675
Liu, Hualong 1363, 1597
Liu, Jia 751
Liu, Jiahong 3042
Liu, Jiaxin 2343, 2369, 2426
Liu, Jinjie 1114
Liu, Jinjun 751
Liu, Kevin 1274
Liu, Liming 1746
Liu, Ming 2521
Liu, Sijia 2369, 2426
Liu, Wen-Chin B. 315

Liu, Wentao 1090
Liu, Xiaosen 1544, 2556, 2675
Liu, Xiaoshan 429
Liu, Y. 1451
Liu, Yan-Fei 1673, 1892
Liu, Yang 2675, 3321
Liu, Yifu 3129
Liu, Yongjie 1501, 3042
Liu, Yu-Chen 2179
Liu, Yunting 1179
Liu, Zeguo 1687
Liu, Zengyang 2488
Liu, Zhan 2521
Liu, Zhanlei 278, 2482
Liu, Ziheng 171, 225
Locher, Fabrice 2495
Locke, William 3141
Lodge, Finlay 3167
Logi, Sean 880
Long, Haihong 651, 2981
Long, Teng 1439, 2597
Loparo, Kenneth A. 2698
Lope, Ignacio 2468
López, Abraham 1959
Lopez-Torres, Carlos 1774
Lu, Che-Yu 1967, 2900
Lu, Fengwang 98
Lu, Guo-Quan 586, 2228
Lu, Lucas 416
Lu, Mowei 1754, 2846, 3206
Lu, Wei 2117
Luan, Shaokang 1034
Lucía, O. 147
Lucía, Óscar 3109
Luckett, Benjamin 919
Luise, Claudio 2640
Lukic, Srdjan 2527
Lumod, Phen 1159
Luo, Fang 2576, 2582
Luo, Tianming 2035
Lv, Jianwei 2343, 2369, 2426
Ma, D. Brian 2541
Ma, Dingkun 2375
Ma, Guangji 2070
Ma, Hangxiao 978
Ma, Tianlu 3181
Ma, Zhedong 1173, 2603
Ma, Zhiyuan 1786
Ma, Zhuxuan 27
Maaz, Syed Mohammad 1267
Mabuchi, Yuuichi 2681
MacFadyen, Martin 3167

Madadi, Mehrnaz 370
Maddela, Avinash 1715
Maekawa, Sari 2924
Mahbub, S. Tahmid 157
Maheshwari, Anuj 3089
Maji, Sounak 1640
Major, Joshua 1824
Mak, Pui-In 978
Maksimoviæ, Dragan 1, 54, 334, 682, 2764
Malannino, Claudia 252
Mallik, Ayan 34, 1135
Mallik, Ranajay 658
Mandrile, Fabio 472
Manjrekar, Madhav 2883
Mannan, Tahmid Ibne 1420, 1858
Manos, Konstantinos 1274, 1693
Mansour, Mahmoud 719
Mantooth, H. Alan 1844, 2407
Manzoni, Stefano 958
Marcault, Emmanuel 1396
Marellapudi, Aniruddh 164, 3334
Marianne, Julien 828
Marin, Brandon 491
Marquardt, Rainer 2960
Martin, Alexander 1211
Martin, Sébastien 828
Martin, Trent 1, 54
Martinez, Wilmar 238, 258, 436, 2167, 2752, 3124
Martinez-Limia, Alberto 1051
Martins, João R.R.O. 1889
Martins, Rui P. 978
März, Martin 1759, 1767
Mather, Barry 645, 3059
Mathieu, Frédéric 2495
Mathúna, Cian Ó. 285
Matiushkin, Oleksandr 1622, 2173
Matsumori, Hiroaki 1237, 3096
Matsumoto, Hirokazu 579
Matsumoto, Yohei 2681
Matsuo, Takayoshi 2932
Mattavelli, P. 2014
Mattavelli, Paolo 2667
Maureira-Riquelme, Ángel 629
Mauromicale, Giuseppe 1070
Mavencamp, Dan 2157
Mazariegos, Pablo 1427
Mazzer, Simone 1444, 2254, 2260
McDonald, Brent 42
McGrew, Tyler 1557
Mekhilef, Saad 746
Mendes, Arthur 2992
Mercier, Patrick P. 315

Metwly, Mohamed Y. 919
Meyer, Stefan 1034
Miao, Honglei 2059
Michelis, Stefano 479
Milivojeviæ, Nikola 1, 54
Min, Hao 1090
Min, Hyungki 1629
Min, Run 2109
Minato, Yuichiro 2913
Mirafzal, Behrooz 2461
Mirkoviæ, Nikola 788
Mishima, Taichi 3248
Mishra, Santanu K. 2213
Mitcheson, Paul D. 1653
Mitrovic, Vladimir 409
Mitsui, Koji 1299
Miyamae, Masaki 2681
Miyanjou, Kazuki 1977
Miyazaki, Tatsuya 1797
Mo, Liping 795
Mo, Xianghao 1375
Mohammad, Mostak 1635
Mohammadi, Sajjad 2474
Mohseni, Parham 2173
Moniruzzaman, Md 943, 951, 1420
Montejano, Misael 637
Monticone, Francesco 1646
Montoya-Acevedo, Diego 629
Moon, Gun-Woo 1899
Moon, Jinyeong 1473, 2413
Moorthy, Radha Sree Krishna 2797
Morris, Lauryn 1311
Moschopoulos, Gerry 1972
Motoori, Shuichiro 1977
Motto, Eric 2356
Mou, Di 2857
Mou, Shin 2419
Mounesi, Reza 1791
Moursi, Mohamed Shawky El 2871, 3064
Mousavi, Mahdi S. 2290
Mu, Qiang 1388, 2790, 3328
Mu, Wei 2597
Mu, Xuchu 978
Muduli, Utkal Ranjan 1940, 2871, 3064
Mueller, Lukas 538
Muenz, Ulrich 1184
Mühlethaler, Jonas 2495
Mujica, Gabriel 868, 3298
Mukhopadhyay, Anwesha 3267
Mukunoki, Yasushige 2356
Müller, Kilian 3241
Mulumudi, Guru Abhilash 1135

Munk-Nielsen, Stig 175, 357, 1034
Murakami, Haruhiko 1569
Muravleva, Ekaterina 1850
Murillo-Yarce, Duberney 1959
Murray, Samantha K. 1905
Murukesan, Karthick 180
Muscat, Isaac 449
Musolino, Francesco 738
Mustakin, Zaheen 1388, 2790
Na, Woonki 2973, 3000, 3006, 3012
Nabila, Kashfia Tajmim 2877
Nabizadah, Ahmad 3160
Nag, Kumar Joy 990, 997
Nagahara, Teruaki 1569
Nagai, Yoshiyuki 423
Nagano, Masanori 285
Nagar, Anshul 2973
Nagasawa, Shinobu 2610
Nagayoshi, Kenichi 1102, 3096
Nakagaki, Akito 2654
Nakagawa, Shigeki 1797
Nakamura, Hirokazu 1728
Nakamura, Keisuke 1237
Nakano, Satoshi 2551
Nakashima, Junichi 2356
Nakata, Yosuke 1383
Nakata, Yuki 21, 2913
Nam, David 2992
Namadmalan, Alireza 2474
Namba, Akira 1797
Namburi, Krishna 1294
Naradhipa, Adhistira M. 498
Narasimhan, Sneha 1866
Narumanchi, Sreekant 1824
Nasiri, Adel 1489, 1791
Nassaji, Abolfazl 2290
Nassar, Rajaie 586, 2228
Nations, Mark 552
Naval, Sourav 1012
Navarro-Rodríguez, Ángel 1935
Neal, Adam T. 2419
Nelms, R.M. 1953, 2703
Nelson, Blake 395, 1167
Nelson, Tolen M. 207
Nelson, Tolen 185
Ng, Wai Tung 983, 1108
Ngo, Khai D. T. 2228
Ngo, Khai 586
Ngo, Minh 111
Nguyen, Allen T. 840
Nguyen, Calvin 1274
Nguyen, Duy T. 231, 1121

Nguyen, Hien 1, 54
Nguyen, Kien 3248
Nguyen, Tung-Tan 389
Ni, Chuan 2117
Nielsen, Morten Rahr 357
Nikmaram, Behnam 2290
Ning, Guangdong 809
Ning, Guangfu 2096
Ning, Shangxian 2660
Nishijima, Kimihiro 1977
Nishimura, Keigo 3096
Nishio, Haruhiko 1108
Nishizawa, Shin-Ichi 2551
Nitta, Honami 1797
Noesges, Brenton A. 2419
Noguchi, Koichiro 1569
Noh, Young-Seok 518
Norman, Patrick 3167
Notake, Koki 1299, 1414
Núñez, Guillermo 1197
Nuzzo, Jeremy 2387
O'Driscoll, Seamus 285, 2009
Oberdieck, Karl 1051, 2589
Oboreh-Snapps, Oroghene 1311
Ochiai, Yuki 3261
Ohi, Toshi 2821
Ohno, Takashi 21
Ohodnicki, Paul R. 544, 2516, 3119
Ohodnicki, Paul 370, 1326
Okamoto, Takahiro 321
Olalla, David 3100
Olimmah, Marshal 395
Onar, Omer C. 1635
Onishi, Hiroyuki 2431
Onuma, Naoto 2681
Opificius, Julian 401
Orabi, Mohamed 775, 906, 2327
Orlando, Tailan 342
Orr, Allison 1211
Oruganti, V.S.R.Varaprasad 801
Ota, Hiroaki 3248
Ou, Shuyu 1501, 3042
Ouyang, Ziwei 246, 252, 1810
Pahlevani, Majid 616, 2953
Pakala, Sriharsh 505
Palani, Praveenkumar 62
Pallantla, Manikanta 2708
Palmal, Manas 1874
Pan, Ci 192
Pan, Qishan 2207
Panja, Pijush Kanti 821
Paplham, Tyler W. 2516

Parashar, Sanket..1347
Paredes-Camacho, Alejandro 1774
Park, Junhyeong...3187
Park, Sung-Bum..3187
Parkhideh, Babak 1388, 2790
Parreiras, Thiago M. 1355, 1615
Pasupuleti, Sai Sushma...................................3316
Patle, Nagesh 2805
Paul, Sayan 1, 54, 334
Paulino, Glaucio H...1274
Pavone, Mario Giuseppe....................................738
Peña-Alzola, Rafael 1935, 3167
Peng, Fang Z. 761
Peng, Hongjie 171, 225
Peng, Xiaochuan 1090
Penof, David 1519
Pereira, Joao...637
Pereira, Lucas 1388, 2790
Pereira, Thiago Antonio 118
Peretz, Mor Mordechai 594
Pérez, Fernando 868, 3298
Pérez, Sara 1197
Perez-Farre, Quirc 1774
Perreault, David J. 132
Perreault, David 124
Petriæ, Ivan Z. 157
Petriæ, Ivan 2764
Petrillo, Gaia 266
Petucco, Andrea 2667
Pfost, Martin 573, 1129, 1576, 2772
Philippe, Antoine 1396
Phukan, Ripun 2082, 2296
Phung, Thanh Hai 195
Picot-Digoix, Mathis...2718
Piel, Joshua J. 2419
Pietrini, Giorgio 670
Pigott, J. 1451
Pilawa-Podgurski, Robert C. N.......... 151, 157, 291, 558,
....2276, 2805
Pillonnet, Gaël 315
Pirson, Nicolas 258
Pizzuto, Matteo 1096
Plum, Thomas 1919
Pong, Man-Hay 389
Pool-Mazun, Erick 1459
Popoviæ, Zoya 682
Porras, David A..27
Porter, Matthew 1020, 1564
Pou, Josep 1927
Pourjafar, Saeid 2173
Prabhakar, Siva 69
Pradhan, Rachit...670

Prakash, Surya..1940
Preindl, Matthias 272, 1255
Prodiæ, Aleksandar 307, 990, 997
Punjabi, Shobhana 1159
Qahouq, Jaber A. Abu............................2779, 2840
Qi, Nianzun 357, 1034
Qian, Ting 2117
Qian, Yijie 524
Qiblawey, Yazan..3071
Qin, Yuan 1564
Qin, Zian 609
Qiu, Tian 1040, 2400
Queiroz, Samuel S...............................1334, 1341
Quenette, Vincent..1889
Rabenold, Elizabeth 2249
Radhakrishnan, Kaladhar 491, 1681
Radici, Christian.........................1403, 1512, 1781
Rafiq, Aamir... 395
Rahman, Md Rashedur...........................943, 951
Rahman, Mohammad Dehan....................1844, 2407
Rahouma, Ahmed.. 27
Rajagopal, Narayanan 1836
Rajpurohit, Chirayu..2764
Raju, Soniya 378
Rallabandi, Vandana...............................1635, 3174
Ram, Achala 2920
Ramasubramanian, Deepak 1184
Ramirez, Juan 1211
Ramkumar, S. 2708
Ramos, Gabriel V. 1355, 1615
Ramos, Regina 1197, 1375
Ran, Li 1832
Rana, Dilip 1040
Rana, Mandeep S. 2213
Rao, Yifan 1274
Rashid, Syed Saeed 1640, 1646
Rathore, Vikas Kumar 594
Raval, Vishwam 727, 3054
Ravichandran, Krishnan 1681
Rawat, Shubham 1347
Raychowdhury, Arijit 518
Reddy, Narsimha 3054
Redondo, Alejandro...............................868, 3298
Reinotas, Jurgis 1754
Ren, Linhao 2343
Ren, Sheng 3181
Ren, Xufu 1439
Restrepo, Carlos 629
Rettner, Cornelius..1759
Richardeau, Frédéric 2718
Rikiishi, Yasuhiro...2284
Ripamonti, Giacomo 479

Ristic-Smith, Aleksandar	1529
Rivas-Davila, Juan	363, 2241
Rizkalla, Maher	2029
Rizzolatti, Roberto	1444, 2254, 2260
Roberts, Gianluca	307
Rodgers, Aidan	1242
Rodriguez, Ezequiel Ramos	1927
Rodriguez, Fernando	3100
Rodriguez, José	746, 895, 2290, 2327
Rodríguez, Juan	1701, 1959
Rogers, Daniel	1529
Rogers, Michael	1569, 2356
Ronanki, Deepak	2647
Rong, Mingzhe	3220
Rong, Zhenshuai	1439
Rosa, Bruno M.G.	1653
Round, Simon	1803
Roy, Soham	1121
Rubinic, Jaksa	416
Rueß, Manuel	3241
Ruiz, Juan M.	2002
Ruiz, Juan	442, 2082, 2296
Ruoff, Dominik Alexander	2589
Ruppert, Daniel	1759
Russo, Andrea	252
Ruszczyk, Adam	1803
Sa, Satyam	103, 449
Saberi, Sajad	2840
Sadasivan, Arya	2461
Sadilek, Tomas	401
Sado, Kerry	2833
Saeedifard, Maryam	2051, 3071, 3321
Saelens, Jonathan	1311
Saggini, Stefano	479, 1444, 2260
Saha, Subrata	1237
Saha, Tarak	3174
Sahoo, Subham	696, 912, 1501
Sai, Ranajit	285
Saiga, Kazuma	2551
Saito, Shoji	1569
Saito, Wataru	2551
Sakai, Hiroto	3077
Salari, Omid	3083
Salehi, Maryam	2883
Samanta, Akash	3141
Sambo, Haifah B.	291
Sandoval, Rolando	1459
Sangwongwanich, Ariya	738, 1501, 3042
Sanjakdar, Omar	1396, 2562
Santi, Enrico	2833, 2865
Santos, Ion Leandro Dos	342
Santos Jr., Euzeli Cipriano Dos	2029
Sanusi, Bima Nugraha	246, 1810, 2035
Saraf, Pushkar	77
Sarajian, Ali	895
Sarda, Radhika	62, 1927
Sarlioglu, Bulent	3174
Sarnago, H.	147
Sarnago, Héctor	3109
Sarofim, Seif	449
Sati, Shraf Eldin	622
Sato, Yuji	1383
Sato, Yuki	579
Satterlee, Ryan	3133
Satyamsetti, Vijayakrishna	1403
Sauter, Bailey	2764
Sayed-Ahmed, Ahmed	2932
Sba, Baher Abu	2453
Sbabo, Paolo	2667
Schanen, Jean-Luc	2562
Scheideler, William	1590
Scherer, Yohannes Amilcar Tekle	342
Sebastián, Javier	1959
Sebata, Kohei	1977
Sekiya, Hiroo	3248
Selvarasu, Uthandi	761, 1465
Sen, Paresh C.	1892
Sen, Tanuj	139, 349, 1882
Sengstock, Jonathan	1242
Sengupta, Arkadeb	90, 118, 1148
Seo, Gab-Su	602, 645, 3059
Seo, Seoktae	1629
Sethupandi, Abishek	62
Seugnet, Léo	2718
Shadmand, Mohammad B.	3071
Shafei, Ahmad El	1326
Shah, Shreyas B.	670
Shahbazi, Reza	1179
Shahsavar, Tala Hemmati	1622
Shang, Shuye	3227
Shao, Hang	2138
Shao, Linbo	1020, 1564
Sharma, Mohit	3141
Shen, Andy	3129
Shen, Xiaobing	436, 2167
Shi, Guannan	1564
Shillaber, Luke	2597
Shimada, Takae	2681
Shimosako, Shumei	3077
Shin, Se-Un	834
Shivdikar, Saumil	2400
Shoji, Tomokazu	2821
Shrestha, Niranjan	801
Shu, Wenze	524

Siddiquee, Ashraf 1294, 1602
Silveira, Hector Bessa 342
Sim, Dong Hyeon 3200
Sim, Si Yuan 505
Singh, Anurag 1, 54, 334
Singh, Prashant 1026
Singla, Rishabh 3054
Siraj, Ahmed 1040, 2400
Sitta, Alessandro 1070
Smith, John 637
Solecki, Alex 1242
Solomentsev, Michael 77
Son, Gibong 1741
Song, Chen 2075
Song, Keqi 1507
Song, Minwoo 3012
Song, Qihao 202
Song, Xiaoqing 1844, 2407
Song, Yubo 696
Song, Zhihao 327
Sönmez, Ertuðrul 1051
Soundararajan, Soundhariya G. 238
Souri, Naser 2202
Sowers, Elizabeth A. 2419
Sozer, Yilmaz 1294, 1602, 2698
Spiazzi, Giorgio 2667
Spieler, Matthias 1495, 1551
Sridhar, Sundaramoorthy 299
Sriram, Vaisambhayana B. 62, 1927
Srivastava, Shubham 2213
Starke, Michael 637, 703
Stauth, Jason T. 1590, 1715
Steiner, Mark 2356
Stella, Fausto 1408
Steyaert, Bernard 1255
Steyn-Ross, Alistair 378, 874
Stillwell, Andrew 1242
Stokowski, Nicole 1242
Strache, Sebastian 2569
Strathman, Sophia A. 1311
Streit, Jochen 1051
Strezelecki, Ryszard 2173
Stricula, Justin 401
Sturdivant, Maurice 544
Su, Gui-Jia 1635
Su, Mei 2096
Sugie, Hisashi 2730
Sui, Qingcheng 436, 2752
Sukita, Yohei 1102
Sullivan, Charles R. 840, 1816
Sun, Bosheng 1990
Sun, Kai 2488

Sun, Lingwei 1108
Sun, Peiyuan 2375
Sun, Ruize 192
Sun, Weifeng 524
Sun, Xiuhu 3328
Sun, Zhen 3275
Sun, Ziang 2981
Sund, Jade 971
Sune, Joseph Benzaquen 164, 3334
Suntharalingam, Piranavan 670
Suzuki, Asamira 1728
Swaminathan, Madhavan 2101
Sweet, Mark 3167
Swoboda, Philipp 2348
Syed, Hadiuzzaman 1051
Szczublewski, Austin M. 185
Tadakuma, Toshiya 2610
Taguchi, Koichi 2356
Taha, Wesam 3147
Tajima, Shin 782
Takahashi, Keita 2610
Takahashi, Yoshiaki 1414
Takamiya, Makoto 1084, 1102, 2551
Takamura, Yota 1797
Takayama, Naoki 2681
Takeda, Ryo 2821
Takeuchi, Kosuke 21
Takeuchi, Toshiro 2828
Takishima, Kenta 423
Takizawa, Sota 2924
Tan, Matthew 1882
Tan, Siew-Chong 3227
Tanaka, Kenichiro 579
Tanaka, Ryota 423
Tanaka, Shinsaku 2284
Tanaka, Toshiyuki 2828
Tang, Ho-Tin 1507, 1582
Tang, Wenyuan 1363, 1597
Tang, Yi 3220
Tant, Mike 1824
Tariquzzaman, Md. 3036, 3048
Tarutani, Masayoshi 1383
Tatetsu, Riku 1977
Tayebi, Milad 854
Teng, Fei 2527
Teng, Yiyina 757
Terauchi, Naoya 285
Terzija, Vladimir 757
Thacker, Thimothy 2992
Then, Han Wui 1681
Thevar, Madasamy Palavesha 62
Thike, Rajendra 1279

Thirumoorthi, Sathya Rupan	1866
Thurlbeck, Alastair P.	1602
Tian, Fanghao	2167, 3124
Tian, Jiachen	2327
Tian, Xiaoyang	1754
Tingbari, Vincent Masabiar	2973
Tomey, Hala	1211
Tomioka, Shohei	579
Tong, Junhong	1786
Tong, Qiaoling	2109
Torres, Javier	1197
Torres, Renato Amorim	1495
Touhami, Mustapha	1012
Tran, Ngoc Ho	2569
Trescases, O.	1451
Trescases, Olivier	103, 449, 676, 1905
Tripathi, Anshuman	62, 1927
Tsai, Chieh-Ju	664, 2131, 2143, 2725, 2735, 2741
Tschanz, James	1681
Tseng, Chien-Hao	2725
Tsou, Ming-Chang	887
Tsuchida, Takayuki	285
Tuzizila, Jeremie	401
Uchida, Yasuo	48
Uddarraju, Praneeth	1311
Uddin, Muhammad Fasih	2453
Uegaki, Shin	1383
Uematsu, Takeshi	3248
Ulrich, Burkhard	1707, 2303, 2502
Umanand, L.	1610, 2785
Umar, Jamil	2973
Umar, Muhammad F.	3071
Umetani, Kazuhiro	321, 2654
Ursino, Mario	1444, 2254, 2260
Uzum, Alper	1294, 2698
Vagnon, Eric	2562
Vanderwegen, Wout	238
Varadarajan, Kamal	180
Vasiæ, Miroslav	788, 1375, 1803
Vedula, Inder	1, 54
Vergès, Gaël	1905
Vico, Enrico	1408
Vines, Peter	1403, 1512, 1781
Vinnac, Sébastien	2718
Vitale, Gianni	3133, 3304
Vohl, Kenny	2757
Wagner, Tomas	3100
Walters, Andrew	951
Wang, Cheng Feng	103, 449
Wang, Daming	2369
Wang, Haiyan	3312
Wang, Haoyu	485, 983, 1544, 2207, 2556, 2675, 2857

Wang, Hongjie	719, 2393
Wang, Huai	912, 2380, 3042
Wang, Jin	1203, 2986
Wang, Jinyan	171, 225
Wang, Jun	538, 1850, 2088
Wang, Kaiyuan	3227
Wang, Kejia	505
Wang, Kun	2088
Wang, Kunrong	2162
Wang, Laili	2375
Wang, Lei	2162
Wang, Liang	983
Wang, Libing	3174
Wang, Lichong	757
Wang, Linguo	2070
Wang, Lisheng	1248
Wang, Liwei	1153
Wang, Maojun	225
Wang, Meng	3312
Wang, Mengqi	2624, 3155
Wang, Pinhe	246, 2035
Wang, Qiong	498
Wang, Rudy	9, 1318
Wang, Rui	1063
Wang, Shaozhe	1459
Wang, Shukai	566, 1882, 2438
Wang, Shumeng	761, 1465
Wang, Shuo	1173, 2603
Wang, Sunqing	3018
Wang, Wei	2692
Wang, Xiao	219
Wang, Xiaohua	3220
Wang, Xiaosheng	795
Wang, Xiaoting	3030
Wang, Xiaoyu	416
Wang, Xinlin	809
Wang, Xiongfei	1615
Wang, Xuan	1791
Wang, Xuliang	1544, 2556, 2675
Wang, Yan	1544, 2556, 2675
Wang, Yao	3275
Wang, Yibo	795, 3181
Wang, Yicheng	3147
Wang, Yiju	1368
Wang, Yulei	219
Wang, Yunxin	2675
Wang, Zijian	1537
Wang, Ziyao	485
Wang, Zuoshuai	3018
Watabe, Kiyoto	2551
Watanabe, Hiroki	21, 48
Watanabe, Kenichi	1102, 3096

Wehr, Erik 2757
Wei, Anran 1983
Wei, Bo 327
Wei, Jinxiao 1832
Wei, Xing 3042
Wei, Xuanjing 1403
Wei, Yuxin 2713
Weihs, Leon 2757
Weiser, Mathias C.J. 2387, 3241
Weng, Sheldon 1681
Wens, Mike 195
Wheeler, Patrick 895
Wicht, Bernhard 848
Wick, Lukas 2380
Williamson, Sheldon............801, 1224, 2322, 3114, 3141
Wilson, Marcus 378
Winkler, Joseph 848
Wipprecht, Lukas 2303
Wojewoda, Leigh 491
Wong, Andy 2833
Wouters, Hans........................ 238, 258, 2167
Wright, Jason 401
Wu, Alan 307
Wu, Chih-Chiang 2687
Wu, Hsiang-Kai 2687
Wu, Shang-Syun 2179
Wu, Taotao 1090
Wu, Teng 2634, 2660
Wu, Tsai-Fu 16, 900
Wu, Xin................................... 651, 2981
Wu, Xinke 1995
Wu, Yang 139, 349
Wu, Yanqing 1995
Wu, Yingzhe 1248
Wu, Yue 1114, 3220
Wu, Yuxuan 2582
Wunderlich, Andrew 2865
Wunderlich, Ralf 2757
Xi, Zichen 1020
Xia, Xiaoyi 2022
Xiang, Zhangwei 429
Xiao, Junjie 609
Xie, Biyun.. 919
Xu, Dehong................... 651, 2894, 2981
Xu, Guo 2096
Xu, Haoran...................................... 2109
Xu, Huangsheng 2207
Xu, Limei.. 2075
Xu, Shen 524
Xu, Wentao 1012
Xu, Wenzhe..................................... 2426
Xu, Xinmiao 1433

Xu, Yun 1114
Xu, Ziyang 2521
Xue, Hui 2117
Xue, Yuxiang 1248
Yabuta, Shigenori 2821
Yagielski, John 3174
Yamaguchi, Koji.................... 1299, 1414
Yamamoto, Keisuke 3194
Yamamoto, Masayoshi 2431
Yamanaka, Kimito 1797
Yan, Decheng 3334
Yan, Yiyang 2343, 2426
Yan, Zhaoheng 1114
Yan, Zhixing 357, 1034
Yang, Bowen 602
Yang, Garam 3000
Yang, Hélène T.W. Ma 983
Yang, Juchen 1203
Yang, Liu 2369
Yang, Qichen 860
Yang, Qiuzhe 1537
Yang, Robert 180
Yang, Xin 1020, 1564
Yang, Xingyu 1953
Yang, Xinliang 409
Yang, Yirui 1173, 2603
Yang, Yongheng 3024
Yang, Yun 3227, 3275
Yang, Zineng 1020
Yao, Wenxi 327
Yao, Yuzhou 1203
Yasko, Mohamed.............................. 3124
Yato, Shinji..................................... 3077
Ye, Liang 285
Ye, Zhengyu 2117
Yeo, Howe Li 62
Yeo, Sungku 2547, 3281
Yi, Lifang 2413
Yi, Zheyuan 2488
Yin, Shan 1248, 2075
Yin, Tianxiang 2535
Yoneyama, Rei 2356
Yoshimoto, Kantaro 423
Yoshimura, Yuto 2654
You, Longxiang 3018
Youssef, Mohamed Z. 3083
Yu, Hao 2343, 2426
Yu, Jingshu 1681
Yu, Ruiyang..................................... 854
Yu, Sheng-Han 2131
Yu, Sheng-Yang 42, 1990
Yu, Wensong 2220

Yu, Xiang 1892
Yuan, Hao 2117
Yuan, Huan 3220
Yuan, Jiaqi 670
Yuan, Jingyi 966
Yuan, Song 1114
Yuan, Tianlong 429
Yuan, Tianshu 2375
Yun, Dam 1659, 2268
Zaabi, Omar Al 1940
Zade, Aditya 719, 1004, 2194
Zaitsu, Toshiyuki 1977, 2654
Zaizen, Shohei 2551
Zaman, Mohammad Shawkat 676
Zan, Xin 3232
Zane, Regan 719, 1004, 2194
Zeineldin, Hatem H. 622
Zekorn, Tobias 2757
Zeng, Hank 2138
Zeng, Jia-En 1967
Zeng, Wenliang 510
Zeng, Zheng 219
Zhan, Cao 278, 2482
Zhang, Ben 3181
Zhang, Bing 2070
Zhang, Bo 192
Zhang, Bohua 573
Zhang, Boran 2675
Zhang, Boyi 1850, 2296
Zhang, Cheng 1512, 3212
Zhang, Chenghui 2713
Zhang, Chi 9
Zhang, Desheng 2109
Zhang, Fuxing 2059
Zhang, Haijin 3312
Zhang, Hely 1286
Zhang, Heng 2369
Zhang, Hengbin 1248
Zhang, Hong 2375
Zhang, Honglang 2075
Zhang, Jiazheng 2628, 2968
Zhang, Jincheng 978
Zhang, Jinfeng 1439
Zhang, Li 2535
Zhang, Qingzheng 2894
Zhang, Renjie 3220
Zhang, Shengke 214
Zhang, Shiqi 757
Zhang, Tianyi 2066
Zhang, Weihang 978
Zhang, Xiangrong 3275
Zhang, Xin 505, 1143, 3018

Zhang, Xiong 2343, 2426
Zhang, Yi 2380
Zhang, Yichi 2380
Zhang, Yifan 2343, 2369, 2426
Zhang, Yifu 2746
Zhang, Yingjie 2603
Zhang, Yuanxin 2857
Zhang, Yuhao 202, 1020, 1564
Zhang, Yuxin 2973
Zhang, Zhe 2907
Zhang, Zhenbin 2846, 3206
Zhang, Zheyu 1040, 1153, 2400
Zhang, Zhi Jin 3321
Zhang, Zhining 1203
Zhang, Zichen 1850
Zhao, Delin 3220
Zhao, Fangzhou 1615
Zhao, Hongbo 357, 1034
Zhao, Shuang 2369, 3227
Zhao, Shuofeng 1824
Zhao, Tiefu 1388, 2790, 3328
Zhao, Tuo 1274
Zhao, Wending 1995
Zhao, Yifan 2521, 2846, 3206
Zhao, Yue 27, 2453
Zheng, Zexiang 2343, 2369
Zhou, Daniel H. 1693
Zhou, Daniel 566, 1274, 2438
Zhou, Dao 2380
Zhou, Fei 2541
Zhou, Feng 2624
Zhou, Jiale 1388, 2790, 3328
Zhou, Kunxiao 809
Zhou, Lufan 1803
Zhou, Mingde 2207
Zhou, Wenqi 2589
Zhou, Xigen 2157
Zhou, Yan 1767
Zhou, Yi 651
Zhou, Yuan 2535
Zhou, Yuebin 2369
Zhou, Zongjie 1047
Zhu, Jiaqi 3312
Zhu, Jinli 761, 1465
Zhu, Junjie 2070
Zhu, Lingyu 278, 2482
Zhu, Liyan 1020
Zhu, Yicheng 558, 2276
Zhu, Zhenhai 1995
Zhuo, Fang 2327
Zolfi, Pouya 1230
Zou, Huanghaohe 1786

Zou, Jiaao..2066
Zou, Jiarui................................... 558, 2276, 2805
Zou, Mingrui..219
Zou, Xudong..2066
Zou, Xuecheng...2109
Zufferli, Kevin 1444, 2260
Zuo, Yu 258, 436, 2167, 2752
Zuo, Zhiling...2070
Zynger-Capaverde, Betina.................................2562